Handbook of Heavy Construction

OTHER McGRAW-HILL HANDBOOKS OF INTEREST

Handbook of Heavy Construction

Edited by

JOHN A. HAVERS

Professor of Construction and Engineering Management
School of Civil Engineering, Purdue University
Lafayette, Indiana
Assistant to R. B. De Mars, President and Chairman of the Board,
Geupel De Mars, Inc., Indianapolis, Indiana

and

FRANK W. STUBBS, JR.

Late Professor Emeritus of Civil Engineering
Purdue University, Lafayette, Indiana

SECOND EDITION

McGRAW-HILL BOOK COMPANY

New York St. Louis San Francisco Auckland Düsseldorf Johannesburg
Kuala Lumpur London Mexico Montreal New Delhi Panama
Paris São Paulo Singapore Sydney Tokyo Toronto

Sponsoring Editors Harold B. Crawford/Daniel N. Fischel
Director of Production Stephen J. Boldish
Editing Supervisor Lila M. Gardner
Designer Naomi Auerbach
Editing and Production Staff Gretlyn Blau, Teresa F. Leaden, George E. Oechsner

HANDBOOK OF HEAVY CONSTRUCTION

 Library of Congress Catalog Card Number 77-107297

07-027278-6

4 5 6 7 8 9 KPKP 7 9 8 7

Dedicated To FRANK W. STUBBS, JR., 1898–1967

Contributors

ROBERT W. ABBETT *Partner, Tippetts-Abbett-McCarthy-Stratton, New York, N. Y.* SECTION 2, CONSTRUCTION CONTRACTS.

CYRIL S. ADAMS *President, Cyril S. Adams, Inc., Houston, Texas.* SECTION 23, STEEL CONSTRUCTION: OXYACETYLENE CUTTING AND JOINING.

R. L. BAUER *Staff Engineer, Caterpillar Tractor Company, Aurora, Ill. (formerly Assistant Chief Engineer, Construction Equipment Division, Harnischfeger Corporation, Milwaukee, Wis.).* SECTION 17, CRANES AND HOISTS.

R. W. BEATY *Chief Paving Engineer, Barber-Greene Company, Aurora, Ill.* SECTION 25, PAVING: BITUMINOUS PAVEMENTS.

DAVID G. BORG *Technical Representative, Explosives Division, Atlas Chemical Industries Inc., Wilmington, Delaware.* SECTION 19, ROCK EXCAVATION: EXPLOSIVES (CO-AUTHOR WITH E. G. BROWN).

GEORGE F. BOWDEN *Vice-President, Research and Development, Symons Manufacturing Company, Des Plaines, Ill.* SECTION 21, CONVENTIONAL CONCRETE: STATIONARY FORMS.

E. A. BRAKER *(deceased), formerly Divisional Product Consultant, Construction Equipment Division, International Harvester Company, Melrose Park, Ill.* SECTION 11, LOADERS: FRONT-END LOADERS.

GEORGE C. BRALYÉ *Manager, Operations Engineering, Fluor Utah Engineers & Constructors, Inc., San Mateo, Calif.* SECTION 19, ROCK EXCAVATION: PLANNING AND DEVELOPMENT (CO-AUTHOR WITH P. E. SPERRY).

E. G. BROWN *Market Planning Coordinator, Explosives Division, Atlas Chemical Industries, Inc., Wilmington, Delaware.* SECTION 19, ROCK EXCAVATION: EXPLOSIVES (CO-AUTHOR WITH DAVID G. BORG).

R. Y. BUSH *Consulting Engineer, Fullerton, Calif.* SECTION 29, CONSTRUCTION DEWATERING.

J. F. CAMELLERIE *Vice-President, Construction Implementation and Procedures, Inc., Huntington, N. Y.* SECTION 21, CONVENTIONAL CONCRETE: SLIPFORMS.

J. R. CHALLIS *Sales Manager, Kolman Division, Athey Products Corporation, Sioux Falls, S. D.* SECTION 11, LOADERS: BELT LOADERS.

HERBERT K. COOK *Vice-President, Engineering, Master Builders, Cleveland, Ohio.* SECTION 21, CONVENTIONAL CONCRETE: MATERIAL SELECTION AND MIX PROPORTIONING.

E. A. COX *Manager, Construction Services, H. B. Zachry Company, San Antonio, Texas.* SECTION 7, EQUIPMENT ECONOMICS.

JOHN A. CRAVENS *Vice-President, Sales, Euclid, Inc., Cleveland, Ohio.* SECTION 13, HAULING UNITS (CO-AUTHOR WITH DUNCAN MACLAREN).

R. D. EVANS *Civil Engineer, Intercontinental Sales, Caterpillar Tractor Company, Peoria, Ill.* SECTION 6, ENGINEERING FUNDAMENTALS: MATERIAL CHARACTERISTICS; EQUIPMENT CHARACTERISTICS; JOB CHARACTERISTICS. SECTION 8, TRACTORS, BULLDOZERS, AND RIPPERS; SECTION 9, SCRAPERS.

E. H. FAUTH *Dean of Career Education and Community Services, Waubonsee Community College, Sugar Grove, Ill. (formerly Group Vice-President, Products, Barber-Greene Company, Aurora, Ill.).* SECTION 11, LOADERS: WHEEL LOADERS.

BEN C. GERWICK, JR. *Executive Vice-President, Santa Fe–Pomeroy, Inc., San Francisco, Calif.* SECTION 28, COFFERDAMS AND CAISSONS.

E. J. HAKER *Product Development Engineer, Rex Chainbelt, Inc., Milwaukee, Wis.* SECTION 16, SOIL SURFACING EQUIPMENT (CO-AUTHOR WITH C. F. RIDDLE).

HANS I. HANSEN *(deceased), formerly Consulting Engineer, Iowa Manufacturing Company, Cedar Rapids, Iowa.* SECTION 20, AGGREGATE PRODUCTION.

CLIFTON H. HUBBELL *Chief Engineer, Sauerman Brothers, Inc., Bellwood, Ill.* SECTION 14, CABLEWAYS AND DRAG SCRAPERS: SLACKLINE CABLEWAYS; DRAG SCRAPERS.

JOHN K. HUNT *Chief Engineer, Johnson Operations, Road Division, The Koehring Company, Champaign, Ill.* SECTION 21, CONVENTIONAL CONCRETE: BATCHING, MIXING, PLACING, AND CURING.

JOHN C. KING *Manager, Federal and State Engineering Services, Master Builders, Cleveland, Ohio.* SECTION 22, SPECIAL CONCRETES AND MORTARS.

J. F. LINDELL *Chief Engineer, Parsons Division, The Koehring Company, Newton, Iowa.* SECTION 12, TRENCHERS.

LAURENCE L. LYLES *General Manager, W. M. Lyles Company, Fresno, Calif.* SECTION 26, PIPELINES: UNDERGROUND UTILITY CONSTRUCTION.

DUNCAN MACLAREN *District Manager, Southwest Region, Terex Division, General Motors Corporation, Dallas, Texas.* SECTION 13, HAULING UNITS (CO-AUTHOR WITH JOHN A. CRAVENS).

ROBERT McLEAN *Vice-President, Construction Data Systems Corporation, Stanford, Calif.* SECTION 4, CONSTRUCTION PLANNING AND SCHEDULING.

E. O. MARTINSON *Martinson Engineering, Milwaukee, Wis. (formerly Vice-President, Research Development and Manufacturing, The Koehring Company, Milwaukee, Wis.).* SECTION 10, EXCAVATORS.

FRANK T. MATTHIAS *Vice-President, Transportation Projects, Kaiser Engineers, Oakland, Calif.* SECTION 3, CONSTRUCTION SERVICES.

HOWARD P. MAXTON *(retired), formerly Vice-President, The Offshore Company, New York, N. Y.* SECTION 1, CONTRACTOR'S ORGANIZATION.

ROBERT S. MAYO *President, Mayo Tunnel and Mine Equipment, Inc., Lancaster, Pa.* SECTION 31, TUNNELING: TUNNELS IN ROCK.

R. W. MUELLER *Construction Safety Services Consultant, Employers Insurance of Wausau, Wausau, Wis.* SECTION 5, CONSTRUCTION SAFETY.

HAROLD C. PRICE *Chairman, H. C. Price Company, Bartlesville, Okla.* SECTION 26, PIPELINES: TRANSMISSION PIPELINES.

JEAN J. REVELT *Manager of Marketing Services, The Lincoln Electric Company, Cleveland, Ohio.* SECTION 23, STEEL CONSTRUCTION: ELECTRIC ARC WELDING.

C. F. RIDDLE *Supervising Design Engineer, Rex Chainbelt, Inc., Milwaukee, Wis.* SECTION 16, SOIL SURFACING EQUIPMENT (CO-AUTHOR WITH E. J. HAKER).

WARREN N. RIKER *Vice-President, Walsh Construction Company, Inc., New York, N. Y.* SECTION 27, PILES AND PILE DRIVING.

J. M. SCOTT *Manager—Construction Equipment, American Bridge Division, U.S. Steel Corporation, Ambridge, Pa.* SECTION 23, STEEL CONSTRUCTION: STEEL ERECTION.

JOHN F. SEIFRIED *Consulting Engineer, Evanston, Ill.* SECTION 21, CONVENTIONAL CONCRETE: REINFORCING STEEL.

E. V. SEMONIN *Development Engineer, Off-the-Road Tire Development, The Goodyear Tire and Rubber Company, Akron, Ohio.* SECTION 6, ENGINEERING FUNDAMENTALS: TIRES FOR CONSTRUCTION EQUIPMENT.

P. E. SPERRY *Construction Engineer, Fluor Utah Engineers & Constructors, Inc., San Mateo, Calif.* SECTION 19, ROCK EXCAVATION: PLANNING AND DEVELOPMENT (CO-AUTHOR WITH GEORGE C. BRALYÉ).

JOHN R. TABOR *President, Mining Equipment Manufacturing Corporation, Racine, Wis.* SECTION 31, TUNNELING: TUNNELS IN EARTH.

JOSEPH E. VOLLMAR, JR. *President, Vollmar Brothers Construction Company, St. Louis, Mo.* SECTION 18, MARINE EQUIPMENT.

RAYMOND M. WELSH *Chief Engineer, Cableway Division, Lidgerwood Manufacturing Company, New York, N. Y.* SECTION 14, CABLEWAYS AND DRAG SCRAPERS: TAUTLINE CABLEWAYS.

WILLIAM G. WESTALL *Consulting Engineer, Largo, Florida.* SECTION 25, PAVING: PORTLAND CEMENT CONCRETE PAVEMENTS.

RUSSELL P. WIBBENS *Director of Marketing, American Institute of Timber Construction, Englewood, Colo.* SECTION 24, TIMBER CONSTRUCTION.

LYMAN D. WILBUR *Vice-President, Morrison-Knudsen Company, Inc., Boise, Idaho. and Chairman of the Board, International Engineering Company, Inc., San Francisco, Calif,* SECTION 30, RIVER DIVERSION.

A. T. YU *Consulting Engineer, Kinnelon, N. J. (formerly Vice-President—Operations, Robins Engineers and Constructors, Hewitt-Robins, Inc., a Division of Litton Industries, Totowa, N. J.).* SECTION 15, BELT CONVEYORS.

Preface

The "Handbook of Heavy Construction" was originally conceived by Frank W. Stubbs, Jr., as a comprehensive reference source for persons who are actively involved in heavy construction. The same objective has been adhered to during the preparation of the second edition. The contents of the book are directed primarily to those contractors who, in recognition of the increasing scope and complexity of their work, accordingly seek to extend their knowledge in allied areas of construction. The book will also be a useful reference source for equipment manufacturers, materials producers, owners' representatives, and students of the construction industry.

The second edition of the "Handbook of Heavy Construction" reflects the best of construction practice in the United States, as of the date of its preparation. Its 47 contributing authors are experts in their particular fields of practice. Their numbers include practicing contractors, equipment specialists, and consulting engineers. Each of them is familiar with the practical aspects of construction, and has prepared his Handbook material accordingly.

The book, as revised, consists of three major parts. The first of these, Construction Management, includes sections on construction organization, contracts for construction, supporting services, and job planning and scheduling. These basic topics have been grouped together because they are applicable to all types of construction.

The Construction Equipment part of the book begins with two general sections, Engineering Fundamentals and Equipment Economics. These important topics provide a rational background for equipment selection. Each of the remaining sections then proceeds to discuss a major class of construction equipment. Examinations of more specialized equipment types, comparisons of operating characteristics, and guidelines for production estimating are included as subtopics within these sections.

The third part of the book, Construction Applications, identifies major construction tasks and analyzes alternative ways for performing them. The major emphasis in this part of the book is on the selection of the best construction method. The equipment needed to perform each task is now discussed in terms of actual job applications, with ample cross-referencing to the material supplied in the Construction Equipment sections. Where supplementary equipment information is required, it is included within the Construction Applications discussions. Each section in this third part of the book thus contains a thorough examination of a major construction application.

Many persons have contributed to this book in the earnest hope that, by so doing, they will have aided the construction industry of which they are a part. The untimely death of Frank W. Stubbs, Jr. in 1967 and my consequent assignment as editor-in-chief placed additional demands on several of those contributors. I would like to express my deep appreciation for their continuing cooperation and unsparing efforts. I would also like to express my appreciation to Alice Tobin, secretary to the President of Geupel De Mars, Inc., who generously typed many hundreds of manuscript pages during the preparation of the book.

John A. Havers

Contents

PART 3 CONSTRUCTION APPLICATIONS

Index follows Section 31.

Handbook of Heavy Construction

Part One

Construction Management

Section 1

Contractor's Organization

HOWARD P. MAXTON

Vice-president (Retired), The Offshore Company,
New York, N.Y.

INTRODUCTION

Top executives, no matter how competent, cannot function to full effectiveness without a sound plan of organization. A well-conceived, long-range organization program permits changes to be made in the right direction as opportunities arise. In the absence of such a plan, changes must be made on the basis of expediency, organization errors are often perpetuated, and fundamental improvements are seldom realized.

Every phase of a company's organization plan should be questioned and tested from a wholly objective viewpoint, without being influenced by present pattern or personnel, precedent or tradition. From such an analysis a plan of organization can be developed which will best meet the current and future requirements of the business.

A good organization chart for the company as a whole, with auxiliary charts for each major division, is an essential first step in the analysis, clarification, and understanding of any organization plan. The process of charting the organization is one good test of its soundness, as any organization relationship which cannot be readily charted is likely to be illogical and therefore confusing to those working under it.[1]*

ORGANIZATIONAL PLANNING

The importance of organization planning grows with the increasing size of the company. The small-size contractor has few organizational problems; he is likely to handle most or all of the bidding, ordering of equipment and materials, and other office work himself or with the help of several assistants. His construction operations require simple organization and coordination techniques, but the usual problems of organizational management generally do not develop. As the company grows in size, the contractor-owner is forced to employ and depend on field superintendents to run the construction end of the business, and he necessarily spends more thought and effort in the bidding, sales, and the administrative end of the business. Gradually additional personnel are added, functions are divided, and finally several types of work formerly handled by one person may require large groups of workers for each type. Then the staff specialist is brought in, and there must be a determination of his place in the organization and an acceptance of his services by the other members of the line and staff.

Organizational planning is concerned with the establishment of relationships so as to integrate and coordinate the activities of individuals or of several groups or divisions. It necessarily involves defining and evaluating the functions of each individual or group of individuals. This careful analysis of the activities of each person or group of persons may result in redefining duties and responsibilities and tends to delineate and coordinate the work of the supervisors and the employees of the firm. The resulting organization chart presents an overall picture of the firm and indicates the chain of command and the lines of communication. Once developed, an organization plan or its resultant chart may tend to point out the weaker links and to provide for replacements due to promotions, reassignments, or retirements.

It is a well-accepted practice in industry to train and develop the younger men from within the organization to provide for replacements and inherent growth or expansion. This policy assures a constantly rising corps of well-trained and proved people to step into the more important supervisory and management jobs of the company and is at the same time a morale builder and tends to lessen the rate of turnover. A lower rate of turnover is especially significant as it applies to the more capable and usually more ambitious type of personnel.

Organizational planning may also reveal duplication of work and effort and reduce confusion of operations within an organization. It provides improved coordination and decreases friction among the personnel and, finally, as a consequence, results in more efficient operations and lower overhead costs.

The need to plan an organization, which means the need to define the functions and duties of the personnel, arises at the time the business begins to grow out of the so-called "one-man" organization. This planning need may not be recognized until the firm has gained considerable stature and the requirements of proper planning are so obvious that they become a necessity. Also, as an organization grows and new personnel is brought in, the lack of clear definitions of duties and responsibilities sometimes creates personnel problems that are difficult to overcome. The "older hands" are sometimes disturbed by the bringing of new personnel into the expanding organization who appear to assume functions and responsibilities formerly handled by the older employees, often unconsciously or consciously discouraging cooperation and teamwork and tending, if not controlled or stopped, to be disruptive and very costly to the growing organization. Therefore, as an organization expands it is important that the determination of work loads be under constant scrutiny and evaluation, and a shifting of these work loads among employees, both old and new, should be a matter that is

* Superior numbers refer to the list of references at the end of this section.

given considerable attention, frequent review, and clear definition. Each employee is entitled to know his sphere of activity and the coordination that is required of him from the various members of the organization. *To the extent authority is required to accompany his responsibility, he should be given that authority.*

Organizational planning is a continuous, long-term proposition requiring farsighted judgment and readiness to make shifts in objectives and personnel to meet changing conditions. Many objectives require advance planning because of problems involving finances, trained personnel, caliber of operations, etc. Desirable changes in personnel may require proper timing if the full benefits are to be achieved from the change. Too often the executive is so engrossed in current worries and problems that he cannot give the necessary time and thought to the broad, long-range objectives that may enable his company to expand its activities and enjoy greater prosperity. It is best to have an immediate plan and a long-range goal. The immediate improvements can be accomplished within the framework of the long-range objective, and each successive move will be a right move in the direction of the final goal.

In the development from the small one-man construction business to the large company, major organizational problems arise. These may be catalogued as follows:

(1) Formulation of Objectives: Division of Work
(2) Delegation of Responsibility: The Accommodation of Personalities
(3) Delegation of More Management Functions: Span of Control
(4) Reducing the Executive's Burden: The Staff Assistant
(5) Establishing a New Function: The Staff Specialist
(6) Coordination of Management Functions: Group Decision Making
(7) Determining the Degree of Delegation: Decentralization[2]

Some organizations have combined under one supervisor the functions of bidding on a particular job, planning its construction-plant equipment, and finally the responsibility for supervising it in the field. That is, if the company is the successful bidder on a job, the supervisor who was instrumental in preparing the bid follows through and goes into the field and runs the job. He is in this way responsible for proving that his bid was adequate. This method of operation has considerable merit in that the estimator prepares his bid based upon his own field experience and knowledge. Also, it eliminates the possibility of the superintendent being held responsible for overruns in the field, when actually the bid prepared by someone else may have been too low.

Elements of Organization It has been said that there are three elements of organization. These are:

1. A common purpose
2. A communication system
3. Willingness to cooperate

These three elements must be present if there is to be an organization. Without a common purpose, there is no point or reason for people to work together. There must be some goal or some objective which a group recognizes and finds desirable if it is to be expected to contribute to its attainment.

There must be a communications system so that information may pass quickly from one to another in order for objectives to be promptly presented.

Only those willing to cooperate will contribute to the work at hand to accomplish any proposed objective. In discussing the willingness to cooperate, it is necessary to consider the question of incentives. Incentives are of two types: personal and financial.

Many objectives are of such a nature that individuals will enjoy certain personal satisfactions in the accomplishment. They will work toward the objective, and the personal values are so sufficient that no financial reward need be involved. Such is the case in all social, civic, fraternal, religious, and other similar types of organizations.

In the case of industrial (construction) organizations, it is necessary that money be added as an incentive; however, money alone does not determine the worth of a job. The personal values must also be considered, and these definitely have a monetary value.

An organization comes into being whenever two or more people get together to accomplish something that one could not accomplish by himself. Additional people may be added as the work requires additional participation and as people are found who are willing to contribute to the accomplishment of the stated objectives. So long as the contributors report to one supervisor, we have a unit organization—the simplest form of organization. This is shown in Fig. 1-1.

As soon as the work grows to the point where one of the original subordinates needs a helper, a new relationship is established. A new unit organization has been created, and we now have a complex organization. This is shown in Fig. 1-2.

Figure 1-3 represents a typical organization chart showing a situation where each supervisor may have five subordinates reporting to him.

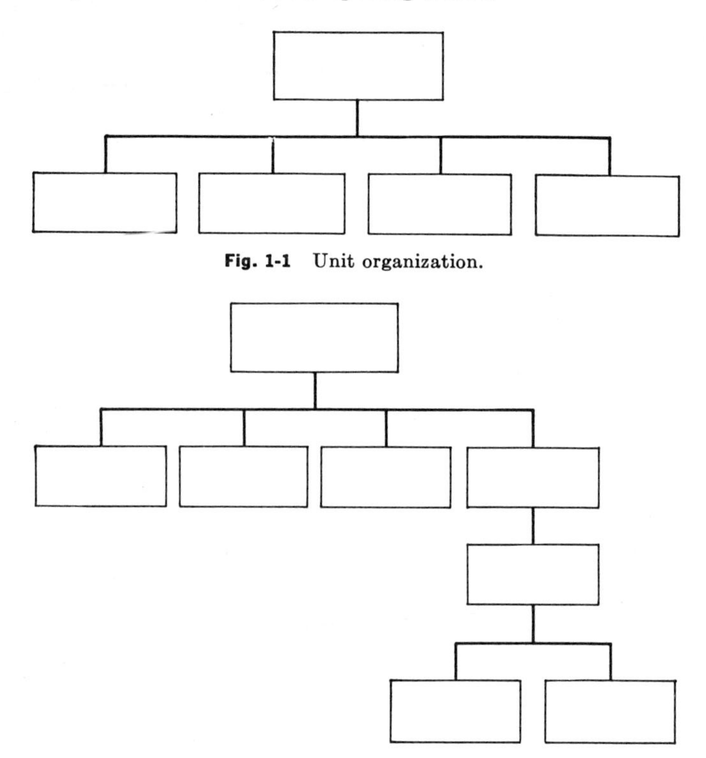

Fig. 1-1 Unit organization.

Fig. 1-2 Complex organization, early stage.

Delegating Responsibility With the growth of the company, the number of employees reporting to the owner-contractor increases. He has an increasing number of management functions requiring long-range planning, policy making, execution, and control. From the very first, when he employed his first helper or assistant, he found it necessary to delegate functions of a minor nature. In all but the smallest company it is necessary to delegate responsibility, and the growth of the firm demands that the owner divest himself of minor functions and responsibilities as quickly as possible. As his organization expands, he finds it necessary to delegate increasingly important functions and responsibilities, and as his company enters the ranks of "big business" within the industry, the head of the firm may find that the determination of policies, the evaluation and handling of problems among his key personnel, and the demands of his "opposite numbers" in responsible positions within the industry and among his clientele take up most or all of his time.

The degree to which he can accomplish this removal of responsibilities from his own shoulders will have a positive effect on the progress and expansion of his company's activities. This is not to say that he should forgo *control* of operations, but simply that he must instruct subordinates how to carry out the time-consuming detail work and brief him as to results.

As the growth of the company requires the owner to delegate greater responsibilities to capable personnel, his organization takes form and breaks down into work groups or departments. During this early growth period it is well to begin to build into the organization necessary control checks and balances. These take form in written reports, cost statements and cost analyses, periodic reports of work done, comparison studies, staff meetings, etc.

The owner or chief executive cannot, however, delegate responsibility for the success or failure of the business. He must multiply his own productivity by delegation of duties, but the crucial question is *how* he delegates the work and *who* his choice is. If there is a weak link or a breakdown in the organization, it is his responsibility to recognize it and apply corrective measures.

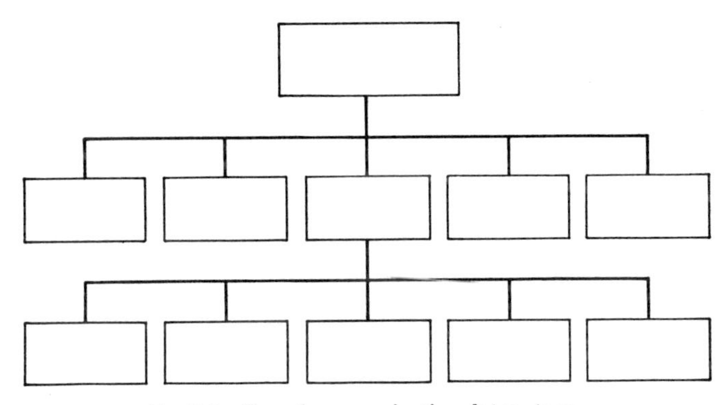

Fig. 1-3 Complex organization, later stage.

At this point, the owner finds that the successful operation of his business, including perhaps the preparation of bids and most certainly including the supervision of his projects, as well as other important functions such as purchasing, selection of plant and equipment, etc., either has become the complete responsibility of others, or that at least he must be largely guided by the advice and recommendations of others with respect to these activities. Experience has shown that five or six primary or key subordinates, each in charge of a department or function, is about the maximum number that the executive can supervise and coordinate and still have time to devote to the long-range planning of the organization.

> In organizational theory, the optimum "span of control"—that is, the number of subordinates who can be effectively supervised by one man—is generally set at between three and six.[2]
>
> The number of subordinates whose tasks are interdependent who can be directed immediately and effectively by one individual is strictly limited. It should not exceed five or six. The ideal number of subordinates for all superior authorities appears to be four.[3]
>
> The average human brain finds its effective scope in handling three to six other brains.[4]

The need to delegate responsibilities is one of the first major organizational problems that arises with growth. Sometimes the rising or expanding owner-contractor finds this problem extremely hard to overcome. He may be reluctant to delegate functions, or if he does he may be reluctant to delegate the necessary authority. Or, having delegated the function and authority, he may spend as much time worrying about the other fellow or insisting that the subordinate follow the contractor's predetermined line of operation, with the result that the contractor spends as much time following the subordinate's footsteps as if he were doing the work himself. Often the owner-contractor may feel that nobody can do the work as well as he, but nevertheless, having delegated both the responsibility and the authority to the subordinate, he may find improvement in the method of doing things as well as improvement in the rate of output. Delegation of responsibilities often brings out in subordinates better qualities and work than were known to exist. Pride in accomplishment is known to be as rewarding to employees as an addition to the pay envelope.

In large organizations, the various functions should be assigned so as to obtain the greatest advantage from the division of labor. Work should be so divided that the incumbent of a position should be able to become a specialist and increase his knowledge on the particular job assigned to him. Then, his special abilities should be used to the fullest.

It has been said the executive job is to organize, visualize, energize and supervise.[5] Management must also select and train a good team of competent people who thoroughly understand and are able to accept and discharge the obligations of their positions. If the executive of a large firm has accomplished the points enumerated above, he may be said to have fulfilled his primary responsibility. Alone and unsupported, even the most competent executive can do only a certain amount of work in a given time. When he breaks up his organization into natural and controllable parts and places competent personnel in charge of many functions, the chief executive has taken a great step toward the successful operation of his business. However, the delegation of work load alone does not ensure the success of business. Heads of departments and their alternates must not only be competent and able but they must integrate and coordinate their work with other individuals and departments of the organization. The executive also must plan ahead and visualize the expansion of his business and the consequent expansion of his organization. He must not only supervise, but he must energize and inspire each subordinate in the team to do a job to the extent of his ability.

Finally, the executive cannot hope to attract and hold competent and ambitious young men in the organization unless they are permitted to share not only in the work load but also in the stimulation and feeling of authority that is rightfully theirs as a result of their contribution to the company's success.

Selecting and Training Subordinates The selection of people requires the utmost care, and the training of employees requires time and patience. In the process it may be discovered that an employee is a misfit in his particular job, or an ambitious young fellow is working under too low a ceiling in his particular line of work, or an employee has hidden talents and can become more valuable to himself and the company in another operation. In all these cases, careful evaluation of the employee's potential may indicate shifts within the organization which, if carried out, will develop better people and a stronger firm. Men of limited capacities and qualifications have their place too and can often be used in other important positions, but they should not be placed in key administrative positions where their administrative deficiencies will adversely affect the entire organization.

Too much care cannot be exercised in the selection of subordinates. Of course, the executive cannot take the time to interview and select all new employees, but if he has selected his key personnel carefully and successfully, they will insist on competent and high-class lesser employees. Some of these same employees will be the key people and managers a few years hence.. Therefore successful selection of both high and low personnel not only pays off immediately in efficiency of operation but sets the stage for expansion and for the replacement of older employees.

There is much to be said about the executive's responsibility to guide and stimulate subordinates. Futhermore, subordinates require coordination of effort and sometimes careful and wise curbing, as it is almost inevitable that ambitious young men going places sometimes "get in each other's hair." An observant and understanding executive can often team up this conflicting manpower and thus develop valuable assets; whereas this same manpower, unguided or uncurbed, can wreck or at least slow down the management machine. Such teaming-up requires careful thought and execution and a thorough understanding of the several personalities involved.

The Staff Assistant An important office that is commonly found in large organizations is that of the staff assistant. This person often is delegated heavy responsibilities by the executive and may be an extremely capable man, sometimes on a par with the other key personnel, although he operates as an individual and does not have any part of the organization under his direct supervision. His duties may consist of transmitting orders and instructions from the chief executive to department heads and other key personnel, receiving and coordinating their reports, and sometimes acting on behalf of or representing the chief executive. It is possible that his duties and influence may earn for him the title of "assistant president." Caution should be exercised, however, that he not be permitted to assume authority that he does not possess or that he does not unconsciously or intentionally attempt to step into his boss's shoes. Nevertheless, his close relationship to the chief executive brings to him a knowledge and understanding of the intimate matters of management and, if he is capable, may in time lead to his advancement to head up new and important projects or divisions of the organization.

The Staff Specialist The staff specialist furnishes the necessary technical aid to the organization. The small or medium-size construction firm will not require the full-time services of such specialists, which may include lawyers, insurance experts, or engineers specializing in some particular field, and therefore outside assistance is employed when needed.

Sooner or later, however, as the company attains growth, it is both economical and feasible to employ full-time specialists in certain fields. These specialists generally work independently and may be available to anyone in the organization for consultation or advice.

Other types of specialists are sometimes appointed to handle functions which were previously under the control of an untrained individual. For example, employee relations may have been handled on an informal hit-or-miss basis. Employee problems, accentuated by growth of the firm and requirements imposed by the Wage Hour Law and other governmental regulations, ultimately become so serious as to force the decision to employ a full-time expert or specialist to handle all personnel and employee affairs.

DIVIDING THE WORK

The final major criterion for dividing the work of the organization is the size of the company. The importance of the chief problems faced by the top management varies as the company grows. Hence the major functions exercised and supervised by the chief executive are likely to change also. This may be illustrated by the work table which the great French industrialist Henri Fayol drew up:[6]

Relative Importance of Requisite Abilities of Personnel in Industrial Concerns

	Requisite abilities, %						
	Managerial	Technical	Commercial	Financial	Security*	Accounting	Total evaluation
One-man business	15	40	20	10	5	10	100
Small firm	25	30	15	10	10	10	100
Medium-sized firm	30	25	15	10	10	10	100
Large firm	40	15	15	10	10	10	100
Very large firm	50	10	10	10	10	10	100
State enterprise	60	8	8	8	8	8	100

* Safeguarding property, avoiding social disturbances in the broad sense, and opposing any influence which endangers the life of the business.

From this table the following conclusions may be drawn:

1. The most important ability of the head of the small industrial company is technical ability.

2. As one goes up the chain of command, the relative importance of managerial ability increases and that of technical ability declines. Equilibrium between these two obtains in medium-sized companies.

3. The most important ability on the part of heads of large companies is managerial skill, and the more important the company the greater the place occupied by this ability.

4. Commercial and financial ability play relatively more important parts in the case of heads of small and middle-sized companies than they do in the case of larger companies.

5. As one goes up the scale of industrial concerns the managerial coefficient increases at the expense of the rest, which tend to even out, approximating up to one-tenth of the total evaluation.

It is clear that the larger the size of the business the greater the emphasis on broad managerial functions, such as planning, forecasting, organizing, commanding, coordinating and controlling.

It is therefore concluded that the most important criterion for the division of work is that of economic efficiency. This should lead to specialization, full utilization of abilities and homogeneity between groups.

Where this criterion is paramount, the basic functions (i.e., those supervised by the chief executive) are those which make the greatest contribution toward profitability. However, the economic criterion, it should be remembered, must usually be modified in the light of noneconomic needs. Both need to be fitted to the particular stage of the growth and the special requirements of the company.[2]

The Chain of Command The chain of command, as described so well by Fayol, is

> the chain of superiors ranging from the ultimate authority to the lowest ranks. The line of authority is the route followed—via every link in the chain—by all communications which start from or go to the ultimate authority. This path is dictated both by the need for some transmission and by the principle of unity of command, but it is not always the swiftest. It is even at times disastrously lengthy in large concerns, notably in governmental ones. Now, there are many activities whose success turns on speedy execution, hence respect for the line of authority must be reconciled with the need for swift action.
>
> Let us imagine that section F has to be put into contact with section P in a business whose scalar chain is represented by the double ladder G-A-Q.

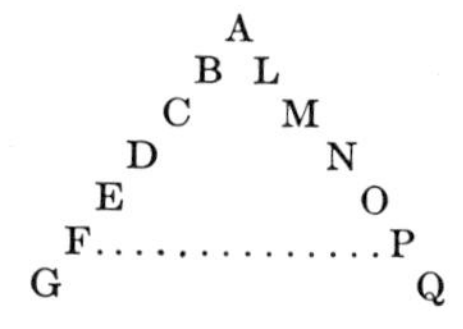

> By following the line of authority the ladder must be climbed from F to A and then descended from A to P, stopping at each rung, then ascended again from P to A, and descended once more from A to F, in order to get back to the starting point.

Fayol continues,

> It is much simpler and quicker to go directly from F to P by making use of FP as a "gang plank" and that is what is most often done. The chain of command will be safeguarded if managers E and O have authorized their respective subordinates F and P to treat directly, and the position will be fully regularized if F and P inform their respective superiors forthwith of what they have agreed upon. So long as F and P remain in agreement, and so long as their actions are approved by their immediate superiors, direct contact may be maintained.[6]

Decentralization It is most logical that some large engineering and construction companies and perhaps some of the medium-size ones have decentralized operations.

In some instances the decentralization was planned, while in others decentralization was a consequence of performing a contract in another section of the country, necessitating the establishment of a branch office. During or following completion of the job, the local management successfully bid on other work, with the result that the branch organization developed permanency.

From an organizational standpoint, decentralization achieves many benefits by breaking down the work load geographically. In many instances, economy of operation may be obtained by local management because of its greater familiarity with conditions in its area, including labor supply and conditions of employment, material markets, and prospective clientele.

Decentralization almost inevitably leads to delegating responsibility, although some firms exercise closer control and supervision over decentralized activities than others. It cannot be achieved merely by delegating responsibility and authority. The unit management must be permitted to organize. It must be able to discharge its obligations.

Even in cases where jobs are local, some contractors establish their field organizations on an autonomous basis with *assists* from the central office. On the other hand, other contractors control construction operations or jobs directly from the home office, including engineering, purchasing, payroll making, etc., and visit the jobs daily, if in the area, to observe progress and give instructions to the superintendents. In each case, however, the central office receives reports of progress, cost, and operations and should be in a position to determine how successfully the job is being constructed. Should progress be slow or costs high, remedial measures can be applied while it is still possible to benefit from corrections taken or changes made.

In achieving decentralization, it is essential that management not sacrifice coordination and control. But the operating subdivision should comprise a complete functional and business entity, one for which the unit manager can be held fully responsible.

Management Philosophy

In many companies the nature of top management philosophy has a profound influence on the nature of decision-making. Over the years an increasing number of companies have adopted decentralization of decision-making as a basic method of organization. These companies include very large units like General Motors, Ford, Sears Roebuck, du Pont, International Harvester, General Electric, and others, and medium-large companies such as Sylvania Electric Products, Inc., Sperry Corporation, and American Brake Shoe Company. The top-management philosophy of these financially successful companies has done much to convince other executives that the adoption of decentralization may be helpful to the economic position of their company. Typical of this thinking is the statement of General Wood [formerly] Chairman of the Board of Sears, Roebuck and Co.:

> We complain about government in business, we stress the advantages of the free enterprise system, we complain about the totalitarian state, but in our industrial organizations, in our striving for efficiency, we have created more or less of a totalitarian organization in industry—particularly in large industry. The problem of retaining efficiency and discipline in these large organizations and yet allowing our people to express themselves, to exercise initiative and to have some voice in the affairs of the organization is the greatest problem for large industrial organizations to solve.[2]

THE ORGANIZATION CHART

The duties of the executive include establishing objectives, forming channels of communications, delegating authority and responsibility, and designing adequate controls. An organization chart is nothing more than a graphic representation of the channels of communication in the formal organization. It is very important that the channels of communication be known and that they be followed.

Many organizations have tried to keep organization charts confidential. This leads to little more than confusion. People are expected to perform without knowing exactly what their job is or to whom they report. In order that duties and responsibilities may be effectively delegated along with commensurate authority, it is necessary that people know where they fit in the organization and to whom they report.

The organization chart gives a very helpful picture of the component parts of an organization joined together to make an integrated whole. It indicates lines of command and of responsibility, from the chief executive downward. Reading upward, it shows the obligations of subordinates. The chart shows lines of communication and tends to clarify functions and authority. The lines of communication are horizontal as well as vertical, and it is usually understood that the lines have arrows on both ends.

The chart presents a formal view of the organization but does not always show the informal relationships that exist among executives and individuals—relationships which are, in actual practice, sometimes stronger and more effective than the formal relationships.

Time brings changes in the organization, sometimes because of expansion or contraction of activities and sometimes because of changes in personnel. Therefore the chart must be subject to periodic review and change.

The construction of the chart requires a thorough knowledge of the natural components of the company's business. There is then a determination of the major and minor subdivisions of the whole organization to achieve the smoothest and most economical coordination and operation. In many instances, an outline of each job's content is shown in a box of the chart, together with the name of the person exercising the responsibility.

Figures 1-4 to 1-7 outline the functions and the divisions for a large construction organization. The functions and organizational requirements are about the same regardless of whether the business is individually owned, a partnership type of company, or a closely held or publicly owned corporation. In certain instances, corporate officers, such as vice-president or treasurer, are indicated by boxes in the charts. These titles are shown to indicate the degree of responsibility attached to the position and do not mean that a vice-president necessarily heads up the operation in question. Likewise, for example, the engineering department is headed by a chief engineer, but it is equally true that the chief engineer, in some cases, may also carry the title of vice-president.

The charts show a natural grouping or breakdown of functions such as may be found in any contractor's organization. Every organization will have variations in the groupings due to the processes of natural growth, to personalities, and to local conditions, management requirements, and environments.

It is hardly possible that everyone will completely agree with the grouping of the functions on all these charts because each person is guided more or less by his own experience. Therefore, there is no quarrel to be had with any rearrangement of any of the functions, although it is anticipated that in most of their principal elements the charts will remain quite unchanged. As an example, many contractors may place their estimating under their chief engineer; the cost engineer often operates independently; and scheduling may be done in the estimating department. Many large contracting organizations have project managers who are in charge of the various field jobs; in other instances, the top field man is given the title of superintendent. Many major construction companies employ full-time engineers and other personnel in research and experimentation, and this work may come under the engineering department or it may be established as a separate activity.

The purpose of Fig. 1-4 is to represent in three primary divisions the activities or functions found in the average engineering and construction company. These three primary groups are:

1. Engineering and design
2. Field construction
3. Administration and control

All items or functions can logically be placed in one of the three groups.

In Fig. 1-5 the functions are arranged into subdivisions for logical administration and control. Figure 1-6 shows a typical field organization which is named Project *A*, being one of the projects identified on Fig. 1-5. This project or job is headed by a project manager or a general superintendent.

Here again, in the field organization, it is unlikely that any contractor will fully agree as to the distribution of functional controls. That is not too important because it is well understood that, for certain reasons due to experience, personalities, or local practice, functions will be handled and controlled in a manner different from those shown on the chart. For example, it is possible that in some instances personnel and safety work may come directly under the control of the superintendent, timekeeping and paying may come under the chief accountant, and purchasing and expediting may be a separate and distinct staff function headed by a purchasing agent (or director of

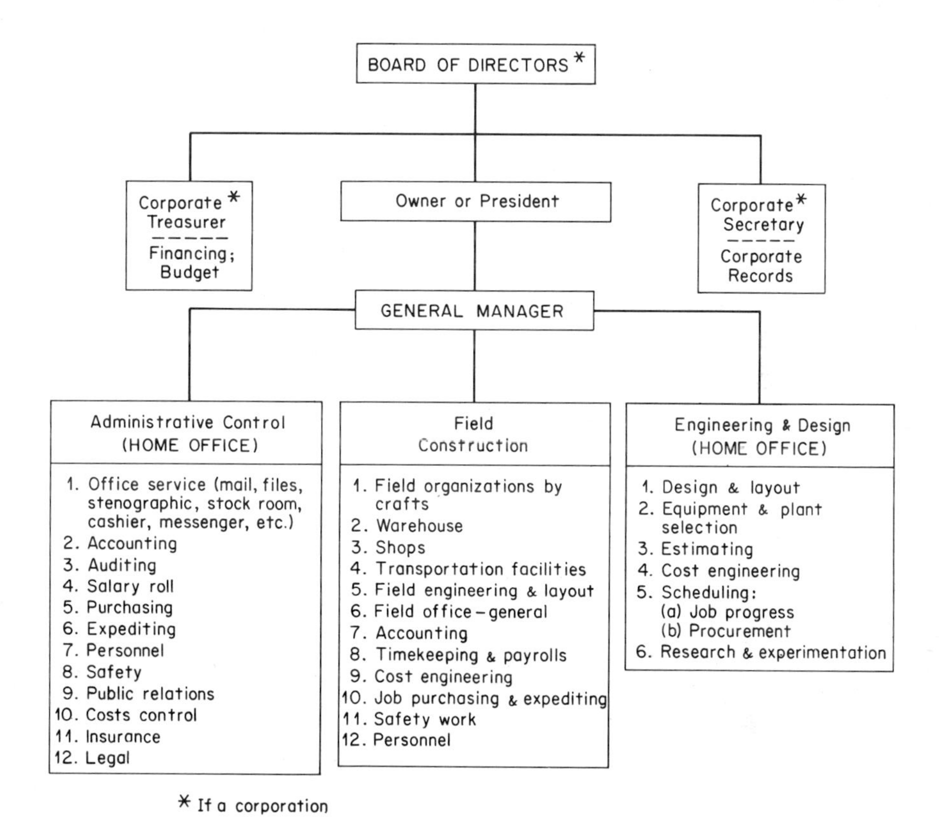

Fig. 1-4 Organizational functions grouped into three basic divisions.

procurement). In other cases, it is possible that the construction plants are laid out or designed entirely in the home office, and there is, therefore, no need for that function in the field organization. The same comment may be made with respect to estimating and inventorying.

On smaller projects or jobs, two or more of the functions shown on the charts may be handled by one individual. On very small jobs all those functions on the entire organization chart which are pertinent or apply in some shape or form might be handled by one man. It is true that on such a job some of these activities assume little or no importance or would appear to take no time or attention; but they probably exist, even though they are not recognized.

Many of the major organizations, operating on a sectional or national basis, have an intermediate office between the home office and the field organization. This office is

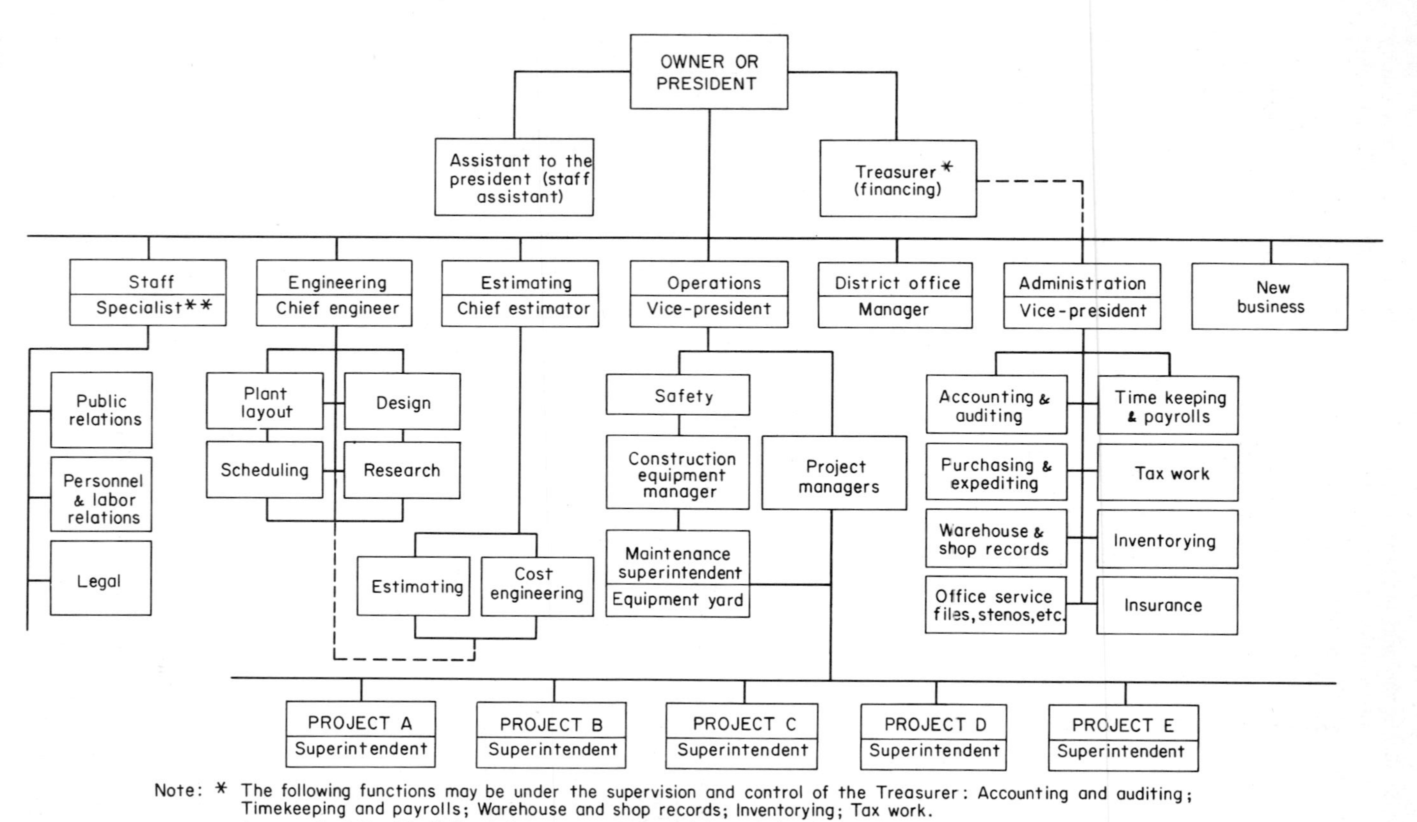

Note: * The following functions may be under the supervision and control of the Treasurer: Accounting and auditing; Timekeeping and payrolls; Warehouse and shop records; Inventorying; Tax work.

** Any or all of the Staff Specialists may report administratively to the Vice-president, Administration.

Fig. 1-5 Home-office organization.

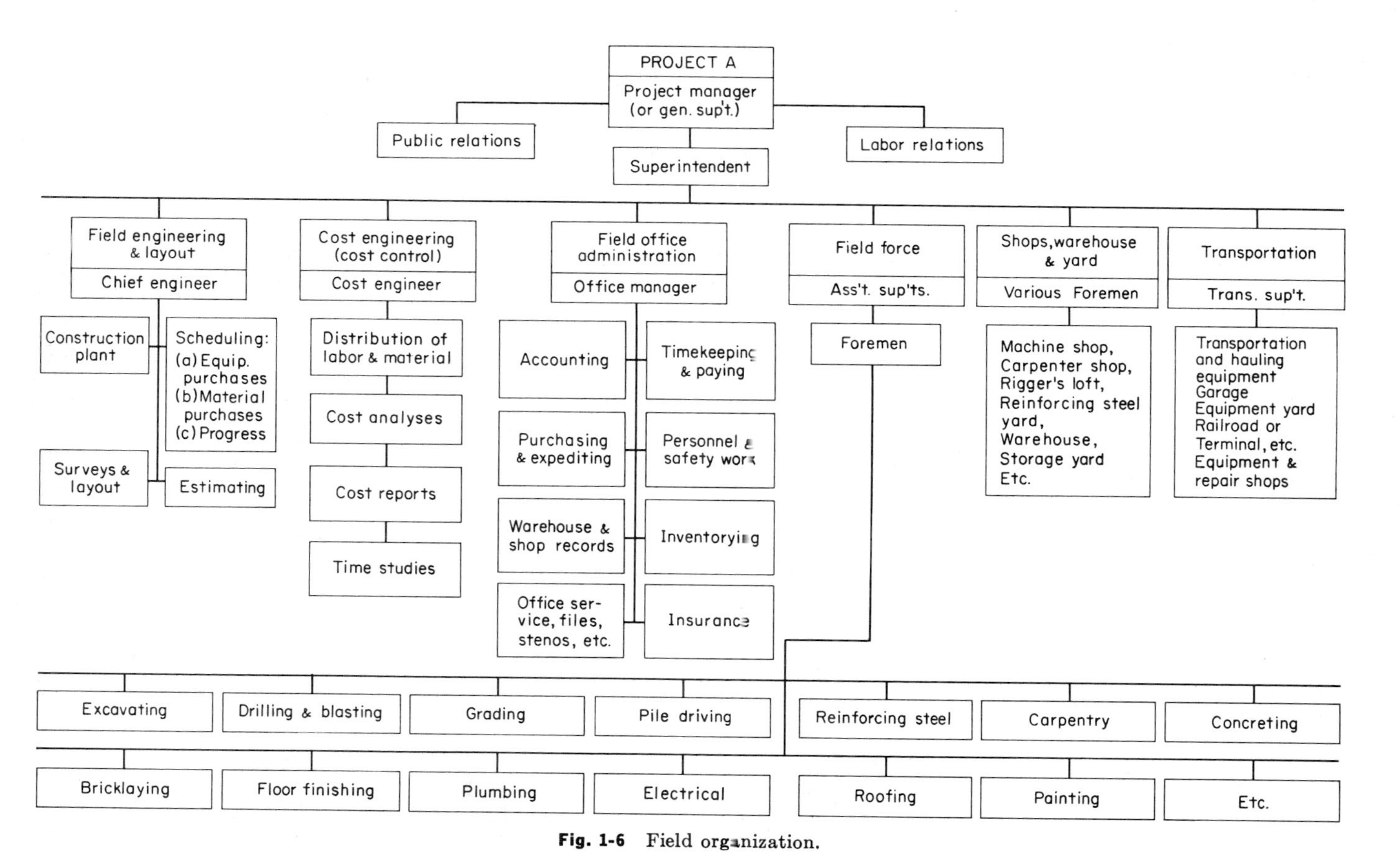

Fig. 1-6 Field organization.

sometimes referred to as the "district office" or "branch office," and might be established along the lines shown in Fig. 1-7. The method and degree of activities and the type of control exercised over and by these offices vary greatly from company to company. Some district offices may function largely as new business or "sales" offices, whereas other district offices operate almost entirely independently, carrying out the same functions in the district that the home office exercises for the company as a whole.

This decentralization has many advantages in that the district office is much closer to the work and can watch the pulse of business activities in the district much better than can the home office. Furthermore, the success or failure of a district office unit can largely be identified as being the responsibility of the district manager and his principal assistants, and the district manager can be charged with the success or failure of the district office. However, in the decentralized office, there may be a certain

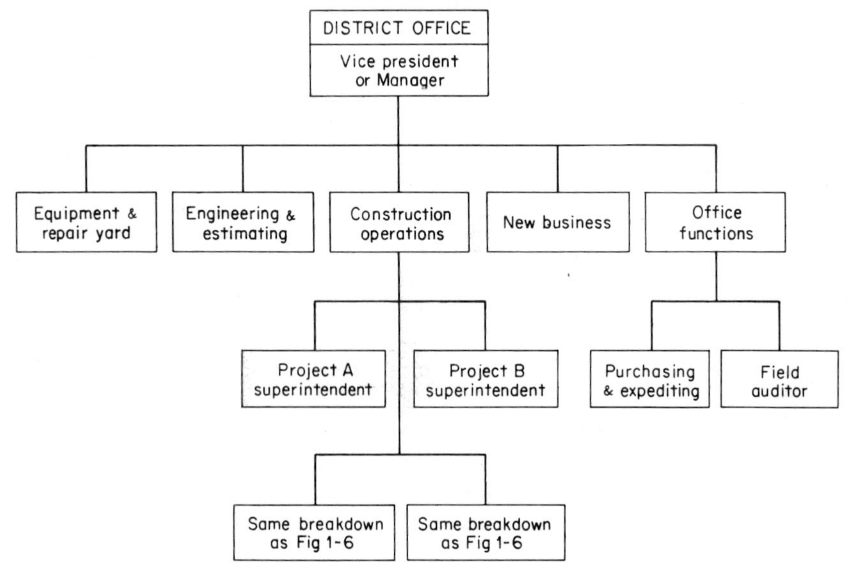

Fig. 1-7 District-office organization.

amount of duplication of organization and effort, which might be more costly as an overhead item. Furthermore, the district manager may sometimes tend to operate independently and consciously or unconsciously refuse the experience and help offered by the home office. These advantages and disadvantages of central authority versus decentralization must be analyzed and weighed in the light of conditions and personalities within each organization. The decision that might be good or advantageous for one organization might have exactly the opposite effect from the standpoint of control and economies in another organization.

The present trend in large companies is toward decentralization, both as to administration and from an operational standpoint.

MANAGEMENT COMMITTEES

Many large organizations have found it advantageous to form various topside management committees that operate on the advisory or sometimes on the policy-making level. Such committees might be named the Executive Committee, the Operating Committee, the Finance Committee, etc. These committees are generally composed of top-echelon officers and may include the chairman of the board or the president as an ex-officio member.

The committees usually function within broadly prescribed powers, often advisory only, and are established for the purpose of setting policy of the company, advising the president and other top management, and handling important matters such as the company's financing and budgeting problems.

An interesting example of the use of the committee is found in the case of the Lukens Management Committee, set up by Lukens Steel Company. This management committee was established because the individual formulation of sales and operating policies by the two subsidiaries and the parent company was causing difficulties. It had become obvious that increased and broader cooperation was necessary in establishing the policies and practices of the parent company and its subsidiaries. The establishment of the committee made it possible to reorganize the company and to allocate fewer responsibilities, especially in operations, thus shortening the span of control of the president.

> From the committee's experience to date, the following major advantages are considered to justify its establishment and continuance:
>
> It offers consultative supervision. This results in uniformity of direction throughout the organization; it helps to provide coordination of long-term and short-term programs toward established objectives; it allows for flexibility in emergency situations, should they arise, without interference with basic schedules. Finally, it provides broader experience for executives and greater interchangeability of management personnel.
>
> Free discussion by members of this group concerning proposed general policies and practices makes for better understanding, improved coordination, and closer cooperation among divisions and departments of the company.
>
> When a recommendation to the board of directors is made by the management committee, the proposal carries much more weight than if it were submitted by an individual executive. The directors know that the recommendation has come to them only after it has been examined and approved by management committee members representing all the important divisions of the company.
>
> The management committee at Lukens has functioned primarily as an advisory body. Line responsibility of division heads is retained for decisions that affect only one division. Decisions affecting more than one division are developed by the committee as a whole.[2]

The American Management Association reports that group action is increasingly characteristic of American management, especially in the larger corporations. There would seem to be no reason why group action cannot also be used to advantage in certain instances in the medium-size companies and organizations. Of the 150 companies surveyed by the American Management Association, 110 reported that they have one or more committees meeting regularly. More than half the companies studied have a general management committee to discuss problems of overall company significance. In most cases, this general committee consists of the working directors of the board and the heads of the major functions of the business. In addition to the general management committees, many companies have functional committees with limited authority over a particular aspect of the business, such as production, sales, or personnel. In addition, frequent informal group meetings are held.

The American Management Association suggests that the major requirements for success of committee operation may be summarized as follows:

1. Work of committees should justify their costs.
2. The principles of effective group action should be applied.
3. Committee mechanics should be arranged so that meetings will not be hampered by procedural difficulties.
4. Only subjects that can be handled better by groups than by individuals should be selected for committee discussion.

REFERENCES

1. Holden, Fish, and Smith, *Top-management Organization and Control*, p. 5, McGraw-Hill Book Company, New York, 1941.

2. Ernest Dale, *Planning and Developing the Company Organization Structure*, p. 14, American Management Association, 1952.
3. Lt. Col. L. Urwick, Executive Decentralization with Functional Coordination, *Management Rev.*, pp. 356, 359, December, 1935.
4. General Sir Ian Hamilton, *The Soul and Body of an Army*, p. 229, Edward Arnold & Co., London, 1921.
5. Lounsbury Fish, *Organization Controls and Executive Compensation*, p. 16, General Management Series No. 142, American Management Association, 1948.
6. Henri Fayol, *General and Industrial Management*, pp. 10–11, Sir Isaac Pitman & Sons, Ltd., London, 1949. Translator Constance Storres.

Section 2

Construction Contracts*

ROBERT W. ABBETT

Partner, Tippetts-Abbett-McCarthy-Stratton, New York, N.Y.

INTRODUCTION

A contract may be defined as an agreement enforceable at law. Only the parties to the agreement are bound by its terms, and no right or liability can accrue directly to a person who is not a party. However, if the contract is made expressly for the benefit of a third person, as for instance in a life insurance contract, such benefit is enforceable at law in a suit initiated by the third person. Indirect or indefinite benefits to a third person are not included in this category and are not necessarily enforceable. Third persons may also acquire rights under a contract, subsequent to its execution, by assignment, succession in office, receivership, and inheritance. Contract laws vary from state to state, and the law governing a specific contract will be determined by the

* The text for this section was adapted largely from Robert W. Abbett, *Engineering Contracts and Specifications*, 4th ed., chap. 11, John Wiley & Sons, Inc., New York, 1963.

place where the contract was made or by the place of residence of the parties unless expressly provided otherwise.

In order to be enforceable at law, a contract must contain the following essential elements:

1. There must be a real agreement or "meeting of the minds."
2. The subject matter must be lawful.
3. There must be a valid consideration.
4. The parties must be legally competent.
5. The contract must comply with the provisions of the law with regard to form.

The absence of any one of these elements is sufficient to void a contract.

TYPES OF CONSTRUCTION CONTRACTS

Like all types of contracts, those for construction work may be written in any one of many different forms, varying from a simple "offer and acceptance" to lengthy legal documents setting forth all the details and technicalities of the work. Standardization is desirable in the interests of economy, and a great deal has been accomplished along this line by professional societies, trade organizations, and various government agencies. As a result of the studies of these organizations, the general form and content of construction contracts are fairly well established for the conditions most likely to arise.

Construction contracts may be classified in two general types: namely, competitive-bid contracts and those which are negotiated with selected contractors. Competitive-bid contracts are usually drawn on a fixed price basis, whereas negotiated contracts usually provide for the owner to pay the actual cost of the work plus compensation to the contractor to cover his overhead expense and profit. Each type is designed to meet the requirements of particular conditions, and there are variations within each classification that should be clearly understood by the engineer.

Competitive-bid Contracts For public-works construction, all who are so inclined may submit a bid, and the contract will be awarded to the lowest responsible bidder. This is required by law for United States government work, and most state and local governments have similar restrictions. Responsibility in this sense usually means that the bidder must have ample working capital, insofar as it can be determined at the time; that he can furnish the required sureties and guarantees; and that his record is free from defaulted contracts and unjustified litigation. Some evidence of previous experience is desirable but is not always required under existing laws. Private construction may be handled in a more informal manner in that the list of bidders may be selected by the owner or recommended by the engineer, and the contract may be awarded to other than the lowest bidder if this appears to be in the best interests of the owner.

Competitive-bid contracts are of two common types. The *lump-sum* form provides for the compensation of the contractor on the basis of a total amount to cover all work and services required by the plans and specifications. The *unit-price* contract includes a breakdown estimate of the number of units of each type of construction and a price for each unit. Inasmuch as the estimate of quantities usually will not be exact at the time the contract is executed, provision should be made for payment to the contractor on the basis of actual quantities, measured in place when the work is completed, at the unit prices specified in the contract. For the purpose of comparing bids, an approximate total cost is determined from the estimated quantities and the unit prices in the bid. This approximate total cost is considered as a lump sum in the determination of the low bidder. With the exception of the method of payment to the contractor, the wording of the lump-sum and unit-price contracts may be similar. Important differences are found, however, in the methods of bidding and accomplishing the work under the two types and also in the wording of the specifications.

The lump-sum contract is indicated where the types of construction are largely standardized and where a variety of operations are required, making it impracticable to break down the work into units. An important percentage of all building construction is accomplished under this form of contract. As a prerequisite, the plans and specifications should be comprehensive and should show in complete detail the requirements of the work. Changes and extra work orders after the contract is signed are expensive and lead to controversies and disputes. Furthermore, when the plans are indefinite, the contractor is forced to gamble on the uncertainties or to increase his bid to cover the worst conditions to be expected. Consequently, high costs to the owner will result if complete information is not furnished. If these hazards are avoided, the owner has the advantage of knowing in advance the exact cost of the work and of having the assurance that the contract will be completed at the earliest practicable date since efficiency and speed in completion will usually tend to increase the contractor's profit.

The unit-price contract has many advantages when the work requires large quantities of relatively few types of construction and the volume of work cannot be exactly determined in advance. It is flexible in that reasonable variations may be made in the amount of required work without formal change orders as long as the changes are restricted to the bid items covered by the contract. The plans and specifications must show the nature and details of the work, but its limits may be left more or less indefinite; the magnitude and scope of the work are approximately indicated by the engineer's estimate. Under these conditions the contractor is not forced to gamble on uncertain conditions, such as depths of underground foundation work, for example, as he would under lump-sum contracts. Most heavy-construction work is accomplished under this type of contract.

Combined lump-sum and unit-price contracts combine the significant features of both lump-sum and unit-price bidding. In general, this type of contract will be indicated when the work includes items the details of which may be defined and broken down into units but which are somewhat indefinite as to quantities. For work of this character, unit-price payment is advantageous. If the same project includes specialty items or structures involving a large number of types of construction, the requirements of which are exactly known and for which the contractor can prepare an accurate estimate and bid, payment on a lump-sum basis will be simpler for everyone concerned. Thus, the contractor's bid will include lump-sum amounts for some items and estimated quantities with corresponding unit prices for others. The combined type is well suited for building construction, for instance. Foundations can be covered satisfactorily on a unit-price basis, whereas the superstructure is well adapted to lump-sum bidding.

Contract documents are prepared to comply with the procedure for setting up a contract, which is rigidly defined by law for public-works projects. The same procedure is recommended for privately sponsored projects, although some of the formalities may be omitted. The prescribed procedure for both lump-sum and unit-price contracts requires the preparation of the following documents:

1. *Advertisement,* or *notice to contractors,* which gives a brief description of the work and invites bids from contractors for its accomplishment.
2. *Instructions to bidders,* which furnish detailed information required for the preparation of the bid or proposal.
3. *Form of bid* or *proposal.* This document is desirable in order that all bids will be uniform and prepared on a common understanding.
4. *Contract,* the legal document binding on the signing parties, which states the work to be done, the compensation for the contractor, the time of completion, and the other legal and business features of the project.
5. *Specifications,* which cover the engineering or technical requirements of the work and describe the details of construction.
6. *Plans,* which show the physical details of the work.

The contract documents, excepting drawings, may be typed, mimeographed, or printed. It is standard practice to bind them all into one volume which, together with the plans for the work, is available to contractors for bidding and construction purposes.

Negotiated Contracts In contrast to competitive-bid contracts in which the award is the result of open competition, negotiated contracts are awarded to selected contractors after a study of their qualifications, previous experience, and facilities. Since competitive bidding usually is required by law for public-works projects, negotiated contracts are normally restricted to private construction. They have many advantages under certain conditions and, during periods of national emergency or war, legal restrictions against negotiated government contracts may be removed by special legislation in order that their advantages may be available for the construction of defense facilities.

The fact that a contract is negotiated has little bearing on its form. That is to say, lump-sum and unit-price contracts may be negotiated as readily as any other type. In general, however, negotiated contracts are based on the premise that the owner will pay the actual cost of the work plus compensation for the services, facilities, and technical knowledge of the contractor. It is only in the provision for the contractor's compensation that the various types of cost-plus contracts differ materially. This arrangement places the contractor more or less on a professional basis, where he represents the owner in matters relating to the work. Under these conditions, the contractor's compensation is of the nature of a professional fee rather than the profit anticipated in a business transaction.

Under the common forms of negotiated contracts, the contractor furnishes the working capital required to finance the construction. Later he is reimbursed in the full amount of the actual cost of the construction plus his compensation for the management, coordination, and accomplishment of the work. Therefore the terms of the contract must anticipate methods for the control of expenditures and for determining the actual cost. This is particularly true of certain items of the contractor's overhead and indirect general expenses in his central office. These are difficult to determine exactly, and an agreement as to some approximate method for their distribution should be made. As an alternative, they may be eliminated altogether as items of cost and the contractor's compensation may be increased a reasonable amount to cover them. This would be a matter for negotiation, which is probably the most satisfactory method for handling the problem.

In order to control expenditures, provision should be made for the engineer to approve all payrolls and purchases prior to the placing of orders. Salaries of key men and executives should also be subject to approval. Wage rates will be those prevailing in the community and this will be stipulated in the contract. The determination of the actual cost of the work can be made only through an adequate cost accounting system, and the accounting methods of the contractor should be subject to the approval of the owner or the engineer.

Cost-plus-percentage-of-cost contracts are undoubtedly the oldest form of the negotiated type of contract and provide for the contractor's profit on the basis of a fixed percentage of the actual cost of the work. In common with all negotiated contracts, this type permits the beginning of construction before the plans are completely developed. This can result in an important saving of time in the completion of urgent projects, and the owner may make any desired changes in the plans and specifications as the work progresses. It has the disadvantage to the owner that the contractor's compensation is increased by an increase in construction cost. Therefore, there is no incentive for the contractor to economize during construction.

Cost-plus-fixed-fee contracts provide for payment to the contractor of the cost of the work plus a fixed amount as a fee for the accomplishment of the work. The amount of the fee is determined from a consideration of the character and scope of the work and its estimated cost. Thereafter the fee remains fixed, although the actual cost of the work may vary considerably from the estimate. In order to negotiate such a contract it is necessary for the scope of the work to be clearly defined and for both parties to agree on the amount of the estimate. This type has had wide usage on government work during war emergencies and has also been used extensively in the construction of large, privately owned buildings for which accurate bidding estimates are difficult to prepare.

From the point of view of the relation between the owner, the engineer, and the contractor, the cost-plus-fixed-fee contract more nearly approaches the ideal construction contract than any yet devised. The contractor is selected on the basis of merit and paid a fixed amount for completing specified work. There is no incentive for him to inflate the construction cost, since his fee would not be affected, and he is free in every respect to act in the best interests of the owner.

On the other hand, there is no specific incentive for the contractor to exert his best efforts toward efficiency, and the cost of the work may be increased thereby. This is to be watched, particularly in the latter stages of a large construction project when frequently there is a tendency to string out the final operations and the completion of the contract. When this happens, excessive overhead costs are to be expected.

Cost-plus-fixed-fee contract with a profit-sharing clause provides an added incentive to the contractor to keep the cost of the work at a minimum. Ordinarily the amount of the fixed fee is based on a preliminary estimate of the cost of the work. The profit-sharing provision allows the contractor to receive a share of any saving if the actual cost should be less than the original estimate. The contractor's share of the amount saved is in addition to the fixed fee.

Cost-plus-fixed-fee contract with a bonus clause may be used when the completion of the work is urgently desired by the owner. A bonus provision may then be written into the contract as an incentive to the contractor to reduce the time of construction to a minimum. This bonus is usually in the form of a fixed amount to be paid to the contractor, in addition to his fee, for each day on which the owner has full use of the completed work before the originally estimated date of completion. The contract may also provide for the assessing of a fixed amount as liquidated damages against the contractor for each day after the originally estimated date of completion required by the contractor to finish the work.

Cost plus a sliding scale of fees is another variant of the cost-plus-fixed-fee type of contract and has been widely used for construction work. It provides for the contractor's fee to change proportionately to the actual cost of the work, in accordance with a sliding scale of fixed fees. As in the profit-sharing type, this may be used as an incentive to the contractor to reduce the cost of the work by allowing the fee to be increased in fixed amounts for various increments of cost less than the original estimate and decreased in fixed amounts for various increments of cost greater than the original estimate. In the latter case it is customary to arrive at a minimum fee which would remain unchanged with further increases in the cost of the work, thus guaranteeing a minimum profit to the contractor.

This type of contract also may be considered a compromise between the cost-plus-fixed-fee and cost-plus-percentage forms by allowing the contractor's compensation to increase with the cost of the work in accordance with the desired sliding-scale rates for various increments of cost in lieu of a constant percentage of the cost.

Cost-plus contract with a guaranteed ceiling price places a maximum limit on the cost of the work to the owner and thus removes one of the principal arguments against cost-plus types of contracts. The owner now knows in advance what the maximum cost of his work will be. The contractor is reimbursed for the actual cost of the work plus his fixed fee, provided that the total amount does not exceed the maximum limit established in the contract. If the total amount should exceed the maximum limit, the contractor is held responsible for the excess and receives no compensation over the guaranteed ceiling price. This type of contract removes some of the uncertainties from the ordinary cost-plus contract but requires that the plans and specifications for the work be sufficiently developed to permit the establishment of a reasonable ceiling price.

Management contracts are used when the contractor is retained in a managerial capacity only. The entire work is then supervised by the contractor, including the purchase of materials, hiring of labor, and letting of subcontracts. He may or may not elect to do parts of the work with his own forces. If management services only are intended, it should be so stated in the contract.

Management agreements are not to be confused with so-called "brokerage contracts" in which the contractor, without the owner's prior knowledge, enters into a contract with the intention of accomplishing all or most of the work by subcontracts with the view of collecting a profit on the work of each subcontractor. These are regarded with disfavor and should not be permitted. Management contractors, to the contrary, are selected on a professional basis for their honesty, technical skill, and administrative ability. Distinction also should be made between management and agency in this connection. Under an agency stipulation the owner is legally bound by the acts of the contractor, whereas the acts of an independent contractor in connection with the work are his own responsibility. Under an independent contractor relationship the owner does not assume responsibility for the work until it is completed and formally accepted. The contract should be explicit in this connection.

Miscellaneous Contracts There are some forms of contract which may be either negotiated or competitively bid, or which may combine both features within a single contract. A few of these contracts will be briefly discussed.

Combined engineering and construction contracts are employed when an owner who contemplates a construction project desires to deal with only one party for all services, both engineering and construction, in connection with the work; this is a so-called "turn-key" or "package" job. This procedure is, in fact, standard practice in most European and South American countries. Some United States companies also operate in this manner. The contract may be drawn either on a firm-price or cost-plus basis, and all planning, design, plans, specifications, and construction services are included under one contract.

Under a combined engineering and construction contract the engineer is automatically allied with the contractor. The engineer should have it clearly understood that he will be guided by the principles of engineering practice and professional ethics in carrying out his duties under the contract. This should be done irrespective of whether he expects to work as a coventurer or as an employee of the contractor.

Joint-venture contracts are used when an extraordinarily large construction project is to be accomplished under a general contract and requires a greater concentration of financial, administrative, and technical resources than can be mobilized by one construction company. This has led to the development of the joint-venture type of contract in which several firms combine their assets, plant, and personnel to undertake such a project. A joint venture is similar to an ordinary partnership or corporation in that each party in the combination shares in the work, risks, and profits or losses of the contract in accordance with the terms of the joint-venture agreement. It differs from a partnership or corporation in that it is usually restricted to one contract and is not a continuing relationship.

A joint venture of several firms may enter into any of the usual types of construction contracts acceptable to the owner. As a prerequisite there should be an independent written agreement between the joint venturers which gives a clear arrangement for the financing and management of the work under the contract and the manner in which the risks and profits or losses are to be shared. The joint-venture agreement should be subject to the approval of the owner and may be made a part of the construction contract if desired.

The joint-venture type of contract came into prominence in the construction of large dam projects, notably Hoover Dam, and has been increasing in popularity. Joint ventures are now common practice in the construction industry in the United States.

Incentive-type contracts for work outside the United States have been used to make specific projects outside the continental limits of the United States attractive to American contractors. There are many uncertainties in this class of work and, therefore, the contracts are usually of the cost-plus type. Typical of the hazards to be encountered in overseas work are unstable foreign currencies, remote and frequently primitive living conditions for American personnel, limited and untrained native labor supply, and difficulties of transportation and communication. Work under these conditions is not susceptible of the efficient administrative control to be expected of contractors on similar work in the United States. Unless a strong incentive factor is provided in the contract in addition to the fee, construction costs are likely to be relatively high.

To give the contractor additional incentive to operate the project at maximum efficiency and minimum cost and to penalize him for inefficient work under these conditions, a cost-plus-base-fee contract with an incentive provision has been developed. Also, an element of competitive bidding has been incorporated which is helpful in selecting the most competent contractor for the work. It also determines the amount of the fee and establishes the basis of the incentive adjustment of the fee.

In practice, prospective bidders are invited to submit their qualifications to the engineer, and those found unsuitable for the work are eliminated from the bidding. Prequalification of bidders places a heavy responsibility on the engineer, and great care is necessary in this step of the procedure. Among the factors involved in establishing contractors' qualifications are experience in the type of construction required, experience in the locality where the work is to be performed, financial resources, quality of personnel and organization, and construction equipment available.

Contractors are instructed to prepare bids upon the basis that all costs will be reimbursed as for the ordinary cost-plus-fixed-fee contract. Each bidder prepares his own estimate of the cost of the work, and this is called the *target estimate.* The bidder also proposes the base fee for which he offers to do the work.

If the actual cost of the work turns out to be less than the target estimate of the successful bidder, his fee is increased. If the actual cost of the work exceeds the target estimate, the base fee of the contractor is reduced.

In analyzing the bids and awarding the contract, weighted consideration is given to the contractor's qualifications, job analysis, target estimate, and proposed base fee. The form of the contract is identical to the ordinary cost-plus-fixed-fee type excepting the target estimate, base fee, and fee-adjustment clauses. In addition, provision must be made for adjustment of the target estimate to recognize changes in wage rates, material prices, and other factors beyond the control of the contractor during the life of the contract.

This type of contract is unnecessarily complex and speculative for projects within the United States, and therefore it is not recommended for domestic work.

IMPLEMENTATION OF THE CONTRACT

Selection of Type of Contract Though the selection of the contract form is, strictly speaking, a function of the owner, it usually devolves on the engineer to furnish recommendations on which the owner may base his decision. In making a decision the fundamental differences between competitive-bid and negotiated contracts should be clearly understood and kept constantly in mind. A competitive-bid contract is a straightforward business transaction in which the owner takes his chances in an open market. He agrees to pay a specific price for a definite service, and constant inspection and supervision are required to make certain that the service is completed in the manner prescribed in the contract. In contrast, under the usual types of negotiated contracts, the owner and the contractor are partners with a common objective, namely, to complete the work in accordance with the owner's requirements as quickly and as economically as possible. In this case emphasis is on the control of expenditures which are reimbursable to the contractor, as well as on inspection and supervision of his workmanship.

It will be advantageous to the owner to adopt a competitive-bid contract when sufficient time is available to work out the plans and specifications in detail, thus minimizing costly changes and extra work orders at a later date; when activity in the construction industry is at a low ebb, thus ensuring keen competition and lower bids; and when some control can be exercised on the quality and character of the bidders.

A negotiated contract will be indicated when it is desirable to begin work before the completion of detail plans and specifications, thus assuring earlier completion of the work; when the requirements of the project cannot be determined definitely in advance of the early phases of construction; or when the nature of the project is such that an accurate estimate cannot be made for bidding purposes.

When a decision has been made between a competitive-bid and a negotiated contract, an analysis of the unique features of the project will usually indicate the most suitable variation to adopt. This selection will be as significant as that determining the general type.

When a competitive-bid contract is selected, a decision must be made between the lump-sum and unit-price types. This decision will be based upon the type of construction involved. In general, the lump-sum contract is used for building construction, equipment, and machinery. The unit-price form is used for most kinds of engineering construction, such as bridges, dams, highways, and foundations. Local conditions may lead to exceptions to these rules, however.

When the owner decides upon the use of a negotiated contract with a selected contractor, the choice of the type of contract will usually be restricted to one of the variations of the cost-plus-fixed-fee form, in view of the disfavor in which the cost-plus-percentage type is held by most engineers. The particular variant to be adopted will depend on local conditions or personal preference. It should be noted, however, that practically all these variations introduce additional speculative elements into the contract as compared with the straightforward fixed-fee basis and in most instances complicate the payment provisions of the contract unnecessarily. An exception to this rule is found in construction work to be performed in remote and primitive localities. For these conditions an incentive-type contract is indicated.

It should be noted that a contract may be negotiated on a lump-sum or unit-price basis as well as on a cost-plus basis. Ordinarily, however, this would require relatively complete plans and specifications, and in most instances it would be to the owner's advantage to obtain competitive bids.

Subcontracts Generally all subcontracts for specialized phases of construction work are awarded on a competitive-bid basis irrespective of the type of the prime contract, although some general contractors deal with favored subcontractors on a negotiated basis. Subcontracts may be of the lump-sum type, the unit-price type, or a combination of the two types, as in the case of competitive-bid prime contracts. Subcontracts are the responsibility of the prime contractor, and the basis on which they are drawn is of no concern to the engineer and the owner under competitive-bid prime contracts except that all subcontracts are subject to approval by the engineer. Under cost-plus prime contracts, however, all subcontracts are items of reimbursable cost, and therefore their amounts and terms should be closely scrutinized.

Changes and Extra Work Frequently, in the course of construction work, it becomes necessary to make changes in the plans and to introduce additional work which was not contemplated at the time of the award of the contract. In a similar manner, it may become desirable to omit certain parts of the project which were included in the contractor's bid. In either case it is necessary to modify the contract by the issuance of a change order. This should always be done in writing, and no changes should ever be undertaken on the basis of verbal orders except in an emergency. The change order may be in letter form and should state explicitly the work required and the price to be paid to the contractor if the change involves any additional cost or the credit to be allowed the owner if the change involves a reduction in cost.

The method of arriving at the cost of changes and extra work should be given careful consideration. If practicable, an agreement with the contractor as to the amount should be made before he is authorized to proceed, in order to avoid disputes after the work is completed. The following are the usual methods of payment for changes:

1. A lump-sum amount to cover all costs and profit to the contractor
2. Unit prices to cover all costs and profit to the contractor
3. The actual cost of the change plus a fixed fee determined in advance of the work on the basis of the estimated cost
4. The actual cost of the change plus a percentage of the cost to cover the contractor's overhead and profit

Methods for the determination of the costs of changes may be established in the contract on the basis of the contractor's actual expenses plus a specific percentage, or procedures for arriving at the costs by negotiation may be specified as described above. When changes are anticipated, unit prices for additions and deductions to the quantities in the engineer's estimate may be obtained in the original bid.

The adjustment to which a contractor is entitled when a change order is issued includes an appropriate change in the time for completion of the contract, and the extension or reduction in time should be covered in the negotiations for the change in price. This is important, particularly when liquidated damages are involved.

The amounts involved in changes and extra work are contractual matters to be negotiated between the owner and the contractor when not provided for in the original contract, inasmuch as it is usually impracticable to request bids from others.

As an alternative to the issuance of a change order, a contract may be modified to cover changes and extra work by the negotiation of a supplemental agreement. The differences between the two methods are not always clear. Most contracts permit the owner to order changes in the work at will, and the contractor is obliged to execute such changes when properly ordered and to accept equitable adjustments in the contract price if indicated. On the other hand, courts have held that a change order may not be used to modify the essential nature of the work, such as, for example, the change to an earth dam under a contract which originally called for the construction of a concrete dam. A contractor would not be obliged to accept such a change except under conditions satisfactory to him under a supplemental agreement.

It is essential that every construction contract shall permit the owner to order extra work or make changes by altering, adding to, or deducting from the work at any time without invalidating the contract or the contractor's bond. All such changes should be executed under the conditions of the original contract except that the cost of the extra work or changes and any claims for extension of time caused thereby should be adjusted at the time the change is ordered.

Claims and Disputes One of the most troublesome aspects of construction contracting practice is the matter of claims for extra costs and extensions of time by the contractor and the disputes that result therefrom. Along with changes and extra work orders, the most frequent causes of controversies and disputes are changed conditions at the site, delays due to acts of the owner and other contractors, delays due to events beyond the control of the contractor, and conflicting or ambiguous statements in the contract documents. Much can be accomplished in avoiding disputes by anticipating the common sources of claims and providing for their resolution in the contract.

Latent or subsurface conditions frequently cannot be determined accurately before the beginning of construction. On private work the contractor is usually obliged to accept full responsibility for the conditions encountered, it being stated that the information contained in the contract documents relative to these conditions is the best obtainable; but the owner accepts no responsibility as to its accuracy, and the contractor is to use it at his own risk. United States government contracts, on the other hand, allow an equitable adjustment in the contract amount if actual subsurface conditions are found to be materially different from those indicated on the plans.

The contract should be explicit as to the responsibility for delays to the contractor caused by acts of the owner or by other contractors employed on the work. Typical of such conditions are the failure of the owner to make the site or access to the site available to the contractor in accordance with the anticipated schedule and the failure of another contractor employed on the site to perform his work in accordance with the anticipated schedule. It is usually required that the contractor be responsible for the coordination of his own work with the requirements of the owner and other contractors, and any delays that result from his failure to do so constitute normal hazards of the contract. Accordingly, such delays are not sufficient grounds for an extension of time, and the contractor is required to bear any increased costs resulting from such delays. Many contractors protest that this is an unfair provision and maintain that a contractor is entitled to an extension of time without assessment of liquidated damages when he is delayed or otherwise suffers loss because of acts of the owner or another contractor which are beyond his control. There is much to be said for this argument, and it appears that the most equitable solution is to permit extensions of time when the contractor suffers unreasonable delay because of acts of the owner or other contractors. While courts will usually hold the owner fully responsible if he should be guilty of some act of negligence or willful misconduct which results in delays and damages to the contractor, all contractors engaged on the site should be required to hold the owner harmless from any damages resulting from their failures to coordinate their own work schedules.

The contractor should be entitled to an extension of time to cover delays to the work caused by events beyond his control, such as unavoidable delays in transportation, strikes, floods, and like events which fall in the general classification of "acts of God." Bad weather normally does not justify an extension of time unless it can be shown conclusively that such weather is unprecedented and could not have been anticipated as a normal hazard of the contract. Fires and accidents may indicate an extension of time when they are beyond the control of the contractor and his own negligence is not involved.

Claims arising from conflicting and ambiguous statements in the contract documents are particularly hazardous to the owner. Courts generally favor the contractor in such disputes, inasmuch as the owner is responsible for the preparation of plans and specifications.

Mistakes and omissions in the plans and specifications, impossible requirements in the specifications, and errors in placing line and grade stakes usually lead to claims by the contractor for which the owner is responsible.

The accurate determination of equitable damages in connection with claims is always difficult, and it is never possible to foresee all conditions which may lead to disputes in construction work. When methods for their resolution are not provided in the contract, claims are best settled by direct negotiations between the owner and the contractor with the assistance of the engineer. If negotiations fail, the only alternatives are litigation and arbitration.

Renegotiation of Contracts and Subcontracts During World War II, a congressional act provided for the renegotiation of any contract or subcontract on government war projects amounting to more than $100,000 to determine whether the profits or costs were excessive. If found to be excessive, the contractor was required to return the excess to the government. Renegotiation proceedings were held before a board of experts and were initiated at the request of the government. When renegotiation was initiated in connection with one war contract, costs and profits under all contracts between that contractor and other government agencies were considered in combination and the total profits from all work were considered in arriving at the amount of the excess. This procedure is contrary to the principles of contract law according to which the court does not pass on the sufficiency of a bargain. However, it was enforceable during the emergency on the grounds that no one should be allowed to profit excessively because of the war. In 1950 the act was reestablished on a continuing and more or less permanent basis. This law is restricted to defense contracts and does not apply in normal practice.

SUGGESTED FORM OF CONTRACT, ASCE-AGC, FOR USE IN CONNECTION WITH ENGINEERING CONSTRUCTION PROJECTS*

Prepared jointly by and reproduced with the permission of th
American Society of Civil Engineers and
The Associated General Contractors of America, Inc.

1966 EDITION

FOREWORD

These Agreement Forms and the Standard General Conditions of Contract which accompany them pertain to Engineering Construction Contracts between Private or Public

Owners and Contractors. They are intended as a guide in the preparation of Contract Documents for such construction and are subject to change or modification to suit particular conditions.

Ordinarily, a complete set of Contract Documents consists of the following:

(*a*) Advertisement, or Notice to Contractors
(*b*) Instructions to Bidders
(*c*) Form of Bid or Proposal
(*d*) Contract (Agreement)
(*e*) General and Special Conditions of Contract
(*f*) Specifications
    1. General
    2. Special
(*g*) Drawings

The first three items are essential to all contracts that are bid competitively. However, the last four are those essential to the actual Contract. The Forms and Standards contained herein pertain to Items (*d*) and (*e*).

The Contract Forms cover three alternative bases of payment to the Contractor:

1. A Lump Sum Basis
2. A Unit Price Basis
3. A Cost-Plus Basis

The Contracting Parties will select the desired form.

The Contract Form for a Cost-Plus Basis of Payment is based upon a payment consisting of Cost Plus a Percentage Fee. Two other types are commonly used: Cost Plus a Fixed Fee and Cost Plus a Percentage or Fixed Fee with a guaranteed cost limit. The text of Articles III and IV can easily be adapted to either of the other two types of Cost-Plus bases of payment. Sometimes, in a Cost-Plus Contract with a guaranteed cost limit, provision is made for dividing equally between the Owner and Contractor any savings below the cost limit that may be achieved in the execution of the work.

Form of Agreement for Engineering Construction

Unit Price Basis

This Agreement, made on the-------------- day of ------------------------, 19----,

by and between --

party of the first part, hereinafter called the OWNER, and --------------------------

----

----

party of the second part, hereinafter called the CONTRACTOR.

It is understood ENGINEER representing Owner shall be -------------------------

----

Witnesseth, That the Contractor and the Owner, for the considerations hereinafter named, agree as follows:

Article I—*Scope of the Work*

The Contractor hereby agrees to furnish all of the materials and all of the equipment and labor necessary, and to perform all of the work shown on the drawings and described in the specifications for the project entitled -------------------------------------

----

----

all in accordance with the requirements and provisions of the following Documents which are hereby made a part of this Agreement:

(*a*) Drawings prepared for same by____

numbered____

and dated________________________________, 19____.

(*b*) Specifications consisting of:

1. "Standard General Specifications" issued by___________________________________

______________________________, _________________________________Edition

2. "Special Conditions" as prepared by_______________________________________

dated______________________________.

3. The "General Conditions of Contract for Engineering Construction"—1966 Edition.

4. Addenda

No.________________________________ Date______________________________

ARTICLE II—*Time of Completion*

(*a*) The work to be completed under this Contract shall be commenced within___________ calendar days after receipt of notice to proceed.

(*b*) The work shall be completed within___________calendar days after receipt of notice to proceed.

(*c*) Failure to complete the work within the number of calendar days stated in this Article, including extension granted thereto as determined by Section 19 of the General Conditions, shall entitle the Owner to deduct from the moneys due to the Contractor as "Liquidated Damages" an amount equal to $_______________________for each calendar day of delay in the completion of work.

(*d*) If the Contractor completes the work earlier than the date determined in accordance with Paragraph (*b*), and the Engineer shall so certify in writing, the Owner shall pay the Contractor an additional amount equal to $_______________________for each calendar day by which the time of completion so determined has been reduced.

ARTICLE III—*The Contract Sum*

(*a*) The Owner shall pay to the Contractor for the performance of the work the amounts determined for the total number of each of the following units of work completed at the unit price stated thereafter. The number of units contained in this schedule is approximate only, and the final payment shall be made for the actual number of units that are incorporated in or made necessary by the work covered by the contract.

Item No.	*Classification*	*Estimated No. of Units*	*Unit*	*Unit Price Bid*	*Total for Item*

(*b*) Should the number of units of completed work of any individual item of the above schedule vary by more than________% from the number of units stated in such schedule of units, either the Owner or the Contractor may request a revision of the unit price for the item so affected, and both parties agree that under such conditions an equitable revision of the price shall be made.

(*c*) Changes in the work made under Section 18 of the General Conditions, and not included in Article I, that cannot be classified as coming under any of the Contract units may be done at mutually agreed-upon unit prices, or on a lump sum basis, or under the provisions of Article V "Extra Work."

ARTICLE IV—*Progress Payments*

The Owner shall make payments on account of the Contract as follows:

(*a*) On not later than the fifth day of every month the Contractor shall present to the Engineer an invoice covering the total quantities under each item of work that has been completed from the start of the job up to and including the last day of the preceding month, and the value of the work so completed determined in accordance with the schedule of unit prices for such items together with such supporting evidence as may be required by the Engineer. This invoice shall also include an allowance for the cost of such material required in the permanent work as has been delivered to the site but not as yet incorporated in the work. Measurements of units for payment shall be made in accordance with the Special Conditions of the Contract.*

(*b*) On not later than the 15th of the month, the Owner shall pay to the Contractor 90 percent of the amount of the invoice—less previous payments made. The 10 percent retained percentage may be held by the Owner until the value of the work completed at the end of any month equals 50 percent of the total amount of the Contract after which, if the Engineer finds that satisfactory progress is being made, he shall recommend that all of the remaining monthly payments be paid in full. Payments for work, under Subcontracts of the General Contractor, shall be subject to the above conditions applying to the general Contract after the work under a Subcontract has been 50 percent completed.

(*c*) Final payment of all moneys due on the contract shall be made within 30 days of completion and acceptance of the work.

(*d*) If the owner fails to make payment as herein provided, or as provided in Article V(*d*), in addition to those remedies available to the Contractor under Section 25 of the General Conditions, there shall be added to each such payment daily interest at the rate of 6 percent per annum commencing on the first day after said payment is due and continuing until the payment is delivered or mailed to the Contractor.

ARTICLE V—*Extra Work*

If the Engineer orders, in writing, the performance of any work not covered by the Drawings or included in the Specifications, and for which no item in the Contract is provided, and for which no unit price or lump sum basis can be agreed upon, then such extra work shall be done on a Cost-Plus-Percentage basis of payment as follows:

(*a*) The Contractor shall be reimbursed for all costs incurred in doing the work, and shall receive an additional payment of________% of all such cost to cover his indirect overhead costs, plus________% of all cost, including indirect overhead, as his fee.

(*b*) The "Cost of the Work" shall be determined as the net sum of the following items:

1. Job Office and all necessary temporary facilities such as buildings, use of land not furnished by the Owner, access roads and utilities. The costs of these items include construction, furnishings and equipment, maintenance during the period that they are needed, demolition and removal. Salvage values agreed on or received by the Contractor shall be credited to the Owner.

2. All materials used on the work, whether for temporary or permanent construction.

3. All small tools and supplies; all fuel, lubricants, power, light, water and telephone service.

4. All plant and equipment at specified rental rates and terms of use. If the rental rates do not include an allowance for running repairs and repair parts needed for ordinary maintenance of the plant and equipment, then such items of cost are to be included in the Cost of the Work.

5. All transportation costs on equipment, materials and men.

6. All labor for the project and including the salaries of superintendents, foremen, engineers, inspectors, clerks and other employees while engaged on the work but excluding salaries of general supervisory employees or officers, who do not devote their full time to the work.

7. All payroll charges such as Social Security payments, unemployment insurance, workmen's compensation insurance premiums, pension and retirement allowances, and social insurance premiums, vacation and sick-leave allowances applicable to wages or salaries paid to employees for work done in connection with the contract.

* In addition to advance payment for materials delivered to site, wording should indicate that, where applicable, advance payment may be in order for materials in storage away from the site, for field plant and equipment, access roads, etc.—details to be spelled out in the Special Conditions.

8. All premiums on fire, public liability, property damage or other insurance coverage authorized or required by the Engineer or the Owner, or regularly paid by the Contractor in the conduct of his business.

9. All sales, use, excise, privilege, business, occupation, gross receipt and all other taxes paid by the Contractor in connection with the work, but excluding state income taxes based solely on net income derived from this contract and Federal income taxes.

10. All travel or other related expense of general supervisory employees for necessary visits to the job excluding expenses of such employees incurred at the Home Office of the Contractor.

11. All Subcontracts approved by the Engineer or Owner.

12. (Insert other costs proper for inclusion in this Contract.)

    a. --

    b. --

    c. --

13. Any other cost incurred by the Contractor as a direct result of executing the Order, subject to approval by the Engineer.

14. Credit to the Owner for the following items:

    a. Such discounts on invoices as may be obtainable provided that the Owner advances sufficient funds to pay the invoices within the discount period.

    b. The mutually agreed salvage value of materials, tools or equipment charged to the Owner and taken over by the Contractor for his use or sale at the completion of the work.

    c. Any rebates, refunds, returned deposits or other allowances properly credited to the Cost of the Work.

(*c*) The cost of the work done each day shall be submitted to the Engineer in a satisfactory form on the succeeding day, and shall be approved by him or adjusted at once.

(*d*) Monthly payments of all charges for Extra Work in any one month shall be made in full on or before the 15th day of the succeeding month. Those payments shall include the full amount of fee earned on the cost of the work done.

IN WITNESS WHEREOF the parties hereto have executed this Agreement, the day and year first above written.

----------------------------------OWNER

WITNESS:

-------------------------------------- By: -----------------------------------
Title

------------------------------CONTRACTOR

WITNESS:

-------------------------------------- By: -----------------------------------
Title

FORM OF AGREEMENT FOR ENGINEERING CONSTRUCTION

Cost-plus Basis

THIS AGREEMENT, made on the-------------day of-----------------------, 19----,

by and between--

party of the first part, hereinafter called the OWNER, and____________________________

____

____

party of the second part, hereinafter called the CONTRACTOR.

It is understood ENGINEER representing Owner shall be___________________________

____

____

WITNESSETH, That the Contractor and the Owner, for the considerations hereinafter named, agree as follows:

ARTICLE I—*Scope of the Work*

The Contractor hereby agrees to furnish all of the materials and all of the equipment and labor necessary, and to perform all of the work shown on the Drawings and described in the specification for the project entitled______________________________________

____

____

all in accordance with the requirements and provisions of the following Documents which are hereby made a part of this Agreement:

(*a*) Drawings prepared for same by____

numbered____

and dated________________________________, 19____.

(*b*) Specifications consisting of:

1. "Standard General Specifications" issued by_________________________________

_________________________, ____________________________________Edition.

2. "Special Conditions" as prepared by_____________________________________

___________________________ dated_____________________________________

3. The "General Conditions of Contract for Engineering Construction"—1966 Edition.

4. Addenda

No.________________________________ Date______________________________

ARTICLE II—*Time of Completion*

(*a*) The work to be completed under this Contract shall be commenced within____________ calendar days after receipt of notice to proceed.

(*b*) The work shall be completed within____________calendar days after receipt of notice to proceed.

(*c*) Failure to complete the work within the number of calendar days stated in this Article, including extension granted thereto as determined by Section 19 of the General Conditions, shall entitle the Owner to deduct from the moneys due to the Contractor as "Liquidated Damages" an amount equal to $________________________for each calendar day of delay in the completion of work.

(*d*) If the Contractor completes the work earlier than the date determined in accordance with Paragraph (*b*), and the Engineer shall so certify in writing, the Owner shall pay the Contractor an additional amount equal to $________________________for each calendar day by which the time of completion so determined has been reduced.

ARTICLE III—*The Contract Sum*

(*a*) The Owner shall pay the Contractor for the performance of the work an amount equal to the actual "Cost of the Work" as defined below, plus-------% of all costs to reimburse the Contractor for indirect overhead and general supervision, plus-------% of all costs, including indirect overhead and general supervision, as commission or profit.

(*b*) The "Cost of the Work" shall be determined as the net sum of the following items:

1. Job Office and all necessary temporary facilities such as buildings, use of land not furnished by the Owner, access roads and utilities. The costs of these items include construction, furnishings and equipment, maintenance during the period that they are needed, demolition and removal. Salvage values agreed on or received by the Contractor shall be credited to the Owner.
2. All materials used on the work whether for temporary or permanent construction.
3. All small tools and supplies; all fuel, lubricants, power, light, water and telephone service.
4. All plant and equipment at specified rental rates and terms of use. If the rental rates do not include an allowance for running repairs and repair parts needed for ordinary maintenance of the plant and equipment, then such items of cost are to be included in the Cost of the Work.
5. All transportation costs on equipment, materials and men.
6. All labor for the project and including the salaries of superintendents, foremen, engineers, inspectors, clerks and other employees while engaged on the work but excluding salaries of general supervisory employees or officers, who do not devote their full time to the work.
7. All payroll charges such as Social Security payments, unemployment insurance, workmen's compensation insurance premiums, pension and retirement allowances, and social insurance premiums, vacation and sick-leave allowances applicable to wages or salaries paid to employees for work done in connection with the contract.
8. All premiums on fire, public liability, property damage or other insurance coverage authorized or required by the Engineer or the Owner, or regularly paid by the Contractor in the conduct of his business.
9. All sales, use, excise, privilege, business, occupation, gross receipt and all other taxes paid by the Contractor in connection with the work, but excluding state income taxes based solely on net income derived from this contract and Federal income taxes.
10. All travel or other related expense of general supervisory employees for necessary visits to the job excluding expenses of such employees incurred at the Home Office of the Contractor.
11. All Subcontracts approved by the Engineer or Owner.
12. (Insert other costs proper for inclusion in this Contract.)

    a. --

    b. --

    c. --

13. Any other costs incurred by the Contractor as a direct result of executing the Order, subject to approval by the Engineer.
14. Credit to the Owner for the following items:

    a. Such discounts on invoices as may be obtainable provided that the Owner advances sufficient funds to pay the invoices within the discount period.

    b. The mutually agreed salvage value of materials, tools or equipment charged to the Owner and taken over by the Contractor for his use or sale at the completion of the work.

c. Any rebates, refunds, returned deposits or other allowances properly credited to the Cost of the Work.

ARTICLE IV—*Progress Payments*

The Owner shall make payments on account of the Contract as follows:

As early as possible after the first day of each month the Contractor shall present to the Owner a statement of all costs incurred on account of the work involved in this Contract during the preceding month. This statement shall be accompanied by copies of supporting invoices and such copies of payrolls, or totalization and distribution of same, as may be required; lists of plant and equipment used, with rates for same, together with any other information necessary to allow the Engineer to verify the accuracy of the statement. To the total of the costs incurred there shall be added the full amount of the percentages earned as set up in Article III, and the full amount shall be paid to the Contractor within 10 days after the receipt of the statement.*

IN WITNESS WHEREOF the parties hereto have executed this Agreement, the day and year first above written.

------------------------------------OWNER

WITNESS:

------------------------------------ By: ------------------------------------

Title

------------------------------CONTRACTOR

WITNESS:

------------------------------------ By: ------------------------------------

Title

FORM OF AGREEMENT FOR ENGINEERING CONSTRUCTION

Lump Sum Basis

THIS AGREEMENT, made on the--------------day of------------------------, 19----,

by and between--

party of the first part, hereinafter called the OWNER, and-------------------------

----

----

party of the second part, hereinafter called the CONTRACTOR.

It is understood ENGINEER representing Owner shall be-------------------------

----

WITNESSETH, That the Contractor and the Owner, for the considerations hereinafter named, agree as follows:

ARTICLE I—*Scope of the Work*

The Contractor hereby agrees to furnish all of the materials and all of the equipment and labor necessary, and to perform all of the work shown on the Drawings and described in the specifications for the project entitled--------------------------------------

----

----

* Here may be inserted a provision for a retained percentage if desired.

all in accordance with the requirements and provisions of the following Documents which are hereby made a part of this Agreement:

(*a*) Drawings prepared for same by____

numbered____

and dated_______________________________, 19____.

(*b*) Specifications consisting of:

1. "Standard General Specifications" issued by__________________________________

_____________________________, _________________________________Edition.

2. "Special Conditions" as prepared by______________________________________

__________________________ dated____

3. The "General Conditions of Contract for Engineering Construction"—1966 Edition.

4. Addendum

No.________________________________ Date________________________________

ARTICLE II—*Time of Completion*

(*a*) The work to be completed under this Contract shall be commenced within____________ calendar days after receipt of notice to proceed.

(*b*) The work shall be completed within____________calendar days after receipt of notice to proceed.

(*c*) Failure to complete the work within the number of calendar days stated in this Article, including extension granted thereto as determined by Section 19 of the General Conditions, shall entitle the Owner to deduct from the moneys due to the Contractor as "Liquidated Damages" an amount equal to $________________________for each calendar day of delay in completion of the work.

(*d*) If the Contractor completes the work earlier than the date determined in accordance with Paragraph (*b*), and the Engineer shall so certify in writing, the Owner shall pay the Contractor an additional amount equal to $________________________for each calendar day by which the time of completion so determined has been reduced.

ARTICLE III—*The Contract Sum*

(*a*) Except for adjustments as provided herein, the Owner shall pay to the Contractor, for the performance of the work, the lump sum of $________________________which amount shall be known as the Contract Sum.

(*b*) The Contract Sum shall be equitably adjusted to cover changes in the work ordered by the Engineer, but not shown on the Drawings or required by the Specifications. Such increases or decreases in the Contract Sum shall be determined by agreement between the Owner, or the Engineer, as his representative, and the Contractor. If it is impracticable to arrive at a pre-agreed-upon amount, the work so ordered may be done under the provisions of Article V.

ARTICLE IV—*Progress Payments*

The Owner shall make payments on account of the Contract as follows:

(*a*) On not later than the fifth day of every month the Contractor shall present to the Engineer an invoice covering the percentage of the total amount of the Contract which has been completed from the start of the job up to and including the last day of the preceding month, together with such supporting evidence as may be required by the Engineer. This invoice shall also include the cost of such material required in the permanent work as has been delivered to the site but not as yet incorporated in the work.* Where provision is made for payment for materials delivered to the site or elsewhere,

* In addition to advance payment for materials delivered to the site, wording should indicate that, where applicable, advance payment may be in order for materials in storage away from the site, for field plant and equipment, access roads, etc.—details to be spelled out in the Special Conditions.

evidence that the Contractor has paid suppliers should be required in order to establish that the Owner is paying for property to which the Contractor has clear title.

(*b*) On not later than the 15th of the month, the Owner shall pay to the Contractor 90 per cent of the amount of the invoice—less previous payments made. The 10 per cent retained percentage may be held by the Owner until the value of the work completed at the end of any month equals 50 per cent of the total amount of the Contract, after which if the Engineer finds that satisfactory progress is being made, he shall recommend that all of the remaining monthly payments be paid in full. Payments for work, under Subcontracts of the General Contractor, shall be subject to the above conditions applying to the general Contract after the work under a Subcontract has been 50 per cent completed.

(*c*) Final payment of all moneys due on the Contract shall be made within 30 days of completion and acceptance of the work.

(*d*) If the owner fails to make payment as herein provided, or as provided in Article V(*d*), in addition to those remedies available to the Contractor under Section 25 of the General Conditions, there shall be added to each such payment daily interest at the rate of 6 per cent per annum commencing on the first day after said payment is due and continuing until the payment is delivered or mailed to the Contractor.

ARTICLE V—*Extra Work*

If the Engineer orders, in writing, the performance of any work not covered by the Drawings or included in the Specifications, and for which no item in the Contract is provided, and for which no unit price or lump sum basis can be agreed upon, then such extra work shall be done on a Cost-Plus-Percentage basis of payment as follows:

(*a*) The Contractor shall be reimbursed for all costs incurred in doing the work, and shall receive an additional payment of--------% of all such cost to cover his indirect overhead costs, plus --------% of all cost, including indirect overhead, as his fee.

(*b*) The "Cost of the Work" shall be determined as the net sum of the following items:

1. Job Office and all necessary temporary facilities such as buildings, use of land not furnished by the Owner, access roads and utilities. The costs of these items include construction, furnishings and equipment, maintenance during the period that they are needed, demolition and removal. Salvage values agreed on or received by the Contractor shall be credited to the Owner.

2. All materials used on the work whether for temporary or permanent construction.

3. All small tools and supplies; all fuel, lubricants, power, light, water and telephone service.

4. All plant and equipment at specified rental rates and terms of use. If the rental rates do not include an allowance for running repairs and repair parts needed for ordinary maintenance of the plant and equipment, then such items of cost are to be included in the Cost of the Work.

5. All transportation costs on equipment, materials and men.

6. All labor for the project and including the salaries of superintendents, foremen, engineers, inspectors, clerks and other employees while engaged on the work but excluding salaries of general supervisory employees or officers, who do not devote their full time to the work.

7. All payroll charges such as Social Security payments, unemployment insurance, workmen's compensation insurance premiums, pension and retirement allowances, and social insurance premiums, vacation and sick-leave allowances applicable to wages or salaries paid to employees for work done in connection with the contract.

8. All premiums on fire, public liability, property damage, or other insurance coverage authorized or required by the Engineer or the Owner, or regularly paid by the Contractor in the conduct of his business.

9. All sales, use, excise, privilege, business, occupation, gross receipt and all other taxes paid by the Contractor in connection with the work, but excluding state income taxes based solely on net income derived from this contract and Federal income taxes.

10. All travel or other related expense of general supervisory employees for necessary visits to the job excluding expenses of such employees incurred at the Home Office of the Contractor.

11. All Subcontracts approved by the Engineer or Owner.

12. (Insert other costs proper for inclusion in this Contract.)

    a. ---

    b. ---

    c. ---

13. Any other costs incurred by the Contractor as a direct result of executing the Order, subject to approval by the Engineer.

14. Credit to the Owner for the following items:

    a. Such discounts on invoices as may be obtainable provided that the Owner advances sufficient funds to pay the invoices within the discount period.

    b. The mutually agreed salvage value of materials, tools or equipment charged to the Owner and taken over by the Contractor for his use or sale at the completion of the work.

    c. Any rebates, refunds, returned deposits or other allowances properly credited to the Cost of the Work.

(*c*) The cost of the work done each day shall be submitted to the Engineer in a satisfactory form on the succeeding day, and shall be approved by him or adjusted at once.

(*d*) Monthly payments of all charges for Extra Work in any one month shall be made in full on or before the 15th of the succeeding month. Those payments shall include the full amount of fee earned on the cost of the work done.

IN WITNESS WHEREOF the parties hereto have executed this Agreement, the day and year first above written.

---------------------------------- OWNER

WITNESS:

-------------------------------------- By: ------------------------------------
Title

---------------------------------- CONTRACTOR

WITNESS:

-------------------------------------- By: ------------------------------------
Title

General Conditions of Contract for Engineering Construction

Index

1. Definitions
2. Execution and Correlation of Documents
3. Design, Drawings and Instructions
4. Copies of Drawings Furnished
5. Order of Completion
6. Ownership of Drawings
7. Familiarity with Work
8. Changed Conditions
9. Materials and Appliances
10. Employees
11. Royalties and Patents
12. Surveys
13. Permits, Licenses and Regulations
14. Protection of the Public and of Work and Property
15. Inspection of Work
16. Superintendence
17. Discrepancies

18. Changes in the Work
19. Extension of Time
20. Claims
21. Deductions for Uncorrected Work
22. Correction of Work Before Final Payment
23 Suspension of Work
24. The Owner's Right to Terminate Contract
25. Contractor's Right to Stop Work or Terminate Contract
26. Removal of Equipment
27. Responsibility for Work
28. Partial Completion and Acceptance
29. Payments Withheld Prior to Final Acceptance of Work
30. Contractor's Insurance
31. Surety Bonds
32. Owner's Insurance
33. Assignment
34. Rights of Various Interests
35. Separate Contracts
36. Subcontracts
37. Engineer's Status
38. Engineer's Decisions
39. Arbitration
40. Lands for Work
41. Cleaning Up
42. Acceptance and Final Payment

SEC. 1—*Definitions*

(*a*) The Contract Documents shall consist of Advertisement for Bids or Notice to Contractors, Instructions to Bidders, Form of Bid or Proposal, the signed Agreement, the General and Special Conditions of Contract, the Drawings, and the Specifications, including all modifications thereof incorporated in any of the documents before the execution of Agreement.

(*b*) The Owner, the Contractor and the Engineer are those named as such in the Agreement. They are treated throughout the Contract Documents as if each were of singular number and masculine gender.

(*c*) Wherever in this Contract the word "Engineer" is used it shall be understood as referring to the Engineer of the Owner, acting personally or through assistants duly authorized in writing by the Engineer.

(*d*) Written notice shall be deemed to have been duly served if delivered in person to the individual or to a member of the firm or to an officer of the corporation for whom it is intended, or to an authorized representative of such individual, firm, or corporation, or if delivered at or sent by registered mail to the last business address known to him who gives the notice, with a copy sent to the central office of the Contractor.

(*e*) The term "Subcontractor" shall mean anyone (other than the Contractor) who furnishes at the site, under an Agreement with the Contractor, labor, or labor and materials, or labor and equipment, but shall not include any person who furnishes services of a personal nature.

(*f*) Work shall mean the furnishing of all labor, materials, equipment, and other incidentals necessary or convenient to the successful completion of the Contract and the carrying out of all the duties and obligations imposed by the Contract.

(*g*) Extra work shall mean such additional labor, materials, equipment, and other incidentals as are required to complete the Contract for the purpose for which it was intended but was not shown on the Drawings or called for in the Specifications, or is desired by the Owner in addition to that work called for in the Drawings and Specifications.

(*h*) Dispute shall mean lack of agreement between any parties that have any obligations, duties, or responsibilities under the terms of the Contract, Drawings, or Specifications.

SEC. 2—*Execution and Correlation of Documents*

The Contract Documents shall be signed in duplicate by the Owner and the Contractor.

The Contract Documents are complementary and what is called for by any one shall be as binding as if called for by all. In case of conflict between Drawings and Specifications, the Specifications shall govern. Materials or work described in words which so applied have a well-known technical or trade meaning shall be held to refer to such recognized standards.

SEC. 3—*Design, Drawings and Instructions*

It is agreed that the Owner will be responsible for the adequacy of design and sufficiency of the Drawings and Specifications. The Owner, through the Engineer, or the Engineer as the Owner's representative, shall furnish Drawings and Specifications which adequately represent the requirements of the work to be performed under the Contract. All such Drawings and instructions shall be consistent with the Contract Documents and shall be true developments thereof. In the case of lump-sum Contracts, Drawings and Specifications which adequately represent the work to be done shall be furnished prior to the time of entering into the Contract. The Engineer may, during the life of the Contract, and in accordance with Section 18, issue additional instructions by means of Drawings or other media necessary to illustrate changes in the work.

SEC. 4—*Copies of Drawings Furnished*

Unless otherwise provided in the Contract Documents, the Engineer will furnish to the Contractor, free of charge, all copies of Drawings and Specifications reasonably necessary for the execution of the work.

SEC. 5—*Order of Completion*

The Contractor shall submit, at such times as may be reasonably requested by the Engineer, schedules which shall show the order in which the Contractor proposes to carry on the work, with dates at which the Contractor will start the several parts of the work, and estimated dates of completion of the several parts.

SEC. 6—*Ownership of Drawings*

All Drawings, Specifications and copies thereof furnished by the Engineer shall not be reused on other work, and, with the exception of the signed Contract, sets are to be returned to him on request, at the completion of the work.

SEC. 7—*Familiarity with Work*

The Owner shall make known to all prospective bidders, prior to the receipt of bids, all information that he may have as to subsurface conditions in the vicinity of the work, topographical maps, or other information that might assist the bidder in properly evaluating the amount and character of the work that might be required. Such information is given, however, as being the best factual information available to the Owner. The Contractor, by careful examination, shall satisfy himself as to the nature and location of the work, the character of equipment and facilities needed preliminary to and during the prosecution of the work, the general and local conditions, and all other matters which can in any way affect the work under this Contract.

SEC. 8—*Changed Conditions*

The Contractor shall promptly, and before such conditions are disturbed, notify the Owner in writing of: (1) Subsurface or latent physical conditions at the site differing materially from those indicated in this Contract; or (2) previously unknown physical or other conditions at the site, of an unusual nature, differing materially from those ordinarily encountered and generally recognized as inherent in work of the character provided for in this Contract. The Engineer shall promptly investigate the conditions, and if he finds that such conditions do so materially differ and cause an increase or decrease in the cost of, or the time required for, performance of this Contract, an equitable adjustment shall be made and the Contract modified in writing accordingly. Any claim of the Contractor for adjustment hereunder shall not be allowed unless he has given notice as above required; provided that the Engineer may, if he determines the facts so justify, consider and adjust any such claims asserted before the date of final settlement of the Contract. If the parties fail to agree upon the adjustment to be made, the dispute shall be determined as provided in Section 39 hereof.

SEC. 9—*Materials and Appliances*

Unless otherwise stipulated, the Contractor shall provide and pay for all materials, labor, water, tools, equipment, light, power, transportation and other facilities necessary for the execution and completion of the work. Unless otherwise specified, all materials incorporated in the permanent work shall be new and both workmanship and materials shall be of good quality. The Contractor shall, if required, furnish satisfactory evidence as to the kind and quality of materials.

SEC. 10—*Employees*

The Contractor shall at all times enforce strict discipline and good order among his employees, and shall seek to avoid employing on the work any unfit person or anyone not skilled in the work assigned to him.

Adequate sanitary facilities shall be provided by the Contractor.

SEC. 11—*Royalties and Patents*

The Contractor shall pay all royalties and license fees. He shall defend all suits or claims for infringement of any patent rights and shall save the Owner harmless from loss on account thereof except that the Owner shall be responsible for all such loss when a particular process or the product of a particular manufacturer or manufacturers is specified, unless the Owner has notified the Contractor prior to the signing of the Contract that the particular process or product is patented or is believed to be patented.

SEC. 12—*Surveys*

Unless otherwise specified, the Owner shall furnish all land surveys and establish all base lines for locating the principal component parts of the work together with a suitable number of bench marks adjacent to the work. From the information provided by the Owner, the Contractor shall develop and make all detail surveys needed for construction such as slope stakes, batter boards, stakes for pile locations and other working points, lines and elevations.

The Contractor shall carefully preserve bench marks, reference points and stakes and, in case of willful or careless destruction, he shall be charged with the resulting expense and shall be responsible for any mistakes that may be caused by their unnecessary loss or disturbance.

SEC. 13—*Permits, Licenses and Regulations*

Permits and licenses of a temporary nature necessary for the prosecution of the work shall be secured and paid for by the Contractor. Permits, licenses and easements for permanent structures or permanent changes in existing facilities shall be secured and paid for by the Owner, unless otherwise specified. The Contractor shall give all notices and comply with all laws, ordinances, rules and regulations bearing on the conduct of the work as drawn and specified. If the Contractor observes that the Drawings and Specifications are at variance therewith, he shall promptly notify the Engineer in writing, and any necessary changes shall be adjusted as provided in the Contract for changes in the work.

SEC. 14—*Protection of the Public and of Work and Property*

The Contractor shall provide and maintain all necessary watchmen, barricades, warning lights and signs and take all necessary precautions for the protection and safety of the public. He shall continuously maintain adequate protection of all work from damage, and shall take all reasonable precautions to protect the Owner's property from injury or loss arising in connection with this Contract. He shall make good any damage, injury or loss to his work and to the property of the Owner resulting from lack of reasonable protective precautions, except such as may be due to errors in the Contract Documents, or caused by agents or employees of the Owner. He shall adequately protect adjacent private and public property, as provided by Law and the Contract Documents.

In an emergency affecting the safety of life, of the work, or of adjoining property, the Contractor is, without special instructions or authorization from the Engineer, hereby permitted to act at his discretion to prevent such threatened loss or injury. He shall also so act, without appeal, if so authorized or instructed by the Engineer.

Any compensation claimed by the Contractor on account of emergency work, shall be determined by agreement or by arbitration.

SEC. 15—*Inspection of Work*

The Owner shall provide sufficient competent personnel, working under the supervision of a qualified engineer, for the inspection of the work while such work is in progress to ascertain that the completed work will comply in all respects with the standards and requirements set forth in the Specifications. Notwithstanding such inspection, the Contractor will be held responsible for the acceptability of the finished work.

The Engineer and his representatives shall at all times have access to the work whenever it is in preparation or progress, and the Contractor shall provide proper facilities for such access, and for inspection.

If the Specifications, the Engineer's instructions, laws, ordinances, or any public authority require any work to be specially tested or approved, the Contractor shall give the Engineer timely notice of its readiness for inspection, and if the inspection is by an authority other than the Engineer, of the date fixed for such inspection. Inspections by the Engineer shall be made promptly, and where practicable at the source of supply. If any work should be covered up without approval or consent of the Engineer, it must, if required by the Engineer, be uncovered for examination and properly restored at the Contractor's expense, unless the Engineer has unreasonably delayed inspection.

Re-examination of any work may be ordered by the Engineer, and, if so ordered, the work must be uncovered by the Contractor. If such work is found to be in accordance with the Contract Documents, the Owner shall pay the cost of re-examination and replacement. If such work is not in accordance with the Contract Documents, the Contractor shall pay such cost.

SEC. 16—*Superintendence*

The Contractor shall keep on his work, during its progress, a competent superintendent and any necessary assistants. The superintendent shall represent the Contractor, and all directions given to him shall be binding as if given to the Contractor. Important directions shall immediately be confirmed in writing to the Contractor. Other directions shall be so confirmed on written request in each case. The Contractor shall give efficient superintendence to the work, using his best skill and attention.

SEC. 17—*Discrepancies*

If the Contractor, in the course of the work, finds any discrepancy between the Drawings and the physical conditions of the locality, or any errors or omissions in Drawings or in the layout as given by survey points and instructions, he shall immediately inform the Engineer, in writing, and the Engineer shall promptly verify the same. Any work done after such discovery, until authorized, will be done at the Contractor's risk.

SEC. 18—*Changes in the Work*

The Owner may make changes in the Drawings and Specifications or scheduling of the Contract within the general scope at any time by a written order. If such changes add to or deduct from the Contractor's cost of the work, the Contract shall be adjusted accordingly. All such work shall be executed under the conditions of the original Contract except that any claim for extension of time caused thereby shall be adjusted at the time of ordering such change.

In giving instructions, the Engineer shall have authority to make minor changes in the work not involving extra cost, and not inconsistent with the purposes of the work, but otherwise, except in an emergency endangering life or property, no extra work or change shall be made unless in pursuance of a written order by the Engineer, and no claim for an addition to the Contract Sum shall be valid unless the additional work was so ordered.

The Contractor shall proceed with the work as changed and the value of any such extra work or change shall be determined as provided in the Agreement.

SEC. 19—*Extension of Time*

Extension of time stipulated in the Contract for completion of the work will be made when changes in the work occur, as provided in Section 18; when the work is suspended as provided in Section 23; and when the work of the Contractor is delayed on account of conditions which could not have been foreseen, or which were beyond the control of the Contractor, his Subcontractors or suppliers, and which were not the result of their fault or negligence. Extension of time for completion shall also be allowed for any delays in the progress of the work caused by any act (except as provided elsewhere in these General Conditions) or neglect of the Owner or of his employees or by other contractors employed by the Owner, or by any delay in the furnishing of Drawings and necessary information by the Engineer, or by any other cause which in the opinion of the Engineer entitled the Contractor to an extension of time, including but not restricted to, acts of the public enemy, acts of any government in either its sovereign or any applicable contractual capacity, acts of another contractor in the performance of a contract with the Owner, fires, floods, epidemics, quarantine restrictions, freight embargoes, unusually severe weather, or labor disputes.

The Contractor shall notify the Engineer promptly of any occurrence or conditions which in the Contractor's opinion entitle him to an extension of time. Such notice shall be in writing and shall be submitted in ample time to permit full investigation and evaluation of the Contractor's claim. The Engineer shall acknowledge receipt of the Contractor's notice within 5 days of its receipt. Failure to provide such notice shall constitute a waiver by the Contractor of any claim.

SEC. 20—*Claims*

If the Contractor claims that any instructions by Drawings or other media issued after the date of the Contract involve extra cost under this Contract, he shall give the Engineer written notice thereof within ________ days after the receipt of such instructions, and in any event before proceeding to execute the work, except in emergency endangering life or property, and the procedure shall then be as provided for changes in the work. No such claim shall be valid unless so made.

SEC. 21—*Deductions for Uncorrected Work*

If the Engineer deems it inexpedient to correct work that has been damaged or that was not done in accordance with the Contract, an equitable deduction from the Contract price shall be made therefor, unless the Contractor elects to correct the work.

SEC. 22—*Correction of Work before Final Payment*

The Contractor shall promptly remove from the premises all materials and work condemned by the Engineer as failing to meet Contract requirements, whether incorporated in the work or not. The Contractor shall promptly replace and re-execute his own work in accordance with the Contract and without expense to the Owner and shall bear the expense of making good all work of other contractors destroyed or damaged by such removal or replacement.

If the Contractor does not take action to remove such condemned materials and work within 10 days after written notice, the Owner may remove them and may store the material at the expense of the Contractor. If the Contractor does not pay the expense of such removal and storage within ten days' time thereafter, the Owner may, upon ten days' written notice, sell such materials at auction or at private sale and shall pay to the Contractor any net proceeds thereof, after deducting all the costs and expenses that should have been borne by the Contractor.

SEC. 23—*Suspension of Work*

The Owner may at any time suspend the work, or any part thereof, by giving________ days' notice to the Contractor in writing. The work shall be resumed by the Contractor within ten (10) days after the date fixed in the written notice from the Owner to the Contractor so to do. The Owner shall reimburse the Contractor for expense incurred by the Contractor in connection with the work under this Contract as a result of such suspension.

If the work, or any part thereof, shall be stopped by notice in writing aforesaid, and if the Owner does not give notice in writing to the Contractor to resume work at a date within________days of the date fixed in the written notice to suspend, then the Contractor may abandon that portion of the work so suspended and he will be entitled to the estimates and payments for all work done on the portions so abandoned, if any, plus________% of the value of the work so abandoned, to compensate for loss of overhead, plant expense, and anticipated profit.

SEC. 24—*The Owner's Right to Terminate Contract*

If the Contractor should be adjudged a bankrupt, or if he should make a general assignment for the benefit of his creditors, or if a receiver should be appointed as a result of his insolvency, or if he should be guilty of a substantial violation of the Contract, then the Owner, upon the certificate of the Engineer that sufficient cause exists to justify such action, may, without prejudice to any other right or remedy and after giving the Contractor and his Surety seven days' written notice, terminate the employment of the Contractor and take possession of the premises and of all materials, tools, equipment and other facilities installed on the work and paid for by the Owner, and finish the work by whatever method he may deem expedient. In such case the Contractor shall not be entitled to receive any further payment until the work is finished. If the unpaid balance of the Contract price shall exceed the expense of finishing the work, including compensation for additional managerial and administrative services, such excess shall be paid to the Contractor. If such expense shall exceed such unpaid balance, the Contractor shall pay the difference to the Owner. The expense incurred by the Owner as herein provided, and the damage incurred through the Contractor's default, shall be certified by the Engineer.

SEC. 25—*Contractor's Right to Stop Work or Terminate Contract*

If the work should be stopped under an order of any court, or other public authority, for a period of more than three months, through no act or fault of the Contractor or of anyone employed by him, or if the Engineer should fail to issue any estimate for payment within seven days after it is due, or if the Owner should fail to pay the Contractor within seven days of its maturity and presentation any sum certified by the Engineer or awarded by arbitrators, then the Contractor may, upon seven days' written notice to the Owner and the Engineer, stop work or terminate this Contract and recover from the Owner payment for all work executed, plus any loss sustained upon any plant or materials plus reasonable profit and damages.

SEC. 26—*Removal of Equipment*

In the case of termination of this Contract before completion from any cause whatever, the Contractor, if notified to do so by the Owner, shall promptly remove any part or all of his equipment and supplies from the property of the Owner, failing which the Owner shall have the right to remove such equipment and supplies at the expense of the Contractor.

SEC. 27—*Responsibility for Work*

The Contractor assumes full responsibility for the work. Until its final acceptance, the Contractor shall be responsible for damage to or destruction of the work (except for any part covered by partial acceptance as set forth in Sec. 28). He agrees to make no claims against the Owner for damages to the work from any cause except negligence or willful acts of the Owner, acts of an Enemy, acts of war or as provided in Sec. 32.

SEC. 28—*Partial Completion and Acceptance*

If at any time prior to the issuance of the final certificate referred to in Section 42 hereinafter, any portion of the permanent construction has been satisfactorily completed, and if the Engineer determines that such portion of the permanent construction is not required for the operations of the Contractor but is needed by the Owner, the Engineer shall issue to the Contractor a certificate of partial completion, and thereupon or at any time thereafter the Owner may take over and use the portion of the permanent construction described in such certificate, and may exclude the Contractor therefrom.

The issuance of a certificate of partial completion shall not be construed to constitute an extension of the Contractor's time to complete the portion of the permanent construction to which it relates if he has failed to complete it in accordance with the terms of this Contract. The issuance of such a certificate shall not operate to release the Contractor or his sureties from any obligations under this Contract or the performance bond.

If such prior use increases the cost of or delays the work, the Contractor shall be entitled to extra compensation, or extension of time, or both, as the Engineer may determine, unless otherwise provided.

SEC. 29—*Payments Withheld Prior to Final Acceptance of Work*

The Owner, as a result of subsequently discovered evidence, may withhold or nullify the whole or part of any payment certificate to such extent as may be necessary to protect himself from loss caused by:

(a) Defective work not remedied.
(b) Claims filed or reasonable evidence indicating probable filing of claims by other parties against the Contractor.
(c) Failure of the Contractor to make payments properly to Subcontractors or for material or labor.
(d) Damage to another contractor.

When the above grounds are removed or the Contractor provides a Surety Bond satisfactory to the Owner which will protect the Owner in the amount withheld, payment shall be made for amounts withheld, because of them.

No moneys may be withheld under (b) and (c) above if a payment bond is included in the Contract.

SEC 30—*Contractor's Insurance*

The Contractor shall secure and maintain such insurance policies as will protect himself, his Subcontractors, and unless otherwise specified, the Owner, from claims for bodily injuries, death or property damage which may arise from operations under this Contract whether such operations be by himself or by any Subcontractor or anyone employed by them directly or indirectly. The following insurance policies are required:

(a) Statutory Workmen's Compensation.
(b) Contractor's Public Liability and Property Damage—
    Bodily Injury:
        each person.............. $__________
        each accident............ $__________
    Property Damage:
        each accident............ $__________
        aggregate................ $__________
(c) Automobile Public Liability and Property Damage—
    Bodily Injury:
        each person.............. $__________
        each accident............ $__________
    Property Damage:
        each accident............ $__________

All policies shall be for not less than the amounts set forth above or as stated in the Special Conditions. Other forms of insurance shall also be provided if called for by the Special Conditions.

Certificates and/or copies of policy of such insurance shall be filed with the Engineer, and shall be subject to his approval as to adequacy of protection, within the requirements of the Specifications. Said certificates of insurance shall contain a 10 days' written notice of cancellation in favor of the Owner.

SEC. 31—*Surety Bonds*

The Owner shall have the right, prior to the signing of the Contract, to require the Contractor to furnish bond covering the faithful performance of the Contract and the payment of all obligations arising thereunder, in such form as the Owner may prescribe in the bidding documents and executed by one or more financially responsible sureties. If such bond is required prior to the receipt of bids, the premium shall be paid by the Contractor; if subsequent thereto, it shall be paid by the Owner. The Owner may require additional bond if the Contract is increased appreciably.

SEC. 32—*Owner's Insurance*

The Owner shall secure and maintain insurance to 100% of the insurable value thereof against fire, earthquake, flood, and such other perils as he may deem necessary and shall name the Contractor and Subcontractors as additional insured. Such insurance shall be upon the entire work in the Contract and any structures attached or adjacent thereto. He shall also secure and maintain such insurance as will protect him and his officers, agents, servants, and employees from liability to others for damages due to death, bodily injury, or property damage resulting from the performance of the work. The limits of such insurance shall be equal to the amounts stated in subparagraphs (b) and (c), of Section 30.

SEC. 33—*Assignment*

Neither party to the Contract shall assign the Contract or sublet it as a whole without the written consent of the other, nor shall the Contractor assign any moneys due to him or to become due to him hereunder, except to a bank or financial institution acceptable to the Owner.

SEC 34—*Rights of Various Interests*

Whenever work being done by the Owner's or by other contractor's forces is contiguous to work covered by this Contract, the respective rights of the various interests involved shall be established by the Engineer, to secure the completion of the various portions of the work in general harmony.

SEC. 35—*Separate Contracts*

The Owner reserves the right to let other contracts in connection with this project. The Contractor shall afford other contractors reasonable opportunity for the introduction and storage of their materials and the execution of their work, and shall properly connect and coordinate his work with theirs.

If the proper execution or results of any part of the Contractor's work depends upon the work of any other contractor, the Contractor shall inspect and promptly report to the Engineer any defects in such work that render it unsuitable for such proper execution and results.

SEC. 36—*Subcontracts*

The Contractor shall, as soon as practicable after signing of the Contract, notify the Engineer in writing of the names of Subcontractors proposed for the work.

The Contractor agrees that he is as fully responsible to the Owner for the acts and omissions of his Subcontractors and of persons either directly or indirectly employed by them, as he is for the acts and omissions of persons directly employed by him.

Nothing contained in the Contract Documents shall create any contractual relation between any Subcontractor and the Owner.

SEC. 37—*Engineer's Status*

The Engineer shall perform technical inspection of the work. He has authority to stop the work whenever such stoppage may be necessary to insure the proper execution of the Contract. He shall also have authority to reject all work and materials which do not conform to the Contract and to decide questions which arise in the execution of the work.

SEC. 38—*Engineer's Decisions*

The Engineer shall, within a reasonable time after their presentation to him, make decisions in writing on all claims of the Owner or the Contractor and on all other matters relating to the execution and progress of the work or the interpretation of the Contract Documents.

SEC. 39—*Arbitration*

Any controversy or claim arising out of or relating to this Contract, or the breach thereof which cannot be resolved by mutual agreement, shall be settled by arbitration in accordance with the Rules of the American Arbitration Association, and judgment upon the award rendered by the Arbitrator(s) may be entered in any Court having jurisdiction thereof.

SEC. 40—*Lands for Work*

The Owner shall provide as indicated on Drawing No._________and not later than the date when needed by the Contractor the lands upon which the work under this Contract is to be done, rights-of-way for access to same, and such other lands which are designated on the Drawings for the use of the Contractor. Such lands and rights-of-way shall be adequate for the performance of the Contract. Any delay in the furnishing of these lands by the Owner shall be deemed proper cause for an equitable ajustment in both Contract price and time of completion.

The Contractor shall provide at his own expense and without liability to the Owner any additional land and access thereto that may be required for temporary construction facilities, or for storage of materials.

SEC. 41—*Cleaning Up*

The Contractor shall remove at his own expense from the Owner's property and from all public and private property all temporary structures, rubbish and waste materials resulting from his operations. This requirement shall not apply to property used for permanent disposal of rubbish or waste materials in accordance with permission of such disposal granted to the Contractor by the Owner thereof.

SEC. 42—*Acceptance and Final Payment*

(a) Upon receipt of written notice that the work is substantially completed or ready for final inspection and acceptance, the Engineer will promptly make such inspection, and when he finds the work acceptable under the Contract and the Contract fully performed or substantially completed he shall promptly issue a certificate, over his own signature, stating that the work required by this Contract has been completed or substantially completed and is accepted by him under the terms and conditions thereof, and the entire balance found to be due the Contractor, including the retained percentage, less a retention based on the Engineer's estimate of the fair value of the claims against the Contractor and the cost of completing the incomplete or unsatisfactory items of work with specified amounts for each incomplete or defective item of work, is due and payable. The date of substantial completion of a project or specified area of a project is the date when the construction is sufficiently completed in accordance with the Contract Documents as modified by any change orders agreed to by the parties so that the Owner can occupy the project or specified area of the project for the use for which it was intended.

(b) Before issuance of final payment, the Contractor, if required in the Special Conditions, shall certify in writing to the Engineer that all payrolls, material bills, and other indebtedness connected with the work have been paid, or otherwise satisfied, except that in case of disputed indebtedness or liens, if the Contract does not include a payment bond, the Contractor may submit in lieu of certification of payment a surety bond in the amount of the disputed indebtedness or liens, guaranteeing payment of all such disputed amounts, including all related costs and interest in connection with said disputed indebtedness or liens which the Owner may be compelled to pay upon adjudication.

(c) The making and acceptance of the final payment shall constitute a waiver of all claims by the Owner, other than those arising from unsettled liens, from faulty work appearing within the guarantee period provided in the Special Conditions, from the requirements of the Drawings and Specifications, or from manufacturer's guarantees. It shall also constitute a waiver of all claims by the Contractor, except those previously made and still unsettled.

(d) If after the work has been substantially completed, full completion thereof is materially delayed through no fault of the Contractor, and the Engineer so certifies, the Owner shall, upon certificate of the Engineer, and without terminating the Contract, make payment of the balance due for that portion of the work fully completed and accepted. Such payment shall be made under the terms and conditions governing final payment, except that it shall not constitute a waiver of claims.

(e) If the Owner fails to make payment as herein provided, there shall be added to each such payment daily interest at the rate of 6 per cent per annum commencing on the first day after said payment is due and continuing until the payment is delivered or mailed to the Contractor.

Section 3

Construction Services

FRANKLIN T. MATTHIAS

Vice-president—Transportation Projects,
Kaiser Engineers, Oakland, Calif.

Type of Job Construction projects are too different to permit the development of a formula that can specifically answer the needs of planning any one project. Wide ranges of equipment and plant requirements exist among such projects as heavy industrial plants, concrete dams, earth dams, highway construction, harbor facilities, power transmission lines, tunnels, hydroelectric projects, and thermal power plants. Even within these classifications, wide variations exist. For example, a wide-river, low-head dam calls for different equipment, different plant, and different construction operations plans than would be suitable for a high-head, canyon type of dam. Services must be designed to fit the specific needs of the job at hand.

Construction Methods and Equipment The design of effective and adequate construction services requires a detailed knowledge of the proposed construction methods, the equipment and construction plant, and the service expected of those facilities. Continuous, trouble-free operation of service facilities is a "must" on a job that is planned to operate three shifts continuously. Where service demands are not so rigorous, some items of lesser quality may be permitted if they represent a reasonable degree of economy. Extra maintenance is to be expected, but this usually can be done during idle shift periods in construction operations. The designer of services must evaluate the costs of service interruptions in terms of the cost of suspended construction work. Delays might force the job into winter conditions or other unfavorable climatic conditions, and this possibility also must be evaluated. In general, money spent for a reasonably high degree of dependability of services is as well invested as that spent for dependable and productive plant and equipment.

Layout Overall economies can result from careful and thorough coordination in planning of services and construction plant. Such coordination will frequently lead to modifications both in plant facilities and in layout of services. The contractor is fortunate if he has a site which, without heavy grading, has sufficient room so that construction plant facilities and services can be located to meet the anticipated requirements economically. When heavy site grading is required, the new topography should be designed to meet plant and service needs with maximum economy, considering both initial and maintenance costs.

Logistics A substantial part of the total supplies required by a construction project is devoted to the needs of the job services. Particularly in remote locations, the cost of transportation of supplies can be a heavy financial burden to the project. Since modern mobile equipment is nearly all diesel powered,* the planner may fail to consider other fuels to provide power for furnishing compressed air, electricity, water supply, and other services. Any economic alternatives of using locally available wood, coal, or natural gas, or a temporary hydroelectric installation for energy production and use of other local materials that can be incorporated into services facilities, should be explored thoroughly. It may take more planning and more engineering to design and build service facilities of a type not normally used, but decisions must be made on the basis of the maximum ultimate economy to the construction program. Planning and engineering costs generally are very small in proportion to installation and operating costs. Reduction in the transportation load results in savings in indirect costs of supervision and administration as well as the direct costs of freight handling. Where there are no local energy-producing materials, the objective should be to utilize fuels of the same type as those needed for the mobile construction equipment and construction plant facilities in order to reduce the numbers of items requiring separate storage and handling facilities. Interchangeability of repair parts for services equipment and facilities will also reduce logistic demands.

Labor In industrial plants which are expected to operate indefinitely, it can be considered a fair trade to invest the capitalized value of one employee's wage costs (possibly $150,000 or more) in equipment which is capable of eliminating the need for the employee. Labor-reducing investments of this magnitude are seldom justified in construction projects. The term of each job is limited, and the equipment may have little value unless it is adaptable to the next job. However, the designer of construction services might begin by estimating the initial expenditure that would balance the

* See Sec. 6, Engineering Fundamentals.

cost of supporting one employee for a particular job. This information would be useful when evaluating the possible use of automatic devices, controls, centralization of equipment, and other items that will reduce manpower. The evaluations must take into account the working rules of those labor unions which may be involved as well as any applicable requirements which may be imposed by regulatory agencies. Most labor unions and most state governments have rules which are designed for personnel protection and require full-time attendants for compressors or compressor plants, diesel-electric stations, steam boilers, and similar installations. The possible use of expensive special controls and automatic devices to reduce labor should be examined with this forced use of attendants in mind.

Fig. 3-1 A shop job for compressed air is to operate the forging hammers that produce shanks on the drill rods. (*Aluminum Company of Canada, Ltd.*)

Safety The best safety is that which is designed and built into plant layout and equipment. Effective safety precautions and education are stressed strongly in modern construction planning and execution. Contractors have learned by experience that safety pays, both in improved employee morale and in lower job costs, because of reductions in lost time and compensation for accidents. Safety designed into physical facilities generally represents a lower investment with respect to safety-record improvements than a continuing program of personnel safety training.

COMPRESSED-AIR SERVICES

Planning Modern construction operations require the use of compressed air in large quantities. Compressed air has almost entirely replaced steam on heavy construction projects as a power source for such equipment as small hoists and pile drivers. It continues as the almost universal source of power for rock-drilling equipment and is coming into increasing use to power shop tools, impact wrenches, drills, hammers, etc. (See Figs. 3-1 and 3-2.) Compressed air can transport materials in pipes effectively. For construction, air is frequently used to transport cement and is common as the medium for placing concrete in tunnels. It is necessary for sandblasting, for placing Gunite, and for ordinary spray painting. An effective cleaning ing tool, much used in construction, is a combined compressed-air and water jet or combined air and steam jet, where the mixing of air with water or steam can be controlled by valves on each feed line. Effective job planning of compressed-air services requires knowledge of the great variety of uses in which compressed-air power provides advantages over other power sources. It is not uncommon for a job superintendent to add powered equipment as the job progresses until compressed-air demands are substantially beyond the originally estimated needs. This tendency must be considered in planning.

Fig. 3-2 Detachable drill bits are sharpened on air-motor-driven grinder. (*Aluminum Company of Canada, Ltd.*)

Air-compressor equipment discussed here will be confined to the types normally used in construction and will not include the very specialized types used in industry, such as high-pressure units for accumulators, hydroelectric governor units, air-blast circuit breakers, and other special purposes. The compressed-air plants built for the Angeles Tunnel Project are identified in Table 3-1.

Air-powered Equipment Types commonly used in construction fall into four general classifications:

Piston-actuated Hammers. This type of equipment probably consumes a greater amount of air on construction jobs than all other types. It includes chipping hammers, hand-held power shovels, riveting hammers, most rock drills, pile-driving hammers, tampers, and similar equipment, including many types of impact wrenches. Air pressure applied to a piston furnishes motive power and is transmitted to a hammer or tool holder through a piston rod. Most of these tools are double acting; i.e., compressed air is supplied and exhausted alternately to both sides of the piston. Steam can be used without disadvantage to power certain large items of this type, such as pile hammers, but it is not practical for hand-held tools. Virtually no other source of power but steam competes with air for this type of equipment.

Air Motors. The two most common types are reciprocating motors driving a shaft through crankshaft linkage and the rotary type in which a shaft is turned by passage of compressed air through the vanes mounted on the shaft. Rotary air motors are generally used on small shop tools, and reciprocating air motors are used where higher power output is required. Popular for hand-held shop tools, the rotary motors are used for drilling, reaming, grinding, polishing, sawing, etc. Rotary air tools are generally light in weight, are controllable both in speed and torque by simple valves, will stall without damage when overloaded, are readily reversible, carry no electrical shock hazard, are not affected by water even when immersed, and present no spark hazard when used in explosive atmospheres. Larger air motors are largely used for hoists, where again compressed air has almost entirely supplanted steam as a motive force on construction hoists. Large hoists, 50 hp and over, are usually electrically driven. Air motors frequently are used as starting motors for internal-combustion engines. They are the power source for many water pumps, especially of the small, portable, immersible sump-pump type.

TABLE 3-1 Compressed-air Plant for Angeles Tunnel Project*

Similar central-station compressed-air plants were built to supply the North Adit, the Osito Adit, and the South Adit.

Central Plant:†	
3—2,000-cfm stationary compressors, electric powered	6,000 cfm
Loads:	
8—Heavy-duty rock drills at 500 cfm	4,000 cfm
Pumps	1,000 cfm
Shop services	300 cfm
Miscellaneous services	700 cfm

* Angeles Tunnel Project is an element of the California Water Plan to supply water to Southern California. The tunnel was enlarged from the size required for water conveyance, and a surge tank was included to take advantage of the pumped storage potential offered by reservoirs at both ends. The tunnel is 37,774 ft long, excavated to 35 ft 6 in. round to produce a 30-ft diameter of finished concrete lining. The surge tank includes a 65-ft-high riser, 28-ft finished inside diameter, and a chamber 235-ft-high, 120-ft finished inside diameter. Major tunnel excavation was from three adits with a short section of the south end driven from the south portal.

† Compressed-air plant at each adit was sized to support drilling at one heading. Drilling from each adit alternated between the two headings operated from each adit.

Air Cylinders. These operate like jacks and are used for that purpose as well as for power-actuated controls, brakes, and similar purposes. Compressed air for controls and brakes is usually produced by individual compressors which are driven by the engine or motor of the equipment on which they are used. Air cylinders are often used as special-purpose jacks in tunnels, where they usually get their air supply from the distribution system that supplies air for drills and other equipment common to tunnel jobs. Air cylinders or air chambers can suppress heavy shocks in such uses as water systems, special shop equipment, mobile equipment, and the like.

Transporters of Materials. High-velocity air or pressure applied by air finds wide use in construction for material transporting and/or placing. Bulk cement unloading and transport through pipes to concrete-plant silos or other storage has become common. The pipe is usually loaded by a combination of mechanical agitation and air suction. Some systems employ air jets to aerate the cement so that it can be picked up readily by the air suction and blown to the storage silo or point of use. Standard equipment is available.

Compressed-air concrete placers, commonly used in placing tunnel concrete, operate by air pressure building up in a chamber in which mixed concrete has been placed. Pressure then moves a "slug" of concrete through a pipe to the forms, sometimes a distance of several hundred feet. Essential to the success of this type of concrete placing is a reliable source of compressed air and an air receiver capacity sufficient to level out the sudden heavy demand as the "slug" of concrete moves rapidly up the pipe. The air volume used varies widely with the size of concrete-delivery pipe, normally 6- or 8-in. diameter, and with distance and frequency of cycles. This same system is used for special types of cement-grout placing where pressures must have a positive limitation that normally cannot be furnished with a displacement pump. Some types of paint sprayers suck the paint into the moving airstream, but it is air velocity that carries it to the spray nozzle. Moving air is utilized in sandblasting, in guniting, and most common of all, as a jet to clean rock areas and shop equipment and to blow away other foreign substances. Many modern carpenter shops employ air to collect sawdust and other wood and fine metal waste by suction, then transport it by a pipeline to a disposal point. Industrial vacuum cleaners find many uses around construction jobs and are frequently powered by a general pressure air-distribution system.

The air-slide conveyor is another air-actuated transporting device. Finely ground materials travel down a slightly inclined, fixed porous belt. A cushion of air, which is forced through the belt, carries the material without the help of any moving parts. It has not been used much on construction, but there seems no reason why it should not meet certain construction needs. Air-lift pumps and air-cushion vehicles may also have applications in construction.

Air Compressors A general understanding of the basic principles of the various types of compressors is important to the planner and designer of a compressed-air service. Accordingly, the fundamental types commonly used are described here. The reader is referred to available literature sources for more detailed information.[1,2,3]*

Ratings. The normal capacity of air compressors is usually given as the cfm of free air that the unit can discharge as compressed air at a stated pressure. This is not a consistent, precise rating for all conditions as it refers to free air at the point of use or rating test. The pressure, humidity, and temperature of the free air affect the amount of compressed air delivered and the power consumed for any given compressor. For the types of compressors usually employed in construction, a delivery pressure of 100 psi is used to qualify the standard rating. Most construction air tools demand about 100 psi for efficient operation, and most compressors intended for construction services are designed for efficient delivery of air at this pressure or slightly higher. There is a trend toward use of higher-pressure air for rock drills to increase drilling speed and efficiency,† and some drilling units are now available with an integrally mounted air compressor that will supply up to 250 psi to the drill. Higher pressures are required when feed lines to air-operated equipment are long or when several units are pumping into a common distribution system. For such cases, it is common to regulate the compressors to about 125 psi, either through motor controls or by safety-release valves or by both. Especially with reciprocating units of the positive-displacement type, it is essential to provide and maintain safety-release valves that exhaust air from a system subjected to more than a safe pressure. Manufacturers of air compressors have gone to great lengths to identify performance through widely different variables of elevation, pressure, humidity, and temperature and should be consulted if operations are to be conducted at extreme altitudes or extreme temperatures.

Types. Four common types of compressor equipment are available. Of these, the reciprocating and rotary units are the most frequently used and the most adaptable to construction. Both portable and stationary units are available through a wide range of capacities and power drives. The distinction between portable and stationary units is possibly more important to the construction planner than is the type of compressor. Portable units, up to 1,500-cfm capacity and powered by internal-combustion engines, are much used because of their mobility. The capacity of standard stationary units runs up to 6,000 cfm or more, although it is rare to use larger than about 2,300-cfm units in construction work. Stationary units commonly are selected for the central compressor station, with a distribution-system type of installation

* Superior numbers refer to the list of references at the end of this section.
† See Sec. 19, Rock Excavation.

justified where compressed-air needs are heavy and the period of need is long. This condition exists on large tunnel jobs, heavy quarrying operations, large dams, and similar operations. Stationary units are rarely used for industrial building, road building, bridges, etc., where the mobility of the portable unit is especially desirable.

The *reciprocating compressor* is by for the most common type in use. Compression is obtained by use of a piston reciprocating in a cylinder. Valves admit pressure air into a receiver or distribution system, then they close while intake air is drawn in as the piston moves back. These compressors can be single or double acting and have one, two, or more stages, depending on the compression ratio and delivery pressure desired. For the 100- to 125-psig range, reciprocating compressors are usually two-stage and have intercooling between stages. Reciprocating compressors above 5,000 cfm are rare except for occasional single-stage, low-pressure units which are used as blowing engines or vacuum pumps. Speeds of reciprocating compressors run from 1,700 rpm for small units to 75 rpm for very large units. Normal for, say, 1,000 cfm, 100 psi would be about 300 or 360 rpm. Control of compression is usually obtained by an on-off starter in the case of very small units and by an unloading valve that allows the compressor to idle without compressing for larger units. Efficiency and heat-dissipation requirements usually dictate two-stage compression above about 60-psi delivery pressures.

Reciprocating compressors are built with many types of models, such as Y, X, duplex, tandem, radial, and semiradial. Many of these are intended for special applications which are rarely encountered in construction service.

Until a few years ago, virtually all portable compressors were reciprocating, but *rotary compressors* are now more generally used, particularly in the larger sizes. Rotary compressors can be classed as those units that employ one or more rotating elements in a fixed housing to develop air pressure. Types included in this category are screw, sliding vane, rotary lobe, and liquid piston. The rotary-lobe and rotary liquid-piston-type units are rarely used in construction and will not be described.

The *rotary sliding-vane compressor* has longitudinal vanes which slide radially in a rotor mounted eccentrically in a cylinder. Air or gas flows into the space between the upper vanes and, when the shaft is rotated, the vanes are pushed down in the slides by the eccentricity of the stator casing. As the compressed air reaches the discharge opening, it escapes through the outlet port. No valves are required, but the vanes require adequate lubricating to ensure that they do not stick in the slides. Moderate amounts of the lubricating oil will then be picked up and carried into the discharge lines. This type of compressor has found wide use as a portable unit in applications where the presence of oil in the discharge line is unobjectionable or even desirable, as in the case of air-operated tools. Rotary compressors are generally quieter in operation than reciprocating units, and lower maintenance cost is claimed since moving parts in contact are minimized. On an equal capacity-pressure basis, efficiencies are comparable with reciprocating units. Sliding-vane rotary compressors are used in the range of 50 to 2,000 cfm and from 5 to 125 psig. Two-stage units are commonly used above 60 psig and are usually intercooled by a water-cooled heat exchanger.

The *rotary-screw compressor* delivers compressed air to the discharge ports of the compression chamber through interacting rotors which are closely fitted but not in contact. Pressure is developed through displacement rather than by the fan-type action of the sliding vane. Speed of rotation of the rotors is generally high (from 3,000 to 12,000 rpm), although some units, particularly in the portable class, match rotor speed to diesel engine speed in the 2,000-rpm range. Most manufacturers now offer the screw-type units in the larger range of sizes of portable compressors.

The *axial-flow compressor* may be visualized as a sequence of propeller-type fans enclosed in a conical pipe and arranged so that the air delivery from each blade passes to the entrance side of the successive blades. Thus, by dynamic action, the series of blades imparts velocity and pressure to the air. The axial-flow compressor is relatively inefficient for high pressures or when the overall compression ratio is greater than about 8. Axial-flow units are adaptable to very large volumes of air, in the range from 10,000 to 1,000,000 cfm, and would not be suitable for a general plant air-service installation. Axial-flow units are high-speed machines, usually operating at 3,600 to

20,000 rpm depending on size and pressure, and are rarely used to produce compressed air for construction service.

The *centrifugal compressor* compresses air or gas by the dynamic action of rotating impellers. Centrifugal force is imparted to the air by the rapidly rotating blades in the impeller casting. Air velocity and pressure are built up by each successive impeller in a multistage unit, and the air is fed tangentially from the last stage to the delivery outlet. Single-stage units are used for delivery pressures of a few psig. The actual design requirements will determine the choice of impeller diameter and number of stages, both of which are parameters of compressor efficiency. Centrifugal compressors are usually not efficient compared to other conventional types for air supplies below 10,000 cfm or above about 150 psi because of the design problems. They operate at high speeds in the range of 2,000 to 10,000 rpm and at volumes from 10,000 to 100,000 cfm.

Multistage units may be designed with integral interstage cooling (diaphragm cooling) or with external interstage cooling. The efficiency of intercooling and the design of air passageways will determine the overall compression efficiency of these units. Steam- or gas-turbine drives are common, but electric drives with gear-speed increasers are also used. Although construction projects rarely demand the large amount of compressed air that this type of compressor delivers in its efficient size range, it occupies little space and perhaps deserves consideration for supplying very large air requirements.

Power Sources. Power units furnished by manufacturers vary widely in terms of horsepower per unit capacity. This is particularly true in portable units, where gasoline or diesel engines are not "tailor-made" for the units and standard sizes of engines may not fit the power needs with any degree of exactness. Also, a substantial percentage of power may be absorbed by auxiliary services such as cooling (both the engine and the air compressor), starter-battery chargers, and oil pumping. The stationary compressor units, particularly in the 1,000- to 2,000-cfm class, are usually electrically driven, and standard motor sizes can be matched fairly closely to the power need. In general, compressors require about 1 hp to compress 5 cfm to 100 psi. Rated power for standard stationary compressors is usually less than the 1 to 5 ratio and for portable units is usually higher.

When selecting portable compressors, the power unit should be considered as carefully as the compressor unit. Most portable units experience frequent load changes, and the engine usually operates on low load much of the time. If the service expected of a compressor is not continuously at full capacity, less than normal engine capacity can be permitted. Normally, the engine should operate at its best efficiency at about full compressor load, and this point should be checked.

Stationary Compressors. Stationary compressors in construction service are virtually limited to being diesel-engine or electrically driven. Diesel engines usually are direct connected or driven through V belts. Electric motors deliver power through V belts or are direct connected.

Portable Compressors. Portable compressors generally cover the range from 100 to 1,200 cfm and are gasoline- or diesel-engine driven. At least one manufacturer offers a 1,500-cfm unit, but this is beyond the normal portable size. Diesel drive is normal for the larger units, but gasoline is still the favored power fuel for the smaller compressors. Most manufacturers supply standard gasoline-powered units only up to 315-cfm capacity but cover the entire range with diesel units. The average characteristics of portable compressors are shown in Table 3-2.

Central-station Plants. In contrast to the portable compressors, the selection of central-station compressed-air plants will require careful engineering studies to identify the necessary auxiliary facilities, equipment, and housing. It is usually good economy to filter the intake air, and this is a "must" if the intake air is dusty. Head losses at the intake pipe and filter should be kept low, as these will adversely affect both power and output capacity of the compressor. Intercoolers between stages of compression and aftercoolers at the discharge side of the compressors will require cooling water. Except where it is economical to have new water continually supplied, an outside cooling tower or other heat-dissipation device must be used.

The size of the cooling unit is greatly influenced by temperature and humidity, and the advice of the supplier of the compressors should be sought if climatic conditions are unusual. Safety valves must be provided in cooling chambers and in any section of pipe that can be closed by a valve and still receive pressure buildup from the compressor. Aftercoolers and air receivers require condensate traps. Under conditions of extreme cold, reheaters are frequently employed to reduce the relative humidity

TABLE 3-2 Representative Characteristics of Portable Compressors

Delivery, cfm at 100 psi	Compressor, rpm	Engine horse-power*	Weight, wheel mounted, lb	Length, in.	Width, in.
		Reciprocating			
125	1,370	41	2,650	116	65
250	1,350	77	5,445	117	65
375	1,600	160	9,250	136	74
600	1,600	216	10,400	139	78
		Rotary			
125	1,850	46	3,080	129	69
250	1,800	85	4,610	94	70
365	1,750	122	6,320	118	76
600	1,750	177	9,700	156	72
900	1,775	260	12,640	166	85
1,200	1,800	392	16,600	174	96
		Rotary-screw†			
150	2,150	70	3,910	145	64
150	7,200	59	2,260	110	68
250	1,850	27	6,390	119	66
250	3,300‡	97	5,050	102	70
365	1,800‡	115	5,800	131	69
365	3,900‡	120	7,515	126	75
600	1,850	196	10,146	145	72
600	4,730‡	196	8,800	135	77
900	1,750	274	13,300	162	88
900	2,560‡	274	11,000	151	82
1,200	1,850	410	17,865	184	80
1,200	3,385‡	410	14,280	159	84

* Most manufacturers offer gasoline and diesel up to about 250-cfm size and diesel only for larger sizes.

† Screw type separately listed because of the current trend toward this type, especially in larger sizes.

‡ Compressor speed higher than engine speed.

and reduce freezing. Careful attention should be given to slope lines toward condensate traps. In certain cases, usually in cold weather, the aftercooler may be bypassed advantageously, and this provision should be designed into the station piping. Where dry, warm air is required by equipment in cold weather, reheaters can be placed in the line near points of use. This increases the air volume or air pressure or both; thus, much of the heat applied adds to the available air energy. In cold weather, the combination of use of the aftercooler bypass, condensate traps along the line, and reheaters can produce good and efficient results. In this use, the pipeline itself acts as an aftercooler and must be carefully watched to avoid buildup of frozen condensate.

Figures 3-3 and 3-4 illustrate central-plant compressor installations on construction projects.

Air Distribution Air-distribution systems require design attention which is equivalent to that devoted to water-distribution lines. Unlike water-distribution systems and because air is compressible, air receivers in a system near points of heavy demand will tend to average the surges of air demand, reduce pipe losses, and maintain more constant pressures on the equipment.

Design of Air-distribution System. The distribution system must be designed to supply enough air, at steady and adequate pressures, to meet the needs of all phases of

Fig. 3-3 Construction power generators and air compressors were housed together at Kemano, B.C., to service the west end of the tunnel, the underground powerhouse, and penstock construction. (*Aluminum Company of Canada, Ltd.*)

the construction job. This statement, academic as it may seem, is important because systems often are inadequate, and revamping is an expensive job. First, the air needs of the job must be evaluated and points of use determined. A sample analysis is shown in Table 3-3 (page 3-13). Then, a flow diagram should be developed to indicate pipe lengths and service-point loads to be supplied at those points. Loop systems are useful, as they are in water systems, and should be used where practicable, but normally the job layout does not permit extensive use of this device. Next, pipe sizes should be selected, losses calculated, and pipe sizes revised if losses are unacceptable. Then, the layout should be studied in detail, refinements made to reduce losses, and the entire system checked again.

Pipe sizes, T's and angles, and air receivers all need special attention to keep losses down. Special attention to avoid and repair leaks in lines during operation will be profitable.

There are readily available tables of friction losses for air flowing in pipes and in various types of fittings.[1,2,4] These data are required to determine (1) proper sizes in relation to volume to be handled and (2) length of pipe. At 100 psi, a 6-in.-diameter straight pipe will carry 3,000 cfm a mile with a pressure loss of about 12 psi. A 4-in.-diameter straight pipe will carry 1,000 cfm a mile with about the same pressure loss. Pipes of 8- or 10-in. diameter are commonly used in long headings of large tunnels where 15 to 20 heavy drifter drills, with air-driven pumps, hoists, and other auxiliary equipment, may demand up to 4,000 cfm. A good rule for selecting pipe hose sizes is, "When in doubt, choose the larger."

Fig. 3-4 The portal of the Brushy Canyon Adit to Ralston Tunnel was situated in a narrow canyon with precipitous walls and no natural benches for construction plant facilities. Diesel-electric generators and compressor plant were located on a widened bench of the access road about 50 ft above the adit portal. American River Project in California. (*Kaiser Engineers.*)

The use of standard T's and pipe angles is a common but wasteful practice in compressed-air systems. Sharp angles can dissipate an amount of pressure equivalent to hundreds of feet of straight pipe. Bending large-diameter pipe to a long radius normally is beyond the capacity of a job shop, and the need for these items must be established early so that they can be purchased in time.

Pipe joints, hose connections, and any other connections in a system are potential sources of leaks that can dissipate energy from a system. The use of badly worn equipment that has developed clearances will excessive waste substantial amounts of air. These losses are not obvious, but they are real and can represent a heavy cost toll on the job.

Pipes and Fittings. All elements producing, transmitting, or using compressed air are under pressure, and state regulations (usually "Industrial Safety Code") will commonly specify minimum standards for such equipment. In general, such regulatory codes are classified under the title of "unfired pressure vessels." Some codes have definite specifications of acceptable standards; some are so general that a user must either rely on precedence as to acceptable articles or must prepare plans and refer them to the appropriate regulatory body for approval. In any case, it is essential for the planner to determine the code of the jurisdiction in which the job is to operate and to investigate acceptable practices in that jurisdiction.

Construction distribution systems and fittings should be of a type easy to assemble and disassemble without damage to the joints. Permanent systems ordinarily use threaded or welded joints. Construction services ordinarily use one of the several makes of clamp connections that permit easy disassembly and reuse on subsequent

TABLE 3-3 Typical Compressed-air Demands for Air-operated Equipment in Good Operating Condition*

	cfm at 100 psi
Heavy-duty pile hammers—standard makes	480–1,800
Cement silo aeration	200
Mobile cement unloaders (usually 40 to 50 psi)	160–350
Single grout pump, average size	250
Drifter drills with pneumatic feed:	
5½-in. diameter cylinder	600
5-in. diameter cylinder	460
4½-in. diameter cylinder	350
4-in. diameter cylinder	250
3½-in. diameter cylinder	225
3-in. diameter cylinder (jackleg)	155
2½-in. diameter cylinder (jackleg)	115
Jackhammer:	
Heavy	130
Medium	90
Light	60
Paving breaker, average size	40
Riveting hammer	25–50
Impact wrenches	15–90
Sump pump, including submersibles	80–170

* These are representative demands and may vary widely for different makes and models. They may be used for preliminary planning but should be checked specifically with suppliers for items actually ordered. Wear in cylinders and pistons adds materially to air consumption. Drifter-drill air consumption is based on wet drilling of horizontal holes, such as tunnel drilling. In dry, quarry-type vertical drilling, blowing out deep holes may demand up to double the air requirement listed. The tabulated information was obtained from job experience and manufacturers' published data.

jobs. A large cost differential exists between standard units and valves and fittings of noncorrosive materials with built-in, long-term dependability. While permanent installations may be able to afford the expensive type, the use of standard units for a construction job will usually result in overall economy.

Most clamp-type fittings can accommodate a limited degree of accidental or deliberate misalignment. This permits lines to be installed with long, sweeping bends that are much more efficient than sharp bends. For pipe sizes under 3 in., standard threaded couplings and fittings are usually the most economical.

The last link connecting distribution lines to air power equipment is usually of flexible hose. Heavy losses occur in right-angle hose-fitting takeoffs from a distribution pipe. A more efficient arrangement is to branch off from a supply line with a manifold which is two or three times the diameter of the supply line and set up a bank of hose-connection fittings. This device reduces air losses and also provides air receiver capacity at the point where it is most effective. Losses in air hoses are about equivalent to losses in straight pipe of the same diameter. Because they are usually short and because compressed-air hose in large sizes is expensive, the hose size selected is commonly slightly smaller than the pipe size used for an equivalent amount of air. Short runs of hose, 50 ft or less, demand ½-in. hose to feed a chipping hammer or light shop tool and at least 1¼ in. for a heavy rock-drifter drill.

Concentrated intermittent loads, such as a battery of rock drills or an air-operated concrete placing "gun," require an air receiver capacity close to the load point. The size of air receivers should be at least 4 to 6 percent of the volume of free air (in cubic feet per minute) delivered through them.

A special development in aluminum pipe is particularly useful in remote areas or extremely rugged topography. A rubber-sealed coupling device will couple pipes

Fig. 3-5 Air for drilling holes 4 in. by 24 ft long for a 3-in. suspension cable was provided by a 315-cfm portable compressor on valley floor 1,000 ft below precipitous cliff. Three-inch aluminum pressure line carried air to drill. (*Aluminum Company of Canada, Ltd.*)

that have end fittings welded to them by the supplier. These fittings make a coupling that is slightly flexible and will permit laying pipe in a snaky configuration. The very light weight of the aluminum pipe and fittings makes it possible for men to carry and handle the pipe and place it in position without the aid of hauling or handling equipment. A particularly difficult demand for compressed air for drilling and grouting suspension-cable anchorages in the mountains of British Columbia was met by the Saguenay-Kitimat Company by the use of 3-in. aluminum pipe with these clamp couplings. Pipe sections and clamps were delivered by helicopter at a bench in the precipitous mountainside near the anchorage sites. A crew of four men was able to assemble about 500 ft of this pipe by attaching sections on the upper end and feeding the assembly down the steep mountain slope. The lower end was then connected to a portable compressor on a rock bench that was accessible from the valley floor. See Fig. 3-5.

Safety The planner must appreciate the need to design safety into construction facilities. Compressed air is dangerous; a failure of any pressure vessel can be explosive. All practical precautions should be taken to provide pressure-relief valves on any part of a system that can be isolated and where pressures might be built up by compressors or heat. A rigid operating routine is needed to check these relief valves frequently to be sure they will work. There should be no possibility of closing the main line between a compressor and an air receiver.

ELECTRIC POWER

Most modern heavy-construction jobs use electric power in a large variety of ways. Electricity has almost entirely supplanted steam for heavy-duty hoists, such as used in cableways and fixed derricks. Concrete aggregate and mixing plants, and associated conveyor equipment, are nearly all operated by electric power. Ease of transmission of electricity and availability of a wide variety of electric motor sizes and types have caused carpenter-shop, machine-shop, plumbing- and piping-shop, and other shop equipment items to be each powered by its own electric motor. This almost universal use of electricity as the primary power source in construction makes it essential for the construction planner to devote intelligent and expert attention to the supply and distribution of electricity.

Planning Where electricity can be supplied from an existing utility system by construction of a moderate length of transmission line, this is nearly always the best and most economical solution. The planner must investigate any such supply thoroughly to determine that the capacity of the supply system and the voltage characteristics will be satisfactory.

The main requirement, assuming the capacity and voltage control are adequate, is dependability. If the tie to an existing system is too long and expensive, or if the supply is likely to be unsatisfactory, a central plant to generate electricity must be provided. If a central generating plant is to be provided, then the planned use of electricity for the entire project should be reviewed carefully. Diesel-driven electric generators in central stations are commonly used for construction. If diesel sets are used to supply power, then diesel-engine drives should be used instead of electric drives for all uses to which diesel is readily and economically adaptable. A compressed-air plant can be driven by diesel power as readily as by electric power. If a concrete plant is far from other centers of activity, possibly a special diesel-electric generator should be operated nearby to save electric transmission costs. In general, the cost of power for a modern construction job is a substantial item and deserves careful planning.

In most cases, the permanent plant or facility under construction will need an electric power supply for its continued operation. A wise contractor will review the ultimate need with his client to explore the possibility of mutual benefits that may be realized by putting a permanent installation in immediately so that it can serve construction needs. See Figs. 3-6 and 3-7.

Electric-powered Equipment Virtually every power-operated tool or piece of fixed equipment can be readily powered electrically. The principal exceptions are tools of the piston-actuated type driven by compressed air.

Nearly all electric motors employed on construction use alternating current. Direct-current motors offer easy and precise control of speed related to load, but ac motors have been improved to the point where they are usually satisfactory for construction equipment. About the only important exceptions are the trolley-battery locomotives and electric-powered shovels for underground work. Direct-current motors for very large hoists, 400 hp or over, will also require extensive starting and control equipment to provide control characteristics which are equivalent to those of dc motors. In some cases, this may make adoption of dc drive the most favorable solution. An ac motor or diesel engine coupled to a dc motor offers the advantages of dc control and is still used in special cases.

The characteristics of the motors selected for construction power are important in the design of power services as well as in the selection of control and starting devices.

Fig. 3-6 Twelve 1,000-kw diesel-electric generators similar to this unit were installed to furnish construction power for the Chute-des-Passes project. (*Aluminum Company of Canada, Ltd.*)

Fig. 3-7 Construction-style switchyard and transformer installation receives power from temporary hydroelectric station or diesel-electric station or both and feeds 44,000-volt distribution lines that serve working areas of Chute-des-Passes project. (*Aluminum Company of Canada, Ltd.*)

Both generation and transmission facilities must be able to carry heavy starting loads, if motors on large units demand them, without damage to other motors in the system. The possible advantage to a system of using synchronous motors with leading power factor should not be neglected. Such motors are readily adapted for powering compressors.

A typical analysis of construction power requirements for a large tunnel and underground powerhouse project is shown in Table 3-4.

Power Production Power needs in settled areas normally can be met by purchase of power from a utility company and construction of a transmission line to deliver electricity to the jobsite. In remote areas, generating equipment will have to be obtained. In virtually all cases, the economic choice is diesel-electric motor-generator units. The choices are either to provide small generator units that can be moved readily to serve different job areas or to set up a central station with transmission and distribution to all points of the job. For large jobs with work-area centers widely separated, a combination of both may be required for the best economy. In the rare case, local opportunities may exist for a hydroelectric plant to serve construction.

TABLE 3-4 Summary of Analysis of Electric Loads For Chute-des-Passes Project in Quebec*

Facility	Connected load, kw				
	Original estimate				Revised total
	Adit 1	Adit 2	Adit 3	Total	
Compressed air†	900	900	1,340	3,140	3,600
Ventilation fans	1,340	1,340	1,340	4,020	4,050
Concrete plant, Cement unloading plant and aggregate plant	.....	535	.....	535	1,100
Camp lighting and service	200	190	255	645	1,780‡
Tunnel lighting	180	150	200	530	550
Shops	100	100	100	300	500
Boiler plant	100	100	100	300	750
Pumps, dewatering	150	150	150	450	470
Water supply	100	50	150	300	400
Power shovels	150	150	225	525	525
Misc. yard lighting	60	55	70	185	225
Tailrace tunnel	.....	.....	230	230	250
Total	3,280	3,720	4,160	11,160§	14,200¶

* The Chute-des-Passes project was built for the Aluminum Company of Canada, Ltd., by Perini-McNamara-Quemont, a joint venture. The project included an intake from an existing reservoir; 6 mi of pressure tunnel 35 ft in diameter; underground powerhouse housing five 200,000-hp turbine generators; a surge tank; and a tailrace tunnel 9,000 ft long and 50 ft in diameter. Access and construction tunnels totaled over 10,000 ft.

† At each of three adits, one additional 720-cfm compressor was diesel powered to provide emergency service in case of electrical-power outage.

‡ Increase due to decision to heat family quarters electrically.

§ Capacity of generating plant set at twelve 1,000-kw units to provide spare units after allowing for diversity factor.

¶ Decision to build hydroelectric plant for construction power gave surplus generating capacity even after load increases. Plan was to use diesel plants for peak loads.

This is not usual, but the possibility should always be checked. In remote areas, where transportation is expensive, when the construction period is several years, and where the hydroelectric potential is at hand, the generally higher first cost of the hydroelectric plant may be much more than offset by fuel savings.

For the Chute-des-Passes hydroelectric project in Quebec, which was completed in 1960, 12 diesel-electric units of 1,000 kw each were purchased to get the job going. Near the principal working area was a dependable hydroelectric potential of about 12,000 hp that could be developed through available used turbine-generator equipment that had produced construction power for a nearby development. Careful cost studies indicated that the savings in fuel oil would be substantially more than the cost of the hydroelectric-system installation, and so it was built. The diesel station was located adjacent to the hydroelectric site so that both plants could deliver power into the common switchyard that serviced the entire job. Some of the diesel units were retained for "peaking."

Initial costs of steam generating plants are high compared with diesel electric. Where suitable local fuel is available, usually coal or natural gas, and where the client may have permanent need for a standby plant on the job, it may prove advantageous to build such a plant early enough to serve construction needs. Even then, some diesel-electric generation probably will be needed to build the permanent plant.

Diesel-electric generators are made as standard units from 15 to 1,000 kw. Units from about 300 kw down usually are available from manufacturers' stock. Larger units can be obtained, but for construction service it is usually better to have a number of smaller units than one or two very large units. For fractional system loads, the plant with smaller units can operate only as many as are needed and thus obtain better operating efficiency. A failure of one of many small units is not the disaster that failure of one of two or three large units might be. Where a project has more than one work center with a separate diesel-electric power supply for each, the possibility should be explored of obtaining a portable turbine-driven generator that can be moved to support any of the generating centers as required. Units in the range of 500-kw size are available, and the trend in development is toward higher efficiencies and lower fuel costs. It is not uncommon for large projects to require generation and distribution facilities to handle 10,000 kva or more.

Diesel-electric sets up to about 350 kw normally generate three-phase power at 220/440 volts. Larger units normally generate at 2,300 volts, which is the normal voltage of motors above 50 hp. Motors of 50 hp or lower normally require 440 or 550 volts down to the range of size of motors operating small shop equipment, which normally take 220 volts. Fractional-horsepower motors on hand-held tools are supplied with a 110-volt current and should have a direct-ground-wire connection to minimize danger to personnel. Motors working in areas that are constantly wet, even if above the 50-hp range, should operate at 440 volts or 550 volts rather than 2,300. Good examples are in cofferdam pumping and in tunnels. While end-use voltages are not important factors in the selection of generating equipment, they are important considerations in the planning and design of the switchyard and transformer equipment.

Power Distribution Design of a distribution system for a large construction job requires detailed analysis of the location of power-demand centers, the voltages required at each, and the extent of the lighting load at each. While some diversity can be expected, the general service lines should be designed for 90 percent or more of the expected connected load in any one area. Most motors used in construction are induction motors and tend toward a lagging power factor that can throw severe recirculating current loads on a distribution system and the transformers. Distribution-system loads should be evaluated in terms of kilovolt-amperes, taking into account power factors anticipated for both full and fractional loads on equipment. In general, conductor sizes and transformer sizes should be selected on the high side of calculated needs. The tendency of a construction force is to use more rather than less power and lighting than was originally estimated.

Figure 3-8 shows the preliminary load study and line diagram of generation and distribution for Adit No. 3 of the Chute-des-Passes project. Figure 3-9 shows a line diagram of the entire system as built to serve operations at the three adit areas.

The extent of electricity needs for lighting is commonly underestimated. Particularly in outside work areas, a job cannot afford to have work done under the severe handicap of poor illumination. It takes a lot of kilowatts to illuminate the working area of a dam or to light up miles of tunnels.

General-area distribution lines and transformers should be located along the edges of the area to keep them away from work activities. Transformer units of 500 kva or less are relatively easy to move and reinstall and, in general, it is better to use two 500-kva units than one 1,000-kva unit. Standardization of transformer sizes has frequently proved to be extremely valuable for construction service. Even if some transformers are larger than their intended load requires, the ability to interchange them freely may make it possible to avoid the shutdown of a construction operation.

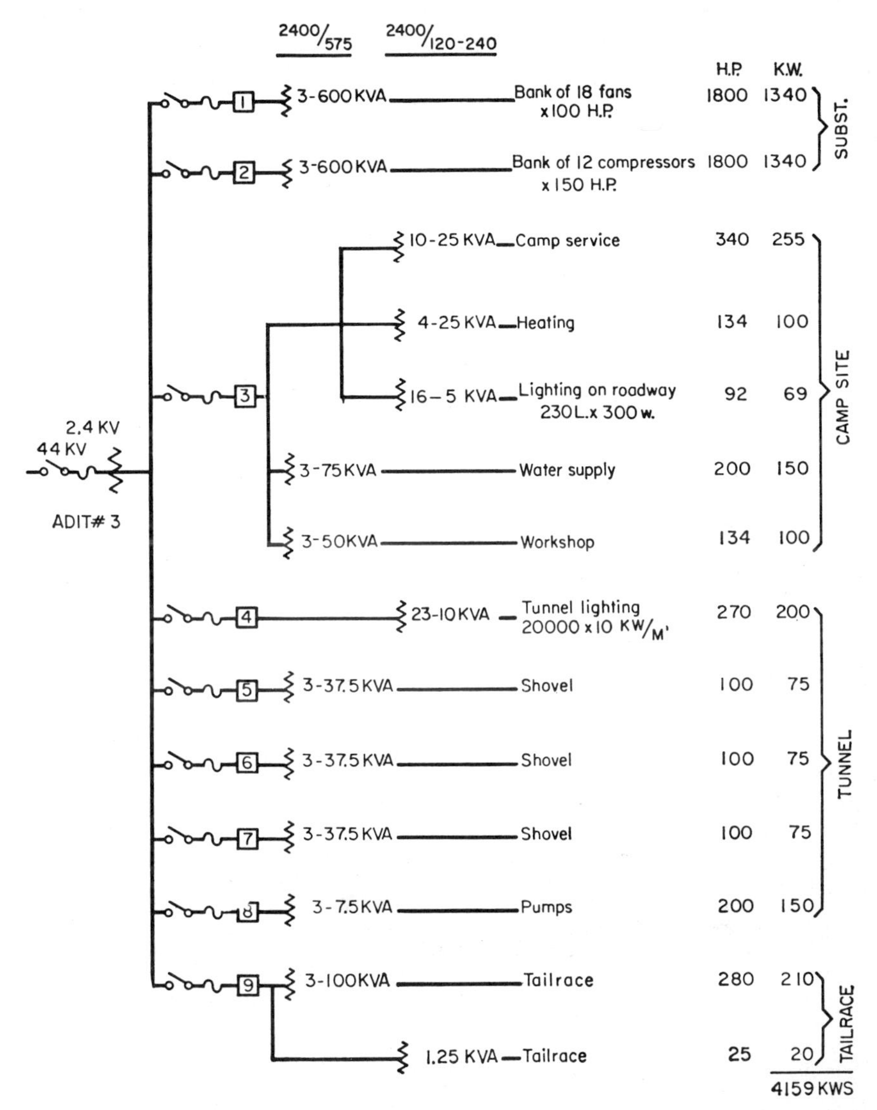

Fig. 3-8 Preliminary load study and line diagram of generation and distribution for Adit No. 3 of the Chute-des-Passes hydroelectric project. The underground powerhouse, 50-ft-diameter tailrace tunnel, penstock tunnels, and part of the pressure tunnel were constructed from this adit. Similar studies were made for Adits 1 and 2 which gave access to the rest of the project. (*Aluminum Company of Canada, Ltd.*)

Power lines are extremely hazardous to mobile cranes, trucks with dump bodies up, fork-lift trucks with high loads, and other high-clearance construction equipment. Although the transmission-line designer will reverse the statement just made, it is a fact that many accidents result from having lines too low—accidents to personnel, to lines, and to equipment. A contractor cannot afford to take undue risks of this kind. In timber country, ample right-of-way clearing must be performed and combustible material must not be allowed to accumulate under the conductor or supporting poles.

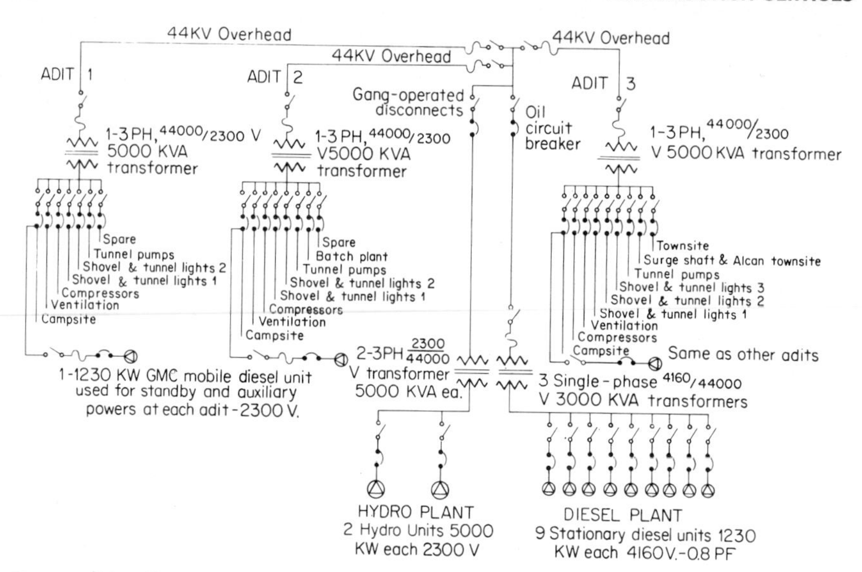

Fig. 3-9 Line diagram of Chute-des-Passes project generation and distribution system as built. (*Aluminum Company of Canada, Ltd.*)

Fig. 3-10 Mobile, dry-type transformer and master switch in Kemano, B.C., tunnel used to supply electric power to mucking machine. (*Aluminum Company of Canada, Ltd.*)

Electrical facilities must be dependable, adequate, and safe if an economically successful job is to result.

Auxiliaries Equipment controls, protective relays, and interlocking controls are more a part of equipment design than of the electric-power distribution system design. The function and type of such auxiliaries must be known to the system designer. They may have important effects on the power-supply operation, even though they may actually consume only minute amounts of power. See Fig. 3-10.

System protection should be built into the original design of the electrical services. Bus lines should exist only at main substations, with circuit-breaker protection on each line which is attached to that bus. Disconnects should permit isolation of circuits for maintenance purposes. Service lines to camps, offices, mess halls, and shops should have independent circuit-breaker protection whenever they feed from circuits that supply heavily powered equipment.

Safety The importance of building safety precautions into the electrical supply system must be appreciated. It is better to avoid the chance of exposure to accidents than to try and train men consciously to avoid doing something that might lead to an accident. The design should make it easier for men to work safely than for them to take risks. Three-wire circuits, for example, will provide a more positive ground for hand-held tools than will a short circuit through a man. Safety needs to be ever present in the mind of the designer.

WATER-SUPPLY SERVICES

Planning Construction projects require supplies of both domestic and industrial water. Domestic water, some or all of which must be potable, is required for offices, shops, job areas, construction camps, etc. Normal job demands include industrial water for cooling electric generating plants and compressor plants, general job services, fire protection, rock drilling, dust prevention in quarries and tunnels, crushing and screening plants, concrete mixing, rock cleanup, concrete curing, grouting, feed water for boiler plants, and refrigeration of water for concrete. Services that require industrial water on special projects include excavating and/or transporting material by sluicing, moisture control of earth dam fill, transporting materials by dredge pipelines, and sluicing rock fill.

The needs for both domestic and industrial water should be analyzed in the same manner that has been outlined for compressed air or electric power. It should be appreciated that most people are careless of water use, and here construction men are no exception. Thus, it is of particular importance in the case of water supply to make generous estimates of needs, with ample provisions to meet those needs.

Water can be an expensive commodity to a construction project, and available sources and quality should be determined in the initial phases of job planning. Water needs to be tested for potability, hardness, scale-forming characteristics in boilers, suitability for concrete, and other similar characteristics. If treatment is needed for a potable domestic supply, the bigger industrial loads should certainly be isolated and supplied through separate lines. For the limited amount of water that does not have to be potable and which is used in camps, offices, shops, and for general purposes around a job, it is usually cheaper to use treated water than it is to put in a dual system of distribution lines.

Careful consideration should be given to the storage requirement, which commonly is a function of the dependability of the supply. In nearly all cases, some gravity-feed water storage should be provided to handle heavy surges of demand and for fire protection. It is not uncommon to need up to 100,000 gal of storage to supply a large construction project.

Particularly in the northern areas, weather protection is of extreme importance. As is the case with other services, the supply must be dependable at all times; the freezing of a major supply line is just as disastrous to a job as the failure of a pumping system. In extreme climates, pipes carrying water may have to be buried as much as 7 ft deep to get below the frost line. Care should be taken to get greater than normal coverage where water lines cross under roads, walks, or yard areas which may be cleared of snow, thus permitting the frost line to go deeper than normal. Permafrost may be encountered in extremely cold climates and will require an entirely different treatment of water distribution. Successful use has been made of pipes laid in boxes on the surface, surrounded by insulating material and carrying either electric heating around the pipe or a steam line in the same box. The return line of a steam-heating system will usually provide enough heat in an insulated box trough to prevent freezing. The usual precautions of insulating the pipe connections from the distribution main to an office, shop, or house should not be forgotten. It is far better to provide ample protection initially than to have to repair or thaw out frozen service lines under conditions that require the use of jackhammers or chipping hammers to excavate frozen ground. A satisfactory and effective method of thawing out frozen pipes is by attaching the leads from a welding machine on either side of the zone that is frozen. The possibility of this should be anticipated in critical areas so that the pipe is accessible.

Water-distribution lines can be of cast iron, steel, cement asbestos, or wood-stave pipe. If there is any possibility of major advantage to the client in putting in permanent lines initially, this should not be overlooked; in this case cast iron is probably best because of its strength, long life, and resistance to corrosion. Care should be taken to avoid high spots in the pipeline that might entrap air and seriously reduce the water-carrying capacity of a pipeline. Where high points cannot be avoided, air-relief valves should be installed and access to them provided from the ground surface.

Sources and Treatment *Sources.* When surface water can be obtained nearby in sufficient quantity and of good quality, this is by far the cheapest method of getting water. It may also be more economical to pump water several miles from a suitable surface supply rather than to develop wells or to treat water from closer sources. The topography should be investigated to find natural reservoir sites which are located high enough to provide gravity feed to the job. If the available water supply consists of a small stream, very thorough investigation should be made of the dependability of its flow. Studies of its stream-flow records, if available, or those of adjoining streams, and investigation with local inhabitants of the area should determine whether or not the supply will be adequate. In northern areas, it should be remembered that winter runoff from small streams can be almost completely stopped during extended periods of cold weather.

If surface water is not available, probably the next most desirable source is groundwater from shallow wells. If an open river or lake contains water that is not sufficiently clean or safe, under some conditions satisfactory water can be obtained from shallow wells excavated in gravel banks near the shore. Pervious natural gravel does a fair job of natural filtering. Wells of this type should be subjected to several days of continuous test pumping to prove out the supply.

A third standard source of water is from deep wells. This is usually the most expensive, as it not only requires expenditures to drill the wells but also demands a continual, heavy pumping load. Water from deep wells commonly contains more minerals than surface water and should be carefully checked to determine its tendency to develop boiler scale. Where the steam-plant demand for water is high, the tendency to develop boiler scale may have to be corrected by using a closed system to condense the steam and reuse the water. Thus, the effect of careful planning of water services may lead to substantial changes in the type of boiler-plant facilities that will improve dependability.

Projects involving tunnel construction may in some cases produce their own water supply by collecting the water from tunnel leakage. This cannot be anticipated beforehand, but the possibility as the job goes along should not be ignored. Water is needed at a tunnel heading to operate the drills and to wet down the blasted rock to clear the working area of dust. Frequently, a separate pumping system from a sump in the tunnel floor can provide enough water to take care of this need. The construction man and the planner need to be alert to take advantage of any such opportunities that develop.

Treatment. Domestic water may require chlorination, filtration, coagulation and settling, aeration, or almost any combination of these treatments. Government health agencies usually will test water for potability on request, and their tests will usually command higher legal respect than those made by private laboratories. It is not safe to use a water supply without adequate tests for bacteria content. It should be remembered that large concentrations of men are involved in construction projects. Poor water can lead to serious local epidemics and result in heavy liability against the contractor.

Various types of chlorine dosers are available for treating water. Where chlorination is used, frequent checks should be made to ensure that small amounts of free chlorine are residual in the water at all points in the system. Long dead-end pipes should be avoided, as tests for free chlorine in those lines may lead to overdosing the water at other points. In tropical or semitropical areas, chlorination may not be sufficient to eliminate certain types of bacteria and filtration may also be required. Filtration may be required in temperate and northern areas to clear the water of suspended solids, but this is not usually a biological consideration.

Pumps and Pumping Most pumps on a construction project are electric driven, particularly those used to service normal distribution. Pumps are available in a wide variety of sizes and are capable of pumping against wide ranges of head. For ordinary ranges of head, impeller pumps directly driven by electric motors are popular. For extremely high heads, displacement pumps are common in permanent installations but are rarely used for construction purposes. Several stages of impeller pumps and the use of boosters will be adequate for almost any head. The water supply must be dependable, and every large construction job should be provided with some emergency portable pumps, gasoline or diesel driven, which should be started frequently to ensure that they will be in working order should emergencies arise. Particularly where fire protection is involved, spare pumping capacity of not less than one pump should always be provided.

Distribution *Domestic Water.* The domestic-water distribution for a construction project must meet all the design and health requirements which are normally established for cities or other communities. It is advisable to observe these regulations rigorously to avoid possible indefensible liabilities. In general, a potable-water system which is separate from the normal domestic-water system is not desirable except where treatment is extremely expensive. The system should be designed to carry ample water, as an increase in capacity of 20 to 40 percent above estimated needs is usually not expensive. The need for and use of water are likely to grow as the job develops, and amounts required are difficult to estimate precisely before the fact. For domestic water, there should be sufficient storage under average conditions to supply something like a day's use, and it is extremely desirable to have that storage feed the system by gravity. Especially for systems with long lines and dead ends, careful consideration should be given to valves and fittings that are not quick closing, as water-hammer shocks can be just as serious on temporary systems as they are on permanent systems.

Industrial. In general, the design requirements of an industrial water-supply system are the same as those for any other supply system. The principal difference is that industrial water is apt to be in demand in large amounts for short periods, and this leads to a need for substantially larger pipes. It is generally impractical to store the very large volumes that would be required to carry the industrial supply for a day or two, but some dependable storage is essential, if only for the purpose of fire protection. It normally is not feasible to separate the water supply for fire protection from the supply for industrial purposes, but special precautions must be taken to ensure that storage water is always available for fire protection.

In a plant area, a liberal number of T's and taps should be built into the initial lines. This is easy and cheap to do initially, but expensive and difficult to do after the system is in operation. Aggregate screening plants usually operate with wet screens, and a 6- to 8-in. pipeline is generally required to handle the large amounts of water needed. Concrete plants require less water, but it is customary to supply them with a 3- or 4-in. line. In tunnels and quarries, an ample supply of water should be provided to wet down the blasted rock and reduce the dust produced during drilling. This is a health precaution to prevent silicosis and is frequently subject to governmental regulation and inspection.

Fire Protection Fire hydrants tapped into the main distribution system should be provided in shop areas, in camp areas, and around the fringes of a job where forest fires might be a hazard. For forest areas remote from the central camp or working areas in which men might be active, 2- to 3-in. portable gasoline pumps with suction hose and hose discharge lines should be strategically placed and operated frequently enough to ensure their good working condition. These pumps will draw from streams or ponds or any other place that water is available and can pump up to 1,000 ft quite effectively. In camp buildings, particularly two-story wooden camps, automatic sprinkler protection is desirable. Sprinkler or deluge water protection should also be provided in areas where highly combustible materials are stored. Automatic fire protection can be ensured, even in an unheated space, by having the protective system filled with air which will be released if a fire triggers the release valves. Water will then fill the pipes behind the exhausted air. A fire engine with gasoline-driven pumpers and hose should be provided for every major construction project to develop effective fire-nozzle pressures beyond the system pressures. In addition to water protection, chemical fire extinguishers must be strategically placed and regularly serviced.

HEATING

Heating requirements on construction projects cover a variety of needs in construction camps, offices, and shops. If these are located reasonably close together and if the job is of reasonably long duration, a central boiler plant with steam distribution is usually the most effective and economical method. Central boiler plants may be fired by fuel oil, natural gas, wood, or coal, and the decision should be made on the basis of the type which is available locally at the lowest cost.

Low-pressure oil-fired boilers are now manufactured in a wide range of sizes. They can operate without the constant attendance usually required by law for high-pressure plants, since they can be controlled by thermostats and pressure-relief valves which operate automatically. These boiler units are growing in popularity for heating mess halls, barracks, and offices, where a single unit will usually handle one or two buildings. Heating can be accomplished through standard radiators or baseboard units. In addition, hot-water systems which will also operate without attendants can be provided.

For large shops and warehouse areas, the favorite heating system uses overhead heaters which blow air through steam radiators or electric heating coils. These units are also convenient, where steam is readily available, for heating areas of fresh concrete in cold weather.

Fuel-oil or kerosene-fired space heaters are low in cost and should receive consideration, especially for short-term jobs. They are a greater fire hazard than central systems for buildings or for areas, and this fact must be considered. Electric space-heater units are relatively firesafe and are available as convector units either with forced draft or using natural thermal circulation. They are also available as radiant units or as low-temperature baseboard units. The standard residential types of forced hot-air, oil-fired furnaces can be used to advantage in camp buildings.

In hot, dry climates, consideration should be given to heating offices and camp buildings with circulating air. This offers the possibility of blowing the air through steam radiators in the winter for heating and using the same setup to blow air through wet filters for cooling in the summer.

Heat is frequently required temporarily in job areas, such as where concrete is placed in the winter and a substantial amount of heat must be provided to protect it from initial freezing. The use of salamanders for this purpose has largely disappeared. Modern heating units which use fuel oil with a powered fan to circulate the hot air have just about supplanted all other types of heaters for remote areas where temporary heat is required. The modern units are far safer than the old-style salamanders, as they blow in large volumes of warm air and there is no high heat concentration anywhere near them.

SEWAGE DISPOSAL

Most construction jobs are of relatively short duration, and complete domestic sewage treatment is not normally justified. The dumping of untreated sewage is usually not permitted or else is sufficiently opposed by public opinion as to be an unacceptable practice. The alternative is treatment in septic tanks, with the effluent discharged to gravel dispersal beds or directly into natural drainage channels. Septic-tank treatment is effective through a wide range of sizes, and the facilities are not expensive. Thus, in a big job, sewage collection will probably be minimized by locating several septic tanks near the sources of sewage.

Industrial wastes resulting from construction are small. Used lubricating oil does a fair job of dust prevention around shop, camp, and plant areas. Shop scrap is mostly solids, which can be burned or buried along with garbage.

VENTILATION*

Underground work requires forced ventilation. This, for long tunnels or underground powerhouses, represents a heavy power load. A large tunnel heading needs 20,000 to 50,000 cfm of air to clear out blasting fumes within a reasonable time. Reversible fans should be used so that the concentration of fumes immediately after a blast can be sucked out for a few minutes, then the fans can be reversed to blow fresh air to the heading so that work can go on. Booster fans generally are installed about every 1,000 ft in long runs of ventilation pipe to make up friction loss and avoid subjecting the usually light, thin-wall ventilation pipe to high pressure and suction loads. Fans are usually electrically driven and are sized to be integral parts of the ventilation pipe system. The fans at the tunnel portal or adit portal may be diesel driven, but the boosters in the tunnel should be electrically driven to avoid adding to underground exhaust fumes. For the Kemano, British Columbia, 25-ft tunnel (1952–1954), each heading used two diesel-driven 20,000-cfm blowers, each with two electrically driven booster fans. At the Chute-des-Passes tunnel (1956–1960), blowers and boosters were electrically driven. Each heading had a 48-in. blower rated at 52,000 cfm and powered by a 100-hp motor. The ventilation power load was 4,050 kw, or almost 30 percent of the total project power demand. At the Angeles Tunnel, near Los Angeles (1967–1969), each heading was served by a 125-hp 48-in. blower supplying about 75,000 cfm with boosters every 1,000 ft.

In most states, industrial safety codes specify minimum requirements for ventilation air in underground work or excavation areas in terms of the operation to be performed. The applicable code should be carefully checked in planning and designing ventilation facilities. For example, if diesel power is allowed in underground work areas, there is generally a minimum specified amount of air supply per maximum horsepower of diesel power generated in the underground area. The design of an adequate ventilation system requires both an understanding of code requirements and a detailed knowledge of the construction methods and equipment to be employed. Sharp turns and abrupt changes in pipe cross section should be avoided, as they greatly increase friction losses as compared to longer radius turns and gradual cross-section transitions. Information is readily available for selecting blowers and pipe sizes and for identifying losses in vent-line bends and fittings.[4]

A quick approximation of ventilation air being supplied to a tunnel heading can be made by measuring smoke-drift velocity toward the tunnel exit and applying that velocity to the tunnel cross-section area.

REFERENCES

1. Ingersol Rand Co., *Compressed Air Handbook.*
2. T. Baumeister and L. S. Marks (eds.), *Standard Handbook for Mechanical Engineers,* 7th ed., McGraw-Hill Book Company, 1967.
3. Literature of manufacturers of compressed-air equipment.
4. Joy Manufacturing Co., *Fan Selection Tables.*

* See Sec. 31, Tunneling.

Section 4

Construction Planning and Scheduling

ROBERT C. McLEAN

Vice-president, Construction Data Systems Corporation, Stanford, Calif.

INTRODUCTION

Construction planning and scheduling has not been, is not now, and never will be a science. It is, in every sense of the word, an art. The function of planning and scheduling of construction work has been carried on in the past by talented individuals, each of whom has had a great capacity for details and an ability to see in his own mind how a specific project would be built. As projects have become more complex, these individuals have had to seek or devise various planning aids to help them to more fully comprehend, evaluate, and remember the numerous ramifications of these more complicated construction undertakings.[1]*

This section will present information on the critical path method (CPM) and explain how it can be implemented by a construction organization. CPM is a systematic approach to planning, scheduling, and control.

Importance of Preplanning Planning can be defined as the devising of a scheme for doing, making, or arranging something; a project; a program. When the prefix "pre" is added, "preplanning" implies prior arranging before the work being planned is under way. In other words, it goes farther than just deciding what work will be accomplished in the next day or two. It delves into the sequencing of work and the detailed planning of work cycles, well in advance.

Unfortunately, there is no accurate way of measuring the value of good planning. Most construction projects are "one of a kind," and it is not feasible to compare a project constructed after a nominal amount of planning with an identical undertaking where planning has been performed in great detail prior to construction. Lacking this basis for direct comparisons, it is difficult to place a dollar value on the thinking that goes into preplanning a job.

CPM is an important tool in the planning of construction work of any size. Essentially, the project is broken down into manageable work items, called activities, that are required to produce the desired result. This allows a separate analysis of how these various activities may be accomplished and what construction methods will be employed. Each activity thus considered will require a certain period of time to perform and will cost a given amount. If either the method or the speed of performance is changed, there will be an associated change in the cost. An activity may be such that it could be accomplished at a relatively low cost, but the time required might then be so long as to interfere with other work on the project. In other words, it probably is not possible or practical to perform each individual activity in the most economical way. Certain activities will have to be expedited so that the overall project efficiency is as high as possible.

Once the project has been taken apart, so to speak, into a number of separate steps, these must all be considered collectively to show how they interrelate. This task is accomplished by preparing a diagram or model on paper to indicate the relationships between activities. Using this model, or "network" as it is called, as a basis for further study, one can evaluate how different plans and schedules will affect the overall project.

Intuition or good judgment alone is hardly enough to depend on in planning any construction project, large or small. There are too many work items that must be integrated to rely or depend on the human mind to encompass them all. For this reason, even the most well-regarded construction managers and planners should strive to systematize their work so that the details are handled routinely, leaving the mind free to think in terms of the overall picture.

* Superior numbers refer to the list of references at the end of this section.

Uses for Detailed Job Coordination The critical path method provides a methodical system for keeping track of detailed activity interrelationships, and it forms the basis for meaningful analysis of the overall project. CPM can be used to a lesser or greater extent to provide information for planning, scheduling, and control as follows: [2–6]

1. It provides a clear identification of the activities which determine the completion time for the overall project. These critical items can then receive detailed attention. This is management by exception, because the less critical activities do not need to be controlled as closely.

2. It gives a quantitative evaluation of the amount of scheduling leeway that any noncritical activity possesses, usually referred to as "float" or "slack." This establishes the time limits for any shifting of starting time or for slowing down the work to facilitate the schedule without detrimental effect to the project duration. In the case of "free float," it establishes the shifting limits so as to not delay any other activity.

3. It systematically provides data, based on project restraints, on the earliest date that each activity may begin and on the latest date that it may start without interfering with the determined completion time. If an activity is commenced as of this latest start, it will become a critical activity.

4. It provides quantitative information to be used in a factual consideration of both time and cost in choosing methods, equipment, work force, and the most economical project completion date.

5. It provides a tool for logically evaluating alternate strategies that may be used in the execution of the work. An entire project may be planned several different ways on paper at a very low cost. The data from these possible procedures can then be rationally compared to pick the best plan.

6. It offers a procedure for assessing the effect of a delay or scheduling variation in one activity on the overall project and on other individual activities. It can be used to evaluate how change orders and extra work orders affect the performance of other activities farther downstream where there is no direct relationship.

7. It can provide a basis for a cost control system that relates to the cost of work items as shown on a CPM network rather than to unit quantities. Each activity can be assigned one or more cost codes.

8. It can, if used properly, be a valuable tool in timing the purchases of materials and in coordinating the prime and subcontractors' work on an overall project basis. There can be a real meeting of the minds on how the work of the various subcontractors dovetails together and must be coordinated. It can help the general contractor perform his function as project coordinator.

9. In the last few years, CPM has become an accepted tool throughout the construction industry. Contractors who are using it have been able to do a better job of planning and scheduling and have realized a greater profit through better management and control.

10. CPM is a valuable tool during the bidding stage of any large project, since it provides a basis for the detailed planning of financial requirements. It can further help in determining the probable length of the job and the time that supervision and equipment will be needed.

Three-phase Application The critical path method can be subdivided into three phases. Phase I encompasses the preparation of a detailed job plan in the form of a network diagram. This diagram displays the interrelationships of the various activities that make up the project and is a valuable tool in itself; it also forms the basis for the other two phases. Probably 60 to 70 percent of the time spent on CPM will be used in the preparation of the network, and the task deserves the full support and effort of top managers of the project being planned. This point cannot be overstressed. If management is not behind the undertaking, and if the information available from a network is not used, a great deal of effort can be wasted. In this phase, the representation of the interplay of activities is the most important idea.

Phase II introduces time as a parameter of job planning. A realistic estimate of the time required by each activity must be prepared. From these time data and the sequential relationships which were diagrammed during Phase I, certain *time elements* or time control data can then be calculated. This second phase points out the critical activities that control the completion of the project and provides a quantitative measure of the float of all noncritical activities. The mechanics of Phase II are extremely simple. Since they are also fairly routine, this phase lends itself to the use of digital computers on networks of significant size or where frequent updating is warranted or required.

Phase III takes into account the possibility of variations in scheduling in order to expedite a project at least cost or to schedule manpower and equipment resources to best advantage. Many individuals and groups have attempted to program algorithms which will automatically generate an "optimum" schedule for this phase of CPM. The writer is of the opinion that a great deal of judgment control is lost when sole reliance is placed on the computer for this purpose. Simplified assumptions must almost invariably be made for computer implementation of these algorithms. Users of CPM, and of Phase III particularly, should realize that the network analysis is only a tool for presenting detailed information on which to base job decisions.

HISTORICAL DEVELOPMENT OF NETWORK METHODS

The actual theory and mechanics of critical path and network techniques have been developed since the mid-1950s. It is interesting to note that two independent research groups, each working on the problem of scheduling and control of projects, came up with separate schemes to improve management's ability to comprehend and analyze complicated production problems. These two schemes both relied on some sort of a picture to show the relationships between the activities of an overall project.[3,4]

One of these two schemes, which has come to be known as the critical path method, was developed in a study made by the engineering services department of E. I. du Pont de Nemours and Co. About the same time, the Special Projects Office of the Bureau of Ordnance, U.S. Navy, set up a task force to study management control techniques. This group developed the system known as PERT (program evaluation and review technique).

These two systems took different views of the basic concept of scheduling, and for very good reason. CPM was developed by an owner-builder company for use in planning, scheduling, and controlling its own work. Background knowledge was available within the company to aid in diagramming a realistic, sequenced plan of actions and in assigning accurate time estimates. Therefore, CPM has been characterized as a "deterministic" approach.

PERT, on the other hand, was developed to coordinate and control a project which consisted essentially of research and development. For most of this work, there was no available backlog of experience on which to base realistic time estimates. Logically, therefore, the PERT system incorporated a "probabilistic" approach in estimating activity and project times. However, in recent years the probability aspects of PERT have essentially been abandoned, and so the two names CPM and PERT have become almost synonymous.

Although each of these systems uses a network as its basis, another of the differences between early PERT and CPM was in the diagramming notation. This difference will be covered in more detail in the subsequent discussion of network techniques.

In 1958, at Stanford University, a study of possible cost reduction techniques for construction was started. Following an independent effort to develop scheduling organization by a system similar to what is now referred to as the "time-cost tradeoff technique," the early work on CPM and PERT, using computers, came to the attention of the researcher.

At this point an investigation was launched to see if this new tool could be successfully and simply applied by contractors in general, without the use of expensive computers. This work, carried out by John W. Fondahl of the Civil Engineering Department, departed from the conventional arrow diagramming techniques and proposed instead to adapt some of the process charting techniques, commonly used in industrial engineering, to CPM networking. From this study came the report entitled "A Non-

Computer Approach to the Critical Path Method for the Construction Industry,"[7] which was made available to contractors in late 1961. A condensed version of the report also appeared at about the same time in *The Constructor*, the official publication of the AGC.

Work has continued at Stanford in developing this system in more detail, and others working with the same diagramming technique have also contributed to its use. The system has been used extensively by the writer and has gained wide acceptance in the construction industry. Much of the material following is based on this technique, which is referred to as the "precedence" diagramming notation.

BASIC NETWORK TECHNIQUES

The use of any of the so-called "critical path" planning systems is dependent on the formulation of a logical sequencing of the various work activities to be planned. Several methods have been devised for this, but most of them require that the plan be put down on paper in the form of a logical sequence diagram. Among the various names which have been devised for these diagrams are "arrow diagrams," "logic diagrams," "flow charts," "precedence networks," or just plain "networks."

Three Types of Network Diagrams As has been mentioned, there have been three areas of development in the use of critical-path-type planning. CPM, PERT, and the variation on CPM that has been advocated at Stanford each has its own diagramming system. CPM and PERT both use, with only minor differences, what is widely known as an "arrow diagram." Since these diagrams are basically the same, the mechanics of arrow diagramming will be presented in general, and then the differences between the two usages will be pointed out. On the other hand, precedence

Fig. 4-1 Arrow and event nodes.

diagramming uses a system of networking that is entirely different from arrow diagramming. This system, too, will be explained and contrasted with arrow networks.

The basic component of an arrow diagram is an arrow with a circle or node at each end, as shown in Fig. 4-1. The arrow "head" is at node j and the "tail" is at i. The arrow represents an activity that requires time to complete, and the two nodes are the beginning and ending *events* of this activity. Usually the nodes are numbered, so that any arrow may be referred to by its two event numbers; e.g., activity i-j. Alternatively, an activity can be identified by a single letter rather than by its pair of event numbers. The number pairs are favored for identification of activities when computers are to be used in processing the time element data.

Since the purpose of a network diagram is to show the activities that make up a project and the relations between them, a number of arrows can be combined into a single diagram. The arrow head of one activity must then meet the arrow tail of succeeding activities, and this makes the diagram preparation difficult at times.

Figure 4-2 represents a plan such that, after activity 1-2 is completed, activities 2-3 and 2-4 can be started. Activity 3-5 can follow after 2-3; 4-5 and 4-6 can follow 2-4, etc. Event 5 is the moment in time when both of the two preceding activities have been completed and activity 5-7 can start.

The diagram must represent explicitly the activities being planned. That is to say, each arrow in a diagram must represent a separable piece of work or allotment of time. One cannot say that activity B will start after activity A is half completed. Instead, activity A must be broken up into two parts, A_1 and A_2, and must be represented as shown in Fig. 4-3. An example of this treatment of overlapping activities is supplied in Fig. 4-4, which illustrates a plan for forming a long wall. Identified activities could include erecting outside form, placing reinforcing, and buttoning up the wall form.

Certain key questions will aid in the formulation of diagrams. These will be identified later in this section, as they are applicable to all diagramming systems.

One other important point must be discussed before we leave arrow diagrams and go on to the precedence diagram system. Figure 4-5 reproduces the diagram formerly shown as Fig. 4-2, but with letters now used to identify activities. One new arrow, labelled d_1, has also been added. This new arrow is dashed rather than solid. It is called a *dummy* or restraint arrow, and it indicates that activity H cannot be started until after D, E, and F are all completed. If D, E, and F were instead terminated at a common node, this would indicate that G depended on the completion of F. Since this

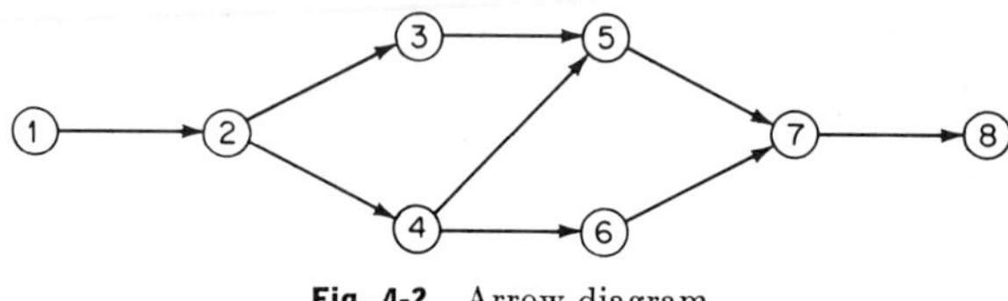

Fig. 4-2 Arrow diagram.

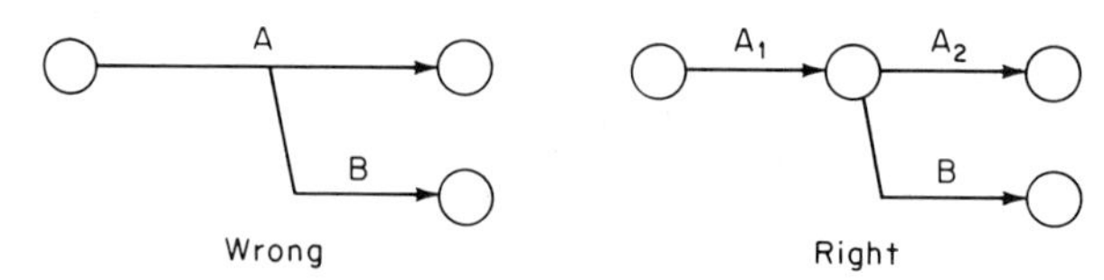

Fig. 4-3 Arrow diagram of overlapping activities.

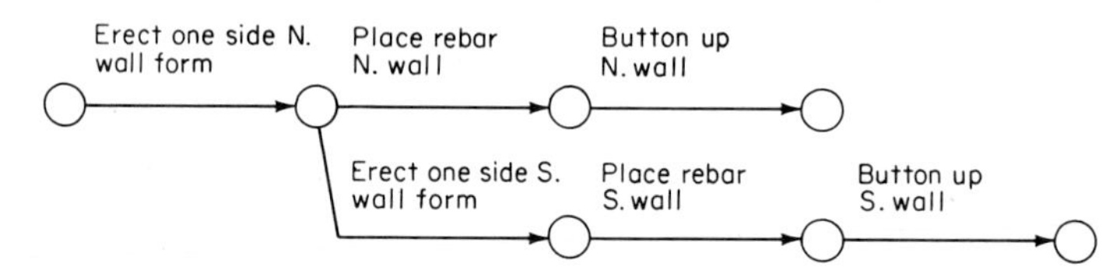

Fig. 4-4 Arrow diagram showing overlapping work on two wall forms.

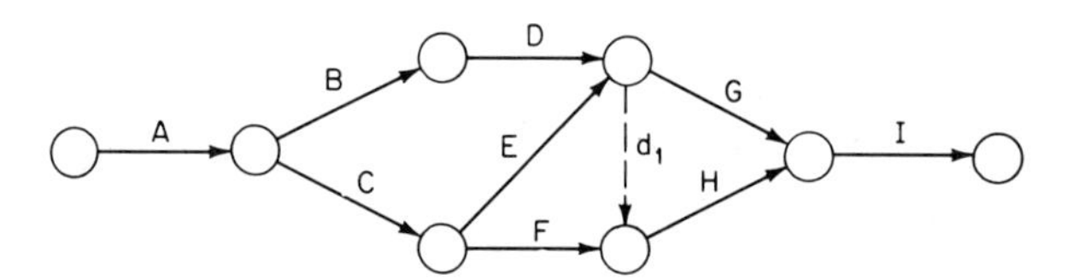

Fig. 4-5 Arrow diagram with dummy activity.

is not a requirement, the dummy is necessary to indicate the proper activity sequences.

A dummy arrow does not have any time or work associated with it as other arrows do. It is merely introduced in the diagram to indicate the proper sequence for the activities, and this concept caused a great deal of anguish for early CPM users. It is quite easy to identify activities that take up time or effort, but a dummy arrow does not represent this type of activity. It is merely an artificial device to indicate a sequence restraint between activities that cannot be shown in a more straightforward manner. As will be shown subsequently, the precedence diagram system does not require its use.

The diagram supplied as Fig. 4-5 indicates the network relations which are listed in the following table:

Activity	*Precedes activities*
A	*B* and *C*
B	*D*
C	*E* and *F*
D	*G* and *H* (via d_1)
E	*G* and *H* (via d_1)
F	*H*
G	*I*
H	*I*
I	

The last few paragraphs have discussed arrow diagramming in general. The customary CPM and PERT approaches both use this type of diagram, although the two systems begin to deviate when actual labels are placed on the diagrams. The arrows used with conventional CPM are labeled with the name of the activity, as shown in Fig. 4-4. In most PERT diagrams, on the other hand, the events are labeled as shown in Fig. 4-6.

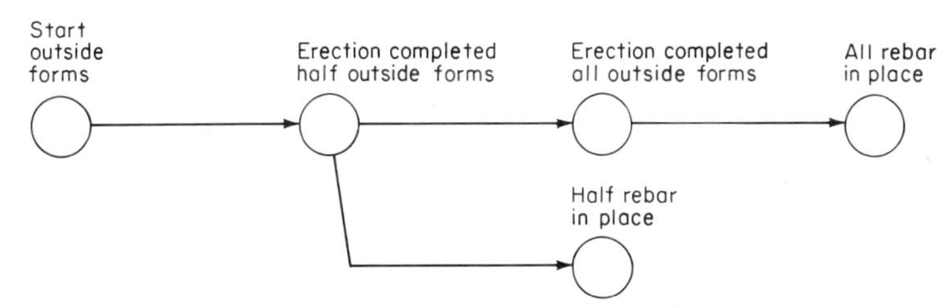

Fig. 4-6 PERT-type diagram with event nodes labeled.

Fig. 4-7 Precedence diagram activities and sequence line.

The descriptions supplied in Fig. 4-6 give an event, or a moment in time, as an actual job reference. This type of diagram is very good for research and development work, where each event may represent the completion of a distinct phase of a study. However, most construction people contend that the labeling of the activities is a much better system where tangible and clearly defined tasks are being considered.

The last diagramming system to be presented is the precedence diagramming notation which has been stressed at Stanford.[7] In this notation the activities are represented by geometric shapes, usually boxes or circles. Activities can be plotted on a work sheet in random order, and the interrelationships are then shown by drawing *sequence lines* between the activity symbols. This eliminates the problem of matching arrow heads and tails that makes arrow diagrams difficult to draw. A simple example of two sequential activities, forming and pouring, is shown in Fig. 4-7.

A sequence of jobs such as forming the outside of a wall (*A*), placing reinforcing steel (*B*), buttoning up forms (*C*), placing concrete (*D*), curing (*E*), and stripping (*F*) would be represented as shown in Fig. 4-8. Arrow heads may be added to show direction of flow, but they are not generally used.

The arrow diagram shown earlier as Fig. 4-5 had nine activities and required one dummy. Using the precedence diagram notation, it would be drawn as shown in Fig. 4-9. The relations between *D* and *H* and *E* and *H* are shown explicitly by connecting these activities with sequence lines. A typical precedence diagram will have as many or more sequence lines as a corresponding arrow diagram has arrows. The advantage is in the ease of placing and relating activities on the diagram without having to worry about the physical placement of activities within the network.

Many arguments can be given pro and con for arrow or precedence diagrams.[8] Suffice it to say that the precedence diagram system is gaining in popularity with CPM users. It has been found to be much easier to use, to explain to field people, and to refine and update as a project progresses. It is obviously applicable to small projects that do not require the use of computers, and it is also being used for large computer applications.

Figures 4-10 and 4-11 present two diagrams for the same project. The first makes use of the arrow method while the second uses the precedence method.

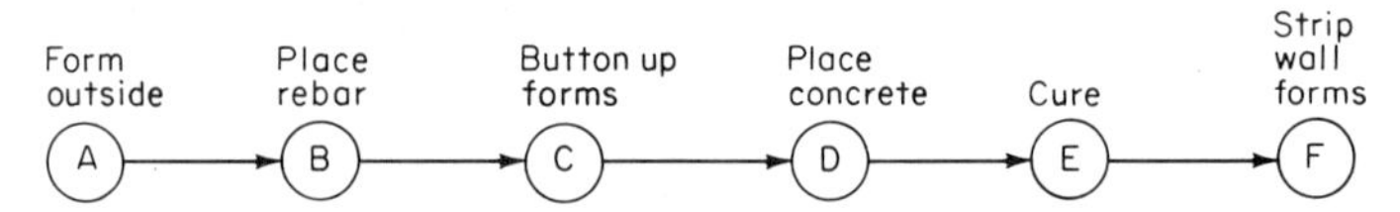

Fig. 4-8 Precedence diagram for activities in sequence.

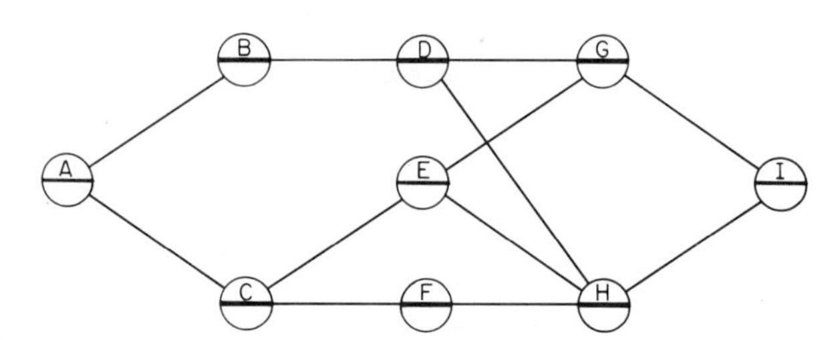

Fig. 4-9 Precedence diagram.

Key Questions in Preparing Diagrams In making the network diagrams for any construction project, there are three questions that will help in setting down the proper logic. They are the same questions that were formerly used to prepare bar charts, but they now refer to specific activities rather than to a major operational breakdown. These questions may be asked in relation to a specific activity:

1. What activities *can* start as soon as this one is completed?
2. What activities *must* be completed before this can start?
3. What other activities *can* be going on simultaneously with this one?

These three questions will pinpoint the relationships between activities as each is added to the diagram. Sequence lines are used to connect activities to show these relationships on the network. The questions will also serve as a checking procedure to ensure that all important items have been included.

Preparation of Actual Diagrams There are as many different sets of procedural steps in making up a network diagram as there are people using CPM. The writer has been involved in using CPM on many projects and has found some of the following ideas to be helpful.

The most important factor is to be sure that the people preparing the diagram know construction and also have specific knowledge of the job being planned. They should understand the basic principle behind network diagramming and be anxious to benefit from its application or, at the very least, be willing to try it to see what value it can

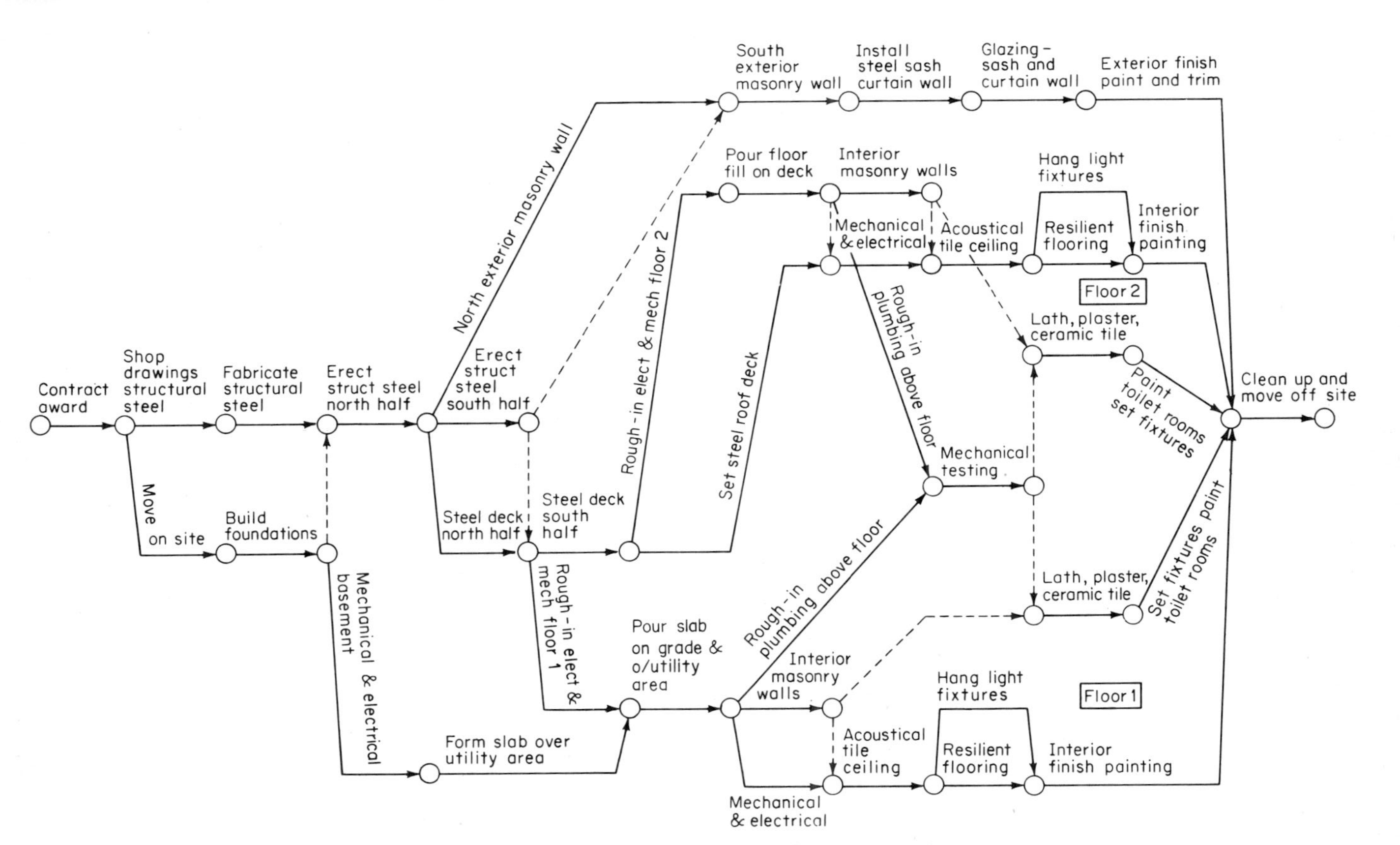

Fig. 4-10 CPM network for small office building (arrow diagram).

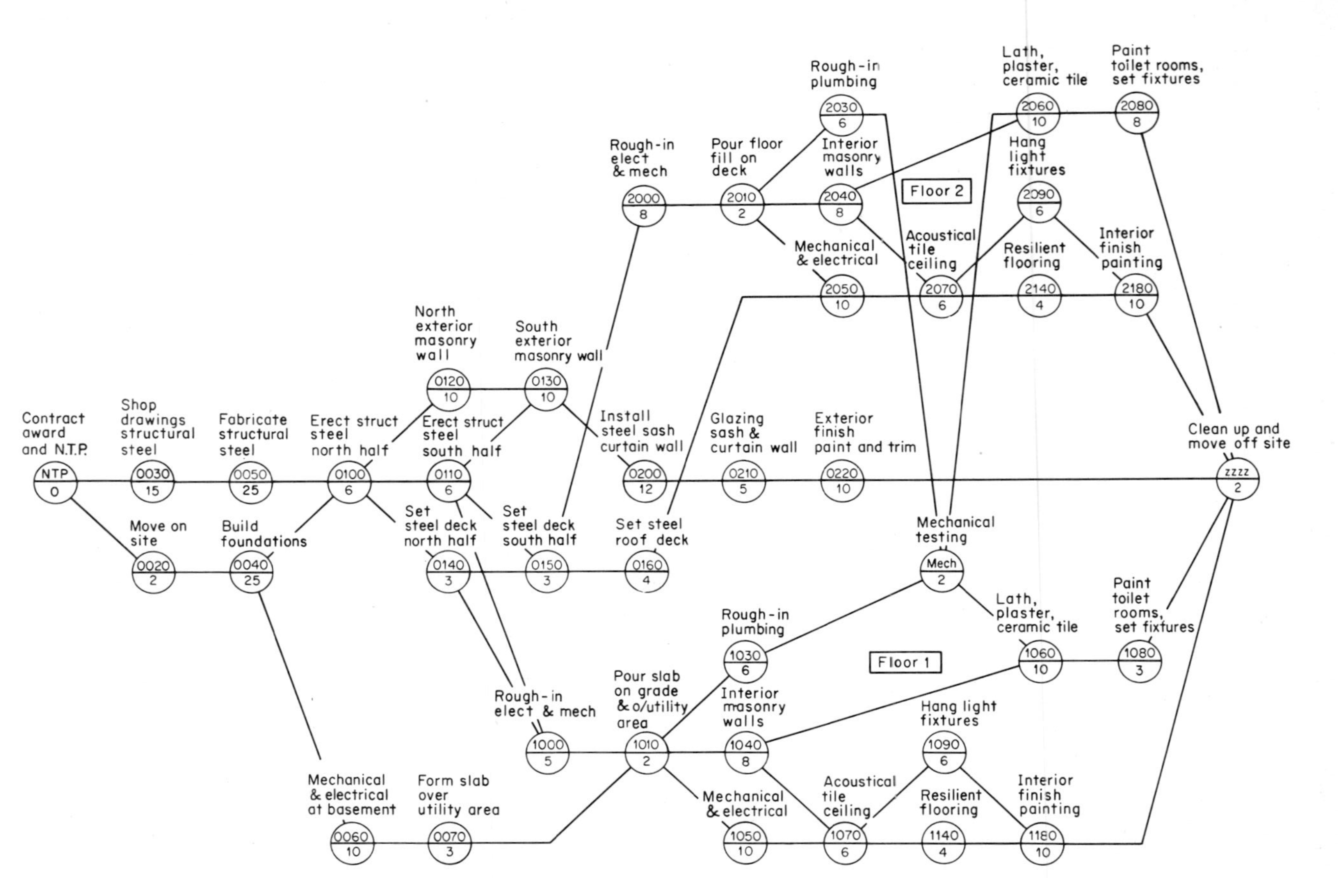

Fig. 4-11 CPM network for small office building (precedence diagram).

offer. Without an open-minded attitude toward CPM, the possible benefits that can be gained from its use will be reduced.

The diagram should be drawn on tracing paper so that it can be reproduced. Typically, several revisions will be made before the "final" diagram will be acceptable to all the people working on the planning. At this point it should be reemphasized that construction scheduling of any sort must be considered as a dynamic effort, subject to change and revision. A CPM diagram is a plan based on the best information and ideas available at the time it is prepared. It must be continually updated as better information becomes available. *This is not a fault of the technique; rather, it is a fact of life on any construction project.*

As a starting point, it is worthwhile to make up a list of all the subcontracts and major purchases that must be transacted during the project. This serves as a checklist to ensure that important procurement items and restraints are not left out.

For most construction work, the working day is probably the best basic time unit. Jobs that take less than a day can be combined, so that a full day's work is included in a given activity. It is not always convenient to do this, but to work in terms of hours or half days implies a precision which is not really practical in long-range planning. The work unit that is set down as an activity can be any time-consuming task that is meaningful to the project management. While the degree of project breakdown will depend on the level of detail at which CPM is being applied, a few common factors are as follows:

1. Responsibility for work, such as subcontract work versus that of the general contractor
2. The type of work and predominant labor classification involved
3. The structural elements involved, such as walls, columns, and slabs
4. Location of the work when it will be performed by different crews
5. The amount of lead time, if any, that will be required for parallel activities
6. Breakdown for payment or cost-control purposes
7. The size of a manageable work task

To begin the actual networking of a project, all the activities that can begin on the first day are put down. The answer to the question "What activities can start as soon as this one is done?" will dictate the next group of activities to be diagrammed. As each activity is included in the network, the questions "What activities must be completed before this one can start?" and "What other activities can be going on simultaneously with this one?" will help in properly tieing it into related activities. In general, the diagram will fan out from a few starting activities into an array of activities, just as the work force builds up on the actual job.

From time to time, say after each group of 50 activities has been added to the diagram, it will be worthwhile to go back over those just added and read back from the diagram exactly what has been represented there. Once the networking gets started, it will often progress so fast that subtle but important relationships between activities will be neglected. By interpreting the plan back from the diagram, people who are learning the system will come to understand how certain relationships must be included in the future.

If the project includes a fair amount of work to be performed by subcontractors, these persons should be consulted to establish how their work must and can fit into the overall picture. It would not be practical to have representatives of the various subcontractors at preliminary planning meetings, but the enlistment of their cooperation in the overall planning of the project will make the coordination in the field much easier. Their work items should be included in the diagram as it is being made up. By the use of the three key questions, they can quickly tell if their work has been tied in properly.

Who Prepares Networks? It is very important to involve the right people in setting up the network diagram. The use to which the planning will be put will determine those persons who should be included in this group. In general, it should be made up of the people who will be involved in the execution of the project. Making up the diagram does force long-range planning, and this effort will be worthwhile even if this is as far as the system is carried.

At the time the job is getting under way, the person who probably is most familiar with it is the man who prepared the estimate. If he will also be involved in the field management, this knowledge will not be lost. If he will not have this field involvement, he can still contribute a great deal in the early stages of a job while this knowledge is fresh in his mind. Very little has been written about the value of CPM as a tool for the transfer of knowledge—for example, from an estimator to a field superintendent—but this application should not be discounted.

Of course, the project superintendent and the project manager will be the two people who will use the information provided to the greatest extent. They are the ones who will make the decisions as to how the job will be run, and the network should reflect their thinking. The activity breakdown should be in work units that will have significance to them and to their field personnel. Since the system can be used to best advantage on a day-to-day basis, they should have a good understanding of the networking methods and procedures.

THE CONCEPT OF TIME ELEMENTS

The actual preparation of the network, described in the preceding pages, is only the beginning of the application of the critical path method to construction work. The network depicts the sequence of activities that must be carried out on the job to complete a given project. The planning value obtained from this effort makes it worthwhile even if this is as far as the technique is carried.[3,4,9,10]

Sequential relationships indicate how the various activities fit together in a project, but sequence is not the only important function in planning. The timing of the various phases of the work is also important. The inclusion of time as a parameter of CPM permits the user to establish the proper timing of the activities represented in the network. It is this latter feature that has made the CPM tool so very useful.

Time Elements These timing data for the activities have been referred to as the *time elements* of a network.[1] The person who is planning a project is interested in knowing three important timing facts which can be obtained from a CPM network:

1. The earliest time that an activity can start
2. The latest time that an activity may be completed without delaying the project completion
3. The leeway available in scheduling an activity

The concept of time elements will be explained in reference to a bar chart, a tool that most readers will be familiar with. As an example, refer to the precedence network in Fig. 4-12*a*, which has seven work items, *A* through *G*. Figure 4-12*b* has these same items plotted as they might appear on a bar chart. It can be seen from this chart that activities *B* and *D* can begin as soon as *A* is finished, and activity *C* can then follow *B*. Note, however, that activity *E*, which follows both *C* and *D*, cannot be started until *D* is completed. In this case it is said to be restrained by the completion of *D*. Activity *G* follows both *E* and *F* but is restrained by the completion of activity *E*. In the bar chart we have shown all the activities as starting immediately after their predecessors when all restraints are considered. Each of these activities is thus plotted at its *early start*. If a daily time scale were added to the drawing, then we could read off the day that a given activity could start. For example, if activity *A* started on day 1 and would require three days to complete, then the early start of both *B* and *C* would be day 4. The *early finish* of activity *A* would be at the end of day 3.

At this point it would be well to define what is meant by a time scale and how it relates to the early start and other time elements. The writer has found through experience that most users of CPM prefer to have a time scale that is referenced to the beginning of a workday. With this convention, when we say that activity *A* has an early start at day 1, we mean that it can start at the beginning of the first day. The early finish is referenced to the beginning of the workday following its actual completion.

Some time scaling systems are referenced to the end of the workday, and activity *A* would then have an early start of zero. This time scaling places the emphasis on the completion of activities rather than on their start.

If we take the bar chart shown and plot all activities as far downstream toward the finish as possible while holding the final activity in its place, it will look as shown in Fig. 4-12*c*. Whereas the first bar chart showed every activity starting as early as possible, this chart shows each of them delayed as long as possible. In other words, the finish of every activity is plotted at the latest time that it can be accomplished without delaying the overall completion of the project. If we scaled off the finish of each activity, we would get what is referred to as the *late finish*. This is one of the three important timing factors we are interested in.

The other time factor of interest is a measure of scheduling any leeway or *float* time that each activity has. It will again be useful to use a bar chart to show this concept. This time Fig. 4-12*b* and *c* will be superimposed as Fig. 4-13 to show the early and late

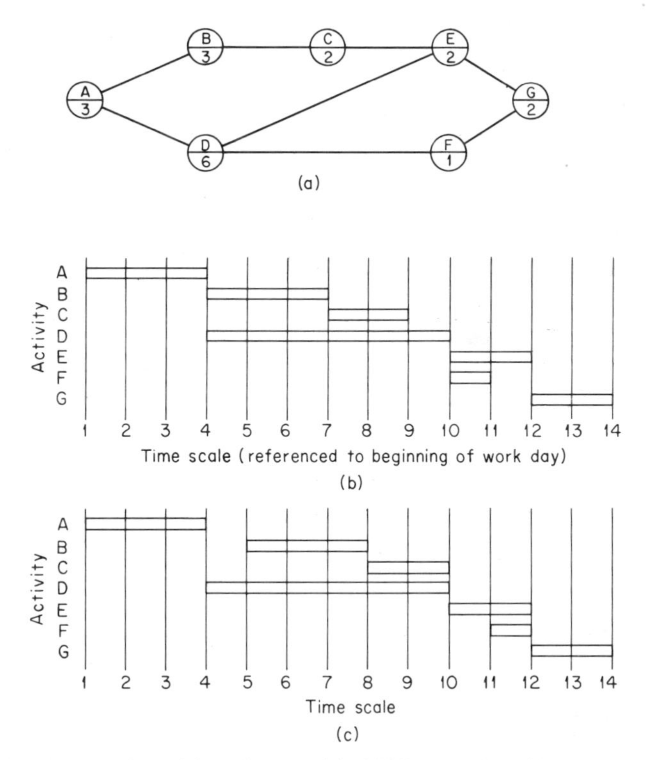

Fig. 4-12 CPM network and bar charts. (*a*) CPM network. (*b*) Bar chart on early start. (*c*) Bar chart on late finish.

charts together. The late bars will be shown shaded and the early bars will be blank. From this diagram we can see that activities *A*, *D*, *E*, and *G* have not been allowed to move at all. Activities *B*, *C*, and *F*, on the other hand, could be delayed by some amount before they would restrain the completion of the overall project. In other words, they have some float or leeway. Actually, two different measures of leeway can be obtained from the diagram.

Free float is defined as the amount of time any activity can be delayed without adversely affecting the early start of any following activity. Referring to the diagram, we can see that activities *C* and *F* have such leeway, designated C_{ff} and F_{ff}. Activity *B*, on the other hand, has no free float because the early start of *C* is actually the same time as the early finish of *B*. In other words, the free float is the least difference between the early finish of an activity and the early start of all following activities.

The other type of float is called *total float*. It is defined as the amount of time that an activity can be delayed without adversely affecting the overall time for project completion. The total float has been designated in Fig. 4-13 as B_{tf}, C_{tf}, and F_{tf}. In this example, if B is delayed up to the amount of time B_{tf}, it will delay the start of C but will not change the completion time for the project. Likewise, if C starts on time but is not completed in the given time, its completion may be delayed as much as C_{tf} without holding up the completion of the project.

It is important to note that the total float shown on the bar chart is equal for both B and C. However, if B falls behind schedule, in reality it uses up some of the total float that is shared by all activities in the chain between A and E via B and D. This should be clearly understood, since the fact that a given activity has a considerable amount of total float can often cause an unwarranted complacency on the job. If this total float is "used up," so to speak, early in a project, it makes the scheduling of later activities on the chain much less flexible. Extreme caution must be used when talking about how much total float a given activity has, because it must be remembered that this float is shared with other activities. Free float, on the other hand, is not shared with other activities. It provides a true measure of how much an activity can be delayed or extended.

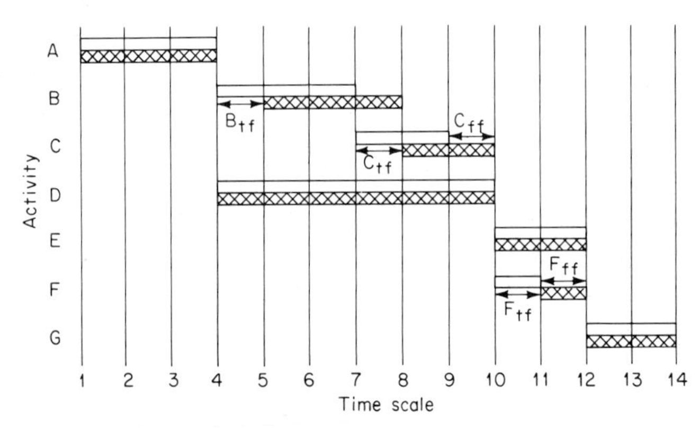

Fig. 4-13 Combined bar chart on early start and late finish showing free and total float.

The time elements that can be obtained for each activity in a network have now been discussed. These are the early start, late start, early finish, late finish, total float, and free float.

Estimating Durations In order to arrive at values for the time elements, an estimated duration must be assigned to each activity. This time should be the number of days, based on the estimator's best judgment, that it will take to perform the activity. Some articles on the subject talk of a "normal" activity time which is associated with least activity direct cost, and a "crash" activity cost which is associated with least activity duration. These concepts are applied in time-cost trade-off procedures. It is felt by the writer that the use of a realistic value for the time that each activity is expected to take, considering other work concurrent and resources available, will yield the best results in basic applications. These time estimates can and will be revised as the planning and scheduling progress.

All time estimates are made in working days, which eliminates the problem of checking on weekends and holidays. When estimating a procurement item that will take several weeks, the time value assigned should be 5 days per week times the required number of weeks. This will keep the times consistent.

In addition to assigning each activity an estimated duration, it is often useful to assign it a label so that it can be referred to easily. It is quite convenient to make up networks with large enough circles so that an activity number can be placed in the top half and the estimated duration in the bottom half. The job description is placed beside the circle.

Since the original network is bound to change, and in fact should be changed as job conditions indicate that its assumptions were not completely realistic, the labeling system should facilitate such revisions. A very good method is to use numbers and label the activities by tens so that additional activities may be subsequently added. If an activity is deleted from the network, its number can be dropped because there is no need for sequential numbering.

Some computer programs for CPM calculations will allow combinations of letters and numbers. This is a good feature because it allows for flexibility. For example, on a dam job, labels related to the work descriptions are helpful. With high-rise buildings, on the other hand, the two high-order digits can be associated with the floor number. Some of these ideas are illustrated in Fig. 4-11.

CPM Time Calculations Various systematic methods have been found convenient in calculating the time elements for each of the activities in a network. Two of these methods will be discussed briefly.

It is pointed out in many references on CPM that only simple arithmetic is involved in determining the various time elements for each of the activities represented on a network.[2,4,9] Although this is true, the work is quite tedious and can be very time consuming. But, compared to the benefits that can be derived, it is well worthwhile.

The idea behind all the calculations has already been shown in the bar chart representations of the time elements. The basis for these calculations will now be set down explicitly in written and mathematical forms. Since this description is intended mainly for applications, certain liberties have been taken in presenting the basic equations.

Figure 4-14 shows a portion of a typical network. The activity in the middle is the one in which we are interested and for which we will explain the calculations. It will be designated as A_i, for the activity of interest. The two activities to the left represent any activities that immediately precede A_i, and will be designated as A_p. The two activities to the right represent any activities that follow A_i, and will be designated A_f.

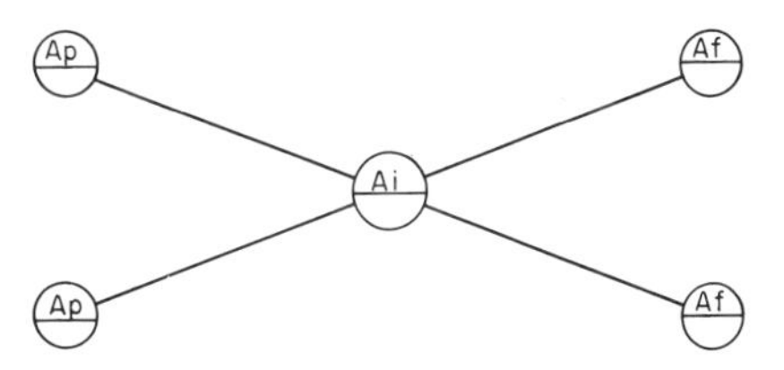

Fig. 4-14 Precedence diagram showing typical relationships to activity of interest for time-element calculations.

The calculations are made for each activity in a network. Therefore each activity will, in turn, be a preceding activity for those that follow it; it will be an activity of interest; and it will be a following activity for those that precede it.

Early Start. The early start (ES) of any activity is either the initial day of the project (if it has no predecessor) or the latest of the early finishes of all activities that immediately precede it. It must be remembered that all time elements are referenced to the beginning of a workday.

$$\text{ES}(A_i) = \max\ (\text{EF}_{\text{all}(A_p)}) \tag{4-1}$$

Early Finish. The early finish (EF) of any activity is its early start plus its estimated duration.

$$\text{EF}(A_i) = \text{ES}(A_i) + \text{duration}\ (A_i) \tag{4-2}$$

The duration of the entire project is the early finish of the last activity.

Late Finish. The last finish (LF) of any activity is either equal to the duration of the project (if it is the final activity) or equal to the earliest of the late starts of all activities that immediately follow it.

$$\text{LF}(A_i) = \min\ (\text{LS}_{\text{all}(A_f)}) \tag{4-3}$$

Late Start. The late start (LS) of any activity is its late finish less its estimated duration.

$$\text{LS}(A_i) = \text{LF}(A_i) - \text{duration}\ (A_i) \tag{4-4}$$

Total Float. The total float (TF) of any activity is the difference between its earliest and latest start times.

$$\mathrm{TF}(A_i) = \mathrm{LS}(A_i) - \mathrm{ES}(A_i) \tag{4-5}$$

Free Float. The free float (FF) of any activity is the least of the differences between its early finish and the early start of all activities that immediately follow it.

$$\mathrm{FF}(A_i) = \min\,(\mathrm{ES}_{\mathrm{all}(A_f)} - \mathrm{EF}_{(A_i)}) \tag{4-6}$$

These six relationships form the basis for all the basic CPM time element calculations. In order to "process" or make calculations for a network, two methods will be suggested.

Tabular Calculations The calculations may be made on a tabular form that can be set up as standard for any job. There are 10 columns in Table 4-1 for the following information: (1) activity number for easy reference; (2) activity duration; (3) critical

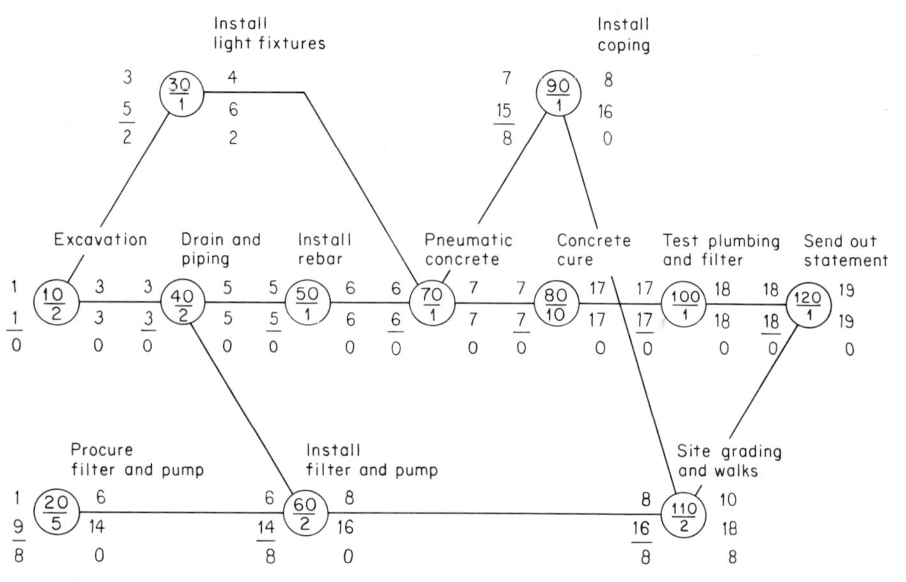

Fig. 4-15 Simplified CPM diagram of swimming pool showing calculations of time elements on the diagram.

indicator; (4) activity description; (5) early start; (6) late start; (7) early finish; (8) late finish; (9) total float; (10) free float. These calculations will be illustrated for the network shown in Fig. 4-15.

This small network depicts 12 activities that might be involved in the construction of a swimming pool. The time estimates are in working days. Note that item 8, concrete cure, normally might require 14 days but is set up for 10 days in recognition of the fact that it will continue over two weekends.

The calculations are actually prepared in three separate "passes" of the network. As each activity is encountered on the table on which columns 1, 2, and 4 have been filled in, the network must be checked to establish what activities precede or follow the activity of interest.

The first pass, or *forward pass* as it is called, is to find the early time elements. Equations 4-1 and 4-2 are used to find the early start and early finish of each activity in the network. The results of these calculations are shown in Table 4-1.

The second pass, or *backward pass*, starts by setting the late finish of the final activity equal to the early finish. The late time element for each activity is then calculated. Equations 4-3 and 4-4 are used find the late finish and late start of each activity by working backward through the diagram. The sequential relationships are obtained from the diagram. The results of this pass are shown in Table 4-2.

In order to calculate the free floats for the network, a third pass, or *float pass*, is made taking into account the previously calculated early finishes and early starts. Using Equations 4-5 and 4-6, the total and free float may be found. The results of this pass are shown in Table 4-3.

TABLE 4-1 CPM Example—Swimming Pool

1	2	3	4	5	6	7	8	9	10
Activity	Duration	Critical	Description of activity	ES	LS	EF	LF	TF	FF
10	2		Excavation	1		3			
20	5		Procure filter and pump	1		6			
30	1		Install light fixture	3		4			
40	2		Install drain and piping	3		5			
50	1		Install rebar	5		6			
60	2		Install filter and pump	6		8			
70	1		Place pneumatic concrete	6		7			
80	10		Concrete cure	7		17			
90	1		Install coping	7		8			
100	1		Test plumbing and filter	17		18			
110	2		Site grading and walks	8		10			
120	1		Send out statement	18		19			

TABLE 4-2 CPM Example—Swimming Pool

1	2	3	4	5	6	7	8	9	10
Activity	Duration	Critical	Description of activity	ES	LS	EF	LF	TF	FF
10	2		Excavation		1		3		
20	5		Procure filter and pump		9		14		
30	1		Install light fixture		5		6		
40	2		Install drain and piping		3		5		
50	1		Install rebar		5		6		
60	2		Install filter and pump		14		16		
70	1		Place pneumatic concrete		6		7		
80	10		Concrete cure		7		17		
90	1		Install coping		15		16		
100	1		Test plumbing and filter		17		18		
110	2		Site grading and walks		16		18		
120	1		Send out statement		18	19 →	19		

Calculations on Drawing A second method, and one that is recommended on projects where tabular calculations are not required, is to make the calculations right on the network. This can be done very quickly on a print which is reproduced from a master network made on tracing paper. The time element data can be spaced around the circle which represents the activity. This might be done as shown in Fig. 4-16.

An alternate method is to use preprinted triacetate symbols that can be placed on tracing paper after the activity label, description, and duration have been typed on them. The symbol has a box in each corner that can be used when filling in the time elements. This type of symbol is used by the Guy F. Atkinson Company on all its precedence networks in place of the circles which have been used in the other examples herein. (See Fig. 4-17.)

When working on the diagram, the same three passes through the network are required. An ES equal to 1 is assigned to all activities which start on the first day.

TABLE 4-3 CPM Example—Swimming Pool

1	2	3	4	5	6	7	8	9	10
Activity	Duration	Critical	Description of activity	ES	LS	EF	LF	TF	FF
10	2	CP	Excavation	1	1	3	3	0	0
20	5		Procure filter and pump	1	9	6	14	8	0
30	1		Install light fixture	3	5	4	6	2	2
40	2	CP	Install drain and piping	3	3	5	5	0	0
50	1	CP	Install rebar	5	5	6	6	0	0
60	2		Install filter and pump	6	14	8	16	8	0
70	1	CP	Place pneumatic concrete	6	6	7	7	0	0
80	10	CP	Concrete cure	7	7	17	17	0	0
90	1		Install coping	7	15	8	16	8	0
100	1	CP	Test plumbing and filter	17	17	18	18	0	0
110	2		Site grading and walks	8	16	10	18	8	8
120	1	CP	Send out statement	18	18	19	19	0	0

The EF is simply the ES plus the activity duration. This EF is then carried forward to all following activities as a possible ES. The ES of an activity will be the greatest of all the preceding EFs which have been carried forward.

The backward pass is just the opposite. The LF of the final activity is filled in from its EF. The LS is simply the LF minus the duration, and is then carried back to all preceding activities. The LF of an activity will be the least of all following LS times which have been carried back.

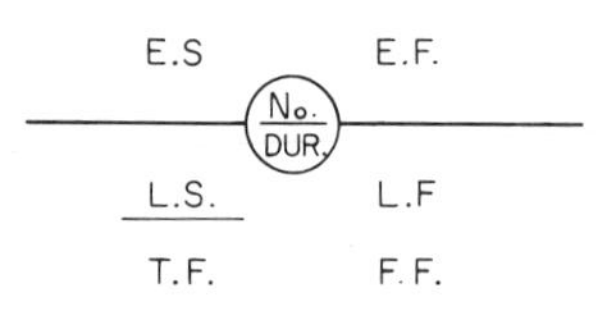

Fig. 4-16 Locations of calculated time elements around activity node when work is done on the precedence diagram.

Total float is merely the difference between the LS and ES from the first two passes. The free float is the least difference between the EF of an activity and the ES of each of the following activities.

The network for the swimming pool, as illustrated in Fig. 4-15, shows time elements which have been calculated directly on the diagram. All three passes have been performed. This system of working on the diagram is a good one for many applications. All the data are available on a single diagram, and one does not have to refer back and forth between the diagram and the time element table.

THE CRITICAL PATH

In previous paragraphs it has been pointed out that some of the activities in the network have a certain amount of scheduling leeway or float. Nothing has been said so far about the activities that do not have any float. At least one chain or sequence of activities from the present time to the end of the job will have zero float, and this chain is referred to as the *critical path.*

Significance of the Critical Path The critical path is often denoted on the network drawing by using a color on the sequence lines which connect the critical activities. It is this sequence of work items that determines the earliest possible finish time of the project. If the start or finish of any of the critical items is delayed, the completion of the entire project will be moved back unless some later critical activity can be speeded up.

The use of CPM on hundreds of construction projects over the last few years has shown that typically only about 10 to 20 percent of the activities shown on a network are actually critical.[3,4] The first project on which Du Pont used CPM was planned and scheduled with traditional methods to permit a comparison between the effectiveness of the two systems. Over one hundred and fifty items were considered to be critical by the traditional methods but, when CPM was used, it was shown that only seven items formed the chain of work which determined the project finish time Only four of these seven activities had been identified by the previous methods.

This Du Pont job was the first of many to show that a great number of activities that had previously been thought of as being of prime importance actually do not control project completion. Such items as a turbine or large boiler are often furnished by the owner but must still be installed in sequence; their procurement has been shown to be critical in many cases, which means that the completion date is not always entirely under the control of the contractor. CPM is a good tool to help point this out to an owner.

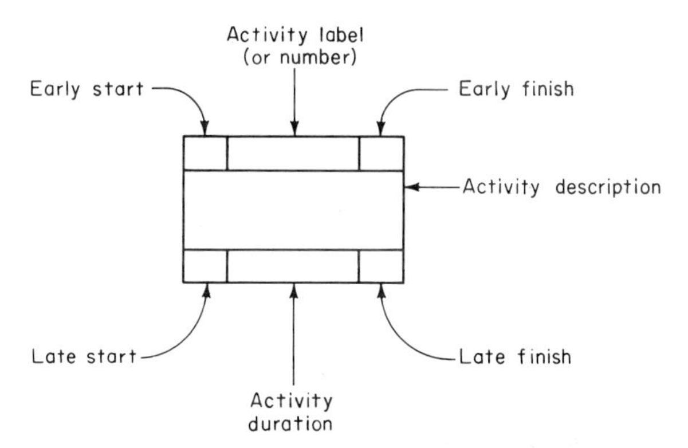

Fig. 4-17 Example of preprinted symbol for developing networks and for calculations of time elements.

The ability to pinpoint the critical sequence or path is a very significant advantage of CPM. For many years, whenever a project fell behind the proposed schedule, the whole project would be put on a crash or overtime schedule. CPM can show graphically that only a few activities need to be replanned and speeded up to put the project back on target. This has often been referred to as *management by exception.* The manager knows, assuming he is satisfied with the network representation of his plan, that only a small percentage of the activities need to be watched closely on a day-to-day basis. The others can be rescheduled as fill-in work or to help level the crew requirements on the job. The ability to make a rational decision as to which items need to be speeded up and which can be slowed down, if necessary, can produce a more economical job plan.

Subcritical Chains The critical path has been defined as a chain of zero-float activities from the present time to the end of the project. Any slippage that occurs in one of the critical activities will delay the completion of the entire project. On many projects, in addition to a critical chain, there will be groups of activities that form subchains, or subnetworks, which include many days of float. There is actually a chain of activities through each subnetwork that determines the completion of the work thus represented. However, even though the item in question may not be needed for its intended purpose until some time after its earliest possible completion, it still does not make sense to consider it unimportant. The men and equipment involved may be needed elsewhere on the overall project. Where the subnetwork is not on the critical path for the overall project, another criterion of job priority is used. That sequence of activities within each subnetwork which has the least amount of total float is viewed as the most critical sequence within this subnetwork.

On many projects, a subcritical chain may tie into the critical path from time to time. In the case of a multistory concrete structure which is formed and poured in two sections, the critical path may run up through the work on one half of the building. The other half of the building may show one or two days' float on every item. Items that are this closely related to critical activities or that have this small amount of float are essentially critical themselves. They should be watched as closely as if they had zero float and were marked as being on the critical path.

Dangers of Overemphasis In many applications of CPM, the concept of the critical path has been overemphasized. It may be a case where a little knowledge is a bad thing, but at any rate an overdependence on being shown the critical items can prove to be a problem. If the project network is not updated to reflect the current state of the project or if the network was not realistic to begin with, the information provided will likewise be unrealistic.

CPM is simply a planning and scheduling tool. It is no more and no less effective than the people who are using it. In the discussion of networking it was stated that the ultimate responsibility for preparing a network diagram must fall to the man or men who have control of the project. If the networking is left completely to an outsider, it will not realistically represent the plan of action in the field. When this occurs, the first inclination is to say it is the system that is poor, when in reality it is the fault of the man in control for not taking time to let this tool be a help to him.

The two biggest dangers of critical path work occur when a network is accepted that is not as close to reality as possible and when too much attention is paid to the colored line on the diagram which shows the original critical path. There is far too much continual change on a project to support the belief that a network which is made up at the beginning of a project can remain effective for the entire project duration. As events occur that were not anticipated, or as delays cause problems in noncritical areas, this information should be added to the network so that its effect can be shown. This is called *updating* the network, and a new set of computations of time elements should be made after every major revision to the network. All too often a good deal of work goes into an original project plan, but then the plan is not used effectively as the job progresses. The mechanics of CPM insert "as of today" facts into the calculations so that one can project the effect of recent actual change. The network should be continually revised as changes occur and as more current information becomes available.

PROJECT PLANNING, SCHEDULING, AND CONTROLLING

Project Planning There are three very important and interrelated phases to the management of any project in which the use of CPM techniques can play an important role. These phases, which will now be discussed, are planning, scheduling, and controlling. The three phases are very distinct, but they are interrelated to such an extent that in a final analysis they must be used in combination to achieve the most beneficial results.

Planning is the working out of a fairly detailed program of how a given project will be constructed. All too often this is done only in the mind of the person who is to run the job. When this is the case, a great deal of the value is lost because others, not knowing exactly what he is thinking, cannot dovetail their work to best advantage. A good many large projects in the past were planned, so to speak, on the back of an envelope or on a napkin over a cup of coffee.

To work to the best advantage, planning should be set down in some form so that all concerned are presented with a clear and detailed outline of how the project is to proceed. Not only will this make it much easier for all the work to be coordinated but it will give a basis for discussion on how the plan might be improved. This is not to say that the man in charge of the project cannot adequately plan the job himself, but rather to imply that two heads are better than one only if they can communicate. Quite often it is very hard to put into words a detailed sequence of how a great many construction operations fit together, but a simple diagram of some sort will alleviate this problem.

A CPM network diagram is a very effective and systematic device for putting the plan into a useful form on paper. Quite often the sequencing of concrete placements in a building will be determined by other factors than just one pour above another. In a parking garage, for example, where there are ramps and parking levels that are cantilevered over each other, the placing sequence is not determined by the floor below but by the adjacent level. This type of problem can be worked out very well with a network planning system.

Making up a diagram, in effect, forces planning in a degree of detail which was seldom achieved before the introduction of CPM. Quite often people will be heard to say that the amount of time spent on the diagram was worthwhile even if this is as far as CPM is carried. The thought that goes into making up the diagram will probably point out many problems that might not have been identified by conventional planning methods. Since these problems are anticipated early in the project, any redesign or material changes that are required will have sufficient lead time so that crash and overtime charges can be kept to a minimum.

About 60 to 70 percent of the work involved in using CPM, except in work where it is used also as a payment control system, goes into the making up of the diagram. Probably 50 to 60 percent of the benefits occur from the more detailed planning and analysis of the whole job that is required to make up a realistic diagram.

CPM at Bidding Stage. In most projects that are built today, completion time is an important factor. Very often this relates to either gained or lost revenues because of early or late punch-out on the project. The idea of a substantial bonus for early completion is becoming more widespread, particularly in the areas of power generating stations, production plant construction, and home and apartment building. In many cases the bonus is of such magnitude that it is worth planning toward a very early completion in order to realize the added compensations.

This anticipated revenue from bonuses may have to take the form of lower bids because other contractors using CPM will be anticipating the same speedups. In projects where substantial savings or bonuses are possible because of early completion, CPM in the bidding stages offers a systematic way to analyze a general working plan and thus gain a more realistic evaluation of possible speedups. A good estimator and project manager who have a sound working knowledge of CPM can rough out a prebid network for a multimillion dollar project in a matter of a few hours. Their time may be very valuable in putting together a bid, but the time taken out to set up a diagram will more than pay for itself by giving them a better feeling for how the job can progress. From this information they will have a better idea as to the alternative plan of action that will be the most competitive and still yield a fair profit.

On many projects, bids are submitted, and the completion date is one of the bid items. In this case a contractor who can make up a good general CPM diagram in order to establish a feasible finish date has an advantage over the contractor who makes an unrealistic "guess," for he may be held to his submitted completion date.

Checking Alternatives. Quite often, after a job has been awarded, there are several different plans of attack that may be followed by the contractor. For example, a highway job might lend itself to being started at either end, or it might be started in the middle and progress both ways. A housing tract might be started close to a highway or at the most remote spot and then be worked back toward the road. A complete set of steel forms instead of handcrafted wooden forms may prove to be a key item on the job. The method of accomplishment is usually up to the contractor, and he should evaluate the various alternatives that are available to him.

The two important parameters in this evaluation are the costs and the times involved. The mechanics of CPM for determining the time elements for a network, as described previously, are very easy to apply. When there are several alternatives to consider, a network can be drawn for each and the time elements can be calculated to compare the time relationships of the various plans.

Actually, more often than not, the alternatives are not entirely different plans. They are, instead, variations of a general plan. In such circumstances it is not necessary that the entire network be redrawn, but, rather, small sections can be changed to represent the alternate plans. In some cases the network will remain the same, indicating that the same sequences will still hold, but the estimated durations for certain activities will be raised or lowered to indicate a planned slowdown or speedup of the work represented.

When the cost of various alternatives is to be considered in addition to time, it is important to be practical in using CPM as a comparison tool and not let the amount of work required outweigh the benefits. For example, one approach that was used early in the development of CPM concepts was to assign a cost or cost function to every activity and then compare entire networks as alternatives. The writer considers this to be not only impractical but nearly impossible in the time that is generally available for project planning. A system which is much better is to consider only the small sections of a network that need be changed to represent the alternate plan. Very often this is an investment in some time- and labor-saving equipment compared to the less mechanized way of accomplishing a given task. Using the principles of engineering economy and considering the timing of expenses (data that are available from a CPM analysis), a logical and valid decision can be made for each alternative. Cost estimates need be prepared only for the activities that will be directly affected by an alternate plan. In many cases planners with years of experience will make this type of analysis informally on many items as they develop an overall network for a project.

Project Scheduling The second phase of the management of any project in which CPM techniques can play an important role is that of scheduling. When using CPM, the functions of planning and scheduling are separated, at least in the early stages.

Scheduling is the assigning of starting dates and completion dates to the various activities that go into the makeup of a project. The critical path method provides a basis for logical decisions for this scheduling. It will be remembered that, after time estimates have been prepared for every activity, the mechanics of CPM provide data on the earliest and latest job times and the measure of float available. This information is far more detailed than that obtainable from any previous scheduling system.

One of the first things to do when preparing the working schedule is to prepare a conversion table which relates working days, taken consecutively, to calendar dates (including allowances for weekends and for legal holidays). This table can be set up as shown in Table 4-4.

After the time elements have been computed for all the activities in the network, they should be analyzed in the scheduling phase to ensure that they are realistic. In other words, this is also a checking phase on the network representation.

In making up a diagram, the tendency is to plan work to be sequential. Replanning may be required so that activities are planned concurrently. In many cases, several work items can be going on at the same time, but each one requires a small portion of one of the others to be done before it is started. That is to say, a small amount of lead time is required on an activity before a second activity that will be operating concurrently can be started. Generally, the theory of CPM requires that one activity be completed before its successor is started. However, there are so many cases where this overlapping is the actual case that special provisions have been built into many computer programs to handle it. The writer prefers to refer to the increment of time between the start of an activity and the start of a somewhat delayed concurrent activity as a *lead-time factor*, although some of the computer programs refer to this as a *lag factor*. Other lag factors than a start-to-start relationship are available to the experienced planner, but it is not recommended that they be employed until one has a thorough understanding of networks. An example of overlapping activities is given in Fig. 4-18.

In this general summary network of a dam project, several of the activities can be overlapped. For example, the SO24 above the sequence line connecting activity 2, "Common Excavation," and activity 4, "Rock Excavation," indicates that "Rock

Excavation" can begin 24 working days after the start of "Common Excavation." By using this lead-time factor, it is not necessary to break the first activity down into two parts to indicate the actual lead-time relationship. It needs to be pointed out that this device is in direct contradiction to strictest CPM theory, and its use should not be attempted until one has a good working knowledge of networking techniques. Only then can shortcuts like this be safely employed.

After all the inconsistencies in the diagram have been corrected, the actual scheduling and replanning can begin. In most applications of CPM to construction projects, the work will be planned to progress as fast as possible. Since the network diagram represents the relationships between activities, the time elements which are obtained will indicate the earliest that each activity can be commenced. These early starts make up the first approximation of the project schedule.

TABLE 4-4 Sample Conversion of Working Days to Calendar Dates

SAMPLE CPM NETWORK SMALL OFFICE BUILDING

STANDARD CONVERSION CALENDAR

CAL DAY	WRK DAY	CAL DATE	
1	1	4 JAN 1971	
2	2	5 JAN 1971	
3	3	6 JAN 1971	
4	4	7 JAN 1971	
5	5	8 JAN 1971	
6		9 JAN 1971	SATURDAY OR SUNDAY
7		10 JAN 1971	SATURDAY OR SUNDAY
8	6	11 JAN 1971	
9	7	12 JAN 1971	
10	8	13 JAN 1971	
11	9	14 JAN 1971	
12	10	15 JAN 1971	
13		16 JAN 1971	SATURDAY OR SUNDAY
14		17 JAN 1971	SATURDAY OR SUNDAY
15	11	18 JAN 1971	
16	12	19 JAN 1971	
17	13	20 JAN 1971	
18	14	21 JAN 1971	
19	15	22 JAN 1971	
20		23 JAN 1971	SATURDAY OR SUNDAY
21		24 JAN 1971	SATURDAY OR SUNDAY
22	16	25 JAN 1971	
23	17	26 JAN 1971	
24	18	27 JAN 1971	
25	19	28 JAN 1971	
26	20	29 JAN 1971	
27		30 JAN 1971	SATURDAY OR SUNDAY
28		31 JAN 1971	SATURDAY OR SUNDAY

In most cases it is not practical to schedule work in detail for more than a month or two in advance. The project will change too much in the meantime. This is not to say that the diagram should not be prepared for the whole job; rather, it is suggested that detailed scheduling of work, carried too far into the future, is often wasted effort. A practical, workable plan would be to prepare a detailed schedule once a month for the work which is to be done in the succeeding two months. This provides an overlap of one month and will still not be getting so far into the future that the whole plan will be changed by then.

In detailed project scheduling, one is interested in the exact date that certain work items will be starting. If the network has been set up to correctly represent the job, then most activities will be scheduled to begin on their early start dates. Of course, since the critical activities must follow each other in sequence, their schedule is already set up.

In the case of noncritical activities, the amounts of total and free float give a measure

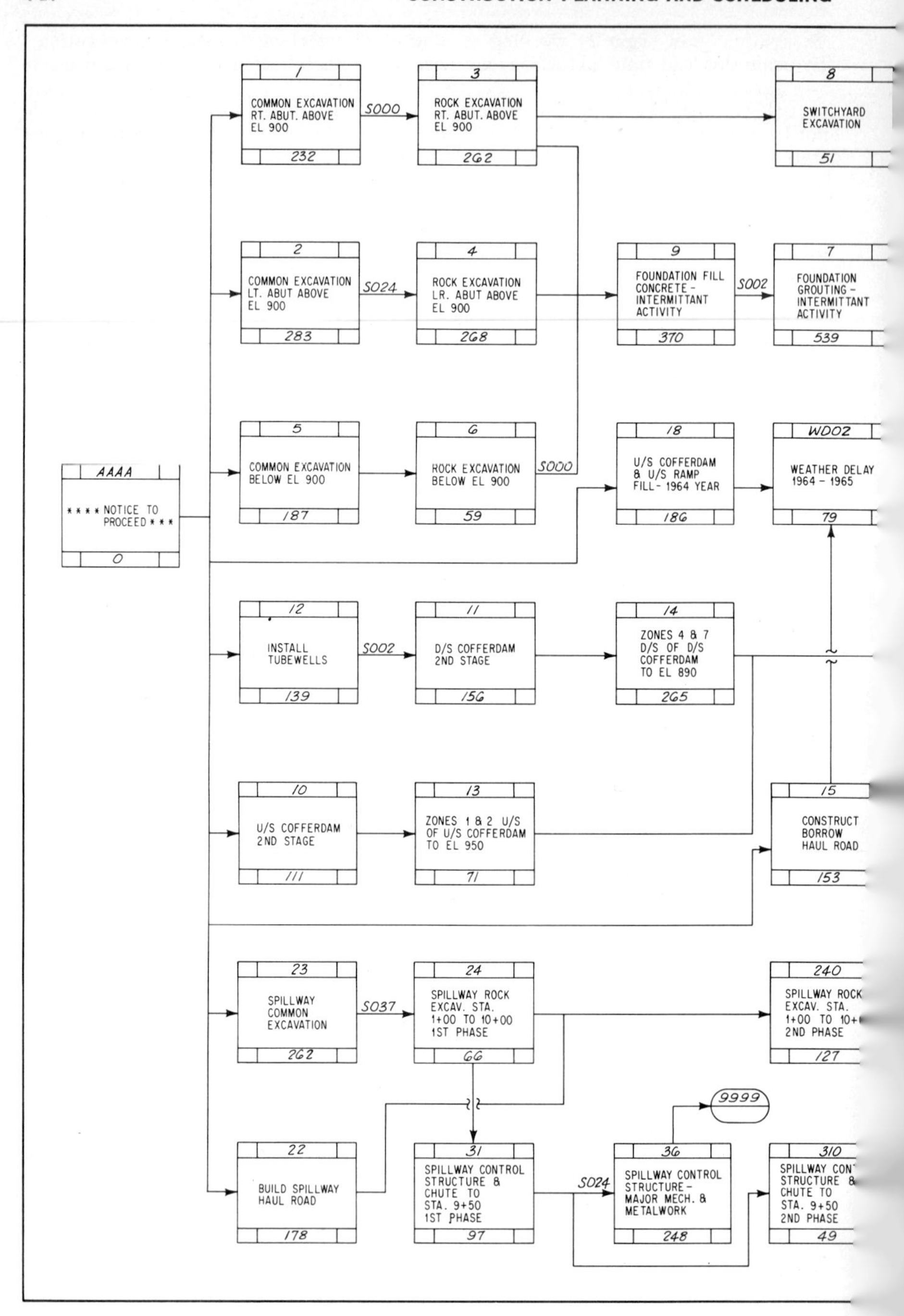

Fig. 4-18 Method for showing overlapping concurrent activities using

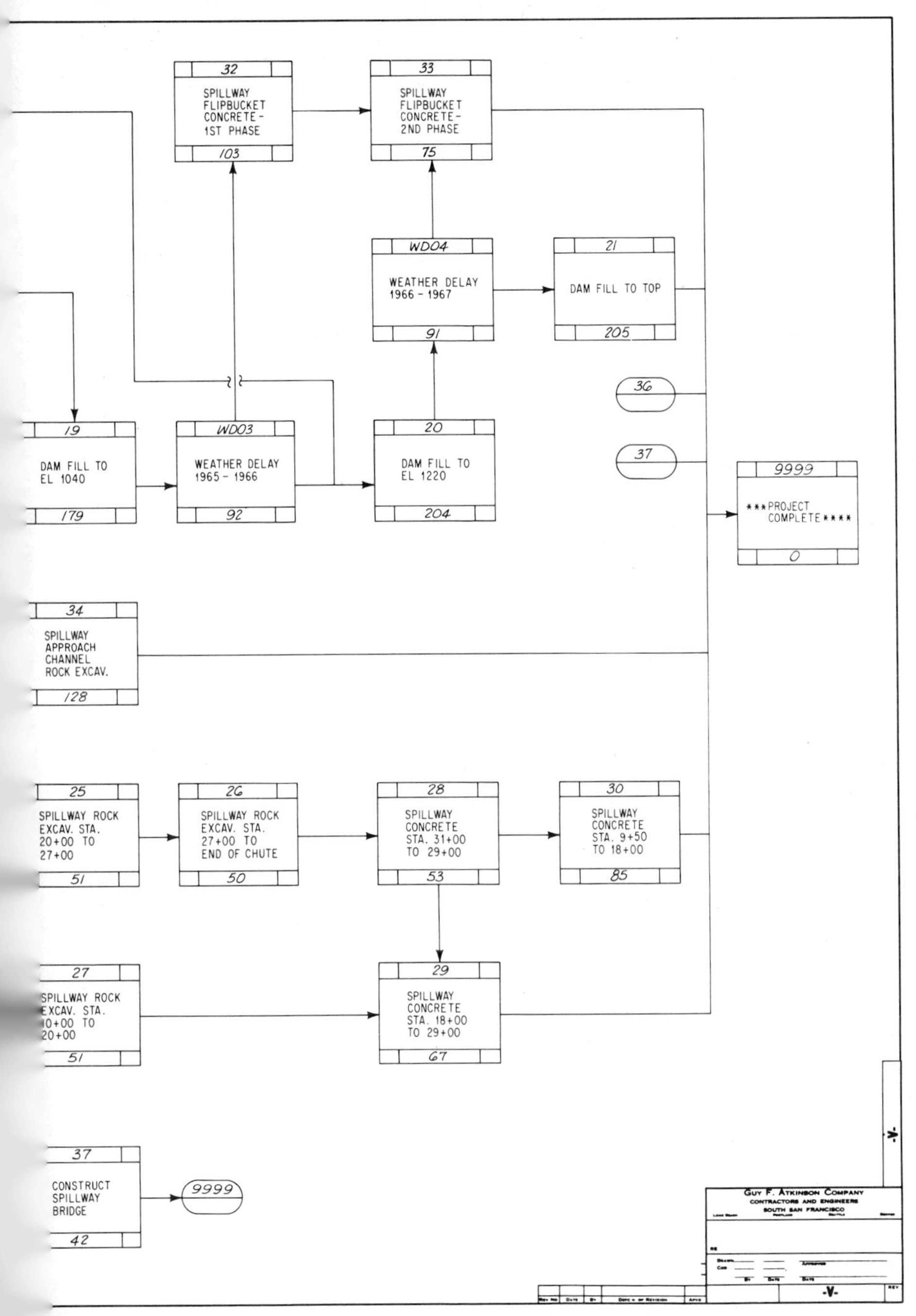

lead-time factors. *(Guy F. Atkinson Company, South San Francisco, Calif.)*

of the scheduling leeway. In many cases, these items should be started as soon as possible because other work follows immediately behind. This will be indicated by a value of zero for the free float. When necessary, however, noncritical activities can be scheduled to start at a time other than their early start. It will be up to the project planner and to supervision to decide at what pace these activities will proceed.

It is not sufficient to merely make up a network diagram and calculate the time elements. Unless decisions are then made based on the information thus provided and the results are communicated to those who need them, much of the value of the network will be lost. There are usually key dates on which the owner and subcontractors will want to base their planning. For example, the owner should be informed of the dates on which the builder expects to order materials and equipment. If the owner must decide on a color or pattern, he will then be on notice as to when his decision must be made.

Subcontractors are usually involved in projects at many locations and for many general contractors. To permit the subcontractors to effectively plan their work on several jobs, the dates that they are expected to come in to perform certain tasks must be given to them well in advance. Table 4-5 is a schedule that might be given to a subcontractor to establish when work is to be done on each floor of a high-rise building. This schedule gives the start and finish dates that the general contractor is using in his planning. Since change is inevitable, revisions to their respective schedules are sent out to subcontractors whenever necessary or on a regular monthly basis.

Some contractors are reluctant to provide this detailed information to others, reasoning that it will be used against them in the event that the job falls behind. This may be the case in certain instances, and it will be up to the people in charge to correctly inform those to whom the schedules are sent. The recipients should understand that this detail is supplied as an aid to their overall work planning and that the schedule will be updated at various intervals. This type of information, if properly understood and used, could be of real help to subcontractors.

Another important scheduling item is shop drawings. The time required to prepare these drawings and have them approved will often be a critical item, particularly for an activity like reinforcing steel that is needed for the foundations. It is up to the contractor to schedule these approvals through the architect's office so that work on fabrication will not be delayed. If a schedule is prepared for this, the design organization will be put on notice by indications of how this work fits into the overall production on the project.

The scheduling of deliveries can be a real problem at a jobsite with little working space. In most cases, material deliveries should be scheduled to arrive a short time prior to the time the materials will be needed. These material delivery dates may be "backed in" from the information available from CPM. An activity designating the procurement of the material should be shown on the network. Its duration will include the time for placing of the order, for fabrication, and for any special handling that will be necessary. This activity will be followed on the network by the one on which the material will be needed. Once the date the item is required is established, then the start time of the preceding item can be arrived at with more accuracy than just a "guesstimate." This is accomplished by scheduling the start of procurement to be some reasonable amount of time earlier than the early start of the activity where it will be required less the estimated duration of procurement and delivery. This technique is not new to any contractor, but CPM provides a much clearer picture of the time relation than was available previously.

Project Controlling The third project management function in which CPM can be of value is in the control or monitor phase. In order for any system to work, it must involve a set of checks and balances to ensure that it is operating properly. This phase of project management is a continual one, and it will indicate when additional work must be done in the other two areas: planning and scheduling.

Control is the comparing of the actual on-the-job progress with that which was anticipated from the previous scheduling. It includes the process of making job decisions based on this comparison, as required to maintain a predetermined schedule. Control also includes keeping current the information that is required for rescheduling.

TABLE 4-5 Subreport of Job Schedule to Be Given to Individual Subcontractor

CONSTRUCTION DATA SYSTEMS CORP. 4/ 6/69

HIGH RISE BUILDING CONSTRUCTION SCHEDULE

THIS REPORT IS SORTED ACCORDING TO EARLY START ORDER

SUB-REPORT NO. 1

ACTIVITY LABEL	DUR.	CR	ACTIVITY DESCRIPTION	EARLIEST START	EARLIEST FINISH	LATEST START	FREE FINISH	FREE FLOAT
00010	60	*	ERECT TOWER STEEL FTG TO 3RD	MAY 1/69	JUL 28/69	MAY 1/69	JUL 28/69	0
00020	60	*	BOLT & WELD STEEL TOWER FTG TO 3RD	MAY 19/69	AUG 13/69	MAY 19/69	AUG 13/69	0
03010	17	*	ERECT TOWER STEEL 3RD TO 7TH	JUL 28/69	AUG 20/69	JUL 28/69	AUG 20/69	0
03020	17	*	BOLT & WELD STEEL TOWER 3RD TO 7TH	AUG 13/69	SEP 8/69	AUG 13/69	SEP 8/69	0
07010	17	*	ERECT TOWER STEEL 7TH TO 11TH	AUG 20/69	SEP 16/69	AUG 20/69	SEP 16/69	0
07020	17	*	BOLT & WELD STEEL TOWER 7TH TO 11TH	SEP 8/69	OCT 2/69	SEP 8/69	OCT 2/69	0
11010	17	*	ERECT TOWER STEEL 11TH TO 15TH	SEP 16/69	OCT 9/69	SEP 16/69	OCT 9/69	0
11020	17	*	BOLT & WELD STEEL TOWER 11TH - 15TH	OCT 2/69	OCT 27/69	OCT 2/69	OCT 27/69	0
15010	17	*	ERECT TOWER STEEL 15TH TO 19TH	OCT 9/69	NOV 3/69	OCT 9/69	NOV 3/69	0
15020	17	*	BOLT & WELD STEEL TOWER 15TH - 19TH	OCT 27/69	NOV 19/69	OCT 27/69	NOV 19/69	0
19010	17	*	ERECT TOWER STEEL 19TH TO 23RD	NOV 3/69	NOV 26/69	NOV 3/69	NOV 26/69	0
19020	17	*	BOLT & WELD STEEL TOWER 19TH - 23RD	NOV 19/69	DEC 16/69	NOV 19/69	DEC 16/69	0
23010	17	*	ERECT TOWER STEEL 23RD TO 27TH	NOV 26/69	DEC 23/69	NOV 26/69	DEC 23/69	0
23020	17	*	BOLT & WELD STEEL TOWER 23RD - 27TH	DEC 16/69	JAN 12/70	DEC 16/69	JAN 12/70	0
27010	17	*	ERECT TOWER STEEL 27TH TO 31ST	DEC 23/69	JAN 19/70	DEC 23/69	JAN 19/70	0
27020	17	*	BOLT & WELD STEEL TOWER 27TH - 31ST	JAN 12/70	FEB 4/70	JAN 12/70	FEB 4/70	0
31010	22	*	ERECT TOWER STEEL 31ST TO ROOF	JAN 19/70	FEB 18/70	JAN 19/70	FEB 18/70	0
31020	22	*	BOLT & WELD STEEL TOWER 31ST - ROOF	FEB 4/70	MAR 9/70	FEB 4/70	MAR 9/70	0
40000	10	*	REMOVE DERRICK USED FOR STEEL EREC	MAR 9/70	MAR 23/70	MAR 9/70	MAR 23/70	0

Two checklists of work can be of value in the control of a project. The first of these is an early or scheduled start file which lists all the activities that are to get under way on any given day. It serves two purposes: it allows the job superintendent to look ahead to work that is scheduled to begin in the future, and it is a checklist to ensure that activities are started on schedule. In the preparation of the project network, detailed sequences will have been worked out well in advance. This will prove much more beneficial in keeping the job moving than will planning on a day-to-day basis on the job. Therefore, having a predetermined list of activities to be started will aid in keeping the project on schedule.

The second checklist is a late-finish file of all activities. This is sort of a tickler file. If jobs are not completed as of their late-finish time, they have become critical to the completion of the entire project. As will be recalled, the late finish of any activity is the latest that it can be completed without interfering with the planned completion of the overall project.

Samples of these two checklists are shown in Tables 4-6 and 4-7. These portray portions of the information for the network of a small office building, Fig. 4-11. Note that the actual calendar dates have been recorded to aid the superintendent in using the lists. Table 4-6 is a partial listing of activities by early start. Table 4-7 shows a similar listing of activities by late finish. As each job is completed, it can be crossed off both lists so that the job record is kept up to date.

Another tool in the checking of the actual progress with the planned schedule is a time-scale network. This network depicts the job plan in much the same way that a bar chart does, with the additional advantage of showing all the interrelationships of the various activities. As was true for detailed job scheduling, it is not practical to think in terms of an up-to-date time-scale network for the whole project. The network will be just as useful as a control tool if it is corrected only for the next month or two whenever the overall network is updated. It is quite often just as easy to prepare the "final" preliminary network on a time scale after initial analysis and replanning has been done. This type of diagram has proved very worthwhile in communicating with people who are unfamiliar with the detail of CPM. With its help, they can visualize the time relationship of the flow of work.

In the case of precedence diagrams, the circle or node is usually plotted as of the early start. The length of the sequence line to the next activity in sequence then indicates, roughly, the time that is available for the activity.

The matter of keeping the schedule up to date is a part of the control function. When the actual project progress is checked against the planned schedule, information is available to show when updating of the schedule would be required or be beneficial. There are two criteria that determine when updating would be advisable. The first is when the job falls behind schedule. That is to say, when one or more critical activities have started late or when a noncritical activity is commenced after its late start. In this latter case, the completion of the overall job has changed and different activities may have become critical. In some cases an entirely new critical path will be established; in others only a subchain will be added to the previous critical path.

The second criterion is based on a review of the network diagram. From time to time on the project, certain sections of work will have to be replanned to expedite their progress. When this occurs, or when a major change of planning comes about, the network no longer is a true representation of the job. Therefore, the data available from the time elements may be totally unrealistic. In other words, whenever the project changes in such a way as to render the network invalid, then the diagram should be changed and the schedules updated. The network is the basis for all CPM calculations and decisions, and it must be maintained to give a realistic model of the job at all times.

Prior to the introduction of the critical path techniques, whenever a few activities fell behind whatever schedule had been set up, the tendency was to put the entire job on a crash basis to make up for "lost" time. The time elements of total and free float now allow a rational evaluation of work slippage, right on the job, as part of the control function. If a certain activity does not commence on time or if it requires more time than the estimated duration, it may or may not adversely affect the over-

TABLE 4-6 Listing of Activity Items by Early Start

CONSTRUCTION DATA SYSTEMS CORP. 2/15/69

SAMPLE CPM NETWORK SMALL OFFICE BUILDING

THIS REPORT IS SORTED ACCORDING TO EARLY START ORDER

ACTIVITY LABEL	DUR.	CR	ACTIVITY DESCRIPTION	EARLIEST START	EARLIEST FINISH	LATEST FREE START	LATEST FREE FINISH	LATEST START	LATEST FINISH	FREE FLOAT	TOTAL FLOAT
NTP	0	*	CONTRACT AWARD AND N.T.P.	JAN 4/71	JAN 4/71	JAN 4/71	JAN 4/71	JAN 4/71	JAN 4/71	0	0
0020	2		MOVE ON SITE	JAN 4/71	JAN 6/71	JAN 4/71	JAN 6/71	JAN 21/71	JAN 25/71	0	13
0030	15	*	SHOP DRAWINGS STRUCTURAL STEEL	JAN 4/71	JAN 25/71	JAN 4/71	JAN 25/71	JAN 4/71	JAN 25/71	0	0
0040	25		BUILD FOUNDATIONS	JAN 6/71	FEB 10/71	JAN 6/71	FEB 10/71	JAN 25/71	MAR 2/71	0	13
0050	25	*	FABRICATE STRUCTURAL STEEL	JAN 25/71	MAR 2/71	JAN 25/71	MAR 2/71	JAN 25/71	MAR 2/71	0	0
0060	10		MECHANICAL & ELECTRICAL AT BASEMENT	FEB 10/71	FEB 25/71	FEB 10/71	FEB 25/71	MAR 16/71	MAR 30/71	0	23
0070	3		FORM SLAB OVER UTILITY AREA	FEB 25/71	MAR 2/71	MAR 22/71	MAR 25/71	MAR 30/71	APR 2/71	17	23
0100	6	*	ERECT STRUC STEEL NORTH HALF	MAR 2/71	MAR 10/71	MAR 2/71	MAR 10/71	MAR 2/71	MAR 10/71	0	0
0110	6	*	ERECT STRUCT STEEL SOUTH HALF	MAR 10/71	MAR 18/71	MAR 10/71	MAR 18/71	MAR 10/71	MAR 18/71	0	0
0120	10		NORTH EXTERIOR MASONRY WALL	MAR 10/71	MAR 24/71	MAR 10/71	MAR 24/71	MAR 16/71	MAR 30/71	0	4
0140	3		SET STEEL DECK NORTH HALF FLOOR 2	MAR 10/71	MAR 15/71	MAR 15/71	MAR 18/71	MAR 15/71	MAR 18/71	3	3
1000	5		ROUGH-IN ELECT & MECH FLOOR 1	MAR 18/71	MAR 25/71	MAR 18/71	MAR 25/71	MAR 26/71	APR 2/71	0	6
0150	3	*	SET STEEL DECK SOUTH HALF FLOOR 2	MAR 18/71	MAR 23/71	MAR 18/71	MAR 23/71	MAR 18/71	MAR 23/71	0	0
0160	4		SET STEEL ROOF DECK	MAR 23/71	MAR 29/71	MAR 31/71	APR 6/71	MAR 31/71	APR 6/71	6	6
2000	8	*	ROUGH-IN ELECT & MECH FLOOR 2	MAR 23/71	APR 2/71	MAR 23/71	APR 2/71	MAR 23/71	APR 2/71	0	0

TABLE 4-7 Listing of Activity Items by Late Finish

CONSTRUCTION DATA SYSTEMS CORP. 4/ 6/69

SAMPLE CPM NETWORK SMALL OFFICE BUILDING

THIS REPORT IS SORTED ACCORDING TO LATE FINISH ORDER

ACTIVITY LABEL	DUR.	CR	ACTIVITY DESCRIPTION	EARLIEST START	EARLIEST FINISH	LATEST START	FREE FINISH	LATEST START	LATEST FINISH	FREE FLOAT	TOTAL FLOAT
NTP	0	*	CONTRACT AWARD AND N.T.P.	JAN 4/71	JAN 4/71	JAN 4/71	JAN 4/71	JAN 4/71	JAN 4/71	0	0
0020	2		MOVE ON SITE	JAN 4/71	JAN 6/71	JAN 4/71	JAN 6/71	JAN 21/71	JAN 25/71	0	13
0030	15	*	SHOP DRAWINGS STRUCTURAL STEEL	JAN 4/71	JAN 25/71	JAN 4/71	JAN 25/71	JAN 4/71	JAN 25/71	0	0
0040	25		BUILD FOUNDATIONS	JAN 6/71	FEB 10/71	JAN 6/71	FEB 10/71	JAN 25/71	MAR 2/71	0	13
0050	25	*	FABRICATE STRUCTURAL STEEL	JAN 25/71	MAR 2/71	JAN 25/71	MAR 2/71	JAN 25/71	MAR 2/71	0	0
0100	6	*	ERECT STRUC STEEL NORTH HALF	MAR 2/71	MAR 10/71	MAR 2/71	MAR 10/71	MAR 2/71	MAR 10/71	0	0
0110	6	*	ERECT STRUCT STEEL SOUTH HALF	MAR 10/71	MAR 18/71	MAR 10/71	MAR 18/71	MAR 10/71	MAR 18/71	0	0
0140	3		SET STEEL DECK NORTH HALF FLOOR 2	MAR 10/71	MAR 15/71	MAR 15/71	MAR 18/71	MAR 15/71	MAR 18/71	3	3
0150	3	*	SET STEEL DECK SOUTH HALF FLOOR 2	MAR 18/71	MAR 23/71	MAR 18/71	MAR 23/71	MAR 18/71	MAR 23/71	0	0
0120	10		NORTH EXTERIOR MASONRY WALL	MAR 10/71	MAR 24/71	MAR 10/71	MAR 24/71	MAR 16/71	MAR 30/71	0	4
0060	10		MECHANICAL & ELECTRICAL AT BASEMENT	FEB 10/71	FEB 25/71	FEB 10/71	FEB 25/71	MAR 16/71	MAR 30/71	0	23
1000	5		ROUGH-IN ELECT & MECH FLOOR 1	MAR 18/71	MAR 25/71	MAR 18/71	MAR 25/71	MAR 26/71	APR 2/71	0	6
0070	3		FORM SLAB OVER UTILITY AREA	FEB 25/71	MAR 2/71	MAR 22/71	MAR 25/71	MAR 30/71	APR 2/71	17	23
2000	8	*	ROUGH-IN ELECT & MECH FLOOR 2	MAR 23/71	APR 2/71	MAR 23/71	APR 2/71	MAR 23/71	APR 2/71	0	0
2010	2	*	POUR FLOOR FILL ON DECK AT FLOOR 2	APR 2/71	APR 6/71	APR 2/71	APR 6/71	APR 2/71	APR 6/71	0	0

all project. If the extension is on a critical item, the time will either have to be made up on a later activity or the job will run over the planned time. In order to make up this time, only the future critical activities need be considered as candidates for expediting.

If the slippage occurs in a noncritical activity, the overall project may not be adversely affected but the start of other work following may be delayed. If the activity has an amount of free float, this gives a quantitative value for the length of time that its finish can be delayed without delaying the start of future work. The project management should understand this concept. If the activity has only total float, some following activities will also be delayed. In other words, total float is shared by the activities on a given chain. The whole string of work items can be shifted to a certain extent, but when one is shifted or delayed, this reduces the amount of scheduling leeway for other future work.

Whenever an activity falls behind its scheduled finish date, a check should be made to evaluate the effect on the project. If future work will be delayed, notice should be sent out to all concerned so that the workers involved will not arrive on the job too soon.

ROLE OF THE COMPUTER

To many people, CPM and PERT techniques are synonymous with the use of digital computers to produce project schedules automatically. Even though CPM is often implemented with a computer-oriented system, it can also be handled manually. The idea that the costly ownership or leasing of a computer is automatically involved has scared many construction people away from CPM. Actually, for many small- to medium-sized jobs, manual CPM is a better system and one which can be handled rather simply with a contractor's own personnel. In any event, a manual approach is a desirable intermediate step in learning to use CPM effectively.[7]

The computer, when properly used, can be a very useful tool in the implementation of CPM. The important thing to remember is that a computer is simply a data processing device. It can handle all the calculations involved in a 1,000-activity network in a matter of seconds, but in doing so it is nothing more or less that a very fast and accurate clerk. The actual calculations involved in basic CPM are very simple but tedious. This makes the computer a "natural" to help set up a CPM schedule on any large project.

A word of caution is in order here. A computer can be a wonderful help or it can be the downfall of a CPM system. The important item is not the "electronic brain," but rather the people who are to use the system. In many cases the network is made up by people in the home office. The computer print-out is just another stack of paper that has to be dealt with. The total value of CPM is lost because the key to its success is an acceptance and a good basic understanding of the background of the system by the man actually making use of it and running the job. It is well to repeat that the value gained from the system will be directly proportional to its acceptance by the project management.

On large jobs, the use of a digital computer can be a real asset. Even though it is conceivable that networks of any size could be analyzed using a completely manual method, there seems to be some point at which it becomes desirable to convert to a computer. This cutoff between manual and computer calculations depends on several things, and the size of the project is only the first item on the list. Other factors to be considered are how the data can best be presented for use as a control tool; how often, if at all, the information will be updated; how much time will be needed to check hand calculations to be sure that no errors have been introduced through a simple, careless mistake; and, of course, the requirements of the specifications.

As a general rule of thumb, the writer has found that computer implementation will be worthwhile if a project will be over six months in duration, will require several updating revisions, and is over 300 activities in size. On projects over this size, the amount of effort involved in running through the time-element calculations by hand increases because of the sheer volume of computations that must be made. The chance for errors is also correspondingly greater.

Whether the computations are to be made by hand or with a computer, the same amount of work is involved in preparing the diagram. If the calculations are to be handled on a computer, each of the activities will have to be identified by an activity label for easy reference. In the case of precedence diagrams, this requires only one label (which can be a numeric or alphabetic label); whereas on arrow diagrams the events at the head and tail of each arrow result in a dual numbering system.

Once the diagram has been numbered properly, the information must be coded onto data sheets for keypunching onto computer cards. Most current programs allow for random numbering of activities, which makes the job of numbering the network much easier. It is well to have this coding performed by someone who knows construction terms, since the coded job descriptions which are used on the diagram may have to be expanded in order to make them more meaningful.

After the coding forms are completed, the information must be keypunched onto computer data cards. When the cards have been punched and sorted into the proper sequence, they can be read into the computer along with a CPM program. This program instructs the computer on how to handle the data and what form the output should take.

The time-element data are calculated extremely rapidly and accurately. Another important feature of computer solutions is the feasibility of having the data presented in several different forms, for improved convenience to the data users. In a manual solution, the time-element data are supplied either in tabular form or else they are recorded on the diagram itself. A lot of recopying is then required to prepare partial schedules and schedules which are listed on different work orders.

With a computer application, the output data may be listed in order of early starts to show the sequence in which jobs may be started. They can also be listed in order of late starts to show the latest dates by which jobs must have been commenced. If the input data are coded properly, a separate schedule can be listed for each work area or for each subcontractor, outputting only scheduling information that relates to each specific work breakdown.

At the same time that the information is being printed out of the computer, the working days can be correlated with a job calendar so that the data are printed out with the time elements in calendar days. This makes the output meaningful to anyone examining it without the necessity of referring to a conversion chart each time a date is to be checked.

Updating Necessity One fact is certain in construction, and that is that things will change. No matter how well a job is planned at the beginning, there will be activities that start late or that do not follow properly in sequence. This is not the fault of the job, the planner, or the CPM system. It is just a fact that realistically must be accepted.

When activities on the job do fall behind schedule, CPM offers a means of evaluating the effect on the overall project and of deciding what the best corrective procedure will be. The first step will be to revise the network diagram so that it realistically represents the project at the time the data are updated. Activities which have been completed should be noted. Activities that are in progress should have their time estimates revised to reflect the amount of work remaining. The network need not be completely redrawn; in fact, it usually is not. The changes can be added with a colored pencil. A line from the top to the bottom, cutting through all activities in progress, will be the cutoff line showing the date of the revision.

The second step is to revise the input data cards. The cards of all those activities that have been completed must be flagged. The new estimated durations for activities in progress must then be punched onto the respective cards in place of the previous estimates. Any additional revisions that are made to the diagram to reflect new ideas or restrictions must also be incorporated into the revised data cards.

The third step is to rerun the cards on the computer to obtain a new set of time-element data. The important idea is that only a relatively small number of cards must be changed for each updating run. This means that although a fairly long time was required originally to code all the data cards, this work is done only once. Subsequent revisions are relatively easy to make because the bulk of the work has already been done.

In some cases, on jobs where CPM is a contract requirement, updating will be done on a biweekly or monthly basis. This may not be a completely logical time interval. The network and time elements should be updated whenever changes that render the relations shown on the network invalid occur on the job or when several activities are started or completed beyond their late starts or finishes. The latter situation indicates that these activity sequences are now critical and that new priorities have been introduced.

Simulation of Alternatives One very useful application of the critical path method that has not been fully utilized in the past is to provide a means of simulating several alternative construction plans or methods to arrive at the best one. A general network can be made to show the overall plan; then a process similar to that for updating can be used to compare minor or major revisions to this general plan. Typically, only a few activities would have to be changed to reflect a new scheme. With the aid of a computer to quickly and accurately recalculate the time-element data, several plans can be analyzed in a short time and at a relatively small cost.

Table 4-8 shows the computer output for the small building project whose network diagram is given in Fig. 4-11. The information is presented in a readable and usable form. The activity dates have been correlated and the time-element data are presented in the form of a dated master report.

IMPLEMENTING A CPM SYSTEM

The critical path method is a very powerful tool for the construction planner. The important point which is stressed in this section is that CPM is just like any other tool in the construction manager's bag of tricks. If used properly and with a sense of balance, the results will be meaningful and rewarding.

The networking phase of CPM is fairly easy to master. Any number of firms have started using network planning after having read only a short article about it in a magazine. The network is the base on which the rest of CPM is built, and as such it is the key to any application. The developers of CPM have listed some of the benefits of network diagramming other than providing the model for critical-path time-element calculations and analysis:

1. Provides a disciplined basis for planning a project. It actually forces long-range thinking.
2. Provides a picture of the scope of a project that is easy to understand and can be clearly read.
3. Provides a vehicle for simulating alternatives.
4. Directs attention to the activities that determine project duration.
5. Shows the interrelationships between activities and pinpoints the responsibilities of the various subcontractors involved.
6. Provides an excellent vehicle for training project personnel.

In the time-element phase, data are available that will be far more accurate for scheduling than data that have been obtained from any other system. The actual calculations are not complicated and can be handled with either a computer or noncomputer approach. Working right on a copy of the diagram has proved to be a fairly workable method, even on fairly large projects. Time-element data provide information that will be extremely useful for coordinating the work of prime and subcontractors as well as providing realistic data on delivery dates and reasonable project completion requirements.

The critical path, or chain of activities that determines the project duration, is defined. This allows the manager to manage by exception. In other words, it pinpoints the activities that must start at a set time in order that the project may be completed as scheduled. The float or scheduling leeway of every activity is available from this phase also. It provides a very useful ranking of criticality among the various activities within a project.

TABLE 4-8 Computer Output for CPM Network

CONSTRUCTION DATA SYSTEMS CORP. 4/ 6/69

SAMPLE CPM NETWORK SMALL OFFICE BUILDING

THIS REPORT IS SORTED ACCORDING TO CARD INPUT ORDER

ACTIVITY LABEL	DUR.	CR	ACTIVITY DESCRIPTION	EARLIEST START	EARLIEST FINISH	LATEST START	FREE FINISH	LATEST START	LATEST FINISH	FREE FLOAT	TOTAL FLOAT
NTP	0	*	CONTRACT AWARD AND N.T.P.	JAN 4/71	JAN 4/71	JAN 4/71	JAN 4/71	JAN 4/71	JAN 4/71	0	0
0020	2		MOVE ON SITE	JAN 4/71	JAN 6/71	JAN 4/71	JAN 6/71	JAN 21/71	JAN 25/71	0	13
0030	15	*	SHOP DRAWINGS STRUCTURAL STEEL	JAN 4/71	JAN 25/71	JAN 4/71	JAN 25/71	JAN 4/71	JAN 25/71	0	0
0040	25		BUILD FOUNDATIONS	JAN 6/71	FEB 10/71	JAN 6/71	FEB 10/71	JAN 25/71	MAR 2/71	0	13
0050	25	*	FABRICATE STRUCTURAL STEEL	JAN 25/71	MAR 2/71	JAN 25/71	MAR 2/71	JAN 25/71	MAR 2/71	0	0
0060	10		MECHANICAL & ELECTRICAL AT BASEMENT	FEB 10/71	FEB 25/71	FEB 10/71	FEB 25/71	MAR 16/71	MAR 30/71	0	23
0070	3		FORM SLAB OVER UTILITY AREA	FEB 25/71	MAR 2/71	MAR 22/71	MAR 25/71	MAR 30/71	APR 2/71	17	23
0100	6	*	ERECT STRUC STEEL NORTH HALF	MAR 2/71	MAR 10/71	MAR 2/71	MAR 10/71	MAR 2/71	MAR 10/71	0	0
0110	6	*	ERECT STRUCT STEEL SOUTH HALF	MAR 10/71	MAR 18/71	MAR 10/71	MAR 15/71	MAR 10/71	MAR 18/71	0	0
0120	10		NORTH EXTERIOR MASONRY WALL	MAR 10/71	MAR 24/71	MAR 10/71	MAR 24/71	MAR 16/71	MAR 30/71	0	4
0130	10		SOUTH EXTERIOR MASONRY WALL	MAR 24/71	APR 7/71	MAR 24/71	APR 7/71	MAR 30/71	APR 13/71	0	4
0140	3		SET STEEL DECK NORTH HALF FLOOR 2	MAR 10/71	MAR 15/71	MAR 15/71	MAR 18/71	MAR 15/71	MAR 18/71	3	3
0150	3	*	SET STEEL DECK SOUTH HALF FLOOR 2	MAR 18/71	MAR 23/71	MAR 18/71	MAR 23/71	MAR 18/71	MAR 23/71	0	0
0160	4		SET STEEL ROOF DECK	MAR 23/71	MAR 29/71	MAR 31/71	APR 6/71	MAR 31/71	APR 6/71	6	6
0200	12		INSTALL STEEL SASH CURTAIN WALL	APR 7/71	APR 23/71	APR 7/71	APR 23/71	APR 13/71	APR 29/71	0	4

In the case of scheduling variation, use is made of the network and the time-element data to replan the project from the standpoint of least-cost expediting and resource allocation. These two concepts actually go hand in hand and can be treated informally to yield reasonable benefits in return for a fairly small investment in time.

When the project duration must be shortened, the list of critical items in the first few months of the project should be looked at. A good estimator and project manager should be able to determine which activities can be expedited at least relative cost.

When certain types of resources or manpower are scarce, planning the best usage of these key items can be very beneficial. In many cases a project manager will build this right into his schedule, interrelating the activities to show the flow of work which is done by the various crews. If this work flow is not shown explicitly, resource loading curves will be a help in defining problem areas so that corrective planning can be done.

The following is a suggested approach to the use of the critical path method within a construction organization.

1. Pick a small, relatively simple project for the first attempt. It is best to become acquainted with the system gradually and expand its use as knowledge is gained. Every organization should develop its own special methods of implementation to fit its needs and the abilities of its personnel.

2. Draw a network diagram for the project that is as accurate as possible. Realize that all of CPM is based on this network model and will only be as good as the diagram information. It is suggested that activity interrelationships first be identified on the basis of the physical work sequences which must be followed. The planner can then go back through the network and include the probable field scheduling restrictions.

3. Prepare an estimated duration for each activity. These estimates should represent reasonable times which are consistent with overall project speed and complexity. They would be the times used to plot a bar chart under conventional planning procedures.

4. Perform the time element calculations for this network to determine the critical path, float of noncritical activities, and project duration.

5. Analyze the data from step 4. Examine the activities that were shown to be critical to see it they form a logical sequence through the project. One should be sure that the network is as realistic as possible before progressing further. Also examine the activities having very small total float, since this indicates that they are nearly critical. Now that the key items are identified, a further breakdown for some of these activities may be desirable.

6. If any major changes are made to the diagram, prepare new estimates for the changed activities and repeat step 4.

7. If the projected completion time given by the time elements is longer than desired, consider the possibility of decreasing the length of the job by revising just the critical items. Now that the importance of these items to project completion is identified, the planner may devote more effort to arriving at faster ways to perform them. Knowledge of noncritical work going on at the same time may suggest speedups by transferring men to the more important activities.

8. Check with subcontractors to see that the time estimates for their work are as accurate as possible. Cooperation between all people involved on a project in setting up the system will yield great benefits. Check key dates with suppliers to see that procurements will not hold up the work.

9. After the "final" preliminary schedule is set up for the project, it will be helpful to prepare a time-scale diagram of the whole project for field use in daily planning and for office use for control. The entire project should be drawn in this manner only once. After each future updating, time-scale networks covering the next 30 to 60 days of operations will be quite valuable and will be fairly quick and easy to prepare.

10. Update the schedule whenever the job falls behind to the extent that the critical relationships may have changed or when it becomes apparent that the network is not a true representation of the way the project is actually being executed. Information based on an unrealistic network is of questionable value. As personnel become

acquainted with the system, reasonable decisions as to the desirability of updating can be made.

11. Review the benefits gained by the use of CPM planning with all key personnel from time to time. Keep in mind that it is a tool. It is a better tool than any previously available for this work, but it is only as effective as the people using it.

This section has not attempted to present a complete, detailed course on the use of CPM in construction. The intent has been to set down basic concepts and to include some ideas as to philosophy behind the proper use of this tool. CPM, properly applied and used, can be a great help to any construction organization.

Probably the most meaningful idea to retain from this brief treatment is that the benefits realized from a CPM system are proportional to the effort that is put into it. It is very important that the project management personnel participate in the development of the basic network diagram. This diagram must reflect their best thinking and thereby be as realistic as possible. It should also be remembered that most networks need to be reviewed and updated at frequent intervals. This is not a fault of the system but rather a fact of life on any complex construction project.

REFERENCES

1. Robert C. McLean, *Implementation of the Critical Path Method through Resource and Multi-Project Planning and Scheduling,* The Construction Institute, Dept. of Civil Engineering, Stanford University Press, Stanford, Calif., 1964.
2. Joseph Horowitz, *Critical Path Scheduling,* The Ronald Press Company, New York, 1967.
3. Joseph J. Moder and Cecil R. Phillips, *Project Management with CPM & PERT,* 2d ed., Van Nostrand-Reinhold Company, New York, 1970.
4. James J. O'Brien, *CPM in Construction Management—Scheduling by the Critical Path Method,* McGraw-Hill Book Company, New York, 1964.
5. *A Manual for Applying the Critical Path Method to Highway Department Engineering and Administration,* E. S. Preston & Associates, Ltd., Washington, D.C., 1963.
6. Russell D. Archibald and Richard L. Villoria, *Network Based Management Systems (PERT/CPM),* John Wiley & Sons, Inc., New York, 1967.
7. John W. Fondahl, *A Non-Computer Approach to the Critical Path Method for the Construction Industry,* 2d ed., The Construction Institute, Dept. of Civil Engineering, Stanford University Press, Stanford, Calif., 1962.
8. John W. Fondahl, "Let's Scrap the Arrow Diagram," *Western Construction,* vol. 43, no. 8, August, 1968, pp. 29–33.
9. *CPM in Construction, A Manual for General Contractors,* The Associated General Contractors of America, Inc., Washington, D.C., 1965.
10. George E. Deatherage, *Construction Scheduling and Control,* McGraw-Hill Book Company, New York, 1965.

Section 5

Construction Safety

ROGER W. MUELLER

Construction Safety Services Consultant,
Employers Insurance of Wausau, Wausau, Wis.

INTRODUCTION

This section has one fundamental purpose: to motivate those in the construction industry to contribute toward a common goal of reducing the intolerable waste of human life and human capacity which results from occupational injuries and disease.

This section seeks to introduce construction management to broader management concepts of loss control and to encourage the application of effective loss-control techniques. It is directed in particular to those management personnel who, by their involvement in loss control, can make the most meaningful and immediate contributions to this common goal.

Construction Safety—A Dilemma Attention is continually directed at the universal problem of occupational safety and health. Current and proposed legislation at the local, state, and national level has spotlighted accidents and occupational disease as major sources of economic loss. This waste amounts to billions of dollars yearly, and the accompanying human suffering and misery have no price tag. Unless industry's loss-control efforts produce acceptable results, restrictive legislation and enforcement that are undesirable to industry may be enacted.

The construction industry, with its rapidly expanding technology, faces the challenge of demonstrating by action its desire and ability to improve the construction safety record. Proof should be in the form of decreasing accident rates and subsequent reduction in losses.

The construction industry has consistently had a high frequency-severity rate. This is confirmed by *Accident Facts*,[1]* an annual publication of the National Safety Council. *Accident Facts* lists the three-year (1967–1969) injury rate for the construction industry as approximately double that for all industries reporting to the National Safety Council. *Accident Facts* also reveals that member companies of the National Safety Council in general have better safety programs and lower injury frequency rates than nonmember companies.

The estimated cost of accidents and occupational disease in the construction industry is staggering. Specific cost statistics continually vary, but there is general agreement that the losses (including those from fire) are in the billions of dollars and increasing. These costs are both real and substantial, and they must ultimately be paid by the contractors' clients in the form of increased construction costs.

Concentrated industry effort in the field of construction safety can produce immediate and favorable results. This has been verified by the application of technical loss control resources to specific problem areas of occupational safety and health. Resultant frequency-severity rates are well below the industry average.

Annually, outstanding safety accomplishments of individual contractor firms are given national recognition by the Associated General Contractors of America. Many other examples of exceptional safety accomplishments are available from various trade associations, governmental agencies, and insurance companies. They are testimony that losses can be controlled.

Safety versus Loss Control Safety has been given many labels. Therefore, no attempt will be made to define safety, since there likely would be considerable disagreement. But where does safety start? Where does it end? What about the old expression, "Safety First"? Is this a misnomer?

The accomplishment of any task involves the assumption of some degree of risk. The degree of risk usually can be reduced or virtually eliminated by enacting and enforcing appropriate controls. To establish a meaningful and objective discussion in this section, the term "loss control" will encompass the entire subject of construction safety.

Loss control includes the control of losses of any kind that are incurred as a result of occupational accident or disease. Losses may be one or a combination of the following: loss of life; bodily injury; loss of earnings; loss of skills; loss of services; loss of loved ones; property damage, including fire loss; and many others. Some losses have specific dollar values, and others cannot be measured in terms of dollars. Real-

* Superior numbers refer to the list of references at the end of this section.

istically, the contractor must also include adverse publicity, poor public relations, and loss of profits.

NECESSITY FOR LOSS CONTROL

There are fundamental factors which provide construction management with the need, motivation, and interest to exert maximum influence in construction loss control. The law-safety aspect will continue to become more involved. A construction firm that does not demonstrate by actions and deeds the application of modern techniques of loss control could be openly susceptible to liability litigation.

Self-preservation There is common agreement among safety professionals that long-term loss control cannot be successful without top management involvement. What should motivate construction management to become involved in loss control? A business undertaking usually has a primary desire to continue to exist and function as an identifiable entity. It has been repeatedly demonstrated in the business world that severe losses resulting from occupational accidents and diseases can be a definite contributing factor to the success or failure of a firm. Therefore, effective loss control is a vital factor affecting the continuance of a successful contractor's business.

Profit Under the private enterprise system, profit is a necessity for the contractor's survival. Many factors contribute to a contractor's profit. These include the application of good management techniques, management skills, good labor relations and, as emphasized before, the control of losses and occupational accidents and disease.

Statutory, Regulatory, and Advisory Standards The legitimate contractor has legal obligations to himself, to his employees, to his clients, and to the public. His legal responsibilities are many. They may be specifically defined by laws, standards, codes, regulations, and accepted good work practice in the absence of any specific requirements.

When considering the specifics of loss control, management in the construction industry has a legal obligation to know, observe, and enforce those requirements applicable to the work being done. The requirements are usually based on a need resulting from prior experience. The implications of these legal obligations for loss control are evidenced by the increased number and dollar size of liability claims awarded to workers in the construction industry. A major source for such claims is the "safe place to work" law which is applicable in more than half of our states. This type of law expressly places general responsibility for occupational safety on the employer.

Many states assess fines, jail sentences, or both as penalties for the violation of safety orders. These offenses are usually labeled as misdemeanors unless another penalty is specified. Some states include penalties in their workmen's compensation laws, oftentimes labeled "serious and willful misconduct" penalties. The penalty is assessed against the employer when it is proven that known unsafe conditions caused or contributed to an employee injury. The penalty is in the form of additional compensation to the injured employee, usually limited to specific amounts, uninsurable by law, and based on the severity of the injury. In some states reduced compensation may be awarded to an employee whose actions, as specifically defined, are considered contributory to the disability. This provision is seldom applied, as sympathy for the injured employee and dependents, if any, prevails.

An example of the trend in state legislation in the field of industrial safety is found in California. In 1963, fines and the lengths of jail sentences were increased for employers whose gross negligence in on-the-job safety resulted in the death of an employee. The provisions of this legislation are applicable except where still more severe penalties are provided in the Labor Code. The maximum limit for imprisonment in the county jail became one year, and fines were increased to not less than $1,000 and to not more than $5,000.

Also in California, the Business and Professions Code was amended to reaffirm and strengthen the authority of the Department of Professional and Vocational Standards. This department can suspend the licenses of employers (which includes all contractors) within its licensing jurisdiction who fail to take required measures for the safety of workmen. Any violation of safety orders of the Division of Industrial Safety that results in death or serious injury becomes "a cause for disciplinary action" by the department.

In the absence of specific safety laws, codes, or regulations, there is an increased tendency to determine employer liability based on established standards and universally accepted good practice procedures for a specific situation.

Owner Requirements—Contractual Liability Owners and contracting agencies are expressing increased interest in loss control. They have recognized that contractors' losses resulting from unsafe work conditions and poor management controls will be reflected in increased construction costs. Therefore, owners are adopting methods to keep these losses to a minimum. Owners also need protection from liability suits resulting from or attributed to the contractor's work or actions. As a result, it is increasingly popular for safety and loss-control features to be incorporated by owners into the contract requirements. Probably the best-known examples of owner-agency requirements for loss control are the Corps of Engineers' *General Safety Requirements*[2] and the Bureau of Reclamation's *Construction Safety Standards.*[3] These manuals officially establish the health and safety requirements for construction by contract. The requirements are considered as contractual obligations, and the government agency personnel enforce them accordingly.

Many private agencies and industries have established their own safety requirements and provisions for contract work. Applicable safety standards may be incorporated into the contract documents by reference. Specific safety requirements may be included as special conditions in the specifications.

The increased use of indemnification clauses in construction contracts carries with it the potential of heavy losses for contractors. These clauses are commonly referred to as "hold-harmless agreements" and, when included in a contract, impose someone else's legal liability on the person who assumes the contractual obligation for performance of a task. The following paragraph is an outstanding example of this broad form of indemnification. Such a clause could be disastrous if the contractor were not fully aware that he was assuming the cost for injuries or damages not caused or contributed to by his employees or subcontractors.

> The Contractor agrees to indemnify and save harmless the Owner, Architect, and Engineer, their agents and employees, from and against all loss or expense, including cost and attorneys' fees, by reason of liability imposed by law upon the Owner, Architect, or Engineer for damages because of bodily injury, including death at any time resulting therefrom, sustained by any person or persons or on account of damage to property, including loss of use thereof, whether caused by or contributed to by said Owner, Architect, or Engineer, their agents, employees or others.

A hold-harmless agreement in the contract specifications may be lengthy or as brief as one sentence, and the latter type may be more dangerous than a comprehensive list of demands transferring liability from one party to another. The effectiveness of such a clause can only be determined by court test, and even then there is no assurance that a subsequent case will be ruled in the same way. It is extremely important that contractors seek competent advice and legal counsel as to the possible implications of such clauses. This will provide guidance to needed insurance protection and to high-risk activities where an added emphasis on loss controls may be warranted.

Legislative Actions The contractor should keep himself informed as to proposed and pending safety legislation which may require additional dollar outlays and impose ineffective restrictions and controls. Contractors should express their interest as individuals and citizens, as businessmen, and as part of an industry. This interest can be demonstrated by active participation in legislative hearings and proposals. Contractors should be active in their industry and trade associations and should encourage others to participate similarly and express opinions and ideas.

Industry and trade associations are only as active and effective as their members' interests dictate. They can provide constructive guidance and influence on industry problems if their efforts are properly directed and endorsed. For example, if a contractor firm indicates disinterest by nonparticipation and the industry performance reflects disinterest, the public may support unfavorable legislation.

Public Relations The results of loss control have a direct effect on the construction industry's public relations. Some news media have the facilities to publicize losses from occupational diseases and accidents. News of a serious loss, and the resultant unfavorable publicity to the parties involved and to the industry, can be carried to millions of people in a matter of hours. The contributing factors may never be known to the public, who are on the receiving end of biased and unqualified opinions. Public sympathy and sentiment can result in the enactment of costly and ineffective governmental loss controls which may not be based on sound economic and moral considerations.

Each construction firm must maintain good public relations within its own operational sphere of influence. Prospective clients are expressing an increased interest in a firm's record of losses and loss-control activities. This loss history may be an indicator of the application of management techniques to production and loss control. An outstanding record of loss control should be reflected in efficient production and in low overhead costs for accident losses.

Poor loss-control techniques, in addition to contributing to a record of costly losses by a firm, may have a direct effect on company personnel attitudes. An unfavorable record of serious occupational accident and disease losses by individual firms could undermine the confidence of employees in company management and ability. It may also strain labor union relationships.

Community public relations can be adversely affected if poor work practices are consistently permitted. Some examples could include unsightly housekeeping on jobsites exposed to the public; poor traffic control and high-hazard situations; nuisances created by construction activity, such as dust, noise, and vibrations; open-excavation hazards; and unscheduled interruptions of normal business activities in a congested area. Good public relations are a necessity for the construction industry.

Moral Aspects The moral aspects of loss control are emphasized last, but not least. There is a wealth of available material which expounds the moral aspects of safety. One of the best-known applications is to highway safety, where the visible results are insignificant. There is complete agreement that any approach to effective loss control cannot be successful without the inclusion of moral considerations.

APPLICATION OF MANAGEMENT PRINCIPLES TO LOSS CONTROL

The factors which have established the need for loss control in construction also serve to establish the importance of applying sound management principles to the techniques of loss control.

Management Principles of Loss Control The basic management principles for controlling, coordinating, and directing the use of personnel, skills, time, material, and money to accomplish a given task can be found in many textbooks. These management principles are planning, organizing, directing, and controlling for profit. If we agree that loss control is a management responsibility, then these same principles must be applied to the technique of controlling losses.

Management must fully integrate loss-control activities into production activities at every level. By doing this, it will put an end to the failures that result from trying to isolate loss control as a separate and distinct activity.

Management Planning for Loss Control Planning is a basis for management action and control. Many of the ingredients of management planning will involve loss control directly or indirectly. Included are a firm's objectives, its organizational design to attain or exceed those objectives, policies, procedures, programs, schedules, and standards. Not to be forgotten is budgeting as a vital planning ingredient which has significant meaning to loss control.

Cost of controls may be an important factor: for instance, providing for safety nets or for a loss-control specialist. Insurance costs must be included in budgeting. Executive leadership should be constantly aware of its responsibility to include loss control in its planning. This responsibility cannot be delegated. The goals and objectives for loss control should be practical and attainable. For example, "Every effort will be made to reduce this firm's losses from accidents and injuries by 50 per cent in one year." The inclusion of loss control in planning will give it purpose and direction through all levels from management and supervision to the employee. A management loss-control policy will provide the impetus for action.

Basically, a loss-control policy states that top management expects maximum effort from everyone to control losses. The policy may be as simple or as detailed as management desires. It should state management's acceptance of loss control responsibility and should fix and define authority and accountability within the company's line organization. It should point the direction that management wants to go. The policy should be a guide for decision making and individual action. Any effective policy should identify acceptable loss-control standards as a basis for future effort and emphasis.

At this point a distinction should be made between loss-control policy and the commonly used term "loss control" or "safety" program. The terms are often incorrectly used interchangeably. A safety program will be considered herein as a strategy to be followed or actions to be taken to achieve or exceed safety objectives. It places emphasis on the company's loss-control activities by establishing and requiring the application of specific safety rules and procedures to work assignments. The safety program should not be a substitute for management policy. A safety program cannot be effective without a safety policy; the program is an action tool for loss control, and it usually results from a management policy.

The complete extent and effectiveness of the policy makeup will be dependent on management's desire to be directly involved in loss control. Guidance in the preparation of a contractor's management policy is offered in the *Manual of Accident Prevention in Construction*[4] published by the Associated General Contractors of America. The National Safety Council also has a data sheet available on management safety policies.[5] Without a loss-control policy, written or implied, all efforts at permanent and effective loss control will be haphazard at best.

Management Organization for Loss Control Management must organize for effective loss control. The ultimate objective of such organization is to integrate the principles of loss control and then proceed on the basis of predetermined management goals. Organization should be within the existing line and staff structure of the company. The functions and network responsibilities, authority, and accountability for loss control at all levels should be defined. Specific areas in which loss controls are to be focused should be identified. The following are eight basic areas for management's consideration when organizing for loss control.

Responsibility for loss control is an important and much misunderstood term. The manager who delegates work will assign duties, grant authority, and create accountability in subordinates. In this process he retains his own obligations and remains accountable for the results. Accountability can never be delegated. These fundamental management concepts must be understood and observed, to preserve the chain of command and to achieve effective management.

Each responsible level of management and supervision should know its obligation for loss control. Job descriptions, in addition to organization charts, are one of the best methods of defining and assigning supervisory positions. The job descriptions can specifically state job requirements and limitations, even at the basic employee level. The position of the safety or loss-prevention specialist within the organization is strengthened when his duties, responsibilities, authority, and accountability are specifically defined. The role of outside assistance in loss control should be identified.

Control of safe working conditions must be job-oriented and is best accomplished by integrating job loss control into job planning. Job planning in construction is a continuous process from the conception of a project to its completion. Construction work activities are scheduled and coordinated by planning, which is motivated by profit. The planning stage is the best time to include loss control as an integral part of production. Loss-control activities should be considered and adopted as basic parts of the work procedure rather than as extra work or cost items which are tacked on later when the work is in progress. Job planning for loss control is not new and revolutionary. It has long been with us and has been known by many names. "Prejob safety conferences," "job safety orientations," and "preplanning for safety" are some names we have heard. Call it what we will, but we must include loss control in every phase of job planning.

The preconstruction planning of a high-rise building or structure may be taken as an example. Along with hundreds of other important details, the job manager should reflect on the inherent "vital few" hazards peculiar to this type of construction. His first thought should focus on the major exposure of falls, which in multistory construction account for more than 25 percent of the injury costs. For openers, the fall exposure may be evaluated for necessary controls:

Where will safety belts be needed?
Will safety nets be required?
What about scaffold design and ladder use?
Will perimeter guarding be needed for floor edges?
How will floor openings be protected?
How will materials and personnel be safely transported to different levels?
How will the public be protected inside and outside the jobsite?
What are the provisions for supervisory and maintenance inspections to maintain necessary loss controls?

These questions are representative of the type of thinking needed to control all potential loss exposures common to construction. But it should not be forgotten that the major exposures anticipated should be only the *minimum* considerations in any job planning. How will these minimum considerations be determined? As stated earlier, the necessity for providing controls and their extent should be based on applicable governing codes, regulations, standards, and accepted good work practices for the work being done. Costs of necessary loss controls should be determined and included as part of the overall job cost. If costs are a governing factor, top management should have a clear understanding with job management and supervision of its willingness to permit supervision to "play the odds." An example would be a management decision as to whether or not excavation shoring is to be in accordance with applicable standards or lesser requirements to meet competition.

Management can place emphasis on the concept of job loss-control planning by directly participating in it. The existing system of job planning should be reviewed to evaluate how loss-control planning can be incorporated. If a company uses any of the systems methods for job planning, such as the Critical Path Method,* management may include an examination of major safety exposures and needed controls in the prejob analysis.[6]

Inspections are a most effective method of detecting specific problems and exposures peculiar to each job. Routine inspections should be on a daily basis or oftener, as determined by the type and severity of exposures. Job inspections should be made by immediate supervision to detect and eliminate hazards and to initiate and enforce controls. Maintenance inspections of equipment of all types by designated personnel are a necessity and should be made on a regularly scheduled basis.† The responsibility for job inspections, types, frequency, reporting of results, and follow-up action should be specifically defined and assigned by responsible management.

Management inspections on a periodic basis are a means of expressing management's involvement in loss control. Any time that top management is on the job, loss control should be given the same emphasis as other management functions. No other action of top management can be as demoralizing to good loss control as an apparent disregard and tolerance of unsafe work conditions on the job. Management should constantly encourage supervisors and all employees to plan for loss control as a part of the job.

* See Sec. 4, Construction Planning and Scheduling.
† See Sec. 7, Equipment Economics.

Another effective method of achieving near total participation in job loss control by all employees is the requirement by top management that "tool-box" or "tailgate" safety meetings be conducted by the immediate supervisor (usually the foreman) on a regular basis. These meetings should be short and held at least weekly, or oftener if a severe exposure exists. A group's activity should furnish the source of all or almost all the topics for its discussions. Examples of such topics might be recent accidents or injuries either on the job or off the job, losses related to current activities, and safe operation and hazards of job equipment. Reviews of probable job loss exposures and hazards, with emphasis on control and corrective actions, make excellent topics. The choice of appropriate topics would seem to be endless, and participation by everyone should be encouraged. The State of California *Construction Safety Orders*[7] require that "supervisory personnel shall conduct 'tool-box' or 'tailgate' safety meetings—or equivalent—with their crews at least every ten working days, or as needed, to emphasize safety."

The prime contractor can and should specify the requirements for loss control that will govern work performed under his control, including work done by subcontractors. The application of this principle of prime contractor responsibility is complicated by various legal technicalities that may be involved. An example would be the use of hold-harmless clauses in subcontracts to avoid liability. Job management should be familiar with these technicalities so that appropriate loss controls can be initiated and enforced.

Governmental projects, in particular, may involve more than one prime contractor. As a result it is not uncommon for the areas of responsibility for loss control to become fuzzy or obscure. It is the basic responsibility of the owner or contracting agency, unless otherwise stated, to resolve any differences and coordinate loss-control activities. Without such definition, serious overlapping loss-control problems may remain unsolved.

Selection and placement of employees is a problem in the construction industry because of the type of work and a relatively high employment turnover rate. Often the employer has little or no choice in the selection and placement of employees. The practice of effective loss control makes it desirable that employees be physically qualified for their job and that they possess needed skills.

Management should establish selection and placement procedures dependent on job needs and employee capabilities. For example, procedures can include interviews and screening of prospective employees, checking references and previous work history, a determination of the physical condition of a prospective employee, job physical requirements, job descriptions, job transfer procedures, and procedures for evaluating the skills of equipment operators. Specific procedures may be influenced by the type of work and contractual requirements.

Management frequently entrusts equipment valued at many thousands of dollars to an equipment operator. Management should consider evaluating operator skills for operating the respective equipment by requiring a qualifying operator's test or license. There may be a need for the imposition of certain physical requirements, such as those for a tunneler working under compressed air, a diver, or a crane operator. Certain government agencies have included in their contractual agreements a requirement that operators of certain types of equipment and machinery be given physical examinations to determine if they are physically qualified to perform their assigned duties without danger to themselves or others. The fulfilling of such requirements may call for the cooperation and agreement of organized labor, which can complicate the process.

Training and supervision of the employee are vital to loss control. How can an employee make the maximum contribution to successful loss control if he is not instructed, directed, and guided in the application of accepted good work methods and procedures? The supervisor of that employee is the most immediate influence. Supervisory training and experience, motivated and encouraged by management, are loss-control techniques of utmost importance. How can job managers and job supervision maintain effective job loss controls if they are not knowledgeable in loss-control techniques and governing requirements for their type of work? Top management should require that job management and supervisors obtain copies of applicable safety codes, regulations, and work standards. If supervisory skills of supervisors and managers are weak, loss control will most certainly suffer, as will all other profit aspects of the job.

Management should consider "built in" training programs for all levels of employees. Such training may be either formal or informal and must clearly identify good safety practices as a part of doing a job right. A highly desirable beginning is a safety indoctrination of the new employee. This may be conducted by the office manager, the personnel manager, the project manager, or the foreman for whom the employee will work. If it is to be effective, it should include the entire management team to some degree. Its objective should be to instill in the new employee a feeling of responsibility for loss control to his employer, his fellow workers, the public, and to himself.

Supervisory training should be continuous. The rapidly expanding technology of the construction industry embraces new materials, methods, machines and equipment, and manpower applications. This, in itself, is sufficient reason for management to constantly update supervisory capabilities. Today's supervisor may be tomorrow's manager. The capable supervisor will demand effective loss controls. He will set the example.

Management has many sources of outside assistance for safety training. Educational institutions countrywide offer educational courses and seminars. Many state agencies conduct one-day or longer safety conferences and discussions. The Associated General Contractors of America offers a correspondence course on construction safety for supervisors. The National Safety Council continuously offers specific safety training courses. The United States Department of Labor has assistance available in the field of occupational safety. Insurance carriers may have specialized safety courses and information for employees and supervisors. Most of the specialized trade associations sponsor and promote safety activities and distribute safety materials and statistics regarding new association problems to their members.

*Equipment selection, application, and maintenance** are also vital to loss control. There is virtually no segment of the construction industry that does not utilize, directly or indirectly, some type of equipment or machine in the performance of its specific task. Many catastrophic losses can be attributed to equipment failure of one type or another.

Management must exercise its control in this area. A specific procedure should be established for equipment selection, application, and maintenance. Equipment should be selected on the basis of what it is expected and designed to do. The selection process should consider the available operator safeguards, such as roll-over protection, noise mufflers, operator protection from heat and dust, ease of operation, adaptability of the machine, the exposure of operator to moving parts, manufacturer's recommendations for operation, training available for operators and maintenance personnel, machine durability, and other factors as they might apply. Control should be maintained over the modification and redesign of equipment for special purposes, a common practice of contractors in the construction industry.

Maintenance procedures are a necessity whenever equipment is involved. Regular maintenance inspections can reveal serious machine deficiencies and prevent equipment failures. Qualified maintenance personnel must be available, and maintenance personnel training must be a continuous process. Files of manufacturers' operation and maintenance manuals should be maintained and kept where they are readily available for reference. Maintenance facilities should be adequate for the need, and a maintenance record system is a necessity. These records may be kept manually, although many large contractors are making effective use of electronic data processing. Loss control and safety are concrete by-products of good equipment selection, application, and maintenance.

Safety education, motivation, and communication are broad topics of interest, and close examination will reveal considerable overlapping into other areas of loss control. Management involvement in safety education, motivation, and communication will deal with all relevant activities not included earlier under Training and Supervision. These activities will generally be considered supplementary to the basic loss-control

* See Sec. 7, Equipment Economics.

efforts as described thus far. Included in them are group meetings of employees, supervision, and management personnel which are devoted to safety and loss control. Periodic management and supervisory meetings can provide a review of current and past loss-control records, problems, and activities. Plans for the future can be discussed. Employee meetings may be utilized to discuss specific employee aspects of loss control.

Safety material such as leaflets on specific subjects for employees and supervisors, the use of safety posters, safety contests, safety film showings, safety bulletin boards, letters from management, payroll inserts, safety incentives and awards, a house safety publication, accident facts on construction operations, and the promotion of off-the-job safety all may supplement other loss-control activities. A wealth of this type of material is available from insurance carriers, trade associations, governmental agencies and jurisdictions, and from the National Safety Council.

There is a need for this type of supplementary activity to maintain a level of employee interest that is consistent with the loss-control objectives and goals of management. All too often safety materials are adopted as substitutes for on-the-job loss-control performance. Whatever is done should be done with a purpose and with meaning. The timing of such educational activities can be critical and can either increase or decrease their effects. A negative example would be the distribution of literature cautioning against nail punctures when this actually was a very rare and minor exposure for the work underway. A positive example would be the showing of a film to employees on the prevention of falls when falls are a severe exposure.

Management should constantly evaluate its need for company safety education, motivation, and communication. This evaluation should be made with an understanding of its complementary purpose and application. Any attempt by management and supervision to utilize these activities and materials as a substitute for their own efforts and shortcomings should be quickly and positively discouraged.

Health services and care of the injured are frequently neglected areas in the construction industry. The treatment of an injured person from the time of the injury to the time of supervised medical treatment can greatly influence the seriousness of the injury and the length of the recovery period, and thereby it can directly affect all the costs associated with an injury. The knowledge that qualified personnel are available to give emergency first aid is also a tremendous morale booster to all employees. It is claimed that supervisors who are qualified to provide emergency first aid are more safety conscious and have fewer accidental injuries occurring under their supervision than untrained supervisors.

Top management should be the prime motivator to ensure adequate care for the injured on the job. Job management should be responsible for establishing the standards for personnel and facilities for a given project. These standards should include, for example, the emergency equipment needed, number of qualified personnel, ambulance service available, availability of physicians and hospital emergency services, and the procedures for the care and handling of injured persons. The amount of detail will be determined by the needs of the project.

The U.S. Army Corps of Engineers[2] and the Bureau of Reclamation[3] have established minimum standards for first-aid medical facilities as a contractual requirement. Specific requirements are based on the total number of employees on a given project. The Bureau of Reclamation requires that every contractor foreman possess a Bureau of Mines or American Red Cross first-aid certificate. The contractor is required to provide for the first-aid instruction.

Top management should be aware of any contractual requirements for first-aid and medical facilities. It is becoming common practice to include minimum requirements in the contracts for large construction projects, private and governmental. Small- and medium-size contractors have the same need for first-aid procedures on the job. Someone in even the smallest work crew should have a working knowledge of how to administer emergency first aid and utilize essential supplies. Management should grasp every opportunity to motivate supervisors and employees to be proficient in administering emergency first aid.

Construction management desiring information regarding minimum job requirements should refer to the pamphlet entitled *A Guide for Medical Services for Construction Projects.*[8] This contains recommendations prepared by the Committee on Medical Service for Construction Projects of the Council on Industrial Health, American Medical Association. Assistance also is available from insurance carriers, industrial physicians, private consultants, and governmental agencies.

Another source of potential serious loss in the construction industry is the rapidly expanding list of occupational health hazards. These hazards are generally identified as those environmental factors or stresses that may seriously affect or endanger the health of the employee. Some examples of these hazards in construction are airborne dust such as silica, which causes silicosis; fumes from solvents; heat exhaustion; dermatitis; excessive noise (hearing loss is compensable in some states); toxic substances; and radiation.

Construction management is responsible for the prevention and control of industry occupational health hazards. Management should initiate the action necessary to prevent and control anticipated hazards for the type of work being done. Such action may be based on known hazards and previous experience, or it may anticipate new hazards from unfamiliar substances and processes. Supervision should be alert to question any suspected exposures, seeking qualified and competent assistance where advisable. The evaluation, prevention, and control of occupational health hazards have become a highly specialized technical science. Assistance as needed should be sought from industrial hygienists, industrial physicians, safety engineers and consultants, manufacturers of substances, insurance carriers, and governmental agencies, including state boards of health.

Management should become familiar with loss-control requirements. Specific legal requirements vary by state, but the absence of such requirements is no excuse to ignore sound principles of prevention and control. A suspected exposure to possible occupational hazards can be extremely costly and seriously detrimental to employee morale and labor relations. The prompt and qualified evaluation of a hazard and the immediate establishment of necessary controls are needed. As in other areas, a greater extension of federal controls is foreseen unless individual states strengthen their specific requirements for the control of serious health hazards. Emphasis by management on the planning for the care of the injured and protection from occupational disease will be a significant contribution to total loss control, both in terms of dollars and in the achievement of good public and employee relations.

Accident and injury records are necessities for effective management loss-control action. The record system should be simple and should provide basic information which is readily retrievable. This information should have uniformity and accuracy, giving a clear description of facts and findings. Records should be adequate to meet industry and insurance requirements and to satisfy the pertinent workmen's compensation laws in the states. All records should be confidential and accessible only to authorized personnel. The detail of the record system is based on management's desires and needs and on job requirements. Top management should identify the responsibility for record keeping, indicating whether this is to be performed on a main-office level or on a project or job level. Management's involvement in record keeping should assist with the development of basic information needed for effective loss control.

Management Direction for Loss Control Top management must accomplish its objectives of effective loss control. As with other management responsibilities, it is desirable that one person in top management assume the leadership and responsibility for this task. The person so designated will direct the preparation and implementation of approved safety plans.

Management direction is the technique of "doing" and will include *delegation.* Specific responsibilities and authority for loss control should be assigned at every level, so that everyone can apply his abilities to the utmost. Top management may assign the project manager the responsibility and authority for loss control on his project as determined and outlined by a management loss-control policy. The project manager then exercises his responsibility and authority within the limits of the firm's loss-control policy, the job requirements, time, money, materials, and personnel under his control and direction.

Top management should *supervise* job management by continued instruction, guidance, and coaching to see that loss control is implemented and carried out. Inspiration and encouragement from top management are needed to *motivate* the job manager toward the successful accomplishment of loss control. This can take many forms. Success may be recognized by salary increases, personal promotion opportunities, or internal and public recognition of accomplishments. Any recognition should reflect the results of loss control on an equal basis with other production objectives.

Training and counseling with job management are excellent activities to provide direction. The job manager may need training or assistance to learn the techniques of loss control and their application. Attendance should be encouraged at special seminars offered by educational institutions, trade associations, governmental agencies, insurance carriers, and the National Safety Council. Training may be within the firm's own organization, by group meetings and discussions with special emphasis on loss-control objectives and problems. Top management can counsel a job manager in specific problems.

Staffing for loss control should be the responsibility of top management and can be included in the function of directing. Top management should determine the need for the special skills of a loss-control specialist either for the overall organization or on a job basis or both. Such a need may result from a contractual agreement which specifies that the services of a full-time or part-time qualified safety professional must be provided on a project, subject to the approval of the contracting agency. Titles by which staff loss-control personnel are known include those of "safety director," "safety manager," "safety engineer," "loss prevention manager," and others.

Top management should determine the qualifications and job description for such a position. Assistance in establishing qualifications, standards, and job descriptions is available from organizations such as the American Society of Safety Engineers[9] and the National Safety Council. The loss-control specialist as a staff adviser usually plans, advises, and helps the line organization with loss control. Occasionally he may be involved with the firm's insurance activities. The position and status of a loss-control specialist in a firm's organizational structure are as varied as the number of firms. Management is responsible for employing a loss-control specialist as an adviser to make recommendations in a specialized area. A loss-control specialist can gain acceptance within the chain of command through his status in the organization, his abilities as an expert in loss control, the endorsement of his loss-control recommendations by top management, and the influence that he may exert on rewards.

The duties of the loss-control specialist in larger organizations may include assisting and advising top management on policy, planning, training of personnel for loss control, specific loss-control programs, and selection of personnel for job loss control. In smaller organizations, his duties may include on-the-job inspections, determining first-aid needs, keeping accident and loss records, and assisting with accident investigations. He may determine the need for the service of other specialists, such as industrial hygienists, chemists, and industrial nurses. The job safety engineer usually reports directly to the job or project manager, and for maximum effectiveness the extent of his responsibility, authority, and accountability should be clearly defined. The lack of such definition is a well-known source of irritation and "wheel-spinning" on many a construction job.

Top management should be alert to a need for specialized staff assistance with loss control. Top management should also coordinate the firm's efforts at loss control. Needs should be kept in their proper perspective as determined by their relative importance, and major problems should receive major attention. This reaffirms the principle of accomplishing immediate, significant results with the least amount of effort through the knowledgeable direction of that effort.

Management Control Management needs measurements of progress toward its planned objectives and a means of initiating corrective action. This introduces the management principle of control. As applied here, "control" includes the establishment of standards, measuring performance on the basis of these standards, and taking appropriate action to correct unacceptable deviations from them. The depth and intensity of control are dependent on the organization, the specific needs of the work undertaken, and the wants and capabilities of management at various levels. All these are fully applicable to loss control.

Records of accidents, injuries, and resultant losses, as stated earlier, are a necessity for rational management action. Management control of losses cannot be complete or effective without records. Management should require summaries of desired information on a monthly, yearly, or similar basis.

Records should include an analysis of accident data abstracted from accident investigations to obtain the information desired by management. Analysis may be by department, body part, type of injury, type of accident, causes, serious or costly accidents, lost time, costs, or by time (of day, of month, of year, etc.). Separate records may be kept for foremen, supervisors, job managers; by project, job, or type of work. The amount of detail should be determined by specific management needs.

Records can provide management with direction for loss-control effort by revealing that certain loss causes produce the majority of losses in dollars. If a firm's losses are not great enough to reveal this information, supplemental statistics available from industry or other sources may help to identify major loss causes. The concept of concentrating control effort on major loss causes is in accordance with sound management principles and with both humanitarian and dollars-and-cents wisdom. This technique of loss control can help management to achieve immediate and significant results with the least amount of effort.

Management can also utilize information from records to sell loss control to itself, its supervision, and its employees. Current accident information will be an attention factor to focus a firm's loss-control efforts on specific loss problems. It can provide a keen spirit of competition between jobs, departments, supervision, and workmen. Accomplishment in loss control, verified by records and publicized, is an excellent method of maintaining good public relations.

Finally, management's progress in controlling losses can best be evaluated through record analysis. Management should decide the method of measurement or combination of methods. Examples of measurements would be by frequency and severity rates, by cost, by the days of lost time, by the number of lost-time injuries, or by a combination of all accidents. There are many adaptations of measurement systems developed for specific needs. Another effective measurement may be the ratio of losses to the respective insurance premium cost for a specified period, commonly referred to in the insurance industry as the "loss ratio." Assistance with records is readily available from the National Safety Council, trade associations, insurance carriers, and governmental and private agencies.

Records in themselves are meaningless. It is a management responsibility to determine the extent of records, assign the duties for record keeping, and ensure the maximum utilization of the resulting information. Management's emphasis on record keeping, analysis, and evaluation, with resultant control actions, will reaffirm management control and involvement.

Costs of employee injuries should be handled and controlled like any other production costs. Occupational injuries and fatalities increase a company's operating cost, as do the expenses of raw materials, parts, and labor. The total operating cost should include property damage losses, not forgetting those from fire, and hidden overhead costs to the extent possible and reasonable. In general, production costs are controlled by:

1. *Budgeting:* setting up a budget based on a level of cost that is expected and acceptable
2. *Costing:* calculating, recording, and allocating current costs for the guidance of management
3. *Accountability:* holding the supervisor of each company unit accountable when his cost exceeds his budget

The application of the above steps to a loss-control costing system is not simple, but it can reveal many problems. For example, the allocation of injury costs on a current basis is complex due to the wide variations in compensation benefits from state to state, the difficulty in accurately determining current costs for injuries, and the influence of the number of dependents on the cost of fatalities. These problems have been approached in various ways.

A method of "current costing" of occupational injuries[10] overcomes the variable factors which complicate the measurement of safety performance based on actual injury costs. This system uses average costs for categories of injuries, and it has proved highly accurate when applied to a large number of cases. A costing system applied by management to controlling occupational injuries is an effective technique of loss control. The National Safety Council, in its *Accident Prevention Manual for Industrial Operations*,[11] has a section entitled "Estimating Accident Costs" which presents another method of obtaining accurate estimates of the total cost of a company's work accidents. A method of estimating uninsured costs is also explained.

Construction management's application of costing to loss control will require record keeping in detail. Costing may be detailed or may consist of a general evaluation of total costs. Costs may be allocated to specific categories, such as injury costs, property damage losses, losses from auto fleet accidents, and general liability losses. Some companies make actual dollar allocations for hidden overhead costs based on their own experience analysis. Consideration might be given to allocating costs by project, operating division, on a companywide basis, or as a cost per man-hour. Management should choose the application of a costing system which will best fit its mode of operation. The validity of the system used will be dependent on the accuracy and completeness of records and the degree of care with which costs are developed and allocated. The application of an effective costing system will enable a company to bring employee injury costs under the direct control of management. Interest by supervision and management will be keen when losses are directly related to production costs and profit.

Frequency and severity of injuries provide a basis for comparing industry performance. The best-known and most widely accepted rating system[12] is supplied by the American National Standard Z16.1. This standard outlines the proper stipulations, defines terminology, and interprets procedures and application. It also describes the standard procedure for keeping necessary records. Formulas based on a million-man-hour unit are provided for calculating disabling injury and severity rates. The rates provide a company with a basis for comparing its injury experience with that of similar companies or with the injury experience of the industry of which it is a part. A guide for accident analysis and records is provided by the American National Standard Z16.2.[13]

Comparisons can also be made with a company's own previous experience. Member companies of the National Safety Council report their frequency and severity rates to the National Safety Council annually. The National Safety Council tabulates the results of these reports by industry classification and publishes annual summaries and comparisons in *Accident Facts*.[1] Trends in injury performance, as related to the frequency and severity of injuries, can be detected and evaluated to institute needed safety controls. The rates also indicate comparative performance within the industry or in relation to other industries. Many safety awards recognize a firm's outstanding loss-control accomplishments on the basis of low frequency and severity injury rates as compared to industry or national rates.

Accident investigation is the responsibility of management, and effective loss control cannot be a reality unless it is performed. Its primary objective is fact finding, not fault finding. The principal purposes for accident investigation are to:

1. Determine the accident causes
2. Evaluate those causes to initiate loss controls that will prevent a recurrence
3. Inform all fellow workers and others concerned with similar situations, so that they may recognize and prevent future accidents
4. Direct the attention of responsible supervision to its responsibilities for loss control
5. Find and record facts that may assist with the determination of legal liability

Management should assert its responsibility for accident investigation by establishing the basic requirements. Major items that should be considered are:

1. What circumstances and conditions determine which accidents should be investigated? Management should provide guidelines. Considerations to include are the seriousness of the loss or of the injury, the frequency of recurrences of certain types of accidents or injuries, and the loss potential in specific operations.

2. Who will be responsible for the investigation? Usually the responsible and immediate supervisor is the designated person. Assistance may be available from a safety engineer or inspector, a management representative, the safety committee, an insurance company representative, or a governmental agency. Participation by management in the investigation of serious losses is a proved tool of loss control.

3. When should the investigation be made? As soon as possible! The construction industry is unique because, in a very short time, evidence of an accident may disappear due to cleanup and continuance of construction work. Fellow workers may leave the job and go to work elsewhere. Job supervision may change. Weather conditions may seriously affect the conduct of an investigation. Consequently, evidence of contributory factors is quickly lost.

4. What type of report will be required? Accident investigation reports may be simple or detailed. Management should provide guidelines for the information desired. The basic purposes of an investigation should be satisfied.

5. Who will receive the report? Management should also make this determination dependent on needs. It is essential that responsible management review all reports of fatal injuries and losses deemed to be serious. Management review and follow-up will reassert management's interest and involvement in loss control.

6. What follow-up action will be taken? Needs will determine management's actions. Action may be limited by authority and economic considerations, or it may be a necessity due to legal requirements. Responsible management and supervision should be involved with recommendations for corrective action. Procedures should be considered for requesting outside assistance if needed.

7. Who will initiate and enforce the action decided upon? Management should define this responsibility. The usual practice is to stay within the organizational structure of management and supervisory responsibility.

It is noteworthy that the *Manual of Accident Prevention in Construction*,[4] published by the Associated General Contractors of America, describes the makeup and application of a "board of inquiry." The function of this board is to conduct an investigation as soon as possible after the occurrence of each serious or fatal accident. The sole objective of this investigation is to prevent that type of accident from recurring in the work of the group members.

SYSTEMS SAFETY

The construction industry is constantly involved with new techniques and complex technology. Loss exposures are multiplied accordingly, but the modern systems approach meets this challenge. The great value of the systems approach to construction loss control is (1) its efficiency in examining and evaluating the complex relationships involved, (2) its effectiveness in highlighting problem boundaries, and (3) its economy in reviewing significant aspects of the problems.

System safety is not new, since it had its origin in aircraft safety systems. Continued applications in the aerospace industry have introduced "systems safety analysis."[14] The National Safety Council is actively engaged in providing basic background information on the application of systems to industrial safety by publishing articles and conducting seminars on systems safety. The application of the systems concept to loss control in the construction industry has unlimited possibilities.

The System An accepted definition of a system that embraces the total area of loss control is "an orderly arrangement of components that are interrelated and act and interact with one another to perform a task or function in a particular environment." Continuing with the application to loss control, another definition for a system should be considered: "a group of things (man, machine, environment) which are related to one another in some dependent manner so that collectively they represent a whole and accomplish a task."

Figure 5-1 illustrates a simple systems representation of the basic, interrelated variables which are involved in the process of loss control for a specific construction task. It portrays the principle that a system is task oriented. The interfaces between man, tool, environment, time, and management represent the interrelationship of the components of performance for task accomplishment. The task may be simple or complex. Systems analysis examines and evaluates a task by breaking the complex task into simpler tasks which can be considered subsystems of the original system. Subsystems are then analyzed in terms of the component parts of the entire system needed to perform the allotted task of the subsystem. The combinations are seemingly endless and provide for a careful and methodical evaluation of many variables. The choice of the basic system is dependent on the problem to be solved.

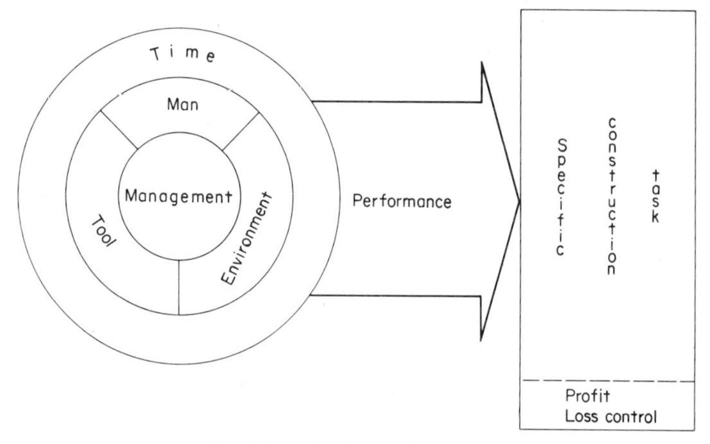

Fig. 5-1 A basic systems model.

Since construction is task oriented, the application of system safety analysis to construction loss control is logical. The construction task is usually divided into specific work segments for design, planning, and accomplishment. System safety analysis provides the desired degree of detail in examining the people involved at any given time, the tools or equipment utilized or affected, the environment in which the task is done, and the level of management and supervision involved.

A specific application is the evaluation of a simple but deadly construction situation—planning a lifting operation with a crane near a power line. Applying the man/tool/environment systems analysis concept described above permits critical examination of the interrelated factors in needed detail. Many questions arise during job planning concerning the variable factors that should be evaluated to establish effective loss controls.

TASK: Is the task clearly defined? Is loss control a part of the task?

TIME: How long will the task take? When will the task be done?

MAN: Will there be constant supervision of the task at all times? Is the operator physically qualified and skilled to operate the specific type of equipment? Will a qualified signalman assist? Has everyone concerned with the task been given adequate instructions?

ENVIRONMENT: What are the requirements for power-line clearances? What are the power-line voltages involved? Should the power company be notified of any hazardous conditions? Will it be necessary to have the power disconnected? What will be the weather and lighting conditions?

TOOL: Is the equipment in good mechanical condition? Is the capacity of the equipment adequate? Has accessory equipment been inspected and is it adequate?

MANAGEMENT: Is it necessary to involve management in the planning? Does job management approve the plan?

These questions suggest the type of analysis which becomes possible by examining and evaluating the basic, interrelated factors of a system. Questions are limited only by the specific operation and the problems anticipated. They are an essential part of work planning. This type of analysis also demonstrates the systems approach to hazard recognition.[15]

Hazard Recognition Hazard recognition has always been basic to loss control. How can hazards be identified, evaluated, and controlled unless they can be recognized? If construction management applies hazard recognition at the job level, unlimited possibilities for effective loss control result. Hazard recognition applies to managers, supervisors, and to the employees. It is applicable to present job exposures and to anticipated future exposures. The recognition of hazards is a by-product of experience, training, and skills.

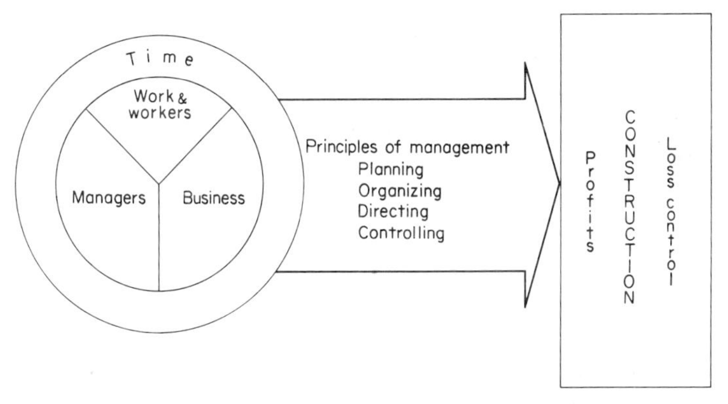

Fig. 5-2 An integrated systems approach to construction safety.

The application includes a working knowledge of the legal requirements for loss control applied to the work, the creation of an awareness in workers of loss-producing combinations and accepted methods of control, development of qualified supervisors, assistance from loss-control specialists, and establishing the requirements to plan for loss control. The list is endless.

Hazard recognition results when interacting variables are evaluated for loss-producing combinations that may exist within a system. The systems concept provides an orderly and comprehensive method of identifying the variables for a specific task. Hazard recognition is embodied in the concept of systems safety analysis.

Network Analysis Systems Management should consider the application of system safety analysis as a loss-control technique to construction projects using a network analysis system for construction job planning and scheduling.* The best-known examples of network analysis systems used in the construction industry are the Critical Path Method[6] (CPM) and the Program Evaluation and Review Techniques[16] (PERT). Frequently, the use of a network analysis system is a contract requirement to control job planning and scheduling on large construction projects. Applying systems safety analysis to network analysis will facilitate the detection and evaluation of major loss exposures.

Network analysis provides for easy identification of known and anticipated hazardous activities, including those of subcontractors. It will reveal the approximate time

*See Sec. 4, Construction Planning and Scheduling.

and duration of those activities and may indicate the number of employees involved in each of them. Hazardous exposures to the public and to employees become readily apparent. Network analysis provides a means of integrating the needed loss controls into production as a part of the work rather than as an extra cost item later.

Systems and Management Loss Control Figure 5-2 is a basic system diagram which illustrates the principles of management and interacting components involved in construction loss control. It provides a simplified look at the entire construction task, which includes loss control. The systems concept applied to construction loss control is a "mind stretching" technique for construction management. Applications may vary from company to company and from project to project. Systems safety analysis has flexibility of application at every level of operations. The volume of published material available on systems safety analysis and industrial applications is rapidly expanding. It has had notable applications in the area of product safety. The technique of systems safety analysis should have a promising future when applied by construction management to the construction industry problem of effective loss control.

INSURANCE

It is a necessity for management to be familiar with insurance and its application, since the cost of insurance may represent from 5 to 15 percent of the contract price for a construction project. Insurance costs are directly related to the loss experience of a firm and of the total industry within a specific workmen's compensation insurance classification. Management should require that the administration of the company's insurance program be coordinated with loss-control efforts.

Insurance as applied to the construction industry is a contract wherein an insurance company agrees to defend the policy holder and pay sums of money in his behalf in the event a claim develops from the covered hazard. There are many varieties of basic hazard and insurance coverages available. Their availability to a contractor is dependent on the amount of premium involved, loss history, the type of work, the professional skill of the insurance representative, the choice of insurance sources available, and other variables. Management should seek the counsel and advice of qualified insurance professionals to ensure adequate and applicable insurance protection. Two basic types of coverages are briefly described here: workmen's compensation and comprehensive general liability. An excellent presentation and explanation of insurance coverages for the construction industry is available in a publication entitled *Insurance for Contractors*.[17]

Workmen's compensation insurance is a basic coverage in the construction industry and is generally a state statutory requirement for most types of construction work. Several states have monopolistic state funds which provide all workmen's compensation insurance in those states. This insurance provides compensation and medical benefits to injured employees in accordance with the provisions of the applicable state workmen's compensation laws. The types of requirements vary widely. Rates are developed in each state for separate work classifications based on a unit of $100 of payroll. These rates are influenced considerably by the state loss experience reported by insurance companies within each specific classification. Rates may be adjusted upward or downward annually, according to loss experience and other variables. A firm's net insurance cost of this coverage is directly influenced by the workmen's compensation loss experience record of that individual firm.

Certain states have provisions for an employer to assume his own workmen's compensation risk. This is known as "self-insurance," and specific state requirements must be satisfied by the employer before this can be done. Self-insurance by a contractor is a venture that requires expert professional assistance and counsel.

Comprehensive general liability coverage is extremely important. It can protect the contractor from major losses which could easily threaten the firm's financial soundness. The general liability policy lists the divisions of coverage, conditions, limitations, and exclusions which restrict rather than extend coverage, including what will be defended by the insurance company. Management should consider the printed policy as a starting point from which its insurance representative can tailor coverages and limits to suit specific requirements by endorsements.

Premium rates are influenced considerably by previous loss experience. High or low insurance premium rates may make the difference between winning or losing a contract award. The availability of certain coverages may be reasonable or prohibitive, depending upon the previous loss history of the contractor. Therefore, successful loss control can contribute to a firm's success in bidding for work.

Management should consult with its insurance representative to determine an acceptable loss ratio. The term "loss ratio" applied to insurance, as stated earlier, is simply a ratio of insured losses to the insurance premium paid for a given time period. An acceptable loss ratio—the break-even point for the insurance carrier—may vary with the type of coverage.

Efficient management should endeavor to control losses to stay below the break-even point. Losses are averaged over a 3- to 5-year period and, if they consistently exceed this break-even point, management should expect to pay higher insurance premiums for the desired coverages. A reliable insurance company should keep responsible management regularly informed of current loss costs. This information can provide management with yet another meaningful tool to emphasize effective loss control.

Insurance may pay for the direct cost of an accident, but indirect costs which are meaningful losses to the contractor are frequently overlooked. Through various combinations of circumstances, the indirect accident cost can easily exceed four to five times the actual insurance-reimbursed loss. This multiple cost factor is easily exceeded by accidents that may occur in specialized high-hazard types of work such as tunneling. A typical analysis of certain uninsured costs is presented in the National Safety Council's *Accident Prevention Manual for Industrial Operations.*[11] Some contractors, based on their own experience, assign estimated dollar costs to some of the hidden overhead costs dependent upon the specifics of an accident.

It is common to have insurance requirements specified as a part of contract agreements. Usually the requirements for workmen's compensation coverage and general liability coverage with specific dollar limits will be stated. Indemnification clauses, commonly referred to as "hold-harmless" agreements and described earlier, may be included. Requirements for railroad protective coverages are included when work is done within railroad jurisdictions.

Evidence of adequate insurance coverages may be required of all subcontractors involved by specifying that "certificates of insurance" be submitted prior to beginning any work. It is management's responsibility to evaluate or obtain assistance in evaluating any contractual requirements so that necessary insurance protection will be acquired. The insurance requirements may also provide the experienced loss-control specialist with clues as to special hazards that are anticipated or may be encountered. Any loss-control measures which have been identified in the job planning stage may then be considered as alternatives to higher insurance costs. This is preventive rather than corrective loss control.

Construction management involved in loss control should choose a qualified insurance carrier. A contractor's insurance protection represents a sizable outlay of money. For this he can rightly expect outstanding service from the representatives of his insurance carrier. Loss control should then be a joint venture of the contractor and his insurance carrier. Management must acknowledge that insurance is a controllable item of expense rather than a fixed one. Profits can be improved through selective use of insurance as part of a loss-control program.

CONCLUSION

Many sources of assistance are available to construction managers in performing their task of loss control. These sources include the Associated General Contractors of America, the National Safety Council, related industry associations, equipment manufacturers, governmental agencies, insurance carriers, educational institutions, and private loss-control consultants. Assistance may be in the form of publications on specific subjects, literature, educational programs, and available personal services. Equipment manufacturers, for example, may provide training programs for equipment operators and maintenance personnel. Construction management should utilize outside loss-control assistance whenever desirable and advantageous. Loss-control problems may be complex, requiring the application of all available resources to obtain optimum solutions.

If construction management is to meet the challenge of effective loss control for the construction industry, the total responsibility of loss control must be accepted as a management function. Applying the accepted principles of management to loss control provides the needed framework for success. The integration of loss control into construction design, planning, and production will demonstrate management's involvement and participation in the construction industry joint venture—total loss control and maximum profits.

REFERENCES

1. *Accident Facts*, National Safety Council, 425 North Michigan Avenue, Chicago, Ill., 60611, 1970.
2. *General Safety Requirements*, EM 385-1-1, U.S.Army Corps of Engineers, U.S. Government Printing Office, Washington, D.C., 20401, March 1, 1967.
3. *Construction Safety Standards*, Bureau of Reclamation, U.S. Department of the Interior, Denver, Colo., 80225, June, 1968.
4. *Manual of Accident Prevention in Construction*, Associated General Contractors of America, Inc., 1957 E Street, NW, Washington, D.C., 20006, 1970.
5. *Management Policies on Occupational Safety*, Data Sheet 585, National Safety Council, 425 North Michigan Avenue, Chicago, Ill., 60611, 1968.
6. *CPM in Construction*, Associated General Contractors of America, Inc., 1957 E Street, NW, Washington, D.C., 20006, 1965.
7. *Construction Safety Orders*, State of California, Division of Industrial Safety, 455 Golden Gate Avenue, San Francisco, Calif., 94102, 1965.
8. *A Guide for Medical Services for Construction Projects*, Division of Occupational Health, American Medical Association, 535 North Dearborn Street, Chicago, Ill., 60616, 1958.
9. American Society of Safety Engineers, 850 Busse Highway, Park Ridge, Ill., 60068.
10. *Three Giant Steps to Management Control of Injury Costs: Costing, Budgeting, Accountability*, Employers Insurance of Wausau, Wausau, Wis., 54401, 1967.
11. *Accident Prevention Manual for Industrial Operations*, 6th ed., National Safety Council, 425 North Michigan Avenue, Chicago, Ill., 60611, 1969.
12. *Method of Recording and Measuring Work Injury Experience*, American National Standard Z16.1, American National Standards Institute, Inc., 1430 Broadway, New York, N.Y. 10018.
13. *Method of Recording Basic Facts Relating to the Nature and Occurrence of Work Injuries*, American National Standard Z16.2, American National Standards Institute, Inc., 1430 Broadway, New York, N.Y. 10018.
14. J. L. Recht, Systems Safety Analysis: An Introduction, *National Safety News*, December, 1965, National Safety Council, 425 North Michigan Avenue, Chicago, Ill., 60611.
15. E. Levens, Systems Concept—A Powerful Tool in Hazard Recognition, *Journal of the American Society of Safety Engineers*, April, 1969, The American Society of Safety Engineers, 850 Busse Highway, Park Ridge, Ill., 60068.
16. Pertinent Facts About Pert, *National Safety News*, pp. 44–47, September, 1968, National Safety Council, 425 North Michigan Avenue, Chicago, Ill., 60611.
17. W. T. Derk, *Insurance for Contractors*, Fred S. James Company, 1 North LaSalle Street, Chicago, Ill., 60602, 1966.

Part Two

Construction Equipment

Section 6

Engineering Fundamentals

R. D. EVANS

Civil Engineer, Intercontinental Sales, Caterpillar Tractor Company, Peoria, Ill. (*Material Characteristics, Equipment Characteristics, and Job Characteristics*)

E. V. SEMONIN

Development Engineer, Off-the-road Tire Development, Goodyear Tire & Rubber Company, Akron, Ohio (*Tires for Construction Equipment*)

A. Introduction

There are basic engineering fundamentals that affect the selection of equipment for handling bulk materials. Most important and significant of these are: (1) material characteristics, (2) equipment characteristics, and (3) job characteristics. There are also important interactions between these characteristics. Tires, although actually an equipment characteristic, will be discussed as a separate topic in this section.

B. Material Characteristics

CLASSIFICATION

In the sense used in this section, the term "material" can refer to earth, rock, processed aggregates, and other substances which are handled in bulk. Rock and earth (dirt) are materials of dominant interest in many areas of heavy construction. In bid documents, excavation is classified as "common," "rock," or "unclassified." "Common" refers to earth excavation, while "unclassified" is a term which reflects the lack of a clear distinction between soil and rock.

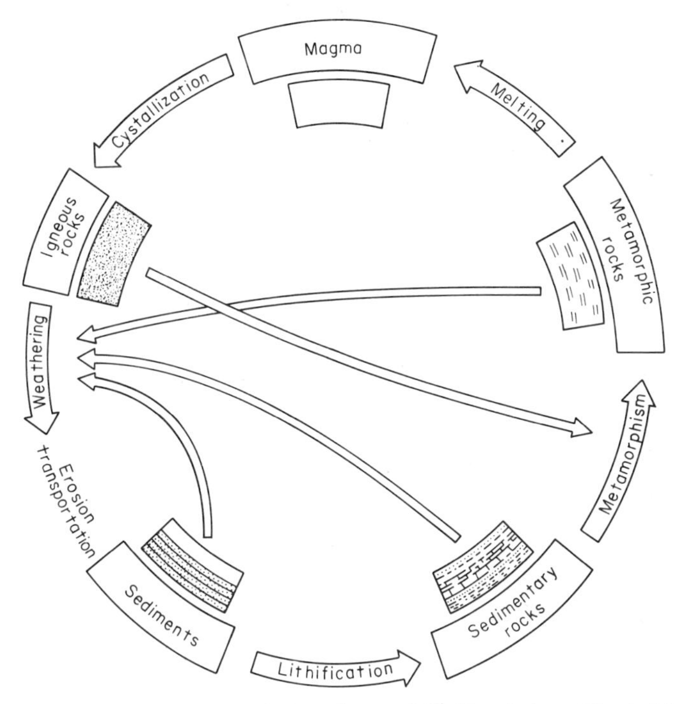

Fig. 6-1 The rock cycle. *(From L. Don Leet and Sheldon Judson, Physical Geology, 3d ed., 1965. Reprinted by permission of Prentice-Hall, Inc.)*

"Geology" is defined as the science of the earth, and "physical geology" deals with the nature and properties of the materials composing the earth, their distribution, formation, alteration, transportation, and distortion. A review of some basic geology will help in understanding the term "material" as applied to rock and earth handling.

Rock is considered by geologists to be the solid part of the earth's surface; it includes both consolidated and unconsolidated deposits. The rock cycle which traces the relationship among the rock classes is illustrated in Fig. 6-1, where the outer circle represents the complete cycle.[1,*] If uninterrupted, this cycle will progress from magma through igneous rocks, sediments, sedimentary rocks, metamorphic rocks,

* Superior numbers refer to the list of references at the end of this section.

and back again to magma. The cycle may also be interrupted at various points, as indicated schematically by the arrows within the circle.

Rocks can be classified according to their mode of origin as igneous, sedimentary, or metamorphic. Table 6-1 describes commonly occurring rocks which have been classified on this basis.[2]

TABLE 6-1 Common Types of Rock*

Type	Name	How formed	Constituents	Texture	Characteristics
Igneous-Intrusive	Granite	Cooled slowly, within the earth's surface	Quartz, feldspar	Coarse crystals	Gray, white, pink, yellow-brown
	Syenite		Feldspar	Coarse crystals	Darker and less mottled than granite
	Diorite		Feldspar	Coarse crystals	Darker but similar to granite
	Gabbro		Feldspar, pyroxene	Coarse-to-medium crystals	Greenish tinged
	Peridotite		Pyroxene, olivene	Coarse-to-medium crystals	Dense, dark green
Igneous-Extrusive	Diabase	Cooled quickly at the earth's surface	Feldspar	Fine grained	Lathlike feldspar, crystals
	Porphyry		Feldspar grains	Crystals embedded in fine matrix	Lathlike feldspar, crystals
	Felsite		Quartz, feldspar	Fine grained, porous, layered	Light gray, yellow, pale or deep red
	Obsidian		Silica granite	Glassy	Gray to black streaked
	Basalt		Pyroxene, olivene	Fine grained	Black cooled lava
Sedimentary	Sandstone	Clastic	Cemented quartz	Sand grains, coarse to fine	White, gray, yellow, dark red
	Shale	Clastic	Clay	Layered	Black to white, green, dark red
	Limestone	Organic chemical	Calcium carbonate	Fine grained	White to gray, soft to hard
	Dolomite	Chemical	Magnesia carbonate	Fine grained	Brown—resembles limestone
	Breccia	Clastic	Pebbles matrix	Coarse	Angular rock fragments cemented
Metamorphic	Slate	Pressure	Clay	Lustered mica flakes	Resembles shale
	Schist	Heat pressure	Mica	Crystallized shale	Micaceous banding
	Gneiss	Heat pressure	Sand, shale	Layered	Gray to white—resembles granite
	Quartzite	Pressure	Sandstone	Sand grained	Resembles sandstone but harder
	Marble	Heat pressure	Limestone, dolomite	Fine grained	Formed from limestone and dolomite

* From A. Brinton Carson, *General Excavation Methods*, p. 285, McGraw-Hill Information Systems Company, New York, 1961.

Igneous rocks are formed by the cooling of molten magma within the earth (intrusive) or by the cooling of lava on the earth's surface (extrusive). Igneous rocks rarely contain fossils, often may be identified by glass or mineral content, and seldom have the stratified, banded, or foliated characteristics of other rocks. Granites, basalts, traprock, pitchstone, and pumice are igneous rocks commonly encountered on construction jobs.

Sedimentary rocks are derived from preexisting rock by direct erosion or solution. Weathering, transportation, and redistribution by streams, glaciers, wind, groundwater, waves, and organisms are the basic formation processes. Typical clastic or mechanical sediments are gravel, conglomerate, sand, sandstone, silt, clay, and shale. Typical nonclastic or chemical and organic precipitates are limestone, dolomite, chert, salt, gypsum, and coal. A prominent characteristic of many sedimentary deposits is stratification or bedding. Sedimentary rocks are the primary parent materials for soil formation. Although 95 percent of the outer 10 mi of the earth consists of igneous rocks or metamorphosed igneous rocks, 75 percent of the exposed rocks are of sedimentary origin.

Metamorphic rocks owe their distinctive characteristics to the transformation or recrystallization of preexisting igneous or sedimentary rocks. These rocks may be changed in mineral composition, texture, or both. The agents which cause metamorphism in rocks are shearing stresses and pressure; chemical action, or liquids and gases; and temperature. Common metamorphic rocks are gneiss, quartzite, schist, marble, anthracite, and slate. All are found on or near the earth's surface.

Soil is the natural surface material which results from the weathering of rock and the decomposition of organic materials. Its formation is influenced by organisms, climate, topography, parent material, and time. Its composition varies with depth, and exposure reveals a series of zones or horizons called a "soil profile." The three major horizons which are shown in Fig. 6-2 can be described as follows:[1]

1. Zone of leaching
2. Zone of accumulation
3. Partially decomposed parent material

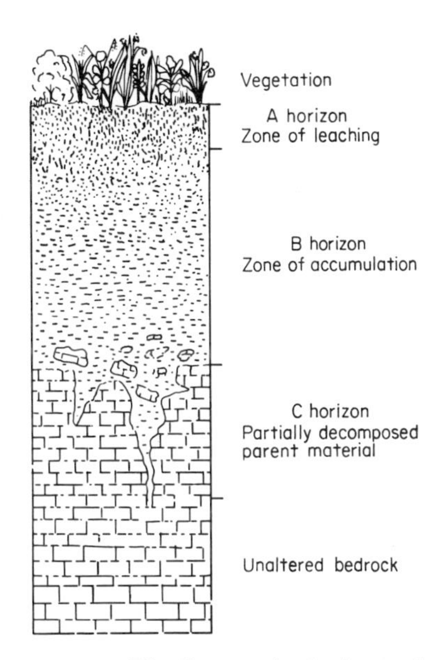

Fig. 6-2 The three major horizons of a soil. (*From L. Don Leet and Sheldon Judson, Physical Geology, 3d ed., 1965. Reprinted by permission of Prentice-Hall, Inc.*)

The three soil horizons are all developed from the underlying parent material. When this material is first exposed at the surface, its upper portion is subjected to intense weathering and decomposition proceeds rapidly. As the depth of decomposed material increases, downward-percolating water begins to leach out some of the minerals and to deposit them lower down. Gradually, the A horizon and the B horizon build up. Weathering of the underlying parent material continues, giving rise to the C horizon. With the passage of time, the C horizon reaches deeper and deeper into the unweathered material and the B horizon keeps moving downward. The A horizon, in turn, encroaches on the upper portion of the B horizon. Finally, a "mature" soil profile is developed. Table 6-2 shows two soil classification systems which are widely accepted by authorities.[2]

Influence of Material on Equipment Selection The construction man reserves the term "rock" for material which must be removed by blasting or some equivalent method. In construction work, bedrock may be encountered in surface outcrops or in excavations which extend below the depth of soil cover. The influence of bedrock on equipment selection is related to rock type, bed thickness, and extent. Generally speaking, the user of quarry rock must drill and shoot, load out with some type of loader, and haul directly to a fill or to a crusher for further processing. Those procedures require special skills, i.e., proper spacing of blastholes for fragmentation, compatible explosives, and correct bench height, to mention a few.* Heavy-duty tractors equipped with rippers are also used to rip or fracture many rocks to a degree

* See Sec. 19, Rock Excavation.

TABLE 6-2 Soil Classification Systems*

Airfield classification system (Casagrande)		Dry strength	Principal classification tests, disturbed samples	Expansion shrinkage	Drainage characteristics	Compaction characteristics	Weight per cu ft, lb	Public Roads Administration (PRA) classification	
Soil groups and typical names	Symbols							Symbols	Description
Well-graded gravel and gravel-sand mixtures; no fines	GW	None	Sieve analysis	Almost none	Excellent	Excellent	125	A–3	Clean gravels; glacial gravels
Well-graded gravel-sand-clay mixtures; excellent binder	GC	Medium to high	Sieve analysis, liquid and plastic limits on binder	Very slight	Practically impervious	Excellent	130	A–1	Well-graded gravels
Poorly graded gravel-sand mixtures; little or no fines	GP	None	Sieve analysis	Almost none	Excellent	Good	115	A–2	Poorly graded gravel and sand
Gravel with fines, very silty gravel, clayey gravel; poorly graded	GF	Slight to high	Sieve analysis, liquid and plastic limits on binder	Almost none to slight	Fair to impervious	Good if closely controlled	120	A–2	
Well-graded sands and gravelly sands; little or no fines	SW	None	Sieve analysis	Almost none	Excellent	Excellent	120	A–3	Well-graded sands; beach sands
Well-graded sand-clay mixtures	SC	Medium to high	Sieve analysis, liquid and plastic limits on binder	Very slight	Practically impervious	Excellent	125	A–1	Sand, clays
Poorly graded sands; little or no fines	SP	None	Sieve analysis	Almost none	Excellent	Good	100	A–3	Sand—no fines
Sand with fines, very silty sands, clayey sands; poorly graded	SF	Slight to high	Sieve analysis, liquid and plastic limits on binder	Almost none to medium	Fair to impervious	Good if closely controlled	105	A–2	
Silts and fine sands, silty or clayey fine sands	ML	Slight to medium	Sieve analysis, liquid and plastic limits on binder	Slight to medium	Fair to poor	Good to poor, close control	100	A–4	Silty soils
Clays, sandy clays, silty clays, lean clays	CL	Medium to high	Liquid and plastic limits	Medium	Practically impervious	Fair to good	100	A–4 A–6	Colloidal clays Sandy clays
Organic silts and organic silt-clays of low plasticity	OL	Slight to medium	Liquid and plastic limits, natural condition after drying	Medium to high	Poor	Fair to poor	90	A–4 A–7	Micaceous sandy loam
Micaceous or diatomaceous fine sandy and silty soils, elastic silts	MH	Slight to medium	Liquid and plastic limits	High	Fair to poor	Poor to very poor	100	A–5	Elastic silts
Clays (inorganic) of high plasticity, fat clays	CH	High	Liquid and plastic limits	High	Practically impervious	Fair to poor	90	A–6 A–7	Expansive clays
Organic clays of medium to high plasticity	OH	High	Liquid and plastic limits, natural condition after drying	High	Practically impervious	Poor to very poor	100	A–7 A–8	Plastic clays
Peat and other highly organic swamp soil	PT	Readily identified	Texture, consistency	Very high	Fair to poor	Not practical		A–8	Muck and peat

* From A. Brinton Carson, *General Excavation Methods*, p. 26, McGraw-Hill Information Systems Company, New York, 1961.

where they can be dozed to a final position, loaded into trucks and hauled, or loaded and hauled by conventional scraper units.

As stated earlier, "common" excavation consists predominantly of soil or earth. Its removal will generally not require the use of explosives, athough tractors equipped with rippers can be used to loosen tight, consolidated earth and thus ease its handling by shovels, scrapers, or dozers.

An "unclassified" excavation may consist of a combination of rock and soil and thus present the excavation problems of both material types. A careful analysis should then be made in order to select the proper equipment.

WEIGHT-VOLUME RELATIONSHIPS

Material weights must be considered in all phases of construction. For example, weight influences the choice of horsepower for acceptable performance of transporting equipment and may establish the basis for payment for hauling. Material weight per unit of volume is also a criterion in determining the acceptability of compacted fills.

Contractors are interested in the volumes occupied by rock and soil materials during the processes of excavation, transportation, and placement. These volumes are influenced by methods of excavation. For example, a shovel excavates an earth bank and loads out a truck, which in turn transports the load to a fill where the material is compacted: bank yard—loose yard (shovel)—loose yard (truck)—compacted yard.

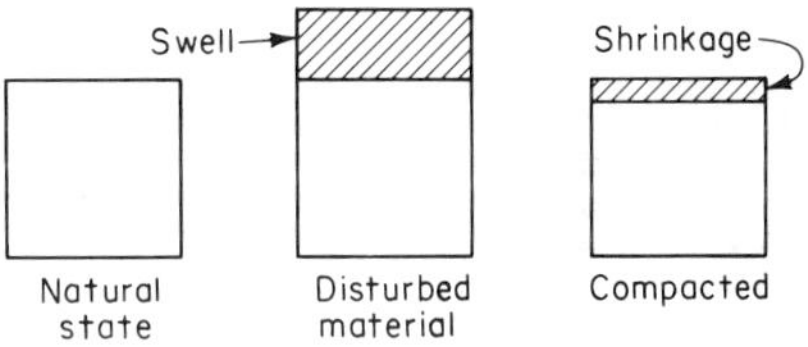

Fig. 6-3 Volume changes in soils.

Volume Measure Soil volume is defined according to its state in the earth-moving process. For simplicity, the following discussion will be confined to English system units, although their metric system equivalents may be directly substituted. The three commonly used measures of soil volumes are illustrated in Fig. 6-3 and are described as follows:

Bank cubic yard (bcy)—1 cu yd of material as it lies in the natural state.

Loose cubic yard (lcy)—1 cu yd of material after it has been disturbed by loading, with consequent swelling relative to its natural state.

Compacted cubic yard (ccy)—1 cu yd of material after it has been compacted. This usually results in shrinkage relative to both its natural and loose states.

For example, assume that 1 bcy of material weighs 3,000 lb and that, by virtue of its material characteristics, it swells to 1.3 cu yd when loaded. The resulting 1.3 lcy still weigh 3,000 lb, although the soil now has 30 percent swell relative to its bank condition. On the other hand, if either 1 bcy cubic yard or 1.3 lcy is compacted, its volume may be reduced to 0.8 ccy. Its weight still remains unchanged at 3,000 lb.

Generally, earth-moving jobs are calculated on the basis of bank cubic yards. Thus, in order to estimate production, the relationships between bank measure, loose measure, and compacted measure must be known. Typical values for a range of materials are listed in Table 6-3. Values are shown in pounds and kilograms for bank and loose states only, as compacted yards would vary widely depending upon compaction effort.

Load Factor The ratio between bank measure and loose measure is called "load factor" (LF):

$$\text{LF} = \frac{\text{bank cubic yards (bcy)}}{\text{loose cubic yards (lcy)}}$$

If the percent of swell of the material is known, the load factor may also be obtained by using the relationship:

$$\text{LF} = \frac{100\%}{100\% + \%\ \text{swell}}$$

TABLE 6-3 Approximate Weight of Materials*

Material	lb/lcy	kg/loose cu m	lb/bcy	kg/bank cu m
Bauxite	2,400	(1,425)	3,200	(1,900)
Caliche	2,100	(1,250)	3,820	(2,265)
Cinders	960	(575)		
Carnotite, uranium ore	2,770	(1,630)	3,680	(2,185)
Clay:				
Dry excavated	1,840	(1,090)	2,560	(1,520)
Wet excavated	3,080	(1,825)	4,280	(2,540)
Dry lumps	1,820	(1,070)	2,530	(1,500)
Wet lumps	2,700	(1,600)	3,750	(2,225)
Natural bed	2,130	(1,265)	2,960	(1,755)
Clay and gravel:				
Dry	1,940	(1,150)	2,290	(1,360)
Wet	2,220	(1,255)	2,620	(1,555)
Coal:				
Anthracite, raw	2,000	(1,190)	2,700	(1,600)
washed	1,850	(1,050)	2,500	(1,485)
Bituminous, raw	1,600	(950)	2,160	(1,280)
washed	1,400	(830)	1,890	(1,120)
Decomposed rock:				
75% R–25% E†	3,300	(1,955)	4,720	(2,800)
50% R–50% E	2,900	(1,715)	3,860	(2,290)
25% R–75% E	2,660	(1,585)	3,320	(1,970)
Earth:				
Loam, dry excavated	2,100	(1,250)	2,620	(1,555)
Moist excavated	2,430	(1,440)	3,040	(1,805)
Wet excavated	2,700	(1,600)	3,380	(2,005)
Dense, packed	3,100	(1,840)	3,880	(2,300)
Flowing mud	2,920	(1,730)	3,650	(2,165)
Packed, dry	2,560	(1,520)	3,200	(1,900)
Granite:				
Broken	2,780	(1,650)		
Solid	.....	....	4,720	(2,805)
Gravel:				
Loose, dry	2,560	(1,520)	2,870	(1,700)
Pit run	3,240	(1,920)	3,630	(2,155)
Dry, ¼–2 in.	2,840	(1,680)	3,180	(1,685)
Wet, ¼–2 in.	3,380	(2,005)	3,790	(2,250)
Sand and clay:				
Loose	2,700	(1,600)	3,380	(2,005)
Compacted	4,050	(2,405)		
Gypsum:				
Broken	3,050	(1,810)	5,340	(3,170)
Crushed	2,700	(1,600)	4,720	(2,800)
Solid	.....	....	4,720	(2,800)
Hematite, iron ore	4,150	(2,465)	4,900	(2,905)
Limestone:				
Broken	2,620	(1,550)		
Solid	.....	....	4,400	(2,605)
Magnetite, iron ore	4,680	(2,785)	5,520	(3,275)
Peat:				
Dry	680	(400)	1,140	(675)
Moist	1,350	(800)	2,250	(1,335)
Wet	1,890	(1,120)	3,160	(1,875)
Pyrite, iron ore	4,340	(2,580)	5,120	(3,040)
Sandstone:				
Broken	2,550	(1,505)	4,260	(2,525)
Solid	3,920	(2,305)	6,550	(3,885)

TABLE 6-3 Approximate Weight of Materials* (Continued)

Material	lb/lcy	kg/loose cu m	lb/bcy	kg/bank cu m
Sand:				
Dry, loose	2,400	(1,440)	2,690	(1,595)
Slightly damp	2,850	(1,680)	3,190	(1,895)
Wet	3,120	(1,860)	3,490	(2,070)
Wet, packed	3,120	(1,860)	3,490	(2,070)
Sand and gravel:				
Dry	2,920	(1,730)	3,240	(1,920)
Wet	3,380	(2,005)	3,750	(2,225)
Slag:				
Broken	2,970	(1,760)	4,960	(2,945)
Solid	3,670	(2,105)	6,130	(3,635)
Snow:				
Dry	220	(130)		
Wet	860	(515)		
Stone, crushed	2,700	(1,600)	4,510	(2,675)
Taconite	4,050–5,400	(2,405–3,200)	7,090–9,450	(4,208–5,600)
Topsoil	1,620	(960)	2,320	(1,375)
Traprock:				
Broken	2,950	(1,745)	4,420	(2,620)
Solid	4,870	(2,880)	7,300	(4,320)

* Varies with moisture content, grain size, degree of compaction, etc. Tests must be made to determine exact material characteristics.

† R, rock, E, earth.

Load factors and percent of swell for various materials are listed in Table 6-4.

Payload Maximum payload, measured in terms of weight, is the weight limitation on the net load which can be carried by a unit of equipment. The actual payload carried by a unit of known tare weight can readily be obtained by weighing. It is also possible to express payload in terms of bank yards, loose yards (in the unit), or compacted yards. The capacity of an equipment unit is then stated in terms of the volume of loose material, either struck or heaped, which it can contain. "Struck capacity" is the volume level-full, while "heaped capacity" is the volume of the unit with load heaped above the top of the body. For scrapers, the slope of the heaped portion of the load is often taken as 1:1. For trucks and trailers, it is variously taken as 1:1, 2:1, or 3:1.

TABLE 6-4 Correlation between Swell Percentage and Load Factor

Swell (percent)	*Load factor*	*Swell (percent)*	*Load factor*
5	0.952	80	0.556
10	0.909	85	0.541
15	0.870	90	0.526
20	0.833	95	0.513
25	0.800	100	0.500
30	0.769	5.3	0.95
35	0.741	11.1	0.90
40	0.714	17.6	0.85
45	0.690	25.0	0.80
50	0.667	33.3	0.75
55	0.645	42.9	0.70
60	0.625	53.8	0.65
65	0.606	66.7	0.60
70	0.588	81.8	0.55
75	0.571	100.0	0.50

To estimate the machine payload in bank cubic yards, the actual payload volume in loose cubic yards is multiplied by the load factor:

$$\text{Load (bcy)} = \text{lcy} \times \text{LF}$$

Manufacturers' specification sheets supply load and volume ratings for their hauling equipment. In handling the lighter materials, heaped capacity may be reached before the permissible payload in pounds is attained; in such a case, sideboards might be added to increase the capacity.

Shrinkage Factor The ratio between compacted measure and bank measure is called "shrinkage factor" (SF):

$$\text{SF} = \frac{\text{compacted cubic yards (ccy)}}{\text{bank cubic yards (bcy)}}$$

Shrinkage factor is either estimated or obtained from job plans or specifications which show the conversion from compacted measure to bank measure.

EXAMPLE: Construct a 10,000-cu yd bridge approach of dry earth with a shrinkage factor of 0.75. Haul unit is rated 14 cu yd struck and 20 cu yd heaped.

How many bank yards are needed? How many heaped loads are required, assuming 25 percent swell? (LF of 0.80 from Table 6-4.)

$$\text{bcy} = \frac{\text{ccy}}{\text{SF}} = \frac{10{,}000}{0.75} = 13{,}333$$

$$\text{Load (bcy)} = \text{lcy} \times \text{LF} = 20 \times 0.80 = 16.0$$

$$\text{Number of loads required} = \frac{13{,}333}{16} = 833$$

Soil Density This is defined as the weight of a unit volume of the soil. "Dry density" refers to the weight of dry solids in this unit volume, and "wet density" is the weight of the dry solids plus the weight of water in a unit volume.

Soil density is dependent on the relative volumes of the solid particles and the void spaces. It is also dependent on the specific gravity of the solids and, in the case of wet density, on the amount of water in the soil. The relative volumes of solid particles and void spaces are dependent upon the type, nature, and extent of compactive effort. For a given set of conditions, the maximum wet density increases with increasing water content until the voids are almost entirely filled.

The specific gravity of the soil solids is the true specific gravity of the soil; it is equal to the weighted average of the specific gravities of all the mineral particles. Although the specific gravity of minerals varies widely, soils normally contain a preponderance of quartz and quartzlike minerals. This narrows the usual range of true specific gravity of soils to values between about 2.55 and 2.75. Within limited geographical areas, the actual range of variation may be much less.

Soil Compaction Most earth structures require compaction to ensure a strong, uniform foundation. In the case of large earth fills, the material is usually spread in controlled layers and rolled or compacted by specialized equipment. Cuts also must be compacted; this is accomplished by first excavating below grade and then building up and compacting them. In either case a foundation or subgrade is provided so that other structures such as buildings, pavement, or airstrips can be constructed.

The nature of soils is such that, as moisture is added to a dry sample, the density to which the sample can be compacted by a given compactive effort increases until an optimum moisture is reached. The density then decreases as more moisture is added, and the sample eventually turns into a plastic mud. Some soils are far more critical with respect to moisture content than others; that is, certain soils will obtain their maximum densities over a fairly wide range of moisture content, whereas others lose stability rapidly as moisture is increased beyond the optimum point.

In order to predict the amount of moisture that should be in the soil at the time of compaction, a series of laboratory standard compaction tests is run with progressively increasing quantities of moisture. The results are plotted on a graph, with the density recorded along the vertical axis and the percentage of moisture recorded along the horizontal axis. A smooth curve can then be drawn through the plotted points, and the optimum moisture point with its corresponding density is read off this curve as its peak (Fig. 6-4). This optimum moisture content is valid for the type and amount of compactive effort employed in the test. Under other conditions, such as those experienced in field compaction, a different optimum moisture content might be applicable.

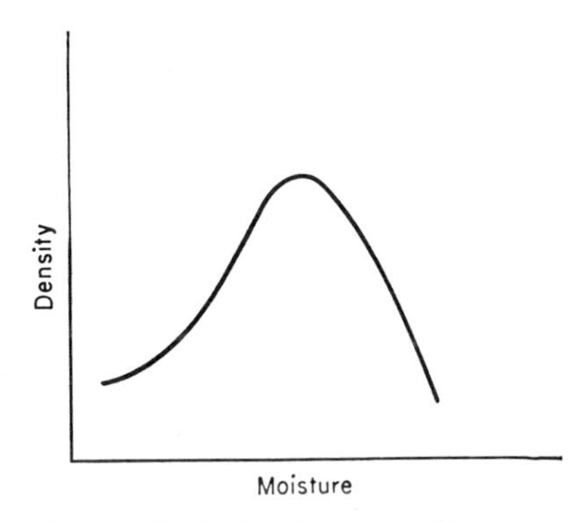

Fig. 6-4 Typical moisture-density curve.

Material Investigations The values listed in Tables 6-2 and 6-3 are consensus figures and are representative of commonly occurring soils and rocks. There are many instances where careful analysis of readily available data will permit more accurate estimating and help to minimize bidding uncertainties. For example, careful study of soil profiles on the plans will yield information as to specific soil types, location, and extent. In some cases, moisture content and densities of soils in place may appear on the plans or may be obtained from the contracting agency. In addition, agricultural soil maps, air photos, geologic maps, and well logs are available and useful.

The contractor may also find it advisable to perform supplementary soil studies. Soil moisture and density measurements at shallow depths are readily and easily carried out and are often of major value. Careful examination of all available information, a minimum of exploration, and some testing may yield a surprising amount of information at a low cost.

C. Equipment Characteristics

POWER UNITS

Gasoline and diesel engines have been used for a considerable period of time as basic power sources for heavy-construction equipment. Gas-turbine engines have been more recently introduced as power sources and are used in conjunction with electric generators in much the same way as diesel-electric units. Electric power is generated by the primary power source and is used to power electric motors. These motors then convert the electricity back into mechanical power. Both gas-turbine and diesel-electric generator units up to the 1,000-hp class are now used in large off-highway trucks, and evaluation is under way to determine their economic status.

Gas and Diesel There are many points in common between diesel- and gasoline-fueled engines. If this discussion is confined to reciprocating-piston 4-stroke-cycle engines of the type used today, the principal components of both engines are quite similar. (2-stroke-cycle engines will be discussed subsequently.) Such differences as exist are largely in the method of introducing fuel and in the process by which it is ignited. Nevertheless, these differences are important in the use and selection of engines.

As early gasoline engines were developed and improved, high compression ratios were tried to obtain increased power and economy. It was soon learned that higher pressures were limited by preignition. This is a form of self-ignition which takes place when heat of compression, combined with heat passed to the air-fuel mixture by hot engine parts, becomes great enough to cause premature firing. Preignition limited early engines to a compression ratio of approximately 4:1; these engines were heavy, expensive, and uneconomical. Only about 7 percent of the fuel's potential heat energy was transformed into work, about the same as for a steam engine.

Improved gasolines and additives (i.e., tetra ethyl), with better ignition characteristics and impeded self-ignition, now permit the use of higher compression pressures. The upper limit on gasoline-engine compression ratio is presently about 12:1, although many engines operate in the range of 7:1 to 9:1 in order to burn cheaper lower-octane gasoline.

Largely because of the inherent limitations of the gasoline engine, the diesel engine was developed. Dr. Diesel applied the principles, outlined by earlier theoreticians, of a cylinder in which air is compressed and thereby heated. When atomized fuels are subsequently introduced into this cylinder, the resulting combustion is more complete and efficient as a result of the higher compression pressures. Compression ratios in diesel engines range up to 22:1, but 16:1 is a commonly used value.

At this point it might be well to consider the fuels which are used in gasoline and diesel engines. Both engines are adaptable to a wide range of fuels. Gas engines basically use gasoline but can be adapted with fair success to use natural gas, liquefied petroleum gases, manufactured gas, sewage gas, and light distillates. Diesel engines have a broad appetite for all kinds of diesel fuels, fuel oils, and "crude"; and they can be converted readily, with an ignition system, to burn natural gas. The use of the term "octane" in connection with gasoline and the term "cetane" in connection with diesel can be explained as follows:

"Octane rating" is a measure of gasoline antiknock quality. A gasoline with a high octane rating can be used in a higher compression engine, and generally it is a better fuel. Originally, gasolines were 50 octane, but now automobiles use 70 to 95 octane and aviation fuels exceed 100.

"Cetane rating" is a measure or grading of diesel fuels according to ignition lag and length of time between injection and ignition. The quicker-burning fuel has the higher cetane number. In diesel fuels the range is from 30 to 60.

Most gasoline engines employ a carburetor, which is a device to mix air and gasoline in the proper amounts. Approximately 15 lb of air are required for the combustion of 1 lb of gasoline; expressed on a volumetric basis, this is roughly the equivalent of 9,000 cu ft of air for each cubic foot of gasoline. These proportions are influenced by engine speeds, engine loads, ambient air pressure, temperature, and humidity. It is a difficult task to maintain optimum air-to-fuel ratios over the usual wide range of operating conditions. The system of mechanically metering and injecting fuel, as employed by a diesel, is generally more precise than a carbureted system.

Ignition is accomplished in gas engines by means of a spark plug which conveys a heat impulse to the air-fuel mixture. The spark plug is backed up by a battery to supply electrical energy; a coil to raise the voltage; a breaker to furnish high voltage; a condenser to provide a quick, clean electrical break; a distributor to time the spark impulses; a wire to complete the electrical path from the condenser to frame; and a ground to act as a return path for circuits.

The system is quite complex in comparison to the diesel-engine system wherein a single, cam-operated plunger pumps fuel to an injector. The injector then sprays the fuel into the cylinder, as timed from a camshaft. There are various types of diesel systems—direct injection, precombustion chambers, separate or combined pumps and nozzles—but all are simple compared to the gas system. A precise air-fuel ratio is not required in diesel engines and, in fact, there is always excess air which is furnished by naturally aspirated, supercharged, turbocharged, and/or aftercooled systems. The excess air which is available in diesel engines, particularly turbocharged and aftercooled versions, accounts for lower combustion and exhaust gas temperatures and is an advantage in satisfying cooling requirements.

Temperatures of combustion vary widely. Naturally aspirated and turbocharged diesels will experience about 4300 to 4700°F, while aftercooled diesels may be in the neighborhood of 4000 to 4100°F. A typical gasoline engine will have peak temperatures of 5000 to 5200°F, at least 10 percent higher than naturally aspirated or turbocharged engines and a full 25 percent higher than aftercooled diesels. Exhaust gas temperatures are also higher in gasoline engines.

Peak cylinder pressures of a modern gasoline truck engine will run about 750 to 900 psi, somewhat higher than for the naturally aspirated, precombustion-chamber diesel. Turbocharged prechamber diesels have peak cylinder pressures of about 1,000 to 1,700 psi, depending on the degree of turbocharging. Naturally aspirated direct-injection diesels run to about 1,700 psi, and when turbocharged to about 300 psi higher. Diesel engines are built more heavily than similarly powered gasoline engines and can compress the charge more highly while keeping internal temperatures lower. Since they work at higher pressures, diesels extract a greater percentage of the available fuel energy than do gasoline engines.

Diesel fuel has more heat energy to begin with than does gasoline, and diesel engines use a higher percentage of the available fuel energy. Fuel consumption of a gasoline engine at full load will approximate 0.6 lb per bhp-hr while a comparable medium-duty diesel will average less than 0.4 lb per bhp-hr. At idling speeds the difference is even more marked, with fuel-consumption rates of gasoline engines running three times those of diesels. Diesel fuel also costs less than does gasoline, although the differential per gallon may vary widely according to the locality.

Although it might seem from this discussion that there are no areas in which the heavy-duty gasoline engine excels, such is not the case. Gasoline engines, being lighter in weight, are favored where minimum weight is an overriding criterion. The light weight of their moving parts permits them to accelerate quickly, although this advantage is largely offset by the greater torque lugability of the diesel. Since the gasoline engine works at lower pressures, it can be built to wider tolerances. Coupled to its light weight, this results in a low purchase price.

Gasoline engines, because of their wider tolerances, will run on fuel and oil too dirty for a modern diesel. Since less mass is involved, working at more liberal tolerances and at much lower cranking pressures, gasoline engines will start more easily at low temperatures than will diesels. However, this situation is exactly reversed if high humidity has caused severe water condensation on the electrical ignition system of the gasoline engine.

The ignition and carburetion problems experienced by gasoline engines, especially as they become older, are a major source of downtime which is not shared by diesels. Although service for gasoline engines is widely available, not every automobile dealer is able to adequately service the heavy-duty gasoline engine. On the other hand, as diesels have become more and more common, excellent service facilities can readily be located.

The diesel is designed to give many more trouble-free miles of service than its gasoline-engine counterpart. Top overhauls, especially, are more frequent on gasoline engines since the high combustion and exhaust temperatures lead to warped or pitted valves and valve seats, top ring wear, and piston cracking. This high heat load is further compounded by the fact that the lightly built (so-called "heavy-duty") gasoline engine operates at a BMEP* of 100 to 125 psi at full load, and from 140 to 160 psi at maximum torque. This compares with a range of 65 to 95 psi at full load and 90 to 120 psi at maximum torque for the medium-duty diesel.

Because of the volatility of the fuel, gasoline-powered vehicles are less safe in a collision. Gasoline stored either in the vehicle or in storage tanks is prone to fire and explosion hazards, and this is reflected in insurance rates. Many contractors are finding that diesel power gives them greater security.

Generally speaking, the heavy-duty modern diesel engine has virtually supplanted gasoline engines for most classes of construction equipment. Its former image as a huge, slow-speed engine no longer fits. Medium-duty direct-injector diesel engines are now competing with heavy-duty gasoline engines. Comparing a medium-duty diesel (up to 250 hp) and a popular heavy-duty gasoline engine today, we find that the latter weighs only 400 lb less.

A final advantage of diesels, one sure to become increasingly important in a society growing ever more alert to air pollution, is the lower emission of carbon monoxide. Not only in city traffic, but anywhere there is a high engine density or low air circulation, the much lower carbon monoxide value in exhaust from diesels will constitute a major advantage.

In summary, the diesel offers:

1. Less need for servicing
2. Longer life
3. Lower fuel consumption
4. Lower-priced fuel
5. Low fire hazard
6. Low CO emission

2-stroke Cycle and 4-stroke Cycle Both 4-stroke cycle and 2-stroke cycle engines are available in construction equipment. Although gasoline 2-stroke cycle engines are widely used in small sizes and for special purposes, this discussion will be confined to 2-cycle engines of the diesel type and more particularly to the uniflow system.

First, a review of the two engine types will establish the operation and functions of each. In the 4-stroke cycle, air intake occurs on the first stroke, compression on the second, power on the third, and exhaust on the fourth. The crankshaft has revolved twice and intake starts at TDC (top dead center). A stroke is defined as the travel of a piston from TDC to BDC (bottom dead center).

In the 2-stroke cycle, all these operations require only two strokes. Air is supplied by a blower and at BDC the piston uncovers ports in the cylinder wall. The compressed air enters to provide fresh air and blow out or scavenge the exhaust air. As the piston moves up, the cylinder intake ports are covered and compression takes place. Fuel is injected and burned at TDC, which starts the piston down on its power stroke. Near the bottom, just before the intake ports are uncovered, the exhaust valves open and lower the pressure so that compressed air can again enter and repeat the cycle. The arrangement of intake ports in the bottom of the cylinder and exhaust ports in its top permits the air to flow in one direction, leading to the name "uniflow" (Fig. 6-5).

Since the 4-cycle engine requires four strokes for its basic function, it follows that

* BMEP, or "brake mean effective pressure," is a product of the mean effective pressure by mechanical efficiency, and is expressed in pounds per square inch.

the 2-cycle engine will have twice as many power strokes as the 4-cycle when operating at the same engine speed. In view of this, it might be assumed that the 2-cycle engine will develop twice the horsepower, per unit of displacement, of the 4-cycle engine. The 2-cycle engine actually develops somewhat less than this, ranging from 1.6 to

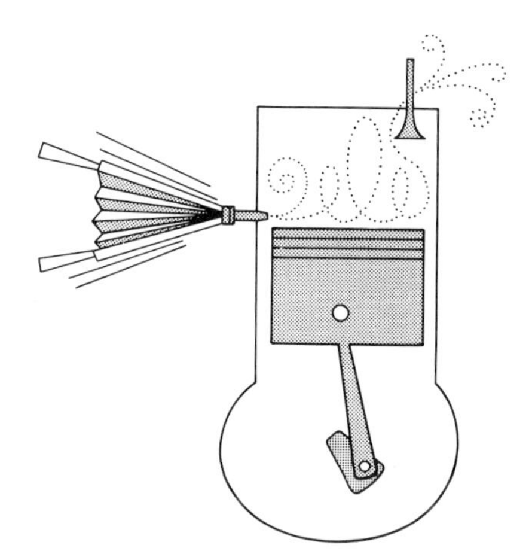

Fig. 6-5 Uniflow system.

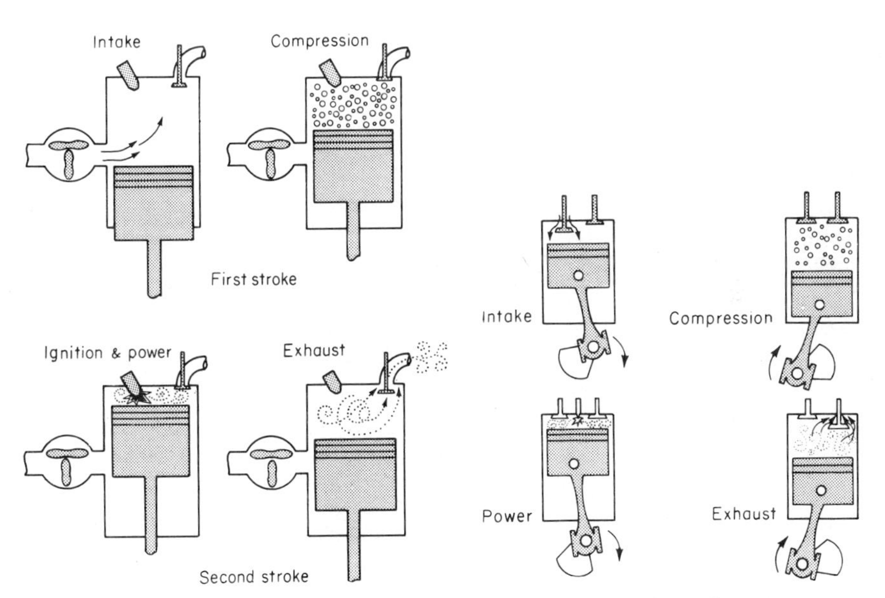

Fig. 6-6 2-stroke cycle system.

Fig. 6-7 4-stroke cycle system.

1.9 that of the 4-cycle engine, because of its shorter effective stroke and the added load required to operate the blower. Air intake for the 2-cycle engine may also be a limiting factor because of the short time which is available to charge the cylinder. Figure 6-6 shows a 2-cycle system and Fig. 6-7 a 4-cycle system.

The major characteristics of 2-stroke and 4-stroke engines can be summarized as follows:

2-STROKE:

1. Power stroke every cycle, but does not produce twice the power of 4-cycle.
2. Good flexibility, permitting power-package combinations with fewer cylinders than 4-stroke engines of comparable displacement because of power stroke for each crank revolution.
3. Engine difficult to maintain in top operating condition because of port rings and dry liners.
4. High noise level.
5. Greater fuel consumption than 4-cycle.
6. Rpm limited to about 2,000 due to valve mechanism.
7. Smaller in size and lighter in weight than the 4-cycle.

4-STROKE:

1. Generally easier to maintain, giving longer trouble-free service.
2. Requires less fuel than 2-cycle.
3. Low noise level.
4. Longer effective stroke.
5. Costs more to manufacture.
6. Operates at higher pressure but at about the same temperature as 2-cycle.

Gas Turbines History records many uses of turbinelike machines to harness the energy of moving fluids and thus perform work. An obvious example is the windmill,

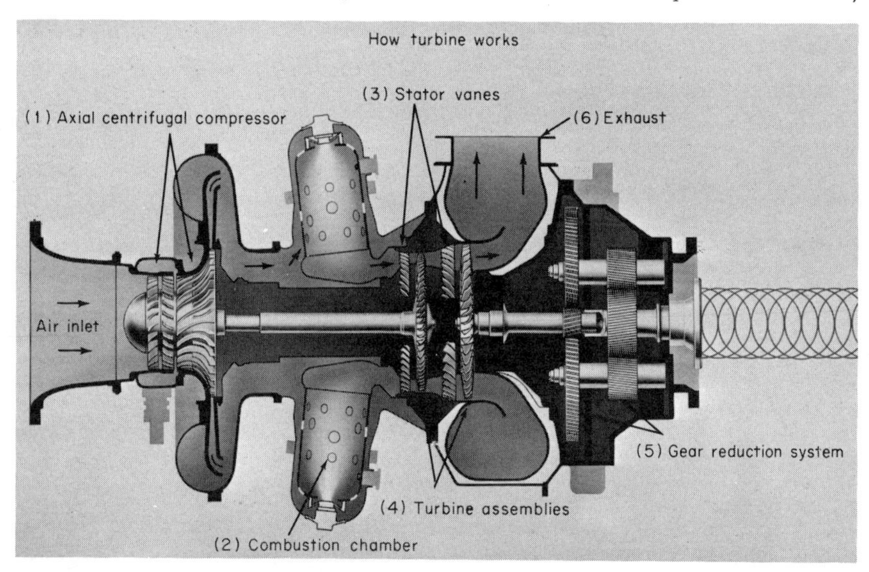

Fig. 6-8 Gas turbine system.

which is powered by moving air. A modern turbine, simply stated, is a rotary engine actuated by reaction of a current of fluid. Air (fluid) is compressed, heated, and expanded over a bladed turbine assembly to produce power. The basic turbine cycle consists of four phases—compression, heat addition, expansion, and exhaust.

One type is shown in Fig. 6-8 and works as follows: Atmospheric air is drawn into the turbine and compressed by a rotating axial-centrifugal flow compressor (1). Compressor increases density and temperature of intake air. After compression, air is directed to combustion chambers (2), where burning of fuel produces heat. Hot gases—directed by stator vanes (3)—expand rapidly over turbine assemblies (4), applying force to blades and causing rotation. Gas-producer turbine wheel drives inlet air compressors; power turbine wheel drives gear reduction system (5). Finally, gases are ducted back to atmosphere (6).

Advantages of gas turbines are:

1. Lighter and smaller than heavy-duty reciprocating engines of comparable hp, both diesel and gas.
2. No radiator or jacket water systems.
3. Fewer moving parts—no pistons, rings, valves, rocker arms, crankshafts, etc.
4. Split shaft units (Fig. 6-8), with one connected to compressor and the other an output section, are linked together only by gas flow. As load is applied to the output shaft and power turbine, causing lug, the gas-producer turbine wheel maintains full speed and power. At output-shaft stall the torque multiplies 2.2 times and thus provides a torque-converter performance.

Disadvantages of gas turbine are:

1. High fuel consumption.
2. High initial price.
3. Components are subjected to high temperatures and stresses.

Many engineers predict a bright future for the gas turbine, based on the premise that its light weight and compact size will offset the high initial price and high fuel consumption.

Table 6-5 compares a gas turbine with a diesel unit.

TABLE 6-5 Comparative Data for Gas Turbine and Diesel Engine

	Turbine	Diesel*
Hp, continuous	300	300
Weight, lb	450	6,500
Weight/hp ratio	1.5	21.7
Length, in	40	95
Height, in	27	57
Width, in	25	40

* Weight includes torque converter and radiator.

Load Factor The horsepower output of an engine is set by its manufacturer to ensure good service life provided the engine is used for the purpose for which it was designed. The engine will consume a given quantity of fuel when developing this rated horsepower (full load) and a lesser quantity at partial load. The load factor, expressed either as a decimal or as a percentage, reflects the ratio between actual and full load.

For example, an engine which is used on a pumping application which requires its full horsepower continuously will have a load factor of 1.0, or 100 percent. If the same engine is used on applications such as dozing or excavating, where it seldom works continuously at full load, its load factor will be less than 100 percent and its fuel consumption may be appreciably reduced.

A unit which is set to consume 10 gal per hr at full load will consume only 8 gal per hr at 80 percent load factor. Actual load factors depend upon type of service, and they range from virtually zero up to 1.0. A tractor will have a high load factor, perhaps up to 0.9, when ripping or pulling plows at full throttle. A medium load factor of 0.75 is typical when dozing, pulling scrapers, or push loading, while a load factor of 0.5 might represent traveling with light loads or idling for long intervals.

POWER TRANSMISSION AND CONTROLS

Transmissions Some form of transmission of power—mechanical, hydrodynamic, hydrostatic, or electrical—is used in practically every type of modern construction equipment. The possible variations are manifold as well as sophisticated.

The following recommended practice, "Automotive Transmission Terminology—SAE J645a" in Ref. 3, enumerates and defines various kinds.

AUTOMOTIVE TRANSMISSION, TERMINOLOGY—SAE J645a

SAE Recommended Practice

Report of Transmission Committee approved April 1956 and last revised November 1962

The following listed definitions are intended to establish suitable terminology for the various kinds of automotive transmissions and related components currently used in this country.

Transmission is a device for transmitting power at a multiplicity of speed and torque ratios.

Mechanical Transmission—A transmission in which the speed and torque ratios are obtained by gears or other mechanical elements.

Countershaft Transmission—A mechanical transmission in which gears are mounted on parallel shafts.

Common types of countershaft transmissions are:

Sliding Gear Countershaft Transmission—A countershaft transmission in which ratios are obtained by axial relocation of gears to mesh with other gears.

Constant Mesh Countershaft Transmission—A countershaft transmission in which gears are constantly in mesh and ratios are obtained by engagement of positive or friction clutches.

Synchronized Countershaft Transmission—a constant mesh countershaft transmission in which means are provided for synchronizing the speeds of engaging elements.

Planetary Transmission—A mechanical transmission using planetary gears.

Hydrodynamic Transmission—A transmission using a hydrodynamic drive unit (fluid coupling or torque converter).

Hydrostatic Transmission—A transmission using fluid placed under pressure by a positive displacement pump to drive a positive placement motor.

Automatic Transmission—A transmission in which normal ratio changes are effected automatically without manual assist.

Semiautomatic Transmission—A transmission in which some of the functions of normal ratio changes are effected automatically.

Power Shifted Transmission—A transmission in which manually selected ratio changes are accomplished through application of power.[1]

Power Assisted Shift Transmission—A transmission in which the effort of manual ratio change is assisted by the application of power.[1]

Manually Shifted Transmission—A transmission in which all ratio changes are effected manually.

Interrupted Drive Transmission is a transmission in which ratio changes cannot be effected without interruption of drive.

Continuous Drive Transmission—A transmission in which ratio changes are effected without interruption of drive.

Split Torque Drive—Comprises two or more parallel paths of torque transmission, one of which is mechanical and one or more hydrodynamic.

[1] Normally hydraulic, pneumatic, or electric.

Electric-driven units are not included in the definition, but the power-shifted and power-assisted types use electric power in addition to hydraulic and pneumatic linkages to accomplish their functions.

Table 6-6 enumerates the principal types of transmissions used in construction equipment units. Application and owners' preference are major factors in transmission selection, and manufacturers are sensitive to users' requirements. This is evidenced by the options offered, although in some instances manufacturing economics will restrict the available options to the predominant use. It should be pointed out that the choices tabulated are not related to percent of total use. For example, 90 to 95 percent of the on-highway tractors (trucks) are equipped with stick shift, countershaft constant-mesh transmissions, whereas the remainder are hydraulic torque converters. In each case economics generally dictates the selection.

Figures 6-9 through 6-12 show various types of transmissions and/or controls.

Controls Mechanical, fluid, and electric linkages are used to control the basic operations of construction equipment. Combinations are also in evidence, such as

TABLE 6-6 Equipment Transmission and Controls Classification by Principal Use*

Equipment	Transmission			Controls		
	Mechanical	Hydraulic	Electric	Hydraulic	Mechanical	Electric
Tractors, on-highway (conventional truck)	1	2	NA	NA	NA	NA
Tractors, off-highway:						
Crawler type	2	1	3	NA	NA	NA
Wheel type	2	1	3	NA	NA	NA
Scrapers:						
Wheel type, self-propelled	2	1	3	1	3	2
Elevating	2	1	3	1	NA	2
Multiple	2	1	3	1	3	2
Excavators, backhoes, and draglines, full-revolving:						
Crawler type	1	2	3†	2	1	NA
Truck-mounted	1	2	NA	2	1	NA
Wheel type, self-propelled	1	2	NA	2	1	NA
Wide-bucket excavator, full-revolving:						
Crawler type	2	1	NA	1	2	NA
Truck-mounted	2	1	NA	1	2	NA
Wheel type, self-propelled	2	1	NA	1	2	NA
Bucket excavator, nonrevolving:						
Crawler type	2	1	3	1	NA	NA
Rubber-tired, including articulated	2	1	3	1	NA	3
Cranes and clamshells, full-revolving:						
Crawler type	2	1	NA	2	1	NA
Truck-mounted	2	1	NA	2	1	NA
Wheel type, self-propelled	2	1	NA	2	1	NA
Loaders:						
Towed	NA	2	NA	NA	1	NA
Self-propelled	1	2	NA	NA	1	NA
Wheel excavators	2	3	1	NA	1	NA
Trenchers:						
Ladder	1	2	NA	NA	NA	NA
Wheel	1	2		NA	NA	NA
Hauling units, on-highway conventional truck, rubber-tired (rear-, side-, bottom-dump, Lo-Boy, etc.)	1	2	NA	NA	2	3
Hauling units, off-highway (rear-, side-, bottom-dump)	2	1	3	1	2	3
Cableways	2	NA	1	NA	1	NA
Belt conveyors	1	NA	2			
		Hydrostatic				
Compactors, self-propelled:						
Sheepsfoot and tamping foot	2	1	3	1	NA	2
Grid	2	1	NA	1	NA	NA
Vibratory	2	1	NA	1	NA	NA
Pneumatic	2	1	NA	1	NA	NA
Cranes:						
Tower	1	2	3	NA	1	NA
Climbing	1	2	3	NA	1	NA
Gantry	1	2	3	NA	1	NA
Hoists	1	3	2	NA	1	NA
Trimming and spreading:						
Motor grader	2	1	NA	1	2	NA
Trimming subbase, subgrade canals	1	2	NA	1	2	NA
Tractor attachments:						
Dozers	NA	NA	NA	1	2	3
Winches	1	2	NA	NA	1	NA
Rippers	NA	NA	NA	1	2	3
Scrapers, towed	NA	NA	NA	1	2	3
Booms, pipelayers	NA	NA	NA	2	1	NA
Rakes, sweeps, etc	NA	NA	NA	1	2	3

* Not in relation to percent of total. Performance preference indicated numerically—1, 2, and 3. NA—not applicable or use insignificant.

† On sizes above 4 cu yd.

hydraulic control of a crane boom for length and radius of load and cable control for lifting the load. In self-loading elevating scrapers, the ladder can be operated by a hydrostatic or electric motor and the remaining functions can be operated by hydraulic rams. Figure 6-13 shows a hydrostatic motor on a ladder and Fig. 6-14 shows an electric motor.

Fig. 6-9 Typical mechanical sliding-gear countershaft transmission.

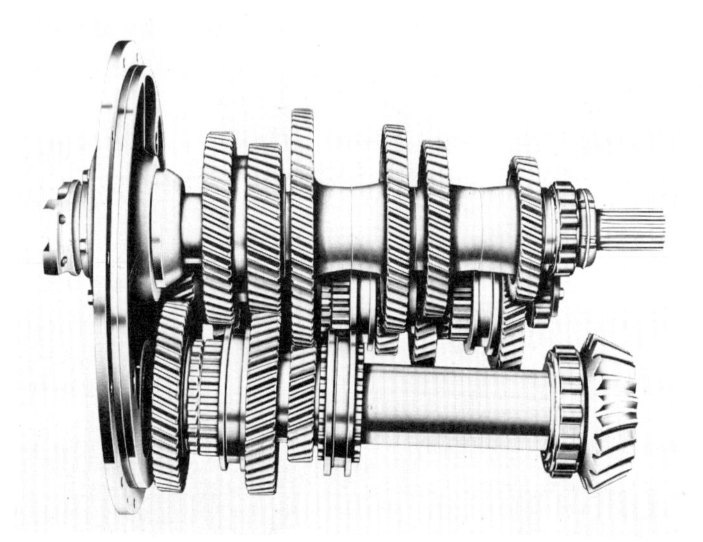

Fig. 6-10 Typical mechanical constant-mesh countershaft transmission.

A complete hydraulic system for controlling a wide-bucket excavator is shown in Fig. 6-15. Figure 6-16 shows a basic self-adjusting vane-type pump and Fig. 6-17 is a cutaway view of filter system, controls, and spool valves.

A complete cable system is shown in Fig. 6-18. Illustrated is a double-drum rear-mounted unit with fair leads for cable operation of scrapers, rippers, or dozers. The unit is driven from the transmission and controlled from the operator's seat.

The configurations and combinations for control systems are numerous and the final choice, as in transmissions, is made by the user based on his judgment of job conditions and economics. Table 6-6 attempts to point out the preferred selection.

EQUIPMENT MOUNTING

Construction equipment, viewed in terms of its mobility, can be classed either as stationary or mobile. Equipment in the first of these categories can be permanently installed, as in a commercial aggregate-producing plant, or it can be transportable

Fig. 6-11 Typical power shift, planetary transmission.

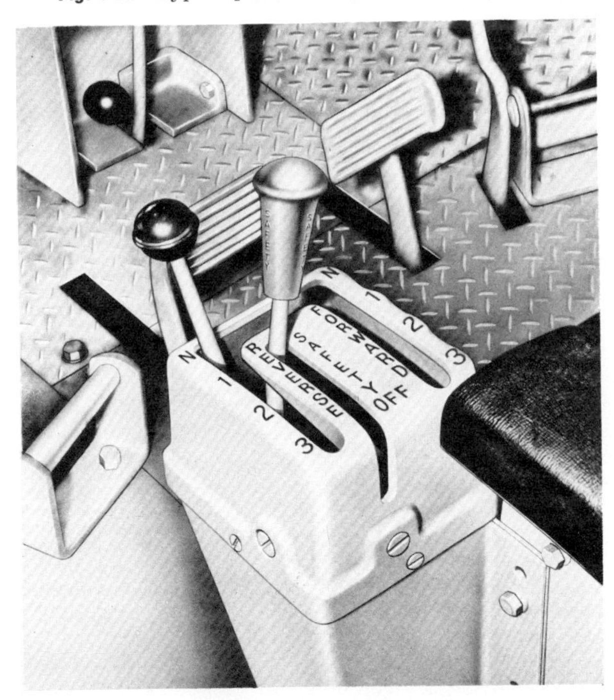

Fig. 6-12 Typical power-shift control console.

either as an entity or by disassembling it. Equipment in the mobile category can be either self-propelled or capable of being towed. The construction industry, by reason of its inherent mobility, places a major emphasis on mobile and transportable equipment.

Fig. 6-13 Hydrostatic motor.

Towed units such as scrapers, compressors, water wagons, and disk plows, to mention a few, rely on a separate prime mover for transporting purposes and in some instances for control, e.g., scrapers. Most common of the prime movers are track-type tractors, wheel-type tractors (2- or 4-wheel drive), wheel- or track-type loaders, motor graders, and off-highway trucks.

Fig. 6-14 Electric motor.

For the most part, the prime mover units can provide drawbar facilities irrespective of their basic design function. Track- and wheel-type tractors can be equipped with dozers or shovels and still tow other equipment. If a prime mover can be used in several applications, there is a reduction in the need for specialized equipment. This flexibility becomes very important when selecting equipment for overseas projects.

Fig. 6-15 Complete hydraulic system.

Self-propelled units can be broadly classified as:

1. Mobile prime mover with demountable attachments such as wide-bucket excavator, dozer, backhoe, pipelayer, etc.
2. Integrated unit such as a power shovel where the power unit (engine), operating through proper gearing and a power mechanism, both propels the unit and enables it to function for its designed purpose. See Fig. 6-19.
3. Power shovel, as mentioned above, which is truck mounted. In this case separate power units (engines) are used—one for the truck and the other for the power shovel. See Fig. 6-20.

Fig. 6-16 Vane-type pump—hydraulic.

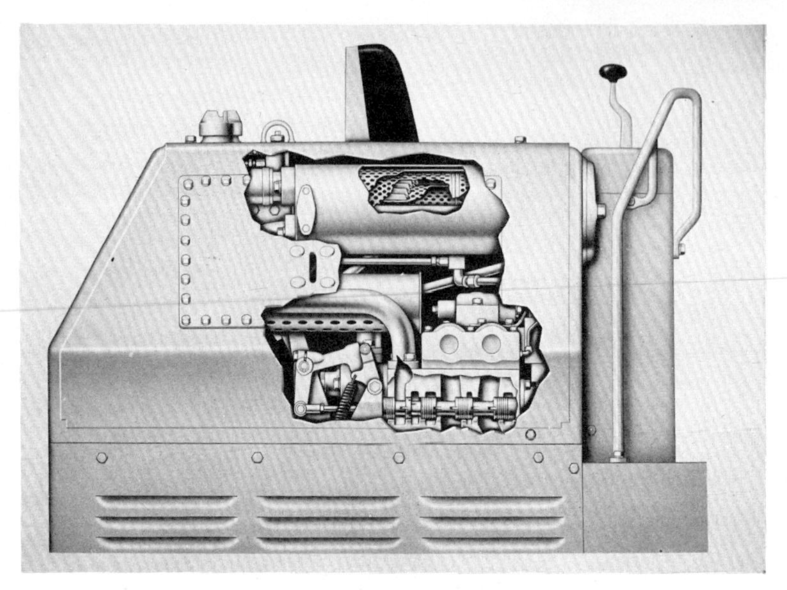

Fig. 6-17 Spool valve controls.

Running Gear Mobile construction equipment can generally be equipped with either tracks or wheels as its running gear. Each type has its advantages and disadvantages, and the choice for a particular job should be the one which obtains the desired production at the least cost. The basic factors which influence this choice will now be discussed.

Fig. 6-18 Cable control system.

Fig. 6-19 Power shovel mounted on crawler-type carrier.

When speed and mobility are important considerations, rubber-tired running gear has its biggest advantage over crawlers. Equipment that spends much of its workday shuttling between jobs (such as a utility machine) or on a job that does not call for a lot of traction (a dozer spreading fill, for example) probably should be on rubber. Speed and mobility become relatively less important when the unit is given a permanent job in a limited area.

Where traction requirements are high, the crawler running gear can offer significant advantages over rubber tires. Consider, for example, a tractor whose effective pulling force is limited by the tractive reaction which can be mobilized. In most materials of interest to contractors, tracks can develop more traction than rubber tires. As a consequence, a crawler-tractor will develop a greater effective drawbar pull than a wheel-type tractor of similar weight and horsepower. Thus, traction is the crawler's

Fig. 6-20 Power shovel mounted on truck-type carrier.

ally for jobs where high drawbar pull requirements exert more influence on the cost per yard than does speed.

The tractive limitation on pulling power is the weight of the machine itself. No mobile unit of existing design can exert more pounds pull than the weight on its driving wheels or tracks. For instance, if a conventional automobile has only 40 percent of its total weight on the rear driving wheels, then the maximum pulling force which it can exert is limited to 40 percent of its total weight. Remember, this is a maximum figure. Underfoot conditions can make it considerably less, as anyone who has ever had an automobile stuck in a snowbank will verify.

In steep terrain, where high speeds are not obtainable, maximum machine gradeability may be important. This gradeability is also related to developable tractive reaction; hence steep grades will tend to favor the use of crawler running gear.

"Rolling resistance" refers to the resistance encountered when a wheeled vehicle traverses a surface; particular importance is attached to the situation where this surface is a soil of low bearing capacity. Tires penetrate a soft surface, leading to increased rolling resistance. Tracks "build their own road," with rolling resistance essentially constant. The superior "flotation" of tracks will favor their use in unstable soils.

POWER RATINGS

Horsepower is defined as "an expression of rate of work output," and all equipment manufacturers can supply specification sheets which show the rated power of their units. While these ratings may be entirely correct, they do not necessarily indicate the "working horsepower" in the sense that this term is interpreted by the contractor. Various factors can influence a horsepower rating, and the manufacturers have not adopted uniform rating procedures.

Basically, there are two types of vehicle horsepower ratings: one based on *performance* and the other on *capacity*. Performance horsepower ratings report the actual work output of the engine as applied to the vehicle. Capacity horsepower ratings are generally described as "maximum" and tell the potential of the basic engine. Both are valid ratings, but they give the engine output under entirely different conditions and thus are not comparable. The four basic factors involved in determining engine horsepower are sea-level correction, rating speed or rpm, accessory load, and load duration.

A sea-level correction allows for the theoretical effect on engine rating caused by differences in ambient temperatures and barometric pressures. The air density where an engine is rated varies according to the location at which the test is performed. By correcting ratings to sea-level standard conditions, a basis forcomparis on is provided. The ratings, however, show only potential outputs under the specified conditions.

"Rating speed," or rpm, indicates the engine speed at which the horsepower rating is produced. It is not necessarily the same rpm at which the engine is set for use in a particular vehicle. While the engine might have the same capacity rating in all uses, its actual output and rpm will be set to correspond with the power needed for a particular unit.

"Accessory load" defines how the engine is equipped for a given rating. Accessories (fan, generator, and air compressor) use up power, and a higher horsepower rating can be obtained by excluding them in rating calculations. The fan is the largest power consumer, in some cases taking as much as 30 hp from an engine.

"Load duration" specifies how long the engine must hold the load to establish the horsepower rating. Extreme output can be taken from an engine only for a short time. A maximum rating generally states that the load is held for a period of 5 min. A flywheel rating is based on unlimited operation of a vehicle when used for the purpose for which it was designed. Throughout the industry, six general rating descriptions are used:

1. Maximum: the bare engine, sea-level and temperature correction, extra rpm (beyond what will be applied in the vehicle), and 5 min of load

2. Maximum: bare engine, sea-level and temperature correction, vehicle rpm, and 5-min load
3. Maximum: bare engine, sea-level and temperature correction, vehicle rpm, and engine's rated load
4. Flywheel: only fan excluded, sea-level and temperature correction, vehicle rpm, and vehicle load (comparable to service conditions)
5. Flywheel: with accessories, sea-level and temperature correction, vehicle rpm, and vehicle load
6. Flywheel: with accessories, factory conditions, vehicle rpm, and vehicle load

The problem that arises now is trying to compare unlike horsepowers when choosing machinery for various jobs. To illustrate the effect of these rating definitions, let us take as an example an engine set at 100 hp and define this horsepower as "flywheel with accessories" (description No. 6). Correcting this particular engine to sea-level standard conditions—the No. 5 flywheel definition—could then bring the horsepower rating to 104. Moving to flywheel definition No. 4 and removing only the fan could raise the horsepower rating to 114.

The next alternative, No. 3, is a capacity-type rating. The bare engine is corrected to sea level at the engine manufacturer's rated load, which could result in a horsepower rating of 120. The No. 2 maximum rating is still based on the vehicle operating speed, but the short-duration load could result in a possible 135-hp rating. The ultimate of capacity ratings, No. 1, might reach as high as 150. Thus, the same engine with a flywheel-horsepower output of 100 could, by these various definitions, have a maximum horsepower as high as 150 for rating purposes.

As another aspect of horsepower ratings, the intended use of a vehicle and the conditions under which it will normally work can result in different flywheel-horsepower settings for essentially the same engine—with equal service life anticipated. The same engine, but with different rpm settings, is used in an industrial wheel tractor rated at 360 flywheel horsepower and a rear-dump truck rated at 400 flywheel horsepower.

Though set at a lower horsepower, the wheel-tractor engine will work harder than the truck engine because of job conditions. The wheel tractor uses power while loading, travels on rough haul roads with high rolling resistance, uses power when dumping on the fill, and usually returns on a road with a high rolling resistance. The machine also weighs over 10 tons more than the truck, both empty and loaded. The off-highway truck does not require power for loading, usually travels on well-maintained haul roads with low rolling resistance, needs only a small amount of power while dumping, and returns on a road with relatively low rolling resistance.

Maximum horsepower ratings reflect capacity, or capability, and flywheel ratings indicate the actual output performance. Flywheel-horsepower output of an engine is set by the manufacturer to give a vehicle a good service life. Thus, no limitations are made on the operation of the vehicle as long as it is used for the purpose for which it was designed and built. It becomes the manufacturer's responsibility to set the engine output to match intended usage and still give good service life. Contractors who know the differences in horsepower ratings are able to buy or rent construction machinery with a better understanding of what it can do on the job.

As used in construction equipment, rated flywheel horsepower must be further corrected for transmission and other losses in order to predict available power for useful work. Transmission losses can be 10 to 20 percent, with minimum losses in manual transmission and final reduction assemblies and with maximum losses in hydraulic transmissions. Electrical generator units which power electrical traction motors provide highly efficient transmission systems with losses of 10 percent or less.

POWER USABLE

In the preceding discussion of power ratings it was indicated that power is energy in action—a force performing work at a given rate. The power setting for the engine will establish the theoretical "power available" for a given machine. It is then necessary to establish whether this power available is, in fact, power usable. The machine is adequately powered for its intended application if "power usable" is at least equal to the power requirement.

The power potential in all classes of equipment is limited by many factors. Among some of the important factors are altitude, temperature, wear, and inadequate maintenance. In addition, traction is a limiting factor which applies to transportation equipment and will be discussed later.

Altitude and Temperature Internal combustion engines use oxygen from the air in the combustion process. At low altitudes, the pressure of the atmosphere is relatively great and a larger volume of air flows into the cylinders of internal combustion engines on the suction stroke. Due to the fact that the atmosphere is also compressed, the oxygen content per unit of volume of air is increased, so power output varies with density or weight of air surrounding the engine. Likewise, temperature affects power as the atmosphere expands with heat and reduces the amount of oxygen content per unit of volume of air.

From these statements, it can be concluded that oxygen content of intake air on a volumetric basis varies directly with an increase in atmospheric pressure (low altitude) and inversely with a rise in atmospheric temperature. Relative humidity also has an effect, as the water in moisture-laden atmosphere takes up volume that could otherwise be occupied by oxygen.

Since the above factors affect the performance of internal combustion engines, a common basis of comparison is provided by the Society of Automotive Engineers Test Code (SAE J816). A correction factor converts observed data to specified standard conditions, as indicated below:[3]

1. Barometric pressure—29.38 in. Hg
2. Temperature—85°F
3. Vapor pressure—0.38 in. Hg

Reduction in rated horsepower thus occurs at the higher altitudes and temperatures and must be recognized in order to estimate actual performance. Approximate corrections can be made by the following "rules of thumb" for naturally aspirated engines:

ALTITUDE: For gasoline and 4-stroke-cycle diesel engines, deduct 3 percent from sea-level rating for each 1,000 ft of altitude above sea level.

For 2-stroke-cycle diesel engines, deduct 1 percent for each 1,000 ft of altitude above sea level.

TEMPERATURE: Deduct 1 percent of rated power at 85°F for each 10°F temperature rise and add 1 percent for each 10°F temperature drop.

The use of turbochargers and aftercoolers reduces loss of power at higher altitudes and temperatures, but the number of variables precludes any convenient formula for predicting the net change in power. It is necessary to consult engine manufacturers for derating data based on specific operating conditions.

The wide variety of engines available today permits a sophisticated approach to the selection of the best engine for a particular application. For example, one manufacturer uses a naturally aspirated engine which has flywheel horsepower ratings ranging from 65 at 1,680 rpm to 105 at 2,200 rpm. The net horsepower at the flywheel of the vehicle engine is measured when operating under SAE standard ambient temperature and barometric conditions—85°F (29°C) and 29.38 in. (746 mm) of mercury. Vehicle engine equipment includes fan, air cleaner, water pump, lubricating-oil pump, fuel pump, and generator. The engine will maintain full horsepower up to 2,500 ft (760 m) altitude. The same basic engine if turbocharged can be rated at 115 flywheel horsepower and maintain this horsepower up to 10,000 ft without derating.

Wear and Inadequate Maintenance These are present, in varying degrees, in most equipment units. They can further reduce "power usable" below the rated value of "power available." For example, ring wear can result in blow-by and a loss of power which is recognized by a tractor operator as a reduction in gradeability. A fouled fuel system can prevent the flow of sufficient fuel for combustion, with a consequent loss of power. Dirty filters, clogged oil coolers, and air leaks in suction lines are additional causes for loss of power. These problems should be recognized and corrected as soon as possible. A well-planned preventive maintenance and inspection program is required to maximize equipment performance.

Traction The ability of tracks or wheels to grip the ground is termed "traction." It is dependent upon the weight on the tracks or driving wheels and the type of surface upon which the vehicle operates. Tracks will pull better on firm earth than on concrete, and rubber tires will pull better on concrete than on firm earth. Traction conditions, therefore, may limit the usable drawbar pull or rimpull.*

A specific type of underfoot condition can be expressed in terms of a "coefficient of traction" (see Table 6-7). This coefficient is a function of the surfaces in contact. It is determined by experiment, and is then applied to the total weight on the driving wheels or tracks. For instance, if we say that the coefficient of traction for rubber tires on ice is 0.12, this means that wheel slippage can be expected if the required pounds pull exceeds 12 percent of the machine weight on the driving wheels. This represents only one limiting condition, since the actual power available in the form of potential rimpull may be less than the rimpull at which slippage occurs.

TABLE 6-7 Coefficients of Traction

Materials	Rubber tires	Tracks
Concrete	0.90	0.45
Clay loam, dry	0.55	0.90
Clay loam, wet	0.45	0.70
Rutted clay loam	0.40	0.70
Dry sand	0.20	0.30
Wet sand	0.40	0.50
Quarry pit	0.65	0.55
Gravel road (loose surface)	0.36	0.50
Packed snow	0.20	0.25
Ice	0.12	0.12
Firm earth	0.55	0.90
Loose earth	0.45	0.60
Coal, stockpiled	0.45	0.60

The usable pounds pull at which slippage is imminent can be determined as follows:

Traction limitation on usable pounds pull
= coefficient of traction × weight on drivers or tracks

EXAMPLE: What is the traction limitation on the maximum usable drawbar pounds pull of a track-type tractor pulling a compactor in loose earth? Weight of tractor: 32,600 lb

Usable pounds pull = total weight of tractor × coefficient of traction
= 32,600 lb × 0.60
= 19,560 lb pull

POWER REQUIRED

This depends on the characteristics of the unit, its functions, and the job conditions. Power is required for each equipment function, but these functions may not be performed concurrently. A mobile unit, for example, must overcome rolling resistance and grade resistance in order to achieve mobility. An elevating scraper, in addition to its function of mobility, also requires power to excavate and load material. A power shovel requires power for mobility and for loading, but these functions are not performed concurrently.

Rolling Resistance This can be defined as the resistance to movement of mobile equipment at constant speed over level ground (Fig. 6-21). The following major

* Rimpull is the pulling force, measured at the ground contact point, which the engine delivers to the tires.

factors determine the rolling resistance for rubber-tired units:

1. Ground penetration
2. Flexing of tire sidwalls
3. Speed
4. Wheel bearing friction
5. Total weight on tires

Factors other than ground penetration can be approximated, for all surface conditions, by a single value of 40 lb for each ton of gross vehicle weight (GVW) which is carried on the tires. It is also possible to relate the tire penetration on a given

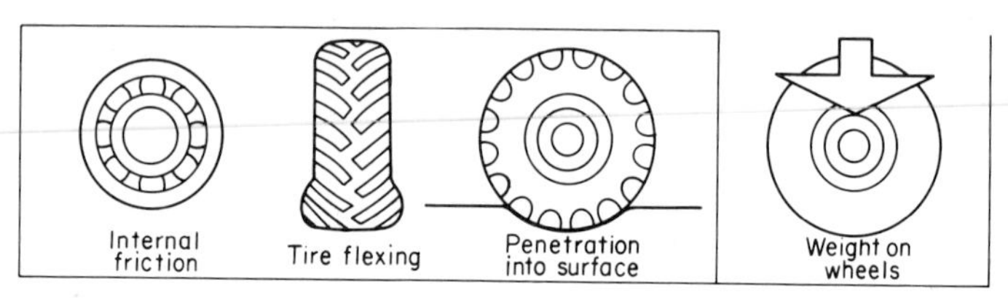

Fig. 6-21 Factors affecting rolling resistance.

ground surface, approximately at least, as a function of this same GVW. By combining factors 2, 3, and 4 with a specified ground condition or surface (factor 1), the rolling resistance per unit of GVW can be stated as a single number, or rolling-resistance factor. It can also be expressed as a percentage of the GVW.

Table 6-8 lists typical rolling-resistance factors. Since these are functions of tire penetration, they can vary considerably for different segments of the same haul road. At a given point on the haul road, the rolling resistance may also change each time a unit passes. Rolling resistance is a very important factor and will frequently determine the maximum speed at which a given piece of equipment can operate.

TABLE 6-8 Typical Rolling-resistance Factors for Rubber-tired Equipment

Description of haul road	Lb/ton	Percent of GVW
Hard, smooth, stabilized, without penetration under load	40	2
Firm, smooth, flexing slightly under load	65	3.25
Rutted dirt, flexing considerably under load	100	5
Rutted dirt, no stabilization, somewhat soft under load	150	7.5
Soft, rutted mud or sand, deep penetration under load	200–400	10–20

To calculate the force or rimpull necessary to overcome the rolling resistance of rubber-tired hauling units, the entire weight of the unit, including both tractor and the loaded wagon or scraper, must be considered. For a tractor-scraper combination of this type, loaded with a gross weight of 56 tons and with a rolling resistance on the haul road of 100 lb per ton, the power required to overcome rolling resistance when moving at constant speed over level ground would be

$$56 \text{ tons} \times 100 \text{ lb/ton} = 5{,}600 \text{ lb rimpull}$$

or

$$112{,}000 \text{ lb} \times 0.05 = 5{,}600 \text{ lb rimpull}$$

To estimate the force or drawbar pull* required to overcome the rolling resistance offered by a track-type tractor towing a wheeled scraper, it is necessary to consider only the gross weight of the scraper unit. The rolling resistance of a track-type tractor is essentially a constant value, independent of ground penetration, since the

* Drawbar pull is the pulling force, measured at the drawbar of a track-type mobile unit, which the engine delivers to the tracks.

unit lays its own traveling surface as it advances. The values of drawbar pull which are supplied by tractor manufacturers have already been corrected for rolling resistance.

The following example illustrates the drawbar pull which is required to overcome the rolling resistance of a track-type tractor which pulls a wheeled scraper. Assume a tractor weight of 20 tons, with scraper and load weighing 44 tons, for a total weight of 64 tons. Assume also that the combination will be operating on a level haul road with a rolling resistance of 150 lb per ton. Since the rolling resistance of the track-type tractor has already been recognized in establishing its drawbar pull rating, it is only necessary to calculate the rolling resistance of the loaded scraper. The power required to move the unit at constant speed over level ground would then be:

$$44 \text{ tons} \times 150 \text{ lb/ton} = 6{,}600 \text{ drawbar pounds pull}$$

Grade Resistance Grade resistance (GR) is the force of gravity which must be overcome when going uphill. It is a function of the total weight of any vehicle, regardless of whether it is a track- or wheel-type machine.

In earth-moving work, grades are most frequently measured in *percent* slope, which is the ratio between vertical rise (or fall) and the horizontal distance in which the rise (or fall) occurs. For instance, a vertical rise of 5 ft in 100 ft horizontal distance would be a 5 percent grade.

When the grade is uphill or *adverse*, the effect is a demand for more power. Grade resistance then is a hindering force. If the grade is downhill or favorable, the effect is a helping force, tending to produce additional pounds pull to propel the vehicle. This downhill effect is commonly called *grade assistance* (GA). Regardless of whether the terrain is uphill, downhill, or level, *rolling resistance* (RR) is always present and must be considered.

When traveling uphill, a vehicle must overcome rolling resistance *plus* grade resistance.

When traveling over level terrain, a vehicle must overcome only rolling resistance.

When traveling downhill, a vehicle must overcome rolling resistance *less* grade assistance.

Both grade resistance and grade assistance are estimated in the same way. A rule of thumb states that each 1 percent of grade produces a hindering or helping force of 20 lb per ton of vehicle weight. This is in addition to rolling resistance. The formula can be expressed as:

$$\text{GR (or GA)} = (\text{total equipment weight} + \text{load}) \times 20 \text{ lb/ton} \times \text{units of \% grade}$$

EXAMPLE: For a loaded track-type tractor-scraper combination weighing 64 tons (tractor, 20 tons; scraper and load, 44 tons) operating on a haul road with an adverse grade of 10 percent, the required drawbar pull to overcome grade resistance would be:

$$64 \times 20 \times 10 = 12{,}800 \text{ lb}$$

The maximum gradient (gradeability) for a mobile equipment unit may be fixed either by available power (drawbar or rimpull) or by tractive reaction which can be developed before slippage occurs.

$$\text{Maximum gradeability (\%)} = \frac{\text{usable drawbar or rimpull (lb)}}{\text{GVW (tons)} \times 20 \text{ lb/ton}}$$

$$= \frac{\text{usable drawbar or rimpull (lb)} \times 100}{\text{GVW (lb)}}$$

$$\text{GR drawbar (rimpull) hp} = \frac{\text{GVW (tons)} \times 20 \text{ lb/tons} \times \text{\% grade} \times \text{mph}}{375}$$

Other Requirements Mobile construction equipment may have either a singular function or multiple functions. The "pure" hauler is an example of a single-function unit; and a power shovel, a self-loading scraper, and a front-end shovel are examples of multiple-function units. Power must be adequate for all functions.

In the case of the "pure" hauler, the function is transportation of material, and the power requirement in terms of rimpull can be calculated by the formulas in the preceding paragraphs. In other cases, power requirements are complex and we must rely on observed results and experience. This is known as the "state of the art."

Controlling Requirements It is evident that equipment functions require power that is compatible with application and job conditions. The controlling power requirement for acceptable performance can be a combination of rolling resistance and grade resistance, or it can be within the category of "complex" requirements mentioned previously.

Neglecting the force required to overcome wind resistance and to provide acceleration, the total resistance to the motion of a mobile equipment unit can be taken as

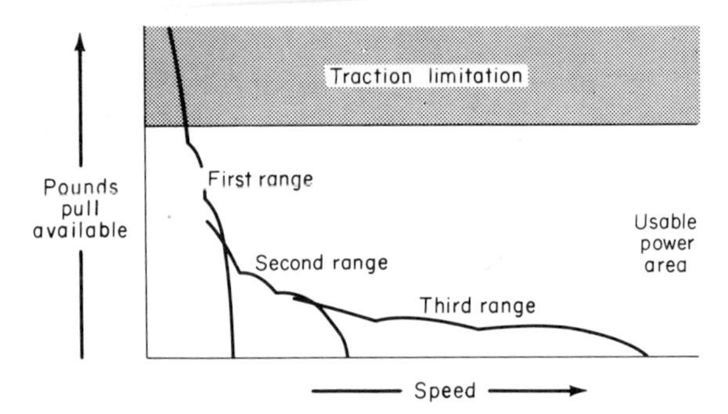

Fig. 6-22 Traction limitation on usable power.

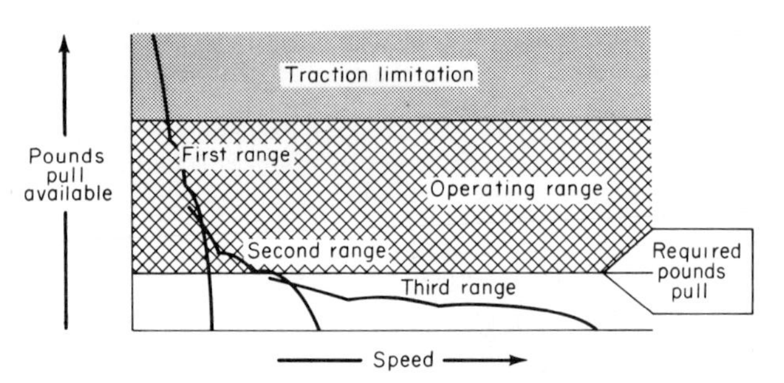

Fig. 6-23 Operating range of usable power.

the sum of rolling resistance and grade resistance. The rolling resistance of a track-mounted self-propelled unit is considered to be constant, as was explained earlier, and has already been accounted for in the drawbar pull rating which is supplied by its manufacturer. The rolling resistance of a wheel-mounted mobile unit is a function of the particular ground surface and the weight which is carried on rubber tires.

If the entire weight of the mobile unit is carried on tires, as for a towed scraper or a wheel tractor, its rolling resistance and its grade resistance can be combined into a single figure for ease in calculation. This combined resistance can be expressed as required pounds pull per ton of GVW on the assumption that 1 percent adverse grade is equivalent to a 20 lb per ton rolling resistance. Alternatively, it can be expressed as an equivalent adverse grade by increasing the actual gradient by 1 percent for each 20 lb per ton of rolling resistance.

Graphically, these power considerations can be shown as follows: Figure 6-22 is a typical rimpull-speed graph (rimpull or pounds available on the vertical axis and speed on the horizontal). The usable power area is limited by the applicable traction factor, which is superimposed at the top of the graph.

Figure 6-23 shows the same chart, but with the power required (total resistance) now superimposed at its bottom. The operating range thus falls between this "power required" and the "power usable," which was shown to be limited by traction.

In the case of the power shovel that moves to a new location only when material is out of operating range, rolling and grade resistances are of little importance. Engine horsepower becomes significant in order to provide power for digging and crowding into a bank and for swinging to load a truck or otherwise dispose of the material. Basic assumptions must be made by the designer regarding line pulls, line speeds, and swing speeds in order to produce a unit with acceptable digging power and rates. Clutches, brakes, cable sizes, shafts, gears, and machine stability must be adequate and contribute to a total design. A working balance of these elements depends upon years of design and operating experience.

D. Tires for Construction Equipment

PRINCIPLES OF TIRE ENGINEERING

A brief summary of some basic engineering concepts of the pneumatic tire should prove helpful in understanding the practical aspects of tire performance.

Fundamentally, a tire is a nonrigid torus. Its main structural members are high-tensile cords of textile fibers or steel cable. These cords are wrapped around wire bead foundations which fasten the entire structure to the rim wheel under compressive forces.

The uninflated tire is not capable of carrying any load beyond that which can be supported by the inherent stiffness of its envelope. In this state the cord and bead structure are in a comparatively relaxed condition. When the tire is inflated but unloaded, its cord structural members are put in tension. The inflation pressure exerts a force acting radially outward, and the tire envelope must contain this force. The cords, now under equal tension, exert a pulling force on the bead wires. The bead wires are in turn put in tension. This enables the tire structure to transmit bending and compression loads, which it could not do when uninflated.

When the tire is loaded, the compressive load stresses appear mostly as a relief of the tension stresses within the deflected contact area of the tire. The vertical components of tension in the tire walls above the ground are relaxed. Excluding the small increment of load supported by the tire structure itself, the total load on the tire exactly equals the reduction in tension within the walls. These reactions are shown in Fig. 6-24.

As indicated in Fig. 6-24, the loaded tire deflects at the ground-tire interface, resulting in a contact zone. The area of this ground contact and the pressures which are created at the ground-tire interface are of critical importance to tire life and performance. The area of the ground contact equals the applied load divided by the sum of the tire pressure and the pressure component exerted by the tire stiffness. This relationship can be expressed as

$$A = \frac{W}{P + P_1}$$

where A = gross contact area of tread-ground contact, sq in.
W = applied load, lb
P = tire inflation pressure, psi
P_1 = tire stiffness component, psi

The relevance of tire load, pressure, and contact area to overall tire function is readily apparent. The load on a tire determines the percent of deflection at a particular pressure. These factors, with appropriate allowance for the tire stiffness component, determine the contact area. Obviously, holding inflation pressure constant, the higher the load on a tire the greater the amount of deflection. A greater amount of deflection increases the contact area and at the same time increases the work-load cycle of the tire (Fig. 6-25). This is of crucial importance because the hysteresis varies with the rates and magnitudes of the stress reversals which occur when a moving tire deflects under load. The greater the energy loss, the higher the temperature increase. Stress relaxation, heat generation, and other factors are manifestations of basic molecular processes which occur when a rubber structure adapts to various stresses.

Briefly, a small tire can carry a greater load by an increase in pressure. However, increasing inflation pressure beyond the recommended amount increases the carcass stresses. It also increases the average contact pressure on the tread and the side thrust on the rim flange. The accepted industry concept is, instead, to increase tire

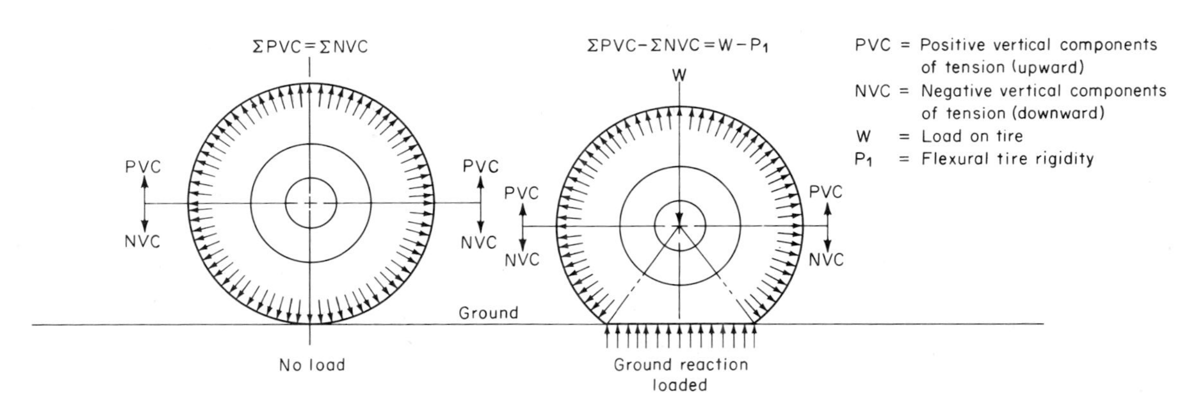

Fig. 6-24 Tire load reactions. The sum of the positive vertical components (above the axle) minus the sum of the negative vertical components (below the axle) is equal to the tire load minus the increment of flexural tire rigidity.

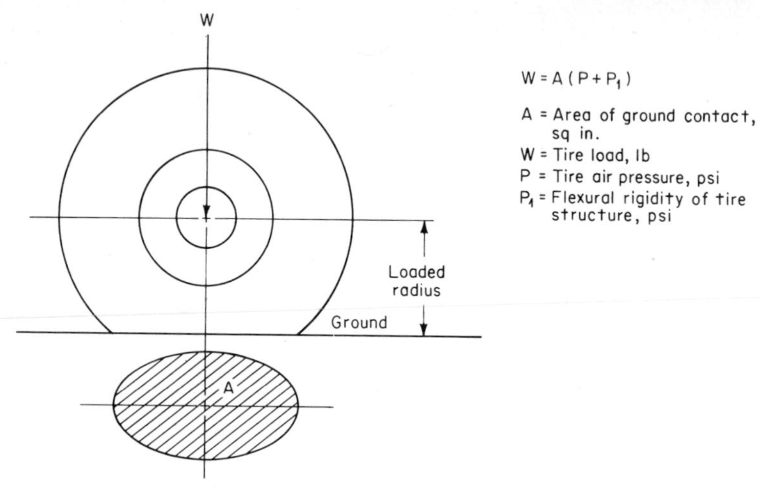

Fig. 6-25 Load, pressure, ground contact.

size when loads are increased. This results in a larger contact area while maintaining a moderate tire inflation pressure, and both tread and carcass stresses are kept within acceptable limits.

The load and rating tables shown in the tire selection portions of this section should be as rigorously followed as structural-steel tables or any other authoritative tables of strength properties.

MECHANICAL FUNCTIONS OF TIRES

Because of its unique combination of air support and air cushioning properties, the tire used for construction equipment performs the following mechanical functions effectively and efficiently:

Support The tire furnishes static and dynamic support while allowing comparatively free motion of the vehicle through rolling. On uneven surfaces the tire uses

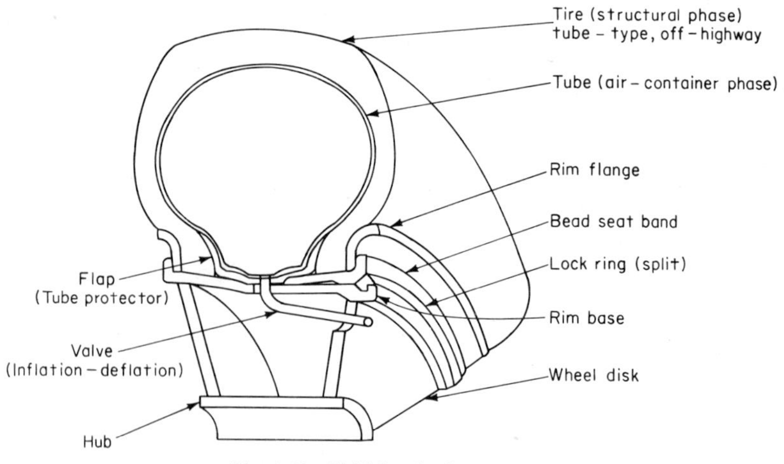

Fig. 6-26 Vehicle wheel assembly.

a minimum of power since it envelops or deflects over many obstacles rather than actually lifting the axle, as required by a rigid wheel. The compressed-air inflation medium furnishes fast, springlike recovery, thus avoiding energy loss due to "impact feedbacks."

Propulsion The rolling radius of the tire acts as a large, flexible, and continuous lever to furnish effective thrust as a moment about the axle. The tire both provides and resists massive work reactions, as required by the machine's engine and drive capabilities.

Traction By virtue of its ground-adhesion capabilities, the tire develops substantial traction, steering, and braking reactions. It also supplies lateral and directional stability. The tire has the ability to exert side thrust when the direction of travel is changed, and it thus does the work of changing the direction of a vehicle. The wide rubber tread of the construction-type tire is advantageous; studies indicate that the total slipping resistance of a tire is equal to the basic frictional resistance between the tread and the ground surface plus a mechanical interlocking between these two surfaces.

Load Distribution The tire distributes the weight of the vehicle over a substantial area of contact. This avoids excessive strains on the wheel and on the ground. Where mobility over weak surfaces is required, ground-contact pressures can be kept low. A tire can develop a wide variation in contact areas as it penetrates into a particular ground medium.

Absorbent Spring The tire acts as a linear spring to cushion shocks to the vehicle and to absorb impact loads. Unfortunately, the tire is not an effective damping device. Eliminating the adverse effects of bounce magnitude and frequency remains a function of the vehicle suspension system.

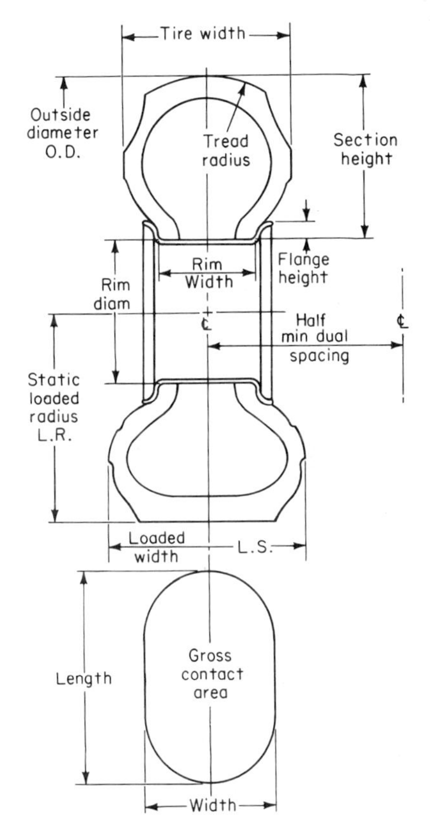

Fig. 6-27 Tire dimensions.

TIRE FORM, STRUCTURE, AND COMPONENTS

These are parts of the vehicle wheel-assembly system. The omega tire shape (Ω) is enclosed at its bottom by the rim section, which actually is a closure for and a part of the air chamber. It is essential from both structural and wear considerations that the recommended rim be used. Figure 6-26 is a simplified diagram of a tube-type assembly which is attached at the hub. The tire and rim assembly can also be demountable at the spoke ends of the wheel or, in the case of planetary wheel assemblies, the rim base can be part of the gear case. Figure 6-27 indicates the major dimensions of a tire and rim assembly.

The major and minor components of a tubeless tire are illustrated in Fig. 6-28. The basic carcass or body shown in this figure is a bias-ply or conventional type. Figure 6-29 diagrams the fundamental differences in carcass construction between the conventional-ply and radial-ply tire.

ON-HIGHWAY TIRES

The following paragraphs will discuss tires intended primarily for highway service, with GVW in the 4,000- to 80,000-lb range. Vehicles under 4,000 lb are usually equipped with passenger-type tires. Proper attention and adherence to the tire-selection data which follow have important implications for tire efficiency and economy as well as for tire and vehicle safety.

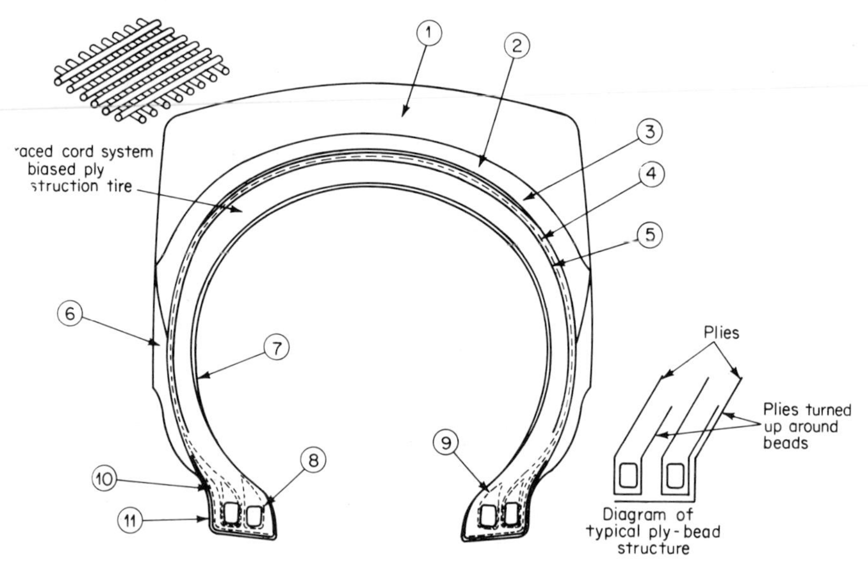

Code	Name	Function	Usual material
1	Tread	Furnishes traction and braking properties—protects tire body	Rubber compound—good wear qualities
2	Undertread	Tread foundation—aids tread adhesion to tire body. May be reinforced with wire filaments.	Rubber compound—good adhesion properties
3	Breaker	Impact resistance—distributes shock loads over tire body	Coated textile cord fabric; may also be coated steel cable
4	Ply	Tension member—transmits bending forces and contains inflation pressure	Coated textile cord with resilient rubber compound
5	Cushion or squeegee	Shear insulation during bias cord pantographing under tire deflection	Resilient rubber compound
6	Sidewall	Abrasion protection—may be reinforced with wire filaments	Rubber compound with flexibility
7	Innerliner	Air seal	Diffusion-resistant rubber compound
8	Bead	Tension member—secures plies to rim (entire tire region which engages rim often referred to as "tire bead")	Insulated high-carbon steel wire
9	Flipper	Relates beads into ply structure	Coated textile cord or square-woven fabric
10	Fabric chafer	Protects tire bead area in region of rim flange	Coated textile cord or square-woven fabric
11	Rubber chafer	Abrasive-resistant air seal	Semicured rubber compound

Fig. 6-28 Conventional tire components.

Size and Load Rating The main criterion for the selection of a particular tire size and load-range rating is the dynamic loading which the tire must support. This design loading should be based on the highest individual wheel load within the GVW distribution for the equipment units in question. The maximum load per tire for this design condition should not exceed the recommended standards for the proposed inflation pressure, load range, and usage.

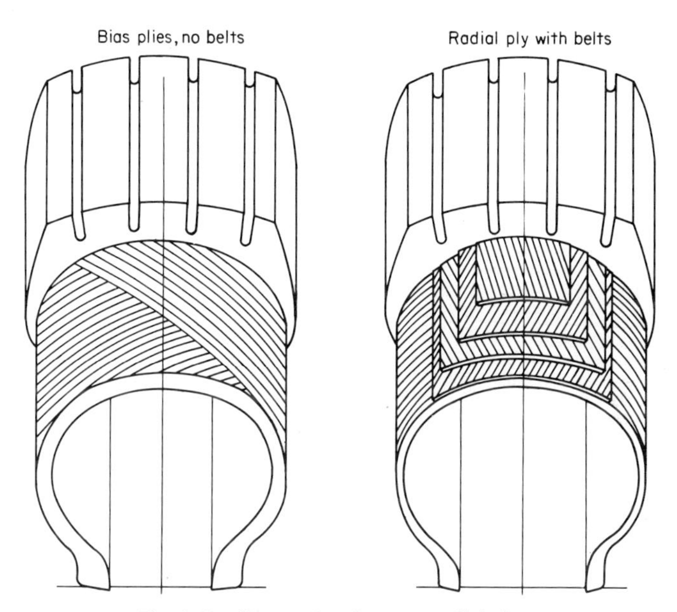

Fig. 6-29 Conventional versus radial tires.

At this point the tire standards established by the Tire and Rim Association, Inc., should be briefly explained, since all of the load tables supplied herein have been formulated by this association. The Tire and Rim Association (T&RA) Standards are technical specifications based on engineering principles and approved practices. The basic standards of this association are recognized in the United States and in practically all countries of the world. (Regional organizations may have some variations.)

The former method of classifying tire strength by "ply rating" has largely been replaced by a "load range" designation. Table 6-9 provides a comparison between the two systems.

TABLE 6-9 Conversion of Ply Rating to Load Range Designation

Load range	*Replaces ply rating*
A	2
B	4
C	6
D	8
E	10
F	12
G	14
H	16
J	18
L	20
M	22
N	24

Maximum tire loadings are established by the Tire and Rim Association as functions of vehicle service, tire size, load range, and inflation pressure. The recommendations for on-highway service are listed in Tables 6-10 to 6-19 inclusive; the heading for each table describes the type of vehicle service. In general, the tire-size designation refers to the nominal tire width (not the actual section width) and the

TABLE 6-10 Conventional Tires for Trucks, Buses, Trailers, and Multipurpose Passenger Vehicles Used in Highway Service—Bias and Radial Ply—Tires Used as Duals
(Tire and Rim Association Standard)

Tire identification		Tire load limits at various cold inflation pressures													
Size	Load range	30	35	40	45	50	55	60	65	70	75	80	85	90	95
LT (light truck) tires mounted on 5° tapered-bead seat rims															
6.00-16 LT	C	990	1,080	1,170	**1,255**										
6.50-16 LT	C	1,120	1,225	1,320	**1,420**										
6.70-15 LT	C	1,060	1,170	1,265	**1,355**										
7.00-13 LT	C	880	960	1,035	**1,110**										
7.00-13 LT	D	880	960	1,035	1,110	1,180	1,250	**1,315**							
7.00-14 LT	C	910	995	1,075	**1,155**										
7.00-14 LT	D	910	995	1,075	1,155	1,225	1,300	**1,365**							
7.00-14 LT	E	910	995	1,075	1,155	1,225	1,300	1,365	1,430	1,495	**1,555**				
7.00-15 LT	C	1,190	1,310	1,420	**1,520**										
7.00-15 LT	D	1,190	1,310	1,420	1,520	1,620	1,715	**1,800**							
7.00-16 LT	C	1,260	1,365	1,475	**1,580**										
7.00-16 LT	D	1,260	1,365	1,475	1,580	1,685	1,780	**1,870**							
7.10-15 LT	C	1,160	1,270	1,370	**1,470**										
7.50-15 LT	D	1,370	1,500	1,620	1,735	1,850	1,965	**2,060**							
7.50-15 LT	E	1,370	1,500	1,620	1,735	1,850	1,965	2,060	2,160	2,250	**2,350**				
7.50-16 LT	C	1,430	1,565	1,690	**1,815**										
7.50-16 LT	D	1,430	1,565	1,690	1,815	1,930	2,040	**2,140**							
7.50-16 LT	E	1,430	1,565	1,690	1,815	1,930	2,040	2,140	2,245	2,345	**2,440**				
8.25-16 LT	D	1,740	1,900	2,050	2,200	**2,340**									
8.25-16 LT	E	1,740	1,900	2,050	2,200	2,340	2,480	**2,600**							
9.00-16 LT	D	1,980	2,165	2,340	2,505	**2,665**									
9.00-16 LT	E	1,980	2,165	2,340	2,505	2,665	2,820	**2,965**							
Tires mounted on 15° tapered-bead seat drop-center rims															
7-14.5	D	1,035	1,125	1,220	1,305	1,390	1,470	**1,550**							
7-14.5	E	1,035	1,125	1,220	1,305	1,390	1,470	1,550	1,625	1,695	1,755	**1,830**			
7-14.5	F	1,035	1,125	1,220	1,305	1,390	1,470	1,550	1,625	1,695	1,755	1,830	1,895	1,960	**2,025**
8-14.5	E	1,190	1,300	1,405	1,505	1,595	1,690	1,780	1,865	1,950	2,030	**2,110**			
8-14.5	F	1,190	1,300	1,405	1,505	1,595	1,690	1,780	1,865	1,950	2,030	2,110	2,180	2,255	**2,330**
7-17.5	C	1,260	1,375	1,485	**1,590**										
7-17.5	D	1,260	1,375	1,485	1,590	1,690	1,790	**1,880**							
8-17.5	C	1,445	1,575	1,700	**1,820**										
8-17.5	D	1,445	1,575	1,700	1,820	1,935	2,050	**2,155**							

NOTE 1: Restrictions listed in text apply to this table.
NOTE 2: Sizes 7.00–15 LT through 7.10–15 LT are furnished in both tubeless and tube types; sizes 7.50–15 LT through 9.00–16 LT are tube type; all other sizes are tubeless.

nominal rim diameters. The recommendations supplied in Tables 6-10 to 6-17 inclusive are subject to the following restrictions:

1. Selection of the tire size shall be based on meeting the requirements of the maximum tire load.
2. Maximum tire load must not exceed the limits indicated by the boldface values in the tables, except that these values may be increased by 9 percent (by 7 percent for tires described in Tables 6-16 and 6-17) when the tire is to be used for intermittent highway operation on a vehicle which is designed for a maximum speed of 50 mph.

3. Minimum recommended cold inflation pressures for various loads must conform to those supplied in the tables except as noted herein:
   a. Cold inflation pressures for sustained driving over 60 mph (over 65 mph for tires described in Tables 6-10 and 6-11) must be increased by 10 psi above those specified in the tables for the loads being carried, subject to a maximum of 100 psi.

TABLE 6-11 Conventional Tires for Trucks, Buses, Trailers, and Multipurpose Passenger Vehicles Used in Highway Service—Bias and Radial Ply—Tires Used as Singles
(Tire and Rim Association Standard)

Tire identification		Tire load limits at various cold inflation pressures													
Size	Load range	30	35	40	45	50	55	60	65	70	75	80	85	90	95
LT (light truck) tires mounted on 5° tapered-bead seat rims															
6.00-16 LT	C	1,130	1,230	1,330	**1,430**										
6.50-16 LT	C	1,270	1,390	1,500	**1,610**										
6.70-15 LT	C	1,210	1,320	1,430	**1,530**										
7.00-13 LT	C	1,000	1,090	1,170	**1,260**										
7.00-13 LT	D	1,000	1,090	1,170	1,260	1,340	1,420	**1,490**							
7.00-14 LT	C	1,030	1,130	1,220	**1,310**										
7.00-14 LT	D	1,030	1,130	1,220	1,310	1,390	1,470	**1,550**							
7.00-14 LT	E	1,030	1,130	1,220	1,310	1,390	1,470	1,550	1,620	1,700	**1,770**				
7.00-15 LT	C	1,350	1,480	1,610	**1,720**										
7.00-15 LT	D	1,350	1,480	1,610	1,720	1,830	1,904	**2,040**							
7.00-16 LT	C	1,430	1,560	1,680	**1,800**										
7.00-16 LT	D	1,430	1,560	1,680	1,800	1,910	2,030	**2,130**							
7.10-15 LT	C	1,320	1,440	1,560	**1,670**										
7.50-15 LT	D	1,560	1,710	1,840	1,980	2,100	2,220	**2,330**							
7.50-15 LT	E	1,560	1,710	1,840	1,980	2,100	2,220	2,330	2,450	2,560	**2,660**				
7.50-16 LT	C	1,620	1,770	1,930	**2,060**										
7.50-16 LT	D	1,620	1,770	1,930	2,060	2,190	2,310	**2,440**							
7.50-16 LT	E	1,620	1,770	1,930	2,060	2,190	2,310	2,440	2,560	2,670	**2,780**				
8.25-16 LT	D	1,980	2,160	2,330	2,500	**2,660**									
8.25-16 LT	E	1,980	2,160	2,330	2,500	2,660	2,820	**2,960**							
9.00-16 LT	D	2,250	2,460	2,660	2,850	**3,030**									
9.00-16 LT	E	2,250	2,460	2,660	2,850	3,030	3,210	**3,370**							
Tires mounted on 15° tapered-bead seat drop-center rims															
7-14.5	D	1,175	1,280	1,385	1,485	1,580	1,670	**1,760**							
7-14.5	E	1,175	1,280	1,385	1,485	1,580	1,670	1,760	1,845	1,925	2,005	**2,080**			
7-14.5	F	1,175	1,280	1,385	1,485	1,580	1,670	1,760	1,845	1,925	2,005	2,080	2,155	2,230	**2,300**
8-14.5	E	1,350	1,475	1,595	1,710	1,815	1,920	2,020	2,120	2,215	2,305	**2,395**			
8-14.5	F	1,350	1,475	1,595	1,710	1,815	1,920	2,020	2,120	2,215	2,305	2,395	2,485	2,565	**2,645**
7-17.5	C	1,430	1,565	1,695	**1,815**										
7-17.5	D	1,430	1,565	1,695	1,815	1,925	2,040	**2,145**							
8-17.5	C	1,640	1,790	1,940	**2,075**										
8-17.5	D	1,640	1,790	1,940	2,075	2,205	2,335	**2,455**							

NOTE 1: Restrictions listed in text apply to this table.
NOTE 2: Sizes 7.00–15 LT through 7.10–15 LT are furnished in both tubeless and tube types; sizes 7.50–15 LT through 9.00–16 LT are tube type; all other sizes are tubeless.

   b. Cold inflation pressures for special operating conditions may be increased up to 10 psi above those indicated in the tables, with no increase in loads.
   c. The cold inflation pressures shall not be increased by more than 10 psi above those indicated in the tables for the loads which they support, and shall not exceed 100 psi.
4. Manufacturers should be consulted to determine the availability and engineering specifications of rims and wheels for the applications identified in these tables.

Table 6-18 applies to dual tires used in 20-mph maximum-speed service on low-platform trailers and straddle trucks. Table 6-19 applies to single tires used in similar applications.

Tube Type versus Tubeless Type On-highway tires are available in either tube-type or tubeless construction. The tire size designation will usually indicate whether the tire is tube type or tubeless and, in any case, all tubeless tires are so marked by the

TABLE 6-12 Conventional Tube-type Tires for Trucks, Buses, and Trailers Used in Highway Service—Bias and Radial Ply—Tires Used as Duals
(Tire and Rim Association Standard)

Tire identification			Tire load limits at various cold inflation pressures									
Size	Load range	40	45	50	55	60	65	70	75	80	85	90
6.50-20	C	1,630	1,750	**1,860**								
6.50-20	D	1,630	1,750	1,860	1,970	2,070	**2,170**					
7.00-17	C	1,640	1,760	**1,870**								
7.00-17	D	1,640	1,760	1,870	1,980	2,080	**2,180**					
7.00-18	D	1,710	1,830	1,950	2,060	2,170	**2,270**					
7.00-20	D	1,840	1,980	2,100	2,220	2,340	**2,450**					
7.00-20	E	1,840	1,980	2,100	2,220	2,340	2,450	2,560	2,660	**2,760**		
7.50-15	E	1,700	1,820	1,940	2,050	2,160	2,260	2,360	2,460	**2,550**		
7.50-17	D	1,850	1,990	2,110	2,230	2,350	**2,460**					
7.50-17	E	1,850	1,990	2,110	2,230	2,350	2,460	2,570	2,680	**2,780**		
7.50-18	D	1,920	2,060	2,190	2,310	2,430	**2,550**					
7.50-20	D	2,070	2,220	2,350	2,490	2,620	**2,750**					
7.50-20	E	2,070	2,220	2,350	2,490	2,620	2,750	2,870	2,990	**3,100**		
8.25-15	F	2,030	2,170	2,310	2,440	2,570	2,700	2,810	2,930	3,040	3,150	**3,260**
8.25-17	E	2,200	2,360	2,510	2,650	2,790	2,930	3,060	**3,180**			
8.25-18	E	2,280	2,450	2,600	2,750	2,900	3,040	3,170	**3,300**			
8.25-20	E	2,460	2,640	2,800	2,960	3,120	3,270	3,410	**3,550**			
8.25-20	F	2,460	2,640	2,800	2,960	3,120	3,270	3,410	3,550	3,690	3,820	**3,950**
9.00-15	F	....	2,590	2,760	2,920	3,070	3,210	3,360	3,490	3,630	**3,760**	
9.00-18	E	....	2,910	3,100	3,270	3,440	3,610	**3,770**				
9.00-20	E	....	3,120	3,310	3,510	3,690	3,870	**4,040**				
9.00-20	F	....	3,120	3,310	3,510	3,690	3,870	4,040	4,200	4,360	**4,520**	
10.00-15	F	....	....	3,140	3,320	3,490	3,660	3,830	**3,980**			
10.00-15	G	....	....	3,140	3,320	3,490	3,660	3,830	3,980	4,130	4,280	**4,430**
10.00-18	F	....	....	3,510	3,710	3,910	4,090	4,280	**4,450**			
10.00-20	F	....	....	3,760	3,970	4,180	4,380	4,580	**4,760**			
10.00-20	G	....	....	3,760	3,970	4,180	4,380	4,580	4,760	4,950	5,120	**5,300**
10.00-22	F	....	....	4,000	4,230	4,450	4,660	4,870	**5,070**			
10.00-22	G	....	....	4,000	4,230	4,450	4,660	4,870	5,070	5,260	5,450	**5,640**
10.00-24	F	....	....	4,240	4,490	4,720	4,950	5,170	**5,380**			
11.00-20	F	....	....	4,100	4,330	4,560	4,780	4,990	**5,190**			
11.00-20	G	....	....	4,100	4,330	4,560	4,780	4,990	5,190	5,390	5,590	**5,780**
11.00-22	F	....	....	4,350	4,600	4,840	5,080	5,300	**5,520**			
11.00-22	G	....	....	4,350	4,600	4,840	5,080	5,300	5,520	5,730	5,940	**6,140**
11.00-24	F	....	....	4,620	4,890	5,140	5,390	5,630	**5,860**			
12.00-20	G	....	....	....	4,930	5,190	5,440	5,680	5,910	**6,140**		
12.00-22	G	....	....	....	5,240	5,510	5,780	6,030	6,280	**6,520**		
12.00-24	G	....	....	....	5,550	5,840	6,120	6,390	6,650	**6,910**		

NOTE: Restrictions listed in text apply to this table.

manufacturer. The wide-base tires shown in Tables 6-16 and 6-17 may replace duals for a particular operation or service, and Table 6-20 gives equivalent single-tire replacements for some of the common dual sizes. The load range of the single wide-base tire should equal the combined load range of the two conventional tires, whether tube type or tubeless.

A summary of these tables and illustrations shows that on-highway tires are resolved into three major form divisions. Referring to Table 6-20, for example, a 10.00-20 dual tube-type assembly is similar to a 11-22.5 dual tubeless assembly, and either of these arrangements may be replaced by a single wide-base tire—the 18-19.5. Each of these three major forms has its characteristic advantages. The 10.00-20 is the standard form and has the advantage of greater interchangeability—more sources of supply and greater choice of tread designs. The 11-22.5, being tubeless, requires no tube or flap. The single wide-base tire provides more flotation for an

TABLE 6-13 Conventional Tube-type Tires for Trucks, Buses, and Trailers Used in Highway Service—Bias and Radial Ply—Tires Used as Singles
(Tire and Rim Association Standard)

Tire identification		Tire load limits at various cold inflation pressures										
Size	Load range	50	55	60	65	70	75	80	85	90	95	100
6.50-20	C	1,860	2,000	**2,120**								
6.50-20*	D	1,860	2,000	2,120	2,250	2,360	**2,470**					
7.00-17	C	1,870	2,010	**2,130**								
7.00-17*	D	1,870	2,010	2,130	2,260	2,370	**2,490**					
7.00-18	D	1,950	2,090	2,220	2,350	2,470	**2,590**					
7.00-20	D	2,100	2,260	2,390	2,530	2,670	**2,790**					
7.00-20*	E	2,100	2,260	2,390	2,530	2,670	2,790	2,920	3,030	**3,150**		
7.50-15*	E	1,940	2,070	2,210	2,340	2,460	2,580	2,690	2,800	**2,910**		
7.50-17	D	2,110	2,270	2,410	2,540	2,680	**2,800**					
7.50-17*	E	2,110	2,270	2,410	2,540	2,680	2,800	2,930	3,060	**3,170**		
7.50-18	D	2,190	2,350	2,500	2,630	2,770	**2,910**					
7.50-20	D	2,360	2,530	2,680	2,840	2,990	**3,140**					
7.50-20*	E	2,360	2,530	2,680	2,840	2,990	3,140	3,270	3,410	**3,530**		
8.25-15*	F	2,310	2,470	2,630	2,780	2,930	3,080	3,200	3,340	3,470	3,590	**3,720**
8.25-17	E	2,510	2,690	2,860	3,020	3,180	3,340	3,490	**3,630**			
8.25-18	E	2,600	2,790	2,960	3,140	3,310	3,470	3,610	**3,760**			
8.25-20	E	2,800	3,010	3,190	3,370	3,560	3,730	3,890	**4,050**			
8.25-20*	F	2,800	3,010	3,190	3,370	3,560	3,730	3,890	4,050	4,210	4,350	**4,500**
9.00-15*	F	....	2,950	3,150	3,330	3,500	3,660	3,830	3,980	4,140	**4,290**	
9.00-18	E	....	3,320	3,530	3,730	3,920	4,120	**4,300**				
9.00-20	E	....	3,560	3,770	4,000	4,210	4,410	**4,610**				
9.00-20*	F	....	3,560	3,770	4,000	4,210	4,410	4,610	4,790	4,970	**5,150**	
10.00-15	F	....	....	3,580	3,780	3,980	4,170	4,370	**4,540**			
10.00-15*	G	....	....	3,580	3,780	3,980	4,170	4,370	4,540	4,710	4,880	**5,050**
10.00-18	F	....	....	4,000	4,230	4,460	4,660	4,880	**5,070**			
10.00-20	F	....	....	4,290	4,530	4,770	4,990	5,220	**5,430**			
10.00-20*	G	....	....	4,290	4,530	4,770	4,990	5,220	5,430	5,640	5,840	**6,040**
10.00-22	F	....	....	4,560	4,820	5,070	5,310	5,550	**5,780**			
10.00-22*	G	....	....	4,560	4,820	5,070	5,310	5,550	5,780	6,000	6,210	**6,430**
10.00-24	F	....	....	4,830	5,120	5,380	5,640	5,890	**6,130**			
11.00-20	F	....	....	4,670	4,940	5,200	5,450	5,690	**5,920**			
11.00-20*	G	....	....	4,670	4,940	5,200	5,450	5,690	5,920	6,140	6,370	**6,590**
11.00-22	F	....	....	4,960	5,240	5,520	5,790	6,040	**6,290**			
11.00-22*	G	....	....	4,960	5,240	5,520	5,790	6,040	6,290	6,530	6,770	**7,000**
11.00-24	F	....	....	5,270	5,570	5,860	6,140	6,420	**6,680**			
12.00-20	G	....	....	....	5,620	5,920	6,200	6,480	6,740	**7,000**		
12.00-22	G	....	....	....	5,970	6,280	6,590	6,870	7,160	**7,430**		
12.00-24	G	....	....	....	6,330	6,660	6,980	7,280	7,580	**7,880**		

NOTE 1: Restrictions listed in text apply to this table.
NOTE 2: The asterisk indicates that the widest specified rim for this tire must be used if clearances permit; otherwise use the measuring rim. If the measuring rim is used, determine its adequacy for the intended service by consulting the rim and wheel manufacturer.

on-highway truck that is required to make off-highway deliveries. The single assembly also reduces the inventories of such components as spacer bands, rims, and disk wheels. The three major forms then consist of the standard tube-type "round" tire—example, the 10.00-20; the tubeless version of the "round" tire—example, the 11-22.5; and the wide-base tubeless single which can replace duals—example, the 18-19.5.

Rim Types Rims are separate but become integrated with the tire once the tire is mounted and inflated. Rims are designed to withstand the air pressure of the tire, the

TABLE 6-14 Conventional Tubeless Tires for Trucks, Buses, and Trailers Used in Highway Service—Bias and Radial Ply—Tires Used as Duals
(Tire and Rim Association Standard)

Tire identification		Tire load limits at various cold inflation pressures										
Size	Load range	40	45	50	55	60	65	70	75	80	85	90
		Tires mounted on 15° tapered-bead seat drop-center rims										
7-22.5	C	1,630	1,750	**1,860**								
7-22.5	D	1,630	1,750	1,860	1,970	2,070	**2,170**					
8-19.5	C	1,850	1,990	**2,110**								
8-19.5	D	1,850	1,990	2,110	2,230	2,350	**2,460**					
8-19.5	E	1,850	1,990	2,110	2,230	2,350	2,460	2,570	2,680	**2,780**		
8-22.5	D	2,070	2,220	2,350	2,490	2,620	**2,750**					
8-22.5	E	2,070	2,220	2,350	2,490	2,620	2,750	2,870	2,990	**3,100**		
9-22.5	E	2,460	2,640	2,800	2,960	3,120	3,270	3,410	**3,550**			
9-22.5	F	2,460	2,640	2,800	2,960	3,120	3,270	3,410	3,550	3,690	3,820	**3,950**
10-22.5	E	....	3,120	3,310	3,510	3,690	3,870	**4,040**				
10-22.5	F	....	3,120	3,310	3,510	3,690	3,870	4,040	4.200	4,360	**4,520**	
11-22.5	F	....	....	3,760	3,970	4,180	4,380	4,580	**4,760**			
11-22.5	G	....	....	3,760	3,970	4,180	4,380	4,580	4,760	4,950	5,120	**5,300**
11-24.5	F	....	....	4,000	4,230	4,450	4,660	4,870	**5,070**			
11-24.5	G	....	....	4,000	4,230	4,450	4,660	4,870	5,070	5,260	5,450	**5,640**
12-22.5	F	....	....	4,100	4,330	4,560	4,780	4,990	**5,190**			
12-22.5	G	....	....	4,100	4.330	4,560	4,780	4,990	5,190	5,390	5,590	**5,780**
12-24.5	F	....	....	4,350	4,600	4,840	5,080	5,300	**5,520**			
12-24.5	G	....	....	4,350	4,600	4,840	5,080	5,300	5,520	5,730	5,940	**6,140**

NOTE: Restrictions listed in text apply to this table.

TABLE 6-15 Conventional Tubeless Tires for Trucks, Buses, and Trailers Used in Highway Service—Bias and Radial Ply—Tires Used as Singles
(Tire and Rim Association Standard)

Tire identification		Tire load limits at various cold inflation pressures										
Size	Load range	50	55	60	65	70	75	80	85	90	95	100
		Tires mounted on 15° tapered-bead seat drop-center rims										
7-22.5	C	1,860	2,000	**2,120**								
7-22.5	D	1,860	2,000	2,120	2,250	2,360*	**2,470***					
8-19.5	C	2,110	2,270	**2,410**								
8-19.5	D	2,110	2,270	2,410	2,540	2,680	**2,800**					
8-19.5	E	2,110	2,270	2,410	2,540	2,680	2,800	2,930*	3,060*	**3,170***		
8-22.5	D	2,360	2,530	2,680	2,840	2,990	**3,140**					
8-22.5	E	2,360	2,530	2,680	2,840	2,990	3,140	3,270	3,410*	**3,530***		
9-22.5	E	2,800	3,010	3,190	3,370	3,560	3,730	3,890	**4,050**			
9-22.5	F	2,800	3,010	3,190	3,370	3,560	3,730	3,890	4,050	4,210*	4,350*	**4,500***
10-22.5	E	....	3,560	3,770	4,000	4,210	4,410	**4,610**				
10-22.5	F	....	3,560	3,770	4,000	4,210	4,410	4,610	4,790	4,970	**5,150**	
11-22.5	F	....	....	4,290	4,530	4,770	4,990	5,220	**5,430**			
11-22.5	G	....	....	4,290	4,530	4,770	4,990	5,220	5,430	5,640	5,840	**6,040**
11-24.5	F	....	....	4,560	4,820	5,070	5,310	5,550	**5,780**			
11-24.5	G	....	....	4,560	4,820	5,070	5,310	5,550	5,780	6,000	6,210	**6,430**
12-22.5	F	....	....	4,670	4,940	5,200	5,450	5,690	**5,920**			
12-22.5	G	....	....	4,670	4,940	5,200	5,450	5,690	5,920	6,140	6,370	**6,590**
12-24.5	F	....	....	4,960	5,240	5,520	5,790	6,040	**6,290**			
12-24.5	G	....	....	4,960	5,240	5,520	5,790	6,040	6,290	6,530	6,770	**7,000**

NOTE 1: Restrictions listed in text apply to this table.
NOTE 2: The asterisk indicates that the widest specified rim for this tire must be used if clearances permit; otherwise use the measuring rim. If the measuring rim is used, determine its adequacy for the intended service by consulting the rim and wheel manufacturer.

TABLE 6-16 Wide-base Tubeless Tires for Trucks, Buses, Trailers, and Multipurpose Passenger Vehicles Used in Highway Service—Bias and Radial Ply—Tires Used as Singles
(Tire and Rim Association Standard)

Tire identification		Tire load limits at various cold inflation pressures									
Size	Load range	30	35	40	45	50	55	60	65	70	75
8.00-16.5	B	**1,360**									
8.00-16.5	C	1,360	1,490	1,610	**1,730**						
8.00-16.5	D	1,360	1,490	1,610	1,730	1,840	1,945	**2,045**			
8.00-16.5*	E	1,360	1,490	1,610	1,730	1,840	1,945	2,045	2,145	2,240	**2,330**
8.75-16.5	B	**1,570**									
8.75-16.5	C	1,570	1,720	1,850	**1,990**						
8.75-16.5	D	1,570	1,720	1,850	1,990	2,110	2,240	**2,350**			
8.75-16.5*	E	1,570	1,720	1,850	1,990	2,110	2,240	2,350	2,470	2,570	**2,680**
9.50-16.5	B	**1,860**									
9.50-16.5	C	1,860	2,030	2,190	**2,350**						
9.50-16.5	D	1,860	2,030	2,190	2,350	2,500	2,650	**2,780**			
9.50-16.5*	E	1,860	2,030	2,190	2,350	2,500	2,650	2,780	2,920	3,050	**3,170**
10-16.5	C	1,840	2,010	2,170	**2,330**						
10-16.5	D	1,840	2,010	2,170	2,330	2,480	2,620	**2,750**			
10-17.5	C	1,910	2,095	2,265	**2,425**						
10-17.5	D	1,910	2,095	2,265	2,425	2,580	2,730	**2,870**			
10-17.5*	E	1,910	2,095	2,265	2,425	2,580	2,730	2,870	3,010	3,140	**3,270**
12-16.5	D	2,370	2,590	2,800	**3,000**						
12-16.5*	E	2,370	2,590	2,800	3,000	3,190	3,370	**3,550**			
14-17.5	D	3,210	3,500	3,790	**4,060**						
14-17.5*	E	3,210	3,500	3,790	4,060	4,320	4,570	**4,800**			

NOTE 1: Restrictions listed in text apply to this table.
NOTE 2: The asterisk indicates that the widest specified rim for this tire must be used if clearances permit; otherwise use the measuring rim. If the measuring rim is used, determine its adequacy for the intended service by consulting the rim and wheel manufacturer.

TABLE 6-17 Wide-base Tubeless Tires for Trucks, Buses, and Trailers Used in Highway Service—Bias and Radial Ply—Tires Used as Singles
(Tire and Rim Association Standard)

Tire identification		Tire load limits at various cold inflation pressures							
Size	Load range	45	50	55	60	65	70	75	80
14-19.5	F	4,350	4,620	4,890	5,140	5,390	**5,630**		
15-19.5	G	5,140	5,470	5,790	6,090	6,380	6,660	**6,930**	
15-22.5	G	5,640	6,000	6,340	6,670	6,990	7,300	**7,600**	
16.5-19.5	H	6,030	6,410	6,780	7,130	7,480	7,810	8,130	**8,440**
16.5-22.5	H	6,590	7,010	7,410	7,790	8,170	8,540	8,890	**9,230**
18-19.5	H	6,700	7,130	7,540	7,930	8,310	8,680	**9,040**	
18-22.5	H	7,310	7,780	8,220	8,650	9,070	9,470	**9,860**	
19.5-19.5	J	7,900	8,400	8,890	9,350	9,800	10,240	**10,650**	

NOTE: Restrictions listed in text apply to this table.

bending stresses due to radial loads, and the side thrusts due to moments about the axle which result from cornering.

The truck tires used for light vehicles are mounted on drop-center or semi-drop-center rims, Fig. 6-30. In some cases the 5° types are interchangeable for the same tire. The drop-center rim includes a well which enables one side of the tire bead to be gradually buttonholed over the rim flange. The depth of the well and the height

TABLE 6-18 Tires for Trucks and Trailers Used in Restricted-speed Highway Service—Bias and Radial Ply—Tires Used as Duals
(Tire and Rim Association Standard)

Tire size designation	Tire load limits at various cold inflation pressures										
	50	55	60	65	70	75	80	85	90	95	100
	Tires mounted on Type I, II, and III rims										
7.00-15	2,270	2,400	2,520	2,640	2,760	2,870	2,990	3,090	3,200	**3,300**(F)	
7.00-20	2,770	2,920	3,080	3,220	3,370	3,510	**3,640**(E)				
7.50-15	2,550	2,700	2,840	2,980	3,110	3,240	**3,360**(E)	3,480	3,600	**3,710**(F)	
7.50-18	2,880	3,050	3,200	**3,360**(D)	3,510	3,650	3,790	3,930	4,060	4,190	**4,320**(G)
7.50-20	3,110	3,290	3,460	**3,630**(D)	3,790	3,950	**4,100**(E)	4,250	4,390	**4,530**(F)	
8.25-15	3,050	3,220	3,390	3,550	3,710	3,860	4,010	4,160	**4,300**(F)	4,430	**4,570**(G)
8.25-20	3,690	3,910	4,110	4,310	4,500	**4,680**(E)	4,860	5,040	**5,210**(F)		
9.00-15	3,630	3,840	4,040	4,240	4,430	4,610	4,780	**4,960**(F)	5,130	5,290	**5,450**(G)
9.00-20	4,370	4,620	4,860	5,100	5,320	5,540	5,750	**5,960**(F)			
10.00-15	4,140	4,380	4,610	4,830	5,040	**5,250**(F)	5,450	5,650	**5,840**(G)		
10.00-20	4,950	5,230	5,500	5,770	6,030	**6,270**(F)	6,510	6,750	**6,980**(G)		
11.00-15	4,530	4,790	5,040	5,280	5,510	5,740	5,960	6,170	6,390	6,590	**6,790**(H)
11.00-20	5,390	5,710	6,000	6,290	6,570	**6,840**(F)	7,100	7,360	**7,610**(G)		
12.00-20	6,140	6,500	6,830	7,160	7,480	7,790	**8,090**(G)	8,380	8,670	**8,940**(H)	
13.00-20	7,150	7,560	7,950	8,340	8,710	9,060	9,410	**9,750**(H)			
14.00-20	8,450	8,930	9,400	9,850	10,290	10,710	11,120	**11,520**(J)	11,920	12,290	**12,670**(L)
14.00-24	9,430	9,980	10,500	11,000	11,490	11,960	12,420	**12,870**(J)			
	Tires mounted on 15° drop-center rims										
7–14.5	1,780	1,880	1,980	2,080	2,170	2,260	2,340	**2,430**(E)	2,510	2,590	**2,670**(F)
8–14.5	2,150	2,280	2,400	2,510	2,620	2,730	2,830	**2,940**(E)	3,040	3,130	**3,230**(F)
9–14.5	2,490	2,640	2,770	2,910	3,040	3,160	3,280	**3,400**(E)	3,520	3,630	**3,740**(F)

NOTE 1: Maximum speed is 20 mph.
NOTE 2: Letters in parentheses denote load range for which boldfaced loads and inflations are maximum.
NOTE 3: Sizes mounted on 15° drop-center rims are tubeless; all others are tube type.

TABLE 6-19 Tires for Trucks and Trailers Used in Restricted-speed Highway Service—Bias and Radial Ply—Tires Used as Singles
(Tire and Rim Association)

Tire size designation	Tire load limits at various cold inflation pressures								
	60	65	70	75	80	85	90	95	100
	Tires mounted on Type I, II, and III rims								
7.00-15	2,590	2,740	2,870	3,010	3,150	3,270	3,410	3,520	**3,650**(F)
7.00-20	3,160	3,330	3,510	3,670	3,840	4,000	**4,150**(E)		
7.50-15	2,910	3,080	3,240	3,400	3,550	3,690	**3,830**(E)	3,970	**4,100**(F)
7.50-18	3,280	3,480	3,650	**3,830**(D)	4,000	4,160	4,320	4,480	**4,630**(G)
7.50-20	3,550	3,750	3,940	**4,140**(D)	4,320	4,500	**4,670**(E)	4,850	**5,000**(F)
8.25-15	3,480	3,670	3,860	4,050	4,230	4,400	4,570	4,740	**4,900**(F,G)
8.25-20	4,210	4,460	4,690	4,910	5,130	**5,340**(E)	5,540	5,750	**5,940**(F)
9.00-15	4,140	4,380	4,610	4,830	5,050	5,260	5,450	**5,650**(F)	**5,850**(G)
9.00-20	4,980	5,270	5,540	5,810	6,060	6,320	6,560	**6,790**(F)	
10.00-15	4,720	4,990	5,260	5,510	5,750	**5,990**(F)	6,210	6,440	**6,660**(G)
10.00-20	5,640	5,960	6,270	6,580	6,870	**7,150**(F)	7,420	7,700	**7,960**(G)
11.00-15	5,160	5,460	5,750	6,020	6,280	6,540	6,790	7,030	**7,280**(H)
11.00-20	6,140	6,510	6,840	7,170	7,490	**7,800**(F)	8,090	8,390	**8,680**(G)
12.00-20	7,000	7,410	7,790	8,160	8,530	8,880	**9,220**(G)	9,550	**9,880**(H)
13.00-20	8,150	8,620	9,060	9,510	9,930	10,330	10,730	**11,120**(H)	
14.00-20	9,630	10,180	10,720	11,230	11,730	12,210	12,680	**13,130**(J)	**13,590**(L)
14.00-24	10,750	11,380	11,970	12,540	13,100	13,630	14.160	**14,670**(J)	
	Tires mounted on 15° drop-center rims								
7-14.5	2,030	2,140	2,260	2,370	2,470	2,580	2,670	**2,770**(E)	**2,860**(F)
8-14.5	2,450	2,600	2,740	2,860	2,990	3,110	3,230	**3,350**(E)	**3,470**(F)
9-14.5	2,840	3,010	3,160	3,320	3,470	3,600	3,740	**3,880**(E)	**4,010**(F)

NOTE 1: Maximum speed is 20 mph.
NOTE 2: Letters in parentheses denote load range for which boldfaced loads and inflations are maximum.
NOTE 3: Sizes mounted on 15° drop-center rims are tubeless; all others are tube type.

TABLE 6-20 Wide-base Single Replacements for Dual Tires

Wide-base single		Dual-tire replacement	
Size	Load range	Size	Load range
10-16.5	C	6.00-16	C
10-16.5	D	6.50-16	C
10-17.5	E		
10-17.5	F		
12-16.5	D	7.00-15, 7.00-16	C
12-16.5	E	7.00-15	D
12-16.5	E	7.50-16	C
14-17.5	D, E	8-19.5	C
15-19.5	G	8.25-20, 9-22.5	E
15-22.5	G, H		
16.5-19.5	H	9.00-20, 10-22.5	E
16.5-22.5	H	9.00-20	E
18-19.5	H	10.00-20, 11-22.5	F
18-22.5	H	10.00-20	F
18-22.5	J	10.00-22, 11-24.5	F
19.5-19.5	J	11.00-22, 12-22.5	F

of the rim flange must be carefully designed to make this possible. The semi-drop-center rim is recommended for 5° bead tires in the higher load ranges. This rim can use a shallow well and still permit tire mounting, since it features a split side flange which is removable. As shown on Fig. 6-30, drop-center rims are of one-piece construction.

The tires listed at the bottoms of Tables 6-10 and 6-11 are mounted on the drop-center rims with 15° bead seats which are used for tubeless tires. The tires listed in Tables 6-14 through 6-16, as well as the last four sizes in Tables 6-18 and 6-19, are also mounted on rims of this type.

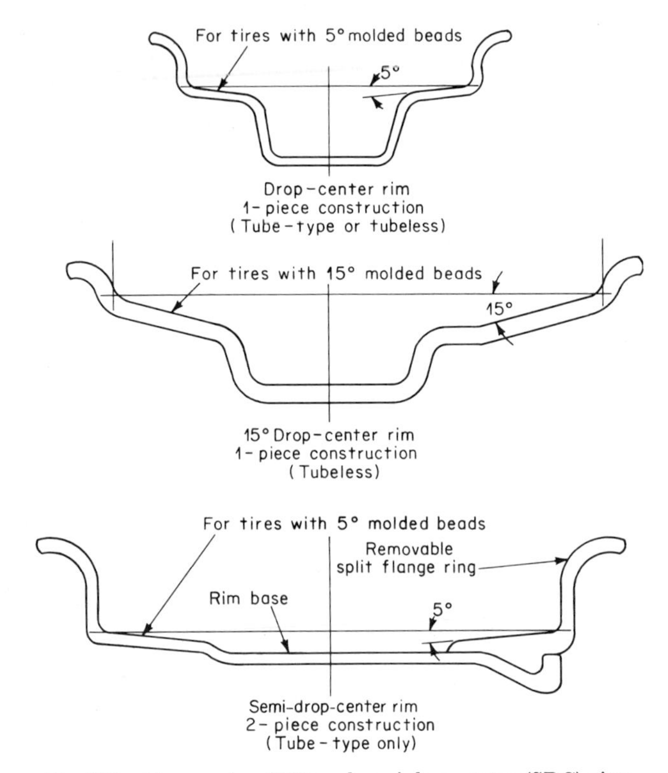

Fig. 6-30 Drop-center (DC) and semi-drop-center (SDC) rims.

The tube-type tires listed in Tables 6-12 and 6-13 are commonly found in highway service. They are generally mounted on rims designed for flat-bead tires. These rim types are shown on Fig. 6-31. The other rim types as shown on Fig. 6-31 are of two- or three-piece construction, with either the side flange or lock ring split, depending on the number of components.

Since the rim is actually a continuation of the tire form, it is of prime importance that the recommended (or optional) rim be used. Each tire is designed for a predetermined radial- and lateral-compression fit on its respective rim, which must match with the tire size. Comprehensive rim data may be found in the *Year Book* of the Tire and Rim Association and the publications of tire manufacturers.

Tread Types The most important factors affecting on-highway tire performance are tire load, size, pressure, and applicable load range. The various types of tread designs in no way change any of the factors mentioned—but they do affect wear and carcass penetration.

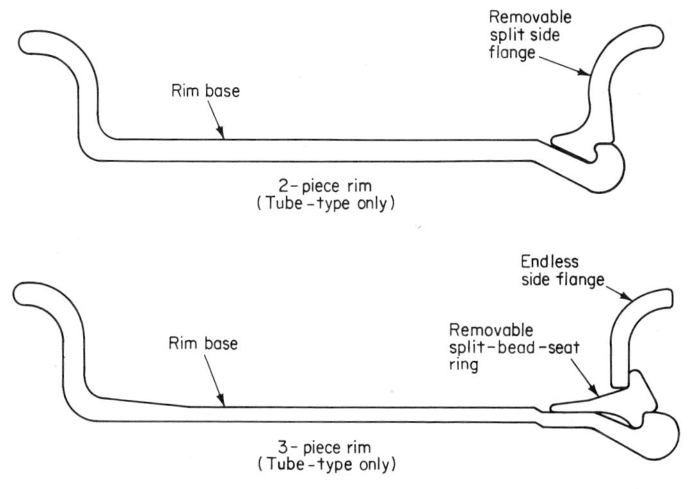

Fig. 6-31 Rim types for flatbead tires.

Some 650 various tread designs are available today for any given size of standard truck tire. These designs are based on six fundamental types, illustrated in Fig. 6-32. In addition to basic elements of groove and ribs, many types and designs with thin grooves or "cuts" are used. The depth of the tread design generally matches up with the type of service and the tire size.

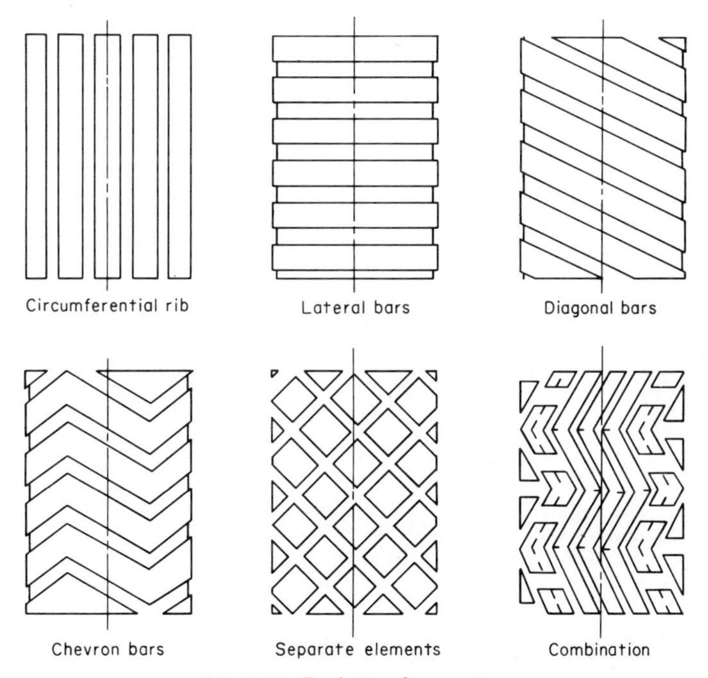

Fig. 6-32 Basic tread geometry.

In the selection of the most applicable tread, the type of highway service and also the wheel positions should be considered. The broad service in which on-highway tires are used includes pickup and delivery trucks (mostly local); heavy-duty, long-distance hauling (interstate); mud and snow; and on- and off-highway travel. (This latter classification will be covered under Off-highway Tires.) Consideration should be given to wheel position—steering, drive, trailing, or all wheel. The all-wheel tire may be mounted in any position. Figure 6-33 illustrates a typical tread type

Fig. 6-33 Light-vehicle tire.

Fig. 6-34 Mud and snow tire.

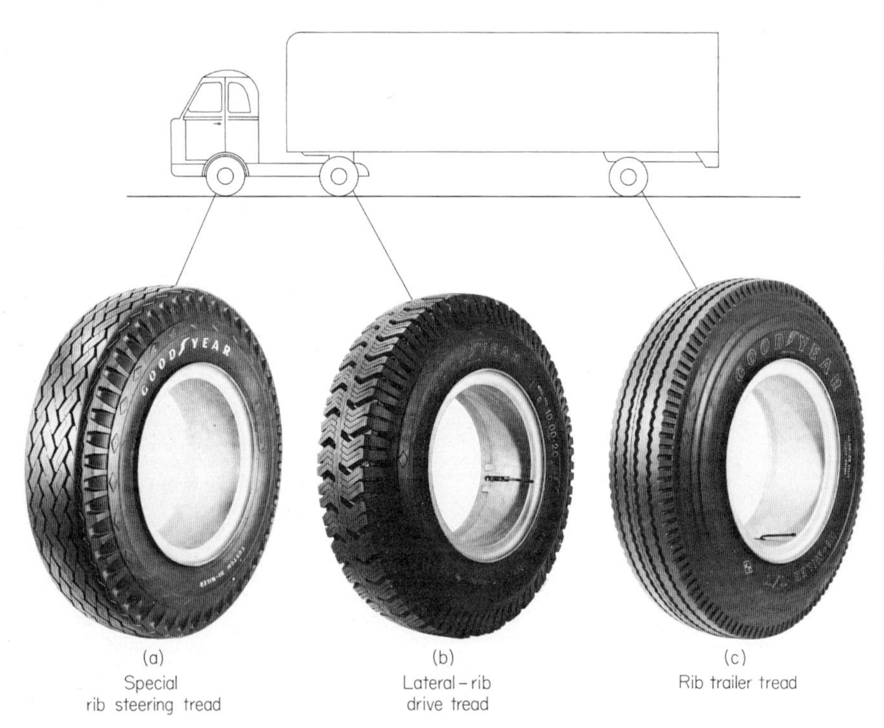

Fig. 6-35 Tractor-trailer tires.

for lighter vehicles and pickup trucks which is used on all-wheel position. Figure 6-34 illustrates a mud and snow type which can be used on the rear axle for increased traction. Figure 6-35 indicates location and tread types for a long-distance hauling unit. A special-rib steering tread is shown on the front tires, with lateral-rib treads on the drives and special-rib trailers on the rear. In actual practice, as drive tires become worn, they may be moved to the trailer position and then ultimately recapped as a trailer tire.

Economic Evaluation For maximum tire economy, tire size and load range must be accurately matched to the maximum axle loads. A tire may be constructed of various textile cords or of steel cords, but the applicable load range for a particular size of tire is independent of the cord structure used. The size and load values establish the structural capacities of the tire.

The various carcass materials and tread designs establish the wear and protective-durability factors for the tire. The basic cord geometry will affect tire performance; the conventional bias-angle tire provides a maximum-strength bead area and sidewall area, while the radial-cord geometry will furnish maximum tread performance. Performance factors are also affected by the tread and sidewall gauges and by the type of rubber compounds used in the tire. Table 6-21 lists tread gauges at the center line for some representative sizes of on-highway tires. A minimum undertread of about 40 percent is recommended for adequate tread-element foundation and as a good base for any retread work. The width and flatness of the tread will also affect tire wear.

TABLE 6-21 Comparative On-highway Tire Tread Gauges

	Tread gauges		
Tire section	Standard	Heavy	Mud and snow
7.00	0.43	0.50	0.67
8.25	0.47	0.59	0.74
9.00	0.49	0.68	0.79
10.00	0.51	0.73	0.81
11.00	0.53	0.77	0.84
12.00	0.55	....	0.88

NOTE: Gauges listed are theoretical molded depths at the tire centerline. They are primarily related to tire width, and are not affected by changes in tire diameter.

All tire components except steel elements deteriorate due to heat. The maximum internal temperature that contemporary tires can tolerate is in the 225 to 250°F range. This refers to the temperature at the tire's "hot spot"—usually in the shoulder area of the tire, under the tread. The critical temperature of a tire cannot be determined by measuring the external tread temperature or by calculating the contained air temperature. It can only be measured by a thermocouple needle inserted to the precise depth of the tire's "hot spot."

Overload, underinflation, high ambient temperatures, and continuous operation all contribute to heat buildup. The result of high heat buildup in the tire is shown by ply separation, blowouts, and short tire life. Proper load, inflation, and load range plus observance of legal speed limits and good driving habits will reduce hazards due to heat.

The following basic procedures are recommended as part of a tire evaluation plan aimed at lower tire costs:

1. Correlation of truck manufacturer's specification weight with actual truck weights to determine if added equipment has resulted in any tire overload. Truck weights should be checked in both unloaded and loaded conditions. Tires should be upgraded as to load range or size if overloads are indicated.

2. Tire inflation should be checked regularly. Checks should be made with an accurate air gauge after the tires have cooled to approximate ambient temperatures.

3. All valves must have valve caps.

4. Any mechanical problems affecting tire performance must be corrected. Misalignments and mismating of dual tires reduce overall tire life.

5. Proper tire rotation and regular tire inspection should be maintained so that tires are not worn irregularly or worn beyond the retread stage. In no case should a tire be worn to the carcass stage. Many tires include indicators which show the safe level of tread wear.

6. A tire record system should be maintained and should include an analysis of tires that have been removed.

A good tire record system will reveal, for a particular operation, the most appropriate tire for maximum economy. The following formula may be used in conjunction with the tire record system:

$$\frac{\text{new tire price} + \text{price of repair and retreading}}{\text{total miles of tire service}} = \text{tire cost per mile of service}$$

The original tread gauge of the tire should be recorded and subsequent gauge readings taken at intervals. Mileage in terms of each $\frac{1}{32}$ in. of wear may then be calculated as a means of estimating the projected tread life.

About two-thirds of the new-tire cost is in the carcass or body of the tire and about one-third in the tread. The operator should first choose a tire carcass with an adequate load range. He should then endeavor to extend tire life by using an extra tread type and/or retreading. If permitted to wear to the smooth stage, a typical tire is good for about 50,000 mi before retread.

The final tire cost per mile of service will ultimately decide, for a particular operator, the tire and tread which result in minimum tire cost for his operation.

OFF-HIGHWAY TIRES

This category will include tires for off-highway operation and for alternating on- and off-highway usage.

Service Requirements The tire service class described as "off-highway" includes a variety of widely differing operations. At the one extreme, vehicles such as logging trucks or dump trucks hauling aggregate may combine on-highway and off-highway service. Maximum speeds for such vehicles would be quite high, but wheel loads would be relatively low because of highway load restrictions.

Next, consider an off-highway truck or a motor scraper engaged in off-highway earth hauling. Loads can be much higher than for the first example, since highway load limitations no longer apply. Speeds are relatively high, with maxima perhaps in the 30-mph range, but they are significantly lower than for on-highway service. Average hauls are relatively short by on-highway standards but may still be measured in miles.

Next, there are equipment classes where travel speeds and travel distances are both decidedly restricted. Motor graders, wheel tractors, and pneumatic-tired compactors all fall within this class. Finally there are wheel-mounted units such as compressors and generators which might be considered as essentially stationary.

These varying service requirements for wheel load, travel speed, and haul distance have necessitated different types of tires. The general principles concerning tire engineering and selection which have been outlined in the preceding discussions still apply, but additional guidelines and structural emphasis are warranted because of the severe service conditions associated with off-highway activity.

Tires which are intended for alternating on-highway and off-highway service are comparable in size to tires designated solely for on-highway operation. For off-highway service, loads may be much larger and tire sizes are correspondingly increased. Large off-highway tires, known as A-2 tires, are those with cross-sectional widths of 16.00 in. and larger. These are tires whose size, construction, and compounds are designed for service other than highway. The two most important facts regarding this class of tires are:

1. Heavier treads, undertreads, and tire bodies combined with heavier loads and increased deflections result in serious heat buildups. The heavier gauges prevent effective heat dissipation.

2. Tires must be selected on the basis of the severity of service. Impacts, cuts, and tears can lead to greatly decreased service life.

Off-highway tires of the conventional type (16.00, 18.00, etc.) are used as duals. The wide-base tires (26.5, 23.5, etc.) are used as singles. Each conventional tire has a wide-base equivalent; for example, the 24.00-29 versus the 29.5-29. The wide-base tires have the same diameters as the conventional tires but are wider in section and use wider rims. Pressures are somewhat less, and the wide-base tires generally have better flotation qualities. With the larger-size tires and equipment units, the initial drive, axle, and wheel configurations usually preclude subsequent changeovers to dual/single or single/dual arrangements.

The choice between singles and duals is dependent upon the basic vehicle design and on the intended type of service. Some of the main advantages of each of these tire systems are as follows:

DUAL-TIRE ARRANGEMENT:

1. Cooler running.
2. Reduced loss hazard. (The replacement costs of duals are less than for a large single tire—also the "width of exposure" to a damaging object is less for each tire of a dual pair than for a wide-tread single tire.)
3. Easier repair and retreadability. (Dual tires are simpler to mount and dismount—are easier to handle and transport both on the jobsite and at the retread plant.)
4. Lower rolling resistance on firm roads.

SINGLE-TIRE ARRANGEMENT:

1. Better flotation and mobility.
2. Better cut-and-bruise resistance due to enveloping power.
3. Lower rolling resistance on loose material.
4. Better ride.

Figure 6-36 illustrates examples of various off-highway tread types. The B and C rock types both have the same external appearance, but the C type has less undertread and a lighter carcass. The rock-type treads are available in the four tread-type variations in order to combine maximum rock-service ability with the minimum heat generation characteristics.

Table 6-22 lists representative tread-gauge comparisons between some of the sizes of off-highway tires. Tread gauges for other sizes will be in approximate proportion.

TABLE 6-22 Representative Off-highway Tire Tread Gauges

Size	Tread type (with deviation)	Tread gauge
18.00-25	E-3	1.24
18.00-25	E-4	1.13
27.00-49	E-3	1.66
27.00-49	E-4	2.50
33.5-39	E-3 (A)	1.67
33.5-39	E-5 (B)	1.37
33.5-39	E-6 (C)	1.37*

* With less undertread gauge.

NOTE: Gauge dimensions shown are theoretical molded depths at the tire center line.

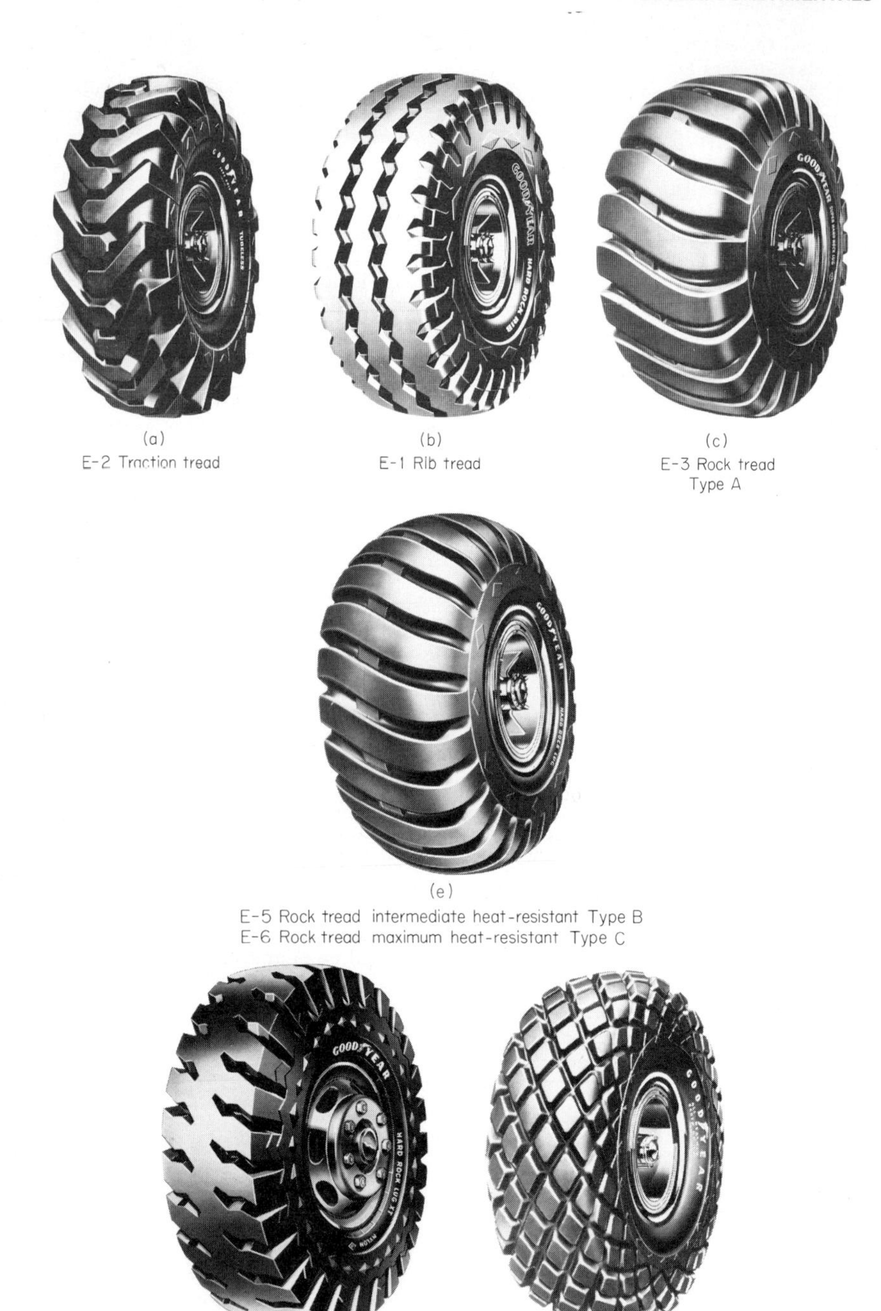

(a)
E-2 Traction tread

(b)
E-1 Rib tread

(c)
E-3 Rock tread
Type A

(e)
E-5 Rock tread intermediate heat-resistant Type B
E-6 Rock tread maximum heat-resistant Type C

(d)
E-4 Rock deep tread

(f)
E-7 Flotation tread

Fig. 6-36 Typical treads for off-highway tires.

Off-highway rims must withstand very high stresses. Moments about the axle and all other forces acting upon the off-highway tires and rims are proportionately higher than for their on-highway counterparts. Figure 6-37 illustrates typical sections of grader and off-highway rims. Note that side flanges on the heavy rims are removable to facilitate tire dismounting. It is important that the specified rim-flange height be used with its proper tire-section size.

Tire Construction Tires intended for alternating on-highway and off-highway service are, for the most part, of tube-type construction. Large off-highway tires,

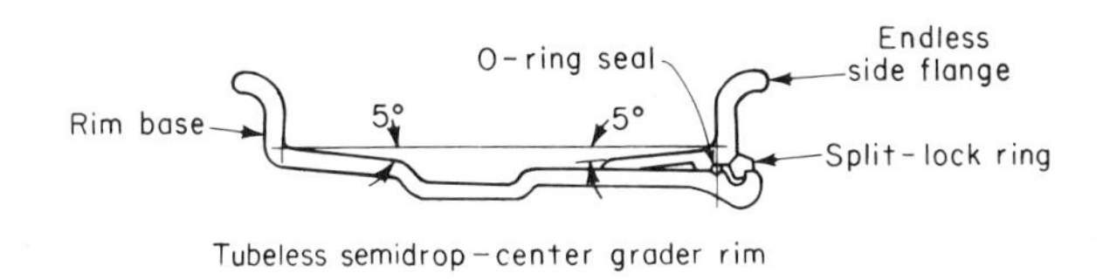

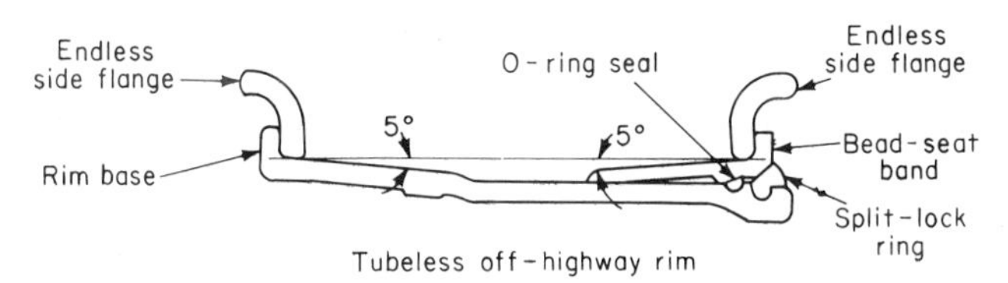

Fig. 6-37 Typical grader and off-highway rims.

however, are practically all tubeless. In these A-2 sizes, tubeless tires have the following advantages:

1. The elimination of heavy tubes and flaps decreases the overall insulating gauge by a substantial amount. This reduces both contained air and internal tire temperatures.
2. The tubeless tire/rim assembly represents a much simpler structure for assembly and disassembly. It also permits the valve to be installed at another place than the rim centerline.
3. Service problems of tubes and flaps are eliminated, thus greatly reducing downtime.
4. Small penetrations through the tire frequently lead to blowouts in a tube-type tire but usually result only in repairable leaks for the tubeless tire.
5. Tubeless tires may be retreaded and repaired by qualified tire-service personnel.

The conventional construction for the off-highway tire body, as for the on-highway tire, is textile bias ply. The load-carrying capacity of this type of off-highway tire is still expressed in terms of "ply rating" rather than the newer "load range" designation which has been accepted for on-highway tires. However, the universal relationships between load and speed remain essentially unchanged regardless of the rating system which is applied: the faster the speed, the lower the allowable load. The reason for this, of course, is heat buildup in the tire and its consequent damaging effects.

The conventional tire may use the following special constructions (added to the basic textile-cord body):

1. Shredded wire undertread—standard in many earth-moving (E-2, E-3, E-4, E-5) and dozer-and-loader (L-2, L-3, L-4, L-5) tire types. Shredded wire reinforces the basic compound and resists small to medium cutting. If the tread is cut, the shredded-wire compound aids in prevention of cut growth. This type of steel reinforcement consists of short lengths of fine carbon-steel wire milled directly into the rubber compound.

2. Shredded-wire sidewall—this reinforced-steel construction may be extended into the tire sidewalls down to the rim flanges.

3. Shredded-wire tread—the entire tread and sidewall may be composed of this compound where hazardous operations are involved.

4. Steel breakers—two narrow plies of steel cable may be added in the crown area of the tire to protect the textile carcass against large cuts and impacts.

5. Cap plies—the entire textile carcass can be encased by steel-cable plies from rim flange to rim flange.

6. Various combinations of the steel cable piles and shredded wire may be, and are, used.

Figure 6-38 diagrams some of the aforementioned constructions.

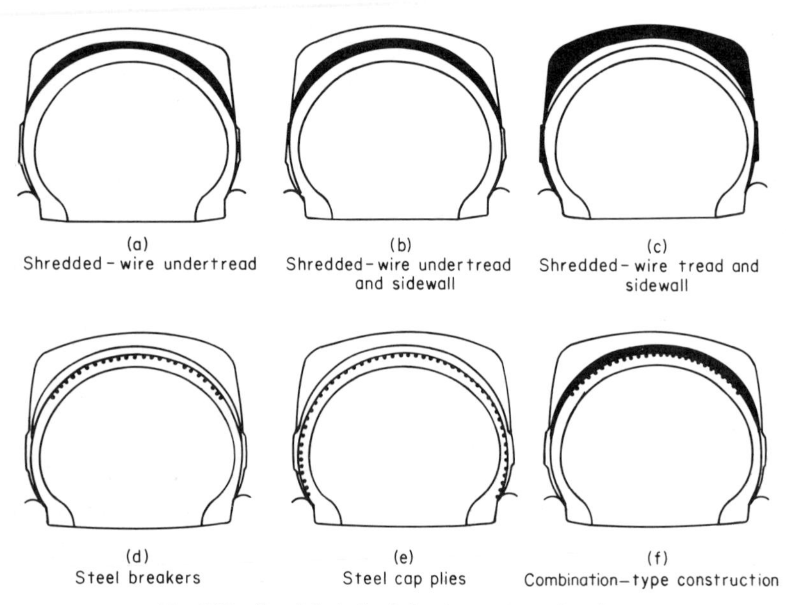

Fig. 6-38 Special steel reinforcing systems for tires.

The radial-belted tire is a more recent type of construction for off-highway tires. Its carcass consists of a single steel ply encircled by a number of near-circumferential steel breakers (belts) which constrict the tread of the tire. Textile carcasses have also been used in conjunction with the steel belt, and shredded wire can be utilized in the radial tire construction. The radial-ply tires have the following advantages:

1. They run cooler because the single-ply steel body dissipates heat faster than the thicker-fabric bias tire.

2. They provide an easier ride, since flexing is concentrated in the sidewall.

3. Normal tread wear life is longer because of the lack of squirm beneath the tread.

4. Some fuel savings may be expected.

Radial-ply tires will merit consideration under the following conditions:

1. High-speed scraper operations in dirt where heat buildup is a factor, contributing to early failure.

2. Long-haul, large-capacity truck service over good haul roads where the loading and dumping areas are relatively free of rocks.

3. Projects where tire maintenance is good and monthly performance records are available.

Table 6-23 is a representative radial-ply lineup. These tires are rated on the load-range rather than the ply-rating system, and in some cases one load range satisfies two or three existing ply ratings.

Mining and Logging Service These tires are an extension of the on-highway tires into off-highway service. This is a more rugged and hazardous type of work than on-highway operation. The mining and logging tires, designated as "ML," usually are constructed of heavier, stronger tire bodies with lug-type or ribbed treads; small tread elements and blades are generally eliminated. Figure 6-39 illustrates some typical mining and logging tires. These tires might be used, for example, on logging trucks which haul the load from the woods over minimal dirt roads and then continue

TABLE 6-23 Radial-ply Off-highway Tires

Conventional size	Load range	Replaces ply rating
18.00-25	L	12-16-20
18.00-25	N	24
18.00-25	S	28-32
18.00-33	S	28-32
21.00-35	W	36-40
21.00-49	W	36-40
24.00-49	X	30-36-42
Wide-base size	**Load range**	**Replaces ply rating**
17.5-25	H	12-14-16
20.5-25	H	12-16
23.5-25	H	12-14-16
26.5-25	L	14-16-20
26.5-29	P	22-26
29.5-25	Q	22-28
29.5-29	Q	22-28
29.5-35	T	28-34
33.25-35	S	32
33.5-33	V	32-38
33.5-39	V	32-38
37.5-33	X	36-42
37.5-39	Y	36-44

NOTE: Available in limited tread types only.

over well-maintained public roads. They may also be used for slow-speed service—cranes, straddle trucks, etc.; some sizes are also used in static service.

Mining and logging (ML) service tires are classified by ply rating rather than by the newer load-range rating. Table 6-24 lists ML tire sizes and load ratings for intermittent highway service at 50 mph maximum speed. Table 6-25 lists similar ratings for off-highway service at a speed of 30 mph. (This latter is not an officially published Tire and Rim Association table but is recognized generally in the tire industry.)

Large-capacity Hauling Units Large-capacity trucks in off-highway hauling service, together with scrapers of all types, present the most critical and demanding service requirements for tires. In selecting tires for these vehicles, the ton mile per hour (tmph) rating system is used to supplement the usual load rating systems (ply range or load range) of the Tire and Rim Association. The Tire and Rim Association rating indicates the load that the tire can carry, and the tmph rating predicts the service performance of the tire.

The tmph rating system (average load multiplied by average speed) has been developed to achieve optimum performance from earthmover tires. It is a way of calculating load, speed, and time factors within safe tire temperature limits. The use of the tmph formula will aid in the prevention of premature tire failures due to excessive heat. Earthmover haul jobs operate at various tmph rates depending on

Fig. 6-39 Mining and logging tires (ML). (*a*) Lug-type tread with center groove. (*b*) Rock-type tread. (*c*) Lug-type tread with center rib. (*d*) Four-rib-type tread.

the load, distance, and time involved in round trips. Thus, each job's haul rate must be determined to compare it with each tire's tmph rate limit.

$$\text{tmph job rate} = \text{average tire load} \times \text{average speed for the day}$$

$$\text{Average tire load} = \frac{\text{“empty” tire load} + \text{“loaded” tire load}}{2}$$

$$\text{Average speed} = \frac{\text{round-trip distance in miles} \times \text{number of trips}}{\text{total hours (from start of first shift to end of last shift)}}$$

EXAMPLE:

"Empty" vehicle tire load = 20,000 lb or 10.0 tons
"Loaded" vehicle tire load = 34,000 lb or 17.0 tons

The earthmover is worked two 8-hr shifts per day—8:00 A.M. to 4:30 P.M. and 6:00 P.M. to 2:30 A.M. (includes 1½ hr of downtime between shifts). Each shift hauls 15 loads 4½ mi each way.

$$\text{Average tire load} = \frac{10.0 \text{ tons} + 17.0 \text{ tons}}{2} = 13.5 \text{ tons}$$

$$\text{Average speed} = \frac{9 \text{ mi round trip} \times 15 \text{ trips} \times 2 \text{ shifts}}{18.5 \text{ hr (total from start of first shift to end of last shift)}} = 14.6 \text{ mph}$$

$$\text{Job rate} = 13.5 \text{ tons} \times 14.6 \text{ mph} = 197 \text{ tmph}$$

Tires used for this job should have tmph ratings of at least 197.

Table 6-26 is a comparison chart relating tmph ratings and tread types for bias-ply tires. The tmph rating for any given type and size of tire is not a fixed value as in the Tire and Rim Association ply rating system and may change if the tire construction and materials are adjusted or improved.

TABLE 6-24 Off-highway Tires Used in Intermittent On-highway Service
(Tire and Rim Association Standard)

Tire size	Tire load limits at various cold inflation pressures											
	40	45	50	55	60	65	70	75	80	85	90	95
7.50-20 ML	2,070	2,220	2,350	2,490	2,620	2,750	2,870	2,990	**3,100**(10)	3,220	3,330	**3,430**(12)
8.25-20 ML	2,460	2,640	2,800	2,960	3,120	3,270	3,410	**3,550**(10)	3,690	3,820	**3,950**(12)	
9.00-20 ML	.....	3,120	3,310	3,510	3,690	3,870	**4,040**(10)	4,200	4,360	**4,520**(12)		
10.00-20 ML	.....	.....	3,760	3,970	4,180	4,380	4,580	**4,760**(12)	4,950	5,120	**5,300**(14)	
10.00-22 ML	.....	.....	4,000	4,230	4,450	4,660	4,870	**5,070**(12)	5,260	5,450	**5,640**(14)	
10.00-24 ML	.....	.....	4,240	4,490	4,720	4,950	5,170	**5,380**(12)	5,590	5,790	**5,990**(14)	
11.00-20 ML	.....	.....	4,100	4,330	4,560	4,780	4,990	**5,190**(12)	5,390	5,590	**5,780**(14)	
11.00-22 ML	.....	.....	4,350	4,600	4,840	5,080	5,300	**5,520**(12)	5,730	5,940	**6,140**(14)	
11.00-24 ML	.....	.....	4,620	4,890	5,140	5,390	5,630	**5,860**(12)	6.090	6,310	**6,520**(14)	
12.00-20, 21 ML	.....	.....	.....	4,930	5,190	5,440	5,680	5,910	6,140	6,360	6,580	**6,790**(16)
12.00-24, 25 ML	.....	.....	.....	5,550	5,840	6,120	6,390	6,650	6,910	7,160	7,410	**7,640**(16)
13.00-24, 25 ML	.....	.....	.....	.....	.....	7,090	7,400	7,700	8,000	**8,290**(16)		
14.00-20, 21 ML	.....	.....	.....	.....	.....	.....	7,800	8,110	8.430	**8,730**(18)		
14.00-24, 25 ML	.....	.....	.....	.....	.....	.....	8,700	9,050	9,400	**9,740**(18)		

NOTE 1: The tabulated load ratings are based on 50-mph maximum speed.
NOTE 2: Figures in parentheses denote ply ratings for which boldface loads and inflations are maximum.

TABLE 6-25 Off-highway Tires for Mining and Logging Service, Short Hauls

Tire size	Tire loads at various inflation pressures														
	30	35	40	45	50	55	60	65	70	75	80	85	90	95	100
7.50-20	2,070	2,270	2,450	2,630	2,790	2,950	**3,110**(10)	3,260	**3,400**(12)	3,540	3,675	3,810	3,940	4,060	4,190
8.25-20	2,460	2,690	2,910	3,110	3,310	3,500	3,680	3,860	4,030	**4,200**(12)	4,360	4,520	4,670	4,820	4,970
9.00-20	2,910	3,180	3,440	3,680	3,920	**4,140**(10)	4,360	**4,570**(12)	4,770	4,970	5,160	5,340	5,530	5,700	5,880
10.00-20	3,300	3,610	3,900	4,180	4,440	4,700	**4,940**(12)	5,180	**5,140**(14)	5,630	5,850	6,060	6,270	6,470	6,660
10.00-22	3,510	3,840	4,150	4,450	4,730	5,000	**5,260**(12)	5,510	**5,760**(14)	5,990	6,220	6,450	6,670	6,880	7,090
10.00-24	3,720	4,070	4,400	4,720	5,020	5,310	**5,580**(12)	5,850	**6,110**(14)	6,360	6,610	6,840	7,080	7,300	7,525
11.00-20	3,590	3,930	4,250	4,560	4,850	**5,120**(12)	5,390	**5,650**(14)	5,900	6,140	6,380	6,610	6,830	7,050	7,270
11.00-22	3,820	4,180	4,520	4,840	5,150	**5,440**(12)	5,730	**6,000**(14)	6,270	6,520	6,780	7,020	7,260	7,490	7,720
11.00-24, 25	4,050	4,440	4,780	5,140	5,470	**5,780**(12)	6,080	**6,370**(14)	6,650	6,930	7,190	7,450	7,710	7,960	8,200
12.00-20, 21	4,090	4,470	4,830	5,180	5,510	5,830	**6,130**(14)	6,430	**6,710**(16)	6,990	7,260	7,520	7,780	8,030	8,270
12.00-24, 25	4,600	5,030	5,440	5,830	6,200	6,560	**6,900**(14)	7,230	**7,550**(16)	7,870	8,170	8,460	8,750	9,030	9,310
13.00-24, 25	5,320	5,820	6,280	6,740	7,170	7,590	7,980	8,370	**8,750**(18)	9,110	9,460	9,800	10,130	10,460	10,780
14.00-20, 21	5,620	6,140	6,630	**7,100**(12)	7,550	8,000	**8,400**(16)	8,820	9,200	**9,580**(20)	9,940	10,300	10,660	11,000	11,300
14.00-24, 25	6,270	6,850	7,400	7,920	8,430	8,920	**9,400**(16)	9,830	10,280	**10,690**(20)	11,100	11,510	**11,900**(24)	12,250	12,610

NOTE 1: The tabulated load ratings are based on 30-mph maximum speed.
NOTE 2: Figures in parentheses denote ply ratings for which boldface loads and inflations are maximum.

The data needed to compute an accurate tmph rating for a particular tire size, type, and service must be obtained from the tire manufacturer or the equipment manufacturer; hence the tmph ratings in the last column in Table 6-26 are very approximate. No attempt has been made in this table to identify a particular size with a specific rating. The radial-ply tire has a higher tmph rating than its equivalent size in a bias-ply tire, since heat is dissipated more rapidly from its single-ply steel body. Depending upon size and type, the tmph range for the radial-ply tire is approximately 300 to 530.

Tables 6-27 and 6-28 indicate permissible loadings based on Tire and Rim Association ratings for tires used in off-highway service with large-capacity hauling units. The proper size and ply rating for a tire can be found by comparing service tire loads with the maximum permissible tire loads listed in these tables. The 30 mph limitation for the listed tire loads refers to maximum speed, not average speed, and applies to round-trip hauls of 3 mi or less. For round trips in excess of 3 mi, rated cold inflation should be increased by 5 lb. It may be noted that Table 6-27 applies to conventional tires which are an extension in size from regular truck size, while tires in Table 6-28 are the wide-base equivalents of these conventional sizes. The wide-base tires require

TABLE 6-26 Off-highway Bias-ply Tires, Treads, and TMPH Ranges

Code designation (conventional and/or wide base)	Tread type†	Tread-type deviation	Approximate tmph range*
E-1 (Fig. 6-36*a*)	Rib		150–220
E-2 (Fig. 6-36*b*)	Traction		145–270
E-3 (Fig. 6-36*c*)	Rock	A (wide base only)	125–390
E-4 (Fig. 6-36*d*)	Rock, deep tread		100–215
E-5 (Fig. 6-36*e*)	Rock, intermediate heat resistant	B (wide base only)	234–424
E-6 (Fig. 6-36*e*)	Rock, maximum heat resistant	C (wide base only)	240–515
E-7 (Fig. 6-36*f*)	Flotation		170–290

* Varies for particular sizes of tires.
† Tread types are identified by a label on the tread. A, B, C deviations are stamped on the tire sidewall.

less pressure and can support a slight load increase due to their greater section width. The conventional tires are used on end-dump trucks in single and dual applications, and in some cases they may be used on bottom dumps and scrapers. The wide-base tires are always used as singles and are the normal tires for scrapers.

The proper type of tire as to tread and/or construction has to be determined by an analysis of the particular job operation in which the vehicle is to be used. The Tire and Rim Association ratings are more concerned with load, and the tmph ratings are more concerned with speed.

Table 6-29 indicates the recommended types of tires for off-highway, large-capacity hauling units. After a satisfactory tire has been identified, an approximation of its tmph rating can be obtained by referring back to Table 6-26. The required tmph rating can be calculated for the proposed service conditions, as described earlier.

Once the tmph rating is satisfied, the most serviceable tread type should be used. In other words, if there happen to be several suitable tires whose tmph ratings meet or exceed the job tmph requirement, the tire whose tmph rating comes closest to the tmph requirement should always be selected. In this way an optimum tire will be chosen—the best tire for wear and durability as well as heat resistance.

Tread type and tire body or carcass are usually matched as to type of service. For example, an E-4 deep-tread rock will have a heavier plied body. In addition,

TABLE 6-27 Conventional Tires for Earthmoving, Mining, and Logging Service for Short Hauls—Tires Used as Singles or Duals
(Tire and Rim Association Standard)

Tire size	Tire load limits at various cold inflation pressures											
	25	30	35	40	45	50	55	60	65	70	75	80
12.00-20, 21	....	4,090	4,470	4,830	5,180	5,510	5,830	**6,130**(14)	6,430	**6,710**(16)		
12.00-24, 25	....	4,600	5,030	5,440	5,830	6,200	6,560	**6,900**(14)	7,230	**7,550**(16)		
13.00-24, 25	....	5,320	5,820	6,280	6,740	7,170	7,590	7,980	8,370	**8,750**(18)		
14.00-20, 21	5,040	5,620	6,140	6,630	**7,100**(12)	7,550	8,000	**8,400**(16)	8,820	9,200	**9,580**(20)	
14.00-24, 25	5,630	6,270	6,850	7,400	7,920	8,430	8,920	**9,400**(16)	9,830	10,280	**10,690**(20)	11,100(24)*
16.00-21	6,670	7,430	8,120	8,780	**9,410**(16)	10,010	10,580	**11,130**(20)				
16.00-25	7,370	8,210	8,980	9,710	**10,400**(16)	11,060	11,700	**12,310**(20)	12,900	**13,480**(24)	14,070	**14,560**(28)
18.00-25	9,520	**10,610**(12)	11,590	**12,530**(16)	13,430	**14,290**(20)	15,110	**15,900**(24)	16,660	**17,400**(28)	18,110	**18,810**(32)
18.00-33	11,040	12,300	13,440	14,530	15,570	16,560	17,520	**18,430**(24)	19,320	20,180	21,000	**21,810**(32)
18.00-49	13,950	15,540	16,980	18,360	19,680	20,930	22,140	**23,290**(24)	24,410	**25,500**(28)	26,540	**27,560**(32)
21.00-25	12,250	**13,650**(16)	14,920	**16,130**(20)	17,290	**18,390**(24)	19,450	**20,460**(28)				
21.00-29	13,170	14,680	16,040	**17,340**(20)	18,590	**19,770**(24)						
21.00-35	14,530	**16,190**(16)	17,700	**19,130**(20)	20,510	**21,810**(24)	23,070	**24,270**(28)	25,440	**26,570**(32)	27,650	**28,710**(36)
21.00-49	17,600	19,610	21,430	**23,170**(20)	24,830	**26,410**(24)	27,940	**29,390**(28)	30,810	**32,170**(32)	33,490	**34,770**(36)
24.00-25	15,830	17,630	**19,270**(18)	20,840	**22,330**(24)	23,750	**25,130**(30)					
24.00-29	16,950	18,890	20,640	22,320	**23,920**(24)	25,440	**26,910**(30)	28,300	**29,670**(36)			
24.00-35	18,590	20,720	22,640	24,480	26,230	27,900	**29,520**(30)	31,050	**32,550**(36)	33,990	**35,380**(42)	
24.00-49	22,310	24,860	27,170	29,380	31,480	33,480	35,420	37,250	**39,050**(36)	40,790	**42,450**(42)	44,080(48)†
27.00-33	22,310	24,860	27,170	**29,380**(24)	31,480	**33,480**(30)	35,420	**37,250**(36)				
27.00-49	27,300	30,410	33,240	35,940	38,520	40,970	43,340	**45,580**(36)	47,780	**49,900**(42)	51,940	**53,930**(48)
30.00-33	27,330	30,450	**33,280**(28)	35,980	**38,560**(34)	**41,020**(40)						
36.00-41	43,070	47,990	52,450	56,710	60,770	64,640	68,380	**71,920**(48)				
36.00-51	47,720	53,170	58,100	62,820	67,330	71,610	**75,750**(42)	79,680	**83,530**(50)	87,230	**90,790**(58)	

* Maximum **11,900** at 90 psi inflation.
† Maximum **45,680** at 85 psi inflation.
NOTE 1: The tabulated load ratings are based on 30-mph maximum speed. For round trips in excess of 3 miles, increase rated cold inflation pressures by 5 lb.
NOTE 2: Figures in parentheses denote ply ratings for which boldface loads and inflations are maximum.

TABLE 6-28 Wide-base Tires for Earthmoving Service for Short Hauls—Tires Used as Singles (Tire and Rim Association Standard)

Tire size	Tire load limits at various cold inflation pressures						
	25	30	35	40	45	50	55
20.5-25	8,650	**9,640**(12)	10,540	**11,390**(16)	12,210	**12,990**(20)	
23.5-25	**11,150**(12)	12,420	**13,580**(16)	**14,680**(20)	15,730	**16,730**(24)	
26.5-25	14,190	**15,810**(16)	**17,280**(20)	18,680	**20,020**(24)	21,300	**22,530**(28)
26.5-29	15,170	**16,900**(18)	18,470	**19,970**(22)	21,400	**22,770**(26)	
29.5-25	**17,830**(16)	19,870	**21,710**(22)	23,480	**25,160**(28)		
29.5-29	**18,990**(16)	21,150	**23,120**(22)	25,000	**26,790**(28)	28,500	**30,140**(34)
29.5-33	**20,110**(16)	22,410	**24,490**(22)	26,480	**28,380**(28)		
29.5-35	20,670	23,030	**25,170**(22)	27,210	**29,160**(28)	31,020	**32,810**(34)
33.25-35	**24,600**(20)	27,410	**29,950**(26)	32,390	**34,710**(32)	36,920	**39,050**(38)
33.5-33	25,310	28,200	**30,820**(26)	33,320	**35,710**(32)	37,980	**40,180**(38)
33.5-39	27,270	30,390	**33,210**(26)	35,910	**38,480**(32)	40,930	**43,290**(38)
37.5-33	30,740	**34,250**(24)	37,430	**40,470**(30)	**43,370**(36)	46,130	**48,800**(42)
37.5-39	32,990	36,750	**40,170**(28)	43,430	**46,540**(36)	49,510	**52,370**(44)
37.5-51	37,310	41,570	45,430	49,120	**52,640**(36)	55,990	**59,230**(44)

NOTE: The tabulated load ratings are based on 30 mph maximum speed. For round trips in excess of 3 miles, increase rated cold inflation pressures by 5 lb. Figures in parentheses denote ply ratings for which boldface loads and inflations are maximum.

this type of tire will normally have a heavier undertread. Gauging the depth of the tread pattern is only a visible criterion of possible tire longevity. Inner tire construction and compound durability are critically important.

Dozer and Loader Service The tires discussed in the previous paragraphs are generally classed as "transport or haulage" tires. Radial load, speed, and distance as exemplified in the Tire and Rim Association systems are critical factors in their selection. The dozer and loader (D&L) tires, in contrast, are classed as "work tool"

TABLE 6-29 Recommended Tires for Off-highway Trucks and Scraper Service

Type of service	Material	Recommended tire type	
		Normal	If heat is factor
Earthmoving, excavation, fill	Mixed soils, some rock	E-1 (front) E-3	E-5 E-6 (if heat is a severe factor)
Earthmoving, excavation, fill	Rock, shale	E-4	E-3 E-5 (if heat is a severe factor)
Open-pit, strip mining	Rock, shale, coal ore	E-4	E-3
Earthmoving, excavation, fill—maximum traction	Sod, earth, very little rock	E-2	E-2 (with heat-resistant construction)
Earthmoving, excavation, fill—minimum traction	Sod, earth, very little rock, sand	E-7	E-7 (with heat-resistant construction)
Long hauls, maximum speed	Improved roads	None	E-1 (with heat-resistant highway rib)

(a) L-2 Traction tread

(b) L-3 Rock tread

(c) L-4 Rock deep tread

(d) L-5 Rock extra-deep tread

Fig. 6-40 Typical treads for dozer and loader tires.

tires. These have special design and service characteristics as follows:

1. Heavy undertreads and/or tread depths
2. Heavy carcass and sidewalls (if required)
3. Various types of steel protective systems
4. Designed for slow-speed service—5 mph and static
5. Maximum stability with minimum deflection
6. Special abrasive-resistant compounds
7. Heat failure problems normally not critical as in "transport" earthmoving tires

Since D&L tires are not designed to be inherently heat resistant, it is most important that they never be interchanged with earthmoving (truck and scraper) tires. This requires care, since most D&L sizes duplicate other earthmoving sizes. Except for being branded "D&L" or a similar designation, they look exactly like other earthmoving tires.

Table 6-30 supplies the Tire and Rim Association loading recommendations for conventional D&L tires, and Table 6-31 lists similar recommendations for wide-base tires. Figure 6-40 illustrates treads for D&L tires, and Table 6-32 supplies typical tire service recommendations. Representative tread gauges for dozer and loader

TABLE 6-30 Conventional Tires for Fork-lift Trucks, Mobile Cranes, Shovels, Mining Cars, Front-end Loaders, and Dozers
(Tire and Rim Association Standard)

Tire size	Tire load limits at various cold inflation pressures													
	35	40	45	50	55	60	65	70	75	80	85	90	95	100
7.00-15	2,410	2,600	2,790	2,970	3,140	3,300	3,460	3,610	3,760	3,910	4,050	4,190	4,320	**4,450**(12)
7.00-20	2,940	3,070	3,410	3,630	3,840	4,030	4,230	4,420	4,600	4,770	4,950	5,120	**5,280**(10)	**5,440**(12)
7.50-10	1,960	2,120	2,270	2,410	2,550	2,690	2,810	2,940	3,060	3,180	3,290	3,400	**3,510**(10)	**3,620**(12)
7.50-15	2,710	2,940	3,150	3,350	3,540	3,720	3,900	4,080	4,240	4,410	4,560	**4,720**(10)	4,870	**5,020**(12)
7.50-18	3,060	3,310	3,540	3,770	3,990	4,190	4,400	**4,590**(8)	4,780	4,960	5,140	**5,320**(10)	5,490	**5,660**(12)
7.50-20	3,300	3,570	3,830	4,070	4,310	4,530	4,750	**4,960**(8)	5,160	5,360	5,550	**5,740**(10)	5,920	**6,110**(12)
8.25-15	3,240	3,500	3,750	3,990	4,220	4,440	4,660	4,860	5,060	5,260	5,450	5,630	5,810	**5,990**(12)
8.25-18	3,650	3,940	4,220	4,490	4,750	5,000	5,240	5,470	5,700	**5,920**(10)				
8.25-20	3,930	4,240	4,550	4,840	5,120	5,380	5,640	5,890	6,130	**6,370**(10)	6,600	6,830	7,040	**7,260**(12)
9.00-10	2,630	2,850	3,050	3,240	3,430	3,610	3,780	3,950	4,110	**4,270**(10)	4,420	4,580	4,720	**4,860**(12)
9.00-15	3,860	4,170	4,470	4,750	5,030	5,290	5,550	5,790	6,030	6,260	6,490	**6,710**(12)		
9.00-20	4,640	5,020	5,380	5,720	6,050	6,370	6,680	6,970	7,260	7,530	7,810	**8,070**(12)		
10.00-15	4,400	4,750	5,100	5,420	5,730	6,030	6,320	6,600	6,870	7,140	7,390	7,650	**7,890**(14)	
10.00-20	5,260	5,690	6,090	6,480	6,860	7,210	7,560	7,890	8,220	8,530	**8,840**(12)	9,140	**9,430**(14)	
11,00-20	5,740	6,200	6,650	7,070	7,480	7,870	8,250	8,610	8,970	**9,310**(12)	9,650	**9,980**(14)		
12.00-20	6,540	7,070	7,570	8,060	8,520	8,960	9,400	9,810	10,210	10,610	**10,990**(14)	11,370	11,730	**12,080**(16)
*12.00-24	**7,240**(6)	7,830	8,390	**8,920**(8)										
*13.00-24	**8,510**(6)	9,200	**9,860**(8)	10,490	11,100	11,670	12,240	**12,780**(12)	13,300	13,810	14,310	**14,800**(16)		
14.00-20	8,980	9,710	10,400	11,070	11,710	12,310	12,910	13,480	14,030	14,570	15,100	15,610	**16,110**(18)	**16,600**(20)
*14.00-24	10,020	**10,830**(8)	11,610	12,350	13,060	13,740	**14,400**(12)							
14.00-24	10,020	10,830	11,610	12,350	13,060	13,740	14,400	15,040	15,650	16,250	16,840	17,420	17,970	**18,520**(20)
*16.00-24	13,670	14,780	15,840	**16,850**(12)										
16.00-25	13,120	14,180	15,200	16,170	17,100	17,990	18,860	19,690	20,500	21,280	22,050	22,810	**23,530**(24)	
18.00-25	16,940	**18,320**(12)	19,630	20,880	**22,090**(16)	23,230	24,360	**25,440**(20)	26,470	**27,490**(24)				
21.00-25	21,810	23,580	25,270	**26,880**(16)	28,430	**29,900**(20)	31,350	**32,740**(24)	34,070	35,380	**36,660**(28)			
24.00-25	28,170	30,450	**32,640**(18)	34,720	36,720	38,620	**40,490**(24)	42,280	44,010	**45,700**(30)				
24.00-29	30,170	32,620	**34,960**(18)	37,180	39,330	41,370	**43,370**(24)	45,290	47,140	**48,950**(30)				
36.00-41	76,650	82,880	88,820	94,470	99,930	105,110	110,190	115,070	119,770	124,370	**128,870**(48)			
36.00-51	84,920	91,820	98,400	104,670	110,720	116,450	122,080	127,490	**132,690**(42)	137,790	**142,780**(50)	147,660	152,330	**15,7000**(58)

* These tires used on semi-drop-center rims.

NOTE 1: The tabulated load ratings are based on 5-mph maximum speed. For 10-mph maximum speed, they must be reduced 13 percent at the same inflation pressures.

NOTE 2: For stationary service conditions, the above loads may be increased up to 57 percent with no increase in inflation.

NOTE 3: For fork-lift truck tires in off-highway service, this table only applies to 8.25-18 and larger sizes.

NOTE 4: Figures in parentheses denote ply ratings for which boldface loads and inflation pressures are maximum.

TABLE 6-31 Wide-base Tires for Fork-lift Trucks, Mobile Cranes, Shovels, Mining Cars, Front-end Loaders, and Dozers
(Tire and Rim Association Standard)

Tire size	Tire load limits at various cold inflation pressures											
	25	30	35	40	45	50	55	60	65	70	75	80
15.5-25	7,460	8,320	9,090	**9,830**(8)	10,530	11,200	**11,850**(12)					
17.5-25	9,010	10,040	10,980	11,870	12,720	**13,530**(12)						
20.5-25	10,890	12,130	13,260	**14,330**(12)	15,360	**16,340**(16)	17,280	18,180	**19,060**(20)	19,900	20,720	**21,510**(24)
23.5-25	14,610	16,270	**17,780**(12)	19,230	**20,610**(16)	21,920	**23,190**(20)	24,390	25,570	**26,700**(24)		
26.5-25	18,600	**20,720**(12)	22,640	**24,480**(16)	26,240	**27,910**(20)	29,520	**31,050**(24)	32,550	**33,990**(28)		
26.5-29	19,880	22,150	24,210	26,180	**28,050**(18)	29,840	**31,570**(22)	33,200	**34,800**(26)			
29.5-25	23,370	26,030	**28,450**(16)	30,760	32,970	**35,070**(22)	37,100	39,020	**40,900**(28)	42,710	44,460	**46,160**(34)
29.5-29	24,880	27,710	**30,290**(16)	32,750	35,100	**37,330**(22)	39,490	41,530	**43,540**(28)	45,470	47,330	**49,140**(34)
29.5-35	27,080	30,170	**32,970**(16)	35,650	38,210	**40,640**(22)	42,990	45,220	**47,400**(28)	49,500	51,520	**53,500**(34)
33.25-35	32,400	36,090	39,450	**42,650**(20)	45,710	48,620	**51,430**(26)	54,090	**56,710**(32)	59,220	61,640	**64,000**(36)
33.5-33	33,160	36,950	40,380	**43,660**(20)	46,790	49,770	**52,650**(26)	55,380	**58,050**(32)	60,630	**63,100**(38)	
33.5-39	35,730	39,810	43,510	**47,040**(20)	50,410	53,620	**56,720**(26)	59,660	**62,540**(32)	65,310	**67,980**(38)	
37.25-35	39,550	44,070	48,160	52,070	55,810	59,360	**62,790**(30)	66,040	**69,230**(36)	72,300	**75,250**(42)	
37.5-33	40,280	44,870	49,040	**53,030**(24)	56,830	60,450	**63,940**(30)	67,250	**70,500**(36)	73,360	**76,630**(42)	
37.5-39	43,220	48,160	52,630	56,910	60,990	64,870	68,620	72,170	**75,660**(36)	79,010	82,240	**85,400**(44)
37.5-51	48,890	54,470	59,530	64,360	68,980	**73,370**(28)	77,610	81,630	**85,570**(36)	89,370	93,010	**96,580**(44)

NOTE 1: The tabulated load ratings are based on 5 mph maximum speed. For 10 mph maximum speed, they must be reduced 13 percent at the same inflation pressure.
NOTE 2: For stationary service conditions, the above loads may be increased up to 57 percent with no increase in inflation.
NOTE 3: Figures in parentheses denote ply ratings for which boldface loads and inflations are maximum.

TABLE 6-32 Recommended Tires for Dozer and Loader Service

Type of service	Material	Recommended tire type
Loading, maximum traction.........	Mixed soils, sand, some gravel, smooth materials	L-2 Fig. 6-40*a*
Loading, mild dozing, some transport	Mixed soils, smooth aggregates, some rock	L-3 Fig. 6-40*b*
Loading, dozing..................	Rock, rough materials, sharp aggregates	L-4 Fig. 6-40*c*
Severe loading and dozing..........	Sharp rock, abrasive materials of all types	L-5 Fig. 6-40*d*

tires are listed in Table 6-33. Equivalent proportions will generally apply to other tire sizes.

Grader Service Road-grader tires are a separate class of tires, with different widths and overall diameters than the regular off-highway tires. Tires for "rock service" graders (G-3) are similar in dimension to other off-highway tires, but their bead areas are designed to fit the standard grader rim. Loads, pressures, ply ratings, and trends are all developed for optimum motor-grader performance.

Table 6-34 applies to grader tire service. The front-wheel tires (sizes 7.00-20 through 10.00-24) are used on standard truck rims. The remaining tires, except the 7.50-20 and 9.00-24, are generally used on semi-drop-center rims.

Figure 6-41 illustrates the basic grader-tire types, and Table 6-35 lists typical applications for these tires.

Compactor Service Today there exist sizes of tires with a variety of ground-pressure ratings to satisfy most compacting service requirements. Table 6-36 is a generally accepted industry table of loads and inflations. The 7.50-15 tires in 4- and 6-ply ratings are known as "smooth rollers." As far as rims and service are concerned, they are not interchangeable with the 7.50-15 tires with 10-, 12-, and 14-ply ratings.

(a) G-3 Rock tread (b) G-1 Rib tread (c) G-2 Traction tread

Fig. 6-41 Typical treads for grader tires.

TABLE 6-33 Representative Tread Gauges for Dozer and Loader Tires

Tire size	Tread type	Tread gauge
29.5-29	L-2	1.50
29.5-29	L-3	1.50
29.5-29	L-4	2.25
29.5-29	L-5	3.75
33.25-35	L-2	1.67
33.25-35	L-3	1.67
33.25-35	L-4	2.50
33.25-35	L-5	4.18

NOTE 1: Not all tread types are available in all tire sizes.
NOTE 2: Gauge dimensions are theoretical molded depths at tire center line.

The former are molded to fit 5° semi-drop-center rims, while the latter are designed for heavy earthmover-type rims. Compactor service tires with rated pressures exceeding 100 psi require special compactor-type rims or other rims approved for these high pressures.

Figure 6-42 illustrates examples of industry compactor-type treads.

TABLE 6-34 Tires for Road Graders
(Tire and Rim Association Standard)

Tire size	Tire load limits at various cold inflation pressures								
	20	25	30	35	40	45	50	55	60
Front wheel									
7.00-20	1,340	1,530	1,700	1,860	2,010	2,150	**2,290**(8)	2,420	**2,550**(10)
7.00-24	1,540	1,750	1,950	2,130	2,300	2,470	2,620	2,780	**2,920**(10)
7.50-24	1,720	1,960	2,180	2,380	2,580	2,760	2,940	3,110	**3,270**(10)
8.25-24	2,040	2,320	2,580	2,820	3,050	3,270	3,480	**3,680**(10)	
9.00-24	2,400	2,730	3,040	3,320	3,590	3,850	**4,090**(10)		
10.00-24	2,710	3,080	3,430	3,750	4,060	**4,350**(10)			
Rear wheel									
7.50-20	1,510	1,720	1,910	**2,090**(6)					
9.00-24	2,400	2,730	3,040	3,320	3,590	3,850	**4,090**(10)		
10.00-20	2,410	2,740	3.050	**3,330**(8)					
10.00-24	2,710	3,080	3,430	**3,750**(8)					
11.00-24	2,910	3,300	3,680	**4,020**(8)	4,350	**4,660**(10)			
12.00-24	**3,300**(6)	3,750	**4,180**(8)						
12.00-28	**3,670**(6)								
13.00-24	3,460	**3,930**(8)	**4,380**(10)						
13.00-24	**3,870**(6)	**4,410**(8)	**4,910**(10)	**5,365**(12)					
13.00-32	4,730	**5,390**(8)							
14.00-20	4,090	4,650	5,180	**5,660**(12)	**6,120**(14)				
14.00-24	4,570	**5,190**(8)	**5,780**(10)	**6,320**(12)					
14.00-28	5,050	**5,730**(8)							
14.00-32	5,510	**6,260**(8)							
16.00-20	5,630	6,400	7,130	**7,780**(14)					
16.00-24	6,230	7,090	**7,900**(12)						
18.00-26	**8,325**(10)								

NOTE 1: Figures in parentheses denote ply ratings for which boldface loads and inflations are maximum.
NOTE 2: For maintenance work on established highways, inflation pressures may be increased 50 percent if desired.
NOTE 3: The tabulated load ratings are based on 25-mph maximum speed.

(a)
C-1 Smooth tread

(b)
C-2 Grooved tread

Fig. 6-42 Typical treads for compactor tires.

Static Tire Service Tires used in mobile-crane or similar service are rated as shown in Table 6-37. The construction of these tires is generally either a mining and logging (ML) type or one of the heavier highway types. For static tire service, it is essential to have rims and wheels which are of adequate strength for the tire loads and pressures.

TABLE 6-35 Recommended Tires for Road Grader Service

Type of service	Material	Recommended tire
Normal road maintenance, blading . .	Mixed earth, some sand, sod, gravel	G-1 Front (Fig. 6-41*b*) G-2 Rear (Fig. 6-41*c*)
Basic road construction, blading.	Soft earth, gravel, some rock	G-2 (Fig. 6-41*d*)
Rough clean-up, severe blading, shovel and pit maintenance	Rock; rough, coarse aggregates; ores; mining products	G-3 (Fig. 6-41*a*)

Economic Evaluation The factors affecting the economics of off-highway tires are both numerous and complex. Tires represent a major expense item for wheel-type construction equipment and must be analyzed along with the expenses of ownership, repairs, labor, and fuel. Just as fuel furnishes motive power to an engine, tread wear develops the traction force which enables a mobile unit to function. A tire consumes its tread in much the same way as an engine consumes fuel. Both fuel consumption and tire wear are directly related to increased horsepower, braking forces, and loads.

There are two main elements in cost evaluation for heavy-construction tires. The first element involves an estimate of tire service life and the second element requires the recording of actual tire service. The estimated hourly costs of tires can be expressed as follows:

$$\frac{\text{tire replacement and maintenance costs}}{\text{estimated tire life in service hours}} = \text{tire cost per service hour}$$

TABLE 6-36 Tires for Compactor Vehicles

Inflation pressure	Tire size									
	7.50-15	7.50-15	9.00-20	10.00-20	11.00-20	13.00-24	16.00-21	18.00-25	21.00-25	30.00-33
35	**2,860**(4)	2,940	5,020	5,650	6,220	9,400	13,120	18,800	24,000	49,500
40	3,090	3,190	5,420	6,110	6,730	10,180	14,200	20,300	26,000	53,500
45	3,310	3,420	5,810	6,540	7,210	10,900	15,200	21,800	27,800	57,300
50	3,520	3,640	6,180	6,960	7,670	11,600	16,190	23,200	29,600	61,000
55	3,740	3,840	6,540	7,360	8,110	12,270	17,130	24,500	31,400	64,400
60	**3,935**(6)	4,040	6,870	7,740	8,530	12,900	18,000	25,800	33,000	67,800
65		4,240	7,200	8,120	8,950	13,510	18,890	27,050	34,600	71,000
70		4,440	7,530	8,480	9,340	14,100	19,700	28,300	36,100	74,200
75		4,610	**7,830**(10)	8,820	9,730	14,700	20,500	29,400	37,600	77,200
80		4,790	8,130	9,160	**10,100**(12)	15,240	21,300	30,500	39,000	80,100
85		4,950	8,420	**9,490**(12)	10,460	15,800	22,100	31,600	40,400	83,000
90		**5,130**(10)	**8,720**(12)	9,820	**10,820**(14)	16,360	22,800	32,750	41,800	86,000
95		5,290	9,000	10,130	11,160	16,860	23,550	**33,800**(24)	43,200	88,600
100		5,450	9,260	**10,440**(14)	11,510	**17,370**(18)	24,250	34,800	44,500	91,400
105		5,620	**9,550**(14)	10,740	**11,840**(16)	17,900	25,000	35,800	45,700	94,000
110		**5,770**(12)	9,800	**11,040**(16)	12,170	18,400	25,660	36,800	47,000	96,500
115		5,920	10,060	11,330	12,490	18,870	26,300	37,800	48,300	99,000
120		6,070	**10,310**(16)	11,620	**12,800**(18)	19,350	27,000	38,800	49,500	101,600
125		6,210		**11,890**(18)	13,110	**19,800**(22)	**27,700**(28)	**39,650**(32)	50,700	104,000
130		**6,350**(14)		12,180	**13,420**(20)	20,300	28,300	40,600	51,900	106,600
135				12,440	13,720	20,700	28,900	41,500	53,000	109,000
140				**12,710**(20)	14,000	21,200	29,600	42,400	54,200	111,200
145				12,970	**14,300**(22)	21,600	30,100	43,200	55,300	113,500
150				13,230		**22,050**(26)	30,800	44,100	**56,400**(44)	**115,900**(64)
155				**13,490**(22)			31,400	**45,000**(40)		
160							**31,950**(36)			

NOTE 1: The tabulated load ratings are based on 5-mph maximum speed.
NOTE 2: Figures in parentheses denote ply ratings for which boldface loads are maximum at indicated inflation pressures.

SPECIMEN
TIRE APPLICATION STUDY FORM

Requested by Contractors Sales Made by Russell Date 1-3-69

Job or Project Name U.S. Dam Location Texas

Contractor Smith Dealer Jones Tire Co.

Job Type & Volume Concrete and earth-filled dam 3,000,000 cu. yd.

No. of Vehicles, Size & Type 12 35 Ton

Tire Size & P.R. Front 18.00-25 32 PR Drive 1800-25 32 PR Rear

Optional Size & P.R. — — —

Haul Road Material & Condition Dry hard Caliche - Gravel top Amb. F° 50°-100°

Haul		Return		Job Hours					Tire Loads			
Ft.	%GR	Ft.	%GR	Shift	Work	Lunch	Bet.	Total		Front	Drive	Rear
1500	5.0	200	0	1st	10	1/2	—	10 1/2	Empty	14,100	7090	—
1200	-.5	1500	-8.	2nd	10	1/2	1 1/2	12	Loaded	20,940	22400	—
800	4.											
200	0	2200	0	3rd					Total (L)	35,040	29490	—
Round trip (A) 1.442 miles				Total	20			(B) 22 1/2	L/4000 (AL)	8.76 T	7.37 T	— T

Cycle Time	Actual	or Calc.
Load		3.2
Haul		2.7
Dump		.8
Return		1.3
Total		8.0
Calc. Cycle hrs.		(C) .1333

IF using a CALCULATED TIME (C), then (A) ÷ (C) = Ave. Speed (AS) 10.8 mph
1.442 ÷ .1333

Approx. Eff. = 90% for Trucks, 80% for Scrapers, therefore (AS) x (AL) x Eff. = 85 TMPH
10.8 x 8.76 x .90

IF using an ACTUAL COUNT of Cycles per day (CD) = ____, then (CD) x (A) = Daily Miles (DM) ____

(DM) ÷ (B) = Ave. Speed (AS) ____ mph

(AS) x (AL) x 100% = ____ tmph

JOB TMPH RATE, FRONT 85 Drive 72 Rear —

TIRE SIZE & TYPE RECOMMENDED AND TIRE TMPH RATE

FRONT: 18.00-25 32PR HRL-XT 120 TMPH

DRIVE: 120 TMPH

REAR: ____ TMPH

Fig. 6-43 Specimen tire application study form.

Tire life may be estimated from the chart which is supplied as Table 6-38. This chart assumes 2,510 "actual-use hours" as the base for mean tire life. Service factors which affect actual service life are identified and adjustment factors are supplied. Another factor of importance is "operator attitude," since the care which the operator exercises will have a marked effect on tire life. The tire-application study form shown as Fig. 6-43 can be helpful in selecting the most economical tire for a particular job.

The Tire Analysis Form shown as Fig. 6-44 has been designed for use in a fleet or job survey. Records of periodic tire surveys are of significant use in selecting the best type of tire for a construction application. Such records, if continued over a period of several months, will determine the facts about a particular tire's alleged economic advantage.

Proper tire selection is the basis for optimum tire economy. Obtain all of the details of the job as accurately as possible, then base the selection of tires on these facts. Proper tire care is also an important factor in reducing tire costs.

TABLE 6-37 Tires for Static Service on Mobile Cranes

Tire size	Ply rating	Maximum	
		Load, lb	Inflation, psi
7.00-15	12	6,875	100
7.00-20	10	8,405	100
	12	8,405	100
7.50-10	10	5,720	100
	12	5,720	100
7.50-15	10	7,815	100
	12	7,815	100
7.50-18	8	7,750	80
	10	8,810	100
	12	8,810	100
7.50-20	8	8,365	80
	10	9,510	100
	12	9,510	100
8.25-15	12	9,235	100
8.25-18	10	9,800	90
8.25-20	10	10,520	90
	12	11,170	100
9.00-10	10	7,230	90
	12	7,685	100
9.00-15	12	11,025	100
9.00-20	12	13,255	100
10.00-15	14	12,465	100
10.00-20	12	14,445	95
	14	14,915	100
11.00-20	12	15,465	90
	14	16,445	100
12.00-20	14	18,140	95
	16	18,725	100
13.00-20	16	22,225	100
14.00-20	18	26,230	100
	20	26,230	100
14.00-24	20	29,260	100
16.00-25	24	38,250	100
18.00-25	24	49,680	100

NOTE 1: For inflation pressures in excess of 100 psi, consult tire and rim manufacturers.
NOTE 2: When these loads are used, the rims and wheels must be checked to determine if they are of sufficient strength.

TABLE 6-38 Estimated Tire Life (Based on 2,510 Mean Hours of Actual Use)

Conditions	*Factor*	*Conditions*	*Factor*
I Maintenance		V Loads (see note)	
Excellent	1.090	T & RA recommended	1.090
Average	0.981	20% overload	0.872
Poor	0.763	40% overload	0.545
II Speeds (maximum)		VI Curves	
10 mph	1.090	None	1.090
20 mph	0.872	Medium	0.981
30 mph	0.763	Severe	0.872
III Surface conditions		VII Grades (drive tires only)	
Soft earth—no rock	1.090	Level	1.090
Soft earth—some rock	0.981	6% max	0.981
Well-maintained gravel road	0.981	15% max	0.763
Poorly maintained gravel road	0.763	VIII Other miscellaneous combinations (see note below)	
Blasted—sharp rock	0.654	None	1.090
IV Wheel positions		Medium	0.981
Trailing	1.090	Severe	0.872
Front	0.981		
Driver:			
Rear dump	0.872		
Dozer and loader	0.763		
Bottom dump	0.763		
Self-propelled scraper	0.654		

Application of this table: Using 2,510 hr as a base, multiply by each succeeding factor to obtain approximate estimated hours as the final product. If miles are used as a base, use 25,100.

Condition VIII is to be used when overloading is present in combination with one or more of the first four conditions—maintenance, speeds, surface conditions, or curves. The combination of these conditions with an overload will create a new and more serious condition which will contribute to early tire failure to a larger extent than will the individual factors of each condition.

TIRE ANALYSIS FORM Page ________

Job ________ Location ________ Contractor ________

Type Unit ________ Tire Serviced By ________

Unit # ________ Date ________

Misc. Comments Eqpt. ________

	1	2	3	4	5	6	7	8	9	10
Position										
Size										
Ply Rating										
Make										
Type										
Serial										
Special Identification										
Non-skid										
Inflation Pressure COLD										
Inflation Pressure HOT										
Miles ☐ Hours ☐										

Comments: ________

Other Misc. Comments ________

Tire Maintenance (Check one)

FAIR	GOOD	V. GOOD

Road Maintenance (Check one)

FAIR	GOOD	V. GOOD

Fig. 6-44 Tire analysis form.

E. Job Characteristics

Each construction job can be viewed as a quantity of work which must be performed, generally with restrictions on its permissible time and cost. The emphasis in this section is primarily directed to the selection of the equipment which is needed to perform the job. As work requirements are analyzed it will be found that each job, in greater or lesser degree, is associated with its singular set of job characteristics. The influence of these job characteristics on equipment selection will vary considerably, and each major job will merit a detailed study.

The following paragraphs will survey the job characteristics which may require consideration in equipment selection, placing particular emphasis upon their engineering fundamentals.

GEOGRAPHIC LOCATION

An initial and thorough inspection of the jobsite generally sets the stage for the concept of how the job will be accomplished, both for domestic and foreign work. Preliminary planning must identify the construction services for the job (power, water, air, heat, transportation) and correlate these services with equipment selection. The task is complex but essential. The following factors may merit particular attention.

Labor In the so-called "developed countries," the availability of labor and adequate levels of skills can usually be assumed. This is not intended to oversimplify the problems that may actually be encountered, but at least the basic capabilities are present. On the other hand, the underdeveloped countries may pose many problems with respect to construction labor, with consequent effects on equipment selection and productivity.

Equipment operators must usually be trained, and many different nationalities have adapted themselves to this phase. Standardization of equipment and procedures is strongly recommended but, even so, productivity will vary widely. Based on 100 percent for an experienced and proficient operator, indigenous operators may produce at only a 60 to 65 percent rate. Productivity can be improved through incentive competition by placing one experienced operator in a group of four to eight local operators. A lack of physical stamina may necessitate relief operators if 8- to 10-hr shifts are contemplated. In this case, one relief operator for 8 to 10 locals will help to maintain production.

In developed countries, due to high pay rates and fringe benefits, economics will frequently justify a high investment in equipment—sometimes up to $150,000—to eliminate one employee. Such action is not necessarily warranted on overseas projects. Labor rates may be much lower and equipment costs, per unit of output, may be much higher.

Parts and Service Availability Aggressive manufacturers and equipment dealers provide parts, service, and operator manuals as well as training and consulting services. Equipment dealers have recently offered an exchange-component plan under which contractors can replace a worn part with a rebuilt and guaranteed assembly. For example, worn engines, transmissions, track chains, etc., can be sent in for repair, and rebuilt units are immediately installed to minimize downtime. The cost is based on the expense of rebuilding the replaced assembly.

Parts and service support varies widely, particularly for work in other countries. A thorough investigation should be made to determine those parts and service facilities which should be provided by the contractors to supplement those provided by dealers. Standardization is important, and each major equipment manufacturer has a family engine that can be used in various pieces of equipment. For example, in-line 4s, 6s, 8s and V-6s, V-8s, V-16s use common parts—pistons, liners, rods, bearings, etc.—

to simplify parts stocks and repair and maintenance procedures. These family engines are sold to other equipment manufacturers to further encourage standardization.

Accessibility Transportation to jobsites in remote areas can impose limitations on equipment selection. As an example, a contractor on a project in Peru at the headwaters of the Amazon River had two choices of access routes: (1) Load out of a United States port, proceed down the east coast of South America to the Amazon, and then go up the Amazon 2,800 mi to the project. (2) Load out of a United States port, proceed to Lima, Peru, and truck the equipment 450 mi over a winding, narrow road which crosses the Andes Mountains at altitudes up to 16,000 ft. Narrow bridges, with a maximum load limit of 15 tons, and small-bore tunnels were also involved. The first route did not impose any restriction as to equipment size, while the second presented many obstacles as to size, weight, need for disassembly, weather, etc.

Logistics play an important part in the supply of equipment and material. Delivery dates must be kept by manufacturers in order that job mobilization and progress can be maintained. In some areas, rental equipment is available (at added job cost) to fill in; but when it is not, job overhead can skyrocket. This further points out the need for planning, so that adequate inventory is maintained and interrelated equipment systems are not interrupted or shut down.

Other Factors Presently, contracting and labor groups are striving to find ways and means to reduce the so-called "seasonality" in construction work. Construction methods that will permit year-round work will help to stabilize the construction industry. Some obstacles to extending the construction season are:

1. Lack of knowledge of cost factors and techniques to make year-round work attractive or feasible
2. Reluctance to change
3. The problem of financing any added cost

Altitude, temperature, rolling resistance, and grade resistance are additional job factors to be considered.

TYPE OF JOB

Construction jobs can be broadly classified as one or more of the following operations:

Transportation (vertical and/or horizontal)
Loading and unloading (includes excavation and filling)
Processing or fabricating

Many items of construction equipment perform more than one of these operations (e.g., an elevating scraper loads, transports, and unloads), but others are highly specialized. The choice is between equipment which is highly efficient for a specialized operation or multipurpose equipment which is reasonably effective for several operations.

In general, equipment systems have been developed for each category of construction job. An acceptable system can usually be identified on the basis of experience and general practice within the industry. The selection of a particular type of equipment within a given equipment system presents more difficulty and involves a consideration of other job characteristics (quantity of material, available space, etc.) in addition to the job type. Attempts have been made to correlate the job types with equipment types as a basis for preliminary equipment selection; that correlation proposed by A. B. Carson for jobs in the "general excavation" category provides an excellent example:[2]

The scope of operations depends on: the total quantity of earth to be moved; the location of the excavation site; its relative width, breadth, and depth; the type of soil to be moved; and the final disposition to be made of the excavated material.

Excavation falls into seven basic types, depending on the factors noted above—bulk-pit excavation; bulk wide-area excavation; loose bulk excavation; limited-area, vertical excavation; trench excavation; tunnel excavation; and dredging. As with most classifications, there are no neat limits to each division and some excavations fall as well within one type as another. Nevertheless, a general grouping will aid us in the initial selection of equipment.

Bulk-pit excavation is primarily excavation of considerable depth, as well as of substantial volume or bulk that must be hauled from the site of operations. In this kind of excavation, the equipment used is operated against the face of a bank from its lower level, and the excavated material is loaded into vehicles at the bottom of the pit to be hauled away from the site. The resulting pit has vertical or nearly vertical walls because of site limitations, such as adjoining streets or buildings.

Bulk wide-area excavation may also have considerable depth and total volume and may also require that the material be hauled away. In this kind of excavation, however, there is complete access to the site from many directions, and the excavation banks can be sloped flatly on two or more sides. It is usually shallower in depth, but larger in area. The excavated material is hauled a shorter distance and deposited over an area similar to that from which it is dug.

Loose bulk excavation differs from the previous two types in the character of the materials to be excavated—wet clays and silts in unconsolidated formations, the bulk of which is *not* hauled away but simply cast into a new position. (Excavations for canals and embankments are of this general type.) Moreover, this excavation is usually performed from the surrounding ground rather than from inside the pit.

In limited-area, vertical excavation, the bulk or volume of the excavation is not the decisive factor. This method of excavation is used in loose or wet soils—unconsolidated formations—where the banks must be supported by shoring or sheathing. The material must of necessity be lifted out vertically.

In trench excavation, the width of operations—and, generally, the depth—is limited. Trenching may be performed in any soil, and will sometimes fall into the category of limited-area, vertical excavation.

Tunnel excavation is limited in width, as in trenching, and in depth or height; it is carried out completely underground. Tunnel excavation, or tunneling, is not generally considered under the heading of "general excavation" and is therefore beyond the scope of this volume.

Dredging is the removal of soils from under water, using the water as a means of transportation to convey the soils to final positions. It might be considered a variant of loose bulk excavation, and, without the presence of water, the excavation might be handled by similar methods.

With his job classification system once established, Carson then supplies a tabulated guide to the selection of basic equipment for each category of excavation task. The guide is reproduced herein as Table 6-39.

PRODUCTION REQUIREMENTS

The quantity of material, the quantity of work for equipment, and the time available for completion dictate the production requirements and thus influence equipment selection. Production planning as related to job requirements will determine the size, type, and number of equipment units. It is also important to analyze the equipment system for the total construction job instead of concentrating exclusively upon the selection of individual units.

Construction operations generally follow a sequence in which one machine performs a phase of the total work and other types of equipment perform subsequent phases. Production characteristics should be balanced for optimum results. For example, the pattern of quarry blasting controls fragmentation, which establishes the size of the loading and hauling equipment, which in turn will influence the sizing of the crusher. Any mismatch or imbalance in the interrelated sequences will affect the job economics adversely. Other examples are pusher-scraper and shovel-truck combinations; concrete delivery, placing and finishing; and pipe-laying operations.

Each machine has a theoretical production capacity for a specific job under ideal conditions. Assuming this theoretical output as 100 percent, the actual capacity or practical performance on the job is somewhat less. The production output is dependent upon the combined efficiency of the machine, the operator, and management. It is necessary to evaluate all three in order to establish a satisfactory balance.

The following material, equipment, and job characteristics should be evaluated

TABLE 6-39 Guide to Selection of Basic Excavation Equipment (P = power shovel, S = scraper, C = clamshell, and D = dragline.)

Class of material	Symbol	Water content	Bulk pit	Bulk wide area	Loose bulk	Limited area vertical
Gravel—well graded	GW	Dry	P	S	P	? C
		Wet	P	S	D	C
Gravel with clay	GC	Dry	P	?	P	? C
		Wet	P	?	D	C
Gravel—poorly graded	GP	Dry	P	S	P	C
		Wet	P	S	D	C
Gravel with fines	GF	Dry	P	S	? D	? C
		Wet	P	D	D	C
Sand—well graded	SW	Dry	P	S	D	C
		Wet	P	S	D	C
Sand with clay	SC	Dry	P	? S	? D	? C
		Wet	P	D	D	C
Sand—poorly graded	SP	Dry	P	S	D	C
		Wet	C	D	D	C
Sand with fines	SF	Dry	P	S	D	C
		Wet	C	D	D	C
Silt and fine sand	ML	Dry	P	S	D	C
		Wet	C	D	D	C
Clay	CL	Dry	P	S	D	? C
		Wet	C	D	D	C
Organic silt	OL	Dry	P	S	D	C
		Wet	C	D	D	C
Fine silts	MH	Dry	P	S	D	C
		Wet	C	D	D	C
Highly plastic clays	CH	Dry	P	S	D	C
		Wet	C	? D	? D	? C
Organic clays	OH	Dry	P	S	D	C
		Wet	C	D	D	C
Peat	PT	Dry	P	S	D	C
		Wet	C	D	D	C

From A. Brinton Carson, *General Excavation Methods*, p. 33, McGraw-Hill Information Systems Company, 1961.

during equipment planning and selection:

1. Characteristics of the project site
2. Weather conditions expected during the construction period
3. Characteristics of the materials
4. Requirements of the plans and specifications
5. Time allotted for the completion of the project
6. Nature, amount, and distribution of work on the job
7. Interference with or dependency of each operation on other operations
8. Location and nature of water supply
9. Transportation facilities and costs for moving to and on the job
10. Working space available
11. Labor cost and availability
12. Availability and relative cost of fuel and electric power
13. Parts supply, repair, and service facilities
14. Types, sizes, capabilities, and limitations of available machines
15. Balance and interdependence of equipment
16. Equipment already owned by the contractor
17. Future work anticipated and the type of equipment needed for that work

18. Initial investment and subsequent operating and maintenance costs of equipment
19. Investment limit of contractor

References

1. L. D. Leet and Sheldon Judson, *Physical Geology*, 3d ed., Prentice-Hall, Inc., Englewood Cliffs, N.J., 1965.
2. A. B. Carson, *General Excavation Methods*, McGraw-Hill Information Systems Company, New York, 1961.
3. *SAE Handbook*, 1969, Society of Automotive Engineers, Inc.

Section 7

Equipment Economics

E. A. COX

**Manager, Construction Services,
H. B. Zachry Company, San Antonio, Texas**

INTRODUCTION

The economics of equipment is a fascinating challenge to our prediction and decision-making abilities. Arkright, Smith, Savage, Taylor, Gantt, Emerson, Gilbreth, and many others have emphasized the utilization of resources for greater productivity.[1]* They have advocated the transfer of skill; that mechanizing the task increases production through greater endurance and reliability. The mechanization which brought about the industrial revolution and subsequent technological advancement is still accelerating rapidly. By extending the effective length of man's arm, we can accomplish much of our work while enjoying a level of comfort and safety that is second to none.

The construction industry takes pride in having held the cost of moving a yard of earth at approximately the same level for the past 30 years. The gain in productivity which has thus offset the increases in labor and equipment costs is directly attributable to the industry's willingness to invest in larger, faster equipment and to the ingenuity of persons within the industry who design the new machines and make them work.

As equipment grows in size and becomes more complex, contractor-management must improve its ability to evaluate and utilize the new hardware. To survive financially, highway and heavy contractors must be knowledgeable in acquiring, utilizing, maintaining, and trading equipment. On most highway contracts today, the cost of equipment allocated against the job will exceed the cost of job labor. On some heavy contracts that are either high-volume, high-production jobs or predominantly rock excavation jobs, the cost of equipment may be double or triple the cost of labor.

This section covers the general concepts of evaluating equipment needs and is limited to the fundamentals of equipment economics. Other sections are devoted to specific construction requirements and include the economics of alternate methods of performing the work. Up-to-date cost and production data for estimating may also be obtained from manufacturers' publications and other referenced information sources.[2–10]

ACQUISITION

Forecasting Equipment Requirements A strategy for building a specific construction capability requires good communications between those responsible for financial planning, bidding, and equipment use and maintenance and those who are responsible for construction management. Market surveys are needed to identify the potential work for the years ahead. Planned projects are evaluated as to soil conditions and equipment requirements. Forecasts based upon the probability of obtaining projects are translated into future equipment needs by type and size of equipment. For example, a smaller highway contractor may conclude that he will obtain work for only two categories of scrapers; i.e., paddle wheels and a spread of the Caterpillar 631 size. A larger contractor may conclude that he can justify a spread of high-production scrapers such as the Caterpillar 666s in addition to the other two spreads. The number of units in each type of spread will depend more upon the contracts already obtained and projects being advertised in the very near future.

When the contractor must acquire additional equipment to accomplish work under contract, the decision to purchase, lease with option to purchase, or to rent without purchase option depends upon the probability of obtaining sufficient work for the equipment to amortize the cost. The decision also depends upon whether the equipment is normally available on a lease or rental basis when needed. Standard equipment is usually readily available on a rental basis, but it is often purchased due to the high probability of obtaining work for it. Special-purpose equipment such as a ringer for a Manitowoc crane often entails greater shipping expense when obtained on a rental basis, but this may be justified by job needs.

Based upon his market forecast, a contractor may elect to phase out of a type of work such as pipeline construction by allowing his spread to become noncompetitive. When market conditions change, he may choose to reenter the field with a major investment in new equipment. Under these conditions, he contemplates a capital investment decision which is much like a proposed investment in any other property. An analysis is required of the anticipated cash flows and earnings at varying levels of

* Superior numbers refer to the list of references at the end of this section.

use. Occasionally a piece of equipment will pay for itself on work already under contract and there is no question of profitability. Alternatives should always be evaluated.

In order for the larger contractor to keep track of his equipment commitments and maintain high utilization, some system of displaying the planned use of equipment is required. For the smaller contractor, a schedule board may be used for a Gantt chart form of presentation. For the larger contractor, such a display may be automated through use of a computer, as illustrated in Fig. 7-1. Such a system is of value only if accurate, up-to-date information can be obtained as to the location and condition of the equipment and as to all known requirements for its uses, including projected overhaul times. Marketing, in the heavy-construction sense of the word, is the matching of plant and personnel requirements of prospective jobs to the available resources of the contractor.

Selection of Equipment Most new projects provide some opportunity to consider the selection of new equipment. The first consideration in any endeavor is getting the job done, and the selection of construction equipment should rightfully be based primarily upon production considerations. In order to meet competition, however, and achieve productivity gains required for our rising standard of living, cost reduction is essential. The contractor must select that equipment which may be used most effectively over its life, taking into account its effect upon the utilization of the other equipment which he owns.

Production estimates should always be based upon the entire equipment system, with provision for move-in and move-out as well as for steady-state production rates. Allowances may then be made for requirements such as additional storage bins and for loss of production during startup and shutdown. One of two approaches may be used. First, if a fixed time for accomplishing the operation has been specified, the contractor must size his equipment to meet the time requirements. Second, if there is no stipulated time requirement, the contractor may make the most effective use of his existing equipment and personnel. The first condition is the most prevalent, and the usual approach is to compute the average production rate required and the peak production rate required. The net working time is estimated by allowing for normal delays such as move-in time and normal weather loss, plus a reserve for contingent losses due to abnormal weather, labor strikes, and supplier delays. The quantity to be produced divided by the net working time in hours will give the effective production rate required. For example, 60 days may be allowed for excavating a channel containing 350,000 bank yards. Based upon a 40-hr week, there is a total of 343 hr available, and the average production rate required is 1,020 bcy per hr. Estimating a loss of 15 percent or 51 hr for weather and another 5 percent or 17 hr for contingent delays, the net working time becomes 275 hr and the effective production rate required becomes 1,273 bcy per hr. Allowing for 83 percent working efficiency, the peak production rate becomes 1,273 divided by 0.83 or 1,534 bcy per hr. The allowance for working efficiency provides time for personal delays, repositioning, minor servicing, unusual material characteristics, operator fatigue, and cleanup. In earthmoving estimates, this job efficiency is often recognized by basing production forecasts on a 50-minute hour rather than on a 60-minute hour. In a new or unique operation such as erecting a new type of structural component, there also may be a significant decrease in initial performance rates while the crew is learning the process. In such cases, it may be necessary to reduce the peak production rate not only by an efficiency factor but also by a learning factor in order to arrive at the required average production rate.

Fluctuations in production rates are absorbed in many ways. Aggregates are stockpiled to ensure an uninterrupted supply for the batch plant. On a pipeline, sufficient open ditch is maintained to ensure that the stringing and bending crews do not catch up with the ditching machine. Enough roadway is fine-graded before starting to place base material to ensure that the operation will be able to continue at its maximum production rate. Prior to starting a slab crew, a sufficient number of beams will be put in place to enable the men to keep working without subsequently waiting for the placement of beams. Unless sufficient time and work are allowed between crews to minimize interference, efficiency may be lowered because equipment breakdowns and

MAJOR EQUIPMENT PROJECTION
AS OF 1 JUL 1969

PAGE 4

DESCRIPTION	EQUIP NR	JOB NR	STATUS	REMARKS	JUL AUG SEP OCT NOV DEC JAN FEB MAR APR MAY JUN JUL AUG SEP OCT NOV DEC
CMI AUTO GRADE SP30 66	329501	1562	RENT		1562***************
GR3301					
A C DD MOTOR GRADER 69	330104	1509	AVAIL		1509***
GR3305					
CAT 112 GRADE 99HP D 65	330501	1557	AVAIL		
J D 570 MOTOR GRADER 68	330502	0024	RENT		0024***
A C DD MOTOR GRADER 69	330503	1545	RENT		1545********1568++++++++
GR3306					
NO. 12 MOTOR GRADER		1584		01	1584+++
CAT 12 GRADE 115HP D 60	330629	1530	RENT		1530********
CAT 12 GRADE 115HP D 63	330639	1545	RENT		1545********1568++++++++
CAT 12 GRADE 115HP D 65	330640	1576	STORE		1576***************************
CAT 12 GRADE 115HP D 65	330641	0024	RENT		0024***
CAT 12 GRADE 115HP D 66	330642	0006	AVAIL		
CAT 12 GRADE 115HP D 66	330643	1545	RENT		1545********1568++++++++
CAT 12 GRADE 115HP D 66	330644	1566	RENT		1566
CAT 12 GRADE 115HP D 66	330645	0024	RENT		0024***
CAT 12 GRADE 115HP D 68	330646L	1561	RENT		1561
GR3309					
NO. 14 MOTOR GRADER		1584		01	1584+++++++++++++++ (from SEP of the following year)
CAT 14 GRADE 150HP D 63	330908	1450	RENT		1450
CAT 14 GRADE 150HP D 63	330911	0006	AVAIL		
CAT 14 GRADE 150HP D 64	330913	1559	RENT		1559******
CAT 14 GRADE 150HP D 65	330915	1530	RENT		1530********
CAT 14 GRADE 150HP D 66	330916	1529	RENT		1529********
CAT 14 GRADER 150 HP 66	330917	1530	STORE		
CAT 14 GRADER 150 HP 66	330918	1487	RENT		1487
CAT 14 GRADE 150HP D 67	330919	1555	OTHER		
CAT 14 GRADE 150 HP 68	330920L	1561	RENT		1561
CAT 14 GRADE 150 HP 69	330921L	1566	RENT		1566
CAT 14 GRADE 150HP D 69	330922L	1562	RENT		1562*************************
GR3310					
NO. 16 MOTOR GRADER		1584		01	1584+++
CAT 16 GRADE 225HP D 65	331001	1581	1530		
CAT 16 GRADE 225HP D 66	331002L	1559	RENT		1559******
CAT 16 GRADE 225HP D 67	331003L	1566	RENT		1566
CAT 16 GRADE 225HP D 67	331004L	1553	AVAIL		
CAT 16 GRADE 225HP D 68	331005L	1576	RENT		1576***************************
CAT 16 GRADE 225HP D 68	331006L	1566	RENT		1566
CAT 16 GRADE 225HP D 68	331007L	1562	RENT		1562*************************

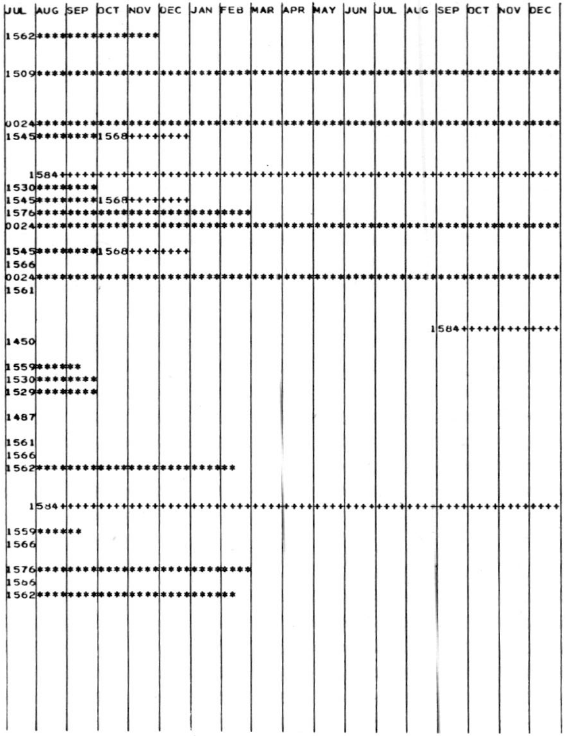

Fig. 7-1 Projected equipment backlog of work. (*H. B. Zachry Company.*)

similar delays will affect a greater portion of the project. When this condition exists, equipment must be sized for a higher peak rate in order to obtain the required average rate. Contractors are becoming more cognizant of the need to be able to estimate the variations in production rate as well as the average production rate. Where job conditions vary significantly, the production rates will vary accordingly.

Most equipment selection is based upon production rates which are taken from textbooks, manufacturers' handbooks, or contractors' records. Manufacturers and their dealers are also performing equipment selection studies for contractors. These studies are usually made on the larger projects where contractors have a greater opportunity to amortize a large percentage of the cost of new equipment. Table 7-1 contains the results of a typical study. Computer programs for computing cycle times are available from equipment manufacturers if contractors prefer to use them in their own studies. The production figures are computed from machine payload, machine weights, rimpull curves, rotating-mass constants, average fixed times, rolling resistances, grade resistances, length-of-haul segments, and tire loading in tons per tire. Other algorithms include engine specifications, converter specifications, final

TABLE 7-1 Typical Equipment Production Study

			Caterpillar 666		Caterpillar 660		Caterpillar 650		Caterpillar 633	
Haul	Quantity, bcy	Distance one way, ft	Cycle, min	Production, bcy/hr	Cycle, min	Production, bcy/hr	Cycle, min	Production, bcy/hr	Cycle, min	Production, bcy/hr
A	1,897,965	5,700	6.26	409	8.36	306	8.01	260		
B*	210,885	8,200	8.11	316	11.15	230	10.81	192		
C	4,218	700	2.25	1139	2.67	958	2.32	897	2.70	534
D	4,702,735	5,967	6.47	396	8.66	296	8.31	250		
E	4,246,818	5,533	6.14	417	8.18	313	7.82	266		
F*	1,202,044	8,200	8.11	316	11.15	230	10.81	192		
G	421,770	6,700	7.00	366	9.48	270	9.13	228		
H	632,655	2,165	3.56	720	4.41	580	4.03	516	4.34	332
I*	3,078,921	13,050	11.68	219	16.54	155	16.26	128		
J	1,370,572	1,000	3.00	855	3.95	648	3.67	566	3.75	384
K	2,319,735	1,000	3.00	855	3.95	648	3.67	566	3.75	384

* NOTE: Due to high ambient temperatures and long haul distances, the ton-mile-per-hour rating is exceeded when using the 666 scraper on the 8,200- and 13,050-ft hauls. Alternatives include reducing the average speed or payload of the machine. Tires are available for the other machines that are within the tmph ratings. It is recommended that the contractor consult with his tire distributor or manufacturer for guidance on tire selection.

drive ratio, engine brake horsepower and rpm, and other data for computing performance.[2] Production may be based on either a 50- or 60-minute hour, but no allowance is made for downtime; operator efficiency; haul-road conditions other than length, grade, and rolling resistance; and other job efficiency factors. The studies generally try to take into account the more significant job conditions, and earthmoving studies are often based upon a seismic analysis of subsurface conditions.

These more sophisticated computer simulation techniques are being used by the larger contractors as well as by equipment manufacturers. As an example, Table 7-2 analyzes the earthmoving requirements for a highway project. Because of the cost of this type of analysis, a mathematical model is more likely to be employed where a given job situation will exist for a relatively long time.[11] The major deterrent to its use is the cost of obtaining accurate input information as to elemental cycle times and the variation in cycle times.

Regardless of the method of estimating production rates for a particular project, the prudent contractor falls back upon his previous experience to verify his answers. For this reason, most estimators have bench-mark production rates for equipment which provide a cross-check on the job under consideration. These bench-mark

TABLE 7-2 Typical Earthmoving Estimate Using Computer Simulation

H.B. ZACHRY COMPANY — EARTHWORK ESTIMATE

JOB 1472 EL PASO HIGHWAY — MARCH 08, 1968

	HAUL STATIONS	DIST(FT) GRADE(%)	QUANTITY(CY) FIN AREA(SY)		NO.	CLASS	DAILY RATE	HOURLY RATE	HRS/ DAY	MIN /HR	PROD/ UNIT	CY/ HOUR	UNIT COST	TOTAL COST	TOTAL HOURS
CUT	333+94.84 – 336+ 0.0	1869.6	29213.1	HAUL	3	HU64570	25.90	19.94	10.0	50	243	731	0.111		
FILL	315+63.47 – 316+71.95	–4.9	13124	LOAD	2	TA73500	17.74	13.84	10.0	50	365	731	0.051	4734.40	39.9
CUT	322+36.31 – 324+ 0.0	642.2	6040.9	HAUL	3	HU64570	25.90	19.94	10.0	50	300	901	0.090		
FILL	316+71.95 – 316+98.26	–4.1	7147	LOAD	2	TA73500	17.74	13.84	10.0	50	450	901	0.042	794.47	6.7
CUT	318+ 0.0 – 319+ 0.0	111.4	2144.9	HAUL	2	HU64570	25.90	19.94	10.0	50	339	679	0.080		
FILL	316+98.26 – 318+ 0.0	–4.2	8544	LOAD	2	TA73500	17.74	13.84	10.0	50	339	679	0.055	288.96	3.2
CUT	321+ 0.0 – 322+36.31	245.4	1947.0	HAUL	3	HU64570	25.90	19.94	10.0	50	327	983	0.082		
FILL	319+ 0.0 – 321+ 0.0	–3.5	14650	LOAD	2	TA73500	17.74	13.84	10.0	50	491	983	0.038	234.81	2.0
CUT	330+ 0.0 – 333+94.84	604.2	46155.9	HAUL	3	HU64570	25.90	19.94	10.0	50	297	893	0.091		
FILL	324+ 0.0 – 330+ 0.0	–7.4	25508	LOAD	2	TA73500	17.74	13.84	10.0	50	446	893	0.042	6126.26	51.7
CUT	371+29.92 – 373+61.08	3292.7	247816.9	HAUL	5	HU64570	25.90	19.94	10.0	50	186	931	0.145		
FILL	336+ 0.0 – 343+ 3.33	–7.0	44341	LOAD	2	TA73500	17.74	13.84	10.0	50	465	931	0.040	45910.31	265.9
CUT	349+ 0.0 – 353+ 0.0	455.0	58771.9	HAUL	3	HU64570	25.90	19.94	10.0	50	306	919	0.088		
FILL	343+ 3.33 – 349+ 0.0	–9.4	31573	LOAD	2	TA73500	17.74	13.84	10.0	50	459	919	0.041	7583.29	63.9
CUT	365+ 0.0 – 371+29.92	1138.7	639819.9	HAUL	3	HU64570	25.90	19.94	10.0	50	244	733	0.111		
FILL	353+ 0.0 – 365+ 0.0	–13.0	87848	LOAD	2	TA73500	17.74	13.84	10.0	50	366	733	0.051	103458.00	872.5

SOURCE: H. B. Zachry Company.

production rates have been derived from prior time studies or other measurements of production and are generally applicable to well-defined job conditions. Often a new project can be classified as harder than one of the bench-mark jobs but easier than another. Since obsolescence is shortening the economic life of much of the heavy equipment, it is important that good bench-mark production rates be obtained as early as possible in the life of a unit.[12]

After determining that an operation may be performed at a satisfactory production rate by two or more types of equipment, the cost of the alternatives must be evaluated. In comparing the return on investment from two or more alternatives, the additional factors of inflation, taxes, resale value, and the time value of money must be considered, together with dealer service and the equipment's reliability, adaptability, obsolescence risk, standardization potential, and other factors.[13] The most important factors to be analyzed are the probability of future work at a profitable price and the accuracy of forecast maintenance cost.

A study of the used equipment being auctioned off, or a walk through a large contractor's equipment yard, will invariably show evidence of misjudgments as to the utilization of equipment. Part of the equipment will reflect the high risk of obsolescence. Other equipment will reflect the unstable market conditions of the construction industry. The continued trend toward larger equipment has increased the pressure on contractors to keep it busy. Utilization is the most important economic factor for high-cost, low-maintenance equipment such as cableways and fixed cranes.

Maintenance cost of some equipment may more than double in severe operating conditions. As an example, the *Caterpillar Performance Handbook* shows a repair factor of 0.04 for a wheel loader operating in a free-flowing, low-density material.[3] When loading shot rock, the repair factor jumps to 0.09, an increase of 125 percent. The economics of whether to use a shovel or a wheel loader for rock excavation depend primarily upon the projected maintenance costs under these service conditions.

The analysis of the design characteristics of each manufacturer's equipment to determine its suitability to job conditions is an important aspect of equipment selection. Design criteria vary between manufacturers, and component options provide a wide range of cost and ruggedness. A balance between reliability, investment cost, and operating cost should be selected, since a management policy of selecting the lowest-priced equipment can often lead to higher costs. Consideration of design reliability should include criteria such as:

1. Reputation of the vendor
2. Warranty or guarantee
3. Service organization of the vendor
4. Use of reputable standard components
5. Adequacy of power train or drive mechanisms
6. Structural design at connecting points

Another consideration in the selection of equipment is the benefits to be derived by standardization.[14] Cost savings due to family standardization may range from 5 to 15 percent of the purchase price of the equipment. The advantages of standardization come from flexibility in scheduling, reduced labor cost for maintenance and repairs, lower spare-parts-inventory cost, and greater purchasing power associated with a larger volume from a single supplier. The disadvantages of standardization include the following: (1) Loss of operational efficiency—due to timing of new models, small gains in efficiency often will be sacrificed to stay with one make. Also, the manufacturer may fall behind competitors in technological improvements. (2) Loss of bargaining power—the loss of competitive activity may lead the manufacturer to give lower trade-in allowances or permit service to deteriorate.

In summary, a list of factors to be considered in comparing equipment alternatives should include:

1. Applicability of the equipment to the planned operation and project conditions
2. Reliability of the equipment and predicted maintenance requirements
3. Service support by the vendor and manufacturer
4. Flexibility and utilization potential of the equipment
5. Investment costs

6. Fuel consumption and operating requirements
7. Operator acceptability and training requirements
8. Safety features
9. Supervisor acceptability
10. Standardization considerations

Equipment Replacement Success in attaining the optimum investment policy is fundamental to the economic growth of a contractor and, in an industry as competitive as heavy construction, the consequences of an unwise equipment investment or replacement policy can easily jeopardize the financial position of the company. Replacement of equipment is an investment and must be strictly considered as such.

When considering replacement or other equipment investment alternatives, it is important to remember that the past may be used only as a guide to the future. The price paid for equipment and the depreciation taken to date are of interest only as they pertain to taxes on capital gains. Investment decisions must be based upon present circumstances and the best estimate of future conditions. Many variables will affect the return on investment. Some of these may be defined mathematically, while others, such as employee morale and safety, cannot as yet be accurately defined mathematically and must be judged by construction management.

A distinction between replacement of equipment due to its physical condition and displacement of the equipment due to obsolescence should be noted. Each new project, each new equipment model, and even each improvement to existing equipment models raises the issue of an investment opportunity by displacement. Contractors generally consider equipment replacement when downtime becomes excessive or when the time for a major overhaul approaches. Some owners review their equipment condition when awarded a new job and make replacement decisions at that time. Other contractors may review their equipment at year end and, based upon tax position and available capital, make their replacement decisions then. However it may be solved, the problem of replacing equipment is a frequently recurring one.

Several general rules for replacement have been offered. One is to replace when the anticipated operating and overhaul costs plus the decrease in salvage value during the next period of use are the same as or greater than the operating and fixed-charge costs for a new piece of equipment.[15] A similar rule is "as long as the average cost is greater than the marginal cost of extending the life of equipment by one additional year, do not replace—as soon as the marginal cost of one additional years' service exceeds the average cost, the asset should be replaced."[16] The Caterpillar Tractor Company states in simpler terms that "whether cumulative cost per hour becomes progressively higher or lower with added machine hours is the key to the replacement decision."[17]

In practice, each time a piece of equipment is repaired, the cumulative cost per service hour is increased. The question becomes one of whether the repaired equipment can then earn enough to provide economic justification for the repair. The decision may be based upon either previously established guidelines or upon a study of the specific piece. Guidelines may be used as a matter of policy; for example, previous replacement studies may have consistently shown that the type of equipment in question should be replaced immediately prior to the second overhaul. By using such a policy, the losses are restricted to the dispersion of individual cases from the average. Records will generally show that the magnitudes of the losses warrant individual consideration. For example, the condition of tractors of equal age may vary considerably due to varying degrees of job severity and maintenance expertise. The Navy has published tables for construction equipment which set forth an accumulated maximum economical repair limit as a percentage of replacement cost and a maximum one-time repair limit as a percent-of-replacement cost by each year of life.[18]

Mathematical models to describe the replacement situation have been in use since the early 1920s. They are constantly being improved by increasing the number of parameters considered in the estimates, and some are very comprehensive.[19] Equipment replacement models will generally contain the following parameters:

1. *The value of money.* Interest rates are used for determining the cost of maintaining an investment in equipment and also for adjusting cost to a common time value. When comparing interest rate with rate of return after taxes, the relationship

"rate of return equals interest rate times one minus the tax rate" may be used. The interest rate or rate of return after taxes should always be greater than the present inflation rate. If the inflation rate is considered separately in the study, however, it should not be included in the interest rate also.

2. *Inflation.* The inflation rate used should be applicable to the cost adjusted. There has been a significant difference between increased prices for new equipment and inflation for mechanics' wages.

3. *Taxes.* The method used should convert costs to a common tax basis before comparing them.[20,21]

4. *Salvage values.* If a trade-in offer is more than the cash value of a piece of equipment, the difference is actually a reduction in the price of new equipment.

5. *Utilization.* In comparing alternatives, the cost should be based upon the amount of work that will be accomplished by the alternative rather than on expenditures.

6. *Standby uses.* Alternative uses of present equipment, such as standby uses, should be weighed against keeping the old equipment on the job or replacing it with new.

Figure 7-2 shows the expenditures incurred to date for a specific D-8 tractor. It also shows the projected costs for a new D-8 tractor, based upon the current purchase price and the average operating cost experience, after adjustment for the time value of money. An overhaul for the old D-8 is estimated at $19,000. An approach to making the decision whether to overhaul or replace is as follows:

Old Tractor with $19,000 Overhaul

Operating cost per hour next 3,000 hours:

Repairs	$2.00
Downtime loss*	1.00
Productivity loss†	0.75
Total	$3.75

Amortization of overhaul:

Hours worked following overhaul	Total overhaul cost, $/hr		Trade-in gain on overhaul, $/hr		Overhaul amortization, $/hr		Operating cost, $/hr		Relative cost, $/hr
1,000	19.00	−	8.00	=	11.00	+	3.75	=	14.75
1,500	12.67	−	4.00	=	8.67	+	3.75	=	12.42
2,000	9.50	−	2.00	=	7.50	+	3.75	=	11.25
2,500	7.60	−	1.00	=	6.60	+	3.75	=	10.35
3,000	6.33	−	0.50	=	5.83	+	3.75	=	9.58
3,500	5.43	−	0.25	=	5.18	+	3.75	=	8.93
4,000	4.75				4.75	+	3.75	=	8.50
4,500	4.22				4.22	+	3.75	=	7.97

* Downtime costs to the contractor are those incurred by idle men, equipment, and other resources while waiting for a piece of equipment which is being repaired.

† Productivity loss is the difference in performance between the old tractor and a new one of the same type and is caused by wear, clogged fuel injectors, and similar factors affecting engine horsepower of the old tractor as well as improvements to the new tractor.

The alternatives under consideration are to increase any remaining unamortized investment in the old tractor by a $19,000 overhaul or to invest in a new tractor. For either alternative, the average hourly costs can be projected over periods of use ranging up to the expected life of the investment. Figure 7-2 indicates that the projected cost curve for the new tractor has a minimum slope at approximately 10,000 hr of use and an hourly cost of $11.50. Since the projected hourly costs for continued use of the old tractor provide for amortization of overhaul costs only, with no allow-

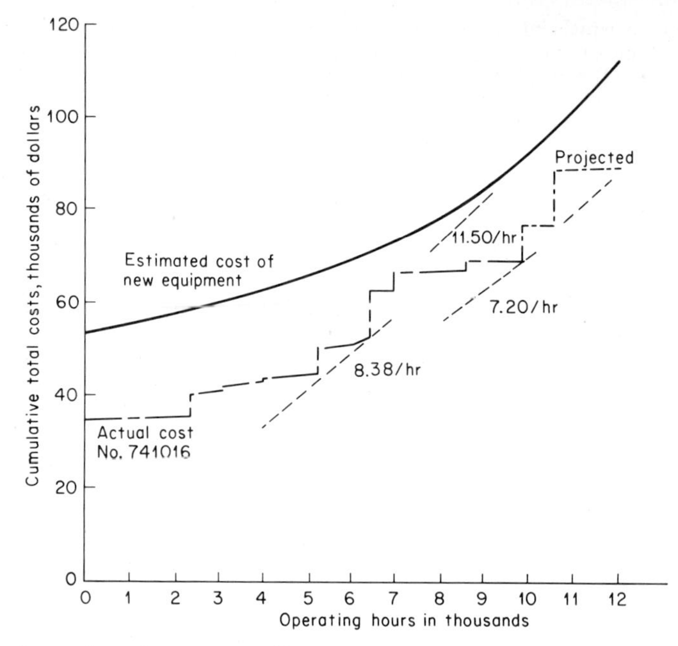

Fig. 7-2 Comparison of actual expenditures for a D-8 tractor with projected expenditures for a replacement unit.

ance for any residual value of the tractor prior to the overhaul, any trade-in allowance for the old tractor can be treated as an effective reduction in the acquisition cost of the new tractor. The projected minimum cost of $11.50 per hr for the new tractor will be correspondingly lowered by approximately $0.50 for each $5,000 of trade-in allowance.

New Tractor (life assumed to be 10,000 working hours)

With trade-in of, $	Relative cost, $/hr	With trade-in of, $	Relative cost, $/hr
0	11.50	15,000	10.00
5,000	11.00	20,000	9.50
10,000	10.50	25,000	9.00

If a trade-in of $15,000 is offered, the new tractor will cost about $10 per hr of use and the old tractor must work approximately 2,800 hr after overhaul to recover the cost of the repair at the $10 per hr rate. If the trade-in amount were increased to $25,000, the new tractor would cost about $9 per hr, and the old tractor would need to work about 3,400 hr at $9 per hr to recover the overhaul cost.

Figure 7-3 shows a break-even chart for the overhaul or trade decision. As an example, if a trade-in of $20,000 were offered, the overhaul would have to be amortized over a 3,000-hr period in order to be justified.

The method just described is a general one for construction equipment and may not be applicable to other types of equipment or to equipment operated under extreme conditions. The most important aspect of replacement is a good analysis of the difference in productive capability. It is also important to make a good appraisal of the value left in a piece of equipment; there is sometimes a tendency to forget that equipment expenditures do not become costs until they are consumed at some later date. For example, an expenditure for a new generator adds value that becomes a part of the cost of using the equipment in the future. The new generator may have substantially more value to the contractor as a new piece of hardware with many hours of life remaining than as a piece of hardware to be immediately traded in or sold. The overall value of many replaced parts in terms of additional useful life is often underestimated.

Although it is possible to allocate some downtime cost to the piece of equipment causing it, this would be very impractical with most accounting systems presently in use. Also, the average downtime costs for a D-8 tractor could have no relationship to the future work planned for the unit. A policy of down-grading the use as a tractor gets older is frequently followed.

All successful replacement systems have the common requisite that a contractor must know his equipment condition and cost in order to be able to intelligently make the best trades for it.

Financing If the investment in equipment represents old, unreliable, noncompetitive equipment, the contractor may find it impossible either to bid jobs profitably with

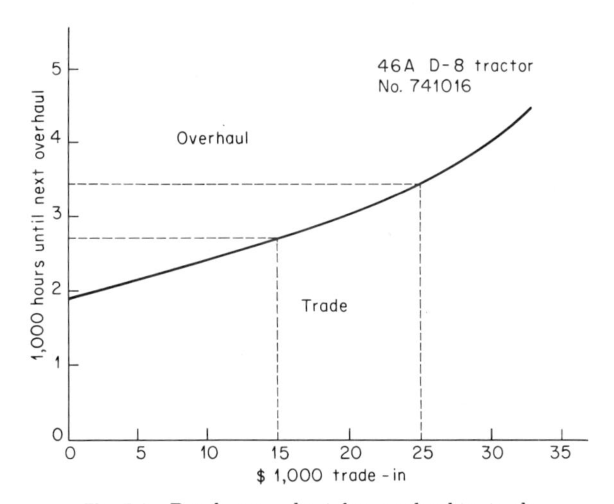

Fig. 7-3 Break-even chart for overhaul or trade.

existing equipment or to borrow funds to purchase new equipment. The company with modern, high-production equipment may still require protection against the unforeseen loss of contracts or other contingencies that are inherent in the construction industry. Additional financial leverage may also be desired to provide the necessary cash flow for a profitable expansion. When financing of equipment becomes necessary, there are four common alternatives. These are dealer financing, leasing, bank financing, and financing through a commercial finance company.

Most dealers have convenient, uncomplicated financing plans available at reasonable interest rates. They are often in a better position to require a smaller down payment. Financing through the dealer does not give as much freedom to shop the market as do other plans, however.

Leasing offers several attractive advantages and has become a very popular form of financing. There may be a tax advantage by expensing monthly rentals as they are incurred. Whether this will be allowed by the tax authority depends upon the nature of the lease and, specifically, the provisions for capturing title to the equipment after expiration of the lease. Personal property taxes may favor leasing. Manufacturer's warranty may also be passed along to the contractor by the leasor. Equipment from several manufacturers may be combined under one lease and, with a firm commitment from a lessor in hand, a contractor may shop as though he were a cash purchaser. Usually no down payment is involved, so leasing may appeal to companies short on cash since it will enable them to present stronger balance sheets.

Table 7-3 shows three basic leasing plans with two terms each.[22] Since the cost of money under leasing plans is generally higher than interest under traditional financing, the proper selection of a plan is important and the benefits must be carefully considered.

A contractor's commercial bank will generally extend additional credit on a secured basis at interest rates more favorable than those available from commercial finance and leasing companies. Also, it will traditionally make more favorable unsecured loans based upon the strength of a company's balance sheet than will other lenders.

TABLE 7-3 Equipment Leasing Alternatives

I. 5-year lease—(available investment tax credit taken by lessor)
Selling price—$1,600,000
Rate equivalent to 7.25% simple per annum
60 rentals of $31,870
Total rentals—$1,912,200
Option to purchase at conclusion of rentals—fair market value

II. 10-year lease—(full 7% investment tax credit taken by lessor)
Selling price—$1,600,000
Rate equivalent to 5.5% simple per annum
120 rentals of $17,365
Total rentals—$2,083,800
Option to purchase at conclusion of rentals—fair market value

III. 5-year lease—(investment tax credit passed on to lessee)
Selling price—$1,600,000
Rate—8.29% simple per annum
60 rentals of $32,667
Total rentals—$1,960,020
Purchase commitment—lessee must purchase equipment within 45 days of final rental payment for 5.7% of original selling price of $1,600,000 or $91,200

IV. 10-year lease—(investment tax credit passed on to lessee)
Selling price $1,600,000
Rate 7.91% simple per annum
120 rentals of $19,334
Total rentals—$2,320,080
Purchase commitment—lessee must purchase equipment within 45 days of final rental payment for 28% of original selling price of $1,600,000 or $448,000

V. 5-year lease—(investment tax credit passed through to lessee)
Selling price—$1,600,000
Rate—9.5% simple per annum
60 rentals of $33,544
Total rentals—$2,012,640
Option to purchase equipment—$1

VI. 10-year lease—(investment tax credit to lessee)
Selling price—$1,600,000
Rate—9.1% simple per annum
120 rentals of $20,283
Total rentals—$2,433,960
Option to purchase equipment—$1

SOURCE: L. E. Schaffer and W. C. Kraft, *Rock Products*, Chicago, April, 1969.

The commercial banks often limit the amount they will lend to an individual contractor, and it is possible that equipment financing will restrict the amount available for unsecured loans.

One of the large commercial finance companies may finance a contractor's equipment as a chattel mortgage loan, or it may purchase the dealer's installment sales contract. Under a chattel mortgage loan the title passes to the purchaser and debt collateral often includes greater assets than the specific piece. The interest as well as the default risk is generally higher than for a title retention plan, and generally the latter is preferred.

The extent to which favorable cash flow can be maintained and taxes are prudently managed compounds the significance of having acceptable debt alternatives available. Financing is certainly one of the most important aspects of equipment economics and should be accorded careful investigation based upon competent advice.

MAINTAINING AVAILABILITY

The importance of adequate equipment maintenance continually increases. The trend toward larger, more sophisticated machines keeps the cost of construction equipment constantly increasing and, along with the greater production rates, there is a greater loss whenever a machine cannot be producing. Improved reliability has been engineered into machinery, and better materials and more sophisticated controls are now being used. However, even the best equipment can be misused by the operator who tries to maintain production under adverse job conditions after a bad night off the job. Good operating practices are overlooked, breakdowns occur, and valuable time and resources are wasted. Downtime has been reduced considerably during the past 15 years, but a compilation of time utilization data for representative construction equipment gives an indication of the magnitude of the problem.[23,24] HRB Committee Report No. 23, dated May, 1953, lists maintenance and repair delays of 15 min

TABLE 7-4 Percentage Distribution and Average Duration of 585 Maintenance and Repair Delays of 15 Min and Over Experienced by Rubber-tired Tractor and Scraper Combinations on Highway Grading Jobs

Type of delay	Percentage of total maintenance and repair delay time	Average duration of each delay, min
Wait for repair parts	20	528
Tractor motor, transmission and drive assembly	40	245
Tractor—other	12	100
Power control unit	11	136
Scraper control system—cable, hydraulic lines, etc	6	56
Scraper—other	6	66
Greasing, oiling, and refueling	5	26
Total	100	137

SOURCE: Highway Research Board Committee Report No. 23, May, 1953.

or more each, as normally experienced by rubber-tired tractor and scraper combinations on active highway grading jobs. These will account for a total downtime to each unit of about 1.9 hr out of every 10-hr workday. Table 7-4 uses data from this report to show the type and extent of the delays encountered. The study showed an average frequency of interruptions for service of once every 12.2 hr and an average delay duration of 137 min.

A contractor's records will probably indicate a lower downtime percentage today, but the importance of minimizing downtime has increased even more.

Reducing Downtime It is a project manager's prime responsibility to make a profit on his project while exercising reasonable prudence in the care and utilization of his resources. To get the most out of the equipment dollar, it is essential that construction management appreciate the relative costs of maintenance and of downtime. An overemphasis upon production costs may influence contractors to push their equipment to obtain maximum production performance. For example, sideboarding scrapers and trucks is still common practice.

The additional equipment costs resulting from adverse job conditions are often underestimated or ignored. If a load on a bearing is increased to twice the design load, its expected life is decreased to one-tenth its design life. The life of a bearing varies inversely with the load raised to the 3⅓ exponential power. Reduction in gear life under excess loads is even greater than bearing-life reduction. The effect of speed on bearing life is almost a straight-line relationship. If speed is increased by 20 percent, then bearing life is reduced by 20 percent. Abuse is sometimes justifiable when an unusual project or circumstance develops, but construction management needs to be cognizant of the consequent price that is paid.

Figure 7-4 shows a concept that has been receiving increasing attention recently. As construction processes become larger, more automated, and more highly integrated into well-balanced systems, the adverse effect of downtime is more pronounced. The use of larger assemblies or standby units reduces both downtime and the skill level required of field mechanics. At one extreme, no spare parts may be carried on the project. At the other extreme, for example, a standby central-mix plant could be kept at the jobsite. Somewhere between these extremes is the most economical level of standby inventory.

Most construction projects lend themselves to segmentation, where an interruption on an early operation will not affect a later operation unless the interruption is of an unusually long duration. For example, sufficient aggregate can be stockpiled so that a crusher breakdown can be repaired before the central-mix plant runs out of aggre-

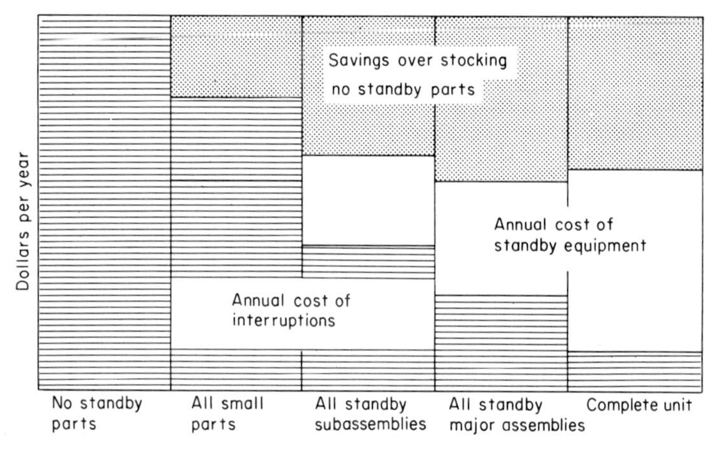

Fig. 7-4 Concept of optimum inventory component level.

gate. Shutdown of the central-mix plant, however, stops the hauling units and paving train as well. As the continuous-process plant becomes more common in construction, analysis of the economics of standby parts or components may be desirable.[25]

Consider a Euclid belt loader pulled by two D-8 tractors servicing a balanced hauling spread of 10 Euclid wagons. An analysis of maintaining standby loader components may be approached as follows:

Assume:

Cost of standby loader	$35,000	
Less salvage	$ 7,000	
Annual cost at 8 years		$3,500
Debt service (40 %)(10 %)(56 %)(35,000)		750
Maintenance		2,000
Total annual standby cost		$6,250
Spread downtime cost per hour	$150/hr	

2—D-8 (loader)
10—LDT34 bottom dump
2—No. 16 motor grader
1—834 compactor
2—D-8 dozer (cut and fill)
1—Water wagon
Miscellaneous

	Without standby	With standby
Engine		
100-hr delivery—6-hr installation		
Major repair 1.5 times per yr		
Annual downtime	159 hr	9 hr
Belt		
30-hr delivery—16-hr installation		
10 times per yr		
Annual downtime	460 hr	160 hr
Other		
Frequency—75 times per yr		
Average duration—2 hr		
Annual downtime	150 hr	

Alternatives:

Ci = Cost of interruption caused by component failure

Ci = Annual frequency × avg hr duration × downtime cost per hr

1. Complete standby loader	
Annual cost of standby unit*	$ 6,250
Ci = (86.5)(0.3)(150)	3,892
Total downtime costs	$ 10,142
2. Both belt and engine standby	
Cost of standby belt	$ 550
Cost of standby engine	1,575
Ci belt = (10)(16)(150)	24,000
Ci engine = (1.5)(6)(150)	1,350
Ci other = (75)(2)(150)	22,500
	$ 49,975
3. Standby belt only	
Cost of standby belt	$ 550
Ci belt = (10)(16)(150)	24,000
Ci engine = (1.5)(106)(150)	23,850
Ci other = (75)(2)(150)	22,500
	$ 70,900
4. Standby engine only	
Cost of standby engine	$ 1,575
Ci belt = (10)(46)(150)	69,000
Ci engine = (1.5)(6)(150)	1,350
Ci other = (75)(2)(150)	22,500
	$ 94,425
5. No standby	
Standby cost	-0-
Ci belt	$ 69,000
Ci engine	23,850
Ci other	22,500
	$115,350

* It is assumed that 0.3 hr will be required to shift to the standby loader in the event of a breakdown of the primary loader.

A major project involving precast components may warrant consideration of the length of time a piece of equipment such as a gantry crane could be down for repairs. It may be possible and desirable to build an inventory of the precast components which require the use of the crane so that construction can be continued by using up the inventory when the crane must be taken out of service for repairs.

To be complete, an evaluation of an acceptable level of downtime must take into account the economics of field repairs versus home shop repairs versus dealer repairs. It must consider the level of parts a dealer may consign to a job as opposed to a purchased inventory. For the more complex assemblies, the specialized training of the dealer must be weighed against the training of the contractor's mechanic.

Current practice for establishing inventories is to select the percentage of original equipment cost that will be maintained and then develop a list of spare parts to equal this amount. On domestic work where a fast part supply is available, the average inventory level is around 3 percent of original cost. Most major manufacturers are able to locate parts if available anywhere within the United States and have them on

TABLE 7-5 Laboratory Report Showing Oil Contaminants

Laboratory Report

RECEIVED FROM:
H. B. ZACHRY COMPANY
P. O. BOX 21130
SAN ANTONIO, TEXAS 78221

OIL SAMPLING FILE

PAGE NO. 13

WRIGHT OIL CO. LABORATORIES
301 S. MEDINA ST.
SAN ANTONIO TEXAS 78207

DATE PROCESSED: 08-05-69 JOB NO: JOB LOCATION:

UNIT NUMBER	SYS CODE	EQUIPMENT DESCRIPTION	SAMPLE DATE	TOTAL ENG. HRS./MILES	ANTI FRZ	MOIS-TURE	FUEL DIL	VISC @ 210° F	FLASH POINT	SAE GRD.		SPECTROGRAPHIC ANALYSIS							
										FIELD	LAB	BORON	LEAD	IRON	ALUM	COPPER	CHROME	TIN	SILIC
453308	A1B	CAT 944 LOAD 2YD 63	04-08	2284				062	440	000	000	011	007	006	004	000	001	015	004
453308	A1B	CAT 944 LOAD 2YD 63	04-26	2401				069	440	000	000	003	018	049	012	000	014	021	005
453308	A1B	CAT 944 LOAD 2YD 63	05-24	2528				000	067	000	425	002	013	034	007	000	006	015	004
453308	A1B	CAT 944 LOAD 2YD 63	06-14	2586				067	420	000	030	002	010	015	005	000	004	015	004
453311	A1B	CAT 944 LOAD 2YD 64	09-08	5811				073	443	030	030	001	007	017	007	000	005	023	004
453311	A1B	CAT 944 LOAD 2YD 64	09-21	5885				072	465	030	030	003	007	017	006	001	003	018	004
453311	A1B	CAT 944 LOAD 2YD 64	10-05	5951				000	000	030	000	002	007	021	008	001	009	027	004
453311	A1B	CAT 944 LOAD 2YD 64	11-09	6097				065	435	030	000	002	007	022	006	000	008	023	004
453311	A1B	CAT 944 LOAD 2YD 64	11-23	6146				071	450	030	040	003	007	017	004	002	009	023	004
453311	A1B	CAT 944 LOAD 2YD 64	12-14	6204				068	445	030	000	002	007	014	004	001	005	029	004
453311	A1B	CAT 944 LOAD 2YD 64	01-04	6228				064	430	030	000	001	008	017	004	002	004	015	004
453311	A1B	CAT 944 LOAD 2YD 64	02-01	6267				065	430	030	000	003	007	031	007	001	007	022	004
453311	A1B	CAT 944 LOAD 2YD 64	04-26	6415				067	440	030	000	002	008	017	006	001	000	013	004
453401	A1B	EUC L20 LOAD 2 1/2Y 63	02-12	1485				053	350	030	000	087	016	118	024	005	005	033	028
453402	A1B	EUC L20 LOAD 2 1/2Y 67	02-06	3489				066	440	030	000	095	007	070	013	000	000	010	018
453403	A1B	CAT 950 LOAD 2 1/2Y 68	09-28	1075				000	000	030	000	005	015	077	016	006	011	028	005
453403	A1B	CAT 950 LOAD 2 1/2Y 68	03-29	1821				071	470	000	000	005	008	038	013	000	007	022	004
453403	A1B	CAT 950 LOAD 2 1/2Y 68	04-26	1924				067	445	000	000	003	007	038	008	000	005	018	004
453403	A1B	CAT 950 LOAD 2 1/2Y 68	05-24	1987				064	395	000	030	003	007	022	005	000	000	015	004
453403	A1B	CAT 950 LOAD 2 1/2Y 68	06-14	2063				065	420	000	030	003	007	032	007	000	002	024	004
453505	A1B	CAT 966 LOAD 2 3/4Y 65	11-09	2002				072	460	030	000	091	011	046	016	000	016	014	016
453505	A1B	CAT 966 LOAD 2 3/4Y 65	12-14	2034				066	440	030	000	012	007	014	004	001	002	025	004
453505	A1B	CAT 966 LOAD 2 3/4Y 65	01-18	2084				063	000	030	000	007	008	030	009	000	005	021	010
453505	A1B	CAT 966 LOAD 2 3/4Y 65	01-31	2093				063	435	030	000	006	009	008	004	000	000	014	004
453505	A1B	CAT 966 LOAD 2 3/4Y 65	05-24	6654				066	435	000	030	004	009	012	003	000	000	012	004
453505	A1B	CAT 966 LOAD 2 3/4Y 65	06-14	6735				066	415	000	030	003	008	011	005	001	000	015	004
		LOAD 2 3/4Y 65	11-23	2018				073	470	030	040	084	007	017	006	000	001	008	004
		LOAD 2 3/4Y 65	11-09	1707				071	460	030	000	068	007	026	010	000	001	018	005

SOURCE: H. B. Zachry Company.

the jobsite within 24 hr. On overseas work, the inventory level will vary more, depending on the urgency in completing the project. Public works projects will require inventory levels in the 6 to 10 percent range. A defense installation may require 10 to 18 percent or higher.

Inventories for individual equipment models may be maintained at any desired level up to and including a complete standby unit. The factors to be considered include (1) the amount of construction that would be stopped by the failure of the machine; (2) the time required to obtain the necessary part at the jobsite; (3) the probability of failure considering hours of use since new or last overhaul as well as severity of use; (4) the complexity of the unit or assembly; and (5) the disposition of excess spare parts after project completion.

Servicing The costs and benefits of preventive maintenance programs are difficult to evaluate, and practice varies widely. An inadequate lubrication program can cause quick deterioration of wearing parts and loss of performance through contamination. Breakdown repair cost and downtime can quickly become excessive. A blown gasket permitting glycol antifreeze to contaminate the oil can rapidly necessitate an engine overhaul.

A good preventive maintenance program prolongs equipment life, reduces cost of downtime, and minimizes lubrication cost. It can mean the difference between 85 and 95 percent availability. This 10 percent increase in availability may represent an effective reduction in equipment costs of about 3 percent. Standardization of lubricants can reduce purchase cost and inventory, and oil drain intervals may be extended through analysis of oil samples. Testing may be accomplished by differential infrared analysis, membrane filtration analysis, spectrographic analysis, spot tests, and other laboratory procedures.[26] Many of the oil distributors can provide a testing service at moderate cost.

Table 7-5 shows a computer-produced report from a system employing spectrographic analysis. The permissible amount of contaminant in the oil depends upon the type of engine and the oil used. Gasoline engines will have lead from the fuel, where diesel engines will not. Heavy-duty S-1 oil, recommended for some equipment, has a high boron content as opposed to S-3 oil, recommended for other types of equipment. The effective use of a sampling program requires frequent consultation with equipment dealers to diagnose the possible causes of high contaminant levels.

Repairs With the best of care, equipment breakdowns will still occur. The contractor must then seek the most economical level of maintenance capability to meet his needs. The economics of an optimum balance between number of mechanics, parts inventory, operator training, movement of equipment for repairs, and dealer arrangements are very complex and intangible. Linear programming techniques have been proposed[15] to plan and schedule maintenance work so as to best utilize the facilities available and thus minimize repair costs. For example, a contractor with more than one shop will frequently be able to do particular maintenance jobs more economically in one shop than in another. A dealer may also do certain repairs that the contractor cannot perform. The formal programming techniques attempt to schedule the repairs to the best shop, but their effective application is hampered by their exacting requirements for accurate input data.

The establishment of repair priorities and the effective scheduling of repair jobs will reduce repair costs and improve the utilization of equipment. The success of such efforts will depend upon knowledge of the construction schedule and the condition of working equipment.

OWNERSHIP AND OPERATING COSTS

The cost of using a given piece of equipment may vary widely due to the influence of the many relevant variables. These variables include local weather conditions, timing of the purchase, obsolescence due to the introduction of new equipment, and other difficult-to-predict factors. Recommended procedures for estimating costs vary from simple, straightforward forms that do not attempt to relate cost to the users' specific applications to very sophisticated models whose requirements for input data are well beyond the ability of the average contractor to provide. The following paragraphs will present an overview of equipment economics, and subsequent paragraphs will identify accounting information needs and outline procedures for estimating and monitoring equipment costs. From the viewpoint of establishing the right price for a given project, the contractor is interested in the equipment investment which must be committed to a project and in the equipment value that will be consumed in order to complete the project. He is also concerned with the administrative costs of owning and managing the equipment.

Value Accumulation By the time a contractor has moved a piece of equipment onto a project, he has already made a substantial investment. This includes the unamortized cost of the equipment and any major modifications to it, plus the miscellaneous costs of ownership. If it is not a new piece of equipment, his investment may include a prior overhaul plus preventive and field maintenance costs.

The acquisition cost of a piece of equipment is normally regarded as the machine price including attachments, the sales tax if any, and the freight charges to the contractor's yard or initial project. In the event of a trade-in, the equipment acquisition cost may be regarded as the purchase price adjusted by the difference between the trade-in allowance and the expected price the contractor could receive after selling expenses on the open market.

Miscellaneous costs of ownership include insurance, ad valorem or personal property taxes, and the cost of special service equipment. Interest is generally included as a miscellaneous ownership cost but cannot be properly considered without relating it to the overall capitalization policy of the company. If a contractor elects to finance half his equipment so that he may invest his capital in a new nonconstruction venture, the cost of interest would not be an equipment cost. On the other hand, if financing becomes necessary in order to obtain larger or more modern equipment, the financing cost would rightfully be allocated to the equipment.

Major modification costs should be regarded in the same manner as investment in new equipment. Equipment modification may serve a need that is not satisfied by available standard equipment, or it may be a method of realizing greater value than could be obtained by trading in old equipment or selling it outright.

A major overhaul restores equipment value that may be realized either through additional use or through a higher sale price. In either event, the overhaul cost represents an additional investment in equipment value to be realized at some future date.

Preventive and field maintenance also restore equipment value which is subsequently consumed through equipment use. A broken hydraulic line, fouled spark plug, or even contaminated oil will make the equipment less valuable, whether to the project or to a different user. Replacing the parts or the contaminated oil increases the value of the equipment to the project or to a subsequent purchaser. In order to simplify accounting and also to better control expenditures, field maintenance costs are often carried as project costs in a contractor's accounts. It is difficult to define precisely where a field repair ends and an overhaul begins, and it is more desirable to think of all repairs as adding or restoring value. Since fuel and lube are immediately consumed, it is appropriate to treat them as project costs. Move-in costs are also immediately consumed and fall within this latter category.

Generally, any equipment expenditure that will not be used up on the first project should be regarded as an additional investment. In return, the contractor expects to receive a number of standard hours of usage during which the equipment must be earning a profit.

Loss of Value As equipment becomes more sophisticated, with an increased number of moving parts and more complex systems for control, the consumption of ownership and maintenance value becomes more difficult to predict. The factors contributing most to the loss of value of a particular piece of equipment to a contractor include (1) use of equipment causing wear on moving parts; (2) operational obsolescence; (3) time-dependent physical deterioration; (4) risk cost of ownership; (5) utility cost of ownership; and (6) miscellaneous ownership costs.

The ranges of ownership and operating costs for different types of equipment under varying job conditions are shown in Table 7-6. Figure 7-5 shows the influence of job conditions and utilization on D-8 tractors. Under unusual job conditions, the variance may be greater. As the estimated repair cost approachès the upper value of the ranges given, management should actively seek other methods of doing the job. Equipment repair costs under "severe" and "very severe" working conditions are difficult to estimate because the type of breakdown becomes less predictable. Frame failures, as an example, occur in a random fashion. The rule of thumb that a 100

TABLE 7-6 Equipment Repair Costs as Percentages of Initial Cost Less Tires

Equipment	Type of service		
	Easy	Medium	Severe
Track-type tractors	70	90	130
Tractor-drawn scrapers	30	40	60
Pipe layers	20	30	40
Wheel-tractor scrapers	20	90	130
Wheel-tractor bottom dumps	40	50	70
Off-highway trucks	60	80	110
Wheel-type tractors	40	60	90
Track-type tractors	70	90	130
Wheel loaders	40	60	90
Motor graders	30	50	70

SOURCE: Caterpillar Performance Handbook.

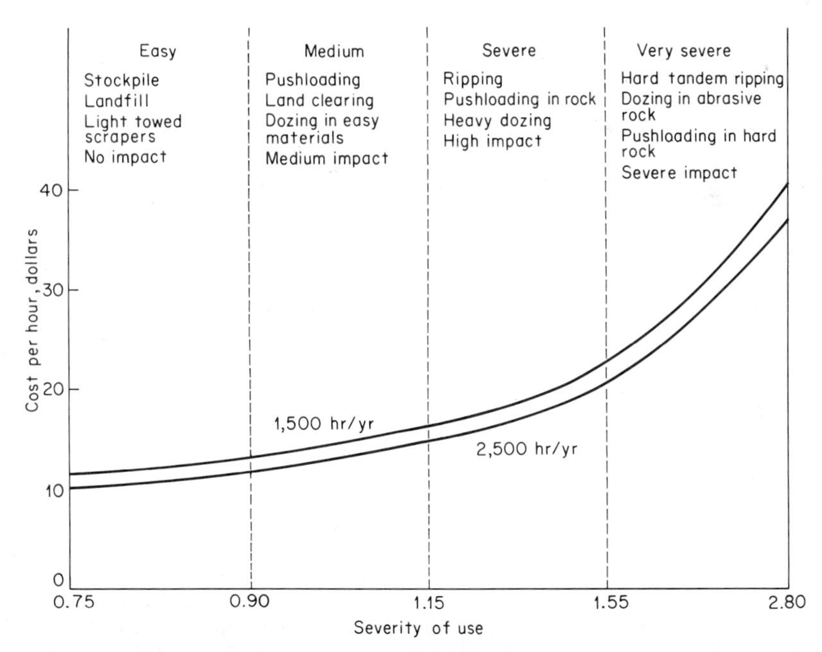

Fig. 7-5 Hourly cost of D-8 tractors under varying job conditions.

percent overload on a bearing results in a 90 percent reduction in bearing life should be remembered.

A survey of used equipment in large contractors' yards will inevitably disclose pieces of equipment in usable condition that cannot be economically used or sold for more than scrap value. The displacement of towed sheepsfoot rollers by self-propelled rollers and tamping wheels for tractors is an example of obsolescence that has caused the former to lose value. A survey of production rate history indicates that the trend toward larger-sized equipment has been geometric rather than linear. As a long-term average effect, the growth in productivity in earthmoving has approximately equaled the growth in inflation. In the period from 1940 to 1960, this growth has averaged around 3½ percent per year. As the rate of technological change increases, the rate of obsolescence and the rate of inflation have increased also.

Physical deterioration takes place even when equipment is not working. Tires deteriorate from age, exposed metals rust, seals dry out and leak, seat coverings and upholstery become brittle and develop cracks, and theft occurs. These costs continue during the times when equipment is not being utilized, and appropriate allowances must be made for them.

The use of equipment by contractors normally carries a higher risk than does its use in other industries. Dam and channel construction exposes equipment to flooding. Work around steep grades increases the likelihood of runaways and complete wreckage of equipment. Booms and power lines are a disturbing mixture. The costs that may result from these factors must be borne as accident losses or as insurance costs.

Unlike other industries, construction does not involve a fixed plant. When work at one project runs out, the equipment loses some of its value as an income-producing asset. Additional moving cost must be incurred to relocate it to another income-producing location. It may also lose some value in its present configuration, and expenditures may be required to change or readjust attachments for the next available project. Since the magnitudes of these costs depend upon the nature of the next project, they are normally estimated as job costs and expenditures are charged directly to the project. In evaluating equipment cost and technological obsolescence, the cost of teardowns, moves, and setups may be significant, however, and should be considered in any replacement study.

Administrative and Overhead Cost Certain administrative and overhead costs must be borne by the contractor who uses heavy equipment. These costs, while not adding to or taking away from the value of his equipment, are generally proportional to the size of his equipment pool. They include accounting expenditures, storage facilities for idle equipment, nonallocable taxes and insurance, and maintenance facility cost. These costs should be applied as a burden, along with the hourly ownership and operating cost.

Shop cost poses a particular problem, and care should be taken not to distort cost records by the manner in which overhead is handled. A common practice is for contractors to estimate and charge field shop cost directly to the project, while home-office shop costs are applied to the equipment through an applied burden charge. Under this policy, however, costs are distorted to the extent that major overhauls are done in the field shop. If field shop costs are to be handled as job overhead, it is preferable that central shop cost be treated as a general overhead item and budgeted accordingly.

The preferred way of allocating both central shop and field shop overhead to the equipment is by applying a burden to all repair labor and part charges. Shop overhead cost will generally be about 20 percent of labor and part charges. Normally, a central shop will be slightly more efficient than a field shop, and the percentage applicable to it will be slightly less than the percentage applicable to field shops.

Some taxes and insurance are difficult to charge directly to an equipment unit. A blanket liability insurance policy on vehicles is an example. For the average contractor, the broad-coverage insurance and taxes will amount to about 5 percent of total equipment cost. Insurance on specific pieces of equipment should either be charged to the equipment as a miscellaneous ownership cost or charged to the project, whichever is more applicable.

The cost of warehousing equipment parts and supplies can best be allocated as a handling charge on the warehouse issues. A small percentage will normally cover warehouse cost.

INFORMATION REQUIREMENTS FOR FINANCIAL AUDIT

Actions by the federal government to regulate spending for planned improvements through tax legislation are felt immediately by the construction industry, and a thorough understanding of the economic effect is essential to profitable operation. Adequate equipment records are needed not only for tax requirements but also for establishing a financial capability. Public contracts and most private projects require performance bonds that may be obtained only by demonstrating a financial position satisfactory to a surety. Good equipment records go beyond these minimum requirements so as to aid in the analysis of tax decisions and to provide a measure of the contractor's performance in equipment management.

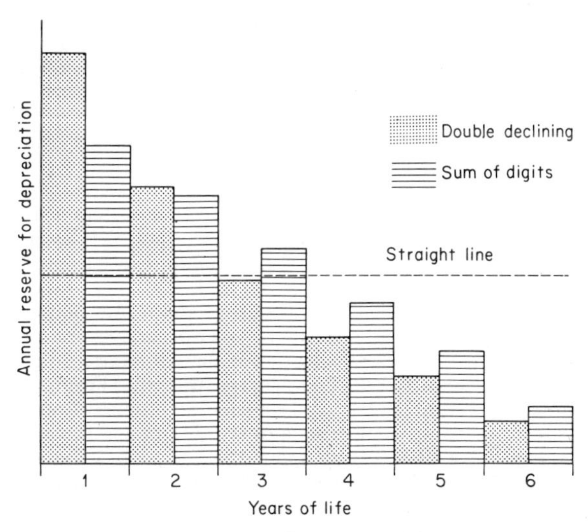

Fig. 7-6 Methods of depreciation.

Information for Taxes[20,21] The contractor is generally most concerned with the influence of income tax law and of the investment tax credit. Other taxes include sales, property, ad valorem, state income, etc.

Studies of the influence of income tax law show that profits are materially affected by the way in which equipment depreciation is handled for tax purposes. The Internal Revenue Code of 1954 limited the use of accelerated methods of depreciation to new equipment which was acquired after December 31, 1953, and which had a useful life of greater than three years. Until January 1, 1962, any gain on the sale of equipment above the book value was treated as capital gains and taxed accordingly. The contractor was charged with holding his depreciation to a reasonable allowance by the depreciable life and salvage value selected. A shorter depreciable life and lower salvage value increased the amount of capital gain that was taxable at the lower rate. The legislation changing the gain on an equipment sale to ordinary income removed the advantage of the lower capital gains rate, but the expanding contractor may still defer taxes by the accelerated methods. Figure 7-6 shows a comparison of the three most common depreciation methods.

The Revenue Act of 1962 provided an incentive to increase the expansion rate by allowing an investment tax credit. The Investment Tax Credit Law provided a credit for the purchase of new equipment with a depreciable life greater than 4 years. Equipment with at least 8 years' life qualified for a 7 percent credit. A life between 6 and 8 years may receive two-thirds of the 7 percent, and a life between 4 and 6 years qualified for one-third of the full credit. Used equipment must be valued at less than $50,000 to qualify for this credit. Other provisions of the act limited the amount of the credit to a percentage of the tax liability.

The effect of the investment tax credit has been to offset the incentive to use shorter depreciable lives without discouraging new investment. With any gain on sales now taxed as ordinary income, contractors are motivated to consider longer lives than previously contemplated so as to take advantage of the investment credit. In 1966, in order to reduce inflationary pressures, Congress suspended the investment credit for investments of more than $25,000. With the suspension, the inflationary pressure eased and the incentive was reinstated after a 5-month suspension. At present (1969), the President has again asked for termination of the investment tax credit, and it is expected that it will be discontinued.

The general effect of the investment credit has been to lengthen the economic life of equipment when other factors are not significant. The 5-year life of a piece of equipment is more likely to be lengthened to 6 years if the machine can be retained the additional time without a major overhaul and without significant increase in cost due to downtime or the introduction of new equipment creating obsolescence.

Equipment parts consumed upon state and federal projects are normally exempt from state sales tax. The laws vary to such an extent that generalizations may be misleading and local advice should be obtained. In any event, state and local legislation may require additional record keeping.

Auditing Requirements Some contractors apply their depreciation and maintenance charges directly to the project to which the equipment is assigned. The common practice in the larger companies is to use a standard rate approach and allocate the cost of equipment to the individual contract and often to an item within the contract. A system combining a calendar time charge and a use (hours worked) charge has been shown to have substantial advantages over other systems. The most obvious advantage of separate time and use charges is that estimates of utilization are not as sensitive to error as they are in a single charge system. From a practical standpoint, usage cannot be accurately recorded on small equipment and, where measurement of usage is impractical, a time charge only must be used. The calendar time charge to the project starts when the equipment is assigned to the job and continues until it is released for transfer to the next project. In the H. B. Zachry Company, all equipment operating costs are normally allocated as use charges. Ownership costs are allocated in equal proportions on a time charge and on a use charge. This method further reduces the effects of inaccurate estimates of utilization. By allocating the ownership charges 50 percent on a time basis and 50 percent on a use basis, the problem of second-shift work is handled in an acceptable manner. In a few of the larger companies, different rates may be used under different conditions, and age is often a basis for using a reduced rate.

Where a usage charge is applied, the equipment costing system is usually similar to a payroll system with labor distribution. Usage is reported on time cards, the appropriate rates are applied, and an equipment usage register and equipment distribution journal are prepared for each project.

Two uncommon features of the Zachry system include a double allocation of the time charge and the use of an adjustment factor for severe conditions. Figure 7-7 shows 50 percent of the ownership cost allocated to a project overhead account. The charges are then reallocated to direct cost items. Table 7-7, taken from an equipment distribution journal, shows the use of a severity factor and a utilization factor. Equipment costs are allocated by daily and hourly rates to distribution codes within a project. The daily rate is converted to an hourly rate by the use of a utilization factor. This factor is estimated for each item and identified by a distribution code at the time the project is bid. The severity factor reflects an estimated deviation from the average condition upon which the standard rates are based. The cost difference between the estimated utilization and the utilization actually reported is charged to a project overhead account. Therefore, the overhead account receives a credit if the equipment is used more than the estimated amount and a debit if the utilization is below the estimated. This system allocates all the cost to the items where the cost was estimated and makes provision for some operations to be worked on a double shift while others are worked single shift. The use of the two factors has resulted in a system in which estimated costs and allocated costs are very comparable.

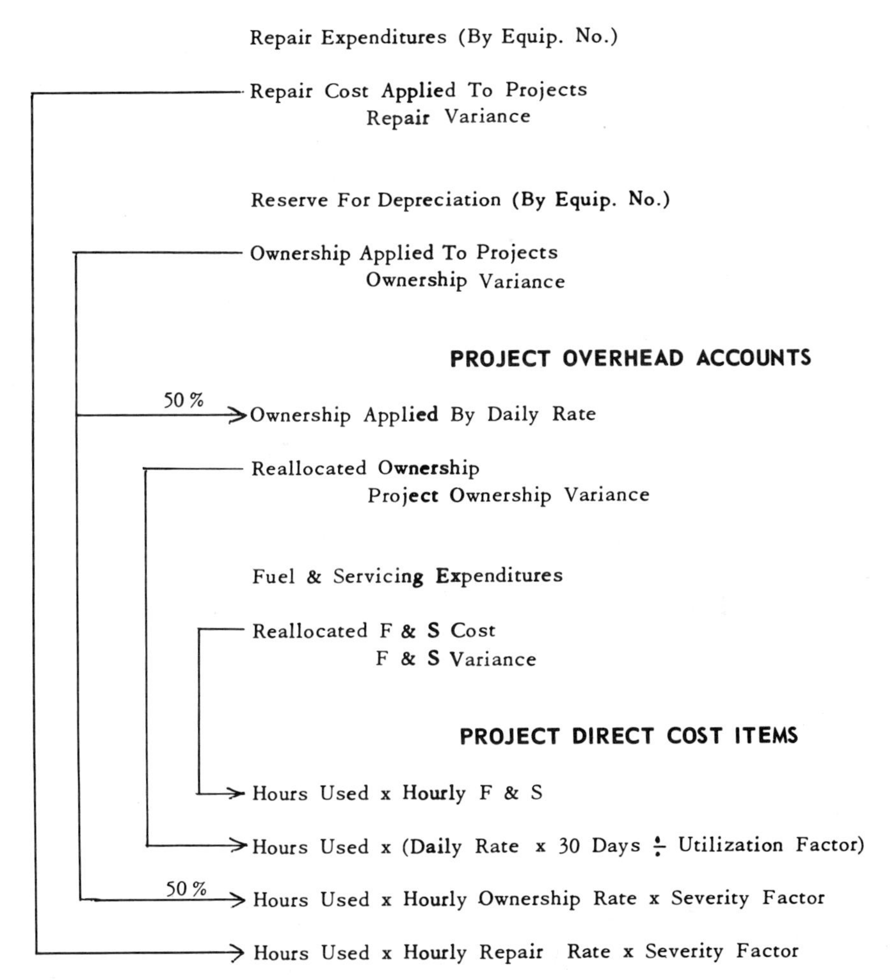

Fig. 7-7 A method of allocating equipment costs to project items. Utilization factor is the estimated hours used per month on item. Severity factor is the estimated ratio of item equipment severity to average company severity. (*H. B. Zachry Company.*)

INFORMATION FOR OPERATING MANAGEMENT

Taylor recognized that a good system was as valuable an asset as a piece of equipment. Certainly, in the management of equipment, facts rather than intuition should be the basis for decisions. The information system should provide the communications necessary to attain the management goals of achieving the optimum levels of (1) the most suitable equipment; (2) in the best condition economically feasible; (3) where needed, when needed; (4) with adequate support to provide acceptable service; and (5) at lowest cost consistent with company objectives. Management from the top is

TABLE 7-7 Equipment Distribution Journal

EQUIPMENT DISTRIBUTION JOURNAL — PAGE

PERIOD ENDING: 04/27/69 — G/L OR JOB 1450 DE CORDOVA DAM

G/L OR JOB	DISTRIBUTION CODE	SUB-DISTRIBUTION CODE	EQUIPMENT NUMBER	D H	CHARGEABLE DAYS & HOURS		UTIL. FACTOR	SEVER. FACTOR	RATES			RENTAL AMOUNT
					RENTAL DAYS	PROD HOURS			DAILY	HOURLY	HOURLY ALLOCA.	
400	831	831	74017	H		58	130	1.00	2.72	2.33	.63	172
						58*						172
400	841	841	900049	H		6	130	1.00	.28	.42	.06	2
400	841	841	901408	H		16	130	1.00	.90	.76	.21	14
						22*						16
400	1025	1025	151298	H		5	130	1.00	.05	.66	.01	3
400	1025	1025	151304	H		9	130	1.00	.05	.66	.01	6
400	1025	1025	151308	H		7	130	1.00	.05	.66	.01	5
400	1025	1025	151309	H		10	130	1.00	.05	.66	.01	7
400	1025	1025	270206	H		10	130	1.00	.56	.20	.13	3
400	1025	1025	270512	H		5	130	1.00	1.23	1.15	.28	6
400	1025	1025	270608	H		9	130	1.00	1.64	1.73	.38	18
400	1025	1025	600401	H		5	130	1.00	6.87	4.95	1.59	32
400	1025	1025	645287	H		5	130	1.00	10.62	4.35	2.45	34
400	1025	1025	645806L	H		5	130	1.00	26.19	14.70	6.04	104
400	1025	1025	647001	H		10	130	1.00	29.92	22.17	6.90	291
400	1025	1025	647002	H		50	130	1.00	29.92	22.17	6.90	1454
400	1025	1025	732705	H		10	130	1.00	7.69	4.48	1.77	63
400	1025	1025	741019	H		5	130	1.00	15.16	9.74	3.50	67
400	1025	1025	741609	H		9	130	1.00	22.80	13.45	5.26	168
400	1025	1025	741611	H		7	130	1.00	22.80	13.45	5.26	132
400	1025	1025	767408	H		5	130	1.00	15.80	12.46	3.65	81
400	1025	1025	770806L	H		5	130	1.00	19.83	14.35	4.58	95
400	1025	1025	886503	H		2	130	1.00	11.00	12.72	2.54	30
						173*						2599
400	1140	1143	151298	H		45	130	1.00	.05	.66	.01	29
400	1140	1143	151304	H		41	130	1.00	.05	.66	.01	27
400	1140	1143	151308	H		43	130	1.00	.05	.66	.01	28
400	1140	1143	254733	H		30	130	1.00	1.56	2.17	.36	76
400	1140	1143	255002	H		40	130	1.00	1.11	1.81	.26	83
400	1140	1143	270512	H		45	130	1.00	1.23	1.15	.28	65
400	1140	1143	270608	H		41	130	1.00	1.64	1.73	.38	86
400	1140	1143	515505	H		2	130	1.00	3.87	3.35	.89	9
400	1140	1143	549402	H		5	130	1.00	1.60	1.14	.37	8
400	1140	1143	591509	H		33	130	1.00				
400	1140	1143	600401	H		4	130	1.00	6.87	4.95	1.59	26
400	1140	1143	645287	H		45	130	1.00	10.62	4.35	2.45	306
400	1140	1143	645805L	H		45	130	1.00	26.19	14.70	6.04	934
400	1140	1143	645806L	H		45	130	1.00	26.19	14.70	6.04	934
400	1140	1143	741019	H		45	130	1.00	15.16	9.74	3.50	597
400	1140	1143	741609	H		41	130	1.00	22.80	13.45	5.26	768
400	1140	1143	741611	H		43	130	1.00	22.80	13.45	5.26	804
400	1140	1143	767408	H		45	130	1.00	15.80	12.46	3.65	726
400	1140	1143	770603	H		33	130	1.00	16.72	12.20	3.86	529
400	1140	1143	770702	H		27	130	1.00	19.83	14.35	4.58	512
400	1140	1143	770805L	H		45	130	1.00	19.83	14.35	4.58	852
400	1140	1143	770806L	H		45	130	1.00	19.83	14.35	4.58	852

SOURCE: H. B. Zachry Company.

essential for attaining these goals and requires proper purchasing decisions, strong maintenance policies, and the establishment of clear standards for performance. Equipment cost estimates must be made with reasonable accuracy. Reliable communications must be maintained between field mechanics and top management. A systematic analysis of meaningful records is essential, and information must be communicated to those responsible for equipment management.

Estimating Costs Effective management is essential in contracting, but even the most efficient contractor must obtain work at the right price to achieve long-term success. Since equipment expenditures are consumed over a long period of time and under unforeseeable conditions, cost will not be exactly as anticipated. In order to facilitate intelligent bidding and to ensure the maintenance of a profitable average, however, cost must be forecast with reasonable accuracy. How close the estimate needs to be depends upon the facts and circumstances of each individual project. Where job conditions are very uncertain, national averages may be the only pertinent data available. The Associated Equipment Distributors compile annually a reference book of nationally averaged rental rates.[9] Equipment manufacturers' handbooks normally show average cost as well as adjustment factors for tailoring a rate to the predicted job conditions. These costs and adjustments are continually being refined through additional testing and measuring procedures. Manufacturers are particularly progressive in determining component life under differing conditions.

It is more desirable to base a particular estimate upon specific costs than to depend upon national averages. Wage rates, training opportunities, and many other circumstances influence operating cost. The use of national averages may lead to the loss of projects that could have been constructed profitably or to the awarding of contracts at an insufficient price to meet equipment cost. Accuracy will be improved if careful consideration of life is given to major components of the larger units. For example, it may be desirable to consider a diesel engine and pump set as two separate pieces of equipment with separate lives and cost.

Figure 7-8 shows a typical format for estimating the average hourly rate for owning and operating a piece of equipment. For a given model of equipment, the contractor may make only one estimate based upon his average or standard job conditions and equipment use. The estimate may then be adjusted for individual project conditions.

Equipment cost estimates are used most extensively for bidding, project planning, allocation of cost to projects, and budgetary control of maintenance. The format shown in Fig. 7-8 is generally satisfactory for these purposes. It should be used with the recognition that factors such as inflation must be considered. The delivered price, line item 1, should include freight as well as attachments. Freight may be approximated at $2.50 per hundredweight. Overall company debt policy should be considered when using item 7 to ensure that interest, insurance, and taxes are not duplicated anywhere in the estimating or accounting system. If overhead cost of a central shop and administrative cost are to be included within a standard rate used by estimating and accounting, they may be added under item 11—Special Items. It is often more convenient to exclude the operator's hourly wage from equipment cost and estimate it as a separate figure.

As company policies change to adapt to different project mixes, the cost of maintaining equipment will change. If a larger percentage of repair work is returned to the dealers, equipment overhead ratios may change. Equipment cost is constantly changing, and the only way to keep accurate estimates of it is by collecting reliable equipment records.

Equipment Records To be meaningful, records must be more than listings of expenditures and more than just averages. The records must be pertinent to the day-to-day decisions and must be reliable enough to instill confidence in their use. They must be where they are needed at the time they are needed. They must be acquired at reasonable cost and convenience. The information system must be simple and flexible enough to survive in the construction-project environment. All these features are necessary, but it is also of primary importance that the information system be wanted and used by those who need the facts.

The diversity of the locations at which contractors build complicates the record-keeping task. Transferring records between jobsites as the equipment is transferred is sometimes awkward. Documents that should be transferred with each machine include its maintenance jacket, preventive maintenance records, maintenance instructions, lubrication charts, parts catalog, tire records, and any operating instructions, Certain parts that are carried in inventory should be transferred with the equipment, and inventory records should be adjusted accordingly.

To facilitate the analysis of equipment records, more contractors are turning to centralized files and data processing systems. The interest in utilizing computers has stimulated more communication between groups using the hardware, and considerable progress has been made in the past decade toward reconciling the conflicting information needs of the different functions within contractors' organizations. Part of the information is often retained in machine-readable form on a combination of magnetic tape and disk files. Table 7-8 shows typical contents and use of files retained by

No.	Item					DATE	
	Machine or Component				D7E PS	82-30 Dozer (Euclid)	
	DEPRECIATION VALUE						
1.	Delivered Price (including Attachments)				45,543.00	48,352.00	
2.	Less Tire Replacement Costs:						
	Front						
	Drive						
	Rear						
3.	Delivered Price Less Tires						
4.	Less Resale Value or Trade-in						
5.	NET VALUE FOR DEPRECIATION				45,543.00	48,352.00	
	OWNERSHIP COSTS						
6.	Depreciation: Net Depreciation Value / Life in Hours	45,543 / 10,000	48,352 / 10,000		4.56	4.84	
7.	Interest, Insurance, Taxes:						
	Annual Rates: Int. 9 % Ins. 2 % Taxes 2 % = 13%						
	Estimated Annual Use in Hours 2,000						
	Yrs. Life: 3, 4, 5, 6, 7, 8, 9, 10						
	% Invest.: 66.7, 62.5, 60.0, 58.3, 57.1, 56.3, 55.6, 55.0						
	Hourly Cost = Price x % Invest x Tot. Ann. Rate / Annual Use in Hours				1.78	1.89	
8.	TOTAL HOURLY OWNERSHIP COSTS				6.34	6.73	
	REPAIR COSTS (Allocated by Standard Rate)						
9.	Tires: Replacement Cost / Estimated Life in Hours						
10.	Repairs: % x Del. Price - Tires / Life in Hours	90%	82%		4.10	3.96	
11.	Special Items:						
12.	TOTAL HOURLY REPAIR COST				4.10	3.96	
	SERVICE COSTS (Direct Job Charge)						
	Unit Price x Consumption						
13.	Fuel: .15 x	6.7	10.0		1.01	1.50	
14.	Lubricants, Filters, Grease:						
	Unit Price x Consumption						
	Engine 1.10 x	.04	.05		.04	.05	
	Transmission 1.10 x	.03	.04		.03	.04	
	Final Drives 1.20 x	.02	.03		.02	.03	
	Hydraulics 1.10 x	.03	.03		.03	.03	
	Grease .20 x	.05	.06		.01	.01	
	Filters x	.16	.23		.16	.23	
15.	Service Labor 4.50 x	.07	.07		.32	.32	
16.	TOTAL SERVICE COST				1.62	2.21	
17.	Operator's Hourly Wage				4.50	4.50	
18.	TOTAL COST PER HOUR				16.56	17.40	

Fig. 7-8 Equipment cost estimate form.

equipment management. Additional information retained may include trends of cost, utilization, and various ratios to be used as a follow-up on management policies. A file of equipment specification data may be retained for simulation of equipment operations. Table 7-9 shows the approximate computer cost of using equipment records with an IBM 360 Model 30 computer with tape and disk storage. Electronic data

TABLE 7-8 Equipment Information Retained by a Large Contractor

File	Contains	Use
Equipment category	Utilization estimates Cost estimates Shipping data	Estimating new work Allocating cost to projects (work breakdown) Controlling maintenance expenditures Selection of categories for maintenance and scheduling system
Equipment by individual piece	Depreciation data Location Availability status Mechanical condition Usage Last maintenance check date Financing or leasing information	Depreciation schedules Allocation of cost to projects Control of utilization Control of diagnostic checks Project equipment investment
Cost history (active equipment)	Cost history Utilization history	Maintenance cost control Establishment of cost estimates Comparison of equipment types
Usage transactions. . .	Usage history	Control of equipment policies Analysis of severity cost
Diagnostic file.	Oil contaminant amounts	Indicates equipment malfunction or excessive wear
Cost history (inactive equipment)	Cost history Utilization history	Special studies such as resale value analysis

SOURCE: H. B. Zachry Company.

processing cost for routine processing of equipment information should not exceed four-tenths of 1 percent of equipment expenditures.

It is impossible at present to diagnose the amount of wear on all equipment parts, and most wear must be judged from the life of the part and knowledge of job conditions. Therefore, it is essential to retain records of part replacements for review when considering future overhauls. Minimum information retained pertaining to a repair

TABLE 7-9 The Use of Data Processing Systems for Equipment Information

File	Pieces of equipment (or classes)	Character per piece of equipment	Disk cylinders required	Annual cost to maintain, $	Time to search for individual piece, sec	Time to display data for individual piece, sec	Time to scan file, min	Cost to scan file, $
Equipment category. . .	717	300	10	575	0.2	1	0.5	0.42
Equipment by individual piece.	3,300	400	45	1,125	0.3	1	2.6	2.16
Diagnostic file (oil sampling).	500	750	15	900	0.2	1	0.75	0.65
Cost history (active equipment).	2,750	3,100	240	2,700	0.4	5	16	13.32
Usage transactions.	2,750	5,100	550	5,850	1.0	5	30	25.00
Cost history (inactive equipment).	8,500	3,100	750	8,737	1.5	5	50	41.70

SOURCE: H. B. Zachry Company.

TABLE 7-10 The Cost History of a Piece of Equipment

EQUIP. NO. 211602
DESCR. BUC 54B CCD 21/2YD 55
COST 65,300
RESERVE 59,292
BK. VALUE 6,008

EXCEPTION REPORT OF UNITS WITH RELATIVELY HIGH REPAIR COSTS INCURRED TO DATE OR DURING THE 30 DAY PERIOD ENDING 10-30-68

ESTIMATED LIFE 60
TO DATE LIFE 162
REMAINING LIFE 102 –
*** CURR MONTH REPAIR COSTS ARE EXCESSIVE.

				PRODUCTIVE			MAINTENANCE COSTS					REPAIR	OWNERSHIP		VARIANCE		
COMMENTS	YR	MO	SEV	DAYS	HOURS	LABOR	PARTS	TIRES	REPAIRS	OTHER	TOTAL	CREDITS	COSTS	CREDITS	REPAIR	OWNER	TOTAL
CURRENT TREND	68	10	100	0	0	3100	7572	0	0	0	10672	0	0	0	10672–	0	10672–
BY MONTHS	68	09	100	14–	63–	857	675–	0	0	0	182	378–	0	426–	196	426	622
	68	08	100	35–	191–	187	311	0	27	29	554	1146–	0	1168–	592	1168	1760
	68	07	100	28–	158–	118	548	0	0	6	672	943–	0	950–	276	950	1226
	68	06	100	42–	229–	55	418	0	17	28	518	1374–	0	1400–	856	1400	2256
	68	05	100	21–	89–	84	1686	0	19	0	1789	534–	0	621–	1255–	621	634–
	68	04	100	28–	78–	1118	4360	0	337	17	5832	468–	0	701–	5364–	701	4663–
	68	03	100	18–	60–	2061	5538	0	47	142	7788	360–	0	482–	7428–	482	6946–
	68	02	100	0	0	1312	4875	0	77	12	6276	0	0	0	6276–	0	6276–
PRIOR YEARS	67		100	206–	872–	1325	1016	0	21	4	2366	5232–	0	6087–	2866	6087	8953
	66		100	356–	1438–	414	1741	0	0	3	2158	8628–	26	8168–	6470	8142	14612
	65		100	242–	985–	691	2165	63	192	0	3111	5910–	0	5547–	2799	5547	8346
	64		100	276–	717–	985	1775	0	954	0	3714	4302–	611–	6329–	588	6940	7528
	63		100	352–	1846–	1234	2677	0	389	0	4300	11076–	719	8072–	6776	7353	14129
	62		100	364–	1279–	2488	5651	0	3594	4	11737	7674–	1199	8348–	4063–	7149	3086
	61		100	323–	1059–	3227	7486	0	739	9271	20723	6352–	1998	7408–	14371–	5410	8961–
	60		100	242–	1173–	2498	6959	0	994	8818	19269	7038–	2330	5550–	12231–	3220	9011–
	59		100	359–	1694–	2144	1891	0	709	854	5598	6946–	3840	8233–	1348	4393	5741
	58		100	316–	1721–	0	0	0	0	3583	3583	8865–	6582	10405–	5282	3823	9105
	57		100	365–	2230–	0	0	0	0	6135	6135	11489–	10970	13487–	5354	2517	7871
	56		100	25–	122–	0	0	0	0	2439	2439	629–	18284	738–	1810–	17546–	19356–
	55		100	237–	1760–	0	0	0	0	323	323	9067–	19590	10644–	8744	8946–	202–
TOTALS TO DATE				3849–	17764–	23898	55994	63	8116	31668	119739	98416–	64927	104764–	21323–	39837	18514

SOURCE: H. B. Zachry Company.

should include the equipment number, date, service meter reading, and a list of parts installed. One successful system is to file invoices as they are paid, for easy reference by equipment number. This centralized equipment folder then provides a history of all charges made to that equipment number and includes a copy of the work order that accompanies an invoice. The file, therefore, contains an itemized list of parts installed on the machine. The file needs to be easily accessible to the equipment manager who makes replacement decisions.

Monitoring Costs Equipment cost estimates should continually be monitored and maintained to provide a measure of performance for project managers and to permit budgetary control over equipment cost. The level at which budgetary control of maintenance is attempted varies significantly between companies. Most companies record their cost to the individual piece of equipment by type of expenditure. A typical breakdown of the types of expenditure used is (1) labor, (2) parts, (3) tires and tubes, (4) lubrication and preventive maintenance, (5) accident repairs, and (6) special modifications. Along with the maintenance expenditures, depreciation and other ownership costs are normally charged to the individual equipment numbers. In increasing numbers, construction companies are extending their budgetary control to a more detailed level by equipment system. A typical breakdown of systems will include (1) engine, (2) electrical and ignition, (3) fuel, (4) cooling, (5) undercarriage, (6) steering, (7) power train, (8) hydraulic, (9) brake system, (10) body, and (11) accessory equipment. The optimum level depends upon the variability of the equipment's use. With more uniform job conditions, greater detail can be effectively utilized. Cost may be summarized and compared by type of equipment, model, and size (i.e., Caterpillar 660 scrapers), or, at a still higher level, by functional type of equipment (i.e., hauling units).

Cost for a given period has little meaning when taken by itself. At minimum, an inception-to-date cost and estimate should be compared, and it is very desirable to maintain cost in such a manner that trends are apparent. Table 7-10 shows a cost history maintained in machine-readable form. A unit cost per hour provides the best control for most situations. Production quantities may be appropriate units of cost for some applications and for very stable job conditions, but are unsatisfactory for most contracting work. They are no longer allowed by the Internal Revenue Service for depreciation purposes.

A maintenance cost system is necessary to protect a contractor's high capital investment in construction equipment. It permits the control of good service procedures and the prevention of excessive operating costs due to overmaintenance or inadequate maintenance. It should also enable the contractor to price his work more intelligently. The fulfillment of these objectives should govern efforts to develop effective information systems.

As long as construction projects are of shorter duration than the life of equipment, the control of equipment costs will be a major challenge.

REFERENCES

1. L. P. Alford, *Principles of Industrial Management*, rev. ed., The Ronald Press Company, New York, 1951.
2. *WABCO Equipment Manual*, Westinghouse Air Brake Co., Construction Equipment Division, Peoria, Ill.
3. *Caterpillar Performave Handbook*, Caterpillar Tractor Co., Peoria, Ill.
4. *Euclid Equipment Handbook*, Euclid Equipment Co., Hudson, Ohio.
5. *Internatinal and Hough Construction Equipment*, International Harvester Co., Construction Equipment Division, Mclrose Park, Ill.
6. NADA, *Official Used Car Guide*, National Automobile Dealers Used Car Guide Co., Washington, D.C., January, 1969.
7. *Rental Rate Blue Book for Construction Equipment*, Equipment Guide-Book Co., Palo Alto, Calif.
8. *The Handbook of New and Used Construction Equipment Values* (Green Guide), Equipment Guide-Book Co., Palo Alto, Calif., vols. I and II.
9. *Nationally Averaged Rental Rates and Reference Data for Construction Equipment*, Associated Equipment Distribution, Oak Brook, Ill., annual publication.

10. *Operating Cost Guide*, Technical Bulletin No. 2, Power Crane and Shovel Association, Chicago, Ill., 1965.
11. Paul M. Teicholz, "A Simulation Approach to the Selection of Construction Equipment," University Microfilms, Inc., Ann Arbor, Mich., 1963.
12. James F. Douglas, *Obsolescence as a Factor in the Depreciation of Construction Equipment*, Bureau of Yards & Docks, U.S. Navy, NBy 45818, The Construction Institute, Stanford University Press, Stanford, Calif., 1963.
13. H. B. Maynard, *Industrial Engineering Handbook*, McGraw-Hill Book Company, New York, 1956.
14. Francis D. Koster, *The Economics of Heavy Construction Equipment Standardization*, Bureau of Yards & Docks, U.S. Navy, NBy 45818, The Construction Institute, Stanford University Press, Stanford, Calif., 1964.
15. Richard F. Perdunn, *Managing Automotive Equipment More Efficiently*, Industrial Education Institute, Boston, April, 1967.
16. William T. Morris, *The Analysis of Management Decisions*, rev. ed., Richard D. Irwin, Inc., Homewood, Ill., 1964.
17. *Planned Equipment Replacement for Excavating Contractors*, Market Division, Caterpillar Tractor Co., Peoria, Ill.
18. *Administration and Operation of Transportation Equipment*, Bureau of Yards & Docks, Washington, D.C., July, 1957.
19. James F. Douglas, *Construction-Equipment Policy: The Economic Life of Equipment*, Facilities Engineering Command, U.S. Navy, NBy 45818, The Construction Institute, Stanford University Press, Stanford, Calif., July, 1966.
20. Carlos A. Herrero, *The Influence of Investment Credit on the Economic Life of Construction Equipment*, Technical Report no. 85, The Construction Institute, Stanford University Press, Stanford, Calif., 1957.
21. James F. Douglas, *Construction-Equipment Policy: Influence of Income Tax Law on Economic Life*, Technical Report no. 69, The Construction Institute, Stanford University Press, Stanford, Calif., September, 1966.
22. L. E. Schaffer and W. C. Kraft, What You Should Know about Financing Equipment, *Rock Products*, Chicago, April, 1969.
23. *Construction and Maintenance Equipment*, Highway Research Board Special Report 68, National Academy of Sciences—National Research Council Publication 959, Washington, D.C., 1962.
24. Average Production Rates for Major Equipment, *Highway Research Record no. 278*, National Academy of Sciences—National Academy of Engineering Publication 1655, Washington, D.C., 1969.
25. Elmer Adrain Cox, "The Economical Use of Standby Equipment for a Typical Kraft Process Pulp and Paper Mill," master of science thesis, A & M University of Texas, College Station, Texas, July, 1955.
26. Willett A. Snook, Used Engine Oil Analysis, *Lubrication*, vol. 54, no. 9, 1968.

Section **8**

Tractors, Bulldozers, and Rippers

R. D. EVANS

Civil Engineer, Intercontinental Sales, Caterpillar Tractor Co., Peoria, Ill.

INTRODUCTION

Both on-highway and off-highway prime movers are used extensively on heavy construction projects. By way of example, they serve as the tractive units for power shovels, cranes, rear dumps, bottom dumps, and mobile drills. Most of these equipment types will be discussed in detail elsewhere in this book. It is the purpose of this section to distinguish briefly between on-highway and off-highway units and to discuss in detail the "self-contained" track type and wheel type. The characteristics and uses of bulldozers and rippers, commonly used accessories for the off-highway prime mover, will also be described.

Fig. 8-1 Rear-dump unit truck.

Fig. 8-2 Concrete mixer mounted on unit truck.

ON-HIGHWAY TRACTORS

Classification On-highway units are generally classed either as straight (or unit) trucks or as truck-tractors. The former class constitutes the integral mounting for conventional rear-dump bodies (Fig. 8-1) and mixers (Fig. 8-2), while the latter class is used in conjunction with low beds (Fig. 8-3), bottom dumps (Fig. 8-4), rear dumps

(Fig. 8-5), and side-dump bodies. Another system of truck classification, particularly favored in military terminology, is by wheels and drivers and is given in that order. For example, a 6 × 4 (Fig. 8-6) has six wheels and four drivers, a 6 × 6 (Fig. 8-7) has six wheels and six drivers, a 4 × 2 (Fig. 8-8) has four wheels and two drivers, etc.

Fig. 8-3 Low-bed truck-tractor.

Fig. 8-4 Bottom-dump truck-tractor.

Fig. 8-5 Rear-dump truck-tractor.

Nominal tonnage rating, which should not be interpreted as an actual load rating, is another method of classifying trucks. This method is used in some states for licensing purposes. Tonnage rating for payload capacity is in general use in Europe, in addition to gross vehicle weight (GVW) and gross combination weight (GCW).

In the United States, the most widely used method of truck classification is based on GVW and GCW.

All the states have rules and regulations which establish vehicle size and weight limits.[1]* The maximum permissible on-highway width is 8 ft in most states, with

Fig. 8-6 Kenworth 6 × 4 unit.

Fig. 8-7 Oshkosh 6 × 6 unit.

a few permitting 8.5 ft. Length, height, tandem-axle spacing, and speeds are other limiting factors which vary from state to state.

Regulations for on-highway trucks are manifold. Since all classes of highways are involved—interstate, primary, secondary, rural, city, etc.—multiple levels of government (federal, state, county, township, municipal) may exert their jurisdictional

* Superior numbers refer to the list of references at the end of this section.

authorities. Various highway types must also be considered—reinforced concrete, asphaltic, gravel, dirt, etc.—each of which may have different load-bearing characteristics.

In view of those factors, it is not surprising that nationwide conformity in truck regulations simply does not exist. This lack primarily affects the operation of freighters and cartage trucks, although all trucks which operate on a public highway are subject to applicable legal controls. Our concern, however, is limited to "semis"

Fig. 8-8 Diamond T 4 × 2 unit.

and tandem units, or "doubles," which may be used to haul earth, base material, or aggregates. These trucks are primarily on-highway trucks, although most of them have been designed and manufactured to be used in off-highway applications as well.

Equipment Characteristics Four basic types of trucks are used for bulk material handling in on- and off-highway service. Table 8-1 identifies these types and indicates

TABLE 8-1 Truck GVW (GCW) Ratings and Payload Capacities

Basic type of truck	Manufacturers' gross vehicle rating, lb		Payload capacity, tons	
	Usual	High	Usual	High
Two-axle	23,000	30,000	7.5	10
Three-axle	44,000	60,000	15.5	21
Four-axle	68,000	74,000	25	25
Tractor-trailer	72,000	80,000	22	26

the usual ranges of gross vehicle ratings and payloads.[2] The basic comparison has been limited to these two items because of the wide selection of engines, transmissions, front and rear axles, and tires. By way of example, Table 8-2 lists manufacturers'

TABLE 8-2 Specification Ranges for Typical Three-axle, Single-unit Dump Trucks

	Usual range	High range
Engine (diesel) hp	225–275	225–275
Transmission	10 speed	10 speed
Rear bogie, lb	32,000	38,000
Speed, mph	55–60	55–60
Tires	10.00–20	11.00–24
Body—cu yd, struck	12–14	16–18
Payload—tons	12–16	20–24
GVW, lb	30,000–45,000	46,000–70,000

specifications within the "usual" and "high" ranges (Table 8-1) for typical dump trucks in the three-axle, single-unit class.[2]

It is frequently necessary for a contractor to select his dump trucks from within the "usual" 30,000–45,000-lb range of GVW in order to comply with highway load restrictions. The same three-axle truck can be equipped with a larger engine, more capacity in its rear axle, tires with a higher load rating, and a body with a greater capacity. Thus, in those states where a 70,000 GVW is permissible by special permit, the actual payload can be increased up to 24 tons. This larger payload then permits the single unit to compete on an economic basis with "semis" and "doubles."

Production Estimating The performance of on-highway units is geared to a repetitive basic work cycle. The time required to perform each work cycle can be analyzed in terms of its fixed elements and its variable elements. It is frequently convenient to view as fixed elements the times required for loading, acceleration, gear shifts, braking, turns, and dumping. Variable time elements then consist of travel times on the haul and the return. Wait times resulting from loading delays and traffic delays can either be identified separately or treated as variable time elements. An efficiency factor can also be introduced to compensate for minor and routine delays which are not specifically identified in the basic time cycle. The production rate in cubic yards or tons per hour can then be expressed by the equation:

$$\text{Production, cu yd (tons) per hr} = \frac{60\ (\text{min})}{\text{fixed} + \text{variable time (min)}} \times \text{payload (yd or tons)} \times \text{percentage efficiency}$$

EXAMPLE

Single-unit Dump Truck

Payload capacity, tons	15
Fixed time, min	10
Variable time, min	5
Efficiency, percent	80
Tons/hr = $^{60}\!/_{15} \times 15 \times 0.80 = 48$	

OFF-HIGHWAY TRACTORS

Classification Off-highway tractors or prime movers fall into two general classes—crawler or track-type (Fig. 8-9), and rubber-tired or wheel-type (Fig. 8-10). The latter class will be viewed as consisting of two-axle units with either 2- or 4-wheel drive. This excludes the single-axle tractor shown in Fig. 8-10*a* which requires another unit such as a single-axle scraper or trailer for support and stabilization; i.e., a two-axle motor scraper. Scrapers of various types, including motor scrapers, are discussed in Section 9, Scrapers. Truck-type hauling units are discussed in Section 13, Hauling Units.

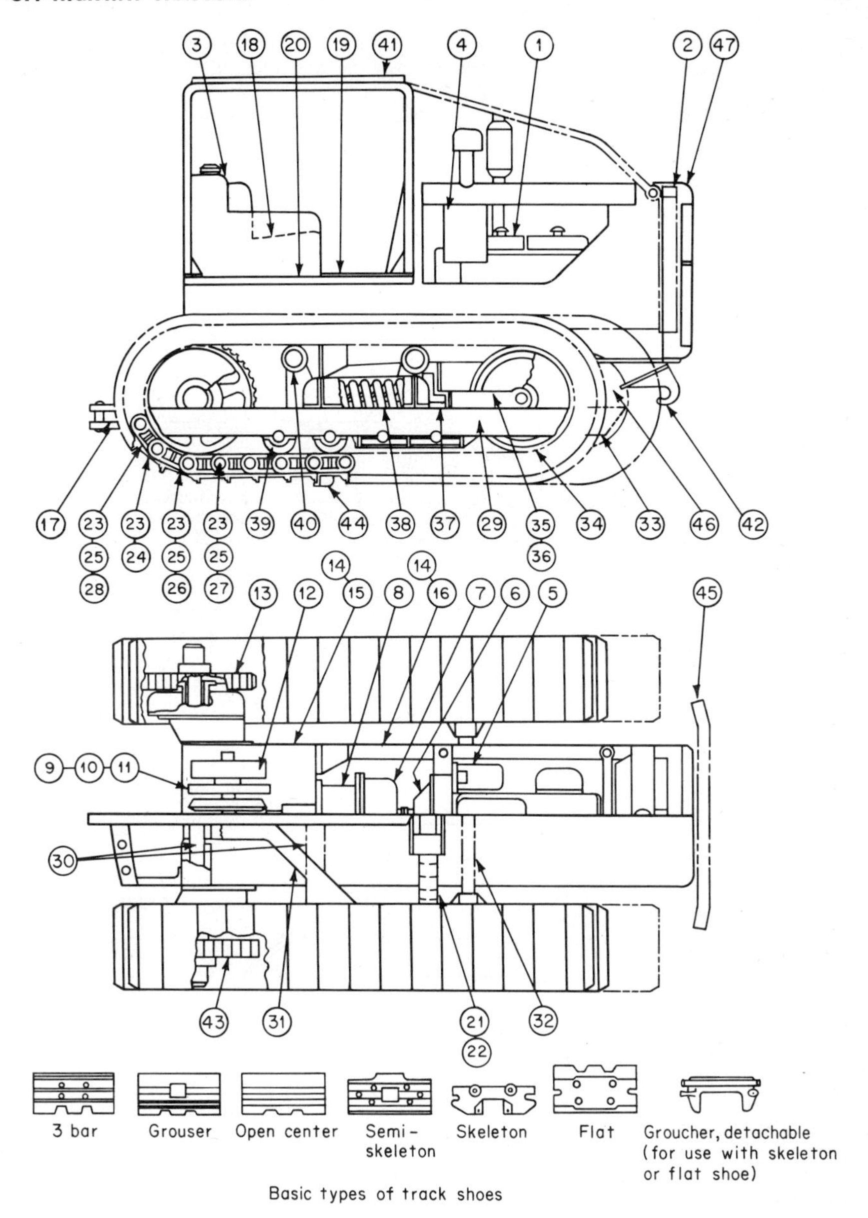

Fig. 8-9 Nomenclature—track-type tractors. (*SAE Handbook, 1969.*)

Figures 8-11 and 8-12 illustrate typical off-highway units.[3] These are self-contained prime movers which provide tractive power for drawbar work. Such work includes bulldozing and push-loading scrapers as well as towing scrapers, sheepsfoot and rubber-tired rollers, dirt and rock wagons, plows and disks, rippers, logging arches, and skid pans.

These prime movers also function as mobile mountings for various attachments and accessories. Prominent among these attachments are blades (straight, angle, tilt, push cups, brush, etc.), rippers, pipelayers, towing and logging winches, backfillers, snowplows and wings, and push blocks.

Equipment Characteristics The major characteristics of off-highway prime movers are discussed briefly below. Reference should also be made to Equipment Characteristics in Section 6, Engineering Fundamentals.

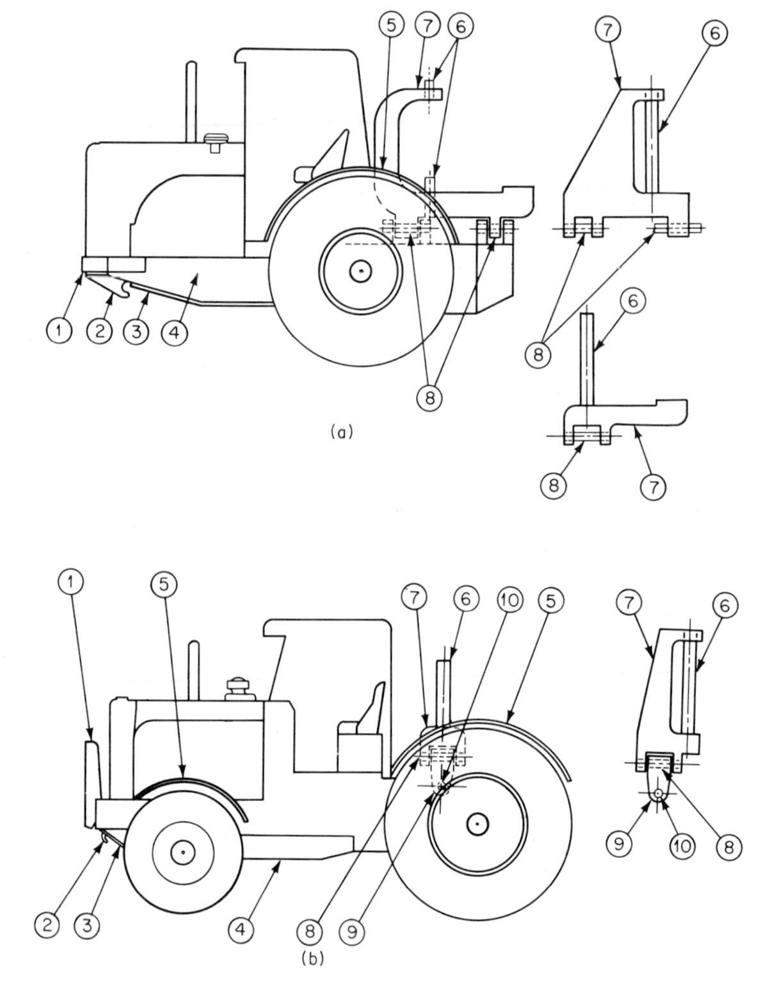

Fig. 8-10 Nomenclature—wheel-type tractors. (*a*) Single axle; (*b*) two axle. (*SAE Handbook, 1969.*)

Prime Movers. Specification sheets for prime movers provide the best means to evaluate equipment characteristics. They indicate the type and horsepower rating of the engine, the type of transmission, final drive, axles, steering, brakes, suspension, tires or tracks, weight, and other pertinent information. Equally important are the drawbar-speed and rimpull-speed curves for track-type and wheel-type tractors, respectively. Figure 8-13 shows a Caterpillar D9G track-type tractor curve and Fig. 8-14 shows a Caterpillar 824 wheel-type tractor curve. Comparative specifications for representative track- and wheel-type tractors,[2] together with data for bulldozers and rippers, are listed in Tables 8-3 and 8-4.

Running Gear. Prime movers for modern off-highway construction equipment are usually available with either tracks or wheels as their running gear. Track-type tractors and power shovels (excavators) have been used since the turn of the twentieth

Fig. 8-11 Caterpillar D8H (PS) with 8S blade and hydraulic control. (*Manufacturer's literature.*)

Fig. 8-12 Caterpillar 834 wheel tractor equipped with blade. (*Manufacturer's literature.*)

century. Wheel-type (rubber-tired) tractor-scrapers were introduced in the late 1930s and were followed by rubber-tired drawbar units, dozers, and loaders. All the latter types have counterpart units mounted on tracks. For any job, a choice of running gear type must be made in order to properly utilize the prime movers and obtain optimum production at the least cost.

When speed and mobility of the prime mover are of major importance, wheels are

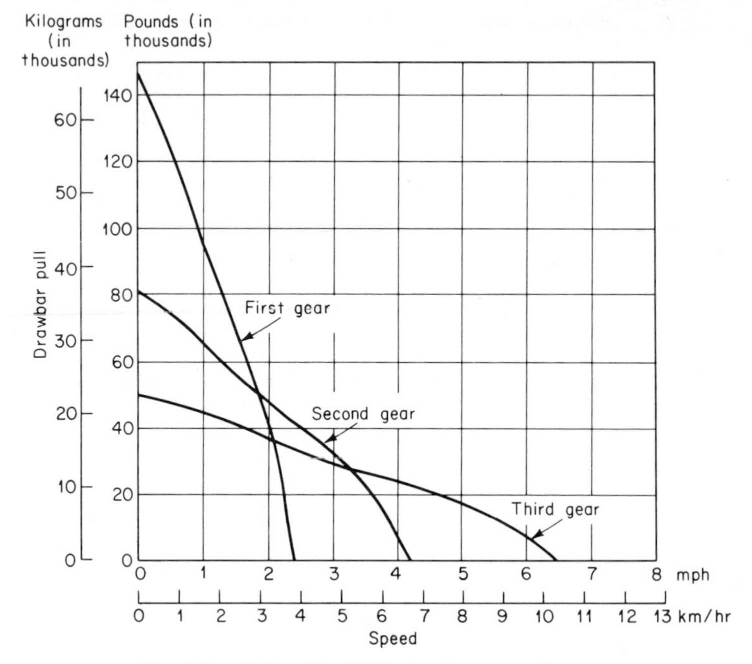

Fig. 8-13 Caterpillar D9G drawbar—speed curve.

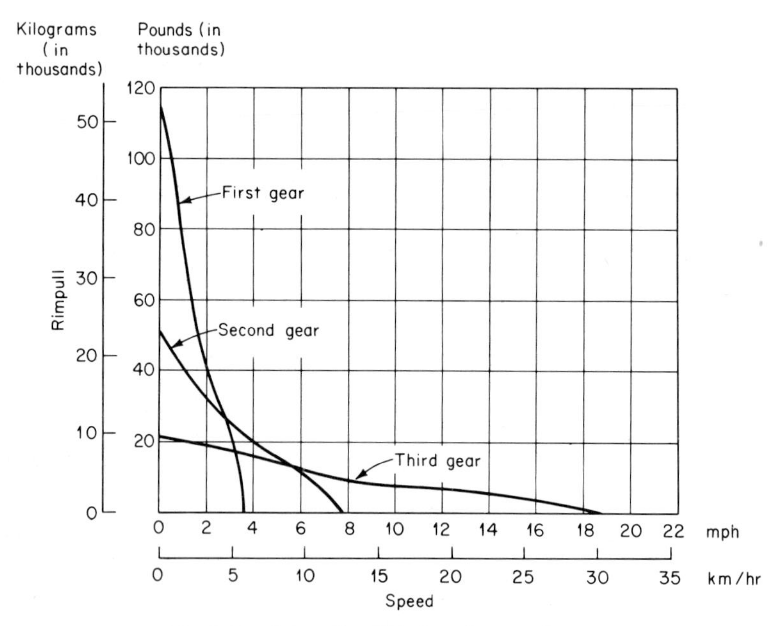

Fig. 8-14 Caterpillar 824 rimpull—speed curve.

TABLE 8-3 Comparative Specifications for Off-highway Track-type Tractors, Bulldozers, and Rippers

Manufacturer	Model	Advertised flywheel hp	Bare tractor weight, lb	Speed range, mph	Bulldozer weight, lb	Ripper weight, lb
Allis-Chalmers	HD21 (PS)	268	50,000	0–8.0	8,870	7,795
	HD16DP (PS)	173	35,700	0–7.2	6,820	5,146
	HD16D (DD)	160	34,500	1.4–5.8	6,820	4,700
	HD11EP (PS)	140	23,000	0–5.4	4,325	3,840
	HD11E (DD)	122.5	21,700	1.4–5.7	4,325	3,840
Caterpillar Tractor Co.	DD9G (PS)	770	160,900	0–6.5	15,140C	NA
	D9G (PS)	385	67,700	0–6.5	16,950	14,900
	D8H (PS)	270	49,700	0–6.5	11,300	10,526
	D8H (DD)	270	48,700	1.6–6.7	11,300	10,526
	D7E (PS)	180	32,000	0–6.0	7,180	5,900
	D7E (DD)	180	31,400	1.5–5.9	7,180	5,900
	D6C (PS)	120	23,700	0–6.4	4,655	3,170
	D6C (DD)	120	23,200	1.5–5.9	4,655	3,170
General Motors Corp. (TEREX)	82-80 (PS)	425	73,000	0–6.9	9,960	16,920
	82–40 (PS)	275	51,300	0–6.1	8,710	12,020
	82–30 (PS)	225	44,500	0–7.3	8,010	12,020
International Harvester Co.	TD25C (PS)	285	54,000	0–6.0	9,965U	10,970
	TD25C (DD)	285	53,600	1.5–4.9	9,965U	10,970
	TD25B (PS)	230	50,300	0–7.1	8,660	8,440
	TD25B (DD)	230	49,870	1.5–5.2	8,660	8,440
	TD20B (PS)	160	30,875	0–6.0	5,260	7,840
	TD15B (PS)	120	22,693	0–5.6	4,290	3,560
	TD15B (DD)	120	22,600	1.5–5.2	4,290	3,560

Information has been obtained from authoritative sources and is based upon information obtained at the time material was prepared for printing. It is believed to be correct; however, it is not guaranteed and is subject to change.

Bulldozer blades above are hydraulic straight blades except C = cushion blade, and U = U blade.

Weight of bare tractors can be increased up to 10 percent depending upon attachments and accessories mounted.

the obvious choice for the running gear. Speed and mobility become relatively less important when traction requirements are high. In these circumstances, crawler running gear can prove greatly superior to rubber tires, since it can normally develop more traction.* The crawler tractor can then normally exert more usable drawbar pull than can a wheel-type tractor of similar weight and horsepower. This situation favors the use of crawler running gear on jobs where high drawbar-pull requirements exert more influence than speed on the cost per yard.

The coefficient of traction (maximum drawbar pull or rimpull at imminent slippage divided by gross vehicle weight) is generally considered to have a maximum value of 0.6 for rubber-tired units operating on all surfaces excluding concrete. For track-type tractors, this maximum figure is about 0.9. The curves in Fig. 8-15 match the rimpull of a typical dozer-equipped wheel-type tractor with the drawbar pull of a comparable track-type machine. Detail *A* in the graph examines more closely the usual working range for both types. The shaded areas represent the ranges of developable traction for soil types varying from dry clay loam to loose earth—0.6 to 0.9 traction coefficients for tracks and 0.4 to 0.6 coefficients for wheels.

* See Sec. 6, Engineering Fundamentals.

TABLE 8-4 Comparative Specifications for Off-highway Wheel-type Tractors and Bulldozers

Manufacturer	Model	Advertised flywheel hp	Operating weight, lb	Speed range, mph	Tire size	Articulated Yes	Articulated No
Caterpillar Tractor Co.	824	300	62,400	0–18.5	29.5–29	x	
	834	400	77,100	0–20.4	29.5–35	x	
Hough (International Harvester)	D120C	285	68,000	0–26.7	29.5–29	x	
	D500	532	145,000	0–26.7	37.5–33	x	
R. G. LeTourneau, Inc.	T450A	475	95,000	0–10	98 in. OD	x	
	T600	635	136,000	0–10	98 in. OD	x	
Michigan-Clark Equipment Co.	180	170	40,000	0–23.6	23.5–25		x
	280	318	66,200	0–31.2	29.5–29	x	
	380	475	97,500	0–27.9	33.5–33	x	
	480	635	105,000	0–28.6	33.5–33		x
MRS Mfg. Co. . .	1–90	175	30,000	0–23.5	23.5–25	x	
	1–105	238	46,000	0–28.6	26.5–25	x	
	1–110	260	55,000	0–30.7	29.5–25	x	
	1–120	432	70,000	0–26.7	29.5–25	x	
WABCO.	C	218	34,600	0–17.2	26.5–25	x	

Information has been obtained from authoritative sources and is based upon information obtained at the time material was prepared for printing. It is believed to be correct; however, it is not guaranteed and is subject to change.

Weight of bare tractors can be increased up to 10 percent depending upon attachments and accessories mounted.

The crawler-mounted tractor of Fig. 8-15, for example, weighs 82,000 lb equipped with dozer and a control unit. At 0.9 coefficient of traction, it can develop 73,800 lb of pull before its tracks start to slip. The rubber-tired unit, working weight 91,000 lb, can produce 54,600 lb of pull at its maximum coefficient (0.6). The working speeds for the two machines, when operating under these conditions, will not be substantially different.

As machine weight increases, available pounds of pull go up. Mounting an 11,000-lb ripper on the crawler might increase its drawbar pull by 9,900 lb. A 75 percent fill of calcium chloride solution (11,000 lb) in the wheel units could produce 6,600 lb more rimpull.

To look at the weight–drawbar-pull relationship another way, the weight of machine required to develop a specified usable drawbar pull is directly related to the applicable coefficient of traction. Accordingly, a rubber-tired machine must usually be heavier than a crawler tractor if it is to develop an equivalent amount of drawbar pull. Of course, the heavier machine will also have to have more horsepower to keep its power-to-weight ratio within reasonable limits. Operating in a material which corresponds to the maximum coefficients of traction for each type of running gear, a wheeled machine needs half again as much weight as a crawler ($0.9/0.6 = 1.5$) to deliver the same drawbar pull.

On wet ground, the coefficient of traction for rubber falls off more rapidly than for tracks. As a result, the wheel tractor must be even more than 1½ times the weight of a track-type tractor for comparable drawbar pull during rainy weather. At the other end of the tractive scale—in loose sand, where traction coefficients for both types are about 0.3—a 91,000-lb wheel unit, with proper flotation, might develop a slightly greater drawbar pull than an 82,000-lb crawler.

Maximum drawbar pull, however, does not tell the full story of the work capabilities of two types of machines. Track-type tractors perform best at speeds up to about 3 mph, while wheel tractors are designed to work best between 2 and 5 mph. It can be seen in Fig. 8-15 that the two curves almost match between 2 and 4 mph. Applications such as fill spreading and high-speed production dozing show off the wheel tractor's rimpull-speed relationship to best advantage.

Wheeled machines, unlike their track-type counterparts, have to expend some of their rimpull in overcoming rolling resistance. Since rolling resistance varies so much,

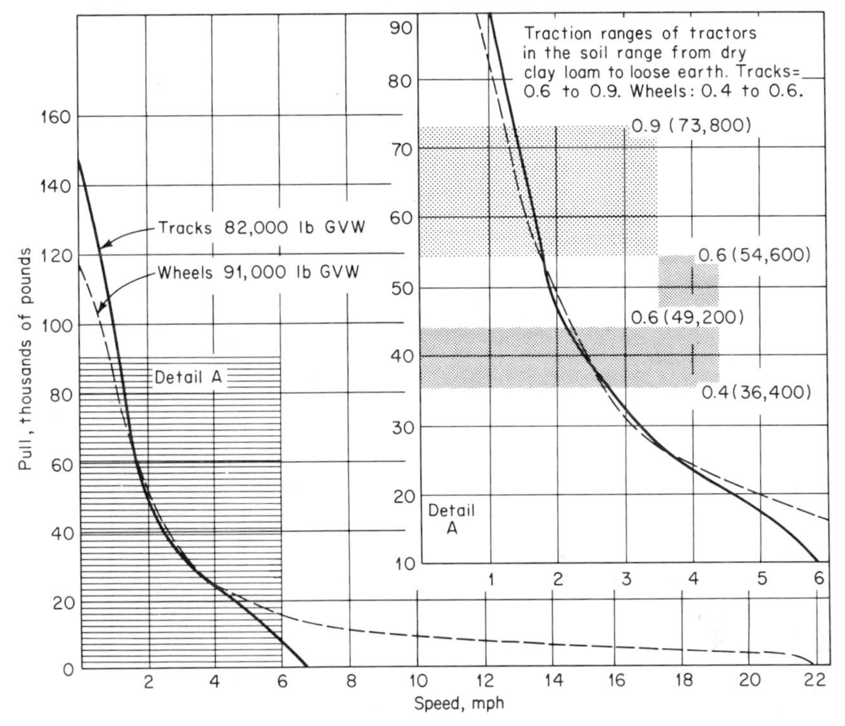

Fig. 8-15 Comparison of wheel-type tractor rimpull with the drawbar pull of track-type (power shift) tractors.

no attempt has been made to allow for it in the curves. An actual rimpull curve for the wheel tractor at work could be somewhere below that shown in Fig. 8-15.

Since a wheel tractor needs more weight and power than a crawler of similar drawbar-pull capabilities, it will generally be more expensive. The higher price, when reflected as an hourly depreciation rate, contributes to a higher hourly owning and operating cost and thus to a higher price per pound of usable drawbar pull. In some applications the faster operating speed of the wheel machines will more than make up the difference, but contractors will want to observe units working side by side before concluding that wheels will give a lower cost just because they are faster.

One cost item deserves special attention—that of tires and tracks. When wheel tractors first became available, relatively high track costs were a telling argument in favor of wheels. The later years have seen major advances in both metallurgy and design which, in most cases, have extended track wear life more than track prices have increased. For example, owners of machines in many applications can expect 50 percent or more longer life today than they could from 1955 components in the same working conditions. Consumer prices for track parts in that time have increased only about 25 percent.

On the other hand, tires for construction equipment have also been improved. Further, they can usually be recapped one, two, and sometimes three times before the carcass must be discarded. All in all, the relative costs of tracks and tires have little effect in choosing between wheel- and track-type tractors. In fact, the most influential element in this decision cannot be any *one* contributing item but has to be the final figure—cost per yard.

The relative advantages and disadvantages of wheel and track running gears can be summarized as follows:

	Wheel advantages	Track advantages
Working conditions	Firm clay, concrete Abrasive soils but no sharp edges Level or downhill work Poor in wet weather	Variety of soils Sharp edges not so destructive Any terrain Good in all weather
Effect on ground	Good compaction; variable with counterweight and ballast	Good flotation; low ground pressure with different shoe sizes and types
Application	Long distances Loose soils Fast return speeds, 8–12 mph Moderate blade loads and long, thin cuts when dozing Highly mobile	Short distances Tight soils Slow return speeds, 4–7 mph Large blade loads and short, heavy cuts when dozing Restricted mobility

Power Transmission and Controls. Present-day track- or wheel-type prime movers are equipped with hydraulic, cable, or electrical controls. Initially cable controls dominated, but hydraulic systems have improved and now afford unexcelled simplicity, capacity, and flexibility. Individual preference carries a lot of weight in selecting controls, but today the choice follows the above order—hydraulic, cable, electric.

Differences in initial price and in cost of operation of the various systems are no longer significant—in fact, they favor hydraulic systems. An additional factor in favor of the hydraulic system is its versatility. The basic hydraulic system has valves—one, two, three or more—which can be used to operate dozers, tilt cylinders, scrapers, rippers, front-end shovels, agricultural implements, and toolbars. Combinations can also be arranged, such as dozer-ripper or dozer-tilt-ripper.

Hydraulic and electrical systems provide down pressure readily for many applications. Cable controls are still required for some types of dozers and for those towed scrapers which lack hydraulic components.

Most track- and wheel-type tractors are offered with a choice of direct-drive or power-shift transmission. If we consider two tractors with identical engine flywheel horsepower, any difference in performance characteristics will be due to the direct-drive transmission having a little more drawbar pounds pull in a narrow speed range compared with the power-shift transmission having slightly less drawbar pounds pull in a wider speed range. In a dozing or drawbar application with a variable load, the direct-drive transmission will provide more drawbar pounds pull with more frequent gear shifting compared with the faster and less frequent shifting of the power-shift transmission which generates slightly less drawbar pounds pull. It is now generally accepted that the power-shift tractor will give superior production in an application which has variable load as compared with the superior production of the direct-drive transmission in an application with a constant load. Hence, a tractor equipped with a dozer will give better production if it utilizes a power-shift transmission, whereas the agricultural tractor will be better applied with a direct-drive transmission.

Most wheel tractors are offered with power-shift transmissions as their principal uses are fill dozing, push loading, utility work, and jobs with variable loads. They are suited best where mobility and high speeds are required.

Production Estimating It should be noted that off-highway prime movers serve as tractive units for a wide variety of equipment. As tractive units only, they do not lend themselves to a production analysis. A production analysis becomes possible only when they do useful work such as dozing, compacting, ripping, pulling scrapers, etc. Tractors used in the aforementioned applications are discussed in other sections of this book, and reference should be made to the appropriate section for production information. For example, bulldozers and rippers are covered in subsequent portions of this section.

In general, off-highway units follow a basic cycle, as discussed previously in this section under On-highway Tractors.

BULLDOZERS

The bulldozer is widely used and is probably the most versatile of any of the earth-moving tools. Figure 8-16 illustrates two basic dozer units—straight and angled—and includes names of major components peculiar to bulldozers and controls.

A bulldozer is a frame-mounted unit with a blade, curved in its vertical section, extending in front of the tractor. Blades range in width from some 6½ ft to around 24 ft and in height from about 5 ft down to approximately 2½ ft.

Blade Types There are four basic blade types: straight, angle, U, and cushion. The track-type tractor can have any of the four basic blades mounted on it, whereas the wheel-type tractor is generally restricted to either the S or C blade or a lightweight U blade for the handling of light materials such as coal and wood chips.

Straight Blades. The S blade attaches by push arms and braces to the tractor frame. It can be adjusted or tilted so that one corner will be approximately 2 ft below ground level on the large blades and 1 ft below ground level on the smaller units. Tip (forward or backward) adjustment can be made to a maximum of 10°.

The S blade is primarily a heavy-duty blade used in excavation and pioneering work. It is also suitable for drifting material and for trenching and backfilling. Since it can be tilted, it makes an excellent penetrating tool in hard excavation and can be successfully used to dislodge large boulders. By changing the pitch (tip), light materials can be drifted or hard soils may be dug, depending upon conditions.

Angling Blades. An A blade is generally wider and heavier than the straight blade for the same tractor, but it is lighter in cross section. It mounts on a C-shaped frame, each end of which is attached to the tractor frame. As its name implies, the blade can be angled to the right or left of a normal S-blade position, up to a maximum angle of 25°. Angling blades can also be tilted in the same manner as the straight blade. However, due to the mounting on the C frame, the blade cannot be tipped. For comparable machines, the angling blade will be some 1 to 2 ft wider than the straight blade.

A blades are most effectively employed in side casting, as in the construction of a sidehill road or in backfilling a trench.

U Blades. Most manufacturers build a U blade for larger machines. This blade has the same cross-sectional strength as the straight blade and can be tilted and tipped in the same manner. The U blade is wider than the straight blade and its outside edges are canted forward 25° to minimize spillage in loose material.

U blades perform much the same duty as the S blade. Their use is more suited to the noncohesive soils or rocks, whereas the S blade is primarily a cohesive-soil tool. The U blade is a good match on a ripping tractor and can also excel in stockpiling coal, drifting material into a grizzly, pioneering, and clearing.

Cushion Blades. The C blade is mounted to the inside of the tractor main frame. It is narrower than the S blade and is used mainly to push-load scrapers. Because the blade is compact, the tractor can maneuver quickly in the cut area to contact scrapers. The reduced blade length also minimizes the chances of cutting tires during

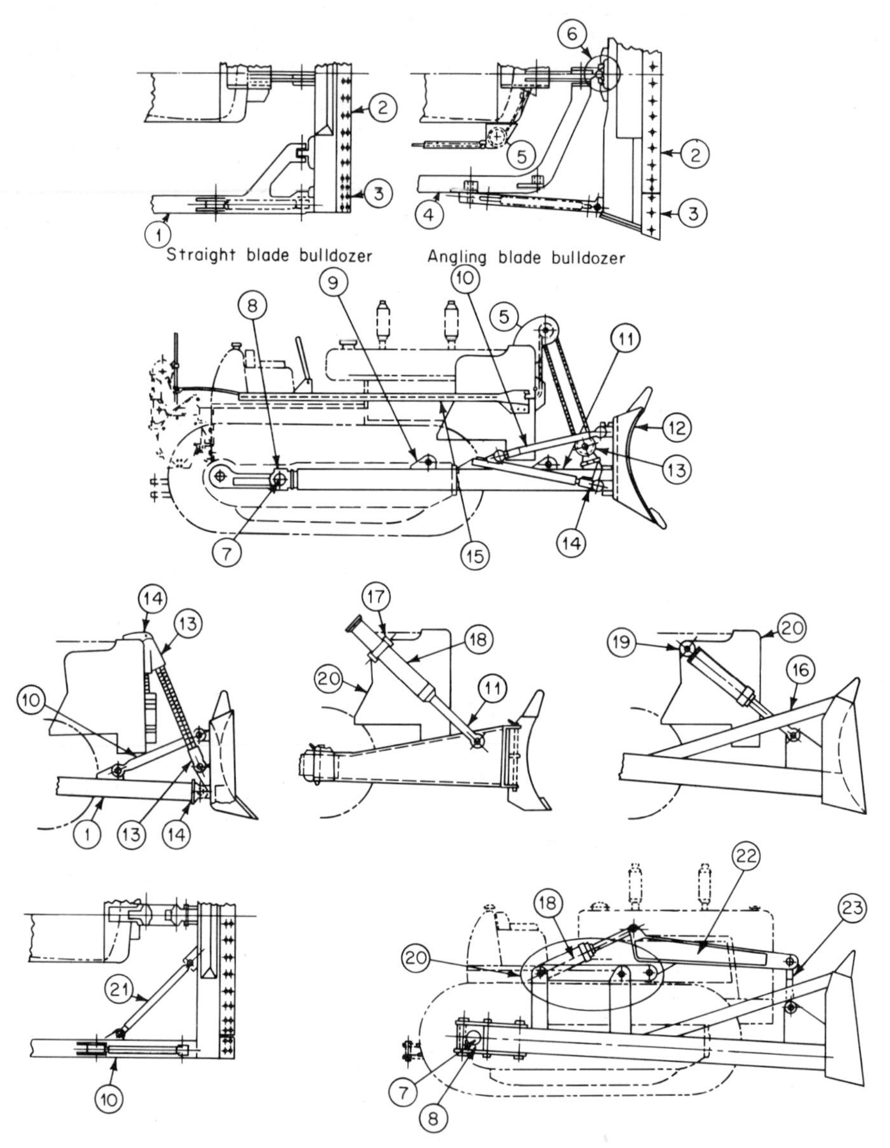

Fig. 8-16 Bulldozer types and nomenclature. (*SAE Handbook, 1969.*)

the push cycle. It is used instead of a push cup because it can also clean up in the cut and thus help to maintain high productivity. The blade is a utility one and not a production tool from the bulldozing aspect.

The blade mounting is usually through a series of rubber and steel disks onto the push arms. Those disks absorb the shock when the blade contacts a scraper at relative speeds up to 3 mph. The C blade cannot be tilted, tipped, or angled as can the S and U blades.

Special Blades. There are many special blades available from tractor or associated equipment manufacturers. Figures 8-17 and 8-20 show some of the more popular blade arrangements.

Fig. 8-17 Fleco brush rake on Caterpillar D8.

Fig. 8-18 Rome KW blade on Caterpillar D7.

Tip, Tilt, and Angle To change the tilt position of S or U blades, the tilting braces are rotated in opposite directions. The end result is that one corner of the blade will be considerably lower than the opposite corner. Some manufacturers have replaced one of the braces by a hydraulic cylinder (called a "tilt" cylinder) which performs the tilting operation. Some tilt controls are separate from the dozer controls (raise, lower) and are hand or foot operated. Others are incorporated with the dozer control lever so that all movements can be made with one lever. In either case, the dozing operation is not interrupted.

To change the tip position of an S or U blade, the tilting braces are rotated in the same direction. They expand to move the angle of tip of the blade forward and contract to move the angle of tip of the blade backward. This adjustment can be manual, hydraulic, or even electric. In the hydraulic adjustment, both braces would be replaced by hydraulic cylinders.

To change the angle of an A blade in relation to the tractor center line, the blade is pivoted about its central connection to the C frame. The blade braces are then pinned into the proper holes on the C or blade-support frame.

Applications *Land Clearing.* Generally speaking, a bulldozer is the first piece of equipment on the jobsite. One of its first tasks will be that of land clearing. Brush and small trees are removed by lowering the blade a few inches into the ground to strike and cut the roots. This is usually done in the lower speed ranges, and backing up may occasionally be necessary to clear the blade so that it can always cut cleanly. Medium trees and brush 4 to 10 in. in diameter usually require more than one pass.

Fig. 8-19 Bladerson bowl blade on Caterpillar D9.

Fig. 8-20 U blade on Caterpillar D7.

About 1 to 3 min is required for this operation, using tractors in excess of 100 drawbar horsepower.

For large trees 12 to 30 in. in diameter, more care and more time are required. First, contact should be made with the blade high and centered for maximum leverage. This contact should be made gently, pushing the tree at half throttle while inspecting for dead limbs which may break and fall. A bulldozer should not charge into trees at full throttle. The direction of fall, usually the direction of lean, is determined. Then a cut is made on the opposite side to a depth sufficient to cut some of the larger roots. The roots on both adjacent sides are cut in a similar manner. After building an earth ramp on the initial side of the tree to obtain greater leverage, the tree is pushed. As the tree starts to fall, the tractor should be reversed quickly to avoid incurring damage as the roots are pulled out of the ground. When stumps are removed, the holes should be filled and the area leveled to prevent water from accumulating. The time consumed for this operation will be 5 to 20 min per tree.

Stripping. Stripping consists of the removal of topsoil that is not usable as fill material or as stable subgrade. Stripping operations should be planned so as to minimize haul distance. In dozer work, the maximum haul distance is approximately 300 ft, and for maximum economy the material should be moved only once. In rough or steep terrain where hillsides and steep ravines must be stripped, hauls are usually longer. For these conditions, stripping should be done so that the topsoil is dozed downgrade to scrapers working at the bottom of the grade. The scrapers then perform the necessary hauling. All stripping operations should be planned so that other earthmoving units can begin excavation as soon as possible.

Pioneering and Sidehill Cuts. Sidehill cuts should always be started or pioneered from the top and then worked downward. Working downhill gives the advantage of gravity. It may be necessary to reach the starting point by climbing up a more gradual slope on the opposite side of the hill. In exceptionally steep or rocky terrain, a block hung on a tree or anchor point may be used to pull the pioneering tractor back up the hill, either by another tractor or by a winch on the pioneering tractor. Very steep areas can be worked by hanging on the line until a bench is well started.

On average terrain, cuts are started by working straight down the hill, making short passes to bench out an area large enough so that the tractor can eventually turn and work parallel to the road. Pioneering cuts should slope into the uphill side for maximum earthmoving efficiency. Short, swinging passes should be made as the dozer works downhill to drift the material over the side of the cut. Pioneering cuts should be wide enough so that the scrapers that follow have room to maneuver.

Backfilling. Dozers are the best equipment for backfilling because material may be pushed directly ahead of the machine over embankments into ditches or directly against a structure. Angling dozers are excellent for backfilling ditches because they can drift material into the trench while maintaining forward motion. With straight dozers, best results are obtained by approaching at an angle, ending up each pass by swinging in toward the trench, culvert, or structure. When backfilling culverts, crawler tractors should not cross the structure unless there are at least 12 in. of solid material on top.

Dozing Rocks and Frozen Ground. Rocks are generally removed by using one corner of the blade, as the full power of the tractor can then be applied to a short section of the blade. Best penetration can be obtained by tilting the corner of the blade. Large single rocks can be removed by using the blade to lift the rock, simultaneously applying power to the tracks. When a large rock is embedded in a group of smaller ones, the smaller ones should be removed before starting to move the large one.

Formations such as shale and sandstone are found in the inclined position. The operator should dig under the outcrop rather than attempt to penetrate from the opposite side. The formation can then be crushed by driving the tractor over it, following which it can easily be removed.

When working in frozen ground, the blade must be tilted for penetration. Once the area is penetrated, the blade will work against the exposed edge and, by pushing and lifting, the ground will be broken.

Ditching. Although the motor grader is better suited for ditching work, a bulldozer can be used effectively for rapid construction of rough ditches. The rough ditch is constructed with a straight blade by working at right angles to the length of the ditch.

Spreading. When trucks, scrapers, or wagons are used for hauling, bulldozers are ideal spreading tools at the fill. The blade should be kept in the straight position so that the material is drifted directly under the cutting edge. Depth of spread is usually set in the job specifications to obtain the desired compaction. Tractor-dozers are also excellent tools for pulling compaction equipment on the fill.

Downhill and Slot Dozing. Whenever possible, dozing should be done downhill for greatest production. In downhill work, it is not necessary to travel down with each load. Several loads can be piled up at the brink of the hill and pushed to the bottom with one pass.

Slot dozing uses spillage from the first few passes to build a windrow on each side of the dozer's path. This forms a trench, preventing spillage on subsequent passes. Cut sections, where possible, should be alternated as "slots" between narrow uncut sections. These uncut sections can then be removed in normal dozing. With favorable grades and soil conditions, the increase in production may amount to as much as 50 percent.

Blade-to-blade Dozing. This gives increased output for haul distances of 50 to 300 ft. At less than 50 ft, the extra yardage obtained is offset by the extra time required to maneuver the second dozer into position. Two or more dozers may be used effectively in blade-to-blade dozing. However, there are limitations to this method. Any delay which affects one machine will simultaneously affect the other machines, with a consequent decrease in production.

Bulldozing Tips.

1. Do not back up further than necessary, and do not push earth greater distances than required.
2. Know where the next pass is going to be made and where material will be placed. Always plan the operation thoroughly.
3. Alternate the tilt and angling adjustments periodically to balance wear on the blade and steering clutches.
4. Keep a tight line on the cable control. Excessive slack in the line during operation tends to cause kinks and uneven spooling, resulting in premature cable breakage or flattening of the cable on the drum.
5. Always replace broken cable with the cable size specified by the tractor and bulldozer manufacturer.
6. When dropping down a steep hill or over the side of a fill, use the blade for a brake.
7. When traveling, carry the blade low or about 14 in. above the ground; this practice will protect the underside of the tractor.

Production Estimating Many bulldozer applications such as ditching, grubbing, stumping, pioneering, and stripping do not lend themselves to an exact production analysis in terms of volume of material moved, but in most cases an approximation can be made of the total machine hours required to complete the job. The work can then be bid as a lump sum, or in some cases on a per acre or lineal foot basis, respectively, for clearing or ditching. In contrast, production dozing on short hauls or feeding belt loaders can generally be analyzed to give reliable production figures. Table 8-5 indicates typical bulldozer production under these conditions. In general, the economical hauling distance for a dozer ranges from 25 to 300 ft.

More detailed estimates of bulldozer production can be obtained by analyzing the working cycle. Consideration must then be given to material type, blade capacity, travel and return distances and speed, and working efficiency. Blade capacities for various sizes of track-type tractors and for different blade types are listed in Table 8-6.

Production in loose cubic yards (lcy) per 60-minute hour is figured in the following manner:

$$\text{Yd/hr} = \frac{60 \text{ min} \times \text{dozer capacity}}{\text{fixed time} + \text{haul time} + \text{return time}}$$

where fixed time = 0.05 min when only forward and reverse lever is required to shift (PS)
= 0.10 min when both forward and reverse lever and gear selector are required to shift (DD)

Haul time and return time are computed from the travel distance and tractor speeds.

Dozing will generally be done at slow speeds from 100 to 170 fpm forward and 220 to 400 fpm in reverse. For actual tractor speeds, see manufacturers' specification sheets.[2]

TABLE 8-5 Estimated Track-type Tractor-Bulldozer Production*
(loose cubic yards per 60-minute hour with power-shift (PS) tractor and S blade)

Tractor flywheel hp	One-way haul distance, ft					
	50	100	150	200	300	400
275–450	1,070	557	379	352	268	216
200–275	940	482	328	305	232	188
125–200	685	355	240	225	171	138
75–125	500	259	176	164	125	101
50–75	295	153	104	97	74	60

For A blade use 75 percent of values for S blade.
For U blade use 125 percent of values for S blade.
For DD (direct drive) use 80 percent of values for power shift (PS).
The table is based on the following conditions:

0 grade { Dozing speed 160 fpm; Return speed 220 fpm up to 150 ft; Return speed 400 fpm beyond 150 ft }

No slot or trench.
Material is in the bank state (no stockpile).

Each of the following will increase production as indicated:
Doze on stockpile, add 15–20 percent.
Doze downhill on 10–20 percent grade, add 20–40 percent.
Use of slot dozing technique, add 15–20 percent.

* Consult manufacturers for production of rubber-tired dozers as tractive effort varies widely and depends upon total tractor weight (counterweight, hydroinflation of tires, tire size, etc.) and coefficient of traction applicable.

TABLE 8-6 Blade Capacities for Track-type Bulldozers
(loose cubic yards measure)

Tractor flywheel hp	A blade	S blade	U blade
275–450	7.4	10.5	13.5
200–275	6.5	9.1	11.8
125–200	4.7	6.7	8.7
75–125	3.4	4.9	7.2
50–75	2.0	2.9	N.A.

EXAMPLE

Consider a 200 to 275 flywheel hp tractor, straight blade, capacity 9.1 lcy, 50 ft travel distance, haul and return in second gear.

$$\text{Fixed time} = 0.05 \text{ min}$$

$$\text{Travel time} = \frac{50 \text{ ft}}{160 \text{ fpm}} = 0.31 \text{ min}$$

$$\text{Return time} = \frac{50 \text{ ft}}{225 \text{ fpm}} = 0.22 \text{ min}$$

$$\text{Production} = \frac{60 \text{ min} \times 9.1 \text{ lcy}}{0.05 \text{ min} + 0.31 \text{ min} + 0.22 \text{ min}}$$

$$= 940 \text{ lcy/hr}$$

To estimate the expected production, load factor and job efficiency should be included. The production in loose cubic yards can be converted to bank cubic yards (bcy) by multiplying lcy by a load factor whose value depends on the type of material.* Assuming a 0.80 load factor and a 50-minute-hour efficiency, the expected production would be

$$940 \text{ yd/hr} \times 0.80 \text{ load factor} \times {}^{50}\!/_{60} = 626 \text{ bcy/hr}$$

The next step is to compute the unit cost of dozing the material in question. In order to accomplish this, it is necessary to obtain an estimate of the hourly owning and operating cost for the unit. This cost can be obtained from the user's records or estimated from data such as that supplied in Table 8-7. Assuming an owning and operating cost of $20 per hour for the bulldozer in the example, we have

TABLE 8-7 Relative Hourly Ownership and Operating Cost Factors for Bulldozers (Operator Wage Excluded)

Tractor flywheel hp	*Relative cost factors*
275–450	5.00
200–275	3.50
125–200	2.50
75–125	1.75
50–75	1.0

$$\text{Unit production cost} = \frac{\$20}{626 \text{ bcy/hr}} = \$0.032/\text{bcy}$$

RIPPERS

The tractor-mounted ripper is finding increasing use in construction, mining, and quarry operations. Due to the greater weight and higher horsepower of track-type tractors as well as to advances in ripper design, this combination can now rip material which could only have been blasted a few years ago. Ripping production and costs vary from material to material, and all rock formations cannot be ripped. However, if a rock can be ripped with a minimum production of 150 to 200 cu yd per hr, the costs are generally lower than for drilling and blasting.† In some cases, a cost reduction of 30 to 70 percent has been recorded. In addition, ripped material can be scraper loaded to effect greater savings.

Types Various types of rippers are illustrated in Fig. 8-21.

Towed Units. These are virtually obsolete, having been replaced by the integral types—heavy-duty hinge and parallelogram. The towed units are cumbersome, hard to maneuver, and do not use tractor weight effectively compared to the integral units. The towed unit is usually cable operated while the integral ripper is hydraulically operated and, in conjunction with the tractor on which it is mounted, provides a more compact and efficient unit. The tractor itself is usually equipped with crawlers, thus avoiding the excessive damage to tires that might be expected in rocky conditions. Figure 8-21*a* illustrates a towed ripper.

Miscellaneous Units. One type of ripper, generally referred to as a "scarifier," is mounted on motor graders and loaders as shown in Fig. 8-21*b*. It can also be mounted on a small crawler tractor.

Backrippers are mounted on dozer blades, as shown in Fig. 8-21*c*. As the dozer proceeds forward, the back rippers pivot in the mounting brackets and drag over the ground. Then, as the tractor backs up to its original dozing point, the rippers engage in the ground. They are used in less severe conditions than the heavy-duty units. The shanks and points can be removed or pinned in the brackets above the blade base when not in use.

* See Material Characteristics in Sec. 6, Engineering Fundamentals.
† See Sec. 19, Rock Excavation.

Recently several manufacturers have introduced a bulldozer-ripper combination. It is essentially a bulldozer with a hydraulically adjustable ripper shank mounted at each end of the blade. A penetration depth up to 12 in. can be obtained, and the tilting feature can be used as desired. This arrangement permits production dozing

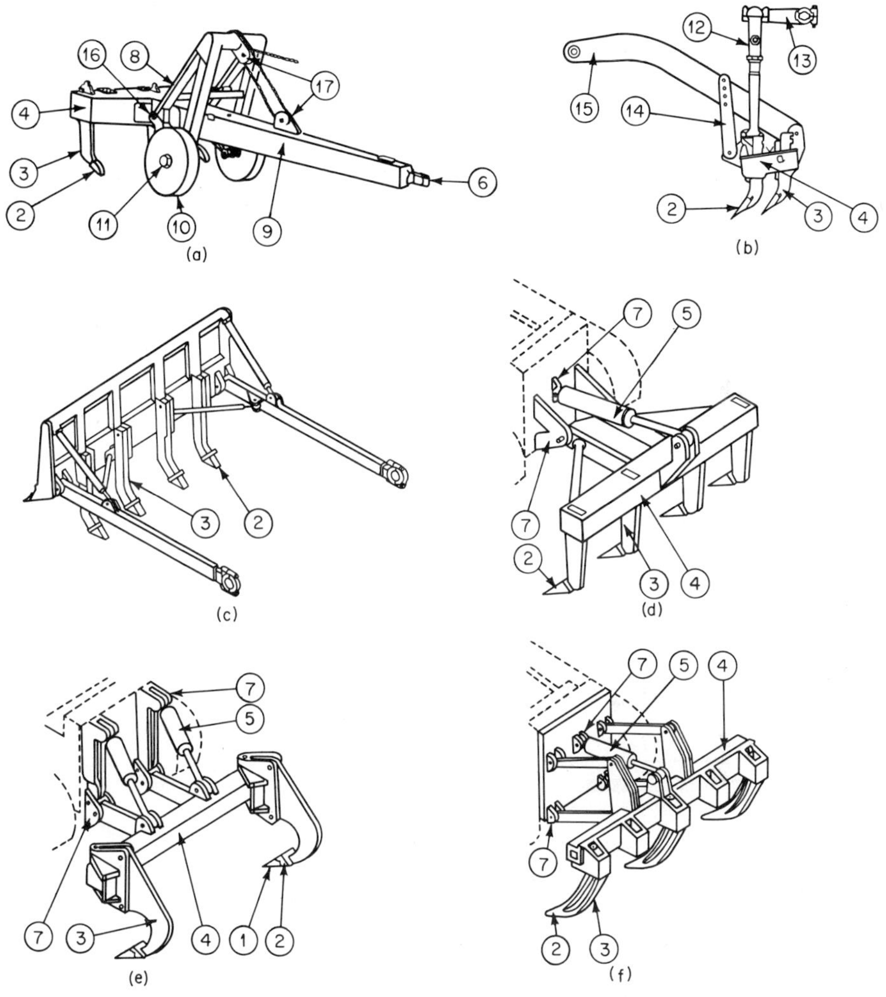

Fig. 8-21 Types of rippers. (*a*) Towed ripper, hinge type; (*b*) ripper, motor-grader mounted; (*c*) back ripper, dozer mounted; (*d*) hinge-type ripper, tractor mounted; (*e*) hinge-type ripper, swivel shanks, tractor mounted; (*f*) parallelogram-type ripper, tractor mounted. (*SAE Handbook, 1969.*)

and increases the dozer effectiveness in severe applications where blade penetration is a serious problem.

Pipeline rippers are similar to heavy-duty types in appearance but are heavier in weight and construction and penetrate deeper—up to 84 in.

One manufacturer makes a ripper in which the shank rests against large rubber blocks on movable brackets. When rock formations are encountered, the shank is compressed against the blocks. It is claimed that rock fragmentation will occur when this built-up energy exceeds the rock strength.

Hinge and Parallelogram Units. The two basic types in general use today are the hinge and parallelogram rippers shown in Fig. 8-21*e* and *f*. The beam of the hinged ripper pivots on link arms to raise or lower the ripper shank. This causes the angle between the ripper point and the ground to vary as the point enters the ground, which causes a penetration problem. The parallelogram ripper linkage allows the angle between the ripper point and the ground to remain constant as the point enters the ground, which improves penetration capability. Shanks are fixed or swivel and will penetrate up to 50 in.

Operation Generally, slow speeds—1 to 1.5 mph—should be used for most ripping operations, as point and track wear increases very rapidly with only small increases in speed. In most cases, a ripping job is begun with only one shank. If the material is easily penetrated and fractures into small pieces, a second shank can then be used. In tough conditions, however, two shanks cause greater track slippage and impose

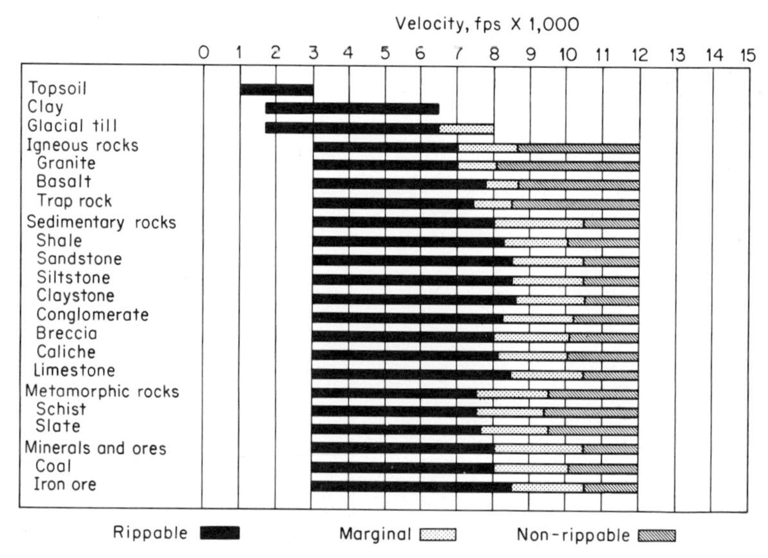

Fig. 8-22 Ripper performance as related to seismic wave velocities. (385 horsepower track-type tractor with parallelogram single-shank ripper.)

severe off-center loads on both tractor and ripper. It is normally desirable to rip as deep as possible, but if stratification is encountered it may be best to remove the material in its natural layers.

Tandem ripping may prove profitable under some conditions. Adding a second tractor to push the first tractor-ripper combination will extend the useful service range of the ripper into harder materials. Many times only a small part of the total rock on a job is too tough for a single tractor and ripper to handle, and in these cases tandem ripping can prove more economical than bringing in a drill crew. In very hard rock it is not uncommon for tandem ripping to increase production three or four times over that obtained with a single tractor, even though the cost may double.

Some rock that is too difficult to rip can be preblasted with a light charge of explosives and then ripped successfully. Experience with this ripping technique is limited, and a careful cost comparison should be made before applying it.

Production Estimating With the advent of the refraction seismograph, it is possible to investigate the rippability of various rock formations. The seismograph measures the velocities of seismic waves through subsurface materials. These velocities can be compared to data obtained from previous tests, wherein rippability and seismic velocity have been correlated for typical materials and for selected machines. An example of such correlation for a 385-hp track-type tractor with a parallelogram single-shank ripper is shown in Fig. 8-22. For tractors in the 200- to 275-hp class, the rippability ranges would be reduced by 15 to 25 percent.

Ripping production can also be estimated from seismic velocities. Figure 8-23 shows production for 385- and 275-hp crawler-tractors with parallelogram single-shank rippers, operating as single units. For example, this figure indicates that the single 275-hp tractor can rip approximately 1,000 bcy per hr in a rock whose seismic velocity is 4,000 fps. Figure 8-22 provides data for preliminary selection of ripping units over a wide range of seismic conditions. The final selection of equipment should be based on job production requirements.

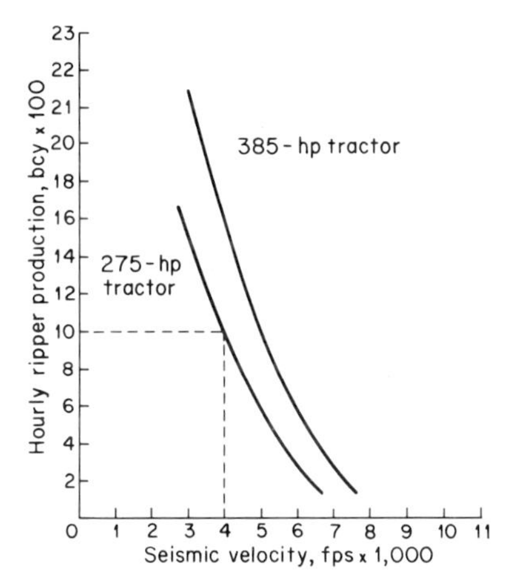

Fig. 8-23 Ripper production. Considerations: power shift tractors; single-shank rippers; production figures based on 100 percent working efficiency (60-minute hour). Chart is for ripping in all classes of material under ideal conditions. For igneous rocks with seismic velocities of 6,000 fps or higher, the production figures should be reduced by 25 percent. If thick laminations, vertical laminations, or other adverse factors exist, the indicated production figures should be derated appropriately.

Ripping ownership and operating costs will vary from 25 to 100 percent above bulldozing costs for comparable tractor sizes. Assuming the owning and operating cost for the 275-hp tractor (including operator's wages) to be $20 per hour for bulldozing and 50 percent greater for ripping, then for the example above the unit ripping cost would be $30/1,000 = $0.03 per bcy. Since the production rate is based on a 60-minute hour, an appropriate adjustment should be made to obtain the expected production cost.*

REFERENCES

1. *Commercial Car Journal, Fleet Reference Annual*, sec. 2, Rules and Regulations, Vehicle Size and Weight Limits, pp. 172–176, April, 1969.
2. Manufacturers' handbooks, literature, and specification data.
3. Society of Automotive Engineers Handbook, 1969.

* See Sec. 7, Equipment Economics.

Section 9

Scrapers

R. D. EVANS

Civil Engineer, Intercontinental Sales, Caterpillar Tractor Company, Peoria, Ill.

INTRODUCTION

The scrapers used in modern earthmoving and heavy construction include towed wheel scrapers, Fig. 9-1; wheel tractor-scrapers, Fig. 9-2; elevating scrapers, Fig. 9-3; and multiple scrapers, Fig. 9-4. All scrapers are either powered or towed by a prime mover which can be (1) a track-type tractor, (2) a two-axle wheel-type tractor, (3) a single-axle wheel-type tractor, or (4) a twin-engine unit. This last, illustrated in

Fig. 9-1 Caterpillar D8H and 463E towed scraper (18/26).

Fig. 9-2 Caterpillar 631 wheel tractor-scraper (21/30).

Fig. 9-3 Caterpillar 613 elevating scraper (—/11).

Fig. 9-5, is identical to types (2) and (3) except that an additional engine is mounted on the scraper to drive the scraper axle. The basic configurations of prime movers other than the twin-engine units are discussed in Section 8, Tractors, Bulldozers, and Rippers, within the Off-highway Tractor topic.

CONVENTIONAL SCRAPERS

Towed wheel scrapers and wheel tractor-scrapers (both two axle and three axle) are frequently referred to as "conventional scrapers." Figure 9-6 shows the components and standard nomenclature for both towed and wheel tractor-scrapers. The towed unit has a truck (component 10) and gooseneck (component 2) arrangement which is

Fig. 9-4 R. G. Le Tourneau L90 multiple scraper (90).

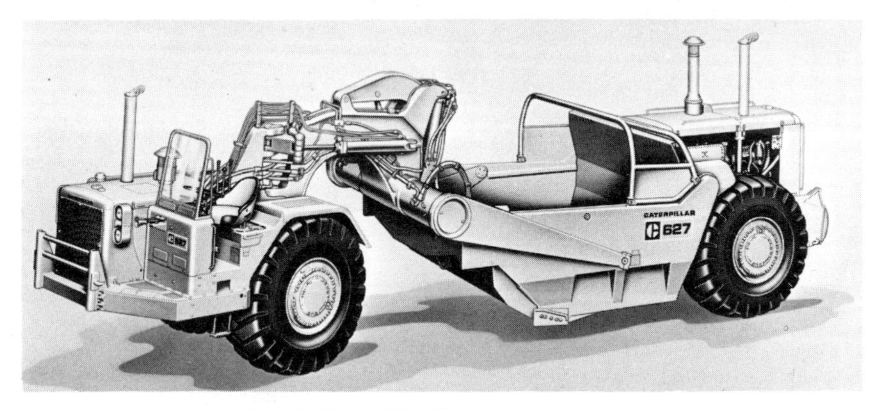

Fig. 9-5 Caterpillar 627 twin-engine scraper.

suitable for towing, whereas the wheel tractor-scraper has a gooseneck only (component 2) which attaches to the tractor portion to make an integrated, self-propelled unit. The principal components of conventional scrapers are the apron, cutting edge, bowl, ejector, push frame and plate, draft frame, hitch, and running gear.

The primary functions of the conventional scraper are to load (generally with assistance from a pusher), haul, spread, and return to a loading point. These basic operations comprise the materials-handling working cycle. The conventional scraper can also be used for light clearing, rough or fine grading, bank sloping, and ditching.

Capacity Ratings Conversion of yardage ratings for conventional scrapers into capacities in terms of pay yards is required for each specific project. Scraper yardage ratings are generally determined on the basis of SAE J741a[1]* as follows:

1. Capacities shall be rated on the basis of cubic yards. Standard ratings shall include both struck and heaped capacities (see Fig. 9-7).

* Superior numbers refer to the list of references at the end of this section.

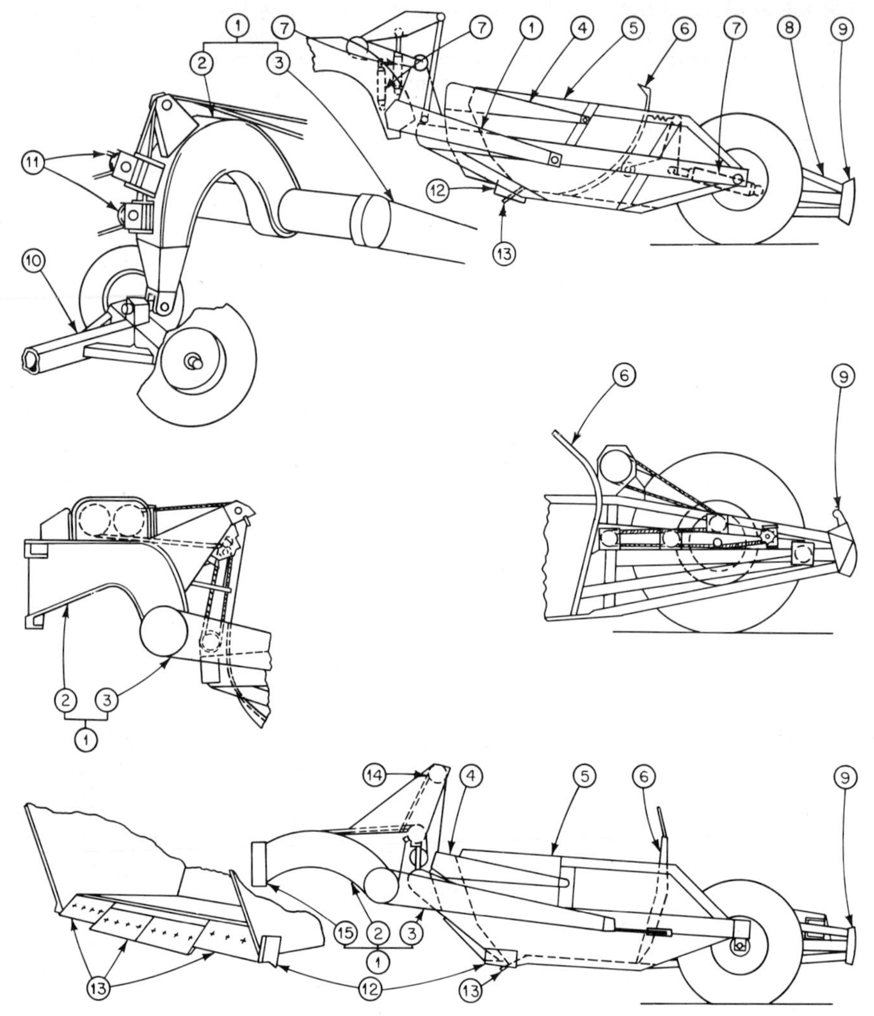

Fig. 9-6 Components and nomenclature for conventional scrapers. (1) Frame, draft, (2) gooseneck, (3) yoke, (4) apron, (5) bowl, (6) ejector, (7) cylinder, (8) frame, push, (9) plate, push, (10) truck, front, (11) fairlead, (12) bit, side, (13) cutting edge, (14) sheave, (15) housing, kingpin.

2. Struck capacity of a scraper shall be the actual volume enclosed by the bowl and apron, struck off by a straight line passed along the top edge of the side plates or adjacent load-carrying mechanism or extensions thereof. Struck capacity shall be given to nearest 0.1 cu yd.

3. When the top of the front apron, in the closed position, is below the top edge of the side plates, the capacity shall be limited, either by a plane from the top edge of the apron to the forward corners of the side plates, or by a plane at a slope of 1:1 extending from the top edge of the front apron to the plane formed by the top edge of the side sheets, whichever gives the smaller capacity.

4. To determine the 1:1 slope for the limiting plane, the scraper shall be set in its normal carrying position with the apron closed.

5. Top extensions of the ejector above the side plates shall not be included in the determination of struck capacity.

6. The volume occupied by apron arms, sheave frames, or other internal projections shall be disregarded in calculating the struck capacity.

7. Heaped capacity of a scraper shall be the sum of the struck capacity and the volume enclosed by the four planes at a 1:1 slope extending upward and inward from the top of the solid portion of the front apron, from the top of the solid portion of the ejector or rear plate, and from the top edges of the side plates. Small barred or screened openings in the apron may be ignored in determining the solid top line.

8. The scraper shall be set in the same carrying position as used for determining struck capacity.

9. For scrapers of less than 12 cu yd struck capacity, the heaped capacity shall be given to the nearest ½ cu yd; for scrapers of 12 cu yd struck capacity and larger, the heaped capacity shall be given to the nearest 1.0 cu yd.

10. If the top edge of the side plate (or extension thereof), front apron, or ejector is not a straight line, a mean line through its configuration shall be used to establish the base line of the plane enclosing the heaped capacity.

11. The possible interference of overhead structures, such as sheave guides and cables, with the heaped capacity shall be ignored.

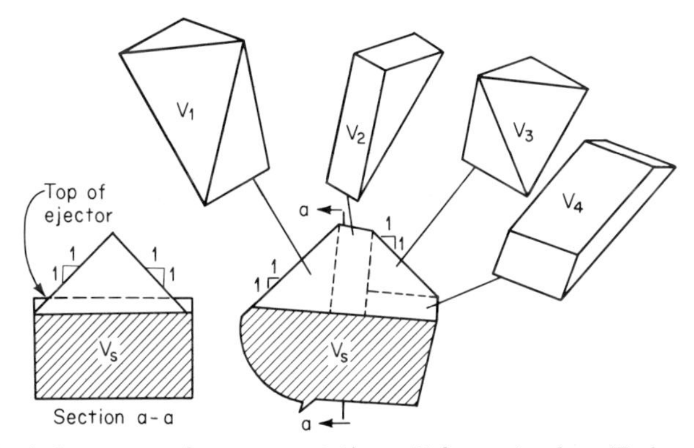

Fig. 9-7 Typical scraper volume computation. Volume struck = V_s (express to nearest 0.1 cu yd). Volume heaped $= V_H = V_s + (V_1 + V_2 + V_3 + V_4)$. If less than 12 cu yd struck, express to nearest 0.5 cu yd. If 12 cu yd struck or more, express to nearest whole cubic yard.

TABLE 9-1 Specifications for Representative Towed Wheel Scrapers

Manufacturer	Model	Capacity, cu yd		Control	Empty weight, lb
		Struck	Heaped		
Caterpillar Tractor Co.	435E	12	17	Cable	22,900
	435F	14	18	Cable	24,900
	463E	18	26	Cable	34,350
	463F	21	28	Cable	34,635
	491C	27	35	Cable	47,500
WABCO (Construction Equipment Division)	DT	8.3	11.5	Cable	15,900
	CT	14	20.0	Cable	25,350
Young Corp.	KS-700	7	9	Cable	15,650
	KS-800	8.2	11	Cable	19,560
	KS-1000	10	13	Cable	20,650
	KS-1200	12.4	16	Cable	25,500
	KS-1500	15	19	Cable	33,500
	KS-1800	18	23	Cable	35,000

12. Both the struck and heaped capacities as defined are, with minor reservations, definitely measurable quantities providing a standard method of comparing the volume of scrapers.

Equipment Specifications Specification details for representative scrapers of the conventional type are summarized[2] in Table 9-1 (Towed Wheel Scrapers) and Table 9-2 (Wheel Tractor-scrapers).

Methods of Operation The materials-handling working cycle of the conventional scraper will be described in terms of an earthwork application. The scraper is also finding increasing usage in handling fragmented rock; this application is described in Section 19, Rock Excavation.

Loading. Figure 9-8*a* shows the loading attitude for the conventional scraper. The

TABLE 9-2 Specifications for Representative Wheel Tractor-scrapers

Manufacturer	Model	Capacity, cu yd		Horse-power	Speed range, mph	Empty weight, lb
		Struck	Heaped			
Allis-Chalmers Mfg. Co.	260 Series A	15	20	320	5.3–33	51,100
	460B	24	33	353	4.2–29	78,500
	460C	24	33	456	5.7–34	84,100
Caterpillar Tractor Co.	621	14	20	300	3.2–30	51,400
	627	14	20	450	3.2–32	63,700
	631C	21	30	415	5.5–32	77,600
	637	21	30	640	5.5–33	86,900
	641B	28	38	550	4.9–31	107,500
	650B	32	44	550	7.2–43	101,700
	651B	32	44	550	5.7–33	117,400
	657B	32	44	950	5.7–33	137,800
	660B	40	54	550	7.6–43	110,300
	666B	40	54	950	7.6–43	130,800
General Motors Corp. (TEREX)	S7	7	9	148	4.2–25.5	26,500
	S24	24	32	432	4.5–29	80,450
	S32	32	43	520	4.8–31.5	105,550
	SS24	24	32	475	6.2–36	80,110
	SS40	40	52	475	5.2–37	102,765
	TS14	14	20	320	3.2–23	53,300
	TS24	24	38	659	5.2–29.6	91,000
	TS32	32	43	865	5.0–31.5	121,400
International Harvester Co.	270	14	18	260	5.7–34	42,200
	295B	24	32	420	5.2–32.5	74,300
R. G. LeTourneau, Inc..	LC35	27	30	950	0–15	75,000
Michigan-Clark Equipment Co.	210	15	20	290	4.3–29.4	49,600
	310	32	40	475	4.7–31	84,300
MRS Mfg. Co.........	1-80S	9	11	186	0–27.6	33,425
	1-90S	11	14	186	0–27.4	35,425
	250-D	43	53	635	0–37	140,340
	250-D	48	58	635	0–37	149,340
Westinghouse Air Brake Co. (WABCO)	D	7.3	9	148	5.1–30.1	23,070
	C229F	15	21	318	4–28	52,600
	B339F	25	34	475	3.7–32.7	86,500

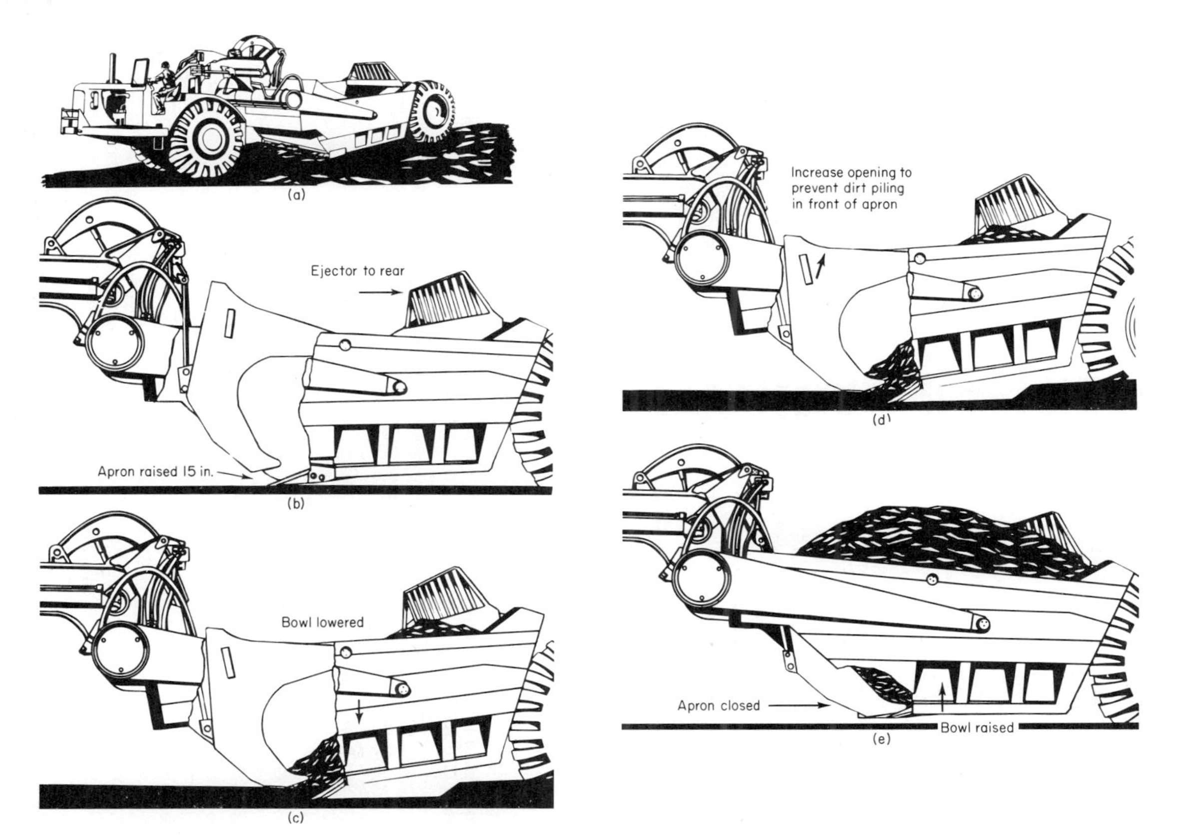

Fig. 9-8 Loading a conventional scraper. (*a*) Loading attitude. (*b*) Approaching the cut. (*c*) Making the cut. (*d*) Apron opening during loading. (*e*) Finishing the cut.

operator approaches the cut with the ejector to the rear and the apron raised approximately 15 in. (Fig. 9-8*b*). The bowl is then lowered to the desired depth of cut (Fig. 9-8*c*) and the tractor transmission shifted to L load range to keep engine speed high. For work other than finish grading operations, available power and length of cut are factors to consider in determining the optimum depth of cut.

Positive apron control permits the operator to choose the correct opening when loading (Fig. 9-8*d*). Increasing the opening will prevent dirt from piling up in the front of the lower apron lip, and decreasing the opening will keep dirt from rolling out of the bowl. In loose material, the bowl should be raised and lowered rapidly; this can be repeated as necessary to pump material into the bowl for maximum loading.

When the bowl is full, the apron is closed and the bowl is then raised (Fig. 9-8*e*). In loose, dry material, the bowl should only be raised slightly, then the apron closed partially and the bowl again raised.

Traveling Loaded. Figure 9-9*a* illustrates the traveling attitude for the loaded scraper. If the scraper controls are of the hydraulic type, as will usually be the case, a bowl-carry check valve will operate automatically in the bowl-lift hydraulic circuit (Fig. 9-9*b*). This valve locks the bowl cylinders hydraulically so that the bowl cannot drop in case a large hose should break.

The bowl should be raised to provide sufficient clearance for high-speed travel (Fig. 9-9*c*). The apron should be fully closed to prevent loss of the material, and the ejector should remain in the rear position (Fig. 9-9*d*).

Unloading. The unloading scraper attitude is illustrated in Fig. 9-10*a*. The bowl should be positioned to spread the material to the desired depth (Fig. 9-10*b*). A partial opening of the apron during the initial unloading process will assist in maintaining an even spread of loose material (Fig. 9-10*c*). If the material is wet and sticky, the apron can be raised and lowered repeatedly until the material behind it is loosened and drops out of the bowl (Fig. 9-10*d*). When the material has fallen from the back of the apron, the ejector is moved forward to push the remaining material out of the bowl at a uniform rate (Fig. 9-10*e*). It is recommended that the apron be raised completely before moving the ejector forward in any type of material.

In some hydraulically controlled scrapers, an automatic two-speed ejector-control system moves the ejector forward slowly when considerable force is required to push the material out of the bowl. As this required force is reduced, the control system automatically increases the speed of ejector movement.

Returning to Cut. The traveling attitude of the unloaded scraper while returning to the cut is illustrated in Fig. 9-11.

ELEVATING SCRAPERS

In contrast to the conventional scraper, which usually depends upon a pusher tractor for loading, the elevating scraper is considered to be self-loading. It is similar to a conventional scraper except that its apron has been replaced by an elevator or "ladder." This elevator is made up of heavy bicycle-type chain with flights or paddles bolted to it. The chain is rotated by a power source which is independent of the forward travel of the scraper. It generally has variable speed capability and can be reversed. A conventional scraper depends upon pusher power to force material over the cutting edge and then backward into the bowl. The elevating scraper must only force the dirt over the cutting edge, since the ladder will carry it from there into the bowl.

Self-loading elevating scrapers were first considered as utility tools—grading between concrete forms, hauling backfill to structures, and the like. Their production capabilities have now been recognized, and they compete with the other scraper classes for large enbankment jobs. They load well in most materials, with the following exceptions: shot rock, ripped rock, boulders, "bony" material that is too large to pass between the cutting edge and the elevator flights, and cohesive material with high moisture content that tends to ball up and stick to flights. The cost of a pusher and its operator is eliminated for the elevating scraper because of its self-loading capability. However, due to the dead weight of its loading mechanism, it has a reduced efficiency in the traveling portion of its working cycle.

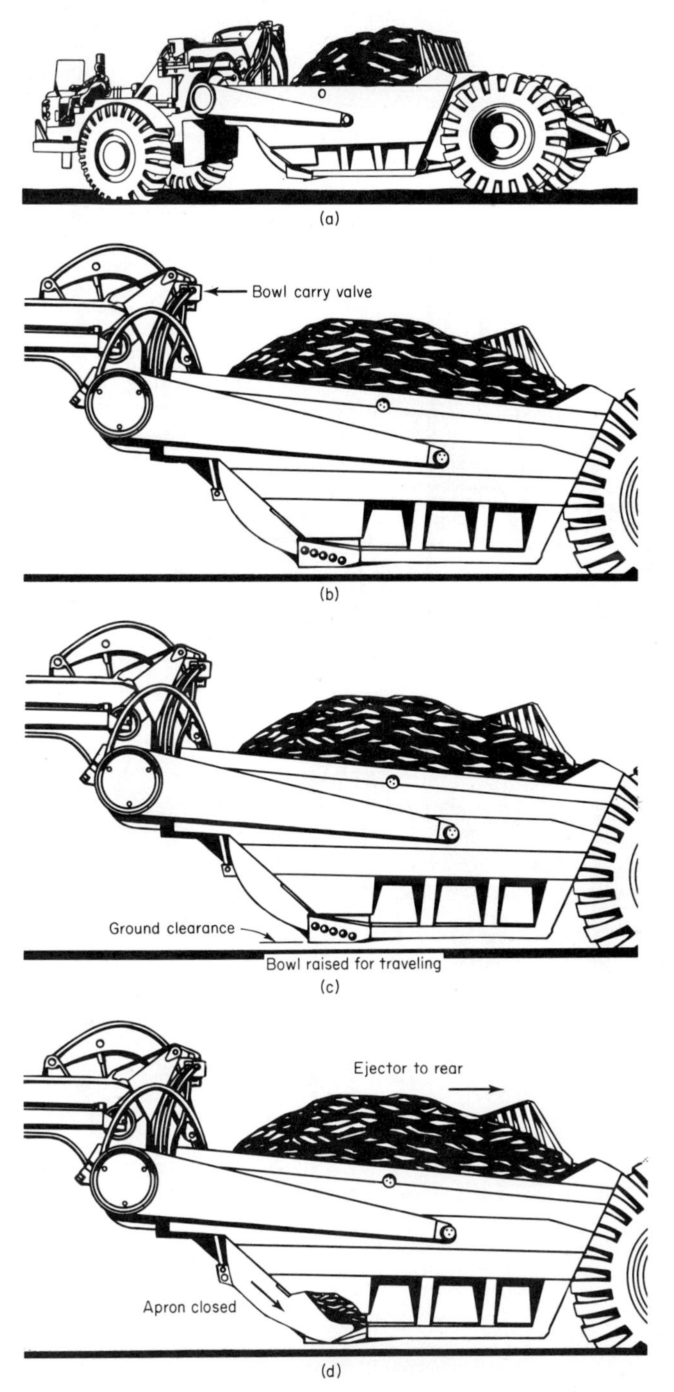

Fig. 9-9 Traveling loaded with a conventional scraper. (*a*) Traveling attitude. (*b*) Bowl-carry check valve. (*c*) Bowl position. (*d*) Apron and ejector position.

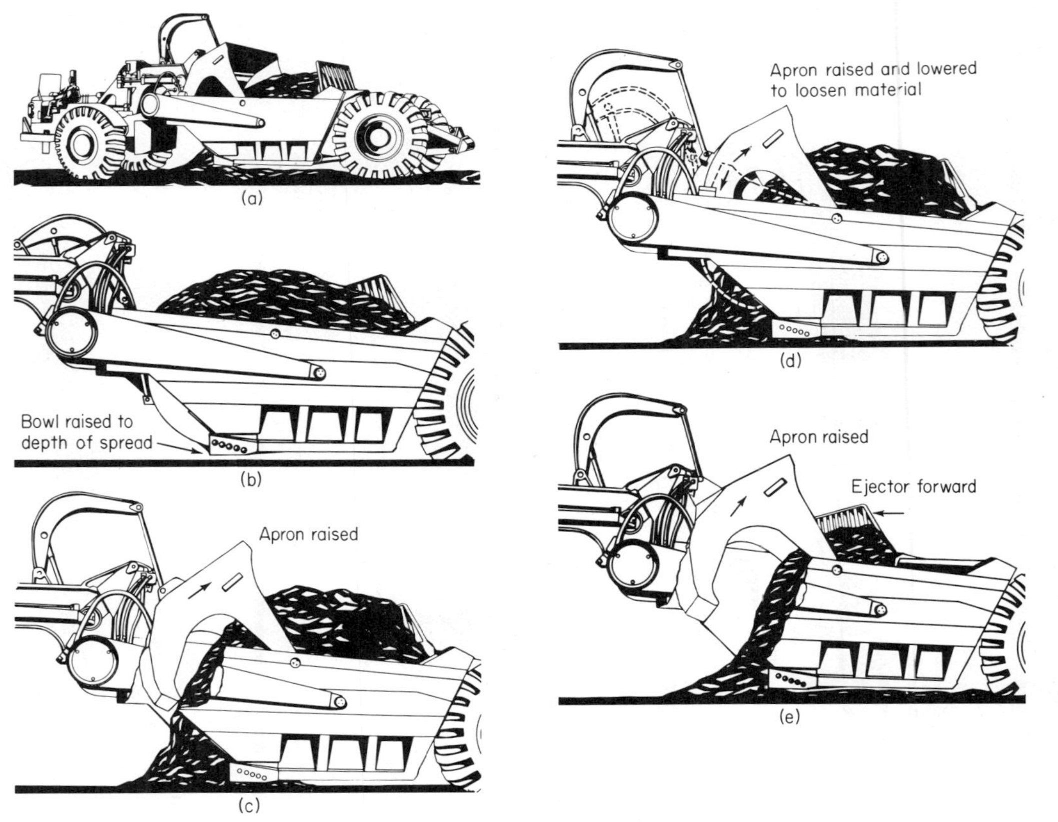

Fig. 9-10 Unloading a conventional scraper. (*a*) Unloading attitude. (*b*) Bowl position. (*c*) Apron position (loose material). (*d*) Apron position (sticky material). (*e*) Ejector movement.

Fig. 9-11 Returning to cut.

Capacity Ratings Yardage ratings for self-loading scrapers are generally determined on the basis of SAE J957, as follows:[1]

1. Capacity shall be rated on the basis of cubic yards.
2. The standard rating shall be the total volume of the heaped capacity.
3. The elevating scraper bowl shall be set at its normal carrying position when calculating its standard rated capacity.
4. To determine the standard rating, it will be necessary first to determine the struck capacity.
5. The struck capacity of an elevating scraper shall be the actual volume enclosed by the bowl and elevator, struck off by a straight line passed along the top edge of the bowl sides or adjacent load carrying mechanism, or extensions thereof.
6. If the top edge of the bowl sides, or extensions thereof, is not a straight line, a mean line through its configuration and parallel to the ground shall be used.
7. The forward edge of the bowl shall be the plane, or planes, formed by the upward straight path, or paths, of the inner edge of the flights on the elevator adjacent to the load.
8. The elevator shall be located in the manufacturer's recommended or normal operating position.
9. The boundary of the lower forward bowl will be a line normal to the cutting edge and passing through the centerline of the elevator bottom idlers, or a line from the cutting edge to the nearest point of the elevator flight path.
10. The heaped capacity shall be the sum of the struck capacity and the volume enclosed by four planes as follows:

a. Planes extending upward and inward at a 1:1 slope from the top edges of the bowl sides.

b. A continuation of the uppermost plane formed by the upward path of the inner edge of the flights on the elevator.

c. A plane extending from the top of the solid portion of the ejector drawn tangent to the path of the outer edge of the flights at the top of the elevator and stopping at the intersection with the plane established by item *b*.

11. The volume occupied by flights, levers, or other internal projections shall be disregarded in calculating the standard rated capacity.
12. For scrapers of less than 12 cu yd, the standard rated capacity shall be given to the nearest 0.5 cu yd; for scrapers of 12 cu yd and larger, the standard rated capacity shall be given to the nearest whole cubic yard.
13. The heaped capacity as defined, with minor variations, is a definitely measurable quantity which provides a standard method of comparing the volume of elevating scrapers. (See Fig. 9-12.)

Equipment Specifications Specification details for representative elevating scrapers[2] are summarized in Table 9-3.

Methods of Operation The working cycle of the elevating scraper is quite similar to that of the conventional scraper.

Loading. In making the cut, the bowl is first lowered to a depth that will permit the elevator and tractor to operate at a high, constant engine speed. Shallow cuts, as illustrated in Fig. 9-13*a*, are a necessity. Deep cuts will tend to force material into the bowl, and this will slow or stall elevator flights. The elevator flights sweep the mate-

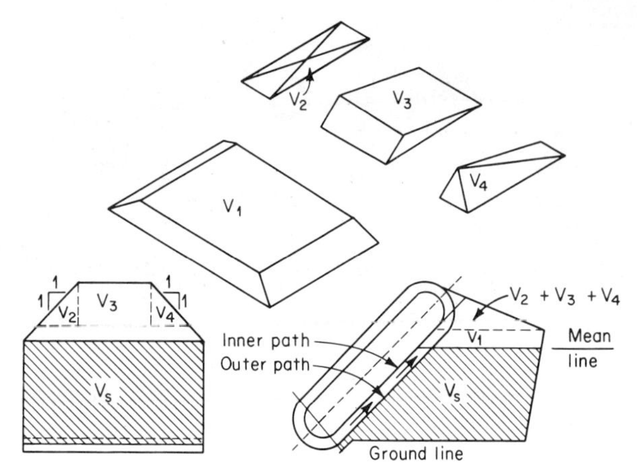

Fig. 9-12 Elevating scraper volume computation. Volume struck = V_s. Standard rating volume heaped = $V_H = V_s + (V_1 + V_2 + V_3 + V_4)$.

TABLE 9-3 Specifications for Representative Elevating Scrapers

Manufacturer	Model	Capacity, cu yd, heaped	Horse-power	Speed range, mph	Empty weight, lb
Allis-Chalmers Mfg. Co.	260E Series A	23	320	3.7–28	59,300
Caterpillar Tractor Co.	613	11	150	0–26	27,200
	J621	21.5	300	5.8–30	63,100
	633	32	415	6.1–33	86,200
Deere & Co.	JD760	9	143	2.6–26	30,650
General Motors Corp. (TEREX)	S7	12	148	4.2–25.5	33,370
	S35E	35	495	4.8–31.5	98,900
Hancock	282G	9	115	0–22	27,000
	292B	11	160	1.5–28	33,300
International Harvester Co.	E200	9	135	2.9–24	26,200
	E270	21	260	3.4–26.5	49,000
	E295	32	420	5.2–32.5	88,060
Michigan-Clark Equipment Co.	110-12	12	178	2.8–26	37,850
	110-14	14	238	3.5–31	43,000
	210H	23	335	3.3–33	56,800
	310H	31	475	3.8–33	93,900
MRS Mfg. Co.	I-80S	12	186	0–27.4	36,400
	I-95S	17.5	250	0–27.4	57,000
	I-100S	20.5	290	0–31.5	63,980
	I-105S	23	337	0–30	73,080
	I-110S	25	389	0–32	78,000
Westinghouse Air Brake Co. (WABCO)	D111A	11	160	3.6–25.4	30,300
	C222F	21	318	4.0–28	58,200
	BT333F	32	475	3.7–32.7	94,285

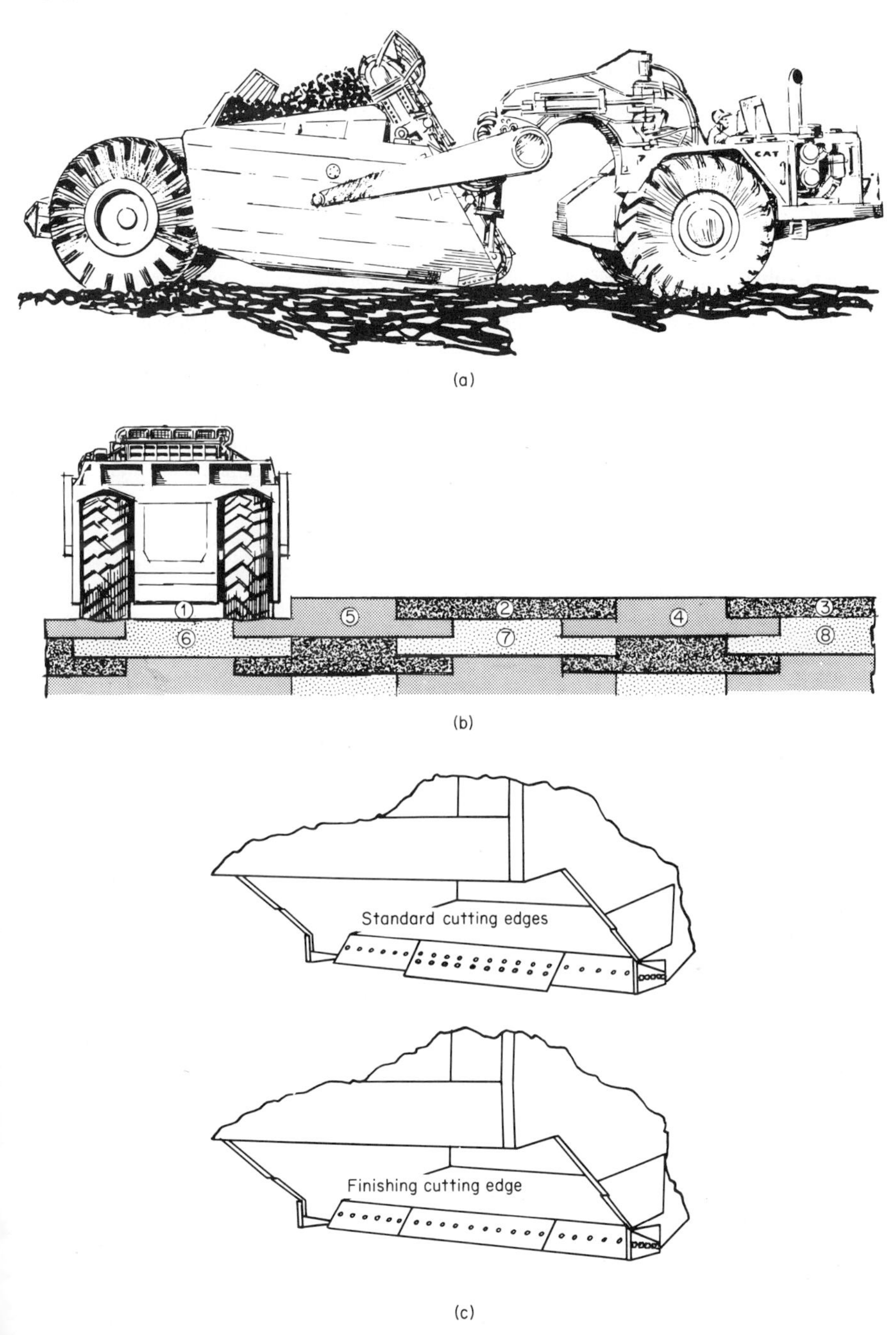

Fig. 9-13 Loading an elevating scraper. (*a*) Shallow cuts to assure elevator efficiency. (*b*) Straddle loading. (*c*) Cutting edges.

rials into the bowl, and any attempts to force material into the bowl will actually increase the time required to load.

The elevator has four speeds forward and one reverse. Materials such as sand, silt, and topsoil, which are easy to penetrate, are loaded at high speeds. This keeps the cutting edge swept clean and prevents the material from being pushed along in front of the cutting edge. If the operator repeatedly lifts and lowers the bowl during the loading operations, the advantage of high-speed elevator operation will be lost. Low speeds are used in loading tough materials, such as hard-packed clay, gumbo, etc. Low elevator speed will lessen elevator bounce and permit the elevator flights to sweep material into the bowl.

If possible, ridges 5 to 6 ft wide should be left between successive loading passes. Then, on the following passes, the operator can straddle load to pick up the ridges at a depth below previous cuts. This will leave other ridges, thus permitting the cycle to be repeated (Fig. 9-13*b*).

In addition to a straight cutting edge, an offset cutting edge or stinger is available to assist in the loading cycle (Fig. 9-13*c*). The center section of this cutting edge is extended as a stinger bit for normal loading. For hard loading, protruding ripping teeth can be bolted to the cutting edge. For doing finish work, loading windrowed material, or doing cleanup work, a finishing center section should be used. When the

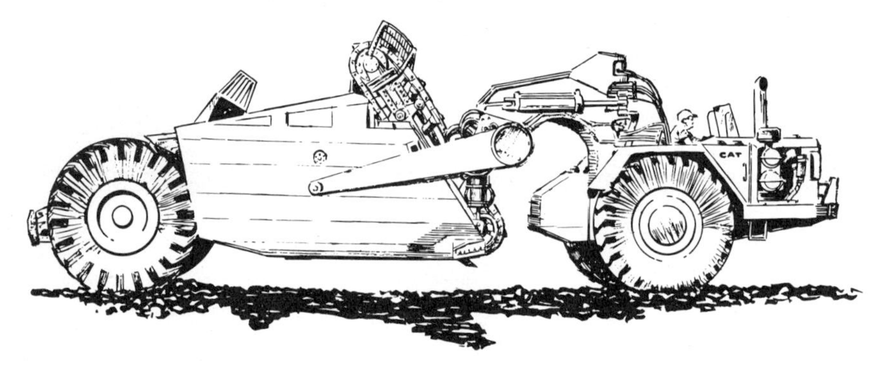

Fig. 9-14 Travel attitude with proper bowl clearance.

bowl is fully loaded, the elevator should be stopped. If this is not done, material will be swept from the face of the load and spilled over the sides of the bowl.

Traveling Loaded. When the scraper is loaded, the elevator should be stopped and the bowl raised until it clears the ground by 1 to 2 in. Travel with the bowl in this position should be continued until any loose mound of material is smoothed out. This will leave a smooth cut in the loading area. The bowl should be raised sufficiently for clearance when traveling (Fig. 9-14).

Unloading. The bowl should be lowered to permit the desired depth of spread. Condition of fill and depth of spread will change the speed and power requirements. Throughout the unloading cycle, the tractor should be operated at full engine speed and without allowing the engine to lug.

With the machine in motion, the ejector floor is opened. The material in the bowl will then begin to unload itself, and the leading edge of the ejector floor will strike off the unloaded material in a smooth, even layer.

Returning to Cut. With unloading completed, the bowl can be raised to the desired height for the haul-road conditions. The ejector floor is closed, and the scraper returns to the cut.

MULTIPLE SCRAPERS

The first scrapers in this category emerged as single-bowl units with power on two axles. In most cases they were push loaded, except in extremely easy loading materials and/or when favorable grades were present to provide additional power (gravity). More recently there has been a trend towards multiple scraper units (as many as three but generally two) with several engines and driving axles. Their combined horsepower and the high tractive effort which is compounded as the scraper bowls are successively loaded places these multiple units in the self-loading category.

Configurations for the multiple scrapers vary depending upon the manufacturers. Some multiple units are permanently coupled and require only one operator station, while others have disconnect features which permit independent use of the individual units as desired. In the latter case, individual operator's controls are overridden by a central control station when operating as a multiple unit.

Loading and spreading are generally done in sequence—first bowl, second bowl, etc. The all-wheel drive units, like the elevating-type units, eliminate pusher and extra operator cost as well as wait time in the cut. However, they have increased maintenance, repair, and tire costs in comparison with conventional scrapers. They load well in all stable materials but may be limited in their traction capabilities when loading in poor underfoot conditions. All wheel-drive units—single, twin, triple—present a complex problem as far as productivity estimating is concerned. It is advisable to consult their manufacturers for production and cost information.

Fig. 9-15 Caterpillar 627 push-pull scrapers.

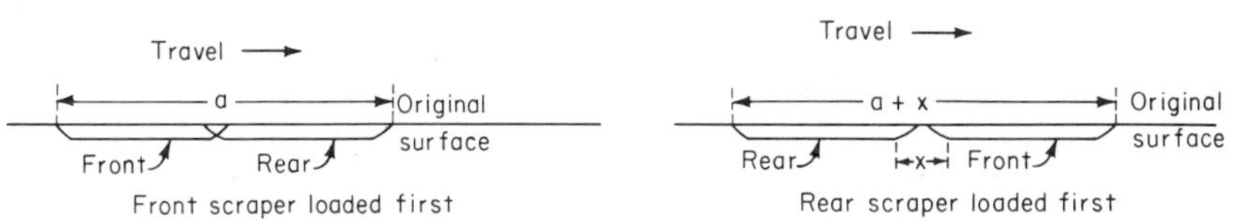

Fig. 9-16 Push-pull loading technique.

A push-pull concept (Fig. 9-15) provides a self-loading feature which is comparable to that of multiple units. Here two units assist each other only during the loading cycle, while on all other portions of the work cycle they operate as individual haul units. The first unit to arrive in the cut becomes the front unit and is loaded first. This takes advantage of the partial load the machine will get while positioning, as most operators self-load as much as possible. In addition, loading the front unit first permits a shorter cut length and results in a smoother cut. The rear unit can eliminate a possible gap by overlapping the first cut slightly, as illustrated in Fig. 9-16.

CONTROL SYSTEMS

Scrapers are either cable controlled or hydraulically controlled. In all cases, power for the scraper control is furnished by the engine of the prime mover. Cable-controlled scrapers are operated either by a mechanical cable-control unit or by an electric system which consists of a generator and electric motors. Hydraulically controlled scrapers are operated by hydraulic pumps and jacks.

Hydraulically controlled scrapers dominate the field. The typical hydraulic system illustrated in Fig. 9-17 consists of a hydraulic tank (7), pump (6), and control valves (4) mounted on the tractor; and lines (1), bowl cylinders (2), apron cylinder (3), and ejector cylinder (5) mounted on the scraper.

The hydraulic tank is mounted in front of the right fender of the tractor. Two elements are used in the filter, which is located in the hydraulic tank. A bypass valve which opens at 15 psi permits the oil to pass directly to the tank sump from the return line in the event that the filter elements should become clogged.

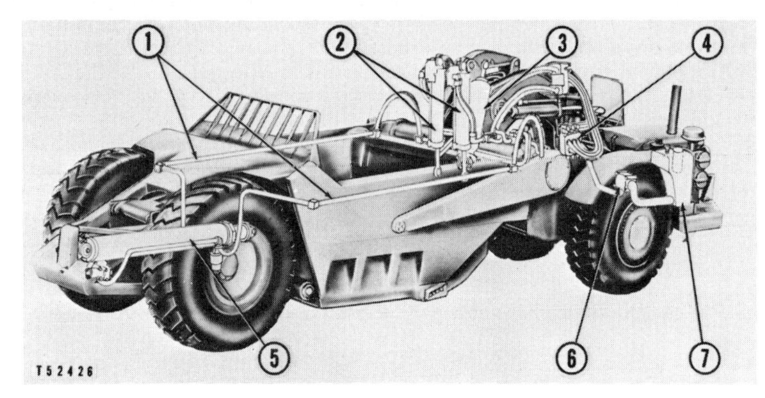

Fig. 9-17 Typical scraper hydraulic system.

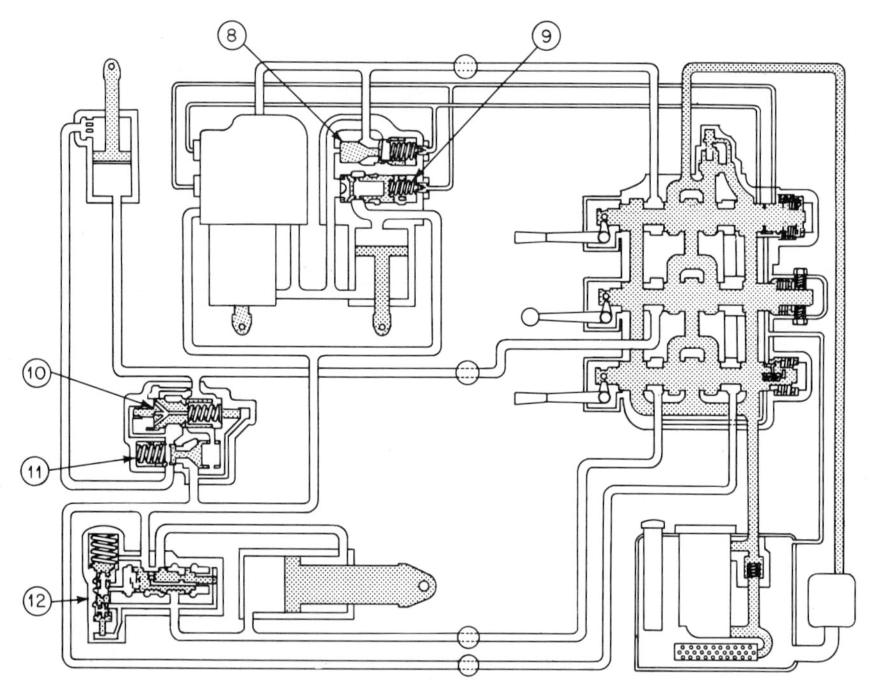

Fig. 9-18 Schematic hydraulic system.

The control valve is located to the right of the operator's compartment. Three control levers are mounted vertically at the right of the operator and control the bowl, apron, and ejector. All high-pressure hoses on the scraper are four-wire spiral wrap with permanent end fittings. The vent lines to the quick-drop valves are smaller, two-wire-braid hoses. The cylinders which are used to raise and lower the bowl are double acting to provide positive down pressure for the cutting edge.

Figure 9-18 supplies a schematic picture of a typical hydraulic system. A check valve and quick-drop valve are mounted on the top of each bowl cylinder. The bowl-carry check valve (8) acts as a safety valve in case a large hose should break or become damaged and prevents the bowl from dropping while traveling. It blocks the oil passages when the bowl control lever is in "hold" position and traps oil in the bowl cylinders. The bowl is then held by the blocked oil, rather than by system pressure, which eliminates the need for high pressure in the hydraulic lines while the machine is traveling. The quick-drop valve (9) permits faster down movement of the bowl while loading. The piston rod ends of both bowl cylinders are connected by a vent hose to equalize the pressure between the cylinders.

One double-acting cylinder is used to open and close the apron. When the apron is moved down, the system relief valve (10) controls the pressure in the piston rod end. A check valve (11) prevents the apron from floating up when the apron lever is in "hold" position and the bowl or ejector is operated.

A double-acting cylinder is used on the ejector. An automatic speed-change valve (12) is used in the ejector circuit to provide a slow and fast speed movement for the ejector. The valve changes from slow to fast, depending on the resistance to the ejector created by the material in the bowl. The fast speed is obtained by automatically supplying the oil exhausted from the piston-rod side of the piston to the other side of the piston along the pump flow.

BASIC CYCLE CONSIDERATIONS

In the basic working cycle, regardless of the scraper type, there are certain efficiency considerations that affect production and unit costs. The following illustrations use hypothetical data and are for comparative purposes only. Table 9-4 supplies basic owning and operating costs for the equipment used in these examples.

TABLE 9-4 Equipment Selection Data

Unit	Flywheel horsepower	Scraper capacity, lcy, struck/heaped	Hourly ownership and operating costs, including operator's wages
Pusher, track tandem or dual	770	....	$49.13
Pusher, track	385	....	23.19
Pusher, track	270	....	15.73
Wheel scraper, twin power	900	40/54	47.48
Wheel scraper	500	40/54	38.01
Wheel scraper, push pull	1,800	64/88	46.25
Wheel scraper, twin power	900	32/44	45.19
Wheel scraper	500	32/44	33.76
Wheel scraper	500	28/38	31.55
Wheel scraper, elevating	400	/32	30.77
Wheel scraper, three axle	400	21/30	25.84
Wheel scraper, two axle	400	21/30	25.82
Wheel scraper, elevating	300	/21.5	19.76
Wheel scraper	300	14/20	16.63
Pusher, wheel	400	....	23.99
Motor grader	150	....	13.73
High-speed compactor	400	....	23.02

The Cut When loading in the cut, material enters the scraper bowl rapidly at first and slows down as the loading progresses. Figure 9-19 is a typical load-growth curve

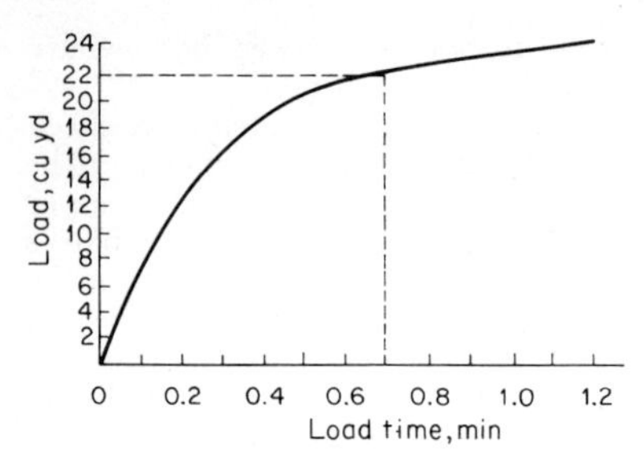

Fig. 9-19 Load growth curve.

which relates the time that a scraper spends in the cut to the load obtained. Note that the load finally reaches an area of the curve where further loading consumes excessive time, and the scraper should then pull out of the cut. Table 9-5 is an example of a 400-hp wheel tractor-scraper, working on an approximate 3,000-ft haul, which is loaded first with 22 and next with 24 cu yd of material. Notice that the load time in the cut must be doubled to pick up the last 2 cu yd of the 24-cu yd load, resulting in a production loss of 240 cu yd per hr.

Adequate pushers are necessary for efficient operation in the cut. Table 9-6 shows the production capability of four 400-hp wheel tractor-scrapers with one pusher.

TABLE 9-5 Controlling Load Time

Payload, bcy	22	24
Load time, min	0.75	1.5
Total cycle time, min	5	6
Trips/60-min hr	12	10
Scraper production, bcy/60-min hr	264	240
Pusher cycle time, min	1.5	2.25
Scrapers loaded/60-min hr	40	26.7
Max pusher production, bcy/60-min hr	880	640
Lost production, bcy/60-min hr	......	240

TABLE 9-6 Adequate Pushers

*4 scrapers, 21/30**		*1 pusher, 385 hp*	
Payload	22 bcy	Payload	22 bcy
Load time	0.75 min	Load time	0.75 min
Scraper cycle time	5 min	Pusher cycle time	1.5 min
Unit production	264 bcy/hr	Loads per hour	40
Potential spread production	1,056 bcy/hr	Maximum pusher production	880 bcy/hr
Potential lost production			176 bcy/hr

Loading cost

4 scrapers	$103.28	4 scrapers	$103.28
1 pusher	23.19	2 pushers	46.38
Total	$126.47	Total	$149.66

$$\frac{\text{Cost}}{\text{bcy}} = \frac{\$126.47}{880\ \text{bcy}} = \$0.144 \qquad \frac{\text{Cost}}{\text{bcy}} = \frac{\$149.66}{1{,}056\ \text{bcy}} = \$0.142$$

* Capacity, loose cubic yards: struck/heaped = 21/30.

Although the cost per hour for the equipment spread is increased by adding another pusher, increased production lowers the cost per cubic yard. The lowest cost per yard occurs when full advantage is taken of the maximum potential of the hauling units.

Tandem pushing may pay off with large scrapers of 28-cu yd and greater capacity when working in difficult loading material such as loose sand and tough clay. Poor traction in the cut, long hauls, and rock cuts are conditions which also warrant tandem pushing. Cushioning devices and power-shift transmissions, introduced in recent years, have made tandem pushing increasingly practical and popular.

Most scraper tire wear occurs in the cut. Tire slippage of up to 20 percent can occur on a tractor-scraper without detection by either the operator or a ground observer. Assume that under a normal condition (a minimum of tire spinning) two scraper tires and four drive tires wear out in approximately 5,000 hr, amounting to 15 percent of hourly owning and operating costs. This costs the contractor at least

$$6 \text{ tires} \times \$2{,}500 \text{ per tire} = \$15{,}000 \text{ or } \frac{\$15{,}000}{5{,}000 \text{ hr}} = \$3 \text{ per hr}$$

Excessive spinning of drive tires while loading can cut tire life in half. Tire costs then become about 30 to 35 percent of owning and operating costs, with tire consumption over the same 5,000-hr period increased to eight drive tires and two scraper tires. This costs the contractor at least

$$10 \text{ tires} \times \$2{,}500 \text{ per tire} = \$25{,}000 \text{ or } \frac{\$25{,}000}{5{,}000 \text{ hr}} = \$5 \text{ per hr}$$

Ripping has gained widespread acceptance in recent years. It improves loading characteristics of many materials, allows scraper loading of rock, reduces wear on haul units, and lowers cost per yard.

There are a few additional factors which can adversely affect cut production:

Poor spotting of the pusher
Rough cut caused by raising the scraper bowl too fast
Poor boost of the scraper by the pusher
Taking a deep cut, loading thick slabby material which is difficult to unload and compact on the fill

The Haul Road A large part of a haul unit's time is spent on the haul road. A poorly maintained haul road causes extreme bounce which results in load loss, machine and operator fatigue, and excessive downtime. The slower speed at which the haul unit must be operated lowers production and raises cost per yard.

Table 9-7 analyzes a four-scraper spread which is operated on a haul road in different conditions of maintenance. With only one motor grader available to work the haul road, the low spread production results in a cost of $ .201 per yd. Adding a second motor grader increases the production, and despite the additional $13.73 per hour cost for equipment, reduces cost per yard to $ .168. Haul-unit speeds under good conditions can easily average 20 mph on a 3,000-ft haul, although even a few obstacles may cut this average speed in half and thus double haul and return time. The resulting loss in hourly production may amount to several hundred cubic yards.

The Fill Fill production cannot exceed compactor production, the critical operation on the fill. In Table 9-8, scrapers are hauling material to the fill area where it is compacted by two tractors pulling sheepsfoot rollers. Because of the slow compaction, production is held to 920 cu yd per hr at a cost of $ .231 per cu yd. Adding another sheepsfoot roller will reduce the cost per yard to $ .218 and allow maximum production to be maintained. However, replacing the three slow sheepsfoot compactors with one high-speed compactor (Table 9-9) reduces cost per yard to $ .177 and minimizes congestion on the fill. Unfortunately, there are still a few states in which "method" specifications on compaction exclude the use of high-speed compactors. In states with "end result" specifications, high-speed compaction has been found to be a very satisfactory method.

Other Factors Finally, here is a group of other important factors which will affect job efficiency:

INFORMED OPERATORS: An operator will function better if he knows what is expected of him and how his job fits into the overall operation.

MATCHED SPREADS: New high-speed units can be slowed down by one slow unit in the spread.

TABLE 9-7 Haul-road Maintenance

Production

Scrapers, 21/30;* pushers, 385 hp; graders, 150 hp

	Excellent road	Poor road
Load time, min	0.75	0.75
Turn, dump, etc., min	0.75	0.75
Haul and return, min	3.5	5
Total cycle time, min	5	6.5
Trips/60-min hr	12	9.2
Unit production, bcy/60-min hr	264	203
Spread production, bcy/60-min hr	1,056	812

Costs

Poor road		Excellent road	
4 scrapers	$103.28	4 scrapers	$103.28
2 pushers	46.38	2 pushers	46.38
1 grader	13.73	2 graders	27.46
Total	$163.39	Total	$177.12

$$\frac{\text{Cost}}{\text{bcy}} = \frac{\$163.39}{812\ \text{bcy}} = \$0.201 \qquad \frac{\text{Cost}}{\text{bcy}} = \frac{\$177.12}{1{,}056\ \text{bcy}} = \$0.168$$

* Capacity, loose cubic yards: struck/heaped = 21/30.

TABLE 9-8 Slow-speed Compaction

Scraper and compactor production

4 scrapers, 21/30*		*2 compactors*, 285 *hp tractors and* 5 × 5 *sheepsfoot rollers*	
Load	22 bcy	Depth per layer	6 in.
Cycle time	5 min	Passes per layer	8
Trips per hour	12	Avg working speed	3 mph
Unit production	264 bcy/hr	Production	735 cu yd/hr compacted or 920 bcy/hr
Spread production	1,056 bcy/hr		

Costs

2 Compactors		*3 Compactors*	
4 scrapers	$103.28	4 scrapers	$103.28
2 pushers	46.38	2 pushers	46.38
2 graders	27.46	2 graders	27.46
2 rollers	35.40	3 rollers	53.10
Total	$212.52	Total	$230.22

$$\frac{\text{Cost}}{\text{bcy}} = \frac{\$212.52}{920\ \text{bcy}} = \$0.231 \qquad \frac{\text{Cost}}{\text{bcy}} = \frac{\$230.22}{1{,}056\ \text{bcy}} = \$0.218$$

* Capacity, loose cubic yards: struck/heaped = 21/30.

TABLE 9-9 High-speed Compaction Costs

4 scrapers, 21/30*	$103.28
2 pushers, 400 hp	46.38
2 graders, 150 hp	13.73
1 high-speed compactor	23.02
Total	$186.41

$$\frac{\text{Cost}}{\text{bcy}} = \frac{\$186.41}{1{,}056\ \text{bcy}} = \$0.177$$

* Capacity, loose cubic yards: struck/heaped = 21/30.

STANDARDIZATION OF EQUIPMENT: Mechanics and operators can become more familiar with the equipment, parts inventories can be lowered, and overall operation simplified.

PREVENTIVE MAINTENANCE: Eliminate needless downtime.

JOB SAFETY: A safe job is a productive job.

EQUIPMENT SELECTION

This discussion will also make use of the cost figures shown in Table 9-4. These will not have the validity of those obtained from a contractor's own cost records and systems, since they are based on the standard industry practice for determining machine hourly owning and operating (O&O) costs. They are valid for comparison purposes, although it should be understood that they are direct machine costs only—including operator wages. No other business or operating expenses are included.

The Push-loading Concept More construction dirt moving is done today with push-loaded fleets of scrapers than with all other systems combined. The economics of the push-loading system are such that its profitable application spans the greatest breadth of the actual conditions encountered in the field. The key to a pushed-scraper spread's economy is the fact that both pusher and scraper share in the work of obtaining the load. The resulting loading costs are lower than for almost any form of top-loading system and for other self-loading systems—elevating or multiple units —providing the pusher cost can be prorated over three to four scraper units.

A top-loading hauling unit will generally have a lower tare weight than its scraper counterpart, and thus it can carry a higher payload. However, the savings which are realized in the haul cost must be sufficient to offset the lower loading cost of the scraper spread before the "pure" haul system becomes economical. The heavier weight of an elevating scraper penalizes it on the haul portion of the cycle, and this weight penalty must be overcome through the elimination of the pusher before the economic performance of the elevating scraper can equal or exceed that of the scraper-pusher combination. The interrelated load-and-haul cost of the scraper-pusher system has proved to be an economical choice for a wide range of haul distances, underfoot conditions, grades, and materials.

In average, common earth materials, a single 385-hp pusher will typically load out 35 to 40 loads of 22.5 bcy per hr. That is a pit production of 800 to 900 bcy per hr with an investment in loading equipment of approximately $80,000. With a pusher cost of $23 to $25 per hr, it translates into a direct loading cost of $ .025 to $ .030 per bcy. A typical top-loading system to deliver this production would require an investment of $150,000 to $250,000, with a corresponding direct loading cost of $ .05 to $ .10 per bcy. While hauling costs with the scraper will be slightly higher than with a pure hauler, the cycle must become quite long before this cost will offset the difference in loading costs. Additionally, the scraper offers an unmatched ability to spread its load in an even lift on the fill. Frequently this can result in substantially lower fill-processing costs than can be obtained with other types of haulers.

With the advent of special-application scrapers, which are designed expressly for rock handling, the push-loaded scraper has one of the widest material appetites of any system. It has been used economically to handle materials ranging from blow sand to gumbo to shot granite. This versatility can be a key factor in picking a system where a variety of materials is to be encountered. For these reasons, push-loaded scraper fleets are the backbone of the earthmoving industry today.

Since push-loaded scrapers occupy such a dominant position in earthmoving systems, the science of matching particular types of scraper equipment to the job has become an important one. There are proper applications for both towed scrapers and wheel tractor-scrapers, two-axle and three-axle rigs, various size scrapers behind the same basic power unit, and for different size units of the same basic type.

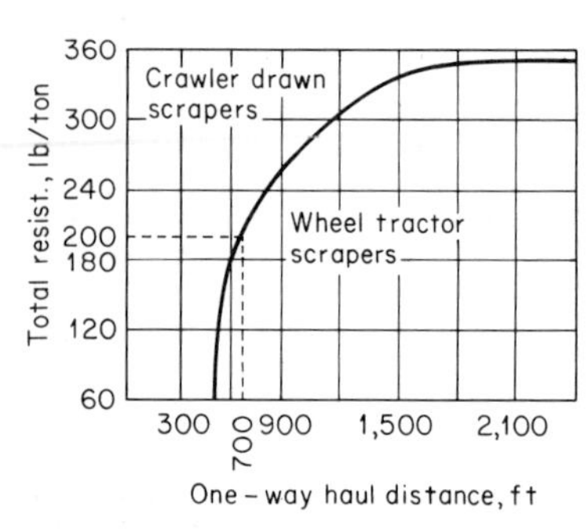

Fig. 9-20 Application zones—crawler versus wheel scrapers.

Towed Scrapers versus Wheel Tractor-scrapers The advantageous economics of the crawler-drawn scraper (towed) as compared to the wheel tractor-scraper stem from its:

- Ability to minimize the adverse effects of high rolling resistance by laying down its own road
- Superior tractive effort in mud
- Ability to handle severe grades
- Ability to economically self-load some materials if the tractor is matched with a proper size of scraper
- Typically lower undercarriage costs in severe rock work when compared to tire costs on a wheel tractor-scraper

The limiting factor of the system is speed—normally a maximum of 4 to 5 mph. Thus, this method must be classed as a short-haul system—the maximum economical haul distance increases as total resistance increases.

Figure 9-20 roughly defines the economic areas, comparing crawler-drawn units with wheel tractor-scrapers of equivalent capacity. As the total resistance increases, the economical haul distance for crawler-drawn scrapers is extended, so that over a 200 lb per ton haul road the crawler spread would be economical out to approximately 700 ft. This relationship will vary somewhat, depending on the specific units considered.

Computerized Vehicle Simulation While side-by-side analysis in the field remains the most conclusive way of answering equipment selection questions for a specific job, computerized vehicle simulation programs are a most valuable tool. In such analyses, the performance capabilities of the various machines are programmed into a digital computer system. Data describing one or a number of specific haul profiles in the field

TABLE 9-10 Loading Summary Data

Scraper capacity, lcy, struck/heaped	Pusher	Avg load time, min	Avg payload, bcy	Avg load factor*
14/20	1–270 hp	0.68	16.0	0.80
/21.5	....	0.83	16.0	0.74
21/30	1–385 hp	0.87	22.5	0.74
/32	....	0.83	22.4	0.75
28/38	2–385 hp	0.68	30.1	0.80
32/44	2–385 hp	0.81	35.0	0.79
32/44	2–385 hp	0.79	36.2	0.82
64/88	....	1.90	68.2	0.77
40/54	2–385 hp	0.95	39.8	0.74
40/54	2–385 hp	0.90	42.1	0.78
Overall avg load factor	....	....	....	0.77

* Ratio of avg payload in bcy to heaped capacity in loose cubic yards.

are then fed into the computer, which in a matter of seconds will print out the productive capability of each machine studied for each haul profile.

Basic input data, especially concerning loadability, are derived primarily from extensive weight and time-study work performed in the field. Table 9-10 is a summary of overall average loading performance recorded in the field on various machines. The data include over 300 field studies made since 1962.

Average load factor as used in Table 9-10 is defined as the ratio of actual payload (in bank cubic yards) to rated heaped-scraper capacity. The tabulated averages were obtained in materials which ranged from 2,400 to 3,600 lb per bcy and included the clays, silts, sands, gravels, and combinations which are normally considered as "scraper dirt."

Table 9-11 is an example of the input data describing the haul-profile segment

TABLE 9-11 Computer Input Data Sheet

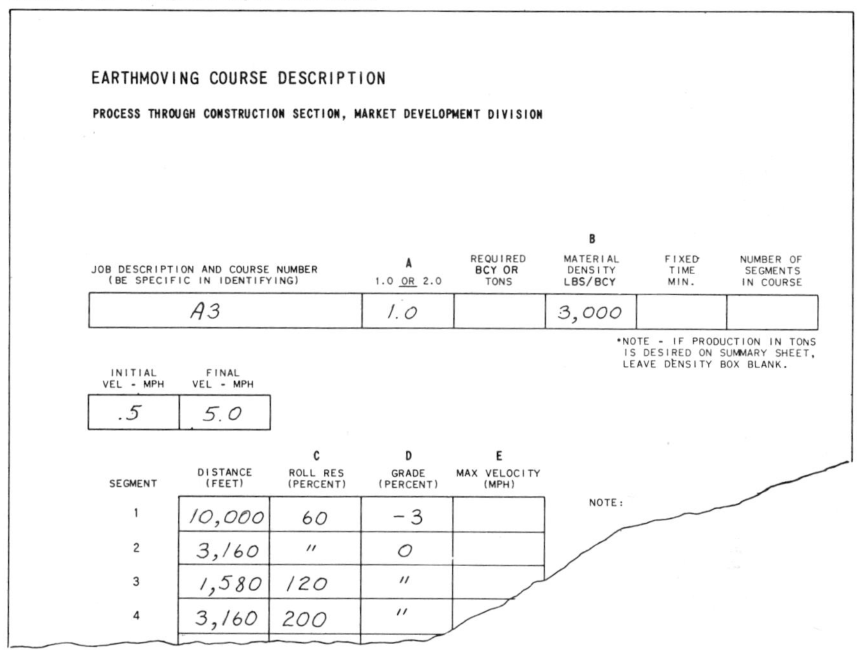

EARTHMOVING COURSE DESCRIPTION

PROCESS THROUGH CONSTRUCTION SECTION, MARKET DEVELOPMENT DIVISION

JOB DESCRIPTION AND COURSE NUMBER (BE SPECIFIC IN IDENTIFYING)	A 1.0 OR 2.0	REQUIRED BCY OR TONS	B MATERIAL DENSITY LBS/BCY	FIXED TIME MIN.	NUMBER OF SEGMENTS IN COURSE
A3	1.0		3,000		

*NOTE - IF PRODUCTION IN TONS IS DESIRED ON SUMMARY SHEET, LEAVE DENSITY BOX BLANK.

INITIAL VEL - MPH	FINAL VEL - MPH
.5	5.0

SEGMENT	DISTANCE (FEET)	C ROLL RES (PERCENT)	D GRADE (PERCENT)	E MAX VELOCITY (MPH)
1	10,000	60	−3	
2	3,160	"	0	
3	1,580	120	"	
4	3,160	200	"	

NOTE:

lengths, grades, rolling resistance, limiting speeds, fixed times, bank densities, and initial and final velocities.

The answer is printed out by the computer on a sheet such as that illustrated in Table 9-12. Data on the print-out, which is a summary sheet, include: payload

TABLE 9-12 Computer Print-out Sheet

PAGE 08 6/05/69 10:58 AM VEHICLE SIMULATION-SALES 0115 RPH-01

COMPUTING DIVISION-RESEARCH DEPARTMENT-CATERPILLAR TRACTOR- 73- 0

VEHICLE - 769 - B (1800x33)
ENGINE - 415 HP W-FAN
TRANSMISSION - 9 SPEED
EMPTY WEIGHT - 57800 LBS
PAYLOAD - 70000 LBS
TIRE SIZE - FRONT DRIVE REAR
1800X33 1 1800X33 1

	HAUL DIST FEET	HAUL TIME MIN	RETURN DIST FEET	RETURN TIME MIN	FIXED TIME MIN	CYCLE TIME MIN	TRIPS PER HOUR	PRODUCTION	TOTAL COURSE MATERIAL	BANK DENSITY LBS/CU YD	TOTAL MACHINE HOURS
MINE 40	2600	1.82	2600	.93	4.0	6.75	8.89	311.27 TONS/HR	0 TONS	0	0
MINE 60	2600	2.05	2600	.96	4.0	7.02	8.55	299.27 TONS/HR	0 TONS	0	0
MINE 80	2600	2.47	2600	.98	4.0	7.45	8.06	282.03 TONS/HR	0 TONS	0	0

(in pounds), haul distance and time, return distance and time, fixed time, cycle time, trips per hour (assumes a 60-minute hour), yardage per trip, production in bank cubic yards per hour, required yardage, and machine hours necessary.

Accuracy of the computer analysis method has been checked against many actual jobs. The sketch and numerical data in Table 9-13 refer to the Bastrop Dam job in

TABLE 9-13 Computer Simulation and Actual Field-study Comparison

Haul profile:

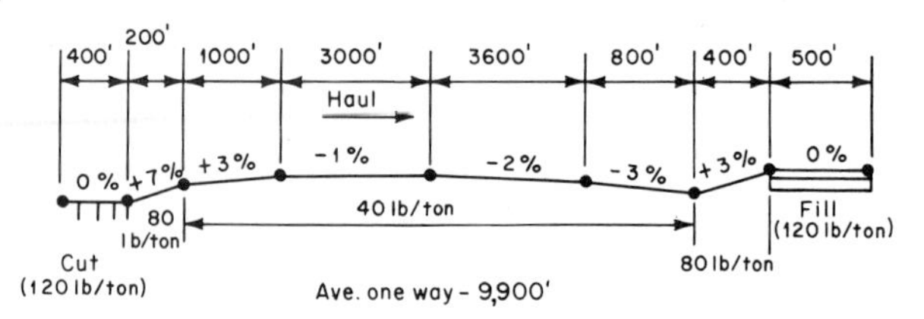

Scrapers: Cat 660-650 combinations
Pushers: Tandem D9G's
Material: Sandy clay—2,800 lb/bcy

Production	Computerized estimate	Actual from field study
Avg load time, min	0.81	0.90
Dump and turn time, min	0.50	0.40
Wait—pusher, min	0.20	0.30
Total fixed time, min	1.51	1.60
Haul time, min	5.07	4.85
Return time, min	4.88	4.95
Total cycle, min	11.46	11.40
Trips/60-min hr	5.24	5.25
Avg payload, bcy	35.0	36.5
Production/scraper, bcy/60-min hr	183	192
Scrapers—balanced spread	7	7
Spread production, bcy/60-min hr	1,281	1,344

Texas. Figures in the first column of this table were obtained from computer analysis, while those in the second column were taken on the job with scales and stopwatches. Note that the actual figures show a slightly higher production rate than the computerized estimate. This is due to a 36.5-yd average payload achieved, compared to the figure of 35.0 which was used in the estimate. However, each machine would reflect this same relative difference, so comparisons would still be valid.

Many contractors now have their own computers and are extending their use to include vehicle simulation in their bidding and machine selection. Computerized vehicle simulation can shed some light on various questions raised concerning wheel tractor-scrapers, such as the following:

Two or three axle?
Large or small?
Single or tandem power?
Size of scraper?
Selection of pusher power?
Track-type pusher versus wheel-type pusher?

Number of Axles Which is needed—traction and maneuverability, or speed? Two-axle scrapers can climb grades that other machines cannot and, with hydraulic steering, they can "duck walk" through mud. However, the better riding qualities of three-axle units permit higher top speeds (40 versus 30 mph) which can pay off on long, flat runs where climbing and "mudding" abilities are secondary. Computer analysis can help pin down the point at which the balance swings one way or the other.

TABLE 9-14 How Many Axles?

Haul profile:

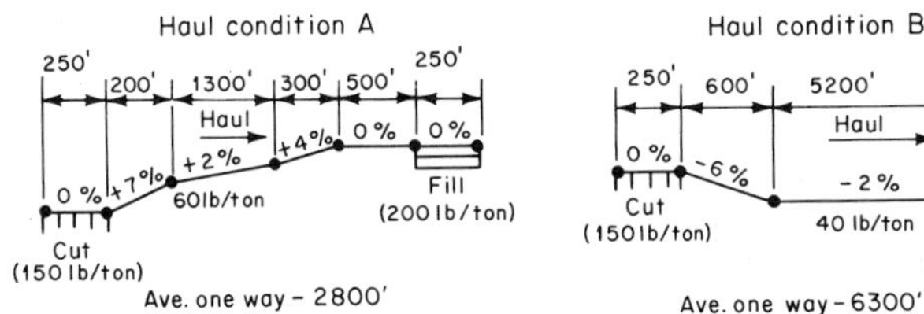

Pusher	1-385 hp		1-385 hp	
Production: Wheel scraper (400 hp)	2-axle	3-axle	2-axle	3-axle
Fixed time, min	1.50	1.50	1.50	1.50
Haul time, min	3.49	3.65	2.86	2.51
Return time, min	1.42	1.44	3.08	2.74
Total cycle, min	6.41	6.59	7.44	6.75
Trips/60-min hr	9.36	9.10	8.06	8.88
Avg payload, bcy	22.5	22.5	22.5	22.5
Avg production/scraper, bcy/60-min hr	211	205	181	200
Scrapers—balanced spread	4	4	4	4
Spread production, bcy/60 min hr	844	820	724	800
Cost of production				
Total hourly spread costs, dollars	126.47	126.55	126.47	126.55
Load and haul cost, dollars/bcy	0.150	0.154	0.175	0.158

In this example, the computer compares the performance of balanced spreads of two-axle and three-axle 400-hp wheel scrapers on two different hauls. The scrapers, identical in capacity, fare about the same on a 2,800-ft run with moderate adverse grades. Undoubtedly, the traction consideration would favor the two-axle unit under these conditions. On the other hand, where the haul is 6,300 ft with favorable grades, load and haul costs for the three-axle rig show a clear superiority of 10 percent.

The advantages of the three-axle machine are gradually disappearing as improved two-axle design features are introduced. With the advent of snubbing devices, for example, the "loping" characteristic of the two-axle machine is minimized. Shock absorption and damping systems can provide the two-axle machine with essentially the same ride characteristics as three-axle units. The third axle, of course, still results in a better tire-weight distribution which provides more tmph (ton-mile per hour) capacity and allows higher workday-average speeds.*

Field tests show production gains up to 14 percent for the snubbed machine, depending on the condition of the haul road and on the cut and fill surfaces. The advantage of the damping feature is not confined to long, high-speed hauls, since the maximum production advantage actually came on the shortest haul (on a rough,

* See Sec. 6, Engineering Fundamentals.

1,000-ft, severe road surface). In some rough areas, primarily the cut and fill, the improved unit was able to maintain twice the speed of the conventional machine. The reduction in the stress levels of the snubbed machine also results in lower repair and maintenance costs.

Weight-to-horsepower Ratio Where scrapers of different sizes are available with the same engine power, other choices are possible. The larger load can mean more production per horsepower hour if underfoot conditions and grades do not greatly hamper cycle times, but rough going or adverse grades may demand lower weight-to-horsepower ratios. Here again, analysis is helpful in viewing the picture.

In Table 9-15, the 28-yd (struck) scraper is matched against the 32-yd (struck) scraper. Both have 500-hp engines. On the job with moderate adverse grades, analysis shows that the 28-yd scraper has a slight cost advantage. On the job with a longer haul and more favorable grades, the larger scraper closes the cycle-time gap and gains the edge with respect to unit costs.

TABLE 9-15 Weight-to-horsepower Ratio?

Haul profile:

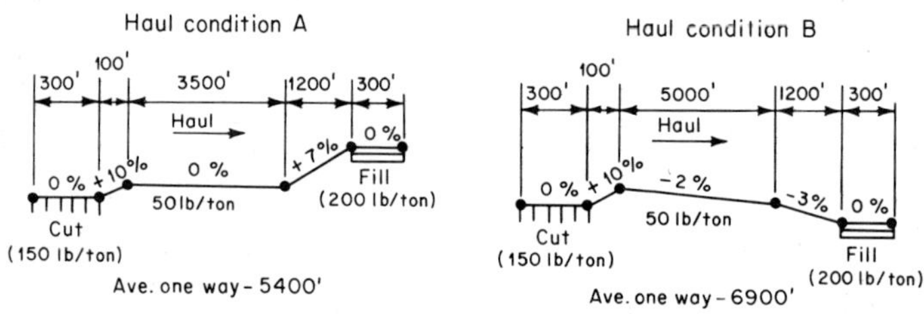

Pushers: Tandem	2-385 hp		2-385 hp	
Production: Wheel scraper (500 hp)	28/38	32/44	28/38	32/44
Fixed time, min	1.30	1.50	1.30	1.50
Haul time, min	5.24	6.30	3.48	3.53
Return time, min	2.33	2.27	3.70	3.97
Total cycle, min	8.87	10.07	8.48	9.00
Trips/60-min hr	6.76	5.96	7.08	6.67
Avg payload, bcy	30.1	35.0	30.1	35.0
Avg production/scraper, bcy/60-min hr	203	209	213	233
Scrapers—balanced spread	6	6	6	6
Spread production, bcy/60-min hr	1,218	1,254	1,278	1,398
Cost of production				
Total hourly spread costs, dollars	235.68	248.94	235.60	248.94
Load and haul cost, dollars/bcy	0.194	0.199	0.184	0.178

Number of Engines Scrapers with engine power on two axles (twin power) can climb adverse grades of 50 percent or more, fully loaded. They can also outperform other units in extremely soft, muddy footing. However, they require an investment up to 25 to 27 percent larger than their single-drive axle counterparts. Thus the question is: When will they produce enough additional yardage to justify the extra cost?

Computer analysis of two typical jobs, as illustrated in Table 9-16, provides a means for comparison.

On the haul involving only moderate grades, Table 9-16 indicates that the spread of single-powered units achieves a direct load-and-haul cost per yard of $ .138, as compared with $ .145 for tandem engine units. However, where grades run 10 to 15 percent adverse, the tandem power units have the edge. This margin will, of course, widen as the grade or rolling resistance becomes more adverse.

TABLE 9-16 Tandem or Single-engine Power?

Haul profile:

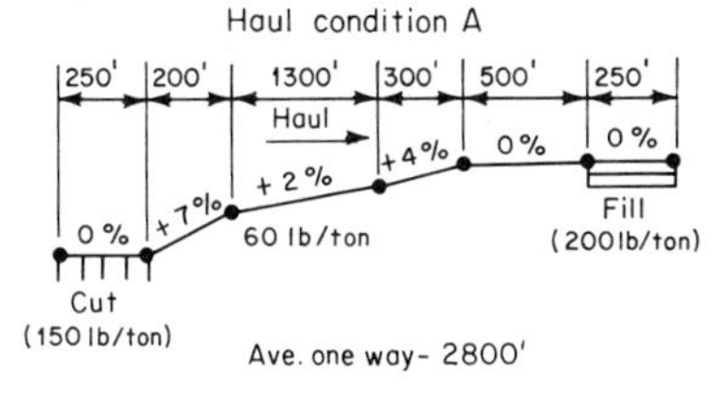

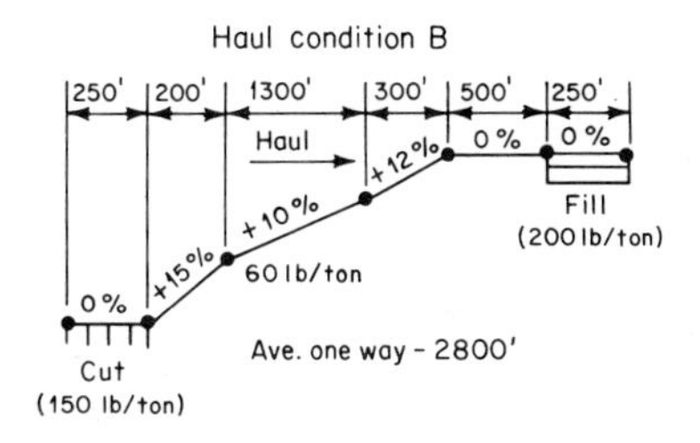

Pushers: Tandem	2-385 hp		2-385 hp	
Production: Wheel scraper (32/44)	500 hp	900 hp	500 hp	900 hp
Total fixed time, min	1.45	1.34	1.46	1.34
Haul time, min	3.03	2.08	5.85	3.54
Return time, min	1.92	1.79	1.99	1.87
Total cycle, min	6.41	5.21	9.30	6.75
Trips/60-min hr	9.36	11.25	6.45	8.89
Production/scraper, bcy/60-min hr	35.0	36.2	35.0	36.2
Scrapers—balanced spread	4	3	6	5
Spread production, bcy/60-min hr	1,312	1,251	1,356	1,610
Cost of production				
Total hourly spread cost, dollars	181.42	181.95	248.94	272.33
Load and haul cost, dollars/bcy	0.138	0.145	0.184	0.169

Many still consider the tandem-powered scraper as a specialized tool—for opening up jobs, extreme adverse grades, short hauls, etc. Today, however, these units are being used in more and more "average conditions" work.

Size of Scraper In general, unit costs of production are lower with large equipment than with smaller machines. Machine cost rises at a slower rate than machine capacity, and operator wages will generally remain constant regardless of machine size. However, if a large machine is to pay its way, its productivity must be fully utilized in a balanced spread. Large units also require more pushing power, which adds costs; and job layouts must provide adequate maneuvering room. To permit a size comparison, data on 21/30 cu yd (400 hp) and 28/38 cu yd (500 hp) scrapers working in balanced spreads on the same typical job were programmed into the computer. These two units have nearly the same weight-to-horsepower ratios (21/30—189 lb per hp, 28/38—192 lb per hp) and provide a good basis for comparison.

The results are shown in Table 9-17. Cycle times increase with size, but not as sharply as productivity. Load and haul costs per yard are \$ .144 for the small unit and \$ .139 for the large unit. These cost differences appear to be moderate, but they can be highly significant when evaluating annual production and capital investment requirements.

TABLE 9-17 What Size Scraper?

Haul profile:

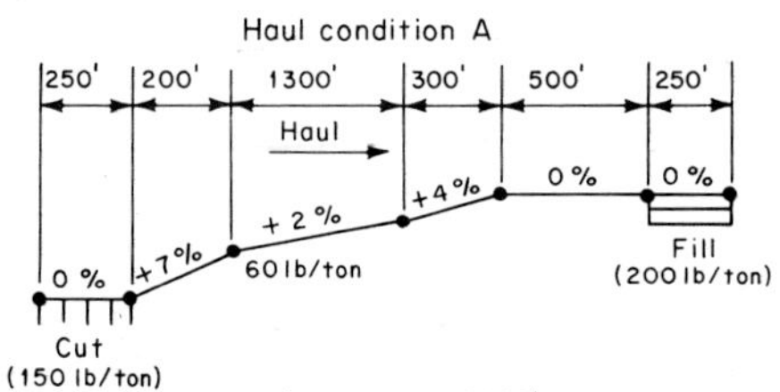

Pusher(s)	1-385 hp	Tandem 2-385 hp
Production: Wheel scraper	21/30 (400 hp)	28/38 (500 hp)
Fixed time, min	1.50	1.30
Haul time, min	2.82	2.67
Return time, min	1.89	1.85
Total cycle time, min	6.21	5.82
Trips/60-min hr	9.72	10.29
Payload, bcy	22.5	30.1
Production/scraper, bcy/60-min hr	219	310
Scrapers—balanced spread	4	4
Spread production, bcy/60-min hr	876	1,240
Cost of production		
Total hourly spread costs, dollars	126.47	172.58
Load and haul cost, dollars/bcy	0.144	0.139

Power for Push Loading Pusher selection can be the most important single element of a properly applied scraper spread, since the entire spread's output is dependent on it. Two general conclusions which can be drawn regarding a properly equipped cut are:

1. There must be a close balance between the hourly load-count capabilities of the pusher equipment and the total loads which can be hauled per hour.
2. Excess capacity of either loading or hauling equipment cannot be fully utilized and detracts from the system's potential unit-cost economy.

The 1-385-hp unit pushing a 21/30 (400 hp) wheel tractor-scraper on the haul profile described in Table 9-18 may serve as an example.

The conclusions from the data in Table 9-17 are:

1. A 3:1 ratio is the closest practical balance—it delivers lowest unit cost.
2. Either 2:1 or 4:1 ratios result in a significant cost penalty.
3. An improvement could be made in 4:1 ratio economy by lowering load time and carrying smaller load. This still results in some penalty, however.

Balance between loading and hauling equipment is essential for economic applica-

TABLE 9-18 Scraper-pusher Matching

One-way haul distance: 2,500 ft — Grade: 0%
Material: avg clay loam—3,000 lb/bcy — Avg R.R.: 80 lb/ton
Scrapers: 21/30 (400 hp) — Pusher: 1-385 hp
Optimum load time for max scraper production: 0.80 min
Avg payload: 22.5 bcy
Avg pusher cycle (load) 0.80 + (return) 0.40 × 0.80 + (boost and shift) 0.25 = 1.37 min
Avg scraper cycle (from computer): 4.49 min

$$\text{Ideal number of scrapers per pusher: } \frac{4.49 \text{ min (scraper cycle)}}{1.37 \text{ min (pusher cycle)}} = 3.28$$

Spread production and cost

Number of scrapers	2	3	4	4 (adjustable load time)
Production limited by	Scrapers	Scrapers	Pusher	Pusher
Load time, min	0.80	0.80	0.80	0.60
Avg payload, bcy	22.5	22.5	22.5	19.0
Spread production at 100% efficiency, bcy/hr	600	900	988	1,040
Hourly spread cost, dollars	74	99	124	124
Load and haul cost, dollars/bcy	0.123	0.110	0.126	0.119

tion for any interdependent load and haul system. Normal matching of push power to scraper size is as follows:

a. 14/20—270 hp usually—180 hp in good material—385 hp in very hard, severe conditions

b. 21/30—385 hp usually—270 hp in good material—tandem 2-385 hp in very hard, severe conditions

c. 28/38—tandem 2-385 hp usually—385 hp in good material

d. 32/44 and up—tandem 2-385 hp

e. Tandem 32/44—occasionally can use 385 hp, but usually tandem 2-385 hp

f. 40/54 and up—never less than tandem 2-385 hp, sometimes triple pushers

A dual 385-hp unit can also be ideal pusher equipment where tandem 2-385-hp capacity is called for, as typified by the results of a field loading study of 32/44 (500 hp) units (see Table 9-19). The dual-pusher configuration is illustrated in Fig. 9-21, and consists of two interconnected tractors which are operated by one set of controls and one operator. A tandem configuration consists of two separate tractors, one behind the other and each with its own operator.

Fig. 9-21 Dual D9G Caterpillar tractors.

TABLE 9-19 Pusher Cost Summary, Dual versus Tandem

	Dual	Tandem
Loads/60-min hr	46.1	40.0
Average load, bcy	33.3	32.0
Production, bcy/60-min hr	1,535	1,280
Hourly O&O costs, dollars	49.13	46.38
Loading costs, dollars/bcy	0.032	0.036
Production advantage, %	20	
Cost advantage, %	13	

Users go to duals when they feel they have reached maximum pit or cut efficiency with tandems. They are looking for that extra five to ten loads per hour which can be gained with duals under the proper set of conditions—primarily with a good pit and good pit supervision. (Jobs have been observed where the loading rate exceeded 90 loads per hour.) The duals have the capability of:

Faster load times
- Full power at the instant of contact
- No wait time for positioning second tractor
- Greater pushing effort
- Faster ground speed and faster loading rate

Quicker boost
Faster return
Less approach and maneuver time

All these factors contribute to a faster cycle time and thus more loads per hour. The economy in using duals results from more loads per hour, not from eliminating one operator.

Track-type versus Wheel-type Pusher The big advantage of wheel-type running gear is mobility—speeds to 20+ mph. In general, wheel pushers are justifiable when this mobility can be used, such as where cuts are thin and the spread moves from cut to cut several times daily.

Basically, the picture is this. A 385-hp track-type unit equipped for pushing weighs around 85,000 lb. With a coefficient of traction of 0.9 (normally applied to tracks), the machine is capable of a tractive effort of some 76,000 lb.* For a wheel tractor to provide a comparable pushing effort, it would need to weigh over 125,000 pounds—assuming a coefficient of traction of 0.6, which is a good figure for rubber in

* See Sec. 6, Engineering Fundamentals.

TABLE 9-20 Pusher Comparison—Wheels versus Tracks

Location: Centerville, Ala.
Material: Sandy clay—3,200 lb/bcy

Scrapers, 21/30 (400 hp)	Wheel pusher, 400 hp	Track pusher, 385 hp
Optimum load time, min	0.74	0.70
Pusher cycle time, min	1.29	1.18
Loads/50-min hr	38.7	42.3
Avg payload, bcy	21.2	22.1
Avg pusher production/50-min hr, bcy	822	938
Avg hourly cost, dollars	23.99	23.19
Loading cost, dollars/bcy	0.029	0.025

most conditions. A 400-hp wheel-type unit with full counterweights weighs about 7,000 lb more than the track unit. The net result is a 15,000- to 20,000-lb differential in drawbar pounds pull—favoring the tracks. For underfoot conditions which are less favorable for the wheel unit, the tractive advantage of the track-type unit could be considerably greater. Table 9-20 makes a comparison of the two types of pusher units under typical field conditions.

This field comparison points up the following:

Wheels push longer—have a longer cycle time—and will load out fewer loads per hour than will tracks. In addition, there is a slightly larger payload in the scraper pushed by tracks. The average hourly cost also favors the track-type machines.

Conclusions

1. Wheels will seldom equal track production for comparable units.
2. Hourly cost of investment is generally higher for wheel pushers.
3. Loading cost per cubic yard favors tracks.

There can be exceptions to these generalizations. For example, rubber-tired dozers may be justified in extreme cases of abrasive sand on the basis of tire costs versus track costs. However, the case must be truly an extreme one to offset the basic economics of the track-type unit.

Using a wheel and two single-track units in tandem offers an economic possibility. The mobility of the wheel unit permits it to push one track unit and then quickly pick up another in a shuttle-type operation. In this manner three pushers (two track and one wheel) can handle a spread instead of the customary four pushers.

Self-propelled, Self-loading Scrapers The self-loading scraper concept has become an earthmoving system of growing importance as contractors continue to strive for increased efficiencies by minimizing the mismatch and bunching effect inherent in push-loaded scraper fleets. Self-loading to date has been achieved with two basic methods—elevating design or multiple-bowl, all-wheel-drive units.

Elevating Scraper. This unit, in itself, is a "balanced spread" and is capable of productivity approaching that of a push-loaded scraper on short hauls. Being a more complex machine, it weighs some 8,500 to 11,000 lb more than its conventional counterpart and represents a $10,000 to $15,000 higher investment to attain essentially the same payload capacity. Yet the elevating scraper has proved very economical for utility work—cleanup and finishing behind production spreads—and is a natural choice for small jobs where productivity of a single unit is all that is needed. One of its most significant advantages is balanced productive capacity that can be bought in $100,000 increments, compared to a minimum balanced-spread investment of $260,000 for two comparable conventional scrapers and a pusher. The concept also can be an economical production earthmoving system under the right conditions.

In short, the basic considerations in selecting elevating scrapers are:

No pusher required
Not vulnerable to mismatch
Minimum balanced spread investment
No rock appetite
Elevator is dead weight when hauling

When examining the economics of elevating scrapers, it is simply a case of the elevator working for you ±1.00 min, during loading only. For the remainder of the cycle, the additional 10,000 lb of elevator weight works against you. Thus, on short hauls in good conditions, if the elevator works for you 1.20 min and against you only 2.00 min, it may look attractive. On hauls of greater length, on extreme grades, or with high rolling resistance, when you find it working for you for 1.20 min and against you 4 to 5 min, it is probably less economical.

Where does it fit economically? Table 9-21 explores a typical situation. In this example, the elevating unit is economical because the elevator worked for you 1.10 min and against you only 2.49 min. Also, there was an unavoidable mismatch in the pushed spread. Total scraper cycle time of 3.17 min divided by pusher cycle time of 1.23 min indicates an ideal scraper-to-pusher ratio of 2.57, rather than the actual

TABLE 9-21 Elevating Scraper Economics, Typical Example No. 1

One-way haul distance: 1,500 ft Grade: level
Rolling resistance: 80 lb/ton

Production	Elevating /32	Conventional 21/30
Avg load time, min	1.10	0.70
Avg dump and turn time, min	0.40	0.50
Haul and return time, min	2.09	1.97
Total cycle time, min	3.59	3.17
Trips/60-min hr	16.7	18.9
Avg payload, bcy	22.4	22.5
Scrapers/pusher	....	2
Scraper production, bcy/60-min hr	374	850
Cost of production		
Total hourly spread costs, dollars	30.77	74.83
Load and haul cost, dollars/bcy	0.082	0.088

ratio of 2.0. The elevating scraper would be more economical as the haul becomes shorter.

Table 9-22 examines a second situation, where the elevating unit is not economical because the elevator worked for you 1.10 min and against you 4.62 min. Also, the conventional spread had a good match of scrapers to pushers (4.3 ideal versus 4 actual). The relative advantage of the conventional scraper would widen as the cycle becomes longer.

TABLE 9-22 Elevating Scraper Economics, Typical Example No. 2

One-way haul distance: 2,500 ft Grade: level
Rolling resistance: 160 lb/ton

Production	Elevating /32	Conventional 21/30
Avg load time, min	1.10	0.70
Avg dump and turn time, min	0.40	0.50
Haul and return time, min	4.22	4.10
Total cycle time, min	5.72	5.30
Trips/60-min hr	10.50	11.30
Avg payload/bcy	22.4	22.5
Scrapers/pusher	....	4
Scraper production, bcy/60-min hr	235	1,016
Cost of production		
Total hourly spread costs, dollars	30.77	126.47
Load and haul cost, dollars/bcy	0.131	0.124

Figure 9-22 supplies a graphic explanation of the approximate areas of economical application of each type of machine. Note that along the break-even limit for elevating scrapers, the haul distance gets shorter as the resistance increases—the elevator works against you a greater portion of the time. Any mismatch between regular scraper and pusher would move the practical intersection somewhat to the right (in favor of the elevator). This relationship identifies the elevator as an economical short-haul system.

Another way of looking at the economics of elevating scrapers is in terms of the optimum ratio of scrapers to pushers. Some feel that elevating scrapers are more economical until this ratio approaches 4:1. At this point, the cost of the pusher is spread over four scrapers and is more than offset by the shorter load time and the lower haul cost of the conventional scraper—thus resulting in a favorable economic situation for the push-loaded scraper fleet.

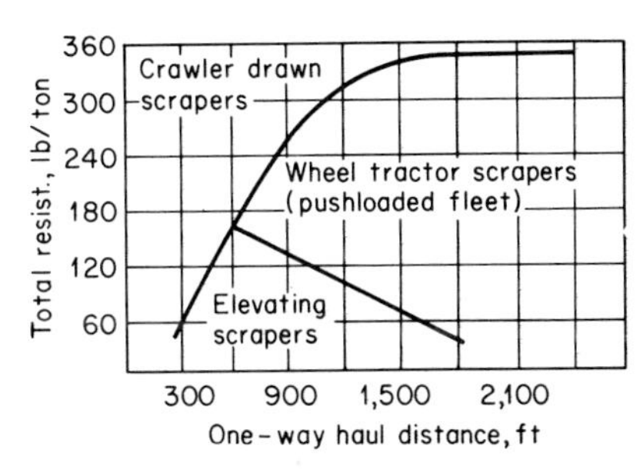

Fig. 9-22 Economical application zones.

The operational range of the elevating scraper may be extended in actual practice. Many contractors do not find it convenient, for one reason or another, to add a scraper every time the haul lengthens out and unbalances the scraper-to-pusher ratio. Instead, they will frequently work with basic spreads of 3:1, 4:1, etc., and the economic range of elevating scrapers is thus extended past that which might be calculated on paper.

Multiple Units. As pointed out previously, the multiple scraper and push-pull concepts are relatively new. Several large contractors now use the multiple scraper system or are appraising it in order to obtain cost information for future bidding, but at present there is not a definite trend in this direction. In contrast, the push-pull concept is rapidly gaining favor and recognition. This is mainly because of its overall versatility and its advantages over the permanently connected multiple-bowl units. These advantages include the following:

- Modification for coupling is relatively inexpensive, less than $10,000.
- There is less vulnerability to complete loss of production—the remaining unit is still capable of operating if one unit is down for repair.
- It has a broader materials appetite—it can be push loaded in hard-to-load materials and cuts.
- It has a higher practical top speed, which the operator is more likely to utilize when conditions permit.
- It has versatility, since it can operate in a spread as a conventional unit.

There are, of course, some limitations to this system. These include:

- The cut must be of sufficient length, generally about 200 ft or more.
- It is difficult to open up cuts.
- Slippery underfoot conditions limit traction and rimpull.
- Rock cuts will result in excessive tire wear.

REFERENCES

1. Society of Automotive Engineers (SAE) Handbook 1969, Standards and Recommended Practices.
2. "Specifications for Your Files," *Construction Methods and Equipment*, November, 1968.

Section **10**

Excavators

E. O. MARTINSON*

Martinson Engineering, Milwaukee, Wis.

INTRODUCTION

The excavators covered in this section are primarily the full-revolving, transportable digging machines. These include power shovels, backhoes, draglines, and clamshells. Front-end loaders, belt loaders, and wheel loaders are discussed in Section 11; wheel

* Formerly Vice-president, Research, Development and Manufacturing, Koehring Company, Milwaukee, Wis.

and ladder trenchers are discussed in Section 12; and drag scrapers and slackline cableways are discussed in Section 14.

The Power Crane and Shovel Association (PCSA), with offices in Milwaukee, Wis., has published a number of bulletins which pertain to the excavators discussed in this section. These PCSA technical bulletins are:

No. 1: *Functional Design, Job Applications, and Job Analysis of Power Cranes and Shovels*
No. 2: *Operating Cost Guide*
No. 3: *Proper Sizing of Excavators and Hauling Equipment*
No. 4: *Cable-Controlled Power Cranes, Draglines, Hoes, Shovels, Clamshells—Mountings, Attachments, Applications*
No. 5: *Power Crane Application in Industrial Plants Operator Manual—125 Ways to Better Power Shovel-Crane Operation*
Hand Signal Cards—For Shovel Crane Operation
PCSA Standard No. 1—*Mobile Power Crane and Excavator Standards* (1968)

The last publication also covers applicable Society of Automotive Engineers (SAE) Test Standards and USA Standards Committee B30 Hand Signals.

The SAE technical committees are continuing work on mobile-crane test standards. The standards are published in the annual SAE handbooks and also as separate bulletins such as J-987, *Crane Structures—Method of Test* (1967). The U.S. Commerce Department has published *Commercial Standard* CS90-58 (1958), which applies to excavators and cranes. Manufacturers' specifications for all makes of excavators are supplied by Equipment Guide Book Co. of Palo Alto, Calif.

MOUNTINGS

Excavators such as full-revolving backhoes, power shovels, draglines, and clamshells are mounted on specially designed carriers. Crawler types are by far the most common excavator mountings, with truck-type or self-propelled rubber-tired running gear being used to some extent on small cable-operated or hydraulic machines. Excavators may also be barge or rail mounted, and the large walking dragline rotates on a structural base or "tub." Small, 180°-swing backhoes can be mounted on industrial and farm tractors.

The three types of mounting used for the "commercial" ⅜- to 4-cu yd range of excavators and for the electric mining shovels are shown schematically in Fig. 10-1. The crawler characteristics are as follows:

1. Can run on soft or wet ground
2. Can operate on sharp rocks and under other adverse conditions
3. Climbs steep grades up to 40 percent
4. Turns in a short space
5. Relatively short length means very little interference with digging operations
6. Low travel speed of about 1 mph
7. Requires job-to-job transportation on trailers or railway cars

Truck-type carriers with rubber-tired running gear have the following characteristics:

1. High road speed and mobility, up to 50 mph.
2. Lower stability over side than over rear.
3. Can be equipped with quick-setting outriggers for certain excavators.
4. High ground pressures from tires or outriggers require firmer and smoother operating locations.
5. Require more operating space.
6. May require removal of some parts such as counterweight or outriggers to attain legal travel weight.
7. Have two cabs and two engines and may require two operators.

The characteristics of the self-propelled carrier with rubber-tired running gear are similar to those of the truck type except that the former has:

1. Medium travel speed, usually 6 mph max but occasionally up to 20 mph
2. Single engine and is controlled completely from one cab with one operator
3. Shorter wheelbase than truck type

EXCAVATOR SIZE RATINGS

The size classification of excavators has traditionally been based on shovel-dipper rating, regardless of which interchangeable attachment is actually used. Backhoe dippers usually have the same capacity ratings as the shovels unless they are deliberately kept smaller for narrow trenches. The dragline bucket and clamshell bucket can usually be larger in capacity than the machine's "shovel-rating."

The various manufacturers formerly used similar flywheel horsepower for their units and thus ended up with comparable machine weights for each yardage rating. Recently there has been a trend to drop all reference to shovel-dipper capacity and,

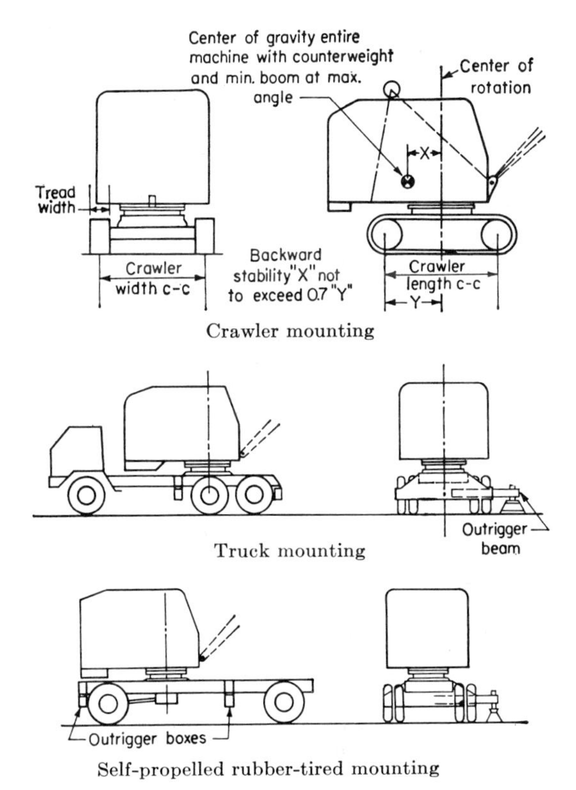

Fig. 10-1 Typical mountings for excavators. Crawler mounting mostly used for excavators except for telescopic grading hoes and small hydraulic hoes to ½ cu yd which may also be mounted on a truck-carrier mounted or a self-propelled rubber-tired carrier.

instead, to simply identify each excavator by a model number. This makes it increasingly difficult to obtain valid comparisons between the excavators of several manufacturers. The introduction of hydraulic-actuated models has resulted in additional variations in excavator specifications. However, the best method for comparing machine sizes still seems to be by the nominal shovel-dipper rating. Large, single-purpose machines, such as walking draglines, are rated according to their actual average bucket size.

CABLE-ACTUATED EXCAVATORS

The large mining and stripping excavators are designed as special-purpose units, with no provision for interchangeability of attachments. In the commercial size ranges of cable-actuated excavators, however, the basic machine can be optionally equipped with a shovel, backhoe, clamshell, crane, pile driver, or dragline.

Shovels The controlled motions which are involved in shovel digging are shown schematically in Fig. 10-2. Shovels are less popular as excavating tools than formerly, largely because of the improved capabilities of the front-end loader.* Nevertheless, there are many applications where the shovel continues to dominate. Hard digging in high banks is a typical example.

A cable-actuated shovel, mounted on a special-use carrier equipped with track-type running gear, is illustrated in Fig. 10-3. The relative dipper sizes and reaches for the smaller sizes of cable-actuated shovels are shown in Fig. 10-4.

The main hoist line for a cable-actuated shovel operates at a nominal speed of 180 fpm with the clutch fully engaged. Shovel dippers are rigged with two-part reeving, resulting in a maximum dipper hoist speed of about 90 fpm. The crowd and retract actions of the dipper stick are controlled by slipping the secondary hoist clutch. Some machines employ a chain drive to a shipper shaft which is geared to the racks of the double dipper stick. Other shovels use a cable crowd with a single dipper stick. The crowd clutch is usually set loose so that the operator will not waste power by

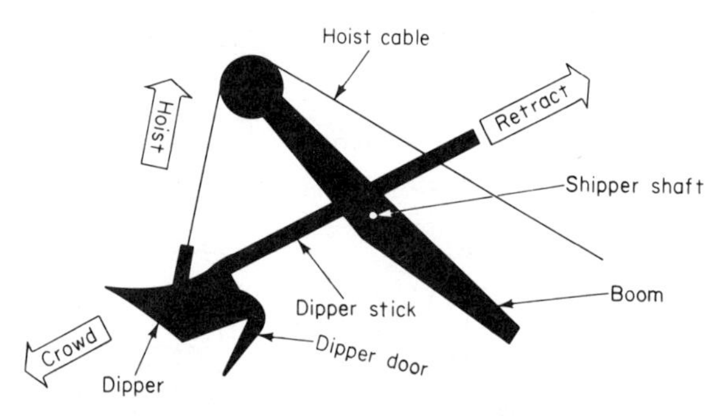

Fig. 10-2 Shovel digging action. Hoist clutch fully engaged during digging. Crowds by slipping clutch. Retracts by releasing clutch and brake.

overcrowding. Another method of generating crowd force is to dead-end the dipper hoist cable at the crowding mechanism. This reduces the energy wasted in a slipping crowd clutch at the expense of possibly slowing down the hoist and complicating the control.

Shovel booms and dipper sticks must be rugged, since they are subjected to severe twisting and bending stresses. In heavy digging, the loads are near the engine stalling point at every cycle. As a consequence, fatigue loads are high on many parts of a shovel. Overloads are prevented when the machine starts to tip, since the operator must then reduce the crowd force.

Backhoes Figure 10-5 shows a cable-actuated backhoe which is mounted on a full-revolving carrier similar to that for the shovel just described. The backhoe attachment is increasing in popularity and is widely employed in trenching and in building-foundation excavating. It has about the same digging force at the dipper lip as the shovel, and it operates at about the same average hoisting speed of 90 fpm. The actual bucket digging force changes considerably during its travel, being lowest when the arm is extended and highest when it is almost completely wound in. When digging at maximum depth with the arm fully extended, the penetrating force of the backhoe is further reduced because the upward pull of the digging lines has a tendency to lift the bucket out of the trench. Typical digging ranges for four popular sizes of hoes are shown in Fig. 10-6.

* See Sec. 11, Loaders.

Hoe time cycles are approximately 20 percent longer than those of shovels because the hoisting distance is greater and because the dipper must be dumped by fully extending the dipper arm. Like shovels, the backhoe dippers can be accurately spotted and controlled during the digging stroke. Backhoe booms and dippers experience high shock and fatigue loads in normal digging use. The backhoe can be abused by an improper chopping action or by side loading when the dipper is dropped into a trench while still swinging at high speed. Hoist brakes can be prematurely worn by permitting long, free drops which are abruptly checked in the last few feet.

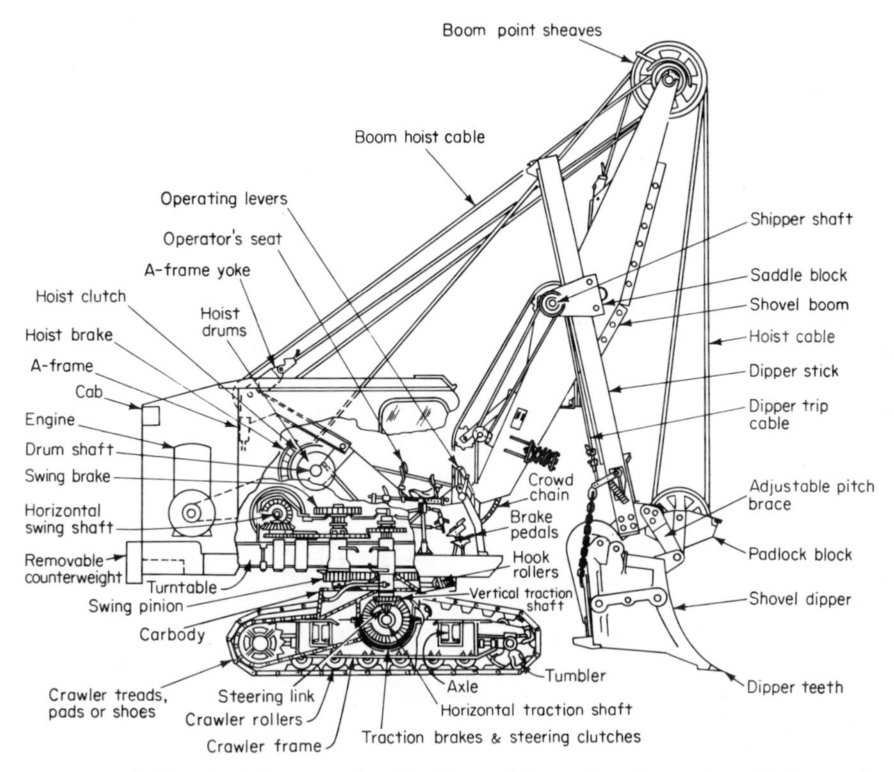

Fig. 10-3 Cable-actuated power shovel. Nomenclature of major parts. Chain-crowd type shown. Basic machinery is same for several other excavator attachments and cranes.

Minor adjustments in hoe dipper width can be obtained through the use of various side cutter teeth or by shimmed, adjustable, hinged side cutters. Greater width variations will require different sizes of dippers.

Draglines and Clamshells The dragline attachment and the clamshell attachment to the cable-actuated excavator are illustrated in Fig. 10-7. Both of those attachments require the length of crane boom which is frequently used for mobile cranes in the medium-lift class. However, an excavator with the dragline or clamshell attachment will require additional counterweights and longer and wider crawlers than will shovels and backhoes.

Clamshells can dig vertically at considerable depths and, when necessary, through small passages. They also have the ability to lift their loads to appreciable heights above grade. Light rehandling buckets with two- to four-part reeving are used for filling overhead bins with loose material from pits, stockpiles, or railway gondola cars. General-purpose buckets, with four to six parts of line reeved in the closing line, are used in excavating shafts or trenches where clearances are restricted by shoring and cross bracing. Still heavier buckets may be needed for underwater digging.

Clamshells are usually reeved to have the same line speeds for closing and for holding, so that both lines will stay tight during hoisting. Should there be any difference in these line speeds, the closing line should be at the higher speed. The clutch should then be set loose enough so that it will not raise the loaded bucket on the closing line alone. A bucket should not be dropped too rapidly, because a sudden stop at the bottom of the excavation can damage its hinge stops. The bucket should also be fully open when lowered, to prevent bending the lip.

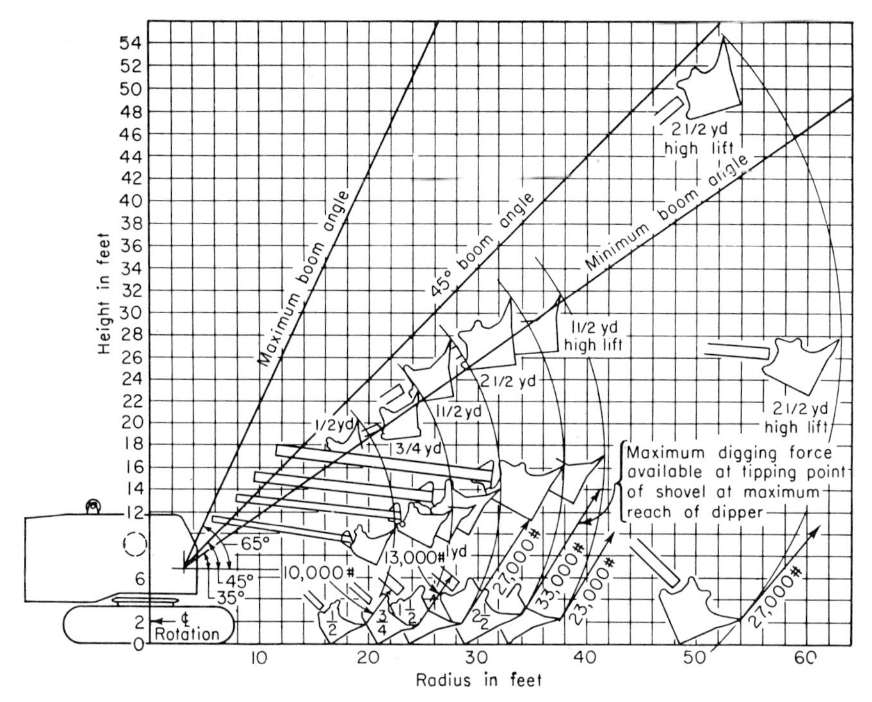

Fig. 10-4 Shovel digging range for sizes ½ cu yd to 2½ cu yd. High-lift range is for special long booms.

A dragline is capable of digging at a distance away from its mounting which is about equal to its boom length and to a depth which is equal to one-half of its boom length. Where necessary to clear high banks or spoil piles, the bucket can be hoisted to elevations of one-half the boom length less the dump height indicated in Table 10-1. Although the dragline bucket cannot be spotted as accurately and positively as can shovels and hoes, a good operator can use it to load hauling units. The dragline is especially useful where the area to be excavated is large, too soft for other machines, or under water.

Clamshell-bucket rated capacity is equal to the actual struck or plate-line inside volume in cubic yards, measured from a surface across the top edge of the side plates and back plate. The dragline-bucket rated capacity is 90 percent of the inside struck volume with the front edge of the lip (not teeth) being the line of the vertical front measuring plane. The 10 percent deduction is to allow for angle of repose of material at the front. Another way of calculating this rated dragline-bucket yardage is to divide the "vertical front" volume in cubic feet by 30.

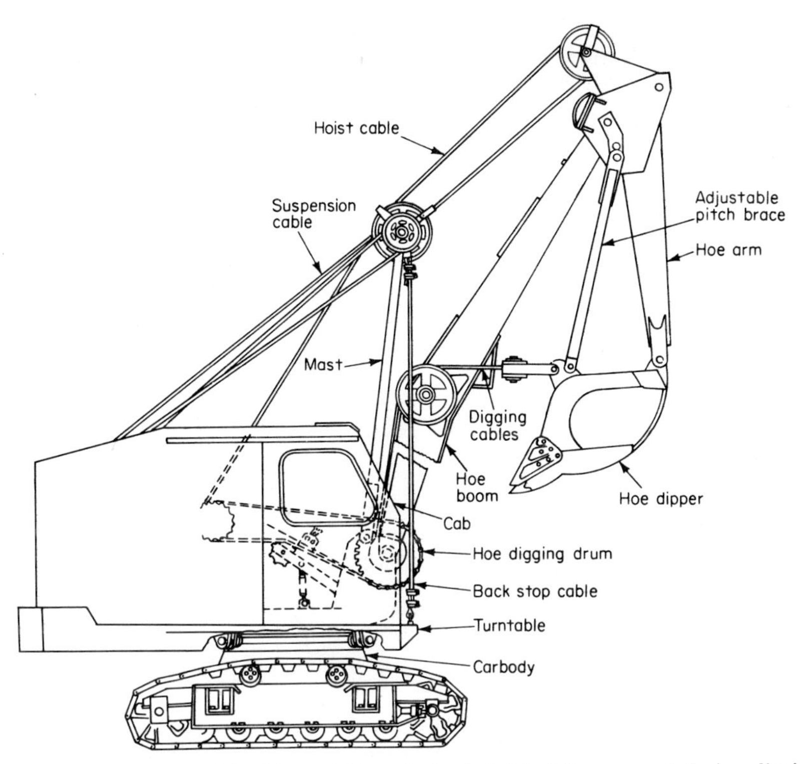

Fig. 10-5 Cable-operated backhoe. Dig or in-haul clutch fully engaged during digging, and cutting load regulated by releasing hoist brake.

Draglines and clamshells are manufactured in several types and weight ranges. The weight of a given size and type of bucket varies between manufacturers. Weight variation is not necessarily in direct proportion to size, even for buckets built by the same manufacturer. A composite average of the bucket weights for several types and a range of sizes is shown in Fig. 10-8.

Long booms for lift cranes can be safely designed with a relatively low factor of safety of 1.6, based on a critical column buckling load which acts in conjunction

TABLE 10-1 Vertical Dumping Heights Required by Clamshells and Draglines

Bucket size, cu yd	Clamshell, vertical height	Dragline dumping height below boom point
½	9 ft	11 ft 9 in.
¾	10 ft	12 ft 9 in.
1	10 ft 6 in.	13 ft 9 in.
1½	11 ft	16 ft
2	11 ft 6 in.	18 ft
2½	12 ft	19 ft
3	12 ft	21 ft
4	13 ft	22 ft

with a 2 percent side load. The crane will operate only a limited number of cycles at peak load, and fatigue is not a significant design consideration. Draglines, in contrast, will operate a large number of cycles at their full lifting loads and with heavy side loads. It follows that draglines require shorter and heavier booms than the lightweight, tubular-alloy, heat-treated booms which are suitable for the high-lift cranes. The single boom-point sheave of the dragline is designed with a wide-angle groove to accept the wide side deflections due to accelerating the swing or holding the bucket against a side slope.

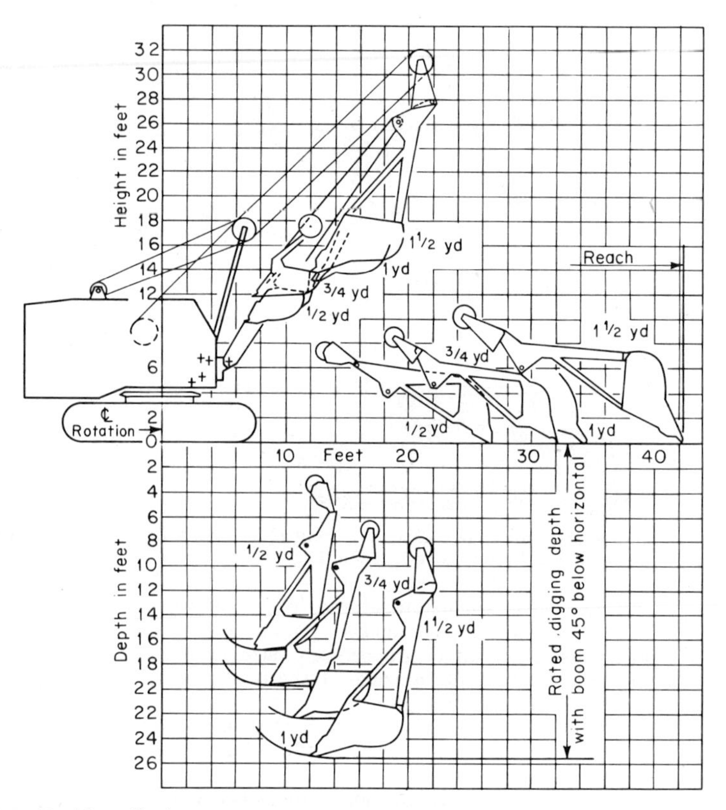

Fig. 10-6 Backhoe digging ranges ½- to 1½-cu yd sizes showing maximum depth, reach, and dump clearance.

Hoisting is performed at about 180 fpm. The drag speed is slightly reduced from this, usually to 160 fpm, by means of a smaller drum. The hoist action is partially opposed by the drag-cable pull, which is slowly released by the drag brake until dumping position is reached. Operating experience indicates that the weight of the loaded drag bucket should be limited to between 35 and 50 percent of the rated maximum engine line pull. Clamshell loaded weight can be somewhat greater, ranging between 45 and 60 percent of this maximum power pull.

Drag inhaul cables are usually one size larger than the hoist and are of the Seale pattern. This has the larger wires on the outside of the cable, even though the smaller ratio of drum-to-cable diameters might suggest small-wire strand cable construction. The operator should prevent spoil from piling up in front of the machine, since otherwise the drag cable may abrade unnecessarily against rocks. Drag buckets must be gently lowered into the ground as the digging pull starts. Draglines, with their long booms and heavy counterweight, have high swing inertias which can cause large heat inputs to swing clutches. This can be a problem if the operator tries to accelerate or stop the swing motion too quickly.

Practical boom lengths and operating ranges for various sizes of clamshells and draglines are indicated in Fig. 10-9. The hoisting capacities of several sizes of excavators are shown in Fig. 10-10. Dragline and clamshell bucket sizes are shown in the

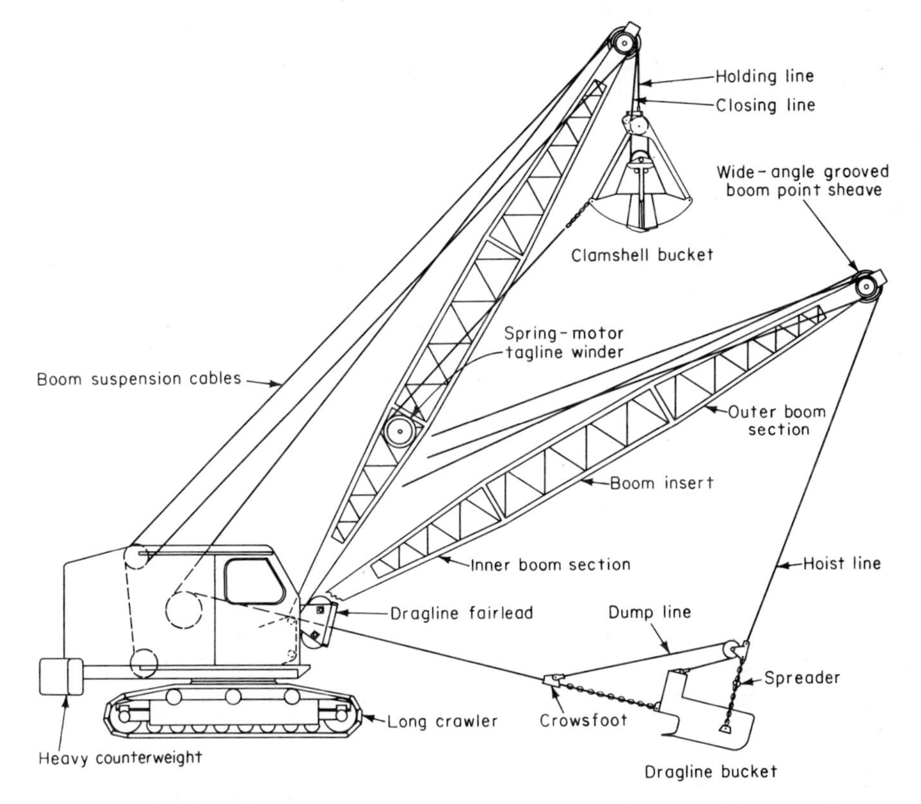

Fig. 10-7 Dragline and clamshell attachment arrangement diagram with nomenclature shown.

rectangles on this chart; the position of each rectangle indicates the practical radius at which the excavator size which it represents might be used. The chart uses an encircled "E" to indicate the usual maximum line pull for each excavator. The encircled "1" on each line size indicates the continuous working line pull which corresponds to one-quarter of the breaking strength of the cable size shown. It should be noted that the line loading for the largest clam in each excavator size comes near this point.

HYDRAULIC-ACTUATED EXCAVATORS

Hydraulic-actuated excavators have rotating superstructures which are somewhat different from those of the cable-actuated machines. Their lower bases may be identical to those of the cable-type excavators, or they may be fully hydraulic with a separate motor powering each crawler. Attachments for the hydraulic-actuated excavator include backhoes, straight-line telescopic hoes, and telescopic boom lift cranes.

The various digging actions of these excavators are powered by hydraulic cylinders and motors and are controlled by hydraulic valves or variable-stroke pumps. Hydraulic excavators have become increasingly popular in the forms of conventional backhoes and straight-line telescopic boom hoes, since they have the advantage of a more maneuverable bucket with a wrist action. In addition, when the hydraulic excavator is engaged in heavy digging, one motion can stall without stopping other actions or killing the engine. This feature permits the operator to maintain a stalling

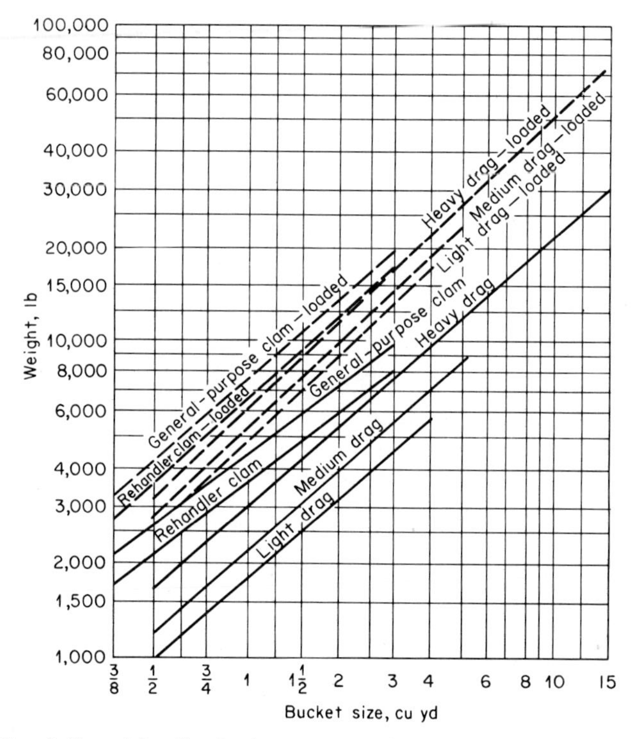

Fig. 10-8 Clamshell- and dragline-bucket average weights. Solid lines for empty buckets. Dashed lines for rated volume at 110 pcf or 3,000 lb per cu yd.

force momentarily while manipulating his other controls, whereas in a standard cable machine he must release the overloaded clutch to avoid killing the engine. If the cable machine were equipped with a torque-converter drive, the engine would not stall under these circumstances, but all machine motions would cease. The hydraulic-actuated backhoe has the further advantage that its digging forces do not tend to lift the bucket out of a deep trench, as does the cable pull on a cable-powered hoe.

Hydraulic pumps are slightly less efficient than gear and chain drives. Because of this, plus line and valve pressure drops, a hydraulic-type excavator requires a slightly larger engine to do the same work as a cable machine. Hydraulic backhoes, for example, require about 25 percent more installed horsepower than comparable cable-type backhoes. Maximum bucket digging forces are limited through relief valves to the design pressure for the hydraulic system, whereas cable machines can develop high instantaneous forces through flywheel inertia or torque-converter multiplication.

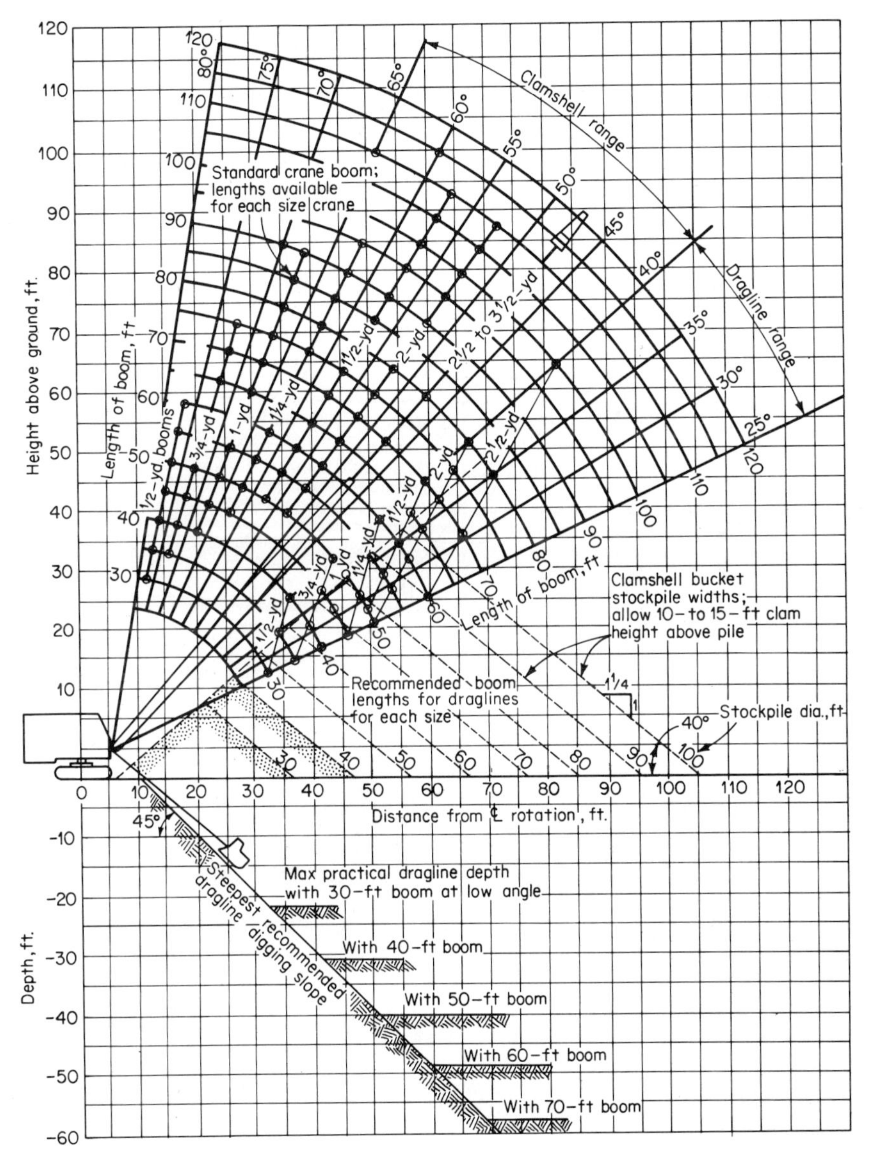

Fig. 10-9 Dragline and clamshell working ranges. Commonly used boom lengths shown for each class of work for graphically determining required boom length to clear stockpiles or hoppers and the working radius or reach.

Cable machines experience wear and heating of clutches, brakes, and wire rope. Hydraulic machines are troubled by gradual erosion of pumps and valves from fine dirt or by sudden failures due to large metal particles in the oil flow; bursting of hydraulic line and hose because of excessive bending, fatigue, and heat; and packing wear and failure.

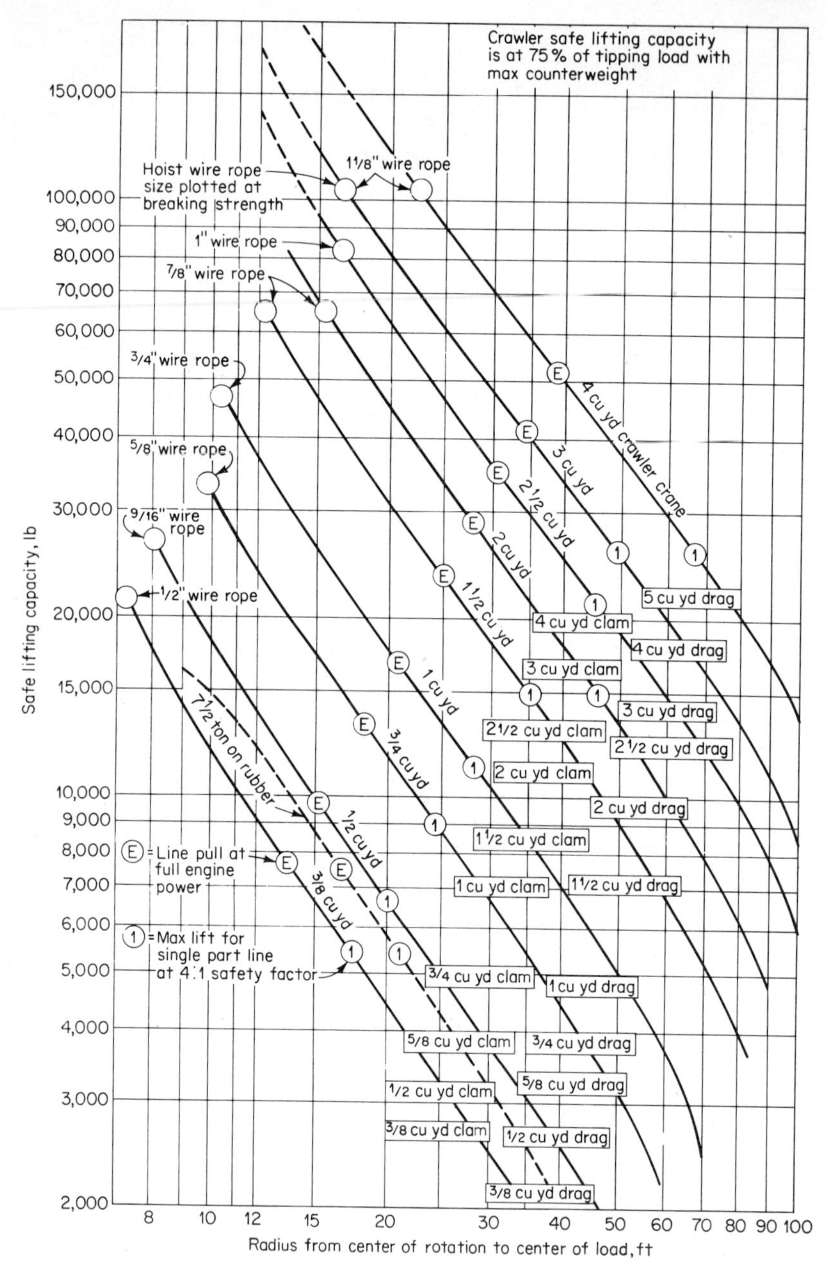

Fig. 10-10 Clamshell and dragline selection for safe load against tipping for nine sizes of crawler cranes. Loaded-bucket weights are plotted to show size of machine required. The maximum operating radius is the intersection of the bucket rectangle with safe-load capacity of one or two sizes for each bucket size.

Oil filtration is very important, and it should be recognized that not all filters perform as well as may be claimed. Filters can be overpressured either by cold oil, which cannot bypass, or by buildup with dirt, causing small holes or cracks which then allow dirt to pass through. Oil should be periodically checked by some form of optical analysis to determine how good the filtration actually is. Foam or air must not be pulled into the pumps because air in minute quantities can soon destroy a pump. If filters are used in the suction line and cause too much resistance, pumps will cavitate (generate vacuum voids) in certain areas and will fail prematurely due to metal erosion.

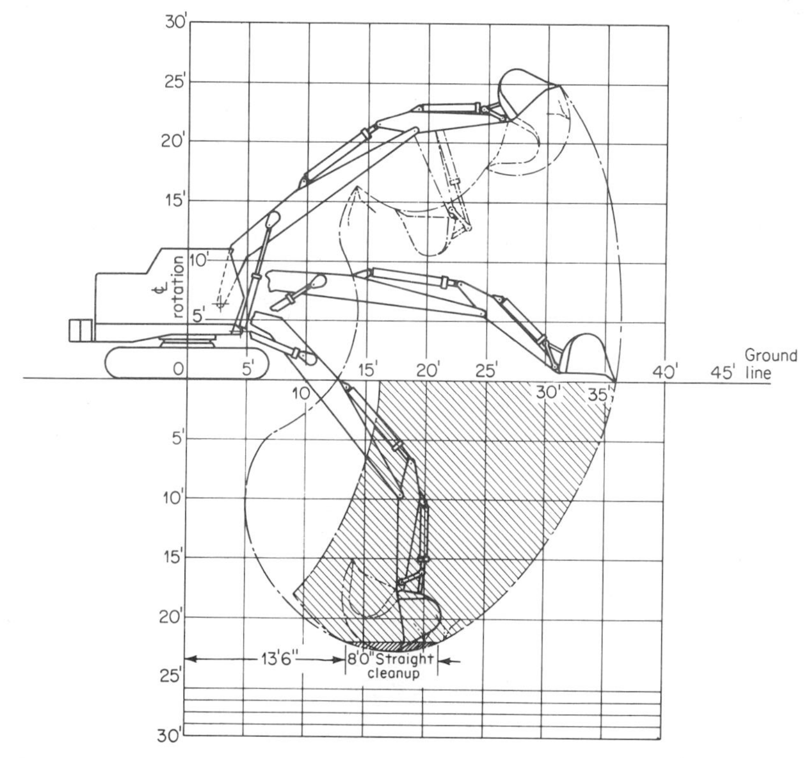

Fig. 10-11 Hydraulic backhoe, 1-cu yd dipper size-range diagram. All boom, arm, and dipper motions hydraulic-cylinder actuated. Swing and traction might be by either hydraulic motors or clutch mechanical drive or a combination of both.

Operators may move control valves too suddenly, causing unnecessarily high surge pressures which occur too fast to be completely absorbed by the relief valves. The result is high stress in hoses and mechanical parts. The slamming of pistons against the end of the cylinder also shortens the life of parts, and cylinders have cushion stops at their ends to lessen this danger. High forces can be generated even with these cushion stops, and the operator should avoid repeated bumping. Although cylinder rods which are exposed to weather are usually chrome plated, they should be coated with grease if the machine is not operated for extended periods. If this is not done, the rods may rust through the chrome. Exposed rods which may be damaged by falling stones or carelessly used tools should be hardened.

Backhoes Nearly all manufacturers use a hydraulic hoe geometry which is similar to that illustrated in Fig. 10-11. The current trend is to use higher operating pressures and several pumps so that two or more motions may be actuated at the same time. Cycle times are similar to those for cable backhoes and are approximately 20 percent longer than for comparable shovels.

Wide-bucket Excavators A hydraulic-actuated wide-bucket loader with a parallel crowd is shown in Fig. 10-12. This attachment is interchangeable with the hydraulic backhoe and is mounted on a full-revolving, track-type carrier. The geometry of the linkage permits the bucket to crowd forward parallel to the ground without requiring manipulations of the hoist or bucket angle. The attachment provides 7 to 10 ft of crowd, and when necessary the excavator may be traveled forward during the first part of this crowd. The brakes set automatically when travel is stopped, and the hydraulic crowd is continued.

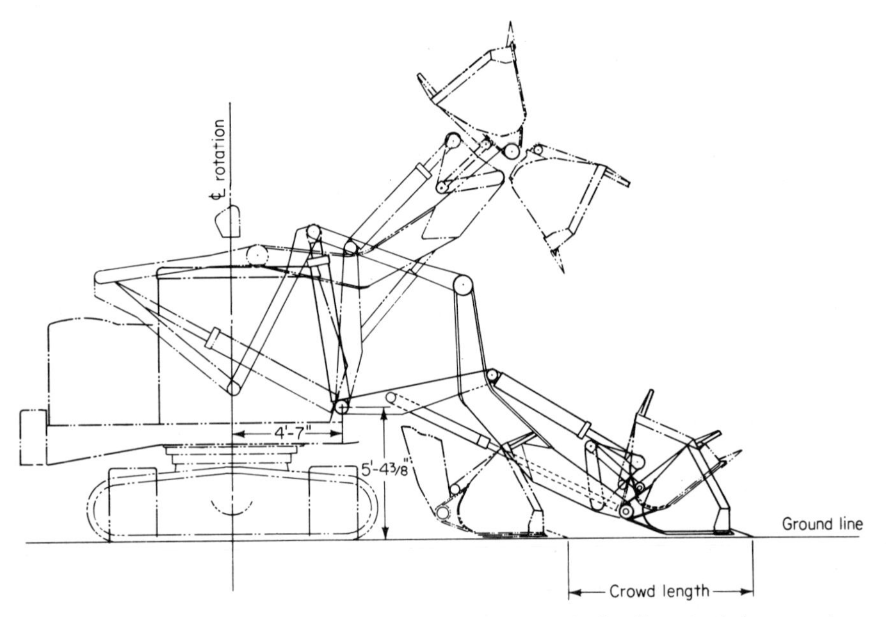

Fig. 10-12 Level crowding wide-bucket full-revolving hydraulic-actuated excavator. Bucket size twice capacity of shovel dipper for similar size machine. Complete cycle time in excavating work averages 30 sec. Small operating space ideal for underground and tunnel excavation.

Since its maximum forward reach is less than that of a shovel, the bucket size can be approximately twice that of a shovel or backhoe of the same weight and power. Its average loading-cycle time in heavy digging is about 1½ times that for a shovel of the same weight and power. It can compete with the wheel-type loaders where excessive tire wear occurs for the latter units, as in handling some classes of rock. In close-quarter operations, such as tunnels or underground loading, the wide-bucket excavator can operate where there is insufficient space for shovels or track-type loaders.

Telescopic Grading Hoes The telescopic grading hoe shown in Fig. 10-13 is a specialized hydraulic-actuated excavator. It can grade slopes and flat areas more readily than other attachments, since it has a straight-line action even when the digging line is not parallel to the boom axis. Various buckets, blades, and hooks are also available for this machine.

The bucket of the hydraulic grading hoe has both a wrist action and a rotation right and left about the longitudinal digging action. These actions make it useful in trimming as well as in excavating. Its carrier can be equipped either with crawler-type

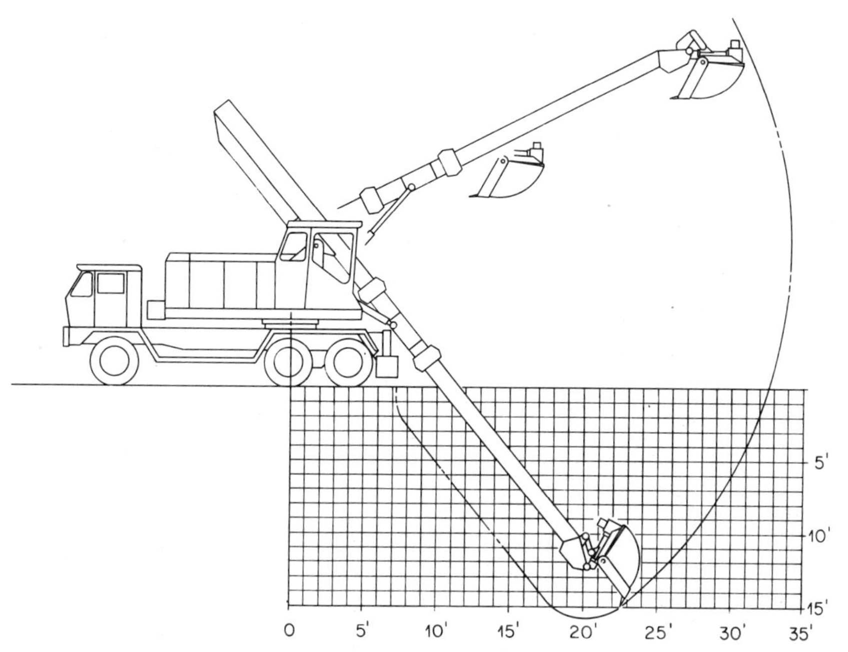

Fig. 10-13 Hydraulic telescopic boom grader. The 12- to 15-ft straight-line action and long reach useful in trimming of slopes and ditches, slab lifting and loading, and general excavating.

or wheel-type running gear. Truck carriers will sometimes include quick-set outriggers for added lifting capability.

SINGLE-PURPOSE EXCAVATORS

Mining Shovels Next in size above the commercial-size shovel is the two-crawler electric mining shovel of the type shown in Fig. 10-14. Shovels in this class are generally similar to those in the smaller size range, but they are not convertible to other attachments. They are built in sizes from 5 to 25 cu yd, and all except the smallest of them are electrically driven. Some of these excavators have a main ac motor which drives the main hoist through eddy-current clutches and the other motions by dc Ward-Leonard control, while other units use dc motors on all motions. Mining shovels are used for direct loading into hauling units and occasionally into the receiving hopper of a conveyor belt.

Walking Draglines These draglines are supported on circular bases or tubs which bear directly on the ground, as shown in Fig. 10-15. A pair of long shoes is suspended from the excavator sides, and these shoes are lowered to the ground when the machine is to be moved. The entire machine and its base are then partially lifted and dragged backward away from the boom direction. The walking shoes have a lift and travel of several feet per stroke and can move the machine at 15 to 20 fpm or 0.2 mph. The ground-contact pressure is around 10 to 18 psi under the base and approximately twice this under the shoes.

Walking draglines are made with drag buckets in sizes from 12 to 220 cu yd and with booms from 170 to 300 ft long. All units except the smallest are electrically driven. Power distribution is generally at 4,160 volts on the smaller machines and 7,200 volts on the larger ones. The peak load can be four times the average load or perhaps two times the installed horsepower of the excavator, and the power supply and distribution system must be sized accordingly.

Walking draglines are generally employed in coal mine stripping and mineral mining. They can dig one-half to two-thirds of the boom length below the machine base, can lift the spoil above base level about two-fifths of the boom length, and have a reach of approximately the boom length from their centers. On this basis, the vertical distance from the bottom of the deepest pit to the peak of the spoil bank could be about the same as the boom length.

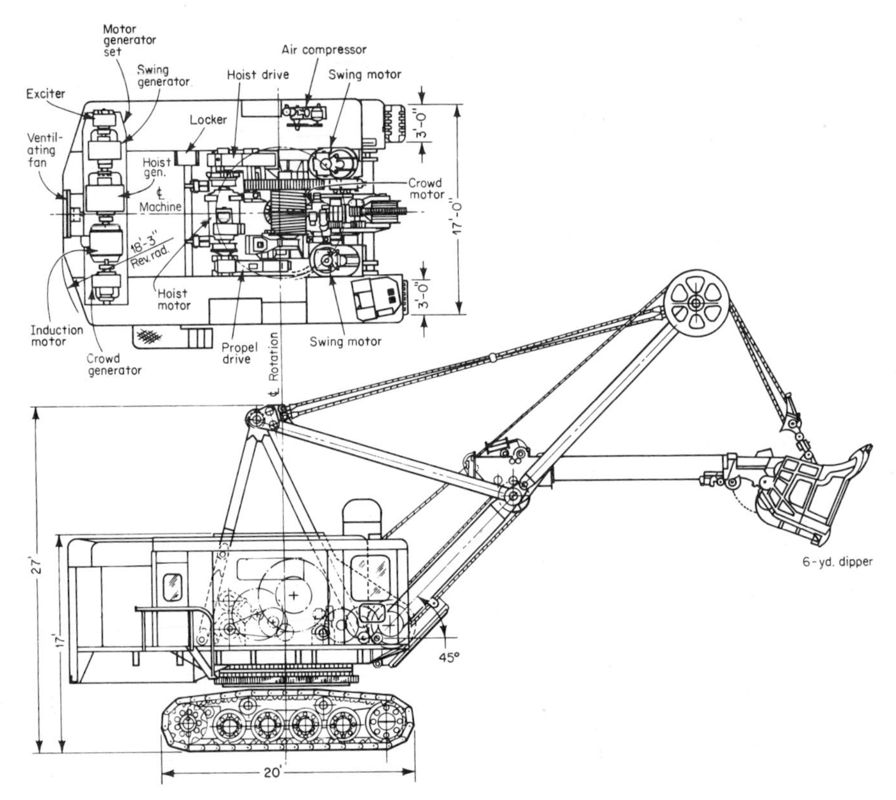

Fig. 10-14 Electric mining shovel, two-crawler machines in sizes from 4- to 25-cu yd dipper capacity. Power supply by trailing electric cable.

The cycle time for the walking dragline is on the order of 60 sec, with hoist speeds of 400 to 700 fpm and drag speeds of 300 to 500 fpm. The larger sizes with longer booms have the higher speeds, and thus they require a disproportionately large increase in power in comparison with the smaller sizes.

Stripping Shovels Like walking draglines, these excavators are used in permanent (10 to 40 years) operating situations in stripping coal overburden. The stripping shovel shown in Fig. 10-16 is mounted on a carrier with four sets of crawlers, one set at each corner of its base. Stripping shovels are presently built in sizes from 40- to 140-cu yd dipper capacity. They cannot lift spoil as high as a walking dragline, and their dumping height is limited to seven-tenths of the boom length. Both classes of machines have similar weight, cost, and power for a given size of bucket. The shovel has a faster cycle and the ability to dig harder material, while the dragline has greater reach and lifting height.

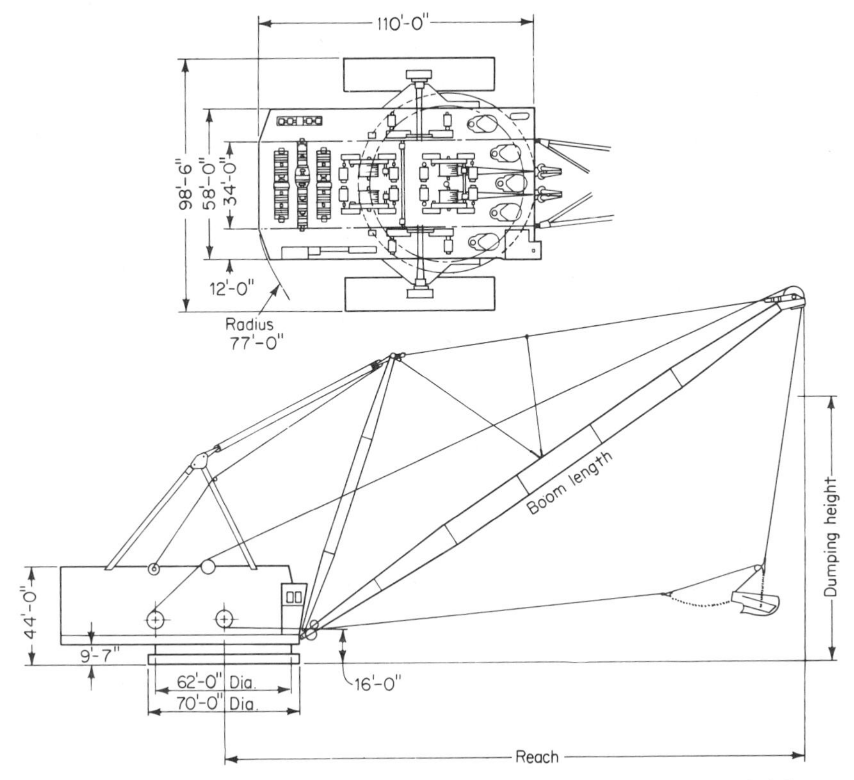

Fig. 10-15 Electric walking dragline, 8- to 200-cu yd bucket capacity, 200- to 300-ft booms. 70-cu yd 6,000,000-lb weight shown.

The stripping shovel crawlers have much less bearing area than the dragline supports, with contact pressures on the order of 60 psi. These shovels can travel at 0.15 to 0.2 mph.

PRODUCTION ESTIMATING

The most practical way to predict the normal hourly production rate for an excavator is to first compute its theoretical peak capacity. A working cycle can usually be identified for the machine, based upon its intrinsic method of operation (excavate, swing, dump, etc.) and the particular conditions under which it is used (angle of swing, etc.). The time elements for this theoretical cycle can be obtained by direct observation, or estimates may be based on average values available in the product literature. Then, after estimating the quantity excavated during each cycle, the peak production rate for the unit can be computed.

This theoretical production rate will not be achieved in actual practice. Instead, appropriate allowances must be made for machine move-ups, ground cleanup, occasional incomplete filling of the bucket, operator's personal lost time, and all other "minor" delays of the types shown in Fig. 10-17. The combined effect of these minor delays can be expressed in the form of a working efficiency factor which, when multiplied by the theoretical peak production rate, will yield an estimate of the "normal" production rate. A value of 0.80 is frequently used for this working efficiency factor.

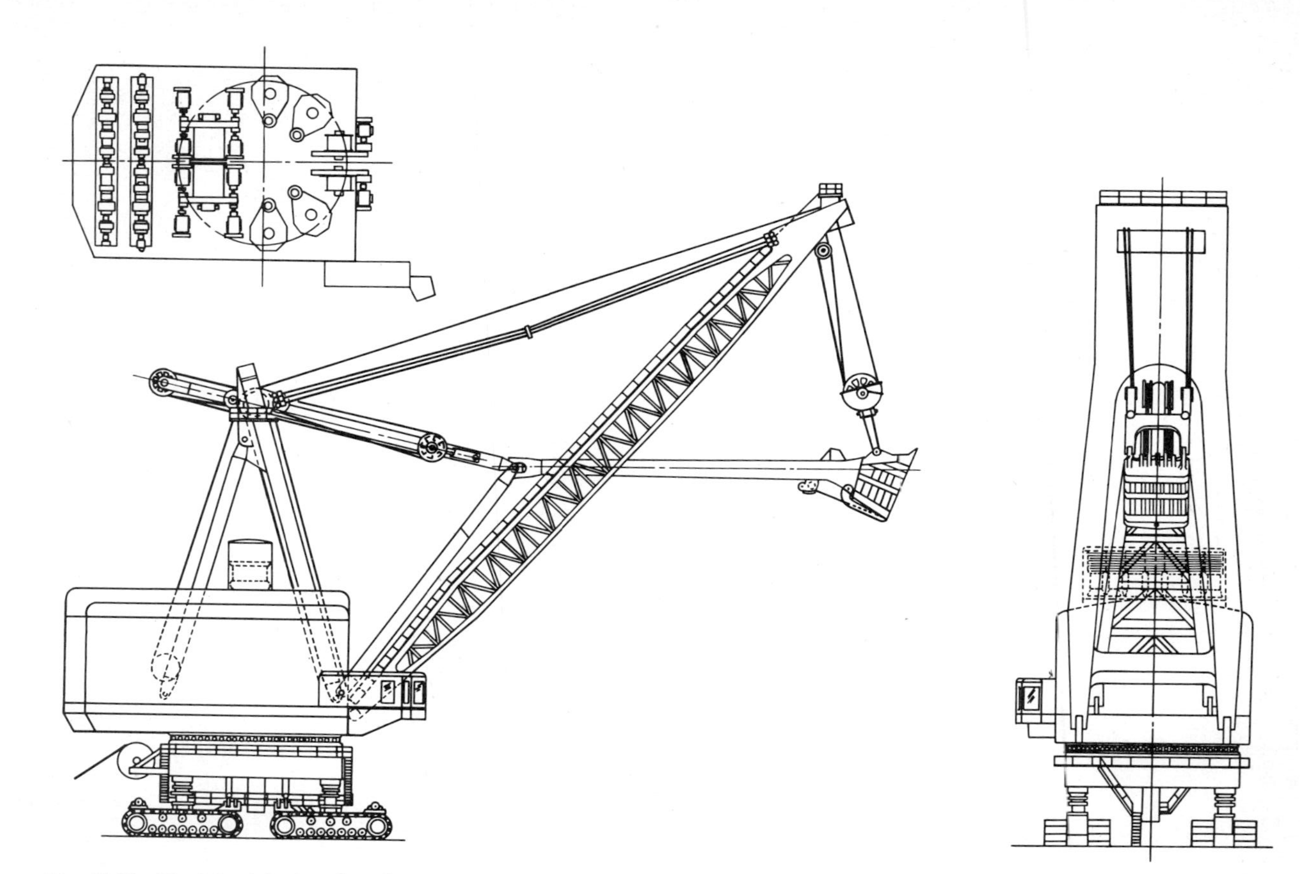

Fig. 10-16 Electric stripping shovel, mounted on four hydraulically leveled crawler sets. Digs harder material than other types. Built in 40- to 140-cu yd dipper sizes.

The normal production rate is approximately the average yardage which can be expected in a well-managed, continuous-production operation such as would be carried on in gravel pits, quarries, mineral mining, and coal stripping and loading, where weather conditions and waiting for trucks would not normally delay operations. However, on jobs such as highway work, there are several factors which may reduce the average production rate to well below the normal production rate. These factors can include waiting for blasting, maintenance during operating hours, bad-weather shutdowns, and excessive breakdowns caused by unusually difficult operating conditions. Experience on highway jobs has indicated that the normal production rate should be multiplied by a contingency efficiency whose value may be on the order of 0.50 (see Major delays, Fig. 10-17) in order to obtain a valid estimate of the average production rate.

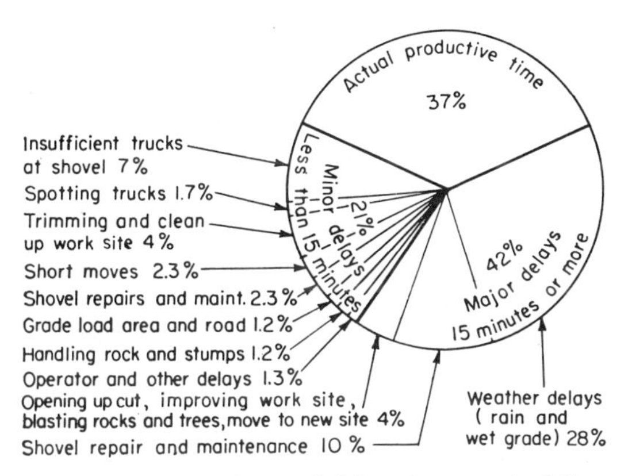

Fig. 10-17 Power shovel productive time and delays in percent of theoretical best time. Recorded average over several months from 16 shovels on 10 highway jobs; data from Highway Research Board 1947–1949 studies which indicate the nature of delays which may be encountered on some jobs.

Figure 10-18 shows the relationships between hauling-unit size, size of shovel or dragline, and required loading time. The chart is based on a working efficiency factor of 0.90, which experience has indicated can be obtained under these conditions. An angle of swing of 90° has been assumed and, if it is felt that the additional refinement is warranted, the indicated outputs could be increased by 20 percent for a 45° average swing angle and decreased by 25 percent for a 180° average swing angle. These values indicate the importance of planning the operation to use a minimum swing angle.

Figure 10-19 shows the reach and height-of-dump capabilities of a wide range of dragline and shovel sizes. In addition to having a greater reach, the larger machines can dig harder materials and handle larger boulders. It can be seen from Fig. 10-19 that draglines have much greater lift height and reach than do comparable sizes of shovels.

Draglines have about a 25 percent longer cycle time than shovels, resulting in 80 percent as many loads per hour. However, dragline buckets are usually larger than the shovel rating of the machine. As a result, the hourly output of a dragline could be the same as for a shovel. The size of bucket which can be used depends on the operating radius needed as well as on the loaded weight of the bucket, as shown in Fig. 10-9. When the operating radius has been established, the boom length can be determined from Fig. 10-8.

The control lever arrangement as it appears in the operating cab of most commercial-range excavators is illustrated schematically in Fig. 10-20. These levers

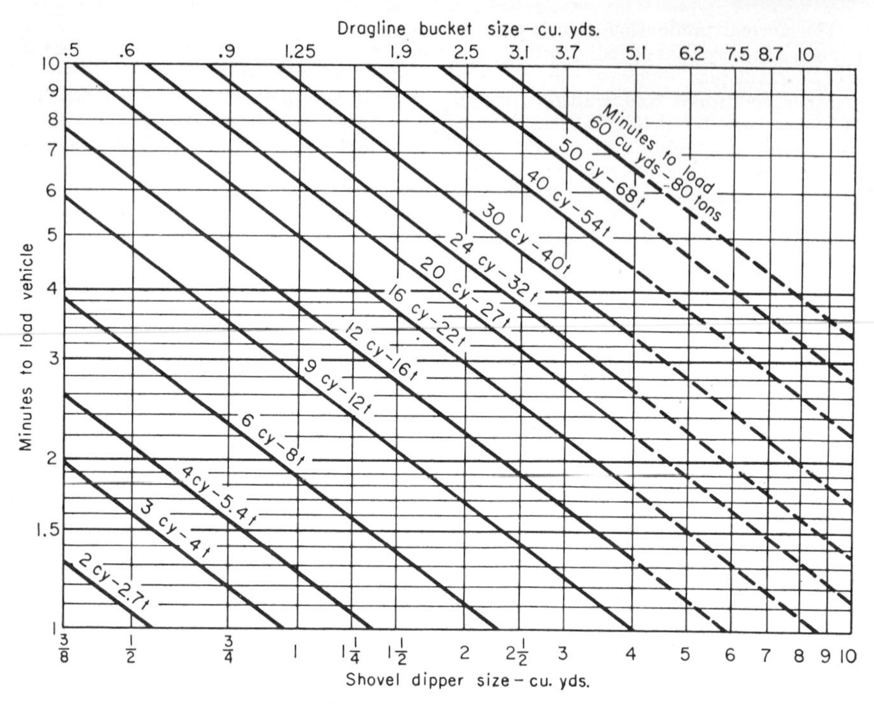

Fig. 10-18 Time required to load various sizes of hauling units with shovels. Dragline size with slower cycle time shown at top of chart. Based on bucket filled every pass, normal digging, and excavator operating at 90 percent of theoretical peak rate during the hauler-loading time interval. Hauler tonnage figures are based on 100 pcf material.

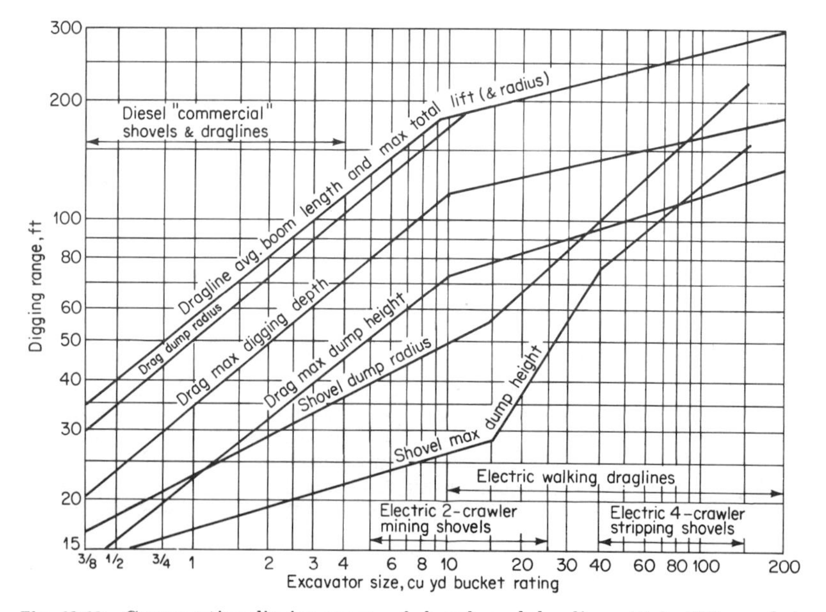

Fig. 10-19 Comparative digging ranges of shovels and draglines, ⅜- to 200-cu yd size.

control the various excavator attachments as they progress through their working cycles. The control levers in small cable machines are mechanically linked to the clutches and brakes. Because of heavier operating effort, air- or hydraulic-powered controls or mechanical power-assisted (servo) controls are used on larger machines. Hydraulic-actuated excavators are controlled by hydraulic valves, but these may be located remotely from the operator and require mechanical linkages also. The electric

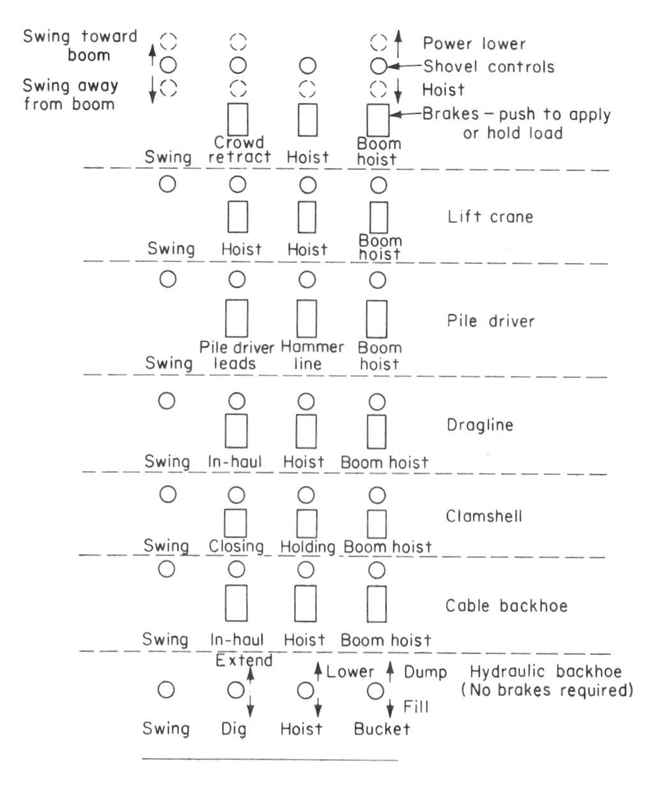

Fig. 10-20 Control lever position arrangement per SAE Standard J-983. Looking forward. Cab may be on right or left side of excavator. Some boom hoists have both "hoist" and "lower" on hand lever and also a foot brake. A pawl or ratchet for boom hoist, engine clutch control, engine throttle, swing lock or brake, crawler steering control, crawler forward and reverse (if independent from swing control), and miscellaneous control locations are not standardized.

mining and stripping shovels and walking draglines have electric control devices in the cab, where the hand lever regulates both hoisting and braking and foot pedals regulate the swing. The electric controls, therefore, have a special arrangement.

Figure 10-21 shows excavator cycle-time elements as functions of excavator type and size. In preparing this figure, an average swing angle of 90° was assumed. Similar information, now converted to theoretical working cycles per hour for each excavator type and size, is presented in Fig. 10-22. These cycles per hour, when multiplied by the rated struck-bucket capacity of an excavator, give its theoretical peak production in loose cubic yards (lcy) per hour. By assuming a swell factor of 1.33 to convert loose cubic yards to bank cubic yards (bcy) and a working efficiency factor of 0.80 to convert peak production rate to normal production rate, the output yardages shown in Fig. 10-23 were obtained. As indicated earlier, it may then be necessary to apply a contingency efficiency factor to this normal output rate in order to obtain a realistic estimate of average output.

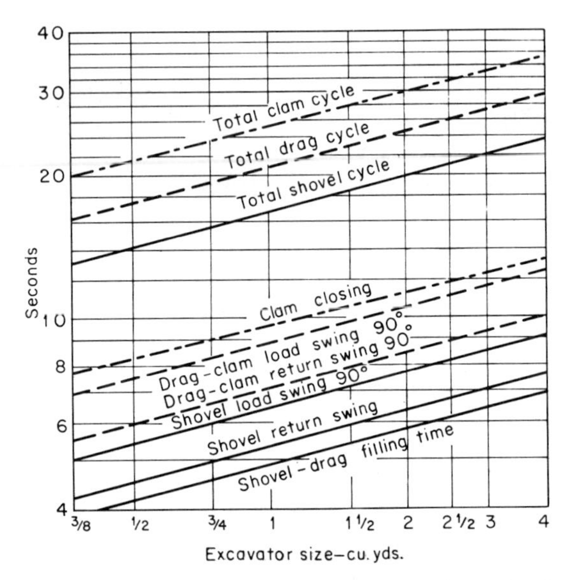

Fig. 10-21 Excavator cycle elements. For 90° swing, minimum bucket travel to fill. Shovels dumping into trucks, backhoes, and draglines to embankments, and clamshells from stockpiles to hoppers.

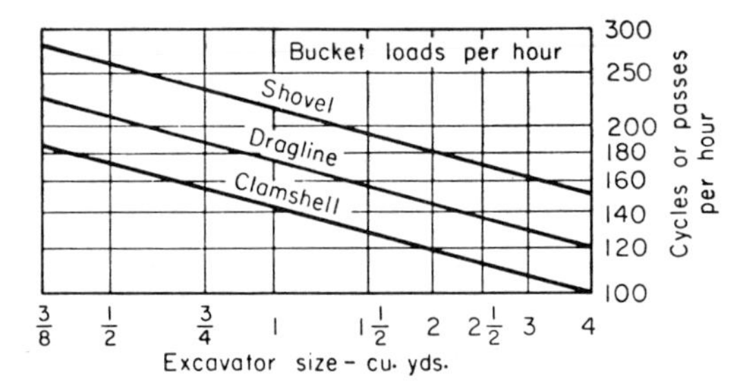

Fig. 10-22 Theoretical cycles per hour for shovels, backhoes, draglines, and clamshells for 90° swing and average range. Derived from Fig. 10-21. No allowance made for move-ups or delays of any kind.

Power Requirement The comparative digging effectiveness of various types of excavators can be compared by relating the installed flywheel horsepower rating for each unit to its estimated production in bank cubic yards per hour. The production rates indicated in Fig. 10-24 were derived in the same manner as those of Fig. 10-23, and the corresponding horsepower ratings are average values as obtained from the product literature. The wheel and ladder trenching machines have the highest output per installed horsepower, but they can dig only under reasonably favorable conditions.* These units work best where long, continuous trenches are produced, and lineal footage of excavated trench may then be of greater interest than the total volume of excavation.

The cable-operated excavators—shovel, backhoe, dragline, and clamshell—have the same engine for each application and are fully convertible machines. Since the backhoe cycle is typically slower than shovels, Fig. 10-24 shows backhoes as having about 25 percent less output per horsepower than the shovel attachment. Dragline cycles are also slower than shovel cycles, but with short boom lengths the dragline

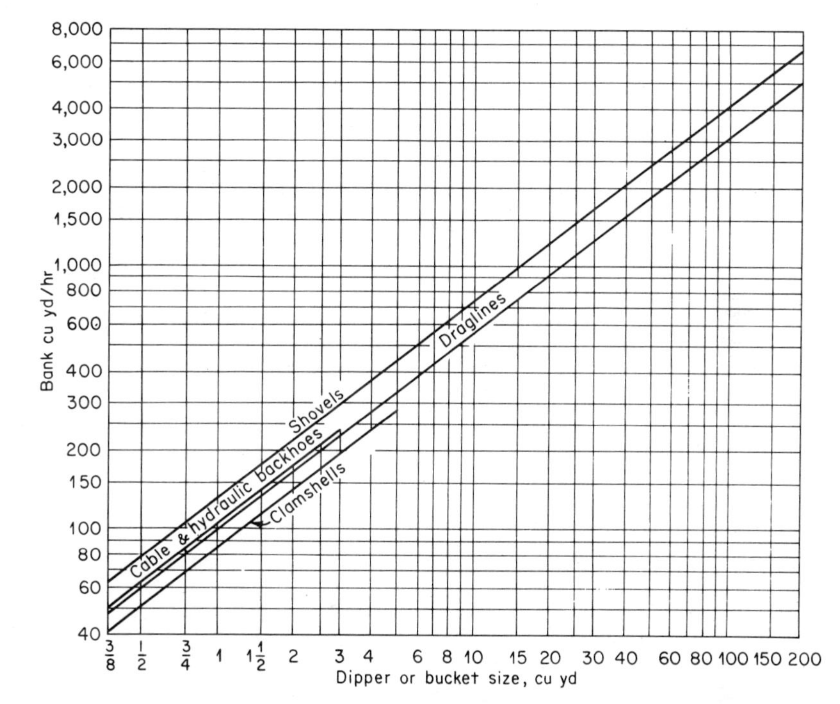

Fig. 10-23 Excavator output for shovels and draglines ⅜- to 200-cu yd size and backhoes ⅜ to 3 cu yd. Output is based on bank (in place) cubic yards with a 1.33 swell factor and the machine operating at 80 percent of its theoretical cycles per hour (Fig. 10-22) with 90° swing, noninterrupted normal digging and nominal digging face or depth.

bucket can be larger than the shovel rating. For these conditions, the shovel and the dragline will have a similar output per horsepower. The hydraulic backhoe, on the other hand, will usually require about 30 percent more installed horsepower per unit of output than will an equivalent cable backhoe.

The backhoe mounted on an industrial or farm tractor is provided with still more power for each cubic yard per hour of excavating capacity, but the load factor would be less since the engine would not be used at its maximum power while digging. Slackline cableways and drag scrapers† would also require more power, since they perform a transporting function in addition to excavating. Each of these machines finds its optimum use under a specific set of job conditions, hence Fig. 10-24 can only provide general guidance as to relative outputs.

* See Sec. 12, Trenchers.
† See Sec. 14, Cableways and Drag Scrapers.

Even though factors other than power requirements will frequently determine the construction method, there are times when a choice of excavating equipment can be made between two or three alternatives. When this situation exists, it is logical to select the unit which will produce at the lowest cost. In trenching, within the range of their depth capabilities, wheel trenchers will excavate at the lowest cost per cubic yard and can dig harder material than a ladder trencher. Trenches whose depths exceed the practical limits for wheel trenchers will require a ladder

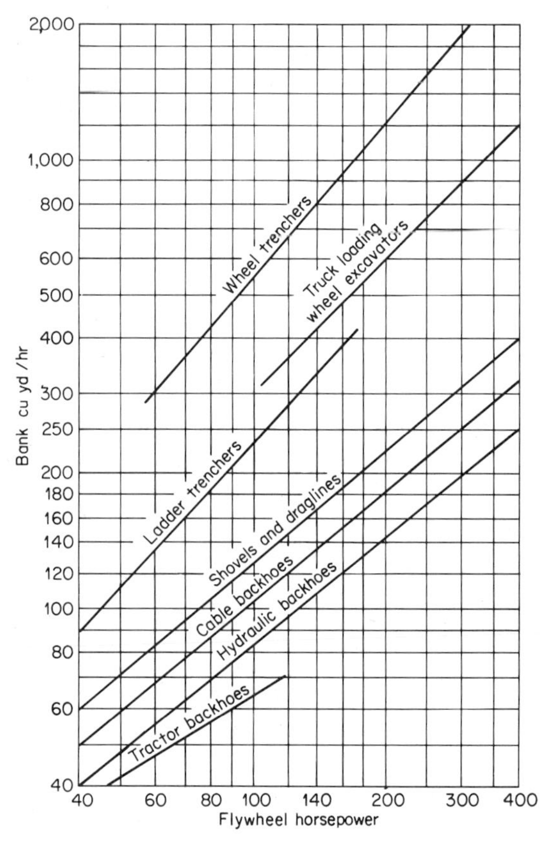

Fig. 10-24 Comparative excavating efficiency of several excavator methods related to the flywheel horsepower rating of the engine usually used. Output is based on bank (in place) cubic yards, easy digging, and 80 percent of theoretical best machine cycle.

trencher. Hard, variable digging with occasional large rocks will require a backhoe. Also, in street work where trucks are loaded, the hoe would be used rather than trenchers. If it is necessary to dig inside of light to medium bracing or around other obstructions, the hydraulic backhoe would be the best choice because of its greater digging flexibility. A clamshell with still lower horsepower productivity than the backhoe might be needed where heavy bracing is used.

As another example, a wheel-type excavator* could load out trucks at three or four times the rate of a shovel (or dragline) of comparable horsepower. The wheel

* See Sec. 11, Loaders.

excavator is a specialized single-purpose machine and would be a suitable choice if the job was big enough to justify its use. The pit in which it works must then be laid out so that little time is lost in turning or backtracking, and there must be no large boulders. If the haul from the pit is short, drag scrapers, slackline cableways, or bulldozers might also be considered. Their output is lower than that of the excavators just discussed, but their use would eliminate the need for separate hauling units.

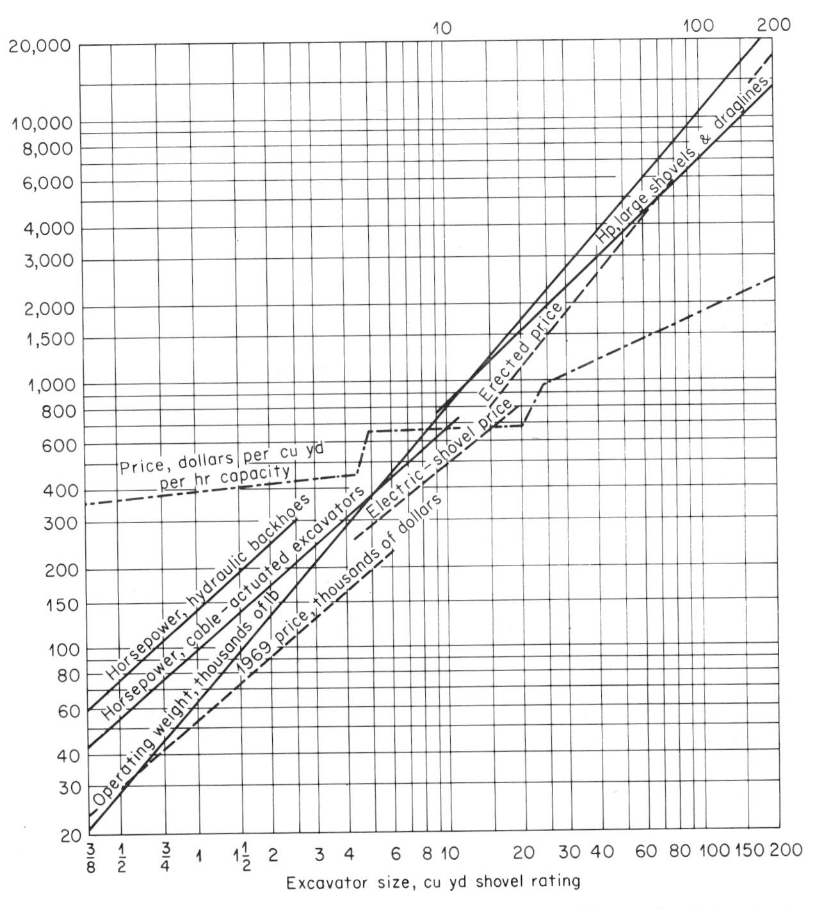

Fig. 10-25 Excavator weight and flywheel horsepower, ⅜ to 200 cu yd. Prices for commercial-range electric mining shovels and large, field-erected electric stripping shovels and walking draglines. Price per cubic yard of hourly capacity derived from data of Fig. 10-23.

Hauling-unit sizes* must be studied in conjunction with loading-unit sizes, since a small shovel may delay a truck too long in loading. It may be desirable to keep the loading time down to 3 min or so on long-duration jobs in quarries and mines. The overall economics of the job will normally dictate the best balance of equipment. The relative flywheel horsepowers for selected excavators in a wide range of sizes from ⅜ to 200 cu yd are shown in Fig. 10-25. These are the manufacturers' power ratings at the governed speed less fan and generator accessories, and the peak power

* See Sec. 13, Hauling Units.

at a pull down in engine speed is considerably higher. If the load is applied suddenly on a cable-actuated machine, as when a bucket is suddenly stopped by a large rock or reaches the end of its travel, the line pull can reach 300 percent of the rated engine power-line pull. This can break the wire rope unless the clutch compensating or overload springs are carefully adjusted to just barely carry the full-power shaft torques. Torque converters will cushion this peak line pull but will still develop from 200 to 500 percent of rated engine full-speed torque before stalling occurs. Electric machines are difficult to rate, but the peak power output for a few seconds at a time may be double the flywheel horsepower shown in Fig. 10-25.

The average horsepower of excavators, calculated from the amount of the fuel used, or from the kilowatt hours metered in the case of electric machines, seems to be about two-thirds of the rated flywheel horsepower. The load factor is thus approximately 67 percent, a figure which is frequently used in sizing diesel engine radiators for excavator service.

Economic Evaluation The operating weights shown in Fig. 10-25 are straight-line averages for all classes of excavators. The weights in the commercial size range apply primarily to shovels and backhoes, and the heavier counterweights and longer crawlers required for dragline- and clamshell-equipped machines may increase the overall weights of these machines by an additional 10 percent. Also, depending on the manufacturer, the weights of larger shovels in the commercial range and in the electric mining-shovel range may be up to 20 percent less than the charted values.

The 1969 prices of excavators varied from $1.10 per lb of operating weight on the smallest shovels and backhoes to $.75 at the upper end of the size range. Excavators up to 6 cu yd in size, when equipped with a dragline or clamshell attachment, cost about 10 percent less than when equipped with a shovel. Electric mining shovels averaged from $.85 to $.78 per lb, while the field-erected price of the large electrics varied from $.65 per lb at the lower end of their range to $.80 at their upper end. It is probable that continuing inflation will increase these prices. For example, if the average annual price increase were to amount to 3 percent, the price would double between 1969 and 1992.

The prices per cubic yard of hourly capacity are also shown in Fig. 10-25. These vary from $350 on small shovels to $2,500 per cu yd per hr for the largest 200-cu yd walking draglines. This sevenfold increase in capital investment per unit of excavating output occurs because the large machines must lift the spoil so much further, requiring more weight to handle the long booms and greater horsepower. A 200-yd bucket has 400 times the volumetric capacity of a ½-yd size but, since the time cycle of the larger bucket is three to four times that of the smaller, the ratio of their production capacities appears to be only about 100:1 (see Fig. 10-23).

The initial equipment cost is the largest factor in the total cost of any excavator operation. This is a "fixed" cost in that it continues each year, regardless of hours of operation. Maintenance, fuel, and operating labor are variable costs, frequently treated as proportional to the hours used. The annual capital cost is computed by amortizing the excavator investment through equal annual payments continued over the expected service life. Each payment includes both capital recovery and interest on the unamortized balance, with the relative balance between these components gradually shifting in favor of capital recovery. Table 10-2 shows total annual fixed cost per dollar of initial investment, both as an annual (average) total and as a cost per operating hour for various usage policies. The factors in column two are based on a 7 percent annual rate of interest and zero salvage value. Similar factors are readily available in standard financial tables and can be applied to other interest rates and service lives.

Composites of 1969 excavator prices plus capital costs taken from Table 10-2 are used to develop the hourly costs of Fig. 10-26. Although the curves of this figure are based on the assumption that each excavator would have no value at the end of the 20-year period, such is not necessarily the case. An excavator's resale price will depend on its condition, on market demand, and on the amount of inflation which has occurred since its initial purchase. Unusual, special, and difficult-to-move excavators may have only scrap value.

Maintenance costs for excavators are difficult to estimate and will vary widely. The average hourly maintenance cost is estimated at 1/20,000 of the original net cost ($1.50 per hr) on the small ⅜- and ½-cu yd units, to 1/200,000 of cost ($100 per hr) on the largest stripping shovels. Operating labor is basically the same for all sizes at $4 to $5.50 per hr except that the largest machines require a full-time oiler or assistant operator. Their use also involves more overtime, since a 7-day-week operation is desirable for a 200-cu yd machine whose fixed capital costs run $250 per hr. Supervision and fringes, other than overtime, are included in operating labor at 20 percent. Repair labor during downtime, lubricants, repair parts, and perishable parts are included in the estimated maintenance cost. Diesel fuel cost is based on the engine fuel consumption of ½ lb per hp per hr, with fuel at $.21 per gal or $.03 per lb and an engine load factor of 65 percent (average power or fuel compared to full load rating). For excavators over 5 cu yd, electric power costs including demand charges, transformer, and line-distribution costs are taken as $.02 per kilowatthour for the 5-cu yd unit down to $.01 per kilowatthour for the largest 200-cu yd strippers. Electrical factor, or average power, is assumed at 40 percent of the installed power shown in Fig. 10-24. The power costs for the larger machines are

TABLE 10-2 Cost of Amortizing Capital Equipment on the Basis of an Equal Cost Per Year (includes taxes at 3 percent and interest at 7 percent)

Depreciation term or life, yr	Yearly payment to retire debt in depreciation term	Property tax and insurance at 3% of average value	Total average annual fixed cost	Fixed costs per operating hour per dollar investment			
				7,200 hr/yr (360 days at 20 hr)	4,000 hr/yr (two shifts)	2,000 hr/yr (one shift)	1,000 hr/yr
40	0.075009	0.015	0.090009	0.000013	0.000022	0.000045	0.000090
20	0.094393	0.015	0.109393	0.000015	0.000027	0.000055	0.000109
15	0.109795	0.015	0.124795	0.000017	0.000031	0.000062	0.000125
10	0.142378	0.015	0.157378	0.000022	0.000039	0.000078	0.000157

higher per cubic yard excavated because the material is moved in the order of ten times as far and is lifted five times as high.

The fixed capital costs are shown in Fig. 10-26 for five different hours per year of operation. These hours of use affect the actual hourly costs more than any other single factor. For example, the total operating cost of a 1-cu yd shovel varies from $10 per hr when used 4,000 hr per year over a 20-year life to $20 per hr if used an average of 500 hr per year for 20 years. A 100-yd walking dragline would cost $250 per hr at 20 hr per day, 360 days per year of operation; and this cost would increase to nearly $600 per hr if the unit were used only 2,000 hr per year.

The least certain of the estimated costs will be the maintenance cost. However, this item will amount to only about one-sixth of the total cost, assuming annual usages ranging from 2,000 hr on small machines to 7,200 hr for the largest. Therefore, even if actual maintenance costs vary considerably from the assumed values, the total hourly costs will not change much.

Figure 10-27 summarizes all the economic factors involved in the selection of excavator size and method. The unit costs per cubic yard for this figure were derived by dividing the total hourly costs of Fig. 10-26 by the estimated hourly yardages of Fig. 10-24. Since all the calculations underlying Fig. 10-27 include the same 80 percent of optimum operating efficiency and same basis of cost, the figure furnishes an approximate guide to the relative costs of all excavator sizes under various yearly hours of use. If the machine is to be used an average of 500 to 1,000 hr per year,

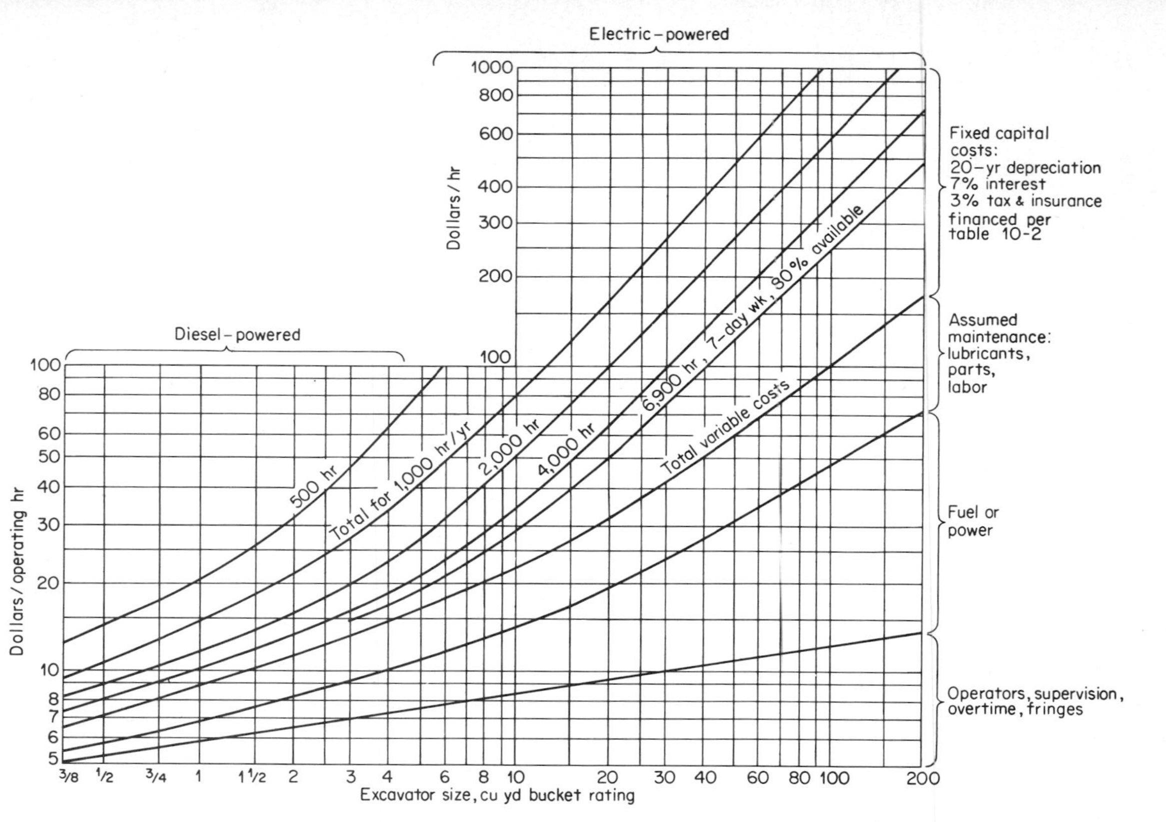

Fig. 10-26 Hourly operating cost relationships for conventional crawler, full-revolving, cable-type excavators such as shovels, backhoes, and draglines show relative importance of cost elements and how they are affected by machine size. Solid lines show total hourly costs and how the annual usage affects this cost. Fixed cost is based on amortizing the purchase price of the machine with 20 years of equal payments to cover the principal and 7 percent interest, plus 3 percent of the average investment (one-half of original cost) to cover taxes and insurance per Table 10-2.

the lowest unit excavation costs are for the 1½- to 2½-cu yd sizes. Larger sizes may still be required in order to obtain the desired digging range or, alternatively, transporting problems may dictate the use of smaller machines. Also, where output is limited by conditions other than excavator capacity, there is no point in using a larger excavator.

A special coal stripping operation with a large shovel or dragline might require the removal of 60 ft of overburden in order to reach a 3-ft seam of coal. This operation then involves moving 20 cu yd, or 36 tons, of overburden per ton of coal. Figure 10-27 indicates a cost of about $.05 per cu yd of overburden moved or $1 per ton of coal uncovered if the coal weighs 2,250 lb per cu yd in its natural state. The cost of loading this coal into trucks with a smaller shovel is shown to be about $.06 per cu yd in place, or $.055 per ton.

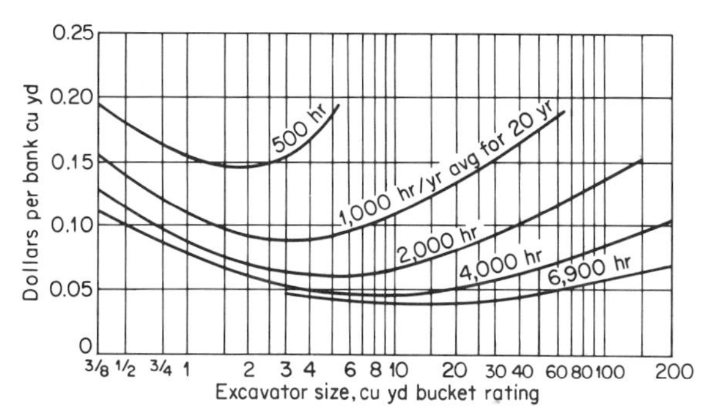

Fig. 10-27 Unit cost per cubic yard for hourly costs of Fig. 10-26 at production rate for all shovels and large draglines shown in Fig. 10-23.

However, if the terrain is such that the largest machine is required to handle deep overburden needing long casting distance and high lift, the stripping cost is shown to rise from $.05 per bank yard for a 60-yd machine to $.07 or 40 percent more for a 200-yd machine. Even though the absolute cost may be argued, the relative cost relationships of Fig. 10-27 are quite realistic and should be most useful in making the size selection.

Section 11

Loaders

E. A. BRAKER*

Divisional Product Consultant, Construction Equipment Division, International Harvester Company, Melrose Park, Ill. *(Front-end Loaders)*

E. H. FAUTH†

Dean of Career Education and Community Services, Waubonsee Community College, Sugar Grove, Ill. *(Wheel Loaders)*

J. R. CHALLIS

Sales Manager, Kolman Division, Athey Products Corporation, Sioux Falls, S.D. *(Belt Loaders)*

* Deceased.
† Formerly Group Vice-president, Products, Barber-Greene Company, Aurora, Ill.

A. Front-end Loaders

Front-end loaders are versatile, self-propelled machines mounted either on crawler- or wheel-type running gear. They are regularly equipped with a front-mounted general-purpose bucket with which they dig, scoop, lift, carry, and dump into hauling units, bins, hoppers, conveyors, and stockpiles. They also transport, spread, and compact fill material. In addition, a variety of buckets, grapples, blades, and other front- and rear-mounted attachments enables them to do such work as doze, scrape, clamp, grub, forklift, rip, ditch, trench, and winch.

CRAWLER-TYPE LOADERS

Except for their integrally mounted equipment, crawler-type loaders are closely related to crawler tractors. On all makes of loaders except one, the track frames are rigidly held to the chassis instead of oscillating. Loader tracks generally extend farther forward for better bucket lifting and carrying stability. Track gauge width between track centers is sometimes greater than on tractors with equal power. Overlapping flat, low-profile, triple-grouser track shoes provide for easier turning and permit some movement over pavement. On many crawler loaders, steering control may be either by hands only or feet only instead of by both hand levers and foot pedals.

The fabricated steel loader frame and the tractor front frame are often integral. Double-acting hydraulic cylinders on each side of the frame actuate the loader lift and bucket mechanism. The hydraulic system includes the hydraulic pump on a live drive from the engine, a pressure-relief valve, an oil reservoir, filters, and loader lift and bucket controls. Additional controls are available to actuate a rear-mounted ripper and/or the clam of a multipurpose bucket.

A power-shift transmission with torque-converter drive makes bucket crowding and lifting easy without killing the engine. It also permits effortless forward-reverse shifts and speed changes. The transmission includes a power takeoff drive for rear-mounted equipment, such as a winch.

Figure 11-1 illustrates the arrangement of drive components in a typical power-shift, crawler-type loader. Table 11-1 summarizes the comparative specifications for makes and models down to 1-cu yd standard capacity, as current in 1969. Crawler-type loaders have the advantages of good flotation and traction on soft or uneven terrain. They operate over sharp objects which would be destructive to rubber tires, and they have compact design for close-quarter maneuverability. However, on-highway transport equipment such as a truck and tilting trailer is needed to move them from job to job over public roads and pavements.

WHEEL-TYPE LOADERS

In 1948, Frank G. Hough began production of the original 4-wheel-drive integral front-end loader. This was equipped with rear-mounted engine, up-front operator, and a 1½-cu yd bucket. Numerous manufacturers have since entered the market with 4-wheel-drive loaders of up to 15-cu yd regular rated capacity. Table 11-2 summarizes comparative specifications of makes and models down to 1¼-cu yd standard capacity, as current in 1969. The net flywheel horsepower tabulated is that published by the loader manufacturers for fully equipped diesel engines at rated governed rpm as regularly installed in the loaders, with power deductions for fan, generator or alternator, air cleaner, water pump, and other standard engine accessories.

The published net flywheel horsepower may be that developed under normal operating conditions at 500 ft or more altitude; or it may be corrected to sea level, which adds about 5 percent to the horsepower normally developed. Some loader specifications give only maximum or bare engine horsepower, which is not comparable to net flywheel horsepower.

Diesel engines power most of the loaders. Some loaders are available with optional makes of diesels, and gasoline engines are also available for the smaller-size loaders. The drive is through a torque converter, power-shift transmission, front- and rear-axle differentials, and planetary gear reductions in all wheels. Differentials are generally of the limited-slip, torque-transferring type in either front, rear, or both axles. The steel main frames of the loaders may be either rigid or articulated. Figure 11-2 illustrates a typical rigid-frame loader with rear-wheel power-assist steering and a general-purpose bucket of 1½-cu yd SAE rated capacity.

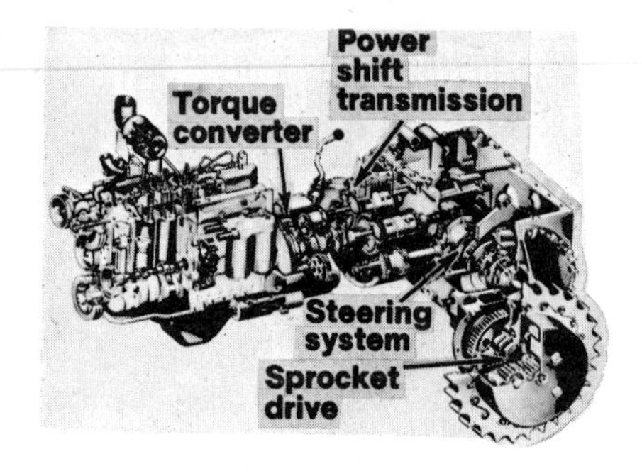

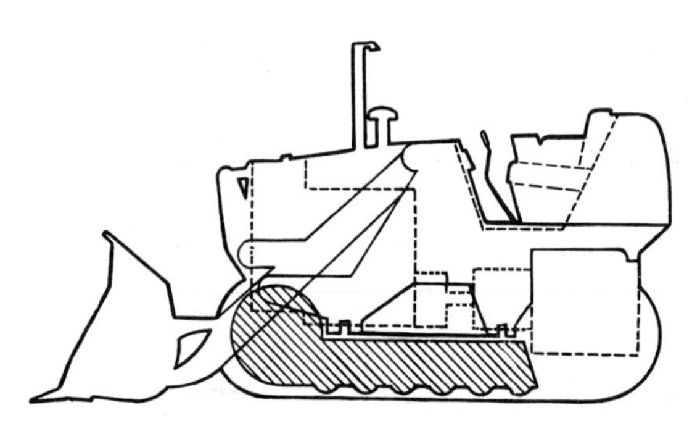

Fig. 11-1 Crawler-type loader.

Figure 11-3 shows the arrangement of drive components in a wheel loader of 10-cu yd SAE rated capacity. The frame of this loader is hinged (articulated) between its front and rear axles. The loader steers by swinging through an angle of 40° to the right or left of center, actuated by hydraulic cylinders. In the illustration, the operator's compartment is mounted on the front frame and turns so that the operator faces the bucket and work area. However, this compartment can also be placed on the rear frame to swing with the engine. There is a trend toward articulated-frame, rubber-tired loaders for heavy-construction work because of faster swing cycles, shorter turning radius, and a longer wheelbase for better roadability.

The hydraulically actuated loader lift mechanism is located ahead of the operator for safety. Loader lift control provides the conventional raise, hold, lower, and float positions. Bucket control has roll-back, hold, and dump positions. Additional hydraulic controls are furnished for optional front- or rear-mounted equipment.

Most loaders have an automatic boom lift kickout which is adjustable for various dumping heights up to the maximum clearance in inches from the ground to the bucket cutting edge at a 45° dumping angle as shown in the comparative specifications. The corresponding reach at the maximum dumping height is measured from the tires or front-most part of the tractor. An automatic bucket positioner that can be adjusted to the desired bucket digging or scooping angle may be either optional or regular equipment.

Some loaders have two foot pedals for applying their 4-wheel power brakes. One of the brake pedals disengages the transmission, allowing the operator to divert full engine power to the live-driven hydraulic pump or pumps for load prying and lifting.

TABLE 11-1 Comparative Specifications of Diesel Crawler-type Loaders

Make	Model	SAE rated capacity, cu yd (3,000 lb/cu yd material)	Shipping weight, lb	Net flywheel horsepower	Speed range, mph		Bucket fully raised 45° dump, in.	
					Forward	Reverse	Clearance	Reach
Allis-Chalmers	21G	4	74,400	268	0–7.6	0–7.0	123	52
	12G	2¾	42,700	195	0–4.7	0–5.3	118⅜	40⅜
	7G	1¾	27,000	116	0–5.1	0–4.7	105	34
	6G	1½	21,000	72	1.5–5.5	2.0–4.1	109	32
	HD-4	1	12,400	NP	0–4.9	0–5.6	98	29
J. I. Case......	1150	1¾	25,515	85	0–6.0	0–7.1	101	42¼
	850	1⅜	19,420	66	0–5.7	0–6.2	100½	35¾
	450	1	12,990	51	0–5.8	0–6.4	98	28
Caterpillar Traxcavators	983	4½	66,200	275	0–6.4	0–7.4	144	55
	977K	2½	41,000	170	0–5.8	0–7.0	126	50
	955K	1¾	28,000	115	0–5.8	0–7.0	114	46⅝
	951B	1½	22,100	85	0–6.0	0–7.3	108	40½
	941	1¼	19,700	70	0–5.7	0–6.9	102	40
John Deere....	450	1⅛	15,690	57	1.8–6.7	1.7–6.4	103	33
International...	250B	2½	41,210	160	0–4.9	0–5.4	117⅜	49¾
	175B	2	27,765	120	0–5.2	0–6.2	107	47¼
	150 PS	1½	19,720	75	0–3.8	0–5.1	102	34
	150 GD	1½	19,520	71	1.5–5.2	1.7–3.5	102	34
	125 PS	1¼	17,700	69	0–5.5	0–6.4	103	37
	125	1¼	17,900	63	1.4–6.5	1.9–6.2	103	37
Massey-Ferguson	MF-500	2⅛	27,890	136	0–4.0	0–4.0	101	35½
	MF-3366	1½	22,420	75	0–3.9	0–4.6	100½	38½

NP = not published.

Wheel-type loaders have much higher speeds than the crawler type and operate best on firm surfaces. They can be run on pavement and moved from job to job under their own power. A wider variety of sizes is available to match job requirements. Their maintenance cost is lower where wet, sandy soils would be highly abrasive to crawler tracks and undercarriages. However, wheel loaders normally have less traction than crawler-type loaders of equal weight, and this limits the usable power which they can develop for heavy digging and grade climbing. Their load-handling capacity is also reduced on boggy ground.

FRONT-END LOADER RATINGS

The Society of Automotive Engineers, Inc. (485 Lexington Ave., New York, N.Y. 10017), has established and published industry SAE standard ratings for bucket volume capacity and tipping load weight as well as a recommended operating load rating

TABLE 11-2 Comparative Specifications of Diesel 4-wheel-drive Loaders

Make	Model	Frame	SAE rated capacity, cu yd (3,000 lb/cu yd material)	Shipping weight, lb	Net flywheel horsepower	Max. speed, mph				Bucket fully raised 45° dump, in.	
						Forward		Reverse			
						Low	High	Low	High	Clearance	Reach
Allis-Chalmers	745	A	3½	35,150	185	2.4	20.9	2.7	23.0	116	33
	645	A	2¾	25,158	154	3.7	22.7	3.8	23.5	110¼	35¾
	545H	A	2	17,880	96	3.6	21.8	3.5	22.5	108½	33⅞
	545	A	1¾	17,230	96	2.7	24.0	3.5	22.5	111½	32¾
Baldwin-Lima-Hamilton	140	A	3½	29,500	180	3.8	28.2	5.1	10.3	119¼	41⅞
	100	R	2½	22,600	137	3.1	24.2	3.2	25.2	110	37½
	80	R	2	19,200	105	3.0	24.2	3.1	25.0	109	32¼
	60	R	1½	15,494	90	3.0	22.4	4.0	8.0	104¾	32
	50	R	1¼	12,060	72	2.8	21.4	3.8	7.6	104	30¼
J. I. Case	W-26	A	3	28,790	160	2.4	21.0	3.3	7.7	115¾	40
	W-24	A	2½	22,900	125	2.7	23.0	3.7	8.5	104	41
	W-10C	R	2	19,350	95	3.2	20.2	3.2	20.2	100½	34¼
	W-9C	R	1¾	18,300	95	3.2	20.2	3.2	20.2	103	33
	W-8C	R	1½	17,905	95	3.1	19.6	3.1	19.6	104¼	32½
	W-7C	R	1¼	12,865	72	3.0	22.0	3.0	22.0	99	27½
Caterpillar Tractor Co.	992	A	10	109,400	550	4.4	22.8	4.8	24.6	179	68
	988	A	6	66,220	325	4.0	22.1	4.0	22.1	130	50
	980	A	4	44,000	235	4.0	26.1	5.1	16.6	121	47
	966C	A	3	32,000	170	4.3	21.1	5.1	25.0	116½	35⅜
	950	A	2¼	22,870	130	4.1	20.7	4.9	24.6	107¾	33⁵⁄₁₆
	930	A	1¾	17,700	100	4.1	25.0	5.0	14.6	111½	32⅝
	920	A	1½	15,400	80	4.0	25.0	4.8	14.2	108	29
John Deere	JD644	A	2¼	22,458	131	7.0	23.0	...	9.0	110	37¾
	JD544	A	1½	17,540	94	7.0	23.0	...	9.0	106	35½

General Motors Corp. (Terex)	72-81	A	9	101,000	438	4.9	15.0	5.9	17.0	159	58
	72-51	A	3½	35,690	191	3.1	22.0	4.3	8.5	122	38
	72-41	A	3	30,590	151	3.2	26.8	3.3	27.5	120	41
	72-31	A	2½	27,620	134	2.9	24.0	3.9	8.8	120	27
	72-21	A	2	21,160	107	2.4	20.3	3.2	7.0	105	24
Hough Payloaders	H-400B	A	10	116,000	500	6.3	15.5	6.3	15.3	156	75
(International Harvester Co.)	H-120C	A	5	54,000	296	4.2	29.4	4.2	29.4	128½	46
	H-100C	A	4¼	43,000	285	4.7	32.7	4.7	32.7	123	61
	H-90E	A	3½	37,000	230	4.6	32.0	4.6	32.0	118	43½
	H-80	A	3	31,900	160	3.9	23.7	4.7	28.4	118	46
	H-65C	A	2¼	23,135	141	4.2	25.8	5.1	31.0	109	43
	H-60B	A	1¾	17,500	95	3.5	21.6	4.3	25.9	103½	47
	H-50C	R	1½	15,640	83	4.2	25.4	5.0	30.5	103	37¼
	H-30B	R	1¼	11,380	73	3.6	21.3	4.2	25.7	94	31½
International	3850	R	1¼	12,900	79	5.3	21.0	3.3	7.1	102	42
KW-Dart Truck Co.	D-600	A	15	151,000	635	3.6	22.0	3.6	22.0	186	90
Massey-Ferguson	MF-55	A	2½	25,380	138	3.0	21.0	3.0	21.0	109	38
	MF-44	R	2	18,420	93	4.0	24.0	4.0	24.0	110	30¾
	MF-33	R	1½	14,940	74	4.0	24.0	4.0	24.0	109½	29⅜
Michigan	475	A	12	137,200	554	3.8	28.2	5.0	28.2	171	66
(Clark Equipment Company)	275	A	6½	60,600	305	4.0	22.5	4.0	22.5	127	53½
	175	A	4½	40,800	261	4.0	23.1	4.0	23.1	115	49
	125	A	3½	35,370	198	4.5	25.5	4.5	25.5	118½	46¼
	85	A	3	31,750	177	3.7	25.4	3.7	25.4	115½	37
	75	A	2½	23,050	150	5.2	36.2	5.2	36.2	106	36
	75	R	2	17,050	102	4.4	26.4	4.4	26.4	109	31
	55	R	1½	13,715	72	4.3	24.3	4.3	24.3	99¼	32¾
Nelson (N. P. Nelson, Inc.)	400B	R	4	37,100	NP	3.5	24.5	3.5	24.5	124	44
	250DC	R	2⅝	25,400	NP	3.5	27.6	3.4	26.7	114	33
	200C	R	2¼	21,350	NP	3.4	27.0	3.3	26.2	114	26
	150C	R	1¾	16,700	NP	3.5	24.5	3.5	24.5	108	24
Pettibone-Mulligan	PM-350	R	3½	32,700	NP	3.9	27.0	3.9	27.0	135	36
	PM-285	R	2¾	25,500	NP	3.4	27.0	3.5	28.0	112	34
	PM-245	R	2	21,500	NP	7.0	27.0	. . .	9.5	112	30

TABLE 11-2 Comparative Specifications of Diesel 4-wheel-drive Loaders (Continued)

Make	Model	Frame	SAE rated capacity, cu yd (3,000 lb/cu yd material)	Shipping weight, lb	Net flywheel horsepower	Max. speed, mph				Bucket fully raised 45° dump, in.	
						Forward		Reverse			
						Low	High	Low	High	Clearance	Reach
Scoopmobile (WABCO)..........	1200	A	10	106,000	478	3.5	26.2	3.5	26.2	162	72
	500	A	5	64,000	294	5.0	20.0	5.0	20.0	120	33
	400B	A	4	43,100	192	4.0	25.0	4.0	25.0	120	24
	350	A	3½	38,860	192	4.0	25.0	4.0	25.0	120	26
	LD-7A	A	2½	24,860	110	3.0	20.0	3.0	20.0	109	29
	LD-7	A	2	21,600	93	3.0	20.0	3.0	20.0	109	29
	150B	A	1½	17,300	93	4.0	20.0	4.0	20.0	110	28
Thew-Lorain (Koehring Co.)......	500A	A	5½	58,850	286	2.2	25.0	2.4	27.3	134½	42½
	400	A	4	40,000	228	2.0	23.2	2.2	25.4	122¼	45¼
	ML-350	R	3½	32,260	NP	2.2	22.7	3.0	8.3	122	43
	ML-275	R	2¾	26,800	NP	1.8	20.8	2.6	7.7	118	43
	ML-250	R	2½	23,870	NP	1.8	20.0	2.5	7.4	115	42
	ML-225A	R	2¼	22,510	NP	1.7	20.0	2.4	7.3	110	39
	ML-200A	R	2	21,490	NP	1.7	20.0	2.4	7.3	110	39
Trojan (Eaton Yale & Towne)....	8000	A	8	102,500	475	3.9	22.1	3.9	22.1	146	50
	6000	A	5½	65,800	335	3.8	21.8	3.8	21.8	131	47
	4000	A	4½	44,000	247	3.9	27.0	3.9	27.0	120	39
	3000	A	3	30,500	185	3.3	23.1	3.3	23.1	115	45
	2000	A	2½	24,700	150	3.5	24.0	3.5	24.0	114	39
	204A	R	2¼	21,550	120	4.6	31.7	4.6	31.7	106	38
	164A	R	1¾	18,040	115	4.4	30.5	4.4	30.5	105	30
	134A	R	1½	16,100	90	3.9	27.0	3.9	27.0	105	29
	124	R	1¼	14,400	90	3.4	21.6	3.4	21.6	107	24

A = Articulated frame.
R = Rigid frame.
NP = Not published.

for front-end loaders. Selection of the proper bucket capacity for the material being handled involves these ratings, which may be briefly described as follows.

Bucket Capacity, SAE Rating SAE rating (nominal heaped) is described in SAE Standard J742b. The volume in cubic yards (or cubic meters) is based on physical

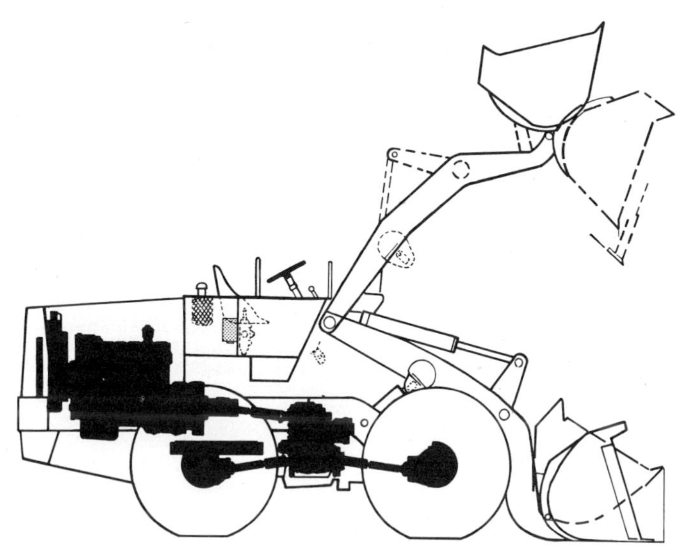

Fig. 11-2 Rigid-frame 4-wheel-drive loader.

dimensions of the bucket only, without consideration of the bucket action on any specific machine. It applies to a bucket with opening oriented in an upright, level position and filled with material heaped at a 2:1 angle of repose.

For bucket capacities of from ¾ to 3 cu yd, the SAE standard rating interval is ⅛ cu yd; for buckets over 3 cu yd, the interval is ¼ cu yd. If the calculated heaped capacity falls below a given rating interval by more than 2 percent, the next lower

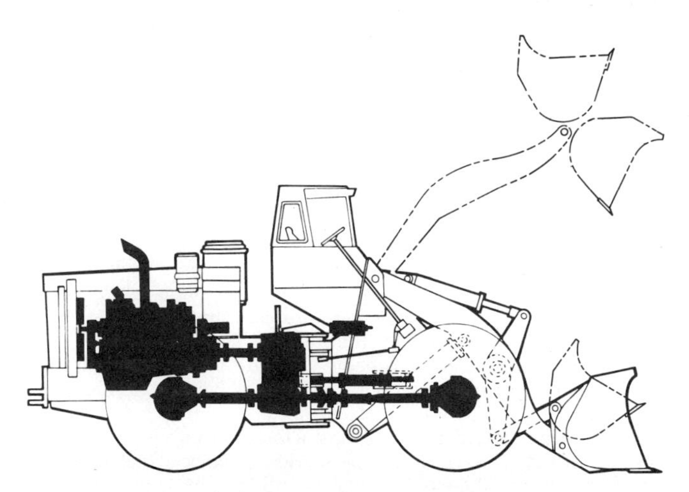

Fig. 11-3 Articulated-frame 4-wheel-drive loader.

interval is deemed to be the SAE rating. Struck capacity is shown decimally to three significant figures when given in addition to the SAE rating.

Tipping Load, SAE Rating This is defined in SAE Standard J732c. It is based on the loader being at operating weight and stationary on hard, level ground. Operating weight with specified bucket is that of a fully serviced loader, including a full fuel tank; counterweight in amount specified, if any; and a 175-lb operator. The bucket is rolled fully back with the center of rated-load gravity at the maximum forward position in the raising cycle. Tipping load is the minimum weight in pounds or kilograms that causes the front rollers of crawler-type loaders to clear the tracks or the rear wheels of wheel-type loaders to clear the ground. Articulated-steering loaders should be rated in full turn position.

Operating Load, SAE Rating To conform with SAE recommended practice J818a, the SAE rated operating load should not exceed 50 percent of the SAE rated tipping load for wheel-type front-end loaders or 35 percent for the crawler type. The higher figure is permissible for wheel-type loaders because they normally operate on harder surfaces and smoother terrain than the crawler type.

The SAE rated capacity in cubic yards of the bucket attached to the loader should not exceed the loader's SAE operating load rating divided by the weight per cubic yard of scooped material. Because of material swell, this weight may be less than that in-bank but more than the loose weight, depending on the digging, crowding, and lifting action of the bucket. For efficient, long-life performance of the loader, it is important to know the loader application and the unit weight of the material to be handled in order to select the right type and capacity bucket.

LOADER BUCKETS AND ATTACHMENTS

General-purpose buckets are available for each model loader in a range of SAE rated yardage capacities approved for handling specified unit-weight materials without exceeding the SAE operating-load rating of the loader. However, unless otherwise ordered, a standard general-purpose bucket is usually supplied for material weighing at least 3,000 lb per cu yd of SAE rated capacity. Therefore, the comparative specifications in Tables 11-1 and 11-2 are based on SAE ratings for material with a minimum weight of 3,000 lb per cu yd. General-purpose buckets are of one-piece all-welded steel construction with a straight cutting edge to which digging teeth can be added if required.

An optional multipurpose or four-in-one bucket can be used as a dozer, scraper, and clamshell in addition to functioning as a conventional shovel bucket. It consists of two major segments. The rear segment is essentially a dozer blade with a straight cutting edge. The front segment is a hydraulically operated clam with a straight cutting edge at the front, and it is usually equipped with digging teeth. Operation of the clam requires a third hydraulic control valve. The multipurpose bucket is much heavier than the standard-capacity general-purpose bucket, and a loader equipped with it will generally need more counterweight for the same operating load rating.

A one-piece bucket of heavy-duty construction with a protruding V-shaped cutting edge is also available. This is known as a "spade-nosed rock bucket" and is used for prying out and scooping up shot rock. Other special buckets include a quarry type with stub teeth to load fragmented shot rock and a skeleton rock bucket to sift out dirt and undesirable small pieces of rock when loading mine-run stock into hoppers or trucks. A side-dump bucket can be used for handling loose, stockpiled material and eliminates turning of the loader to dump into a hauling unit.

Straight or angling dozer blades, a grubber blade for land clearing, logging and lumber grapples, and other front-end attachments for special applications are available in place of buckets on many loaders.

A rear-mounted ripper-scarifier or a towing winch not only provides counterweight but increases the utility of a crawler-type loader. Quick-attachable, rear-mounted backhoes are available for the smaller wheel- and crawler-type loaders for ditching and trenching.

Canopies, cabs, cab pressurizers, heaters and air conditioners, rollover protection devices, safety belts, and backup warning lights are among the attachments available for operator comfort and safety. Instrument panel covers, fuel and oil caps, and hood side doors all of which lock into place are good investment features to protect against vandalism. Tires other than standard are available to suit operating conditions.

MAJOR LOADER APPLICATIONS

Front-end loaders are regularly used for loading, hauling, excavating, clearing, and cleanup. Ability to carry, compact, and spread cover material for sanitary landfill makes them excellent units not only for air- and water-pollution control but also for profitable land making.

Loading This is a major application in which operating conditions are often favorable for wheel-type loaders. It consists of scooping, lifting, turning, and dumping materials such as sand, gravel, and crushed or shot rock from stockpile, bank, or construction area into hauling units. The larger, articulated wheel-type loaders,

TABLE 11-3 Estimated Times (minutes) to Load Hauling Units by V-method from In-bank with Crawler Units and from Stockpiles or Free-running Banks with Rigid-frame Wheel Loaders

Loader standard capacity, cu yd	Hauling-unit nominal struck/heaped capacity, cu yd, and tonnage rating						
	4/5 cu yd, 6 tons	6/7.5 cu yd, 9 tons	8/10 cu yd, 12 tons	10/12 cu yd, 15 tons	12/15 cu yd, 18 tons	14/17 cu yd, 21 tons	16/20 cu yd, 24 tons
1¼	2.0	3.0	4.0				
1½	1.5	2.5	3.3	4.0			
1¾	...	...	3.0	3.5	4.5		
2	...	2.0	2.5	3.0	4.0		
2½	...	1.5	2.0	2.5	3.0	3.5	
3	...	...	...	2.0	2.5	3.0	3.5
4	...	...	...	1.5	2.0	2.0	2.5

because of their mobility and high productivity, are particularly satisfactory for loading shot rock into off-highway hauling units.

Figure 11-4 shows efficient methods for loading loose material from bank or stockpile into hauling units. To determine the normal number of loader passes for average-weight loose materials, divide the 2:1 heaped capacity rating of the hauling unit by the nominal heaped capacity of the loader bucket. A poor match of the loader bucket to the hauling unit either means a partially loaded bucket and a decrease in loading efficiency or a partially loaded hauling unit and a decreased hauling efficiency. For heavy materials, the number of loader passes may have to be restricted to the hauling-unit struck capacity to avoid exceeding the hauling-unit tonnage rating.

Time for each pass depends on kind of material; size and type of loader; bucket filling, lifting, and dumping conditions; and the proficiency of the loader and hauler operators. Tables 11-3 and 11-4 give estimated times in minutes under average conditions to load the material, including shot rock, into hauling units of various struck/heaped capacities with power-shift loaders of 1¼- to 15-cu yd SAE bucket capacity. Table 11-3 allows a minimum of 0.5 min per pass for in-bank loading with crawler units and for stockpile or free-running bank loading with rigid-frame wheel loaders.

Table 11-4 starts with a basic time of 0.42 min for swing-loading loose bank or stockpiled material. It adds 0.20 to 0.28 min for prying out, traveling with, and loading shot rock into off-highway haulers. The times given in Tables 11-3 and 11-4 are subject to modification from job experience and consultation with loader manufacturers' representatives. Tonnage ratings for hauling units are based on material averaging 3,000 lb per cu yd of struck hauling-unit capacity.

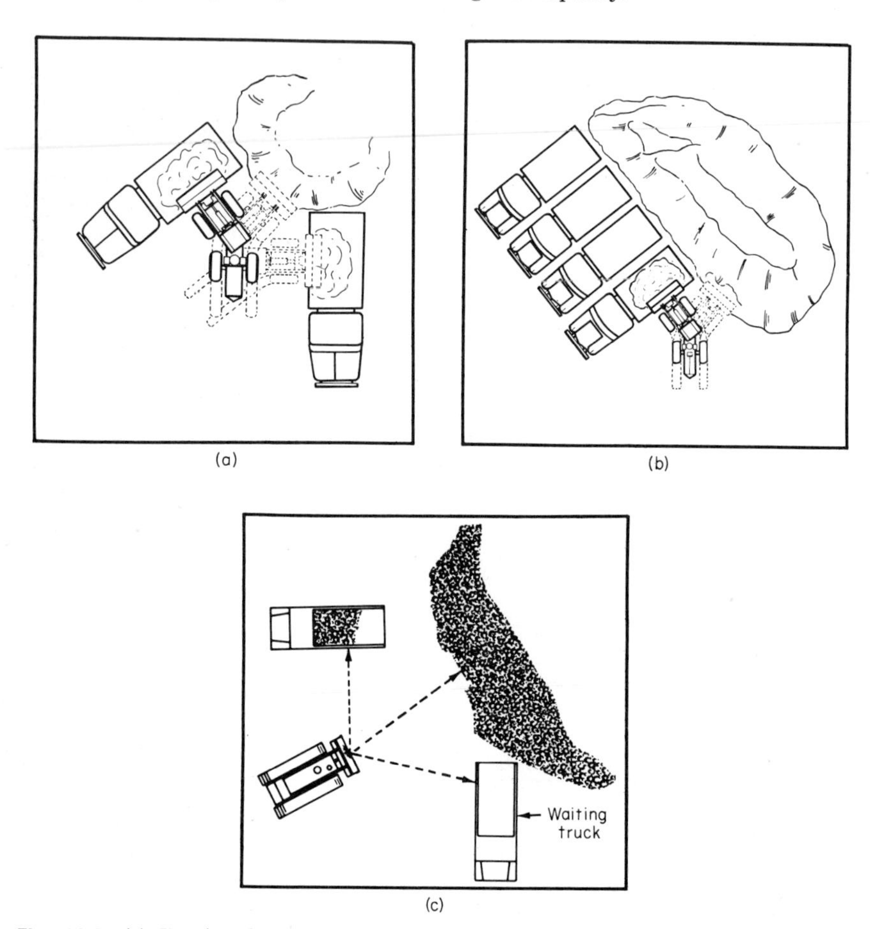

Fig. 11-4 (*a*) V-swing loading with articulated-frame loader; (*b*) step loading with articulated-frame loader; (*c*) V-system loading with crawler or rigid-frame wheel-type loader.

Hauling Rubber-tired loaders are excellent for moving loose materials over short distances to dump into hauling units, hoppers, conveyors, bins, or where needed on construction sites. Production in loose cubic yards at 100 percent working efficiency is approximately the SAE rated capacity of the bucket in cubic yards multiplied by 60 minutes and divided by the cycle time in minutes.

Cycle time consists of two elements: (1) fixed time for loading, lifting, and dumping the bucket, four changes of direction and two turns; and (2) travel time for load carrying and empty returning. High reverse speeds enable faster time cycles when turns are not over 90°, all made in the bucket loading area.

A fixed time allowance of 0.30 min is generally adequate. However, if 180° turns must be made at each end of the haul, an additional 0.06 to 0.10 min may be required. Travel time depends on the loader's average forward and reverse speeds over the

TABLE 11-4 Estimated Times (minutes) to Load Loose Materials into Hauling Units by V-swing or Step Method with Articulated-frame Wheel Loaders

Loader standard capacity, cu yd	Hauling-unit nominal struck/heaped capacity, cu yd, and tonnage rating								
	8/10 cu yd, 12 tons	10/12 cu yd, 15 tons	12/15 cu yd, 18 tons	15/18 cu yd, 22 tons	20/24 cu yd, 30 tons	24/30 cu yd, 36 tons	30/36 cu yd, 45 tons	33/40 cu yd, 50 tons	40/50 cu yd, 60 tons
Stockpile:									
2	2.1	2.5	3.3	3.8					
2½	1.7	2.1	2.5	3.0	4.2				
3	...	1.7	2.1	2.5	3.4	4.2			
3½	...	...	1.8	2.2	3.0	3.8			
4	...	1.3	1.6	2.0	2.6	3.3	3.8		
4½	...	...	...	1.7	2.4	2.9	3.4		
Shot rock:									
5	...	...	1.9	2.5	3.1	3.7	4.5		
6	...	...	...	1.9	2.5	3.1	3.8	4.5	
8	...	...	...	...	2.0	2.6	3.2	3.4	
9	...	...	...	1.3	2.0	2.6	2.7	3.4	3.9
10	...	...	...	1.3	2.0	2.0	2.7	2.8	3.5
12	...	...	...	...	1.4	2.0	2.1	2.7	2.8
15	...	...	...	...	...	1.4	...	2.1	

distance and terrain involved. Allowance must be made for acceleration and braking as well as for the maximum speed at which full bucket loads can be safely carried without spilling.

Table 11-5 records the estimated production per 60-minute hour on level terrain

TABLE 11-5 Estimated Production Capabilities of an Articulated Wheel Loader with 3½ cu yd SAE Rated Bucket (3,000 lb/cu yd material)

One-way distance, ft	Average speed, mph	Cycle time, min	Production per 60-minute hour		
			Work cycles	Cu yd (3,000 lb/cu yd material)	Tons
25	2.9	0.496	121	424	636
50	4.3	0.564	106	371	556
75	5.1	0.635	94	329	493
100	5.8	0.691	87	304	456
150	7.0	0.786	76	266	399
200	8.0	0.868	69	241	361
300	8.8	1.07	56	196	294
400	9.4	1.27	47	164	246

for a loader with a $3\frac{1}{2}$-cu yd SAE rated capacity bucket. The loader is assumed to have four power-shift speeds forward and reverse, as follows:

Gear	1st	2d	3d	4th
mph	0–4.6	0–8.4	0–17.5	0–32.0

Hauls begin in first gear and change to second after the first 25 ft; this is the top speed to carry a full load safely without spillage. Returns over the longer distances are made with third gear reverse as top speed. These top speeds are reduced by percentages to compensate for the accelerating and braking times required.

For example, the load-carrying average speed for the 400 ft haul distance is reduced to about 90 percent of the 8.4 mph second speed, or say 7.6 mph. To reach the top third-gear return speed requires more acceleration and braking time, but the average speed should be at least 70 percent of 17.5 mph, or say 12.3 mph. The average speed for the 800-ft total travel distance then is

$$\frac{2 \times 7.6 \text{ mph} \times 12.3 \text{ mph}}{7.6 \text{ mph} + 12.3 \text{ mph}} = 9.4 \text{ mph}$$

Total cycle time for the 800-ft travel distance and production per 60-minute hour are then computed as follows:

$$\begin{aligned}
800 \text{ ft travel} \div 9.4 \text{ mph} \times 88 \text{ fpm} &= 0.97 \text{ min}\\
\text{Fixed time, loading, dumping, etc.} &= \underline{0.30}\\
\text{Total cycle time} &= 1.27 \text{ min}\\
\text{Cycles per 60-minute hour} = 60 \div 1.27 &= 47 \text{ cycles/hr}\\
\text{Production in 3,000 lb/cu yd material} = 3\tfrac{1}{2} \times 47 &= 164 \text{ cu yd/hr}\\
\text{Production in tons} = 164 \times 3{,}000 \text{ lb}/2{,}000 \text{ lb} &= 246 \text{ tph}
\end{aligned}$$

Job conditions normally result in a 48- to 50-minute productive hour, corresponding to working efficiencies of 80 to 83 percent. To convert loose yards to in-bank yards, multiply by the material swell factor.

Excavating Crawler and heavy-duty wheel-type loaders are excellent for many excavation jobs. Figure 11-5 shows a typical basement digging operation arranged so that the loader can be driven forward out of the excavation on a moderately inclined ramp. Loaders have advantages over dozers of comparable size because they can be used to lift excess excavated material and dump it into trucks or over the edge of the excavation into stockpiles.

Fig. 11-5 Basement excavation.

A rear-mounted ripper or scarifier can be used on a crawler loader to loosen such materials as compacted earth, hardpan, shale, slate, and decomposed rock that do not readily yield to direct bucket loading. The ripper can also be used to tear up old brick, asphalt, or broken concrete pavement. One operator can then both dig and load the material for removal.

The amount of material that can be excavated per hour depends on the kind of material and the distance it is to be moved. Figure 11-6 gives travel time in minutes

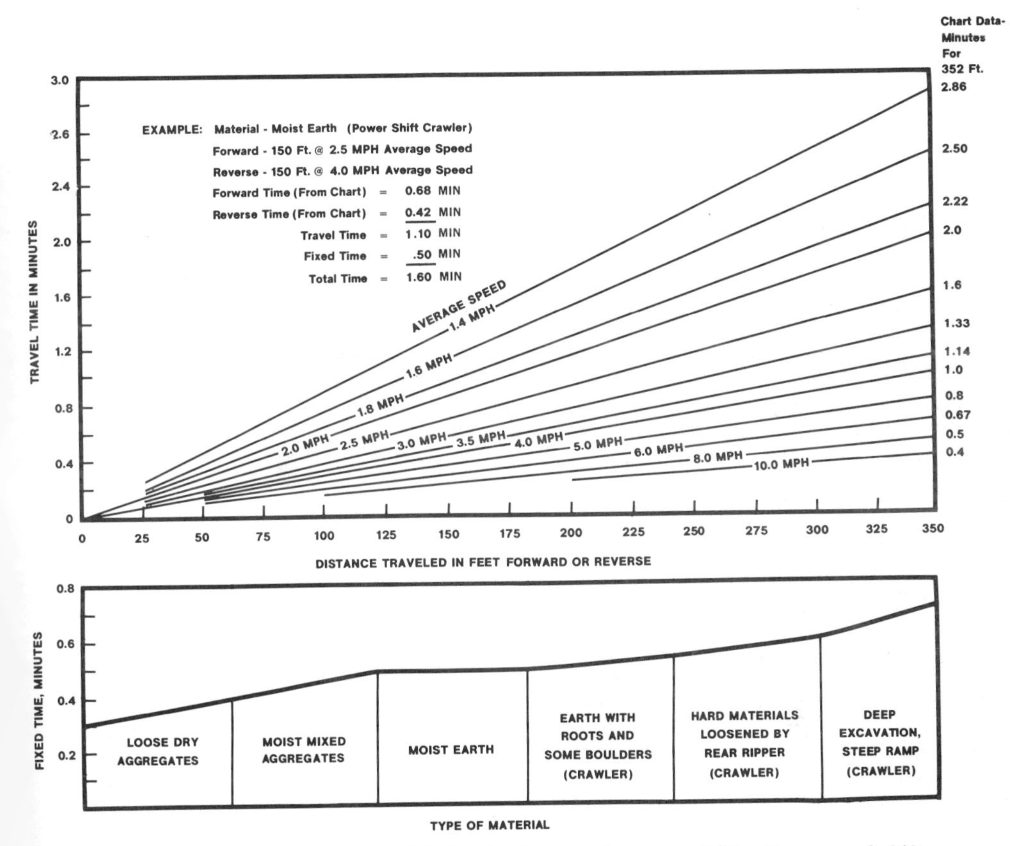

Fig. 11-6 Loader cycle times. (Tabulated values are for power shift. For manual shift, add 0.05 min for each gear change.)

for distances to 350 ft and the fixed time allowance for filling the bucket, dumping, four power-shift changes of direction, and a full load cycle of hydraulics. Although higher peak speeds can be realized on firm, level ground, it is seldom that average travel speeds for crawler-type loaders exceed 2 to 2.5 mph forward and 4 mph reverse when used for excavation. Confined work areas, slippage, and steep ramps tend to reduce actual travel speeds.

Clearing and Cleanup On urban renewal projects, loaders scoop up and load the debris of demolished buildings into hauling units. They also knock down small buildings, and they clean up after snowstorms, tornadoes, floods, and other disasters. Crawler loaders with multipurpose buckets clear construction sites of trees, rocks, brush, and trash, often using a rear-mounted winch to speed the process. A special

TABLE 11-6 Estimated Hourly Owning and Operating Costs for Front-end Loaders Working under Average Conditions*

Loader type	Diesel crawler		Diesel articulated 4-wheel drive				
Capacity, cu yd (3,000 lb/cu yd material)	1¾–2	2½	2½	3	3½	5	10
Ownership costs:							
Depreciation, 10,000 hr, 5 yr	$2.70	$3.90	$2.82	$3.31	$ 4.32	$ 5.65	$10.50
Interest, insurance, taxes at $0.03 per $1,000 total price	0.81	1.17	0.90	1.08	1.44	1.95	3.60
Total ownership costs	$3.51	$5.07	$3.72	$4.39	$ 5.76	$ 7.60	$14.10
Operating costs:							
Fuel at $0.15 per gal	$0.72	$0.93	$0.63	$0.72	$ 0.78	$ 1.11	$ 1.95
Engine oil at $1.00 per gal	0.08	0.11	0.07	0.09	0.10	0.12	0.14
Transmission oil at $1.00 per gal	0.04	0.05	0.01	0.01	0.01	0.02	0.02
Final drive lube at $1.00 per gal	0.02	0.03	0.01	0.01	0.01	0.02	0.02
Hydraulic oil at $0.80 per gal	0.02	0.03	0.02	0.02	0.02	0.02	0.03
Chassis lube at $0.18 per lb	0.02	0.02	0.01	0.01	0.01	0.02	0.02
Filters	0.10	0.11	0.10	0.12	0.13	0.14	0.18
Tires, 3,000-hr life	......	......	0.60	0.97	1.60	2.84	5.00
Repairs	2.43	3.51	1.69	2.05	2.88	3.39	6.30
Total operating cost	$3.43	$4.79	$3.14	$4.00	$ 5.54	$ 7.68	$13.66
Total O & O costs without operator	$6.94	$9.86	$6.86	$8.39	$11.30	$15.28	$27.76
Data used to estimate costs:							
Approx. list price, f.o.b. factory	$27,000	$39,000	$30,000	$36,000	$48,000	$65,000	$120,000
Approx. tire replacement cost	....	....	$ 1,800	$ 2,900	$ 4,800	$ 8,500	$ 15,000
Depreciation price (less tires)	....	....	$28,200	$33,100	$43,200	$56,500	$105,000
Fuel, gal per hr	4.8	6.2	4.2	4.8	5.2	7.4	13.0
Engine oil, gal per hr	0.080	0.110	0.068	0.090	0.100	0.115	0.140
Transmission oil, gal per hr	0.040	0.045	0.007	0.009	0.009	0.015	0.016
Final drive lube, gal per hr	0.023	0.029	0.007	0.009	0.009	0.020	0.020
Hydraulic oil, gal per hr	0.024	0.033	0.002	0.020	0.020	0.023	0.034
Chassis lube, lb per hr	0.080	0.100	0.050	0.060	0.060	0.070	0.100
Repairs, percent of depreciation	90	90	60	60	60	60	60

* Do not include operator's wages, supervision, overhead, or profit. Figures must be adjusted to suit job conditions and actual purchase price. Tire life varies considerably and may be 1,500 hr or less in shot rock.

grubber blade is available for use in place of the bucket for large-scale clearing and piling.

Loaders are not only the first on the site to prepare it for building and construction operations but also the last in order to backfill, spread, level, and top with selected soil for grass and landscaping.

OWNING AND OPERATING COSTS

Table 11-6 gives estimated owning and operating cost per hour, exclusive of operator's wages, for representative size ranges of crawler- and wheel-type loaders.

B. Wheel Loaders

The wheel excavator or loader can be considered as a competitive substitute for the other types of loaders discussed in this section as well as for the various excavators which are discussed in other sections. It is important not to limit one's thinking when considering loaders and loading applications.

Wheel excavating and/or loading should be considered whenever one or more of the following job characteristics are present:

1. When the terrain of the loading point is level so that the material must be elevated in order to be top-loaded.
2. When the distance between loading and discharge points is large (above ½ mi) so that top-loaded haul units provide an economic advantage.
3. When material size is small on the average (up to 24 in. for the larger wheels and to 15 in. for smaller units), and when relatively few massive rocks are anticipated.
4. When the required rate of moving material from loading to dumping point is relatively high (above 1,000 bcy per hr) and when total job size warrants bringing in a single-purpose loading unit. The latter breakpoint will depend upon the wheel-loader size, but it will be in the 100,000-bcy range for the smaller contractor-type units.
5. When it is desirable to blend materials in order to distribute moisture or different materials evenly.
6. When particle size can be influenced by loading technique and is important to ease of compaction. (This will apply to materials such as gumbos, which would be discharged in large chunks by a scraper.)
7. When reclaiming materials, such as salt, coal, etc., which tend to agglomerate in the stockpile and to resist flow into under-pile conveyor systems.
8. When high water tables or job economics preclude the installation of under-pile conveyor systems.
9. When moving material into processing in a continuous flow, rather than by an intermittent loading process, is desirable.

DEVELOPMENT

The wheel excavator had its origin in Europe in the 1920s. It was used in the stripping and mining of brown coal, where the objective was the loading of material at a relatively high rate and at a minimal unit cost. This was accomplished by mounting a cutting wheel with digging buckets at the end of a boom which provided both horizontal and vertical traverse to the cutting elements. The buckets discharged onto a digging ladder conveyor which in turn discharged onto a stacker conveyor, both of which were mounted on the excavator chassis. The same general approach was followed in all the European wheel excavators as well as in most of the excavators which were exported to other countries or built under license in other parts of the world.

This type of wheel excavator, sometimes called the Kolbe wheel after its German inventor, is manufactured by a number of German companies. Among them are Friedkrupp GMBH Maschine und Stahlbau, Rheinhausen; Weserhutte Otto Wolfe GMBH in Bad Oeynhausen; and Demag-Lauchhammer of Dusseldorf. Rights to manufacture and sell the Demag-Lauchhammer machine in the United States have been granted to the Wellman Engineering Company, an affiliate of McDowell Company, Inc., of Cleveland, Ohio (see Fig. 11-7). Similar machines have been designed, manufactured, and sold in the United States by the Bucyrus-Erie Company of Milwaukee (see Fig. 11-8) and by Mechanical Excavators, Inc., of Los Angeles. Both the German and American machines will be described in more detail in later paragraphs.

Fig. 11-7 Bucket wheel excavator used in stripping brown coal. *(Demag-Lauchhammer and McDowell-Wellman Engineering Co.)*

Figure 11-9 illustrates a somewhat different approach which has been followed in the design of wheel excavators by Barber-Greene Company of Aurora, Ill. A digging wheel, complete with buckets, teeth, etc., is mounted centrally in a structural-steel main frame and suspended on a set of three crawlers or wheels. The elevation and slope of the wheel relative to the plane of the suspension are adjustable hydraulically. The digging elements discharge material onto a conveyor which is rigidly

Fig. 11-8 Bucyrus-Erie model 1054 WX wheel excavator.

Fig. 11-9 Barber-Greene model XL-50 excavator.

mounted inside the wheel and perpendicular to its plane. This conveyor discharges onto a second conveyor which is colinear with the first but can be adjusted vertically to accommodate hauling units with varying heights. A third conveyor, again colinear with the other two, can be used to guide the flow of material to side-by-side haul units without halting the flow of material.

The wheel approach is particularly useful in loading materials which tend to agglomerate in the stockpile and to resist flow into under-pile conveyor systems. Examples of these materials are salt, fertilizer, chemicals, coal, etc. For materials with a low angle of repose, a unit very similar to the wheel excavator is used (see Fig. 11-10). For materials with a high angle of repose, which tend to be undercut with consequent danger to machine and operator, a boom with a rotary cutter element is attached to the digging wheel. This increases capacity by guiding material flow down the pile and directly onto the conveyor, thus reducing the amount of material which must be lifted and discharged onto the belt by the bucket wheel.

TYPES

The two general types of wheel excavators are described below. To simplify this discussion, the two types are designated as "boom-mounted wheel excavator" and "integral wheel excavator" respectively.

Fig. 11-10 Barber-Greene double-wheel reclaimer handling iron-ore pellets.

Boom-mounted Wheel Excavator Units of this type can be further subdivided into two categories for ease of understanding. The first category includes the relatively small machine, designed for the use of the contractor, which is transportable as a unit. The second category includes the large machine which may either be assembled on the job from the ground up or created by adding components to a conventional large-shovel chassis.

Figure 11-11 shows a plan and an elevation of a small contractor machine. The digging wheel includes eight buckets and can be adjusted vertically by means of hydraulic rams. It is mounted on a boom, as is the conveyor belt onto which it discharges. This boom, complete with digging wheel and conveyor, can swing through a complete 360° circle by rotating on the crawler-mounted undercarriage.

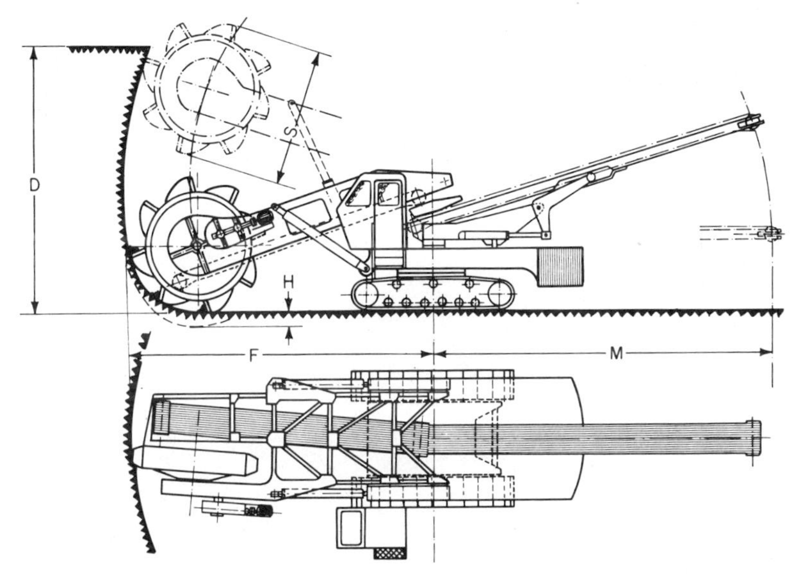

Fig. 11-11 Weserhutte boom-mounted wheel excavator (plan and elevation views).

A second conveyor receives material from the first conveyor. Although the discharge point of the first conveyor is fixed relative to the loading point of the second conveyor, the second conveyor can be hydraulically slewed through an arc of 180° and can be raised and lowered hydraulically to accommodate receiving vehicles of varying heights.

Power to operate the digging wheel, each of the conveyors, the crawlers, and to rotate the superstructure is furnished by individual electric motors. These motors, in turn, are driven by the output of a motor-generator set mounted on the chassis or by centrally generated power which is delivered to the machine through cables. The single operator controls all functions of the excavator.

The machine just described is typical of the several small, contractor-type, boom-mounted wheel excavators (see also Fig. 11-12). On these machines, the material which discharges from the wheel is customarily guided by a chute to the receiving conveyor belt. One major variation is to attach the buckets to a cone in such a way that the material discharges directly to the belt and in the direction of travel of the belt (see Fig. 11-13). The machine may also be mounted on a rubber-tired chassis instead of on crawlers.

Although larger versions of the boom-mounted wheel excavators are massive machines (see Fig. 11-14), they operate in a manner similar to the smaller units. The large machines are often used in coal-stripping operations, frequently in conjunction

Fig. 11-12 Demag-Lauchhammer type 430 bucket wheel excavator stripping overburden in phosphate mine.

Fig. 11-13 Mechanical Excavator, Inc., boom-mounted wheel excavator.

with a shovel. The wheel then strips the softer, unstable, upper overburden, and the shovel concentrates on the harder material underneath and just above the coal seam. Advantages claimed for this approach are as follows:

1. Overstripping of the softer, unstable, upper overburden by the wheel excavator develops a comparatively wide bench beyond the shovel high wall. This eliminates most of the sliding hazard of the excavator high wall.
2. Spoil-bank sliding tendencies are greatly reduced when the wheel excavator casts the softer, more unstable material on top of the harder shovel spoil and places it one cut or more back.
3. A shovel and wheel excavator combination results in a wider pit than is attainable with either a shovel or a large dragline working alone.
4. Use of a wheel excavator permits the shovel to work more efficiently against lower overburden depths. When the shovel must work close to its maximum reach and height, more care is necessary in spotting to dump and usually greater swing

angles are involved, all resulting in longer cycle times and reduced shovel production rates.

5. Where the more difficult shovel spoil requires shooting, increased blasthole drilling efficiency can be expected when the wheel removes the easier upper overburden. The shelf created by the wheel permits drills to start vertical blastholes directly at the surface of the more difficult layer. This also reduces the problem of soft material at the top sloughing into holes already drilled.

6. Wheel-excavator operation will facilitate restoration of the soil for agricultural use. The topsoil and upper burden are cast on top of the spoil bank, and the distributed pattern in which the spoil is deposited leaves lower peaks and smaller ridges to be leveled off.

The large wheel excavators are also used on major earthmoving projects such as earth-fill dams. The operating crew of such a machine typically consists of two operators and two oilers. One operator in a cab on the central tower controls all functions except truck loading. The second operator is in a cab on the stacking ladder and controls the truck-loading function. In coal-stripping operations, the operating crew can be reduced to two men—an operator and an oiler.

Fig. 11-14 Bucyrus-Erie model 684 WX boom-mounted wheel excavator.

Integral Wheel Excavator Two elevation views of this type of wheel excavator are shown in Fig. 11-15. The digging wheel is mounted integrally with the structural-steel main frame, which is in turn mounted on three hydraulically driven crawlers. The elevation and transverse slope of the digging elements are controlled by hydraulically adjusting the relative elevation of two of the three crawlers. The digging wheel is driven from the main power unit, through a torque converter and mechanical power train, to a pin-and-segment type drive unit on the wheel rims. Forward speed of the machine is infinitely variable as a result of the hydraulic drive to the crawlers, and it can thus be adjusted to correspond exactly with the rate at which the material is excavated. The single operator in the cab is in complete control of the excavator.

The buckets discharge material to the first belt conveyor, which is rigidly mounted relative to the wheel. The second conveyor is vertically adjustable to compensate for variable vehicle heights. The third conveyor, called the "transfer conveyor," is used to permit instantaneous transfer of material flow between two loading stations. This permits truly continuous loading without the need to interrupt flow for positioning haul units. All three conveyors are hydraulically driven and positioned.

A modified version of this type of wheel excavator, which has been adapted to reclaim materials from stockpiles, is shown in Fig. 11-16. Note the boom with rotating cutters which is used to prevent undercutting of the pile and caving and which guides the material flow directly to the conveyor belt to minimize the load on the

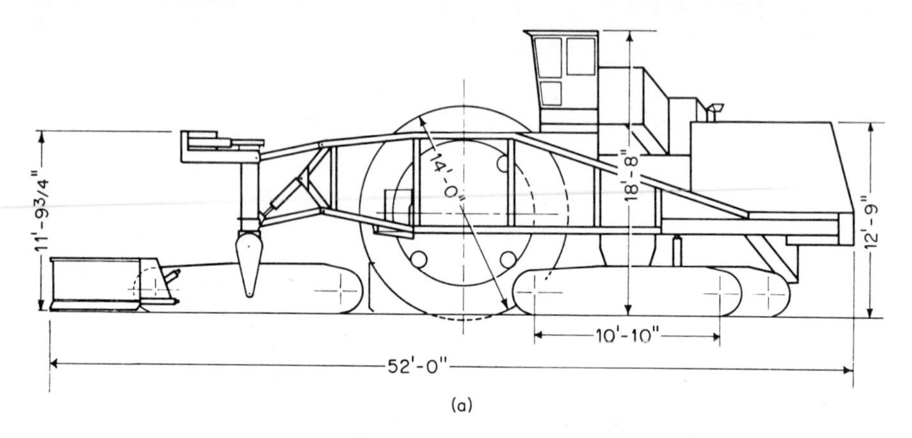

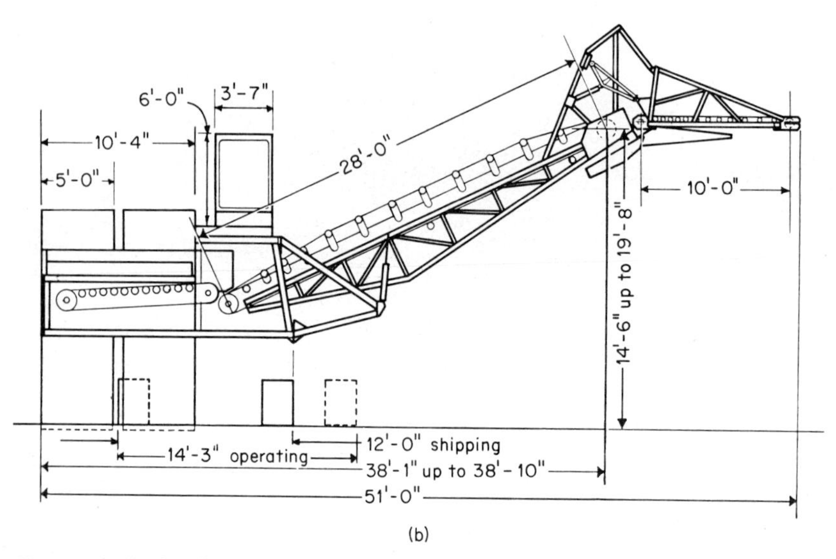

Fig. 11-15 Barber-Greene model XL-50 integral wheel excavator. (*a*) Elevation view No. 1; (*b*) elevation view No. 2.

digging wheel. The operation of this machine is otherwise very similar to that of the integral wheel excavator.

PERFORMANCE CHARACTERISTICS

Boom-mounted Wheel Excavator Specifications for the Weserhutte wheel excavators, contractor-type machines, are as listed in Table 11-7. The Mechanical Excavator machines have specifications as listed in Table 11-8. Models 1000, 2000, and 3000 are larger than the normal contractor-type unit, but their specifications are included for convenient reference.

The larger boom-mounted wheel excavators are special-purpose units, and their specifications are not standardized. The Bucyrus-Erie wheel excavator which was used on San Luis Dam, Los Banos, Calif., had specifications as in Table 11-9.

Fig. 11-16 Barber-Greene model XA-50 single-wheel reclaimer.

Integral Wheel Excavator Specifications for the Model XL-50 integral wheel excavator manufactured by Barber-Greene Company are supplied in Table 11-10.

PRODUCTION ESTIMATING

The customary rating procedure for conventional excavating equipment, such as shovels and draglines, is related to the bucket size in cubic yards. Capacity ratings then incorporate the number of digging cycles per unit of time to estimate the output in cubic yards per hour.

The capacity rating for a wheel excavator is somewhat different, since the theoretical maximum capacity is directly proportional to the number and size of buckets on the wheel and to the angular velocity of the wheel. This theoretical maximum must be discounted to compensate for the "diggability" of the material, different degrees of bucket filling, and the horsepower available to the wheel. Table 11-11 lists the peak loading rates which have been projected by the Barber-Greene Company for its XL-50 wheel excavator.

The actual output from an equipment spread which is built around a wheel excavator is dependent upon many factors. It can be expected that the spread production rate will be significantly less than the peak loading rate for the excavator. The

TABLE 11-7 Weserhutte Wheel Excavator Specifications

Machine	Main measurements*					Number and size of buckets	Approximate output in heavy material, cu m/hr	Approximate output in light material, cu m/hr	Approximate output of installed motor, kw
	D	F	H	M	S				
AR 45 5.5/0.2	5.5 m (18 ft)	6.7 m (22 ft)	0.2 m (8 in.)	6.8 m (22⅓ ft)	2.75 m (9 ft)	8–45 liters	100	150	40
AR 100 7.5/0.3	7.5 m (24½ ft)	9 m (29½ ft)	0.3 m (1 ft)	9 m (29½ ft)	3.7 m (12 ft)	8–100 liters	200	300	90
AR 200 7.5/0.3	8.5 m (28 ft)	10.3 m (33¾ ft)	0.3 m (1 ft)	10.5 m (34½ ft)	4.3 m (14 ft)	8–200 liters	400	600	150

* See Fig. 11-11 for locations of indicated measurements. All measurements in parentheses are approximate.

TABLE 11-8 Mechanical Excavator, Inc., Specifications for Wheel Excavators

Specifications	Model number					
	300	500	700	1,000	2,000	3,000
Overall length	45 ft 0 in.	45 ft 0 in.	45 ft 3 in.	79 ft 0 in.	84 ft 0 in.	90 ft 0 in.
Overall width	11 ft 8 in.	11 ft 8 in.	12 ft 8 in.	14 ft 0 in.	17 ft 0 in.	25 ft 0 in.
Overall height	11 ft 8 in.	11 ft 8 in.	13 ft 6 in.	28 ft 0 in.	29 ft 0 in.	31 ft 0 in.
Wheel diameter	11 ft 6 in.	11 ft 6 in.	11 ft 6 in.	19 ft 6 in.	20 ft 0 in.	25 ft 0 in.
Number of buckets	6	6	6	6	6	6
Normal wheel speed	12 rpm	12 rpm	12 rpm	8 rpm	9 rpm	8 rpm
Swing speed at cutting radius	0–90 fpm	0–90 fpm	0–90 fpm	0–90 fpm	0–90 fpm	0–90 fpm
Max theoretical capacity	480 cu yd/hr	720 cu yd/hr	1,080 cu yd/hr	2,100 cu yd/hr	3,600 cu yd/hr	4,800 cu yd/hr
Ladder belt:						
Width	30 in.	30 in.	36 in.	42 in.	60 in.	72 in.
Length	16 ft 0 in.	16 ft 0 in.	16 ft 0 in.	31 ft 6 in.	31 ft 6 in.	33 ft 6 in.
Speed	350 fpm	350 fpm	450 fpm	450 fpm	500 fpm	500 fpm
Discharge belt:						
Width	30 in.	30 in.	36 in.	42 in.	60 in.	72 in.
Length	26 ft 0 in.	26 ft 0 in.	26 ft 0 in.	36 in. 9 ft	40 ft 0 in.	40 ft 0 in.
Speed	400 fpm	500 fpm	500 fpm	500 fpm	550 fpm	550 fpm
Engine (wheel drive)	GM6-V-71	GM8-V-71	GM8-V-71	GM8-V-71	GM12-V-71	GM16-V-71
Engine (generator drive)	....	....	....	GM8-V-71	GM12-V-71	GM16-V-71
Drive motors:						
Ladder belt	15 hp	25 hp	35 hp	50 hp	100 hp	150 hp
Discharge belt	15 hp	25 hp	35 hp	50 hp	100 hp	150 hp
Propel speed	0–20 fpm	0–20 fpm	0–88 fpm	0–88 fpm	0–88 fpm	0–88 fpm
Cutting radius	10 ft 0 in.	11 ft 0 in.	20 ft 9 in.	39 ft 0 in.	40 ft 0 in.	45 ft 0 in.
Cutting height	20 ft 0 in.	20 ft 6 in.	11 ft 6 in.	20 ft 0 in.	20 ft 0 in.	25 ft 0 in.
Total weight (approx.)	41 tons	41 tons	45 tons	115 tons	160 tons	250 tons

TABLE 11-9 Specifications for Bucyrus-Erie Model 684 WX Wheel Excavator

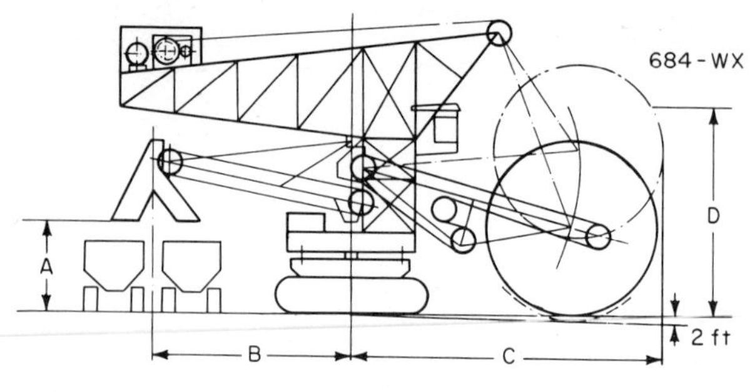

General

Number of buckets	10
Rated bucket capacity	2½ cu yd
Wheel diameter over lips	30 ft 0 in.
Working weight, approximate	1,625,000 lb
Ballast required (furnished by purchaser)	125,000 lb
Width of crawler treads	72 in.
Overall length of mounting	32 ft 3⅜ in.
Total effective bearing area	345 sq ft
Maximum stacker swing	180°
Width of belts	84 in.
Length of digging ladder	38 ft 0 in.

Working ranges

A: Dumping height, approximate	14 ft 4 in.
B: Dumping radius, approximate	63 ft 0 in.
C: Cutting radius, approximate	60 ft 0 in.
D: Cutting height, approximate	40 ft 0 in.

Electrical equipment

Wheel drive motor (blown)	(1)	375 hp at 230 volts dc 75°C continuous 750 hp at 460 volts dc 75°C continuous
Propel motor	(1)	300 hp at 230 volts dc 75°C intermittent 600 hp at 460 volts dc 75°C intermittent
Swing motors	(2)	25 hp at 230 volts dc 75°C continuous 50 hp at 460 volts dc 75°C continuous
Digging ladder conveyor	(1)	300 hp 440-3-60 ac 55°C continuous
Digging ladder hoist motor (blown)	(1)	187½ hp at 230 volts dc 75°C continuous 375 hp at 460 volts dc 75°C continuous
Stacker conveyor motor	(1)	300 hp 440-3-60 ac 55°C continuous

TABLE 11-10 Specifications for Barber-Greene Model XL-50 Wheel Excavator

Capacity	1,000–3,000 bcy/hr
Overall height	13 ft 0 in.
Overall length	43 ft 0 in.
Overall width (with conveyor)	36 ft 0 in.
Overall width (less conveyor)	17 ft 6 in.
Minimum turning radius	42 ft 0 in.
Wheel diameter	14 ft 0 in.
Wheel speed	6–11 rpm
Wheel width (both)	10 ft 6 in.
Bucket capacity (each)	½ cu yd
Conveyor	60 in. × 25 ft 0 in.
Conveyor speed	0–600 fpm
Discharge height (with flop gate)	8 ft–12 ft 6 in.
Machine travel speed	0–2½ mph
Power unit	Model 343 (403 hp) Caterpillar diesel engine with torque converter

TABLE 11-11 Projected Peak Loading Rates for Barber-Greene Model XL-50 Wheel Excavator

Material	Expected peak loading rates in bcy/hr (where horsepower is the limiting factor)*			Theoretical peak bucket-capacity loading rates in bcy/hr (if horsepower were not the limiting factor)†						
					24-bucket pattern			18-bucket pattern		
	Loading-rate range	Tooth pattern‡	Bucket pattern§	Percent of swell	1st gear	2d gear	3d gear	1st gear	2d gear	3d gear
Loam, dry	2,400–3,000	1 or 2	24	35	3,240	3,940	4,640	2,430	2,950	3,460
Wet sand	2,500–3,300	1 or 2	24	18	3,720	4,500	5,300	2,780	3,370	3,970
Moist sand	2,700–3,600	1 or 2	24	20	3,650	4,430	5,220	2,730	3,310	3,900
Gravel $\frac{1}{4} \times 4$ in.	1,800–2,800	1 or 2	24	25	3,510	4,250	5,010	2,620	3,180	3,750
Sand and dry clay	2,300–3,300	1	24	42	3,070	3,720	4,380	2,300	2,780	3,280
Dry clay and gravel	1,400–2,700	1	24	42	3,070	3,720	4,380	2,300	2,780	3,280
Clay, tough moist (heavy)	1,400–2,600	2	18 or 24	45	3,020	3,670	4,320	2,260	2,740	3,240
Clay, abrasive, hard	1,100–1,900	2	18	48	2,940	3,560	4,200	2,130	2,670	3,140
Caliche	1,000–1,500	1	18 or 24	60	2,720	3,300	3,880	2,030	2,470	2,910
Soft shale	1,300–2,100	1	18	60	2,720	3,300	3,880	2,030	2,470	2,910
Soft shale and rock	1,200–2,300	1	18	60	2,720	3,300	3,880	2,030	2,470	2,910
Soft sand rock, lightly shot	1,500–2,000	2	18 or 24	60	2,720	3,300	3,880	2,030	2,470	2,910
Soft sand rock, well shot	1,700–2,300	2	18	65	2,630	3,190	3,760	1,970	2,390	2,810
Bituminous coal	1,300–1,800	1	18 or 24	48	2,940	3,560	4,200	2,130	2,670	3,140

* Horsepower loading rate: This rate is based on the limitations of horsepower, wheel torque, or tooth penetration. Rates are based on optimum depth of cut, 10 to 12 ft.

† Bucket-capacity loading rate: This shows the theoretical loading rates obtainable with material swell percentages if horsepower were not the limiting factor. The bucket capacity is rarely the loading rate limitation except in sticky material, where these bucket-capacity loading rates are subject to reduction due to bucket buildup.

‡ Tooth pattern: Type 2 with a wider tooth spacing provides penetration for digging hard materials. Type 1 with closer tooth spacing is used to protect against bucket wear where material is less difficult to penetrate.

§ Bucket pattern: Maximum rock size discharged from the bucket is generally limited to 9 in. in diameter on the 24-bucket pattern and 15 in. in diameter on the 18-bucket pattern. The 18-bucket pattern is sometimes preferable in cemented materials.

following equation is suggested as a basis for estimating the production rate on a given job:

Production rate = production efficiency × expected peak loading rate

where production efficiency = $A \times B \times C \times D \times E$

- A = haul unit availability = (1 − time waiting for haul units/total shift time)
- B = excavator turnaround efficiency = (1 − time spent in runback/total shift time)
- C = work schedule efficiency is a general factor reflecting contractor's job organization, maintenance, etc.
- D = excavator availability = (1 − time lost due to downtime during shift/total shift time)
- E = rock delay factor = (1 − time spent bypassing occasional oversize material/total shift time)

The values of the factors which affect overall job efficiency will vary and must be estimated for each job by each contractor.

It is important to keep the following points in mind when considering the production rate of a wheel excavator or reclaimer:

1. The total cost of moving material at a given rate between two points is the overriding consideration. This includes the cost of loading, hauling, and dumping.
2. The wheel excavator, when used for top-loading an economical hauling device such as a bottom-dump truck, can frequently reduce the total cost of moving materials below that incurred by a combined loading-hauling unit such as a scraper. In such cases it is misleading to evaluate the competitive position of the wheel by considering only its unit loading cost.
3. Where the wheel is competing directly against another loading device such as a belt loader and dozers or a front-end loader, one may more reasonably compare direct loading costs of the competitive units.

SELECTION OF UNIT

The first step in evaluating the suitability of a unit for a given project is, of course, to estimate the overall production rates desired for the project. A wheel excavator

EXAMPLE

Hourly Owning and Operating Costs for a Wheel Loader

LOADING RATE: 2,800 bcy/hr

PRODUCTION RATE: 2,800 loading rate × 66% production efficiency = 1,840 bcy/hr

OWNING AND OPERATING COST:

		Per production hour
Owning cost		$33.30
Operating cost per loading hour		
Fuel	$ 3.60	
Lube and filters	0.61	
Belt and flashing liners	1.20	
Teeth and buckets	0.99	
Rebuilding and major repairs	18.90	
Total per loading hour	$25.30	
Production efficiency 66% × total $25.30		16.70
Labor cost:		
Operator		5.28
Spotter		5.28
Maintenance $\left(\frac{3.14 \text{ man-hours} \times \$8.61 \text{ labor rate}}{10 \text{ production hours per shift}}\right)$		2.70
Total cost per production hour		$63.26

COST PER BCY:

$\frac{\$63.26 \text{ cost per production hour}}{1{,}840 \text{ bcy/hr production rate}}$ .... $ 0.035 per bcy with an adequate loading rate can then be selected, keeping in mind that the peak loading capability of the wheel must be suitably discounted on the basis of anticipated production efficiency if the specified job production rates are to be met.

Upon selecting the wheel excavator to be used, the appropriate number and type of haul units can also be selected. This step is particularly important when the cost of moving material by the wheel-excavator technique is to be compared with the use of a combined loading-hauling unit such as a scraper. The unit cost of loading material with the selected unit may be estimated by the procedure indicated in the example on page 11-30. The values used are only for illustrative purposes, actual values for a given unit can readily be substituted into this format. Overall costs of a wheel-excavator–haul-unit spread may be estimated by applying a similar format to the cost of owning and operating the haul units and then adding loading and hauling costs.

C. Belt Loaders

WHY CONSIDER BELT LOADERS?

This question can be answered in one word—production! Belt loaders, or top-loaders, as they are sometimes called, provide maximum production at a minimum cost with capacities up to 1½ cu yd per sec (see Fig. 11-17).

Belt loaders, like other earthmoving tools, are limited in their realm of application. Many types of earthmovers are available, each of which has particular features which make it a desirable tool to utilize on specific types of jobs. Belt-loader manufacturers claim several advantages for their units in addition to high production rates.

Fig. 11-17 40-ft by 72-in. portable belt loader loading glacial moraine into rear-dump wagons. (*Kolman Division, Athey Products Corporation.*)

First, the "pouring" of material eliminates body damage to and maintenance problems with haul units, difficulties which normally accompany the impact and shock of "dump" loading (see Fig. 11-18).

Second, conventional scraper loading will often produce voids in the load and at corners, and little heap will be realized. The top-loading of scrapers ensures the elimination of voids and guarantees heaped loads with every cycle (see Fig. 11-19).

In addition, the utilization of a mixed fleet of haul units which includes scrapers will allow continuing production with the scrapers whenever the belt loader must be moved to a new setting.

Finally, and probably most important, is the reduction in cycle time. To fully utilize the capabilities of high-speed rubber-tired haul units, the utmost attention must be paid to the haul road between the cut and fill. Ideally, grades are kept to a minimum, especially on the haul, with the best situation being one in which a slight downgrade can be utilized. If hard-surfaced roads are not practical, a motor grader should be used to maintain the road. By placing the belt loader at the bottom of the cut, adverse grades and the time lost when scrapers climb them are eliminated. Also,

Fig. 11-18 Top loading eliminates body damage and maintenance problems.

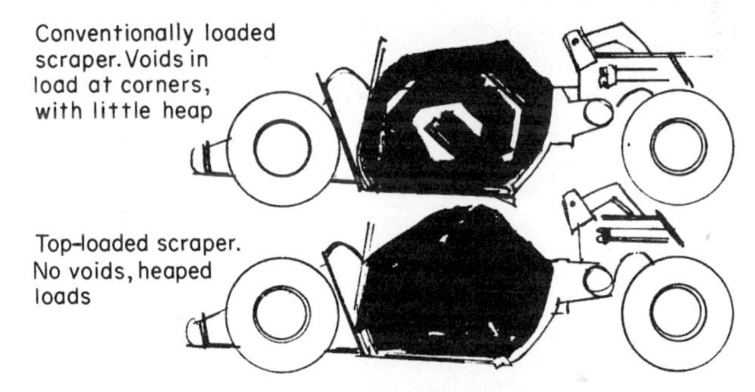

Fig. 11-19 Top loading eliminates voids and ensures heaped loads every trip.

Fig. 11-20 42-ft by 60-in. portable belt loader loading dirt and rock. (*Barber-Greene Company.*)

Fig. 11-21 50-ft by 60-in. portable belt loader with 5- by 12-ft single-deck vibrating grizzly. (*Kolman Division, Athey Products Corporation.*)

the replacement of worn parts on self-loading units is reduced substantially. Tire life is increased because of the prepared haul roads.

Since most belt-loader manufacturers claim the ability to top-load virtually any material—dirt (see Fig. 11-20); boulder-strewn soil (see Figs. 11-17 and 11-21); shot rock (see Fig. 11-22); sand, gravel, clay, gumbo, glacial moraine (see Fig. 11-23); etc.

Fig. 11-22 41-ft by 48-in. portable rock loader with 5- by 10-ft single-deck vibrating screen. (*Pioneer Division, Portec Incorporated.*)

—more flexibility is available for the various job applications. Models range in weight from 38,000 to 80,000 lb and portability is emphasized by all manufacturers. Two types of undercarriage are available and will be discussed subsequently.

When material size restrictions are involved on a particular job, the belt loader offers another advantage. By mounting a vibrating screen or vibrating grizzly at the discharge end of the loader, oversize material is eliminated at the loading point. Consequently, the costly rehandling of rejected or undesirable oversize material at the fill becomes unnecessary. (See Figs. 11-21 and 11-22.)

Fig. 11-23 25-ft by 60-in. belt loaders, 100-ft by 48-in. shuttle conveyors, and 15,000 ft by 66 in. stationary conveyor moving 60 million yd of glacial moraine from deposit to dam site. (*Kolman Division, Athey Products Corporation.*)

Experience has shown bulldozers to be the most economical means of charging belt loaders. Three to six dozers, utilizing gravity to its maximum, can keep material flowing continuously to the belt-loader feed track.

Since their inception, belt loaders have taken a definite position in the field of high-capacity earthmoving tools. When applicable, their use will continue to provide contractors with the cost-per-yard savings needed in the complex and competitive field of heavy construction.

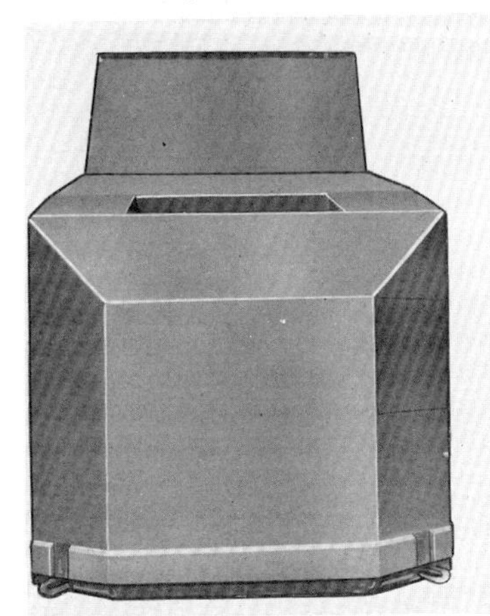

Fig. 11-24 Belt loader–dozer trap incorporating a reciprocating-plate feeder. (*Kolman Division, Athey Products Corporation.*)

Fig. 11-25 Direct-feed belt loader–dozer trap with adjustable backboard and trap "vari-plates" for variable trap opening. (*Kolman Division, Athey Products Corporation.*)

EVOLUTION OF BELT LOADERS

By 1957, many contractors had realized the advantages of a dozer-fed trap and belt conveyor for top-loading, and they were building their own units. Examination of the market potential and of the "homemade" models already in use encouraged equipment manufacturers to enter the field.

In 1959, the first production model was introduced by Kolman Manufacturing Company of Sioux Falls, S.D. This 60-in. belt loader was tested on a 6½ million cu yd job where it reduced the previous number of scrapers by 75 percent without loss of production. Further developments by all manufacturers have allowed the belt loader to take its place with the other available and necessary earthmoving tools.

Understandably, each manufacturer claims that his particular model has unique advantages. (See Table 11-12.) However, in general, there are two basic types or designs. One incorporates a feeder built into a steel dozer-trap enclosure at the tail end and is used primarily for free-flowing materials (see Fig. 11-24). The other design also has a steel dozer trap at the tail end but does not include a feeder (see Fig. 11-25). The latter trap design allows direct feed of material onto the conveyor belt and is used primarily for sticky, hard-to-handle material.

The conveyor frame design on both types is quite similar. All conveyors with the exception of the Pioneer 4841 Rock Loader (which will be reviewed below) have a heavy-duty lattice-frame design which is reinforced for the rugged applications to which these units are subjected. Although most manufacturers offer 50- and 60-ft lengths, the 50-ft model predominates and can be set up to top-load virtually any size haul unit with little or no excavation.

The steel dozer-trap housing on both types of units is also constructed similarly by all manufacturers. The trap is enclosed with heavy steel plate and is rigidly crossbraced for maximum support. The feeder-dozer trap combination incorporates a reciprocating-plate feeder built within the trap structure. Material enters the feed track through an opening on top of the trap and, in turn, is fed at a controlled rate to the conveyor belt. The feeder, normally driven from the conveyor tail shaft, controls the material fed to the conveyor belt and thus minimizes belt wear.* Capacity of the feeder is controlled through an adjustable feed-control gate. Most manufacturers provide a self-relieving type of gate which can be forced open by oversize material. Once the oversize material has passed the feeder, this gate will drop to its preset position (see Figs. 11-26 and 11-27). In addition, some manufacturers provide adjustment to the feeder pitch and stroke length to compensate for changes in material characteristics. One manufacturer offers a compound-action feeder which, according to his claims, will minimize the chance of "bridging." Feeder clutches are usually available to allow disengaging the feeder drive while the main conveyor belt continues to operate. Most feeders have adjustable machined edges for a close fit that will minimize leakage of fine material over the sides of the feeder plate.

Some manufacturers provide replaceable feeder wear plates. These make it unnecessary to replace the entire feeder plate when wear occurs.

It is not feasible to employ a feeder when handling extremely sticky types of material which will not flow. The most suitable trap design for this type of material allows direct feed to the conveyor belt. An extra-long slotted opening is provided on top of the trap, with a minimal drop distance onto the conveyor belt. This design reduces the chance of material bridging over the trap opening and allows continuous material flow. Since many semisticky or free-flowing materials could flood this type of trap, some manufacturers offer a variable-opening trap which can be adjusted according to the predominant material characteristics. (See Fig. 11-25.)

Although many manufacturers sanction the use of variable-opening traps in shot rock or material containing a percentage of rock, wear life of the belt and other components is lessened because of the direct belt feed. In addition, the controlled rate of feed realized from the model with a feeder is also lost.

A third type of belt loader has been developed by Pioneer Division, Portec Incorporated, and is designated as the Pioneer 4841 Rock Loader (see Fig. 11-22). This

* See Sec. 15, Belt Conveyors.

TABLE 11-12 Belt-loader Specifications*

BELT LOADER SPECIFICATIONS

	Manufacturer and Model		Belt Width	Conveyor Length	Max. Rated Capacity YPH	Belt Speed FPM	Undercarriage Wheel or Skid	Standard Diesel Power Unit	Horsepower Net Rating	Idler Specifications: Roll Dia., Number of Rolls, Trough Angle	Feeder: RPF = Reciprocating Plate Feeder, DF = Direct Feed	Discharge Height	Loading Clearance (Approx.)	Overall Height	Overall and Towing Width	Overall Length	Towing Height	Towing Length	Approx. Weight	
1	BARBER–GREENE	PL–90	48"	42'	2,000	360	W	IHC	130	6" dia., 3–roll, 20°	RPF	13'–10"	10'–0"	18'–6"	11'–0"	49'–0"	13'–6"	46'–6"	48,000	1
2			60"		3,600			GMC	252			14'–1"		19'–6"	12'–0"	50'–1"	14'–6"	47'–7"	64,500	2
3		PM–90	48"	51'	2,000						DF	17'–3"	12'–0"	19'–6"	10'–6"	56'–8"	14'–6"	54'–0"	56,000	3
4			60"		3,600							17'–6"			11'–6"				66,000	4
5	HEWITT–ROBINS	450	48"	50'	2,400	350	W	GMC	170	7" dia., 5–roll, 40°	RPF	15'–9"	12'–0"	18'–6"	10'–0"	53'–6"	13'–0"	51'–0"	57,000	5
6		460		60'								19'–0"	15'–6"	21'–0"		63'–6"	13'–6"	61'–0"	62,900	6
7		550	60"	50'	3,600				227			15'–9"	12'–0"	18'–6"	11'–0"	53'–6"	13'–0"	51'–0"	60,850	7
8		560		60'								19'–0"	15'–6"	21'–0"		63'–6"	13'–6"	61'–0"	69,100	8
9	KO–CAL	4845–R	48"	45'	2,800	370	W	GMC	105	6" dia., 3–roll, 20°	DF	13'–8"	11'–0"	15'–0"	9'–4"	48'–0"	14'–0"	47'–10"	38,000	9
10		4860–R		60'					154			20'–2"	17'–6"	23'–3"		61'–6"	13'–6"	54'–6"	51,300	10
11		6045–R	60"	45'	3,600		S	A–C	220			14'–6"	12'–4"	15'–10"	10'–0"	49'–0"	K–D	K–D	63,920	11
12		6050–R		50'								16'–0"	13'–6"	19'–6"		55'–0"			67,220	12
13		6060–R		60'								19'–0"	16'–6"	22'–6"		64'–0"			73,820	13
14		4845–S	48"	45'	2,800	220 or 370	W	GMC	105		RPF	12'–5"	10'–11"	15'–0"	9'–4"	49'–0"	13'–6"	49'–0"	48,000	14
15		4860–S		60'					154			19'–0"	17'–6"	23'–0"		63'–6"		56'–3"	58,500	15
16		6045–S	60"	45'	3,600		S	A–C	220			14'–4"	12'–4"	18'–0"	10'–0"	50'–9"	K–D	K–D	76,250	16
17		6050–S		50'								16'–0"	13'–5"	19'–6"		56'–0"			81,350	17
18		6060–S		60'								18'–5"	16'–2"	23'–0"		65'–3"			86,000	18
19		6050		50'		370	W	GMC	154			16'–8"	13'–0"	21'–8"		65'–6"	13'–6"	49'–3"	45,000	19
20	KOLBERG	348–50	48"	50'	2,000	350	W	GMC	130	6" dia., 5–roll, 40°	RPF	13'–0"	10'–0"	19'–6"	10'–6"	58'–0"	13'–6"	58'–0"	49,300	20
21		348–60		60'					154			15'–3"	12'–0"	21'–6"		68'–0"		68'–0"	55,250	21
22		360–50	60"	50'	3,600				227			13'–0"	10'–0"	19'–6"	11'–6"	58'–0"		58'–0"	61,138	22
23		360–60		60'								15'–3"	12'–0"	21'–6"		68'–0"		68'–0"	68,958	23
24		448–50	48"	50'	2,000				154		DF	13'–0"	10'–0"	19'–6"	10'–6"	55'–6"		47'–6"	55,000	24
25		448–60		60'								15'–3"	12'–3"	21'–6"		65'–6"		57'–6"	60,000	25
26		460–50	60"	50'	3,600				227			13'–0"	10'–0"	19'–6"	11'–6"	55'–6"		47'–6"	61,000	26
27		460–60		60'								15'–3"	12'–3"	21'–6"		65'–6"		57'–6"	68,000	27
28		472–50	72"	50'	4,800							13'–8"	11'–0"	19'–10"	12'–0"	58'–0"		48'–4"	76,000	28
29	KOLMAN	303	48"	50'	2,400				130		RPF	13'–5"	10'–0"	15'–1"	10'–0"	55'–0"	12'–6"	52'–0"	59,500	29
30				60'					154			15'–7"	12'–6"	17'–3"		65'–0"		62'–0"	65,000	30
31			60"	50'	3,600				227			13'–5"	10'–0"	15'–1"	11'–0"	55'–0"		52'–0"	70,000	31
32				60'								15'–7"	12'–6"	17'–3"		65'–0"		62'–0"	76,000	32
33		404	48"	50'	2,400				154		DF	13'–5"	10'–6"	15'–1"	10'–0"	51'–0"		48'–0"	55,860	33
34				60'								15'–7"	12'–8"	17'–3"		61'–0"		58'–0"	63,550	34
35			60"	50'	3,600				227			13'–5"	10'–6"	15'–1"	11'–0"	51'–0"		48'–0"	58,965	35
36				60'								15'–7"	12'–8"	17'–3"		61'–0"		58'–0"	67,040	36
37			72"	50'	4,800							14'–0"	11'–0"	20'–0"	12'–0"	59'–0"	13'–6"	49'–6"	80,000	37
38	PIONEER	4841	48"	41'	2,000	0 to 80	W	Electric	100	APRON CONVEYOR	DF	17'–0"	15'–0"	26'–11"	10'–2"	46'–7"	13'–6"	44'–7"	74,950	38

Barber–Greene Company
400 N. Highland Avenue
Aurora, Illinois 60507

Hewitt–Robins
Division of Litton Industries
4603 W. Mitchell Street
Milwaukee, Wisconsin 53214

Ko–Cal – C. S. Johnson
Division of Koehring Company
P. O. Box 3067
Champaign, Illinois 61820

Kolberg Manufacturing Company
West 21st. Street
Yankton, South Dakota 57078

Kolman Division
Athey Products Corporation
5100 W. 12th Street
Sioux Falls, South Dakota 57101

Pioneer Division
Portec Incorporated
3200 Como Avenue S. E.
Minneapolis, Minnesota 55414

* From "Specifications for Your Files," *Construction Methods and Equipment*, November, 1968.

unit features a hydrostatic-drive apron feeder rather than a conveyor belt and reciprocating-plate feeder and, according to the manufacturer, is designed for loading and/or separating (when a screen or grizzly is added) extremely large, abrasive materials.

As previously mentioned, all belt-loader manufacturers claim portability as a feature of their units. Two basic types of undercarriages are available. The first type involves a skid plate mounted under the conveyor frame and dozer trap which makes it possible to skid the entire unit when relocation within the pit is necessary. Normally, when this type of undercarriage is furnished, units must be knocked down or put on a lowboy trailer for long hauls. The second and most common type of under-

Fig. 11-26 Self-relieving, adjustable, radial feed-control gate with remote hydraulic control. (*Kolman Division, Athey Products Corporation.*)

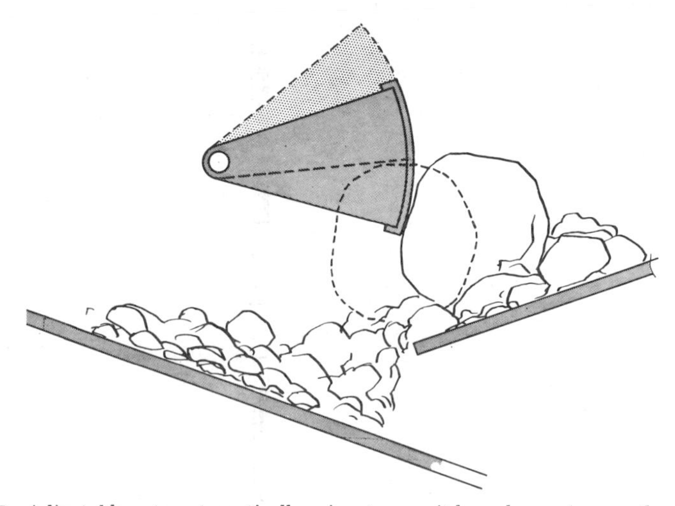

Fig. 11-27 Adjustable gate automatically raises to permit large lumps to pass through, then returns to preset position. (*Barber-Greene Company.*)

carriage involves mounting the entire unit on pneumatic tires. Movement can then be accomplished with relative ease by using dozers to push and pull the machine around the pit. The dozer trap is skid-mounted on most models, with pneumatic tires used for longer movements. A fifth-wheel hitch is available and can be bolted

Fig. 11-28 Portable belt loader with pneumatic tires, skid plate under dozer trap, and fifth wheel being towed onto the jobsite by a tractor. (*Kolman Division, Athey Products Corporation.*)

to the bottom of the trap skid plate for tractor towing over the road on long hauls (see Fig. 11-28).

Relocation within the pit can normally be accomplished by an experienced crew in less than 2 hr. It is necessary to plan hours in advance so that all is ready when the

Fig. 11-29 Pit relocation of 50-ft by 60-in. portable belt loader utilizing two dozers for skidding unit to new cut. (*Kolman Division, Athey Products Corporation.*)

actual move is started. First, one of the feed dozers will be utilized to open the new cut at the next location while the other dozers continue to feed the loader. Just prior to relocation, the dozers clean any material away from the top and sides of the trap. The unit can then be pulled out of the bank by two dozers, aided by a cable pull-out assembly if the trap is equipped with one. Dozer push-blocks protrude from some trap models, and two dozers can push against these to position the trap in the new cut. Dozers then begin to push material over the trap and resume normal operation (see Fig. 11-29).

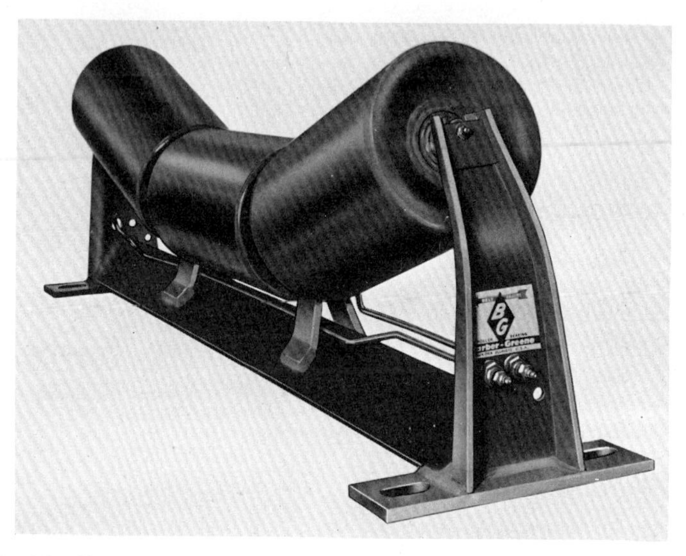

Fig. 11-30 6-in.-diameter grease-type, roller-bearing, three-roll, 20° troughing idler with ¼-in.-diameter shaft. (*Barber-Greene Company.*)

Two basic troughing-idler designs are utilized.* One is a three-roll idler with a 20° trough angle (Fig. 11-30). The other is a five-roll design with a 40° trough (Fig. 11-31). Both are available with steel rolls (Fig. 11-30) or rubber impact rolls (Fig. 11-31). The rubber impact rolls are normally used at the loading point, with the steel-roll type used on the remaining area. Troughing-idler spacing will vary from

Fig. 11-31 6¼-in.-diameter, sealed ball-bearing, five-roll, 40°, rubber-impact troughing idler with vulcanized rubber ribs and 1¼-in.-diameter shaft. (*Kolman Division, Athey Products Corporation.*)

approximately 12 in. at the loading point to 24 in. on the main conveyor. According to manufacturers of the five-roll troughing idler, additional load support is realized by using ten bearings to support the material load rather than using six bearings as in

* See Sec. 15, Belt Conveyors.

the conventional three-roll design. In addition, the deeper trough tends to minimize the chance of material spillage.

In addition to the hydrostatic drive described for the Pioneer unit, three basic drives are utilized by the other major manufacturers. The first, used exclusively by Barber-Greene Company on one of its two models, incorporates a totally enclosed torque-arm reducer mounted on the head shaft and driven by V belts from a diesel engine (Fig. 11-20).

The second type, used exclusively by Ko-Cal Division, Koehring Company, features a dual drive to both sides of the head pulley. Power is transmitted from the power unit through V belts to a propeller shaft and into a heavy-duty differential. Then, by chain, power is transmitted to both sides of the head pulley to drive the conveyor belt. An optional countershaft is provided and, when incorporated in the drive train, reduces the belt speed. According to Ko-Cal, the slower belt speed can be utilized when handling extremely large oversize material and, by simply bypassing the countershaft through sprockets provided, belt speed can be increased to provide a maximum rate of production in other materials.

The third type, introduced by Kolman Manufacturing Company and used by the remaining manufacturers, incorporates V belts from the power unit to a primary countershaft. Through a roller-chain drive enclosed in an oiltight chain case, power is transmitted to the secondary countershaft. Power from the secondary countershaft to the head-pulley shaft is through roller chain normally equipped with a drip-type oiler. The basic drive can be supplemented by adding a roller-chain drive from the extended secondary countershaft to the screen jackshaft. The additional drive can then be used to power a vibrating screen or grizzly. A head-pulley clutch mounted on the secondary countershaft allows disengagement of the drive to the head pulley, thus providing stopping and starting of the conveyor belt while the screen or grizzly (if included) continues to vibrate. In the two basic drive types which were previously described, however, an entirely separate drive would be required for these ancillary functions.

Each belt-loader manufacturer believes that his model has certain advantages or refinements which make it particularly desirable. It is, of course, up to the user to establish what machine he feels will best do his job, with flexibility for future applications. He should closely review all current models and accessories which are available before deciding on any given unit. For example, an enclosed operator's cab (available with air conditioning and heater) is sometimes advantageous. Some units are furnished with remote hydraulic controls which allow complete control of the conveyor, feeder, and feed-control gate from the operating platform in full view of the entire operation. Hydraulically controlled vibrating screen or grizzly deflector plates, horn signals, special belting, and operating lights with generator sets are a few more of the numerous optional accessories.

Some manufacturers, looking to the future, are developing reversing gates and reversible conveyors which can be mounted at the discharge end of the belt loader. These will make it possible to direct uninterrupted material flow into haulers parked side by side, eliminating lag time between haul units and increasing the effective hauling capacity. Also, where terrain restricts the use of any type of haul unit, belt loaders are being used as primary feed units to long conveyor systems. These minimize or completely eliminate the need for any haul whatsoever. (See Fig. 11-23.)

JOB ANALYSIS AND UNIT SELECTION

The key to belt-loader application is *volume*. Whenever the belt loader can be set up for loading numerous haulers, its quick-loading feature substantially reduces the unit cost of handling material. The following general job factors can be evaluated to determine whether or not a belt loader should be used on a given job.

First, as previously stated, the volume to be moved is most important. When belt loaders were first introduced, many earthmoving experts felt that a minimum of 1 million cu yd should be involved before considering a belt loader. This opinion is still held by many. However, many contractors now have these units as part of their equipment fleet and use them profitably on jobs containing as little as 250,000 cu yd. Consequently, the possible application of a belt loader to future jobs should also be considered if the original job is less than 1 million cu yd.

Second, material depths are equally important. In order for dozers to use gravity to its maximum, the greater the face height, the better. Obviously, if the depth of cut is less than 15 ft, scrapers can handle it far better than belt loaders. Most manufacturers believe that a minimum bank height of 20 ft is necessary to make top-loading economically feasible.

Third, although haul distances and haul roads were mentioned earlier, continuing study in these two areas is warranted. Cycle time on long hauls can be cut sharply by fast loading. Since the belt loader is positioned at the bottom of the cut, maintained haul roads and elimination of adverse grades will increase haul-unit efficiency and reduce tire wear. Also, the reach of belt loaders has paid off on short hauls where haulers or scrapers have trouble negotiating poor grades or footing directly at the cut.

Fourth, when job conditions dictate a closing distance between the dozers and belt-loader trap in excess of 500 ft, belt loading becomes uneconomical.

Although the size and volume of most jobs may make this point incidental, other equipment in the fleet should also be considered in the final evaluation. Belt loaders may require as many as six dozers, depending on the dozer size, pushing capacity, length of the push, output capacity of the belt loader, and capacity of the haul units. Conversely, they can load a mixed fleet of haulers ranging from on-highway trucks to scrapers. Consequently, haulers which might have otherwise been idle can be utilized.

Material size classification also deserves study. A vibrating screen or grizzly mounted on the belt-loader head end can scalp off oversize rock at the borrow area far more cheaply than conventional rock rakes can clear the fill area. This could be a determining factor in selecting a belt loader for a specific job.

Once the application is apparent, the proper type of loader can be selected on the basis of the information which has been presented in this section. The unit selection is primarily dependent on the material characteristics, and the unit's capacity should suffice to satisfy the production requirements. Although all 60-in. belt loaders have an average rated capacity of 1 lcy per sec, the lag time incurred during a haul-unit exchange makes the 3,600 lcy per hr capacity rating unrealistic. In order to get a more realistic hourly production rating, let us assume that loading of a 45-cu yd haul unit requires 45 sec and that a 15-sec time lag occurs during haul-unit exchange.* On this basis, the estimated productive rate would be 45 cy yd per min or 2,700 lcy per hr. By applying an efficiency factor of 80 percent (approximately equivalent to a 50-minute hour), a conservative hourly production rating of 2,160 lcy is obtained.

The number of haul units required will vary with the haul distance. Ideally, when a full hauler pulls out from under the belt loader, another empty hauler should be standing by to pull under. The following formula may be used to determine the required number of haul units·

$$\text{Number of haul units} = \frac{\text{total cycle time}}{\text{load time}}$$

The number of dozers required can be computed by utilizing the following formula:

$$\text{Number of dozers} = \frac{\text{total belt loader capacity (lcy/hr)}}{\text{capacity (lcy/hr) of a given dozer at a given pushing distance}}$$

The unit cost of loading material with the selected unit may be estimated by the procedure indicated in the following example. The values used in this example are only for illustrative purposes. Actual values for a given unit can readily be substituted into this format.

* See Sec. 13, Hauling Units.

EXAMPLE

Hourly Owning and Operating Costs for Kolman 50 ft × 60 in. Model 303 Portable Belt Loader

DATA:	
Selling price, f.o.b. factory	$ 65,000
Add freight	2,000
Total delivered selling price	67,000
Less belt price	4,500
Total depreciable value (less belt)	$ 62,500
Working conditions	Average
Depreciation:	
Years	6
Total hours	10,000
Hours per year	1,665
Estimated belt life, lcy	1,000,000
Estimated average hourly capacity, lcy/hr	2,160
HOURLY OWNERSHIP COSTS:	
Depreciation: $\frac{\$62{,}500 \text{ delivered price less belt}}{10{,}000 \text{ hr (6 yr at 1,665 hr/yr)}}$	$ 6.25
Interest: $\frac{8\%/\text{yr} \times \$67{,}000 \times 6 \text{ yr}}{10{,}000 \text{ hr}}$	3.22
Taxes and insurance: $\frac{2\%/\text{yr} \times \$67{,}000 \times 6 \text{ yr}}{10{,}000 \text{ hr}}$	0.81
Total hourly cost of ownership	$10.28
HOURLY OPERATING COSTS:	
Fuel (avg 10 gal/hr × 19 cents per gal)	$ 1.90
Lubricating oil, grease, and filters	0.24
Maintenance, minor repairs, and overhaul (avg estimated 5 percent of total depreciable value per year):	
$\frac{5\%/\text{yr} \times \$62{,}500}{1{,}665 \text{ hr/yr}}$	1.88
Belt: $\frac{\$4{,}500/\text{belt} \times 2{,}160 \text{ lcy/hr}}{1{,}000{,}000 \text{ lcy/belt}}$	9.72
Operator's wages per hour*	5.00
Total hourly cost of operation	$18.74
Total hourly cost of ownership and operation†	$29.02

* When material calls for selective dozer feeding or where union requirements prevail, a second operator or a spotter on the dozer trap or an oiler may be required.

† The above owning and operating costs do not include the costs of a scalping grizzly or vibrating screen.

Once the total hourly ownership and operating costs have been established, they can be added to the costs of the other equipment on the job. The final step is to compute the total cost per yard.

$$\text{Total cost/lcy} = \frac{\text{total hourly cost of ownership and operation, \$}}{\text{average hourly capacity (lcy)}}$$

Continuous flow production capacity for the Kolman Model 303, 50 ft × 60 in. belt loader is rated by the manufacturer at 3,600 lcy per hr. The practical production capacity of this loader can be estimated by allowing a 15-sec time lag for switching hauling units and a conservative efficiency rating of 80 percent (approximately equivalent to a 50-minute hour). Because the belt loader cost remains constant, and because the system production is determined by the type and size of push units and haul units, the use of larger haul units will increase system production and lower the loading cost per loose cubic yard. Using the assumed time lag and efficiency factors, and applying the estimated owning and operating cost of $29.02 per hour for the Model 303 belt loader, the following relationships are obtained between hauling unit capacity, system production rate, and belt loader costs per loose cubic yard.

Capacity of haul unit, cu yd	System production, lcy/hr	Belt loader cost, $/lcy
78 (100 tons)	2,415	0.0120
50	2,220	0.0131
40	2,100	0.0138
30	1,920	0.0151
25	1,800	0.0161
20	1,645	0.0176
15	1,440	0.0202
10	1,150	0.0252

Section 12

Trenchers

J. F. LINDELL

Chief Engineer, Parsons Division of the Koehring Company, Newton, Iowa

INTRODUCTION

Trenchers are machines for excavating trenches or ditches of considerable length and variable width, depth, and contour. A trencher consists of a self-propelled tractor or carrier mounting, fitted with either crawler-type or wheel-type running gear, which has the capability of traveling at a creep speed in ranges from zero to 30 fpm or more; an excavating device which consists of several buckets or cutters attached to either the periphery of a wheel or to a chain, with the capability of digging and discharging continuously; and a means of conveying the excavated material to either side of the excavated trench and depositing it in windrows. Trenchers are available in many configurations and sizes but can be broadly classified in two groups:

ladder trenchers and wheel trenchers. The tractor-mounted backhoe, which finds extensive use in many trenching applications, is not included in this section.

LADDER TRENCHERS

The excavating member of a ladder trencher consists of a long boom with a drive sprocket or sprockets on its upper end and an idler sprocket(s) or roller(s) on its lower end. An endless chain with cutters or buckets attached to it connects the upper sprocket and the lower sprocket. The sprocket(s) on the upper end of the boom or ladder are usually connected through a chain drive to a power source on the trencher carrier. As the buckets or cutters travel down the upper side of the boom on the digging chain, they enter the excavation; returning to the drive sprocket(s) on the lower side of the boom, they excavate material from the trench. This material is discharged into the conveying system as the buckets or cutters pass over the upper sprocket(s) on the boom. A typical ladder trencher, showing the relationship between the boom, chain drive, conveyor, and carrier, is illustrated in Fig. 12-1. Ladder trenchers can be subdivided into two groups: utility ladders and construction ladders.

Fig. 12-1 Construction ladder trencher capable of digging trenches to 11 ft deep and 30 in. wide. (*Parsons Division of the Koehring Co.*)

Utility Ladder Trenchers These trenchers are characterized by a single-chain excavator with cutters attached directly to the chain. They are normally powered with air-cooled engines in the range of 9 to 35 hp and are available with either crawler-type or wheel-type running gear. The conveying system which receives spoil from the cutters and discharges it at the edge of the trench is frequently of an auger-type rather than a belt conveyor.

Utility trenchers are available in sizes as small as 1,000 lb (Fig. 12-2) ranging up to 5,000 to 6,000 lb, as illustrated in Figs. 12-3 and 12-4. They are capable of digging trenches as narrow as 3 in. for the smaller machines and as wide as 20 in. with the larger machines, with maximum effective digging depths generally in the range of 5 ft. The machines operate most effectively in clay- or loam-type soils and their production will be considerably reduced in harder materials such as soft sandstone, soft shale, or soft limestone. They are not very effective in soils which contain hard rocks or boulders whose diameters exceed either 10 in. or the trench width, whichever is the lesser. Because of their limitation in discharging material, utility ladder trenchers are generally restricted in the total area of trench face which they can excavate. Thus, when digging at greater widths, the depths at which they can operate are correspondingly reduced.

Fig. 12-2 Small utility trencher being used for installation of a buried residential telephone service. (*Davis Manufacturing, Inc.*)

Utility ladders are used extensively for installation of gas, water, telephone, and electric-power service lines. The larger sizes are also frequently used for gas and water mains and can do an excellent job on footings and foundations on residential and other small buildings. Trencher manufacturers offer a multiplicity of attachments for their utility ladders. Backhoe and dozer attachments are shown in Figs. 12-3 and 12-4, and auger-boring attachments, angledozers, and vibratory plows are also available.

Maintenance costs on the smaller trenchers have been reduced during the last few

Fig. 12-3 Articulated utility trencher. Note single digging chain with cutters directly attached, auger conveyor and backhoe-backfill–blade attachment. (*Vermeer Manufacturing Co.*)

years, largely as a result of improved product quality. Nevertheless, it is still a fact that the trencher digging chain and cutters are high maintenance items. A number of manufacturers do not warranty the digging chain and cutters, considering them expendable parts.

Construction Ladder Trenchers Machines in this class are characterized by a double bucket chain or, less commonly, by a heavy-duty single chain with cutters directly attached. Buckets which are used in the double bucket chain configuration have replaceable teeth or cutters on their leading edges.

Construction ladder trenchers have digging depths ranging from 5 to 25 ft and digging widths ranging from 8 in. to 6 ft. Attachments are also available on the larger

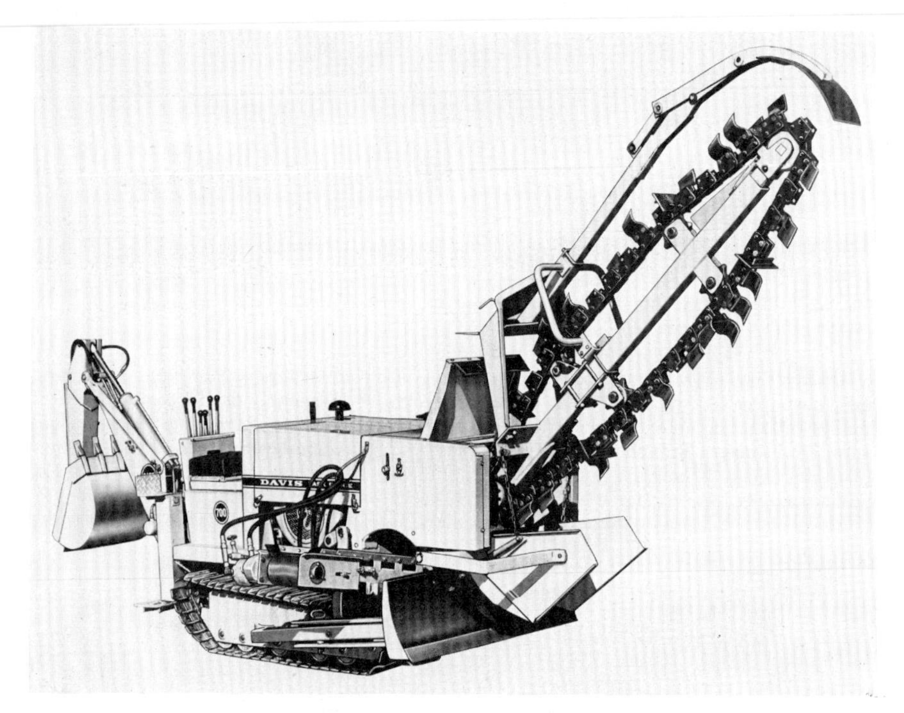

Fig. 12-4 Crawler-mounted utility trencher with backhoe attachment. *(Davis Manufacturing, Inc.)*

machines which can extend the digging widths to as much as 10 ft. Machine weights range from 6,000 lb to 60,000 lb. Power ranges from 40 hp to 160 hp, with liquid-cooled gas engines being predominant in the smaller sizes and diesel engines favored in the larger sizes. Most trenchers in the latter category are crawler mounted.

Construction ladder trenchers are used primarily in urban areas. The smaller machines are suitable for gas, water, power, and communication service lines as well as for small water mains and sewer lines. The larger machines are used principally for large water mains and large sanitary and storm sewers. Figure 12-5 illustrates a medium-sized ladder trencher on a sanitary sewer project. Large ladder trenchers, because of their capability of digging straight trenches with smooth, even walls, lend themselves to the use of support equipment such as portable trench shields and backfill conveying systems. All these can help to lower costs and expedite the job. A project of this type is illustrated in Fig. 12-6.

Fig. 12-5 Ladder trencher on sanitary sewer project. (*Parsons Division of the Koehring Co.*)

Fig. 12-6 Trench shield and backfill conveyor used with large ladder trencher on storm sewer project. (*Parsons Division of the Koehring Co.*)

WHEEL TRENCHERS

The excavating member of a wheel trencher generally consists of a pair of circular rims whose outside diameters are connected by U-shaped buckets or cutters. The wheel which is thus formed is driven by gear segments on the inside diameter of the rims. These gear segments consist of forgings with gear teeth on their faces and roller paths on their inside diameters. The entire wheel assembly is carried in a frame on a pair of gear pinions and two or three pairs of rollers. The wheel is turned by a chain drive which connects a power source on the trencher carrier to a shaft on which the wheel drive pinions are attached. The wheel turns in a direction with the buckets traveling away from the carrier in the top position when discharging material and toward the carrier at the bottom of their travel while digging. The buckets or cutters on the wheel perform their excavating task while traveling upward. As they near the top position of the wheel, the excavated material either drops from the buckets or is forced from them by a cleaner mechanism. A conveyor

Fig. 12-7 Compact wheel trencher excavating trench for a residential water main. (*Cleveland Trencher Co.*)

receives the spoil, passes it through the wheel, and deposits it onto a spoil pile alongside the trench.

Wheel trenchers can generally excavate material at almost twice the rate of a ladder machine with equivalent horsepower. However, judgment must be exercised when equipping them for the specific type of work and soil in which they are to be used. Knowledgeable selections must be made from the various teeth, buckets, bucket backs, bucket cleaners, and other miscellaneous items which are available. The largest standard models of wheel trenchers in current production will dig to a maximum depth of approximately 8½ ft although still larger units have beem custom built for specific job applications.

Wheel trenchers are used primarily in four broad categories: utility distribution, farm drainage, transmission pipelines, and irrigation.

Utility Distribution* Figure 12-7 illustrates a wheel trencher which is suitable for utility work. These trenchers are generally rather compact machines that are capa-

* See Underground Utility Construction in Sec. 26, Pipelines.

ble of digging trenches with depths of 4 to 6 ft and widths up to a maximum of 24 in. Their crawler-type running gear is usually equipped with flat shoes to minimize damage to streets and lawns. If a machine is equipped with a crumbing shoe, this shoe is generally power-actuated so that it can be retracted to permit digging to the desired depth in the shortest forward travel distance. The crumbing assembly is attached to the rear of the wheel frame. It consists of a post extending downward to a shoe which rides the bottom of the ditch, and a plate extending in an arc downward from the upper part of the crumber post. The plate is shaped to parallel the path of the excavator buckets, with adequate clearance from it, and its lower end is attached to the crumbing shoe. The crumber assembly keeps loose spoil from falling into the trench and thus produces a clean, smooth trench bottom. When a particularly clean trench is not desired the crumbing shoe may be omitted or the machine may be operated with the power-actuated crumbing shoe retracted.

Fig. 12-8 Shiftable wheel trencher with wheel offset and tilted, leaving trench behind right crawler with vertical side walls. (*Barber-Greene Co.*)

Shiftable wheel trenchers are becoming increasingly popular for utility work, as they can excavate closer to obstructions and further out on roadsides than the conventional machines. They can also dig a vertical trench while operating on a side slope. Hydraulic controls are used to shift or tilt the wheel. Figure 12-8 illustrates a machine of this type operating in a residential area. The shiftable wheel trencher is generally equipped with a power-actuated crumbing shoe.

Farm Drainage Wheel trenchers used for this type of work will generally have digging depths ranging from 6 to 7 ft. Usual digging widths are 20 to 24 in. and occasionally as wide as 36 in. As these machines are often working in soft, sticky soil, their crawlers are generally equipped with high-grouser-type crawler shoes. The shoe widths are seldom less than 20 in. and may be as much as 36 in.

The wheel for a farm-drainage trencher will frequently have buckets which are equipped with gumbo cutters rather than with teeth. A gumbo cutter is a lip, attached to the bucket, which has a sharp, smooth leading edge. The type of bucket back will vary with the type and condition of soil. A backless bucket is used during extremely sticky digging, tine backs represent an intermediate case, and solid backs will be used during the dry season. Farm drainage trenchers are invariably equipped with bucket cleaners and/or rim cleaners when either tine-back buckets or backless buckets are used. The crumbing shoe behind the wheel is equipped with a trench shield to prevent spoil from falling back into the ditch. Automatic tiling devices are in common usage today; these are chute-type attachments which position the tile precisely in the bottom of the trench as the trencher progresses. Figure 12-9 shows a machine so equipped. Tile as large as 24 in. in diameter has been placed with an automatic layer, but the more common sizes are 10 to 12 in. for main drainage lines and 4 to 8 in. for laterals.

Wheel trenchers lend themselves admirably to farm drainage work since they can readily maintain a very precise grade. The most commonly used method is to pre-position grade targets alongside the line where the trench is to be dug. The trencher is equipped with a sight bar which is mounted on the wheel frame and referenced to the axial center line of the wheel. The operator then aligns this sight bar with the pre-positioned targets, either ahead of or behind the machine. There has also been some promising development work on automatic grade-control systems, including the use of

Fig. 12-9 Farm drainage trencher equipment with automatic tile layer. Accurate grade is maintained with string line.

lasers. At this time, the costs are somewhat prohibitive but are expected to come down.

Transmission Pipelines* Every area of the country is now serviced with pipelines which transport natural gas, gasoline, fuel oil, and even anhydrous ammonia for fertilizer. The construction of these transmission pipelines generally involves large excavation quantities and many miles of trenching with few interruptions. The wheel trencher is particularly well suited for this type of work. One common procedure is to distribute a number of trenchers over the job. If three trenchers were used, for example, the first could dig the first mile, the second trencher the second mile, the third trencher the third mile, the first trencher the fourth mile, and so on. Under good digging conditions, a production rate of 1 mi per day per trencher is not unusual.

Trenchers used for pipelines are generally heavy-duty machines which are equipped with diesel power, wide crawler tracks with high grouser shoes, heavy-duty buckets and wheel equipment, and trench slopers when required. The various models of wheel trenchers which are available from manufacturers for pipeline service have digging capabilities which encompass widths from 24 in. to 60 in. and depths from 6 ft to

* See Transmission Pipelines in Sec. 26, Pipelines.

8½ ft. The required size of trencher for a specific job is determined primarily by the pipe diameter. A common specification for the trench width is the outside diameter of the pipe plus 16 in. The trench depth requirement is the outside diameter of the pipe plus the specified cover. This depth of cover is most often in the range of 3 to 4 ft. Figure 12-10 shows a medium-size pipeline trencher opening trench for a gas transmission line.

Irrigation A conventional irrigation distribution system consists of trapezoidal ditches, either lined or unlined, or of pipes which operate under partially full conditions of flow. Since this irrigation system is gravity operated, accurate grade control for the trenching operation is required. This control becomes even more essential when paving machines follow the trenchers and specified lining or wall thickness must be maintained. The wheel trenchers used for irrigation work are, for the most part, the same type of trenchers as are used in pipeline work. However, the wheel equipment may have to be modified to cut the trench section which is specified for a particular project. Two types of irrigation ditches are in common usage, one of which has an open trapezoidal section which may or may not be concrete lined. Variables in this

Fig. 12-10 Pipeline trencher. (*Buckeye Division of Gar-Wood Industries.*)

type of ditch are the depth, bottom width, and side slopes. Wheel trenchers equipped to dig such ditches have special buckets which are shaped to cut the ditch bottom and, sometimes, a short portion of the side slope. These trenchers are also equipped with rotary, conical-shaped cutters that cut the side slope and deliver the spoil to the trencher buckets for removal. (See Fig. 12-11.) A crumber attachment fits the trapezoidal section precisely, in order to leave a finished ditch ready for paving. Because of the variables in a trench of this type, the attachments must generally be custom-made to specification.

The second type of irrigation trench in common use is constructed with vertical sides and an accurately shaped semicircular bottom. Using this shaped bottom as a pattern, concrete is cast in place to form a continuous circular conduit. Trenchers for this type of work are equipped with special round-bottom buckets of very precise dimension which must usually be custom made. Another system of irrigation replaces the gravity-flow methods of water distribution with pressurized flow in conduits. The method is becoming increasingly popular, since a pressure system is quickly installed, requires less land area, and can be used in conjunction with sprinkler heads to service land that is not uniformly sloped.

Either polyvinyl or concrete pipe is used for the distribution system, and this is

Fig. 12-11 Wheel trencher equipped to cut trapezoidal section. Note buckets, rotary side cutters, and crumber. Section is finished and ready for concrete lining. (*Barber-Greene Co.*)

installed in much the same manner as utility pipelines. Conventional wheel trenchers are then employed with no special attachments required.

PRODUCTION

Many elements influence the production of trenchers. Factors which pertain to the machine itself will include horsepower, bucket capacity, digging-wheel speeds or bucket-line speeds, and availability of a creep speed or other speeds suitable for the soil type and type of digging. In addition, of course, there are factors such as hardness, rock content, moisture content, etc., which are related to the type of soil which is being excavated. Finally, job conditions such as trench widths and depths, length of trench between obstructions, provisions for spoil disposal, etc., must be evaluated.

Under ideal digging conditions, assuming adequate power and the availability of a suitable creep speed, the production of a trenching machine is in direct proportion to the bucket capacity and to the wheel and bucket-line speeds. Actual production is reduced by the presence of large rocks or boulders which make it necessary to reduce wheel or bucket-line speeds in order to prevent damage to the machine. Sticky soils also tend to reduce production, since it is virtually impossible to clean all such material out of the digging buckets. The net effect of this, of course, is to reduce effective bucket capacity.

Trenchers can excavate extremely hard materials such as caliche, sandstone, shale, coral or oolite, and soft limestone. However, production may then be less than 10 percent of that which would be obtained under ideal digging conditions. Successful operation in these materials requires a heavy-duty machine which is equipped with the proper type of digging teeth and, in general, it will be found that wheel trenchers can do a better job than ladder trenchers. The light-duty trenchers, since they are equipped only with cutters attached to a chain, have a cutting action which is more comparable to that of a saw than that of an excavator. As a result, their production depends to a great extent on their chain speed and cutter configuration. They do have the particular advantage of being able to dig much narrower trenches than machines equipped with buckets. In these circumstances, the volume of material is usually of less importance than the footage of trench. Figure 12-12 is a chart illustrating the capabilities of various types and sizes of trenchers as compared to the capabilities of several sizes of tractor-mounted backhoes.

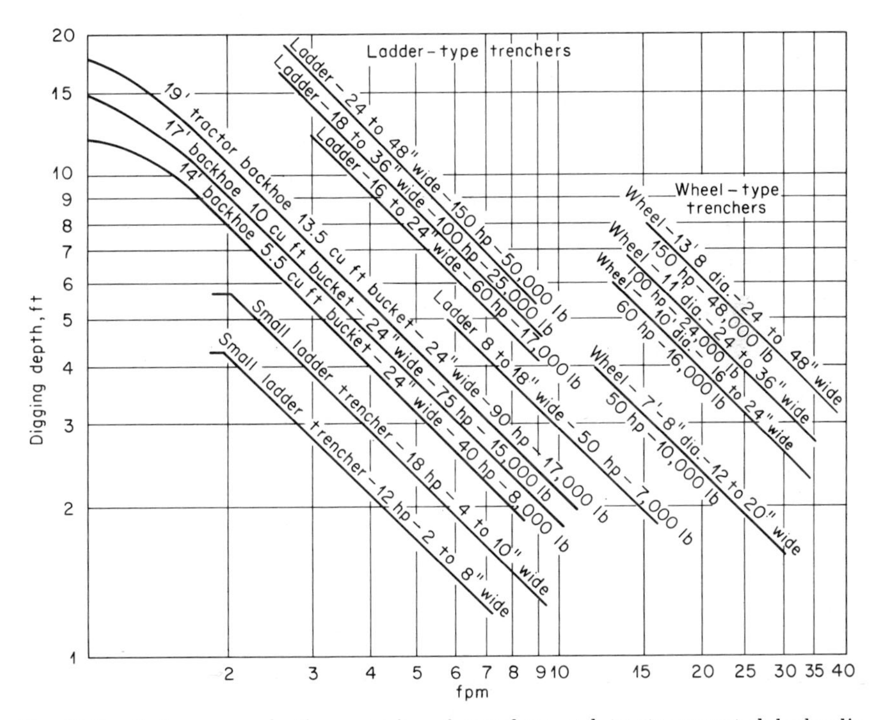

Fig. 12-12 Optimum productive capacity of trenchers and tractor-mounted hydraulic backhoes in easy digging (sandy loam) with expert operators. For hard digging, derate up to 90 percent. *(Parsons Division of the Koehring Co.)*

OPERATION

A trencher is perhaps one of the easiest pieces of construction equipment to operate, since it is slow moving and to a large degree automatic. As is true for all types of construction machinery, the operator of a trencher must still be continuously alert to the hazards involved in its use. There is always the risk of a cave-in when trenching in unstable soil; this endangers not only the machine and its operator but also any workmen who may be nearby. Another danger is the possibility of striking underground utility lines such as those for gas, water, power, or communications. Apart from the damage to the lines themselves and the consequent disruption of utility service, this can damage the trencher unless its operator immediately stops the machine. It is essential that the trencher operator be at the controls at all times. He must remain alert and conscious of the possible hazards to personnel working around the machine and of possible damage to the machine itself.

Trenching operations should begin with the machine fully serviced and ready to operate before its engine is started. The operator then maneuvers it in line with the trench which he proposes to cut. When the machine is in position, he puts it into creeper speed and engages the brakes. He then lowers the digging member to a position just above the surface of the ground, starts the digging member and the conveyor, and slowly begins cutting into the earth.

It is possible to leave a ladder trencher stationary while lowering its boom to the desired digging depth. A wheel trencher which is equipped with a fixed crumber must be advanced slowly to permit the entry of the crumber into the trench. If the trencher is equipped with a power crumber, this can be raised and the wheel lowered with the trencher remaining stationary. After digging to depth, a creep speed is selected which is suitable to the soil conditions, power availability, and bucket capacity. A trencher equipped with a hydraulically driven creep system is particularly advantageous because its creep speeds are infinitely variable.

While digging, the conveyor should be extended sufficiently to fully discharge the excavated spoil at the desired distance from the trench. The operator must keep the machine digging along the prescribed alignment for the trench, which sometimes involves following a string line. The required depth or grade must also be maintained with the aid of the grade-control system with which the machine is equipped.

MAINTENANCE

The following discussion should not be construed as a maintenance guide for trenchers. Rather, it points out a few areas where particular attention is desirable to prevent premature failure of important components and systems on the trencher. It is recommended that careful study be given to the maintenance manual which is supplied with the trencher.

Each type of construction machinery has its own particular area of high maintenance costs. The part of the trenching machine which requires the greatest attention is, of course, its digging member. In ladder trenchers this involves cutting teeth, excavator chain, head sprockets, and idler rollers. In wheel machines, high maintenance items in addition to the teeth and buckets will include the wheel-gear segments, wheel rollers, and drive pinions. It is very important that trencher teeth be replaced before they become sufficiently worn so that any portion of the trencher bucket, chain, or wheel contacts the side of the trench or the face of the cut. Further, it is desirable to keep sharp teeth in use as the digging effort is thus considerably decreased. The cost of tooth replacement is usually more than offset by fuel savings and reduction of maintenance on other parts of the machine.

In contrast to crawler tractors, the running gear of a trenching machine is a relatively long-life item and requires little maintenance other than routine lubrication. This can be attributed to the basic pattern of operation for the trencher. While it is digging, it is moving at a relatively slow speed under a steady load, resulting in little wear and shock on its running gear.

Most trenchers are now equipped with hydraulic systems for operating various functions of the machine. As for all hydraulically driven equipment, it is exceedingly important that the hydraulic fluid be of the proper type and viscosity for the class of service and the prevailing ambient temperatures. It is also exceedingly important that the machine be properly equipped with suitable strainers and filters, and that these be maintained as recommended by the manufacturer. When replenishing or replacing hydraulic oil or replacing filters, extreme care should be taken not to introduce dirt or contaminants into the hydraulic system.

With the coming of hydraulically driven conveyors and increased conveyor speeds, the rates of wear for conveyor belts have increased greatly. This can be of particular importance when working in dry, abrasive materials. Conveyor belt maintenance can be reduced by insisting that the operator refrain from running the conveyor belt any faster than is necessary to properly discharge the spoil and locate the spoil bank. Belt life will be extended if the operator is instructed to shut off the conveyor at the same time that he stops the bucket line. Most hydraulic conveyors are independently driven and will continue to operate unless positive action is taken to shut them down.

OPERATIONAL COSTS

The initial cost of a trencher is approximately the same as for other construction machinery of equivalent size, horsepower, and complexity. Assuming that a useful life of 7,000 to 8,000 operating hours is distributed within a 5-year depreciation period, it can be expected that the average trencher will have maintenance and repair costs which are approximately equivalent to its depreciation over the 5 years. This equivalence has frequently been applied when estimating maintenance costs for machines in the 15,000- to 25,000-lb weight category. While the very small utility trenchers will have much higher maintenance costs in relation to their depreciation, they are generally not used as many hours per year as some of the larger trenchers.

Labor costs, of course, vary tremendously in different parts of the country. The contractor must make his own determination for these, giving appropriate consideration to the location and particular circumstances of his job. The operator is usually the only workman required for a trencher, although the labor contracts in a few areas demand an oiler in addition. This is particularly likely to be required for the larger machines.

Because of the tremendous differences in types and sizes of trenchers, differences in soil types and conditions, labor rates, and other possible variables, it is not meaningful to present any general formula for predicting the cost of trencher operation. Most trencher manufacturers provide their dealers with specific data on the productivity, maintenance cost, and depreciation cost for trencher operation in various soil conditions. It is suggested, therefore, that the contractor contact his equipment dealer when he requires such information to supplement that in his own records.

Section 13

Hauling Units

JOHN A. CRAVENS

Vice-president, Sales, Euclid, Inc., Cleveland, Ohio

DUNCAN MacLAREN

District Manager, Southwest Region, Terex Division,
General Motors Corporation, Dallas, Texas

INTRODUCTION

There are today some sixteen manufacturers of "on-highway" hauling units; some twenty-two manufacturers produce vehicles classified as "off-highway" hauling units. While the contents of this section are applicable to both types of vehicles, it is important to recognize that the terms "on-highway" and "off-highway" refer not only to whether a hauling unit may or may not legally be operated on public highways but also to the construction of the unit itself.

From a legal standpoint, the distinction between on-highway and off-highway hauling units can be made as follows:*

On-highway haulers are vehicles whose design and manufacture are intended to permit their use on the public road, and as such they are subject to motor vehicle registration. In general, they will comply with gross vehicle weights, axle loadings, and width limitations imposed by the states. (See Fig. 13-1.) Off-highway haulers

Fig. 13-1 Typical on-highway rear-dump hauler. (*White Truck Division, White Motor Co.*)

are vehicles whose design and manufacture are intended to preclude their use on public roads, thus they do not comply with on-highway limitations imposed by the various states, and are not subject to motor vehicle registration. Typical characteristics are widths in excess of 8 ft 6 in. and/or axle loads at rated payload in excess of 24,000 lb.

In addition to legal distinctions, there are specific design features which characterize the off-highway hauling units:

1. Body construction is specifically designed to absorb high-impact loading. The floor of the body may be constructed of high-tensile steel (100,000 psi) or aluminum and will generally be fitted with steel wear plates to reduce abrasion damage.

* See Sec. 8, Tractors, Bulldozers, and Rippers.

2. The power train will consist of a diesel engine—often turbocharged—coupled to a torque converter and power-shift or automatic transmission. The transmission will be of the planetary or countershaft design and will often be mounted remotely from the engine. A means is provided for the final gear reduction at the drive wheels. This is usually a planetary-type axle or an internal driving gear and ring gear in the wheel hub or final drive case.

3. The ratio of pounds of gross vehicle weight to horsepower (flywheel or net) is generally lower than for an on-highway hauling unit. Top speed is slower than for a comparable on-highway hauling unit (about 40 to 45 mph), but gradeability under adverse conditions of grade and rolling resistance is from 25 to 40 percent better.

4. In most cases, a rigid-type frame is used to hold all driving components in alignment. Box sections, I beams, or ship channel frame rails—generally separated by cylindrical torque tubes—are used. On large hauling units, the frame may be built

Fig. 13-2 Typical off-highway hauler, 35-ton rated payload. (*Euclid, Inc., subsidiary of White Motor Co.*)

in two sections and joined by means of two or more vertical pins so as to articulate for better maneuverability.

5. Wheel and drive configuration may be 4 by 2, 4 by 4, 6 by 2, or 6 by 4. Tires designed for earthmoving, mining, and logging service are standard equipment.*

In addition, for the off-highway hauler, extra attention is given to oversize air cleaners and filters on all engine openings, vents in the drive-train gear cases, shockproof mounting of all instruments, dustproofing of electrical systems, heavier cabs (although it has only been in the last few years that serious attention has been given to human engineering or "operator-oriented" controls), cushioned mounting for radiators, and heavier mounting of fenders and fuel tanks. It follows that the initial purchase price of a specially designed off-highway hauler is considerably higher than that of a highway-type unit which has been modified for off-highway service. (See Fig. 13-2.)

In construction or mining, it is important for a prospective buyer to analyze the differences between on-highway and off-highway hauling units. It can be prohibitively expensive to misapply a strictly highway-type unit to a rugged off-highway load-and-haul situation. Careful analysis of the job to be done is necessary to make the proper equipment choice.

Generally, the on-highway hauler is found on long, high-speed hauls over roads with low rolling resistance. Loading may be with a power shovel of 1½- to 3-cu yd capacity for reduced-impact loading or by relatively small front-end loaders; or the hauler may be loaded from an overhead hopper.

* See Tires for Construction Equipment in Sec. 6, Engineering Fundamentals.

TYPES OF OFF-HIGHWAY HAULING UNITS

There are several distinct types of off-highway hauling units, each of which has specific design features which enable it to satisfy a particular set of job requirements. Major types of off-highway units include:

1. The rear-dump truck carrying a body on the truck chassis.
2. Bottom-dump tractor-trailer unit.
3. A side-dump truck carrying the body on the truck chassis. This unit may be dumped either by its own integral hoist mechanism or by an independent hoist or "sky hook." Independent hoists are used in places where a fixed dump point is required, such as a hopper or crusher, and one dumping mechanism can thus handle many trucks.
4. Side-dump tractor-trailer unit.
5. Rear-dump tractor-trailer unit.

To select the right size and type of hauling unit, the estimator or equipment engineer must know thoroughly the nature of the job which has to be done. He must also be acquainted with the particular operating characteristics and the advantages and limitations of the various machines which could conceivably do the job. A tentative selection of the type of hauler can then be made, and estimates can be prepared for several suitable types and sizes of machines. Following this, the estimated hauling costs per cubic yard or ton of material can be compared. In selecting a hauling unit, these general criteria are used:

1. The versatility of the hauling unit; can it be efficiently used in other applications?
2. The compatibility between the unit being considered and currently used hauling or loading equipment.
3. Possible restrictions on maneuvering space at the loading or dumping areas or on side and overhead clearances.
4. Possible vehicle weight or width restrictions on haul roads, including bridges.
5. The effect of extreme grades, either favorable or unfavorable, particularly on the loaded haul.

Rear-dump Load-on-back Truck The rear-dump truck with body mounted on the truck chassis is the most popular type of rubber-tired hauling vehicle. It can haul free-flowing material such as earth, sand, and gravel and more bulky material such as blasted rock, ore, shale, and coal. (See Table 13-1 for typical specifications.)

The normal excavating type consists of a body shell with heavy rib reinforcements and an inner liner or wear plate which is usually welded to the body shell (Fig. 13-3). This "double bottom" can take the loading impact of rock dropped from large loading shovels. Special "low profile" bodies have been developed to cater to the lower load-height capability of the wheel loader, which is increasingly popular for loading off-highway haulers (Fig. 13-4).

When the rock is not well blasted, it is frequently necessary for the shovel operator to carefully balance a huge piece of blasted rock on the shovel-dipper teeth before carefully placing it in the truck body. Even at best, the loading impacts are extremely severe in this type of work. For continuous heavy-duty rock service in quarries and in rock excavations, a heavy-duty rock option body is available. This has front and side liners in addition to floor wear plates. Liner plates in larger trucks will run from ¼- to 1-in.-thick alloy-steel plate.

In the colder climates, design precautions can be taken to prevent material from freezing to the rear-dump body. Exhaust gases from the engine are carried through ducts in the front slope sheet, in the bottom of the body, and on a portion of the side walls. In most cases this is adequate to prevent material from freezing in the body. In extremely severe conditions it may be necessary to clean out the forward bulkhead, the corners, and along the junction of the sides and bottom of the body despite the application of heat. Under these circumstances the conventional body can become so clogged with frozen material that it requires cleaning after every second or third trip.

The heated body can also help considerably when handling wet, adhesive clay soils;

bauxite; and other sticky materials. Heat which flows between the liner plate and the outer body shell creates a dried layer of material next to the body shell and facilitates the dumping of wet clay-type soils.

Use rear dumps when:

1. The material to be hauled is large rock, ore, shale, etc., with individual particle sizes over 12 in. in diameter, or when the material involves both free-flowing and bulky components
2. The hauling unit must dump into restricted hoppers or over the edge of a waste bank or fill
3. The hauling unit is subject to severe loading impact while under a shovel, dragline, or hopper
4. Maximum maneuverability—such as a rapid spotting—is required at the loading or dumping areas
5. Maximum gradeability is required

Fig. 13-3 50-ton off-highway hauler with standard body. (*Euclid, Inc., subsidiary of White Motor Co.*)

Fig. 13-4 50-ton off-highway hauler with low-profile body. (*Euclid, Inc., subsidiary of White Motor Co.*)

Bottom-dump Tractor-trailer The bottom-dump tractor-trailer consists of a diesel-powered prime mover or tractor with large earthmover-type single or dual tires on each drive wheel and a semitrailer with drop bottom or clamshell-type doors (see Fig. 13-5). The trailer hopper is generally wider and deeper at the front to put more weight on the drive axle. It is also wider at the top than at the bottom so that the side walls of the trailer are sloped inward. This shape of hopper is only suitable for

TABLE 13-1 Specification for Rear Dumps and Rear-dump Trailers

Vehicle model	Shipping weight, lb	Payload, max, lb	Capacities, cu yd		Overall length	Overall width	Overall height	Loading height	Engine		Tires
			Struck	Heaped					Make and model	Flywheel horsepower	
Allis-Chalmers/Athey:											
260-A2233R	50,360	66,000	22	26	33 ft 2 in.	11 ft 0 in.	12 ft 4 in.	9 ft 7 in.	AC AC260	282 hp at 2,100 rpm	29.5-29 (34)
460-460R	79,900	96,000	30	35	36 ft 2 in.	12 ft 9 in.	13 ft 7 in.	10 ft 11 in.	AC AC460	340 hp at 1,950 rpm	33.5-33 (38)
Caterpillar 769B	57,800	70,000	22.3	35.6*	25 ft 6 in.	11 ft 11 in.	12 ft 10 in.	9 ft 8 in.	Cat D343	415 hp at 1,900 rpm	18.00-25 (32)
Caterpillar/Athey:											
621-PR621	55,765	60,000	20	24	32 ft 10 in.	11 ft 5 in.	12 ft 4 in.	6 ft 7 in.	Cat 621	300 hp at 2,200 rpm	29.5-29 (34)
631B-PR631B	62,300	80,000	24.5	29.0	35 ft 11 in.	12 ft 0 in.	12 ft 10 in.	9 ft 9 in.	Cat 631 B	360 hp at 1,900 rpm	29.5-39 (34)
630B-PR630B	65,300	80,000	24.5	29.0	39 ft 6 in.	12 ft 1 in.	12 ft 9 in.	9 ft 9 in.	Cat 630B	360 hp at 1,900 rpm	29.5-35 (34)
651-PR651	87,600	120,000	40	47	40 ft 5 in.	14 ft 8 in.	13 ft 10 in.	11 ft 9 in.	Cat 651	500 hp at 1,900 rpm	33.5-39 (32)
660-PR660	118,190	120,000	45	53	45 ft 5 in.	16 ft 6 in.	14 ft 5 in.	12 ft 5 in.	Cat 660	500 hp at 1,900 rpm	21.0-49 (36) dual
Challenge-Cook:†											
SDS-35	23,930	70,000	20	27	24 ft 1 in.	10 ft 7 in.	8 ft 8 in.	8 ft 8 in.	....	....	12.00-24 (16)
SDS-50	29,460	100,000	30	37	24 ft 1 in.	10 ft 7 in.	10 ft 3 in.	10 ft 3 in.	....	....	14.00-24 (20)
SDS-70	35,030	140,000	30	40	24 ft 10 in.	13 ft 5 in.	11 ft 0 in.	11 ft 0 in.	....	....	14.00-24 (20)
Euclid:											
R-13	24,200	26,000	8.7	10.9	21 ft 6½ in.	9 ft 1¾ in.	10 ft 7 in.	7 ft 10 in.	GM 4-71N	160 hp at 2,100 rpm	12.0-25 (14)/12.0-25 (18)
R-22	37,000	44,000	14.7	18.3	25 ft 1 in.	11 ft 4 in.	12 ft 4 in.	9 ft 1½ in.	GM 6-71N	239 hp at 2,100 rpm	16.0-25 (16)/16.0-25 (24)
R-22	37,300	44,000	14.7	18.3	25 ft 1 in.	11 ft 4 in.	12 ft 4 in.	9 ft 1½ in.	Cum N855C	250 hp at 2,100 rpm	16.0-25 (16)/16.0-25 (24)
R-35	56,000	70,000	23.3	29.0	27 ft 9 in.	12 ft 7 in.	12 ft 6 in.	10 ft 8 in.	GM 12V71N	434 hp at 2,100 rpm	18.0-25 (32)
R-35	55,500	70,000	23.3	29.0	27 ft 9 in.	12 ft 7 in.	12 ft 6 in.	10 ft 8 in.	Cum NTA 855C	380 hp at 2,300 rpm	18.0-25 (32)
R-50	79,000	100,000	33.3	41.0	33 ft 2½ in.	13 ft 2½ in.	14 ft 3 in.	12 ft 11 in.	GM 16V-71N	635 hp at 2,100 rpm	21.0-35 (28)/21.0-35 (36)
R-50	79,700	100,000	33.3	41.0	33 ft 2½ in.	13 ft 2½ in.	14 ft 3 in.	12 ft 11 in.	Cum VT1710C	635 hp at 2,100 rpm	21.0-35 (28)/21.0-35 (36)
R-105	137,000	210,000	60.0	74.0	34 ft 3 in.	15 ft 6 in.	16 ft 3 in.	15 ft 1 in.	GM 12V149T	1,000 hp at 1,900 rpm	24.0-49 (36)
Euclid/Athey:											
4UOT-E915R	26,980	30,000	9.0	10.3	27 ft 0 in.	10 ft 0 in.	9 ft 6 in.	7 ft 4 in.	GM 4UOT	135 hp at 2,100 rpm	23.5-25 (16)
39LOT-E2740R	70,250	80,000	27.0	32.0	36 ft 5 in.	12 ft 9 in.	13 ft 7 in.	10 ft 3 in.	GM 39UOT	398 hp at 2,100 rpm	33.5-33 (32)
47LOT-E4060R	90,250	120,000	40.0	47.0	39 ft 6 in.	14 ft 9 in.	13 ft 10 in.	11 ft 7 in.	GM 47LOT	498 hp at 2,100 rpm	37.5-33 (42)
49LOT-E4466B	79,130	132,000	44.0	54.0	54 ft 2 in.	12 ft 0 in.	13 ft 1 in.	11 ft 1 in.	GM 49LOT	441 hp at 2,100 rpm	33.5-39 (38)
John Deere/Athey											
JD760-PR760	28,395	30,000	9	10.3	29 ft 1 in.	10 ft 0 in.	9 ft 5 in.	7 ft 4 in.	JD 760	143 hp at 2,200 rpm	23.5-25 (16)
Hardwick:‡											
621-E621	58,370	70,000	20	25§	34 ft 1 in.	11 ft 11 in.	11 ft 6 in.	7 ft 11 in.	Cum D336	300 hp at 2,200 rpm	29.5-29 (28)
630B-E631	78,900	90,000	25	33§	39 ft 9 in.	15 ft 4 in.	12 ft 1 in.	8 ft 6 in.	Cum D343	400 hp at 1,900 rpm	29.5-35 (34)
631B-E631	76,800	90,000	25	33§	36 ft 9 in.	15 ft 4 in.	12 ft 1 in.	8 ft 6 in.	Cum D343	400 hp at 1,900 rpm	29.5-35 (34)
641-E641	98,800	110,000	31	42§	39 ft 4 in.	16 ft 8 in.	13 ft 8 in.	9 ft 3 in.	Cum D346	500 hp at 1,900 rpm	33.5-39 (38)
651-E651	104,500	120,000	34	48§	40 ft 7 in.	16 ft 8 in.	13 ft 11 in.	9 ft 6 in.	Cum D346	500 hp at 1,900 rpm	37.5-39 (36)
660-E660-80	114,860	160,000	48	56§	43 ft 5 in.	17 ft 8 in.	15 ft 10 in.	11 ft 3 in.	Cum D346	500 hp at 1,900 rpm	21.0-49 (36) dual
660-E660-100	123,060	200,000	60	68§	43 ft 5 in.	19 ft 2 in.	16 ft 10 in.	12 ft 3 in.	Cum D346	500 hp at 1,900 rpm	21.0-49 (36) dual
769B-E769	58,400	70,000	22.8	36.8*	25 ft 2 in.	13 ft 2 in.	12 ft 8 in.	9 ft 6 in.	Cum D 343	415 hp at 1,900 rpm	18.0-25 (32)
779-E779	103,200	170,000	54	68	29 ft 4 in.	16 ft 8 in.	13 ft 3 in.	12 ft 4 in.	Cum D 348	960 hp at 2,000 rpm	21.0-49 (36)
International Harvester:											
100	48,150	60,000	18.9	24	26 ft ¼ in.	11 ft 6 in.	11 ft 10 in.	10 ft 5 in.	IH DT1817	435 hp at 2,200 rpm	18.0-25 (20)/18.0-25 (28)
65B	36,000	40,000	13.1	17.3	24 ft 6 in.	10 ft 10 in.	11 ft 2 in.	9 ft 3½ in.	IH DVT573	300 hp at 2,600 rpm	13.0-25 (16)/16.0-25 (24)
65B-PS	36,500	40,000	13.1	17.3	24 ft 6 in.	10 ft 10 in.	11 ft 2 in.	9 ft 3½ in.	IH DVT573	300 hp at 2,600 rpm	13.0-25 (16)/16.0-25 (24)
100B	50,000	70,000	20.9	26	26 ft ¼ in.	11 ft 10 in.	11 ft 10 in.	10 ft 10½ in.	IH DT1817	435 hp at 2,200 rpm	18.0-25 (20)/18.0-25 (32)
180	68,445	100,000	33.3	41	29 ft 2 in.	13 ft 3 in.	12 ft 3 in.	11 ft 6 in.	GM 16V71N	560 hp at 2,100 rpm	18.0-25 (32)

180	63,600	90,000	30	38	27 ft 7 in.	13 ft 3 in.	12 ft 2 in.	10 ft 11 in.	GM 12V71N	456 hp at 2,100 rpm	18.0-25 (32)/18.0-25 (28)
140 PH	60,690	80,000	26.6	33.6	25 ft 1 in.	13 ft 3 in.	12 ft 0 in.	11 ft 3½ in.	IH DT817B	420 hp at 2,200 rpm	18.0-25 (24)
140 PH	59,230	73,200	24	31	25 ft 1 in.	13 ft 3 in.	12 ft 0 in.	10 ft 9½ in.	IH DT817B	420 hp at 2,200 rpm	18.0-25 (20)
I-H/Athey:											
E270-12030R	45,800	60,000	20	25	30 ft 10 in.	11 ft 5 in.	12 ft 2 in.	9 ft 5 in.	IH E270	300 hp at 2,100 rpm	29.5-29 (34)
295-13045	70,345	90,000	30	35	35 ft 1 in.	12 ft 9 in.	13 ft 7 in.	10 ft 11 in.	IH 295	375 hp at 2,100 rpm	33.5-33 (38)
Koehring:											
960	17,470	18,000	5.5	7.0	15 ft 10 in.	8 ft 5 in.	9 ft 10 in.	7 ft 5 in.	GM 4-71N	119 hp at 1,800 rpm	10.0-20 (12)/16.0-25 (16)
1860	35,730	36,000	11.0	13.5	19 ft 10 in.	10 ft 7 in.	11 ft 10 in.	9 ft 3 in.	GM 6-71N	227 hp at 2,100 rpm	13.0-25 (18)/26.5-25 (26)
2460	38,000	48,000	14.0	18.0	23 ft 3 in.	11 ft 0 in.	11 ft 8 in.		GM 8V-71N	318 hp at 2,100 rpm	16.0-25 (24)/16.0-25 (24)
KW-Dart:											
D-2330	50,800	70,000	24	31	25 ft 5 in.	12 ft 2½ in.	12 ft 3 in.	10 ft 9½ in.	Cum NT-380	355 hp	18.0-25 (28)
D-2440	64,700	80,000	26	32§	29 ft 3 in.	12 ft 6½ in.	13 ft 3 in.	11 ft 2 in.	Cum V12-525	485 hp	18.0-33 (32)
D-2541	74,300	100,000	31	39§	29 ft 10 in.	15 ft 1½ in.	13 ft 9½ in.	11 ft 6 in.	Cum V12-525	485 hp	21.0-35 (28)
D-2551	79,000	130,000	43	67	30 ft 5 in.	15 ft 3½ in.	13 ft 9½ in.	12 ft 6 in.	Cum VT12-635	585 hp	21.0-35 (36)
D-2561	88,000	150,000	50	60	31 ft 3 in.	15 ft 2 in.	14 ft 11 in.	13 ft 4 in.	Cum VT12-700	700 hp at 1,000 rpm	24.0-35 (36)
D-2661	96,500	170,000	56	78	31 ft 3 in.	15 ft 3½ in.	15 ft 3 in.	14 ft 8 in.	Cum VT12-700	650 hp	21.0-49 (40)
D-2771	132,000	240,000	65	85	38 ft 1 in.	18 ft 6 in.	16 ft 10 in.	14 ft 1 in.	GM 12V-149	740 hp	27.0-49 (42)
DE-2771	140,000	220,000	65	85	38 ft 1 in.	20 ft 0 in.	16 ft 10 in.	14 ft 1 in.	Cum VT12-700	700 hp at 1,000 rpm	27.0-49 (42)
Mack:											
M-15X	31,400	30,000	11	15½	24 ft 1 in.	11 ft 1 in.	11 ft 1 in.	8 ft 4 in.	Mack END-673P	180 hp at 2,100 rpm	13.0-25 (18)/14.0-25 (20)
M-20X	32,000	30,000	13	17	24 ft 1 in.	11 ft 1 in.	11 ft 1 in.	9 ft 2 in.	Mack END-711	200 hp at 2,100 rpm	13.0-25 (18)/16.0-25 (24)
M-25X	41,800	50,000	16	20	25 ft 2½ in.	11 ft 2 in.	11 ft 11 in.	10 ft 1 in.	Mack END-864	270 hp at 2,300 rpm	16.0-25 (24)/18.0-25 (28)
M-30X	46,800	60,000	20	26	26 ft 11½ in.	11 ft 5 in.	12 ft 1 in.	10 ft 3 in.	Cum NT335CI	335 hp at 2,100 rpm	18.0-25 (24)/18.0-25 (28)
M-32SX	56,000	64,000	22	29	29 ft 0 in.	11 ft 2½ in.	11 ft 9 in.		Cum NT335CI	320 hp at 2,100 rpm	16.0-25 (24)
M-45SX	75,500	90,000	28	36½	31 ft 8 in.	11 ft 11½ in.	12 ft 2 in.	10 ft 10½ in.	Cum V12-525	450 hp at 2,100 rpm	18.0-25 (24)
M-50AX	70,800	100,000	32	39	27 ft 0 in.	13 ft 1 in.	13 ft 8 in.	12 ft 0 in.	Cum V12-525	525 hp at 2,100 rpm	21.0-35 (28)
M-65AX	84,400	130,000	42	51	28 ft 7 in.	15 ft 10 in.	13 ft 9 in.	12 ft 9 in.	Cum VT635	635 hp at 2,100 rpm	21.0-35 (36)
M-70SX	100,200	140,000	44	56	34 ft 5 in.	13 ft 4 in.	13 ft 7 in.	11 ft 8 in.	Cum VT700	700 hp at 2,100 rpm	18.0-33 (32)
Michigan:											
210	48,400	60,000	20	24	32 ft 9 in.	11 ft 3 in.	12 ft 4 in.	9 ft 7 in.	Cum NT855	335 hp at 2,100 rpm	29.5-25 (34)
310	73,250	80,000	27	32	37 ft 5 in.	12 ft 9 in.	13 ft 7 in.	10 ft 3 in.	GM 12V-71N	475 hp at 2,100 rpm	33.5-33 (32)
310	73,250	90,000	30	35	37 ft 5 in.	12 ft 9 in.	13 ft 7 in.	10 ft 11 in.	GM 12V-71N	475 hp at 2,100 rpm	33.5-33 (32)
Michigan/Athey:											
210-M2030R	48,400	60,000	20	24	32 ft 9 in.	11 ft 3 in.	12 ft 4 in.	9 ft 7 in.	Cum NT855	290 hp at 2,100 rpm	29.5-29 (34)
310111-M2740R	73,250	80,000	27	32	37 ft 5 in.	12 ft 9 in.	13 ft 7 in.	10 ft 3 in.	GM 12V-71N	475 hp at 2,100 rpm	33.5-33 (32)
310111-M3045R	74,250	96,000	30	35	37 ft 5 in.	12 ft 9 in.	13 ft 7 in.	10 ft 11 in.	GM 12V-71N	475 hp at 2,100 rpm	33.5-33 (38)
Unit Rig & Equipment:											
M-85	114,000	170,000	54	78	32 ft 4 in.	14 ft 7 in.	16 ft 8 in.	14 ft 6 in.	Cum VT-12-700	700 hp	21.0-49 (40)
M-100	118,000	200,000	60	86	32 ft 4 in.	15 ft 4 in.	16 ft 10 in.	15 ft 6 n.	Cum VT-12-700	700 hp	24.0-49 (42)
M-120	130,000	240,000	76	115	32 ft 4 in.	17 ft 2 in.	17 ft 1 in.	15 ft 8 in.	Cum VT-12-700	700 hp	27.0-49 (42)
M-200	228,500	400,000	110	148	43 ft 4 in.	24 ft 0 in.	20 ft 7 in.	17 ft 0 in.	EMD 645-E4	1,650 hp	36.0-51 (50)
Wabco:											
35	51,120	70,000	23.3	29	24 ft 4 in.	12 ft 4 in.	11 ft 11 in.	10 ft 6 in.	Cum NT 855C	380 hp at 2,300 rpm	18.0-25 (28)
50	74,890	100,000	30.7	38.2	28 ft 10 in.	13 ft 5 in.	12 ft 11 in.	11 ft 8 in.	Cum V12-525	525 hp at 2,100 rpm	21.0-35 (32)
65B	85,500	130,000	42	54	31 ft 6 in.	15 ft 11 in.	13 ft 6 in.	12 ft 10 in.	Cum VT12-635	635 hp at 2,100 rpm	24.0-35 (30)
75B	91,500	150,000	45	57	31 ft 6 in.	15 ft 11 in.	13 ft 6 in.	13 ft 3 in.	Cum VTA-1710-C	700 hp at 2,100 rpm	24.0-35 (42)
100A	143,800	200,000	56	72	35 ft 1 in.	18 ft 7 in.	15 ft 8 in.	15 ft 0 in.	GM 16V-71NT	665 hp at 2,100 rpm	27.0-49 (36)
120A	155,800	240,000	63	80	35 ft 1 in.	18 ft 7 in.	16 ft 0 in.	15 ft 8 in.	GM 12V-149T	920 hp at 1,900 rpm	27.0-49 (48)
75B/120RD	154,440	240,000	71.6	86	65 ft 5 in.	20 ft 5 in.	14 ft 5 in.	14 ft 6 in.	Cum VTA-1710-C	700 hp at 2,100 rpm	24.0-35 (42)/27.0-49 (36)

* SAE 1:1 rating.
† Trailers only, no engines.
‡ Hydraulic rear ejection.
§ SAE 2:1 rating.

TABLE 13-2 Specifications for Bottom Dumps and Bottom-dump Trailers

Vehicle type	Shipping weight, lb	Payload, max, lb	Capacities, cu yd		Overall length	Overall width	Overall height	Loading height	Engine		Tires
			Struck	Heaped					Make and model	Flywheel horsepower	
I-H/Athey:											
295-13045B	67,523	90,000	30	35	41 ft 2 in.	12 ft 8 in.	11 ft 4 in.	10 ft 3 in.	IH 295	375 hp at 2,100 rpm	33.5-33 (32)
PH100-13248B	66,300	96,000	32	38	51 ft 5 in.	12 ft 7 in.	11 ft 6 in.	9 ft 10 in.	IH PH100	375 hp at 2,100 rpm	29.5-35 (34)
PH180-15075B	105,100	150,000	50	60	58 ft 0 in.	12 ft 6 in.	12 ft 8 in.	12 ft 1 in.	IH PH180	580 hp at 2,100 rpm	18.0-49 (36) dual
PH180-166100B	117,912	200,000	66	75	61 ft 10 in.	16 ft 8 in.	14 ft 1 in.	13 ft 6 in.	IH PH180	580 hp at 2,100 rpm	21.0-49 (36) dual
PH180-1120-100H	109,332	200,000	120	131	62 ft 5 in.	13 ft 6 in.	14 ft 3 in.	13 ft 1 in.	IH PH180	580 hp at 2,100 rpm	21.0-49 (36) dual
KW-Dart:											
D-4655	139,500	240,000	74	85	65 ft 7 in.	15 ft 9 in.	13 ft 10 in.	13 ft 0 in.	Cum VTA-1710-C700	650 hp	21.00-35 (28)
D-4651	130,300	240,000	120	180	68 ft 6 in.	15 ft 1 in.	13 ft 10 in.	10 ft 9 in.	Cum VT12-700	650 hp	21.00-35 (28)
Michigan 310	72,100	90,000	30	35	43 ft 10 in.	12 ft 7 in.	11 ft 4 in.	9 ft 9 in.	GM 12V-71N	475 hp at 2,100 rpm	33.5-33 (38)
Michigan/Athey 310111-M3045B	72,100	90,000	30	35	43 ft 10 in.	12 ft 7 in.	11 ft 4 in.	9 ft 9 in.	GM 12V-71N	475 hp at 2,100 rpm	33.5-33 (38)
Wabco:											
90	77,650	180,000	106	118	55 ft 7 in.	12 ft 7 in.	12 ft 8 in.	....	Cum VT-12-635	600 hp at 2,100 rpm	18.0-33 (32)/18.0-49 (32)
90	77,650	180,000	106	118	55 ft 7 in.	12 ft 7 in.	12 ft 8 in.	....	Cum V-12-525	525 hp at 2,100 rpm	18.0-33 (32)/18.0-49 (32)
90	77,650	180,000	106	118	55 ft 7 in.	12 ft 7 in.	....	....	GM 12V-71	475 hp at 2,100 rpm	18.0-33 (32)/18.0-49 (32)
100CH	88,810	200,000	114	126	57 ft 7 in.	14 ft 2 in.	13 ft 3 in.	13 ft 3 in.	Cum VT-1710C	600 hp at 2,100 rpm	18.0-33 (32)/18.0-49 (32)
75B/120CH	128,800	240,000	132	150	65 ft 2 in.	15 ft 11 in.	14 ft 11 in.	13 ft 3 in.	Cum VT-1710C	665 hp at 2,100 rpm	24.0-35 (36)/21.0-49 (36)
Caterpillar/Athey:											
621-PW621	54,000	60,000	20	25	40 ft 10 in.	11 ft 10 in.	10 ft 0 in.	9 ft 5 in.	Cat 621	300 hp at 2,200 rpm	29.5-29 (28)
631B-PW631B	62,100	82,000	27	32	44 ft 10 in.	12 ft 1 in.	11 ft 5 in.	9 ft 8 in.	Cat 631B	360 hp at 1,900 rpm	29.5-35 (34)
630B-PW630B	65,500	96,000	32	38	48 ft 1 in.	12 ft 1 in.	10 ft 6 in.	9 ft 10 in.	Cat 630B	360 hp at 1,900 rpm	29.5-35 (34)
650-PW650	82,100	132,000	44	51	51 ft 11 in.	12 ft 6 in.	12 ft 6 in.	11 ft 0 in.	Cat 650	450 hp at 1,900 rpm	37.5-39 (32)
PH630BS	65,000	102,000	50	56	50 ft 5 in.	12 ft 0 in.	11 ft 3 in.	10 ft 10 in.	Cat 630B	360 hp at 1,900 rpm	29.5-35 (34)
660-PW660	101,270	200,000	50	59	54 ft 6 in.	15 ft 6 in.	13 ft 2 in.	11 ft 5 in.	Cat 660	500 hp at 1,900 rpm	21.0-49 (36) dual
PH630S	64,780	102,000	64	69	50 ft 0 in.	12 ft 0 in.	11 ft 3 in.	10 ft 10 in.	Cat 630B	360 hp at 1,900 rpm	29.5-35 (34)
PH660	115,500	200,000	120	131	59 ft 8 in.	13 ft 6 in.	13 ft 9 in.	12 ft 9 in.	Cat 660	500 hp at 1,900 rpm	21.0-49 (36) dual
Challenge-Cook:*											
E-2500	13,420	50,000	18	22	38 ft 7 in.	7 ft 11 in.	7 ft 1 in.	7 ft 1 in.	....	....	11.0-22.5 (12)
E-3000	14,570	60,000	20	24	38 ft 7 in.	7 ft 11 in.	7 ft 5 in.	7 ft 5 in.	....	....	11.0-25.5 (12)
E-3000L	12,420	60,000	20	24	36 ft 10 in.	7 ft 11 in.	8 ft 0 in.	8 ft 0 in.	....	....	11.0-25.5 (12)
E-3500	16,000	70,000	20	24	38 ft 7 in.	7 ft 11 in.	7 ft 9 in.	7 ft 9 in.	....	....	10.0-20 (14)
E-3500L	12,860	70,000	20	24	36 ft 10 in.	7 ft 11 in.	8 ft 0 in.	8 ft 0 in.	....	....	10.0-20 (14)
E-4000	18,000	80,000	27	31	40 ft 11 in.	7 ft 11 in.	8 ft 5 in.	8 ft 11 in.	....	....	11.0-22 (14)
E-5000‡	22,240	100,000	30	34	40 ft 11 in.	9 ft 1 in.	8 ft 7 in.	9 ft 1 in.	....	....	12.0-24 (16)
E-5000G	26,670	100,000	31	35	39 ft 9 in.	9 ft 1 in.	8 ft 7 in.	9 ft 7 in.	....	....	12.0-24 (16)
E-6000	27,170	120,000	33	37	40 ft 11 in.	9 ft 1 in.	9 ft 3 in.	9 ft 9 in.	....	....	14.0-24 (16)

E-6000G	30,670	120,000	34	38	39 ft 9 in.	9 ft 1 in.	9 ft 3 in.	10 ft 3 in.	....	....	14.0-24 (16)
E-7000‡	32,000	140,000	43	52	43 ft 6 in.	11 ft 10 in.	10 ft 5 in.	10 ft 5 in.	....	....	14.0-25 (24)
E-7000G	36,000	140,000	43	52	40 ft 8 in.	11 ft 10 in.	10 ft 11 in.	10 ft 11 in.	....	....	14.0-25 (24)
E-8500	52,370	170,000	50	60	48 ft 0 in.	11 ft 10 in.	11 ft 10 in.	11 ft 10 in.	....	....	18.0-25 (20)
CH-7500‡	38,900	150,000	80	95	43 ft 0 in.	13 ft 8 in.	11 ft 9 in.	11 ft 9 in.	....	....	14.0-25 (24)
Euclid:											
B-63	77,375	126,000	42	54	57 ft 6 in.	12 ft 7½ in.	11 ft 1 in.	10 ft 0 in.	GM 12V-71N	475 hp at 2,100 rpm	16.0-25 (20)/37.5-33 (36)
B-70	79,000	140,000	47	60	57 ft 6 in.	12 ft 7½ in.	11 ft 1 in.	10 ft 8 in.	GM 12V-71N	475 hp at 2,100 rpm	16.0-25 (20)/37.5-33 (42)
B-100	113,900	200,000	60	78	63 ft 10 in.	15 ft 4 in.	12 ft 5 in.	10 ft 6 in.	GM 16V-71N	635 hp at 2,100 rpm	18.0-25 (20)/21.0-49 (28) dual
B-110	116,900	220,000	67	85	63 ft 10 in.	15 ft 4 in.	12 ft 5 in.	11 ft 2 in.	GM 16V-71N	635 hp at 2,100 rpm	18.0-25 (20)/21.0-49 (32) dual
Fruehauf:*											
D2-M20/28	10,625	50,000	20	22	28 ft 3 in.	7 ft 11½ in.	9 ft 7 in.	9 ft 7 in.	....	....	10.0-20 (12)
X2-M22	11,800	60,000	22.75	29.4	36 ft 8 in.	7 ft 7 in.	8 ft 10¼ in.	8 ft 9 in.	....	....	11.0-22.5 (12)
EI & B2 M280†	13,490	30,000	10.4	11.9	46 ft 8½ in.	7 ft 11½ in.	8 ft ½ in.	8 ft ½ in.	....	....	10.0-22 (12)
International Harvester 180PH	103,720	170,000	50	59.6	....	....	....	11 ft 10 in.	GM 12V-71N	456 hp at 2,100 rpm	18.0-25 (28)/18.0-25 (29)

* Trailers only, no engines.
† Train: semi and 4-wheeler.
‡ Train model available with doubled capacities.

use with relatively free-flowing materials, since large pieces of rock or shale might jam in the smaller bottom opening when the vehicle is dumping. The tractor may be of the 4-wheel type, which is particularly well suited to long high-speed hauls; or it may be an overhung-engine type of prime mover without a steerable front axle. This latter type is particularly adapted to working in soft soil conditions or around structures where very good maneuverability is required. (See Table 13-2 for typical bottom-dump specifications.)

Fig. 13-5 100-ton bottom dump being loaded by belt loader. (*Euclid, Inc., subsidiary of White Motor Co.*)

The bottom dump has the following characteristics:

1. Adaptable to free-flowing materials such as earth, sand, gravel, crushed stone, and the like.
2. When fitted with large single tires having air pressures of 30 to 50 psi, it offers better flotation than the rear-dump unit. It can thus haul through soft oils or sand better than the rear dump.
3. Adaptable to long, level, high-speed hauls, particularly when fitted with large-diameter dual tires. The machine is generally geared for higher travel speed than a rear dump, resulting in reduced gradeability and performance characteristics. The rear dump is built for pulling up steep grades; the bottom dump is built for longer, level hauls.
4. Spreads its load in windrows by dumping while in motion. It can also dump in drive-over hoppers, in which case it stops during the dumping operation.
5. Ability to pull on steep grades is restricted because of unfavorable power-weight ratio compared to rear dumper and less weight on drive wheels, which limit traction in soft soils or wet weather. In general, long, adverse grades should not exceed 5 percent for best application.
6. Low cost per ton of carrying capacity, and mechanical simplicity of operation and maintenance.

When equipped with large, single, flotation-type tires, and to a lesser degree when fitted with large-diameter duals, the bottom dump offers a significant advantage over dual-tired rear-dump equipment when the bearing capacity of the soil is low. In general, the lower the air pressure in the tire, the better the flotation. While there are certain exceptions to this statement, the inflation pressure of a tire can be used as a rough index of its flotation ability. In recent years, the so-called "wide base" tire has come into vogue.* This tire has a wider cross section or "bulge" and, as it sinks further into soft ground, its contact area increases exceptionally rapidly. This adds greatly to the flotation effect. Generally, the dual-tire rear-dump trucks have

* See Tires for Construction Equipment in Sec. 6, Engineering Fundamentals.

tire inflation pressures of 60 to 80 psi, whereas the single-tired bottom-dump equipment uses tires inflated to 30 to 50 psi.

Use bottom dumps when:

1. The material to be hauled is relatively free-flowing.
2. The hauling unit has unrestricted dumping into a drive-over hopper or the load is to be spread in windrows.
3. The haul is relatively level, thus permitting high-speed travel.
4. Long grades on the loaded haul do not exceed 3 to 5 percent. This is recommended as a general rule for maximum unit efficiency, but it is not the measure of maximum gradeability of a bottom dump.

Side-dump Load-on-back Truck Side-dump trucks are widely used in quarry-rock hauling and stripping operations. Many of the cement plants and large stone quarries which formerly used rail equipment had crushers and hoppers which were adapted to side-dump railroad cars. It was a logical evolution that, in the changeover from rail to truck, the side-dump vehicle should be considered (see Fig. 13-6 and Table 13-3).

Fig. 13-6 20-ton Euclid chassis with Easton side-dump body. (*Easton Car & Construction Co., Easton, Pa.*)

Because of the necessity of dumping over the drive tires and the design requirement for stability, the predominant type of chassis used with side-dump bodies is the tandem-axle or 6-wheel variety, especially where an integral hoist is required. The 4-wheel type of truck has been used in quarry operations where independent hoists are used at the crusher or hopper. Side-dump bodies are used in quarry stripping to dump the overburden "on the fly," whereas the rear-dump truck must stop, reverse, and dump its load over a waste bank.

For example, an engineer selecting equipment for a stripping operation might have a choice of a tandem-axle vehicle with side-dump body carrying 20 tons payload or a 4-wheel vehicle with single-drive axle mounting a rear-dump body carrying 20 tons payload. He would weigh the merits of side dumping versus end dumping and would also analyze the differences in the two hauling vehicles in traction, complexity of maintenance, performance characteristics, tire capacities, dumping angle of the body, and other pertinent features. These factors are often more important than the question of side-dump or rear-dump method of getting rid of the load.

Load-on-back side dumps are custom built to suit specific applications, so "stock" models are not offered. Easton Car & Construction Co., Easton, Pa., is a prime supplier of such bodies for use on any manufacturer's rear-dump chassis.

Side-dump Tractor-trailer The side-dump tractor-trailer is of heavy-duty construction and is able to handle both free-flowing material and rock. The dump angle is approximately 55°, which is adequate for all but the most adhesive materials.

TABLE 13-3 Specifications for Side Dumps

Vehicle type	Shipping weight, lb	Payload, max, lb	Capacities, cu yd		Overall length	Overall width	Overall height	Loading height	Engine		Tires
			Struck	Heaped					Make and model	Flywheel horsepower	
Caterpillar/Athey:											
PD630	72,000	90,000	30	34	48 ft 11 in.	13 ft 3 in.	10 ft 2 in.	10 ft 10 in.	Cat 630B	360 hp at 1,900 rpm	29.5-35(34)
PD660	128,660	144,000	48	53½	56 ft 3 in.	17 ft 6 in.	11 ft 4 in.	8 ft 9 in.	Cat 660	500 hp at 1,900 rpm	21.0-49(36) dual
Euclid/Easton:											
203LDT/TD3650H	94,860	100,000	30	36	54 ft 4 in.	13 ft 10 in.	11 ft 0 in.	9 ft 9 in.	GM 12V-71N	475 hp at 2,100 rpm	18.0-49(24) dual

There are two kinds of side-dump trailers, both of which are of the "tub" type. One has fixed, flared sides and the other has a downfolding gate on one side so that there is no obstruction to the material when dumping. The first type of body is more often seen in quarry installations where there is a firm roadbed at the dumping point and the side of the body can extend down into the hopper area when dumping. The downfolding-side trailer is commonly used in stripping operations, where the side of the body chutes the load away from the trailer tires (see Fig. 13-7). The

Fig. 13-7 Euclid tractor with 50-ton Easton side-dump trailer.

unit moves forward while dumping. In some designs the body can dump either to the left or the right, depending upon operating requirements. (See Table 13-3 for typical specifications.)

Use side dumps when:

Bottom-dump criteria apply but where dumping is restricted to one direction (e.g., over a bank, where other hauling units cannot maneuver).

Rear-dump Tractor-trailer (Rocker) The overhung-engine type of prime mover which pulls a rear-dump semitrailer has some unusual and interesting operating

Fig. 13-8 Terex tractor with Athey rear-dump trailer. (*Athey Products Corp., Raleigh, N.C.*)

characteristics (see Fig. 13-8). This type of machine has excellent maneuverability because the overhung-engine tractor can turn 90° either to the right or to the left without moving forward. In some designs, the wheelbase can be shortened for turning by raising the body to the dump position. This moves the trailer wheels forward and closer to the prime-mover drive wheels, shortening the wheelbase and making it possible to obtain an extremely short turning diameter. This is of particular importance in tunnel work or pioneering on narrow mountain roads. Both the prime mover and the trailer are equipped with large single tires with rock treads. This gives the hauling unit better flotation characteristics than for a dual-tired rear dump. (See Table 13-1 for typical specifications.)

The tractor-trailer rear dump does not have the overall gradeability or traction of a rear-dump truck. It generally has a greater gross weight per horsepower than the load-on-back truck. On steep grades, the tractor-trailer unit has a weight transfer off the drive axle and back to the trailing axle. This limits its traction on wet or slippery haul roads.

The prime mover for the tractor-trailer rear dump is also designed to haul a rubber-tired scraper. It is thus possible to use the prime mover with a rear-dump trailer for rock hauling; then, when "scraper dirt" is encountered, the rock hauler can be parked, its wheels and tires removed, and a scraper bowl installed. The prime mover can then continue to operate, hauling a scraper rather than the rear-dump trailer. The interchangeability of the prime mover is of particular interest in regions where there is both rock and dirt to be hauled yet insufficient volume to keep a specialized rock hauler or earthmoving scraper consistently at work.

Use rocker-type dumpers when:

1. Distance from loading area to dumping area does not exceed 500 ft.
2. The maximum grade on the loaded haul does not exceed 10 percent. This is not a measure of the maximum gradeability of rocker-type dumpers, but it is a practical limit for efficient operation.

HAULER PERFORMANCE

Performance and Retarder Charts The performance potential or gradeability of a hauling unit can be expressed by means of performance and retarder charts. Once the variables of rolling resistance and grade resistance have been identified for a given haul road, these charts specify the maximum attainable speed for a given hauling unit when operating in each transmission range or gear. Performance and retarder charts for a 35-ton hauling unit are shown in Figs. 13-9 and 13-10. Each chart is actually two separate graphs placed side by side. The right-hand graph relates the power provided by the unit's power train, expressed as rimpull, to the vehicle speed in miles per hour. The left-hand graph relates the two major factors which dictate the required rimpull at any specified vehicle speed; these factors are vehicle weight and total resistance* (grade resistance plus rolling resistance) to forward motion. The procedure for reading the chart is as follows:

1. Determine the vehicle specifications. Select the appropriate chart showing the correct engine, transmission, gear ratio, and tire size.
2. Establish the vehicle weight both empty (net vehicle weight—NVW) and loaded (gross vehicle weight—GVW). If the actual NVW and GVW are different from the "rated" NVW and GVW, draw the correct lines on the graph.
3. Estimate the total resistance (rolling resistance plus grade resistance). Rolling resistance is always a positive number; grade resistance is positive for a vehicle traveling uphill and negative for a vehicle traveling downhill. Remember that rolling resistance and grade resistance are measures of the rimpull required to keep a vehicle in motion under particular haul-road conditions. The hauler must develop rimpull equal to this total resistance in order to maintain a constant speed. Total resistance is expressed—in these charts—as a percent of the vehicle's weight.
4. When the total resistance as a percent of vehicle weight has been established, locate it on the vertical scale at the far left side of the chart. Read down the appropriate slanted line. Stop at the vertical GVW line if the hauler is loaded, or at the vertical NVW line if the hauler is empty.
5. Extend the point of intersection horizontally to the performance curve. The intersection of this horizontal line and the performance curve gives the vehicle's speed.

* See Sec. 6, Engineering Fundamentals.

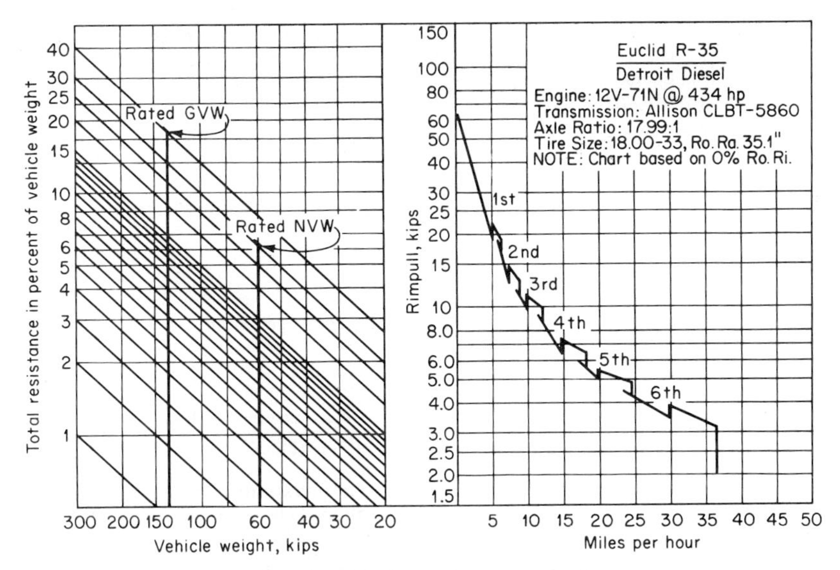

Fig. 13-9 Hauler performance chart.

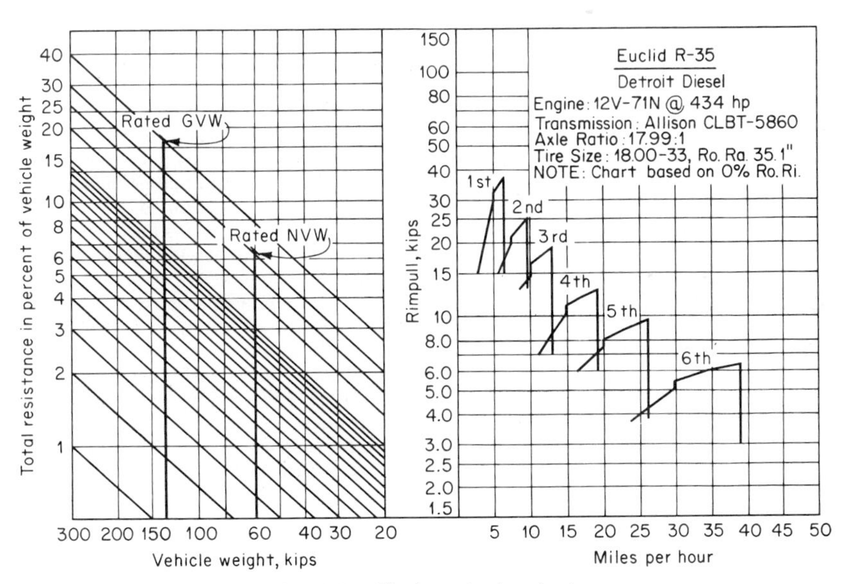

Fig. 13-10 Hauler retarder chart.

6. When the horizontal "required rimpull" line intersects both the converter and lock-up (i.e., "direct drive") rimpull curve in a given gear range, two interpretations could be made. For example, as illustrated in Fig. 13-11, the speed selection could be either 14.3 mph in converter or 16.7 mph in lock-up. As a rule of thumb, select the lock-up speed range when accelerating and select the converter speed range when slowing down.

Retarder charts are read in a similar manner to a performance chart, although the values shown in the total resistance column are actually negative numbers. Thus, for example, a negative 13 percent grade with a 3 percent rolling resistance is read as 10 percent total negative resistance. The resulting horizontal line (from step 5, above) may cross two or more retarder rimpull curves. The retarding speed range falls between the first and last points of the intersection which the horizontal line makes with the curve.

Any of the speeds which fall within the maximum and minimum required rimpull values may be used by selecting the appropriate transmission range, by varying the amount of oil in the retarder, or by accelerating the engine above the speed at which it is driven by the other members of the power train.

In the event detailed specifications are not available when a preliminary performance check is being made, vehicle speed in miles per hour can be determined when the total payload and net weight of the vehicle as well as the horsepower of the engine are known. The following formula is used:

$$\text{Vehicle speed in mph} = \frac{\text{net horsepower at drive wheels} \times 375}{\text{GVW in tons} \times [40 + (20 \times \text{percent of grade})]}$$

The net horsepower at the drive wheels equals the maximum engine horsepower at governed speed times 0.81. This assumes a 10 percent loss in efficiency due to the power requirements of driving accessories such as fan, air compressor, generator or alternator, and hydraulic pumps. Another 10 percent is deducted for friction losses in the hauler chassis.

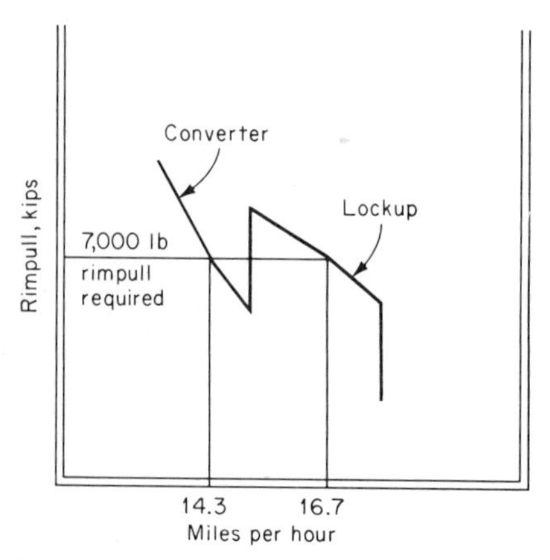

Fig. 13-11 Converter speed versus lock-up speed.

Each percent of grade is approximately equivalent to 20 lb of rimpull or tractive effort per ton of gross weight. The above formula includes a figure of 40 lb per ton for rolling resistance, which is representative for hauling equipment which is operated over well-maintained haul roads. The hauler is thus working against the equivalent of a 2 percent grade because of road resistance even when hauling on the level. While this figure is somewhat empirical and will vary according to the haul roads and basic vehicle design, it is typical for both on-highway and off-highway service under well-maintained conditions.

Applying this formula to a 35-ton hauler with a gross weight of 65.25 tons and a 434-hp engine rating, the results for a 4 percent grade are as follows:

$$\text{Speed} = \frac{(434 \times 0.81) \times 375}{65.25 \times [40 + (20 \times 4)]} = 17 \text{ mph}$$

Now change the rolling resistance from 40 lb per ton to 80 lb per ton. This would be a typical value for a soft or poorly maintained dirt surface:

$$\text{Speed} = \frac{(434 \times 0.81) \times 375}{65.25 \times [80 + (20 \times 4)]} = 12.5 \text{ mph}$$

Poor haul roads can drastically reduce the hauling speed. It is common practice in construction and mining projects to use motor graders for haul-road maintenance where high production is required. Graded and well-drained haul roads, permitting two-way traffic and providing a wide radius and good visibility on curves, are important to lowering hauling costs.

Operating Conditions In the tables used for estimating hauler production and costs, the data are frequently grouped under the headings "favorable," "average," and "unfavorable." These terms assist in further identifying the hauling job.

Favorable conditions:

1. Material being excavated and hauled:
   Topsoil
   Loam/clay mixture
   Compacted coal
   "Tight" earth (no rock)
2. Loading area (cut or borrow):
   Unrestricted in length or width
   Dry and smooth (or under constant maintenance)
3. Rolling resistance:
   Under 4 percent
4. Supervision:
   Constant at both the loading and dumping areas
5. Weather:
   Little rainfall or other weather delays
   Ambient temperatures of 40° to 60°F

Average conditions:

1. Material being excavated and hauled:
   Clay with some moisture
   Soft or well-ripped shale
   Loose sand with some binder
   Mixtures of different earths
   Sand/gravel mixture
2. Loading area (cut or borrow):
   Some restrictions in length or width
   Dry, with some loose material
3. Rolling resistance:
   4 to 8 percent
4. Supervision:
   Intermittent at both the loading and dumping areas
5. Weather:
   Moderate rainfall or other weather delays
   Ambient temperatures of 60° to 100°F

Unfavorable conditions:

1. Material being excavated and hauled:
   Heavy (dense) or wet clay
   Loose sand with no binder
   Coarse gravel (no fines)
   Caliche or unripped shale
   Frequent boulders or rock outcroppings
2. Loading area (cut or borrow):
   Tight restrictions in length or width
   Wet, slippery, and/or soft (not maintained)
   Units load uphill or on a side slope
3. Rolling resistance:
   Above 8 percent
4. Supervision:
   None (e.g., no spotters) at either the loading or dumping area
5. Weather:
   Excessive rainfall or other weather delays
   Ambient temperatures in excess of 100°F or below 0°F

Seldom is an entire job—considering individual segments of the haul cycle and equipment costs—given a blanket description of "favorable," "average," or "unfavorable." Weather may be a consideration only if the hauling job must be completed within a fixed time period.

EXAMPLE: A contractor is required to move 400,000 bcy of rock in 6 calendar months, or 180 days.

$$6 \text{ calendar months} \times 24 \text{ working days per month} = 144 \text{ working days}$$

The estimator determines that 312.5 bcy per hr can be moved.

$$312.5 \text{ bcy per hr} \times 10 \text{ hr} = 3{,}125 \text{ bcy per day}$$
$$400{,}000 \text{ bcy} \div 3{,}125 \text{ bcy per day} = 128 \text{ working days}$$

The estimator has provided approximately 11 percent margin between the "days available" and the "days required." This margin could provide for weather delays and other unforeseen events. It may also encompass equipment breakdown—poor machine availability. By proper job supervision, driver education, and other improvements, the actual hourly hauling rate might well be increased beyond the estimated value.

Machine availability is the percentage of the scheduled working hours during which a unit's mechanical condition is such that the unit is able to haul. Preventive maintenance, regular inspection, and servicing of hauling units are very important in maintaining high unit availability. Proper driver training to prevent misuse of hauling equipment is also important.*

Table 13-4 lists average availabilities for hauling units as functions of hauling unit type and anticipated life. These values may be considered as adequate for estimating purposes, assuming good preventive-maintenance practices, "average" operating conditions, and operation for 50 percent or less of any 24-hour period.

TABLE 13-4 Availabilities of Hauling Units

Types of unit	Years of anticipated life					
	1	2	3	4	5	6
	Average availability, percent*					
Rear dumps	90	89	88	86	84	83
Bottom dumps, side dumps	92	90	89	88	86	86
Rocker-type dumpers	88	86	83	80	78	78

* Factors such as severe operating conditions, multishift operations, and unsatisfactory preventive maintenance practices can reduce machine availability by as much as 20 percent below the levels considered "average." Similarly, favorable conditions may increase first-year values by as much as 5 percent and can show an increasing advantage over average conditions and subsequent years.

One method of allowing for maintenance and repair time is to include spares or standby units in the hauling fleet. Spare units are an inexpensive form of production "insurance." A standby unit does not incur any operating costs and should, therefore, be charged to the job at its ownership cost only. The hauling cost per ton or yard will be increased only slightly by having spare units on standby. On the other hand, the lack of spare units can result in costly production losses. To calculate the required total fleet, operating haulers plus spares, the following simple formula may be used:

$$\text{Required fleet} = \frac{\text{number of operating haulers} \times 100}{\text{percent machine availability}}$$

* See Sec. 7, Equipment Economics.

Assume that an estimate calls for the continuous use of 10 trucks and that their type, age, and condition indicate 85 percent machine availability. A fleet of 11.8 or 12 trucks will then be required to satisfy the production requirements. Two spares would be provided to ensure that 10 haulers are actually in operation at all times.

ESTIMATING HAULER PRODUCTION

Hauler production under optimum conditions can be estimated by means of the formula:

$$\begin{array}{l}\text{Hauler production}\\ \text{per 60-minute hour}\\ \text{(tons or cubic yards)}\end{array} = \frac{\text{60 minutes}}{\text{time to complete one cycle (min)}} \times \begin{array}{l}\text{hauler's payload}\\ \text{per cycle}\\ \text{(tons or cubic yards)}\end{array}$$

Payload The term "rated payload" refers to the maximum recommended load for a hauling unit as established by the manufacturer. Either the volumetric capacity or the rated payload will limit the allowable load, depending on the characteristics of the material which is to be hauled. To determine a hauler's payload per trip, the bank (in place) and loose (excavated) weights per cubic yard of the material should first be established.* A check can then be made to determine whether the rated payload is exceeded when the hauling unit is filled to its heaped capacity with loose material. Whichever situation governs, heaped capacity or rated payload, will establish the maximum payload. The payload in pounds of a hauling unit should not be exceeded, since such overloading could adversely affect the life and performance of the unit.

EXAMPLE: The maximum capacity of a typical 35-ton rear-dump unit is as follows:

Struck capacity (SAE)	23.3 cu yd
Heaped capacity (3:1)	27.0 cu yd
Heaped capacity (2:1 SAE)	29.0 cu yd
Rated payload	70,000 lb

This specification indicates that the hauler can be loaded (with the material heaped at 2:1 on the body) with 29 cu yd of material providing that the 29 cu yd weigh no more than 70,000 lb.

Assume that the unit is used to haul material with a unit weight of 90 pcf in its bank condition and a unit weight of 77.8 pcf in a loose state. The rated payload of 70,000 lb for the hauling unit will not be exceeded until the payload amounts to 33.3 lcy. However, this hauling unit is not equipped with any devices (e.g., top extensions) for increasing its volumetric capacity beyond its rated capacity. Therefore, the maximum payload which can be carried without excessive spillage is estimated at 29.0 cu yd, loose measure. The hauler's payload per cycle or trip can also be expressed as 60,900 lb, as 30.45 tons, or as 25.0 bcy.

Cycle Time In preparing an estimate of hauler production, each cycle is considered to be made up of the following parts:

$$\begin{array}{l}\text{Time to complete}\\ \text{one cycle}\end{array} = \begin{array}{l}\text{loading}\\ \text{time}\end{array} + \begin{array}{l}\text{haul}\\ \text{time}\end{array} + \begin{array}{l}\text{dump-and-turn}\\ \text{time}\end{array} + \begin{array}{l}\text{return}\\ \text{time}\end{array} + \begin{array}{l}\text{spot-and-delay}\\ \text{time}\end{array}$$

EXAMPLE: A 35-ton rear dump, being loaded with shale by a 9-cu yd wheel loader, might have the following cycle time:

Load: 3 passes at 0.90 min per pass	2.70 min
Haul: 2,700 feet at 18.0 mph avg	1.70 min
Dump and turn	1.30 min
Return: 2,700 feet at 21.0 mph avg	1.46 min
Spot and delay	0.25 min
Total time to complete one cycle	7.41 min

Loading Time A hauling unit may be loaded by a power shovel, dragline, clam, belt loader, front-end loader (either rubber-tired or track-type), or by a gravity hopper.

* See Material Characteristics in Sec. 6, Engineering Fundamentals.

Loading time is simply the hauler's calculated payload divided by the loading rate (e.g., pounds per minute).

*Power Shovels and Draglines.** Tables 13-5 and 13-6 show typical output rates, in loose cubic yards per 60-minute hour, for power shovels and draglines. Both of these tables are based on data published by the Power Crane and Shovel Association. The tabulated production figures are based on 90° swing (shovel) and 120° swing (dragline) from the pile to the hauler.

TABLE 13-5 Power-shovel Output [in loose cubic yards per 60-minute hour (90° swing)]

Bucket capacity, lcy	Easy digging (0.85–1.00 dipper factor)	Medium digging (0.80–0.90 dipper factor)	Hard digging (0.70–0.80 dipper factor)	Extreme digging (0.40–0.60 dipper factor)
$\frac{3}{4}$	141–166	111–125	82–94	47–76
1	182–214	142–160	107–122	62–100
$1\frac{1}{4}$	222–262	171–193	131–150	75–122
$1\frac{1}{2}$	261–307	200–225	155–177	89–144
$1\frac{3}{4}$	298–351	226–254	178–204	102–165
2	334–393	253–285	200–228	114–185
$2\frac{1}{2}$	408–480	303–341	241–275	137–222
$2\frac{3}{4}$	446–525	328–369	260–297	149–242
3	478–563	354–398	280–320	160–260
$3\frac{1}{2}$	546–643	406–456	318–363	182–296
4	609–717	451–507	356–407	203–330
$4\frac{1}{2}$	672–791	498–561	391–447	223–363
5	735–865	546–615	428–489	245–398
$5\frac{1}{2}$	793–934	592–666	466–532	266–433
6	853–1,005	638–718	502–573	287–467
$6\frac{1}{2}$	914–1,075	683–767	540–617	308–501
7	973–1,145	730–821	575–657	329–535
8	1,087–1,280	826–930	646–738	369–600
9	1,210–1,425	919–1,032	718–821	410–666
10	1,340–1,580	1,018–1,145	788–900	450–731
11	1,462–1,720	1,108–1,246	855–976	505–821
12	1,580–1,860	1,198–1,348	923–1,057	528–857
13	1,700–2,000	1,288–1,450	990–1,133	565–918
14	1,820–2,140	1,378–1,550	1,060–1,210	597–982
15	1,945–2,290	1,465–1,650	1,123–1,283	643–1,045

EXAMPLE: Determine the load time of a 35-ton rear dump working under a 5-cu yd shovel in shale. Digging is classified as "hard."

$$\text{Loading time} = \frac{29 \text{ lcy}}{489 \text{ lcy/hr}} \times 60 \text{ min per hour} = 3.56 \text{ min}$$

As a check on the hourly output of a shovel or dragline, loading time can also be computed as follows:

$$\text{Loading time} = \frac{\text{number of shovel or dragline}}{\text{passes required}} \times \frac{\text{shovel or dragline cycle time}}{\text{for one pass}}$$

The number of required passes is determined as follows:

$$\text{Number of passes required} = \frac{\text{payload of hauling unit (lcy)}}{\text{shovel dipper or bucket capacity (lcy)} \times \text{shovel dipper or bucket factor}}$$

Shovel dipper and dragline bucket factors for various digging conditions are listed

* See Sec. 10, Excavators.

in Table 13-7. As a rule of thumb, the normal cycle time for one 90° pass of a shovel can be approximated as:

Easy digging conditions.................. 0.30 min
Medium digging conditions............... 0.40 min
Hard digging conditions.................. 0.50 min
Extreme (rock) digging conditions......... 0.70–1.00 min

Normal cycle time for one 120° pass of a dragline bucket can be approximated as:

Easy digging conditions.................. 0.40 min
Medium digging conditions............... 0.50 min
Hard digging conditions.................. 0.60 min

Cycle times for bucket dipper swings other than 90° and dragline swings other than 120° should be adjusted appropriately.

TABLE 13-6 Dragline Output [in loose cubic yards per 60-minute hour (120° swing)]

Bucket capacity, lcy	Easy digging (0.85–1.00 dipper factor)	Medium digging (0.80–0.90 dipper factor)	Hard digging (0.80–0.90 dipper factor)
¾	102–120	82–93	63–71
1	130–153	105–118	81–91
1¼	155–183	126–142	98–110
1½	177–208	146–164	116–131
1¾	198–233	164–185	132–149
2	208–257	181–204	147–166
2½	253–298	212–238	175–197
3	280–330	234–264	195–219
3½	306–360	256–288	216–243
4	332–391	280–315	238–268
4½	361–426	305–344	262–295
5	389–458	330–372	284–320
5½	416–490	355–399	310–349
6	447–526	381–429	334–376
6½	475–559	406–468	359–404
7	509–599	433–487	385–433
8	579–681	489–550	436–480
9	648–763	546–614	493–555
11	784–923	655–737	586–659
13	920–1,082	761–856	683–768
15	1,056–1,242	864–972	778–875

EXAMPLE: Consider the same 35-ton rear dump being loaded by a 5-cu yd shovel in shale (2,100 lb per lcy).

$$\text{Number of passes required} = \frac{29 \text{ lcy}}{5 \text{ lcy} \times 0.80 \text{ dipper factor}} = 7.25 \text{ (use 7 passes)}$$

$$\text{Loading time} = 7 \text{ passes} \times 0.50 \text{ min per pass} = 3.50 \text{ min}$$

Generally, three to six passes of a shovel or dragline to fill a hauling unit are considered a good balance between the hauler and the loading device. The fewer the passes the better, of course, providing that the size of the hauler body is not too small in comparison to the throat opening of the shovel dipper or dragline bucket (so that spillage and impact of material on the hauler are minimized). Figure 13-12 illustrates recommended shovel capacity ranges.

TABLE 13-7 Shovel Dipper and Bucket Factors for Various Digging Situations

Condition	Shovel dipper factor	Dragline bucket factor	Type of material	Specific materials
Easy digging . . .	0.85–1.00	0.85–1.00	Loose, soft, free-flowing materials. Materials which will fill the dipper or bucket to capacity and frequently provide a heaped load. Overload compensates for swell of material.	Bituminous coal Cinders or ashes Clay—with loose gravel Clay—sandy Earth—loose Gravel—small Loam Muck Sand—dry or moist Very well-blasted material
Medium digging	0.80–0.90	0.80–0.90	Harder materials that are not difficult to dig without blasting but that break up and swell, causing voids in the dipper or bucket.	Anthracite coal Clay—wet or dry Clay gravel—packed Earth—packed Gravel—coarse
Hard digging . . .	0.70–0.80	0.80–0.90	Materials that require some breaking up by blasting or shaking. More bulky and somewhat hard to penetrate, causing voids in the dipper or bucket.	Clay— heavy, wet, sticky Gravel—cemented Gravel—with large boulders Gumbo—heavy, wet Limestone—well broken Ore formations—not of rock character, but requiring some blasting Sand rock—other blasted rock Shale—blasted
Extremely hard digging	0.40–0.60	Not applicable	Blasted rock, hardpan, or other bulky materials which cause considerable voids in the dipper and are difficult to penetrate.	Caliche Clay—tough, rubbery, that shaves from the bank Conglomerate Granite Limestone Sandstone Taconite Shale—hard, tough Traprock Any of these, blasted to large pieces, mixed with fines and dirt

*Belt Loaders.** These are used to excavate material, elevate it to a specified height by means of a continuous belt or series of buckets, and discharge it into a hauling unit. Belt loaders, as well as overhead hoppers, are particularly advantageous when the material is free-flowing. Loading rates vary according to the discharge capacity of the belt or hopper throat opening. A Super BV belt loader, for example, has a maximum discharge capacity of 7,785 lcy per 60-minute hour.

* See Belt Loaders in Sec. 11, Loaders.

*Front-end Loaders.** In loading a hauling unit with a rubber-tired or track-type front-end loader, maximum efficiency can be obtained by placing the hauling unit as close to the loader as possible, so as to reduce travel distance between the two. In matching a hauling unit with a front-end loader, it is important to consider whether the loader's specified dump height (the distance from the tip of the bucket to the ground when the bucket is tilted forward to its maximum extent) is compatible with the hauler's loading height. Such consideration ensures that the tip of the loader bucket will clear the top of the hauler body. It is also important to check the front-end loader's specified reach (the horizontal distance from the tip of the bucket to the front of the tires when the loader bucket is in the maximum dump-height position).

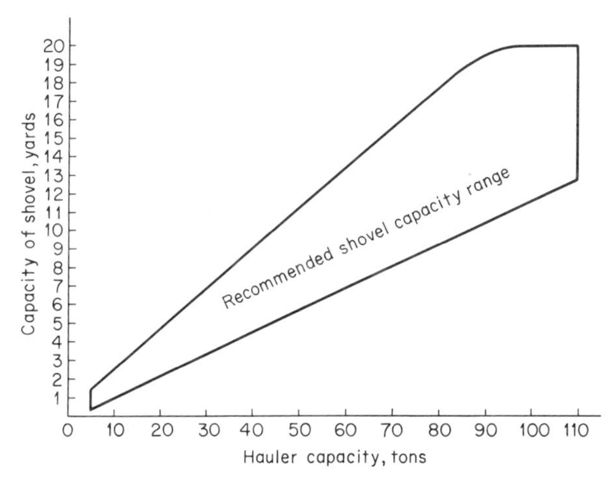

Fig. 13-12 Match between size of hauler and size of shovel.

The reach should be such that the load will be properly distributed over the width of the hauler body.

For estimating purposes, assuming minimum haul distances (e.g., between material pile and hauler) and average digging conditions, typical cycle times for rubber-tired front-end loaders are supplied in Table 13-8.

TABLE 13-8 Typical Cycle Times for Front-end Loaders When Loading Hauling Units

Rated bucket capacity, lcy	*Loader cycle time, min/pass*
Less than 4	0.40–0.60
4–10	0.50–1.00
10–20	0.80–1.50

Travel Time Whether the hauler is loaded and proceeding to the dump area or is empty and returning to the loading site, travel time is determined as follows:

$$\text{Travel time (min)} = \frac{\text{length of road section (ft)}}{\text{average speed over road section (mph)} \times 88}$$

The factor 88 is 5,280/60 and is used to convert feet and miles per hour into minutes.

The determination of a hauler's maximum speed over a given section of road where both rolling resistance and grade resistance are known has already been discussed under Hauler Performance. However, the hauler will not operate at its maximum speed over the entire length of a road section. The theoretical maximum speed must therefore be reduced to a practical average to compensate for such items as lower

* See Front-end Loaders in Sec. 11, Loaders.

speeds during the acceleration or deceleration of the vehicle. This reduction can be accomplished by introducing a speed factor whereby

$$\text{Average speed (mph)} = \text{maximum speed (mph)} \times \text{speed factor}$$

A speed factor of 1.00 means that the vehicle will continuously maintain its theoretical maximum speed along the road section, while a speed factor less than or greater than 1.00 indicates that the average vehicle speed will be slower or faster, respectively, than this theoretical maximum speed. Several variables affect the speed factor: the weight-to-horsepower ratio for the hauling unit, the length of the road section, and the initial or end speed of the hauler. The weight-to-power ratio is determined as follows:

If the hauler is loaded:

$$\text{Weight-to-power ratio} = \frac{\text{gross vehicle weight}}{\text{flywheel or net horsepower}}$$

If the hauler is empty:

$$\text{Weight-to-power ratio} = \frac{\text{net vehicle weight}}{\text{flywheel or net horsepower}}$$

A hauler with a higher weight-to-power ratio will have, relatively, less power available to the drive wheels to overcome the resistance of the grade and the haul-road surface than will a unit with a lower weight-to-power ratio. Consequently, the higher a hauling unit's weight-to-power ratio, the longer it will take to build up to its maximum speed and the lower will be its average speed over a given haul-road section. While acceleration and momentum affect only the travel speeds at the beginning and end of a given road section, the length of the road section also is important. The longer the section, the more time will be spent at maximum speed and the nearer to 1.00 will be the speed factor. A hauler which enters or leaves a road section at or near the maximum speed will have a speed factor closer to 1.00 than a unit which starts from stop or must slow down at the end of the road section.

The effect of a vehicle's momentum is evident when a hauler enters a road section which has a different grade than the previous section. If the vehicle enters a grade traveling at a faster speed than its theoretical maximum speed on that grade, then its average speed on that grade will be greater than the maximum speed (as determined from the performance charts) and the speed factor will be greater than 1.00. Likewise, if the hauler enters a grade traveling at a slower speed than the maximum attainable speed on that grade, then the average speed on the new grade will be less

TABLE 13-9 Factors for Conversion of Maximum Speed to Average Speed—Unit Starting from Stop

Haul-road length, ft	Downgrade*			Level*			Upgrade*		
	1	2	3	1	2	3	1	2	3
0–250	0.45	0.40	0.35	0.38	0.34	0.30	0.30	0.25	0.20
251–500	0.54	0.48	0.42	0.44	0.40	0.36	0.36	0.31	0.26
501–750	0.63	0.56	0.50	0.50	0.46	0.42	0.42	0.37	0.32
751–1,000	0.72	0.64	0.56	0.56	0.52	0.48	0.47	0.44	0.38
1,001–1,250	0.80	0.72	0.62	0.62	0.58	0.54	0.55	0.50	0.44
1,251–1,500	0.87	0.80	0.70	0.68	0.64	0.60	0.60	0.55	0.50
1,501–1,750	0.94	0.85	0.75	0.75	0.70	0.65	0.65	0.60	0.55
1,751–2,000	1.00	0.90	0.80	0.85	0.80	0.70	0.75	0.70	0.60

* Weight-to-power ratios: 1 = 300 lb per hp and under; 2 = 301–400 lb per hp; 3 = 401 lb per hp and over.

TABLE 13-10 Factors for Conversion of Maximum Speed to Average Speed—Unit in Motion When Entering Road Section

Haul-road length, ft	Downgrade*			Level*			Upgrade*		
	1	2	3	1	2	3	1	2	3
0–250	0.70	0.62	0.54	0.65	0.58	0.50	0.60	0.54	0.46
251–500	0.80	0.70	0.60	0.75	0.66	0.56	0.68	0.62	0.52
501–750	0.90	0.78	0.66	0.85	0.74	0.62	0.75	0.70	0.58
751–1,000	1.00	0.86	0.72	0.95	0.82	0.68	0.85	0.78	0.64
1,001–1,250	1.10	0.94	0.78	1.05	0.90	0.74	0.95	0.86	0.70
1,251–1,500	1.20	1.00	0.84	1.15	0.98	0.80	1.00	0.95	0.76
1,501–1,750	1.30	1.10	0.90	1.25	1.05	0.85	1.05	1.00	0.80
1,751–2,000	1.40	1.20	0.96	1.35	1.15	0.90	1.10	1.05	0.84

* Weight-to-power ratios: 1 = 300 lb per hp and under; 2 = 301–400 lb per hp; 3 = 401 lb per hp and over.

than the maximum speed and the speed factor will be less than 1.00. These effects are due to the momentum that the hauler accumulates on the previous road section.

Speed factors for varying road lengths and conditions are supplied in Tables 13-9 and 13-10. When the length of a road section and the average travel speed are known, the travel time can be obtained directly from Table 13-11.

TABLE 13-11 Conversion of Travel Distance (ft) and Average Travel Speed (mph) to Travel Time (min)

Speed, mph	Travel distance, ft									
	100	200	300	400	500	600	700	800	900	1,000
3	0.379	0.757	1.136	1.515	1.893	2.272	2.652	3.030	3.409	3.788
4	0.284	0.568	0.853	1.136	1.420	1.705	1.989	2.273	2.557	2.841
5	0.227	0.454	0.681	0.908	1.136	1.363	1.591	1.818	2.045	2.273
6	0.189	0.378	0.568	0.757	0.946	1.136	1.325	1.515	1.705	1.894
7	0.162	0.324	0.487	0.649	0.811	0.974	1.136	1.299	1.461	1.623
8	0.142	0.284	0.426	0.568	0.710	0.852	0.994	1.136	1.278	1.420
9	0.126	0.252	0.378	0.505	0.631	0.757	0.883	1.010	1.136	1.263
10	0.113	0.227	0.341	0.454	0.568	0.681	0.795	0.909	1.023	1.136
12.5	0.091	0.182	0.273	0.363	0.454	0.545	0.636	0.727	0.818	0.909
15	0.075	0.151	0.227	0.303	0.378	0.454	0.530	0.605	0.681	0.757
17.5	0.065	0.129	0.194	0.259	0.324	0.389	0.454	0.519	0.584	0.649
20	0.057	0.113	0.170	0.227	0.284	0.341	0.397	0.454	0.511	0.568
22.5	0.050	0.101	0.151	0.202	0.252	0.303	0.353	0.404	0.454	0.505
25	0.045	0.090	0.136	0.181	0.227	0.272	0.317	0.363	0.408	0.454
27.5	0.041	0.082	0.124	0.165	0.206	0.248	0.289	0.330	0.371	0.412
30	0.038	0.076	0.113	0.151	0.189	0.227	0.265	0.303	0.341	0.379
32.5	0.035	0.070	0.104	0.139	0.174	0.209	0.244	0.279	0.314	0.349
35	0.032	0.065	0.097	0.129	0.162	0.194	0.227	0.259	0.291	0.324

Turn-and-dump Times The time required to dump the payload and to maneuver the hauling unit into position for the return route is dependent on the destination of the material, the type and maneuverability of the hauling unit, the type and condition of the material, and the road maintenance in the dumping area. The prediction of conditions at the dumping area is based on the estimator's personal experience and familiarity with either the hauling unit or the job. Representative times are listed in Table 13-12.

TABLE 13-12 Typical Turn-and-dump Times (min)

Operating conditions	Rear dumps	Bottom dumps	Side dumps
Favorable	1.0	0.4	0.7
Average	1.3	0.7	1.0
Unfavorable	1.5–2.0	1.0–1.5	1.5–2.0

Spot Time The time for the hauling unit to be spotted at the loading unit in a position ready for loading is dependent on the type of approach (straight ahead or back in), the absence or presence of a spotter to guide the driver, and the time interval between hauling units. Typical spot times are supplied in Table 13-13.

TABLE 13-13 Typical Spot Times (min)

Operating conditions	Rear dumps	Bottom dumps	Side dumps
Favorable	0.15	0.15	0.15
Average	0.30	0.50	0.50
Unfavorable	0.80	1.00	1.00

Delay Factors and Job Efficiency There are frequently found to be hazards or obstructions in the haul road for which time allowances should be made when computing the cycle time. In addition, conditions resulting in "continuous" delays may exist, which again increases cycle time. Typical items which initiate intermittent or continuous job delays, thereby reducing cycle efficiency, are identified in Table 13-14.

TABLE 13-14 Typical Delay Factors

Intermittent factors (Estimate delay time for each item)	*Continuous factors* (Estimate delay time over the entire haul and return route)
One-way haul roads	Extremely high, or variable, rolling resistance
Delay at passing points	Wet or slippery haul roads
Multiple curves or switchbacks	Unskilled operators
Blind corners	Long, downgrade hauls
Bridges	Inexperienced management and/or supervision
Underpasses	
Railroad crossings	
Cross traffic	

Obviously, an estimate which is too optimistic in predicting the hourly production ability of each hauling unit will prove to be unrealistic. It will be found that the specified production cannot be maintained and that the number of haulers assigned to the job is insufficient. An estimate must be based on a realistic evaluation of the sustained or average hauling production over a long period of time.

The "effective" working hour is a simple device which makes allowance for the unavoidable delays which are encountered on all operations. The 50-minute hour, which assumes that delays will average 10 minutes per clock hour, is frequently used as a basis for estimating. This is equivalent to an 83 percent overall job efficiency. The hauling cycle time, computed as explained earlier in this section and expressed in minutes, is then divided into 50 minutes to determine the average number of trips which will be completed each clock hour. An assumption of a 50-minute hour is probably adequate for average conditions. For those jobs considered to be "favorable" from the standpoint of possible delays, a 55-minute hour may be used, which is equivalent to a 92 percent overall efficiency. Conversely, projects which are subject to unusual delay and efficiency problems should be estimated on the basis of a 45- or even a 40-minute working hour, equivalent to overall efficiencies of 75 or 67 percent.

ESTIMATING HAULING COSTS

If an estimate of hauling costs is to be made with existing equipment on which complete and accurate cost records have been maintained, then, of course, such cost records should be used. The "Estimated Hourly Ownership and Operating Cost" form supplied as Fig. 13-13 has two uses: (1) it allows the estimator to calculate costs for a known machine and a specific application; (2) it allows the estimator to calculate machine costs for a new and, perhaps, untried hauling unit.

Ownership Costs* Depreciation is a major item in this category. Other items include interest, taxes, insurance, and storage.

TABLE 13-15 Annual Hours of Hauler Operation, by Type of Service

Type of service	*Hours operated per year*
Contracting—rear dump, bottom dump, side dump, belt loader, scraper, rocker-type rear dump	2,000
Industrial—rear dump, bottom dump, scraper	3,000
Mining—rear dump	3,500

Depreciation. The total amount to be depreciated equals the delivered price of the hauler (purchase price, optional equipment, and freight) less the original or "replacement" value of the tires (including tires, tubes, and federal excise or state taxes). This deduction is made because tire costs are carried as an operating expense.

If, at the end of a specified depreciation period (e.g., 10,000 hr), there will still be a residual value—for salvage or resale—in the unit, then such value will be deducted from the delivered price before computing the schedule of depreciation charges.

TABLE 13-16 Years to Complete Write-off for Haulers, AGC Guide*

Type of equipment	*Years to complete write-off*
Rear dumps, off-highway:	
5–12 cu yd (7.5–18 tons)	5
12–18 cu yd (18–27 tons)	6½
18–28 cu yd (27–42 tons)	6
28–35 cu yd (42–52 tons)	5½
35–74 cu yd (52–111 tons)	5
Bottom dumps, side dumps	5
Shuttle-type dumpers	5

* From *Contractors Equipment Ownership Expense*, Associated General Contractors of America, Washington, D.C.

Here is a method which may be used for cost estimating, although it is not meant to designate any legal depreciation for tax purposes. To determine the depreciation period, first select the "hours operated per year" from Table 13-15. Second, select the number of years until complete write-off from Table 13-16.

An alternate method of selecting the number of years to complete write-off is supplied in the "Depreciation Guidelines and Rules" prepared by the Internal Revenue

* See Sec. 7, Equipment Economics.

Service of the U.S. Treasury Department. This establishes equipment life based on specific applications, as indicated in Table 13-17.

TABLE 13-17 Years to Complete Write-off for Equipment, I.R.S. Guide*

Application	*Years to complete write-off*
General contract construction	5
Logging	6
Mining	10
Marine contract construction	12

* From I.R.S. Bulletin, July, 1962.

The depreciation period is the hours operated per year times the number of years to complete write-off. The hourly depreciation cost is the total amount to be depreciated divided by the depreciation period.

35-ton rear-dump hauling unit
Model

OWNERSHIP COST

1. DEPRECIATION: (No allowance made for salvage or resale value)

Purchase Price	58,300 lb		$ 75,820
Extras			
Severe application kit	4,000 lb	$ 2,695	
Optional tires (see below)	2,530 lb	$ 3,155	
	lb	$	
	lb	$	
	lb	$	
	lb	$	
	lb	$	
	lb	$	
	lb	$	
	lb	$	
Total Cost of Extras			$ 5,850
Freight – 64,830 lb @ $ 2.00 hundred pounds			$ 1,297
Delivered Price			$ 82,967
(Less) Original Value of Tires (Includes Tires, Tubes, and Taxes)			
Front Tires: 18.00-33 (32) rock-2 reqd.		$ 3,632	
Drive Tires: 18.00-33 (32) rock-4 reqd.		$ 7,264	
Trailer Tires:		$	
Total Value of Original Tires			$ 10,896
Total Amount to be Depreciated			$ 72,071

Depreciation Period

Hours Operated per Year:	2,000
Number of Years to Complete Write-off:	5
Depreciation Period (Hours):	10,000

Hourly Depreciation Cost . . . . . $ 7.21

2. INTEREST, TAXES, INSURANCE, and STORAGE:
(13%–Interest 9%, Taxes 2%, Insurance and Storage 2%).

(13%) X Average Yearly Investment $ 49,780
——————————— (OR: .039 X 82,967) . . . . . $ 3.24
Hours Operated per Year: 2,000

A. TOTAL *HOURLY* OWNERSHIP COST . . . . . $ 10.45

Fig. 13-13 Estimated hourly ownership and operating cost for rubber-tired equipment.

Interest, Taxes, Insurance, Storage. The hourly cost of these items will depend on the prevailing rates and on the total hours and number of years over which the hauling unit is depreciated. The hourly cost is calculated as follows:

$$\text{Hourly cost of interest, taxes, insurance, storage} = 13\% \times \frac{\text{average yearly investment}}{\text{hours operated per year}}$$

This 13 percent, of course, may vary throughout the different sectors of the hauling-equipment market. It is comprised of 9 percent for interest, 2 percent for taxes, and 2 percent for insurance and storage. The 9 percent interest rate is the simple-interest equivalent of the 6 percent add-on interest rate commonly used in equipment financing. The average yearly investment is the delivered price of the hauling unit times the appropriate percent factors from Table 13-18.

EXAMPLE: A five-year depreciation period is equal to 20 percent write-off per year. Therefore, the residual value is decreased by 20 percent of the original depre-

OPERATING COST

3. TIRE REPLACEMENT COST: (Includes Tires, Tubes, and Taxes.)

Original Value of Tires: $ 10,896 / Estimated Tire Life: 2,5000 hours .... $ 4.36

4. TIRE REPAIRS: at 12 % of Tire Replacement Cost .... $ .52

5. GENERAL REPAIRS: (Including Parts and Labor)

(45 %) X *Hourly* Depreciation Cost: $ 7.21 X Depreciation Period: 10,000 Hours / 10,000 Hours .... $ 3.24

6. FUEL COST:

(Est. Consumption: 9.0 Gal/Hr) X ($ 0.14 per Gallon) .... $ 1.26

7. SERVICE COST: (Oil, Grease, Including Greasing Labor)

(Est. at 1 3 of Fuel Cost .... $ .42

8. CUTTING EDGE REPLACEMENT COST: (Scraper Only)

($ ____ per Section) X (4) / 1,000 Hours .... $ ____

9. OPERATOR: (Including Fringe Benefits)

($ 6.00 per Hour) .... $ 6.00

B. TOTAL *HOURLY* OPERATING COST .... $ 15.80

C. TOTAL ESTIMATED *HOURLY* OWNERSHIP AND OPERATING COST .... $ 26.25

DOES NOT INCLUDE PROFIT, OVERHEAD, AND SUPERVISION

Fig. 13-13 (*Continued*).

TABLE 13-18 Average Yearly Investment versus Years to Complete Write-off

Years to complete write-off	*Average yearly investment as a percent of delivered price*
1	100
2	75
3	66.6
4	62.5
5	60
6	58.3
7	57.1
8	56.2
9	55.5
10	55

ciable amount for each year of service until, at the end of the fifth year, the machine is completely depreciated:

Year	*Value at beginning of year, percent of first cost*
1	100
2	80
3	60
4	40
5	20
	300

Average yearly investment = 300 ÷ 5 = 60 percent of delivered price.

For simplified estimating, the hourly cost of interest, taxes, insurance, and storage can be taken directly from Table 13-19.

TABLE 13-19 Hourly Cost of Interest, Taxes, Insurance, and Storage per $1,000 of Delivered Price (9 percent interest, 2 percent taxes, 2 percent insurance and storage)

Years to complete write-off	Depreciation period, hr				
	7,500	10,000	15,000	20,000	25,000
1	$0.017	$0.013	$0.009	$0.007	$0.005
2	0.026	0.020	0.013	0.010	0.008
3	0.035	0.026	0.017	0.013	0.010
4	0.043	0.033	0.022	0.016	0.013
5	0.052	0.039	0.026	0.020	0.016
6	0.061	0.046	0.030	0.023	0.018
7	0.069	0.052	0.035	0.026	0.021
8	0.078	0.059	0.039	0.029	0.023
9	0.087	0.065	0.043	0.033	0.026
10	0.095	0.072	0.048	0.036	0.029

EXAMPLE: A hauling unit with a delivered price of $80,000 is to be depreciated over a six-year period, with a total depreciation period of 10,000 hours. The hourly costs of interest, insurance, taxes, and storage are

$$\$80{,}000 \times \frac{\$0.046}{\$1{,}000} = \$0.68 \text{ per hr}$$

Operating Costs* These include the costs of tire replacement and repair, general repairs, fuel, and miscellaneous servicing.

Tire Replacement Cost.† Worn tires can be replaced with a complete set of new ones; or a certain percentage of them—measured over the life of the unit—can be recapped. According to industry averages, 75 percent of all tires can be recapped if this is done before the tread is worn down into the casing. Recapping costs average approximately 50 percent of the original value of the tire. Recap life is computed, again as an average, at 75 percent of the original tread life. When exact recapping cost and life are known or can be accurately projected, the following formula can apply:

$$\text{Hourly tire replacement cost} = \frac{\text{replacement cost} + \text{recap cost}}{\text{original life} + \text{recap life}}$$

Tire Repairs.† In estimating total tire costs for a hauling unit, it is customary to include an allowance for tire repair cost. This can be expressed as a percentage of hourly tire replacement cost, as indicated in Table 13-20.

TABLE 13-20 Tire Repair Factors Expressed as Percentages of Hourly Tire Replacement Cost

Operating conditions	*Percent of tire replacement cost*
Favorable	10
Average	12
Unfavorable	15

General Repairs. The hourly cost of general repairs for hauling units, including parts and labor, can be estimated as a percentage of hourly depreciation costs. Selection of the percentage to be applied is a matter of judgment. The combined effect of operating conditions, operator skill and training, method of loading and degree of overload, etc. should all be considered. Table 13-21 lists typical values for repair factors.

TABLE 13-21 Repair Factors for Hauling Units Expressed as Percentages of Hourly Depreciation Costs (Based on a depreciation period of 10,000 hr)

Type of equipment	Operating conditions		
	Favorable	Average	Unfavorable
Rear dumps	37	45	60
Bottom dumps	30	35	45
Shuttle-type dumpers	40	45	50
Scrapers	42	50	62

The repair factors in Table 13-21 are based on a depreciation period of 10,000 hr. For depreciation periods of shorter or longer duration, appropriate adjustments should be made. The expression

$$\frac{\text{Depreciation period, hr}}{10{,}000\ \text{hr}}$$

which appears following "General Repairs" in Fig. 13-13 should also be revised in such cases.

EXAMPLE: A bottom dump has an hourly depreciation cost of $8.00. The depreciation period is 10,000 hr. The unit will work under generally favorable conditions.

* See Sec. 7, Equipment Economics.

† See Tires for Construction Equipment in Sec. 6, Engineering Fundamentals.

The hourly repair cost is

$$30 \text{ percent} \times \$8.00 \times 10{,}000/10{,}000 = \$2.40$$

Fuel Costs. These are based on local rates per gallon and on engine fuel consumption. Engine fuel consumption depends on the gross horsepower of the engine and on the duty cycle, or application, of the hauling unit. One method of estimating fuel consumption for diesel engines up to 700 hp is to multiply the gross horsepower by the appropriate factor from Table 13-22.

TABLE 13-22 Equipment Fuel Consumption (gallons per hour per horsepower)

Type of equipment	Operating conditions					
	Favorable		Average		Unfavorable	
	Factor	Duty, percent	Factor	Duty, percent	Factor	Duty, percent
Rear dumps	0.015	25	0.021	35	0.027	45
Bottom dumps	0.018	30	0.024	40	0.030	50
Side dumps	0.018	30	0.024	40	0.030	50
Scrapers, self-loaded	0.027	48	0.033	55	0.039	65
Scrapers, push-loaded	0.024	40	0.030	50	0.036	60

The factors in Table 13-22 are based on a normal consumption rate of 0.42 lb of fuel per brake horsepower-hour, or 0.06 gal per horsepower-hour multiplied by the duty cycle. The duty cycle is the percent of working time that the engine in the

TABLE 13-23 Hauler Servicing Costs as Percentages of Hourly Fuel Costs

Operating conditions	*Service cost as a percent of hourly fuel cost*
Favorable	20
Average	33
Unfavorable	50

hauling unit is operating at maximum output. Diesel engines rated at greater than 700 hp, and gas turbines, may have higher fuel consumption rates than those supplied in Table 13-22, and revised factors should then be established.

TABLE 13-24 Estimated Hourly Costs for Rear-dump Haulers in Current Use*

Capacity of rear-dump hauler, tons	Hourly ownership cost, dollars	Hourly operating cost, dollars	Hourly ownership and operating cost, dollars
13	2.87	7.70	10.57
22	4.31	9.43	13.74
35	6.61	13.04	19.65
50	8.89	16.49	25.38
105	12.59	21.97	34.56

* Supplied by Euclid, Inc., Cleveland, Ohio, in 1970.

EXAMPLE: A 35-ton rear-dump hauling unit has a gross engine horsepower of 434. Under average operating conditions, fuel consumption will be 434 × 0.021, or approximately 9 gal per hr. Hourly fuel cost is \$0.14 × 9 = \$1.26.

Service Cost. This is the cost of oil, grease, filters and the labor involved in performing minor end-of-shift servicing tasks. The amount charged will vary according to the frequency at which this servicing is done, and also upon the labor rate.

For hauling units, the hourly service cost can be estimated on the basis of the size of the unit's engine. Since the engine size has already been correlated with fuel consumption, it follows that the service cost can be expressed as a percent of fuel costs. This is illustrated in Table 13-23.

TABLE 13-25 Estimated Hourly Costs for Bottom-dump Haulers in Current Use*

Capacity of bottom-dump hauler, tons	Hourly ownership cost, dollars	Hourly operating cost, dollars	Hourly ownership and operating cost, dollars
63	9.34	16.63	25.97
70	9.93	17.05	26.98
100	12.03	20.74	32.77
110	12.45	22.01	34.46

* Supplied by Euclid, Inc., Cleveland, Ohio, in 1970.

Comparative Hauler Costs Rear-dump haulers in common use have estimated hourly ownership and operating costs as listed in Table 13-24. These values are based on depreciation periods of 15,000 hr.

Bottom-dump haulers in common use have estimated hourly ownership and operating costs as listed in Table 13-25. These values are also based on depreciation periods of 15,000 hr.

Section **14**

Cableways and Drag Scrapers

RAYMOND M. WELSH

Chief Engineer, Cableway Division, Lidgerwood Manufacturing Company, New York, N.Y. (*Tautline Cableways*)

CLIFTON H. HUBBELL

Chief Engineer, Sauerman Brothers, Inc., Bellwood, Ill. (*Slackline Cableways and Drag Scrapers*)

A. Tautline Cableways

Aerial transportation equipment, and particularly a tautline cableway, is a custom-engineered system designed especially to meet certain requirements. The reader should realize that a cableway or any ropeway is a complex system and cannot be classified in standard sizes—like an ordinary crane, excavator, or derrick.

An accurate contour map of the surrounding property is essential to lay out and design a cableway properly, and it requires experienced and skilled technicians to study the location to be served. The system must be arranged so that the load may be raised or lowered at any point under the span.

A number of simple cableways, in capacities of 5 to 10 tons and most frequently driven by steam-powered hoisting engines, were built by Lidgerwood prior to 1900. Hundreds more of these cableway plants were supplied by Lidgerwood in the early part of the twentieth century, with the capacities increased to 20 tons around 1920. The modern cableway, however, is different in design and construction from these earlier installations. It has been developed in response to the demand for a heavy-duty, high-efficiency hoisting and hauling device, and its principal use is in the construction of large dams for hydroelectric development, water storage, irrigation, or flood control. Such construction projects involve the pouring of large quantities of concrete, in some cases in excess of 5,000,000 cu yd, and offer a tremendous challenge to operate the equipment at the highest possible speeds.

The high-speed characteristic of the modern cableway is possible only as a result of new features that have been devised and added to the cableway systems over the years. To incorporate these changes, the cost of the equipment has increased considerably. The standard or conventional cableway and the Travelift are still practical and should be considered for lesser projects, since their initial costs are much less.

DESCRIPTION

A typical tautline cableway can be described simply as a system for hauling, hoisting, and lowering material by means of a carriage which runs on a cable suspended between two towers. One tower is known as the "head tower," and the hoist and haul machinery is located near it. The other tower is known as the "tail tower." Either tower or both towers may travel, depending upon the work area to be covered. A haul rope attached to the carriage hauls it along the suspension cable, and the corresponding drive is called the "haul drive." A second rope to hoist the load is brought to the carriage, down to the fall block, and then back to the carriage. This rope is dead-ended either at the tail tower or at the carriage, depending on the tautline cableway system employed.

A tautline cableway should not be confused with the slackline cableway which is used for excavating. The latter machine digs its load, lifts it free of the excavation, conveys it, and automatically dumps it at any desired point along the span. The digging bucket is suspended from a trolley carrier which travels on an inclined track cable. This cable is slackened off to lower the bucket into the excavation and tightened to lift the bucket into the air for inhaul to the dumping point. The principles of design and operation of the slackline-cableway excavator are explained in a subsequent portion of this section.

Although a cableway is sometimes referred to as a "tramway," there is a distinct difference. A tramway conveys material but does not hoist and lower its load. In brief, a tautline cableway is a cable crane working on a single span, while a tramway is a conveyor which can work over multiple spans and great distances.

APPLICATIONS

When the work performed by the early plants furnished by Lidgerwood is analyzed, it becomes apparent that other machines now perform certain tasks which formerly were considered to be cableway applications. Large portable cranes, with booms of great length, have replaced cableways on many projects. Many of the early cableways were supplied for logging purposes, but today the logging industry is able to take out timber with huge diesel tractors that were nonexistent around the turn of the century.

Cableways have been applied to work such as long cuts and adjacent fills, graving docks, quarries, sewage disposal plants, bridge building, penstocks, isolated high structures, and the building of relatively high dams in narrow canyons. However, the most common use of the modern cableway is for the placement of concrete in dam construction. Its second most prevalent use in recent years has been for building bridges. Frequently highway or railway bridges of relatively short span length must be constructed across streams where it is difficult or hazardous to construct falsework to support the structure during erection. The placement of concrete in the piers and the erection of steel may be handled readily, and in many instances economically, by a tautline cableway.

On any project where a reinforced-concrete dam and a powerhouse are to be constructed, a cableway should always be evaluated as a possible scheme for materials handling. Cableways leave a clear, unobstructed space between the towers over the whole area, and this can be a major advantage during construction. In addition, cableways are practically immune to the hazards of high water, flash floods, and ice runs. The modern cableway can be operated at much higher speeds than whirly cranes and, thanks to the flexible nature of its track cable, loads are lifted smoothly and deposited gently.

CABLEWAY COMPONENTS

A cableway consists of three main parts:

1. The track cable, on which the carriage runs, and its supports
2. The power application system—electrical equipment, hoists and ropes
3. The load carriage and rope carriers

The heart of any ropeway system is the main track cable. This may be attached to movable towers, stationary towers or masts, stationary anchors, or a combination of these. It is generally of locked-coil construction, which is produced in sizes up to 4-in. diameter with a breaking strength of over 900 tons.

The power application system has two main functions: (1) to move the carriage on the track cable, and (2) to lift and lower the load. Power is applied by one of several different reeving systems to move the carriage along the track cable and to raise or lower the hook load. All cableways for high output are electrically powered, but a self-contained diesel power unit is often used to drive a cableway for low-speed operation—such as bridge construction, etc. The various rope reevings, types of hoists, and electrical power systems will be described subsequently. The operating ropes, which are the real workhorses in a tautline cableway, are expendable items and will be discussed under Maintenance.

The load carriage is made as light and simple as possible and must be compatible with the fall-rope carrier system. It contains the wheels which run on the track cable and usually includes two large sheaves for the hoisting rope. The track wheels are mounted in bogies or equalizer frames to distribute the load as evenly as possible on the track cable.

With the exception of the Travelift system, it is necessary to use a set of fall-rope carriers for spans in excess of 1,000 ft. The carriers prevent sag in the hoist rope between the carriage and towers.

TYPES OF CABLEWAYS

Based on the tower combination which it employs, a cableway may be classified into any of these types: fixed, luffing, radial traveling, and parallel traveling.

Fixed Cableways A cableway which has two stationary or fixed towers is classified as a fixed cableway. The towers may be needle-type masts; guyed, four-legged towers; stable towers; or simple A frames. As they cannot be moved laterally, the area that can be served by the cableway is necessarily confined to a straight line.

Luffing Cableways These are a development of the fixed cableway and are useful where a long, narrow strip has to be served. The luffing masts are fixed in position but are mounted on ball-pivot bases so that they are able to sway sideways and at right angles to the line of the main cable. The amount of luff at each side of the vertical is usually 15 percent of the mast height. A cableway with a 100-ft mast can then serve a 30-ft width for the length of its span. The luffing movement is controlled by winches which work through tackle connected to the side anchor cables.

Radial-traveling Cableways In this system, a movable tower radiates about a stationary tower. The angle of travel subtended by a radial cableway is usually less than 40°. If a relatively small angle is employed, a single track-cable backstay will suffice. If the angle approaches 20° or more, two backstays will be required. The reason for this is obvious: as the traveling tower moves toward either end of its circular trackway, the line of force revolves about the radial tower and the lateral horizontal component of force at the top of the tower increases. When it is considered that on a 25-ton cableway at full load the horizontal component of tension in the track cable may exceed 500 kips, the importance of the direction of the line of force becomes apparent.

In radial cableways the track cables must be connected at the stationary tower in such a manner that the track cable may follow the path of the traveling tower without imposing bending at the socket or its connections. On modern heavy-duty installations, this is usually accomplished by means of suspended universal trunnions.

Parallel-traveling Cableways This term applies to cableways where both terminal towers are movable. It is then essential that these towers move on straight and parallel trackways. Modern heavy-duty cableways of the traveling type are invariably electrically powered, with towers of fabricated structural steel. The movable tail tower is similar to the movable head tower, except that a machinery deck is not required for the former and the head-frame arrangement is slightly different due to the change in sheave requirements.

CABLEWAY SELECTION

If a decision to employ cableways is made, a layout involving span, tower heights, and tower locations is required. The physical characteristics of the cableway installation will be governed by the terrain, and by the shape of the dam in the case of hydroelectric projects. Generally, more economical coverage is obtained by head and tail towers traveling on parallel tracks. These tracks, of course, may be used for more than one cableway.

The height of towers is governed by the span and the height of the dam structure. A clearance of about 50 ft is required between the low point of the trajectory of the carriage wheels and the dam crest. The distance from the chord line to this low point of the trajectory, known as the "sag," is then calculated. In this way the location of the chord line at the center of the span is determined. The alternative elevations of the ends of the chord line may be juggled to adjust for suitable structures. A movable head tower should be at least 75 ft high, since this height is necessary to reduce the fleet angle of the operating ropes leading from the hoist drum to the crown sheaves. It is also necessary to provide an adequate deck area for mechanical and electrical equipment.

Spans At present, tautline cableways are seldom used for spans shorter than 1,000 ft or longer than 3,000 ft. Normal spans will range from 1,500 to 2,500 ft, as indicated in Table 14-1. Although in a few instances spans in excess of 3,000 ft have been used, they are not recommended or considered practical for heavy loads at high speeds.

Capacities Construction cableways are designated by their hook-load capacity, either in cubic yards of concrete or in tons. While smaller cableways are supplied occasionally, accepted standard capacities are 2, 4, 6, and 8 cu yd, with a maximum hook-load capacity of 25 tons for the latter. When rated in tons, cableways are clas-

TABLE 14-1 Typical Concrete-handling Cableway Installations

Year	Project and size	Number and capacity	Span, ft	Track cable diam, in.	Type	Operating speeds, fpm		Particulars
						Haul	Hoist	
1968	Dworshak Dam, Idaho 6,700,000 cu yd	three 25 tons	2,920	3½	Parallel	2,200	950	Separate hoist and haul drives
1967	Bullards Bar Dam, California 2,700,000 cu yd	two 25 tons	2,765	3½	Radial	2,200	950	Separate drives
1965	Mossy Rock Dam, Washington 1,270,000 cu yd	one 25 tons	2,400	3¼	Radial	2,100	950	Separate drives
1964	Guri Dam, Venezuela 1,200,000 cu yd	one 25 tons one 12.5 tons	2,900 2,575	3½ 2¼	Radial Fixed	1,600 1,500	700 375	Separate drives Conventional
1963	Rapel Dam, Chile 1,650,000 cu yd	one 25 tons	1,755	3	Parallel	1,500	600	Separate drives
1962	Yellowtail Dam, Montana 1,460,000 cu yd	two 25 tons	2,100	3¼	Radial	1,800	900	Separate drives
1959	Shihmen Dam, Formosa 1,700,000 cu yd	two 25 tons	2,750	3¼	Radial	1,400	350	Single motor conventional
1958	Flaming Gorge Dam, Montana 1,000,000 cu yd	two 25 tons	1,900	3¼	Radial	1,200	400	Separate drives
1957	Glen Canyon Dam, Arizona 6,000,000 cu yd	two 50 tons one 25 tons	1,800/2,050 1,800	4 3	Parallel Parallel	1,500 1,400	650 350	Separate drives Separate drives
1955	Greers Ferry Dam, Arkansas 1,000,000 cu yd	one 25 tons	2,100	3¼	Radial	1,600	650	Separate drives

sified as being of 6-, 12-, 18-, or 25-ton capacity. The 25-ton cableways are generally rated as good for a 25 percent increase for occasional loads under careful handling. Regardless of the standards cableway manufacturers devise, the variations in the service requirements will usually necessitate major deviations from these standards, and the end result is a custom-built system. Although construction cableways of 40- to 50-ton capacity using 12-cu yd buckets have performed satisfactorily on spans of about 2,000 ft, the practical limit of capacity when spans approach 3,000 ft is 25 tons. This is being demonstrated on recent projects where huge masses of concrete are being handled.

Installations Table 14-1 contains a list of the important cableway installations used in the construction of dams since 1955. Most of these installations had separate hoist and haul drives which enabled the operators to place the maximum amount of concrete in a given time.

Table 14-2 lists major cableway installations for the period from 1930 to 1955. The comparable operating speeds are not given, but in most cases the speed is 1,200 fpm on the haul and 300 to 400 fpm on the hoisting. An exception is the Pine Flats Cableway in 1951, which had separate drives and operated at a haul speed of 1,800 fpm and a hoist speed of 500 fpm. Also, in the case of the Travelift-type cableways built in the early 1950s, some time was saved by the overlapping of the hoist and haul movements. This applies to Lookout Point and the other Travelift installations.

To underscore the wide possibilities of cableway usage, mention should be made of the Hoover Dam service cableway. This Lidgerwood cableway was installed permanently for handling heavy equipment from the rim of the canyon down to the powerhouse levels on either side of the river. It is normally rated at 160 tons capacity

TABLE 14-2 Typical Heavy-duty Cableway Installations, 1930–1955

Year	Project and size	Number and capacity	Span, ft	Track cable diam. in.	Head tower		Tail tower		Particulars
					Height, ft	Fixed or traveling	Height, ft	Fixed or traveling	
1954	Table Rock Dam, Arkansas, 1,200,000 cu yd	One 8 cu yd	2,600	3¼	398	F	163	T	Travelift formerly at Lookout Point. Standard reeve with 200-hp a-c motor
1952	Peribonka Dam, Canada, 275,000 cu yd	One 4 cu yd One 8 cu yd	2,285 2,069	2¾ 3¼	125 175	T T	203 200	F F	Travelift 500-hp a-c hoist motor. Four 50-hp motors driving tower. Single-compartment air-dump buckets
1951	Chief Joseph Dam, Washington, 969,000 cu yd	Two 8 cu yd	2,500	3	2–110	T	1–240	F, mast type	Both standard reeving. Single-compartment air-dump buckets
1951	Lookout Point Dam, Oregon, 860,000 cu yd	One 8 cu yd	2,600	3¼	398	F	163	T	Travelift 500-hp a-c drive. Two 75-hp motors driving tower. Air-dump buckets
1951	Pine Flats Dam, California, 2,200,000 cu yd	Two 8 cu yd	2,340	3½	2–65	T	2–45	T	One cableway standard reeve with 500-hp motor. One cableway English reeve with four 250-hp d-c motors driven by two 900-hp motor-generator sets. Two 75-hp motors driving each tower. Air-dump buckets
1950	Detroit Dam, Oregon, 1,500,000 cu yd	Two 8 cu yd	1,983	3¼	2–75	T	2–45	T	Travelift 600-hp a-c hoist motor. Two 75-hp motor on each tower. Double-compartment air-dump buckets
1950	Conemaugh Dam, Pennsylvania, 300,000 cu yd	One 6 cu yd	1,420	3	87	T	39	T	Travelift 350-hp a-c hoist motor. Two 40-hp motors driving head tower. Two 30-hp motors driving tail tower. Double-compartment air-dump buckets
1950	Hungry Horse Dam, Montana, 2,900,000 cu yd	Three 8 cu yd	2,415	3	1–200	F	3–52	T	Standard reeve with dump line. Buckets two-line bottom-dump type. Three cableways from Shasta Dam
1950	Philpott Dam, Virginia, 320,000 cu yd	One 6 cu yd One 8 cu yd	2,415 1,700	3 3	1–102 110	T F, mast type	1–85 52	F T	Standard reeve, 500-hp a-c hoist-motor equipment from Allatoona Dam. One 100-hp motor driving tail tower
1949	Tignes Dam, France, 850,000 cu yd	Two 8 cu yd One 4 cu yd	2,000 2,000	3 2½	2–117 1–158	T T	....	F F	Standard reeve 500-hp a-c hoist motor
1949	Narrows Dam, Arkansas, 300,000 cu yd	One 8 cu yd	1,662	3	90	F	110	T	Standard reeving equipment assembled from other jobs

TABLE 14-2 Typical Heavy-duty Cableway Installations 1930–1955 (Continued)

Year	Project and size	Number and capacity	Span, ft	Track cable diam, in.	Head tower		Tail tower		Particulars
					Height, ft	Fixed or traveling	Height, ft	Fixed or traveling	
1947	Allatoona Dam, Georgia, 500,000 cu yd	One 8 cu yd	1,710	3	110	F	52	T	Standard reeving
1947	Des Joachims, Canada 489,000 cu yd	One 10 ton	2,175	2¾	105	T	159	F	Parts assembled mainly from other jobs. Converted to Travelift
1947	Fall River Dam, Kansas, 150,000 cu yd	One 15 ton	1,475	2¾	103	T	150	F	Assembled from other jobs
1947	Wolf Creek Dam, Kentucky, 1,300,000 cu yd	Two 8 cu yd	2,300	3	.....	T	372	F	Head towers and mechanical equipment rebuilt from Norris. New tail tower
1945	Chastang Dam, France, 327,000 cu yd	Two 8 cu yd	1,146	3	2–80	T	2–60	T	Standard reeve with 400-hp a-c drive
1941	Shasta Dam, Washington, 6,000,000 cu yd	Seven 8 cu yd	Up to 2,670	3	1–460	F	Up to 125	T	All cableways standard reeve. Three units d-c drive; four units a-c drive. Hoists three-drum type for two-line bottom-dump buckets
1935	Norris Dam, Tennessee, 1,000,000 cu yd	Two 6 cu yd	1,925	3	2–75	T	2–110	T	Standard reeving with 400-hp a-c drive
1932	Boulder Dam, Arizona, 4,000,000 cu yd	Five 8 cu yd	2–2,575 2–1,405 1–1,365	3 3 3	2–90 2–75 1–98	T T F	2–90 2–42 1–16	T T T	All cableways standard reeve with 500-hp a-c hoist motors

and will handle up to 200 tons. The stationary span is 1,256 ft and the cable trackway consists of six 3½-in.-diameter steel cables.

REEVING SYSTEMS

There are several different reeving systems for the operation of the carriage and hook load. The simplest of these is the English reeve, which is satisfactory for short spans where fast speed is not required. The most prevalent system at present is the modern, high-capacity unit with separate hoist and haul drives. The most successful installations have employed the new Lidgerwood-Pohlig differential type of carrier, which enables the operator to work at high speeds over long spans without long delays and with safety.

English Reeve This simple reeving is illustrated in Fig. 14-1. It is usually a light-capacity system and is sometimes made up by cableway riggers on the job. There have been cases where it was advantageous to use such a rig during the installation of the prime cableways for a major project.

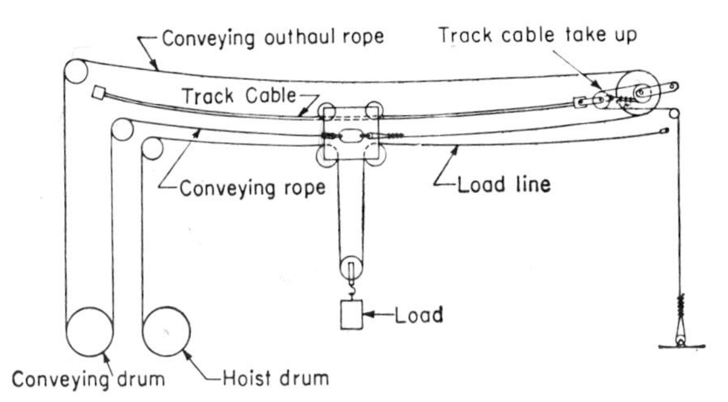

Fig. 14-1 Simple English reeve.

The English cableway carriage consists of a simple frame enclosing track wheels at the top and two rope sheaves at each end of the lower part. The haul rope is attached to each end of the frame and runs endlessly through sheaves positioned at each end of the track cable. The outhaul portion runs back above the track cable, over a sheave at the head tower, down to the gypsy or elliptical drum of a double-drum hoist. The hoist rope runs off the second drum through a tower lead sheave, then passes over the nearest carriage sheave, down to the load block, and up over the second carriage sheave to a dead anchor at the far end of the track cable.

Sag in the hoist rope between the carriage and the towers is outhauled by an extra weight on the fall block. This system is practical only for relatively short spans and light loads. When applied to long spans, the actual weight of the cable develops such a sag that the uplift on the hook becomes unmanageable and the sagging hoist rope may interfere with other ropes or foul the structure below.

Standard Lidgerwood This reeving involves the support of the hoist or load line between the carriage and head tower by fall-rope carriers (see Fig. 14-2). The hoist rope itself is dead-ended on the carriage after multiple parts of reeving between the fall block and carriage. Tandem carriages are used for the larger-capacity cableways to distribute the load on more wheels and also to provide separation of the fall lines. The fall-rope carriers are spread at given intervals by a button line. As the carriage travels out and back on the track cable, the carrier frames are alternately removed from the horn and buffer board on the one end of the carriage and picked up by it. The operating lines run over rollers inserted in the carrier units. As the carriage moves outward from the head tower, the buttons on the button line pull the carriers off, and as the carriage returns the horn picks up the carriers as the buffer board on the carriage runs into these units progressively.

The button-stop fall-rope carrier system, so far as the method is concerned, was developed by Lidgerwood and first used about 1890. These early carriers had rigid button rings which caused a very high impact and shock and limited the travel speeds. Just after the turn of the century, the Miller patented shock-absorbing feature was added to the carrier ring assembly. These new carriers operated successfully on thirteen 6-ton traveling cableways with dc motors which were sold to the United States government in 1908 for use in the construction of the Panama Canal. This

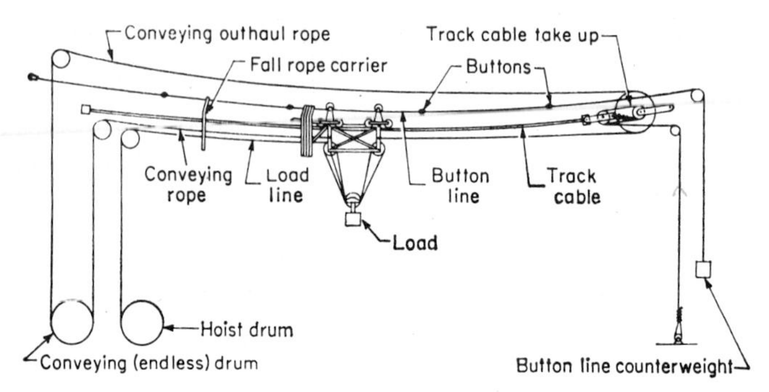

Fig. 14-2 Standard Lidgerwood reeve.

improvement enabled the operators to travel in excess of 1,200 fpm; however, it was Lidgerwood's practice to develop the system to operate at a normal speed of 1,200 fpm, since the operators will find means to overspeed anyway.

This system is operated by a double friction-drum hoist with a single motor drive. The hoist contains one cylindrical drum large enough to accommodate the hoisting or load line and a second endless or spool-type drum for the haul line. The friction and brakes are usually electropneumatically controlled. The frictions are set by direct air-thrust cylinders, and the brakes are weight-set and air-released.

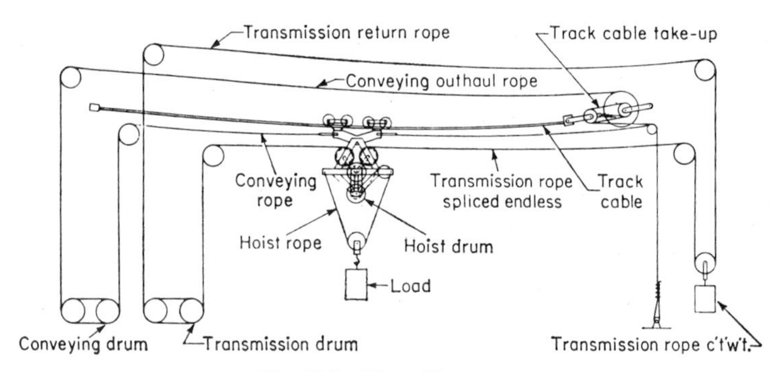

Fig. 14-3 Travelift reeve.

The standard Lidgerwood system, sometimes referred to as a "conventional" cableway, has been the principal system used in heavy-duty cableways for three-quarters of a century.

Travelift Figure 14-3 illustrates the first successful attempt to incorporate the hoist mechanism of a cableway system in the load carriage. The technique involves the use of an endless rope circulating over the span and arranged to transmit the hoisting power from a driving hoist at the head tower to a transmission and separate rope drum in the underframe of the carriage. Thus the hoist rope between carriage and fall block is completely isolated from the transmission rope which drives it. The transmission rope, which is spliced and really endless, is reeved through traction sheaves; these in turn drive a hoist drum through a chain drive. Therefore a two-part hoist line which is dead-ended on the carriage and on the drum may hoist or lower the load, depending upon the direction that the transmission rope is driven through the carriage.

Both the haul lines and transmission line can be operated in the same manner with the Travelift system. These tractive ropes must be activated either by an elliptical drum or a more elaborate system involving tandem grooved drums. Most of the Travelift installations had the operating ropes reeved around the drive and idler drums of the main hoist. These hoists had disk clutches on the high-speed-motor extension shaft and band brakes on the drum flange, brakes weight-set and air-released, and friction clutches air-set.

A sheave block with counterweight is placed in the spliced transmission-rope system at the tail tower. This is necessary to assure proper tension for tractive effort in the

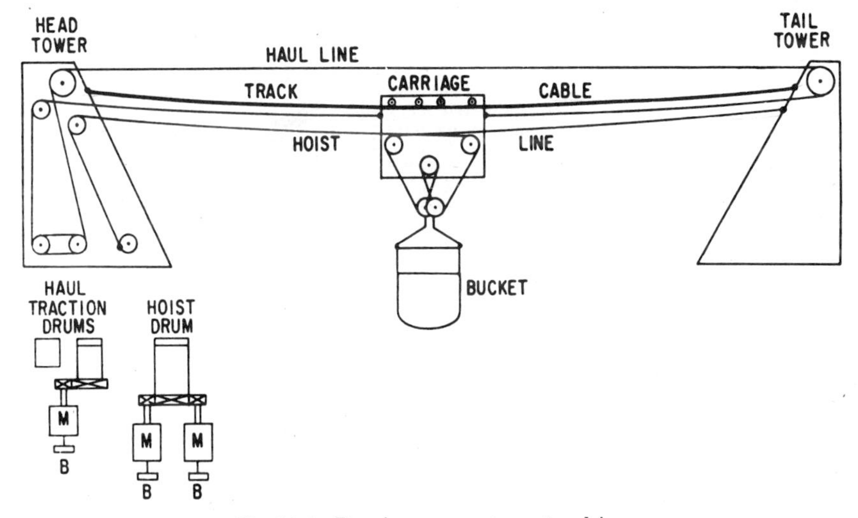

Fig. 14-4 Reeving—separate motor drive.

three transmission sheaves to hoist the load. Since the hoist operating ropes are both under controlled tension, the endless lines eliminate the use of fall-rope carriers. The hoisting or transmission drum is geared for two speeds, and it is possible to hoist or lower while hauling. This feature enables the operator to hoist and haul simultaneously with a single motor drive.

Dual-motor Drives Separate hoist and haul drives are favored today for cableways on major construction projects. They are high-powered, high-speed rigs with dc adjustable voltage controls. The reeving is simple, but each driving drum is separately powered. To handle the operating ropes on long spans at high speeds, a system of differential-type fall-rope carriers is necessary.

The reeving for dual drive is illustrated in Fig. 14-4. The haul drive consists of a powered drum with several grooves, ahead of which is mounted a similar idler drum. The haul line passes around these drums several times to give traction to the rope. The hoisting or load line is the same as that part of the English reeve system except that in this case, due to the high rope speed, the sheaves and rope leads are placed so as to eliminate reverse bends. Since the drive motors are coupled directly to the hoists for electrical control, no clutches or frictions are required.

The success of the Lidgerwood-Pohlig differential carrier has contributed to the higher-speed operation of the separate drive system. Figure 14-5 illustrates one of the carriers. Spans of 2,000 ft require ten carriers, five on each side of the carriage. Spans of 2,500 to 3,000 ft have fourteen carriers to a set, or seven on each side. As the carriage moves toward the head tower, each fall-rope carrier also moves in the same direction. Although each carrier has a speed of its own, these speeds are coordinated so that the carriers will all arrive at the head tower simultaneously. As the carriage moves toward the tail tower, a similar arrangement applies. The carriers are spaced by a separate rope, which is driven by the movement of the carriage and is looped around the pulleys on both sides of the carriers. Depending on the direction of travel, the rope on the one side drives the carriers while the return rope on the opposite side acts as a reactive force to hold the carrier system in step.

Fig. 14-5 Lidgerwood-Pohlig differential carrier.

The separate hoist and haul drives eliminate the friction clutches and associated manually controlled brakes, add to the ease with which these rigs can be handled, and thus contribute to safe operation. One master switch completely controls the hoisting and lowering, and a second master switch controls the forward and reverse movement of the carriage. In stopping, regenerative braking slows the load down smoothly to a very low speed before electric brakes set automatically.

CABLEWAY DESIGN

The notations and formulas which follow will apply to spans of fixed length. It has already been explained that the term "cableway" applies only to a single-span unit whose two ends are anchored. One or both of these anchors may be the top of a movable tower, the top of a stable tower, or a ground anchor. Where either or both ends of a cableway track cable are anchored to movable or stable towers, the span length will decrease as the tower deflects horizontally. This factor can be neglected for short towers, but for towers of considerable height it might cause sufficient additional deflection to affect clearances. However, such a special condition does not come within the scope of this section, and in all cases it will be assumed that cables are anchored to rigid supports.

The first step is to select the size, construction, and grade of main cable with a proper factor of safety and then to determine the maximum tension in the cable due to dead and live loads. It is then necessary to erect the cable at such a deflection that the maximum safe working tension will not be exceeded when the load is applied. These steps can be accomplished, for a given set of initial conditions, by referring to

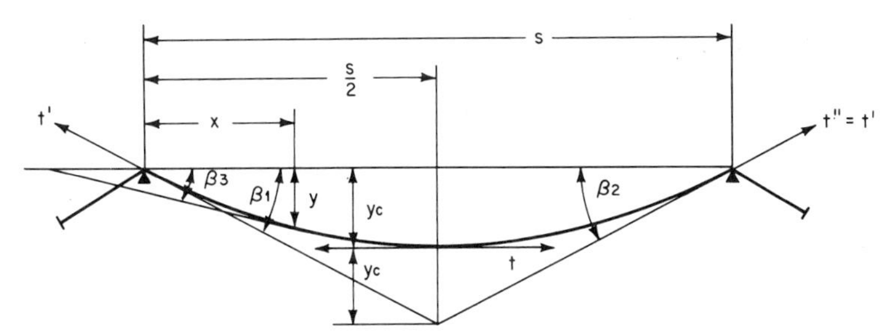

Fig. 14-6 Level span—uniformly loaded, anchored.

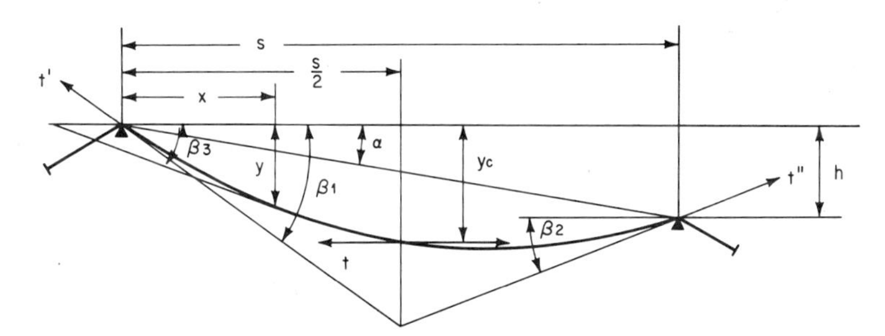

Fig. 14-7 Inclined span—uniformly loaded, anchored.

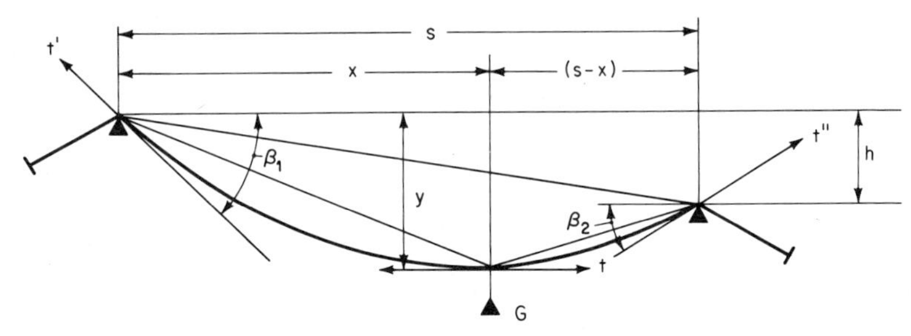

Fig. 14-8 Inclined span—single load at any point, anchored.

Figs. 14-6 to 14-8. From these data most problems encountered in the preliminary design and the selection of a cableway may be solved.

Notation

A = net cross-sectional area of cable, sq in.
E = modulus of elasticity, psi—for locked-coil cable use 19,000,000
G = weight of individual concentrated load, lb
G_1 = concentrated load, including impact, lb
h = vertical difference in elevation of supports, ft

k = ratio of deflection to span = y/s for level spans
L = length along cable under load between supports, ft
L_1 = total length of cable when cable only is supported, ft
L_0 = length along cable at zero tension, ft
L_e = length along cable at erection tension, ft
P = change in total length of cable per pound of tension = L/AE, ft/lb
s = horizontal distance between supports, ft
t = horizontal component of cable tension, lb
t' = maximum cable tension at left support, lb
t'' = maximum cable tension at right support, lb
t_e = erection tension, lb
w = weight, lb, per horizontal foot of span length for uniformly distributed load
x = horizontal distance from left support to a point on the cable, ft
y = vertical deflection from left support to the same point on the cable, ft
y_c = vertical deflection from support at center of span, ft
α = angle between horizontal and a chord between supports, deg
β_1 = angle between horizontal and a tangent to a cable curve at left support, deg
β_2 = angle between horizontal and a tangent to a cable curve at right support, deg
β_3 = angle between horizontal and a tangent to a cable curve at any point in span, deg

Formulas A cableway span of uniform weight and section will form a catenary, but, for practical purposes, a parabola coincides very nearly with a catenary. Since the parabolic equations are much simpler, they therefore are universally used. For purposes of this section, computations are based on the assumption that uniform loading is distributed horizontally and that the cable will form a parabolic arc.

For a Level Span, Uniformly Loaded. When the horizontal component of cable tension is known, the center deflection is

$$y_c = \frac{ws^2}{8t} \tag{14-1}$$

The deflection at any point in the span is

$$y = \frac{wx(s - x)}{2t} \tag{14-2}$$

When the center deflection is known, the horizontal component of cable tension is

$$t = \frac{ws^2}{8y_c} \tag{14-3}$$

For an Inclined Span, Uniformly Loaded. When the horizontal component of cable tension is known, the center deflection is

$$y_c = \frac{ws^2}{8t} + \frac{h}{2} \tag{14-4}$$

The deflection at any point in the span is

$$y = \frac{wx(s - x)}{2t} \pm x \tan \alpha \tag{14-5}$$

$$\tan \beta_1 = \frac{ws}{2t} + \tan \alpha \tag{14-6}$$

$$\tan \beta_2 = \frac{ws}{2t} - \tan \alpha \tag{14-7}$$

$$\tan \beta_3 = \frac{w}{t}\left(\frac{s}{2} - x\right) \pm \tan \alpha \tag{14-8}$$

When the center deflection is known, the horizontal component of cable tension is

$$t = \frac{ws^2}{8y_c - 4h} \tag{14-9}$$

The low point of the inclined span occurs when $\tan \beta_3 = 0$. Therefore

$$x = \frac{s}{2} + \frac{t}{w} \tan \alpha \tag{14-10}$$

When deflection at any other point is known,

$$t = \frac{wx(s - x)}{2(y - x \tan \alpha)} \tag{14-11}$$

$$t' = t \sec \beta_1 \tag{14-12}$$

$$t'' = t \sec \beta_2 \tag{14-13}$$

When the deflection is known, the length of cable in an inclined span is

$$L_1 \text{ or } L = \sqrt{s^2 + h^2}\left(1 + \frac{8}{3} k^2\right) \tag{14-14}$$

For an Inclined Span with a Concentrated Load at Midpoint. When a cableway carriage is suspended at point C midway between towers A and B, two subchords AC and BC are formed. The cable then assumes two catenary arcs which intersect at C. Proceeding on the assumption that the difference between the catenary and the parabolic curves is negligible, the center deflection is found from

$$y_c = \frac{s(2G + ws)}{8t} + \frac{h}{2} \tag{14-15}$$

$$t = \frac{s(2G + ws)}{8yc - 4h} \tag{14-16}$$

$$t' = t \sec \beta_1 \tag{14-17}$$

$$t'' = t \sec \beta_2 \tag{14-18}$$

$$\tan \beta_1 = \frac{G + ws}{2t} + \frac{h}{s} \tag{14-19}$$

$$\tan \beta_2 = \frac{G + ws}{2t} - \frac{h}{s} \tag{14-20}$$

For an Inclined Span with a Single Concentrated Load at Any Point. The deflection may be found from the formula

$$y = \frac{x(ws + 2G)^2(s - x)}{2t[ws^2 + 4G\sqrt{x(s - x)}]} + \frac{hx}{s} \tag{14-21}$$

This formula is used for two purposes: (1) for calculating the maximum tension in a cable by measuring the deflection at any given point, and (2) for calculating the trajectory of the moving load.

In the erection of a cableway, it is necessary to know the erection deflection and hence the erection tension of the unloaded span. For this condition, the values of L_1 and t_e must satisfy the equation

$$L - L_1 = \frac{(t - t_e)L}{AE} = P(t - t_e) \tag{14-22}$$

and, by substitution,

$$Pt_e + L - (Pt + s) = \frac{w^2 s^2}{24 t_e^2} \tag{14-23}$$

In any cableway system, it is well to calculate the unloaded tensions of the track cable as well as the loaded tensions. The determination of the unloaded tensions of the track cable provides a guide for the inherent stresses in the haul rope before it actually does its own job of hauling the carriage along the track cable. This involves the use of the modulus of elasticity in tension for the particular construction of cable selected. For locked-coil or smooth-coil cables, the modulus of elasticity is taken as 19,000,000 psi. This value will increase if the cable is prestressed or after rolling loads have passed over it for some time.

Calculations It has been established that most cableway spans will be between 1,000 and 3,000 ft in length, and the most popular size for a hydroelectric project of importance is the 25-ton capacity.

The uniform quality of track cable today permits the use of a working factor of safety of 2.75. The deflection is generally in the range of 5 to 7 percent, and the ideal is 5½ to 6 percent to achieve a desirable life of the main cable.

It is necessary to assume a size and grade of cable for the calculations, and the average cableway designer would guess that 3¼-in.-diameter special-grade locked-coil cable could be satisfactory for a 25-ton rig operating over a medium length of span.

EXAMPLE

It is required to find the suitability of 3¼-in.-diameter locked-coil cable for a 25-ton-capacity level cableway. We will assume a span of 2,000 ft and specify that the track cable is to have a safety factor of not less than 2.75.

The carriage weighs 12,000 lb, load block 2,500 lb, and the suspending hoisting rope 1,500 lb. Therefore

$$G = 12{,}000 + 2{,}500 + 1{,}500 + \text{maximum load } 50{,}000 \text{ lb} = 66{,}000 \text{ lb}$$
$$G_1 = 1.10G = 72{,}600 \text{ lb}$$

Since s and G are known, we will assume 3¼-in.-diameter track cable weighing 25.6 lb per ft (w) and having a nominal breaking strength of 1,160 kips. Then

$$t' = \frac{1{,}160{,}000}{2.75} = 421{,}800 \text{ lb, say 422 kips}$$

Since $t' = t \sec \beta_1$, then $t = t'/\sec \beta_1$.

To obtain a value for β_1 we will assume y_c to be a desirable deflection of 6 percent of span length $= 0.06 \times 2{,}000 = 120$ ft; then, from formula,

$$t = \frac{s(ws + 2G_1)}{8y_c} = \frac{2{,}000[(25.6 \times 2{,}000) + (2 \times 72{,}600)]}{8 \times 120} = 408 \text{ kips}$$

$$\tan \beta_1 = \frac{(G_1 + ws)}{2t} = \frac{72{,}600 + 51{,}200}{816{,}000} = 0.1510$$

$$\beta_1 = 8°35'$$

Then

$$t' = t \sec 8°35' = 408 \times 1.0113 = 413{,}000 \text{ lb}$$
$$t' = t''$$

therefore

$$t'' = 413{,}000 \text{ lb}$$

It is apparent that the 3¼-in. diameter was the proper selection. To determine the least deflection y_c, use the maximum allowable tension $t' = 422$ kips.

$$t = t'/\sec \beta_1 = 422{,}000/1.0113 = 415 \text{ kips}$$

$$y_c = \frac{s(ws + 2G_1)}{8t} = \frac{2{,}000(51{,}200 + 145{,}200)}{8 \times 415{,}000} = 118.5 \text{ ft}$$

Therefore, the least operating deflection at center of span under full-load capacity will be 118.5 ft or 5.925 percent of span length.

Most of the 8-cu yd cableways, although rated 25 ton, are only required to handle a nominal load of 22 to 23 tons at the hook. The weight of 8 cu yd of concrete at 4,000 lb per yd is 32,000 lb, and the average bucket weighs about 12,000 lb, resulting in a total hook load of 44,000 lb. Therefore, the maximum load of 25 tons used in the above calculations will normally provide a margin of safety. Also, the 1,160-kip

TABLE 14-3 Typical Cableway Characteristics and Duty Cycle

Load and machinery data	
Power supply	2,300 volts, 3 phase, 60 cycles
Weight of bucket	12,000 lb
Weight of concrete (8 cu yd)	32,000 lb
Fall block and spreader	3,000 lb
Suspended load, full bucket	47,000 lb
Suspended load, empty bucket	15,000 lb
Weight of carriage	14,000 lb
Hauling loaded	61,000 lb
Hauling empty bucket	29,000 lb
Normal haul speed	1,500 fpm
Hoist or lower loaded	600 fpm
Hoist or lower empty	700 fpm
Average haul distance	500 ft
Average hoist at load point	40 ft
Average lower at dump	300 ft
Time to load and dump bucket	25 sec
Grade at pickup point	17.3 percent
Weight of hoist rotating parts	21,000 lb
Overall mechanical efficiency	85 percent
Span	1,750 ft

Operation	Time, sec	Hoist, ft	Haul, ft	Time with hoist and haul overlapped, sec
Load bucket	15.0			15.0
Accelerate hoist	7.0	34.0		
Hoist (loaded) 600 fpm	0.15	1.5		8.0
Decelerate hoist	0.85	4.5		
Accelerate haul out	3.4		39	
Haul out (loaded) 1,500 fpm	15.2		380	
Decelerate haul out	7.0		81	32.5
Accelerate lower	0.95	4.7		
Lower (loaded) 600 fpm	27.45	274.5		
Decelerate lower	4.1	20.8		
Dump	10.0			10.0
Accelerate hoist	2.05	11.5		
Hoist (empty) 700 fpm	24.2	281.0		
Decelerate hoist	1.25	7.5		27.5
Accelerate haul in	2.8		31	
Haul in (empty) 1,500 fpm	17.0		425	
Decelerate haul in	3.9		44	
Accelerate lower	1.4	8.0		
Lower (empty) 700 fpm	1.8	21.0		5.0
Decelerate lower	1.8	11.0		

If conditions are perfect, time = 98 sec

Add 35 to 40 percent leeway (terminal and control delays) +37

Practical average cycle = 135 sec

Assume 3,600/135 = 27 trips per hour, approximately

posted strength of 3¼ in locked-coil cable is conservative, and under actual test the strength exceeds 1,200 kips.

Duty Cycle In a typical concrete-placing project, an estimate must be made of the rate of placement required. The profile of the dam site should then be studied to determine the average distance from pickup point to pour point. If we arrive at the approximate center of gravity of the huge mass of concrete and settle on the point at which material will be picked up, we can establish average distances for hook movement, both vertically and horizontally.

Knowing the average distances of travel, and with load and machinery data such as shown in Table 14-3 available, an average duty cycle can be developed. The duty cycle shown in this table is for an 8-cu yd capacity rig with separate drives. The hoist and haul units are powered by 750- and 500-hp motors respectively. It is not feasible or practical to expect perfect conditions to exist; therefore a certain amount of tolerance should be considered when assuming an average duty cycle.

CABLEWAY FEATURES

The parts that make up cableway systems are similar but will vary with the different reeving and the type of cableway. A brief description of the main items, with some of the essential features, is supplied below.

Towers There are cableway towers of various forms, the simplest of which is the stationary type of steel construction. This may consist of needle-type masts; rocking-type or guyed four-legged structures; or it may be a stable structure.

The needle-type mast has a universal base with a ball-and-socket joint at the bottom, and it is guyed at the top. Such masts are often equipped with luffing devices which enable the masts to be leaned laterally.

The rocking-type towers are four-legged towers so designed that the rear legs have a free uplift but are restrained horizontally, while the entire vertical component of the superimposed load will be carried in compression by the front legs.

The guyed, four-legged tower has the base of each leg fastened firmly to the foundation and is commonly employed when the track cable and backstays are connected to a universal trunnion. This assures that the vertical component of the superimposed loads falls within the four legs.

Stable towers are not commonly used, but there are cases where they have served well. As the name implies, there are no guys to offset the horizontal components of the loads that are imposed. A tower of this type which serves several cableways will require a generous foundation of sufficient weight to overcome the uplift of the back legs.

The traveling tower is more commonly used in modern practice, particularly in dam construction. Traveling towers for radial- or parallel-type cableways normally range up to 200 ft high, although higher towers are practical and have been used. The heavy towers are fabricated from structural steel and are mounted on trucks for mobility. These trucks contain standard-size car wheels, and they are arranged in pairs and connected to equalize the load at each corner. The trackways to accommodate the trucks on most installations of 25-ton capacity are standard railroad gauge front and rear with 132-lb rails. The front tracks are inclined to oppose the horizontal component of pull against the tower. To equalize the overturning moments of the towers, reinforced-concrete blocks as ballast are placed on a platform around the rear vertical legs. These features are illustrated in Fig. 14-9.

A reversible, motor-driven winch is installed in the tower. An endless rope is driven by the spool drum and, with tackle blocks anchored at either end of the trackway, the tower can be moved satisfactorily. However, when the trackways are put down to ARA specifications and are common to several cableways, the traction drive is more practical and is almost universally used today. The motorized gear reducers are connected to the wheels by chain and sprockets. See Fig. 14-10 for a typical traction drive.

The towers on heavy-duty installations are driven by a minimum of eight wheels; this is necessary to minimize slippage when the rails are wet. On a 125-ft-high tower,

Fig. 14-9 Typical head tower—radial cableway.

Fig. 14-10 Typical traction drive—gear head motor and chain.

the counterweight may equal 550 tons and the rolling weight would be in excess of 700 tons; therefore the wheel loads might run as high as 40 tons.

Main Cable On minor installations where slower speeds are permissible and the required length of service is not excessive, a regular wire rope will be used as a track cable. The smooth- or round-wire track strand can also be used where operating requirements are minimal. However, for all major high-speed, long-service installations, locked-coil track cable must be considered. (See Fig. 14-11.) The smooth exterior surface of this cable eliminates impact loads on the outer wires, and the outer wires are prevented from protruding by the locked construction.

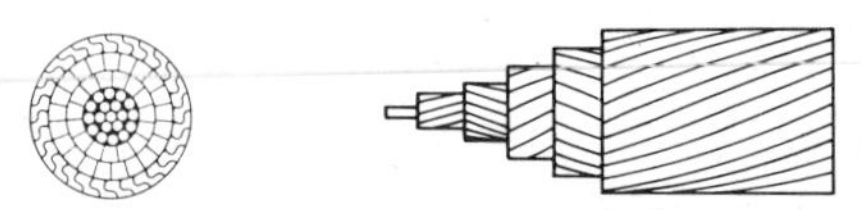

Fig. 14-11 Locked-coil track cable.

The cable is socketed at each end in a forged steel open socket which is arranged for attaching to the eyebars. These eyebars are attached to a pin at one tower and to the main cable-adjusting takeup at the other end. The takeup consists of heavy bars and pins, together with the steel sheaves for multiple purchase of a separate wire rope. This takeup is required to regulate the tension in the main cable.

Hoists A small, job-built cableway will probably be adapted to use whatever type of contractor's hoist is available. Such a hoist, if nonreversing, will require three friction drums. One drum will be used for hoisting, with the other two used for inhaul and outhaul.

The most common arrangement is the double-tandem friction-drum hoist. This is used with a conventional cableway driven by an electrical motor with reversible control. Even though these hoists have loose drums, the load is lowered under power regeneratively and can be retarded by plugging the motor.

With the usual arrangement of the conventional cableway, the hoist drum friction is engaged and the hoist rope is wound on the drum when hauling in toward the head tower. Conversely, the hoist drum will pay out rope when the carriage is hauled out. This is necessary in order that the hook may travel parallel to the main cable. To accomplish this, the diameter of the lagging on the haul drum is made the same as the

Fig. 14-12 Standard double-drum hoist conventional cableway.

average diameter of the several layers of rope on the hoisting drum. The bucket is normally moved on a square cycle—haul out, lower, dump, hoist, and haul in. All hoist and haul movements are performed separately. The drum frictions and brakes are air operated, and both drums are powered by a single motor. A typical, standard, conventional two-drum hoist is shown in Fig. 14-12.

Although transmission of power to a hoist on the carriage by wire rope is not a new idea, it is refined in the Travelift system. Operating speeds are increased somewhat, and fall-rope carriers are eliminated. A cable-powered traction sheave arrangement, operating a chain-driven load hoist, is actually mounted on the carriage. The driving winch is of the single-motor type, with drums for hoist-rope transmission and the haul drives. The friction clutches and brakes on the winch allow the operator to hoist and haul simultaneously within the limits of the power of a single motor. Thus, travel and lift (Travelift).

Fig. 14-13 Typical high-speed main hoist 2,400/3,000 fpm.

It is interesting to note that the conventional or standard Lidgerwood and the Travelift systems are powered by single 400- to 600-hp electric motors to do the work singly or together. Recent separate-drive rigs go from 800 to 1,000 hp on the haul and 1,000 to 1,250 hp on the hoist. The operating speeds and the torque of these larger motors require large and sophisticated hoists which are similar to the units used for mine and shaft work. Figure 14-13 shows the arrangement favored; two heavy-duty mill motors are generally used for each drive. With these amplidyne-controlled drives, the current is limited to give smooth, even acceleration. Hoists are equipped with helical gears and antifriction journals throughout.

Cable Carriages Depending on the reeving employed, the carriages will vary to suit the operating conditions. A 25-ton-capacity carriage for the standard, conventional reeving will generally have 16 wheels and is made in two sections. The duplex carriage allows the hoisting rope to be spread over a greater angle, if required. On deep drops, this avoids twisting and counteracts the pendulum action which can be caused by rapid deceleration after traveling at high speed. A duplex arrangement with fall block and bucket attached is illustrated in Fig. 14-14.

Experience has proved that the life of the main cable can be prolonged by the use of

Fig. 14-14 Typical duplex carriage—25-ton cableway.

a greater number of track wheels. This distributes the load and reduces the pressure under the wheels. Figure 14-15 shows a 12-ton carriage with eight track wheels. The average 6-ton-capacity unit will contain four track wheels.

The carriage for the dual-motor drives has the load line of the lifting hoist reeved the same as for the English reeve, with the rope dead-ended at the tail tower and not at the carriage. Therefore, while traveling without hoisting or lowering, the load line merely moves through the carriage.

The Travelift carriage is illustrated in Fig. 14-3; it consists of bogie-connected track wheels arranged so as to provide equal wheel loads. A trio of direct-connected traction sheaves and a hoist drum plus a separately mounted fleeting sheave complete the undercarriage assembly.

Slack-rope Carriers The most important factor in the efficient operation of a cableway is the method of supporting the hoisting and conveying ropes. When these ropes are slack, they must of necessity be supported at intervals in an efficient manner, particularly when the span is long.

It was pointed out in the discussion of reeving systems that slack carriers are not

Fig. 14-15 Typical single carriage—12-ton cableway.

necessary for the English reeve when employed on very short spans or for the Travelift system where the transmission line (as well as the haul line) is installed with back tension. For the conventional cableways and for the modern dual drives with spans often in excess of 2,500 ft, an approved method to prevent sagging is important.

The carriers mentioned in the earlier description of standard Lidgerwood reeving are shown very clearly in Fig. 14-16. In this type of carrier, the eye which is picked up by the button is hinged, and the head of the carrier is also hinged. This shock-absorbing carrier is very durable, as has been proved in usage at high speeds, since the hinged construction of eye and head materially reduces the shock of impact.

For continual operation at speeds in excess of 1,500 fpm, all cableway people have found it is necessary to use a differential type of carrier. For many years attempts were made to develop this type of carrier, but with little success. Finally, about 1955, Karl Sommer of Germany invented a carrier system which has made it possible to operate at higher speeds satisfactorily. This carrier is shown in Fig. 14-5 and is known as the Lidgerwood-Pohlig carrier.

Fig. 14-16 Typical Lidgerwood shock-absorbing type of button-stop carrier.

Sheaves The proper selection and the rigging of the sheaves in a cableway system are very important. Even with the best arrangement, the replacement of the operating ropes will still be one of the greater costs of the cableway operation. Rope life will be increased by the use of as large a sheave as appears to be practical. Sheaves of large diameter do two things: they increase the usable strength in a rope and they increase its life. If the sheave (or drum) diameter is at least five hundred times the diameter of the outer wires, the loss in strength due to bending the rope around the sheave will not then be more than 4 percent.

It is also important to machine the groove in the steel sheave so it will support the rope properly. The fleet angle of tower sheaves should be kept to a minimum, since this will reduce wear on sheaves and rope. In the case of radial-type cableways, the sheaves in the tower must be nested in fair-lead brackets. The rope speeds encountered in cableway operation make it necessary that the fair-lead bracket rotate easily and follow the movement of the traveling tower. Therefore the bearings should be of the antifriction type.

POWER SUPPLY

The smaller cableways, such as those used for bridge construction, may be powered by gasoline or diesel engines, but the heavy-duty cableways without exception use electric power. The power line to the site will generally be 2,300 to 4,160 volts. Usually, economy dictates that the potential for the hoist motor be at least 2,300 volts. Separate feeders at different voltages are needed to supply power for movement of towers and for auxiliary equipment like air-operated brakes and clutches. A careful evaluation of the horsepower requirements should be made, taking into account the weight of revolving parts and the acceleration of the operating ropes and carriage. These must be considered as well as the hook load, particularly on a high-speed installation.

Electrical Equipment Early cableways generally used wound-rotor motors with refined secondary control. Some of the refinements included current-limit acceleration, plugging, and overspeed slowdown. Most of these cableways had single-motor drives, with the drive motor clutched to the drums for hoisting and hauling as required. A few dc drives were also used with the standard-reeved single-motor systems.

The more recent dual-motor drives use dc adjustable-voltage hoist drives and both dc and ac haul drives. These hoist drives are selected to meet the duty cycle of the job, but in general the limiting factor is not the rms rating but rather the commutation ability of the motor at maximum speed. When load is being lowered and placed at the pour site, the hoist drive must have adequate margin to provide a relatively short deceleration time without exceeding the permissible commutation limits of the dc motor.

The dc hoist motor is shunt wound and electrically connected to the armature of an adjustable-voltage generator. This generator and exciter are direct driven by a synchronous motor. Since the generator output can be varied from zero to a maximum in either polarity, the speed of the motor will follow the voltage. The dc motors are of the most rugged mill type, operated normally at approximately double voltage. In most cases, they include blowers for maximum ventilation.

Control For a drive that is to operate continuously at constant load and speed, there would be no object in interposing a motor-generator set between the load motor and the electric power supply. However, in a cableway drive where accelerations and decelerations are continuously repeated, power is saved in the following manner with the dc amplidyne-controlled adjustable-voltage drive:

1. During each acceleration with an ac wound-rotor drive, a good percentage of the power drawn from the line is wasted in the secondary resistors. With the adjustable-voltage drive, practically all the input power is used to accelerate the load.

2. Power must be taken from the line to plug an ac motor and its load to rest during each deceleration. The adjustable-voltage drive, instead of requiring power to stop its load, transfers the energy in the moving load back through the mechanical and electrical systems to drive the synchronous motor of the motor-generator set as a generator and pump power back into the line.

The control of most cableways requires considerable jogging where the motor is not accelerated to full speed. The power demand of a dc system while jogging is much less than in the case of an ac system, because a large value of current is obtainable from the adjustable-voltage generator at a very low voltage.

CABLEWAY OPERATION

When a cableway is used in the placement of concrete in a large dam structure, it is important to locate the operator so as to give him as clear a view of the job as possible. However, the pickup area is the only point where visual operation by the operator is important. When the batch of concrete is run out to the pour point, the signalman in that area takes over regardless of whether or not the operator can see that location clearly. In a wide dam, most of the work, even when in the operator's line of vision, is so far away that he cannot really see what is happening in the form. On modern cableway projects the operator maintains contact with the signalman in the pour area, who "talks" the load into position for dumping by means of two-way radio. The cableway operator on occasion has also used closed-circuit television to monitor an instrument panel at the machinery house. The television camera focuses on dials of instruments that register the haul, hoist, and lowering speeds and also indicate the position of the bucket relative to the carriage and the head tower. This operating information, together with the simplicity of a single lever for each of the hoist and haul controls of the modern separate-motor-drive cableway, adds to the confidence of the operator to operate at extremely high speeds. They also add to the safety of the operation.

The concrete is brought to the cableway in transfer cars which run on tracks for the full traverse of the cableways. The concrete is then dumped into buckets, which rest on a platform extending the length of the transfer tracks. This arrangement permits a bucket to be filled at any point along the cableway platform. To cause the bucket to swing clear of the transfer tracks, the carriage is run a few feet toward the center of the span just before hoisting.

Maintenance The largest recurrent repair item for a tautline cableway system is the replacement of the operating ropes. These ropes are inspected periodically or every few days and are greased about once a week on the average job. Records of the durability of these ropes on different installations vary to such an extent that it is difficult to predict a useful life expectancy. However, it is fair to say that the hoist and haul ropes on a well-designed rig should handle up to 300,000 cu yd.

A track cable of locked-coil construction will usually transport between 1½ million and 2 million tons of payload. Actually, the smooth surfaces of lubricated track cables do not wear out, but the outer wires do fail from fatigue. Consequently the wearing surface may be destroyed when too many wires break in the same area.

The hoist, carriage, and sheaves are similar to any other piece of equipment in that they demand oil and grease. However, the cableway is perhaps the most expensive and important piece of equipment on the job and therefore requires particular attention.

Costs Some will argue that the high-powered heavy-duty cableway is too costly an investment. This may be correct for an ordinary or relatively small project, but where topography is favorable and where the volume of concrete to be handled is of substantial proportions, comparisons have shown that there is no more economical way to construct a dam.

All cableways require expert erectors, operators, and maintenance crews. Contractors who have had experience with cableways and who have organizations of riggers and other personnel skilled in the handling of cableways are prepared to make an appraisal of the physical and economic characteristics of the job.

It is impossible to estimate the cost of a cableway as a prime piece of construction equipment in the same way as one might estimate the price of a power shovel. The cost of the shovel may vary only with bucket size, width of tracks, and the manufacturer, while the cost of a cableway will not only depend on the bucket capacity, but also on the span, height of towers, size of motors, and whether the control is alternating or direct current, etc.

Choice of cableways is generally based on a consideration of the span required, first cost, salvage value, and operational cost. By way of example, consider a representative modern cableway of 25-ton capacity, 2,000-ft span, parallel-traveling type with 70-ft head tower and 30-ft tail tower. The separate drives with 750-hp dc hoist and 500-hp ac haul and modern differential carriers would class this as a moderately high output rig, and its approximate cost should run about $625,000. Preparation and laying of trackways and erection and rigging costs would, for average conditions, run about $95,000.

The operation costs are functions of both production and time, the same as for any other construction equipment. Labor costs for an 8-yd installation including the operator, electrician-oiler, signalman, and part-time rigging and track-maintenance crews would amount to about $25 per hour. The cost of ropes and mechanical maintenance and replacement would be about $ .10 per cu yd.

B. Slackline Cableways

DESCRIPTION

The slackline cableway is a long-range excavator that digs material and transports it for distances up to 1,000 ft. The slackline cableway digs its load, lifts it free of the excavation, conveys it to the discharge point, and dumps automatically. It is recommended for deep digging, long hauls, and delivery of material to an elevated point such as a stockpile or hopper.

The slackline cableway can dig from ground level to 125 ft below water. Handling capacity is governed by haul distance and digging depth. This allows fast, short hauls when maximum tonnage is needed to meet demands. Conversely, when time

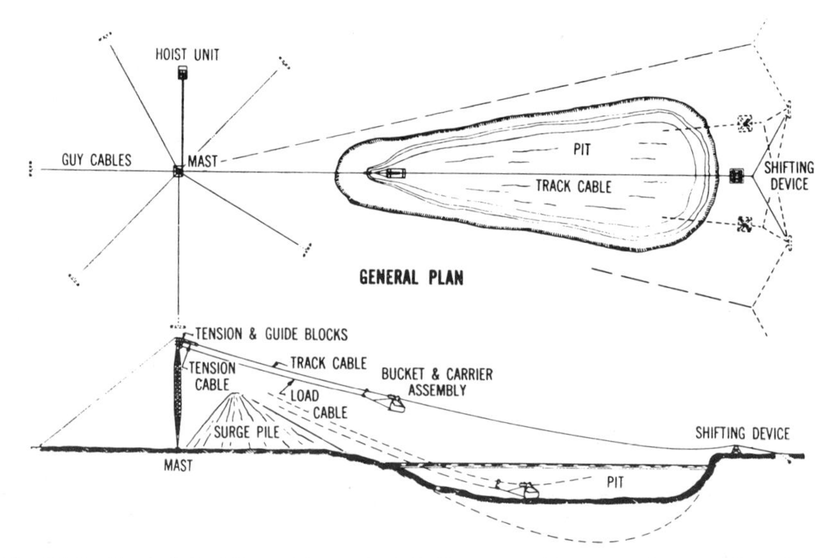

Fig. 14-17 Typical slackline cableway installation excavating an underwater deposit and hauling to surge pile. (*Sauerman Bros., Inc.*)

permits, the reach of the slackline can be stretched to the extreme limits to rebuild or create a reserve pile.

Slacklines are used to excavate sand and gravel, clean out settling ponds and reservoirs, deepen rivers and channels, and reclaim industrial wastes.

One operator handles all the digging, conveying, and stockpiling operations. Slacklines are, therefore, economical excavating and conveying machines. They are built in sizes from ½ to 3½ cu yd.

The slackline cableway consists of a bucket and carrier assembly, a track cable with tensioning cable, a load cable, blocks, a guyed mast to provide height at the head end or dump end, an anchorage at the tail end which is usually movable so that an area can be covered, and a hoist. The hoist is specifically designed for slackline operation to provide a slow digging speed and a fast conveying speed. It can be powered by an electric motor or diesel power unit with a torque converter. Figure 14-17 shows a typical slackline cableway installation.

The frequency of movement of the tail end determines the type to be used. This frequency depends upon depth and span of excavation, handling capacity of the machine, and nature of the material. The typical tail end uses a short tower to provide some elevation of the track cable, which is anchored to a bridle cable. The line of operation is shifted by moving the tower and its track cable connection on the bridle cable by auxiliary equipment.

Another type of tail end is a rapid-shifting system using two towers with a bridle system between them. The track cable is anchored to a bridle frame on the bridle cable, which can be shifted by a motor-powered winch controlled by the operator from his station near the hoist. Movable towers, usually rail mounted, are also used to anchor the track cable. Noncaving materials require more frequent movement of the tail end than free-caving materials.

Since the bucket returns to the digging point by gravity, it is necessary that the head end be high enough in relation to the tail end. Usually the difference between the elevations of the head and tail ends is one-seventh of the horizontal span between the two ends. If the bucket is to discharge into an elevated hopper, care must be taken in the hopper design and location to allow sufficient dumping clearance for the bucket.

FUNCTION

The operating principle of the slackline cableway is simple. The digging bucket is suspended from a carrier which travels on the inclined track cable. The track cable is slackened to lower the bucket into the excavation and tightened to lift the bucket so it can be conveyed to the dumping point. (See Fig. 14-18.)

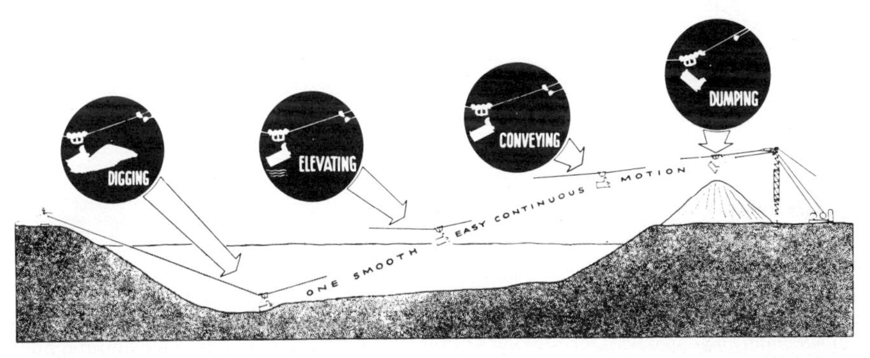

Fig. 14-18 The slackline cableway digs, elevates, conveys, and dumps in one continuous motion. (*Sauerman Bros., Inc.*)

The motive power of the cableway is a two-drum hoist. The load cable, attached to the front chains of the bucket, leads over a guide block at the top of the mast down to the front drum of the hoist. With the bucket resting on the material to be excavated, a forward pull on the load cable causes the bucket to dig and fill to capacity.

After the bucket is loaded, the track cable is raised by taking up the tension cable on the rear drum of the hoist. This raises the loaded bucket clear of the unexcavated material. The loaded bucket is then conveyed quickly along the track cable to the dumping point by the front drum of the hoist.

The bucket dumps automatically, its dumping position along the track cable being determined by a stop button fastened to the track cable. The button may be moved if it is desired to change the dumping point of the bucket. After dumping, the bucket returns at high speed by gravity down the track cable to the digging point. Figure 14-19 shows a slackline excavating a deep deposit.

Slackline cableways have much in common with drag scrapers. Both types of machines use a bucket for digging and transporting material, cables for pulling the bucket and raising it or changing its line of operation, blocks for directing the cables, supports for the blocks, and a hoist for powering the cables. The bucket and cables are the only parts of the machines to contact the material being handled.

The main difference between the two machines is that the slackline uses a bucket which lifts the material whereas the drag scraper makes use of a bottomless scraper which drags the material rather than lifting it. Therefore, the slackline is used to advantage where the haul is long, making its high conveying speed effective, and where the ability to lift the load is desirable or the material is "soupy."

Fig. 14-19 Slackline cableway excavating an underwater deposit of sand and gravel, as viewed from the tail end. (*Sauerman Bros., Inc.*)

PERFORMANCE AND PRODUCTION CHARACTERISTICS

Table 14-4 shows specifications for slackline cableways. It also lists slackline handling capacities in cubic yards per hour for typical average hauls when digging free-caving sand and gravel at a depth of 30 ft below the mast. Allowances should be made in the handling capacities for digging at greater depths, time required for maintenance and shifting the tail end, and abnormal digging conditions.

SELECTION OF UNIT

In selecting a slackline cableway machine, the following information should be available:

1. Description of the work
2. Type of material, weight per cubic foot, size, free-caving or noncaving
3. Elevation of delivery point in relation to digging point
4. Depth of excavation
5. Length and width of deposit and height of bank above water level
6. Location of plant in relation to deposit
7. Whether material is to be delivered to stockpile or hopper
8. Desired hourly capacity and number of hours of operation of the machine per day
9. Whether hoist is to be electric or diesel powered

The owning and operating costs for a 2-cu yd slackline cableway equipped with a torque converter diesel and set up on an 800-ft span are calculated as follows. On an average haul of 300 ft and a depth of 30 ft, this machine will produce 110 cu yd of free-caving sand and gravel per hour. Figures used are for the year 1969.

Owning Costs:

Delivered price.................................... $78,000.00

$$\text{Depreciation} = \frac{\text{delivered price}}{\text{life in hours}} = \frac{\$78,000.00}{20,000} \ldots\ldots = \$\ 3.90$$

Interest, insurance, taxes

Annual rates: interest, 9%; insurance, 2%; taxes, 2%.................... 13%

Estimated annual use in hours.. 2,000

Year's life = 10; % investment... 55.0

$$\text{Hourly cost} = \frac{\text{price} \times \text{investment} \times \text{total annual rate}}{\text{annual use in hours}}$$

$$= \frac{\$78,000.00 \times 0.55 \times 0.13}{2,000} \ldots\ldots = \$\ 2.79$$

Total hourly ownership costs.................................... $ 6.69

Operating Costs:

Fuel, 8 gal at $0.15 per gal.................................... $ 1.20

Maintenance at $0.025 per cu yd (110 cu yd per hr)................ 2.75

Operator.. 5.00

Total hourly operating costs.................................... $ 8.95

Total Owning and Operating Costs per Hour......................... $15.64

$$\text{Cost per cu yd} = \frac{\$15.64 \text{ per hr}}{110 \text{ cu yd per hr}} \ldots\ldots \$\ 0.142$$

TABLE 14-4 Specifications and Handling Capacities of Slackline Cableways

Kind of power	Size of bucket, cu yd	Length of span, ft	Handling capacity in cubic yards per hour when digging 30 ft below mast on an average haul of						Approximate shipping weight, lb
			150 ft	200 ft	250 ft	300 ft	400 ft	500 ft	
Electric...	½	400	25	22	20	17	...	...	16,040
Diesel....	½	400	25	22	20	17	...	...	17,390
Electric...	¾	500	39	37	36	33	...	...	22,795
Diesel....	¾	500	57	54	51	46	...	...	23,020
Electric...	1	600	51	50	46	43	37	...	38,005
Diesel....	1	600	83	80	70	65	55	...	39,255
Electric...	1½	700	...	79	76	73	66	57	54,740
Diesel....	1½	700	...	97	91	84	72	61	53,970
Electric...	2	800	...	...	94	90	80	70	64,930
Diesel....	2	800	...	...	114	110	98	84	67,330
Electric...	2½	900	...	...	113	112	105	95	92,200
Diesel....	2½	900	...	...	150	147	125	107	93,600
Electric...	3½	1000	...	...	...	156	140	129	154,940
Diesel....	3½	1000	...	...	...	182	161	143	157,840

Diesel engine equipped with torque converter and tail shaft governor. The handling capacities stated above are for operations in free-caving materials, such as ordinary sand and gravel. Figures in the third column show the ordinary length of span for each size of machine. Operating cables can be provided for longer spans up to the maximum capacity of each hoist, but excessively long spans should be avoided. Electric motors are 40-deg continuous rated. Diesel engines are suitable for operating these machines up to 3,000-ft altitude.

SOURCE: Sauerman Bros., Inc.

C. Drag Scrapers

DESCRIPTION

The drag scraper is fundamentally a simple machine used to dig material and move it for several hundred feet. The basic components are a bottomless crescent-shaped scraper, cables to pull the scraper, blocks to guide the cables, supports for the blocks, and a hoist to provide motive power and spool the cables. Additional equipment is often added to change the line of operation or raise the scraper on the return trip. This equipment can be additional drums on the hoist or separate winches and movable towers, trolleys, or bridle frames.

Drag scraper machines are used to excavate materials in the dry or under water. They will dig to depths of 100 ft or from banks over 200 ft high; operate on unstable

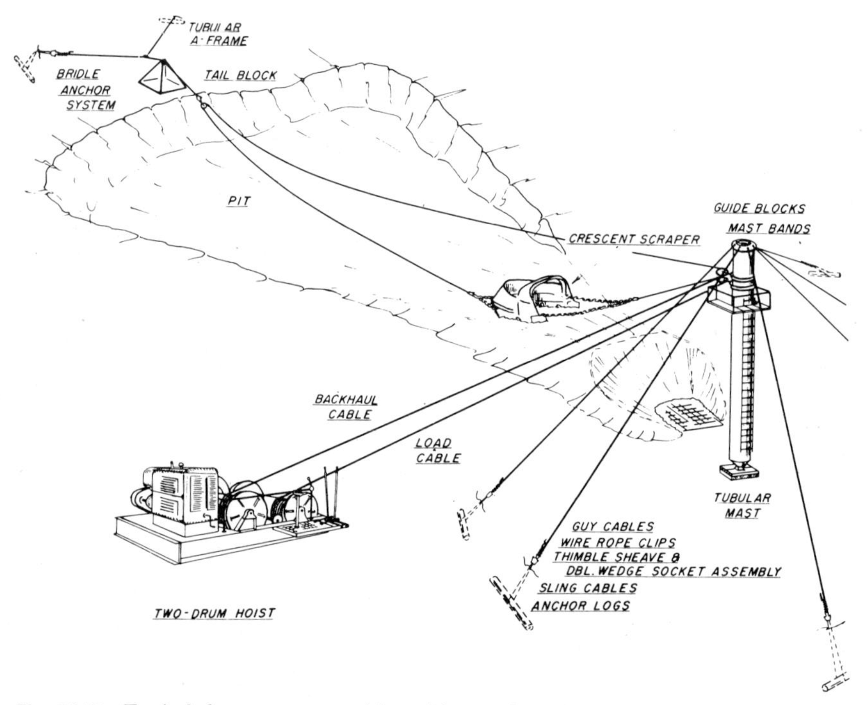

Fig. 14-20 Typical drag scraper machine with two-drum hoist. (*Sauerman Bros., Inc.*)

ground that will not support mobile equipment; handle noncaving materials; move hot, hazardous, or dusty materials; and operate where there is low headroom. They find use in the construction, sand and gravel, mining, cement, steel, coke, fertilizer, marine, aluminum, food, paper, and power industries. The key to the success of the drag scraper machine is the bottomless crescent scraper. When properly designed, this scraper has the ability to dig effectively until loaded, then stop digging and ride its load.

Drag scrapers are built in a range of sizes from ⅓ to 20 cu yd. They are furnished without teeth for easy digging and with teeth for difficult digging. They are light in weight in relation to their volumes when compared to buckets which lift their loads. The scraper and cables are the only components of the machine which contact the material being handled, so maintenance is low. Also, because of this, there is little dead weight being moved with the material. The ratio of payload to total weight (payload plus dead load) is 70 to 80 percent for drag scrapers as compared to 15 to 55 percent for mobile equipment in which the power plant and operator are moved with the material.

Drag scraper machines are of two general types: straight haulage machines and track cable machines. In a straight haulage machine, the scraper travels at grade on both inhaul and backhaul phases of the operating cycle. Simple machines are arranged to operate in one line, as shown in Fig. 14-20. Machines can also be arranged

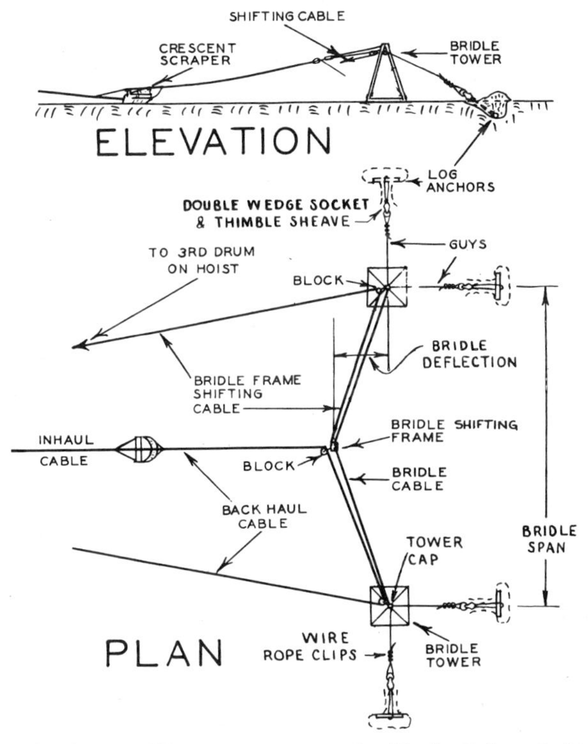

Fig. 14-21 Rapid-shifting bridle arrangement used with three-drum hoist. (*Sauerman Bros., Inc.*)

to operate over an area by shifting the line of operation. Figure 14-21 shows a bridle arrangement with shifting of the line of operation provided by a third drum controlled by the operator. Movable towers or trolleys running on trackways are also used for this purpose.

Straight haulage machines are usually built in sizes from ½ to 5 cu yd with hauls up to 500 ft. They can be operated by a single operator at the hoist or from a remote location. They can be arranged for automatic operation or semiautomatic operation under the supervision of an attendant. Figure 14-22 shows a drag scraper in operation.

The second type of drag scraper is the track cable machine. This differs from a straight haulage machine in that the scraper is raised up by a track cable and is returned to the digging point through the air. This reduces the wear on the scraper, provides positive placement of the scraper, and permits a high return speed. Track cable machines require height to produce the necessary uplift from the track cable to lift the empty scraper. Guyed masts or towers, either stationary or movable, provide the height. Track cable machines are built in sizes up to 15 cu yd and operate on spans up to 1,000 ft.

Fig. 14-22 4-cu yd drag scraper machine with rapid-shifting bridle, as viewed from the tail end. (*Sauerman Bros., Inc.*)

Track cable machines can use either a two-drum or a three-drum hoist. With a two-drum hoist, one drum is used to pull the scraper with the inhaul cable and the second drum is used to raise the scraper with the track cable. The scraper is returned to the digging point by gravity. For long spans, a three-drum hoist is generally

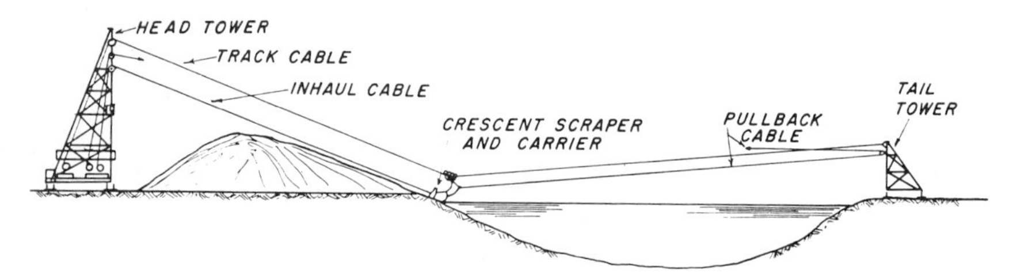

Fig. 14-23 Track-cable drag scraper with pullback cable. (*Sauerman Bros., Inc.*)

used. The third drum is used to pull the scraper back to the digging point through the air with the pullback cable. (See Figs. 14-23 and 14-24.)

FUNCTION

The principle of operation of the straight haulage machine is quite simple. The scraper is pulled from the digging point to the delivery point with the inhaul or load cable by one drum of the hoist. The scraper is then pulled back to the digging point with the backhaul cable by a second drum of the hoist. The backhaul speed is generally faster than the inhaul speed. The operator varies the digging point to suit conditions by returning the scraper to different points along the span.

The capacity of the scraper is greater for the shorter hauls, and thus the operator can take short hauls when high capacities are required and longer hauls when handling rates can be reduced. The line of operation is shifted as required to cover an area by moving the bridle frame, tower, or trolley, depending on the equipment used.

Different arrangements can be used for automating straight haulage machines. These machines are provided with electropneumatic control of clutches and brakes. They can be powered with electric motors or with diesel power units with torque converters. The simplest automatic arrangement is to provide continuous operation of the scraper between inhaul and outhaul limits that are set and changed by the attendant as required.

Another arrangement provides for an automatically varying haul distance by increasing the length of haul each trip of the scraper. The machine can be arranged so that the line of operation is periodically changed. The operation of the scraper

Fig. 14-24 5-cu yd track-cable drag scraper machines building a dike with fill excavated from a lake bottom deposit. (*Sauerman Bros., Inc.*)

machine can also be regulated by timers, counters, bin-level indicators, and other such devices. These serve to coordinate the scraper operation with other operations such as plant production, conveyors, or trucks.

In the track cable machine, the scraper is suspended from a carrier which rides on the track cable. The track cable is slackened to lower the scraper at the digging point until it rests on the material. The scraper is then pulled into the delivery point by the inhaul cable while riding the material. The track cable is taken up or let out as required as the scraper is inhauled.

At the delivery point the track cable is tensioned, lifting the scraper into the air and dumping the material. With the track cable tensioned, the scraper is returned to the digging point by gravity or pulled back with the pullback cable. The track cable is then slackened off to repeat the cycle.

PERFORMANCE AND PRODUCTION CHARACTERISTICS

Table 14-5 shows handling capacities in cubic yards per hour for straight haulage machines in sizes from ½ to 5 cu yd when handling free-caving material. These machines are powered with diesel units equipped with torque converters.

Table 14-6 shows handling capacities for track cable machines in sizes from 6 to 15 cu yd when handling free-caving material. These machines are powered with electric motors.

Either electric motors or diesel power units can be used with either straight haulage

TABLE 14-5 Rated Handling Capacities of Drag Scraper Machines from ½ to 5 cu yd

Machine size, cu yd	Capacity in cubic yards per hour at average haul of				
	100 ft	200 ft	300 ft	400 ft	500 ft
½	40	24	18	14	
¾	62	34	25	20	16
1	82	48	35	27	21
1½	130	75	53	42	33
2	172	98	72	55	44
3	293	172	120	93	76
4	391	230	161	123	100
5	488	288	200	155	126

Handling rates are based upon normal digging, free-caving material with an average weight of 3,000 lb per cu yd. Drag scraper machines 1½ through 5 yd show capacities when equipped with torque converter diesel power.
SOURCE: Sauerman Bros., Inc.

TABLE 14-6 Rated Handling Capacities of Track Cable Machines from 6 to 15 cu yd

Length of haul, ft	Capacity in cubic yards per hour for indicated size of drag scraper				
	6 cu yd	8 cu yd	10 cu yd	12 cu yd	15 cu yd
100	420	560	700	840	1,050
200	282	375	470	564	705
300	215	280	350	420	525
400	168	224	280	336	420
500	138	184	230	276	345
600	120	160	200	240	300
700	108	144	180	216	270
800	96	128	160	192	240
900	...	...	140	168	210
1,000	...	...	130	156	195

SOURCE: Sauerman Bros., Inc.

or track cable machines. Handling capacities are usually greater for the diesel units with torque converters than for corresponding electric motors, as higher speeds are obtained.

Allowances should be made in the handling capacities for maintenance, shifting the line of operation, and difficult digging conditions.

SELECTION OF UNIT

The following information is needed for selecting a drag scraper machine:

1. Description of work
2. Type of material, weight per cubic foot, size, free-caving or noncaving
3. Elevation of delivery point in relation to digging point
4. Depth of excavation

5. Length and width of area to be worked and height of bank above water level
6. Location of plant in relation to area worked
7. Whether material is to be delivered to stockpile, hopper, or trucks
8. Desired hourly capacity and number of hours of operation of the machine per day
9. Whether hoist is to be electric or diesel

The owning and operating costs for a 5-cu yd straight-haulage drag scraper machine equipped with a torque converter diesel and set up on a 500-ft span are as follows. On an average haul of 300 ft this machine will produce 200 cu yd per hour of free-caving material. Figures used are for the year 1969.

Owning Costs:

Delivered price.................................... $95,000.00

$$\text{Depreciation} = \frac{\text{delivered price}}{\text{life in hours}} = \frac{\$95{,}000.00}{20{,}000} \ldots\ldots = \$\ 4.75$$

Interest, insurance, taxes

Annual rates: interest, 9%; insurance, 2%; taxes 2%.................... 13%
Estimated annual use in hours.. 2,000
Year's life = 10; % investment....................................... 55.0

$$\text{Hourly cost} = \frac{\text{price} \times \text{investment} \times \text{total annual rate}}{\text{annual use in hours}}$$

$$= \frac{\$95{,}000.00 \times 0.55 \times 0.13}{2{,}000} \ldots\ldots = \$\ 3.40$$

Total hourly ownership costs.................................. $ 8.15

Operating Costs:

Fuel, 14 gal at $0.15 per gal.................................... $ 2.10
Maintenance at $0.025 per cu yd (200 cu yd per hr)................ 5.00
Operator.. 5.00

Total hourly operating costs..................................... $12.10

Total Owning and Operating Costs per Hour........................ $20.25

$$\text{Cost per cu yd} = \frac{\$20.25}{200 \text{ cu yd per hr}} \ldots\ldots \$\ 0.101$$

Drag scrapers are similar in many ways to slackline cableways. The drag scraper handles material by dragging rather than lifting it. This means that, for a given size of bucket, the loads are lighter and mechanical parts and power units smaller.

SCRAPER USED WITH DRAGLINE

The use of a bottomless scraper in combination with a revolving dragline or crawler crane increases the range of operation and usefulness of these units. The reach of a dragline can be extended by using a track cable with a scraper and carrier assembly. So equipped, the machine's range is limited usually only by the spooling capacity of the drag cable drum. It can reach farther, dig deeper under water, and take material out of soft areas without the nuisance of mats and the hazard of undermining. The dragline is generally used as the inhaul or load cable.

If it is desired to excavate material and pile it in front of the dragline, the cable is reeved through a guide block in an elevated position in a boom support. The hoist line is used as the track cable. It is reeved through a block on the boom tip to an anchorage which may be fixed or movable. (See Fig. 14-25.)

A scraper of the efficient crescent design can be larger than the dragline bucket it replaces because the scraper is much lighter, size for size. The bottomless scraper hauls at ground level and deposits its load automatically when lifted. Thus, the crane is required to raise only the empty scraper. Large machines can usually double their rated bucket size with a crescent-type scraper. Smaller units can use about a half size larger.

The method of operation is as follows. The track cable is slackened off while the scraper is digging and bringing in a load. When it arrives at the dumping point, the operator pulls the track cable taut by a few revolutions of the drum which lifts the scraper into the air. The scraper, being bottomless, deposits its load automatically when lifted. Release of the drag brake allows the load cable to unwind from the drag drum, and the carrier, impelled by gravity, carries the empty scraper down the track cable.

Draglines equipped with crescent scrapers are used for such jobs as cleaning out rivers and ponds, deepening channels, digging and backfilling trenches for underwater pipelines, excavating marshes and peat bogs, constructing and repairing levees, building marinas, excavating sand and gravel, and grading beaches.

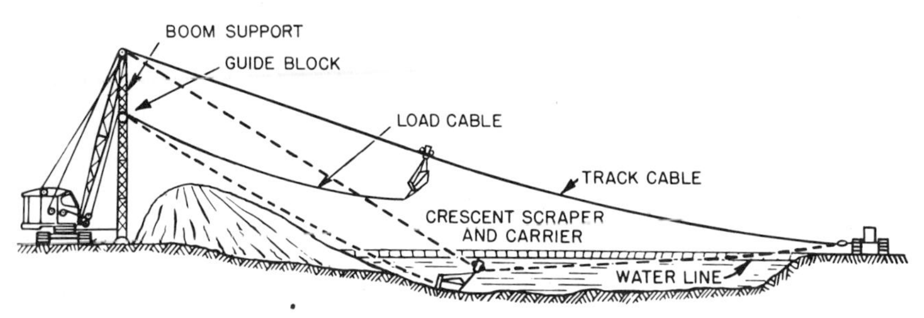

Fig. 14-25 Crescent scraper used with dragline with boom support and track cable. (*Sauerman Bros., Inc.*)

The span between the dragline and the tail end is usually a maximum of about seven times the boom height. The distance that the scraper returns by gravity depends for the most part on the spooling capacity of the drag drum. If this distance is less than the width of a river or pond, the machine can be used to dig from the middle to one bank, then it can be relocated on the other bank to finish the excavation. The size of the scraper that can be used depends upon the available line pulls in the drag and hoisting cables of the dragline and the overturning moment which the machine can resist. The size of the scraper can often be increased by using a boom support or boom back guys to stabilize the machine.

A scraper can also be used in place of the dragline bucket in a casting operation. Here again, a larger scraper can be used. With this arrangement the scraper can not only excavate at grade level or below grade but can also pull down banks or piles. The scraper in a casting operation can be equipped with a lifting hitch so that a partial load can be lifted. This load is about 50 percent of the rated load of the scraper.

Section 15

Belt Conveyors

A. T. YU

Vice-president—Operations, Robins Engineers & Constructors,
Hewitt-Robins, Inc., Division of Litton Industries, Totowa, N.J.

INTRODUCTION

One of the noteworthy developments in the modern construction industry is the increasingly frequent use of automated belt-conveyor systems to handle earth, aggregates, rocks, and other bulk material. The growing popularity of long belts for overland bulk haulage is particularly impressive.

The principal advantages of a belt-conveyor system over other means of overland haulage are its low maintenance, ability to cross adverse terrain, low labor requirement, high reliability, and excellent safety record. When the total tonnage to be carried is sufficient to justify the higher initial investment, a belt-conveyor system nearly always results in the lowest overall per ton-mile cost.

As early as the 1920s, long belt-conveyor systems were successfully used for earth movement. Five million cubic yards of dirt were moved for more than a mile from Denny Hill, through the business district in Seattle, onto barges.[1*] This was followed over a period of several decades by a series of large earth and dam projects.[2,3]

Modern advances in technology have greatly enhanced the capabilities of belt-conveyor systems. The use of belt with a steel-cable core has made possible single-flight center distances[4] of over 15,000 ft. High speeds (1,200 fpm) and wide belts (120 in.) have upgraded capacities[5] to over 20,000 tons per hour (tph). Sophisticated controls and devices such as closed-circuit TV have drastically reduced operating costs and, at the same time, have increased systems reliability. For the design and operation of the systems, electronic data processing systems have emerged as an efficient and sometimes indispensable tool. Developments in related systems and components such as portable crushers, screens, etc., have also been instrumental in making the belt-conveyor system more versatile and adaptable to heavy construction.

Volumes have been written on the subject of belt conveyors and related equipment.[6–10] The reader is referred to these for an in-depth treatment of the subject. This section is limited to a general outline of the belt-conveyor system's application in heavy construction.

BELT-CONVEYOR SYSTEMS AND SUBSYSTEMS

To attain its optimum utilization, a belt conveyor must be viewed as an integral part of an overall system. Only after the objective and function of an entire project are clearly analyzed and defined can the most suitable and economical belt conveyor or belt-conveyor system be selected. Thus the adequacy of a belt-conveyor system, including its subsystems and components, must be weighed in terms of the overall system which it serves and in terms of today's as well as tomorrow's needs. For example, a 36-in. belt system may require considerably less investment than an equivalent 42-in. network and will probably operate at a lower per-ton cost. Yet the 42-in. system may be the more desirable choice if its use could result in substantially faster completion of the project and/or savings in the number of required shifts in the quarry, thus leading to lower overall project and unit costs.

The need for a systems approach has become further accentuated by the increased demand for automation. Without complete synchronization with the overall system which it serves, automatic control of a belt-conveyor system alone has no meaning.

A typical belt-conveyor system used for an earth dam project is shown in Fig. 15-1. Here, many million yards of rock and earth are simultaneously required to top out a large earth dam and to be barged out for a causeway. Various grades of material from a nearby borrow pit are excavated, stockpiled, and then moved to the construction site via an overland belt-conveyor system. Reclaim conveyors at the site then route the respective materials to truck bins or to a barge loader.

In the system illustrated, while the proper selection and sizing of each of the individual components is important, the overriding consideration is still the most economical yard of earth or rock excavated and hauled to the destination. A costly, wide belt may not be necessary if the bulk of the material to be hauled is alluvial, and a

* Superior numbers refer to the list of references at the end of this section.

crusher can easily handle the occasional large boulders. If space is a problem at the dam site, a larger stockpile near the excavating pit may be the solution. If the terrain is rugged and steep downhill grades are involved, an overland belt system not only will minimize investment but may even generate power for running other equipment.

This illustrative system may be divided into several subsystems according to the specific functions each performs; they are analyzed and described as follows:

Plant Feed Systems Since a belt conveyor is a continuous bulk-handling device, the system will attain its highest productivity and efficiency when it can be loaded uniformly at its maximum design rate. Thus the key to maximizing the utilization of a belt-conveyor system is its feed. A good plant feed system must be capable of adapting to the working cycles of equipment and converting intermittent random feed into steady, uniform flow.

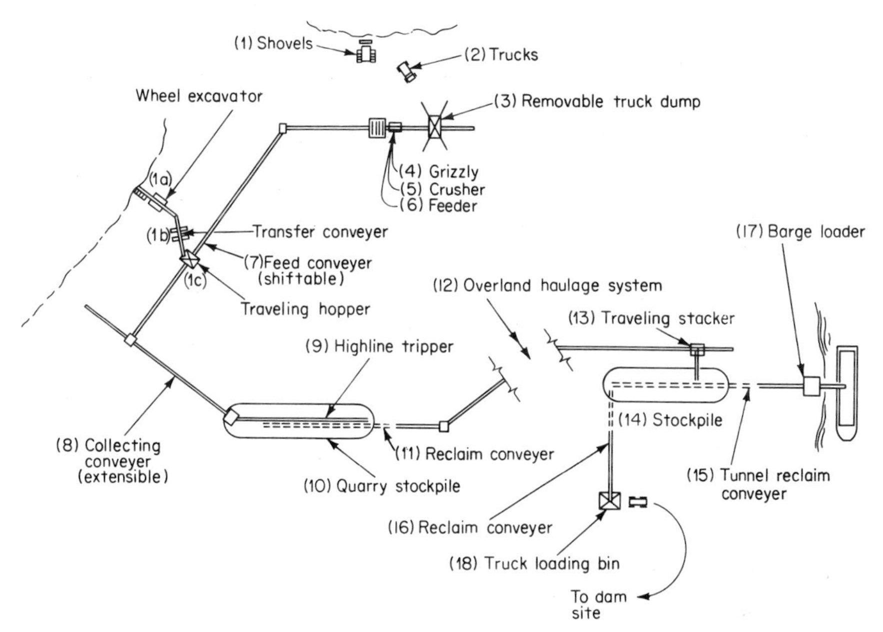

Fig. 15-1 A typical belt-conveyor system in a construction project.

In the earthmoving project example, alluvial or relatively easy digging material is dug by a bucket wheel excavator which is itself a continuous loader. Through a crawler-mounted portable transfer conveyor and a traveling receiving hopper, material is fed onto the feed conveyor nearly parallel to the face of the borrow pit. This conveyor should be of shiftable design to enable it to move sideways toward the pit with the advance of excavation.

For the rock zones, where large shot rock must be loaded by shovels, a different feed system is required. Here trucks are unloaded into portable or semiportable truck dumps, over a grizzly, prior to being crushed to sizes that are economical for belt-conveyor handling. Surge hoppers are provided when needed, and mechanical feeders receive the combined product to deposit it onto the long feed conveyor. The truck dump may be designed to accommodate rear-, bottom-, or side-dump trucks with greater than 100-ton capacity (see Fig. 15-2).

Table 15-1 summarizes the major advantages and disadvantages of the most widely used types of feeders. Several of these feeders are illustrated in Fig. 15-3. On occasion, if the reclaim conveyor is relatively short, it can also serve as a feeder. It

Fig. 15-2 Removable truck dump in bolted construction is readily dismantled and reerected as excavation progresses.

TABLE 15-1 Feeders

Type	Major advantage	Disadvantage
Vibrating feeder (mechanical, magnetic, or pneumatically actuated)	Versatility for a variety of materials	Inclination adjustment may be difficult
Reciprocating plate feeder	Simplicity in operation and maintenance	Not readily adjustable
Apron feeder	Low cost, relatively rugged construction	Cleanup, maintenance
Manganese feeder	Rugged construction for heavy-duty service	Cleanup, maintenance, high cost
Belt feeder	Economical, simple, and clean operation	Not sufficiently rugged for hard lumps
Plow feeder	Positive, continuous flow	Relatively limited capacity—tunnel expense
Gate feeder	Extremely simple and economical operation	Difficult to regulate feed, poor wear resistance
Chain feeder	Simplicity	Regulation difficulty
Dozer trap-belt loaders	Mobility, simple operation	Difficulty in handling boulders

then receives material directly from the hopper bottom by gravity. This is possible only if the muck contains enough fines to cushion the impact of the large lumps. Even then, regulating the surge may be a problem.

Storage-Reclaim Systems Since fluctuation in input and output of a system or a plant is unavoidable, a storage-reclaim system provides the flexibility and reserve needed for a continuous, smooth operation. It is insurance against breakdowns, strikes, weather, maintenance, and other unexpected contingencies.

How much insurance to provide depends on the requirements of the overall system. One must consider the normal rate of plant input and all the factors which might affect it plus the normally required rate of reclaim and the possible consequences if this is not

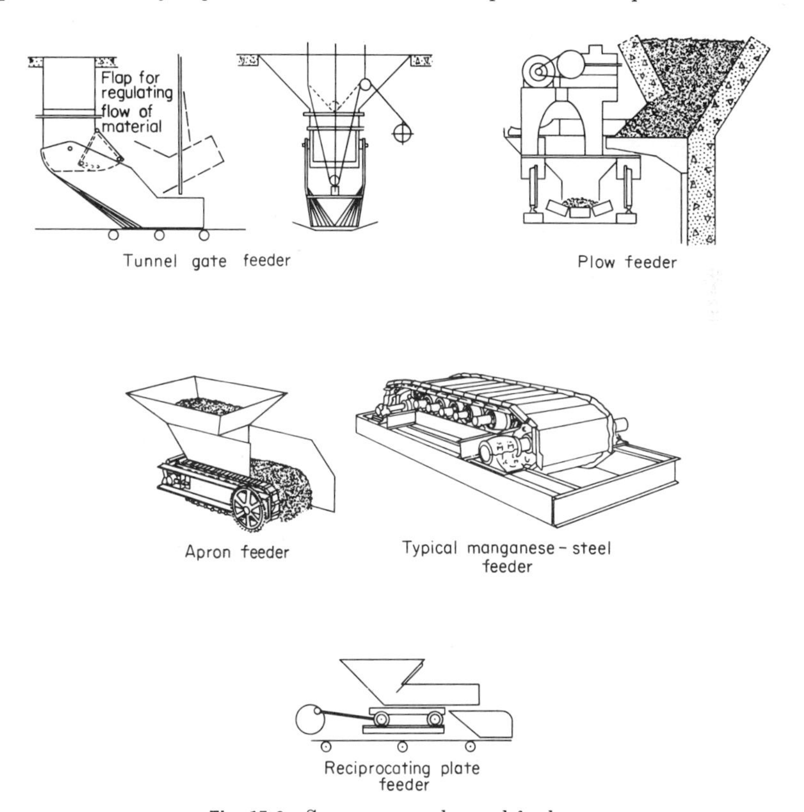

Fig. 15-3 Some commonly used feeders.

maintained. The characteristics and types of material to be stored, the necessity for blending or shelter, and the most suitable equipment to use are additional factors. The cost of the storage itself should be included in this all-encompassing evaluation. Here one must consider the available space, terrain, soil and foundation requirements, water table, and other engineering factors.

When the quantity of material to be stored is relatively small, storage bins are often used. The major advantages of bin storage are cleanliness of operation and the ease of material reclaim. Simple gates under the bins are operated mechanically, pneumatically, or hydraulically and can feed trucks or rail cars directly and quickly. Prefabricated storage bins of up to 5,000-cu yd capacity are commercially available. For larger quantities, yard storage would normally be required.

The capacities of outdoor stockpiles can vary from several thousand tons to several million tons. Figure 15-4 gives the storage capacities of materials with different angles of repose. When a tunnel-gravity reclaim system is used, only the part of the stockpile reclaimable by gravity is considered *live storage* (see Fig. 15-5). The dead storage is normally regarded as a safety reserve to be used for emergencies.

For outdoor yard storage, if neither degradation nor dust is a problem, high-line

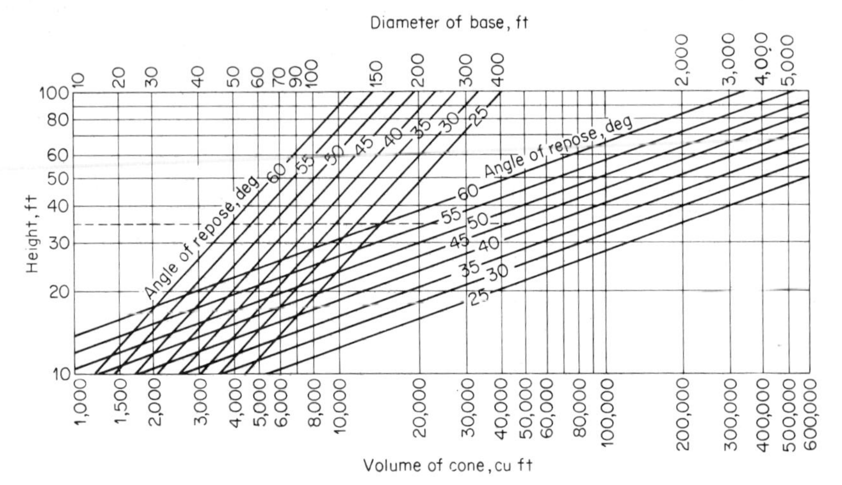

Fig. 15-4 Capacities of conical stockpiles. Note: To find diameter of base and volume of a cone, enter chart at the left with height of cone. Follow across until reaching the first set of lines "Angle of repose" of the material to be stored; directly above, at the top of the chart, read the diameter of the base. Then resume reading along the height of the cone until intersecting with second set of lines "Angle of repose;" directly below, at the bottom of the chart, read volume (cubic feet) of the given cone. The heavy line example indicates that a cone 35 ft high, built up of material with a 45° angle of repose, will have a base diameter of 70 ft and a volume of 45,000 cu ft. The operations may be reversed to find height of a cone to give any required volume. Frustums of cones may also be calculated by finding the volume of the full cone and of the upper section, and subtracting the latter from the former.

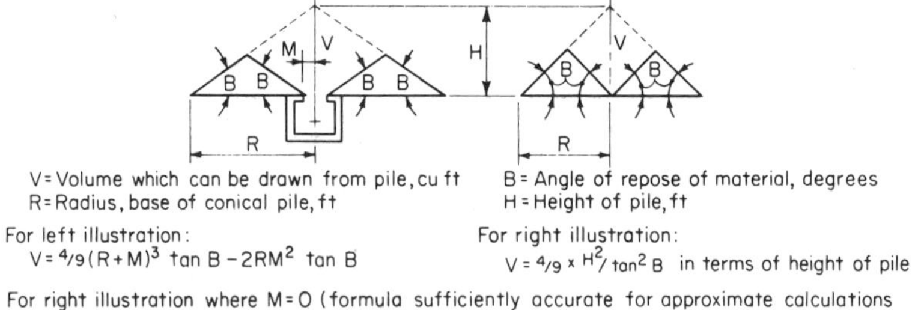

V = Volume which can be drawn from pile, cu ft
R = Radius, base of conical pile, ft
B = Angle of repose of material, degrees
H = Height of pile, ft

For left illustration:
$V = 4/9(R+M)^3 \tan B - 2RM^2 \tan B$

For right illustration:
$V = 4/9 \times H^3/\tan^2 B$ in terms of height of pile

For right illustration where M = 0 (formula sufficiently accurate for approximate calculations left hand illustration also) $V = 4/9\ R^3 \tan B$

Fig. 15-5 Method of determining the amount of material which can be withdrawn by a tunnel conveyor from a conical pile.

trippers and shuttle conveyors may be used to form long, rectangular stockpiles. When dust or weather becomes a problem, the stockpile may be housed. Traveling stackers or stacker towers can build stockpiles of various shapes on the ground. A double-wing, track-mounted stacker (see Fig. 15-6) forms two longitudinal stockpiles, one on each side of the track. A track-mounted "radial" stacker builds a kidney-shaped pile. A rubber-tire mounted stacker may be over 100 ft in length. The crawler-mounted stacker, equipped with its own propelling drives, usually has the greatest flexibility. Throwers, or slingers, are occasionally installed at the end of the stackers to increase their reach. If dust is a problem, extensible telescopic chutes may be installed at the heads of the stackers, or the stackers may be fitted with luffing booms to decrease drop.

The most commonly used method for reclaiming yard stockpiles has been gravity feed to reclaim conveyors housed in a tunnel. The tunnel runs under the stockpile and is built of reinforced concrete, timber, or corrugated steel culvert. Openings in its roof are provided with feed gates or mechanical feeders to withdraw the material and feed it at a uniform rate onto the tunnel conveyor.

The bucket-wheel-type reclaimer, either rail- or crawler-mounted, is rapidly gaining in popularity. When the controlled blending of material is required, a bridge-mounted bucket wheel which is a combination reclaiming-blending machine is also available. The wheel-mounted buckets dig into the storage pile and discharge, as

TABLE 15-2 A Belt-conveyor System Compared with Other Commonly Used Haulage Systems

Haulage system	Advantages	Disadvantages
Railroad	Rugged equipment—resistance to abuse Same installation may be used to haul equipment, supplies, and personnel No need for power transmission nor large power plant if diesel equipment is used	High construction costs, especially in rugged terrain Inability to negotiate steep grades, consequently requiring much longer run Maintenance High accident rate in difficult terrain
Trucks	Low investment cost Highest flexibility Can carry equipment, supplies, and personnel	High maintenance—road and equipment Require a team of operators and shop personnel High operating cost High accident rate
Belt conveyors	Low operating cost Comparatively lower investment cost for high tonnage and rugged terrain Requires least number of operators Reliability and safety Noiseless—quieter operation	Normally cannot be used for hauling supplies, equipment, or personnel Rubber belt is vulnerable to damage unless all design precautions are exercised and operating restrictions are properly policed

they reach the highest or topmost position, onto an adjacent conveyor belt. The capacities of these machines vary from 150 to 4,000 tph at a cost of $250,000 to $1 million installed.

Overland Haulage Systems Because of the large quantities of materials involved in heavy construction, the cost of hauling each yard of earth or rock can materially affect the profit of an entire project. Hence, all factors must be carefully weighed before the most economical haulage system is finally chosen. A belt-conveyor system is compared with other commonly used haulage systems in Table 15-2.

In addition to the systems identified in this table, there are other means of long-distance haulage for special applications which are, or are becoming, available. For example, the overhead cableway has been used very successfully for concrete dams and in mining operations. Cableways are particularly advantageous when the construction is in a deep gorge where employment of other equipment is not practical.*

The cable-belt system utilizes two independent steel cables, one on each side of the rubber belt, to support the belt and to carry tension. Since the belt is no longer required to transmit drive force, its cost tends to be lower than that of equivalent

* See Sec. 14, Cableways and Drag Scrapers.

belting for conventional conveyors. The speed of this system is comparatively lower, and cable wear has been a problem in some installations.

Table 15-3 compares the costs of alternative haulage systems with that for a 6.5 mi, six-flight conveyor system with a 36-in.-wide belt. This system rises 700 ft and then plunges 1,700 ft prior to arriving at the terminal. This rugged terrain can be traversed with only 6.5 mi of belt-conveyor system as compared to 14 mi of rail line or 10.8 mi of truck road.

Examples of long belt-conveyor haulage systems used for heavy construction and other industries are shown in Table 15-4. Other well-known examples are Grand Coulee Dam (48 in., 5,000 ft, 1930); Shasta Dam (36 in., 9.6 mi, 1940); Bull Shoals Dam (30 in., 7 mi, 1948); Great Salt Lake Crossing (54 in., 2.5 mi, 1955); Trinity Dam (42 in., 3 mi, 1958); George Washington Bridge approach (42 in., 2 mi, 1961); Orovillc Dam (54-, 72-, 96-in., 5 mi—plant and land, 1964).

TABLE 15-3 Relative Cost Comparison—Various Haulage Systems*

	Conveyor	Railway	Cableway	Trucks
Length required (miles)	6.5	14.0	6.5	10.8
Relative cost per ton-mile	1.00	0.58	2.29	1.30
Relative cost per ton	1.00	1.26	2.29	2.16
Relative capital cost	1.00	1.30	0.81	0.97

* ADAPTED FROM: Robert C. Temps, Reynolds' Overland Conveyor in Jamaica, *Skillings' Mining Review*, Feb. 3, 1968.

TABLE 15-4 Examples of Long Belt-conveyor Systems

Width, in.	System length, ft	Max single-flight length, ft	Speed, fpm	Capacity, tph	Material	Location	Year	Note
47	12,140	12,140	492	1,760	Excavated earth	Kobe, Japan	1964	Move mountain to sea through culvert and tunnel
66	15,000	15,000	1,100	12,000	Excavated earth	Portage Mtn. Dam, Vancouver, Can.	1965	Overland
42	54,048	10,591	492	1,200	Stone	Nagato, Japan	1964	Overland/tunnel
36/42	35,756	6,450	600	800/1,200	Coal	Greene City, Pa.	1965	Overland/underground
36	50,300	16,400	590	2,240	Iron ore	Marcona, Peru	1966	Overland
42	33,377	10,950	500	....	Bauxite	Jamaica	1967	Overland

Ship and Barge Loaders When excavated material must be routed over wide stretches of water, ship or barge loaders equipped with belt conveyors serve as a link between water and land.

The simplest loader is a fixed-tower type. The tower supports an overhanging boom-conveyor which reaches out to trim the ship or the barge transversely. Lengthwise trimming is accomplished by moving the vessel with its own gear, a towboat, or winches on shore.

A traveling loading tower affords the added flexibility of not requiring movement of the vessel for trimming. The tower itself is fed by a trailing tripper, and both move along the pier together.

The capacity of a modern ship- or barge-loading system may vary from several hundred tons per hour to over 10,000 tph.[11] Before loader size, capacity, and type can actually be determined, consideration must be given to the size of vessels and the frequency of their arrivals, the loading cycle time, and climatic and water conditions at the pier (wind, tide, current, surf, waves, and ice). A ship- or barge-loading system is, in essence, a large batch-handling subsystem. A storage-reclaim system becomes a key to converting it into a smooth, continuous bulk-handling operation.

Portable Systems Portable conveyor systems are usually lighter, lower in cost, and easily fitted to a job's needs. Since virtually no erection costs are involved, they may readily be moved to the most desirable locations to handle assignments. In practice, the components and subsystems are unitized in a system or plant and broken down into logical module units such as feeders, screens, crushers, and stackers.[12] Depending on how they are supported, the portable systems may be classified in five groups:

1. Rubber-tire running gear
2. Wheels running on rails
3. Crawler-type running gear
4. Skid mounted
5. Semiportable

The systems which travel on rubber tires are the lightest and are readily towed around the jobsite by dozers or tractors. A heavy unit can be loaded onto a truck-trailer

Fig. 15-6 Typical double-wing rail-mounted stacker.

for long-distance highway movement. The tires may be left on when the unit arrives at the new site, or they may be temporarily removed and the plant jacked up and supported on timber or concrete blocks to provide a firm foundation. Power is furnished by trailing cables, portable diesel generators, or directly connected gasoline or diesel engines. Because of its relatively low cost, the system with rubber-tire running gear is by far the most widely used. Among other applications, it has been used in ship and barge loading.[13]

Machines which travel on rails are generally used in more permanent installations where a stable, ballasted rail bed can be prepared. The movement of these machines is confined to the rail line limits. Typical machines in this category are hoppers, stackers (Fig. 15-6), shuttle conveyors, trippers, blending-reclaim machines, feeders, ship and barge loaders, etc.

Machines which travel on crawler tracks have the greatest flexibility by virtue of their ability to move around in rough terrain. They are usually large, sturdy machines, especially those designed for excavation. The wheel-on-boom bucket excavators are equipped with their own boom conveyors to receive and discharge excavated material onto a connecting conveyor network. A transfer conveyor which moves on crawlers adds to the system flexibility.

A crawler-type machine is usually backed up by a shiftable feed-collecting conveyor. This shiftable conveyor has unitized deck stringer sections mounted on rail ties. Specially designed lugs enable the conveyor to "snake" into new positions without causing damage to belt, machinery, or structures. Shifting is accomplished by specially designed rigs mounted on dozers. A 60-in.-wide, 1,000-ft conveyor can be moved laterally up to 6 ft and realigned in a matter of hours.

When frequent movement is not required, semiportable plants are used. These plants are built with simple bolted construction, readily erected on light foundations. Upon completion of a project or a phase of it, they are easily dismantled and moved to another location.

BELT-CONVEYOR COMPONENTS

Belts In a belt conveyor, the most important and usually most costly component is the rubber belt. An oversimplified description of the construction of a conveyor belt is a carcass imbedded in rubber. The carcass carries tension and absorbs impact. It is made up of one or many plies or layers of fabric (cotton or synthetic), bonded together by a friction and/or skim coat of rubber. When the carcass is of single-layer construction, it could be fabric yarns or steel-wire stranded cables, the latter being the newest type of carcass with the highest tension rating. The rubber compound may be natural, synthetic, or a blend. *Breakers* with cord or leno (an open mesh fabric) construction are sometimes placed between the cover and the carcass to increase the adhesion as well as the impact and puncture resistance of the belt.

The conveyor-belt industry has rated various types of belt rubber in three grades according to their cut, tear, and abrasion resistance:

Cover grade	Cut and tear resistance	Abrasion resistance	Application
1	Excellent	Excellent	Large-sized rock, sharp cutting materials, extremely rugged service
2	Good	Excellent	Heavy-duty service—sized materials with limited cutting action, primarily abrasion
3	Low	Good	General light-duty service—small, sized material

Allowable working tension varies according to the type of carcass employed. It may be as low as 27 lb per in. per ply for a low-cost cotton fabric carcass and as high as 240 lb per in. per ply for a high-tensile synthetic construction (see Table 15-5). A belt with a steel wire-rope core can carry up to 4,000 lb per in. of belt width, using ⅜-in. stranded cables.

In selecting a conveyor belt, in addition to the quality of the rubber, one must also consider the tension requirements, troughability over idlers, longitudinal flexibility as a function of the pulley diameter, number of fabric plies, impact resistance, and cover thicknesses in order to find the most appropriate and economical choice.

Conveyor belts are usually shipped in rolls, to be made endless at the job site. Vulcanized splices are considered superior to mechanical splices in view of their higher strength and longer service life as well as the resultant cleanliness. On the other hand, when high initial costs are not justified and time does not allow for extensive vulcanizing procedures, mechanical fasteners such as Flexco metal links have been used successfully in many installations. A simple mechanical splice may be made in a matter of minutes, compared to many hours for making and curing a vulcanized splice. Mechanical fasteners have been lifesavers for quick repairs of cuts and rips in conveyor belting in an emergency. A compromise between the two methods of splicing is the so-called "cold patch," such as the one manufactured by Rema Co. This method requires the same stripping and gluing procedure as used in the vulcanized splices, but without the application of heat after the splices are made. This greatly reduces the time required for the completion of a splice and still has many of the advantages inherent in a vulcanized splice.

Pulleys and Idlers Pulleys are usually located at the terminals of a belt conveyor, where they support the belt and transmit driving power. Between the terminals, the conveyor belt is supported by troughing idlers on its loaded upper half and by return idlers on its normally empty lower half.

A conveyor pulley is usually of welded steel construction. Its continuous rim is supported by steel end disks which are fitted with compression hubs on shafting.

TABLE 15-5 Tension Ratings of Conveyor Belts with Fabric Carcass (lb/ply/in.)*

Fabric identification, RMA†	Normal mechanical fastener splice	Normal vulcanized splice
35	27	35
43	33	43
50	40	50
60	45	60
70	55	70
90	...	90
120	...	120
155	...	155
195	...	195
240	...	240

* ADAPTED FROM: *Belt Conveyors for Bulk Materials*, Conveyor Equipment Manufacturers Association, 1966.

† Rubber Manufacturers Association.

Some pulleys are crowned at the center to help the training of the conveyor belt. Lagging of rubber or other material is used to increase the coefficient of friction between the belt and the drive pulley. It also helps to reduce wear on the pulley face and eliminate material buildup. Wing-type pulleys of slatted construction also help to eliminate material buildup on the pulley face. Pulley outfits with shop-assembled shafting and bearings are usually shipped as complete assemblies for ease of field erection.

Troughing idlers are commonly of the three-roll type, with the center roll horizontal and the two side rolls inclined to form a trough (Fig. 15-7*a*). The standard troughing angle for side rolls is 20°, but advances in belting technology are responsible for a trend toward *deep trough* idlers. These idlers permit greater belt capacity without increasing belt width. Their troughing angles may vary from 25° to 35° or even 45°. Transition idlers (Fig. 15-7*b*) with gradually varying troughing angles are required to properly support the belt as it approaches the flat terminal pulleys.

The idler rolls are usually made up of steel tubing, with welded end disks which house antifriction bearings. Two systems of lubrication are available. The so-called "sealed for life" idlers use factory lubricated and sealed rolls and can only be relubricated by disassembling the seals in shops. The regreasable idlers may be lubricated on site through grease fittings. Some of these idlers have internal or

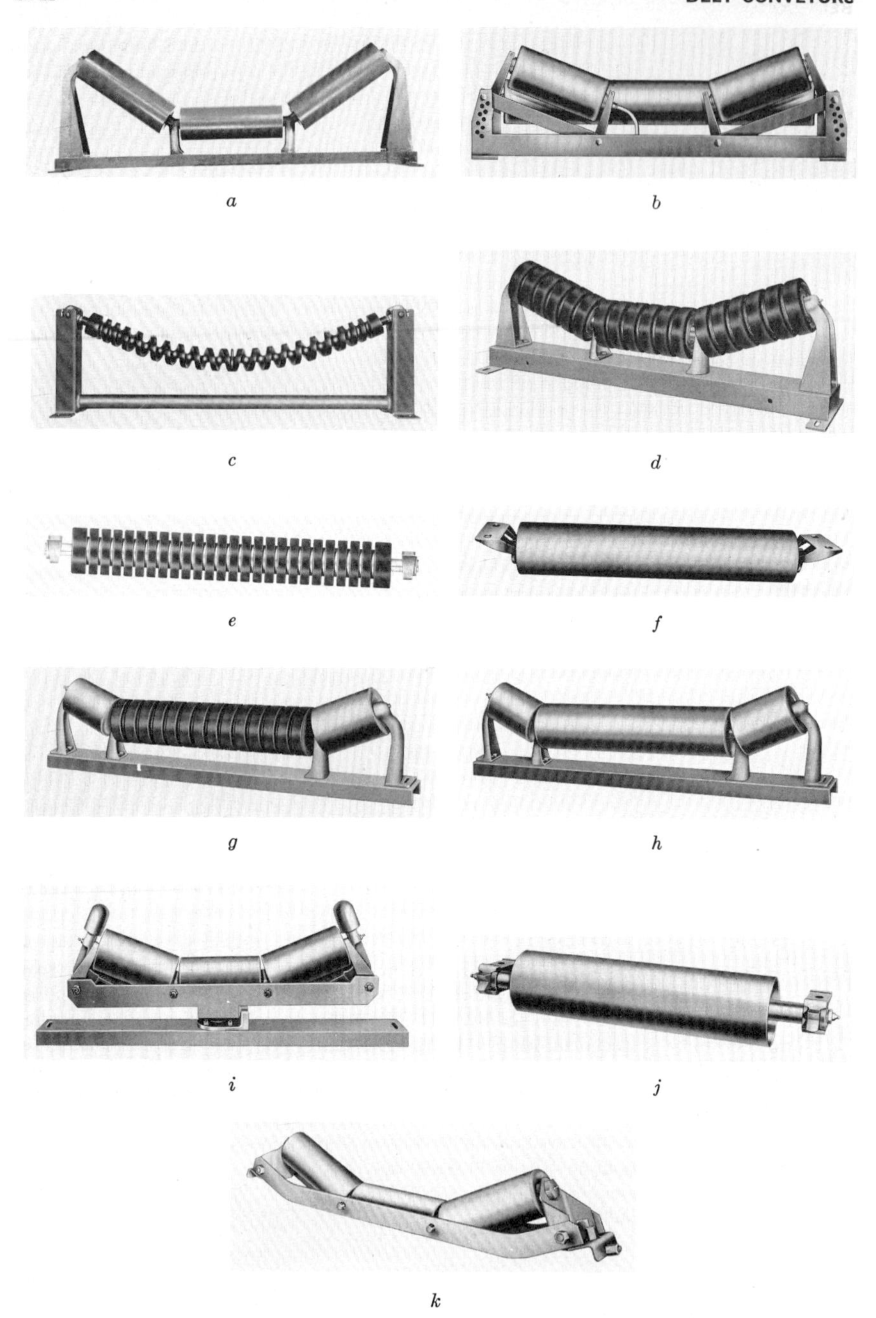

Fig. 15-7 Troughing idlers. (*a*) Standard troughing idler. (*b*) Transition troughing idler. (*c*) Spiral catenary idler. (*d*) Rubber-disk impact idler. (*e*) Rubber-disk return idler. (*f*) Flat-roll return idler. (*g*) Picking idler (impact). (*h*) Picking idler. (*i*) Troughing-training idler. (*j*) Return-training idler. (*k*) Wire-rope stringer idler.

external piping connecting all the bearings to one grease fitting for one-shot lubrication.

For lighter duties, catenary troughing idlers (Fig. 15-7*c*) may be substituted for the three-roll type. Individual rubber rolls, or a continuous spiral assembly, are molded on a flexible catenary member (usually a stranded cable). The assembly rotates on fixed bearings at the ends of the catenary member. The rolls may also rotate on individual bearings supported by a fixed, flexible catenary member. The spiral type may be made up of a continuous steel coil spring, which also helps to absorb impact.

The spacing of troughing idlers is dictated by the permissible belt sag between the idlers and thus is a function of belt tension and the load carried on the belt. Impact idlers (Fig. 15-7*d*) at loading points are spaced as close as physical dimensions permit, still allowing room for cleanup. The spacing on the return idlers (Fig. 15-7*e* and *f*) is usually a function of weight of the belt only and varies anywhere from 8 to 12 ft. For the economical design of long conveyors in which the tension varies substantially from one terminal to another, the spacing between troughing idlers may be graduated from around 3 ft at the low-tension sector to over 10 ft where high belt tension can support the load between the idlers through catenary action.

To cushion impact on the belt at feed points, the rubber-disk impact idlers illustrated in Fig. 15-7*g* are used. To facilitate hand picking, the length of the troughing side rolls may be reduced and the length of the center flat roll increased (Fig. 15-7*h*). This enables the belt to assume a relatively flat yet troughed configuration. Rubber-disk return idlers and helical or spiral return idlers are often used for very sticky material. These idlers have a self-cleaning feature to minimize excessive buildup on the return belt or on the return idlers or pulleys. Sections of rubber sleeves vulcanized on flat return idlers may achieve the same results.

Various types of training idlers are available for carrying (Fig. 15-7*i*) as well as return belt runs (Fig. 15-7*j*). These idlers induce auxiliary side forces on the belt if it begins to run out of line. A sophisticated installation may even employ a hydraulically actuated training idler.

Drives Nearly all belt conveyors are driven by electric motors.* Speed reduction may be achieved by chain and sprockets, V belts, or gearing. The most common device consists of reduction gears which are housed in a box as an entirely shop-assembled unit. The shaft-mounted reducer, with ratings up to 1,200 hp, has the advantage of ease of replacement. The parallel-shaft base-mounted units have the greatest load capacity. Right-angle helical and worm-gear reducers provide added flexibility in machinery arrangement and layouts. These speed reducers may be directly coupled to drive pulley shafts or used in conjunction with chain sprockets and other speed reduction devices to achieve further reduction.

Flexible couplings between drive and driven shafts allow for minor misalignments. To prevent a loaded, inclined belt conveyor from running backwards, a backstop can be used. This allows free motion of the drive pulley in the forward direction but automatically prevents rotation of the drive pulley in reverse. The most common types of backstops are ratchet and pawl, differential band brake, and overrunning clutch; sophistication and costs run in the same order.

To prevent a loaded, declined conveyor from running out of control, a brake is required. This may be in the form of a disk brake on the motor or a shoe brake on a brake-type coupling. Eddy-current brakes exert a retarding force through a magnetic field without direct mechanical connection.

Drives are usually located at the discharge ends of inclined conveyors. For stackers and booms, where weight at the discharge end is critical, the drive is normally placed internally near the feed end. The drive for a declined conveyor is usually located at the tail end, where the conveyor is fed.

Motors and Controls The most commonly used motors for conveyor applications are the squirrel-cage and wound-rotor types. On rare occasions, dc motors are also used. A motor for a conveyor-belt drive must have sufficient torque to overcome the starting friction of the system. At the same time, it should not impose exces-

* Gasoline engines are used where electric power is not readily available. Generally, they are not as economical and require higher maintenance.

sive stress and strain on the belting or on its associated mechanical and structural components.

The wound-rotor motor, with external resistance connected through a controller to the secondary winding, provides stepped-torque control between maximum and minimum limits. The number of steps is increased as the torque limits are moved closer together.

Because of its relatively simple and rugged construction, with low initial cost and maintenance, the squirrel-cage motor is by far the most popular in conveyor drive application. There are a number of standard designs of the squirrel-cage motor, and NEMA (National Electrical Manufacturers Association) classifies them into five categories according to speed-torque characteristics. The most commonly used is the NEMA-B design. When higher starting torque is required, the NEMA-C design can provide a minimum of 200 percent of the full-load torque at start.

The variable-speed motor provides added flexibility in conveyor drives. Speed control may be infinitely variable, or the motor may be of a definitive multispeed type. Torque control is also obtained with oil- and magnetic-type couplings.

Since conveyor applications frequently involve a dusty atmosphere, most of the motors used are the totally enclosed, fan-cooled type. Occasionally, open drip-proof motors are used to minimize initial cost.

To achieve the highest degree of automation, nearly all the modern conveyor systems are sequentially interlocked. It is generally so arranged that a conveyor cannot start until the conveyor which it feeds has reached its rated speed. To avoid pileups at the junctions during normal stopping and to clear the belt of material, a reverse sequential-stop system with appropriate time delays assures that no conveyor will stop until after its feed conveyor has stopped and it has completely cleared of material.

Transfer Chutes Transfer chutes are provided at belt conveyor junctions to confine, guide, and feed material. These are usually custom-designed devices for a specific type of material and a specific application. A properly designed chute minimizes the impact on the belt it feeds. A properly designed *stone box* can achieve this without eliminating the beneficial velocity component in the direction of the belt travel. Especially in the case of an angular junction, the stream of material should be guided so that only the velocity component in the direction of the receiving belt is kept intact. This component helps to minimize abrasion as the belt accelerates the material to the same speed at which it travels. On the other hand, the transverse component of the stream should be minimized, if not completely eliminated, to avoid its tendency to push the belt out of line and to reduce the wear of the belt cover.

Occasionally, a fixed grizzly is provided to screen out the fines so that they fall on the belt first. They then form a cushion for the lumps to fall on and thereby minimize possible belt damage. Fishplates may be provided at the bottom of the chute to help guide slabby or lumpy material in the direction of the belt travel (see Fig. 15-8). Adjustable curtains may be used to guide the material and to reduce its impact and the resultant wear.[14] Skirtboards, in lengths of 10 ft or more, guide and contain the material after it has fallen on the belt.

A wide variety of wear plates are available. These are usually bolted inside the chutes with countersunk bolts for ease of replacement. Special rubber liners work well where wear is not accompanied by gouging and sharp cuts. For areas of extreme impact and abrasion, used manganese crusher jaw plates, if available, can outwear any other abrasion-resistant material.

Supporting Structures Supporting structures for a belt-conveyor system consist essentially of the terminal supports and the intermediate sections. At the terminals, structural framework supports the pulleys, belt tension takeups, drives, chutes, screens, feeders, and other associated equipment. The intermediate structures consist of stringers made up of steel channels or angles to support the troughing and return idlers (see Fig. 15-9). Occasionally, deck plates are provided to protect the return belt and, at the same time, provide lateral strength to the supporting structure. Short posts, in turn, support these stringers and anchor them either directly on the foundations or on elevated trusses. These stringers may serve the dual purpose of supporting the idlers and functioning as the top chord of a deck truss.

For belt widths under 42 in., a service walkway on one side of the belt may be sufficient. To adequately service wider belts, walkways on both sides of the conveyor may be necessary.

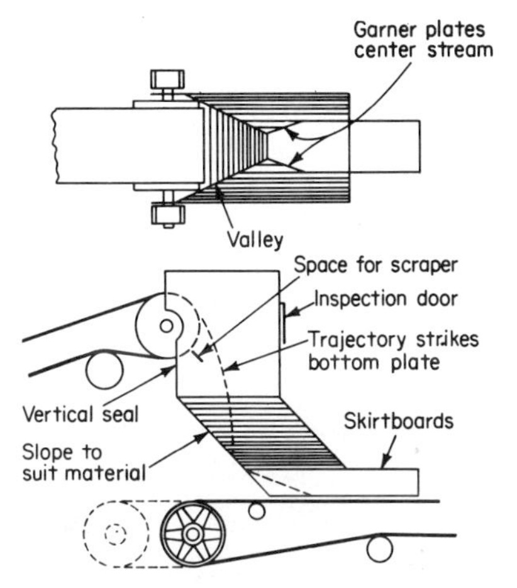

Fig. 15-8 Typical transfer chute and fishplate (or Garner plate).

Fig. 15-9 Typical intermediate sections—troughing idlers supported on steel channels and angle posts.

For conveyor widths up to 48 in., preengineered and prefabricated trusses of standard construction with a wide choice of spans are available. Their use can result in considerable savings, both in fabrication and erection costs.

Wire ropes, which replace steel stringers, have gained popularity for their ease of construction and lower investment cost. They are used with specially designed idlers and supporting stands (Fig. 15-10).

Accessories *Takeups* are used to ensure proper belt tension. The simplest and most positive is the gravity type, which utilizes a counterweight to maintain belt tension. For shorter conveyors, screw-type takeups may be adequate.

Electrically controlled takeups may be used where exceptionally high belt tension necessitates sophisticated monitoring and control. Where belt vibrations become a problem, a hydraulic takeup may be the solution.

Scales. Various standard belt scales for continuous weighing are available. These are usually precalibrated and are capable of registering capacity rates in terms of tons per hour. They are set within a limited percentage range of the conveyor's rated capacity. Certain belt scales (such as Merrick's Type E Weightometer) can offer an accuracy of 0.5 percent over a range from one-third to full load. The sum total of tonnage handled is registered on a totalizer. Electronic remote-controlled devices enable monitoring of tonnage readings in a control center.

Since catenary action of the belt between idlers can materially affect the accuracy of the scales, it is mandatory to maintain the tension in the conveyor belt as constant as possible. Hence, counterweighted takeups are preferred and the scale should be located at the low tension sector along the conveyor.

Fig. 15-10 Belt conveyor supported by wire-rope stringer.

Belt Cleaners. These devices are provided near the discharge end of the conveyor, immediately after the mainstream of material has been unloaded. Counterweighted or spring-loaded belt scrapers with high durometer rubber inserts are the simplest devices. They are usually located under or behind the discharge pulley so that the scrapings can be collected on the conveyor fed by the same transfer chute.

Rubber or bristle brushes are available for special applications. They may be driven independently or through the conveyor drive. Where water is abundant and drainage presents no problem, strong jet spray does a good job in cleaning the belt.

Rocks which accidentally have fallen onto the return belt can become wedged between the belt and the tail pulley and cause severe damage. A simple belt plow, riding on top of the return belt and located in front of the tail pulley, is a simple and effective safeguard.

Safety Devices. For open long-conveyor installations, it may be necessary to protect the belt from unwanted detraining forces caused by side wind. A vertical steel plate along the belt conveyor provides the simplest wind shield. A more elaborate installation may have a hood cover over the belt. When the presence of wind is sporadic, wind hoops spaced at 30- to 40-ft intervals may be sufficient to prevent the belt from being lifted by gusts.

A rattrap may be installed where occasional boulders tend to roll back on steep conveyors. A rattrap is essentially a simple one-way swing gate which permits the boulders to pass through in the direction of material flow. Any boulders which run backwards are stopped at the gate where they are accumulated and wedged together until they move forward again en masse.

Low-cost *wire-mesh guards* are versatile and have been used extensively around drives and moving machinery as well as under conveyor trusses. Pull-cord *emergency switches* are often installed along the entire length of a conveyor. The cord is mechanically tied to switch arms, and the conveyor drive will stop instantaneously when the cord is pulled. The electrical circuits are usually designed so that the conveyor cannot restart until the cause for stoppage is corrected and the switch arm is mechanically reset.

Other safety devices include overload cutout switches, belt-penetration switches (to avoid extensive belt damage), etc. These provide added protection to the system.

ENGINEERING DESIGN

Material Characteristics A thorough study of the characteristics of the materials to be handled is a prerequisite for a well-engineered conveyor system with properly selected components. Heading the list of pertinent items is size, not only the maximum lump size but also the percentage breakdowns of the various size ranges. A reliable and representative screen analysis will provide this information.

TABLE 15-6 Maximum Safe Inclinations of Troughed Belt Conveyors for Handling Various Bulk Materials

Material	Inclination, degrees	Rise in ft per 100 ft
Cement—loose	22	40.4
Clay—fine dry	23	42.4
Clay—wet lump	18	32.5
Concrete—wet	15	26.8
Earth—loose	20	36.4
Gravel—bank run	18	32.5
Gravel—screened	15	26.8
Gypsum—powdered	23	42.4
Limestone	18	32.5
Rock—fine and crushed	22	40.4
Rock—mixed	20	36.4
Rock—sized	18	32.5
Sand—dry	15	26.8
Sand—damp	20	36.4
Sand—tempered and foundry	24	44.5

High percentages of fines may create a dust problem, especially at transfer junctions. When moisture is present cohesive particles tend to cling to belting and machinery. In addition to being a nuisance this buildup can cause detraining of the belt. On the other hand, if the material contains many large lumps, wear becomes the major concern. Rounded boulders have a tendency to roll back on a steeply inclined belt, causing damage to machinery and injury to personnel. Table 15-6 indicates maximum safe inclinations of troughed belt conveyors for handling various bulk materials.

The origin of the material can often provide a hint about its ultimate behavior on a belt conveyor. For example, crushed aggregates are usually angular and will result in more belt-cover wear than normally rounded alluvial material. Well-screened and well-sized lumps, without the benefit of the fines to cushion the impact, cause substantially more wear on the belt and wear plates.

TABLE 15-7 Belt-conveyor Capacities [for 20° and 35° troughing idlers—capacities in tons (2,000 lb) per hr]

Belt width, in.	Cross sectional area, sq ft		Speed, fpm (material weight 100 pcf)																	
			100		200		400		500		600		700		800		900		1,000	
	20°	35°	20°	35°	20°	35°	20°	35°	20°	35°	20°	35°	20°	35°	20°	35°	20°	35°	20°	35°
24	0.369	0.451	111	135	221	271	443	542	553	677	664	812	774	948	885	1,083	996	1,219	1,106	1,354
30	0.599	0.733	180	220	360	440	719	880	899	1,100	1,079	1,320	1,259	1,540	1,438	1,760	1,618	1,980	1,798	2,200
36	0.886	1.083	266	325	531	650	1,063	1,300	1,328	1,625	1,594	1,950	1,860	2,275	2,125	2,600	2,391	2,925	2,657	3,250
42	1.228	1.502	368	450	737	901	1,473	1,802	1,841	2,252	2,210	2,703	2,578	3,153	2,946	3,604	3,314	4,054	3,683	4,505
48	1.625	1.988	488	596	975	1,193	1,950	2,385	2,438	2,982	2,925	3,578	3,413	4,174	3,900	4,771	4,388	5,367	4,875	5,963
54	2.078	2.542	624	763	1,247	1,525	2,494	3,050	3,118	3,813	3,741	4,576	4,365	5,338	4,988	6,101	5,612	6,863	6,235	7,626
60	2.587	3.164	776	949	1,552	1,899	3,105	3,797	3,881	4,746	4,657	5,696	5,434	6,645	6,210	7,594	6,986	8,543	7,762	9,493
72	3.772	4.613	1,132	1,384	2,263	2,768	4,527	5,536	5,659	6,919	6,790	8,303	7,922	9,687	9,054	11,071	10,185	12,455	11,317	13,839
84	5.180	6.334	1,554	1,900	3,108	3,800	6,216	7,601	7,770	9,501	9,324	11,401	10,878	13,301	12,432	15,201	13,986	17,102	15,540	19,002
96	6.811	8.327	2,043	2,498	4,086	4,996	8,173	9,992	10,216	12,491	12,259	14,989	14,302	17,487	16,345	19,985	18,388	22,483	20,432	24,981
108	8.664	10.592	2,599	3,178	5,198	6,355	10,396	12,711	12,996	15,889	15,595	19,066	18,194	22,244	20,793	25,422	23,392	28,599	25,991	31,777
120	10.740	13.130	3,222	3,939	6,444	7,878	12,888	15,756	16,109	19,695	19,331	23,634	22,553	27,573	25,775	31,512	28,997	35,451	32,219	39,390

NOTE: Table follows essentially CEMA formulation based on 25° surcharge angle and 100-pcf material density (maximum edge distance being 5.5 percent belt width plus 0.9 in.); for other density, tph (listed above for 100 pcf) $\times \frac{\text{(known weight/cu ft)}}{100}$.

The minimum width of a conveyor belt is usually governed by the maximum lump size of the material carried. A general rule is to have the belt width two to three times the size of a maximum lump. Minimum belt width is more critical for well-sized material than for unsized. In analyzing the maximum lump size, one must pay particular attention to the fracture pattern of the rock. For instance, some rocks tend to be slabby, with one dimension many times smaller than the others. When slabs occur frequently, a conveyor designer must provide sufficient belt width and transfer space to handle them.

Capacity, Speed, and Sizing of Equipment The carrying capacity of a belt conveyor is expressed in terms of tons per hour (tph). This capacity is a function of belt width, speed, and material density. Thus, a conveyor of a given width, running at a fixed speed, has its theoretical capacity cut in half when loaded with a material whose bulk density is one-half that of the material for which the conveyor was designed.

Table 15-7 lists belt-conveyor carrying capacities in short tons per hour for material weighing 100 pcf, belt widths up to 120 in., speeds varying from 100 to 1,000 fpm, and idler troughing angles at 20° and 35°. As can be seen, the use of 35° deep trough idlers can increase the carrying capacity of a belt conveyor by as much as 25 percent.

TABLE 15-8 Suggested Minimum Belt Width in Relation to Lump Size

Material	Maximum lump size, in.								
	Ratio of belt width to lump size	Belt width, in.							
		24	30	36	42	48	54	60	72
Quarry	2¼	11	13	16	19	21	24	27	32
Occasional lump	3	8	10	12	14	16	18	20	24
Crusher run (no slabs)	3½	7	9	10	12	14	15	17	21
Crusher run, fines removed	4	6	8	9	11	12	13	15	18
Sized material from screening operation	4½	5	7	8	9	11	12	13	16

A belt conveyor can theoretically attain as high a capacity as desired by merely increasing its speed, and belt conveyors carrying overburden have been run as fast as 1,200 fpm.[15] In normal practice, however, belt speeds are limited by many factors. Among these are the character of the materials conveyed, expected operating and maintenance conditions, and vibration of the supporting structures.

Extremely abrasive materials, sharp-edged rocks, and very large lumps require lower speeds to reduce belt cover wear. Fine granular materials will spread out at high speeds as the belt passes over idlers, particularly at low tension. Lumps then tend to be forced toward the edges, causing spillage. If dust is a problem, high speeds usually aggravate it.

Conditions at the feed point are important for a high-speed conveyor. When material is loaded on a slope, the tendency to roll back is more pronounced on a faster-running belt. This condition is aggravated by the presence of large numbers of round boulders (see Table 15-8 for lump size limitations).

Generally, higher speeds may be allowed if belt-to-belt transfers are in line. Angular transfers introduce a transverse velocity component which causes chute or belt wear. The wear is intensified as the belt speed increases. At extremely high speeds, vibrations of the belt and structures should be carefully examined.

To avoid possible pileup at the junctions, the conveyor which receives the material should never have lower speed than the preceding feed conveyor. Table 15-9 lists recommended maximum belt speeds in feet per minute for various widths of conveyors and types of material.

Design Calculations A belt conveyor involves a wide variety of components which must function together as a mechanical complex. The belt itself is made up of materials of various degrees of elasticity and strength. With the many variables involved, an exact mathematical analysis of horsepower and tension is virtually impossible. Empirical design formulas based on limited field and laboratory tests are available and have been satisfactory for most applications.

These formulas are derived on the assumption that the power required to drive a conveyor may be subdivided into two principal parts, namely: (1) power to overcome friction, and (2) power to lift the load.* For certain transfers, additional power is required to accelerate the load.

The power required for lifting the load is directly proportional to the lift and capacity rate in tons per hour.† Power to overcome frictional resistance concerns all

TABLE 15-9 Maximum Recommended Belt Speeds (fpm) for Standard Service

Material	Maximum speed*	Belt width, in.							
		24	30	36	42	48	54	60	72
Granular fines									
Minus 1-in. lump	900	600	700	800	900	900	900	900	900
Occasional lump 10% belt width	800	550	600	700	750	800	800	800	800
Half maximum-sized lump									
Rounded pieces	700	500	600	650	700	750	750	750	750
Abrasive and sharp	650	450	550	600	650	650	650	650	650
Maximum-sized lump									
Rounded pieces	650	400	500	550	600	650	650	650	650
Abrasive not sharp	600	400	450	500	550	600	600	600	600
Abrasive and sharp	550	350	400	450	500	550	550	550	550

* Higher speeds may be used under special conditions and impact forces must be duly considered.

of the moving parts such as belts, idlers, pulleys, takeups, drives, and other accessories. The interaction of all these components, in turn, will vary depending on the configuration of the system. Furthermore, their friction coefficients vary as a function of the temperature, load, tension in the belt, and operating and climatic factors.

In addition to horsepower requirements, a complete belt-conveyor analysis must take into consideration the belt tension. This information is required for the selection of the belt and for the design of the mechanical components, such as pulleys,

* For a declined conveyor which delivers material below the elevation at which it is received, the load on a belt actually tends to generate power. When the decline is steep and the power derived from lowering the load exceeds the frictional horsepower, the system becomes "regenerative." The motor becomes a generator and has been frequently used to feed power back to the system for other uses.

† A simple rule of thumb for estimating lift horsepower is lift in feet multiplied by capacity in 1,000 tph. Thus, 60 hp would be required for a conveyor with a net rise of 20 ft handling 3,000 tph.

shafting, reducers, couplings, structures, and other accessories. Empirical formulas, in slightly different forms, are published in various manufacturers' handbooks.[6-10] The most widely accepted are those listed in the CEMA handbook.[6] Various graphical solutions for these formulas have also been developed.[8,16]

Electronic data processing has substantially aided the engineering design of complex belt-conveyor systems. Various horsepower, tension, and component design programs have been written and are used extensively for general applications. The

PROGRAM E0140 24FEB69
BELT TENSIONS CID 1156

OUTPUT DATA INQUIRY NO

CONTR SYS. NO. CUSTOMER NAME ENGR. NO.

CONVEYOR P1

SUMMARY OF BELT TENSIONS 1

PROFILE	MAXIMUM TENSIONS FOR ANY LOADING CONDITION				
POINT	ACCELERATE	DECELERATE	RUNNING NF	RUNNING RF	BREAKAWAY
1	10117.	6343.	6648.	6436.	6996.
2	10215.	6342.	6689.	6437.	7080.
3	11576.	6439.	7946.	6591.	8585.
4	20454.	8796.	12659.	10842.	14496.
5	22608.	10161.	15245.	13357.	17242.
6	23590.	9965.	15513.	13535.	17757.
7	30629.	14450.	27565.	25274.	30529.
7	7765.	8687.	8472.	8667.	8255.
8	7842.	8618.	8513.	8600.	8404.
9	7843.	7995.	7956.	7992.	7920.
10	7700.	7700.	7700.	7700.	7700.
11	8076.	7931.	7975.	7934.	8016.
12	7972.	7045.	7163.	7066.	7288.
13	8448.	7052.	7196.	7086.	7355.
14	8453.	6724.	6886.	6766.	7069.
15	9792.	6289.	6551.	6377.	6861.
16	9920.	6287.	6553.	6377.	6865.
TMAX	30629.	14450.	27565.	25274.	30529.
TMIN	7700.	6287.	6551.	6377.	6861.
TE	22863.	5763.	19092.	16606.	22274.
ACMAT	98213.	87062.	98213.	98213.	98213.
WK2	508.	450.	508.	508.	508.

Fig. 15-11 Typical computer printout for belt-conveyor design program.

solution of a complex belt-conveyor problem may now be derived in a matter of minutes on a computer, as compared to hours or days of hard manual labor. Figure 15-11 shows a typical computer printout.

Other Parameters Any belt-conveyor system should be designed with the operator and maintenance men in mind. Belt operation and maintenance are discussed in a subsequent portion of this section; this material should be studied in conjunction with that which immediately follows.

Before any capacity requirements and component selections are finalized, the designer must always raise the question of adequacy of the system to accommodate future expansion. More often than not, little or no additional investment is required to make provisions for future expansion if the system is properly planned in advance. Layouts can then be arranged so that minimal alteration to machinery or structures will be required when the future stages of expansion are implemented. Since fieldwork usually is costly and plant downtime is equally hard to come by, early recognition of these requirements reflects sound engineering judgment. In the same connection, standardization of components rarely costs more at the initial stage. Even if some of the components are slightly oversized at a small added cost, overall savings may still result from a reduced spare parts inventory and lower maintenance costs.

Rubber belting is usually the most vulnerable and, at the same time, the most costly component in an entire belt-conveyor system. A well-designed belt-conveyor system is one which pampers the rubber belt to the limit. All the remaining components in a well-engineered system must be selected on the basis of maximum possible belt life.

Once the material is settled on the belt, very little can happen unless it is again disturbed. It follows from this that transfer points are where troubles usually develop. Each transfer merits careful study by an experienced designer. It should

Fig. 15-12 A long overland conveyor follows the relatively undisturbed land contour.

be so engineered that it will accommodate the anticipated range of material characteristics without major field alterations.

If the terrain and right-of-way conditions permit, long conveyors with the least number of transfers will minimize drives, chutes, and other junction accessories as well as their inherent maintenance. Minimizing the number of transfer points will also reduce belt wear and degradation of the materials hauled.

Longer conveyors may be subjected to high belt tensions. These may dictate structural requirements for the belt and its associated machinery which are different from those suitable for shorter conveyors elsewhere in the system. In such a situation, in the interests of standardization and minimizing spares inventory, a designer may choose to deliberately limit the length of some conveyors in the system. In so doing, he will have to cope with the undesirable consequence of a corresponding increase in the number of transfers.

To minimize cut and fill as well as bridge work, a long overland belt conveyor will approximately follow the existing land contours (Fig. 15-12). The consequent savings in cost should be weighed against any increases in complexity of the design.

One or a number of the components in a belt-conveyor system may become critical as various combinations of design conditions are investigated. For example, when the majority or all of the declined sections of a roller-coaster conveyor are loaded, the conveyor may tend to run away. Conversely, if all the inclined sections are loaded, the system would require the maximum driving force to restart after an accidental or emergency stop. A long conveyor system possesses high inertia, especially when loaded, and its tendency to coast or drift could result in serious pileups at junctions. In addition to a properly designed braking system, flywheels at succeeding drive stations may be needed to ensure the continued flow of material in the proper sequence.

OPERATION AND MAINTENANCE

A belt-conveyor system, if properly designed, will give the user years of low-maintenance, easy operation. If automation is carried out to the highest degree, an entire conveyor system representing millions of dollars of investment may be safely operated by one man in an air-conditioned control center. This operator merely pushes the proper buttons on an illuminated panel, using indicator lights to check on the flow sequencing. The operator of any belt-conveyor system must be keenly aware of the capabilities for which the plant was designed. Most important of all, he must carefully police the plant feed to ensure that foreign material or materials far in excess of design specifications do not enter the system. Violation of this simple rule may result in plugging the key transfer chutes or cutting long conveyor belts, with costly damages as well as the resultant downtime.

Overloading is another common cause of unsatisfactory plant performance. Any component has its design limit and safety margin, and repeated overstress will lead to premature fatigue failures and consequent downtime. Overloading at transfer points causes plugging of chutes and possible damage to the belt, as well as spillage on the conveyors.

It is the operator's prime responsibility to obtain high production. He can best achieve this by making sure that the downtime is minimized, rather than by resorting to periodic overloads which nearly always defeat his purpose. He must constantly remind himself that a belt-conveyor system is a continuous bulk-handling device. As such, it must be kept in continuous operation in order to give the maximum output. If the simple rules of operation are not abused and if the maintenance instructions are followed, a belt-conveyor system will be a faithful, trouble-free workhorse.

In maintenance, the center of attention should again be the rubber belt itself. An acceptable preventive maintenance program must include at least the following:

Belt Alignment. Alignment of the belt's supporting structures must be periodically inspected to ensure that there is no ground settlement or damage from accidents or other causes. A belt conveyor on a canted support cannot be expected to run centered. Alignment of head and tail pulleys, drives, and idlers should be periodically checked for the same purpose.

Transfers. Transfer junctions and chutes should be inspected to see that excessive wear has not changed the flow pattern. The wear plates may have worked loose, causing them to fall on the belt. This also applies to the skirtboards or other guiding devices. Loose bolts which have escaped inspection have proved to be one of the most frequent starters of belt damage.

Cleanup. Accumulation of dripping on rolling parts or under the belt can result in a pileup that makes the training of the belt impossible. It may also cause direct abrasion and wear of machinery or the belt. Plows and scrapers should be inspected periodically to see that excessive wear has not neutralized their effectiveness. If trolley-mounted magnetic devices are used for tramp iron removal, they must be discharged prior to being used again. Oil and grease are also detrimental to rubber belting and should be removed if inspection shows them to be present.

Climatic Considerations. In extremely dry climates, the periodic removal of dust from machinery and electrical components is a necessary part of the preventive maintenance program. Ozone coupled with intensive heat is injurious to rubber; if a long, open conveyor is to be exposed to intensive sunlight for any length of time, it should be run periodically to avoid possible damage to the section of the belt that is continually exposed to intensive heat and sun. The use of the wrong grease for the

```
BASIC ENGINEERING DATA

CONTR.  RUN        CUSTOMER NAME                         CUSTOMER ENGR     DATE    INQUIRY
SYS.NO. NO.                                                 NO.    NO.               NO.
   0     1 DEMONSTRATION CONVEYOR                                        10DEC67

CONVEYOR  NO  77.0
   54.  IN  WIDTH,    3000.0  FT  LENGTH,    19.9  FT  LIFT,   4520.  NET  TONS/HR
   584.  FT/MIN  ACTUAL,  0.95  LOAD RATIO,  776.  MAX  ALLOWABLE  SPEED
   612.  FT/MIN  DESIGN,    4.4  IN EDGE DIST,   12.0  MAX  ALLOWABLE  LUMP

MATERIAL
   3.0  IN  MAX  LUMP,    130.  LB/CU.FT,  7.  MTL  CLASSIFICATION,  5.  LUMP  DIST

BELT TENSIONS  -  1000  LBS
   37.5  RUN  FULL  MAX,    26.5  TE  MAX,  11.0  T2RF,  11.0  TU,  37.5  RUN  INT  MAX
   62.1  ACCEL,  10.7  RUN  MIN,  11.51  ACTUAL  AVAIL  WF  T1/T2
   35.7  MAX  DECEL,    47.2  TE  BRKWY

BELT
  40824.  AWT  LBS    ,  1.05  T-TOL  S.F,

HG  TAKE-UP
   31.0  FT  NET  TRAVEL

HORSEPOWER
  CALC  AT       515.8  MOTOR  467.9  BELT  23.3  FRINGE
  515.8  INT  RUN  MAX,     170.1  INT  RUN  MIN  498.9  AVG  ACCEL  USED
  515.8  ABS  MAX  LOADING  OR  UNLOADING

WR2  LB-FT2       532.8  CONV  2162.4  FULL  LOAD,    2162.4  INT  LOAD

1ST  DRIVE
  1. MOTOR,  400.0  HP,  400.0  ALT  HP,  0.  ALLOWABLE  STARTS/HR

  REDUCER  498.4  ME  HP,  446.0  TH  HP  1.25  INPUT  S.F.,  1.78  ACTUAL  S.F.
                35.  LBS  RTD  O.H.L.,  37.83  RATIO,  0.0  REGEN  HP
          0.953  PERCENT  VARIATION  OF  REQD  TO  STD  RATIO
   3.0  FT-KP  BACKSTOP  TORQUE  REQD  AT  L.S.S.

2ND  DRIVE
  1. MOTOR,  300.0  HP,  125.0  ALT  HP,  0.  ALLOWABLE STARTS/HR

  REDUCER  329.2  ME  HP,  338.0  TH  HP  1.25  INPUT  S.F.,  1.56  ACTUAL S.F.
                25.  LBS  RTD  O.H.L.,  38.04  RATIO,  0.0  REGEN  HP
          0.953  PERCENT  VARIATION  OF  REQD  TO  STD  RATIO
   2.1  FT-KP  BACKSTOP  TORQUE  REQD  AT  L.S.S.

TIME - SECONDS
  14.3  MAX  ACCEL,  16.3  DRIFT  FULL  LOAD

VERTICAL CURVES
  PT  CONDITION   RADIUS  TANGENT  PI  DIST  TENSION

  1  RUNNING       -51.1      0.7      0.0     21041.
  1  BREAKAWAY       0.0      0.0      0.0         0.
  1  ACCEL,          0.0      0.0      0.0         0.
  1  DECEL,          0.0      0.0      0.0         0.
  1  INPUT,          0.0      0.0      0.0         0.

  2  RUNNING      1505.2     35.1      0.4     23080.
  2  BREAKAWAY    2409.1     56.2      0.7     36944.
  2  ACCEL,       3649.7     85.2      1.0     55967.
  2  DECEL,       1144.7     26.7      0.3     17552.
```

Fig. 15-13 Sample computer printout of Example 1.

idler bearings has been a common cause for many conveyors' inability to start in cold weather. This situation may be further aggravated when the equipment is new and the idlers have not yet been run in.

ESTIMATING COSTS

To illustrate the methods of estimating capital and operating costs, two examples are given, one for a 3,000-ft belt conveyor and another for a 4-mi overland haulage system. The procedures and calculations are only intended to serve as a guide, and the values

ENGINEERED ESTIMATE

CONTR. SYS. NO.	RUN NO.	CUSTOMER NAME	CUSTOMER NO.	P.E. NO.	DATE	INQUIRY NO.
0	1	DEMONSTRATION CONVEYOR			10DEC67	

CONVEYOR NO. 77.0

NO. OF UNITS	DESCRIPTION		WEIGHT LBS. LOT	LIST PRICE EACH	LIST PRICE LOT
	IDLERS	CODE 21			
749	54 IN 3600	4.00 FT SPG, BM 389	107856.	92.40	69208.
250	54 IN 3610	12.00 FT SPG, BM 3472	19250.	47.30	11826.
4	54 IN 3600	1.33 FT SPG, BM 389	576.	92.40	370.
	OTHER RCD	CODE 29			
1	HGTU FRAME OUTFIT FOR 30 FT TRAVEL		1030.	649.00	649.
1	IDLER CARRIAGES		475.	310.00	310.
2	6 IN X 57 IN SPEC BEND PLY 1-7/16 SFT		180.	59.10	118.
2	54 IN DOUBLE BLADE SPRING BELT SCRAPER		220.	131.95	264.
	BELT	CODE 41			
6276	FT, 54 IN 6 PLY RN130 AJAX 6/32 TC INCL 1 INCL OP 2/32 BC INCL 0 INCL OP INCL 51 FT FOR 6 SPLICES		105852.	28.30	177611.
	OTHER HRD	CODE 49			
1	ITEM RUBBER & CEMENT FOR 6 VULC SPLICES		120.	80.00	480.
	PARALLEL SHAFT REDUCERS	CODE 61			
1	29D , STYLE 1 46.3 OUTPUT RPM 37.83 RATIO FURNISH BASE MOUNT MOTOR MOUNT 2 COUPLINGS MOUNT COUPLING GUARDS		14200.	17902.00	17902.
1	24D , STYLE 1 46.0 OUTPUT RPM 38.04 RATIO FURNISH BASE MOUNT MOTOR MOUNT 2 COUPLINGS MOUNT COUPLING GUARDS		10400.	12111.00	12111.
	OTHER GPD	CODE 69			
1	COUPLING GUARDS		50.	115.00	115.
1	COUPLING GUARDS		175.	300.00	300.

Fig. 15-13 (*Continued*).

derived are approximate. Operating and maintenance conditions differ, and the actual costs of installing and operating a conveyor system may vary considerably.

Example 1 shows the use of the computer technique to obtain some of the basic design and cost data.

EXAMPLE 1 Basic Data and Production Estimates

OPERATION: 70 million cu yd to be moved in 5 years, 340 working days per year. Two 8-hr shifts per day.

CONFIGURATION: Overland 3,000 ft with net lift of 19.9 ft.

MATERIAL: Dense excavated earth with lumps up to 3 in., abrasive and sharp.

DENSITY: 130 pcf.

CAPACITY, SPEED, AND SIZING:

$$70{,}000{,}000 \text{ cu yd/5 yr/340 days/16 hr} = 2{,}570 \text{ yd per hr} \times \frac{27 \times 130 \text{ pcf}}{2{,}000 \text{ lb}}$$
$$= 4{,}520 \text{ tph}$$

Table 15-9 recommends belt speed range of 600 to 650 fpm for belt widths 36 in. and up (half maximum-sized lump, abrasive and sharp).

From Table 15-7, at 600 fpm using regular 20° trough, choose 54-in. belt with a maximum rated capacity of 4,863 tph.*

DESIGN AND COST DATA BY COMPUTER:

With the above data fed as input to the computer, all the belt tension, horsepower requirement, equipment selection, and cost information is obtained in the form of programmed printouts. An excerpt of the detailed printout is shown in Fig. 15-13. The cost of the 6-ply, 54-in. belting is $177,600 and the complete conveyor cost, exclusive of structures, is $351,800.

Capital and Operating Costs

Capital Costs:

Machinery and belting	$351,800	
Structural support at $15/ft × 3,000 ft	45,000	
Electrical wiring and controls	20,000	
Total equipment	$416,800	
Freight (5%)	20,800	
Erection (25%)	104,200	
	$541,800	
Contingency (2%)	10,800	
Total investment	$552,600	
Assume no salvage value		
Depreciation over 5 yr at	$110,500/yr	
Interest, insurance, and tax†	$ 33,200/yr	
Total annual capital cost		$143,700
At $\frac{70{,}000{,}000 \text{ yd}}{5 \text{ yr}}$ = 14,000,000 yd/yr		$.0103/yd
or, $\times \frac{27 \times 130}{2{,}000}$ = 24,570,000 ton/yr		$.0059/ton

Operating Costs:

Power:		
Installed hp 700 × usage factor 0.8 × 0.746		
× $0.01/kwh = $4.18/hr		
× 16 hr/day = $66.88/day		
× 340 days/yr	= $ 22,700/yr	
Maintenance:		
Machinery—5% of $194,200 (mechanical and electrical)	= $ 9,700/yr	
Belt—20% of $177,600	= $ 35,500/yr	
Operating labor: (2 shifts × 1 man)		
2 man-shifts per day × $30 = $60, $60/day × 365 days	= $ 21,900/yr	
Total annual operating cost		$ 89,800
At 14,000,000 yd/yr		$.0064/yd
or 24,570,000 ton/yr		$.0037/ton

Summary:

	Capital costs	*Operating costs*	*Total*
Per year	$143,700	$89,800	$233,500
Per cubic yard	$.0103	$.0064	$.0167
Per ton	$.0059	$.0037	$.0096
Per ton-mile (equiv.—prorated by multiplying by $\frac{5{,}280 \text{ ft}}{3{,}000 \text{ ft}}$ = 1.76)	...	...	$.0169

* Table 15-7 lists all capacities on basis of 100-pcf density. The 4,863 tph is obtained by modifying capacity rating with density variation, thus, 3,741 × 130/100 = 4,863 tph.

† Average investment (5 + 1)/(2 × 5) × $552,600 = $331,600. Assume 10 percent interest, insurance, and tax = $33,200 per year.

EXAMPLE 2 Basic Data and Production Estimates

OPERATION: 40 million cu yd to be moved in 3 yr, 330 days/yr, two 8-hr shifts per day.

CONFIGURATION: Principally downhill, overland a distance of 4 mi.

MATERIAL: Excavated earth and rocks, passed through crusher. Maximum lumps 14 in., abrasive, not sharp.

DENSITY: 100 pcf.

CAPACITY, SPEED, AND SIZING:

$$40{,}000{,}000 \text{ cu yd}/3 \text{ yr}/330 \text{ days}/16 \text{ hr} = 2{,}530 \text{ yd per hr} \times \frac{27 \times 100 \text{ pcf}}{2{,}000 \text{ lb}}$$

$$= 3{,}420 \text{ tph}$$

Table 15-8 suggests minimum belt width of 48 in. Table 15-9 recommends belt speed of 600 fpm for belt widths 48 in. and up.

From Table 15-7, at 600 fpm, choose 35° deep trough idlers, use 48-in. belt with a maximum rated capacity of 3,578 tph. Note that if 20° trough idlers were used, one would have to use the next size, 54-in. belt.

DESIGN: Assume an installed total horsepower of 1,800 (actually a comprehensive analysis is required on the basis of an exact survey of the land contour).

Capital and Operating Costs

Capital Costs:

Item	Amount	Total
Machinery, structures, electrical:		
Terminals at \$150 per hp × 1,800 hp	\$ 270,000	
Intermediate sections at \$0.75 per in. belt width per ft conveyor length × 48 in. × 21,120 ft	760,300	
Belting at \$0.60 per in. belt width per ft length × 48 in. × 21,120 × 2	1,216,500	
Total equipment	\$2,246,800	
Freight (5%)	112,300	
Erection (25%)	561,700	
	\$2,920,800	
Contingency (2%)	58,400	
Total investment	\$2,979,200	
Salvage value (20% of equipment)	−449,400	
Depreciable value	\$2,529,800	
Depreciation over 3 yr at	\$ 843,300/yr	
Interest, insurance, taxes*	\$ 168,600/yr	
Total annual capital cost		\$1,011,900
At $\frac{40{,}000{,}000}{3}$ = 13,333,000 yd/yr		\$.0759/yd
or, × $\frac{27 \times 100}{2{,}000}$ = 18,000,000 ton/yr		\$.0562/ton

Operating Costs:

Item	Amount	Total
Power:		
Installed hp 1,800 × usage factor 0.8 × 0.746 × \$0.01/kwh = \$10.74/hr × 16 hr/day = \$172/day × 330 day/yr	= \$57,000/yr	
Maintenance:		
1.5% of \$2,979,200 (investment)	= \$44,700/yr	
Operating labor and supervision: (2 shifts with 4 men/shift)		
8 man-shifts per day × \$30 = \$240/day × 365 days	= \$87,600/yr	
Total annual operating cost		\$189,300/yr
At 13,333,000 yd/yr		\$ .0142/yd
or 18,000,000 ton/yr		\$ .0105/ton

* Average investment: (3 + 1)/(2 × 3) × \$2,532,100 = \$1,686,500. Assume 10 percent interest, insurance, taxes = \$168,600.

Summary:

	Capital costs	*Operating costs*	*Total*
Per year	$1,011,900	$189,300	$1,201,200
Per cubic yard	$.0759	$.0142	$.0901
Per ton	$.0562	$.0105	$.0667
Per ton-mile	$.0141	$.0026	$.0167

REFERENCES

1. New Developments in Earthmoving, *Construction Methods*, March, 1965.
2. Move a Mountain to Span a Sea, *Elastomers Notebook*, E. I. du Pont de Nemours & Co., August, 1958.
3. Oroville Dam, A New Concept in Earthmoving, *Construction Methods*, October, 1964.
4. R. W. Stensrud, Marcona's Cross Country Conveyor System, *Mining Engineering*, November, 1968.
5. A. T. Yu and R. K. Quinn, A New Dimension in Iron Ore Transportation on the Great Lakes, *Skillings' Mining Review*, Feb. 15, 1969.
6. Conveyor Equipment Manufacturers Assoc., *Belt Conveyors for Bulk Materials*, Cahners Publishing Co., Inc., Boston, 1966.
7. Conveyor Belt Engineering, *Hewitt-Robins Bulletin No. 175*, 1962.
8. *Handbook of Belting*, The Goodyear Tire & Rubber Co., Inc., Akron, Ohio, 1953.
9. Materials Handling and Processing Equipment, *Link-Belt Catalog 1000*, 1958.
10. F. V. Hetzel and R. K. Albright, *Belt Conveyors and Belt Elevators*, John Wiley & Sons, Inc., New York, 1941.
11. A. T. Yu, Recent Trends and Advances in Shiploading, *Skillings' Mining Review*, Apr. 27, 1968.
12. A. T. Yu, Portable Module Systems for Open Pits, *Engineering and Mining Journal*, June, 1967.
13. New for Ships—A "Roll Away" Loading System, *Modern Materials Handling*, September, 1964, pp. 50–51.
14. A. T. Yu, After 1¾ Million Tons of Iron Ore, Long Belt Shows Hardly Any Wear, *Engineering and Mining Journal*, March, 1964.
15. E. R. Traxler, Conveyor Belt Horsepower and Operating Distances Continue to Increase, *Engineering and Mining Journal*, August, 1961.
16. A. T. Yu, Graphical Determination of the Power Requirements for Long Belt Conveyors with Different Gradients, *Fordern und Heben*, vol. 10, 1964.

Section **16**

Soil Surfacing Equipment

E. J. HAKER

Product Development Engineer, Rex Chainbelt Inc., Milwaukee, Wis.

C. F. RIDDLE

Supervising Design Engineer, Rex Chainbelt Inc., Milwaukee, Wis.

This section covers four classes of construction equipment relating to the fields of spreading, stabilization, compaction, and trimming. Included in each of these classes of equipment is a general discussion of the subject, followed by a categorized list of the equipment being manufactured. Specifications are shown for each of the categories. Additional data on equipment selection and average production to be expected are also included.

SPREADING EQUIPMENT

General The art of soil spreading can be divided into two basic types: spreading the earth fill in embankment work and spreading highly graded granular materials on base course and shoulder work. Although there may be some resemblance between the two operations, they are actually separate and distinct. Each has its own procedures and problems.

On a typical embankment operation, earth fill is brought in from the borrow area in bulk haul units such as scrapers or rear- or bottom-dump trucks. In most cases this haul dirt must be spread to a uniform and controlled lift thickness. In this way it becomes possible to achieve a high uniform density during the compaction operation.

In base course or shoulder spreading, the granular soil is brought in, usually in rear-dump units, and placed on the finish subgrade. Base courses are generally placed in one or more lifts, each of which is compacted to the specified requirements. After compaction, final exact trimming removes all the excess material. Careful control is required in spreading each lift of base course, both to ensure compaction uniformity and to minimize the quantity of material which must subsequently be removed during the trimming operation.

Equipment Types The category of soil spreading equipment encompasses many varieties of construction equipment. The more commonly used types for each of the two areas of soil spreading are as follows:

EARTH FILL SPREADING:

Bulldozers
Scrapers (pan and elevating)
Motor graders
Large units

BASE COURSE AND SHOULDER SPREADING:

Motor graders
Tractor-mounted box spreaders
Automatic trimmer type of equipment

Bulldozers. Crawler tractors with rugged, front-mounted blades have for years been the workhorses in earth-fill spreading. They can also provide secondary services on the fill as push help for unloading haul vehicles. Bulldozers are manufactured in a wide range of sizes from 5,000 to 160,000 lb gross weight and from 42 to 770 hp.*

Scrapers. Although basically haul units, scrapers can be generally considered to have a spreading function on the fill. Scrapers are manufactured in both the self-propelled and towed versions. The self-propelled version is available in two styles, the pan type and the elevating type.†

Large Spreading Units. These are being developed in an attempt to keep pace with the increasing volume of the haul units, and they have created a great deal of interest among contractors. The theory behind the large units is that one large spreading unit on the fill creates much less congestion than several smaller units. One manufacturer has introduced a large, 4-wheel drive, rubber-tired unit with a center-mounted blade. (See Fig. 16-1.) The earth leveling capacity of this new unit has been estimated by its manufacturer to be in the area of 1,500 to 4,000 cu yd per hr. Spreading units of this size are economically limited to only the very large type of fill sites.

Tractor-mounted Box Spreaders. This is a popular type of spreader for spreading granular type soils in base course or shoulder work. Capacities to 1,000 tph are estimated by the manufacturers. (See Fig. 16-2.)

Automated Base Trimmers. These large, automatically controlled units are very versatile. Among their many capabilities is that of spreading base course soils.

* See Sec. 8, Tractors, Bulldozers, and Rippers.
† See Sec. 9, Scrapers.

Although it is one of the newer types of equipment, the automated base trimmer is seeing increased usage as a spreader because of its fast, full-width operation. These fine graders are subsequently discussed in more detail under Trimming Equipment.

Motor Graders. The most versatile tool in the category of spreading equipment is unquestionably the motor grader illustrated in Fig. 16-3. This unit is capable of effectively spreading most types of soils in addition to performing many other duties.

Fig. 16-1 Large spreading unit. (*Raygo Inc.*)

Fig. 16-2 Tractor-mounted box spreader. (*Jersey Spreader Company.*)

(See Fig. 16-4.) Originally developed some eighty odd years ago, it has consistently maintained its popularity and improved its performance level through the years. In addition to spreading, the motor grader can handle the following operations effectively:

1. Fine finish grading
2. Ditching of all types (see Fig. 16-5)
3. Haul-road maintenance (earthmoving and logging)
4. Ripping and scarifying
5. Bank cutting
6. Snow removal

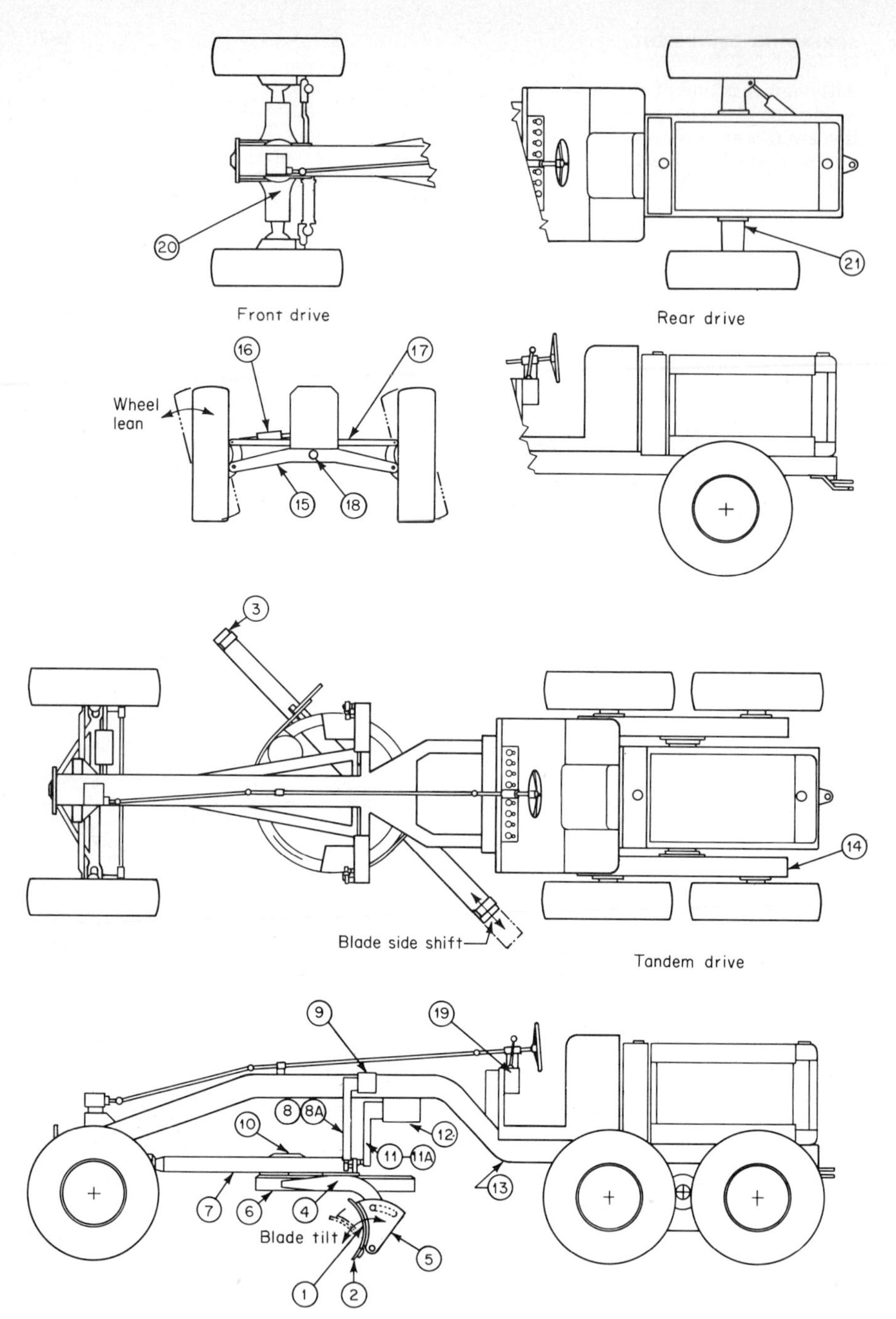

Fig. 16-3 Motor grader nomenclature. (1) Blade, (2) cutting edge, (3) bit, end, (4) arm blade, (5) bracket, tilt, (6) circle, (7) drawbar, (8) link, blade lift, (8a) cylinder, blade lift, (9) lift, blade, (10) circle reverse, (11) link, circle side shift, (11a) cylinder, circle side shift, (12) circle side shift, (13) frame, main, (14) drive, tandem, (15) axle, front, (16) wheel lean, (17) tie bar, wheel lean, (18) pin, axle pivot, (19) control, power, (20) drive, front, (21) drive, rear. (*SAE Handbook, 1969.*)

The movement of the blade and attachments is generally hydraulically powered and is actuated by controls which are within easy reach of the operator.

Specifications for the various sizes of motor graders are shown in Table 16-1.

Production Several procedures have been advanced for estimating the production from a piece of spreading equipment. Among these are formulas that pertain to the motor grader. One of the best accepted of these formulas is the following:*

$$\text{Time to complete a job (hr)} = \frac{\text{no. of passes} \times \text{distance in miles}}{\text{avg speed in mph} \times \text{efficiency factor}}$$

Fig. 16-4 Motor grader spreading earth fill. (*Caterpillar Tractor Company.*)

Fig. 16-5 Motor grader on ditch and bank work. (*Caterpillar Tractor Company.*)

The effects of variations in operator performance and in the type of soil can usually be compensated for in the efficiency factor. This factor has to be estimated on the basis of good judgment, but it should generally fall into the range from 0.70 to 0.90. Production estimation for other types of spreading equipment should be handled similarly, using the above formula as a general guide.

* Caterpillar Tractor Company.

TABLE 16-1 Specification Data for Motor Graders

Size classification, gross weight, lb	Rating, hp	Wheel-base, in.	Tread, in.	Max speed, mph	Blade size		Manufacturers
					Length, ft	Height, in.	
6,500–10,000 (Includes tractor maintainer)	65–75	114–170	60–68	18–21	9–10	13½–16½	Allis-Chalmers Galion Huber
10,000–22,000	70–110	205–230	74–84	18–25	10–12	18–25	Austin-Western Caterpillar Cleveland DRMCO John Deere Galion WABCO
22,000–28,000	100–170	225–240	76–84	18-25	12–13	24–27	Allis-Chalmers Austin-Western Caterpillar Cleveland DRMCO Galion Huber Pettibone-Mulliken WABCO
28,000–35,000	150–220	230–245	76–84	19–26	12–13	27–29	Austin-Western Caterpillar Cleveland DRMCO Galion Huber Pettibone-Mulliken WABCO
35,000–50,000	200–230	250–270	86–98	28–31	14	31–32	Caterpillar WABCO

STABILIZATION EQUIPMENT

General Soil stabilization, perhaps one of the least familiar aspects of highway and airfield construction, is considered to be one of the more important steps. Stabilization, as the name implies, is intended to make the soil structure more stable or less subject to change from outside influences. Mechanical stabilization can be accomplished by equipment which mixes the soil, using one or more rotors, to produce a more homogeneous material. In its simplest form, stabilization can merely consist of blending the concentrations of sand and aggregate which are naturally present in a soil so as to obtain a more homogeneous state. In the more practical usage, however, stabilization involves the mixing of an additive with the soil before blending the combination into a uniform mixture. This additive can be a complementary type of soil or it can be a manufactured agent such as portland cement, asphalt, or lime. Table 16-2 lists the more commonly used stabilization agents.

Stabilized soils are used in the base, subbase, and embankment layers. In order to carry traffic, they must be covered with either an asphaltic or concrete surface course to prevent abrading. The minimum surface treatment should be an asphalt-chip seal coat.*

Equipment Types The equipment which is used to stabilize soils is of two basic types: the in-place mobile mixer and the central-mix plant. The in-place mixer can be

* See Sec. 25, Paving.

further divided into three separate categories: the multiple-pass units, the single-pass units, and the road-mix units. Only self-propelled stabilizers are covered here; although some tow-type stabilizers are still being manufactured, they represent a very small and declining part of today's market.

Multiple-pass Stabilizer. This is the most popular type of stabilization equipment in use today. It consists basically of a carrier whose large engine drives an attached rotor. This rotor is rotating within an enclosure called a "hood" or "mixing chamber." (See Fig. 16-6.)

The carrier consists of a frame mounted on four rubber-tired wheels, with a single engine providing power for both the rotor drive and propulsion. The rotor drive is mechanical through a two- or three-speed transmission. Hydrostatic power transmission is preferred for the propulsion drive, since it provides an infinite range of working speeds as well as dynamic braking control.

TABLE 16-2 Common Stabilizing Agents

Agent or additive	Form used	Suitable soils	Reaction with soil	Construction area
Asphalt	Cutback liquid or aqueous emulsion	Sandy and granular types	Physical binder	Base course
Lime	Fine powder or aqueous slurry	Plastic (plasticity index higher than 10)	Chemical reaction to reduce plasticity, also a physical binder	Subbase and subgrade
Portland cement	Fine powder	Granular (plasticity index lower than 10)	Physical cementation	Base course
Salts (calcium chloride, sodium chloride)	Granules	Sands and gravels	Moisture retention	Base course

The rotor consists of a series of connected disks. These hold the cutting teeth or tines at their outer periphery and generally rotate in a forward down direction at speeds of from 100 to 250 rpm. The cutting teeth or tines are mostly of the quick-change variety to minimize the downtime when replacing a worn set. The rotor, which can be raised or lowered, is covered on all sides except the front with a steel-plate enclosure called the "hood" or "mixing chamber." The rear surface is slanted and has a hinged lower portion called a "tailboard," which regulates the release of the mixed soil. (See Figs. 16-7 and 16-8.)

A fluid system is available on all units to facilitate the introduction of a liquid stabilizing additive such as water, asphalt, or lime slurry. This fluid system basically consists of a pump driven by a separate engine, a meter, a spray bar mounted on the hood, hoses connecting these components, and a means of regulating the flow of the additive.

Single-pass Stabilizer. This is now being manufactured by only one firm in the United States. It consists basically of a carrier with a large engine driving three or four rotors, mounted within an attached hood. The machine functions in a similar manner to that described for the multiple-pass unit. The carrier is track mounted, and its engine provides power to drive both the rotors and the crawler tracks.

Either three or four counter-rotating rotors, depending on the model, are mounted within a single hood. Each rotor performs a specialized function. The first rotor provides the cutting function, the second rotor blends the soil into a homogeneous mixture, and the remaining rotors accomplish a transverse pug-mill type of mixing. The single hood or mixing chamber functions in a similar manner to the hood in a multiple-pass stabilizer. (See Fig. 16-9.)

The fluid handling system is also comparable to that for the multiple pass stabilizer.

Road-mix Stabilizer. This type of stabilizer, which is presently being produced by only one manufacturer, can be described as a traveling mixer. Its function is confined to mixing windrowed soils, and it accomplishes this by straddling the windrow. It

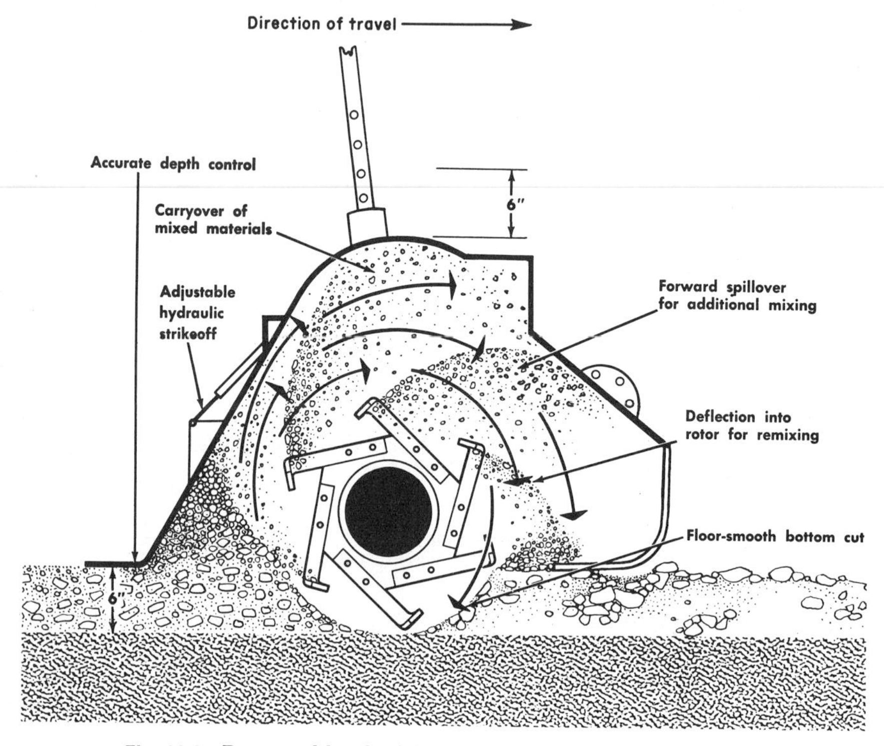

Fig. 16-6 Rotor and hood mixing action (multiple-pass stabilizer).

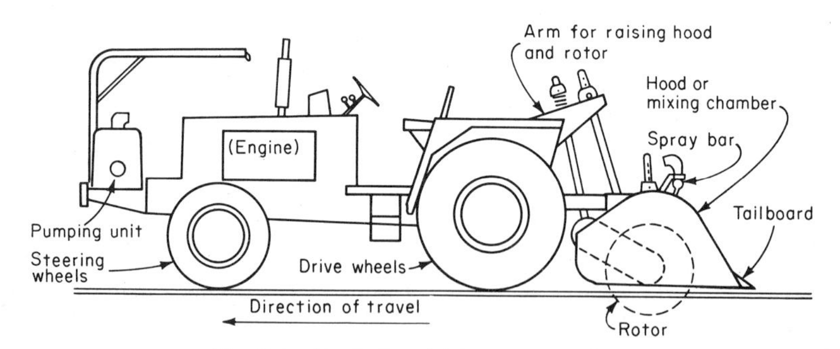

Fig. 16-7 Typical multiple-pass stabilizer.

consists primarily of a large, self-propelled carrier with an underslung mounted longitudinal pug mill.

The carrier is formed of a long frame upon which is mounted a single large engine. This engine provides the power for the mixing rotor and the propulsion drive. The entire unit is mounted on six rubber-tired wheels, four of which are driving in the smaller models and six in the larger ones.

A single pug-mill-type drum provides the mixing action. Its mixing drum is equipped with spray nozzles for introducing the asphalt or water additive. Side guide plates are utilized to better funnel the windrow soil into the mixing drum area. A horizontal blade is also mounted to the rear of the mixing drum to partially level the processed windrow soil. Figure 16-10 illustrates this type of stabilizer.

Fig. 16-8 Large-size multiple-pass stabilizer. (*Rex Chainbelt Inc.*)

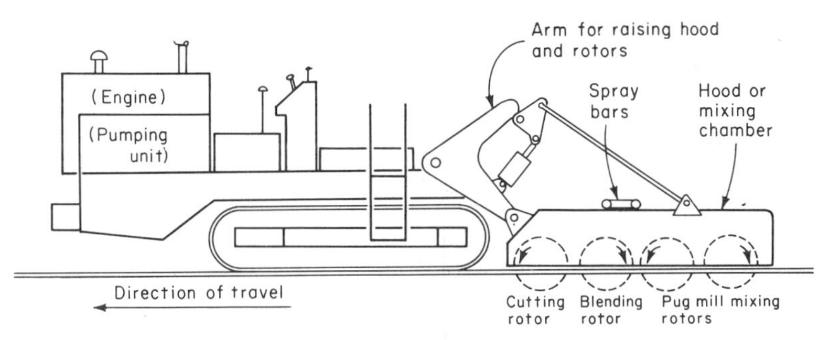

Fig. 16-9 Single-pass stabilizer.

Table 16-3 shows the characteristics of the equipment used for in-place stabilization.

Central-plant Stabilization. Soil stabilization is also accomplished in a central mixing plant which usually is located at the source of the soil or aggregate to be used. The central mix method, popular in some areas of the country, has been largely confined to portland-cement stabilization and, to a smaller degree, asphalt stabilization. It has been estimated that as much as 40 percent of the soil cement used in the United States is being mixed in a central plant.

Two basic types of central-mix plants are in use today: the batch type and the continuous-flow type. The continuous-flow plant appears to be the more popular, due to its higher production capabilities. Production of an average continuous-flow plant is about 400 to 600 tph, with the largest producing about 1,000 tph.

The typical central mix plant includes a conveyor with a loading hopper, a cement storage tank with an accurate feed mechanism, and a pug mill with a water supply. A front-end loader or clam bucket is needed to keep the hopper loaded with the soil or aggregate, so as to ensure even distribution on the conveyor. The cement feed is timed to the conveyor speed. Cement is added to the material on the conveyor and the water is added in the pug mill. Figure 16-11 illustrates a typical central-mix plant.

In the continuous-flow plant, the pug mill has a holding hopper to facilitate more continuous production of the plant.

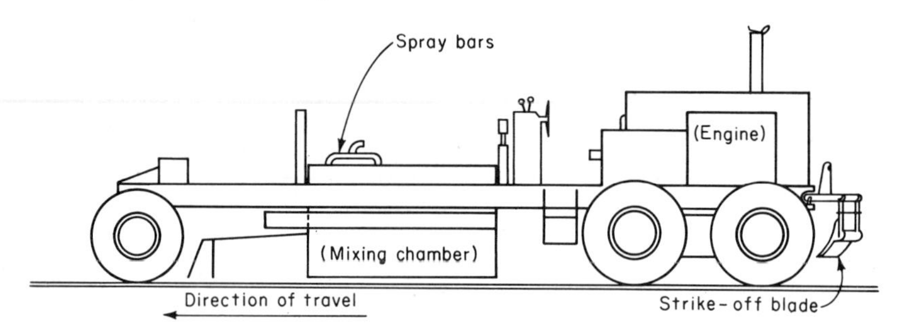

Fig. 16-10 Road-mix stabilizer.

TABLE 16-3 Equipment for In-place Stabilization

Type	Cutting or mixing width, ft	Cutting depth, in.	Mixing depth, in.	Working speed, fpm	Working weight, lb	Rotor diameter, in.	Average production, cu yd per hr	Manufacturers
Multiple pass, standard unit	6–7	4–9	6–16	0–200	12,000–18,000	28–35	100–200 (two passes)	Bros Buffalo-Springfield Pettibone-Wood Rex Seaman
Multiple pass, large unit	7–8	6–14	9–24	0–200	24,000–40,000	33–54	200–350 (two passes)	Bros Buffalo-Springfield Pettibone-Wood Raygo Rex Seaman
Single pass...	5–11	7–8	7–8	0–50	32,000–60,000	28–30 (three or four rotors)	200–350 (one pass)	Buffalo-Springfield
Road mixer...	(windrow mixing)	....	....	0–50	26,000–40,000	42–54 (pug mill)	100–400 (one pass)	Pettibone-Wood

Haul times from the central-mix plant to the jobsite can be critical, especially with soil cement and 30 to 45 min is now considered to be the limit. An open dump-type of truck with a canvas cover is generally used as a haul vehicle. After the stabilized soil has arrived at the jobsite, it is quickly spread and compacted.

Equipment Selection Several factors are involved in the selection of the best type of equipment for the job:

1. Type of job
2. Size of job
3. Contractor economics
4. Specifications of governing agency

General guidelines when considering these factors are as follows.

Type of Job. Individual considerations here include type of soil, type of additive, equipment function, depth of cut or mixing, and soil location. These considerations are discussed here, together with a general analysis of each.

TYPE OF SOIL: Soil can vary from tough cutting, high plasticity index (PI) clay to easily mixed sandy soil. Clay generally requires a multiple-pass stabilizer, preferably a larger, higher horsepower unit. Easily mixed sandy soils can be handled by most types of equipment, including the road mix and central plant.

TYPE OF ADDITIVE:

Portland cement, asphalts: Most types of equipment. Central-plant and road-mix units perform well.

Lime: Generally a multiple-pass stabilizer, preferably the larger sizes.

Salts: In-place stabilizers generally used.

EQUIPMENT FUNCTION: This refers to whether a pulverizing or mixing operation is involved.

Pulverizing: Generally a multiple-pass unit.

Mixing: Most types of equipment. Central-plant and road-mix units perform well.

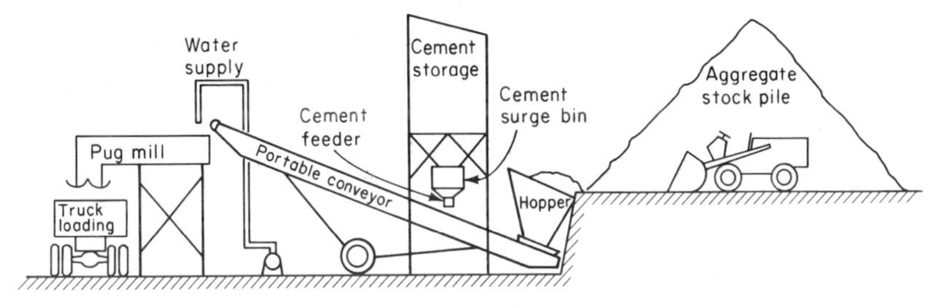

Fig. 16-11 Typical central-mix plant (soil cement).

DEPTH OF CUT: This generally determines size of unit, as the larger units tend to cut deeper.

SOIL LOCATION: Soil location determines whether in-place mixing or a central plant will be used. If soil must be hauled in, some consideration should be given to central-plant mixing.

Size of Job. Although the size of the job and production expected per hour should provide some control as to the type of equipment selected, other factors far outweigh this one. Production can generally be increased by adding to the fleet.

Contractor Economics. This important factor generally refers to the capital investment that a contractor feels he should have in stabilization equipment. Initial cost and expected production must be weighed here, and a contractor's previous experience is perhaps the strongest influence. For a contractor first entering the stabilizing field, the selection of a standard size of multiple-pass stabilizer is generally the safest way to start. Experience gained with this relatively inexpensive unit can better help him to further evaluate the purchase of the larger equipment in the future.

Specifications of the Governing Agency. Some agencies clearly spell out the type of equipment to be used for a particular job. Other agencies may generally bracket around several types.

COMPACTION EQUIPMENT

General Compaction might be considered to be the most important phase in the construction of any soil structure. If the compaction is inadequate this is generally reflected in the failure of any subsequent construction.

Compaction can be defined as the process of densifying or increasing the unit weight of a soil mass through the application of static or dynamic forces, with the resulting expulsion of air and, in some cases, moisture. The effects of compaction on the soil are twofold. First, it reduces future settlement by rearranging the particles to form a more compact soil mass. Secondly, it builds strength into the soil structure.

Static forces are considered to be those which are produced by a roller or compactor which accomplishes soil densification primarily by its heavy weight. Dynamic forces are those which utilize a combination of weight and energy to produce a vibratory or tamping effect on the soil. Compactors utilize both static and dynamic forces to achieve the required soil density.

Equipment Types Today's compaction equipment is represented by many large, highly responsive, and versatile self-propelled units. These can produce a high degree of compaction with only a minimum of equipment breakdown problems. Present-day compactors and rollers can be divided into four major classes: rubber tired, segmented

Fig. 16-12 Medium-size (21-ton) rubber-tired compactor. *(Bros Division—American Hoist.)*

pad or tamping foot, vibratory, and steel wheel. Each of these classes can further be subdivided into two or more type or size classifications. As an example, rubber-tired rollers are being manufactured in three distinct size categories, with tire diameter being the primary identification.

Each of the four major classes of compactors is discussed separately in the following paragraphs.

Rubber-tired Compactor. This is frequently referred to as a "pneumatic" compactor, and has been called the "universal" compactor. The last name may exaggerate the facts, but it suggests the broad field of usage for rubber-tired compactors. (See Fig. 16-12.) Although one of the newer types, rubber-tired compactors have achieved excellent results in almost every type of compaction. They first appeared in the early 1930s as towed wagons with a rock-filled box, and they have now graduated to the level where they are being manufactured in three distinctly different tire sizes. In addition, the towed wagons have almost completely given way to the modern, powerful, self-propelled versions. Table 16-4 lists pertinent data for the rubber-tired compactors.

The compaction tools are the rubber tires, and these are specially designed to provide a specific ground contact pressure (GCP) on the soil. The magnitude of GCP is controlled by both wheel loading and tire inflation pressure. A national standardization of GCP values was achieved early in 1967, and each currently manufactured model of rubber-tired compactor will carry a decal which identifies its maximum allowable wheel load and the ground contact pressures which can be attained with the various combinations of wheel loadings and tire inflation pressures. Figure 16-13 shows a typical decal covering the 7.50 × 15 tires.

Rubber-tired compactors have seen considerable usage in rolling asphaltic concrete surfaces. Here they perform the function of an intermediate roller, immediately following the breakdown steel-wheel roller. In most instances, the rubber-tired roller is followed by finish steel-wheel rolling. When used as an intermediate roller, the rubber-tired roller contributes to compaction uniformity and to the orientation of aggregate particles in their most stable positions.

TABLE 16-4 Rubber-tired Compactors

Basic size class	Number of wheels	Tire size	Weights: Shipping, lb	Weights: Max ballasted, ton	Max wheel loading, lb	Rolling width, in.	Rated hp	Speeds, mph: Av working	Speeds, mph: Max travel	Manufacturers
Light; 15-in.-diam wheels	9 or 11 or 13	7.50 × 15	6,000–19,000	10–16	2,500–4,000	68 (9-wheel), 96 (11-wheel), 99 (13-wheel)	40–110	3–7	14–21	Austin-Western Bros Buffalo-Springfield Ferguson Galion Grace Huber Hyster Ingram Rex Rosco Seaman Tampo
Medium; 20-in.-diam wheels	7 or 9 or 11	9.00 × 20, 10.00 × 20, 11.00 × 20	13,000–21,000	21–30	4,500–6,000	74 (7-wheel), 77–96 (9-wheel), 96 (11-wheel)	80–130	4–10	13–16	Bros Buffalo-Springfield Ferguson Huber Hyster Ingram Michigan Raygo Tampo
Heavy; 24-in.-diam wheels	7	13.00 × 24	20,000–24,000	30–35	9,000–10,000	85–96	100–150	4–10	12–17	Bros Ferguson Grace Huber Ingram Tampo

Rubber-tired compactors are designed to provide uniform coverage and equal wheel loading even on rougher terrain. Uniform coverage is possible because the rear wheels are staggered to roll the spaces left by the front wheels. Equal wheel loading on uneven terrain is provided by means of their articulated axles.

The feature of changing tire inflation pressures "on the go" has caused a great deal of interest among contractors. This is a very direct means of varying the ground-contact pressures; the higher the tire pressure, the higher the ground-contact pressures, assuming the ballasted weight has remained constant.

CERTIFIED MAXIMUM GROUND CONTACT PRESSURES

ISSUED BY
BITUMINOUS EQUIPMENT MANUFACTURERS BUREAU
UNDER THE SPONSORSHIP OF
CONSTRUCTION INDUSTRY **CIMA** MANUFACTURERS ASSOCIATION
135 S. LaSalle St. Chicago, Illinois 60603
FOR 7.50 x 15 SMOOTH TREAD COMPACTOR TIRES

TIRE PLY		4 PLY	6 PLY			10 PLY					12 PLY						14 PLY							
TIRE PRESSURE		35	35	50	60	35	50	60	70	90	35	50	60	70	90	110	35	50	60	70	90	110	120	130
WHEEL LOAD		GROUND CONTACT PRESSURES and CONTACT AREAS																						
1000	GCP	37	37	44	49	38	44	47	51	57	38	44	47	51	60	65	46	50	54	56	61	68	71	74
	CA	27	27	23	20	26	23	21	20	18	26	23	21	20	17	15	22	20	19	18	16	15	14	14
2000	GCP	43	43	50	55	46	52	56	60	67	46	53	56	60	69	75	54	59	62	65	72	78	82	86
	CA	47	47	40	36	43	38	36	33	30	43	38	36	33	29	27	37	34	32	31	28	26	24	23
2500	GCP	45	45	52	58	49	56	59	64	71	50	57	60	65	74	78	57	63	66	70	76	83	87	90
	CA	56	56	48	43	51	45	42	39	35	50	44	42	38	34	32	44	40	38	36	33	30	29	28
3000	GCP	47	47	55	61	53	60	65	67	75	53	60	64	69	77	83	60	66	70	73	80	87	91	94
	CA	64	64	55	49	57	50	46	45	40	57	50	47	43	39	36	50	45	43	41	38	34	33	32
3500	GCP			57	63		62	67	71	80		64	67	71	81	86		68	73	76	83	90	94	98
	CA			61	56		56	52	49	44		55	52	49	43	41		51	48	46	42	39	37	36
4000	GCP				65			68	73	82			70	75	84	89			75	79	86	94	98	101
	CA				62			59	55	49			57	53	48	45			53	51	47	43	41	40

GCP—Ground Contact Pressure
CA—Ground Contact Area

MAXIMUM ALLOWABLE WHEEL LOAD THIS ROLLER 3500

PERFORMANCE FIGURES HAVE BEEN APPROVED, SUBJECT TO TIRE MANUFACTURERS NORMAL TOLERANCE BY
FIRESTONE TIRE & RUBBER CO.
GOODYEAR TIRE & RUBBER CO.
GOODRICH TIRE & RUBBER CO.
U.S. RUBBER TIRE CO.
GENERAL TIRE CO.

Fig. 16-13 Typical decal showing ground-contact pressures for rubber-tired compactors.

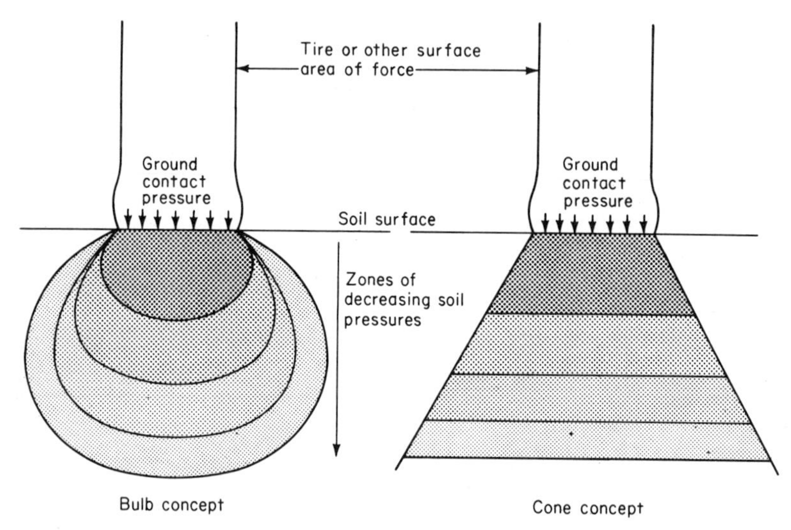

Fig. 16-14 Theories of soil compaction.

Two theories have been advanced to attempt to explain soil behavior beneath a loaded contact surface such as a rubber tire. These theories are referred to as the "bulb concept" and the "cone concept." (See Fig. 16-14.) Attempts have been made through the years to assign values to the forces in the various zones of each concept.

Segmented-pad or Tamping-foot Roller. This self-propelled roller represents a new breed of compactor, filling the need for faster compaction and higher production. Initially developed in the mid-1950s with a rolling type of pad on the wheels, today's units feature the tamping pad or the tamping type of foot. The inherent dynamic beat developed by these tamping pads or feet can in many cases produce a high degree of compaction. (See Fig. 16-15.)

The need for these high-production earth-fill compactors has been generated by the steadily increasing capacity of the hauling units. Statistics have been developed to show that an average scraper today has the potential to haul five times the volume that an average scraper hauled 10 years ago. Factors such as increased haul yardage, higher speeds, and faster loading and unloading cycles have paced this increased volume in fill hauling.

The modern segmented-pad or tamping-foot compactor has satisfied the need for faster compaction through its higher speeds, greater maneuverability, and increased pulling power. Its use will reduce grade congestion, especially on the more crowded fill areas. Some units feature full-width compaction in one pass. A front-mounted blade can be attached to increase the versatility of the compactor.

Fig. 16-15 Segmented-pad compactor (32,000 lb). (*Rex Chainbelt Inc.*)

The bulb and cone concepts as discussed under the rubber-tired compactors also apply to the segmented-pad or tamping-foot compactor. However, the pressures exerted by the pads or feet are of much greater magnitude than the pressures exerted by the rubber tires.

"Walking out" is a term normally associated with this type of compactor. It refers to a condition of deep penetration of the pads or feet on the first pass, with less penetration on each subsequent pass. The wheels in effect ride higher with each pass; hence the term "walking out."

The tow-type sheepsfoot is increasingly being replaced by the large, self-propelled units. However, despite the fact that its popularity has decreased greatly, the towed unit still fills a compaction need. With its slower speed and lower cost, it can provide economical compaction in the smaller types of fill areas.

Today's large, self-propelled units are basically manufactured in three distinct sizes. Table 16-5 shows the various characteristics of each size.

Vibratory Compactors. These units combine static weight with a generated, cyclic type of force. The dynamic type of compaction is relatively new and really came into its own during World War II. It was introduced to satisfy the need for higher densities in granular types of soils.

Two basic types of vibratory compactors are available, the roll type and the plate type. The roll type, being the more popular, is manufactured in numerous styles and sizes. The plate type is generally available only in a self-propelled version. It should also be noted that the roll types are available with either a smooth or a foot type of roll.

Today's roll type of vibratory compactor has come a long way from the models which were available only a decade ago. (See Fig. 16-16.) Many units now offer a

TABLE 16-5 Segmented-pad and Tamping-foot Compactors

Basic size class	Working weight, lb	Rolling width, ft	Drum or wheel diam, in.	Rated hp	Working speeds, mph	Estimated production capability, tph	Manufacturers
Light, self-propelled	30,000–36,000	7–9	60–70	130–250	3–12	700–900	Buffalo-Springfield FWD Wagner Rex Pactor
Medium, self-propelled	40,000–55,000	8–10	70–80	250–350	3–12	900–1,200	Bros Buffalo-Springfield Caterpillar FWD Wagner Hyster Michigan Rex Pactor
Heavy, self-propelled	60,000–100,000	12–14	75–85	400–600	3–12	1,600–2,500	Caterpillar R. G. LeTourneau Rex Pactor
Tow types.	. . . .	6–10	48–80	. . . .	2–5	300–600	(Numerous)

Fig. 16-16 Self-propelled vibratory compactor—roll type. (*Raygo Inc.*)

system which will indicate the vibrating amplitude of the roll, and this information is a guide to the vibration frequency at which the roller and the soil are in resonance. There is some evidence to indicate that vibration at this resonant point will produce faster compaction. The frequency can be varied in most vibratory compactors. Many vibratory rollers are also available in a suitable version to compact deep-lift asphalt bases, complete with spray system and smooth tires.

Although there is little standardization in the vibratory compactors which are being manufactured today, an attempt is made here to classify them according to size and force developed. Table 16-6 indicates the range of units available.

The generally accepted rating for vibratory compactors is "total applied force." This is expressed in pounds or tons and is the numerical sum of the dynamic force plus the static weight of the vibrating surface on the soil.

TABLE 16-6 Vibratory Compactors

Type	Roll or plate width, in.	Dynamic force, lb	Static weight, lb	Engine hp	Max frequency, vibrations/min	Manufacturers
Roll, self-propelled	60–75	15,000–20,000	12,000–15,000	65–80	1,600–1,800	Ferguson Tampo Vibro-Plus
	80–85	24,000–30,000	15,000–20,000	80–130	1,500–1,800	Bros Raygo Rex Seaman Tampo Vibro-Plus
	90–102	35,000–75,000	25,000–50,000	100–150	1,400–1,800	Bros Pettibone-Wood Raygo Tampo
Roll, towed	60–75	15,000–25,000	6,000–10,000	25–40	1,400–2,200	Bros Buffalo-Springfield Essick Ferguson Hyster Jackson Michigan Rex Rosco Tampo Vibro-Plus
	76–84	25,000–60,000	10,000–25,000	60–75	1,300–1,600	Bros Essick Ferguson Jackson Michigan Tampo Vibro-Plus
Roll, tandem, self-propelled	30–36	4,500–8,000	3,000–4,500	25–35	2,500–3,000	Essick Raygo Vibro-Plus
Plate, self-propelled	150–160	30,000–60,000 (six plates)	8,000–13,000	50–75	2,100–4,200	International-Vibration Jackson Lima

A brief discussion of the basic terminology used in discussing vibratory compactors is presented here. (See Fig. 16-17.)

1. *Vibrating mass:* The total entity that vibrates as a unit.

2. *Surcharge:* The nonvibrating mass that surrounds and is affixed to the vibrating mass through the isolators. As a general rule of thumb, the greater the surcharge weight, the greater the depth of compaction.

3. *Vibration isolator:* The connection between the vibrating mass and the surcharge or nonvibrating members; generally of an elastic rubber type of compound.

4. *Frequency:* The number of vibrations in a unit of time (usually 1 min). Frequency generally relates to production; other factors being equal, the greater the frequency, the greater the production.

5. *Dynamic force:* The force generated, usually by centrifugal means, and generally described in pounds or tons.

6. *Amplitude:* The vertical movement of the vibrating mass from its position at rest to a maximum in one direction. Total vertical movement is actually double amplitude or displacement. Amplitude is not controlled by the vibratory compactor alone. It is the result of a vibrating system composed of the compactor's vibrating mass and the soil being compacted. Soil conditions such as type, moisture content, and degree of compaction affect amplitude considerably.

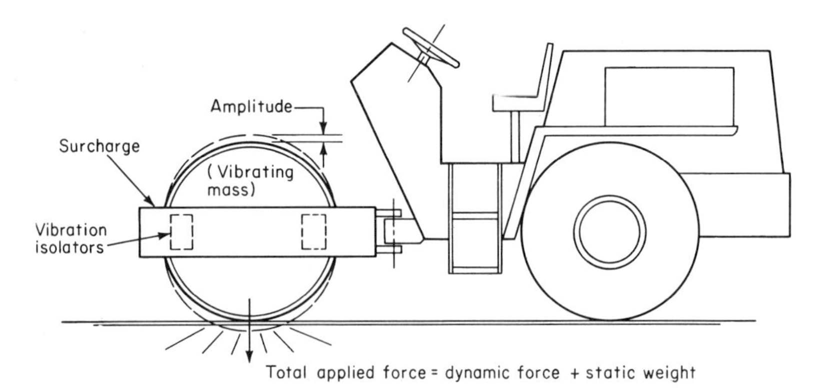

Fig. 16-17 Vibratory compaction terminology.

Steel-wheel Rollers. As their name implies, these are rollers in addition to being compactors. They are designed primarily for pavement course rolling, and their rolls are machined to provide a smooth, concentric surface.

There are two basic styles of steel-wheel rollers: the tandem with two rolls of equal width, and the three-wheel with one wide guide roll and two narrow drive rolls. In each of these styles there are three size categories. There is also a unique version of the tandem, called the "three-axle" tandem, which is similar to the tandem in appearance except that it has two guide rolls on a walking-beam type of arrangement. The three-axle tandem is generally the heaviest of the steel-wheel rollers in total ballasted tonnage. The walking beam allows the forces exerted by the guide rolls to be varied.

Present-day features in steel-wheel rollers include hydrostatic drive, hydrostatic steering complete with steering wheels, a pressure-type fog-spray system for wetting the rolls, and self-contained portability wheels on the 4- to 6-ton tandems to facilitate towing without the use of a trailer. Hydrostatic drive, introduced in 1961, has proved to be the single biggest improvement in recent years. It provides infinitely variable speed control together with dynamic braking action. (See Fig. 16-18.)

Steel-wheel rollers through the years have become fairly well standardized as far as basic dimensions are concerned. This standardization also applies to the weight distribution on the rolls themselves. Table 16-7 indicates the weight proportioning for both the tandem and the three-wheel rollers being manufactured.

Table 16-8 shows the characteristics for each of the size categories.

Equipment Selection Soil, the relatively thin crust covering the surface of our planet, actually provides the reason for compaction. Naturally occurring soil does not normally possess the strength required to support heavy loads, but this strength can frequently be built into the native soils through compaction. As man continues to build his vehicles larger and heavier, soil strength requirements also become increasingly greater. In order to move these vehicles, it becomes necessary to construct highways and airfields with adequately strong foundations. The strength required in these foundations is built in by means of compaction.

Soils suitable for highway or airfield construction are actually of two basic types: the granular or nonplastic soils exemplified by sands and gravels, and the fine-grained or plastic soils such as the clays. Naturally occurring soils are generally mixtures of these two types, resulting in an almost endless variety of soils. Each of these must be analyzed for the best methods of compaction. In order to simplify the identification

Fig. 16-18 Tandem steel-wheel roller. (*Rex Chainbelt Inc.*)

TABLE 16-7 Weight Proportioning—Steel-wheel Rollers

Type of roller	Weight percentage	
	Guide roll	Drive roll
Tandem	35–40	60–65
3 wheel	30	70

and classification of these numerous soil mixtures, various agencies concerned with highway and airfield construction have developed soil classification systems. These systems are in general agreement and classify the granular-type soils as best and the plastic-type soils as poorest for construction.

The type of soil determines to a large extent the type of compactor that should be selected for efficient compaction. Other factors such as machine availability and soil moisture are also important, but the type of soil appears to be the biggest factor. Figure 16-19 shows a rule-of-thumb relationship between type of compactor and type of soil.

Each of the four types of compactors also has its own individual characteristics, including the method of rating its compaction forces. Table 16-9 indicates the rating method applied to each type of compactor.

TABLE 16-8 Steel-wheel Rollers

Type	Size class, max ballasted weight, tons	Rolling width, in.	Com-pression roll width, in.	Overall length, ft	Rated hp	Max rolling speeds, mph	Roll compressive forces, lb/lineal in.		Suitable applications	Manufacturers
							Com-pression roll	Guide roll		
Tandem	5–6	42	42	11–13	40–65	4–8	155–190	85–120	Asphalt patching and small jobs	Buffalo-Spring-field, CH & E, Ferguson, Galion, Huber, Ingram, Michigan, Rex
Tandem	8–10	50	50	14–16	75–125	4–8	190–265	120–160	Asphalt break-down and finish rolling	Buffalo-Spring-field, Ferguson, Galion, Huber, Ingram, Michigan, Rex, Tampo
Tandem	12–14	54	54	16–17	85–150	4–8	265–340	160–210		Austin-Western, Buffalo-Spring-field, Ferguson, Galion, Huber, Hyster, Ingram, Rex, Tampo
3-axle tandem	20	54	54	22–23	75–120	5–8	320–325	180–185	Asphalt finish rolling	Buffalo-Spring-field, Galion, Ingram
3-wheel	6	69	18 (two rolls)	14–15	60–80	4–7	200–250	110–140	Asphalt break-down rolling	Ingram
3-wheel	8–10	72–76	20 (two rolls)	16–17	75–120	4–7	250–380	140–170		Austin-Western, Galion, Ingram, Michigan, Tampo
3-wheel	12–14	76–84	24 (two rolls)	17–18	75–120	4–7	380–540	170–200		Austin-Western, Buffalo-Spring-field, Ferguson, Galion, Huber, Ingram, Michigan

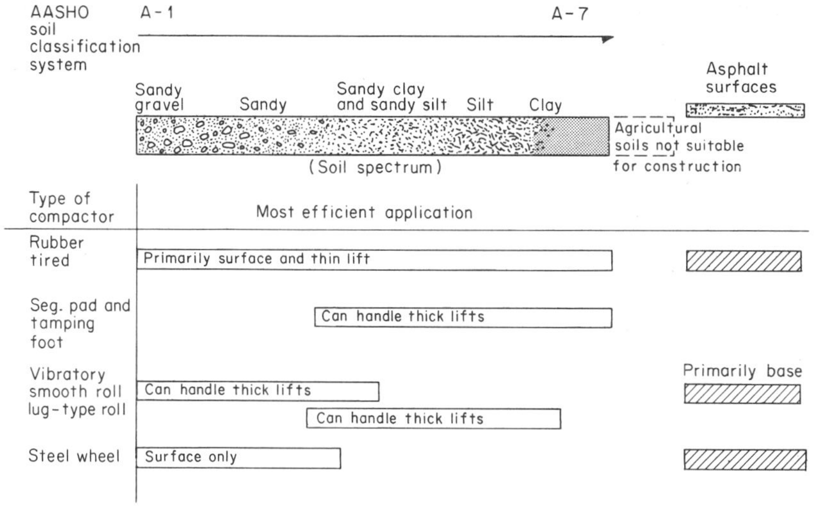

AASHO soil classification system A-1 A-7

Sandy gravel · Sandy · Sandy clay and sandy silt · Silt · Clay

(Soil spectrum)

Agricultural soils not suitable for construction

Asphalt surfaces

Type of compactor | Most efficient application

Rubber tired: Primarily surface and thin lift

Seg. pad and tamping foot: Can handle thick lifts

Vibratory smooth roll: Can handle thick lifts

lug-type roll: Can handle thick lifts

Primarily base

Steel wheel: Surface only

Fig. 16-19 Compactor suitability.

TABLE 16-9 Compactor Rating Methods

Type of compactor	Method of rating	Terms used
Rubber-tired.......	Tire pressure on soil surface per unit area	Ground contact pressure (GCP) expressed in psi
Segmented pad and tamping foot	Pad or foot pressure on soil surface per unit area	
Vibratory.........	Total force applied to soil	Total dynamic force generated plus static weight on roll expressed in pounds
	Frequency	Expressed in cycles or vibrations per minute
	Dynamic force	Force generated expressed in pounds
Steel wheel........	Roll pressure on pavement surface per unit of roll width	Expressed in pounds per lineal inch of roll width (pli)

TRIMMING EQUIPMENT

General We have entered into a new era in trimming equipment, in which the old methods and machines are rapidly giving way to the new. It was not too many years ago that the motor grader was the only equipment used for trimming. Today we have a variety of large, highly specialized, but extremely versatile machines to do much of the trimming. The result is more accuracy and greater production.

The actual transition into this new era started when various attachments were adapted to the old workhorse, the motor grader, to convert it into a semiautomatic trimming machine. This type of equipment change was the result of two factors: the tightening up of tolerances and the economic squeeze for higher production. The closer tolerances in turn resulted in better yields and consequently greater profits for the contractor.

Another significant factor which provided stimulus for the new automatic trimmers is the steadily decreasing number of highly skilled motor-grader operators. Each year there is a greater shortage of operators who can precisely control the motor grader blade and thus achieve the necessary high degree of trimming accuracy.

Equipment Types The years from 1955 to 1965 saw the evolution from the conventional motor grader to the large, automatic, full-width trimmers. The chronology of this transition was as follows:

Mid–1950s: Canal trimmers were developed.

1960: Attachment developed for conventional motor grader permitted semiautomatic trimming of one lane at a time.

1961: First fully automatic surface profile trimmer developed using a wire control system. Steering, however, was manual as the unit was towed by a crawler tractor. This unit trimmed a single lane at a time.

1962–1965: Several companies independently developed and introduced fully automatic trimmers for both surface profile and steering using wire control systems. These units were basically full-width machines, trimming two lanes in a single pass.

Fig. 16-20 Fully automatic trimmer (rear view). (*CMI Corporation.*)

Today's automatic trimming equipment centers around the control system in which a wire or a plastic line is very tightly stretched between stakes offset from the grade. The wire anchors on these stakes are adjustable, allowing very accurate alignment of the wire to the design grade. Steel wire was used in the early designs, but nylon cord has become the standard for present-day usage. The tightly stretched wire or cord provides a base plane for trimming the grade and subsequent paving operations. A horizontal movable arm, mounted on the trimming machine and spring-tensioned to this wire, is connected to an electric switch and provides the means of control. If the rough grade is low, the arm will lift and produce a signal to raise the trimming blade, thus controlling the profile of the trimmed surface. A vertical movable arm, traveling along the wire, controls steering in a similar manner.

Trimming equipment can be of several types, the more common of which are discussed below.

Automatic Trimmers. These large, self-propelled units are available in either a single- or double-lane version. They feature extremely precise trimming of the surface profile combined with high production, and they are accurately controlled by means of a "wire" guidance system. A typical two-lane unit will have a frame measuring about 30 ft wide and 30 ft long, each corner of which is mounted on a crawler track through a hydraulically controlled vertical member. The vertical movement in each of these members provides the means for accurately trimming the surface profile. The actual cutting of the grade is accomplished by a series of cutting teeth mounted on a full-width rotor. This rotor is followed by an adjustable strike-off blade, which in turn is followed by an auger for directing the spoil and then by the final trim blade. (See Fig. 16-20.) Through setup adjustments, the spoil can be cast to either or both sides.

One trimmer unit of this type can be converted into a portland-cement concrete paver, base-course spreader, or full-width asphaltic concrete paver by removing and adding components.

Motor Grader. This unit has long been the standard for trimming and, with a skilled operator, it can perform this task very well. Due to the pressures of higher production and the decreasing supply of skilled operators, the large, automatically controlled trimmers have now taken over much of its job. However, the motor grader still maintains a place as a trimmer on the secondary roads and on the smaller type of job.

Canal Trimmers. These operate on similar principles to the automatically controlled base and subbase type of trimmers. (See Fig. 16-21.)

Fig. 16-21 Canal trimmer in operation. (*R. A. Hanson Company.*)

New Specialized Equipment. A recent development is a large, bidirectional motor grader which can be automatically controlled by use of a reference "wire." This unit features eight large, rubber-tired wheels, all driving; and an engine on each end to develop a combined total of 450 hp. The machine is about 40 ft long and weighs 65,000 lb.

Other new trimming units from several manufacturers are either in the planning or development stages.

Equipment Selection The two major units used in trimming today are the large, automatically controlled trimmers and the conventional motor grader of average size. As a rule of thumb, it will take from four to six motor graders to equal the production of one automatic double-lane unit. It should also be noted that the automatic units will tend to trim to closer tolerances than the motor grader. It is apparent that the large, automatic units are suited to the large, main-line type of job, while the motor grader is more suited to the smaller type of project.

Type of unit	*Average trimming production*
Automatically controlled units (single or double lane)	1 mi in 4 hr
Conventional motor grader (with skilled operator)	
Single lane	1 mi in 8–12 hr
Double lane	1 mi in 16–24 hr

Section 17

Cranes and Hoists

R. L. BAUER*

Staff Engineer, Caterpillar Tractor Company, Aurora, Ill.

INTRODUCTION

This section will cover hoisting devices where the load is freely suspended on a hook or shackle rather than raised into place.† Such devices as elevators, fork lifts, excavating equipment, or jacks (rams) will not be included.

Historically, man's muscle, along with domesticated animal power, has been the means of raising loads large and small. With the advent of winching machinery, larger loads have been handled faster and more easily. This has permitted construction projects of magnitudes and speeds not otherwise possible. Lifting capacities for hoisting units normally range up to 800 tons; but there appears to be no maximum limit to future loads, to the height they can be raised, or to the distance through which they can be moved. As the requirement arises, the appropriate equipment will be developed.

* Formerly Assistant Chief Engineer, Construction Equipment Division, Harnischfeger Corp., Milwaukee, Wis.

† See also Steel Erection, Sec. 23, Steel Construction.

Each style of hoisting device has evolved to satisfy a special need within the construction industry. The stiffleg derrick in current use is a unit which, apart from its modern prime mover, is not too different from the derricks used in ancient Egypt. The mobile crane is the same basic structure with the added feature of mobility, so that it can be readily moved from one work site to another. The development of mobile cranes closely parallels that of the excavator industry, as the machinery used is similar and often common.* The hydraulic crane was developed to provide superior mobility and a reduced setup time in comparison with the mechanical crane, although it sacrifices maximum length of boom and some load-picking capacity in exchange for these advantages. Cranes of the gantry class, particularly those equipped with wheel-type running gear, can move their loads for distances which are beyond the reach of other styles of cranes. The tower crane was developed to facilitate the rebuilding of large areas of apartments and office buildings which were destroyed during World War II, and it has subsequently been used on a variety of urban projects. The original tower crane designs were European and, mobile tower cranes excluded, most of the units used in this country are still imported.

Although major manufacturers for each type of equipment will be identified, these lists are not intended to be complete or up to date. Equipment specifications are constantly changing as models are added or obsoleted. Once the equipment parameters for a proposed use have been identified, equipment dealers can readily furnish current specifications. Publications such as *Construction Methods and Equipment*, McGraw-Hill, or *Construction Equipment and Materials*, Conover-Mast, can also be consulted for reference material. Federal, state, and/or municipal codes or recommendations will frequently apply to the use of crane and hoist equipment.

MOBILE MECHANICAL CRANES†[1,2,3]

This type of crane has a mechanical, rigid drive from its power plant through transmissions, chains, gears, shafts, friction clutches, and cables. The basic crane motions are raising and lowering the load, swinging the load by rotating the crane body on its carrier mounting, and raising and lowering the crane boom. Other auxiliary functions may be provided as needed.

The power plant in the crane body is usually a gasoline or diesel engine, but occasionally an electric motor is employed. Power from the prime mover is transmitted through a clutch to a single- or multiple-speed transmission or to a torque converter. The clutch provides a means of disconnecting the drive train for ease of starting and for maintenance work. A multiple-speed transmission allows the selective use of high line speeds for rapid hoisting with reduced loads and lower line speeds for heavier loads or precision control of hoisting.

Torque converters can be substituted for multiple transmissions to provide comparable line-speed control without the requirement of shifting (see Fig. 17-1). Several variations of torque converters and converter controls are available, and an appropriate selection can be made to match the specific operating requirements. A standard single- or three-stage converter will normally have two governors, one at the engine to control the throttle setting and one at the output shaft to control the winching-machinery speed. The engine governor can be a variable control which is used to regulate engine speed, or it can be set to act only as an emergency control shutoff and thus protect against overspeeding.

If the engine governor is a variable one, the machinery or output-shaft governor could either be of the fixed-setting type or eliminated entirely. However, it is general practice to provide a variable-control governor for the output shaft, regardless of the type of engine governor. The variable-control output governor makes it possible to limit machinery speed for precision slow-speed hoist control of loads while still permitting swing or second load lines to be engaged without loss of line pull up to the capacity of the engine. When dual variable-governor controls are installed on a converter, the one calling for the least speed at the engine will govern.

* See Sec. 10, Excavators.

† Superior numbers refer to the list of references at the end of this section.

Split drives, at least one of which will be equipped with a converter, are used to transmit engine power at separate speeds to the different crane functions. Speed and torque output from the converter can remain essentially constant at any one setting, despite some variation in engine speed. This allows one speed control, say on the load hoist line, and a separate speed control on the boom swing motion. The direction of one drive can even be reversed through the converter so it is possible to lower the hoist line without affecting the speed or direction of the swinging motion.

The load hoist motion is controlled with a friction clutch and brake. These components can be actuated mechanically or with power air, power hydraulics, static hydraulics, or a combination such as air over hydraulics. Although each system has its specific merits, the mechanical one is usually less expensive. It also permits the

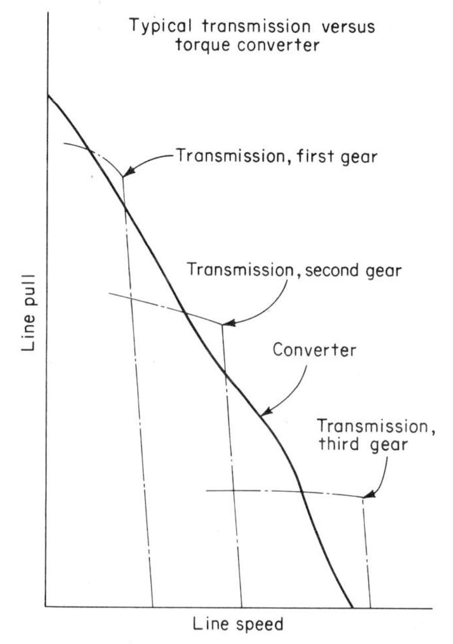

Fig. 17-1 Line pull versus line speed, multiple-speed transmission and torque converter.

operator to "feel" the reaction of the load more directly than with a power system. The static hydraulic system has similar characteristics. Power air requires rather large components, since it operates at 80 to 120 psi, but leakage is not critical. Excessive atmospheric moisture can be troublesome, although it can normally be controlled with alcohol evaporators and moisture ejectors. Power hydraulics can be operated with pressures to 3,000 psi, so its components are smaller than for power air. Any leakage in a power hydraulic system is critical, but this is not a problem when proper components are selected.

The crane boom is a latticed structure consisting of three or four main chords which are connected with lacings to form a beam. The chords and lacings are either rolled angle or tubular sections and are usually made of alloy or high-strength steels to minimize their dead weight. The boom foot pivots on, and is supported by, a revolving frame. The boom hoist system is connected to the boom point and thereby supports the boom. It is also used to vary the boom angle and, as a consequence, the load radius. Maximum boom angle is 80° above the horizontal. The maximum rated crane radius is usually equal either to the boom length or to its projected length when raised 10 to 15° above the horizontal. There are several variations of boom tips available—a heavy-duty tip which accommodates the maximum number of parts of hoist-rope reeving and thus allows the handling of maximum crane-rated loads; a light-duty tip which permits a longer overall boom, although this will not have enough sheaves to allow the maximum reeving for the maximum load; a hammerhead or offset tip to permit reduced overhead height and facilitate the handling of bulky loads; and container-handling tips which are designed primarily for ship loading of large containers.

A lighter structural section, known as a "jib," can be mounted on top of the boom if so desired. Although the maximum lift capability of the crane is thereby reduced, the combined length of boom and jib will be greater than can be obtained by the boom alone. An offset jib for the boom will also permit larger load radii than an equivalent length of boom and can be advantageous when working close to a high structure.

Boom hoisting and lowering are controlled in much the same way as the load-hoist motion. The boom hoist brake is spring set and normally, but not always, it remains set for the raising motion and is released only for lowering. Lowering must be with power, as a "live boom" is never permitted. A safety ratchet or pawl is engaged to the boom hoist drum at all times except when lowering the boom.

TABLE 17-1 Major Producers, Crane Ratings, and Boom Lengths for Crawler-type Mobile Cranes

Company	Range of ratings, tons	Maximum boom plus jib length, ft
American	9–300	290 + 80
Bucyrus-Erie	13–75	210 + 40
Hanson	7–15	40
Insley	9–45	120 + 30
Koehring	10–85	170 + 30
Lima	19–150	220 + 50
Link Belt	8–140	230 + 70
Little Giant	10–16	50
Lorain (Div. of Koehring)	16–40	110 + 40
Manitowoc	35–220	220 + 60
Marion	31–175	120 + 30
Northwest	13–100	170 + 50
P & H	17–110	210 + 50
Schield Bantam (Div. of Koehring)	9½	60 + 15
Unit	9–12½	60

The swing motion of the crane can be controlled with friction clutches, electromagnetic clutches, hydraulic pump and motor, or an electric motor.

A mobile crane can be mounted on a crawler-type unit, a conventional truck, or a self-propelled wheel-type carrier. The crawler unit is propelled by power which is transmitted from the crane body through a shaft at the center of crane body rotation. Positive locks or friction brakes are used to steer the crawler unit and also serve to hold the crane in position while it is in use. The friction brake can be manually operated, or it can be automatically spring set and released when torque is applied from the engine.

The ability of the crawler-type crane to lift loads without tipping is almost always greater over the end than over the side, and caution is required when swinging a load to a less stable position or when luffing it out to a greater radius. Various widths and lengths of tracks are available for most crawler-type carriers to allow for variations in the maximum crane load. A larger track also permits use of larger counterweights, so the gain in stability is twofold. Transporting a large crawler unit can be a problem, and some models have the capability of increasing or decreasing the width between their tracks. This track separation is fully extended for maximum stability when the crane is handling maximum loads. Crane load ratings are reduced when the tracks are moved closer together, but the task of transporting the crane or moving it on the jobsite is made easier.

Major producers of crawler cranes, together with representative crane ratings and crane boom lengths (including jibs, where available), are listed in Table 17-1.

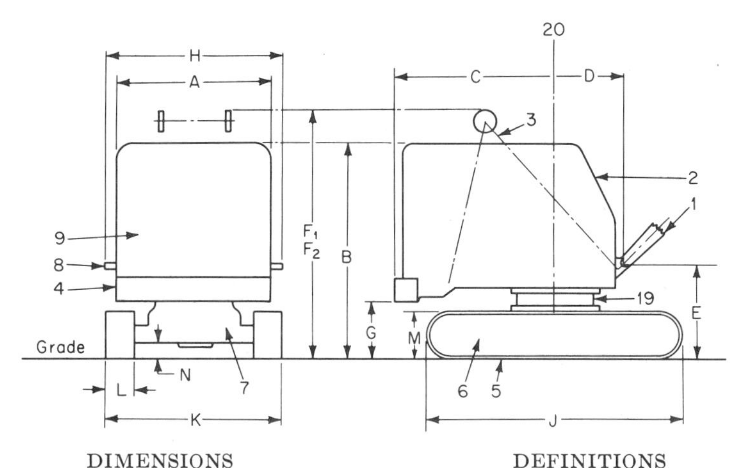

DIMENSIONS

A. Width of cab
B. Maximum height of cab above grade
C. Swing clearance (radius of rear end from axis of rotation)
D. Distance of boom foot pin to axis of rotation
E. Height of boom foot pin above grade
F_1. Gantry height above grade when in lowered position
F_2. Gantry height above grade when in raised operating position
G. Distance under counterweight to grade
H. Overall width when running boards are used
J. Overall length of crawler
K. Overall width of crawler
L. Width of crawler tread shoes
M. Height of crawler tread belt at center of end tumblers
N. Minimum clearance under crawler base to grade

DEFINITIONS

1. Front end attachment
2. Cab
3. Gantry or A frame
4. Counterweight
5. Crawler tread belt
6. Crawler side frame
7. Carbody or crawler base
8. Running board
9. Revolving superstructure
19. Swing circle or roller path
20. Axis of rotation

Fig. 17-2 Crawler-type crane mounting. For ground clearance dimensions see SAE J894. (*Society of Automotive Engineers.*)

Figures 17-2 and 17-3 illustrate the components and nomenclature for crawler-mounted cranes and accessory equipment as established by the Society of Automotive Engineers (SAE).

Trucks used for crane mounting are 6 by 4 or 6 by 6 for the smaller crane sizes, 8 by 4 for intermediate sizes, and 12 by 6 for the larger sizes.* (See Fig. 17-4.) Other combinations are also available for special requirements. Power plants and transmissions are generally matched to give a creeping speed of 1 mph and travel speeds of up to 40 or 50 mph. The carrier power must be adequate for the required travel speed and gradeability. Outriggers which are manually extended and retracted

* See Sec. 6, Engineering Fundamentals.

will normally be supplied as standard equipment, with hydraulic power actuation commonly available as an option on most models. Outriggers, outrigger housings, counterweights, and front-end equipment can be removed prior to traveling, to the extent necessary to satisfy local highway limitations as to total load, axle loads, and

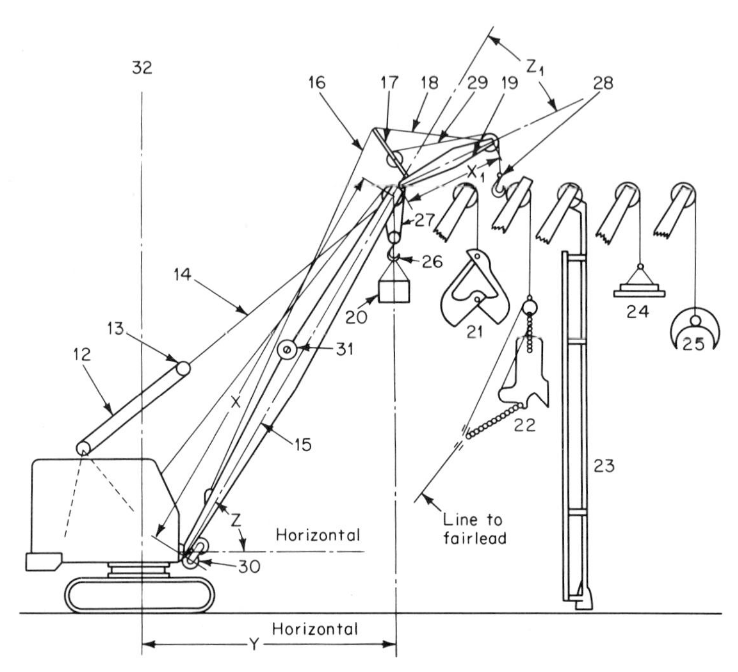

DIMENSIONS

X. Boom length from boom foot pin to boom head sheave pin
X_1. Jib length from jib foot pin to jib head sheave pin
Y. Radius of load (also applies to jib hook load)
Z. Boom angle
Z_1. Offset angle of jib (also can be given as an offset dimension)

DEFINITIONS

12. Derricking or live boom hoist rope
13. Floating harness or bridle
14. Pendants, guys or boom backstays
15. Crane boom
16. Jib backstay lines
17. Jib mast
18. Jib front stay lines
19. Jib
20. Concrete bucket
21. Clamshell bucket
22. Dragline bucket
23. Pile driver leads
24. Magnet
25. Grapple
26. Main lift-hook block
27. Main hoist line
28. Jib or whipline hook
29. Jib or auxiliary hoist line
30. Dragline fairlead
31. Tagline winder or magnet take-up reel
32. Axis of rotation

Fig. 17-3 Boom equipment for mechanical crane. (*Society of Automotive Engineers.*)

load per inch of tire width. The weights of these components are available from the manufacturer. In general, the published lifting capabilities for truck-mounted cranes will be based on outriggers set, over side and over rear; and without outriggers set, over side and over rear. The over-side ratings are usually the lowest.

The self-propelled rubber-tired carrier, as illustrated in Fig. 17-5, is similar to a truck in configuration. It can be powered from the crane body in much the same way as described for the crawler-type machine, or it can be powered from a separate prime mover which is located in the carrier and remote controlled from the crane body. Crane units which are mounted on self-propelled rubber-tired carriers may or may not have extensible outriggers. The front axles of the carrier will usually have lock-out devices to eliminate any pivotal effects and to improve over-the-side stability when operating without outriggers.

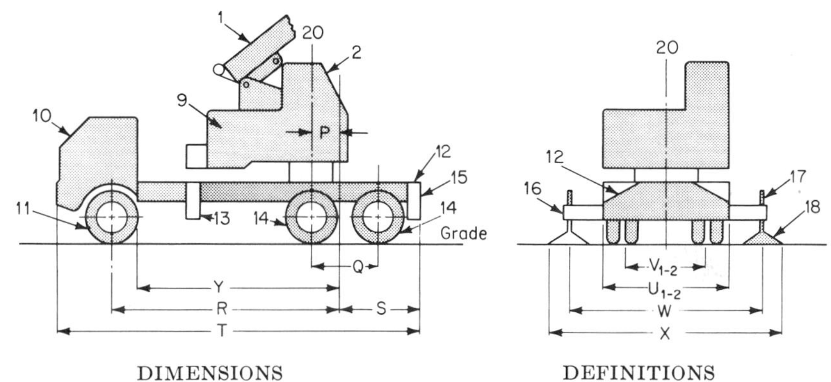

DIMENSIONS

P. Distance from center of rear axle or bogie to axis of rotation
Q. Distance between centers of axles of tandem axle bogie
R. Wheelbase (wheelbase for tandem front axle is measured to tandem center pivot point)
S. Distance from center of rear axle or bogie to rear end of frame
T. Overall length of carrier
U_1. Maximum overall width with retracted outriggers (floats removed)
U_2. Max. overall width with retracted outriggers (floats attached)
V_1. Track or tread width, rear axle
V_2. Track or tread width, front axle
W. Effective length of extended outriggers
X. Overall width over floats with outriggers extended
Y. Distance from back of carrier cab to center of rear axle or bogie (known as CA distance in trucking industry)

DEFINITIONS

1. Boom
2. Cab
9. Revolving superstructure
10. Carrier cab
11. Front axle
12. Carrier frame
13. Front outrigger box
14. Rear axle
15. Rear outrigger box
16. Outrigger beam
18. Outrigger float
19. Swing circle or roller path
20. Axis of rotation

Fig. 17-4 Truck-type crane mounting (6 by 4 or 6 by 6 shown). (*The Power Crane and Shovel Assoc.*)

Major producers of wheel-type mobile mechanical cranes, together with representative crane ratings and crane boom lengths (including jibs, where available), are listed in Table 17-2.

The upper body of a mobile mechanical crane can rotate through a full 360° about its point of attachment with the lower carrier mounting. This rotation is accomplished through hook rollers with live rollers, hook and house rollers, hook rollers only, or antifriction swing circle bearings. The latter type of unit has the lowest friction so provides the most precise swing control. However, the other types can approach this degree of control if they are properly designed.

Special auxiliary equipment for mobile mechanical cranes is available when desired. This includes foot throttle control, modified hand throttles, and combination throttle controls in addition to the dual throttle controls or torque converters which were discussed earlier. Hoist-drum rotation feelers or indicators can be applied so the crane operator can detect slight drum motions without looking at the hoist drum or hoist line. Power load lowering is available on the hoist drum for easing loads into place. Its use will reduce the reliance otherwise placed on the hoist brake. Dual hoist brakes will reduce brake heating by increasing brake area and are an asset for long-distance lowering or fast-cycle operation. Partially setting or dragging swing brakes can help control swing motion when handling loads in winds or on slopes.

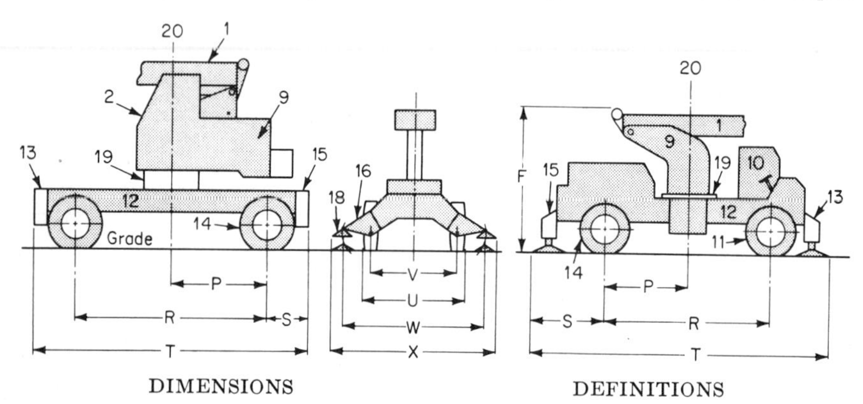

DIMENSIONS

F. Maximum height of attachment when in travel position
P. Distance from center of rear axle to axis of rotation
R. Wheelbase
S. Distance from center of rear axle to rear end of frame
T. Overall length of carrier
U. Maximum overall width with retracted outriggers
V. Track or tread width
W. Effective length of extended outriggers
X. Overall width over floats with outriggers extended

DEFINITIONS

1. Boom
2. Cab
9. Revolving superstructure
10. Carrier cab
11. Front axle
12. Carrier frame
13. Front outrigger box
14. Rear axle
15. Rear outrigger box
16. Outrigger beam
18. Outrigger float
19. Swing circle or roller path
20. Axis of rotation

Fig. 17-5 Self-propelled rubber-tired crane mounting. (*The Power Crane and Shovel Assoc.*)

Inching devices have been applied to hoist and swing motions for fine control. Special brake linings can also modify and improve load control. Boom-angle indicators, mechanical or electrical, are an aid to the operator in spotting loads and in using the crane-capacity chart. Electronic or hydraulic load indicators can also be installed. Although they are not accurate load-weighing devices, they do aid the operator in keeping loads within the crane's capacity when either rated load or rated load radius is approached.

Overload indicating systems are available. These indicate an approach to an unstable condition or to any other predetermined load limitation. These devices utilize mechanical, hydraulic, or electronic means to sense the load and load radius or boom angle and thereby determine the amount of overturning moment applied by the live load. Visual and audio signals aid in preventing accidental overload of a machine when this equipment is applied.

Cold-weather starting aids can be supplied, and heaters can be provided in either or both upper and lower operator stations. Lights can be powered through individual lighting plants or through increased capacity of the electric system on the engine. Accessory equipment which is normally intended for highway trucks can be applied, such as special mirrors, clearance lights, etc. Counterweight removal devices, magnets, and magnet generators are available. The crane body can also be equipped with various types of digging attachments.* Concrete-handling buckets in various configurations and sizes are available, as are load hooks, sheave blocks, etc.

TABLE 17-2 Major Producers, Crane Ratings, and Boom Lengths for Wheel-type Mobile Cranes

Company	Range of ratings, tons	Maximum boom plus jib length, ft
American	18–175	290 + 100
Bay City (Manitowoc)	45	160 + 60
Bucyrus-Erie	25–110	220 + 50
Hanson	15–25	50
Insley	15–45	140 + 40
Koehring	20–65	170 + 40
Lima	25–140	200 + 50
Link Belt Speeder	15–82	180 + 60
Little Giant	12½–20	80
Lorain (Div. of Koehring)	20–110	220 + 60
Manitowoc	65–140	220
Northwest	26–60	170 + 50
P & H	15–250	320 + 80
Schield Bantam (Div. of Koehring)	15	70 + 30
Unit	25–35	120 + 30

MOBILE HYDRAULIC CRANES[1,2,4]

The working motions of the hydraulic crane are powered and controlled through a pressurized hydraulic system. Pumps are used to supply the desired pressures, and the versatility of the crane may be enhanced as the number of pumps is increased. If a single pump is used to control more than one operating motion, then one motion must necessarily have priority over the others. True independence of motion is not obtained unless each motion is controlled by a separate pressure system.

The pumps supply hydraulic oil at pressures up to 6,000 psi to actuate cylinders or hydraulic motors and thus power the required crane motions. Each pump, motor, and cylinder is sized to give the end force and speed desired. The maximum power which can be supplied to each crane component is limited to that which can be provided by the pump which services it. The structural design of each hydraulic crane component can thus be based upon the limiting load which its particular hydraulic subsystem can support. In mechanical cranes, on the other hand, it is possible to direct full crane power to each operating function, and component design must be based upon this possibility.

Use of variable-volume pumps and/or motors can provide the hydraulic crane with operating characteristics which are similar to those obtained with mechanical cranes equipped with torque converters. Hoist motion is accomplished through a hydraulic-motor power winch with a built-in gear reducer or a special low-speed control hydraulic motor. The winch is mounted on the boom or in the vicinity of the boom pivot point.

The crane boom consists of two or more telescoping boxes which are fabricated from steel plate or special latticed designs (see Fig. 17-6). These boxes are extended or retracted by the action of one or more hydraulic cylinders. Some boom designs combine mechanical control with hydraulic control; for example, the upper part of the

* See Sec. 10, Excavators.

boom is extended mechanically and locked in position, then further extension is accomplished through a hydraulic system. The boom will work in a range from the horizontal to 80° above horizontal. Jibs can be applied to the tip of the boom in order to gain reach at the expense of a reduced maximum load.

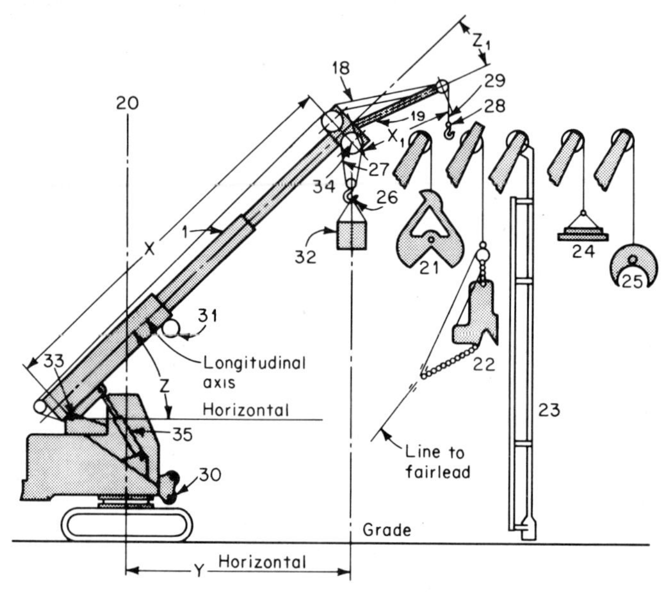

DIMENSIONS

- *X*. Boom length pin
- X_1. Jib length from jib foot pin to jib head sheave pin
- *Y*. Radius of load (also applies to jib hook load)
- *Z*. Boom angle
- Z_1. Offset angle of jib (also can be given as an offset dimension)

DEFINITIONS

1. Crane boom
18. Jib front stay lines or brace
19. Jib
20. Axis of rotation
21. Clamshell bucket
22. Dragline bucket
23. Pile driver leads
24. Magnet
25. Grapple
26. Main lift-hook block
27. Main hoist line
28. Jib or whipline hook
29. Jib or auxiliary hoist line
30. Dragline fairlead
31. Tagline
32. Concrete bucket
33. Boom pivot pin
34. Load hoist sheave pin
35. Boom hoist cylinder

Fig. 17-6 Boom equipment for hydraulic crane. (*The Power Crane and Shovel Assoc.*)

Boom hoist is powered through hydraulic cylinders, which are locked through check valves and act as brakes to hold the boom in the raised position. Swing motion is accomplished with a hydraulic pump and motor with a gear reduction. Several variations of fixed and variable displacement pump and motor are used to obtain the desired characteristics.

The carrier for the mobile hydraulic crane can be fitted with crawlers in much the same way as described for the mobile mechanical crane. Forward motion is then accomplished through a hydraulic motor and gear reduction, located in the crane carrier. A swivel at the center of the crane body rotation carries the hydraulic fluid to this motor and returns it to the tank after use. The hydraulic motor can be used as a partial brake but, due to internal leakage, an external friction brake is used in conjunction with it.

Trucks and other wheel-type carriers can also be utilized, just as they are on mobile mechanical cranes. One variation has a single engine in the carrier, with pump drives to supply hydraulic power to the upper crane body. This eliminates the cost of a separate power plant in the upper crane body but requires a complicated swivel and a more elaborate control system.

Where applicable, the auxiliary equipment which has been described for the mechanical cranes can also be attached to hydraulic machines. This includes boom-angle indicators, load-weighing devices, and load-moment systems. One system peculiar to hydraulic machines is the crossover of pump flow to allow two-speed application on hoist winches and propel motors.

Major producers of mobile hydraulic cranes, together with representative crane ratings and crane boom lengths, are listed in Table 17-3.

TABLE 17-3 Major Producers, Crane Ratings, and Boom Lengths for Mobile Hydraulic Cranes

Company	Range of ratings, tons	Maximum boom length, ft
BLH Austin-Western	5–20	81
Bucyrus-Erie	5–70	230
Drott	1–12½	40
Galion	8–12½	81
Grove	6–55	165
Hanson	5	17
Hein-Werner	8	35
Hy-Dynamic	10–12½	75
Hy-Hoe	20	83
Link-Belt Speeder	5–9	37
Lorain (Div. of Koehring)	15–25	103
P & H	9½–65	105
Pettibone	5–55	140
"RO" Mobile Crane	5	29
Schield Bantam (Div. of Koehring)	12½–17½	80
Silent Hoist & Crane	2½–15	18
Warner & Swasey	15–45	125

TOWER CRANES[1,2]

There are three basic types of tower cranes: (1) mobile, or a modified crane attachment of tower and jib boom on a mobile crane; (2) rail mounted; and (3) climbing or self-erecting. Each of these will do basically the same job, but in a different fashion. The first and second types stay on the perimeter of the working area and reach inside it, while the third type works from a fixed location within the facility which is being constructed. The mobile tower crane has the versatility of quick setup on a job and easy conversion to other modes of operation. The rail type is less versatile, but for large, long-range projects in a confined area it has a lower initial cost. However, it is not adaptable to short-term projects. The climbing tower crane is initially the most economical but can service only a relatively small area.

The mobile tower crane has a jib boom at or near the top of a latticed tower structure which is fixed to a rotating crane body. Height is limited by the structural strength of the tower and by the feasibility of raising the structure off the ground.

The rail-mounted tower crane is similar in function to the mobile tower crane just described except for its propulsion mechanism and a simplified machinery arrangement. It is not convertible to other functions such as crane, dragline, etc.

The climbing tower crane has a fixed tower which is intermittently supported to some other structure. Rotation of the jib boom is accomplished at the top of the

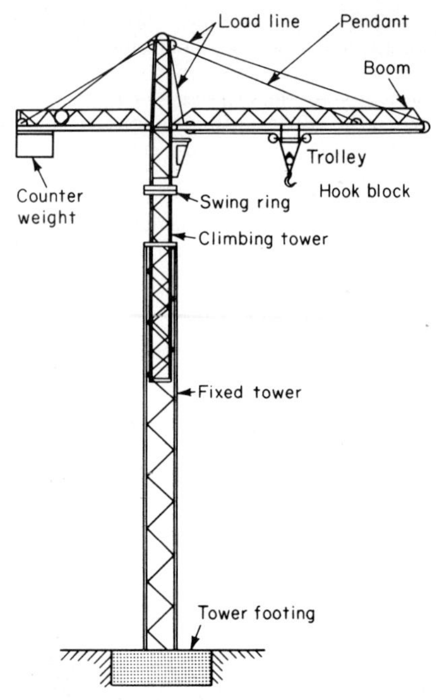

Fig. 17-7 Tower crane.

tower, as illustrated in Fig. 17-7. Load radius is varied by luffing the boom or with a trolley traversing the boom used as a horizontal beam.

Companies producing mobile tower cranes are:

Bucyrus-Erie Co.
Harnischfeger
Lima
Link Belt Speeder
Lorain (Div. of Koehring)
Manitowoc

Companies producing rail-mounted and climbing tower cranes are:

Bell-Liebherr (distributed by Bell, built by Liebherr)
Favelle Industries Pty. Ltd.
Heed Linden
Manitowoc-Potain (distributed by Manitowoc, built by Potain)
Sarnia Commander Climbing Crane

GANTRY CRANES[1,2]

The gantry crane is a structure designed to straddle a load and raise it on a bridge-type structure (see Fig. 17-8). A trolley can move the load crosswise to the structure, similar to an industrial shop crane. Cranes for construction purposes normally move on rubber tires, although rail-mounted gantries are common for industrial uses. The gantry crane can be used for industrial yard work, shipping (rail, truck, or ship) facilities, or on construction sites. These cranes are tailor-made to specifications for specific applications.

Power for the crane is supplied by a gasoline or diesel engine in conjunction with hydraulic pump and motor systems or an electric drive. Steering can be conventional or the crane can be moved sidewise, slewed, or turned pivotally on its own axle. This permits considerable versatility in placing loads such as steel beams for specific types of construction. Two or more gantry cranes can be used to handle large loads without sacrificing the versatility of slewing or sidewise travel for load placement.

Gantry cranes with rotating crane bodies (sometimes referred to as "whirlys") combine some features of the gantry crane with those of the mobile mechanical crane.

Fig. 17-8 Gantry crane. (*Harnischfeger Corp.*)

Recommended reference publications are:

*USA Standard B*30.2—1943 and *USA Standard B*30.2.0—1947, The American Society of Mechanical Engineers, United Engineering Center, 345 East 47th Street, New York, N.Y. 10017

Companies producing gantry cranes are:

Clark (Van Carrier)
Cook Bros. (Straddle Trailer)
Drott (Travelift)
Gerlinger Carrier Co.
Harnischfeger
R. G. LeTourneau
Renner Mfg. Co. (Comporter)
Shaw Box Crane and Hoist
Towmotor Co.

DERRICKS[1,2]

The derrick is a nonmobile, permanently anchored structure. Figure 17-9 illustrates its mast and boom, with their accompanying rigging and supports. A stiffleg derrick has its mast supported by nonrigid guy lines. The mast and boom rotate freely on the base or sill through a bearing or ball-socket arrangement. This rotation can be either a full 360° or partial, depending on the guy or stiffleg design in relation to the boom length. The boom is luffed in much the same fashion as on a mobile crane. The power source is either a gasoline or diesel engine or an electric motor. Power transmission is much the same as for a mobile mechanical crane.

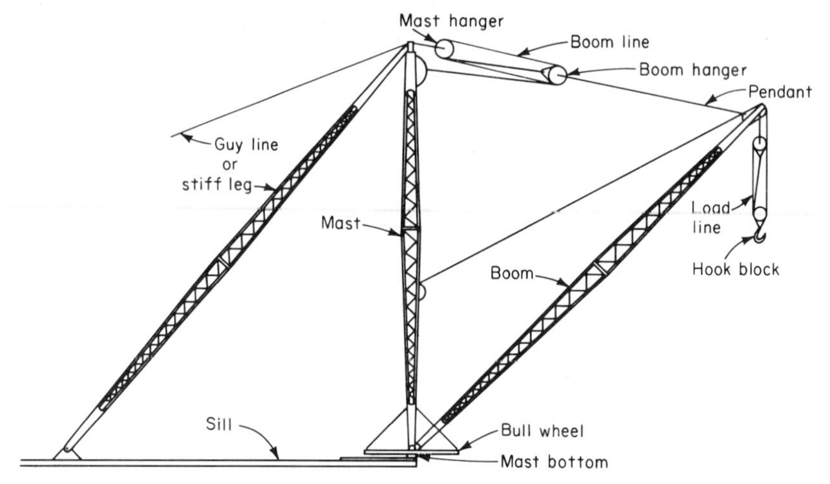

Fig. 17-9 Derrick.

Recommended reference publications are the same as for gantry cranes. Companies producing derricks are:

American Hoist and Derrick
Clyde Iron Works
Manitowoc Engineering Co.
Star Iron Works
Washington Iron Works

HELICOPTERS

Lifting and transporting structures or structural material by helicopter is a relatively new concept. Lifts onto existing high structures beyond the reach of normal cranes and material transportation into isolated areas without roads which are capable of supporting mobile cranes are types of service to be considered for this new concept of material handling.

This construction tool has definite limitations in its use. Atmospheric conditions and work delivery areas greatly influence what can or cannot be done, more so than for conventional hoisting procedures. Costs are relatively high and must be carefully analyzed, although convenience is sometimes an offsetting factor.

Companies producing helicopters are:

Bell Hellicopter Co. (Div. of Textron)
Boeing Corp.
Hughes Aircraft Co.
McDonnell Aircraft Corp.
Piasecki Aircraft Corp.
Sikorsky Aircraft (Div. of United Aircraft Corp.)

EQUIPMENT SELECTION

The selection of equipment to do a job or a series of expected jobs is influenced by several factors. For a variety of small lifts, a hydraulic or mechanical crane with adequate capacity when operating at the maximum expected range might be desirable. A machine rated at 15 to 20 tons would be able to handle 6 tons at a 20-ft radius and 50-ft height, but a larger machine would be required if the same 6-ton load is to be handled at 200-ft radius and 250-ft height. If tight quarters are expected, such as a high building and narrow streets, a mobile tower crane may be more adaptable than a standard crane. If most jobs are to be relatively light lifts, a small machine can be purchased and larger machines rented for the fewer expected heavy or long-range lifts. Adaptability to other equipment such as dragline, shovel, or hoe can be a factor if there will not be enough crane work to utilize the full time of the equipment. Roadability at the job or en route to the job can influence the choice of equipment. Economics of initial costs, maintenance cost, expected downtime, taxes, and amortization versus rental charges must be considered. The ability to sell, trade, or rent the equipment to others will be an additional factor to consider. Speed of load handling

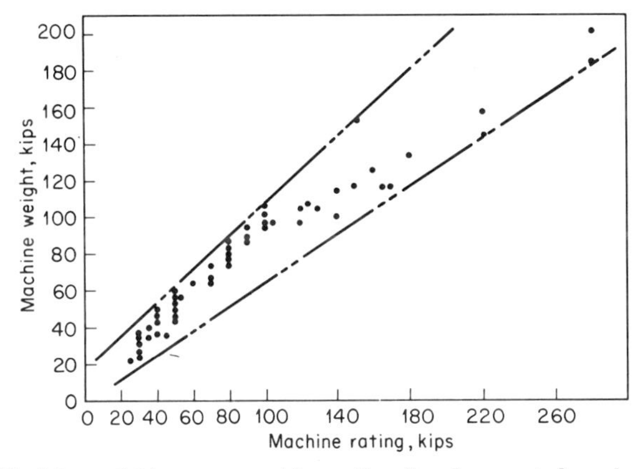

Fig. 17-10 Machine weight versus machine rating (truck-mounted mechanical cranes).

and power ability can be factors, and ease of operation and smoothness of load control will influence the selection. Safety features such as operator visibility, adequate guarding of machinery, and clutch and brake size should be investigated.

There are several methods of comparing various pieces of equipment. Figure 17-10 is a plot of machine ratings versus machine weight. Similar graphs can be constructed to compare power, speed, line pull, cost, ratings at various radii, etc. The information for these comparisons should be compiled at the time of an expected purchase or rental in order to have current data. All in all, each job will require some compromises in the choice of lifting equipment. Where alternatives exist, the selection should favor the larger units for improved capabilities and prolonged equipment life. Specialty equipment, except for long-term jobs, should be sought on a rental basis so as to preserve capital for more universally used tools.

REFERENCES

1. *USA Standard B30.5; Safety Code for Cranes, Derricks, Hoists, Jacks and Slings, Crawler, Locomotive and Truck Cranes*, American Society of Mechanical Engineers, New York, N.Y.
2. *S.A.E. Construction and Industrial Machinery Handbook Supplement HS* 39, Society of Automotive Engineers, New York, N.Y.
3. *Mobile Power Crane and Excavator Standards, PCSA Std. No.* 1, Power Crane and Shovel Association, Milwaukee, Wis., 1968.
4. *Mobile Hydraulic Crane Standards, PCSA Std. No.* 2, Power Crane and Shovel Association, Milwaukee, Wis., 1968.

Section 18

Marine Equipment

JOSEPH E. VOLLMAR, JR.

President, Vollmar Bros. Construction Company, St. Louis, Mo.

INTRODUCTION

When one begins to design or plan for the purchase of marine equipment, it is wise to start with consideration of the hull—that essential and costly element of all floating derrick barges, dump scows, pile drivers, and tugs. All equipment that floats must be mounted in or on a hull.

The hull must conform to the job which is to be done. A hull that is too large for the equipment will be costly to maintain and awkward to use. If it is too small, it will probably render poor service and may even be dangerous. It must be of steel construction, which is expensive, and its cost will be determined by its shape, tonnage, size, and special equipment.

With each year of its life, floating equipment becomes more expensive to maintain. It frequently develops a "suicide complex" of leaks and other problems.

BARGES AND SCOWS

Selection The problems confronting a prospective lessee or purchaser of a barge or scow are many. For example, is it of sufficient size to remain stable when loaded? Will its framing support the proposed deck load? In addition to ascertaining its physical shape and value, the questions of how long, how deep, how wide, is it the type needed, all must be answered as parts of the seek, find, and inspect routine.

The nomenclature of barges and scows should be explained. These two names are interchangeable, and the choice between them depends largely upon the location of use. Scows are referred to as "barges" on the inland waterways; conversely, barges are referred to as "scows" on the Great Lakes, Gulf, and coastal waters.

Many of the older scows on the coastal waters are of the keel type or model bow construction. However, most barges today, whether oceangoing or for use on the inland waterways, have scow-type rakes (on bow and stern) unless they are to be used in a matched or integrated tow. This is especially true of contractors' barges, which generally are limited for ease of operation to a length not exceeding 150 ft. Oceangoing barges are usually shorter than river barges, since a long barge suspended between two waves will have a tendency to buckle. Consequently, the depth-to-length ratio is usually greater for ocean and lake barges.

Many specialized forms of floating transporting equipment are available to the contractor. Of these, side-dump or bottom-dump barges will frequently merit his consideration. Bottom-dump barges are generally in short supply on the inland rivers, being more easily leased for service on the coastal waters or the Great Lakes. Side-dump barges are used in shallow water but, because of their specialized nature, are not easily located or leased.

Inspection To ascertain the seaworthiness of a barge, it should be removed from the water by dry dock or on marine ways. Once out of the water, the hull below the waterline can be inspected. Holes should be drilled to determine the plate thickness in any areas which are badly pitted. The greatest degree of pitting usually occurs in the vicinity of the light (unloaded) waterline. If the barge is old and of riveted construction, which was the standard method of construction until shortly after 1930, a close inspection should be made for rivets which are loose or which have their heads rusted off. Rivets on the exterior of the hull should be inspected for wear, as they have a tendency to wear off from rubbing against lock walls, other barges, etc.

On all barges used on construction work, the knuckles (where side and bottom meet) and rakes (underside of sloping bow and stern) are subject to considerable abuse from hitting the bottom and working off rock banks, etc. Some old barges which have had excessively hard service may have corrugated sides; that is, the side plates are indented between the frames. This condition does not usually impair the strength of the barge unless the framing is badly bent. However, if a barge has corrugated sides, the outside of the side plate should be inspected to see if it is worn thin from excessive rubbing. If so, the worn area can be improved by welding a strap over it.

Barges which have been used in brackish, salt, or acid waters, such as on the Gulf, the oceans, or the Monongahela River, may show evidence of excessive deterioration in the form of bad pitting around the light waterline. They may also be more rusted inside than a barge used in fresh water, and barnacles may be present. Salt or mud deposits on the inside of the hull bottom are evidence of previous leaks, as are wooden plugs or concrete patches. In brackish, salt, or acid waters, an anticorrosive paint may increase the life of the floating equipment.

Money spent for expert advice before purchasing a barge may save considerable and needless later expense. The American Bureau of Shipping, which can be contacted in almost every major river city, is usually able to furnish inspectors at a reasonable fee. Normally, the marine insurance companies will recommend or perhaps assist in finding an inspector if they are to carry the hull or cargo policy. Regardless of the nature of the barge, scow, or hull, it is well to remember the Latin phrase *caveat emptor*, which in effect states, "Let the buyer beware." The value of expert counsel and advice cannot be overstressed.

TUGS AND TOWBOATS

Selection Without a doubt, the most useful piece of floating equipment on a marine construction project is a towboat or tug. These units are similar; the tug has a model bow with a keel running from stem to stern, while the towboat has a scow bow similar to that of the rowboat. Modified scow bows are those in which the designer has attempted to add streamlining.

For contractors' work on the inland waterways, it is difficult to surpass the scow-bow towboat because of its ability to work close to the bank or in shallow water. A towboat requires less draft than a tug of similar size, although a tug can take heavier weather on open waters because its additional freeboard increases its tolerance to rolling. The type of stern affects the ease of shallow-water operation, and it is usually advisable to secure a towboat with a semi-tunnel. The underside of the after bottom is then reduced in depth to accommodate the screws. With twin engines, the propeller shafts come out of the hull on the sides and are held steady at the propellers by means of struts. In the case of the tug, the propeller shaft comes through the keel.

The steering system for contractors' boats is usually of the manual type; i.e., the wheel is connected to the rudder by means of gears and shafts or by cables and drums. If hydraulic steering has been installed, the system should be capable of moving the rudder from hard over to hard over in 12 sec or less. All single-engine, hand-steered boats will give harder rudder response in one direction than the other; this should not be the case with a twin-engine boat if its engines are synchronized properly.

The typical operating crew on a contractor's boat consists of a pilot, an engineer, and either one or two deckhands depending upon the type of towing or fleet shifting to be done. Boats whose engines and gears are not pilot-house controlled will require the services of a full-time engineer.

Bunks, galleys, and crew's quarters are fine sundry items to have on a towboat, as are radar, ship-to-shore telephones, and depth-indicating systems. However, these can be expensive if they are not needed for a particular operation. A pair of good searchlights is a wonderful aid for shifting equipment at night. Proper coast-guard lights are a definite must, as are life jackets. It is well to remember that the boat may have to pass a U.S. Coast Guard inspection.

Inspection When inspecting a boat for possible lease or purchase, several factors should be considered. It is helpful to go to a reputable shipyard for assistance; if this is not possible, the first thing to check is the foundation of the engine. The framing and engine supports should be heavy and sound. Notice the type, spacing, and thickness of the framing; this is a key to the general condition and design. If the beams are small, the channels are light, and the welding is poor—beware! Next, the soundness of the hull should be examined. Its exterior should be looked at carefully, especially at the waterline. If this area is heavily pitted, the buyer will probably have a "leaker" on his hands. Also, the interior of the hull should be checked for wooden plugs, plate patches, or a new section of hull plating.

The engines are key items, and their satisfactory operation requires good service facilities at every port. A shutdown for lack of parts is always expensive. Engines manufactured by either Caterpillar or General Motors are very satisfactory, as excellent service facilities are always available. Repair parts for other makes must usually come from the factory. Beware of "bastard" boat design; a small boat "souped up" with big engines usually means low freeboard in addition to other troubles. "Rudderless" boats can also be troublesome; they are usually someone's exclusive dream, and parts for repair will be hard to find. Also be careful of boats with excessive "roll" when the rudder is hard over in a tight turn.

DREDGING

Hydraulic Dredge Hydraulic dredging is the ultilization of power with water to transfer solid material. The main parts of a hydraulic dredge are the hull, the gantries, the spud keepers and spuds, suction, the ladder with the suction line and intake, the cutter head, the dredge pump and engine, the hoisting equipment (usually five drum), and sundry equipment such as light plants, service pumps, compressor, etc. Figure 18-1 illustrates typical dredge components.[1,*] The average dredge is constructed on a barge or on a hull which is designed for this particular service.

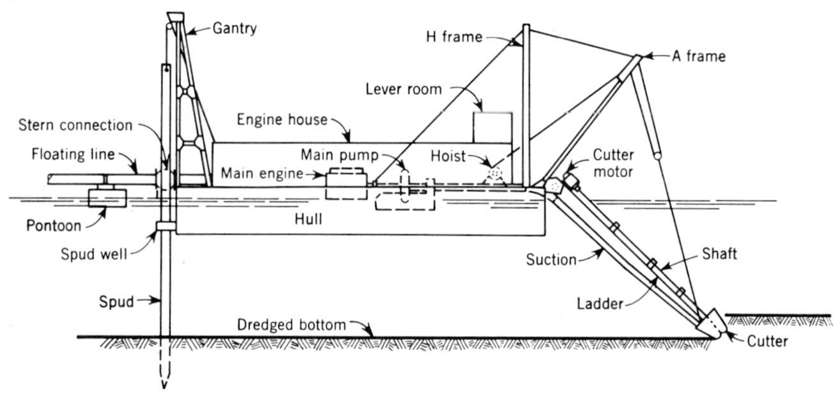

Fig. 18-1 Dredge components. *(From J. Huston, Dredging Fundamentals, J. Waterways Harbors Div., Am. Soc. Civil Engrs., no. WW3, paper 5390, August, 1967.)*

Hydraulic dredges are rated by the size of their discharge lines, and they vary in size from midgets to mammoths. A small dredge equipped with a 6-in. dredge pump driven by a gasoline engine can be built onto a hull only 30 ft long and 14 ft wide, with a 2-ft draft. It is ideally suited for dredging work in most lakes or ponds where deposits, such as mud and gravel, are to be removed in limited quantities. In contrast, a dredge with a 20-in. pump connected directly to a 1,300-hp engine requires a hull 137 ft long and 40 ft wide, with a draft of 10 ft. Such a dredge is capable of handling gravels and boulders through pipelines over a mile long. Its discharge pipe must be of sufficient size to accommodate the spoil. Table 18-1 lists typical specifications for dredges.

For free-flowing material such as sand or gravels, a plain suction nozzle without cutter or agitator is used. The suction force of the pump causes the solid material from the bottom to be picked up. Where this material is firm and compacted or does not disintegrate and flow readily, the suction must be provided with a revolving cutter. This will break up the material in the immediate vicinity of the suction end so that it can be picked up and transported. There are several types of cutter heads; some have cast blades and others have removable teeth (see Fig. 18-2). The motor and drive shaft for the cutter head are mounted on a structural steel framework called the "dredge ladder."

The ladder length, usually at a 45° inclination, determines the maximum dredging depth. Although ladder lengths of up to 225 ft have been employed,[2] lengths will customarily range between 20 and 140 ft. The suction pipe is carried within the ladder and is connected to the suction pump aboard the dredge by a ball joint or length of rubber suction hose. This connection permits the adjustment of the suction ladder to the desired depth of dredging. The cutter and ladder must be of sufficient strength

* Superior numbers refer to the list of references at the end of this section.

TABLE 18-1 Typical Specifications for Five Sizes of Dredges[1]

Item	Size of dredge discharge, in.				
	12	16	20	24	28
Length, ft	100	120	140	160	175
Beam, ft	35	40	45	50	50
Depth, ft	8	9	10	12	15
Displacement, tons	560	840	1,200	1,850	3,000
Pump power, bhp	570	1,000	1,500	2,700	5,000
Pump speed, rpm	500	400	350	325	300
Cutter power, bhp	150	200	400	750	1,000
Cutter speed, rpm	5–30	5–30	5–30	5–30	5–30
Spud length, ft	55	60	70	90	100
Ladder length, ft	50	55	60	70	80
Maximum pipeline, ft	2,500	4,000	5,000	7,000	9,000
Maximum width of cut, ft	160	200	220	270	325
Minimum width of cut, ft	50	60	70	90	90
Maximum digging depth, ft	35	40	45	50	60
Minimum digging depth, ft	4	5	6	8	12

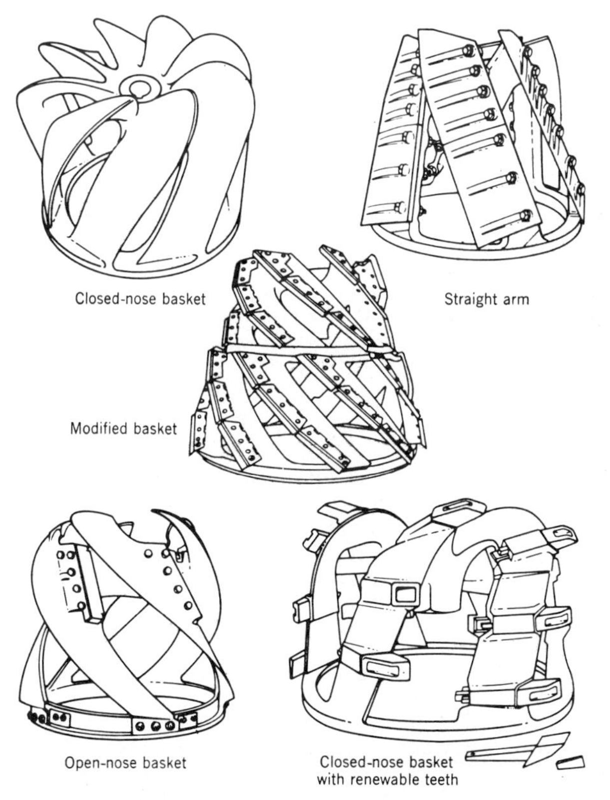

Fig. 18-2 Various cutter types. *(From J. Huston, Dredging Fundamentals, J. Waterways Harbors Div., Am. Soc. Civil Engrs., no. WW3, paper 5390, August 1967.)*

to withstand the sudden shocks and strains incurred when encountering boulders, compact layers, etc. Cutter speeds vary between 5 and 40 rpm on most dredges, and the power applied to the cutter may be as high as 3,000 hp.

The dredge ladder is suspended by a block and tackle from the forward end of the bow gantry. This bow gantry, sometimes called the "A frame," is usually pin connected to the bow of the forward deck. The stern, or spud, gantry is made high enough so that a spud can be hoisted until its bottom is level with the bottom of the hull. The spud keepers are located at the deck and near the lower knuckles of the stern. Spuds are usually cylindrical, with their diameters depending upon the size of the dredge, and their lengths are contingent upon the depth of the water in which the dredge is to be used. It is by means of the spuds that the dredge can be anchored or maneuvered to "walk" along the river bottom. Power for raising and lowering the spuds is usually derived from the main hoist engine.

Swing lines are installed in order to swing, or walk, the dredge. These lines lead down the ladder to sheaves near the suction end of the ladder, and then to anchors placed on either side of the dredge and at 90° to its center line. The swinging pull is thus directed nearest the cutter head where the heaviest load is applied.

The size of the dredging pump is determined by the output required, the conditions of the material to be excavated, and the lengths of the floating discharge and shore lines. The pump must develop sufficient suction to lift the cuttings, and it must discharge at sufficient velocity to keep the dredged material in suspension. A velocity of 12 to 14 fps is usually adequate, although large dredge pumps may operate at higher velocities. In the case of an extremely long floating discharge line or long shore lines, additional booster pumps may be necessary to gain maximum efficiency of the dredge.

The required brake horsepower for the pump can be computed as follows:

$$\text{bhp} = \frac{QHS}{3960E}$$

where Q = pump discharge, gpm
H = total head, ft of fresh water
S = specific gravity of dredged material
E = pump efficiency, expressed as a decimal

The total head is the sum of suction and discharge heads, including both static and dynamic components. Specific gravity of the suspension is approximately as follows: 10 percent solids = specific gravity of 1.1; 20 percent solids = specific gravity of 1.2, etc. The hydraulic dredged material should run between 15 to 20 percent solids for good production. The efficiency of the dredge pump depends upon its make and design but normally varies between 0.50 and 0.60.

The discharge pipe of the dredge is usually constructed of relatively light, spiral-welded material. It is subject to considerable wear, the extent of which is directly proportional to type, quantity, and velocity of the material which passes through it.

Each piece of floating pipeline is mounted on a float, with the simplest type of float consisting of either two or four steel barrels or drums. On the inland waterways, the Corps of Engineers uses small barges (15 ft wide by 30 or 40 ft long) for this purpose. A circular track mounted on the barge permits the dredge pipe to rotate. Shore pipe is usually fitted up with flanges, and in the majority of cases does not have to be shifted as often as the river discharge line. When equipping a dredge with discharge pipe, the pipe nearest the pump should be the heaviest, the floating line next in weight, the shore pipe the lightest.

The actual setup of the dredge usually is performed in the following order. The dredge is towed to the desired starting location, the spuds are lowered, and the swing-line anchors are run. The pontoon line is then set into position at the stern of the dredge and connected to the dredge discharge. Each piece of line is individually added until the desired length is obtained. Bank lines are then installed to the point of discharge.

If a hydraulic fill is to be constructed, provision must be made for the runoff of the dredge water. Small dikes can be arranged to form a shallow pond, and the pump water is then directed into this pond so that settling of the dredge material will take place. Small levees must be kept built up as the hydraulic fill proceeds, and this is best done by using a small bulldozer or a loader.

The number of employees needed to man a dredge project depends largely upon the size of the dredge. The responsibility for the dredge and its operation rests with the dredge superintendent (dredge captain). It is his job to know the dredge, its equipment and capacity, and to be familiar with the engine, its repair and maintenance. Serving under the dredge captain are the pumpers, one for each 8-hr shift. The pumper operates the hoist and cutter head, calls for pump engine speeds, and controls the amount of material being dredged and the swing of the dredge. Working under the pumper is usually one mate per shift. Through the mate's direction such shift personnel as deck laborers (for changing the floating lines, running cables, and swing anchors) and the shift engineers receive instructions. The shift engineer is responsible for engine speeds, the maintenance of the engines, hoists, etc. A single oiler per shift is usually employed to assist the shift engineer.

The operation of hoists, cutter head, swing controls, and marine signals is usually performed from the control house. This house can be mounted on top of the machinery housing so as to be in full view of the entire operation. The pumper is stationed within this house, where he operates the hoist for controlling the ladder, spuds, and swing lines. He is also responsible for displaying and giving the proper marine signals to other vessels.

Ladder or Bucket Dredge This is a specialized piece of dredging equipment which is adapted for excavating submarine trenches and for use in sand and gravel production. It consists of a ladder-mounted endless chain whose buckets discharge onto a conveyor. The ladder dredge, because of its nature, is not able to convey its material beyond its point of bucket discharge except as made possible by supplementary means. The ladder is also vulnerable to shifting of the barge caused by currents, passing vessels, or rough weather.

Dipper Dredge These dredges have distinct advantages when excavating certain types of material and are especially useful for breaking up ledge rock or excavating blasted rock. The dipper dredge usually operates with three spuds: two forward and a crowding spud at the stern. Prior to starting work, a large percentage of the dredge's weight is hoisted onto the two forward spuds; this makes a solid foundation from which the dipper can operate. The crowd spud (at an angle of 20 to 30°) prevents the dipper stick from pushing the dredge backward. The dipper bucket is mounted at the end of the dipper stick, where its operation is similar to that of a power shovel. The boom which supports the dipper stick swings, and the bucket discharges into a waiting scow barge or deck barge. As the dipper dumping radius is extremely limited, the barge must constantly be moved while the loading operations are taking place.

Clamshell Dredge This is exceptionally useful for deep digging where it is impractical for any other equipment to operate. When spuds are not used, anchor lines must be placed to keep the dredge in position while the dredging operation is being performed. Standard clamshell buckets may be used, including the orange-peel type. The actual selection of bucket size is dependent upon the size of the equipment and the type of material to be excavated. Deep digging demands proportionately longer bucket ropes.

It is not uncommon to anchor a mobile crane on the deck of a flat-deck barge and use it as a clamshell dredge. This is perhaps the most economical method of excavating relatively small volumes of material, such as cleaning out powerhouse intakes or removing silt from the face of a dock. The clamshell dredge has an advantage over the dipper dredge inasmuch as it can work in closer areas and, because of the longer reach of its boom, can load barges which are farther away.

FLOATING CRANES, DERRICKS, AND PILE DRIVERS

Cranes and Derricks Until a successful "sky hook" is developed, a floating crane or derrick will be one of the indispensable tools of the marine contractor. There are many ways of approximating this proverbial "hook" aboard a hull, the most economical of which is by the installation of a fixed boom, or A frame, on the bow of a barge. If heavy structural members and large cable are used, extremely heavy loads can be lifted. These fixed booms do not provide versatility in handling loads, since the entire hull or barge must be shifted to deposit the load where it is desired. Steel stiffleg derricks are available in rated capacities of 3 to 800 tons. They are extremely popular and can be equipped for hook work or dragline operation.

Whether on pile driving operations, material handling, or steel erection, equipment which will revolve through 360° is a great asset. Barge-mounted revolving cranes are expensive and, as noted earlier, a contractor may place a crawler-type crane aboard a barge when a clamshell operation or relatively light lifting is to be performed. The crawler equipment should then be securely lashed to the deck to prevent its sliding overboard. The operator should give careful consideration to the load and to the radius at which it is to be lifted. Unfortunately, once the crane is firmly anchored to the deck, the operator can no longer "feel" the load. Many boom failures have occurred for this reason.

For capacities larger than can be handled with a barge-mounted crawler crane, it will be necessary to procure a crane similar to those manufactured by the American Hoist and Derrick Co. These cranes range in capacity from 7 tons at 80 ft to 30 tons at a 95-ft radius. Many older machines, despite the fact that they may be steam powered, are still extremely serviceable. They should not be overlooked when equipping a project on a limited budget.

As noted earlier, a barge-mounted stiffleg derrick provides a very economical lifting unit. Such a derrick, with its engine, will be considerably less expensive than a crawler crane of comparable capacity. Several new and efficient designs have been introduced; for example, the number of barge connections required to fasten the derrick to the hull has now been reduced to three main points. A 180° swing can also be accomplished. Through the use of the bull wheel and swing engine, the capacity of the derrick barge is comparable to that of the revolving crane.

Before attempting to mount any type of lifting equipment aboard a barge or hull, whether A frame, crawler crane, or deck-mounted revolving crane, a study must be made to ascertain whether the deck beams and framing will carry the load which is to be imposed. If the framing of the barge is deemed satisfactory, then planking such as mats and additional I beams should be employed. The planking should spread the load over a satisfactory number of frames so that undue strain will not be imparted onto the deck framing system. Assuming that the barge has sufficient capacity to carry the equipment which is to be mounted aboard, a study should be made by a qualified marine engineer to ascertain whether the loads to be handled at any given radius will cause the hull to become unseaworthy and capsize.

One of the insatiable requirements of "progress" seems to be that all aspects of a business must grow. A greater amount of capital is required, manpower needs rise, and equipment also increases in size and capacity. The search for oil in the tidelands is an excellent example. This operation, in contrast to land drilling, requires the construction of large marine structures and the use of heavy floating equipment. Today projects are being undertaken that would not have been attempted five years ago. Sunken barges, with cargo intact, are being successfully salvaged where they would have been abandoned in the past. To perform these services, derrick and crane barges of colossal size are being built. The undisputed giant of the crane barges is 400 ft long, 100 ft wide, and 78 ft deep. It can work in the open sea in any part of the world, has a boom 233 ft in length, and is capable of lifting a 500-ton load and swinging it in a full circle of 360°. This gigantic crane, which is owned by J. Ray McDermott and Co., is shown in Fig. 18-3.

Auxiliary equipment for this mammoth "hook" includes pile driving apparatus which will deliver 120,000 ft-lb to drive steel piles up to 6 ft in diameter. The rig is also equipped for laying large-diameter pile frames under water. Air-conditioned quarters are provided for the entire crew, and a potable water supply and adequate storage space make it possible to maintain the operation for an extended period without refueling or reprovisioning.

Fig. 18-3 500-ton revolving crane barge.

Fig. 18-4 500-ton derrick barge.

Another "500 tonner" is located on the Eastern seaboard. This is a derrick of the stiff-legged A frame type, especially designed for the transferring of heavy equipment from dockside to floating vessels. It is owned by the Raymond International Co., and is illustrated in Fig. 18-4.

By utilizing slings and strongbacks, these behemoths of the construction industry can be teamed up so as to lift heavy loads beyond the capacity of either one (see Fig. 18-5). Great care must be exercised when using two or more cranes for a "lift." To prevent damage to the equipment and to reduce the likelihood of accidental injuries to the operating personnel, strain gauges should be used on all lines where overload may occur.

Pile Drivers Pile-driving operations occur on marine projects more frequently than any other type of operation. The federal government is attempting to improve the channels of our inland waterways, and much of this work entails pile driving. With adaptation, crawler or fixed-barge derricks may be employed for pile-driving work. Most contractors prefer to use a pile driver with a fixed set of leads and,

Fig. 18-5 Two barge-mounted stiffleg cranes, combined lifting capacity of 800 tons.

using this equipment in conjunction with hoisting equipment, piles varying in length from 30 to over 100 ft can be driven depending upon the water depths.

A popular pile hammer for driving wood piles is the No. 2 Vulcan, single-acting hammer delivering 7,260 ft-lb of energy per stroke.* When driving steel or concrete piles, a McKiernan Tierney Model 10-B-3, Double Acting Vulcan Model 30-C, or a Vulcan No. 1 may prove suitable. Experience and actual trial will best dictate the preferred size and type of hammer. If the hammer is steam operated, the boiler must be of sufficient size to permit continuous operation of the pile hammer.

The operational crew of a floating pile driver usually consists of a foreman, three deckhands, two engineers, a fireman, and a leadsman. Through the incorporation of a series of staging into the leads, the pile is hoisted until vertical. It is then brought into the leads through the coordination of the deck gang and the man on one of the

* See Sec. 27, Piles and Pile Driving.

upper stages. The leadsman gives the order to lower the hammer onto the piles and start the driving operation. When the pile has been driven to grade, the operation is repeated on the next pile. A stern engine, in addition to the main three-drum hoisting engine, is frequently utilized for the control of the anchor lines.

MISCELLANEOUS EQUIPMENT

Floating equipment of a specialized nature, such as compressor barges, pump barges, tool-storage barges, office barges, or equipment for submarine divers, is usually assembled in accordance with the needs and demands of the project. These requirements should be established by experienced and competent engineers.

Auxiliary Diving Equipment When assembling a compressor barge to supply air for a pneumatic caisson, the size of the compressors must be established by someone thoroughly familiar with this type of project. The required compressor capacity will depend upon the size of the caisson and the frequency of use of the air locks.

Submarine divers are usually "problems" to the project superintendent. The diver, being himself an artisan, will insist (and rightfully so) that his compressors, lines, and tender have an area assigned to them where they will be undisturbed. In order to accomplish this, it is usually advisable to reserve for his use a low barge or pontoon which may easily be shifted about. This has the additional advantage of being readily moved out of the way should the diver's work be relatively intermittent. There should also be adequate shelter for the comfort of the diver while changing gear during cold or inclement weather. A tool-house barge with a coal-fired cannonball stove is quite suitable or, if this is not available, a lean-to can be constructed using 2 by 4s and tarpaulins. Shelter should also be provided for the diver's tender, who will operate the diver's communication system.

It is usually advisable to have a diver's ladder constructed. The diving suit, together with its ballast and weighted shoes, becomes extremely heavy when out of water. For this reason, the rungs on a diver's ladder must be more substantial than those on an ordinary ladder. These rungs can be constructed of 2 by 4s and must be close together, usually 10 in. riser top to riser top. The ladder is placed so that it extends 7 or 8 ft into the water and approximately 5 or 6 ft above the deck of the barge. In this manner the diver can bring himself to the surface by inflating his suit, be pulled to the ladder by his tender, and then walk up the ladder until he is flush with the barge deck. His handlers can assist him in stepping from the ladder onto the deck. When working in deep water, it is advisable to provide a decompression chamber to be used in the event that the diver makes an emergency rise or is stricken with the bends. The operation of the decompression chamber should be undertaken only by a thoroughly qualified person.

Transportation The type of transportation to be supplied for a project depends upon the number of men to be employed and the distance over which they must be transported. If the project is near a landing, the job tug or towboat will easily suffice. If it is in back bayou country, motor launches which are capable of fairly moderate or fast speeds can be employed. The project superintendent should be cognizant of the fact that boats which are used for transporting crews should not be overloaded, and the rules of the U.S. Coast Guard should be observed. Because of the automobile, quarters boats are now very rarely employed; when they are, they are used only on the most inaccessible and distant marine projects. Today, men live ashore and assemble at a landing to await transportation to the project.

Communications The key to effecting a saving on many marine projects is the ability of the supervisory personnel to coordinate the work with subcontractors, the home office, and other construction gangs. This communication can best be performed by mobile-type radio telephones. On a large project the initial cost of a base station and several portable units is small in comparison with the benefits realized. Units such as walkie-talkies or transceivers can be installed in the superintendent's automobile or aboard the towboat or other floating equipment. The mobile service departments of the Bell Telephone Co., Motorola Corp., and the Radio Corporation of America can give excellent advice and supply reliable equipment for this type of communication.

MARINE RISK INSURANCE

The contractor whose work requires the use of floating equipment is subjected to hazards peculiar to the nature of marine construction. There is risk of financial loss resulting from damage to or destruction of such equipment as he may own; he is also responsible for injury to the person or the property of others for which he may be held accountable at law. A qualified insurance broker can negotiate the necessary insurance coverages with an underwriter to eliminate the greater share of these possibilities of loss. Because of its special nature, marine insurance is a field of insurance unto itself. It is extremely important before attempting to move any hulls or equipment that a broker well versed and thoroughly qualified (of whom there are relatively few) be contacted and proper insurance arranged.

Marine insurance, in general, pertains to water-borne traffic on oceans, lakes, rivers, canals, etc. This type of insurance protects against loss or damage to the hull, liability for injury or damages to persons or property, or loss of cargo.

Hull Insurance This protects the insured for specified damages to his marine equipment and is frequently written with a deductible clause. It covers the hull, fittings, machinery, and fixtures of a vessel for loss against the perils of navigation, such as collision, stranding, sinking, heavy weather, fire, and explosion. This insurance also protects against loss from salvage charges. The hull policy developed for river marine hazards is a modified form of that used for ocean marine.

Most policies contain the Inchmaree clause covering latent defect and negligence of the master, officers, and hands. This latent-defect clause does not cover loss due to the negligence of the owners or managers, nor does it cover the cost of the part in which the latent defect existed. In addition, collision liability known as the "running down" clause insures against liability for damage to another vessel, its freight and cargo. Usually 20 percent of the insurance can apply to property separately stored ashore. Port-risk hull coverage is a variation of hull insurance which is designed for vessels laid up for a long time. On larger and more costly vessels, a premium savings can be made by covering part under a hull policy and the remainder under a disbursements insurance policy. The latter policy then comes into effect in the event of a total loss.

Protection and Indemnity Insurance This coverage is usually obtained by endorsement to a hull policy. Its terms are similar to the protection afforded by an automobile liability and property damage policy, broadened to include employer's liability. It includes the insured's legal liability as owner of the insured vessel for loss of life and injury to employees, including the master and the crew. It also embraces those not employees and the insured's liability for damage to property in so far as these risks are not covered by the running-down clause of the hull policy.

Workmen's Compensation Insurance This is designed and written as prescribed by the workmen's compensation statute of the state in which the insured is domiciled or operating. It normally excludes coverage for masters of vessels and crew members, although these excluded employees may be brought under a workmen's compensation policy by a voluntary compensation endorsement. The insurance company will then pay such amounts as an injured employee would have received had the act been automatically applicable.

An employee who is included in a compensation policy by this type of endorsement is not obligated to accept the compensation payment set forth in the act, but may instead elect to bring an action in Maritime Law under the Jones Act. Should this occur, the insurance company will protect and defend the employer against loss under the employer's liability section of its policy. It is important that all owners carry these policies properly endorsed, especially in view of the fact that the various phases and full effect of the applicable laws have not been rendered certain by court decisions.

REFERENCES

1. J. Huston, Dredging Fundamentals, *J. Waterways Harbors Div., Am. Soc. Civil Engrs.*, no. WW3, paper 5390, August, 1967.
2. O. Erickson, Latest Dredging Practice, ASCE *Transactions*, vol. 127, pt. iv, 1962.

Part Three

Construction Applications

Section 19

Rock Excavation

GEORGE C. BRALYÉ

Manager, Operations Engineering, Construction Division, Fluor Utah Engineers & Constructors, Inc., San Mateo, Calif. *(Planning and Development)*

P. E. SPERRY

Construction Engineer, Fluor Utah Engineers & Constructors, Inc., San Mateo, Calif. *(Blast Design)*

E. G. BROWN

Market Planning Coordinator, Explosives Division, Atlas Chemical Industries, Inc., Wilmington, Del. *(Explosives)*

DAVID G. BORG

Technical Representative, Explosives Division, Atlas Chemical Industries, Inc., Wilmington, Del. *(Explosives)*

A. Introduction

The uninterrupted escalation of the labor component of construction costs for the past 20 years has necessitated a continuing examination of construction methods. Efforts have been made to offset these increased labor costs by obtaining the maximum benefits from improved equipment, better materials, and the utilization of scientific methods in the planning of operations. Rock excavation, particularly, has been subject to such system optimization. The drilling methods and equipment of 10 years ago are today economical only in special applications; the explosives formerly popular have been replaced with cheaper, more powerful, and easier to handle blasting agents; improved ripping equipment has led to the mechanical excavation of grades of rock which were formerly drill-and-shoot applications; scientific methods of blast design now permit economic comparisons of alternate blasting geometrics and, while not yet fully developed, these methods must eventually replace trial-and-error practice in blast design.

Open-cut rock excavation requires practical solutions to the problems of working in material which is not of the contractor's choosing, which is arbitrarily located, and which must be removed within a limited time period. Quarry rock excavation involves these problems to a lesser extent, since it is performed in preselected material and usually at locations which have been chosen because of their topographic and geologic advantages.

Rock excavation in open cuts and quarries continues to be a developing science. The material in the following pages is presented with this in mind, although it is emphasized that the gap between theory and field performance has been significantly shortened. A meeting on the common ground of standard practice can be reasonably anticipated in the near future.

B. Planning and Development

GEOLOGY AND ROCK MECHANICS

The art of rock excavation necessarily requires a thorough knowledge of rock structure and its reaction when subjected to external forces. Both geologic knowledge and engineering technology must be applied to the problems of rock removal. The science of geology is of particular importance since it locates and identifies the material, predicts its behavior when subjected to the required construction processes, and recommends the basic approach to the stress and stress-relief process which is the nature of rock excavation.

Rock Types and Structures The various types of rock may be grouped into three categories:

1. Igneous, intrusive—granite, diorite; dykes
   Igneous, extrusive—volcanic deposits
2. Metamorphic—quartzite, marble, slates, schists, etc.
3. Sedimentary—limestone, sandstone, etc.

In construction open cuts, any type of rock may be encountered, whereas the types of rock are limited in quarry operations. Fault zones, folds, tilt, and attitude of the rock formation will influence the operational planning for either category of rock excavation. General stability and possible water problems must be evaluated and, as the detailed planning proceeds, consideration must also be given to joints, fracturing, bedding, dip and strike, massivity, grain, and texture. All these characteristics of rock structure affect blasting and handling methods.

TABLE 19-1 Hardness Chart (Mohs' Scale) for Various Minerals and Rocks

Mineral or rock	Hardness	Scratch test
Diamond	10.0	
Carborundum	9.5	
Sapphire	9.0	
Chrysoberyl	8.5	
Topaz	8.0	
Zircon	7.5	
Quartzite	7.0	
Chert	6.5	
Traprock	6.0	
Magnetite	5.5	
Schist	5.0	Knife
Apatite	4.5	Knife
Granite	4.0	Knife
Dolomite	3.5	Knife
Limestone	3.0	Copper coin
Galena	2.5	Copper coin
Potash	2.0	Fingernail
Gypsum	1.5	Fingernail
Talc	1.0	Fingernail

NOTE: Hardness is the resistance of a *smooth* plane surface to abrasion and is measured in terms of the above scale. Approximations may be reached by scratching with fingernail, copper coin, and knife. Scratch test is not applicable to the harder materials.

Percussion drilling is practical within the 2.5 to 7.5 range; rotary drilling from 1.0 to 5.0.

There are three principal types of rock structures: bedded structures (sedimentary), block structures (igneous), and sheet structures (igneous). All these structures are jointed, and some sedimentary and all the igneous rocks have planes of easy fracture. Although the spacing of joints is measured in feet, these planes of easy fracture may be measured in inches. A bedded structure consists of bedding planes and two sets of joints, a block structure has three sets of joints, and a sheet structure has one set of joints. In quarry operations, the spacing of the bedding planes and joints determines the largest piece of stone that can be quarried.

Rock Characteristics and Mechanics The characteristics of the rock within these structures determine its suitability for use when specified products are required, as in quarrying; and its mechanical action under stress dictates the excavation procedure. Porosity and weathering are influential in the site selection; fragmentation and gradation resulting from optimum blasting limit the usable products; specific gravity determines the amount to be produced (for example, a 20 percent decrease in specific gravity would increase the volume of breakwater stone by 50 percent); durability, hardness, and strength affect the cost, speed, and methods of drilling and blasting. Table 19-1 supplies a widely used scale (Mohs' scale) which ranks the hardness of various minerals and rocks. Average unit weights for rock materials, both in the solid state and after blasting, are listed in Table 19-2.

TABLE 19-2 Average Weight in Pounds per Unit of Volume of Different Materials Blasted

Material	Specific gravity	Solid		Broken	
		lb/cu ft	lb/cu yd	lb/cu ft	lb/cu yd
Basalt	3.01	188	5,076	122	3,299
Coal, anthracite	1.3 –1.84	98	2,546	64	2,255
Coal, bituminous	1.2 –1.5	84	2,268	55	1,474
Diabase	2.6 –3.03	176	4,752	114	3,089
Colomite	2.8 –2.9	181	4,887	117	3,177
Gneiss	2.62–2.92	179	4,833	116	3,141
Granite	2.55–2.86	169	4,563	110	2,966
Gypsum	2.3 –3.28	174	4.698	113	3,054
Halite (rock salt)	2.1 –2.56	145	3,915	94	2,545
Hematite	4.5 –5.3	306	8,262	199	5,430
Limestone	2.35–2.87	163	4,401	107	2,861
Limonite	3.6 –4.0	237	6,399	154	4,159
Magnetite	4.9 –5.2	315	8,505	205	5,528
Marble	2.08–2.85	154	4,158	100	2,703
Mica, schist	2.5 –2.9	168	4,536	109	2,948
Porphyry	2.5 –2.6	159	4,293	103	2,790
Sandstone	2.0 –2.78	149	4,023	97	2,615
Shale	2.4 –2.8	162	4,374	105	2,843
Slate	2.5 –2.8	171	4,617	111	3,001
Talc	2.56–2.8	167	4,509	108	2,931
Traprock	2.6 –3.0	174	4,698	113	3,054

Overburden and Surface Weathering A distinction is made in the nature of overburden and weathering between sedimentary and igneous rock formations. In the case of sedimentary rock there is a marked transition from soil overburden to rock, and the amount of overburden is usually readily determinable. The surface topography permits easier, less costly removal since it either parallels the bedding planes or is more or less inclined where the region has been subject to earth movements. On the other hand, the topography of an igneous formation tends to be much more irregular. The distinction between overburden and weathered rock is then less clear, the overburden is uncertain in amount and location, and its removal can require special excavation in pockets and seams. These characteristics must be taken into consideration when predicting the yield of useful product from a potential quarry site. An improper evaluation of the overburden and weathering of a given formation can, for example, result in an uneconomical low-bench operation since the top material may have to be removed and wasted.

EVALUATION AND PLANNING

Planning a rock excavation project has two broad limitations: one is imposed by specifications and local laws, while the other concerns the nature and location of the rock.

Specifications and Local Laws Laws protect populated areas or existing installations from vibration disturbances, excessive noise levels, and dust nuisances. Specifications control the grading of the blasted product (fragmentation), the height and slope of the excavated cuts (number of benches), the quality tolerances of the product, and the duration of the operation. Some specifications elaborate on the local laws as they pertain to vibration and noise levels and go so far as to stipulate formulas for computing the weight of explosive for each delay in a shot. In some cases, they also specify the actual blasthole diameter to be drilled. Quarry locations can be chosen to minimize the effects of these controls, but construction open cuts may be required anywhere. The excavation of open cuts, therefore, can be subject to any or all of these restrictions.

Geologic and Engineering Evaluation A complete geologic and engineering evaluation is essential for the proper planning of a rock excavation task. Geologic exploration, testing, and interpretation serve to identify, catalog, and evaluate rock structure, characteristics, mechanics, weathering, and overburden features at a potential excavation site. This information is vital to the design engineer and to the party who must perform the work of removing, processing, and utilizing the rock.

The overall terrain must be physically examined and a surface map should be prepared which indicates the principal geologic features. Aerial photographs are a valuable guide in the interpretation and evaluation of the observed physical features. The body of the formation may be examined by core drilling to obtain samples or by the excavation of test pits and trenches to reveal surface conditions and by seismic tests to determine soundness. These explorations yield geologic samples for testing and evaluation, suggest methods of efficient use of the terrain for development purposes, indicate favorable access routes, and confirm the existence of other natural features such as groundwater, site clearing, and weather conditions which would affect excavation operations.

The samples taken are subjected to standard tests, such as ASTM or British Standard, to determine tensile, compressive, rupture, and shear characteristics. The design engineer then uses these results in his design. The contractor is advised concerning those characteristics of the rock which will affect his plan of operation; his principal concerns are the structure of the rock and its hardness.

Site and Equipment Factors The nature of the rock and the topography of the site are described in the findings of the geologic and engineering evaluation. Each site is unique, and it is unwise to assume that previously successful operating methods will apply in apparently similar situations. The principal distinction is whether the rock is sound or fissured. This characteristic, together with the topography, influences the selection of drilling equipment and the choice between large or small drills.

The hardness of the rock determines the type of drill and explosives, the required production dictates the drill size and number, and the structure of the rock is the basis for the design of the blasting round. Topography limits the overall approach to the excavation, including the location and the grades of access and haul roads, the face development, and the drainage requirements. System optimization requires that the selection of equipment be based upon an evaluation of all the previously mentioned features. In many cases, however, a contractor may already have a substantial investment in rock-handling equipment. In these circumstances it is frequently to his economic advantage to use less efficient units which he already owns rather than to buy new, ideally suited units. Compromises are thus made with certain of the geologic and engineering recommendations, with the result that construction rock excavations are rarely made using ideal methods and equipment.

When new equipment is being selected for a rock job, the emphasis should be placed upon the loading and hauling units and the primary crusher, if one is to be used. The investment for these items will greatly exceed that for drilling equipment. The two operations, drill-shoot and load-haul, meet on the common ground of fragmentation. While it may be possible to drill and shoot rock quite economically when producing a certain maximum size, it may not be economical to purchase and operate loading and hauling equipment which can handle that size. An accommodation is made when the drilling and shooting method is revised, usually at a higher cost, to produce the fragmentation required.

Operations Planning Usually, the first step in the development of the site is the removal of the overburden and unsuitable (weathered) materials located over the rock. In quarries, after the initial opening up, this is usually a continuous, concurrent operation involving the excavation of low benches above and ahead of the rock excavation. For construction excavations, this operation may be completed for the entire excavation area before rock removal is started. In either case, scrapers, dozers, and dozer-rippers can be efficiently utilized. Overburden removal for quarries is normally

Fig. 19-1 Hard ripping using single tooth. (*Caterpillar Tractor Co.*)

minimal, since the ratio of stripping to usable rock is taken into consideration when selecting the site. In construction cuts the ratio of weathered material to firm rock can be quite high, and rippers are increasingly used to excavate this incompetent rock.

With the advent of the heavy-duty integral hydraulic ripper and larger, more powerful tractors, the range of rippable materials has been extended considerably into what were once drill-and-shoot operations. Figure 19-1 shows the use of a single tooth in hard ripping; a push block could be used if additional power were required. Medium-hard ripping, using two teeth, is illustrated in Fig. 19-2. For still easier ripping, a third tooth could be added. The excavated rock can subsequently be loaded out and hauled to disposal with scrapers. This ripper-scraper operation has obvious economic advantages over a drill-shoot-shovel-truck spread. Where the required fragmentation permits, the use of rippers is now an accepted excavation means to reduce the volume of rock which must be drilled and shot.

Rippability of subsurface material can now be predicted with remarkable accuracy by the refraction seismograph, although final answers must still be obtained by actually trying to rip the formation. Figure 19-3 illustrates the principle of the refraction seismograph; this device measures the velocity of seismic waves when traveling through different kinds of subsurface materials.[1,*] The speed of the wave in a hard,

* Superior numbers refer to the list of references at the end of this section.

tight rock is fast, up to 20,000 fps, while in loose soil it is as slow as 1,000 fps. Thus, by measuring the speed of the seismic wave through various layers of material, the degree of consolidation can be determined. From this information it is possible to plan the excavation method and equipment.

Rock, as previously stated, can be classed into three main categories; igneous, sedimentary, and metamorphic; rippability generally follows this grouping. Figure 19-4 correlates the observed performance of a heavy-duty ripper with the rock type and the measured seismic velocity.

Fig. 19-2 Medium-hard ripping using two teeth. (*Caterpillar Tractor Co.*)

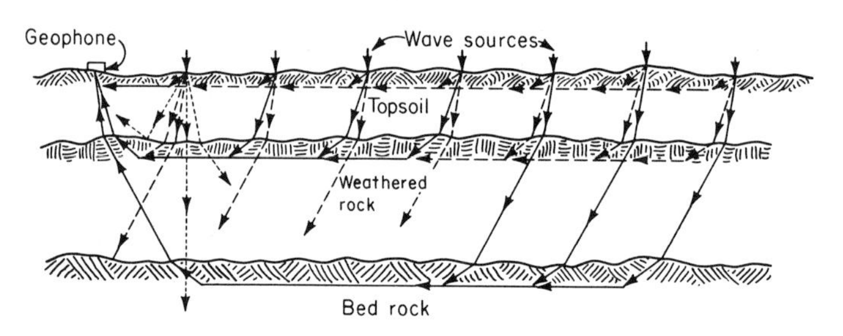

Fig. 19-3 Operation of the refraction seismograph. (*Caterpillar Tractor Co.*)

For the particular equipment combination to which it applies, this performance chart indicates that many materials having seismic velocities up to 8,500 fps can be ripped. This appears to be the current economic limit, since costs generally rise sharply when ripping production falls below 400 bcy per hour. Factors other than seismic velocity can affect rippability; for example, extremely thick laminations (18 in. or more) affect penetration, lift-out, and material breakage.

After the overburden and rippable materials have been removed, the drilling and shooting requirements for further development of the excavation are more clearly defined. A limited amount of experimental drilling and shooting, preferably opening up a face at least 20 ft high, should confirm and supplement the geologic report and enable the final planning of the excavation procedure. Contour maps and layouts are used to determine the optimum method of working the production face or faces.

Factors to be considered at this stage should include:

1. *Length of face (quarries).* The shorter the face, the easier it may be to obtain the required volume of rock from a location which can take the best advantage of local topographic features.

2. *Fragmentation required.* This is dictated by specifications when working quarries. In construction cuts, it should be the optimum for the loading and hauling equipment which is to be used.

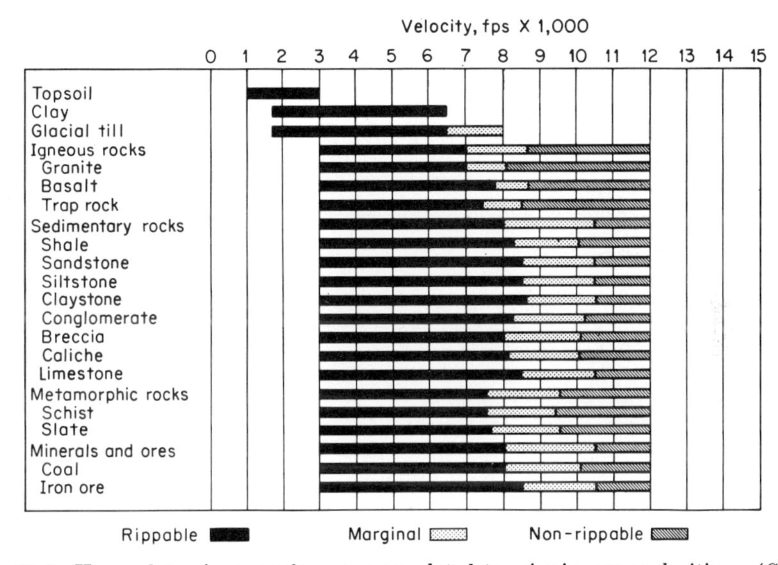

Fig. 19-4 Heavy-duty ripper performance as related to seismic wave velocities. (*Caterpillar Tractor Co., Handbook of Ripping.*)

3. *Drill size.* This selection is based on the rock structure and hardness, the fragmentation required, and the type and amount of explosive to be used.

4. *Number of benches.* The selection of equipment is the primary factor in determining the number of benches to be worked. An economic comparison should be made between using a large drill to work one bench versus using smaller drills on two or more benches, with additional haul roads. Such a comparison could consist of evaluating the following situation:

Drill size, in.	Air consumption, cfm	Number of holes	Powder required, lb	Number of benches	Number of drills	Number of haul roads
Small 3½.....	1,800	2,000	60,000	2	3	2
Large 8.......	900	150	70,000	1	1	1

5. *Production required.* The overall selection of equipment dictates the excavation method which, in return, corresponds to a certain production. This production must be examined to verify that the work will be performed within the permissible time and at reasonable unit costs, including adequate allowances for equipment depreciation. High-volume, long-term projects, such as commercial quarries, can economically

justify the purchase of bigger, better, and more efficient equipment; whereas short-term, low-volume construction excavations cannot.

6. *Site condition upon termination.* The selected method of operation must also provide for the ultimate treatment of the excavated site. Emphasis is now being placed on increased amounts of site restoration for purposes of safety and aesthetics. The excavation method should facilitate the subsequent performance of such work as backfilling haul roads, grading disposal areas, trimming slopes, and filling to avoid the accumulation of water.

With the overall method of operation once selected, the details of the rock removal can be planned. These will include drilling equipment, explosives, and blast design.

DRILLING EQUIPMENT

There is a definite need for various kinds of drilling equipment and drilling methods to economically overcome the problems encountered when the endless combinations of rock types, rock structures, and formations must be drilled. The successful operator investigates each possible drilling method in the continuing effort to improve drilling production to lower costs.

Drilling equipment is grouped into three broad categories: (1) percussion, (2) rotary, and (3) abrasion drills. The percussion and rotary drills are the production tools of the industry. Abrasion drills such as diamond-and shot-core drills are special application tools and are limited in use. The thermal-jet piercing drill, another special application tool, is finding increased usage as its technique and equipment design improve.

Percussion Drills The percussion drill is the most versatile of all rock drills and can economically drill rock over a wide range of hardness or abrasiveness. Percussion drills employ a reciprocating air-operated hammer (piston) which strikes a rotating bit or a string of rotating drill rods with a bit on its bottom. An air or water system is used to remove stone cuttings from the drill hole. The rotating action of the rod and bit is supplied by one of several devices: a rifle bar, an integral rotation motor, or an external rotation motor. (See Fig. 19-5.)

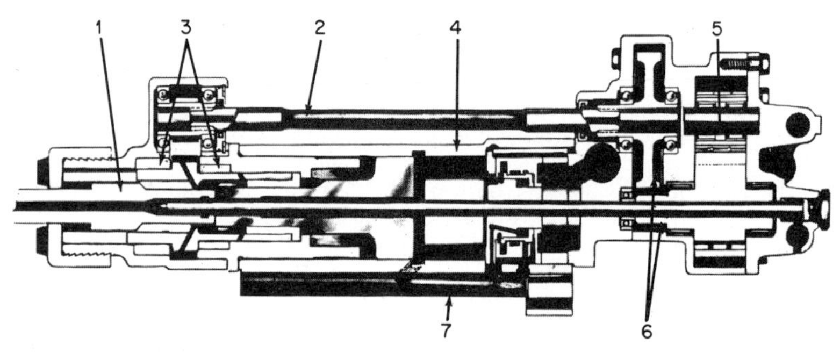

Fig. 19-5 Independent rotation rock drill. Provides a combination of rotation, speed, impact force, feed pressure, and hole blowing for efficient penetration in any type of rock. Piston hammer and rotation motor operate as independent units, permitting control of rotation and percussion to meet operating conditions: light impact and fast rotation for soft ground and heavier percussion and slower rotation for hard rock drilling. *(Gardner Denver Company.)*

The rock is broken at the bottom of the hole by the impact of repeated blows from the compressed-air-operated hammer, and the stone cuttings are blown out of the hole by air applied through the hollow drill rod. Bits with four sharpened rectangular carbide inserts are normally used and, for maximum production, are resharpened frequently on the job. Button bits with multiple cylindrical carbide inserts have also

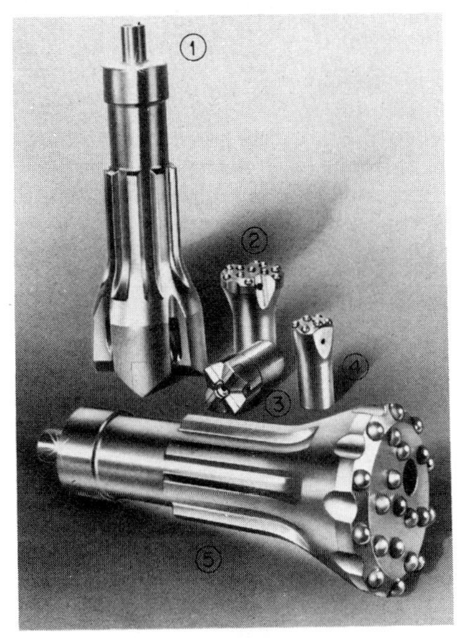

(1) 6-in. down-the-hole drill bit
(2) 3½-in. button bit for crawler drill
(3) 2½-in. threaded bit for crawler or jackhamme
(4) 1⅞-in. taper socket bit
(5) 8-in. down-the-hole button bit

Fig. 19-6 Drill bits, tungsten carbide inserts. (*Ingersoll Rand Company.*)

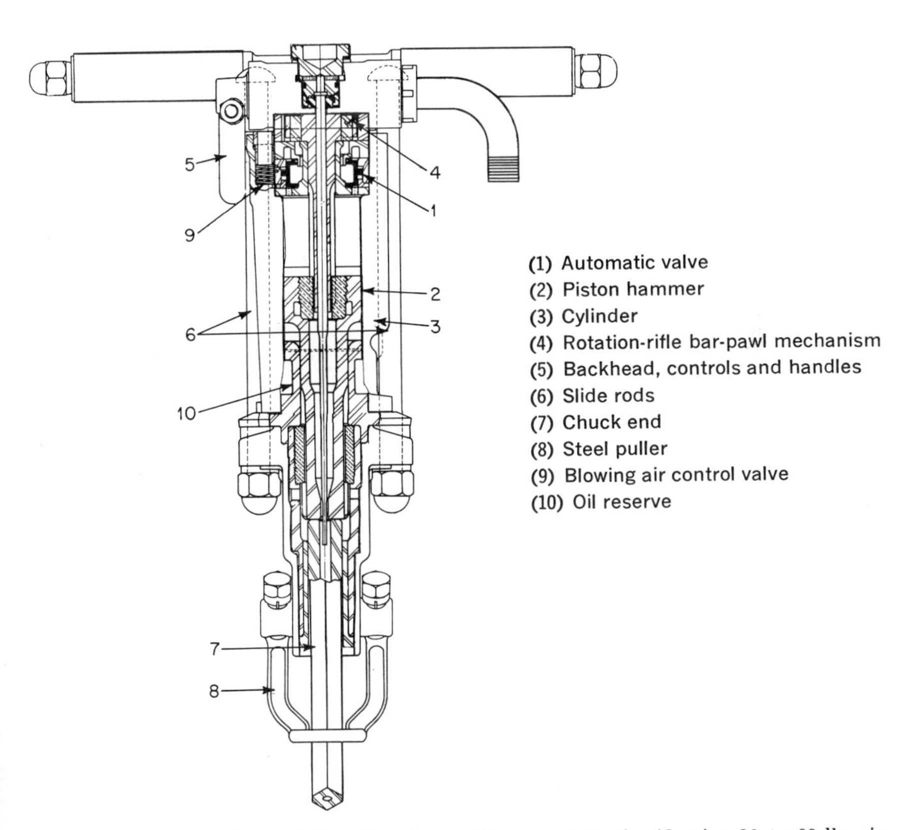

Fig. 19-7 Typical sinker drill. Bore 2⅜ to 2⅝ in.; weight classification 30 to 60 lb; air consumption 70 to 130 cfm; hole size 1¼ to 2 in. (*Gardner Denver Company.*)

been introduced. Some rock button bits are self-sharpening and have extremely long life. Figure 19-6 illustrates several types of rock bits.

Surface percussion drills are the most widely used type of percussion drills. The drill operates above the surface of the ground, and the drill rod string is placed between the drill and the bit. Surface drills have the advantage of being light in weight and highly maneuverable. Sinkers (jackhammers), stopers, and feed leg drills are examples of smaller surface drills that can be used in close quarters by one man with no drill mounting required. Percussion drifters do require mountings, but these are small in comparison with those for rotary drills, which must support as much as 90,000 lb of thrust. Surface percussion drills operate at any angle through a complete circle and thus can be used for horizontal, up-hole, and angle drilling to adapt to adverse rock formations.

Fig. 19-8 Medium-duty crawler drill. (*Ingersoll Rand Company.*)

Hand-held drills (jackhammers) used in open-cut excavation can drill small holes up to 20 ft deep; however, they are seldom used beyond a 10-ft depth. Jackhammers are not used on production work except in special cases, but they are restricted to drilling oversize muck and for spot work in reducing high grade. Jackhammers are classed by weight; the heaviest, 65 lb, is the only one used for hole drilling. Bits smaller than 2 in. and light hexagonal drill steel in 2-ft multiples are then employed. Jackhammers consume up to 150 cfm of compressed air at 85 psi and average between 10 and 15 ft of small-diameter hole per hour. Figure 19-7 illustrates the major components of a jackhammer.

Heavy percussion drills which are mounted on track-type carriers are commonly referred to as airtracks; drills of this type are illustrated in Figs. 19-8 and 19-9. These drills are used for all construction work except for the largest excavations, and they are also used for some quarry drilling. Airtracks are capable of drilling 2- to 5-in. holes up to 100 ft deep, but they are normally used for 2½- to 3½-in. holes up to 30 ft deep. The airtrack carriers are highly maneuverable and can tow their air compressors. Air motors are used for drilling and for advancing or retracting the drill rod in the feed shell; hydraulically operated booms are used to support and position the drill. Airtracks can be classified by drill size. The smaller, more common drills have 4½- to 5-in.-diameter hammers and require 600 cfm of air at 90 psi (at sea level); the large drills have 5- to 6-in. hammers and operate on 1,200 cfm at 90 psi. The smaller machines weigh 4 to 5 tons, while the larger ones weigh as much as 10 tons.

The feeds of the smaller airtracks are long enough to use 12-ft steels, while the larger

Fig. 19-9 Heavy-duty crawler drill. *(Gardner Denver Company.)*

machines use 12- to 20-ft lengths. A 12-ft steel is the most a man can handle, so mechanical changers are used with 20-ft steels. Holes deeper than the original steel length are drilled by coupling on additional pieces of steel. These couplings dictate the minimum bit size, since the drilled hole must be ¼ to ½ in. larger than the coupling diameter.

Drill steel is available in a wide range of combinations of shank, rod section, threading, lengths, and bit ends, with steel sizes from ⅞-in. diameter to 4 in. Significant progress has been made in sectional drill steel, and advanced technology in this field has permitted the development of the larger, more efficient drilling units being used today. Figures 19-10 through 19-12 illustrate typical drill steels.

Variations in drill steel are described as follows:

For drifters (airtracks)

Shank:
- Round lugged, cup seal
- Round splined, cup seal

Rod sections:
- 1¼-in. hex or 2½-in. tube rod
- 1½-in. hex or 3-in. tube rod
- 1¼- or 1½-in. continuously threaded (rope thread) or 3-in. tube rod
- 1¾-in. hex or 4-in. tube rod
- 1⅞-in. hex

Bit end:
- Bottom drive threaded

For sinkers (jackhammers)

Shank:
- Hexagon collared
- Quarter hexagon collared

Rod sections:
- ⅞-in. hex
- 1-in. hex

Bit end:
- Bottom drive threaded
- Shoulder drive threaded
- Tapered

Airtrack production is from 30 to over 60 fph dependent upon the hole size and depth, air pressure, rock type, job characteristics, and frequency of bit sharpenings. The following ranges of useful life may be expected from the airtrack accessories:

Bits	400 to 4,000 ft
Steel	800 to 3,000 ft
Couplings	400 to 1,200 ft
Striking bars	1,000 to 3,000 ft

(1) Cup seal shank
(2) Reverse buttress thread
(3) Coupling
(4) Sectional rod, carborized and shot peened
(5) Bit

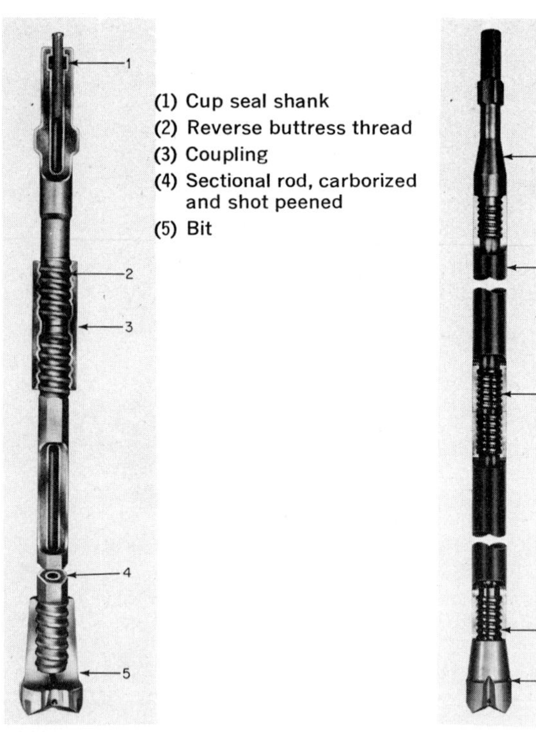

Fig. 19-10 Sectional drill steel. *(Gardner Denver Company.)*

(1) Shank
(2) Tubular drill rod
(3) Internal coupling
(4) Bit adapter
(5) Bit

Fig. 19-11 Tube-rod sectional drill steel. *(Gardner Denver Company.)*

In operation, the striking bar transmits the blow from the hammer to the drill steel, and much of the energy of the airtrack is dissipated in the drill steel and couplings. As holes are drilled deeper, the air pressure used to remove drill cuttings drops, with a consequent slowdown in the operation. These factors limit the economical hole depth for the airtrack drill and have led to the development of the down-the-hole drill.

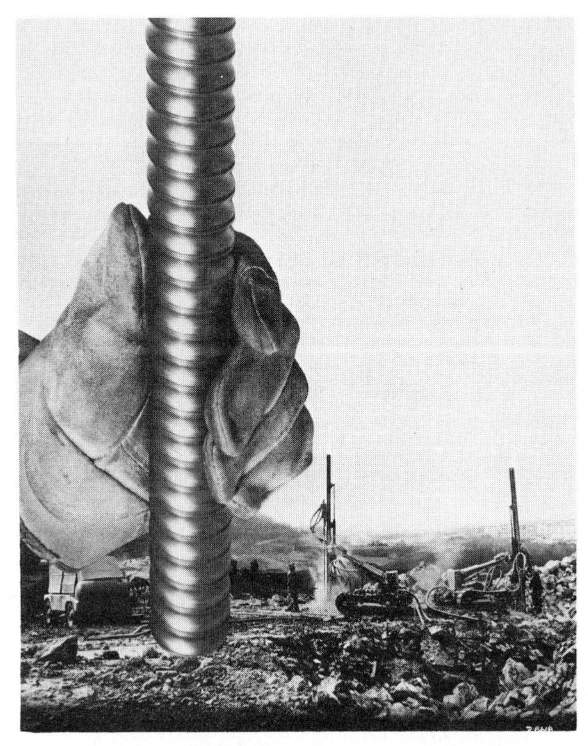

Fig. 19-12 Continuously threaded drill steel. (*Ingersoll Rand Company.*)

The nine United States manufacturers of self-propelled percussion drills produce machines having specifications that fit within the ranges listed in Table 19-3.

TABLE 19-3 Specification Ranges for Self-propelled Percussive Drills

Weight	7,100–8,874 lb	Boom lift	
Width	6 ft 3 in.–7 ft	Above horizontal	40°–65°
Minimum height	4 ft–4 ft 8¾ in.	Below horizontal	0°–42°
Pad width	10–11 in.	Max. horizontal drilling height	8–11 ft
Pad contact area	1,010–1,510 sq in.	Propelling motor, each track	6–11.5 hp
Ground clearance	9½–12 in.	Drifter bore	4½–5 in.
Traveling speed	2.5–3 mph	Average penetration	
Boom swing	8 ft 7 in.–17 ft	(Barre granite)	16–30 in./min

Below-surface drills or down-the-hole drills operate at the bottom of a rotating drill string directly above the bit. The outside diameter of the drill is smaller than the hole and thus enables the drill to follow the bit down the hole. Since the drill string is above the drill, the footage life of the drill rods is very good and the energy loss between drill and bit is held at a constant minimum.

In dry drilling, cleaning is less of a problem with the down-the-hole drill since the exhaust air from the drill provides some of the hole-cleaning air requirements. Drill noise is less of a problem, too, because after the drill enters the hole the noise level drops to a low figure even at close range.

The principal disadvantage of the down-the-hole drill has been the risk of operating a percussion device below the surface. If there is a rock fall or if mud collars form above the drill, withdrawal becomes a problem and the drill is sometimes lost in the hole. Recent lubrication improvements and water injection systems have practically eliminated the mud collar problem; however, some of the risk remains. Down-the-hole drills do not operate satisfactorily if subsurface water in great volume or of high head is encountered. Water pressure exceeding 100 psi could prevent effective drilling.

Fig. 19-13 Down-the-hole drill. (*Ingersoll Rand Company.*)

Down-the-hole drills are available as either low- or high-pressure drills (100 to 250 psi). The high-pressure-drill operation may be limited by the effectiveness of the available compressors. Table 19-4 supplies a guide to drilling speeds for down-the-hole drills. The tabulated drilling speeds include allowances for setup and steel changes when using 25-ft steel to drill holes 40 to 80 ft deep for DHD-275 and DHD-1060A, and 50 to 125 ft deep for DHD-1070 and DHD-1090A.

Churn drills or well drills are of the reciprocating percussion type. They are used for large blastholes from 3 to 8 in. or more in diameter. The bit is attached to a heavy bar which is lifted by a line hoist and dropped into the hole, producing chipping and crushing by the weight of the blow.

Drilling progress is dependent largely on the weight of the drill tools, and for this reason a hole of less than 5 in. in diameter is not usually practical. A 6-in. hole can be driven through limestone at rates of 3 to 6 fph and through traprock or granite at rates of 1 to 1.7 fph.

Churn drills have given way to rotaries and heavy airtracks in the construction industry. They are no longer considered modern production tools.

TABLE 19-4 Drilling Speeds for Selected Down-the-hole Drills*

Drill	Air pressure, psi	Blows per minute	Free air, cfm	Carset bit gauge, in.	Hole-cleaning velocity, fpm	Drilling speed, fph
DHD-275	100	875	220	5	4,530	12–15
DHD-1060A	100	1,025	420	6	3,760	20–24
DHD-1070	100	1,026	575	7	8,350	10–24
DHD-1090A	100	1,025	680	8	3,900	20–24

* Material: granite (Mohs' scale 4.0).
SOURCE: Ingersoll Rand Company.

Rotary Drills Rotary drills provide their drilling action by the application of heavy down pressure on a roller-cone bit of the type illustrated in Fig. 19-14. Separate variable-speed motors, hydraulic and electric, create the pulldown pressure and the drill rotation. A proper combination of drill rotation and drill pressure is used to accommodate the material being drilled. Rotary drills may be either diesel or electrically powered. Figure 19-15 illustrates the use of a rotary rig in an Arizona copper mine to drill 9-in. holes to a depth of 50 ft. Figure 19-16 shows a rotary drill which is fitted with an auxiliary boom for angle drilling.

The derrick frame of a rotary drill provides storage space for four or five drill pipes, each of which can vary in length from 20 ft in the smaller drills to 35 ft in the larger drills. It also serves as the headframe and guide for the hydraulic pulldown motor and the hoisting drum. The drill pipes can be handled mechanically, with the result that adding or removing a section can be done in a few minutes. Rotary drills are usually mounted on crawler-type carriers and weigh from 25 to 85 tons. Pulldown pressures vary from 30,000 lb for a 6- to 7-in. drill hole to 90,000 lb for a 15-in. hole machine.

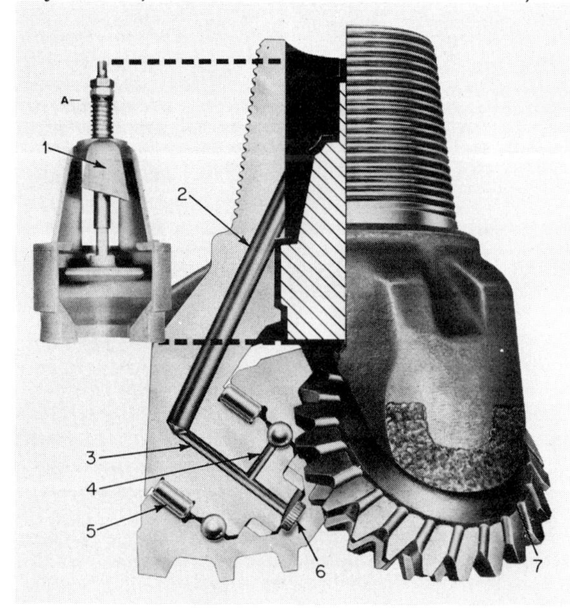

(1) Arrestor value, spring loaded. Prevents slurry or cuttings from entering the bit body and plugging air passages
(2) Air inlet to bearings
(3) Drilled port
(4) Two ports to ball race on nonload-bearing side of journal
(5) Air pressure regulator
(6) Air slot for air to friction bearings
(7) Shirt tail, tungsten carbide band to prevent exposure of bearings by abrasive wear.

Fig. 19-14 Rotary blasthole drill bit. (*G. W. Murphy Industries, Reed Drilling Tools Division.*)

The rotary drilling method was limited for many years to soft formations, such as clays and soft and sandy shales, because of the limitations of available drilling bits. With the introduction of roller-cone bits and drag bits with tungsten carbide inserts, much harder formations can be drilled by rotary drills. This has been demonstrated by the economical operation of rotary drills in the Minnesota taconite mining industry.[4]

Air compressors remove cuttings (bailings) from the hole as drilling progresses, and they can also be used if the machine is used for percussion down-the-hole drilling. Compressor capacity varies from 600 cfm on the smaller machines to 2,500 cfm on the largest drill. The use of air compressors for bailing out the cuttings from suitable formations has revolutionized rotary drilling. This practice eliminates the expense of delivering water to the drill hole and has also increased rotary bit life and penetration rates. To provide efficient rotary drilling performance, bailing velocity must be adequate to remove the cuttings. The return velocity of the hole-cleaning air varies according to drilling conditions: i.e., weight of cuttings, hole depth, amount of moisture encountered, etc. The consensus of most authorities is that a return air velocity of

5,000 fpm represents the minimum acceptable value. The effectiveness of this bailing velocity is subject to variation, dependent upon the weight of the cuttings to be removed.

Bailing velocity in feet per minute can be determined by the formula

$$V_B = \text{cfm} \times 144/\text{area of annulus (in.}^2)$$

The area of the annulus is the hole area minus the OD drill pipe area.

Failure to provide enough air to clean the hole results in regrinding of the cuttings under the bit. This slows down penetration rate and increases bit cost. Figure 19-17 is a nomograph relating bailing velocity, hole diameter, cfm air requirement, and drill pipe size. As the result of improved drill design, better quality of materials, and advances in bit bearing capacity and tooth design, modern rotary drills are capable of drilling a 10-in. hole in hard limestone or granite at a penetration rate of 20 to 30 fph (at 100 percent availability). Rotary drills can be adapted to drill at an angle, a developing procedure in certain rock formations which reduces backbreak, toe, and secondary drilling. Some rotary drills can also be used as down-the-hole drills.

Fig. 19-15 BE model 45-R rotary drill. (*Bucyrus-Erie Co.*)

The required bit pressure is the most important factor to be considered in the selection of rotary drill equipment. This bit loading varies widely for different types of rock. In the harder rocks it also varies with the bit diameter (the larger the bit, the greater its bearing capacity and the higher the unit loadings permissible). It is for this reason that rotary drilling in hard and extremely hard rock with a bit smaller than 9 in. becomes uneconomical. When a 6- to 7½-in. hole in hard rock is to be drilled, the percussion down-the-hole drill comes into its own.

Rocks have already been classified on the basis of their origin as sedimentary, igneous, and metamorphic. In the application of blasthole drills it is also expedient to classify rock as soft, medium, hard, and extremely hard. Table 19-5 shows this type of rock classification.[5] A different type of rotary bit is recommended for each of these rock hardness classifications, with the major variations being in the cutter tooth design. In the case of extremely hard formations, a carbide button insert bit is used. This type of bit produces a crushing action for rock fracture rather than the chipping and abrading which occur with the steel tooth bits. Four types of rock bits are illustrated in Fig. 19-18.

Table 19-6 lists typical performance data for rotary and down-the-hole drills when operating in each rock classification.[5] These data illustrate the wide differences in performance and bit life. Generally, it is economical to use rotary drills with 6- to 12-in. bits in soft and medium rock, with 7- to 12-in. bits in hard rock, and with 10- to 15-in. bits in extremely hard rock. Down-the-hole drills are not economical in soft or medium rock but become effective when used in hard rock with bit sizes ranging from 6 to 12 in.

Abrasive Drills *Diamond-core drills* are used principally for exploratory drilling when rock samples are required for physical examination and testing. The core barrels of these drills have standard diameters, EX, AX, and NX ($1\frac{7}{16}$ in., $1\frac{27}{32}$ in., and $2\frac{15}{16}$ in.). Larger diameters have proven to be uneconomical because of the abrasive (diamond) cost, and shot-core drills are used to obtain large-diameter cores. A $1\frac{1}{2}$-in. drill, which can take a 3-in. core, requires seven to eight diamonds of about

Fig. 19-16 Rotary drill with auxiliary boom for angle drilling. (*Gardner Denver Company.*)

TABLE 19-5 Classification of Rocks by Degree of Hardness

Soft	Medium	Hard	Extremely hard
Shale Clay Soft limestone	Limestone Sandstone Sandy shales Porphyries Iron ore	Siliceous Limestone Dolomite Granite Iron ore Quartzite	Iron formation Quartzite Hard igneous rocks

two carats each. Bortz, which is sharper and harder than black diamond but is less durable and tough, can also be used as the abrasive. It costs about 10 percent as much as a diamond bit; however, most drillers prefer the diamond drill since its longer bit life more than offsets its higher cost. The lower investment for a bortz bit becomes an advantage if a drill bit is lost.

Shot-core drills (calyx drills) utilize chilled shot as the abrasive material. The shot breaks up under the bit, and the resulting sharp, angular particles are the cutting

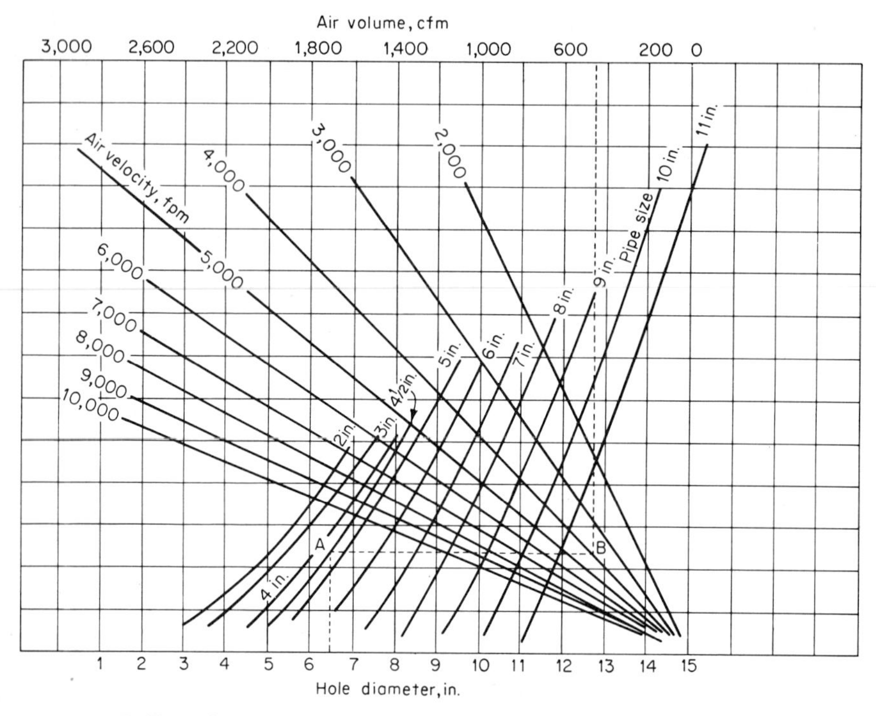

Fig. 19-17 Bailing velocity chart. *(G. W. Murphy Industries, Reed Drilling Tools Division.)*

TABLE 19-6 Drilling Performance

Rock classification	Bit loading, lb/in. bit diam	Penetration rate, ft/hr (100% availability)	Bit life, ft of hole, 9⅞-in.-diam bit
Rotary drilling			
Soft	2,000–3,000	70–150	5,000–20,000
Medium	3,000–4,500	30–70	2,000–5,000
Hard	4,000–5,500	20–30	500–2,000
Extremely hard	5,000–7,500	10–30	400–1,500
Down-the-hole drilling			
Hard	500–1,000*	15–30	600–1,200
Extremely hard	500–1,000	15–30	300–500

* For percussion drilling, down pressure is only required to hold the bit on the bottom since the air hammer provides the drilling action. In deeper holes it is sometimes necessary to hold back on the drill rod to control the weight on the bit.

medium for grinding the rock. Drilling speeds are slower than for diamond drilling, but the bit and abrasive costs are substantially less.

Shot drills are used to obtain cores in diameters ranging from $3\frac{1}{2}$ to 72 in. Cores 6 ft in diameter can be driven and recovered up to 100-ft depths; 4 ft to 200 ft; 1 ft to 600 ft; and 5- or 6-in. cores can be drilled to 2,500-ft depth. Calyx cores are principally

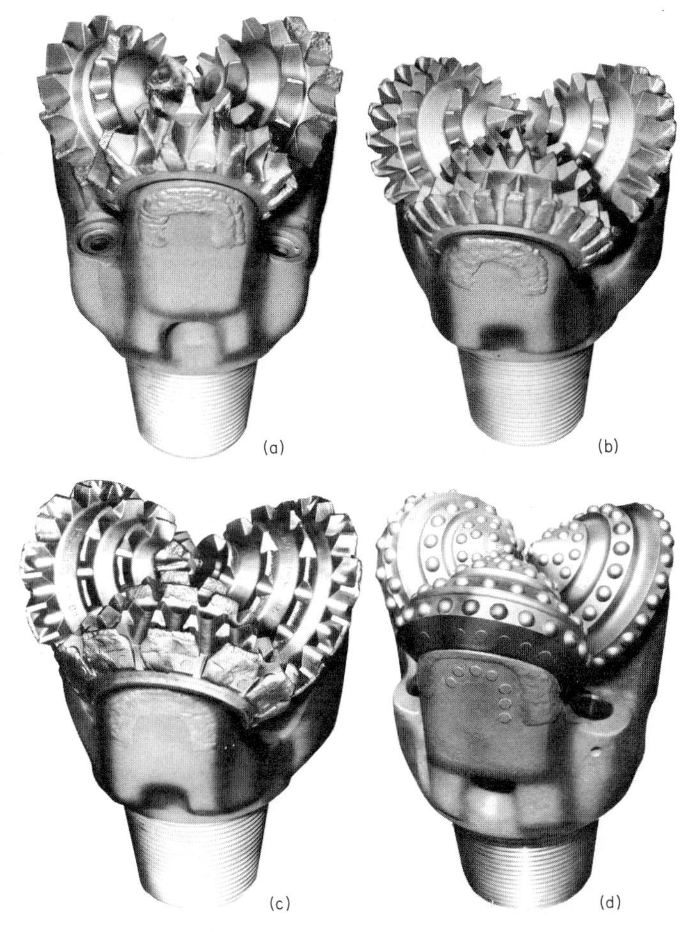

Fig. 19-18 Rotary bits. (*a*) Type A, for soft formations, bit loading 1 to 3,000 lb/in. of bit diameter, rotary speeds 120 to 85 rpm. (*b*) Type B, for medium formations; loadings 2 to 5,000 lb/in. speeds 100 to 60 rpm. (*c*) Type C, for hard formations, 4 to 7,000 lb/in. speeds 80 to 40 rpm. (*d*) Type D, for extremely hard formations, loadings 3 to 9,000 lb/in. speeds 80 to 40 rpm, tungsten carbide button inserts. (*G. W. Murphy Industries, Reed Drilling Tools Division.*)

taken in dam foundations, where large-diameter holes make it possible to visually inspect the rock formation by lowering the observer into the hole.

Special Application Drills *Thermal piercing drills* (jet drills) have been used effectively to drill blastholes in hard rock such as granite, quartzite, syenite, sandstone, and magnetic taconite as well as other hard, spallable formations.

The thermal process disintegrates or spalls the rock by utilizing a rocket-type burner which consumes a mixture of oxygen and fuel oil. The burner is a single-orifice high-velocity type consisting of an atomizer section, a combustion section, and a face cap. The kerosene and oxygen are atomized and burned under pressure in the combustion chamber. The hot burning gases are ejected at supersonic speed (6,000 fps) from the orifice in the face cap. Water is emitted on the periphery at the blowpipe in a series of jets at a point just above the atomizer section.

The jet piercing action is continuous; the force of the burning gases, plus steam formed from the cooling water which flows to the burner, carries the spalled particles out of the hole. Jet piercing has the potential to drill blastholes at a much faster rate—under certain conditions ten times faster—than rotary or percussion drilling. Increased acceptance of this method of drilling can be expected as technique and equipment improvements enable it to enter the production field of blasthole drilling.

Drills of the future, employing novel or advanced rock-cutting techniques, are being studied today. Several of these have the potential of delivering destructive energy more efficiently to the rock. Techniques are being advanced in the mechanical, thermal, chemical, and fusion fields.[6]

Processes which presently have significant research priority include ultrasonic, fluid erosion, spark, high-velocity flame, softeners, plasma, dissolvers, and laser. Others less potentially efficient include the electron beam, cryogenic, electron disintegration, pellet, and water cannons.

BLAST DESIGN

There is an optimum expenditure for the drilling and blasting processes which are required to remove rock and to yield a desired product. This product must be such that all subsequent handling can be performed at an acceptable cost. For example, minimum drilling and blasting cost will yield poor fragmentation, with the result that shovel-truck production is greatly reduced and shovel maintenance costs are severely increased.

Criteria for Design *Fragmentation* must be compatible with the type and kind of loading equipment available. Larger shovels can handle coarse fragmentation; shovels can dig in a tighter muckpile than front-end loaders; belt loaders require medium-sized material; scrapers require very good fragmentation as well as proper disposition of the muck. Rock to be used for riprap is normally shot with low powder and drilling factors to produce the large size required; rock to be used in embankments may have maximum specified sizes; material to be crushed should be small enough to minimize crusher delays and maintenance (slightly improved fragmentation can increase crusher production greatly). Experience indicates that it is usually cheaper to shoot all the rock to less than maximum size rather than risk the added costs of secondary shooting or sorting of material on the fill, which is necessary when oversize material is produced.

Bench height is principally determined by the drill equipment capabilities. In general, benches should be as high as practical in order to decrease the number required and thereby minimize equipment movement and bench cleanup for drilling. The height, however, should not exceed the economic drilling depth nor should it result in a hazardous face or muckpile. With airtrack drills, the optimum bench height is generally 20 to 30 ft; with the larger drills, 30 to 60 ft. The hole depth should be a multiple of the drill-steel length; because adding a third steel is uneconomical, optimum drilling costs will result if the hole is only two drill rods deep. The required amount of subdrilling must be added to the hole depth when designing the bench height; i.e., two 12-ft drill rods and 2 ft of subdrilling will result in a face of 22 ft, not 24 ft. The larger drills are often equipped with high masts so that a bench up to 50 ft high can be drilled without adding steel.

Much of the energy of the above-the-hole percussion drill is absorbed by the drill steel and couplings, and its penetration rate may decrease as much as 20 to 30 percent for every drill steel added. If the rock is so abrasive that the bit must be resharpened before the hole is bottomed out, consideration should be given to reducing the bench height and changing bits after each hole is drilled. Drilling high benches in some rock formations may cause the bit to wander, thus producing an inaccurate drill pattern with poor fragmentation and grade control.

Bench heights are also dictated by the loading equipment to be employed. Large shovels can economically work higher rock cuts, but there is no rule of thumb to determine the best height. This is probably because different types of rock have varying propensities to "run" to the shovel. A front-end loader cannot safely work as high a face as a shovel; also, the muckpile for a loader operation may be spread out and lowered by using more powder in the shot.

Geology and topography often determine bench height. A rock formation that must be shot against the dip may have loose material on the bench face and be unsafe to work at more than a minimum height. A formation which is shot with the dip may also present this problem, but the conditions can be alleviated by means of angle drilling. It is usually undesirable to subdivide high benches because of difficult access for loading and hauling equipment, but it becomes necessary when production must be increased by working more than one bench. Generally, a 30-ft-wide bench would be considered the minimum, with widths of 50 to 60 ft being preferable. A balance between the bench widths and heights must be maintained so that the entire cut slope is not too steep. The principal reason for the popularity of low benches is the elimination of the hazards of working under a high, hard-to-scale rock face.

Hole size is often dictated by the available drilling equipment, but it is generally true that the most economical hole size is that which can be drilled the fastest. The hole size should increase with the bench height, and suggested ranges are shown in Table 19-7. However, many 60-ft benches have been drilled with 3-in. bits, and

TABLE 19-7 Relationship between Bench Height and Suggested Hole Diameter

Bench height, ft	*Hole diameter, in.*
Less than 10	2–3
10–20	2¼–4
20–40	4–10
40–60	6–12

35-ft benches have been drilled with 12-in. bits. In very blocky and seamy rock, small holes may break the rock best because the powder is more uniformly distributed. Where suitable, large-diameter holes are becoming more popular in the construction industry for reasons of economy. The mining industry, which is able to conduct more long-term experiments, has demonstrated that large-diameter holes loaded with AN/FO (ammonia nitrate–fuel oil) are more economical than smaller holes shot with the higher-energy but more expensive slurries.

Final blast-design criteria, including the selection of hole size, should be made only after careful evaluation of all costs involved in drilling, shooting, loading, hauling, fill treatment, and crushing.

Selection of Explosives Of the many kinds of explosives mentioned in the Explosives portion of this section, present heavy-construction practices tend to use three kinds: ammonia nitrate–fuel oil (AN/FO); its high density and bulk-strength counterparts, slurries; and nitroglycerin (NG) dynamites for loading out of water. The first two of these explosives are free flowing and fill the hole completely, thus ensuring efficient breakage by transmitting all the explosive force to the rock. Their high degree of coupling with the rock contrasts with the wasted explosive stress which often occurs in holes loaded with poorly tamped stick powder.

Both AN/FO and slurry are readily loaded into the hole by pouring from a bag or by blowing or augering from a bulk truck. Loading stick powder, especially into deep holes, is a slow and costly operation. The cost per pound for AN/FO is 10 to 20 cents less than for stick powder and 5 to 15 cents less than for slurries. The ready acceptance of AN/FO by the industry is due to its low cost, good coupling, and ease of loading. Its principal unfavorable characteristic is that it is not water resistant. This has caused many users to select the more costly slurries which, in addition to good water resistance, have the same coupling abilities and ease of loading as AN/FO. The greater strength of the slurry may also offset its higher material cost by permitting a wider drilling pattern or the use of smaller holes.

Research has shown that explosives which have a good impedance match to the rock (relation between the density and the propagation velocity of the explosive and the density and acoustical velocity of the rock) will transfer a greater percentage of available energy to the rock. Thus AN/FO usually results in lower costs when used with the plastic-acting rocks such as shale, soft and medium limestone, and sandstone. For the same reasons, slurry performs best in the harder, more brittle rocks such as hard limestone, quartzite, gneiss, and all igneous rocks. Slurries are generally economical whenever the drilling cost per foot of hole is more than explosive cost per foot of hole; that is, the harder the rock the more economical slurry becomes. In wet holes, slurries can be loaded as a bottom charge, with NG dynamite up the hole to above the water level and AN/FO used to top out the charge.

The use of explosives in modern heavy-construction open-cut excavation can be summarized as follows:

STICK POWDER: Used principally for presplitting, as primers for secondary shooting, and for loading out of water in wet holes

AN/FO: Used in dry holes in soft, plastic rock and as top loads in hard rock (wet or dry)

SLURRIES: Used as bottom loads in wet holes suitable for AN/FO or in wet or dry holes in hard, brittle rock

Blasthole Design An optimum blasthole design and spacing exists for every different combination of rock, explosive, and blasthole geometry. These variables influence the results of blasting to such an indeterminable extent that until recently all blast designs depended upon trial-and-error methods using "powder factors" (pounds of powder per cubic yard of blasted rock). These powder factors were selected by persons experienced in the art, after giving appropriate consideration to the rock type and natural jointing, fragmentation required, blasthole geometry, delay-cap pattern, number of free faces, and kind of explosives to be used. Order-of-magnitude figures for the required pounds of powder per bank cubic yard of rock are as follows: for average rock and fair fragmentation, ¾ to 1 lb; for easily broken rock and poor fragmentation, ½ lb; for hard-to-break rock and good fragmentation, 1¼ lb or more.

The pressures from rising labor costs at home and competition abroad have forced our mining industry to pioneer the application of scientific methods to the problem of rock excavation by explosive means. Three principal objectives for this research have been (1) determination of the mechanics of rock breakage, (2) improvement of the empirical powder-factor method of design by correlating it with blast geometry, and (3) development of scientific methods of blast design. This work has now progressed to the point where large economies can be realized when the trial-and-error system is replaced by more rational methods of design. The construction industry is already applying these new techniques with considerable success.

Mechanics of Rock Failure. Various studies have generally agreed on the mechanics of rock failure—that it occurs in three stages, and what each stage consists of—but they have not agreed on the relative importance of each stage to the failure process. Nevertheless, an attempt is made below to combine the presently accepted concepts of the various stages of rock breakage.

The detonation releases enormous chemical energy which transforms the explosive material into a gas under tremendous pressure. At the wall of the blasthole this pressure generates a shock wave that radiates outward into the rock in all directions. Crushing of the walls of the blasthole and radial cracking are the immediate effects of this shock wave; since it travels at 10,000 to 15,000 fps, these effects are completed in several microseconds. As the shock wave moves outward, its velocity and strength are greatly reduced by the damping capacity of the rock as well as by each fissure or other discontinuity in the rock. If a free face is not encountered before the wave is damped to below its capacity to break the rock, little apparent breakage occurs. When the shock wave reaches a free face, however, it is reflected back into the rock as a tension wave. If at some distance back into the rock this reflected wave exceeds the tensile strength of the rock, a slab is broken off and propelled from the face. This slabbing action is increased by any natural rock jointing near the free face.

The gas generated by the explosion, after creating the initial shock wave, continues to expand and acts against the inertia of the rock mass. The strong compression wave which is propagated induces high stress in the rock and causes the initially formed radial cracks to extend further. At some point the two failure processes meet and vent each other, and the slabbing ceases. The highly stressed rock bursts when stress is relieved and the expanding gases, if their energy has not been depleted by the stressing action, are explosively released and impart additional throw to the bursting rock. In a properly designed blast, a maximum amount of the gas energy has been utilized for rock stressing and the air blast and fly rock are minimized.

The principal area of disagreement in the failure theories is in the relative amounts of rock breakage that can be attributed to stress-relief fracturing and to tension

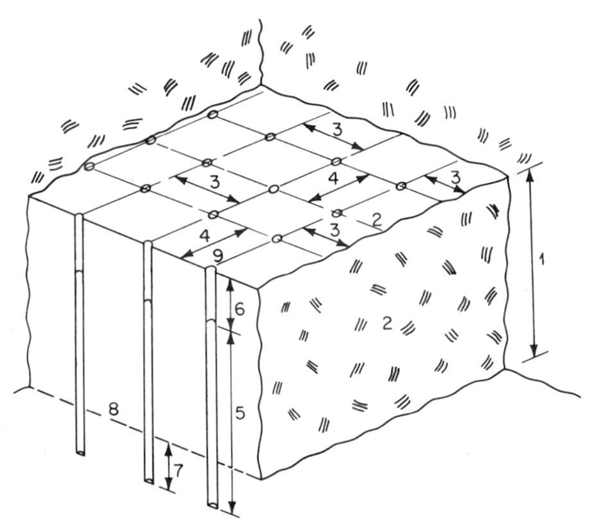

Fig. 19-19 Blasting round.

slabbing. Early laboratory experiments in homogeneous material led researchers to conclude that most of the breakage was caused by slabbing, but recent field measurements have indicated that less than 5 percent of the explosive energy is carried by the shock wave toward the free face.[7] It has been postulated from these studies that the shock wave is not necessary to produce rock breakage and that stress relief is the controlling factor. Present thinking seems to favor this concept, particularly so when consideration is given to the attenuating effect on the shock wave of the natural discontinuities in rock and to the amplifying effect of these same rock characteristics on the radial cracking and stress-relief bursting of the rock. Additionally, it has been demonstrated that, in the softer plastic rocks which have been well blasted, the tension-slabbing effect is almost totally suppressed by the damping action of these materials.

Empirical Powder-factor Design. Blast designs using powder factors have long been popular, and extensions of this method have been explored as the basis for more economical design. This work was directed toward overcoming the principal drawbacks of the empirical methods for designing a blasting round: the lack of established relationships between powder type, hole size, bench height, burden, spacing, subdrilling, stemming, and the number of free faces (see Fig. 19-19).

Tables were compiled listing the experience gained with various bench heights and rock types in conjunction with certain hole sizes and spacings. These experience tables provided an excellent starting point for blast design and demonstrated the fact that optimum burden and spacing vary with rock type and bench height; however, the tables did not include information on the types of explosives or powder-factor values.

Subsequent listings which included powder factors as well as successfully used blasthole geometrics were more useful and provided helpful guidance when similar blasting situations were encountered. A listing of this type is supplied in Table 19-8. Such tables also demonstrated how blasthole geometry might change with various hole diameters, bench heights, powders, powder factors, and delay-cap patterns.

Five "standard blasting ratios" have been developed by Ash.[8] These provide a range of geometric relationships with which a trial-and-error blast design can be started; modifications for different explosives and rock densities are also indicated.

Burden.	20 to 40 × explosive diameter
Hole depth.	1.5 to 4.0 × burden
Subdrilling.	0.3 (minimum) × burden
Stemming.	0.5 to 1.0 × burden
Spacing.	1 to 2 × burden

The burden standard is modified to 20 to 25 for light-density explosives and to 40 for slurries. Burdens on the low side of the standard range are recommended for heavy, massive rock, and burdens on the high side are used for light or highly fractured rock. These standards have the attractive feature of supplying numerical values. Unfortunately, the many variables which must be considered will permit a wide range of possible answers. Prudence then dictates that selection of the values actually to be used should be made only by experienced personnel.

Scientific Methods of Blast Design. Formulas which have been developed by Langefors and Kihlstrom[9] can be used with an experimentally determined rock constant to calculate the weight of powder required to break a specified burden of a single hole in a certain height of bench. A separate formula modifies the powder weight according to the degree of fixation at the bottom of the hole, the strength of the powder, and the hole spacing. Additional formulas are used to compute the amount of additional powder required if multiple row patterns are used or if greater fragmentation is needed. One-third of the burden distance is added to the hole depth for subdrilling, and the full burden distance is used for stemming. The diameter is that which provides the space needed to load the required powder weight into the length of the hole (less stemming), which conforms to the bench height. Rock excavation for Morrow Point Dam was accomplished using this method.[10]

Scientific inquiries into blast design were sponsored by the United States government during the 1940 to 1950 period. Some of the experimental work was conducted by C. W. Livingston, who developed a method of blast design based on cratering.[11] This was one of the first scientific methods employed; it consisted of observations of the behavior of rock when single spherical charges were detonated at varying depths below the rock surface. The Iron Ore Company of Canada[12] has applied this method to reduce its blasting costs by 40 percent and increase the yield of rock per foot of drilling by 30 percent. Another mining operation obtained a 61 percent increase in rock yield by using the cratering method of blast design.[13] The heavy-construction industry is also beginning to use this method successfully.

Blast design by the crater method requires small test blasts in the actual material which is to be removed. These tests use the same type of explosive that is to be employed in the production shots and seek to identify its optimum utilization in terms of yield of fragmented rock. The test results are used to select the weight of powder for each production blasthole. Two basic formulas are applied to calculate the burden and to verify the breakage at the collar. Spacing, subdrilling, and stemming are usually determined empirically.

An example of blast design with the crater method is given below. Since the method is applicable to holes which are drilled horizontally as well as to those which are

TABLE 19-8 Successfully Used Blasthole Geometrics and Blasting Agents

Rock	Hole diam, in.	Bench height, ft	Subdrill, ft	Burden, ft	Spacing, ft	Collar, ft	Type of explosive	Powder factor, lb/bcy
Andesite	2½	16	1	5	6	4	AN/FO	1.4
Fresh granite	2½	27	3	6	6	2	Gelatin	1.4
Fresh granite	2½	50	5	8	8	4	AN/FO	0.9
Traprock (basalt)	2½	37	3	7	8	8	Gelatin	0.7
Blocky granite	3	27	3	7	7	8	NCN	1.0
Fractured granite	3	40	5	7	7	10	Ammonia	0.8
Massive granite	3	20	3	6	6.3	5	½ Slurry, ½ AN/FO	2.0
Granite	3	30	4	7	8	7	AN/FO	1.1
Blocky dolomite	3	30	4	7	8	5	AN/FO	0.9
Massive limestone	3	20	P†	7	7	8	Gelatin	0.9
Blocky limestone	3	24	2	7	8	7	AN/FO	1.0
Metamorphosed limestone	3	40	4	7	9	6	Slurry	1.7
Schist	3	16	2	6	7	4	NCN	1.5
Hard shale	3	18	2	8	8	7	Ammonia	1.1
Hard gneiss	3½	40	4	8½	10	6	AN/FO	1.0
Sandstone	3½	18	2	8	8	8	AN/FO	1.0
Sandstone	3½	23	2	10	10	7	AN/FO	0.8
Jointed quartzite	4	37	3	12	12	10	AN/FO	0.7
Sandy shale	5⅝	14	2	10	10	9	AN/FO	1.2
Sandstone	5⅝	23	2	8	10	11	AN/FO	1.3
Jointed limestone	6	33	3	10	15	D*	NCN	1.0
Sandstone	6¼	37	3	18	18	12	AN/FO	1.2
Blocky basalt	6½	40	5	18	18	7	AN/FO	1.0
Dolomitic limestone	6½	47	3	15	16	D*	Slurry	0.5
Gneiss	6½	45	5	18	20	14	Slurry	1.3
Dense basalt	7	40	4	17	18	16	½ Slurry, ½ AN/FO	0.9
Highly faulted limestone	7⅜	25	5	15	15	D*	AN/FO	0.8
Blocky limestone	8	20	P†	14	14	D*	Gelatin	0.7
Massive limestone	8½	55	5	21	24	21	NCN	1.1
Jointed basalt	9	36	4	24	24	12	AN/FO	0.8
Sandstone	9⅞	40	4	22	24	5	AN/FO	1.2

* D = Deck loaded.

† P = Parting plane at bottom of hole, therefore no subdrilling.

drilled vertically, the terms "depth" and "distance" are used synonymously in the following notation:

N = critical charge depth. This is measured normal to the primary free face, extending from the blasthole collar to the explosive center of gravity, and is expressed in feet. It is the maximum distance at which at least one rock slab is broken out at the primary free face.
d_o = optimum charge depth. This is measured in the same manner as described for the critical charge depth, and is also expressed in feet. It is the charge distance which produces the maximum volume of crater and corresponds to the most efficient use of the explosive in terms of quantity of rock removal per unit weight of explosive.
d_c = depth of charge. This is measured in the same way as the critical and optimum charge depths, and is stated in the same units. It is the distance between the blasthole collar at the primary free face and the explosive center of gravity.
d_b = blasthole depth. This is the distance between the blasthole collar and the bottom of the blasthole.
B = burden distance between a secondary free face and the explosive center of gravity. It is measured normal to the secondary free face and is expressed in feet.
W = weight of the explosive charge in each blasthole, lb.
E = $N/W^{1/3}$, strain energy factor for the given rock and explosive charge.
S = spacing between explosive centers of gravity of adjacent blastholes. It is expressed in feet.
V = volume of crater produced in a given rock by a given explosive charge W at a distance d_c from the primary free surface. It is expressed in cubic feet.
Δ_o = d_o/N, optimum depth/critical depth. This is the optimum depth ratio.
Δ_B = B/N, burden depth/critical depth. This is the depth ratio in the burden direction.
Δ_c = d_c/N, depth of charge/critical depth. This is the depth ratio in the collar direction.

Assume that the production plan calls for the use of a slurry as the explosive. The material to be excavated is hard granite, and 3½-in.-diameter holes are specified in order to utilize the available excavating and drilling equipment. These holes will be drilled vertically downward from the rock surface (primary free face) to produce a 20-ft-high bench (secondary free face). For these conditions, Table 19-9 indicates average values for Δ_o and E as follows: $\Delta_o = 0.50$, $E = 4.50$.

The test blasts should be made in sound rock that has not been previously damaged by blasting. The largest practicable holes are desirable for testing in order to minimize the influence of charge shape and the effects of local variations in geology. We will assume the use of 4½-in. test holes, each charged with 10 lb of explosive (including the primer). From this, together with the approximate value for the strain energy factor obtained from Table 19-9, a trial value for the critical charge depth can be obtained.

$$N = EW^{1/3} = 4.50 \times \sqrt[3]{10} = 9.7 \text{ ft}$$

Since 10 lb of the slurry (specific gravity = 1.45) will fill 1 ft of the 4½-in.-diameter test hole, the center of gravity of the charge will be located 6 in. above the bottom of the hole. Based on the trial value of $N = 9.7$ ft, select a hole depth $d_b = 10$ ft for the first trial ($d_c = 9.5$ ft). A test hole is drilled to this depth and loaded with a measured 10 lb of slurry; the remainder of the hole is carefully filled with stemming to its collar.

Assume that several slabs are formed at the hole collar by this shot, indicating that the 9.5-ft charge depth is less than the critical charge depth. Thus, for the next trial, the hole is drilled 11 ft deep ($d_c = 10.5$ ft) and loaded and stemmed as before. The resulting blast is observed to barely crack the primary free face, indicating that N is somewhere between 9.5 and 10.5 ft. Accordingly, it is estimated that $N = 10.2$ ft.

The strain energy factor E can now be calculated using this experimentally obtained value of N:

$$E = \frac{N}{W^{1/3}} = 10.2/\sqrt[3]{10} = 4.75$$

This value of E will be used in the subsequent calculations.

Table 19-9 lists $\Delta_o = 0.5$ as an average value of the optimum depth ratio for this rock type and explosive. Applying the relationship $d_o = \Delta_o N$, the trial value of the optimum charge depth d_o is $0.5 \times 10.2 = 5.1$ ft. Accordingly, the next test hole is drilled to a depth of 5.5 ft and loaded as before ($d_o = 5.0$ ft). The shot is observed to produce neither flyrock nor noise, and it is concluded that this charge depth is greater than the optimum charge depth. A second hole is then drilled to 5.0-ft depth ($d_c = 4.5$ ft), loaded, and fired. It is observed to produce minor flyrock and noise, indicating that the charge depth is now slightly less than the optimum charge depth. Interpolating, it is estimated that $d_o = 4.7$ ft. The optimum depth ratio is then calculated as $\Delta_o = d_o/N = 4.7/10.2 = 0.46$. This experimentally obtained value checks fairly closely with the average value of $\Delta_o = 0.50$ which was supplied in Table 19-9.

TABLE 19-9 Approximate Values of Δ_o and E for Use in the Crater Method of Blast Design

Type of explosive	Design factor	Type of rock: Plastic — Permafrost, soft shale, soft limestone	Type of rock: Intermediate — Sandstone, hard shale, hard limestone	Type of rock: Brittle — Granite, quartzite, basalt
Slurry	Δ_o	0.85	0.70	0.50
	E	2.50	3.50	4.50
	$\Delta_o E$	2.10	2.45	2.25
AN/FO	Δ_o	0.95	0.85	0.50
	E	2.00	3.00	4.00
	$\Delta_o E$	1.90	2.55	2.00

The test program just described has produced acceptable blast design factors with four blasts requiring 33 ft of drilling; more accurate values could be obtained with an extended program. For the best determination of Δ_o, the volume of each test crater should be ascertained and V/W plotted against Δ_o. Optimum depth would be the high point on such a curve.

The production blast design for the 20-ft bench makes use of the test data. For the first trial computation, assume that there will be 3 ft of subdrilling and 9 ft of stemming (subdrilling is usually ⅓ to ½ of the burden, and stemming is at least equal to the burden). With these assumptions, a 3½ in vertical borehole in the 20-ft bench will accommodate a powder column 14 ft (20 + 3 − 9) in length. The explosive charge W in each borehole will then amount to 84 lb, based on 6 lb per ft of powder column for a 3½-in. borehole. Using the experimentally obtained value of $E = 4.75$, the computed critical depth for this explosive charge will be $N = EW^{1/3} = 4.75\sqrt[3]{84} = 20.8$ ft.

To ensure uniform fragmentation from the production blast, it is necessary to evaluate the charge depths to the two free faces in terms of the critical charge depth ($\Delta_B = B/N$ and $\Delta_c = d_c/N$). A suitable burden distance for this type of rock excavation is $B \leq d_o$ and, by definition, $d_o = \Delta_o N$. From the test data, $\Delta_o = 0.46$ and $N = 20.8$ ft. Substituting, $B = d_o = 0.46 \times 20.8 = 9.6$ ft. To allow for the lower breaking efficiency of the elongated production charge, we will reduce the design burden to 9 ft. The design ratio in the burden direction is then $\Delta_B = B/N = 9/20.8 = 0.43$.

Experience has indicated that the depth ratio in the collar direction, Δ_c, should be about 0.9 for optimum fragmentation and backbreak. For our design the distance from the primary free face (surface) to the center of gravity of the charge is 9 ft (stemming) plus 7 ft (half the column load), or 16 ft, and $\Delta_c = d_c/N = 16/20.8 = 0.77$. This depth ratio will produce excellent fragmentation at the surface but will also result in an undesirable amount of throw and backbreak. To adjust Δ_c closer to the desired value of 0.9, the decision is made to increase the subdrilling and also to reduce the weight of powder. Increasing the subdrilling to 4 ft and decreasing the powder weight to 75 lb (12½ ft of powder column), new values are obtained as follows:

$$N = 4.75 \times \sqrt[3]{75} = 20.0 \text{ ft}$$

and

$$d_o = 0.46 \times 20.0 = 9.2 \text{ ft (use } B = 9.0 \text{ ft)}$$

TABLE 19-10 Typical Blasthole Layout Card

Table for Pit Foremen

Material: Iron formation — Explosive: Hydromex

$E = 4.26$ $\Delta_o = 0.53$ — Hole diameter: 9⅞ in.

Bench height, ft	Burden, ft	Spacing, ft	Depth, ft	Weight, lb	Column, ft	Collar, ft
21	17	24	25	506	10	15
24	18	25	28	603	12	16
26	19	26	30	710	14	16
28	20	28	32	825	17	16
31	21	29	35	958	19	16
33	22	31	37	1,102	22	16
36	23	32	40	1,258	25	15
39	24	34	43	1,426	29	14

The 75 lb of slurry will occupy 12.5 ft of 3½-in. hole, hence the stemming distance is $20 + 4 - 12.5 = 11.5$ ft. The distance from the collar of the hole to the center of the explosive is

$$d_c = 11.5 + (0.5 \times 12.5) = 17.8 \text{ ft}$$

The depth ratio $\Delta_c = 17.8/20.0 = 0.89$, which is satisfactory.

The selection of the spacing, S, for the blastholes is needed to complete the design. Test blasts could also be used to determine the optimum hole spacing but, in this case, we will resort to empirical standards which set the hole spacing at 1.1 to 1.4 times the burden distance. We will choose a hole spacing of 11 ft, which results in

$$\frac{S}{B} = \frac{11}{9} = 1.22$$

The first production blast will thus be laid out with 3½-in. vertical holes on a 9- by 11-ft pattern. Each hole will be drilled 24 ft deep and loaded with 75 lb of slurry. The resultant blast should be checked for fragmentation, flyrock, throw, noise level, and complete breakage at grade. Minor adjustments can then be made and a blasthole layout card, similar to the one used by Iron Ore Company of Canada (Table 19-10), may be prepared for the pit foreman.[14]

The crater method of blasthole design can be the basis for an accurate cost comparison of various hole sizes, bench heights, and powder types. Further experience with this method is needed before it can completely replace the trial-and-error application of the improved powder-factor methods, but recent work indicates that the crater method of design will soon be at the nomograph stage. This will encourage its use in even the smallest rock excavation project.

Blasting Round Design Once the blasthole design has been formulated, a layout can be made of a number of blastholes in one or more rows to produce the blasting "round" illustrated in Fig. 19-19.

Single versus Multiple Rows. The choice between single and multiple rows will be influenced by the rock, the production rate desired, and the length and height of bench. The layout and timing of the blasting rounds must be such that shot rock will always be available for the loading equipment (see Table 19-11). One row of holes produces a lower muckpile than multiple rows, reduces backbreak, and provides some assurance that the blast is shot to grade before the next row is shot. A hole which does not pull to grade can cause a major delay in any cut or quarry operation. The high grade must then be redrilled and shot to prevent a progressively adverse condition from developing. A hole not shot to grade may have been caused in several ways: it may have been drilled short; the powder may not have been placed to the bottom of the hole; the bottom charge may have been inadequate; the hole may have been drilled inaccurately causing the burden to be greater than designed; the charge may have misfired; or the burden-spacing relationship may have been wrong.

Former quarry practice favored shooting one row at a time in order to avoid these difficulties. With the introduction of reliable delay caps and the application of improved blast design and layout, multiple-row rounds have now increased in popularity. They have the advantages of better fragmentation and fewer shutdowns for shooting. Heavy-construction blasting is invariably accomplished with multiple rows because of production requirements.

Layout. Blasting rounds may be laid out in either a square, rectangular, or staggered (diamond) pattern. The square or rectangular pattern is preferred, since it is less confusing to drill and the effect of a staggered pattern can still be obtained by delaying the detonation across the corners of the pattern. When possible, rock excavation should be planned so that the bench face is parallel to the major vertical jointing. This permits shooting across the planes of greatest weakness, producing maximum fragmentation and minimizing high grade and hole cutoffs. It also allows maximum hole spacing and results in a more nearly vertical bench face for the next round. In open-jointed formations that dip steeply, the bench should run across the major jointing to minimize hole cutoffs. The round should then be staged by delays so that breakage is across the face as much as possible.

Angle drilling may be used to overcome the undesirable backbreak that occurs when blasting a face parallel to a steeply dipping joint system. The drawback of angle drilling is that it requires more drilling and drill-rig maneuvering. Greater care is needed to ensure that the planned burden-spacing relationship is maintained with angled holes, whereas it is easy to drill plumb holes accurately. Extreme drilling inaccuracies can result when drilling at as little as 20° off vertical. Table 19-12 lists the deviations and losses, either deliberate or accidental, in hole depth due to angle drilling.

Relief and Buffers. To minimize drilling and powder, a blasting round should have two free faces (top and face of bench) to which the broken rock may swell and move. Sidehill and through cuts make use of the natural free faces, and a vertical free face may be developed in flat terrain by means of a sinking cut. This type of cut requires larger, more closely spaced vertical holes and about twice the amount of powder. The excessive vibration which it usually produces can be minimized by using lifters (horizontal holes at the bottom of the cut) combined with the increasingly deeper vertical holes.

Buffer shooting is popular in rock that fragments with increased confinement and that breaks readily to grade. This method relies on shooting into the muckpile left from the previous shot, using this pile as a buffer to reduce the normal relieving effect

TABLE 19-11 Material in the Solid Displaced at Different Drill-hole Spacings

Distance between face and drill-hole centers, ft	Distance between drill-hole centers, ft, and volume of material displaced, cu yd per ft of hole depth*																		
	7	8	9	10	11	12	13	14	15	16	17	18	19	20	21	22	23	24	25
7	1.81	2.00	2.33	2.70	2.85	3.11													
8	2.00	2.37	2.65	2.96	3.26	3.55													
9	2.33	2.65	3.00	3.33	3.66	4.00													
10	2.70	2.96	3.33	3.70	4.10	4.44	4.81	5.18	5.55	5.92									
11	.....	3.26	3.66	4.10	4.48	4.88	5.30	5.70	6.11	6.52									
12	.....	.....	4.00	4.44	4.88	5.33	5.77	6.22	6.66	7.11	7.55								
13	.....	.....	.....	4.81	5.30	5.77	6.26	6.74	7.22	7.70	8.18								
14	.....	.....	.....	5.18	5.70	6.22	6.74	7.26	7.77	8.30	8.81	9.33	9.85	10.37					
15	.....	.....	.....	5.55	6.11	6.66	7.22	7.77	8.33	8.88	9.44	10.00	10.55	11.11	11.66	12.23	12.78	13.33	13.88
16	.....	.....	.....	.....	.....	7.11	7.70	8.30	8.88	9.48	10.07	10.66	11.30	11.85	12.44	13.03	13.63	14.22	14.81
17	.....	.....	.....	.....	.....	7.55	8.18	8.81	9.44	10.07	10.70	11.33	11.96	12.59	13.22	13.85	14.48	15.11	15.74
18	.....	.....	.....	.....	.....	8.00	8.66	9.33	10.00	10.66	11.33	12.00	12.66	13.33	14.37	14.66	15.33	16.00	16.66
19	.....	.....	.....	.....	.....	.....	9.15	9.85	10.55	11.30	11.96	12.66	13.37	14.07	14.77	15.43	16.18	16.88	17.60
20	.....	.....	.....	.....	.....	.....	9.63	10.37	11.11	11.85	12.59	13.33	14.07	14.81	15.55	16.30	17.03	17.77	18.51
21	.....	.....	.....	.....	.....	.....	.....	.....	11.66	12.44	13.22	14.37	14.77	15.55	16.33	17.11	17.88	18.66	19.44
22	.....	.....	.....	.....	.....	.....	.....	.....	12.22	13.03	13.85	14:66	15.48	16.30	17.11	17.92	18.74	19.55	20.40
23	.....	.....	.....	.....	.....	.....	.....	.....	12.78	13.63	14.48	15.33	16.18	17.03	17.88	18.74	19.59	20.44	21.39
24	.....	.....	.....	.....	.....	.....	.....	.....	13.33	14.22	15.11	16.00	16.88	17.77	18.66	19.55	20.44	21.33	22.22
25	.....	.....	.....	.....	.....	.....	.....	.....	13.88	14.81	15.74	16.66	17.60	18.51	19.44	20.40	21.39	22.22	23.14

* Average conversion: To convert cubic yardage into tons, multiply the figures given above by 2.3 for granite; 2.27 for limestone; 1.55 for sandstone; 2.18 for shale; and 2.53 for traprock.

of the free bench face. It has the advantage of protecting equipment from flyrock, thus eliminating shovel move-out. It also builds a high muckpile, which is a great advantage when the shooting is confined to low benches. If a front-end loader is the loading tool, the higher and more compact muckpiles are not desirable and buffer shots should not be used. This does not reduce the effectiveness of loaders, since they are so mobile that the time required for them to move out is unimportant.

Delay Patterns. Blastholes are detonated either in rows or in controlled sequences to lower the drilling and powder factors, increase fragmentation, produce a more easily handled muckpile, and reduce backbreak and vibration. Almost all multiple-row blasting is now detonated with millisecond (or short-period) delays, either electric or nonelectric.

Inherent variations in the firing time of caps range from ±20 percent in the lower delays to ±10 percent in the higher delays. This, in itself, essentially ensures that each hole will be detonated at a different time, even when all caps in a row have the same nominal delay period. Caps whose firing times can vary enough to cause overlap (i.e., a No. 5 firing before a No. 4) should not be used together unless the round is

TABLE 19-12 Deviation and Loss per 100 ft of Depth Resulting from Angle Drilling

Deviation, degrees	Deviation, ft	Loss of depth, ft
1	1.75	0.02
2	3.49	0.06
3	5.23	0.14
4	6.98	0.24
5	8.72	0.38
6	10.45	0.55
7	12.19	0.75
8	13.92	0.97
9	15.64	1.23
10	17.36	1.52
15	25.94	3.37
20	34.20	6.03
25	42.25	9.38

deliberately planned so that this will not be detrimental. Important considerations when planning the delay sequences are:

1. Provide proper relief (minimum 90° breakage angle) for every hole.
2. Make the pattern as simple as possible to minimize confusion during loading.
3. Shoot across major jointing.
4. Plan the "throw" to produce a muckpile for optimum loading conditions.
5. Maintain the planned burden and spacing.
6. If possible, orient the delay pattern to accommodate the existing jointing of the rock. Examples of such orientation are illustrated in Fig. 19-20.

DRILLING AND SHOOTING

Bench Preparation Overburden removal in preparation for a drilling and blasting operation has already been discussed. It is worthwhile to spend considerable effort to leave the surface reasonably level and smooth so as to facilitate the hole layout and the equipment moves between holes. Leveling shots, either short down-holes or horizontal lifters, can be used for localized leveling of the drilling surface. A level surface is particularly important where large rotary or down-the-hole drills are used; in fact, in situations where the surface of a proposed cut is unusually rugged, it may be economically preferable to use the lighter airtrack drills and thus avoid the cost of leveling the surface for the operation of rotaries.

Blasthole Drilling The hole geometry and depth must be accurately controlled if the drilling and shooting is to produce the desired results. It is good practice to establish a baseline and grade at a sufficient distance back from the face and then to locate holes for the next shot from this baseline. Since the top surface of the rock is usually uneven in spite of reasonable attempts to keep it level, the bottom elevation of each hole should be determined from an established bench mark instead of simply measuring the depth of drilling at each hole location.

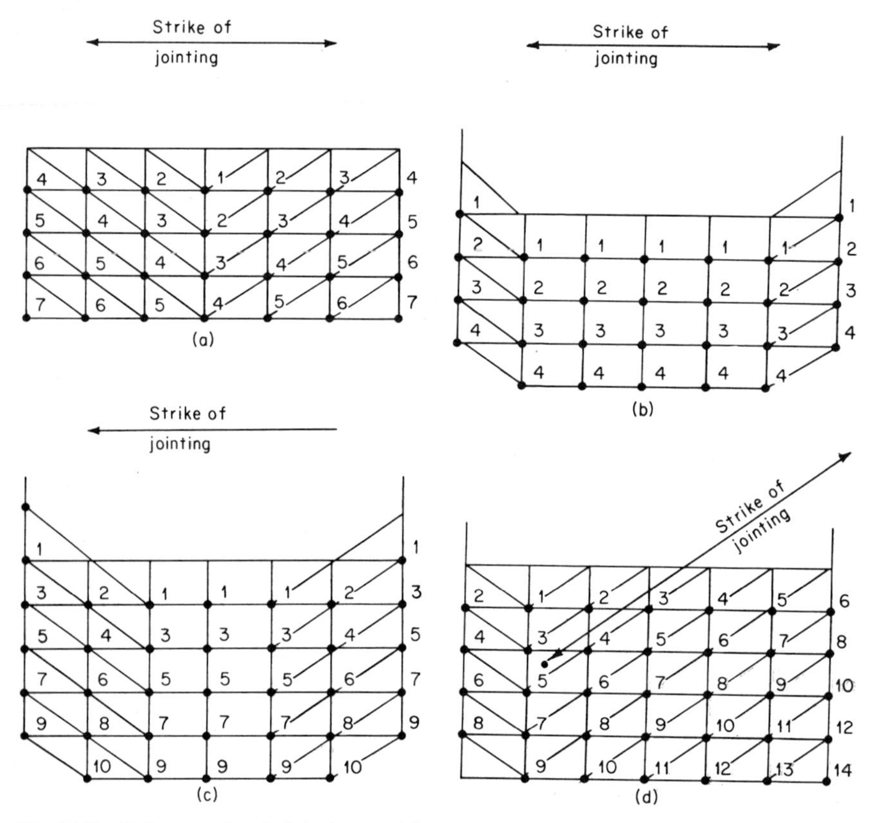

Fig. 19-20 Influence of rock jointing on delay pattern. (*a*) Poor delay pattern for shooting across jointing parallel to face. Muckpile will be tight and backbreak may be excessive. (*b*) Preferred delay pattern for shooting across parallel jointing. Results in better muckpile and less backbreak. (*c*) Alternative delay pattern for shooting across jointing. Results in best fragmentation and easiest to handle muckpile. (*d*) Pattern for shooting across jointing not parallel to the face. Muckpile will be to the left.

In large quarry operations, a log of each hole drilled should be kept as a permanent record. This log should list the thickness of strata, seams, pockets, presence of water, deviation of hole, depth, diameter, and location. Such information may subsequently be of value if it becomes necessary to revise the blast design. Possible modifications could include the type and kind of explosive and the location of the stemming, based on the rock conditions revealed by the drilling. A blasting log, when kept as a permanent record, can also serve to refute the validity of damage claims against the contractor or quarry operator.

Drilling Production. When the blastholes have been located, drilled, and loaded accurately, all in conformance with the blast design, the theoretical requirements to produce the desired rock fragmentation have been satisfied. The actual amount of rock produced will depend upon the rates at which these tasks can be performed.

Drilling production is principally affected by the compressed air delivered to the tool, the sharpness of the bit, the operating condition of the drill and, in rotary drills, by the full development of the down pressure specified. In percussion drills, the weak link is usually the compressor size; this component should never be undersized, and proper consideration must always be given to operating location (at or above sea level) and operating conditions. A regular and complete preventive maintenance program for the drill and compressor is the best assurance of obtaining the designed performance.

Competent supervision and skilled and conscientious operators are required to perform the work so that the blast design is closely followed and the variances kept to a minimum. There are enough variations inherent in rock geology that, for the sake of an efficient blast design, all other variances should be held to a minimum.

TABLE 19-13 Theoretical Drilling Cost Comparison by Hole Sizes—Rotary Drilling*

Hole diameter, in.	7⅜	9	10⅝	12¼	15
Bucyrus-Erie drill model	30-R	40-R	50-R	60-R	61-R
Labor—two men at $6.00/hr.	$12.00	$12.00	$12.00	$12.00	$12.00
Maintenance, repair labor, supplies, oil	$3.50	$4.30	$6.50	$8.60	$10.75
Power at $0.015/kwh	....	$1.80	$3.00	$3.00	$5.75
Diesel fuel at $0.15/gal	$1.50				
Total operating cost/hr.	$17.00	$18.10	$21.50	$23.60	$28.50
Drilling rate, fph	70	80	90	100	110
Operating cost/ft	$0.243	$0.226	$0.239	$0.236	$0.260
Bit life, ft	6,000	7,000	8,000	9,000	10,000
Bit cost/lin ft	$0.026	$0.027	$0.031	$0.033	$0.053
Total drilling cost/ft	$0.269	$0.253	$0.270	$0.269	$0.313
Ratio—hole areas	1.0	1.5	2.1	2.8	4.2
Hole spacing—center to center, ft	20	25	30	34	41
Cu yd/ft of hole	15	23	32	42	63
Drilling cost/cu yd	$0.0179	$0.0110	$0.0084	$0.0064	$0.0050
Production per 7¼-hr shift:					
Feet drilled	508	580	653	725	798
Cu yd prepared	7,620	13,340	21,549	30,450	50,274

* Materials: Typical Midwest coal-mining overburden.

Drilling Economics. Table 19-13, prepared by Bucyrus-Erie, demonstrates the theoretical effect on the cost per cubic yard of excavated material by the use, when possible, of larger-diameter blastholes in similar material. The increased drilling rate and wider hole spacing used with a 15-in. hole will more than double the rate of rock breakage obtained with a 10⅝-in. hole.

Table 19-14 shows actual drilling costs for two different-sized holes in separate but similar pits. Since drilling depth for the larger holes is greater than normal, this comparison favors the larger holes to a greater degree than could usually be expected.

When the desired hole size and depth and the characteristics of the material to be shot fall within the capabilities of several kinds of drills, an economic comparison should be made to indicate the best selection. Two such comparisons follow.[15]

CASE 1

Cost Comparison—Rotary and Percussion Drills

TYPE OF ROCK: Medium (3.0 to 4.0 on Mohs' scale)

DAILY PRODUCTION REQUIRED: 5,000 tons

HOLE SIZE: 4 in.

HOLE DEPTH: 30 ft

TABLE 19-14 Actual Drilling Costs by Hole Size, Rotary Drilling

Item	Pit 1	Pit 2
Stripping shovel, cu yd	45	65
Average depth of overburden, ft	47.7	73.4
Type of overburden	Limestone	Limestone
Hole diameter, in	9	10⅝
Hole pattern, ft	27 × 30	40 × 40
Drill model	50-R	50-R
Bit life, ft	6,000	8,000
Drilling rate, fph	97	88
Cu yd drilled/ft	29.9	60.3
Powder factor, lb/cu yd	0.27	0.25
Costs:		
Operating costs, includes drill labor, dozer charges	$0.1348*	$0.1339*
Maintenance	$0.0792	$0.0597
Miscellaneous supplies	$0.0075	$0.0035
Bit cost	$0.0380	$0.0343
Total cost/ft	$0.2595	$0.2314
Total cost/cu yd	$0.0087	$0.0038

* Labor component based on $3.50 per hour.

HOLE SPACING: 8 ft × 10 ft

EQUIPMENT:

Rotary: 1 unit, 15,000-lb pulldown class, 250-cfm compressor. Tungsten carbide drag bit

Percussion: 2 units, airtrack with 4½-in. bore drill, independent rotation, 1½-in. steel, and 600-cfm compressor

Cost per Foot Drilled

Item	Rotary	Percussion
Capital equipment (5-yr amortization)	$0.030	$0.070
Drill steel, couplings, and shanks	none	0.051
Kelly bar and adaptor	0.016	none
Bits	0.075	0.015
Bit grinding	0.003	0.004
Rock drill oil, coupling grease	0.002	0.008
Compressor, airtrack, and drill maintenance	none	0.056
Rotary-drill maintenance	0.028	none
Fuel	0.011	0.056
Labor	0.020	0.040
Total cost per ft of hole	$0.185	$0.300
Cost per ton	$0.030	$0.048

CONCLUSION: Rotary drilling will reduce drilling costs approximately 38 percent in some *medium* rock quarries.

CASE 2

TYPE OF ROCK: Hard (4.0 to 5.5 on Mohs' scale)
DAILY PRODUCTION REQUIRED: 3,500 tons
HOLE SIZE: 4½ in.
HOLE DEPTH: 30 ft

HOLE SPACING: 9 ft × 12 ft
EQUIPMENT:
Rotary: 2 units, 15,000-lb pulldown class, 250-cfm compressor, roller-cone bits
Percussion: 1 unit, heavy airtrack with 5-in. bore drill, independent rotation, 1¾-in. hexagonal steel, and 900-cfm compressor

Cost per Foot Drilled

Item	Rotary	Percussion
Capital equipment (5-yr amortization)	$0.120	$0.110
Drill steel, couplings, and shanks	none	0.102
Kelly bar and adaptor	0.020	none
Bits	0.340	0.101
Bit grinding	none	0.001
Rock drill oil, coupling grease	0.002	0.018
Compressor, airtrack, and drill maintenance	none	0.046
Rotary drills maintenance	0.026	none
Fuel	0.045	0.067
Labor	0.080	0.040
Total cost per foot of hole	$0.633	$0.485
Cost per ton	$0.072	$0.055

CONCLUSION: Percussion drilling will reduce drilling costs approximately 24 percent in some *hard* rock quarries.

These studies were made using the hardness of the rock as a basis for comparing the drill performances. However, rock hardness is only one factor which must be considered when selecting a drill. Any rock formation, hard or soft, can be highly abrasive. When drilling in such formations, penetration rates and the holding of the drill gauge become serious problems. Many formations of relatively low hardness can be rotary drilled at a rapid rate, but the abrasiveness of the rock raises the cost per foot in favor of the percussion or down-the-hole drill. Experience shows that the latter drills give longer bit life in abrasive rock. Since bit cost is a major component in rotary drilling costs, percussion or rotary down-the-hole drills are usually more economical in abrasive rock drilling.

Loading The loading operation includes the hole preparation, placing of the charge in the desired manner with one or more primers in readiness for firing, and the confinement of the charge by stemming. Loading of blastholes should not be attempted by anyone who is not thoroughly experienced and skilled in all phases of the procedure. Safety precautions should be observed, not only during the loading process but also before the work is started.

If a thunderstorm approaches during loading, work should be stopped at once and all personnel should leave the danger zone. When using electrical initiation, a lightning storm several miles away may be dangerous, and a strike nearby is likely to detonate part or all of the blasting caps in the charge. Lightning alarms are now available and can detect thunderstorms which are close enough to be hazardous.

Before loading, the condition and depth of all blastholes should be checked. The holes must be cleaned out in order to utilize the available space for explosives and to facilitate loading in general. Cleaning is usually done by compressed air or water or a combination of the two.

Cartridge Loading. Dynamite cartridges are available in 8-, 12-, 16-, or 24-in. lengths. The primer and the next cushioning cartridge should be lowered carefully into the hole, but the remaining cartridges may be dropped. The primer cartridge itself is never tamped, but in tight holes a wood tamping pole may be used to ensure contact from cartridge to cartridge. A tamping pole made from straight-grained hardwood is preferred. The tamping process can be facilitated by slitting the cartridge in two longitudinal cuts, one on each side and about 4 or 5 in. long. These cuts allow the cartridge to collapse easily in the hole under a relatively light blow from the tamping pole. Care should be taken to see that the fuse or cap leg wires do not become damaged by kinking or pinching between the pole and the wall of the hole.

Bulk Loading. Free-running blasting agents such as AN/FO and slurries can be poured or mechanically pumped directly into the prepared blasthole. Bulk loading of blastholes with water gels and bulk AN prill-fuel mixtures is economical where large quantities are used per hole and per shot. There are three basic types of water gel loading trucks that are capable of placing products in the blasthole at rates from 400 to 750 lb per min. They are (1) a finished-product pumper which consists of a truck containing a tank and a pump that places the gel in the hole, (2) a batch pumper containing a mixing unit that batches a tankful of finished product on site, and (3) a

Fig. 19-21 Bulk loading of water gels. (*E. I. duPont de Nemours & Co., Inc.*)

continuous pumper that carries all the ingredients that are mixed as needed at the blasthole. Trucks range in capacity from a few thousand to 30,000 lb. Generally, the product is pumped in a low-viscosity form, with an ingredient added just prior to delivery to the blasthole to accomplish the final gelling in the hole. Figure 19-21 illustrates the bulk loading of water gels.

The handling and loading of bulk AN/FO has evolved because of the savings in cost between bulk and packaged products and the field-labor savings in loading blastholes with mechanical equipment. Standard conveying equipment such as belts, buckets, or augers incorporated in bulk delivery trucks may be used to handle prills (see Fig. 19-22). Additionally, many simple but effective means have been used to load bulk prills into blastholes: modified concrete mixers, dump bodies with stationary or swivel chutes, shoveling from a standard flat-bed truck, etc. Oil can be sprayed onto prills as they fall into the hole; the mixture is not normally uniform.

The capacity of bulk-delivery bodies varies from 1 to 10 tons, and one type of bulk-loading truck uses several augers to deliver the mixture to the hole. Oil is metered into the prill stream at a point far enough back from the auger discharge end to provide a uniform mixture by the time it reaches the blasthole. Portable bulk loaders are adaptable to many operations, and delivery rates of 300 to 500 lb of prill–fuel-oil mix per minute are available. Oil is metered accurately to provide a 94 percent prill to 6 percent oil ratio.

Tamping is not normally required with bulk loading, although in the smaller-diameter holes some agitation with a small pole may be advisable to prevent bridging. The performance of all blasting agents depends to a great extent on sufficient priming, regardless of the hole diameter. In small-diameter holes, a single primer is considered adequate to ensure detonation of short columns on the order of 15 ft or less; for longer charges, additional primers should be placed not over 15 ft apart. In large-diameter blastholes, multipoint priming is recommended with a detonating cord downline. Generally, a primer every 10 or 20 ft in large-diameter deep holes is adequate. These primers should have satisfactory diameter length, and performance characteristics. In large-diameter holes where there is little or no chance of an interruption of the AN/FO column, multiple priming with unsatisfactory primers will actually be detrimental.

Fig. 19-22 Loading bulk prills into borehole. (*E. I. duPont de Nemours & Co., Inc.*)

When a top load of free-flowing blasting agent is used above a bottom load of cartridge agent in water, it is recommended that a plug of water-resistant product be used above the water level. Unless such a plug is used, the free-running material will fill the space around the cartridges and, in so doing, will displace the water and raise the water level. This, in turn, can desensitize a portion of the upper charge. The net effect is equivalent to an undesirable separation in the powder column. A slurry product in a plastic film cartridge is sometimes used for the plug.

Stemming and Deck Loading. It is generally advisable, in the interest of safety and efficiency, to confine explosives by a suitable inert material. This stemming protects the loaded charges from accidental ignition or detonation, but its most important function is to confine the charge and thereby obtain the best possible efficiency. Conventional stemming materials, listed in decreasing order of their effectiveness, are (1) a mixture of sand and plastic clay, (2) clay, (3) sand, (4) loam, and (5) water for dry holes. In wet holes, crushed stone chips are preferred.

Stemming material may be used in the column of the blasthole as a spacer for deck loading purposes or as a means of concentrating the powder charge where a greater explosive effort is required. For example, if the drilling log indicates that the hole will cut across a horizontal seam of soft material, the powder in this vicinity can be omitted and stemming substituted. When the powder column is broken for this reason, additional primers will be required.

The required weight of powder for each hole is determined by the blast design. If this calculated charge is insufficient to provide a column load which extends up to the desired height, then deck loading should be used. With this procedure the main charge in the bottom is supplemented by one or more smaller charges, properly located higher in the hole and separated from each other by stemming. If the hole is to be fired electrically, it is necessary to prime each charge, generally by placing two caps in the bottom load and at least one in each of the decks.

Coyote Tunnel Blasting Coyote blasting is most suited to formations which yield the desired fragmentation by displacement only. Breakage must be provided through the movement of the burden by a highly concentrated charge. Additionally, coyote blasting is sometimes the only practical method of shooting down a high face because of the rough terrain and difficulty of access to the top of the formation. It is especially adaptable to the diced basalt formations of the Northwest. These rocks are highly checked and cracked, and disintegrate readily when disturbed.

The face to be shot should be between 75 and 100 ft high for best results. A system of tunnels is driven from the elevation of the quarry floor; these tunnels are usually horizontal and about 4 by 5 ft in cross section. This section area is just large enough to provide work space and is usually ample to contain the explosive charge. It is important that the formation to be shot be accurately surveyed and that the tunnels be driven to the desired line and grade using instrument control. The simplest layout is a T configuration whose stem consists of a main adit driven perpendicular to the hillside or quarry face. A single wing, or crosscut, is located at the far end of this stem and is driven at 90° to the left and to the right. The length of the adit is generally less than the height of the face.

For vertical burdens of over 100 ft, more than one set of wings is required. When working very high faces, it may be an advantage to supplement the coyote tunnels with at least one row of large vertical drill holes. This row is placed parallel to the rear wing and 20 to 30 ft back from it. The holes will help to improve breakage and trim down the crest so as to minimize hazardous overhangs. As a practical matter, such top drilling is difficult to accomplish because it requires access to the area by large drill rigs.

Blasting agents such as AN/FO are replacing dynamites in coyote shooting because of their outstanding safety and physiological (non-headache) properties. When loading a coyote shot, normal practice is to divide the total charge for each wing into units of a calculated size. These units are then placed at 20- to 25-ft centers, leaving a 12- to 15-ft space between the last charge in each wing and the adit intersection to prevent rifling of the blast through the adit when the blast is fired. Detonation is normally by double lines of Primacord, well protected and cross-tied.

The charges nearest the adit should be protected with a sandbag bulkhead so that they will not be disturbed by the stemming operations in the intersection and the adit. Stemming can consist of any heavy, inert material (tunnel muck is often used). It is also good practice to stem between the separate charges in the wings, even though some operators do not do so. The material chosen for a coyote shot will probably contain weak areas or seams not easily discernible which, if unprotected, could cause dissipation of the blasting energy or blowouts. Stemming is certainly advisable for confining charges where either the horizontal or vertical burdens vary greatly from one part of the wing to another or where the wing intercepts an open seam or an obviously soft spot. Stemming has been placed by blowing, tramming with the mucker, and by hand placing sandbags.

The loading ratios (pounds of explosive per cubic yard of rock) must be varied to suit the type of rock and the layout. The charge is usually calculated on the basis of the total tonnage or yardage in the "square" shot area. The yardage of interest is the rock volume bounded by vertical planes passed through the back wings, through both ends of the cross cuts, and through the face. The loading ratio for a layout consisting of a long adit and short wings should be heavier than that for a short adit and long wings since it is easier to move out the burden in a blast of the latter proportions.

Tunneling economics favor layouts having longer adits and shorter wings; in some cases, it is cheaper to make two layouts with short wings than to make one layout with long wings. Coyote shots have been employed to blast a face 1,000 ft in length, using six adits for two L-shaped layouts on the ends and four T-shaped layouts evenly spaced between.

Secondary Breakage In any major blasting operation it is inevitably necessary to blast oversize rock pieces and boulders to facilitate their loading, crushing, or placement in fill. This operation is termed "secondary shooting," and it is a very costly aspect of rock excavation and handling. Extremely large chunks must be redrilled, loaded, and shot (block holing). Smaller pieces may be broken by the external detonation of dynamite cartridges; the explosive is laid on the surface of the rock and contained with a heavy pack of mud (mud capping). There is always a tendency to overload in block holing, and mud capping is inherently an inefficient use of explosives since a large portion of the blast energy is expended into the air (with an attendant loud noise which tends to create poor public relations in populated areas). Depending upon the rock, two to six sticks of 1¼- by 8-in. dynamite per cubic yard may be required to break rock when using this method.

An alternate method uses a "headache ball" or "dropball" attached to a crane. The ball is made of cast semisteel or nickel alloy, weighs 500 to 8,000 lb, and is dropped from varying heights depending upon the impact required. Jacking and wedging and, more recently, hydraulic tools are also used effectively to reduce oversize shot fragments. These latter methods require drill holes.

Although all these methods are expensive, there is usually no alternative but to reduce the oversize fragments. This need is particularly urgent if their presence is delaying and disrupting the loading-out operations of the shovels and trucks.

Overbreak Control The object of overbreak control is to reduce and better distribute the explosive charge and, in so doing, to minimize stressing and fracturing of the rock beyond the neat excavation line. Blasting techniques to reduce the amount of overbreak are used when it is desirable to preserve the natural strength of rock walls, as in dams, open highway cuts, and structure foundations. Control of overbreak is a practical concern in construction projects, since the natural material which is blasted away or damaged outside the neat line must usually be replaced (at no compensation) by comparable construction materials, such as concrete and compacted backfill. Such added costs are the penalty that must be paid for indifferent or careless blasting practices.

Methods of blasting to control overbreak have become increasingly scientific, but they still place a major reliance on trial-and-error adjustments. This is not surprising when we consider the number of geologic variables involved. For many years, line drilling of unloaded holes was the only method used for overbreak control. Other techniques were developed as operators continued to seek ways to load some of these relief holes and thus cut down on the total amount of drilling. These techniques have been given such names as "cushion blasting," "presplitting," "preshearing," "smoothwall blasting," and "sculpture," "perimeter," and "contour" blasting. Their chief departure from the original line-drilling principle is that some or all of the holes are loaded with relatively light, well-distributed charges of explosives. The firing of these charges tends to shear the rock between the holes and thus permits wider hole spacing, less drilling, and consequently less drilling cost.

The controlled blasting techniques can be discussed as three major methods: (1) line drilling, (2) presplitting (preshearing), and (3) trim blasting (smoothwall, cushion, sculpture, perimeter, contour). The principles, applications, advantages, and limitations of these methods are described in the following paragraphs:

Line Drilling

PRINCIPLE: Precise drilling of a single row of small-diameter, closely spaced holes which create a plane of weakness along the neat line.

APPLICATION: Best suited for homogeneous formations where natural planes of weakness are at a minimum. Optimum results are obtained when primary excavation is removed to within one to three rows of holes from the neat line and when drilling can be done perpendicular to the geologic strike of the formation. Drill holes are generally

2 to 3 in. in diameter and are spaced from 2 to 4 diameters. The primary blastholes adjacent to the line drill holes are loaded more lightly and are located at 50 to 75 percent of the primary burden and spacing distances from the line holes.

ADVANTAGES: Can produce a fairly smooth surface in formations where other controlled methods, even with light shooting, may cause damage beyond the neat line. When used between loaded holes, shearing is promoted and results are improved, with less powder being required.

LIMITATIONS: Can be unpredictable except in very homogeneous formations. Drilling costs are high because of close spacings. Requires close control of drilling operations.

Presplitting

PRINCIPLE: Drilling, loading, and firing a single row of holes along the neat line before the adjoining main excavation area is blasted. When two charges are shot simultaneously in adjacent holes under optimum loading and spacing conditions, linear cracking produces a fractured zone between the holes. With a balanced spacing and loading, the fractured zone between the holes will be a narrow sheared area. The subsequent production blast can break at this boundary, producing a smooth wall with little overbreak and reducing damage to the finished surface.

APPLICATION: Presplit holes are string loaded with full or partial cartridges 1 to 2 in. in diameter or with special continuous columns of ⅞- and 1-in.-diameter charges. Usually, all holes are stemmed completely around and between charges. It is also desirable to increase the charge in the first few feet of bottom charge to two or three times that used in the upper portion. This promotes shearing at the bottom of the hole where it is most difficult to obtain. Holes are fired simultaneously with a detonating cord downline. In some formations, results may be improved by using line drill guide holes between loaded holes. These will promote shearing and give better results without increasing the powder charge per hole. Table 19-15 shows

TABLE 19-15 Diameters, Spacing and Loading—Presplitting Average Rock

Hole diameter, in.	Spacing, ft	Explosive charge,* lb/ft
1½–1¾	1–1½	0.08–0.25
2–2½	1½–3	0.08–0.25
3–3½	1½–3	0.13–0.50
4	2–4	0.25–0.75

* Cartridges should be no larger than half the diameter of the hole.

spacings and loadings which can be used as guides for average rock conditions.[16]

The depth that can be presplit at one time is dependent upon the driller's ability to maintain good hole alignment. Deviation greater than 6 in. will give inferior results; when using 2- to 3½-in. holes this usually limits the maximum depth to 50 ft.

Presplit operations can proceed in advance of the primary excavation, but prudence dictates that they should not advance too far into unknown rock because the rock characteristics may change. In order to expose the results of the presplit operation sooner and permit any necessary adjustment, the presplitting may be accomplished simultaneously with the primary shot. Earlier-period electric blasting caps are then used to initiate the presplit 50 or more milliseconds ahead of the production shot.

ADVANTAGES:

1. Before production shot. Easier to coordinate drilling and loading of presplitting with production shot. Deep cuts can be presplit in one shot, with production shots blasted later in two or more benches. Increased hole spacing with reduced drilling costs. It is not necessary to return to blasted slopes or walls after the primary excavation has been made.
2. Simultaneously with the production shot. Results can be seen sooner; entire

blast may be drilled, loaded, and shot in one operation, and this may reduce costs. Maintains all operations in one area for better control.

LIMITATIONS:

1. Before production shot. Results cannot be observed until production shot is made. Can possibly disrupt rock formation on sidehill cuts, which can increase drilling difficulties for production shots. Requires clearing of working area well ahead of normal operation. This may increase logistic and supervisory problems.

2. Simultaneous shooting. Difficulty in coordinating the drilling and loading of the two types of shots.

Trim Blasting

PRINCIPLE: Similar to presplit except that the main burden is removed first. A relatively narrow rock mass is left as a buffer zone between the final row of burden holes and the neat-line limits. Drilling hole sizes can range from 2 to 6 in., and explosive charges will consequently be heavier than in presplitting.

APPLICATION: Cushion blasting, which is a form of trim blasting, is drilled similarly to presplit drilling except that every third or fourth hole is of larger diameter. Only these large-diameter holes are loaded, and the intermediate guide holes are fully stemmed. Since larger-diameter holes can produce better alignment for greater depths, cushion blasting has been successfully accomplished at 90-ft depths. Cushion blasting was originally developed using dynamite cartridges which were carefully placed and located in large holes. The cartridge dynamite is now being replaced by modern ⅞- to 1-in. continuous-column charges which are placed in smaller holes. Table 19-16 may be used as a guide for loads and patterns for cushion blasting in average rock conditions.

TABLE 19-16 Suggested Loads and Patterns for Cushion Blasting[12]

Hole diameter, in.	Spacing, ft	Burden, ft	Explosive charge, lb/ft
2–2½	3	4	0.08–0.25
3–3½	4	5	0.13–0.50
4–4½	5	6	0.25–0.75
5–5½	6	7	0.75–1.00
6–6½	7	9	1.00–1.50

To promote shearing at the bottom, a large bottom charge is used as described for presplitting. For maximum cushioning, the charges should be placed as close as possible to the excavation side of the hole. Cartridges are either taped to the down line or drop-loaded between applications of stemming at predetermined locations. In poorly consolidated formations, the increased use of unloaded, small-diameter guide holes is recommended. Where only the top portion of a formation is weathered, the guide holes need be drilled only to the depth of this condition. In highly competent rock, the stemming in the loaded holes can sometimes be omitted with good results.

Smooth-wall blasting, a form of trim blasting, is used principally in underground work. It is particularly applicable in tunnel construction.

ADVANTAGES: Since the primary excavation is accomplished first, full advantage of the geology revealed may be taken when loading trim-blasting shots. This is not possible in other controlled techniques.

In cushion blasting, using larger-diameter holes, better drilling alignment is possible. Better results are obtained in unconsolidated formations, and these results are ascertainable after the first shot.

LIMITATIONS: Trim blasting requires that the drilling, loading, shooting, and excavation spread be moved back into the working area twice: once to perform the production excavation, and again to do the trim or cushion blasting. This normally makes the process more costly than presplitting.

The primary blast may remove part or all of the planned buffer which was to be used in the trim blasting. This could require complicated loading in the individual holes or for the whole shot.

Site conditions may make it very difficult or unsafe to move drilling equipment back onto narrow berms above high blasted faces. It is not practical for cutting around curves or corners without also using line-drill or presplit techniques.

Overbreak control methods are obviously most successful in solid, homogeneous rock. In order of usual preference, the controlled methods to be used in poor to good rock are line drilling, cushion blasting, and presplitting. This sequence could be altered or reversed for other economic considerations. It is always necessary to evaluate each system in terms of geology, physical site conditions, and the skill of the labor available.

Noise and Vibration Control Vibration and noise from blasting are lessening as public nuisances. This is partly because of the operator's successful attempts to exercise closer control over the nature and timing of his blasting and partly because decreased vibration and noise are natural side benefits from the use of scientific blast designs. A shot which utilizes the explosives to obtain a maximum of breakage will usually minimize vibration.

The public becomes aware of blasting by two physical sensations—noise and vibration. These nuisances have occasioned a number of claims for damage, real or imagined. As a result, the search for improved blasting techniques continues.

Noise. Blasting vibrations are lessened by the use of the modern millisecond blasting cap, which detonates successive holes at intervals as short as 0.005 to 0.0025 sec. Unfortunately, a similar reduction in sound does not usually occur. Since sound travels in air at a relatively low velocity, the noise generated by each detonation in a delay shot is merged in some or all directions. Weather conditions affect the intensity of the noise from a blast as it is heard at a distance, since the variables that affect sound propagation are wind velocity and temperature as a function of altitude. However, the generally accepted idea that noise from a blast is reflected from the lower surface of clouds is mistaken. A cloud layer or flat sheet of clouds is often a sign of temperature inversion, and it is this condition that actually causes an increase in noise level.

The relation of the sound wave propagation to the blast site is shown in Fig. 19-23 for three basic weather conditions. Generally favorable conditions are clear to partly cloudy skies, with fleecy clouds and relatively warm daytime temperatures. Cloudy days with rapidly changing winds, perhaps with showers, would also be favorable. Late evening or night blasting usually results in noise disturbances, since temperature inversions increase during that part of the day.

Unfavorable conditions consist of relatively still air, usually foggy, hazy, or smoky. When smoke from a nearby stack fans out horizontally, it is a good indication that an unfavorable inversion exists.

Although there are no established noise thresholds which must be observed by law or regulation, it is the operator's obligation not to create a noise nuisance. Unfortunately, nuisance can be defined in as many ways as there are people hearing the noise.

Vibration. Vibration, unlike sound, can be measured, graded, computed, and classified. Seismic vibrations can be and have been studied extensively, and it is possible to derive standards to apply to the vibrations which are generated in quarry and construction blasting.

The peak particle velocity in the seismic wave train is the measured parameter which best correlates with structural damage. Research investigations[17] by Langefors of Sweden, Edwards of Canada, and the U.S. Bureau of Mines have indicated that the structural soundness of nearby structures will not be adversely affected if the peak particle velocity is limited to 2 in. per sec (see Fig. 19-24). The state of New Jersey, one of the leaders in adopting blasting regulations, has placed this critical velocity at 1.92 in. per sec.

Peak particle velocity is determined by seismic measurements. The instruments

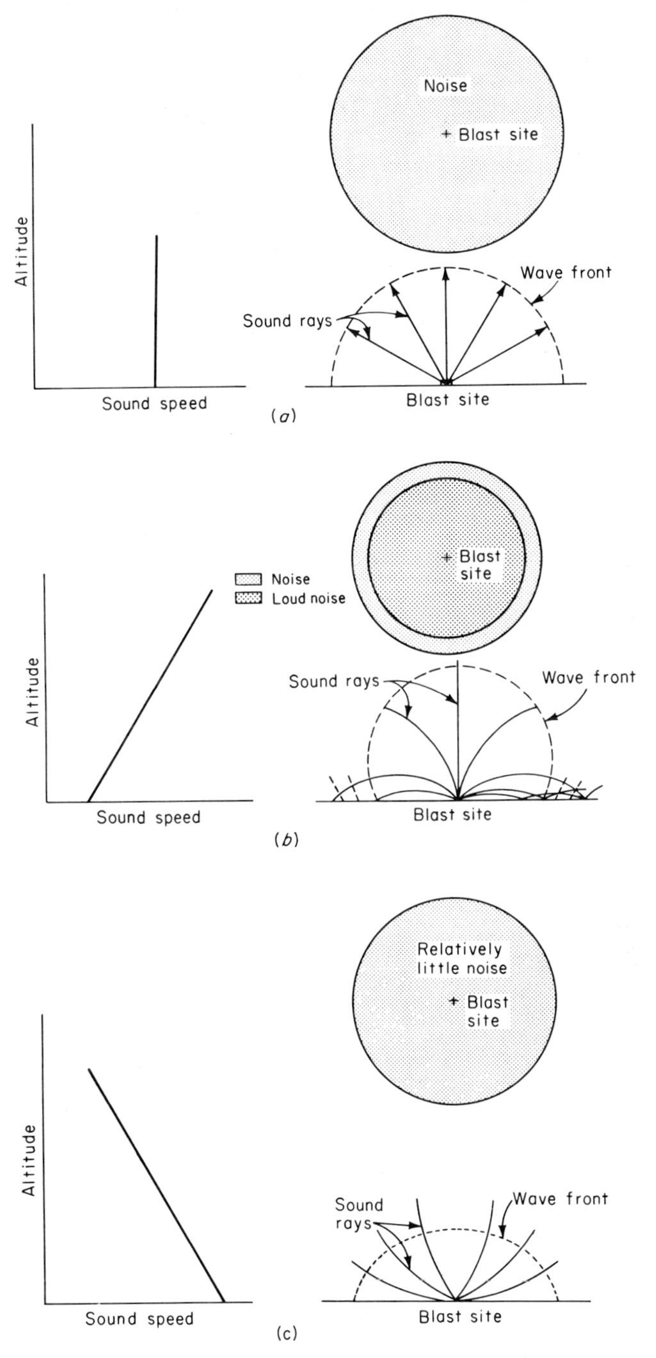

Fig. 19-23 Influence of weather conditions on sound propagation from a blast site. (*a*) Sound propagation from an isothermal condition (sound speed constant with altitude). (*b*) Sound propagation from a positive thermal gradient inversion (sound speed increases with altitude). (*c*) Sound propagation from a negative thermal gradient (sound speed decreases with altitude).

used in the field to measure blast vibrations will record either particle velocity versus time or displacement versus time. In the latter case, both displacement in inches and frequency in cycles (amplitude and frequency) can be obtained.

In the event that no seismic measurements can be taken, various agencies have compiled "distance-loading" tables to regulate blasting. Table 19-17, used by the state of New Jersey, is an example. Such tables do not take into consideration the specific nature of the terrain (overburden) nor the effect of the different underlying materials in damping out or absorbing the energy of the elastic vibration waves.

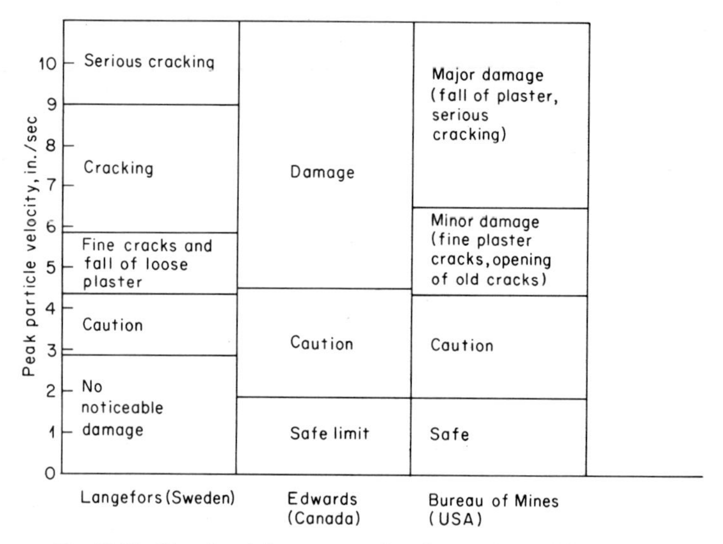

Fig. 19-24 Structural damage as related to peak particle velocity.

TABLE 19-17 Quarry Charge versus Safe Distance to Nearest Structure

Distance to nearest structure, ft	*Maximum quantity of explosives per instantaneous shot or per delay shot, lb*
100	75
200	92
300	116
400	140
500	176
600	209
700	259
800	330
900	403
1,000	495
1,200	770

The terrain damping coefficient may vary widely. Assume, for example, that the vibration produced by a wave at a given point on the surface of a rock mass is assigned a coefficient of 1. The measured displacement at this same point for the same wave would be possibly ten times as much if an unconsolidated soil overlies the rock. This assumes that the overburden is of an average thickness, perhaps one-quarter to one-half the wavelength. Abnormal thicknesses of overburden (half a wavelength or more) of water-soaked sand, dry sand, gravel, or loam may be subject to thirty times the displacement. Seismic measurement is the most reliable method of determining vibration effects and hence the optimum distance-loading relationship.[18]

The energy ratio is another method of expressing the vibration amplitude–peak velocity effect. In the energy ratio theory, a ratio value of 1 corresponds to the 2-in.-per-sec limitation on peak velocity. The energy ratio is expressed as

$$ER = (3.29FA)^2 = 1 \text{ for critical locations}$$

where F = frequency of vibration, cycles per sec
A = amplitude (displacement) of vibration, in.

This formula was applied on the Dworshak Dam project, where it predicted that the explosive charge should be limited to 9,000 lb per delay when shooting foundation rock in areas where the overburden had been stripped. The maximum charge was reduced to only 2,400 lb per delay in areas where the overburden remained in place.

Powder-loading guidelines, derived from a ratio called "scaled distance," have also been recommended to minimize damage to nearby structures and nuisance to the

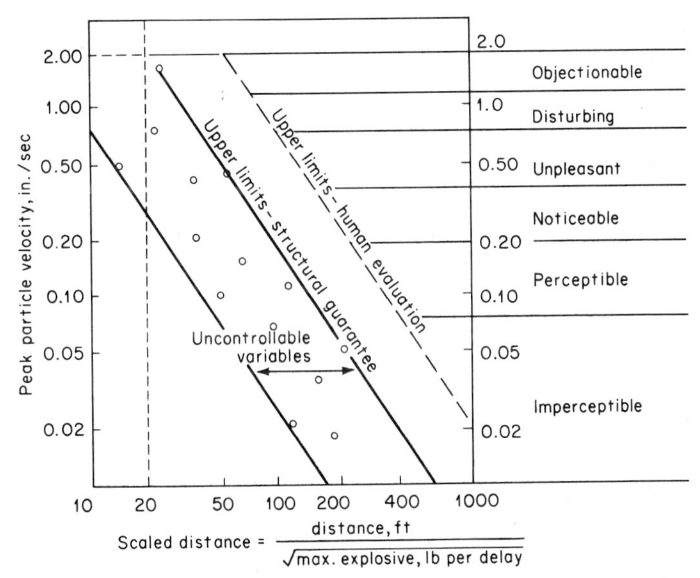

Fig. 19-25 Scaled distance as a function of peak particle velocity. To avoid cracking of plaster, the scaled distance would be limited to 20 (at 1,000 ft, use 2,500 lb of explosive per delay) at this particular site.

public.[17] The scaled distance is determined by the ratio of the blast's distance and the square root of its explosive weight. Two blasting situations, one involving a distance of 100 ft from a 1-lb blast and the other 1,000 ft from a 100-lb blast, would therefore have the same scaled distance (and thus the same predicted vibration). Figure 19-25 indicates the measured vibrations from a quarry operation and relates peak particle velocity to scaled distance.

Human Factors. The operator can have reasonable assurance that his powder loadings are not causing structural damage wherever particle velocities are kept to 2 in. per sec or less. However, he may still find it necessary to operate with much lower loadings because of the adverse reaction of his neighbors.

Human response to vibrations is much more sensitive than structural response and is a real problem that deserves proper consideration. It has become apparent that human evaluation is based not only on the amplitude of motion but also on its frequency. The higher the frequency, the less the motion that will be tolerated. Figure 19-25 correlates peak velocities and scaled blast distances in terms of observed human response to blasting operations. To operate with the fewest complaints, the peak velocity should not exceed 1 in. per sec, a scaled distance of 75. For the example given on this figure, at a distance of 1,000 ft from the blast, the weight of explosive should be limited to 170 lb per delay to avoid adverse human reactions. This limit would be applicable for a short-term construction operation; for a permanent quarry operation it should be lower still.

The charge limitation suggested in this example is a self-imposed one, based upon generalized relationships between particle velocity, scaled blast distance, and human response. It is entirely conceivable that seismograph tests after operations have begun will permit the limit to be raised considerably. For example, if seismic tests indicate that a scaled distance of 20 rather than 50 is acceptable, this will permit the shooting of almost six times as much powder per delay, with resulting economies. Seismic testing is another example of the successful application of scientific methods to blasting operations.

LOADING AND HAULING

The modern rock-moving contractor is no longer limited to the traditional method of loading and hauling shot rock by shovel and truck. There now are various systems to evaluate as well as a multitude of individual machines in each system category. Consequently, the trend today is to undertake more prejob studies in order to identify the optimum system and its optimum equipment. Computer analyses are now routine and are capable of furnishing immediate answers to the questions of the theoretical economic advantage of one system over another.

In the following paragraphs, a review is made of the major factors which affect this system and equipment selection. The operating characteristics of the equipment are also examined.

Loading The classic method of loading shot rock has been by power shovel. However, large wheel-type front-end loaders have now taken over a sizable share of the top-loading work. The selection of equipment for loading shot rock must usually be considered from one of two viewpoints: (1) a quarry operation where the requirements are specific, unchanging, long-term, and of high volume; and (2) the construction open cut where the equipment requirements may be for multipurpose use, short duration, and limited volume.

In each case, the selection will be based on these basic considerations:

1. The cost per cubic yard of material loaded. *Factors:* Total volume and rate of production required; depreciation, taxes, insurance; repair and maintenance; labor and materials; operating costs and estimated escalation.

2. Operating conditions. *Factors:* Size and weight of material; geometry of cut; primary crusher size; height, depth, and tightness of muckpile; height of face; size of hauling units; weather conditions; optimum rate of loading (to reduce number of haul units required); availability of attachments for conversion of the loading unit for other uses; transportation limitations (state regulations may limit size for highway moves, overall dimensions may limit rail transportation); capability to move in and out as required for blasting operations.

3. Existing and forecast economic conditions. *Factors:* Favorable or unfavorable for resale of equipment (influences the depreciation charged to work, and hence unit costs); limitations in available capital may make the delivered cost of the unit the primary consideration, regardless of the cost per cubic yard or operating conditions.

None of the above factors can be arbitrarily selected, but instead the entire system must be evaluated. For example, the cost of drilling and shooting to produce the desired fragmentation and muckpile characteristics may prove to be too costly because of the rock structure and hardness. When the drilling and shooting are revised to fall within economic limits, the fragmentation may have changed to such an extent that the previously selected shovel is no longer suitable. Again, if the haul characteristics are found to be such that only a certain type of haul unit can be used economically, a shovel or loader size which has been selected for other reasons may not be compatible with this particular haul unit. The use of loaders may be curtailed for this reason, since their top-loading capability is limited. Or, once the optimum fragmentation, shovel size, and hauling units are selected, it may be found that the geometry of the cut will not permit an efficient loading operation. As a result, some units of the system must then be changed.

*Shovels.** The shovel is still the recognized method of loading rock in heavy construction and quarry excavations. It is then the principal link in the production chain, and its selection must be given careful consideration since this can often mean the difference between profit and loss.

When a shovel unit is being selected, proper consideration should be given to those features which would affect its anticipated production for the proposed application. There are three principal features to be evaluated: (1) convertibility to other uses, (2) kind of power (diesel, diesel-electric, or electric), and (3) length of crawlers (long or short). A fourth feature to consider is the kind of mechanism used for the machine swing.

If the specific application is loading rock in a quarry, we are not interested in the shovel's convertibility characteristics. However, we will want short crawlers, electric power (if an energy source can be supplied economically), and a fast, powerful swing mechanism. For construction purposes we will normally prefer a universal machine which can be readily converted to a dragline, crane, or backhoe. We might then be prepared to sacrifice rock-excavating ability by selecting a machine with long crawlers, a mechanically operated swing, and diesel or diesel-electric power. It must be borne in mind, however, that on modern, large-quantity construction projects, the rock excavation requirements for a shovel may be of such a magnitude that quarry standards in equipment selection may be economically applicable. In such a case, with short tracks being selected, the alternative use of the machine as a dragline may be limited by stability problems and its hoisting capabilities as a crane would be reduced. The type of swinging mechanism for the shovel should also be selected on the basis of the specific requirements. The choice between mechanical, electric, hydraulic, or air mechanism is of particular importance if conversion to a crane or dragline is contemplated.

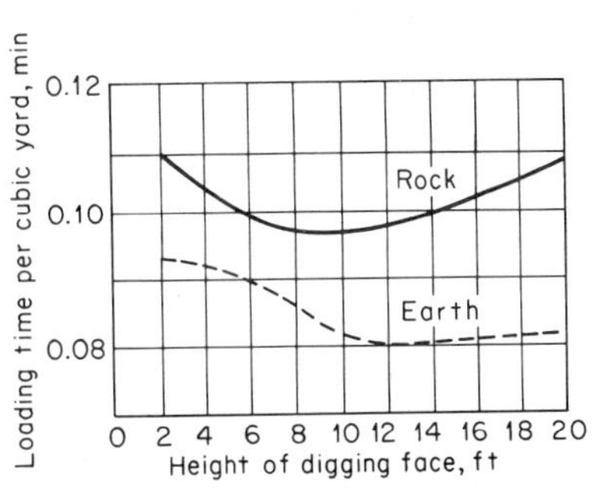

Fig. 19-26 Shovel production as affected by height of digging face.

Present-day competition in standard shovel models is keen, and the buyer gets what he pays for. All shovels are quite similar in their principal features, having enclosed drives running in lubricant, metered air controls, torque converters, and the same selection of motors, i.e., GMC, Caterpillar, and Cummins. All makes offer high-speed independent boom hoists, power load lowering, and basic safety features.

Shovels are best suited for close-range work where they have good control of the digging; they are speedy with little lost motion when the muckpile has the proper characteristics. As previously discussed under Blast Design, the muckpile disposition is the result of many factors, but its height is important to the extent that it permits the shovel bucket to be filled in a single loading pass. A digging face of 6 to 12 ft will usually produce near-optimum shovel production when loading rock with no oversize material. If oversize material is limited to approximately 5 percent, the bucket loading curve will still have the same general characteristics.

Figure 19-26 indicates that the loading time per cubic yard will increase for faces less than 6 ft and will reach a maximum at a face height of 2 ft. Face heights greater than the optimum 6- to 12-ft range will also increase loading time. It is not always practical or possible to use explosives in such a way as to produce a muckpile height which yields minimum bucket-loading costs and, when the face is low, second or even

* See Sec. 10, Excavators.

third bucket passes may be justified. The working cycle for the shovel consists of loading, swinging, dumping, and returning. The larger the percentage of bucket loading time in this complete work cycle, the less desirable it is to make additional passes.

Good operation in rock dictates that the shovel be kept close to the working face of the muckpile. When the shovel is in this position, the shovel bucket has a tendency to work its way into the loose rock as it is lifted. This saves much hard work for both shovel and operator by keeping the use of the crowd mechanism to a minimum.

The swinging time and the spotting of the bucket in its dumping position over the hauling unit are prime factors in production. Every attempt should be made to plan the operation so that the shovel swing between loading and the hauling unit is 90° or less. Where space permits, much shovel time may be saved by having hauling units spotted on both sides of the shovel. This eliminates shovel waiting time while the loaded hauling unit pulls out and the empty unit maneuvers into loading position. Under certain conditions, where access is tight or visibility is poor, a spotter may be

Fig. 19-27 Caterpillar M210-8-4 wheel loader–truck operations. (*Caterpillar Tractor Co.*)

used to direct the hauling units under and away from the shovel and to regulate the traffic in the loading area.

The spotting of hauling units can be simplified and repair and maintenance costs appreciably reduced if the access area around the shovel is kept clean and free of rock spillage from the loading operation. This service, which is usually performed periodically by a dozer or wheel loader, is necessary to ensure an uninterrupted loading operation.

*Wheel-type Front-end Loaders.** Within a certain range of operating conditions, a choice is possible between the traditional shovel and the modern, large, rubber-tired front-end loaders (see Fig. 19-27). For example, Caterpillar's 988 loader with a 6-cu yd rock bucket can equal the production of a 3½- to 4-cu yd shovel if operating conditions are favorable. Table 19-18 supplies the results recorded[19] by a Montana contractor when loading shot blue-gray basalt weighing 4,700 lb per bcy.

Another study involved a highway job in Pennsylvania which required the loading and hauling of rock weighing 2,800 lb per bulk cu yd. Here an International H-400 front-end wheel loader equipped with a 10-cu yd bucket and a 71B shovel with a 4-cu yd bucket were compared when loading out Model 180 Payhauler trucks. The

* See Sec. 11, Loaders.

TABLE 19-18 Comparison of Front-end Loader and Shovel Costs in Loading Shot Basalt

Unit	Caterpillar 988 loader	4½-cu yd shovel	2½-cu yd shovel
Truckloads per shift	150	170	100
Yardage per shift, bank cubic yards—solid state	1,900	2,160	1,000
Average hourly production:			
Bcy/hr	238	270	125
Tons/hr	560	635	294
Estimated average hourly machine cost*	$25.60†	$29.50	$22.31
Per bank cubic yard	$ 0.108	$ 0.109	$ 0.178
Per ton	$ 0.045	$ 0.046	$ 0.075

* Hourly cost computed by standard industry methods, including operator wages in force on this contract.

† Hourly cost for 988 reflects severe conditions.[19]

following results were observed:[20]

1. Each truck could be loaded by three to five passes with the H-400 or eight to ten passes by the 71B.
2. When loaded with the H-400, the trucks carried 6.6 percent (3.2 tons) more than when loaded with the 71B.
3. The average load time for the H-400 was 2.55 min. The corresponding loading rate of 12.7 tons per min was 58 percent faster than the shovel loading rate.

Further comparison can be made from an estimate published by the Caterpillar Tractor Co. which pertains to power-shovel and front-end-loader operations.[19] (See Table 19-19.) The power-shovel prices, ownership and operation costs, and production figures used in this estimate were taken from Power Crane and Shovel Association publications. The owning and operating costs for the loader are based on severe operating conditions, and the loader production is averaged from the results of several job studies. The equipment prices were current at the time that the estimate was prepared, so the production cost figures are comparable.

The power shovel of today thus has significant competition from the large front-end loaders under certain conditions. There are, of course, certain limitations to the loader's operation. For example, it would be hazardous to work it under a high face. The loader is only capable of loading loose rock, since it does not have the digging capabilities of the large shovels; the muckpile must also be fashioned to the loader's capability, and this may require higher blasting costs or the assistance of a feeder dozer. These conditions for loader operation add to the overall costs.

There are also benefits associated with the loader operation which can wholly or partially offset these costs. The loader's mobility can reduce the move-out time prior to shooting and the time for moving back immediately after. A movement

TABLE 19-19 Power Shovel—Front-end Loader, Estimated Production and Cost

Item	Power shovels					988 loader
Unit, cu yd	2½	3	3½	4	5	6
Estimated price	$117,210	$126,215	$127,140	$154,225	$218,925	$76,715
Depreciation hours	27,000	27,000	27,000	28,000	28,800	10,000
Estimated O&O	$24.24	$24.62	$26.24	$27.75	$35.89	$28.79
Production:						
Tons/hr	440	512	584	656	872	625
Cost/ton	$0.056	$0.048	$0.045	$0.042	$0.041	$0.046

between cuts which can be made in 10 min with a loader might take 1 or 2 hr with a shovel. The loaders can load out from tight quarters and work either side of the cut, and they can back off quickly should a slide occur. The rock buckets on large loaders can handle big boulders which would have to be redrilled and shot to pass through a shovel bucket. Loaders are multiple-purpose machines which can perform their own cleanup; their move to the next job can be made over the highway under their own power or on a single lowboy.

*Belt Loading.** Under some circumstances, it is practicable for ripped material and certain kinds of shot rock to be dozed to a conveyor belt for top loading of haul units. The topography and work layout must be suitable if this loading system is to be competitive, and the material must include sufficient fines. Although this is not the best system for many job conditions, it merits consideration whenever possible. The belt loader is a high-production unit, and a very large investment in shovel equipment would be needed to match the loading capacity of a 60- or 72-in. belt.

Scrapers are also used to load out ripped and shot rock, and specially strengthened models are available for this application. This developing usage will be discussed under Hauling.

Hauling As the primary considerations for a haulage system, the hauling unit must be strongly constructed, with adequate speed and power, and the haul roads must be the best that economic restrictions warrant. Even with matched loading and hauling units, it may still be found that the system is inefficient because of poor haul-road characteristics. For a stipulated production, the haul-road conditions (rolling resistance, alignment, grades, etc.) will dictate the required type and number of haul units as well as their operating and maintenance costs. In any planned system the haul road is the one item, more than any other, which the contractor can directly influence. It is good practice, therefore, to devote considerable time and money to the planning, construction, and maintenance of such roads.

Rapid economic comparisons between various haul units operating on various haul roads can be made by using a high-speed digital computer. To obtain the predicted result, the haul road as built and maintained should have equal or better characteristics than the theoretical road used in the system selection. Grades should be held to a minimum, good surfacing and drainage provided, all curves widened and straightened if possible, and visibility improved by clearing brush and trees so that the highest speed safely attainable can be used.

Trucks.† Most haul units in use today are off-highway rear-dump trucks. Due to continuous improvements over the years, these trucks now have better weight-to-horsepower ratios, improved bodies and suspension systems, and larger payload capacities. Less than 15 years ago the standard construction size was 22 tons. Since then contractors have moved into the 40- to 50-ton class as normal practice, and indications are that even the larger units now available will be put to increasing use (Tarbela Dam contractors will use 110-ton bottom-dump trucks).

Although the mine and quarry operators have led the way in the use and development of the larger haul units, the heavy-construction contractor, faced with the ever-increasing volumes of material to be hauled in modern large projects, has been quick to utilize the largest units available. In some cases, he has had new and larger units developed for a specific job. Oroville, San Luis, and Mangla Dams, all of which were successfully completed in the period between 1960 and 1965, required the combined handling of over 232 million cu yd of material at a rate of approximately 3.5 million cu yd per month. It is obvious from these figures that considerable emphasis must have been placed on the development and use of large-size hauling units. A hauling system which includes these high-capacity units should be carefully planned for their effective use. It is no advantage for a truck to have improved transporting ability if an inordinate amount of time is required to load it. As the ability to transport materials at a faster rate increases, the capability to load it faster must also be provided. The loading unit must be matched with the hauling unit.

* See Sec. 11, Loaders.
† See Sec. 13, Hauling Units.

*Scrapers.** When the haul distances are short, it is often economical to use the scraper both as the rock loading unit and as the hauling unit. (See Fig. 19-28.) Scrapers are adapted for this purpose by strengthening the bowl bottom (both top and bottom plates), cutting edge supports, router bit supports, and the apron lip. Their physical use is dependent upon the geometry of the cut and the loadability of the rock; their economic use is dependent upon their maintenance costs, availability (downtime), and tire costs.

Scraper loadability is limited to material having 1½- to 2-ft-size fragments or boulders, with enough smaller pieces to provide lubrication for good loading action and to fill the voids. When working in this kind of material, such as well-ripped limestone, the scraper load factor is about 0.65 compared to 0.75 for dirt.

Sedimentary rocks such as shale, sandstone, and well-laminated limestone will generally break up with sufficient fines to make scraper loading feasible. Solid metamorphic rocks such as granite and basalt will break into blocks of relatively constant size with few fines and are extremely difficult to load.

Fig. 19-28 Scraper loading rock; push unit assisting. (*Caterpillar Tractor Co.*)

The cost factors for loadability, repair and maintenance, machine availability, and tire life will vary significantly when different job conditions prevail. The following descriptions indicate the classes of material which may be encountered:[21]

Condition	*Description*
A	*Favorable—Common Earth.* Common earth materials with adequate cohesive properties for good loading; no rock particles; smooth, well-maintained cut.
B	*Unfavorable—Common Earth.* Clay, hardpan, or lightly cemented granular materials; relatively hard to load without ripping; no sizable rock particles; average cut maintenance.
C	*Moderately Severe Earth-Rock Mixture.* Well-graded mixture of earth and sandstone, shale, caliche, or other rippable materials; maximum normal particle size, 8 to 12 in.; adequate fines to fill voids and produce fair loading conditions; average cut conditions.
D	*Severe Earth-Rock Mixture.* Poorly graded earth-rock combinations. Predominantly rock particles; moderate percentage of fines; maximum normal particle size, 12 to 18 in.; somewhat rough, rocky cut and fill conditions.
E	*Extremely Severe—Ripped or Shot Rock.* Sharp, block-type fracturing; minimum percentage of fines; very hard loading due to interaction of rock particles; maximum normal particle size, 18 to 24 in.; rough rock cut and fill with average maintenance.

* See Sec. 9, Scrapers.

Using these criteria, hundreds of field studies have made it possible to evaluate scraper performance when loading and hauling rock. Figure 19-29 illustrates the results which can be expected when operations are closely supervised, skillful operators are employed, and a thorough and effective preventive maintenance program is in effect.[21]

Loading and Hauling Economics A realistic study of the overall economic picture should be made before choosing between a scraper and shovel-truck spread. A typical job situation can be analyzed as shown in the following example.[19]

The job, which until recently would have automatically been assigned to a shovel-truck spread, requires the moving of 750,000 bcy of a severe earth-rock cut (condition D). This material weighs 3,600 lb per bcy. The haul road is 2,500 ft in length, with a rolling resistance of 60 lb per ton and a favorable grade of 6 percent for 1,200 ft. The cut is 150 ft long with a rolling resistance of 100 lb per ton. The fill is 300 ft long and has a resistance of 180 lb per ton.

Two balanced spreads are chosen for comparisons. Spread A consists of three Caterpillar 631 scrapers, specially adapted for rock, with a Cat D-9 pusher. Spread B consists of a 5½-cu yd shovel and two Caterpillar 769 hauling units.

Spread costs (Job condition D assumed)	Spread A, scrapers	Spread B, shovel-trucks
Average load factor (from Fig. 19-29)	0.65	70,000 lb (rated)
Repair factor (from Fig. 19-29)	100%	85%
Average availability (from Fig. 19-29)	87%	92%
Average tire life (from Fig. 19-29), hr	1,500	2,500
Estimated average hourly cost/hauling unit*	$38.33	$23.79
Total hourly cost of haul units	$114.99	$47.58
Estimated hourly pusher/shovel cost	$24.40	$38.89
Spread cost per hour	$139.39	$86.47

* Estimated by standard industry method. Scraper costs for Condition D include repair reserve of $6.83, and tire reserve of $6.32. Operators' wages of $4.00 and oilers' wages of $2.50 per hour are assumed. The costs for the 769 truck are developed for average rock-loading conditions with the truck equipped with extra tread-depth tires.

Spread production (Machine performance calculated by digital computer)	Spread A	Spread B
Average load times, min	0.80	2.0
Average loaded travel time in cut, min	0.10	
Average positioning time at dump, min	....	0.15
Average dump time, min	0.25	0.30
Average position and wait time, min	0.25	0.25
Total fixed time, min	1.40	2.70
Haul time, min	1.16	1.01
Return time, min	1.37	1.10
Total cycle time, min	3.93	4.81
Trips per 60-minute hour	15.27	12.48
Average payload (0.65 load factor), bcy	19.5	19.5 (35.1 tons)
Unit production per 60-minute hour, bcy	298	243
Total hourly spread production, bcy	894	486

The above production figures assume 100 percent working efficiency and 100 percent machine availability, and must be adjusted as follows:

	Spread A	Spread B
Calculated time to complete job with working efficiency and machine availability each taken at 100%, hr.	839	1,543
Estimated time at 75% working efficiency, 87% scraper availability, and 92% truck availability, hr.	1,286	2,235
Estimated total for load and haul costs.	\$179,256	\$186,555
Estimated production cost/bcy.	\$0.239	\$0.249

This comparison indicates that substitution of scrapers for the shovel-truck system will result in a cost savings of \$0.010 per bcy and will reduce the total working time by 949 hr. This answer is obtained under conditions where each scraper is charged an hourly repair reserve over 50 percent higher than for a truck (\$8.80 versus \$5.31); an hourly tire cost almost triple that of the truck (\$8.02 versus \$2.54); and more than 60 percent greater downtime (13 percent versus 8 percent).

While the production cost per bcy is usually the basic criterion for comparing spreads, the capital investment required for each bcy per hour of productive capacity is another important factor which should be considered. Comparing the spreads of our example we find that the investment per bcy per hour of production capacity is \$760 for

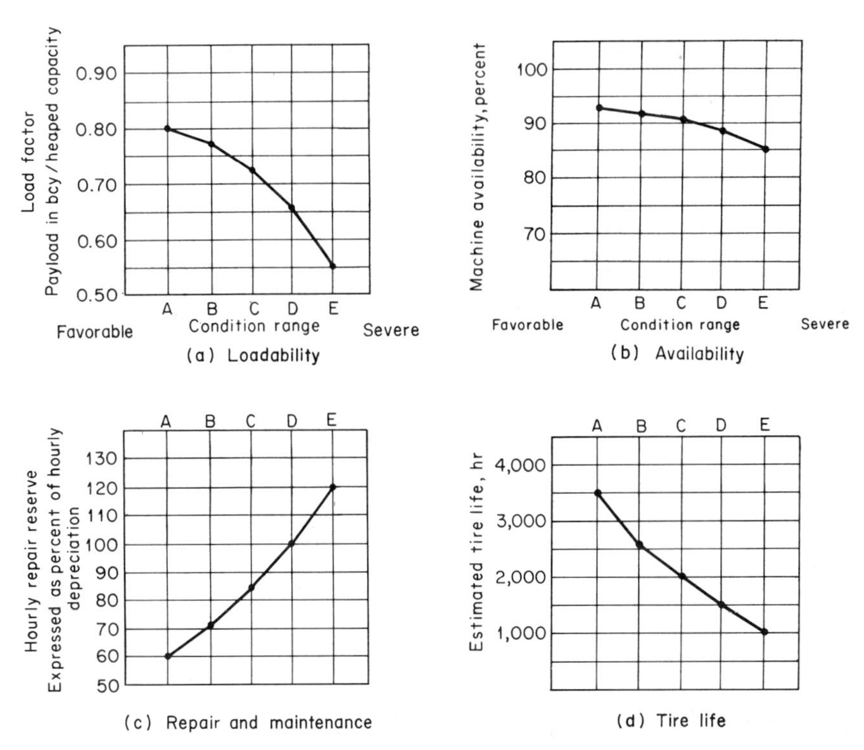

Fig. 19-29 Wheel tractor performance when loading and hualing shot rock under varying conditions.

spread B and $427, or 78 percent less, for spread A. The anticipated return on investment is a very important factor in any equipment selection decision. Obviously, from this viewpoint and by reason of its lower projected cost per bcy, spread A is the favored choice.

A scraper which is used to haul shot or ripped rock to an embankment can provide still another possible cost saving. It can spread the material uniformly in the dumping process, and this lessens the requirements for spreading dozers or blades as compared to the spot-dump unloading by trucks.

This example, however theoretical, should alert the contractor to the real possibilities of scraper applications in rock loading and hauling. Pusher-scraper equipment, due to its basic efficiency and versatility, can frequently overcome the economic handicaps of reduced payloads, greatly accelerated repair and tire costs, and lower availability. In such cases, it offers an economical method of moving rock materials under severe conditions.

C. Explosives

This subsection describes briefly the basic types of explosives commonly used in the construction industry, their applications, and their characteristics. It also discusses loading methods and initiation systems. Because of differences between similar products produced by the several suppliers, much of the information in this subsection is supplied on a generalized basis. For more precise information about a particular type of product, the supplier must be consulted.

CHARACTERISTICS OF EXPLOSIVES

Types Various types of explosives are available for use in construction applications. The selection of the best type of explosive for a particular use, as noted earlier under Blast Design, will involve an evaluation of the properties of the explosive and of other properties pertaining to the rock itself. The major types of explosives are briefly discussed below.

Ammonia gelatins are powerful, plastic, high-density explosives which generate very high detonation pressures, borehole pressures, and detonation velocities. Having excellent water resistance, they are often used as a bottom charge in wet holes. They are also used as a bottom charge in either wet or dry holes when there is a tendency to leave a high bottom. The lower grades are used as a column charge in rock which is extremely hard to fragmentize or where water conditions are unusually severe. The 75 percent and higher grades are sometimes used as primers for AN/FO and slurries, but, because of an inherent characteristic of ammonia gelatins to initiate at low-order detonation velocity and build up to high order, 2 ft of column height may have to be used to ensure that high-order detonation velocity is attained. There are several ammonia gelatins available, such as the Atlas Power Primer, that are specially formulated to initiate immediately at high-order detonation velocity. These products, therefore, are preferred as primers for the various types of AN/FO and slurry.

Ammonia dynamites differ from ammonia gelatins because the nitroglycerine is absorbed into other ingredients in the mixture rather than being gelatinized. The resultant product is less powerful and water resistant and has a lower velocity and density. Ammonia dynamites are available in varying grades and in either high- or low-density formulations. They are cap sensitive and are frequently used as a column charge below the water level in wet holes. However, because they are less water resistant than the ammonia gelatins, it is recommended that one or more cartridges of the latter explosive be used as a primer at the bottom of a wet hole. The grade of cartridged ammonia dynamite should be one which can be expected to blast just as well as poured AN/FO would if the hole were dry.

Semigelatins are classed between ammonia gelatins and ammonia dynamites because they combine some of the physical advantages of the former with some of the high-cartridge-count advantage of the latter. They are more powerful, plastic, cohesive, and water resistant than ammonia dynamites, although they are not the equal of ammonia gelatins in these respects. They are frequently used in tunnels and other underground work. On surface operations where holes are wet, they are sometimes used as a column charge above a bottom charge of ammonia gelatin until the column is out of the water. In easily broken rock or shallow holes, they are sometimes used as a bottom charge.

Ammonium nitrate–fuel oil (AN/FO) is available in several product types. The simplest of these is a mixture of explosive-grade ammonium nitrate prills and No. 2 diesel fuel oil. This mixture is available in moistureproof bags from which it is poured into either vertical holes or a pneumatic loader. In addition, there are a number of types of specially designed bulk-loading trucks available. The adaptability of a particular job to such mechanization and the type of equipment best suited to job conditions should be discussed with the explosive supplier.

A more powerful explosive mixture is *crushed AN/FO* in which some of the prills have been crushed prior to mixing. This provides better explosive characteristics and a higher loading density. Crushed AN/FO is normally supplied in moistureproof bags.

In those instances where it is desirable to eliminate decking in the borehole, a *low-density AN/FO* is available in which a portion of the ammonium nitrate is replaced by a lightweight, inert material. This product, which is available in moistureproof bags, has explosive properties which are quite weak.

A *high-density AN/FO* is available in waterproof bags for loading in wet boreholes. It is packaged in 4-in.-diameter and larger sizes and attains sinking density. When loaded directly into wet holes, it is entirely dependent on the package for its water resistance.

Two more recent developments are *aluminized crushed* and *aluminized high-density AN/FO*. These products provide more blasting energy than their corresponding nonaluminized mixtures and are useful for column loading where good blasting results are rather difficult to attain. Aluminized crushed AN/FO is available in moistureproof bags for pouring into the dry portion of a borehole. Aluminized high-density AN/FO is usually supplied in dimensional water-resistant cartridges. These cartridges are normally 4 in. or more in diameter, but some success has been achieved with cartridges as small as 2½ in. in diameter.

The various AN/FO mixtures generate more efficient explosive reactions as the column diameter increases. For instance, the following detonation velocities have been recorded in borehole measurements:

Borehole diameter, in.	*Detonation velocity, fps*
3½	12,300
6½	13,500
7	13,900
10⅝ and above	14,600

Not only does the volume of AN/FO per lineal foot of borehole increase as the diameter of the hole increases, but the efficiency of explosive reaction per cubic foot of AN/FO also increases. The effects are additive.

If job conditions require that relatively small-diameter boreholes be drilled and if it is difficult to obtain the desired breakage, higher costs for drilling and handling may more than offset the low selling price of AN/FO. Under such conditions, an evaluation of overall costs may show a clear economic advantage to using a more powerful explosive.

In order to obtain the full potential energy from any AN/FO product, proper priming is essential. Extensive laboratory and field trials have shown that a good primer must meet the following three requirements:

ADEQUATE SIZE: Primer diameter must be as close to the diameter of the AN/FO column as possible. Primer length must be sufficient to ensure that the primer is detonating at high-order velocity when it initiates the AN/FO.

GOOD EXPLOSIVE PROPERTIES: High-order detonation velocity must be achieved in a short distance.

HIGH DETONATION PRESSURE

In holes 1½ to 4 in. in diameter, an ammonia gelatin specially formulated to achieve high-order detonation velocity in a short distance is the most efficient and economical primer. In 5-in.-diameter or larger holes, consideration should be given to the use of slurries for priming or boostering AN/FO products. The slurry must meet the three requirements of a good primer that are described above. If the slurry is not cap sensitive, it must be initiated with a primer, preferably a cartridge 2 in. in diameter or larger of one of the higher grades of ammonia gelatin.

Slurries have been available for the last several years. They are essentially mixtures of ammonium nitrate, water, and a sensitizer such as TNT, smokeless powder, aluminum, etc. The ability of slurries to completely fill the diameter of a borehole permits maximum loading density per lineal foot of hole, with maximum blasting forces applied to the rock. In many instances this permits expanded burden and spacing dimensions and results in lower drilling costs per cubic yard of rock. Slurries vary from excellent to fair in their resistance to water, but many are capable of prolonged exposure to water. Their explosive properties also vary over a wide range, and care must be taken to ensure that the properties of a particular brand or grade will fit the job conditions.

Packaged slurries are generally used in boreholes that are 4 in. or larger in diameter. Several slurries can be poured or pumped, and these are generally used in 2- to 4-in.-diameter boreholes. Slurries are used as a bottom charge or as an up-the-column booster for AN/FO. The more powerful brands are particularly useful as a bottom charge where toe problems are difficult to overcome. Slurries require the use of an adequate primer and, just as with AN/FO, this primer must attain high-order detonation velocity immediately. It must also be large enough in diameter and length to initiate the slurry efficiently.

Properties Explosives are generally classified either as "high explosives" (HE) or "nitro-carbo-nitrates" (NCN). Those products which will not explode when subjected to the blasting action of a No. 8 strength blasting cap or the impact action of a high-power rifle bullet are generally classified as NCN. Those products that do explode when subjected to either of these actions are generally classified as HE. These classifications may have an effect on the type of storage that must be provided. It is necessary to know the content and meaning of all pertinent regulations so as to meet those conditions that apply to a particular jobsite. Table 19-20 lists the properties of typical HE and NCN explosives.

Cartridge strength is an expression of the comparative borehole pressure developed by equal volumes of explosives. It is a useful measuring stick because the loading crew in reality loads a volume rather than a weight of explosives into a borehole. It should be understood that explosives such as a slurry or poured AN/FO (which completely fill the borehole diameter) produce more borehole pressure than an untamped cartridge of an explosive of the same cartridge strength. The untamped cartridge fills only a portion of the borehole diameter, not all of it.

Weight strength is an expression of the comparative borehole pressures developed by equal weights of explosives. As has just been stated, the loading of boreholes is actually a volumetric process. Explosives vary widely in density, and the applicability of a particular weight strength to a given set of blasting conditions must also be judged on the basis of the density of the explosive in question. Cartridge strength takes into account both weight strength and density.

Confined velocity provides an indication of the type of blasting action which can be expected from an explosive. A high velocity normally indicates that an explosive will provide a sharp, breaking action. This characteristic is needed to fragmentize rock at the bottom of a borehole or any vein of hard, brittle rock. A low velocity normally indicates a slower, heaving action, which is needed in softer, more resilient rock. The lower-velocity grades are frequently adequate for up-the-hole loading.

Detonation pressure is a blasting property which has only been measurable in recent years. Explosives produce two useful and sequential forms of energy. Shock energy occurs immediately, and borehole pressure is developed shortly thereafter. Shock energy can be represented as a curve starting from zero and building up very rapidly to a peak, then decreasing somewhat more slowly to zero. Detonation pressure is the measure of the peak value attained during this action and is customarily expressed in kilobars (one thousand times normal atmospheric pressure, or about 14,700 psi). In breaking hard, brittle rock, an explosive with a high detonation pressure frequently provides finer and more uniform breakage. This, in turn, results in lowered costs in handling and crushing. It should be understood that an explosive, such as a slurry or AN/FO, that completely fills the borehole diameter transmits all its shock energy and detonation pressure to the surrounding rock because the explosive is directly coupled to the borehole wall. An untamped, firmly cartridged explosive cannot do so because the air gap between the explosive column and the borehole wall absorbs a substantial portion of the energy.

TABLE 19-20 Properties (Approximate) of Commonly Used High Explosives and Nitro-carbo-nitrates

Grade	Classification*	Cartridge strength, %	Weight strength, %	Confined velocity, fps, in thousands, 1½ × 8 in.	Detonation pressure, kilobars	Density		Water resistance*	Fumes*
						Cartridge count, 1¼ × 8 in./50 lb	Specific weight, grams/cc		
Ammonia gelatin, 75%	HE	75	63	17.5	135	100	1.42	E	G
Ammonia gelatin, 60%	HE	60	51	16.0	120	97	1.46	E	G
Ammonia gelatin, 40%	HE	40	33	12.0	75	90	1.58	E	G
Semigelatins†	HE	31–52	65	10.5–15.0	35–65	150–110	0.95–1.29	G	G
High-density ammonia dynamite, 60%	HE	50	60	13.0	50	110	1.29	F	G
High-density ammonia dynamite, 40%	HE	36	40	11.0	40	105	1.35	F	G
Low-density ammonia dynamites†	HE	26–47	65	9.2–11.0	34–46	160–118	0.88–1.60	F	G
Presplit powders†	HE	13–52	33–65	9.0–15.0	16–75	....	0.19–0.81¶	E–F	G–P
Slurries†	HE or NCN	30–50	40–60	9.0–18.0 (4 in.)	20–95	....	1.20–1.65	E–F	G
AN/FO‡	NCN	23	65	12.0 (3 in.)	16	....	0.81	P	P
AN/FO, crushed‡	NCN	25	65	12.5 (3 in.)	27	....	0.84	P	P
AN/FO, low density‡	NCN	7	65	10.0 (3 in.)	......	....	0.55	P	P
AN/FO, high density§	NCN	30	65	13.0 (5 in.)	34	....	1.10	F	P
Aluminized AN/FO, crushed‡	NCN	33	70	12.5 (3 in.)	28	....	0.95	P	P
Aluminized AN/FO, high density§	NCN	43	70	13.0 (5 in.)	35	....	1.12	F	P

* HE, high explosive; NCN, nitro-carbo-nitrate; E, excellent; G, good; F, fair; P, poor.
† Range of properties for products available.
‡ Product loaded to fill borehole.
§ Packaged in water-resistant package.
¶ Density in pounds of explosive per lineal foot.

Density of explosives is commonly stated either as the cartridge count (number of 1¼- by 8-in. cartridges packed in a 50-lb case) or as a specific weight (grams per cubic centimeter). A cartridge count of 140 is approximately equal to a specific weight of 1.0. Water has a specific weight of 1.0; therefore any explosive having a cartridge count greater than 140 or a packaged specific weight of 1.0 or less will not sink in water.

Water resistance is a self-defining term. The gelatins and some slurries are virtually impervious to water, while AN/FO dissolves in water. In many instances a product which deteriorates in water is packaged to provide water resistance for as long a time as the package remains intact. Careful handling of such products is necessary in order to ensure that the package does remain intact.

Fumes is a term which relates to toxic gases produced in the explosive reaction. Explosives are usually formulated to produce a minimum of toxic fumes. Carbon monoxide is formed if the explosive reaction creates a deficiency of oxygen, while an excess of oxygen results in the formation of oxides of nitrogen. The fume characteristics shown in Table 19-20 are generalized so as to relate to products made by all manufacturers. The fume rating of an explosive is normally not a matter of primary importance for surface work, but any explosive used under ground should preferably carry a "Fume Class 1" rating. This means that it produces less than 0.16 cu ft of toxic gases per 1¼- by 8-in. cartridge. If fume rating is an important characteristic, the explosive supplier can furnish information about a particular grade.

Vibration and Air Blast As the population grows, vibration and air blast from shooting become of greater importance in overall planning of a job. If the work is in a populated area, consideration should be given to the hiring of a vibration consultant prior to the selection of haulage or drill equipment or prior to the determination of the amount of explosive to use per blast or per delay period. The close proximity of structures may require small blasts, thereby dictating small-diameter holes and, in some cases, shorter holes. The latter will increase the number of lifts. When estimating vibration levels, the following relationship is frequently used to determine the maximum recommended amount of explosive per delay period:

$$W = \left(\frac{d}{50}\right)^2$$

where W = maximum recommended weight of explosive, lb per delay
d = distance from blast site to point of interest, ft

When vibration problems cause inconveniences, such as small-sized blasts with the need for more frequent moving of equipment, costs will increase. These additional costs can be minimized if care is taken to use the maximum number of millisecond delay periods. This practice will permit the use of a larger, more economical blast. In order to keep vibrations at the desired level, it may be necessary to use a different delay period in each deck. When decking is used, however, the amount of powder per hole is reduced. This means that burden and spacing must be reduced in order to maintain the desired powder factor. Should it become necessary or desirable to shoot a weight of explosive which is in excess of that indicated by the approximate formula just stated, or if any possibility of damage exists even though this recommended limit is not exceeded, a consultant should be called in.

Under wet conditions, some grades of explosives may propagate the blast from one hole to another. So, when working under wet conditions where a vibration problem would result if all holes were to detonate simultaneously rather than by delay period, the explosive supplier should be consulted for selection of the proper explosive.

Air blast creates as many complaints as do vibrations. Careful shot design with adequate stemming is essential to noise control. Bottom-hole initiation with electric blasting caps is desirable. If detonating cord is used, the trunk lines, millisecond-delay (MS) connectors, and the exposed portion of the down line must be covered to keep the noise at a minimal level. Weather conditions can affect the transmission of sound, and in certain cases shooting times must be scheduled when conditions permit.

Weather conditions affecting sound transmission are:

Wind velocity and direction
Barometric pressures
Atmospheric temperatures
Temperature inversions

In any event, where vibration or air blast may be a problem, a specialist should be consulted before the job begins so as to avoid imposition of unnecessary restrictions by the local community and as protection in the event of legal complications. (See also Figs. 19-23 to 19-25.)

Safety Recommendations The Institute of Makers of Explosives, being aware of the conditions that might create an accident, has published the following safety pamphlets:

Prevention of Accidents in the Use of Explosives (commonly referred to as "Do's and Don'ts")
Standard Storage Magazines (Pamphlet No. 1)
American Table of Distances (Pamphlet No. 2)
Rules for Storing, Transporting and Shipping Explosives (Pamphlet No. 5)
Safety in the Handling and Use of Explosives (Pamphlet No. 17)
Radio Frequency Energy—A Potential Hazard in the Use and Transportation of Electric Blasting Caps (Pamphlet No. 20)
How to Destroy Explosives (Pamphlet No. 21)
Explosives in Agriculture (Pamphlet No. 11)

Every blasting supervisor should have a complete set of these publications, be fully aware of the information contained therein, and carefully manage his operation so as to comply with the suggestions made. Every member of a blasting crew should be aware of and instructed in the procedures contained in "Do's and Don'ts." In order that "Do's and Don'ts" may be readily available, explosive manufacturers normally package one or more copies in every case of explosives and caps.

In addition to the suggestions contained in the publications cited above, there are federal, state, and local laws and regulations pertaining to the transportation, storage, and use of explosives. Supervisory personnel should become well acquainted with these and comply with them.

MECHANICAL LOADING METHODS

AN/FO and various slurries are relatively insensitive, and their nature permits them to be poured into a borehole. With their increasing use, it was only a matter of time until mechanical systems to handle them were developed. The labor savings therefrom can be substantial, and the use of bulk AN/FO in surface operations is now commonplace. Even small jobs, formerly not considered large enough, are modifying bulk systems to meet their needs. In this manner, contractors enjoy savings both in labor and in the initial cost of the product.

Systems used to handle bulk AN/FO fall into two types: pneumatic and auger. Each type has been used to load holes ranging from 2½ to 10⅝ in. in diameter. Each job must be considered separately in relation to the type of equipment to be used, because job variations are too numerous to enable the establishment of any meaningful blanket recommendations. Factors affecting equipment selection are expected consumption, location, terrain, diameter and depth of hole, and state and local regulations.

Under dry conditions, AN/FO products have found extensive use in underground work through pneumatic loading. Pneumatic loaders fall into three broad categories:

Venturi loaders
Pressure pots
Combination loaders

Venturi loaders, although imparting a high velocity to the AN/FO and a good borehole loading density, have relatively slow loading rates of 12 to 18 lb per min. Pressure pots, on the other hand, impart a low velocity to the AN/FO and yield a lower borehole loading density but can result in higher loading rates of 30 to 50 lb per min. Combination pneumatic loaders attempt to combine the advantages of both the venturi and pressure-pot types; an air bypass system leads to a venturi located at either the outlet of the pressure pot or near the end of the hose. Combination loaders provide a medium borehole loading density and loading rate.

Job characteristics of the underground operation will determine which pneumatic system to use. Generally speaking, smaller jobs will tend to use venturi loaders while the bigger jobs will employ the pressure pot or combination loader. If the job is large enough to warrant them, specially designed pressure pots with multiple hoses equipped with venturi boosters should be considered.

When pneumatically loading ammonium nitrate products either from bulk trucks or underground pneumatic equipment, consideration must be given to the generation and accumulation of static electricity. Numerous investigations since the original introduction of pneumatic loading have led to the conclusion that static electricity is always generated. However, if certain favorable conditions exist, this static electricity will probably be drained away before it accumulates to a hazardous level. In general, conditions which are necessary to prevent static accumulation are:

Ambient relative humidity of 50 percent or greater.

A properly grounded pneumatic loader.

Use of a semiconductive loading hose approved for this service.

Electrical resistance of the earth between the grounding device, which is a portion of the pneumatic loader, and the borehole should be a maximum of 1,000,000 ohms (1 megohm).

The operator must not wear gloves when loading, and the use of nonconductive liners in the borehole should be prohibited.

If AN/FO is to be loaded pneumatically, tests prior to and periodic checks during the operation should be made. This would include testing for grounding resistance, static voltage, and relative humidity. Additional information can be obtained from most explosive suppliers and the U.S. Bureau of Mines.

Bulk slurry loading has been for the most part limited to jobs large enough to justify the installation of special and costly support equipment such as field plants and slurry pump trucks. While this equipment is normally supplied by the slurry manufacturer, its cost must be passed on in the price of the slurry. Pumped slurry has certain basic advantages, such as filled boreholes. If this feature is desirable on jobs which cannot support a bulk slurry system, special slurries which can be poured into the hole from cartridges are available.

INITIATION SYSTEMS

Initiation systems fall into two broad categories, electric and nonelectric. In the first category, electric blasting caps include instantaneous, millisecond-delay, and long-period-delay caps. The nonelectric category includes detonating cord with millisecond delay connectors, Primadets, and fuse caps with safety fuse.

While each type of system offers certain unique advantages, it is generally agreed that initiation of the shot by means of electric blasting caps gives more advantages and greater versatility. This method of initiation provides a precisely timed sequence of blasts which produces more uniform fragmentation and a muckpile shape that is easily handled. It provides a greater number of delay periods, which can offer greater flexibility in the size of a shot when firing in populated areas where vibration and noise must be controlled. It permits bottom-hole initiation even when cap-sensitive powders are used. These are distinct advantages because they nearly always provide better blasting results.

Electric Initiation Systems An electric blasting (EB) cap consists of a base charge, a flash or ignition charge, and an electrically activated ignition device sealed in a waterproof shell assembly. It may or may not contain a delay element. (See Fig. 19-30.)

The EB cap without a delay element is generally referred to as a "No. 6" or an "instantaneous EB" cap. The term "No. 6" refers to the strength designation, which is standard for the construction and most other industries. The No. 6 strength is also used in millisecond-delay and long-period-delay EB caps.

The basic difference between instantaneous EB caps and delay EB caps is the column of delay powder in the cap between the ignition device and the flash charge. It is important to remember that when sufficient electrical current is applied to a blasting circuit containing delay caps, all caps activate at the same instant but a delay in firing time occurs because of the time required to burn through the delay powder before the flash and base charges are activated. For this reason, the EB cap ignition

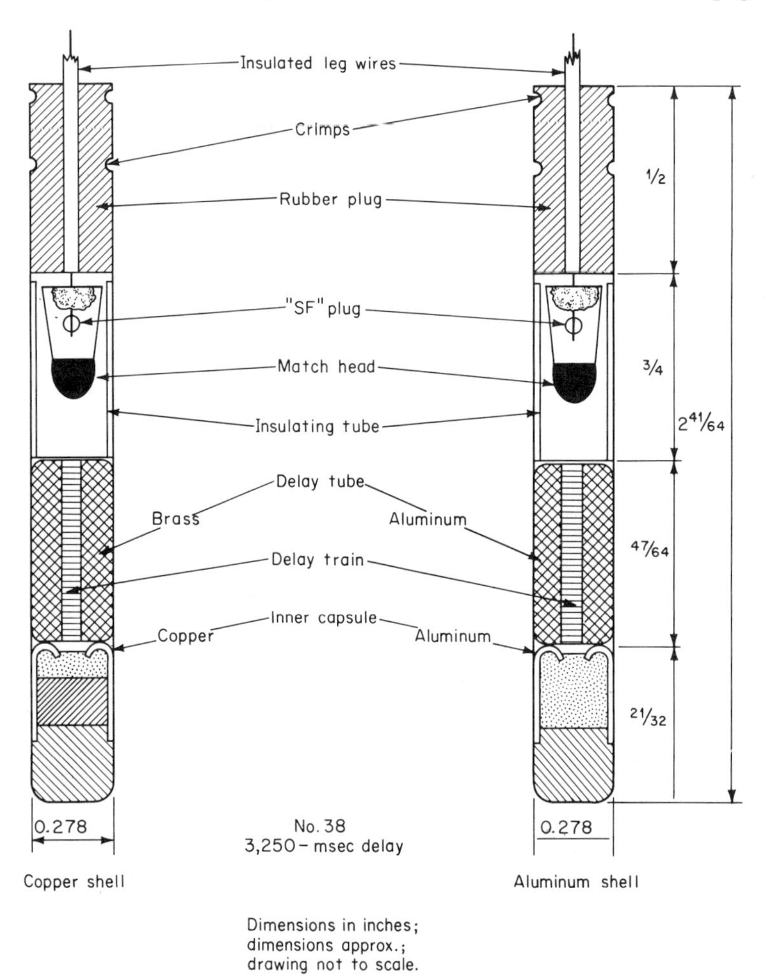

Fig. 19-30 Typical SF delay electric blasting cap (cross-sectional view).

system is not subject to cap cutoffs from rock shifts or other outside factors. This is a distinct advantage.

Delay EB cap systems are of two general types—millisecond delay and long-period delay. Timings of typical millisecond-delay and long-period-delay series are supplied in Table 19-21. Millisecond-delay caps are used primarily in surface work. The proper use of millisecond delays can result in considerable savings through improved fragmentation as well as better control of flyrock, air blast, and vibration.

The drill pattern must be selected concurrently with the millisecond-delay EB cap pattern. Figure 19-31 shows a staggered-row drill pattern for a road cut with throw of blasted rock directed toward the center. This drill pattern could be used with either of the cap patterns illustrated in Fig. 19-31*a* and *b* and would normally be employed in shallow cuts with fairly easy-breaking rock. A slightly different staggered-row type of drill pattern is shown in Fig. 19-32. This could be used in shooting to the open end of a cut, using the cap patterns shown in Fig. 19-32*a* and *b*. The cap pattern extending in a straight row into the cut shown in Fig. 19-31*a* and the cap pattern extending in a straight row across the cut shown in Fig. 19-32*a* would result in greater

TABLE 19-21 Timings for Typical Delay Caps

(*a*) Atlas Rockmaster EB caps (millisecond delays)

Delay No.	Average time, msec	Delay No.	Average time, msec
1	8	20	1,000
2	25	21	1,125
3	50	22	1,250
4	75	23	1,375
5	100	24	1,500
6	125	25	1,625
7	150	26	1,750
8	175	27	1,875
9	200	28	2,000
10	250	29	2,125
11	300	30	2,250
12	350	31	2,375
13	400	32	2,500
14	450	33	2,625
15	500	34	2,750
16	550	35	2,875
17	650	36	3,000
18	750	37	3,125
19	875	38	3,250

(*b*) Atlas Timemaster EB caps (long-period delays)

Delay No.	Average time, msec	Delay No.	Average time, msec
1	500	9	4,500
2	1,000	10	5,000
3	1,500	11	5,500
4	2,000	12	6,000
5	2,500	13	6,500
6	3,000	14	7,000
7	3,500	15	7,500
8	4,000		

movement than the alternating cap patterns used in Figs. 19-31*b* and 19-32*b*. These latter patterns would produce less movement but would result in better and more uniform fragmentation.

In certain cases, it may be desirable to use a V-cut type EB cap pattern to obtain the desired breakage or adequate control of the shot. In using the V-cut cap pattern, the holes should be drilled on the square with equal burden and spacing. When the V-cut cap pattern is applied to this drill pattern, the true burden and spacings which occur during the blast are as shown in Fig. 19-33.

Figure 19-34 demonstrates the importance of considering rock conditions prior to

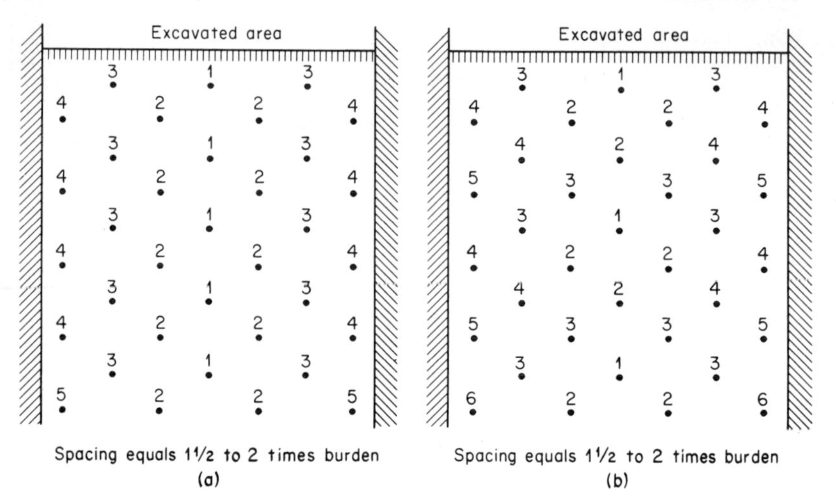

Fig. 19-31 Drill and cap patterns for throw toward center of cut.

selection of drill and EB cap patterns. The rock surrounding the holes to the right of delay 1 in the cap pattern will have a tendency to ride up along the bedding planes, leaving a high bottom. The pitch of the bedding planes in the rock to the left of delay 1 will help to keep this side of the cut to the desired grade.

When pronounced bedding planes, either horizontal or pitched, are encountered,

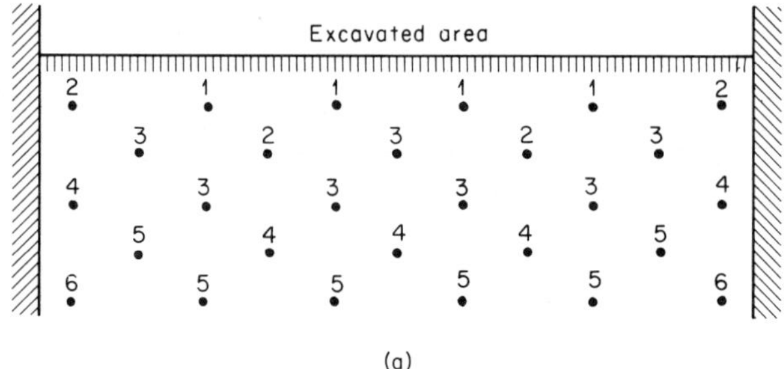

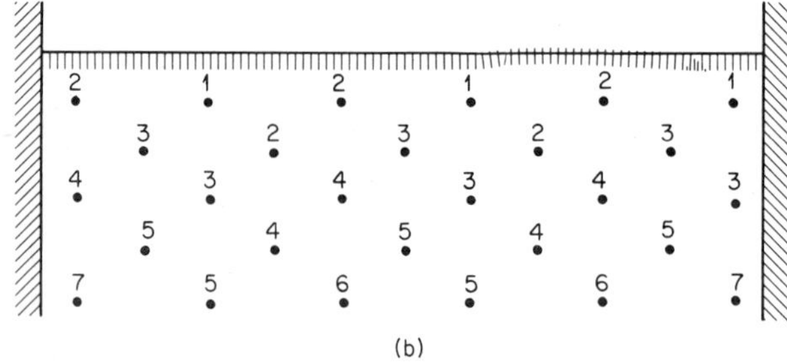

Fig. 19-32 Drill and cap patterns for throw toward open end of cut.

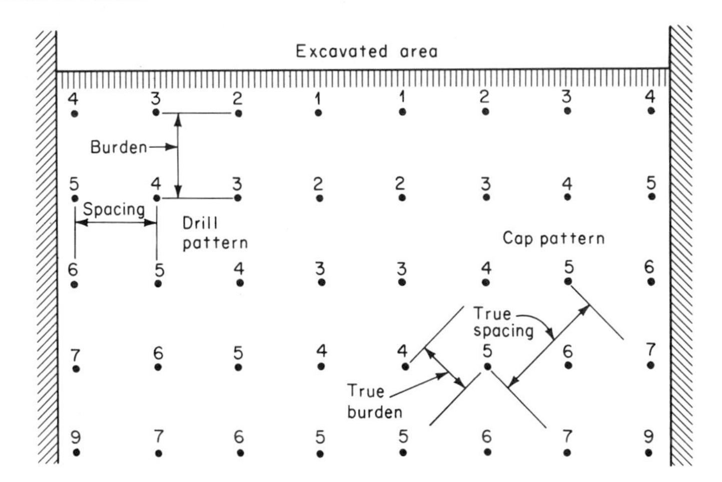

Fig. 19-33 Drill and cap pattern for V cut.

cutoffs in the powder column can be caused by a shift of rock when using delay-cap patterns. In Fig. 19-34, this may occur in the rock to the left of delay 1. In order to prevent the possibility of column cutoff, an additional delay EB cap with primer should be used near the top of the hole. This top cap is usually one delay period longer than the period used in the priming cartridge at the bottom of the hole.

The selection of drill patterns and EB cap patterns is of prime importance. Geological factors such as bedding, dip, jointing, and seams must be considered, as well as width of cut, depth, control of shot, and type of equipment used for loading and haulage. Normally, patterns such as in Fig. 19-31*a* and *b* are used in conjunction with the maneuverable front-end loaders because the rock is apt to be scattered further and the muckpile will be lower. A V-cut EB cap pattern is used where it is desirable to keep the rock in a higher pile for faster digging with the much less maneuverable shovels.

The long-period delays are sometimes used on the surface in shooting ditches, but they are most commonly employed for tunneling or underground mining. In underground work, the longer time interval has the advantage of allowing the rock blasted by one delay period to move out of the way before the next period fires. It also reduces air blast and leaves a higher muckpile close to the face for faster digging, with little scattered rock.

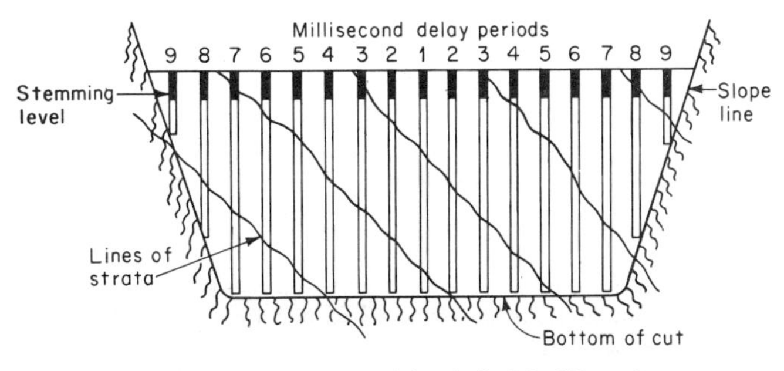

Fig. 19-34 Cap pattern used in pitched bedding planes.

The time interval between delays in the long-period caps will vary between manufacturers. There is no significant difference in performance as long as there is no overlap in firing time and the time interval between periods is at least 125 msec. Time intervals shorter than 125 msec will tend to give the effect of millisecond delays, that is, greater air blast and greater scattering of the muckpile. This is not to say that millisecond caps should not be used underground. They can, under certain conditions, offer savings in reduced powder factor or increased fragmentation when the powder load per hole is tailored to conditions and strictly enforced. In extremely blocky ground where cutoffs of the powder column are a problem if long-period delay caps are used, millisecond delays may be a solution. In using millisecond delays underground, it is important to skip periods in the first part of the series to allow enough time for the rock in the cut to move before the next hole fires.

Both Atlas and Du Pont have handbooks which give in detail the proper procedure to use when wiring-in various types and sizes of shots. When a blasting circuit is completed, it is of prime importance to check it carefully with a blasting galvanometer or a specially designed volt-ohmmeter. These instruments are equipped with silver chloride cells which produce an electrical current well below the minimum required to shoot an EB cap. Under no circumstances should a silver chloride cell in either type of instrument be replaced by any other electrical power source. This is an absolute safety requirement. Detailed instructions on the use of these instruments are available from the manufacturer.

The three sources of electric energy generally used in the blasting industry are condenser-discharge blasting machines, mechanically operated blasting machines, and power line circuits. Of the three systems, condenser-discharge blasting machines are most widely used. They are preferred because of their simplicity of operation and their assurance of providing adequate firing current for almost any size of shot encountered. Although several sizes of blasting machines are available, a high-capacity model which is capable of firing 1,000 to 1,200 EB caps is well worth the small additional cost. Such a machine can eliminate misfires which might otherwise occur under adverse conditions such as extremely wet conditions or when poor practice has been used in tying in a circuit. All machines carry the manufacturer's recommendations on use and servicing. These should be followed closely.

Mechanical-type blasting machines range in size from the hand-held 10-hole twist machine to the commonly used rackbar machine. These machines are entirely reliable when used as recommended. In actual practice, however, the size of a shot is frequently larger than originally anticipated and the shooter ends up exceeding the recommended number of caps per circuit. In addition, mechanical-type blasting machines have the disadvantage of being dependent on the skill of the operator.

Power line circuits are normally used in tunnel work and occasionally on surface work. It is essential that a power line circuit, as constructed, be capable of delivering to the blasting circuit the electrical energy needed to fire the caps reliably. To ensure this, consideration must be given to the capacity (kilovolt-amperes) of the transformer as well as proper circuitry to supply the necessary voltage and amperage. To prevent arcing inside a delay cap, a time limiting switch which is voltage sensitive, such as Atlas' Arcmaster switch, should be used.

Nonelectric Initiation Systems Fuse blasting caps are nonelectric caps containing an explosive charge and are available in either No. 6 or No. 8 strength. They are initiated by the fire that spits from the end of the safety fuse which is crimped into the cap. Safety fuse contains a train of black powder, supported and enclosed in wrappings of textile and waterproofing materials.

In the construction industry, cap-and-fuse blasting is normally used for secondary blasting. In practice the primer assembly is first made up according to recommendations available from the supplier and is then inserted into the hole in the boulder along with the desired amount of explosive. In multiple-cap shots, the preferred method of igniting safety fuse is by Quarrycord, an Ensign-Bickford product. First, all holes are loaded and stemmed, then the ends of safety fuse are joined to a length of Quarrycord with a special connector. The Quarrycord, when lit, burns at a rate of approximately 1 fps and ignites each length of safety fuse in turn. In shots consisting of no more than a few holes and in an area where exiting of personnel is unhampered, it is possible to rapidly ignite each fuse end in turn with hot fuse lighters, pull-wire fuse lighters, or lead spitters and then get out of the danger area.

The cap-and-fuse ignition system is sometimes used in small tunnels. In this application, the preferred method of achieving a timed sequence is through the use of the relatively slow-burning Ignitacord, an Ensign-Bickford product. Ignitacord is a wire-bound flexible cord with a core of thermite. It burns with a short, intense flame and should be used only where this flame does not impose a safety hazard.

Primadet Delays, an Ensign-Bickford product, are nonelectric delay blasting caps for in-hole initiation of explosives. They consist of three components: a length of Primaline, a 4-grain-per-ft detonating cord; a No. 6 strength blasting cap with a delay element; and a connector to facilitate attaching the end of the Primaline to a 25-grain-per-ft detonating cord trunk line. In principle the trunk line ignites the Primaline, which in turn ignites the delay column in the cap. When the delay column burns through, it ignites the explosive charge in the cap which in turn detonates the explosive in the hole. Primadets, which are available in 19 different millisecond and 10 different long-period delays, are designed for use where there is a definite hazard from extraneous electricity. It is necessary to use primers and column loads that are insensitive to the 4-grain-per-ft detonating cord; this excludes the use of most cap-sensitive grades of explosives. Primadets are used primarily with AN/FO and in underground work, but they have also been used on the surface when loading and rock conditions are not too severe.

Detonating cord is a high-energy, linear detonating device which is used primarily to initiate commercial explosives. It consists of a strong, flexible outer cover with a core containing an explosive. When initiated with a blasting cap, it detonates along its length at a velocity of approximately 21,000 fps. Detonating cord is classified according to the number of grains of explosive per linear foot. Detonating cords with core loads as low as 18 to as high as 400 grains per ft are available, but those most commonly used in the construction industry are from 25 to 60 grains per ft. All strengths are capable of initiating cap-sensitive explosives at any point along the length of the cord.

Detonating cord can be used as:

A supplement in a blasting system initiated by an EB cap

A primary initiating device

A presplitting explosive

In formations that contain seams, caves, or highly jointed conditions which may interrupt the explosive column, detonating cord is used for the entire length of the explosive column in order to ensure propagation. Where it is necessary to use decks, detonating cord can be used along the length of the hole to initiate the various decks. In both instances, EB caps, usually at the bottom of the hole, are used to initiate the shot. Fifty-grain detonating cord is frequently used for such rock conditions, but savings can be realized by the use of 25-grain cord in medium-sized holes if conditions are not too severe.

For secondary shooting, for some forms of ditching, or when it is desirable to load behind the drill and leave the shot sitting for some time, it may be advantageous to use detonating cord as the initiating system for a shot. In this case detonating cord is used both as a down line and as a trunk line running from hole to hole. Initiation of the trunk line is by means of an EB cap or a cap-and-fuse assembly. The strength of the down line is dictated by job conditions, but the trunk line could be 25 grains per ft. For secondary shooting, an 18 grains per ft detonating cord should be effective for the down lines.

If delay firing is desired when using detonating cord, it can be done in one of three ways:

Direct priming in the hole with millisecond EB caps

Attachment of millisecond EB caps to exposed detonating cord at the collar of the hole just prior to firing of the shot

Use of millisecond-delay connectors

The use of millisecond-delay EB caps in accordance with the first of these two methods was discussed earlier. Millisecond-delay connectors are available in intervals of 5, 9, 17, and 25 msec, and other intervals can be obtained by combining two or more periods in series. Millisecond-delay connectors consist of a molded plastic sleeve with a delay element inside a copper tube which has been inserted into the center portion of the plastic sleeve. Each sleeve end is fashioned so that the detonating cord can be looped and locked in place with a tapered pin. When laying out the trunk line, sufficient slack should be allowed to accommodate the connectors. The trunk line is cut at the location where the connectors are to be inserted. The connector should be placed reasonably close to the hole that is to be delayed in order to reduce the possibility of damage by surface disturbance from holes fired previously. Trunk lines which are used with or without millisecond-delay connectors should be interconnected so that detonation can reach each hole from an alternate direction in case of a break in the line.

In presplit work, detonating cord is used as a down line for initiating the presplit grades of high explosives. Most conditions warrant the use of 50-grain detonating cord. However, in those cases where the higher-strength presplit powders are used, it may be possible to use 25-grain cord. Where noise is not a factor, the holes are tied together with a detonating cord down line. If noise is a factor, EB caps are used for initiation in the hole.

In certain cases, detonating cord may be used as a presplit powder. Many factors influence the number of grains to use per foot of borehole, but a good starting point in solid rock would be 100 grains per ft of borehole for each foot of spacing between holes. The use of detonating cord as a presplitting powder is particularly useful when it is necessary to presplit a hard cap rock. Detonating cord is used in the top portion of the hole, with normal presplitting powder in the bottom portion. Sand stemming is of prime importance in this instance.

As with other explosive products, certain precautions and techniques should be followed in the use of either electric or nonelectric initiation systems. Detailed information is readily available from the supplier or manufacturer.

REFERENCES

1. *Handbook of Ripping*, Caterpillar Tractor Co., Peoria, Ill., 1966.
2. K. McGregor, *The Drilling of Rock*, C. R. Books, Ltd., London W.C. 2, England.
3. E. H. Dickerson, *Rock Drill Data*, Ingersoll-Rand Company, 1960.
4. W. G. Piper, Large Diameter Drill Holes in Overburden Blasting, *Mining Congress Journal*, November, 1963.
5. *Blast Hole Drill Application*, Bucyrus-Erie Company, South Milwaukee, Wis., 1961.
6. Wm. C. Maurer, *Novel Drilling Techniques*, Pergamon Press, Oxford, England, 1968.
7. Melvin A. Cook, Behavior of Rock During Blasting, Seventh Symposium on Rock Mechanics, Penn State University, 1965.
8. Richard L. Ash, The Mechanics of Rock Breakage, *Pit and Quarry*, August–November, 1963.
9. V. Langefors and B. Kihlstrom, *Rock Blasting*, John Wiley & Sons, Inc., New York, 1963.
10. Darrel W. Hanen, Drilling and Blasting Techniques for Morrow Point Powerhouse, Ninth Symposium on Rock Mechanics, Colorado School of Mines, Golden, Colo., 1967.
11. C. W. Livingston, *Explosives in Ice*, U.S. Army Snow, Ice and Permafrost Research Establishment, Corps of Engineers, Wilmette, Ill., 1960.
12. Alan Baurer et al., How IOC Puts Crater Research to Work, *Engineering and Mining Journal*, September, 1965.
13. Alan Baurer, Application of the Livingston Theory, *Quarterly of the Colorado School of Mines*, vol. 56, no. 1, Golden, Colo., 1961.
14. L. C. Lang, Mining Itabirite on the Bong Range, Liberia, *Canadian Mining Journal*, 1965.
15. Product literature, Gardner Denver Company, Quincy, Ill., 1968.
16. *Blasters Handbook*, E. I. duPont de Nemours & Co., Wilmington, Del., 1966.
17. John F. Wiss, Effects of Blasting Vibrations on Buildings and People, Presentation at Fifteenth Annual National Highway Conference ARBA, Detroit, Mich., Sept. 20, 1967.
18. Don L. Leet, *Vibrations from Blasting Rock*, Harvard University Press, Cambridge, Mass., 1960.
19. *Rock Handling Systems*, Caterpillar Tractor Co., Peoria, Ill., 1966.
20. Field Report, International Harvester Company, Construction Equipment Div., Melrose Park, Ill., 1966.
21. *Economics of Moving Rock with Scrapers*, Caterpillar Tractor Co., Peoria, Ill., 1965.

Section 20

Aggregate Production

HANS I. HANSEN*

Consulting Engineer, Iowa Manufacturing Company, Cedar Rapids, Iowa

* Deceased.

INTRODUCTION

The production of aggregate is a major industry, with a tremendous impact on the national economy. In 1968 the total amount of aggregate produced from stone in quarry operations was estimated at 800 million tons valued at $1.3 billion,*[1] and from sand and gravel in pit operations a total of 923 million tons valued at $1.02 billion.[2] The aggregate was produced in an estimated 3,500 stationary stone plants and 6,200 stationary sand and gravel plants, as reported by the Bureau of Mines, in addition to an estimated 4,600 individual portable plants. Tables 20-1 and 20-2 list the production of commercial crushed stone, sand, and gravel plants for the years 1963 and 1964.

The construction industry is the major user of this aggregate. Table 20-3 indicates that 660 million tons of aggregate, approximately 37 percent of the 1,774 million tons produced from all sources, was used in the year 1966 for highway construction and maintenance.[3] It is estimated that the total United States production of aggregate will increase 50 percent in each decade, as it did in the period 1956 to 1966 (see Fig. 20-1). Estimated aggregate production for each state in the year 1965 is shown in Figs. 20-2 and 20-3.

With the increased demand for aggregate, there has been a trend toward more rigorous specification requirements. Compliance with these requirements is controlled by random sampling of the aggregate and is reflected in more exacting processing procedures. In spite of this and its associated effects on equipment and labor costs, Fig. 20-4 shows that the increase in the per-ton price of aggregate has been moderate.

Several common aggregate terms are defined below, beginning with the aggregate itself. In the broadest usage, *aggregate* can be defined as a mineral material such as sand, gravel, crushed stone, slag, shells, or combinations thereof which, together with a binding medium or alone, forms an end product.[4,5]

Sand is the granular material passing the ⅜-in. sieve and almost entirely passing the No. 4 sieve, also predominantly retained on the No. 200 sieve, which results from natural disintegration and abrasion of rock or from processing of completely friable stone.

Gravel is the granular material predominantly retained on the No. 4 sieve which results from natural disintegration and abrasion of rock or from processing of weakly bound conglomerate.

Crushed stone is the granular product which results from the processing of rock.

Blast-furnace slag is the nonmetallic product, consisting essentially of silicates and aluminosilicates of lime and of other bases, which is developed in a molten condition simultaneously with iron in a blast furnace. The production of slag in 1968 was reported to have been 29 million tons.[6]

Coarse aggregate is defined as that portion of an aggregate retained on the No. 4 sieve, and the *fine aggregate* as that portion of an aggregate passing the No. 4 sieve.

* Superior numbers refer to the list of references at the end of this section.

TABLE 20-1 Number and Production of Commercial Crushed-stone Plants in the United States by Size of Operation, 1963–1964

Annual production, short tons	1963				1964			
	Number of plants	Production		Cumulative total, thousand short tons	Number of plants	Production		Cumulative total, thousand short tons
		Thousand short tons	Percent of total			Thousand short tons	Percent of total	
Less than 25,000	871	7,253	1.2	7,253	987	8,121	1.2	8,121
25,000 to 50,000	307	11,188	1.8	18,441	345	12,408	1.9	20,529
50,000 to 75,000	242	14,994	2.4	33,435	259	16,019	2.4	36,548
75,000 to 100,000	195	16,669	2.7	50,104	193	16,753	2.6	53,301
100,000 to 200,000	535	76,363	12.4	126,467	497	71,317	10.9	124,618
200,000 to 300,000	251	61,290	9.9	187,757	267	64,989	9.9	189,607
300,000 to 400,000	165	57,451	9.3	245,208	176	60,942	9.3	250,549
400,000 to 500,000	129	57,601	9.3	302,809	129	57,622	8.8	308,171
500,000 to 600,000	89	48,777	7.9	351,586	103	55,448	8.5	363,619
600,000 to 700,000	64	41,262	6.7	392,848	69	44,283	6.7	407,902
700,000 to 800,000	46	34,211	5.5	427,059	36	26,635	4.1	434,537
800,000 to 900,000	33	27,985	4.5	455,044	41	34,978	5.3	469,515
900,000 and over	96	163,380	26.4	618,424	111	186,605	28.4	656,120
Total	3,023	618,424	100.0	618,424	3,213	656,120	100.0	656,120

SOURCE: W. R. Barton and P. G. Cotter, *Stone*, Bureau of Mines.

TABLE 20-2 Number and Production of Domestic Commercial Sand and Gravel Plants by Size of Operation, 1966–1967

Annual production, short tons	1966				1967			
	Plants		Production		Plants		Production	
	Number	Percent of total	Thousand short tons	Percent of total	Number	Percent of total	Thousand short tons	Percent of total
Less than 25,000	2,544	40.8	26,162	3.8	2,609	41.3	26,032	3.9
25,000 to 50,000	970	15.6	35,259	5.2	949	15.0	35,164	5.3
50,000 to 100,000	957	15.3	69,457	10.2	1,016	16.1	74,142	11.1
100,000 to 200,000	841	13.5	120,711	17.8	804	12.7	115,065	17.2
200,000 to 300,000	382	6.1	94,052	13.3	400	6.3	96,901	14.5
300,000 to 400,000	190	3.0	65,048	10.1	201	3.2	69,570	10.4
400,000 to 500,000	116	1.9	52,737	7.8	101	1.6	45,129	6.8
500,000 to 600,000	62	1.0	32,929	4.8	67	1.1	36,928	5.5
600,000 to 700,000	41	0.7	26,945	4.0	32	0.5	21,053	3.2
700,000 to 800,000	25	0.4	18,979	2.8	36	0.6	26,852	4.0
800,000 to 900,000	24	0.4	20,658	3.0	25	0.4	21,300	3.2
900,000 to 1,000,000	15	0.2	14,187	2.1	20	0.3	19,056	2.9
1,000,000 and over	69	1.1	102,473	15.1	55	0.9	80,409	12.0
Total	6,236	100.0	679,597	100.0	6,315	100.0	667,601	100.0

SOURCE: James Paone, *Sand and Gravel*, Bureau of Mines.

TABLE 20-3 Aggregate Production versus Highway Usage

Production data through 1966 from *Minerals Yearbook* of the Bureau of Mines, U.S. Department of the Interior; projected for 1975 and 1985 on basis of continued growth rate equal to average for 1956–1961 and 1961–1966 periods, except as noted.

Consumption in highway construction and maintenance through 1966 based on AASHO surveys; projected for 1975 and 1985 on basis of highway needs reported by AASHO to Congress.

	Production, millions of tons				Highway construction and maintenance	
Year	Sand and gravel	Crushed stone	Iron blast-furnace slag*	Total	Millions of tons	Percent total production
1956	625	504	32	1,161	443†	38
1961	752	613	27	1,392	555	40
1966	934	811	29	1,774	660	37
1975	1330	1260	33‡	2,623	1199	46
1985	1970	2060	37‡	4,067	1447	36

* Excludes expanded slag lightweight aggregate used almost exclusively in building construction.

† Assumes maintenance added 10 percent to construction use of 403 million tons.

‡ Estimates projected on basis of moderate increase in iron and steel production.

SOURCE: AASHO–ARBA Joint Committee Report, 1968.

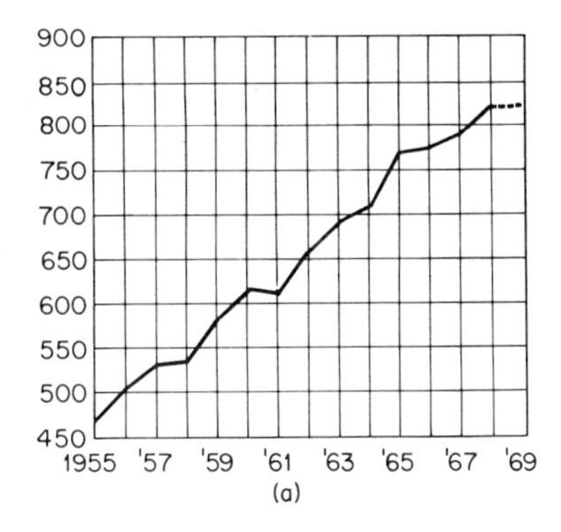

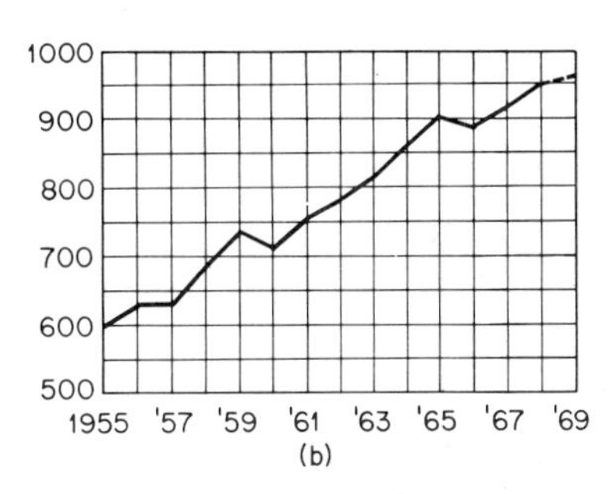

Fig. 20-1 Annual United States aggregate production, 1955 to 1969. (*a*) Crushed stone production in millions of tons. (*b*) Sand and gravel production in millions of tons.

AGGREGATE SOURCES

Sand and gravel pits are found in every state. The Bureau of Mines reports indicate that California is the leading producer, followed by Michigan, New York, Ohio, and Minnesota. The combined sand and gravel production of these five leading states represents approximately 32 percent of the total production in the United States.[7] In quarry operations, approximately 71 percent of the output is limestone, with Pennsylvania being the leading producer among the states. Oregon, Washington, and New Jersey are the major producers of the basaltic types of crushed stone, while granite quarries are mainly found in Georgia and North Carolina.[8]

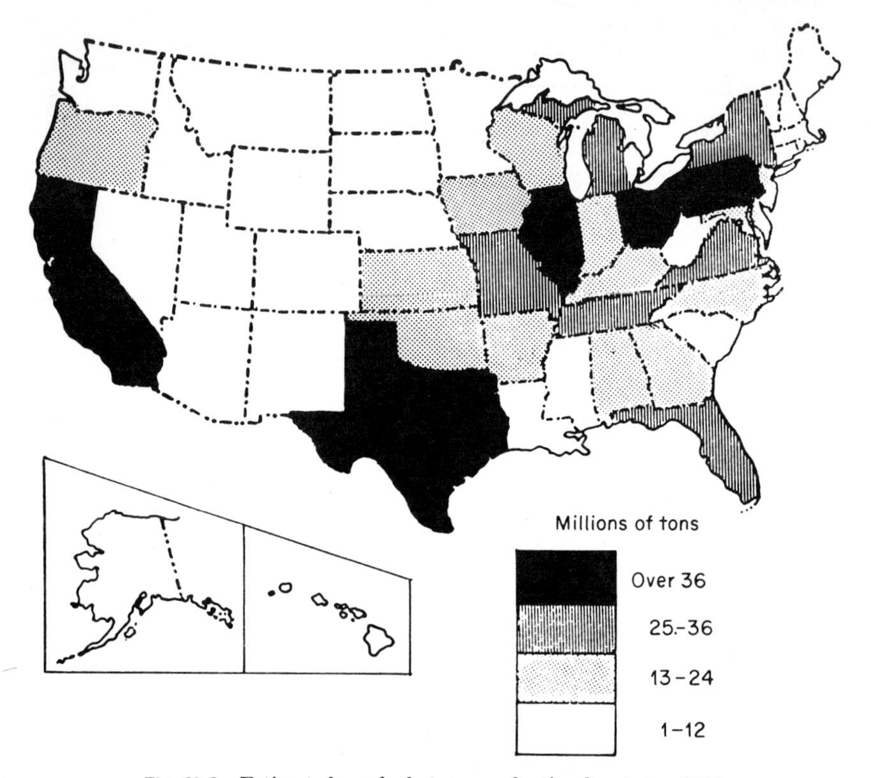

Fig. 20-2 Estimated crushed stone production by states, 1966.

Specification Requirements Most major contracts covering the production of aggregates will include requirements which are based on the published standards of one or more of the following:

1. American Association of State Highway Officials[9]
2. American Society for Testing and Materials[10]
3. Corps of Engineers[11]
4. Federal Specifications[12]

These specifications will include details which are of importance to the producer as well as to the user, and they will cover the range from raw material to finished aggregate. Substantially the same subjects are covered by each issuing agency, and subsequent portions of this section indicate the pertinent specification designations and describe the scopes of applicable specifications.

Classification of Rocks The general classification of the natural rock covers three groups: igneous, sedimentary, and metamorphic. Each of these classes is again divided as to type and family. Table 20-4 lists the general classification of rocks which are used for aggregates, while Table 20-5 indicates the types and chemical composition of rock-forming minerals. Table 20-6 lists the primary mineral composition of the more common rocks used for aggregates.[13]

Igneous Class. This class is characterized by absence of fossils, presence of glass, uniformity of structure, and interlocking crystals. Rocks of this class were formed by cooling and crystallization of molten material. The rate of cooling determined the size of the crystals: coarse-grained crystals occur in the intrusive igneous type, and fine-grained crystals are found in the extrusive type. The igneous class includes rocks such as granite, diorite, basalt, and diabase.

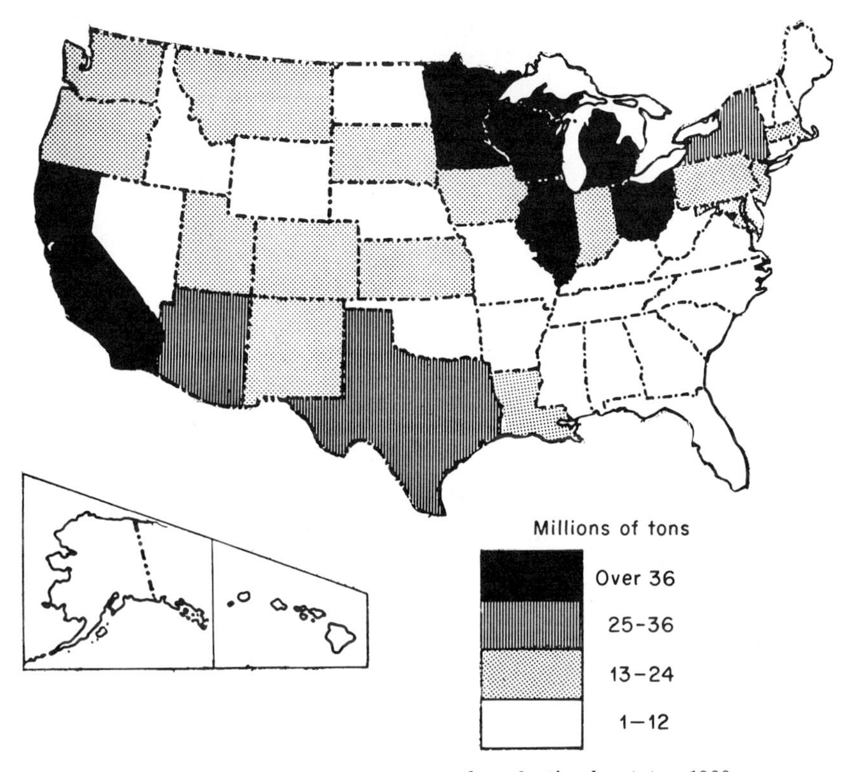

Fig. 20-3 Estimated sand and gravel production by states, 1966.

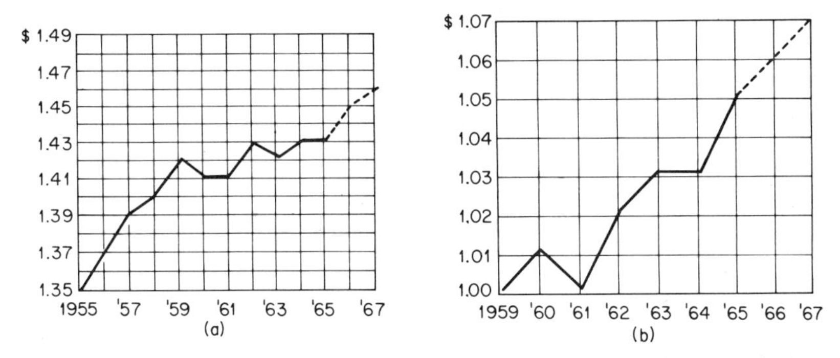

Fig. 20-4 Average prices for processed aggregates, 1955 to 1967. (*a*) Crushed stone prices in dollars per ton. (*b*) Sand and gravel prices in dollars per ton. Also includes contractor and government produced material.

Sedimentary Class. This class is characterized by rounded grains, presence of fossils, stratification in relatively thick layers, and abrupt changes in color from layer to layer. These rocks were largely formed by the deposition of water- and wind-transported rock grains. Two rock types within this class are separated on the basis of main mineral components. "Calcareous" includes limestone and dolomite, while "siliceous" includes shale and chert.

TABLE 20-4 General Classification of Rocks

Class	Type	Family	
Igneous	Intrusive (coarse-grained)	Granite* Syenite* Diorite* Gabbro	Peridotite Pyroxenite Hornblendite
	Extrusive (fine-grained)	Obsidian Pumice Tuff Rhyolite*†	Trachyte*† Andesite*† Basalt* Diabase
Sedimentary	Calcareous	Limestone Dolomite	
	Siliceous	Shale Sandstone Chert	Conglomerate‡ Breccia‡
Metamorphic	Foliated	Gneiss Schist	Amphibolite Slate
	Nonfoliated	Quartzite Marble Serpentinite	

* Frequently occurs as a porphyritic rock.
† Included in general term "felsite" when constituent minerals cannot be determined quantitatively.
‡ May also be composed partially or entirely of calcareous materials.
SOURCE: Bureau of Public Roads.

TABLE 20-5 Rock-forming Minerals

Name	*Composition*
	Primary minerals
Quartz	Silicon dioxide
Feldspar:	
Orthoclase	Silicate of potassium and aluminum
Microcline	Silicate of potassium and aluminum
Plagioclase	Silicate of sodium, calcium, and aluminum
Pyroxene:	
Augite	Silicate of calcium, iron, magnesium, and aluminum
Amphibole:	
Hornblende	Complex silicate principally of calcium, iron, magnesium, and aluminum
Mica:	
Muscovite	Hydrous silicate of potassium and aluminum
Biotite	Hydrous silicate of potassium, magnesium, iron, and aluminum
Magnetite	Iron oxide
Rock glass	Variable
Garnet	Silicate of aluminum, iron, and calcium
Olivine	Silicate of magnesium and iron
	Secondary minerals
Calcite	Calcium carbonate
Dolomite	Calcium and magnesium carbonate
Kaolin	Hydrous silicate of aluminum
Chlorite	Hydrous silicate of iron, magnesium, and aluminum
Epidote	Hydrous silicate of calcium, aluminum, and iron
Limonite	Hydrous iron oxide
Opal	Hydrous silicon dioxide

SOURCE: Bureau of Public Roads.

TABLE 20-6 Mineral Composition of Rocks

Name of rock	Number of samples tested	Essential mineral composition, percent*													Remainder
		Quartz	Orthoclase microcline	Plagioclase	Augite	Hornblende	Mica	Calcite	Dolomite	Chlorite	Kaolin	Epidote	Iron ore	Rock glass	
Igneous rocks															
Granite	165	30	45	(8)	...	...	6	...	...	...	(6)	...	...	...	5
Biotite granite	51	27	41	9	...	...	11	...	...	...	(7)	...	...	...	5
Hornblende granite	20	23	34	12	...	13	4	...	...	...	(10)	...	...	...	4
Augite syenite	23	(4)	52	7	8	...	4	...	...	(3)	(11)	(3)	(4)	...	4
Diorite	75	8	7	30	...	27	(4)	...	...	(3)	(8)	(5)	(3)	...	5
Gabbro	50	...	...	44	28	9	...	...	...	(3)	(6)	...	...	...	10
Rhyolite	43	32	45	(3)	...	...	(5)	...	...	(4)	(3)	...	(4)	...	4
Trachyte	6	(3)	42	...	...	6	...	(3)	...	(3)	(14)	(8)	(7)	9	5
Andesite	67	...	...	48	14	3	...	...	...	(6)	...	(3)	(8)	12	6
Basalt	70	...	...	36	35	...	...	...	...	...	...	...	(3)	21	5
Altered basalt	196	...	...	32	31	...	...	...	...	(9)	(4)	...	(4)	12	8
Diabase	29	...	...	44	46	...	...	...	...	...	...	...	(4)	...	6
Altered diabase	231	...	...	35	26	...	...	...	...	(15)	(9)	...	(4)	...	11
Sedimentary rocks															
Limestone	875	(6)	...	...	...	...	...	83	8	...	...	...	...	...	3
Dolomite	331	(5)	...	...	...	...	...	11	82	...	...	...	...	...	2
Sandstone	109	79	(5)	...	...	...	...	...	...	...	(4)	...	(9)	...	3
Feldspathic sandstone	191	35	26	...	...	...	...	(3)	...	(3)	(22)	...	(4)	...	7
Calcareous sandstone	53	46	(3)	...	...	...	...	42	...	...	...	...	(3)	...	6
Chert	62	93	...	...	...	...	...	...	...	...	...	...	...	...	7†
Metamorphic rocks															
Granite gneiss	107	37	35	(3)	...	...	18	...	...	...	...	...	...	...	7
Biotite gneiss	62	31	35	(5)	...	...	21	...	...	...	...	...	...	...	8
Hornblende gneiss	18	10	16	15	(3)	45	(4)	...	...	...	...	...	...	...	7
Mica schist	42	37	16	...	...	...	38	...	...	...	...	...	...	...	9
Biotite schist	17	34	13	(3)	...	...	41	...	...	...	...	...	...	...	9
Chlorite schist	23	11	...	10	...	(5)	...	...	...	39	...	28	(4)	...	3
Hornblende schist	68	10	(3)	12	...	61	...	...	...	...	...	(7)	...	...	7
Amphibolite	22	(3)	...	8	...	70	...	...	...	...	...	12	...	...	7
Slate	71	29	(4)	...	...	...	55	...	...	...	...	...	(5)	...	7
Quartzite	61	84	(3)	...	...	...	(4)	...	...	...	...	...	...	...	9
Feldspathic quartzite	22	46	27	...	...	...	(7)	...	...	(3)	(10)	...	...	...	7
Pyroxene quartzite	11	29	19	15	24	...	...	...	...	...	...	...	(5)	...	8‡
Marble	61	(3)	...	...	...	...	...	96	...	...	...	...	...	...	1

* Values shown in parentheses indicate minerals other than those essential for the classification of the rock.
† Includes 3 percent opal.
‡ Includes 3 percent garnet.
SOURCE: Bureau of Public Roads.

TABLE 20-7 Physical Properties of United States Rocks

No.	Name of rock specimen	Location found	Hardness Shore number	Coefficient of thermal expansion, room temp to 212°F	Max compressive stress, psi	True specific gravity	Apparent specific gravity	Wt, lb/cu ft	Absorption, %	Pores, %	Solids, %
	Igneous Rocks										
	Rhyelite-granite series										
1	Felsitic rhyolite	Mojave, Calif.	86.0	41×10^7	17,250	2.59	2.49	155.2	1.67	4.13	95.9
2	Rhyolite breccia	Animas Forks, Colo.	101.4	45	42,500	2.65	2.62	163.8	0.41	1.02	98.9
3	Granite porphyry	Winchester, Mass.	78.7	34	17,700	2.67	2.65	165.2	0.27	0.70	99.3
4	Binary granite	Concord, N.H.	94.4	49	25,800	2.68	2.57	160.4	1.55	3.98	96.0
5	Biotite granite	Picton Island, N.Y.	103.7	36	34,500	2.64	2.61	163.0	0.36	0.94	99.1
6	Biotite granite	Cripple Creek, Colo.	97.0	19	25,900	2.60	2.57	160.7	0.48	1.19	98.9
7	Biotite granite	Westerly, R.I.	91.8	47	20,650	2.63	2.60	162.4	0.39	1.00	99.0
8	Biotite granite	Westerly, R.I.	99.1	*	19,930	2.65	2.63	164.2	0.34	0.88	99.1
9	Biotite granite	Barre, Vt.	95.2	52	28,500	2.65	2.63	164.2	0.34	0.88	99.1
10	Biotite granite	Llano Co., Tex.	91.1	39	23,500	2.66	2.64	165.0	0.20	0.44	99.5
11	Biotite granite	Spring Creek, Colo.	84.6	*	8,830	2.61	2.63	158.2	1.13	2.87	97.1
12	Biotite granite (granitite)	Woodbury, Vt.	98.9	38	25,380	2.65	2.63	164.0	0.38	0.97	99.0
13	Biotite muscovite granite	Georgetown, Colo.	97.4	66	27,500	2.68	2.56	166.6	0.21	0.58	99.4
14	Biotite muscovite granite	Peekskill, N.Y.	85.1	41	18,000	2.60	2.58	161.2	0.25	0.64	99.3
15	Biotite muscovite granite	Mount Airy, N.C.	96.9	45	24,700	2.62	2.60	162.1	0.38	0.98	99.0
16	Alkali granite	Quincy, Mass.	75.4	29	19,200	2.67	2.64	164.8	0.36	1.01	99.0
17	Alkali granite	Quincy, Mass.	100.5	58	22,600	2.64	2.62	163.6	0.33	0.88	99.1
18	Aplite	Boulder Co., Colo.	71.3	50	20,750	2.61	2.50	156.1	1.67	4.11	95.8
19	Fluorite granite	Clinton Co., N.Y.	50.9	56	13,250	3.04	2.99	186.0	0.58	1.67	98.3
20	Spring Creek granite	Cripple Creek, Colo.	85.6	37	22,200	2.60	2.54	158.5	0.92	2.63	97.7
21	Hornblende granite	Fredericksburg, Tex.	83.1	37	25,800	2.63	2.61	162.8	0.36	0.89	99.1
22	Amphibole granite	Hurricane Island, Maine	100.5	60	23,700	2.64	2.62	163.5	0.32	0.79	99.2
	Trachyte-syenite series										
23	Quartz syenite	Ticonderoga, N.Y.	70.1	37	21,600	2.67	2.63	164.1	0.62	1.54	98.4
	Dacite-quartz-diorite series										
24	Granodiorite	St. Cloud, Minn.	98.8	42	29,500	2.71	2.70	168.5	0.19	0.50	99.5
25	Dacite	San Luis Obispo, Calif.	99.6	28×10^7	23,970	2.55	2.46	153.6	1.44	3.50	96.4
	Andesite-diorite series										
26	Hornblende andesite	Mount Shasta, Calif.	73.7	23×10^7	13,610	2.49	2.22	138.8	4.86	10.77	89.2
27	Andesite porphyry	Newton, Mass.	69.4	25	22,400	2.81	2.79	174.3	0.29	0.74	99.2
28	Andesite porphyry	Boulder Co., Colo.	82.7	39	32,700	2.71	2.69	167.8	0.27	0.70	99.3
29	Andesite	San Juan Co., Colo.	90.2	57	28,750	2.72	2.71	169.3	0.05	0.10	99.9
30	Obicular gabbro diorite	Davis Co., N.C.	66.1	*	25,000	3.02	2.97	185.2	0.57	1.67	98.3

TABLE 20-7 Physical Properties of United States Rocks (Continued)

No.	Name of rock specimen	Location found	Hardness Shore number	Coefficient of thermal expansion, room temp to 212°F	Max compressive stress, psi	True specific gravity	Apparent specific gravity	Wt, lb/cu ft	Absorption, %	Pores, %	Solids, %
	Igneous Rocks (*Continued*)										
	Basalt-gabbro series										
31	Hornblende basalt	Chaffe Co., Colo.	89.0	26	30,300	2.78	2.77	172.9	0.17	0.44	99.5
32	Olivine basalt	Jefferson Co., Colo.	77.2	22	23,950	2.75	2.74	171.0	0.10	0.22	99.8
33	Olivine basalt	Mt. St. Helens, Wash.	75.7	33	6,540	2.84	2.27	138.2	9.97	22.06	77.9
34	Basalt porphyry	Lake Co., Ore.	75.2	26	18,750	2.81	2.76	172.2	0.66	1.76	98.2
35	Dolerite porphyry	Cape Ann, Mass.	87.5	35	23,600	2.87	2.34	177.5	0.37	1.03	98.9
36	Fine-grained diabase	Somerset Co., N.J.	96.4	35	38,000	2.96	2.95	184.4	0.06	0.17	99.8
37	Diabase	Somerville, Mass.	71.6	31	23,250	2.85	2.82	176.0	0.38	1.00	99.0
38	Greenstone	Winton, Minn.	53.8	50	*	2.70	2.69	167.7	0.19	0.46	99.5
39	Bytownite gabbro	Duluth, Minn.	80.2	30	*	3.00	3.00	187.3	0.00	0.00	100.0
40	Orthoclase gabbro quartz	Wichita Mts., Okla.	88.8	20×10^7	18,600	2.73	2.72	169.6	0.25	0.62	99.3
41	Orthoclase gabbro	Duluth, Minn.	76.5	*	26,400	2.87	2.86	178.4	0.13	0.29	99.7
	Sedimentary Rocks										
	Breccias and conglomerates										
42	Syenite breccia	Boulder Co., Colo.	83.6	30×10^7	27,550	3.12	3.10	193.3	0.27	0.78	99.2
43	Chert breccia	Cherokee Co., Kans.	104.5	67	*	2.60	2.56	159.8	0.66	1.65	98.3
44	Limestone breccia	Boulder Co., Colo.	27.3	52	5,960	2.80	2.28	142.1	8.26	18.73	81.2
	Sandstones										
45	Argillaceous sandstone	Portageville, N.Y.	57.2	47	13,900	2.65	2.48	155.1	2.48	6.10	93.8
46	Red (ferruginous) sandstone	Potsdam, N.Y.	61.3	57	18,350	2.63	2.49	155.6	2.12	5.26	94.7
47	Gray standstone (berea grit)	Berea, Ohio	42.0	51	5,840	2.66	2.13	132.7	9.36	19.84	80.1
48	Gray sandstone	Keeseville, N.Y.	86.5	37×10^7	33,350	2.62	2.58	160.9	0.66	1.62	98.3
49	Sandstone	Jordan, Minn.	22.0	56×10^7	4,920	2.60	1.91	119.5	13.80	26.40	73.6
50	Sandstone	Medina, N.Y.	72.7	63	18,270	2.60	2.46	153.7	2.09	5.15	94.8
51	Brownstone	Somerset Co., N.J.	51.3	55	14,100	2.52	2.35	146.8	2.92	6.75	93.2
52	Calcareous sandstone	Socorro, N.M.	60.3	65	18,810	2.62	2.31	144.5	5.14	11.85	88.1
	Limestones and dolomites										
53	Coquina	St. Augustine, Fla.	*	*	220	2.73	1.19	74.0	47.80	56.70	43.3
54	Limestone	Boulder Co., Colo.	56.5	40	19,780	2.78	2.67	166.6	1.56	4.10	95.8
55	Limestone	Onondaga Co., N.Y.	64.2	24	26,900	2.73	2.72	169.7	0.12	0.27	99.7

TABLE 20-7 Physical Properties of United States Rocks (Continued)

No.	Name of rock specimen	Location found	Hardness Shore number	Coefficient of thermal expansion, room temp to 212°F	Max compressive stress, psi	True specific gravity	Apparent specific gravity	Wt, lb/cu ft	Absorption, %	Pores, %	Solids, %
56	Coral limestone	LeRoy, N.Y.	58.5	28	16,780	2.70	2.67	166.4	0.52	1.38	98.6
57	Coral limestone	Jeffersonville, Ind.	57.9	49	14,220	2.68	2.66	166.1	0.31	0.75	99.2
58	Gray limestone	Valcour Island, N.Y.	58.1	17	19,980	2.71	2.69	167.8	0.25	0.63	99.3
59	Gray limestone	Ruth, Nevada	43.0	48	8,470	2.70	2.68	167.4	0.26	0.64	99.3
60	Encrinal limestone	Lockport, N.Y.	36.8	68	12,280	2.71	2.66	166.1	0.70	1.85	98.1
61	Encrinal limestone	Trenton Falls, N.Y.	44.1	56	10,400	2.70	2.67	166.4	0.49	1.27	98.7
62	Argillaceous limestone	Rochester, N.Y.	66.4	51	23,350	2.80	2.74	170.9	0.80	2.19	97.8
63	Pale gray limestone	Concrete, Colo.	45.4	24	18,500	2.66	2.54	158.9	1.73	4.36	95.6
64	Cherty limestone	Buffalo, N.Y.	96.2	*	*	2.65	2.65	165.4	0.08	0.15	99.8
65	Oolitic limestone	Batesville, Ark.	48.0	31	14,520	2.66	2.61	163.2	0.70	1.82	98.2
66	Oolitic limestone	Bedford, Ind.	27.9	43	8,210	2.66	2.29	143.2	5.97	13.66	86.3
67	Black oolite (fossiliferous)	Milton, Pa.	59.3	30	20,900	2.72	2.71	169.4	0.14	0.30	99.7
68	Dolomitic limestone	Gouverneur, N.Y.	46.6	58	9,230	2.74	2.73	170.3	0.12	0.28	99.7
69	Dolomitic limestone	Rochester, N.Y.	59.8	58	20,100	2.76	2.65	165.4	1.49	3.89	96.0
70	Birdseye limestone	Watertown, N.Y.	64.1	37	18,670	2.72	2.70	168.4	0.27	0.71	99.3
71	Chocolate "Tenn. marble"	Near Knoxville, Tenn.	45.7	53	14,290	2.69	2.68	167.6	0.09	0.19	99.8
	Siliceous rocks										
72	Diatomaceous silica	Santa Barbara, Calif.	17.5	15	477	1.93	0.48	30.3	154.60	74.90	25.1
73	Chert	Joplin, Mo.	97.9	64	86,300	2.59	2.48	155.1	1.69	4.10	95.8
	Chemical precipitates										
74	Travertine (onyx marble)	Suisun, Calif.	41.1	62	14,220	2.68	2.63	164.3	0.75	1.94	98.0
75	Travertine (onyx marble)	Great Salt Lake, Utah	42.1	44×10^7	16,200	2.65	2.64	165.2	0.09	0.21	99.7
	Metamorphic Rocks										
	Gneisses and crystalline schists										
76	Biotite gneiss	Uxbridge, Mass.	79.6	34×10^7	12,950	2.74	2.68	167.1	0.84	2.23	97.7
77	Biotite gneiss	Baltimore, Md.	87.2	44	15,640	2.66	2.65	165.8	0.12	0.30	99.7
78	Biotite gneiss	Cripple Creek, Colo.	95.7	13	14,950	2.68	2.65	165.6	0.47	1.16	98.8
79	Granitoid gneiss	Salisbury, N.C.	99.4	42	29,400	2.62	2.61	162.9	0.12	0.32	99.7
80	Gabbro gneiss	Albemarle Co., Va.	71.4	43	27,200	3.15	3.12	195.0	0.30	0.86	99.1

TABLE 20-7 Physical Properties of United States Rocks (Continued)

No.	Name of rock specimen	Location found	Hardness Shore number	Coefficient of thermal expansion, room temp to 212°F	Max compressive stress, psi	True specific gravity	Apparent specific gravity	Wt, lb cu ft	Absorption, %	Pores, %	Solids, %
	Quartzites and slates										
81	Baraboo quartzite	Ableman, Wis.	93.5	61	33,800	2.65	2.64	164.8	0.12	0.34	99.7
82	Quartzite	Dell Rapids, S.D.	97.2	60	23,400	2.63	2.61	162.8	0.25	0.65	99.4
83	Quartzite	Sioux Falls, S.D.	77.8	*	43,800	2.68	2.67	166.6	0.13	0.40	99.6
84	Red slate	Granville, N.Y.	75.3	49	19,100	2.79	2.78	173.6	0.13	0.29	99.6
85	Gray slate	Bangor, Pa.	59.1	45	*	2.76	2.71	169.3	0.67	1.84	98.2
86	Green slate	Pawlet, Vt.	58.8	49	17,700	2.77	2.77	173.0	0.00	0.00	100.0
87	Catlinite	Pipestone Co., Minn.	30.2	35	18,300	2.88	2.78	173.7	1.30	3.62	96.4
	Crystalline limestone and dolomites										
88	Crystalline limestone	Rutland, Vt.	35.7	27	6,540	2.71	2.68	167.3	0.40	1.14	98.8
89	Napoleon gray marble	Phenix, Mo.	42.5	51	13,180	2.71	2.66	165.8	0.77	2.02	98.0
90	White (yule) marble	Marble, Colo.	38.3	38	*	2.72	2.70	168.7	0.19	0.45	99.5
91	Dolomitic marble	Lee, Mass.	46.0	50	24,900	2.88	2.86	178.7	0.29	0.81	99.2
92	Pink and gray banded marble	Hewitts, N.C.	55.0	38	15,420	2.74	2.73	170.2	0.14	0.38	99.6
93	French gray marble	Plattsburg, N.Y.	52.5	48	*	2.52	2.51	156.5	0.26	0.65	99.4
94	St. Lawrence marble	Gouverneur, N.Y.	41.1	45	14,400	2.53	2.52	157.5	0.13	0.31	99.7
95	Pittsford Valley marble	Florence, Vt.	42.8	45	*	2.50	2.49	155.5	0.19	0.48	99.5
96	Variegated dolomitic, marble	Swanton, Vt.	66.1	44	32,450	2.83	2.82	175.7	0.13	0.40	99.6
97	Verd antique	Pyrenees Mts., France	71.4	34×10^7	13,500	2.75	2.73	170.1	0.29	0.76	99.2

* Data not obtained because of lost or defective specimen.
SOURCE: Iowa State University.

Metamorphic Class. This class is characterized by separation of crystals into approximately parallel layers and by the ability to break readily into thin slabs. Rocks of this class were formed from another type of rock by the great heat and pressure developed during the shrinkage and buckling of the earth's crust. Gneiss, a foliated type, was formed from granite; marble, a nonfoliated type, was formed from limestone.

Physical Properties of Rocks The properties of rocks within the same class can vary greatly, as shown in Table 20-7. The data in this table summarize the test results for 97 rock specimens which were obtained at specific locations.[14]

The suitability of a rock for use as coarse or fine aggregate in construction is dependent in large part upon its ability to resist the breakdown of traffic, weathering, and cyclic changes in temperature from freezing to thawing. Standard tests are performed on rock samples in order to predict the behavior of the rock when used as a construction aggregate. Table 20-8 shows the geographical distribution of more than 13,000 rock samples obtained throughout the 50 states, and Table 20-9 lists average values for the physical properties of the principal rocks thus sampled. Table 20-10 shows average values for the compressive strength of rock types commonly used as aggregates.[15]

Identification of Common Rocks A great deal can be gained by the producer and by the engineer from a knowledge of the identification of rock. In addition to their varying intrinsic properties, rocks may be affected in different ways by the aggregate processing itself. Some rock types, after processing, will have better finished shapes than others. Granite has a good quality for aggregate and yields a good finished product, although it is the source material for less than 8 percent of quarried aggregate. Approximately 11 percent of all quarried rock is basalt, diabase, and gabbro, a group generally known as "traprock." These are dark, heavy rocks with a tendency to crush to flat pieces in the finer sizes. Limestone and dolomite are the most widely found common quarry rocks and account for approximately 71 percent of the total. These high-calcium stones usually crush to yield well-shaped aggregate and lend themselves well to processing by impact breakers and hammer mills. Sandstone may vary greatly in its physical properties and must be checked carefully before it is used as an aggregate. Approximately 4 percent of quarried aggregate is produced from this type of rock.[16]

It is becoming increasingly important to include detailed petrographic examination in the evaluation of any raw material for use as aggregate. Such an analysis as a quality test by qualified personnel is both quick and reliable. In order to make a petrographic examination of rock samples for aggregate, it is necessary to be able to identify the various rock types. Table 20-11 outlines a system for such identification. The basis for this table is the appearance and character of newly fractured surfaces of unweathered rock.

Descriptive Nomenclature of Constituents of Natural Mineral Aggregates:

ASTM Designation........................ C294-67
Corps of Engineers Designation............ CRD-C139-56

Scope. The purpose of this nomenclature is to provide brief, useful, and accurate descriptions of some of the more common or more important natural materials found as constituents of mineral aggregates. The descriptions provide a basis for common understanding when they are used to designate aggregate constituents. It should be emphasized that many of the materials described will frequently occur in particles that do not display all the characteristics given in the description. Further, most of these materials grade from varieties which meet one description to varieties which meet another, with all intermediate stages being found. These descriptions are not adequate to permit the accurate identification of the natural constituents of mineral aggregates. In many cases, this identification can only be made by a qualified geologist, mineralogist, or petrographer using the apparatus and procedures of these sciences. References to these descriptions may, however, serve to indicate or prevent gross errors in identification.

Identification of the constituent materials in a mineral aggregate may facilitate the recognition of its properties, but identification alone, however accurately it may be

TABLE 20-8 Geographical Distribution of Samples

State	Number		State	Number	
	Coarse aggregate	Fine aggregate		Coarse aggregate	Fine aggregate
Alabama	158	56	Nebraska	20	8
Arizona	60	9	Nevada	1	0
Arkansas	173	130	New Hampshire	68	14
California	159	25	New Jersey	147	22
Colorado	71	6	New Mexico	43	14
Connecticut	110	20	New York	295	98
Delaware	53	22	North Carolina	808	188
District of Columbia	96	248	North Dakota	9	44
Florida	67	99	Ohio	625	294
Georgia	398	54	Oklahoma	121	78
Idaho	21	0	Oregon	56	7
Illinois	182	11	Pennsylvania	836	36
Indiana	265	15	Rhode Island	82	9
Iowa	35	3	South Carolina	149	206
Kansas	70	13	South Dakota	60	11
Kentucky	112	19	Tennessee	261	132
Louisiana	33	57	Texas	310	341
Maine	154	123	Utah	30	0
Maryland	444	225	Vermont	61	3
Massachusetts	433	42	Virginia	1,407	404
Michigan	179	54	Washington	237	2
Minnesota	54	24	West Virginia	366	40
Mississippi	106	110	Wisconsin	178	70
Missouri	81	17	Wyoming	19	7
Montana	39	16	Total	9,742	3,426

SOURCE: Bureau of Public Roads.

TABLE 20-9 Average Values for the Physical Properties of the Principal Kinds of Rocks

Kind of rock	Bulk specific gravity		Absorption		Loss by abrasion				Hardness		Toughness	
					Deval test		Los Angeles test					
	No. of tests	Avg	No. of tests	Avg %	No. of tests	Avg %	No. of tests	Avg %	No. of tests	Avg	No. of tests	Avg
Amphibolite	81	3.02	81	0.4	87	3.9	30	35	56	16	70	14
Basalt	229	2.86	228	0.5	203	3.1	24	14	192	17	203	19
Breccia	22	2.57	22	1.8	25	6.4	...	...	22	17	17	11
Chert	74	2.50	74	1.6	78	8.5	6	26	29	19	29	12
Conglomerate	25	2.68	26	1.2	27	10.0	...	...	11	16	11	8
Diabase	332	2.96	309	0.3	340	2.6	63	18	253	18	285	20
Diorite	59	2.92	57	0.3	60	3.1	...	...	45	18	48	15
Dolomite	668	2.70	667	1.1	708	5.5	134	25	586	14	612	9
Eclogite	8	3.11	8	0.1	9	2.5	...	...	6	18	6	21
Epidosite	22	3.03	22	0.4	20	3.7	10	20	19	18	18	17
Felsite*	149	2.66	147	0.8	150	3.8	9	18	118	18	127	17
Gabbro	46	2.96	46	0.3	45	3.0	4	18	38	18	42	14
Gneiss	419	2.74	424	0.3	602	5.9	293	45	365	18	386	9
Granite†	662	2.65	666	0.3	718	4.3	174	38	589	18	703	9
Limestone	1,695	2.66	1,673	0.9	1,677	5.7	350	26	1,209	14	1,315	8
Marble	184	2.63	162	0.2	175	6.3	41	47	162	13	188	6
Peridotite	10	3.31	9	0.3	7	4.1	...	...	5	15	7	9
Quartzite	208	2.69	204	0.3	233	3.3	119	28	146	19	161	16
Sandstone	716	2.54	707	1.8	699	7.0	95	38	613	15	681	11
Schist	297	2.85	296	0.4	314	5.5	136	38	180	17	212	12
Serpentine	20	2.62	20	0.9	19	6.3	13	19	11	15	13	14
Slate	84	2.74	78	0.5	71	4.7	7	20	57	15	69	18
Syenite	39	2.74	38	0.4	31	4.1	14	24	26	18	32	14

* Including andesite, dacite, rhyolite, and trachyte.
† Including granodiorite, pegmatite, and unakite.
SOURCE: Bureau of Public Roads.

TABLE 20-10 Compressive Strength of Types of Rock Commonly Used as Aggregates

Type of rock	Number of samples*	Compressive strength, psi		
		Avg†	After deletion of extremes‡	
			Max	Min
Granite	278	26,200	37,300	16,600
Felsite	12	47,000	76,300	17,400
Trap	59	41,100	54,700	29,200
Limestone	241	23,000	34,900	13,500
Sandstone	79	19,000	34,800	6,400
Marble	34	16,900	35,400	7,400
Quartzite	26	36,500	61,300	18,000
Gneiss	36	21,300	34,100	13,600
Schist	31	24,600	43,100	13,200

* For most samples, the compressive strength is an average of 3 to 15 specimens.

† Average of all samples.

‡ Of all the samples tested, 10 percent with the highest or lowest values have been deleted as not typical of the material.

SOURCE: D. O. Woolf, *ASTM, Spec. Tech. Pub.* No. 169, 1955.

TABLE 20-11 Preliminary Classification of Rocks

Group I: Glassy, wholly or partly
Group II: Not glassy; dull or stony; homogeneous; so fine-grained that grains cannot be recognized
Group III: Distinctly granular
Group IV: Distinctly foliated; no effervescence with acid
Group V: Clearly fragmental in composition; rounded or angular pieces or grains cemented together

Group I: Glassy Rocks

1. Glassy luster; hard; conchoidal fracture; colorless to white or smoky gray; generally brittle. *Quartz*
2. Solid glass; may have spherical inclusions; brilliant vitreous luster; generally black. *Obsidian*

Group II: Dull or Stony, Very Fine-grained Rocks

Subgroup II A. Not scratched by fingernail, but readily scratched with knife.
1. Particles almost imperceptible; dull luster; homogeneous; clay odor; little if any effervescence with acid; laminated structure; breaks into flakes. *Shale*
2. Little if any clay odor; brisk effervescence with acid. *Limestone*
3. Little if any clay odor; brisk effervescence with acid only when rock is powdered or acid is heated. *Dolomite*
4. Soapy or greasy feel; translucent on thin edges; green to black; no effervescence. *Serpentinite*

Subgroup II B. Not scratched with the knife or scratched only with difficulty; no effervescence with acid
1. Light to gray color; clay odor possible; may have a banded flow structure. *Felsite*
2. Very hard; pale colors to black; no clay odor; conchoidal fracture; waxy or horny appearance *Chert.* If dark gray to black, *Flint*
3. Heavy; dark color; may have cellular structure, may contain small cavities filled with crystalline minerals. *Basalt*

Group III: Granular Rocks

Subgroup III A. Easily scratched with the knife
1. Brisk effervescence with acid. *Limestone* or *marble*
2. Brick effervescence only with warm acid or with powdered rock. *Dolomitic marble*

Subgroup III B. Hard; not scratched with knife or scratched with difficulty; grains of approximately equal size
1. Mainly quartz and feldspar; usually light colored, sometimes pinkish. *Granite*
2. Mainly feldspar; little quartz (less than 5 percent); light colors of nearly white to light gray or pink. *Syenite*
3. Feldspar and a dark ferromagnesian mineral
   a. Major constituent feldspar; rock of medium color. *Diorite*
   b. Ferromagnesian mineral equal to or in excess of feldspar; rock of dark color
      (1) Grains just large enough to be recognized by the unaided eye *Diabase*
      (2) Coarse-grained rock. *Gabbro*
4. Mainly ferromagnesian minerals; generally dark green to black
   a. Predominant olivine with pyroxene or hornblende. *Peridotite*
   b. Predominant hornblende. *Hornblende*

TABLE 20-11 Preliminary Classification of Rocks (Continued)

5. Mainly quartz
   a. Fracture around grains. *Sandstone*
   b. Fracture through all or through an appreciable percentage of grains. *Quartzite*

Subgroup III C. Hard; not scratched with knife or scratched with difficulty; large distinct crystals in finer ground mass

1. Crystals of feldspar and quartz with some of a ferromagnesian mineral (generally biotite) in a light-colored ground mass (of feldspar and quartz). *Granite porphyry*
2. Crystals of feldspar and usually a ferromagnesian mineral in a light colored ground mass (of feldspar). *Syenite porphyry*
3. Crystals of ferromagnesian minerals or of striated feldspar, or both, in a medium-colored ground mass (of feldspar and ferromagnesian minerals). *Diorite porphyry*
4. Crystals of quartz or feldspar, or both, generally with a ferromagnesian mineral, in a predominant fine-grained ground mass of light color. *Felsite porphyry*
5. Crystals of feldspar or of a ferromagnesian mineral, or both, in a fine-grained dark or black heavy ground mass, *Basalt porphyry*

Group IV: Foliated Rocks

1. Medium to coarse grain; roughly foliated. *Gneiss*
2. More finely grained and foliated. *Schist*
   a. Consists mainly or largely of mica with some quartz. *Mica schist*
   b. Meduim green to black; consists mostly of a felted or matted mass of small bladed or needlelike crystals arranged in one general direction. *Hornblende schist or amphibolite*
   c. Glassy or silky luster on foliation surfaces; splits readily into thin pieces. *Sericite schist*
   d. Soft, greasy feel; marks cloth; easily scratched with fingernail; whitish to light gray or green. *Talc schist*
   e. Smooth feel; soft; glimmering luster; green to dark green. *Chlorite schist*
3. Very fine grain; splits easily into thin slabs; usually dark gray, green, or black *Slate*

Group V: Fragmental

1. Rounded pebbles embedded in some type of cementing medium. *Conglomerate*
2. Angular fragments embedded in a cementing medium. *Breccia*
3. Fragments of volcanic (fine-grained or glassy) rocks embedded in compacted volcanic ash. *Volcanic tuff* or *volcanic breccia*
4. Quartz grains, rounded or angular, cemented together. *Sandstone*
5. Quartz and feldspar grains cemented together to resemble the appearance of granite. Arkose (feldspathic sandstone)

SOURCE: Bureau of Public Roads.

accomplished, cannot provide an adequate basis for predicting the behavior of aggregates in service. Mineral aggregates composed of constituents of any type or combination of types may perform well or poorly in service depending upon the exposure to which they are subjected, the physical and chemical properties of the matrix in which they may be embedded, their physical condition at the time they are used, and upon other factors.

Method of Petrographic Examination of Aggregate for Concrete:

ASTM Designation........................ C295-65
Corps of Engineers Designation............ CRD-C127-64

Scope. This method outlines procedures for the petrographic examination of representative samples of materials proposed for use as aggregates in concrete. The specific procedures employed in the petrographic examination of any sample will largely depend on the purpose of the examination and the nature of the sample.

Source Evaluation It has become increasingly difficult to obtain the natural rock from which most aggregate is produced, and a new project may involve prospecting for a source of raw material. The geologic history of the general location then becomes of prime importance. Sources of information are manifold and should be consulted.[17] The following is a partial listing of such sources:

1. "Geologic Map of the United States" by the United States Geologic Survey. Scale 1 to 2,500,000
2. State geologic maps as published by the federal or state governments
3. Reports of various state topographic, hydrologic, and geologic surveys, many covering in detail core drillings and excavations
4. Information from local sources such as operators of existing pits and quarries, also well diggers and drillers
5. Aerial surveys and maps, including records of geophysical explorations

Aggregate (clean separation)	Estimated tons of separated aggregate				Total estimated tons separated aggregate	Estimated tons required
	Between elevations 3,240 and 3,200	Between elevations 3,200 and 3,160	Between elevations 3,160 and 3,120	Between elevations 3,120 and 3,100		
Sand	135,000	596,000	2,240,000	981,000	3,952,000	1,500,000
Coarse aggregate $\frac{3}{16}$–$\frac{3}{4}$ in.	231,000	1,081,000	1,728,000	451,000	3,491,000	1,100,000
Coarse aggregate $\frac{3}{4}$–$1\frac{1}{2}$ in.	202,000	816,000	1,367,000	408,000	2,793,000	1,100,000
Coarse aggregate $1\frac{1}{2}$–3 in.	168,000	732,000	1,192,000	396,000	2,488,000	1,100,000
Coarse aggregate 3–6 in.	103,000	552,000	681,000	174,000	1,510,000	800,000

(*a*)

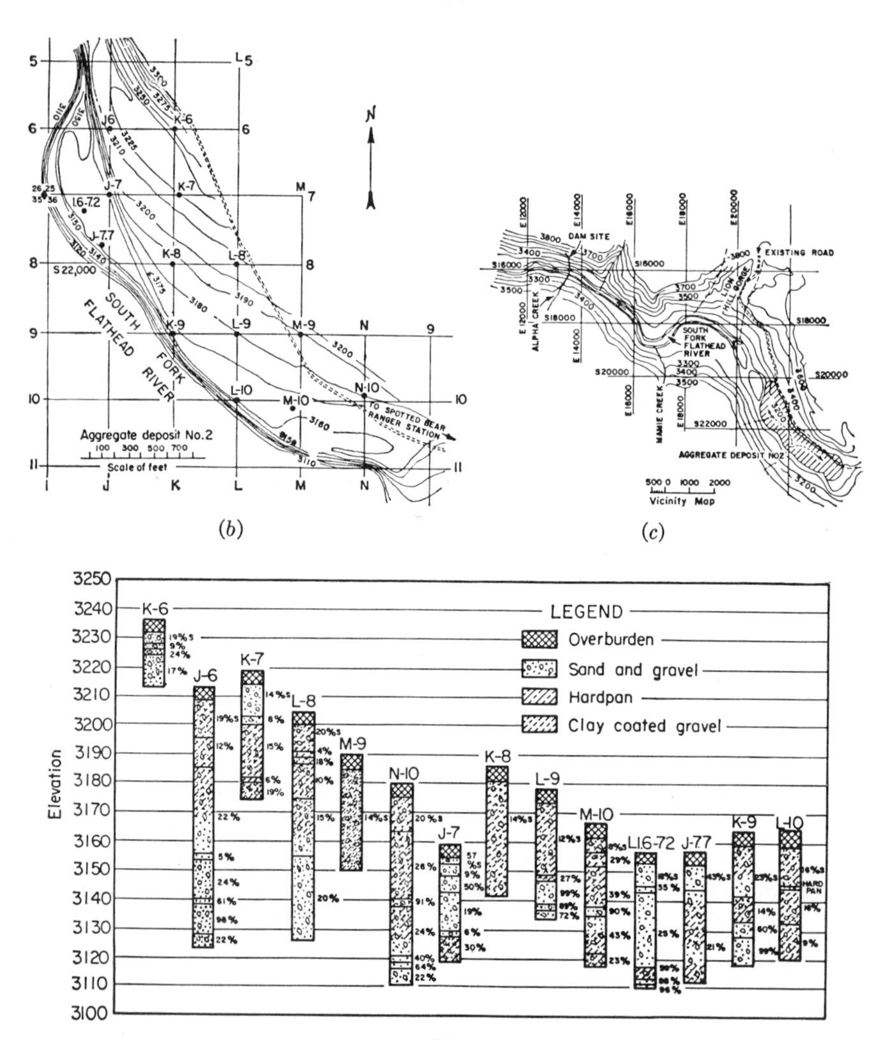

(*b*) (*c*)

(*d*)

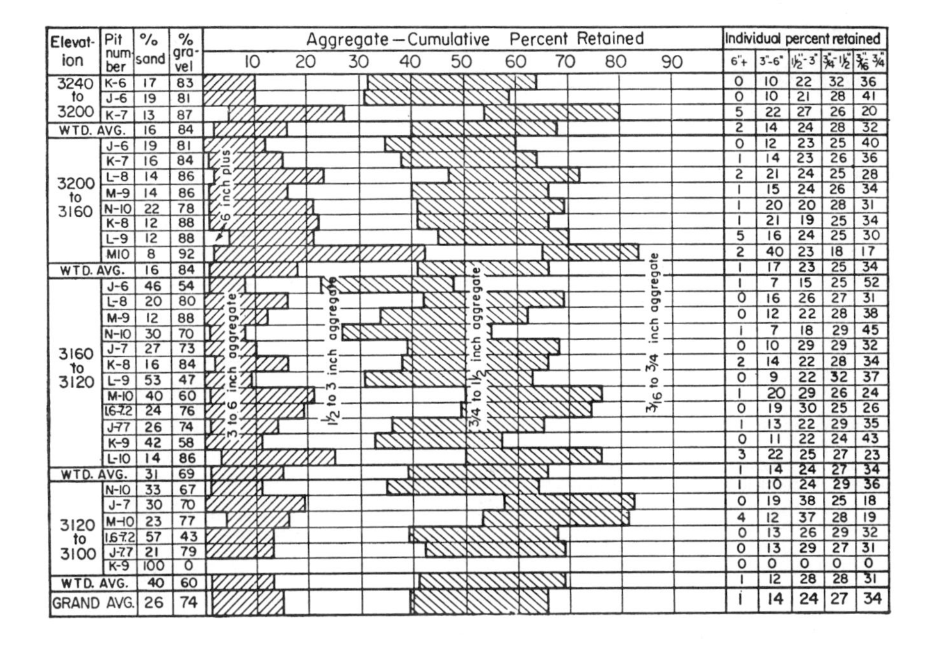

Elevation	Pit number	% sand	% gravel	6"+	3"-6"	1½"-3"	¾"-1½"	3/16"-¾"
3240 to 3200	K-6	17	83	0	10	22	32	36
	J-6	19	81	0	10	21	28	41
	K-7	13	87	5	22	27	26	20
WTD. AVG.		16	84	2	14	24	28	32
3200 to 3160	J-6	19	81	0	12	23	25	40
	K-7	16	84	1	14	23	26	36
	L-8	14	86	2	21	24	25	28
	M-9	14	86	1	15	24	26	34
	N-10	22	78	1	20	20	28	31
	K-8	12	88	1	21	19	25	34
	L-9	12	88	5	16	24	25	30
	M10	8	92	2	40	23	18	17
WTD. AVG.		16	84	1	17	23	25	34
3160 to 3120	J-6	46	54	1	7	15	25	52
	L-8	20	80	0	16	26	27	31
	M-9	12	88	0	12	22	28	38
	N-10	30	70	1	7	18	29	45
	J-7	27	73	0	10	29	29	32
	K-8	16	84	2	14	22	28	34
	L-9	53	47	0	9	22	32	37
	M-10	40	60	1	20	29	26	24
	I.6-7.2	24	76	0	19	30	25	26
	J-7.7	26	74	1	13	22	29	35
	K-9	42	58	0	11	22	24	43
	L-10	14	86	3	22	25	27	23
WTD. AVG.		31	69	1	14	24	27	34
3120 to 3100	N-10	33	67	1	10	24	29	36
	J-7	30	70	0	19	38	25	18
	M-10	23	77	4	12	37	28	19
	I.6-7.2	57	43	0	13	26	29	32
	J-7.7	21	79	0	13	29	27	31
	K-9	100	0	0	0	0	0	0
WTD. AVG.		40	60	1	12	28	28	31
GRAND AVG.		26	74	1	14	24	27	34

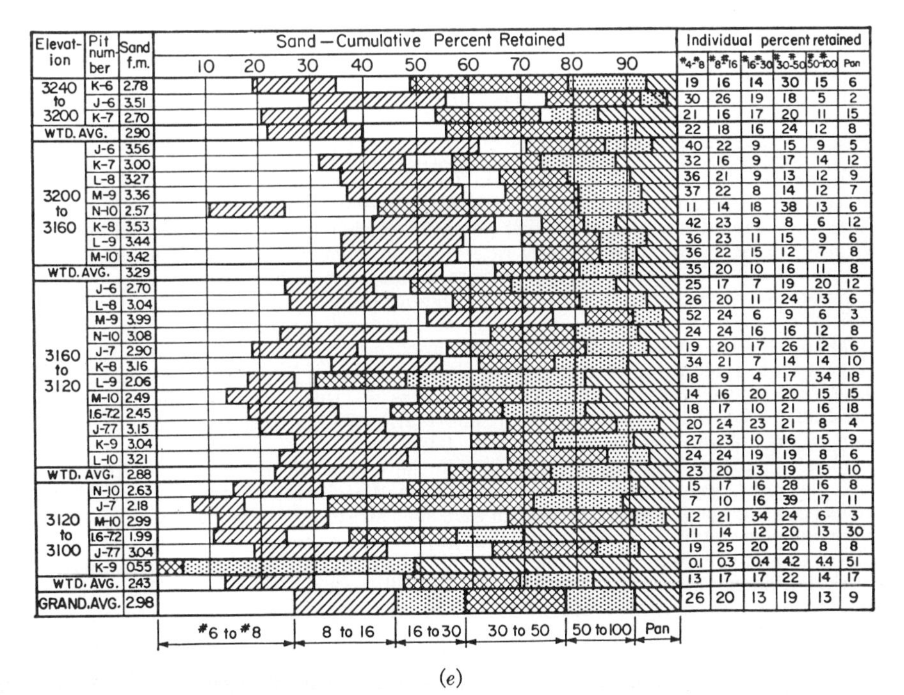

Elevation	Pit number	Sand f.m.	#4-#8	#8-#16	#16-#30	#30-#50	#50-#100	Pan
3240 to 3200	K-6	2.78	19	16	14	30	15	6
	J-6	3.51	30	26	19	18	5	2
	K-7	2.70	21	16	17	20	11	15
WTD. AVG.		2.90	22	18	16	24	12	8
3200 to 3160	J-6	3.56	40	22	9	15	9	5
	K-7	3.00	32	16	9	17	14	12
	L-8	3.27	36	21	9	13	12	9
	M-9	3.36	37	22	8	14	12	7
	N-10	2.57	11	14	18	38	13	6
	K-8	3.53	42	23	9	8	6	12
	L-9	3.44	36	23	11	15	9	6
	M-10	3.42	36	22	15	12	7	8
WTD. AVG.		3.29	35	20	10	16	11	8
3160 to 3120	J-6	2.70	25	17	7	19	20	12
	L-8	3.04	26	20	11	24	13	6
	M-9	3.99	52	24	6	9	6	3
	N-10	3.08	24	24	16	16	12	8
	J-7	2.90	19	20	17	26	12	6
	K-8	3.16	34	21	7	14	14	10
	L-9	2.06	18	9	4	17	34	18
	M-10	2.49	14	16	20	20	15	15
	I.6-7.2	2.45	18	17	10	21	16	18
	J-7.7	3.15	20	24	23	21	8	4
	K-9	3.04	27	23	10	16	15	9
	L-10	3.21	24	24	19	19	8	6
WTD. AVG.		2.88	23	20	13	19	15	10
3120 to 3100	N-10	2.63	15	17	16	28	16	8
	J-7	2.18	7	10	16	39	17	11
	M-10	2.99	12	21	34	24	6	3
	I.6-7.2	1.99	11	14	12	20	13	30
	J-7.7	3.04	19	25	20	20	8	8
	K-9	0.55	0.1	0.3	0.4	4.2	4.4	51
WTD. AVG.		2.43	13	17	17	22	14	17
GRAND. AVG.		2.98	26	20	13	19	13	9

(*e*)

Fig. 20-5 Exploration of a pit source. (*a*) Estimate of quantities of gravel in deposit and quantities required for construction. (*b*) Plan of aggregate deposit. (*c*) Vicinity map. (*d*) Log of test pits. (*e*) Aggregate depths, grading data, and screen analysis. (*Bureau of Reclamation.*)

Once a potential source of raw aggregate material has been identified, its suitability must be evaluated. This analysis might be confined solely to the project for which the aggregate is going to be used, or it may be extended to include commercial possibilities beyond the project. For either case, appropriate consideration should be given to such items as the title holder to the land on which the deposit is located, the size and type of deposit, and its location in relation to existing road systems or railroad sidings. For a construction project of a definite size, the most important question will be the hauling distances between the material source and the points of use. Suitable deposits should be covered by options, including details as set out below:

THIS OPTION, made and entered into by and between ________________________, party of the first part, and ________________________, party of the second part, owner of the property in ________________________ county in the State of ________________________,

Witnesseth:

THAT WHEREAS, the party of the first part is contemplating crushing stone at or near the premises of the second party in ________________________ county in the State of ________________________.

WHEREAS, the second party has land convenient and thought to be suitable for the opening of a pit/or quarry and the crushing of aggregates.

NOW THEREFORE, in consideration of the sum of ____________ receipt of which is hereby acknowledged and other good and valuable consideration it is mutually agreed as follows:

(1) The second party hereby grants and gives unto the party of the first part an option for 120 days to lease the second party's property described as follows:

(Insert description of property.)

(2) When and if the party of the first part furnishes aggregate for the planned construction and exercises this option, the party of the first part agrees to pay to the party of the second part a royalty of ____________ per ton.

(3) This option is conditioned as follows:

(a) That second party's property is free of previous interfering leases or options.

(b) That the party of the first part actually furnishes aggregate for the planned construction.

IN WITNESS WHEREOF, the parties hereto set their hands this ________________________ day of ________________________, 19____ as full acceptance of the above terms and agreement.

The option should be witnessed and notarized and is best handled by an attorney familiar with the specific local, state, and federal requirements to be met in each case.

Factors to be evaluated for a commercial operation, in addition to those just described, should include the market potential, anticipated cost of operation and its relation to the competitive picture, rate of output required to satisfy the consumer, estimated raw material reserve, and future expansion possibilities.

Sand and Gravel Pits. In general, the digging of test holes is the surest and the quickest way to get a reliable picture of the material within a pit deposit which is available for aggregate production. This applies both to small and large deposits. A thorough exploration of a pit source is of particular importance because of the frequent requirements for specified percentages of crushed material in the finished product. Figure 20-5 shows pertinent information for the explored areas of a gravel deposit, as obtained by the Bureau of Reclamation.[18]

Stone Quarries. The proper sampling of the site is of the greatest importance. Samples must provide an average representation of each rock stratum and should also reveal any significant variations within each such stratum. Core drilling of the site is invaluable to obtain the necessary rock samples. It will also give information as to the size of the deposit. A drilling log should be maintained, and the final evaluation of the site should be based on the cores and the drilling log. Where there is an existing quarry face, representative samples should be taken from freshly exposed faces.

A typical cross section of a rock face in a limestone deposit is shown in Fig. 20-6. This rock is a part of the Laurel (Silurian) formation. The chemical composition of one sample in each of the five ledges is shown in Table 20-12, together with physical test results based on AASHO and ASTM standard methods of test.

Sample Requirements. The taking of reliable samples includes the detailed recording of pertinent observed facts. Careful sampling is of major importance, since an

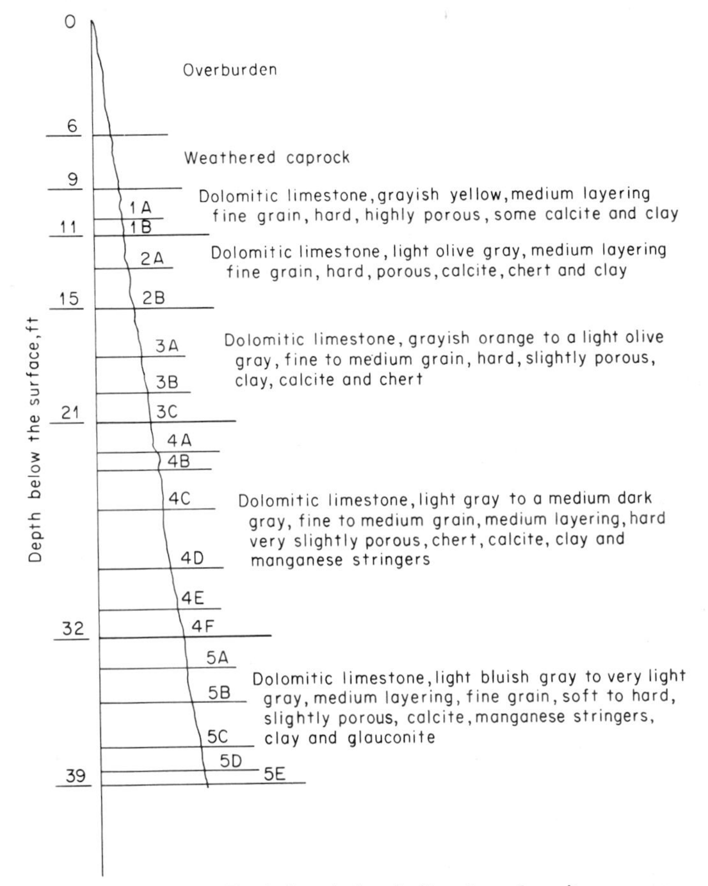

Fig. 20-6 Typical rock face in limestone deposit.

unreliable sample can cause costly mistakes in bid estimates and in projections of equipment requirements. The sampling of rock to be used for aggregate is covered by a number of specifications, of which the following are the more important:

AASHO Designation	T2-60
ASTM Designation	D75-59 (1968)
Corps of Engineers Designation	CRD-C100-64
Federal Specification Designation	Method 101.01

Scope. The methods cover the sampling of stone, slag, sand, gravel, and stone blocks for the purpose of a preliminary investigation of the source of supply, acceptance or rejection of such a source, or inspection of materials. Table 20-13 shows the sample requirements as outlined by the Corps of Engineers in CRD-C100-64.

TESTS OF ROCKS AND AGGREGATES[19]

Physical and chemical tests are of value in assessing the suitability of rocks and aggregates. The physical tests will show the extent of compliance with whatever

TABLE 20-12 Chemical and Physical Test Results from Limestone Deposit Shown in Fig. 20-6

Laurel (Silurian) formation

Ledge No.	1-A	2-B	3-C	4-D	5-E
Insolubles	5.00	6.76	4.72	1.94	5.58
Silicon dioxide	3.99	4.67	3.66	1.49	4.53
Aluminum oxide	0.85	1.36	0.63	1.07	1.03
Iron oxide	0.64	0.70	01.12	0.77	2.22
Calcium carbonate	54.83	58.83	54.19	62.11	57.11
Magnesium carbonate	39.54	32.11	39.42	33.96	33.61
Calcium carbonate equivalent (CCE)	101.88	97.04	101.10	102.52	97.11
Loss on ignition	43.76	43.85	44.70	45.12	42.95

Ledge No.	Thickness of ledge, ft	Percent of wear, Los Angeles abrasion	Percent of loss, 5 cycles, Na_2SO_4 soundness	Percent of water absorption, 24 hr	Specific gravity	
1	2	26.2	12.8	2.1	Apparent	2.78
					Bulk	2.69
2	4	24.3	6.4	1.9	Apparent	2.78
					Bulk	2.70
3	6	25.4	9.3	1.4	Apparent	2.80
					Bulk	2.73
4	11	26.7	5.5	0.9	Apparent	2.79
					Bulk	2.75
5	7	31.3	6.7	1.2	Apparent	2.80
					Bulk	2.74

specifications are to be met. The chemical analysis will give important specification information and very necessary guidance for equipment selection, such as an indication of the expected abrasiveness of the material. For concrete aggregates, especially, the chemical composition of the aggregate is very important.

In general, the tests of rock will cover the following characteristics:

1. Abrasion resistance
2. Toughness
3. Hardness
4. Specific gravity
5. Absorption
6. Compressive strength
7. Soundness
8. Other properties

Abrasion Resistance Two testing procedures have been devised to measure the resistance of rock to abrasive wear. The Los Angeles abrasion test first came into use in 1932, and it is now a requirement in most of the states. The Deval test was developed by the French in 1878 and subsequently modified, with a final change from circular screen openings to square openings in 1937. It has been found to be of less value than the Los Angeles test as an indication of the behavior of aggregate in actual use.

Abrasion of Coarse Aggregate (Los Angeles Machine):

AASHO Designation	T96-60
ASTM Designation	C131-66
Corps of Engineers Designation	CRD-C117-46
Federal Specification Designation	Method 208.11

Scope. The specification covers the Los Angeles procedure for the testing of crushed gravel, rock, and slag or uncrushed gravel for resistance to abrasion. Ledge rock, when hand broken into approximately cubical fragments of the specified sizes and tested in this manner, has been found to have a loss of approximately 85 percent of that for crushed rock of the same quality.

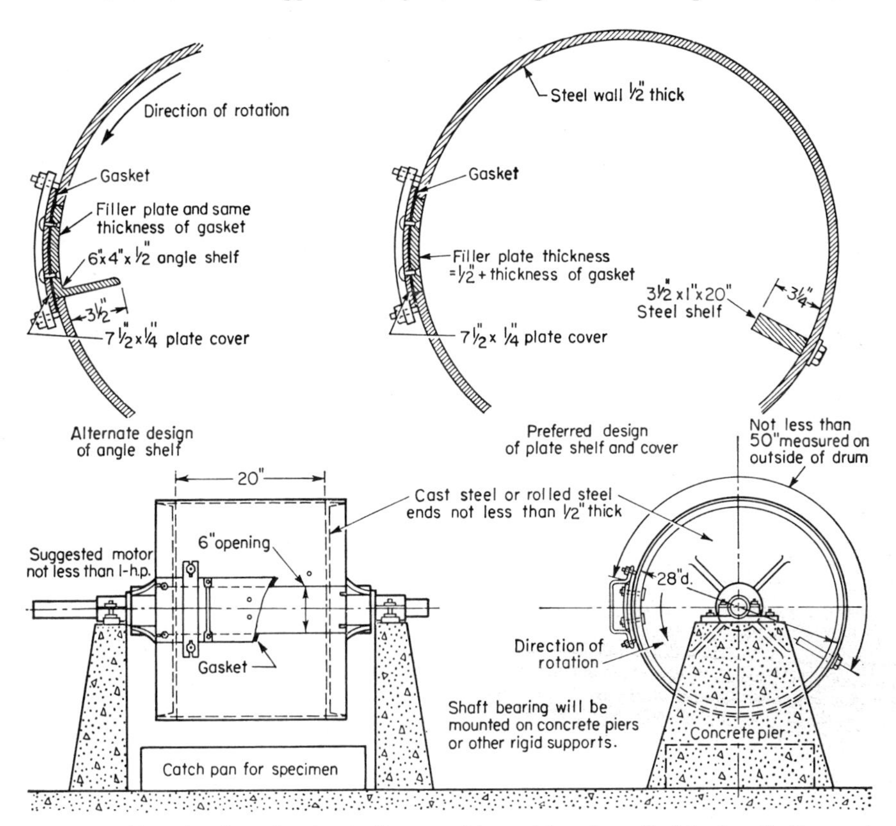

Fig. 20-7 Los Angeles abrasion-testing machine. (*American Society for Testing and Materials.*)

Apparatus. The Los Angeles abrasion testing machine shown in Fig. 20-7 consists of a steel cylinder having an inside diameter of 28 in. and a length of 20 in., with a 3½-in. shelf. The cylinder is supported for rotation on stub shafts, and an opening with a removable cover is provided for loading the sample together with an abrasive charge of a maximum of twelve 1⅞-in.-diameter cast-iron or steel spheres. Material is tested by rotating the machine from 500 to 1,000 revolutions, depending on the sample, at a speed of 30 to 33 rpm. The sample size should be 10,000 g for the coarser grading and 5,000 g for the finer. The abrasive wear is reported as the difference between the retained weight on the designated screen and the weight of the original sample, expressed as a percentage of the original weight.

TABLE 20-13 Corps of Engineers Requirements for Rock Sampling (CRD-C100-64)

Purpose and source	Size	Amount
Petrographic examination		
Core samples	4-in. diam	All recovered
Pieces broken from each ledge or bed of a face	At least 1 lb each	At least 5 lb
Stockpiles of processed crushed stone*	Fine aggregate	50 lb
	Coarse aggregate:	
	No. 4 to ¾ in.	50 lb
	¾ to 1½ in.	100 lb
	1½ to 3 in.	200 lb
	3 to 6 in.	1,000 lb†
Sand and gravel (processed or pit run)*	Sand	50 lb
	Coarse aggregate:	
	No. 4 to ¾ in.	100 lb
	¾ to 1½ in.	200 lb
	1½ to 3 in.	400 lb
	3 to 6 in.	1,000 lb†
Elementary tests only		
Fine aggregate:		
Processed, natural or manufactured	...	300 lb
Unprocessed sand, or rock for sand making	...	500 lb
Coarse aggregate:		
Processed	No. 4 to ¾ in.	500 lb
	¾ to 1½ in.	500 lb
	1½ to 3 in.	1,000 lb
	3 to 6 in.	1,500 lb
Unprocessed	No. 4 to ¾ in.	1,000 lb
	¾ to 1½ in.	1,500 lb
	1½ to 3 in.	1,500 lb
	3 to 6 in.	2,000 lb
Elementary tests and freezing and thawing		
Fine aggregate:		
Processed, well graded	...	1,500 lb
Poorly graded or unprocessed sand, or rock for manufacturing fine aggregate	...	2,000 lb
Coarse aggregate:		
Processed	No. 4 to ¾ in.	1,500 lb
	¾ to 1½ in.	1,500 lb
	1½ to 3 in.	1,000 lb
	3 to 6 in.	1,500 lb
Unprocessed	No. 4 to ¾ in.	3,000 lb
	¾ to 1½ in.	3,000 lb
	1½ to 3 in.	1,500 lb
	3 to 6 in.	2,000 lb

NOTE: Preliminary evaluation of massive rock may be made on AX (1⅛-in.) core or of complex or thin-bedded rock on NX (2⅛-in.) core.

* Obtained by compositing and reducing larger samples.

† Or not less than 300 pieces, whichever is larger.

TABLE 20-13 Corps of Engineers Requirements for Rock Sampling (CRD-C100-64) **(Continued)**

Purpose and source	Size	Amount
Mixture proportioning studies (per mixture)‡		
¾-in. aggregate mixture	Fine aggregate	800 lb
	No. 4 to ¾ in.	1,200 lb
	Cement	4 bags
1½-in. aggregate mixture	Fine aggregate	1,000 lb
	No. 4 to ¾ in.	1,000 lb
	¾ to 1½ in.	1,000 lb
	Cement	4 bags
3-in. aggregate mixture	Fine aggregate	2,000 lb
	No. 4 to ¾ in.	1,500 lb
	¾ to 1½ in.	1,000 lb
	1½ to 3 in.	2,000 lb
	Cement	5 bags
6-in. aggregate mixture	Fine aggregate	3,000 lb
	No. 4 to ¾ in.	2,000 lb
	¾ to 1½ in.	1,500 lb
	1½ to 3 in.	2,500 lb
	3 to 6 in.	3,000 lb
	Cement	7 bags
Airfield pavement mixtures¶	Fine aggregate	3,000 lb
	Coarse aggregate§	(§)
	Cement	16 bags
Unprocessed rock samples (for tests other than petrographic examination)		
6-in. core for processing to fine or coarse aggregate and elementary tests only	Per ledge or zone	20 ft
36- to 40-in. calyx core or quarry or ledge rock for processing to fine and coarse aggregate for elementary and freezing-and-thawing tests	Per aggregate	25 tons

‡ Processed materials complying with applicable specifications for grading for mixture proportioning studies only.

§ Coarse aggregate minimum requirements:

Amount (in pounds) Passing Sieve Sizes

Maximum size, in.	No. 4 to ¾ in.	No. 4 to 1 in.	¾ to 1½ in.	1 to 2 in.	1½ to 2½ in.
1	.....	5,000			
1½	3,000	.....	3,000		
2	.....	3,000	.....	3,000	
2½	2,500	.....	2,500	.....	2,500

¶ These quantities represent minimum amounts needed to establish optimum mixture proportions using three cement factors or water-cement ratios, and molding sufficient 6- by 6-in. beams for nine flexural strength tests at each of three test ages from mixtures representing each cement factor or water-cement ratio. Should investigation of additional mixtures or additional tests be desired, the quantities must be increased proportionally. These quantities must also be increased proportionally if the grading of the samples submitted is such that reprocessing at the laboratory is required.

SOURCE: Corps of Engineers, U.S. Army.

The test is required at the present time by 39 states with a specification range of 25 to 65 percent, and it is generally required on government projects where wear values from 40 to 50 percent are commonly specified. Refer to Table 20-9 to obtain average wear values for the principal types of rock.

Abrasion of Coarse Aggregate (Deval Machine):

AASHO Designation	T3 and T4-35
ASTM Designation	D2-33 (1968) and D289-63
Corps of Engineers Designation	CRD-C141-56
Federal Specification Designation	Method 208.0

Scope. The specification covers the Deval procedure for the testing of crushed gravel, rock, and slag or uncrushed gravel for resistance to abrasion.

Apparatus. The Deval abrasion-testing machine consists of a hollow iron cylinder, 20 cm in diameter and 34 cm deep inside, closed at one end and with a cover at the other end. The cylinder is mounted on a shaft set at an angle of 30°. The usual sample weight is 5,000 g; for a material with a specific gravity below 2.20 it is 4,000 g. The material is tested by rotating the machine 10,000 revolutions at a speed of 30 to 33 rpm with an abrasive charge of six cast-iron spheres, each with a diameter of 1⅞ in. and weighing about 430 g. The difference between the retained weight on the designated screen and the weight of the original sample, expressed as a percentage of the original weight, shall be the loss by abrasion.

The test is required at the present time by nine states, with a specification range of 3 to 8 percent. Refer to Table 20-9 to obtain average wear values for principal types of rock.

Toughness The toughness test is made to determine the resistance of a rock to impact.

Toughness of Rock:

Corps of Engineers Designation	CRD-C132-53
Federal Specification Designation	Method 207.0

Scope. The specification covers the procedure for determining the toughness of rock. The toughness is defined as the resistance to fracture under impact, expressed as the final height of blow required to fracture a test specimen 24- to 25-mm in diameter and 25-mm high resting on a 50-kg cast-iron anvil. The blow is delivered by a 2-kg hammer on a 1-kg steel plunger, with a spherically shaped lower end 1 cm in radius.

The average height of blow in centimeters of three test specimens shall be reported as detailed in the specification. Table 20-9 gives average values of toughness of the principal kinds of rock.

Hardness Mohs' scale indicates graded values of mineral hardness, as listed below. It was developed in 1812 and is now being used successfully for correlating physical and technological properties of rocks.

Talc	1	Calcite	3	Orthoclase	6	Corundum	9
Gypsum	2	Fluorite	4	Quartz	7	Diamond	10
		Apatite	5	Topaz	8		

Approximately 70 percent of the crushed rock produced in the United States is in the range of 3 to 4 on this scale.[20]

The hardness of rocks, on a broad range, can be covered by the following four categories:

Soft	Medium	Hard	Very hard
Asbestos rock	Limestone	Granite	Taconite
Gypsum rock	Dolomite	Quartzite	Granite
Slate	Sandstone	Iron ore	Felsite
Talc		Traprock	Traprock
Limestone			

Unit Weight, Specific Gravity, Absorption, and Moisture Content These tests are necessary to obtain quantitative design data for the subsequent use of the aggregate. They also provide some indication of the probable behavior of the material.

Unit Weight of Aggregate:

AASHO Designation........................ T19-56
ASTM Designation......................... C29-68
Corps of Engineers Designation.............. CRD-C106-57
Federal Specification Designation............ Method 201.0

Scope. This method of test covers the procedure for determining the unit weight of fine, coarse, or mixed aggregates.

Specific Gravity and Absorption of Fine Aggregates:

AASHO Designation........................ T84-60
ASTM Designation......................... C128-68
Corps of Engineers Designation.............. CRD-C108-60
Federal Specification Designation............ Method 209.1

Scope. (1) This method of test is intended for use in making determinations of bulk and apparent specific gravity and absorption (after 24 hr in water at room temperature) of fine aggregate. The bulk specific gravity is the value generally desired for calculations in connection with portland-cement concrete. (2) This method determines directly the bulk specific gravity as defined in the Standard Definition of Terms Relating to Specific Gravity (AASHO Designation M132), or the bulk specific gravity on the basis of weight of saturated surface-dry aggregate, or the apparent specific gravity as defined in the Standard Definitions M132.

Specific Gravity and Absorption of Coarse Aggregates:

AASHO Designation........................ T85-60
ASTM Designation......................... C127-68
Corps of Engineers Designation.............. CRD-C107-60
Federal Specification Designation............ Method 209.0

Scope. (1) This method of test is intended for use in making determinations of bulk and apparent specific gravity and absorption (after 24 hr in water at room temperature) of coarse aggregate. The bulk specific gravity is the value generally desired for calculations in connection with portland-cement concrete. (2) This method determines directly the bulk specific gravity as defined in the Standard Definition of Terms Relating to Specific Gravity (AASHO Designation M132), or the bulk specific gravity on the basis of weight of saturated surface-dry aggregate, or the apparent specific gravity as defined in the Standard Definitions M132.

Method of Test for Absorption by Aggregates (Field Method):

Corps of Engineers Designation............ CRD-C109-48

Scope. This method of test is intended for use in making an approximate determination of absorption (of moisture) by fine or coarse aggregate.

Surface Moisture in Fine Aggregate:

AASHO Designation...................... T142-48
ASTM Designation....................... C70-66
Corps of Engineers Designation............ CRD-C111-48

Scope. This method of test covers a procedure for determining, in the field, the amount of surface moisture in fine aggregate by displacement in water. The accuracy of this method depends upon accurate information on the bulk specific gravity of the material in a saturated surface-dried condition. The same procedure, with appropriate changes in the size of the sample and the dimensions of the container, may be applied to coarse aggregate.

Method of Test for Surface Moisture in Fine Aggregate (Field Method):

Corps of Engineers Designation............ CRD-C112-48

Scope. This method of test covers a rapid procedure for determining the percentage of surface moisture in fine aggregate by a displacement method. The method as set forth was modified from one proposed by C. R. Waters, District Engineer, New York State Department of Public Works, Buffalo, N.Y. The present method was published in 1945.

Method of Test for Moisture Content of Aggregate by Loss in Weight by Drying:

Corps of Engineers Designation............ CRD-C113-60

Scope. This method of test covers the procedure for determining the percentage of evaporable moisture in and on a sample of aggregate by loss in weight by drying.

Voids in Aggregate for Concrete:

AASHO Designation....................... T20-42
ASTM Designation........................ C30-37 (1964)
Corps of Engineers Designation............ CRD-C110-48

Scope. This method of test outlines the procedure for determining voids in aggregate.

Compressive Strength This is of special importance for the evaluation of an aggregate in connection with crusher selection. See Table 20-10 for typical values.

Method of Test for Compressive Strength of Natural Building Stone:

ASTM Designation............ C170-50 (1958)

Scope. This method covers the sampling, preparation of specimens, and procedure for determining the compressive strength of natural building stone.

Soundness The general scope of the soundness test is covered by various specifications. See Table 20-9 for typical test results.

Method of Testing Stone for Resistance to Freezing and Thawing:

Corps of Engineers Designation............ CRD-C144-63

Scope. This method covers a procedure for determining the resistance of stone to freezing and thawing. Information developed by use of this method may be applicable in the evaluation of stone for use as slope protection, as concrete aggregate, or for other purposes.

Soundness of Aggregate by Freezing and Thawing:

AASHO Designation............ T103-62I

Scope. This method describes the procedure to be followed in testing aggregates to determine their resistance to disintegration by freezing and thawing. It furnishes information helpful in judging the soundness of aggregates subjected to weathering action. However, in selecting aggregates, principal dependence should be placed on service records of the materials when exposed to actual weathering conditions. In the absence of such information, the test method affords a useful guide. Because of the limited amount of information concerning the significance of the test results and because of the lack of information as to the uniformity of tests made in different laboratories, the test method should not be used as an arbitrary basis of rejection; it should be used only to furnish information to indicate whether or not the materials require further investigation of their soundness.

Soundness of Aggregate by Use of Sodium Sulfate or Magnesium Sulfate:[21]

AASHO Designation....................... T104-57
ASTM Designation........................ C88-63
Corps of Engineers Designation.............. CRD-C137-62 and C115-55
Federal Specification Designation............ Method 203.01

Scope. This method covers the procedure to be followed in testing aggregates to determine their resistance to disintegration by saturated solutions of sodium sulfate or magnesium sulfate. It furnishes information helpful in judging the soundness of aggregates subject to weathering action, particularly when adequate information is not available from service records of the material exposed to actual weathering conditions. Attention is called to the fact that test results by the use of the two salts differ considerably, and care must be exercised in fixing proper limits in any specifications which may include requirements for these tests.

Potential Reactivities of Aggregates (Chemical Method):

ASTM Designation........................ C289-60
Corps of Engineers Designation............ CRD-C128-62

Scope. This method of test covers a chemical method for determining the potential alkali reactivity of aggregates and their potentiality for producing abnormal expansion in concrete when used with high-alkali cement. It is based on the amount of reaction of the aggregate with a sodium hydroxide solution under controlled laboratory conditions of test.

Potential Volume Change of Cement-Aggregate Combinations:

ASTM Designation............C342-67

Scope. (1) This method of test is intended to determine the potential expansion of cement-aggregate combinations by measuring the linear expansion developed by the combinations in mortar bars subjected to variations of temperature and water saturation during storage under prescribed conditions of test. (2) This method is regarded as having particular applicability to certain cement-aggregate combinations common in the central part of the United States (Oklahoma, Kansas, Nebraska, and Iowa).

Other Properties Several miscellaneous tests are briefly summarized below. All are of interest in aggregate selection and application.

Amount of Material Finer than No. 200 Sieve in Aggregate:

AASHO Designation........................ T11-60
ASTM Designation......................... C110-67
Corps of Engineers Designation.............. CRD-C105-57
Federal Specification Designation............ Method 202.11

Scope. This method of test outlines the procedure for determining the total quantity of material finer than a standard No. 200 (74-micron) sieve in aggregates.

Production of Plastic Fines in Aggregates:

AASHO Designation.............. T210-64I
California Designation............ 229-C

Scope. This method describes the procedure for determining the durability factor of aggregates. The durability factor is a value indicating the relative resistance of an aggregate to producing detrimental clay-like fines when subjected to the prescribed mechanical methods of degradation.

Percentage of Particles of Less than 1.95 Specific Gravity in Coarse Aggregates:

AASHO Designation............ T150-49

Scope. This method of test covers a procedure for the approximate determination of shale in coarse aggregate. The test separates, along with the shale, other particles of low specific gravity or relatively high surface area such as iron oxides, soft particles, and other small materials.

Clay Lumps in Aggregates:

AASHO Designation........................ T112-64I
ASTM Designation......................... C142-67
Corps of Engineers Designation.............. CRD-C142-65
Federal Specification Designation............ Method 205.0

Scope. This method of test covers the procedure for the approximate determination of clay lumps in the routine examination of aggregates.

Method of Test for Clay Lumps and Friable Particles in Aggregates (Slaking Test Method):

Corps of Engineers Designation............ CRD-C118-55

Scope. This method of test covers the procedure for determination of clay lumps and friable particles in aggregates by slaking in water.

Coal and Lignite in Sand and Lightweight Pieces in Aggregates:

AASHO Designation........................ T113-63I
ASTM Designation......................... C123-66
Corps of Engineers Designation.............. CRD-C122-64
Federal Specification Designation............ Method 236.0

Scope. (1) This method of test covers a procedure for the approximate determination of coal and lignite in the routine laboratory examination of sands. The method separates, along with the coal and lignite, other particles of low specific gravity, such as small pieces of wood, vegetable matter, etc. (2) This method of test covers a procedure for the determination of the approximate percentage of lightweight pieces in aggregate by means of sink-float separation in a heavy liquid of suitable specific gravity.

Soft Particles in Coarse Aggregates:[22]

AASHO Designation........................ T189-63I
ASTM Designation......................... C235-68
Corps of Engineers Designation.............. CRD-C130-57
Federal Specification Designation............ Method 228.0

Scope. This method of test covers the procedure for determining the quantity of soft particles in coarse aggregates on the basis of scratch hardness. It is to be used to identify materials which are soft, including those formed of a soft material and those which are so poorly bonded that the separate particles in the piece are easily detached from the mass. The test is not intended to identify other types of deleterious materials in aggregate.

Note. A study of tests for the determination of soft pieces in aggregate gave as a conclusion that the scratch-hardness test using a hard yellow brass scribe was the only one considered suitable for laboratory and field use.

Organic Impurities in Sand for Concrete:

AASHO Designation........................ T21-60
ASTM Designation......................... C40-66
Corps of Engineers Designation.............. CRD-C121-63
Federal Specification Designation............ Method 235.0

Scope. This method of test covers the procedure for an approximate determination of the presence of injurious organic compounds in natural sands that are to be used in cement mortar or concrete. The principal value of the test is to furnish a warning that further tests of sand are necessary before they are approved for use.

Measuring Mortar-making Properties of Fine Aggregates:

AASHO Designation........................ T71-63I
ASTM Designation......................... C87-68
Corps of Engineers Designation.............. CRD-C116-58
Federal Specification Designation............ Method 206.01

Scope. This method of test covers the procedure for measuring the mortar-making properties of fine aggregate for concrete by means of a compression test on specimens made from a mortar of plastic consistency and gauged to a definite water-cement ratio. Its principal use is the determination of the effect of organic impurities revealed by the colorimetric test.

Method of Calculation of the Fineness Modulus of Aggregate:[23]

Corps of Engineers Designation.............. CRD-C104-48
Federal Specification Designation............ Method 204.0

Definition. The fineness modulus is an empirical factor obtained by adding the total percentages of a sample of an aggregate retained on each of a specified series of sieves and dividing the result by 100. The sieve analysis is expressed as cumulative percentages coarser than each of sieves No. 100, 50, 30, 16, 8, and 4, and ⅜ in., ¾ in., and 1½ in. and larger, increasing in the size ratio of 2 to 1 for each of these sieves and no others, when determining the fineness modulus of either fine or coarse aggregate. The sieve analysis shall be made in accordance with CRD-C103.

Range of Values. The same value of fineness modulus may be obtained from several different gradations. In general, a small value indicates a fine material while a large value indicates a coarse material. The value for sands commonly ranges from 2.00 to 3.00 and for coarse aggregate from 6.50 to 7.50 when all the material is finer than the 1½-in. sieve. Combinations of fine and coarse aggregate have intermediate values.

Note. The fineness modulus as a measure of size distribution is a most helpful tool for arriving at a well-balanced design of concrete mixtures.

Method of Test for Flat and Elongated Particles in Aggregate:

Corps of Engineers Designation, Coarse............ CRD-C119-53
Corps of Engineers Designation, Fine.............. CRD-C120-55

Scope. The specifications cover the procedure for the determination of flat and elongated particles in aggregate. A flat particle is defined as one having a ratio of width to thickness greater than 3. An elongated particle is one defined as having a ratio of length to width greater than 3. The length l, width w, and thickness t, are, respectively, the greatest, intermediate, and least dimensions of any particle as measured along mutually perpendicular directions. They should be considered as the principal dimensions of the circumscribing rectangular prism.

Apparatus. Any suitable equipment by which aggregate particles may be tested for compliance with the specifications is acceptable. Figure 20-8 shows a detail drawing of a proportional caliper device as developed by the Concrete Research Division, Waterways Experiment Station, Corps of Engineers, U.S. Army. The unit has three axis positions by the use of which openings in the ratio of 1 to 2, 1 to 3, and 1 to 5, respectively, may be obtained. When percentages are based on weight, a balance or scale sensitive to 0.5 percent of the weight to be weighed is required. For fine aggregates, a stereoscopic microscope equipped with a 10× ocular and 0.66× and 2.0× objectives to give magnifications of 6.6 and 20 diameters is required. In addition, dissecting needles and sharp-nosed forceps are needed to turn and pick up fine aggregate grains.

Procedure for Coarse Aggregate. A representative sample of the supply of each size of coarse aggregate to be tested shall be selected, sieved, and reduced by quartering and/or splitting until approximately 100 particles are obtained of each sieve size larger than the ⅜-in. sieve present in the amount of 10 percent or more of the sample. Each of the particles in the sieve-size sample shall be tested, using the testing device, and segregated into one of three groups: (1) flat, (2) elongated, (3) not flat or elongated.

Test for Elongation (Fig. 20-9*a*). Set the larger opening equal to the length of the particle. If the width of the particle can be placed within the smaller opening, the particle is elongated.

Test for Flatness (Fig. 20-9*b*). Set the larger opening equal to the width of the particle. If the thickness of the particle can be placed within the smaller opening, the particle is flat.

Procedure for Fine Aggregate. Each particle shall be examined, using the microscope as outlined in the specification, and assigned by inspection to the category of flat, elongated, or not flat or elongated. Particles which fall on the borderline between two categories shall be assigned half to each category.

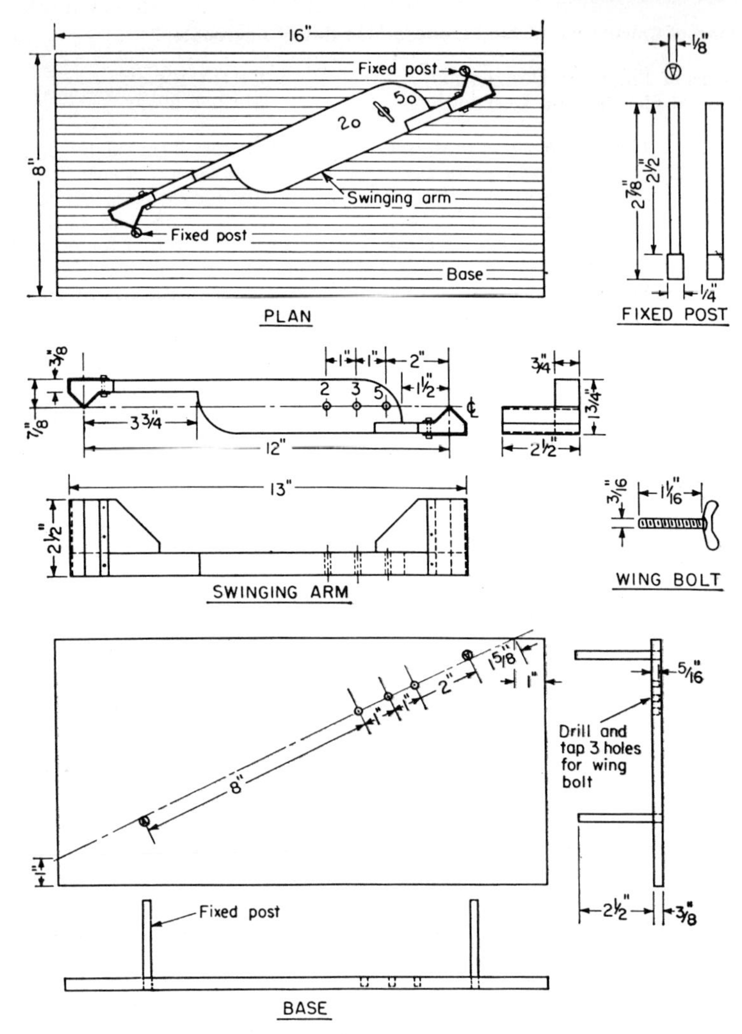

Fig. 20-8 Proportional caliper device for checking flat and elongated particles. (*Waterways Experiment Station, Corps of Engineers.*)

Method of Test for Sand Equivalent:[24]

California Designation............ C217E

Scope. This test is intended to serve as a rapid field test to show the relative proportions of detrimental fine dust or clay-like material in soils or in fine aggregate.

Sieve Analysis of Fine and Coarse Aggregates:

AASHO Designation........................ T27-60
ASTM Designation......................... C136-67
Corps of Engineers Designation.............. CRD-C103-60
Federal Specification Designation............ Method 202.01

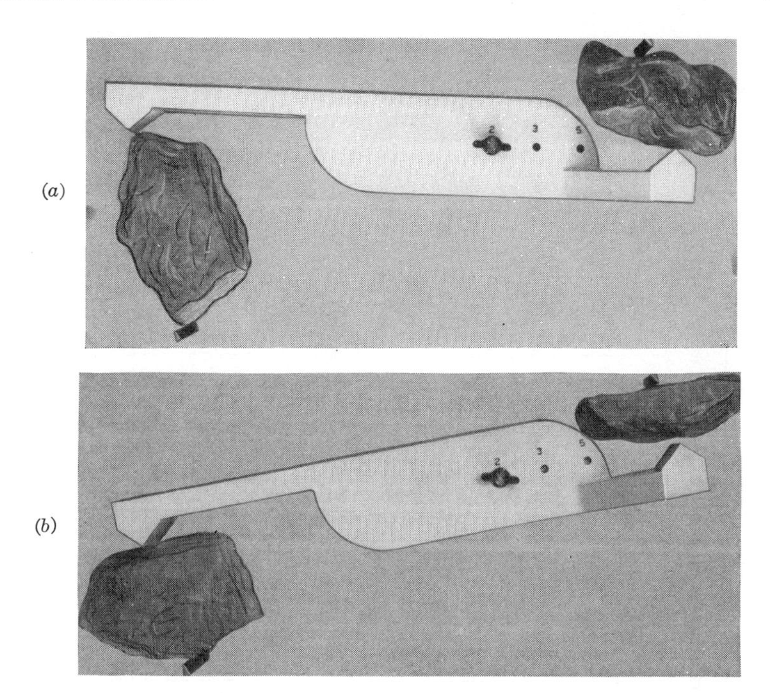

Fig. 20-9 Use of a proportional caliper. (*a*) Test for elongation. (*b*) Test for flatness. (*Iowa Manufacturing Company.*)

Scope. This method of test covers a procedure for the determination of the particle-size distribution of fine and coarse aggregates using sieves with square openings. This method is also applicable to the use of laboratory screens with round openings. It is not intended for use in the sieve analysis of aggregates recovered from bituminous mixtures or for the sieve analysis of mineral fillers.

AGGREGATE GRADATIONS

In order to reduce construction cost, a great deal of the activity of various organizations has been concentrated on the standardization and, not the least, on the uniform application of aggregate gradations. In connection with the Interstate Program, the Bureau of Public Roads made a special study of the problems involved in standardizing aggregate gradations and of the economic gains therefrom. Subsequently, in 1961, a set of specifications was issued as a guideline for the various states in connection with the federal highway projects.[25] The American Association of State Highway Officials (AASHO), through its committee on construction, made a complete review of all state specifications for highway construction (excluding bridges). The intent was to develop AASHO standard construction specifications which would be sufficiently broad to permit their adoption by all of the states. These AASHO Guide Specifications for Highway Construction were published in 1964.[26]

The Bureau of Public Roads has outlined the goal of standardization of aggregate gradation specifications as follows:

1. To develop a minimum number of standard aggregate gradations that can be uniformly adopted nationwide for general usage, while at the same time recognizing the need for some variations by special provisions to fit locally available materials.
2. To achieve uniformity in the number and sizes of sieves to be used in specifying aggregate gradations.
3. To develop and adopt a simple and uniform system for identification of the standard aggregate gradations.[27]

TABLE 20-14 Sizes of Coarse Aggregates

SPR No.	Nominal size square openings*	Amounts finer than each laboratory sieve (square openings), percent by weight														
		4	3½	3	2½	2	1½	1	¾	½	⅜	No. 4	No. 8	No. 16	No. 50	No. 100
1	3½–1½	100	90–100	......	25–60	......	0–15	......	0–5							
1-F†	3½–2	100	90–100	......	......	0–10	0–2									
2-F†	3–1½	...	100	90–100	......	......	0–10	0–2								
2	2½–1½	...	......	100	90–100	35–70	0–15	......	0–5							
24	2½–¾	...	......	100	90–100	......	25–60	......	0–10	0–5						
3	2–1	...	......	......	100	95–100	35–70	0–15	......	0–5						
357	2–No. 4	...	......	......	100	95–100	......	35–70	......	10–30	......	0–5				
4	1½–¾	...	......	......	......	100	90–100	20–55	0–15	......	0–5					
467	1½–No. 4	...	......	......	......	100	95–100	......	35–70	......	10–30	0–5				
5	1–½	...	......	......	......	......	100	90–100	20–55	0–10	0–5					
56	1–⅜	...	......	......	......	......	100	90–100	40–75	15–35	0–15	0–5				
57	1–No. 4	...	......	......	......	......	100	95–100	......	25–60	......	0–10	0–5			
6	¾–⅜	...	......	......	......	......	......	100	90–100	20–55	0–15	0–5				
67	¾–No. 4	...	......	......	......	......	......	100	90–100	......	20–55	0–10	0–5			
68	¾–No. 8	...	......	......	......	......	......	100	90–100	......	30–65	5–25	0–10	0–5		
7	½–No. 4	...	......	......	......	......	......	......	100	90–100	40–70	0–15	0–5			
78	½–No. 8	...	......	......	......	......	......	......	100	90–100	40–75	5–25	0–10	0–5		
8	⅜–No. 8	...	......	......	......	......	......	......	......	100	85–100	10–30	0–10	0–5		
89	⅜–No. 16	...	......	......	......	......	......	......	......	100	90–100	20–55	5–30	0–10	0–5	
9	No. 4–No. 16	...	......	......	......	......	......	......	......	......	100	85–100	10–40	0–10	0–5	
10	No. 4–0‡	...	......	......	......	......	......	......	......	......	100	85–100	.....	.....	.....	10–30
G1§	1½–No. 50	...	......	......	......	......	100	80–100	......	50–85	......	20–40	15–35	5–25	0–10	0–2
G2§	1½–No. 8	...	......	......	......	......	100	65–100	......	35–75	......	10–35	0–10	0–5		
G3§	1½–No. 4	...	......	......	......	......	100	60–95	......	25–50	......	0–15	0–5			

* In inches, except where otherwise indicated. Numbered sieves are those of the United States Standard Sieve Series.
† Special sizes for sewage trickling filter media.
‡ Screenings.
§ The requirements for grading depend upon percentage of crushed particles in gravel. Size G1 is for gravel containing 20 percent or less of crushed particles; G2 is for gravel containing more than 20 percent but not more than 40 percent of crushed particles; G3 is for gravel containing crushed particles in excess of 40 percent.
SOURCE: Simplified Practice Recommendation R163-48, National Bureau of Standards.

TABLE 20-15 Typical Uses for Sizes Given in Table 20-14

Use	SPR size number and nominal size*																						
	1	1-F	2-F	2	24	3	357	4	467	5	56	57	6	67	68	7	78	8	89	9	10	G1	G2	G3
	3½–1½	3½–2	3–1½	2½–1½	2½–¾	2–1	2–No. 4	1½–¾	1½–No. 4	1–½	1–⅜	1–No. 4	¾–⅜	¾–No. 4	¾–No. 8	½–No. 4	½–No. 8	⅜–No. 8	⅜–No. 16	No. 4–No. 16	No. 4–0	1½–No. 50	1½–No. 8	1½–No. 4
Water-bound macadam:																								
Coarse aggregate	x	...	...	x	...	x																		
Filler	...	...	...	...	...	...	...	...	...	...	...	...	...	...	...	...	...	...	...	...	x			
Bituminous macadam, penetration method:																								
Coarse aggregate	x	...	...	x	...	x																		
Choke	...	...	...	...	...	...	...	...	...	x	x	...	x	x	...	...	x							
Seal	...	...	...	...	...	...	...	...	...	...	...	...	...	...	...	...	...	x	x	x				
Bituminous plant mixes, base or surface courses:†																								
Base, open mix	...	...	...	...	x	x	...	x	...	...	x													
Base, closed mix	...	...	...	...	...	...	x	...	x	...	...	x	...	x										
Binder course	...	...	...	...	...	...	...	...	...	...	...	x	...	x										
Surface course, coarse grading	...	...	...	...	...	...	...	...	...	x	x	x	...	x										
Surface course, fine grading	...	...	...	...	...	...	...	...	...	...	...	...	...	...	...	...	x	x	x	x				
Seal	...	...	...	...	...	...	...	...	...	...	...	...	...	...	...	...	x	x	x	x				
Bituminous road mix:																								
Mixing course	...	...	...	...	...	...	...	...	...	...	x	...	...	x										
Choke	...	...	...	...	...	...	...	...	...	...	...	...	...	...	...	...	...	x	x	x				
Seal	...	...	...	...	...	...	...	...	...	...	...	...	...	...	...	...	...	x	x	x				
Drag leveling course:																								
Leveling course	...	...	...	...	...	...	...	...	...	...	...	...	...	x	x	...	x							
Seal	...	...	...	...	...	...	...	...	...	...	...	...	...	...	...	...	...	x	x	x				
Bituminous surface treatment	...	...	...	...	...	...	...	...	...	x‡	x‡	...	x	x	x	x	x	x	x	x				
Seal for airport construction	...	...	...	...	...	...	...	...	...	...	...	...	...	...	...	...	...	...	x	x				
Portland-cement concrete	x	...	...	x	...	x	x	x	x	...	...	x	...	x	...	x								
Railroad ballast:																								
Stone or slag	...	...	...	...	x	x	...	x	...	...	x	x												
Gravel	...	...	...	...	...	...	...	...	...	...	...	...	...	...	...	...	...	...	...	...	...	x	x	x
Roofing	...	...	...	...	...	...	...	...	...	...	...	...	...	...	...	...	...	x						
Sewage trickling filter media	...	x	x																					

* In inches, except where otherwise indicated. Numbered sieves are those of the United States Standard Sieve Series.
† For plant mixes the aggregate should consist of appropriate sizes selected from Table 20-14 combined with suitably graded fine aggregate.
‡ Bottom course of multiple surface treatment.
SOURCE: Simplified Practice Recommendation R163-48, National Bureau of Standards.

This subject was the basis of a cooperative study by the National Sand and Gravel Association (NSGA), the National Crushed Stone Association (NCSA), the National Slag Association (NSA), and the major producers and users of aggregate. The Joint Technical Committee proposed a set of gradation specifications which was approved by the National Bureau of Standards and published in 1948 by the Department of Commerce as a Simplified Practice Recommendation.[28] Table 20-14 lists the recommended gradations for coarse aggregate, and Table 20-15 indicates typical construction uses for these gradations. The SPR system does not cover the complete gradings for portland-cement concrete or bituminous paving mixtures because it does not include sands or mineral fillers. These latter gradations are specified under AASHO and ASTM standards.

The AASHO and ASTM standard gradations which are recommended for general use by all State Highway Departments are referenced in the following paragraphs, together with the corresponding Corps of Engineers and Federal Designations. Where correlation exists between these standard gradations and the SPR sizes listed in Table 20-14, each gradation is keyed by a combination of upper-case and lower-case letters to a corresponding column in Table 20-16. By referring to this table, the equivalent SPR (Simplified Practice Recommendation) sizes can be immediately identified.

TABLE 20-16 Correlation between SPR Sizes and Standard Gradations Proposed in AASHO, ASTM, Corps of Engineers, and Federal Specifications

SPR No.	Reference key to AASHO, ASTM, Corps of Engineers, and Federal Specifications									
	A	B		C		D	E	F	G	H
	a	a	b	a	b	a	a	a	a	a
1	x		x				x	x		
1-F										x
2-F										x
2	x		x				x	x		x
24	x								x	
3	x	x	x	x	x	x	x	x	x	
357	x	x	x	x	x	x				
4	x	x	x	x	x	x x			x	
467	x	x	x	x						
5	x			x	x	x			x	
56	x				x					
57	x	x	x	x	x	x				
6	x			x		x				
67	x	x	x	x	x	x				
68	x			x		x				
7	x	x	x	x		x				
78	x			x		x				
8	x		x	x	x	x				
89	x									
9	x				x					
10	x						x	x		
G1									x	
G2									x	
G3									x	

A Standard Size of Coarse Aggregate for Highway Construction:

a. AASHO Designation............ M43-49
a. ASTM Designation............. D448-54 (1968)

B Coarse Aggregate for Portland-cement Concrete:

a. AASHO Designation......................... M80-51
b. ASTM Designation.......................... C33-67
b. Corps of Engineers Designation.............. CRD-C131-55
CRD-C133-65
b. Federal Specification Designation............ SS-A-281b

Fine Aggregate for Portland-cement Concrete:

Y. AASHO Designation......................... M6-51
Z. ASTM Designation.......................... C33-67
Z. Corps of Engineers Designation.............. CRD-C131-61
CRD-C133-65
Z. Federal Specification Designation............ SS-A-281b

Gradations as follows:

Sieve size, U.S. Standard	Total passing, percent by weight	
	Y	Z
⅜ in.	100	100
No. 4	95–100	95–100
No. 8	......	80–100
No. 16	45–80	50–85
No. 30	......	25–60
No. 50	10–30	10–30
No. 100	2–10	2–10

C Coarse Aggregate for Bituminous Paving Mixtures:

a. ASTM Designation......................... D692-63
b. Federal Specification Designation............ SS-C-731a

Grading requirements as shown in Table 20-17A.

Fine Aggregate for Bituminous Paving Mixtures:

ASTM Designation............ D1073-63
Gradings No. 1, 2, and 3 in Table 20-17B.

D Hot-mixed, Hot-laid Asphalt Paving Mixtures:

a. ASTM Designation............ D1663-67
Table 20-17C shows the composition of representative paving mixtures.

Mineral Filler for Bituminous Paving Mixtures:

ASTM Designation............ D242-64
Grading requirements as shown below:

Sieve size, U.S. Standard	*Total passing, percent by weight*
No. 30	100
No. 50	95–100
No. 100	90–100
No. 200	70–100

TABLE 20-17A Grading Requirements for Coarse Aggregate in Bituminous Paving Mixtures

SPR No.	Nominal size (sieves with square openings)	Amounts finer than each laboratory sieve (square openings), percentage by weight									
		2½ in.	2 in.	1½ in.	1 in.	¾ in.	½ in.	⅜ in.	No. 4	No. 8	No. 16
3	2 in. to 1 in.	100	90–100	35–70	0–15	......	0–5				
357	2 in. to No. 4	100	95–100	......	35–70	......	10–30	......	0–5		
4	1½ in. to ¾ in.	...	100	90–100	20–55	0–15	......	0–5			
467	1½ in. to No. 4	...	100	95–100	......	35–70	......	10–30	0–5		
5	1 in. to ½ in.	...	......	100	90–100	20–55	0–10	0–5			
57	1 in. to No. 4	...	......	100	95–100	......	25–60	......	0–10	0–5	
6	¾ in. to ⅜ in.	...	......	......	100	90–100	20–55	0–15	0–5		
67	¾ in. to No. 4	...	......	......	100	90–100	......	20–55	0–10	0–5	
68	¾ in. to No. 8	...	......	......	100	90–100	......	30–65	5–25	0–10	0–5
7	½ in. to No. 4	...	......	......	......	100	90–100	40–70	0–15	0–5	
78	½ in. to No. 8	...	......	......	......	100	90–100	40–75	5–25	0–10	0–5
8	⅜ in. to No. 8	...	......	......	......	......	100	85–100	10–30	0–10	0–5

SOURCE: Bureau of Public Roads.

TABLE 20-17B Grading Requirements for Fine Aggregate in Bituminous Paving Mixtures*

Sieve size, U.S. Standard	Amounts finer than each laboratory sieve (square openings), percent by weight		
	Grading No. 1	Grading No. 2	Grading No. 3
⅜ in.	100	......	100
No. 4	95–100	100	80–100
No. 8	70–100	95–100	65–100
No. 16	40–80	85–100	40–80
No. 30	20–65	65–90	20–65
No. 50	7–40	30–60	7–40
No. 100	2–20	5–25	2–20
No. 200	0–10	0–5	0–10

* It is recognized that for certain purposes satisfactory results may be obtained with materials not conforming to these specifications. In such cases the use of fine aggregate not conforming to the grading requirements of these specifications may be authorized only under special provisions based on field experience or laboratory studies of the possibility of designing a mixture of materials to be used on the job that will yield bituminous paving mixtures equivalent in quality to the job mix requirements.

SOURCE: American Society for Testing and Materials.

TABLE 20-17C Composition of Asphalt Paving Mixtures

Sieve size	Nominal maximum size of aggregates							
	2-in.	1½-in.	1-in.	¾-in.	½-in.	⅜-in.	No. 4	No. 8
	Asphalt concrete						Sand asphalt	Sheet asphalt
	Grading of total aggregate (coarse plus fine, plus filler if required): amounts finer than each laboratory sieve (square opening), percentage by weight							
2½ in.	100							
2 in.	90–100	100						
1½ in.	....	90–100	100					
1 in.	60–80	....	90–100	100				
¾ in.	....	60–80	......	90–100	100			
½ in.	35–65	....	60–80	......	90–100	100		
⅜ in.	....	....	......	60–80	....	90–100	100	
No. 4	15–50	20–55	25–60	35–65	45–70	60–80	80–100	100
No. 8*	10–40	10–40	15–45	20–50	25–55	35–65	65–100	95–100
No. 16	....	....	......	......	....	......	40–80	85–100
No. 30	....	....	......	......	....	......	20–65	70–95
No. 50	2–15	2–16	3–18	3–20	5–20	6–25	7–40	45–75
No. 100	....	....	......	......	....	......	3–20	20–40
No. 200†	0–4	0–5	1–7	2–8	2–9	2–10	2–10	9–20
	Asphalt cement, percentage by weight of total mixture‡							
	3½–7½	3½–8	4–8½	4–9	4½–9½	5–10	7–12	8½–12
	Suggested coarse aggregates, SPR sizes							
	3 and 57	4 and 67	5 and 7 or 57	67 or 68 or 6 and 8	7 or 78	8		

* In considering the total grading characteristics of an asphalt paving mixture, the amount passing the No. 8 sieve is a significant and convenient field control point between fine and coarse aggregate. Gradings approaching the maximum amount permitted to pass the No. 8 sieve will result in pavement surfaces having comparatively fine texture, while coarse gradings approaching the minimum amount passing the No. 8 sieve will result in surfaces with comparatively coarse texture.

† The material passing the No. 200 sieve may consist of fine particles of the aggregates or mineral filler, or both. It shall be free from organic matter and clay particles and shall be nonplastic when tested by the method of test for liquid limit of soils (ASTM Designation D423), and the method of test for plastic limit and plasticity index of soils (ASTM Designation D 424).

‡ The quantity of asphalt cement is given in terms of percentage by weight of the total mixture. The wide difference in the specific gravity of various aggregates, as well as a considerable difference in absorption, results in a comparatively wide range in the limiting amount of asphalt cement specified. The amount of asphalt required for a given mixture should be determined by appropriate laboratory testing or on the basis of past experience with similar mixtures, or by a combination of both.

SOURCE: Bureau of Public Roads.

The mineral filler shall consist of limestone dust, portland cement, or other suitable mineral matter.

E Crushed Stone, Crushed Slag, and Crushed Gravel for Dry-bound or Water-bound Macadam Base Courses:

a. ASTM Designation............ D694-62 (1968)

F Crushed Stone and Crushed Slag for Water-bound Surface Course:

a. AASHO Designation............ M77-64I

Aggregate for Masonry Grout:

ASTM Designation............ C404-61
For grading requirements, see Table 20-18.

TABLE 20-18 Grading Requirements for Masonry Grout

Sieve size, U.S. Standard	Amounts finer than each laboratory sieve (square openings), percent by weight				
	Fine aggregate			Coarse aggregate	
		Size No. 2			
	Size No. 1	Natural	Manufactured	Size No. 8	Size No. 89
½ in.	......	......	......	100	100
⅜ in.	100	......	......	85–100	90–100
No. 4	95–100	100	100	10–30	20–55
No. 8	80–100	95–100	95–100	0–10	5–30
No. 16	50–85	60–100	60–100	0–5	0–10
No. 30	25–60	35–70	35–70		0–5
No. 50	10–30	15–35	20–40		
No. 100	2–10	2–15	10–25		
No. 200	......	......	0–10		

SOURCE: American Society for Testing and Materials.

Aggregate for Masonry Mortar:

ASTM Designation............ C144-66T
For grading requirements, see Table 20-19.

Mortar Sand:

AASHO Designation............ M45-42
Gradation as shown below:

Sieve size, U.S. Standard	*Total passing, percent by weight*
No. 8	100
No. 50	15–40
No. 100	0–10
No. 200	0–5

TABLE 20-19 Grading Requirements for Masonry Mortar

Sieve size, U.S. Standard	Total passing, percent by weight	
	Natural sand	Manufactured sand
No. 4	100	100
No. 8	95–100	95–100
No. 16	70–100	70–100
No. 30	40–75	40–75
No. 50	10–35	20–40
No. 100	2–15	10–25
No. 200	......	0–10

G Crushed Stone, Crushed Slag, and Gravel for Railroad Ballast:

a. Federal Specification Designation............ SS-C-743
See special study on properties of sand and gravel for railroad ballast.[29]

H Crushed Stone, Crushed Slag, and Gravel for Sewage Trickling Filter Media:

a. Federal Specification Designation............ SS-744a

Sieves for Testing Purposes:

AASHO Designation........................ M92-42
ASTM Designation......................... E11-61
Corps of Engineers Designation.............. CRD-C101-55
CRD-C102-63
Federal Specification Designation............ Method 301.0

Scope. These specifications cover woven-wire cloth sieves, round-hole screens (sieves), and square-hole perforated-plate screens (sieves) for precision testing in the classification of materials according to size (mechanical analysis, fineness, and particle-size determinations). The sieves covered by these specifications are intended for general use. See Tables 20-20 through 20-23. To get reliable results from testing, great care must be taken in the selection of representative samples. Poor procedures are the greatest cause of inconsistency in end results.[30]

The International Standards Organization (ISO) is endeavoring to establish an International Test Sieve Series. The ISO Standard includes 19 sieve sizes, all of which are contained in the sieve series of AASHO M92-42 and ASTM E11-61 specifications. The proposed ISO sieves are identified in Table 20-20.

A number of nations have adopted sieve standards which follow very closely the openings of the U.S. Standard Series sieves. A comparison of several such standards is supplied in Table 20-24.

Specifications which govern the aggregate for construction under the auspices of a state highway department can be obtained from the department headquarters. Aggregate gradations for highway purposes, even though they are in the main based on the Simplified Practice Recommendations (Tables 20-14 and 20-15), will vary in designation from state to state. Test and other specifications will, in general, refer to the AASHO and ASTM Designations. See Tables 20-25 and 20-26 for the Michigan State Highway Department Standard.

TABLE 20-20 Nominal Dimensions, Permissible Variations, and Limits for Wire Cloth of Standard Sieves (U.S. Standard Series, $\sqrt[4]{2}$ ratio)

Sieve designation		Sieve opening		Permissible variations in average opening, percent	Permissible variation for not more than 5 percent of openings, percent	Permissible maximum variation in individual openings, percent	Nominal wire diameter[c]	
Standard	Alternate	mm	in. (approximate equivalents)				mm	in. (approximate equivalents)
107.6 mm	4.24 in.	107.6	4.24	±3	+4	+5	6.40	0.2520
101.6 mm	4 in.[b]	101.6	4.00	±3	+4	+5	6.30	0.2480
90.5 mm	3½ in.	90.5	3.50	±3	+4	+5	6.08	0.2394
76.1 mm	3 in.	76.1	3.00	±3	+4	+5	5.80	0.2283
64.0 mm	2½ in.	64.0	2.50	±3	+4	+5	5.50	0.2165
53.8 mm	2.12 in.	53.8	2.12	±3	+4	+5	5.15	0.2028
50.8 mm	2 in.[b]	50.8	2.00	±3	+4	+5	5.05	0.1988
45.3 mm	1¾ in.	45.3	1.75	±3	+4	+5	4.85	0.1909
38.1 mm	1½ in.	38.1	1.50	±3	+4	+5	4.59	0.1807
32.0 mm	1¼ in.	32.0	1.25	±3	+4	+5	4.23	0.1665
26.9 mm	1.06 in.	26.9	1.06	±3	+5	+6	3.90	0.1535
25.4 mm	1 in.[b]	25.4	1.00	±3	+5	+6	3.80	0.1496
22.6 mm[a]	⅞ in.	22.6	0.875	±3	+5	+6	3.50	0.1378
19.0 mm	¾ in.	19.0	0.750	±3	+5	+6	3.30	0.1299
16.0 mm[a]	⅝ in.	16.0	0.625	±3	+5	+6	3.00	0.1181
13.5 mm	0.530 in.	13.5	0.530	±3	+5	+6	2.75	0.1083
12.7 mm	½ in.[b]	12.7	0.500	±3	+5	+6	2.67	0.1051
11.2 mm[a]	7/16 in.	11.2	0.438	±3	+5	+6	2.45	0.0965
9.51 mm	⅜ in.	9.51	0.375	±3	+5	+6	2.27	0.0894
8.00 mm[a]	5/16 in.	8.00	0.312	±3	+5	+6	2.07	0.0815
6.73 mm	0.265 in.	6.73	0.265	±3	+5	+6	1.87	0.0736
6.35 mm	¼ in.[b]	6.35	0.250	±3	+5	+6	1.82	0.0717
5.66 mm[a]	No. 3½	5.66	0.223	±3	+5	+10	1.68	0.0661
4.76 mm	No. 4	4.76	0.187	±3	+5	+10	1.54	0.0606
4.00 mm[a]	No. 5	4.00	0.157	±3	+5	+10	1.37	0.0539
3.36 mm	No. 6	3.36	0.132	±3	+5	+10	1.23	0.0484
2.83 mm[a]	No. 7	2.83	0.111	±3	+5	+10	1.10	0.0430
2.38 mm	No. 8	2.38	0.0937	±3	+5	+10	1.00	0.0394
2.00 mm[a]	No. 10	2.00	0.0787	±3	+5	+10	0.900	0.0354
1.68 mm	No. 12	1.68	0.0661	±3	+5	+10	0.810	0.0319
1.41 mm[a]	No. 14	1.41	0.0555	±3	+5	+10	0.725	0.0285
1.19 mm	No. 16	1.19	0.0469	±3	+5	+10	0.650	0.0256
1.00 mm[a]	No. 18	1.00	0.0394	±5	+7½	+15	0.580	0.0228
841 μ	No. 20	0.841	0.0331	±5	+7½	+15	0.510	0.0201
707 μ[a]	No. 25	0.707	0.0278	±5	+7½	+15	0.450	0.0177
595 μ	No. 30	0.595	0.0234	±5	+7½	+15	0.390	0.0154
500 μ[a]	No. 35	0.500	0.0197	±5	+7½	+15	0.340	0.0134
420 μ	No. 40	0.420	0.0165	±5	+12½	+25	0.290	0.0114
354 μ[a]	No. 45	0.354	0.0139	±5	+12½	+25	0.247	0.0097
297 μ	No. 50	0.297	0.0117	±5	+12½	+25	0.215	0.0085
250 μ[a]	No. 60	0.250	0.0098	±5	+12½	+25	0.180	0.0071
210 μ	No. 70	0.210	0.0083	±5	+12½	+25	0.152	0.0060
177 μ[a]	No. 80	0.177	0.0070	±6	+20	+40	0.131	0.0052
149 μ	No. 100	0.149	0.0059	±6	+20	+40	0.010	0.0043
125 μ[a]	No. 120	0.125	0.0049	±6	+20	+40	0.091	0.0036
105 μ	No. 140	0.105	0.0041	±6	+20	+40	0.076	0.0030

TABLE 20-20 Nominal Dimensions, Permissible Variations, and Limits for Wire Cloth of Standard Sieves (Continued)

Sieve designation		Sieve opening					Nominal wire diameter[c]	
Standard	Alternate	mm	in. (approximate equivalents)	Permissible variations in average opening, percent	Permissible variation for not more than 5 percent of openings, percent	Permissible maximum variation in individual openings, percent	mm	in. (approximate equivalents)
88 μ[a]	No. 170	0.088	0.0035	±6	+20	+40	0.064	0.0025
74 μ	No. 200	0.074	0.0029	±7	+30	+60	0.053	0.0021
63 μ[a]	No. 230	0.063	0.0025	±7	+30	+60	0.044	0.0017
53 μ	No. 270	0.053	0.0021	±7	+30	+60	0.037	0.0015
44 μ[a]	No. 325	0.044	0.0017	±7	+30	+60	0.030	0.0012
37 μ	No. 400	0.037	0.0015	±7	+30	+60	0.025	0.0010

[a] These sieves correspond to those proposed as an International (ISO) Standard. It is recommended that wherever possible these sieves be included in all sieve analysis data or reports intended for international publication.

[b] These sieves are not in the fourth root of 2 series, but they have been included because they are in common usage.

[c] The average diameter of the warp and of the shoot wires, taken separately, of the cloth of any sieve shall not deviate from the nominal values by more than the following:

Sieves coarser than 595 μ .... 5 percent
Sieves 595 μ to 125 μ .... 10 percent
Sieves finer than 125 μ .... 15 percent

NOTE: All measurements of openings and wire diameters shall be made on the completed sieve.
SOURCE: American Society for Testing and Materials.

TABLE 20-21 Thickness of Plates for Round-hole Screens

Screening area, sq in.	Diam of opening, in.	Thickness of plate, in. Min	Max
Under 100	All sizes	0.049	0.066
100 and over	1/16 and 1/8	0.049	0.066
	1/4 to 2 1/2	0.060	0.100
	3 and 3 1/2	0.075	0.130
	4 and 5	0.105	0.160
	6 and 8	0.120	0.175

Where square-hole plate screens are specified for use, the openings shall be the same as the openings of woven wire-cloth sieves, but in other respects, except for the arrangement of the openings, they shall conform to the requirements for round-hole screens.
SOURCE: American Society for Testing and Materials.

AGGREGATE PRODUCTION

The develoment of the aggregate source will follow a study of the results of the exploration and test programs and a thorough evaluation of the economic factors. Each phase of aggregate production should be analyzed prior to source development, including utilities, haul roads, flowsheet, equipment selection, opening of deposit, and the availability of competent people for an efficient operation. A National Sand

and Gravel Association (NSGA) report covering 258 aggregate plants with a total production of 143,617,991 tons for 1965 indicated that labor requirements averaged 0.10 man-hours per ton of sand and gravel produced. Reported man-hour requirements varied considerably, with the extreme high and low values being 0.36 and

TABLE 20-22 Spacing of Openings in Plates for Round-hole Screens

Nominal diameter of opening, in.	*Nominal width of metal between adjacent openings, in.*
1/16	3/64
1/8	3/32
1/4	1/8
3/8	3/16
1/2	3/16
5/8	3/16
3/4	1/4
7/8	1/4
1	3/8
1 1/4	3/8
1 1/2	1/2
2	5/8
2 1/2	3/4
3	3/4
3 1/2	3/4
4	3/4
5	1
6	1
8	1

The openings shall be arranged so that their centers lie at the vertices of triangles which are approximately equilateral within the limits given by the permissible variations in width of metal and diameter of opening.

SOURCE: American Society for Testing and Materials.

TABLE 20-23 Approximately Equivalent Round- and Square-opening Testing Screens

Square openings, in.	*Round openings, in.*	*Square openings, in.*	*Round openings, in.*
No. 8	1/8	1 1/4	1 1/2
No. 4	1/4	1 1/2	1 3/4
3/8	1/2	1 3/4	2
1/2	5/8	2	2 3/8
5/8	3/4	2 1/4	2 3/4
3/4	7/8	2 1/2	3
7/8	1	3	3 1/2
1	1 1/4	3 1/2	4 1/4
1 1/8	1 3/8	4	4 3/4

This table shows the sizes of round openings that are approximately equivalent to the stated sizes of square openings in testing sieves. Numbered sieves are those of the U.S. Standard Sieve Series.

SOURCE: Simplified Practice Recommendation, R163-48, National Bureau of Standards.

0.01 respectively. Although the production labor requirements will vary from pit to pit and from quarry to quarry, the efficiency of an aggregate processing operation is not necessarily a function of the size of a plant or its total output.

The aggregate production can be divided into two distinct areas, namely, the pit or quarry operation and the plant operation.

Pit Development The extent of site preparation at a pit will vary. It may be as simple as starting on a gravel bar in a dry river bed, or it may call for extensive clearing and grubbing plus the removal of overburden. Site preparation may involve the use of bulldozers, scrapers, draglines, and, at times, power shovels.*

The handling of the raw material from pit to aggregate plant can be done in various ways. Use may be made of wheel loaders or front-end loaders and hauling units; or bulldozers, belt loaders, and hauling units; also front-end loaders and conveyors.† Where a portable plant is being operated in the pit, front-end loaders and draglines can be used to recover the sand and gravel and place it directly into the plant feed hopper. Where the pit excavation is below water level, a dredge can be used to excavate the raw material and transport it directly to a pump box at the plant. For handling sand and gravel from bars in rivers or underwater diggings, drag scrapers come into use.‡ In order to achieve an efficient operation, it is of the utmost importance to size the equipment in such a way that the pit excavation and transporting capacity is in line with the processing-plant capacity.

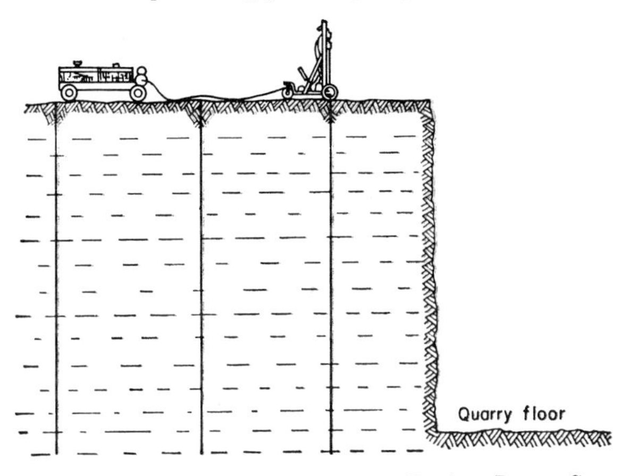

Fig. 20-10 Cross section of a quarry face. *(Gardner-Denver Company.)*

Quarry Development The clearing and grubbing for a quarry development, including the removal of any overburden, will follow the same general approach and require the same type of equipment as has been described for the pit. A quarry can be worked either in a single bench or in multiple benches, and an operations plan is very important for the best development. A single-bench operation has the advantage of maintaining the entire quarry floor at one level. When the rock formation has horizontal stratification because of natural separations or clay seams, these should determine the quarry floor. Otherwise, the depth of the face is limited only by the capabilities of the drilling equipment and could go as high as 100 ft, although generally it is 20 to 50 ft. A multiple-bench operation is used when the face is too deep for a single shot or when there are natural seams. It is also used when the material is found to vary in quality from seam to seam, as indicated by samples from the core drilling. This type of operation permits more continuity than in a single-bench operation.

The drilling and blasting in the quarry is in reality the first aggregate processing step. It is recognized that size reduction of rock by explosives is the cheapest way; hence the spacing of the drill holes and the blasting procedure should be such that the best fragmentation of the rock is obtained.§ Typical drill patterns for 6½-in.-diameter holes in an average limestone formation may range from a spacing of 20 by 14 ft for 30- to 50-ft faces to a spacing of 24 by 16 ft for 50- to 70-ft faces. Subdrilling to a depth of 3 to 6 ft below the quarry floor is an average requirement and is frequently supplemented by an additional 1 to 2 ft of hole to accommodate drill cuttings and sludge (see Fig. 20-10).

* See Sec. 8, Tractors, Bulldozers, and Rippers; Sec. 9, Scrapers; and Sec. 10, Excavators.
† See Sec. 11, Loaders; Sec. 13, Hauling Units; and Sec. 15, Belt Conveyors.
‡ See Sec. 14, Cableways and Drag Scrapers.
§ See Sec. 19, Rock Excavation.

TABLE 20-24 Comparison of United States, Tyler, Canadian, British, French, and German Standard Sieve Series

United States[a]		Tyler[b]	Canadian[c]		British[d]		French[e]		German[f]
Standard	Alternate	Mesh designation	Standard	Alternate	Nominal aperture	Nominal mesh No.	Opening, mm	No.	Opening
107.6 mm	4.24 in.								
101.6 mm	4 in.								
90.5 mm	3½ in.								
76.1 mm	3 in.								
64.0 mm	2½ in.								
53.8 mm	2.12 in.								
50.8 mm	2 in.								
45.3 mm	1¾ in.								
38.1 mm	1½ in.								
32.0 mm	1¼ in.								
26.9 mm	1.06 in.	1.05 in.	26.9 mm	1.06 in.					
25.4 mm	1 in.	....	....	....	....	...	.....	..	25.0 mm
22.6 mm	⅞ in.	0.883 in.	22.6 mm	⅞ in.					
19.0 mm	¾ in.	0.742 in.	19.0 mm	¾ in.	....	...	.....	..	20.0 mm
16.0 mm	⅝ in.	0.624 in.	16.0 mm	⅝ in.					18.0 mm
13.5 mm	0.530 in.	0.525 in.	13.5 mm	0.530 in.	....	...	.....	..	16.0 mm
12.7 mm	½ in.	....	....	....	....	...	.....	..	12.5 mm
11.2 mm	7/16 in.	0.441 in.	11.2 mm	7/16 in.					
9.51 mm	⅜ in.	0.371 in.	9.51 mm	⅜ in.					10.0 mm
8.00 mm	5/16 in.	2½	8.00 mm	5/16 in.	....	...	.....	..	8.0 mm
6.73 mm	0.265 in.	3	6.73 mm	0.265 in.					
6.35 mm	¼ in.	....	....	....	....	...	.....	..	6.3 mm
5.66 mm	No. 3½	3½	5.66 mm	No. 3½					
4.76 mm	4	4	4.76 mm	4			5.000	38	5.0 mm
4.00 mm	5	5	4.00 mm	5	....	...	4.000	37	4.0 mm
3.36 mm	6	6	3.36 mm	6	3.35 mm	5			
2.83 mm	7	7	2.83 mm	7	2.80 mm	6	3.150	36	3.15 mm
2.38 mm	8	8	2.38 mm	8	2.40 mm	7	2.500	35	2.5 mm
2.00 mm	10	9	2.00 mm	10	2.00 mm	8	2.000	34	2.0 mm
1.68 mm	12	10	1.68 mm	12	1.68 mm	10	1.600	33	1.6 mm

1.41 mm	14	12	1.41 mm	14	1.40 mm	12			
							1.250	32	1.25 mm
1.19 mm	16	14	1.19 mm	16	1.20 mm	14			
1.00 mm	18	16	1.00 mm	18	1.00 mm	16	1.000	31	1.0 mm
841 μ	20	20	841 μ	20	850 μ	18			
							0.800	30	800 μ
707 μ	25	24	707 μ	25	710 μ	22			
							0.630	29	630 μ
595 μ	30	28	595 μ	30	600 μ	25			
500 μ	35	32	500 μ	35	500 μ	30	0.500	28	500 μ
420 μ	40	35	420 μ	40	420 μ	36			
							0.400	27	400 μ
354 μ	45	42	354 μ	45	355 μ	44			
							0.315	26	315 μ
297 μ	50	48	297 μ	50	300 μ	52			
250 μ	60	60	250 μ	60	250 μ	60	0.250	25	250 μ
210 μ	70	65	210 μ	70	210 μ	72			
							0.200	24	200 μ
177 μ	80	80	177 μ	80	180 μ	85			
							0.160	23	160 μ
149 μ	100	100	149 μ	100	150 μ	100			
125 μ	120	115	125 μ	120	125 μ	120	0.125	22	125 μ
105 μ	140	150	105 μ	140	105 μ	150			
							0.100	21	100 μ
88 μ	170	170	88 μ	170	90 μ	170	.....	..	90 μ
							0.080	20	80 μ
74 μ	200	200	74 μ	200	75 μ	200			
									71 μ
63 μ	230	250	63 μ	230	63 μ	240	0.063	19	63 μ
									56 μ
53 μ	270	270	53 μ	270	53 μ	300			
							0.050	18	50 μ
44 μ	325	325	44 μ	325	45 μ	350	.....	..	45 μ
							0.040	17	40 μ
37 μ	400	400	37 μ	400					

[a] U.S. Sieve Series—ASTM Specification E11-61.
[b] Tyler Standard Screen Scale Sieve Series.
[c] Canadian Standard Sieve Series 8-GP-1b.
[d] British Standards Institution, London BS-410-62.
[e] French Standard Specifications, AFNOR X-11-501.
[f] German Standard Specification DIN 4188.
SOURCE: The W. S. Tyler Company.

TABLE 20-25 Grading Requirements for Coarse Aggregates and Surfacing Aggregates, Michigan State Highway Department

Michigan series No.	Class	Material†	Intended use (for the guidance of the user and not a part of the specifications)	Total percent passing (square sieve opening—U.S. Standard Sieve Series)										Loss by washing
				2½ in.	2 in.	1½ in.	1 in.	¾ in.	½ in.	⅜ in.	No. 4	No. 8	No. 30	
4	A	Gravel, stone, slag	Concrete—pavement, base course	100	95–100	65–90	10–40		0–20	0–5				‡
6	AA	Gravel, stone, slag	Concrete—structures (when specified), prestressed beams			100	95–100		30–60		0–8			‡
	A	Gravel, stone, slag	Concrete—pavement, structures, base course, sidewalk, curb, etc.			100	95–100		30–60		0–8			‡
9	A	Stone, slag	Bituminous base and binder courses—sheet asphalt, bituminous concrete				100	60–80		0–25	0–10			‡
				(Gravel material before crushing shall be retained on a 1¼-in. sieve.)										
17	A	Gravel, stone, slag	Concrete—structures				100	90–100	50–75		0–8			‡
18	A	Gravel, stone, slag§	Cover for surface treatment				100	90–100	50–80		0–25	0–15		‡
19	A	Stone	Aggregate resurfacing, aggregate base course					100		45–70		15–40		
				(Gravel material before crushing shall be retained on a 1¼-in. sieve.)										
20	A	Gravel, stone, slag, or blends	Bituminous aggregate surface course, hot plant mix (4.11)					100		60–85		40–60	20–35	0–7¶
	B	Gravel, stone, slag, or blends	Bituminous aggregate surface course, road mix (4.08) and plant mix (4.09)					100		60–90		40–65	20–40	2–10¶
21*	AA	Gravel, stone, slag	Aggregate base course			100	85–100		40–75			20–40		3–7
	A	Gravel, stone, slag	Aggregate base course			100	85–100		40–75			20–40		3–7
	AA	Gravel, stone, slag	Aggregate base and surface courses, aggregate shoulders, aggregate resurfacing				100	90–100		65–85		30–50		3–7
	A	Gravel, stone, slag	Aggregate base and surface courses, aggregate shoulders, aggregate resurfacing				100	90–100		65–85		30–50		3–7
	B	Gravel, stone, slag	Aggregate base and surface courses, aggregate resurfacing				100	90–100		65–85		25–45		3–7

22*	C	Gravel, stone, slag	Aggregate base and surface courses, aggregate resurfacing				100	90–100		60–80		20–40		3–7
	D	Gravel, stone, slag	Aggregate base and surface courses, aggregate resurfacing				100	90–100		60–80		20–40		3–7
	E	Gravel, stone, slag	Aggregate base and surface courses, aggregate resurfacing				100	90–100		60–80		20–40		3–7
23	A	Gravel, stone, slag	Aggregate shoulders, aggregate approaches				100			60–85		25–55		7–15
24	A	Gravel, stone, slag	Selected subbase				100			60–85		30–55		3–7
25	A	Stone, slag§	Bituminous concrete wearing course, cover for surface treatment					100	98–100	60–90	10–25	0–12		‡
				(Gravel material before crushing shall be retained on a 1-in. sieve.)										
	B	Gravel, stone, slag§	Cover for surface treatment, small patches					100	98–100	60–90	10–25	0–12		†
	C	Gravel, stone, slag	Cover for surface treatment					100	98–100	60–90	10–25	0–12		‡
26	A	Stone, slag§	Cover for surface treatment						100	98–100	10–30	0–10		‡ ‡
				(Gravel material before crushing shall be retained on a 1-in. sieve.)										
	B	Gravel, stone, slag§	Cover for surface treatment						100	98–100	10–30	0–10		‡
	D	Gravel, stone, slag§	Cover for surface treatment						100	98–100	10–30	0–10		‡
31	AA	Stone, slag	Bituminous concrete wearing course						100	98–100	35–65	5–25		‡
				(Gravel material before crushing shall be retained on a 1-in. sieve.)										
	A	Stone, slag	Bituminous concrete wearing course						100	98–100	35–65	5–25		‡
				(Gravel material before crushing shall be retained on a 1-in. sieve.)										
	C	Stone	Bituminous patching mixture						100	98–100	35–65	5–25		‡
				(Gravel material before crushing shall be retained on a ⅞-in. sieve.)										

* Natural sand or screenings may be uniformly added to increase the fines content.
† Stone material shall be either crushed rock, crushed boulders, crushed stone, or crushed gravel.
‡ See loss by washing in Table 20-26.
§ Slag shall not be used for surface treatment unless specifically provided.
¶ Passing No. 200 sieve.
SOURCE: Michigan Department of State Highways.

TABLE 20-26 Physical Requirements for Coarse Aggregates and Surfacing Aggregates, Michigan State Highway Dept.

Michigan series No.	Class	Gravel and stone[a]											Blast-furnace slag						
		Wear, Deval abrasion, AASHO: Uncrushed[b]	Wear, Deval abrasion, AASHO: Crushed	Wear—Los Angeles abrasion	Soundness	Loss by washing	Soft particles	Sum of soft particles, chert, and hard absorbent particles	Incrusted particles	Thin or elongated pieces	Freeze-thaw durability	Crushed material	Soundness	Loss by washing	Sum of coke and coal particles	Sum of coke, coal, and soft particles	Sum of coke, coal, iron, and soft particles	Thin or elongated pieces	Freeze-thaw durability
		Percent maximum										Percent minimum	Percent maximum						
4	A			36	12[c]	0.8[d]	2.5	8	[f]	15	[g]		6	1.5	1		5	15	[g]
6	AA			36	9	0.8[d]	2[e]	4	[f]	15	[g]		6	1.5	1		4	15	[g]
	A			36	12[c]	0.8[d]	2.5	9	[f]	15	[g]		6	1.5	1		5	15	[g]
9	A			36		3	5	10[h]		15		100	6	3	1	3		15	
17	A			36	12[c]	0.8[d]	3	10	[f]	15	[g]		6	1.5	1		5	15	[g]
18	A			40		3	3			15		50		3	1		5	15	
19	A			40								100	Not permitted						
20	A	20	30									25							
	B	20	30																
21	AA	20	30									100							
	A	20	30									25							
22	AA	20	30									50							
	A	20	30									25							
	B	25	35									25							
	C	20	30			[i]						50							
	D	30	40									25							
	E	40	50									25							
23	A	40	50																
24	A	40	50																

TABLE 20-26 Physical Requirements for Coarse Aggregates and Surfacing Aggregates, Michigan State Highway Dept. (Continued)

Michigan series No.	Class	Gravel and stone[a]											Blast-furnace slag						
		Wear, Deva abrasion, AASHO		Wear—Los Angeles abrasion	Soundness	Loss by washing	Soft particles	Sum of soft particles, chert, and hard absorbent particles	Incrusted particles	Thin or elongated pieces	Freeze-thaw durability	Crushed material	Soundness	Loss by washing	Sum of coke and coal particles	Sum of coke, coal, and soft particles	Sum of coke, coal, iron, and soft particles	Thin or elongated pieces	Freeze-thaw durability
		Uncrushed[b]	Crushed									Per-cent mini-mum							
		Percent maximum											Percent maximum						
25	A			36	12[c]	3	3	6[h]		15		100	6	3	1	3		15	
	B			36		3	5			15		50		3	1		5	15	
	C			40		3	5			15				3	1		5	15	
26	A			36	12[c]	3	3	6[h]		15		100	6	3	1	3		15	
	B			36		3	5			15		40		3	1		5	15	
	D			40		3	5			15				3	1		5	15	
31	AA[i]			36	12[c]	3	3	6[h]		15		100	6	3	1	3		15	
	A			36	12[c]	3	3	6[h]		15		100	6	3	1	3		15	
	C			40		3	5	10[h]		20		100	Not permitted						

[a] Stone material shall be crushed rock, crushed boulders, crushed stone, or crushed gravel.
[b] For each 10 percent crushed particles, one percent additional is allowed above the maximum wear allowance for uncrushed materials.
[c] Aggregates with a higher percentage of loss may be used provided a satisfactory service record of at least 5 years' duration under similar conditions of service and exposure shall have been demonstrated.
[d] Loss by washing of 1.5 percent permitted for material produced entirely by crushing ledge rock, traprock, or boulders.
[e] Clay-ironstone particles shall not exceed 1.0 percent.
[f] The aggregate shall not contain more than 10 percent of particles which are incrusted for more than one-third of their surface area. At least 60 percent of the material shall be entirely free from incrustations.
[g] Aggregates intended for use in exposed concrete will be required to demonstrate, to the satisfaction of the engineer, adequate freeze-thaw resistance for the particular use, either by means of extended field record in similar concrete, similarly exposed, or by accelerated laboratory freeze-thaw tests, or both.
[h] Chert material shall not be included.
[i] Aggregate produced from limestone quarries will not be permitted. Aggregate produced from gravel sources shall not contain more than 70 percent carbonate rock particles.
[j] The sum of "loss by washing" and shale material shall not exceed 10 percent. In no case shall the material contain more than 7 percent "loss by washing" at the time of placing in the stockpile.
SOURCE: Michigan Department of State Highways.

Contract drilling by an outside source is very common in some parts of the country, with current (1969) prices varying from quarry to quarry and ranging up to $2 per ft of drill hole. The average yield in tons of rock per pound of explosive will be in the range from 2 to 4 tons for a limestone quarry, while in granite and traprock it may be 2 to 3 tons. A high quarry cost is incurred when large rocks left as by-products from the initial blasting must be further reduced in size in order to meet the requirement of the primary crusher in the plant. Either secondary blasting or a drop ball must be employed for this purpose, and there is an old quarry man's saying, " It is cheaper to roll them down the mountain." Where power shovels are used for loading shot rock into the hauling unit, the dipper will serve as an approximate gauge to the maximum size of the rock which can reach the primary crusher. (See Table 20-27.)

TABLE 20-27 Standard Openings of Manganese Dippers

Capacity, cu yd	Dipper opening, in.		Capacity, cu yd	Dipper opening, in.	
	Height	Width		Height	Width
½	28½	28½	2¼	45	49½
¾	29½	36	2½	45	52
1	33	38¼	3	46	59
1¼	35¾	41½	3½	53	59
1½	38	43½	4	51	67½
1¾	40	46¼	4½	57	67½
2	40	49½	5	55½	72½

SOURCE: Amsco Division, Abex Corporation.

Few records of the gradation of shot material as obtained in the quarry are available. In 1938, the Bureau of Mines investigated the relation of gradation to primary crushing,[31] and Table 20-28 indicates the findings pertaining to quarry-run material. In 1964, the Atomic Energy Commission sponsored the experimental use of nuclear explosives to break rock and thus manufacture crushed stone aggregate. This experiment, conducted by the Lawrence Radiation Laboratory, yielded interesting and worthwhile information on the gradation of material following the explosion[32] (see Table 20-29). This type of information is needed in order to get a better picture of the gradation and the fragmentation which result from blasting. It is also helpful when there is an interest in obtaining large particles within a specified size range for use as riprap.

For a quarry operation, in much the same way as for a pit operation, it is necessary to have the proper balance between the capacities of the individual plant units. Without such a balance, there will be low equipment utilization and poor production efficiency. A special attempt should be made to obtain a satisfactory haul way out of the quarry in order to reduce maintenance costs for the hauling units.

Processing Plant The production of aggregate calls for a variety of equipment, ranging in size from the very small to the very large. The individual pieces of equipment can be arranged, frequently with the aid of a flow diagram, into an equipment system which meets the aggregate processing needs of each producer. Capacity of the plant equipment is always stated in tons per hour (tph) and is normally based on a material weight of 100 pcf. All tons are short tons of 2,000 lb.

TABLE 20-28 Gradation of Quarry-run Material

Sieve opening, in.	Cumulative percent passing		
	Vanport limestone	Granite	Pickaway limestone
34	.....	100.00	
24	100.0	93.70	100.00
17	93.7	67.68	60.75
12	73.1	54.08	36.80
8.5	58.8	40.38	23.94
6.0	48.2	32.87	15.84
4.25	37.9	27.92	8.10
3.00	29.2	27.82	3.67
2.12	20.8	18.33	1.63
1.50	14.8	15.20	1.00
1.05	10.1	11.91	0.54
0.742	7.6	10.33	0.37
0.525	5.1	8.49	0.22
0.371	4.1	7.71	0.16
0.263	3.0	6.53	0.13
0.185	2.3	5.80	0.11
0.131	.....	4.90	0.09
Pan			

SOURCE: Bureau of Mines.

The plant units can be divided into the following main categories:

- Crushers
  - Primary
  - Secondary
  - Tertiary
  - Others
- Screens
  - Grizzly
  - Scalping
  - Finishing
- Feeders
  - Vibrating
  - Apron
  - Reciprocating
  - Others
- Conveyors
- Washing equipment
- Special equipment

Crushers Most information on the crushing of rock and the energy required is empirical. The early work identified with crushing was performed by Rittinger, who stated in 1867 that the energy required for crushing is proportional to the areas of the new surfaces formed. Kick stated in 1885 that the energy required to produce any desired reduction in volume of a rock particle is a constant, and is not related to the original size of the particle.[33,34] In 1951, Bond presented his "Third Theory of Comminution," which stated that the work input in rock crushing varies as the new crack length produced or as the square root of one-half the new surface area produced.[35] This theory is the basis for the work index, W_i, which has been found in laboratory tests to vary from 15.1 for gravel in Michigan to 39.2 for gravel in Arizona. For limestone this index varies from 5.0 to 22.2, for traprock from 13.1 to 56.0, and for granite from 6.67 to 26.5.[36]

The horsepower requirement for a crusher is usually given by the manufacturer of the equipment. It has been found from field tests that a jaw crusher with a 200-hp motor can reduce 40-in.-maximum limestone rock to minus 8 in. at a rate of 700 tph. It has also been found that a hammer mill with a 500-hp motor can reduce 2½-in. stone to 100 percent passing a No. 8 mesh at a rate of 200 tph.

TABLE 20-29 Summary of Fragment Size Distribution from Underground Explosions in Rock

Danny Boy* (Basalt)		Pre-Schooner Delta† (Basalt)		Hardhat‡ (Granodiorite)	
Sieve size	Percent passing	Sieve size	Percent passing	Sieve size	Percent passing
6 ft	100	6 ft	100	6 ft	100
5 ft	88	5 ft	100	4 ft	89.4
4 ft	83	4 ft	92	2 ft	61.1
3 ft	75	3 ft	74	1 ft	43.9
2 ft	63	2 ft	57	6 in.	36.1
1 ft	43	1 ft	38	3 in.	29.0
6 in.	30	6 in.	28	2 in.	25.0
4 in.	24	4 in.	25		
3 in.	19	3 in.	23		
2 in.	15	2 in.	20		
1½ in.	13	1½ in.	18		
1 in.	11	1 in.	16		
¾ in.	9	¾ in.	14		
½ in.	8	½ in.	12		
⅜ in.	7	⅜ in.	11		
No. 4	5	No. 4	8		

* U.S. Army Engineer Waterways Experiment Station, Corps of Engineers, *Investigations of Manufacture of Rip-Rap and Aggregate by Nuclear Methods*, PNE-5003, unpublished.

† U.S. Army Engineer Waterways Experiment Station, Corps of Engineers, Project Pre-Schooner, *Geologic Investigations and Engineering Properties of Craters*, PNE-505, unpublished.

‡ Estimate based on (1) field observation; (2) visual estimates published by Boardman, Rabb, and McArthur (1963); and (3) size determinations by Rodean (1964) from photographs taken of Hardhat rubble.

SOURCE: U.S. Atomic Energy Commission.

Primary Crushers. Types generally used include the gyratory, the jaw (swing or overhead eccentric), and the impact breaker (single impeller or double impeller). Single-roll crushers and special heavy-duty hammer mills also see occasional use as primary crushing units for aggregate production. The ratio of reduction in a crusher is defined as the ratio of the maximum size of the feed to the maximum particle size of the product. This ratio varies from 9 to 1 for the pressure type of crusher up to 15 to 1 for the impact type of crusher.

The gyratory crusher illustrated in Fig. 20-11 is used in the larger stationary plants for high-capacity crushing. Its size is designated by the maximum cross dimension of its moon-shaped feed openings, one of which is located on each side of the spider. It is a pressure-type crusher whose head has a gyratory motion without rotation. The cavity is of a form which could be obtained by rotating a jaw crusher section around a vertical axis. There should be adequate clearance above the crusher to permit maintenance, and the available hoisting equipment should have sufficient lifting capacity to handle spider, mantle, and eccentric shaft. Table 20-30 lists sizes and production rates for one type of gyratory crusher.

Gyratory and jaw crushers are classed as pressure-type crushers. Their capacities are dependent upon a number of variables, ranging from cavity nip angles, speed, and shaft eccentricities to density and friability of feed. There is very little factual information available, since few study programs have covered the entire reduction from quarry size to finished product.[37] The nip angle is defined as the angle formed by

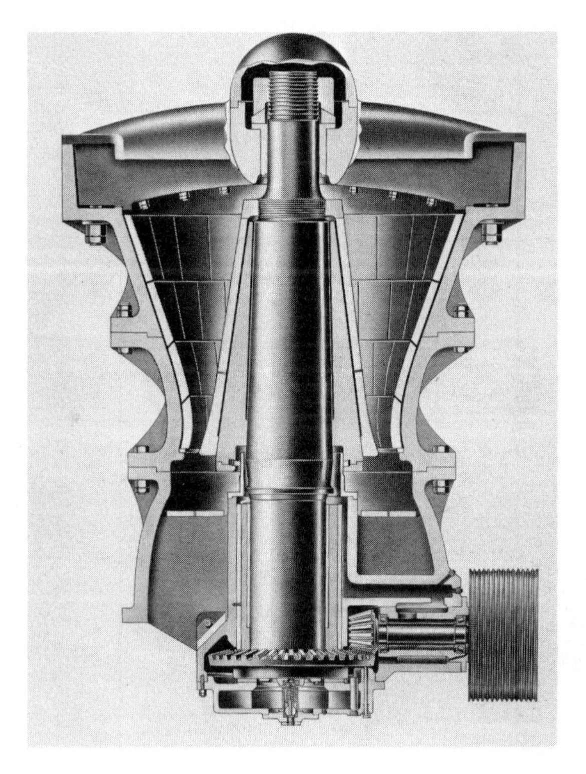

Fig. 20-11 Primary gyratory crusher. *(Fuller Company, General American Transportation Corp.)*

the tangential lines to jaw faces or cavity sections. It normally varies from 21 to 25°, but may run to 27 or 30° for curved surfaces.

The swing-jaw crusher is a stationary type used for heavy crushing in hard rock. It requires less head room than a gyratory crusher and is generally more economical. It has the drawback of requiring a manually controlled feeder for its operation. Figure 20-12 illustrates the Blake type of swing-jaw crusher, and Table 20-31 lists typical sizes and production rates.

The overhead eccentric-jaw crusher illustrated in Fig. 20-13 is the most commonly used primary unit for the average plant. Crusher sizes range up to the 48- by 60-in. unit of this type which was introduced in 1967. The overhead eccentric-jaw crusher, due to its intrinsic design, has a much higher production than the swing-jaw crusher as measured in tons per hour of processed aggregate per pound of crusher weight.

TABLE 20-30 Capacities of Bulldog Gyratory Crushers, Type TC*

Size of opening, in.	Size of each receiving opening, in.	Approximate capacities in tph based on materials weighing 100 pcf when crushed — Discharge opening—closed side												Max rpm driving pulley	Horse-power required†	Approximate shipping weight, lb
		2 in.	2½ in.	3 in.	3½ in.	4 in.	5 in.	6 in.	7 in.	8 in.	9 in.	10 in.	11 in.			
30	30 × 118	345	420	495	560	625	765	910	.....	.....	.....	.....	.....	380	200	175,000
36	36 × 136	...	505	595	665	750	915	1,070	1,220	.....	.....	.....	.....	340	250	255,000
42	42 × 153	...	...	675	770	860	1,045	1,220	1,395	1,550	.....	.....	.....	340	250	285,000
48	48 × 166	...	...	...	945	1,050	1,380	1,500	1,710	1,910	2,100	.....	.....	310	350	450,000
54	54 × 190	...	...	...	...	1,250	1,520	1,790	2,015	2,250	2,500	2,700	.....	310	400	500,000
60	60 × 210	...	...	...	...	.....	2,150	2,500	2,925	3,200	3,450	3,800	4,100	285	500	750,000

* Capacities are based on the rpm tabulated. To save power, crusher should be operated at proportional speeds to suit the production desired, but never at less than 60 percent of the rpm tabulated.

† The horsepower varies with the size of the product, the capacity, and the hardness and toughness of the material.

SOURCE: Fuller Company, General American Transportation Corp.

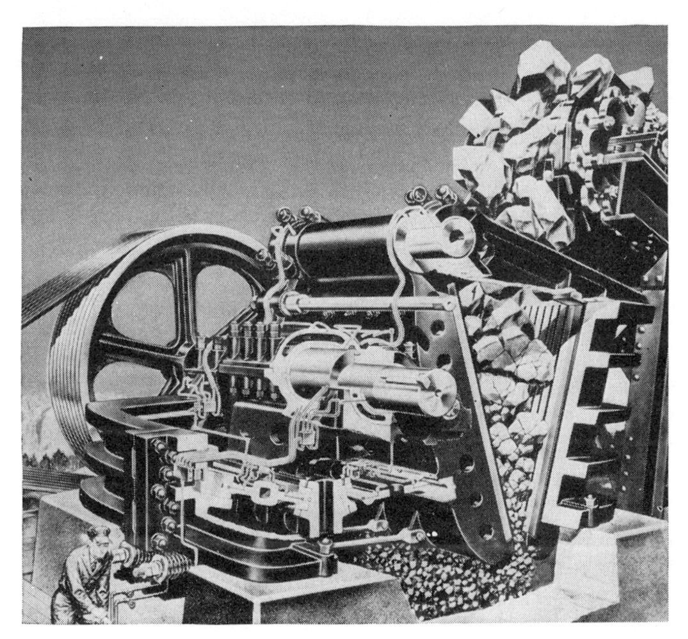

Fig. 20-12 Swing-jaw crusher, Blake type. (*Kennedy Van Saun, subsidiary of McNally Pittsburg.*)

TABLE 20-31 Capacities of Swing-jaw Crushers, Blake Type

Size of crusher receiving opening, in.	Approx. crusher weight, lb	Approx. speed, rpm	Required hp	Capacity at setting, tph	Setting, in.	Capacity at setting, tph	Setting, in.
7 × 10	6,000	275	7	7–10	2½		
10 × 24	18,500	275	15	25–30	2½		
14 × 24	25,000	275	25	45–55	4		
18 × 36	51,000	250	40	70–90	5		
24 × 36	80,000	200	75	150–175	5	180–210	7
30 × 42	115,000	200	100	190–230	6	240–270	8
36 × 48	160,000	200	125–150	230–270	6	280–320	8
48 × 60	350,000	175	175–200	400–475	6	525–600	8

SOURCE: Kennedy Van Saun, a subsidiary of McNally Pittsburg.

Table 20-32 lists sizes and capabilities of the overhead eccentric-jaw crusher, and Table 20-33 lists typical gradations for the crushed product.

The twin-jaw crusher, with two overhead eccentric jaws, will have less wear on the jaw plates than a single-jaw crusher because the jaw timing reduces slippage (see Fig. 20-14). The capacity of the twin-jaw crusher is also higher than that of the single-jaw crusher. Table 20-34 shows capacities for typical twin-jaw crushers.

To get the best possible crushing performance from the overhead eccentric-jaw crusher, a change can be made from a straight to a bellied jaw (see Fig. 20-15). The bellied jaw will be advantageous when crushing gravel with a large percentage of

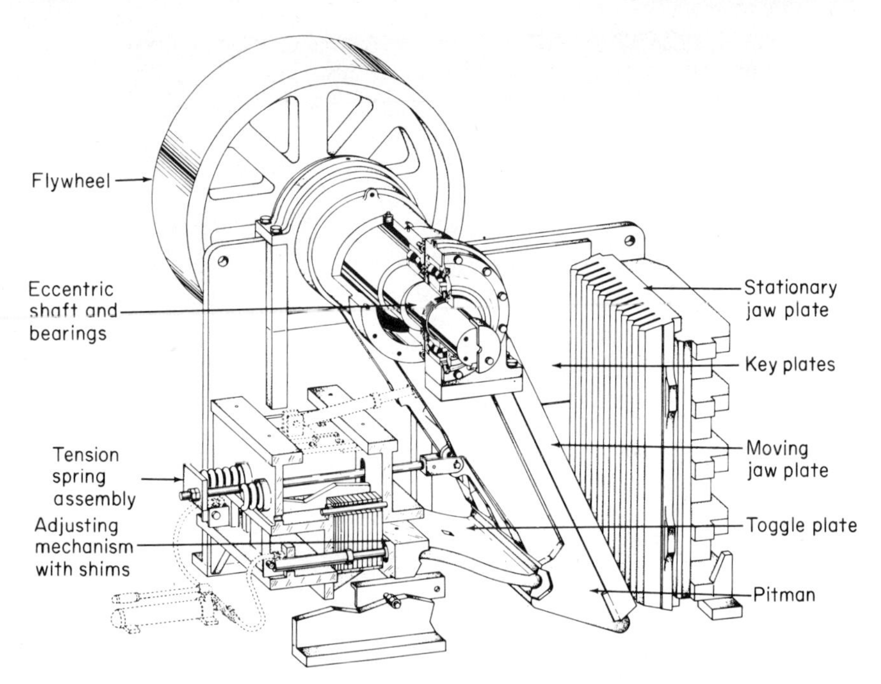

Fig. 20-13 Typical overhead eccentric-jaw crusher.

Fig. 20-14 Twin-jaw crusher. (*Iowa Manufacturing Company.*)

sizes which are close to the discharge setting. The material will then be dispersed higher in the crushing cavity, with less concentrated jaw wear, and more of the discharged material will be close to the discharge setting. The normal speed for the jaw crushers will be found in the range 250 to 300 rpm. Gravity and cavity shape will influence the optimum speed for the crusher, and there is a point beyond which capacity does not increase with speed. At the higher speeds, it will also be found that more of the discharged material will be close to the jaw setting. The overhead eccentric-type jaw can also be used as a secondary crusher. For this use the jaw length will generally be shorter and the feed opening somewhat narrower.

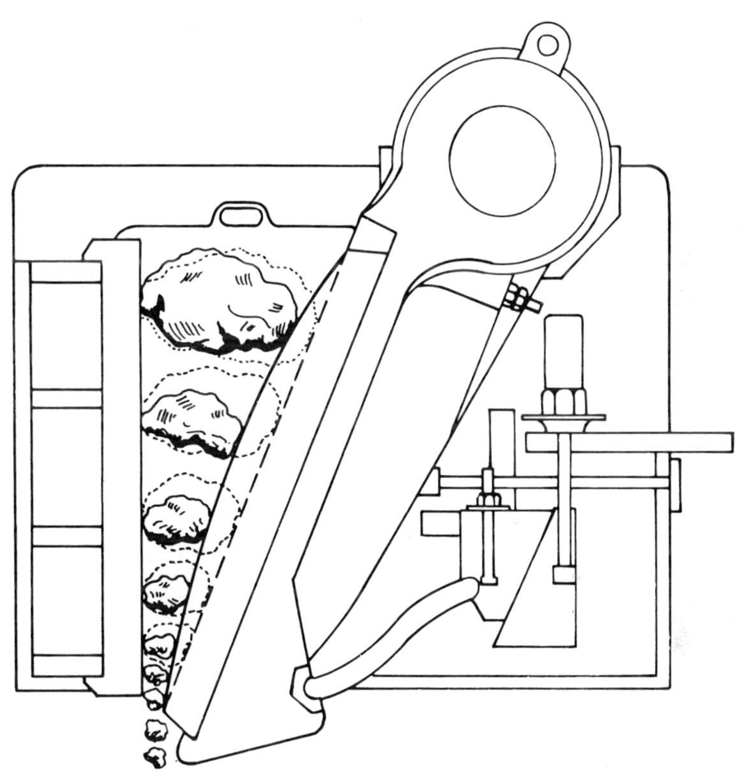

Fig. 20-15 Straight- and bellied-jaw crusher cavity.

The double-impeller impact breaker is a high-capacity primary unit with a large ratio of reduction. It is used mainly for crushing limestone. To evaluate the expected manganese wear, the percentages of SiO_2, Fe_2O_3, and Al_2O_3 are added to 25 percent of the $MgCO_3$ content. The larger this total is, the higher will be the abrasiveness factor and the maintenance cost. For an abrasiveness factor of 10, this cost will vary from 4 to 8 cents per ton of material crushed.

There are two distinct types of double-impeller breakers. One has its impellers running in the same direction, and the reduction is accomplished mainly by the impact of the stone with the impeller bar and with the stationary bar structure. The other type has its impellers running in the opposite direction from each other, which gives a reactivation of the material and an impact against incoming stone (see Fig. 20-16). The feed opening on this type of crusher is large, which reduces the quarry problems

TABLE 20-32 Capacitics of Overhead Eccentric-jaw Crushers

Size of crusher, in. (feed opening)*	42 × 48	30 × 42	25 × 36	18 × 32	15 × 38	14 × 24	10 × 36	10 × 30	10 × 21	10 × 16
Net weight crusher, lb, about	105,800	53,000	38,000	22,000	20,500	10,000	12,420	9,900	5,650	4,950
Weight, crated, lb, about	106,700	54,500	39,000	23,000	21,300	10,500	13,000	10,350	6,000	5,200
Cubical contents, crated for export, cu ft, about	1,616	900	650	370	360	165	185	170	130	115
Electric motor required, hp†	150–200	125–150	75–100	50–60	50–60	20–30	20–30	15–25	15–20	10–15
Drive pulley:										
Diameter, in.	72	60	60	48	48	38	38	33	33	
Face, in.	17	14½	14½	12½	12½	10½	10½	10½	8½	8½
rpm	220	250	250	275	275	320	300	320	350	350
Capacity, tph‡										
½-in. discharge opening	....	....	....	....	....	....	....	....	5–7	4–6
¾-in. discharge opening§	....	....	....	....	....	....	18–27	13–20	7–10	6–8
1-in. discharge opening	....	§	....	....	§	17–25	23–33	17–25	9–13	8–11
1½-in. discharge opening	§	....	....	....	38–57	25–35	29–43	23–34	15–20	10–15
2-in. discharge opening	....	....	§	40–60	48–72	30–45	36–54	29–43	19–26	14–20
2½-in. discharge opening	....	....	....	50–75	57–86	37–55	43–65	35–52	22–33	17–25
3-in. discharge opening	....	....	90–130	60–90	67–100	43–65	50–75			
3½-in. discharge opening	....	140–220	100–150	70–105	76–114					
4-in. discharge opening	....	160–240	113–170	80–120	....	....	....	....	....	§
5-in. discharge opening	300–450	190–285	143–215	100–150	....	....	§			
6-in. discharge opening	333–500	220–330	173–260							
7-in. discharge opening	366–550	257–380								
8-in. discharge opening	406–610									
9-in. discharge opening	467–670									
10-in. discharge opening	480–720									
11-in. discharge opening	520–780									
12-in. discharge opening	560–840									

* To secure the capacities specified, all feed to primary crushers should be smaller than feed opening of crusher in at least one dimension.

† Horsepower required varies with the size of product being made with the crusher, the capacity, and the hardness of the rock or ore.

‡ The capacities given are in tons of 2,000 lb and are based on crushing limestone weighing loose about 2,600 lb per cubic yard and having a specific gravity of 2.6. No crusher, when set at any given discharge opening, will produce a product all of which will pass a screen opening of the same dimensions as the given discharge opening. The amount of oversize will vary with the character of the rock. Capacities given are with the jaws in closed position. For close settings, all undersize material should be removed from the feed so as to eliminate packing and excessive wear on the jaw dies. Wet, sticky feeds will tend to reduce crusher capacities.

§ Where no rating is specified in the capacity table for a certain discharge opening, the crusher cannot be operated economically at that opening.

SOURCE: Smith Engineering Works Division of Barber-Greene.

TABLE 20-33 Estimated Percentage Product Gradation for Overhead Eccentric-jaw Crusher

Product size, U.S. Standard	Crusher closed-side setting, in.																	
	3/4	1	1 1/4	1 1/2	1 3/4	2	2 1/4	2 1/2	2 3/4	3	3 1/2	4	5	6	7	8	9	10
+ 10 in.	.....	.....	.....	.....	.....	.....	.....	.....	.....	.....	.....	.....	.....	6.0	18.0	27.0	34.0	40.0
− 10 in. + 9 in.	.....	.....	.....	.....	.....	.....	.....	.....	.....	.....	.....	.....	.....	6.0	6.0	5.0	5.0	5.0
− 9 in. + 8 in.	.....	.....	.....	.....	.....	.....	.....	.....	.....	.....	.....	.....	7.0	8.0	7.0	7.0	6.0	5.0
− 8 in. + 7 in.	.....	.....	.....	.....	.....	.....	.....	.....	.....	.....	.....	.....	10.0	8.0	7.0	7.0	6.5	6.0
− 7 in. + 6 in.	.....	.....	.....	.....	.....	.....	.....	.....	.....	.....	.....	10.0	9.0	9.0	8.0	6.5	6.0	5.5
− 6 in. + 5 in.	.....	.....	.....	.....	.....	.....	.....	.....	.....	4.0	13.0	12.0	10.0	9.0	7.0	6.5	6.5	6.5
− 5 in. + 4 in.	.....	.....	.....	.....	.....	.....	.....	5.0	10.0	12.0	13.0	13.0	10.0	8.0	7.0	7.0	6.0	5.0
− 4 in. + 3 1/2 in.	.....	.....	.....	.....	.....	.....	7.0	8.0	9.0	8.0	8.0	7.0	6.0	5.0	4.5	3.5	3.5	3.5
− 3 1/2 in. + 3 in.	.....	.....	.....	.....	.....	7.0	9.0	9.0	9.0	9.0	8.0	6.0	5.0	4.5	4.0	3.5	3.0	3.0
− 3 in. + 2 3/4 in.	.....	.....	.....	.....	4.0	5.0	4.5	4.5	4.5	4.5	4.0	3.5	3.0	2.5	2.0	2.0	1.7	1.5
− 2 3/4 in. + 2 1/2 in.	.....	.....	.....	4.0	5.0	6.0	5.5	5.5	5.0	4.5	4.0	3.5	3.0	2.5	2.5	2.0	1.8	1.5
− 2 1/2 in. + 2 1/4 in.	.....	.....	.....	3.0	6.0	6.0	5.5	5.0	4.5	4.5	3.5	3.5	3.0	2.5	2.0	1.7	1.7	1.5
− 2 1/4 in. + 2 in.	.....	.....	5.0	6.0	6.0	7.0	6.5	6.0	5.0	4.5	4.0	3.5	3.0	2.5	2.0	1.8	1.8	1.6
− 2 in. + 1 3/4 in.	.....	2.0	5.0	7.0	7.0	7.0	6.0	5.0	5.0	5.0	4.0	3.5	3.0	2.5	2.5	2.0	1.8	1.8
− 1 3/4 in. + 1 1/2 in.	.....	6.0	8.0	9.0	10.0	7.5	7.0	7.0	6.0	5.5	4.5	4.0	3.5	3.0	2.5	2.5	1.7	1.6
− 1 1/2 in. + 1 1/4 in.	3.0	6.0	7.0	8.5	7.0	6.5	6.0	5.0	5.0	4.5	4.0	3.5	2.5	2.5	2.1	1.8	1.7	1.4
− 1 1/4 in. + 1 in.	7.0	13.0	13.0	10.5	9.0	8.0	7.0	6.5	6.0	5.5	5.0	4.5	3.5	3.0	2.5	2.0	1.8	1.7
− 1 in. + 7/8 in.	4.0	7.0	6.0	5.5	4.5	4.0	3.5	3.5	3.0	2.5	2.5	2.4	2.0	1.5	1.5	1.2	1.1	0.9
− 7/8 in. + 3/4 in.	11.0	11.0	9.0	7.5	7.0	5.5	5.0	4.5	4.5	4.0	3.5	2.8	2.5	2.0	1.7	1.5	1.4	1.2
− 3/4 in. + 5/8 in.	8.0	5.5	4.5	3.8	3.5	3.3	3.0	2.7	2.5	2.5	2.0	1.8	1.5	1.2	1.1	0.9	0.7	0.6
− 5/8 in. + 1/2 in.	11.0	8.0	7.0	5.4	5.0	4.2	3.5	3.4	3.0	3.0	2.2	2.2	1.7	1.6	1.3	1.1	1.0	0.9
− 1/2 in. + 3/8 in.	14.0	10.5	8.5	7.3	6.5	5.5	5.2	4.8	4.3	3.8	3.6	3.1	2.6	2.2	1.9	1.7	1.4	1.2
− 3/8 in. + 5/16 in.	4.0	3.0	3.0	2.5	1.9	1.8	1.6	1.4	1.4	1.4	1.2	1.1	0.8	0.7	0.7	0.5	0.4	0.3
− 5/16 in. + 1/4 in.	6.5	5.0	4.0	3.0	2.8	2.7	2.3	2.0	2.0	1.6	1.4	1.3	1.1	1.0	0.8	0.7	0.6	0.5
− 1/4 in. + No. 4	7.5	5.5	5.0	4.2	3.6	3.0	2.8	2.7	2.3	2.3	2.0	1.9	1.5	1.3	1.0	0.9	0.7	0.6
− No. 4 + No. 8	10.5	7.6	6.5	5.5	4.8	4.3	3.9	3.6	3.4	3.1	2.8	2.5	2.0	1.6	1.4	1.1	1.0	0.7
− No. 8 + No. 10	2.1	1.6	1.3	1.1	1.0	0.9	0.8	0.8	0.7	0.6	0.6	0.5	0.4	0.4	0.3	0.2	0.2	0.2
− No. 10 + No. 16	4.6	3.3	2.8	2.4	2.1	1.9	1.7	1.6	1.5	1.4	1.2	1.1	0.9	0.7	0.7	0.5	0.3	0.3
− No. 16 + No. 30	3.0	2.1	1.9	1.6	1.4	1.2	1.1	1.0	1.0	1.0	0.8	0.7	0.6	0.5	0.4	0.3	0.3	0.2
− No. 30 + No. 40	0.9	0.8	0.6	0.5	0.5	0.4	0.4	0.3	0.2	0.2	0.2	0.2	0.2	0.2	0.1	0.1	0.1	0.06
− No. 40 + No. 50	0.6	0.5	0.5	0.4	0.3	0.3	0.2	0.2	0.2	0.2	0.2	0.1	0.1	0.1	0.1	0.1	0.1	0.04
− No. 50 + No. 100	1.0	0.6	0.6	0.6	0.5	0.5	0.5	0.5	0.5	0.4	0.4	0.4	0.3	0.2	0.2	0.2	0.1	0.1
− No. 100	1.3	1.0	0.8	0.7	0.6	0.5	0.5	0.5'	0.5	0.5	0.4	0.4	0.3	0.3	0.2	0.2	0.1	0.1
Total %	100.0	100.0	100.0	100.0	100.0	100.0	100.0	100.0	100.0	100.0	100.0	100.0	100.0	100.0	100.0	100.0	100.0	100.0

EXAMPLE: To find the percent of each size in the product of a crusher set at 2 1/2 in. closed setting, find 2 1/2 in. at the top of the chart. Follow the column down until it intersects with the horizontal range size column. The first reading is 5 in. and 5 percent, indicating that 5 percent of the material will be over 5 in. and 95 percent passing. By following down the column you will find:

Over 2 1/2 in.:	32 percent	1 3/4 to 2 in.:	5 percent	1 to 1 1/4 in.:	6.5 percent	5/8 to 3/4 in.:	2.7 percent
2 1/4 to 2 1/2 in.:	5 percent	1 1/2 to 1 3/4 in.:	7 percent	7/8 to 1 in.:	3.5 percent	to 5/8 in.:	22.8 percent
2 to 2 1/4 in.:	6 percent	1 1/4 to 1 1/2 in.:	5 percent	3/4 to 7/8 in.:	4.5 percent	Total:	100 percent

SOURCE: Iowa manufacturing Company.

with oversize rocks. The unit can be used as a secondary crusher for the preparation of cubical material and for elimination of flats. Tables 20-35A and 20-35B list the characteristics and capacities of double-impeller impact breakers, and Fig. 20-17 shows typical gradation curves for the crushed product.

TABLE 20-34 Capacities of Twin-jaw Crushers

Model of crusher	Horsepower required	Crusher capacity, tph,* with closed-side settings, in.					
		1	1¼	2	3	4	5
1216	20–40	16–26	19–32	28–47	42–70		
1236	75–125	.....	36–63	54–90	84–139	115–176	
1836	75–125	.....	.....	48–80	70–135	96–176	160–220

* Based on 100 pcf material.
SOURCE: Iowa Manufacturing Company.

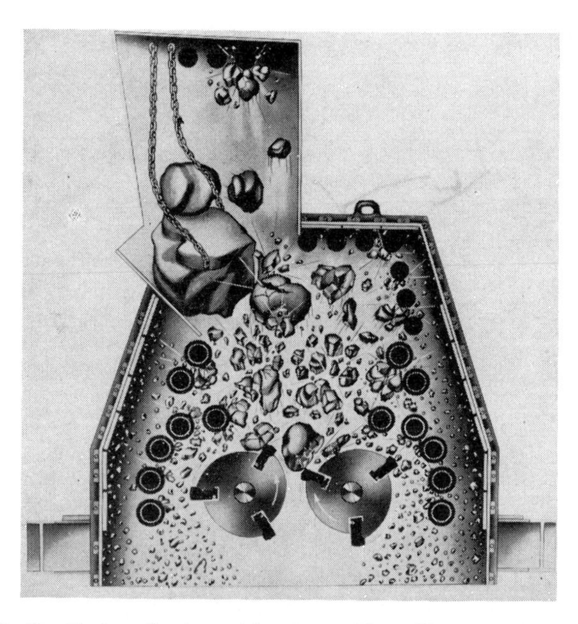

Fig. 20-16 Double-impeller impact breaker. (*Iowa Manufacturing Company.*)

The single-impeller impact breaker is also a high-capacity primary unit. It is used for crushing limestone, usually in open-circuit operation. Table 20-36 lists characteristics and capacities for units of this type, and Fig. 20-18 illustrates their operation. Typical gradation curves for the products from single-impeller impact breakers are shown in Fig. 20-19.

TABLE 20-35A Double-impeller Impact Breakers: Characteristics

Model No.	2222	3042	3645S	3645H	4350S	4350H	5360H
Overall dimensions:							
Height	7 ft 4⅞ in.	8 ft 0 in.	8 ft 6 in.	8 ft 7½ in.	11 ft 6³⁄₁₆ in.	11 ft 8½ in.	14 ft 3 in.
Length	10 ft 6 in.	10 ft 6 in.	13 ft 0 in.	13 ft 0 in.	17 ft 0 in.	17 ft 0 in.	20 ft 0 in.
Width	5 ft 2 in.	6 ft 5 in.	7 ft 4 in.	7 ft 7 in.	8 ft 4 in.	9 ft 0 in.	10 ft 0 in.
Inlet opening	22 in. × 22 in.	30 in. × 42 in.	36 in. × 45 in.	36 in. × 45 in.	43 in. × 50 in.	43 in. × 50 in.	53 in. × 60 in.
Outlet opening	20 in. × 96 in.	30 in. × 96 in.	36 in. × 88 in.	36 in. × 88 in.	44 in. × 135⅜ in.	43 in. × 141¾ in.	53 in. × 176 in.
Weight, lb, with skid‡	21,900	26,600	39,700	41,800	75,360	92,800	172,500
Mounting skid length	Variable	Variable	13 ft 0 in.	13 ft 0 in.	17 ft 0 in.	17 ft 0 in.	20 ft 0 in.
Mounting beams	12 in. WF	12 in. WF	14 in. WF	14 in. WF	16 in. WF	18 in. WF	24 in. WF
Impellers:							
Total weight with shaft and bars, lb (ea.)	2,930	4,400	6,300	6,725	10,160	10,360	17,440
Diam outside bars	35½ in.	35½ in.	37½ in.	37½ in.	42¼ in.	42¼ in.	49¼ in.
Core (material)	Cast nickel steel	Cast nickel steel	Cast nickel steel	Cast nickel steel	Cast nickel steel	Cast nickel steel	Cast nickel steel
Weight, lb (ea.)	1,900	2,900	4,500	4,600	7,525	7,450	12,570
Impeller bars	Manganese	Manganese	Manganese	Manganese	Manganese	Manganese	Manganese
Weight, lb (ea.)	265	400	450	430	530	530	720
Number required	6	6	6	6	6	6	6
Wearing surfaces	4	4	4	1	1	1	1
Shafts:							
Material	Chr-Ni-Moly steel	Chr-Ni-Moly steel	Chr-Ni-Moly steel	Chr-Ni-Moly steel	Chr-Ni-Moly steel	Chr-Ni-Moly steel	Chr-Ni-Moly steel
Diam through impeller	4¹⁄₁₆ in.	4¹⁄₁₆ in.	5 in.	6⅛ in.	6¹⁄₁₆ in.	7½ in.	10 in.
Diam at ends	3⁷⁄₁₆ in.	3⁷⁄₁₆ in.	3⁷⁄₁₆ in.	4¹⁵⁄₁₆ in.	4¹⁵⁄₁₆ in.	5¹⁵⁄₁₆ in.	7 in.
Key at ends	⅞ sq in.	⅞ sq in.	⅞ sq in.	1¼ sq in.	1¼ sq in.	1½ sq in.	1½ sq in.
Weight, lb (ea.)	185	235	375	625	750	1,030	2,200
Bearings	Heavy-duty antifriction, spherical self-aligning						
Breaker bars:							
Material	Mang-ARS	Mang-ARS	High carbon steel	High carbon steel	High carbon steel	High carbon steel	High carbon steel
Diameter	5 in.	5 in.	5 in.	5–6½ in.	6½ in.	6½ in.	8 in.
Sleeves:							
Material	....	....	Manganese	Manganese	Manganese	Manganese	Manganese
Diameter	....	....	7 in.	7–9 in.	9 in.	9 in.	12 in.
Liner plates:							
Material	Mang-ARS	Mang-ARS	Mang-ARS	Mang-ARS	Mang-ARS	Mang-ARS	Mang-ARS
Thickness	1 in.	1 in.	1–1¼ in.	1–1¼ in.	1–1¼ in.	1½ in.	1½ in.
Side plates:							
Material	H. R. S.	H. R. S.	H. R. S.	H. R. S.	H. R. S.	H. R. S.	Cast steel
Thickness	1 in.	1 in.	1½ in.	1½ in.	1½ in.	2 in.	3–4½ in.
Hp required:							
Electric*	(2) 40–50	(2) 60–75	(2) 100–125	(2) 100–125	(2) 150	(2) 150	(2) 200
Diesel†	(1) 110–140	(1) 150–190	(1) 230–290	(1) 230–290	(2) 160–175	(2) 160–175	(2) 210–230
Rpm	350–900	350–900	350–900	350–900	350–675	350–675	350–575
Practical minimum product size, closed circuit	1½ in. minus	1½ in. minus	1½ in. minus	1½ in. minus	2½ in. minus	2½ in. minus	4 in. minus
Capacities	Dependent upon type and characteristics of the material processed						
Tons per hour	75–125	125–250	250–500	250–500	400–800	400–800	600–1,200

* If new frame motors are used, increase electric hp requirement by 25 percent.
† Diesel hp requirements are continuous rating.
‡ Weights do not include feed chute, chain curtain, discharge chute, or drives.
SOURCE: Iowa Manufacturing Company.

TABLE 20-35B Double-impeller Impact Breakers: Output Gradations for Primary Crusher in Open-circuit Operation

Sieve size	Percent passing		
	Soft limestone	Hard limestone	Hard dolomite
6 in.	100	100	72
4 in.	100	85	52
3 in.	94	75	40
1½ in.	71	40	25
¾ in.	52	22	16
No. 4	27	10	5

SOURCE: Iowa Manufacturing Company.

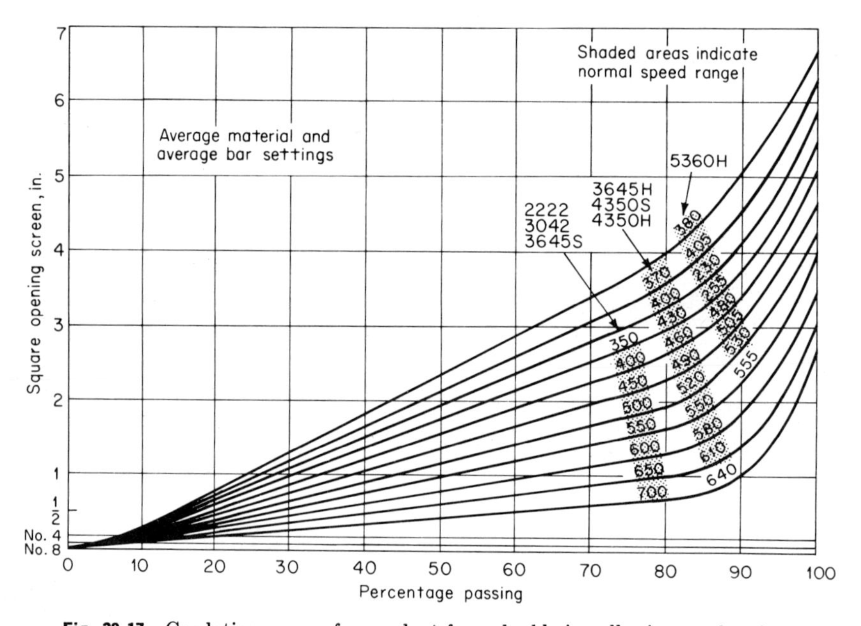

Fig. 20-17 Gradation curves for product from double-impeller impact breaker.

Secondary Crushers. Types generally used include the gyratory, cone, roll, and hammer mill.

The secondary type of gyratory crusher, illustrated in Fig. 20-20, has a small eccentric throw. This will give a finished product in which no more than 10 to 20 percent of the material is larger than the closed setting of the unit. See Tables 20-37 and 20-38 for sizes and capacities of gyratory crushers. The hydraulic adjustment and the safety release can be combined so that the reset of the crusher is automatic after any tramp iron passes through the crushing cavity. Figure 20-21 illustrates the principle of the hydraulic adjustment, which provides a simple method of maintaining a desired crusher setting.

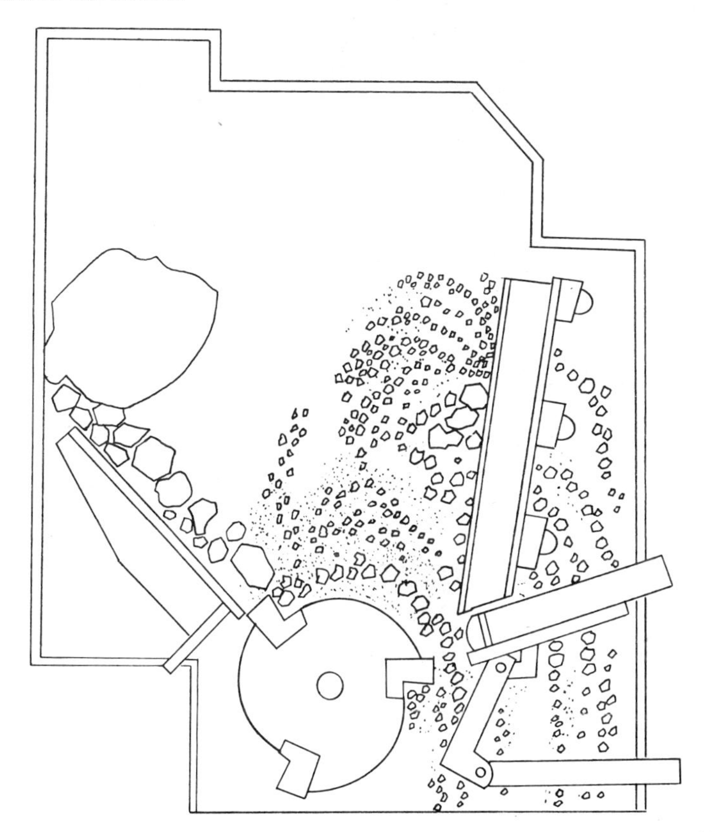

Fig. 20-18 Single-impeller impact breaker. *(Universal Engineering Corp., subsidiary of Pettibone Mulliken Corp.)*

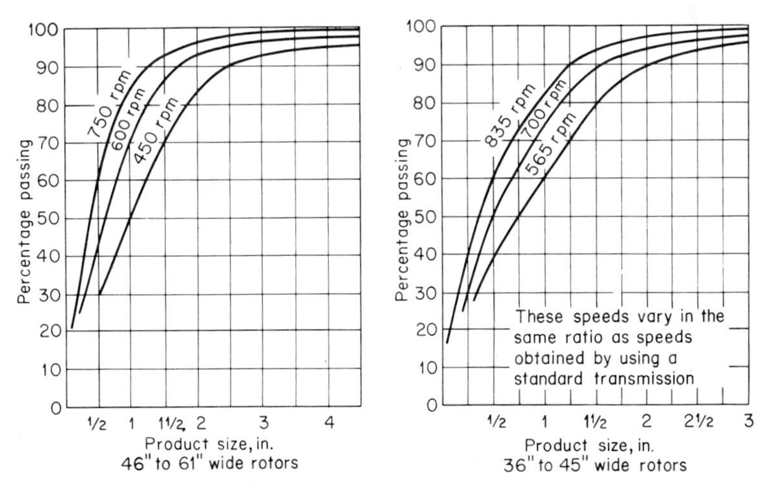

Fig. 20-19 Gradation curves for product from single-impeller impact breaker.

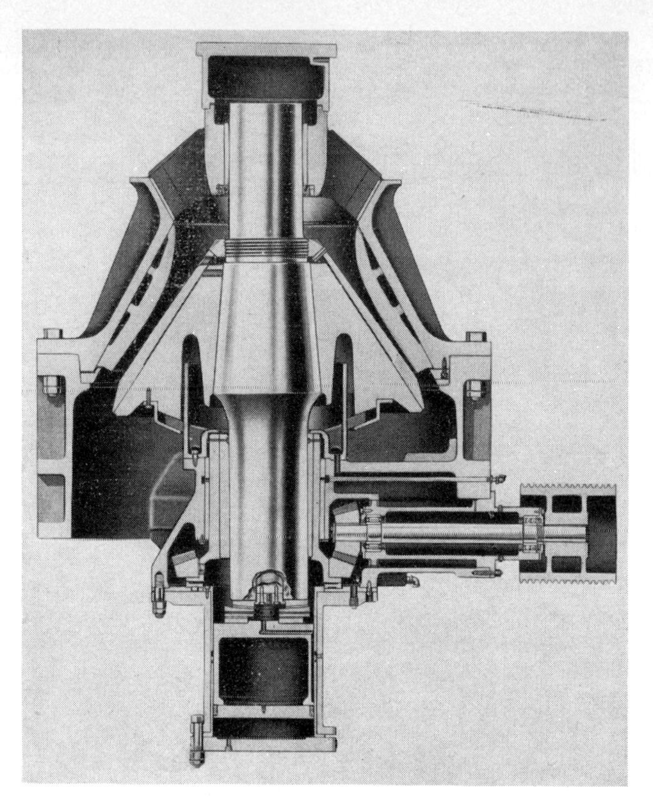

Fig. 20-20 Secondary gyratory crusher. (*Allis-Chalmers.*)

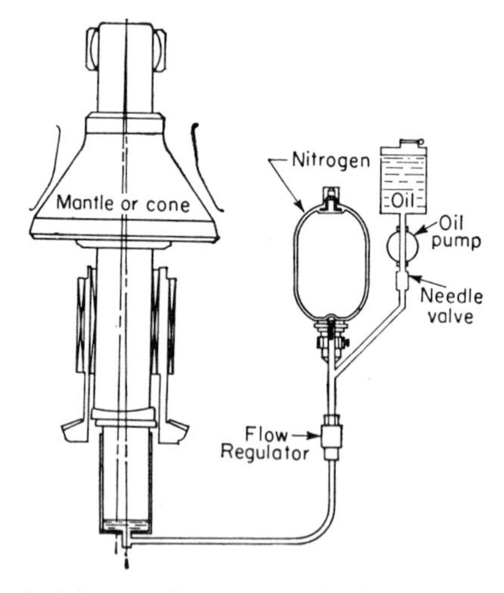

Fig. 20-21 Hydraulic principle of adjustment and safety release on gyratory crusher. The main assembly, including the crushing cone, is supported on a hydraulic jack. When oil is pumped into or out of the jack, the main shaft assembly is raised or lowered, changing the crusher setting. When tramp iron enters the crushing chamber, pressure on oil in the hydraulic jack becomes greater than the gas pressure in the accumulator. This forces oil from the hydraulic system to the accumulator, allowing the crushing cone to lower until tramp iron is discharged. After discharge, it returns to the original setting. (*Allis-Chalmers.*)

TABLE 20-36 Single-impeller Impact Breakers

Model	3642	4442	5042	6060
Speed, rpm	450–900	400–850	425–900	450–600
Hp required	150–200	200–350	250–400	300–500
Liners	Special abrasive-resistant steel			
Rotor:				
Material	Fabricated steel			
Width, in	36	44	50	60
Outside diam hammer circle, in	37½	42	42	52
Weight with 3 hammers (approx.), lb	6,100	11,575	13,145	26,875
Hammers:				
Material	Manganese steel			
Number	3	3	2,3 or 4	3 or 4
Hammer width, in	36	44	50	60
Weight, each (approx.), lb	328	400	455	625
Shafts:				
Material	Special alloy steel			
Diam at core, in	6 15/16	7½	7½	10½
Diam at sheaves, in	4 7/16	5 5/16	5 5/16	6 15/16
Bearings	Antifriction bearings (self-aligning)			
Capacity, tph:				
Primary crusher, nominal 3 in. minus	Up to 400	Up to 600	Up to 750	Up to 1400
Secondary crusher, nominal 1 in. minus	Up to 175	Up to 250	Up to 400	Up to 450
Weight (approx.), lb	24,759	42,520	59,890	87,200

SOURCE: The Universal Engineering Corp., subsidiary of Pettibone Mulliken Corp.

TABLE 20-37 Capacities of Gyratory Crushers*

Crusher size, in.	Wt., lb	Motor hp	Feed opening, in.	Max. feed size, in.	Crusher capacity, tph, with closed-side settings, in.												
					5/16	3/8	7/16	½	5/8	¾	7/8	1	1¼	1½	1¾	2	2½
28	12,000	40/60	4	3	20	25	30	35	40	45	50	60					
			6	4½	...	...	30	35	40	45	50	60	70	85			
40	28,000	75/125	4	3	...	60	70	85	95	105							
			6	4½	...	...	70	85	95	105	115	125					
			8	6	...	...	...	85	95	105	115	125	135	150	170		
52	45,000	125/200	4	3	...	...	...	140	160	180							
			8	6	...	...	...	140	160	180	200	220	240				
			11	8	...	...	...	...	160	180	200	220	240	270	300		
70	90,000	200/300	4	3	...	...	...	195	220	245							
			8	6	...	...	...	...	220	245	270	300	330				
			12	9	...	...	...	...	...	245	270	300	330	360	400		
			15	11	...	...	...	...	...	...	270	300	330	360	400	440	500

* Capacities based on closed circuit operation (6:1 reduction ratio is recommended) and 100 pcf crushed material.

SOURCE: Kennedy Van Saun, a subsidiary of McNally Pittsburg.

TABLE 20-38 Capacities of Hydrocone Gyratory Crushers*

Crusher size, in.	22	30	36	45	51	60	84
Approx. total weight, lb	6,600	13,500	18,400	36,500	44,750	72,800	205,500

22- and 30-in. Hydrocones

Crusher designation	Max. hp	Eccentric throw, in.	Crusher capacity, tph, with closed-side setting of discharge opening, in.								
			1/8	3/16	1/4	5/16	3/8	7/16	1/2	5/8	3/4
322 coarse chamber†	20	3/16	....	9.5	10	10.5	11	12	13		
	20	1/4	....	....	14	14.5	15	16	17		
	25	5/16	....	....	....	16	17	20	22		
	25	3/8	....	....	....	....	22	25	27		
	30	1/2	....	....	....	....	...	...	36		
222 intermediate chamber‡	20	1/4	7	7.5	8	9					
	25	3/8	....	12.5	13	14					
	30	1/2	....	....	19	20					
122 fine chamber§	30	1/2	10	11	13						
530 coarse chamber†	30	1/4	....	....	20	21	22	23	24	25	27
	30	5/16	....	....	....	28	29	30	32	33	35
	40	3/8	....	....	....	....	35	36	38	40	42
	50	1/2	....	....	....	....	...	...	51	53	55
	50	5/8	....	....	....	....	...	...	...	66	68
330 intermediate chamber‡	30	1/4	11.5	12	12.5	13	14	15	16		
	40	3/8	....	18	20	22	24	26	28		
	50	1/2	....	....	25	26	28	30	32		
	50	5/8	....	....	....	36	38	39	41		
1 1/2 30 fine chamber§	50	1/2	14	18	21	25	28				
	50	5/8	....	20	24	28	31				

Max. feed sizes up to 4 7/8 in. for coarse chamber, 2 7/8 in. for intermediate chamber, and 1 3/8 in. for fine chamber.

36-in. Hydrocones

Crusher designation	Max. hp	Eccentric throw, in.	Crusher capacity, tph, with closed-side setting of discharge opening, in.												
			1/8	3/16	1/4	5/16	3/8	7/16	1/2	5/8	3/4	7/8	1	1 1/8	1 1/4
736 coarse chamber†	60	1/2	...	...	...	...	...	...	67	69	72	75	78	81	84
	75	5/8	...	...	...	...	...	...	...	88	92	96	100	104	107
	100	3/4	...	...	...	...	...	...	...	...	111	115	120	125	130
	100	7/8	...	...	...	...	...	...	...	...	...	139	143	148	152
	125	1	...	...	...	...	...	...	...	...	...	...	160	168	175
	125	1 1/8	...	...	...	...	...	...	...	...	...	...	...	178	198

TABLE 20-38 Capacities of Hydrocone Gyratory Crushers* (Continued)

36-in. Hydrocones

Crusher designation	Max. hp	Eccentric throw, in.	Crusher capacity, tph, with closed-side setting of discharge opening, in.												
			⅛	3/16	¼	5/16	⅜	7/16	½	⅝	¾	⅞	1	1⅛	1¼
436 intermediate chamber‡	60	½	...	...	38	39	40	41	43	45	48				
	75	⅝	...	...	...	48	49	51	52	57	59				
	100	¾	...	...	...	...	58	60	62	67	71				
	100	⅞	...	...	...	...	...	69	71	77	82				
	125	1	...	...	...	...	...	...	80	87	94				
	125	1⅛	...	...	...	...	...	...	...	95	105				
236 fine chamber§	60	½	22	24	26	28	29	31	32	36					
	100	¾	...	36	38	40	42	44	45	50					
	100	⅞	...	...	43	45	47	49	52	56					
	125	1	...	...	46	49	52	55	57	63					
	125	1⅛	...	...	...	53	56	59	64	70					

Max. feed size 7 to 2 in.

45-in. Hydrocones

Crusher designation	Max. hp	Eccentric throw, in.	Crusher capacity, tph, with closed-side setting of discharge opening, in.													
			5/32	3/16	¼	5/16	⅜	½	⅝	¾	⅞	1	1⅛	1¼	1½	2
945 coarse chamber†	100	⅝	...	...	...	...	...	...	98	101	105	109	113	117	125	141
	125	¾	...	...	...	...	...	...	...	126	130	135	139	143	152	169
	125	⅞	...	...	...	...	...	...	...	...	155	160	165	169	178	197
	150	1	...	...	...	...	...	...	...	...	...	185	190	195	205	225
	150	1⅛	...	...	...	...	...	...	...	...	...	...	216	221	232	253
	200	1¼	...	...	...	...	...	...	...	...	...	...	...	247	259	281
	200	1⅜	...	...	...	...	...	...	...	...	...	...	...	...	286	310
645 intermediate chamber‡	100	⅝	...	...	...	70	71	74	76	79	81	84	87			
	125	¾	...	...	...	...	86	89	92	96	99	102	105			
	125	⅞	...	...	...	...	...	105	109	112	116	120	124			
	150	1	...	...	...	...	...	121	125	129	133	138	142			
	150	1⅛	...	...	...	...	...	...	141	146	151	156	161			
	200	1¼	...	...	...	...	...	...	157	163	168	174	179			
	200	1⅜	...	...	...	...	...	...	...	180	186	192	198			
345 fine chamber§	100	⅝	29	30	33	36	39	44	50	55						
	125	¾	...	37	41	44	47	54	61	67						
	125	⅞	...	...	48	52	56	64	72	80						
	150	1	...	...	56	60	65	74	83	92						
	150	1⅛	...	...	...	69	74	84	94	105						
	200	1¼	...	...	...	77	82	94	106	117						
	200	1⅜	...	...	...	...	91	104	117	130						

Max. feed sizes up to 8⅞ in. for coarse chamber, 5⅞ in. for intermediate chamber, and 2⅞ in. for fine chamber.

TABLE 20-38 Capacities of Hydrocone Gyratory Crushers* (Continued)

51- and 60-in. Hydrocones

Crusher designation	Max. hp	Eccentric throw, in.	Crusher capacity, tph, with closed-side setting of discharge opening, in.													
			3/16	1/4	5/16	3/8	7/16	1/2	5/8	3/4	7/8	1	1 1/4	1 1/2	1 3/4	2
1051 coarse chamber†	75	1/2	...	...	...	...	...	90	95	100	105	110	117	125	132	
	100	5/8	...	...	...	...	...	...	125	130	135	140	145	154	163	
	125	3/4	...	...	...	...	...	...	...	160	165	170	185	195	200	
	150	7/8	...	...	...	...	...	...	...	...	190	195	205	210	220	
	175	1	...	...	...	...	...	...	...	...	...	235	245	250	260	
	175	1 1/8	...	...	...	...	...	...	...	...	...	...	270	275	280	
	200	1 1/4	...	...	...	...	...	...	...	...	...	...	305	315	325	
	200	1 3/8	...	...	...	...	...	...	...	...	...	...	...	330	340	
651 intermediate chamber‡	75	1/2	...	70	71	73	74	76	80	84	88	92	100			
	125	3/4	...	...	...	110	115	120	125	130	140	150	165			
	175	1	...	...	...	...	...	145	155	160	170	175	185			
	200	1 1/4	...	...	...	...	...	...	185	205	220	235	255			
3 1/2 51 fine chamber§	75	1/2	38	40	50	56										
	125	3/4	...	55	63	70	77	85								
	175	1	...	75	80	90	100	110	120							
	200	1 3/8	...	...	95	105	115	125	140	160						
1260 coarse chamber†	150	3/4	...	...	...	...	...	...	...	200	210	220	230	240	250	260
	175	7/8	...	...	...	...	...	...	...	...	235	245	255	265	275	285
	200	1	...	...	...	...	...	...	...	...	...	265	275	285	295	305
	225	1 1/8	...	...	...	...	...	...	...	...	...	...	315	325	335	345
	250	1 1/4	...	...	...	...	...	...	...	...	...	...	350	365	380	395
	275	1 3/8	...	...	...	...	...	...	...	...	...	...	...	415	430	445
	300	1 1/2	...	...	...	...	...	...	...	...	...	...	...	460	475	490
760 intermediate chamber‡	150	3/4	...	...	...	130	140	150	160	170	180	190	210	230		
	200	1	...	...	...	...	...	190	200	210	220	230	250	270		
	250	1 1/4	...	...	...	...	...	...	240	250	260	270	280	300		
	300	1 1/2	...	...	...	...	...	...		300	310	320	330	350		
460 fine chamber§	150	3/4	...	80	90	100	105	115	125							
	200	1	...	85	95	105	115	125	135	150						
	250	1 1/4	...	...	110	120	135	145	155	175						
	300	1 1/2	...	...	...	135	150	160	185	205						

Max. feed size to 12 in.

* Stepped line indicates recommended minimum and maximum setting for each throw.

† *Coarse Crushing Chamber* The coarse crushing chamber affords the maximum feed opening for a given size of crusher. Crushers fitted with a coarse chamber can be choke fed provided that product size material in the feed is removed. Approximately 60 percent of the discharged crushed product from 22- and 30-in. Hydrocones will be smaller than the closed-side setting.

The coarse chamber has a relatively short parallel zone and is designed to be operated at a closed-side setting equal to or greater than the eccentric throw. For example, a crusher with a 3/8-in. eccentric throw should be operated at a 3/8-in. (or more) closed-side setting and, therefore, a 3/4-in. open side setting. Optimum capacity and product will result when operated under these conditions, as well as most economical wear on the mantalloy crushing surfaces.

TABLE 20-38 Capacities of Hydrocone Gyratory Crushers* (Continued)

84-in. Hydrocones

Crusher designation	Max. hp	Eccentric throw, in.	Crusher capacity, tph, with closed-side setting of discharge opening, in.												
			1/4	5/16	3/8	7/16	1/2	5/8	3/4	7/8	1	1 1/4	1 1/2	1 3/4	2
1784 coarse chamber†	300	1	...	...	...	...	...	...	...	...	440	450	460	470	480
	350	1 1/4	...	...	...	...	...	...	...	...	...	555	565	575	585
	400	1 1/2	...	...	...	...	...	...	...	...	...	...	690	700	710
	450	1 3/4	...	...	...	...	...	...	...	...	...	...	...	870	885
	500	2	...	...	...	...	...	...	...	...	...	...	...	...	1005
1084 intermediate chamber‡	250	3/4	...	...	180	190	200	210	220	230	240	260	280	300	320
	300	1	...	...	...	...	275	285	295	305	315	335	355	375	395
	350	1 1/4	...	...	...	...	...	345	355	365	375	385	405	425	445
	400	1 1/2	...	...	...	...	...	...	440	450	460	470	490	510	530
	450	1 3/4	...	...	...	...	...	...	...	555	570	585	600	630	660
	500	2	...	...	...	...	...	...	...	...	655	675	695	715	735
584 fine chamber§	250	3/4	120	123	126	128	131	136	141	146	151				
	300	1	153	157	161	165	169	177	185	193	201				
	350	1 1/4	...	200	205	209	214	224	234	244	254				
	400	1 1/2	...	...	250	256	262	273	284	295	306				
	450	1 3/4	...	...	...	302	308	321	333	345	358				
	500	2	...	...	...	...	348	363	378	393	408				

Max. feed size to 17 in.

One-way dimension (slot size) of the feed to a crusher fitted with a coarse chamber should not exceed two-thirds to 70 percent of the feed opening. The maximum feed size to a 736 Hydrocone crusher would therefore be about 5 in. one-way dimension.

The use of a wobble plate feeder, furnished as optional equipment, is recommended if the feed size is relatively large, if the crusher is to be operated in closed circuit, or if the feed to the crusher is irregular.

‡ *Intermediate Crushing Chamber* The intermediate crushing chamber has a feed opening somewhat less than a coarse crushing chamber, but because of its longer parallel zone, it is designed to be operated at a closed-side setting equal to or greater than half the eccentric throw. For example, with a 3/4-in. eccentric throw, the minimum closed-side setting would be 3/8 in. Approximately 68 percent of the discharged crushed product from 22- and 30-in. Hydrocones will be smaller than the closed-side setting.

Crushers fitted with this type of chamber can be choke fed provided that product size material in the feed is removed ahead of the crusher. The one-way dimension or slot size of the feed to a crusher should not exceed approximately half the receiving opening. A 436 Hydrocone crusher with a 5/8-in. eccentric throw could be operated at a 5/16-in. closed-side setting and feed size should not exceed 2 in. one-way dimension.

The wobble plate feeder, although not required under most circumstances, is recommended if the feed is irregular or if the crusher is operated as a recrusher, at a relatively close setting, or in closed circuit.

§ *Fine Crushing Chamber* The fine crushing chamber has the longest parallel zone and therefore the smallest feed opening for any given size of crusher. It can be operated at ratios of eccentric throw to closed-side setting of up to 4 to 1. With a 3/4-in. throw, for example, a 236 Hydrocone crusher could be operated at 3/16 in. on the close side.

Because of their design, crushers with fine crushing chambers cannot be choke fed but must be equipped with the wobble plate feeder. The maximum one-way dimension of the feed approaches the crusher feed opening. A 236 Hydrocone crusher can be fed with material up to 2-in. one-way dimension. Approximately 75 percent of the discharged crushed product from 22- and 30-in. Hydrocones will be smaller than the closed-side setting.

SOURCE: Allis-Chalmers.

Fig. 20-22 Standard cone crusher. (*Nordberg Manufacturing Company.*)

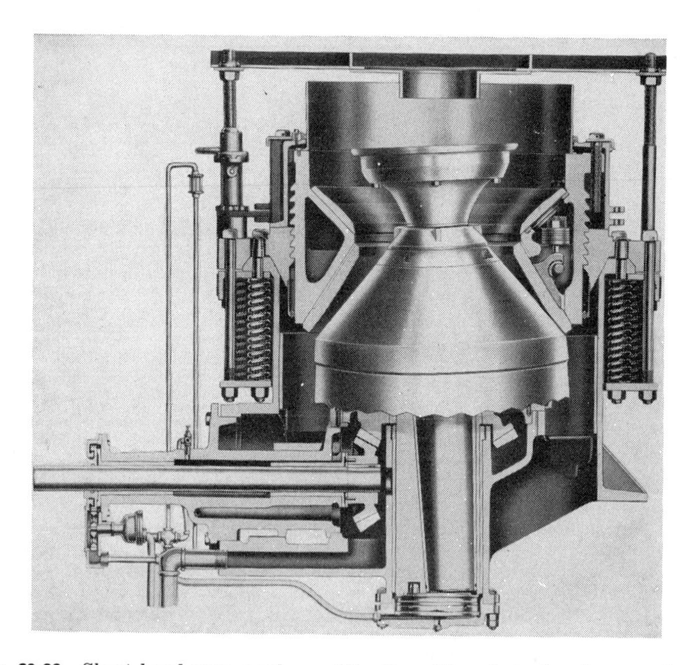

Fig. 20-23 Short-head cone crusher. (*Nordberg Manufacturing Company.*)

The cone crusher has a far greater motion of the crushing head than does the gyratory crusher (see Figs. 20-22 and 20-23). As a result, the product contains approximately 40 percent material larger than the crusher setting. The crushing cavity design varies according to the required feed size, and the gravitational movement of the material through this cavity is timed to ensure multiple crushing action on each particle.

The cone crusher can be used both in open and closed circuit, depending upon the flowsheet requirements. Tables 20-39 and 20-40 list characteristics of typical cone crushers, including their capacities in open- and closed-circuit operation. The estimated gradation of the product from a cone crusher operating in open circuit is supplied in Table 20-41. As a rule, the top size of product from a cone crusher in open-circuit operation is approximately twice the closed-side setting. In closed-circuit operation, the top size is determined by the size of the screen cloth opening, which is balanced with the crusher setting. (See Table 20-42.)

The roll crusher is in general use as a finishing crusher for all types of rock. Its ratio of reduction is determined by its shell configuration, being a maximum for serrated shells and a minimum for smooth shells (see Fig. 20-24). A frequently used double-roller combination consists of a smooth shell and a smooth beaded or corrugated shell. The diameter and spacing of the rolls determine the maximum feed size, and the roll width will then determine the capacity for a given peripheral speed. (See Tables 20-43 and 20-44.) On the newer types of roll crushers, the finger gears

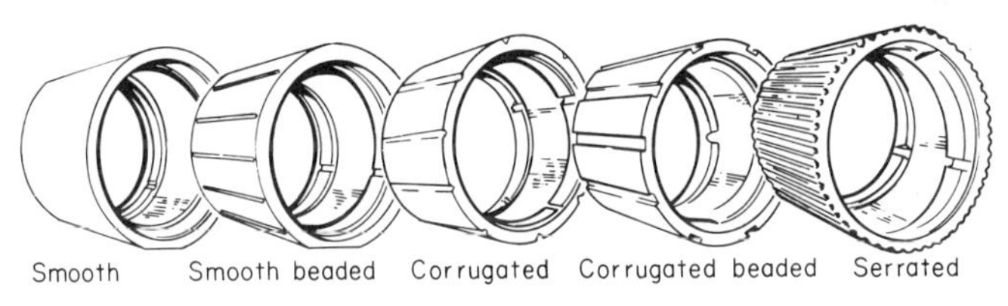

Fig. 20-24 Typical shells for roll crushers.

Fig. 20-25 Triple-roll crusher. (*Portec Inc., Pioneer Division.*)

TABLE 20-39 Capacities of Cone Crushers, Open Circuit

Size	Type of cavity	Recommended hp (electric)	Full-load rpm	Weight, lb	Recommended minimum discharge setting A	Feed opening with minimum recommended discharge setting A	
						B Closed side	B Open side
2 ft (610 mm)	Fine	30	575	9,900 (4,491 kg)	¼ in. (6 mm)	2¼ in. (57 mm)	2¾ in. (70 mm)
	Coarse				⅜ in. (10 mm)	3¼ in. (83 mm)	3¾ in. (95 mm)
	Extra coarse				½ in. (13 mm)	3½ in. (89 mm)	4 in. (102 mm)
3 ft (914 mm)	Fine	60	580	22,000 (9,979 kg)	⅜ in. (10 mm)	3⅜ in. (86 mm)	4⅛ in. (105 mm)
	Coarse				½ in. (13 mm)	4⅞ in. (124 mm)	5 11/16 in. (144 mm)
	Extra coarse				¾ in. (19 mm)	6⅞ in. (175 mm)	7½ in. (191 mm)
4 ft (1,219 mm)	Fine	100	485	37,100 (16,828 kg)	⅜ in. (10 mm)	5 in. (127 mm)	5⅝ in. (143 mm)
	Medium				½ in. (13 mm)	6⅛ in. (156 mm)	6¾ in. (171 mm)
	Coarse				¾ in. (19 mm)	7⅜ in. (187 mm)	8¼ in. (210 mm)
	Extra coarse				¾ in. (19 mm)	8 15/16 in. (227 mm)	9¾ in. (248 mm)
4¼ ft (1,295 mm)	Fine	150	485	47,100 (21,364 kg)	½ in. (13 mm)	4½ in. (114 mm)	5⅜ in. (137 mm)
	Medium				⅝ in. (16 mm)	7⅜ in. (187 mm)	8¼ in. (210 mm)
	Coarse				¾ in. (19 mm)	9 in. (229 mm)	10 in. (254 mm)
	Extra coarse				1 in. (25 mm)	10⅜ in. (264 mm)	11¼ in. (286 mm)
5 ft (1,524 mm)	Fine	200	485	91,600 (41,549 kg)	⅝ in. (16 mm)	6¾ in. (171 mm)	7½ in. (191 mm)
	Medium				¾ in. (19 mm)	8¾ in. (222 mm)	9¼ in. (235 mm)
	Coarse				⅞ in. (22 mm)	9¾ in. (248 mm)	10½ in. (267 mm)
	Extra coarse				1 in. (25 mm)	11½ in. (292 mm)	12¼ in. (311 mm)
5½ ft (1,676 mm)	Fine	200	485	92,600 (42,003 kg)	⅝ in. (16 mm)	7⅛ in. (181 mm)	7¾ in. (197 mm)
	Medium				⅞ in. (22 mm)	8⅝ in. (219 mm)	9½ in. (241 mm)
	Coarse				1 in. (25 mm)	9⅞ in. (251 mm)	10⅞ in. (276 mm)
	Extra coarse				1½ in. (38 mm)	13½ in. (343 mm)	14½ in. (368 mm)
7 ft (2,134 mm)	Fine	300	435	148,500 (67,359 kg)	¾ in. (19 mm)	10 in. (254 mm)	11 in. (279 mm)
	Medium				1 in. (25 mm)	11½ in. (292 mm)	12¾ in. (324 mm)
	Coarse				1¼ in. (32 mm)	13½ in. (343 mm)	14⅞ in. (378 mm)
	Extra coarse				1½ in. (38 mm)	16¾ in. (425 mm)	18⅛ in. (460 mm)

SOURCE: Nordberg Manufacturing Company.

have been replaced by rubber tires. Another improvement is the replacement of the compression spring with a gas-hydraulic-type spring or a nest of preloaded disk springs. The triple-roll crusher (Fig. 20-25) has two crushing stages, which increase its overall reduction in comparison with a twin-roll crusher. Table 20-45 compares the capaci ties of twin- and triple-roll crushers. Table 20-46 lists the recommended horsepower for each type and Table 20-47 supplies capacity multipliers for split feed versus a single feed to a triple-roll crusher.

To get the best possible performance from a roll crusher, the feed size should be such that there is no slippage. Where the size of feed is too large, the manganese wear will be increased out of proportion to any gain in production. Estimated product gradation for closed-circuit operation was supplied in Table 20-42, and comparable data for open-circuit operation is listed in Table 20-48. The oversize material in the product from a roll crusher can result from irregularities in the shells due to welding or overloading.

Capacities, tph, at indicated discharge setting A										
¼ in. (6 mm)	⅜ in. (10 mm)	½ in. (13 mm)	⅝ in. (16 mm)	¾ in. (19 mm)	⅞ in. (22 mm)	1 in. (25 mm)	1¼ in. (32 mm)	1½ in. (38 mm)	2 in. (51 mm)	2½ in. (64 mm)
15	20	25	30	35	40	45	50	60		
...	20	25	30	35	40	45	50	60		
...	...	25	30	35	40	45	50	60		
...	35	40	55	70	75	80				
...	...	40	55	70	75	80	85	90		
...	...	...	...	70	75	80	85	90		
...	60	80	100	120	135	150	170	180		
...	...	80	100	120	135	150	170	180		
...	...	...	...	120	135	150	170	180	185	
...	...	...	...	120	135	150	170	180	185	
...	...	100	125	140	150	175	190	220	250	
...	...	...	125	140	150	175	190	220	250	
...	...	...	...	140	150	175	190	220	250	
...	...	...	...	...	...	175	190	220	250	
...	...	...	145	175	200	230	250	275	300	
...	...	...	...	175	200	230	250	275	300	
...	...	...	...	...	200	230	250	275	300	
...	...	...	...	...	...	230	250	275	300	
...	...	...	160	200	235	275	320	365	430	450
...	...	...	...	...	235	275	320	365	430	450
...	...	...	...	...	...	275	320	365	430	450
...	...	...	...	...	...	...	...	365	430	450
...	...	...	...	370	400	500	620	750	1,000	1,050
...	...	...	...	...	400	500	620	750	1,000	1,050
...	...	...	...	...	...	...	620	750	1,000	1,050
...	...	...	...	...	...	...	...	750	1,000	1,050

Hammer mills are of two types, one with the rotor running in the same direction at all times and the other with the direction reversible (see Figs. 20-26 and 20-27). The high-speed rotor is provided with two to four rows of hammers, each swinging freely around an end fastened to the rotor. Provisions are made for adjustment by shifting the hammers outward and by reversing them. The reduction of the rock takes place by impact and by attrition, and the size of the product is controlled by the spacing of the bars in the grate bar structure. Hammer mills are normally used only in limestone or similar material with a low abrasion value. Because of the large reduction in the mill and the extensive work performed, the horsepower requirements are somewhat high. Tables 20-49 and 20-50 supply characteristics and capacities for nonreversible and reversible hammer mills, respectively. Typical product gradation from a nonreversible hammer mill is given in Table 20-51.

Tertiary and Other Crushers. Substantially the same types of crushers which are used as secondary crushers can also be used as tertiary crushers. To meet the

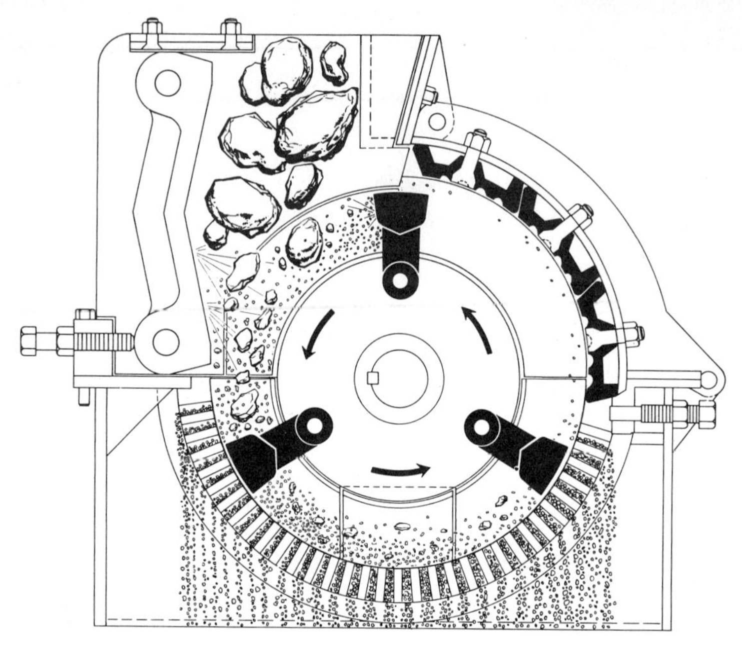

Fig. 20-26 Nonreversible hammer mill. (*Rogers Manufacturing Division, subsidiary of A. M. Byers Co.*)

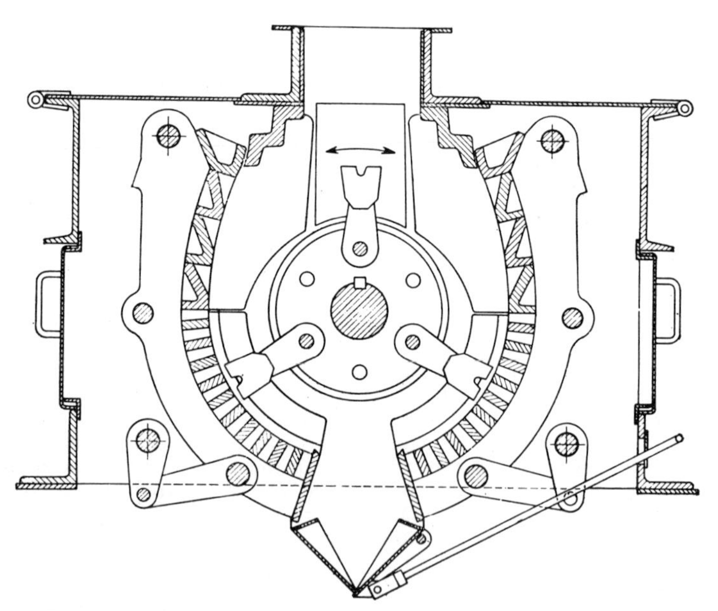

Fig. 20-27 Reversible hammer mill. (*Pennsylvania Crusher Corporation, subsidiary of Bath Industries, Inc.*)

TABLE 20-40 Capacities of Cone Crushers, Closed Circuit

Size	Type of cavity	Recommended hp (electric)	Full-load rpm	Weight, lb	Recommended minimum discharge setting C	Feed opening with min. recommended discharge setting C		Capacities, tph, at indicated discharge setting C						
						D Closed side	D Open side	⅛ in. (3 mm)	3/16 in. (5 mm)	¼ in. (6 mm)	⅜ in. (10 mm)	½ in. (13 mm)	⅝ in. (16 mm)	¾ in. (19 mm)
2 ft (610 mm)	Fine	30	575	10,100 (4,581 kg)	⅛ in. (3 mm)	¾ in. (19 mm)	1⅜ in. (35 mm)	6	8	10	14	20		
	Coarse				3/16 in. (5 mm)	1½ in. (38 mm)	2 in. (51 mm)	...	8	10	14	20		
3 ft (914 mm)	Fine	75	580	23,200 (10,523 kg)	⅛ in. (3 mm)	½ in. (13 mm)	1⅝ in. (41 mm)	15	20	30	40	50	60	65
	Medium				⅛ in. (3 mm)	1 in. (25 mm)	2 in. (51 mm)	15	20	30	40	50	60	65
	Coarse				¼ in. (6 mm)	2 in. (51 mm)	3 in. (76 mm)	...	...	30	40	50	60	65
4 ft (1,219 mm)	Fine	150	485	47,400 (21,500 kg)	⅛ in. (3 mm)	1⅛ in. (29 mm)	2½ in. (64 mm)	20	35	50	75	100	125	150
	Medium				¼ in. (6 mm)	1⅝ in. (41 mm)	3 in. (76 mm)	...	35	50	75	100	125	150
	Coarse				5/16 in. (8 mm)	2¾ in. (70 mm)	4 in. (102 mm)	...	...	...	75	100	125	150
	Extra coarse				⅝ in. (16 mm)	4¾ in. (121 mm)	5½ in. (140 mm)	...	...	...	...	...	125	150
5 ft (1,524 mm)	Fine	200	485	92,900 (42,139 kg)	3/16 in. (5 mm)	1 in. (25 mm)	2½ in. (64 mm)	...	50	75	108	142	174	206
	Medium				¼ in. (6 mm)	1¾ in. (44 mm)	3¼ in. (83 mm)	...	...	75	108	142	174	206
	Med. coarse				⅜ in. (10 mm)	2 9/16 in. (65 mm)	3⅞ in. (98 mm)	...	...	...	108	142	174	206
	Coarse				⅜ in. (10 mm)	3¼ in. (83 mm)	4⅝ in. (117 min)	...	...	...	108	142	174	206
	Extra coarse				½ in. (13 mm)	4⅛ in. (105 mm)	5⅝ in. (143 mm)	...	...	...	...	142	174	206
5½ ft (1,676 mm)	Fine	200	485	93,900 (42,593 kg)	3/16 in. (5 mm)	1⅜ in. (35 mm)	2¾ in. (70 mm)	...	65	90	135	175	210	245
	Medium				¼ in. (6 mm)	2⅛ in. (54 mm)	3½ in. (89 mm)	...	...	90	135	175	210	245
	Coarse				⅜ in. (10 mm)	3¾ in. (95 mm)	5¼ in. (133 mm)	...	...	...	135	175	210	245
	Extra coarse				½ in. (13 mm)	6 in. (152 mm)	7¼ in. (184 mm)	...	...	...	...	175	210	245
7 ft (2,134 mm)	Fine	300	435	154,600 (70,126 kg)	3/16 in. (5 mm)	2 in. (51 mm)	3¾ in. (95 mm)	...	120	160	240	300	360	420
	Medium				⅜ in. (10 mm)	3⅞ in. (98 mm)	5¾ in. (146 mm)	...	...	...	240	300	360	420
	Coarse				½ in. (13 mm)	5 in. (127 mm)	7 in. (178 mm)	...	...	...	...	300	360	420
	Extra coarse				⅝ in. (16 mm)	6 5/16 in. (160 mm)	8¼ in. (210 mm)	...	...	...	...	...	360	420

SOURCE: Nordberg Manufacturing Company.

TABLE 20-41 Estimated Percentage Product Gradation for Cone Crusher in Open Circuit

Product size, U.S. Standard	Closed-side setting of crusher, in.										
	2½	2	1½	1¼	1	⅞	¾	⅝	½	⅜	¼
+4½ in.	2										
−4½ in. + 4 in.	4										
−4 in. + 3½ in.	6	3									
−3½ in. + 3 in.	12	5									
−3 in. + 2½ in.	16	13	4								
−2½ in. + 2¼ in.	9	10	5	2							
−2¼ in. + 2 in.	7	10	7	4							
−2 in. + 1¾ in.	8	10	11	6	3						
−1¾ in. + 1½ in.	7	10	13	12	6	4					
−1½ in. + 1¼ in.	7	9	14	16	12	7	4				
−1¼ in. + 1 in.	5	8	13	17	19	17	12	6			
−1 in. + ⅞ in.	2	3	6	7	11	12	11	6	3		
−⅞ in. + ¾ in.	2	3	5	7	9	12	13	12	6		
−¾ in. + ⅝ in.	2	3	4	7	10	11	14	16	12	4	
−⅝ in. + ½ in.	11	2	4	5	8	10	12	17	19	12	
−½ in. + ⅜ in.	...	11	3	4	6	8	12	14	20	24	8
−⅜ in. + 5/16 in.	...	...	11	2	3	4	4	7	10	14	13
−5/16 in. + ¼ in.	...	...	...	11	2	3	4	5	8	12	19
− ¼ in. + No. 4	...	...	...	...	11	12	2	4	6	12	20
− No. 4 + No. 8	...	...	...	...	...	...	12	3	4	7	12
− No. 8 + No. 10	...	...	...	...	...	...	...	10	12	6	11
− No. 10 + No. 16	...	...	...	...	...	...	...	...	...	9	4
− No. 16 + No. 30	...	...	...	...	...	...	...	...	...	...	13
− No. 30											
Total %	100	100	100	100	100	100	100	100	100	100	100

SOURCE: Nordberg Manufacturing Company.

TABLE 20-42 Estimated Percentage Product Gradation, Roll Crusher and Cone Crusher in Closed Circuit

Product size, U.S. Standard	Closed-side setting of crusher or top size of product, in.			
	¾	⅝	½	⅜
+⅝ in.	4			
−⅝ in. + ½ in.	12	6		
−½ in. + ⅜ in.	24	19	8	
−⅜ in. + 5/16 in.	14	15	13	4
−5/16 in. + ¼ in.	12	16	19	12
−¼ in. + No. 4	12	15	20	24
− No. 4 + No. 8	7	11	16	24
− No. 8 + No. 10	6	7	12	21
− No. 10 + No. 16	2	3	3	4
− No. 16 + No. 30	7	8	3	4
− No. 30	...	...	6	7
Total %	100	100	100	100

TABLE 20-43 Theoretical Roll-crusher Capacities*

Roll setting (clear opening), in.	Roll width, in.															
	16		18		20		22		24		25		30		36	
	Cu yd/hr	Tph	Cu yd/hr	Tph	Cu yd/hr	Tph	Cu yd/hr	Tph	Cu yd/hr	Tph	Cu yd/hr	Tph	Cu yd/hr	Tph	Cu yd/hr	Tph
5	...	...	...	...	...	...	...	...	...	...	...	.....	1,158	1,653	1,390	1,984
4	...	...	...	...	...	...	...	...	...	...	770	1,010	924	1,245	1,109	1,494
3½	...	...	...	...	...	...	...	...	648	876	675	895	810	1,092	972	1,310
3	...	...	...	...	...	...	510	672	555	750	580	770	693	936	832	1,123
2¾	...	...	...	...	420	564	465	630	507	675	531	705	636	858	763	1,030
2½	...	...	348	468	381	513	423	573	462	612	483	639	579	780	695	936
2¼	276	375	315	423	378	468	381	516	414	558	432	585	519	702	623	842
2	246	333	279	378	309	417	339	459	369	498	384	519	462	624	554	749
1¾	216	291	246	327	270	363	297	399	324	438	339	456	405	548	486	658
1½	186	249	210	282	231	312	255	342	279	375	291	390	348	468	418	562
1	123	168	138	186	156	210	171	231	186	249	192	261	231	312	277	374
¾	92	126	105	141	117	156	126	171	138	186	144	195	174	234	209	281
½	63	84	69	93	78	105	84	114	93	126	96	129	117	156	140	187
¼	30	42	36	45	39	51	42	57	45	63	48	66	57	78	68	94

* Based on material passing the opening of the rolls and not a square screen opening. The peripheral speed of the roll used for the tabulation is 500 fpm, with a uniform flow of material and a weight of 100 pcf. For actual capacity of roll crusher, use a factor depending on type of material. For average crushing conditions, use 50 percent of values in table. For extreme crushing conditions, use 35 percent of values in table.

TABLE 20-44 Maximum Size of Feed for Roll Crushers*

Crusher setting, in.	Equipped with two smooth shells					Equipped with one smooth and one smooth-beaded shell					Equipped with two smooth-beaded shells or one smooth-beaded and one corrugated shell‡				
	Roll shell diameter, in.					Roll shell diameter, in.					Roll shell diameter, in.				
	16	24	30	40	55	16	24	30	40	55	16	24	30	40	55
1/4†	5/8	3/4	3/4	1	1										
			7/8	1 1/4	1 1/2										
3/8†	3/4	7/8	1 1/8	1 1/4	1 1/4										
				1 3/8	1 3/4										
1/2†	7/8	1 1/8	1 1/4	1 1/2	1 5/8	7/8	1 1/8	1 1/4	1 1/2	1 5/8					
					1 7/8	1 1/8	1 3/8	1 1/2	1 3/4	2					
5/8†	1	1 1/4	1 3/8	1 5/8	2	1 1/4	1 3/8	1 5/8	1 3/4	2					
							1 1/2	1 3/4	2	2 1/4					
3/4†	1 1/8	1 3/8	1 1/2	1 3/4	2 1/8	1 3/8	1 5/8	1 7/8	2	2 3/8	1 5/8	2	2 1/4	2 1/2	2 7/8
									2 1/4	2 5/8					3 1/8
7/8	1 1/4	1 1/2	1 5/8	1 7/8	2 1/4	1 1/2	1 3/4	2	2 1/4	2 3/4	1 3/4	2 1/4	2 3/8	2 3/4	3 1/4
1	1 3/8	1 5/8	1 3/4	2	2 3/8	1 3/4	2	2 1/4	2 1/2	3	2	2 3/8	2 1/2	2 7/8	3 1/2
1 1/4	1 5/8	1 7/8	2	2 1/4	2 5/8	2	2 3/8	2 1/2	2 3/4	3 1/4	2 3/8	2 3/4	2 7/8	3 1/4	3 3/4
1 1/2	1 7/8	2 1/8	2 1/4	2 1/2	2 7/8	2 3/8	2 5/8	2 3/4	3 1/8	3 5/8	2 5/8	3 1/8	3 1/4	3 5/8	4 1/8
1 3/4	2 1/4	2 3/8	2 1/2	2 3/4	3 1/4	2 3/4	3	3 1/8	3 1/2	4	3 1/4	3 1/2	3 5/8	4	4 3/4
2			2 3/4	3 1/8	3 5/8			3 1/2	4	4 1/2			4	4 1/2	5 1/4
2 1/2			3 1/4	3 5/8	4 1/8			4 1/4	4 5/8	5 1/8			4 3/4	5 1/4	5 7/8
3				4 1/4	4 3/4				5 3/8	5 7/8				6 1/8	6 5/8

* Based on regular shaped particles with relatively smooth surfaces, such as average gravel. Highly friable material, irregular shaped particles, and/or particles with relatively rough surfaces permit somewhat larger feed sizes. However, size or volume of feed must be regulated so that discharge volume does not exceed 70 percent of theoretical maximum (solid ribbon) volume, which could occur at the closer settings.

† Top row of figures is recommended size; bottom row is maximum particle. Recommended particle sizes are within the prescribed reduction ratio. Maximum particle sizes are the permissible occasional oversize particles that will be gripped under normal conditions.

‡ Shells can be with or without beads as dictated by material characteristics.

SOURCE: Iowa Manufacturing Company.

TABLE 20-45 Capacities of Twin- and Triple-roll Crushers in Open Circuit Operation*

Twin-roll crusher roll settings†

Crusher designation	¼ in.	½ in.	¾ in.	1 in.	1¼ in.	1½ in.	2 in.	2½ in.	3 in.
2416	16	31	47	63	79	94			
3024	33	66	100	133	166	200	266		
4022	34	69	103	138	172	207	276	344	414
4030	53	106	160	213	266	320	426	532	640
5424	44	87	131	175	228	262	350	437	525
5536	65	130	195	261	326	390	522	652	782

Triple-roll crusher roll settings‡

Crusher designation	¼ in.	½ in.	¾ in.	1 in.	1¼ in.	1½ in.	2 in.	2½ in.
3024	52	104	156	208	260	312		
4022	58	117	176	234	292	350	468	584
4030	79	159	238	318	398	476	636	796
5424	65	131	198	262	328	392	524	655
5536	97	195	293	391	489	586	782	977

* For closed circuit operation, use 85 percent of the values from the above table. Capacity may vary as much as ±25 percent. The capacity at a given setting is dependent on hp, slippage, type of shells, and feed size.

† Capacity in tons per hour. Based on 75 percent of theoretical ribbon of material, single feed.

‡ Capacity in tons per hour. Based on 50 percent of theoretical ribbon of material.

TABLE 20-46 Recommended Horsepower for Twin- and Triple-roll Crushers

Crusher designation	Twin-roll crusher		Triple-roll crusher		Twin roll weight, lb	Triple roll weight, lb
	Electric	Diesel, cont. hp	Electric	Diesel, cont. hp		
2416	50	75	. . .	175	8,500	
3024	125	175	150	200	11,200	15,200
4022	150	200	200	275	20,700	29,600
4030	250	325	300	400	30,700	42,900
5424	250	325	300	400	38,000	51,300
5536	350	475	450	600	. . . . . .	80,800

SOURCE: Portec Inc., Pioneer Division.

TABLE 20-47 Capacity Multipliers for Open Circuit Operation of a Triple-roll Crusher, Twin Feed versus Single Feed

Feed split ratio, coarse/fine	*Crusher capacity multiplier*
20/80	0.83
30/70	0.97
40/60	1.13
50/50	1.35
60/40	1.66
67/33	2.00
70/30	1.95
80/20	1.75
90/10	1.55

EXAMPLE: 4030 triple-roll crusher, single feed, 1-in. setting, has capacity of 318 tph. Twin feed with 60/40 feed split, the capacity is 318 × 1.66, or 528 tph.

SOURCE: Portec Inc., Pioneer Division.

TABLE 20-48 Estimated Percentage Product Gradation for Roll Crusher, Open Circuit (for closed circuit, see Table 20-42)

Product size, U.S. Standard	Crusher closed-side setting, in.																
	¼	⅜	½	⅝	¾	⅞	1	1¼	1½	1¾	2	2¼	2½	2¾	3	3½	4
+5 in.	.....	.....	.....	.....	.....	.....	.....	.....	.....	.....	.....	.....	.....	.....	4.0	13.0	22.0
−5 in. + 4 in.	.....	.....	.....	.....	.....	.....	.....	.....	.....	.....	.....	.....	5.0	10.0	12.0	13.0	13.0
−4 in. + 3½ in.	.....	.....	.....	.....	.....	.....	.....	.....	.....	.....	.....	7.0	8.0	9.0	8.0	8.0	7.0
−3½ in. + 3 in.	.....	.....	.....	.....	.....	.....	.....	.....	.....	.....	7.0	9.0	9.0	9.0	9.0	8.0	6.0
−3 in. + 2¾ in.	.....	.....	.....	.....	.....	.....	.....	.....	.....	4.0	5.0	4.5	4.5	4.5	4.5	4.0	3.5
−2¾ in. + 2½ in.	.....	.....	.....	.....	.....	.....	.....	.....	4.0	5.0	6.0	5.5	5.5	5.0	4.5	4.0	3.5
−2½ in. + 2¼ in.	.....	.....	.....	.....	.....	.....	.....	.....	3.0	6.0	6.0	5.5	5.0	4.5	4.5	3.5	3,5
−2¼ in. + 2 in.	.....	.....	.....	.....	.....	.....	.....	5.0	6.0	6.0	7.0	6.5	6.0	5.0	4.5	4.0	3.5
−2 in. + 1¾ in.	.....	.....	.....	.....	.....	.....	2.0	5.0	7.0	7.0	7.0	6.0	5.0	6.0	5.0	4.0	3.5
−1¾ in. + 1½ in.	.....	.....	.....	.....	.....	.....	6.0	8.0	9.0	10.0	7.5	7.0	7.0	6.0	5.5	4.5	4.0
−1½ in. + 1¼ in.	.....	.....	.....	.....	.....	7.0	6.0	7.0	8.5	7.0	6.5	6.0	5.0	5.0	4.5	4.0	3.5
−1¼ in. + 1 in.	.....	.....	.....	.....	10.0	11.5	13.0	13.0	10.5	9.0	8.0	7.0	6.5	6.0	5.5	5.0	4.5
−1 in. + ⅞ in.	.....	.....	.....	8.0	4.0	6.5	7.0	6.0	5.5	4.5	4.0	3.5	3.5	3.0	2.5	2.5	2.4
−⅞ in. + ¾ in.	.....	.....	8.0	6.9	11.0	10.0	11.0	9.0	7.5	7.0	5.5	5.0	4.5	4.5	4.0	3.5	2.8
−¾ in. + ⅝ in.	.....	4.0	4.0	9.0	8.0	6.6	5.5	4.5	3.8	3.5	3.3	3.0	2.7	2.5	2.5	2.0	1.8
−⅝ in. + ½ in.	.....	6.0	10.0	10.5	11.0	9.5	8.0	7.0	5.4	5.0	4.2	3.5	3.4	3.0	3.0	2.2	2.2
−½ in. + ⅜ in.	3.0	12.0	20.0	15.6	14.0	12.2	10.5	8.5	7.3	6.5	5.5	5.2	4.8	4.3	3.8	3.6	3.1
−⅜ in. + 5⁄16 in.	5.0	5.6	5.0	4.5	4.0	3.5	3.0	3.0	2.5	1.9	1.8	1.6	1.4	1.4	1.4	1.2	1.1
−5⁄16 in. + ¼ in.	13.0	11.5	10.0	8.3	6.5	5.7	5.0	4.0	3.0	2.8	2.7	2.3	2.0	2.0	1.6	1.4	1.3
−¼ in. + No. 4	20.0	15.2	10.5	9.0	7.5	6.5	5.5	5.0	4.2	3.6	3.0	2.8	2.7	2.3	2.3	2.0	1.9
− No. 4 + No. 8	26.0	20.3	14.5	12.5	10.5	9.1	7.6	6.5	5.5	4.8	4.3	3.9	3.6	3.4	3.1	2.8	2.5
− No. 8 + No. 10	5.0	3.8	2.7	2.4	2.1	2.1	1.6	1.3	1.1	1.0	0.9	0.8	0.8	0.7	0.6	0.6	0.5
− No. 10 + No. 16	11.0	8.4	5.9	5.2	4.6	4.0	3.3	2.8	2.4	2.1	1.9	1.7	1.6	1.5	1.4	1.2	1.1
− No. 16 + No. 30	7.6	5.9	4.2	3.6	3.0	2.5	2.1	1.9	1.6	1.4	1.2	1.1	1.0	1.0	1.0	0.8	0.7
− No. 30 + No. 40	2.4	1.8	1.3	1.1	0.9	0.8	0.8	0.6	0.5	0.5	0.4	0.4	0.3	0.2	0.2	0.2	0.2
− No. 40 + No. 50	1.6	1.2	0.9	0.8	0.6	0.6	0.5	0.5	0.4	0.3	0.3	0.2	0.2	0.2	0.2	0.2	0.1
− No. 50 + No. 100	2.2	1.8	1.3	1.1	1.0	0.8	0.6	0.6	0.6	0.5	0.5	0.5	0.5	0.5	0.4	0.4	0.4
− No. 100	3.2	2.5	1.7	1.5	1.3	1.1	1.0	0.8	0.7	0.6	0.5	0.5	0.5	0.5	0.5	0.4	0.4
Total %	100.0	100.0	100.0	100.0	100.0	100.0	100.0	100.0	100.0	100.0	100.0	100.0	100.0	100.0	100.0	100.0	100.0

To find percent of each size in the product of a roll crusher set at ½ in., find ½ in. at the top of the chart, follow the vertical column down until it intersects with the horizontal range size column. The first figure is ⅞ in. and 8 percent indicating that 8 percent of the material will be over ⅞ in. and 92 percent passing. The figures between the heavy lines indicate the amount of oversize that can be expected at a given setting. By following down the vertical column we find the following:

Over ½ in.:	22 percent	5⁄16 to ¼ in.:	10 percent	No. 8 to No. 10:	2.7 percent
½ to ⅜ in.:	20 percent	¼ to No. 4:	10.5 percent	to No. 10:	15.3 percent
⅜ to 5⁄16 in.:	5 percent	No. 4 to No. 8:	14.5 percent	Total:	100 percent

SOURCE: Iowa Manufacturing Company.

TABLE 20-49 Capacities of Nonreversible Hammer Mills

	Mill size—diameter by length of rotor								
	30 × 18 in.	30 × 24 in.	30 × 30 in.	30 × 36 in.	34 × 20 in.	34 × 30 in.	34 × 40 in.	48 × 54 in.	48 × 62 in.
Motor horsepower	60–125	75–150	100–200	150–250	50–100	100–150	150–250	350	400
Mill rpm	900–1,600	900–1,600	900–1,600	900–1,600	600–1,200	600–1,200	600–1,200	600–1,200	600–1,200
Feed opening, in.	8 × 18	8 × 24	8 × 30	8 × 36	15 × 22½	15 × 32½	15 × 42½	25 × 54	25 × 62
Capacity, tph:									
Agricultural lime	30	40	50	60	10–25	20–45	40–65	70–115	80–130
¼ in.	35	47	58	70	20–45	35–65	45–85	80–150	90–170
⅜ in.	40	53	67	80	30–55	45–70	55–90	95–155	110–180
½ in.	45	60	75	90	45–65	55–75	65–95	115–165	130–190
¾ in.	55	74	92	110	55–75	65–105	85–125	150–185	170–250
1 in.	65	87	108	130	55–75	80–125	105–145	185–255	210–290
1½ in.	80	107	133	160	75–95	105–145	145–185	255–325	290–370
2 in.	95	127	158	190	85–105	115–165	165–205	290–360	330–410
Weight, lb	6,500	8,500	10,500	12,500	8,500	10,200	12,950	22,500	25,250

SOURCE: Rogers Manufacturing Division, subsidiary of A. M. Byers Co.

TABLE 20-50 Capacities of Reversible Hammer Mills

Crusher designation	Capacity, tph*	Motor size, hp	Total weight, lb	Weight of rotor, lb
1–24	10–20	15–30	4,000	670
2–24	20–30	30–50	5,000	680
3–24	30–40	50–60	6,000	690
3–30	40–50	60–75	8,500	1,230
4–30	50–60	75–100	10,000	1,500
5–30	60–75	100–125	11,500	1,800
7–36	75–100	125–150	15,000	1,740
9–36	100–150	150–250	18,000	2,265
100–42	150–200	250–300	23,000	3,875
102–42	200–250	300–400	26,000	5,180
104–42	250–300	400–500	29,500	6,485
13–48	250–350	400–550	45,000	7,800
15–48	350–450	500–700	55,000	10,555
17–54	450–600	600–900	67,000	13,310

* Based on secondary reduction of average limestone or cement rock to a nominal ¾ in. and under product. For reduction of other materials, or for crushing to finer or coarser products, capacities and horsepower requirements change correspondingly.
SOURCE: Pennsylvania Crusher Corporation, subsidiary of Bath Industries, Inc.

TABLE 20-51 Typical Product Gradation for Nonreversible Hammer Mill*

Screen size, U.S. Standard	Cumulative percent passing for rotor speeds of			
	1,200 rpm			750 rpm
2 in.	100.0			
1¾ in.	99.0	.....	.....	100.0
1½ in.	97.0	100.0	.....	96.7
1¼ in.	91.0	98.0		
1 in.	83.0	93.0	100.0	86.2
¾ in.	69.0	85.0	99.0	77.7
½ in.	57.5	74.5	96.5	62.0
⅜ in.	50.0	68.0	93.0	49.4
No. 4	36.0	53.0	77.0	
No. 8	25.0	40.0	52.0	16.9

* Grate bars changed to make 1½ in., 1¼ in., ¾ in., 1½ in. top size respectively. The 1,200-rpm test was with hard limestone and the 750-rpm test was with soft limestone.

specification requirements for fine aggregate, a Gyradisc or cone-type fine crusher comes into use (see Fig. 20-28). This crusher can be used with all types of rock, operating in closed circuit with a screen. Well-graded feed of minus ¾-in. size, in no event exceeding 1½-in. maximum, will give crusher outputs[38] as listed in Tables 20-52 and 20-53. Fine aggregate specifications and requirements for agricultural liming material produced from limestone can be met by employing a high-velocity lime mill (Fig. 20-29, Tables 20-54 and 20-55).

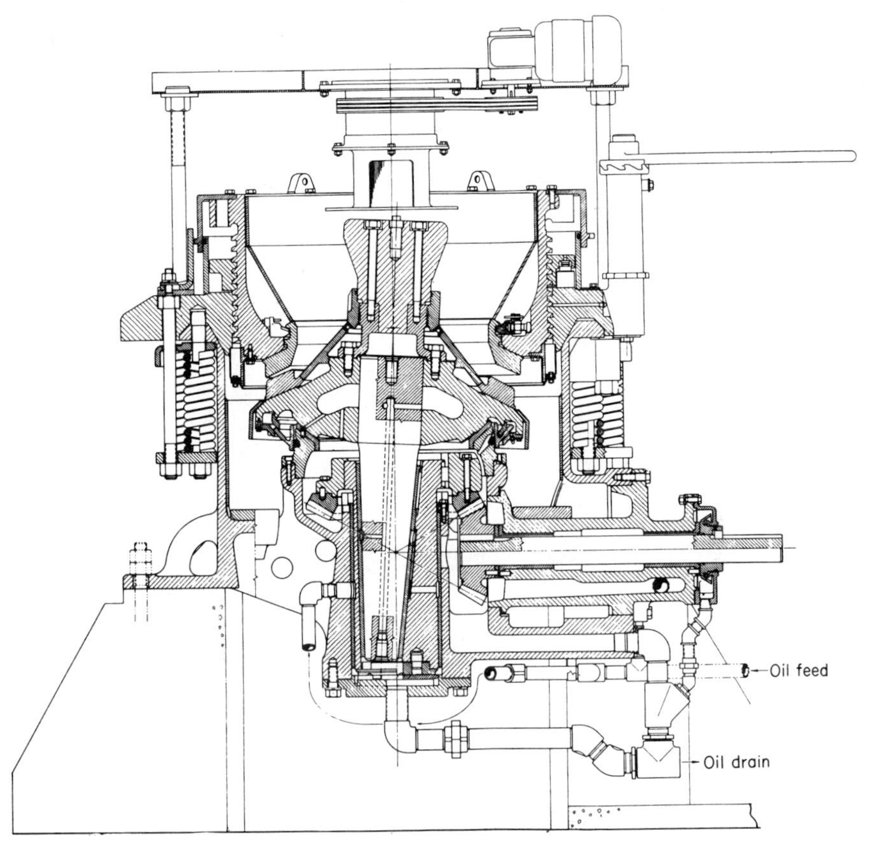

Fig. 20-28 Gyradisc crusher. (*Nordberg Manufacturing Co.*)

TABLE 20-52 Capacities of Gyradisc Crushers*

Crusher size, in.	Tph through crusher	Finished product (screen undersize), tph							
		3/8 in.	1/4 in.	No. 4	No. 6	No. 8	No. 12	No. 16	No. 20
36	75	60	50	35	30	23	18	15	10
48	120	105	80	55	45	40	30	25	17
54	140	125	90	65	50	45	35	30	20
84	280	250	180	130	100	90	70	60	40

* Capacities are based on results secured in actual practice. The figures apply to material weighing 100 pcf and a properly graded feed. A feed size of minus 3/4 in. is typical, but a minus 1½-in. feed can be readily accepted. Optimum capacity is obtained with a closed-side setting ranging between 1/4 in. and 7/16 in. The setting is not necessarily related to the closed-circuit screen opening.

SOURCE: Nordberg Manufacturing Co.

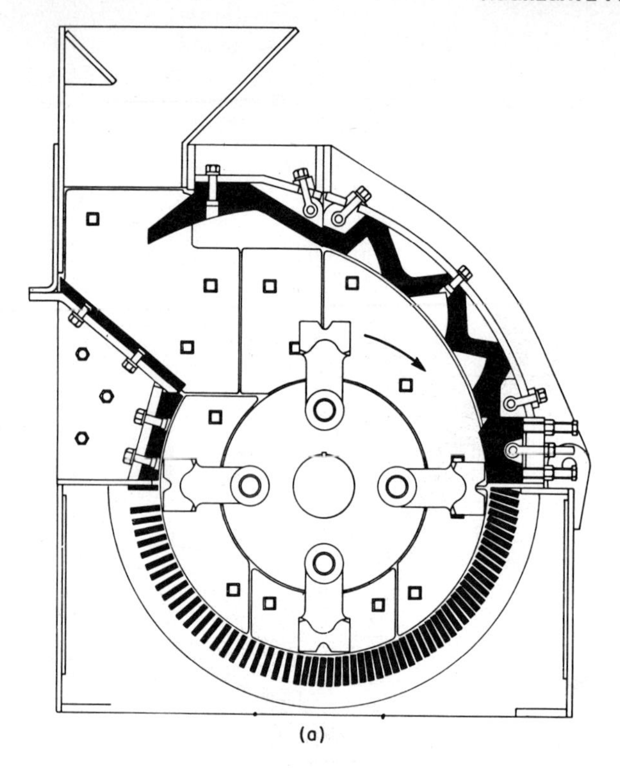

(a)

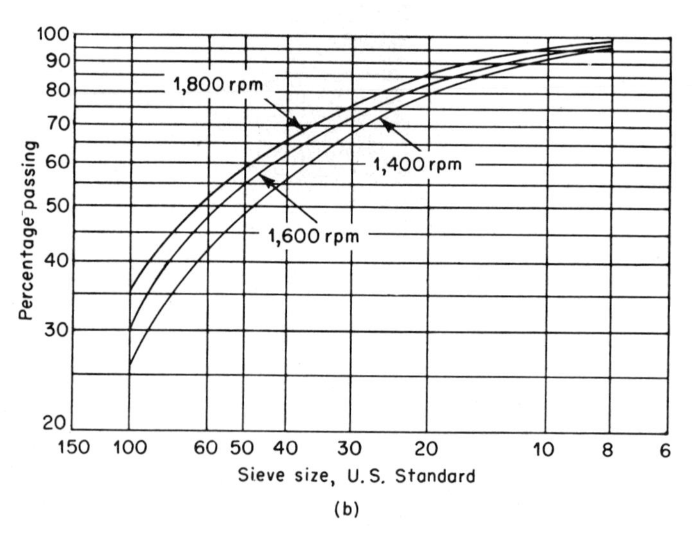

(b)

Fig. 20-29 High-velocity lime mill. (*a*) Sectional view. (*b*) Output in medium limestone. (*Iowa Manufacturing Company.*)

TABLE 20-53 Performance of 54-in. Gyradisc at Cooley Gravel Co.

Screen size, U.S. Standard	Percent passing					
	Minus No. 4 pit run sand	Gyradisc Feed	Gyradisc Crusher product	Screened through No. 10 Hummer screen	Concrete sand after blending	Specs. ASTM C-33
3/8 in.	......	100	100	......	100	100
No. 4	100	84.5	92	......	99.0	95–100
No. 8	74.5	42.6	69	100	93.0	80–100
No. 10	......	29.7	60	99.5		
No. 16	46.0	10.2	43	90.5	70.0	50–85
No. 30	27.0	4.7	28	60.5	40.6	25–60
No. 40	......	.....	21	50.2		
No. 50	11.0	3.4	18	37.0	18.0	10–30
No. 80	......	.....	12	26.5		
No. 100	1.5	.....	10	20.5	8.0	2–10
No. 200	......	.....	6	11.4	3.0	0–5
Fineness modulus	3.40	.....	3.36	1.96	2.70	

SOURCE: National Sand and Gravel Association.

TABLE 20-54 Capacities of High-velocity Lime Mills*

Model	Capacity, tph	Required hp	Mill rpm	Weight, lb
20	20–40	175–250	1,400–1,800	6,300
30	40–60	250–400	1,400–1,800	8,200
40	60–85	400–700	1,400–1,800	11,000

* Based on average condition in limestone. Recommended feed size 2 to 3 in. Product 40 to 75 percent passing 40 mesh and 20 to 50 percent passing 100 mesh.
SOURCE: Iowa Manufacturing Company.

TABLE 20-55 Capacities of "6,000" Bradley Hercules Mills*

Capacity, tph	Screen sizes, U.S. Standard				
	No. 10	No. 20	No. 6	No. 100	No. 200
30–35	...	100	76	67	55
35–40	100	90	61	51	40

* Based on an average hard and dry limestone. Feed size minus 2½ in. Power required, approximately 300 hp. Weight, approximately 62,500 lb.
SOURCE: Bradley-Fish Inc., subsidiary Bradley Pulverizer Company.

For producing specification sand, a rod mill or ball mill is used (Fig. 20-30). These units are heavy and require approximately 4 hp per ton produced per hour. The capacity will vary with the diameter and length of the unit. The control of the gradation and the fineness modulus is accomplished in wet grinding by increasing the pump dilution to obtain fewer fines. Similar control is accomplished for both wet and dry grinding by increasing the rate of flow or by reducing the rod charge. The overall metal cost is estimated to be about $0.32 per ton of sand (Table 20-56). There is normally a requirement for fineness-modulus control in manufactured sand, and in a wet operation this is accomplished by the use of a water scalping tank. For a dry operation, the control is obtained by the use of an air separator. For fine crushing and grinding, use is also made of multiple and single cage disintegrators, vertical and horizontal impactors of special construction, and grinding mills of various designs.

Fig. 20-30 Rod mill. (*The Mine and Smelter Supply Co.*)

Screens A stationary grizzly can be used to remove sizes less than the crusher setting before the feed reaches the primary crusher. The grizzly can also be a part of the feeder, in which case it will have either a vibrating motion or a reciprocating motion, depending upon the feeder type. A vibrating, heavy-duty grizzly with either one or two decks is frequently used in conjunction with an apron-type feeder. The purpose of the bottom deck, where this arrangement is used, is to remove quarry fines and dirt. The material thus removed becomes waste or, on occasion, is used as a nonspecification base material.

Scalping screens can be either inclined or horizontal. They are of the heavy-duty type and are capable of handling pit or quarry materials in sizes up to 15 in. Scalping screens are mainly single- or double-deck screens, and a rough separation is more important than a high efficiency. The finishing screen can be any one of a variety of types, ranging from revolving to centrifugal screens. The most important factor is adequate screening capacity under all circumstances. The close tolerances required by modern gradation specifications will often necessitate more screening area than is available in most aggregate processing plants.

TABLE 20-56 Center Peripheral Discharge Rod Mills

Characteristics

Mill size, ft	Approx weight without rods, lb	Capacity, tph*	Horsepower: To run	Horsepower: Of motor	Rod charge, tons†	Mill rpm
2 × 6	7,900	1–2	7–9	10	1.7	38.0
3 × 8	23,500	4–6	25–30	30	5.9	35.0
4 × 10	40,600	10–13	50–60	60	11.6	30.0
5 × 12SD	50,300	15–20	85–90	100	17.5	28.0
5 × 12	63,000	22–28	110–115	125	21.0	26.5
6 × 12	90,500	30–38	180–197	200	29.7	23.0
7 × 12	110,000	40–45	200–215	225	34.9	21.5
8 × 12	122,000	65–75	325–350	350	52.9	17.5
9 × 12	147,500	80–98	415–450	450	67.0	15.5
10 × 14	260,000	120–140	595–640	700	99.7	14.0
11 × 14	310,000	160–190	735–780	800	116.7	12.7
12 × 14	380,000	190–230	880–925	1,000	138.9	11.7

Gradation of products, percent

U.S. Standard mesh	Kaiser aggregate plant	Mt. Morris Dam	Buggs Island Dam
+ No. 4	.....	.....	0.1
– No. 4 + No. 8	.....	3.7	0.9
– No. 8 + No. 16	19	14.6	5.8
– No. 16 + No. 30	30	27.8	18.3
– No. 30 + No. 50	19	24.6	24.5
– No. 50 + No. 100	9	15.7	25.6
– No. 100	.....	.....	24.8
– No. 100 + No. 200	6	8.8	
– No. 200	17	4.8	
Total, %	100.0	100.0	100.0

* The capacities are given in dry tons per hour based on minus ¾-in. plus No. 4 screened feed of medium hard gravel. Mill discharge is generally less than 5 percent plus No. 4 in wet open-circuit operations. For dry grinding work reduce the capacities by approximately 30 to 50 percent.

† 45 percent mill volume, new liners and rods.

SOURCE: The Mine and Smelter Supply Company.

The simplest screen is the revolving screen, which in a wet operation gives a good scrubbing action as the material tumbles slowly under a water spray. Although this type of screen has a low capacity, as indicated in Table 20-57, its rugged design and low operating cost make it a favorite in many localities. Inclined screens are also in common use and are available with two different vibration mechanisms. The first of these, wherein the screen is vibrated by electromagnetic means, is particularly suited for screening very fine materials. Coarser materials can be screened more efficiently with a screen which is vibrated by mechanical means, giving it a circular

TABLE 20-57 Capacities of Revolving Screens

Capacity of screens is based on diameter and per foot of length of screen. Screen length total in increments of 4 ft as follows:

36-in. diam, 8 to 16 ft
48-in. diam, 12 to 20 ft
60-in. diam, 12 to 20 ft
72-in. diam, 16 to 32 ft

Capacities and speeds

Screen diameter, in.	36	48	60	72
Average feed, tons/hr of stone or gravel	40	85	125	150
Peripheral speed, fpm	175	190	190	190
Revolutions per minute	19	15	12	10
Horsepower per foot of length	0.65	0.8	1.0	1.4

Capacities and Lump Sizes

Screen diam, in.	Max ring size of feed, in.	Tons/hr per foot of length for stone or gravel												
		Ring size of product, in.												
		½	¾	1	1¼	1½	1¾	2	2½	3	3½	4	5	6
36	1	3.2	4.1	4.7										
	2	1.9	2.5	3.2	3.6	4.1	4.4	4.8						
	3	1.3	1.9	2.3	2.8	3.2	3.4	3.8	4.2	4.8				
	4	1.1	1.5	1.9	2.2	2.5	2.9	3.2	3.6	4.1	4.4	4.8		
48	1	4.2	5.4	6.3										
	2	2.5	3.4	4.2	4.8	5.4	5.9	6.3						
	3	1.8	2.5	3.1	3.7	4.2	4.6	5.0	5.7	6.3				
	4	1.4	2.0	2.5	3.0	3.4	3.8	4.2	4.8	5.4	5.9	6.3		
	5	1.1	1.6	2.1	2.5	2.9	3.2	3.6	4.2	4.7	5.2	5.6	6.3	
60	1	5.2	6.7	7.8										
	2	3.1	4.2	5.2	6.0	6.7	7.3	7.8						
	3	2.2	3.1	3.8	4.6	5.2	5.7	6.2	7.1	7.8				
	4	1.7	2.5	3.1	3.7	4.2	4.7	5.2	6.0	6.7	7.3	7.8		
	5	1.4	2.0	2.6	3.1	3.6	4.0	4.5	5.2	5.9	6.5	7.0	7.8	
	6	1.2	1.7	2.2	2.7	3.1	3.5	3.9	4.6	5.2	5.8	6.3	7.1	7.8
72	1	6.3	8.1	9.4										
	2	3.7	5.1	6.3	7.2	8.1	8.8	9.4						
	3	2.7	3.7	4.6	5.5	6.3	6.9	7.5	8.5	9.4				
	4	2.1	3.0	3.7	4.5	5.1	5.7	6.3	7.2	8.1	8.8	9.4		
	5	1.6	2.4	3.1	3.7	4.3	4.8	5.4	6.3	7.0	7.8	8.4	9.4	
	6	1.4	2.1	2.7	3.3	3.7	4.3	4.7	5.5	6.3	7.0	7.5	8.6	9.4

SOURCE: Link-Belt Company.

throw. The inclined screen is normally used without a feedbox, and the material to be screened is deposited directly on the screen cloth. The inclination on the screen will vary up to a maximum of 20°.

Horizontal screens can be any one of a variety of types. Their motion can be straight, substantially straight, or elliptical. They can have from one to five decks, can be with or without spray bars, and can be provided with devices to facilitate fine material screening. Figure 20-31 illustrates an auxiliary screening device which can be attached to a horizontal vibrating screen. The maximum amount of fine

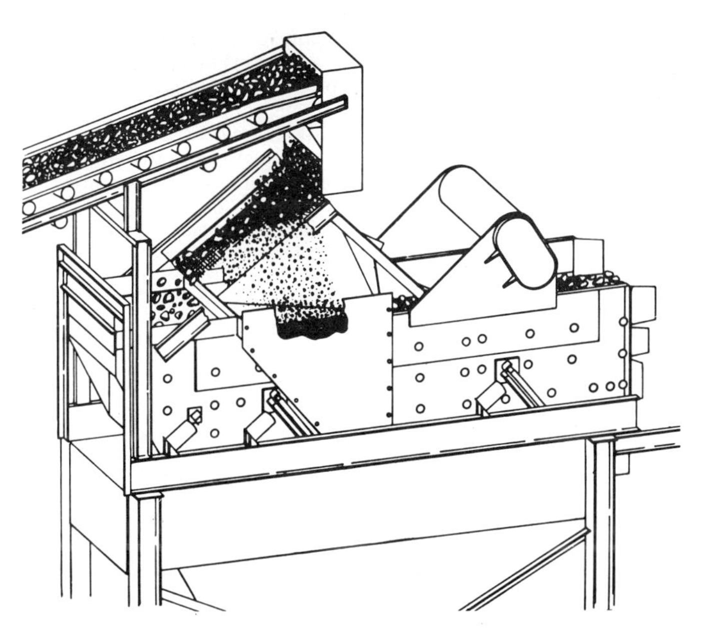

Fig. 20-31 Auxiliary screening device. Flow diagram illustrates how material falling through screen wire is diverted to chute on outside of screen box. The material can be added to material passing bottom deck or taken off separately. (*Iowa Manufacturing Company.*)

material is passed around the bottom deck of the screen, giving added effective capacity for fine material screening.

The centrifugal screen, with its vertical and cylindrical screening surface, is mainly used for fine screening from No. 8 size and down.[39] The screen capacity, based on a nominal 36 sq ft of screening surface, is given for each opening size in Table 20-58.

Screening capacity can be increased by the use of spray bars on the screens to 1.5 to 2.0 times that for dry screening. Water required for a washing operation should be at a rate of 5 to 10 gpm per ton of material per hr. As an example, 200 tph would require 1,000 to 2,000 gpm of water. In an aggregate plant where the material is

TABLE 20-58 Capacities of Centrifugal Screens (nominal 36-sq ft screening surface)*

Size separation, U.S. Standard Series	*Feed to screen, tph*
No. 8	60–80
No. 10	55–70
No. 12	50–60
No. 14	45–55
No. 18	35–40
No. 20	30–35
No. 30	20–30
No. 40	15–20
No. 60	12–15
No. 80	7–10
No. 100	5–8

* The tabulated capacities are for dry screening of materials weighing 100 pcf with approximately 50 percent oversize and 20 percent half size.

SOURCE: AIME, 1965.

brought to the screening units by pumping into a splash box above the screen, a 4- by 12-ft horizontal double-deck screen can, as a rule, handle a maximum of approximately 3,000 gpm of water and about 175 tph of sand and gravel, of which approximately 100 tph is sand. The amount of sand will be the determining factor.

The screen deck can be either punched plate, special rubber, or wire cloth. For abrasive and coarse material, rubber and punched plate have the advantage of longer life. Round, hexagonal, or square holes should be punched to the equivalent size of the testing screen used to check the gradation of the finished aggregate. (Refer to Tables 20-21 through 20-23.) Screen cloth should be selected with the finest wire diameter consistent with the anticipated screening service. The various screen users and manufacturers have agreed on the recommended wire diameters in relation

TABLE 20-59 Clear Openings, Wire Diameters, and Open Areas in Production Screens*

	Light		Standard light		Standard heavy		Heavy	
Clear opening, in.	Wire diameter, in.	Open area, percent	Wire diameter, in.	Open area, percent	Wire diameter, in.	Open area, percent	Wire diameter, in.	Open area, percent
4	1/2	79.0	5/8	74.8	3/4	70.9	1	64.0
3 1/2	7/16	79.0	1/2	76.6	5/8	72.0	3/4	67.8
3	7/16	76.2	1/2	73.5	5/8	68.5	3/4	64.0
2 3/4	3/8	77.4	7/16	74.4	1/2	71.6	5/8	66.4
2 1/2	3/8	75.6	7/16	72.4	1/2	69.4	5/8	64.0
2 1/4	3/8	73.4	7/16	70.1	1/2	66.9	5/8	61.2
2	5/16	74.8	3/8	70.9	7/16	67.3	1/2	64.0
1 3/4	5/16	71.9	3/8	67.8	7/16	64.0	1/2	60.5
1 1/2	1/4	73.4	5/16	68.5	3/8	64.0	7/16	59.9
1 3/8	1/4	71.5	5/16	66.5	3/8	61.6	7/16	57.5
1 1/4	1/4	69.4	5/16	64.0	3/8	59.2	7/16	54.8
1 1/8	0.225	69.6	1/4	67.0	5/16	61.0	3/8	55.7
1	0.225	66.6	1/4	64.0	5/16	58.0	3/8	52.9
7/8	0.207	65.3	0.225	63.3	1/4	60.5	5/16	54.3
3/4	0.192	63.4	0.207	61.4	1/4	56.3	5/16	49.8
5/8	0.177	60.7	0.192	58.5	0.225	54.0	1/4	51.0
1/2	0.162	57.1	0.177	54.5	0.192	52.2	0.207	49.8
7/16	0.148	55.8	0.162	53.2	0.177	50.7	0.192	48.3
3/8	0.135	54.1	0.148	51.4	0.162	48.7	0.177	46.1
5/16	0.120	52.2	0.135	48.8	0.148	46.0	0.162	43.4
1/4	0.105	49.6	0.120	45.6	0.135	42.2	0.148	39.4
3/16	0.080	49.1	0.092	45.1	0.120	37.2	0.135	33.8
1/8	0.054	48.7	0.072	40.2	0.092	33.4	0.105	29.5

* Based on Simplified Practice Recommendation R147-42, Wire Diameters for Mineral Aggregate Production Screens.
SOURCE: U.S. Department of Commerce.

to the job requirements and opening size (Table 20-59). It can be seen from Table 20-59 that, for a given clear opening, the percent of open area increases as the wire diameter decreases. As an example, the percentage of opening for a 1/4-in. double-crimp square-opening screen cloth with a wire diameter of 0.105 in. is 49.6 percent; for 0.120-in. wire diameter, 45.6 percent; for 0.135-in. wire diameter, 42.2 percent; and for 0.148-in. wire diameter, 39.4 percent. In other words, on a 4- by 12-ft screen deck with wire cloth having a 1/4-in.-square opening and 0.105-in. wire diameter ("light" in Table 20-59), the total free area is 23.8 sq ft. The same wire cloth with a 0.148-in. wire diameter ("heavy" in Table 20-59) has a total free area of 18.9 sq ft or only about 80 percent of that available with the fine wire screen. The increased screen efficiency from finer wires should be carefully weighed against the increased replacement of the wire cloth because of wear, in order to obtain the best results and the lowest cost.

The blinding of a screen is frequently a problem. In crushed material it is caused by elongated particles of such a shape that they hang up in the openings. Vibrating screens with sharp reversing action of the stroke will have less tendency to blind than those with circular action. Readjustment of the crusher to get less material of a size close to the screen opening will help. In fine screening, wet or dry, floating chains on screen cloth will frequently eliminate a blinding problem. The chains on flat-top screen cloth, with a 1¼- to 2½-in.-square opening and a width of 6 in. less than the deck width, should be anchored outside the screen so that they float on top of the deck. For screening of fine material such as agricultural lime, some screens are provided with a rubber cylinder or a ball deck under the screen deck (see Fig. 20-32). The screen action causes the balls to bounce against the screen cloth and thus keep it free from blinding. Using the screen wire cloth as an electric-resistance heating element and holding the wire temperature at approximately 100°F under load and 160 to 180°F under no-load conditions is another way of avoiding blinding in fine screening operations. Where moisture is present and No. 4 screen openings or smaller are involved, a stainless-steel wire cloth will stay clean longer because of its freedom from rust.

Inefficient screening can be caused by a number of factors. Unless the screening area is adequate to handle the load at all times, including surges, there will be a carry-

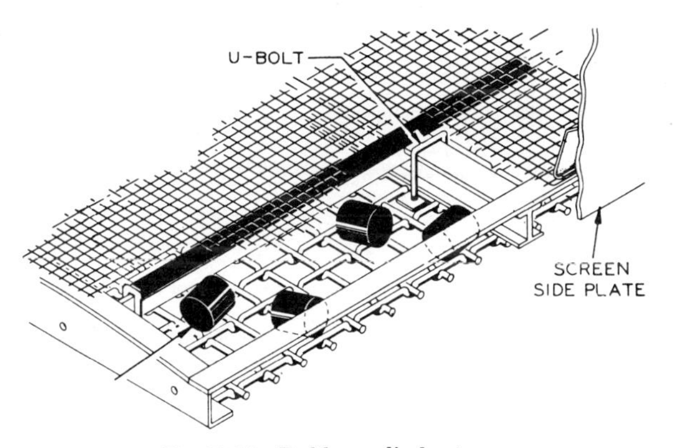

Fig. 20-32 Rubber cylinder tray.

over of undersize material. In closed-circuit operation this will increase the circulating load, throwing extra and unnecessary work on all other equipment in the circuit. In the finished aggregate screening operation, it will mean that the material will not meet specifications. The method of feeding the material onto the screen is of prime importance and, for screens without a vibrating feedbox, a rock box should be provided to spread the material properly over the full width of the top end of the screen. A vibrating feedbox as an integral part of the screen will give the best distribution and a better utilization of the screen deck as it acts as an extension of the top deck.

The capacity of a screen will vary with a number of factors inherent in its design. Screen capacity will also vary with the material type and gradation. There are two functions performed by a screen, the first of which is the stratification of the material being conveyed over the screen deck. The coarse material in the feed will move to the topmost layer, leaving the finer material at the bottom layer for better screening. Second, there is the actual separation of the material at the screen deck. To get the optimum in efficiency for the screening operation, the depth of bed at the discharge end should be less than three to six times the size of the screen-deck opening.[40] Because of the meager reports on tests of screens, empirical equations must be used to predict screen capacities, and their application must be tempered with experience and

TABLE 20-60 Screen Capacity Factors

Basic capacity B*

Screen size, U.S. Standard	B	Screen size, U.S. Standard, in.	B	Screen size, U.S. Standard, in.	B
No. 20	0.65	7/8	5.10	2 3/4	8.80
No. 10	0.99	1	5.50	2 7/8	9.00
No. 8	1.11	1 1/8	5.80	3	9.25
No. 7	1.24	1 1/4	6.10	3 1/8	9.50
No. 6	1.39	1 3/8	6.30	3 1/4	9.75
No. 5	1.57	1 1/2	6.50	3 3/8	10.00
No. 4	1.80	1 5/8	6.80	3 1/2	10.25
1/4 in.	2.50	1 3/4	7.00	3 5/8	10.50
5/16 in.	2.75	1 7/8	7.25	3 3/4	10.70
3/8 in.	3.20	2	7.50	3 7/8	10.90
7/16 in.	3.50	2 1/8	7.70	4	11.20
1/2 in.	3.80	2 1/4	7.90	4 1/8	11.40
9/16 in.	4.20	2 3/8	8.20	4 1/4	11.60
5/8 in.	4.50	2 1/2	8.40	4 3/8	11.80
11/16 in.	4.70	2 5/8	8.60	4 1/2	12.10
3/4 in.	4.80				

* Capacity in tph/sq ft fed to deck. Wire cloth has square openings with 50 percent open area; feed is 40 percent half-size, 25 percent oversize; weight of screened material is 100 pcf. Efficiency, 90 percent. If a 75 percent screening efficiency is acceptable, such as in scalping, multiply the tabulated values by 1.5.

Condition factor, K

Tenacity and/or surface moisture condition	K
Wet, muddy, or otherwise sticky rock; gravel, sand, etc.	0.75
Surface-wet quarried or mined material; material from stockpiles with surface moisture greater than 6 percent but nonhygroscopic	0.85
Dry pit-run material; dry, lumpy, crushed rock. Surface moisture less than 4 percent	1.00
Naturally dry materials, uncrushed; materials that have been dried prior to screening; or materials screened while hot	1.25

Shape factor, P

Elongated particles, percent	P	Elongated particles, percent	P
5	1.00	40	0.75
10	0.95	50	0.70
15	0.90	60	0.65
20	0.85	70	0.60
30	0.80	80	0.55

The first column represents the percentage of elongated feed particles that have more than a 3-to-1 ratio of length to width, and that have a width larger than half the aperture width but smaller than one and a half times the aperture width.

Factor S		Factor D	
Angle of incline, degrees	S	Deck	D
Horizontal	1.20	Top	1.00
5	1.15	Second	0.90
10	1.05	Third	0.80
15	1.00	Fourth	0.70
20	0.95		

Oversize factor, V; half-size factor, H

Percent of feed*	Oversize factor, V	Half-size factor, H	Percent of feed*	Oversize factor, V	Half-size factor, H
0	0.91	0.40	50	1.18	1.20
5	0.92	0.45	55	1.25	1.30
10	0.93	0.50	60	1.33	1.40
15	0.95	0.55	65	1.42	1.50
20	0.97	0.60	70	1.55	1.60
25	1.00	0.70	75	1.75	1.70
30	1.03	0.80	80	2.00	1.80
35	1.06	0.90	85	2.60	1.90
40	1.09	1.00	90	3.40	2.00
45	1.13	1.10	95	4.30	2.10

* For V factor, percent of feed not passing opening. For H factor, percent of feed less than half the screen opening size.

Slot factor, T

For square openings, the factor is 1.00. For long-slot screens and round openings, the factors are:

Slot length 6 or more times width	1.60
Slot length 3 to 6 times width	1.40
Slot length 2 to 3 times width	1.10
Round openings	0.80

Weight factor, W

For screened material weighing 100 pcf, the factor is 1.00. For other weights, factor is proportional, e.g., 0.50 for 50 pcf; 1.25 for 125 pcf, etc.

Open area factor, O

For 50 percent open area, the factor is 1.0; for 40 percent, it is 0.8; for 45 percent, 0.9; for 60 percent, 1.2; and so on.

Wet screening factor, M

Size of opening, in.	M	Size of opening, in.	M
1/32	1.25	5/16	1.9
1/16	1.5	3/8	1.75
1/8	1.75	1/2	1.5
3/16	1.9	3/4	1.0
1/4	2.0		

Use factor M when water is added to the material at a rate of 5 to 10 gpm for each tph.

NOTE: For feed sizes larger than 1/2 in., wet screening becomes much less effective. For size 20 mesh and smaller, wet screening poses problems.

field correlations. The following formula is the one most commonly used. See Table 20-60 for basic screen capacities and values of the various factors.

$$A = B \times S \times D \times V \times H \times T \times K \times P \times W \times O \times M$$

where A = Permissible total feed (oversize plus undersize) to screen deck, in tph per sq ft
B = Basic capacity of screen deck, tph per sq ft
S = Factor for angle of screen incline, measured in degrees
D = Deck factor
V, H = Oversize and half-size factors
T = Slot factor
K = Condition factor
P = Shape factor
W = Weight factor
O = Open area factor
M = Wet screening factor (omit for dry screening)

The following example covers the necessary calculations for a finishing screen in a typical screening operation:

The plant is receiving material from a gravel pit and is producing 250 tph of minus 1-in. material. Since the bottom deck of the double-deck screen is the critical one, the calculations for the top deck will be omitted. Including material returned from the secondary crusher, the feed for the bottom deck is approximately 309 tph. Size distribution for this feed is 19 percent of plus 1 in. sizes and 43 percent of minus ½ in. sizes. The material is natural rounded gravel weighing 125 pcf, with 5 percent flats from crushing and a surface moisture greater than 6 percent. Table 20-59 indicates that the open area for a standard light 1-in. square-opening screen cloth is 64 percent. Actual capacity of a square foot of horizontal screen deck is estimated as follows:

$$A = 5.50 \times 1.20 \times 0.90 \times 0.97 \times 1.06 \times 1.0 \times 0.85 \times 1.00 \times 1.25 \times 1.28 = 8.3 \text{ tph per sq ft}$$

With 309-tph feed the screening area required is approximately 37 sq ft. A 4- by 10-ft screen would be a natural choice. Should the screen cloth on the bottom deck be changed to ½-in. square opening, and assuming that 25 percent of the feed to it is of minus ¼ in. size, approximately 72 sq ft of screen area would be required. A 5- by 16-ft screen would then provide some reserve capacity for contingencies.

The experienced plant operator will keep an eye on the screen loading. As the manganese wears on the crushers, there will be a slow buildup in the circulating load to the screen. The material which is approximately the same size as the screen opening is the most difficult to separate, and its presence will affect the efficiency of the screening operation. Gradation specifications are written in such a way that a certain amount of undersize is permitted in the oversize or finished material and, to the user of the end product, the actual percentage of undersize will define the efficiency of the screening operation.

In closed-circuit operation there will be an additional screen loading to take into consideration, over and above the loading for an open-circuit operation.[41] The output from the various types of crushers will contain a certain percentage which is above the crusher setting, the amount of which will vary from approximately 20 percent for a roll crusher to approximately 40 percent for a cone crusher. Table 20-61 lists maximum values of the screen-loading rate S for a screen which receives the output from a crusher in a closed-circuit system. Here R is the circulating load expressed in percent of the primary feed to the crusher; e is the screen efficiency; O_F is the percentage of oversize in the feed; r is the percentage of oversize in the crusher product. The table clearly shows the influence of the crusher setting on the total loading to a screen in closed-circuit operation.

With the increased emphasis on compliance with gradation specifications and with the application of random sampling control, the screening operation takes on more importance. The extra cost of an ample screening area for the plant is usually a sound

TABLE 20-61 Steady-state Rates of Screen Loading S for Crusher and Screen Operating in Closed Circuit. Values of S Are Expressed as Ratios of Raw Feed Rate F

Case 1: Crusher receives the raw feed F, plus the load R recirculated from the screen; the screen receives the discharge from the crusher.

r, percent oversize in crusher product	Screen efficiency e, percent						
	100	95	90	85	80	75	70
0	1.00	1.05	1.11	1.18	1.25	1.33	1.43
5	1.05	1.11	1.17	1.24	1.32	1.40	1.50
10	1.11	1.17	1.24	1.31	1.39	1.48	1.59
15	1.18	1.24	1.31	1.38	1.47	1.57	1.68
20	1.25	1.32	1.39	1.47	1.56	1.67	1.79
25	1.33	1.40	1.48	1.57	1.67	1.78	1.91
30	1.43	1.50	1.59	1.68	1.79	1.91	2.04
35	1.54	1.62	1.71	1.81	1.92	2.05	2.20
40	1.67	1.75	1.85	1.96	2.08	2.22	2.38
45	1.82	1.91	2.02	2.14	2.27	2.42	2.60
50	2.00	2.11	2.22	2.35	2.50	2.67	2.86
60	2.50	2.63	2.78	2.94	3.13	3.33	3.57
70	3.33	3.51	3.70	3.92	4.17	4.44	4.76
80	5.00	5.26	5.56	5.88	6.25	6.67	7.14

Examples

Case 1

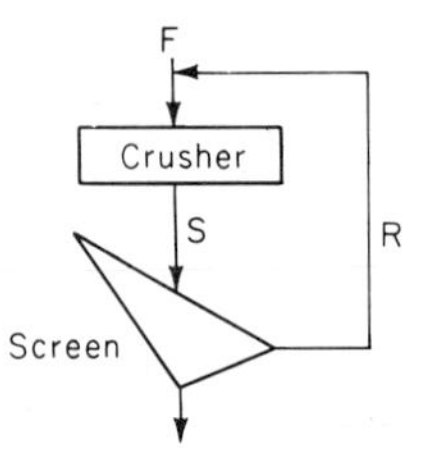

$$S = \frac{10{,}000}{e(100 - r)}$$

$$R = \frac{10{,}000}{e(100 - r)} - 1$$

e = 90 percent, screen efficiency
r = 20 percent, oversize in crusher product

$$S = \frac{10{,}000}{90(100 - 20)} = \frac{10}{7.2} = 1.39$$

Case 2

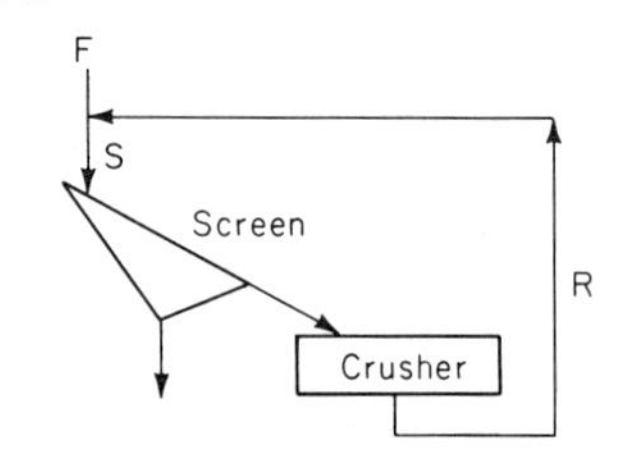

$$S = \frac{10{,}000 + e(O_F - r)}{e(100 - r)}$$

$$R = \frac{10{,}000 - e(100 - O_F)}{e(100 - r)}$$

e = 90 percent, screen efficiency
r = 20 percent, oversize in crusher product
O_F = 30 percent, oversize in raw feed F

$$S = \frac{10{,}000 + 90(30 - 20)}{90(100 - 20)} = \frac{10.9}{7.2} = 1.51$$

investment. The additional area will eliminate problems when specifications do change, while still maintaining the required efficiency.

Feeders The feeding method varies with the type of operation and with the type of primary crusher. In plants with a gyratory primary, the hauling units can "choke feed" the crusher by dumping directly on top of it. Quarry operations where a jaw crusher or impact breaker is used as a primary unit will require a feeder, which will usually be either a vibrating grizzly or the apron type. The skill of the operator will determine to a large extent the life of the feeder and the output of the primary unit. The feeder should be provided with a rock box of sufficient capacity to hold approximately two loads from the hauling units, so that rock is being dumped upon rock. End dumping is the preferable loading method, and the rock box should be designed especially for side dumping if this method is to be used. An apron feeder can be shorter for shovel-fed portable units, and it is frequently inclined up to 10° to get less of a lift for the shovel. The capacity of the feeder should be balanced with the loading and the capacity of the primary crusher (see Table 20-62).

TABLE 20-62 Capacities of Apron Feeders

Width of feeder, in.	Capacity, tph,* at speeds (fpm) of					Length of feeder, ft	Hp required at approx 35 fpm
	10	20	30	40	50		
18	20–35	35–70	55–105	70–140	90–175	3–6	3–5
24	30–60	60–120	90–180	120–240	150–300	4–8	5–7½
30	50–95	95–190	145–285	190–380	240–475	5–10	5–10
36	70–135	135–270	205–405	270–540	340–675	6–12	7½–15
42	95–185	185–370	275–555	360–720	465–925	8–14	10–20
50	130–260	260–520	390–780	520–1,040	650–1,300	12–16	15–25
60	185–370	370–740	555–1,110	740–1,480	925–1,850	16–18	20–30

* The capacity is based on material depth on feeder of approximately half the width of the feeder and will vary with the type and size of material. The operating speed is generally 30 to 40 fpm. The feeder should not be inclined more than 10° to the horizontal, with the high end toward the crusher.

The feeder illustrated in Fig. 20-33 is a vibrating type which includes a grizzly as an integral part. This grizzly is used either to bypass material around the primary crusher or to remove quarry fines and dirt. Table 20-63 lists capacities of typical vibrating grizzly feeders.

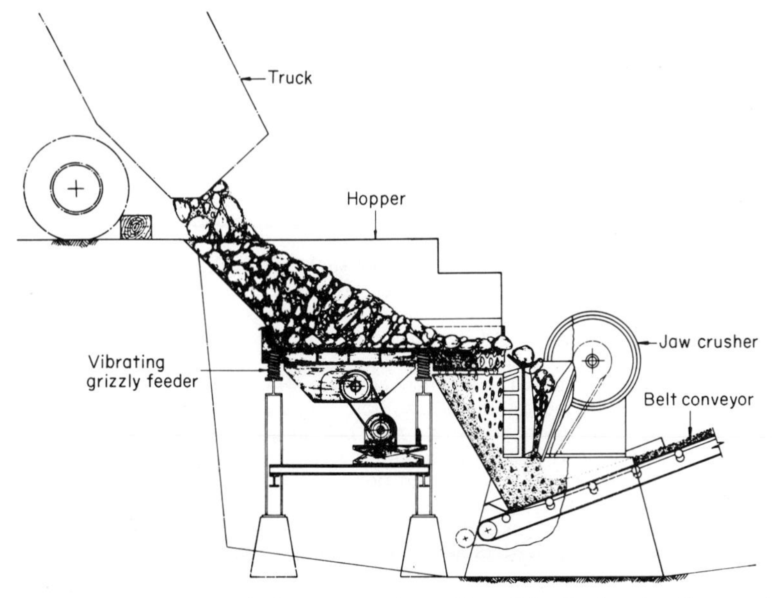

Fig. 20-33 Vibrating grizzly feeder operation. (*Simplicity Engineering Company.*)

TABLE 20-63 Capacities of Vibrating Grizzly Feeders

Model	Motor required, hp	Standard widths, in.	Standard lengths, ft	Pan depth, in.		Average capacity, tph
				Standard	Max	
OA-10-KA	1–3	12–24	4–10	6	9	50
OA-10-JA	2–5	18–36	5–18	8	12	150
OA-10-N	5–10	26–50	8–18	9	13	300
OA-10-A	7½–15	26–50	8–20	9	13	600
OA-10-MA	10–30	38–74	10–24	9	24	1,000

RATE OF TRAVEL: Installed on a horizontal plane, dry, free-flowing material will travel at approximately 30 fpm; at 5° decline, 60 fpm; at 10° decline, 120 fpm.

MATERIAL CHARACTERISTICS: Dry, free-flowing material will feed at the highest rate of travel. Adhesive or moist material will have a tendency to create friction, resulting in a lower rate of travel.

The chart below shows average capacities based on a rate of travel of 60 fpm (5° decline), and free-flowing material with a weight of 100 pcf (loose). If feeder is to be installed on a horizontal plane, reduce capacities shown by one-half. If existing space imposes a restriction on pan width, the feeder may be installed at 10° decline, which will double the capacity figures shown by the chart.

The lines on the chart represent material bed depth on the feeder pan. Capacities for material beds of 6, 8, 9, or 10 in. may be derived from those shown.

If material weight per cubic foot is, for instance, 35 lb, multiply capacities shown by 0.35; if weight is 125 lb per cu ft, multiply by 1.25.

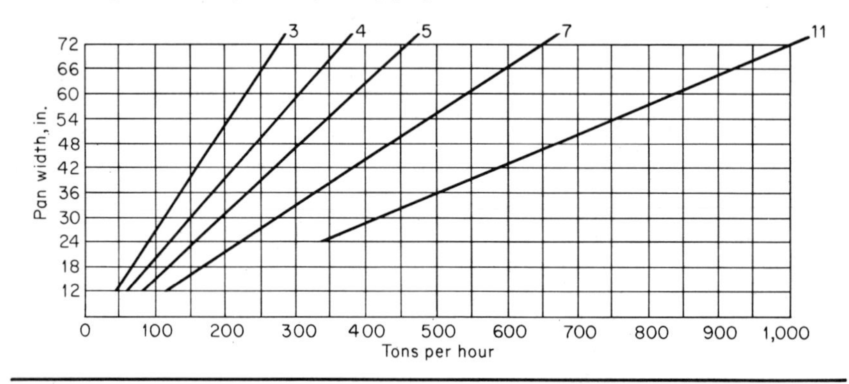

For pit operation where the material is of such a size that a reciprocating plate feeder can handle it, a trap over the feeder can be employed and dozers utilized to push the material into the trap. The feeder is usually provided with an adjustable swinging gate which is preloaded by a spring or weight to allow an occasional large rock to pass through without jamming. The gate is adjustable by a rope pull which is controlled by the operator at the primary crusher, often 50 to 100 ft away. Heavy-duty reciprocating plate feeders are used to feed primary crushers in quarry operations. The reciprocating plate feeder requires a rock box, as do the vibrating grizzly and the apron feeder, and it is the most commonly used feeder for portable tandem-type crushing plants in sand and gravel operations. Capacities of reciprocating feeders are supplied in Table 20-64.

For controlled feeding of secondary or tertiary crushers or for conveyor feeding, electromagnetic vibratory feeders are used. Table 20-65 shows characteristics and capacities for this type of feeder. To a lesser degree, belt feeders also find use for controlled feeding of finished material.

Conveyors The most frequently used type of equipment for the handling of aggregates within a plant is the belt conveyor.[42]* A belt with sufficient strength to stand the pull should be selected, and the rubber should be heavy enough to resist the impact and wear of the aggregate. The conveyor should be inclined no more than will permit the handling of the material without sliding, normally 15 to 22°. There is a definite relationship between the width of the belt and the maximum lump size of material that the belt will handle. The normal operating speed for a belt conveyor will vary between 250 to 350 fpm for an 18-in.-wide belt and 400 to 550 fpm for a 48-in.-wide belt. The wider belt operates at a higher speed.

TABLE 20-64 Capacities of Reciprocating Feeders

Width of feeder, in.	Capacity, tph*	Hp required
18	25–150	2–3
24	50–300	3–5
30	75–450	5–7½
36	100–600	7½–10

* Capacities are given for sand and gravel. Capacity for crushed rock will be approximately 75 percent of the tabulated value.

TABLE 20-65 Capacities of Electromagnetic Vibratory Feeders

Model	Size, in.	Capacity, tph*	Power consumption, watts	Shipping weight, lb
F-22	10 × 36	30	250	550
FH-22	18 × 30	50	250	600
F-33	18 × 42	65	450	1,050
FH-33	24 × 42	100	1,000	1,350
F-44	30 × 60	200	1,000	2,450
F-45	36 × 60	300	1,500	3,300
FH-45	42 × 60	425	1,500	4,300
F-55	48 × 60	600	2,000	7,400
F-66	54 × 72	800	2,000	8,400
F-86	60 × 84	1,000	2,500	9,600
F-88	72 × 84	1,250	4,600	11,000

* Capacity is based on handling damp sand or coarser material weighing 100 pcf with trough sloped 6° downgrade.

SOURCE: Syntron, a division of FMC Corp.

A belt conveyor with ample capacity should always be selected, giving full consideration to surge loads. For simplification of maintenance problems, a few standard sizes should be selected to cover all the plant handling requirements. In selecting sizes and determining horsepower, it should be remembered that there is an increased horsepower requirement whenever skirt boards are used. Belt life can be prolonged

* See Sec. 15, Belt Conveyors.

by frequent checking on alignment, and neglect of this one factor is considered to be the largest contributor to edge wear.

To prevent tramp iron from entering the crushers and causing trouble, an electronic metal detector can be used in conjunction with the belt conveyor. This device will detect nonmagnetic metal, such as manganese dipper teeth, in addition to magnetic

TABLE 20-66 Capacities of Fine-material Washers

Single-screw fine-material washer-classifier-dehydrator

Screw diameter, in.	20	20	24	24	30
Tub length, ft.	22	25	22	25	22
Capacity, tph*	30	30	50	50	75
Maximum material size, in.	$\frac{3}{8}$	$\frac{3}{8}$	$\frac{3}{8}$	$\frac{3}{8}$	$\frac{3}{8}$
Hp (normal) required (electric)†	5	5	$7\frac{1}{2}$	10	10
Water required, gpm at 25 psi	30–195	30–195	30–235	30–235	40–275
Screw speed, rpm (normal)	38	38	32	32	26
Weight of washer, lb	6,650	7,350	6,100	6,800	8,300
Loaded weight, lb	16,850	18,000	17,850	18,900	23,750

Screw diameter, in.	30	36	44	54	66
Tub length, ft.	25	25	32	34	35
Capacity, tph*	75	100	175	275	400
Maximum material size, in.	$\frac{3}{8}$	$\frac{3}{8}$	$\frac{3}{8}$	$\frac{3}{8}$	$\frac{3}{8}$
Hp (normal) required (electric)†	15	15	25	40	60
Water required, gpm at 25 psi	40–275	40–360	60–740	60–950	60–1,350
Screw speed, rpm (normal)	26	21	17	14	11
Weight of washer, lb	9,000	10,400	17,900	27,500	40,200
Loaded weight, lb	24,900	29,800	69,900	86,500	131,000

Double-screw fine-material washer-classifier-dehydrator

Diameter of flights, in.	36	44	54	66
Tub length, ft.	25	32	34	35
Capacity, tph*	200	350	550	800
Maximum material size, in.	$\frac{3}{8}$	$\frac{3}{8}$	$\frac{3}{8}$	$\frac{3}{8}$
Hp (normal) required (electric)†	30	50	75	120‡
Water required, gpm at 25 psi	105–620	125–1,400	125–1,700	125–2,050
Screw speed, rpm (normal)	21	17	14	11
Weight of washer, lb	17,650	34,400	55,200	77,500
Loaded weight, lb	54,150	123,200	164,200	249,500

* Capacity ratings are stockpiled materials and are to be used as a guide only. They are based on average amount of fines in material feeds. Reduced screw speed and capacity require less horsepower.

† Horsepower requirements listed are for electric motors. When replacing with gasoline or diesel engine, multiply the electric motor horsepower by $1\frac{1}{2}$ and select an engine with a corresponding continuous horsepower—if continuous horsepower is not given, use 80 percent of maximum horsepower given for the engine.

‡ Two 60-hp motors.

SOURCE: Eagle Iron Works.

metal such as alloy drill bits. The unit is sensitive and can be adjusted to the belt speed.

Washing Equipment In order to meet specification requirements for fine and coarse aggregate, it is often necessary to include a washing operation. In addition, the fineness-modulus control requirement can best be handled this way.[43] Where the pit material is such that there is a hump in its gradation curve, the surplus size or sizes can be removed by splitting the material and properly reblending it. The same concept holds true for manufactured sand, where the terminal velocity or settling speed of the sand particles in relation to the velocity of the overflow of water will determine the size retained. The capacities of fine-materials washers, based on an average sand, are indicated in Table 20-66. Where the percentage of material passing the No. 50 sieve increases, it is necessary to slow down the speed of the screw. Compensation for the resulting decrease in capacity and water-handling ability must then be made in the unit size selection (see Fig. 20-34 and Table 20-67).

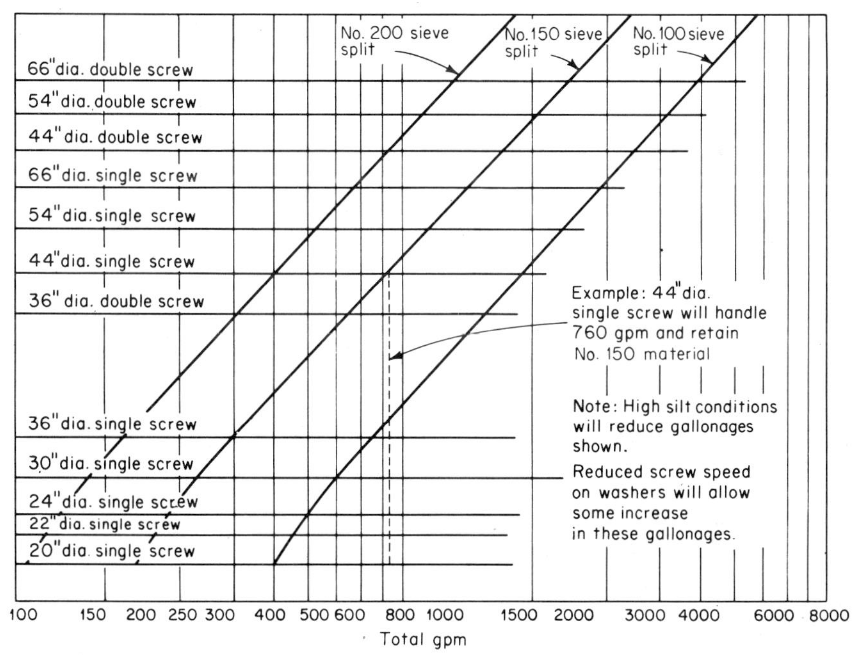

Fig. 20-34 Capacities of fine-material washers in relation to water volume. Select a diameter, either single or double screw, along the vertical edges of the chart: use the horizontal line beneath the diameter selected to enter the chart. Follow this horizontal line across the chart until it intersects a diagonal line which sets forth the desired mesh of the retained product. Drop a vertical line from this intersection and read the water capacity in gallons per minute.

A simple way to handle a large volume of water and to control the gradation is to add a water scalping tank to the plant equipment. Table 20-68 lists capacities of such tanks. As an example, a 10-in. pump will discharge to a screen about 3,000 gpm of water with 175 tph of sand and gravel. Without provisions for handling the volume of water, the minus No. 100 material will be lost. Coarse material from the upper decks of the screen can be further washed and dewatered in a coarse-material washer (see Table 20-69). Fines in overflow water can be recovered in a cyclone (see Tables 20-70 and 20-71).

TABLE 20-67 Recommended Screw Speed Reductions for Fine-material Washers

Identification of material	Percent passing U.S. Standard sieve size			Screw speed reduction	
	No. 50	No. 100	No. 150	Percent of normal	Resultant peripheral, fpm
No. 1	15	2	0	100	200
No. 2	20	5	0	75	150
No. 3	30	10	3	50	100
No. 4	40	20	7	25	50
No. 5	85	40	20	16	32

SOURCE: Eagle Iron Works.

TABLE 20-68 Capacities of Water Scalping Tanks

Length of single tank, ft	Type of tank	Maximum allowable gpm of slurry at low silt content to save fine sand retained on U.S. Standard sieve size			Loaded weight, lb, with 3-cell flume	Loaded weight, lb, with 3-cell lined flume
		No. 100	No. 150	No. 200		
20 ft 0 in.	Water Mizer	1,150	550	325	28,200	30,700
20 ft 0 in.	Standard	2,300	1,200	700	55,150	57,300
24 ft 0 in.	Standard	2,800	1,400	800	66,600	69,150
28 ft 0 in.	Standard	3,200	1,600	900	77,900	81,000
24 ft 0 in.	Wide	3,500	1,800	950	91,950	93,900
28 ft 0 in.	Wide	4,100	2,100	1,100	108,150	110,400
32 ft 0 in.	Wide	4,700	2,400	1,250	124,800	127,450
36 ft 0 in.	Wide	5,300	2,700	1,400	141,000	144,050
40 ft 0 in.	Wide	5,900	3,000	1,550	157,100	160,550
Length of double tank, ft	**Type of tank**	**Maximum allowable gpm of slurry at low silt content to save fine sand retained on U.S. Standard sieve size**			**Loaded weight, lb, with 3-cell flume**	**Loaded weight, lb, with 3-cell lined flume**
		No. 100	**No. 150**	**No. 200**		
28 ft 0 in.	Standard	6,500	3,200	1,800	156,400	162,600
24 ft 0 in.	Wide	7,000	3,600	1,900	184,100	188,700
28 ft 0 in.	Wide	8,200	4,200	2,200	226,050	231,400
32 ft 0 in.	Wide	9,400	4,800	2,500	249,150	255,250
36 ft 0 in.	Wide	10,600	5,400	2,800	281,150	288,050
40 ft 0 in.	Wide	11,800	6,000	3,100	313,500	320,700

SOURCE: Eagle Iron Works.

Where clay or other unwanted constituents are present, special washers are required. These take such forms as revolving scrubbers, combinations of washers with revolving or vibrating screens, or paddle log washers. The capacity of a log washer will vary with the type and size of material. The type and toughness of the material to be removed from the aggregate will determine the length of the unit. The slope of the log washer, which can be adjusted from level to 3 in. per ft, will also change the retention time and hence the performance (see Table 20-72).

TABLE 20-69 Capacities of Coarse-material Washers

Single-screw coarse-material washer-dewaterer

Screw diam, in.	22	22	24	24	30	36
Tub length, ft	15	18	15	18	18	18
Capacity tph	45–55	45–55	60–75	60–75	100–125	150–175
Max. material size, in.	2	2	2	2	2½	2½
Hp required (electric)*	10	15	15	15	20	30
Water required (gpm at 25 psi)	250–350	250–350	300–400	300–400	350–450	400–600
Screw speed, rpm	40	40	40	40	35	32
Weight of washer, lb	5,250	6,050	6,150	7,100	9,450	12,150
Loaded weight, lb	9,550	10,650	13,850	15,350	22,250	30,350

Double-screw coarse-material washer-dewaterer

Screw diam, in.	30	36
Tub length, ft	18	18
Capacity, tph	200–250	300–350
Maximum material size, in.	2½	2½
Hp required (electric)*	40	50
Water required (gpm at 25 psi)	600–800	700–900
Screw speed, rpm	35	32
Weight, lb	15,750	19,200
Loaded weight, lb	33,950	47,000

* Power requirements are based on electric motors and on handling gravel. Ample horsepower is provided to handle the maximum-sized material set forth in the specifications. If the washer is handling a high percentage of this maximum-sized material, or if the washer is handling crushed stone or ore, the capacity of the unit will be reduced from that set forth in specifications. Installation of paddles in place of screw flights will also reduce capacity slightly, and if more than six sets of paddles are employed on each conveyor shaft, the larger required motors will be used. When replacing the electric motor with gasoline or diesel engine, multiply the electric motor horsepower by 1½ and select an engine with a corresponding continuous horsepower—if continuous horsepower is not given, use 80 percent of maximum horsepower given for the engine.

SOURCE: Eagle Iron Works.

Special Equipment For removal of undesirable lightweight material in an aggregate, such as shale and other deleterious material, a heavy-media operation is effective.[44] The process utilizes water plus a mixture of magnetite and ferrosilicon, with an average specific gravity of approximately 2.5. Material with specific gravity below this value will therefore float when placed in the mixture. The required equipment includes launderers, rinsing screens, and magnetic separators. Acceptable aggregate has been

TABLE 20-70 Capacities of Sand Cyclones

	10 in.	14 in.	21 in.
Max valve opening, in.	2	3	4
Inlet diameter, in.	3	4	6
Overflow, in.	4	6	8
Gpm	100–300	150–450	300–1,000
Tph (max)	10	35	90
Separation (U.S. Std. sieve size)	100–270	50–200	30–150
Empty weight, lb	400	760	1,400

SOURCE: Eagle Iron Works.

TABLE 20-71 Capacities of Sand Clones

Series	*gpm*
1000	Up to 100
1500	100–400
2000	400–600
2120	600–800
2130	800–1,000
3000	700–2,100
3100	1,700–2,500

SOURCE: Process Machinery Division, Arthur G. McKee & Company.

TABLE 20-72 Capacities of Log Washers

12-in. logs			10-in. logs		
Length, ft	Double weight, lb	Hp	Length, ft	Double weight, lb	Hp
25	42,500	75	20	23,000	50
30	47,500	100	25	27,000	60
35	52,500	125	30	31,000	75

Length of washer depends on amount and toughness of clay. Maximum size of feed, 4 in.

12-in. Washmaster log washer capacity: sand and gravel, 150 tph; crushed limestone, 100 tph.

10-in. Washmaster log washer capacity: sand and gravel, 125 tph; crushed limestone, 80 tph.

Water required: 50 to 500 gpm at 25 psi.

Standard operating speed, 28 rpm.

SOURCE: McLanahan Corporation.

produced in heavy-media plants with reported costs of operation (less preliminary crushing and screening) of 8 cents per ton in 1966. This cost is based on an average plant capacity of 100 tph. (See Fig. 20-35.)

For dry control of fines in fine aggregates, an air separator of the type illustrated in Fig. 20-36 comes into use. The separation process can control fineness modulus to some extent and can remove fines for agricultural lime or mineral fillers. Capacities of typical air separators are supplied in Table 20-73.

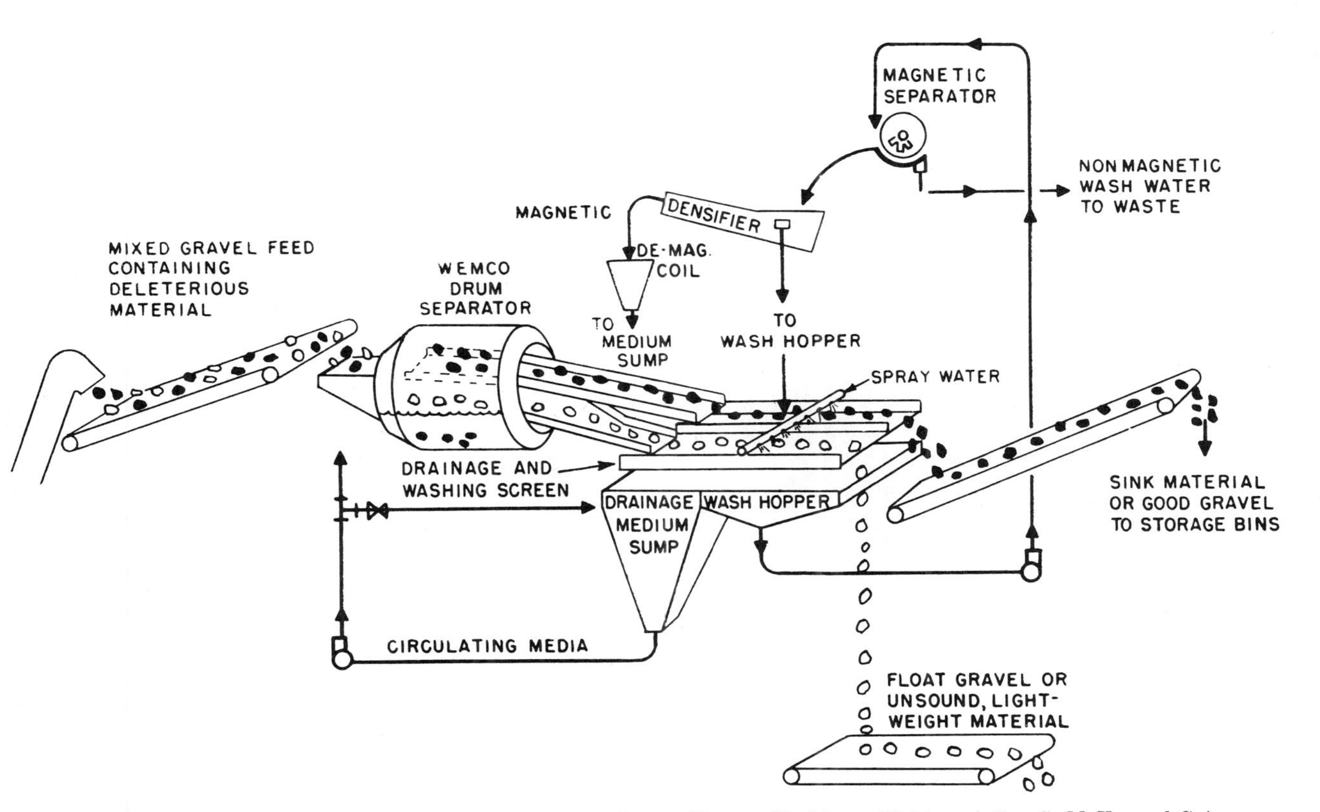

Fig. 20-35 Diagram of heavy media separation plant. (*Process Machinery Division, Arthur G. McKee and Co.*)

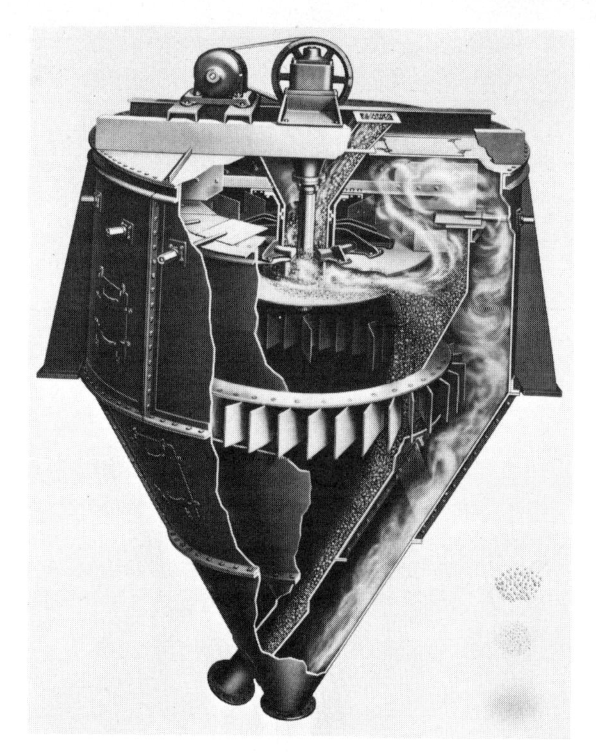

Fig. 20-36 Whirlwind air separator. (*Sturtevant Mill Co.*)

TABLE 20-73 Capacities of Whirlwind Air Separators

Size, ft	Approx. motor hp (3 phase, 60 cycle)	Motor rpm	Drive shaft rpm	Approx. intake, tph	Domestic shipping weight, lb
3	7½	1,800	1,100	up to 3	1,000
4½	10	1,800	590	up to 6	1,900
6	15	1,200	600	up to 15	3,800
8	20	1,200	575	up to 40	9,500
10	30	1,200	550	up to 80	13,000
12	40	1,200	600	up to 120	18,500
14	75	1,200	590	up to 160	21,000
16	125/150	1,200	585	up to 400	31,000
18	250/300	1,200	510	up to 700	48,000
20	400/500	1,200	514	up to 1,400	70,000

SOURCE: Sturtevant Mill Company.

PLANT FLOWSHEETS

Regardless of the flowsheet set up for the processing of aggregates, the plant is at all times a combination of individual and known units. The combination of these units will vary from location to location, since no two quarries are alike and only rarely will two sand and gravel pits have the same pit analysis.

In general, portable plants will have capacities of up to 1,200 tph and will include the same types of units as will stationary plants. Innumerable combinations are possible in the usage of portable plant units. A simple sand and gravel plant, as an example, could consist of a reciprocating feeder, a feed conveyor, a screen, a jaw crusher, a roll crusher, an under-crusher conveyor, and necessary products conveyors; all these units would be combined on a structural-steel chassis with rubber-tired wheels.

Fig. 20-37 Commander portable crushing and screening plant. (*Iowa Manufacturing Co.*)

TABLE 20-74 Capacities of Commander Portable Crushing Plants

Model	Jaw crusher	Roll crusher	Screen size	Capacity, tph,* when crushing to a top size of					
				1½ in.	1¼ in.	1 in.	¾ in.	⅝ in.	½ in.
111	1016SJ	1616	30 in. × 10 ft	....	....	60–96	44–84	36–72	28–60
211	1216TJ	1616	30 in. × 10 ft	....	....	80–120	60–104	44–88	32–72
322	1024SJ	2416	42 in. × 10 ft	172–280	140–240	120–192	92–144	64–112	44–84
332	1024SJ	3018	42 in. × 10 ft	192–300	160–260	136–212	104–160	76–132	48–96
422	1036SJ	2416	42 in. × 10 ft	224–360	184–324	144–252	84–168	68–140	56–112
432	1036SJ	3018	42 in. × 10 ft	248–400	204–360	160–280	100–184	80–152	68–124
443	1036SJ	3025	48 in. × 10 ft	300–520	248–416	200–320	160–264	120–220	100–172
522	1236SJ	2416	42 in. × 10 ft	224–360	184–324	144–252	84–168	68–140	56–112
532	1236SJ	3018	42 in. × 10 ft	248–400	204–360	160–280	100–184	80–150	68–124
543	1236SJ	3025	48 in. × 10 ft	300–520	248–416	200–320	160–264	120–220	100–172
544	1236SJ	3025	48 in. × 12 ft	300–520	248–416	200–320	160–264	120–220	100–172
555	1236SJ	3030	48 in. × 14 ft	400–620	320–480	240–360	200–300	160–240	120–180
565	1236SJ	4026	48 in. × 14 ft	360–600	292–456	224–344	185–285	145–228	108–176
575	1236SJ	4130	48 in. × 14 ft	460–700	365–550	275–410	220–340	184–275	135–195
645	1236TJ	3025	48 in. × 14 ft	480–700	380–540	280–400	232–332	184–264	140–200
765	1636SJ	4026	48 in. × 14 ft	360–600	292–456	224–344	185–285	145–228	108–176
855	1242SJ	3030	48 in. × 14 ft	375–650	325–475	250–375	220–300	160–250	120–190
865	1242SJ	4026	48 in. × 14 ft	400–650	340–500	270–380	230–310	170–252	128–182

* All capacities listed are approximate and based on normal crushing conditions where 25 percent of the material coming from the pit will be crushed and material weighing 100 pcf. Capacities must be based on pit analysis, type, size, and hardness of material and even distribution of material between the jaw crusher and roll crusher. To find approximate capacity when 50 percent of the material is to be crushed, divide any above figure by 2. For 100 percent, divide by 4.

SOURCE: Iowa Manufacturing Company.

A typical portable plant is illustrated in Fig. 20-37. Such a plant configuration can be set up to handle either small or large capacities, as indicated in Table 20-74.

Approximately 68 percent of all aggregate processing plants are of the stationary type. By referring to Tables 20-1 and 20-2, it can be seen that annual production is less than 400,000 tons for the majority of these plants. However, the present trend is in the direction of larger plants with increased capacities. In developing the flowsheet for a stationary plant, consideration should be given to possible expansion and to the handling of contingencies. A portable plant is an excellent way of supplementing the capabilities of a stationary plant and could also be desirable when it

TABLE 20-75 Preliminary Cost of Aggregate Plant

Quantity	Description	Price
	Basic portable plant—300 tph	
1	35- by 46-in. primary jaw-crusher unit, scalping screen, 48 in. delivery conveyor, diesel power and dolly	$ 135,000
1	Ground-level heavy-duty feeder, 48 in. × 45 ft	36,000
7	5- by 16-ft screen units at $25,000	175,000
2	36-in. by 50-ft conveyors at $8,000	16,000
23	24-in. by 50-ft conveyors at $6,000	138,000
1	5530 roll crusher with delivery conveyor and power or 4030 roll (minus $7,000)	72,000
1	4¼ standard unit with delivery conveyor and power	80,000
1	4-ft scale house unit with delivery conveyor and power	80,000
2	Portable tunnels with feeders	20,000
1	Portable screw	14,000
	Subtotal	$ 766,000
	Erection of individual units at 1 percent	8,000
	Total	$ 744,000
	Other items to make basic 300-tph portable plant operational	
1	Fresh water supply and sludge disposal system with necessary pumps and pipe (swag)	$ 50,000
1	Electrical system—includes all necessary starters, wiring, trailers, etc., less generators	90,000
1	Site preparation	10,000
1	Foundations	5,000
1	Freight	30,000
1	Office trailer and storage trailer and scale house	15,000
1	Scale	8,000
	Subtotal	$ 208,000
	Miscellaneous plus contingencies at 25 percent of variables	52,000
	Total	$ 260,000
	Mobile equipment	
2	6-yd front-end loader	$ 130,000
1	4-yd front-end loader or 6-yd front-end loader (plus $20,000)	45,000
2	35-ton quarry trucks at $65,000	130,000
1	25-truck crane	50,000
1	1½-yd self-propelled rubber-tired crane	40,000
2	Flat-bed trailers at $6,500	13,000
3	Generator capacity	90,000
		$ 498,000
	Summary	
	Basic portable plant at 300 tph	774,000
	Other items	260,000
	Mobile equipment	498,000
	Grand total	$1,532,000

SOURCE: AIME, 1968.

becomes necessary to produce an aggregate size which is not presently made by the stationary plant. A portable plant is also useful to balance the production of various sizes or to test a market for a new product.[45,46]

To complete the layout and arrangement for an aggregate plant, whether stationary or portable, consideration should be given to the traffic pattern. The flow of traffic will be determined by the location of the source of material in relation to the plant proper as well as by the location of bins, stockpiles, and scale house. Having a well-arranged site with adequate loading equipment will eliminate bottlenecks and decrease the handling cost. It is also important to the overall public relations program.

For a general consideration of the investment involved, the cost of the crushing-plant equipment will vary between $1,500 to $2,500 per tph of plant capacity. In addition there will be the cost of the plant site itself and of the necessary drills, shovels, draglines, and other equipment for hauling, loading, and auxiliary purposes. The final investment may well be double the cost of the plant units. Frequently the raw material is obtained on a royalty basis, which may vary from 7 to 15 cents per ton of finished aggregate over the scale. Such an investment justifies careful checking and sample taking prior to proceeding with the plant development.

Many problems are encountered by the producer before an aggregate processing plant is finally producing material and earning a return on its investment. It is rare to find a published report of actual problems and costs, but the record supplied as Table 20-75 indicates a total cost of $1,532,000 for a 300-tph stationary and movable plant.[47] This total was approximately twice the cost of the plant itself.

The following paragraphs cover the analysis of some typical plant flowsheets.

Portable Sand and Gravel Plant, 250 tph A highway job calls for 100,000 tons of selected subbase, Michigan State Specification 24A. This quantity is to be produced in forty 10-hr working days. There is no permanent market for this particular material, but there are other job locations, so the plant will have to be portable. To meet the requirement, its capacity should be 250 tph minimum. (See Fig. 20-38.)

The pit analysis shows approximately 60 percent gravel and 40 percent sand, with gradation as follows:

Product size, U.S. Standard	*Percent*
−8 in.	100
−8 in. + 3 in.	6.5
−3 in. + 1½ in.	15.5
−1½ in. + ¾ in.	18.0
−¾ in. + No. 4	19.3
− No. 4 + No. 8	5.7
− No. 8 + No. 30	15.0
− No. 30	20.0

The specification calls for 100 percent minus 1-in. material. Therefore, the material to be crushed is the plus 1-in. material which, based on the pit analysis, is approximately 34 percent of 250 tph, or 85 tph. With a top size of 8 in., a typical two-crusher portable plant can be used. Table 20-74 identifies a No. 645 tandem plant with a capacity of 280 to 400 tph of minus 1-in. material based on a 25 percent crushing requirement. Should 100 percent of the material require crushing, the plant capacity would be in the 70- to 100-tph range. Since only 34 percent or 85 tph of crushing is actually required, with a production capacity of 250 tph, it appears that this plant has the desired capacity. The plant units consist of a 1236 twin-jaw crusher (Table 20-34), a 3025 roll crusher (Table 20-45), and a 4- by 14-ft horizontal screen. The jaw setting will be 1¼ in., the roll setting will be ¾ in., and the screen will be fitted with 2-in. square-opening screen cloth on its top deck and with 1-in. cloth on its bottom deck. Standard heavy wire will be employed for the screen cloth, resulting in 67.3 percent open area for the 2-in. mesh and 58.0 percent open area for the 1-in. mesh (Table 20-59). The pit gradation can then be rewritten to show the size distribution in the feed and in the output from each processing unit. Size percentages for the final product are also listed. Pertinent data have been taken from Tables 20-33 and 20-42.

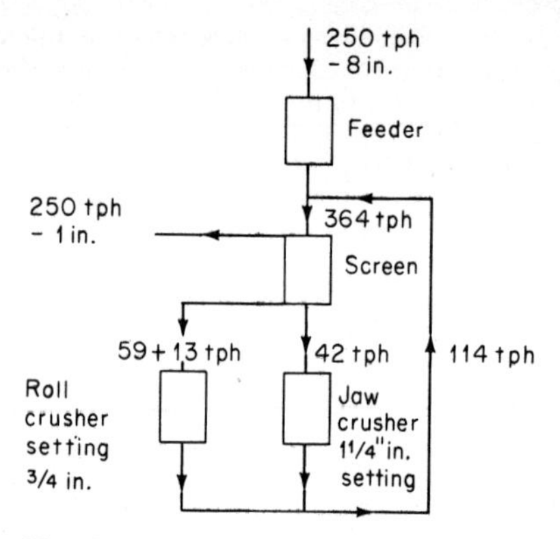

Fig. 20-38 Flowsheet, portable sand and gravel plant, 250 tph.

Size range	Total feed		Jaw, tph	Roll, tph	Total product	
	Percent	Tph			Tph	Percent
+2 in.	16.8	42.0				
−2 in. + 1 in.	17.2	43.0	16.0			
−1 in. + ⅜ in.	18.9	47.5	14.7	23.6	85.8	34.3
−⅜ in. + No. 8	12.1	30.3	7.8	26.5	64.6	25.8
−No. 8	35.0	87.2	3.5	8.9	99.6	39.9
	100.0	250.0	42.0	59.0	250.0	100.0

Sieve size, U.S. Standard	Cumulative passing, percent	
	Product	Specification
1 in.	100.0	100
⅜ in.	65.7	60–85
No. 8	39.9	30–55

The calculated requirements are for 42 tph of material to be crushed in the jaw crusher and 59 tph to be crushed in the roll crusher. The plant capacity is seen to be adequate to meet both of these requirements. The screen will be checked only for capacity on the finishing or bottom deck, since the capability of this deck will determine the screen adequacy. The calculated capacity per sq ft of this screen deck (see Table 20-60) is:

$$A = B \times S \times D \times V \times H \times T \times K \times P \times W \times O \times M$$

With 59 tph primary feed to the roll crusher, the circulating load for 25 percent oversize (Table 20-48) and 90 percent screen efficiency is 38.7 percent (Table 20-61) or

13 tph. This will give a total of 72 tph passing over the screen deck and 250 tph passing through the screen deck, or a total feed of 322 tph.

B = 5.50 tph per sq ft of screen
S = 1.20 for horizontal deck
D = 0.90 for the second deck
V = 0.98 for 22.4 percent ($^{72}\!/_{322} \times 100$) oversize
H = 1.20 for approximately 50 percent half size
T = 1.00 for square screen openings
K = 0.75 for wet gravel
P = 1.00 for no more than 5 percent elongated particles
W = 1.00 for 100 pcf material
O = 1.35 for 67.3 percent open screen area
M = disregarded for dry screening

$$A = 5.5 \times 1.2 \times 0.9 \times 0.98 \times 1.2 \times 1.0 \times 0.75 \times 1.0 \times 1.0 \times 1.35 = 7.1 \text{ tph}$$

The required screening area is 322/7.1 or 45.4 sq ft. A 4- by 14-ft screen is included in the plant tentatively selected, and this is adequate. By changing the screen cloth mesh and readjusting the crusher setting, other aggregate specifications can be

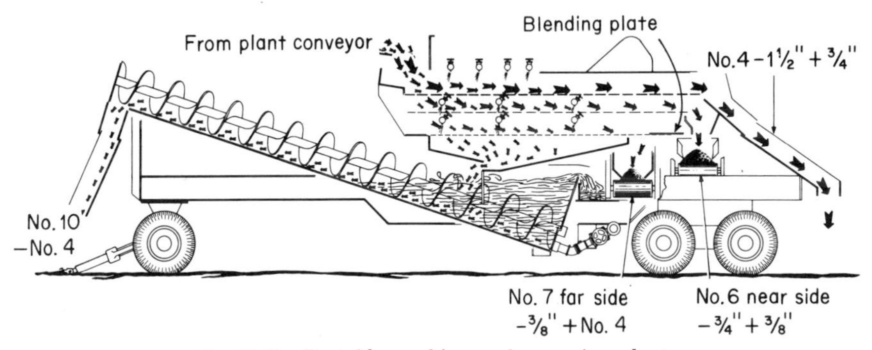

Fig. 20-39 Portable washing and screening plant.

met, but each such combination will change the plant capacity. A portable bin unit of 20-ton capacity is provided for truck loading of the finished material.

Should the same plant be required to produce 300 tph in another pit with a specification requirement for washing of the aggregate, a washing and screening plant will be necessary (see Fig. 20-39). Based on the pit gradation plus crushing of the oversize to minus 1½ in., the four materials can be split into SPR sizes (Table 20-14) to yield the following quantities and gradations:

50 tph of size No. 4	1½ in. to ¾ in.
60 tph of size No. 6	¾ in. to ⅜ in.
70 tph of size No. 7	⅜ in. to No. 4 (modified top size is ⅜ in.)
120 tph of size No. 10	minus No. 4

The plant includes a 5- by 16-ft triple-deck horizontal screen. The fine-material washer is a 44-in. by 32-ft single screw (Table 20-66) with a listed capacity of 175 tph. This unit, when slowed to 75 percent of normal speed, will handle the 120 tph of minus No. 4 and will save material retained on a No. 100 screen. Total water required is approximately 1,500 gpm. Figure 20-34 indicates that the unit can handle this amount of water. With a basic screen capacity of 1.80 tph and a wet screening factor of 1.9 (Table 20-60), the No. 4 screen has the capacity to handle the 190-tph feed to the third deck.

Stationary Quarry Plant, 300 tph The plant is to be stationary, with a fixed market demand of 500,000 tons per year and approximately a 9-month operating season. A production rate of 300 tph will meet this requirement. The market is mainly covered by four SPR sizes of aggregate, Nos. 4, 6, 7, and 10, with the sale about even for each size (see Table 20-14). The sizes are as follows:

No. 4	1½ in. to ¾ in.
No. 6	¾ in. to ⅜ in.
No. 7	⅜ in. to No. 4 (modified top size ⅜ in.)
No. 10	minus No. 4

The following listing of equipment is based on a rough evaluation of its capacity when arranged according to the flowsheet supplied as Fig. 20-40. Reference has been made to pertinent equipment capacity tables within this section.

3042 jaw crusher primary
Vibrating feeder with grizzly
4- by 10-ft double-deck inclined and scalping screen
4¼-ft secondary cone crusher, open circuit
5- by 16-ft double-deck secondary screen
4-ft tertiary cone crusher, closed circuit
6- by 20-ft, 15° inclined triple-deck finishing screen
Necessary conveyors and bins

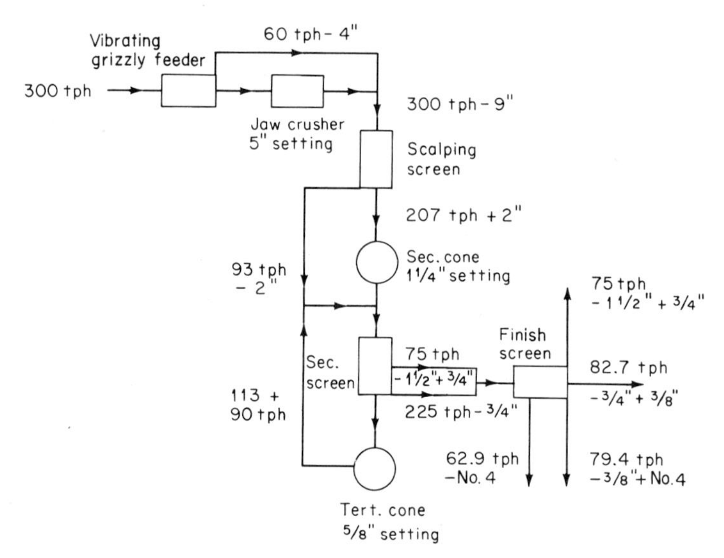

Fig. 20-40 Flowsheet for stationary quarry plant, 300 tph.

The amount of quarry material of less than 4-in. size, which corresponds to the spacing of the grizzly bars on the feeder, is estimated from Tables 20-28 and 20-29 to be a minimum of 20 percent or 60 tph. The 3042 jaw crusher with a closed setting of 5 in. has a listed capacity of 190 to 285 tph (Table 20-32), which will meet the crushing requirement of 240 tph. The scalping screen is provided with 4-in. square-opening punched plate for the top deck. This provides protection for the bottom deck, which has 2-in. square openings in heavy wire screen cloth with an open area of 64 percent. Table 20-33 indicates that there will be 69 percent of plus 2-in. material in the output from the jaw crusher; hence 207 tph will go to the 4¼-ft secondary cone crusher with coarse cavity set at 1¼ in. Table 20-39 indicates a crusher capacity of 190 tph at this setting in open-circuit operation. Since adjustments in setting can be made to balance the circuit once the system is in operation, the unit is adequate. The expected crusher product gradation, based on Tables 20-33 and 20-41, is as follows:

Product size, U.S. Standard	Primary jaw, 5-in. setting, tph	Secondary cone, 1¼-in. setting, tph	Total product, tph
+4 in.	138.0		
−4 in. + 2 in.	69.0		
−2 in. + 1½ in.	19.5	54.0	73.5
−1½ in. + ¾ in.	31.5	83.0	114.5
− ¾ in. + ⅜ in.	17.4	37.0	54.4
− ⅜ in. + No. 4	10.2	17.2	27.4
−No. 4	14.4	15.8	30.2
	300.0	207.0	300.0

Approximately 75 tph is required of each size category; hence 75 tph will be removed from the 114.5 tph of the passing 1½ in., retained on ¾ in., material which bypasses the tertiary crusher. The primary feed to the tertiary cone crusher in closed circuit operation will be 73.5 + (114.5 − 75) = 113.0 tph. A 4-ft cone crusher with coarse cavity and a ⅝-in. setting has a capacity of 125 tph in closed circuit (Table 20-40), which meets the requirements. The gradation of the product, as estimated from Table 20-42, is as follows:

	Feed, tph	Tertiary cone, ⅝-in. setting, tph	Total product, tph	
To tertiary	113.0			
−1½ in. + ¾ in.	75.0		75.0	No. 4
− ¾ in. + ⅜ in.	54.4	28.3	82.7	No. 6
− ⅜ in. + No. 4	27.4	52.0	79.4	No. 7
−No. 4	30.2	32.7	62.9	No. 10
	300.0	113.0	300.0	

The bottom deck will be the most critical on all the screens. Accordingly, the selection of the screen size for each case will be based on calculations for this deck. Reference will be made to Tables 20-60 and 20-61.

Scalping Screen, a 15° Inclined Screen. Based on the material breakdown following the jaw crusher discharge, the total feed to the bottom deck is 162 tph. Of this, approximately 69 tph or 42 percent is oversize and 53 tph or 33 percent is half size.

B = 7.50 tph per sq ft of screen
S = 1.00 for 15° inclination
D = 0.90 for the second deck
V = 1.10 for 42 percent oversize
H = 0.86 for 33 percent half size
T = 1.0 for square screen openings
K = 0.85 for surface-wet quarried material
P = 1.00 for no more than 5 percent elongated particles
W = 1.00 for 100 pcf material
O = 1.28 for 64 percent open screen area

Applying the formula, the calculated capacity per sq ft of screen is:

$$A = 7.5 \times 1.0 \times 0.9 \times 1.1 \times 0.86 \times 1.0 \times 0.85 \times 1.0 \times 1.0 \times 1.28 = 7.0 \text{ tph}$$

The required screening area is $^{162}/_{7.0}$ or approximately 23 sq ft. The 4- by 10-ft screen is adequate.

Secondary Screen. This is a 5- by 16-ft horizontal screen with $\frac{3}{4}$-in. standard light wire on its bottom deck, thus 61.4 percent open area. The primary feed to the cone is 113 tph; and the circulating load, for 90 percent efficiency and 40 percent oversize, is 80 percent of this feed or 90 tph. This results in a total feed to the bottom screen deck of 300 + 113 + 90 − 73.5 (the plus $1\frac{1}{2}$-in. material) = 429.5 tph. Of this, 75 + 39.5 + 90 = 204.5 tph goes over the screen and 225 tph goes through the screen. This last is combined with 75 tph of material which passes $1\frac{1}{2}$ in. and is retained on $\frac{3}{4}$ in., making a total of 300 tph for the finishing screen. The actual capacity A per sq ft of screen can be calculated as before.

B = 4.80 tph per sq ft of screen
S = 1.20 for a horizontal screen
D = 0.90 for the second deck
V = 1.16 for 47.7 percent (204.5/429.5 × 100) oversize
H = 0.86 for 33.2 percent (142.3/429.5 × 100) half size
T = 1.0 for square screen openings
K = 0.85 for surface-wet quarried material
P = 1.00 for no more than 5 percent elongated particles
W = 1.00 for 100 pcf material
O = 1.25 for 61.4 percent open screen area

$$A = 4.8 \times 1.2 \times 0.9 \times 1.16 \times 0.86 \times 1.0 \times 0.85 \times 1.0 \times 1.0 \times 1.25 = 5.5 \text{ tph}$$

The required screening area is 429.5/5.5 or 78 sq ft, and the 5- by 16-ft screen is adequate.

Finishing Screen. This is a 6- by 20-ft inclined triple-deck screen with No. 4-mesh screen on the bottom deck, light wire, and 49.1 percent open area. The total feed to the deck is 79.4 tph + 62.9 tph = 142.3 tph, with 79.4 tph or 55 percent going over the deck and approximately 36 tph or 25 percent being half size.

B = 1.80 tph per sq ft of screen
S = 1.00 for 15° inclination
D = 0.80 for the third deck
V = 1.25 for 55 percent oversize
H = 0.70 for 25 percent half size
T = 1.0 for square screen openings
K = 1.00 for dry crushed rock
P = 1.00 for no more than 5 percent elongated particles
W = 1.00 for 100 pcf material
O = 1.00 for 49.1 percent open screen area

$$A = 1.8 \times 1.0 \times 0.8 \times 1.25 \times 0.7 \times 1.0 \times 1.0 \times 1.0 \times 1.0 \times 1.0 = 1.26 \text{ tph}$$

The required screening area is 142.3/1.26 or 114 sq ft, so the 6- by 20-ft screen is adequate.

The main conveyors are each 30 in. wide with a rated capacity of 450 tph. The conveyor for closing the circuit on the tertiary cone crusher is 24 in. wide, as are the conveyors to the loading bins. The four bins each have a capacity of 50 tons. Stockpiling of material is by truck.

Stationary Sand and Gravel Plant, 150 to 175 tph Figure 20-41 shows the flowsheet for a sand and gravel plant with a capacity of 150 to 175 tph. The material for the feeder passes over a grizzly with a 3-in. bar spacing to a 25- by 36-in. jaw crusher (Table 20-32), then to a 4- by 12-ft double-deck horizontal washing screen. The plus $1\frac{1}{2}$-in. material is crushed to minus $1\frac{1}{2}$ in. in a roll crusher which operates in closed circuit with the screen. The minus ¼-in. material and the water are flumed to a water scalping tank (W.S.T.) (Table 20-68) and split three ways; mason sand and concrete sand, each of which is dewatered in a fine-materials screw (Tables 20-66 and 20-67 and Fig. 20-34), and waste, which is flumed to a sludge pond. The passing 1½-in., retained on ¼ in., material is passed to a log washer for scrubbing and discharged to a 3- by 8-ft triple-deck screen for final sizing.

Sand and Gravel Pumping Operation, 150 to 175 tph Figure 20-42 shows the flowsheet of a plant with a capacity of 150 to 175 tph. The 10-in. pump discharges about 175 tph of sand and gravel and 3,000 gpm of water in a pump box above the 4- by 12-ft triple-deck screen. The plus 1½-in. material is crushed to minus 1½-in.

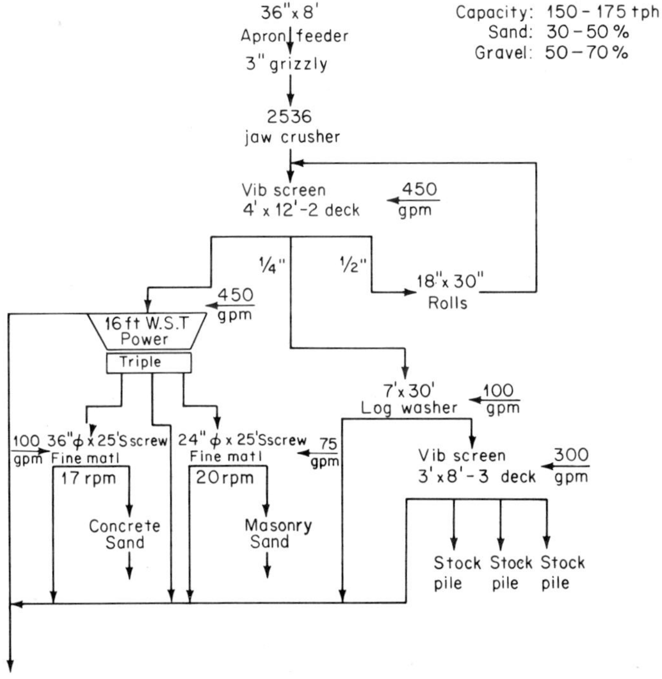

Fig. 20-41 Flowsheet for a 150- to 175-tph stationary crushing, screening, and washing plant for sand and gravel. (*Eagle Iron Works.*)

sizes in a jaw crusher which operates in closed circuit with the screen. The minus ¼-in. material and water are flumed to the water scalping tank (Table 20-68) and split three ways. The mason sand and concrete sand are each dewatered in a fine-materials screw (Tables 20-66 and 20-67 and Fig. 20-34), and the waste is flumed to a sludge pond. The passing 1½-in., retained on ¼ in., material is passed to a log washer for removal of clay balls and other deleterious material, then it is screened over a double-deck screen to produce two sizes of gravel, passing 1½ in. and retained on ¾ in., and passing ¾ in. and retained on a No. 4 mesh. The water addition after the triple-deck screen is made for rinsing purposes.

Stationary Sand and Gravel Plant, 300 tph Figure 20-43 shows the flowsheet of a plant with a rated capacity of 300 tph. The material is brought from the pit to the plant by a belt conveyor and deposited on a 4- by 12-ft triple-deck screen for washing. The plus 1¾-in. material, approximately 5 tph, goes to a bin. The passing 1¾-in. and

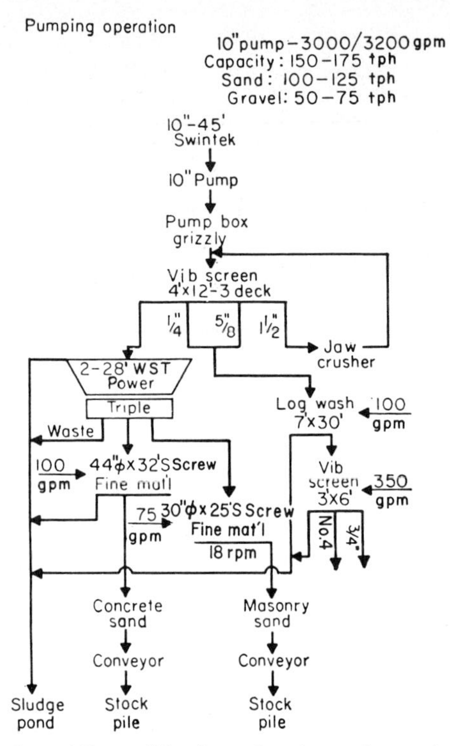

Fig. 20-42 Flowsheet for a 150- to 175-tph sand and gravel pumping operation. (*Eagle Iron Works.*)

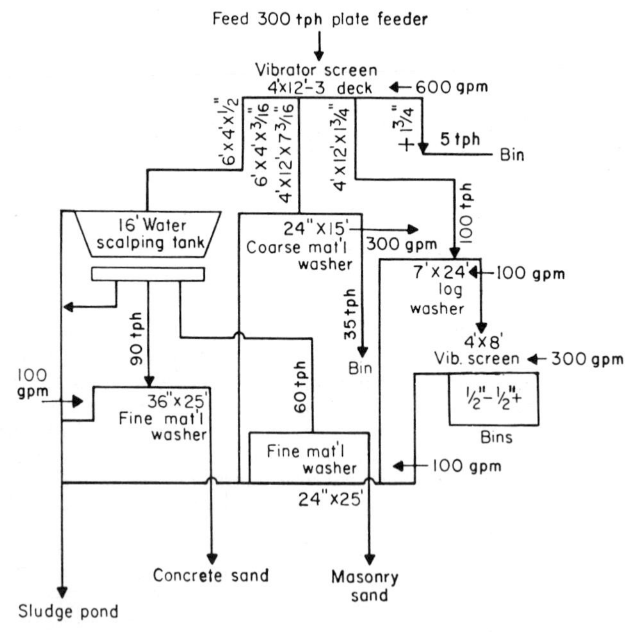

Fig. 20-43 Flowsheet for a 300-tph stationary sand and gravel plant. (*Eagle Iron Works.*)

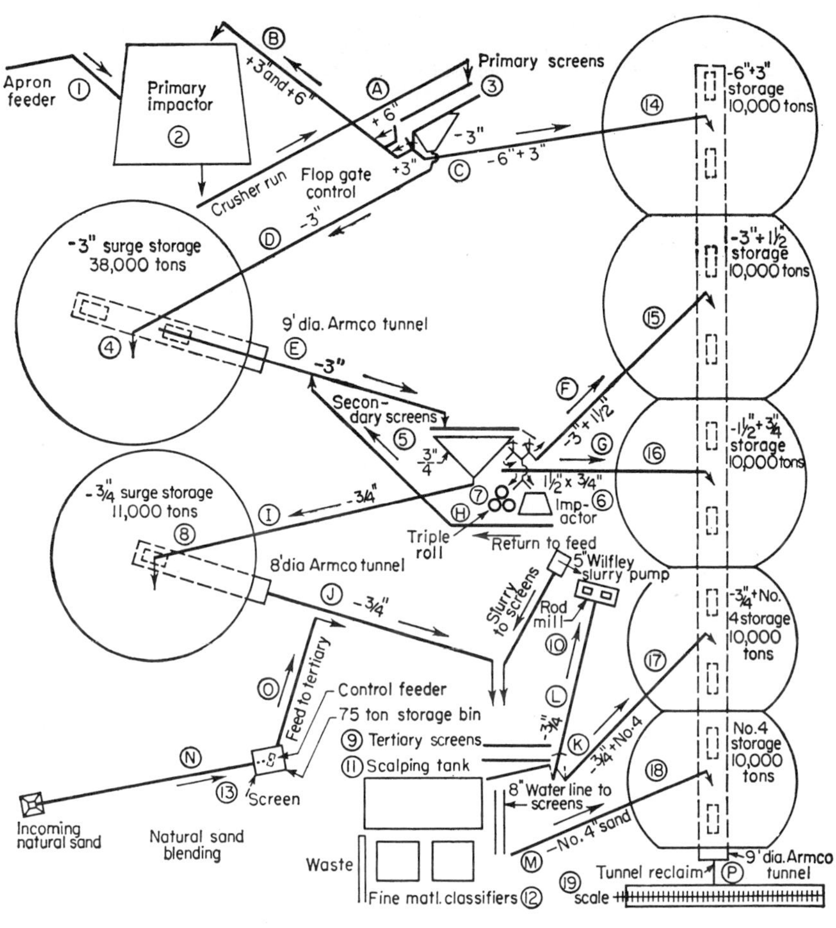

Conveyors*			
A	48″ × 190′	Crusher run	Impactor to primary screens
B	36″ × 185′	+6″ and +3″	Primary screens to impactor
C	30″ × 189′	−6″ + 3″	Primary to finished storage
D	42″ × 218′	−3″	Primary to surge pile
E	36″ × 178′	−3″	Surge to secondary
F	30″ × 175′	−3 + 1½	Secondary to finished storage
G	24″ × 163′	−1½ + ¾	Secondary to finished storage
H	24″ × 86′	Crusher run	Secondary to secondary feed
I	30″ × 218′	−¾	Secondary to surge
J	30″ × 155′	−¾	Surge to tertiary
K	24″ × 145′	−¾ + No. 4	Tertiary to finished storage
L	24″ × 75′	−¾	Tertiary to rod mill
M	24″ × 212′	−No. 4 sand	Tertiary to finished storage
N	24″ × 125′	Natural sand	Incoming to bin
O	18″ × 80′	Natural sand	Bin to tertiary feed
P	42″ × 503′	All	Reclaiming

* Twelve Barber-Greene, three Pioneer; two Link Belt; all Goodyear belting.

Plant Equipment		
1	50″ × 14½′	Cedarapids heavy-duty apron feeder
2	5360H	Cedarapids impactor
3	5′ × 12′	Pioneer double-deck screen (two)
4	48″ × 60″	Syntron feeder (two)
5	4″ × 12′	Cedarapids double-deck screen (three)
6	3030	Cedarapids impactor
7	5424	Pioneer triple roll crusher
8	42″ × 60″	Syntron feeder
9	4′ × 14′	Cedarapids double-deck screen (four)
10	8′ × 12′	Marcy rod mill
11	32′	Eagle water scalping tank
12	36″ × 25′	Eagle double fine matl. classifier (two)
13	4′ × 12′	Pioneer screen
14	48″ × 72″	Syntron F55 feeder (two)
15	48″ × 72″	Syntron F55 feeder (two)
16	48″ × 60″	Syntron F55 feeder (two)
17	48″ × 60″	Syntron F55 feeder (two)
18	48″ × 60″	Syntron F55 feeder (two)
19	100-ton	Howe track scale

Fig. 20-44 Flowsheet for a 1,000-tph stationary crushing plant for dolomitic limestone. (*Construction Methods and Equipment, January, 1956.*)

retained on 7⁄16-in. material, a total of 110 tph, goes to a log washer and then to a 4- by 8-ft single-deck vibrating screen for a split. The passing 1¾-in. and retained on ½-in. fraction is directed to one bin and the minus ½-in. fraction to another. The material passing the 7⁄16-in. screen and retained on the No. 4 mesh, a total of 35 tph, goes to a coarse-materials washer (Table 20-69) and then to a bin. The 150 tph of minus No. 4 material is flumed to a water scalping tank (Table 20-68), with a split of two sizes of sand, and is dewatered in two fine-materials washers (Tables 20-66

TABLE 20-76 Aggregate Specifications for 1,000 tph Stationary Crushing Plant (See Fig. 20-44)

Crushed-stone aggregate

Sieve size, U.S. Standard	Percent passing (by weight)			
	Aggregate size, in.			
	3⁄16–¾	¾–1½	1½–3	3–6
7 in.	......	......	......	100
6 in.	......	......	......	90–100
4 in.	......	......	100	20–45
3 in.	......	......	90–100	0–15
2 in.	......	100	20–55	
1½ in.	......	90–100	0–15	
1 in.	100	20–55		
¾ in.	90–100	0–15	0–5	
⅜ in.	20–55	0–5		
No. 4	0–5			

Manufactured sand

Sieve size, U.S. Standard	Percent passing (by weight)
No. 4	95–100
No. 8	75–90
No. 16	50–70
No. 30	30–50
No. 50	15–30
No. 100	5–10
No. 200	Max. 60 % of −100

Fineness modulus: 2.50–2.90

and 20-67 and Fig. 20-34), with 90 tph of concrete sand going to one bin and 60 tph of masonry sand to the other. Water additions are made at various points to ensure clean material, and the total water requirement is 1,500 gpm.

Stationary Quarry Plant, 1,000 tph Figure 20-44 shows the flowsheet for a stationary crushing plant with a capacity of 1,000 tph. The coarse aggregate ranges from 6 in top size to a No. 4 mesh; the fine aggregate, a manufactured sand, must meet a fineness modulus value of 2.50 to 2.90. The size requirements are typical for heavy-construction jobs involving concrete construction (Table 20-76), and the flowsheet is basic in its application. For basalt or granite as raw material, the impact crusher will be replaced with gyratory crushers (Table 20-30) or swing-type jaw crushers (Table 20-31); the secondary and tertiary crushers will be gyratory crushers (Tables 20-37 and 20-38) or cone crushers (Tables 20-39 and 20-40).

For this project the well-shot material (about 3 tons per pound of powder) is hauled from the quarry to the primary crusher, a 5360H double-impeller impact breaker (Table 20-35A), and reduced to minus 6-in. material. This is scalped over a 5- by 12-ft double-deck screen, the minus 3-in. material going to a 38,000-ton stockpile and then to the secondary crushers, a 5424 triple-roll crusher (Table 20-45) and a 3042 impact breaker (Table 20-35A) operating in closed circuit with three 4- by 14-ft screens. The minus ¾-in. material goes to an 11,000-ton stockpile for manufacture of sand. This last is a wet operation in closed circuit with an 8- by 12-ft Marcy

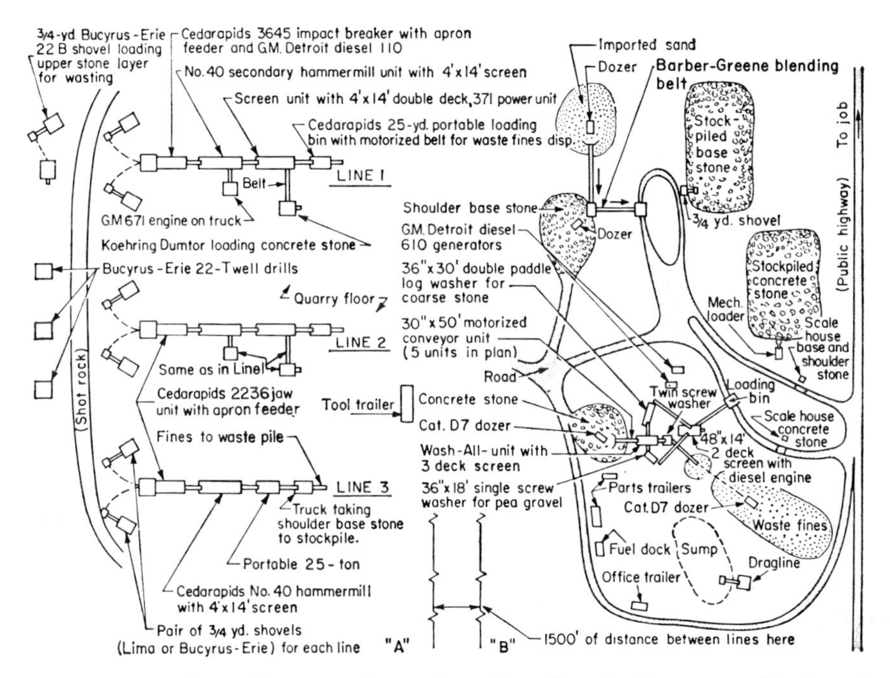

Fig. 20-45 600-tph crushing, screening, and washing plant for limestone. (*Roads and Streets, September, 1956.*)

rod mill (Table 20-56), washing screens, a 32-ft water scalping tank (Table 20-68), and 36-in. by 25-ft fine-material washers (Tables 20-66 and 20-69 and Fig. 20-34). Aggregates can be blended by the use of vibratory feeders (Table 20-65) to the 42-in. belt which runs in a tunnel under the stockpiles. The reclaiming belt feeds a loading conveyor and chute system, which discharges the material into trucks or railroad cars for transportation to the jobsite.

Portable Quarry Plant, 600 tph Figure 20-45 shows a typical solution to the material requirements for a highway project. The approach is a basic one, utilizing portable units which will be removed at the completion of the project. The type of equipment will vary with the characteristics of the raw material, but otherwise the flowsheet will change only to accommodate the material specifications. The portable crushing, screening, and washing plant has a rated capacity of 600 tph. It produces four types of specification aggregates, with a top size of 1½ in. Drilling and blasting is done with a single 17-ft face, using an 8- by 10-ft drill pattern and shooting twice a week for a two-shift loading operation. Because of deep fissuring and undesirable material, as well as specification requirements, waste and excess fines amount to approximately 30 percent. Each of the unitized plants consists of a primary crusher and a secondary crusher operating in closed circuit with a screen. There is also a screening plant for lines No. 1 and No. 2 with two unloading points, one for waste material and one for concrete aggregate. The screening plant is omitted on line No. 3. The rock is handled directly from the quarry face to the feeder by two ¾-yd shovels for each primary. The material to be washed is hauled approximately 2,000 ft to a portable washing plant. The control of the material sizes is mainly by screen and crusher adjustment and by grate bar selection for the hammer mills. The finished aggregate is stockpiled in 4-ft lifts, each of which is set back to avoid sliding and segregation.

PLANT OPERATION

Factors which are of great importance in the plant operation include the federal and state code requirements pertaining to safety. These have been very much strength-

TABLE 20-77 Typical Equipment Sound Levels (decibels)

Equipment	Decibels
Trucks (15-ton up)	89–101
Shovels (diesel)	91–107
Shovels (electric)	83–91
Bulldozers	102–106
Cranes	88–99
End loaders	95–97
Road graders	91–96
Locomotives (diesel)	88–100
Primary crushers	88–91
Secondary crushers	91–101
Vibrating screens	98–109
Rod and ball mills	89–104
Pneumatic drills:	
Surface	97–98
Underground	112–121
Diamond drill	101

SOURCE: Mining Congress Journal, 1967.

TABLE 20-78 State of California Noise Exposure Norm*

Frequency band, cps	American Standard preferred frequencies for acoustical measurements	Octave band sound pressure level, decibels (re: 0.002 dynes/sq cm)
20–75	63	110
75–150	125	102
150–300	250	97
300–600	500	95
600–1,200	1,000	95
1,200–2,400	2,000	95
2,400–4,800	4,000	95
4,800–10,000	8,000	95

*If an employee is exposed to noise for 5 or more hours per normal workday, the levels shown are those at and above which the wearing of hearing protectors is mandatory. For employees whose exposure to occupational noise is less than 5 hours per day, the noise levels may be 3 db higher for each halving of the exposure time; e.g., for an exposure of 2½ hr, the noise levels encountered may be 3 decibels higher in all frequency ranges than the values shown; if the exposure time is 1¼ hours per day, the noise levels may be 6 decibels higher than indicated here before the ear protectors must be worn.

SOURCE: State of California Code, 1962.

ened in the last few years, and a good safety program is a must.* Environmental control as this pertains to noise, water, and air pollution is also assuming greater importance, following the Air Quality Act of 1967.[48]

The expected noise levels in and around an aggregate plant are shown in Table 20-77.[49] The State of California in November, 1962, established the noise exposure standards listed in Table 20-78.

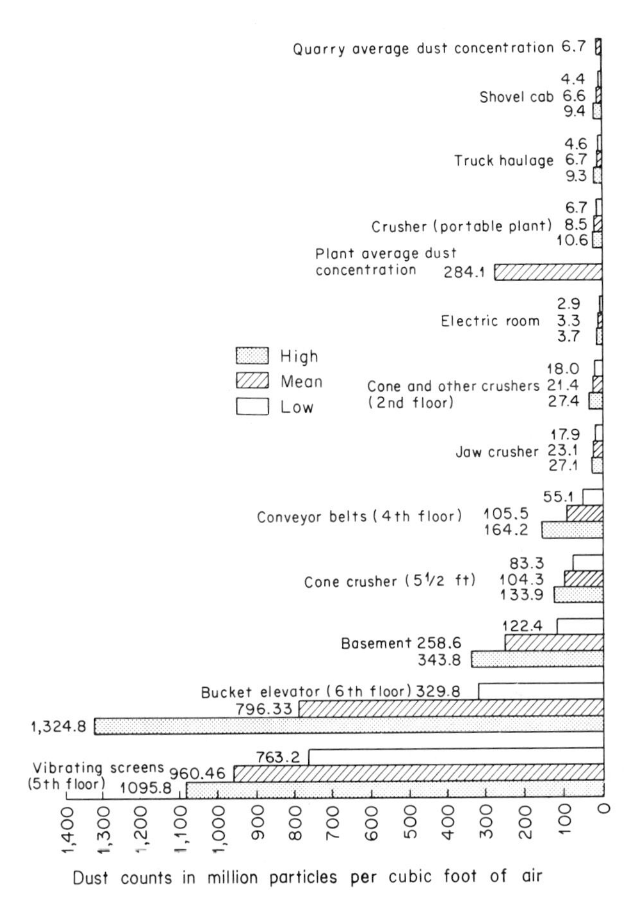

Fig. 20-46 Observed dust counts for representative construction activities. (*Bureau of Mines.*)

Air pollution control requirements are rapidly being enacted in every state. The dust emission around quarry operations can be of such a magnitude that it becomes both a health hazard and a safety hazard (see Fig. 20-46). The National Crushed Stone Association (NCSA) has published a special report on a dust monitoring system with an initial cost of less than $800.[50]

For all types of plant operation, both stationary and portable, it is desirable to keep the dust down on haul roads and other unpaved roads around the plant. It is estimated that this can be accomplished with calcium chloride for 15 cents per sq yd per year. With used motor oil or other available low-cost oils, the estimated cost

* See Sec. 5, Construction Safety.

with a monthly application is approximately 10 cents per sq yd per year. The control of dust in this way will increase safety and decrease the problem with dust contamination in stockpiles.

In stationary plants, the use of a wetting agent will help reduce the dust problem.[51] Exhaust systems for both portable and stationary plants should be designed to meet

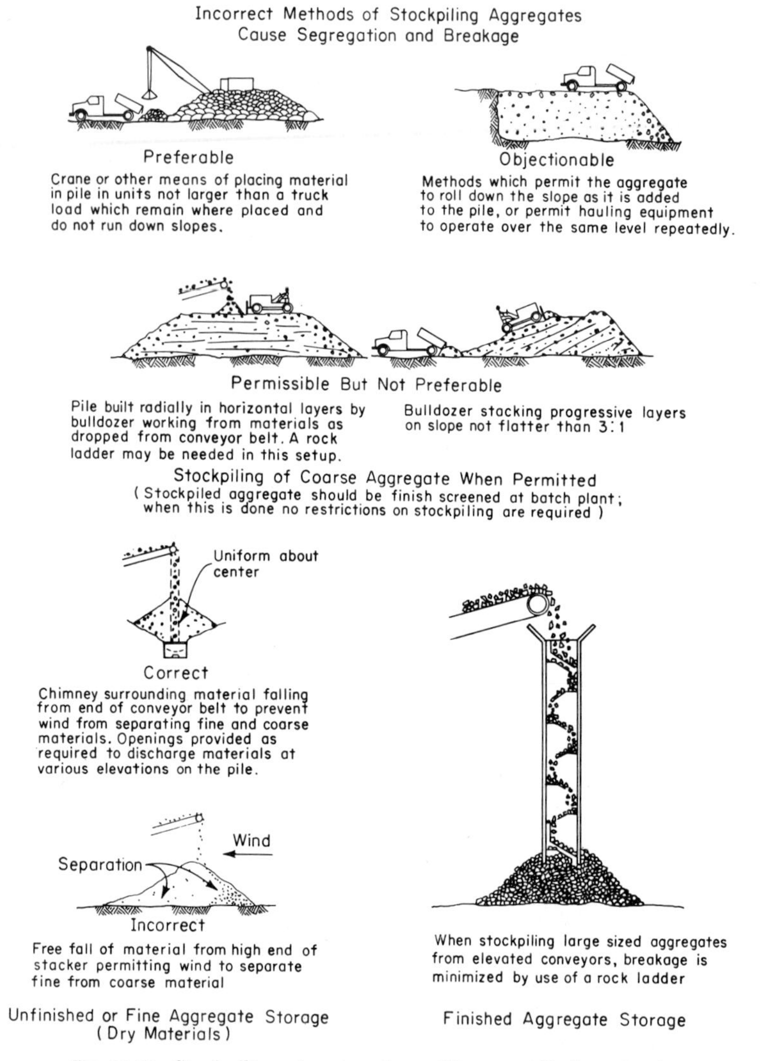

Fig. 20-47 Stockpiling of aggregates. (*Bureau of Reclamation.*)

the special requirements for the plant and its surroundings. Hoods should be provided for all conveyor transfer points and for screens and crushers. The duct velocity to collectors should be from 3,000 to 4,000 fpm. The air volume required will depend upon design details at each point; a belt transfer point requires an air volume of approximately 550 cfm per ft of belt width, and a screen requires an air volume of 75 cfm per sq ft of screen-box area. Crushers require from 1,000 to 10,000 cfm of exhaust air, depending on the type of crusher and the material. The highest exhaust requirement is for impact-type crushers and limestone. The collector itself can be dry or wet electrostatic, filter type, or cyclone. The cost will range from $1 per cu ft for complete cyclone-type collectors to $3 per cu ft for bag-filter-type installations. The efficiency requirement has become so stringent that bag filters will be the best solution to the problem.

There is an increased interest in the methods for stockpiling of the finished aggregate. Some specifications will call for a definite method in order to avoid segregation and breakage (see Fig. 20-47). In estimating the total job requirements, the loss due to stockpiling should be included; this can amount to 5 to 10 percent of the stockpile size.

To maintain gradation control, to check for the best possible operation of the plant, and for inspection of the finished aggregate, it is necessary to have testing equipment. The testing equipment includes the testing sieves and the mechanical equipment for shaking the samples down to the full specification range; also sample pans, sample splitter, scales for weighing the sample, and a hot plate for drying the fine aggregate fraction. A well-located test house with adequate room for the test equipment will also be of great value. Necessary adjustments of crushers to compensate for wear can be anticipated by studies of test samples. In this way, overloading of any single crusher or screen unit can be avoided.

The sample taking is a very important step.[52] Small quantities should be taken at various points from a stockpile, both top and bottom. A good location for sample taking is at the end of the finished-material conveyor, where a pail or pan can be used to obtain a complete cross section of the discharged material. The recording of the test result can be made on standard graph sheets for comparison with the specification limits.

The variation within each stockpile and the overall problem of random sampling for specification control have been the subjects of a great deal of investigation.[53] It is evident from the reports that a flat stockpile made with truck-dumped material and reclaimed with a front-end loader will ensure adequate control of degradation with minimum sampling. A portable conveyor and front-end (FE) loader combination, a frequently used arrangement, will require 35 samples to give a confidence level of 95 percent (Tables 20-79 and 20-80).

TABLE 20-79 Summary of Stockpile Tests

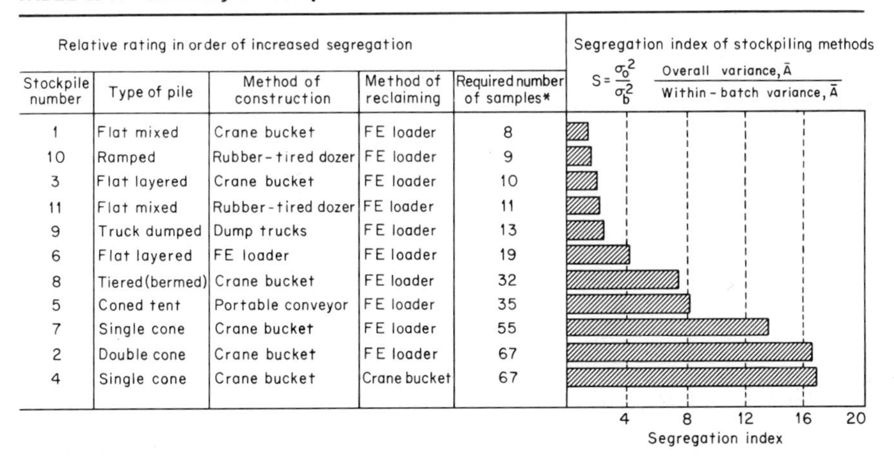

Relative rating in order of increased segregation				
Stockpile number	Type of pile	Method of construction	Method of reclaiming	Required number of samples*
1	Flat mixed	Crane bucket	FE loader	8
10	Ramped	Rubber-tired dozer	FE loader	9
3	Flat layered	Crane bucket	FE loader	10
11	Flat mixed	Rubber-tired dozer	FE loader	11
9	Truck dumped	Dump trucks	FE loader	13
6	Flat layered	FE loader	FE loader	19
8	Tiered (bermed)	Crane bucket	FE loader	32
5	Coned tent	Portable conveyor	FE loader	35
7	Single cone	Crane bucket	FE loader	55
2	Double cone	Crane bucket	FE loader	67
4	Single cone	Crane bucket	Crane bucket	67

* NOTE: The required number of samples was calculated at a 95 percent confidence level and a desired accuracy of $\pm 0.05\ \bar{A}$. This column shows that the number of samples must be increased as the variability of the aggregate gradation increases.
SOURCE: Highway Research Board.

TABLE 20-80 Comparative Stockpile Costs

Stock-pile No.	Pile type	Construction method	Construction time, hr	Total cost	Cost per ton*	Segregation index†
1	Flat-mixed	Crane bucket	10.25	$ 758	$0.50	1.35
2	Double cone	Crane bucket	10.25	758	0.50	16.48
3	Flat layered	Crane bucket	6.75	500	0.33	1.96
4	Single cone	Crane bucket	10.25	758	0.50	16.86
5	Coned tent	Portable conveyor	11.00	817	0.54	8.10
6	Flat layered	Front-end loader	13.00	959	0.64	4.05
7	Single cone	Crane bucket	11.00	817	0.54	13.36
8	Tiered (bermed)	Crane bucket	13.00	959	0.64	7.37
9	Truck dumped	Dump trucks	7.00	517	0.34–0.00‡	2.30
10	Ramped	Rubber-tired dozer	11.00	808	0.54	1.59
11	Flat-mixed	Rubber-tired dozer	9.25	683	0.45	2.10
			112.75	$8,334		

* Assuming 1,500 tons per stockpile.
† Based on $\bar{A}$.
‡ Essentially zero cost if aggregate is delivered to stockpile site in trucks.
SOURCE: Highway Research Board.

The maintenance of plant and equipment is of prime importance. The shop should be such that it can handle the day-to-day repairs and support a sensible preventive maintenance program. When possible, major overhauls should be spaced to coincide with the off season. Spare parts should be carried in quantities large enough to ensure continued operation. Manganese and screen cloth should always be on hand to meet the week-to-week requirements. The upkeep of hammers for hammer mills, breaker bars for impact breakers, and other wearable manganese parts can best be done by welding.*

REFERENCES

1. Stone, *Mineral Industry Surveys*, Bureau of Mines, 1968.
2. Sand and Gravel in 1968, *Mineral Industry Surveys*, Bureau of Mines.
3. AASHO–ARBA Joint Committee Report, Task Group on Aggregates, 1968.
4. *A Glossary of the Mining and Mineral Industry*, Bureau of Mines.
5. *Definition of Terms Relating to Concrete and Concrete Aggregates*, ASTM Designation C125-58, American Society for Testing and Materials.
6. Review and Forecast, 1968–1969, *Pit and Quarry*, January, 1969.
7. Sand and Gravel, *Minerals Yearbook*, Bureau of Mines, 1964.
8. Stone, *Minerals Yearbook*, Bureau of Mines, 1964.
9. Standard Specifications for Highway Materials and Methods of Sampling and Testing, part I and part II, AASHO.
10. ASTM Standards, including Tentatives, part 10, ASTM, 1964.
11. *Handbook for Concrete and Cement*, part III, Corps of Engineers.
12. *Road and Paving Materials, Method of Sampling and Testing*, Federal Specification.
13. *The Identification of Rock Types*, rev. ed., Bureau of Public Roads, 1960.
14. *Physical Properties of Typical American Rocks*, Bulletin 131, Iowa State University.
15. *Results of Physical Tests of Road Building Aggregate*, Bureau of Public Roads, 1953.
16. Nonmetallic Minerals—Facts and Figures, *Pit and Quarry*, 1964.
17. *Prospecting for Natural Aggregates*, NSGA Circular No. 94, and *Rock Products*, May–June, 1965.
18. *Concrete Manual*, 7th ed., Bureau of Reclamation.
19. *Annotated Tables of Strength and Elastic Properties of Rocks*, University of Illinois, AIME, 1956.
20. *The Concept of Hardness as Applied to Mineral Aggregates*, Highway Research Board, 1967.
21. *Soundness and Deleterious Substances*, ASTM Tech. Pub. No. 169A; and NSGA Circular No. 97.

* See Electric Arc Welding and Oxyacetylene Cutting and Joining in Sec. 23, Steel Construction.

22. Methods for the Determination of Soft Pieces in Aggregate, *ASTM Proc.*, vol. 47; and NSGA Circular no. 44.
23. Extending Application of Fineness Modulus, *ACI Journal Proc.*, vol. 43, December, 1947.
24. *Investigation of Concrete Sand*, NSGA Tech., Info. Letter No. 266.
25. *Standard Specifications for Construction of Roads and Bridges on Federal Highway Projects*, Bureau of Public Roads.
26. *Guide Specifications for Highway Construction*, AASHO.
27. *Aggregate Gradation for Highways*, Bureau of Public Roads.
28. *Coarse Aggregates (Crushed Stone, Gravel and Slag)*, Simplified Practice Recommendation, R163-48, U.S. Department of Commerce.
29. *Properties of Sand and Gravel for Railroad Ballast*, NSGA Circular No. 62.
30. *Manual on Test Sieving Methods*, STP447, ASTM, 1969.
31. *Primary Crushing—Report of Investigations*, Nos. 1, 2, and 3, Bureau of Mines, 1938.
32. *Aggregate Production with Nuclear Explosives*, Lawrence Radiation Laboratory, UCRL-12180.
33. Carl Naske, *Zerkleinerungsvorrichtungen und Mahlanlagen*, 1921.
34. Fred C. Bond, *Crushing and Grinding*, Bulletin 402, Bureau of Mines.
35. *The Third Theory of Comminution*, AIME Paper TP3308B, 1951.
36. *How to Determine Crusher and Grinding Mill Sizes Accurately*, Allis-Chalmers.
37. D. H. Geisenkieng, *Jaw Crusher Capacities*, AIME, 1951.
38. C. G. Cooley, *Production of Fine Sand from Gravel*, NSGA, 1963.
39. J. D. Grace, *Screening by Centrifugal Force*, AIME, 1965.
40. *Vibrating Screens and Their Operation*, NCSA, 1969.
41. Samuel E. Gluck, Screens in the Chemical Process Industries, *Chemical Engineering*, 1965.
42. John H. Berstrom, Belt Conveyor Roundup, *Rock Products*, 1963.
43. C. E. Golson, *Modern Classification Methods Applied to Fine Aggregates*, AIME, 1959.
44. C. E. Golson, HMS Plants for Low Grade Deposits, *Western Construction*, September, 1959.
45. Evans Wheaton, *Economics of Portable Plants*, NCSA, 1967.
46. Max F. Oleschlaeger, *Practical Application of the Use of Portable Plants*, NCSA, 1966.
47. Frederick C. Moore, John W. Matthews, and Edwin E. Dotter, *Problems Encountered in the Design and Construction of a New Aggregate Plant*, AIME, 1968.
48. Hans I. Hansen, *Dust and Noise Control—A Problem*, APAO, Canada, 1968.
49. James M. Botsford, Control of Mining Noise Exposure, *Mining Congress Journal*, August, 1967.
50. F. A. Renninger, *The Problem of Fugitive Dust in the Highway Construction Industry*, Highway Research Board, 1967.
51. Montague Hankin, Jr., *Various Methods of Dust Collection at Stone Plants*, NCSA, 1966.
52. C. W. Matthews, Stockpiling of Materials, *Rock Products*.
53. *Effect of Different Methods of Stockpiling and Handling of Aggregates*, Highway Research Board, Program Report 46; and Warren B. Warden, M-W Tech, Paper No. 90.

Section **21**

Conventional Concrete

HERBERT K. COOK

Vice-president, Engineering, Master Builders, Cleveland, Ohio (*Materials Selection and Mixture Proportioning*)

JOHN K. HUNT

Chief Engineer, Johnson Operations, Road Division, The Koehring Company, Champaign, Ill. (*Batching, Mixing, Placing, and Curing*)

JOHN F. SEIFRIED

Consulting Engineer, Evanston, Ill. (*Reinforcing Steel*)

GEORGE F. BOWDEN

Vice-president, Research and Development, Symons Manufacturing Company, Des Plaines, Ill. (*Stationary Forms*)

J. F. CAMELLERIE

Vice-president, Construction Improvements and Procedures, Inc., Huntington, N.Y. (*Slip Forms*)

A. Materials Selection and Mixture Proportioning

The major ingredients in concrete are water, cement, and aggregates. In an increasing number of cases, admixtures are also used in concrete. The most widespread use of admixtures is for entraining air in concrete to increase its durability and workability. Admixtures for reducing water content and for the control of set are also in common use.

The type of cement, the source of aggregates, the type or types of admixtures, and the mix proportions for concrete for use in heavy construction have usually been thoroughly explored and final selection has been made before the construction contract has been awarded. This is particularly true for large government jobs where it is customary for the job specifications to include the type of cement, the designation of acceptable sources of aggregate, and the requirement that the concrete will be proportioned in accordance with mixture designs provided to the contractor by the responsible government agency. Similar specifications and information are usually provided by the owner or his engineer for private construction projects. Nevertheless, the contractor may occasionally find himself in the position where such information is not supplied, and in any event it is his responsibility to produce and place concrete of the specified quality. To meet this responsibility, he must have at least a general knowledge of the types of cement available, including their special characteristics and intended uses. He should know something about the requirements for and properties of satisfactory aggregates and admixtures as well as the properties of those which are not satisfactory, and their effect on the concrete both before and after placement and hardening. Most particularly, he should know the effect of variations in the proportions of the ingredients in the concrete upon its properties and placing characteristics.

CEMENT

The greatest proportion of the cement used in the United States is manufactured to meet specifications prepared by either the American Society for Testing and Materials or the General Services Administration of the Federal government.

The American Society for Testing and Materials' *Standard Specification for Portland Cement*[1]* (ASTM Designation C 150) provides for five types. Type I is intended for use in general concrete construction when the special properties specified for the other types are not required. It is the type most widely used and will automatically be provided when no other type is stated in the purchasing agreement. Type II is for use in general concrete construction exposed to moderate sulfate action or where a moderate heat of hydration is required. Type III is intended for use when high early strength is required. Type IV is for use when a low heat of hydration is required. Type V is for use when high sulfate resistance is required. It should be noted that Types IV and V are not usually carried in stock and, in advance of specifying their use, purchasers should determine whether they are or can be made available.

ASTM's *Standard Specification for Portland Cement* also provides for three types of cement which contain an air-entraining agent.[1] These cements are identified as Types IA, IIA, and IIIA. They are similar to the corresponding three types without air except that their strength requirements are somewhat lower and they are required to entrain air in a standard mortar within specified limits. There are no ASTM specifications for air-entraining Types IV and V.

Federal Specification[2] SS-C-192g includes requirements for the same five types of

* Superior numbers refer to the list of references at the end of this section.

non-air-entrained cements and the three types of air-entrained cements as the ASTM specifications.

The ASTM and the federal government also have specifications for a number of special cements[1,2] such as natural cement, portland-pozzolan cement, portland blast-furnace slag cement, and slag cement. Somc of these cements are used as replacements for or additions to portland cement and some are used in place of portland cement. Natural cement and slag cement are normally viewed as partial replacements for portland cement. Portland blast-furnace slag cement and portland-pozzolan cement may be considered as combinations of replacement materials and portland cement, preblended at a cement plant. The reasons for using these blends, either at the concrete batching plant or preblended prior to delivery, may include economy, reduction of heat generation in mass concrete, improvement of impermeability or watertightness, and improvement of workability. The manner in which they are to be used will be detailed in the job specifications. Because of this and because their general use does not approach the widespread use of portland cement, lengthy discussion of their special applications does not appear to be warranted herein.

Cement manufactured to meet a given ASTM type will usually meet the corresponding federal type and vice versa[3] unless some of the special options, such as low alkali content and heat of hydration, are made applicable. Cements for which these options are applicable usually are made to order and, in advance of specifying their use, the purchaser should determine the manufacturer's ability to produce them. The individual test procedures used to determine compliance with the requirements of the ASTM and Federal Specifications[1,2] are normally of interest primarily to the testing agency, but they become of interest to the purchaser or contractor if there is a question relative to noncompliance.

AGGREGATES

Aggregate is often regarded as graduated fragments of any hard, inert material which can be mixed with water and a cementing material to form concrete. Many of the problems that occur in the production of concrete stem from the fact that some of the materials proposed for use as aggregate turn out in practice not to be adequately hard, adequately inert, or suitably graduated.

That portion of the aggregate which will pass through a No. 4 sieve is called fine aggregate (or sand), and that which is retained on this sieve is called coarse aggregate. It is also customary and frequently mandatory that the coarse aggregate be divided into several sizes, depending upon the maximum size to be used in the concrete. The primary reason for this is to prevent segregation in grading during stockpiling and handling. Standard practice is to separate the coarse aggregate into No. 4 to ¾ in.; ¾ to 1½ in.; 1½ to 3 in.; and 3 to 6 in. On some large projects, particularly for very lean concrete such as is used in gravity dams, it is sometimes required that the fine aggregate also be separated into two sizes and that these be batched separately. The workability of such concrete is very critical and it is mandatory that the grading of the fine aggregate be very closely controlled to ensure that the concrete can be transported and consolidated in the forms in an efficient and uniform manner.

Contract requirements for aggregates, and whether or not acceptable sources are designated, vary more widely than do the requirements for cement. The reason for this may be attributed, in part, to the geological differences in deposits from one region to another, to differences in the degree of exposure of the concrete, and to differences in strength and other requirements for the particular structure.

Most specifications for aggregate are based on ASTM[4] Designation C 33 or its equivalent Federal Specification[2] SS-A-281b. The requirements of these specifications are sometimes modified, either because local materials cannot meet some of the requirements or because special properties dictated by unusual construction or exposure conditions are called for. These standard specifications are suitable for structural concrete and paving. For such structures as large gravity dams, it is customary to prepare more detailed specifications designed to fit more closely the requirements of the individual job, its location, and the suitability of the local aggregates.

The Corps of Engineers is perhaps unique because its specifications for aggregates[5] for its civil works construction have no limits except with respect to gradation. The purpose of such specifications is to obtain the best materials that are locally and economically available. Initially, a thorough investigation is made by the Corps of Engineers of all known sources of aggregate within economic range of the proposed project. Sources which are satisfactory from the standpoint of quality and economy are then listed in the invitations for bids as "acceptable sources." The successful bidder is given the opportunity to submit samples from one source other than those listed. If this source is equal in quality to the designated acceptable sources and is adequate in quantity, he is permitted to use it in the project; otherwise, he must obtain his aggregate from one of the designated "acceptable sources." Local producers, if any, will usually be listed as acceptable sources; the contractor can then obtain quotations from these suppliers and bid accordingly. In remote locations where he must produce his own aggregate, the contractor must base his bid price on his estimated cost of obtaining material from one of the acceptable sources. If, after having been awarded the contract, he can locate a satisfactory source of his own and can produce from it at a lower price, he is in an excellent position to make a better profit.*

In any event, on a large project the contractor will want to assure himself of the quantity and quality of the available aggregate sources before he bids on the job. Where he selects the source himself, he will usually be required to submit samples to a designated testing laboratory or sometimes to a laboratory of his own choice.

The rapidly increasing use of both insulating grades and structural grades of lightweight aggregate warrants some discussion of these materials. Lightweight aggregates for concrete for insulation are intended for use in concrete not exposed to the weather. The prime consideration is the thermal insulating value of the resulting concrete. Such aggregates are generally of two types. They may be prepared (1) by expanding materials such as perlite or vermiculite or (2) by expanding, calcining, or sintering products such as blast-furnace slag, diatomite, fly ash, shale, or slate. Requirements for acceptability are contained in ASTM Designation C 332, *Standard Specifications for Lightweight Aggregates for Insulating Concrete.*[4] On the other hand, for lightweight aggregates intended for use in structural concrete, the prime considerations are lightness in weight and sufficient compressive strength in the concrete for structural purposes. They, also, are of two general types which are prepared (1) by expanding, calcining, or sintering materials such as blast-furnace slag, clay, diatomite, fly ash, shale, or slate or (2) by processing natural materials such as pumice, scoria, or tuff. Requirements for acceptability are contained in ASTM Designation C 330, *Standard Specifications for Lightweight Aggregate for Structural Concrete.*[4]

Both of these types of lightweight aggregate are essentially manufactured materials. The selection of suitable and economic sources of supply is therefore a matter of finding appropriate manufacturing plants close to the jobsite or of otherwise arranging for the supply of materials of suitable quality at a reasonable shipping cost. The number of aggregate manufacturing plants and their geographic locations are such that these materials can generally be made available for most jobs.

WATER

Water is a most important material on any heavy-construction job. It is used in the concrete as mixing water and on the surface of the concrete for curing. Considering the attention that is devoted to the testing, inspection, and control of the other ingredients of the concrete, it may seem unusual that so little attention is given to this very important material.

Fortunately the general criterion applies that water which is good enough to drink

* See Sec. 20, Aggregate Production.

is satisfactory for use in concrete. The converse is not necessarily true, since water that is not potable may often be quite satisfactory for use as mixing water. For example, except where corrosion of imbedded items might be a problem, seawater can be and has been used quite satisfactorily as mixing water in concrete. However, such use is not recommended if satisfactory fresh water is available. In the event of doubt about a source of water, there are more reliable criteria of its acceptability than its potability. If the pH of the water is between 6.0 and 8.0 and if it is free from organic matter, it may be considered to be safe for use. Also, if the 7- and 28-day strengths of mortar cubes made with the questionable water are within 90 percent of similar mortar cubes made with distilled water, the water may be considered acceptable.[4]

Water is essential to hydration, the chemical interaction between the water and the cement, and it is also required to make the plastic concrete sufficiently workable so that it can be handled and consolidated in the forms. Approximately 2.5 to 3 gal of water per sack of cement are required for the chemical hydration process; and this proportion, in theory, will produce maximum strength. However, as a practical consideration such concrete will be completely unworkable for normal consolidation procedures. It is therefore necessary to add extra water to produce the required workability. The key words are "required workability," and the addition of water beyond this point will rapidly reduce the strength of the concrete as well as all its other desirable properties.

Water is also used for curing the concrete, particularly in the case of large structures such as dams and bridges where ample supplies of water are almost always readily available. The primary consideration for the acceptability of water for use in curing is that it be nonstaining, especially for structures exposed to public view and where aesthetics or architectural appearance are important. Rather minor amounts of staining materials in the water may cause objectionable surface staining over the extended curing period. The principal objectionable materials in such water are iron and organic matter. The potential staining characteristics of these impurities cannot necessarily be evaluated on the basis of a chemical analysis. They are best evaluated by a performance type of test where a small concrete slab is actually subjected to curing with the water in question. An example of such an evaluation is Test Method[6] CRD C 401 of the Army Corps of Engineers.

ADMIXTURES

The use of admixtures has helped to improve the quality and economy of concrete as a construction material. The acceptance of proved materials has grown steadily, and admixtures of one type or another are used in approximately 80 percent of the concrete placed in the United States and Canada. They are selected for specific purposes and have provided qualities which in most cases could not have been obtained as effectively or as economically by any other means.

A definition of terms is always desirable, and this seems to be particularly true with respect to admixtures. In the concrete industry an admixture is defined as any material other than water, aggregates, and hydraulic cement used as an ingredient of a concrete or mortar mixture and added to the batch immediately before or during mixing. When such materials are interground or blended with the hydraulic cement, they are described as "additions" to cement. For example, an air-entraining agent which is added to a cement is properly described as an "addition." When the same air-entraining agent is batched directly into the concrete, it is properly described as an air-entraining "admixture." Admixtures, rather than additions, are the materials under discussion in this section.

In addition to the admixtures which entrain air, there are many others which may be used to procure certain benefits. The most widely used are chemical admixtures to control setting time and/or to reduce water. Others are accelerators, retarders, plasticizers, cementitious and pozzolanic materials, waterproofers, and materials used to inhibit alkali-aggregate expansion.

Air-entraining Admixtures The most commonly used admixtures are those employed to entrain air.

Occasionally concrete may be placed where weight is an important factor, but the engineer who specifies density is usually endeavoring to obtain concrete which will be resistant to the passage of water and to the effects of corrosive solutions and which will withstand severe weathering. It is almost universally true that these objectives can be met most readily by concrete which contains entrained air.

Consider, first of all, the action that takes place in a non-air-entrained concrete mix. When the concrete is first placed the aggregates are uniformly dispersed within a more or less fluid medium. Since they are heavier than the fluid, they tend to settle, even in fairly stiff mixes, and this settling action will force the light ingredients in the concrete to the surface. These materials are the water and weak fines, such as silt. Thus, in the zone where maximum resistance to wear and weathering is needed, there is a layer of the weakest materials in the concrete. In addition, as the water moves to the surface it creates connected channels and voids. These provide easy access for moisture after the concrete has cured. If the aggregate particles continue to settle until they are in intimate contact with one another, they have little opportunity to move with the cement paste as it sets. The paste shrinks as it dries out after curing; and if movement of the aggregate is restricted or prevented, some bond may be lost. When subjected to weathering, such as freezing moisture, this condition may result in pop-outs.

In air-entrained concrete, on the other hand, the minute, disconnected bubbles of air which are dispersed uniformly throughout the mortar act much like water wings, buoying up the heavier pieces of aggregate. Since settlement of the heavier ingredients is reduced, bleeding of water and the light fines to the surface is held to a minimum. Some bleeding does occur, but the channels are fewer and, because of the interference of the air bubbles, they are disconnected.

Because the connected channels and voids are less frequent, permeability and absorption of the hardened concrete are reduced. The penetration of moisture into the concrete in cold weather is retarded, and the force exerted by the moisture when it freezes is absorbed and cushioned by the small bubbles of air which are a permanent part of the concrete. In the plastic state, the air also acts as a cushion that prevents the pieces of aggregate from making close contact. As a result, they are more free to follow the slight movement of the cement paste as it shrinks. Bond between the paste and the aggregates, especially at the surface, is improved. So long as the air is entrained properly and the amount is controlled within optimum limits, the concrete surface no longer consists of high water content and weak fines, and it will have greater resistance to both weathering and moderate wear. Resistance to freezing and thawing, particularly scaling caused by the heavy application of deicing salts, is very markedly reduced.

One disadvantage of the normal type of air-entraining admixture is the possible loss of strength in medium and rich mixes. This disadvantage can frequently be overcome by the use of a water-reducing admixture in conjunction with the air-entraining admixture. Water-reducing admixtures will be discussed in more detail later, but it is pertinent to mention them here. Some water-reducing admixtures will concurrently entrain air, others will not but will greatly reduce the amount of air-entraining admixture required to produce a specified amount of air in the concrete. Water-reducing admixtures in each of the above categories will reduce or completely overcome the strength loss that is common in medium or rich mixes when only the normal types of air-entraining admixtures are used. Such a combination may be of interest to the contractor who is required to meet mix specifications for both strength and air content. In lean mixes, the water reduction produced by normal air-entraining admixtures can often be sufficient to counteract the effect of air on strength.

Accelerators Calcium chloride, the least expensive of the known accelerators, is also widely favored because of simplicity of use and ease of control. However, the use of calcium chloride or of any other accelerator does not remove the need for good cold-weather concreting practice.[7] Calcium chloride should not be used where stray dc electric currents might exist, where embedded items of different metals are in contact or in close proximity, or where the concrete is in contact with stressed wires or cables. These prohibitions apply to the use of calcium chloride in the normally used amounts of 1 or 2 percent by weight of the cement. They also apply to equivalent amounts of any other chloride that may be added to or be present in the materials used in the concrete.

With the possible exception of prestressed wires, on which there is a lack of definitive information, it is generally agreed that the presence of up to 0.1 percent calcium chloride by weight of cement or its equivalent of other chlorides will cause no corrosion problem. There is a tendency on the part of some specifiers to require that "the admixture shall contain no calcium chloride." This is unrealistic from several standpoints. First, within the maximum limit of 0.1 percent there is no demonstrable evidence of any corrosion problem due to calcium chloride. Next, the requirement does not preclude the use of another chloride. For example, small amounts of chloride are essential ingredients of other types of admixtures. In reality, except for reagent-type chemicals, detectable amounts of chloride can be found in almost any chemical or concrete material. If it is actually desired that no chloride be present in the concrete, it is not sufficient to exclude chloride from an admixture which is used in a very small amount compared to the other ingredients of the concrete. The reasons for and the questionable benefits of such highly restrictive requirements should be carefully considered before specifying.

Other accelerating admixtures are available, but in almost every case it would not be practical to use them alone. This may be because of their physical form, because the small amounts required would make dispensing hazardous, because of their expense, or because of their unpredictable nature when used in too small or too large amounts. Those that are used in concrete are generally combined with other materials into a multicomponent admixture. Many of these have the added advantage of water reduction, and some have none of the disadvantages of calcium chloride.[8] They should meet the requirements of ASTM C 494, *Chemical Admixtures for Concrete*,[4] for Type C if water reduction is not required and for Type E if water reduction is required or desired—which it normally is.

Retarders The demand for retarding agents has been stimulated by the fact that concrete without retardation may set too fast to permit achievement of the desired design characteristics of the structure. Examples are bridge decks or other concrete structures placed over cambered steel where the dead load of the concrete must be in place before the concrete sets or where it is necessary to prevent cold joints in mass concrete or other types of construction. Another stimulus has been the difficulty experienced at times with false-setting cement. Sucrose, starches or cellulose materials, gypsum, and some organic acids and their salts are effective as retarders. Their use requires very precise dispensing and careful inspection and control on the job to prevent serious damage to the concrete, such as prolonged delay of set or excessive expansion of the concrete.

Some retarding admixtures which contain carbohydrate derivatives have been developed. These have been found quite satisfactory in controlling or slowing down the set of concrete as long as they are used in accordance with the manufacturer's instructions. They are capable of moderate water reductions, an important benefit in itself. Some may tend to increase bleeding of the concrete, which is often considered objectionable.

One form of salts of sulfonated lignin has established an excellent record as a retarder as well as a means for controlling the evolution of heat in mass concrete. Under most circumstances the quantity required for almost exact control of setting time can be determined with minimum preliminary tests. The substance also entrains air in the concrete, thus providing a high degree of durability and reducing water requirements with consequent gain in ultimate strength. These properties assure the contractor that it will have no detrimental effect when properly used.

ASTM Special Technical Publication No. 266 describes the use of water-reducing admixtures and set-retarding admixtures and their effects on the properties of concrete.[9] Additional information on the use and control of retarders may be found in an excellent paper by Tuthill and Cordon.[10]

Water-reducing, Set-controlling Admixtures[4] Admixtures in this category are widely used. ASTM C 494 includes requirements for five types. These are:

Type A: Water-reducing admixtures
Type B: Retarding admixtures
Type C: Accelerating admixtures
Type D: Water-reducing and retarding admixtures
Type E: Water-reducing and accelerating admixtures

Type A admixtures are primarily designed to reduce the water content of concrete without significantly affecting setting time. The other types, as their names imply, are for controlling the setting time with or without water reduction.

Plasticizers Air-entraining agents, some accelerators, and the retarders have a plasticizing action in concrete. With slump or workability held constant, these materials permit reductions in the water content as compared to a plain mix. By the same token, they permit placement of concrete in heavily reinforced or complicated form work without increasing the water content of the mix.

Finely divided materials, inert fillers, and cementitious and pozzolanic products might also serve as plasticizing agents in mixes which are deficient in fines. Many of these plasticizers, however, will have no special benefit in mixes containing properly graded aggregates. Adding plasticizers to such mixes may even increase the water demand.

Cementitious and Pozzolanic Materials This group includes natural cements, slag cements, hydraulic lime, fly ash, volcanic ash, diatomaceous earth, and other materials. Their presence in concrete otherwise lacking fines may be quite beneficial. On jobs where prolonged wet curing is possible, some of them may be used to distinct advantage as a replacement for part of the cement, thus reducing the heat of hydration. The cements and limes usually have service records which the contractor can examine when assessing their value. The others in this group may require investigation and specific testing under job conditions to establish their merit.

Waterproofers All construction men know that there is no such thing as waterproof concrete. Agents sold for waterproofing purposes are advertised as helpful in "reducing permeability" or "reducing absorption." Any admixture which reduces water, entrains air, or reduces the voids in the hardened concrete is more or less effective in reducing both absorption and permeability. The extent of its influence as a waterproofer can probably be related directly to its ability to reduce water and bleeding. Those with the lowest water requirements and with the lowest bleeding will produce concrete that is more nearly watertight.

Water repellents, such as fatty acids, are often used to coat the pores and capillaries of the concrete. They help to resist the ingress of water in walls above grade or in slabs laid on the ground, but their value in concrete below grade and subject to water pressure is questionable. If sizable pressures are involved, they may actually be detrimental. Below-grade concrete is waterproofed better and more economically by the use of water-reducing and air-entraining agents, and the best results above grade or on the ground are procured by combining these with one of the water repellents.

Inhibitors of Alkali-Aggregate Reaction One of the greatest headaches to construction men in some parts of the United States has been the reaction between the alkalis in cement and certain aggregates, causing an expansion which results in cracking of the concrete and occasionally in what might be described as a blowout rather than a pop-out. Unfortunately, research on the subject is not conclusive enough at this time to designate any specific materials as certain cures. It is known that the use of low-alkali cement is indicated, and there seems to be evidence that some pozzolanic materials are quite helpful in counteracting the reaction. There is a wealth of literature on the phenomenon of alkali-aggregate reaction, means of detecting it, and proposed ways of controlling it.

OTHER RELATED MATERIALS

There are two other types of materials which, although they do not enter into the categories previously discussed and are not actually used in the concrete, are frequently a part of a large concrete project, particularly for industrial construction.

Special Treatments for Concrete Floors Concrete floors subject to heavy-duty wear require a quality that might be termed "toughness." Much depends on the type of traffic or usage of the floor. If abrasion and impact are involved, the hard, brittle concrete aggregates and cement paste that produce such excellent results elsewhere may prove to be a handicap. For areas subject to impact and abrasion, a softer but tougher material is desirable. Iron is one such material.

A number of commercial products applied as "shakes" are available and contain reasonably well-graded iron particles as the aggregate. In selecting the one to be employed, due consideration should be given to the gradation of the iron (for the densest surface, the grading should be similar to that of good concrete sand), the cleanliness and freedom from oil of the particles, and the absence of any detrimental materials such as aluminum or zinc. It is also helpful if workability agents are present so that the shake can be incorporated into a concrete of low water content. As a general rule, the use of air-entraining admixtures in the base concrete will not permit the incorporation of sufficient shake material nor allow proper finishing of the surface.

Shake materials are available in a variety of colors for enhancing the appearance of concrete floors, defining traffic aisles, and for other purposes. They are available with natural aggregates for use where heavy traffic is not a problem or with iron aggregates for heavy industrial applications.

Grouting Materials for Machinery and Bridge Plates This application of concrete or mortar involves peculiar problems and requires the use of special admixtures. Some persons advocate dry packing of grouts, but subsequent dismantling of such installations has sometimes shown as little as 15 percent contact between bed plate and grout. The use of flowable grouts has proved more satisfactory for two reasons: (1) there is greater assurance of filling the grouting space, and (2) there is a substantial saving in labor.

When flowable grouts are employed, an expanding cement or an admixture is used to counteract shrinkage of the mortar. The common admixtures contain either aluminum powder or iron aggregate combined with a catalyst. The aluminum powder generates a gas which causes an expansion. In the iron type, the catalyst causes a controlled shrinkage-correcting action which affects both settlement and drying.

Use of aluminum involves certain precautions which, for successful results, make it mandatory to have a competent concrete technician on hand to supervise its use. With the iron-grouting compound there are fewer factors which affect its efficiency, and the desired result can be obtained by average workmen following the directions of the manufacturer.

PROPORTIONING CONCRETE MIXTURES

The proportioning of a concrete mixture is the process of determining the most economical and practical combination of materials that will produce a mixture which in the plastic state can be readily handled and consolidated in the forms and which in the hardened state will develop the required strength, durability, volume stability, and watertightness for the job under consideration.

A discussion of all the considerations involved in the proportions of concrete mixtures is not possible in this section, but an attempt is made to provide sufficient information to enable the reader to design an adequate and economical mix to meet all but the most stringent job requirements. Free use has been made of the information contained in ACI 613, *Recommended Practice for Selecting Proportions for Concrete,* published by the American Concrete Institute.[11]

Contract specifications for concrete are generally based on one of three basic approaches:

1. Mixture proportions designed and controlled by the owner
2. Minimum cement content, maximum water-cement ratio, and range of slump specified
3. Concrete strength specified, usually at an age of twenty-eight days

For contracts under condition 1, the contractor has only to comply with the instructions of the owner with respect to the mixture proportions provided. Under condition 2, the contractor may prepare his own mixture designs but the cement, water, and slump requirements must be within specification limits. Under condition 3, the contractor has complete freedom in the selection of mixture proportions, subject only to the requirement that the concrete attain the specified strength at the age or ages indicated. The responsibility of the contractor is obviously considerably greater under condition 3 than under conditions 1 and 2. However, in this situation, he also has greater control over the cost of concrete production.

Basic Physical Properties of Materials The detailed specification requirements for cement and the general considerations relative to the selection of aggregate sources from a quality standpoint have been discussed earlier. The specific properties which must be known for computation purposes or which have a bearing on the proportioning of concrete mixtures are reviewed below. Accepted methods for determining the properties of the materials are contained in references 1 and 4 at the end of this section.

Cement. The physical and chemical characteristics of the cement influence the properties of the hardened concrete. For this reason, cement from the same source and of the same type as that which will be used in actual construction should be used in the laboratory concrete tests. However, the only property of the cement directly concerned in the computation of concrete-mix proportions is specific gravity. For blends of portland cement with other hydraulic materials, such as natural cement or slag cement, the specific gravity of the individual cements should be determined and the specific gravity of the blend to be used on the job should be calculated by direct proportion. For normal portland cements, both plain and air-entraining, the specific gravity may be assumed to be 3.15 without introducing appreciable error in the mix computations.

Aggregates. Gradation, specific gravity, absorption, and moisture content of both fine and coarse aggregate are physical properties essential for mix computations. The dry-rodded weight of the coarse aggregate is also required. Other tests which may be desirable for large or special types of work include petrographic examination and tests for chemical reactivity, soundness, durability, and resistance to abrasion, plus tests for various deleterious substances. All such tests provide information of value in judging the ultimate quality of the concrete and in selecting appropriate proportions.

Aggregate gradation or particle-size distribution is a major factor in controlling unit water requirements, the proportion of coarse aggregate to sand, and the cement content of concrete mixes for a given degree of workability. While numerous "ideal" aggregate grading curves have been proposed, experience and individual judgment must continue to play important roles in determining acceptable aggregate gradings. Additional workability realized by use of air-entrainment permits, to some extent, the use of less restrictive aggregate gradations, but caution should be used in depending on this means of "improving" the gradation.

Undesirable sand grading may be corrected to desired particle-size distribution by separating the sand into two or more size fractions and recombining in suitable proportions, increasing or decreasing the quantity of certain sizes to balance the grading, or reducing excess coarse material by grinding. Undesirable coarse aggregate grading may be corrected by crushing excess coarser fractions, wasting excess material in other fractions, supplementing deficient sizes from other sources, or a combination of these methods. To the extent that grading limitations and economy in use of cement permit, the proportions of various sizes of coarse aggregate should be held closely to the grading of available materials. Whatever processing is done in the laboratory should be practical from the standpoint of field economy and job operation. Samples of aggregates for concrete-mix tests should be representative of aggregate selected for use in the work. For laboratory tests, the coarse aggregate should be cleanly separated into the required size fractions to provide for uniform control of mix proportions.

Elements of the Trial Mix The first step toward the selection of the proportions of materials to be used in the concrete is to estimate, on the basis of the best information available on the materials and job conditions, the trial mix which will require the least final adjustment. The following basic properties of the materials must be determined by actual laboratory tests before the trial mixes can be made:

1. Sieve analysis, specific gravity, absorption, and moisture content when batched for both the fine and the coarse aggregate.
2. Dry-rodded unit weight of the coarse aggregate.
3. Whether or not the cement is air-entraining, and its specific gravity. The specific gravity can generally be taken as 3.15 without the introduction of appreciable error in the computations.

After determination of the above properties of the materials, the initial trial proportions are established by following the seven steps indicated below and explained in detail immediately thereafter.

1. Select the water-cement ratio from established relationships to meet the specified requirements for durability and strength.
2. Select the limits of slump that will permit proper handling and consolidation of the concrete under job conditions.
3. Determine the maximum available size of coarse aggregate which is suitable for use under the job conditions.
4. Estimate the minimum percentage of sand that will provide adequate workability.
5. Estimate the amount of water per cubic yard of concrete necessary to meet the requirements for slump, maximum aggregate size, and percent sand, as established in steps 2, 3, and 4.
6. If an admixture is used, select the kind and amount required to comply with the conditions established in steps 1 to 5.
7. Calculate the trial-mix proportions that will meet the above conditions.

Water-Cement Ratio. This should be selected on the basis of the durability and strength required in the structure. The durability requirements will be governed by the field-exposure conditions, and the strength will be governed by the design loads. Where the contract specifications state a minimum cement factor and a maximum permissible water-cement ratio, these requirements must, of course, be met.

Entrained air is of great benefit in ensuring durable concrete, and should be used when exposure to weathering is expected to be severe. It is also of great benefit in increasing the workability of the concrete, particularly for lean mixes for massive sections. Table 21-1 indicates the maximum permissible water-cement ratios in terms of gallons of water per bag of cement for various types of structures and degrees of exposure. This table guides the user in selecting the water-cement ratio required for adequate durability.

The maximum water-cement ratio or minimum cement factor to produce the required strength can best be determined by making and testing specimens with the concrete proportioned in accordance with the trial mix or mixes and using the same materials, including cement, that will be used in the job. Table 21-2 indicates the approximate strengths that can be expected for a range of water-cement ratios using materials of good quality in a well-designed mix. A corresponding table for flexural strength is not given because the flexural strength varies over an excessively wide range with given proportions of different materials. Where flexural strength is specified, the required proportions should be determined by laboratory tests using job materials to ensure that the required strengths will be attained. The compressive strength should also be verified by laboratory tests, but the water-cement ratios shown in Table 21-2 should normally approximate the required strengths.

Strengths for air-entrained concrete for a given water-cement ratio (Table 21-2) are indicated as being 20 percent lower than for non-air-entrained concrete. This reduction applies only when the water-cement ratio is the same for both the plain and the air-entrained concrete. If the cement content and consistency are kept constant, less mixing water is required for the air-entrained concrete than for the plain; the resulting decrease in water-cement ratio increases the strength, which partially or entirely offsets the reduction in strength caused by air entrainment.

Slump and maximum aggregate size should now be selected. Concrete proportions should be such that the concrete has the stiffest consistency (lowest slump) that will allow it to be placed efficiently and thoroughly. In the selection of the slump for concrete for heavy construction, full advantage should be taken of the reductions in slump that can be attained by the consolidation of concrete with large heavy-duty vibrators, which are a requirement for such construction. Table 21-3 shows the recommended slumps for various construction conditions.

Within the limits of economy and form and reinforcing spacing, the largest permissible maximum size of aggregate should be used. The use of large aggregate permits a reduction in water and cement requirements because the paste content of the concrete is decreased. The maximum sizes of aggregate recommended for various types of construction are shown in Table 21-4. Aggregate larger than 6 in. in size is rarely used because the requirements for processing and handling equipment rapidly approach uneconomical proportions.

TABLE 21-1 Maximum Permissible Water-Cement Ratios for Different Types of Structures and Degrees of Exposure (in gallons per bag)

Type of structure	Exposure conditions*					
	Severe wide range in temperature or frequent alternations of freezing and thawing (air-entrained concrete only)			Mild temperature rarely below freezing, or rainy, or arid		
	In air	At the water line or within the range of fluctuating water level or spray		In air	At the water line or within the range of fluctuating water level or spray	
		In fresh water	In sea water or in contact with sulfates†		In fresh water	In sea water or in contact with sulfates†
Thin sections, such as railings, curbs, sills, ledges, ornamental or architectural concrete, reinforced piles, pipe, and all sections with less than 1 in. concrete cover over reinforcing	5.5	5.0	4.5‡	6	5.5	4.5‡
Moderate sections, such as retaining walls, abutments, piers, girders, beams	6.0	5.5	5.0‡	§	6.0	5.0‡
Exterior portions of heavy (mass) sections	6.5	5.5	5.0‡	§	6.0	5.0‡
Concrete deposited by tremie under water	...	5.0	5.0	...	5.0	5.0
Concrete slabs laid on the ground	6.0	...	....	§		
Concrete protected from the weather, interiors of buildings, concrete below ground	§	...	....	§		
Concrete which will later be protected by enclosure or backfill but which may be exposed to freezing and thawing for several years before such protection is offered	6.0	...	....	§		

* Air-entrained concrete should be used under all conditions involving exposure and may be used under mild exposure conditions to improve workability of the mixture.

† Soil or groundwater containing sulfate concentrations of more than 0.2 percent.

‡ When sulfate-resisting cement is used, the maximum water-cement ratio may be increased by 0.5 gal per bag.

§ Water-cement ratio should be selected on basis of strength and workability requirements.

Unit water content can be estimated from Table 21-5, which indicates the approximate water requirements for different slumps and maximum sizes of aggregates. It

TABLE 21-2 Compressive Strengths of Concrete for Various Water-Cement Ratios*

Water-cement ratio, gal/bag of cement	Probable compressive strength at 28 days, psi	
	Non-air-entrained concrete	Air-entrained concrete
4	6,000	4,800
5	5,000	4,000
6	4,000	3,200
7	3,200	2,600
8	2,500	2,000
9	2,000	1,600

* These average strengths are for concretes containing not more than the percentages of entrained and/or entrapped air shown in Table 21-5. For a constant water-cement ratio, the strength of the concrete is reduced as the air content is increased. For air contents higher than those listed in Table 21-5 the strengths will be proportionally less than those listed in this table.

Strengths are based on 6- by 12-in. cylinders moist-cured under standard conditions for 28 days. See Method of Making and Curing Concrete Compression and Flexure Test Specimens in the Field (ASTM Designation C 31).

TABLE 21-3 Recommended Slumps for Various Types of Construction*

Types of construction	Slump, in.†	
	Maximum	Minimum
Reinforced foundation walls and footings	5	2
Plain footings, caissons, and substructure walls	4	1
Slabs, beams, and reinforced walls	6	3
Building columns	6	3
Pavements	3	2
Heavy mass construction	3	1

* Adapted from Table 4 of the 1940 Joint Committee Report on Recommended Practice and Standard Specifications for Concrete and Reinforced Concrete, ACI.

† When high-frequency vibrators are used, the values given should be reduced by about one-third.

TABLE 21-4 Maximum Sizes of Aggregate Recommended for Various Types of Construction

Minimum dimension of section, in.	Maximum size of aggregate,* in.			
	Reinforced walls, beams, and columns	Unreinforced walls	Heavily reinforced slabs	Lightly reinforced or unreinforced slabs
2½–5	½–¾	¾	¾–1	¾–1½
6–11	¾–1½	1½	1½	1½–3
12–29	1½–3	3	1½–3	3
30 or more	1½–3	6	1½–3	3–6

* Based on square openings.

is well to keep in mind the importance of keeping the unit water content to the absolute minimum. The following statement appears in the fifth edition of the *Concrete Manual* of the Bureau of Reclamation: "For a given set of materials and water-cement ratio, the unit water content (water required per cubic yard of concrete) is the most important basic factor affecting the quality of concrete."[12] The quantity of water per unit volume of concrete required to produce a mix of the desired consistency is influenced by the maximum size, particle shape, and grading of the aggregate and by the amount of entrained air. Within the normal range of mixes, it is relatively unaffected by the quantity of cement.

The quantities of water given in Table 21-5 are sufficiently accurate for estimating initial proportions. They are the maxima which should be expected for the indicated maximum sizes of fairly well-shaped but angular aggregates, graded within the limits of conventional specifications. If otherwise suitable aggregates have higher water requirements than indicated in Table 21-5, it is probable that they are less favorably shaped or are not so well graded as normally may be expected. Unless otherwise indicated by laboratory tests, the proportions of such aggregates should be adjusted or the cement increased, or both, to maintain the desired water-cement ratio.

Cement requirements for a given concrete can be computed using the maximum permissible water-cement ratio selected from Table 21-1 or 21-2 and the water requirements from Table 21-5. The cement factor is obtained by dividing the gallons of

TABLE 21-5 Approximate Mixing Water Requirements for Different Slumps and Maximum Sizes of Aggregates*

Slump, in.	Water, gal/cu yd of concrete for indicated maximum sizes of aggregate							
	⅜ in.	½ in.	¾ in.	1 in.	1½ in.	2 in.	3 in.	6 in.
	Non-air-entrained concrete							
1–2	42	40	37	36	33	31	29	25
3–4	46	44	41	39	36	34	32	28
6–7	49	46	43	41	38	36	34	30
Approx amount of entrapped air in non-air-entrained concrete, %............	3	2.5	2	1.5	1	0.5	0.3	0.2
	Air-entrained concrete							
1–2	37	36	33	31	29	27	25	22
3–4	41	39	36	34	32	30	28	24
6–7	43	41	38	36	34	32	30	26
Recommended avg total air content, %.............	8	7	6	5	4.5	4	3.5	3

* These quantities of mixing water are for use in computing cement factors for trial batches. They are maxima for reasonably well-shaped, angular, coarse aggregates graded within limits of accepted specifications.

If *more* water is required than shown, the cement factor, estimated from these quantities, *should* be increased to maintain the desired water-cement ratio except as otherwise indicated by laboratory tests for strength.

If *less* water is required than shown, the cement factor, estimated from these quantities, *should not* be decreased except as indicated by laboratory tests for strength.

mixing water required per cubic yard by the water-cement ratio in gallons per bag of cement. If a minimum cement factor is specified, the corresponding water-cement ratio for estimating strength can be computed by dividing the gallons of water per cubic yard by the cement factor in bags per cubic yard. Selection of proportions for concrete should be based on whichever of the limitations specified (durability, strength, or cement factor) requires the lowest water-cement ratio.

Aggregate requirements can now be examined. The minimum amount of mixing water and the maximum strength will result for given aggregates and cement when the quantity of coarse aggregate used is the largest consistent with adequate placeability and workability. The quantity of coarse aggregate that can be used increases with the maximum size of aggregate. The quantity of coarse aggregate can be determined most effectively by laboratory investigations of the materials, subject to later adjustment as necessitated by field conditions. An estimate of the best proportions can be made on the basis of established relationships for aggregates graded within conventional limits. Table 21-6 shows these relationships, with the quantities of coarse aggregate expressed as dry-rodded bulk volumes per unit volume (cubic yard) of concrete.

Concrete of comparable workability can be expected with aggregates of comparable size, shape, and grading. This will be true if a given percentage of dry-rodded coarse aggregate is used per unit volume of concrete and if the volume of mortar remains constant, even though the solid volumes of cement, water, air, and sand may be interchanged. In the case of different types of aggregates, particularly those with different particle shapes, the use of a fixed, dry-rodded, solid volume of coarse aggregate automatically makes allowance for differences in mortar requirements as reflected by the void content of coarse aggregate. For example, angular aggregates have a higher void content and therefore require more mortar than rounded aggregates. A fuller discussion of the proportioning of concrete mixes on the basis of the dry-rodded volume of coarse aggregate may be found in reference 13 at the end of this section.

TABLE 21-6 Volume of Coarse Aggregate per Unit of Volume of Concrete*

Maximum size of aggregate, in.	Volume of dry-rodded coarse aggregate per unit volume of concrete for different fineness moduli of sand			
	2.40	2.60	2.80	3.00
3/8	0.46	0.44	0.42	0.40
1/2	0.55	0.53	0.51	0.49
3/4	0.65	0.63	0.61	0.59
1	0.70	0.68	0.66	0.64
1 1/2	0.76	0.74	0.72	0.70
2	0.79	0.77	0.75	0.73
3	0.84	0.82	0.80	C.78
6	0.90	0.88	0.86	0.84

* Volumes are based on aggregates in dry-rodded condition as described in Method of Test for Unit Weight of Aggregate (ASTM Designation C 29). These volumes are selected from empirical relationships to produce concrete with a degree of workability suitable for the usual reinforced construction. For less workable concrete such as required for concrete-pavement construction, they may be increased about 10 percent.

Another method of proportioning fine and coarse aggregate is to compute the total solid volume of aggregate in the concrete mix and multiply this volume by the recommended percentage of sand. The percentage of sand in the concrete mix has been used extensively as a means of expressing the proportions of sand and coarse aggregate; the sand is expressed as a percentage of the total solid volume of the entire aggregate. Table 21-7, taken from the Bureau of Reclamation *Concrete Manual*,[14] indicates the approximate air and water contents per cubic yard of concrete and the proportions of fine and coarse aggregate. It duplicates much of the information supplied in Table 21-6, but contains additional data such as the recommended percentages of sand for each maximum size of coarse aggregate.

Computation of Mix Proportions In the computations that are given below, examples are shown for computing mix proportions both on the basis of estimating the quantity of coarse aggregate from its dry-rodded volume (Example 1) or on the basis of the computed total solid volume of aggregate multiplied by the recommended percentage of sand (Example 2). Either method is satisfactory and under normal conditions will produce approximately the same proportions. Maintaining the solid volume of coarse aggregate constant in concrete has certain advantages: (1) mix adjustments for changes in air content and cement content are made automatically, and (2) in mixing-plant operation, batch-setting changes for coarse aggregate become unnecessary for changes in other proportions.

The computation of mix proportions and the application of the foregoing discussion can best be explained by showing the calculations for specific examples. For these examples the following design criteria and mix materials will be assumed:

1. Type I non-air-entraining cement will be used in Example 1 and Type II non-air-entraining cement in Example 2. The specific gravity of each cement will be taken as 3.15 and the weight of a bag as 94 lb.

2. The coarse aggregate has a specific gravity, bulk dry, of 2.68 and an absorption of 0.5 percent.

3. The fine aggregate has a specific gravity, bulk dry, of 2.64; an absorption of 0.7 percent; and a fineness modulus of 2.8.

4. Both the coarse and fine aggregates are of satisfactory quality. The coarse aggregate is stockpiled in four size groups: No. 4 to ¾ in., ¾ to 1½ in., 1½ to 3 in., and 3 to 6 in.

TABLE 21-7 Approximate Air and Water Contents per Cubic Yard of Concrete and the Proportions of Fine and Coarse Aggregate (For concrete containing natural sand with a fineness modulus of 2.75 and average coarse aggregate and having a slump of 3 to 4 in. at the mixer)

Max size of coarse aggregate, in.	Actual weight of coarse aggregate per unit volume of concrete as a percentage of dry-rodded unit weight	Air-entrained concrete			Non-air-entrained concrete		
		Recommended air content, %	Avg. water content, lb/cu yd	Sand, % of total aggregate by solid volume	Approx. percent entrapped air	Avg. water content, lb/cu yd	Sand, % of total aggregate by solid by volume
⅜	41	8	322	59	3.0	352	61
½	52	7	306	50	2.5	336	53
¾	62	6	283	42	2.0	316	45
1	67	5	267	37	1.5	300	41
1½	73	4.5	245	33	1.0	280	36
2	76	4	229	30	0.5	266	33
3	81	3.5	204	28	0.3	242	31
6	87	3	164	24	0.2	210	28

Adjustment of values for other conditions

Changes in conditions stipulated	Effect on values: Unit water content, %	Sand, %	Dry-rodded coarse aggregate, %
Each 0.1 increase or decrease in fineness modulus of sand	....	±0.5	±1
Each 1-in. increase or decrease in slump	±3		
Each 1% increase or decrease in air content	±3	±0.5–1.0	
Each 0.05 increase or decrease in water-cement ratio	....	±1	
Each 1% increase or decrease in sand content	±1	....	±2
For angular coarse aggregate	+7 to 10	+3 to 5	
For lower slump concrete as in pavements	−3	−3	+6

NOTE: If aggregates are proportioned by percent sand method, use the first and second column; if by dry-rodded coarse aggregate method, use the first and third columns.

Example 1. Concrete is required for an architectural reinforced-concrete wall having a minimum thickness of 5 in. The wall will be subjected to wide ranges in air temperature in a northern climate. A 28-day compressive strength of 3,500 psi is specified for at least 80 percent of the strength tests. The concrete will be consolidated with high-frequency vibrators.

The trial-batch proportions are computed as follows:

1. Because of the severity of the exposure, air-entrained concrete will be used. Reference to Table 21-1 indicates that the water-cement ratio should not exceed 5.5 gal per bag.

2. Table 21-2 indicates that a water-cement ratio of 5 gal per bag, with air-entrained concrete, will produce a compressive strength at 28 days of 4,000 psi. Previous experience with job control[15] indicates that approximately this average strength is required to ensure that at least 80 percent of the strength tests will be above the 3,500 psi specified. For this reason the lower value of 5 gal per bag will be used.

3. Tables 21-3 and 21-4 show that a maximum slump of 4 in. and a maximum aggregate size of ¾ in. should be used.

4. The approximate quantity of mixing water for a 4-in. slump in air-entrained concrete with ¾-in. aggregate is found in Table 21-5 to be 36 gal per cu yd. This table also indicates that the air content should be 6 percent. The entrained air will be obtained by use of an admixture added at the mixer as a solution. The volume of solution added should therefore be included in the computations as part of the volume of mixing water. The quantity of solution added should be sufficient to entrain the required amount of air. The amount recommended by the manufacturer will, in most cases, produce the desired air content. In this example it is assumed that 1 qt of air-entraining solution per bag of cement is required to provide 6 percent air.

5. From items 2 and 4 the required cement content is computed as follows:

$$\frac{36 \text{ gal water/cu yd}}{5 \text{ gal water/bag of cement}} = 7.2 \text{ bags/cu yd}$$

6. From Table 21-6 it is found that, with a fine aggregate having a fineness modulus of 2.8 and a ¾-in. coarse aggregate, 0.61 cu ft of coarse aggregate on a dry-rodded basis will be used in each cubic foot of concrete. The quantity in a cubic yard will then be $27 \times 0.61 = 16.47$ (16.5 cu ft). The dry-rodded weight was determined to be 100 lb per cu ft; so the weight of coarse aggregate per cubic yard of concrete will be 100×16.5, or 1,650 lb.

7. Having established the quantities of cement, water, coarse aggregate, and air, the sand content is calculated as follows:

$$\text{Solid volume of cement} = \frac{7.2 \text{ bags/cu yd} \times 94 \text{ lb/bag}}{3.15 \text{ (sp gr)} \times 62.4 \text{ lb water/cu ft}} = 3.44 \text{ cu ft}$$

$$\text{Volume of water} = \frac{36 \text{ gal/cu yd}}{7.5 \text{ gal/cu ft}} = 4.80 \text{ cu ft}$$

$$\text{Solid volume of coarse aggregate} = \frac{1{,}650 \text{ lb/cu yd}}{2.68 \text{ (sp gr)} \times 62.4} = 9.87 \text{ cu ft}$$

Volume of air $= 0.06 \times 27$	= 1.62 cu ft
Total solid volume of ingredients except sand	= 19.73 cu ft
Solid volume of sand required $= 27 - 19.73$	= 7.27 cu ft
Required weight of dry sand $= 7.27 \times 2.64$ (sp gr) $\times 62.4$	= 1,198 lb

8. The estimated batch weights per cubic yard of concrete are:

Cement $= 7.2$ bags $\times 94$ lb/bag	= 677 lb
Water $= 36$ gal $\times 8.33$ lb/gal	= 300 lb
Sand (dry basis)	= 1,198 lb
Coarse aggregate (dry basis)	= 1,650 lb

9. It should be noted that the above batch weights are for 1 cu yd of concrete on a dry basis, no correction having been made for moisture content. The fine and coarse aggregate used in the field will almost always carry some free moisture. The moisture

will also vary from day to day or even from hour to hour. The determination of moisture contents and the correction of batch weights based on the determinations are therefore among the most important operations in the daily production of concrete. To illustrate the manner in which such corrections are made, let us assume that the moisture content of the sand has been found by test to be 5.0 percent and the coarse aggregate, 1.0 percent. The weight of dry sand was found to be 1,198 lb. The amount of moist sand to be weighed out must therefore be 1,198 × 1.05 = 1,258 lb. On the same basis, the weight of moist coarse aggregate will be 1,650 × 1.01 = 1,667 lb.

10. The water in the aggregates in excess of the amount which they will absorb must be considered as part of the mixing water and the weight of the mixing water corrected accordingly. The absorption of the sand was previously stated as being 0.7 percent. The amount of free water which the sand contains is therefore 5.0 − 0.7 = 4.3 percent. Correspondingly, the free water on the coarse aggregate will be 1.0 − 0.5 = 0.5 percent.

The weight of mixing water contributed by the sand will be 0.043 × 1,198 = 52 lb, and for the coarse aggregate it will be 0.005 × 1,650 = 8 lb. These two weights (52 + 8 = 60 lb) must be subtracted from the indicated weight of mixing water (300 − 60). The weight of water to be batched will then be 240 lb, assuming that the air-entraining admixture is weighed with the mixing water. If the air-entraining admixture is added separately, the weight of mixing water should be reduced correspondingly. In this instance, 1 qt or 0.25 gal of solution is required per bag of cement and the total weight of the solution is 0.25 gal/bag × 7.2 bags × 8.33 lb/gal = 15 lb. The net weight of mixing water per cubic yard of concrete is then 240 − 15 = 225 lb.

11. If a 5-yd batch is required, each of the batch weights will be multiplied by 5, and for batches of other sizes the individual batch weights will be multiplied by the number of yards required. Occasionally, particularly for mixers of fractional-yard capacities, the batch size is specified in terms of the number of bags of cement required for the batch; for example, if it is desired to use a three-bag batch in a 16-cu-ft mixer, the total volume of concrete will be 27 × 3/7.2 = 13.3 cu ft, and each batch weight of the materials will be 3/7.2 of that given above.

Example 2. In the first example, since coarse aggregate of ¾-in. maximum size was used, it was in accordance with good practice to batch the coarse aggregate as a single ingredient. Now assume that concrete is required for a massive footing which will be below ground level and will not be exposed to weathering or other attack. Structural-design requirements call for a 28-day compressive strength of 2,000 psi. No reinforcing is required.

The trial batch proportions are calculated as follows:

1. While the structure will not be exposed to weathering, it is decided to use an air-entraining admixture to provide added workability and to reduce the water content. The water-cement ratio will be established solely on the basis of the required strength. Table 21-2 indicates that a water-cement ratio of 8 gal per bag will produce a strength of 2,000 psi. Since the specifications require only a nominal strength of 2,000 psi with no minimum requirement with respect to tests, this water-cement ratio will be acceptable.

2. Table 21-3 shows that a slump of 1 to 3 in. should be used. Consolidation will be accomplished with heavy-duty vibrators of normal frequency. The mix will be proportioned for an average 2-in. slump.

3. Table 21-4 recommends the use of 6-in. maximum-size coarse aggregate, which is available.

4. Table 21-5 indicates that, for 6-in. aggregate and a 1- to 2-in. slump, the approximate mixing water requirement will be 22 gal per cu yd at a recommended air content of 3 percent.

5. From items 1 and 4 the cement content is found to be 22 gal per cu yd/8 gal per bag = 2.8 bags per cu yd.

6. Because it is difficult to determine accurately the dry-rodded unit weight of 6-in. aggregate, the aggregate proportions will be established on the basis of percent sand. Previous experience (see Table 21-7) indicates that approximately 24 percent sand should be satisfactory.

7. On the basis of the above established values, the aggregate volumes for 1 cu yd are calculated as follows:

$$\text{Solid volume of cement} = \frac{2.8 \times 94}{3.15 \times 62.4} = 1.34 \text{ cu ft}$$

$$\text{Volume of water} = \frac{22}{7.5} = 2.93 \text{ cu ft}$$

Volume of air = 0.03 × 27 = 0.81 cu ft
Solid volume of all ingredients except aggregate = 5.08 cu ft
Total solid volume of aggregates = 27 − 5.08 = 21.92 cu ft
Solid volume of sand = 21.92 × 0.24 = 5.26 cu ft
Solid volume of coarse aggregate = 21.92 − 5.26 = 16.66 cu ft

8. The estimated batch weights per cubic yard of concrete are:

Cement = 2.8 × 94 = 263 lb
Water = 22 × 8.33 = 183 lb
Sand (dry basis) = 5.26 × 2.64 × 62.4 = 867 lb
Coarse aggregate (dry basis) = 16.66 × 2.68 × 62.4 = 2,786 lb

9. The coarse aggregate is stockpiled in four size groups and will be weighed in four batches. It is therefore necessary to calculate a combined grading and to determine the percent of each size group and, from this, the batch weight for each group.

A number of approaches to the optimum combined grading of aggregates have been developed and discussed in the literature. However, it is believed that combining the various sizes on the basis of maximum density is the most widely used and accepted method. The procedure discussed below is based on the work of Talbot and Richart,[16] modified by Bolomey,[17] and is the procedure used by the Corps of Engineers.[6] The relative proportions of the size groups are determined by taking a complete grading combination from the coarse-aggregate grading curves shown in Fig. 21-1. These curves start at the No. 4 sieve and are derived from the following formula:

$$p = \frac{\sqrt{d} - \sqrt{0.1875}}{\sqrt{d_{\max}} - \sqrt{0.1875}} 100$$

where d = sieve opening, in.
$d_{\max}$ = maximum size of aggregate, in.
p = cumulative percent passing

$$\sqrt{0.1875} = 0.433013$$

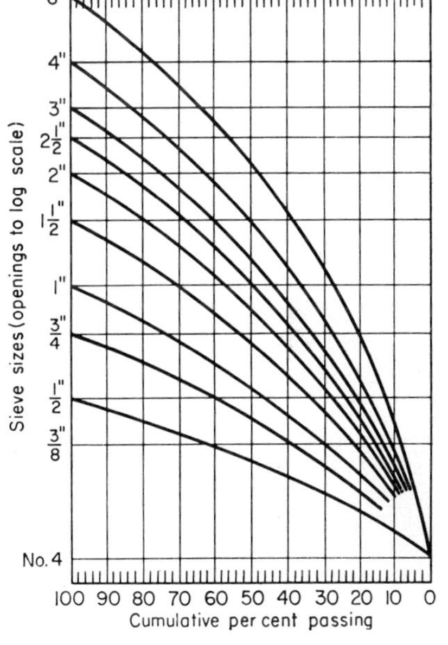

Fig. 21-1 Grading curves for coarse aggregate. (*Corps of Engineers.*)

Values derived from this equation are also given in Table 21-8. The objective in this example is to arrive at the combination of the four sizes of coarse aggregate that will approximate the maximum density curve for 6-in. maximum-size aggregate shown in Fig. 21-1 or listed in the second column in Table 21-8. Combined gradings using varying percentages of each of the four sizes of coarse aggregate are computed, and the proportions resulting in the combined grading most closely approximating the

TABLE 21-8 Proportions of Aggregate of Various Sizes

Sieve size, in.	Maximum size, in.									
	6	5	4	3	2½	2	1½	1	¾	½
	Cumulative percent passing									
6	100									
5	89.4	100								
4	77.7	86.9	100							
3	64.4	72.0	82.9	100						
2½	56.9	63.7	73.3	88.4	100					
2	48.7	54.4	62.6	75.5	85.5	100				
1½	39.3	43.9	50.5	60.9	69.0	80.7	100			
1	28.1	31.4	36.2	43.6	49.4	57.8	71.6	100		
¾	21.5	24.0	27.6	33.3	37.7	44.1	54.7	76.4	100	
½	13.6	15.2	17.5	21.1	23.9	27.9	34.6	48.3	63.3	100
⅜	8.9	9.9	11.4	13.8	15.6	18.3	22.6	31.6	41.4	65.4
No. 4										

TABLE 21-9 Combined Gradings

Sieve retained on	Grading of individual size groups, % retained				36%	26%	16%	22%	Recombined cumulative, %		Grading curve 6 in. max, %	
	3–6 in.	1½–3 in.	¾–1½ in.	No. 4–¾ in.	3–6 in.	1½–3 in.	¾–1½ in.	No. 4–¾ in.	Ret.	Pass.	Ret.	Pass.
	(1)	(2)	(3)	(4)	(5)	(6)	(7)	(8)	(9)	(10)	(11)	(12)
6 in.	1.0	0.0	.....	.....	0.4	0.0	....	....	0.4	99.6	0.0	100.0
3 in.	91.0	5.0	.....	.....	32.8	1.3	....	....	34.5	65.5	......	64.4
2 in.	8.0	56.0	.....	.....	2.8	14.6	....	....	51.9	48.1	51.3	48.7
1½ in.	0.0	33.0	0.0	.....	....	8.5	0.0	....	60.4	39.6	......	39.3
1 in.	.....	5.0	67.0	0.0	....	1.3	10.7	0.0	72.4	27.6	71.9	28.1
¾ in.	.....	1.0	28.0	4.0	....	0.3	4.5	0.9	78.1	21.9	......	21.5
½ in.	.....	0.0	3.0	28.0	....	....	0.5	6.2	84.8	15.2	86.4	13.6
⅜ in.	.....	.....	2.0	25.0	....	....	0.3	5.5	90.6	9.4	91.1	8.9
No. 4	.....	.....	0.0	39.0	....	....	....	8.6	99.2	0.8	100.0	0.0
Smaller than No. 4	.....	.....	.....	4.0	....	....	....	0.8	100.0	0.0		
Total	100.0	100.0	100.0	100.0	36.0	26.0	16.0	22.0				

grading taken from the curve are used. Table 21-9 illustrates the method of calculation. Columns 1 to 4 represent the grading of the four sizes of aggregate available at the job. From the procedure described above it is found that the proportions which best approximate the maximum density curve are 36 percent of the 3- to 6-in. aggregate, 26 percent of the 1½- to 3-in., 16 percent of the ¾- to 1½-in., and 22 percent of the No. 4 to ¾-in. The percentages of material retained on each sieve for each of the four size groups when used in these proportions are shown in columns 5 to 8. Columns 9 and 10 show the combined gradings, and columns 11 and 12 show the gradings taken from the curve.

It was determined in item 8 that a total weight of 2,786 lb of dry coarse aggregate was required for 1 cu yd of concrete. The dry weights of each of the stockpiled sizes are determined as follows:

$$2{,}786 \times 0.36 = 1{,}003 \text{ lb of 3- to 6-in. size}$$
$$2{,}786 \times 0.26 = 724 \text{ lb of } 1\tfrac{1}{2}\text{- to 3-in. size}$$
$$2{,}786 \times 0.16 = 446 \text{ lb of } \tfrac{3}{4}\text{- to } 1\tfrac{1}{2}\text{-in. size}$$
$$2{,}786 \times 0.22 = 613 \text{ lb of No. 4 to } \tfrac{3}{4}\text{-in. size}$$

10. As in Example 1, the batch weights for Example 2 are on a dry basis. This is not the condition in which they will be in the field. Further, the moisture contents of the four sizes of coarse aggregate may not be identical. To illustrate the calculations of corrected batch weights, let us assume the following conditions, all other properties of the materials being as previously stated:

Moisture content of sand = 5.0 percent
Moisture content of 3- to 6-in. aggregate = 0.5 percent
Moisture content of 1½- to 3-in. aggregate = 0.5 percent
Moisture content of ¾- to 1½-in. aggregate = 1.0 percent
Moisture content of No. 4 to ¾-in. aggregate = 1.5 percent
Air-entraining admixture at the rate of 0.15 gal per bag of cement is required to entrain 3 percent air; it is batched separately from the mixing water

Batch weights of aggregates in the field:

Sand = 867 × 1.05 = 910 lb
3- to 6-in. aggregate = 1.003 × 1.005 = 1,008 lb
1½- to 3-in. aggregate = 724 × 1.005 = 728 lb
¾- to 1½-in. aggregate = 446 × 1.01 = 450 lb
No. 4 to ¾-in. aggregate = 613 × 1.015 = 622 lb

Corrections in amount of mixing water:

Free water in aggregates:

Sand = (0.05 − 0.007) × 867 lb = 37 lb
3- to 6-in. aggregate = (0.005 − 0.005) × 1003 = 0 lb
1½- to 3-in. aggregate = (0.005 − 0.005) × 724 = 0 lb
¾- to 1½-in. aggregate = (0.01 − 0.005) × 446 = 2 lb
No. 4 to ¾-in. aggregate = (0.015 − 0.005) × 613 = 6 lb
45 lb

(*Note:* The absorption of 0.5 percent is the same for each of the four sizes of coarse aggregate)

Water in admixture:

0.15 gal/bag × 2.8 bag × 8.33 lb/gal = 3 lb

The total amount of water to be subtracted from the original amount of mixing water is therefore 45 + 3 = 48 lb. The amount of mixing water to be batched in the field is 183 − 48 = 135 lb.

The following table compares the computed batch quantities for Example 2 with those which would actually be used in the field.

Ingredient	Material quantities, lb		
	Computed	Used in field	
	For 1 cu yd	For 1 cu yd	For 4 cu yd
Cement	263 (2.8 bags)	263 (2.8 bags)	1,052
Water	183 (22 gal)	135 (20 gal)	540
Admixture	3 (0.4 gal)	3 (0.4 gal)	12 (1.6 gal)
Sand	867 (dry)	910 (moist)	3,640 (moist)
3- to 6-in. coarse aggregate	1,003 (dry)	1,008 (moist)	4,032 (moist)
1½- to 3-in. coarse aggregate	724 (dry)	728 (moist)	2,912 (moist)
¾- to 1½-in. coarse aggregate	446 (dry)	450 (moist)	1,800 (moist)
No. 4 to ¾-in. coarse aggregate	613 (dry)	622 (moist)	2,488 (moist)

11. Adjustments of batch weights for batches greater or smaller than 1 cu yd are made on a proportional basis in the same manner as discussed in item 11 for Example 1.

Concrete-mix Tests The values listed in the tables of this section may be used for establishing preliminary trial mixes. However, they are based on averages obtained from a large number of tests and do not necessarily apply exactly to the materials being used on a particular job. It is therefore excellent policy to make a series of concrete tests to establish the relationships needed for selection of the proper proportions of the materials actually to be used on the project. An example of a series of such tests is shown in Table 21-10. The first mix of the series was a computed trial mix, arrived at in the manner previously discussed and shown in the examples. The second mix was adjusted to increase slump, but it appeared to be oversanded and to contain insufficient coarse aggregate. In the third mix, the amount of coarse aggregate was increased to an amount which was estimated to be the maximum which would still produce a mix of satisfactory workability. The sand content was correspondingly reduced. Three additional mixes were then made, with the water-cement ratio varying over a range of 5 to 7 gal per sack. From these mixes the relationship between water-cement ratio, cement content, and strength was established for the job materials. The field mixes could then be selected directly from Table 21-10.

TABLE 21-10 Typical Minimum Mix Series to Establish the Properties of Concrete Made with Field Materials

Mix No.	Net water-cement ratio, gal/bag	Water content, gal/cu yd	Cement content, bags/cu yd	Aggregate content, lb/cu yd		Slump, in.	Per cent air	28-day strength, psi		Workability		
				Sand	Coarse aggregate			Compression	Flexure	Segregation	Rodability	Finish
1	5.6	32.0	5.70	1,186	1,940	1	4.5	3,500	555	None	Good	Good
2	5.6	34.0	6.07	1,112	1,940	3¼	4.5	3,500	555	None	Excellent	Excellent
3	5.6	33.5	5.98	1,064	2,008	2½	4.5	3,500	555	None	Excellent	Very good
4	5.0	33.5	6.70	1,008	2,008	3	4.5	4,000	600	None	Excellent	Excellent
5	6.0	33.5	5.58	1,096	2,008	2¾	4.5	3,200	525	None	Excellent	Very good
6	7.0	33.5	4.79	1,158	2,008	3	4.5	2,600	450	None	Very good	Good

Mix No. 1—Low on slump.
Mix No. 2—Oversanded, increased coarse aggregate and lowered water for Mix No. 3.
Mix No. 3—Workability satisfactory.

In laboratory tests it seldom will be found, even by experienced operators, that the desired adjustments will develop as smoothly as indicated in Table 21-10. Also, it should not be expected that field results will check exactly with laboratory results. An adjustment of the selected trial mix on the job is usually necessary and will be discussed later. Closer agreement between laboratory mixes and field mixes will be assured if machine mixing is employed for making the laboratory mixes. This is especially desirable if air-entraining admixtures are used, since the type of mixing influences the amount of air entrained. Before mixing the first batch, the laboratory mixer should be "buttered" or the mix "overmortared" as described in ASTM Designation C 192, because a clean mixer will retain a percentage of the mortar. Similarly, any processing of materials in the laboratory should simulate as closely as practical the corresponding treatment in the field.

The minimum series of tests illustrated in Table 21-10 may be expanded as the size and special requirements of the work require. Alternate aggregate sources, aggregate gradings, types and brands of cements, admixtures, maximum sizes of aggregate, and considerations of concrete durability, volume change, temperature rise, and thermal properties are some of the variables that may require a more extensive program.

Adjustments for Field Use Having established by laboratory test the mix proportions that appear to produce the results desired for job conditions, the next step is to translate these proportions to the batch weights to be used in the job mixer. Most large jobs will use mixers of 4-cu yd capacity. Items 10 and 11 of Example 1 and items 10 and 11 of Example 2 explained how the batch weights per cubic yard were corrected for moisture content and adjusted to the batch weights for the size batch desired. In this case, for a 4-cu yd mixer, the corrected batch weights for a 1-cu yd batch are simply multiplied by 4. Let us assume that the batch material quantities tabulated at the end of item 10, Example 2, were found by laboratory tests to be satisfactory for job requirements. The batch weights for a 4-cu yd batch will then be as shown in the last column of the tabulation.

It will be noted that, in the figures for the 4-cu yd batch, the number of bags of cement and the number of gallons of water have not been indicated in parentheses, whereas the number of gallons of admixture have been shown. This is because, while it is customary to indicate the number of bags of cement and the gallons of water in a 1-cu yd batch, in large batching operations it is almost universally required that the cement and water be batched by weight. On the other hand, because of the small amount used, the admixture may be batched either by weight or by volume. It should also be noted that while the weights for 4 cu yd have been indicated to the nearest pound, the quantities will actually be weighed to the nearest scale division. The 4,032 lb of 3- to 6-in. coarse aggregate will probably be weighed to the nearest 50 lb.

After trying a few full-size field batches on the job, it is found that the water content per cubic yard must be increased from 22 to 23 gal to provide adequate workability for consolidation. Consequently, the cement factor should be increased to 23 gal/cu yd ÷ 8 gal/bag = 2.9 bags/cu yd. The batch quantities should be recomputed accordingly, following the procedures demonstrated in Example 2.

If the field use indicated that less water could be used, it should be reduced. However, as previously recommended, no consequent reduction in cement factor should be made unless laboratory tests indicate that such reduction can be made. It is necessary, however, that some adjustment be made in the batch quantities to compensate for the loss in volume due to the reduced volume of water. This may be done by increasing the solid volume of sand by an amount equal to the volume of the reduction in water. For example, assume that 21 gal of water is required instead of the 22 gal for the concrete in Example 2. Then 21/7.5 is substituted for 22/7.5 in computing the volume of water in the batch, or the solid volume of water per cubic yard becomes 2.80 instead of 2.93 cu ft. The difference in solid volume of 0.13 cu ft is added to the sand, making it 5.39 instead of 5.26 cu ft per cu yd, and this is converted to a weight basis as indicated in the example.

The percentage of air in concrete can be measured directly with an air meter, or it can be computed from theoretical and measured unit weights in accordance with ASTM methods.[4] For any given set of conditions and materials, the amount of air entrained is roughly proportional to the quantity of air-entraining agent used. Increasing the cement content or fines, decreasing the slump, or raising the temperature of the fresh concrete usually decreases the amount of air entrained for a given quantity of agent. The grading and particle shape of the aggregates, particularly the sand, also have an effect on the amount of air entrained. The job mix should not be adjusted for minor fluctuations in water-cement ratio or air content. A variation in water-cement ratio of ±0.25 gal per sack of cement, resulting from maintenance of constant workability, is considered normal. A variation of ±1 percent in air content is also considered normal.

SPECIAL PROPORTIONING REQUIREMENTS

Six-inch Aggregate Concrete All the procedures used here in arriving at the estimated mix proportions, in making laboratory trial mixes and tests, and in the adjustment of the final mix in the field are as applicable to concrete containing 6-in. aggregate as to concrete made with aggregate of smaller size. However, limitations in the size of laboratory testing equipment introduce special problems in the selection of mix proportions and the making of laboratory and routine control tests when 6-in. aggregate is used.

It will be noted that the aggregate proportions in Example 2 were selected from Table 21-7 instead of Table 21-6 because of the difficulty of obtaining an accurate dry-rodded weight for 6-in. aggregate. Actually, ASTM Method C 29 for determining the unit weight of aggregate is recommended for use only with aggregate of up to 4-in. size.[4] The size of the unit weight measure, the capacity of the scales required, and the physical difficulties of moving a full unit weight measure, to say nothing of the technical difficulties of adequately consolidating the aggregate in the measure, make it impractical to determine the unit weight of 6-in. aggregate or concrete containing such aggregate.

Similar difficulties are encountered in the determination of slump, in the making of compressive- or flexural-strength specimens, and in the determination of air content of the concrete by either the unit-weight or pressure method. It is generally accepted that any test specimen should be at least five times the diameter of the largest aggregate used. A concrete cylinder for compressive-strength tests of concrete containing 6-in. aggregate should therefore be 30 in. in diameter and 60 in. tall. Such a cylinder obviously could not be hand-rodded. It would weigh approximately 3,675 lb and, if it developed the relatively low strength of 2,000 psi, would require a testing machine with a minimum capacity of 1,500,000 lb. Several methods of circumventing these difficulties are in common use, although all of them are compromise solutions to the problem. They are satisfactory for job-control purposes if their limitations are recognized and compensated for accordingly.

The most common method for determining slump, strength, and air content is to wet-screen the concrete over a 1½-in. sieve and to determine these properties on the concrete which passes the sieve. It is obvious that the slump, for example, of the concrete containing the 1½-in. aggregate is not the same as for the concrete with the full-size aggregate, but the relationship of this slump to adequate workability in the forms is usually sufficient for routine control purposes. Many organizations depend upon good control of aggregate grading, moisture content, and batching operations, together with close observation of the workability of the concrete in the forms, and do not make slump tests. The Kelly ball test described in references 4 and 18 is another means of determining concrete workability. Strength tests are customarily made on 6- by 12-in. cylinders containing the 1½-in. aggregate concrete. The strength of the job concrete is sometimes verified by testing cores drilled from the structure.

Testing for the air content of the job concrete is one of the more important routine control procedures. Determination of air content on the minus 1½-in. portion of the concrete has been used almost universally on large jobs. Extensive investigations have been made of the reliability of and the relationship between the air content of concrete which has been wet-screened over a smaller sieve and the air content of the original concrete. The results of investigations performed by the Corps of Engineers are described in reference 19.

Lightweight Aggregate Concrete The basic considerations for the proportioning of concrete using normal weight or natural aggregates also apply to the proportioning of concrete using lightweight aggregate. However, excluding those aggregates which have sealed or coated surfaces, most lightweight aggregates have capacities for absorption which may require several days to satisfy. This makes determination of their absorption and specific gravity values impractical, and lightweight concrete mixes are usually established by a series of trial mixes on a cement-content basis at the required consistency.[20]

The pertinent properties of lightweight aggregates will vary widely from one manufacturing source to another. For example, because some lightweight aggregates are rounded with smooth surfaces, the water requirements to produce a 2-in. slump in non-air-entrained concrete can vary from 300 to 450 lb per cu yd for different aggregates. This wide range in water requirements is reflected in a corresponding range in cement content to produce a given strength. The structural strength of different lightweight aggregates also has an important effect on the cement requirement, particularly for higher-strength concretes. Because of the specialized nature of the design of mixture proportions for structural-grade lightweight-aggregate concrete and the special techniques involved in mixing and placing, it is suggested that the contractor and the owner either place their reliance on the recommendations of the lightweight-aggregate manufacturer or consult with others who have a sound knowledge of this kind of concrete. This does not imply that there is anything particularly difficult about the use of lightweight concrete, but it is a specialized field of concrete technology that the average contractor or owner will not have the time or the funds to explore adequately.

B. Batching, Mixing, Placing, and Curing

MATERIAL HANDLING AND STORAGE

One of the more difficult goals in the production of concrete is to have each batch as nearly like each other batch as possible. This means that not only must the same number of pounds of each material be used in each batch but also that the characteristics of the materials used should not vary from batch to batch or from day to day.

Material Characteristics If the gradation changes within a given aggregate size range, then the concrete will be different. One way to reduce segregation and maintain proper gradation is to have more size ranges, with each range covering a narrower span of sizes. Finish-screening immediately before the aggregate goes into the batching storage bin reduces the effects of segregation and degradation, but this can only screen the materials available and cannot overcome poor gradation.

If all aggregate comes from a single source, the variation in intrinsic strength will probably be small; but if several sources are used, the variations may become significant and must be considered in the concrete design.

A uniform moisture content in all aggregate materials as batched is very desirable in reducing variations in consistency or slump. Stockpiles of sand may need to drain for 24 to 48 hr to achieve a stable moisture content. Moisture meters are very useful for detecting a change in moisture content, but they should be checked and calibrated fairly often and particularly any time a different type of sand from a different source is to be used.

The strength of concrete varies considerably depending on the actual cement used to produce the concrete. It has been established that the compressive strength of concrete produced using cement from the same source fluctuates considerably over a period of one year. If cement from more than one source is used, the variations in the compressive strength become much higher. Concrete must be designed based on the cement that is actually to be used in the concrete.[21,22]

Admixtures must be handled carefully, because a very small change in the amount of a given admixture may have adverse effects on the concrete to an extent far exceeding the actual percentage of error. Since many admixtures are being used in a highly concentrated form, it is extremely difficult to batch them within the necessary tolerances. This creates even more troubles.

Plant Types Concrete batching plants may be divided into types based on several criteria. They may be either *low-profile* on the one hand or *stack-up, gravity,* or *tower* plants on the other, depending on the flow of material through the plants or their physical configuration. They may be classified as *central-mix, transit-mix,* or *dry-batch* plants depending on the function. They may be *mass-concrete, paving, ready-mix,* or *concrete-products* plants depending on the ultimate use for which the plant is intended. They may also be classified on the basis of the plant's mobility as *permanent, portable,* or *mobile.*

A stack-up, gravity, or tower plant is one in which the materials are elevated to the storage bin and flow by gravity into the weigh hoppers and into a central plant mixer or directly into a hauling unit. There is no generally accepted definition of a low-profile plant, but if the materials are elevated in the flow from the storage bin to the ultimate receiving unit then the plant is normally considered to be low-profile. In the most common of these low-profile plants the material is elevated into the storage bin, discharged by gravity into the weigh hoppers, then discharged onto a belt conveyor and elevated to a central plant mixer or into a transit-mix or dry-batch truck. Another type of low-profile plant elevates the materials to the overhead storage bins, discharges from the storage bin to a belt conveyor, and elevates the materials to the weigh hoppers. From the weigh hoppers the materials may be discharged by gravity into a central plant mixer or into a transit-mix or dry-batch truck.

A plant which has a central plant mixer is called a central-mix plant. A transit-mix plant is one where the aggregate, cement, and usually water are batched and then discharged into the transit-mix truck for mixing enroute to the jobsite. In some cases only the aggregate and cement are batched at the plant, and water is added either from the tank of the transit-mix truck or from a supply at the jobsite. A plant which is used for batching only the aggregate and cement, placing these materials into compartments of trucks for hauling to the job, is normally referred to as a dry-batch plant (Fig. 21-2).

Fig. 21-2 Mobile one-stop dry-batch paving plant. (*C. S. Johnson Operations, Road Divison, Koehring Company.*)

Mass-concrete plants are normally large, permanent, central-mix stack-up plants (Figs. 21-3 and 21-4). They are designed to produce a large amount of concrete at a given installation, so ease of transporting the plant is not an important criterion. There are usually between two and six mixers in a mass concrete plant. The mixers may be either 2 or 4 cu yd each. There are many types of paving plants available today. The most recent ones are utilizing central-mix plant features and eliminating the paver mixer at the jobsite. The paving plant may be either stack-up or low-profile; portable or mobile; and either central-mix, dry-batch, or transit-mix. The most common paving plant today is a low-profile, mobile, central-mix plant. With the large quantities of concrete that can be produced and paved, there is a definite need for a high-capacity, central-mix, low-profile mobile plant for the paving industry (Figs. 21-5 and 21-6). Ready-mix plants may be almost any combination of the various types. Many ready-mix producers are utilizing small low-profile mobile plants (Fig. 21-7) to move into a job which is large enough to support a separate small plant. This reduces the travel time and therefore also the number of trucks necessary to handle the job. In some cases a ready-mix producer will use a low-profile mobile plant (Fig. 21-7) to move into an area to sample the market. If the market demands are such that it would appear profitable, then the ready-mix producer may install a large permanent plant (Figs. 21-8 and 21-9) and move the mobile plant to another location. Concrete-products plants are normally designed for one particular location and are normally permanent installations. They may be either low-profile or stack-up plants. They are almost invariably central-mix plants because the material must be mixed in the plant and is usually processed within the same plant area.

Permanent plants quite often take a considerable amount of time for erection. This time may involve several weeks because pieces have to be bolted together. There may be preassembled sections which can be erected in large bulk so that there are relatively few pieces to put together and relatively few connections to make, thus reducing erection time and cost. Portable plants can be easily transported or hauled, but several days may still be required to prepare for a move and to set up after being

Fig. 21-3 Large mass-batching and mixing plant. (*C. S. Johnson Operations, Road Division, Koehring Company.*)

moved. Mobile plants are furnished with their own wheels, axles, and a fifth wheel so that a tractor may be connected to the plant to tow it along the road from one site to another. These are usually used where the plant must be moved frequently and the moving time must be cut to an absolute minimum. Sometimes only a few hours are necessary to get a plant ready to move and another few hours may be necessary before the plant is ready to be put into operation after it has been moved.

Material Handling Concrete batching, mixing, and placing may be considered essentially a problem in material handling. The raw materials—aggregate, cement, water, and admixture—must be delivered to the plant site; either stored and then reclaimed to the batching bins or delivered directly to the batching bins; measured; delivered to a plant mixer, transit mixer, or to a dry-batch truck and then to a paving mixer; and the mixed concrete delivered to its final destination by a truck, transit mixer, concrete bucket, buggy, conveyor belt, chute, or some other means. Many different material handling methods must be used in the course of getting these different materials through the different phases involved in the production of concrete. All aspects of the material handling problem, from the arrival of the raw material at the batch-plant site to the final placement of the finished concrete, are important, and any one of them may become the bottleneck for the entire job. Many problems of quality control are inherent in some handling methods and steps must be taken to ensure that proper safeguards are taken, proper methods are used, and proper equipment is on hand to satisfactorily perform each step in this process.

It is essential that the plant be arranged and the delivery of the various materials be scheduled in such a way that the different functions of the material handling do not interfere with each other. The trucks or railroad cars delivering the raw materials must not interfere with the equipment used to transport the finished product from the batch-plant site. This requires a considerable amount of preplanning, forethought, and constant supervision to ensure that there are no interfering processes and that all phases of the work proceed according to plan. Close cooperation and good coordination between the material suppliers and the batch-plant superintendent are essential.

Since the materials used in the production of concrete have widely different characteristics, several different methods of transporting the raw materials to the batch-plant site must be used. The aggregate may be delivered to the batch-plant site in barges, railroad cars, or trucks. Cement may be delivered to the batch-plant site in railroad cars or trucks. Water is either available at the site or pumped to the site. Admixtures are normally delivered either premixed in containers or in dry bulk form.

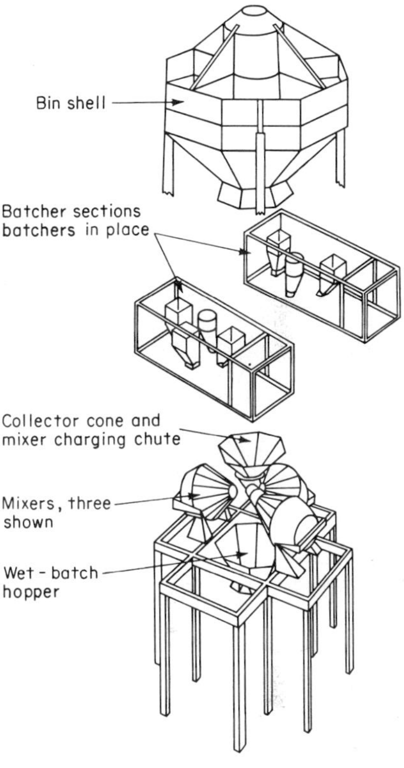

Fig. 21-4 Schematic diagram of a large batching and mixing plant for a dam construction project. (*Concrete Construction Handbook.*)

Each of these materials requires different handling methods to place it into the proper storage facilities. The aggregate may be put into a stockpile or directly into a storage bin by the use of belt conveyors, bucket elevators, or clamshell buckets. Cement is usually placed into a watertight storage silo or bin by pneumatic pumping or by a screw conveyor and bucket elevator. Water and the liquid admixtures are pumped to a storage vessel.

The different materials also require different methods of feeding into the measuring devices. Either a single- or double-clam gate is used for aggregate; a rotary plug valve, a vane feeder, or a screw conveyor is used to feed cement; water may be fed by gravity into a weigh hopper to be measured by weight or may be pumped through a meter for volume measurement; and admixtures may be pumped through a meter, blown into a volumetric measuring sight glass, fed by gravity flow through a calibrated orifice based on time, or gravity-fed into a weigh hopper.

After all the ingredients are measured, the means used depend on whether the plant is a central-mix, shrink-mix, transit-mix, dry-batch, or concrete-products plant. In a central-mix plant the material is discharged into a central plant mixer, completely mixed, then discharged into some vehicle for transportation to the final placing site.

This vehicle might be an open-top agitating unit, an open-top nonagitating unit, a dump truck, a concrete bucket, or a belt conveyor.

If a shrink-mix plant is used, the materials are fed into a central plant mixer directly after weighing, partially mixed or shrink mixed, and then fed into a transit-mix truck to be mixed for some additional time while enroute to the jobsite.

Fig. 21-5 Large, mobile, low-profile central-mix paving plant with two tilting mixers. (*C. S. Johnson Operations, Road Division, Koehring Company.*)

Fig. 21-6 Large, unitized, mobile, low-profile central-mix paving plant for one mixer shown in travel position. (*C. S. Johnson Operations, Road Division, Koehring Company.*)

In a transit-mix plant the materials are put into a transit-mix truck directly after being weighed and are mixed entirely in the truck. Some specifications require that all mixing be done at the plant site and that the truck be operated at agitating speed while enroute to the job. Other specifications require that all mixing take place on the jobsite after the transit-mix truck has arrived there. Still other specifications allow any combination of these two methods, or they may allow mixing while enroute to the jobsite. When the transit mixer has arrived at the jobsite, some further material handling is necessary. The concrete must be unloaded from the transit-mix truck and directed to the final resting place. Chutes are normally provided with the truck. The concrete may be directed immediately through the chutes into forms or transported further by means of belt conveyors or concrete pumps.

If a dry-batch plant is used, the material is normally discharged from the weigh hoppers directly into dry-batch hauling units. These may have from two to seven separate compartments, each of which holds one complete batch of aggregate and cement. The dry-batch trucks then transport the material to a paver, where it is mixed with water and discharged into forms or in front of a slip-form paver.

In a concrete-products plant, the material is usually discharged into a mixer and then discharged into forms, a concrete-block machine, or a concrete-pipe machine.

Fig. 21-7 Small, unitized, mobile, low-profile ready-mix plant. *(C. S. Johnson Operations, Road Division, Koehring Company.)*

The type of plant and the type of material handling equipment vary widely depending on the final end use of the concrete being produced, the quantity of concrete being produced in a given time, the specifications governing the production of concrete, the geographic location of the plant, the actual physical location of the plant with respect to land surrounding it, and the amount of room available at the plant site.

Aggregate may be delivered to the plant site by rail, truck, or barge. The method of rehandling aggregate depends on the method of delivery to the plant site. Aggregate may be delivered by rail in bottom-dump cars which may be discharged into a hopper, fed onto a belt conveyor, and then delivered directly into the bin or onto a stockpile. Clamshell buckets may be used to unload the railroad cars and place the material either into the bin or onto stockpiles.

If the aggregate is delivered by truck, it may be placed directly onto stockpiles or into a loading hopper which feeds either a belt conveyor or a bucket elevator. If the batch plant is located at the side of a hill, arrangements may be made so that the truck can drive on an open grill directly out over the storage bin and dump into the compartments of the bin. Barges are usually unloaded with clamshell buckets.

When a stockpile is built up using a clamshell bucket, the material should be placed, not thrown, in layers in order to reduce segregation and degradation. The stockpile may also be made by using an inclined belt conveyor. In order to build a longer stockpile, either a radial stacker or a parallel stacker is quite commonly used

Fig. 21-8 Large stack-up central-mix ready-mix plant. *(C. S. Johnson Operations, Road Division, Koehring Company.)*

Fig. 21-9 Large low-profile central-mix ready-mix plant. *(C. S. Johnson Operations, Road Division, Koehring Company.)*

(Figs. 21-10 and 21-11). A reversing shuttle conveyor may also be used in conjunction with an inclined conveyor to build long stockpiles (Figs. 21-12 and 21-13). The parallel stacker builds a stockpile in a straight line and the radial stacker builds a stockpile on a radius. The radial stacker produces somewhat more dead storage if the

Fig. 21-10 Radial stacker with raising and lowering boom. (*Atlas Conveyor Company.*)

Fig. 21-11 Parallel stacker with raising and lowering boom. (*Barber-Greene Company.*)

stockpile is over a tunnel conveyor. Either rock ladders or some provision for lowering or raising the end of the conveyor should be used if the stockpile is to be very high. This is to reduce immediate degradation.

Aggregate may be delivered to the batching plant storage bin by an inclined conveyor fed directly from a hopper (Fig. 21-14) or by a tunnel conveyor (Fig. 21-15) under the stockpile. If a double-clam gate is used to reclaim onto a tunnel conveyor, the material should be discharged into the center of the belt. If a single-clam gate is used, the material should be discharged in the direction of the belt travel. This eliminates the lateral force on the belt. The number of gates to be used under each stockpile will be determined by the height of the stockpile, the length of the stockpile, and the amount of dead storage that can be tolerated.

Aggregate bucket elevators may be used to charge the aggregate storage bin. These are normally used only when space is not available for a belt conveyor. This is due to the additional maintenance and high-wear characteristics of the bucket elevator when used for aggregate.

Fig. 21-12 Inclined belt conveyor with reversing shuttle belt conveyor. (*Barber-Greene Company.*)

Fig. 21-13 Parallel stacker, shuttle belt conveyor, and tunnel belt conveyor. (*Atlas Conveyor Company.*)

The aggregate bin may also be charged with a clamshell bucket, either directly from barges, railroad cars, or reclaimed from a stockpile. When a single inclined conveyor or a bucket elevator is used to feed the storage bin, some means must be provided to direct the material to the proper compartment. When the compartments are in tandem (in a straight line), a shuttle conveyor mounted at the top of the bin may be used. If the charging height of the storage bin is low, either a radial stacker or a parallel stacker may be used to charge the bin directly. A pivoted distributor (turnhead) may be used to direct the material to the proper compartment when the compartments are arranged so they can be fed from a single point. In many cases more than one belt conveyor is used to charge the bin. For high-production plants this is very feasible, particularly if the height of the bin is such that the conveyors are not excessively long. Some low-profile plants can be charged directly with front-end loaders.

In many cases today, some form of automatic bin filling is utilized. In a low-profile plant, for example, one belt conveyor may be used to feed each compartment. The belt conveyor is normally fed from a hopper, which in turn is fed by a front-end loader or clamshell bucket. The belt conveyors are kept fully loaded and start and stop at a signal given by a bin-level indicator in the aggregate compartment.

Fig. 21-14 Inclined belt conveyors to plant storage bin. (*Atlas Conveyor Company.*)

Fig. 21-15 Tunnel belt conveyor. (*Barber-Greene Company.*)

When a tunnel conveyor, inclined conveyor, and pivoted distributor are used, the automatic bin filling becomes somewhat more complicated. The belt conveyors usually run constantly whether any material is being fed to the bin or not. When the material in one bin compartment reaches a predetermined low level, the pivoted distributor rotates to that compartment and the tunnel gate for the correct material opens and begins filling. When the material level reaches a point near the top of the bin, the system is told to stop and the tunnel gate closes. After enough time has elapsed to clear the tunnel conveyor and the inclined conveyor, the pivoted distributor is free to move to another compartment to begin filling it. The system should be interlocked so the tunnel gate cannot open unless the pivoted distributor is at a compartment which indicates a need for material and the tunnel conveyor and the inclined conveyor are running; the pivoted distributor should not rotate until the entire system has had sufficient time to clear itself.

A bucket elevator could obviously be used instead of the inclined conveyor. If the bin compartments are in tandem, a shuttle conveyor could be used instead of the pivoted distributor. If there are only two compartments, a flop chute or two-way gate could replace the pivoted distributor.

Plant Storage Concrete-batching-plant storage bins are available in almost every conceivable size and shape. They may be furnished with aggregate only or cement only; with aggregate storage compartments around the outside and a cement storage compartment in the middle; with aggregate storage on both ends and cement in the middle; with aggregate on one end and cement on the other; or with aggregate in the center and cement on both ends. The storage bins may be square, rectangular, hexagonal, octagonal, or round. The compartments may be arranged in a tandem fashion so that the materials are side by side or they may be arranged so that the compartments are in a square. There may be as few as one or as many as eight, twelve, or even sixteen or more aggregate compartments. There may be only one or as many as two, three, four, six, or more compartments for holding different types of cement. Capacities of the storage bins vary from a few cubic yards of aggregate up to several hundred cubic yards and from a few up to one or two thousand barrels of cement.

Large central-mix paving plants have three or four aggregate storage compartments and one or two cement storage compartments. The aggregate bin is usually arranged for 5 to 20 min of storage. Normally some system of automatic bin charging is utilized. The cement is normally delivered in pneumatic unloading trucks. Auxiliary storage is commonly furnished with 1,000-barrel mobile units. As many as two to six trucks may be unloading at one time. The production capacity varies from about 200 cu yd per hr up to about 1,000 cu yd per hr, depending on the number of mixers, the size of the mixer, the mixing time, and whether the plant is equipped with individual single-material aggregate batchers or a cumulative aggregate batcher.

Dry-batch paving plants may have one or two separate plants for cement. Sometimes several sets of batchers are used under these plants so that more than one compartment of the hauling truck may be filled in one cycle. There is usually a very small amount of aggregate storage in the bin. Cement may be delivered by pneumatic unloading trucks or by railroad cars with a screw conveyor and elevator to charge the cement storage facilities. Up to 630 cu yd per hr is possible from a dry-batch paving plant. This would require from five to seven 34E dual-drum pavers, depending upon the required mixing time. The 630 cu yd per hr is based on having four complete sets of single-material batchers charging dry-batch trucks with seven compartments.[23]

Mass-concrete plants are commonly furnished with five aggregate compartments and with facilities for storing cement and pozzolan. The aggregate bin should have sufficient overhead storage for one to two hours' operation at the rated capacity using the maximum-size aggregate. This will ensure continuity of the pour in case there is a breakdown of the material handling equipment. Normally from three days' to one week's supply of aggregate is kept on hand in stockpiles and from three days' to one week's supply of cement is kept on hand in large steel silos. The production capacity depends on the number of mixers, the mixing time, and the size of the mixers. Up to 480 cu yd per hr should be produced from six 4-cu yd mixers.

Ready-mix plants may have from three to twelve or even sixteen aggregate compartments and from one to six or even more cement compartments. Since ready-mix plants are commonly called upon to furnish concrete for many different purposes, they require a number of different sizes and types of aggregate and cements. The storage capacity of the ready-mix bin normally depends to a large extent on the peak concrete production in cubic yards per hour (see Table 21-11). The production capacity of ready-mix plants varies up to 300 cu yd per hr, with a few plants exceeding this. Very few ready-mix concrete plants produce at the peak capacity for more than a short time in any one day. The peak capacity is normally needed only during the start-up time in the morning when all the trucks must be loaded in a short time. After this, the trucks are usually staggered, so that the production rate drops considerably.

Concrete-products plants normally have provisions for four aggregate compartments and sometimes more if specialty items or ornamental blocks are being produced. Usually one and sometimes two cement compartments are used. The production capacity is normally very small compared to other types of plants. Aggregate and cement storage depends largely on the availability of the materials.

TABLE 21-11 Typical Storage Capacity for Commercial Ready-mix Plants[23]

	Aggregate storage, cu yd				Cement storage, bbl*		
		Fixed plant				Fixed plant†	
Peak concrete production, cu yd/hr	Mobile plant, bin	Bin with 3 to 5 compartments	Bin with 6 or more compartments	Total storage	Mobile plant total storage	Overhead storage	Total storage
10	0–15	...	...	110	165‡		
25	0–30	30	40	275	165‡	70	350‡
50	35	50	65	550	165‡	130	650
100	40	100	125	1,100	350‡	260	1,300
150	45	225	270	1,650	525	390	1,950
200	50	400	480	2,200	700	520	2,600
300	60	650	780	3,300	1,050	780	3,900

* Cement-storage capacity is shown at 4 cu ft per bbl; cubic feet of storage space required is four times the figure shown. Because of fluffing in handling, actual capacity may be 70 to 85 percent of the storage capacity shown.

† Overhead and total cement storage should be increased about 20 percent if there are three or more types of cement.

‡ 165 bbl is the minimum storage for cement delivered in 125-bbl pneumatic trucks. 450 bbl is the minimum storage for cement delivered in 400-bbl rail hopper cars.

Special Processing After screening to produce the proper gradation of aggregate, it is sometimes necessary to rescreen the materials to reduce the effects of degradation and segregation. This rescreening procedure can only reject undersize material and cannot correct for poor gradation in the original material supplied.* The rescreen may be located on the ground and the material elevated to the storage bin after rescreening. It is usually better to screen on the top of the plant to get as close to the actual batching process as possible. This eliminates one step in the material handling process after the rescreening is finished. Most specifications require checking at the batcher discharge for a determination of correct gradation. If the screens are located on top of the bin, it is necessary to have the system carefully designed to reduce vibration. Excessive vibration will be transmitted to the lower part of the bin and may have an adverse effect on the controls and scales. Normally the material is sorted into the different sizes and then blended together before it is run through the screen.

On large installations, two double-deck screens, one above the other, are usually adequate. Some cases may require two sets of screens operating side by side. One triple-deck screen with the bottom deck used for two sizes may be adequate for small installations. Either a sand bypass chute or a separate sand conveyor should be

* See Sec. 20, Aggregate Production.

provided. The sand takes about one third of the total loading time, so a separate sand conveyor increases the effective screen capacity by about 50 percent.

Heating of some of the materials in cold weather is necessary in order to be able to place and protect concrete before it freezes.

Hot water is the most common and simplest method of raising the temperature of concrete. If the water is too hot, there is a danger of flash set when it contacts the cement. Hotter water can be used if the water and aggregate are mixed before the cement is introduced into the mixer. Sparger nozzles in a water reservoir may be used with a steam valve controlled by a thermostat to regulate the water temperature. If boiler feed water does not require treatment, this is the most economical method of heating water. Steam coils may be used in the water reservoir to heat water without loss of boiler water except that lost through leaks in the system. In-line water heaters have steam coils built into a section of the water line. This method heats the water as it passes into the reservoir and requires less area of heating coils than the coils in a tank because both the steam and water are moving.

Frozen aggregate must be heated above 32°F to make warm concrete, but it should not be heated above 100°F because of the danger of flash set. If the temperature of the aggregate can be kept above 32°F, additional heating will be reduced considerably. Live steam jets that run directly into aggregate storage bins can heat the aggregate rapidly. This also increases the moisture content in the aggregate. The material near the jets may be overheated and not enough heat may be transferred to some of the material which is away from the jets. Steam coils, hot-water coils, and hot-oil coils tend to heat the aggregate slowly because the aggregate is a poor heat conductor. One advantage of these methods is that they stabilize the moisture content of the aggregate. These coils should be located near the center of the bin in the line of the aggregate flow, which will also reduce heat loss. These coils can be used to heat stockpiles.

Electric heating units attached to the bottom of the bin can prevent freezing of the aggregates. Heaters to heat the area below the bin bottom also prevent freezing. This does not normally produce enough heat in the storage bin to thaw out the material for any great distance into the bin.

When placing concrete in hot weather, it is often necessary to cool some or all of the materials that are used in the concrete. Using the coldest water available, shading and sprinkling the aggregate stockpiles, and painting storage and handling equipment white to reduce heat absorption from the sun all help to reduce the additional cooling.

If the aggregate stockpiles are sprinkled, enough water must be used to keep the aggregate wet, and there must be more than enough water to replace the moisture evaporated. Usually from ⅛ to ½ in. per hr is required to keep the piles wet. A very coarse spray should be used since a fine spray increases relative humidity and decreases the water temperature, both of which reduce evaporation and cooling.

The use of ice in cooling concrete is widespread and very effective, but the amount that can be used is limited by the amount of moisture in the aggregates. About 80 percent of the added mix water may be ice, but some water is needed to dilute the admixtures.

Ice may be purchased in blocks and ground before use in the plant, or it may be produced in flake, cube, chunk, or crushed form. Flake ice melts faster in the mixer. Ice is usually elevated to the batching floor by a bucket elevator, stored in an insulated bin and delivered to the weigh hopper by a screw conveyor. A weigh hopper very similar to an aggregate weigh hopper may be used for ice, but the discharge gate must be very large.

Coarse aggregate may be cooled by a vacuum process which removes the air and vapor to maintain a pressure corresponding to the desired temperature, usually about 34°F. This may be done in the overhead storage bin or on the ground before the material is elevated to the storage bin.

Circulating cold air at about 38° through coarse-aggregate bins can cool the aggregate to about 40°F. Aggregate smaller than about ¾ in. has such high air resistance that it is not feasible to use this method.

Inundation of coarse aggregates in about 35°F water has been used to cool the aggregate to about 40°F. Some cold water (about one third of the tank) is pumped into the tank, which is then filled with the coarse aggregate. Cold water is pumped in the bottom of the tank and flows out over weirs near the top. The water is then rechilled and reused.

Sand can be cooled to 34°F by a vacuum cooling process similar to that described for coarse aggregate. Since the sand has a high resistance to the flow of vapor, it must either be cooled quickly in small amounts or circulated to expose the material surfaces to the vacuum.

Screw-conveyor coolers are available for cooling sand. In these the screw conveyor is surrounded by a jacket with 35 to 40°F water or 20 to 30°F brine circulating through it. Sand may be cooled to about 50°F with water and to about 40 to 50°F with brine.

Cement and pozzolan cooling is difficult and expensive because both of these materials lose heat slowly. Coolers similar to the screw-conveyor cooler for sand are sometimes used.

Selecting the best combination of cooling processes for a particular application is a difficult problem. Many variables are involved, and in some of the systems certain components may lend themselves to other uses.

BATCHING

After a trial mix has been established, it is important that the same proportions be maintained in every batch of concrete. The volumes of aggregate, cement, and water are the most important factors in producing consistent batches. For several reasons it is not convenient to measure the amount of materials on a volumetric basis. The aggregates tend to bulk or change volume depending upon the moisture content, method of handling, and the size and shape of the particles; the cement volume varies greatly with the amount of aeration in the cement; and the volume of water varies depending on the temperature. It is also not possible to measure as accurately and rapidly using volumetric methods as by weight. Since the specific gravity of most

TABLE 21-12 Typical Composition of Highway and Airport Paving Concrete[23] (Values in lb per cu yd of resultant concrete based on saturated surface-dry aggregates)

	Lowest observed value	Second centile	Median	Ninety-eighth centile	Highest observed value
Paving mixes, all mixes:					
Cement	385	442	557	674	677
Water	157	150	231	312	316
Sand	975	955	1,185	1,415	1,580
Total coarse aggregate	1,660	1,790	2,030	2,300	2,380
Total aggregate	3,020	3,040	3,230	3,500	3,560
Wet concrete	3,906	3,896	4,018	3,140	4,142
Paving mixes, 2 aggregates:					
Sand	1,066	1,043	1,180	1,359	1,342
Coarse	1,660	1,775	2,013	2,193	2,163
Paving mixes, 3 aggregates:					
Sand	975	950	1,190	1,430	1,580
Coarse, smaller quantity	360	520	930	1,220	1,190
Coarse, larger quantity	956	910	1,160	1,410	1,740
Coarse, smaller aggregate	360	520	1,000	1,300	1,274
Coarse, larger aggregate	683	750	1,100	1,410	1,740
Paving mixes, 4 aggregates:					
Sand	986	.....	.....	.....	1,311
Coarse, smallest quantity	143	.....	.....	.....	528
Coarse, middle quantity	733	.....	.....	.....	992
Coarse, largest quantity	850	.....	.....	.....	1,203

TABLE 21-13 Typical Composition of Commercial Ready-mix Concrete[23] (Values in lb per cu yd of resultant concrete based on saturated surface-dry aggregates)

	Lowest observed value	Second centile	Median	Ninety-eighth centile	Highest observed value
All mixes except pavement and lightweight mixes:					
Cement	188*	330	552	785	877
Pozzolan	0	.....	.....	.....	75
Cement + pozzolan	230	330	522	785	877
Water	210	209	269	372	383
Sand	985	1,019	1,342	1,596	1,670
Total coarse aggregate	1,454	1,524	1,888	2,150	2,267
Total aggregate	2,755	2,990	3,233	3,410	3,442
Wet-concrete weight	3,809	3,845	4,011	4,148	4,180
1½ in. max, 2 aggregates:					
Cement	370	352	490	655	658
Water	210	208	246	302	292
Sand	1,018	950	1,310	1,550	1,550
Coarse aggregate	1,780	1,750	1,950	2,150	2,143
Total aggregate	3,017	3,000	3,265	3,450	3,442
Wet-concrete weight	3,844	3,820	3,990	4,165	4,165
1½ in. max, 3 aggregates:					
Cement	380	426	533	640	750
Sand	985	960	1,200	1,440	1,410
Fine aggregate	750	660	1,010	1,400	1,362
Coarse aggregate	607	650	1,010	1,250	1,155
Total coarse aggregate	1,736	1,690	2,010	2,300	2,267
Total aggregate	3,040	3,000	3,220	3,440	3,410
1½ in. max, 4 aggregates:					
Cement	432	.....	.....	.....	658
Water	321	.....	.....	.....	339
Sand	1,122	.....	.....	.....	1,311
Fine aggregate	201	.....	.....	.....	202
Medium aggregate	1,167	.....	.....	.....	1,170
Coarse aggregate	644	.....	.....	.....	646
Total coarse aggregate	2,012	.....	.....	.....	2,018
Total aggregate	3,134	.....	.....	.....	3,329
Wet-concrete weight	4,086	.....	.....	.....	4,113
¾ to 1 in. max, 2 aggregates:					
Cement	188*	305	505	705	750
Water	225	225	270	345	347
Sand	1,040	1,080	1,360	1,630	1,670
Coarse aggregate	1,660	1,670	1,860	2,140	2,170
Total aggregate	3,021	3,040	3,235	3,430	3,440
Wet-concrete weight	3,809	3,860	4,005	4,150	4,131
⅜ to ⅝ in. max, 2 aggregates:					
Cement	393	360	585	910	877
Water	267	250	325	400	383
Sand	1,000	1,210	1,430	1,650	1,625
Coarse aggregate	1,454	1,400	1,680	1,920	1,885
Total aggregate	2,755	2,710	3,160	3,320	3,300
Wet-concrete weight	3,960	3,970	4,020	4,065	4,180
Lightweight insulating or fill concrete:					
Cement	318	300	530	770	700
Water	401	360	460	560	509
Lightweight aggregate	214	180	460	2,350	1,900
Wet-concrete weight	989	850	1,600	3,050	2,265
Lightweight structural concrete:					
Cement	461	415	590	900	852
Water	250	220	365	510	473
Sand	250	100	1,200	1,410	1,380
Lightweight aggregate	710	640	850	1,680	1,530
Total aggregate	1,592	1,450	2,030	2,270	2,260
Wet-concrete weight	2,410	2,310	3,000	3,340	3,274

* Plus 45 lb pozzolan.

materials is relatively constant, proportioning by weight is the most accurate method of producing consistent concrete. Notable exceptions are cinders, expanded slag, and other lightweight materials where the specific gravity changes drastically.

Most specifications require that cement be weighed on a scale separate from the other materials for greater accuracy. Water is usually weighed on a scale separate from other materials, although often, for economy, it is measured volumetrically through a meter. The different sizes of aggregate may each be measured on a separate scale, or they may be weighed "cumulatively" on a single scale. For cumulative batching, the fill gate for one aggregate material is opened and the desired amount of that material is put into the weigh hopper. The gate is closed and then a second gate is opened and the desired amount of the second material is put into the weigh hopper. This procedure is followed until all the aggregate materials have been weighed.

Typical ranges of material quantities useful for establishing batching-equipment capacities are shown in Tables 21-12 through 21-15. These are based on a large

TABLE 21-14 Typical Concrete Mixes for Specific Uses[23] (lb per cu yd)

Use	Cement or cement plus pozzolan	Water	Sand	Total coarse aggregate	Total aggregate	Wet-concrete weight
Grout	590–2,360	318–730	316–2,700	....	316–2,700	3,406–3,645
Pumped concrete	500–580	280	1,555	1,485	3,040	3,820–3,900
Nuclear shielding	564	300	900	3,860	4,760	5,624
Prestressed concrete	658–752	267–287	1,125–1,200	1,920	3,045–3,120	4,045–4,064
Lightweight prestressed concrete	846	459	180	1,293	1,473	2,780
Regular concrete block	455	177	1,425–1,950	1,425–1,950	3,375	4,007
Lightweight concrete block	340	280–560	....	....	1,750–2,200	2,490–2,940
Concrete pipe	640–830	250–420	1,120–2,230	985–1,860	2,960–3,330	4,115–4,315
Porous concrete	596	213	....	2,835	2,835	3,644

number of batches from many locations. Two formulas out of one hundred taken at random can be expected to be less than the second centile value and two can be expected to be more than the ninety-eighth centile value. Half of a large number of formulas can be expected to be above and half below the median. Slag aggregates have been excluded but would be about 90 percent of the aggregate weights shown.[23]

Aggregate Aggregates are weighed in a container which is usually suspended from a scale system and fed by a fill gate or fill valve. This fill gate may be either a single- or a double-clam gate. The weigh hopper must be of sufficient capacity to hold the material to be batched so that it does not touch the fill gate. The volume of the hopper should be at least 38 cu ft per cu yd of rated capacity. A typical fill-gate size is 18 by 18 in., with about 10 by 17 in. net opening. This allows about 2.2 cfs for coarse gravel and about 3.6 cfs for fine sand. Considerably larger fill gates are often used for larger aggregate material up to a maximum of 3 in., 4 in., or even 6 or 8 in. Larger fill gates are also commonly used on high-speed plants where a batcher of more than 6-cu yd capacity is used. If the gate opening is too large and the material is flowing too fast, the gate opening must be restricted by welding a bar across the gate or by an adjustment on the air cylinder. For special high-speed production, two gates may be used to feed one aggregate material. Unless special care is taken when this is done, the accuracy of weighing may suffer.

Cement Cement and cementitious materials are particularly difficult to control. When they are well aerated they flow very rapidly and will spurt out through a small pinhole through which water will barely pass. If they are dead (have low air content), it is almost necessary to pull the material from the storage bin. This creates many problems in the accurate measurement of these materials. Obviously the storage

TABLE 21-15 Typical Composition of Mass Concrete [Weights in lb per cu yd of resultant concrete based on saturated surface-dry aggregates. w/c or w/(c + p) indicates water-cement ratio or water cement + pozzolan ratio by weight][23]

	Lowest observed value	Second centile	Median	Ninety-eighth centile	Highest observed value
Grout mixes:					
Cement	590	.....	810	.....	1,020
Pozzolan	0	.....	.....	.....	260
Cement + pozzolan	590	.....	.....	.....	1,020
Water	318	.....	325	.....	550
w/(c + p), %	32	.....	39	.....	78
Sand	2,065	.....	2,350	.....	2,700
Wet weight	3,465	.....	3,600	.....	3,645
¾-in. max mixes:					
Cement	318	410	610	740	711
Pozzolan	0	.....	.....	.....	105
Cement + pozzolan	423	410	610	740	711
Water	225	270	307	355	342
w/(c + p), %	45	44	51	63	59
Sand	991	970	1,255	1,445	1,450
No. 4 to ¾-in. aggregate	1,355	1,685	1,775	1,835	2,080
Total aggregate	2,801	2,765	3,025	3,155	3,280
Wet-concrete weight	3,794	3,790	3,920	4,090	4,043
1½-in. max mixes:					
Cement	282	375	590	805	765
Pozzolan	0	.....	.....	.....	100
Cement + pozzolan	374	375	590	885	852
Water	163	160	270	310	310
w/(c + p), %	25	20	45	64	60
Sand	825	795	990	1,180	1,180
No. 4 to ¾-in. aggregate	830	810	1,050	1,210	1,576
¾- to 1½-in. aggregate	444	990	1,110	1,330	1,372
Total coarse aggregate	1,844	1,960	2,180	2,330	2,300
Total aggregate	2,996	2,990	3,145	3,300	3,397
Wet-concrete weight	3,823	3,820	3,990	4,190	4,277
3-in. max interior mixes:					
Cement	188	160	280	550	500
Pozzolan	0	.....	.....	.....	89
Cement + pozzolan	255	240	320	550	500
Water	152	140	185	290	252
w/(c + p), %	50	49	57	79	74
Sand	748	680	875	1,050	997
No. 4 to ¾-in. aggregate	642	590	800	1,010	938
¾- to 1½-in. aggregate	745	685	830	980	927
1½- to 3-in. aggregate	750	840	1,010	1,160	1,115
Total coarse aggregate	2,490	2,430	2,660	2,870	2,769
Total aggregate	3,330	3,250	3,540	3,850	3,766
Wet concrete	3,980	3,930	4,090	4,250	4,193

bin and silo must have no holes for the cement to leak from and must be watertight to prevent water from going into the bin or silo.

The hoppers may be either conical or rectangular and should have 50 to 60° net bottom slopes. The volume of the weigh hopper should be 9 cu ft per cu yd of rated batch capacity plus 3 cu ft to allow for fluffing and variations in the material. The dust seal between the fill gate and the weigh hopper should be flexible or arranged in

TABLE 21-15 Typical Composition of Mass Concrete (Continued)

	Lowest observed value	Second centile	Median	Ninety-eighth centile	Highest observed value
3-in. max exterior mixes:					
Cement	245	250	400	550	625
Pozzolan	0	.....	.....	.....	82
Cement + pozzolan	286	275	425	575	625
Water	160	145	210	280	266
w/(c + p), %	42	41	49	64	61
Sand	685	650	845	1,140	1,061
No. 4 to ¾-in. aggregate	395	300	760	980	910
¾- to 1½-in. aggregate	675	650	840	980	927
1½- to 3-in. aggregate	750	700	1,030	1,450	1,370
Total coarse aggregate	2,445	2,420	2,610	2,920	2,884
Total aggregate	3,210	3,140	3,470	3,840	3,750
Wet-concrete weight	3,980	3,940	4,100	4,260	4,235
6- to 8-in. max interior mixes:					
Cement	148	150	205	400	400
Pozzolan	0	0	56	110	118
Cement + pozzolan	198	200	255	410	400
Water	94	108	165	305	309
w/(c + p), %	36	40	64	85	86
Sand	714	705	825	1,210	1,170
No. 4 to ¾-in. aggregate	410	415	565	725	721
¾- to 1½-in. aggregate	333	380	580	770	913
1½- to 3-in. aggregate	675	650	815	1,140	1,096
3- to 6- or 8-in. aggregate	433	700	950	1,150	1,150
Total coarse aggregate	2,560	2,540	2,900	3,260	3,211
Total aggregate	3,542	3,520	3,750	3,980	3,925
Wet-concrete weight	3,914	3,910	4,150	4,390	4,306
6- to 8-in. max exterior mixes:					
Cement	247	247	307	441	441
Pozzolan	0	0	50	105	106
Cement + pozzolan	247	247	368	441	441
Water	130	132	175	232	226
w/(c + p), %	41	41	49	56	56
Sand	710	680	790	1,030	1,022
No. 4 to ¾-in. aggregate	280	260	500	650	679
¾- to 1½-in. aggregate	201	460	575	670	675
1½- to 3-in. aggregate	635	620	790	1,060	1,069
3- to 6- or 8-in. aggregate	764	760	950	1,290	1,230
Total coarse aggregate	2,500	2,420	2,885	3,090	3,055
Total aggregate	3,440	3,410	3,630	3,970	3,913
Wet-concrete weight	4,036	4,040	4,185	4,310	4,290

some manner so that it will not affect the accuracy of weighing. There must also be an air vent in the weigh hopper to prevent the buildup of air pressure which would also affect the scale system. Normally a vibrator is attached to the weigh hopper to ensure that all cement is discharged from the weigh hopper. If cement is held in the weigh hopper for more than a minute or so it will probably lose air and will be extremely difficult to discharge.

The cement fill gate may be a rotary vane feeder, a rotary plug valve, or a screw conveyor. The discharge from the weigh hopper may be any of the above types or a conical valve, which is an inverted cone which pulls up into the bottom of the weigh hopper to close the gate opening. Another discharge valve in common use is a flexible tube which is squeezed together to close the discharge opening. An opening is normally provided in the cement weigh hopper to allow an overload to be corrected, but these overload ports are very rarely used in practice. The flow of cement may be controlled by adjusting the pressure of aeration into the storage bin or silo or by adjusting the amount of gate opening.

Water and Admixtures Water and liquid admixtures are weighed in hoppers of light-gauge corrosion-resisting alloy or galvanized steel. A quick-acting valve should be used to feed the weigh hopper. Many admixtures are viscous and sticky, and steeper sides on the weigh hopper are necessary for better discharge of the material. The admixture hopper often has some provisions for running water through the valves to clean them if one end of the valve is open to the air. No provision is normally made to remove overload from these weigh hoppers since the hopper top is open. Water and admixtures are often measured volumetrically in hoppers, tanks, and sight glasses. Since the volume varies with temperature, it is necessary to correct for this difference in volume when there is a considerable difference in temperature, such as when hot water is measured volumetrically. Neglecting this temperature difference in volume results in a 1.6 percent error in the amount of water if 140°F water is used and if the volume was calculated based on 60°F. If water is weighed, this correction does not need to be made.

Water meters are commonly used and within certain ranges are generally accurate within ±1 percent. The repeatability is usually better than the accuracy of these meters. If hot water is to be used in a meter, then a hot-water disk must be furnished. Cold water can be used through a hot-water meter with somewhat less accuracy, but hot water should never be used through a cold-water meter.

A wide variety of admixtures are now commonly used with concrete. The quantities of these materials may vary from as little as ¼ oz to as much as 2 qt per sack of cement. The extremely small dosage rates provide problems in accurate dispensing. Some of the admixtures are not compatible with each other, so it may be necessary to add one to the water batcher, one to the sand, and one or more directly into the mixer or mixer truck. It may be possible to dispense them in sequence, so that one goes into the water line and when it is finished another may be put into the water line.

One of the simplest, least expensive, but least satisfactory methods of dispensing admixture is by using an adjustable orifice which is set so that a specific amount of admixture flows through the orifice in a given length of time. A timer is then set for the number of seconds required for the proper amount of admixture. This method gives no assurance that any admixture flows through the orifice. For example, if the supply tank is empty or if the orifice is clogged up, then no admixture will flow. The height of admixture in the storage tank over the orifice will vary to some extent and this will cause a difference in the flow rate of the admixture, thus producing an error in the amount put into the concrete.

A closed chamber with positive displacement may be calibrated and used to insert admixtures. With this method, admixture is forced into a section of the chamber by an air cylinder. The valve is closed, the air cylinder is operated, and the admixture is forced out into the water line. This method is superior to the timer but still leaves the possibility of an empty tank or a valve which is stuck. No air is allowed into this chamber at any time so there is no real danger that air will be mixed with the admixture to dispense an erroneous quantity of admixture.

When meters are used to dispense admixtures, many of the above problems are eliminated (Fig. 21-16). These meters usually have an electrical impulse device in the head which is calibrated from ½ oz to 2 qt per pulse. These pulses are counted by a preset counter in the control system. When the proper number of pulses has been counted, the meter is stopped. If no admixture goes through the meter, the control system can refuse to discharge the batch or can give the operator an indication that he received no admixture.

Sight glasses can be equipped with a pulsing device (Fig. 21-17) which may be calibrated from 0.1 oz to 1 qt per pulse. Interlocks may be provided to indicate zero level, to prevent the bottle from being discharged unless the quantity is within tolerance, and to prevent overfilling the bottle.

Fig. 21-16 Meter for dispensing admixture. (*Protex Industries, Inc.*)

Weigh hoppers are sometimes used for measuring admixtures. This is by far the most foolproof, most positive, and most accurate system, but it is also the most expensive. It requires a scale system and a complete set of controls very similar to those used for cement, aggregate, or water. There may be almost any number of admixtures used in a given plant and it is not feasible to weigh all these materials on one set of scales. The quantities vary from 1 oz to 20 qt, and a scale system for admixtures should not be used below about 10 percent of its capacity if it is to give the accurate results needed for properly dispensing admixtures. The number of scales needed depends on the range of quantities of the admixtures used.

Scales Almost all the scales presently used in concrete batching plants are the lever type which are similar to the old steelyard scale. These use the principle of pivoting a lever near one end so that a relatively small weight can balance a considerably larger weight. By combining several levers together in one system in this fashion, it is possible to have a large-capacity weigh hopper suspended from one end of the scale system balanced by an indicating device at the other end of the scale system which requires a load of only 5 to 25 lb.

The hopper is usually suspended by four hanger rods from two main levers (Fig. 21-18). One main lever consists of two simple levers connected by a torque tube to transmit the load from the short lever in the back to the longer lever in the front. The two main levers are connected together by a gathering shackle, and the load from both main levers is transferred to the first extension lever. Since the load from both of the main levers is combined before it is transferred, the two main levers act as a single simple lever. More levers may be used to reduce the load as required by the final indicating device.

The ratio of a lever is the ratio of the weight on the load knife-edge and the weight on the power knife-edge when the lever is balanced. This is the ratio of the distance

Fig. 21-17 Sight glass with pulsing unit for dispensing admixture. (*Protex Industries, Inc.*)

from the load knife-edge to the fulcrum knife-edge and the distance from the power knife-edge to the fulcrum knife-edge. The ratio is normally expressed as 2:1, 3:1, or 6-2/3:1.

The ratio of a scale lever system is the product of the ratios of the individual levers. If the ratio of the main levers is 5:1, the ratio of the first extension lever is 4:1, and the ratio of the second extension lever is 4:1, then the ratio of the lever system is $5 \times 4 \times 4$ or 80:1. If the system is balanced and 800 lb is put in the hopper, 10 lb must be added to the power knife-edge to return the system to balance.

The levers are constructed with the knife-edges positioned as accurately as possible. An error of 0.001 in. in the location of a single knife-edge may cause an error that would

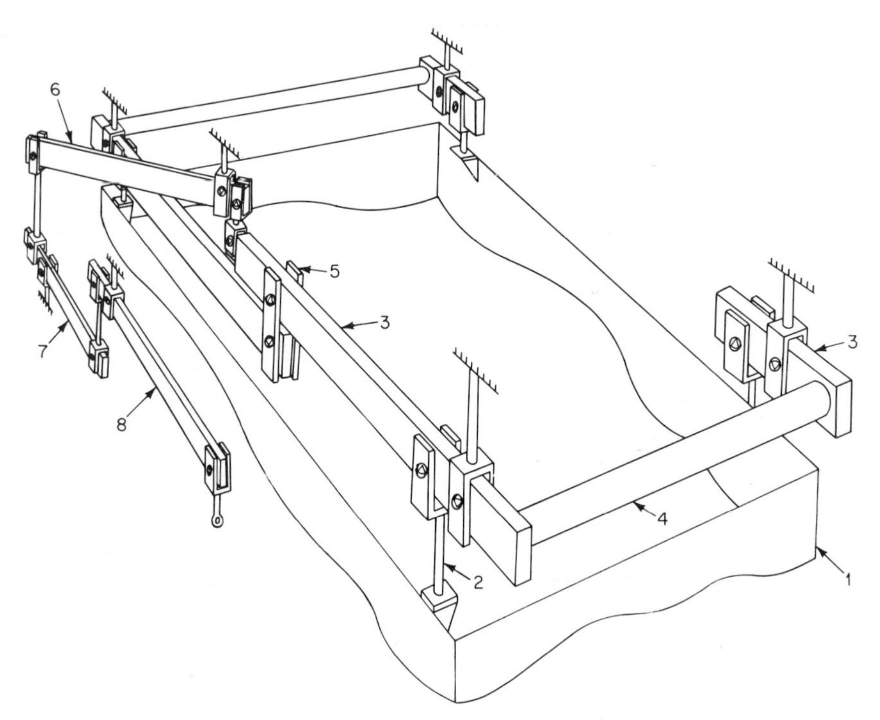

Fig. 21-18 Typical batcher scale lever system. (1) Weigh hopper, (2) hanger rod, (3) main lever, (4) torque tube, (5) gathering shackle, (6) first extension lever, (7) second extension lever, (8) third extension lever.

result in the condemnation of the scale system. For this reason there must be some adjustment built into the lever system. A nose iron at the junction point of the main levers provides for changing the ratio of the main levers to compensate for any errors in the actual location of the knife-edges throughout the scale lever system. Adjustable knife-edges are also used to adjust the ratio of the scale lever system. These are mounted in such a way that they can be rotated in their holes to change the location of one knife-edge with respect to the other knife-edges in that lever.

Either a beam scale or a dial scale may be used as an indicating device with the lever system.

A dial scale consists of a pendulum weight, a rack, a pinion, a dial pointer, and a dial chart. The pendulum weight produces a counterreaction to the lever system and drives the dial pointer by means of the rack and pinion. The dial scales provide a continuous indication of weight from zero to full scale capacity, allow quick changes of material quantities, are easier to adapt to graphic or digital recording, and are easier to use with electronic controls.

The beam scale is usually mounted in a beam box (Fig. 21-19) and consists essentially of a tare beam or balance beam with an indicator attached plus one or more weigh beams. Each weigh beam has two poise weights which may be moved to set the desired weight. The beam scales are simpler, more rugged, easier to maintain, easier to calibrate, have smaller weight divisions when used for cumulative weighing, and always indicate balance in the same place so it is easier and quicker to determine if the proper weight is on the scale.

Either scale should incorporate an adjustable dash pot for damping the scale system. This dash pot may be adjusted by changing the viscosity of fluid in it or by changing the size of the holes through which the fluid must pass. The dash pot must be so constructed that it produces no friction when the system is in balance or not

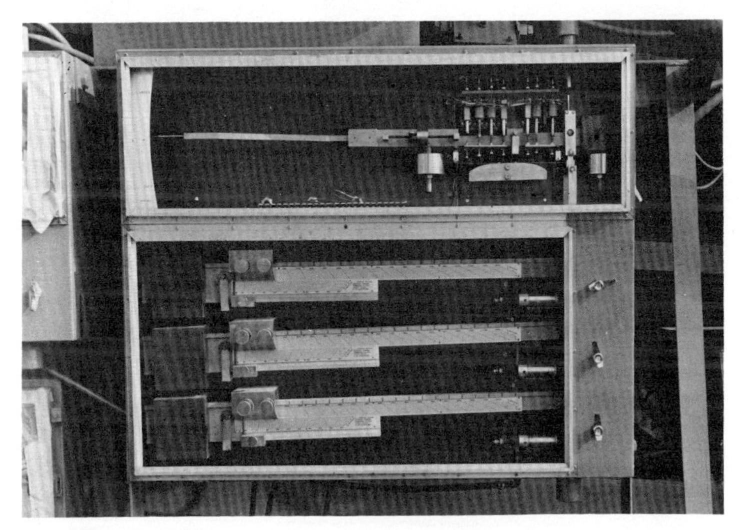

Fig. 21-19 Beam box with three weigh beams, transmission beam, and balance indicator. (*C. S. Johnson Operations, Road Division, Koehring Company.*)

moving, but it must produce a considerable damping effect at any time that the scale is in motion. This reduces the oscillations of the scale to a minimum so that it will settle down quickly.

On a multiple-beam scale, one which is used to weigh more than one material cumulatively, there should be one separate weigh beam for each material to be weighed. This allows setting the net weight of each material on a separate weigh beam. At the beginning of the cycle, one weigh beam is dropped on the scale and the material is put into the scale until the balance beam indicates balance again. Then a second beam is dropped onto the scale and another gate is opened and material put into the batcher until the scale is again in balance. The third weigh beam is dropped on the scale and another gate is opened and this continues until all the materials have been weighed. It is necessary to leave each weigh beam on the scale while the later ones are added so that at the end of the cycle all the weigh beams are on the scale. All the weigh beams should be removed from the scale when the material is discharged, and the scale will indicate balance when the weigh hopper is empty.

When a dial scale is used to weigh materials cumulatively, it is only necessary to watch the dial pointer to determine when the correct amount of material is in the weigh hopper. The net weights of the materials must be added together to determine the desired scale reading.

There are many different specifications dealing with the static load tolerances that are applied when a scale is checked with calibrated test weights. The most restrictive of these requires ±0.05 percent of full scale capacity as a maximum tolerance for new scales. The most liberal allows ±0.5 percent of full scale capacity as the tolerance for the scale system. The most common tolerance for scales is ±0.1 percent of full scale capacity.

Controls Control systems vary from the simple hand-lever type to the highly sophisticated electronic-type controls with which the operator merely selects the mix design, either by a switch or by picking a punch card, selects the number of cubic yards

Fig. 21-20 Hand-lever aggregate and cement batcher with separate scale system for aggregate and cement. *(C. S. Johnson Operations, Road Division, Koehring Company.)*

desired, and pushes the start button. This control system then produces the materials required for the concrete and records the weights of each material as well as the time, date, mix identification number, truck number, and any other information which is desired. Obviously the more complicated systems require a higher degree of maintenance, more maintenance, and a much more advanced method of troubleshooting.

The manual controls operate by means of mechanical hand levers that are physically attached to the gates. When the operator pulls on a hand lever, the gate opens (Fig. 21-20). By watching the scale, the operator can determine when to close the gate by pushing on the hand lever. The primary disadvantage of this system is that all quality control is in the hands of the operator who is pulling the levers. Operator fatigue is one of the big problems. An operator who is fresh at the beginning of the day can produce batches rapidly with considerable accuracy, but he may be doing a relatively poor job by late afternoon. If the primary operator is sick or on vacation, his substitute must be able to operate as efficiently and as accurately or the quality of the concrete suffers. One of the advantages of this manual system of operation is that it allows almost an infinite variety of control functions which are not possible with even the most sophisticated electronic equipment today. For example, the operator can control the sequence of batching and the sequence of discharging, hold back some of the water, control the discharge rate of any of the materials to match the acceptance rate of transit-mix trucks, control the rate of flow from the fill gates of the materials depending on how the materials react on a given day, weigh an extremely small batch in a large hopper with a fair degree of accuracy by reducing the feed rate to a minimum, and also maintain a fast cycle when large batches are required by leaving the fill gate open wide until the batch is almost complete. Thus the main disadvantage, leaving the quality control in the hands of one operator, is also its greatest advantage because the operator can exercise discretion depending on circumstances. This is still beyond the scope of even the most sophisticated control systems on the market.

Manual controls with air assist are usually considered the next step. Air rams are attached to the fill gates and discharge gates of the batchers. The operator merely has to push a button until the proper weight is reached, then release it. This eliminates the operator fatigue problem to a considerable extent. It also takes away part of the flexibility of the completely manual system. The operator can no longer hold the fill gates to some specific opening, because they open wide or not at all. He can, however, alternately open and close the gates to produce a jogging action when he nears final weight, but this is somewhat slower than the manual operation and it requires constant operator attention and action.

Semiautomatic batching is achieved when the operator can preset a weight on the scale and actuate the charging device or fill valve by pulling a lever or pushing a button. When the correct weight is in the weigh hopper, the fill gate will close. This system, of course, frees the operator from the need to watch the batcher closely, but it also introduces a possibility of error when the rate of material flow changes because of a gradation change, a change in moisture content, or any other factor. In these cases, the cutoff points should be adjusted to produce the correct weights with the new set of conditions. A semiautomatic control system may or may not have interlocks to prevent the discharge of the batcher if the materials are not within the desired tolerances.

Many types and styles of automatic control systems are available. Some of the systems may be used with either beam scales or dial scales. Some of the systems may be arranged with mix selection and others are designed to control only one mix design. The basic cutoff element which tells the system that the scale has reached the correct weight may consist of microswitches, reed switches, photoelectric cells, solenoid cutoff units, or electronic devices.

Automatic controls normally produce more consistent and more reliable results than the other types. The operator must select the amount of material desired and push on a button to start the cycle. The control system automatically batches each material, checks to see that it is within the required tolerances, and tells the operator that he may discharge it.

Since the automatic system is designed to relieve the operator of some of the decisions, it is important that some interlocks be built into the system to be sure an incorrect batch is not discharged. There should be an interlock to prevent the start of a new batch unless the scales are at zero. The fill gate should not be able to open unless the discharge gate is completely closed. The batcher should not be able to discharge unless all materials are within the required tolerance, the fill gates are closed, the mixer is empty, and the mixer is in the correct condition to receive a new batch. The mixer should not be able to discharge until the mixing time has elapsed.

Some provision must be made to take care of the column of material in the air just before the final weight is reached. There are at least two methods of doing this. In one case the aggregate gate is alternately opened and closed, or jogged, when the weight approaches the final cutoff point so that material comes into the weigh hopper in short spurts. This reduces the column of material in the air so that the error is negligible. With cement, the gate may be jogged or the opening may be cut down so that the cement feeds in slower. A second approach to this problem is to set the final cutoff point so that it is some number of pounds below the final cutoff weight. Then the gate starts to close before the cutoff point is reached. This is normally referred to as a "midair" or "material in suspension" adjustment. One of the difficulties is that the material flow is not always constant and the amount of this compensation must necessarily change. The amount of this compensation must also change if a different quantity of material is to be weighed because the column of material will be of a different height and therefore a different weight. When materials are weighed cumulatively

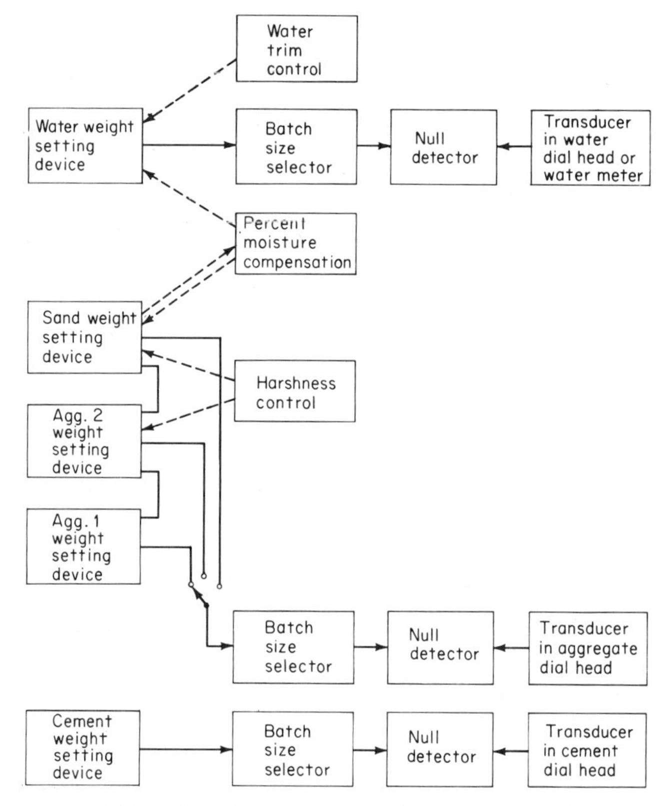

Fig. 21-21 Block diagram of an electronic control system.

in the same batcher, the first material will have a relatively tall column but the next material which weighs on top of it will have a relatively shorter column and the two settings could not be the same. This problem is aggravated when the batch size changes. Anything that changes the weight of this column of material requires the operator to adjust the amount of compensation for it, so the responsibility for accuracy is in the hands of the operator.

A special type of automatic control system is the electronic control. This term refers to those systems which use some type of electric transducer attached to the scale system which feeds back information to the basic control unit where desired weights for each material are set by some adjustable electrical device. The electronic control system consists essentially of three parts: a transducer which is mounted on the scale, a weight-setting device which is mounted in the controller at the operator's station, and a null detector which is also in the controller (Fig. 21-21). Typical examples of the transducer are potentiometers, linear variable differential transformers, special inductive devices, and pulse generators. Typical examples of the weight-setting devices are potentiometers (Fig. 21-22), punch cards (Fig. 21-23), push cards, and thumb-wheel switches (Fig. 21-24). For a null detector, either vacuum tube or transistor amplifiers, magnetic amplifiers, or sensitive meters may be used.

An electrical signal from the weight-setting device representing the amount of material to be batched and an electrical signal from the transducer at the scale representing the amount of material actually being weighed are fed into the null detector. When these two signals are equal, the system is told to stop batching. With the punch card or push card, any number of preset mixes may be kept on hand so that a wide variety of mixes is available at any time. By punching a new card, a new mix can be established. There may be several sets of potentiometers or thumb-wheel switches, so that different mixes may be preset and then selected as needed by the operator. Some of the systems have mix-selection units which are built as plug-ins so that any number may be preset, stored, and plugged into the unit as they are needed. Where several presets are available, all but one are usually set for the mixes to be most used during a particular day and one preset section is left for the odd or unusual mixes which will always come in during the day.

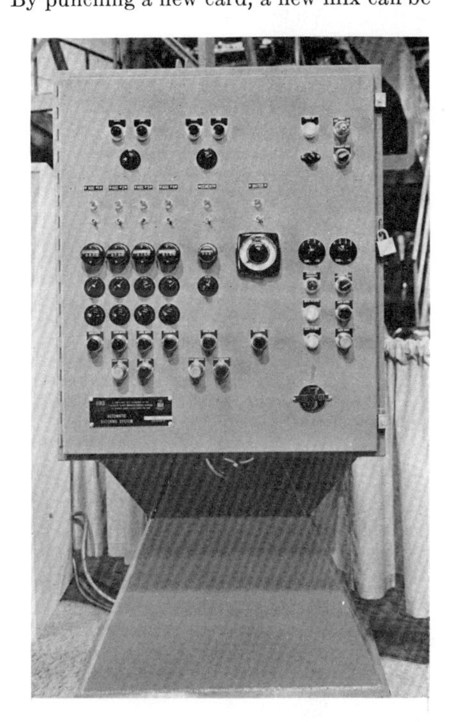

Fig. 21-22 Electronic control system with potentiometer weight selection. (*C. S. Johnson Operations, Road Division, Koehring Company.*)

One of the primary advantages of an electronic control system is that a switch may be incorporated which can be set to the number of cubic yards desired. The weights can be programmed into this system as needed for a 1-cu yd batch, and the batch-size selector switch may be used to select the number of cubic yards desired for a particular batch. This is obviously of special interest and benefit to a ready-mix producer. For a paving contractor who normally uses the same size batch and the same mix design for an entire day or week, it would have relatively little value. Quite frequently the batch-size selector switch is calibrated in ¼-cu yd increments. This switch may also be set to give repeat batches by calibrating the switch to call for more cubic yards than can be batched at one time. For example, if a 4-cu yd batcher is used, the switch can have settings up to 12 cu yd. If the switch is set for 6 cu yd, two batches of 3 cu yd each would automatically be produced. If the switch is set for 12 cu yd, three batches of 4 cu yd each would be produced.

Since the moisture content usually varies in the fine aggregate, it is convenient to have some fairly simple method of adjusting both the fine aggregate and water weights to compensate for the amount of moisture in the fine aggregate. In most electronic control systems a switch may be installed to make this adjustment. It is relatively simple to design a circuit to adjust the sand weight, since the percent moisture is a direct percentage of the sand weight. Adjusting the amount of water to be called for is more difficult because the number of pounds to be corrected is not a direct percentage of the water weight. The control system must determine the number of pounds of water to be deducted, convert this number of pounds into some electrical signal, and then subtract this signal from the amount of water to be batched. This problem is

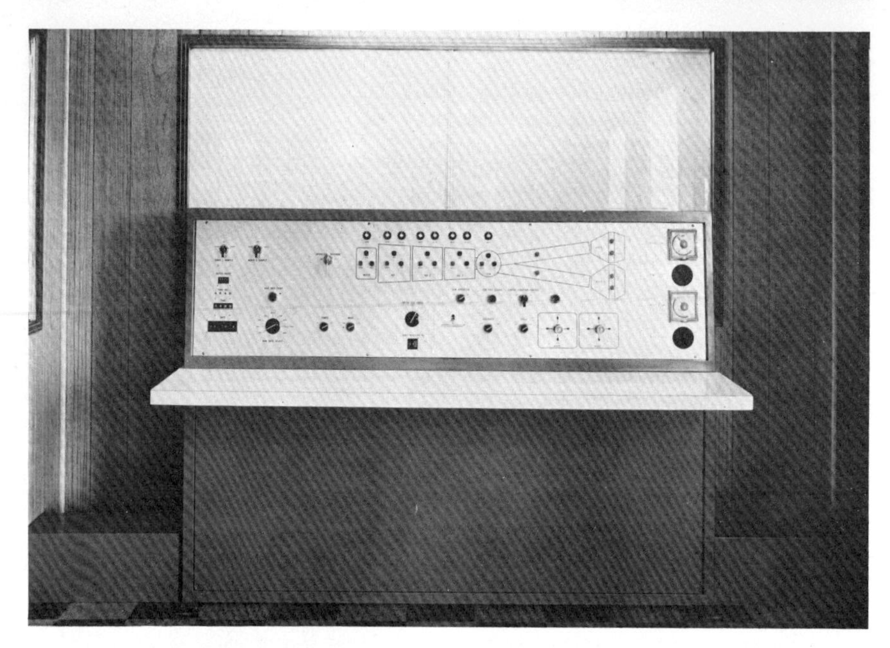

Fig. 21-23 Electronic control system with punch-card weight selection. (*Auto Controls, Inc.*)

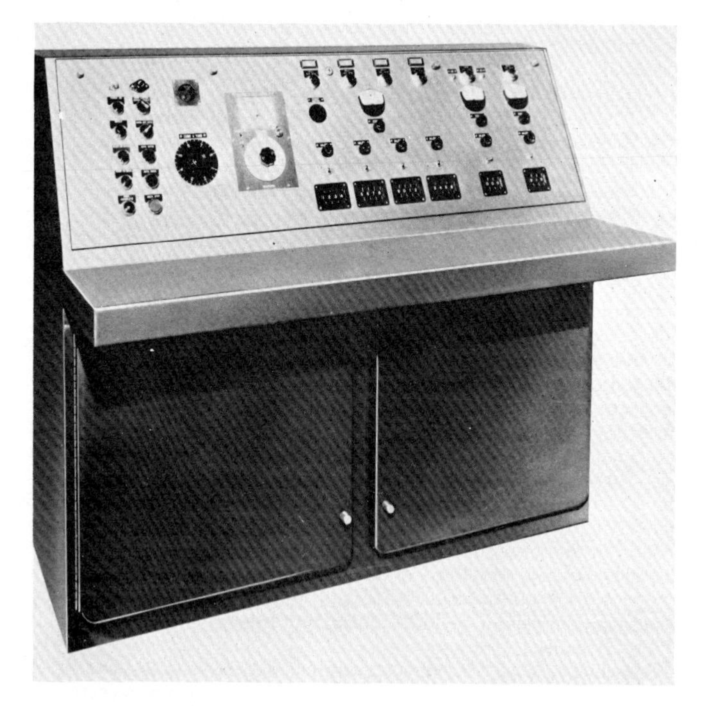

Fig. 21-24 Electronic control system with thumb-wheel weight selection. (*C. S. Johnson Operations, Road Division, Koehring Company.*)

complicated by the fact that the actual amount of water to be compensated is usually a considerable portion of the actual water weight.

A control which can be and frequently is added to the electronic type control systems is a "slump" control, or more accurately a "water trim" control. This merely gives the operator a quick and easy method of changing the amount of water called for without changing his preset or punch card. Another control is a "harshness" control which increases the amount of sand and decreases the amount of one aggregate or increases the amount of one aggregate and decreases the amount of sand to control the harshness of the mix while retaining the same total weight of aggregate plus sand.

The electronic control systems are especially adaptable to remote operation of the complete plant. Since it is not necessary to go to the batcher to change the amount of each material required or the number of cubic yards, the batcher may indeed be several hundred feet from the control system, so that the operator is not bothered by the harsh sounds of the gates opening and closing and the aggregate hitting the steel weigh hoppers. This requires that the control systems be at least as fully interlocked as the normal automatic systems, if not more so. The operator also should have some indicating lights or some form of remote dial indicator to tell him what is going on at the batcher.

The electronic control systems are considerably more complicated and more complex than most other types of automatic controls and therefore require a higher caliber of maintenance and troubleshooting personnel. Since every control function requires additional pieces or parts in the system and since the possibility of failure increases as the number of parts increases, one must anticipate a correspondingly greater amount of trouble from the highly sophisticated, complex systems than from those which are relatively simple.

Tolerances Since it is not reasonable to expect absolutely perfect batching from any plant or any system, some tolerance must be established to determine the maximum allowable errors permitted during the batching operation. The delivery tolerance for cement and water varies from ±0.5 to ±1 percent of the net weight desired. The most common tolerance is ±1 percent. The normal tolerance for admixture is ±3 percent of the net amount desired, but the tolerance is often stated as ±3 percent or ±1 oz, whichever is greater. This creates a very dangerous possibility when extremely small dosage rates and small-size batches are used. If the dosage rate is ¼ oz per sack of cement, if a 4 bag mix is used, and if only 1 cu yd is being batched, then the total amount of admixture desired is 1 oz. With a tolerance of ±1 oz, all the admixture could be omitted from the batch or twice the required amount could be put into the batch and it would still be within the allowable tolerances. One solution to this problem would be to specify a minimum dosage rate per bag of cement based on the accuracy and repeatability of available dispensing equipment. Another solution would be to have the specifications rigidly enforced as ±3 percent of the required amount and force someone to design equipment capable of dispensing admixtures within ±0.03 oz. Note that with a 4-cu yd batch under the same conditions, the allowable error on admixture is ±1 oz in a total of 4 oz, which is ±25 percent of the desired amount.

The delivery tolerance for aggregate varies from ±0.5 to ±3 percent. The most common tolerance is ±2 percent for fine aggregate and ±3 percent for the larger aggregate from 3 to 8 in. When a separate scale is used for each size aggregate, the tolerance is a percentage of the net weight desired. When aggregates are weighed in a cumulative batcher, the tolerances present a very confusing and usually misunderstood situation. There are at least three different methods currently used for specifying tolerances on cumulative aggregate batchers. The first method allows ±2 percent of the net weight of each material (Figs. 21-25 and 21-26). The second method allows ±2 percent of the desired scale reading at the completion of batching of each material (Fig. 21-27). The third method allows ±2 percent of the total batch weight as a tolerance for each individual material (Fig. 21-28). Note that all these methods guarantee that the total batch will be within ±2 percent. If the materials are weighed in ascending or descending order of size, then the error in one size material will be taken up by the next size material. There is more variation in gradation of the aggregate than in the errors produced by any of these three methods.

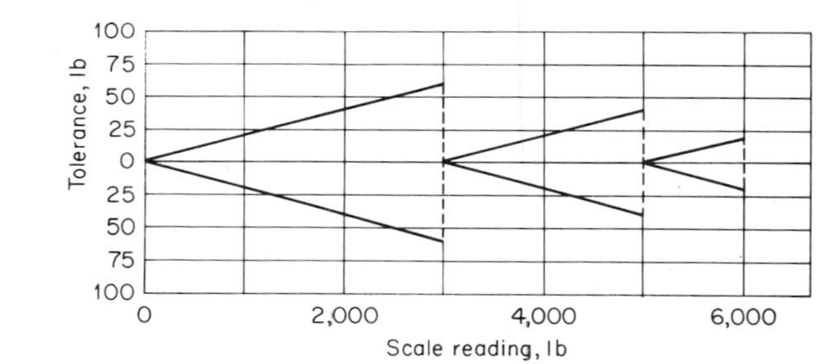

Fig. 21-25 First method of cumulative batching tolerance.

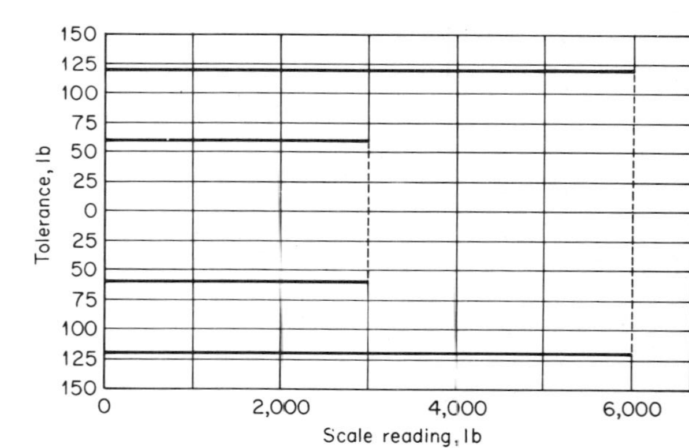

Fig. 21-26 First method of cumulative batching tolerance.

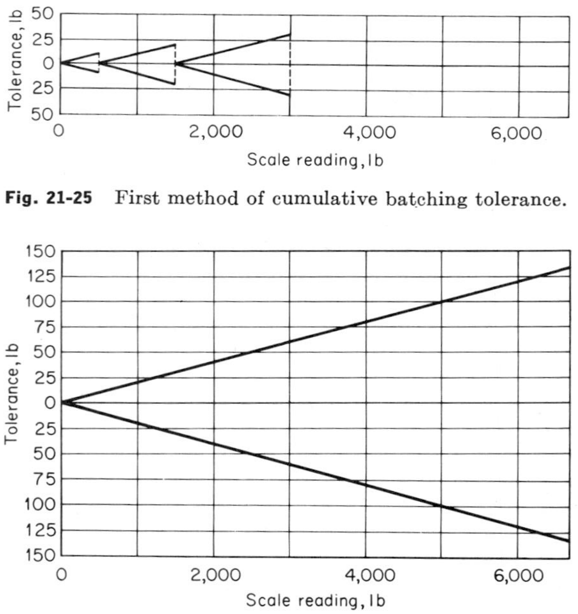

Fig. 21-27 Second method of cumulative batching tolerance.

Fig. 21-28 Third method of cumulative batching tolerance.

There are many types of tolerances which apply to a concrete batching plant. Since the different tolerances, their interpretation, and their enforcement determine to a large extent the original cost of concrete batching-plant control equipment and the scales to be used, a realistic approach to the entire situation should be taken. The most reasonable way to do this is to compare all the tolerances to see the relative effect that each has on the ultimate quality of the concrete. Any other factors which will affect the quality of the concrete should also be considered at the same time.

The ultimate use of the batching plant, whether it is to be used as a paving plant, ready-mix plant, mass-concrete plant, or concrete-products plant, must be taken into consideration in this evaluation of tolerances. For example, a paving plant should normally be expected to produce batches which are near the capacity of the plant.

TABLE 21-16 Typical Tolerances for Batch Operations

Item	*Tolerance*
Test weight	±0.01% of net weight
Scale check	±0.1% of scale capacity (or ±1 scale graduation)
Moisture compensation	±0.5% of sand batch weight
Delivery tolerances:	
Cement	±1.0% of batch weight
Aggregate	±2.0% of batch weight
Water	±1.0% of batch weight

One exception to this would be when the load limit for the hauling unit requires a lower maximum batch than the plant is designed to produce. Most mass-concrete plants usually produce the maximum size batch possible. With some exceptions, a concrete-products plant normally would use the maximum-size batch. Ready-mix plants would necessarily produce batches ranging from the maximum size possible down to the minimum batch which can be produced with sufficient accuracy to meet the applicable specifications. Thus, the ready-mix plant is the only one which should be very much concerned about producing a very small batch.

Assume an 8-cu yd batcher with a 6,000-lb cement dial scale with 5-lb gradations, a 30,000-lb aggregate dial scale with 25-lb gradations for weighing aggregate cumulatively, and a 2,500-lb water dial scale with 2.5-lb gradations. The tolerances for test weights vary considerably depending on which specifications are to be used, but assume that the tolerance is ±0.01 percent of the net weight of each test weight. This is equivalent to 0.005 lb, or 0.08 oz, or about 22.7 grams for a standard 50-lb test weight. The static test of the scales using the calibrated test weights is quite commonly required to be within ±0.1 percent of the scale capacity, which is usually ±1 scale graduation. The actual delivery tolerances which are most common are: cement, ±1 percent of the batch weight; aggregate, ±2 percent of the batch weight; and water, ±1 percent of the batch weight. Moisture compensation is another factor to be considered. Usually ±0.5 percent of the sand batch weight is used as a tolerance for the amount of moisture present in the sand. These tolerances are listed in Table 21-16.

Figure 21-29 shows the relative effect of these tolerances for cement on an 8-cu yd

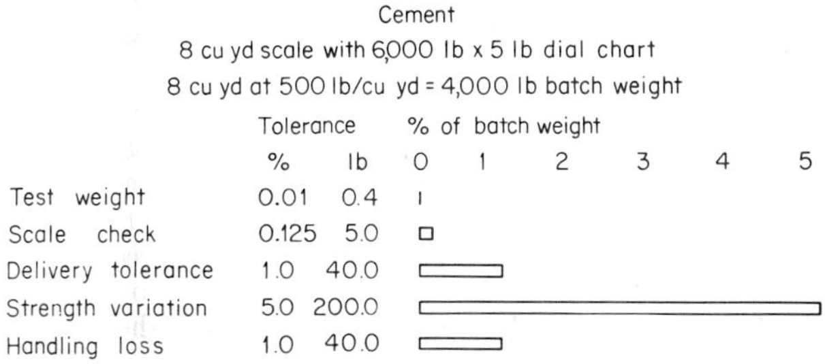

Fig. 21-29 Typical tolerances for an 8-cu yd batch of cement.

batch of concrete assuming 500 lb per cu yd or 4,000 lb for the entire 8-cu yd batch. One additional item has been added to this graph. Since the strength of concrete produced varies due to the physical and chemical differences in the cement used in the concrete, a factor of 5 percent variation is shown here. This is not inconsistent with a number of studies which have been made over a period of years.[21,22] Obviously there is not a 200-lb variation as shown in the graph, but the 5 percent variation of strength is what would be expected if the cement were consistent but if the amount of cement varied by ±5 percent in the batch. The 5 percent figure used here is conservative and in many cases will be considerably more than 5 percent.

Figure 21-30 shows the effect of the various tolerances on an 8-cu yd batch of aggregate assuming 1,000 lb per cu yd or 8,000 lb for the complete batch. Sand is shown

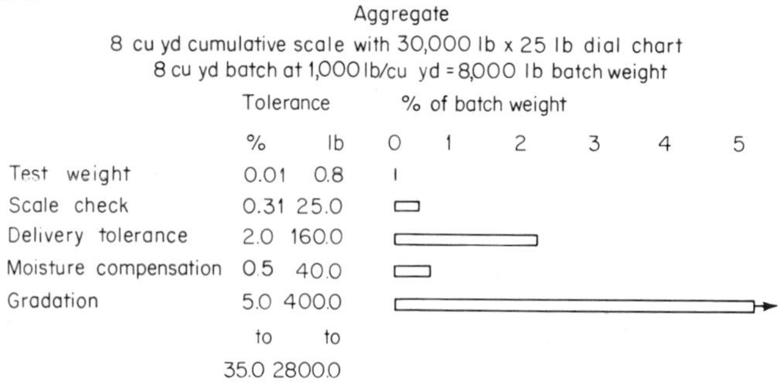

Aggregate

8 cu yd cumulative scale with 30,000 lb x 25 lb dial chart

8 cu yd batch at 1,000 lb/cu yd = 8,000 lb batch weight

	Tolerance %	Tolerance lb
Test weight	0.01	0.8
Scale check	0.31	25.0
Delivery tolerance	2.0	160.0
Moisture compensation	0.5	40.0
Gradation	5.0 to 35.0	400.0 to 2800.0

% of batch weight: 0 1 2 3 4 5

Fig. 21-30 Typical tolerances for an 8-cu yd batch of sand in a cumulative batcher.

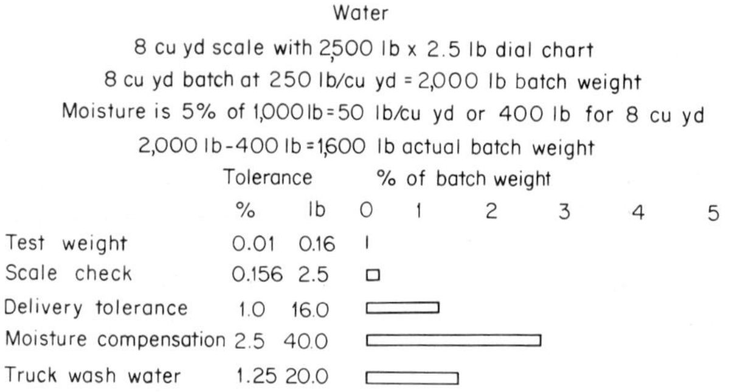

Water

8 cu yd scale with 2,500 lb x 2.5 lb dial chart

8 cu yd batch at 250 lb/cu yd = 2,000 lb batch weight

Moisture is 5% of 1,000 lb = 50 lb/cu yd or 400 lb for 8 cu yd

2,000 lb - 400 lb = 1,600 lb actual batch weight

	Tolerance %	Tolerance lb
Test weight	0.01	0.16
Scale check	0.156	2.5
Delivery tolerance	1.0	16.0
Moisture compensation	2.5	40.0
Truck wash water	1.25	20.0

% of batch weight: 0 1 2 3 4 5

Fig. 21-31 Typical tolerances for an 8-cu yd batch of water with 5 percent moisture in the sand.

here so that we may see the effect of the error in moisture compensation. The ±0.5 percent error in determination of moisture content is shown as 40 lb, and of course this assumes an exact compensation, so that there is no allowance made for errors in actual compensation. The entire error shown is due to the inaccuracy of moisture-content measurement. Since gradation of the aggregate obviously affects many of the qualities of concrete and since there is and must be some latitude in the actual gradation and in the consistency of gradation, a ±5 to ±35 percent variation due to gradation changes has been indicated. This would correspond to a change of from 400 to 2,800 lb of sand in the batch. It would normally be compensated for by 400 to 2,800 lb of some different size of material, thus keeping the yield essentially the same.

Figure 21-31 shows the effect of the various tolerances on a batch of water, assuming 250 lb per cu yd or 2,000 lb total amount of water in the batch. The 40-lb error in moisture compensation is due entirely to the error in determining the amount of moisture in the sand shown in the graph for the aggregate. This again assumes a perfect actual compensation for the amount of moisture which is in the sand. If we assume that there is about 5 percent moisture in the sand, then the amount of water to be deducted from the water batch weight would be 5 percent of 1,000 lb or 50 lb per cu yd. The actual batch weight of water would be 200 lb per cu yd for a total of 1,600 lb of water for the 8-yd batch. The ±1 percent delivery tolerance is almost always applied to the actual amount of water to be batched rather than to the total amount of water to be included in the batch. Therefore we have ±1 percent of 1,600 lb or ±16 lb as a delivery tolerance. As the graph shows, however, the 0.5 percent error in determining the actual moisture content of the sand produces an error of 40 lb in the water batch weight, which is 2.5 percent of the 1,600 lb of water to be batched. Another item of concern has been added at the end of this graph. In ready-mix plants it is extremely difficult for the truck driver to remove all wash water from the truck, and it is even more difficult for a driver to determine exactly how much water

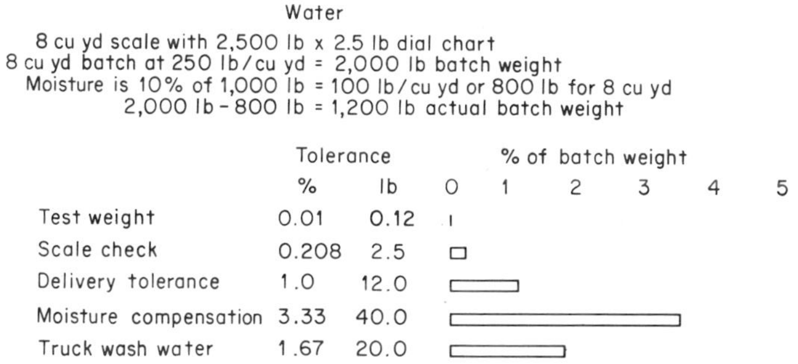

Water

8 cu yd scale with 2,500 lb x 2.5 lb dial chart
8 cu yd batch at 250 lb/cu yd = 2,000 lb batch weight
Moisture is 10% of 1,000 lb = 100 lb/cu yd or 800 lb for 8 cu yd
2,000 lb - 800 lb = 1,200 lb actual batch weight

	Tolerance %	Tolerance lb
Test weight	0.01	0.12
Scale check	0.208	2.5
Delivery tolerance	1.0	12.0
Moisture compensation	3.33	40.0
Truck wash water	1.67	20.0

(% of batch weight: 0 1 2 3 4 5)

Fig. 21-32 Typical tolerances for an 8-cu yd batch of water with 10 percent moisture in the sand.

he still has in the truck. A conservative estimate of about 2½ gal of water has been shown here as a variation from the amount of wash water presumed to be in the truck as opposed to the actual amount of wash water in the truck. This would give an error of about 20 lb or 1.25 percent of the 1,600 lb of water to be batched. Since the errors introduced by the moisture content in the sand and by the truck wash water both exceed the delivery tolerance, and since these three errors greatly exceed the scale-check tolerance—and even this is considerably more than the test-weight tolerance—it is difficult to understand why so much emphasis is placed many times on these relatively smaller errors while the larger ones are, many times, completely ignored or overlooked.

In these illustrations it is possible to see an analogy to the old story of the man who measured one mile to the nearest thousandth of an inch by driving the first 0.9 mile in an automobile and checking the distance with his odometer, then pacing off almost 0.1 mi, measuring about 10 ft with a tape measure, and then measuring the last ¾ in. with a micrometer. In this fashion he had measured a complete mile with an accuracy of 0.001 in. The moisture compensation, the strength variation due to differences in the cement, and the gradation variations in aggregate are comparable to the driving of the automobile and measurement with the odometer. The delivery tolerance in each case can be compared to the pacing of the second portion of the mile. The scale check would correspond to the micrometer measurement. Obviously this story is ridiculous, but it does have a counterpart in real life in the ultraprecise measurements required in many concrete batching plants throughout the country today.

The situation becomes even worse when the moisture content increases, and Fig. 21-32 indicates the situation we have when the moisture content of the sand is assumed to be 10 percent. The delivery tolerance is based on the actual amount of water to be batched and not on the total amount of water in the batch (which would include the moisture in the sand). The percentage errors due to the moisture compensation and the truck wash water therefore increase when the amount of water to be batched is decreased.

As mentioned previously, many factors affect the minimum-size batch which can reasonably be produced from a given concrete batching plant. Since this problem applies basically to a ready-mix concrete plant, a 3-cu yd batch has been shown in Fig. 21-33. It is noticeable that the percent error of the scale check tends to increase as the size of batch decreases. Also the percent error due to the truck wash water, which is a fixed number of pounds, increases as the batch size decreases. There are many excellent plants now operating that are producing at less than one-third of their capacity and doing a fine job, but in general it should not be assumed that a plant can consistently produce batches within these tolerances if the number of pounds required is less than about one third or one fourth of the scale capacities.

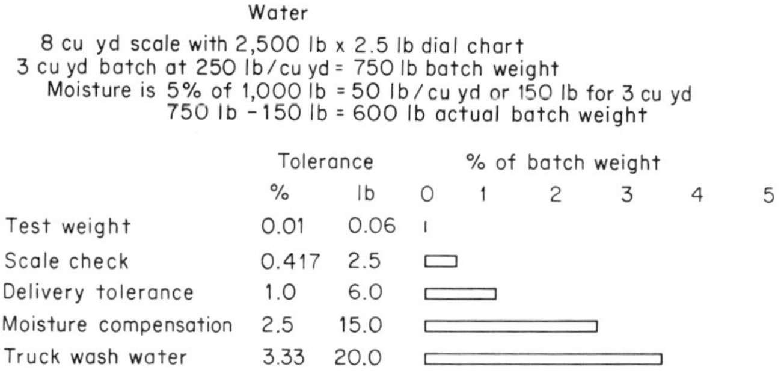

Water

8 cu yd scale with 2,500 lb x 2.5 lb dial chart
3 cu yd batch at 250 lb/cu yd = 750 lb batch weight
Moisture is 5% of 1,000 lb = 50 lb/cu yd or 150 lb for 3 cu yd
750 lb - 150 lb = 600 lb actual batch weight

	Tolerance %	Tolerance lb
Test weight	0.01	0.06
Scale check	0.417	2.5
Delivery tolerance	1.0	6.0
Moisture compensation	2.5	15.0
Truck wash water	3.33	20.0

% of batch weight: 0 1 2 3 4 5

Fig. 21-33 Typical tolerances for a 3-cu yd batch of water with 5 percent moisture in the sand.

Recorders The batch weights of materials in a concrete batching plant have been recorded for many, many years on large mass-pour plants. Several reasons are usually given when the recording of batch weights is required. Where many different mix designs are being used, it is important to know that the correct mix design was used for a particular batch of concrete. It is also essential to know that no material was inadvertently omitted from the batch. Another reason for recording is to ensure that the correct amount of each ingredient was actually put into the batch of concrete.

A ready-mix plant uses many different mix designs throughout one day's operation. Recording for this application should be primarily to ensure that the correct mix design was used for each specific batch. A batch plant that is used only for paving would normally use one mix design throughout any given day with only minor changes in water content and perhaps some small changes in the amount of sand. The only reasons for recording on such a plant would be to have a record that no batch omitted an ingredient and to indicate that the correct amount of each ingredient was put into each batch. An automatic control system with proper and complete interlocks would ensure that the correct amount of each material was put into the batch and would ensure that no material was omitted.

There are basically three types of recording being used in the concrete industry today. Graphic recorders have a pen which moves on a lined chart and indicates the amount of material placed in the weigh hopper. Digital recorders produce digital representations either on a continuous tape or on individual tickets. Photographic recorders are merely cameras that take photographs of the dial charts or of the beam boxes. All the recorders can be arranged to record the time, date, sequence batch number, truck number, mix type, and other miscellaneous information.

Mechanically Coupled Graphic Recorders (Figs. 21-34 and 21-35). The simplest and most direct type of recording is the mechanically coupled graphic recorder. This device involves a pen arm attached directly to the dial scale system so that the pen records in conjunction with the pointer indication on the dial. The method has several advantages:

1. It gives a complete history of the batch from zero weight to final weight and then back to zero weight again.
2. There is no time delay involved in getting the recording.
3. It is very difficult for an operator to tamper with the recording to make a false or incorrect reading.
4. Accuracy is usually very good and may often be in the neighborhood of ± 0.05 percent of full scale reading.
5. It is very reliable because it is a simple, directly coupled system.

Fig. 21-34 Mechanically coupled graphic recorder with 20-in.-wide recording charts. (*C. S. Johnson Operations, Road Division, Koehring Company.*)

There are also some disadvantages:

1. The pens may clog or run out of ink and produce no record at all.
2. There can be an appreciable error if the pens are not adjusted properly.
3. Readability is sometimes not good unless a very wide chart is used. (Chart widths vary from 4 to 20 in.)
4. Duplicate records are not easy to obtain.

Electrically Coupled Graphic Recorders. These recorders are similar to the mechanically coupled graphic recorders, but they are driven electrically from a signal from the primary scale system. Some of the advantages are the same as those of the mechanically coupled recorders:

1. They give a complete history of the batch from zero weight to final weight and then back to zero weight again.
2. It is very difficult for an operator to tamper with the recording to make a false or incorrect reading.

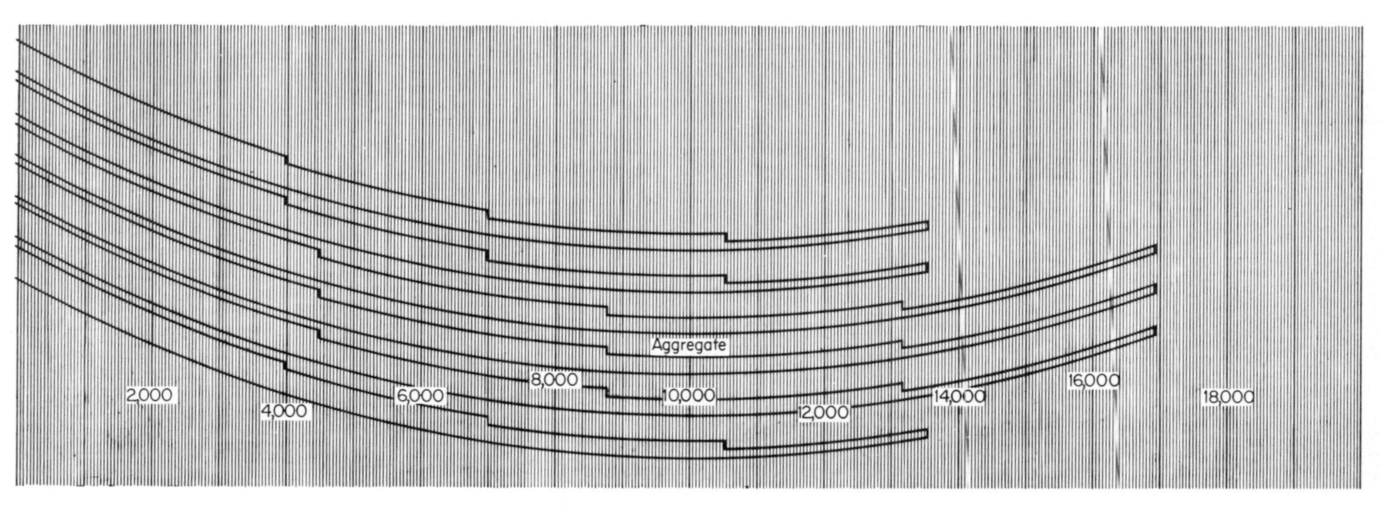

Fig. 21-35 Graphic record of several batches in a cumulative aggregate batcher.

3. These devices are very reliable, although less so than the mechanically coupled recorders.

There are also some disadvantages:

1. The pens may clog or run out of ink and produce no record at all.
2. There can be an appreciable error if the pens are not adjusted properly.
3. The accuracy is not as good as that of the mechanically coupled recorder because of the extra step of transmitting the signal from the primary scale system. There can be a large error if this signal is not sent or received properly.
4. Readability is sometimes not good unless a very wide chart is used.
5. There is usually some delay because the driving system must be some distance behind the primary scale system.
6. Duplicate records are not easy to obtain.

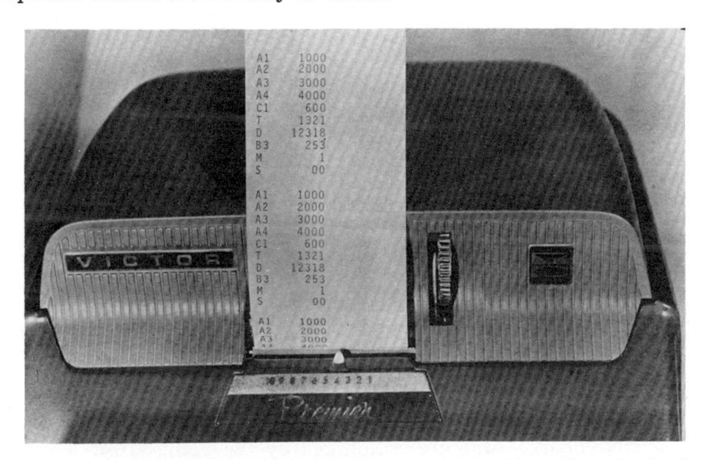

Fig. 21-36 Digital printer with continuous tape. (*Auto Controls, Inc.*)

Digital Recorders (Fig. 21-36). The digital recorder may record on either ticket or continuous tape. There are three primary advantages:

1. It has a high accuracy, usually ± 0.05 percent of full scale.
2. It has very good resolution, usually ± 0.1 percent of full scale.
3. Duplicate records are immediately available.

There are several disadvantages:

1. It does not present a complete history of the batch.
2. There is a necessary time delay while the printer completes its printing. In some cases the recording device which is attached to the scale may have a memory built into it, so the actual delay is in setting up the memory and not in the printing itself.
3. A return to zero may or may not be printed.
4. It is much easier for the operator to tamper with the system and to cause a number to print which would reflect the correct weight as desired but would not correspond to the actual weight.
5. Since the printer itself is a fairly complicated mechanism, and since there is by necessity a transmitting unit coupled with the scale, the digital recorder is not as reliable as the graphic recorder.
6. The digital recorder requires a much higher degree of sophistication in maintenance and troubleshooting.
7. A maintenance contract with the actual manufacturer of the digital printing equipment should be secured so that the units are cleaned and checked periodically to reduce downtime and reduce necessary repair work.

Photographic Recorders. These devices involve a camera which is mounted so that it can produce on film a picture of all dial scales when each batch is complete. It must be arranged to photograph a clock and something to represent the date, sequence number, batch type, etc., if this is required. There are several advantages:

1. The basic equipment is relatively inexpensive.
2. It is simple to operate.
3. Accuracy can be quite good depending on the quality of the camera and the film used.
4. Duplicate records are available.

There are also some disadvantages:

1. Care must be taken to reduce parallax errors, which could be very high.
2. If the dials are widely separated, it may be necessary to use more than one camera.
3. The record may not be available for several days and may be lost completely if the film is defective, if an error is made in processing the film, or if the operator neglected to put film in the camera.
4. There is a time delay to operate the camera.
5. A continuous record of the batch is not made, and it is possible for the operator to hold the scales (manually) at the correct reading to obtain an acceptable record.
6. Zero weights are not always recorded.

MIXING

A concrete mixer may be defined as a device or machine used to combine portland cement, water, aggregate, and other ingredients in a homogeneous mixture to produce concrete. The number, size, shape, angle, and arrangement or location of blades in a mixer are largely empirical. Mixers may be divided into three general types. Plant mixers are usually mounted as a stationary part of a concrete batching plant. Paving mixers are mounted on wheels or tracks; they are designed to mix as they move along the roadway so that the concrete can be placed as soon as it is mixed. Transit mixers or truck mixers are designed to mix while they are enroute from the batching plant to the site where the concrete will be used. Specifications and standards for concrete-plant mixers are established by the Plant Mixer Manufacturers Division of the Concrete Plant Manufacturers Bureau. The Truck Mixer Manufacturers Bureau publishes specifications and standards for truck mixers and agitators.

Plant mixers are either tilting or nontilting. Two types of tilting mixers and three types of nontilting mixers are commonly used in concrete plants (Table 21-17).

TABLE 21-17 Standard Sizes and Types of Concrete-plant Mixers*

Mixer type	*Size and rated mixing capacity, cu yd*
Single compartment, two-opening, nontilting type	1, 2, 3½
One-opening tilting type, mixing angle of drum 15° with horizontal	2, 4½, 6, 8, 9, 10, 12, 15
Two-opening front- or rear-charge and front-discharge tilting type	2, 4½, 6, 8, 9, 10, 12, 15
Vertical-shaft type	1, 1½, 2, 3½, 4½

* These sizes are established by standards of the Plant Mixer Manufacturers Division, Concrete Plant Manufacturers Bureau, 900 Spring Street, Silver Spring, Md. 20910.

Tilting Mixers Tilting mixers have a rotating drum which discharges when the drum is tilted from 50 to 60° downward. The drum may either be horizontal or tilted slightly upward during charging and mixing operations. The inside of the drum is usually weld-coated or supplied with replaceable, abrasion-resistant liner plates.

One-opening tilting mixers with the mixing angle of the drum at 15° with the horizontal usually have a large opening for fast charge and fast discharge (Figs. 21-37 and 21-38). These mixers have been successfully used with almost every type of concrete and for almost every application for ready-mix concrete, mass concrete using large aggregate, cement base course, and pavement concrete.

Two-opening rear-charge and front discharge tilting mixers have the mixing axis of the drum horizontal (Fig. 21-39). The mixer is charged through the rear opening, which is relatively large for fast charging, and discharged through a smaller front opening. These mixers may be used with pavement concrete, ready-mix concrete, or low-slump concrete.

Fig. 21-37 One-opening tilting mixers, mixing angle 15° with horizontal. (*C. S. Johnson Operations, Road Division, Koehring Company.*)

Fig. 21-38 Self-erecting one-opening tilting mixer, mixing angle 15° with horizontal, shown in travel position. (*C. S. Johnson Operations, Road Division, Koehring Company.*)

Nontilting Mixers Nontilting mixers have either a rotating drum or rotating blades in a stationary drum. These mixers may discharge through gates in the mixing compartment or by chutes that direct the concrete from the rotating drum.

Single-compartment two-opening nontilting mixers have a rotating drum. These mixers are normally charged through the rear opening by means of a chute which projects into the mixer. The mixing blades pick up the concrete and spill it over onto the concrete in the bottom. These blades direct the concrete toward the front of the mixer, where a chute is hinged so that it can be rotated into a position to discharge

Fig. 21-39 Two-opening rear-charge and front-discharge tilting mixer, horizontal drum. (*Rex Chainbelt, Inc.*)

material as received from the blades. This mixer is normally used for ready-mixed concrete and high-slump concrete applications. The difficulties with cleaning the mixer and its relatively slow discharge have contributed to its lack of use in recent years.

Vertical-shaft mixers (turbine, pan, or compulsory mixers) have a mixing compartment with rotating blades (Fig. 21-40). There may be one or more rotating vertical

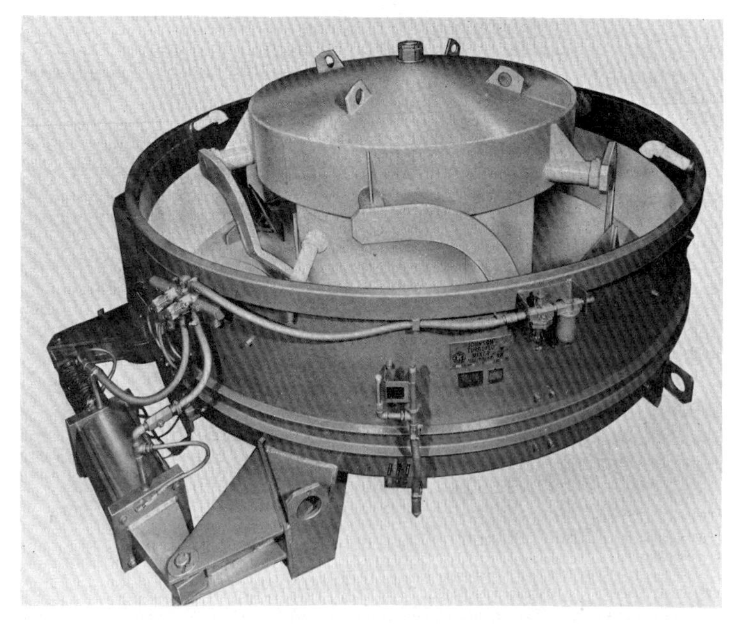

Fig. 21-40 Vertical-shaft mixer. (*C. S. Johnson Operations, Road Division, Koehring Company.*)

shafts with blades attached. In some mixers there are rotating blades and a rotating mixing compartment. The mixer has a high wear factor because of the rapid movement of concrete against the walls and against the blades or paddles. The power requirements are quite high compared to other mixers. These mixers are used for a wide variety of slump requirements, but excessive power is required for low slump and the capacity of the mixer may be reduced for very low-slump applications. These mixers are most commonly used in concrete-products plants, but some have been successfully used in paving operations and ready-mix concrete plants. Concrete is discharged through one or more horizontal doors in the bottom of the mixing compartment. A mixer load indicator is normally used with these mixers to give an indication of the actual loading of the mixer motor and gears. This helps to alert the operator if he is overloading the mixer. These mixers are sometimes very sensitive to sequence of feeding or charging of material. The aggregate and cement should be blended before reaching the mixer, or they should be fed into the mixer together.

Fig. 21-41 Horizontal-shaft mixer. (*Besser Company.*)

The water should be introduced rapidly with the cement and aggregate, and it should be placed well inside the mixer. If water dribbles down the side, the wall may be lubricated and the concrete will slide with the blades around the mixer.

Horizontal-shaft mixers have a stationary, horizontal, cylindrical mixing compartment and one or more horizontal rotating shafts with blades or paddles attached (Fig. 21-41). They are ideally suited to mixing very low-slump concrete and are normally used in concrete-products plants. The blades work the material longitudinally and radially over the material in the bottom of the mixer. The concrete is usually discharged through a door mounted low on one side of the mixer. Liner and blade wear are relatively high. Horizontal-shaft mixers are commonly rated in cubic feet of loose material per batch rather than in cubic yards of concrete. Because of the high absorption, many lightweight aggregates used in these mixers require prewetting before cement is added to the mixer. Water may be added to the aggregate and mixed from 15 sec to several minutes before the cement is introduced.

Paving Mixers Paving mixers are mounted on tracks and move as they are mixing (Fig. 21-42). There are usually two or three drums mounted in one single shell. The cement and aggregate are charged into the first drum from a skip on which the dry-batch trucks dump the material. The water and admixture are put into the mixer from containers mounted on the mixer. The water and admixtures are measured volumetrically at the mixer. The mixing time usually starts when the skip is at its top position. The material is mixed in the first drum and then transferred to the second drum. When all the material has been transferred from the first drum to the second, a new batch may be placed in the first drum. When the material has been mixed, it is discharged into a bucket which is supported by a boom on the mixer. This bucket is then moved out laterally and the concrete is dumped on the ground.

Fig. 21-42 Tri-batch paver. (*Koehring Division of Koehring Company.*)

A mixer timer with a mechanical interlock on the bucket is a part of the mixer. The batch size varies from 1.26 to 1.51 cu yd.

Truck or Transit Mixers (Table 21-18) Truck mixers or transit mixers either mix or agitate concrete in a drum mounted on a moving truck or trailer (Fig. 21-43). There are three types: inclined-axis revolving drum, horizontal-axis revolving drum, and open-top revolving blade or paddle mixers. The inclined-axis type is the most

TABLE 21-18 Capacities of Truck Mixers*
(In cubic yards)

Mixing capacity	*Agitating capacity*†
6	7¾
6½	8½
7	9¼
7½	9¾
8	10½
8½	11¼
9	12
10	13¼
11	14¾
12	16
13	17½
14	19
15	20¼
16	21¾

* These sizes are established by standards of the Truck Mixer Manufacturers Bureau, 900 Spring Street, Silver Spring, Md. 20910.

† Unless otherwise restricted on the manufacturer's data plate on the mixer. Some mixers will agitate this amount only when the drum opening is closed.

common. This mixer discharges by reversing the drum rotation so that the blades tend to screw the concrete back to the drum opening. Truck mixers may be used to completely mix the concrete. This may be done in the yard before the mixer goes to the job to place the concrete, it may be done at the jobsite, or it may be done enroute to the jobsite. Truck mixers are also used to complete mixing when a central plant mixer is used to shrink-mix the concrete. When the truck mixers are used to transport central-mixed concrete, they are usually operating as agitating units. Specifications allow the truck mixer to haul about 10 percent more shrink-mix or central-mix concrete than transit-mix concrete.

Mixing speed is not less than 4 and not more than 12 rpm of the drum or blades. Agitating speed is not less than 2 and not more than 6 rpm of the drum or blades. Drums are often operated at higher speeds while charging or discharging to handle the material faster. Concrete should be mixed between 70 and 100 revolutions at mixing speed after all ingredients are in the drum. The mixing water should also be in the drum before the mixing begins. If any water is added after the concrete has been mixed, the mixer should be rotated a minimum of 30 additional revolutions at mixing speed. Any additional rotation of the drum should be at agitating speed. The overall total number of revolutions should not exceed 250.

Fig. 21-43 Truck mixer. (*Challenge-Cook Bros., Inc.*)

Mixing Time One of the most important factors in the time necessary for proper mixing is the sequence of charging material into the mixer. If the materials are blended well as they enter the mixer, the mixing time may be substantially reduced. Many low-profile central-mix plants use a conveyor belt to charge the mixer, and the materials are well blended or premixed on the belt before they enter the mixers in most of these plants. The large, mass-concrete plants are arranged so that the material is batched in single material batchers which feed into a collecting hopper and from there to the mixer. The materials are blended in the collecting hopper before they enter the mixer.

Mixing times for tilting mixers are shown in Table 21-19. These mixing times are primarily for mass pour work. Tilting mixers used for highway paving projects normally require mixing times of 40 to 90 sec. It is quite common to require 75 sec mixing time without mixer performance tests and to reduce this time to 60 sec or less if performance tests are made. In a ready-mix plant using a mixer for shrink mixing, 30 sec is usually adequate. The turbine-type mixers which are charged adequately will mix in about 30 sec, but sometimes 45 sec is necessary to produce a consistent mix. The horizontal-shaft mixers used in concrete-products plants have a mixing time from 2 to 5 min. Mixing times for paving mixers range from 50 to 90 sec, with 60 sec being the most common. The transfer time between drums is usually considered part of the mixing time. Truck or transit mixers require from 8 to 15 min mixing time.

TABLE 21-19 Tilting-mixer Production Rates—Mass Concrete[23,*]

Nominal or rated mixer capacity, cu yd	Typical mixing time, min	Typical cycle time, min	Batches/hr	Cu yd/hr
2	**1½**	**2**	**30.0**	**60.0**
3	2	2½	24.0	72.0
3½	**2¼**	**2¾**	**21.8**	76.4
4	2½	3	20.0	80.0
4½	**2⅝**	**3⅛**	**19.2**	**86.4**
	2¾	**3¼**	**18.5**	**83.1**
5	2¾	3¼	18.5	92.3
	3	3½	17.1	85.7
5½	**2⅞**	**3⅜**	**17.8**	97.8
	3¼	**3¾**	**16.0**	88.0
6	3	3½	17.1	**102.9**
	3½	4	15.0	**90.0**
7	**3¼**	**3¾**	**16.0**	112.0
	4	**4½**	**13.3**	93.3
8	**3½**	**4**	**15.0**	**120.0**
	4½	**5**	**12.0**	**96.0**

*Above production rates are for mixers with a batch equal to the nominal capacity shown. Mix times shown are typical for specification work in the absence of mixer-performance tests. Mixer sizes in boldface type are rated capacities of Plant Mixer Manufacturers Division standard mixers. They are guaranteed to mix rated capacity when slump is between 1½ and 3 in. and aggregate size is not over 3 in. Before 1966, mixers were rated in cubic feet, with 28 cu ft of rated capacity per cubic yard of nominal capacity, and guaranteed to mix a 10 percent overload. For mass concrete, a batch of the nominal capacity shown is usually mixed in these obsolete-size mixers, and batches of 4 or 8 cu yd are common in 4½- and 8½-cu yd mixers because of large aggregate and because of concrete bucket, crane, and cableway limitations.

Overmixing should be avoided because it increases fines and thus requires more water to maintain consistency, drives out entrained air, and tends to increase the concrete temperature.

PLACING

Concrete should always be described as being placed rather than poured. The expression "poured" comes from the early concrete work where it was necessary to have very wet and sloppy concrete in order to get it into the forms. This was prior to the introduction of high-frequency vibration for the proper placing of concrete.

It is very important to prevent separation of the coarse aggregate from the concrete during placing operations. This separation occurs primarily because the materials in concrete differ greatly in particle size and specific gravity. If the concrete is allowed to flow, the coarse aggregate separates from the mortar. If the concrete is confined laterally, the heavier materials tend to settle and the lighter materials tend to rise toward the top. Separation of the aggregate from the concrete occurs most frequently at the ends of chutes and conveyor belts, at hopper gates, at a discharge point from a mixer, and during transportation.

Concrete should be placed as near as possible to its final position and should not be allowed to flow horizontally or on a slope in forms. It should not be placed in separate piles and the piles then leveled and brought together. Concrete should not be placed in big piles and allowed to run or be worked over a long distance to its final position. Concrete should be placed into the face of previously placed concrete, not away from it. When concrete is placed in layers, each succeeding layer should be placed while the one below it is still plastic, and the top portion of the lower placement should be vibrated to ensure that the concrete is continuous and without separation.[24] It is important to avoid the entrapment of air within partially enclosed spaces which are to be filled with concrete. In wall and column placements, the slump should be reduced as the level of concrete rises in order to offset water gain which will weaken the upper portion of the concrete and make it less durable. The slump should be the minimum that can be vibrated well into the critically exposed portions of the work. When placing concrete in walls, curbs, and slabs, work should proceed from the corners and ends of the forms toward the center, rather than toward the corners and ends, thus avoiding accumulation of mortar and wetter concrete in those parts of structures where exposure is most severe.

The handling equipment and methods used should be selected because they can handle the mix design appropriate to the job at hand; it would be a mistake to design a mix to suit the placing methods to be used rather than the job to be done. The batching, mixing, transporting, and placing should be done at such a rate that the concrete placing may proceed without serious interruptions. Cold joints must be avoided.

Concrete should be placed as soon as possible after mixing. The actual amount of time allowable varies with conditions such as the weather and the ultimate use of the concrete. Forms should be clean, tight, adequately braced, and constructed of materials that will impart the desired texture to the finished concrete. Reinforcing steel should be clean and free of loose rust or mill scale at the time concrete is placed. Any coatings of hardened mortar should be removed from the steel.

Before beginning placement, the contractor should be certain that all transporting, placing, vibrating, finishing, and curing equipment is clean and in proper repair and that it is adequate and properly arranged so that placing may proceed without undue delays. Sufficient personnel should be available to handle all phases of the operation. Concrete placing should not be started when there is a probability of freezing temperatures unless adequate facilities for cold-weather protection have been provided.

Chutes One of the most important problems in handling or transporting concrete is the segregation or separation of the coarse aggregate from the mortar. Concrete should drop vertically and in the clear regardless of the type of equipment used. If drop chutes or elephant trunks are used, the upper sections may be at an angle to facilitate placing of the concrete, but the lower section should always be vertical. Concrete should not be dropped through reinforcement steel or other objects which tend to separate it, nor should it be directed against the forms.

Unless chutes are constructed and used properly, segregation and loss of slump will result. The chutes must be deep enough to handle concrete in such a way that it will move smoothly, and the chutes' slope should be such that concrete of the required slump will slide, not flow. End control should be provided so that the concrete will drop vertically and without segregation from the end of the chute. Two sections of metal drop-chute elephant trunk will serve to control end segregation. A mere baffle is not adequate. Open chutes which are exposed to wind and the sun tend to accelerate slump loss and must be avoided. Enough chutes should be used, or the chutes should be sufficiently portable, that the concrete may be placed directly into its final position instead of being forced to move horizontally by gravity or through vibration. If the concrete is placed in piles, the coarse aggregate which runs down the side should be shoveled back onto the pile and mixed into the concrete.

Buckets Buckets are a most satisfactory means for handling and placing concrete when they are designed for the job conditions and properly operated. They should not be used where they will have to be hauled so far by truck or railroad that there will be noticeable separation or bleeding due to settlement or a loss of slump greater than 1 in. Bucket capacities for handling concrete range from 1 and 2 cu yd for structural work up to 12 cu yd for the larger mass-concrete projects. Each bucket should have a capacity of at least one batch of concrete as mixed to avoid splitting of batches in loading buckets. Such splitting invariably results in segregation. Buckets should be capable of prompt discharge of low-slump or lean-mix concrete. The dumping mechanism should permit discharge of a relatively small portion of concrete

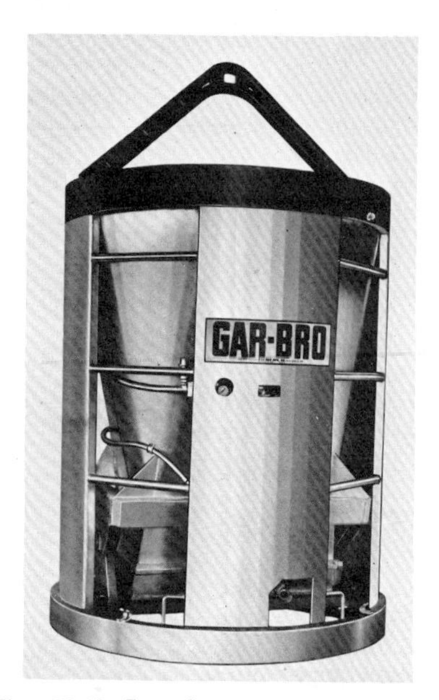

Fig. 21-44 Low-slump concrete bucket. (*Gar-Bro Manufacturing Company.*)

Fig. 21-45 General-purpose concrete bucket (*Gar-Bro Manufacturing Company.*)

Fig. 21-46 Lay-down concrete bucket. (*Gar-Bro Manufacturing Company.*)

in one place. Also, the discharge should be controllable so that it will cause no damage to or misalignment of the forms. Buckets should be filled and discharged without noticeable separation of coarse aggregate.

If the discharge is too slow or if the bucket is too low, or if the discharge is not vertical, the low-slump concrete will stack in a cone and the coarse aggregate will separate and roll down the cone. Rock pockets usually are formed as a result.

Low-slump or mass-concrete buckets (Fig. 21-44) have bottom slopes of 60 to 70° and large gate areas to handle coarse aggregate up to 8 in. They are usually operated with compressed air.

General-purpose buckets (Fig. 21-45) are usually manually operated and handle from 2- to 6-in. slump and about 2½- to 3-in. maximum aggregate.

Lay-down buckets (Fig. 21-46) are commonly filled from a transit-mix truck where low filling height is required.

Buggies Either manual or power-operated buggies (Fig. 21-47) may be used to transfer concrete for limited distances. The hand-operated carts or buggies usually have capacities up to about 6 or 8 cu ft. The power-operated buggies have capacities up to ½ cu yd. The buggies need a relatively level runway or path. Depending on conditions, each hand-operated cart should be able to transport up to about 5 cu yd concrete per hr, and each power-operated buggy should be able to transport up to about 20 cu yd per hr.

Pumps Concrete pumps were originally designed and used to place concrete in tunnels where the space available for placing equipment was very tight and large equipment could not be used. The first models of pumps could not handle lightweight aggregates or harsh mixes and were most effective with high-slump mixes containing large proportions of cement and sand. Today, most or all of these limitations have been overcome and concrete pumping is usually just as fast or faster than other types

Fig. 21-47 Power-operated buggies. (*PCM Division of Koehring Company.*)

of placing. Normally, fewer people are needed on a pumping job. A concrete-pump system requires one pump operator and two or three men handling the hose. Concrete pumps may be used in tandem to obtain higher pours, and one pump may feed into another for this purpose. Several pumps may be used side by side for higher production rates.

At least three different methods of pumping concrete are available. In a piston type which is used by many pump manufacturers (see Figs. 21-48 to 21-50), a valve on the hopper opens and closes to let concrete into the line and a valve on the line opens and closes to prevent concrete from being pulled back when the piston is retracted. This sucks the concrete in through the hopper and pumps it out through the line. Some pumps use compressed air for blowing concrete through the lines. One type of concrete pump (Fig. 21-51) uses a squeezing action which compresses the hose as it rotates and forces the concrete through the hose. This creates a suction which pulls more concrete into the line to make a continuous flow available.

Concrete pumps may easily be utilized where space is at a premium and there is no room for a crane. There is no interference, such as would be encountered with a crane, with any overhead equipment. The pipeline or hose offers minimum disturbance to other operations around the construction site. In at least one instance concrete has been pumped vertically 500 ft. Some pumping units are equipped with booms (Fig. 21-52) which facilitate placing of the concrete in areas where it is difficult to place a pipeline. The boom carries the first 50 or 60 ft of line.

A concrete pump can deliver up to 100 cu yd per hr through lines which may range in size from 2 to 8 in. Pipe is used wherever possible, including changes in direction, since it has less friction than flexible hose. However, the last 25 or 30 ft of a line is usually hose to make it easier to adjust the placing position.

One of the problems of pumping lightweight concrete is the pressure which exists in the pumping line. This tends to force the mixing water into the highly absorptive lightweight aggregates. Prewetting the lightweight aggregates should reduce this problem.

Fig. 21-48 Concrete pump. (*Norton Construction Products Division, Clipper Manufacturing Company, Inc.*)

Fig. 21-49 Concrete pump. (*PCM Division of Koehring Company.*)

Friction is one of the major problems in a concrete-pump installation, and it is advisable to run some grout through the pipe before starting the pumping operation. This produces a thin coating of grout on the pipe and helps to reduce friction. If there is too much water in the mix, the grout coating will be washed away. If sharp, irregular rocks are used, these are likely to cut through the grout coating and cause a blockage of the pipe. Good gradation of the aggregates is a requirement for good, pumpable concrete. Natural round aggregate is best, but crushed blocky aggregate may also be used satisfactorily. Irregular and elongated crushed aggregate is very poor for pumping because it causes particle interference and the aggregates will tend to jam and cause a plug in the pipe. Some of the concrete pumps are equipped with a reversing system so that the operator can pump material back into the hopper. This helps to unblock jams in the line without disconnecting the lines.

Belt Conveyors* Belt conveyors for placing concrete have been used for many, many years, but their use has increased considerably in recent years. The belt conveyors are able to move the concrete around or over obstacles which would present

Fig. 21-50 Concrete pump. (*Whiteman Manufacturing Company.*)

Fig. 21-51 Truck-mounted concrete pump. (*Challenge-Cook Bros., Inc.*)

serious problems with many other methods of placing. Some units are self-propelled (Fig. 21-53) so that they may be used in one location for a short placement, then quickly and easily moved to another location for a subsequent placement.

Many of these conveyors may be used in series to move concrete over a considerable span. The system of belts may be made a variable length by running each conveyor up over the next one (Fig. 21-54). With provisions for moving each belt conveyor with respect to the next, the contractor may start placing concrete at either end of a job and extend or retract the belt system to continue placing concrete without inter-

* See Sec. 15, Belt Conveyors.

Fig. 21-52 Truck-mounted concrete pump with boom. (*Challenge-Cook Bros., Inc.*)

Fig. 21-53 Portable belt conveyor for placing concrete. (*Morgen Manufacturing Company.*)

ruption of the pour. A swinging belt conveyor may be used at the discharge end for versatility in placing as needed.

A side discharge may be used to enable concrete to be placed at any point along the conveyor, which greatly facilitates the placing of concrete over a wider area.

A scraper of rubber or other suitable material should be used to prevent loss of mortar on a return belt. Long belts should be protected from the rain or hot sun by suitable covers.

Belt speeds normally run up to 900 ft per min, and concrete placing at rates up to 300 cu yd per hr has been done successfully.

A special application is a belt conveyor which is mounted on a truck mixer (Fig. 21-55). This enables the contractor to reach over walls, ditches, or other obstacles at the jobsite and eliminates rehandling of concrete in many cases. It also enables the truck to stay a safe distance from forms and excavations. The belt conveyor is

Fig. 21-54 Series of belt conveyors for placing concrete. (*Oury Engineering Company, Marion, Ohio.*)

Fig. 21-55 Belt conveyor mounted on a truck mixer. (*Challenge-Cook Bros., Inc.*)

adjustable in both horizontal and vertical directions so the concrete may be placed wherever it is needed within the reach of the belt conveyor.

Trucks Transit-mix trucks are used as agitating units to transport central-mixed concrete or as mixing units to mix as they are enroute to the jobsite. Some open-top hauling units have specially shaped bodies with rotating blades for agitating central-mixed concrete and chutes to discharge the concrete. Other open-top hauling units are nonagitating and may be either side-dump (Figs. 21-56 and 21-57) or end-dump. Sometimes regular dump trucks are used as hauling units for central-mixed concrete.

Consolidation The objective of consolidating concrete is the elimination of voids within the concrete. Well-consolidated concrete is free of rock pockets and bubbles of entrapped air, and it is in close contact with the forms, reinforcement, and other embedded parts. The best method of securing consolidation of freshly placed concrete is thorough vibration. This permits placing concrete of lower slump than would be possible with handworking and therefore produces a better-quality concrete.

Fig. 21-56 Open-top side-dump concrete hauling unit. (*Maxon Construction Company Inc.*)

Fig. 21-57 Open-top side-dump concrete hauling unit. (*T. L. Smith Company.*)

Internal vibration is usually the most effective, and immersion-type vibrators should be inserted vertically at points 18 to 30 in. apart and slowly withdrawn. A short amount of vibration at this spacing is much better than longer periods of vibration at wider spacing. Systematic spacing of the points of vibration is desirable to ensure that no part of the concrete is missed.[14,25]

When concrete is placed in layers, the entire depth of a new layer should be vibrated. The vibrator should penetrate several inches into the layer below to ensure a thorough union of the layers. It is essential that there be good consolidation in an area where a newly placed section of concrete adjoins other work.

High-slump concrete should not be vibrated too much because water and fine material will be brought to the surface, ultimately leading to scaling or dusting. Concrete should not be moved horizontally by vibration because this results in segregation.

Construction Joints A construction joint is the plane of contact between two increments of concrete, with the second increment placed after the first has hardened. A construction joint may be horizontal, as in a wall or column; or it may be vertical, as in a slab or adjacent blocks of a dam. Reinforcing steel is normally continuous across a construction joint, or dowels are provided. Some designs call for shear keys in the joint.

One of the best methods of cleaning a construction joint is wet sandblasting and washing. The area should then be completely dried immediately prior to placement of fresh concrete. This method is very simple, dependable, and economical.[14]

Special Placing Techniques *Cold-weather concreting* normally means placing, finishing, and curing at temperatures which are at or below freezing.[26] Provisions must be made to prevent damage to concrete from freezing and thawing before it has had time to harden properly. After the concrete has hardened, the temperature should be allowed to drop gradually, since sudden changes in the temperature of new concrete will promote cracking and deterioration of strength and durability. All surfaces which are to be in contact with the new concrete should be raised to a temperature as close as possible to the temperature of the concrete to be used. All ice, snow, and frost should be removed.

Heating of some or all of the concrete materials so that the resultant concrete is between 40° and 60°F is recommended. Quite often the heating of the mixing water alone will be sufficient to accomplish this. If necessary, the aggregate can also be heated. The finished concrete should be kept from freezing for a sufficient period of time to allow for proper hardening and curing.

Placing concrete in hot weather creates a number of problems.[27] The higher ambient temperature causes the concrete to set or harden at a faster rate. The time between final mixing and hardening of the concrete becomes shorter, therefore the time during which the concrete may be worked is shorter and in some cases is entirely too short. This problem, of course, is made worse by the heat of hydration produced by the concrete.

One solution is to decrease the temperature of the concrete to offset the detrimental effects of the higher temperature. This may be done by reducing the temperature of the water, adding ice in place of some of the mixing water, cooling the sand and aggregates, and cooling the cement. Any one or all of these measures may be necessary in a given application. Another simple, partial solution is to paint all the storage bins, mixers, trucks, or other hauling units white to reflect the sunlight instead of absorbing it.

The temperature of the water may be reduced by selecting the coolest water source available or by chilling the water. Mixing time should be kept to a minimum during hot weather because excessive mixing increases the temperature of the concrete. Water-reducing retarders extend the setting time of concrete and in some cases reduce the amount of mixing water necessary to obtain a given amount of slump. Cooling coils are often run in mass concrete to reduce the temperature.

It is best to schedule the start of a placement very early in the morning when the temperature is relatively low and try to arrange to stop before the highest temperatures are reached late in the afternoon. It is also well to have all equipment that will be in contact with the concrete as cool as possible. The batchers, mixers, hoppers, buckets, and trucks should be washed with cold water. It is essential that the concrete be placed as soon as possible after delivery. The necessary curing procedure should be started as soon as possible after the concrete has been placed.

Placing concrete under water should be avoided if at all possible. When it becomes necessary, from ½ to 1 sack of cement per cu yd should be added to the batch. A wet, plastic mix is required since working or vibrating of the concrete is seldom possible. The slump should be 5 or 6 in. A tremie is usually the best method of placing concrete under water. This is a pipe consisting of sections joined together, with a funnel-shaped section at the top to receive the concrete. The tremie discharge end should be free so that it can be moved to the desired point of discharge. The bottom of the pipe should be kept continuously immersed in the concrete, and the tremie should be kept full of concrete at all times. Flow is regulated by raising or lowering the tremie. The concrete should be placed in approximately horizontal layers, and there should be no interruption of the placing until it has been completed. It is not necessary to dewater the foundation area when a tremie is used.*

A watertight bucket can also be used to place concrete under water. The bucket must discharge from the bottom, and the gate should not be opened until the bucket comes in contact with the foundation or previously placed concrete. After the concrete has been discharged, the bucket must be raised slowly until it is well away from the concrete.

FINISHING AND CURING

Finishing The specifications for a job should describe the required finish, since concrete surfaces can be finished in almost limitless variations. The actual finish will be determined by the characteristics of the form or form liners, by any work done on the concrete after the forms are removed, or by the treatment of unformed surfaces. The texture and appearance of the surface will depend on the rubbing, grinding, or sandblasting of the surface and may range from very rough to satin smooth. The specified finish should be consistent with the purpose of the concrete and the use to which it will be put.

Curing Concrete hardens because of the chemical reaction, called hydration, between portland cement and water. The object of curing is to prevent or replenish the loss of necessary moisture during the early, relatively rapid stage of hydration. Optimum curing is defined as the act of maintaining controlled conditions for freshly placed concrete for some definite period following the placing or finishing operations to assure the proper hydration of the cement and the proper hardening of the concrete. Five requirements for proper curing are:[28]

1. Preservation of adequate water content in concrete
2. Maintenance of a fairly constant temperature above freezing
3. Preservation of a reasonably uniform temperature throughout the whole body of concrete
4. Protection from damaging mechanical disturbances
5. Passage of sufficient time for hydration of cement and hardening of concrete

Sufficient water may be maintained by building a small dam around the perimeter and keeping the concrete covered with water. Periodic sprinkling may also be used, but the concrete must not be allowed to dry out between sprinklings. Sand, burlap, canvas, or straw may be placed over the concrete and kept continuously wet.

The concrete may be covered with waterproof paper or liquid, membrane-forming curing compounds to prevent water from escaping. No addition of water is then necessary. The curing compounds should be sprayed on as soon as the concrete has been finished and there is no free water left on the surface.

The time necessary for curing depends on the cement, mix proportions, required strength, size and shape of the concrete mass, weather, method of curing, and future exposure conditions. This time may vary from a few days for very rich mixes in special applications to a month or more for mass concrete.

* See Sec. 28, Cofferdams and Caissons.

C. Reinforcing Steel

The fact that the thermal coefficients of expansion for concrete and steel are approximately equal, i.e., 0.0000065 per degree Fahrenheit, makes possible the use of reinforced concrete as a structural medium. Concrete may be reinforced with plain or deformed steel bars, or with steel wires, steel-wire cables, or steel-wire mesh. In general, the steel resists the tensile stress within a member of the structure, while the concrete resists the compressive stress. Occasionally, steel bars are also required to assist the concrete in compression. Concrete and main reinforcing steel act in combination to resist shear stresses, although supplementary steel may also be supplied for this purpose.

DEFORMED REINFORCING BARS

The bond, or adherence of the concrete to the steel bars, must be sufficient to ensure composite action between the steel and the concrete when the reinforced-concrete member is loaded or stressed. Deformed bars (ASTM Designation A 305) cost about the same as plain bars but develop much greater bond stress resistance; they are used almost universally in the United States to reinforce concrete.

Sizes and Grades Established measurements, types of deformations, and maximum tolerances for deformed reinforcing bars[29] are prescribed in ASTM Designation A 305. Figure 21-58 illustrates several bar types. Table 21-20 lists the 11 standard sizes of deformed bars which have been adopted by the American Society for Testing and Materials. The former No. 2 plain bar is now classed as wire reinforcement.

Deformed bars are rolled from new billet steel, rerolled rail steel, and axle steel. Bars in common use are of three minimum yield-strength levels: 40,000 psi, 60,000 psi, and 75,000 psi. These are designated as Grade 40, Grade 60, and Grade 75, respectively. Rail-steel bars, which are in limited use, are supplied in Grade 50 rather than Grade 40. Most fabricators stock only new billet steel, and this only in Grade 40.

Table 21-21 lists ASTM physical requirements, as of 1968, for standard deformed reinforcing bars.

With the development of higher-strength concrete, high-strength reinforcing steel merits serious consideration. Properly selected combinations of these two materials will permit the design of economical columns for structures 20 stories or more in height. Instead of lapping large-size column bars, couplings can be utilized as illustrated in Fig. 21-59. Alternatively, the bar ends can be milled and butt-welded to save steel and reduce congestion of the reinforcement. High-strength steel bars and wires are also used in prestressed and poststressed concrete structures, resulting in greatly increased bridge and building spans.

Bar Supports Reinforcing steel should be accurately positioned in the forms and securely held in place during the pouring of concrete. This can be accomplished by using prefabricated bar supports and wire ties. When constructing footings, concrete bricks or precast concrete blocks are adequate to support the mats of bars. Heavy foundation mats of reinforcing steel will require special steel bar supports, with base plates resting on the subgrade.

Available bar supports for constructing concrete slabs, joists, beams, and girders include factory-made wire, strip steel, and plastic units. Metal bar supports which are exposed to the weather should be galvanized or cadmium-plated to prevent rust on concrete surfaces. When constructing walls, the reinforcing steel should be securely tied to the wall form ties.

WIRE REINFORCEMENT

Cold-drawn steel wire is used for reinforcing applications such as the following:

1. Spirals for concrete columns
2. Welded wire fabric for concrete slabs and pavements

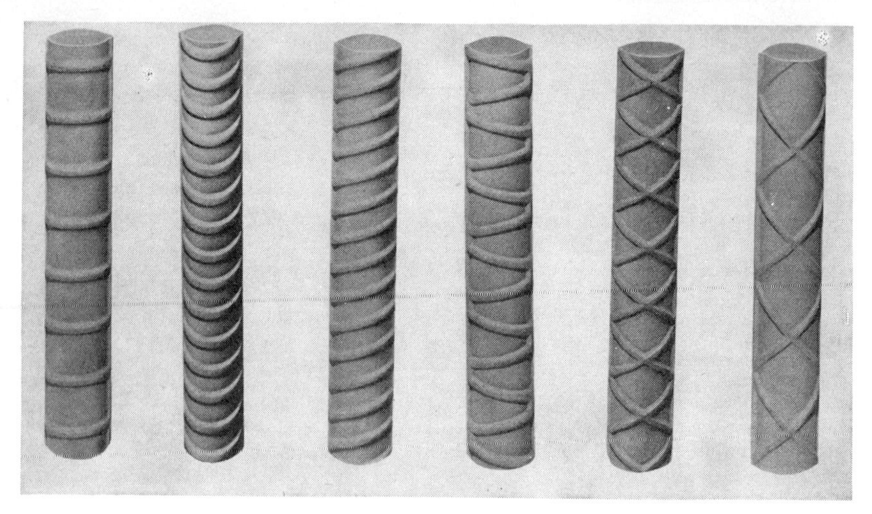

Fig. 21-58 Six types of A305 reinforcing bars.

TABLE 21-20 ASTM Standard Reinforcing Bars

Bar size designation No.	Weight, lb/ft	Nominal dimensions—round sections		
		Diameter, in.	Cross-sectional area, sq in.	Perimeter, in.
3	0.376	0.375	0.11	1.178
4	0.668	0.500	0.20	1.571
5	1.043	0.625	0.31	1.963
6	1.502	0.750	0.44	2.356
7	2.044	0.875	0.60	2.749
8	2.670	1.000	0.79	3.142
9	3.400	1.128	1.00	3.544
10	4.303	1.270	1.27	3.990
11	5.313	1.410	1.56	4.430
14*	7.65	1.693	2.25	5.32
18*	13.60	2.257	4.00	7.09

* Sizes No. 14 and No. 18 are large bars which may not be carried in regular stock. Advance arrangements should be made with the supplier.

3. Precast-concrete pipe
4. Prestressed concrete
5. Bar supports
6. Tying wire

The applicable ASTM Designations are A82 for plain wire and A496 if the wire is deformed.[29] The wire must have a minimum yield strength of 70,000 psi and a minimum ultimate strength of 80,000 psi. Plain wire gauges range from 0000000 to 14 (0.4900- to 0.0800-in. diameter). Gauges in deformed wire range from D-1 to D-31 (0.113- to 0.628-in. diameter).

Spirals Cold-drawn wire or plain round rods can be coiled and assembled with spacers to enclose the core of a concrete column and thus increase its load capacity.

TABLE 21-21 Physical Requirements for Standard ASTM Deformed Reinforcing Bars

Type of steel and ASTM specification No.	Size Nos. inclusive	Grade	Tensile strength min, psi	Yield min, psi*	Minimum elongation in 8-in. length, %†	Cold bend test‡
Billet steel A615	3–11, 14, 18	40	70,000	40,000	1,100,000 / tensile strength min 12 %	Under size No. 6 90°, $d = 3t$ Nos. 6, 7, 8 90°, $d = 4t$ Nos. 9, 10, 11 90°, $d = 5t$
						Nos. 14, 18 None
	3–11, 14,18	60	90,000	60,000	1,000,000 / tensile strength	Under size No. 6 90°, $d = 4t$ Nos. 6, 7, 8 90°, $d = 5t$ Nos. 9, 10, 11 90°, $d = 6t$
					7	Nos. 14, 18 None
	11, 14, 18	75	100,000	75,000	Varies with bar size, 5–7½ %	No. 11 90°, $d = 8t$ Nos. 14, 18 None
Axle steel A617	3–11	40	70,000	40,000	1,100,000 / tensile strength min 12 %	Under size No. 6 90°, $d = 3t$ Nos. 6, 7, 8 90°, $d = 4t$ Nos. 9, 10, 11 90°, $d = 5t$
	3–11	60	90,000	60,000	1,000,000 / tensile strength	Under size No. 6 90°, $d = 4t$ Nos. 6, 7, 8 90°, $d = 5t$ Nos. 9, 10, 11 90°, $d = 6t$
Rail steel A616	3–11	50	80,000	50,000	1,000,000 / tensile strength min 4.5–5 %	None
	3–11	60	90,000	60,000	1,000,000 / tensile strength min 5 %	None

* Yield point or yield strength. See specifications.
† For base sizes of deformed bars. See specifications for adjustment for small and large sizes and for values for plain bars.
‡ d = diameter of pin around which specimen is to be bent, and t = nominal diameter of specimen. Values shown are for deformed bars.

Spirals are fabricated from ⅜-in. round, ½-in. round, and ⅝-in. round wire or rods to a clear pitch of not less than 1⅜ in. or more than 3 in. Cold-drawn wire should have a minimum yield strength of 70,000 psi (ASTM A82). Plain round rods should conform to ASTM Grade 40, 50, or 60 with minimum yield strengths of 40,000, 50,000 or 60,000 psi respectively.

Welded Wire Fabric Welded wire fabric is commonly called "mesh" when used for concrete reinforcement. It is manufactured from cold-drawn wire in square or rectangular patterns and is resistance-welded at all intersections. It must conform to ASTM Designation A185 if made of smooth wire. Should deformed wire be required, ASTM Designation A497 is applicable.[29] The designation WWF 6 × 6, 8/8 denotes that both longitudinal and transverse wires are spaced at 6 in. and that both wires are 8 gauge. This type of mesh is a two-way fabric, while WWF 4 × 12, 4/8 is a one-way fabric. Welded wire fabric with wires lighter than 10 gauge is usually available only in rolls, and WWF with wires heavier than 6 gauge is usually available only in sheets. Wires of either type which are 11 gauge or lighter should be galvanized. Welded wire fabric in sheets may be furnished to order, with sizes limited only by shipping restrictions. The common uses of welded wire fabric are temperature reinforcement, reinforcement in highway or street pavements, main reinforcement in comparatively short slab spans, and concrete-pipe reinforcement. Table 21-22 lists typical styles of welded wire fabric.

TABLE 21-22 Typical Styles of Welded Wire Fabric

Style designation	Spacing of wires, in.		Size of wires, AS & W gauge		Sectional area, sq in./ft		Weight per 100 sq ft, lb
	Longitudinal	Transverse	Longitudinal	Transverse	Longitudinal	Transverse	
(a) One-way (rectangular) types*							
24-1414†	2	4	14	14	0.030	0.015	16
212-04	2	12	0	4	0.443	0.040	169
212-15	2	12	1	5	0.377	0.034	144
212-26	2	12	2	6	0.325	0.029	124
212-37	2	12	3	7	0.280	0.025	107
212-48	2	12	4	8	0.239	0.021	91
212-59	2	12	5	9	0.202	0.017	77
212-610	2	12	6	10	0.174	0.014	66
212-711	2	12	7	11	0.148	0.011	56
312-04	3	12	0	4	0.295	0.040	119
312-15	3	12	1	5	0.252	0.034	102
312-26	3	12	2	6	0.216	0.029	87
312-37	3	12	3	7	0.187	0.025	75
312-48	3	12	4	8	0.159	0.021	64
312-59	3	12	5	9	0.135	0.017	54
312-610	3	12	6	10	0.116	0.014	46
312-711	3	12	7	11	0.098	0.011	39
312-812	3	12	8	12	0.082	0.009	32
412-26	4	12	2	6	0.162	0.029	69
412-37	4	12	3	7	0.140	0.025	59
412-48	4	12	4	8	0.120	0.021	51
412-59	4	12	5	9	0.101	0.017	43
412-610	4	12	6	10	0.087	0.014	36
412-711	4	12	7	11	0.074	0.011	31
412-810	4	12	8	10	0.062	0.014	27
412-812	4	12	8	12	0.062	0.009	25
412-912	4	12	9	12	0.052	0.009	22
412-1012	4	12	10	12	0.043	0.009	19
412-1112*	4	12	11	12	0.034	0.009	16
412-1212*	4	12	12	12	0.026	0.009	13
48-711	4	8	7	11	0.074	0.017	33
48-812	4	8	8	12	0.062	0.013	27
48-912	4	8	9	12	0.052	0.013	23
48-1012	4	8	10	12	0.043	0.013	20
48-1112*	4	8	11	12	0.034	0.013	17
48-1212*	4	8	12	12	0.026	0.013	14
48-1214*	4	8	12	14	0.026	0.008	12
612-3/04	6	12	000	4	0.206	0.040	91
612-2/04	6	12	00	4	0.172	0.040	78
612-00	6	12	0	0	0.148	0.074	81
612-03	6	12	0	3	0.148	0.047	72
612-11	6	12	1	1	0.126	0.063	69
612-14	6	12	1	4	0.126	0.040	61
612-22	6	12	2	2	0.108	0.054	59

TABLE 21-22 Typical Styles of Welded Wire Fabric (Continued)

Style designation	Spacing of wires, in.		Size of wires, AS & W gauge		Sectional area, sq in./ft		Weight per 100 sq ft, lb
	Longitudinal	Transverse	Longitudinal	Transverse	Longitudinal	Transverse	
(a) One-way (rectangular) types* (*continued*)							
612-25	6	12	2	5	0.108	0.034	52
612-33	6	12	3	3	0.093	0.047	51
612-44	6	12	4	4	0.080	0.040	44
612-66	6	12	6	6	0.058	0.029	32
612-77	6	12	7	7	0.049	0.025	27
(b) Two-way (square) types ‡							
2 × 2—10/10	2	2	10	10	0.086	0.086	60
2 × 2—12/12†	2	2	12	12	0.052	0.052	37
2 × 2—14/14†	2	2	14	14	0.030	0.030	21
2 × 2—16/16†	2	2	16	16	0.018	0.018	13
3 × 3—8/8	3	3	8	8	0.082	0.082	58
3 × 3—10/10	3	3	10	10	0.057	0.057	41
3 × 3—12/12†	3	3	12	12	0.035	0.035	25
3 × 3—14/14†	3	3	14	14	0.020	0.020	14
4 × 4—4/4	4	4	4	4	0.120	0.120	85
4 × 4—6/6	4	4	6	6	0.087	0.087	62
4 × 4—8/8	4	4	8	8	0.062	0.062	44
4 × 4—10/10	4	4	10	10	0.043	0.043	31
4 × 4—12/12†	4	4	12	12	0.026	0.026	19
4 × 4—13/13†	4	4	13	13	0.020	0.020	14
4 × 4—14/14†	4	4	14	14	0.015	0.015	11
6 × 6—0/0	6	6	0	0	0.148	0.148	107
6 × 6—1/1	6	6	1	1	0.126	0.126	91
6 × 6—2/2	6	6	2	2	0.108	0.108	78
6 × 6—3/3	6	6	3	3	0.093	0.093	68
6 × 6—4/4	6	6	4	4	0.080	0.080	58
6 × 6—4/6	6	6	4	6	0.080	0.058	50
6 × 6—5/5	6	6	5	5	0.067	0.067	49
6 × 6—6/6	6	6	6	6	0.058	0.058	42
6 × 6—7/7	6	6	7	7	0.049	0.049	36
6 × 6—8/8	6	6	8	8	0.041	0.041	30
6 × 6—9/9	6	6	9	9	0.035	0.035	25
6 × 6—10/10	6	6	10	10	0.029	0.029	21

* One-way fabrics with longitudinal wires up to and including 7/0 can be furnished. The one-way fabrics tabulated above are used primarily in building construction. The sizes and spacings of transverse wires are selected to meet the requirements for temperature reinforcement.

† Usually furnished only in galvanized wire.

‡ A two-way fabric—for a given size of longitudinal wires—is any style in which the sectional area of transverse steel is greater than the minimum required for proper fabrication by reason of the transverse wires either having a spacing which is less than the permissible maximum or being of larger size than the permissible minimum. Two-way fabrics are not limited to the styles tabulated above but may be produced in a wide range of sizes and spacing. Fabric is available with wire sizes of from ½-in. round to 16 gauge at spacings dictated by design requirements.

Fig. 21-59 Coupling connections for vertical reinforcing steel.

Wire and Wire Strands for Prestressed Concrete These reinforcing materials are initially stressed up to 70 percent of their ultimate strength. Cold-drawn wires, depending upon their diameters, will have ultimate strengths ranging between 220,000 and 250,000 psi. Wire strands are also available, with ultimate strengths between 250,000 and 270,000 psi. The specifications[29] for wire and strand (seven wire), respectively, are supplied in ASTM Designations A421 and A416. The seven-wire strand, uncoated, is most popular because the larger strand and larger wire size cost less per pound of prestress. The most popular wire sizes are ⅜ in., 7/16 in., and ½ in.

Various end anchorages have been designed, and two types are in common use. One of these is based on mechanical holding, such as threaded or button-headed units. The other relies on friction or wedge devices.

Table 21-23 lists the pertinent properties of strand and wire for prestressed concrete.

ESTIMATING AND PRICING

Fabricators of reinforcing steel can be found in the principal cities of the United States. Upon request, when lists of bars with bending details are *not* furnished to them, these fabricators will prepare an estimate of the reinforcing steel required and will submit a lump-sum or an average-unit price quotation.

Working from the plans and specifications which are supplied by the design engineer, all bars are first listed as to size, length, type of bending (heavy, radial, or light) and grade of steel. Weights are then computed and a base price is applied to the total, plus extras for size, type of bending, placing drawings and shop details, and trucking or rail freight. There is usually an extra charge for high-strength grades of steel. If column spirals are required, their weights are computed by size and priced. Bar chairs and spacers are estimated and priced, to be included in the lump-sum bid. Welded wire fabric is quoted separately, with unit prices and estimated areas for each style.

The placing of steel is normally quoted and executed by a contractor other than the fabricator. Bar placing is quoted at a price per ton, and placement of welded wire fabric is quoted at a square-foot price. The tonnage of bars that can be placed in an hour will vary sharply with the type of reinforced-concrete structure. Naturally, a structure requiring heavy bars will cost less per ton of placing than one for which light

TABLE 21-23 Reinforcement for Prestressed Concrete

(a) Prestressed concrete strand, uncoated, stress-relieved*,†

Strand dia, in.	Approx. weight, lb/1,000 ft	Approx. area, sq in.	Minimum strengths, lb: Ultimate	Minimum strengths, lb: Yield	Minimum strengths, lb: At 70% of ultimate	Standard ft/reel	Standard ft/coil
3-wire strand							
1/4	126	0.036	9,000	7,650	6,300	2 × 20,000	
5/16	201	0.058	14,500	12,300	10,150	2 × 12,000	
7-wire strand ASTM A416							
1/4	122	0.036	9,000	7,650	6,300	25,000	
5/16	198	0.058	14,500	12,300	10,150	20,000	
3/8	274	0.080	20,000	17,000	14,000	15,000	22,000
7/16	373	0.110	27,000	23,000	18,900	12,000	16,000
1/2	494	0.144	36,000	30,600	25,200	9,000	12,000
0.600	735	0.218	54,000	45,900	37,800	6,000	8,000
7-wire strand high tensile							
3/8	288	0.085	23,000	19,550	16,100	15,000	22,000
7/16	396	0.117	31,000	26,350	21,700	12,000	16,000
1/2	526	0.154	41,300	35,105	28,910	9,000	12,000
9/16	651	0.192	51,700	43,945	36,230	7,000	9,300

(b) Prestressed concrete wire, uncoated, stress-relieved (ASTM A421 when applicable)‡,§

Wire dia, in.	Wire gauge No.	Approx. area, sq in.	Approx. weight, lb/1,000 ft	Minimum strengths: Ultimate psi	Ultimate lb	Yield psi	Yield lb
0.276	...	0.0598	203.2	235,000	14,050	188,000	11,240
0.250	...	0.0491	166.7	240,000	11,780	192,000	9,430
0.196	...	0.0302	102.5	250,000	7,550	200,000	6,040
0.192	6	0.0289	98.32	250,000	7,225	200,000	5,780
0.177	7	0.0246	83.56	255,000	6,270	204,000	5,020
0.162	8	0.0206	70.00	259,000	5,340	207,000	4,270
0.148	9	0.0173	58.66	263,000	4,550	210,000	3,640
0.135	10	0.0143	48.61	268,000	3,830	214,000	3,060
0.120	11	0.0114	38.73	273,000	3,110	218,000	2,490
0.105	12	0.0087	29.69	279,000	2,430	223,000	1,945

* Approximate modulus of elasticity: 27,800,000 psi.
† Minimum elongation at rupture: 3.5 percent in 24 in.
‡ Approximate modulus of elasticity: 29,000,000 psi.
§ Minimum elongation at rupture: 4 percent in 10 in.

bars are required. Here are some figures compiled by a steel-setting contractor of wide experience in Chicago:

Foundations: 5½ man-hours per ton
Concrete-joist construction: 8 man-hours per ton
Flat-slab construction: 10 man-hours per ton
Slab, beam, and girder construction: 11½ man-hours per ton
Walls, bridge piers, and abutments: 16 man-hours per ton

The efficiency of steel setters varies widely, and the estimator must predict the probable accomplishments of his men in order to quote an average price per ton. His bid includes the placing of the bar chairs and spacers to support the bars, as well as the wire and tying required.

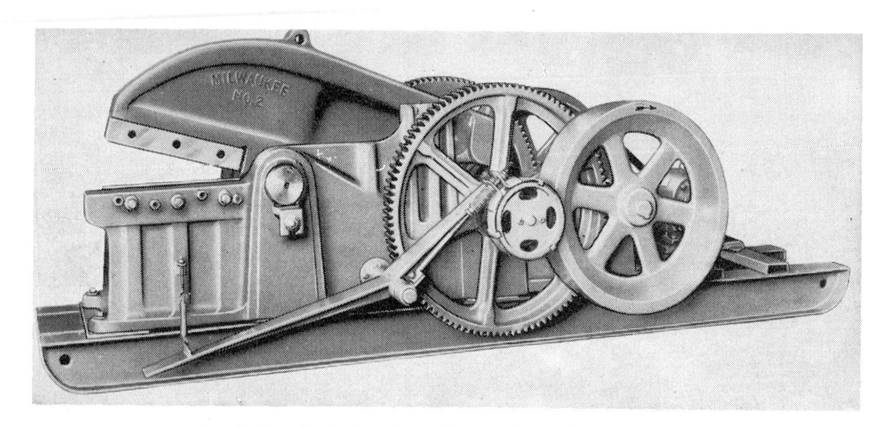

Fig. 21-60 Alligator shear for cutting reinforcing bars.

The cost of placing welded wire fabric with proper laps is related to the speed with which the concrete slab is poured, since the steel fabric must be kept in position. The normal placing cost can be increased by 1¼ to 2½ cents per sq ft by excessive waiting time.

FABRICATION

After an order for reinforcing steel is received—other than lists of bars—the fabricator must usually prepare placing drawings for the architect's or engineer's approval. For some projects, notable state highway structures, the structural plans are completed by the design engineer and become placing plans. The American Concrete Institute's *Manual of Standard Practice for Detailing Reinforced Concrete Structures* is the recognized authority for preparing placing drawings.[30] Upon approval of these drawings, order sheets are prepared and sent to the shop for fabrication.

A reinforcing-bar fabricating plant is usually serviced by one or more railroad spur tracks. Arriving cars will transport stock-length bars and wire products from the producing mill. The bars are unloaded by overhead cranes or hoists and are placed in racks or piles by size and grade.

Electric-powered shears and benders, operated by skilled crews, fabricate the bars in accordance with the shop details. The bars are then bundled and tagged for delivery by trucks or freight cars. Fabrication is normally made in sequence; that is, foundation steel first, then story by story as required.

A modern fabricating shop has bays spanning 65 ft or more which are equipped with cranes or hoists to facilitate the transfer of stock-length bars to and from the storage racks. The first operation is the cutting of the bars to length as listed on the shop-order sheets. This cutting is usually done by an "alligator" shear, electric

Fig. 21-61 Automated shearing line. (*Ducommun Metals & Supply Co.*)

powered and capable of cutting a number of bars at one stroke (see Fig. 21-60). On one or both sides of the shear is a table or series of rollers at least 60 ft long. The table has a scale for manual measurements of length, graduated in feet and inches and with its zero reference at the shear blade. Tables are also manufactured with automatically erected measuring stops which are controlled by labeled pushbutton keys on a panel at the shear (Fig. 21-61). After shearing, one side of the table is power-raised and the sheared bars slide off into the desired low rack.

The shearing process is apt to leave a burr on the bar at the cut. This may be objectionable for smooth or plain dowels which are designed to slip in the concrete. In this case, a special shear can be used to obtain "square" cuts. The use of this shear, as for saw cuts, carries an extra charge.

Bends are made cold, without heating the bars. Several types of electric-powered bending machines are available for this purpose (see Fig. 21-62). These include Wallace, Alamo, Arnold, and Schenck benders for truss bends, angle bends of any degree, and hooks. On at least two units, as many as six bends of one or more bars can be made without changing the mandrels or pins on the revolving table. The benders are used for radial bending by the setting of the mandrels.

Stirrups and ties of ¼- to ⅝-in. diameter are made on a Kardong stirrup machine or a bulldozer. The Alamo spiral machine can coil ¼- to ⅝-in. wire to spirals of the desired diameters. Spacers of cold-rolled channels, with punched-out clips, are then applied to obtain the required pitch of the spiral. Stock coils of wire must occasionally be straightened and cut to length, and this is accomplished on a Shuster wire straightener.

As the steel is delivered, invoices must be issued and accompanied by bar lists. These documeent should be checked by the buyer. The normal terms of payment for steel are ½ percent discount for payment within 10 days of delivery or 30 days net, the amount of the invoice.

PLACING

The placing or setting of reinforcing steel must be strictly in accordance with the placing drawings. The design engineer computes the reinforcing steel requirements for a definite position in each concrete member of the structure. Therefore, it is absolutely necessary that the steel setter be exact in the placing of every bar.

Fig. 21-62 Reinforcing-steel bender. (*Arnold Rebar Bender, Inc.*)

Before the steel is placed, all debris should be removed from the forms. Any work of other trades which involves inserts, sleeves, conduits, ducts, straps, and anchors should also be completed to prevent the subsequent displacement of the reinforcing.

Usually reinforcing steel is sold "f.o.b. trucks jobsite" or "f.o.b. cars nearest siding" and the steel setters must unload the steel and arrange it on the ground according to marks on the tags wired to the bars. The foreman of the steel-setting crew subsequently directs his men in transferring bars from this stockpile to their proper positions in the forms. On large projects he will have cranes and hoists to assist him in moving the steel. He must be an expert in reading structural plans and in planning an efficient and economical placing and tying of the bars.

Foundation bar mats are usually assembled at the jobsite, with the two layers of bars at right angles. Alternate intersections are tied with soft, annealed wire or are secured by clips. Occasionally, welding of alternate intersections is specified. The mats should be supported the required distance above the subgrade by concrete bricks or precast blocks.

Pier and column bars are usually assembled with their ties or spirals and then erected in place. They are positioned so as to maintain the prescribed distance from the forms.

The bar chairs and spacers for slabs, joists, beams, and girders are placed before the reinforcing is set. Spacing should conform to the *Building Code Requirements for Reinforced Concrete,* ACI 318, unless otherwise specified.[31,32] Stirrups are distributed as required. Straight bars are usually placed before placing the bent or trussed bars. Finally, temperature steel consisting of bars or welded wire fabric is placed in its prescribed position. Wire tying follows to prevent displacement of bars during the pouring of concrete. Concreting runways should not rest on or be supported by the reinforcing steel.

Structural iron workers place most of the reinforcing steel in the United States, although this work is claimed by other trades in few localities. In metropolitan New York, for instance, fabrication on the job and the placing are claimed by the lathers.

D. Stationary Forms

Formwork, by definition, is the total system of support for freshly poured concrete. It consists of form sheathing plus all supporting members, hardware, and necessary bracing. Objectives of concrete forming include dimensional accuracy, strength, and economy in the finished product.[33]

In the past, formwork was largely restricted to job-built units. These forms were built in place on the jobsite, used once, and then scrapped. The trend today is toward increasing prefabrication, reuse of forms, and greater mechanization of assembly and erection. As a general policy, the layout and design of the formwork, as well as its construction, are the responsibility of the contractor.

Forms can be classified on the basis of reuse as *single use* or *multiple use*. In terms of the fabricator, they can be classified as *user-built* (job or shop) or *commercially supplied* (prefabricated forms). Forms in the latter category, when designed for reuse, can ordinarily be either rented or purchased outright.

PLANNING FOR FORMWORK

An initial step in formwork planning is to identify the requirements of the owner or architect with respect to the concrete finish. The contractor can then explore methods which will keep forming costs to a minimum. In so doing, he must consider the job requirements and the available materials.

The objective of the contractor, through optimizing all the factors involved in his concrete work, is to produce a satisfactory result at the least possible cost. To do otherwise would be contrary to good engineering and design practices. Reuse of a form has a profound effect on the formwork cost which must be allocated against each unit of formed area. Usually it is desirable to obtain the greatest number of form reuses, but this objective must be weighed against labor to reset the form each time, concrete placement methods, and general job scheduling. It is often poor economy to use a form that, although it has a low initial cost, cannot perform with satisfactory results for the planned number of reuses. On the other hand, one could not hope to purchase one form to cast 200 columns on a specific job. The intermittent labor requirement to strip and reset one form and the cost of placing concrete in individual columns as compared to placing several in a set would have adverse effects on overall job costs.

Specification Requirements It is essential that the owner or architect be explicit as to the finish requirements for the various parts of the building. He should categorize these requirements into specific areas of exposed critical work, areas of exposed semicritical work, and areas that have merely structural requirements. While so doing, the owner/architect should avoid specifying the design of the formwork.

The designer should also avoid specifying work of closer tolerances than those indicated in the current issue[33] of ACI 347, *Recommended Practice for Concrete Formwork*. Otherwise, he runs the risk of unnecessarily increasing the cost of the structure. Poor general finish and tolerance specifications both fall in the same category as specifying 4,000-psi concrete for portions of the building where 3,000-psi concrete would be adequate or requiring structural-steel members which are much larger than those required to support the anticipated loads. These all are examples of poor design.

Job Requirements for Formwork Once the designer has done his part, as evidenced by specifying finish and tolerance limits which are consistent with the intended use of the structure, the contractor can proceed with a detailed analysis of job requirements for formwork. These manifest themselves in terms of formwork timing, number of uses, forming method, quality of performance, and value for salvage. It is exceedingly important not to overemphasize any one of these factors.

Timing. In general, the time of primary interest to the contractor is that required to complete the entire project. Formwork is only a part of the project, and all construction operations must be analyzed in sequence. It would be inappropriate, for example, to adopt a fast method of forming that prevents subsequent trades from doing their work. In some cases, this means an exchange; the formwork itself may cost more, but subsequent trades are able to accomplish their work more quickly. As a result, the total job can be completed at an earlier date. The selection of a particular formwork material can also be governed by whether or not it can be procured without delaying the job schedule.

Number of Uses. The cost of the formwork is a function of the number of uses. Forming costs are normally compared on the basis of dollars per square foot for the area of concrete which is formed. Therefore, the initial costs and subsequent handling costs for the formwork are divided by the number of uses in order to establish the formwork unit cost. A compromise is usually necessary to arrive at the number of reuses which will result in acceptable formwork costs while avoiding conflict with other major job factors. That is, if the structure required 100 columns to be poured, maximum possible reuses would be accomplished with one form being used 100 times. However, this would adversely affect the cost of the concreting and the job scheduling.

Forming Method. The various features of the structure will have bearing on the best forming method to use. Cost comparisons can be made between manual erection and dismantling of small form units versus handling larger sections of forms with cranes or other mechanical devices. The method which is selected must not result in undesirable interference with other construction operations within the specific job environment. For example, the method of placing concrete (chutes, buckets, pumps, tremies) may have a bearing on the requirements for the formwork.

It has sometimes been found that the use of a stronger form, despite its higher initial cost, is justified in terms of job economics by its effect on reducing the labor cost of placing the concrete. For example, substituting a stronger form may permit a crane and a crew of men to increase the rate of pour from 5 to 10 ft per hr. This can result in a significant reduction in the cost of the in-place concrete. To put it another way, the increased cost of formwork in some cases can be more than offset by reduced labor and equipment costs in placing the concrete. Formwork costs and placing costs should be taken under consideration jointly.

Quality of Performance. In considering all the factors involved in timing, number of uses, and methods of forming and placing concrete, the essential requirement is that of acceptable performance. The finished job must satisfy the specifications. Alignment and deflection must be within permissible limits, and the concrete finish must be as specified.

Salvage Value. This factor injects itself into job requirements in four distinct ways. First, form materials such as steel, aluminum, magnesium, etc., can be sold for their scrap value. Second, forming material may be used as a permanent part of the structure. For example, steel floor forms may be left in place as a composite design. Third, there is the possibility that the forming material can be used on future projects after the completion of the current project. Fourth, the user need not pay the full cost of the material or equipment but can instead pay only a fraction of this in the form of rental. Vertical shoring, horizontal shoring, wall forms, and scaffolding are only a few examples of materials which can be obtained on a rental basis. In this way the contractor can reap the benefits of multiple use for a wide selection of formwork equipment without the requirement of ownership and at a fraction of its total cost.

Material Available for Formwork Another factor that the contractor must evaluate is the availability of formwork material. Wood, steel, plastic, magnesium, paper, etc., may all be considered if available. In most cases, the form is made of a combination of several of these materials. Where materials have been salvaged from previous formwork, their possible reuse should be studied. In selecting formwork materials, the contractor should keep in mind that his objective is to construct the facility in accordance with specification requirements, within the overall schedule, and at least total cost.

FORM MATERIALS

There are numerous materials available for forming. Several of those which are frequently used will be described in the following paragraphs.

Lumber* Lumber is a commonly available material and has excellent strength, weight, and cost factors. Characteristics of lumber include a pervious surface with varying texture between the spring and summer grains, leading to differences in color for the concrete surfaces which it contacts. The moisture absorption of form lumber will vary depending on its moisture content at the time of concrete casting and also on the presence and type of parting agent.

The useful life of lumber forms, measured in terms of the number of castings that can be made against them with acceptable results, is dependent on the use of appropriate parting agents or other surface treatments. Lumber can be obtained with smooth, sanded surfaces or it may be left rough-sawn to transfer distinctive textures to the concrete surface. The role of the parting agent will become more critical on the rough surface than on the smooth surface. Without suitable protection, a wooden form surface will deteriorate through alternate wetting and drying and the adhesion of concrete. This deterioration should receive appropriate attention, since its effects can be detected on the finished concrete surface.

Plywood* Plywood is probably the most popular facing material for cast-in-place concrete in the industry today. It is relatively inexpensive and its performance is good provided that it is an exterior grade. Exterior grade means that the adhesive agents in the plywood will have a satisfactory resistance to the excess moisture in the fresh concrete. An unsuitable type of plywood will have a short life and few uses. Any plywood which is selected by a builder should have a history of good performance in the concrete forming field.

Plywoods can be purchased with various types of surface treatments which are intended to promote better performance over a longer period of time. These treatments include light mill-applied coatings, some of which will provide good parting characteristics through several pours. Other surface treatments provide protection for little more than the first cast, and will have to be renewed prior to each subsequent cast. Plywoods can also be obtained with a plastic coating which will extend the usable life of the material to a considerable degree. In most cases, this higher-cost coating will also need some parting agent to supplement its excellent characteristics. Some of these coatings will provide a nearly impervious surface and will eliminate grain raise and its subsequent transfer to the concrete surface. Therefore, if a grain-raise imprint is desirable on the concrete surface, the impervious coatings should be avoided.

Metals Steel, aluminum, and magnesium can be used as forming materials. Steel is a commonly used material for formwork because it has a long usable life and excellent strength and cost factors. The impervious steel surface of the form will impart a uniform color to the concrete. Steel-form fabrication is normally done in a remote fabrication plant. Since field modifications of a steel surface are costly and difficult, all use requirements of the form must be anticipated prior to its fabrication.

The steel skin should be thick enough to carry the concrete load between support members without exceeding permissible deflections. It should also be thick enough to resist damage from vibrator contact or through normal field handling of formwork. A parting agent will be necessary for ease in stripping, to eliminate laitance buildup on the form face, and to minimize the possibility of concrete staining due to rust. These last two items can seriously affect the quality of subsequent casts. When white portland cements are to be used in the concrete, the steel skin should first be

* See Sec. 24, Timber Construction.

pickled or sandblasted. This will remove mill scale which may otherwise cause staining, even when the steel forms are well coated with a parting agent.

Aluminum forming material which is in contact with the concrete will be affected to various degrees by alkalis and other salts which are present in the concrete. Parting agents will help to reduce this reaction. The long-run performance of aluminum forms will vary from one location to the next because of the variations in coatings and concrete and the presence or absence of other metals that may cause galvanic action and lead to pitting of the aluminum surface. Because of its excellent strength-to-weight ratio, aluminum has been used to a great extent for structural portions of the formwork where it is not necessarily in direct contact with the cast concrete.

Magnesium is similar in performance to aluminum. While it is said to be less susceptible to the alkalinity of concrete, its actual performance also varies considerably from job to job. As with aluminum, its behavior is influenced by the nature of the concrete, the parting agents used, and the presence of other materials that can cause a galvanic reaction.

Plastics These are playing a more and more important role in the concrete-forming industry. They have impervious surfaces that usually impart a smooth finish to the concrete without the discoloration which is characteristic of the absorptive-type materials. Plastics come in several categories but can be essentially classified as reinforced or unreinforced.

Reinforced plastic consists of a resin matrix with embedded glass fibers in various forms. The presence of these glass fibers greatly increases the strength of the resin material. Reinforced plastic has found considerable acceptance in custom-made forms for specific job requirements, since the possible shapes and sizes of reinforced plastic forms are almost limitless. An appropriate resin must be used on the form surface in order to assure good performance through a reasonable number of uses. The glass fibers should be kept from contact with the concrete.

Unreinforced plastics can be obtained in sheet form with smooth or textured surfaces. The light-textured surfaces transfer an imprint to the concrete, thus toning down the usual characteristics of an extremely smooth surface. Sheet plastic and those plastics that can be heat-formed into rib designs, etc., need appropriate structural backup to enable them to support the concrete pressure loads. Unreinforced plastics are normally used as liners in a form system that provides all the structural requirements of concrete containment. The role of the plastic liner is merely that of changing the characteristics of the cast surface.

Preformed foam plastics can be used in at least two applications. First, the plastic can be used by itself, usually to form recesses in a wall. Preformed foam planks are often used when conditions preclude their salvage, and they can easily be attached to a basic forming system. For example, they can be used to box out walls for slabs. These slab box-outs usually have reinforcing rods and require no draft, thus eliminating the possibility of their withdrawal. Secondly, foam plastic is used in backing up thin, vacuum-formed, plastic form liners where the concrete pressure would otherwise cause deformation.

Plaster Molds Where custom design of a sophisticated and detailed nature is desired, molds can be made of plaster. The concrete is cast against these molds and the plaster is then broken away from the finished concrete. Obviously, the plaster mold is a one-use form.

Absorptive and Impervious Liners The designer should consider the relative merits of formwork that is impervious and formwork that is absorptive. Each type leaves its characteristic imprint on the concrete surface. The use of an impervious form liner will usually result in a lighter color of concrete and a more uniform appearance. Examples of impervious form liners are steel, plastics, and plywoods which are either of high density or are treated with certain coatings. Absorptive liners, on the other hand, will absorb some water from the fresh concrete as long as the moisture content of the liners is below the saturation point. Such absorption results in a darker color for the hardened concrete and a somewhat softer finish. Variation in concrete color is dependent upon the absorptive capacity of the form. For example, concrete cast against wood will be darker along the spring-growth grains and lighter against the summer-growth grains. Certain parting agents will reduce or nearly eliminate this effect.[34]

FORMWORK REQUIREMENTS

Form Structure *Thickness of Material.* The thickness of the various form materials depends on several major factors: the modulus of elasticity of the material, that is, its stretch per unit load; the span that the material must bridge; the estimated loads, including that of the concrete itself; any deterioration of the physical characteristics of the material which will occur during its usable life; and the maximum deflection that will be acceptable on the finished concrete surface.[35]

Fig. 21-63 High wall bracing.

Bracing. The form face is designed to provide a total envelope to contain the liquid concrete. This form face must maintain the appropriate alignment, whether straight or curved, when the pressure loads of concrete are placed against it. To achieve these requirements, additional members may be placed on the back side of the form face. These members are identified as joists and stringers when they support horizontal sheathing and as studs and wales when they support vertical or near-vertical surfaces. They are usually long, straight members, made of steel or timber, which are capable of holding the form face in proper alignment between its support points. Vertical shores provide the support points for stringers, while wales will receive their support either from external bracing or from concrete ties which act in tension between the wales on opposite sides of a wall. (Various types of ties are discussed under Form Ties on page 21-98.) Figures 21-63 and 21-64 illustrate typical bracing systems.

Alignment. The objective is to place the concrete in whatever final position has been prescribed by the designer. In forming horizontal members, alignment is largely a matter of correct initial positioning of the formwork plus control of formwork deflection. The alignment of wall members involves two distinct considerations. First, as with the horizontal formwork, the concrete pressure must be contained so as to keep the deflection of the loaded form within specified limits. Next, this form alignment must be maintained against external forces such as wind and activity by the workmen.

Bracing to resist these external forces is generally placed on one side of the wall only. Bracing both sides may cause misalignment and in any event is not required.

Form Joints *Prevention of Leakage.* If water with some cement content is allowed to leak from the form, a surface blemish will result. This blemish is characterized by an aggregate-rich surface which is inconsistent with the normal, dense adjacent surfaces. The area will also have a darker color since a lesser amount of water is available for hydration. This aggregate-rich condition penetrates the concrete mass to a considerable depth, and noticeable discoloration may still remain even after relatively deep sandblasting. Grout leakages through fine openings in the formwork must therefore be controlled where appearance is critical. The use of low-slump concrete is helpful in this regard. Form joints can also be made relatively grout-tight by one of several means: (1) placing a lumber batten strip behind plywood joints, (2) wetting wood forms several hours prior to placing the concrete to expand the wood

Fig. 21-64 Wood bracing with adjustable turnbuckle for form alignment.

and help in closing joints, (3) lining forms with a separate face, staggering the lining joints with those of the structural form, (4) installing rubber gaskets between sections of steel forms, and (5) caulking the formwork joints.

Fins. Fins are thin projections of set concrete which extend from a poured surface. They occur where grout has tried to escape through a form joint but has been blocked from complete escape. They are usually not considered objectionable, since they can easily be knocked off with little or no effect on the appearance of the finished surface. For some walls, broad fins are specified by the designer to obtain a pleasing appearance. This effect is accomplished by lining a form face with planks which can either be kept some specific distance apart or randomly placed. The fins can remain as they appear after the forms are stripped, or they can be broken back with special hammers.

Textures and Patterns All forms will have characteristic markings which are normally transferred to the finished concrete surface. Sources of these markings include the following: size of the formwork face (i.e., 4- by 8-ft sheets of plywood); plank widths; various absorptive characteristics of the face; preformed panels; proprietary types of panels with particular perimeter configuration; grain of wooden faces; grain raise due to moisture; number and size of fasteners, etc. Assorted textures and patterns can be produced on the finished concrete surface from these form marks.

A wide variety of textures and patterns for a concrete surface can also be obtained through the use of form liners. This is a very practical approach, since the form facing can be selected and designed entirely apart from structural considerations for the formwork. The contractor then has a great deal of choice as to ways of supplying the necessary backup strength in the forming system.

Wood liners can be used to form feature strips and rustication strips. They can also provide a checkerboard pattern by a 90° change in the direction of grain or planks in adjacent panels.

Plastic liners provide a wide choice of surface textures and designs. Surfaces can be had with smooth or slightly grained (sometimes called "hair-cell") finishes. These grained finishes impart a soft, pleasing effect to the concrete. Certain plastics can be heat-formed into almost limitless variations in design.

Rubber liners may also be considered for relatively shallow textures on the concrete surface. Rubber can be obtained and used either in sheet form or in solid, extruded shapes. It is suggested that the rubber be checked for its resistance to deterioration from oils which are commonly used as parting agents.

Temperature changes can profoundly affect certain liner materials. Metal and plastic liners, if secured in cool weather, can buckle when exposed to a large increase in temperature. This effect usually is not serious because, as the concrete comes in contact with the material, the liner temperature is immediately reduced to near what it was when installed.

Form Ties *Types.* Ties for wall forms will fall generally into one of the following groups: (1) plain coil strapping or wire; (2) preformed strapping, with spacers to ensure specific wall thicknesses and preformed notches to provide breakback in the wall for the desired concrete cover; (3) preformed wire ties for specific wall thicknesses and breakback characteristics; (4) pull ties which are removed completely from the wall after forms are removed; (5) she-bolt ties where an inner male-threaded unit is left in the wall and the outer fastening devices are removed and reused; (6) he-bolt ties where the outer fastening devices are reusable, with an expendable female-threaded unit left in the wall.

Figure 21-65 illustrates various types of commonly used ties. The gang form tie is identical to the standard panel tie except that each of its ends has been extended $2\frac{13}{16}$ in. to permit the connection of a gang form bolt. The 6,000-lb flat tie is used on single-lift forming and is designed for 3-ft spacing. Its use can reduce labor and material handling, both in erection and in stripping, by as much as one-third. It can also be used on higher walls, where its spacing is dependent on job conditions and rate of pour.

Characteristics. Ties are available in almost any desired strength category. Tie strength is usually selected to match the form strength, tie spacing desired, and anticipated pouring rates. The tie spacing must be such that form deflections are kept within acceptable limits in addition to satisfying strength requirements. Each type of tie leaves a characteristic hole in the wall surface. Wire snap ties leave very small holes, usually about $\frac{1}{4}$ in. in diameter and with a normal depth of 1 in. Wood or plastic cones can be used with the snap tie when deeper breakbacks (up to 2 in.) are required. These will increase the size of the hole, usually to 1-in. diameter. Cones are sometimes required to reduce grout leakage at the point where the tie passes through the form.

The characteristic holes of she-bolts depend on the strength category of the tie. Hole diameters usually fall in the range of $\frac{3}{4}$ to $1\frac{1}{2}$ in. He-bolts usually require a cone when it is necessary to provide a concrete cover over the portion of the tie which is left in the wall. Diameters of these cones will range from 1 to 2 in. A pull tie will range in diameter from $\frac{1}{4}$ to $1\frac{1}{2}$ in. and will leave a hole of similar size to the rod diameter; this hole passes completely through the wall, leaving no portion of the tie as expendable. All the above-mentioned ties leave round and relatively clean holes for subsequent patching.

There are two types of ties which leave marks other than those just described. The proprietary flat tie is broken off just below the wall surface by a sharp lateral blow. This causes some slight spalling at the concrete surface, the depth of which is usually limited to about $\frac{1}{2}$ in. Next, a wire tie may have a thin washer at the surface of the wall for spreader action. This washer leaves a shallow impression in the face of the wall around the hole, while the wire itself breaks back at a greater depth.

Form Coatings A form coating is used primarily to prevent the concrete, because of

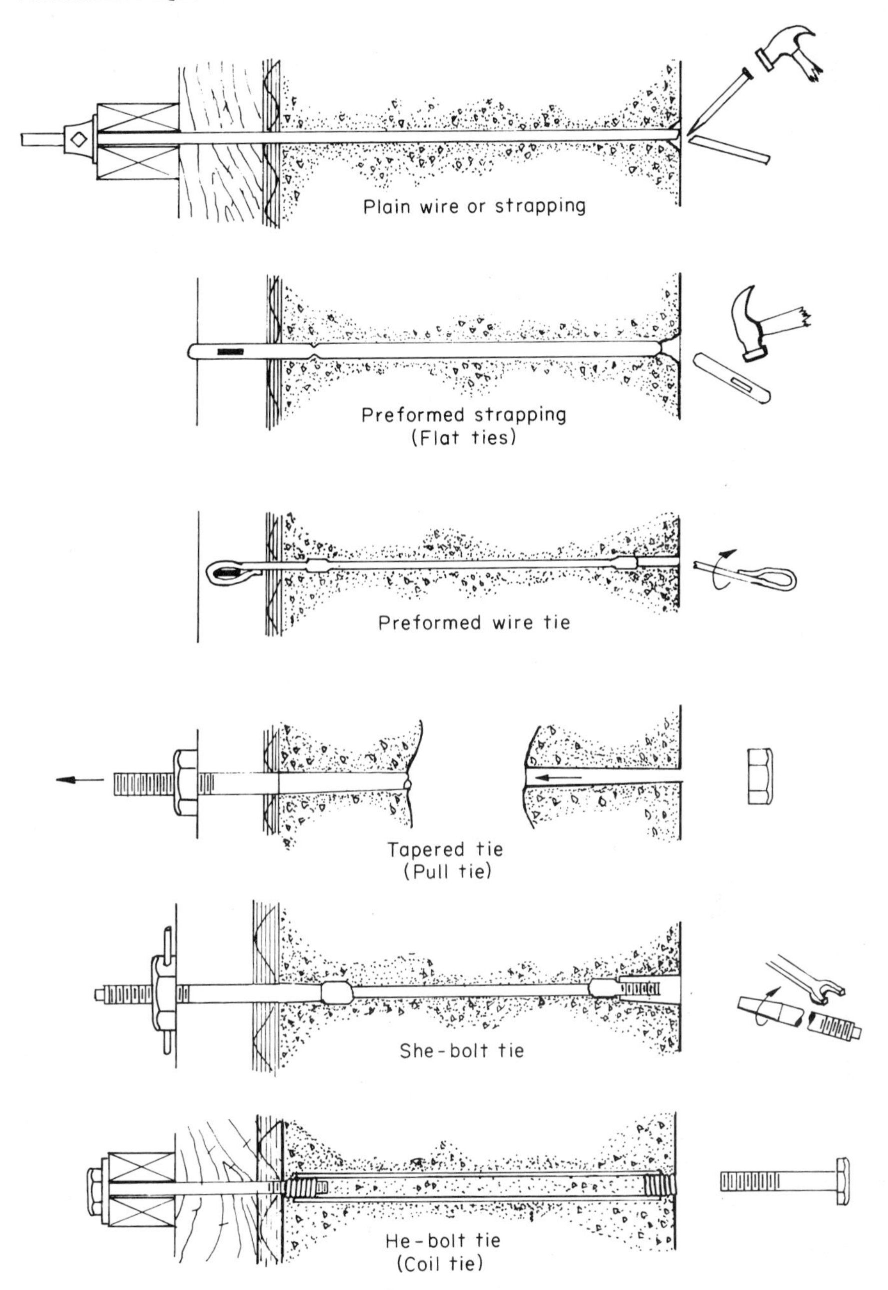

Fig. 21-65 Common types of form ties.

its adhesive characteristics, from bonding to the form face. In using a form coating, the following benefits are achieved: (1) The physical act of pulling the form from the wall is made easier. (2) Spalling of the finished wall is greatly reduced by eliminating any adhesion at vulnerable locations such as corners and edges of feature strips. (3) The form face is protected from the physical deterioration that could come about by its contact with concrete. For example, a good bond breaker will reduce the tendency of the steel form to rust and will keep moisture from entering wood surfaces with subsequent grain raise, fiber raise, loss of strength, etc. (4) When adhesion of the concrete to the form face is prevented, the form face will remain clean and thus provide a consistent surface for subsequent casts. Conversely, if the concrete is allowed to adhere and build up on the form face, subsequent casts will reflect this buildup. An alternative is to incur additional field labor costs for cleaning the form face prior to each use. It is important to note that parting agents, or form oils as they are sometimes called, should have no effect on the concrete surface. Assurance of this should be requested from the supplier of the product.

Shores On all suspended concrete floor and beam systems which are poured in place, various types of shoring are used to support forms and concrete until the concrete has set. Many years ago, dimension lumber was used exclusively for shoring purposes. For the majority of work, 4 by 4s were used. They were cut slightly under the length required, and any adjustment in elevation was made by driving wooden wedges under the shore. The wedges were then toenailed.

Because of variations in story heights and the problems involved in adjusting the lengths of shores, many types of patent shoring systems have been developed. These permit adjustments in height, some even with full load. They also have positive load capacities, ease of erection and stripping, and clamping devices which facilitate the attachment of T heads, stringers, etc. These systems are generally rented for a particular job, but they may also be purchased. Patented shoring systems fall into three main categories, each having unique steel clamping and adjusting devices:

1. Dimension lumber combined with structural-steel shapes
2. Steel pipes
3. Tubular-steel scaffolding

Lumber alone is still used in special cases. Shipping costs, materials on hand, and identical story heights on multistory buildings are factors to consider. However, for the main part, proprietary shoring systems are replacing the lumber shores.

JOB-BUILT FORMS

The initial cost of the materials used in single-use job-built forms for simple concrete members is low, and the materials themselves may frequently be reused on other phases of the job. The size of the job may be so small that the introduction of multiple-use forms would be impractical. Most contractors are familiar with single-use job-built forms, and this can reduce delays which might be occasioned if job superintendents had to acquaint themselves with new techniques.

The job location and type introduce other cost factors. The contractor may have only inferior local materials at his disposal, but rather than pay the costs of freight and rentals for prefabricated forms and incur an undesirable delay while obtaining them, it may be wise for him to use what is locally available. Much so-called "architectural" concrete lends itself best to job-built forms, since the design is often not repeated. The work may also involve an intricate pattern and appearance which is not easily reproduced with factory-made forms. However, if a contractor has the opportunity for several reuses of his forms, it is probable that he will consider using some prefabricated type of forming system.

Forms must be left in place long enough for the concrete to gain sufficient strength to support its own weight and that of any construction load involved. Materials most commonly used in constructing job-built single-use forms are dimension lumber or plywood. The sheathing is formed from 1-in. boards or from plywood in ⅝- or ¾-in. thicknesses. Studs and wales for simple, straight walls where high concrete pressures are not anticipated can consist of 2 by 4s, used singly or in pairs. Heavier pressures require stronger framing, and steel channels may then be required for the wales.

Form Design The design of job-built forms is governed by the job conditions, the job specifications, and the anticipated conditions of loading. (See Table 21-24.)

Wall Forms. The maximum lateral pressures to be anticipated in wall forms are listed in Table 21-25. Fluid concrete exerts a hydrostatic pressure on the forms, the maximum value of which is largely a function of the rate of pour and the length of time required for the concrete to take its set. The values in Table 21-25 apply to structural concrete weighing approximately 150 lb per cu ft which is placed at controlled rates under prescribed conditions of temperature. The tabulated maximum pressures are valid for internal vibration of normal-density concrete placed at 10 ft per hr or less and with a slump no more than 4 in. Depth of internal vibration must be limited to 4 ft below the top of the concrete. Good placement procedures are assumed; i.e., vibration is used for consolidation only and not for lateral movement of the concrete.

Column Forms. The maximum lateral pressures to be anticipated in column forms are listed in Table 21-26.

Slab Forms. In designing forms for slabs, it is essential to consider the following factors: (1) the dead load of concrete and reinforcing steel; (2) a live load, superimposed on the dead load, which may result from the impact of dropping concrete or the weight of temporary heaped concrete, workmen, and equipment such as buggies, runways, etc. (40 to 50 psf is often used to cover these live loads for design consideration); (3) shore spacing, shore safe-load capacity, and shore bearing capacity; (4) the time required for concrete to develop sufficient strength for form removal; (5) the ceiling finish and deflection specifications.

Form Hardware Table 21-27 lists design safety factors and typical uses for various items of form hardware. Snap ties are the common method used to tie and spread user-built wall forms. These ties have guaranteed strengths of from 3,000 to 5,000 lb and built-in spreaders. For loads greater than 5,000 lb, concrete-accessory manufacturers provide tie screws and she-bolts which range in strength from 5,000 to 30,000 lb and even higher for special jobs.

Forms are braced externally where interior tying methods are not practical. External bracing, for example, is indicated in cases of inaccessibility—where there is steel piling or an existing structure on the opposite side of the forms. The bracing serves to keep the forms in alignment and to withstand pressures of poured concrete.

PREFABRICATED FORMS

Commercially supplied forms are frequently referred to as "prefabricated forms." Multiple-use prefabricated forms are normally available either on a purchase or a rental basis, and their merits should be investigated when a number of form reuses is anticipated. These forms are sturdily constructed for a long, usable life, and this factor may reduce their cost per use below that of single-use forms. Additional savings may be realized through lower costs for erecting and stripping, since the prefabricated forms are manufactured in integral units which are still small enough and light enough for one man to handle.

In comparing the costs of user-built forms with those of commercially supplied forms, the contractor should take into consideration the condition of the prefabricated forms, the service provided with them, and the cost of accessories. Form manufacturers will frequently provide engineering layouts at no charge or at a nominal cost, and their field representatives will instruct the contractor's crew in the proper use of the equipment.

Special or extenuating circumstances notwithstanding, prefabricated forming has proved to be consistently beneficial. Because it is preengineered, the risk of on-site or untried design is greatly reduced. The simplicity of its assembly provides uniform results (both in cost and performance) which are little affected by the caliber of the

TABLE 21-24 Form Materials and Strength Data for Design*,†

Item	Principal use	Specification and design data sources
Lumber	Form framing, sheathing, and shoring	"National Design Specification for Stress Grade Lumber and Its Fastenings," National Lumber Manufacturers Association ACI Special Publication No. 4, *Formwork for Concrete* DESIGN DATA: *Wood Handbook*, Forest Products Laboratory, U.S. Department of Agriculture *Wood Structural Design Data*, National Lumber Manufacturers Association *Timber Design and Construction Handbook* (in Canada), Part 4, Section 4.3, "Wood," of the National Research Council National Building Code of Canada
Plywood	Form sheathing and panels	U.S. Product Standard PS 1-66 ACI Special Publication No. 4, *Formwork for Concrete* DESIGN DATA: ASTM, USASI CSA Plywood Manufacturers Association of British Columbia American Plywood Association *Plywood: Properties, Design and Construction*
Steel	Heavy forms and falsework	*Manual of Steel Construction*, American Institute of Steel Construction *Light Gage Cold-Formed Steel Design Manual*, American Iron and Steel Institute Manufacturers' data
	Column and joist forms	Simplified Practice Recommendations R87-32 and R265-63, U.S. Department of Commerce Manufacturers' data DESIGN DATA: *Manual of Standard Practice*, Concrete Reinforcing Steel Institute
	Permanent forms	American Iron and Steel Institute and manufacturers' data

Material	Use	Specification / Design data
	Welding of permanent forms	SPECIFICATION: American Welding Society DESIGN DATA: American Iron and Steel Institute
Aluminum‡	Lightweight panels and framing; bracing and horizontal shoring	Manufacturers' data *Aluminum Construction Manual*, Aluminum Association
Hardboard§	Form liner and sheathing; pan forms for joist construction	Manufacturers' data CSA
Insulating board: Wood fiber Glass fiber Foamed plastic	Permanent forms	Manufacturers' data CSA
Fiber or laminated paper pressed tubes or forms	Column and beam forms; void forms for slabs, beams, girders, and precast piles	Manufacturers' data
Corrugated cardboard	Internal and under-slab voids; voids in beams and girders (normally used with internal "egg crate" stiffeners)	Manufacturers' data
Concrete	Footings Permanent forms Precast floor and roof units Molds for precast units	ACI Building Code, ACI 318-63 ACI Building Code, ACI 318-63 ACI 711-58
Fiber-glass–reinforced plastic	Ready-made column and dome pan forms; custom-made forms for special architectural effects	ACI Special Publication No. 4, *Formwork for Concrete*, and manufacturers' data
Plastics: Polystyrene Polyethylene Polyvinyl chloride	Form liners for decorative concrete	Manufacturers' data
Rubber	Form lining and void forms	Manufacturers' data
Form ties, anchors, and hangers	For securing formwork against placing loads and pressures	Manufacturers' data; see Table 21-27 for recommended safety factors

TABLE 21-24 Form Materials and Strength Data for Design*,† (Continued)

Item	Principal use	Specification and design data sources
Plaster	Waste molds for architectural concrete	Manufacturers' data
Coatings	Facilitate form removal	Manufacturers' data
Steel joists	Formwork support	"Standard Specifications and Load Tables for Open Web Steel Joists," Steel Joist Institute
Steel frame shoring	Formwork support	"Recommended Steel Frame Shoring Erection Procedure," Steel Scaffolding and Shoring Institute; also manufacturers' data
Form insulation	Cold-weather protection of concrete	ACI 306-66 and manufacturers' data

* Adapted from *Recommended Practice for Concrete Formwork* (ACI 347-68) by permission of the American Concrete Institute.

† Since handbooks, standards, and specifications of the type cited here are frequently rewritten or updated, the latest available version should be consulted.

‡ Shall be readily weldable, nonreactive to concrete or concrete containing calcium chloride, and protected against galvanic action at points of contact with steel.

§ Check surface reaction with wet concrete.

TABLE 21-25 Maximum Lateral Pressure for Design of Wall Forms*,† (Based on ACI Committee 347 pressure formulas)

Rate of placement, *R*, ft/hr	*p*, maximum lateral pressure for temperature indicated, psf					
	90°F	80°F	70°F	60°F	50°F	40°F
1	250	262	278	300	330	375
2	350	375	407	450	510	600
3	450	488	536	600	690	825
4	550	600	664	750	870	1,050
5	650	712	793	900	1,050	1,275
6	750	825	921	1,050	1,230	1,500
7	850	938	1,050	1,200	1,410	1,725
8	881	973	1,090	1,246	1,466	1,795
9	912	1,008	1,130	1,293	1,522	1,865
10	943	1,043	1,170	1,340	1,578	1,935

* Reprinted from *Formwork for Concrete*, 2d ed., by permission of the American Concrete Institute.

† Do not use design pressures in excess of 2,000 psf or 150 × height of fresh concrete in forms, whichever is less.

TABLE 21-26 Maximum Lateral Pressure for Design of Column Forms*,† (Based on ACI Committee 347 pressure formulas)

Rate of placement, *R*, ft/hr	*p*, maximum lateral pressure for temperature indicated, psf					
	90°F	80°F	70°F	60°F	50°F	40°F
1	250	262	278	300	330	375
2	350	375	407	450	510	600
3	450	488	536	600	690	825
4	550	600	664	750	870	1,050
5	650	712	793	900	1,050	1,275
6	750	825	921	1,050	1,230	1,500
7	850	938	1,050	1,200	1,410	1,725
8	950	1,050	1,178	1,350	1,590	1,950
9	1,050	1,163	1,307	1,500	1,770	2,175
10	1,150	1,275	1,435	1,650	1,950	2,400
11	1,250	1,388	1,564	1,800	2,130	2,625
12	1,350	1,500	1,693	1,950	2,310	2,850
13	1,450	1,613	1,822	2,100	2,490	3,000
14	1,550	1,725	1,950	2,250	2,670	
16	1,750	1,950	2,207	2,550	3,000	
18	1,950	2,175	2,464	2,850		
20	2,150	2,400	2,721	3,000		
22	2,350	2,625	2,979			
24	2,550	2,850	3,000			
26	2,750	3,000	3,000 psf maximum governs			
28	2,950					
30	3,000					

* Reprinted from *Formwork for Concrete*, 2d ed., by permission of the American Concrete Institute.

† Do not use design pressures in excess of 3,000 psf or 150 × height of fresh concrete in forms, whichever is less.

workmen. Because of this, both production standards and cost standards become predictable. Interchangeable inventory becomes a reality because similar forming equipment can be brought together from a contractor's various jobsites and used without problems of incompatibility.

Maximum salvage is another important benefit. With prefabricated forms, waste, theft, and deterioration of equipment are greatly reduced.

Supplemental equipment for the commercially supplied forming system, under most circumstances, is readily available for rent or purchase. This feature is of considerable value when late decisions are made as to the forming equipment required on a specific project. Prefabricated forming equipment is inventoried in its assembled condition

TABLE 21-27 Design Capacities of Formwork Accessories*,†

Accessory	Safety factor	Type of construction
Form tie	1.5	Light formwork; or ordinary single lifts at grade and 16 ft or less above grade
	2.0	Heavy formwork; all formwork more than 16 ft above grade or unusually hazardous
Form anchor	1.5	Light form panel anchorage only; no hazard to life involved in failure
	2.0	Heavy forms—failure would endanger life—supporting form weight and concrete pressures only
	3.0	Falsework supporting weight of forms, concrete, working loads, and impact
Form hangers	1.5	Light formwork. Design load, including total weight of forms and concrete with 50 psf minimum live load, is less than 150 psf
	2.0	Heavy formwork; form plus concrete weight 100 psf or more; unusually hazardous work
Lifting inserts	2.0	Tilt-up panels
	3.0	Precast panels
Expendable strand deflection devices‡	2.0	Pretensioned concrete members
Reusable strand deflection devices‡	3.0	Pretensioned concrete members

* Reprinted from *Recommended Practice for Concrete Formwork* (ACI 347-68) by permission of the American Concrete Institute.

† Design capacities guaranteed by manufacturers may be used in lieu of tests for ultimate strength.

‡ These safety factors also apply to pieces of prestressing strand which are used as part of the deflection device.

for quick shipment and on-the-job assembly. The user who owns some prefabricated equipment can frequently supplement it with rented units during his periods of peak requirements.

Materials and prices for the various types of commercially available forms will vary over wide ranges. Because of this, it is difficult to find a common denominator for comparison. An evaluation should be made by each contractor in terms of total formwork costs, including both material and labor. This total cost will include the job cost of erecting, stripping, and maintenance; form construction, purchase, or rental cost divided by the number of uses he can expect in a reasonable amount of time; cost of ties, hardware, and other tools per job; form adaptability to different jobs on which he wishes to bid; and the appearance of the finished surface provided by the forms.

Steel-framed forms are designed for long life and minimum maintenance costs. Aluminum frames are excellent for lightweight handling, but their durability must be evaluated. Steel-faced frames are designed for great durability. Many forms today feature plastic-coated plywood faces which will give a very smooth finish and resist abrasion for a great many pours. If the contractor follows the manufacturer's care and cleaning recommendations, he may get as many as 200 reuses from these forms. The amount of plastic coating will vary from manufacturer to manufacturer. Some firms supply plastic-coated plywood only at an extra cost.

Fig. 21-66 Use of semiprefabricated forms.

For heavy construction, a semiprefabricated form is available which uses 25,000-lb-capacity she-bolts and tapered ties. The system consists of 4- by 4-ft steel frames. The contractor supplies his own plywood and bolts it directly to these steel frames. Each tie will support 16 sq ft of formwork and is generally attached 4 ft on centers at the panel intersection. Due to the system's weight, it is engineered primarily for gang forming on heavy-construction jobs such as dams, bridge piers, and large culverts. Figure 21-66 illustrates the use of semiprefabricated forms on the construction of a 40-ft-high retaining wall in New Jersey.

GANG FORMS

Today, the use of gang forms is commonplace. Much of gang forming's success can be attributed to the rapid progress of cranes and other mechanical means of transporting forms. Gang forms are most economical where a crane can work to advantage. For example, on a job requiring high walls and columns, after the crane has set the wall forms it can be put to use setting or stripping ganged column forms. The cost of the crane can then be prorated over the large number of square feet formed.

In contrast, either job-built or prefabricated forms would require costly scaffolding to be set up by hand all along the wall. Generally speaking, the cost of this manual labor would be far greater than the cost of mechanical handling for the larger forms. Another saving would be made for the large panels by assembling them on the ground where the carpenters can work to their greatest efficiency. In setting small forms up on a high wall, productivity is reduced because every item must be brought up to the workmen. This causes much delay and lost motion. A similar situation exists in stripping forms; the giant forms or gang forms would not require any assembly, while small units would have to be assembled and disassembled every time. In addition to their use in forming retaining walls, giant or gang forms have proved economical on bridges, locks, sewage- and water-treatment plants, utility plants, tunnels, sewers, culverts, multistory buildings, and certain unusual jobs.

The advantage of using a large form lies in the savings in erection. A large form area is erected at one time without the handling of individual parts, lumber, and connecting hardware. Weather, rate of pour, and size of wall will govern the dimensions of the lumber used. Normally, plywood facing with 2 by 6 and 2 by 8 studs and/or 2 by 6 or 2 by 8 wales is used. Snap ties, coil ties, and she-bolts with wale extensions are used with large forms. Strengths of individual ties will range from 3,000 to 30,000 lb.

Some prefabricated forms lend themselves well to gang forming, especially with recent advances in manufacturers' gang-form hardware. Then, after the job is completed, the disassembled forms may be put to use on other types of work. With the giant forms, a large section of the work is stripped at one time and the form is then moved by crane or rolling scaffolding to its next setup without further handling. Sizes of these large forms range up to 30 by 50 ft and are limited only by the capacity of the crane or by the difficulties of handling.

It is necessary only to move two large forms into place and secure the ties to be ready for the next pour. Speed of erection, therefore, is much greater with large forms than with individual built-up or prefabricated forms. For architectural purposes, such as on highway retaining walls, various feature strips are easily attached to the gang-form face. Figure 21-67 shows the circular gang forms used to construct the all-circular Church of the Holy Family in Parma, Ohio. Figure 21-68 shows prefabricated forms which have been ganged into 16- by 20-ft sections. These were used to construct battered walls 20 ft high, 18 in. at the bottom and tapering to 12 in. at the top.

FORM APPLICATIONS

Wall Forms Available prefabricated forming equipment for walls includes the following:

Plastic-coated plywood panels ½ in. thick with steel frames
Plywood panels ½ in. thick with steel frames
Steel-faced panels welded to steel frames
Aluminum and magnesium framed panels with plywood faces

In prefabricated wall forms the tie is connected directly to the forms rather than to the wales, thus eliminating the need for a row of wales for every row of ties. The wale is thus used as a means of form alignment rather than as an anchor for the ties. Considerable labor as well as material are saved, and much of the usual hardware is eliminated.

The designs for many types of prefabricated forms are fairly similar, but hardware and ties vary with each manufacturer. Some manufacturers' ties are adaptable to most forming systems, while others can only be used on a particular system. Safe load capacities for ties vary from 1,000 to 6,000 lb. Figure 21-69 shows the insertion of 3,000-lb ties during the construction of a warehouse in New Jersey.

Where built-up forms were once common on complicated walls, prefabs are now often used. Complicated walls may include Y and V walls, straight walls with corbels, curved walls that have a batter in them, and other complicating features. Reuse is a key factor. If the wall or variation is repeated, prefabricated forms may be the more economical method of forming. Figure 21-70 shows prefabricated forms used to construct counterforts on high channel walls for a flood-control project in Arlington, Oregon.

Column Forms Columns occur in all sizes and shapes; e.g., round, square, rectangular, octagonal, oval, etc. The forms for such columns can be made in many ways,

Fig. 21-67 Circular gang forms used for the all-circular Church of the Holy Family in Parma, Ohio.

Fig. 21-68 Battered walls, 20 ft high, 18 in. at the bottom, and tapering to 12 in. at the top, formed by ganging prefabricated forms into 16- by 20-ft sections.

using wood with steel strapping, wood with proprietary column clamps, all steel, paper, aluminum staves, etc. Gang forming is often used in reducing the labor involved in erecting and recycling the forms. Certain proprietary column forms are available on a rental basis.

Beam Forms Beam forms almost invariably are built in place on the job where they are to be used. The soffit or beam bottom is built with 2-in. lumber. For the beam sides, ¾-in. plywood or 1-in. boards reinforced with 1 by 4s are utilized. It is not necessary to use internal tying devices on shallow beams, inasmuch as the beam sides are held in place at the bottom by kickers and at the top by the slab forms.

Fig. 21-69 Insertion of 3,000-lb panel ties during the construction of a warehouse in New Jersey.

Fig. 21-70 Prefabricated forms assembled to form counterforts on high channel walls for a flood-control project in Arlington, Ore.

TABLE 21-28 Shore Spacing under Beams*

Depth of beam, in.	12	15	18	21	24	27	30	33	36	48	54	60
Shore spacing	4 ft 3 in.	4 ft 0 in.	3 ft 11 in.	3 ft 9 in.	3 ft 8 in.	3 ft 7 in.	3 ft 6 in.	3 ft 5 in.	3 ft 4 in.	3 ft 2 in.	3 ft 1 in.	3 ft 0 in.

* Spacing of shores is limited by ⅛-in. deflection of nominal 2-in. bottom. Provision is made for 50 psf construction load.

Spandrel beams, on the other hand, may be as much as 4 to 5 ft in height or depth. Internal tying devices such as snap ties would then be used to contain the concrete pressure. The use of prefabricated forms is feasible in many cases where larger beams are being formed, such as the spandrel beams on a multistory reinforced-concrete parking garage.

Table 21-28 is concerned with the spacing of shores under beams of various depths. The deflection of the nominal 2-in. beam bottom is the basis for the spacings listed in this table. Modifications must be made if the beam widths run in excess of 18 in. The weight of the concrete and various construction loads will then have to be checked

TABLE 21-29 Spacing of Joists, Inches—Simple Spans—Two Supports

Slab thickness, in.	Size of joist, in.	Span of joists																
		4′0″	4′6″	5′0″	5′6″	6′0″	6′6″	7′0″	7′6″	8′0″	8′6″	9′0″	9′6″	10′0″	10′6″	11′0″	11′6″	12′0″
4	2 × 4	24	16	11														
	2 × 6	...	...	...	24	20	14	10										
	2 × 8	...	...	...	...	...	...	24	19	15	11							
	2 × 10	...	...	...	...	...	...	...	...	...	24	19	15	12	10			
	2 × 12	...	...	...	...	...	...	...	...	...	...	...	...	22	19	15	14	10
6	2 × 4	20	13	10														
	2 × 6	...	...	...	22	16	11											
	2 × 8	...	...	...	...	...	...	20	15	12								
	2 × 10	...	...	...	...	...	...	...	...	24	19	15	12	10				
	2 × 12	...	...	...	...	...	...	...	...	...	...	24	21	17	15	12		
8	2 × 4	17	11															
	2 × 6	...	...	24	18	13	10											
	2 × 8	...	...	...	...	...	23	17	13	10								
	2 × 10	...	...	...	...	...	...	...	...	20	16	12						
	2 × 12	...	...	...	...	...	...	...	...	...	...	22	18	14	12	10		
10	2 × 4	15	10															
	2 × 6	...	...	23	16	11												
	2 × 8	...	...	...	...	24	19	14	11									
	2 × 10	...	...	...	...	...	...	...	22	17	13	11						
	2 × 12	...	...	...	...	...	...	...	...	...	24	19	15	12	10			
12	2 × 4	13	8															
	2 × 6	...	...	20	14	10												
	2 × 8	...	...	...	...	23.	17	12										
	2 × 10	...	...	...	...	...	...	24	19	15	12	10						
	2 × 12	...	...	...	...	...	...	...	...	24	21	17	13	11				

Forming surface, ¾-in. plywood.
Deflection limited to ⅛ in.
Construction load, 50 psf.

to see that the total weight will not exceed that which the shores can safely support.

Slab and Beam Forms *Flat Slab.* The formwork for flat slabs is comprised of four main parts: the forming face, joists, stringers or girts, and shores. Three-quarter-inch plywood is usually used for the forming material, with 4 by 4s and 4 by 6s used for joists and stringers, respectively. Other commonly used lumber dimensions are shown in Tables 21-29 and 21-30.

Joists are commonly spaced on either 16- or 24-in. centers, inasmuch as these spacings are suitable for plywood sheets 4 ft wide and 8 ft long. As noted in Tables 21-28, 21-29, and 21-30, any number of combinations may be used in working out the spacing of the stringers and shores. On certain types of flat-slab construction, drop heads, which are additional thicknesses of concrete surrounding a column, may tend to complicate the spacing of stringers and shores.

One manufactured slab-form system permits an average construction speed of a deck a week. The system consists of prefab forms, sliding ledger angles, steel stringers, cross bracing, and adjustable steel shores. The same forms that are used for vertical wall construction are utilized for most decking requirements. Material adaptability, ease of assembly, and speed of erection and stripping are among the many advantages. The deck can readily be adjusted to final grade, and each shore can support up to

TABLE 21-30 Spacing of Joists, Inches—Partially Continuous Spans—Three or More Supports

Slab thickness, in.	Size of joist, in.	Span of joists										
		5'-0"	5'-6"	6'-0"	6'-6"	7'-0"	7'-6"	8'-0"	8'-6"	9'-0"	9'-6"	10'-0"
4	4 × 4	...	24	20	14	11						
	2 × 6	...	...	...	24	18	13	10				
	2 × 8	...	...	...	...	...	...	24	19	15	12	10
	2 × 10	...	...	...	...	...	...	...	...	...	24	20
5	4 × 4	...	24	17	13							
	2 × 6	...	...	...	24	18	13					
	2 × 8	...	...	...	...	...	...	22	17	14	11	
	2 × 10	...	...	...	...	...	...	...	...	24	21	17
6	4 × 4	...	22	16	11							
	2 × 8	...	...	24	19	14	11					
	2 × 8	...	...	...	...	...	24	20	15	12		
	2 × 10	...	...	...	...	...	...	...	...	24	19	16
7	4 × 4	...	21	14	10							
	2 × 6	...	...	24	17	13						
	2 × 8	...	...	...	...	...	23	18	14	11		
	2 × 10	...	...	...	...	...	...	...	...	22	17	14
8	4 × 4	24	19	13								
	2 × 6	...	...	22	16	12						
	2 × 8	...	...	...	...	24	21	16	13	10		
	2 × 10	...	...	...	...	...	...	...	24	20	16	13
9	4 × 4	24	18	12								
	2 × 6	...	24	20	15	11						
	2 × 8	...	...	...	...	...	20	15	12			
	2 × 10	...	...	...	...	...	...	...	23	18	15	12
10	4 × 4	24	16	11								
	2 × 6	...	24	19	14	10						
	2 × 8	...	...	...	...	...	18	14	11			
	2 × 10	...	...	...	...	...	...	24	21	17	14	11
12	4 × 4	21	14	10								
	2 × 6	...	23	16	12							
	2 × 8	...	...	...	...	21	16	12				
	2 × 10	...	...	...	...	...	...	24	19	15	12	10

Forming surface, ¾-in. plywood.
Maximum deflection, ⅛ in.
Construction load, 50 psf.

60 sq ft of forms. Reshoring is not necessary, and the usual "forest of shores" is also eliminated.

Forms are set on sliding ledger angles on both sides of the steel stringers without connecting hardware or clamps. In stripping, the sliding ledger angles are lowered to permit form removal with the stringers and shores remaining in place. The normal reshoring operation is thus eliminated.

Almost any deck-height requirement is possible, and forming equipment is quickly released for reuse. Complete forming of 100 sq ft of deck per man-hour is easily obtained with the system, resulting in a low cost for carpenters and laborers. Figure

Fig. 21-71 Typical slab work.

21-71 shows typical slab work with a prefabricated decking system. Shores from previous pours are left in place, and stripped panels are used for the next deck erection.

Pan construction is the forming of slabs with metal or plastic pans. These are poured monolithically with a thin floor-slab section in a series of light concrete joists. The light joists are supported at the ends by concrete beams which are poured at the same time as the floor and joists. Pans can usually be rented or purchased, and in some localities, companies specializing in pan forming will subcontract the complete forming operation. Table 21-31 is concerned with the dead loads for various pan sizes. The soffits are usually supported by 4 by 6 girts or stringers and shores.

Void-in-slab construction is similar to pan construction in that paper tubes or tiles serve to form the voids between the small concrete joists as well as the underside of the thin slab. The tubes or tiles, however, become permanent parts of the floor structure. It is common practice to set the tile in position on falsework which is similar in construction to flat-slab forming. Large paper tubes are also often used in this same manner. Shoring can consist of 4 by 4s which are wedged to the proper elevation, but these members are slow to install and are awkward to adjust for height. Patented steel and combination steel and wood shores are usually preferred where they are available (see Table 21-32).

CARE OF FORMS

Some suggestions for the proper care of multiple-use forms are as follows:

Forms should be thoroughly cleaned and oiled after every use. The first time forms are erected, a light coat of oil should be sprayed on the reverse side. This procedure should be repeated every five to eight pours. Various form oils are available, and the type should be chosen carefully. The intent is to keep the forms clean, leaving no residue of concrete to affect the surface formed in subsequent casts. Some

TABLE 21-31 Vertical Load for Design of Slab Forms, psf* (Includes weight of concrete and reinforcing steel plus construction live load of 50 psf; weight of formwork not included)

Solid-slab construction

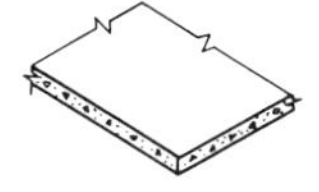

	Slab thickness, in.									
	3	4	5	6	7	8	9	10	11	12
100-lb concrete...........	75	83	92	100	108	117	125	133	142	150
125-lb concrete...........	81	92	102	113	123	134	144	154	165	175
150-lb concrete...........	88	100	113	125	138	150	163	175	188	200

Typical joist-slab construction
Actual weights and dimensions vary slightly from one manufacturer of forming systems to another

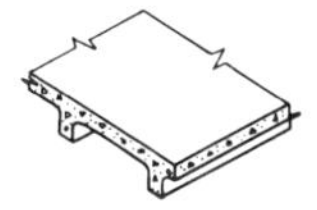

Depth of steel form, in.	20-in.-wide forms				30-in.-wide forms			
	Joist width, in.	2-in. slab	2½-in. slab	3-in. slab	Joist width, in.	2½-in. slab	3-in. slab	3½-in. slab
6	4	89	95	102	4	91	98	
	5	92	98	105	5	93	100	106
	6	94	100	107	6	95	102	108
8	4	95	101	108	4	95	101	
	5	98	104	111	5	98	104	111
	6	101	107	114	6	100	106	113
10	4	100	106	113	4	100	106	
	5	104	110	117	5	102	108	116
	6	108	114	121	6	105	111	118
12	4	107	113	119	4	104	110	
	5	111	117	124	5	107	113	120
	6	116	122	128	6	111	117	124
14	4	113	120	126	5	112	118	124
	5	118	125	131	6	116	122	130
	6	123	130	136	7	120	126	133

* Reprinted from *Formwork for Concrete*, 2d ed by permission of the American Concrete Institute.

TABLE 21-31 Vertical Load for Design of Slab Forms, psf* (Continued)

Typical "waffle" type two-way joist systems
Actual weights and dimensions vary slightly from one manufacturer of forming systems to another

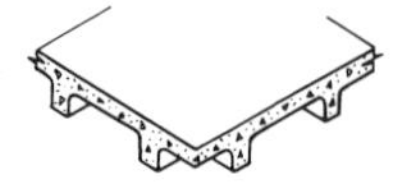

Size of pan form, in.			Slab thickness above form, in.				
Depth	Outside plan	Inside plan	2	2½	3	3½	4½
4	24 × 24	19 × 19	96	103	109	115	
6	24 × 24	19 × 19	109	116	122	128	
8	24 × 24	19 × 19	123	129	135	141	
	36 × 36	30 × 30	111	117	123	130	143
10	24 × 24	19 × 19	129	136	142	148	
	36 × 36	30 × 30	120	128	133	139	152
12	24 × 24	19 × 19	145	152	158	164	
	36 × 36	30 × 30	133	139	145	152	165
14	36 × 36	30 × 30	141	148	154	160	173

TABLE 21-32 Spacing of Shores for Flat Slab Construction*

Slab thickness, in.	Shore spacing										
	5′-0″	5′-6″	6′-0″	6′-6″	7′-0″	7′-6″	8′-0″	8′-6″	9′-0″	9′-6″	10′-6″
6	4-6	4-0	4-0	3-6	3-0	3-0	3-0	2-6	2-6	2-6	2-0
7	4-0	4-0	3-6	3-0	3-0	2-6	2-6	2-6	2-6	2-0	
8	4-0	3-6	3-0	3-0	2-6	2-6	2-6	2-0			
9	3-6	3-0	3-0	2-6	2-6	2-6	2-0				
10	3-6	3-0	2-6	2-6	2-6	2-0					
11	3-0	2-6	2-6	2-6	2-0						
12	3-0	2-6	2-6	2-0							

* Spacings based on stock 4 by 6 lengths, with maximum load (including 50-psf construction load) of 3,000 lb per shore. Maximum fiber stress in shore is 1,500 psi. Deflection of 2 by 10 soffits does not exceed ⅛ in.

oils have a tendency to discolor concrete and otherwise affect the finished surface. Forms should be piled face to face and back to back in trucking and must not be unloaded by dumping. When not in use, the forms should be piled carefully to minimize weathering and to prevent rotting. The pile should have a slight pitch to shed water, and strips should be used between forms to permit evaporation.

E. Slip Forms

The slip-forming method lends itself to the construction of towerlike concrete structures which can be generated by translating a horizontal area vertically upward. In this method, the plan area of the building is laid out on the foundation and the formwork is constructed anywhere from 3½ to 6 ft high. The forms are filled with concrete and slowly moved upward by jacking. As a set of forms moves upward, additional concrete is placed in the top and hardened concrete is exposed at the bottom. This is a continuous process as compared to the conventional method of placing the concrete in lifts. There is no requirement for horizontal joints and slip forming can be carried on around the clock, although it is usually economical to stop the process over weekends. Stops may also be planned on a daily basis or at predetermined elevations.

The plan shape or extent of the structure is not a limitation. It can be circular, rectangular, cruciform, curved, irregular, solid, hollow, or cellular. The important thing is that this area must be projected straight upward for some appreciable height, usually at least 30 to 40 ft, depending on economic and other considerations. Since the forms must slide past the face of the concrete, no projections beyond this face are possible. Floors, for instance, are placed later using keys, beam pockets, weld plates, dowels, or other devices which can be kept flush to the concrete surface. Sections of the concrete may be deleted during the slide, however. Openings may be formed, inserts cast in place, and even columns and girders may be produced by placing formwork and inserts which fit within the confines of the concrete surfaces.

Slip-formed concrete surfaces are almost always vertical, since any variation therefrom results in complications and increased costs. It is possible to slip-form stepped, tapered, or even vertically curved surfaces; but such construction must be justified in terms of overall job economy, time requirements, or aesthetics. Tapered chimneys have been slip-formed to heights of 1,200 ft, and it can be expected that still taller chimneys will be slip-formed in the future.

Some of the major construction uses of slip forms, in addition to bins and silos, are in the high-rise commercial buildings. These include central cores which act as "backbones" and take lateral loading from the structural-steel or precast-concrete framing. The cores usually include elevator shafts, stairwells, toilet facilities, and mechanical runs. Another use of slip forms is in the construction of bearing walls for apartment houses, in combination with cast-in-place or precast concrete slabs. Other uses are water towers, dam intake towers, missile silos, monumental towers (often with restaurants on top), cooling towers, and air traffic-control towers.

DESIGN OF SLIDING FORMS

The forms may be divided into three basic components: sheathing, wales, and yokes. These forms are subjected to both vertical and lateral loading. Figures 21-72 and 21-73 illustrate a typical form, and Fig. 21-74 shows the working deck with a placing deck above and a finisher's scaffold below. The entire weight of all decks and of the finishing scaffolds is carried on the jack rods. The jack rods will be discussed later, but now it can be said that the vertical loads are transmitted from the sheathing and decks to the wales, through the yokes to the jacks, and into the jack rods. The only function of the concrete is to support its own weight and to prevent the jack rods from buckling. In addition to the dead loads, live loads of at least 40 psf for the deck and 50 lb per lineal foot of scaffold must be included in the vertical design loads. Reinforcing steel, forming boxes, and other materials to be stored on the decks must also be considered.

Another important vertical loading on the forms is the "drag force" or the friction of the concrete against the forms as they are raised. The magnitude of this loading is very difficult to determine since it is affected by inherent properties of the concrete (some concretes are more adhesive than others), by the moisture content, workability, rate of set, temperature, form surface, and the condition of the form surface. In order to reduce this force, the forms are given a slight inward batter of about $\frac{1}{16}$ in. per ft of form height. Oiling or plastic treatment of the sheathing is desirable to prevent water absorption by the forms, as excess surface moisture has a beneficial lubricating effect. A drag load of 100 lb per lineal foot of forms is suggested as a good design criterion.

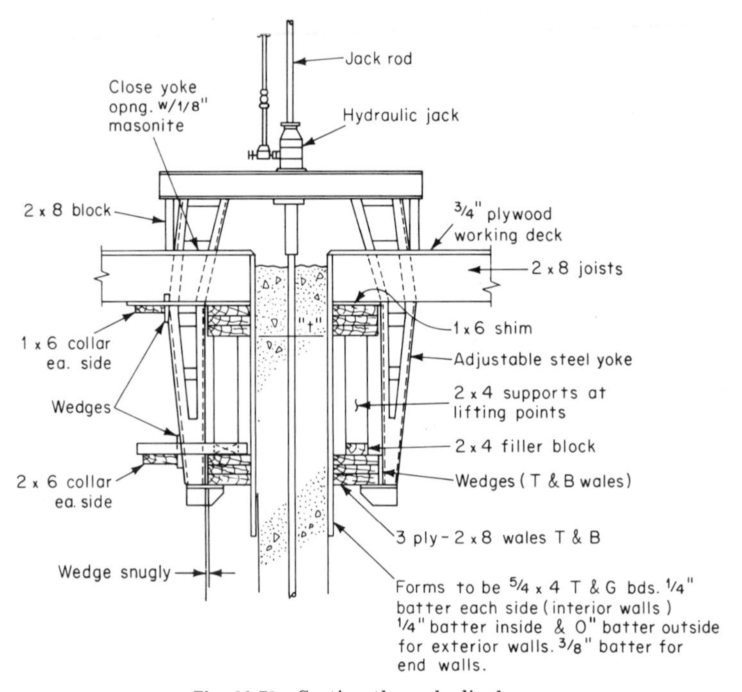

Fig. 21-72 Section through slip form.

The drag forces are picked up directly by the sheathing and transmitted to the wales as a more or less uniform load. The deck loads are applied directly to the top wales by joists and beams, and the scaffold loadings are also applied to the top wales by the scaffold brackets. These loads must be carried by the forms to the supporting yokes. For short spans the wales act as vertical beams, with the sheathing distributing the loads between the upper and lower wales. If the sheathing can take shear, as in the case of steel or plywood, the form will act as a girder. For long spans or heavy loads, the form is trussed to carry the loads.

In addition to the vertical loads, the forms must carry the hydrostatic lateral pressure of the plastic concrete. The sheathing must be designed to support this hydrostatic pressure between the wales and as a cantilever at the ends. The wales must, in turn, carry the pressure as lateral beams between yokes. The only time that the lateral pressure on the lower wales is large is when the empty forms are first filled. Under normal conditions the concrete in this area is not plastic, and the upper wales take the lion's share of the lateral pressure.

In figuring the hydrostatic pressure of the concrete, the formula recommended by ACI Committee 347 may be used. Although this formula gives lower values than that used for designing fixed forms, many years of slip-form design have indicated that its use will give adequate safety against lateral failure. In any case, the hydrostatic head of concrete should never be more than three-quarters of the form height. Table 21-33 shows allowable yoke spans for various sizes of doubled wales. Note that for curved walls allowances must be made for the material lost due to cutting.

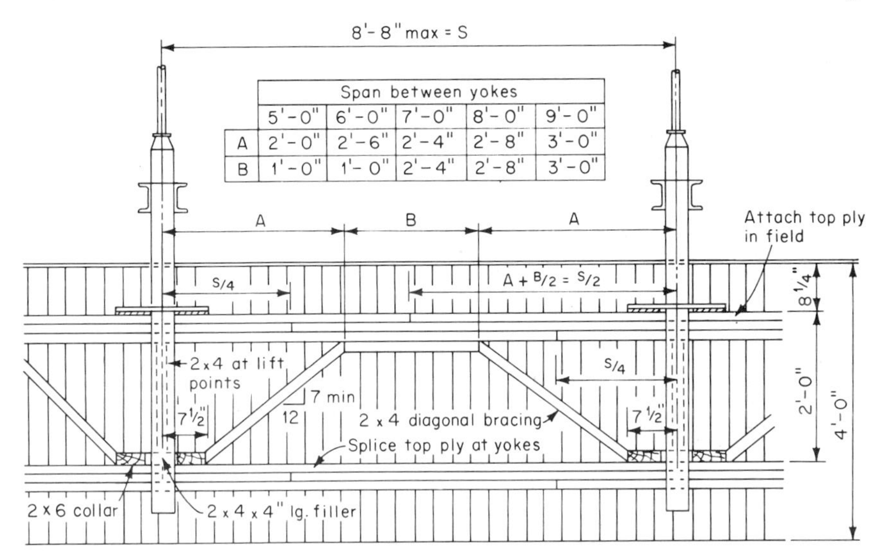

Fig. 21-73 Side view of slip form

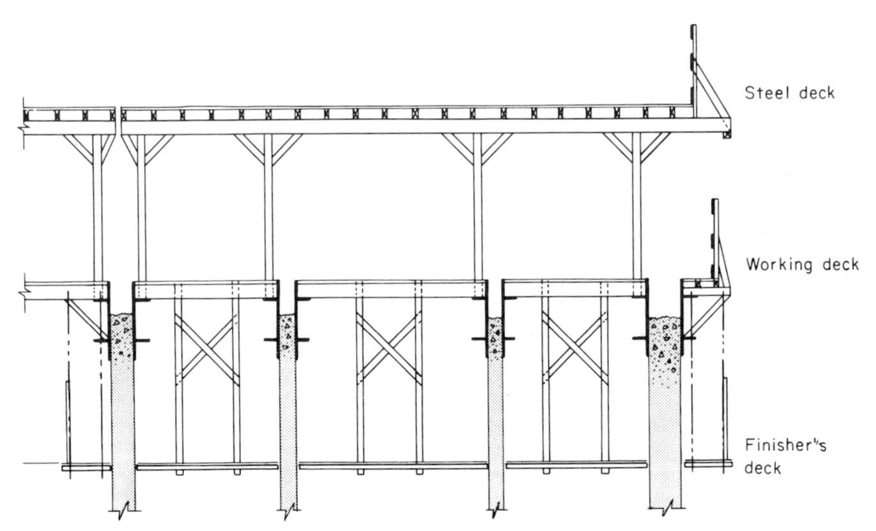

Fig. 21-74 Slip-form decks.

The sheathing for slip forms may be fabricated from any of several materials. The most commonly used material is wood because of its economy and its ability to stand racking, distortion, and abuse without permanent damage. It also lends itself to easy repair and alteration. The wood section most frequently employed is a 1-in stave in 3- to 6-in widths. The staves may be tongue-and-groove or square ended, and of either soft wood or hard wood, depending on the requirements of the job. Three-quarter-inch plywood sheathing is also used, since it has the advantage of reducing the labor required to build the forms. However, plywood forms need much heavier wale systems and bracing than do staves because of a serious tendency to distort in use. Plywood is also limited to straight surfaces or mild curves because of bending difficulties.

TABLE 21-33 Allowable Yoke Spans for Various Sizes of Doubled Wales

	Allowable span, ft											
Rate of slide*....	9 in./hr				12 in./hr				15 in./hr			
Temperature, F..	40°–50°	50°–60°	60°–70°	70°–80°	40°–50°	50°–60°	60°–70°	70°–80°	40°–50°	50°–60°	60°–70°	70°–80°
Top wales												
Double 2 × 4....	5.88	6.29	6.86	7.60	5.00	5.53	6.09	6.86	4.47	4.89	5.53	6.09
Double 2 × 6....	9.12	9.76	10.7	11.8	7.81	8.61	9.43	10.7	6.92	7.61	8.61	9.43
Double 2 × 8....	12.2	13.0	14.2	15.7	10.4	11.5	12.7	14.2	9.27	10.1	11.5	12.7
Double 2 × 10...	15.4	15.5	18.0	20.0	13.2	14.5	16.0	18.0	11.7	12.8	14.5	16.0
Double 2 × 12...	18.6	20.0	21.8	24.1	15.9	17.6	19.3	21.8	14.3	15.5	17.6	19.3
Bottom wales												
Double 2 × 4....	5.15	5.58	6.14	6.91	4.24	4.79	5.37	6.14	3.60	4.12	4.79	5.37
Double 2 × 6....	8.01	8.66	9.56	10.7	6.70	7.45	8.33	9.56	5.56	6.40	7.45	8.33
Double 2 × 8....	10.6	11.5	12.8	14.4	8.94	9.18	11.1	12.8	7.41	8.60	9.18	11.1
Double 2 × 10...	13.6	14.6	16.2	18.2	11.5	12.6	14.1	16.2	10.1	11.2	12.6	14.1
Double 2 × 12...	16.4	17.7	19.6	22.0	13.9	15.2	17.0	19.6	11.3	13.1	15.2	17.0

* Rate of slide is maximun expected at any time.

Steel forms are sometimes used for slip forming. These are several times as expensive as wood but are justified if sufficient reuse is anticipated. They have been used successfully in the construction of underground missile silos which were poured with one-sided forms, the braced sides of the excavation acting as the second form face. Still other types of form material may be used for sheathing as long as they are smooth, strong, somewhat ductile, fairly impervious to water, and not subject to serious changes in properties due to temperature variations.

Wales, except those for steel forms, are usually made of wood and are built up in two or three plies with the joints staggered. Two-ply wales are always built of 2-in.-thick material. Three-ply wales may be built of 2-in. material or a combination of 2- and 1-in. material. Most forms have two wales; some are designed for three wales. For very high forms, perhaps 6 ft in height, additional wales will be required.

The yokes are designed in the shape of an inverted U whose legs are attached to the wales. These legs carry the vertical loads in tension and the lateral loads as cantilever beams. The cross arm of the yoke must be designed as a beam, supported at the center by the jack and subjected to moments from both vertical and lateral loads. Yokes are usually built of either wood or steel and will have a certain capability for adjustment to permit their use on more than one job. Steel yokes must be adjustable over a considerable range as they are used over and over on many jobs.

JACKING SYSTEM

Propulsion and support of the forms is an important aspect of the sliding form operation. This propulsion and support is effected by means of jacks and the rods on which these jacks are supported. Twenty-five years ago practically all slip-form jacking in this country was accomplished using manually operated screw jacks, but today they are rare. Manual operation has been largely replaced by electric, hydraulic, and pneumatic jacking. The jacks are more or less cylindrical in shape, with a hole in the center through which the jack rod passes. They usually have two jaw clutches which alternately grip and raise, so that the jack climbs the jack rod very much like a monkey climbs a palm tree.

The extent of each climb is governed by the stroke of the jack, which is commonly 1 in. The speed of the jack is a function of its stroke length times the number of strokes per hour. Jacks in use at this time are all capable of speeds in excess of 20 in. per hr. Actually, the speed of the sliding form operation is not controlled by the capabilities of the jack but rather by the rate of set of the concrete. If the jacking operation is too rapid, plastic concrete will fall out from the bottom of the forms. If the jacking rate is too slow, the concrete will adhere to the forms and either cause the forms to bind or, what is most probable, the concrete will tear horizontally and lift with the form.

The optimum jacking rate is greatly influenced by concrete properties and temperature. Higher rates of slide are required in hot, dry weather and lower rates in cold, wet weather. Slip-form operations are normally not carried out at slide rates less than 2 in. per hr; 30 in. per hr is the highest speed known to the writer. The normal range of operation is between 6 and 18 in. per hr, resulting in average rates of 8 to 14 in. per hr on a 24-hr basis.

Jack rods are usually either solid steel rods or circular pipe of a fairly high carbon steel. The most common rod diameter is 1 in., although other diameters are used with certain jacks. There are slip-forming systems which use channels outside the concrete instead of jack rods in the concrete, but it is believed that only the pipe and rod jack rods are in use in this country.

Jack rods are normally in axial compression and must be designed to carry this loading without buckling. Where concrete is not present, 4- by 6-in. wood posts are used to give the necessary lateral bracing. The jack rods are fastened to these posts with blocks and jay bolts. One end of the bolt is shaped like the letter J and is wrapped around the jack rod, while its other end passes through a hole in the post. A block of wood is placed between the jack rod and the post, and the bolt is then tightened. Under some conditions, jack rods are suspended from a structure above the top of the slide. This places the jack rods in tension and tends to increase the accuracy of the structure as to vertical plumb. Such a system was used for the underground missile silos, with the support structure resting on the ground over the excavation. It was also used at the World's Fair in the erection of the observation towers for the New York State Exhibit. In this case the jack rods were hung from temporary steel towers.

The normal spacing of jack rods is between 4 and 9 ft, but this may be varied upward or downward as required. The considerations that affect the jack spacing are the following: maximum allowable span of the wales, curvature of the wall, capacity of the jacks, capacity of the jack rods, distribution of loads to obtain uniform loading on all jacks, and placing of jacks to support all corners. It is usually necessary to concentrate jacks at points of heavy loading such as deck beams, concrete hoppers, and bridge landings. The proper layout of the jacking system is of major importance to the success of the slip-form operation. Jacks are almost always set up to operate simultaneously from a central pressure or power source. When a button is pushed, all jacks on the job climb a certain distance up the jack rods. Most of the jacks have excellent accuracy as to the amount of climb, but field conditions and mechanical imperfections make continuous checking necessary. Leveling adjustments must be made as required, although many of the jacks have devices which make them almost completely self-leveling. In any case, the jacks can be individually raised or lowered by hand or by manipulating valves.

DESIGN OF WORKING DECK AND BRACING

Working decks are supported directly on the forms and rise with them. When the span between forms is too great for the deck joists, beams or trusses are used. The deck sheathing and joists should be designed for dead load plus a local live load of 75 psf or concentrated live loads from concrete buggies or other construction equipment, whichever is greater. Power buggies have not been used on slip-form decks because of the very high lateral loadings they impart. Beams and trusses may be designed for a uniform live load of 40 psf.

If the deck is to be used as a slab form at the end of the slide, it must be designed to take the weight of this slab without excessive deflection.

Table 21-34 gives maximum spans for various types of deck sheathing and Table 21-35 gives deck-joist spacings. These tables are based on live loading from nonpowered concrete buggies or 75 psf, whichever is the more critical. The spacings must be modified to take care of any greater loading, if present.

TABLE 21-34 Maximum Spans for Deck Sheathing (Loading from nonpowered concrete buggies or 75 psf, whichever is the more critical)

Type of sheathing	*Maximum span, in., center to center of joists*
1- by 6-in. square edge	12
1- by 8-in. square edge	16
1- by 6-in. tongue and groove	22
1- by 8-in. tongue and groove	24
½-in. plywood	12
⅝-in. plywood	16
¾-in. plywood	22
⅞-in. plywood	24

TABLE 21-35 Floor-joist Spacing for Sliding Deck (Loading from nonpowered concrete buggies or 75 psf, whichever is the more critical)

	Spacing, in., for joists of indicated sizes				
Span, ft	2 × 4	2 × 6	2 × 8	2 × 10	2 × 12
4	38	91	130	164	200
5	20	59	104	131	160
6	12	41	73	110	132
7	...	27	53	85	114
8	...	18	41	65	96
9	...	13	32	52	76
10	...	9	26	42	61
11	...	...	21	34	50
12	...	...	18	29	42
13	...	...	16	25	36
14	...	...	...	21	31
15	...	...	...	19	27
16	...	...	...	16	24
17	...	...	...	...	21
18	...	...	...	...	19
19	...	...	...	...	17
20	...	...	...	...	15

The deck on a slip-form project has the important function of tying all the formwork together so that the structure goes up as a unit. The deck must be designed to maintain the plan dimensions throughout the height of the structure. Distances between walls must remain constant; square corners must stay square; circular arcs must maintain radius. The well-designed and -constructed deck tends to keep itself level. It prevents corners and projections of the structure from moving horizontally. It tends to keep straight lines straight. In order to accomplish these functions and in addition resist wind loads, the decks must be properly braced in the horizontal plane. This bracing may be wood, steel rods, steel plating, trusses, or combinations of any or all of these.

REINFORCING STEEL

Reinforcing steel for slip-form work must be detailed, placed, and inspected in a somewhat different manner than would be the case for conventional forming. In slip forming, the vertical steel is set up in the forms and held in place by templates attached to the deck and moving with it. The steel is lapped and tied to the rod below; it is held at the top by the templates at heights of from 4 to 10 ft above the deck. The higher the template, the longer the rebar that can be used. Laps are usually staggered; this is better structurally and distributes the work load more evenly for the ironworkers. Vertical bars must be limited in their lengths, as considerable "whip" may otherwise develop on a windy day. The length of vertical steel is usually kept between 14 and 20 ft, depending on the size of bars. Bars heavier than No. 6 may be made as much as 5 ft longer than this, but the templates must be high. Bars longer than 20 ft become difficult to handle in the field unless special arrangements are made.

Horizontal steel is set a layer at a time as the work progresses. The bars must fit below the cross arm of the yoke and must be threaded in through the vertical steel and the jack rods. Rebar detailing must verify that it will be physically possible to place the bars, but there is seldom any reason why horizontal bars cannot be detailed so that their placing is very easy. In this system, the rebar men are always working safely and comfortably right at "floor level." The actual lengths of bars may be limited by the design of a particular structure and will seldom exceed 20 ft. Bent bars or bars with hooks must be shorter than straight or hoop steel.

Whenever it is possible for the splices in horizontal bars to be inaccurately placed, such as in a continuous hoop of large diameter, it is advisable to add a few inches to the length of each bar to give the ironworkers a reasonable tolerance. The spacing of horizontal steel should be carefully studied to give the easiest possible placing in the field. Spacings of less than 6 in. tend to keep the ironworkers too busy and make inspection somewhat difficult. Spacings of 10 to 12 in. are normally ideal if this can be arranged. The larger-diameter bars are harder to handle and place, so that it is desirable to use bars of ¾-in. diameter or smaller. Of course, it will not always be possible to keep within the suggested size and spacing limitations and yet furnish the required steel area. The designer must use his judgment and come up with the best possible solution for the particular structure. Spacings as small as 3 in. have been used, and bar sizes up to No. 11 have been employed.

Ties and stirrups must be detailed with particular care so as to ensure easy placement. Hooks bent at 90° are preferred to standard hooks as they can be rotated about the axis of the bar and positioned around a vertical. Care must be taken in placing ties so as not to foul the yokes if a basket-type arrangement is used or if the ties are dropped in from the top. If they cannot be dropped in from the top, the ties must be detailed in pieces that can be easily placed from the sides of the column, pilaster, or wall and yet fully meet the structural requirements.

Inspection of reinforcing steel in slip-form work requires a technique of its own. Large areas of reinforcing cannot be inspected at one time, nor can concreting be delayed until inspection of steel has been completed. The reinforcing is constantly disappearing into the concrete as the forms rise at a rate of 12 in. an hour or more. In order to facilitate placing and inspection, all horizontal steel should be designed in horizontal layers; that is to say, all horizontal steel should be at the same vertical spacing or at least in multiples of the same spacing. Using this system, a reinforcing-steel crew knows that a full set of steel as shown on the drawings must be installed at certain elevations. The inspector, too, can check layer by layer to make sure that no bar is left out of any set. Particular care must be taken to ensure that lighter steel such as hairpins, ties, corner bars, etc., is not omitted.

Positive means must be established to identify the location of each layer of horizontal steel. One method is to mark the layer spacings on vertical bars at several locations by means of saw cuts or tightly wound tying wire. Keel marks have a tendency to get lost if used alone, but they are very good location markers for saw cuts or wire markings. Some firms embed light angle irons in the concrete at one or two locations. These angles are bolted together in sections as the slide progresses and are drilled with holes at the proper location of each horizontal layer. When possible, short bars which do not interfere with the yokes can be tied several courses in advance. They will then furnish excellent guides for the rest of the bars.

Sometimes the steel and the concrete are placed alternately in layers of equal thickness. This sounds like an excellent system, and sometimes it does actually work in the field. However, most of the time it is extremely difficult to maintain, and it tends to slow down both crews and make inspection difficult. A great help to both placing and inspection is to mark the ends of bars right on the deck so that individual measurements are not necessary each time a layer is placed. Since concrete is constantly being placed in the forms, these markers must be of a raised type such as double-headed nails, V notches, steel plates, or wood blocks. At the same time, they must not be a tripping hazard to the men working on the deck.

Tying is often greatly reduced on slip-form work, since the steel is placed directly on the concrete. Some variation in spacing is usually not of structural significance unless the effective steel area is reduced over lengths of 3 or 4 ft, or unless maximum allowable spacings for the steel are exceeded. On the other hand, it is very important to keep proper clearance between the steel and the forms. The inspector must see that this clearance is properly maintained and that sufficient wire ties are used to assure proper cover for the steel. Vertical steel presents no problems in placing or in inspection as long as the lengths and ties, if any, are properly detailed. Bars larger than No. 11 must have welded or mechanical connections, and will seriously delay the slip-form operation.

CONCRETE PLACING

In general, the concrete mixes used for slip-forming are the same as those employed for other methods of concrete construction. Concrete to be slip-formed must usually have a higher slump than concrete that is to be formed by more conventional methods, and slumps between 3 and 5 in. are most commonly specified. Placing concrete in slip forms with a slump below 3 in. becomes very difficult if not impossible. On the other hand, in tropical climates and using certain types of aggregates and cement, slumps up to 7 in. have been required. Vibration, retarders, and workability agents can be used under the proper conditions to retain adequate workability while reducing the slump. The slump actually required is difficult to predict and will vary continuously with changes in the weather. Since improper slump will result in poor concrete, adequate provisions should be made for its careful control. Several batch mixes should be prepared, covering a range of slumps and all meeting the strength requirements. The batching plant should then be alert and ready to switch from one batch mix to another. The superintendent in charge should be authorized to change slumps, as required, by using one of the previously approved batches.

Another point that requires special consideration is the coarse aggregate to be used in the concrete mix. For walls less than 8 in. thick, the maximum size of aggregate must be limited to ¾ in. For most other slip-form work, maximum aggregate size should be limited to 1 in. Crushed stone or gravel is usually specified, but uncrushed gravel has been used successfully. Lightweight and slag concrete present no serious problems.

In slip-form work, concrete is placed in the forms in layers of 6 to 8 in. Each layer is spaded or vibrated as it is placed, and care should be taken that the forms are filled evenly and kept as nearly full as possible. Although systematic filling of the forms is advisable, the system should be set up to avoid inducing torsion in the structure as a whole. For instance, a hollow rectangular tower cannot be filled constantly in a clockwise direction as this will tend to make the forms rotate about the centerline of the structure. Placing in alternately clockwise and counterclockwise directions will solve this problem. If the rate of raising the forms is seriously slowed down for any reason, care must be taken to avoid cold joints. Placing the concrete in thinner layers, say 2 to 3 in., will help reduce the time factor between successive placements.

Vibration of concrete was at one time considered dangerous in slip-form work. Thanks to the work of ACI Committee 609 and to favorable experience with vibration, it is now recommended rather than prohibited. The vibrator may be allowed to penetrate as deeply as it will under its own weight, but it should not be forced deeper into the concrete. Further, it should not be used too long in one place.

The slip-form decks must be kept as nearly broom clean as possible in order to prevent hardened concrete spillings from finding their way into the forms. When cleaning the decks, the sweepings should be directed away from the forms. Clean-out openings must be provided in the deck or some other arrangement made for disposal of debris. When daily stops are made, exposed portions of the forms must be thoroughly cleaned before resuming work.

Admixtures have been successfully used with slip-form work. The use of air-entraining agents and workability agents is quite common. Under normal slip-forming conditions, no ill effects result from the use of these admixtures; in fact, their effects are usually quite beneficial. Naturally, the usual rules of careful use and control apply here as elsewhere. Care must be taken to see that any admixtures are properly dispersed throughout the concrete. If, for instance, a retarder gets concentrated on one side of a structure, the concrete in this area is too plastic to allow the forms to move. The concrete on the other side has not been retarded to the same extent and is setting rather rapidly. This results in a very awkward situation: unless the forms are moved, one side will bind; and if they are moved, the concrete will fall out of the bottom of the other side.

Control of set, workability, moisture content, and temperature are of particular importance in sliding-form operations. In hot weather, or when slide rates must be reduced because of many inserts being placed or for any other reason, admixtures are recommended. Under those conditions, retarding workability agents are usually used to delay the set and to increase the plasticity of the concrete without adding excessive amounts of water. There are some claims that these agents will also improve the ability of the concrete to slide past the forms. This has not as yet been established, as workability and plasticity are not identical with lack of adhesion. In some instances at least, it appears that the use of these workability agents does not completely remove the necessity of increasing the water content of the mix.

A very effective way of controlling the set of concrete during hot-weather placing is by replacing a part of the mixing water with crushed ice. When the necessary ice is available, this method is economical, effective, and entirely safe. The normal precautions of cooling aggregates, painting equipment white, and keeping the equipment in the shade are also in order. In hot weather, in traffic-congested areas, and most especially when these two conditions are combined, the use of plant-mixed concrete becomes dangerous. If this situation exists and a site plant is uneconomical or undesirable, then the trucks should be dry-batched at the plant and the mixing water added at the site. In this way the concrete does not undergo excessive set in the trucks.

At all times, but especially when the temperatures go over 70°F, the number of ready-mix trucks in use must be such that no delay will result in the concrete placing. Even icing and placing the concrete in 2-in. layers will not prevent a cold joint if the concrete supply is interrupted. Auxiliary equipment and parts replacements are "musts" for the concrete-handling system.

In cold weather, accelerators may be required to provide heat at a greater rate and to reduce the setting time. Careful consideration must be given before high-early cement is used on slip-form work. Unless this is really required, the setting time may be excessively reduced and lead to binding of the forms. It may be sufficient to increase the cement factor and thus accelerate the chemical reaction and obtain early strength.

Insulating the forms is very beneficial, at times absolutely necessary. Lightweight enclosures can also be suspended from and attached to the slip forms. In cold weather the minimum enclosure should be a tarpaulin enclosure for the deck railing and the finisher's scaffold. Heating is often required for these enclosures. Salamanders can be placed at intervals on the finisher's scaffold, but they constitute a serious fire hazard. If steampipes can be suspended from the forms and supplied from steam generators, the fire hazard will be eliminated and the concrete curing will be greatly improved.

Inspectors may check the rate of slide by plunging steel rods into the concrete and measuring the depth of hard concrete in the forms. This depth should be 12 to 30 in. and, if it is less than 12 in., the forms must be slowed down until the concrete sets further. If the depth of hard concrete exceeds 30 in., the rate of slide must be increased to prevent binding of the forms. Experienced personnel can judge the rate of slide by standing on the scaffolds below the forms and scratching the green concrete to see how hard it is as it comes out of the forms.

CONTROL AND TOLERANCE

As a slip form rises there is a tendency for it to translate, or possibly even to rotate or to combine translation with rotation. The smaller a structure is in plan, the greater this danger becomes. Fortunately, correction is entirely possible, and those structures which most easily go out of plumb are the ones which can most readily be corrected. Any movements are actually of a small order of magnitude, but they could result in a tall building being several inches out of plumb. Adequate provisions must be made to limit and correct any deviations as slip forming proceeds.

If the deck is not level, the building will naturally "grow" in the direction toward which the deck is tilted. Therefore, the first requirement is that the deck be level at all times unless a tilt is deliberately introduced to make a correction. The forms must be carefully built and leveled before the slide starts. Once the slide starts, levels must be checked continuously and all jacks kept within ½ or ¾ in. of correct elevation. If a jack gets too far ahead of the others, it is made to miss one climb. If it lags behind, it must be brought up to the proper level manually. Even with automatically leveled jacks, human inspection is still necessary to ensure proper operation of the leveling system.

The simplest and most effective way of checking is to cut hacksaw marks in the jack rods at 12-in. intervals. The jacks may then be checked against these saw marks. It is necessary, of course, to make sure that the saw cuts are accurately positioned. "Water level" systems are also used, where a central tank is filled with water and plastic hose is run to the various jacking points. A vertical tube then indicates the water level at each point. Care must be taken to expel all air bubbles from the system or false readings will be encountered.

Periodic checks must be made to determine how close to plumb the various points of the structure are. These checks may be required as often as every 3 hr for buildings having close tolerance and a plan that is particularly subject to "drifting." Checks at 12- to 24-hr intervals are sufficient for slip-form construction which has a stable plan arrangement and which seeks to meet the ACI tolerance of 1 in. per 50 ft of height.

Checks on vertical plumb must be made to detect movement in any direction; readings should give movements in both east-west and north-south directions. Targets are usually painted in several significant positions around the structure. These targets are clearly marked on the forms themselves. Once the form has been raised a foot or so, a transit or a carpenter's level is used to project the line down to the concrete wall. A chisel cut in the concrete can then be used as the lower target. To check the plumb, the cross hairs of a transit are first centered on the form target. The telescope is then rotated through a vertical angle until the horizontal cross hair is on the lower target. The distance between the vertical cross hair and the chisel cut is the amount that the form above is out from plumb in a direction perpendicular to the transit sight. The transit should be plunged and a second shot taken to eliminate transit maladjustments.

Another method is the use of an optical plumb. This is essentially a weighted telescope which hangs freely in a pivoted collar so that the line of sight is always plumb. Target boxes are set at the base of the slide using the optical plumb, and deviation from plumb can be checked by direct reading through the telescope. As many target boxes and mounting collars are required as there are points to be checked, but a single optical-plumb telescope can be moved from position to position. At heights in the vicinity of 150 ft, readings with the optical plumb become difficult because of wind and vibration. When this happens, transit readings must be used for the remainder of the slide or the target boxes must be reset further up.

A word is in order at this time as to acceptable tolerances. ACI Committee 347 allows a variation from plumb of 1 in. per 50 ft of height. With good control it is possible to limit and correct any actual variations, and the following tolerances are suggested in lieu of extending the ACI tolerance as a straight-line function: 1 in. tolerance for structures up to 50 ft, 1½ in. for structures up to 100 ft, and 2 in. for all structures 100 ft high or higher. The designer should keep in mind that the tolerances just stated are minimum ones. Where greater tolerances are permissible, they may save some headaches and result in maximum economy.

FINISHING AND CURING

Slip forming provides an excellent opportunity to finish wall surfaces while the concrete is still quite green. When the concrete is finished below the form it is 4 or 5 hr old, rather than 24 hr or more. It is therefore possible to do excellent patching and to obtain a very dense surface. The most popular and probably the best finishing technique is the float-and-brush finish, which results in a sandy textured, attractive surface. When required, as on the inside of flour bins, a steel-trowel finish may be applied.

Because of the combined effect of the deflection of the forms due to the hydrostatic pressure of the concrete and the taper built into the form, a phenomenon called "shingling" often occurs at levels where the form has become empty to a considerable extent and is then filled. This happens, for instance, at planned stops. The forms must be kept moving once the concreting is completed in order to prevent binding and to allow the slide to start the next day or as planned. This movement usually is continued for 2 to 3 hr after the last concrete is in place, leaving 1 to 2 ft of empty form. This represents $\frac{1}{16}$ in. of draft or a $\frac{1}{16}$-in. gap between the forms and the top of the concrete wall. When concreting is started again, the hydrostatic pressure will increase this gap and an unsightly overpour will occur. This can be minimized to almost nothing if the amount of empty form is kept down to 12 or 15 in.

If aesthetic effects are desired, as for architectural concrete, special care must be taken to prevent shingling and cold joints. It is also desirable to keep the depth of plastic concrete in the form as uniform as possible. Special effects can be obtained by setting masonry or plastic form inserts into the forms on the exposed faces.

The two main methods of curing slip-formed walls are water curing and membrane curing. Water curing is effected by attaching punctured garden hose to the finisher's scaffold so that streams of water constantly impinge on the wall and then flow down the face, hopefully in a continuous curtain. Fog-spray nozzles can also be used and are superior to punctured hose. Unfortunately, difficulties in applying the desired water curing are frequently encountered. Due to clogged holes, variations in water pressure, wind, evaporation, and human indifference, the chances of maintaining a constant curtain of water on the walls for several days, and particularly over the weekends, become somewhat slim. Overly zealous use of the water is not the answer, since this can result in muddy conditions on the ground and in scouring and staining on the walls. Union labor has also been known to walk out in protest.

Membrane curing can be applied from a garden-type pesticide sprayer by a man walking around the structure on the finisher's scaffold. A fugitive dye will help prevent "missed" areas. Once the membrane is on, no further maintenance is required.

CONNECTIONS FOR BEAMS AND SLABS

Adequate connections must be devised for connecting the floor systems to the previously cast concrete walls. These connecting systems should have maximum tolerance and flexibility.

Slabs are usually attached to walls by means of 1⅝- to 4-in.-deep keys. Dowels, if required, can be placed inside the key forms and bent out into the slab after the slide is over. Threaded inserts to take threaded rods, through holes, or hooked connections may also be used.

Concrete beams can be connected in much the same manner as the slabs except that a pocket or through opening is utilized instead of a key. Beam dowels will usually be of larger diameter and greater length than slab dowels. Stubs for welding or mechanical coupling such as Cadwelding will be required.

Attachment of structural-steel beams to concrete walls can be accomplished by means of pockets and vertical anchor bolts or by weld plates cast into the concrete and welding. Precast beams are attached in the same manner.

LABOR CONSIDERATIONS

Slip form construction requires supervision of a high caliber and at least one person "on deck" who is sufficiently experienced in this method of construction; the rest of the crew does not require any more know-how than is required for conventional concrete placing. Local labor has been employed at remote underdeveloped locations on many occasions without ill effect; in this country, in at least one instance, inexperienced college students have been utilized. However, like all other work, slip forming will go better and faster when the workers are skilled in their trades.

A typical slip-form crew will consist of the following:

Superintendent
Assistant superintendent or deck foreman
Jacker (one or more)
Control men (layout)
Foreman for each trade
Carpenters (form watchers and insert placing)
Lathers or ironworkers
Laborers (placing, curing, helping, etc.)
Cement finishers
Standby electrician (complete lighting at night)
Engineers (operate cranes and hoists)

It is obvious that there is a minimum size of crew regardless of the size of the job. A slip-form crew will seldom number less than 12, regardless of how little concrete is being placed. When the rate of concrete placing is low, permission to use a composite crew such as is employed in chimney construction will be very beneficial.

In general, slip-form operations of less than 5 cu yd of concrete per ft of wall height will result in astronomical in-place costs per yard of concrete. The economy of the method improves rapidly and almost in direct ratio to the concrete quantity per foot of slide. At 20 yd of concrete per ft, the economy is quite good; and at somewhere between 20 and 30 yd per ft the cost per yard becomes quite stable.

The most important economic advantage of slip-form construction is its very low cost of formwork per square foot of contact area. There is also a considerable saving in scaffolding. This economy is offset by the necessity of paying overtime premium for labor. The amount of premium pay required will probably determine the economy or lack of economy of a slip-form operation.

In areas where the unions are agreeable to shift work and where the labor supply is ample, the most economical slip-form procedure is three 8-hr shifts on a 24-hr round-the-clock operation, stopping only Sundays or for the weekend. In this case the premium time may be on the order of 10 to 15 percent.

Where unions will not allow shifting and where labor is in short supply, daily slides will be in order. Although no overtime premium will be paid to the crew as a whole, a skeleton crew is required for approximately 2 hr to free the forms. The premium in this case will be about 15 to 20 percent. Where shift operations are accepted by the unions but where there is a shortage of men, a common arrangement is to use two 12-hr shifts. This results in a 20 to 40 percent premium.

References

1. *ASTM Book of Standards*, part 9, "Cement, Lime, Gypsum," latest issue, American Society for Testing and Materials, Philadelphia, Pa.
2. References herein to federal specifications are given by their applicable number—i.e., Fed. Spec. SS-C-192g for portland cement. Copies may be obtained as outlined under General Information in the *Index of Federal Specifications*, which is available in most libraries or from the Superintendent of Documents, U.S. Government Printing Office, Washington, D.C. 20402. The latest issue will apply.
3. J. L. Gilliland, *A Survey of Portland Cement Specifications of the United States of America*, ASTM Special Technical Publication 441, August, 1968.
4. *ASTM Book of Standards*, part 10, "Concrete and Mineral Aggregates," latest issue, American Society for Testing and Materials, Philadelphia, Pa.
5. *Standard Guide Specifications for Concrete, Guide Specifications, Civil Works Construction*, CE 1401.01, Department of the Army, Corps of Engineers, Office of the Chief of Engineers, September, 1965, or latest issue.
6. *Handbook for Concrete and Cement*, Corps of Engineers, U.S. Army Waterways Experiment Station, Vicksburg, Miss.
7. "Recommended Practice for Winter Concreting," ACI-604, *ACI Journal*, Proc. vol. 52, p. 113, October, 1955, revised September, 1965, *ACI Journal*, Proc. vol. 62, p. 1009.
8. "Admixtures for Concrete," ACI 212–63, *ACI Journal*, Proc. vol. 60, p. 1481, November, 1963.
9. *Symposium on Effect of Water-Reducing Admixtures and Set-Retarding Admixtures on Properties of Concrete*, ASTM Special Technical Publication 266, June, 1960.
10. Lewis H. Tuthill and William A. Cordon, "Properties and Uses of Initially Retarded Concrete," *ACI Journal*, Proc. vol. 52, p. 273, November, 1955.
11. "Recommended Practice for Selecting Proportions for Concrete," ACI 613-54, *ACI Journal*, Proc. vol. 51, p. 49, September, 1954, or latest issue.
12. *Concrete Manual*, 5th ed., p. 130, U.S. Bureau of Reclamation, Denver, September, 1949.
13. A. T. Goldbeck and J. E. Gray, *A Method of Proportioning Concrete for Strength, Workability, and Durability*, National Crushed Stone Association, Bulletin 11, November, 1965.
14. *Concrete Manual*, 7th ed., p. 146, U.S. Bureau of Reclamation, Denver, 1963.
15. *Recommended Practice for Evaluation of Compression Test Results of Concrete*, ACI 214-65, January, 1965, or latest issue, American Concrete Institute, Detroit, Mich.
16. A. N. Talbot and F. E. Richart, *Strength of Concrete: Its Relation to Cement, Aggregates and Water*, University of Illinois, Engrg. Expt. Sta., Bull. 137, 1923. See also *Proc. ASTM*, vol. 43, p. 917, 1943.
17. J. Balomey, "Determination de la résistance à la compression des mortiers et bétons," *Bull. tech. Suisse romande*, 11, 14, 15, and 17, 1925.
18. J. W. Kelly and M. Polivka, "Ball Test for Field Control of Concrete Consistency," *ACI Journal*, Proc. vol. 51, p. 881, May, 1955.
19. *Investigation of Field Methods for Determining Air Content of Mass Concrete*, Corps of Engineers, U.S. Army, Waterways Experiment Station, Vicksburg, Miss., Technical Memorandum 6-352, November, 1952.
20. *Recommended Practice for Selecting Proportions for Structural Lightweight Concrete*, ACI 613A-59 or latest issue, American Concrete Institute, Detroit, Mich.
21. D. L. Bloem, *How to Get Good Concrete*, NRMCA Publication No. 121, National Ready Mixed Concrete Association, Silver Spring, Md., 1966.
22. L. D. Long, "Quality Control of Concrete," *Civil Engineering*, November, 1964.
23. Joseph J. Waddell, ed., *Concrete Construction Handbook*, McGraw-Hill Book Company, New York, 1968.
24. *Recommended Practice for Measuring, Mixing, and Placing Concrete*, ACI 614, American Concrete Institute, Committee 304, 1959.

25. *Consolidation of Concrete*, ACI 309, American Concrete Institute, Committee 609.
26. *Recommended Practice for Cold Weather Concreting*, ACI 306, American Concrete Institute, Committee 306, 1966.
27. *Recommended Practice for Hot Weather Concreting*, ACI 605, American Concrete Institute, Committee 305, 1959.
28. *Curing Concrete*, ACI 308, American Concrete Institute, Committee 612, 1958.
29. *ASTM Book of Standards*, part 4, "Structural and Boiler Steel; Reinforcing Steel; Ferrous Filler Metal," latest issue, American Society for Testing and Materials, Philadelphia, Pa.
30. *Manual of Standard Practice for Detailing Reinforced Concrete Structures*, ACI Standard 315-65, American Concrete Institute, Detroit, Mich.
31. *Building Code Requirements for Reinforced Concrete*, ACI 318-63, American Concrete Institute, Detroit, Mich.
32. *Commentary on Building Code Requirements for Reinforced Concrete*, Report of ACI Committee 318, American Concrete Institute, Detroit, Mich.
33. *Recommended Practice for Concrete Formwork*, ACI 347-68, American Concrete Institute, Detroit, Mich.
34. *Concrete Surface Blemishes*, Technical Report of Cement and Concrete Association, London, July, 1964.
35. *Formwork for Concrete*, 2d ed., Special Publication No. 4, American Concrete Institute, Detroit, Mich., 1969.

Section 22

Special Concretes and Mortars

JOHN C. KING

Manager, Federal and State Engineering Services,
Master Builders, Cleveland, Ohio

A. Preplaced-aggregate Concrete*

Preplaced-aggregate concrete[1]† is made by first placing clean, coarse aggregate in the forms. The voids in the aggregate mass are then filled with a fluid cement-sand mortar which is almost always pumped into the aggregate from the lowest point in the mass. The preplaced-aggregate method of producing concrete is also referred to as "grouted-aggregate concrete," "injected concrete," and "prepacked concrete." Prepakt and Colcrete are trade names, the first being used by an association of contractors and the second referring to the use of a special mixer for preparing the grout. The name "Naturbeton" has been applied by some when the method is used to obtain exposed-aggregate surface finishes. Preplaced-aggregate concrete has proved particularly useful in new construction where the placing of conventional concrete would present major difficulties and in concrete repair work where low drying shrinkage is desirable. The preplaced-aggregate method may be used for either plain or reinforced concrete.

MATERIALS‡

Because several of the properties of preplaced-aggregate concrete depend on the materials used, this subject will be treated first.

Coarse Aggregate The coarse aggregate for preplaced-aggregate concrete (PPAC) may be either natural gravel or crushed rock meeting the requirements of ASTM Designation C-33 with respect to quality. It should be graded for least practicable voids and will normally range in size from a minimum of about ¾ in. (with sharp cut-off at ½ in.) to the largest size economically available and convenient to handle without excessive segregation. The maximum size should also be consistent with the spacing of reinforcement, if any, and with the general principle of not exceeding one-fourth of the dimension between forms. In thin sections, say under 12 in., the minimum size may be reduced to ½ or even ⅜ in.

In mass concrete placements, consideration may be given to larger minimum sizes and to top sizes ranging up to "man size" or "derrick stone." The larger channels thus provided for grout flow will permit coarser sands to be used in the mortar, with consequent increase in sand-to-cement ratio.

The void content of the coarse aggregate in place will vary from 33 to 37 percent for a ¾- to 6-in. grading to 38 to 42 percent for a ¾- to 3-in. range. As the size range narrows, the void content will go up over 45 percent for a single size. Using gap grading, voids can be lowered; a void content as low as 25 percent was obtained experimentally with a 50–50 combination of ¾- to 1½-in. and 6- to 10-in. sizes.

The aggregate in place must be free of fines, and the individual particles must be clean. Thus, washing and screening are usually required at or very near the last point of handling before placement. It is desirable that the aggregate be surface wet when placed and so maintained until grouted. Covering the aggregate surface or occasionally fog spraying it will maintain satisfactory dampness when the aggregate must remain in place for long periods before grouting. Hosing in place is highly undesirable because any downward flow of water will wash fines to the bottom, where it is virtually impossible to remove them; the result will be an unbonded, permeable joint or contact.

Grout *Cement* may be any type of portland cement, although Types I and II are most frequently employed. Natural and slag cements have been blended with portland cement on occasion, but experience in this area is insufficient to suggest proportions. All cements should meet appropriate ASTM or other recognized standards.

* See also Sec. 22-B, Grouting.
† Superscript numerals refer to the list of references at the end of this section.
‡ See Material Selection and Mix Proportioning in Sec. 21, Conventional Concrete.

Fly ash and natural pozzolans are frequently used to replace up to 30 percent of the weight of portland cement. Fly ash and natural pozzolans should meet the requirements of ASTM Designation C-618 except that loss on ignition for fly ash should not exceed 5 percent. Either fly ash or natural pozzolan, particularly the former, is favored as a mortar ingredient to improve pumpability, extend grout handling time, and reduce bleeding. Since both materials produce something less than half the heat of hydration which is liberated by an equal weight of portland cement, and since they produce this heat more slowly, their use will also minimize temperature rise. Both can be used in excess of cementing material requirements to make up for missing fines in the sand. Fly ash is frequently less expensive than cement, a cost factor worthy of consideration.

Grout fluidifiers when used in the mortar provide water reduction, retardation of early stiffening, dispersion of the mortar ingredients, reduction of bleeding, and fluid-phase expansion. The expansion is caused by the release of hydrogen gas as a very small amount of aluminum powder reacts with the normal alkalinity of a cement-water combination. The expansion is important for two reasons: (1) to completely overcome the bond-weakening effects of the traces of bleeding that collect under the preplaced-aggregate particles regardless of the care used in selecting, proportioning, and mixing the materials; and (2) to provide a gas-bubble entrainment system for freeze-thaw durability and reduced bleeding. Most commercially available grout fluidifiers are formulated to meet the requirements of CRD-C-566 when used at the rate of 1 percent by weight of the cementing material.

Grout fluidifiers are required in most United States specifications. European practice leans toward the use of high speed shear mixers to hold bleeding to tolerable levels. It has been the writer's experience that even when such mixers are used, the addition of a fluidifier will still provide worthwhile improvements in properties and economies.

Sand, except for its grading, should meet the usual requirements (such as ASTM Designation C-33) for any good conventional concrete. Where the minimum size of coarse aggregate is 1 in. or less, all the sand should pass a No. 8 screen, 95 percent should pass No. 16, and 10 to 25 percent should pass No. 100. The grading should be as uniform as practicable between these limits to minimize bleeding; the fineness modulus should fall between 1.20 and 2.10. Where the coarse aggregate minimum is ⅜ to ½ in., 100 percent of the sand should pass a No. 16 screen.

Coarse aggregate having a minimum size of 2 in. will generally accept a normal, well-graded concrete sand. In general, it will still be desirable to maintain the fraction passing the No. 100 screen in the vicinity of 10 percent.

PROPERTIES AND PROPORTIONING

Properties Concrete made by the preplaced-aggregate method exhibits some properties similar to those of conventional concrete, and others which are of unique special value.[2–4]

Compressive strength (cylinders):

Normal range: 2,500 to 6,000 psi in 28 days without fly ash
2,000 to 6,000 psi in 90 days when 20 to 30 percent of the cement is replaced by fly ash or natural pozzolan
Extreme range: 500 to 13,000 psi at one year

Unit weight of PPAC with coarse aggregate blended for 40 percent voids will be approximately the same as for conventional concrete made of the same materials. However, where the coarse aggregate is graded for lower voids, both the water and cement contents will be reduced (without loss of strength); their volume replacement with coarse aggregate may result in some increase in unit weight.

Drying Shrinkage. The piece-to-piece contact of the preplaced-aggregate reduces drying shrinkage to about half that of conventional concrete made with the same top-size aggregate.

Bond to Existing Concrete. Because fluid mortar penetrates the surface pores and drying shrinkage is so low, the efficiency of bond to properly prepared existing concrete is usually assumed by the writer to exceed 80 percent of the tensile strength of the weaker concrete. This conclusion is based on the observation of a number of cores and the testing of a few, none of which broke along the surface of contact.

Durability. The resistance to alternate cycles of freezing and thawing of PPAC made with an expanding fluidifier in the mortar is approximately equal to that of good air-entrained concrete. In the absence of data to the contrary, the durability of a PPAC which lacks fluidifier or an effective air-entraining agent can be assumed similar to that of non-air-entrained conventional concrete.

Permeability. PPAC made with fly ash at approximately 20 percent of the weight of portland cement and with clean, coarse aggregate has been found to be five to ten times less permeable at 6 months than conventional concrete made with cement alone. On the other hand, the use of dust- and dirt-coated aggregates will increase permeability as well as reduce strength because the coating prevents a tight bond of mortar to aggregate.

Modulus of elasticity appears to be about the same as that of conventional concrete of the same strength when the PPAC coarse aggregate is graded to contain about 40 percent voids. Unpublished data appear to indicate that as the mortar content decreases, the modulus increases, and vice versa. Where this property is critical, it should be determined by tests using proposed job materials

Underwater Placement. Properly prepared and intruded mortar will displace still water from the voids in the aggregate mass without appreciable dilution. PPAC placed under water will equal the quality of that placed above.

Proportioning The strength and other properties of PPAC depend largely but by no means wholly on the proportioning of the mortar, assuming that all the materials in the finished concrete meet the previously stated requirements and suitable placing procedures are followed. The grading limits of coarse aggregate are determined by the dimensions and other details of the construction in which the PPAC is to be used.

Coarse aggregate will normally range in size from ¾ up to 1½ in. for structural beams and columns and up to 4 or 5 in. for bridge abutments and piers. For mass concrete in dams, lock walls, large bridge piers, and the like, the top size may be increased to minimize void content. Whether or not the minimum size is increased will depend on the grading of the sand in the mortar. Conversely, the minimum size of the coarse aggregate should be reduced to ⅜ or ½ in. for structural repairs or thin facings ranging from 3 to 12 in. thick, and a sand should be selected to suit this coarse aggregate grading.

Weight ratios of cementing materials to sand for structural-grade PPAC will usually be in the vicinity of 1:1 for beams, columns, and thin sections and will range up to 1:1½ for aggregate gradings up to 5 in. or so. When the minimum size of coarse aggregate permits the use of coarser sand and appropriate equipment is used for mixing and pumping the mortar, the ratio of cementing materials to sand may be increased to 1:3 or even higher.

Ratio of portland cement to other cementing materials may vary from 100 percent portland cement to 70 percent cement, 30 percent fly ash or natural pozzolan by weight. Within these limits, the amorphous silica in the pozzolanic material will react with the lime liberated by the hydrating portland cement in a secondary reaction to produce insoluble cementitious products. These are similar in strength to the hydrated cement provided suitable curing conditions are maintained. With portland cement alone, a plot of strength gain versus time will be similar to that for conventional concrete. As the pozzolanic proportion is increased, strength matching the 28-day strengths for the 100 percent portland-cement design will occur at progressively later periods. The time for equal strength occurs at about 90 days for 70–30 blends, and the strength curve is then rising at a steeper slope than for the cement-only curve.

For estimating purposes, the quantity of cement or cementing material will be approximately 6 sacks per cu yd for structural concrete and less for mass concrete.

Ratio of water to cementing materials by weight should be in the range of 0.40 to 0.45 (4½ to 5 gallons per sack) for structural-grade mixes and may run as high as 0.62 (7 gallons per sack) for lean, mass concrete.

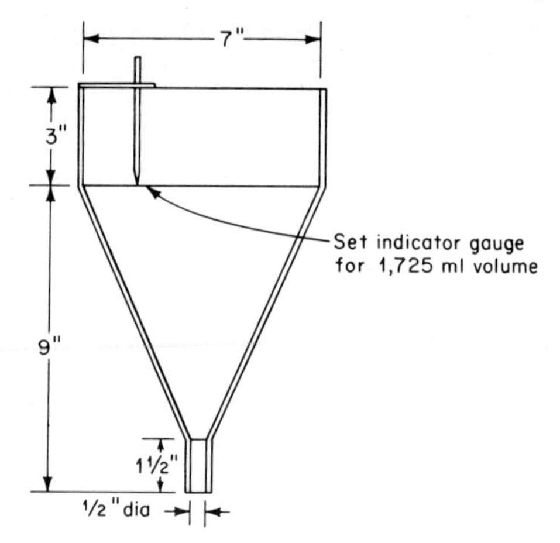

Fig. 22-1 Flow cone. Instructions for field use: (1) Be sure discharge tube is clean and undamaged. A dirty, nicked, or out-of-round tube will not give an accurate reading. (2) Wet cone with water before using. Shake out excess water. (3) Hold cone firmly and vertically. (4) Place finger over discharge tube and fill to proper level with grout. (5) Remove finger and look directly down into the flow cone. Note time required for cone to empty (when you first see light through discharge tube). (6) Wash flow cone thoroughly after use.

Grout fluidifier should be proportioned in accordance with the manufacturer's requirements, modified as necessary to provide:

1. A water reduction of at least 5 percent from the same mix without fluidifier.
2. A minimum expansion in a period of 2 to 3 hr of (*a*) 2 percent for concrete not requiring freeze-thaw durability or (*b*) 5 to 8 percent for concrete exposed to freezing or a combination of expansion exceeding 2 percent and air-entrainment totaling 5 to 8 percent.

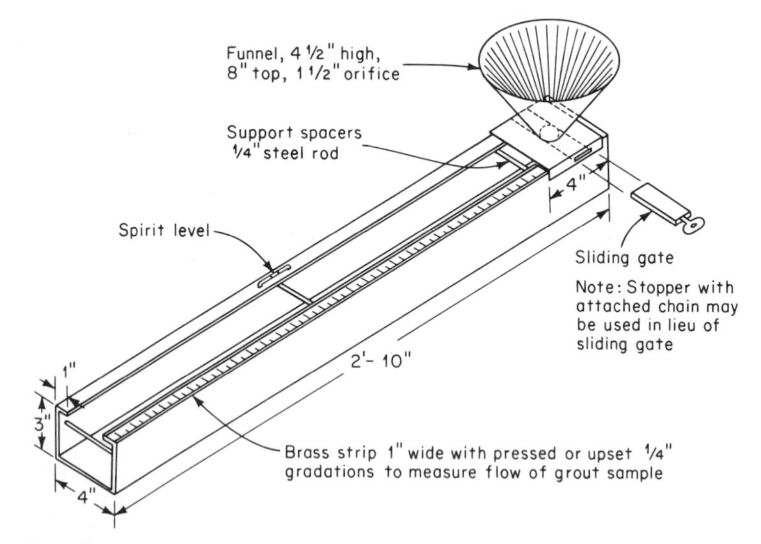

Fig. 22-2 Flowmeter. First fill with 1.20 U.S. quarts of grout. Then open gate and read flow of grout on scale.

3. Bleeding not exceeding one-half the expansion or 1½ percent, whichever is less. The grout proportioning or sand grading may have to be modified to meet this requirement.

Grout consistency for preplaced coarse aggregate having a minimum size of ½ or ¾ in. should be 20 ± 2 sec when measured by the flow cone shown in Fig. 22-1. Using the same sand, i.e., fineness modulus (FM) not over 2.1, the flow-cone consistency may be increased to 30 sec for 1½-in. minimum-sized coarse aggregate and decreased to barely flowable for 6-in. stone. The consistency of thick mortars (over 50 sec) and those containing coarser sands must be controlled by some other means, such as the flowmeter (Fig. 22-2) developed in Europe for coarse-sanded mortars. In this meter, consistency is measured by the distance a standard volume of mortar flows along a rectangular channel. Flows of from 10 to 22 in. are normal.

Having determined the proportions of solids and water to provide the required strength at the desired consistency, the cone or meter is used during the work to maintain the selected water content.

CONSTRUCTION PROCEDURES AND EQUIPMENT

In planning PPAC construction, the contractor can achieve a successful result at minimum cost if he recognizes that, although the method is basically a simple one, there are important differences with respect to conventional concrete procedures. These are described below in the order in which the work is accomplished.

Preliminary Work This will include erection of forms, setting grout pipes and reinforcing steel, and placing the coarse aggregate within the forms.

Forms must be tighter and generally stronger than for conventional concrete.[5] Joints less than 1⁄16 in. between boards, plywood sheets, or steel panels will be self-sealing; wider joints must be caulked with rags or similar water-permeable material, mortared, or stripped on the inside with tape. Where forms are tailored to meet a rock or other rough surface, the sealing should be carefully done and adequately braced to avoid blowing. Grout, water, and air-inflated tubes have been used on occasion.

Forms should be designed to withstand pressures developed by the coarse aggregate plus the fluid head of grout at the level under consideration, usually the bottom. Ketchum's grain-bin formulas or other methods for granular materials have been used for estimating aggregate loads. The mortar pressure will be approximately 1 psi per ft of height of fluid mortar in the forms, and thus will depend upon the rate of pumping and the stiffening time.

Grout pipes and reinforcement may be placed before, during, or after the erection of forms. Grout pipes may also be driven for short distances through in-place coarse aggregate.

Placement of coarse aggregate may be accomplished by any convenient means that will not cause segregation or excessive breakage. Crushed aggregates should not be dropped more than about 5 to 7 ft without prior testing for breakage; hard gravel can generally be dropped 10 to 12 ft without damage. Belts, chutes, bottom-dump buckets, wheelbarrows, and even human hands have been employed. Whatever the method, the aggregate must not be allowed to cone in the forms; i.e., the aggregate must be brought up essentially level.

An exception to the dropping limit occurs when the aggregate is placed through water. This provides a safe terminal velocity for any depth of fall. Segregation is avoided by continuous movement of the placing device.

When placed around blockouts, closely spaced reinforcement, under turbine scroll cases, tunnel linings, and the like, the aggregate may be rodded, shoved, or blown into final position. Vibration for consolidation is seldom employed, the cost of the mortar being usually less than the cost of filling the 1 to 2 percent or so of void reduction gained thereby.

Batching and Mixing Mortar Weight batching with conventional equipment is used for volume work. Since the coarse aggregate fraction of the concrete is already in place and usually accounts for over 60 percent of the total volume, the mortar batching and mixing equipment will be much smaller than that required for ordinary concrete.

Hence, any batching procedure, including hand handling of whole sacks of cement and cubic-foot boxes of sand, is acceptable if suitable controls are set up to maintain accuracy of proportioning.

Mixing for large-volume work is preferably accomplished in horizontal drum mixers with rubber-tipped paddles arranged to sweep the walls of the drum. Capacities ranging from 1 to 6 cu yd have been used. Vertical-shaft paddle mixers have also been used for smaller work. Generally, two mixers or "tubs" are set side by side so that while one is mixing, the other is agitating as it delivers grout by gravity to the funnel of a common pump. In the United States, the mortar has also been batched and mixed at ready-mix plants, or batched there and transit-mixed in conventional concrete trucks enroute to the job.

Materials are added in the order of water, grout fluidifier, and other solids for large mixers; water, fluidifier, cementing materials, and sand for small mixers. The mixing cycle is 3 to 5 min, but grout containing fly ash and a fluidifier may be held at agitating speeds at temperatures below 70°F for as long as 2 hr without harm. Without fly ash, 1 hr of agitating time is a suitable limit for 3,000-psi concrete. The controlling factors are maintenance of pumpability and an excess of expansion over bleeding at the time the grout is intruded. For 4,000-psi concrete and up, the holding limits should either be set at 30 and 15 min, respectively, or determined for the materials actually being used.

A high-speed shear mixer[6] developed in Europe provides an excellent suspension. Water, admixture if any, and cement are first circulated through a specially designed high-speed centrifugal device which produces a smooth grout having some of the characteristics of a colloidal suspension. For small work (one or two sacks of cement per batch), sand is added to the same drum; for a greater production rate, the grout is passed to a second drum for the addition of sand. The larger equipment is reported to have an output rate of some 240 cu ft per hr. Because high-speed mixing requires a lot of power, it is essential that the mixing cycle be limited to 20 to 30 sec per batch. If the grout is not quickly discharged to a holding agitator or tank, it will heat up and thus increase the mixing water requirement.

Transporting the Mortar The mortar may be moved from the mixing plant to the point of use by pumping, trucks, or gravity. Of these alternatives, pumps are most frequently used. In addition to their convenience, they are able to meter the grout, provide controllable pressure and delivery volume, furnish the mortar to otherwise inaccessible locations hundreds of feet above or below the mixers, and provide a good measure of control against the inclusion of air slugs.

For long hauls, ready-mix trucks or tanks with agitators may also be used. When this method is employed, the mortar should be discharged into a holding tank, also agitated, from which pumps or gravity can deliver to the inserts.

In some instances, grout may be delivered advantageously by gravity, such as down a mine shaft or from a mixing plant located well above the use point. Gravity lines should be free flowing (i.e., without valves) and vented at the upper end to ensure against the creation of a strong vacuum which might dry the mortar and cause a plug or collapse a hose. They should discharge into a receiver tank with an agitator to allow air bubbles to escape before the mortar goes to the pumps.

Grout pumps may be of either the piston displacement or screw types. Air-driven piston pumps equipped with rubber pistons and rubber valves or valve seats are preferred over those driven by engine and crankshaft because the man at the insert can shut off the flow without damaging the pump. Air-driven pumps can be selected that will deliver from less than 1 to over 5 cfm, depending on fluid and piston sizes. Their delivery pressures range from 50 to 1,000 psi or more, depending on the diameter ratios of the fluid and air pistons. Engine-driven pumps can deliver from very small quantities to as much as 25 cfm at pressures to over 5,000 psi. The use of dual or triple pistons will minimize pulsation in the grout lines.

Screw pumps also find wide use, especially for the smaller jobs. The screw pump has the advantage of delivering without throb, but rotor and stator replacement becomes a problem with hard, sharp sand.

For high lifts or long distances between base pump and point of use, relay pumping setups may be necessary. The frequency of relays will depend upon available pump capacities. For commonly available equipment, relays at 500 ft vertically or 1,000 to 2,000 ft horizontally are reasonable starting points for planning. A relay station should include a receiving agitator to feed the next set of pumps and thereby eliminate pulsation interference.

Grout lines should be sized to provide for a mortar velocity in the vicinity of 2 to 4 fps. Although properly proportioned grouts will not segregate, this velocity will keep the mortar "alive" over the whole cross section of the conduit and thus minimize grout buildup around its perimeter. Oversized lines encourage slow choking; undersized lines waste pumping effort. For short lines, 100 ft or less, and for small quantities, 1-in. hoses are adequate; for larger volumes and longer lines, pipes or hoses up to 3 or 4 in. should be considered. The writer has dropped mortar as far as 2,500 ft through 2-in. vertical pipe in mining applications and pumped it uphill in large volumes through 4-in. lines at 30° for 1,900 ft (900-ft rise) without difficulty.

All grout should be passed through a ⅜- to ½-in. screen at some point between the mixer and pump to remove lumps, stones, tramp iron, pieces of wood, etc., that always seem to get into the mix sooner or later.

Valves of the plug type, providing unrestricted flow, should be installed on the insert ends of all grout supply hoses. Wyes with a valve on each arm may be used to connect a grout supply line to two inserts. Although mortar will be pumped through only one insert at a time, the wye will permit changing connections without pause in the grout flow. Unions or full-section quick couplings are necessary for making connections to inserts. All grout-line joints and valves, whether at the insert or elsewhere, must be drip tight or plugging will occur.

Grouting Methods Specifications normally require that the mortar be intruded into the preplaced aggregate through pipes extending from the surface to the lowest point in the mass or, for walls, directly by pipe connection to holes in the forms, again commencing at the bottom. The pipes are usually withdrawn (but may be left in place) as the mortar rises, care being taken to ensure that the lower ends are 2 or more ft below the surface of the grout. Horizontal pipes are occasionally used to reach under flat surfaces. They may then be withdrawn through the forms or, alternatively, left in place after being unscrewed from a vertical section to which they may be attached.

Vent pipes are used to bleed air and water from trapped sections such as overbreaks in the crown areas of tunnels. These pipes must be carefully located to reach the high points if such places are to be filled. Pipes for this purpose may be a size or two smaller than the grout pipe. They are left open during pumping until good grout returns from the open end, then closed and left in the concrete. If the visible ends are unsightly, a coupling may be placed 6 in. from the surface and the end unscrewed as the mortar attains initial set.

Grout pipes are usually placed along with the reinforcement before the coarse aggregate, but the writer has found it possible to drive them afterward as much as 20 ft into 5-in. aggregate under water. Grout-pipe spacing will depend on the grading of the coarse aggregate and the configuration of the mass, including height of lift. For ½-in. minimum size aggregate, the spacing should be 5 ft or less; for ¾-in., the spacing may be in the range of 5 to 15 ft, the greater spacing being used in deep masses of aggregate. In an unreported test case, the writer found no loss of strength or other readily observable properties in cores of PPAC taken at 5-ft intervals up to 20 ft from an insert pipe, the limit of the experiment.

Grout pipes may vary in size from ¾ in. for patching; 1 or 1½ in. for medium to fairly large placements; and up to 2 or 2½ in. for very large masses, particularly at depths over 50 ft where the larger diameter is needed for pipe strength. Larger-sized pipe up to 3 or 4 in. is advisable where concrete sand is used in the mortar or where the minimum size of coarse aggregate permits the pumping of very thick (barely flowable) mortars.

The rate of mortar rise in the aggregate mass should be slow enough to minimize cascading, which can occur when rapid rise through a large void permits fallback over smaller ones. This tends to trap air, and encourages dilution of the grout under water. In the writer's opinion, the optimum rate of rise at any point is not over 1 fpm and the maximum permissible is 2 fpm. Production is obtained by keeping the grout pipes 3 to 5 ft below the grout surface and pumping through more than one insert at a time so that mortar rises simultaneously over a wide area.

The grout surface is easily followed by the observation of "color" at seeps in the forms or through the use of sounding wells located at strategic points in the mass. Continuously slotted (unwelded or cut and reamed) pipe is often used for this purpose, the grout surface being located by plumb line and weight. For underwater placements, the weight of the plumb may be adjusted so as to sink through the water but still float on the mortar.

Figure 22-3 shows a turbine scroll case during consolidation of the first lift of preplaced aggregate and illustrates the use of inserts and sounding wells. The smaller pipes with half unions attached are inserts, and a grout hose is connected to one which

Fig. 22-3 Turbine scroll case during mortar pumping operation. (*Prepakt Concrete Co.*)

slopes to reach under the scroll case. The larger, more widely spaced pipes are sounding wells through which a workman is feeling for the grout surface with a light, wooden rod. The open ends of all pipes should be capped, as some still are, during aggregate placement.

When estimating the position of the grout surface between sounding wells, it may be assumed that a 20-sec grout will take a slope of approximately 1:4 or 1:5 above water and 1:8 to 1:10 under water for ½-in. and ¾-in. minimum-sized aggregates, respectively.

Although grouting mortars using coarser sands may not appear quite as fluid, the larger voids in the coarse aggregate grading required for use of such mortars will have a compensating effect. Slopes for these mortars may also be assumed to be in the 1:5 to 1:10 ranges for above- and below-water placements, respectively.

Grouting by gravity delivery through inserts is frowned upon because of the probability that air will enter the inserts between batches of grout and become trapped in the preplaced aggregate. Above water, entrapped air bubbles are the equivalent of honeycomb. The same applies to PPAC which is placed under water, except that the air that does bubble out will cause serious dilution as it breaks through the grout surface.

A second grouting method which is suitable for shallow lifts (5 to 10 ft) and large minimum-sized aggregate (4 to 5 in.) consists of pouring the mortar onto the aggregate. To avoid honeycomb from entrapped air, grouting should start at one corner and continue at that point until the mortar shows at the surface. The pouring point may then be moved forward to maintain an advancing mortar slope. This method cannot be used for underwater placement since grout falling even a short distance through water will distintegrate.

Repairs to Existing Concrete When the PPAC method is to be used to restore a deteriorated concrete beam, column, or other concrete structure, the first step is to remove the weakened material by chipping or sandblasting. This will expose the sound concrete and, where so desired, can be used to roughen its surface before applying PPAC. Space must be provided for the replacement or addition of at least 3 in., preferably 4 in., of new PPAC. Forms are then placed (well anchored to the old concrete), filled with coarse aggregate of proper grading for the thickness being placed, and the grout is intruded. When the forms are filled, a closing pressure of 5 to 10 psi should be held for several minutes to drive out all air and water through a venting section at the highest point. After the forms are removed (24 to 48 hr), the new concrete must be properly cured.

Quality Control The following U.S. Army Corps of Engineers' publications specify procedures for selecting materials and determining the several factors pertaining specifically to preplaced-aggregate concrete:

CRD-C 79 Test for Flow of Grout Mixtures
CRD-C 81 Test for Expansion of Grout Mixtures
CRD-C 82 Test for Setting Time of Grout Mixtures
CRD-C 84 Test for Compressive Strength of Preplaced Aggregate Concrete
CRD-C 85 Selection of Properties for Grout Mixtures
CRD-C 566 Specifications for Grout Fluidifier
CE-1401.03 Guide Specifications for Preplaced Aggregate Concrete

The strength of preplaced-aggregate concrete can be determined reliably only by intruding preplaced-aggregate cylinders or by taking cores from the finished work. Molds consist of the usual 6-in.-diam by 12-in.-long steel cylinders, plus top and bottom plates fastened either to the cylinder or to each other with long bolts. Both plates are drilled and threaded for 1-in. pipe connections. The top plate is also perforated with a number of $\frac{1}{16}$- or $\frac{1}{8}$-in. holes. When the cylinder has been filled with coarse aggregate, with sizes over one-fourth the cylinder diameter discarded, a piece of cloth is placed over its top and the top plate fastened in place. Mortar is intruded slowly from the bottom inlet until good mortar is extruded from the top outlet of the cylinder. The valve on this outlet is then closed, and pumping is continued to build up an internal pressure of approximately 5 psi before closing the valve on the bottom inlet. The cylinder should be tapped lightly during filling to encourage the displacement of air voids. This "babying" is desirable to produce cylinders matching the quality of the more slowly intruded PPAC on an actual job. Cylinders prepared in this manner have been found to match (within ± 200 psi) the strengths of cores taken from work for which the tests were made.

Cylinders made by intrusion through a vertical insert which is withdrawn when the cylinder is full are unreliable, since the space left by the insert is large compared to the diameter of the specimen. The use of mortar cylinders[7] is also unreliable because a bleeding or nonexpanding mortar may show a high strength; preplaced aggregate grouted with this mortar is likely to show a much lower strength because of loss of bond on the undersides of the aggregate particles.

During the course of the work, the consistency of the mortar should be measured frequently by the flow cone or other procedure as a check on water content. Bleeding and expansion tests should be run at least daily to provide warning of serious variations in sand grading and changes in cementing materials characteristics. Sets of cylinders should also be made to provide verification of the strength of the in-place concrete.

The following is a list of some of the more critical precautions that should be observed for the production of acceptable preplaced-aggregate concrete:

1. Voids of adequate size for the intrusion of mortar must exist in the preplaced-

aggregate mass. The few fine particles that cannot be eliminated from the aggregate as it is placed must not be allowed to segregate into pockets.

2. The coarse-aggregate particles must be very nearly dust free to permit the mortar to contact the aggregate surface and establish bond.

3. The mortar must be fluid enough to flow through the voids yet have a low enough water content to provide strength.

4. The mortar must be free of oversized particles that may block void passages.

5. Mortar must be introduced into the preplaced aggregate in such a manner as to completely displace all air and water. This usually requires that the intrusion start at the lowest point in the aggregate mass and that means be established for the free and complete escape of air and water.

6. When preplaced-aggregate concrete is to be produced under water, the water must be clean and nearly still. Silt-contaminated water will coat the aggregate particles and reduce bond. Organically contaminated water can promote the growth of algae or deposit slime, either of which will reduce bond and may prevent the mortar from hardening. When PPAC must be placed in polluted water, the aggregate should be grouted as placed or at least within a very few days.

APPLICATIONS

Brief descriptions of projects constructed by the preplaced-aggregate concrete method may suggest advantageous uses to the engineer.

Bridge Piers The largest structure ever built using the preplaced-aggregate concrete method is the Mackinac Straits Bridge (Figs. 22-4 and 22-5) connecting southern Michigan with its northern peninsula.[8] The bridge is 18,000 ft long over water and

Fig. 22-4 Artist's phantom view of Mackinac Bridge piers. The main tower piers extend 200 ft to bedrock.

rests on 34 piers. It includes a suspended span of 8,000 ft which requires two cable anchorage piers extending 100 ft above water and two main tower piers some 200 ft deep. The total volume of concrete in the piers is approximately 450,000 cu yd.

The preplaced-aggregate concrete method was proposed by the writer to the contractor, Merritt-Chapman and Scott, and was accepted by the Mackinac Bridge Authority and its consulting engineer, Dr. D. B. Steinman, for reasons of economy, quality, and speed of construction. Concrete exceeding the specification requirement of 3,000 psi at 90 days under water was obtained using less than 3 sacks of cement and 150 lb fly ash; a slightly richer mix was used for the 4,000-psi concrete above the water surface. Since fly ash was available in bulk at very low cost, a considerable materials saving was realized. Coarse aggregate (¾ to 5 in.) was brought from a quarry 50 mi distant in 10,000- to 15,000-ton self-unloading boats which discharged the aggregate

Fig. 22-5 Aggregate being deposited from a self-unloading boat to main pier, Mackinac Bridge, and simultaneously mixed and pumped. (*Prepakt Concrete Co.*)

directly into the caissons and sheet-pile forms at rates approaching 2,500 tph. Simultaneously, and also while the aggregate boats were away after new loads, the mortar was mixed and pumped from a relatively small but completely equipped floating plant. This plant, attended only by an open barge containing sand and a covered one holding cement and fly ash, produced sufficient mortar to consolidate the preplaced aggregate at rates up to 250 cu yd per hr. The combined operation resulted in the setting of new world records for underwater concrete placement: 6,250 cu yd in 24 hr, 33,100 cu yd in 7 days, and 103,100 cu yd in 30 days. It is estimated that the contractor saved approximately $750,000 in floating mixing plant and $1,200,000 in cementing material, as well as one season of construction time from his original schedule based on using conventional concrete and tremie placement.

There is another feature which makes the PPAC method attractive for bridges. Where the piers are located in swift water or where violent storms are probable, cofferdams or forms may be quickly stabilized by the immediate placement of a large volume of coarse aggregate.

Dams For mass concrete in dams,[9] PPAC can be produced with one to two sacks of cement plus 100 to 150 lb of pozzolan per cubic yard as rapidly as forms and coarse aggregate can be placed. It is suggested that coarse aggregate be placed by conveyors or by buckets handled from cranes or cableways. When handled by buckets, it should be remembered that aggregate alone weighs about a third less than mixed concrete; hence, the lifting equipment can handle a greater volume of material per trip. Furthermore, placement of inert aggregate may be interrupted at any time without fear of producing a cold joint. Precooling of the coarse aggregate either by inundation in the forms or by chilling before placement is a simple matter.

Heavily Reinforced Concrete Where steel reinforcement is unusually heavy, the placement of coarse aggregate in the inert state offers some advantages. The stone may be rodded into place without danger of honeycomb and without harm to the finished concrete through delays in getting the forms filled. The reinforcement and coarse aggregate may also be placed in lifts while access to tight areas is possible; yet the member or group of members can be made as monolithic as desired by delaying the

Fig. 22-6 Inertia block for forging hammer. (*Prepakt Concrete Co.*)

mixing and pumping of the mortar to an appropriate time. Where heat of hydration is a factor, the aggregate and steel may be precooled in the forms by placing shaved ice on top of the aggregate and allowing the cold water to trickle down through the stone or by inundation of the aggregate with chilled water. In either event, the intruded grout will displace the water without dilution, embedding the aggregate at minimum temperature.

Figure 22-6 illustrates an inertia block for a forging hammer with post-tension cables, reinforcement, and grout pipes installed. The PPAC method was the only feasible technique for placing high-strength concrete in this intensely reinforced structure. Field-made cylinders showed strengths of 4,070 psi at 28 days, 5,850 psi at 28 days, 8,990 psi at 90 days, 10,600 psi at 6 months, and over 13,000 psi at 1 year.[10]

Figure 22-7 shows the use of crushed ice in a very heavily reinforced coking tower to

Fig. 22-7 Melting crushed ice precools coarse aggregate and reinforcement in coking tower structure. (*Prepakt Concrete Co.*)

Fig. 22-8 At AEC's materials testing reactor, barytes coarse aggregate is hand placed in tight spots around piping. (*Prepakt Concrete Co.*)

(*a*)

(*b*)

Fig. 22-9 Repair of viaduct by Prepakt method. (*a*) Before, (*b*) after. (*Intrusion-Prepakt, Inc.*)

cool both coarse aggregate and reinforcement in advance of the pumping of chilled mortar. Ice was also used prior to pumping the inertia block.

Turbine Scroll Cases In large hydroelectric power plants, the PPAC method has been used for embedding large turbine scroll cases (Fig. 22-3) to save both time and cost. Whereas the conventional procedure requires the placement of concrete in 2-ft lifts in not more than two opposing quadrants at a time plus thorough lift-surface preparation and 3-day waits between succeeding lifts, the PPAC method permits total embedment in not more than two lifts. Thus a 30-ft-deep embedment will require at least 45 days for ordinary concrete but not over 2 weeks with PPAC. Both the Corps of Engineers and the Bureau of Reclamation have specified the PPAC method as a permissible procedure for concrete backfill around large turbines.

Nuclear Reactor Containment Vessels Large pressure containment vessels, whether of the spherical or domed-cylinder types, can be assured of full undersurface contact when PPAC is used for support. The space to be backfilled is usually 3 or more ft deep at the center and may be 20 or more ft deep at the perimeter. The diameter of the embedment will vary with the size and shape of the vessel, being on the order of 70 to 150 ft. Since the space to be filled is more horizontal than vertical, care must be taken to plan fully the insert and grout-surface telltale pipe locations and the sequence of the pumping operation in advance of aggregate placement. In general, the mortar should first be brought up to contact with the low point of the sphere, then this contact should be maintained as the remaining aggregate is consolidated. In this type of backfill operation, complete filling with coarse aggregate is desirable mainly from the standpoint of minimizing the quantity of mortar to be used; all open space will be filled with mortar to provide the support.

High-density Concrete for Biological Shielding The PPAC method has been used in many instances for the placement of high-density concrete for biological shielding around nuclear reactors, as shown in Fig. 22-8. Coarse aggregates such as barite, limonite, magnetite, ferrophosphorus, and combinations of these with scrap steel such as chopped waste reinforcement and punchings have produced unit weights of 220 to 290 pcf. Unit weights up to 395 pcf are obtainable if steel is used for both coarse and fine aggregates, but the cost of the materials usually makes it more economical to increase the thickness of the shield.[11,12] Advantages of the PPAC method in this work are:

1. Uniform shielding without segregation is assured.
2. The density may be varied from point to point by changing the coarse aggregate proportions. Injection of grout in a single operation provides a monolithic result.
3. The precise location of instrumentation through the concrete can be checked before pumping the grout and while it is still possible to make corrections.
4. Absence of vibration reduces the possibility of displacement.
5. The characteristics of the grout are such that the complete embedment of instrumentation (pipes, tubes, etc.), including the undersides, is assured.

Structural Concrete Repairs The excellent permanent bond obtainable between preplaced-aggregate concrete and properly prepared conventional concrete makes it possible to restore deteriorated plain concrete or reinforced structural concrete to original condition or enlarge existing members to take additional loads. Figure 22-9 illustrates this application. Masonry and structural concrete repair is one of the widest uses for PPAC.

B. Grouting

Grouting[13,14] is a combination of engineering and art wherein fissures, voids, or cavities in rock or soil masses are filled with a fluid that will harden in place to increase the overall strength and/or impermeability of the mass. Engineering judgment is required to determine the feasibility of the work, the type of grout to be used, and the general conditions of execution, including the establishment of limitations. Successful results are also highly dependent on the experience and skill of the workmen, their immediate supervisors, and the field engineer.

Grouts may be considered under two broad categories. Those which contain finely divided solids are called "solid-suspension" grouts, and those which are pure solutions are known as "chemical" grouts. The two types are occasionally combined, as in the case of chemical grouts which contain inert fillers or, rarely, portland cement.

Solid-suspension grouts containing reactive solids such as portland cement and pozzolans, swelling materials such as bentonite, and/or inert fines such as clays are generally preferred when the flow channels in the void system are large enough to permit the solids to flow through them. In broad terms, cracks narrower than 0.01 in. or granular materials composed of particles finer than ⅛ in. will not readily accept solid-suspension grouts.

Chemical grouts, on the other hand, will flow anywhere that water can be made to move. The rate of movement is then dependent on the viscosity of the grout as well as on the pressure gradient. Chemical grouts usually consist of two chemicals which, when combined, will stiffen to a low- or moderate-strength gel. The hardening time and eventual strength of this gel are controlled by the proportioning of the reactive chemicals and the amount of water (if any), and often by a "modifier" which may be considered as a negative catalyst; i.e., the modifier controls the rate of reaction without being otherwise necessary for the basic reaction. Most chemical grouts are proprietary; the user is advised to study the manufacturer's literature for properties, capabilities, proportioning, and methods of placing.[15] For this reason, the major emphasis in the following paragraphs will be on solid-suspension grouts.

MATERIALS

The basic materials for solid-suspension grouts are cement and water. Other ingredients frequently included are sand, pozzolans, clays, and admixtures. Proprietary materials are also available for special purposes.

Cement If it meets the usual requirements for concrete,* the cement will be suitable for most grouting. For pressure grouting fine seams in hard rock, finely ground or air-separated cements are sometimes required.

Water This should also be suitable for concrete.* A minimum specification should call for potability.

Sand The sand for mortars to match concrete strengths should normally be the same as for PPAC. If lower strengths are acceptable, the sand grading becomes less critical. Where mortars containing high sand contents are to be pumped, the finer fractions are important for pumpability.

Fly Ash and Natural Pozzolans When used in mortar-type grouts where strength is important, these should meet the requirements which have been supplied in this section for PPAC. For most other purposes, the loss-on-ignition requirement for fly ash may be relaxed. Fly ash and natural pozzolans are seldom used in fine-seam grouting.

Bentonite and Other Clays These materials vary so much from one location to another that each should be investigated individually for its suitability in the particular work. Bentonites are given a "barrel rating" in the oil drilling industry; the higher

* See Material Selection and Mix Proportioning in Sec. 21, Conventional Concrete.

this rating, the more "mud" of a given consistency it will produce. For example, one ton of "90 barrel minimum" bentonite, a high classification, will produce 90 barrels (45 gal each) of 15-centipoise mud. Bentonites are used for their swelling and gelling action; clays and silts are used as fillers.

Admixtures These can be highly beneficial in many instances. Water reducers, Types A (normal setting), D (retarding), and E (accelerating), ASTM Designation C-494, will greatly improve the strength and will help with pumpability where concrete strengths are desired. Grout fluidifiers (see Preplaced-aggregate Concrete) will provide water reduction, retardation, and expansion. Calcium chloride is especially effective in speeding up the setting time. Certain proprietary lubricating-suspending admixtures are also available; these will suspend high volumes of sand in relation to cement to provide strengths ranging from over 2,000 psi to as low as 50 psi with virtually no bleeding.

Special-purpose Materials High strength with volume stability can be obtained by using special materials which are composed of iron aggregate, catalysts, and plasticizer for jobsite mixing with cement and sand. Factory-premixed combinations of these materials are also available. Widely accepted examples of the latter are Embeco Grouts and Mortars, premixed by Master Builders, which will provide 8,000 to 12,000 psi at ages ranging from 4 to 90 days, depending on the formulation. Premixed grouts of the gas-expansion type will provide strengths on the order of 5,000 to 7,000 psi combined with a measure of bleeding correction.

PROPORTIONING

Grouts are proportioned[16] to satisfy as well as possible a number of overlapping and, to some extent, conflicting conditions. As a fluid, the grout must be pumpable and be able to penetrate the material into which it is to be pumped. After it is in place, it must harden to a stable condition and provide strength and/or impermeability, all as economically as possible. The characteristics of the voids to be filled, together with the desired end result, determine which of these conditions are paramount and which are secondary. Proportioning will therefore be discussed in connection with the type of void to be filled.

Fine Seams and Cracks These are usually encountered, or at least suspected, in rock under dams and associated structures, around penstocks, and in the bases and walls of deep excavations. The objective of grouting in these areas is to prevent water movement and/or add strength to the rock mass. The openings to be grouted vary from tight discontinuities to spaces an inch or more across, with the narrower widths predominating. This makes penetrability the first objective in proportioning the grout. With this in mind, grouts for such work usually employ only cement and water and start with dilute suspensions on the order of 1 sack of cement to 30 to 40 gal of water. The grout is then gradually thickened by reducing the water-cement ratio. When open cracks are encountered, economy dictates that grouting sand be added in proportions which may range from equal parts sand and cement where high strength is desired to as much as 10 or 12 parts of very fine sand to 1 part cement where only a stable seal is needed. When sand is employed, pumpability demands that the water content be adjusted to provide sufficient consistency to maintain the sand in suspension. Pozzolans in amounts up to 30 to 50 percent by weight of the cementing material may be used with sanded grouts for economy and to improve pumpability. When used in the higher proportions, the pozzolans will delay the hardening of the grout for several hours. Lignosulfonate water-reducing retarders may be used, especially with sand grouts, to reduce the water requirement, control stiffening time, and improve pumpability. Grout fluidifiers will do the same and will add expansion. Retarders and fluidifiers are of little use for water reduction in neat cement grouts when the water content exceeds 6 to 7 gal per sack, but it has been reported that their dispersing and lubricating action increases the take.* Chemical admixtures, including fluidifiers, should be proportioned in the mix in accordance with their manufacturers' recommendations.

* See Reference 15, p. 59.

Cavity Filling The desirability of a combination of economy and pumpability indicates the use of the sanded grouts just described. Pea gravel, 1/4 to 3/8 in., may be suspended in these grouts to help fill space and choke off channels through which water may be flowing. Water-reducing retarders will improve pumpability and suspending properties; calcium chloride in amounts up to 2 to 3 percent of the weight of cement can be employed to overcome the retarding effect of the water reducers and speed up setting characteristics. Grout fluidifiers will provide expansion to hold the grout tightly against roof surfaces while the grout hardens.

Filling Vertical or Steeply Sloping Spaces In general, all the preceding remarks concerning sanded grouts apply. When the space is such that the upper surface of the grout is not restrained, the expansion contributed by grout fluidifiers is of little value since gravity will hold the grout tightly to the cavity sides.

Mixes for Precision Grouting This term includes the placing of grout under machinery, where precise alignment requires both volume stability and strength, and under column bases and the like where strength is the prime factor but volume stability is also desirable. Equal weights of activated ferrous aggregate, cement, and sand mixed with water to a pumpable or pourable consistency are generally specified for this work. If the blending is done at the jobsite, the ferrous aggregate manufacturer's instructions with respect to the selection of components and their proportioning and mixing must be very carefully followed to attain the desired results. The use of factory premixed grouts requiring only mixing with water at the job greatly simplifies this problem. For items requiring less critical support, cement-sand grouts proportioned 1:1 by weight and including a grout fluidifier are occasionally used. Units that are bolted to shims and are to receive their entire support therefrom may be grouted with a plain cement-sand mixture to keep the dirt out and prevent tools and trash from disappearing underneath.

Post-tensioned Cable Mixes Where good bond between the grout and cables is desired and the detailer provides fittings (1/2-in. minimum) adequate to permit the ingress of a good grout, cement and water mixed to a 20- to 25-second flow cone* consistency can be pumped. Water-reducing retarders will greatly improve both the pumpability and strength. Grout fluidifiers will do the same and will also add expansion, which further improves pumpability and provides protection against cable bursting from freezing. When smaller fittings are provided, as is often the case, a thinner (15 ± sec) grout must be prepared; this grout will provide corrosion protection but not much strength.

Grouting in Granular Materials[17] The problem with granular materials is generally one of getting grout to travel as far as possible, so that an area will be consolidated or a grout curtain established from the fewest possible grouting points. This puts a premium on penetrability. To achieve this, dilute grouts are used more frequently than thicker ones; since strength is not generally a controlling factor, clays and silts may be used in amounts several times that of the cement. Where only the arresting of water movement is desired, the cement is sometimes eliminated and clay or bentonite used alone. Water-reducing retarders are occasionally used as lubricants. Proportioning with respect to water-solids ratios can only be determined by field trial and modified as the work progresses. An experienced engineer can estimate the probabilities from a detailed study of grain-size distributions of soil samples and grouting materials.

MIXING AND PUMPING

Mixing Procedures and equipment for mixing solid-suspension grouts are similar to those described for the mortars used for PPAC with the exception that clays are usually handled in a pug-mill-type mixer. When bentonites are used and an immediate thickening effect is desired, it is important to mix the bentonite with the water before adding the cement. A better blend is sometimes obtainable by separately mixing water with the cement and the bentonite, then combining the two slurries. When chemical admixtures are used, they should be batched into the water before the solids.

* See Fig. 22-1 and adjacent text.

Grouts should be screened as they leave the mixer to protect the pumps from stones, pieces of wood, etc., and to minimize premature choking of the seams in rock or voids in granular materials. For thin grouts containing only cement and/or clays, a ¼-in. screen resting on a funnel attached to the pump intake can be sufficient provided someone clears off the screenings as necessary. For thicker grouts, including those containing sand, a ⅜- or ½-in. heavy wire screen is suitable. For volume work, the screen should be mounted on a holding agitator. A typical arrangement is shown in Fig. 22-10.

Chemical grouts should be handled in accordance with manufacturer's instructions.

Fig. 22-10 Grout consistency is continuously checked by the use of a flow cone. Author is shown obtaining a sample as the transit-mix truck discharges grout into the agitator through a vibrating screen. (*Civil Engineering.*)

Pumping For solid-suspension grouts, pumps are almost always of the piston-displacement type. The fluid ends of the pumps should be equipped with case-hardened steel cylinders, rubber piston heads, and rubber valves and/or valve seats, all arranged for easy disassembly and thorough, frequent cleaning. Whether air-piston or engine (crankshaft) driven, the output volume will increase and the output pressure decrease as the fluid-end piston diameter enlarges. To cover a variety of volume and pressure demands, some pumps are available with interchangeable liners and pistons.

GROUTING PROCEDURES

Grouting procedures vary widely, depending on the type of grout and the particular grouting requirement. Because of this, the coverage supplied herein can only be very general.

Hard Rock Grouting Grouting holes are drilled in the rock in a predetermined pattern, usually making use of rotary or percussion drills. Where the split-spacing grouting method is used, a first series of holes will be drilled, washed, and grouted. Successive series of holes are drilled and treated in the same fashion until the curtain or area will accept no more grout. In the progressive grouting method, holes are drilled at a predetermined minimum spacing, then washed and grouted in order from one end to the other. The second method permits better washing from one hole to the next of the seams to be grouted, but it requires somewhat more skill. It may also cost more if grout flows easily from one hole to another because many of the holes may have to be redrilled or several adjacent ones may have to be grouted simultaneously.

Washing and grouting may be done in stages as the holes are drilled, a procedure which strengthens the rock from the top down to permit better control of pressure and grout "take" as the holes get deeper. For the successive depth stages, a grout pipe and packer are inserted in the hole to a point in the zone just completed. These are then used to wash and grout the new stage.

The procedure is sometimes reversed; i.e., the hole is drilled to full depth, with grouting proceeding in stages from the bottom up. Some engineers feel this gives them a better knowledge of the rock and its seams before grouting starts; others are concerned that grout may bypass the packer and prevent it from being withdrawn, or that grout will escape all the way to the surface and limit grouting pressures.

In addition to washing the cuttings from a hole, most grouters also apply water pressure in an attempt to wash fine materials from the cracks, either to the next hole or to some distance from the hole. If circulation from hole to hole is obtained, compressed air may be alternated with water to obtain better cleaning.

Mixing may be accomplished at any convenient location. The grout is pumped to the point of use and back to the mixer in a return line, with desired maximum grouting pressure maintained by throttling at the return end. Grout is tapped off this line to the hole; a valve and pressure gauge at the top of the hole are used to control the actual grouting pressure. Closer grout pressure control and a better knowledge of grouting conditions, including "take," are effected by drawing from the grouting line to an agitator and pump at the grouting point. A mixing and pumping setup on skids close to the hole gives best overall control, since mixes can be thickened or thinned as the occasion demands without the waste and delay that a long circulating line provides.

Pressures are usually limited to 1 psi per foot of depth to minimize the possibility of rupturing the rock. When grouting badly fractured rock, benchmarks may be set and observed for signs of heave during grouting operations.

Grouting Cavities and Open Seams When these are to be filled with sanded grouts, it is only necessary to gain access at the lowest point above which positive filling is required. Pumping then proceeds from this location. Recirculating lines are generally neither desirable nor necessary, although the crew should make sure that the line is kept alive. This can be checked by wasting a bit of grout every 5 or 10 min when there are delays due to changing connections or for other reasons.

Loss of grout in channels through which water is flowing can sometimes be controlled by loading the grout with the largest-size gravel that the channel will accept along with smaller sizes, using larger grout holes for access when possible. Although at first most of the grout may be lost, the gravel will hopefully "hang up" on irregularities to build up a reverse filter. A better procedure for larger cavities, erosion channels under dams, unformable voids under bridge piers, and the like is the recently developed "Bagpipe" Grout Confinement Method which is described on page 2-27.

Precision Grouting Machine and column bases may be grouted with precision by pouring if they are small and flat bottomed (narrowest dimension not over 4 to 6 ft), otherwise pumping is required. For pouring, the vertical space between the concrete and the plate should be at least 1 in. for a width of 2 ft, increasing by 1 in. for each fraction of 2 ft wider; for pumping, the space should be 2 in. or more. Pea gravel, which contributes to economy and will absorb some of the heat of hydration, may be added to grout placed in thicknesses of over 2 in. Any space over 6 in. (8 in. with gravel in the grout) should be filled in two or more stages, with 24 to 48 hr between stages to allow for escape of the heat of hydration.

Prior to setting the bed plate, the concrete should be roughened and the underside of the plate cleaned of grease, dirt, and loose rust. For pouring, the plate should be formed and the grout placed as shown in Fig. 22-11 so that the material flows under the short dimension of the plate. A steel strap should be inserted from the far side and worked back and forth the minimum amount necessary to encourage the grout to flow through and rise at least an inch over the underside of the plate.

If the space is to be pumped, a hose or pipe may be inserted along the center line in the long direction and slowly withdrawn as the area is filled, using straps from both sides to bring the grout to both edges. The side forms should be vertical, with the top edges even with the top of the plate so that short boards about 2 ft long can be nailed over the edge spaces to retain the grout and force it to flow in the desired direction.

Vibrators should never be used to move grout for precision grouting because they encourage settlement of the heavier particles and bleeding of the mixing water. The latter will leave more space between the grout and plate than the shrinkage-correcting properties of the grout can overcome.

Before grouting is started, the concrete should be flooded with water for at least 6 hr so that water will not be drawn from the grout when it is placed. Most of the free water should be blown out before pouring, but minor remaining amounts will be displaced by the grout without harm to the grout.

Procedures for curing, trimming, and protecting the exposed edges of ferrous aggregate grouts vary with the formulation of the grout. The manufacturer's instructions should therefore be carefully studied and followed. Some manufacturers provide free job service with their materials; the user is advised to call for it when available.

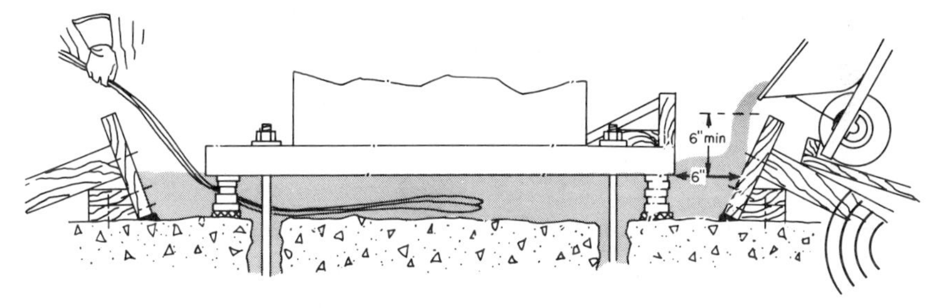

Fig. 22-11 Procedure for forming and grouting under machine bases. *(By permission of Master Builders.)*

Post-tensioned Cable Grouting The pump, hoses, and cable duct are flushed with water; the duct is blown clear; then, with the hose disconnected from the duct, grout is pumped until all water is displaced from the pump and hose. Then the hose is connected to the duct and pumping starts.

Key features for successful work and a smooth operation are:

1. Use of a screened grout so that lumps do not plug the small cable grout inlet.
2. A slow, steady flow of grout, free from pulsations if possible, continuing without interruption from start to finish for each cable duct.

Grout should be pumped until all air and water have been displaced from the duct, as evidenced by an outflow of grout which has the same consistency as that in the mixer and is free of bubbles. The outlet is then closed, followed by closure of the inlet valve with some pressure on the line. If the grout does not contain a gas expanding agent, such as aluminum powder, some engineers prefer a closing pressure of up to 100 psi to compress stray bubbles left in the duct. If the grout does contain an expansion agent, closing pressures over 10 to 20 psi will prevent the desired expansion from occurring.

Grouting in Granular Materials Procedures for grouting granular materials will vary widely with the nature of the material to be grouted and with the end result desired. In most cases, a pipe or casing will be driven to the desired depth. Drilling will be necessary to get through boulders, in which event the hole will reduce in size as each drilled area is passed. An alternative method is to wash the fines from a zone 2 to 10 ft deep under the rock and grout the gravel remaining in the pocket thus created with a grout that will provide sufficient strength to permit continued drilling. Chopping bits and churn drills are used to get through gravels and cobbles; if the gravel contains sufficient fines to retain drilling muds, the muds will both temporarily stabi-

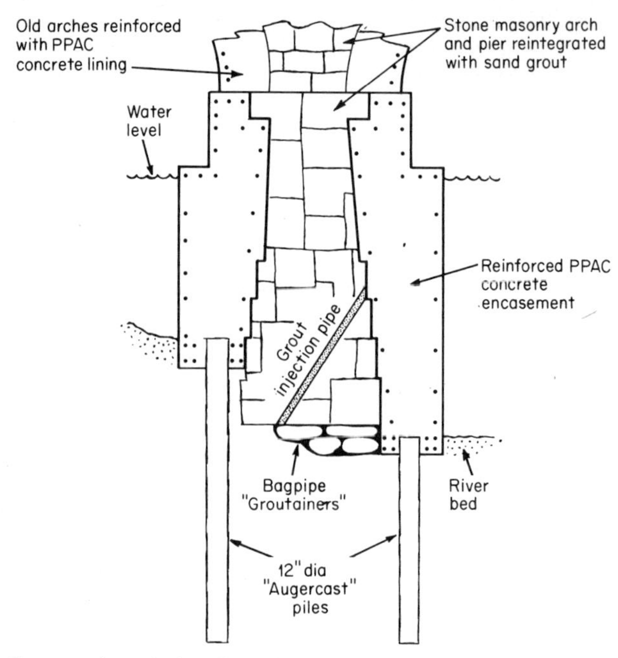

Fig. 22-12 Cross section of pier, Great Northern Railway Bridge, Minneapolis, showing repair procedures. (*Turzillo Corporation.*)

Fig. 22-13 Two piers of Great Northern Railway bridge, Minneapolis, were undercut by record flood in spring of 1965 and settled as much as 14 in. Bridge was saved by application of techniques shown in cross section of pier. (*Turzillo Corporation.*)

lize the walls of the hole and assist in floating out the cuttings. For limited depths in silts, sands, and fine gravels, one can mix grout with the *in situ* material by pumping grout through a rotating shaft having a mixing head on its lower end.

Reintegration of Cracked Structures Old masonry piers, sometimes stone filled, and cracked concrete piers and abutments have often been strengthened and reintegrated by grouting procedures.

In the general situation, the foundation stability of a pier is first checked and corrected where necessary by filling voids under the base with grout and by taking appropriate measures to prevent further scour (see Figs. 22-12 and 22-13). The portion below water may then be encased in conventional or preplaced-aggregate concrete. If the pier is of the masonry type, joints above water are repointed, holes are drilled to the interior, and all voids are filled with a neat cement grout having some "body" or a rich sand grout if the take is sufficiently free.

Cracks in concrete piers and abutments are filled in the same manner. However, if tensile stresses are to be resisted, holes may be drilled perpendicular to the cracks and reinforcing rods grouted in. Should these bars extend clear through the structure, they may be prestressed before grouting for additional efficiency.

SPECIAL APPLICATIONS

A few examples of grouting work will be cited to indicate the broad usage for grouts in construction as well as to illustrate some relatively new methods that may be useful to both the engineer and contractor.

Sand Grout as Concrete As previously indicated, easily pumpable sand grouts can be proportioned with the assistance of appropriate admixtures at water-cement ratios to match the strength and durability properties of conventional concrete. Taking advantage of this property, a steel penstock to be installed in an existing concrete-lined tunnel, for instance, can be enlarged to leave only a 1- to 6-in. annular space for back-

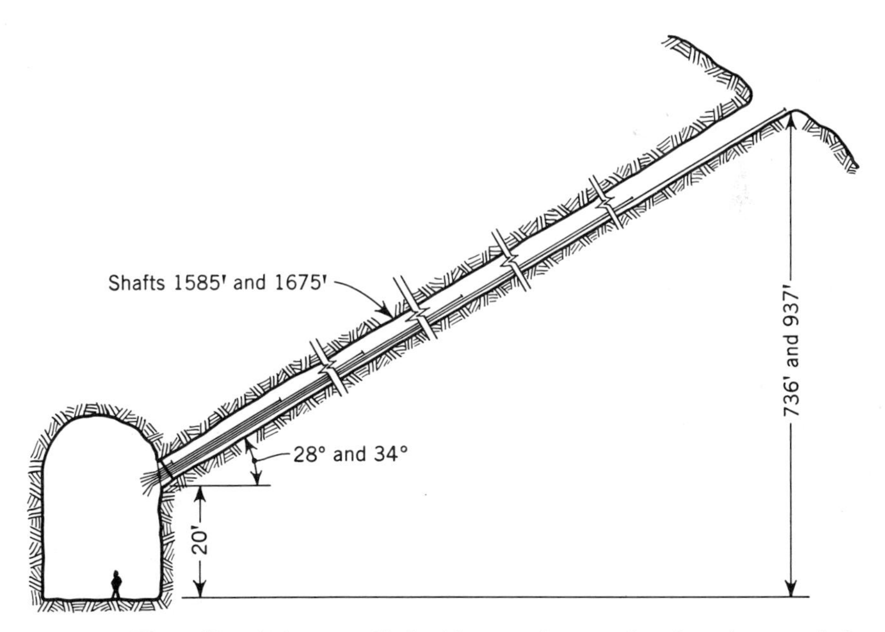

Fig. 22-14 The utility shafts were filled with grout by pumping through successively longer pipes. One shaft, 1,585 ft long, required 1,600 cu yd of grout; the other, 1,665 ft long, took 1,700 cu yd. (*Civil Engineering.*)

filling with a sand grout in contrast to the 12 in. or more normally required for pumped concrete.

In a different situation, at the NORAD* Center,[18] it was necessary to backfill two 5- by 5-ft shafts leading from an underground room to the surface. Each shaft was nearly 1,900 ft long and sloped upward at an angle of approximately 30° from the horizontal, as shown in Figs. 22-14 and 22-15. The rise was about 900 ft. The contractor decided that placing concrete from the surface would be very costly and slow because of extremely difficult access to the adit. Instead, he chose to fill the shafts by pumping a fine-grained concrete—i.e., sand grout—from the bottom, inside the mountain. After placing 4-in. grout pipes extending successively 80 ft, 400 ft, 800 ft, 1,200 ft, and 1,600 ft up each shaft, he constructed bulkheads and pumped 35-ft-long plugs, using a rich mix which attained a strength of approximately 4,000 psi in about 7 days as determined by impact-hammer testing. Pumping for shaft filling was then started

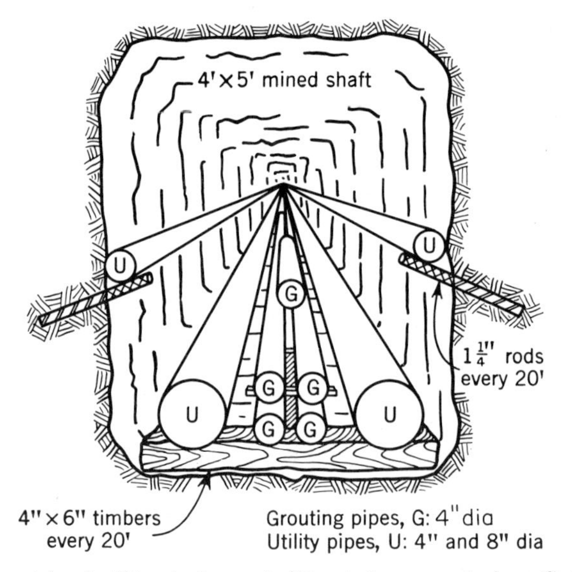

Fig. 22-15 Isometric of utility shaft near bulkhead shows grout pipes, G, through which high-strength grout was pumped to fill the shaft and embed the utility pipes, U. All pipes were strapped to their supports. (*Civil Engineering.*)

by connecting to the shortest pipe and continued until the grout returned through the 400-ft pipe. The grout connection was then transferred to this pipe and pumping resumed until grout discharged from the next-longer pipe. At this time the hose connection was changed again and the process repeated until an inspector at the top reported by phone that the shaft was full. The procedure was repeated on the second shaft.

The grout was batched into 6-yd transit-mix trucks at a ready-mix concrete plant 11 mi away and mixed enroute. At the work site, the grout was discharged through a vibrating screen into an 8-yd agitator. The grout was pumped by a three-cylinder ram pump† having a reported capacity of 60 cu yd per hr at 8,000 psi. The actual rate averaged 25 to 30 cu yd per hr at normal pumping pressures running up to 1,500 psi. The pumping time for each shaft was from 50 to 60 continuous hours.

Grout mixes and strengths are given in Table 22-1. The rich mix was used for the plugs; the lean mix for the lower half of each shaft; and the intermediate mix for the upper sections to reduce pumping pressures.

* Combat Operations Center, North American Air Defense Command.
† Oilwell 48P-HD, 4½- by 8-in. triplex, single acting.

It is believed that two world records for concreting were established on this job. The first is for the maximum height that concrete has ever been pumped in a single lift; i.e., without pumps at intermediate stages. The second is that the columns of concrete in the shafts are the highest and longest lifts of concrete ever placed monolithically.

TABLE 22-1 Results of Compression Tests (on 6- by 12-in. cylinders)

Zone of use	Strength, psi			Mix, nominal
	7 day	28 day	90 day	
Plugs.	. . . . .	5,000	. . . . .	282 lb Ideal Type II cement 75 lb fly ash 400 lb sand 1.00 lb Pozzolith Retarder
Lower half of raise	1,500	3,000	4,700	188 lb Ideal Type II cement 75 lb fly ash 600 lb sand 0.75 lb Pozzolith Retarder
Upper half of raise	1,800	3,500	5,000	188 lb Ideal Type II cement 75 lb fly ash 400 lb sand 0.75 lb Pozzolith Retarder

SOURCE: *Civil Engineering*, October, 1963.

Sand Grout in Special Containers Two procedures utilizing flexible, permeable containers as forms for concrete-quality sand-cement grouts have recently been developed. One of these, the "Bagpipe" Grout Confinement Method,* was conceived initially to plug holes under dams so that normal grouting procedures could be used to reestablish watertightness and structural stability. The method lends itself to underpinning operations and to the filling of larger-size voids and crevices, especially under water, where forming to confine the grout is difficult. In using the process, a porous, flexible bag (a "Groutainer") is placed in the void. Sand grout is then injected and the Groutainer expanded to fill the space. Sufficient grout oozes through the permeable bag to bond and seal the container to the surrounding rock or soil. When the grout hardens, the result is a concrete-filled cavity as illustrated in Fig. 22-12.

The Groutainer may be furnished to the job in various sizes or fabricated at the site to suit the application. To fill large openings, any number of bags may be stacked one atop the other and inflated in order from bottom to top. When swift water is to be stopped, as in a sizable leak under a dam, the Groutainer is placed in a heavy wire cage and the combination inserted in collapsed condition. Inflation expands both; the cage provides a handle to control the depth of insertion and permits higher grout inflation pressures. The reinforcement cage may also be placed inside the grout bag when desired.

A somewhat similar construction procedure involves a process known as the "Fabriform"† concept. The method is essentially one of forcing structural grout or concrete into porous fabric forms under light pressure and maintaining the pressure for a short period of time in order to force out excess mixing water while retaining the

* Patent and trademark owned by Turzillo Corporation, Brecksville, Ohio.

† U.S. patent applications and trade name by Construction Techniques Incorporated, Cleveland, Ohio.

cement, thus reducing the water-cement ratio. This induces rapid stiffening and high early strengths.

Fabriforms are made of a high tensile-strength fiber such as nylon, usually woven so as to limit cement loss to a negligible amount. The forms are fabricated to produce spheres, elongated cylinders, or toroids (doughnut shapes) when inflated with grout at pressures limited by the diameter and strength of the fabric.

The application of flexible forms of either type to construction problems is limited only by the imagination. For instance, a fabric mattress configuration has been laid on a river bank from above high water to a point below scour action, then inflated with grout to provide revetment protection (Fig. 22-16) against bank erosion from currents and waves. Worm-eaten timber piles have been restored by placing a sleeve around the pile to span the weakened section; the sleeve is closed with a zipper and the ends are banded, then filled with a concrete-quality sand grout. It has been suggested that circular bridge piers be formed by lowering toroids inside or outside of temporary guides, stacking them one above the other as they are inflated. When

Fig. 22-16 Fabriform mattress used for bank protection. Quilted configuration provides permeability for relief against uplift. (*Civil Engineering; Construction Techniques, Inc.*)

the grout is hard, the enclosed volume can be filled with tremie or PPAC. The concept is also obviously applicable to lining or plugging tunnels, forming for encasement of bridge piers, and as a jacking device for raising and underpinning structures.

Augered-in-place Piles These may be considered as a grouting procedure in which a column of concrete-strength grout is constructed in the ground by running a full-flight auger to the desired depth (Fig. 22-17), then pumping sand-cement mortar through its hollow shaft to fill the space created as the auger is withdrawn. These piles or grout columns may be reinforced when necessary by inserting a rod or simple cage into the soft grout immediately on removal of the auger. By using an appropriate tip, the piles can be seated several feet into shale and moderately hard rock.

The method was invented* for foundation stabilization for situations where grouting could not provide sufficient strength and the vibration and heaving from driven piles could not be tolerated. When it was found that the capacity of the pile could be estimated from the shear value of the soil in contact with it and the cost was found to be competitive with driven piles, use of the pile spread to new construction including building support and cofferdams (Fig. 22-18).

Grout columns are also used for groundwater cutoffs and for arresting seepage through small earth dams. For this work, the piles are placed tangent to each other as

* Patented procedures used by Turzillo Corporation, Intrusion-Prepakt, Inc., and licensees.

Fig. 22-17 St. Thomas Hospital expansion project, Akron, Ohio. Auger is placing foundation piles in excavation next to an underground service tunnel. Note rough surface of cast-in-place pile wall used as shoring in background. (*Turzillo Corporation.*)

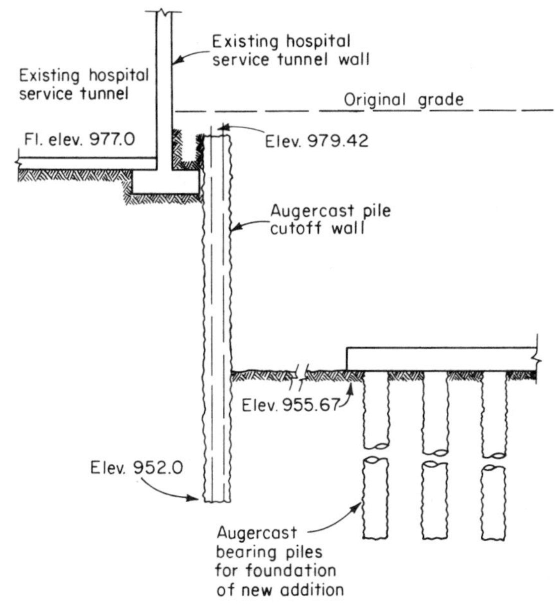

Fig. 22-18 Section showing use of Augercast piles on St. Thomas Hospital expansion project for cutoff wall and for foundation of new hospital addition.

shown in Fig. 22-17. They may also be placed to overlap into each other 2 or 3 in. to form a grout-concrete wall.

Augered-in-place piles 12 to 18 in. in diameter are readily placed to depths of 60 ft or so and have been placed to over 100 ft.

Mixed-in-place Piles The mixed-in-place pile* is produced by forcing a low-water-content cement grout through a rotating shaft to a mixing head which blends it with the material in place, utilizing the native gravel, sand, silt, or clay as the aggregate. The unit strength of the hardened column will vary with the foundation material and the amount of grout pumped: from the equivalent of first-class concrete to as low as 500 to 1,000 psi. Lateral penetration of the grout provides high frictional resistance, giving the pile potentially greater load-carrying capacity than driven piles, particularly in silts and clays. Mixed-in-place piles range in size from 12 to 24 in. and have been placed as deep as 50 ft.

This type of pile derives its economy from the use of the material *in situ,* which reduces aboveground requirements for coarse aggregate and sand. Placement is also quite rapid, only 10 to 15 min being required to place a pile to a depth of 30 ft.

Mixed-in-place piles, tangent to each other or intermixed to overlap 3 to 6 in., have been used as shoring for excavation next to existing buildings and as cutoffs under dams placed on permeable riverbed sands and fine gravel.

References

1. Preplaced Aggregate Concrete for Structural and Mass Concrete, Report by ACI Committee 304, *ACI Journal,* Proc. vol. 66, no. 10, October, 1969 and April, 1970.
2. *Concrete Manual,* 7th ed., U.S. Bureau of Reclamation, Denver, Colo., 1963.
3. Investigation of Resistance of Preplaced-aggregate Concrete to Freezing and Thawing, Misc. Paper C-68-6, U.S. Army Waterways Experiment Station, Vicksburg, Miss.
4. Raymond E. Davis, Prepakt Method of Concrete Repair, *ACI Journal,* Proc. vol. 32, no. 2, August, 1960.
5. ACI Standard 347-68, Recommended Practice for Concrete Form Work, sec. 5.2.
6. D. F. Orchard, *Concrete Technology,* vol. 2, p. 443, John Wiley & Sons, Inc., New York, 1962.
7. John C. King, Strength Evaluation and Quality Control of Prepacked Concrete, *ACI Journal,* Proc. vol. 64, no. 11, November, 1967.
8. R. E. Davis, Jr., and C. E. Haltenhoff, Mackinac Bridge Pier Construction, *ACI Journal,* Proc. vol. 53, no. 31, December, 1956.
9. U.S. Army Engineer Waterways Experiment Station, Investigation of the Suitability of the Prepakt Concrete for Mass and Reinforced Concrete Structures, T.M. No. 6-330, October, 1951.
10. A. M. Klein and J. H. A. Crockett, Design and Construction of a Fully Vibration-controlled Forging Hammer Foundation, *ACI Journal,* Proc. vol. 49, no. 29, January, 1953.
11. Harold S. Davis, Frederick L. Browne, and Harry C. Witter, Properties of High Density Concrete Made with Iron Aggregate, *ACI Journal,* Proc. vol. 52, no. 44, March, 1956.
12. Harold S. Davis, High Density Concrete for Shielding Atomic Energy Plants, *ACI* Journal, Proc. Vol. 54, no. 56, May, 1958.
13. Report on Cement Grouting, Committee on Grouting, *J. Soil Mechanics and Foundations Div., Am. Soc. Civil Engrs.,* vol. 88, proc. paper 3098, April, 1962.
14. Bibliography on Cement Grouting, Committee on Grouting, *J. Soil Mechanics and Foundations Div., Am. Soc. Civil Engrs.,* vol. 89, no. SM4, proc. paper 3575, July, 1963.
15. Guide Specifications for Chemical Grouts, Committee on Grouting, *J. Soil Mechanics and Foundations Div., Am. Soc. Civil Engrs.,* vol. 94, no. SM2, proc. paper 5830, March, 1968.
16. Thomas B. Kennedy, Symposium on Grouting: Research in Foundation Grouting with Cement, *J. Soil Mechanics and Foundations Div., Am. Soc. Civil Engrs.,* vol. 87, no. SM2, proc. paper 2794, April, 1961.
17. John C. King and Edw. G. W. Bush, Symposium on Grouting: Grouting of Granular Materials, *J. Soil Mechanics and Foundations Div., Am. Soc. Civil Engrs.,* vol. 87, no. SM2, proc. paper 3486, April, 1961.
18. J. D. Davenport and J. C. King, "At NORAD Center—New Techniques for Backfilling Shafts," *Civil Engineering,* October, 1963.

* Patent rights owned by Intrusion-Prepakt, Inc.

Section 23

Steel Construction

JAMES M. SCOTT, JR.

Manager—Construction Equipment, American Bridge Division, U.S. Steel Corporation, Ambridge, Pa. (*Steel Erection*)

JEAN J. REVELT

Manager of Marketing Services, The Lincoln Electric Company, Cleveland, Ohio (*Electric-arc Welding*)

CYRIL S. ADAMS

President, Cyril S. Adams, Inc., Houston, Tex. (*Oxyacetylene Cutting and Joining*)

A. Steel Erection

ADVANCE PLANNING

The erection of structural steel can be divided into two separate operations: advance planning and field operations. The success of the field operations depends in large measure on the care and thoroughness of the advance planning. Large bridge and complicated building work necessitates many drawings and layouts to guide the field personnel through each step. The sequence of shipping and erecting must follow

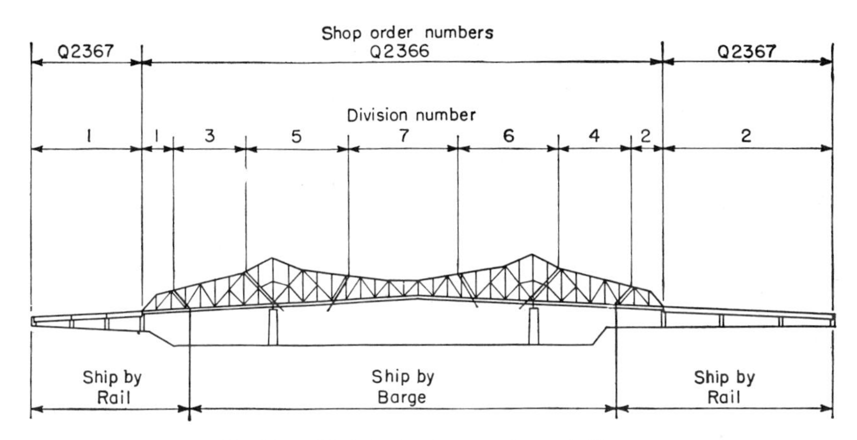

Fig. 23-1 Divisioning for a highway bridge.

specific patterns. The size and weight of individual pieces must be within fabricating capacity, shipping limitations, and erecting equipment capacity.

Divisioning and Shipping A system of subdividing a structure into units called "divisions" is used to schedule the steel from the rolling mill to the erection site. This divisioning is determined by the order in which the structure will be erected. Required shipping dates are then established for each division. The fabricating plant uses this information to schedule material orders, prepare shop drawings, fabricate the material and prepare it for shipment. The size and pattern of divisions will vary with the type of structure. Tier buildings are normally divisioned horizontally. Each tier of columns and beams, usually consisting of two floors, makes up a division. In manufacturing buildings, each aisle with its columns, trusses, and purlins will constitute a division. In bridge work, a division will include all the steel between piers or between falsework bents.

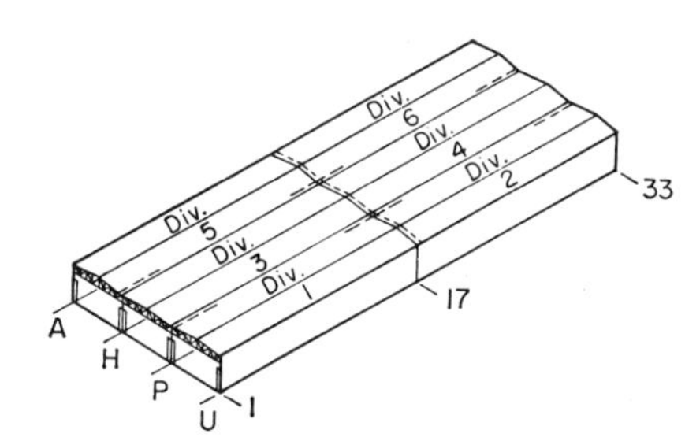

Fig. 23-2 Divisioning for a manufacturing building.

Typical examples of divisioning are shown in Figs. 23-1 to 23-3.

Erection Procedure The type of erecting equipment to be used is determined by the type of structure and the site conditions.* Buildings and bridges that are under 200

* See Sec. 17, Cranes and Hoists.

ft in height can be erected with truck cranes or crawler cranes. Taller structures beyond the reach of crane booms are erected with guy derricks. If the site conditions permit, the lower portion of a tall structure is erected with cranes and the balance is erected with guy derricks. High bridges and those over water are erected with derricks placed on the structure. Derrick boats, cranes on barges, and derricks on barges are also used to erect bridges over water.* Long, heavy railroad girders are erected with derrick cars.

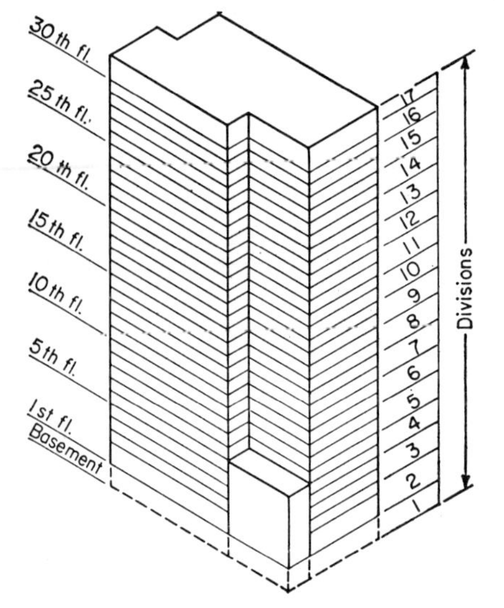

Fig. 23-3 Divisioning for a tier building.

The determination of the type of erecting equipment to be used governs the procedure to be followed in erection. A detailed study of the job is required to make drawings of the steps to be followed in the field. On simple building and bridge construction, these drawings will outline only the major steps. On complicated construction, many drawings are required to show each position of the erecting equipment, specify the erection loads, and give detailed field instructions.

Size and Weight After the type of equipment is determined and the general erection procedure established, the required capacity of the equipment must be determined. The size and weight of heavy members must be checked against fabricating plant capacity, shipping clearances, and erecting equipment capacity. Additional field splices may be required to reduce erection weights or to stay within plant capacity or shipping clearances. Long, slender members designed to act as integral parts of the completed structure must be investigated to determine if they will be stable during shipping and erection. Layouts of complicated joints must be made to see that sufficient clearance is provided for field assembling the members and fastening them.

FIELD OPERATIONS

The operations performed from the receipt of the material to the completed structure can be separated into six general steps:

1. Receiving and unloading
2. Sorting for erection
3. Moving to the site
4. Erecting
5. Aligning
6. Fastening

Receiving and Unloading A receiving area or yard is required to store material until it is required for erection. Few fabricating plants can economically produce structural steel in the exact sequence required for erection. It is, therefore, usually necessary to receive a complete division before starting erection. As shipments can be made by rail, truck, or barge, the location of the receiving area will depend upon the method of shipment. Truck and crawler cranes are used to unload at the majority of jobs. For semipermanent locations, derricks prove very useful. In some metropolitan areas, the railroads will furnish, for an additional tariff, "ground storage" where they will unload and store and then reload according to detail lists furnished by the consignee. Gantry-crane service is also available at many railroad team tracks. The storage space at these team tracks is usually limited, however, and the steel must be removed

* See Sec. 18, Marine Equipment.

promptly. When the fabricating plant and the site are on navigable rivers, steel can be delivered by barge. On the inland waterways, standard 1,000-ton hopper barges are used and carry about 500 tons of fabricated steel.

Sorting for Erection The individual pieces in a structure should be delivered to the erecting crew as they are required. This is accomplished by arranging the steel as it is received into some orderly pattern based on the erection procedure. When this sorting cannot be done at the fabricating plant, it is done at the receiving site. The columns for tier buildings are sorted and stored by tiers; the beams are sorted and stored by bundles, each bundle containing all the beams required for a particular predetermined floor area. For bridges, individual members are stored in the sequence in which they will be erected.

Moving to the Site When streets connect the yard and the site, pole trailers, flatbed trucks, and straddle trucks are used to haul the steel to the site. On some large projects, rail connections are available and local shifting engines are employed. Bridges over water or rugged terrain require a material track over which the steel is moved on small standard-gauge cars. Bridges over navigable water permit loading the steel on scows or barges and delivering it direct to the erecting crew.

Erecting The actual placing of the steel in its permanent location in the structure is performed by a "raising gang." By referring to erecting drawings furnished by the fabricator, the gang determines the particular piece required and how it fits into the structure. The "ground crew" selects the correct piece and directs the erecting crane in hoisting the piece to its proper location. The "top men" or "connectors" guide the member into its exact position, secure it, and release the erecting crane.

Aligning Immediately following the placing of a member in a structure, it is necessary to align it correctly and place sufficient temporary fasteners in each joint to ensure stability. In light building work, the temporary bolts placed by the erectors to secure the member are usually sufficient, and the joints are aligned by the bolting crews which follow. On large tiered buildings, column splices and the floor near the derrick are brought into permanent alignment by "fitting-up crews" which follow the "connectors," placing drift pins and fitting bolts. Another crew places a system of temporary wire-rope bracing which aligns the structure and holds it until the permanent fasteners can be placed. On bridge work, it is usually necessary to place drift pins and fitting bolts in a connection before the member can support itself and the derrick can be released. For this reason the fitting-up crews follow immediately behind the connectors, placing pins and bolts.

Fastening The final operation in steel erection is the placing of the permanent fasteners: rivets, bolts, or welding. The tools and procedures used for field fasteners are described under Erection Tools.

ERECTION EQUIPMENT

The hoisting equipment used to erect steel structures includes derricks, cranes, and variations of these units. This general class of construction equipment is described in Section 17, Cranes and Hoists. The specialized lifting units which are used in steel erection are described herein in greater detail.

Stiffleg Derricks The stiffleg derrick is the most versatile of the derricks used in steel erection. It is essentially a mast and boom, with two sloping fixed legs supporting the mast. The legs are relatively short compared to the maximum boom lengths used and require counterweighting or tying down to resist the uplift caused by the load. Stiffleg derricks are used as yard derricks for unloading and storing material and for bridge erection when mounted on a frame which permits the unit to be moved. Capacities and sizes vary widely, and derricks with booms up to 150 ft in length and with capacities up to 150 tons have been used in steel erection. Stiffleg derricks can also be mounted on barges to be used as floating derrick boats or on a tower for erecting high structures. These derricks are usually powered with gasoline or diesel hoisting equipment. Figure 23-4 shows the basic stiffleg derrick. Figure 23-5 shows the method of determining reactions. Figure 23-6 shows a typical stiffleg derrick mounted on a frame to make a traveler for bridge erection.

Guy Derricks The guy derrick consists of a boom and a mast supported by wire-rope guys. These derricks are used principally in the erection of tiered buildings. They can also be set up to serve as yard derricks and have been used for bridge erection under certain conditions. In building erection the guys are anchored to the steel, while in other setups some form of counterweight or deadman is used to anchor the guys. The boom of the guy derrick must always be shorter than the mast. The boom can then be brought up to the mast and will clear the guys while swinging.

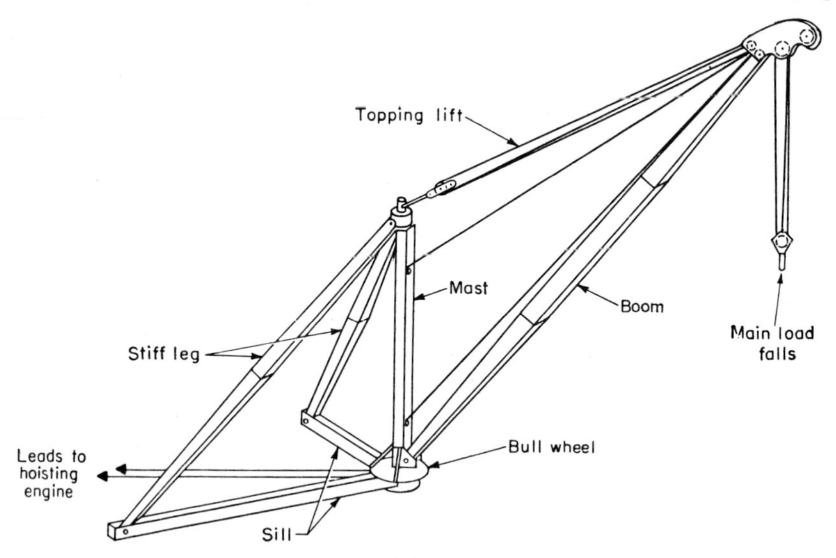

Fig. 23-4 Stiffleg derrick.

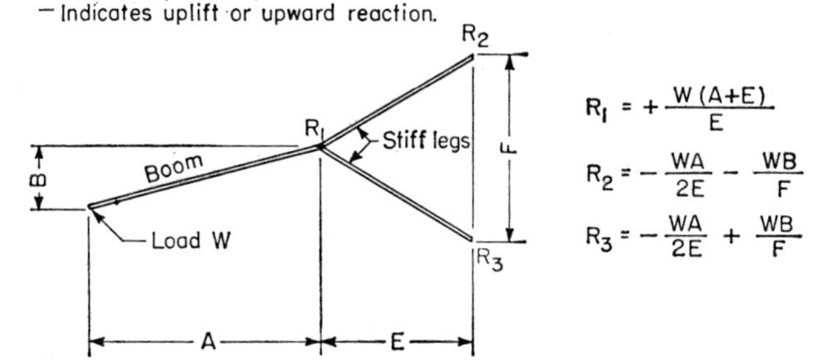

Fig. 23-5 Diagram for computing stiffleg-derrick reactions.

The guy derrick used in steel erection is available with booms up to 150 ft in length and capacities up to 70 tons. Figure 23-7 shows a typical guy derrick as used in tier building erection.

Specialized Units There are many erecting units which are designed for specific applications. The trolley system shown in Fig. 23-8 has been used when bridge erection is begun at both ends and there is difficulty in delivering material to both locations. It has also been used in the erection of arches and viaducts over gorges where the use of conventional erection units was not feasible. Trolley systems up to 1,000 ft in length have been used satisfactorily.

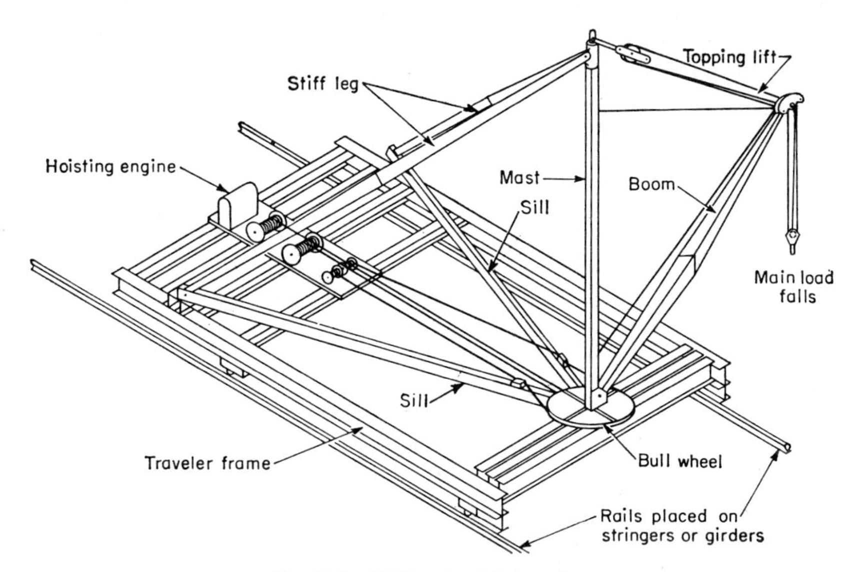

Fig. 23-6 Stiffleg-derrick traveler.

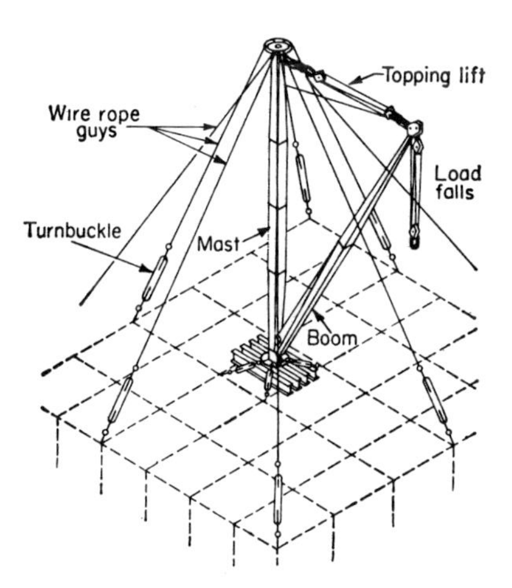

Fig. 23-7 Guy derrick.

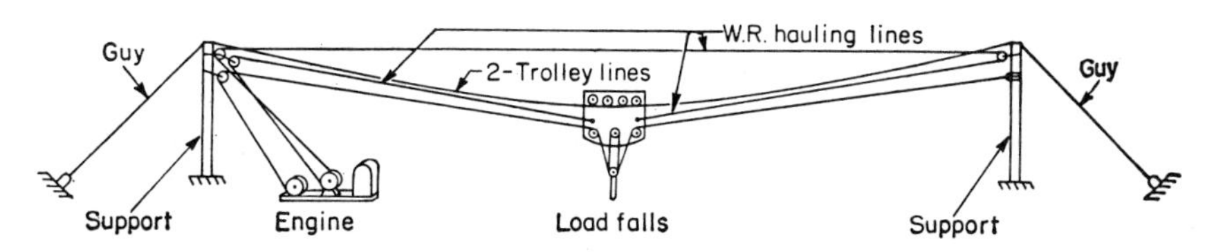

Fig. 23-8 Trolley system.

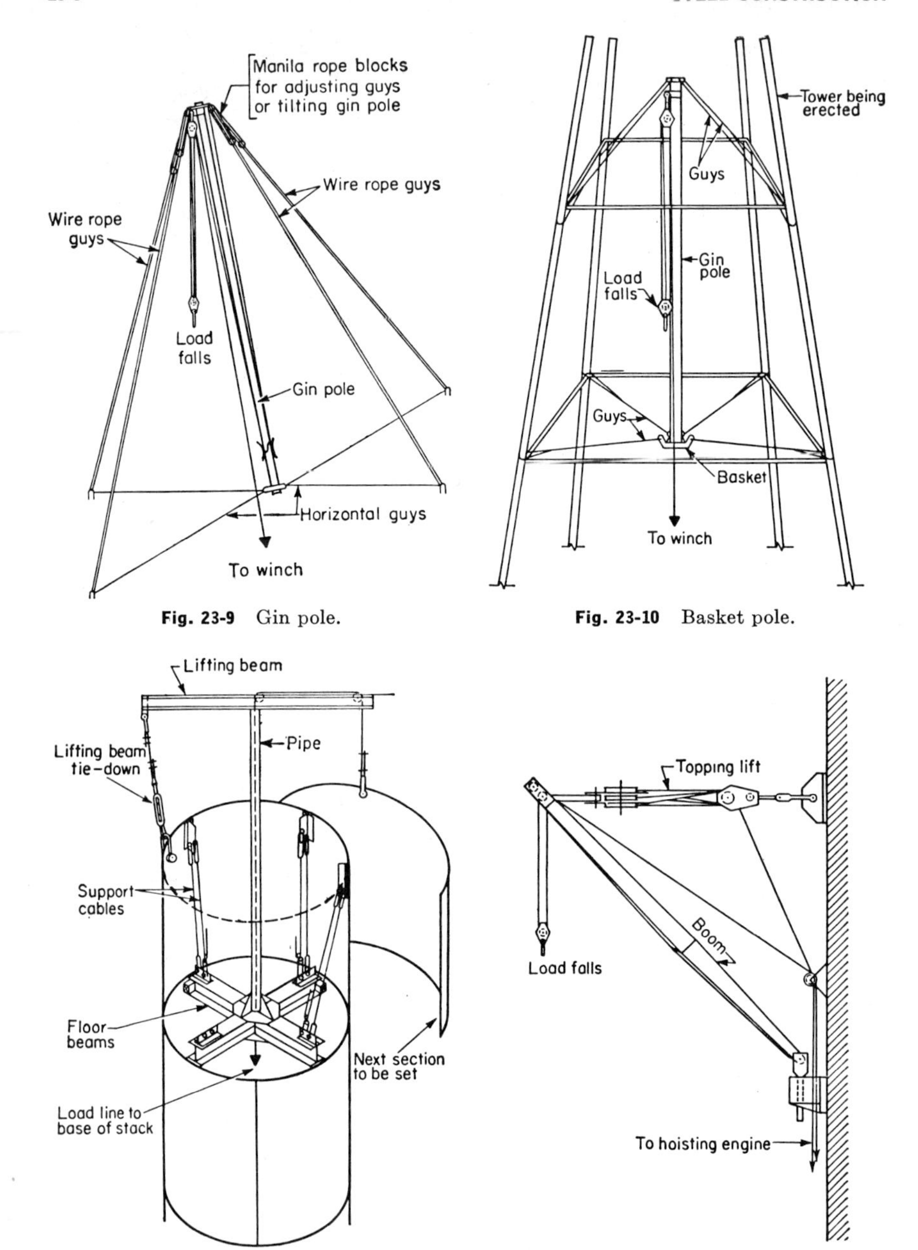

Fig. 23-9 Gin pole.

Fig. 23-10 Basket pole.

Fig. 23-11 Stack rig.

Fig. 23-12 Chicago boom.

The gin pole (Fig. 23-9) is used in the erection of light members when it is not practical to move in a crane or a derrick. Gin poles can be made of wood, pipe, or structural-steel shapes.

The basket pole (Fig. 23-10) is a variation of the gin pole. It is supported by guys at the top and bottom and is used in the erection of towers.

The stack rig shown in Fig. 23-11 is designed for use in erecting steel-plate stacks.

The Chicago boom (Fig. 23-12) is used for erecting and transferring material when some convenient permanent structure is available to act as a mast for attaching the topping lift and is capable of supporting the lower end of the boom.

The telegraph consists of two sets of hoisting cables attached to suitable supports, as shown in Fig. 23-13. By taking up on one cable and letting out on the other, a load may be transferred between the supports.

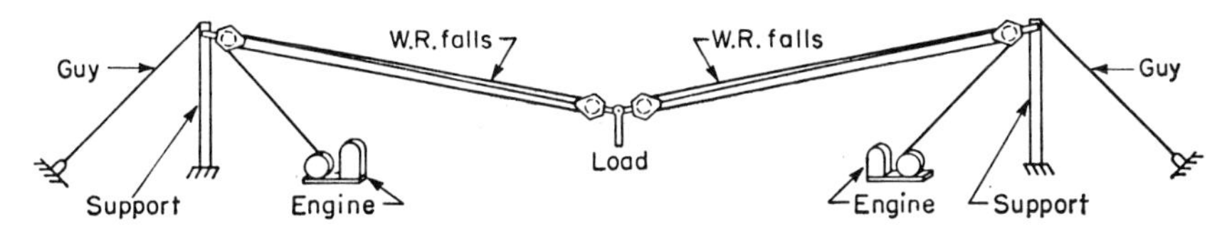

Fig. 23-13 Telegraph.

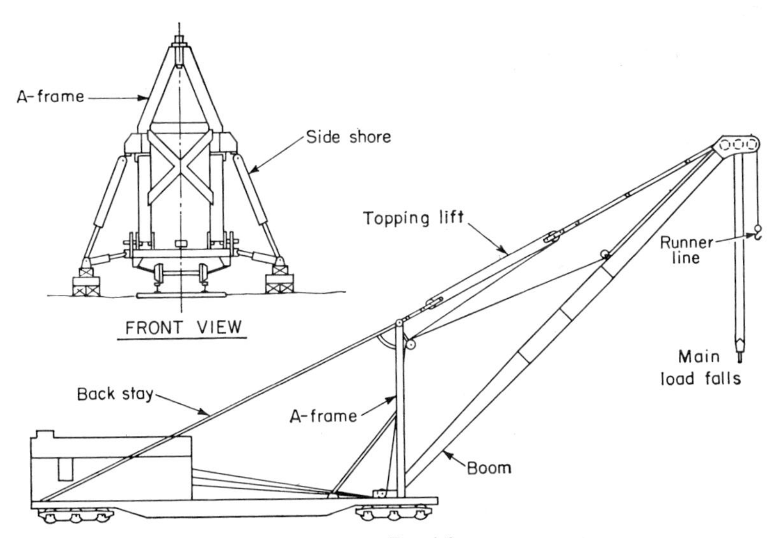

Fig. 23-14 Derrick car.

The derrick car shown in Fig. 23-14 is designed to erect long, heavy girders in railroad work. It is a derrick mounted on a flatcar with a self-propelling power unit. Derrick cars are usually of special design and have been built with booms up to 80 ft long and capacities up to 80 tons. Their maximum lifting capacity is with the boom straight ahead.

ERECTION TOOLS

The small tools used in steel erection vary widely with the type and size of structure. Most of the tools are standard manufacturers' items, but their specific types, sizes, and designs are selected by the individual user. The paragraphs below describe typical erection activities and the corresponding erection tools.

Material Handling To attach a load to the hook of a crane or derrick, a wire-rope sling (Fig. 23-15) is used when the size and weight of the load permit. The basic wire-rope sling consists of a length of rope with an eye spliced in each end. This sling is placed around the member and one end is passed through the opposite eye to "choke" the member, as shown in Fig. 23-15. The sling should be protected when it is bent over sharp edges to prevent cutting individual wires. This protection piece or "soft-

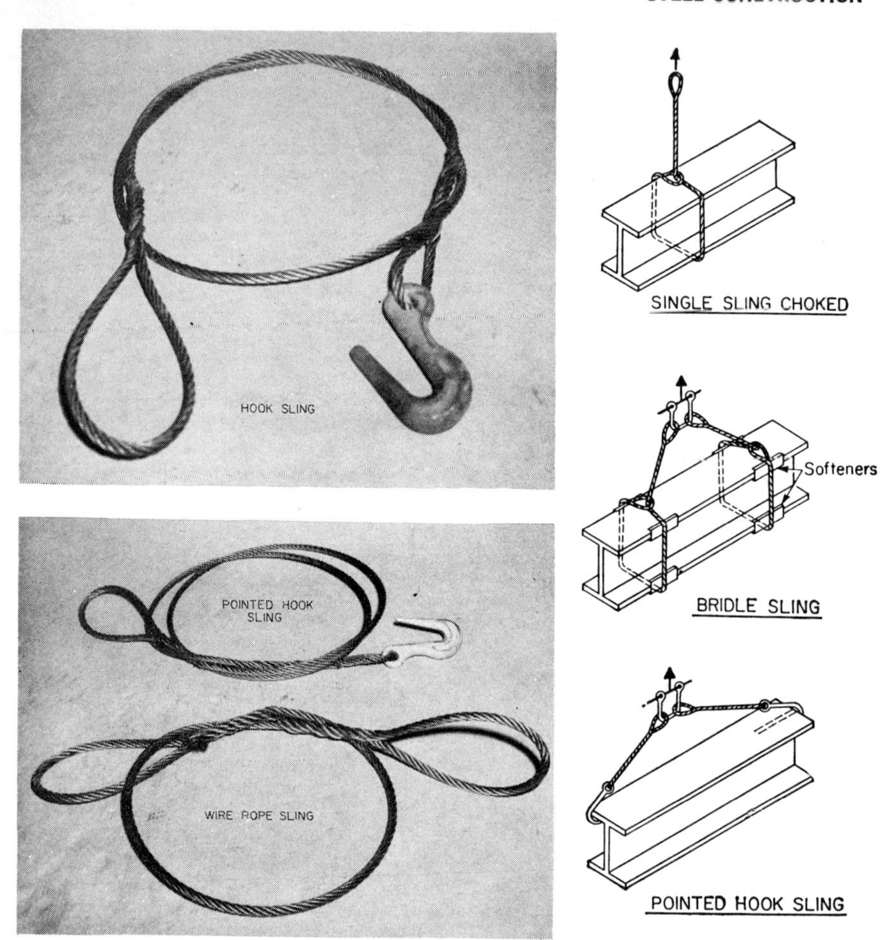

Fig. 23-15 Wire-rope slings.

ener'' can be wood or a short length of split pipe. On long members, two slings are often used together in a bridle hitch. Pointed hook slings are used in pairs when sorting material.

The capacity of a sling varies with the type of wire rope, the factor of safety desired, and the purpose for which it is used. When two slings are used in a bridle hitch, the capacity of each sling is reduced as the angle between them increases. To determine the capacity of each sling in a bridle hitch, a distance should be measured along one sling for the same number of units as the capacity of sling in tons (distance A, Fig. 23-16). The vertical distance from that point to the load (distance B, Fig. 23-16) must then be measured. The number of units in this vertical distance B is the working load for each sling. The unit used may be any convenient one, such as feet or half feet, but the same unit must be used for each measurement.

On heavy girders and chord sections, hitches made from structural sections are used. Figure 23-17 shows various types that are commonly used for 50- to 75-ton pieces. It is the usual practice for the drawing room to calculate the center of gravity of the member, design the hitch, and leave sufficient open holes in the member so that the hitch can be readily bolted to the member in the field.

Connecting The actual connecting of two pieces of steel requires the use of the tools shown in Fig. 23-18. The connecting bar, coffing hoist, and maul are used to bring the members into their permanent positions. The bull pin and fork wrench are used to align the holes in each member. The drift pin, in combination with fitting bolts, is used to keep the pieces in their permanent positions until the permanent fasteners are placed. The number of bolts and pins used in each joint varies with the type of structure. In ordinary building work where no erection, wind, or other load will be applied, only enough pins and bolts are used as are necessary to hold the member in place. In joints where erection, wind, or other loads will be applied, all the holes in a joint are filled with pins and bolts, in a combination of about 40 percent pins and 60 percent bolts.

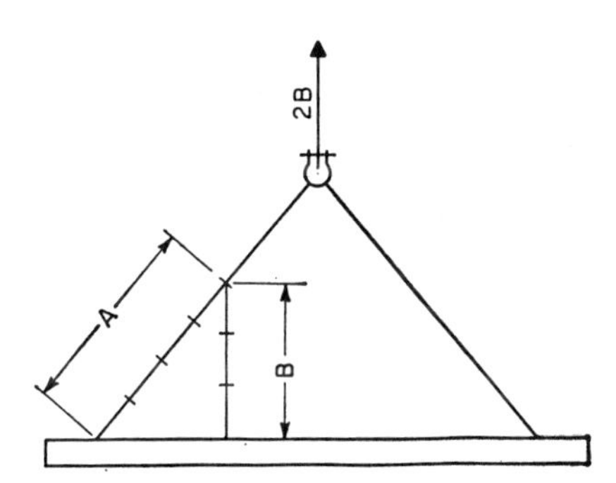

Fig. 23-16 Bridle-hitch capacity diagram.

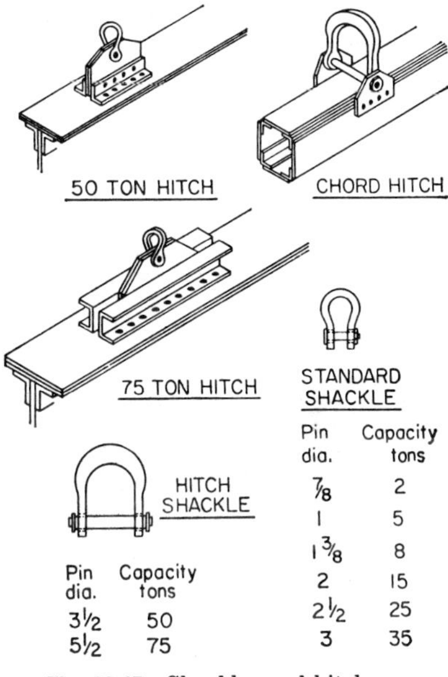

Pin dia.	Capacity tons
3½	50
5½	75

Pin dia.	Capacity tons
⅞	2
1	5
1⅜	8
2	15
2½	25
3	35

Fig. 23-17 Shackles and hitches.

Rigging Wire hoisting rope and steel tackle blocks are used in hoisting loads and actuating the booms of cranes and derricks. The 6 by 19 (6 strands of 19 wires each) rope, fashioned from improved plow steel, is most generally used in hoisting.

Manila rope is used for hand lines, tag lines, hand falls, scaffold lines, and minor uses. It should not be used for hoisting purposes with powered equipment. For hand lines and hand falls, a ¾-in.-diameter manila rope is used. Scaffolds are usually hung with 1-in.-diameter manila rope.

The stress in the rope in a set of tackle blocks is equal to the total load supported divided by the number of parts of rope in the blocks. When the load is moved and the rope passes around a sheave in a tackle block, the stress in the part unwinding from the sheave is greater than in the part winding on the sheave by an amount equal to the resistance at the sheave. This resistance comes from the friction within the rope as it bends around the sheave and the friction between the sheave and the pin upon which it turns. The ratio K of the stress in the unwinding part to that in the winding part can be found by the following equation:

$$K = 1 + C\frac{d}{R} + 2f\frac{r}{R} \tag{23-1}$$

where d = diameter of rope, in.
R = radius of center line of rope at sheave, in.
r = radius of pin on which sheave turns, in.
f = coefficient of friction of sheave on pin
C = constant depending on kind of rope

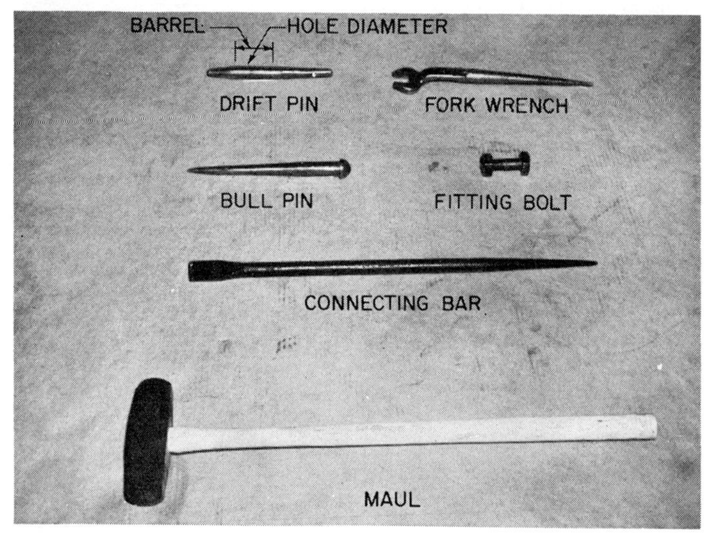

Fig. 23-18 Connecting tools.

Experiments have determined a value for $C = 0.225$ for 6 by 19 hoisting rope and a value for $f = 0.08$ for metalline bushed sheaves.

Where the lead line leads off the fixed block and does not form one of the parts carrying the moving block and load as shown in Fig. 23-19, and when the blocks are rove from the center, the lead-line pull required to lift a load can be found from

$$P = W \frac{K^n(K - 1)}{K^n - 1} \tag{23-2}$$

where P = lead-line pull, lb
W = load lifted, lb
K = ratio of stress in unwinding part to that in winding part [Eq. (23-1)]
n = number of parts of rope from moving block

When the falls are reversed so that the lead line comes off the moving block and serves as one part of the lifting falls as shown in Fig. 23-20, the lead-line pull required can be found from

$$P = W \frac{K^{n-1}(K - 1)}{K^n - 1} \tag{23-3}$$

Table 23-1 lists typical lead-line pulls per 1,000 lb of load for various types of blocks rove with 6 by 19 wire hoisting rope as shown in Fig. 23-19, based on Eq. (23-2).

TABLE 23-1 Lead-line Pull per 1,000-lb Load

Number of parts n	Metalline bushed sheaves, 16-in. OD, 14-in. groove, 2-in. pin diam			Ball or roller bearing sheaves, 20-in. OD, 17-in. groove, 6.25-in. pin diam		
	¾-in. rope $K = 1.045$	⅞-in. rope $K = 1.048$	1-in. rope $K = 1.052$	¾-in. rope $K = 1.022$	⅞-in. rope $K = 1.025$	1-in. rope $K = 1.028$
2	534	536	539	516	519	521
5	228	230	232	213	215	217
10	126	128	131	113	114	116
15	93	95	98	79	81	83
20	77	79	82	62	64	66

If falls are reversed as in Fig. 23-20, divide the values in the tables by K. Multiply the lead-line pull by the corresponding value of K for each outside block that bends the line 180°. If the line bends only 90°, use 75 percent of the increase for 180°. The tabulated values are based on reeving the blocks as shown in Fig. 23-21. If the blocks are laced starting from one side and continuing across the block as shown in Fig. 23-22, the lead-line pulls given in Table 23-1 will be increased by about 25 percent.

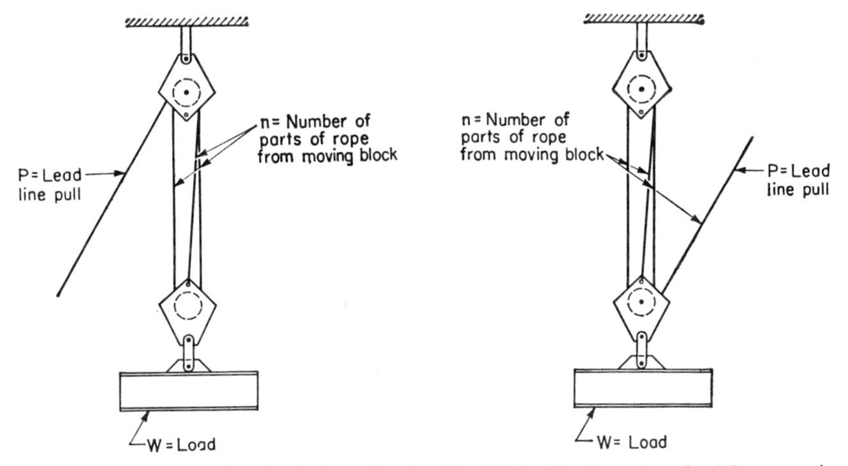

Fig. 23-19 Block reeving, lead from fixed block.

Fig. 23-20 Block reeving, lead from moving block.

The diagrams in Fig. 23-23 indicate the method of reeving a set of blocks when multiple parts are required. The blocks are shown as they would hang when reeved or as lying on the ground in position for reeving. The numbers indicate the order in which the rope passes around the sheaves, beginning at the lead-line end. The location of the number indicates the side of the block on which the rope enters. The numbers shown on the reeved falls in Fig. 23-21 are the same numbers as shown for eight parts in Fig. 23-23. After leaving the sheave at the highest number, the rope goes to the becket. For an added number of parts, where not shown, the diagram for the next-higher number may be used and the end becketed to the block or to the last sheave.

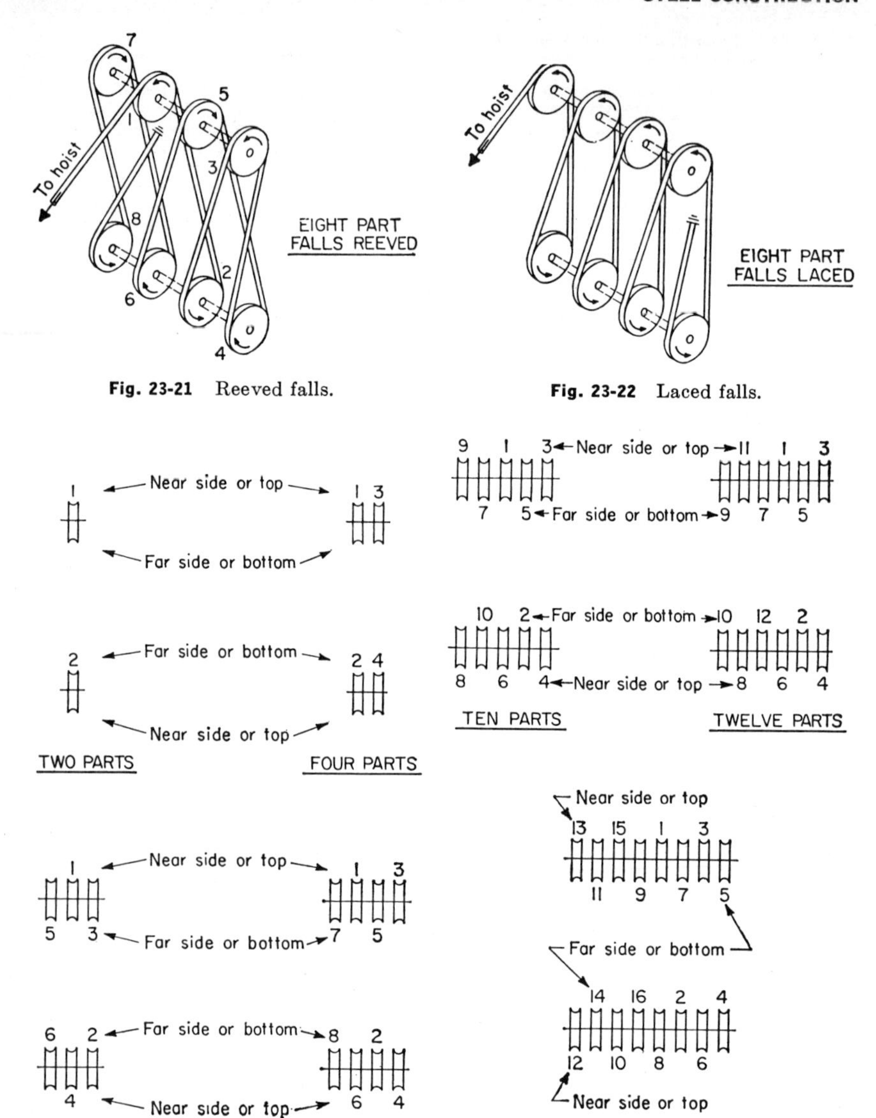

Fig. 23-21 Reeved falls.

Fig. 23-22 Laced falls.

Fig. 23-23 Reeving diagram.

Bolting High-strength carbon-steel bolts as used for field bolting are made in accordance with ASTM Designation A325[1,*] and high-strength alloy-steel bolts are made in accordance with ASTM Designation A490;[2] both types are heat treated. These bolts transfer shear force in a structural joint either by (1) clamping the parts together so tightly that the friction force between them exceeds the external shear force, or (2) through shear stress developed on the area of the bolt. Hardened steel washers are required under the head and under the nut of A490 bolts to reduce galling

* Superior numbers refer to the list of references at the end of this section.

of the connected parts when used to connect material having a specified minimum yield point less than 40,000 psi.

High-strength bolts are placed in accordance with *Specification for Structural Joints Using ASTM A325 or A490 Bolts.*[3] These specifications require that high-strength bolts be tightened to give minimum bolt tensions on completion of the joint, as shown in Table 23-2.

TABLE 23-2 High-strength Bolts

Bolt size, in.	Minimum fastener tension,* thousands of pounds	
	A325 bolts	A490 bolts
½	12	15
⅝	19	24
¾	28	35
⅞	39	49
1	51	64
1⅛	56	80
1¼	71	102
1⅜	85	121
1½	103	148

* Equal to 70 percent of specified minimum tensile strengths of bolts, rounded off to the nearest thousand pounds.

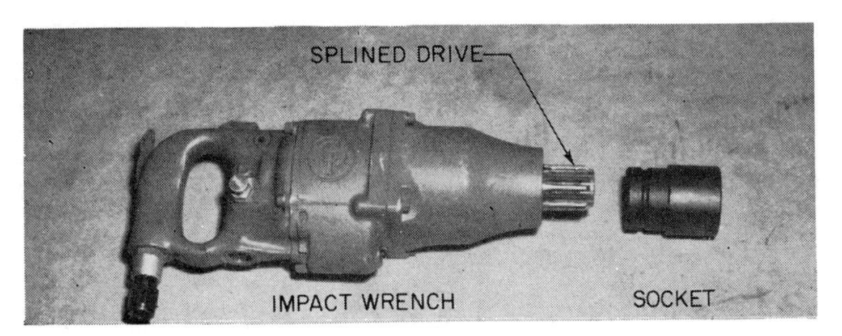

Fig. 23-24 Impact wrench.

The pneumatic impact wrench shown in Fig. 23-24 is used to tighten high-strength bolts. Impact wrenches with sufficient power to rotate the nut a predetermined fraction of a turn are used in the turn-of-nut method. The steps required to tighten bolts by this method are:

1. Bring sufficient bolts in a joint to a "snug-tight" condition to ensure that the several parts are properly compacted; i.e., brought into full contact with each other. Snug-tight is defined as "the tightness attained by a few impacts of an impact wrench or the full effort of a man using an ordinary spud wrench."
2. Place bolts in remaining holes and bring to snug-tight condition.
3. Tighten all bolts additionally by the applicable amount of nut rotation as specified in Table 23-3, with tightening progressing systematically from the most rigid part of the joint to its free edges.

High-strength bolts with ribbed shanks that provide full bearing must be driven

TABLE 23-3 Nut Rotation* from Snug-tight Condition

Disposition of outer faces of bolted parts		
Both faces normal to bolt axis, or one face normal to axis and other face sloped not more than 1:20 (bevel washer not used).		Both faces sloped not more than 1:20 from normal to bolt axis (bevel washers not used).
Bolt length† not exceeding 8 diameters or 8 in.	Bolt length† exceeding 8 diameters or 8 in.	For all lengths of bolts.
½ turn	⅔ turn	¾ turn

* Nut rotation is rotation relative to bolt regardless of the element (nut or bolt) being turned. Tolerance on rotation: 30° over or under.

For coarse thread heavy hex structural bolts of all sizes and lengths and heavy hex semifinished nuts.

† Bolt length is measured from underside of head to extreme end of point.

into the holes by hand or pneumatic hammers. The holes in all adjoining plates must be in perfect alignment to permit the ribbed bolt to be entered. Tightening is performed as described above. There is under development a high-strength bolt that is tensioned by a power tool which then swages a collar into grooves of the bolt shank.

Welding Field welding on structures is done with portable welding machines. These machines can be electric, gasoline, or diesel, depending upon the availability of power on the job. Coated electrodes are designed for use with each type of steel and position of joint to produce efficient and acceptable welds. Automatic and semiautomatic systems using various combinations of wire and shielding materials are available for large, repetitive-type joints in all structures. Refer to Electric-arc Welding on page 23-39 for detailed welding information.

Riveting For most structures, specifications and/or building codes specify either field bolting or field riveting. Since the introduction of high-strength structural bolts in the early 1950s, only a few bridge specifications require field riveting.

Rivets are heated in a hand-operated forge using Pocahontas or smithing coal. They are heated to a light cherry red or to about 1900°F and should be driven before their temperature has dropped below 1000°F. When the hot rivet is inserted in the hole, the manufactured head is held with a bucking-up tool and the stock projecting through the hole is hammered down, forcing material into the hole and forming the driven head.

Figure 23-25 shows the common types of pneumatic riveting hammers, a pneumatic air jack for bucking-up rivets, and the various types of hand bucking-up dollies.

Safety Devices The safety helmet or "hard hat" is now a standard piece of equipment for the steel erector. There are many types available, all designed to protect the wearer from falling objects. Many lives have been saved by the hat deflecting or retarding falling objects.

The safety belt is designed to prevent men from falling. Belts are made of leather or heavy webbing and have a short length of rope attached. This rope, usually an 11-ft length of ½-in.-diameter spun nylon, is tied to the structure. If a workman should lose his footing or a scaffold collapse, his fall will be arrested by the rope. Since the rope restricts the man's movements, it is used mostly while riveting or bolting.

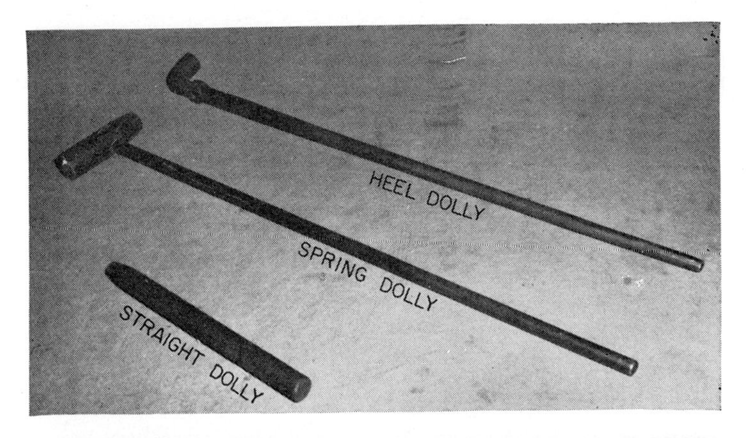

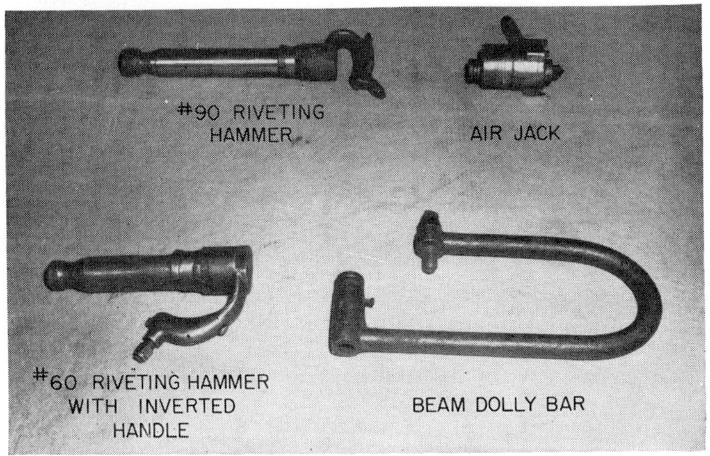

Fig. 23-25 Riveting tools.

Safety goggles, designed to provide protection from smoke, dirt, welding flashes, or impact, are worn when men are working in places that subject them to these hazards. Various types are available for specific types of protection.

Safety vests made of nonsinkable material and life rings are used when men are working over water. In addition, a safety boat manned by an experienced boatman is employed to rescue anyone who falls into the water.

TEMPORARY ERECTION MATERIAL

This temporary material can be divided into two classes. The first class is erection material framing into the structure and acting in conjunction with the structure during erection. The second class is falsework which acts independently of the structure and is used to support the structure during erection.

Erection Material The design drawings for a structure do not, except in special cases, provide for the temporary material required in erection. After the erection procedure is established, the stresses produced in individual members by erection loads are determined and compared with the design stresses. An erection stress which is larger than the design stress but does not produce a reversal of stress, can usually be provided for by increasing the area of the member. Wide-flange chords are increased by substituting a heavier section. Built-up chords can be increased by providing thicker plates or angles. Members that have a reversal of stress during erection can be increased in size, stiffening beams can be added, or their effective lengths can be reduced by providing intermediate points of support.

In erecting bridges, the expansion shoes can be temporarily converted to fixed shoes by placing wedges under the rockers or by welding plates between the rockers and the bedplates. The erection procedure often changes a positive design reaction at the abutment to a negative reaction or uplift. To provide for this temporary uplift condition, counterweights of steel, rock, ingots, concrete, or other materials may be

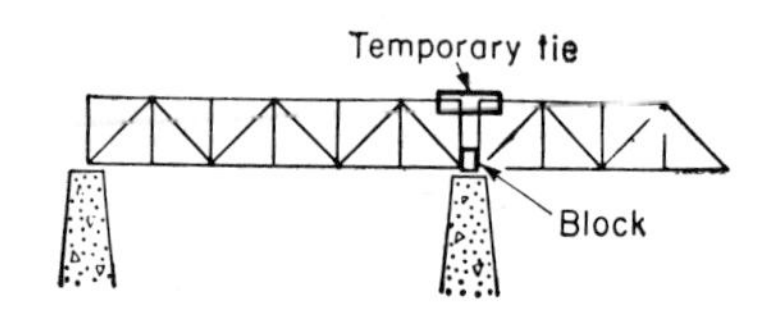

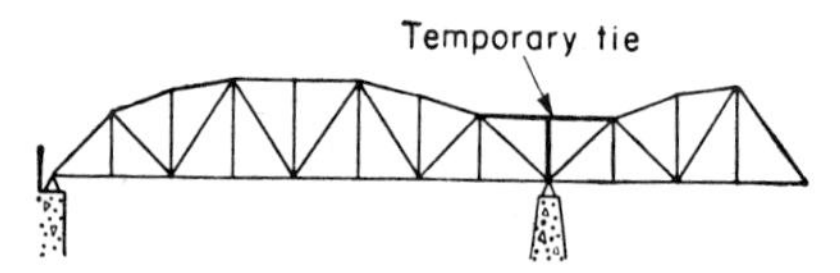

Fig. 23-26 Temporary ties used in bridge.

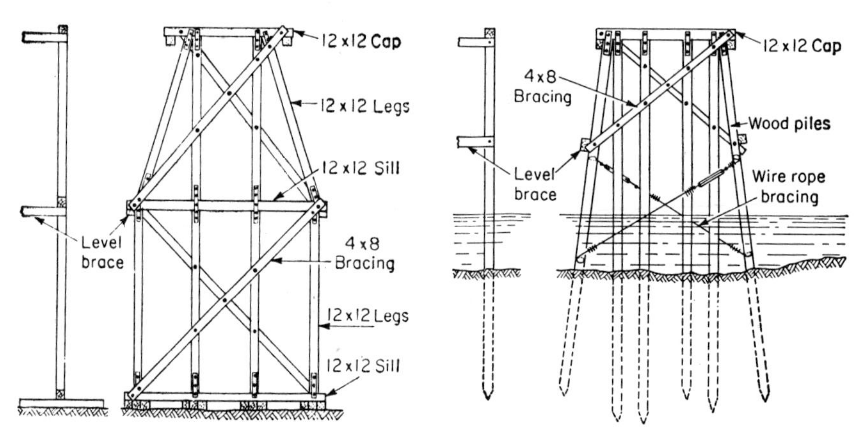

Fig. 23-27 Typical timber falsework bent. **Fig. 23-28** Typical wood-pile falsework.

added to the structure. On large structures, anchorage material is embedded in the foundations and the span is tied to this embedded material with wire rope, temporary steel links, or turnbuckles.

Adjacent deck truss spans are often tied together to act as continuous spans during erection. These temporary ties can be extensions of the gusset plates or additional plates. Adjacent through truss spans are also tied together for erection purposes. These temporary ties are usually similar to the truss chords and tie the spans together at the *U*1 points. Figure 23-26 shows some temporary tie applications.

Falsework The temporary material used to support a structure until it is self-supporting is called "falsework." On spans which are built over land, falsework is constructed of either wood or steel bents placed on temporary spread footings or mud sills. The design of the footings depends upon the load to be carried by the falsework bent and the load-carrying capacity of the soil. Table 23-4 lists common bearing values for various soils and masonry.

Timber bents are commonly built of Douglas fir or southern yellow pine legs 12 in. square, with caps and sills of the same size. The legs, sills, and bracing are fastened

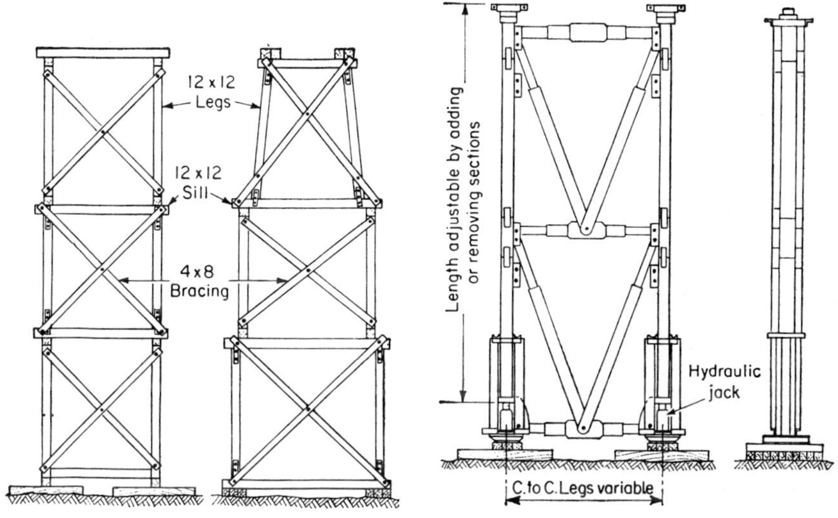

Fig. 23-29 Typical tower timber falsework bent.

Fig. 23-30 Adjustable steel falsework bent.

TABLE 23-4 Common Bearing Values for Soils and Masonry

Soil	*Tons/sq ft.*
Mud, soft	¼–½
Mud, stiff	½–1
Soft clay and loam	½–1
Clay, wet and mixed with sand	1–1½
Clay, wet	1½–2
Clay, permanently dry	4–6
Hard clay and coarse sand	2–4
Firm, coarse sand and gravel	4–6
Cemented sand	4–6
Cemented gravel	8–10
Hardpan	10–12
Masonry	*psi*
Common brick	300
Hard brick	400
Concrete, 1:2:4 mixture	900
Concrete, 1:2½:5 mixture	700
Sandstone masonry	500
Limestone masonry	700
Granite masonry	1,000

together with bolts and splice plates.* For ordinary timber and average workmanship, the load on a 12- by 12-in. column leg should not exceed 25 tons, with a maximum length of 30 ft. Loads up to 36 tons can be safely carried on 12- by 12-in. legs up to 14 ft long when using new timber and above-average workmanship. This load of

* See Sec. 24, Timber Construction.

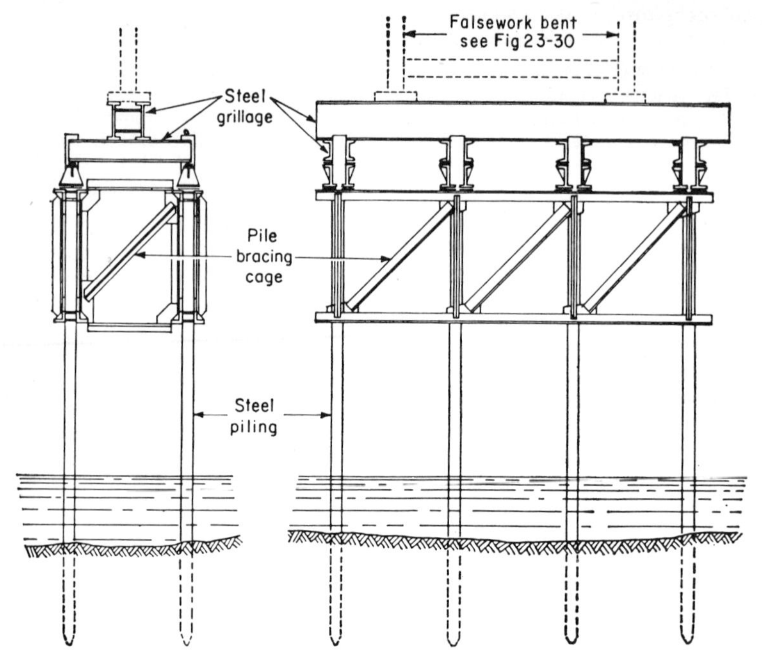

Fig. 23-31 Typical steel-pile falsework bent.

Fig. 23-32 Single-track railroad bridge under construction using falsework.

36 tons is the bearing value of the caps and sills when using 12- by 12-in. timbers and is usually the limiting factor in the design of timber falsework.

In cases where the soil will not support the load carried by the falsework or the bent is built over water, piles are used to furnish the necessary bearing. These piles can be either wood or steel.* Figure 23-27 shows a typical sill-supported single timber bent, and Fig. 23-28 shows a wood-piling bent. If the load is too heavy for a single bent, double bents or tower bents (Fig. 23-29) may be used. Falsework bents of steel construction are designed using current AISC specifications[4] for steel. As indicated in Fig. 23-30, the steel falsework bents are usually detailed to permit adjustments in their height and width, to accommodate varying field conditions. Typical steel-pile bents are shown in Figs. 23-31 and 23-32.

BRIDGE ERECTION

The specific erection procedure to be followed on a particular structure is determined by the erector after a detailed study of the conditions involved. Some of the items

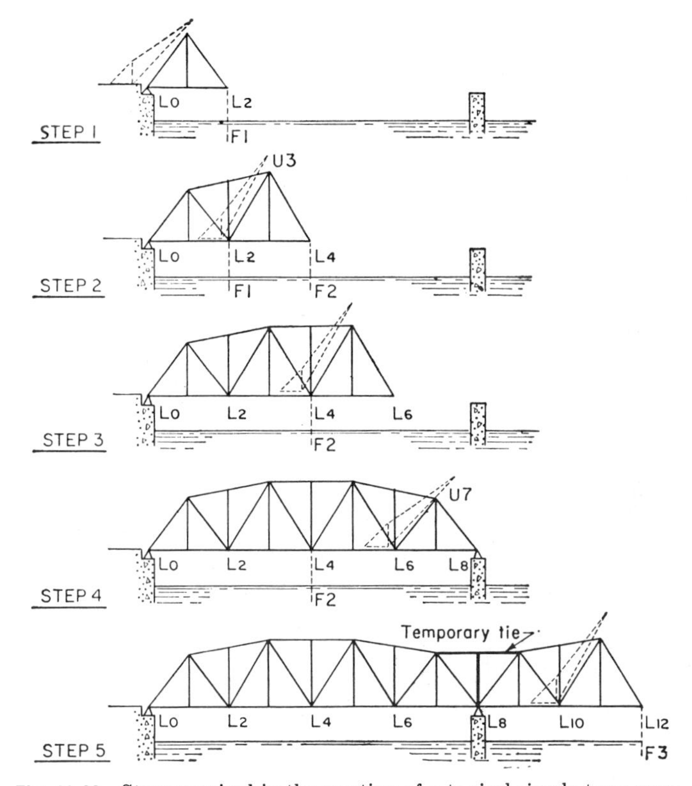

Fig. 23-33 Steps required in the erection of a typical simple-truss span.

to be considered are the length of time required for erection, the equipment available, the site conditions, and the type of structure.

Simple-span Truss Bridges Assuming that a simple-truss span is to be erected with a stiffleg derrick traveler and that two falsework bents are used, the individual steps required are as illustrated in Fig. 23-33.

* See Sec. 27, Piles and Pile Driving.

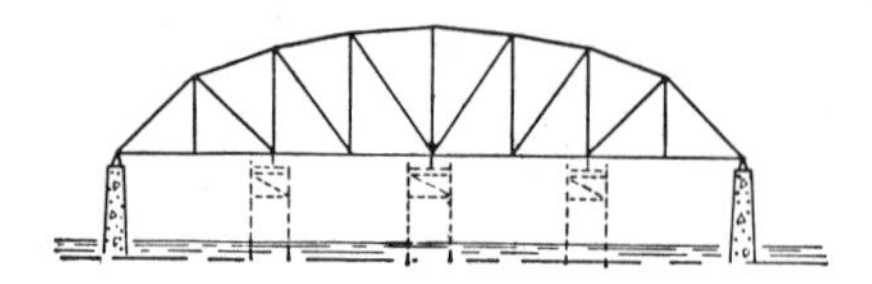

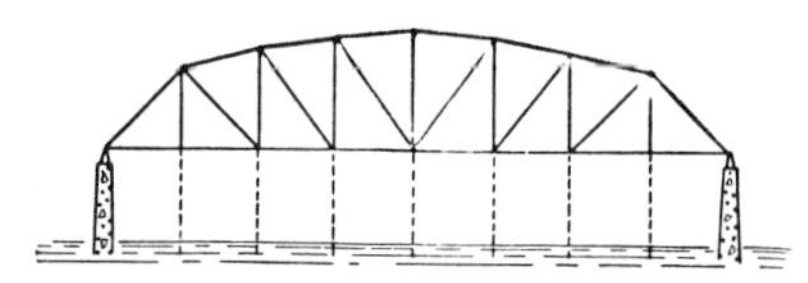

Fig. 23-34 Simple-truss span erection using full falsework.

Fig. 23-35 Simple deck truss highway span being floated into position.

Step 1. Place falsework bent $F1$.
Erect steel between $L0$ and $L2$.

Step 2. Move traveler to $L2$.
Place falsework bent $F2$.
Erect steel between $L2$ and $L4$.
Adjust elevation of falsework bent $F2$ to close joint at $U3$.
Jack up falsework bent $F2$ to release falsework bent $F1$ and remove $F1$.

Step 3. Move traveler to $L4$.
Erect steel between $L4$ and $L6$.

Step 4. Move traveler to $L6$.
Erect steel between $L6$ and $L8$.
Jack up at $L8$ or jack down at falsework bent $F2$ to close joint at $U7$.
Jack down at falsework bent $F2$ and release $F2$.

The above erection procedure applies to simple-truss bridges that fit the site and equipment conditions stated. There are many other methods of erecting simple-truss bridges. Figure 23-34 shows two alternates used where full falsework is required because of special erection conditions.

Fig. 23-36 Cantilever erection of continuous-truss span with two stiffleg derrick travelers.

Fig. 23-37 Top chord jacking point for suspended span of a highway bridge.

When the cost of placing falsework is prohibitive, simple-truss spans have been erected in a temporary position and floated into their permanent location as shown in Fig. 23-35.

A saving in the number of falsework bents can sometimes be made when a series of similar adjacent simple-truss bridges is to be erected. By employing the use of a temporary tie between spans as shown in Fig. 23-33, Step 5, the falsework bent required at $L10$ can be eliminated.

Multiple-span Truss Bridges In multiple-span bridges of the cantilever or continuous type, similar steps are employed for the erection of the anchor arms. The conventional method for erecting the center spans is to continue erection from each side by cantilevering from the anchor arms until the center is reached, as shown in Fig. 23-36. The joining of the two halves of the bridge is then accomplished by jacking. Hydraulic jacks in the chord sections provide the adjustment necessary in cantilever spans.

Figure 23-37 shows a top chord joint in a cantilever bridge that contains the temporary hydraulic jacks used in joining the two cantilevered halves. For continuous spans, jacks located at the abutments and on the main piers can move the span sufficiently to effect a closure at midspan.

One of the alternate methods of erecting multiple-span truss bridges is the balanced-cantilever method, in which erection is started at the main piers (Fig. 23-38) and

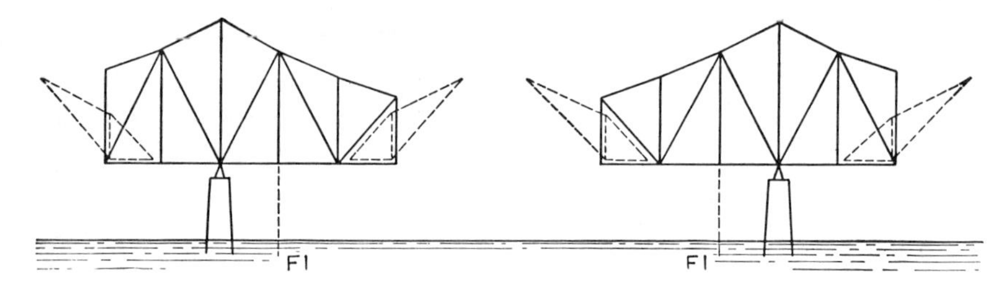

Fig. 23-38 Balanced cantilever erection.

Fig. 23-39 Highway truss span under construction showing balanced cantilever erection.

continues in both directions. In this method it is necessary to keep a close control on the steel erected on each side of the main pier to maintain a balanced condition. Figure 23-39 shows a cantilever bridge being erected using the balanced-cantilever method. On this bridge a temporary falsework span replaces the falsework bents usually used in the anchor arm.

Suspension Bridges The steps required in the erection of a parallel-wire suspension bridge are illustrated in Fig. 23-40.

Step 1. *Tower Erection.* The main towers of a suspension bridge are erected first. The first sections are usually erected with a floating crane or derrick boat which then assembles a creeper traveler. This traveler then erects successive tower sections and moves up the tower legs until the tower is completed. Figure 23-41 shows a suspension-bridge tower being erected by a creeper traveler.

Step 2. *Catwalk and Cables.* A working platform called a "catwalk" is erected in a position a few feet below the final elevation of the main cables. Each catwalk is made up of several large-diameter wire ropes which support a woven wire or wood walkway. A tramway system is then installed from anchorage to anchorage along the catwalk for pulling the bridge wires. At each anchorage a wheel, called "the spinning wheel," is attached to the tramway system. The reels of bridge wire, about $\frac{3}{16}$ in. in diameter, are placed on the anchorages; the end of the wire is fastened to the anchorage and looped over the spinning wheel. The wheels are then pulled along the catwalk and over the towers to the opposite anchorage, the wires are attached to the anchorage, and the procedure is repeated. At regular stations along the catwalk the wires pulled by the spinning wheel are collected and temporarily bound into bundles of about 300 wires each, called "strands." After the required

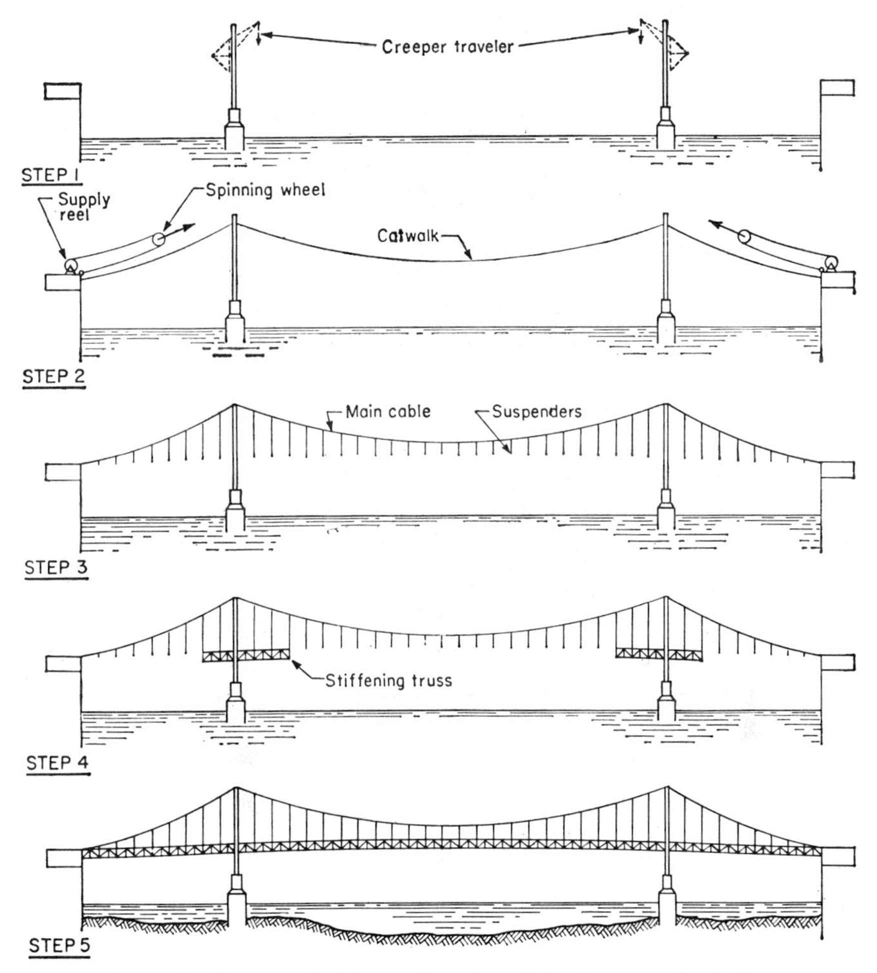

Fig. 23-40 Steps required in erection of parallel-wire suspension bridge.

number of strands has been placed, these strands are compacted into a circular section and bound securely. Figure 23-42 shows a typical cross section of a parallel-wire cable before compaction. Figure 23-43 shows the spinning of the main cables on a suspension bridge. When the strands are preformed before erection, the spinning wheel is eliminated and the strands are pulled from one anchorage over the towers to the opposite anchorage.

Step 3. *Cable Bands and Suspender.* Castings which are designed to grip the main cable are called "cable bands" and are placed around the cable at each point of suspension of the roadway. Wire-rope hangers or suspenders are attached to the cable bands to receive the roadway steel. The main cable between cable bands is then covered with a protective coating.

Step 4. *Stiffening Trusses.* The erection of the stiffening trusses starts at each tower and proceeds in both directions. The trusses and sufficient cross bracing are hung

Fig. 23-41 Suspension-bridge tower under erection with creeper traveler.

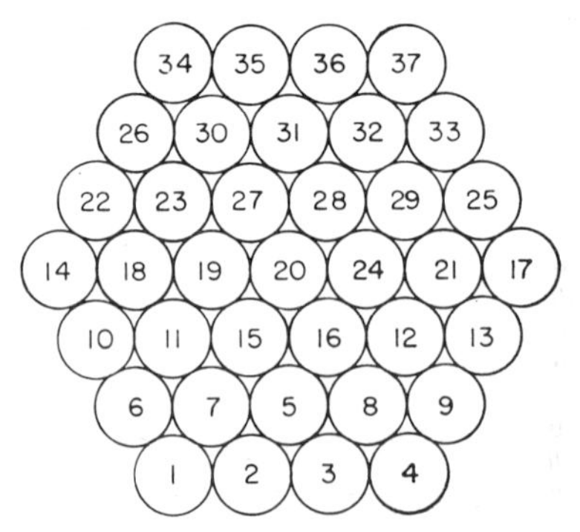

Fig. 23-42 Cross section of one cable of a parallel-wire suspension bridge showing strands before compaction.

off the suspenders at successive points until they meet at the center of the bridge and reach the anchorages at the ends of the bridge.

Step 5. *Floor System*. Following the erection of the stiffening trusses is the erection of the remainder of the floor system and the placing of the roadway surface.

In the erection of the self-anchoring suspension bridge, the steps are reversed. The floor system is placed first, supported on falsework, and followed by the towers. The cables are placed and the suspenders hung from the cable bands. The floor system is attached to the suspenders, its weight transferred to the cables through the suspenders, and the falsework released.

Fig. 23-43 Suspension-bridge cable under construction.

Beam-and-girder Bridges The erection of many beam-and-girder bridges can be performed with crawler or truck cranes operating under the spans. Little or no special planning is then required, other than a determination of the proper size of crane to handle the maximum lift. When the site conditions and the type of structure preclude the use of cranes to erect from below, the erector must devise specific erection procedures for each structure. In Fig. 23-44 a series of long simple-span girders are shown being erected with a stiffleg derrick traveler. The start of the erection of a continuous deck-plate girder span is shown in Fig. 23-45. For structures over navigable water it is often possible to use floating equipment such as derrick boats or cranes mounted on barges. A continuous-beam bridge under erection using a truck crane mounted on a barge is shown in Fig. 23-46. A method combining floating operations and a traveler is illustrated in Fig. 23-47. The girders at the pier

Fig. 23-44 Long simple-span highway girders under construction.

Fig. 23-45 Continuous deck-plate girder highway bridge under construction.

Fig. 23-46 Continuous-beam bridge under construction.

Fig. 23-47 Continuous-girder bridge under construction.

to the left have been assembled on barges and floated into position. After these girders are lowered onto the piers, the traveler erects the remaining girders to complete the span.

Single-track deck-plate girder spans required for railroad work, which cannot be erected by cranes from below, are usually too narrow to use a stiffleg derrick. For these spans a derrick car as shown in Fig. 23-48 is used. These derrick cars can erect

Fig. 23-48 Single-track deck-plate girder bridge being erected with derrick car.

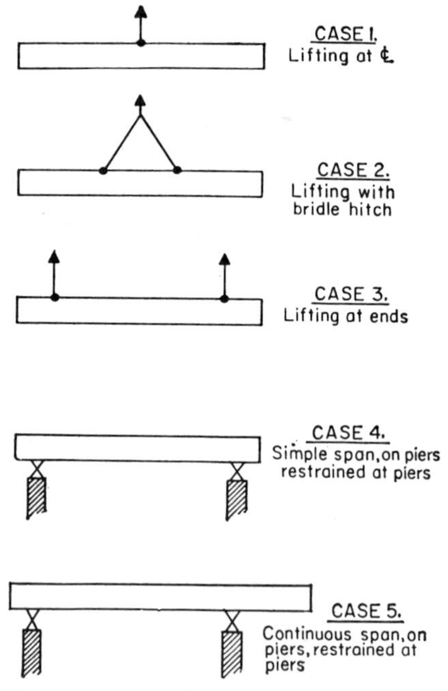

Fig. 23-49 Conditions where lateral buckling of long, slender girders may occur.

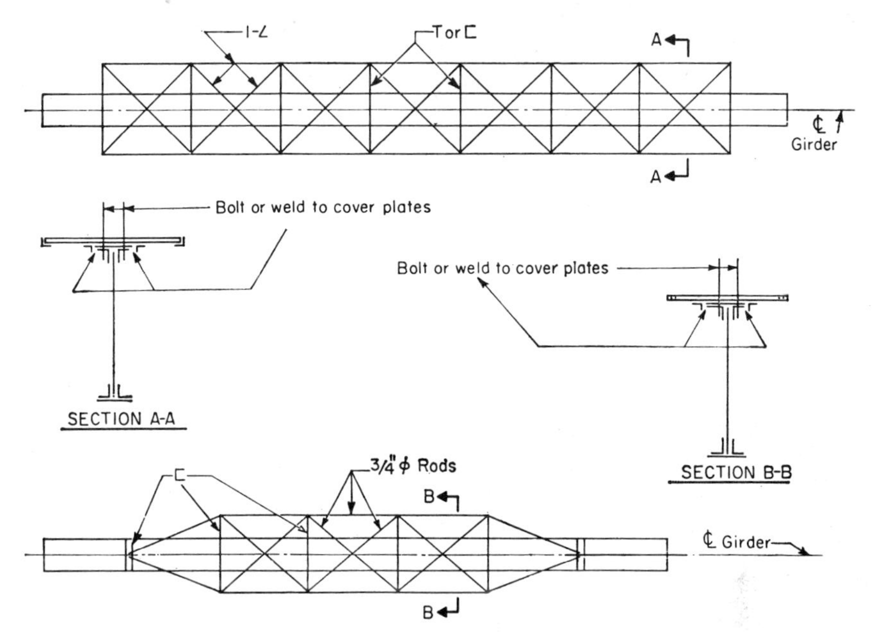

Fig. 23-50 Typical top-flange stiffening trusses for girders subject to lateral buckling.

Fig. 23-51 Erection of long girders with two crawler cranes. Top flange of girder supported by shores to ground.

the girders either singly or in pairs, depending on the capacity of the car and the length and weight of the girders.

The possibility of lateral buckling during erection is prevalent when long, slender girders are involved. Many of these girders are designed to be stable only when connected to adjacent girders or when their top flanges are made a part of the floor system of the bridge. Figure 23-49 indicates three typical cases where lateral buckling may occur during erection and two cases where it may occur after the girder is erected. Experience has shown that the approximate stability can be related to the ratio l/b, where l is the length of girder, in., and b is the width of cover plate, in. (or flange angles if there are no cover plates). As a general rule, girders with an l/b of less than 70 will be stable for all cases. When the l/b ratio is greater than 70, special provision must be made to prevent lateral buckling. One exception

Fig. 23-52 Manufacturing building erection with crawler cranes.

to these general rules is the girder in Case 2, Fig. 23-49, where an l/b ratio up to about 90 is stable. These guide rules are satisfactory for the general run of work and, in exceptional conditions, the permissible unit stress in axial compression due to the dead weight of the girder can be determined by the use of the secant formula.

When lateral buckling is indicated, a top-flange stiffening truss as shown in Fig. 23-50 can be used. These trusses are fastened to the girder while it is on the ground and are removed after the girder is erected and tied into the permanent structure. Although these stiffening trusses will prevent lateral buckling, they are expensive to furnish and place, and the extra weight to be lifted may require heavier hoisting equipment.

Several methods of erection for long, slender girders will eliminate the need for these stiffening trusses. One is to assemble two adjacent girders, with their bracing in place, on the ground and erect the assembled unit in one piece. Another method is to provide shores or guys to hold the top flange until the permanent bracing is erected, as shown in Fig. 23-51. A third method is to hold the girder with the hoisting equipment until an adjacent girder is erected and the bracing connecting the two has been erected.

BUILDING ERECTION

The erection of buildings which are within the reach and capacity of conventional crawler and truck cranes presents few erection-procedure problems. The order of erection follows the obvious pattern of columns, girders, and beams or columns, trusses, and purlins. Figures 23-52 and 23-53 illustrate typical building erection with conventional cranes.

Fig. 23-53 Powerhouse erection with crawler cranes.

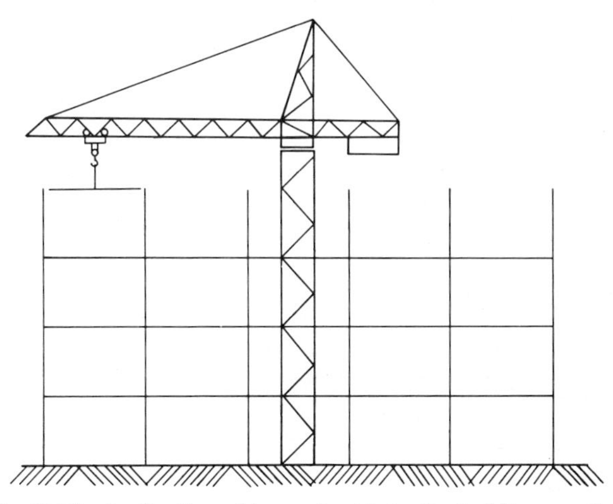

Fig. 23-54 Application of tower derrick to tier building erection.

Fig. 23-55 Tier building under construction with tower truck crane.

Fig. 23-56 Tier buildings being erected with guy derricks.

Tower derricks and tower cranes are also used in building construction, as illustrated in Figs. 23-54 and 23-55. These cranes and derricks permit horizontal reaches, with the unit adjacent to or inside the structure, that are not possible with a conventional boom.

Buildings beyond the reach or capacity of the above equipment are erected with guy derricks as shown in Fig. 23-56 or with stiffleg derrick creepers as shown in Fig. 23-57. The derrick creepers jump themselves up by means of lifting falls attached to the completed steel. Guy derricks are jumped as shown in Fig. 23-58. An explanation of each step follows:

Fig. 23-57 Tier building being erected with stiffleg derrick creepers.

Step 1. Guy-derrick mast is assembled on the ground, with its base in the approximate required location. The mast is tipped up vertically, and guys are anchored to column bases. The boom is inserted, and the topping lift and load lines are reeved. The derrick is ready to operate. The first-tier steel is erected.

Step 2. The boom is removed from the boom seat by picking with topping lift falls, revolved 180°, and placed in a temporary jumping shoe. The top of the boom is guyed off with temporary guys.

Step 3. The load falls are attached to the mast above its center of gravity, the mast guys are moved to the top of the next tier, and the mast is raised to its new position. The mast guys are adjusted and the load falls unhooked.

Step 4. The temporary guys on the boom are removed, and the topping lift falls are used to raise the boom and place it in the boom seat. The derrick is now ready to operate, and the next tier of steel is erected.

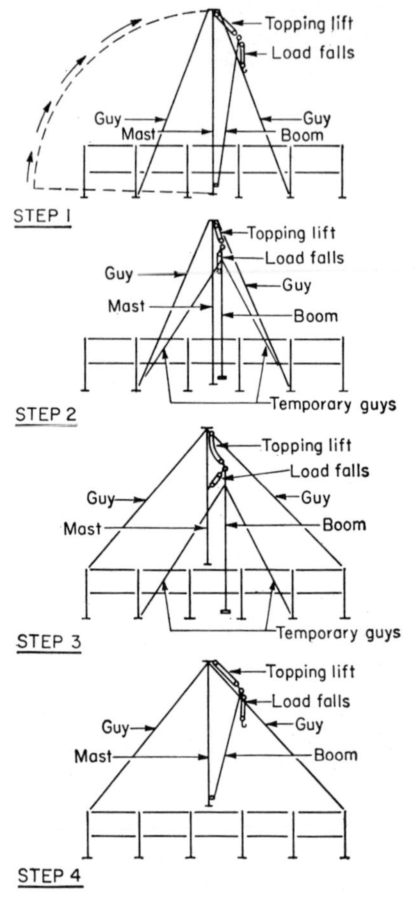

Fig. 23-58 Method of jumping guy derrick during tier building erection.

ERECTION COSTS

The costs incurred in the erection of structural steel can be categorized as direct labor costs, direct expenses, and indirect costs. The items of direct labor cost and direct expense can be defined as those incurred on account of a specific project and are chargeable directly to the project. The accounting practices and operating policy of the individual contractor will determine the indirect costs applicable to a project. These are usually expressed as a percentage of labor costs or direct expenses or both.

Direct Labor Costs These costs will vary with the current wage rates in the area and with the operations involved. The majority of the labor on steel erection is performed by structural ironworkers, boilermakers, and operating engineers. The balance of the direct labor includes supervisory, engineering, and clerical personnel and such incidental common labor as is required by the individual project.

The labor required is usually expressed in terms of man-hours per ton of steel erected. Assuming that a crew can erect 30 tons of steel in an 8-hr day with a crawler crane, the direct labor man-hours will amount to:

1 foreman at 8 hr = 8
1 operating engineer at 8 hr = 8
1 oiler at 8 hr = 8
4 ironworkers at 8 hr = 32
Direct labor man-hours = 56
56 ÷ 30 tons = 1.86 man-hours per ton

Erection with a guy derrick requires two additional ironworkers. The hoisting engine for a guy derrick remains on a lower floor as the derrick is jumped (Fig. 23-58), and one man is required to relay signals to the hoisting engineer. The guy derrick is usually swung manually, requiring another man. Using the same assumptions as above, the direct labor cost will then amount to:

1 foreman at 8 hr = 8
1 operating engineer at 8 hr = 8
6 ironworkers at 8 hr = 48
Direct labor man-hours = 64
64 ÷ 30 tons = 2.13 man-hours per ton

The number of tons of steel that can be erected per day will naturally vary with the type of construction. On structures that have many small, light pieces, the tons that can be erected per day will be less than on a structure that has only heavy pieces. Average tons of steel erected per day are listed in Table 23-5.

TABLE 23-5 Average Tons of Steel Erected per Day for Various Types of Structures

Type of structure	*Average tons of steel erected per day*
Schools	10–25
Warehouse buildings	15–30
Manufacturing buildings	20–35
Tier buildings	20–40
Powerhouses	10–30
Beam bridges	25–75
Girder bridges	20–50
Truss bridges	10–35

This labor cost of placing rivets and bolts can be expressed in man-hours per unit. This cost is influenced by the size of the rivet or bolt. It is also related to the number that it is possible to place at one location, since the cost of moving tools and hanging scaffolds at each point is usually included as part of the rivet or bolt cost. Assuming that a four-man riveting gang can drive 300 rivets in one day and that a two-man bolting gang can place 300 bolts in one day, the per-unit man-hours would be:

RIVETS:
1 foreman ½ time, 4 hr = 4
1 compressor operator ½ time, 4 hr = 4
4 ironworkers 4 × 8 hr = 32
Direct labor man-hours = 40
40 ÷ 300 rivets = 0.133 man-hour per rivet

BOLTS:
1 foreman ½ time, 4 hr = 4
1 compressor operator ½ time, 4 hr = 4
2 ironworkers 2 × 8 hr = 16
Direct labor man-hours = 24
24 ÷ 300 bolts = 0.08 man-hour per bolt

Direct Expense The materials and services which are required during the construction period and consumed on the job are usually considered as a direct cost of steel erection. These direct-expense items include the following typical examples:

Field-office construction and maintenance
Lease of receiving area
Transportation of men
Grease, oil, and fuel
Oxygen and acetylene
Freight on equipment
Local hauling of equipment
Local hauling of steel
Railway and highway flag protection
Demurrage
Miscellaneous stores and supplies
Union welfare and pension contributions

These items of direct cost must be determined for each project. As a general rule, their total will average 15 to 30 percent of the direct labor cost

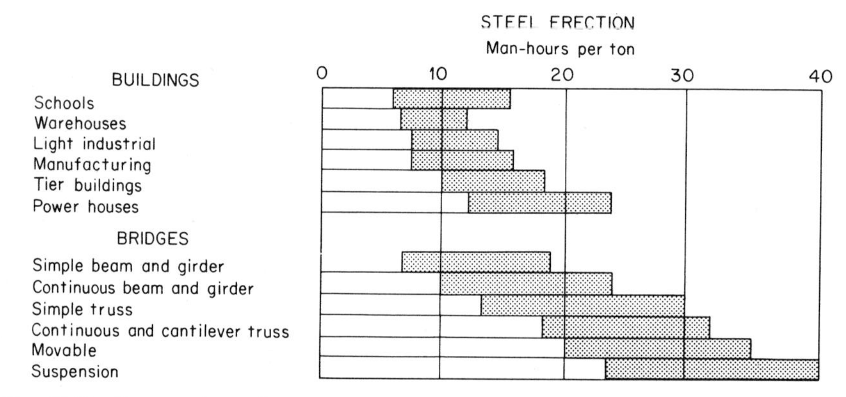

Fig. 23-59 Range of man-hours per ton erected for various types of steel structures.

Indirect Costs The method of costing the expenses which are not directly chargeable to an individual job or which are chargeable to several jobs varies with individual management policy. Some items are considered a part of overhead and are expressed as a percentage of direct labor cost or of direct labor cost and direct expense. For other items the individual job is charged according to the length of time the item is used or according to the quantity of material used.

Average Costs Because of the many variables affecting the field operations in steel erection, erecting prices vary widely depending on the site conditions and the types of structure. In Fig. 23-59 are shown the range of man-hours per ton for the erection of various types of structures.

B. Electric-arc Welding

PRINCIPLES AND PROCESSES

Electric-arc welding employs the heat of an electric arc to melt metals and effect their joining by the intermixing of substance in the joint area. In the type of arc welding used in steel construction, molten filler metal—usually supplied by a covered electrode—mixes with the melted "base" metal of the parts being joined and bridges the parts. As it cools, the molten metal solidifies and the parts become essentially one.

In electric-arc welding, the work to be welded is made part of an electric circuit known as the "welding circuit" (Fig. 23-60). The source of power is a welding generator, transformer, or transformer-rectifier. One cable carrying current from the power source is attached to the work, and another cable is attached to an electrode holder. An arc is established between the electrode and the work when the two are brought in close proximity. The arc is moved along the work, melting the filler and base metal as it progresses. Since the heat of the arc is intense, melting is almost instantaneous.

Various processes employ the electric arc to obtain welding heat. Each process has its advantages, but all have one requirement in common—namely, a need to protect or shield the molten metal from the chemical action of oxygen and nitrogen in the atmosphere. Molten metal has strong affinity for both oxygen and nitrogen, and unless these gases are excluded from the arc area, oxides and nitrides will be formed as impurities that embrittle and weaken the weld. The arc welding processes used in steel construction all include some method of shielding the arc and molten metal from the atmosphere. Welds correctly made by these processes are as strong as the metal of the parts joined, or even stronger.

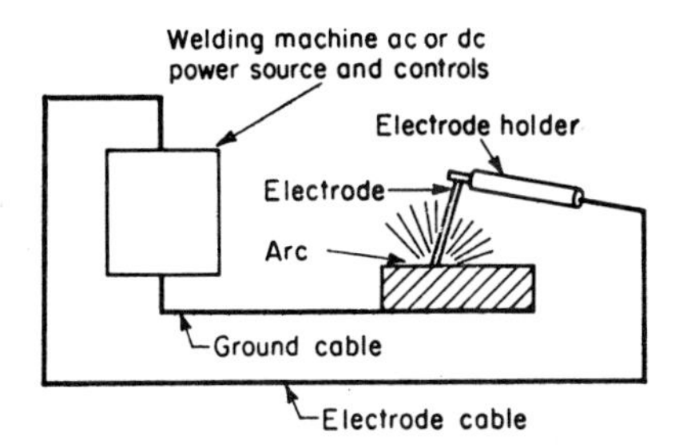

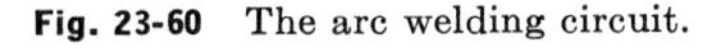

Fig. 23-60 The arc welding circuit.

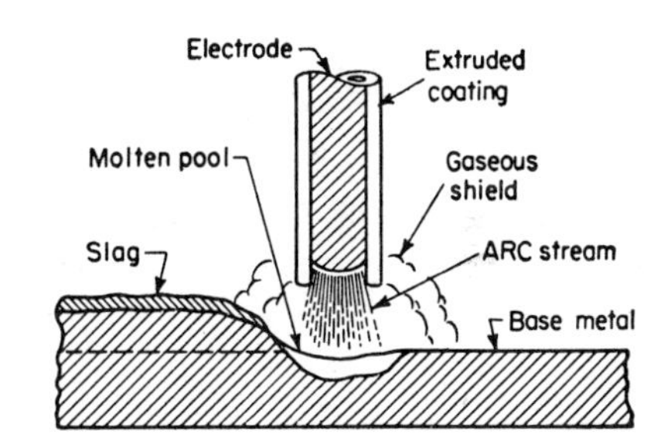

Fig. 23-61 The manual electrode-shielded arc.

Manual Self-shielded Arc Welding This is also called "hand" or "stick-electrode" welding (Fig. 23-61), and it is the most widely used process in the fabrication shop and in the field. A consumable electrode, which is maneuvered by hand along the joint to be welded, supplies filler metal for the weld deposit, produces a gaseous shield around the arc and a slag deposit over the newly solidified metal to prevent atmospheric contamination, and provides scavengers and deoxidizers to refine the grain structure of the weld. The shielding and refining ingredients come from a coating applied to the electrode during manufacture. Manual electrodes range in length from 9 to 18 in. and, for use, are clamped in a device called an "electrode holder." Welding begins by touching the tip of the electrode against the work, then withdrawing it to establish an arc. As the tip of the electrode is consumed, the electrode is progressively lowered to maintain a uniform distance (usually $1/16$ to $3/32$ in.)

between it and the work. Simultaneously, the operator moves the electrode along the work at a rate such that the molten filler metal forms the desired weld size.

Semiautomatic Self-shielded Arc Welding This process, illustrated in Fig. 23-62, is rapidly taking over many of the shop and field welding applications of stick-electrode welding. Self-shielded arc welding provides increased weld deposit rates with cost reductions as great as 75 percent. The wire electrode used with the semiautomatic adaptation of this process is mechanically fed through a welding gun to the arc from

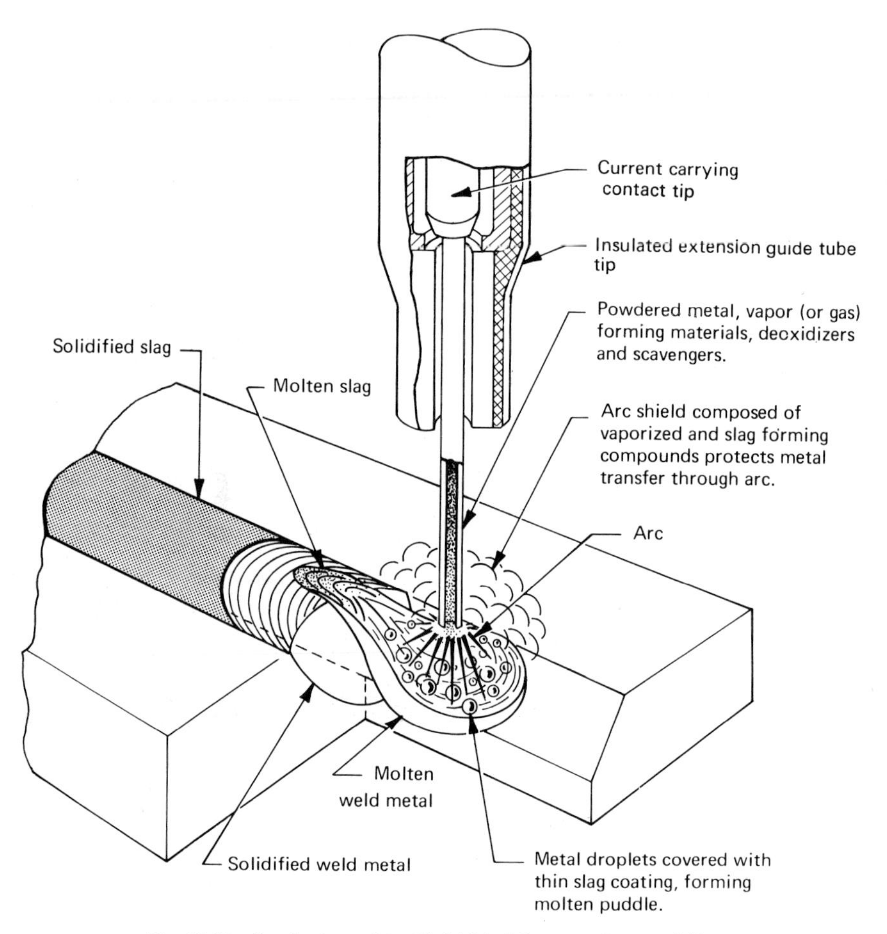

Fig. 23-62 Semiautomatic self-shielded flux-cored arc welding.

a 50-lb coil. Not only is the frequent changing of electrodes eliminated with the semiautomatic gun, thus saving welding time, but higher amperages can be used with it than with the stick-electrode holder. This increases the rate of deposit and permits faster welding speeds.

The construction of the continuous electrode wire used with semiautomatic self-shielded arc welding differs from that of the manual electrode. The filler metal is now on the outside, and the ingredients for shielding, scavenging, deoxidizing, and fluxing are contained in a center core. Such electrode wires are referred to as "flux cored." Operator maneuvering with the semiautomatic gun is similar to maneuvering with the stick electrode. The trigger on the welding gun completes the welding circuit and also initiates wire feed.

Full-automatic Self-shielded Flux-cored Arc Welding The process utilizing flux-cored electrode wire may also be used with full-automatic welding heads. The need for fixtures and manipulators tends to limit the application of full-automatic welding in construction work. It is well suited to the shop fabrication of repetitive parts used on the construction job, such as tapered girders.

Submerged-arc Welding In this process, the arc and the molten metal pool are protected by submerging the arc in a bed of granular, inorganic flux (Fig. 23-63). The granular flux is deposited automatically around the electrode wire as it is fed to the work, and the arc is completely hidden. The flux may be "neutral"—made

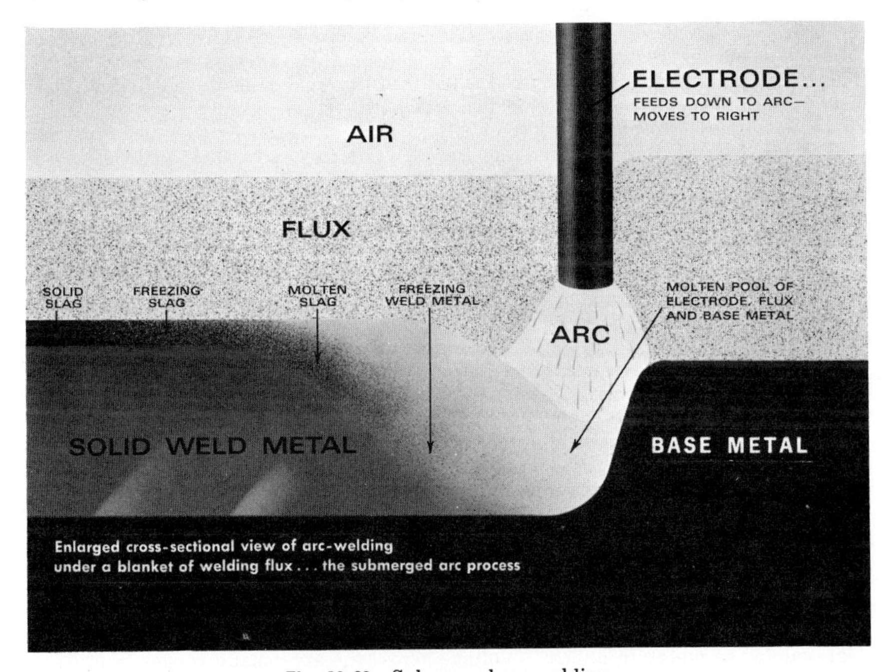

Fig. 23-63 Submerged-arc welding.

up merely of protective shielding, deoxidizing, and slag-forming ingredients—or it may contain alloying elements that enter into the chemical and structural composition of the weld metal. Granules of flux unfused during welding may be collected and reused.

Submerged-arc welding may be performed with either semiautomatic or full-automatic equipment, using a mechanical electrode feeder to advance the electrode wire to the hidden arc. The high currents result in deep penetration, making the process excellent for the fabrication of structural members where deep penetration aids in the development of weld strength. In field welding, the submerged-arc method has limited use, since gravity requires that the bed of granules be laid down on work that is in the flat position. In the fabrication shop, however, flat positioning can be achieved by the use of weldment positioners and fixtures. Here the process finds acceptability not only for its speed, efficiency, and deep-penetration characteristics, but also for the fact that it minimizes distortion.

Gas-shielded Arc Welding This process employs a shield of gas which is carried through a tube to the arc area and released during the welding operation to protect the molten metal from atmospheric contamination. The gases most commonly used are the inert gases, argon and helium, and carbon dioxide (CO_2), a gas nonreactive with molten metal.

CO_2 gas-shielded arc welding is commonly called "CO_2 welding" and is used extensively in steel construction. The electrode may be either a solid wire, a fabricated flux-cored electrode, or a flux-coated electrode. The electrodes or the flux associated with them contain elements to perform scavenging and deoxidizing action in the molten metal pool. Although producing high-quality welds, CO_2 welding has disadvantages in fieldwork because winds are likely to blow away the gaseous shield and because it is awkward for the operator to handle a semiautomatic welding gun to which a gas tube must be attached. For these reasons, self-shielded flux-cored arc welding is generally preferred for field uses such as in the field splicing of high-rise building columns and the joining of beams to columns.

Gas-tungsten arc welding is another variation of the gas-shielded arc welding process. The arc is struck between a virtually nonconsumable tungsten electrode and, if filler metal is required, a welding rod is fed into the weld zone and melted with the base metal. The weld zone is shielded with an inert gas, usually argon or helium, which is fed through the welding torch.

The process is suitable for making fusion welds in nearly all industrial metals, particularly aluminum, aluminum magnesium alloys, other nonferrous metals, and the gauge thicknesses of stainless steel. It is also applicable for welding various combinations of dissimilar metals. Gas-shielded tungsten arc welds, because of positive protection from the atmosphere, tend to be stronger, more ductile, and more corrosion resistant than welds made with other processes. Corrosion due to flux entrapment does not occur, and postwelding cleaning is minimized. The entire welding operation takes place practically without spatter, sparks, or fumes. A major disadvantage of the process is cost, which limits its use in construction welding to assemblies and joints critical enough to warrant the expense.

Other Welding Processes There are nearly two dozen other commercially used welding processes, some of which may have limited application in construction work. Electroslag and electrogas welding are accepted by the American Institute of Steel Construction for the fabrication of steel structures, along with the self-shielded metal arc welding processes, the gas-shielded processes, and the submerged-arc process.

EQUIPMENT FOR ARC WELDING

The equipment for arc welding is highly developed and specialized, with many variations and options to suit the particular job and add to welding efficiency. In general, the equipment is designed to provide a continuous supply of current at a sufficiently high voltage to maintain an arc. Voltage across the arc usually ranges from 17 to 45 and, in operation, is constantly varying because of changes in arc conditions. Currents range from 10 to as high as 1,200 amperes, with the latter applicable to automatic operations. Either direct current or alternating current may be used, although the former is more common for field welding.

Alternating versus Direct Current Satisfactory welding can be accomplished with either alternating or direct current. Each type of current has advantages which produce maximum efficiency under certain conditions.

Alternating 60-cycle current reverses direction 120 times a second. This feature produces both advantages and disadvantages in welding. Because of the current reversal, the magnetic field that is set up in the work metal by the flow of electricity also reverses. This, in effect, prevents the magnetic field in the arc area from building up and deflecting the arc from its intended path. Alternating current is therefore less susceptible to the condition known as *arc blow*, which is experienced with direct current. This means that, if arc-blow conditions exist and an alternating current is used, it is not necessary to reduce current or make other time-losing adjustments that are necessary with direct current. Higher currents and larger electrodes can be used. Alternating current, therefore, is generally faster than direct current for production welding of plate, particularly on short seams with currents above the 250-ampere range.

Certain electrodes, namely those with iron powder in their coatings, have been designed so that they can be used with higher values of alternating current than direct current. Therefore, on downhand production welding jobs, the greatest speeds can be obtained by using these electrodes with alternating current.

The disadvantages of alternating current also stem from its alternating characteristic. Any electrode which cannot be designed to operate in the inherently less stable arc of alternating current must be used only with direct current. Reverse-polarity electrodes, low-hydrogen electrodes, stainless-steel electrodes, hard-surfacing electrodes, and nonferrous electrodes are generally better suited to dc operation. Some of these types can be used but do not perform as well with ac current.

One additional problem with alternating current is the fact that the pulsating arc is less suited to welding thin-gauge metals. The arc is difficult to sustain at low currents and voltages and is less smooth. Metal transfer is pulsating rather than in a steady flow, causing more spatter. The electrode cannot be moved as fast, which tends to cause metal to pile up. In addition, a transformer welding current fluctuates with variations in input voltage, so that burn-through tendency on thin-gauge metals is more difficult to control with alternating current.

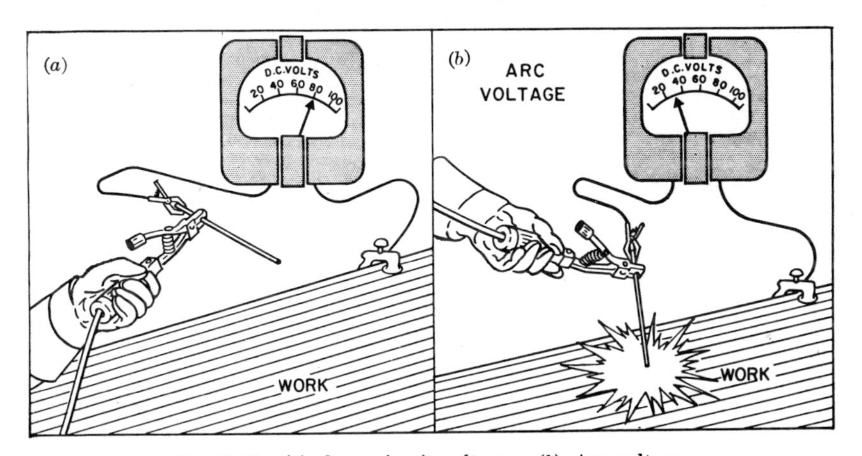

Fig. 23-64 (*a*) Open-circuit voltage. (*b*) Arc voltage.

Direct current is also preferred for out-of-position welding or where the arc must be crowded. The dc arc has less tendency to short out as globules of metal tend to bridge the arc gap.

Direct current permits changing the polarity of the electrode from either positive to negative. This change makes distinct differences in the operation of certain electrodes. For example, E6010 and E6011 electrodes operate best with electrode positive (reverse) polarity when welding heavy plate. For welding thin-gauge metals, such as 14, 16, and 18 gauge, they operate best with the electrode negative (straight) polarity. The change in polarity changes the nature of the metal transfer across the arc.

Summarizing, alternating current is better than direct current for welding where arc blow is encountered and for high-speed downhand welding with iron-powder electrodes. Direct current is better for welding stainless, hard-surfacing, nonferrous, out-of-position, and sheet metal and with electrodes whose operation is sensitive to polarity.

DC Generators These are the most widely used source of welding current for construction work, since they provide maximum arc control and maximum versatility for welding under all conditions. Most of these generators will have dual controls to permit the regulation of both volts and amperes. To understand the need for dual controls, it is necessary to gain an understanding of the effects these controls produce. This, in turn, requires reference to the two types of welding voltages, open-circuit voltage and arc voltage (Fig. 23-64).

Open-circuit voltage is the voltage generated by the welder when no welding is being done. Arc voltage is the voltage drop across the arc during welding. The open-circuit voltage drops to the arc voltage when the arc is struck and can be varied between 50 and 100 volts by means of the voltage-control rheostat. Arc voltage (between 16 and 40 volts) is determined by the design of the electrode and is varied by changing the arc length (Fig. 23-65). If the arc is shortened, the arc voltage decreases. If the arc is lengthened, the arc voltage increases.

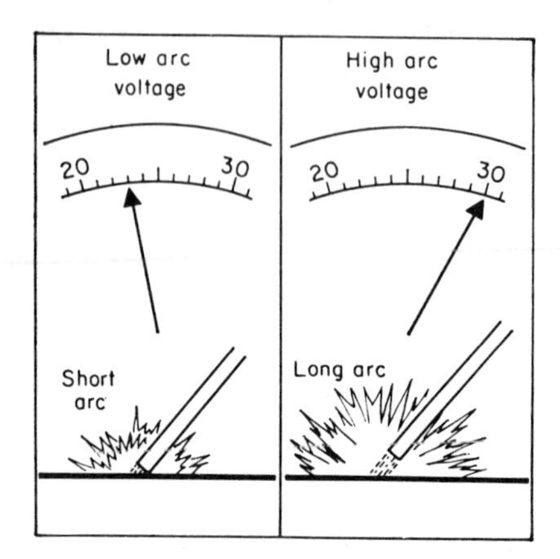

Fig. 23-65 Arc length is related to arc voltage.

Because of differences in metals or welding positions, different welding operations will require different arc characteristics. Adjustable voltage and current controls make possible a choice of different arc characteristics to match different jobs and welding conditions.

A *volt-ampere curve* is a graph that shows all the conditions of output for a welder at one given set of control settings. Consequently, with two continuous controls, an infinite number of volt-ampere curves can be obtained. By varying only the voltage control, the open-circuit voltage can be changed to produce any desired volt-ampere curve, such as those shown in Fig. 23-66. By varying only the current control, the welding current is changed, such as shown in Fig. 23-67.

In the welding operation, the desired welding current is known. Consequently, the important thing to understand is how the current varies while one is welding under different voltage settings. Reference to two volt-ampere curves having the *same normal welding current* but different control settings (see Fig. 23-68) will help explain how current varies with different voltage settings. Curve A, with the high open-

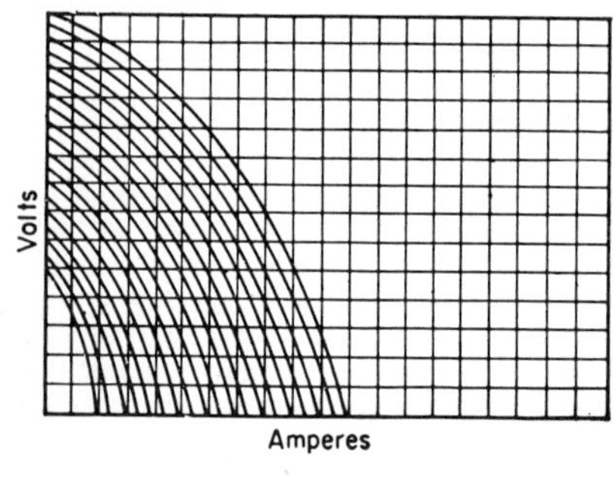

Fig. 23-66 Volt-ampere curves obtained by varying only voltage.

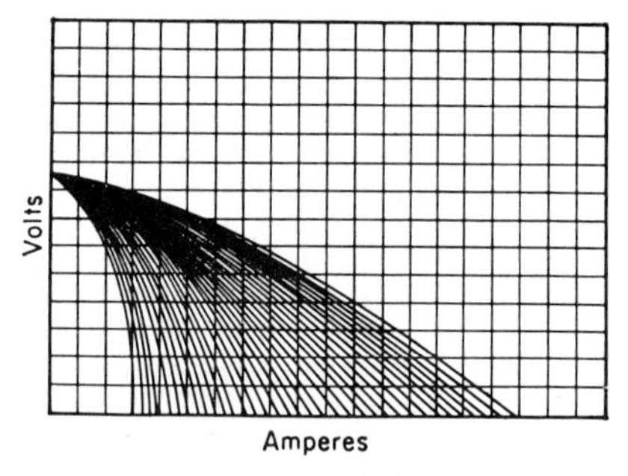

Fig. 23-67 Volt-ampere curves obtained by varying only current.

circuit voltage, is for a voltage setting in a normal operating range. Curve B, with the low open-circuit voltage, is for a vertical and overhead range setting. The normal welding current on *both* curves is 140 amperes at 25 arc volts and is shown by point N. However, if the arc length is varied by the operator, either accidentally or on purpose, then the different curves give different results.

In the case of curve A, shortening the arc to give 20 volts moves the point down to H, which changes the welding current to about 150 amperes. On the other hand, with the controls set to give a curve such as *B*, shortening the arc to 20 volts would move the point to *L* and increase the current to about 180 amperes. When the electrode is shorted against the work (arc voltage = 0), curve *A* shows a current of a little over 180 amperes whereas curve *B* would have a short-circuit current of about 240 amperes. Conversely, lengthening the arc on curve *A* decreases the welding current moderately, but, on curve *B*, it decreases the welding current considerably.

For *downhand welding and sheet-metal applications,* no change of current during welding is desired. Voltage should be set in a normal welding or large electrode range, which gives a medium to high open-circuit voltage setting (curve *A*, Fig. 23-68). Thus, very little change in current will result when an unsteady hand shortens or lengthens the arc.

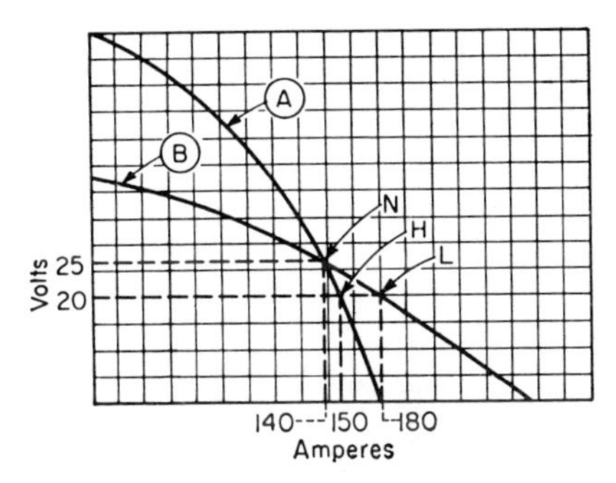

Fig. 23-68 Flat versus steep volt-ampere curve.

When welding in the *vertical or overhead position,* the operator wants to alternate a strong digging arc for penetration with a cool arc which permits the molten metal to solidify. With the welder set for a low open-circuit voltage, as on curve *B* in Fig. 23-68, the operator can control the current merely by changing the length of his arc. If he crowds the electrode into the fillet, it will increase the current and give good penetration and a high deposition rate. As he withdraws the arc away from the work, the current will cut down to a lower value, cooling off the puddle and letting the metal solidify. Thus, in-and-out motion to control the current, combined with a forward and backward "whipping" motion, gives the operator complete control over his puddle in vertical and overhead welding.

The usual procedure for adjusting the machine should be to first set the voltage for the type of application and then adjust the current control to that desired. Practically all current adjustments can be made with the current control. However, if a slight increase or decrease in current is needed, a fine adjustment can be obtained by slightly changing the voltage setting. Such fine adjustments will not materially change the arc characteristics.

Three-phase rectifier welders also produce direct current but do not offer dual control. They usually have a "step type" current control with a limited number of settings. Thus, the voltage control must be used for intermediate current and the number of volt-ampere curves for a specific current setting is limited.

AC/DC Welders The universal combination ac/dc welder, consisting of an ac transformer and a rectifier, is used to good advantage in construction work. This type of welder permits the selection of alternating current with iron-powder type electrodes for fastest downhand welding, then switching to direct current for fastest overhead and out-of-position welding.

Modern machines are designed to operate with an open-circuit voltage that is safe and much smaller than that of older types of equipment. Units are also available with low-voltage contactors that still further reduce the open-circuit voltages and thus reduce the hazard of electrical shock that, in the past, has discouraged the use of ac equipment for field welding.

Since transformer welders operate on single-phase power, a satisfactory single-phase or three-phase power source must be available. Where it can be used, this type of machine will provide the most efficient operation and lowest power cost. Diesel-powered generators are generally not satisfactory for operating, since the line load for this auxiliary power generator must be balanced—a requirement that is difficult to meet with single-phase welders.

Cables Welding current is conducted from the generator or transformer to the arc by an insulated copper or aluminum cable. This cable is made up of numerous hair-like wires enclosed in a durable paper wrapping surrounded by an insulating material, a natural or synthetic rubber. The paper wrap allows the conductor to slip readily within the insulation when the cable is bent. A high-grade insulation contributes to flexibility. Wear resistance is provided by a tough braided-cotton reinforcing and by the special composition of the outer covering.

Cable sizes vary, ranging from No. 4 up to No. 0000. Size selection is based on total cable length, electrode plus ground, and on welder capacity. Table 23-6 gives recommended cable sizes for given machine capacities and distances. Using undersized cables causes an excessive voltage drop in the cable and impairs both welder performance and the ability to produce quality welds. Figure 23-69 is a nomograph useful in determining the voltage drop.

TABLE 23-6 Cable Sizes

Machine sizes, amp	Cable sizes for combined lengths of electrode and ground cable				
	0–50 ft	50–100 ft	100–150 ft	150–200 ft	200–250 ft
200	2	2	2	1	1/0
300	1/0	1/0	1/0	2/0	3/0
400	2/0	2/0	2/0	3/0	4/0
600	2/0	2/0	3/0	4/0	Two 3/0
900*					

* Used for automatic applications only.

Current Measuring Devices Since a proper value of current is essential for best procedure, an accurate method of measuring current is needed. Meters in the welding circuit and calibrated dials on the welding machine are used but have definite deficiencies. Meters in the welding circuit, regardless of their accuracy when new, eventually become inaccurate or broken because of continual oscillations of the pointer under welding service and vibration. Dial calibrations, although quite accurate with normal lengths of cable leads, do not take into account added lengths of cable or poor connections. Another method of current measurement is the use of the clip-on or tong-type ammeter which clamps over the welding cable and indicates the amperage directly.

If meters or dials of known accuracy are not available, the most satisfactory method of determining the current is by measuring the number of inches of electrode melted off in 1 min. Specified melt-off rates are given on procedure sheets. These melt-off rates are approximately those that will be obtained when the recommended current is used. It may be found that a melt-off rate up to 10 percent higher or lower than the value given may be more desirable because of arc blow, fit up, or some other operating condition. In such cases, the melt-off rates are specified as the optimum to be used.

As an example, when the procedure sheet for the joint to be welded recommends $\frac{3}{16}$-in. E6010 electrode at 180 amperes, the desired melt-off rate is found to be approximately $9\frac{1}{2}$ in. per min. To measure the current, one sets the controls of the welding machine at 180 amperes and welds a new length of electrode for exactly 1 min. If less than 8 in. of electrode are melted off, the current setting should be increased until additional test runs give a $9\frac{1}{2}$-in. melt-off. Conversely, if more than $9\frac{1}{2}$ in. are melted off, the current may be adjusted downward. In the latter case, it is not necessary to adjust the current all the way down to a $9\frac{1}{2}$-in. melt-off rate unless problems of poor surface appearance or burn-through are encountered.

Other Items In addition to the power sources, welding machines, and cables, various other items of equipment are used in construction welding. These include electrode holders, semiautomatic guns, wire feeders, welding heads and nozzles, attachments for guns and heads to vary the electrical stick-out of wire-type electrodes, equipment for metering shielding gas, tractors and portable motorized trackless carriages for carrying welding heads and guns along the joint, and many types of full-automatic systems, fixtures, and manipulators for shop fabrication. The manufacturers' literature provides information on their adaptability to the type of welding to be done.

With the increasing use of semiautomatic self-shielded welding on the construction job (Fig. 23-70) to replace slower stick-electrode welding, the automatic wire feeder is becoming an item of equipment almost as standard as the power source. The feeder

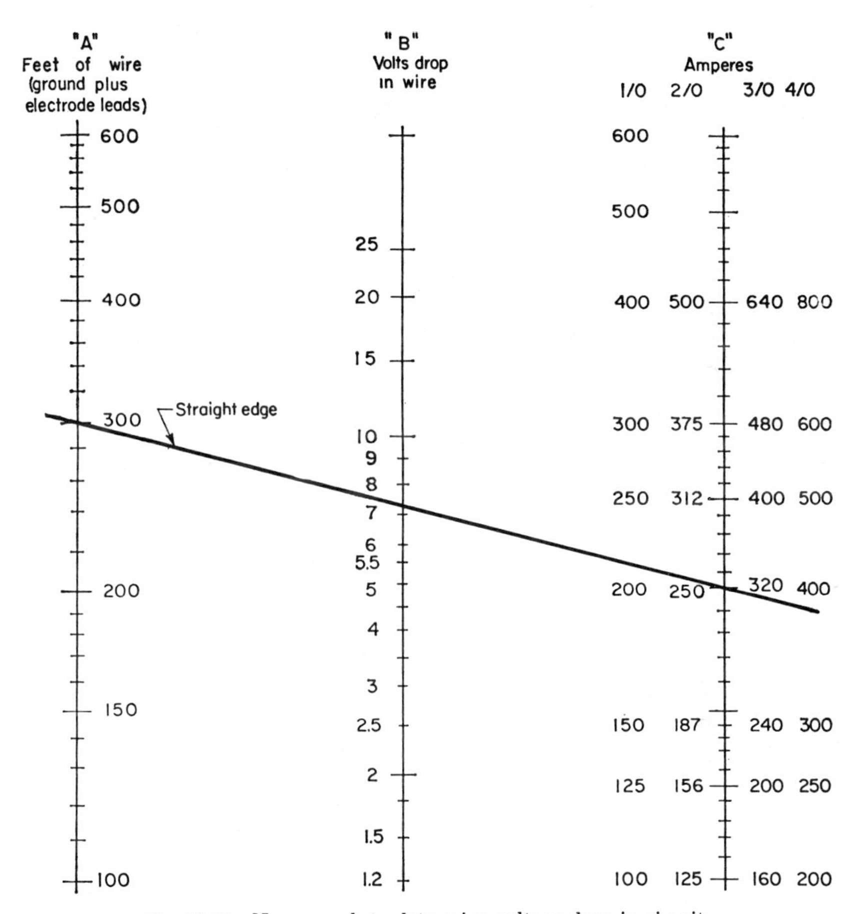

Fig. 23-69 Nomograph to determine voltage drop in circuit.

mechanically pushes or pulls the continuous electrode to the semiautomatic welding gun, depending on where it is located in respect to the coil of electrode wire. Welding current controls are located on the feeder, and speed of feed is adjustable through ranges as great as 32 to 500 in. per min without gear changes. Known as "squirt welders," these feeders are used with both the semiautomatic self-shielded open-arc welding process and the semiautomatic submerged-arc process. Wire feed is actuated with the press of the trigger on the welding gun that completes the welding circuit. With submerged arc, a flux tank is incorporated in the equipment or flux is provided from a separate source as feed to the nozzle surrounding the electrode tip.

To increase the operator's freedom of movement, an extension cable between the coil of electrode wire up to 45 ft in length is provided as an option. In this case, the feeder pulls the electrode 45 ft from the coil and pushes it another 15 ft to the semiautomatic gun. By the mere moving of the small feeder unit as required, the operator thus has a freedom of movement up to 60 ft. This factor is especially important in high-rise building or ship construction, preventing excessive movement of the heavier items of welding equipment.

Fig. 23-70 The wire feeder (left foreground) gives the construction welder substantial freedom of movement. Here, a beam flange-to-column connection is being made with the semiautomatic self-shielded flux-cored process.

Long stick-out extension guides for semiautomatic self-shielded open-arc, semiautomatic submerged-arc, and full-automatic submerged-arc guns and heads are simple devices that enable increased deposition rates without increasing heat inputs. With these guides, deposition rates can be increased up to 50 percent where deep penetration is not a requirement.

ELECTRODE SELECTION

General Electrode selection is based in part upon the steel type—and there are many types of carbon and alloy steels used in construction work. The American Institute of Steel Construction in its 1969 revised *Specifications for the Design, Fabrication and Erection of Structural Steel for Buildings* recognizes 12 ASTM steel types for structural use. These steels are designated by the American Society for Testing and Materials as A36, A53, A242, A375, A441, A500, A501, A514, A529, A570, A572, and A588. Table 23-7 gives the compositions and mechanical properties of these steels.

To make full-strength welds in these structural steels, the electrode must provide a weld metal with a tensile strength matching or exceeding the tensile strength of the specific steel. Thus, an E60XX manual electrode producing filler metal in the range of 62 to 67 ksi (kips per square inch) would be used with A36 steel, which has a minimum tensile strength of 58 ksi, whereas an E70XX electrode producing filler metal of 72 ksi would be suitable for A441 high-strength, low-alloy steel with a minimum tensile strength of 70 ksi.

TABLE 23-7 AISC Recognized Structural Steels

ASTM GRADE and Descriptive Information	Mechanical Properties			Material Shape	Thickness Group Or Grade	Chemical Requirements (Ladle) Percent						
	Tensile Strength† (ksi min.)	Yield Point (ksi min.)	Elongation in 2 in. % min.			C Max.	Mn† Max.	P Max.	Si† Max.	Si †	Cu Min.	V Min.
A36 Structural Steel	58 to 80	36	23	Shapes		0.26	–	0.04	0.05	–	0.20*	
	58 to 80	36	23	Plate	to 3/4 incl.	0.25	–	0.04	0.05	–	0.20*	
	58 to 80	36	23	Plate	over 3/4 to 1-1/2 incl.	0.25	0.80 - 1.20	0.04	0.05	–	0.20*	
	58 to 80	36	23	Plate	over 1-1/2 to 2-1/2incl.	0.26	0.80 - 1.20	0.04	0.05	0.15 - 0.30	0.20*	
	58 to 80	36	23	Plate	over 2-1/2 to 4" incl.	0.27	0.85 - 1.20	0.04	0.05	0.15 - 0.30	0.20*	
	58 to 80	36	23	Plate	over 4 to 8 incl.	0.29	0.85 - 1.20	0.04	0.05	0.15 - 0.30	0.20*	
	58 to 80	36	23	Bars & Bar Shapes	to 3/4" incl.	0.26	–	0.04	0.05	–	0.20*	
	58 to 80	36	23	Bars & Bar Shapes	over 3/4" to 1-1/2"incl.	0.27	0.60 - 0.90	0.04	0.05	–	0.20*	
	58 to 80	36	23	Bars & Bar Shapes	over 1-1/2" to 4" incl.	0.28	0.60 - 0.90	0.04	0.05	–	0.20*	
A53 GRADE B Welded & Seamless Steel Pipe electric resistance or seamless (only chemistry limit is phosphorous)	60 min.	35	**		B			**				
A242 High Strength Low Alloy Structural Steel	70 min.	50	**	Plates & Bars	to 3/4" incl.	0.22	1.25		0.05			
	67 min.	46	**	Plates & Bars	over 3/4 to 1-1/2 incl.	0.22	1.25		0.05			
(other alloying elements may be added.) If Cmax ≤0.15, $Mn_{max.}$ can be raised to 1.40	63 min.	42	24	Plates & Bars	over 1-1/2 to 4" incl.	0.22	1.25		0.05			
	70 min.	50	**	Struc. Shapes	I	0.22	1.25		0.05			
	67 min.	46	**	Struc. Shapes	II	0.22	1.25		0.05			
	63 min.	42	24	Struc. Shapes	III	0.22	1.25		0.05			
A375 High Strength, Low Alloy Hot Rolled Steel Sheet Strip	70 min.	50	22			0.22	1.25		0.05			
When used for welding, the chemistry shall be checked for weldability based on evidence acceptible to the buyer												
A441 High Strength Low Alloy Structural Manganese Vanadium Steel	70 min.	50	**	Plates & Bars	to 3/4" incl.	0.22	.85 - 1.25	0.04	0.05	0.30	0.20	0.02
	67 min.	46	**	Plates & Bars	over 3/4" 1-1/2"incl.	0.22	.85 - 1.25	0.04	0.05	0.30	0.20	0.02
	63 min.	42	24	Plates & Bars	over 1-1/2" to 4" incl.	0.22	.85 - 1.25	0.04	0.05	0.30	0.20	0.02
	60 min.	40	24	Plates & Bars	over 4" to 8" incl.	0.22	.85 - 1.25	0.04	0.05	0.30	0.20	0.02
	70 min.	50	**	Struc. Shapes	I	0.22	.85 - 1.25	0.04	0.05	0.30	0.20	0.02
	67 min.	46	**	Struc. Shapes	II	0.22	.85 - 1.25	0.04	0.05	0.30	0.20	0.02
	63 min.	42	24	Struc. Shapes	III	0.22	.85 - 1.25	0.04	0.05	0.30	0.20	0.02
A500 Cold-Formed Welded and Seamless Carbon Steel Structural Tubing in Rounds and Shapes	45 min.	33	25	Round Structural Tubing	A	0.26		0.04	0.05		0.20*	
	58 min.	42	23	Round Structural Tubing	B	0.26		0.04	0.05		0.20*	
	45 min.	39	25	Shaped Struc. Tub.	A	0.26		0.04	0.05		0.20*	
	58 min.	46	23	Shaped Struc. Tub.	B	0.26		0.04	0.05		0.20*	
A501 Hot-Formed Welded and Seamless Carbon Steel Structural Tubing.	58 min.	36	23			0.26		0.04	0.05		0.20*	
A529 Structural Steel (42 ksi min. yield 1/2 in. max. thickness)	60 to 85	42	19			0.27	1.20	0.04	0.05		0.20*	
A570 GRADES D & E Hot Rolled Carbon Steel Sheets & Strip, Struc. Quality	55 min.	40	**		D	0.25	0.60 - 0.90	0.04	0.04		0.20*	
	58 min.	42	**		E	0.25	0.60 - 0.90	0.04	0.04		0.20*	

* When specified.
** See ASTM Standards for details.
† Where two figures are given, this is a min-max range.

TABLE 23-7 AISC Recognized Structural Steels (Continued)

ASTM GRADE and Descriptive Information	Material Shape	Thickness Or Group	Machanical Properties: Tensile Strength† (ksi min.)	Yield Point (ksi min.)	Elonga. in 2 in. % min.	Grade	Chemical Requirements (Ladle) Percent: C† Max.	Mn† Max.	P Max.	S Max.	Si† Max.	Ni† Max.	Cr† Max.	Mo† Max.	Cu† Max.	V† Max.	Cb Max.	Ti† Max.	Others
A572 High-Strength Low-Alloy, Columbium-Vanadium Steels of Structural Quality	Shapes And Plates		60 min.	42	24	42	0.21	1.35	0.04	0.05	(5)					(2)	(1)(3)		N (4)
			60 min	45	22	45	0.22	1.35	0.04	0.05	(5)					(2)	(1)(3)		(4)
			65 min.	50	21	50	0.23	1.35	0.04	0.05	(5)					(2)	(1)(3)		(4)
			70 min.	55	20	55	0.25	1.35	0.04	0.05	(5)					(2)	(1)(3)		(4)
			75 min.	60	18	60	0.26	1.35	0.04	0.05	(5)					(2)	(1)(3)		(4)
			80 min	65	**	65	0.26	1.35	0.04	0.05	(5)					(2)	(1)(3)		(4)

Alloy content shall be in accordance with one of the following –

(1) Cb 0.005-0.05
(2) V 0.01-0.10
(3) Cb(.05 max)+V 0.02-0.10
(4) N (with V) 0.015

(4) N(.015 max) when added as a supplement to V shall be reported, and the minimum ratio of V to N shall be 4 to 1.
(1)(3) Cb when added either singly or in combination with V unless combined with 0.15 min Si shall be restricted to plate or bar thickness of 1/2" max, and to shapes of Table A, Group 1 of Spec A6.
(5) Si (0.30 max) for shapes and plates to 1-1/2"; for grade 42 plates over 1-1/2" (0.15 – 0.30 max)

A588 High-Strength Low-Alloy Structural Steel with 50 ksi min. Yield Point to 4 inch Thickness

Material Shape	Thickness Or Group	Tensile Strength† (ksi min.)	Yield Point (ksi min.)	Elonga. in 2 in. % min.
Plates And Bars	to 4" incl.	70 min.	50	21
	over 4" to 5" incl.	67 min	46	21
	over 5" to 8" incl.	63 min	42	21
Shapes	Groups 1,2,3,4	70 min	50	19
	Group 5	67 min	46	19

Grade	C† Max.	Mn† Max.	P Max.	S Max.	Si† Max.	Ni† Max.	Cr† Max.	Mo† Max.	Cu† Max.	V† Max.	Cb Max.	Ti† Max.	Others
A	0.10-0.19	0.90-1.25	0.05	0.05	0.15-0.30	–	0.40-0.63	–	0.25-0.40	0.02-0.10	–	–	–
B	0.10-0.20	0.75-1.25	0.04	0.05	0.15-0.30	0.25-0.50	0.40-0.70	–	0.20-0.40	0.01-0.10	–	–	–
C	0.15 Max.	0.80-1.35	0.04	0.05	0.15-0.30	0.25-0.50	0.30-0.50	–	0.20-0.50	0.01-0.10	–	–	–
D	0.10-0.20	0.75-1.25	0.04	0.05	0.50-0.90	–	0.50-0.75	–	0.30 Max.	–	0.04	–	Zr 0.05-0.15
E	0.15 Max.	1.20 Max.	0.04	0.05	0.15-0.30	0.75-1.25	–	0.10-0.25	0.50-0.80	0.05 Max.	–	–	–
F	0.10-0.20	0.50-1.00	0.04	0.05	0.30 Max.	0.40-1.10	0.30 Max.	0.10-0.20	0.30-1.00	0.01-0.10	–	–	–
G	0.20 Max.	1.20 Max.	0.04	0.05	0.25-0.70	0.80 Max.	0.50-1.00	0.10 Max.	0.30-0.50	–	–	0.07	–

A514 High-Yield-Strength, Quenched and Tempered Alloy Steel Plate, Suitable for Welding

Material Shape	Thickness Or Group	Tensile Strength† (ksi min.)	Yield Point (ksi min.)	Elonga. in 2 in. % min.
Plate	to 3/4" incl.	115 to 135	100	18
	over 3/4" to 2-1/2" incl.	115 to 135	100	18
	over 2-1/2" to 4" incl.	105 to 135	90	17

Grade	C† Max.	Mn† Max.	P Max.	S Max.	Si† Max.	Ni† Max.	Cr† Max.	Mo† Max.	Cu† Max.	V† Max.	Cb Max.	Ti† Max.	Others
E	0.10-0.20	0.40-0.70	0.035	0.04	0.20 0.35	–	1.40-2.00	0.40-0.60	0.20-0.40-			0.04-0.10	B .0015-.005
F	0.10-0.20	0.60-1.00	0.035	0.04	0.15-0.35	0.70-1.00	0.40-0.65	0.40-0.60	0.15-0.50	0.03-0.08		–	B .002-.006

† Where two figures are given, this is a min-max range.

Manufacturers' literature often specifies the electrode types that are suitable for a given steel or the steels that may be welded with a particular electrode. When steels or electrodes are identified by their trade names rather than by ASTM, SAE, or other classification designations, the application recommendations of reputable manufacturers will be based on application code requirements.

Welds on low-alloy high-strength steels can often be made with only one or two electrode types. However, for the mild steels used in construction work, a variety of different electrode types may be used to give full-strength welds. The electrode selection should then be based on the requirements of the joint if maximum welding efficiency is to be achieved. Joints may be classified as "freeze," "fill," or "follow" types or a combination of these, and electrode selection can be narrowed to the electrode group that best fills the freeze-fill-follow characteristics of the joint.

Figure 23-71 illustrates various types of freeze, fill, and follow joints and combinations thereof, and it gives the stick-electrode type having performance characteristics suited to the joint. Note that the welding position is also a factor. Thus, a corner fillet would be made with a "fill" type electrode—E7024 or E7028—in the flat position, whereas an E7014 fill-freeze type electrode would be used for the same weld with the work inclined 15° or more. Reference to the AWS numbering system and electrode groups, presented in Table 23-8 and Fig. 23-71, will make clear why the third or fourth digit of the electrode number (depending on whether it is a four- or five-digit number) changes according to the position of the piece which is to be welded. For an inclined piece, an all-position fill-freeze electrode will give optimum performance.

Freeze joints are joints welded vertically or overhead, and the molten metal must solidify quickly to keep it from spilling out. For 3⁄16- to 5⁄8-in. plate, fast-freeze electrodes, EXX10 or EXX11, should be used. With plate thicker than 5⁄8 in., the fill-freeze low-hydrogen electrodes, EXX15, EXX16, or EXX18, are more economical because deposit rates are higher and they can make the desired weld size with fewer, bigger beads.

Fill joints are groove, flat, and horizontal fillets and lap welds in plate over 3⁄16 in. thick. They require fast-fill electrodes with high deposition rates, such as EXX24 or EXX27, to fill in the shortest time. Fast-fill electrodes are adaptable to joints that are presented level or only slightly downhill (less than 15° of incline). More steeply inclined fill joints are best welded with fill-freeze electrodes, such as EXX14.

Follow joints occur with thin-gauge metal (under 3⁄16 in. thick). Here the requirement is for electrodes that weld at high travel speeds with minimal skips, misses, slag entrapment, and undercut. Fillets and lap welds in all positions are best welded with EXX12 or EXX13 fill-freeze electrodes, which have excellent fast-follow characteristics. Other types of follow joints are best welded with fast-freeze electrodes that have good follow-freeze ability.

Manual Self-shielded Arc Welding As indicated by Table 23-8 and suggested by the foregoing discussion of electrodes for freeze, fill, and follow types of joints, the basic stick electrodes can be grouped as fast freeze, fast fill, fill freeze, low hydrogen, and alloy steel. It may be helpful to summarize the characteristics of the groups and the applications of specific electrodes in each group.

The *fast-freeze group* is applicable to welding in all positions for mild steel in general fabrication and maintenance. The electrodes are particularly good for vertical and overhead welding, and such out-of-position welds have x-ray quality. Welds have deep penetration.

E6010 is the basic fast-freeze electrode type for dc welding electrode positive (reverse polarity). Commercial electrodes falling in this category have been especially designed to give the arc stability, light slag, and wash-in that are needed for critical out-of-position applications, such as pipe welding.

E6011 types include electrodes that are similar to the E6010 type but also suited for use with industrial ac welders. If operated electrode negative (straight polarity), these electrodes are excellent for sheet-metal edge, corner, and butt welds. Low-slag electrodes within the designation are excellent for tack welding, especially when the joint is to be filled with weld metal deposited by semiautomatic self-shielded flux-cored wire processes. Such low-slag electrodes are also used for vertical-down welding, making stringer beads, bridging gaps, ac pipe welding, and on applications that require an extra-low silicon deposit. Other electrodes meeting this designation are intended for general shop use with small ac welders having low open circuit voltage.

E7010 electrodes are commonly used for welding high-tensile pipe or other materials requiring a 70-ksi weld joint. Operation characteristics are similar to those of the E6010 electrodes.

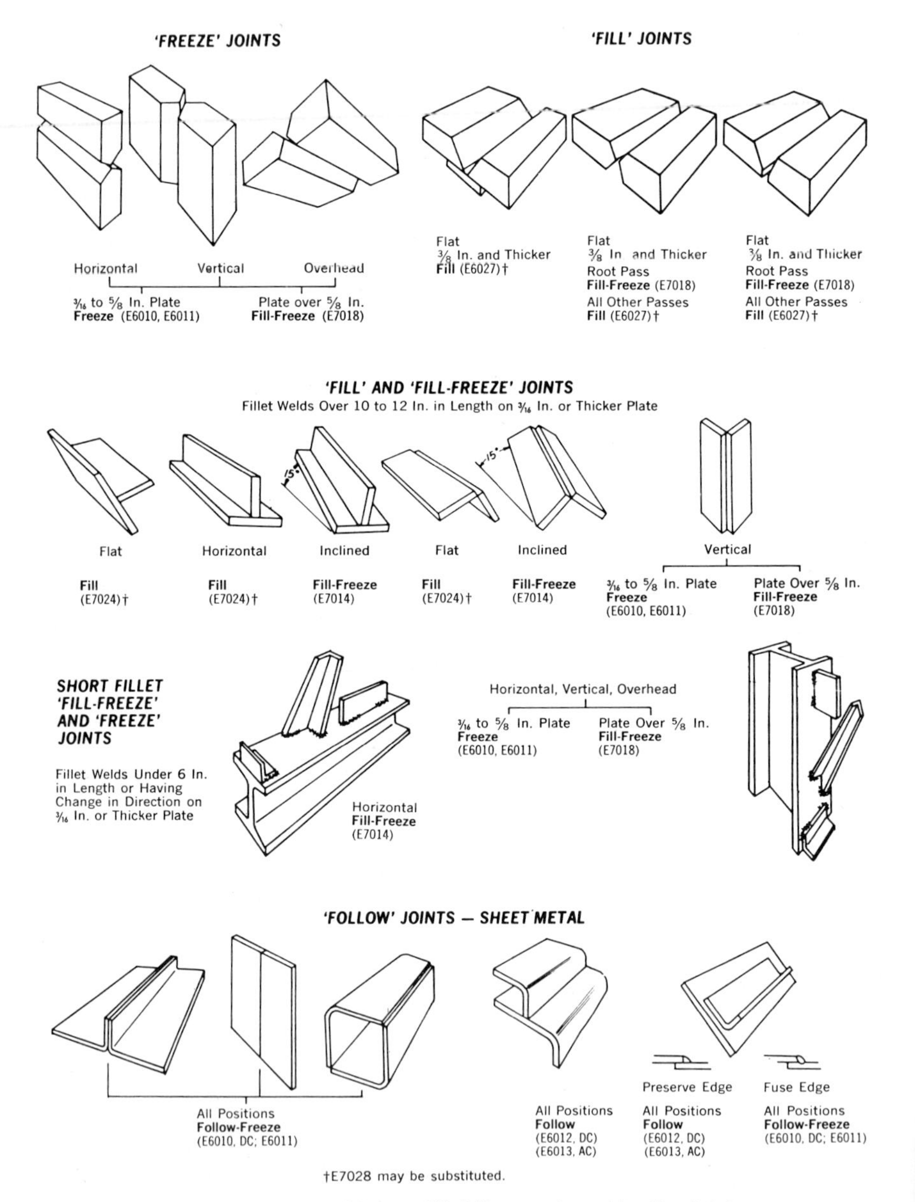

Fig. 23-71 Types of freeze, fill, follow, and combination joints.

TABLE 23-8 The AWS Numbering System and Electrode Groups

AWS Numbering System

a. The prefix E designates arc welding electrode.
b. The first two digits of four-digit numbers and the first three digits of five-digit numbers indicate tensile strength:
E60xx......... 60,000 psi tensile strength
E70xx......... 70,000 psi tensile strength
E110xx........110,000 psi tensile strength
c. The next-to-last digit indicates position:
Exx1x........All positions
Exx2x........Flat position and horizontal fillets
Exx3x........Flat position only
d. The last two digits together indicate the type of coating and the current to be used (see Electrode Groups below).
e. The suffix (example: EXXXX-A1) indicates the approximate alloy in the deposit.
-A1.............. ½% Mo
-B1.............. ½% Cr, ½% Mo
-B2..............1¼% Cr, ½% Mo
-B3..............2¼% Cr, 1% Mo
-C1..............2½% Ni
-C2..............3¼% Ni
-C3..............1% Ni, 0.35% Mo, 0.15% Cr
-D1 & D2........0.25–0.45% Mo
-G...............0.50 min Ni, 0.30 min Cr, 0.20 min Mo, 0.10 min V
Only one element is required.
E11018-M........1.25–2.50 Ni, 0.40 Cr, 0.30–0.35 Mo, 0.05 V

Electrode Groups

The "Fast-freeze" Group:
Exx10.........................Organic, dc only
Exx11.........................Organic, ac or dc
The "Fast-fill" Group:
Exx24.........................Rutile, approx. 50% iron powder, ac or dc
Exx27.........................Mineral and approx. 50% iron powder, ac or dc
The "Fill-freeze" ("Fast-follow") Group:
Exx14.........................Rutile and approx. 30% iron powder, dc or ac
Exx12.........................Rutile (fast-follow), dc or ac
Exx13.........................Rutile (fast-follow), ac or dc
The Low Hydrogen Group:
With "fill-freeze" characteristics
Exx18.........................Low hydrogen and approx. 30% iron powder, dc or ac
With "fast-fill" characteristics
Exx28.........................Low hydrogen and approx. 50% iron powder, ac or dc
The Alloy Steel Group:
Exxx18.......................Low hydrogen and approx. 30% iron powder, dc or ac

The *fast-fill group* contains electrode types for production welds on 3⁄16-in. or thicker mild and alloy steel. These electrodes are applicable to flat and horizontal fillets, laps, and deep-groove butt welds. They may also be used in a slightly downhill position. Fast-fill electrodes have the highest deposition rates of all stick electrodes and give smooth beads with negligible spatter. Penetration, however, is shallow. The 50 percent iron powder used in their coatings is chiefly responsible for their high deposition rates since it, as well as the core wire, contributes to the weld deposit.

E6027 electrodes are used principally for flat, deep-groove joints. Excellent wash-in characteristics and a friable slag make slag removal easy, even in deep grooves.

E7020-A1 types are used in applications similar to those of E6027 when a 70-ksi tensile strength, ½ percent molybdenum deposit is needed.

E7024 electrodes are the basic fast-fill members of the group. Commercial types in the E7024 group are designed to accentuate the deposition rate, slag control, appearance, and other desirable characteristics.

The *fill-freeze group* is made up of electrodes for welding mild steel in all positions. Applications include downhill fillets and laps, irregular or short welds that change direction or position, sheet-metal laps and fillets, and fast-fill joints with poor fit-up. Because some of these electrodes in the 5/64- through 3/16-in. sizes have good fast-follow characteristics, they are excellent for sheet-metal welding. Deposition rates and penetration are medium, and appearance ranges from smooth and ripple free to distinctly rippled.

E6012 electrodes are the basic fill-freeze types for general-purpose welding. They work equally well with ac or dc current.

E6013 electrodes are often used in place of E6012 electrodes for welding sheet steel and plate when appearance and ease in operation are more important than speed.

E7014 electrodes have a high iron-powder content, giving them the best fast-fill ability of the fill-freeze electrodes. Members of this group are frequently used for production welding on short, irregular, or downhill fast-fill joints.

The *low-hydrogen group* contains electrodes which are especially suitable for structural welding on mild and alloy steel members. Low-hydrogen electrodes produce dense, x-ray quality welds with excellent notch toughness and ductility. The low hydrogen content of the deposit reduces the danger of underbead and microcracking on high-carbon and low-alloy steels and on thick weldments. The electrodes are used to resist cracking in medium- to high-carbon steels, hot-short cracking in phosphorus-bearing steels, and porosity in sulfur-bearing steels. They are especially applicable to thick sections and restrained joints in mild and alloy steel plate when shrinkage stresses tend to cause weld cracking and to alloy steels that require 70-ksi tensile-strength deposits.

E7018 low-hydrogen electrodes have fill-freeze characteristics permitting all-position welding. Deposition rates are high for an all-position type electrode as a result of the iron powder in the coating. Bead appearance is flat or slightly convex with distinct ripples. The moderately heavy slag is easily removed.

E7028 low-hydrogen electrodes have fast-fill characteristics. Electrodes of this type perform best on flat fillets and deep-groove joints and also work well with horizontal fillet and lap welds.

The *alloy-steel group* contains electrodes which are designed for the 80- to 110-ksi tensile steels, such as used in bridges, skyscrapers, and other structures that must meet critical strength requirements. The electrodes have low-hydrogen fill-freeze operating characteristics similar to the E7018 electrodes.

E8018-B2 electrodes produce a 1¼ percent chromium, ½ percent molybdenum deposit commonly required for high-temperature, high-pressure piping. They also meet the requirements of E9018-G for welding some 90-ksi steels.

E8018-C3 electrodes produce an 80-ksi tensile-strength weld for general-purpose welding of many high-strength alloys. The 1 percent nickel content of the deposit makes them suitable for welding low-temperature alloys that require good notch toughness down to −60°F. They are also used for fillet welds on 110-ksi quenched and tempered steels, such as A514 and A517.

E8018-C1 electrodes give a 2¼ percent nickel deposit with a notch toughness of 20 ft-lb at −75°F. They are used for welding low-temperature alloys employed in equipment for the storage, piping, and transportation of liquid ammonia, propane, and other gases.

E11018-M electrodes give the 110-ksi tensile-strength deposit needed for full-strength welds on quenched and tempered steels, such as A514 and A517.

Semiautomatic Self-shielded Arc Welding The electrodes used with semiautomatic self-shielded welding are continuous coils of flux-cored wires of varying diameters and operating characteristics. In general, these electrode wires are designed as substitutes for the stick electrodes used in the manual self-shielded processes. The manufacturers' literature presents data on their applications and the procedures to be used for specific joints, and the electrodes must be qualified for the specific job.

Before the 1969 revisions of its building specifications, neither flux-cored arc welding nor gas-metal arc welding was recognized by the American Institute of Steel Construction. Now both are recognized, with A5.20-69 of the American Welding Society establishing performance requirements and electrodes for the flux-cored process and A5.18-69 similarly treating the gas-metal arc-welding processes.

The electrode types for the semiautomatic self-shielded process correspond functionally to those of stick electrodes. There are electrode wires with *freeze-fill* characteristics for:

1. All-position single-pass welds
2. Multiple-pass welding on plate in all positions—flat, vertical, and overhead
3. Roundabouts, short assembly welds, tack welds, seal beads, stringers, and work in confined grooves

The *follow-freeze* electrodes are especially designed for welding 14-gauge through 3/16-in. mild steel in all positions.

The *fast-fill* electrode wires are limited to welds positioned flat or horizontal. They are used extensively in structural welding, and give exceptionally high deposition rates and cost reductions in:

1. Multiple-pass fillet, lap, and deep-groove butt welds
2. Single-pass 1/4- to 1/2-in. fillets and laps, corner welds, and butts where a gap and backup plate are practical

Electrode sizes range from 0.068- to 0.120-in. diameter. They fall under AWS E70T-G and E70T-4 of A5.20-69, *Specifications for Mild Steel Electrodes for Flux-Cored Arc Welding.*

Gas-shielded Arc Welding As noted previously, electrodes for the gas-shielded processes may be solid wire, flux cored, or flux coated. It is beyond the scope of this section to detail the types and their applications. In general, the electrodes are presented in a variety suited to the requirements of the metal to be welded, the joint types, and the positions—much as are the self-shielded electrodes. The manufacturers' literature or reference works on gas-shielded metal arc welding should be consulted as needed when the gas-shielded processes are to be used on the job.

Submerged-arc Welding Electrodes and fluxes for submerged-arc welding merit more discussion because of the unique characteristics of this welding process and its widespread use in structural fabrication and equipment rebuilding.

Both the ferrous and nonferrous electrodes used with submerged-arc welding are bare wires with clean, bright surfaces to facilitate the use of relatively high welding currents. Electrodes are normally sold in the form of coils ranging from 25 to 200 lb. On very high-production jobs, the electrode may be fed from a coil in a drum, in which case the coil may weigh up to 1,000 lb.

As solid wire—without a coating or core—submerged-arc electrodes are offered in lesser variety than electrodes for the other processes discussed. Diameter and chemical composition are the variables. Thus, one producer offers four compositions of submerged-arc wires for carbon steel welding—one a low-cost general-purpose type, one a general-purpose type for higher-strength welds and for rust porosity resistance, one for use on 70-ksi high-tensile steel where restraint-cracking resistance is needed, and one for high travel speeds and rust porosity resistance. Special electrode wires are used for stainless-steel welding and usually contain about 20 percent chromium and 9 percent nickel.

Flux Selection The fluxes used with electrode wires may be "neutral," in which case they do not contribute alloying substances to the weld deposit, or they may be alloy fluxes which are intended to impart certain alloy elements to the weld. Fluxes are granulated, fusible mineral materials, essentially free from substances that would create a large amount of gas during welding. They are made of specified chemical compositions so as to develop particular performance characteristics. A flux has several functions to perform, including prevention of atmospheric contamination of the weld metal and exerting a scavenging-deoxidizing action on the molten metal in the weld crater. The choice of flux depends on the weld procedure to be employed, the type of joint, and the composition of the metal to be welded.

Five basic fluxes are offered by one manufacturer for carbon steel welding. These are listed below with the types of welds for which they are recommended. These may be used to illustrate how flux selection influences applications and performance characteristics with the submerged-arc process.

An All-purpose Flux—High-strength Welds

For about 70 percent of industrial submerged-arc welding. Especially suited for semiautomatic welding because of minimal flash-through.

For 70-ksi-minimum tensile welds when used with a low-carbon wire or 80-ksi welds with a low-carbon wire containing 0.23 percent Si and 1.0 percent Mn.

For roundabouts, 3 o'clock position, and downhill welds.

A High-travel-speed Flux

For all single-pass automatic welds at high travel speeds, using single-arc, twin-arc, or tandem-arc equipment.

For semiautomatic welds on 14-gauge to ¼-in. plate, using $^1/_{16}$-in. high-travel-speed wire.

A Flux for Cracking and Porosity Control

For applications where cracking or porosity is a problem.

For ⅜-in. and larger flat fillets made semiautomatically using a constant-voltage power source.

A Low-cost Flux for Multiple-pass Welds

For multiple-pass welds where good impact strength, low consumption, and low flux cost are important.

A Low-alloy Pickup Flux for Multiple-pass Welds

For multiple-pass welds where maximum impact strength, low alloy pickup, and low flux consumption are important.

For maximum resistance to rust porosity and restraint cracking when used with a designated wire.

For welds with a minimum tensile-strength requirement of 70 ksi.

For welding with alloy electrode wires.

Alloy fluxes contain elements that enter into the chemical constitution of the weld metal, giving it desired mechanical properties. Chromium, nickel, molybdenum, and vanadium are commonly included. Alloy fluxes are usually used with mild steel electrode wires, but they may also be used with alloy wires. When used with mild steel wires, the total of the four alloying elements usually does not exceed 8 percent. If used with an alloy wire, the flux alloy and wire alloy are additive.

There is a cost advantage in using an alloy flux in conjunction with a mild steel electrode to give an alloy weld deposit. The desired metal composition can often be produced at as much as 70 percent lower cost than for an alloy electrode wire to give a similar weld-metal chemistry. When the desired composition of weld metal requires a special blending of alloying elements in the flux, this is readily accomplished and the special blend can be delivered in a short time. Special blends of alloy electrode wire, however, are likely to take a long time for delivery and are obtainable only at a premium price.

TYPES OF WELD JOINTS

Joint design influences the choice of electrode, and it also influences the choice of the welding process. The specification of joint design and sizing is a function of the design engineer and is outside the scope of this section. Nevertheless, knowledge of the nomenclature of weld joints is essential for anyone concerned in any way with construction welding.

The American Welding Society has standardized the nomenclature and symbols for various types of welds, and the terms used in the following paragraphs are in agreement with this standardization. Figure 23-72 is a composite sketch of common welds and provides a good introduction to the subject.

Five basic types of welded joints are shown in Fig. 23-73: butt, corner, edge, lap, and T. For each type of joint, alternative types of welds are applicable. In general, each type of weld will involve a different edge preparation.

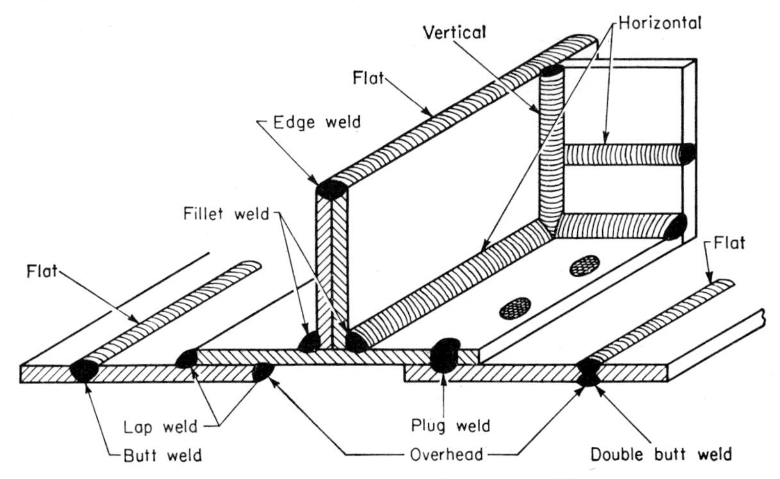

Fig. 23-72 Composite sketch of common weld types.

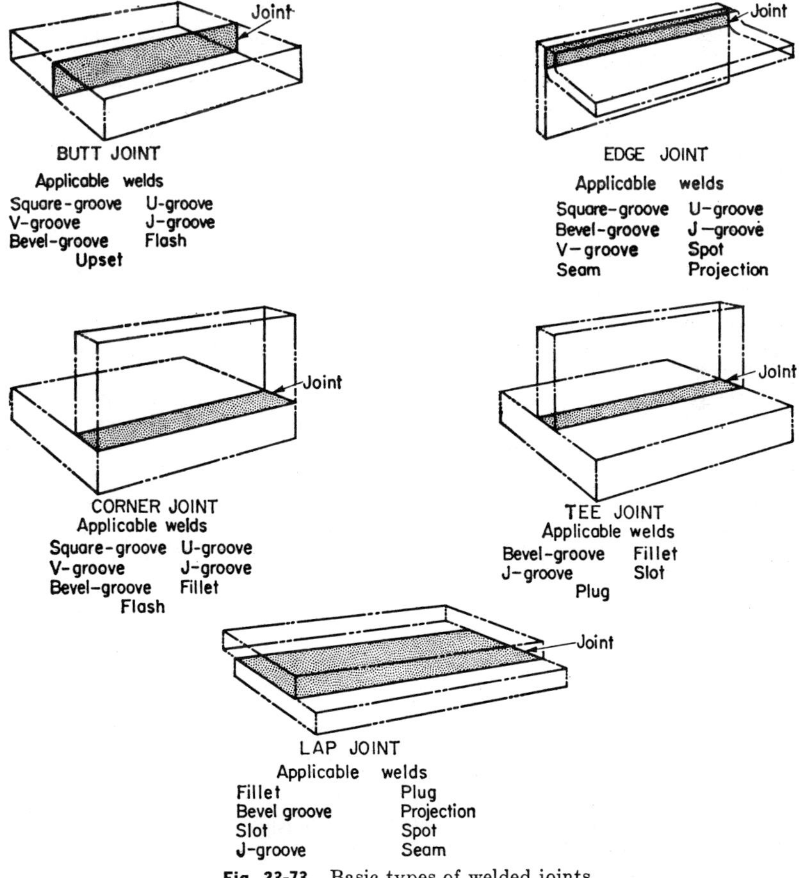

Fig. 23-73 Basic types of welded joints.

BUTT JOINTS: square, single-V, double-V, single-bevel, double-bevel, single-U, double-U, single-J, and double-J grooves

CORNER JOINTS: single-J, double-J, square, single-V, single-bevel, or double-bevel grooves and plain fillet

EDGE JOINTS: single-U, single-J, double-J, or single-V grooves and simple bead

LAP JOINTS: fillet, plug, slot, plus resistance-welding alternatives—spot, seam, or projection

T JOINTS: single-bevel, double-bevel, single-J, or double-J grooves and square

The designer selects the joint according to his study of (1) the load and its characteristics—whether the load is in tension or compression and whether bending, fatigue, or impact stresses in any combination are present; (2) the manner in which the load is applied—steady, variable, or sudden; and (3) the cost of the joint preparation and subsequent welding. If the joint is a connection for a structural member of a building or bridge, it must conform to specifications established by the American Welding Society (D1.0-69-Building; D2.0-69-Bridge), American Institute of Steel Construction (*Specifications for the Design, Fabrication and Erection of Structural Steel for Buildings*), or American Association of State Highway Officials (*Standard Specifications for Highway Bridges*), or it must be submitted to the governing organization for approval. The several types of welded joints are discussed in more detail below.

Butt Joints The square butt joint is suitable for all usual loads. It requires full and complete fusion, particularly when the load is intermittent or involves fatigue (rapidly alternating or pulsating) stresses. Base metal for this joint must be a good weldable steel, since a large portion of this metal is melted during welding. The thickness of plate generally is ⅜ in. or lighter, although thicknesses up to ¾ in. can be handled adequately by automatic submerged-arc welding. Joint preparation is simple, requiring only a matching of the adjoining edges with a gap dependent upon plate thickness. Simple edge preparation means low cost for this type of joint.

The single-V groove weld is generally used for butt joints with plate of ⅜-in. thickness or heavier. It is found occasionally on thinner plate when welding can be done from one side only. Preparation cost is a factor to consider when specifying a single-V groove weld, and more electrode is used in welding than with the square edge.

The double-V groove weld is specified for butt joints on plates thicker than those which can be welded with a single V as well as for work that can be welded from both sides. The cost of preparation is higher than for the single V, but only about half as much electrode is required. Warpage or distortion can be reduced by alternating the beads, welding on one side and then the other to keep the joint symmetrical in the process.

The single-U groove weld is usually found on butt joints in thick material. It is often specified in preference to the single V and double V on heavier plates since it economizes on weld metal. Machining or flame-cutting plates to form a single-U joint reduces the amount of weld metal required for thick plates but increases preparation costs. The joint is welded from one side except for a single bead that is placed on the side opposite the U to give 100 percent penetration.

The double-U groove weld is common to very heavy plates where welding can be done from both sides. The joint takes less weld metal than the single U, but it costs more to prepare.

T Joints The square-T joint corresponds to the square butt joints in that no machining of plates is necessary. Fillet welds are used on one or both sides of a square-T joint. It is adaptable to all plate thicknesses, principally for loads that place the welds in longitudinal shear. More weld metal is required than for other T joints.

The single-bevel weld is employed, in most instances, for joining plates ½ in. and thinner where welding is possible from one side only. It is more costly to machine than the square T, but is lower in electrode cost.

The double-bevel weld is usually specified for T joints in thick plate which can be welded from both sides. Machining costs are obviously higher than for the single bevel, but further reduction in weld metal can usually be accomplished as compared to the square or single bevel.

While the single-J weld may be used for conventional plate thicknesses, it is generally applied to 10-in. plate and heavier. Welding is from one side, although a final finish bead on the opposite side is advisable.

The double-J weld accommodates 1½-in.-thick or heavier plate, where welding can be done from both sides. Preparation cost is higher than for the single J, but less weld metal is needed, thus reducing the cost per joint.

Lap Joints The single-fillet weld has widespread applications in lap joints. Its advantages include practically no plate-edge preparation. However, when fatigue or impact loads are encountered, stress distribution should be studied carefully. If loading is not too severe, the single-fillet lap joint is adaptable to plate of all thicknesses.

A lap joint with a double-fillet weld handles loads much more severe than can be met by the single-fillet lap joint. The two fillets should normally be of the same size, although occasionally one may be smaller than the other. Because of its low preparation cost, the double-fillet lap joint finds widespread application.

Corner and Edge Joints The flush corner joint is acceptable where loads are not severe or when welding 12-gauge and lighter plate. If used for heavier plates, caution should be exercised to ensure that loading will not be excessive.

The half-open joint is generally found on plates heavier than 12 gauge where welding can be done from one side only and where fatigue or impact is not likely to be severe. The contact surfaces of the base metal in this type of joint aid welding by reducing the tendency to burn through the plates at the corner.

The full-open joint is suitable for heavy loads in welding plate of all thicknesses where welding can be done from both sides. When properly made, the joint shape provides good stress distribution, accommodating fatigue or impact loads nicely.

The edge joint is used in joining plates ½ in. or thinner for light loads without impact or fatigue stresses.

Weld Symbols Each type of weld is indicated on drawings by a standard symbol which has been developed by the American Welding Society, A2.0-68. The symbols denote not only the type of weld and its location but also the size and length of bead. They indicate quickly to the designer, draftsman, production supervisor, and welding operator alike the exact welding technique developed for each joint or connection to satisfy all conditions of material strength and service.

The system of symbols for weld specification has been extended to all forms of electric and gas welding. A condensed summary of AWS symbols is presented in Fig. 23-74.

MATERIAL PREPARATION FOR WELDING

Preparation of material for welding includes such steps as forming, cutting, machining, assembly, and positioning. The first three of these are normally shop operations, although flame-cutting work is occasionally handled in the field. Press brakes and rolls can accommodate most rudimentary forming. Cutting operations are accomplished with shears or oxyacetylene torches.

Flame gouging with oxyacetylene equipment is a variant of cutting which is adapted to the preparation of J or U grooves for welding; to the removal of unwanted weld metal, such as tack welds or other temporary joints; and to the preparation of the undersides of groove welds for a backup pass. Air-arc gouging can also be used for these applications. With this technique, metal removal is accomplished by bringing it to a molten temperature by using the intense heat of a carbon arc, then blowing it away with a jet of compressed air.

Mechanical setups for flame cutting or gouging are well suited to plate-edge preparation, particularly for those jobs requiring bevels. In many instances, special cutting arrangements have been devised so that two or four edges of each plate can be beveled simultaneously.

Care in forming, cutting, and handling of shapes to be welded to avoid poor fit-up is a major factor in welding costs. Figure 23-75 illustrates the effect of fit-up upon

AMERICAN WELDING SOCIETY

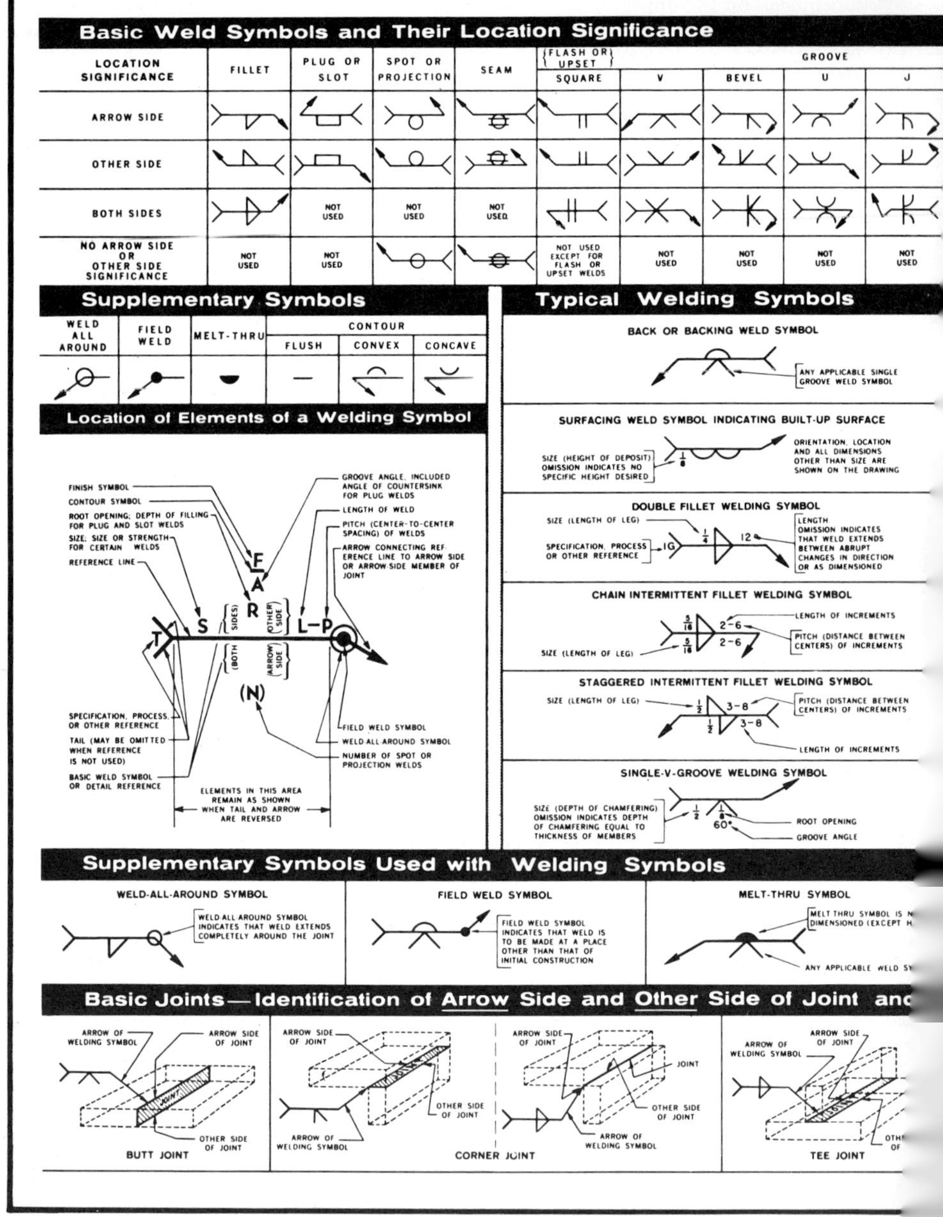

DESIGNATION OF WELDING PROCESSES BY LETTERS

CAW	Carbon-Arc Welding
CW	Cold Welding
DB	Dip Brazing
DFW	Diffusion Welding
EBW	Electron Beam Welding
EW	Electroslag Welding
EXW	Explosion Welding
FB	Furnace Brazing
FCAW	Flux Cored Arc Welding
FOW	Forge Welding
FRW	Friction Welding
FW	Flash Welding
GMAW	Gas Metal-Arc Welding
GTAW	Gas Tungsten-Arc Welding
IB	Induction Brazing
IRB	Infrared Brazing
IW	Induction Welding
LBW	Laser Beam Welding
OAW	Oxyacetylene Welding
OHW	Oxyhydrogen Welding
PAW	Plasma-Arc Welding
PEW	Percussion Welding
PGW	Pressure Gas Welding
RB	Resistance Brazing
RPW	Projection Welding
RSEW	Resistance-Seam Welding
RSW	Resistance-Spot Welding
SAW	Submerged Arc Welding
SMAW	Shielded Metal-Arc Welding
SW	Stud Welding
TB	Torch Brazing
TW	Thermit Welding
USW	Ultrasonic Welding
UW	Upset Welding

DESIGNATION OF CUTTING PROCESSES BY LETTERS

AAC	Air Carbon-Arc Cutting
AC	Arc Cutting
AOC	Oxygen-Arc Cutting
CAC	Carbon-Arc Cutting
FOC	Chemical Flux Cutting
MAC	Metal-Arc Cutting
OC	Oxygen Cutting
PAC	Plasma Arc Cutting
POC	Metal Powder Cutting

STANDARD WELDING SYMBOLS

NONPREFERRED SYMBOLS;
USE PREFERRED SYMBOL WITH PROCESS REFERENCE IN THE TAIL

FLARE-V	FLARE-BEVEL	BACK OR BACKING	SURFACING	FLANGE EDGE	FLANGE CORNER	ARC-SEAM OR ARC-SPOT	RESISTANCE-SPOT	PROJECTION	RESISTANCE-SEAM	FLASH OR UPSET
		GROOVE WELD SYMBOL	NOT USED				NOT USED		NOT USED	NOT USED
		GROOVE WELD SYMBOL	NOT USED				NOT USED		NOT USED	NOT USED
		NOT USED	NOT USED	NOT USED	NOT USED	NOT USED	NOT USED	NOT USED	NOT USED	NOT USED
NOT USED	NOT USED	NOT USED		NOT USED	NOT USED	NOT USED		NOT USED		

SINGLE-V-GROOVE WELDING SYMBOL INDICATING ROOT PENETRATION

SIZE { DEPTH OF CHAMFERING PLUS ROOT PENETRATION

$\frac{1}{4}+\frac{1}{2}$ 0 90°

ROOT OPENING

GROOVE ANGLE

PROJECTION WELDING SYMBOL

PROJECTION WELDING REFERENCE MUST BE USED

SIZE (STRENGTH IN LB PER WELD) DIA. OF WELD MAY BE USED INSTEAD FOR CIRCULAR PROJECTION WELDS

RPW 500 6 (4)

PITCH (DISTANCE BETWEEN CENTERS) OF WELDS

NUMBER OF WELDS

DOUBLE-BEVEL-GROOVE WELDING SYMBOL

OMISSION OF SIZE DIMENSION INDICATES A TOTAL DEPTH OF CHAMFERING EQUAL TO THICKNESS OF MEMBERS

ARROW POINTS TOWARD MEMBER TO BE CHAMFERED

$\frac{1}{8}$ 50°

ROOT OPENING

GROOVE ANGLE

SEAM WELDING SYMBOL

SIZE (WIDTH OF WELD) STRENGTH IN LB. PER LINEAR INCH MAY BE USED INSTEAD

PROCESS REFERENCE MUST BE USED TO INDICATE PROCESS DESIRED

RSEW .30" 3-9

LENGTH OF WELDS OR INCREMENTS OMISSION INDICATES THAT WELD EXTENDS BETWEEN ABRUPT CHANGES IN DIRECTION OR AS DIMENSIONED

PITCH (DISTANCE BETWEEN CENTERS) OF INCREMENTS

WELDING SYMBOLS FOR COMBINED WELDS

$\frac{1}{4}$ $\frac{1}{8}$ 60° $\frac{5}{16}$ T-3 $\frac{3}{8}$ $\frac{1}{8}$ $\frac{1}{4}$

FLASH OR UPSET WELDING SYMBOL

FW

PROCESS REFERENCE MUST BE USED TO INDICATE PROCESS DESIRED

PLUG WELDING SYMBOL

SIZE (DIA. OF HOLE AT ROOT)

INCLUDED ANGLE OF COUNTERSINK

$\frac{1}{2}$ $\frac{1}{2}$ 6 45°

PITCH (DISTANCE BETWEEN CENTERS) OF WELDS

DEPTH OF FILLING IN INCHES OMISSION INDICATES FILLING IS COMPLETE

SQUARE-GROOVE WELDING SYMBOL

OMISSION OF SIZE INDICATES COMPLETE JOINT PENETRATION

SLOT WELDING SYMBOL

DEPTH OF FILLING IN INCHES OMISSION INDICATES FILLING IS COMPLETE

$\frac{5}{8}$

ORIENTATION, LOCATION AND ALL DIMENSIONS OTHER THAN DEPTH OF FILLING ARE SHOWN ON THE DRAWING

FLARE-V-AND FLARE-BEVEL-GROOVE WELDING SYMBOL

SIZE IS CONSIDERED AS EXTENDING ONLY TO TANGENT POINTS

$\frac{1}{4}$ $\frac{1}{16}$ $\frac{1}{8}$ $\frac{1}{16}$

ROOT OPENING

SPOT WELDING SYMBOL

NUMBER OF WELDS (5)

SIZE (DIA. OF WELD) STRENGTH IN LB. PER WELD MAY BE USED INSTEAD

.25" 4 RSW

PITCH (DISTANCE BETWEEN CENTERS) OF WELDS

PROCESS REFERENCE MUST BE USED TO INDICATE PROCESS DESIRED

EDGE- AND CORNER-FLANGE WELD SYMBOLS

$\frac{1}{8}+\frac{1}{8}$

RADIUS $\frac{1}{16}+\frac{3}{32}$

SIZE OF WELD $\frac{1}{16}$

HEIGHT ABOVE POINT OF TANGENCY

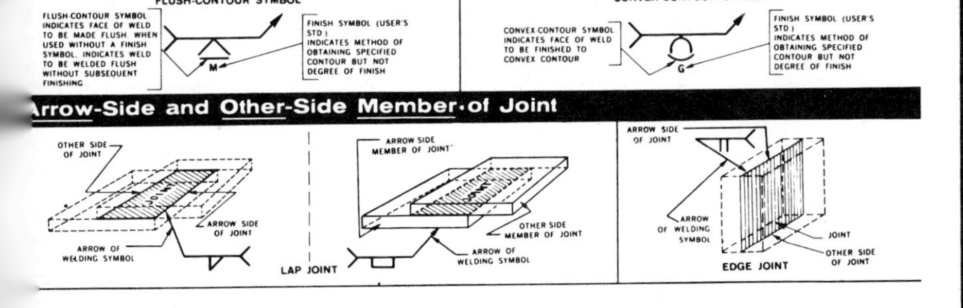

PLUG OR SLOT	SPOT OR PROJECTION	SEAM	FLASH OR UPSET SQUARE	GROOVE V	BEVEL	U	J	FLARE-V	FLARE-BEVEL	BACK OR BACKING	SURFACING	FLANGE EDGE	CORNER

NONPREFERRED SYMBOL—USE PREFERRED SYMBOL WITH PROCESS REFERENCE IN THE TAIL

ARC-SEAM OR ARC-SPOT	RESISTANCE SPOT	PROJECTION	RESISTANCE SEAM	FLASH OR UPSET

Fig. 23-74 Summary of AWS symbols.

welding speed. As the size of the gap increases, speed progressively decreases from the recommended maximum with a gap of $\frac{1}{32}$ in.

The position of the joint has a pronounced effect on the speed and ease of welding. A change from the vertical or overhead to the downhand position can result in an increase in speed of as much as 400 percent. Figure 23-76 shows typical speeds for a $\frac{5}{16}$-in. fillet weld in $\frac{3}{8}$-in. plate.

Optimum positioning of heavy weldments for downhand welding in field operations is often difficult if not impossible. Joints should therefore be designed to permit as

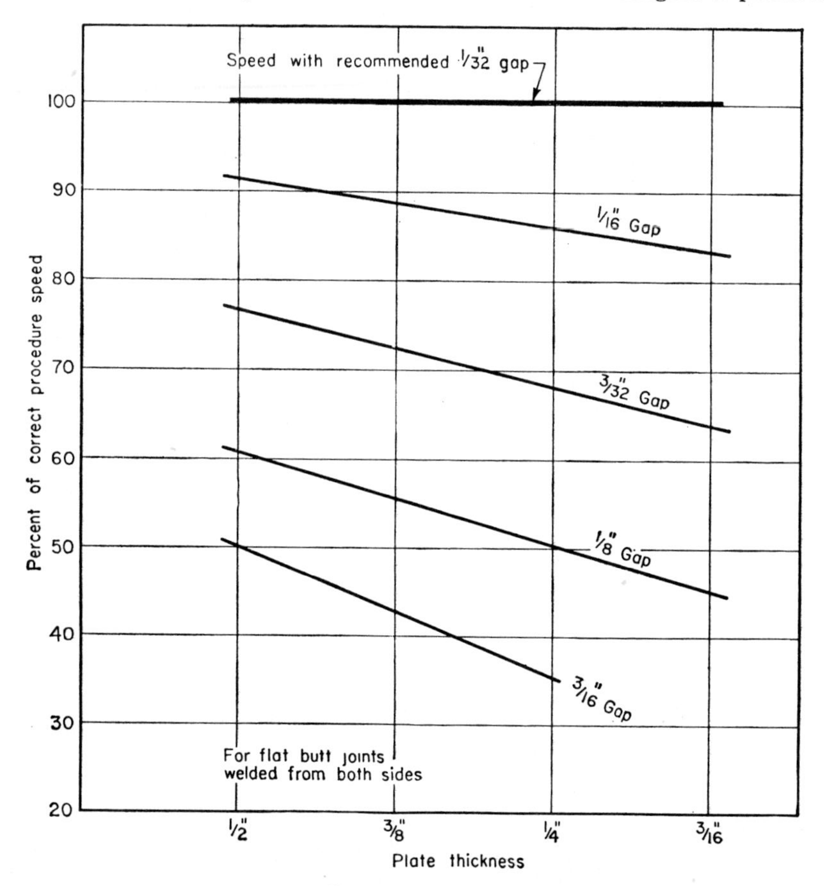

Fig. 23-75 Effect of fit-up on welding speeds.

much downhand welding as possible. It may be possible to break a job down into major subassemblies that can be handled flat in the shop, or at the erection site through the use of positioners, jigs, or fixtures. These subunits can then be transferred to the structure for final erection, where out-of-position welding may be held to a minimum. It is at this point that the ingenuity of the individual contractor will be put to the test, since there are no hard and fast rules to follow. Many special devices have been contrived to position and hold pieces to be welded in the field, even if only for a preliminary tack welding to maintain alignment until finish welds can be made.

Toggle clamps, pneumatic clamps, and hydraulic and screw jacks find frequent application in rigging up for field welding. In the shop, work positioners, both powered and hand-operated types, aid in orienting joints for speedy downhand welding. Assembly jigs and fixtures designed for the specific job at hand also contribute to a higher operating factor, or more feet of welded joint per hour. Cylindrical welded structures can be handled nicely, either in the shop or field, on work rolls that allow turning the weldment under the arc for girth seams and positioning longitudinal seams for welding flat. Such jobs are usually suited to automatic or semi-automatic welding methods.

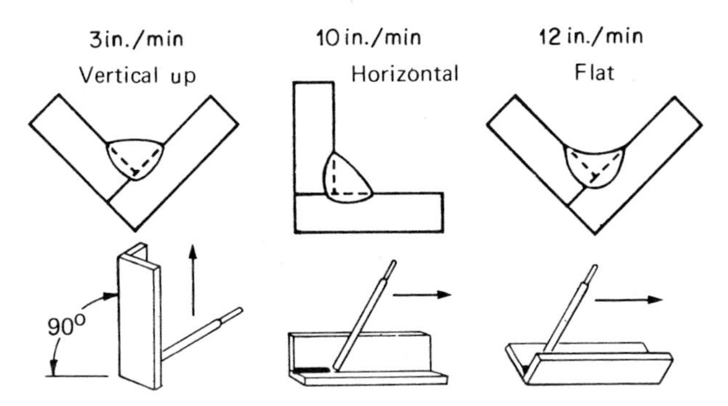

Fig. 23-76 Changing welding position from vertical to flat for this 5⁄16-in. fillet weld can increase speed by almost 300 percent.

WELDING PROCEDURE

Welding procedures have been established for the various types of joints and the processes applicable to them. Each procedure should specify such things as:

The type of joint
The process by which it is to be welded
The mechanics of welding (from one side, from both sides, with backup strip, etc.)
The number of passes and sequence of bead placement
The preparation and fit-up
The electrode and the polarity to be used
The current for various sizes of electrodes and various thicknesses of plate
The melt-off rate and arc speed with different current settings
Other factors that will be helpful in the welding or in estimating costs
Pounds of electrode consumed per foot of weld

Figure 23-77 shows a typical set of procedures—in this case for a single-V butt weld in plate. Note that two passes are made on one side, following which the plate is turned and a back pass is made on the other side. If this is not feasible, the last pass is made overhead with current adjustments. Note also that the feet of joint that could be welded per hour is given at 100 percent operating factor. This assumes uninterrupted welding for the entire hour—a physical impossibility, and therefore requiring adjustment in the light of first-hand experience with the job in consideration. This information is helpful in estimating costs, as is "pounds of electrode per foot of weld."

There are literally thousands of procedures governing various joints and processes. They serve as general guides in establishing the precise procedures for the particular conditions of the job. When recommended procedures are followed, the resulting weld will be equal to the plate in section and strength unless otherwise stated in the procedures.

The American Welding Society has prequalified welding procedures for use in buildings and bridges. These are contained in AWS D1.0-69-Buildings and AWS D2.0-69-Bridges codes. If a procedure is proposed that is not prequalified, it must be qualified by special tests before being used in structures meeting code requirements. As new

Procedures for Butt Welds

Flat Position* **Single V Groove**
Welded from Both Sides

For use where plates can be turned for downhand welding on both sides.*

Preparation:
60° V groove with 1/8-in. shoulder
Fit-up:
Recommended gap 1/32 to 1/16 in.

Electrode and polarity:
E-6027—AC
E-6011—AC

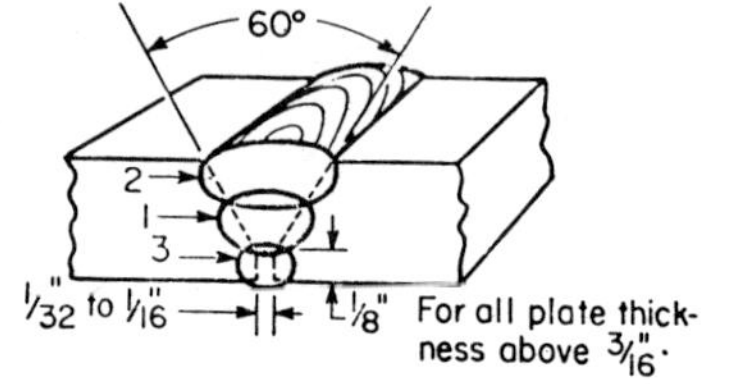

Plate thickness, in.	Electrode size, in.	Alternating current, amp	Electrode melt-off rate, in./min	Arc speed in./min for first pass	Passes or beads	Ft of joint welded per hr (100% operating factor)	Lb of electrode per ft of weld
3/8	3/16 E-6011	175	10.0	9.0	1	....	0.17
	3/16 E-6027	280	12.3	...	2 and back	....	0.34
						20.1	0.51
1/2	1/4 E-6011	275	9.0	8.2	1	....	0.23
	7/32 E-6027	315	10.8	...	2 and back	....	0.52
						17.4	0.75
5/8	1/4 E-6011	275	9.0	7.7	1	....	0.24
	1/4 E-6027	375	10.0	...	2 and 3 back	....	0.80
	7/32 E-6027	315	10.8	...		....	0.23
						13.2	1.27

* Where the plates cannot be turned over, chip into bead 1 and put the last pass in overhead with 3/16 in. E-6010 or E-6011 with the current adjusted to produce an electrode melt-off of approximately 8½ in. per min.

Fig. 23-77 A typical set of welding procedures.

electrodes are developed and innovations made in welding processes, the compendium of procedural data becomes increasingly voluminous.

QUALITY OF WELDS AND WELDMENTS

The quality of welds is assured by inspectors and supervisors who see that the proper types and sizes of electrodes are used and that the current settings follow the established procedures. A check of settings and electrodes, followed with a check of the surface appearance, will tell a qualified inspector enough about the strength of the weld to satisfy normal requirements. Visual inspection is often the only inspection required, although in much welding on structural members, radiographic, ultrasonic, dye-penetrant, and other inspection techniques may be required by either the job specifications or the applicable codes. The AWS publications which contain information on practices for inspecting structural welding are D1.0-69-Buildings and D2.0-69-Bridges.

Surface appearance is an important index of weld quality. Good surface appearance will be characterized by:

No cracks, serious undercuts, overlaps, surface holes, or visible slag inclusions
Uniform ripples and uniform width of bead
Butt welds flush or slightly above the plate surface, without excessive buildup
Fillet welds with legs of equal size on each plate

PRACTICAL WAYS TO MINIMIZE DISTORTION

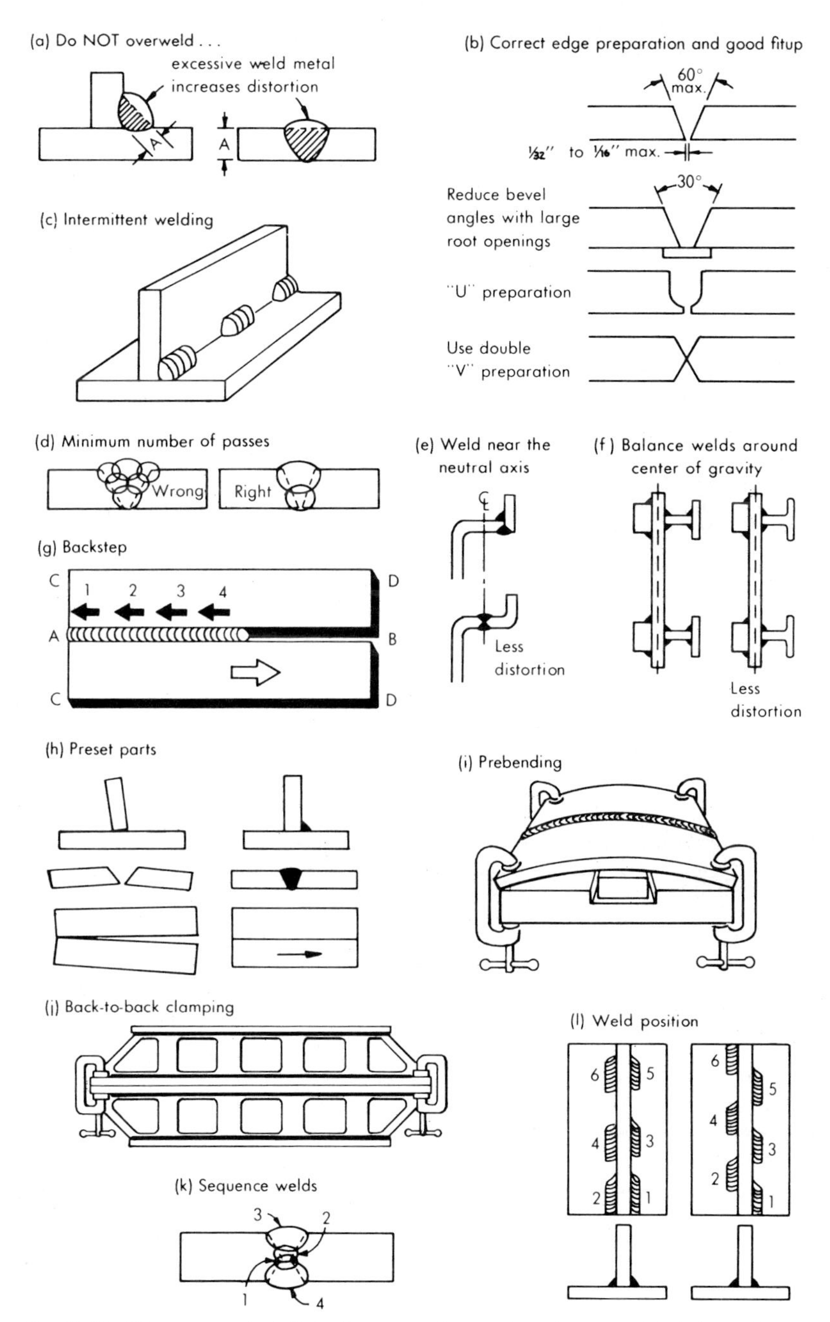

Fig. 23-78 Ways to control distortion or make constructive use of shrinkage forces.

Conscientious welding operators have a tendency to put down more weld metal than called for by the drawings or specifications in the belief that a small amount of overwelding will assure strength in the connection. Unfortunately, this human tendency to give full measure often increases the shrinkage forces and causes distortion, thus giving the owner an inferior rather than a better product. Welds are sized by the designers who determine the sizes necessary to carry the loads, using adequate margins of safety. Any increase in weld size beyond that specified by the designer can only be useless or do harm—and the cost is run up substantially.

Welding sequences are frequently spelled out on drawings or indicated by welding engineers. Again, these must be followed to the letter, since they are the sequences determined as least likely to cause distortion in the weldment. Distortion control is a complex subject involving much study at the design stage—or after the weldment has been made, to determine why it distorts—and a discussion of the theory and art of it is outside the scope of this section. Figure 23-78 graphically portrays some of the practices that either minimize distortion or make constructive use of the shrinkage forces.

HARD-SURFACING

One of the major uses of arc welding in the construction field is the repair and maintenance of equipment. Because of the severe wear conditions encountered, various parts of shovels, dozers, draglines, tractors, and other machinery must be periodically replaced or rebuilt. Many parts of such machinery might be said to be "consumable"—that is, they are worn out on the job during a service life that is known to be finite.* Except in the case of catastrophic failure, the work involved in restoring these worn parts to a serviceable condition is more correctly termed "rebuilding" than "repair." Some companies deliberately schedule the rebuilding of parts rather than waiting for breakdowns or wear beyond the point of easy restoration. Tractor rollers and idlers, dragline bucket teeth and lips, crusher rolls and hammers, shovel tracks, and many types of pins, sprockets, and links on construction machinery are illustrative of parts that are "consumable" in service and subject to frequent rebuilding.

The arc-welding method of rebuilding involves the restoring of a worn part to approximately its original dimensions with a wear- and abrasive-resistant weld deposit. After deposition of sufficient material to form the new surface, machining with carbide tools may be required to bring the part to the desired dimensions.

Rebuilding, as well as the hard-surfacing of new parts, may be accomplished with manual welding methods or mechanized techniques employing the open-arc or the submerged-arc welding processes. The abrasive-resistant weld metal may be deposited with an alloy stick electrode, with a special flux-cored hard-surfacing electrode wire used alone or in conjunction with a neutral flux, or with an alloy flux and a mild steel wire.

One advantage of using an alloy electrode or a special flux-cored hard-surfacing wire is the uniformity of the composition of the deposit despite variations in welding procedures. Conversely, an advantage in using an alloy flux and mild steel wire is the ability to control the chemistry of the final weld metal by variations in procedures. Both methods have their preferred applications, with abrasion resistance, impact, hardness, machinability, corrosion resistance, and the thickness of deposit (number of layers of weld metal) entering into the selection.

* See Sec. 7, Equipment Economics.

C. Oxyacetylene Cutting and Joining

The manually operated oxyacetylene torch for cutting and welding metals has become an indispensable tool of the construction industry. It has reduced the need for a large inventory of expensive equipment and parts and has proved to be a time and money saver in construction work. Jobsite fitting, along with certain jobsite fabrication and the correction of shop-fabrication deviations and errors in steel, can be performed without the expense of returning the materials to the shop. Most of this all-around construction dexterity with the ferrous metals (particularly structural and plate steel) applies to the cutting torch; but often just as important is the ability to weld, braze, gouge, heat, clean, harden, and straighten many metals by employing minor attachments to the basic tool.

Compressed hydrogen (normally used for underwater cutting) and the liquefied petroleum gases (butane and propane) are bottled in many localities, but acetylene compressed into cylinders is universally available and is favored for construction work. Oxygen gas of 99.5 percent or greater purity, with a free volume of 244 cu ft as measured at atmospheric pressure and at 70°F, is compressed into a hollow steel cylinder with a volume of about 1.5 cu ft. Acetylene gas, also of 99.5 percent or greater purity, is similarly compressed from 250 cu ft in volume measured at atmospheric pressure and at 70°F. Acetone is used to absorb this acetylene gas as it is compressed at 300 psi into the steel cylinders. Each cylinder is filled with a special porous and cemented mixture which keeps the acetylene gas stable and in solution with the acetone.

The cylinders for oxygen and for acetylene, as for other compressed gases, are manufactured and regularly tested in strict conformity with the Interstate Commerce Commission's regulations for each specific gas which is to be handled. Cylinders are provided with fusible safety plugs which will blow in the event of excessive internal pressure, such as may be caused by fire. Although sturdily built, the cylinders should be transported and handled with the care and respect normally due any object with explosive properties, no matter how dormant.

CUTTING[5]

Conventional Cutting When iron or steel is heated to a temperature of about 1,600°F (it melts at around 2,700°F), it can be oxidized readily. In the flame-cutting process, a mixture of oxygen and acetylene is passed through small "preheat" orifices in the cutting tip or nozzle of the torch. This flame, with suitable adjustments, is directed onto the metal surface until oxidation has been accomplished. Additional pure oxygen is then passed through the center of the cutting tip, and this combines chemically with the steel or iron to form ferric oxide. The erosive force of the oxygen jet, aided by the chemical reaction, cuts a neat hole in the metal. The cutting tip can be moved along the desired path to yield a slit or kerf which is comparable to that produced by a saw blade. With gas pressures held to constant values, the rate of cutting is a function of tip size and thickness of material.

The theoretical quantity of oxygen needed to oxidize 1 lb of iron is 4.6 cu ft. In normal practice this consumption rate is reduced to 3 cu ft per lb of iron because the ambient atmosphere provides some of the oxygen. In addition, complete oxidation is not required since the oxygen cutting jet actually erodes some of the iron. The chemical structure of the adjacent metal is somewhat altered by the high cutting temperatures, and quick cooling in the atmosphere tends to harden the cut surface. Where the carbon content of the metal does not exceed 0.30 percent, as is true for normal structural and plate steel, these effects are not harmful even though discoloration at the cut surface may signify some tempering. Flame cutting, when neatly done, is sufficient preparation for direct fusion welding by either electricity or flame.

When high-carbon or alloy steels are cut, uniform preheating of an area along each side of the intended cut is required. This will prevent quick cooling and reduce the chemical changes, hardness, and distortion shrinkage. Usual preheating temperatures range from 600 to 1100°F (from a faint red color to almost bright red), with the lower values preferred in order to simplify the cutting operation.

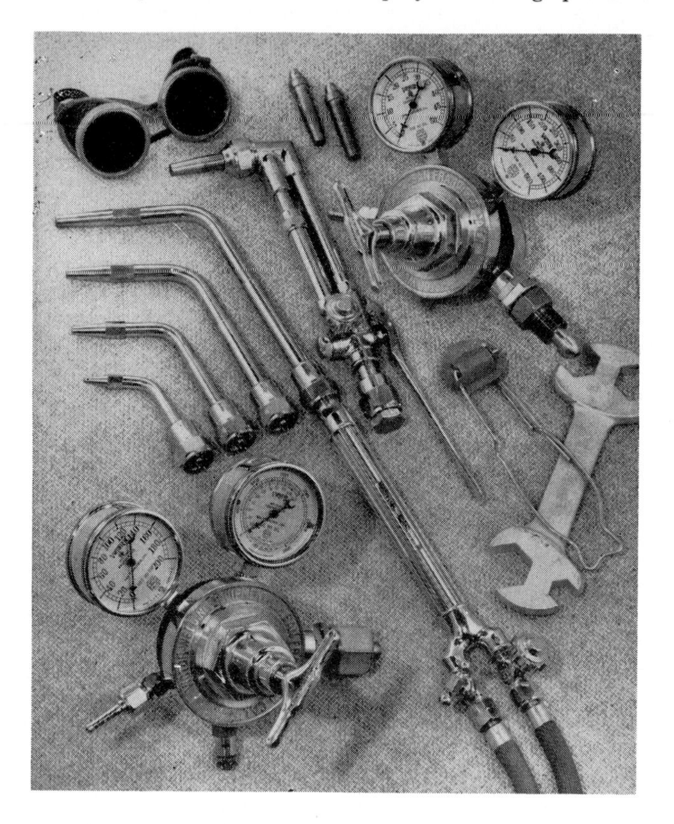

Fig. 23-79 Set of oxyacetylene manual equipment. (*Victor Equipment Co.*)

Figure 23-79 shows a complete set of combination oxyacetylene cutting and welding equipment. The two pressure regulators are used for adjusting cylinder pressures and maintaining constant acetylene and oxygen pressures at the torch. The red and green rubber hoses permit free movement of the torch in an area around the cylinders. The welding torch with assorted heating tips and the cutting attachment with assorted burning tips constitute a complete set of welding and cutting tools. There is a special wrench to fit the union of the set, a friction lighter, and protective goggles for the operator. Replacement tips and hoses are readily available. The tools are made for production and construction use and are designed to stand up under severe service.

Table 23-9 lists the estimated consumption of both oxygen and acetylene, along with other pertinent information for cutting various thicknesses of steel plates. The gas pressures which are recommended for cutting should be measured at the regulators. For each 25 ft of sound ¼-in. hose, add 1 psi at the regulator for hose friction loss. When long hoses must be used, larger-diameter hoses should be substituted to minimize the pressure drop between regulator and torch.

The manual cutting torch attachment, mounted with the desired cutting tip, is fastened to the welding-torch butt with a fixed wrench. The acetylene valve on the assembled torch is opened, the gas ignited with a friction lighter, and the acetylene adjusted until the flame from the preheat holes clears the end of the tip by about 1/8 in. The adjusting oxygen valve (the oxygen valve on the butt is wide open) is then regulated until a neutral flame is established. Equal quantities (volumes) of oxygen and of acetylene are then being consumed and a bright inner cone appears within the flame. The point at which the feathery edges of this flame disappear and a sharp inner cone is visible marks the neutral flow. When the cutting oxygen lever is depressed, a slight amount of oxygen is diverted from the preheat flame to produce a carbonizing flame; accordingly, it is necessary to adjust the oxygen valve in order to produce neutral preheat flames when the cutting lever is down. The torch

TABLE 23-9 Oxyacetylene Cutting Torch Operating Data for Steel

Steel thickness, in.			Oxygen					Acetylene			Cutting speed, in./min	
				Pressure, psi		Consumption, cfm			Consumption, cfm			
Min	Max	Tip size*	Drill jet†	Min	Max	Min	Max	Pressure, psi	Min	Max	Min	Max
1/64	3/16	000	72	7	20	0.5	1.6	5	0.10	0.18	17	35
1/8	3/8	000½	69	8	25	0.7	1.9	5	0.12	0.20	17	35
1/4	1/2	00	65	8	25	1.0	2.1	5	0.15	0.22	17	25
1/4	3/4	0	58	9	30	1.3	2.7	5	0.15	0.26	17	20
1/4	1½	1	54	15	35	1.3	3.8	5	0.15	0.32	14	17
1/2	3	2	52	25	50	1.8	4.8	5	0.18	0.38	11	15
2	5	3	50	30	50	3.1	7.1	5	0.26	0.48	7	14
4	6	4	45	35	60	4.9	8.2	5	0.35	0.54	6.5	9
6	8	5	40	40	70	6.7	12.5	5	0.44	0.75	3.8	6.5
8	10	6	35	45	70	10.0	15.0	6	0.60	0.85	3.5	3.8
10	12	7	28	50	80	15.0	16.0	6	0.70	0.95	3.2	3.5
12	14	8	20	55	80	16.0	19.0	10	0.95	1.10	3.1	3.2
13	15	9	11/64	50	90	16.0	22.0	10	0.95	1.16	3.0	3.2
16	Up	10	3/16	60	100			10				

* Victor Equipment Company size.
† Drill size of oxygen jet.

is held so that the cone end of the preheat flame just touches the metal. The tip is positioned at a 90° angle with the work if a square cut is desired and at a lesser angle if a beveled cut is desired.

The deseaming or the flushing away of welds from joints occasionally becomes necessary for repairs or for the correction of previous fusion-welding work. This is performed with special tips for burning torches which are bent to a deflection angle of some 35°. The operation is similar to cutting except that the metal to be removed is melted and gouged out of its bed with the oxygen jet and the preheat flame.

When it becomes necessary to examine a questionable weld, a special flattened tip can be skillfully applied to flush it away in a thin layer at each pass. This permits the examination of the weld at several levels for slag inclusions, porosity, and other imperfections. At the same time, the joint can be prepared with the cutting torch for partial or complete rewelding to comply with job inspection requirements.

Underwater Cutting The ability to cut off steel piling to grade and perform repairs to hydraulic and marine structures and to floating equipment without expensive use of cofferdams or dry docks makes this process of prime importance. The cutting torch for underwater work is short, enabling the operator to work close to the cut.

In addition to oxygen and fuel gas, compressed air must be used at the cutting tip. The cutting unit is provided with a bubble-forming compressed-air sleeve that also serves as a spacer for setting the correct distance to the work. Underwater cutting requires the services of a skilled diver who is also familiar with the cutting torch. The cutting work, after the diver has located and positioned himself properly, is somewhat similar to cutting in air. The stream of compressed air allows the flame to burn within an air bubble, and this heats the steel more rapidly than if the flame were burning in the water.

The oxygen pressure, fuel gas pressure, and compressed air pressure all must be correspondingly increased as the depth of water in which cutting is to be done increases. Safety requires that the torch (regulator) pressure of oxyacetylene gas not exceed 15 psi. This limits its use to shallow waters, and in deep waters (above 25 to 30 ft) hydrogen is used as a fuel gas. The oxygen pressure must always be slightly more

TABLE 23-10 Approximate Fresh-water Underwater Cutting Pressures
(Pressures in pounds per square inch)

Underwater depth, ft	Water pressure, psi	Cutting pressures, psi		
		Compressed air	Hydrogen fuel gas	Gaseous oxygen
10	4.3	35–60	20–60	80
20	8.6	40–65	25–65	85
30	12.9	50–70	35–70	90
40	17.3	55–80	40–80	100
50	21.6	60–85	45–85	105
60	25.9	70–90	55–90	110
70	30.3	75–100	60–100	115
80	34.6	85–105	70–105	120
90	38.9	90–110	75–110	125
100	43.3	95–115	80–115	135
125	54.1	115–125	100–125	145
150	64.9	130–140	110–140	160
175	75.7	145–155	120–155	175
200	86.6	165–170	135–170	190
225	97.4	180–185	150–185	200

SOURCE: *Welding Handbook*, 5th ed., American Welding Society.

than the compressed-air pressure in order to prevent interference with the oxygen cutting jet. Table 23-10 indicates the approximate gas pressures for flame cutting in fresh waters. For equal depths in sea or salt water, the required water pressures and gas pressures will be higher by some 3 percent (ratio of 64.0 to 62.4).

JOINING

Fusion Welding[6,7] Oxyacetylene fusion welding is a form of fusion welding that uses the heat from the flame to melt and fuse the base metal of the members to be joined. A welding rod is used to supply a filler metal whose metallurgical properties are comparable to those of the metal in the joined members. When necessary, whether on account of the impurities in some metals or to obtain satisfactory fusion, a fluxing agent is employed at the time of welding. The equipment needed for a welding set is shown in Fig. 23-79 along with the cutting attachments. Many sizes of welding tips to meet the heating and welding requirements of various thicknesses of metal are available.

The welding tip is fastened to the torch body and receives a mixture of oxygen and acetylene through the needle valves on the torch. The acetylene needle valve is first opened and the flame ignited with a spark or friction lighter; the valve is then adjusted until the flame has a slight tendency to jump away from the tip. The oxygen needle valve on the torch is now opened until a well-defined white cone appears at the tip, enveloped by a large bluish cone. At this time the flame is neutral, with equal volumes of gaseous oxygen and gaseous acetylene passing from the cylinders through the regulators. If the flow of oxygen is reduced or the acetylene increased, the flame will be carbonizing or reducing. Three flame zones can then be observed instead of the usual two, and the inner cone is no longer well defined. If the flow of oxygen is increased above neutral, the flame will be oxidizing. It is then similar to the neutral flame but contains a shorter and more pointed inner cone whose color has changed from white to almost purple. The neutral flame is normally desired for welding, and at this temperature the molten metal at the weld flows smoothly.

Although the oxyacetylene fusion weld in quantity cannot compete on an economic basis with the electric fusion weld in quantity, there are many occasions when it can be a useful tool on construction work. Electric welding equipment may be troublesome or costly to handle under these conditions, particularly for the small amount of work involved in equipment repairs and alterations. The portability and simplicity of an oxyacetylene welding unit will favor its frequent usage.

Construction welding will be in downhand, vertical, and overhead positions as conditions require. The downhand position is preferred by virtue of its simplicity, but oxyacetylene welding can also be performed in the other positions. Backhand welding, in which the flame is pointed back or away from the direction of travel and the welding rod follows the flame, is in general use since it has been found to produce sound welds at a lesser cost.

In heavy construction, the most common metal encountered is mild steel in rolled plates and shapes, or steel with less than 0.20 percent carbon. Such steels are ductile and are readily weldable by the oxyacetylene process. Since these steels melt at about 2700°F, a great deal of intense heat is required to melt the base metal and filler rod. Allowances for distortion, as well as the development of welding sequences to minimize distortions and residual stresses, are similar to those recommended for electric-fusion welding. Provision for some ⅛ in. of weld shrinkage per lin ft is recommended for long, continuous welds, and the edges of butt welds should be beveled and kept slightly separated in order to secure full penetration of the joint. High-carbon steels can also be readily welded, using a rod which is of comparable high-carbon material and with an excess of acetylene used in the flame. The welded joint should be annealed after the welding has been completed.

To repair broken steel castings, the procedure is similar to that for welding plates and shapes. Steel castings with a carbon content in excess of 0.25 percent should be welded with rods whose strength and carbon content are comparable to those of the parent metal. Large pieces will require preheating in the joint area to permit welding without shrinkage cracks and distortions.

Table 23-11 lists representative operational data for flame welding. The weldable range of thickness for steel plates (or steel thickness) is shown for each size of tip, as are the estimated gas consumptions and the suggested pressures. Inasmuch as welding speed will depend upon the operator to a great degree, the tabulated values for production are only approximate. The gas pressures as recommended for welding are measured at the regulators. For each 25 ft of sound ¼-in. hose, add 1 psi at the regulator to compensate for hose friction loss.

Braze Welding[8] "Oxyacetylene brazing," in its normal interpretation, denotes the process whereby wrought iron, cast iron, steel, copper, bronze, brass, aluminum, and other metals and metal alloys are joined in similar or mixed combinations without melting or fusing the base metal but with the aid of a nonferrous (normally an alloy of copper) filler rod. The several brazing processes all use nonferrous metal or alloy filler rods whose melting points are in excess of 1000°F but are always lower than those of the materials being joined.

Of the various classes of brazing, that known as "braze welding" is the most common to the heavy-construction industry. Braze (or bronze) welding implies that the filler rod is of zinc bronze, melting between 1600 and 1700°F. This material is applied to a prepared joint, which is usually of open-V shape. The preparation is similar to that described for oxyacetylene fusion welding except that in bronze welding the lower temperatures prevent the parent or base metal from melting and fusing with the filler.

This process can be used effectively to make emergency repairs on construction equipment. With a minimum requirement for disassembly, broken castings of iron, steel, or the bronzes can be repaired. Since braze welding does not require the melting of the parent metal, the metallurgical or chemical structure of the casting is left relatively unchanged. With the lower temperatures required, preheating of the casting to ensure sound joints is simple, and local preheating often suffices for the repair work.

TABLE 23-11 Oxyacetylene Welding Torch Operating Data for Steel

Steel thickness, in.					Oxygen				Acetylene					
					Pressure, psi		Consumption, cu ft/hr		Pressure, psi		Consumption, cu ft/hr			
Min	Max	Tip size*	Drill size†	Length of inner flame cone, in.	Min	Max	Min	Max	Min	Max	Min	Max	Filler rod size, in.	Welding speed, ft/hr
	1/32	000	75	7/32	1/2	2	1/2	3	1/2	2	1/2	3	1/16	35
1/64	3/64	00	70	7/32	1	2	1	4	1	2	1	4	1/16	30
1/32	5/64	0	65	3/8	1	3	2	6	1	3	2	6	1/16	26
3/64	3/32	1	60	3/8	1	4	4	8	1	4	4	8	1/8	22
1/16	1/8	2	56	3/8	2	5	7	13	2	5	7	13	1/8	17
1/8	3/16	3	53	3/8	3	7	8	36	3	7	8	36	3/16	11
3/16	1/4	4	49	5/8	4	10	10	41	4	10	10	41	3/16	8
1/4	1/2	5	43	1	5	12	15	50	5	15	15	59	1/4	5
1/2	3/4	6	36	1 1/16	6	14	55	125	6	15	55	127	1/4	3
3/4	1 1/4	7	30	1 1/4	7	16	78	160	7	15	78	152	1/4	2
1 1/4	2	8	29	1 1/4	9	18	90	172	8	15	81	160	5/16	
2	2 1/2	9	28	1 7/16	10	20	100	184	9	15	90	166	5/16	
2 1/2	3	10	27	1 7/16	11	22	106	203	10	15	100	169	3/8	
3	3 1/2	11	26	1 7/16	13	24	118	233	11	15	106	175	3/8	
3 1/2	4	12	25	1 1/2	14	28	126	281	12	15	111	211	3/8	

* Victor Equipment Company size.
† Drill size of opening in tip.

Braze welding is by no means restricted to castings of iron, steel, or bronze; malleable-iron articles can often be repaired by this oxyacetylene process along with low- and high-carbon steels, steel alloys, wrought iron, galvanized steel, copper, copper alloys, nickel, monel, and even dissimilar metals. The low melting point of the filler metal allows almost all common metals to be joined by braze welding.

A successful oxyacetylene braze-welding operation requires proper joint preparation by cleaning all surfaces of oil, grease, dirt, and oxides. It is necessary to provide 45° bevels on all edges 3/16 in. and thicker, and the flux must be correct for the particular bronze rod which is to be used. This flux, usually of the type recommended by the filler-rod manufacturer, is applied in the form of a powder, paste, gas, or rod coating to the prepared surfaces of the parent metal and to the brazing rod. The oxyacetylene flame is used to preheat the area along the joint and induce unequal expansion, following which the joint temperature is raised to the proper brazing level. When this occurs, flux and filler rod are melted into the joint to form the bronze weld. The plasticity of high-zinc bronze filler prevents residual shrinkage stresses in the joint from being excessive.

Properly braze-welded joints can be expected to develop ultimate strengths in the 50,000-psi area in many metals. Such a joint is comparable in strength with fusion welding of mild steel. This high strength factor permits broken gear teeth, flattened chisels, pump casings, worn surfaces, heavy-machinery parts, and similar items to be readily restored to workable condition, often saving costly replacements. The reclaiming of parts is almost limitless in heavy-construction work, as is the ability to reround elliptical holes and to build up and rethread or retap threaded surfaces.

Soldering[9] Soldering, in both the soft and the hard types, is a useful oxyacetylene process for minor repairs, sealing, light or gauge metal fabrication, electrical work, and the joining of small tubes. Soft solders for joining most of the common metals are usually alloys of tin, lead, or tin and lead, and they can be applied with light and careful use of the oxyacetylene flame. All soft solders melt below 800°F and require skill with the torch in order to avoid damage to the base metals as well as to the flux and solder. Fluxes as recommended by the solder producer must be used to clean the surfaces to be joined, to prevent oxidation, and to allow the solder to spread thinly over the joint area. A small welding tip can be used on the torch for soft soldering.

Hard solders are more commonly known as "silver solders" or "silver-brazing alloys" and are used in a form of low-temperature brazing. In silver brazing, many of the common alloys and metals (ferrous and nonferrous) can be securely joined at considerably greater strength than for soft solder. The principal ingredients of the filler metal are silver, zinc, and copper, with other metals occasionally added. As for soft soldering, a suitable flux must be used to prevent oxidation and to assist in the flowing of the alloy. The base metals must be thoroughly cleansed of all dirt, grease, oxides, and scale; mechanical cleaning is often the most satisfactory method for accomplishing this.

RELATED APPLICATIONS

Flame Cleaning The removal of mill scale, grease, and rust from steel surfaces, immediately prior to priming with paint or other coating, is effectively performed with a descaling attachment for the oxyacetylene welding torch. This attachment is normally either 4 or 6 in. wide, permitting a broad, flat area to be covered with few passes. It is equipped with a hard skid or shoe which protects the head from wear and spaces the flame with respect to the work. Circular tips are also available for working around rivet heads and at other than flat surfaces.

Painted surfaces can be cleaned in a similar fashion, since the flame burns the old paint away and cleans the original steel surface. Mill scale and rust expand as the steel surface is flame cleaned and thus break their bond with the base steel. The flame-cleaned surface is ready for immediate painting since the cleaning and descaling remove all surface moisture. It is best to apply prime coating while the newly cleaned surface is still warm; otherwise atmospheric moisture will condense on the cool surface and initiate rusting or oxidation. This drying feature is of fundamental importance along waterways, in humid climates, and in corrosive atmospheres.

Hard-surfacing and Metalizing Various steel parts of heavy-construction equipment are subjected to heavy wear from abrasion, skidding, or rolling. Examples include the cutting blades on bulldozers, scrapers, and graders; and teeth or wearing edges on dragline buckets, shovels, dredges, and drills. Those and similar articles can often be maintained on the job, and with minimum shutdown time and expense, by using the oxyacetylene welding flame to hard-surface the wearing areas with steel-base alloys. Surfacing with applicable alloys can also be used on steel to protect against corrosion and to build up low-friction wearing surfaces (also see Braze Welding).

The advanced development in hard-surfacing alloys for abrasion resistance permits layers as thin as $\frac{1}{32}$ in. or as thick as required to be applied successfully to steel wearing surfaces. Applying the hard-surfacing material with the aid of the oxyacetylene flame enables the two materials to be secured by means of sweating. In this operation the welding-torch flame (see Table 23-11 for tip size; tip should be one size larger than for welding comparable thicknesses of parent metal) is set to burn a slight excess of acetylene. An envelope, three times as long as the inner cone and with the flame at its tip, is thus established immediately around the inner cone. The flame carburizes a thin surface layer of the parent metal, provides a liquid surface onto which the alloy can flow without penetrating the original surface more than 0.01 in., and preheats the base. The original surface must be thoroughly cleaned before the hard-surfacing is applied. Only enough alloy is added to the surface to permit the part to function as originally intended, and all sharp corners of the base metal must be rounded before building up any surface.

The metalizing process also provides a means of protecting steel surfaces, including valves and pipes, against corrosion. A special head or gun receives oxygen and acetylene, as for a welding torch, in addition to compressed air. The metalizing wire is fed at an adjustable rate through the nozzle and into the neutral flame, where it is melted and sprayed onto the prepared surface. A wire-type gun uses compressed air at 60 psi and, depending upon its size, consumes from 10 to 50 cfm of air. The gun weighs 3 to 6 lb and can use metalizing wire from 20-gauge to 3/16-in. diameter, the latter being used for applying metal spray to large surface areas. With very heavy (10 lb) guns it is possible to spray as much as 20 lb of steel or as much as 55 lb of zinc per hour.

Since the bond between the original surface and the sprayed metal is purely a mechanical one (except for high-molybdenum alloys), the surface to be coated must be cleaned and roughened to receive the metalizing spray. Where large surface areas require coating, the cleansing and roughening of the areas are done by blasting with sharp sand or other suitable abrasive. The economy of using sand virtually eliminates other abrasives on field-construction projects.

When used for corrosion protection, the metalizing process is not unlike spray painting. The spray is played slowly over the area to be coated, ordinarily traversing it three to six times with alternate horizontal and vertical passes until the desired thickness of coating is reached. The application quantity can be expressed in pounds of metal wire per unit area of surface and should be calculated in advance for the coating thickness desired. The actual thickness of the coating can also be checked with an electric thickness gauge. The contractor may encounter highly corrosive environments, such as in chemical plants or along salt water. Under these conditions he may wish to use the metalizing process as a supplement to painting for the corrosion protection of his construction equipment.

Heating Processes The heat-treating of minor tools and equipment parts can be performed on the job as a maintenance and repair operation. Flame tempering and flame hardening under field conditions are crude in comparison with metallurgical procedures in the shop, but they can prove of value on construction projects. Articles of ferrous metal which are too brittle or are otherwise in need of softening can be quickly heated beyond their critical temperatures with the aid of an oxyacetylene welding torch. The article is then slowly cooled to ambient temperature, either in still air or packed in an insulating material such as powdered asbestos or lime to retard its rate of cooling. This annealing or normalizing operation softens the brittle steel to a more ductile consistency. Flame hardening of ferrous metals in the field is limited to small tools and parts. These are heated rapidly with an oxyacetylene welding torch until they are above their transformation temperature range and are then quenched in water or oil.

The flame straightening of steel can be used to restore fabricated items that have been bent in handling, or it can be used in correcting buckling that has developed in booms, frames, and other structural units of construction machinery. Each case must be studied separately, since the applying of heat to load-carrying steel members may precipitate further buckling. The member should be fully supported, preferably in a manner such that its own weight will assist in the straightening process. The bent part can then be heated with a large welding tip to a faint red color (800 to 900°F), following which gravity loads or external loads can be applied until it is restored to its original shape. Straightening can also be accomplished by differential cooling of the heated area, using moist rags or water. Several cycles of alternate heating and sudden cooling may be required before the bent member is sufficiently straightened. When straightening has been completed by either method, the working area should be annealed or normalized by heating it to its critical temperature and then allowing it to cool slowly in still air.

It frequently becomes necessary to bend steel reinforcing bars to sharp angles or to small radii. The area which is to be bent should be heated to a fairly bright red in order to prevent rupture or damage during the bending process, and the oxy-acetylene welding torch with heating tip is a simple tool for this heating job. It is also useful for straightening hooks or bends on reinforcing bars as required on the job. Since these bars may have already been damaged by cold bending, it is good practice to heat them to a high temperature when straightening them. Subsequent slow cooling in air will allow the heated portions of the bar to normalize and regain fair structural qualities.

References

1. *High Strength Bolts for Structural Steel Joints, Including Suitable Nuts and Plain Hardened Washers*, ASTM Designation A325-68, American Society for Testing and Materials.
2. *Quenched and Tempered Alloy Bolts for Structural Steel Joints*, ASTM Designation A490-67, American Society for Testing and Materials.
3. *Specification for Structural Joints Using ASTM A325 or A490 Bolts*, approved by the Research Council on Riveted and Bolted Joints of the Engineering Foundation.
4. *Manual of Steel Construction*, American Institute of Steel Construction.
5. Slottman, G. V., *Oxygen Cutting*, McGraw-Hill Book Company, New York, 1951.
6. *Welding Handbook*, 5th ed., American Welding Society, 1964.
7. Potter, M. H., *Oxyacetylene Welding*, 3d ed., American Technical Society, 1956.
8. *Brazing Manual*, American Welding Society, 1963.
9. *Soldering Manual*, American Welding Society, 1959.

Section 24

Timber Construction

RUSSELL P. WIBBENS

Director of Marketing,
American Institute of Timber Construction,
Englewood, Colo.

INTRODUCTION

Wood has always been one of man's most important construction materials. However, its emergence from a status as merely a carpentry material to that of a true engineering material has taken place in relatively recent times. This emergence is due to the development of reliable stress grading of structural lumber, improved mechanical fastenings, new adhesives and glue-laminating techniques, and improved fabricating techniques. One of the developments which has expanded the use of timber in the field of heavy construction is the establishment of a structural timber fabricating industry. This industry, acting through its trade association—the American Institute of Timber Construction (AITC)—has initiated a quality-control and inspection program for structural glued laminated timber and has developed sound engineering practices for structural timber framing.

AITC has published timber design and construction information based on the industry's experience and engineering knowledge. This information, together with technical information published by other organizations in the wood-products industry, has made it possible for those in the field of heavy construction to build reliably, durably, and economically with timber.

Because of the versatility of structural timber framing, wood is used in a great variety of structures such as bridges, towers, electric utility structures, piers and wharves, and other marine structures. Timber piles are used in the foundations of many major structures. Wood formwork and centering is widely used in the construction of large concrete and steel structures.

To assist those involved in the design and construction of such structures, this section presents information on selected timber applications in the field of heavy construction. It includes information on properties of wood, mechanical fastenings, design, and other subjects related to timber construction.

PROPERTIES OF WOOD

Characteristics Wood, unlike most other construction materials, is an organic material with a cellular structure. Aspects of structural design and analysis that are dependent upon the material used are considerably different for wood than for other structural materials.

Trees which supply wood used for construction increase in diameter by adding new layers of cells from the pith, or center, outward. In climates where temperature limits the growing season of a tree, each annual increment of growth is known as an "annual ring." In many species of wood, each annual ring consists of two layers. The light, inner layer is made up of large, thin-walled cells formed in the spring when growth is fastest and is known as "springwood." The darker outer layer is made up of smaller, thicker-walled cells formed later in the year and is known as "summerwood."

The strength of wood is closely related to its density or specific gravity. Summerwood contains more solid wood substance than does springwood and is, therefore, denser and stronger. The proportion of width of summerwood to width of annual ring is sometimes used as a visual measure of the quality and strength of wood.

A living tree develops in the presence of moisture. The amount of moisture in wood is important for all construction purposes. Moisture content is the weight of the water in the wood, expressed as a percentage of the oven-dry weight. The removal of moisture from wood is known as "seasoning." Seasoned wood is stiffer and stronger than unseasoned wood of the same dimensions.

Between zero and approximately 30 percent moisture content, wood shrinks as it loses moisture and swells as it absorbs moisture. Above approximately 30 percent moisture content, there is no dimensional change with variation in moisture content. Wood shrinks or swells most in the direction of the annual rings (tangentially), about one-half or two-thirds as much across these rings (radially), and generally very little along the grain (longitudinally). Also, like most other solids, wood expands on heating and contracts on cooling. Radial or tangential dimensional changes due to temperature variations are relatively small for wood used in construction. Longitudinally, dimensional changes may be important in structures which are subjected to large temperature variations.

The most important electrical property of wood is its resistance to the passage of an electric current. The electrical resistance of wood varies inversely with moisture content, density, and temperature. In addition, wood's electrical resistance is greater across the grain than along it.

An important acoustical property of wood is its ability to dampen vibrations. The sound absorption of wood varies with moisture content, direction of grain, and density. Wood's sound absorption is low; however, this property may be improved by patterning the surface.

Additional information about wood's physical properties and characteristics may be found in the *Wood Handbook*.[1*]

Durability Wood structures, when properly designed and constructed and given reasonable maintenance, will resist damage by moisture, living organisms, chemicals, and fire. When conditions warrant, wood's performance can be further improved by the use of chemical treatments. The control of moisture content in wood structures is important not only from the standpoint of dimensional stability but also for protection from decay and insect attack and from checking.

Decay of wood is caused by fungi, which are low forms of plant life. Decay fungi use organic material, such as wood, for food. For growth they need, simultaneously, food, suitable temperatures (generally 40 to 110°F), sufficient air, and adequate moisture. The elimination of any one of these essentials inhibits the development of fungi. Wood permanently and totally submerged in water cannot decay because air is excluded. Wood will not decay when its moisture content is continuously less than 20 percent. When conditions for decay cannot be controlled, decay can be prevented or stopped by using species of high natural decay resistance or by treating the wood with preservatives. A number of species have heartwood of high natural decay resistance; however, the sapwood of all species is low in decay resistance.

Precautions effective against decay are, for the most part, effective also against damage by wood-destroying insects. These precautions include moisture control and preservative treatments. Marine borers, which attack wood structures in salt or brackish waters, can be controlled by pressure treatment of the wood with heavy retentions of creosote or creosote–coal-tar solutions.

Proper seasoning of wood and the control of the rapid loss of moisture from wood in service are important to prevent excessive checking, or separation of grain. Checking is the result of rapid lowering of surface moisture content combined with differential moisture contents of the inner and outer portions of the piece. The outer cells in a piece of wood try to shrink with a loss of moisture but are restrained by the inner cells, which have a higher moisture content. The more rapid the rate of drying, the greater the differential shrinkage and the higher the shrinkage stresses which cause checking.

Wood is superior to many building materials in its resistance to chemical attack.

* Superior numbers refer to the list of references at the end of this section.

It has excellent resistance to mild acids, particularly at room temperatures, and to organic acids. Wood is seldom used in contact with solutions that are more than weakly alkaline. Oxidizing chemicals and solutions of iron salts, in combination with damp conditions, should be avoided. Acids with pH values above 2 and bases with pH values below 10 have little weakening effect on wood at room temperatures if the duration of exposure is moderate.

Maximum safety to life and safety to property can be achieved by taking advantage of the excellent fire endurance of wood in large cross sections and by close attention to structural details. When exposed to fire, wood forms a self-insulating surface layer of char and thus provides its own fire protection. Although the surface chars, the undamaged wood below the char retains its strength and will support loads in accordance with the capacity of the uncharred section. The fire endurance and excellent performance of heavy timber are attributable to the size of the wood members and to the slow rate at which charring occurs.

Strength Because of the orientation of its cells and the manner in which a tree increases in diameter, wood has different strength properties with respect to its three principal axes: longitudinal (parallel to grain), radial (perpendicular to grain, radial to annual rings), and tangential (perpendicular to grain, tangential to annual rings). The difference in strength properties in the radial and tangential directions is seldom of practical importance; in most structural designs it is sufficient to differentiate only between properties parallel and perpendicular to the grain.

TABLE 24-1 Usual Design Factors for Duration of Loading

Duration of load	*Factor*
Permanent	0.90
Normal	1.00
2 months (as for snow)	1.15
7 days	1.25
Wind or earthquake	1.33
Impact	2.00

Wood used for construction is most commonly in the form of stress-graded lumber, structural glued laminated timber, or plywood. Commercially available wood species in each of these three structural forms are assigned allowable unit working stresses. These allowable unit stresses are then adjusted for the service and use conditions to which the particular member will be subjected. Included among the factors for which adjustments to allowable unit stresses are made are moisture conditions, duration of load, temperature, and treatment.

Higher allowable unit stresses are permitted for wood structural members which, as in most covered structures, remain continuously dry in service. Lower stresses are applicable when members are used in exterior or submerged construction.

Wood has the ability to absorb overloads of considerable magnitude for short periods of time. Allowable unit stresses for wood are given on the basis of "normal" loading duration. This assumes that the full maximum design load will be applied for approximately 10 years and 90 percent of this maximum design load will be applied continuously throughout the remainder of the life of the structure. Recommended adjustments of allowable unit stresses for other durations of loading are given in Table 24-1.

The allowable unit stresses for wood apply to structural members used under ordinary temperature conditions. Some reduction of stresses may be necessary for members which are subjected to elevated temperatures for repeated or prolonged periods of time, especially where the high temperature is associated with a high moisture content in the wood. Tests of wood conducted at about $-300°F$ showed that the important strength properties of dry wood in bending and in compression, including stiffness and shock resistance, were much higher at the extremely low temperature than at normal temperature.

Temperature effect on strength is immediate; its magnitude depends on the moisture

TABLE 24-2 Allowable Unit Stresses for Stress-graded Lumber (normal duration of loading; dry conditions of use)[a]

Grade	Allowable unit stresses, psi					
	Extreme fiber in bending, F_b	Tension parallel to grain, F_t	Compression parallel to grain, F_c	Horizontal shear, F_v	Compression perpendicular to grain, $F_{c\perp}$	Modulus of elasticity E, 10^{-6} psi
POSTS AND TIMBERS: 5 × 5 IN. AND LARGER, WIDTH NOT MORE THAN 2 IN. GREATER THAN THICKNESS						
Douglas fir and larch[b,c]						
Dense Structural	1,750	1,150	1,350	140	455	1.70
Structural	1,600	1,050	1,250	140	415	1.62
Dense No. 1	1,300	900	1.200	140	455	1.53
No. 1	1,150	750	1,000	140	385	1.46
Douglas fir, Coast Region[d,e]						
Dense Select Structural	1,900	1,900	1,650	120	455	1.60
Select Structural	1,750	1,750	1,500	120	415	1.60
Dense Construction	1,500	1,500	1,400	120	455	1.60
Construction	1,200	1,200	1,200	120	390	1.60
Hemlock, West Coast[d,f]						
Construction	1,200	1,200	1,100	100	365	1.40
Hemlock, Western, true firs, and Douglas fir[b,g]						
Structural	1,200	800	1,000	115	245	1.36
No. 1	1,100	700	900	115	245	1.36
Pine, Southern[h,i,j]						
Dense Structural 86	2,200	1,450	1,350	140	315	1.60
Dense Structural 72	1,850	1,250	1,150	120	315	1.60
Dense Structural 65	1,650	1,100	1,000	105	315	1.60
Industrial 86	1,900	1,250	1,150	140	270	1.60
Industrial 72	1,600	1,050	975	120	270	1.60
Industrial 65	1,400	950	875	105	270	1.60
Dense Industrial 58	1,500	1,000	900	110	315	1.60
Industrial 58	1,250	850	775	110	270	1.60
No. 1 Dense SR	1,500	1,000	1,050	110	315	1.60
No. 1 SR	1,250	850	925	110	270	1.60
No. 2 Dense SR	1,300	850	775	95	315	1.50
No. 2 SR	1,100	725	675	95	270	1.40

TABLE 24-2 Allowable Unit Stresses for Stress-graded Lumber (normal duration of loading; dry conditions of use)[a] **(Continued)**

Grade	Allowable unit stresses, psi					Modulus of elasticity E, 10^{-6} psi
	Extreme fiber in bending, F_b	Tension parallel to grain, F_t	Compression parallel to grain, F_c	Horizontal shear, F_v	Compression perpendicular to grain, $F_{c\perp}$	
BEAMS AND STRINGERS: 5 IN. AND THICKER, WIDTH MORE THAN 2 IN. GREATER THAN THICKNESS						
Douglas fir and larch[b,c]						
Dense Structural	1,900	1,250	1,300	140	455	1.70
Structural	1,750	1,150	1,200	140	415	1.62
Dense No. 1	1,600	1,050	1,100	140	455	1.70
No. 1	1,350	900	950	140	385	1.62
Douglas fir, Coast Region[d,e]						
Dense Select Structural	2,050	2,050	1,500	120	455	1.60
Select Structural	1,900	1,900	1,400	120	415	1.60
Dense Construction	1,750	1,750	1,200	120	455	1.60
Construction	1,500	1,500	1,000	120	390	1.60
Hemlock, West Coast[d,f]						
Construction	1,500	1,500	1,000	100	365	1.40
Hemlock, Western, true firs, and Douglas fir[b,g]						
Structural	1,400	900	950	115	245	1.36
No. 1	1,200	800	850	115	245	1.36
Pine, Southern[h]						
Use Southern pine stresses given under Posts and Timbers						
JOISTS AND PLANKS: 2 TO 4 IN. THICK, 6 IN. AND WIDER						
Douglas fir and larch[b,c]						
Dense Structural[k]	2,100	1,400	1,750	155	455	1.91
Structural[k]	1,900	1,300	1,600	155	385	1.81
Dense No. 1[k]	1,750	1,200	1,500	125	455	1.91
No. 1[k]	1,500	1,000	1,300	125	385	1.81
No. 2[k]	1,200	800	1,150	95	385	1.63
Dense Structural[l]	1,950	1,300	1,650	140	455	1.70
Structural[l]	1,750	1,200	1,500	140	385	1.62
Dense No. 1[l]	1,600	1,100	1,450	140	455	1.70
No. 1[l]	1,350	950	1,200	140	385	1.62
No. 2[l]	1,100	750	1,050	115	385	1.46

TABLE 24-2 Allowable Unit Stresses for Stress-graded Lumber (normal duration of loading; dry conditions of use)[a] **(Continued)**

Grade	Allowable unit stresses, psi					Modulus of elasticity E, 10^{-6} psi
	Extreme fiber in bending, F_b	Tension parallel to grain, F_t	Compression parallel to grain, F_c	Horizontal shear, F_v	Compression perpendicular to grain, $F_{c\perp}$	
Douglas fir, Coast Region[d,m]						
Dense Select Structural MC15	2,350	2,350	1,900	135	455	1.76
Dense Select Structural DRY	2,150	2,150	1,700	130	455	1.76
Select Structural MC15	2,150	2,150	1,700	135	415	1.76
Select Structural DRY	2,000	2,000	1,550	130	415	1.76
Dense Construction MC15	2,000	2,000	1,600	135	455	1.76
Dense Construction DRY	1,850	1,850	1,450	130	455	1.76
Construction MC15	1,750	1,750	1,400	135	390	1.76
Construction DRY	1,600	1,600	1,200	130	390	1.76
Standard MC15	1,500	1,500	1,250	110	390	1.76
Standard DRY	1,300	1,300	1,100	105	390	1.76
Hemlock, West Coast[d,m]						
Select Structural MC15	1,750	1,750	1,350	115	365	1.54
Select Structural DRY	1,600	1,600	1,200	110	365	1.54
Construction MC15	1,650	1,650	1,250	115	365	1.54
Construction DRY	1,500	1,500	1,100	110	365	1.54
Standard MC15	1,400	1,400	1,100	90	365	1.54
Standard DRY	1,300	1,300	1,100	85	365	1.54
Hemlock, Western, true firs, and Douglas fir[b,g]						
Structural[k]	1,400	950	1,200	125	245	1.52
No. 1[k]	1,200	800	1,050	100	245	1.52
No. 2[k]	950	650	900	75	245	1.37
Structural[l]	1,300	900	1,150	115	245	1.36
No. 1[l]	1,100	750	1,000	115	245	1.36
No. 2[l]	900	600	850	90	245	1.22

TABLE 24-2 Allowable Unit Stresses for Stress-graded Lumber (normal duration of loading; dry conditions of use)[a] **(Continued)**

Grade	Allowable unit stresses, psi					
	Extreme fiber in bending, F_b	Tension parallel to grain, F_t	Compression parallel to grain, F_c	Horizontal shear, F_v	Compression perpendicular to grain, $F_{c\perp}$	Modulus of elasticity E, 10^{-6} psi
Pine, Southern[h,j,n]						
Dense Structural 86 KD[o]	3,000	2,000	2,350	160	475	2.00
Dense Structural 72 KD[o]	2,500	1,650	2,000	135	475	2.00
Dense Structural 65 KD[o]	2,250	1,500	1,800	120	475	2.00
Dense Structural 86[o]....	2,750	1,850	2,050	150	475	1.90
Dense Structural 72[o]....	2,300	1,550	1,700	125	475	1.90
Dense Structural 65[o]....	2,100	1,400	1,550	115	475	1.90
Industrial 86 KD[p]......	2,550	1,700	2,000	160	405	1.90
Industrial 72 KD[p]......	2,150	1,400	1,700	135	405	1.90
Industrial 65 KD[p]......	1,900	1,300	1,550	120	405	1.90
Industrial 86[p]..........	2,350	1,550	1,750	150	405	1.80
Industrial 72[p]..........	1,950	1,300	1,450	125	405	1.80
Industrial 65[p]..........	1,800	1,200	1,300	115	405	1.80
No. 1 Dense SR KD[q]...	2,000	1,350	1,850	125	475	2.00
No. 1 SR KD[q].........	1,700	1,150	1,600	125	405	1.90
No. 2 Dense SR KD[q]...	1,750	1,150	1,400	105	475	1.80
No. 2 SR KD[q].........	1,500	1,000	1,150	105	405	1.70
No. 1 Dense SR[q].......	1,850	1,250	1,600	120	475	1.90
No. 1 SR[q].............	1,600	1,050	1,350	120	405	1.80
No. 2 Dense SR[q].......	1,600	1,050	1,200	100	475	1.70
No. 2 SR[q].............	1,350	925	1,000	100	405	1.60
No. 1 Dense KD[r].......	2,000	1,350	1,600	105	475	2.00
No. 1 Dense[r]...........	1,850	1,250	1,350	100	475	1.90
No. 1 KD	1,700	1,150	1,350	105	405	1.90
No. 1[r]................	1,600	1,050	1,150	100	405	1.80
No. 2 Dense KD	1,750	1,150	1,400	90	475	1.80
No. 2 Dense[r]...........	1,600	1,050	1,200	90	475	1.70
No. KD[r]..............	1,500	1,000	1,150	90	405	1.70
No. 2[r]................	1,350	925	1,000	90	405	1.60

[a] All dimensions are nominal.

[b] See *Rules for Grading Western Lumber*, Western Wood Products Association, Portland, Ore., July 1, 1968, for grading provisions applicable to use of these stresses. When 2-in.-thick lumber is manufactured at a maximum moisture content of 15 percent and used under conditions where moisture content does not exceed 15 percent, multiply the allowable unit stresses for 2-in.-thick lumber by the following factors:

Thickness	F_b	F_t	F_c	F_v	$F_{c\perp}$	E
2 in.............	1.08	1.08	1.17	1.05	1.00	1.05

Footnotes for Table 24-2 (Continued)

When lumber is used under conditions where moisture content will exceed 19 percent for extended periods of time, multiply the allowable unit stresses by the following factors:

Thickness	F_b	F_t	F_c	F_v	$F_{c\perp}$	E
2 in.	0.84	0.84	0.70	0.97	0.67	0.94
3 and 4 in.	0.87	0.87	0.70	1.00	0.67	0.98
5 in. and over	1.00	1.00	0.91	1.00	0.67	0.98

[c] Stresses are not applicable to Douglas fir lumber manufactured in the states of Arizona, Colorado, Nevada, New Mexico, and Utah.

[d] See *Standard Grading and Dressing Rules Number 15*, West Coast Lumber Inspection Bureau, Portland, Ore., February 15, 1968, for grading provisions applicable to the use of these stresses. When lumber is used under conditions where moisture content is at or above fiber saturation point, as when continuously submerged, multiply the allowable unit stresses by the following factors (applicable to all thicknesses):

F_b	F_t	F_c	F_v	$F_{c\perp}$	E
1.00	1.00	0.90	1.00	0.67	1.00

[e] If pieces are permitted to season to a substantial depth from the surface before full load is applied, use E = 1,760,000 psi.

[f] If pieces are permitted to season to a substantial depth from the surface before full load is applied, use E = 1,540,000 psi.

[g] True firs include California red, grand, noble, Pacific silver, and white fir. Stresses are applicable to Douglas fir lumber manufactured in states specified in footnote *c*.

[h] See *Standard Grading Rules for Southern Pine Lumber*, Southern Pine Inspection Bureau, New Orleans, La., July 1, 1968, for grading provisions applicable to the use of these stresses. When lumber is used under wet conditions or where moisture content is at or above fiber saturation point, as when continuously submerged, multiply the $F_{c\perp}$ values by 0.67 and use the allowable unit stresses listed under "Posts and Timbers" for corresponding grades multiplied by the following factors:

F_b	F_t	F_c	F_v	E
1.00	1.00	0.90	1.00	1.00

[i] Stresses are for lumber 2½ in. and thicker of over 19 percent moisture content.

[j] SR means "stress-rated."

[k] Stresses are for lumber 2 in. thick, 6 in. and wider.

[l] Stresses are for lumber 3 and 4 in. thick, 6 in. and wider (designated as "Heavy Joists, Planks and Studs" by WWPA).

[m] MC15 indicates lumber was dried to 15 percent maximum moisture content before surfacing. DRY indicates lumber was dried to 19 percent maximum moisture content.

[n] KD designates kiln-dried lumber or 15 percent maximum moisture content. Grades not designated KD have 19 percent maximum moisture content.

[o] Stresses are for 2- to 4-in.-thick lumber.

[p] Stresses are for 1- to 4-in.-thick lumber.

[q] Stresses are for 2½- to 4-in.-thick lumber.

[r] Stresses are for 2-in.-thick lumber.

content of the wood and, when the temperature is elevated, on the duration of exposure. Under ordinary atmospheric conditions, wood which is exposed for a short time to temperatures not excessively above normal can be expected to recover essentially all its original strength when the temperature is reduced to normal. Experiments indicate that air-dry wood can probably be exposed to temperatures up to nearly 150°F for a year or more without an important permanent loss in most of its strength properties, but while heated its strength will be temporarily reduced compared to its strength at normal temperature.

When wood is exposed to temperatures of 150°F or higher for extended periods of time, it is permanently weakened, even though the temperature is subsequently reduced and the wood is used at normal temperatures. The permanent or nonrecoverable strength loss depends on a number of factors, including the moisture content and temperature of the wood, the heating medium and time of exposure, and to some extent on the species and the size of the piece.

The allowable unit stresses for wood apply to structural members which have been treated with a preservative when this treatment is in accordance with American Wood-Preservers' Association standard specifications.[2] The effects on strength of other treatments should be investigated.

Stress-graded Lumber. Grades of lumber in which the defects or growth characteristics are limited are known as "structural grades" or "stress grades." Stress-graded lumber falls into any of several lumber classifications, each of which contains grades which are assigned allowable unit stresses. These classifications are defined as follows:

1. *Posts and timbers.* Lumber of square or approximately square cross section, graded primarily for use as posts or columns carrying longitudinal load but adapted for miscellaneous uses in which strength in bending is not especially important.
2. *Beams and stringers.* Lumber of rectangular cross section, graded with respect to its strength in bending when loaded on the narrow face.
3. *Joists and planks.* Lumber of rectangular cross section, graded with respect to its strength in bending when loaded either on the narrow face as a joist or on the wide face as a plank.

Table 24-2 gives allowable unit-stress values for species of stress-graded lumber most commonly used in construction. These stress values apply to sawn lumber under normal duration of loading and under continuously dry service conditions.

For stress-graded lumber used under other than normal durations of loading, the stress values in Table 24-2 must be multiplied by the appropriate adjustment factor from Table 24-1.

For stress-graded lumber used under other than continuously dry conditions of service, the stress values in Table 24-2 must be multiplied by the appropriate adjustment factor as given in the footnotes to the table.

The stress values in Table 24-2 apply for stress-graded lumber which has been preservatively treated by an approved process. For stress-graded lumber which has been pressure impregnated with fire-retardant chemicals, reduce the stress values in Table 24-2 by 10 percent.

Structural Glued Laminated Timber. Structural glued laminated timber is an engineered, stress-rated product of a timber laminating plant. Assemblies of suitably selected and prepared wood laminations are securely bonded together with adhesives, with the longitudinal grain of all laminations approximately parallel. The separate laminations may not exceed 2 in. in net thickness. They may be comprised of pieces end joined to form any length, of pieces placed or glued edge to edge to make wider ones, or of pieces bent to curved form during gluing.

Table 24-3 gives allowable unit stresses for species of structural glued laminated softwood timber included in U.S. Commercial Standard CS 253-63,[3] for which standard structural laminating specifications are available. The stresses apply for horizontally laminated members (load acts perpendicular to wide face of lamination) comprised of four or more laminations. Requirements for slope of grain, type and location of end joints, manufacturing procedures, and other provisions given in the specifications must be met if these allowable unit stresses are to apply.

The stress values in Table 24-3 apply to glued laminated timber used under normal

TABLE 24-3 Allowable Unit Stresses for Structural Glued Laminated Softwood Timber (normal duration of loading)

Combination symbol	Allowable unit stresses, psi				
	Extreme fiber in bending, F_b	Tension parallel to grain, F_t	Compression parallel to grain, F_c	Horizontal shear, F_v	Compression perpendicular to grain, $F_{c\perp}$
DRY CONDITIONS OF USE					
Douglas fir and larch[a,b] E = 1,820,000 psi					
26F	**2,600**	1,900	1,400	195	See footnote[e]
24F	**2,400**	1,800	1,400	195	
22F	**2,200**	1,800	1,300	195	
T-6[c]	2,400	**2,400**	**1,800**	195[d]	
T-5[c]	2,200	**2,200**	**1,600**	195[d]	
T-4[c]	2,100	**2,100**	**1,500**	195[d]	
T-3[c]	2,100	**2,100**	**1,600**	195[d]	
T-2[c]	1,800	**1,800**	**1,400**	195[d]	
T-1[c]	1,500	**1,500**	**1,200**	195[d]	
Douglas fir, Coast Region[a,f] E = 1,800,000 psi					
A	**2,600**	1,600	1,500	165	450
B	**2,400**	1,600	1,500	165	450
C	**2,200**	1,600	1,500	165	450
E	2,600	**2,600**	**2,200**	165	450
F	2,200	**2,200**	**2,100**	165	450
G	1,800	**1,800**	**1,800**	165	385
H	1,200	**1,200**	**1,500**	165	385
Pine, Southern[g] E = 1,800,000 psi					
A-1	2,600	2,600	2,000	200	385
A-2	2,600	2,600	2,000	200	450
A-3	2,600	2,600	2,000	200	450
A-4	2,600	2,400	2,000	200	450
B-1	2,400	2,600	2,000	200	385
B-2	2,400	2,600	2,000	200	450
B-3	2,400	2,600	2,000	200	385
B-4	2,400	2,400	2,000	200	385
C-1	2,200	2,400	1,900	200	450
C-2	2,200	2,600	2,000	200	385
C-3	2,200	2,400	1,900	200	385
D-1	2,000	2,400	1,900	200	385
D-2	2,000	2,200	1,900	200	385
E-1	1,800	2,200	1,900	200	385
E-2	1,800	2,200	1,800	200	385
Redwood, California[a,h] E = 1,300,000 psi					
A	**2,200**	2,000[i]	2,200	125	325
B	**2,200**	2,000[i]	2,000	125	325
C	**2,200**	2,000	2,000	125	325
D	2,200	**2,200**	**2,200**	125	325
E	2,000	**2,000**	**2,000**	125	325
F	1,400	**1,800**	**1,800**	125	325

TABLE 24-3 Allowable Unit Stresses for Structural Glued Laminated Softwood Timber (normal duration of loading) **(Continued)**

Combination symbol	Allowable unit stresses, psi				
	Extreme fiber in bending, F_b	Tension parallel to grain, F_t	Compression parallel to grain, F_c	Horizontal shear, F_v	Compression perpendicular to grain, $F_{c\perp}$
	WET CONDITIONS OF USE				
	Douglas fir and larch[a,b] E = 1,510,000 psi				
26F	**2,100**	1,500	1,000	175	See footnote[e]
24F	**1,900**	1,400	1,000	175	
22F	**1,800**	1,400	950	175	
T-6[c]	1,900	**1,900**	**1,300**	175[d]	
T-5[c]	1,800	**1,800**	**1,200**	175[d]	
T-4[c]	1,700	**1,700**	**1,100**	175[d]	
T-3[c]	1,700	**1,700**	**1,200**	175[d]	
T-2[c]	1,400	**1,400**	**1,000**	175[d]	
T-1[c]	1,200	**1,200**	**900**	175[d]	
	Douglas fir, Coast Region[a,f] E = 1,600,000 psi				
A	**2,000**	1,300	1,100	145	305
B	**1,800**	1,300	1,100	145	305
C	**1,600**	1,300	1,100	145	305
E	2,000	**2,000**	**1,600**	145	305
F	1,800	**1,800**	**1,500**	145	305
G	1,400	**1,400**	**1,300**	145	260
H	950	**950**	**1,100**	145	260
	Pine, Southern[g] E = 1,600,000 psi				
A-1	2,000	2,000	1,500	175	260
A-2	2,000	2,000	1,400	175	300
A-3	2,000	2,000	1,400	175	300
A-4	2,100	2,000	1,400	175	300
B-1	1,900	2,000	1,400	175	260
B-2	2,000	2,000	1,400	175	300
B-3	1,900	2,000	1,400	175	260
B-4	1,900	2,000	1,400	175	260
C-1	1,800	1,800	1,400	175	300
C-2	1,800	2,000	1,400	175	260
C-3	1,700	1,800	1,400	175	260
D-1	1,600	2,000	1,400	175	260
D-2	1,600	1,800	1,400	175	260
E-1	1,400	1,800	1,400	175	260
E-2	1,400	1,800	1,300	175	260
	Redwood, California[a,h] E = 1,200,000 psi				
A	**1,800**	1,600[i]	1,600	110	215
B	**1,800**	1,600[i]	1,500	110	215
C	**1,800**	1,600	1,500	110	215
D	1,800	**1,800**	**1,600**	110	215
E	1,600	**1,600**	**1,500**	110	215
F	1,100	**1,500**	**1,300**	110	215

Footnotes for Table 24-3

[a] Boldface values in F_b column are for members stressed principally in bending. Boldface values in F_t and F_c columns are for members stressed principally in axial tension or axial compression.

[b] See *Standards for Structural Glued Laminated Members Assembled with WWPA Grades of Douglas Fir and Larch Lumber*, Western Wood Products Association, Portland, Ore., 1966, for provisions applicable to the use of these stresses.

[c] Allowable unit stresses for bending, tension, and compression may be increased when slope-of-grain limitations are more restrictive than basic requirements. See reference given in footnote *b*.

[d] Values also apply for members loaded parallel to wide face of laminations when no splits or checks are present at ends of member. When checks or splits are allowed to be present, the allowable horizontal shear stress may not exceed 145 psi for dry conditions of use or 130 psi for wet conditions of use.

[e] Compression-perpendicular-to-grain stresses associated with the lamination that is in the area of bearing vary with rate of growth as follows:

Rate of growth	$F_{c\perp}$ value, psi	
	Dry use	Wet use
Medium grain	385	255
Close grain	410	275
Dense grain	450	300

[f] See *Standard Specifications for Structural Glued Laminated Douglas Fir (Coast Region) Timber*, West Coast Lumbermen's Association, 1963, West Coast Lumber Inspection Bureau, Portland, Ore., for provisions applicable to the use of these stresses.

[g] See *Standard Specifications for Structural Glued Laminated Southern Pine Timber*, Southern Pine Inspection Bureau, New Orleans, La., 1965, for provisions applicable to the use of these stresses. The following limitations on number of laminations are applicable:

Combination symbol	*Number of laminations*	*Combination symbol*	*Number of laminations*
A-1 and B-4	9 or more	B-3	25 or more
A-2	14 to 21	C-1	6 or more
A-3	22 or more	C-2	14 or more
A-4	13 or more	C-3	18 or more
B-1 and E-1	4 or more	D-1 and D-2	10 or more
B-2 and E-2	12 or more		

[h] See *Standard Specifications for Structural Glued Laminated California Redwood Timber*, California Redwood Association, San Francisco, 1965, for provisions applicable to the use of these stresses.

[i] Allowable unit stresses may be increased when slope-of-grain limitations are more restrictive. See reference given in footnote *h*.

TABLE 24-4 Allowable Unit Stresses for Plywood, Group 1 Species (normal duration of loading; dry conditions of use)*,†

Type of stress	Exterior A-A, A-C, C-C, and comparable grades of Overlaid plywood; Structural I A-A; Structural I A-C; Structural I C-C; and Marine, psi	Exterior A-B, B-B, B-C, C-C (Plugged); Plyform Class I and comparable grades of Overlaid plywood; Structural I C-D; Standard Sheathing (Exterior Glue); all Interior grades with Exterior Glue, psi	All other grades of Interior including Standard Sheathing, psi
Extreme fiber in bending, tension. Face grain parallel or perpendicular to span (at 45° to face grain use 1/6)	2,000	1,650	1,650
Compression. Parallel or perpendicular to face grain (at 45° to face grain use 1/3)	1,650	1,550	1,550
Bearing (on face)	340	340	340
Shear in plane perpendicular to plies.‡ Parallel or perpendicular to face grain (at 45° increase 100%)	250	250	230
Shear, rolling, in plane of plies.§ Parallel or perpendicular to face grain (at 45° increase 1/3)	53	53	48
Modulus of elasticity in bending. Face grain parallel or perpendicular to span		1,800,000	

* Conforming to U.S. Product Standard PS1-66 for Softwood Plywood, Construction and Industrial. Group 1 species include Douglas fir, larch, Southern pine and tanoak, Table adapted from *Plywood Design Specification*, American Plywood Association, Tacoma. Wash., November, 1966.

† Where moisture content is 16% or more, multiply the dry location values by the following factors:

For all grades of Exterior and Interior plywood with exterior glue,

Extreme fiber in bending—75%	Modulus of elasticity—89%
Tension—69%	Shear—84%
Compression—61%	Bearing—67%

For all other grades of Interior,

Extreme fiber in bending—69%	Modulus of elasticity—80%
Tension—69%	Shear—84%
Compression—61%	Bearing—67%

‡ Shear-through-the-thickness stresses are based on the most common structural applications where the plywood is attached to framing around its boundary. Where the plywood is attached to framing at only two sides—as in the heel joint of a truss—reduce the allowable shear-through-the-thickness values by 11 percent where framing is parallel to face grain and 25 percent where it is perpendicular.

§ For Marine and Structural I grades use 75 psi. For Structural II use 56 psi.

duration of loading. For other durations of loading, the stress values must be multiplied by the appropriate adjustment factor from Table 24-1.

Table 24-3 is divided into sections for dry-use and wet-use service conditions. Allowable unit stresses for dry-use conditions are applicable when the moisture content in service is less than 16 percent, as in most covered structures. Allowable unit stresses for wet-use conditions are applicable when the moisture content in service is 16 percent or more. This occurs in exterior or submerged construction and in some structures housing wet processes or otherwise having high relative humidities.

The stress values in Table 24-3 also apply for glued laminated timber which has been treated with preservative by an approved process.

Plywood. Plywood is made up of a combination of three or more layers of wood veneer, glued together with the grain of alternate plies at right angles. Plywood possesses more nearly equal strength properties along the length and width of a panel than does solid wood. The two basic types of plywood are Exterior and Interior. Type depends on veneer grade and on whether waterproof (exterior or wet-use) or water-resistant (interior or dry-use) adhesive is used. Softwood plywood is manufactured from some thirty different species, which have been classified into four groups depending on stiffness. The strongest species, such as Douglas fir, larch, and Southern pine, are in Group 1.

Table 24-4 gives allowable unit stresses for softwood plywood. These stresses are for normal duration of loading and dry conditions of service. For other durations of loading, the stresses must be multiplied by the appropriate adjustment factor from Table 24-1. For other service conditions, apply the appropriate adjustment factor given in the footnote to the table.

Sanded grades of softwood plywood, referred to in Table 24-4, are generally identified by the grade of veneer used for the faces. Veneer grades range from A, the highest commonly used; through C, the lowest used in fully Exterior plywood; to D, used only in inner plies and backs of Interior plywood. Unsanded sheathing grades include Structural I, Structural II, and Standard. The Structural grades are intended for maximum efficiency in engineered applications. A designation known as an Identification Index appears in the grade trademarks of Standard sheathing, Structural I, Structural II, and C-C Exterior. It is a set of two numbers separated by a slash. The number on the left indicates recommended maximum spacing in inches for supports when the panel is used for roof decking. The number on the right shows maximum recommended spacing in inches for supports when the panel is used for subflooring.

The stresses in Table 24-4 apply for plywood which has been treated with preservative by an approved process. For plywood which has been pressure-impregnated with fire-retardant chemicals, multiply the stress values in Table 24-4 by 83 percent and the modulus-of-elasticity values by 90 percent.

Section Properties Tables 24-5 through 24-8 give properties of sections for sawn lumber, glued laminated timber, and plywood. Sawn lumber is available in even 2-ft increments of length, with a limit on the maximum length normally available. Glued laminated timber is available in any length; however, length limitations may be imposed by transportation requirements. The standard panel size for plywood is 4 by 8 ft, although other sizes are available.

The use of the sizes given in these tables will result in construction economy, since standard sizes are less costly than special sizes and lengths.

Table 24-5 gives properties of sections for sizes of lumber and timber conforming to Simplified Practice Recommendation 16-53, *American Lumber Standards for Softwood Lumber*. Table 24-6 gives properties of sections for dry sizes of lumber (moisture content 19 percent or less) conforming to Product Standard 20-70, *American Softwood Lumber Standard*,[4] which superseded SPR 16-53, effective March 1, 1970.

Table 24-7 is based on the laminating industry's recommended practice of using lumber of nominal 1- or 2-in. thickness for laminating. Lumber of 1-in. nominal thickness is dressed to ¾ in. before gluing and is generally used for the more sharply curved structural members. Lumber of 2-in. nominal thickness is dressed to 1⅝ or 1½ in. before gluing and is generally used for straight or slightly curved members.

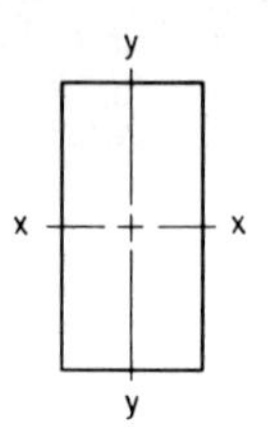

TABLE 24-5 Properties of Sections of ALS Sizes of Lumber and Timber (prior to March 1, 1970)*

Nominal size,† in.	Actual size, S4S,‡ in.	Area, in.²	Section modulus, in.³		Moment of inertia, in.⁴	
			Axis x-x	Axis y-y	Axis x-x	Axis y-y
$b \times d$	$b \times d$					
12 × 20	11½ × 19½	224.25	728.8	429.8	7,106	2,471
18	17½	201.25	587.0	385.7	5,136	2,218
16	15½	178.25	460.5	341.6	3,569	1,964
14	13½	155.25	349.3	297.6	2,358	1,711
12	11½	132.25	253.5	253.5	1,458	1,458
10 × 20	9½ × 19½	185.25	602.1	293.3	5,870	1,393
18	17½	166.25	484.9	263.2	4,243	1,250
16	15½	147.25	380.4	233.1	2,948	1,107
14	13½	128.25	288.6	203.1	1,948	965
12	11½	109.25	209.4	173.0	1,204	822
10	9½	90.25	142.9	142.9	679	679
8 × 20	7½ × 19½	146.25	475.0	182.8	4,625	685
18	17½	131.25	382.8	164.1	3,350	615
16	15½	116.25	300.3	145.3	2,327	545
14	13½	101.25	227.8	126.6	1,538	475
12	11½	86.25	165.3	107.8	951	404
10	9½	71.25	112.8	89.1	536	334
8	7½	56.25	70.3	70.3	264	264
6 × 16	5½ × 15½	85.25	220.2	78.15	1,707	214.9
14	13½	74.25	167.1	68.06	1,128	187.2
12	11½	63.25	121.2	57.98	697	159.4
10	9½	52.25	82.7	47.90	393	131.7
8	7½	41.25	51.6	37.81	193	104.0
6	5½	30.25	27.7	27.73	76.3	76.3
4 × 16	3⅝ × 15½	56.19	145.2	33.95	1,125	61.53
14	13½	48.94	110.1	29.57	743	53.60
12	11½	41.69	79.9	25.19	459	45.66
10	9½	34.44	54.5	20.81	259	37.72
8	7½	27.19	34.0	16.43	127	29.78
6	5½	19.95	18.3	12.05	50.2	21.84
4	3⅝	13.14	7.94	7.94	14.4	14.39
3 × 16	2⅝ × 15½	40.69	105.1	17.80	814.6	23.36
14	13½	35.44	79.73	15.50	538.1	20.34
12	11½	30.19	57.86	13.21	332.7	17.34
10	9½	24.94	39.48	10.91	187.6	14.32
8	7½	19.69	24.61	8.61	92.3	11.30
6	4½	14.43	13.23	6.31	36.4	8.28
4	3⅝	9.52	5.75	4.16	10.4	5.46

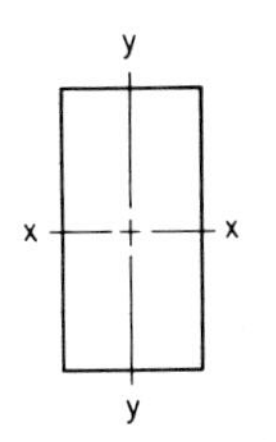

TABLE 24-5 Properties of Sections of ALS Sizes of Lumber and Timber (prior to March 1, 1970)* **(Continued)**

Nominal size,† in.	Actual size, S4S,‡ in.	Area, in.[2]	Section modulus, in.[3]		Moment of inertia, in.[4]	
			Axis x-x	Axis y-y	Axis x-x	Axis y-y
$b \times d$	$b \times d$					
2×16	$1\frac{5}{8} \times 15\frac{1}{2}$	25.19	65.07	6.82	504.3	5.54
14	$13\frac{1}{2}$	21.94	49.36	5.94	333.2	4.83
12	$11\frac{1}{2}$	18.69	35.82	5.06	206.0	4.11
10	$9\frac{1}{2}$	15.44	24.44	4.18	116.1	3.40
8	$7\frac{1}{2}$	12.19	15.23	3.30	57.1	2.68
6	$5\frac{1}{2}$	8.93	8.19	2.42	22.5	1.97
4	$3\frac{5}{8}$	5.89	3.56	1.60	6.45	1.30
2	$1\frac{5}{8}$	2.64	0.72	0.72	0.58	0.58
1×12	$\frac{25}{32} \times 11\frac{1}{2}$	8.98	17.22	1.17	99.0	0.46
10	$9\frac{1}{2}$	7.42	11.75	0.97	55.8	0.38
8	$7\frac{1}{2}$	5.86	7.32	0.76	27.5	0.30
6	$5\frac{1}{2}$	4.30	3.94	0.56	10.8	0.22
4	$3\frac{5}{8}$	2.83	1.71	0.37	3.10	0.14
2	$1\frac{5}{8}$	1.27	0.34	0.17	0.28	0.07

* Sizes conforming to *American Lumber Standards for Softwood Lumber*, Simplified Practice Recommendation 16-53.

† For members over 12 in. in dimension, check availability with local suppliers. Lengths are generally available in 2-ft increments with a limit on maximum length; check with local suppliers for lengths normally available.

‡ Surfaced on four sides.

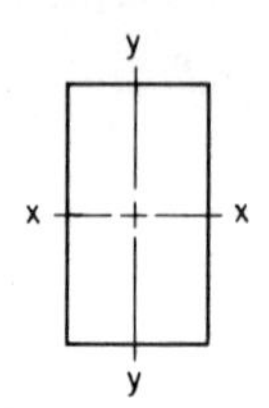

TABLE 24-6 Properties of Sections of ALS Dry Lumber Sizes (effective March 1, 1970)*

Nominal size,† in.	Actual size, S4S,‡ in.	Area, in.²	Section modulus, in.³		Moment of inertia, in.⁴	
			Axis x-x	Axis y-y	Axis x-x	Axis y-y
$b \times d$	$b \times d$					
4 × 16	3½ × 15¼	53.38	135.7	31.1	1,034	54.5
14	13¼	46.38	102.4	27.0	678	47.3
12	11¼	39.38	73.8	23.0	415	40.2
10	9¼	32.38	49.9	18.9	231	33.0
8	7¼	25.38	30.7	14.8	111	25.9
6	5½	19.25	17.6	11.2	48.5	19.6
4	3½	12.25	7.15	7.15	12.5	12.5
3 × 16	2½ × 15¼	38.12	96.9	15.9	739	19.9
14	13¼	33.12	73.2	13.8	485	17.2
12	11¼	28.12	52.7	11.7	297	14.6
10	9¼	23.12	35.6	9.64	165	12.0
8	7¼	18.12	21.9	7.55	79.4	9.44
6	5½	13.75	12.6	5.73	34.7	7.16
4	3½	8.75	5.10	3.65	8.93	4.56
2 × 16	1½ × 15¼	22.88	58.1	5.72	443	4.29
14	13¼	19.88	43.9	4.97	291	3.73
12	11¼	16.88	31.6	4.22	178	3.16
10	9¼	13.88	21.4	3.47	98.9	2.60
8	7¼	10.88	13.1	2.72	47.6	2.04
6	5½	8.25	7.56	2.06	20.8	1.55
4	3½	5.25	3.06	1.31	5.36	0.98
2	1½	2.25	0.56	0.56	0.42	0.42
1 × 12	¾ × 11¼	8.44	15.8	1.05	89.0	0.40
10	9¼	6.94	10.7	0.87	49.5	0.32
8	7¼	5.44	6.57	0.68	23.8	0.25
6	5½	4.12	3.78	0.52	10.4	0.19
4	3½	2.62	1.53	0.33	2.68	0.12
2	1½	1.12	0.28	0.14	0.21	0.05

* Sizes at 19 percent maximum moisture content conforming to *American Softwood Lumber Standard* (see reference 4).

† For members over 12 in. in dimension, check availability with local suppliers. Lengths are generally available in 2-ft increments with a limit on maximum length; check with local suppliers for lengths normally available.

‡ Surfaced on four sides.

TABLE 24-7 Properties of Sections for Glued Laminated Timber

Actual size, in.	Area A, in.^{2}	Section modulus S, in.^{3}	Moment of inertia I, in.^{4}	Volume per lineal foot, cu ft
11 × 66	726	7,990	264,000	5.04
65¼	718	7,810	255,000	4.98
65	715	7,750	252,000	4.97
64½	710	7,630	246,000	4.93
63¾	701	7,450	237,000	4.87
63⅜	697	7,360	233,000	4.84
63	693	7,280	229,000	4.81
62¼	685	7,100	221,000	4.76
61¾	679	6,990	216,000	4.72
61½	676	6,930	213,000	4.70
60¾	668	6,770	206,000	4.64
60⅛	661	6,630	199,000	4.59
60	660	6,600	198,000	4.58
59¼	652	6,440	191,000	4.53
58½	644	6,270	184,000	4.47
57¾	635	6,110	177,000	4.41
11 × 57	627	5,960	170,000	4.35
56⅞	626	5,930	169,000	4.34
56¼	619	5,800	163,000	4.30
55½	610	5,650	157,000	4.24
55¼	608	5,600	155,000	4.22
54¾	602	5,500	150,000	4.18
54	594	5,350	144,000	4.12
53⅝	590	5,270	141,000	4.10
53¼	586	5,200	138,000	4.07
52½	578	5,050	133,000	4.01
52	572	4,960	129,000	3.97
51¾	569	4,910	127,000	3.95
51	561	4,770	122,000	3.90
50⅜	554	4,650	117,000	3.85
50¼	553	4,630	116,000	3.84
49½	544	4,490	111,000	3.78
11 × 48¾	536	4,360	106,000	3.72
48	528	4,220	101,000	3.67
47¼	520	4,090	96,700	3.61
47⅛	518	4,070	95,900	3.60
46½	512	3,960	92,200	3.55
45¾	503	3,840	87,800	3.49
45½	500	3,800	86,300	3.48
45	495	3,710	83,500	3.44
44¼	487	3,590	79,400	3.38
43⅞	483	3,530	77,400	3.35
43½	478	3,470	75,400	3.32
42¾	470	3,350	71,600	3.27
42¼	465	3,270	69,100	3.23
42	462	3,230	67,900	3.21
41¼	454	3,120	64,300	3.15
40⅝	447	3,030	61,500	3.10
11 × 40½	446	3,010	60,900	3.09
39¾	437	2,900	57,600	3.04
39	429	2,790	54,400	2.98
38¼	421	2,680	51,300	2.92
37½	412	2,580	48,300	2.86
37⅜	411	2,560	47,900	2.86

TABLE 24-7 Properties of Sections for Glued Laminated Timber (Continued)

Actual size, in.	Area A, in.²	Section modulus S, in.³	Moment of inertia I, in.⁴	Volume per lineal foot, cu ft
11 × 36¾	404	2,480	45,500	2.81
36	396	2,380	42,800	2.75
35¾	393	2,340	41,900	2.73
35¼	388	2,280	40,200	2.69
34½	380	2,180	37,600	2.64
34⅛	375	2,140	36,400	2.61
33¾	371	2,090	35,200	2.58
33	363	2,000	32,900	2.52
32½	358	1,940	31,500	2.48
32¼	355	1,910	30,700	2.46
11 × 31½	346	1,820	28,600	2.41
30⅞	340	1,750	27,000	2.36
30¾	338	1,730	26,600	2.35
30	330	1,650	24,800	2.29
29¼	322	1,570	22,900	2.23
28½	314	1,490	21,200	2.18
27¾	305	1,410	19,600	2.12
27⅝	304	1,400	19,300	2.11
27	297	1,340	18,000	2.06
26¼	289	1,260	16,600	2.01
26	286	1,240	16,100	1.99
25½	280	1,190	15,200	1.95
24¾	272	1,120	13,900	1.89
24⅜	268	1,090	13,300	1.86
24	264	1,060	12,700	1.83
23¼	256	991	11,500	1.78
11 × 22¾	250	949	10,800	1.74
22½	248	928	10,400	1.72
21¾	239	867	9,430	1.66
21⅛	232	818	8,640	1.61
21	231	808	8,490	1.60
20¼	223	752	7,610	1.55
19½	214	697	6,800	1.49
18¾	206	644	6,040	1.43
18	198	594	5,350	1.38
17⅞	197	586	5,240	1.36
17¼	190	546	4,700	1.32
16½	182	499	4,120	1.26
16¼	179	484	3,930	1.24
15¾	173	455	3,580	1.20
15	165	412	3,090	1.15
14⅝	161	392	2,870	1.12
11 × 14¼	157	372	2,650	1.09
13½	148	334	2,260	1.03
13	143	310	2,010	0.993
12¾	140	298	1,900	0.974
12	132	264	1,580	0.917
11⅜	125	237	1,350	0.869
11¼	124	232	1,300	0.859

TABLE 24-7 Properties of Sections for Glued Laminated Timber (Continued)

Actual size, in.	Area A, in.^{2}	Section modulus S, in.^{3}	Moment of inertia I, in.^{4}	Volume per lineal foot, cu ft
9 × 54	486	4,370	118,000	3.37
53⅝	483	4,310	116,000	3.35
53¼	479	4,250	113,000	3.33
52½	472	4,130	108,000	3.28
52	468	4,060	106,000	3.25
51¾	466	4,020	104,000	3.23
51	459	3,900	99,500	3.19
50⅜	453	3,180	95,900	3.15
50¼	452	3,790	95,200	3.14
9 × 49½	446	3,680	91,000	3.09
48¾	439	3,560	86,900	3.05
48	432	3,460	82,900	3.00
47¼	425	3,350	79,100	2.95
47⅛	424	3,330	78,500	2.95
46½	418	3,240	75,400	2.91
45¾	412	3,140	71,800	2.86
45½	410	3,100	70,600	2.84
45	405	3,040	68,300	2.81
44¼	398	2,940	65,000	2.77
43⅞	395	2,890	63,300	2.74
43½	392	2,840	61,700	2.72
42¾	385	2,740	58,600	2.67
42¼	380	2,680	56,600	2.64
42	378	2,650	55,600	2.62
41¼	371	2,550	52,600	2.58
9 × 40⅝	366	2,480	50,300	2.54
40½	364	2,460	49,800	2.53
39¾	358	2,370	47,100	2.48
39	351	2,280	44,500	2.44
38¼	344	2,200	42,000	2.39
37½	338	2,110	39,600	2.34
37⅜	336	2,100	39,200	2.34
36¾	331	2,030	37,200	2.30
36	324	1,940	35,000	2.25
35¾	322	1,920	34,300	2.23
35¼	317	1,860	32,800	2.20
34½	310	1,780	30,800	2.16
34⅛	307	1,750	29,800	2.13
33¾	304	1,710	28,800	2.11
33	297	1,630	27,000	2.06
32½	292	1,580	25,700	2.03
9 × 32¼	290	1,560	25,200	2.02
31½	284	1,490	23,400	1.97
30⅞	278	1,430	22,100	1.93
30¾	277	1,420	21,800	1.92
30	270	1,350	20,200	1.88
29¼	263	1,280	18,800	1.83
28½	256	1,220	17,400	1.78
27¾	250	1,160	16,000	1.73
27⅝	249	1,140	15,800	1.73
27	243	1,090	14,800	1.69
26¼	236	1,030	13,600	1.64
26	234	1,010	13,200	1.62
25½	230	975	12,400	1.59
24¾	223	919	11,400	1.55
24⅜	219	891	10,900	1.52
24	216	864	10,400	1.50

TABLE 24-7 Properties of Sections for Glued Laminated Timber (Continued)

Actual size, in.	Area A, in.^{2}	Section modulus S, in.^{3}	Moment of inertia I, in.^{4}	Volume per lineal foot, cu ft
9 × 23¼	209	811	9,430	1.45
22¾	205	776	8,830	1.42
22½	202	759	8,540	1.41
21¾	196	710	7,720	1.36
21⅛	190	669	7,070	1.32
21	189	662	6,950	1.31
20¼	182	615	6,230	1.26
19½	176	570	5,560	1.22
18¾	169	527	4,940	1.17
18	162	486	4,370	1.12
17⅞	161	479	4,280	1.12
17¼	155	446	3,850	1.08
16½	148	408	3,370	1.03
16¼	146	396	3,220	1.02
15¾	142	372	2,930	0.984
15	135	338	2,530	0.938
9 × 14⅝	132	321	2,350	0.914
14¼	128	305	2,170	0.890
13½	122	273	1,840	0.844
13	117	254	1,650	0.812
12¾	115	244	1,560	0.797
12	108	216	1,300	0.750
11⅜	102	194	1,100	0.711
11¼	101	190	1,070	0.703
10½	94.5	165	868	0.656
9¾	87.8	143	695	0.609
9	81.0	122	547	0.562
7 × 42	294	2,060	43,200	2.04
41¼	289	1,980	40,900	2.01
40⅝	284	1,920	39,100	1.98
40½	284	1,910	38,800	1.97
39¾	278	1,840	36,600	1.93
7 × 39	273	1,770	34,600	1.90
38¼	268	1,710	32,600	1.86
37½	262	1,640	30,800	1.82
37⅜	262	1,630	30,500	1.82
36¾	257	1,580	29,000	1.79
36	252	1,510	27,200	1.75
35¾	250	1,490	26,600	1.74
35¼	247	1,450	25,600	1.71
34½	242	1,390	24,000	1.68
34⅛	239	1,360	23,200	1.66
33¾	236	1,330	22,400	1.64
33	231	1,270	21,000	1.60
32½	228	1,230	20,000	1.58
32¼	226	1,210	19,600	1.57
31½	220	1,160	18,200	1.53
30⅞	216	1,110	17,200	1.50

TABLE 24-7 Properties of Sections for Glued Laminated Timber (Continued)

Actual size, in.	Area A, in.^{2}	Section modulus S, in.^{3}	Moment of inertia I, in.^{4}	Volume per lineal foot, cu ft
7 × 30¾	215	1,100	17,000	1.49
30	210	1,050	15,800	1.46
29¼	205	998	14,600	1.42
28½	200	948	13,500	1.38
27¾	194	898	12,500	1.35
27⅝	193	890	12,300	1.34
27	189	850	11,500	1.31
26¼	184	804	10,600	1.28
26	182	789	10,200	1.26
25½	178	759	9,670	1.24
24¾	173	715	8,840	1.20
24⅜	171	693	8,450	1.18
24	168	672	8,060	1.17
23¼	163	631	7,330	1.13
22¾	159	604	6,870	1.11
22½	158	591	6,640	1.09
7 × 21¾	152	552	6,000	1.06
21⅛	148	521	5,500	1.03
21	147	514	5,400	1.02
20¼	142	478	4,840	0.984
19½	136	444	4,320	0.948
18¾	131	410	3,840	0.911
18	126	378	3,400	0.875
17⅞	125	373	3,330	0.869
17¼	121	347	2,990	0.838
16½	116	318	2,620	0.802
16¼	114	308	2,500	0.790
15¾	110	289	2,280	0.765
15	105	262	1,970	0.729
14⅝	102	250	1,820	0.711
14¼	99.8	237	1,690	0.693
13½	94.5	213	1,440	0.656
7 × 13	91.0	197	1,280	0.632
12¾	89.2	190	1,210	0.620
12	84.0	168	1,010	0.583
11⅜	79.6	151	859	0.553
11¼	78.8	148	831	0.547
10½	73.5	129	675	0.510
9¾	68.2	111	541	0.474
9	63.0	94.5	425	0.438
8¼	57.8	79.4	328	0.401
8⅛	56.9	77.0	313	0.395
7½	52.5	65.6	246	0.365
5¼ × 36	189	1,130	20,400	1.31
35¾	188	1,120	20,000	1.30
35¼	185	1,090	19,200	1.28
34½	181	1,040	18,000	1.26
34⅛	179	1,020	17,400	1.24

TABLE 24-7 Properties of Sections for Glued Laminated Timber (Continued)

Actual size, in.	Area A, in.^{2}	Section modulus S, in.^{3}	Moment of inertia I, in.^{4}	Volume per lineal foot, cu ft
5¼ × 33¾	177	997	16,800	1.23
33	173	953	15,700	1.20
32½	171	924	15,000	1.18
32¼	169	910	14,700	1.18
31½	165	868	13,700	1.15
30⅞	162	834	12,900	1.13
30¾	161	827	12,700	1.12
30	158	788	11,800	1.09
29¼	154	749	11,000	1.07
28½	150	711	10,100	1.04
27¾	146	674	9,350	1.01
27⅝	145	668	9,220	1.01
27	142	638	8,610	0.984
26¼	138	603	7,910	0.957
26	136	592	7,690	0.948
25½	134	569	7,250	0.930
5¼ × 24¾	130	536	6,630	0.902
24⅜	128	520	6,340	0.889
24	126	504	6,050	0.875
23¼	122	473	5,500	0.848
22¾	119	453	5,150	0.829
22½	118	443	4,980	0.820
21¾	114	414	4,500	0.793
21⅛	111	391	4,120	0.770
21	110	386	4,050	0.765
20¼	106	359	3,630	0.738
19½	102	333	3,240	0.711
18¾	98.4	308	2,880	0.684
18	94.5	284	2,550	0.656
17⅞	93.8	280	2,500	0.652
17¼	90.6	260	2,250	0.629
16½	86.6	238	1,960	0.602
5¼ × 16¼	85.3	231	1,880	0.592
15¾	82.7	217	1,710	0.574
15	78.8	197	1,480	0.547
14⅝	76.8	187	1,370	0.533
14¼	74.8	178	1,270	0.520
13½	70.9	160	1,080	0.492
13	68.2	148	961	0.474
12¾	66.9	142	907	0.465
12	63.0	126	756	0.438
11⅜	59.7	113	644	0.415
11¼	59.1	111	623	0.410
10½	55.1	96.5	506	0.383
9¾	51.2	83.2	406	0.355
9	47.2	70.9	319	0.328
8¼	43.3	59.6	246	0.301
8⅛	42.7	57.8	235	0.296
5¼ × 7½	39.4	49.2	185	0.273
6¾	35.4	39.9	135	0.246
6½	34.1	37.0	120	0.237
6	31.5	31.5	94.5	0.219
5¼	27.6	24.1	63.3	0.191

TABLE 24-7 Properties of Sections for Glued Laminated Timber (Continued)

Actual size, in.	Area A, in.^{2}	Section modulus S, in.^{3}	Moment of inertia I, in.^{4}	Volume per lineal foot, cu ft
3¼ × 24	78.0	312	3,740	0.542
23¼	75.6	293	3,400	0.525
22¾	73.9	280	3,190	0.513
22½	73.1	274	3,080	0.508
21¾	70.7	256	2,790	0.491
21⅛	68.7	242	2,550	0.477
21	68.2	239	2,510	0.474
20¼	65.8	222	2,250	0.457
19½	63.4	206	2,010	0.440
18¾	60.9	190	1,780	0.423
18	58.5	176	1,580	0.406
3¼ × 17⅞	58.1	173	1,550	0.403
17¼	56.1	161	1,390	0.389
16½	53.6	148	1,220	0.372
16¼	52.8	143	1,160	0.367
15¾	51.2	134	1,060	0.355
15	48.8	122	914	0.338
14⅝	47.5	116	847	0.330
14¼	46.3	110	784	0.322
13½	43.9	98.7	666	0.305
13	42.2	91.5	595	0.293
12¾	41.4	88.1	561	0.288
12	39.0	78.0	468	0.271
11⅜	37.0	70.1	399	0.257
11¼	36.6	68.6	386	0.254
10½	34.1	59.7	314	0.237
9¾	31.7	51.5	251	0.220
3¼ × 9	29.2	43.9	197	0.203
8¼	26.8	36.9	152	0.186
8⅛	26.4	35.8	145	0.183
7½	24.4	30.5	114	0.169
6¾	21.9	24.7	83.3	0.152
6½	21.1	22.9	74.4	0.147
6	19.5	19.5	58.5	0.135
5¼	17.1	14.9	39.2	0.118
4½	14.6	11.0	24.7	0.102
3¾	12.2	7.62	14.3	0.085

TABLE 24-8 Effective Section Properties for Plywood (12-in. widths—all plies from same species group)*

Thickness, in.	Approx. weight, psf	Effective thickness for shear: All grades using exterior glue, in.	All grades using interior glue, in.	Properties for stress applied parallel to face grain: Area for tension and compression, $in.^2$	Moment of inertia, $in.^4$	Effective section modulus, $in.^3$	Properties for stress applied perpendicular to face grain: Area for tension and compression, $in.^2$	Moment of inertia, $in.^4$	Effective section modulus, $in.^3$
Rough panels									
5/16 R	1.0	0.318	0.300	2.400	0.026	0.147	1.200	0.002	0.032
3/8 R	1.1	0.375	0.375	2.400	0.048	0.215	1.500	0.003	0.049
1/2 R†	1.5	0.574	0.500	3.600	0.100	0.339	2.400	0.029	0.183
5/8 R	1.8	0.662	0.625	4.586	0.175	0.477	3.000	0.056	0.286
3/4 R	2.2	0.750	0.750	4.600	0.266	0.603	4.500	0.132	0.509
13/16 R	2.4	0.794	0.813	4.605	0.319	0.668	4.500	0.182	0.628
7/8 R	2.6	0.949	0.875	6.900	0.474	0.922	4.500	0.207	0.643
1 R	3.0	1.037	1.000	5.354	0.574	0.976	6.639	0.391	1.020
1 1/8 R	3.3	1.125	1.125	6.840	0.815	1.231	5.250	0.502	1.122
Sanded panels‡									
1/4 S	0.8	0.276	0.240	1.680	0.013	0.091	1.200	0.001	0.027
3/8 S	1.1	0.375	0.375	1.680	0.040	0.182	2.100	0.007	0.079
1/2 S	1.5	0.574	0.500	3.120	0.081	0.277	2.400	0.029	0.183
5/8 S	1.8	0.662	0.625	3.135	0.135	0.367	2.914	0.076	0.345
3/4 S	2.2	0.750	0.750	3.876	0.207	0.470	4.400	0.168	0.597
7/8 S	2.6	0.949	0.875	3.994	0.329	0.639	5.786	0.279	0.811
1 S	3.0	1.037	1.000	5.520	0.498	0.846	6.646	0.454	1.119
1 1/8 S	3.3	1.125	1.125	5.978	0.664	1.003	6.660	0.650	1.385

* All properties adjusted to account for reduced effectiveness of plies with grain perpendicular to applied stress.

† For 1/2-in., three-ply, use the following:

1/2 R	1.5	0.463	0.500	3.000	0.110	0.373	2.000	0.008	0.088

‡ Includes touch-sanded.

SOURCE: *Plywood Design Specification*, American Plywood Association, Tacoma, Wash., November, 1966.

MECHANICAL FASTENINGS

A number of mechanical devices have been developed for fastening wood. The use of readily available mechanical fastenings in construction is more economical than the use of specially designed connecting hardware.

The density of wood is a primary factor in determining the allowable loads for mechanical fastenings. Species groupings based on density and related factors are given in Table 24-9.

Additional information on the design of mechanical fastenings for wood may be found in the *National Design Specification for Stress-Grade Lumber and Its Fastenings*[5] and the *Timber Construction Manual*.[6]

Metal fastenings and connections subject to deterioration by corrosion or by chemical attack should be protected by painting, galvanizing, or plating. In highly corrosive atmospheres, as in chemical plants, it may be desirable to countersink all metal parts into the wood and cover them with hot pitch or tar. In such extreme conditions, the wood should be at or below moisture equilibrium at the time of fabrication to prevent subsequent shrinkage that could open avenues of attack for the corrosive atmosphere.

Iron salts are frequently very acidic and show hydrolytic action on wood in the presence of free water, thus accounting for the softening and discoloration of wood around corroded metal fastenings. This action is especially pronounced in acidic woods, such as oak, and in woods containing considerable tannin and related compounds, such as redwood. It can be eliminated by using zinc-coated, aluminum, copper, or other noncorrosive fastenings where available.

Timber Connectors Timber connectors are metal devices installed in the contact faces of adjacent wood members to transmit loads from one member to another, the joint being prevented from separating by one or more bolts.

TABLE 24-9 Species Load Groups for Fastenings[6]

Species	Specific gravity*	Timber-connector load groups	Lag-bolt and driven-fastening† load groups
Ash, commercial white	0.61	A	I
Beech (American)	0.67	A	I
Birch:			
Sweet	0.71	A	I
Yellow	0.66	A	I
Cedar:			
Alaska	0.46	C	III
Port Orford	0.44	C	III
Cypress, Southern	0.48	C	III
Douglas fir:			
Coast type‡	0.51	B	II
Rocky Mountain type	0.45	C	III
Elm, rock	0.66	A	I
Hemlock, Western	0.44	C	III
Hickory (true)	0.74	A	I
Larch, Western	0.59	B	II
Maple:			
Black (hard)	0.62	A	I
Sugar (hard)	0.68	A	I
Oak:			
Commercial red	0.66	A	I
Commercial white	0.71	A	I
Pecan	0.65	A	I
Pine:			
Norway	0.51	C	III
Southern‡	0.59	B	II
Poplar, yellow	0.43	C	IV
Redwood (old growth)	0.42	C	III
Spruce:			
Red	0.41	C	IV
Sitka	0.42	C	IV
White	0.45	C	IV
Sweetgum	0.53	B	III

* Based on weight and volume when oven-dry.
† Nails, spikes, wood screws, spiral dowels.
‡ When graded for density, these species qualify for group A connector loads.

Split rings are beveled metal rings which are inserted into precut grooves in the contact faces of the wood members. About half the depth of the ring is in each of the two members being joined. A tongue-and-groove "split" in the ring permits simultaneous bearing of the inner surface of the ring against the core left by grooving and of the outer face of the ring against the outer wall of the groove. After the split ring has been inserted in the grooves, the two members being joined are held together with bolts and washers, the bolts being placed in holes that are concentric with the rings. Split rings are manufactured in 2½- and 4-in. diameters. They are a most efficient device for joining wood to wood.

Shear plates are metal plates which are completely embedded in precut daps in the face of a wood member, flush with the surface. They are primarily intended for wood-to-steel connections. When used in pairs, shear plates may be employed for wood-to-wood connections and are useful for demountable structures. The bolt in a shear-plate connection transfers shear stress between shear plates or between a shear plate and a metal side member. Shear plates are manufactured in 2⅝- and 4-in. diameters.

Toothed rings, clamping plates, and spike grids are other forms of timber connectors sometimes used for special applications. Information on the design of these

TABLE 24-10 Allowable Loads for One Split Ring and Bolt in Single Shear (normal duration of loading)

Split ring diam, in.	Bolt diam, in.	Number of faces of piece with connectors on same bolt	Thickness (net) of lumber, in.	Allowable load per connector unit and bolt, lb					
				Loaded parallel to grain*			Loaded perpendicular to grain†		
				Group A‡	Group B‡	Group C‡	Group A‡	Group B‡	Group C‡
2½	½	1	1 min	2,630	2,270	1,900	**1,580** 1,900	**1,350** 1,620	**1,130** 1,350
			1⅝ and thicker	3,160	2,730	2,290	**1,900** 2,280	**1,620** 1,940	**1,350** 1,620
		2	1⅝ min	2,630	2,270	1,900	**1,580** 1,900	**1,350** 1,620	**1,130** 1,350
			2 and thicker	3,160	2,730	2,290	**1,900** 2,280	**1,620** 1,940	**1,350** 1,620
4	¾	1	1 min	4,090	3,510	2,920	**2,370** 2,840	**2,030** 2,440	**1,700** 2,040
			1⅝ and thicker	6,140	5,260	4,380	**3,560** 4,270	**3,050** 3,660	**2,540** 3,050
		2	1⅝ min	4,310	3,690	3,070	**2,490** 3,000	**2,140** 2,570	**1,780** 2,140
			2	4,950	4,250	3,540	**2,870** 3,440	**2,470** 2,960	**2,050** 2,460
			2⅝	6,030	5,160	4,310	**3,490** 4,180	**3,000** 3,600	**2,490** 3,000
			3 and thicker	6,140	5,260	4,380	**3,560** 4,270	**3,050** 3,660	**2,540** 3,050

* Minimum edge distances are 1¾ in. for 2½-in. split rings and 2¾ in. for 4-in. split rings.

† Minimum edge distances (loaded and unloaded) are 1¾ in. for 2½-in. split rings and 2¾ in. for 4-in. split rings. Boldface allowable load values for each thickness of lumber are for minimum edge distances. Other allowable load values for each thickness of lumber are for loaded-edge distances of 2¾ in. or more for 2½-in. split rings and 3¾ inches or more for 4-in. split rings.

‡ See Table 24-9 for species in each group.

types of timber connectors may be found in the *Design Manual for TECO Timber Connector Construction.*[7]

Tables 24-10 and 24-11 give allowable loads for one split ring and bolt and for one shear plate and bolt, respectively. Loads for more than one timber connector unit are the sum of the loads permitted for each connector unit. The tabulated loads apply (1) for normal duration of loading, (2) for connectors installed in wood members that are seasoned prior to fabrication and that remain dry in service, and (3) for loads acting parallel to grain (0° angle between direction of load and longitudinal axis of member) or perpendicular to grain (90° angle between direction of load and longitudinal axis of member).

TABLE 24-11 Allowable Loads for One Shear Plate and Bolt in Single Shear (normal duration of loading; wood side plates[a,b])

Shear-plate diam, in.	Bolt diam, in.	Number of faces of piece with connectors on same bolt	Thickness (net) of lumber, in.	Allowable load per connector unit and bolt, lb[c]					
				Loaded parallel to grain[d]			Loaded perpendicular to grain[e]		
				Group A[f]	Group B[f]	Group C[f]	Group A[f]	Group B[f]	Group C[f]
2⅝	¾	1	1⅝ min	3,370[g]	2,890	2,410	**1,960**	**1,680**	**1,400**
							2,350	2,020	1,680
		2	1⅝ min	2,620	2,250	1,870	**1,520**	**1,300**	**1,090**
							1,820	1,560	1,310
			2	3,190[g]	2,730	2,270	**1,850**	**1,590**	**1,320**
							2,220	1,910	1,580
			2⅝ and thicker	3,370[g]	2,890	2,410	**1,960**	**1,680**	**1,400**
							2,350	2,020	1,680
4	¾	1	1⅝ min	4,750	4,070	3,390	**2,760**	**2,360**	**1,970**
							3,310	2,830	2,360
			1¾ and thicker	5,090[g]	4,360	3,640	**2,950**	**2,530**	**2,110**
							3,540	3,040	2,530
		2	1¾ min	3,390	2,910	2,420	**1,970**	**1,680**	**1,400**
							2,360	2,020	1,680
			2	3,790	3,240	2,700	**2,200**	**1,880**	**1,570**
							2,640	2,260	1,880
			2⅝	4,440	3,800	3,170	**2,580**	**2,210**	**1,840**
							3,100	2,650	2,210
			3	4,830	4,140	3,450	**2,800**	**2,400**	**2,000**
							3,360	2,880	2,400
			3⅝ and thicker	5,090[g]	4,360	3,640	**2,950**	**2,530**	**2,110**
							3,540	3,040	2,530
4	⅞	1	1⅝ min	4,750	4,070	3,390	**2,760**	**2,360**	**1,970**
							3,310	2,830	2,360
			1¾ and thicker	5,090	4,360	3,640	**2,950**	**2,530**	**2,110**
							3,540	3,040	2,530
		2	1¾ min	3,390	2,910	2,420	**1,970**	**1,680**	**1,400**
							2,360	2,020	1,680
			2	3,780	3,240	2,700	**2,200**	**1,880**	**1,570**
							2,640	2,260	1,880
			2⅝	4,440	3,800	3,170	**2,580**	**2,210**	**1,840**
							3,100	2,650	2,210
			3	4,830	4,140	3,450	**2,800**	**2,400**	**2,000**
							3,360	2,880	2,400
			3⅝ and thicker	5,090	4,360	3,640	**2,950**	**2,530**	**2,110**
							3,540	3,040	2,530

[a] For metal side plates, tabulated loads apply except that, for 4-in. shear plates, the parallel-to-grain (not perpendicular) loads for wood side plates shall be increased 18, 11, and 5 percent for groups A, B, and C woods, respectively, but loads shall not exceed those permitted by footnote *c*.

[b] Metal side plates, when used, shall be designed in accordance with accepted metal practices. For steel, the following unit stresses, in pounds per square inch, are suggested for all loadings except wind: net section in tension, 20,000; shear, 12,500; double-shear bearing, 28,125; single-shear bearing, 22,500. For wind, these values may be increased one-third. If bolt threads are in bearing, reduce the preceding shear and bearing values by one-ninth.

[c] The allowable loads for all loadings except wind shall not exceed 2,900 lb for 2⅝-in. shear plates; 4,970 lb and 6,760 lb for 4-in. shear plates with ¾-in. and ⅞-in. bolts, respectively; or, for wind loading, shall not exceed 3,870 lb, 6,630 lb, and 9,020 lb, respectively. If bolt threads are in bearing on the shear plate, reduce the preceding values by one-ninth.

[d] Minimum edge distances are 1¾ in. for 2⅝-in. shear plates and 2¾ in. for 4-in. shear plates.

[e] Minimum edge distances (loaded and unloaded) are 1¾ in. for 2⅝-in. shear plates and 2¾ in. for 4-in. shear plates. Boldface allowable load values for each thickness of lumber are for minimum edge distances. Other allowable load values for each thickness of lumber are for loaded-edge distances of 2¾ in. or more for 2⅝-in. shear plates and 3¾ in. or more for 4-in. shear plates.

[f] See Table 24-9 for species in each group.

[g] Loads exceed those permitted by footnote *c* but are needed for proper determination of loads for other angles of load to grain. Footnote *c* limitations apply in all cases.

For other than normal duration of loading, the tabulated loads should be multiplied by the appropriate adjustment factor from Table 24-1.

For connectors installed in unseasoned lumber which seasons in service, use 80 percent of the tabulated loads. For connectors installed in wood members used under wet-use service conditions, use 67 percent of the tabulated loads.

For connectors installed in wood members pressure impregnated with fire-retardant chemicals and kiln dried after treatment, use 90 percent of the tabulated loads. For connectors installed in fire-retardant treated wood which has not been kiln dried after treatment, use 80 percent of the tabulated loads.

For angles of load to grain (angle between direction of load and longitudinal axis of member) other than 0 or 90°, the allowable load may be determined from the Hankinson formula:

$$N = \frac{PQ}{P \sin^2 \Theta + Q \cos^2 \Theta}$$

where N = allowable load in pounds, or stress in psi, acting at angle to grain
P = allowable load in pounds, or stress in psi, acting parallel to grain
Q = allowable load in pounds, or stress in psi, acting perpendicular to grain
Θ = angle between direction of load and longitudinal axis of member, degrees

Spacing of connectors, edge distance, and end distance must be sufficient to develop the required strength of timber connector joints. Spacing is the distance between centers of connectors measured along a line joining their centers. Edge distance is the distance from edge of member to the center of the connector closest to the edge of member, measured perpendicular to the edge. For members loaded perpendicular to grain, distinction must be made between loaded and unloaded edges. The loaded edge is the edge toward which the load induced by the timber connector acts. The unloaded edge is the edge away from which the load induced by the connector acts. End distance is the distance measured parallel to the grain from center of connector to the nearest end of the member. Table 24-12 gives (1) connector spacings and end distances for members loaded parallel and perpendicular to grain and (2) appropriate percentages of tabulated allowable loads. For spacings and end distances intermediate between minimums and those required for maximum load, the percentages are determined by straight-line interpolation. Connector edge distances are given in Tables 24-10 and 24-11.

The tabulated loads in Tables 24-10 and 24-11 are based upon the assumptions that the wood at the joint is clear and relatively free from checks, splits, and shakes and that the slope of grain at the joint does not exceed 1 in 10. If knots are present within a distance from the critical section of half the diameter of the connector, the area of the knot should be subtracted from the net area at the critical section. The critical section is that section, taken at right angles to the direction of the load, which gives the maximum stress based on the net area. The net area at this section is equal to the full cross-sectional area of the member less the projected area of the connector within the member and the projected area of that portion of the bolt hole not within the connector's projected area. When connectors are staggered, adjacent connectors with parallel-to-grain spacing equal to or less than one connector diameter are considered to occur at the same critical section.

The tension or compression stress at the critical section should not exceed the safe stress of clear wood in compression parallel to grain.

Bolts Standard machine bolts with square heads and nuts are used extensively in wood construction. Since a tight fit requiring forcible driving of bolts is not recommended, bolt holes should be 1/16 in. larger than the diameter of the bolt. Careful centering of holes in main members and splice plates is necessary. Standard cut washers or a metal plate or strap should be inserted between the wood and the bolt head and between the wood and the nut. Nuts should be tightened snugly, but not so tight as to cause crushing of the wood under the washer or plate.

Table 24-14 gives allowable loads for one bolt loaded in double shear. Loads for more than one bolt, of same or varying sizes, are the sums of the loads permitted for each bolt. The tabulated loads apply (1) for normal duration of loading, (2) for bolts installed in wood members that remain dry in service, (3) for loads acting parallel to grain (0° angle between direction of load and longitudinal axis of member) or perpendicular to grain (90° angle between direction of load and longitudinal axis of member), and (4) for bolts installed in three-member joints with wood side members at least half the thickness of the main member.

For other than normal duration of loading, the tabulated loads should be multiplied by the appropriate adjustment factor from Table 24-1.

For bolts installed in seasoned or unseasoned lumber that seasons in place and that remains dry in service, the tabulated loads may be used for a joint with wood side members having a single bolt and loaded parallel or perpendicular to grain; for a single

TABLE 24-12 Timber Connector Spacings and End Distances with Corresponding Percentages of Tabulated Loads

<table>
<tr><th rowspan="2">Connector and diameter</th><th colspan="2">Spacing parallel to grain</th><th colspan="2">Spacing perpendicular to grain</th><th colspan="3">End distance</th></tr>
<tr><th>Spacing, in.</th><th>Percent of tabulated load</th><th>Spacing, in.</th><th>Percent of tabulated load</th><th>Tension member, in.</th><th>Compression member, in.</th><th>Percent of tabulated load</th></tr>
<tr><td colspan="8">Parallel-to-grain loading</td></tr>
<tr><td rowspan="2">2½-in. split rings and 2⅝-in. shear plates</td><td>6¾</td><td>100</td><td rowspan="2">3½ (min)</td><td rowspan="2">100</td><td>5½</td><td>4</td><td>100</td></tr>
<tr><td>3½</td><td>75</td><td>2¾</td><td>2½</td><td>62.5</td></tr>
<tr><td rowspan="2">4-in. split rings and shear plates</td><td>9</td><td>100</td><td rowspan="2">5 (min)</td><td rowspan="2">100</td><td>7</td><td>5½</td><td>100</td></tr>
<tr><td>5</td><td>75</td><td>3½</td><td>3¼</td><td>62.5</td></tr>
<tr><td colspan="8">Perpendicular-to-grain loading</td></tr>
<tr><td rowspan="2">2½-in. split rings and 2⅝-in. shear plates</td><td rowspan="2">3½ (min)</td><td rowspan="2">100</td><td>4¼</td><td>100</td><td colspan="2">5½</td><td>100</td></tr>
<tr><td>3½</td><td>75</td><td colspan="2">2¾</td><td>62.5</td></tr>
<tr><td rowspan="2">4-in. split rings and shear plates</td><td rowspan="2">5 (min)</td><td rowspan="2">100</td><td>6</td><td>100</td><td colspan="2">7</td><td>100</td></tr>
<tr><td>5</td><td>75</td><td colspan="2">3½</td><td>62.5</td></tr>
</table>

row of bolts loaded parallel to grain; for multiple rows of bolts loaded parallel to grain provided each row of bolts has a separate splice plate; and for a joint with steel side members having a single row of bolts parallel to grain in each member and loaded parallel or perpendicular to grain. For other arrangements of bolted joints, use 40 percent of the tabulated loads.

For bolts installed in wood members pressure impregnated with fire-retardant chemicals and kiln dried after treatment, use 90 percent of the tabulated loads. For bolts installed in fire-retardant-treated wood which has not been kiln dried after treatment, the 90 percent value should be reduced in accordance with the preceding paragraph.

For bolts used in joints that are exposed to the weather, use 75 percent of the tabulated loads. For bolts used in joints that are always wet, use 67 percent of the tabulated loads.

For bolts installed in three-member joints with steel side members, increase tabulated parallel-to-grain loads by 25 percent, but do not increase tabulated perpendicular-to-grain loads. For bolts installed in three-member joints with wood side members less than half the thickness of the main member, use tabulated loads given for a main member which is twice the thickness of the thinnest side member. For bolts installed in two-member joints, use half the tabulated load for a piece twice the thickness of the thinner member. For bolts installed in multiple-member joints (other than two or three members), use half the tabulated load for a piece twice the thickness

TABLE 24-13 Spacing and Distance Values for Bolts* [6]

Dimension	Parallel-to-grain loading	Perpendicular-to-grain loading
Spacing between bolts in a row	Minimum of 4 times bolt diameter.	4 times bolt diameter unless the design load is less than the bolt bearing capacity of side members; then spacing may be reduced proportionately.
Staggered bolts	Adjacent bolts are considered to be placed at critical section unless spaced at a minimum of 8 times the bolt diameter.	Staggering not permitted unless design load is less than bolt bearing capacity of side members.
Spacing between rows of bolts	No requirement if net area requirements at critical section are met.	2½ times bolt diameter for l/d† ratio of 2; 5 times bolt diameter for l/d ratios of 6 or more; use straight-line interpolation for l/d between 2 and 6.
	Spacing between rows paralleling a member may not exceed 5 in. unless separate splice plates are used for each row.	
End distance	In tension, 7 times bolt diameter for softwoods and 5 times bolt diameter for hardwoods. In compression, 4 times bolt diameter.	Minimum of 4 times bolt diameter when members abut at a joint.
Edge distance	1½ times the bolt diameter, except that for l/d† ratios of more than 6, use half the row spacing.	Minimum of 4 times bolt diameter at edge toward which load acts.

* These are minimum values for the allowable bolt loads given in Table 24-14.
† Ratio of length of bolt in main member, l, to diameter of bolt, d.

of the thinnest member multiplied by the number of shear planes involved. (For example, in a four-member joint, number of shear planes is three.)

Spacing of bolts, edge distance, and end distance must be sufficient to develop the required strength of bolted joints. Table 24-13 gives recommended spacings and end and edge distances.

For parallel-to-grain loading of lumber seasoned before installation, the net tension area at the critical section must equal 80 percent of the total area in bearing under all bolts at the joint in the member for softwoods and must equal 100 percent for hardwoods. These net tension areas must equal 33 percent and 42 percent, respectively, of that total area in bearing for lumber installed unseasoned and which seasons in place.

The net area at the critical section of a bolted joint is equal to the full cross-sectional area less the projected area of bolt holes at that section. Where bolts are staggered,

TABLE 24-14 Allowable Loads (in Pounds) for One Bolt Loaded in Double Shear (normal duration of loading)[5]

Length of bolt in main member, l, in.	Diameter of bolt, d, in.	l/d	Projected area of bolt, $A = l \times d$ sq. in.	Douglas fir, larch, and Southern pine: Parallel to grain, P	Douglas fir, larch, and Southern pine: Perpendicular to grain, Q	Hemlock, West Coast and Western: Parallel to grain, P	Hemlock, West Coast and Western: Perpendicular to grain, Q	Pine, Eastern white, Ponderosa, sugar, and Idaho white: Parallel to grain, P	Pine, Eastern white, Ponderosa, sugar, and Idaho white: Perpendicular to grain, Q	Redwood: Parallel to grain, P	Redwood: Perpendicular to grain, Q
1⅝	½	3.3	0.8125	1,010	480	850	450	710	380	940	380
	⅝	2.6	1.0156	1,290	540	1,070	510	890	420	1,200	420
	¾	2.2	1.2188	1,550	600	1,280	570	1,070	470	1,450	470
	⅞	1.9	1.4219	1,810	670	1,500	620	1,250	520	1,690	520
	1	1.6	1.625	2,070	730	1,720	680	1,430	570	1,930	570
2	½	4.0	1.00	1,180	590	1,030	550	860	460	1,100	460
	⅝	3.2	1.25	1,560	670	1,320	630	1,100	520	1,460	520
	¾	2.7	1.50	1,910	740	1,580	700	1,320	580	1,780	580
	⅞	2.3	1.75	2,230	820	1,850	770	1,540	640	2,080	640
	1	2.0	2.00	2,550	890	2,110	840	1,760	700	2,380	700
2⅝	½	5.3	1.3125	1,280	780	1,180	730	980	610	1,180	610
	⅝	4.2	1.6406	1,890	880	1,670	820	1,390	690	1,760	690
	¾	3.5	1.9688	2,430	980	2,060	920	1,720	760	2,250	760
	⅞	3.0	2.2969	2,900	1,080	2,420	1,010	2,020	840	2,700	840
	1	2.6	2.625	3,340	1,170	2,770	1,100	2,310	920	3,110	920
3	½	6.0	1.50	1,290	890	1,200	830	1,000	690	1,200	690
	⅝	4.8	1.875	1,980	1,000	1,790	940	1,490	780	1,850	780
	¾	4.0	2.25	2,660	1,120	2,320	1,050	1,930	870	2,470	870
	⅞	3.4	2.625	3,250	1,230	2,760	1,150	2,300	960	3,020	960
	1	3.0	3.00	3,790	1,340	3,170	1,260	2,640	1,050	3,530	1,050
3⅝	½	7.3	1.8125	1,290	1,020	1,200	950	1,000	830	1,200	830
	⅝	5.8	2.2656	2,010	1,210	1,870	1,140	1,560	950	1,860	950
	¾	4.8	2.7188	2,860	1,350	2,600	1,270	2,170	1,050	2,680	1,050
	⅞	4.1	3.1719	3,680	1,490	3,240	1,390	2,700	1,160	3,450	1,160
	1	3.6	3.625	4,430	1,620	3,790	1,520	3,160	1,270	4,120	1,270
4	½	8.0	2.00	1,290	1,040	1,200	980	1,000	890	1,200	890
	⅝	6.4	2.50	2,010	1,330	1,870	1,250	1,560	1,050	1,870	1,050
	¾	5.3	3.00	2,890	1,490	2,680	1,400	2,230	1,160	2,690	1,160
	⅞	4.6	3.50	3,830	1,640	3,430	1,540	2,860	1,280	3,570	1,280
	1	4.0	4.00	4,720	1,790	4,120	1,680	3,430	1,400	4,400	1,400
4½	½	9.0	2.25	1,290	1,020	1,200	960	1,000	900	1,200	900
	⅝	7.2	2.8125	2,010	1,440	1,870	1,350	1,560	1,170	1,870	1,170
	¾	6.0	3.375	2,890	1,680	2,700	1,570	2,250	1,310	2,690	1,310
	⅞	5.1	3.9375	3,920	1,840	3,610	1,730	3,010	1,440	3,670	1,440
	1	4.5	4.50	4,980	2,010	4,450	1,890	3,710	1,570	4,660	1,570
	1⅛	4.0	5.0625	5,980	2,190	5,210	2,050	4,340	1,710	5,560	1,710
5½	⅝	8.8	3.4375	2,010	1,450	1,870	1,360	1,560	1,270	1,870	1,270
	¾	7.3	4.125	2,890	1,940	2,700	1,820	2,250	1,590	2,690	1,590
	⅞	6.3	4.8125	3,940	2,250	3,670	2,100	3,060	1,760	3,670	1,760
	1	5.5	5.50	5,120	2,460	4,670	2,300	3,970	1,920	4,770	1,920
	1⅛	4.9	6.1875	6,440	2,680	5,840	2,510	4,870	2,090	6,000	2,090
6½	⅝	10.4	4.0625	2,010	1,390	1,870	1,300	1,570	1,230	1,870	1,230
	¾	8.7	4.875	2,890	1,940	2,700	1,820	2,250	1,700	2,690	1,700
	⅞	7.4	5.6875	3,940	2,510	3,670	2,350	3,060	2,070	3,670	2,070
	1	6.5	6.50	5,140	2,880	4,810	2,700	4,000	2,270	4,790	2,270
	1⅛	5.8	7.3125	6,500	3,170	6,050	2,970	5,040	2,470	6,050	2,470
7½	⅝	12.0	4.6875	2,010	1,300	1,870	1,220	1,570	1,200	1,870	1,200
	¾	10.0	5.625	2,890	1,880	2,700	1,760	2,250	1,660	2,690	1,660
	⅞	8.6	6.5625	3,940	2,500	3,670	2,340	3,060	2,180	3,670	2,180
	1	7.5	7.50	5,140	3,130	4,810	2,940	4,000	2,590	4,790	2,590
	1⅛	6.7	8.4375	6,500	3,610	6,110	3,370	5,090	2,850	6,050	2,850
9½	¾	12.7	7.125	2,890	1,690	2,700	1,580	2,250	1,570	2,690	1,570
	⅞	10.9	8.3125	3,940	2,350	3,670	2,190	3,060	2,080	3,670	2,080
	1	9.5	9.50	5,140	3,050	4,810	2,860	4,000	2,680	4,790	2,680
	1⅛	8.4	10.6875	6,500	3,830	6,110	3,630	5,090	3,350	6,050	3,350
	1¼	7.6	11.875	8,040	4,590	7,510	4,310	6,260	3,820	7,480	3,820
11½	1	11.5	11.50	5,140	2,850	4,810	2,680	4,000	2,580	4,790	2,580
	1⅛	10.2	12.9375	6,500	3,660	6,100	3,440	5,090	3,260	6,050	3,260
	1¼	9.2	14.375	8,040	4,490	7,510	4,210	6,260	3,950	7,480	3,950

adjacent bolts with parallel-to-grain spacing less than eight times the bolt diameter are considered to occur at the same critical section.

Lag Screws and Lag Bolts Lag screws (or lag bolts) are basically large screws with square heads which are turned with a wrench. They are used in locations where a bolt would be difficult or impossible to install or where a nut on the surface would be objectionable. Standard cut washers or a metal plate or strap should be installed between the wood and the screw head.

Lag screws require prebored lead holes. Lead holes for the shank (unthreaded) portion of the lag screw should have the same diameter as the shank and a depth equal to the length of the shank. Lead holes for the threaded portion should have a depth at least equal to the length of the threaded portion and diameters as follows: for Group I species, 65 to 85 percent of the shank diameter; for Group II species, 60 to 75 percent of the shank diameter; and for Groups III and IV species, 40 to 70 percent of the shank diameter. The larger percentage in each case applies to lag screws of larger diameters. See Table 24-9 for species in each group.

Figure 24-1 gives allowable withdrawal loads for one lag screw. Table 24-15 gives allowable lateral loads for one lag screw. Loads for more than one lag screw are the sum of the loads permitted for each lag screw. The allowable loads apply (1) for normal duration of loading, and (2) for dry-use service conditions.

For other than normal duration of loading, the allowable load should be multiplied by the appropriate adjustment factor from Table 24-1.

For lag screws installed in seasoned or unseasoned lumber that seasons in place and that remains dry in service, the allowable loads may be used for a joint having a single lag screw loaded parallel or perpendicular to grain, for a single row of lag screws loaded parallel to grain, and for multiple rows of lag screws loaded parallel to grain provided each row of lag screws has a separate splice plate. For other types of lag-screw joints in unseasoned lumber, use 40 percent of the allowable loads.

For lag screws installed in wood members pressure impregnated with fire-retardant chemicals and kiln dried after treatment, use 90 percent of the allowable loads. For lag screws installed in fire-retardant-treated wood which has not been kiln dried after treatment, the 90 percent value should be reduced in accordance with the preceding paragraph.

For lag screws used in joints that are exposed to the weather, use 75 percent of the allowable loads. For lag screws used in joints that are always wet, use 67 percent of the allowable loads.

The allowable withdrawal loads determined from Fig. 24-1 are for one lag screw installed in the side grain of a member. If possible, lag screws should not be loaded in withdrawal from end grain. When this condition is unavoidable, use 75 percent of the allowable withdrawal load. The withdrawal load may not exceed the allowable tensile strength of the lag screw at its net (root) section. Penetration of the threaded portion of the lag screw of 7 diameters for Group I species, 8 diameters for Group II species, 10 diameters for Group III species, and 11 diameters for Group IV species will develop approximately the ultimate tensile strength of the lag screw in axial withdrawal. See Table 24-9 for species in each group.

The tabulated lateral loads in Table 24-15 are for one lag screw installed in the side grain of a joint comprised of two wood members of Group II species (see Table 24-9) with load applied either parallel or perpendicular to grain. For lag screws used in a two-member joint with a metal side member up to ½ in. thick, increase the tabulated parallel-to-grain loads by 25 percent, but do not increase the tabulated perpendicular-to-grain loads. For angles of load to grain other than 0 or 90°, the allowable lateral load may be determined from the Hankinson formula.

Spacing of lag screws, edge distances, and end distances must be sufficient to develop the required strength of lag-screw joints. The recommended spacings and end and edge distances given in Table 24-13 for bolts apply also for lag screws with a shank diameter equal to the bolt diameter.

Net section requirements for lag-screw joints are the same as for bolted joints for lag screws having a shank diameter equal to the bolt diameter.

Wood Screws Common wood screws with flat, oval, or round heads are sometimes used in wood construction, although seldom for structural applications.

Wood screws require prebored holes to prevent splitting of the wood. For withdrawal resistance, lead holes should have diameters as follows: for Group I species, 90 percent of the root diameter of the screw; and for Group II, III, and IV species, 70 percent of the root diameter of the screw. For lateral resistance, lead holes should

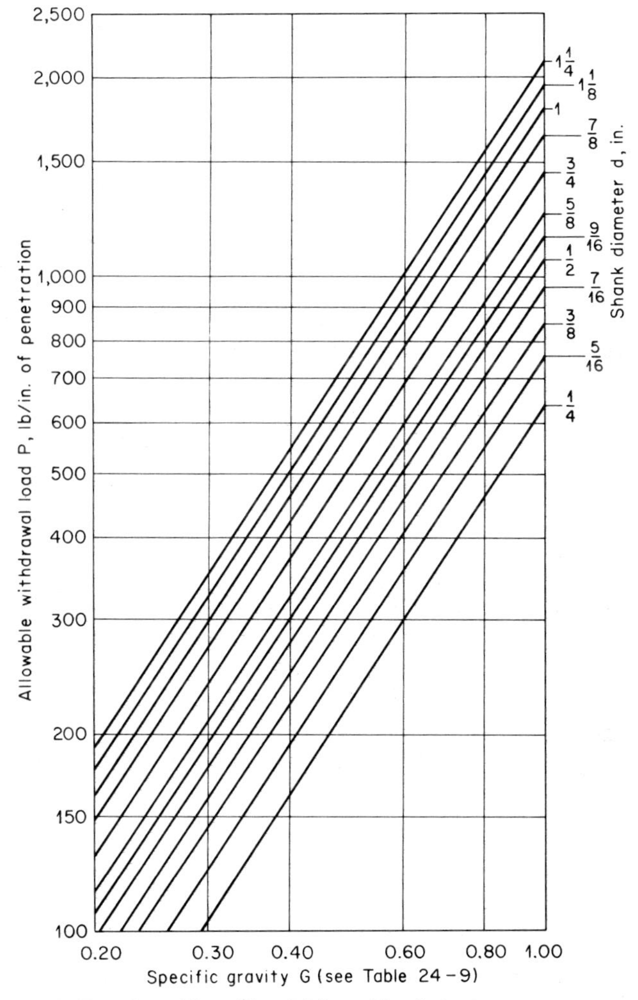

Fig. 24-1 Allowable withdrawal loads for lag screws.

have diameters as follows: for Group I species, the part of the lead hole receiving the shank (unthreaded) portion should have about the same diameter as the shank, and the part of the lead hole receiving the threaded portion should have about the same diameter as the root of the thread; and for Groups II, III, and IV species the diameters of the portions of the lead hole receiving the shank and threaded portions should be about seven-eighths of the diameters of the shank and the root of the thread, respec-

TABLE 24-15 Allowable Lateral Loads for Lag Screws in Group II Species (normal duration of loading; dry-use conditions)*[6]

Lag screw dimension	Thickness of side member															
	1⅝ in.				2⅝ in.				3⅝ in.				5½ in.			
Length, in.	4	5	6	7	6	7	8	9	8	9	10	11	11	12	13	14
Diam, in.	Load acting parallel to grain, lb															
¼	170	190	220	240	170											
5⁄16	210	280	320	350	270											
⅜	240	360	410	450	370	410	460	500	390	410						
7⁄16	270	400	520	560	430	550	610	670	530	550						
½	300	450	610	670	480	650	770	830	640	720	820	910	700	740		
⅝	360	540	710	810	560	750	970	1,120	790	1,000	1,240	1,350	1,010	1,130	1,230	1,290
¾	...	...	...	...	630	860	1,110	1,350	910	1,170	1,450	1,760	1,190	1,460	1,720	1,820
⅞	...	...	...	...	730	970	1,220	1,480	1,010	1,270	1,600	1,940	1,380	1,670	2,000	2,360
1	...	...	...	...	800	1,080	1,380	1,670	1,090	1,390	1,740	2,100	1,550	1,890	2,260	2,620
1⅛	...	...	...	...	...	.....	.....	.....	1,200	1,530	1,870	2,240	1,690	2,080	2,480	2,860
1¼	...	...	...	...	...	.....	.....	.....	1,280	1,620	2,000	2,340	1,800	2,210	2,670	3,100
	Load acting perpendicular to grain, lb															
¼	170	180	210	230	170											
5⁄16	170	240	270	300	230											
⅜	180	270	320	340	280	310	360	380	300	320						
7⁄16	190	280	360	390	300	390	430	470	370	380						
½	190	290	390	440	310	420	500	540	420	470	530	590	450	480		
⅝	210	320	430	490	340	450	580	670	470	600	750	810	610	680	730	780
¾	...	...	...	...	350	470	610	740	500	650	800	970	660	810	950	1,000
⅞	...	...	...	...	380	500	630	770	530	660	830	1,010	720	870	1,040	1,230
1	...	...	...	...	400	540	690	830	550	700	870	1,050	780	950	1,130	1,310
1⅛	...	...	...	...	...	.....	.....	.....	600	770	940	1,120	850	1,040	1,240	1,430
1¼	...	...	...	...	...	.....	.....	.....	640	810	1,000	1,170	900	1,110	1,340	1,550

* See Table 24-9 for species in this group.

tively. About two-thirds of the length of a wood screw is threaded. See Table 24-9 for species in each group.

Figures 24-2 and 24-3 give allowable withdrawal loads and allowable lateral loads, respectively, for one wood screw. Loads for more than one wood screw are the sum of the loads permitted for each screw. The allowable loads apply (1) for normal duration of loading, and (2) for seasoned lumber under dry-use service conditions.

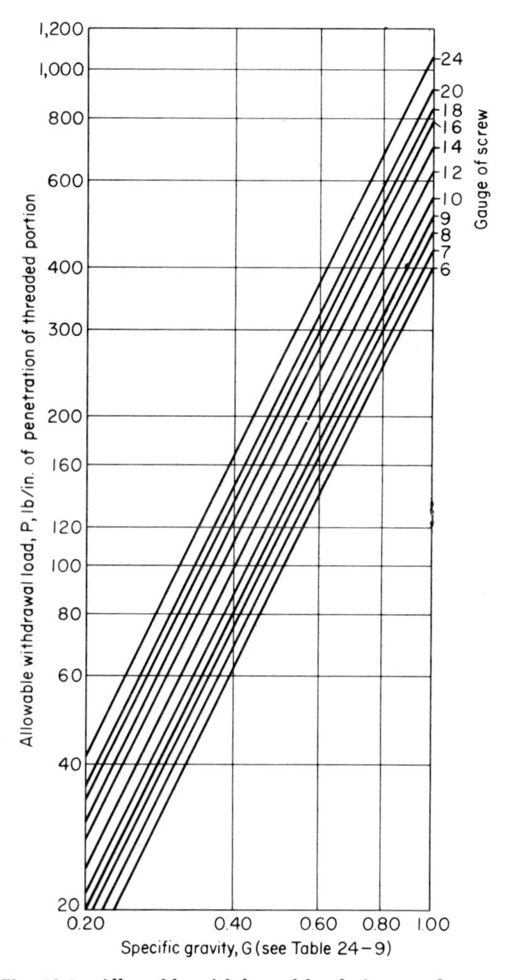

Fig. 24-2 Allowable withdrawal loads for wood screws.

For other than normal duration of loading, the allowable loads should be multiplied by the appropriate adjustment factor from Table 24-1.

For wood screws used in joints that are exposed to the weather and for use in lumber pressure impregnated with fire-retardant chemicals, use 75 percent of the allowable loads. For wood screws used in joints that are always wet, use 67 percent of the allowable loads.

If possible, wood screws should not be loaded in withdrawal. When this condition is unavoidable, the tensile strength of the screw at its net (root) section may not be exceeded. The allowable withdrawal loads determined from Fig. 24-2 are for wood screws installed in the side grain of a member. Wood screws should not be loaded in withdrawal from end grain.

The allowable lateral loads determined from Fig. 24-3 are for one wood screw used in a joint comprised of two wood members, with the screw installed in the side grain, embedded approximately 7 diameters into the member receiving the point and loaded

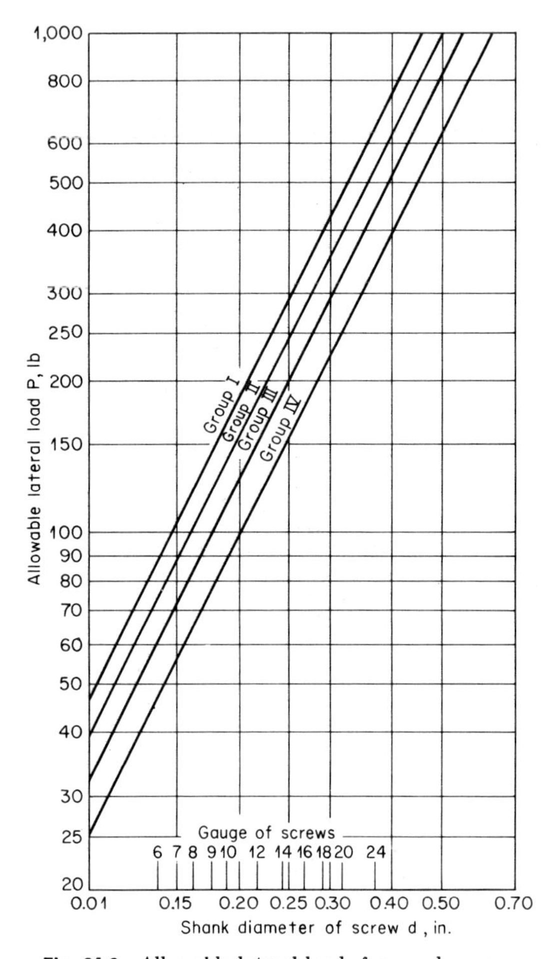

Fig. 24-3 Allowable lateral loads for wood screws.

at any angle of load to grain. If the penetration is less than 7 diameters, the allowable loads should be reduced proportionately; however, a penetration of less than 4 diameters should not be used. For wood screws installed in end grain, use 67 percent of the allowable lateral loads. For wood screws used in a two-member joint with a metal side member, increase the allowable lateral loads by 25 percent.

Spacing of wood screws, edge distances, and end distances must be sufficient to prevent splitting of the wood.

Nails and Spikes Common wire nails and spikes conform to the minimum sizes given in Table 24-16. Hardened deformed-shank nails and spikes are made of high-carbon steel wire and are headed, pointed, annularly or helically threaded, and heat treated and tempered to provide greater strength than common wire nails and spikes. However, the same allowable loads as given for common wire nails and spikes of the corresponding pennyweight size are used for hardened deformed-shank nails and spikes with few exceptions.

When it is necessary to avoid splitting of the wood, a prebored nail hole should be used. The diameter of the hole should not exceed 90 percent of the nail or spike diameter for Group I species and 75 percent of the diameter for Groups II, III, and IV species. See Table 24-9 for species in each group.

TABLE 24-16 Minimum Sizes of Common Wire Nails and Spikes

Pennyweight	Length, in.	Wire diameter, in.
Common wire nails		
6d	2	0.113
8d	2½	0.131
10d	3	0.148
12d	3¼	0.148
16d	3½	0.162
20d	4	0.192
30d	4½	0.207
40d	5	0.225
50d	5½	0.244
60d	6	0.263
Common spikes		
10d	3	0.192
12d	3¼	0.192
16d	3½	0.207
20d	4	0.225
30d	4½	0.244
40d	5	0.263
50d	5½	0.283
60d	6	0.283
5/16 in.	7	0.312
3/8 in.	8½	0.375

Figures 24-4 and 24-5 give allowable withdrawal loads and allowable lateral loads, respectively, for one nail or spike. Loads for more than one nail or spike are the sum of the loads permitted for each nail or spike. The allowable loads apply (1) for normal duration of loading, (2) for dry-use service conditions, and (3) for nails or spikes installed in a joint comprised of two wood members.

For other than normal duration of loading, the allowable loads should be multiplied by the appropriate adjustment factor from Table 24-1. For nails and spikes used in diaphragm construction, increase the allowable loads by 30 percent.

For nails and spikes installed in wood members pressure impregnated with fire-retardant chemicals and kiln dried after treatment, use 90 percent of the allowable loads. For nails and spikes installed in fire-retardant treated wood which has not been kiln dried after treatment, see provisions in the following paragraph.

If possible, nails or spikes should not be loaded in withdrawal. When this condition is impossible, the allowable withdrawal loads determined from Fig. 24-4 apply for nails or spikes driven in the side grain of seasoned wood or in unseasoned wood that will remain wet in service. When driven in unseasoned wood that will season under load, use 25 percent of the allowable withdrawal loads with any nails or spikes except hardened deformed-shank nails, for which the full allowable load may be used. When driven in fire-retardant treated wood which has not been kiln dried

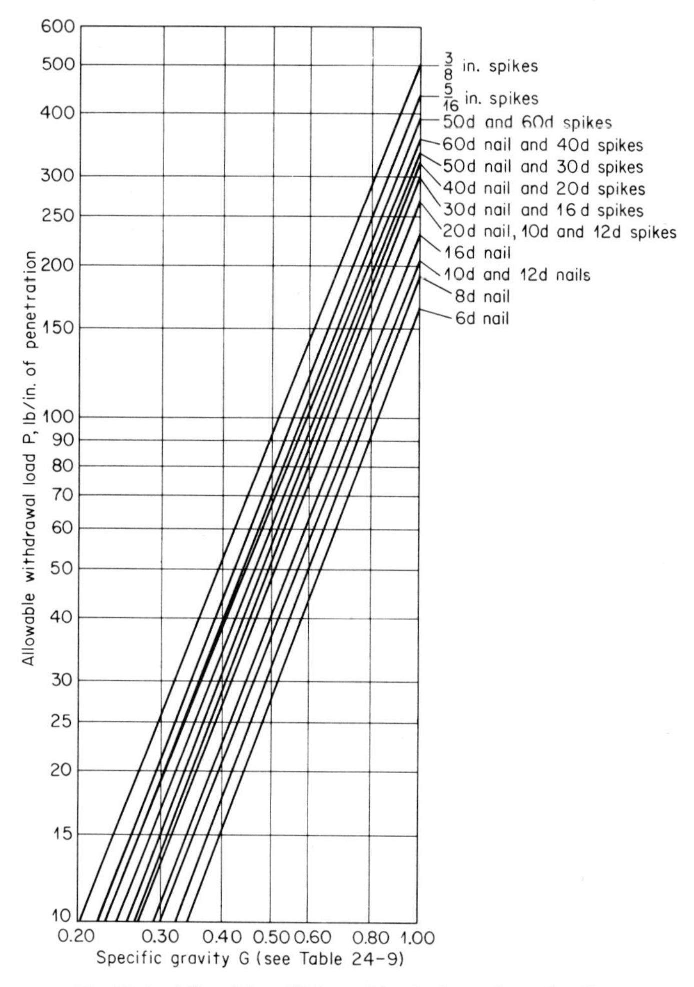

Fig. 24-4 Allowable withdrawal loads for nails and spikes.

after treatment, use 22.5 percent of the allowable withdrawal loads with any nails or spikes except hardened deformed-shank nails, for which 90 percent of the full allowable load may be used. Nails or spikes should not be loaded in withdrawal from end grain.

The allowable lateral loads determined from Fig. 24-5 are for one nail or spike used in a joint comprised of two seasoned wood members of Group III species (see Table 24-9) with the load applied in any lateral direction and with the nail or spike driven in the side grain. The minimum permissible penetration as shown in Fig. 24-5 is one-third the depth of penetration required for full allowable load. For nails or spikes driven in three-member joints and fully penetrating all three members, the allowable lateral loads may be increased by 33 percent when each side member is not less than one-third the thickness of the center member and may be increased by 67 percent when each side member is equal in thickness to the center member. For nails or spikes driven in the side grain of unseasoned wood that will remain wet or will

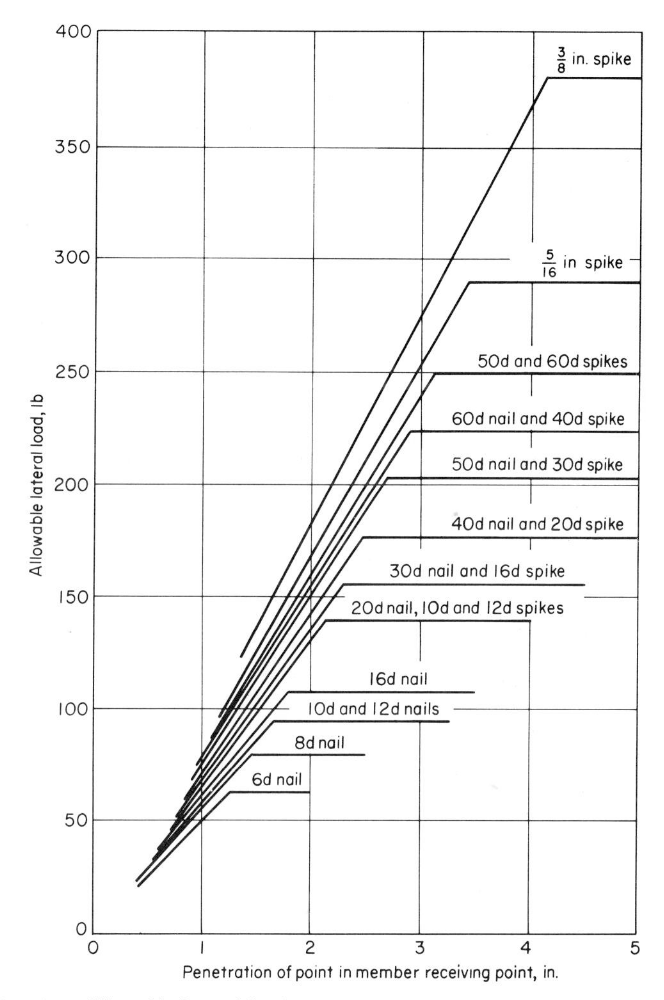

Fig. 24-5 Allowable lateral loads for nails and spikes in Group II species.

be loaded before seasoning, use 75 percent of the allowable lateral loads with any nails or spikes except hardened deformed-shank nails, with which the full allowable load may be used. For nails and spikes installed in fire-retardant treated wood which has not been kiln dried after treatment, use 67.5 percent of the allowable lateral loads with any nails or spikes except hardened deformed-shank nails, with which 90 percent of the full allowable load may be used. For nails and spikes used in joints with metal side plates, increase the allowable lateral loads by 25 percent. For nails or spikes driven in end grain, use 67 percent of the allowable lateral loads.

Spacing of nails and spikes, edge distances, and end distances must be sufficient to avoid unusual splitting of the wood.

The allowable withdrawal loads for toenailed joints, for all seasoning conditions, are 67 percent of the allowable withdrawal loads determined from Fig. 24-4. The allowable lateral loads for toenailed joints are 83 percent of the allowable lateral loads determined from Fig. 24-5. Toenails should be driven at an angle of approximately 30° with the member and started approximately one-third of the length of the nail from the end of the member.

Drift Bolts, Drift Pins, and Spiral Dowels Drift bolts, drift pins, and spiral dowels are metal fasteners used to connect large wood members such as timber caps to pile bents, stringers to caps, and similar assemblies. A drift bolt is a rod with a common bolt head formed on one end. Drift pins are steel rods cut to the desired length. Spiral dowels are twisted, square steel rods with spirally grooved ridges in the entire length, the lead of the spiral thread being sufficient to permit driving instead of turning for installation.

Drift bolts and spiral dowels are available in approximately the same range of diameters as bolts and in lengths as desired. Drift pins can be made from any length or diameter available in steel rods. To prevent splitting of the wood and to develop strength, drift bolts and pins should be driven into prebored holes having a diameter $\frac{1}{16}$ in. less than the drift-bolt diameter. Prebored holes for spiral dowels should have a diameter about 75 percent of the outside diameter of the dowel.

Allowable lateral loads for drift bolts or pins driven into side grain of wood should not exceed and should ordinarily be taken as less than that for bolts of the same diameter. Drift bolts and pins should have greater penetration into members than bolts to compensate for the lack of thread and nut and, in the case of the drift pin, for lack of head and washer. Allowable lateral loads for drift bolts or pins driven into end grain should be 60 percent of the allowable side-grain load perpendicular-to-grain values of equal diameter bolts. To develop this strength, the drift bolt or pin should penetrate at least 12 diameters into the end grain.

Allowable lateral loads for spiral dowels driven in side grain depend upon depth of penetration of the dowel, upon specific gravity of the wood, and upon thickness of the side member. The spiral dowel should penetrate at least 7 diameters into the holding member and the thickness of the side member should be at least five times the diameter of the spiral dowel. The following formula may be used to determine the allowable lateral load, P_L, in pounds, for one spiral dowel driven into the side grain of seasoned wood in a two-member joint and under normal duration of loading.

$$P_L = 3{,}700Gd^2$$

where G = specific gravity (see Table 24-9)
d = outside diameter of spiral dowel, in.

Allowable lateral loads for spiral dowels driven into end grain should be 60 percent of the load determined by the above formula. To develop this strength, the spiral dowel should penetrate at least 12 diameters into the end grain.

Allowable withdrawal loads for drift bolts and pins driven into side grain of seasoned wood in a two-member joint and under normal duration of loading may be determined from the formula

$$P_S = 420Gd^{3/4}$$

where P_S = allowable withdrawal load from side grain, pounds per inch of penetration
G = specific gravity (see Table 24-9)
d = diameter of drift bolt or drift pin, in.

Allowable withdrawal loads for drift bolts or pins driven into end grain of seasoned wood in a two-member joint and under normal duration of loading may be determined from the formula

$$P_E = 195Gd^{3/4}$$

where P_E = allowable withdrawal load from end grain, pounds per inch of penetration and other terms are as defined above.

Allowable loads in withdrawal from side grain for spiral dowels may be determined by increasing the allowable loads in withdrawal from side grain for drift bolts or pins of the same diameter by 45 percent. Allowable loads in withdrawal from end grain for spiral dowels may be determined by increasing the allowable loads in withdrawal from end grain for drift bolts or pins of the same diameter by 80 percent.

For other than normal duration of loading, the allowable loads should be multiplied by the appropriate adjustment factor from Table 24-1.

For drift bolts, drift pins, or spiral dowels installed in unseasoned wood that will season in place, the full allowable lateral loads as determined above may be used. Where the wood will remain unseasoned or is subjected to alternate wetting and drying, use 70 percent of the allowable lateral loads as determined above.

For drift bolts, drift pins, or spiral dowels installed in wood that will remain unseasoned in place, the full allowable withdrawal loads as determined above may be used. If the wood will be subjected to alternate wetting and drying, use 25 percent of the allowable withdrawal loads as determined above.

Framing Anchors Framing anchors are light metal fittings used to provide a more positive connection between wood members than does toenailing. The several types of manufactured framing anchors are right-angled pieces formed from light-gauge galvanized steel. These may be bent to conform with special conditions of use. Nails appropriate to the particular framing anchor should be used in all holes provided in the anchor to develop its full strength.

Allowable load values for framing anchors are generally determined by test, but an estimate may be based on the lateral resistance of the nails used. An increase of one-third is permitted for short-term loading such as wind or earthquake. The manufacturer's data for the particular type of framing anchor used should be followed. Additional design information may be found in the *Timber Construction Manual.*[6]

Hangers Standard and special hangers are used in timber construction for connecting purlins to beams and other similar applications. Stock hangers are available from a number of manufacturers; however, the greater number of hangers are of special design. Allowable loads based on tests or manufacturer's data for the particular hanger used should be followed. Additional design information may be found in the *Timber Construction Manual.*[6]

DESIGN

A great variety of structural systems may be designed to serve particular needs. The relative economies of one system over another will depend upon the specific requirements of the job. In selecting a particular framing system, an important factor which the designer must take into account is its economical span range. The information on economical span ranges herein is intended for preliminary selection purposes only. All structural systems require complete and careful analysis before final selection is made.

Other cost-reducing design considerations include:

1. The detailing of as few and as simple joints as practicable.
2. The avoidance of unnecessary variations in members; i.e., repetitive use of the same member where possible and minimization of the number of different members.
3. Modification of cross section by tapering or building up local areas can be an economical way to control stresses.
4. Judicious use of multiple, cantilever, and suspended-beam systems tends to balance positive and negative bending moments.

More detailed information on timber design may be found in references 5, 6, and 8 to 10 at the end of this section, as well as in other publications.

Columns Wood compression members or columns may be of solid sawn or glued laminated timber or may consist of spaced members. The latter comprise two or more wood compression members with parallel longitudinal axes. The members are separated at the ends and midpoints by blocking and are joined to the end blocking with connectors possessing adequate shear resistance.

The allowable concentric load P, in pounds, on a column is determined from the formula

$$P = AF_c'$$

where A = cross-sectional area of column, sq in.
F_c' = allowable unit stress in compression parallel to grain, in psi, adjusted for slenderness (l/d) ratio

The values for F_c' determined by the following formulas may not exceed the allowable unit stresses in compression parallel to grain (F_c) for the species as given in Tables 24-3 and 24-4, adjusted for service and loading conditions. The values for F_c' determined by the following formulas are subject to the duration-of-loading adjustments given in Table 24-1.

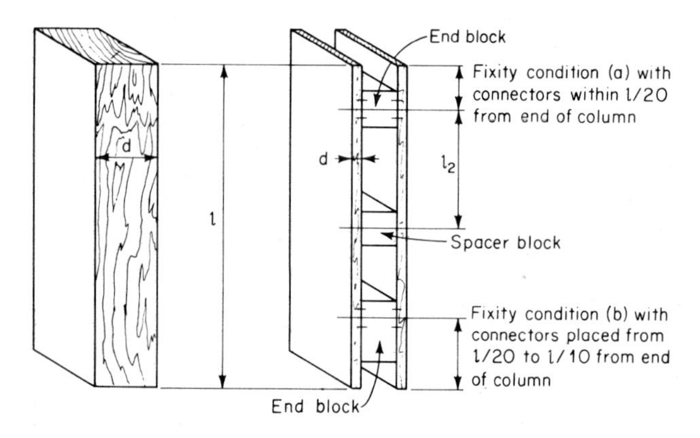

Fig. 24-6 Wood columns.

For rectangular solid or glued laminated timber columns, F_c', in psi, may be determined from the formula

$$F_c' = \frac{0.30E}{(l/d)^2}$$

where E = modulus of elasticity, in psi (see Table 24-3 or 24-4)
l = unsupported length of column, in. (see Fig. 24-6)
d = least dimension of column, in. (see Fig. 24-6)

The ratio l/d may not exceed 50. The formula was derived for pin-end conditions but may also be used for square-cut ends.

For round columns, F_c' may not exceed the value determined for a square column of equal cross-sectional area or that determined by the formula

$$F_c' = \frac{3.619E}{(l/r)^2}$$

where r = radius of gyration of section, in.

For tapered columns, d may be taken as the sum of the least dimension plus one-third the difference between this dimension and the maximum thickness parallel to this dimension.

For the individual members of a spaced column, l/d may not exceed 80, nor may l_2/d exceed 40 (see Fig. 24-6). To obtain spaced column action, end blocks are required when the l/d ratio for the individual members exceeds $\sqrt{0.30E/F_c}$. When l/d ratios for the individual members of spaced columns do not exceed this value, the individual members are designed as solid columns. A factor is introduced in the spaced column design formula which depends on fixity conditions of the column (see Fig. 24-6).

For the individual members of a spaced column, F_c' may be determined from the formulas

$$F_c' = \frac{0.75E}{(l/d)^2} \qquad \text{for fixity condition } (a)$$

$$F_c' = \frac{0.90E}{(l/d)^2} \qquad \text{for fixity condition } (b)$$

The individual members in a spaced column are considered to act together to carry the total column load. Each member is designed separately on the basis of its l/d ratio. Because of the end fixity developed in spaced columns, a greater l/d ratio than that allowed for solid columns is permitted. This fixity is effective only in the thickness direction; the l/d in the width direction is subject to the provisions for solid columns.

Connectors are not required for a single spacer block located in the middle tenth of the column length, l. Connectors are required for multiple spacer blocks, and the distance between two adjacent blocks may not exceed half the distance between centers of connectors in the end blocks. When spaced columns are used as truss compression members, panel points which are stayed laterally are considered as the ends of the spaced column. The portion of the web members between the individual pieces of which the spaced column is comprised may be considered as the end blocks. In the case of multiple connectors in a contact face, the center of gravity of the connector group is used in measuring the distance from connectors to the end of the column (see Fig. 24-6).

The total load capacity determined by using the spaced-column formulas should be checked against the sum of the load capacities of the individual members taken as solid columns without regard to fixity; their greater d and the l between lateral supports which provide restraint in a direction parallel to the greater d should be used.

Spacer and end-block thicknesses may not be less than those of the individual members of the spaced column; nor may thickness, width, and length of spacer and end block be less than required for connectors of a size and number capable of carrying the computed load.

Beams Standard engineering formulas for bending, shear, and deflection may be used to design or analyze solid sawn or glued laminated timber beams. Deflection often governs design, particularly in longer spans; however, shear is likely to control for short, heavily loaded beams.

Figure 24-7 shows the common type of wood beams. The first part of a beam name describes the top surface; the word following the hyphen describes the bottom surface. In general, a tapered beam has a sawn surface and a pitched beam an unsawn surface. Sawn surfaces on the tension side of a beam should be avoided.

Simple-span straight beams are generally most economical for spans ranging from 6 to 40 ft for solid sawn lumber and timber and from 10 to 100 ft for glued laminated timber. Tapered, pitched, and curved beams are most economical for simple spans from 25 to 100 ft. Straight and single and double-tapered straight beams can be furnished in solid sawn or glued laminated timber; but curved or pitched beams can be furnished only in glued laminated timber.

Unit-strength values, computed by the usual engineering methods from test data on wood bending members, decrease as the size of the member increases. Current industry practice takes this decrease into consideration by multiplying the allowable unit bending stress (F_b) for timber beams over 12 in. in depth by a depth effect factor, C_d:

$$C_d = 0.81 \left(\frac{d^2 + 143}{d^2 + 88} \right)$$

in which d = depth of member, in.

Recent research by the U.S. Forest Products Laboratory indicates that this decrease in strength is an effect of size and stress distribution within the member rather than merely a depth effect.

When a curved glued laminated member is subjected to bending, a stress is induced

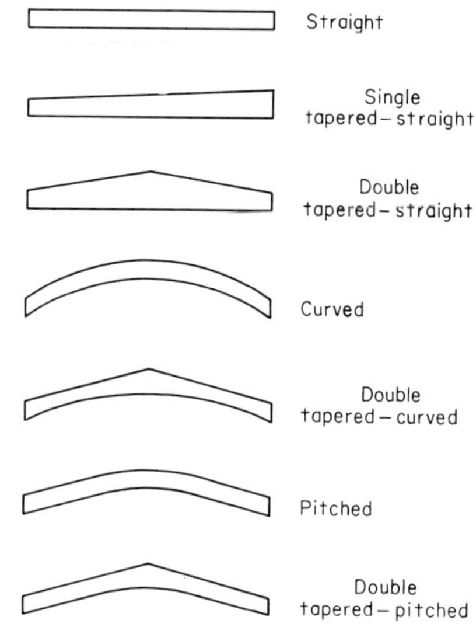

Fig. 24-7 Types of wood beams.

in a radial direction at right angles to the grain. The maximum magnitude of this stress, f_r, occurs at the neutral axis of a rectangle section and is given by the formula

$$f_r = \frac{3M}{2Rbd}$$

where M = bending moment, in.-lb
R = radius of curvature of centerline of member, in.
b = width of member, in.
d = depth of member, in.

According to presently accepted design criteria, when the bending moment (M) is in the direction tending to decrease curvature (increase the radius), this radial stress is tension and is limited to one-third of the value of the allowable unit stress in horizontal shear (F_v). When M is in the direction tending to increase curvature (decrease the radius), the radial stress is compression and is limited to the value of the allowable unit stress in compression perpendicular to the grain ($F_{c\perp}$). Research is presently being conducted in an effort to refine current design criteria for curved structural glued laminated members to resist radial tension stresses.

Stress is induced when laminations are bent to curved forms. The allowable stress in bending (F_b) for curved members must, therefore, be reduced through multiplication by a curvature factor C_c as follows:

$$C_c = 1 - 2{,}000 \left(\frac{t}{R} \right)^2$$

where t = thickness of lamination, in.
R = radius of curvature of inside face of lamination, in.

The ratio t/R may not exceed $1/100$ for hardwoods and Southern pine nor $1/125$ for softwoods other than Southern pine. The curvature factor is not applied to stresses in the straight portion of a member, regardless of curvature in other portions.

Conditions often require that a beam not exceed certain deflection limitations. Table 24-17 gives deflection limits for timber beams as recommended by AITC. Member size as determined by deflection is usually limited by either applied load

TABLE 24-17 Recommended Beam Deflection Limitations

Use classification	Applied load only*	Dead load plus applied load*
Roof beams:		
Industrial	$l/180$	$l/120$
Commercial and institutional:		
Without plaster ceiling	$l/240$	$l/180$
With plaster ceiling	$l/360$	$l/240$
Floor beams:		
Ordinary usage†	$l/360$	$l/240$
Highway bridge stringers	$l/200$ to $l/300$	
Railway bridge stringers	$l/300$ to $l/400$	

* l = span length.

† Ordinary usage classification is intended for construction in which walking comfort, minimized plaster cracking, and the elimination of objectionable springiness are of prime importance.

SOURCE: *Timber Construction Standards*, AITC 100-65, American Institute of Timber Construction, Washington, 1965.

TABLE 24-18 Recommended Minimum Camber for Glued-Laminated Timber Beams

Roof beams*	1½ times dead-load deflection
Floor beams†	1½ times dead-load deflection
Bridge beams‡	
Long span	2 times dead-load deflection
Short span	2 times dead-load + ½ of applied-load deflection

* Roof beams. The minimum camber of 1½ times dead-load deflection will produce a nearly level member under dead load alone after plastic deformation has occurred. Additional camber is usually provided to improve appearance and/or provide necessary roof drainage. Roof beams should have a positive slope or camber equivalent to ¼ in. per foot of horizontal distance between the level of the drain and the high point of the roof in addition to the minimum camber, to avoid the ponding of water.

† Floor beams. The minimum camber of 1½ times dead-load deflection will produce a nearly level member under dead load alone after plastic deformation has occurred. On long spans, a level ceiling may not be desirable because of the optical illusion that the ceiling sags. For warehouse or similar floors where live load may remain for long periods, additional camber should be provided to give a level floor under the permanently applied load.

‡ Bridge beams. Bridge members are normally cambered for dead load only on multiple spans to obtain acceptable riding qualities.

SOURCE: *Timber Construction Standards*, AITC 100-65, American Institute of Timber Construction, Washington, 1965.

only or by applied load plus dead load, whichever governs. Applied load is live load, snow load, wind load, etc. For special uses, such as beams supporting vibrating machinery or carrying moving loads, more severe limitations may be required.

Camber is built into a structural member by introducing a curvature, either circular or parabolic, opposite to the anticipated deflection movement. Camber recommendations given in Table 24-18 vary with design criteria for various conditions of use and, in addition, are dependent upon whether the member is of simple, continuous, or cantilever span; whether roof drainage is to be provided by the camber; and other factors. Reverse camber may be required in continuous and cantilever spans to permit adequate drainage.

Cantilever-beam systems permit longer spans or larger loads for a given size member than do simple-span systems, provided member size is not controlled by compression perpendicular to grain at the supports or by horizontal shear. They are usually more economical than simple spans when the span is over 40 ft. In general, a uniform section should be used throughout the length of a cantilever system. Increasing the depth of section at supports to meet shear requirements causes areas of stress concentration at the point of change and should be avoided. For economy, the negative bending moment at the supports of a cantilever beam should be equal in magnitude to the positive moment.

Continuous-span beams are commonly used in both building and bridge construction to reduce maximum moments, thus reducing the section size required.

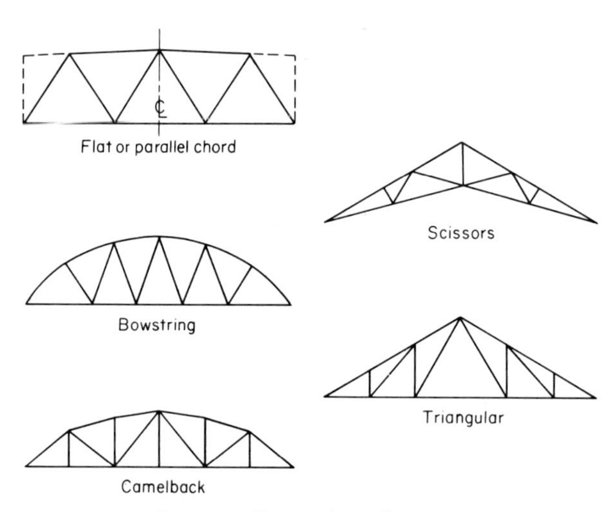

Fig. 24-8 Types of wood trusses.

Combined Loading Members subjected to both flexure and axial loads should be so proportioned that

$$\frac{P/A}{F_t} + \frac{M/S}{F_b} \leq 1 \qquad \text{or} \qquad \frac{P/A}{F_c'} + \frac{M/S}{F_b} \leq 1$$

where P = total axial load, lb
M = total bending moment, in.-lb
A = cross-sectional area of member, sq in.
S = section modulus of member, in.^{3}
F_b = allowable unit stress in extreme fiber in bending, psi
F_c' = allowable unit stress in compression parallel to grain adjusted for l/d of the member under consideration, psi
F_t = allowable unit stress in tension parallel to grain, psi

Trusses Types of wood trusses and arrangement of members may be chosen to suit the shape of structure, the loads, and the stresses involved. The types most commonly built are bowstring, flat or parallel chord, pitched, triangular or A-type, camelback, and scissors (see Fig. 24-8). For most construction other than houses, trusses usually are spaced 12 to 20 ft apart.

Joints are critical in the design of a truss. Use of a specific truss type is often governed by joint considerations.

Chords and webs may be single-leaf (or monochord), double-leaf, or multiple-leaf members. Monochord trusses and trusses with double-leaf chords and single-leaf web system are the commonest arrangements. Web members may be attached to the sides of the chords, or they may be in the same plane as the chords and attached with straps or gussets.

Individual truss members may be solid sawn, glued laminated, or mechanically laminated. Glued laminated chords and solid sawn web members are usually used. Steel rods or other steel shapes may be used as members of timber trusses if they meet design and service requirements.

The bowstring truss is the most popular type. Spans of 100 to 200 ft are common, with single or two-piece top and bottom chords of glued laminated timber, webs of solid sawn timber, and metal heel plates, chord splice plates, and web-to-chord connections. This system is light in weight for the loads that it can carry. It can be shop or field assembled. Attention to the top chord, bottom chord, and heel connections is of prime importance, since they are the major stress-carrying components. Since the top chord is nearly the shape of an ideal arch, stresses in chords are almost uniform throughout the bowstring truss; web stresses are low under uniformly distributed loads.

Parallel-chord trusses, with slightly sloping top chords and level bottom chords, are used less often because chord stresses are not uniform along their length and web stresses are high. Different cross sections are required for successive chords, and web members and web-to-chord connections are heavy. Eccentric joints and tension stresses across the grain should be avoided in truss construction whenever possible, particularly in parallel-chord trusses.

TABLE 24-19 Recommended Timber-truss Depth-to-span Ratios[6]

Bowstring	1/6 to 1/8
Triangular or pitched	1/6 or deeper
Flat or parallel chord	1/8 to 1/10

Triangular trusses and the more ornamental camelback and scissors trusses are used for shorter spans in the range of 50 to 90 ft. They usually have solid sawn members for both chords and webs; and degree of seasoning of timbers, hardware, and connections is of considerable importance.

For joints, split-ring and shear-plate connectors are generally most economical. Sometimes, when small trusses are field fabricated, only bolted joints are used.

Longitudinal sway bracing perpendicular to the plane of the truss is usually provided by solid sawn X bracing. Lateral wind bracing may be provided by end walls or intermediate walls or both. The roof system and horizontal bracing should be capable of transferring the wind load to the walls. Knee braces between trusses and columns are often used to provide resistance to lateral loads.

Horizontal framing between trusses consists of struts at bottom-chord level and diagonal tie rods, often of steel with turnbuckles for adjustment.

Table 24-19 gives recommended depth-to-span ratios for wood trusses.

For ordinary roof loads and truss spacing of 16 to 24 ft, vertical X bracing is required every 30 to 40 ft of chord length. This bracing is placed in alternate bays. Horizontal T-strut bracing should be placed from lower chord to lower chord in the same line as the vertical X bracing for the complete length of the building. If a ceiling is framed into the lower chords, these struts may be omitted.

Joists, spaced 12 to 24 in. center to center, usually rest on the top chords of the trusses and are secured there by toenailing. They may also be placed on ledgers attached to the sides of the upper chords or set in metal hangers, thus lowering the roof line.

Purlins, large cross-section joists spaced 4 to 8 ft center to center, are often set on top of the top chords. They are then butted end to end and secured to the chords with clip angles. Purlins may also be set between the top chords on metal purlin hangers.

Arches Glued laminated timber arches may be two hinged, with hinges at each base, or three hinged, with an additional hinge at the peak or crown. Figure 24-9 illustrates various arch types. Two-hinged arches are generally most suitable for relatively low rise-to-span applications, and three-hinged arches are suitable for higher rise-to-span applications.

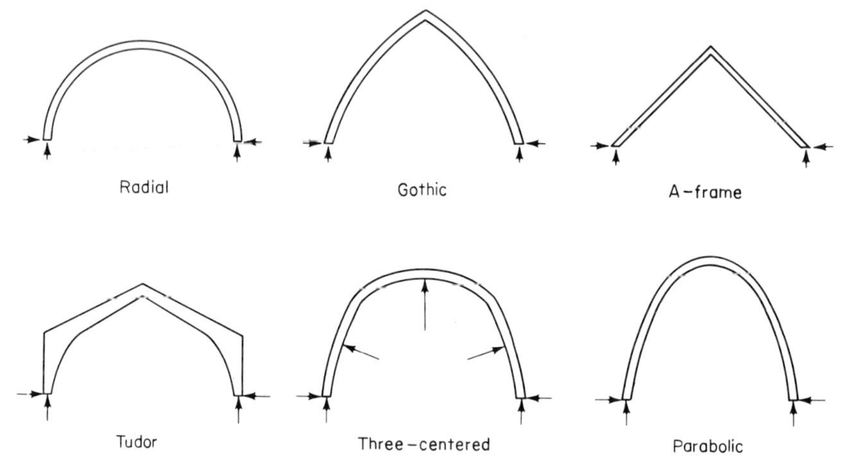

Fig. 24-9 Types of wood arches.

In the design of arches, adequate provision must be made for resistance of horizontal thrust at the supports. This can be supplied by tie rods, buttresses, foundations, or other adequate means. Tie rods may be above the ceiling or below grade.

PROTECTION AND PRESERVATION

Design and construction principles which will assure long service and avoid damage by decay and insects in wood structures include:

1. Positive site and structure drainage to prevent accumulations of water
2. Adequate separation of wood from known moisture sources
3. Ventilation and condensation control in enclosed spaces
4. Cleanup of construction sites to remove stumps, wood debris, stakes, or wood concrete forms before backfilling and pouring of foundations

Additional measures for protecting wood against attacks from termites include mechanical barriers. Termites must maintain contact with a source of moisture such as the ground, and mechanical barriers prevent their passage. Such barriers are usually metal shields used between the foundation and the wood and fitted tightly around pipes and other equipment connecting wood with the ground. All metal joints must be soldered or interlocked to be effective. The metal shields should extend out from the foundation at an angle of 45° for a horizontal distance of at least 2 in. If protection of wood in addition to that obtained by structural methods is needed, the soil adjacent to the foundations should be thoroughly poisoned with a suitable soil poison.

To supplement good design and construction practice, periodic inspection of a structure should be made to ensure that preventive measures are being maintained and that damage is not present.

When conditions favorable to decay or insect damage are unavoidable, wood should be adequately treated with a wood preservative. Wood members are permanent without treatment if located in enclosed buildings where good roof coverage, proper roof maintenance, good joint details, adequate flashing, good ventilation, and a well-drained site assure moisture content of wood continuously below 20 percent. Wood members are also permanent without treatment in arid or semiarid regions where climatic conditions are such that the equilibrium moisture content seldom exceeds 20 percent, and then only for short periods.

Where wood is in contact with the ground or with water, or where there is air and the wood may be alternately wet and dry, a preservative treatment applied by a pressure process is necessary to obtain a long service life. In enclosed buildings where moisture given off by wet-process operations maintains an equilibrium moisture content in the wood of above 20 percent, wood structural members must be preservatively treated. Wood exposed outdoors without protective roof covering and where the wood moisture content can go above 18 to 20 percent for repeated or prolonged periods also requires preservative treatment. Similarly, where wood structural members are subject to condensation by being in contact with masonry, preservative treatment is necessary.

Each type of preservative and method of treatment has certain advantages. The final choice depends on the service expected of the member for the specific conditions of exposure. The minimum retentions shown in Table 24-20 may be increased where severe climatic or exposure conditions are involved.

Creosote and creosote solutions have low volatility. They are practically insoluble in water and thus are most suitable for severe exposure, contact with ground or water, and where painting is not a requirement and a creosote odor is not objectionable.

Oil-borne chemicals are organic compounds dissolved in a suitable petroleum-carrier oil. They are suitable for outdoor exposure, or where leaching may be a factor, or where painting is not required. Depending on the type of oil used, they may result in relatively clean surfaces. While there is a slight odor from such treatments, it is usually not objectionable.

Water-borne inorganic salts are dissolved in water or aqua ammonia, which evaporates after treatment and leaves the chemicals in the wood. The strength of solution is varied to provide the desired net retention of dry salt. These salts are suitable where clean and odorless surfaces are required. The surfaces are paintable after proper seasoning.

To obtain preservatively treated glued laminated timber, lumber may be treated before gluing and the members then glued to the desired size and shape. Alternatively, the already glued and machined members may themselves be treated. When laminated members do not lend themselves to preservative treatment because of size and shape, gluing of treated laminations is the only method of producing adequately treated members.

When treating before gluing is required, water-borne salts, or oil-borne chemicals in mineral spirits, or AWPA P9 volatile solvent[2] are recommended. When treatment after gluing is required or desired, creosote, creosote solutions, or oil-borne chemicals are recommended.

Neither construction materials alone, nor construction features alone, nor detection and fire extinguishing equipment alone can provide the maximum fire safety. There must be a proper combination of these factors to provide the necessary degree of protection for those using the structure and for the structure itself. Some of the more important design considerations include:

1. The required degree of protection as dictated by the occupancy, contents, or operations taking place in or on the structure
2. The installation of automatic alarm and/or sprinkler systems
3. Firestopping and the elimination or proper protection of concealed spaces
4. Separation of areas in which hazardous processes or operations take place
5. Coordination of the number, size, type (such as direct to the outside), and the accessibility of exit ways (particularly stairways) and their distance from each other
6. Enclosure of stairwells and use of self-closing fire doors
7. Proper placement of fire walls and proper protection of openings in them
8. Exposure protection through horizontal separation

The most important protection factors are the prompt detection of a fire, immediate

TABLE 24-20 Recommended Minimum Retentions of Preservatives
(in pounds per cubic foot)

Preservatives	Coastal waters				General use			
	Sawn and laminated timbers		Laminations		Sawn and laminated timbers		Laminations	
	Western woods*	Southern pine	Western woods*	Southern pine	Western woods*	Southern pine	Western woods*	Southern pine
	Creosote or creosote solutions							
Coal tar creosote......	12	20	15	25	8†	8†	10	10
Creosote-coal tar solution............	Not recom.	20	Not recom.	25	8†	8	10	10
Creosote-petroleum solution............	Not recommended				8†	8	10	10
	Oil-borne chemicals							
Pentachlorophenol (5 % in specified petroleum oil).......	Not recommended				0.4‡	0.4	0.5	0.5
Penta (water-repellent) (moderate decay hazard).............	Not recommended				0.2	0.2	0.25	0.25
	Water-borne inorganic salts							
Chromated zinc arsenate (Boliden salt)...............	1.00	1.00	1.25	1.25	0.50	0.50	0.625	0.625
Acid copper chromate (Celcure)...........	1.00	1.00	1.25	1.25	0.50	0.50	0.625	0.625
Ammoniacal copper arsenite (Chemonite).	0.50	0.50	0.625	0.625	0.30	0.30	0.375	0.375
Chromated zinc chloride............	1.00	1.00	1.25	1.25	0.75	0.75	0.94	0.94
Copperized chromated zinc chloride........	1.00	1.00	1.25	1.25	0.75	0.75	0.94	0.94
Chromated copper arsenite (Green-salt-Erdalith)...........	0.75	0.75	0.94	0.94	0.35	0.35	0.44	0.44
Fluor chrome arsenate phenol (Tanalith-Wolman salt).......	0.50	0.50	0.625	0.625	0.35	0.35		0.44
Fluor chrome arsenate phenol (Osmossar-Osmosalt)..........	0.50	0.50	0.625	0.625	0.35	0.35	0.44	0.44

* Douglas fir, Western hemlock, Western larch.
† 10 pcf for timber less than 5-in. thick.
‡ 0.5 pcf for timber less than 5-in. thick.
SOURCE: *Timber Construction Standards*, AITC 100-65, American Institute of Timber Construction, Washington, 1965.

alarm, and rapid extinguishment of the fire. The fact that there are people in the vicinity of a structure is no guarantee that a fire will be discovered promptly. The promptness with which a fire is discovered and the speed with which it is extinguished have an important bearing on life safety and on the extent of damage to the structure.

ERECTION AND MAINTENANCE

Erection The erection of structural timber requires experienced erection crews and adequate lifting equipment to protect lives and property and to assure that the framing is not improperly assembled or damaged during handling. The unloading and storage of structural timber members before erection also demands care and good judgment.

Structural timber framing is subject to surface marring and damage when not properly handled and protected. At the erection site, the following precautions are suggested:

1. Lift members out of railroad cars or roll them on dollies or rollers; do not drag or drop them. Unload trucks by hand or crane; do not dump or drop members.
2. During unloading with lifting equipment, use fabric or plastic belts or other slings which will not mar wood. If chains or cables are used, provide protective blocking or padding.
3. Guard against soiling, dirt, footprints, abrasions, or injury to shaped edges or sharp corners.

If structural timber framing is to be stored before erection, it should be placed on blocks well off the ground. Individual members should be separated by strips so that air may circulate around all four sides. The tops and all sides of storage piles should be covered with moisture-resistant paper. Clear polyethylene films should not be used because wood members are subject to bleaching from sunlight. Individual wrappings should be slit or punctured on the lower side to permit drainage of water.

Waterproof wrapping paper used for the in-transit protection of glued laminated members should be left intact until the members are enclosed. If wrapping has to be removed at certain connection points during the erection, it should be replaced after the connections are made. If it is impractical to replace the wrapping, all of it should be removed.

Trusses are usually shipped partially or completely disassembled and are assembled on the ground at the site before erection. Arches, which are generally shipped in halves, may be assembled on the ground or connections may be made after the half arches are in position. When trusses and arches are assembled on the ground at the site, they should be placed on level blocking to permit connections to be fitted properly and tightened securely without damage. The end compression joints should be brought into full bearing, and compression plates should be installed where specified.

Before erection, the assembly should be checked for prescribed overall dimensions prescribed camber, and accuracy of anchorage connections. The erection should be planned and executed in such a way that the close fit and neat appearance of joints and the structure as a whole will not be impaired.

The accuracy and adequacy of necessary abutments, foundations, piers, and anchor bolts are the responsibility of the buyer. Before erection begins, all supports and anchors should be complete, accessible, and free of obstructions. The weights and balance points of the structural timber framing should be determined before lifting begins, so that proper equipment and lifting methods may be employed. When timber trusses of long span are raised from a flat to a vertical position preparatory to lifting, stresses entirely different from the normal design stresses may be introduced. The magnitude and distribution of these stresses will vary, depending on such factors as the weight, dimensions, and type of truss. A competent rigger will consider these factors in determining how much suspension and stiffening, if any, is required, and where it should be located.

All framing must be true and plumbed. Permanent bracing is bracing so designed and installed as to form an integral part of the final structure. Erection bracing is bracing which is installed to hold the framing in a safe position until sufficient permanent bracing is in place to provide full stability. Proper and adequate temporary erection bracing is introduced whenever necessary to take care of all loads to which the structure may be subjected during erection, including equipment and its operation. This bracing is left in place as long as may be required for safety. Part or all of the permanent bracing may also act as erection bracing.

Erection bracing plumbs the framing during erection and gives it adequate stability to receive purlins, joists, and roofing materials. It may include sway bracing, guy ropes, tying off framing nearest to end walls, steel tie rods with turnbuckle takeups, struts, shoes, and similar items. As erection progresses, bracing is securely fastened in place to take care of all dead load, erection stresses, and normal weather conditions. Excessive concentrated construction loads, such as bundles of sheathing, piles of purlins, roofing, or other materials, should be avoided. Final tightening of alignment bolts should not be completed until the structure has been properly aligned.

The field joining, holding, and welding of steel connections should be performed according to the shop-work requirements for such operations except such requirements as expressly apply to shop conditions only. The specifications of the American Institute of Steel Construction and of the American Welding Society should be complied with.

Unless otherwise specified, all fields cuts should be coated with an approved moisture seal if the member was initially coated. All field framing is done in accordance with the requirements of shop practice except such requirements as expressly apply to shop conditions only. If timber framing has been pressure treated, field framing after treatment should be avoided or, at least insofar as possible, held to a minimum.

All timber framing which requires moisture-content control, whether sawn or glued laminated timbers, should be protected against moisture pickup during erection operations. Any fabricated structural materials which are to be stored for an extended period of time before erection should, insofar as is practicable, be assembled into subassemblies for storage purposes.

After all roofing and permanent bracing are installed, the structure has been properly aligned, and connections and fastenings have been finally tightened, the temporary erection bracing may be removed.

Maintenance Little if any structural maintenance is normally required for properly designed and built timber structures. It is important to distinguish between structural characteristics and seasoning characteristics. Because of shrinkage during seasoning, sufficient looseness may develop in bolts so that tightening will be necessary to keep the faces of the members in reasonably close relationship. Usually no maintenance other than tightening in the first year or so is necessary during the life of the structure.

Inspection for bolt tightness and for the general condition of the joints and the members is usually desirable from 3 to 6 months after erection. For buildings, the check should be made after the first heating season. It is important to remember that most so-called "seasoning defects" in unseasoned material are anticipated and taken care of by adjustments in the design values. The inspector should look primarily for the following defects:

1. General looseness of joints, such as more than $\frac{1}{16}$-in. average opening between members or any one opening approaching $\frac{1}{8}$ in.

2. Excessive deflection, elongation of bolt holes, or other signs of unusual joint movement (a proper joint will seldom move more than $\frac{1}{32}$ in.)

3. Splits or checks of unusual size or length, particularly those tending to run across the face or thickness of a member and out the side or edge, those which the forces in the joints may cause to increase in size or length, and those on bolt centers in short splice plates

4. Twisting or bowing of members or misaligned sections

For normal dead and live load, joints can be tightened by turning up the bolts without jacking the trusses to relieve the load. When washers begin to embed in the wood, the joint is usually as tight as possible; as a rule, there is no need to relieve the load or use oversize washers or separate clamping arrangements. Sometimes grooves that are too shallow prevent complete closure of timber connector joints, but small openings are not serious. Timber connector joints may be checked with a probe, such as a hacksaw blade, to be sure the connectors are installed. A probe check is particularly desirable if there is any sign of trouble. If timber connectors have been omitted, joint reinforcement will be necessary.

Normally one bolt tightening is sufficient unless the framing was totally unseasoned when erected or unless the assemblies are composed of large-size members, such as beam and stringer grades. For such members, complete seasoning may take a year or more. The approximate amount of shrinkage of unseasoned timber can be estimated at about 1/32 in. per inch of thickness or width. Bolts should preferably be tightened during the humid season, so that the timber will not swell subsequently—because of seasonal increases in moisture content—and force the washer into the wood.

Splits and checks are probably the most common cause of concern to the layman. As there are no precise rules, the evaluation of these normal seasoning characteristics is a matter of engineering judgment. The two best criteria for such defects are a knowledge of their cause and of the potential seriousness of an increase in their size or length. A knowledge of the cause will usually indicate whether an increase in size or length can be expected. Checks are usually unimportant. Most splits are the results of seasoning shrinkage, aggravated by connections to other members that prevent a piece from shrinking normally. Internal cross-grain stresses are thus set up between the connections and cause the member to split.

A short split running parallel to the edges or at an angle corresponding to that permitted by the grade is usually not serious. A bolt hole bored at the end of a split will frequently arrest it. Stitch bolts or yokes around the member near the joint help prevent further opening of a split, but they should not be used for closure unless forces in the joint have increased the size of the split beyond that caused by shrinkage. Splits through timber connector joints are not particularly serious. The connector covers a wide area, whereas a split through a bolt hole may greatly reduce the strength of that bolt unless adjacent bolts prevent the wood from spreading under the wedge action of the bolt.

If the need for maintenance is debatable and subsequent inspection is planned, the size and length of a split may be marked on the member, with the date, and checked later. Obviously, if other maintenance of a similar nature is needed near the same area, the doubtful split may be included at little extra cost. A split that has not increased in size or length by the time of a second inspection is usually not serious. Tension members, particularly chord members and their splice plates, are most vulnerable to the effect of splits and should be dealt with accordingly.

Twisting and bowing of members may be a sign of improper design or bracing, joint trouble, or partial member failure due to use of an improper grade of wood for the particular service condition. If none of these causes appears to be responsible, the trouble may stem from seasoning that is in itself not serious, or it may have been present from the time of erection because of inaccurate fabrication. Bowing of single-compression members should usually be corrected by stiffening plates or braces. Bowing of one piece of a two-piece member, on the other hand, requires study and analysis to determine whether the bow should be corrected or merely prevented from increasing. Straightening the bowed member may force it to carry a higher percentage of the load. Such redistribution may be undesirable, as the bow may have properly balanced the stresses.

Deflection is not itself a sign of trouble provided it is not caused by failure of members or connections. Too frequently, laymen think a structure is failing because they find the truss is deflected below horizontal. Often such deflection should have been expected because of an initial neglect to provide camber or to apply it properly. More than average deflection may be expected in structures built of unseasoned material or with less truss depth than that normally recommended (see Table 24-19). A sudden and substantial increase in deflection without change in load, however, is usually a sign of trouble. Frictional forces in joints will sometimes delay normal joint deformation and permit a sudden increase in deflection as the wood shrinks and the forces are finally overcome. Excessive deflection is primarily a result of trouble rather than its cause. Deflection can affect secondary stresses adversely, but such stresses tend to be dissipated over a period of time, particularly in unseasoned lumber.

If the cause of deflection is determined and corrected, or if the deflection gives no sign of increasing or causing other trouble, it is usually not necessary to attempt to recamber the structure. Proper recambering is expensive because it requires that the structure be shored and the lengths of members adjusted. If the structure appears to be satisfactory except for deflection, it is often more economical to adjust roof and ceiling supports to eliminate the low spots than to recamber the members.

Proper timber design will avoid conditions that may lead to undue maintenance. If unseasoned lumber is to be used, details should be designed where possible to permit shrinkage of adjacent pieces without setting up internal stresses that increase the chances of splitting. Thus, in connections, saw kerfs may be provided between rows of bolts, or split splice plates may be used. Additional end distances on bolts reduce the possibility that normal end checks will extend into the joint area. End coating or sealing of the members during fabrication reduces end checking.

For those joints in which seasoning will set up internal drying stresses between connections, stitch bolts may be provided in the original design. These can easily be retightened at the time of general tightening. Stitch bolts are also specified for members loaded at a substantial angle to the grain if three or more bolts with connectors are required. With these safeguards, usually only bolt tightening is necessary for unseasoned or partially seasoned lumber.

LAND STRUCTURES

Formwork and Centering Plywood and lumber are used extensively in the construction of structural and architectural formwork for concrete foundations and structures.* Framing employing either glued laminated or sawn material is often economically used as falsework to support formwork for large poured-in-place concrete structures or as centering for placing precase concrete, stone, and steel structures which are assembled in place.

Falsework is the temporary structure erected to support work in the process of construction. It may be composed of structural units such as columns, beams, trusses, arches, and bracing designed by standard engineering procedures. The type and/or combination of structural units used depends on site conditions, the type and shape of supported structure, and the sequence of operations to be followed.

Centering is specialized falsework used in the construction of arches, shells, space structures, or any continuous structure where the entire falsework must be lowered (struck or decentered) as a unit to avoid introducing injurious stresses in any part of the structure. The lowering of centering is accomplished by the use of jacks, wedges, sand jacks, or some other device which has a controlled rate of travel.

Falsework must be designed to support all anticipated vertical and lateral loads until such time as these loads can be carried by the supported structure. Vertical loads include the weight of the concrete together with its reinforcement and attachments; the dead weight of the falsework and formwork; and the live loads imposed by workmen, equipment, and materials during construction. A concrete weight including reinforcing of 150 pcf is commonly used for design. The American Concrete Institute (ACI) recommends a minimum construction live load of 50 psf of horizontal projection. Many designers use 75 psf for construction with powered concrete buggies. Lateral forces may include the hydraulic pressure of the concrete, wind loads, and equipment loads. Concrete hydraulic pressure varies with method and rate of placement, temperature, consistency and proportion of mix, size and shape of form, and amount and distribution of reinforcing steel.

Deflections at various stages of construction and pouring, and possible movement during curing, are important considerations. Deflection due to dead load of the formwork, falsework, and reinforcing placed before pouring should be compensated for by cambering.

* See Sec. 21, Conventional Concrete.

For most exterior falsework uses, the timber and its connections should be designed for wet-use conditions and normal duration of loading. Glued laminated timber bonded with water-resistant adhesives may be employed if the material is protected by several coats of suitable sealer. The extra expense of waterproof adhesive may be justified by extreme service conditions or by possible salvage value.

In some structures, depending on the job conditions, falsework or centering may be reusable. If such is the case, the designer should provide for ease of reuse in designing the falsework. Reuse conditions include, among others, the amount of disassembly needed for stripping, the weight of the disassembled units in relation to the available handling equipment, the possible provision for traveling falsework (for example, rollers or track arrangements), and stresses imposed by the moving and extra handling. If reuse is not possible, demolition and salvage values should be kept in mind.

Detailed information on formwork, falsework, and centering may be found in *Formwork for Concrete,*[11] ACI Special Publication No. 4.

Towers Wood towers are used to support the wood tanks which provide fire protection and stabilize water pressure for buildings and which provide stored water for railroads. Wood lookout towers are used for forest fire protection and are also used to support outdoor theater screens. For permanent towers, the wood should be treated with a preservative and hardware should be galvanized.

Towers are designed for vertical loads and for horizontal loads of wind and, in some areas, of earthquake. Usually the horizontal loads are critical. Customary tower framing consists of the legs needed to support the load, horizontal struts, and diagonal bracing members. Depending on economic considerations, foundation conditions, available ground area, or other factors, towers are designed to be either freestanding or guyed.

A freestanding, framed tower subject to horizontal forces (usually wind or earthquake) is designed as a cantilever truss. The legs and diagonal bracing members are subject to reversal of stresses. Diagonal web members of either steel rods or wood are frequently double-crossed so that they need be considered as acting only in tension.

The horizontal force used in earthquake design is a percentage of the weight of a structure and of the load that it supports. For example, design stresses in a tank tower will be maximum when the tank is full. Because of the great weight supported at some distance above the base, the horizontal design loads for the weight of the water are likewise applied at the top of the tower. Anchorage to foundations is, therefore, especially critical in tank towers subject to earthquake.

Transmission-line towers have special horizontal load problems if a power line breaks. Such breaks are most apt to be caused by accumulated ice load on the lines. The line tension on the unbroken side of the tower, plus the line dead load and any ice load, may cause torsion in the tower.

In a guyed tower, there is no cantilever action below the guy line. Under either horizontal wind or earthquake forces, the compression stress in the tower legs is increased because of the vertical component of the tension stress in the guy system.

The number of legs needed by a tower will depend on the load it is to carry and the kind of service it is to perform. Some three-legged radio towers have been built, but three-legged towers are not as common as four-legged ones because of the problem of connecting struts and diagonal braces at an angle to square-sided wood members. Four-legged towers are used for all tower purposes except large tank towers. For those towers that must carry heavy loads, it is sometimes advantageous to use three pieces of lumber in an L arrangement at each of the four corners. Loading beams are placed across the top of the legs to distribute the load uniformly to each piece in each leg. The outstanding piece forming the angle leg provides the necessary area for strut and bracing connections and permits concentric joints at panel points. Another method used to simplify connection of struts and braces to a single leg is to offset panel-point joints on adjacent faces of each leg. The offset allows the bolts to pass through the leg, one set above the other. Although this method requires that the usual horizontal struts be placed at a slight slope, each leg and each set of braces will be identical. Fabrication and assembly are therefore simplified.

Tank towers, because of the great weight supported, present special problems. The weight of water produces a vertical load in all the columns, and wind forces produce compression stresses in the leeward columns and tension in the windward columns; therefore, it is necessary to investigate the tower legs for both full and empty tank conditions in order to determine maximum compression and tension stresses. Since wood tanks are generally not fastened to the tops of towers, it is important to keep the ratio of height to diameter such that an empty tank will not be overturned by wind.

To support the bottom of a large wood tank and to distribute the load uniformly to the legs, a series of closely spaced chime joists is used. The chime joists are cut to lengths that fit within the inside diameter of the tank and are placed normal to the plane of the bottom planks. The chime joists also provide a horizontal shear connection between the tank and the tower, because the bottoms of the side-wall staves extend below the bottom of the tank in which the chime joints are fitted. Joists or dunnage beams that bear directly on caps are uniformly spaced normal to the plane of the chime joists. In order to keep the caps from being overly large, legs or columns are usually spaced not more than 8 ft on centers. Horizontal shear stresses higher than those normally allowed are permitted by some designers for the chime and dunnage joists and for the caps.

To increase resistance to overturning by widening the base, the legs of towers are customarily battered. Batter of 1½ in. in 12 in. is common. Some tank towers with multiple legs and a height of not over 20 ft, however, have all legs vertical.

Next to tank towers, towers for outdoor theater screens are the most common type of tower. Critical design stresses are caused by wind pressure and are usually much greater than those for open-sided or skeleton towers. Although screen towers may be guyed, most are freestanding. Primary design is based on wind pressures (or seismic forces) normal to the screen, but it is important that proper resistance to forces parallel to the screen be provided by bracing or by making the screen face self-sheathing. Since ground area is usually not restricted, a maximum base width that is consistent with reasonably sized horizontal struts should be used in order to improve stability, reduce deflection, and permit more economical foundations. If local building codes do not control, some designers permit higher than usual unit stresses for such structures on the assumption that winds are infrequent and not hazardous to life because of the improbability of the theater's being in use under storm conditions.

Bridges Timber bridges are used on highways and railways as well as for pedestrian overpass structures. Timber highway bridges are primarily used on secondary highway systems, on park and forest roads, and for temporary service on construction projects. However, a 290-ft-span glued laminated timber arch bridge was completed in 1968 for use on a United States interstate highway. Timber bridges consist of several basic types, including trestles, girder bridges, truss bridges, and arch bridges. In the design of any of these types, consideration must be given to the moving loads on the bridge. For highway bridges, design loads and their application should be in accordance with the recommendations of the American Association of State Highway Officials. For railway bridges, the recommendations of the American Railway Engineering Association should be followed.

Trestles. The trestle is probably the simplest type of timber bridge. Timber trestles consist of stringers supported by pile or frame bents. The bridge deck is applied to the stringers. Pile and frame bents are capped by timbers 12 by 12 in. or larger, adequately fastened to the tops of the piles or posts.

If pile penetration or height of bent is such that piles longer than those commercially available are required, or if pile bearing values are low and a large number of piles must be driven, posts may be used on top of the pile bents. Frame bents must rest on some type of foundation structure, such as concrete footings or piles. Sway bracing and longitudinal tower bracing, appropriate to the height of the bent, must be provided.

The design of pile bents is largely empirical. Through the years, bent types have become somewhat standardized by the trial and error of experience. For the usual single-track railway trestle, bents commonly contain six piles; those under highway bridges with deck widths up to 26 or 28 ft normally contain four or more piles. Pile spacing may sometimes be controlled by shear or bending in the caps. If practical, spacing for friction piles should be sufficient to prevent overlapping of pressure areas in the soil. If pile heads are from 2 to 3 ft apart at cutoff or cap level, the outer piles are frequently driven at an angle to the vertical, or battered, to provide a greater separation below ground.

Lateral loads from wind or from moving live loads are resisted by sway bracing and by batter piles, usually of 1:12 to 3:12 batter. If they have adequate penetration in firm soil, all piles in a bent act as cantilevers and have some resistance to lateral forces. Battering the outer piles on each side of a bent provides considerable resistance, and if the bent is sway braced, it tends to act as a unit. Bracing should be fastened to the piles as securely as possible, for fastenings are frequently the weakest point in the bracing system. If large lateral forces and considerable vibration are likely, as in railroad trestles, special spike-grid timber connectors and bolts should be used for fastening braces to the piles or posts. Bolts are generally adequate for attaching the sway bracing on highway bridges unless considerable height is involved and large areas are exposed to wind. Even for highway bridges, however, the stronger attachment secured by use of grids is desirable because rigid connections contribute to unit action in the bent.

If horizontal forces act on the structure, piles on opposite sides of the group axis may receive quite different loads. Such pile groups are so designed that the piles on one side of the group will not be overloaded under any combination of forces. Uplift resistance of 40 percent of the friction bearing value is commonly permitted.

Spacing of bents is determined, in part, by the commercially available lengths of stringers, which are fabricated in even-foot increments. The ends of interior stringers are usually lapped and fastened at the bearing on the caps, while exterior stringers are butted at the ends and spliced over the bent caps. At least one end of every other exterior stringer should be adequately fastened to the bent caps.

Stringers are designed as simple span beams. Sizes and spacing are determined by the span and loading conditions. Solid bridging should be provided at the ends of stringers to hold them in line and also to serve as a fire stop. Bridging should also be placed between stringers at midspan or, on long spans, at third points. Fire curtains should be provided at intervals of about 100 ft.

Girder Bridges. Girder bridges consist of glued laminated or solid sawn timber girders supporting a bridge deck. These may be used for spans which exceed the practical limits of timber trestles, for spans less than those economical for truss bridges, or where a truss bridge is not desirable. Substructures similar to those used for timber trestles can be used for girder bridges, the girder being fastened to the bent caps by means of a fabricated steel girder seat. Timber girders are designed as beams. Lateral forces acting on the girder will enter into the design of the lateral bracing system.

Truss Bridges. Truss bridges may be one of two types: deck-truss bridges, in which trusses support the bridge deck and roadway; or through-truss bridges, in which the roadway passes between two parallel trusses forming the bridge structure. The deck-truss type is the more economical, since substructures and lateral bracing are narrower, but restrictions on underclearance may limit its use. Deck trusses may be of the parallel-chord type or of the bowstring type with the truss built up to the level of the floor beams. Through trusses may be one of these truss types, but the bowstring is usually more economical.

Substructures for truss bridges may be similar to those for timber trestles; however, because the vertical loads are greater and are concentrated at the ends of the trusses, the bents must be capable of carrying a greater load, and a system of cribbing is required for bent caps. For longer spans and heavier loads, timber, stone, or concrete piers may be required. Lateral forces are greater on truss bridges, and a carefully designed substructure–sway-bracing system is necessary.

The design of trusses for bridges is similar to that for roof trusses. The length of the truss panel is determined by economical spacing of floor beams, a minimum number of joints, and commercially available lengths of timber. As in roof-truss design, joint design is an important consideration. Bridge truss joints should be designed to eliminate or minimize pockets which may collect moisture.

Arch Bridges. When size conditions are such that considerable height is required between foundation and roadway, an arch bridge may be most economical because of less need for substructure framing. Arch bridges may be of the two-hinged or three-hinged type; two-hinged designs are frequently used on shorter spans, and three-hinged designs are used on long spans.

Glued laminated timber arches may be fabricated to the desired shape and the ends built up to the level of the roadway by means of post bents. Post bents may be connected to the arch by means of steel gusset plates. These should be designed for erection loads, possible stress reversals, and lateral forces as well as for the anticipated bridge loads.

Bridge Decks. The selection of decks for timber bridges is determined largely by density of traffic. Plank decks may be used for light traffic or for temporary bridges. Laminated decks can be used for heavier traffic conditions. Asphaltic wearing surfaces may be applied on the decking, although this is not usually done for plank decks.

Composite timber-concrete decks are sometimes used in timber bridge construction. Composite timber-concrete construction combines timber and concrete in such a manner that the wood is in tension and the concrete is in compression (except at the supports of continuous spans, where negative bending occurs and these stresses are reversed). Composite timber-concrete construction is of two basic types: T beams and "slab" decks.

T beams consist of timber stringers which form the stems and concrete slabs which form the flanges of a series of T shapes. Composite beams of this type are usually simple span bridges. Slab decks use, as a base for the concrete, a mechanically laminated wooden deck made up of planks set on edge. Alternate planks are raised 2 in. to form longitudinal grooves, usually by employing planks of two different widths and alternating them in assembly. This composite type is commonly used for continuous-span bridges and trestles.

In both deck types, a means of horizontal shear resistance and a means of preventing separation are needed at the joint between the two materials. In T beams, resistance to horizontal shear is generally provided by a series of notches cut into the top of the timber stringer, while nails and spikes partially driven into the top prevent vertical separation of the concrete and timber. Other adequate methods may be used. In slab decks, shear resistance is accomplished either by means of notches cut into the tops of all laminations or by patented triangular steel-plate "shear developers" driven into precut slots in the channels formed by the raised laminations. When notches are used, grooves are milled the full length of both faces of each raised lamination to resist uplift and separation of the wood and concrete. When the steel shear developers are used, nails or spikes are partially driven into the tops of raised laminations to resist separation.

In T-beam design, secondary shearing stresses due to temperature changes must be considered in designing for horizontal shear resistance. These stresses are induced by the thermal expansion or contraction of the concrete, both of which are resisted by the wood. It is assumed that the wood is unaffected by normal temperature changes.

Shear connections for temperature change are neglected in slab-deck-type composite construction; however, expansion joints should be provided in the concrete slab. The concrete slab should be reinforced for temperature stresses. In continuous spans, steel sufficient to take care of negative bending stresses is necessary over interior supports.

The dead load of the composite structure is considered to be carried entirely by the timber section. The composite structure carries positive bending moment and, over interior supports in continuous spans, steel reinforcing and the wood act to resist negative bending moment.

Piles Foundation piles may be classified as end-bearing piles or friction piles, according to the manner in which they transfer their loads to the underlying soil.*

End-bearing piles penetrate a layer of low-strength soil before bearing on a firm stratum such as rock. Their load is transmitted through bearing on the point of the pile to the firm stratum. Such piles are designed as timber columns. In all but the very poorest soils, lateral support may be assumed for the portion of the pile in the ground.

* See Sec. 27, Piles and Pile Driving.

Friction piles—which constitute the great majority of foundation piles—support loads through the friction developed between the piles and the soil they are driven into. Timber piles readily lend themselves to support by friction because of the relatively large circumference area of the pile relative to pile cost.

The primary requirement in designing pile foundations is knowledge of the subsurface conditions at the site. Where possible, pile driving or loading tests under actual site conditions should be used. Such tests are the best means of determining the load-carrying capacity of piles. When tests are too expensive or impractical in relation to the size of the job, load-bearing capacity of driven piles may be estimated from a pile-driving formula. Numerous empirical formulas have been developed for this purpose.

Where friction piles are driven in clusters or groups, most building codes require a reduction in the allowable load per pile due to overlapping in the soil of pressure areas of the individual piles. Pile spacing is one of the important factors affecting pile group loads. In general, a minimum spacing of 3 ft on centers is recommended for friction piles.

Wide pile spacing in a group is advantageous from the standpoint of bearing capacity. However, heavier footings are required to distribute the load properly. In some cases, outer piles in a group may be driven at an angle to the vertical, or battered, to provide greater spacing below ground.

Recommendations for the design of timber piles in foundations may be found in *Timber Foundation Pile Study.*[12] The American Society for Testing and Materials, *Standard Specifications for Round Timber Piles,* ASTM D25-58,[13] classifies round timber piles according to use under three general divisions:

CLASS A: Piles suitable for use in heavy railway bridges or other heavy framed construction

CLASS B: Piles suitable for use in docks, wharves, bridges, building or other foundations, and general construction

CLASS C: Piles suitable for use in foundations which will be completely submerged for cofferdams, falsework, or light construction

Table 24-21 lists requirements for the pile classes of ASTM D25-58.

Sheet Piling Wood sheet piling is used in bulkheads and retaining walls to resist earth pressures as well as the surcharge of structures which may be built on the retained earth. Bulkheads for waterfront structures are designed to resist wave action and thus protect land at the shoreline from encroachment by the sea.

Wood sheet piling is usually made up of single wood planks, square edged or tongued and grooved; of mechanically laminated planking spiked together to form lapped or tongue-and-groove joints; or of large timbers, which are milled with tongues and grooves or, more commonly, have dimension lumber spiked to the abutting edges to form a tongue and groove. Mechanically laminated sheet piling in which three planks are spiked together with the middle plank offset to form a tongue on one edge and a groove on the other is commonly known as "Wakefield piling."

The planks in tongue-and-groove sheet piling should either be surfaced on four sides or on one side and one edge. The bottom, or toe, of each section is usually cut on a diagonal with the point on the groove side so that, in driving, the pile will be continuously wedged back against the previously driven pile. These sections are driven with the tongue leading. After driving, the adjacent pieces may be spiked together through the joint which, in effect, gives a solid wall of spiked timber sheeting.

Timber bulkheads are composed of wood sheet piling supported by a line of round vertical piles, to which are bolted two or more lines of horizontal timbers known as "wales." Tie rods provide tension to hold the backfill, and timber struts provide for compression.

The stability of bulkhead retaining walls is dependent upon (1) the thickness and lengths of the sheet piles, which in turn are dependent upon the soils into which the piles are driven and upon the height of the backfill; (2) the size, length, and spacing of the round piles supporting the sheet piles with the attached wales; and (3) the anchorage system.

Tie rods should be of either corrosion-resistant alloy metals or steel rods carefully

TABLE 24-21 Circumference and Diameters of Timber Piles

Length, ft	Class A						Class B						Class C					
	3 ft from butt				At tip, min		3 ft from butt				At tip, min		3 ft from butt				At tip, min	
	Min		Max				Min		Max				Min		Max			
	Circumference, in.	Diameter (approx.), in.	Circumference, in.	Diameter (approx.), in.	Circumference, in.	Diameter (approx.), in.	Circumference, in.	Diameter (approx.), in.	Circumference, in.	Diameter (approx.), in.	Circumference, in.	Diameter (approx.), in.	Circumference, in.	Diameter (approx.), in.	Circumference, in.	Diameter (approx.), in.	Circumference, in.	Diameter (approx.), in.
Douglas fir, hemlock, larch, pine, spruce, or tamarack																		
Under 40	44	14	57	18	28	9	38*	12*	63	20	25	8	38†	12†	63	20	25	8
40 to 50 incl.	44	14	57	18	28	9	38	12	63	20	22	7	38	12	63	20	19	6
55 to 70 incl.	44	14	57	18	25	8	41	13	63	20	22	7	38	12	63	20	19	6
75 to 90 incl.	44	14	63	20	22	7	41	13	63	20	19	6	38	12	63	20	19	6
Over 90	44	14	63	20	19	6	41	13	63	20	16	5	38	12	63	20	16	5
Oak and other hardwoods, cypress																		
Under 30	44	14	57	18	28	9	38*	12*	57	18	25	8	38†	12†	63	20	25	8
30 to 40 incl.	44	14	57	18	28	9	41	13	63	20	22	7	38	12	63	20	22	7
Over 40	44	14	57	18	25	8	41	13	63	20	19	6	38	12	63	20	19	6
Cedar																		
Under 30	44	14	69	22	28	9	38*	12*	69	22	25	8	38†	12†	69	22	25	8
30 to 40 incl	44	14	69	22	28	9	41	13	69	22	25	8	38	12	69	22	25	8
Over 40	44	14	69	22	25	8	41	13	69	22	22	7	38	12	69	22	22	7

* In Class B piles, a minimum circumference of 34 in. or diameter of 11 in. at a point 3 ft from the butt may be specified for lengths of 25 ft and under.
† Class C piles, a minimum circumference of 31 in. or diameter of 10 in. at a point 3 ft from the butt may be specified for lengths of 25 ft and under.
SOURCE: American Society for Testing Materials, *Standard Specifications for Round Timber Piles*, ASTM D25-58 (under the jurisdiction of ASTM Committee D-7 on Wood).

coated with an efficient protective coating. Tie rods may be attached to the wales at one or two points. If single rods are used, attachment is usually at a point two-thirds of the way up the bulkhead height from the bottom. For high bulkheads, attachment is at points one-third and two-thirds up from the bottom.

Poles Wood poles are used in a wide variety of structures such as communication and electric transmission lines, street and highway lighting, outdoor advertising signs, outdoor theater screens, towers, and buildings. These structures employ preservatively treated round poles set into the ground as columns. The ground furnishes both vertical and horizontal support and prevents rotation at the base.

Calculation of required depth of embedment in soil for poles subject to lateral loads can be made by approximate analysis. However, depth of embedment should be checked by tests or against experience in the same type of soil.

Isolated poles such as flagpoles or utility poles are essentially cantilever beams. Column or axial loads are usually small and can be ignored in most such cases.

The American National Standards Institute (ANSI) has established classes of poles based on size. Information on these pole classes is given in *Specifications and Dimensions for Wood Poles,* ANSI 05.1—1963.[14]

In some structures, a bracing system can be provided at the tops of the poles to reduce bending moments at the bases of the poles and to distribute loads. The design of structures supported by poles without bracing requires good knowledge of soil conditions in order to eliminate excessive deflection or side sway.

Bearing values under butt ends of poles should be checked. It is common practice to backfill holes around the pole with well-tamped native soil, sand, or gravel. Backfilling with concrete or soil cement can develop a more effective pole diameter; consequently, it can be used as a means of reducing required depth of embedment. Concrete backfill also increases the area of the pole for skin friction and thus increases the bearing capacity. Skin friction is effective where uplift due to wind may act on a pole through its connections to the roof framing.

In order to increase bearing capacity under pole butts, concrete footings may be used. These should be designed to withstand the punching shear of the pole and the bending moment due to the pole load. Thickness of concrete footings should never be less than 12 in. and may need to be greater, depending on the loads. Concrete footings should be considered even in firm soils such as dry, hard clay; coarse, firm sand; or gravel.

MARINE STRUCTURES

Piers and Wharves Landing places for boats and ships, such as piers and wharves, are commonly built of timber. Timber for these structures offers economy in construction, ability to absorb impact, and ease of change and repair.

Piers and wharves are designed for vertical loads by the usual engineering methods; however, methods of design for the less-well-defined horizontal forces acting on such structures are largely based on experience. Consideration must be given to wave action and to current pressure, as well as to the shock pressure or impact caused by ship maneuvering.

Unless they are temporary, waterfront structures should be preservatively treated for protection from insects, decay, and marine borers. Salt-water air is particularly destructive to unprotected steel and, therefore, corrosion-resistant metals or protective coatings such as galvanizing are recommended for all hardware.

The proper selection of timber for waterfront structures is dependent not only upon the conditions of exposure but also upon a general study of the traffic expected. The amount and kind of traffic will affect the width of structure, the necessary freeboard, and the required depth of water. The last factor, of course, directly influences the length of piles.

The allowable bearing capacity of the piles of a pier or wharf normally determines pile spacing and thus controls the general design. Information on wood piles for land structures also applies for piles used in marine structures.

Columns for covered superstructures are often spaced 20 ft on centers, and cargo doors are detailed to fit into 20-ft bays. It should be noted, however, that separate pilings are driven to support the columns for the superstructure. The column footings do not form a part of the substructure deck supports; instead, they consist of groups of four to six piles each, cut off at mean low-water mark, with timber caps and deck planking that carry the pedestals upon which the shed columns rest.

If additional frictional resistance is required for the outboard piles, it has been common practice to lag the section of long piles below mud line. Lagging consists of bolting two or four 6 by 8 timbers parallel to the piles before driving the piles into place.

Other means of procuring greater bearing from bent-supporting piles consist of spacing piles on 2½-ft centers in the bent or driving double bents on 2½- to 3-ft centers. Local soil conditions and experience, however, sometimes preclude such close spacing. Double bents are capped and decked over with timber to carry the pedestals on which the deck rests.

An important design detail for piers which extend out from the shoreline is corner pile clusters at the outboard corners. At these points, the pier must have great stability and the ability to withstand the wear and abrasion caused by ships being brought into or out of the pier slip. Spacing and number of piles in these clusters will depend upon the character of the harbor bottom, the size of the piling, and the size of the vessels to be accommodated.

Fender systems are employed to absorb the shock of impinging vessels so that neither the pier, which they are designed to protect, nor the ship itself is damaged. Fender systems are almost always made of wood-pile and timber elements. Fenders are attached to the ship-side faces and the ends of the pier or wharf. They are composed of a row of vertical piles driven on 5-ft centers to depths of a few feet into the harbor bottom. They are separated by sawed timbers called "chocks," which are framed into the vertical fender piles at their tops and at points of high and low water. As the fender piles are usually larger in diameter than the chocks, they extend outward from the chocks and thereby absorb the heaviest mechanical wear. Fenders are expendable protective devices that should be replaced as soon as the wear they were designed to absorb becomes excessive. Fender systems are also used as protective devices for bridge piers and similar structures over navigable waters.

The wood which is used for fenders must be high in resistance to mechanical abrasion. Many species are satisfactory, and choice is largely a matter of availability and cost relative to service life. Since decay usually starts at the butt or top ends of fender piles, it is recommended that these ends be protected with wood preservative followed by a bitumastic sealing coat.

The design of pier and wharf decks is dependent upon their design live loads and the spacing of the substructure bents. Decking is usually applied in two layers. The subplanking of 3- to 4-in. thickness is laid transversely to the stringers, which span between caps, and a top layer of 2-in. plank is laid at 90° to the subdeck. The top planking is thereby readily replaceable if abraded by heavy traffic. If it is desired to protect timber decking from mechanical wear, a bituminous or concrete wearing surface may be applied over the wood subplanking. Another form of deck is the composite timber-concrete deck discussed in connection with timber bridges.

Fire hazard in timber piers may be minimized by installing bulkheads, bridging over the bents between stringers, providing sprinkler systems, and using piling encased in concrete or composite piling with concrete above the waterline. Fire-tight bulkheads are often built under piers at spacings not exceeding 150 ft and extending a foot below the mean low-water mark. They help in confining a fire to one section. When deck stringers are used over the bents, the spaces between and over the caps should be filled with solid bridging so that the top of the bent acts as a fire wall. As an extra precaution, automatic sprinkler systems may be installed to give the pier the same protection afforded industrial and commercial buildings.

Rock Cribs Rock cribs are used to support bridges and piers or to protect shorelines against erosion or scouring. They are also used on land for supporting roadway embankments.

Rock cribs are normally composed of a framework of 10 by 10 or 12 by 12 timbers, laid horizontally with log-cabin corners without mortises, and well drift-bolted. Timber connectors can be used at joints to increase the strength of the structure. The space enclosed by the timber framework is filled with local rock or stone. Arrangements should be such that the timber cells do not exceed 10 or 12 ft on a side. If greater dimensions are required, multiple cells should be built.

Bottoms of rock cribs should be embedded well below the surface wash or scour of waves and current. The height should not be much greater than the width unless the top is stabilized laterally.

Since rock cribs are porous, no hydraulic pressure is built up within them. The rock fill produces a lateral pressure on the timber frame.

Dolphins Dolphins are groups of timber piles driven in a cluster, reinforced with timbers, and bound at the top with cables. Their primary purpose is to provide a stable marine element that will protect a pier or wharf from collision and assist in bringing ships to their final mooring place.

The number of piles in a dolphin depends upon the depth of water and consistency of the harbor bottom, the strength of the pile species, and the impact stresses of heavy craft moving at a maximum assumed speed.

A design for a long-lived and stable dolphin consists of a core of pressure-treated vertical piles, separated in the tidal zone by 12 by 12 pressure-creosoted timbers set in two directions and securely wrapped by cable. Around this stable and permanent core, a single line of closely driven, untreated fender piles is driven to take the wear of water craft. If the outer ring of piles becomes so abraded as to require replacement, it can be readily removed and replaced without damage to the permanent central-core cluster.

REFERENCES

1. Forest Products Laboratory, *Wood Handbook*, Handbook No. 72, U.S. Department of Agriculture, 1955.
2. *AWPA Standards*, American Wood-Preservers' Association, Washington, D.C. (revised annually).
3. *Structural Glued Laminated Timber*, Commercial Standard CS 253-63, U.S. Department of Commerce, 1963.
4. *American Softwood Lumber Standard*, Product Standard 20-70, U.S. Department of Commerce, 1969.
5. *National Design Specification for Stress-Grade Lumber and Its Fastenings*, National Forest Products Association, Washington, D.C., 1968.
6. American Institute of Timber Construction, *Timber Construction Manual*, John Wiley & Sons, Inc., New York, 1966.
7. *Design Manual for TECO Timber Connector Construction*, Timber Engineering Company, Washington, D.C., 1962.
8. W. F. Scofield and W. H. O'Brien, *Modern Timber Engineering*, Southern Pine Association, New Orleans, 1963.
9. *Douglas Fir Use Book*, West Coast Lumbermen's Association, Portland, Ore., 1962.
10. Timber Engineering Company, *Timber Design and Construction Handbook*, McGraw-Hill Book Company, New York, 1956.
11. M. K. Hurd, *Formwork for Concrete*, Special Publication No. 4, 2d ed., American Concrete Institute, Detroit, 1969.
12. Dames & Moore, *Timber Foundation Pile Study*, American Wood-Preservers' Institute, Washington, D.C., 1966.
13. *Standard Specifications for Round Timber Piles*, ASTM D25-58, American Society for Testing and Materials, Philadelphia, 1964.
14. *Specifications and Dimensions for Wood Poles*, ANSI 05.1—1963, American National Standards Institute, New York, 1963.

Section 25

Paving

R. W. BEATY

Chief Paving Engineer, Barber-Greene Company, Aurora, Ill.
(Bituminous Pavements)

WILLIAM G. WESTALL

Consulting Engineer, Largo, Fla. *(Portland-cement Concrete Pavements)*

A. Bituminous Pavements

This subsection deals with the construction of bituminous pavements and is specifically directed toward the production and placement of the bituminous-treated materials which are incorporated in the total flexible pavement structure.

PAVEMENT ELEMENTS, MATERIALS, AND TYPES

Pavement Elements[1]* The flexible pavement structure consists of several courses of treated and/or untreated mineral materials placed upon the subgrade or prepared earthwork foundation. The quality (and therefore the cost) of the materials involved will usually increase with each successive course, from the subgrade upward.

Subbase Course. This course consists of an imported material of higher quality than the subgrade, but the term "subbase" usually indicates that it does not meet the quality requirements for a base-course material. It is used in the pavement structure for reasons of economy, as in the case of a locally available granular material. Subbase materials may or may not be treated with additives such as hydrated lime, portland cement, or one of several forms of bituminous materials.

Base Course. This course is placed directly on the prepared subgrade or upon the completed subbase if such a course is included in the pavement structure. Base-course materials are comprised of naturally occurring mineral materials or crushed mineral aggregates of a quality superior to those used in subbase construction. The materials may or may not be treated with additives such as portland cement or bituminous materials. The thickness of the base course will usually range from 4 to 10 in., depending on the types of materials used and the engineering determination of the required structural thickness for the particular pavement design involved. The treatment of the base-course materials with portland cement or bituminous additives results in a reduction in required thickness, thereby reducing the total quantity of materials required for a given design strength. In recent years, the use of bituminous-treated base-course material has been gaining in favor because of its economy and also because its superior binding and waterproofing qualities improve its durability.

Leveling Course. In most cases, the term "leveling course" designates a course of bituminous paving material placed upon the base course or upon an existing surface for resurfacing projects to eliminate surface irregularities. It follows that the leveling course must be of variable depth in order to remove sags and high spots in the underlying surface. In some states, however, the pavement course directly underlying the surface course is called the leveling course. It is then comparable to the "binder course," as next described.

Binder Course. For the high-type bituminous pavements (asphalt-concrete pavements), the binder-course material differs from the surface-course material in that the quality may be somewhat lower. For example, since the binder course is not exposed to the action of traffic, requirements for cohesion and/or skid resistance are not prime considerations. The binder course must have stability at least equal to that of the surface course. The thickness of the binder course is usually on the order of 2 to 3 in., depending on the maximum aggregate particle size used and the overall design thickness requirement.

Surface Course. The functions of the surface course of the flexible pavement are several: (1) to provide a smooth, quiet-riding surface; (2) to provide a surface which is cohesive and durable to resist the wearing stresses imposed by traffic; (3) to provide a surface with a high coefficient of friction between the surface and vehicle tires so that adequate skid resistance is provided; and finally (4) to provide a layer of material

* Superior numbers refer to the list of references at the end of this section.

TABLE 25-1 Principal Uses of Asphalt

Type of construction	Paving asphalts					Liquid asphalts																	
						Rapid curing (RC)				Medium curing (MC)				Slow-curing (SC)				Emulsified¶					In-verted emul-sion
	40–50	60–70	85–100	120–150	200–300	70	250	800	3,000	70	250	800	3,000	70	250	800	3,000	RS-1	RS-2	MS-2	SS-1	SS-1 h	
Asphalt concrete:																							
Highways		x	x	x	x												x						
Airports		x	x	x																			
Parking areas	x	x	x																				
Curbs		x	x*																				
Industrial floors	x																						
Blocks	x																						
Plant mix, cold laid:																							
Graded aggregate												x	x			x					x		x
Mixed-in-place:																							
Open-graded aggregrate								x	x			x	x			x				x			
Dense-graded aggregate											x	x			x	x					x		x
Clean sand							x				x	x										x	
Sandy soil											x				x						x		x
Penetration macadam:																							
Large voids			x					x	x										x				
Small voids				x			x											x					
Surface treatments:																							
Aggregate seal				x	x			x	x			x	x					x	x				x
Sand seal							x				x							x					
Slurry seal																					x	x	
Fog seal																					x†	x†	

Penetration treatment							x				x			x	x								
Prime coat, open surfaces						x	x				x												
Prime coat, tight surfaces										x				x									x
Tack coat						x															x†	x†	
Dust laying														x							x†		
Patching mix:																							
Immediate use							x	x			x	x				x							x
Stock pile											x				x	x							x
Hydraulic structures:																							
Membrane linings, canals, and reservoirs	x‡																						
Hot-laid, graded aggregate mix for groins, dam facings, canal and reservoir linings		x																					
Crack filling						x												x				x§	
Soil treatment, membrane envelope, mixtures	x‡																						
Expansion joints	Blown asphalts, mineral-filled asphalt cements, and preformed joint compositions																						
Undersealing portland-cement concrete	Blown asphalts																						
Roofing	Blown asphalts																						
Miscellaneous	Specially prepared asphalts for pipe coatings, battery boxes, automobile undersealing, electrical wire coating, insulation, tires, paints, asphalt tile, wall board, paper sizing, waterproofing, floor mats, ice cream sacks, adhesives, phonograph records, tree grafting compounds, grouting mixtures, etc.																						

In northern areas where rate of curing is slower, a shift from MC to RC or from SC to MC may be desirable. For very warm climates, a shift to next heavier grade may be warranted.

* In combination with powdered asphalt

† Diluted with water

‡ Also 50–60 penetration blown asphalt and prefabricated panels

§ Slurry mix

¶ Cationic quick-setting grades are comparable in use to RS-1 and RS-2
Cationic mixing grades are comparable in use to MS-2 and SS-1

SOURCE: The Asphalt Institute.

which is relatively impervious to the infiltration of water and air, thus preventing accelerated aging of the bituminous binder material. The thickness of the surface course will vary, but it is usually on the order of 1 to 2 in. However, where durable, nonpolishing aggregates are in short supply and therefore expensive, it is not unusual to find surface courses placed as thin as ½ to ¾ in.

For roads subjected to low volumes of traffic, such as secondary roads, the surface course may consist of a surface treatment of about ¾-in. thickness. This type of surface course, which will be described subsequently, is placed directly on a base course of treated or untreated crushed stone or gravel.

Pavement Materials The materials for a flexible pavement consist of aggregates plus one or more types of bituminous material.[2]

Bituminous Materials. The principal bituminous binder material is petroleum asphalt, although a minor amount of flexible pavement construction utilizes the road tars, usually in the subbase and base courses. Asphalt is used in three different forms: (1) asphalt cements of the several penetration grades, which require heating to relatively high temperatures to achieve the fluidity required for application and/or mixing; (2) cutback asphalts, which are diluted with naphtha or kerosene to render them more fluid for handling (usually, heating to a relatively low temperature is also required to achieve the proper viscosity for their application); and (3) emulsions of paving-grade asphalts and water with emulsifying and stabilizing agents. Emulsified asphalts usually require little or no elevation of temperature for application and/or mixing.

Table 25-1 indicates the uses of the various asphalt materials in accordance with Asphalt Institute suggested criteria.[2]

Aggregates. For bituminous pavement construction, a great variety of aggregate types and gradations are utilized.* Important characteristics of aggregates for flexible pavement uses are resistance to wear, polishing, and weathering (soundness). Table 25-2 shows recommended aggregate gradations for various bituminous mixtures.[2] It can be seen from this that a wide range of aggregate gradations may be used, varying from coarse, open-graded mixes to dense, fine-graded mixes.

In the case of asphalt concrete, the aggregates are heated and dried before mixing with the heated asphalt cement. For mixtures involving use of the cutback asphalts, the aggregates may or may not be heated and dried. For hot-mix cutback asphalt mixtures, the temperature of the aggregates is substantially lower than for asphalt concrete. The same is true for emulsified asphalt-aggregate mixtures.

Pavement Types Major types of bituminous pavements will be identified briefly in the following paragraphs. Discussions of the materials, construction methods, and equipment used in each of these types of pavements will be found in subsequent portions of this subsection.

Bituminous concrete is a composition of asphalt cement and high-quality aggregates, carefully controlled for asphalt content, aggregate gradation, and void content. These parameters are carefully controlled to produce a pavement mixture which meets the specified criteria for stability, cohesion, and impermeability.

Bituminous plant mix is generally similar to asphalt concrete except that the levels of control for production and placement, quality of materials, and design criteria are lower than for bituminous concrete. Methods of production and placement are similar to those for bituminous concrete.

Bituminous penetration macadam is constructed in multiple lifts of coarse, open-graded crushed stone, gravel, or slag. Each lift receives an application of asphalt, after which a succeeding, smaller-sized aggregate is used to produce a well-keyed, stable layer. If the base lift of the macadam pavement is coated with bituminous material in a central mixing plant prior to placement, the pavement is called a "bituminous macadam" type.

Bituminous road mix, as the name implies, is constructed by mixing aggregates and bituminous materials on the roadway, usually from a windrow. It is generally considered to be the lowest-quality type of mixed paving material because of the low level

* See Sec. 20, Aggregate Production.

TABLE 25-2 Compilation of Suggested Mix Compositions*

Mix type	2½ in.	1½ in.	1 in.	¾ in.	½ in.	⅜ in.	No. 4	No. 8	No. 16	No. 30	No. 50	No. 100	No. 200	Percent asphalt
Mix seal														
II a	....	....	....	....	....	100	40–85	5–20	......	......	......	......	0–4	4.0–5.0
II b	....	....	....	....	100	70–100	20–40	5–20	......	......	......	......	0–4	4.0–5.0
Surface														
II b	....	....	....	....	100	70–100	20–40	5–20	......	......	......	......	0–4	4.0–5.0
II c	....	....	....	100	70–100	45–75	20–40	5–20	......	......	......	......	0–4	3.0–6.0
III a	....	....	....	....	100	75–100	35–55	20–35	......	10–22	6–16	4–12	2–8	3.0–6.0
III b	....	....	....	100	75–100	60–85	35–55	20–35	......	10–22	6–16	4–12	2–8	3.0–6.0
IV a	....	....	....	....	100	80–100	55–75	35–50	......	18–29	13–23	8–16	4–10	3.5–7.0
IV b	....	....	....	100	80–100	70–90	50–70	35–50	......	18–29	13–23	8–16	4–10	3.5–7.0
IV c	....	....	100	80–100	....	60–80	48–65	35–50	......	19–30	13–23	7–15	0–8	3.5–7.0
V a	....	....	....	....	100	85–100	65–80	50–65	37–52	25–40	18–30	10–20	3–10	4.0–7.5
V b*	....	....	....	100	85–100	....	65–80	50–65	37–52	25–40	18–30	10–20	3–10	4.0–7.5
VI a	....	....	....	....	100	85–100	....	65–78	50–70	35–60	25–48	15–30	6–12	4.5–8.5
VI b*	....	....	....	100	....	85–100	....	65–80	47–68	30–55	20–40	10–25	3–8	4.5–8.5
VII a*	....	....	....	....	....	100	85–100	80–95	70–89	55–80	30–60	10–35	4–14	6.0–11.0
VIIIa	....	....	....	....	....	....	100	95–100	85–98	70–95	40–75	20–40	8–16	6.5–12.0
Binder														
II c	....	....	....	100	70–100	45–75	20–40	5–20	......	......	......	......	0–4	3.0–6.0
II d	....	....	100	70–100	....	35–60	15–35	5–20	......	......	......	......	0–4	3.0–6.0
III b	....	....	....	100	75–100	60–85	35–55	20–35	......	10–22	6–16	4–12	2–8	3.0–6.0
III c	....	....	....	100	75–100	60–85	30–50	20–35	......	5–20	3–12	2–8	0–4	3.0–6.0
III d	....	....	100	75–100	....	45–70	30–50	20–35	......	5–20	3–12	2–8	0–4	3.0–6.0
IV c	....	....	100	80–100	....	60–80	48–65	35–50	......	19–30	13–23	7–15	0–8	3.5–7.0
Leveling														
III b	....	....	....	100	75–100	60–85	35–55	20–35	......	10–22	6–16	4–12	2–8	3.0–6.0
V b*	....	....	....	100	85–100	....	65–80	50–65	37–52	25–40	18–30	10–20	3–10	4.0–7.5
VI b*	....	....	....	100	....	85–100	....	65–80	47–68	30–55	20–40	10–25	3–8	4.5–8.5
Base														
I a	100	35–70	....	0–15	....	....	....	0–5	......	......	......	......	0–3	3.0–4.5
II d	....	....	100	70–100	....	35–60	15–35	5–20	......	......	......	......	0–4	3.0–6.0
II e	....	100	70–100	50–80	....	25–60	10–30	5–20	......	......	......	......	0–4	3.0–6.0
III d	....	....	100	75–100	....	45–70	30–50	20–35	......	5–20	3–12	2–8	0–4	3.0–6.0
III e	....	100	75–100	60–85	....	40–65	30–50	20–35	......	5–20	3–12	2–8	0–4	3.0–6.0
IV d	....	100	80–100	70–90	....	55–75	45–62	35–50	......	19–30	13–23	7–15	0–8	3.5–7.0
Skip gradations which have been successfully used														
A	....	....	....	100	95–100	....	50–70	30–50	......	......	5–25	......	2–10	4.0–9.5
B	....	100	95–100	....	60–80	....	30–50	20–40	......	......	5–25	......	1–10	4.0–9.5
C	100	95–100	....	60–80	....	....	25–45	15–35	......	......	3–20	......	0–5	4.0–9.5

* May be used for base where coarse aggregate is not economically available.
SOURCE: The Asphalt Institute.

of control of aggregates and bitumen proportioning. This type of pavement is generally regarded as an intermediate-type bituminous pavement.

Surface treatments and seal coats are characterized by an application of liquid asphalt or emulsified asphalt, which then may or may not be covered with a spread of sand, screened gravel, or crushed stone. The term includes such applications as prime coat, tack coat, single or multiple surface treatments of a base course, single seal-coat treatments of an existing pavement surface, fog seal coats, plant-mix seal coats, and special applications for sealing and/or rejuvenating an existing bituminous surface. Single seal treatments applied for the purpose of restoring the integrity and/or skid resistance of an existing surface are sometimes called "armor" coats.

CONSTRUCTION OF SURFACE TREATMENTS, BITUMINOUS PENETRATION MACADAM, AND ROAD-MIXED PAVEMENTS

A discussion of surface treatments, seal coats, macadam-type construction, and road-mix methods is appropriate at this point.[3] The base must be protected from weathering and the effects of traffic. It is entirely possible that engineers may prefer to allow traffic on the completed base for a period of months in order to prove the construction before proceeding with any of the high-type bituminous surfacing methods. Such proving would allow correction for settling and would contribute to a better-riding road.

The construction procedures which will be discussed herein may result in the final pavement surface. They may also comprise one or more of the earlier steps in "stage" construction. In the latter case, with the exception of some types of seal coats and surface treatments, the paving will seldom be the final riding surface for high-type highway construction.

Some types of liquid asphalts which are used in regular seal-coat work have also been employed as corrective treatments on old paving surfaces. These light application treatments are often regarded as regular maintenance operations. There can be several purposes for *seal coating* using bituminous materials, the most important ones being as follows:

To seal the pavement against oxidation and the entrance of moisture
To rejuvenate or enliven a dry, hardened, and weathered surface
To reinforce or build up the pavement and at the same time stop damaging erosion and wear

Light liquid asphalt applications may also be employed as nonskid treatments for a slippery surface, to improve reflective properties, and to mark special traffic lanes. The amount of liquid material applied usually determines the use of a light application of sand, commonly referred to as a "blotter." In fact, this type of maintenance work is sometimes referred to by road engineers as "blotter work." The technique may be used as spot spray patching, depending upon necessity, or it may be used for a complete coverage.

Tack Coat A tack coat is a thin application of bituminous material to a nonabsorptive old road surface—usually asphalt, brick, stone block, or concrete—to provide a bond between the old pavement and the new surface.

In applying the tack coat, the surface is first swept or flushed to free it from dust and foreign material. Cutbacks, tars, or emulsions are then applied with a distributor in the thinnest application that will give complete coverage, seldom over 0.1 gal per sq yd.

Low viscosity or light bitumens should be used for thin coverage, but the residual must produce a hard, tacky surface. If a tack coat is not cured to a tacky condition, it acts as a lubricant and defeats its purpose.

The following bituminous materials are usually used:

RC- 70	RS-1	RT-7
RC-250	RS-2	RT-8

Prime Coat A prime coat is a light bituminous material applied to a stone, gravel, stabilized, or similar absorptive base that is to be given a bituminous surface. The

Fig. 25-1 Base priming with bituminous distributor. *(Littleford Bros.)*

purpose of the prime coat is to act as a bonding agent as well as to seal the joint between the base and the new pavement. The capillary rise of moisture into the surface and surface moisture working through to the base are thus retarded.

Sweeping does not remove all the loose dust and, without a prime coat to provide adhesion, the dust acts as a lubricant between the base and surface. Prime-coat materials are more fluid than tack-coat materials, so that they penetrate into the base a slight amount to help in holding down the dust. The heat of hot-laid bituminous surfaces or the cutback materials in cold-laid surfaces will soften the hardened prime and give the desired adhesion.

With thicker surface courses, priming is not so necessary because of the inherent stability present in any thick mat. However, during the construction of a surface of any thickness, the presence of a prime coat materially aids in waterproofing the base, so that, in the event of rain, it dries off more quickly, and holds lost time to a minimum.

The primer is applied with a distributor at an approximate rate of 0.25 to 0.50 gal per sq yd (see Fig. 25-1). The quantity applied should be absorbed in 24 hr, and the normal drying or curing period is approximately 48 hr. It is advisable to underprime rather than overprime, as any unabsorbed material is often taken up by the new pavement and surface bleeding may result.

As with tack coats, traffic is excluded from the primed surface to prevent loss of any of the primer and to prevent dust from collecting. When traffic cannot be excluded, medium-fine sand is spread over the surface to protect it by blotting up any excess bituminous material. Before the bituminous surface is placed, any loose sand remaining is swept off the base. Figure 25-2 illustrates the sequence of activities.

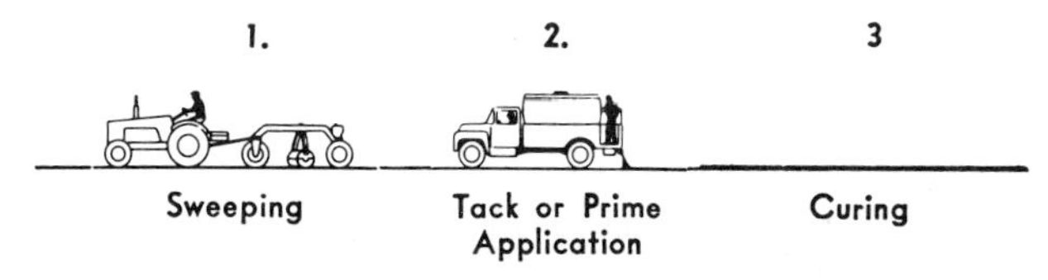

Fig. 25-2 Sequence of operations for priming or tack coating.

Bituminous materials used for priming are listed below. The more viscous materials are used with the coarser bases.

RT-1	SC- 70	MC- 70
RT-2	SC-250	MC-250
RT-3		MC-800

Rapid-curing cutbacks are not desirable because the distillate has a tendency to separate from the asphalt cement and penetrate into the road base, leaving an excess of asphalt cement on the surface.

Fog Seal Coat The preceding terms, "tack" and "prime" coat, are normally associated with resurfacing or new paving projects. A fog seal coat can be considered as a remedial or maintenance treatment for deteriorating surfaces,. sealing and rejuvenating the mat for continued service. It is also used as a sealing coat in some forms of hot plant-mix work or as an intermediate treatment in new construction. A slow-breaking emulsion is diluted with water in ratios of 1:1 to 1:3 and applied at the rate of 0.1 to 0.2 gal per sq yd. Usually no cover aggregate is required.

Slurry Seal Coat When the surface to be treated is unusually rough, weathered, and checked, showing cracks sometimes up to ⅛ in. wide, applications of liquid treatments are inadequate even when used in conjunction with blotter materials. A *slurry seal treatment*, consisting of an emulsified asphalt combined with a mixture of fine

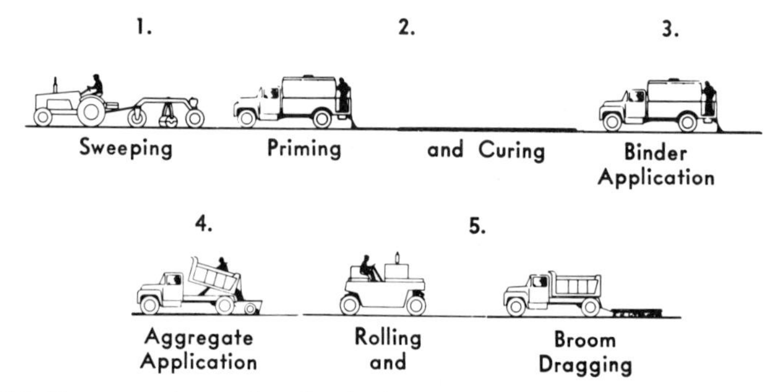

Fig. 25-3 Sequence of operations for single-pass surface treatment road construction.

aggregates, provides a much thicker crack-filling material and is applied with a mechanical squeegee. The use of a slow-setting asphalt emulsion eliminates the necessity for heating the material prior to its application. It also provides a workability period prior to breaking of the emulsion and consequent balling of the mixture, which would prevent adequate squeegee application. The mix should be relatively free flowing and of a creamy consistency, permitting it to be squeegeed into all the cracks so as to promote a smooth surface. Such a treatment will contribute little or no structural strength but is designed to reduce maintenance and patching work and retard further deterioration of the mat. Slurry treatment work often precedes a complete coverage by regular seal-coating methods. The slurry-seal method can be used for special deslicking coatings designed for nonskid performance. In such cases, special types of fines aggregates providing hard, sharp edges are employed.

Single-pass Seal Coats Single-pass seal coats are thin, bituminous-aggregate courses applied to existing bases or surfaces of any type. The sequence of operations, which is illustrated in Fig. 25-3, is essentially as follows:

1. Sweeping: To remove dirt and other foreign matter.
2. Priming and curing: Optional, depending upon condition of base.
3. Binder application: Bituminous material is applied by distributor. See Table 25-3.
4. Aggregate application: Applied by mechanical spreaders towed by dump trucks, by spreaders attached to tail gate, or by self-propelled spreaders with receiving hoppers. Aggregate occasionally is cast by hand from a moving truck or from a windrow.

5. Rolling and broom dragging: Alternate use of roller (pneumatic or steel wheel) and broom drag to smooth and compact the surface.

Multiple-pass Surface Treatments This method of road construction is very similar to the single-pass method except that there are two or more applications of aggregate and bitumen to give a surface having a more substantial load-carrying ability. Larger aggregate particles in the first course are held in place by successive layers of smaller

TABLE 25-3 Quantities of Asphalt and Aggregate for Single Surface Treatments and Seal Coats*

Size of aggre-gate	Pounds of aggre-gate per sq yd†,§	Gallons of asphalt per sq yd†,¶	Hot weather		Cool weather	
			Hard aggregate	Absorbent aggregate	Hard aggregate	Absorbent aggregate
¾–⅜	40–55	0.28–0.35	120–150 RC3000, RS2	RC3000, RS2	RC800, RS2‡	RC800, RS2‡
¾–No. 8	30–45	0.23–0.30	200–300 RC800, RS2	RC800, RS2	RC800 RS2‡	RC800 RS2‡
½–No. 4	25–35	0.20–0.25	200–300 RC250,800 RS1, 2	RC250,800 RS1, 2	RC250,800 RS1‡ RS2‡	RC250,800 RS1‡ RS2‡
½–No. 8	25–35	0.20–0.25	RC250,800 RS1, 2	RC250,800 RS1, 2	RC250,800 RS1‡ RS2‡	RC250,800 RS1‡ RS2‡
⅜–No. 4	20–25	0.20–0.25	RC250,800 RS1, 2	RC250,800 RS1, 2	RC250,800 RS1‡ RS2‡	RC250,800 RS1‡ RS2‡
⅜–No. 8	20–25	0.20–0.25	RC250,800 RS1, 2	RC250,800 RS1, 2	RC250,800 RS1‡ RS2‡	RC250,800 RS1‡ RS2‡
¼–No. 8	15–20	0.15–0.20	RC250,800 RS1, 2	RC250,800 RS1, 2	RC250,800 RS1‡ RS2‡	RC250,800 RS1‡ RS2‡
Sand	10–15	0.10–0.15	RC250,800 RS1, 2	RC250,800 RS1, 2	RC250,800 RS1‡	RC250,800 RS1‡ SS1‡

* These quantities and types of materials may be varied according to local conditions and experience.

† The lower application rates of asphalt shown in the table should be used for aggregate having gradings on the fine side of the limits specified. The higher application rates should be used for aggregate having gradings on the coarse side of the limits specified.

‡ Caution should be exercised when using this material under poor drying conditions.

§ The weight of aggregate shown in the table is based on aggregate with a specific gravity of 2.65. In case the specific gravity of the aggregate used is less than 2.55 or more than 2.75 the amount shown in the table above should be multiplied by the ratio which the bulk specific gravity of the aggregate used bears to 2.65.

¶ Under certain conditions, the heavier grades of MC liquid asphalts may be used in cool weather.

aggregate particles. This method is sometimes designated as "inverted penetration" because the sequence of material application is opposite to that used in penetration macadam construction.

The required steps or operations, as illustrated in Fig. 25-4, are as follows:

1. Sweeping: To remove dirt and other foreign matter.
2. Priming and curing: Optional, depending upon condition of base.
3. First binder application: Bituminous material is applied by distributor.
4. First aggregate application: Large-size aggregate is applied by mechanical spreaders towed by dump trucks, by spreaders attached to tail gate, or by self-propelled spreaders.

5. First rolling and broom dragging: These two steps are alternated to seat aggregate.

6. Second binder application: Same type of bituminous material, again applied by distributor.

7. Second aggregate application: Smaller size and quantity of aggregate is spread evenly to fill surface voids in first course.

8. Final rolling and broom dragging: Alternate use of roller (pneumatic or steel wheel) and broom drag to smooth and compact surface. Broom dragging fills voids for uniform texture.

The number of applications is governed by the desired thickness; there could be as many as three or four.

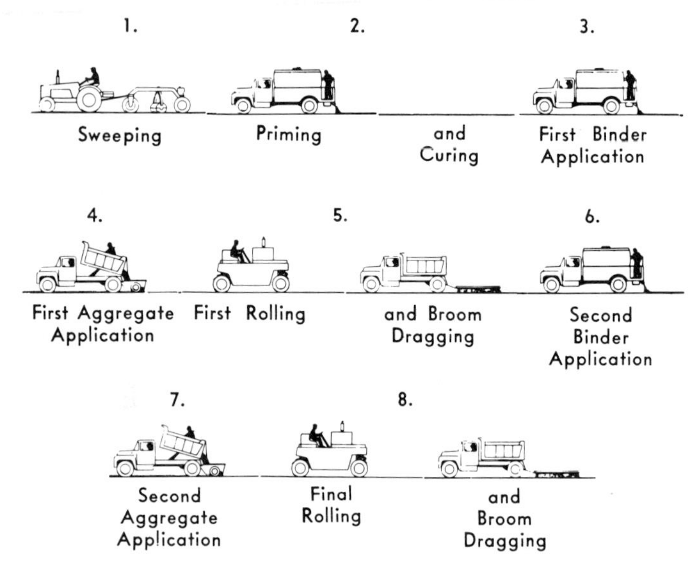

Fig. 25-4 Sequence of operation for inverted penetration or multiple-pass surface treatment road construction.

Plant-mixed Seal Coat An open-graded seal coat which is premixed in a regular mechanical plant and laid with a finishing machine to an approximate depth of ½ in. is frequently used in lieu of the seal coats just described. By mixing in a hot-mix asphalt plant, regular 85–100 or 120–150 penetration asphalts can be used. A tack coat is desirable on a dry, hardened mat.

Penetration Macadam The penetration-macadam method of road construction is older than any other bituminous construction method. The name stems from the use of macadam-type aggregate and the penetration method of applying the binder to the aggregate.

Inverted penetration mats are seldom thicker than 1½ in., but penetration macadam can be 4 in. or more thick. For this reason, and because of the nature of their construction and the size of aggregate used, penetration-macadam roads will generally handle heavier traffic loads and may be base courses rather than surface courses. See Fig. 25-5.

The sequence of operations, as shown in Fig. 25-6, is essentially as follows:

1. Sweeping: To remove dirt and other foreign matter.

2. First aggregate application: Largest-size aggregate applied, usually with spreader box attached to dump truck or with bituminous finisher.

3. First rolling: Compacts coarse aggregate and locks it into place. (Pneumatic or steel wheel roller.)

4. First binder application: Bituminous material is applied by distributor.

5. Second aggregate application: Smaller size and quantity of aggregate, usually called "key" aggregate, is spread evenly to fill voids in first course. Broom drags may be used to aid in distributing keystone.

6. Second rolling: Starts immediately following keystone application, while binder is still warm, to bond and compact aggregate.

7. Second binder application: Same type of bituminous material is applied in smaller quantity.

8. Third aggregate application: A still smaller size and quantity of chips or key aggregate may be immediately applied. Acts as blotter treatment and gives nonskid surface that resists traffic abrasion.

9. Final rolling and broom dragging: Alternate use of roller (pneumatic or steel wheel) and broom drag smooths and compacts surface. Broom dragging fills voids for uniform texture.

The three passes which have been outlined are the usual number. However, many roads are constructed with four, and some with only two.

Fig. 25-5 Finisher equipped with automatic screed controls placing plant-mixed macadam base course.

Road Mix This is a common method of constructing a bituminous road. As the name implies, the aggregate and binder are mixed on the road. The sun and wind remove the moisture from the aggregate.

The three methods of blending the aggregate and binder will be discussed later. The other steps or operations, as shown in Fig. 25-7, are essentially as follows:

1. Sweeping: To remove dirt and other foreign matter.

2. Priming and curing: Optional, depending upon condition of base.

3. Aggregate application: New aggregate is dumped into truck-towed windrow shaper or dumped directly onto road for blade shaping. Old aggregate (from existing road), or combination of old and new aggregate, is blade windrowed. An evener may be dragged over windrow to remove any major irregularities.

4. Aerating: If windrowed aggregate is too wet ($1\frac{1}{2}$ to 2 percent surface moisture by weight is usual limit), it must be aerated with blade, harrow, or some type of tiller to expose individual particles to sun and wind. Following drying, aggregate is usually windrowed again. Windrow may then be spread out by blade prior to mixing operation.

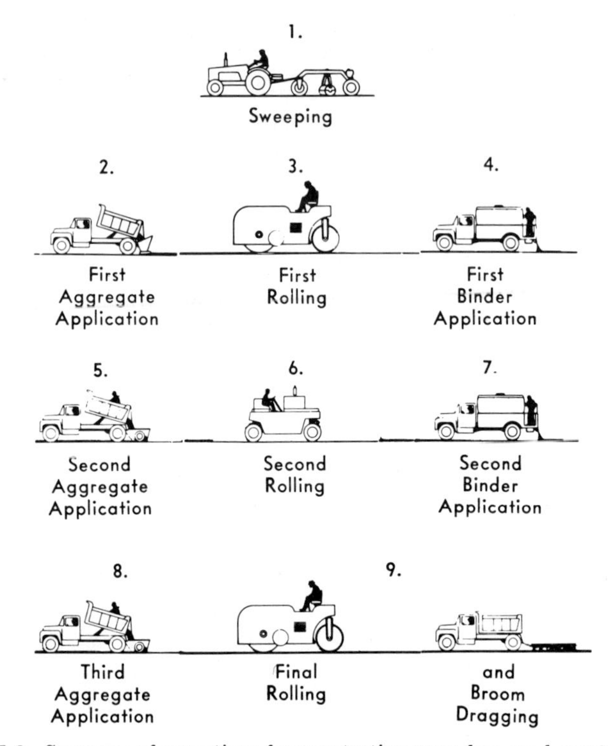

Fig. 25-6 Sequence of operations for penetration macadam road construction.

5. Mixing: The three methods of mixing aggregate and binder are blade, mechanical, and travel-plant mixing.

6. Spreading mix: Blades distribute mix to required loose depth and contour.

7. Rolling: Rolling starts at outside edges and progresses toward center with succeeding passes. Roller may be either smooth steel or pneumatic tired or both, depending upon type of mixture. The period of time between spreading and rolling is a matter of experience and judgment; it varies from job to job, depending upon materials used in mix. On some mixes the roller operates directly behind the blade, while on other mixes it may be necessary to hold the roller back as much as 24 hr. In extreme cases, the roller is kept off the mat for two days.

The three methods of mixing will now be discussed.

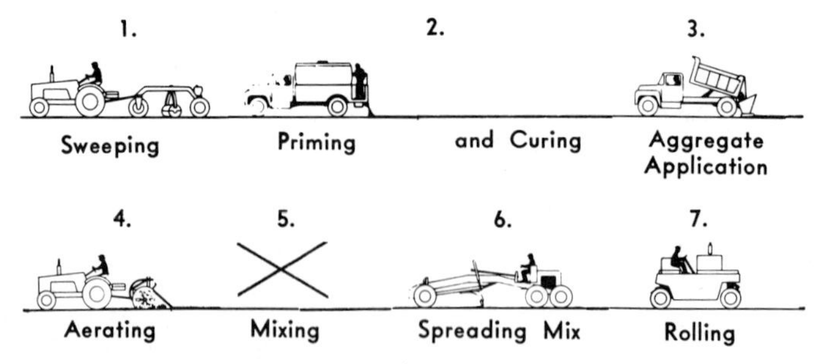

Fig. 25-7 Sequence of operations for road-mix construction.

Blade mixing makes use of blade graders or some type of multiple-blade drag to blend the aggregate and binder together. (See Fig. 25-8.) After the aggregate has been dried and spread uniformly over one lane of the road, the following steps will complete the mixing:

1. Partial application of bitumen: Ordinarily applied in three separate applications by distributor. Alternates with step No. 2.

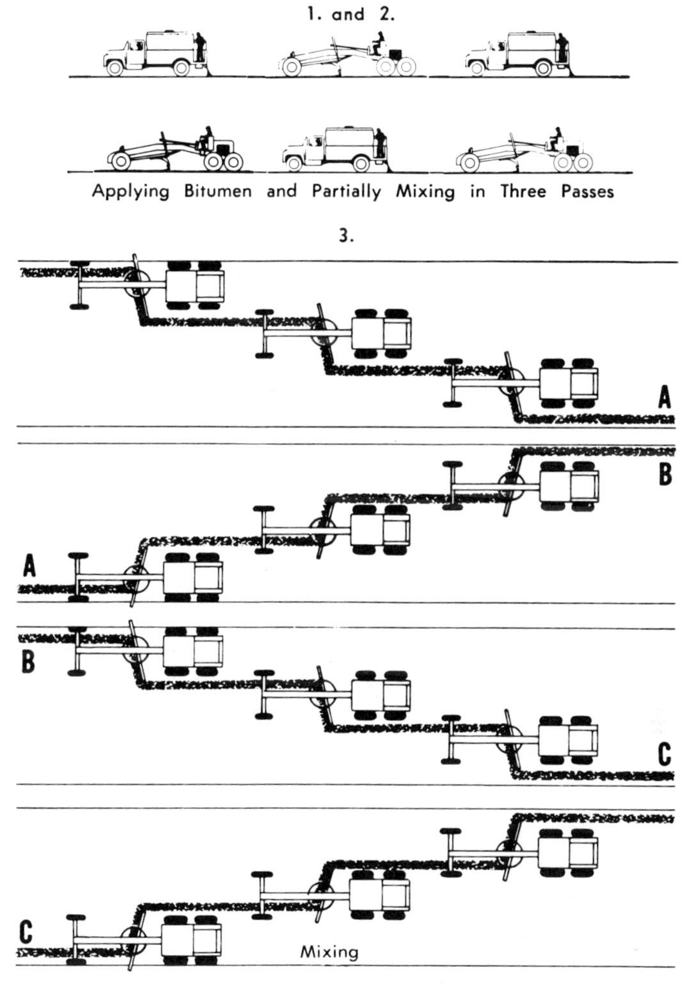

Fig. 25-8 Blade mixing.

2. Partial mixing: Motor grader partially mixes bitumen and aggregate, leaving as little free bitumen as possible on the surface and preventing pools of it from collecting.
3. Mixing: Partially mixed material is formed into a windrow and bladed back and forth across the road until aggregate particles are coated and there are as few rich and lean spots as possible. To reduce the required number of passes, multiple-blade drags are sometimes used instead of graders on open-graded aggregate road mixes.

Mechanical Mixing. The machines that are used to perform "mechanical mixing" have some type of mechanically operated mixing device but do not pick up the aggregate and accurately proportion it with the binder. One type of mixer has no provision for adding the binder, and so can be likened to blade mixing; a distributor operating independently of the mixer applies the binder. Another type adds the binder at the time of mixing by spraying it into the mixing chamber as the machine moves ahead. Some mixers make only one pass; these usually work with windrowed aggregate and leave the mix as a windrow. Other mixers work without windrows and make several passes.

The actual blending by any of the mechanical mixers involves the agitation of the aggregate and bituminous material by some arrangement of tines or paddles attached to a rotating shaft. This shaft is so mounted that it operates

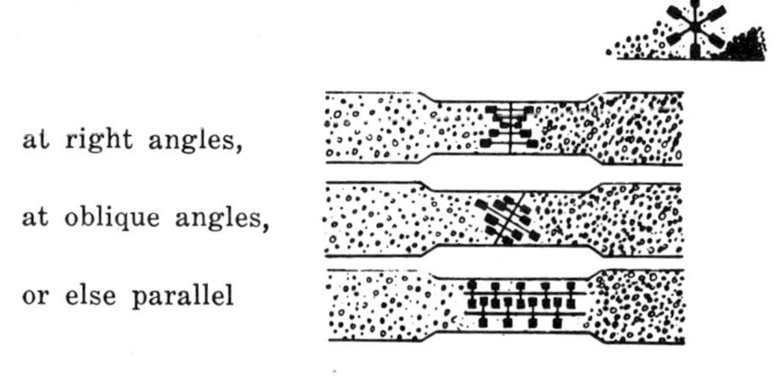

to the direction of travel. Frequently, when the mounting is parallel to the direction of travel, two pug-mill shafts with attached paddles are used as shown in the sketch above.

Travel-plant Mixing. Travel plants are machines which accurately proportion the aggregate and binder—independent of the windrow size, travel speed, or human error. Consequently, travel-plant mixing is a further improvement in road mixing operations.

To achieve accurate proportioning, the windrowed aggregate is elevated by a bucket loader to a hopper. An apron feeder then conveys the aggregate out through a calibrated gate to a pug mill. The aggregate feeder is mechanically interlocked with a bitumen pump, and as the measured quantity of aggregate enters the pug mill it is sprayed and initially coated with the metered binder. After the aggregate and binder have been thoroughly combined in the pug mill, the mix is discharged to a windrow at the rear.

PRODUCTION OF HOT-MIXED BITUMINOUS CONCRETE

Asphalt Plants[4] Figure 25-9 illustrates a modern asphalt mixing plant. The flow of material through all asphalt plants is similar up to the point of measuring and mixing (see Fig. 25-10). Aggregate is fed from storage to the dryer, which removes the moisture and heats the aggregate.

The heated, dried aggregate is continuously fed to vibrating screens which divide it into the specified separations. Each size then flows into its respective hot bin. The next steps are measuring and mixing, and the two different systems used for this phase are referred to as "batch" and "continuous" types.

Aggregate Feeding The first step in achieving maximum plant output is a properly designed cold-feed system. Inadequate feeding methods result in costly delays. Hot-bin starvation, whether caused by bridging of sand or irregular feeding rate of raw aggregate, stops mix production (see Figs. 25-11 and 25-12).

Drying costs, the most expensive item in mix production, are reduced by maintaining a balanced flow of aggregate through the dryer. Only those aggregate sizes which are needed to meet mix requirements should be dried. Every effort should be made to eliminate costly, wasteful drying of surplus materials that will eventually overflow the hot bins and create rehandling costs as well as littering the plant site.

Any blending of minus No. 8 material necessary to achieve the required density and stability in the mix must be done at the cold feed. All sand, fine and coarse, passes through the bottom-deck screen into the No. 1 hot bin. Therefore proportioning of the various sand fractions must be done at the cold feed.

The increasing use of total and proportional controls, many of which are operated from remote panels, has lowered the cost of maintaining a feed which is properly matched to the needs of the final mix. The following points should be considered in selecting a feed system (see Fig. 25-13).

Fig. 25-9 A modern asphalt mixing plant.

Tons per Hour Required. The maximum operating capacity of the plant, as well as the relative proportions of each aggregate size, will set the capacity requirements for the cold-feed systems.

Number of Separations of Materials. Mix design and gradation of available aggregates will determine the number of materials to be handled. This will dictate the number of stockpiles or bins required as well as the method of reclaiming.

Types of Materials to Be Handled. Consideration should be given to the characteristics of the aggregates handled. For example, extremely abrasive aggregates, such as trap rock, or sticky materials that may tend to bridge over bin openings or adhere to the feeder will affect the selection of system components.

Aggregate Source. The source of aggregate plus the frequency of delivery will also affect the choice of system components. Frequent delivery will reduce the amount of

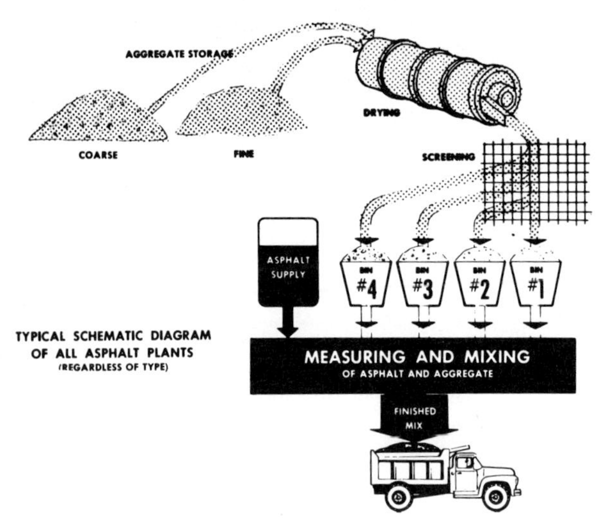

Fig. 25-10 Typical schematic diagram of all asphalt plants.

live storage required at the plant site; less frequent delivery will mean providing sufficient storage and the necessary equipment to stockpile and reclaim this material.

Flexibility Required. The necessary degree of mix flexibility should be considered in the planning. A "drive-in" operation where a variety of mixes will be needed over a short period of time requires a feed system that will permit quick and economical switching of mixes.

Portability. Whether the plant operation is to be portable, semiportable, or stationary will help determine its configuration.

Fig. 25-11 The cold-feed system shown has clamshell-charged field hoppers spaced over a gathering conveyor that feeds the plant's dryer.

Aggregate Drying The most frequently encountered bottleneck in asphalt-plant operation is the dryer. The optimum dryer is the one that will meet the desired performance level at the lowest investment and operating cost. However, the optimum design changes in different geographic areas and under different operating conditions. There are a large number of elements that affect drying, and changing one variable usually causes other factors to vary. The key to optimum dryer performance is proper balance of all these factors to satisfy a particular operating objective.

With the same drum gas velocity* and the same length of dryer, the productive capacity varies in direct proportion to the increase in drum cross-sectional area. This area, of course, is in proportion to the square of the diameter (Fig. 25-14).

Fig. 25-12 A tunnel conveyor system for reclaiming aggregates from stockpiles offers lowest handling costs, gives quick mix-change flexibility, and is ideally suited for total and proportional control feeding from a remote station.

With the same drum gas velocity and the same drum diameter, capacity increases with more length, but not at a proportional rate. For instance, Fig. 25-15 shows that a 50 percent increase in the length of a 20-ft dryer results in an increase of only 20.5 percent in productive capacity.

For a specific size of drum, greater productive capacity can be obtained by increasing the airflow and burner system, although each successive increase in burner and air is of reduced effectiveness. The result, of course, is a consequent increase in the drum gas velocity in that specific drum. Figure 25-16 shows a 29.1 percent capacity increase due to a 50 percent increase in drum gas velocity.

There are practical limits to such an increase. The dust carry-out under higher gas velocities may be one such limitation, depending on the operating conditions of the job (see Fig. 25-17). Combustion-chamber and drum-shell temperatures are not critical, because of the cooling effect of the high-velocity air which passes over them. However, the equipment could be severely damaged in the event of an emergency such as a stoppage of the cold feed.

* Cubic feet of exhaust air is common terminology, but its effect on overall performance is more easily related to the velocity of the gas as it passes through the drum. The drum gas velocity index (DGV) is obtained by dividing the total discharge of exhaust (cfm) by the cross-sectional area of the drum (sq ft).

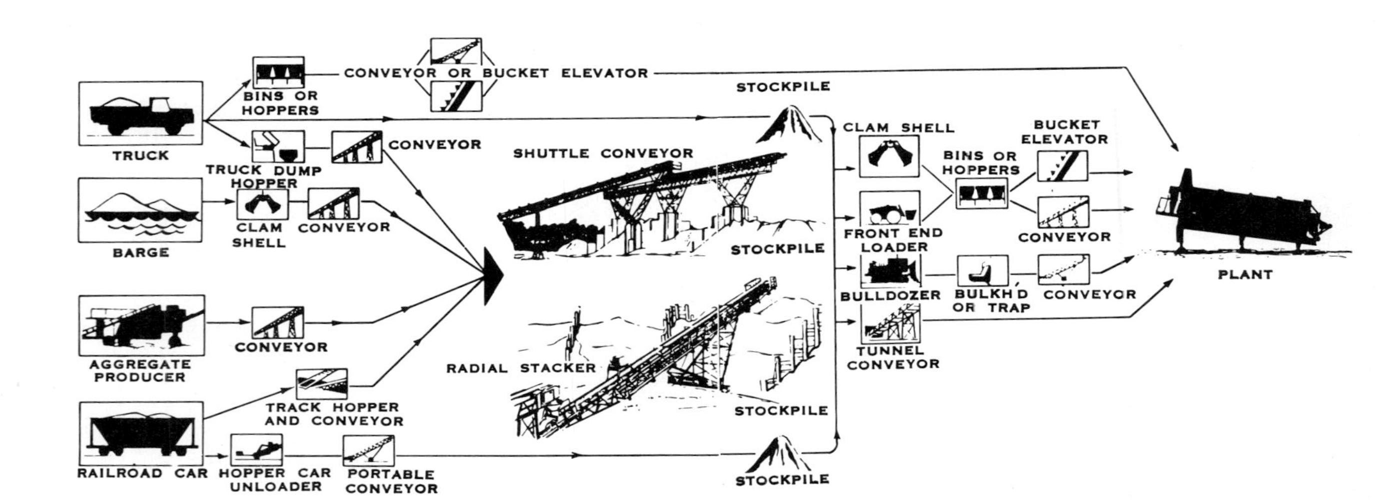

Fig. 25-13 Flow diagram of most of the basic elements employed in typical aggregate feed systems.

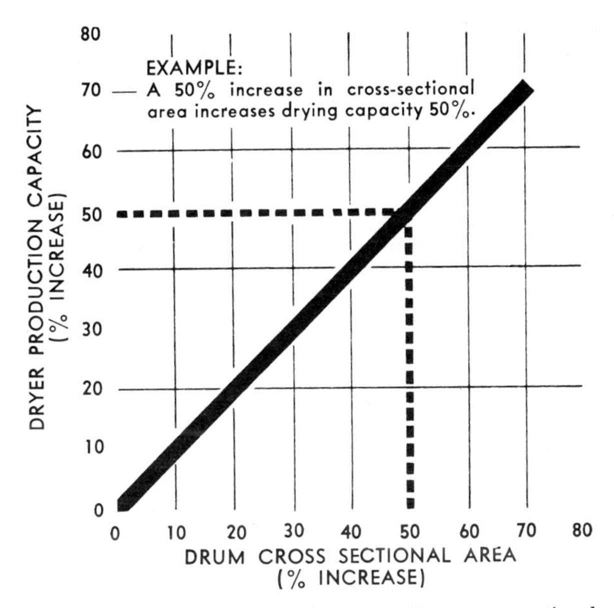

Fig. 25-14 Dryer production capacity versus drum cross-sectional area.

The percentage of moisture to be removed has a major effect on dryer capacity, as indicated in Fig. 25-18. Here dryer capacity is correlated with the moisture to be removed, rather than with total moisture in the aggregate, since the residual amount for different mixes can be variable. To achieve the desired retention time, which can be a key factor in internal moisture removal, loading of the drum can be controlled through flight design, rpm of drum, slope of drum, etc. Because any

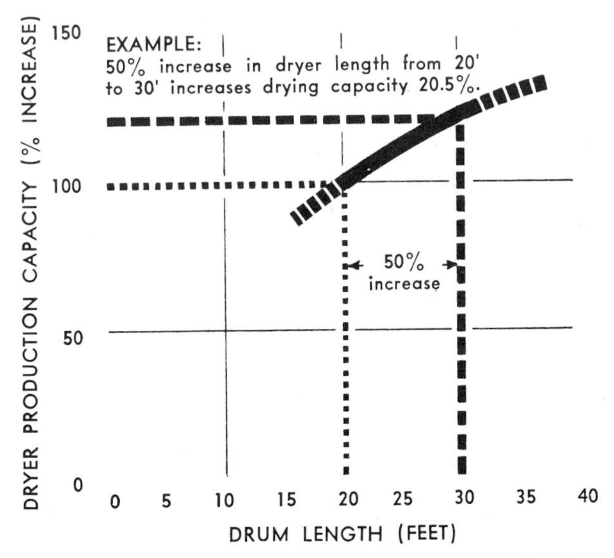

Fig. 25-15 Dryer production capacity versus drum length

length and diameter drum can be balanced out to have the same percent loading, retention time becomes directly proportional to the volume of the drum. However, a dryer should not be loaded up with material beyond 16 percent of the drum volume or there is apt to be erratic flow with material surges.

Determination of the retention time which is optimum for internal moisture removal is too complex a subject to be discussed here. We do not yet have yardsticks in aggregate gradation, coefficient of heat, aggregate internal structure, etc., to either predict it or measure it. Also, we cannot evaluate the adverse effects of internal moisture upon pavement durability.

The most effective heat transfer in drying aggregate is through radiation. On the other hand, material spends only 3 to 5 percent of its retention time in actual veil suspension. The rest of the time it is cascading and soaking at the bottom and in the flights of the drum, during which period there is heat transfer from the smaller

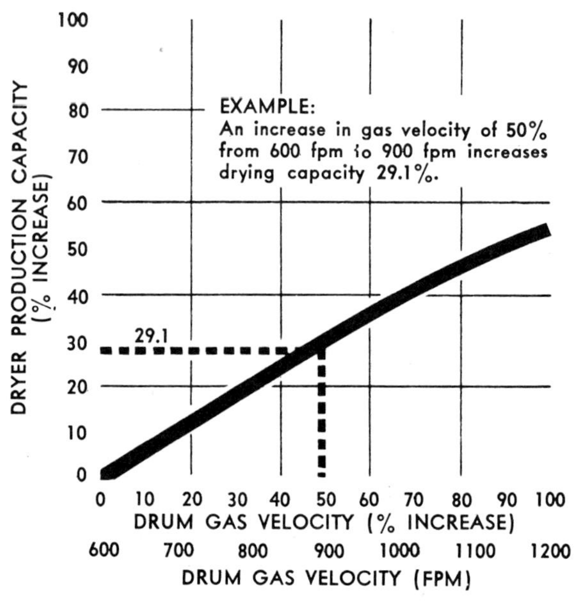

Fig. 25-16 Dryer production capacity versus drum gas velocity.

particles to the larger particles. Two basic types of oil burners are used for asphalt-plant dryers; the low-pressure air-atomizing type with a blower or fan to supply air at 1 to 2 psi to the burner, and the high-pressure steam-atomizing type which requires sizable quantities of dry, hot steam at approximately 125 psi (see Figs. 25-19 and 25-20). The latter type is seldom used on a new installation because of the steam requirement.

A long, narrow flame is desirable to gain maximum radiation effect on the aggregate particles. A short, bushy flame which impinges on the combustion chamber and drum end will usually mean reduced capacity and increased maintenance. Gas burners and combination oil-gas burners are used in many areas where availability and economics permit. The gaseous fuels provide efficient, clean heat, but there should be assurance that the large quantities required for asphalt-plant operations will be available for the length of time needed.

Specifications will normally state the amount of residual moisture permitted in the dried aggregate. They also will specify the temperature of the aggregate for mixing. These limits will vary with the type of mix and asphalt being used. Generally, the moisture requirement will be met* whenever a steady temperature is maintained for the aggregate as it is discharged from the dryer.

Aggregate temperature is measured by either a thermometer or a thermocouple attached to an indicating pyrometer. The latter has a much faster response to changes in temperature and hence is more desirable. Recording devices are available for either type of sensor and can be supplied with extra pens in the event the temperature of a hot bin also must be recorded.

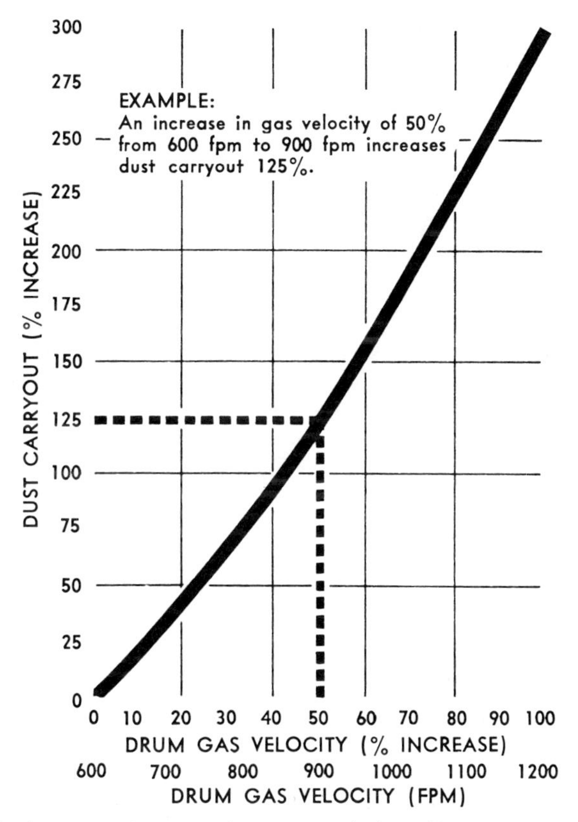

Fig. 25-17 Dust carry-out versus drum gas velocity. Dust carry-out of aggregate increases by the square of the increase in drum gas velocity as illustrated above.

Automatic controls are available to change the rate of fuel to the burner whenever the aggregate discharge temperature varies beyond a preset range. The controls include safeguards that shut off the fuel in the event of flame failure or excessive air temperatures (see Fig. 25-21).

Dust Collection The dust collection equipment must be considered an integral part of the drying system. It plays a major part in providing an even flow of adequate air volume and in collecting and returning a portion of the fine aggregate to the mix. It is important that a proper balance be achieved between the airflow necessary for optimum drying performance and that which will permit proper dust control. The increasing use of aggregates having "borderline" fines content is encouraging greater

* Exceptions to this include (1) some cutback mixes where it is necessary to cool the aggregate after drying, and (2) aggregates causing problems of stripping or foaming because of internal moisture sweating.

concentration on the efficient collection of these valuable fines which otherwise would be lost, requiring expensive replacement with mineral fillers.

The dust collection equipment serves a secondary purpose by reducing nuisance dust in the plant vicinity and minimizing air pollution in the general area.

Dry System. Virtually all primary dust collectors in asphalt plants today are of the cyclone type, in which centrifugal force determines the amount of fines precipitated. Dust-laden air from the dryer has a certain velocity which is induced by the dust-collector fan. When this dust-laden air attempts to follow the sharp curve in the cyclone, centrifugal forces throw the dust out of the airstream and against the wall

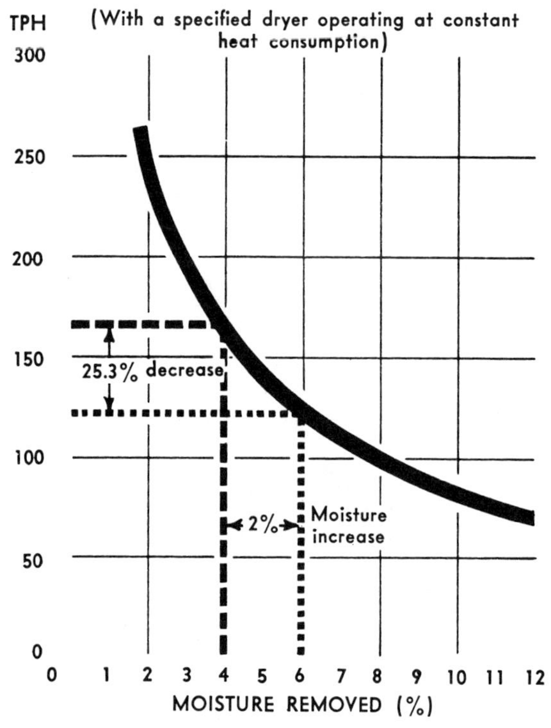

Fig. 25-18 Tons-per-hour capacity at various moisture contents. Moisture content in the incoming aggregate is often guessed, for example, to be between 4 and 6 percent. And yet, there is a 25.3 percent difference in capacity obtainable in this small 2 percent moisture range.

of the cyclone. As the particles of dust strike the wall, their speed is decreased and they gradually slide down the walls of the cyclone to the collecting screw or discharge chute. (See Fig. 25-22.)

The weight of the particle also affects dust collection. Heavy particles, because of their greater momentum, tend to be deposited more rapidly than light particles. Thus, light particles require more time to leave the airstream than is necessary for heavy particles. To give light particles this additional time, the airstream follows a continuously spiraling, downward path rather than taking a straight course through the dust-collector cyclone. The air is finally exhausted through a cylindrical duct located in the center of the cyclone, thus throwing additional dust out of the airstream as it reverses direction.

To prevent clogging in a small cyclone, the dust or fines must be extremely dry. The material for bituminous plant applications is usually dried to under 1 percent moisture content. This is suitable for mixing and application purposes, but not for dust collection in very small cyclones (approximately 6-in. diameter or less). In cyclones of this size, even a small moisture content will cause dust to build up and plug the cyclone. Also, aggregates used for bituminous mixtures often contain twigs, straws, etc., which can quickly cause plugging in a small cyclone. Therefore, although very small cyclones are desirable for dust collection because of their high efficiencies, they are too refined for primary collection in bituminous plant applications. In recent years, however, a trend has developed toward their use as secondary units to reclaim additional fines that otherwise would be lost.

To gain proper airflow capacity to handle larger dryers, more cones are added to the collector instead of increasing diameter. Thus efficiency is maintained.

Fig. 25-19 High-pressure natural gas burner.

Wet System. In metropolitan areas, more stringent pollution codes may force the contractor to reevaluate and improve his secondary dust collection system. A wet collector system is frequently used because of its overall economy and lower first cost.

Industrial wet collectors are not new by any means, but most attempts to modify or adapt them to asphalt-plant work have not been successful. Many of the problems of asphalt-plant operation, such as high volume of airflow, high percentage of fines, high temperatures, etc., are not common to industrial collection.

Most wet collectors work on the principle of wetting the dust particles that remain in the exhaust air after it has been put through the dry cyclone dust collector. The method of wetting the particles and separating the resulting sludge varies appreciably between designs (see Figs. 25-23 and 25-24). There is a very broad range of wet collector efficiencies as a result of these differences.

Fig. 25-20 Low-pressure combination oil and gas burner.

Fig. 25-21 Automatic burner controls are available for aggregate dryers. Maximum and minimum limits are set for the desired temperature of the aggregate. Controls vary burning rate as aggregate temperature approaches high or low limits.

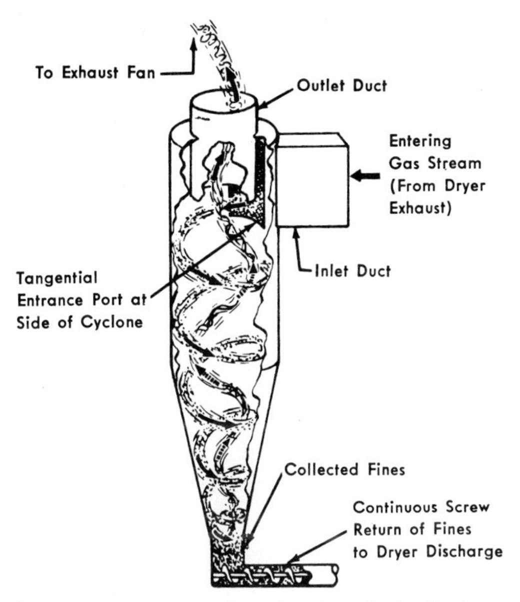

Fig. 25-22 Cross section of cyclone dust collector.

Fig. 25-23 Asphalt plant wet collector, single-nozzle orifice type. This provides high collection efficiency with less than 1 percent makeup water.

Screening* Aggregate gradation variations in an asphalt mix may be traceable to:
1. Stockpile variations
2. Method of reclaiming from stockpile
3. Handling of the material between the cold feeder and the mixer

In high-type mixes, the screening unit is used to reduce the variation and to signal when correction in the aggregate feeding is required. For this reason, a good cold-feed unit is still the most important single contributor to a well-proportioned mix.

Screening Efficiency. While the dryer is the primary potential bottleneck in an asphalt plant, a second major limiting factor of plant capacity is the screen. A screen can pass material at only a limited rate and, when loaded beyond its capacity, material which would normally pass through the screen into the desired hot bin will overrun into the next bin. To control this, standard specifications have established controls on the permissible amount of overrun.

Some specifications simply require a minimum screen efficiency, while others specify a limit on variation within a given bin. One state controls this by specifying

Fig. 25-24 Spray-type wet collector. (*Pioneer Division, Portec Inc.*)

that the bin containing the fine material shall not contain more than 10 percent of material retained on the No. 8 sieve or that sieve size used on the bottom deck. Conversely, the material in all other bins shall not contain more than 10 percent of material passing a No. 8 sieve.

Overrun is difficult to control, since it will vary from test to test under supposedly identical conditions. This, of course, is due to minor changes in material gradation and type. The best test of the aggregate is a composite test of all bins. In this way, one can determine whether, despite the overrun, the aggregate is meeting specifications.

Selecting Screen Sizes. The hot-bin splits can be determined by a plot of the total aggregate specification. The split should be made to obtain approximately equal percentages in each bin. For the three-bin split in Fig. 25-25, the smallest practical screen size is selected (No. 10), and the remaining 64 percent is split 33 to 31 by a 9⁄16-in. screen. For a two-bin split in the same example, a 5⁄16-in. screen is used.

To separate aggregate into specified sieve sizes used in testing, screen cloths having slightly larger openings are used on the asphalt-plant vibrating screen. (See Table 25-4.)

Screen Deck Arrangements. In Fig. 25-26, a 3⅓-deck screen is shown. A full deck screen area is provided for the two smaller screen cloths, but the top deck is split for scalping and No. 3 bin aggregate openings. In the example, the 9⁄16-in. area is greater than the 1⅛-in. area to further balance screening capacity.

* See Sec. 20, Aggregate Production.

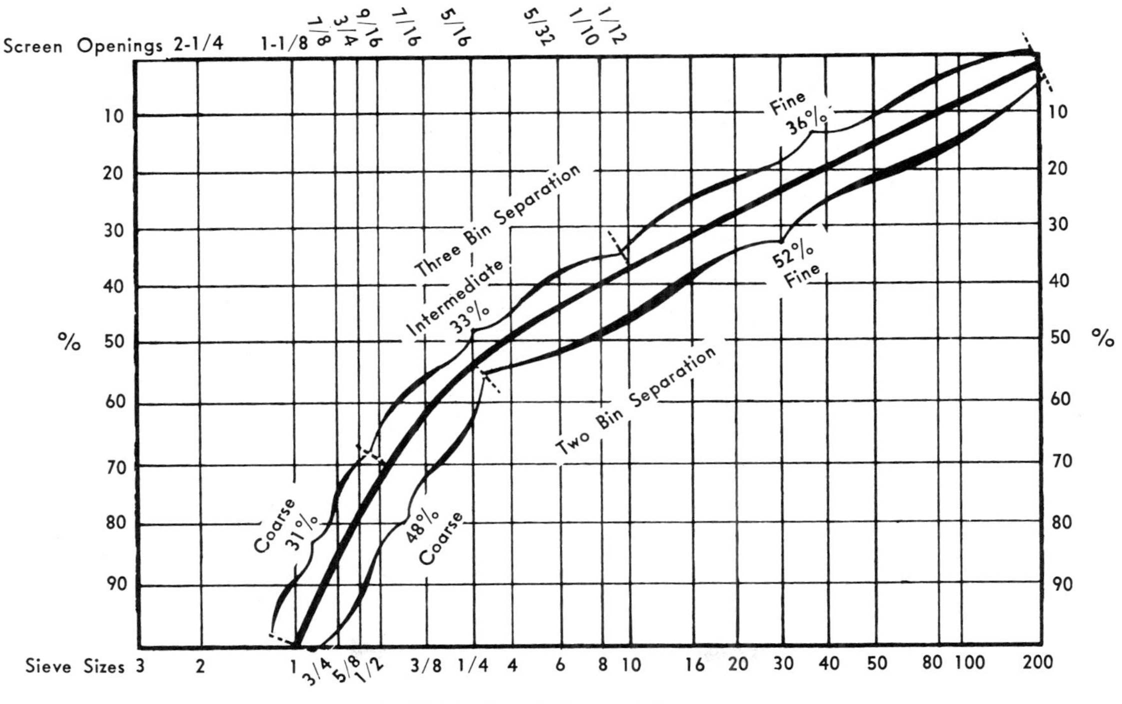

Fig. 25-25 Cumulative gradation curve.

TABLE 25-4 Plant Screen Selection

U.S. Standard Sieve size of bin aggregate, in.	*Actual size of square openings of vibrating screens, in.*
No. 10 (0.079)	1/12
No. 8 (0.094)	1/10
No. 7 (0.111)	1/8
No. 6 (0.132)	5/32
No. 5 (0.157)	3/16
No. 4 (0.187)	1/4
1/4	5/16
5/16	3/8
3/8	7/16
7/16	1/2
1/2	9/16
9/16	5/8
5/8	3/4
3/4	7/8
7/8	1
1	1 1/8
1 1/8	1 1/4
1 1/4	1 3/8
1 3/8	1 1/2
1 1/2	1 5/8
1 5/8	1 3/4
1 3/4	2
2	2 1/4
2 1/4	2 1/2

A typical 2-deck screen arrangement is shown in Fig. 25-27. The bottom deck is split, dividing its area between the No. 1 and No. 2 bins. The fine screen is subjected to heavier loading, since it must carry No. 3 as well as No. 2 bin aggregate.

Vibrating Screen Capacity. The screen capacities listed in Table 25-5 are representative of a broad range of experience and are neither the maximum nor the minimum that may be expected. The capacity of a screen for a particular operation will be influenced by many factors, including:

1. Aggregate moisture content
2. Gradation of the material
3. Screen load versus amount passing
4. Size of screen wire

Where experience indicates that blinding, moisture buildup, or other screening problems will not be encountered, the data in Table 25-5 may be used for dried aggregates. In preparing this table, it has been assumed that approximately 25 percent of the screen load is retained. For example, an indicated 60-tph capacity corresponds to a total load on the screen of approximately 80 tph, with 20 tph being retained and carried to the next bin.

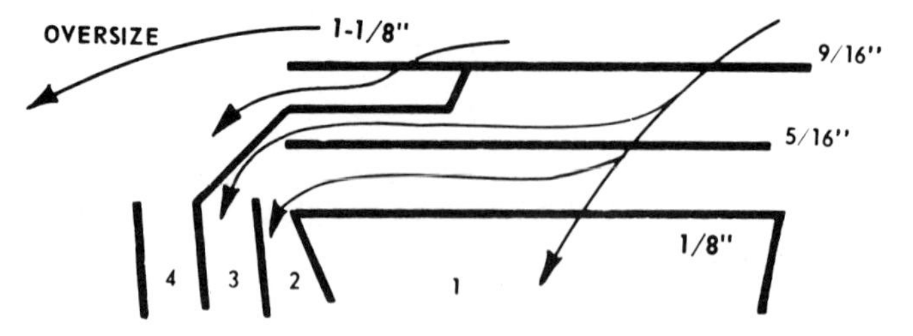

Fig. 25-26 3½-deck screen arrangement.

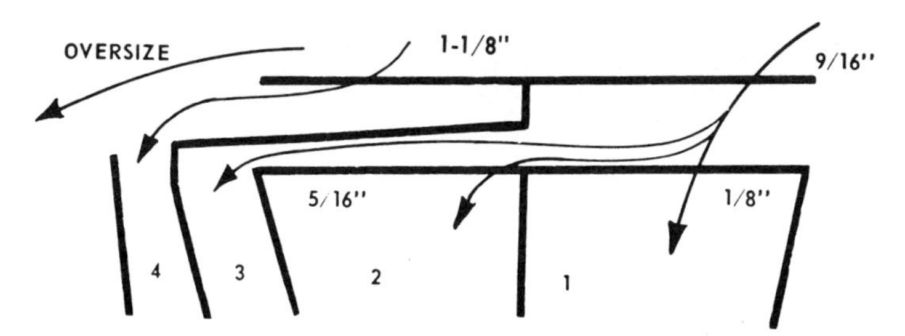

Fig. 25-27 Double-deck screen arrangement.

Measuring and Mixing The batch type and continuous type of plant will be discussed separately. Figure 25-28 illustrates a typical batch plant, and Fig. 25-29 shows a continuous-mix plant.

Batch Type. Total batch weight must be made up of the proper amount of aggregate from each hot bin and the correct percentage of asphalt. In a manually operated batch plant, the operator opens the first aggregate gate while watching a dial scale, then closes the gate as the pointer approaches the required weight for that aggregate size. He progressively opens each succeeding bin gate to accumulate the proper aggregate batch in the weigh hopper. The total aggregate batch is discharged into the pug mill and dry mixed for a prescribed time. The operator then proceeds to weigh another batch. At the completion of a dry-mix cycle, the proper amount of asphalt is dumped or sprayed onto the aggregate. The asphalt may be measured by a meter or by weighing on its own scale.

The operator's controls may be completely manual or may be power operated. Cycling controls may be incorporated to handle the various operations after the aggregate has been weighed and discharged into the mixer. This is often referred to as "semiautomatic" operation.

In recent years there has been a trend toward fully automatic operation of batch plants. Such plants handle the entire measuring and mixing phase automatically and include safeguards to ensure accuracy, batch after batch. This frees the operator to coordinate other plant operations, such as proper bin balancing through remote cold-feed control and regulation of aggregate temperature through remote burner control.

TABLE 25-5 Plant Screen Capacity

Clear square opening of plant screen cloth, in.	tph per sq ft	tph for deck size indicated			
		3 × 8 ft	4 × 10 ft	4 × 14 ft	6 × 14 ft
1/10	1 1/4	30	50	70	105
1/8	1 1/2	36	60	84	126
3/16	2	48	80	112	168
1/4	2 1/2	60	100	140	210
3/8	3 1/4	78	130	180	275
1/2	3 3/4	90	150	210	315
5/8	4 1/4	100	170	240	355
3/4	4 3/4	115	190	265	400
7/8	5 1/4	125	210	295	440
1	5 3/4	140	230	320	485

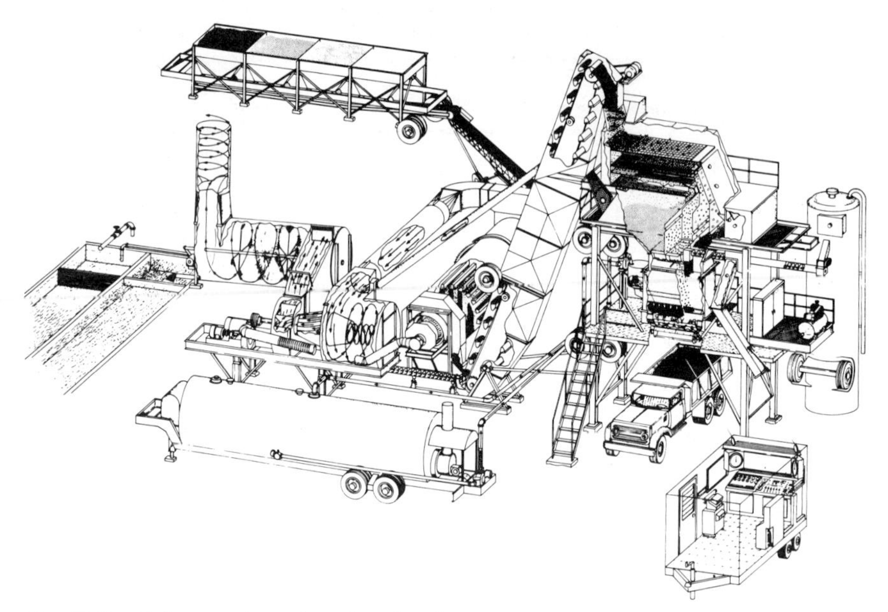

Fig. 25-28 Cutaway view of modern batch-type asphalt plant. (*Iowa Manufacturing Company.*)

Continuous Type. The screen and hot bins in a continuous plant are parts of its gradation control unit. Feeders under the bins deliver measured quantities of the hot aggregate to a bucket elevator, where it is elevated to the pug mill and sprayed with asphalt. The paddles in the pug mill are set to propel the mix toward the discharge end as the mixing action takes place.

Calibrated gates located over apron-type feeders control the amount of aggregate which is drawn from each hot bin. The feeders are interlocked with the asphalt metering pump, and the output of both aggregate and asphalt is controlled through a reference shaft and a revolution counter. The aggregate and asphalt fed to the pug mill during each revolution of the counter may thus be thought of as a small batch. Each batch will contain just the right proportion of each aggregate size and the correct amount of asphalt. In actual operation, these batches come in a continuous stream.

Fig. 25-29 Cutaway view of a modern continuous-mix asphalt plant.

Once the gates and pump are set, the operation becomes automatic and accuracy is assured. Continuous plants are equipped with warning devices or automatic cutoffs which are actuated if the supplies of aggregate or asphalt run short. Samples can be taken at any time to check gradation or rate of feed.

A "single aggregate" type of plant can be used to produce intermediate types of mixes which do not require a split or screening of the aggregate after drying. Any required aggregate blending is accomplished prior to drying, and the heated aggregate is deposited in one bin. Under this bin, a feeder with an adjustable gate is interlocked with the asphalt pump, just as in the highest-type plants previously described. Intermediate or single-aggregate plants offer the advantages of central-plant drying, precise control of asphalt content, and thorough twin-shaft pug-mill mixing. They are economical units for asphaltic mixes intended for secondary roads and other surfaces which do not carry the heavy traffic of primary roads.

Mixing time is the number of seconds necessary to properly combine the aggregate and bitumen and obtain a homogeneous mixture of thoroughly coated particles. The optimum mixing time involves both a rate and an overall cost factor. This optimum time is necessarily influenced by many variables.

A. Environmental
1. Type of aggregate (chemical and structural variables)
2. Design of the mixture
3. Aggregate moisture
4. Humidity
5. Viscosity of the bitumen

B. Pug-mill design
6. Shape of pug mill
7. Peripheral speed of paddle tips
8. Number of paddles
9. Size of paddles
10. Shape of paddles
11. Position of paddles or action produced as a result of paddle arrangement

C. Operational and system
12. Method of introducing bitumen
13. Method and sequence of adding aggregate
14. Condition of pug mill
15. Size of pug mill versus batch size
16. Condition of related equipment
17. Personnel and operating variables

Gradual recognition of improved designs and techniques is evidenced by a trend toward a lowering of mixing times required by specification. A test used to determine the time required for complete mixing has been developed in research carried out by the Barber-Greene Company. This Standard Method of Test for Degree of Particle Coating of Bituminous-Aggregate Mixtures, ASTM Designation D2489, expresses completeness of mixing in terms of the ratio of completely coated coarse particles of aggregate to the total number of coarse particles in the sample.[6]

The total mixing time for batch mixing, as typically specified, is the interval of time between the closing of the aggregate weigh-hopper gate and the opening of the pug-mill gate. This total mixing time is further broken down into a dry-mix period and a wet-mix period. Several studies have indicated that, at most, only a short dry-mix time is needed. In some instances the dry-mix time contributes nothing to achieving a complete mix. Pug-mill design, aggregate gradation, and the relative segregation of the aggregates as they are introduced into the pug mill are factors to be considered.

The required mixing time on a batch-type asphalt plant is one factor in the determination of plant production. The output of the plant and the daily production of the entire operation may be dependent on this mixing time. Assuming sufficient drying and screening capacity, an increase in mixing time means a decrease in production. (See Fig. 25-30.)

The continuous-mix asphalt plant differs from the batch type of intermittent

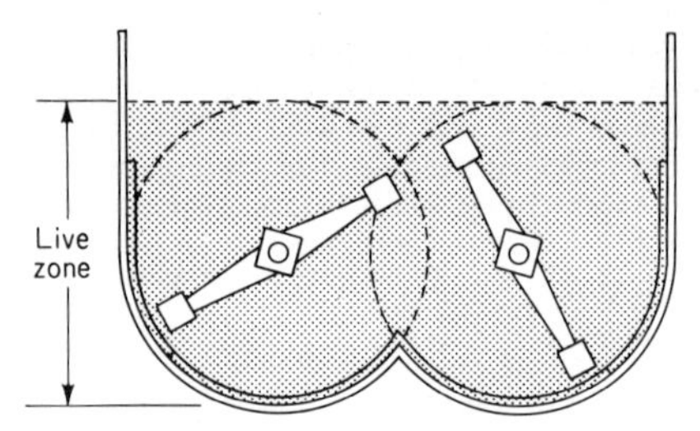

Fig. 25-30 Batch sizes expressed in percent of live zone. Live zone, which offers a uniform method of indicating pug-mill cubics is the net volume below a line extending across the top arc of the inside body-shell radius. Shafts, liners, paddle arms, and tips are all deducted to determine the net cubic feet. Batch sizes may be expressed as a percent of live zone. For example, if a pug mill has a live zone of 150 cu ft, mixing a batch of 7,500 lb (based on 100 pcf aggregate) will utilize 50 percent of the zone cubics.

asphalt plant in that it is possible to change the mixing cycle or mixing time without affecting the output of the plant. The mixing process in a continuous plant begins by blending the graded aggregates as they come from each feeder gate of the gradation unit to the bucket elevator and thence to the mixer. This preblending of the aggregates achieves a thorough "dry mixing." The bitumen is sprayed into the curtain of aggregate as it enters the pug mill, and thus a precoating occurs even before the measured mixing cycle starts. The mixing cycle is completed when the material is discharged from the pug mill.

Most specifications require that mixing time be measured by the following formula:

$$\text{Mixing time, sec} = \frac{\text{pug-mill dead capacity, lb}}{\text{pug-mill output, lb/sec}}$$

Using this formula, the mixing time can be computed by measuring the level of the material below the top of the pug mill in order to determine the dead load.

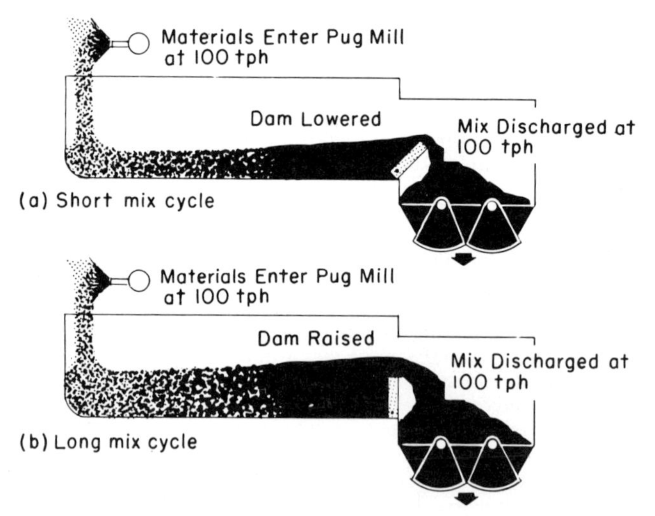

Fig. 25-31 The mixing cycles of the continuous plant. Each cycle can be increased or decreased without affecting the daily tonnage output. In (*a*), note that the right end wall of the pug mill proper is an adjustable dam. The height of the dam is controlled from the operator's platform.

Figure 25-31*a* illustrates an "easy" mix where a short mixing cycle gives adequate coating and mixing. The dam is set in a low position, which causes the material to be propelled by the paddles through the pug mill at a relatively shallow depth. This means that any given particle will pass at a relatively rapid rate through the pug mill and over the dam into the discharge hopper.

Figure 25-31*b* shows a mix requiring a longer mixing cycle, which is accomplished by raising the dam. The mass of material is now relatively deep, and this considerably increases the length of time that any given particle remains in the pug mill.

Note that, in both examples, the materials enter the pug mill at 100 tph and the mix is discharged at 100 tph. The operator has additional control of mix time since any number of paddles can be set to propel or retard. In any case, the rate of discharge per hour is the same as the rate of charging the pug mill, regardless of the means used to lengthen or shorten the mixing cycle.

Plant Production Frequently, a plant's capacity is judged on the basis of its pugmill size alone. An hourly capacity rate is then calculated from the mixer size and required mixing time. However, it is important to recognize the interdependence

TABLE 25-6 Plant Capacity in Relation to Components

Plant capacity in tph (with 5 % asphalt cement)		Cycle times in seconds for batch sizes, lb, of			Minutes of hot bin storage for bin sizes (tons) of			% of total aggregate through ⅛-in. sand screen (capacity from Fig. 25-5)			Dryer capability in % moisture removal (example only)	
									3½ deck			
Total mix	Aggregate only	5,000	6,000	7,000	25	30	50	4 × 12 ft double deck with 4 × 7 ft sand deck	4 × 10 ft	4 × 14 ft	7 × 30 ft	9 × 20 ft
125	119	75	90	97	12	15	25	35	50	70	9.1	13.0
150	143	60	72	78	10	12	21	29	42	59	7.2	10.6
175	166	51	62	67	9	11	18	25	36	50	6.0	9.0
200	190	45	54	58	8	9	16	22	31	44	5.0	7.6
225	214	40	48	52	7	8	14	19	28	39	4.2*	6.6
250	238	36	43	47	6	7	12	17	25	35	3.6*	5.7

* Provided maximum dryer conveying capacity is also sufficient for these tonnages. Actually, top tonnage may be only 200–210 tph for example, regardless of how dry the aggregate is.

of all plant components. Under a given set of environmental conditions, aggregate characteristics, and mix requirements, the limiting factor of plant output may be the dryer, the screens, the mixer, and other items as well. Table 25-6 shows a representative selection of components to achieve the indicated tonnages under varying conditions.

For best results, plant components should be selected on the basis of the anticipated conditions under which the plant will be operated.

Fines Feeding and Handling High-type asphalt plants and stabilization plants often require separate feeding systems for introducing fines materials into the mix. Typical of the materials for these applications are ground limestone, hydrated lime, portland cement, and wet or dry fly ash.

In plant operations where fines usage is high, a system using a storage silo for maintaining several days' supply of fines is often advisable. The ultimate choice of this system is usually dependent on the availability of bulk fines and their price in relation to bagged fines. The bulk-handling system, consisting of receiving hopper, screw conveyor, and dust-tight elevator, is used to charge the storage silo as well as the vane feeder which meters the fines into the plant.

In operations where the volume of fines used does not justify a bulk silo, the plant owner will want to consider a bag-feeding system. The system consists of a ground-mounted feeder, dust-tight elevator, surge hopper, vane feeder or screw conveyor, and an overflow chute. A typical batch-plant bag system is illustrated in Fig. 25-32.

Bulk or bag fines systems are equally adaptable for continuous-mix plants. Final metering of the fines to the mix is accomplished through a variable-speed vane, screw, or belt feeder, depending on the material to be handled and the capacity required. In each case, the fines feed is interlocked with the aggregate and asphalt feed to ensure constant accuracy.

Where excess fines are encountered in the raw aggregate feed, a bypass system can be employed to receive the fines collected by the dust collector. The required amount of fines is then fed back to the mix and any surplus amounts are diverted to a storage bin for disposal or other use.

Plant Accessories *Asphalt storage* should normally be equal to one day's needs. Specific circumstances may render this unnecessary, such as where a plant is very close to a refinery and transport trucks can deliver the asphalt at or near the mixing temperature.

Fig. 25-32 A typical batch-plant bagged-fines feeding system.

Two smaller storage tanks are more desirable than one big one. Should there be only one tank, it will not always be full and at proper temperature in the morning. The dumping of a cold load of asphalt on top of a small quantity of hot binder can make it necessary to shut down until the tank has been brought back to the correct temperature. With two smaller tanks, there is always enough flexibility and extra capacity in the heater to take care of one of the tanks during the daytime.

Sizes of tanks, pipes, and heaters can be selected on a preliminary basis from the data supplied in Table 25-7, but they must be influenced by individual plans and circumstances. Distance, method of transportation, heating facilities, asphalt type, and fuel to be used are all factors.

Plant Arrangement. Many factors are important in selecting a satisfactory plant site. They include sufficient work space, proximity to job or market, traffic considerations, availability of aggregates, good footing and water-table height, and noise abatement and dust nuisance regulations.

Once the site has been selected, the plant should be located with prime consideration given to truck traffic patterns and prevailing winds. If practical, the operator's platform should be located upwind so that any dust is normally blown away from operator and engine locations.

Tanks and heaters should be located to minimize piping and to facilitate the delivery of asphalt and fuel without interfering with plant operations. Representative arrangements are shown in Figs. 25-33 and 25-34.

TABLE 25-7 Guide for Selecting Asphalt Tanks, Pipes, and Heaters

Plant capacity, estimated 10-hr output	30 250	55 500	80 750	105 1,000	130 1,250	155 1,500	180 1,750	205 2,000	250 2,450	tph tons
Asphalt requirements—storage and pipe size										
Asphalt required per day at 6%	15 3,800	30 7,600	45 11,400	60 15,200	75 19,000	90 22,800	105 26,600	120 30,400	147 37,200	Tons Gallons
Storage tanks	1 4,000	2 4,000	2 6,000	2 8,000	2 10,000	3 8,000	3 10,000	3 10,000	4 10,000	No. gallons
Asphalt pipe	2	2	2	3	3	3	3	3	3	Diam, in.
Fuel requirements										
Estimated fuel oil per day	500	1,000	1,500	2,000	2,500	3,000	3,500	4,000	4,900	Gallons
Recommended fuel tank size	2,000	2,000	3,000	4,000	5,000	6,000	8,000	8,000	10,000	Gallons
Recommended supply and return fuel line size	2	2	3	3	3	3	3	3	4	Diam,* in.
Hot oil heater data										
Hot oil pipe size	2	2	2	2	2	2	2	2	2	Diam, in.
Output per hr	450	450	750	800	800	1,200	1,600	1,600	2,000	Units of 1,000 Btu
Estimated fuel per day	70	70	70	90	90	90	125	125	175	Gallons

* If supply tank is located over 40 ft from mixer unit, it would be advisable to increase recommended diameters of supply and return pipes by 1 in.

Heaters, tanks, and piping are selected on the basis of the daily asphalt requirements and the amount needed in storage. In turn, the system directly affects the efficiency of the overall operation of the plant. The required heating capacity is dependent upon the losses encountered in the plant, piping, and tanks and the desired rate of heating for asphalt which is received at less than application temperature.

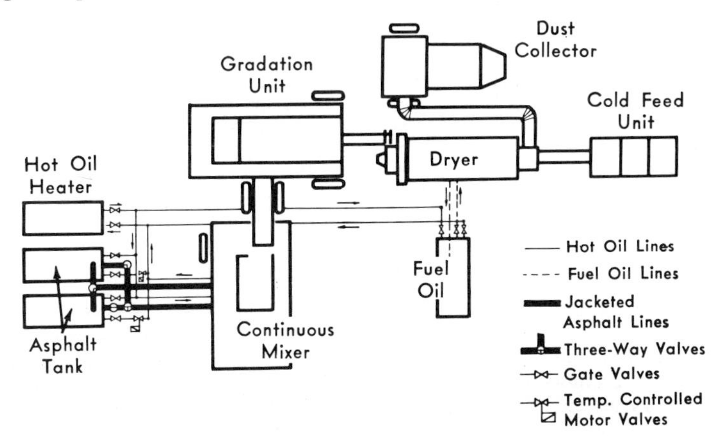

Fig. 25-33 Typical continuous plant and piping arrangement.

The following formula and tables can assist in arriving at total Btu requirements for a particular application. However, it is always advisable to contact the heater manufacturer for a firm recommendation.

I. Heat losses from mixer in Btu/hr

50-ton capacity	100,000
100-ton capacity	180,000
150-ton capacity	265,000
200-ton capacity	350,000

II. Heat losses from pipe

Heat loss per 100 ft of 3-in. diameter pipe with 4-in. jacketed line = 170,000 Btu/hr

III. Heat losses from tanks in Btu/hr for asphalt temperature of 300°F and air temperature of 50°F*

Tank size, ft	Tank capacity, gal	Heat losses from tanks, Btu/hr		
		Uninsulated tank	1-in. insulation	2-in. insulation
8 × 14	5,000	350,000	31,000	18,000
10 × 15	8,000	500,000	43,000	25,000
10 × 17	10,000	540,000	46,000	28,000
11 × 21	15,000	720,000	62,000	37,000
12 × 24	20,000	900,000	76,000	46,000
12 × 32	25,000	1,150,000	96,000	58,000

* Corrections for heat loss for tank temperature other than 300°F:

Asphalt temperature	Multiply values from table by
350°F	1.25
325°F	1.15
250°F	0.75
200°F	0.50

IV. Heat required to raise asphalt temperature

Btu per hour = specific heat × temp. rise per hour × asphalt weight

Example: For a required heat rise of 15°F *per hr*

Btu/hr = 0.5 Btu/lb-°F × 15°F/hr × 8 lb/gal × no. gallons

Gallons in tank	5,000	8,000	10,000	15,000	20,000	25,000
Btu/hr (1,000)	300	480	600	900	1,200	1,500

NOTE: Asphalt supplier can give exact specific heat and weight. Contractor's need or desire may indicate different temperature rise per hour.

V. Safety factor

In calculating total Btu heat loss, unknown factors must be taken into consideration. The heat loss which takes place from fuel tanks, transfer pumps, valves and related equipment, wind velocity, etc., must be allowed for. Since it is next to impossible to pinpoint each of these factors, it is recommended that an additional 10 to 15 percent of the above Btu requirements be added. If the unit is to be located in a very windy area, additional allowances may be in order.

The heater, storage tanks, and plant equipment should be grouped as compactly as possible in order to keep the length of supply and delivery lines to a minimum. Hot transfer oil should be circulated through the system with a velocity of about 7 ft per sec to promote adequate heat transfer. Unusually long supply lines may require the use of an additional circulating pump. It is usually considered good practice to install hot oil delivery and return manifold lines of at least 2 in. in diameter.

Lateral lines from these manifolds to each individual tank and the mixing plant are reduced to about 1 in.

By providing gate valves at the entrance and exit points in each of the various components, hot oil may be stored during transport in these units. It is advisable to provide this for the manifold lines. In addition, these gate valves may be used to balance the oil flow through the entire system, helping to regulate the desired pressure at approximately 30 psi. Hot oil will always seek the path of least resistance. Uniform system resistance may be balanced by the use of valves. It is also advisable to install small air-venting valves at the high points in order to assist in completely filling the system.

Piping for an average asphalt plant layout can be quite costly. Flexible-type couplings have become popular in reducing setup time. Plant components, manifold lines, asphalt-jacketed lines, etc., can be positioned and the entire system connected with a minimum loss of time. One or two plant moves will generally pay for the additional cost of these couplings.

The jumper lines from one jacketed supply line to another can also be of the flexible design and should be at least ¾ in. and preferably 1 in. in size.

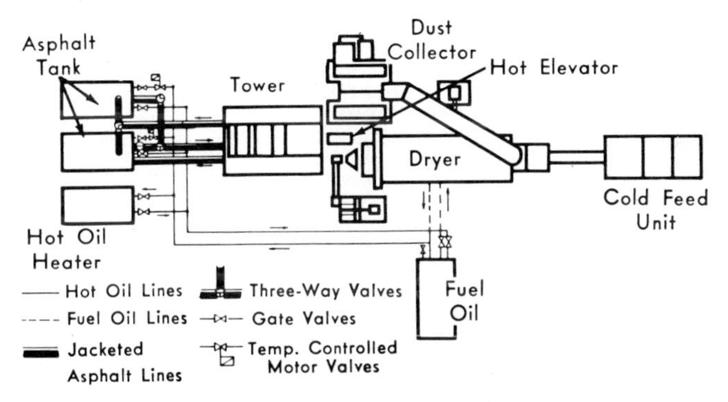

Fig. 25-34 Typical batch plant and piping arrangement.

Heating Coils. The Btu's of heat generated by the heater can be exchanged to the asphalt by circulating the heating fluid through coils in the tanks. The surface area of these coils must be adequate for this purpose. Engineering calculations, verified by experience, indicate that approximately 600 ft of 2-in. standard black steel pipe per each 10,000 gal of asphalt will be necessary in order to raise the asphalt temperature at the desirable rate of 15°F per hr. High wind velocity and low ambient temperatures will materially reduce the heating rate in uninsulated tanks. It is generally agreed that insulation of all the heating equipment, tanks, delivery pipes, etc., will pay for its cost in one or two operating seasons. A minimum of 2 in. of fiber glass or magnesia block-type insulation is recommended.

The configuration and placement of the coils in the tank are very important. Figure 25-35 shows the generally accepted method, which will provide for uniform heating of the entire contents. Circulation of the liquid asphalt, as soon as it becomes pumpable, is often used as a means of increasing the temperature-rise rate.

The selection of tank accessories should receive careful study. The hot-oil motor valves are usually located on the outlet lines of the circulating system, and the thermostat control for these valves must be installed so as to ensure control of the contents at all levels. An indicating thermometer with large dial is desirable, as well as an indicator showing the level of the asphalt in the tank. Delivery and return lines must be installed properly, as well as an adequate tank venting system.

Shutoff cocks at the asphalt outlet are definitely needed. A set of outlets is often provided at the rear of the tank for transfer of the contents to another tank. Adequate cleanout openings and an entry manhole are necessary. Some thought should be given to providing adequate jacklegs or sills for erection as a permanent installation on portable trailer-type frames. Consultation with a tank manufacturer on all given requirements will help to ensure economy and proper operation of the entire system.

The entire heating system, including asphalt and fuel-oil storage tanks, can be installed for economical, fully automatic operation. When insulated tanks are employed, it is advisable to install an electric control to shut down the hot-oil heater operation when no heat is required for maintaining temperature. This is particularly desirable for nighttime operation when the machinery is to be left unattended. Heat loss from uninsulated tanks will probably necessitate continuous heat input but will still require thermostatically controlled motor valves.

Fig. 25-35 Placement of heating coils inside asphalt storage tank. (*Columbian Steel Tank Co.*)

Insulation of the dryer-burner fuel-oil storage tank is desirable, especially if the plant is operated in an area with strong crosswinds or if the ambient temperature is low. This is especially true in high-altitude areas and in parts of the country which experience low nighttime temperatures in early spring and fall.

Experience has shown conclusively that extra care and an investment in adequate equipment for asphalt and fuel-oil storage can be key factors in a continuous and profitable operation.

CONSTRUCTION OF HOT-MIXED, HOT-LAID BITUMINOUS CONCRETE PAVEMENTS[3,5]

Bituminous surfaces have been in use in the United States since before the turn of the century. The major drawback in their application in the early days of road construction was the method of spreading and leveling the material on the roadway. Many methods were used, most of them highly inefficient. Perhaps the oldest method used was that of hand spreading. In this operation the mix was dumped from the trucks onto dump boards and then shoveled onto the road, following which the asphalt was raked smooth to grade and contour by hand. With the increasing cost of hand labor, this method became uneconomical. In addition, a smooth, level, and even-textured surface was very difficult to obtain by this procedure.

Motor graders have been used to spread mix after it has been windrowed or dumped on the road, and this method is still used today for small spreading jobs. However, its disadvantages far outnumber its advantages. Since poor joints frequently accompany blade spreading operations, two graders are usually required to handle the entire width of the road in one pass. There is no compaction gained from blade spreading, and so all compaction of the mat must be obtained from rolling. Segregation of the mix is also a problem in road-grader spreading.

In 1937, after a 7-year period of development, the Barber-Greene Company introduced the tamping-leveling asphalt-finishing machine. This machine lowered the cost and increased the rate of laying asphalt mix.

Fig. 25-36 A modern asphalt finishing machine.

The Finisher A modern asphalt finishing machine must be able to handle all types of asphalt mixes. It must evenly spread and compact these materials, leaving a finished surface ready for final compaction by the steel and pneumatic-tire rollers. It must compensate for minor grade irregularities and must be able to form proper crowns and superelevations for modern-day highways. It must be able to push the large trucks that carry the bituminous mix from the hot-mix plant, and it must be able to operate on surfaces which range from old cracked pavements to newly prepared subbases. Therefore, the finisher must be a highly refined yet ruggedly built piece of construction machinery.

A typical asphalt finisher consists of two main units, the tractor unit and the floating-screed unit (see Fig. 25-36). The tractor unit provides the motive power through crawlers or rubber tires which travel on the road base. The tractor unit includes the receiving hopper, feeders, distributing augers or spreading screws, power plant, transmission, dual controls, and operator's seat.

The screed unit is towed by the tractor unit and rides on the finished surface. The tamper or vibrator units, thickness controls, crown controls, screed heater, and screed plate are all parts of it.

Finisher Operation The plan and side views shown in Fig. 25-37 trace the flow of bituminous mix from the receiving hopper at the front of the finishing machine to the finished pavement behind the screed unit at the rear of the machine. The mix is dumped into the receiving hopper from a truck, which is pushed ahead by the finisher. Rollers mounted on the front of the finisher contact the rear tires of the truck and allow the finisher to push the truck while it is dumping into the hopper.

After receiving the material in the hopper, two independently controlled bar feeders carry the mix back through the control gates to the spreading screws. Each spreading screw is synchronized to its respective feeder, permitting the operator to distribute the mix accurately in front of the screed unit.

The screed unit is attached to the tractor by two long screed arms that pivot well forward on the track casing of the tractor unit. These arms provide no support for

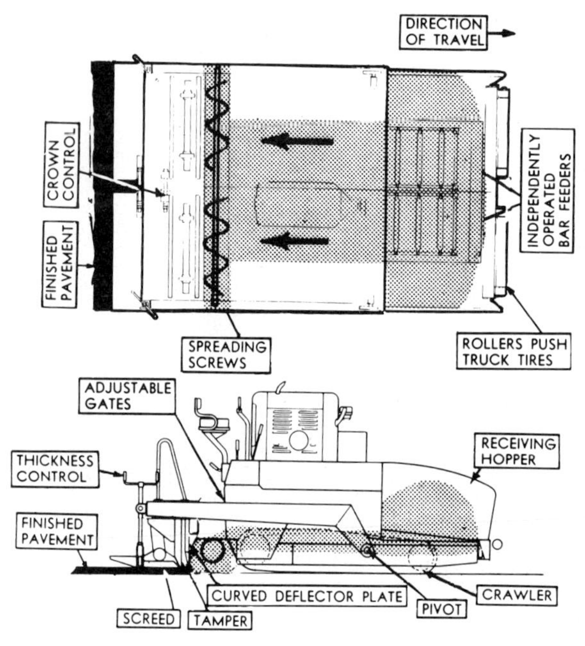

Fig. 25-37 Material flow diagram.

the screed when it is in operating position. Thus the floating action of the screed, as it travels along the road, compensates for surface irregularities which raise or lower the tractor unit. As the tractor pulls the screed into the material, the screed will seek the level where the path of its bottom surface is parallel to the direction of the pull. The screed can also be adjusted for a new thickness of mat by changing the depth-adjustment controls.

The screed unit consists of the tamping bar, which is the compacting medium as well as the strike-off; the screed plate, which gives the ironing action and provides a stable floating action in the mat laid; the thickness controls by which the tilt of the screed plate is changed in order to increase or decrease the thickness of the mat being placed; and the curved deflector plate.

The spreading screws are mounted on the tractor unit, each screw responding automatically to the amount of mix in the screw chamber along with the bar feeder on the same side. The automatic feeder controls can be moved outward as extensions are added, providing an adequate and constant supply of mix at the ends of the screws at all times. Flow-control gates are located at the rear of the tractor unit and have

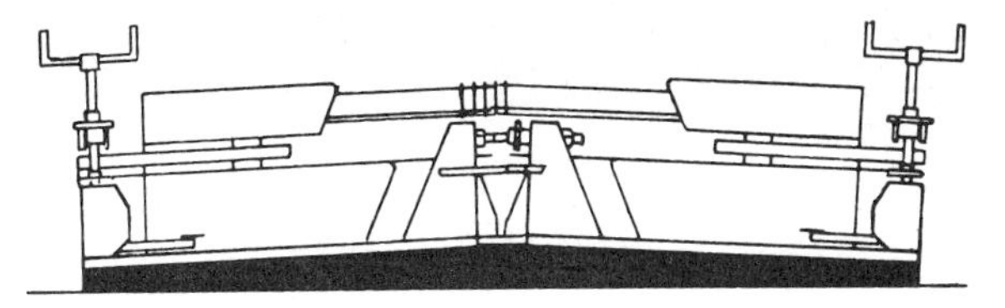

Fig. 25-38 Positive crown.

a crank adjustment within easy reach of the screed man. These gates, properly set, ensure a uniform flow of mix to the screw chamber. This minimizes feeder-clutch actuations and helps to maintain a uniform level of mix ahead of the screed unit.

The screed is also equipped with a heater to prevent the mix from sticking to the screed plate. It is used to heat the screed plate when the air temperature is cold or when starting out first thing in the morning.

The horizontal face on the bottom of the tamper bar strikes off the material so the screed plate can ride smoothly over it. The bevel face of the tamper bar effects the primary compaction of the mix as the paver moves along. The density of the loose asphaltic mix is increased as the material is forced down by the beveled face of the tamper. The horizontal face of the tamper imparts some compaction, but its primary function is in striking off the material.

The horizontal bottom face of the tamper bar will gradually wear to a knife edge with use. There is very little change in its compactive or strike-off action as it wears to this condition.

Vibrating screeds are also used on asphalt pavers requiring compaction and strike-off in the same operation. Strike-off is done by the leading edge of the screed and the compactive effort occurs under the screed plate.

Crown and Superelevation. The crown-control adjustment at the center of the screed provides for the desired contour of the finished pavement. Both crown screws are connected by a chain so that a single lever adjusts the leading and trailing edges of the screed uniformly to the specified crown. The leading edge can also be adjusted independently to achieve the slight additional crown normally required for best surface results. This adjustment is made while paving. It is possible to adjust the crown on the screed from ½ in. negative to 2 in. positive. (See Fig. 25-38.)

The screed can also be adjusted to produce a superelevation by increasing the thickness on one side of the screed through the thickness control as shown in Fig. 25-39.

Leveling Action. Inherent in the basic design of most asphalt finishers is the principle of self-leveling. This feature, properly utilized, can produce a smooth and level mat over a rough subbase or a cracked and uneven roadway. The screed, when pulled into the material, will automatically seek a level where its bottom becomes parallel to the direction of pull asserted by the pivot joint of the finisher. This inherent self-leveling principle depends upon time, distance, and mix stability rather than upon a fixed mechanical ratio. Minor base irregularities are virtually eliminated because of it. Coupled with compaction before strike-off, it assures a smooth-riding mat.

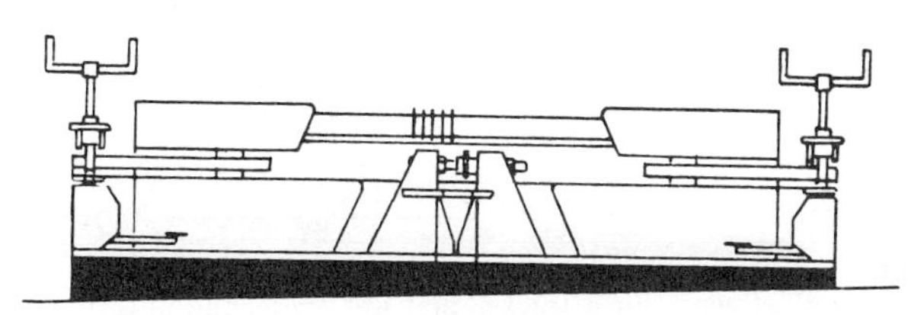

Fig. 25-39 Superelevation.

Fig. 25-40 Finisher maintains a level mat over a level surface—indefinitely.

Thickness Controls. The screed-control operator can also change the thickness of the mat through the use of the thickness controls on the screed. These controls allow the screed to be tilted either up or down. The screed will then either rise or fall until it has reached a new level, with its bottom again parallel with the direction of pull. Figure 25-40 shows the asphalt finisher laying a level mat of uniform thickness over a level base. It will continue to lay such a mat indefinitely without any manual adjustment until (1) the operator changes the thickness control, which causes the machine to automatically lay an ascending or descending ramp until the new thickness has been reached and the screed is again parallel to the direction of the pull (see Fig. 25-41); or (2) an irregularity in the base causes the machine to automatically increase or decrease the thickness of the mat, compensating for the irregularity and producing a level surface over an irregular base. The mat is automatically made thicker to fill in depressions in the base and thinner to smooth out bumps. (See Fig. 25-42.)

The leveling feature of the finisher is further accentuated through the use of the tamper bar, which automatically compacts more material into the depressions in the base. This results in a smooth and level mat after compaction. If paved by a screed not having compaction before strike-off, a depression will reappear after the mat has been rolled and opened for traffic.

Due to the relatively long reaction time of the screed to any change in level of the tractor unit, the level of crawler or tires of the tractor unit may change but the screed will take much longer to come to the new level.

Automatic Screed Controls Control of screed level to established grade lines and matching of adjacent mats or gutters can be accomplished through manual, semiautomatic, or fully automatic devices. These systems control the angle of attack of the screed. Some systems will automatically control transverse slope and provide for manual adjustment to construct superelevations or gradually varying cross slopes. Thus, side-to-side rocking motion of the automobile can be alleviated even though cost considerations may dictate the need for compromise in longitudinal leveling.

Primary methods used for establishing a longitudinal grade reference include string line to grade, joint-matching shoe, and long ski. Following is a brief review of these and other methods.

String Line to Grade. With this system the grade is established by a survey crew and marked by a string line (see Fig. 25-43). The string-line system is used on jobs where leveling to a desired profile is the requirement. For other objectives (i.e., leveling for a smooth-riding surface, joint matching, laying to an existing curb or gutter), there are other reference systems that may be more workable or economical.

Fig. 25-41 When adjustments for a thicker mat are made, the finisher makes a gradual transition to the new mat thickness. Abrupt changes cannot be made.

Fig. 25-42 Finisher automatically compensates for base irregularities, laying level mat without operator attention.

When working with a string-line system, special care should be taken in making the survey and translating it into a string line. A poor survey defeats the purpose of a grade-controlled job before paving starts. After the string line is established, care should be exercised to see that a proper amount of tension is maintained on the line to keep it true. A sagging line will mislead the control system and result in a wavy surface.

Joint-matching Shoe. Where an adjacent line of pavement serves as a grade line, a small joint-matching shoe can be used to sense the surface grade information and pass it on to the control mechanism. The method also finds considerable application in matching curbs, gutters, or in controlling uniform thickness over a surface previously established to a desired contour (see Fig. 25-44).

Long Ski. A semirigid boom or ski is another device that can be used as a grade reference with automatic screed controls. With this system, the ski glides over the high points of the base or old surface, signaling the control mechanism as its midpoint proceeds ahead of the screed. On jobs with serious surface irregularities, a long ski is used (see Fig. 25-45). This long ski spans from peak to peak, sending impulses back to the finisher. A mat is thus laid in the low spots between peaks, with a minimum cover over the high spots to eliminate surface irregularities. Where the base surface is relatively level, a shorter ski can be used.

Other Methods. The traveling string-line method is similar to the long ski. The ski is mounted on sets of wheels, and a string line is tensioned between posts at each end. Usually, the length of the traveling string line is 30 ft. At any given moment, the grade follower on the finisher is following a straight line drawn between the two surface points contacted by the wheels.

Other mobile references include beams supported on multiple spring-loaded "feet" and flexible pipes equipped with tensioned string lines. These methods improve longitudinal grade without the expense of a complete survey and grade setting.

Fig. 25-43 Finisher with automatic screen controls and string line.

Fig. 25-44 Finisher with automatic screed controls using joint-matching shoe. (*Pioneer Division, Portec Inc.*)

Finisher Applications The asphalt finishing machine has been widely used on all types of paving and resurfacing projects. It is chiefly through its introduction that low, intermediate, and high-type asphalt surfaces are economically feasible.

New Construction. As used here, this term refers to bituminous pavements laid on newly constructed and graded roadbeds—usually a well-drained, stabilized gravel or stone base. This type of construction is a simple task for the asphalt finishing machine since subgrades are improved, steep grades are at a minimum, bases are of a prepared material, and obstructions are seldom encountered. However, the asphalt finishing machine may be working on a relatively soft base and will tend to disturb it if proper flotation is not provided.

The "compaction before strike-off" principle of the tamping and leveling finisher enables it to lay a smooth mat over an irregular base which may vary as much as 1 in. from established grade. Irregularities and depressions in the base surface are filled in and properly compacted before the screed unit levels off the surface of the mat. Otherwise, these depressions will again show up after rolling. It is always desirable to

Fig. 25-45 Finisher with automatic screed controls using long ski.

lay at least two courses on any varying base in new construction. Irregularities remaining after the first course can be minimized in the second course.

Resurfacing. Old pavements may be very irregular and full of potholes and other surface failures. By resurfacing this pavement with bituminous material, the life of this roadway can be extended far beyond its original design life.

"Resurfacing" means the placing of a smooth bituminous mat (single or double course) on an old concrete, brick, or bituminous road. On some resurfacing projects, especially in city work, the irregularities or potholes require patching ahead of the finisher. Skip patching or scratch courses are used only in extreme pavement failure cases where holes or depressions are large enough to allow the entire side of the machine to drop into them. Also, if deep potholes are allowed to fill as the machine passes over them, later compaction of this material will in turn cause depressions in the final surface.

On resurfacing where an extreme crown exists it is desirable to first reduce it. This may be done by laying a leveling course consisting of two separate wedge courses on the old road, feathering each course near the center of the crown as illustrated in Fig. 25-46. The most common material for this type of work is fine-graded aggregates. This is an important point, as fine material should be used on all leveling courses to take care of the thin section laid over bumps in the old pavement.

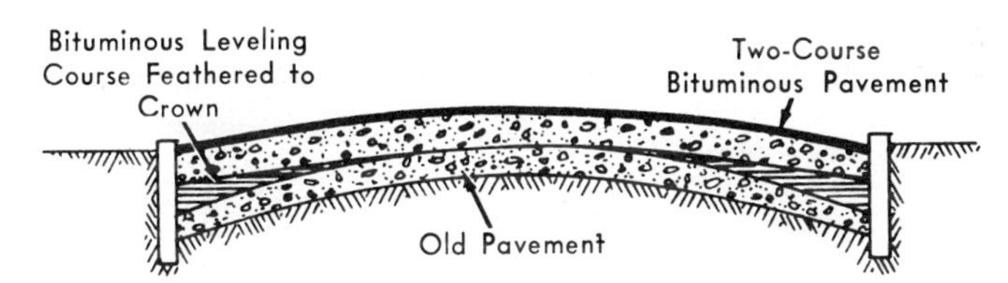

Fig. 25-46 One method of reducing a high crown.

Bituminous layers are often placed on old pavement surfaces to increase the life of the pavement or to produce a nonskid surface. This is usually a simple operation and is covered in the application just discussed. In the resurfacing operation, the finisher is operating on a solid base provided by the old pavement surface. Therefore, traction becomes the important feature of the tractor unit.

Laying Variable Widths. The first step before starting any laying operation is to decide what width of mat to lay. The machine must then be fitted with proper extensions, cutoff shoes, etc. With the use of cutoff shoes, a standard finisher can be set up to lay pavement widths which can vary over a wide range of dimensions. By adding extensions singly or in pairs, the laying width of the machine can be further increased.

In multiple-lane work, the cutoff shoe should always be used opposite the joint-matching side, and the final lane should be at least the width of the machine. Using these simple rules and knowing the total width of the pavement, the proper width and sequence of lanes can be established. Extensions which are added to the finishing machine should perform the same function as the main part of the screed. If the screed is a tamping one, the extension section should also incorporate a tamper bar. If it is a vibratory screed, it is equally important that the extensions also contain the vibration feature. Otherwise, the mat laid under the extension section will be different from the main portion.

Matching Pavements. When matching a joint to a mat previously laid, a small overlap is helpful in preventing any decrease in thickness. If the machine is started out with an overlap of 2 in., a qualified operator should be able to guide the machine with an overlap of not less than 1 in. nor more than 3 in. provided care was exercised in laying a straight first lane.

When matching a joint that has been previously laid and rolled, the depth of the overlapping mat must be sufficient so that additional compaction from the roller will bring the new mat down only to the level of the old mat. Excessive overlap prevents sufficient density of material in the new mat ahead of rolling, and it will also be likely to cause bridging and tearing under the screed.

A recent development in joint matching has been a propane-fueled infra-red heater attached on the side of the finisher next to the previously laid mat. The purpose of this unit is to preheat the existing mat edge just ahead of the screed to achieve a hot joint between the two mats.

Finishers Working in Echelon. This type of operation, illustrated in Fig. 25-47, is frequently used in airport and highway construction. The first finisher usually operates from 50 to 100 ft ahead of the second finisher, which is matching the joint of the first machine. In addition to the high capacity thus obtained, the joint quality is improved because both strips are hot when the joint is made. This assures union of material and a smoother joint with less rolling.

In tandem operations, the roller must keep 6 to 12 in. away from the inside edge on the first strip. The second strip can then be laid the same depth as the unrolled part of the first strip, and the roller can compact the joint while the material is hot.

Fig. 25-47 Finishers working in echelon.

City Street Construction. Paving and repaving operations in city street construction impose some added considerations. It is often desirable to flatten the crown of the pavement at the intersections. This is accomplished by adjusting the synchronized crown control, which reduces the crown in both the leading and trailing edges of the screed by an equal amount.

Manholes require some handwork in asphalt paving operations. On the binder or first course, the screed is lifted over the obstruction, or a ramp is used to make the screed ride up and over the manhole. The location of the manhole is usually marked on the curbs so that the screed of the machine may be run as close as possible to the manhole before it is lifted over it. On the top or surface course, the machine paves over the manhole cover and the excess mix is shaved off afterwards.

There is a danger of jamming the screed into the curb face, making it necessary to raise the screed, reposition the finisher, and clean up the excess mix. So, it is advisable to hold the screed edge about 4 to 6 in. away from the curbing. Excess material is then bled out to the curb, and this narrow strip of material is raked out by hand. It is important to keep the level of the loose, hand-raked material a little higher than the machine-laid material to compensate for compaction produced by the tamper and screed.

Airports and Other Large Areas. Due to the extreme width of airport runways, special precautions must be taken to construct a surface which is free from water pockets. It is good practice to start at the edge of the runway and work toward the center. In this way, if the overlapping joints do not roll smooth, they will not form water pockets. (See Fig. 25-48.)

A method often used to assure a true crown is to check the mat transversely behind the finisher with a straightedge or a crown board upon which is mounted a hand level. This level is adjusted so that when its bubble records on center, the end of the straightedge nearest the center of the runway is higher than the outside end. Similar precautions are necessary to ensure proper drainage when surfacing parking lots or other large areas.

Compaction and Rolling The several layers of a flexible pavement, including the granular or stabilized base and subbase and the asphalt surface itself, must all be adequately compacted and finally rolled to an established contour. Compaction and rolling equipment for this purpose includes steel and pneumatic-tire rollers as well as more specialized types which are designed to produce the desired effect in different materials under varying conditions.

Rolling of the various bituminous layers in the road requires skill and care. A well-engineered and executed construction job which employs the most advanced grading and leveling techniques and the best of materials and workmanship can be canceled entirely by slipshod methods in rolling and finishing. Additional considerations include atmospheric conditions, temperature, and working characteristics of the mix.

It is generally agreed that initial rolling should follow the spreading operation immediately or as soon as possible thereafter. On hot summer days it may be necessary to delay the initial pass for a short time to allow the mixture to cool. A common

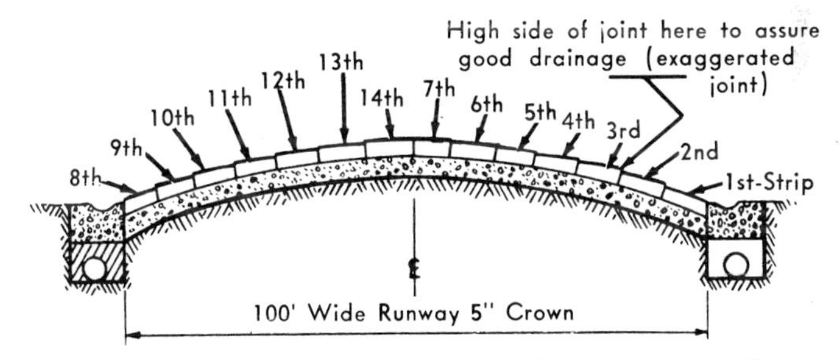

Fig. 25-48 Runway paving technique to eliminate water pockets.

sequence of rolling begins with the steel three-wheel breakdown roller, followed by a pneumatic-tire intermediate rolling operation and a final or finish rolling using a steel-wheel tandem roller. The value of an experienced foreman cannot be disputed. Being familiar with his materials and equipment and cognizant of atmospheric conditions, he can readily determine the timing and sequence of rolling patterns. This can be verified on the first pass of the roller and additional changes can be made if required.

Practice has indicated that the large wheels of the three-wheel breakdown roller exert a great compressive force with little displacement of the mix. This is particularly desirable on longitudinal joint work. A definite rolling pattern should be observed in order to assure a uniformly and correctly compacted mat. (See Fig. 25-49.)

It is universal practice to start rolling at the outside edges of a newly laid mat and roll inward toward the center, or crown. In the case of superelevated curves, rolling is started at the low side and carried to the high side in the same manner. A careful roller operator will observe these requirements, regulating his speed to produce a smooth, finished surface which is free from tears and depressions. Some mixes may be prone to tenderness and will show failure from excessive compaction. A good operator must be aware of these signs and know when to stop the operation.

Immediately following the initial rolling operation, the surface of the mat should be checked with a straightedge for compliance with specifications. The mat, still being warm and relatively soft, can be corrected for variations at this time.

After the initial rolling and any necessary corrections are completed, the pneumatic-tire roller follows. Having noted that traffic will frequently continue to com-

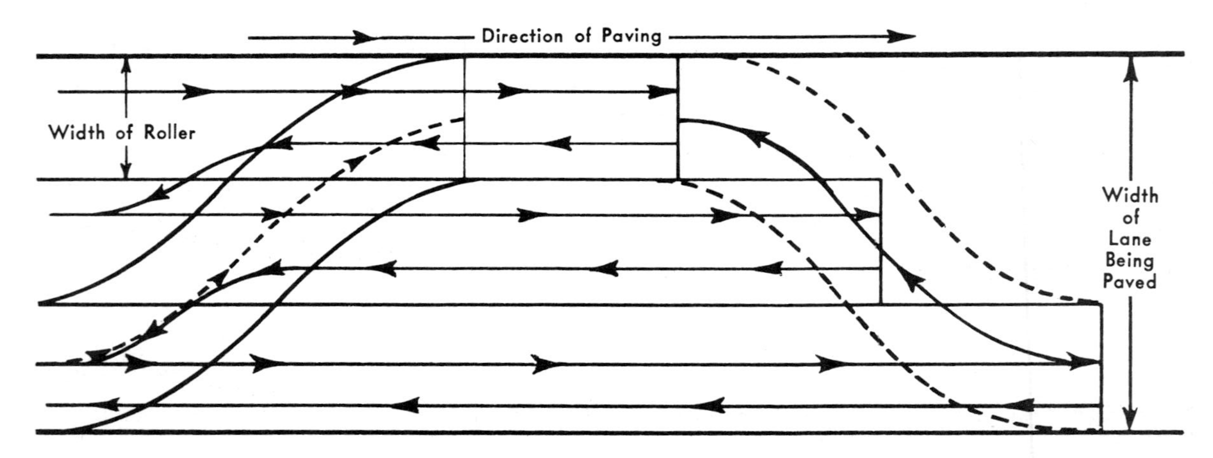

This is a recommended rolling pattern. Every pass of the roller should proceed straight into the compacted mix and return in the same path. After the required passes are completed, the roller should move to the outside of the pavement on cooled material and repeat the procedure. A second method shown is to move to the outside across the ends of the previous passes. This requires turning on the hot mix, however, and the first method is preferred.

Fig. 25-49 Correct rolling pattern.

pact a new pavement, engineers now believe that much of this consolidation should be accomplished during the pavement construction in order to eliminate roughness which will otherwise appear after the first few months of traffic. Pneumatic-tired rollers of the type illustrated in Fig. 25-50 have gone a long way toward accomplishing improved compaction.

The pneumatic-tire roller follows the same pattern as the breakdown roller, starting at the outside edge and working toward the center or crown of the road. As soon as uniform coverage has been accomplished and no further densification is apparently occurring, pneumatic-tire rolling should stop. Further rolling will probably produce little or no further compaction and in some cases may encourage pumping of the asphalt cement to the surface. The flexibility of the pneumatic-tire roller is especially good since its compactive effort is a combination of tire pressures, tire size, and the ballast weight. Engineers feel that pneumatic-tire rollers can be used to reproduce, artificially, the same conditions that would occur from heavy truck traffic.

At the conclusion of the pneumatic-tire rolling operation, the surface of the bituminous mat, although almost totally densified and compacted to the specifications, may still present an unfinished appearance. The application of a tandem steel-wheel

Fig. 25-50 Pneumatic-tire roller. (*Galion Iron Works & Mfg. Co.*)

finishing roller will eliminate any marks left by the pneumatic-tire rolling and will promote a smooth, uniform surface texture. This final rolling must be done before the temperature of the mix has cooled beyond the point permitting such corrections (see Fig. 25-51).

The number and size of the various rolling units are usually specified by the engineer in charge. Pertinent factors include the speed of the paving operation, the nature of the mix being placed, general requirements of the job, and daily atmospheric conditions. The spacing and speed of the three rolling operations will therefore vary, depending on these circumstances. It is not unusual to find the finished rolling operation a considerable distance behind the paving machine during hot summer days. On the other hand, the three operations should be quite closely grouped in cool weather.

Figure 25-52 illustrates the basic types of compactors. The three rolling units just described are most commonly used, but there are other units which offer special advantages. A three-axle tandem steel-wheel roller is frequently used to assist in leveling operations. Some designs of this equipment will permit the automatic shifting of weight or ground-contact pressures to the center wheel when it goes over high spots, thus assisting in flattening these spots for additional leveling. As soon as the high spot flattens out, the roller weight is redistributed uniformly.

A variation of the plain steel-wheel roller employs vibration by eccentric weighted shafts for a combination of static and dynamic compactive efforts. The vibrating

Fig. 25-51 Tandem steel-wheel roller. (*Galion Iron Works & Mfg. Co.*)

force of the eccentric weighted shaft can be varied to suit a variety of situations. In principle, the vibratory forces are transmitted in all directions and to greater depths than static forces. They assist in orienting or relocating particles for greater compaction.

The size and weight of the three rollers most generally used in bituminous work will vary considerably. In general, the three-wheel breakdown roller will vary in weight from 10 to 20 tons on average-sized construction work. The tandem-axle rollers will vary from 8 to 12 tons. Self-propelled pneumatic-tire rollers have an even wider weight range, from about 5 to 35 tons or more. For an individual roller size, the contact pressure may be varied considerably by changing the ballast weight and tire pressures. Some very large pneumatic-tire compactors of the towed type range upward to 50 tons in weight. They are often used for proofrolling, wherein test strips are subjected to high stress for a specified number of loading repetitions.

Fig. 25-52 Types of compactors used in bituminous pavement construction.

Paving Tips Common causes of mat deficiencies are summarized in Table 25-8. Recommendations for paving practice are summarized in the following paragraphs:

Laying

1. Set truck brake slightly before dumping.
2. Operate finisher at a rate only slightly greater than the capacity of the plant which supplies the mix, thus avoiding frequent stops.
3. End strips with square, vertical edge.
4. Vary screed crown to suit contour at intersections.
5. Clean areas over manholes prior to rolling.
6. Allow at least 4-in. clearance between outside of screed and straight curb.
7. On binder, treat gutter flange same as straight curb (4-in. clearance). Screed can overhang gutter on top course.
8. Leave uncompacted, hand-raked mix at a higher level than mat placed under screed.
9. Give careful thought to the width of each strip to be laid so as to minimize longitudinal joints and to permit maximum capacity.
10. Lay cutoff strips prior to final strip.
11. Cutoff shoe should be on opposite side of matching joints.
12. Small overlap (1 to 3 in.) is helpful when matching joints.
13. Brooming in most cases robs a joint of needed material and causes voids or depressions at the joint.

TABLE 25-8 Causes of Mat Deficiencies

Location	Cause	Cracking	Tearing	Wavy mats	Segregation
Mix	Excess No. 200 mesh material	x			
	Too hot or too cold	x			
	Too dry or too rich		x		
	Lack of fines		x		
	Too cold		x		
	Improper ratio of mat thickness to aggregate size		x		
Plant	Mixing temperature fluctuations			x	
	Segregated aggregate stockpile				x
	Poor cold feed				x
	No. 1 hot bin segregation				x
	Insufficient dry mix time				x
Trucks	Truck brake set too hard			x	
	Improper loading of truck				x
Roller	Improper rolling			x	
	Overrolling where base deflects	x			
	Turning too abruptly	x			
	Reversing too abruptly	x		x	
Finisher	Buildup in hopper sides				x
	Flushing of fines				x
	Screed overcontrol			x	
	Overloading spreading screws			x	
	Screed rams holding			x	
	Condition of tamper or screed		x		
	Adjustment of tamper or screed		x		

Rolling

1. Reverse roller slowly and smoothly.
2. Hold roller speed to point where shoving and separation of mat do not occur.
3. On short strips where hot, unrolled joints are to be matched, keep roller 6 in. to 1 ft from edge.
4. Avoid stopping roller in same transverse location after each pass.
5. Do not allow roller to stand on hot mat.

B. Portland-cement Concrete Pavements

This subsection will discuss the employment of modern machines and methods for producing durable concrete pavement with good riding qualities. It will not attempt an exhaustive survey of all the methods that may be used to build an acceptable pavement, nor will it try to provide an inspector's guide covering all the minute details of pavement construction. Rather, it will focus primarily on the high-production capabilities of mechanized devices and the advantages of their use on large paving projects.

Within recent times, the procedures and equipment for construction of concrete pavement have undergone revolutionary changes. The old methods, involving an assortment of manually adjusted machines and a great deal of hand labor, have been almost entirely displaced. A few versatile and sophisticated machines have now eliminated most of the hand labor and have greatly increased the rate of production.

Probably the two most significant developments of recent times have been the central-mix plant and the slip-form paver. These relatively new concepts in concrete pavement technology are practicable for use in most situations and are providing advantages in both economy and productivity for the paving contractor. The development and adoption of these and other mechanized and automated devices, together with greatly improved control procedures, have gone a long way toward eliminating human error. They have made it possible for contractors to do a better job, with greater production and reduced labor costs.

Evidence of the greatly improved construction methods is the rate at which concrete pavement are now being built. Construction of more than a mile a day of 24-ft pavement is commonplace. Some contractors are achieving more than 2 mi a day with a single paving spread.

PREPARATION FOR PAVING*

Preparation for paving will include stabilization of the material that is to provide the pavement foundation, setting forms when they are to be used, and finishing of the bearing surface to grade within the specified tolerances. The foundation material supporting the concrete may, in rare instances, consist of the on-site soil; but it will more generally consist of select materials used to improve stability and provide a uniform bearing capacity. Stabilized subbases, using portland cement, lime, or bituminous materials, have gained rapid acceptance by designers and contractors for several reasons: (1) they not only give additional pavement support and prevent "mud pumping," but they also furnish the contractor with an all-weather working platform; (2) they provide more precise and stable subbase elevations, which facilitate form setting and are absolute requirements for slip-form paving; and (3) they result in low yield loss of the concrete.

Compaction Equipment Certain items of equipment have long been successfully used for compacting subgrade and subbase materials to specified density. Sheepsfoot rollers work well in heavy clay soils, while pneumatic-tired wobble-wheeled rollers work better on gravelly material. Vibratory compactors work well on sandy materials. Tandem-axle steel rollers are well adapted for final compaction. The use of these items of equipment is well known and will not be further dealt with in this discussion.

Automated Earthwork Equipment Final preparation for paving has been greatly facilitated by the use of multipurpose combination machines that can trim the subgrade, spread the subbase materials, and trim the compacted subbase. These

* See also Sec. 16, Soil Surfacing Equipment.

machines offer considerable advantage over older methods that employed a variety of equipment and sizable crews and that required laborious adjustments in order to produce a compacted surface and bring it within specified tolerances.

A typical machine of this new type has a 30-ft wheelbase and is 28 ft wide, traveling on four 16-in. crawler tracks. The 28-ft working width may be extended by the addition of wings that are available for the purpose. The basic components are a single, two-piece, 30-in.-diameter rotary cutter; a 30-in.-diameter helicoid screw spreader with hydrostatic transmission for rotation in either direction; and tandem two-piece moldboard templates. The machine is automatically controlled for elevation by four control systems, one at each corner, which operate from a neoprene string line. Positive ⅛-in. tolerance control is provided by an autolevel control system. Tracers operating from the string lines instantly sense any irregularities from a true plane, and the master control unit immediately makes a hydraulic correction. The machine can trim subgrade and spread subbase material, each at a rate of 75 fpm, or it can trim compacted subbase (fine grading) at a rate of 0.5 mph.

The machine is generally put into operation after the bulk earthwork (cutting, filling, etc.) has been completed and the contractor is ready to start cutting the final

Fig. 25-53 Combination trimmer-spreader preparing subbase in multiple-lane construction.

profile. Recommended procedure is to set the machine to cut 0.3 ft above grade elevation and make an initial pass. If there are low spots, fill material is brought in and another pass is made with the setting 0.1 ft above grade. Any remaining lows are then filled and, if the schedule calls for immediate application of subbase material, a final cut to neat line tolerance is then made.

The machine is particularly adapted to spreading stabilized or natural subbase materials that have been windrowed on the subgrade (see Fig. 25-53). With the previously described controls, the machine spreads the select materials uniformly and discharges excess material from either side or from both sides simultaneously. After final compaction is completed on the subbase, the machine trims or fine grades the completed subbase to within ⅛-in. tolerance so that forms can be set or the slip-form paver operated without any further grading preparations.

Forms and Form Setting Mechanization of paving operations has resulted in heavier and more complex equipment which, in some instances, will operate directly on the forms. This makes it more important than ever that, when forms are used, they be suitable for the loading and properly set for carrying this equipment as well as for providing a pavement that is built to proper line and grade.

Before any forms are set, they should be individually inspected and approved for use on the project. The top of the rail should be straightedged to check compliance with the requirements of the specifications. The braces, locks, and wedges should be in good order, and the welds or rivets fastening the braces should be tight. Twisted, bent, or otherwise defective forms should be removed from the project.

The use of the trimming and fine-grading equipment previously described should greatly simplify the setting of forms to the proper line and grade. A minor amount of work may be required to ensure the firm seating of forms at the proper elevation. The forms should be set accurately to line and grade, form pins driven, and the wedges and form locks made tight. The forms should then be tamped by hand or machine methods. If the tamping is to be effective, it will be necessary to place a small ridge of material to be tamped under the forms along both the inside and outside of the base. Very little tamping will be required if the trimming and fine-grading equipment was properly adjusted.

CONCRETE PRODUCTION

At the beginning of 1969, about 80 percent of the concrete used for highway paving was being produced in central-mixing plants. Such wide acceptance of this relatively new method of concrete production is substantial evidence of the advantages which it offers to the paving contractor. Because of the equipment investment, dry-batching and on-site mixers will no doubt continue to be used for a considerable period of time. This discussion, however, will deal only with central-mix methods of concrete production.

Fig. 25-54 Central-mix plant with twin tilting mixing drums.

Central-plant mixing became possible with the development of highly portable, automated batching and mixing equipment, the parallel development of special spreaders for central mix, and the adoption of more realistic mixing times for large drum mixers. Special truck bodies were also developed for hauling and discharging central-mix concrete—although a considerable amount of the concrete is hauled in regular dump trucks that can be used for other tasks such as hauling aggregates.

A detailed description of central-mix plant operations would serve no useful purpose, since there will be some variations depending on the make or model of the equipment being used. The principal components of the central-mix plant are the aggregate and cement bins, water tank, batch conveyor, mixing drums, and control panel. For transportability, the large components are generally equipped with wheels and built-in electric and braking systems so that they can be moved as "trailers" by suitable truck-tractors. Erection of the plant in a new location can be rapidly accomplished. Manufacturers claim that a four-man crew can set up the plant in about 8 hr, using a crane as erection equipment.

Plant capacity is variable, depending on the mixing time and the size and number of mixing drums. A combination frequently used consists of twin 8-cu yd mixing drums that together are capable of producing up to 600 cu yd of concrete per hour (see Fig. 25-54). Some plants are equipped with one horizontal mixer that is used in com-

Fig. 25-55 Central-mix plant with one horizontal-drum and one tilting-drum mixer.

bination with a tilting mixer, as shown in Fig. 25-55. The horizontal drum mixes for approximately half the required mixing time and then discharges into the tilting drum for completion of the mixing cycle. This arrangement provides 50 to 60 percent more production than a tilting drum alone and avoids the complication of alternately charging two tilting-drum mixers. Tilting mixing drums of 10-cu yd capacity

Fig. 25-56 Central-mix plant with one large tilting-drum mixer.

are also available, and a plant with a single mixer of this capacity may be adequate for one paving site. (See Fig. 25-56.)

Mixing time on central-mix concrete has been established at 50 to 60 sec providing certain conditions are met. Studies made by the U.S. Bureau of Public Roads[7] reveal that minimum mixing time for large central mixers of the tilting-drum type should be determined by plant-mixer performance tests in any case where reasonably possible and in all cases involving 20,000 cu yd or more of concrete. Mixer performance test data should be based on three concrete samples taken at the mixer from each test batch. Samples should be representative of the front, middle, and rear portions of the batch at the time of or immediately prior to discharge. In no event should mixing time be less than 40 sec. When mixer performance tests are not made, minimum mixing time should be 75 sec provided that blending of materials during charging is achieved to the satisfaction of the engineer.

Proper plant adjustment essentially means correctly controlled timing of each ingredient during charging of materials into the mixer and thorough blending of all the batch components (aggregates, cement, water, and any admixtures used).

One of the most significant features of the central-mix plant is its virtually complete automation. A single operator at the control panel performs all the functions of charging, mixing, and discharging. The control panel is usually placed in an air-conditioned trailer which is located to oversee all operations. The slump meter which measures the electric current required to turn the mixer is calibrated to slump tests and gives the operator a reliable check of the consistency of the concrete before it leaves the mixer. Batching uniformity is achieved by push-button electronic and hydraulic operations of weighing and recording each ingredient for each batch. An automatic moisture meter in the sand hopper evaluates the moisture and automatically compensates the total amount of water in each batch—another aid in the maintenance of uniformity.

HAULING CENTRAL-MIXED CONCRETE

Concrete produced in central-mixing plants may be hauled to the paving site in several types of trucks. Special bathtub-type trucks have been developed for the purpose. Both agitating and nonagitating types have been manufactured, and bodies have been designed to dump from the rear or from the side. Side-dump units such as shown in Fig. 25-57 are particularly adaptable to situations where space for truck maneuvering is limited and driving on the subbase is not permitted. These special haul units are produced with bodies having capacities of 6 and 8 cu yd, although the 8-cu yd model is more consistent with the 8-cu yd mixing drums frequently used at the central-mix plant. Specifications may sometimes require that concrete be hauled in agitator units, although there is ample evidence that low-slump air-entrained concrete retains its quality when hauled in nonagitating units.

Central-mix hauling on many projects is being done in ordinary truck boxes. Dump trucks become practical and economical when they can operate on stabilized subbases and discharge into spreader boxes or directly on the subbase as is permitted in some slip-form paving operations.

Some further discussion of haul units will be included in the descriptions of spreading equipment and slip-form paving.

PLACING CONCRETE AND STEEL

Concrete paving operations have advanced to the point where spreading, consolidating, and finishing are almost wholly mechanized operations. Machine methods of placing and finishing have several advantages. First, there are the reduced manpower requirements and lower construction costs. Second, when the equipment is in good mechanical condition, is kept in proper adjustment, and is manned by trained and capable operators, the resulting pavement surface will be smoother.

Fig. 25-57 Side-dump trucks servicing two-lift slip-form project.

Slip-form Paving As previously stated, slip-form paving is doubtless one of the most significant developments in modern concrete pavement technology. It is discussed here because such operations as the hauling and spreading of concrete, as well as the installation of dowels and reinforcing steel, are involved in both slip-form and fixed-form methods of paving.

While the several manufactured models will vary in methods of obtaining results, all acceptable slip-form pavers perform the functions of spreading, vibrating, striking off, consolidating, and finishing the pavement to the prescribed cross section and profile with a minimum of handwork required for an acceptable riding surface. The name "slip form" is derived from the fact that the side forms of the machine, which vary from 16 to 48 ft in length, slide forward with the paver and leave the slab edges unsupported.

The basic operating process on one widely used machine is illustrated in Fig. 25-58 and consists of molding the plastic concrete to the desired cross section and profile under a single, relatively large conforming screed. This is accomplished with a full-width dead screed, 3 to 6 ft in length, which is maintained at a predetermined elevation and cross slope with hydraulic jacks that are actuated through an automatic control system. The control is referenced to offset grade lines preerected parallel to the planned profile for each pavement edge. Concrete is delivered into the forward open-bottom receiving hopper. It is then, in effect, liquefied by intense high-frequency internal vibration so that it flows into and completely fills the space between the sliding side forms, conforming plate, and the underlying subbase as the paver moves forward.

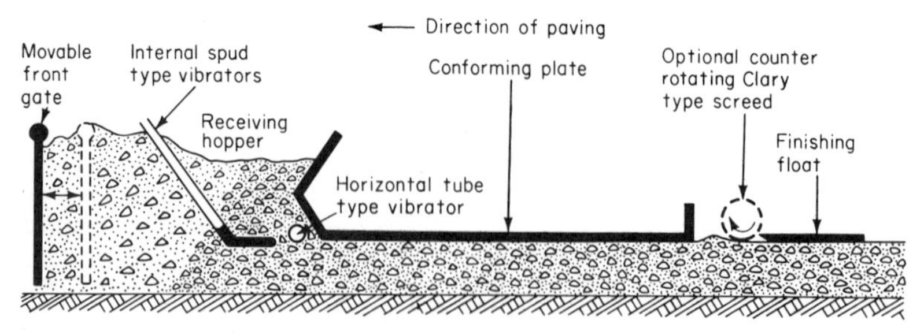

Fig. 25-58 Schematic diagram of slip-form paver.

The rotating screed located behind the conforming screed is maintained at the proper elevation and cross slope by the same controls as the conforming screed. It cuts away any excess concrete and is made to carry a slight excess of grout for filling tears or other imperfections in the surface. Variations in the amount of grout carried by the rotating screed are a signal to the paver operator that mix consistency or pressure has changed enough to cause variations in the surge behind the conforming screed. He can then adjust the frequency of vibration to compensate. Vibrating frequency is decreased to compensate for increasing mortar in front of the rotating screed and is increased when the amount of mortar ahead of the rotating screed is decreasing.

All available models of slip-form pavers have front vibration and some type of strike-off screed as means of obtaining initial consolidation. The machines may have

Fig. 25-59 Slip-form paver with auger spreader.

a vibrating screed and oscillating screeds following the initial process, and at least one model provides a belting action preceding the final float component. Some pavers do not have a receiving hopper, operating instead with concrete deposited directly on the subbase and spread transversely by a paddle or an auger (see Fig. 25-59). All models are equipped with a final float-finisher.

While most slip-form pavers are used to pave the 24-ft paving lanes that are common to two- and four-lane divided highways, machines are available that will place 36- and 48-ft-wide pavements in one pass. These exceptionally wide machines are adaptable to multilane urban expressways and to airport pavements.

Slip-form paving has definitely been proved as a speedy and economical method for producing smooth and durable concrete pavement. There are some attendant problems, however, that should be given consideration in connection with the employment of these machines. It appears as a somewhat formidable undertaking to thoroughly compact a 24-, 36-, or 48-ft-wide slab of concrete, 6 to 12 in. thick, and shape it to final section and profile all with a single machine which is traveling from 6 to 12 fpm. This task, nevertheless, is being accomplished through knowledgeable recognition of its problems.

Although many details are involved, there are three essentials for successful operation of the slip-form paver:

1. A thorough understanding of the machine plus its accurate adjustment and expert operation
2. Construction of the subbase to accurate tolerances for grade and cross section, including an area for the paver tracks just outside each pavement edge that is shaped and compacted to comparable grade tolerances
3. An extremely high degree of batch-to-batch uniformity in the concrete mix delivered at the paver

Certainly any contractor who acquires a slip-form paver will require and obtain manufacturer assistance in training operators in the initial use of the machine. The matter of subbase preparation has already been discussed under Preparation for Paving. Concrete of the essential uniformity is within the capabilities of a modern central-mix plant. The installation of reinforcing steel, also a problem, will be given attention later.

Spreading Concrete The use of centrally mixed concrete created the need for spreading equipment specifically designed to receive concrete from trucks and spread it rapidly and uniformly on the subbase. Such equipment is available in several

Fig. 25-60 Hopper-type spreader being loaded for second lift on formed paving project.

variations. The typical spreader consists essentially of a hopper than can be loaded from either shoulder and that moves transversely back and forth across the paving lane. (See Fig. 25-60.)

A typical, widely used spreader is equipped with a 12-ft hopper that can take the 8 cu yd of concrete hauled by a side-dump truck and deposit it on the subbase in 30 sec. The machine may be equipped with either flanged wheels or crawler tracks, making it interchangeable for either formed or slip-form paving. From a central control panel, one operator controls the machine in its forward or reverse travel, spreading, and strike-off. The hopper is hydraulically controlled and automatically places and strikes off the concrete to a preset depth. The hydraulic strike-off is an extrusion-type bent plate with vertical adjustments for various depths of placement. The spreader's crawler assembly has integral concrete retainers that form the slab edge when the machine is used preceding the slip-form paver. This arrangement permits the placing of concrete at partial depth to accommodate the installation of steel-mesh reinforcement.

Other types of spreaders have been developed that are especially adapted for use with concrete hauled in standard dump trucks. Belt spreaders that can receive concrete from dump trucks operating off the shoulder have been used on some projects.

Box-type spreaders are also available for receiving concrete from conventional dump trucks and spreading it on the subbase.

The subbase should be moistened ahead of the spreading of concrete. Forms must be oiled to prevent concrete from adhering to them.

Steel Reinforcement and Dowels The installation of steel mesh, bar mats, and dowels has long been the source of major problems for the paving contractor. In the past, it was the general practice to construct a formed pavement in two lifts with the reinforcing steel sandwiched between the two layers. Dowels were assembled in a supporting framework and placed at proper elevation on the subbase preceding the first course of concrete. Handling, processing, and installation of the steel components were tedious and expensive procedures and were subject to errors that contributed to pavement defects.

In the quest for more economical and efficient methods of installing dowels and reinforcement, two devices have been developed. These are showing good results, particularly in the construction of fixed-form pavement. Dowel-placing machines, one of the improved devices, were developed several years ago but have been widely used only recently. The dowel placer, one version of which is illustrated in Fig. 25-61, is

Fig. 25-61 Dowel-placing machine.

relatively simple in principle. Dowels are set in a slot device that, when vibrated into the fresh concrete, places the bars in exact position with minimum disturbance of the concrete (see Fig. 25-62). This procedure eliminates the requirement for elaborate supporting framework for the dowels and also ensures that the dowel bars will be uniformly surrounded with concrete. The dowel bars are frequently given a plastic coating so that they will not require painting or greasing prior to insertion into the concrete.

The mechanical mesh placer is a very recent development. It had its beginning when the mesh was taken off the shoulder or back slope where it was spread and placed on a mesh carrier towed by the spreader. The next step was to develop a following device which would force the steel into the concrete. For placing mesh or reinforcing mat between forms, the placing machine is usually mounted on the front end of a long-wheelbase finishing machine. The mesh or mat is depressed in 15-ft-long sections from the slab surface, with the machine standing at rest on the side forms. The placing machine uses vibration in placing the reinforcement, which should eliminate any necessity for other surface vibration prior to finishing the slab surface.

When permitted by the specifications, the dowel placer and the mesh placer may be used in combination for placing dowels and distributed steel on a formed concrete paving project. Since dowel bars are placed at the midpoint of the slab and the distributed steel is usually placed in the top third of the pavement, the mesh placer can

Fig. 25-62 Dowel placer with bars ready for lowering into concrete.

logically follow the dowel placing machine in the construction of formed concrete. Recognizing that the purpose of distributed steel is to prevent surface cracking at long joint intervals, it should be permissible to omit reinforcement immediately adjacent to joints where the internal stresses are minimal. With such discontinuity, the mesh placer will not interfere with the previously installed dowels.

The installation of dowels and distributed steel in pavement built by slip form has usually been done by the two-lift method. Generally this is accomplished by having the mechanical spreader place a 6- or 7-in. layer of concrete at a reduced width. A reduction of about 2 in. on each side allows passage of the slipform paver. This machine will form the completed pavement edge as it places the top layer of concrete above the reinforcing steel. In this operation it is necessary to eliminate internal vibration which would interfere with the steel. The mesh or mats are placed by hand, generally from a mesh cart towed by the spreader. With this method of installation, the dowels will usually be placed on the subbase and supported by the proper framework.

Mechanical dowel and mesh placers have been used with the slip-form paver (see Fig. 25-63). Such use is largely dependent on the resourcefulness and ingenuity of the contractor and the flexibility of job specifications. Space does not permit the enumeration of all equipment combinations that have been successfully employed.

Fig. 25-63 Slip-form paver with mesh placer mounted in front.

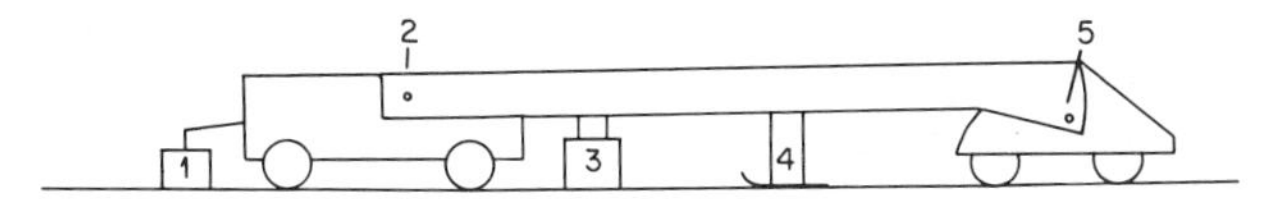

Fig. 25-64 Schematic diagram of finisher-float. (1) Front oscillating screed, (2) pivot point, (3) suspended metering screed, (4) float, (5) pivot point.

A considerable amount of continuously reinforced concrete pavement is being built throughout the country. The general practice, in both formed and slip-form paving, has been to support the steel at the proper height and to place it on the subbase ahead of the paving operation. With machines available for vibrating bar mats into the plastic concrete, it appears probable that this will become the predominant method of installation.

FINISHING CONCRETE

For many years, dating almost back to the beginning of mechanized paving operations, the two-screed finisher and longitudinal float were standard equipment for consolidating and finishing concrete pavement. While these machines did an acceptable job of consolidation, a great deal of handwork had to follow the machines in order to provide a smooth surface.

The combination finisher-float has provided a great improvement over methods formerly used for building fixed-form concrete pavement. This machine furnishes screeding and floating action with one piece of equipment, and it leaves few imperfections to be corrected by hand tools or other methods.

The finisher-float assembly is approximately 24 ft long and consists essentially of an oscillating front screed, a following screed suspended from the central frame, and a pan float following the second screed. Figure 25-64 illustrates the pivoted fore and aft components. This pivoting provides a bogie action which contributes to surface smoothness. The screeds and float are raised and lowered by hydraulic controls handled by a single operator.

Another piece of equipment that is designed to give a final finish to the slab is the

Fig. 25-65 Tube finisher with burlap drag on slip-form slab.

tube finisher illustrated in Fig. 25-65. This self-propelled machine incorporates an 8-in.-diameter by 40-ft-long aluminum finishing tube that is angled diagonally across the pavement. The tube is hydraulically operated and can be raised or lowered and swing-adjusted to either diagonal. It can be operated in either direction and worked back and forth until a satisfactory surface is obtained. The machine is equipped with a water-spray system that can distribute a fine mist over the entire length of the finishing tube when this is desirable. Another component of the finisher is a hydraulically operated burlap drag that can be used with the final pass of the machine. The tube finisher is adaptable for use on either formed or slip-form pavement.

The proper use of the tube finisher should eliminate the need for handwork except for the possible use of straightedges to check the surface during final finishing operations.

JOINTS

The types of joints to be provided in the concrete pavement, their intervals and dimensions (depth and width), will be covered by the job plans and specifications. Because joints are so critical to pavement performance, however, some discussion of their types and purpose should be included here.

The four types of joints used in concrete pavement are (1) construction joints, both longitudinal and transverse; (2) longitudinal center joints, sometimes called "hinged joints," to relieve curling and warping stresses; (3) contraction joints in the transverse direction across paving lanes; and (4) expansion joints which are full pavement depth, with a compressible filler to permit the joint to close as the pavement expands. Regularly spaced expansion joints are seldom provided under modern practice. Tests and experience have shown that, under normal conditions, contraction joints provide adequate space for pavement expansion. Expansion joints are still used adjacent to fixed structures and at asymmetrical intersections.

Jointing arrangement may vary widely, depending on the local construction practices. Dowels and distributed steel are omitted in some localities where cement-stabilized subbases are used. In these instances, contraction joints will be at short intervals, on the order of 15 to 20 ft. The joints may be skewed diagonally across the pavement, or they may be placed at irregular intervals such as 15, 18, and 21 ft, with repetitions of this randomized grouping. Joints placed at such short intervals may be of rather narrow width, such as ⅜ in., since each joint will have little movement in opening and closing. Where distributed steel reinforcement is used, joints will be spaced at much wider intervals, such as 40 to 100 ft, since wider joint spacing is the primary reason for the use of distributed steel. At these longer intervals, contraction joints will be much wider, from ½ in. to as much as 1 in., in order to accommodate a greater amount of movement.

Joint Forming While there are several acceptable ways of forming joints, sawing is by far the most common and doubtless the most economical method. Saw manufacturers have been very active and resourceful and are able to provide power saws that are adaptable to any jointing situation. This equipment has been in use for several years, so it will not require a detailed description here. Something should be said about the timing of sawing, however. Since contraction stresses develop as soon as the concrete hardens, contraction joints must be sawed very early. The exact time of sawing depends upon the type of aggregate, curing method, cement factor, and the weather. Generally, all contraction joints should be sawed as soon as this can be done without damage to the surface. A slight amount of raveling is permissible and desirable because it gives the operator a good gauge of his timing. If there is no raveling at all, the concrete is too hard and cracks may develop ahead of the saw. All joints should be sawed in succession to provide the plane of weakness at the time of maximum contraction and before the slab gains too much strength. This will ensure cracking and uniform opening of all joints. Sawed joints should be thoroughly flushed or blown clean immediately after sawing to remove all residue, which otherwise may cake on the sides of the joint and hinder proper sealing.

When longitudinal center joints are sawed, the timing is not so critical. These joints may be formed at any convenient time before traffic is allowed on the pavement. A common method of forming longitudinal center joints in slip-form paving is by the use of a continuous polyethylene strip. This is automatically placed into the concrete from a roller carried at the rear of the slip-form paver. Center joints in formed pavement may be formed in a similar manner by devices that are available.

Expansion joints, when required, are generally installed by hand methods. The methods of installing expansion joints have long been in use and do not require further elaboration.

Wherever construction joints occur, they should be grooved to a depth sufficient to retain joint sealer. Sawing is generally the most practical method of providing a joint groove. Some hand edging may be required during construction in order to prevent the plastic concrete from overlapping the hardened concrete already in place.

Joint Sealing Materials and methods for sealing joints are covered in detail by the specifications. It is important that the joints be sealed as soon as practicable to minimize the intrusion of foreign materials. Joint openings may be protected by a temporary filler such as a section of rope or a strip of tape over the opening. In any case, the joint opening should be thoroughly cleaned just before sealing.

CURING

The hardening and strength gain of concrete are dependent on the availability of moisture for the hydration processes. The durability and long-term utility of the pavement are greatly dependent on the timely use of adequate curing methods.

The specifications will cover the approved curing and protection methods, and it is essential that the curing be accomplished as required. All equipment and materials required for the application of curing must be available on the job before concrete placement is started each day.

The two most common methods of curing are by the use of waterproof membrane and polyethylene sheets. Equipment is available that is adaptable to the application of white-pigmented curing compound on either formed or slip-form pavement. Polyethylene sheeting may be placed on the pavement in long strips, anchored by loose dirt, and rolled up for further use after the curing period has expired. One advantage of the use of polyethylene sheeting is that it protects the pavement from damage by sudden rainstorms.

PAVING NONSTANDARD SECTIONS

The preceding discussions have dealt with procedures for continuous construction of standard sections of pavement. While most of the constructed pavement will fall in this category, the contractor is frequently faced with the problem of building irregularly shaped sections and short stretches of pavement that are inaccessible to mechanized equipment such as the slip-form paver.

Traffic interchange ramps and bridge approaches are generally bypassed by the slip-form paver. Some contractors have built ramps with the slip form, but such methods appear to need more development. Omitted sections of pavement at bridges result from inability to pave closer than the length of the paver. During slip-form operations, when a bridge is crossed, the contractor places construction joints at the points where operation of the paver is discontinued and then restarted. The resulting gaps must be paved by other methods at a later date.

Ramps are usually paved with forms, using a two-screed finishing machine or a long-wheelbase finisher-float. False forms are used to carry the equipment where the pavement narrows. Bridge approaches are formed, and usually paved, by hand methods with concrete delivered by truck mixers. Frequently a hand-operated vibratory screed is used, followed by hand floating and straightedging.

Most contractors employ either a special crew or a subcontractor to follow the main paving operation and complete the nonstandard sections.

PRODUCTION AND COST DATA

This portion of the discussion deals with slip-form paving and central-mix plant operation.

What are perhaps the most reliable data on cost and production have been included in the report entitled *Slip-Form Paving in the United States*, prepared by L. R. Gillis and L. S. Spickelmire and published in 1967 as a Technical Bulletin of the American Road Builders' Association.[8] The authors, both members of the California Division of Highways, compiled and analyzed data from several states in which the slip-form paver has been widely used. The following pertinent conclusions have been extracted from the report and are presented either as direct quotations or in paraphrased form.

Capital Investment Conventional paving machines, forms and related handling equipment for an average side form project will require an investment in excess of $250,000. The equipment needed to place an equivalent pavement by slip-form methods will require an investment in the order of $125,000 or about one-half.

Because of the lesser amount of equipment required on the project, field repair and maintenance problems are also correspondingly reduced.

Economy Based on their analyses of cost factors in several states, Gillis and Spickelmire concluded that savings of at least $0.50 per square yard can be achieved by using slip-form instead of fixed-form paving methods.

Operational Efficiency One slip-form paver takes the place of three or four conventional concrete paving machines which include such items as spreaders, finishers, float finishers, and belt and burlap drag machines. In addition, an average of 10,000 to 12,000 lineal feet of steel paving forms and the equipment necessary for their continuous rehandling each day are eliminated. It is not difficult to see where slip-form methods also result in a substantial savings in manpower, both skilled and unskilled. The overall savings in labor will average between 20 and 30 men, of which about 30% would be skilled.

Crew Size and Payroll Gillis and Spickelmire estimated the following crew as required for "slip-form paving projects of average complexity under average conditions for State highway construction in California during 1965." The wage scales are, of course, those in effect in California at the time stated.

Estimated Crew

Paving foreman	1 each at $5.50 to $5.60 per hr
Slip-form operator	1 each at $5.50 to $5.60 per hr
Screed men	2 each at $4.10 to $4.20 per hr
Dumpmen	2 each at $3.80 to $3.90 per hr
Grade setter	1 each at $4.80 to $4.90 per hr
Laborers (wire)	2 each at $4.00 to $4.10 per hr
Laborers	3 each at $3.80 to $3.90 per hr
Finishing foreman	1 each at $4.90 to $5.00 per hr
Concrete finishers	2 each at $4.60 to $4.70 per hr
Curing machine operator	1 each at $4.30 to $4.40 per hr
Joint saw operator	1 each at $5.10 to $5.20 per hr

Central-mix Plant It would be extremely difficult to make a valid comparison of the costs of a dry-batch operation and a central-mix plant. Harold J. Halm, executive director of the American Concrete Paving Association, makes this statement which may be considered as authoritative:[9]

> The cost of a central plant set-up is no greater than a conventional batch spread; it may be considerably less. There is usually less plant space required since there is only one stop involved. A further advantage is that several paving sites can be serviced at one time. Where specifications allow, both paving and structural projects can be furnished simultaneously from a single plant.

References

1. K. B. Woods, *Highway Engineering Handbook*, McGraw-Hill Book Company, New York, 1969.
2. *The Asphalt Handbook*, Manual Series No. 4 (MS-4), The Asphalt Institute, College Park, Md.
3. *Bituminous Construction Handbook*, Barber-Greene Company, Aurora, Ill.
4. *Asphalt Plant Manual*, Manual Series No. 3 (MS-3), The Asphalt Institute, College Park, Md.
5. *Specifications and Construction Methods for Asphalt Concrete*, Specification Series No. 1 (SS-1), The Asphalt Institute, College Park, Md.
6. *ASTM Book of Standards*, part II, American Society for Testing and Materials, Philadelphia.
7. F. M. Bozarth and W. E. Grieb, *A Study of Mixing Performance of Large Central Plant Concrete Mixers*, Office of Research and Development, Bureau of Public Roads, U.S. Department of Commerce, paper presented at Highway Research Board Annual Meeting, Washington, D.C., January, 1966.
8. L. R. Gillis and L. S. Spickelmire, *Slip-Form Paving in the United States*, Technical Bulletin No. 263, 1967, American Road Builders' Association, Washington, D.C.
9. Harold J. Halm, *Today's Trends in Concrete Paving*, paper presented at the Ohio Highway Engineering Conference, Columbus, Ohio, April, 1966.

Section 26

Pipelines

LAURENCE L. LYLES

Executive Vice-president and General Manager, W. M. Lyles Company, Fresno, Calif.
(Underground Utility Construction)

and

HAROLD C. PRICE

Chairman, H. C. Price Company, Bartlesville, Okla.
(Transmission Pipelines)

A. Underground Utility Construction

It is quite likely that utility construction is the second oldest type of construction that man has undertaken. Probably the first thing man sought was shelter, where he improvised and improved his living conditions. It is also likely that the second convenience he wanted was ready access to water, a need which he could satisfy by digging a ditch and having the water come closer to his cave. The Greeks and Romans developed sophisticated utilities and had both running water and workable sewer installations in some homes. Lead pipes for carrying water were found in the ruins of Pompeii.

The types of work that will be discussed in this subsection will include all utilities installed underground in a city to deliver energy or service to a business or home. These utilities specifically include water service; sanitary sewers; gas, electricity, and telephone service; and storm sewers. The construction problems involved with each of these utilities are very similar, although there are variations in the actual installation of the different types of pipe and conduit. Therefore, most of the steps in construction will be discussed in terms that will apply to all utilities. Each phase of utility construction will be discussed in its normal sequence.

PRELIMINARY WORK

Planning Planning is the first and most important step in the construction of underground work; it is also the point at which the most money can be made. Although sophisticated planning and scheduling techniques such as the Critical Path Method have not proved particularly successful due to the linear nature of utility work, equally thorough planning effort is necessary for a profitable job. The actual order in which various operations will be performed is rather simple because it is obvious that a trench must be dug before pipe can be installed. The timing—when each operation is to occur and how long it will take—is the key to a smooth and profitable operation. Jobs go well because they are planned well and not because of luck.

The proper execution of an underground job requires a "balanced crew." This means a crew made up of the right equipment and manpower to have all phases of the project progress at the same speed. To obtain a balanced crew, goals or objectives must first be determined: the number of feet of pipe to be installed per day, week, or whatever. Then consideration must be given to the manpower and equipment to be used, to see if they will provide the required production. Invariably there will be one phase of the work that will control the speed of the whole operation. When this is found, the work must be planned in such a way that the controlling phase will progress at the requisite speed. For example, the excavation of the trench with the equipment planned may not give the footage desired. One may then decide to change types of excavators or possibly use more than one machine.

Once the production requirements for the controlling work phase have tentatively been established, other pieces of equipment and some of the manpower on the job that do not have full days of production will invariably be found. For example, the pipe-installation crew may need only 4 hr to install all the pipe that the excavating crew can provide for in 8 hr. To maintain a balanced crew, additional work must be found in their crafts, such as making tie-ins or setting fire hydrants, for those men who would otherwise not have a full productive day. If this is impossible, the possibilities of increasing the speed of the controlling phase of the work must be looked into so that, in the end, full production may be obtained from all the manpower and most of the equipment.

Locating Crosslines Crosslines, existing underground structures that are in the work area, are probably the key item in setting production objectives because of the time required to excavate under them with any type of equipment. The number of crosslines on the job can be determined during the bidding stage; however, their exact location and depth is a job problem. Determining the location of all crosslines should be accomplished before the entire crew is brought on the job because, especially in the old areas of a city, you will get surprises as to what you find.

To do the job properly and safely, owners of all possible crosslines in the area of construction should be contacted. The utility companies, city departments, fire-alarm services, local TV antenna companies, etc., will be just as anxious as you are that no lines be cut and usually will give full cooperation. Many will come to the construction area and mark the location of their lines. Others will have you bring in a map and then indicate their lines on it.

After obtaining all available information, the only sure way to safeguard the existing lines is to have a small crew of men actually pothole and expose the lines at the points where the trench will cross them. The crosslines should be exposed at least one full day in advance of excavation so that they will not cause delays for an expensive crew. This is particularly important where the utility line is to be set to a given grade, because you must be sure that there is sufficient clearance at the crossline for the utility line to be correctly installed.

PAVEMENT REMOVAL

Pavement Breaking Pavement breaking is the shattering or cutting of the pavement into pieces that can be economically removed from the area to be excavated. Techniques vary according to the type and thickness of the material to be broken. Several manufacturers have developed fine, self-propelled machines that break pavement very economically. It goes without saying that the tougher the breaking job, the larger the machine should be. The most commonly used types of machines are the ones where a 1,000-lb or larger weight with a cutting or shattering tool is lifted about 9 ft in the air and allowed to fall free and strike the pavement. Other machines use pneumatic power to accomplish the same result (see Fig. 26-1).

An experienced operator can cut a straight line as marked on the pavement. Generally, the most economical procedure is to cut the pavement about 1 in. outside the planned trench on both sides and then break or shatter the material in between the outside cuts. Applying this procedure of three passes, 4-in. pavement can be broken at the rate of about 110 ft of trench per hour and 6-in. pavement at about 70 ft of trench per hour.

The use of hand pavement breakers (jackhammers) should be limited to jobs where the amount of pavement to be broken is very small. Since it is not possible to determine the efficiency of the operation of pavement breakers by watching or listening, it is advisable to have them tested periodically to determine the impact they are producing as compared to new tools. Such a test will usually show many tools that are only 20 to 40 percent efficient.

By removing any protruding pavement material of 2-in. or larger size after breaking, it is often possible to allow traffic to run over the broken pavement prior to its removal. This permits a better traffic flow in the construction area.

Pavement Removal Pavement removal can be accomplished by the use of a rubber-tired skip loader of a 1½-yd or larger size with a scoop attached to the front of the bucket. This scoop should be smaller than the width of the trench and slightly deeper than the pavement to be removed. If the pavement has been thoroughly shattered, this is an economical method for removing the broken pavement and loading it into dump trucks. As an example, an operator should be able to remove pavement from a 3-ft-wide trench and load it into a dump truck at the rate of about 7 cu yd of pavement per hour. Keeping in mind the balanced crew, a full 8-hr shift may be considerably more than is needed for this particular task. Other duties which can occupy the same men and equipment will then have to be planned, such as removal of excess excavated material, placing base rock, etc.

Sawing Pavement If the pavement is sawed at all, it can be sawed either before or after pavement breaking. The procedure to be followed by the contractor is usually established either by the specifications or by the customary practice in the community where the work is being done. The purpose of sawing pavement is to give an even line to repave against and leave the street looking neat, even though patched.

The advantage of sawing *before* breaking pavement is that the sawing takes the place of cutting the edge with another piece of equipment. As a result, the combined cost of pavement breaking and pavement sawing is reduced.

The disadvantages to sawing ahead are twofold: (1) in old and fractured pavement, pieces of pavement outside the sawed area are likely to become dislodged and break out; (2) the excavating equipment might catch a tooth on the edge of the pavement and break it back outside the sawed area. Both of these can leave you with uneven edges, the very thing that sawing was meant to prevent.

Fig. 26-1 Pavement breaker. (*Arrow Manufacturing Company.*)

The advantage of sawing *after* pavement breaking and just prior to repaving is that you can always leave a neat straight line to repave against. The disadvantages again are twofold: (1) the trench to be repaved is usually wider than it would otherwise have been, and (2) after sawing it is necessary to chip off the narrow piece between the sawed line and the trench. Both of these add cost to the repaving process.

Good, diamond-bladed saws are made by several manufacturers. The deeper the cut and the greater the quantity of sawing that has to be done, the larger and more powerful the saw should be for economical operation.

Using an 8-hp saw, one may expect to saw 1 in. deep in asphaltic concrete at the rate of approximately 165 lin ft per hour, and 2 in. deep at the rate of about 120 lin ft per hour. For sawing concrete, the above rates should be reduced by 40 percent.

EXCAVATING THE TRENCH

Equipment and Methods Excavation of a trench in its simplest form is merely a question of how to remove the earth from the trench for the lowest cost per cubic yard. When choosing the type of equipment to excavate a trench, the contractor will, of course, give first consideration to the equipment he owns. However, he should not be misguided into thinking that it is always cheaper to use a machine that he owns, especially when it is the wrong machine for the job. The types of excavating machines that will be discussed are wheel-type trenchers, ladder-type trenchers, and backhoes.

Where they can be used, wheel-type trenchers are the most economical means of digging a trench.* The limiting factors on wheel trenchers for utility work are the width they will dig (usually a 36-in. maximum) and the depth they will dig (usually 7 ft or less). In addition to these limitations, they leave a large quantity of earth underneath each crossline intersected. As a result, before you can say that a wheel trencher is the best machine for a certain job, the cost of the trenching plus the cost of removing the extra dirt at crosslines must be compared to other means of excavating the trench.

The ladder type of trenching machine is the next most economical method of removing earth from a trench. Generally speaking, the cost per cubic yard for excavation with a ladder trencher is approximately 50 percent higher than with a wheel trencher. Ladder machines come with almost any width and depth of cut and, compared to backhoes, leave an even-sided trench that can be shored economically and safely. In general, where there are few crosslines, the ladder trencher will give more feet of trench per day than a backhoe. Ladder trenching machines are relatively slow in lowering and raising the bucket line and, as a result, jumping crosslines and turning corners becomes expensive. They also leave a sizable amount of earth after jumping a crossline, but this can be reduced if the operator will back the machine so that the ladder digs back under the crossline. The remaining amount of earth will then be less than a wheel trencher leaves.

When using either the wheel or the ladder trencher, the largest buckets and the least amount of side-cutter width should be used within practical application. The reason for this is that the larger the buckets on the machine, the more earth will be excavated every time the wheel or ladder makes a complete revolution.

Backhoes serve several useful purposes in excavating for utilities. First, a backhoe will dig over and around a crossline with a minimum of delay. By straddling the trench already dug with a trenching machine, a backhoe is often used to remove the excess earth left at a crossline. The cost per cubic yard of excavating a complete trench with a backhoe is more than the cost with a trenching machine for the part that a trenching machine can do. However, when the cost of complete backhoe excavation is compared to that of excavating with a trencher plus a backhoe to excavate under numerous crosslines, there are many jobs where the use of a backhoe for all excavation is most economical.

Hydraulically controlled backhoes seem to be the fastest moving and most economical ones on the market. They come in all sizes, from the smallest ones that can straddle an existing trench for the purpose of excavating around crosslines to the large and very large ones. Backhoes can be found to dig almost any depth that a contractor would need for utility work. Where all the excavated material has to be hauled away from the construction site, it is easy to excavate and load the material into dump trucks in the same operation. There are three main disadvantages to a backhoe: (1) it is more expensive in cost per cubic yard to get the earth out of the trench, (2) it is very difficult to shore a trench dug with a backhoe because of the uneven sides of the trench, and (3) the bottom of the trench is left uneven.

Exact production figures for the trench excavation are impossible to give due to the many variables such as soils, working conditions, etc. The following examples are given as guides:

Assuming that the trench to be excavated is 2½ ft wide and 5 ft deep and will not

* See Sec. 12, Trenchers.

be dug to grade, the production in cubic yards per hour will approximate:

Conditions	Wheel trencher	Small backhoe, 44-hp diesel
Good soil—no crosslines.	200	50
Hard soil—no crosslines.	100	20
Good soil—crossline every 50 ft.	95	35
Hard soil—crossline every 50 ft.	60	16

If the above trench were to be excavated to grade, as required for sewer lines, etc., the production would be decreased by about 25 percent.

Assuming that the trench to be excavated is 4 ft wide and 12 ft deep and will be dug to grade, the production in cubic yards per hour will approximate:

Conditions	Ladder trencher	1½-cu yd hydraulic backhoe
Good soil—no c osslines	180	160
Hard soil—no crosslines	100	90
Good soil—crossline every 50 ft.	100	125
Hard soil—crossline every 50 ft.	60	60

Up to this point the discussion of trench excavation has been limited to normal, stable soils. A discussion of some of the problem soils would probably be of value.

Rock Excavation Rock excavation refers to the removal of materials encountered in trenching that are too hard to be removed with available excavating equipment. When rock has to be blasted, there are a few procedures that can maximize safety. First, when there are any buildings within 200 ft of the blasting, special precautions should be taken. The dirt overburden in the trench area should be removed with a backhoe so that the trench thus formed can be used to help minimize the lateral force of the blast. Adequate mats made of hemp, wire, or timbers should be placed over the charge to prevent any rocks from being thrown into the air. The proper type of blasting material should be chosen to minimize the possibility of damage to building foundations. Holes for the blasting material should be drilled about 6 in. below the bottom of the proposed trench. Only experience will tell how close the holes should be drilled to each other. Alternating the holes from one side of the trench to the other will usually produce a trench that is easier to clean out.

After the charge has been set off, the most economical means of removing the shattered rock is with a backhoe. If the overburden that has been removed before drilling and blasting is fairly deep, it may be both safer and cheaper to completely backfill the ditch after blasting and then excavate it with a backhoe.

Dewatering The method used for removal of underground water will, of necessity, vary depending on the amount of water entering the trench, the type of soil, and the hardships or hazards caused by the water. Methods of dealing with these problems will be discussed, starting with the most severe water conditions and then going on to the less serious conditions. In almost every case, it is advisable to remove all the underground water possible before starting excavation.

The most common and most successful method for "drying up" an area is by means of well points.* Since the use of well points is an accepted practice, details that are

* See Sec. 29, Construction Dewatering.

applicable under all conditions will not be repeated. The first consideration for well points is to determine the direction from which the underground water is coming. If it is seeping in from just one side of the proposed trench, then one row of points set to intercept the water will be sufficient. However, if the underground water is a pool or is able to seep into the excavation from more than one direction, it will very likely be necessary to place a row of points on each side of the excavation. The positioning of the well points should give adequate consideration to the space required for the excavator, for storage of the excavated material, and for the men and equipment needed to install the utility lines.

Where the quantity of water to be removed is not great and where the soil is sufficiently permeable so that well pumping can lower the water level for a fairly good distance (25 ft or more), the use of a series of wells with individual pumps has proved economical. This system requires the drilling and graveling or sanding of wells at appropriate intervals along the proposed trench. The wells can be any size from 12-in. diameter up and should have a center casing into which the suction hose can be placed. These wells should be at least 6 or 8 ft below the depth to be excavated because the water level, as it is pumped down, will slope toward the suction inlet in a conical shape. Before the excavation work approaches each well, a pump (usually gasoline powered) should be set up at the well and put into operation with the suction hose near the bottom of the well. The water pumped from the wells should be discharged at a location where it will not flow back into the work area. This system has proved economical where there are short areas to be dewatered and in sandy areas that are easy to dewater.

In some areas the soil may be compact enough to leave vertical banks that, after shoring, are safe even with water entering the trench. Under the circumstances, the trench can be overexcavated by a depth of 6 to 12 in. and the overexcavated section backfilled with coarse gravel. Water will then flow through the gravel to the lowest point of the trench, where it can be pumped out. This system, the "sump method," will work in areas where a relatively small amount of water is entering the trench at one time due to the firm, tight soil through which the trench was excavated.

Shoring Trench shoring should first be examined from the viewpoint of its requirements. Alternatives for shoring installation and removal can then be explored.

Requirements. Recent years have seen the development of aluminum shores, using hydraulic jacks to hold them in place. In general, these have proved to be the most economical means of shoring trenches. They are light in weight, and the man-hours saved in moving and installing them more than offset their high initial cost. They have a long life and can be used repeatedly if properly maintained. In addition, because they can be installed from the surface, they afford greater safety to the installer than wood shoring (see Fig. 26-2).

At present, no standards have been developed as to the design for aluminum shoring. One is thus forced to go back to the old timber requirements and make sure that the aluminum shores are at least as strong. A few of the rules, taken in part from the Construction Safety Orders of the State of California, that can be applied to shoring in *hard* or *compact* soil are:

1. Any trench over 5 ft deep should be shored.
2. Vertical shores for trenches from 5 to 10 ft should be 2 by 8 timbers set not more than 8 ft apart.
3. Horizontal braces between shores in trenches from 5 to 10 ft deep should be screw jacks or 4 by 4 timbers not over 5 ft apart.
4. Vertical shores for trenches over 10 ft deep should be 3 by 8 timbers set not more than 6 ft apart.
5. Horizontal braces between shores in trenches over 10 ft deep should be screw jacks or 4 by 6 timbers not over 5 ft apart. In the case of trenches over 3 ft wide, 6 by 6 timbers should be used.
6. Vertical shores should extend from above the ground to the bottom of the trench, if possible, but in no case end more than 2 ft from the bottom.

For soils that are not hard and compacted, the spacing of the shores must be reduced. In conditions of unstable or running soil, the shoring will have to be solid, and it is then called "sheeting."

Installation. As previously mentioned, the man or men who are installing aluminum shoring with hydraulic jacks can all stay on the surface. If timber shoring is to be installed, the same principles of safety should be applied. All possible work should be done from the surface; i.e., the vertical shores should be placed from the surface and the first horizontal brace installed just below the surface from above. Then, using a ladder, a man should go down into the trench just enough to install the next lower brace or trench jack, etc. By this means, the trench is made safe above him as he descends to install additional horizontal braces. Trench jacks and horizontal braces should never be used by a man as a ladder for getting in or out of a trench, as they are not designed to take a vertical load.

Fig. 26-2 Aluminum-hydraulic trench shoring. *(Allied Steel & Tractor Products, Inc.)*

A trench should never be left open without shoring. Shoring should be installed immediately behind the excavator.

Removal. When planning for the removal of shoring, it is safest to anticipate the possible collapse of the trench sides. The newly installed utility line will then be safeguarded routinely by being covered with loose or compacted fill before the shores are removed. For conditions where the trench will almost certainly cave in on removal of the shores, the trench can also be filled up to the bottom horizontal brace. It is then safe for a man to go down on a ladder and remove this brace, after which the additional trench space can be filled up to the next horizontal brace or screw jack.

If the trench is expected to stand after the removal of the shoring, all the shoring can be removed just ahead of backfilling. This should not be done until all work by men within the trench has been completed and the newly installed utility line has been protected or covered. A workman can then use a ladder to descend to the bottom horizontal trench jack and remove it. The remaining horizontal jacks are removed as he ascends the ladder. Shoring removal is hazardous work, despite the best safety precautions, and one man should never be permitted to work at it by himself.

Sand Excavation Methods of excavation in sand should be carefully studied. Adequate protection must be provided for crosslines.

Methods. Considering both economy and safety, trench excavation in sand is the greatest hazard that a contractor encounters in utility work. If the trench is to be

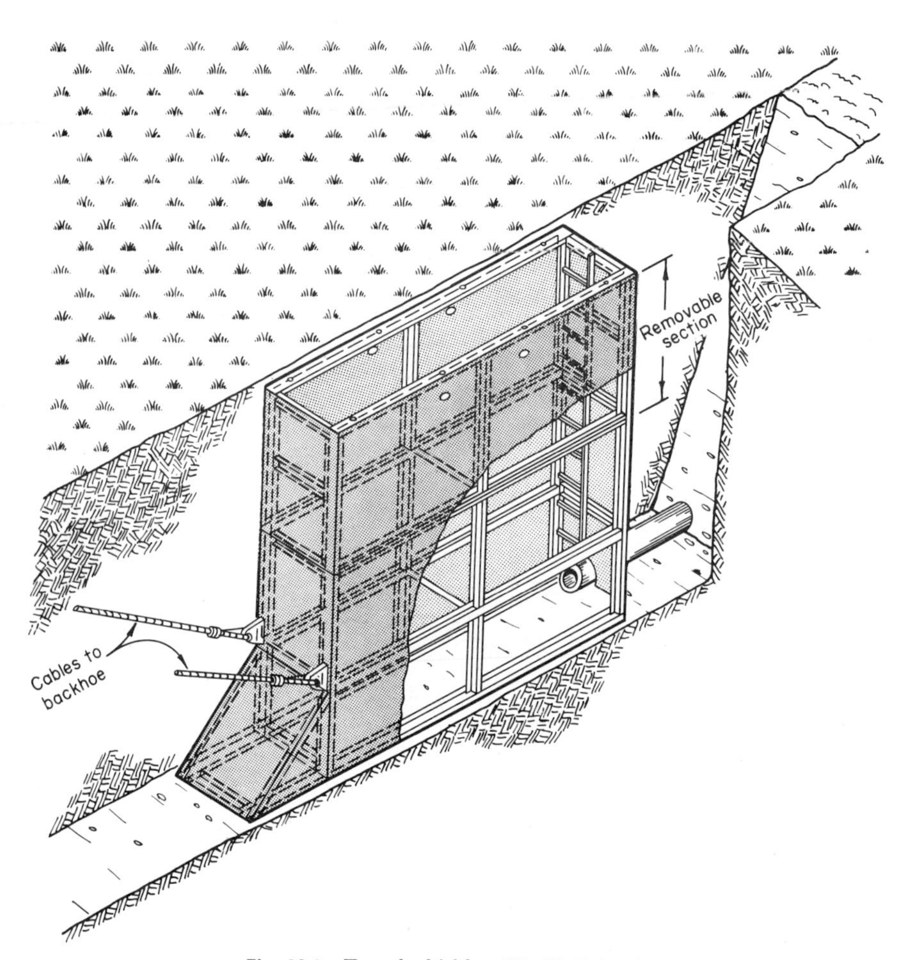

Fig. 26-3 Trench shield. (*W. M. Lyles Co.*)

excavated in a new and undeveloped subdivision or on a beach where the width of the trench is not important, the answer is both simple and relatively economical: a large backhoe or dragline may be used to "bail" out the sand. However, the trench to be excavated is usually located on a city street or highway where there is a limit to the width of the excavation that can be made. It may be necessary to drive steel sheeting prior to the trench excavation. This method has long been used and, under the most severe conditions, is the only safe means of getting the utility line installed. However, two other methods are becoming more popular when they can be applied:

1. The trencher-shield system consists of a steel shield attached to a ladder trencher. The men then have a working area which is completely protected on four sides by steel siding, appropriately braced. The length of this shield must be sufficient to allow for the installation of a length of pipe and its jointing with the last piece laid within the confines of the shield. The back side of the box, away from the trencher, has an opening large enough to clear the pipe that has been installed. As the trenching and pipe laying proceed, the excavated material must be placed in the trench immediately behind the shield by means of a conveyor or a skip loader. Where this system can be used, it is probably the safest and best means available. It has one major disadvantage—the size of the trenching machine ladder and the shield is such that it is very slow and expensive to remove it from the trench, and thus crosslines cannot be economically jumped. However, in many cases, crosslines can be temporarily cut and removed economically to allow the use of the trencher-shield system for excavating sand (see Fig. 26-3).

2. An increasingly popular system for excavating in sand makes use of a similar shield with a backhoe used as the excavator. In this system the shield is not attached to the backhoe but instead rides free on the bottom of the trench as a protective box for the workman. Cutting edges on both of its sides extend forward toward the backhoe, thus minimizing the unbraced space between the shield and the backhoe bucket. To move the shield, the backhoe operator reaches back with the backhoe bucket and hooks a cable onto it. He can then pull the shield toward the backhoe the distance desired. The disadvantage of this system is that the shield is not stable because it is freestanding. The sides of the excavation are rough trimmed, and sufficient caving can occur to knock the shield out of line. In addition, controlling the grade of the pipe and the depth of the shield is difficult. The backhoe-shield has the advantage that, since it is not attached to the backhoe, it can be dragged under shallow crosslines if the sides of the excavation above the shield can be sloped to a safe angle of repose.

Crossline Protection If the crosslines are removed before a trench is excavated in sand, their only protection requirement is for their proper replacement. Here the most important factor is to be sure that all the sand under the reinstalled crossline is thoroughly compacted. If this is not done, the subsequent backfilling and compaction of the trench is sure to depress the replaced line, causing it to be at the wrong elevation or even breaking it.

Where crosslines are not removed and sand is excavated from under them, there is also the problem of safeguarding the crossline while the excavation is open. Long timbers or telephone poles that will span the excavated area can be used as supports from which the crossline can be suspended by means of ropes or cables.

INSTALLATION OF PIPE OR CONDUIT

With the ever-increasing inventiveness of men, we are seeing new materials used every year for pipe and conduits. As a result of the constant change and because the manufacturers go to great lengths to furnish details to all interested parties as to the composition, application, and installation methods for their pipe, information in this section will be confined to the commonly used pipes and conduits.

Water Mains The most commonly used types of pipe for water mains are cast iron and asbestos cement, both of which have many years of successful use behind them. In larger sizes, steel-cylinder concrete pipe is widely used. Plastic pipe is gaining acceptance, and it is likely that reinforced plastic mortar (RPM) pipe will also become accepted.

Due primarily to the ever-increasing cost of labor, almost all types of pipe are now being produced with a rubber or neoprene type of joint. These joints have several advantages in addition to the labor-saving advantage. They allow flexibility at each joint and, in the case of steel pipe, allow the joining of the lengths of pipe without destroying the interior pipe protection by welding.

Generally, the fittings, valves, and fire hydrants for the line can be obtained with the same type joint as the pipe. However, some localities prefer a flanged joint for valves and fire hydrants to make replacement easier.

The use of flexible joints has made the proper installation of thrust blocks even more important. The size of the bearing surface of the thrust block is generally specified. The main items to which attention should be given are (1) that a good bearing surface with the pipe is obtained and (2) that the bearing surface against the trench side is against native soil or well-compacted backfill. In soft, wet, or sandy soils, larger thrust blocks may be needed. If the waterline is to be tested at pressures over 150 psi, extra attention should be given to thrust blocks.

Installation. Manual installation of water pipe is usually the most economical when the sections are 20 ft in length or less and weigh 150 lb or less. The pipe is first strung close to the trench or is rolled within reach of the trench. Two men who are standing in the ditch then reach over and lower the pipe into position. Using this method, with adequate trench prepared ahead and not too many crosslines, 2,000 ft of water main can be installed in an 8-hr day. In most cases, the fittings and valves will be too heavy to install manually. When possible, the best method is to prefabricate fittings and valves and have them located adjacent to the trench where they will be installed. When the laying crew reaches the location for a fitting or valve, they can cut the pipe to length and, with the help of a piece of equipment, lower the fitting into place. This will minimize the installation time for valves or fittings.

The installation of heavier lengths of pipe requires equipment to lift the pipe and lower it into place. For pipes up to 1,200 lb per joint, a swing crane mounted on a small tracklaying tractor has proved economical. Side-boom tractors have proved economical for larger-sized pipe. These two types of equipment have the economic advantage of requiring only one operator and no helper or oiler. Any piece of equipment that requires the employment of two men should be avoided if possible. In addition to the added cost of one man, there invariably seem to be delays involved in the relaying of signals back and forth.

Larger-diameter water mains have to be installed in deeper trenches that invariably require shoring or protection. In open areas, sloping the sides of the trench is the most economical means of getting protection. However, many times sloping is not possible and the trench must be shored. Care must be taken not to remove necessary shoring or, if shoring must be removed, not to have men working in the unprotected trench. The best procedure is to remove only enough shoring to allow one length of pipe to be lowered and then not allow any men in the trench until the pipe is in position and the shoring is replaced. The pipelayers can then go down into the trench to make the necessary joints, etc.

The production attainable for installing large-diameter or heavy pipe is directly related to the number of obstacles encountered. In wide-open laying without shoring, an average of 2,000 ft per day is often maintained. However, the greater the number of crosslines and the greater the amount of shoring, the more slowly the installation will proceed.

Testing. There are several practices that will make the testing of water mains easier and more economical:

1. When installing the pipe, each joint should be gauged with a depth gauge to see that the rubber gasket or ring is positioned properly in the completed joint.
2. When possible, all joints should be left exposed for testing.
3. The line to be tested should be filled to operating pressure and allowed to remain at operating pressure for 1 or 2 days.
4. Testing should not be done against old valves, or, unless it is absolutely necessary, even against new valves. Small amounts of water leaking past valves can give an unsuccessful test when in reality the line is watertight.
5. When testing, the pressure should be raised gradually to the required test pressure.
6. The line should be walked and every exposed joint should be checked visually immediately after operating pressure is put on the line and again after the test pressure is put on the line.

There are several means of bringing a water main up to pressure economically. The simplest and most economical, if it will produce the desired pressure, is the use of a small, conventional water pump. The use of high-pressure water pumps will do the job too; however, these have to be kept in perfect repair and, with infrequent use, they often prove costly. An inexpensive means can be the use of a 200- to 400-gal pressure vessel (such as a butane tank) as a reservoir for the required water and a bottle of nitrogen as the pressure source. Connect the bottom of the pressure tank to the water main by means of tubing of proper strength. Connect the nitrogen bottle to the pressure tank through a pressure regulator. By regulating the pressure of the nitrogen admitted to the pressure tank, an even, controlled pressure can be applied to the water main.

Water Services Building connections generally are installed with copper, galvanized steel, or plastic pipe. When possible, the use of a very small trenching machine or narrow backhoe has proved the cheapest means of excavating the necessary trench.

In areas with good pavement, curbs, gutters, and crosslines, the installation of service lines by boring may prove more economical than open-cut excavation. This can be done by the use of an air motor that rotates a bore pipe (usually ¾ in. or 1 in.) with a cutting head attached. A short trench is excavated, usually in the location where the service will terminate, about 15 ft long and a little deeper than the planned depth of the service. The bore pipe is screwed together and should be at least 20 ft longer than the length of bore desired. It is then aligned in the trench by means of metal guides to point it in the proper direction both horizontally and vertically. The air motor is attached to the rear end of the bore pipe. Two to four men, depending on the hardness of the soil, then start the motor and push the bore pipe in the direction of the desired bore. Water is usually allowed to flow through the bore pipe, thus helping with the cutting by lubricating the hole bored and flushing out the excess dirt. With a soft soil, two men can install a 30-ft bore in less than an hour. Harder soils take more men and a longer time. Where water is not allowed in the boring process, a less satisfactory but still economical result can be obtained by the use of compressed air instead of water.

Sanitary and Storm Sewers These two types of sewers will be discussed together because the installation methods are identical except for the different types of pipe.

The most important item to look into prior to commencing work on a sewer job is the location of the proposed pipe in relation to existing pipes. Because the exact depth of the lines has been predetermined to match the elevation of existing or future lines, it is usually necessary for them to be installed at the elevations indicated. Known crosslines should be exposed prior to the start of excavation to ensure that existing lines will not interfere with the space allocated for the new line.

The second most important item to watch in sewer construction is that the trench be excavated to grade as closely as possible. This will allow quicker and more economical fine grading of the ditch and also give a more uniform base for installing the pipe. Grades for pipe can vary from 0.1 to 6 percent depending on the terrain. Needless to say, careful fine grading is necessary when working with minimum grades. When working with close tolerances and checking to see if pipe is laid to grade, it is important to remember that the flow line of the pipe is the critical point to check, not the top or bell of the pipe. The reason for this is that often the allowable variation in the diameter or straightness of the pipe will be as much as the variation allowed in installing the pipe.

The type of pipe most commonly accepted for sanitary sewers is vitrified clay. It has the characteristics necessary to withstand the chemical action from sewer gases and it is made throughout the country. If the clay pipe has plastic joining surfaces such as "Speed-seal" or "Wedge-lock," a watertight line can easily be installed. Epoxy-lined asbestos-cement pipe is also popular and, with its pliable joint, makes a watertight line.

Storm sewers generally are installed with concrete pipe which, depending on the load conditions, is either reinforced or nonreinforced. In many instances, leakage is not a factor and cement mortar joints are all that is necessary. When a leakproof joint is necessary, a rubber gasket pipe is usually required.

One word of caution is in order concerning testing of installed lines for leaks. With both types of sewers, it is a poor practice to bid on lines with mortar joints that are to be pressure tested. The reason is that the mortar shrinks away from the joint when drying, often cracks if it dries too fast, and will crack if there is any appreciable contraction or expansion of the pipe.

The smaller sizes of pipe weighing less than 150 lb per section can best be laid by hand by two layers. For plastic joints, it may be necessary to use a steel bar to push the pipe "home" and make the joint completely tight. In trenches over 5 ft deep, it is advisable to have the pipe lowered to the layers with a rope by one or two men standing on the surface. Under conditions where the trench can be prepared, a 3- or 4-man crew can install an average of about 2,000 ft of pipe per day. Quality installation requires that the body of the pipe be laid on a uniform foundation, with holes dug to make room for the bells.

A similar method of installation should be used for heavier pipe, except that equipment should be used to lower the pipe into position. In selecting the type of equipment, the key items to consider are (1) equipment that will require only one man to operate, and (2) equipment that can move the pipe from the surface to the trench bottom with a minimum of movements. Extra movements only delay the laying process.

Gas Distribution The gas distribution installation that will be discussed is the medium-pressure, 20- to 50-psi, system generally used for distributing gas within cities. Gas transportation between cities is usually effected by high-pressure lines operating at pressures of several hundred pounds per square inch, and the gas pressure must be reduced to the distribution pressure as it enters the distribution system. The pressure is further reduced at each service connection by means of a regulator to approximately 8 in. of water or a little over 4 oz of pressure.

The most commonly used pipe for gas distribution is black steel pipe, which is usually wrapped or plastic coated to protect the steel from corrosion or electrolysis. Plastic pipe is used under some circumstances; and it is probable that, with ever-increasing improvements, it may find more use in the future. Some areas that have virtually no problems with corrosion or electrolysis use bare black steel pipe. Copper pipe has been used successfully but generally is too expensive.

Pipe sizes for distribution mains will vary with the requirements; however, 2-in. pipe is the most common size. Joining of the lengths of steel pipe is usually done by welding to make the joints permanent and leakproof. A qualified welder and helper can average about 50 welds of 2-in. pipe in an 8-hr day. Pipe is usually supplied in 20- or 40-ft lengths. Welding of distribution mains and house services is almost always done with the pipe above ground. This is possible because the pipe is usually light enough to be lowered in the trench and under crosslines easily. The pipe is rested on skids or other objects to facilitate its alignment and allow for its rotation during welding.

Pipe of 2 in. or less can be welded most economically with acetylene, while 4-in. and larger pipe can be welded most economically with arc welding. Pipe of 3-in. diameter can be done either way, depending on the welders and equipment available. Welds which are properly made are actually stronger than the pipe itself. A utility company will generally require that a welder's competence be demonstrated by means of a test.

Most pipe for distribution is already wrapped when it reaches the job, so the contractor is chiefly concerned with the wrapping of the joints after the welds are made. There are, however, several points that should be stressed concerning wrapped pipe. It should be handled carefully to protect the wrap and should not be dragged. It should not be laid on pavement without padding under it, as small rocks can penetrate the wrapping. Joint wrapping during installation should be equal to or better than the yard wrapping. In many areas where electrolysis and corrosion are a problem, high-voltage electrical current will be used to test the pipe wrapping for defects just prior to the lowering of the pipe into the trench. The bottom of the trench should be cleared of all stones, hard objects, and metal that might penetrate the wrapping. The same care should apply to the earth placed adjacent to the pipe. Where suitable material is not available from the excavation, imported sand is often required to give the necessary protection to the pipe wrapping.

Contact between dissimilar metals in underground piping causes the creation of electrical current. When different types of pipe are joined, such as copper services from a steel main line, an insulated connection should always be used. Distribution systems are usually tested for leaks with a pressure of 100 lb. A recording chart is installed after the pressure has been put in the line, and no leaking or drop in the line pressure is allowable during a 24-hr period. If the line is to be buried before acceptance testing, the preliminary testing of the welds with soapy water before wrapping and while the line is under 100 lb of air pressure is a good precaution.

Excavating the trench for gas distribution lines can usually be most economically done with a wheel-type trencher with narrow buckets. If the trench is about 3 ft or less in depth, a ditch 6-in. wide is sufficient. For trenches more than 3 ft deep, the ditch should be dug 14 in. wide to allow a man to stand in the bottom of the trench to clean out any debris.

Electrical and Telephone The direct-burial methods for installing underground electrical and underground telephone lines are almost identical. The methods and systems of installing these types of underground services for homes in new subdivisions are in a state of constant improvement. There is little question that better and more economical methods will be developed in the near future. There is a trend at present to install both utilities in the same trench, with a distance of 12 in. of compacted fill between the telephone and electrical cables. If there is both a primary and secondary

Fig. 26-4 Cable plow. (*W. M. Lyles Co.*)

electrical cable in the trench, it is usual to have these separated by 12 in. of compacted fill.

Where only one cable is to be buried in a location, plowing the cable in has proved very economical. Plowing in cables in a subdivision is also relatively new, and standard methods have not as yet evolved. However, there are several promising machines in use that are easily moved from one location to another and which have sufficient traction and power to plow in a cable up to 42 in. deep (see Fig. 26-4).

The installation of a duct system for underground electrical distribution has several peculiarities. First of all, there are invariably both primary and secondary circuits to be installed, and these normally are separated by 1 in. of concrete. Secondly, the usual structure for electrical distribution is a multiple combination of round duct conduits of the proper size and in the proper quantity, which are subsequently encased with pea-gravel concrete. Each electrical distribution structure is formed by stacking these ducts with shaped separators between them and then tying the ducts together with a soft steel wire. If this is done properly, with boards of adequate width placed between the structures and the trench sides, there is little chance of the structure moving during the concrete pour. To ensure that the duct structure does not float, a small man can walk on top of it and work the concrete in place with a shovel. The concrete should be just wet enough to work into the voids between the conduits but not so wet as to encourage floating of the ducts.

A variety of materials can be used for the conduits which are to be encased. Plastic, asbestos-cement, and fiber ducts have all been used extensively, and all are easily installed in accordance with the manufacturer's recommendations.

Pull boxes, manholes, and transformer vaults are now being precast and hauled to the job. These offer several advantages over the old cast-in-place method. They allow the street to be opened and closed faster. They are less hazardous to the workers in caving conditions as there is less labor necessary below the surface. And, lastly, by being built under controlled conditions, they usually provide a more economical finished product. The only disadvantage to precasting is that when the casting yard is too far from the job, the hauling cost sometimes offsets the savings.

The installation of telephone conduits and structures is very similar to that just described for electrical structures. Telephone ducts are usually precast or formed in sections that can be hauled to the site and installed. Vitrified-clay multiple duct

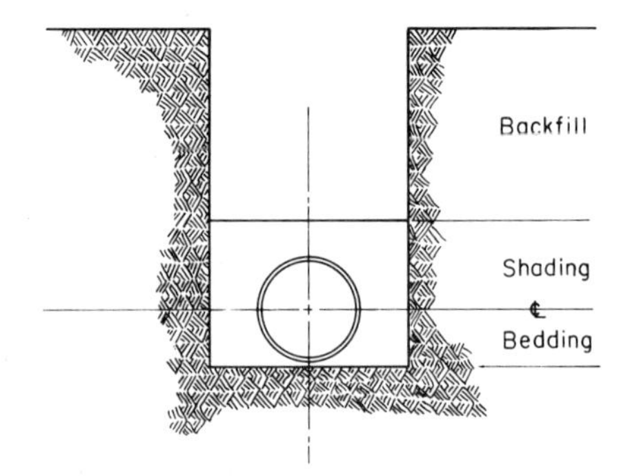

Fig. 26-5 Trench cross section.

was for years the standard for the industry. It is still in use in many locations, although precast-concrete multiple ducts have become more popular. Most multiple-duct sections are too heavy to lay by hand, so equipment is generally used to lower them in the trench. An ice-tong type of clamp that will go around the middle of the structure will allow the installation of the duct in its final position without the necessity of moving it manually. As a rule, the final duct structure for telephone need not be watertight, as the cables that will be installed in the conduits will be waterproof. However, all joints should be tight enough to prevent silt from washing into the duct.

BACKFILLING THE TRENCH

Shading and Bedding There are two distinctly different reasons for shading and bedding, so they are best discussed separately.

Shading, which is the covering of the pipe with loose or compacted soil, is primarily for the purpose of protecting the pipe and preventing its movement during backfilling. Soil can be placed over the pipe by several means. Shading by hand is the old reliable means and is probably the only sure way in conditions where the soil is lumpy or rocky. Hand shading is usually necessary when the material is placed in thin layers for compaction. Shading pipe where the soil is sandy, fine, and not to be compacted in layers can often be accomplished by the use of an angledozer. Damage to or movement of the pipe will not occur if the dirt does not fall directly on the pipe. The angledozer blade should move the dirt so that backfilled material will roll ahead, around and over the pipe. (See Fig. 26-5.)

Bedding usually refers to the support conditions under the bottom half of the pipe. The type of bedding will affect the ability of the pipe to support structural loading without adverse effect. The importance of bedding is then a matter of the additional strength needed by the pipe to fulfill its designed purpose. The larger the pipe, naturally, the greater the need for and the greater the expense of proper bedding. The problem of getting firm dirt under the pipe and the small working space available to work alongside of the pipe complicate proper bedding.

For smaller-diameter pipe, where a good foundation is all that is needed or where a low relative compaction is accepted, flooding the area around the pipe is the best and least expensive means of gaining the desired results. Of course, this method is unsatisfactory with soils that swell when saturated. When using the flooding method, sufficient material should be placed over the pipe to ensure that it will not float when the pipe area is flooded.

When a greater degree of compaction is needed to give support to the pipe, two systems are often used. The first employs mechanical compaction of the soil below the spring line (center) of the pipe. In spite of the fact that this system is often required, it may give only fair results in many soils. No matter how much compactive effort is placed vertically along the sides of the pipe, the earth can only be pushed back under the pipe and compacted for a limited distance. The method has gained acceptance because the degree of compaction at the sides of the pipe can be easily tested; however, this is really not the location that is important for giving support to the pipe. A man using a 33-lb pneumatic tamper should be able to compact about 4 cu yd per hr when properly moistened soil is being placed in layers by others.

The second system which is in use is that of filling the trench to the spring line with sand and then compacting the sand. Dry sand can be compacted mechanically very well and will get well under the pipe, but moist sand compacted mechanically will not be forced under the pipe. The most effective means of compacting moist sand is by saturation and internal vibration with a concrete-type vibrator. By this means, relative compaction values of 80 percent are attainable. Care must be taken to be sure that sufficient water is added to saturate the sand but not so much at any point as to cause the pipe to float.

Equipment and Methods While the methods of backfilling are as unlimited as man's imagination, they do have some common characteristics. Backfilling methods which utilize machines, where they can be employed, are the most economical. The greater the amount of earth to be moved, the larger the machine should be, etc. Where the excavated material is to be used for backfill, a track-type angledozer is generally the most economical unit. Where there is pavement that can be damaged by a tracklayer, rubber-tired equipment should be used. The angledozer has a definite advantage over the straight dozer. It can roll the dirt into the ditch if it can be angled as much as 38°, and it stays closer to the trench as compared to just shoving in a bladeful at a time. There is a definite relationship between the cost of machine backfilling and machine trenching. As a rule of thumb, it will cost about one-third as much to replace the backfill mechanically as it does to excavate mechanically. This factor seems to hold true even in the complicated or deep trenches where the excavation costs can become quite high.

Once satisfactory bedding and shading have been accomplished, the methods of backfilling are largely determined by the compaction requirements. However, there are two warnings that should be given no matter what type of compaction is necessary, and these concern care of exposed crosslines: (1) large quantities of dirt must not be pushed in on unsupported crosslines; and (2) large quantities of uncompacted earth must not be placed under exposed crosslines. In general, great care should be exercised with regard to crosslines while backfilling, or the consequences of breaks, damage, etc., may have to be dealt with at a later date.

Compaction *Requirements.* The required degree of compaction is usually spelled out in the specifications. If it is not, then the best policy where settlement of the trench is undesirable is to obtain a compaction equal to that of the existing ground.

Probably one of the most misused specifications with which contractors must comply is the one requiring 90 to 95 percent relative compaction in trenches where the existing ground has a relative compaction of anywhere from 65 to 85 percent. The result of compliance with such specifications is an island of compacted earth in a sea of relatively soft earth, and no one benefits. However, comply the contractor must, so the following paragraphs will deal with the means of compliance.

There are several precautions that may be taken where future settlement of the trench is undesirable. First, attention should be given to compacting the earth below the spring line of the pipe. Otherwise, voids under the pipe will allow later settlement. Secondly, the depth of trench to be backfilled at one time must be controlled. The depth should be the optimum depth which, for the equipment to be used, will produce the specified compaction. A plan must be formulated for obtaining compaction under crosslines, around structures, and throughout other areas peculiar to the job.

Methods. Flooding has long been the most extensively used method for obtaining compaction. It is probably the best means available in pure sand and may prove acceptable in relatively free-draining soil such as sandy loam. Some future settlement will occur in soils other than pure sand, no matter how good a job of flooding or jetting is done, and relative compaction of greater than 80 percent is unusual. Compaction by flooding may also be acceptable for clay-like materials in farm fields, but it is not satisfactory for work in existing or future streets. As a rule of thumb, trenches should be flooded or jetted in lifts of not over 5 or 6 ft.

Proper moisture is the key to all other types of compaction, since methods of placing and compactive effort will have little effect if the proper amount of moisture is not present. Knowledge of the optimum moisture for the type of soil is essential. If this optimum is not known, it may be necessary to experiment with varying moisture contents on the job. Laboratory tests may be helpful in indicating the probable range of required moisture. They will also show how the earth looks and feels when it contains optimum moisture.

Hand compaction with the use of pneumatic tampers has long been a standard method. It is now generally considered the most expensive means because of labor costs and the necessity of keeping the thickness of the layers of material to be compacted to about 6 in. It will ultimately be used only on very small jobs or in areas where other equipment is not suitable. As an example, a man operating a pneumatic tamper weighing about 33 lb would do well to compact to 90 percent relative compaction more than 5 cu yd per hr.

There are several hand-vibratory compactors on the market which can be used in place of the pneumatic tamper when hand compaction is required. These vibrators use a combination of weight and vibration and can obtain the same results as the pneumatic tamper in much thicker layers. Needless to say, the heavier the machine, the greater the impact and usually the greater the vibrating surface and effectiveness.

Compaction of trenches with machines too heavy to be hand held is becoming more popular as a means of obtaining the desired results economically. Use of vibrating rollers where the trench is wide enough and where the crosslines are infrequent is probably the most efficient method. The sheepsfoot roller is also effective but usually requires more passes.

Since most trenches do not lend themselves to a roller type of compaction, the impact type of compactor (the same machine as that described for pavement breaking) has been developed to compact trenches economically. These machines, which basically drop about a 1,000-lb weight a distance of 9 ft, have proved to be the most economical means of obtaining relative compactions in the 90 percent range.

The depth of material that such machines can compact and the speed at which the compaction can be obtained are dependent on the type of soil and the degree of compaction desired. In sandy loams, compaction in 24- to 36-in. lifts is not uncommon. In clay-like materials, lifts of as little as 8 in. may be necessary. For compacting a suitable soil in 12-in. lifts, an average of 15 cu yd per hr with a machine would be reasonable. As much as 40 cu yd per hr can be compacted if 3-ft lifts are used.

The advantages of such machines are (1) they are highly mobile, (2) they require only one operator, (3) they can easily compact on both sides of a crossline, and (4) good results are obtained.

The disadvantage of these machines is that, if care is not exercised near the pipe or conduit, the installed pipe can be damaged or crushed. Continued tamping in an effort to obtain unrealistic results can actually move the trench sides and even raise the existing pavement on the sides of the trench.

PAVING THE TRENCH

Removing excess material is the first step in trench repaving. When compaction tests are required on the backfill, it is quite common to refill the trench to the road surface so that the street can be used until the results of testing are available. When this is done, it is best to fill the top portion of the trench with the required depth of base material. Temporary paving can be placed over this fill and permanent paving can later be installed with a minimum of effort. On lightly traveled streets, base rock is often left for a week or so without temporary pavement.

The removal of the material necessary for paving the trench can best be done with a loader, usually rubber tired. A scoop that will remove the proper width and depth of material should be installed on the bottom of the bucket edge, and the loader should be used to dig out the material and load it into a dump truck. This same method can be used with a self-loading scraper when the quantity of earth to be removed is sufficient and when the disposal site is near enough to allow for transportation by the scraper.

Cleaning pavement edges is, of course, a "must" if a good bond is to be secured between the old and new pavement. Either a stiff broom or an air jet will serve the purpose. For a really good job, the edges of the pavement should be given special attention just prior to the placing of the new pavement. For asphalt pavement, a bitumal spray or slurry will help provide a good bond. With concrete, wetting the edge just before the concrete is placed will improve the bond.

The types of material to be used will normally be specified, so the placing of materials will be the next item of importance. On very wide trenches the procedures for handling of materials are quite similar to those used in paving a street, so no discussion of them is necessary.

For narrow trenches, a dump truck with a narrow gate built into the tailgate can economically place materials such as base rock and asphaltic pavement. The gate allows the release of the material in a width less than that of the trench and at the same time allows the control of the quantity of material being released. Trench-paving boxes are manufactured which can be pulled behind any dump truck and give the same controls. Both methods work well, with the latter probably being the most economical on trenches of any length.

Production quantities are impossible to give due to the multitude of varieties of work required. Probably the greatest insurance of good production is the use of an experienced crew. Low costs and good quality work can best be obtained by keeping the same men on a crew assigned to do all the trench paving.

B. Transmission Pipelines

The United States today has over 850,000 mi of pipeline for natural gas alone. These pipelines moved over 19 trillion cu ft of gas into 36 million homes and over 3.1 million factories and stores. This volume of gas represents about one-third of the nation's annual consumption of fuel energy. Accordingly, this subsection will deal principally with the construction of the large-diameter pipelines used for natural gas.

The vast network of oil pipelines further illustrates the importance of pipelines. An oil pipeline is present wherever there is, or has been, a productive oil well. Small feeder lines of 2-in. pipe or larger bring oil from the wells in a field to a larger crude line for delivery to a refinery. This may be a short distance or a matter of hundreds of miles. After processing in the refinery, the petroleum products are carried by products pipelines to market centers for distribution.

In the delivery of natural gas from the producing field to the consumer, there are three principal and distinct stages. These are production, transmission, and distribution. The manner in which natural gas is gathered follows an almost identical pattern to that for oil. Field gathering lines take the natural gas into a central plant for removal of impurities and then into the big transmission systems. The long-distance natural-gas transmission arteries are necessarily of larger diameter than those used for oil, now being primarily 22- to 42-in. pipe as compared with the 8- to 36-in. pipe used in the major oil-products lines. Since the transmission phase carries the greatest amount of gas the longest distance, it is the construction of these pipelines that we shall describe.

PLANNING TRANSMISSION LINES

Approximately 85 percent of the proved natural-gas reserves in the United States are located in the Southwest, and the transmission lines which serve the highly industrialized markets of the Northern and Eastern states are necessarily of great length and cross all kinds of terrain. Building such a pipeline is a mammoth task and involves huge sums of money. Before a new transmission line is even put on the drafting board, both assured markets and supplies must be determined. An investment of 160 million dollars or more is involved in the construction of a pipeline of about 1,000 miles in length, and the financing and normal contract arrangements on both the producing and consuming ends are usually set up on a 20-year basis. The planning of such important pipelines must go beyond the initial capacity to satisfy present demand. Allowance must be made for expansion, too, in order to satisfy a foreseeable increase of demand.

Planning by a transmission company generally includes allowance for increasing capacity in two ways. The first is to use a sufficiently large pipeline so that additional compressor stations may be constructed along its route to push the gas in greater quantity at a faster rate. The second is the advance planning for additional sections, called "loops," to be built later to run parallel to the main line. Only after months and often years of planning does actual construction begin.

Once a starting date for construction has been set, the aim is toward the earliest possible completion of the pipeline. It is to the advantage of both the consumers and those who have invested their money in the pipeline to start the gas flowing at the earliest possible moment. In view of this and the need for assured permanence through the quality of construction, there is an absolute necessity for specialization in building pipelines.

Just as proper planning is perhaps the most important single part of developing a new transmission system, so it is with the actual construction of the pipeline. The different terrain over which a pipeline passes demands flexibility, and both equipment and personnel must be geared to immediate and frequent change while building a uniform product.

The proper building of a pipeline requires the advance consideration of a great number of factors. A few of these will be considered, and the main problems which might arise from each will be pointed out (see Fig. 26-6).

Length. It is axiomatic that the larger the project, the lower the overhead in connection with moving equipment, men, and supplies to a new location.

Location. The location of the pipeline right-of-way determines its accessibility. Moving the vast array of equipment is a big job, and it is important to move it by rail as close as possible to the starting point of new work.

The number and adequacy of access roads to the pipeline right-of-way are other important factors determined by location. The more access roads there are, the easier it is to string the fabricated pipe sections along the route as well as to move special equipment onto the right-of-way. Closer supervision, too, is another advantage of a sufficient number of access roads. In mountainous or remote rural areas

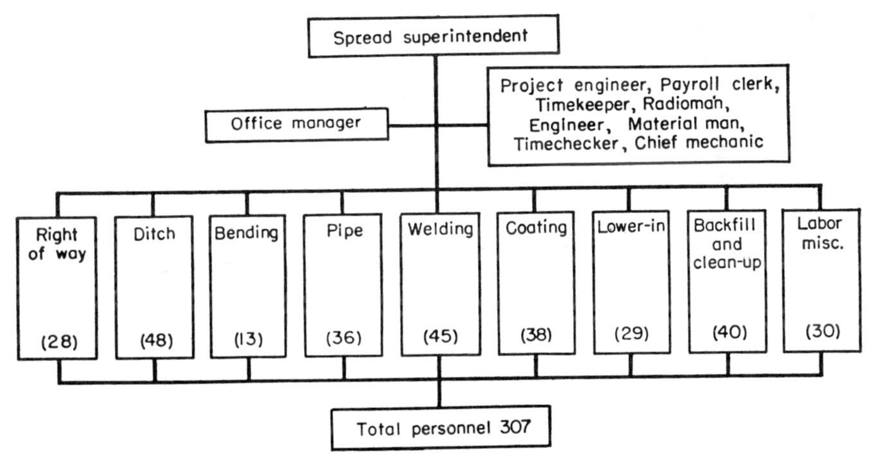

Fig. 26-6 Organization chart for typical pipeline spread.

where the number of such roads is insufficient, additional roads must be built by the pipeline constructor. This, of course, is an important item of cost.

Terrain. The type of land which must be crossed in building a pipeline is one of the biggest cost factors. In cutting a ditch across open farmland or treeless hills, a ditching machine can make relatively swift progress. There is also little problem in clearing the right-of-way to give room and footing for the construction crews to operate. On the other hand, both effort and cost increase sharply where heavily wooded and rocky areas must be crossed. Under the latter conditions, the entire right-of-way must be cleared of all brush, trees, and stumps and the irregular surface smoothed out. For this one task alone, upward of a hundred men may be required.

When rock is encountered, the pipeline ditch must be prepared by blasting and the use of backhoes. Where steep cliffs are encountered, the heavy equipment must often be supported by cables.

Weather. The man who keeps the records of seasonal rainfall is important in planning the rate of progress. A great deal of grief may be avoided if an approximate idea can be obtained of the number of days on which the big equipment will be stalled because of slick underfooting.

Labor Availability. Advance arrangements are necessary to ensure an uninterrupted supply of labor. Building and moving at the same time, a pipeline construction unit is often faced with varying labor conditions with regard to both supply and local union relationships. The simple act of crossing a county line can present a major problem unless advance arrangements have been made.

Right-of-way. Another point of major difference between pipeline construction and other types of construction is in the ownership of the site of construction. The property actually belongs to a third party, the landowner, and right-of-way privileges are purchased by the gas company building the new pipeline. Special consideration of the property is vitally important. Excessive property damage can be costly to both the constructor and the pipeline owner.

CONSTRUCTING TRANSMISSION LINES

Most current pipeline projects involve the construction of loops to existing facilities. Because these loops are necessarily scattered out from the main line, it is not unusual to move men and equipment 350 mi while building only 50 mi of pipeline. Thus, while the right-of-way is being prepared on the most distant loop, welding is being done on the middle loop and pipe is lowered into the ditch on the first loop.

TABLE 26-1 Typical Equipment Spread for Transmission Pipeline Construction

Number required	*Item*	*Number required*	*Item*
2	Passenger cars	4	Welding 200-amp gas drivers
15	Flatbeds	6	Tree saws
20	Pickup	1	Cleaning machine
1	Float	1	Coat-and-wrap machine
2	Lowboy	3	Sets wagon drills
1	Office trailer	1	Backfiller
4	Warehouse vans	2	Ditchers
20	Two-way radios	4	Air compressors—900 cu ft
2	Utility trucks	5	Tar kettles—27 bbl
1	Grease truck	2	Tar kettles—10 bbl
1	Fuel truck	8	3- to 6-in. water pumps
1	Water truck	10	Backhoes and clams
12	Dozers	1	Bending machine
16	Sidebooms	1	Steam cleaner
6	Tow cats	1	Road-boring machine
24	Welding 300-amp diesels	1	Rock ripper
		1	Hot-pass rig
1	Tack rig		

This basic pattern of activity remains unbroken from start to finish in pipeline construction. Wet weather is perhaps the major cause of delay, although other difficulties are common. There may be a turnover in labor, a change in suppliers, or a move by headquarters. To pipeline construction specialists, these things are all in a day's work.

Preliminary Work Long before construction actually starts, the exact nature of the terrain has been determined from a personal inspection of the proposed route by company executives. This inspection is essential for the determination of the men, equipment, and supplies for the job. After meeting with the labor unions of the locality and arranging for housing, the next step is moving in a construction spread. The equipment requirements have already been determined during the construction planning phase. (See Table 26-1, for example.) This equipment will be moved, preferably on railroad flatcars, as close as possible to the proposed starting location.

The office manager of the spread pulls the headquarters office trailer into the first headquarters location. Arrangements have been made in advance for warehousing and yard space to accommodate the equipment and the bustling activity that goes with the start of a new job.

The office manager and spread superintendent immediately set about completing arrangements for an adequate supply of fuel and oil to power the machinery. They also determine the location of the closest repair parts terminals and do other tasks necessary to a smooth-running organization.

Clearing Even before all the men and equipment have been moved into location, the first construction group has already started clearing the right-of-way. Appropriately dubbed the "right-of-way gang," it is its job to clear a smooth highway over the pipeline route for the pipeline building crews that follow. In thick forest country such as found in the mountains of West Virginia, upward of a hundred men may be put to work clearing underbrush, cutting trees, tearing out stumps, and filling treacherous pockets or holes. Speed is the essence of pipelining, and there must be room and footing on which to work.

Cooperating with the landowners, this crew salvages all marketable timber and stacks it neatly alongside the right-of-way for recovery and sale. It then stacks and burns all the underbrush, treetops, and limbs. This part of the work is first cleared with the local forest service wherever possible, and great caution is exercised to see

Fig. 26-7 Ditching machine. (*H. C. Price Co.*)

that no fire gets out of control. The men have back-pack water sprayers, a large water truck stands by for immediate action, and a dozer buries unburnable debris and is available in case it should become necessary to smash a fire line through surrounding woods.

Trenching The second unit that goes out on the right-of-way is the ditching crew. If a 36-in. pipe is to be installed, a huge ditching machine will be used to dig a continuous trench, 52 in. wide and 6 ft deep. The excavated material is piled on one side of the ditch. On a typical spread, the ditch crew has a complement of approximately 50 men, far more than are required simply to operate the ditching machine itself. (See Fig. 26-7.)

Working ahead of the ditching machine, a stake-setting crew sets out markers to show the exact line to be followed, as determined earlier by the engineering department of the gas company. A small bulldozer accompanies the ditching machine to further smooth out any humps or ruts left in the right-of-way, and there is often a tow tractor to pull the ditching machine up steep hills.

In mountainous terrain there is far more to preparing the ditch than the simple job of the trenching machine. Loosening of hard rock often calls for large amounts of explosives. Tandem wagon drills, mounted on steel frames and powered by 900-cfm air compressors, can be used to make the necessary boreholes.

Once the rock is loosened by careful blasting under the direction of the powdermen, backhoes are moved in to put the finishing touches to the tougher sections of ditch. Special auger-boring machines can be used to tunnel under main highways and railroads. In such crossings a section of thick-walled steel pipe of larger diameter than the line itself is put into place for casing as the boring machine cuts through the crossing.

Stringing After the ditch crews have passed along the right-of-way and prepared the bed for the big transmission artery, the sections of steel pipe make their appearance. The "stringers" haul the pipe from where it has been received and stockpiled along a railroad spur and lay it along the ditch side. Most of the large pipeline construction companies now subcontract the stringing to a specialist in this field.

Fig. 26-8 Pipe-bending machine.

Bending With the pipe now in place along the ditch, the next equipment on the route is the bending machine (see Fig. 26-8). Accompanied by a crew of 10 to 15 men with side-boom tractors to lift the pipe into place, the bending machine applies just the right amount of pressure to bend the pipe to the specified degree. Each section of pipe is thus actually form-fitted to the contour of the ditch.

After the bending crew has made a start, the tempo of activity along the line begins to pick up to the fast, continuous pace that is typical of efficient pipeline construction. The pipeline is now starting to take shape.

Welding Perhaps the key job in accomplishing speed in transmission-line construction is up to the next crew. Called the "pipe gang," it features two to four experienced and fast-working welders who are accompanied by two side booms and a battery of tractor-mounted electric welding generators. It is their job to make the first bead on the welds which link the sections of pipe into a continuous line. In dry weather with good footing along the right-of-way, it is not uncommon to see these men actually trotting from joint to joint as they make these first beads. Completion of the welds is left to crews coming along behind.

As each section of pipe is lifted into place, the inside lineup clamp puts the two ends into exact position and the welders start the first welding bead. Lift, lineup, weld! The routine is the same over the entire route, progressing swiftly toward the distant goal of the last tie-in weld at the far end of the line.

A portable machine for down-the-line double-jointing has been developed by an H. C. Price Co. special-projects engineer. This machine, which is adaptable to any type of welding, makes it possible to joint two 40-ft pipe sections on the right-of-way in flat country and thus cut welding and lineup time approximately in half. The use of this sled-mounted machine eliminates the problems encountered in hauling 40-ft lengths of yard-jointed pipe. Such hauling is not only more expensive but is prohibited in some states (see Fig. 26-9).

The pipe gang is followed by two to four more welders, with their normal accompaniment of helpers and buffers to clean slag and flakes from the previous welds. This is the second of three welding units that pass along the line, and the completed weld will contain a total of four or five beads. This second bead is called the "hot pass," and gives additional strength and protection until the welding crews can complete the full weld.

The main welding gang of 10 to 20 additional welders follows the second, or hot-pass, bead. This welding gang, with assisting labor, side booms, and generators, all

Fig. 26-9 Down-the-line double jointing. (*H. C. Price Co.*)

flows out in a long string and works in an atmosphere of feverish activity (see Fig. 26-10).

Welding is perhaps the most important single job in the building of a permanent and strong pipeline. Close checks on each individual weld are made by expert inspectors of the pipeline company and by the contractor's welding foreman. Each welder is thoroughly tested before he is hired, and sample welds are put through rigorous testing. Welded-steel straps are pulled apart by a hydraulic tensile-testing device and, in order for the weld to pass the test, the steel of the pipe itself must pull apart before the weld gives way.

Now, 10 to 15 percent of the joining is done by metal-inert-gas (MIG) welding with an inert-gas machine. This process completes a pipe joint faster than stick welding. Also, there are high-strength wire electrodes for MIG welding that can join pipe of higher tensile strength than can be joined by stick welding. The same testing standards apply to MIG welding.

Cleaning Next the welders are followed along the pipeline by the pipe cleaning machine. This large machine moves along the pipe itself, much as an upright unirail train moves along its track, but with its weight supported by another side-boom tractor. On this machine, several steel-wire brushes move in a circular motion around the pipe, scraping it clean of the dirt and rust that have accumulated between the factory and the field.

As soon as the metal is burnished clean, a primer is flowed evenly over the bright steel by the primer machine in preparation for the coating and wrapping gang which follows.

Coating and Wrapping In well-timed construction, a swift pace is maintained toward completion as the coating and wrapping machine performs its work. This machine applies the final protective coating against corrosion and, in much the same manner as the cleaning and prime-coat machines, it moves along the pipe itself. It is followed closely by tractor-pulled sleds loaded with supplies.

Fig. 26-10 Welding gang. (*H. C. Price Co.*)

As hot tar is cascaded down over the pipe, an assembly on the back of the machine revolves completely around the pipe. This assembly feeds a wide and continuous glass fiber or felt wrapping, or both, from rolls mounted on spindles on the revolving section. As the coat-and-wrap crew moves along, a side boom lifts the pipe ahead of it with the use of a pipe cradle. The side boom moves forward, with the pipe sliding easily over the small wheels in the bottom of the lifting cradle, and the pipe is thus raised to the proper height for the coat-and-wrap machine (see Fig. 26-11).

Tapes applied over a primer are being used by some companies. The application is by much the same principle, except that the need for hot materials and outer wrap is eliminated.

Lowering In After the coat and wrap is completed, the pipe may be lowered back onto carefully stacked wood skids along the ditch where it awaits the next-to-the-last process of lowering in. Alternatively, the pipe may be lowered into the ditch directly behind the coating and wrapping machine. Prior to lowering in by either method, a final inspection is made of the ditch to make sure that it is in proper condition. Where the ditch has been blasted out of solid rock, crews haul soft earth to prepare a soft bed in its bottom and thus avoid any damage to the pipe or its coating.

Once everything is in readiness, side booms in groups of four to eight and spaced approximately 50 ft apart are used to lower the long sections of pipe into the prepared ditch. In a hilly or mountainous area, the contractor will have a greater number of horizontal and vertical bends and the lowering in must be done smoothly and easily. Each section of pipe must fit into the ditch without knocking the pipe sections ahead off their skids or causing unnecessary strain on the pipe.

Normally, the welders have left pipe unjointed about every ½ mi to facilitate lowering in and to ensure an exact fit of the pipe to the ditch contours. Once two sections are in the ditch, end to end, a tie-in weld is made right in the ditch to join the sections (see Fig. 26-12).

Fig. 26-11 Coat and wrap machine. (*H. C. Price Co.*)

Backfilling As much care is taken to see that a soft dirt cover goes directly on top of the pipe as was taken in preparing the ditch bed to receive it. A soft and deep protective pad of earth is first pushed into the ditch and the wide-bladed backfiller then completes the job. In solid-rock areas, this often entails moving the earth from distant locations.

Cleanup By no means the least important job is the cleanup. Major pipeline contractors operate under the theory that a good job of cleanup of the right-of-way is an excellent job of public relations for both their own company and for the transmission company. Bulldozers and hand labor in generous quantity leave a smooth and neat right-of-way at the finish of each job.

Testing Completed sections of the pipelines are pressure tested to ensure the final integrity of the system. Hydrostatic testing is used most often, and water is used more than any other fluid. Sometimes gas or air is used for testing natural gas lines.

The owner company specifies the pressure the contractor is to use in testing, the length of the section to be tested, and the span of time this pressure is to be held in the section of pipe. The pressure is based upon a predetermined percent of the minimum yield strength for which the pipe has been designed.

Fig. 26-12 Side booms lower pipe section into ditch. (*H. C. Price Co.*)

A scraper or "pig" is first pushed through the pipe to clean the inside, and the section of pipe to be tested is then filled with filtered water. Sometimes chemical additives are also required. Pressure is then built up to the percentage of operating pressure which has been designated for the test. This pressure must be maintained for the period of time specified by the owner company.

If the pressure remains within the specified limits during the test period, the section is usually accepted and approved by the owner company. If there has been a reduction in pressure, the location of suspected leaks must be found, the faulty pipe or weld must be cut out, and repairs made or a section of the pipe replaced. After this is done, the test must be repeated before that section of pipe can be accepted by the owner company. This testing is carried out on sections of the pipe until the entire pipeline has been tested and approved as ready for use.

Section 27

Piles and Pile Driving

WARREN N. RIKER

Vice-president, Walsh Construction Company, New York, N.Y.

INTRODUCTION

This section is limited to those items which the contractor will find useful in his business. Theory relative to pile design will not be presented unless it has some direct bearing on the subjects being discussed. Additional material pertaining to the design and behavior of piles can be found in the articles and papers listed at the end of this section.

A pile is a structural member and must be thought of as such. Piles are used to:

1. Transmit the load of a structure through a fluid or stratum of low bearing value to one of more adequate capacity
2. Eliminate and/or control the settlement of a structure where the ground has undesirably high compressibility
3. Protect marine structures, such as bridge foundations, from possible scour by flash floods
4. Consolidate loose, granular soils by the wedging action of high-volume piles
5. Anchor structures against uplift, overturning, and the effects of earthquake stresses
6. Protect riverbanks or provide a bulkhead in the form of sheet piling
7. Serve as moorings or anchorages and as dolphins when driven in clusters and wrapped with cable at the top
8. Act as protective devices in the form of fendering for wharves and piers
9. Form groins for the protection of seacoasts
10. Form permeable dikes or jetties for the control of river flow and maintenance of channel

Piles are conventionally divided into two classes: bearing and sheet piles. Piles in the first class behave somewhat like loaded columns and are stressed axially along their lengths. This loading may be either tension or compression but is more frequently the latter, as is implied by the term "bearing pile." A sheet pile serves as part of a bulkhead, as a fender pile on a wharf, as a part of a groin or permeable dike, etc. It must be adequately sized to resist bending, abrasion, or other stresses which may result from its particular service conditions.

Piles are also classified according to the materials from which they are constructed. The four principal types are wood, concrete, steel, and composites of two or more of these materials. Reinforced-concrete piles, by convention, are included in the "concrete" type instead of being classed in the "composite" group. Figure 27-1 illustrates the classes and major types of piles.

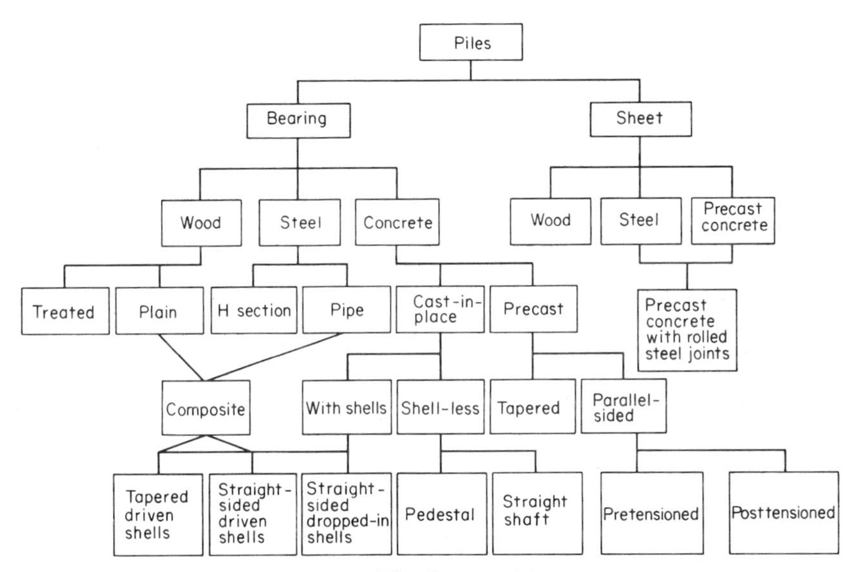

Fig. 27-1 Pile classes and types.

FOUNDATION INVESTIGATIONS

Reliable foundation investigations have a very definite value for the owner, designer, and contractor. Each of these parties should insist that such investigations be performed, if only to protect his own interest in the work. The investigations will indicate the type of foundation which will probably be required, thus providing the owner with guidance as to whether or not the property should be purchased. They will also provide the designer with the information on which the foundation design will be based. The investigations will indicate to the contractor the problems that he can expect to encounter in the construction of the foundation; with this information he can proceed with a check on the foundation design, its practicability under the reported conditions, and the possibility of effecting economies in design and/or construction. The chief values of a thorough foundation exploration are in minimizing the unexpected and in identifying the problems that must be solved. Adequate foundation investigations provide an excellent point of departure for a project, but the owner must still make a judicious selection of his design engineer and his contractor. Not all engineers have the experience to interpret a foundation report and prepare an adequate foundation design, nor do all contractors have the requisite experience to construct a satisfactory foundation.

Ground Survey The engineer who performs the preliminary examination and survey of the site should:[1]*

1. Survey the site and establish property lines
2. Outline the proposed improvement on the plot plan
3. Record the following surface observations on the plot plan:
   a. Nature and extent of filled areas, if any
   b. Indications of peat or organic silt
   c. Locations and types of rock outcroppings
   d. Water courses or ponds
   e. Contour levels
   f. Structures on site and adjacent thereto
   g. Nearby overhead or surface structures
   h. Overhead wires on site or in the vicinity
   i. Railroads and highways
   j. Shorelines and soundings through water (if any) on the site
   k. Character of water over surface of site; salt or fresh, currents, floods, or wave action
   l. Wind loads and effects on piles
4. Investigate subsurface construction, using information from drawings in public offices, utility companies, and owners of adjacent property (with a check in the field as far as practical to detect errors and changes) for size, alignment, grade, and elevation of the following:
   a. Sewer lines
   b. Water and gas mains
   c. Compressed-air, steam, or other piping
   d. Telephone, telegraph, or electric conduits
   e. Foundations on site or on adjoining land
   f. Subways, tunnels, basements, and other subsurface construction
5. Study boring logs and other pertinent records, paying special attention to past and probable future variation in groundwater level; also, examine groundwater and soil for chemicals and carbon dioxide, which may be destructive to concrete, steel, or wood
6. Collect available records of the behavior of structures in the vicinity, and observe the settlements and conditions of these structures in relation to their foundation design

These survey instructions, although written for the guidance of the foundation designer, are also of value to the contractor. Items 3*a*, *b*, *c*, *g*, and *h*, and also item 5, are particularly important, and no contractor should accept a contract without satisfying himself that the type of piling to be driven will be adequate for the anticipated conditions. The lack of adequate survey information prior to bidding has lost many piling contractors large sums of money. Unforeseen conditions can increase pile-driving costs or even lead to failures in piling already driven. In some cases, the piling as originally designed may prove to be inadequate for the conditions actually encountered.

Subsurface Exploration Various methods, each with its own characteristics and limitations, are used to explore subsurface conditions. A thorough foundation investigation may require the use of several of these methods in order to reveal the subsurface conditions fully.

Soundings can be used to explore the ground to the depth proposed for the piling. The usual procedure is to drive, push, or otherwise force a steel rod or pipe, generally ¾ to 1 in. in diameter, into the ground until it can be driven no further. The blows or forces applied to the rod or pipe as it penetrates into the ground up to the depth of refusal are recorded as a part of the test. It is dangerous to rely on this method to determine the bearing power and/or length of a pile unless by previous experience or knowledge of the soil it is known beforehand that the test will give an acceptable answer. This method is rarely used by persons experienced in pile driving, since its results can be grossly misleading.

* Superior numbers refer to the list of references at the end of this section.

Borings are by far the best method of determining the subsurface soil conditions. A common procedure for the making of borings is the Gow method, which uses a 2-in. split-barrel sampling spoon to obtain a partially disturbed 1⅜-in.-diameter soil sample that is representative of the soil strata penetrated. The sampling spoon is driven into the ground by a 140-lb weight falling for a distance of 30 in. As each sample is obtained, the number of blows required to drive the spoon a distance of 12 in. into the soil is recorded. This blow count is used as an indication of the density and shearing strength of the soil, and the soil samples thus obtained are used to classify the character of the soil. Variations of this boring method include the use of larger samplers and heavier drop weights. Seamless thin-wall steel tubing of 2- to 3-in. diameter is sometimes used without the split-barrel sampling spoon.

In the standard Gow method, the sampling spoon is advanced ahead of a 2½-in. pipe casing. This casing is driven by a 300-lb weight, usually falling a distance of 30 in. The number of blows required to advance the 2½-in. pipe casing for a distance of 12 in. into the ground can also be counted, as described for the sampling spoon. The driving of this casing will more nearly approach the driving of a pile, since the frictional resistance of the ground will accumulate along its entire depth of penetration.

Although the logs of blows versus penetration for the sampling spoon and for the casing are both used as a means of predicting the required length of the pile, it takes experience to apply either one successfully for this purpose. The relative densities and the thicknesses of the successive layers of soil must be evaluated for their resistance to penetration and displacement by the pile, and the diameter of the pile must be related to the size of sampling spoon and/or boring casing. The person who predicts pile lengths should possess experience, judgment, and an ability to evaluate the factors of pile shape and load under similar ground conditions. In the end, one man's judgment may be pitted against another's, since no sure method has been developed for predicting pile lengths with acceptable and consistent accuracy. W. W. Moore and Trent R. Dames have described a procedure wherein the shearing strength of each soil stratum penetrated by a friction pile is evaluated by laboratory tests and is subsequently used to predict the required length of the pile.[2] This method has frequently been used to predetermine pile lengths and bearing values, although there are divergent opinions as to its reliability.

Test pits or, on occasion, deeper *test caissons* can be sunk into the ground to permit the direct observation of existing soil conditions. The information thus gained will seldom lead to any decision as to the type or length of pile, but general knowledge is obtained as to the foundation conditions to the depth of the test pit or caisson.

Test piles are the most satisfactory method for determining the required length and safe load capacity of a foundation pile. These test piles are driven in various locations on the site and are then tested by loading. In this way, the probable length of pile required to carry the desired load will be determined under actual conditions. This method, combined with a good boring program, will give the best prejob foundation information.

TIMBER PILES

Wood was the first material used for piling. Evidence of its early usage has been found in the lake region of Switzerland, in Ireland, and in northern Europe. Even today, where suitable woods are available, a large percentage of piling driven is wood. If properly used and not too heavily loaded, wood piles are often the most economical type.*

Species Although every conceivable species of wood has been used as piling, years of usage have indicated those species which are most suitable for each region of the world. In North America, southern yellow pine, Douglas fir, and oak are generally considered to be the best woods for foundation piles.[3,4] In certain areas the woods classified as "mixed hardwoods" are considered equal to the foregoing. Cedar, maple, Norway pine, and woods of similar classifications are less desirable as piling materials. Resistance to marine borers is an important consideration for timber used in marine

* See Sec. 24, Timber Construction.

work, and greenheart and angelique were once considered as ideal materials for this purpose. But they, too, have been found to be affected by marine borers, and improved methods of pressure treating domestic timbers have reversed the trend toward their importation. The use of certified, inspected, pressure-treated piling for marine construction is mandatory under most specifications.[5,6]

A timber pile, to be considered a good structural member, should:

1. Be cut from live, sound trees and be free from fungus or other bacterial infection at the time of cutting
2. Be reasonably straight, so that a line drawn between the centers of the butt and tip lies wholly within the body of the pile
3. Be free of twists, splits, and shakes
4. Have all loose bark removed
5. Be peeled if it is to be creosoted or otherwise treated

Uses Timber piling, because of the ease with which it can be worked, is readily adaptable for use in trestles, wharves, etc. It can be obtained in a wide range of lengths and in many combinations of butt and tip diameters to meet almost any reasonable specification. The strength of timber piling will depend on the species of wood, as will its wearing qualities. Wood is resilient and will bend under load or impact where other materials will break. For this reason, it makes excellent fenders on waterfront structures and dolphins for anchorages. Timber piling is relatively light, handles easily, and will resist driving stresses within its normally accepted range of design loads.

Durability Untreated wood piles are subject to the attacks of termites, carpenter ants, etc. Although an untreated wood pile which is cut off at or below the permanent ground level will retain its structural properties indefinitely, any portion which remains above the groundwater level will rot and ultimately disintegrate. It is considered unwise to pull and later redrive an untreated wood pile since, more often than not, it will split and break under the blows of the hammer.

Pressure-treated wood piles, when used for foundations and marine structures, will obviate most of the problems encountered with untreated wood piles. Treatment of wood piling increases the life of the pile, although it does not extend it indefinitely. Dual treatment of marine piles with jackets of reinforced concrete and/or polyvinyl chloride (PVC) to at least 2 ft below the mud line and to above the high high-water line will also give added life to the piles.[5,6]

CONCRETE PILES

Concrete piles fall into two general categories: precast and cast in place, each of which is further subclassified (see Fig. 27-1). The precast piles are either tapered or parallel sided, and each of these two types may be plain reinforced, pretensioned, or posttensioned. Cast-in-place piles may be the shell-less type, with either pedestal or straight shafts; or the driven-shell type, with either tapered or parallel sides; or the dropped-in-shell type; or combinations of these types. Many of these piles are identified by trade names such as Raymond, Simplex, or Cobi. Each type of concrete pile has certain characteristics peculiar to itself, and all have a common characteristic in that their strength depends on the quality and strength of the concrete.[7]

Precast-concrete Piles Precast-concrete piles can be designed and reinforced to satisfy the particular requirements of each job. They can be cast and cured under controlled conditions to achieve maximum strength, but they are heavy and bulky and may be difficult to handle and drive unless the proper equipment is available. If their lengths are not predetermined before casting by driving test piles, they must include liberal allowances for possible field adjustments. Multiple hair cracks will sometimes develop in plain-reinforced precast piles as a result of the shrinkage of the concrete after curing. Unless these cracks are sealed by a coating of bituminous or plastic material, they can lead to the ultimate deterioration of a pile which is exposed to seawater. Also, multiple-point pickups are necessary for handling plain-reinforced precast-concrete piling if cracking is to be avoided.

Both of these cracking hazards can be eliminated through the use of prestressed piles. To some extent, it might be said that pretensioning (or posttensioning) gives concrete piling the handling characteristics and strength of steel while retaining the durability of concrete. Knowledge of prestressing has developed and matured to a large degree in the last 15 years, particularly as applied to the larger dimensions of concrete piling, and a number of companies have developed materials and methods for performing this work with a high degree of perfection. The Roebling Company manufactures high-strength cables and anchoring devices for use in pretensioned and posttensioned piling. Raymond International manufactures, assembles, and drives sectional posttensioned cylinder piles in diameters from 24 to 54 in., and with wall thicknesses from 4 to 6 in. These piles are characterized by their high structural strength, high load capacity, and flexibility in length up to 200 ft. They are particularly adaptable to marine structures and land causeways, since they can act as the columns supporting the bridge seat.

Cast-in-place Concrete Piles To describe the various types of cast-in-place piles, it is necessary to identify the predominant piles in the shell-less and shell classes and examine the individual characteristics of each. It should be noted that the tasks of placing and curing concrete for cast-in-place piles are frequently performed in close proximity to driving operations. This is somewhat at variance with good engineering practice.

Driven-shell Type. Some shells are of sufficient strength to permit direct driving, while others require a mandrel during driving. Some driven piles have straight sides, others have tapered sides, and still others have combinations of straight and tapered sides.

The *Cobi pile* has a thin-gauge corrugated steel shell, such as Armco Hel-Cor, with parallel sides. A steel-plate boot is welded to its point, and a multileaved mandrel is inflated pneumatically to grip the shell during driving. Shells are available in No. 16, 18, 20, 22, and 24 gauge. Piles can be made in long lengths by increasing the diameter in steps where the initial 40-ft length of shell is exceeded, or by welding two or more 40-ft lengths together. The pile has the advantage of its one-piece shell for most pile lengths and the questionable disadvantage of being driven with a flexible core.

The *Ray-Side pile,* supplied in 8-ft straight-sided sections by the Raymond Concrete Pile Division of Raymond International, Inc., is driven with a core whose mechanical lugs engage driving rings located at each 8-ft joint section of the shell pile. The Ray-Side pile can be combined with other Raymond pile types to produce a pile with a heavy-tapered point and a straight-sided top or one with a step-tapered lower section and a straight-sided upper section.

The *Union Metal pile* has either a fluted and tapered shell or a combination of a tapered shell and a straight-sided shell. The tapered shell has an 8-in.-diameter conical point and is available in three tapers: 1 in. in 2.5 ft, 1 in. in 4 ft, or 1 in. in 8 ft. Lengths of the tapered sections run between 20 and 70 ft, in multiples of 10 ft. Straight-sided sections in 12-, 14-, 16-, and 18-in. diameters and in 10-, 20-, 40-, 70-, and 100-ft. lengths are available for welding on top of each of the three tapered sections. Shells are available in 3, 5, 7, 9, and 11 gauges and are driven directly without the use of a mandrel.

The piles are manufactured and sold by the Union Metal Manufacturing Co. and can be driven by anyone with the proper equipment. They have the advantage of being driven in one piece, with but a single joint between the lower and upper sections. However, shells in the lighter gauges will not absorb and transmit the heavy hammer blows required for normal driving and at the same time have the strength to prevent twisting and lateral collapse.

The *Raymond Standard pile* has an 8-in. point and a uniform taper of 0.4 in. per ft. It is driven in lengths up to 37 ft with the aid of a mechanically operated, collapsible mandrel. Shells are available in lengths of 4 and 8 ft and in gauges of 18, 20, 22, and 24. They are reinforced internally with spirally wound wire of No. 3, 1, 0, or 00 gauge. This pile is used for foundations in loose, granular soils where quick compaction is required. A stub pile is also available, with the same characteristics and uses as the standard pile but with a 10.8-in. point.

The *Raymond Step-Taper pile* is made in 4-, 8-, 12-, 16-, and 24-ft lengths and in

10 shell butt diameters ranging in steps of 1 in. from 8⅝ to 17¼ in. Shell gauges range from No. 12 to No. 24, although all shell sizes are not available in all gauges. Point diameters can be furnished from 8⅝ to 14¼ in. The pile is assembled by screwing successive sections together until the desired length of pile, up to 120 ft, is obtained. A mandrel is used during driving.

The *Raymond Pipe Step-Taper pile* is a combination of pipe for the lower section and shell for the upper section; 10¾- and 12¾-in. outside-diameter pipes in different wall thicknesses are used in lengths up to double-double random lengths. The joint between the shell and pipe is watertight. This pile is classed by some authors as a steel and concrete composite pile.[8]

The *Raymond Ray-Step pile* combines the features of the Raymond Standard pile and the Raymond Step-Taper pile. It has a standard point, tapering from 8 to 14 in. for a total length of 10 ft 9 in. Step-Taper shell sections form the upper portion of the pile and can be furnished in 4-, 8-, 12-, and 16-ft lengths to a butt diameter of 17 in. This pile is used for bearing in granular soils at depths beyond the 37-ft range of the standard pile.

Cuneiform piles are made of sections of Armco spiral pipe assembled in Step-Taper fashion with drive sleeves between the different shell sizes. This pile is made by the Albert Pipe Supply Co., which also manufactures the Cobi tips used with this pile. The Cuneiform pile is driven without a core, and shell thickness must be adequate to take driving stresses.

Dropped-in-shell Type. In this class of pile, an outer casing is driven as for the driven-shell type. After driving, a permanent light shell is placed inside the casing to act as a form for the concrete. At some later stage, the outer casing is withdrawn. As will be discussed subsequently, the bearing value of any pile whose driven shell is extracted is in question unless a load test is subsequently performed.

The *MacArthur Cased-Concrete pile* is formed by driving a heavy casing and close-fitting core into the ground to the desired point elevation. The core is then withdrawn and a light steel shell dropped into the casing. This interior shell is then filled with concrete and the casing is withdrawn.

The *MacArthur Cased-Pedestal pile* is formed by driving a casing and fitted core to the required point elevation, as for the MacArthur Cased-Concrete pile. The core is then withdrawn and a charge of concrete placed in the casing; then the core is reinserted in the casing so that it rests on and compresses the concrete. The casing is next withdrawn to the elevation of the point of the core, and both casing and core are driven into the charge of concrete below to form an enlarged foot or pedestal at the base of the pile.

Again the core is withdrawn. A light-gauge shell is dropped into the casing and forced into the previously compacted concrete, after which withdrawal of the casing is completed. The liner shell is filled with concrete either before or after the casing is withdrawn.

The *Western Button Bottom pile* has an enlarged precast concrete point that is driven by the heavy casing to the desired bearing stratum or resistance. A shell is dropped into the casing, fastened to the point, and filled with concrete. The casing is then withdrawn. This pile is primarily a point-bearing one, although some additional bearing value may be gained by friction after the ground moves back into the annular space left by the withdrawal of the driving casing.

Shell-less Type. In this type of pile, a shell is driven to the desired tip elevation and filled with concrete. The shell is then withdrawn, leaving the uncased concrete to function as a pile. The bearing value of any pile other than the driven-shell type, regardless of whether or not it has a dropped-in casing to contain the concrete, should be viewed as questionable unless a load test is performed. Once the driven casing has been withdrawn, the dynamic formulas for predicting pile capacity are no longer applicable. Also, as noted earlier, driving of pile shells may be proceeding concurrently with concrete placing and/or curing for adjacent piles. The completed pile cannot be inspected; this is of particular concern for the uncased pile, which may be "pinched off" along its length due to the lateral ground pressures produced by the driving of adjacent piles.

The *Simplex pile* has a steel casing with a cast steel, fabricated steel, or precast concrete point. The casing and the point are first driven into the ground to the desired resistance. The casing is then filled with concrete and withdrawn, leaving the point and the uncased concrete in contact with the ground.

The *MacArthur Pedestal pile* is formed in a manner similar to that described for the MacArthur Cased-Pedestal pile, except that no interior shell is used.

The *Franki pile* was developed in Belgium. A steel casing is driven to the bearing stratum and cleaned out, either during or after driving. A quantity of very low slump concrete is then placed in the casing and rammed into place by dropping a heavy cast steel mandrel. When a certain resistance to impact is obtained, the steel casing is slowly withdrawn while more concrete is added. The process of compacting continues, and the resulting shaft of concrete has a pedestal point and a series of bulges along its length where the impacts of the ram have driven the concrete into the surrounding soil. The advantages claimed are the increased bearing of the pile due to the pedestal point and the successive ridges on the shaft, a greatly improved high-strength concrete due to the compaction efforts, and an increase in the density of the soil around the pile. The shaft of the pile can be reinforced with a cage of steel before the concrete is placed and rammed.

Other Types. There are pilelike foundations which are formed by making holes in the ground and filling them with concrete or grout. Although these are not piles in the true sense of the word, since they are drilled in place instead of being driven to a specific bearing value, they are frequently referred to as such. Information regarding them is given in the following paragraphs, but their actual bearing values must be determined by tests.

The *Prepakt pile* is formed by drilling a hole in the soil to the desired bearing stratum, then placing a grout pipe in the center of the hole and filling the annular space with selected gravel or stone. When the stone has reached the cutoff elevation, sand-cement grout is pumped into the grout pipe and the voids of the stone are filled as the grout pipe is slowly withdrawn.*

The *Pali-Radice* or *Root Pattern pile* is a patented development of the Italian Fondedila Company. It is of small diameter and is generally used for underpinning purposes. It is constructed by drilling a small-diameter (4- to 9-in.) hole through the structure to be underpinned and into the ground beneath. This hole is carried to the desired depth, preferably extending to or into rock. A steel bar is then inserted into the hole for its full length and the annular space concreted and/or grouted under pressure. The smaller-diameter shaft is reported to support a load of 10 tons and the larger, 30 tons.

The *Benoto pile* (or caisson) is a double-walled section caisson, driven by means of a machine which simultaneously oscillates it and cleans it out. As the cutting edge advances deeper into the ground, additional sections of casing are added. After bottom is reached, cleaned, and inspected, any required cage reinforcing can be installed. The casing is then filled with concrete progressively as it is rotationally oscillated and withdrawn. The same process is used whether the shaft is of small diameter and called a pile or of large diameter and called a caisson.

STEEL PILES

Steel in various shapes is frequently used for piling.[9] Commonly occurring forms include the H pile, the pipe pile (generally concrete filled), and various combinations of structural-steel sections. Steel piles will take hard driving and will develop high load-bearing capacities when driven to firm bedding in rock or to refusal in other materials. However, portions above the permanent groundwater level will rust and ultimately disintegrate if not protected. In such cases, it is customary to allow $\frac{1}{16}$-in. additional thickness on each exposed face as an allowance for corrosion. Acid soils are particularly destructive to steel piling, and the ground should be analyzed for its pH value before any piles are driven. It may prove necessary to protect the

* See Sec. 22, Special Concretes and Mortars.

pile with special coatings to prevent deterioration. The proposed structure, process, and the surrounding industrial area should also be examined for the possible occurrence of stray electrical currents. If these are present, galvanic or electrolytic protection may be necessary to retard the disintegration of the pile.

H Piles These piles are formed of rolled-steel H sections, which are available in 8- by 8-in., 10- by 10-in., 12- by 12-in., and 14- by 14-in. sizes and in weights ranging from 36 to 117 lb per lineal ft. This piling can be ordered from the mill in lengths up to 80 ft, although extra charges are made for piling over 65 ft, and it can be fabricated into considerably longer lengths by splicing. A standard H section does not have the same section modulus about both axes, but welded flat sections can be added to increase the moment resistance at critical locations.

Steel H piles are of uniform section and strength throughout their length. They are usually considered to have small displacement in granular soils, although this may be questionable, and they are easily integrated into a structure where bracing members must be fastened to the pile. However, they cannot be inspected for damage to the point, which may bend and/or split. They also have a tendency to drive to a greater length for the same bearing value than other piles unless the bearing stratum is rock, in which case all piles will drive to the same depth.

Pipe Piles Steel pipe, with the bottom end either closed or open, is frequently used for piling. The pipe can be left unfilled or filled with sand, concrete, or both. However, common usage reserves the term "pipe pile" for a concrete-filled pipe. A pipe pile has certain advantages over other steel piles. Its section modulus is the same in any direction, and the full length of its interior can be inspected after driving and cleaning. When filled with concrete it is subject to corrosion on its outside only, and when filled with sand or left unfilled it can be sealed at its top to limit corrosion to that resulting from the free oxygen content of the atmosphere within the pile.

Pipe piles are commonly driven in diameters ranging from 6 to 36 in. The moot question for the larger pipe sizes is whether the member should be called a pile or a caisson, with the accepted division between the two occurring in the range of 30 to 36 in. in diameter. Pipe piles of less than 18-in. diameter which are to be filled with concrete or sand can be driven closed-end. Larger pipes are usually driven open-end and then cleaned out before they are filled with concrete.

The closed-end pile either has a flat plate welded to its bottom point or a cast-steel point, conical or concave in shape, which is permanently fixed on the end of the pipe by the act of driving the pile. Experience has proved the advantage of the flat plate over the conical or concave point in tending to push obstructions aside rather than deflecting the pile.

Because of their higher cost, pipe piles are used for high load concentrations. Wide variations in the bearing capacities of pipe piles can be obtained by varying the pipe diameter, wall thickness, and strength of concrete used for filling. A pipe pile can be strengthened in a localized region by increasing its wall thickness. Alternatively, a prefabricated cage of reinforcing steel or a rolled structural H section can be placed at the proper location inside the pipe before the concrete is poured.

Sections of pipe for piles are joined by butt-to-butt welding, or by employing sleeves of cast steel or split sections of pipe which are welded to the pile, or by combinations of the two methods. With modern welding techniques, it is unnecessary to reinforce a welded joint or to use a beaded-spacer backup ring for welding the joint. A flat backup ring behind properly prepared, beveled and flat-joint edges will give a perfect full-welded joint.

Fabricated Steel Section Piles These are frequently fabricated from one or more rolled-steel sections. The possible combinations are limited only by the available sections and the ingenuity of the designer. All these combinations are intended to increase the section modulus of the pile with respect to one or more bending axes, to increase the end area and/or surface area of the pile, or to increase its total displacement of soil. Common types of fabricated piles include three wide-flange sections (Fig. 27-2*a*) and an I-beam section with plates or channels welded to its flanges to increase the section modulus of the whole section or in one direction axially (Fig. 27-2*b*). Occasionally, pairs of wide-flange sections are welded together at the flanges

(Fig. 27-2*c*). A unique fabricated-steel-section pile can be made at low cost from new or secondhand railroad rail (Fig. 27-2*d*) by welding the flanges of three rails together. This welding need not be continuous, but it should equal at least 50 percent of the pile perimeter at each joint. The pile has very low ground displacement, a high section modulus, and a bending resistance that can be varied according to the weight of the rail section used.

Box Piles These are made by forming rolled sections of sheet piles into square, rectangular, hexagonal, or other boxlike shapes. The enclosed area of a box pile

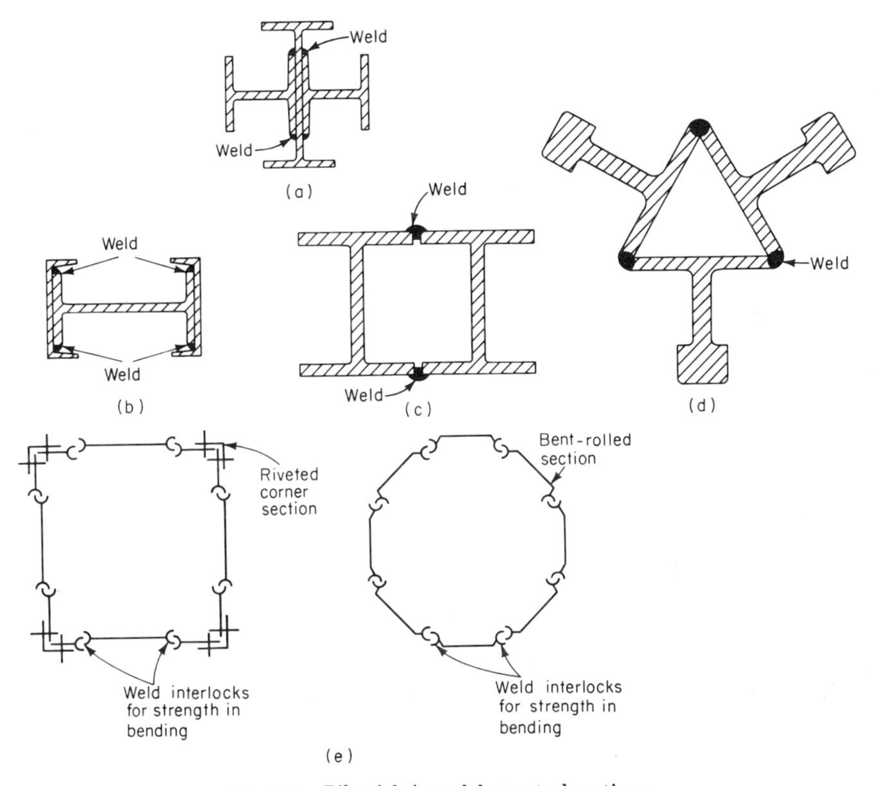

Fig. 27-2 Piles fabricated from steel sections.

may or may not be cleaned out and filled with sand or concrete. The pile is of little value in bending unless the interlocks between adjacent sheets are welded together so that the individual piles act as a unit and not separately (Fig. 27-2*e*). The individual components of the box pile can be driven separately, which reduces the required size for the driving equipment. Considerable skill in pile driving is involved in the proper construction of a box pile, and the loads must be very high to justify its cost.

COMPOSITE PILES

Strictly speaking, any pile that is formed of two or more materials is a composite pile. This definition could include both the reinforced-concrete pile and the steel-pipe pile which is filled with concrete. By common usage, the term "composite pile" is ordinarily reserved for a pile which has its lower part of one material and its upper part of another. The principal feature to be emphasized in all composite piles is the joint between the two materials. This joint should be permanent and mechanically made so that it will not separate under driving or afterward. For a pile carrying direct axial load, it is not necessary that the joint be able to take bending. Where it is possible that lateral forces may introduce bending stresses at the joint, the connection must be designed accordingly.

Types Basically, composite piles can be classified into three types on the basis of the materials from which they are constructed.

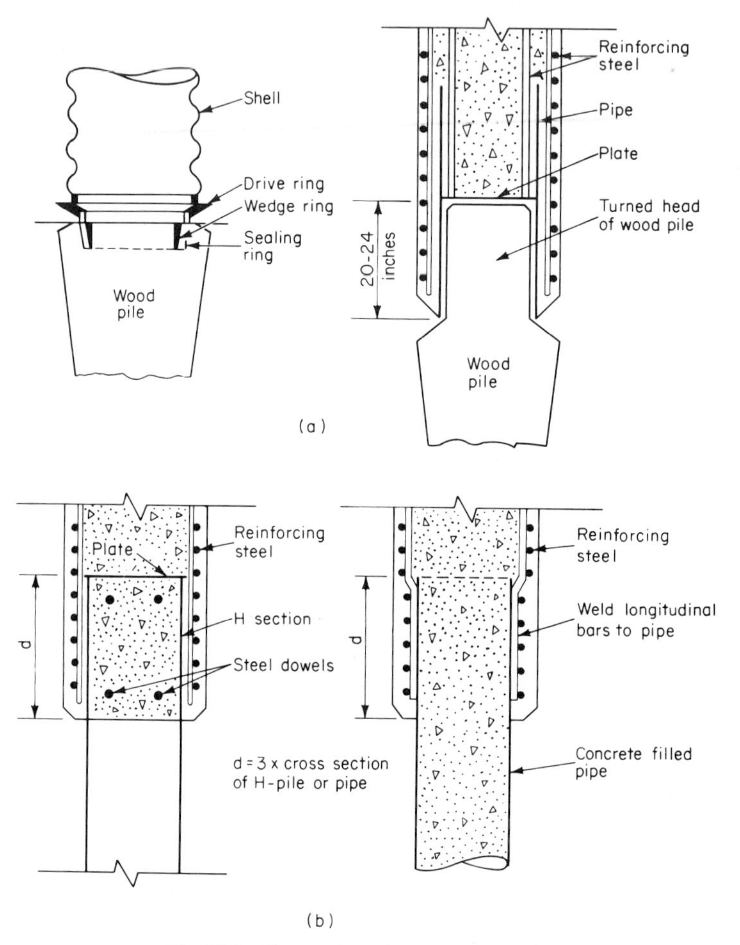

Fig. 27-3 Connections for composite piles.

The *wood-and-concrete composite pile* has a lower section of wood and an upper section of concrete, either precast or cast in place. If provision must be made for bending at the joint, the wood section of the pile should be formed into a tenon and a cage of reinforcing steel placed in the concrete section of the pile. This portion of the pile must then be designed to resist the lateral forces.

Occasionally the upper portion of the composite pile is made of precast concrete, with a socket for the wood pile cast into its lower end. The head of the wood pile must then be sized to fit the socket neatly.

The *wood-and-steel composite pile* can have a wood lower section and a steel upper section, or vice versa. Although the combination of steel and wood is an unusual one for a pile, ground conditions could be such that the customary concrete top could be replaced by steel. The writer has no direct knowledge of this type of composite piling having been used in any permanent installations, but this does not rule out the fact that it is a feasible alternative. If a steel pipe is used for the upper portion of the pile, a flat plate should be welded inside it, approximately 18 in. from its lower end, to transmit the driving forces to the wood pile (Fig. 27-3*a*).

The *steel-and-concrete composite pile* has a lower section of steel and an upper section of concrete. The lower portion can be a rolled-steel H section or a piece of pipe. The upper portion can consist of a cast-in-place concrete section, or concrete pipe, or a precast-concrete section. However, an enlarged concrete section which is cast on the end of a steel H section or pipe to increase its displacement or end bearing is not considered as a composite pile but rather as lagging for the steel pile (Fig. 27-3*b*).

Uses Each type of composite pile has its own characteristic applications.

Wood-and-concrete composite piles are used in the same ways and for the same reasons as wood piles. The concrete portion extends from the permanent groundwater line to the pile cap or point of load, thus replacing that portion of the wood pile that would be expected to rot or disintegrate.

Steel-and-concrete composite piles are used to support high loads where steel in the upper section might disintegrate because of unusual ground conditions, such as chemical reaction in underfill, chemical waste dumps, and occasionally electrolysis.

Sometimes short sections of steel H piling are fastened to the end of cast-in-place or precast-concrete piling to assist in penetrating rock or hardpan. In such cases, the pile is not considered a composite pile. Circumstances will occur where the depth to adequate bearing material is so great that a pile driver cannot handle an all-shell pile of the required length. A pipe may then be added as an extension to the upper end of the shell. Such a combination is regarded as a variation of the cast-in-place pile and not a composite pile.

Wood-and-steel composite piles are so uncommon that their characteristics must of necessity be those of the usage to which they may be adapted.

SHEET PILES

Sheet piling ordinarily takes its name from the material from which it is made, occasionally from its inventor, and sometimes from the manner in which it is most commonly used Sheet piles are made from all the basic pile materials, either plain or combined. Their forms or shapes are such that they interlock to form a wall or "sheet" of piling.* Sheet piling thus finds its place in bulkheads or retaining walls to contain a fill; in a cellular cofferdam to support the adjoining earth and prevent its movement; as a protective cofferdam in the construction of a bridge pier or similar structure; and as a deep enclosure around the base of a waterfront structure to prevent erosion and scour.

Sheet piling retains the basic characteristics of the materials from which it is made. Wood sheet piling can be considered as cheap and temporary. Treated wood is somewhat more durable but still temporary. Concrete sheet piling, if properly made and cured, can be considered the most permanent of all types. Steel sheet piling is more durable than wood but less so than concrete, and allowance must be made for the corrosion that will occur.

Wood Sheet Piling This is made of single wood planks, square edge or tongue and grooved; of laminated planking spiked together to form lapped joints or tongue-and-groove joints, known more commonly as Wakefield sheeting; and of large timbers, which can be milled with tongue and grooves but more commonly have dimensioned lumber spiked to the abutting edges to form a tongue and groove. The jointing is for the purpose of making a tight wall (Fig. 27-4*a*). Timber sheeting has the usual structural limitations of wood; it cannot take hard driving and should be used only where it can be properly braced and supported.

Wood sheet piling, and any other piling without a tension interlock, will produce a tight wall only if it is driven in a manner that forces the joint elements together.

* See also Sec. 28, Cofferdams and Caissons.

The joint must then be held in position by a well-bolted and tied system of wales. The natural tendency of lumber to warp and twist makes tight joints in wood sheet piling difficult to obtain. Wakefield sheeting made of laminated dimension lumber is the best form of wood sheet piling to give tight joints. Slight irregularities and

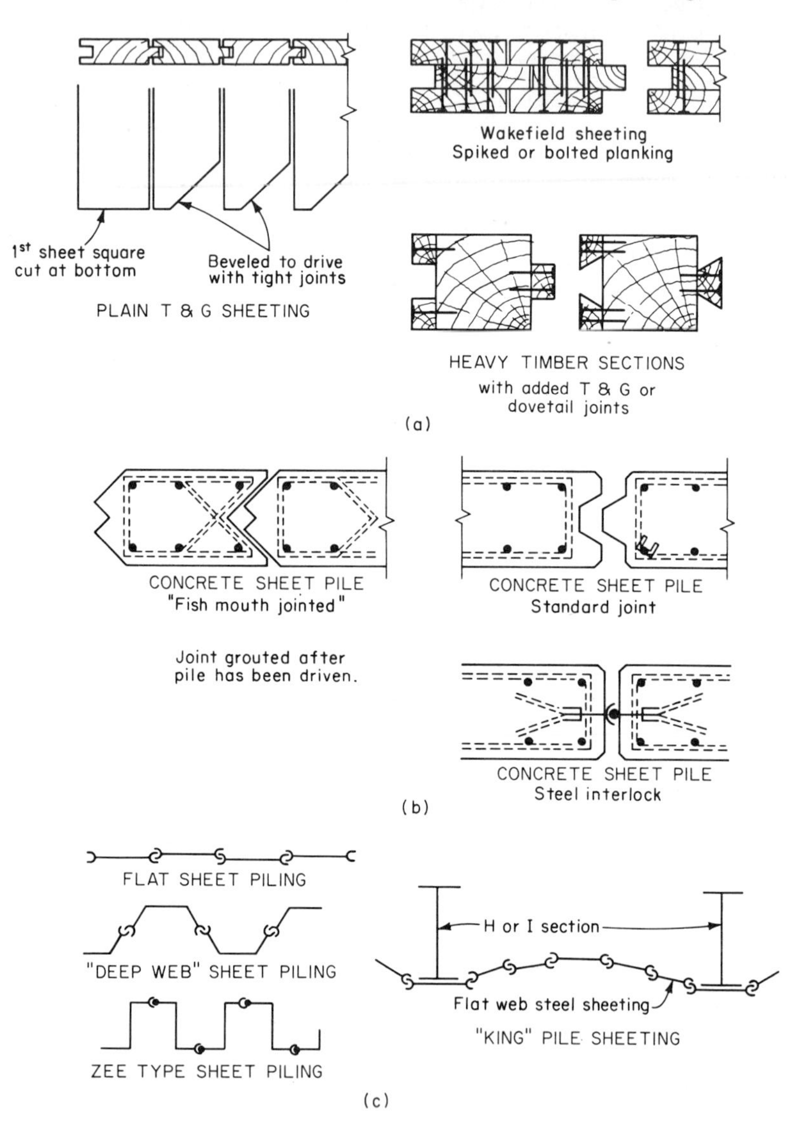

Fig. 27-4 Examples of sheet piling.

warp can be compensated by spiking or bolting pieces of lumber with opposite twists together. The use of "select" grades of lumber will also help to eliminate this problem. The Wakefield tongue and groove can be varied to suit conditions and, after driving, the adjacent pieces can be spiked together through the joint to produce what is effectively a solid wall of timber sheeting.

Concrete Sheet Piling This can be cast to the desired shape and size, with matching grooves for driving alignment and for subsequent pressure grouting. Concrete sheet piling must be initially cast to whatever length is required for driving, since it is difficult to extend it at a later time. Its features are similar to those described earlier for concrete bearing piles, except that the sheet piles are primarily designed to resist lateral forces and bending rather than vertical loads. The principal difficulty with the use of concrete sheet piles is the joint between sections which, unless properly designed, cannot be made watertight. The "fish-mouth" joint illustrated in Fig. 27-4*b* permits the concrete sheet piling to be driven together with a welding action, and afterward the "mouth" can be jetted clean and grouted. If properly done, this type of joint will make a watertight wall. Should joints fail, they can be cleaned out with a jackhammer drill and regrouted.

Steel Sheet Piling This can be rolled in various shapes, weights, and sizes, with maximum mill lengths in the range of 70 ft. Steel sheeting is flexible in that it can be cut, patched, lengthened, and reinforced with comparative ease. Localized increases in its section modulus can be effected by welding plates along its webs and/or flanges.

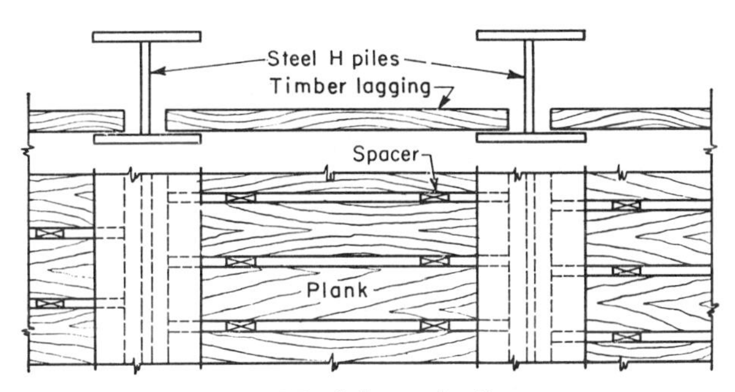

Fig. 27-5 Subway sheeting.

Steel sheet piling has a unique tension-interlock feature and can be made to fit almost any shape of excavation, bulkhead, or cofferdam. It is the only type of sheeting that can be used for cellular cofferdams or similar applications where tension is transmitted via the sheets through the interlocks. "King piles" are sheet piles to which steel I-beam or H-pile sections have been welded or riveted for added strength. The sheet piles arch between adjacent King piles, acting in tension to resist lateral loading. The King piles anchor the sheet pile arches and thus take all of the bending stresses (Fig. 27-4*c*).

Steel H piles are used extensively as "soldier" piles to support the sides of trench excavations by a method known as *subway sheeting*. The H piles are driven on 5- to 8-ft centers in advance of the excavation. Then, as excavation proceeds, timber lagging is placed back of the flanges of the H piles and spaced so as to allow a free draining face (Fig. 27-5).

CHARACTERISTICS OF PILE LOADS

Primary Loading The forces or loads which have to be considered in the design of a pile can be grouped according to their lines of action, as follows:

1. Vertical, either downward or upward, directed axially along the pile length or parallel thereto
2. Horizontal
3. Rotative, tending to impart a twist to the pile
4. Any combination of the above

The downward vertical load is the most common, and the "bearing capacity" of a pile is expressed in terms of this loading. An upward vertical load, simply called "uplift," requires a pile that can take tension. A vertical force (load), which is parallel to the pile axis but acting at some distance from it, will produce bending or limited rotation of the pile. A horizontal load will also cause pile bending, which will be resisted by the reaction of earth or any other material which is in firm contact with the bending portions. The distribution and magnitude of this reaction will depend on the depth and character of the affected strata. Stability requires that reactive forces be sufficient to counteract the pile movements which are initiated by the horizontal force.

A rotative movement is induced in a pile when a horizontal force acts at a distance from the pile axis. The twisting action thus imparted will, in most cases, also cause bending in the pile.

Variable Loading A pile may be subjected to a cyclical form of loading by a reciprocating or oscillating piece of equipment. It may also receive an intermittent impact loading, such as from a hammer or stamping machine. These variable loads can vary in their intensity and/or direction. Piles which are intended to resist such loads must be more carefully driven than simple bearing piles. An alternating downward and upward loading can cause a lightly driven pile to "work" to the point where the bond between it and a surrounding cohesive soil is destroyed. Pile motion in a granular soil will tend to consolidate the adjacent material and contribute to pile and area settlement. Piles which are used to support a hammer or other impact loading may also be driven farther into the soil by the repeated blows. Piles which are subjected to variable loadings should be driven to five or six times their design bearing values provided that the soil will take the load and the piles will take the driving stresses. Where either of these conditions cannot be satisfied, more piles at lower bearing values must be used.

Handling and Storage In addition to sustaining bearing, bending, and twisting stresses that result from its loading as an integral part of a structure, a pile must be capable of withstanding the stresses of handling. Timber piles seldom present undue problems in this respect, since the growing trees from which they are fashioned have already withstood more stress from wind than can be expected during handling. Precast-concrete piles, on the other hand, must be specifically designed and reinforced to withstand these stresses. Design may be based on uniform reactions at the pickup points, or the pickup points can be located so that the bending moments at and between the points of support do not exceed some uniform value. Steel H piling and pipe piling, being of uniform section, should be picked up at points which give uniform bending moments at the pickup points and also between these points.

The magnitudes of bending moments produced in a pile during handling will depend on the method used in lifting and on the location of the pickup point(s). For ease in the handling of the pile, the reactions at the pickup points should be the same. The location of the pickup points for equal reactions on uniform section piles and the maximum moment for each arrangement are shown in Fig. 27-6.

If the lifting points cannot be arranged for equal reactions, the maximum bending moments for each of several lifting arrangements should be calculated. The one with the lowest value of the maximum bending moment can then be used. Alternatively, an arrangement that equalizes the bending moments at supports and between supports may be selected.

All piles should be stored on level ground on timber blocking so that the axis of each pile is maintained in a straight line. Wood piles should be stacked alternately butt to tip, with blocking between layers. Steel H piles should be stacked flange to flange, with a 2-in. space between flanges and with blocking between layers. The 2-in. space facilitates the use of slings for handling, as does the blocking. Pipe piles in single and double random lengths are stored in one of two ways. In the first method, the stack is built in layers with the pipe in each layer touching the next pipe and with blocking between layers. This blocking is not for handling; rather, it keeps the lowest outside pipe from being pushed out under the weight of the stacked pipe, which would lead to the collapse of the entire stack. In the second method, the stack is built in layers, with each successive layer containing one less length of pipe and with no blocking between the layers. The end pipe on the bottom layer must then be blocked or welded to the adjacent pipe to prevent collapse of the stack. Precast-concrete piles must be stored so that their pickup points rest on level blocking, or else the piles will crack. It is also necessary to locate the blocking of successive tiers exactly over the blocking below, or the weight of the upper layers of piling will bend the lower piles; in the case of precast-concrete piles, it will break them.

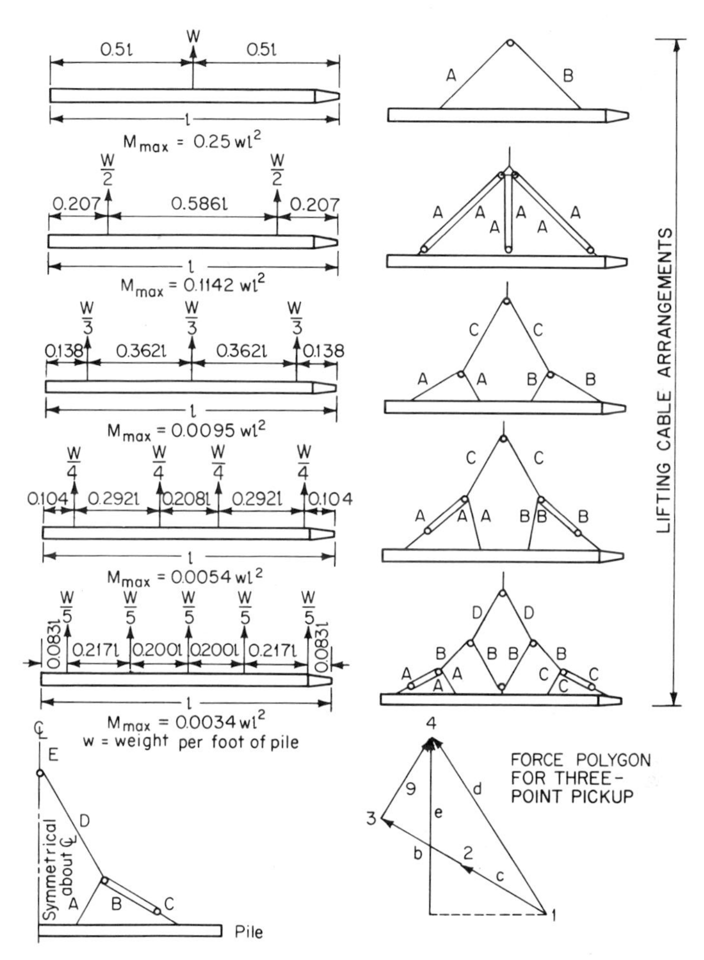

Fig. 27-6 Pickup points for equal loading of uniform-section pile.

Batter Piles These are frequently used in conjunction with vertical piles to resist horizontal forces. The reaction to the horizontal loading is then supplied by forces acting along the longitudinal axis of each pile. This is illustrated in Fig. 27-7, which shows the coplanar resolution of the applied load and the reactive forces. The force B which the batter pile must exert when it acts in conjunction with a bearing pile to resist a horizontal force H is expressed by the formula

$$B = \frac{H}{\sin \alpha} \tag{27-1}$$

where α is the angle between the bearing pile and the batter pile.

The vertical or uplift force V which is then imparted to the bearing pile is

$$V = \frac{H}{\tan \alpha} \tag{27-2}$$

Any downward load which is carried by the vertical pile will help to resist this vertical uplift reaction. If the applied horizontal force is constant and permanent, the uplift component can provide an effective increase in the bearing value of the vertical pile. For example, if we assume that the normal bearing value of a vertical pile is 50 tons, and further assume that a horizontal force imparts a vertical component of 20 tons to this pile, then the actual load that can be placed on the bearing pile is 50 tons plus 20 tons, or 70 tons.

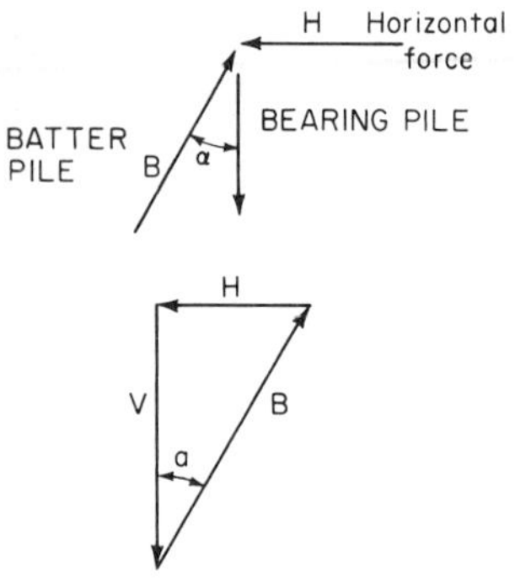

Fig. 27-7 Force diagram.

With proper layout and batter, all forces on the piles become linear and all bending and twisting forces can be eliminated from further consideration in the design. The position of the batter pile with respect to the vertical pile must be such that the pile axes are in line or intersect, otherwise twisting and bending of the vertical pile will occur.

STRUCTURAL CAPACITIES OF PILES

The allowable loading on a pile will normally be limited by the ability of the soil to take the load transmitted by the pile rather than by the ability of the pile to transmit the load. An exception to this may occur when an unsupported length of pile projects above the ground, including any portion of a pile which is in water. This unsupported length must then be designed as a structural member supporting a direct axial load. In the ground, adequate lateral support for the pile is provided by all soils except the very soft or most fluid. Buckling of the pile in that portion below ground can be disregarded in all cases but those noted.

Unless a pile can be set by jetting, the stresses set up in it under the impact of the hammer will usually exceed, by five to six times or more, the stresses which will occur under its allowable static load. Many codes recognize this and stipulate very low allowable static loads. The allowable structural load for a composite pile should be based on the weaker of the pile's two structural parts.

Timber Piles Building codes will usually permit loads up to 30 tons for most wood piles, and up to 40 tons for southern yellow pine, without requiring that loading tests be performed. If there is no applicable building code, and in the absence of loading tests, the structurally allowable axial load R on a round timber pile can be taken as the least of the values given by the following formulas:

$$R = AF_c \tag{27-3}$$

$$R = \frac{3.619AE}{(l/r)^2} \tag{27-4}$$

$$R = \frac{0.30A^2E}{l^2} \tag{27-5}$$

where R = structurally safe axial load on pile, lb

A = cross-sectional area of pile, sq in., taken as the sum of the minimum (point) area and one-third of the difference between the minimum and maximum (butt) areas

F_c = allowable value of unit compression parallel to the grain, psi, with appropriate adjustments for degree of seasoning, service conditions, and duration of loading

E = modulus of elasticity, psi, with appropriate adjustments as described for F_c

l = unsupported length of pile, in.

r = least radius of gyration of section, in.

Values of F_c and E are readily available in published tables.[3,4] These values are frequently based on seasoned wood and continuously dry conditions of service, whereas fresh-cut (unseasoned) piling may actually be used under wet conditions. The tabulated values of F_c and E must therefore be adjusted for the anticipated conditions of service and use.

Concrete Piles *Concrete-filled pipe piles* can be designed for structurally safe loadings according to the following recommendations:[7]

$$R = 0.25f_c'\left(1 - 0.000025\frac{l^2}{r_c^2}\right)A_c + f_sA_s \tag{27-6}$$

The value of f_s is governed by the following relationship when the steel pipe has a yield strength of at least 33,000 psi and a l/r_s ratio equal to or less than 120:

$$f_s = 17{,}000 - \frac{0.485l^2}{r_s^2} \tag{27-7}$$

where R = structurally safe load on pile, lb
f_s = allowable axial unit stress in the steel pipe, psi
f_c' = crushing strength of concrete cylinders at 28 days, psi
A_c = area of concrete, sq in.
A_s = net area of steel pipe in sq in., after deducting 1/16-in. wall thickness for corrosion
r_c = radius of gyration of concrete-filled steel pipe, in.
r_s = radius of gyration of steel pipe, in.
l = unsupported length of pipe, in.

Cast-in-place concrete piles that are entirely embedded in the ground should be designed so that the concrete stress at the level of the bearing stratum does not exceed $0.25f_c'$. Where cast-in-place piles extend above ground, or where the ground above the bearing stratum is fluid or semifluid, the "unsupported length" of the pile should be determined and the allowable pile load gauged as for precast-concrete piles.

Precast-concrete piles, in the absence of building codes, can be designed by the methods recommended for tied or spirally reinforced columns.[7]

For tied columns:

$$R = 0.85A_g(0.25f_c' + f_sp_g) \tag{27-8}$$

For spirally reinforced columns:

$$R = A_g(0.25f_c' + f_sp_g) \tag{27-9}$$

where A_g = gross area of column, sq in.
f_s = nominal working stress in psi for vertical column reinforcement, to be taken as 40 percent of the specification value of the yield point f_s' for the grade of steel used*
p_g = ratio of effective cross-sectional area of vertical reinforcement to gross column area A_g

When the unsupported pile (column) length l is greater than 10 times the smallest lateral dimension d, the allowable axial load R should be reduced to R' by the following formula:

$$R' = R\left(1.3 - 0.03\frac{l}{d}\right) \tag{27-10}$$

Steel Piles Whether made of pipe, H section, or fabricated shapes, steel piles should be designed in accordance with sound engineering principles. Building codes, where applicable, will generally state the permissible structural loading for piles. Lacking such guidance, a steel pile can be viewed as an axially loaded column on the basis of the following formulas:[9]

* See Reinforcing Steel in Sec. 21, Conventional Concrete.

For main compression members,

$$R = Af_s \tag{27-11}$$

$$C_c = \sqrt{\frac{2\pi^2 E}{f_s'}} \tag{27-12}$$

for $Kl/r < C_c$,

$$f_s = \frac{[1 - (Kl/r)^2]f_s'}{\frac{5}{3} + \frac{3Kl/r}{8C_c} - \frac{(Kl/r)^3}{8C_c^3}} \tag{27-13}$$

for $Kl/r > C_c$,

$$f_s = \frac{12\pi^2 E}{23(Kl/r)^2} \tag{27-14}$$

where A_s = net area of steel column, sq in., after deducting 1/16-in. wall thickness for corrosion
E = modulus of elasticity of steel (29,000,000 psi)
K = effective length factor for pile (see Fig. 27-8)
l = unsupported length of pile, in. (see Fig. 27-8)
f_s' = minimum yield strength, psi, for grade of steel employed
f_s = allowable working stress in steel, psi
r = governing radius of gyration for section, in.

DISTRIBUTION OF PILE LOADS TO SOIL

As stated earlier, the allowable pile load is normally governed by the ability of the pile to transfer its load to the soil rather than by the structural capacity of the pile. Various pile-loading formulas can be used to predict this allowable load for a single pile under a specified set of conditions; however, piles in groups may behave quite differently from single piles. The action of piles is also influenced by whether the soil is granular (sands and gravels), cohesive (clays), or alluvial (silts). Basically, a bearing pile can distribute its applied load to the surrounding soil in three different ways (see Fig. 27-9):

1. By end bearing, in which all the load is transmitted through the point of the pile to the underlying material. An example is a pile placed through water with its tip end resting on a clean rock bottom.
2. By friction, in which the load is transmitted by friction between the surface of the pile and the adjacent soil. An example is a pile driven in a clay soil.
3. By a combination of the above.

Practically every pile transmits its load by the last of these three ways. In theory, a pile could be set on a clean rock bottom to develop a simple end-bearing situation. In actual practice, however, a hole would first be drilled into the rock and the pile would then be set into the hole and grouted. This would cause the load to be transferred to the rock along the grouted sides of the pile tip as well as on its end. Even in a homogeneous clay, where a high percentage of the load is transmitted to the ground by the adhesion of the clay to the sides of the pile, there is some load transmission through the pile point. The same applies to a compaction pile where, because of the displacement of the pile, the soil is compacted. Here the load is transmitted to the soil by a combination of bearing and side thrust, but the point of the pile still transmits a portion of the total load.

End-bearing Piles When a pile is used as a column to transmit its load in end bearing to bedrock, hardpan, or some other firm stratum, the structural load is considered to be acting entirely at the end of the pile point. Thus, where a pile is essentially resting on rock, the limiting load on the pile is normally assumed to be governed by the allowable bearing value of the rock in question. This reasoning is valid provided that the tip section can carry the load without crushing.

A single end-bearing pile, in transmitting its load, creates a bulb of pressure under the tip. Contours of equal pressure within this bulb can be computed by applying the Boussinesq equations or similar theoretical relationships.[10] For all practical purposes, any rock will sustain the load transmitted by a pile that has been driven to (or into) it to the required dynamic resistance. The bulbs of pressure beneath

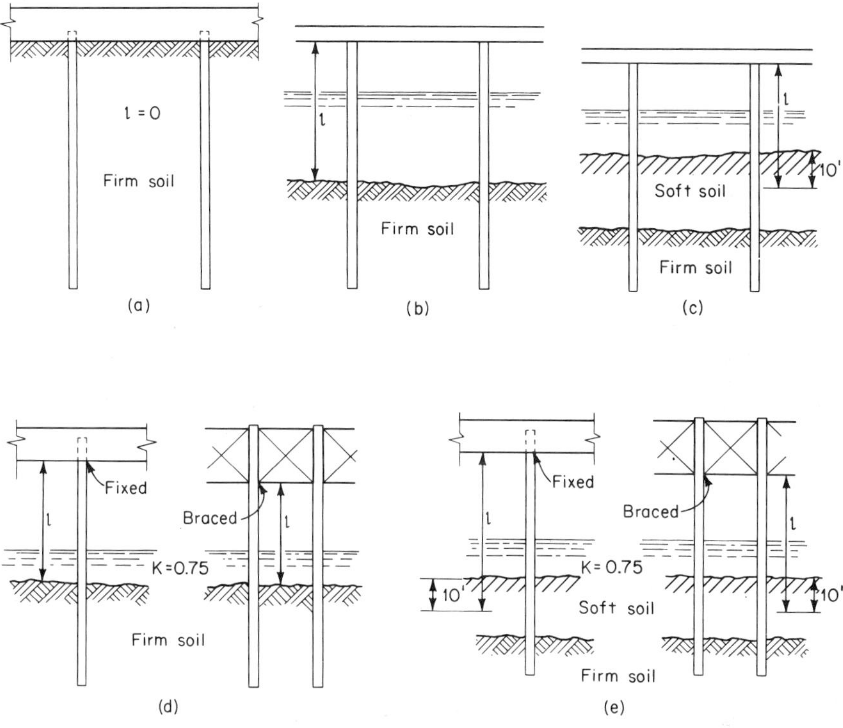

Fig. 27-8 Determination of unbraced lengths for steel piles. (*a*) Entire length embedded in any soil or combination of soils, other than virtually fluid material. In this case the pile is supported throughout its length, and no reduction in load is required because of slenderness ratio. (*b*) Unbraced length above firm soil. The effective length should be taken as the distance from the surface of the firm soil to the nearest connection above the ground. $K = 1.0$. (*c*) Unbraced length above soft soil. The effective length should be taken as the distance from a point 10 feet below the surface of the soft soil to the nearest connection above the ground. $K = 1.0$. (*d*) Unbraced length, top fixed, firm soil. Where the upper end of the unbraced length is fixed by embedment in concrete or is braced in a manner providing the equivalent of continuity, a K factor of 0.75 may be applied to the actual unbraced distance down to the surface of the firm soil, to allow for such restraint. $K = 0.75$. (*e*) Unbraced length, top fixed, soft soil. This condition is the same as (*c*), except that a K factor of 0.75 may be applied to the actual unbraced distance down to a depth of 10 feet below the surface of the soft soil, because of the fixed end condition. $K = 0.75$.

adjacent piles in a pile group will then intersect and overlap,[8] as indicated in Fig. 27-10. The overlap of pressure contours will not cause critical pressures on the bearing stratum if the piles are spaced 3 to 3½ diameters apart.

End-bearing piles may rest in a hard stratum above a soft, compressible material that can be expected to consolidate under load. The piles and pile groups should then be arranged, by battering if necessary, to obtain a distribution of loads such that area settlement of the soft stratum will be uniform and within tolerable limits.

Care should also be used to ensure that the tips of the piles do not pass through the bearing stratum but instead stop above its base. The minimum permissible distance h between the tips of piles and the bottom of the bearing stratum which supports them can be calculated as follows, assuming a pile spacing d:

$$h = \frac{d}{2} \tan 45° = 0.5d \qquad (27\text{-}15)$$

Figure 27-11 illustrates the situation just described where, for practical purposes, the distance h can normally be assumed to be 5 ft. Intensity of loading and the compressibility of the underlying soil must be considered jointly when using this method, since the bearing stratum functions as a raft foundation to ensure a more or less uniform pressure on the soft stratum of soil. Settlement then should be uniform and equal for all parts of the structure.

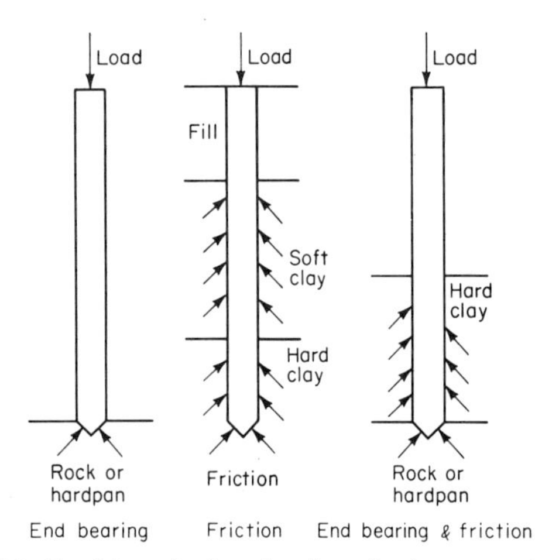

Fig. 27-9 Load transfer from bearing piles to surrounding soil.

Friction Piles Rock or some other firm, unyielding material such as hardpan may not be present within physical or economical reach to permit the use of an end-bearing pile. Instead, the piles may be driven into a deep stratum of nearly uniform soil or through successive layers of different soils until the required load-carrying capacity is obtained by the friction of the surface of the pile against the soil in contact with it. Such piles are called "friction piles." As different soils provide this friction in different ways, it is sufficient to describe the two extremes—the cohesive soils (clays) and the cohesionless soils (sands and gravels). Most of the others are mixtures of these two basic types, with or without the addition of rock and dust (silts).

In the case of cohesive soils or clays, two factors resist the movement of a friction pile: one is the adhesion of the soil (clay) to the pile and the other is the shearing value of the clay itself. In clays, and to a minor extent in some silts, the shearing strength of the soil is less than its adhesion to the pile; failure then takes place by shearing within the soil at some short distance from the surface of the pile. This is clearly demonstrated during the pulling of a pile, which will come out of the ground with a thin layer of clay adhering to its surface.

The ultimate load at failure of a friction pile in clay can be calculated with reasonable accuracy if the shearing strength of the clay is known. This ultimate load calculation can be considered to apply only after any effects from the remolding of the clay during driving have disappeared. The tendency of a pile toward slippage or yielding under load is greatest just after driving and diminishes with time. As a pile is driven in clay, a very thin layer of almost liquid clay is produced along the

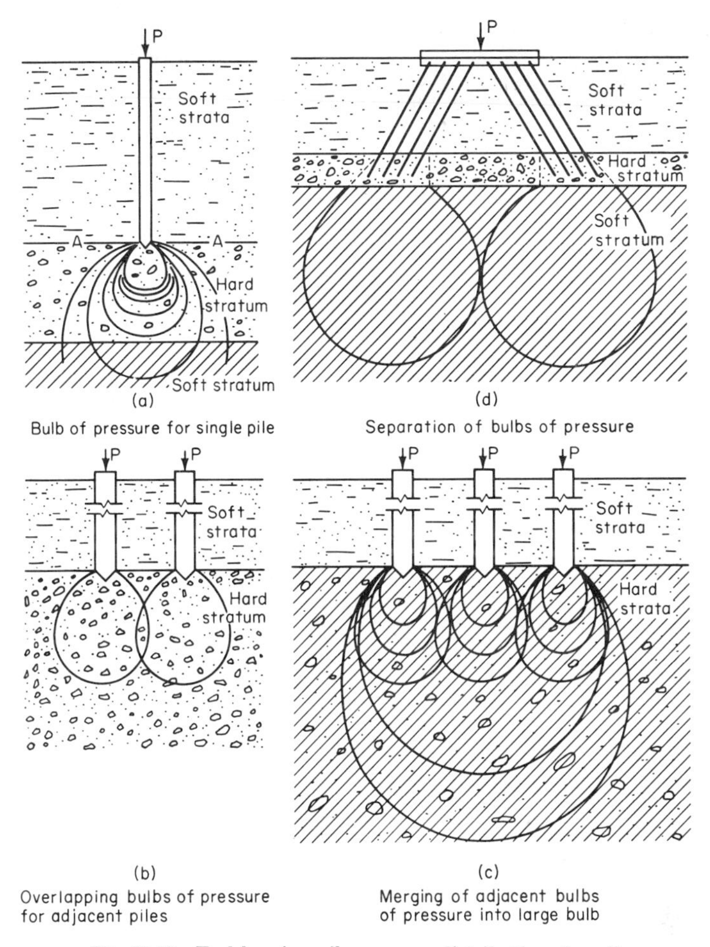

Fig. 27-10 End-bearing-pile pressure distributions in soil.

surface of the pile. For the usual case, it is believed that the clay adjacent to the pile will slowly regain its original (premolding) strength. In fact, there are some indications that the shearing strength of the clay immediately around the pile may actually increase over a period of time because of this remolding process.

The action of a friction pile in a cohesionless soil such as sand is quite different from its action in the cohesive soil just described. The internal friction or shearing strength of a sand varies with the soil density and the extent of the applied pressure. It is also affected by whether the sand is dry or saturated. There are still further factors to consider if shearing strength tests are to be made on the sand for the purpose of determining the probable capacity of a pile: the apparent cohesion of the sand, whether it is cemented or partially cemented in its natural state, and whether the shear tests

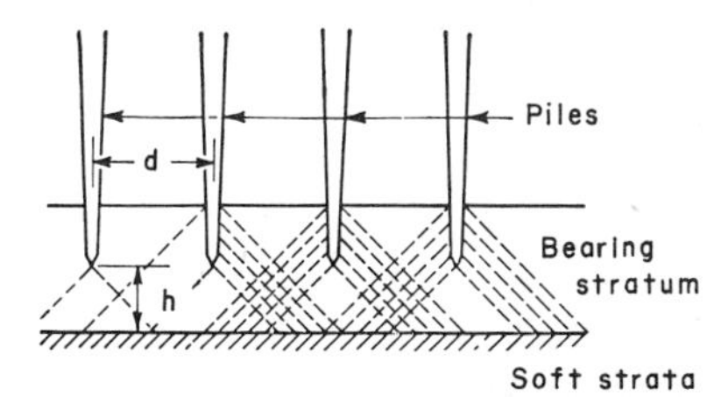

Fig. 27-11 Piles in a bearing stratum overlying a compressible material.

are to be made on the basis of a slow-shear test or on a consolidated, quick-shear test.[10] These concepts will not be pursued further, since the purpose of this section is to dwell on the practical aspects of piles and pile driving.

When a pile is driven into sand, the sand is made more dense (i.e., compacted) by virtue of the fact that the pile displaces its own volume of sand. The vibrations of the pile due to the hammer blows are also transmitted to the soil, causing the grains of the sand to fall more closely together and thus reducing the void spaces between them. This latter type of compaction is more apparent in a saturated sand than in a dry sand. There have been cases where pile driving has consolidated a bed of loose sand to the extent that the entire level of the piled area and the ground adjacent to it are at a lower elevation than before driving commenced. Particular care must be taken in deep beds of sand, whether loose or dense, where piling is used under machinery which will later set up vibrations in the piles and surrounding soil. These transmitted vibrations will cause continued consolidation of the sand and result in settlement, almost always unequal, in the building and adjacent properties.

Area loading is an important factor affecting the behavior of pile foundations in clay soils. This topic is treated extensively in the soil mechanics literature and will only be discussed here in passing. Cases occur where friction piles driven in a clay stratum are adequate to carry a considerable load, but more compressible layers are present beneath the supporting clay stratum and result in area settlement of the total pile structure. In other words, the soil in which a pile is driven may be adequate to prevent a general shear failure, but the compressible layers below the piled area can still permit excessive settlement. It is also important that individual column and footing loads be distributed so as to produce reasonably uniform reactions at the tips of piles within or above a compressible stratum. If this condition is not satisfied, there will be differential settlement of the foundation. This will not be the fault of the pile driving, but is instead the result of inadequate foundation investigations and/or poor foundation design.

Each driven pile develops an area of pressure around it. When the pile load is applied, a bulb of pressure is developed in the surrounding soil and extends below the point of the pile. Figure 27-12 illustrates the configurations of pressure bulbs developed in soils adjacent to single and group friction piles.[8] Note that these pressure distributions are somewhat different from those illustrated earlier for end-bearing piles (Fig. 27-10). Single piles, widely spaced so that the individual bulbs of pressure do not overlap, can take the full load assigned to them without overloading the soil. Groups of piles, however, present a different situation. If there is an overlapping of the pressure bulbs below the piles, the result may be an overloaded soil area and excessive settlement of the structure. Herein lies the fallacy of basing a pile design load on the results of load tests on single piles. Wherever possible, load tests should be made on groups of piles and the design load per pile then determined from the group load.

A friction pile, regardless of the load that it is designed to carry, is simply a structural member that transmits its load to the soil around it. This load transfer continues for each foot of pile length until, in theory at least, there is no load at the pile point. Thus, a friction pile receives a load at its top and progressively transfers this load to lower depths in the ground where less settlement will occur. Chellis, in *Pile Foundations*,[8] points out that:

> Settlement produced by a uniform load increases in proportion to the diameter of the loaded area for cohesive soils, whereas in cohesionless soils the size of the area has little effect. Settlement under a unit load decreases with increasing depth of foundation. However, in addition to depending upon the depth of the foundation, settlement also depends on the ratio of depth to diameter of the loaded area, so that for equal potential settlement reductions, the depth-to-diameter ratios should be kept equal. In cohesionless materials the effect of this ratio on settlement is less than in more cohesive soils. This principle indicates that the value of the piles may be greatly affected by the relation of their lengths to the width of the loaded area. Under a narrow structure every effort should be made to keep the piles longer than the width of the structure, so that a lowered bulb of pressure will occur.

Combination Piles A large number of soil borings indicate the presence of mixed strata conditions, i.e. loose sand over soft clay over sandy clay over hardpan, etc. Piles driven into such soils will fall into the category of combination piles, or part friction and part end bearing, and the possibilities for their behavior under load are

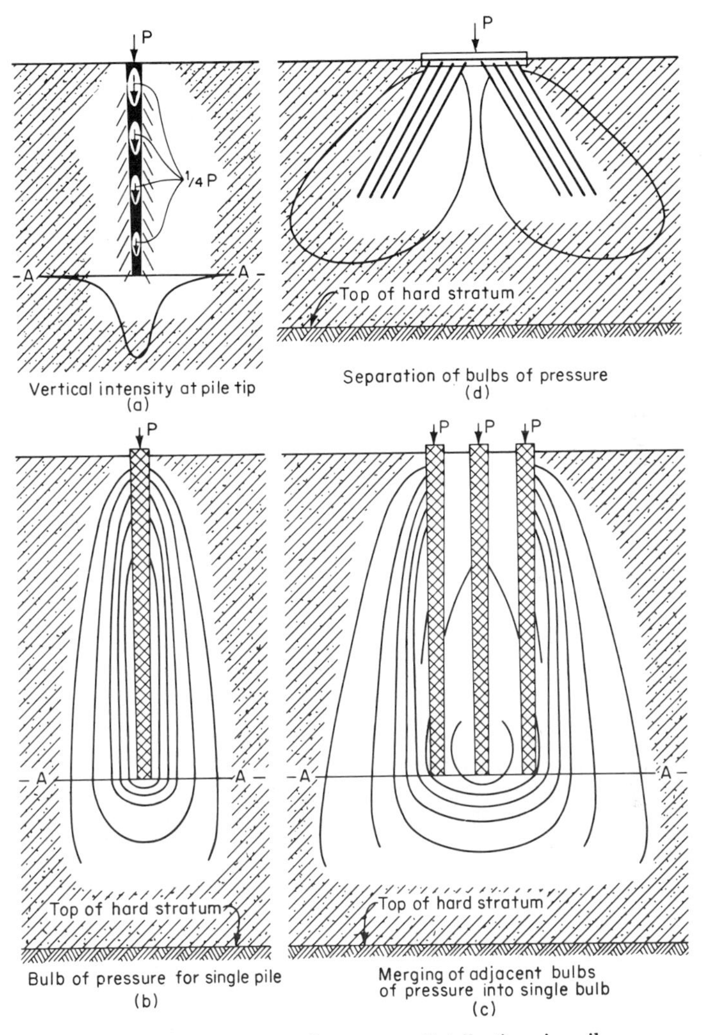

Fig. 27-12 Friction-pile pressure distributions in soil.

limitless. However, the pertinent features of the predominating soil stratum will usually prevail. Its characteristics will then determine the behavior of the pile.

Compaction Piles Although every pile tends to, and does, compact all soils, there are piles which are designed to emphasize this action. A tapered pile, particularly one with a heavy taper such as the Raymond Standard pile, may significantly com-compress and consolidate the soils around it. This effect can be pronounced in a cohesionless soil such as sand but may be almost nonexistent in a cohesive or clay soil. The use of heavy-taper compaction piles in loose granular soils has the effect of creating an island of compacted soil in the area where the piles have been driven. This method can be applied when the more recent development of vibratory compaction of the granular soil cannot be used. The compaction effect of the heavy taper has been put to good use by fixing a heavy-taper section to the point of a straight-sided pile or by casting an oversize heavy-taper concrete section on the end of a length of pipe or H pile.

PILE-DRIVING EQUIPMENT

Contrary to general belief, the driving of piles is an exacting process. To achieve the desired results, the hammer should be correctly sized for the work it is to do and the driver should have the capacity and be of the proper type to handle the pile. Accessory tools such as jets, spuds, and preexcavators should be used when necessary and in a manner to install the pile properly.

Hammers The original tool for driving piles, the drop hammer, has almost disappeared from service; those drop hammers still in use range between 1,000 and 5,000 lb in weight. A drop hammer for the driving of piles has two disadvantages: (1) the rate of driving is low, allowing a temporary set of the pile to occur in some soils; (2) there is a tendency to let the hammer fall from a great height in order to increase the rate of driving; the resulting high-velocity impact stresses often cause fracture of the head of the pile and will, at times, cause the pile to break at the tip or in the middle.

The mechanical hammers that have superseded the drop hammer are of six distinct types: (1) single acting, (2) double acting, (3) differential acting, (4) diesel, (5) hydraulic, and (6) vibratory. Each has different pile-driving characteristics.

Historically speaking, all pile hammers except the newly developed diesel, vibratory, and hydraulic hammers have been operated by steam. However, in the last 10 years there has been a trend toward compressed-air operation, particularly for the lighter hammers and the smaller or intermittent driving operations. The development of diesel-operated compressors, in capacities up to 1,200 cfm at 125 to 150 psi with the addition of heat interchangers, has made the operation of pile driving simpler and more economical. Manufacturers' catalogs now list the hammer requirements for either steam or air operation and, what is more important, the hammer valving has been redesigned to operate under both conditions. Most of the newer hammers will no longer freeze up at the valve due to the cooling effect of the expanding air at the exhaust.

Tables 27-1 through 27-3 are taken from the data sheets of the Vulcan Iron Works, Inc. They give the specifications of comparable sizes of pile hammers, both in current production and discontinued, and of comparable sizes of pile extractors. In the absence of more detailed analyses of the piles and their driving requirements, Table 27-4 provides some guidance as to appropriate hammer sizes for various types of piling. The hammer sizes in this table have been based on the *Engineering News* formula (see subsequent discussion of dynamic pile formulas). It should be noted that the accepted pile-driving formulas are not applicable to the vibrating hammer, nor are the following comments concerning stroke and weight of ram.

The stroke or fall of the hammer ram is a factor that influences the energy delivered by the hammer. In fact, for a single-acting hammer,

$$\text{Energy} = \text{weight of ram} \times \text{height of fall}$$

The slide bar of a modern impact hammer can be adjusted so that the blow (energy) remains constant, or it can be adjusted to increase or decrease (soften) the impact. The slide bar should be fixed to the ram so that the valve action is the same for each stroke, and no hammer should be used with a loose slide bar. Improper adjustment of the slide bar can result in a bouncing hammer if the valve operation is delayed or in a cushioning effect that could, in extreme conditions, negate the blow if the valve is permitted to open too soon.

The weight of ram is an important factor, since a heavy-ram impact hammer working on a short stroke is more effective in driving a pile than a light-ram long-stroke hammer. The weight of the ram, the length and speed of the stroke, and their relation to the weight of the pile are important to the proper driving of the pile. In theory, a pile can be of such a length that all the energy which it receives from a hammer blow is absorbed in its mass. Under these circumstances, a blow of the hammer will not advance the point of the pile. To appreciate this statement, it is necessary to understand what happens when the hammer hits the pile.

A hammer blow causes a temporary compression at the pile head. This compression travels the length of the pile in the form of a wave and will move the pile point forward when it is reached. However, as this wave of energy moves through the pile, it loses some of its strength by a temporary compression of the material of the pile. In a short pile, this effect is negligible and can be disregarded; in a long pile, the energy absorption by the temporary compression of the pile can be considerable and becomes a factor which affects the ultimate driving resistance of the pile. The friction between the ground and the pile obviously resists the downward motion of the pile, and when this resistance is added to the inelastic energy absorption within the pile, the sum of the two can be enough to completely absorb the energy of a small hammer. For this reason, it is absolutely necessary that heavy-ram hammers be used in the driving of long piles.

The size of the ram should be gauged for the work that it has to do. A heavy-ram slow-acting hammer is more effective than a light-ram fast-acting hammer in driving a pile of a given weight, even though the two hammers may have the same rated energy per blow. The heavier-ram hammer will drive the pile deeper and will produce a pile of better bearing value than the equally rated lighter-ram hammer. As a general rule, pile driving should employ the heaviest-ram hammer that will not cause permanent damage to the pile. Except in cases where space or other considerations limit the size of the hammer or where the pile material will not take the hammer impact, the writer recommends that the ram be sized so as to approximate the weight of the pile to be driven, with a minimum ram weight of 5,000 lb. If the ram weight exceeds twice the pile weight, the material should be checked for resistance to impact.

Specifications will frequently attempt to control the permissible ram weights of the hammer by stating that "the ratio of the weight of the hammer to the weight of the pile should not be less than 1/10 or more than 10." This is like saying that anything from a tack hammer to a sledgehammer can be used to drive any nail. There is no accepted formula for sizing the hammer to the pile, but recent tests indicate that pile driving is most effective when the weight of the ram or striking parts of the hammer is between 0.80 and 1.25 times the weight of the pile to be driven. In other words, if the average of these ratios is taken as a guide, the best results should be obtained when the weight of the ram equals or slightly exceeds that of the pile.

The *single-acting hammer* is commonly known as the "Vulcan" type. Steam or air raises the ram, which then drops by gravity. Ram weights range from 550 to 20,000 lb, but the majority of piles are driven by hammers with rams weighing 5,000 and 6,500 lb which, if in good condition, have a blow count of 50 to 55 per min. These are the sizes of hammers which are commonly owned by contractors, but they are not necessarily the right ones for their assigned jobs.

The most important thing to check in the operation of a single-acting hammer is the setting of the valve. Improper timing of the valve can result in the admission of steam to the cylinder below the piston long enough before the end of the downstroke to act as a cushion, thus snubbing the hammer blow. A valve cut off too soon on the upstroke will result in a "short-stroke" hammer. Improperly spaced guide rods will cause a drag on the ram weight.

The *double-acting hammer* uses steam or air, both to raise the ram and to impart additional energy to its downstroke. Since part of the energy per blow is derived from the steam or air pressure, the double-acting hammer will have a lighter ram than the single-acting hammer for equally rated energy per blow. Its action is faster than that of the single-acting hammer, but the blow delivered by each stroke will vary over a wide range unless there is very careful control of the boiler or air pressure.

As in single-acting hammers, the valve setting in double-acting hammers is impor-

TABLE 27-1 Specifications for Comparable Sizes of Currently Available Pile Hammers

Make of hammer	Type	Style	Size	Rated energy, ft-lb	Blows per min	Weight of striking parts, lb	Total weight, lb	Length of hammer	Jaws, in.	Air, cfm	Nominal boiler hp	ASME boiler hp	Steam or air psi req.	Size of hose, in.	¶$\sqrt{E \times W}$ rating
Energy over 100,000 ft-lb															
Vulcan	Sing.- act.	Open	040	120,000	60	40,000	86,000	17′-11″	14¼ × 80	*3,500	‡300	600	120	(2) 3	69,282
Energy 50,000 to 100,000 ft-lb															
Vulcan	Sing.-act.	Open	020	60,000	60	20,000	39,000	15′-0″	11¼ × 37	*1,756	‡130	278	120	3	34,640
Super-Vulcan	Dif'ent'l.	Open	200C	50,200	98	20,000	39,050	13′-2″	11¼ × 37	*1,746	‡120	260	142	3	31,685
M.K.T. Corp.	Sing.-act.	Closed	S20	60,000	60	20,000	§38,650	18′-5″	† × 36	1,720	150	280	150	3	34,640
Energy 30,000 to 50,000 ft-lb															
Vulcan	Sing.-act.	Open	014	42,000	60	14,000	27,500	14′-6″	11¼ × 32	*1,282	‡100	200	110	3	24,248
Super-Vulcan	Dif'ent'l.	Open	140C	36,000	103	14,000	27,984	12′-3″	11¼ × 32	*1,425	‡100	211	140	3	22,449
Vulcan	Sing.-act.	Open	010	32,500	50	10,000	18,750	15′-0″	9¼ × 26	*1,002	‡75	157	105	2½	18,027
M.K.T. Corp.	Sing.-act.	Closed	S14	37,500	60	14,000	§31,600	14′-10″	† × 36	1,260	90	190	100	3	23,000
M.K.T. Corp.	Sing.-act.	Closed	S10	32,500	55	10,000	§22,200	14′-1″	† × 30	1,000	65	140	80	2½	18,027
Energy 20,000 to 30,000 ft-lb															
Vulcan	Sing.-act.	Open	08	26,000	50	8,000	16,750	15′-0″	9¼ × 26	* 880	‡ 60	127	83	2½	14,422
Super-Vulcan	Dif'ent'l.	Open	80C	24,450	111	8,000	17,885	11′-4″	9¼ × 26	*1,245	‡ 80	180	120	2½	13,985
Vulcan	Dif'ent'l.	Closed	8M	24,450	111	8,000	18,400	10′-6″	9¼ × 32	*1,245	‡ 80	180	120	2½	13,985
M.K.T. Corp.	Sing.-act.	Closed	S8	26,000	55	8,000	§18,100	14′-4″	† × 26	850	55	119	80	2½	14,422

Energy 10,000 to 20,000 ft-lb																
Vulcan	Sing.-act.	Open	06	19,500	60	6,500	11,200	13′–0″	8¼ × 20	* 625	‡ 50	94	100	2	11,258
Super-Vulcan	Dif'ent'l.	Open	65C	19,200	117	6.500	14,886	12′–1″	8¼ × 20	* 991	‡ 70	152	150	2	11,201
Vulcan	Sing.-act.	Open	1	15,000	60	5,000	10,100	13′–0″	8¼ × 20	* 565	‡ 40	81	80	2	8,660
Super-Vulcan	Dif'ent'l.	Open	50C	15,100	120	5,000	11,782	10′–2″	8¼ × 20	* 880	‡ 60	125	120	2	8,689
Vulcan	Dif'ent'l.	Closed	5M	15,100	120	5,000	12,900	9′–4″	8¼ × 28	* 880	‡ 60	125	120	2	8,689
M.K.T. Corp	Dbl.-act.	Closed	11B3	19,850	95	5,000	14,500	11′–1″	† × 26	900	60	126	100	2½	9,785
M.K.T. Corp	Dbl.-act.	Closed	10B3	13,100	105	3,000	10,850	9′–4″	† × 24	750	50	104	100	2½	6,269
M.K.T. Corp	Sing.-act.	Closed	S5	16,250	60	5,000	§12,375	13′–3″	† × 24	600	40	84	80	2	9,000
M.K.T. Corp	Compound	Closed	C5	16,000	110	5,000	11,880	8′–9″	† × 26	585	30	56	100	2½	8,944
Union	Dbl.-act.	Closed	0	19,875	110	3,000	14,500	10′–1″	† × 30½	800	50	...	125	2	6,360
Union	Dbl.-act.	Closed	1	12,725	125	1,600	10,000	8′–2″	† × 28	600	40	...	100	1½	4,530
Energy 5,000 to 10,000 ft-lb																
Vulcan	Sing.-act.	Open	2	7,260	70	3,000	7,100	12′–0″	7¼ × 19	* 336	‡ 25	49	80	1½	4,666
Super-Vulcan	Dif'ent'l.	Open	30C	7,260	133	3,000	7,036	8′–11″	7¼ × 19	* 488	‡ 40	70	120	1½	4,666
Vulcan	Dif'ent'l.	Closed	3M	7,260	133	3,000	8,490	7′–11″	7¼ × 24	* 488	‡ 40	70	120	1½	4,666
M.K.T. Corp	Sing.-act.	Closed	S3	9,000	65	3,000	§ 8,800	12′–4″	† × 20	400	25	57	80	1½	5,200
M.K.T. Corp	Dbl.-act.	Closed	9B3	8,750	145	1,600	7,000	8′–2″	† × 20	600	45	85	100	2	3,742
Union	Dbl.-act.	Closed	1½A	8,280	135	1,500	9,200	8′–4″	† × 24	450	35	...	100	1½	3,524
Energy under 5,000 ft-lb																
Vulcan	Dif'ent'l.	Closed	DGH900	4,000	238	900	5,000	6′–9″	See note	* 580	‡ 40	75	78	1½	1,897
Vulcan	Dif'ent'l.	Closed	DGH100A	386	303	100	786	4′–2″	3¼ × 8¼	* 74	‡ 5	8	60	1	196
M.K.T. Corp	Dbl.-act.	Closed	7	3,600	225	800	5,000	6′–1″	† × 21	450	35	63	100	1½	1,697
Union	Dbl.-act.	Closed	3	3,660	160	700	4,700	6′–4″	† × 23½	300	20	...	100	1¼	1,600
Union	Dbl.-act.	Closed	6	445	340	100	910	3′–10″	....	75	....	...	100	¾	210
Union	Dbl.-act.	Closed	7A	320	400	80	540	3′–7″	....	70	....	...	100	¾	160
M.K.T. Corp	Dbl.-act.	Closed	3	356	400	68	675	4′–10″	....	110	....	...	100	1	155

Note: The Vulcan DGH900 is available with three jaw sizes: 6¼ × 18, 7¼ × 19, and 8¼ × 20.

* Adiabatic compression of air assumed.

† This dimension is optional.

‡ 12 sq ft of heating surface per bhp.

§ With flat anvil.

¶E = rated striking energy, ft-lb; W = weight of striking parts, lb.

TABLE 27-2 Specifications for Comparable Sizes of Discontinued Models of Pile Hammers

Make of hammer	Type	Style	Size	Rated energy, ft-lb	Blows per min	Weight of striking parts, lb	Total weight, lb	Length of hammer	Jaws, in.	Air, cfm	Nominal boiler hp	ASME boiler hp	Steam or air psi req.	Size of hose, in.	¶$\sqrt{E \times W}$ rating
Energy 50,000 to 100,000 ft-lb															
Super-Vulcan	Dif'ent'l.	Closed	20000	50,200	98	20,000	40,500	12′-9″	12¼ × 42	*1,746	‡120	260	142	3	31,685
Energy 30,000 to 50,000 ft-lb															
Super-Vulcan	Dif'ent'l.	Closed	14000	36,000	103	14,000	29,444	12′-3″	10¼ × 38	*1,425	‡100	211	140	3	22,449
Vulcan	Sing.-act.	Open	OR	30,225	50	9,300	18,050	15′-0″	9¼ × 26	* 364	‡ 72	140	100	2½	16,765
Energy 20,000 to 30,000 ft-lb															
Super-Vulcan	Dif'ent'l.	Closed	8000	24,450	111	8,000	18,477	10′-3″	9¼ × 32	*1,245	‡ 80	180	120	2½	13,985
Vulcan	Sing.-act.	Open	0	24,375	50	7,500	16,250	15′-0″	9¼ × 26	* 841	‡ 60	128	80	2½	13,485
Energy 10,000 to 20,000 ft-lb															
Super-Vulcan	Dif'ent'l.	Closed	5000	15,100	120	5,000	12,140	9′-1″	8¼ × 28	* 880	‡ 60	125	120	2	8,689
Energy 5,000 to 10,000 ft-lb															
Super-Vulcan	Dif'ent'l.	Closed	3000	7,260	133	3,000	7,250	7′-8″	7¼ × 24	* 488	‡ 40	70	120	1½	4,666
Energy under 5,000 ft-lb															
Vulcan	Compound	Closed	Calif. E	3,800	150	950	3,800	7′-9″	6¼ × 18	* 270	‡ 25	45	70	2	1,899
Super-Vulcan	Dif'ent'l.	Closed	1800	3,600	150	1,800	4,274	6′-5″	6¼ × 20	* 308	‡ 25	45	120	1¼	2,545
Super-Vulcan	Dif'ent'l.	Open	18C	3,600	150	1,800	4,139	7′-8½″	6¼ × 18	* 303	‡ 25	45	120	1¼	2,545
Vulcan	Sing.-act.	Open	3	3,600	80	1,800	3,700	9′-6″	6¼ × 18	* 220	‡ 18	28	80	1¼	2,545
Vulcan	Sing.-act.	Open	4	825	80	550	1,400	7′-0″	4¼ × 14	* 36	‡ 8	15	80	1	673
Vulcan	Dif'ent'l.	Closed	DGH100	386	303	100	786	4′-2″	3¼ × 8¾	* 74	‡ 5	8	60	1	196
Vulcan	Compound	Closed	Calif. G	375	270	100	750	3′-11″	3¼ × 8¾	* 35	‡ 7	12	65	1	193

* Adiabatic compression of air assumed.
‡ 12 sq ft of heating surface per bhp.
¶ E = Rated striking energy, ft-lb; W = Weight of striking parts, lb.

TABLE 27-3 Specifications for Comparable Sizes of Pile Extractors

Make of extractor	Size	Rated energy, ft-lb	Blows per min	Weight of ram, lb	Total weight, lb	Stroke, in.	Overall length	Diam. link pin, in.	Hoist pull, tons	Air, cfm	Nominal boiler hp	ASME boiler hp	Steam or air psi req.	Size of hose, in.	¶$\sqrt{E \times W}$ rating
Energy over 1,500 ft-lb															
Vulcan..........	1200A	1,640	530	1,200	9,200	2	12′-7″	2 7/16	150	*1,020	‡65	140	100	2	1,400
Energy 1,000 to 1,500 ft-lb															
Vulcan..........	800A	1,000	550	800	5,400	2 1/8	10′-8″	2 7/16	100	* 740	‡40	97	100	1½	894
M.K.T. Corp....	E4	1,200	400	400	4,400	3	10′-5″	2 3/8	100	550	35	...	100	1½	632
Energy 500 to 1,000 ft-lb															
Vulcan..........	400A	500	550	400	2,850	2	9′-4″	2 7/16	50	* 342	‡25	49	100	1¼	447
M.K.T. Corp....	E2	700	450	200	2,600	3	8′-4″	2 3/8	50	400	30	...	100	1½	374
Energy under 500 ft-lb															
Vulcan..........	200A	250	550	200	1,500	2	7′-10″	2 7/16	25	* 173	‡18	24	100	1	223

* Adiabatic compression of air assumed.
‡ 12 sq ft of heating surface per hp.
¶ E = Rated striking energy, ft-lb; W = Weight of striking parts, lb.

TABLE 27-4 Suggested Hammer Sizes for Driving Various Types of Piles*

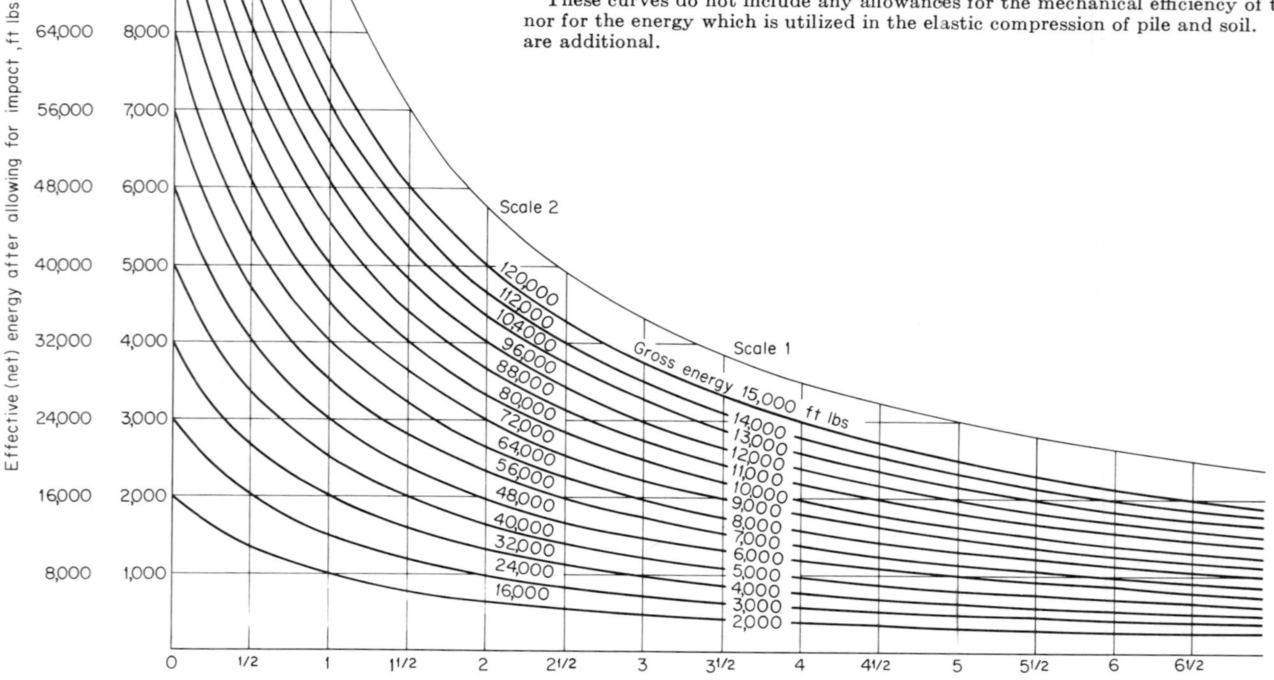

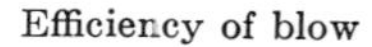

Efficiency of blow

Assuming the material which is struck as being inelastic, the diagram illustrates the energy remaining after impact, referred to as effective (net) energy, for various ratios of pile weight to ram weight and for a selected gross striking energy.

Conversely, knowing the weight of the pile and the effective energy necessary to drive it, then the gross striking energy that the hammer must develop when fitted with different weights of ram can be readily determined.

These curves do not include any allowances for the mechanical efficiency of the hammer nor for the energy which is utilized in the elastic compression of pile and soil. Such losses are additional.

Type of pile		Sheet steel piling†			Timber piling		Concrete piling		H-beam piling		Pipe piling	
		Lt. wt.	Med. wt.	Hvy. wt.	Lt. wt.	Hvy. wt.	Lt. wt.	Hvy. wt.	Lt. wt.	Hvy. wt.	Lt. wt.	Hvy. wt.
Weight, lb/lin ft		20	30	40	30 (Approx. 13″ dia.)	60 (Approx. 18″ dia.)	150 (Approx. 12″×12″)	400 (Approx. 20″×20″)	36 to 53 (8″ to 12″)	74 to 117 (12″ to 14″)	25 to 50 (8″ to 12″)	50 to 116 (12″ to 18″)
Driving through earth, moist clay, and loose gravel; normal frictional resistance												
Length of pile, ft	Depth of penetration	Suggested size of hammer										
25	½					2	2	1		2		2
	Full			2		2	2	1	2	2–1	2	2–1
50	½	2	2	2	2	2	1	06	2	1	2	1
	Full	2	2	2	2	2–1	1	06	2–1	1	2–1	1
75	½		2	2		1		08	1	06	1	06
	Full			2–1		1		08	1	08	1	08
Driving through stiff clay, compacted gravel, and sand; high frictional resistance												
25	½				2	2	2	1	2	2–1	2	2–1
	Full		2	2	2	2	2–1	1	2	1	2	1
50	½	2	2	2	2–1	2–1	1	08	2	1	2	1
	Full		2	2		1		010	2–1	06	2–1	06
75	½		2	2–1		1		010	1	08	1	08
	Full			1		06		010	06	010	06	010

* For use only in the absence of exact data.
† Sizes of hammers listed are based upon driving two sheet piles simultaneously.

tant. In addition, the rings on the pistons must be close fitting (but not tight), since loss of steam or air around the piston will lower the effectiveness of the blow. Double-acting hammers are totally enclosed for underwater operation. A double-acting hammer will drive a given pile faster than will a single-acting hammer with the same ram weight and energy per blow. The faster-acting hammer keeps the pile in motion for a greater portion of the time and thus reduces the opportunity for the pile to "set" between blows.

The *differential hammer* also imparts energy to the falling ram by applying steam or air pressure on the head of the piston, but in this case the piston has two different diameters. The force added to the falling ram is then the unit pressure multiplied by the difference in the two piston areas. The differential hammer is a short-stroke hammer, with blow counts that run 75 percent higher than for the same weight of ram in the single-acting type.

The comments regarding operation of the double-acting hammer apply equally to the differential hammer.

In the *diesel hammer*, the ram is the cylinder and is free falling. To start the hammer, the cylinder is raised by means of a cable to its maximum height. It is then automatically disengaged from the line, allowing it to fall freely. The air in the cylinder is compressed on the downward stroke of the ram-cylinder, and this to some extent cushions the blow. Just prior to the ram striking, diesel fuel is injected into the cylinder, where it explodes. The force of this explosion, combined with the rebound from the ram striking the anvil block, is sufficient to return the ram-cylinder to its maximum height. Thus, after the first blow, the diesel hammer will continue to operate as long as fuel is injected. The timing of the fuel injection is all-important; too early an injection will nullify the blow, and too late an injection will miss the rebound effect needed to return the ram and cylinder to starting position.

The advantage of this hammer is that it eliminates the need for a boiler. Its disadvantages are that it is not self-starting and does not deliver a uniform blow.

The *sonic hammer*, also known as a "vibrating" or "resonant" hammer, is the newest development in driving equipment. Its use was initiated in Europe, where electric vibrators were attached to the sides of steel sheet piling to assist in driving. The success of this experiment led to the development of the sonic hammer.

The sonic hammer utilizes the principle that a forced vibrating system will absorb peak energy from the source of energy when operating at resonant frequency but will absorb very little energy at off-peak frequencies. If the tip of the pile is one-half or a full multiple of the wavelength which is emitted at resonant frequency by the source, it will be kept in a high-velocity reciprocating motion at near-maximum stroke as long as the energy source is maintained at the critical or resonant frequency. At sonic speed, the longitudinal vibration of the tip of the pile is faster than the movements of the nonelastic particles of soil in returning to their original undisturbed positions. The bonds between the particles of soil are momentarily broken, so that the soil is in a fluid state for a miniscule distance from the surface of the pile.

The vibrating pile moves through this "fluid" soil which, because of its inelastic state, does not, to a material degree, transmit the vibrations to the surrounding soil. This explains the absence of shock wave in sonic-hammer driving; in most soils the effects of the vibration can be felt only for a modest radius of 15 to 20 ft from the pile being driven. However, the effects of vibrations on deep beds of granular soils are quite apparent, as consolidation will take place and the area will therefore settle.

The Bodine hammer, developed and used by the Resonant Pile Corporation, employs an oscillating device which is driven by either one or two internal combustion engines. This device, which alternately pushes down and pulls up on the top of the piling, is operated at a frequency which corresponds to the natural resonance frequency of the mechanical assembly of which the pile is a part.

The Foster Vibro-Driver is a vibratory hammer developed in France. It operates on the same resonance principle as the Bodine hammer, but uses an electrical vibratory device to provide the impulse to the pile (see Fig. 27-13).

Pile Drivers These are divided into two general classes, land drivers and floating drivers. There are also special drivers, intended either for service on land or on water, which are built for a "one time" use. These latter drivers are considered to be in a class by themselves.

The purpose of a pile driver is to hold the pile in a fixed position during driving. The rig should be stable and heavy enough to handle both pile and hammer without obvious movement or swaying. The leads should be as long as necessary to handle and drive the pile in one piece. When long, slender piles are driven vertically, and when heavy or flexible piles are driven on a batter, the leads of a pile driver should be fitted with automatic devices to hold the pile in line and to keep it from bending during driving. Where warranted by job conditions, such as limited headroom, piles can be driven or jacked by using short lengths of piling which are successively fastened to the previously driven piece. In so doing, it must be recognized that the resulting pile may not be straight and may be of questionable capacity.

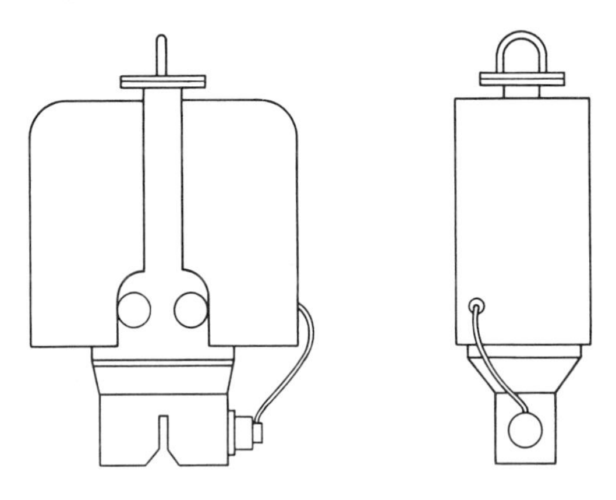

Foster Vibro Driver/Extractor–Model No.	2-17	2-35	2-50
Driving Characteristics:			
Maximum energy rate delivered, ft-lb/sec	18,440	37,970	54,250
Vibration frequency, rpm min.	1,090	890	700
Vibration frequency, rpm max.	1,290	1,120	1,020

Fig. 27-13 Vibratory driver/extractor.

Pipe piles are frequently used when short lengths of piling must be driven and joined. The pipe must have square-cut ends, and each successive piece should be full-welded to the previously driven piece of pipe, using backup rings to ensure a good weld. Welding is the best method of assuring continuity of the piling, watertightness, and reasonable straightness. A sleeve-joint connection is always a second choice and will not give as good results as the welded joint.

There are no rules that can be set down to determine the ideal size of the driver except that it should be big enough to do the job. This is somewhat analogous to the saying, "Don't send a boy to do a man's job."

In *land drivers*, the trend has been toward the mobile crawler-crane type. Today, one can transform any crane into a pile driver by fitting it with a set of leads, a spotter, a hammer, and an auxiliary boiler or compressor. Diesel hammers and those of the sonic and hydraulic variety do not require boilers but do require power in other forms. Some crawler-type drivers are specially designed to drive piles which are vertical or battered in any direction. However, there is still a field for the lightweight skid drivers and the large roller rigs.

A *floating driver* can be assembled by placing a land driver on a suitable barge and adding winch engines for the fore, aft, port, and starboard lines. If the driver is steam operated, additional boiler capacity will be required to operate the auxiliary equipment. For practical operation, two boilers should be provided: at least one for the hammer, depending on its size, and one of 40 to 50 hp for the auxiliary equipment. Where there is a large number of piles to drive or where batters are required, a floating rig which can drive vertical and compound batters and can do any necessary jetting of the piles at the same time will do a better and quicker job. In addition to the regular pile-driving crew, men are required to handle the extra lines and equipment. In order to keep a floating pile driver supplied with piles, at least one pile barge and a tug or large power launch will be needed. More often than not, a second launch will be required to handle the pile-barge lines and anchors and to supply the pile barge with fuel.

Special-use pile drivers can be built to run on railroad tracks for use in building trestles and other "right-of-way" piled structures. Motorized fork-lift trucks and other types of mobile equipment can be fitted with a hammer for close-in work and short-headroom driving. Low-head, short-lead drivers are used for the same purpose.

Accessory Equipment When properly used, each accessory tool used in pile driving serves a definite purpose. Knowledge of when and how to use each tool is as important as the tool itself.

The *driving head* serves a double purpose. It distributes the blow of the hammer uniformly over the head of the pile and, with the cushion block added, softens the sharp impact of the hammer blow which could otherwise damage or destroy the pile. Driving heads, also called "driving caps" or "helmets," are fabricated of forged cast steel and occasionally of structural steel. It is important that the base of the driving head be shaped to fit the particular pile or sheeting being driven, since piles and sheeting can be damaged beyond repair by the use of the wrong driving head. In addition, the proper driving head assures the maximum efficient transmission of the hammer blow to the pile.

The *cushion block* softens the blow of the hammer just enough to prevent injury to the driving head and the head of the pile. Many materials have been used more or less successfully as cushion blocks, with wood being the most frequent selection. Woods vary in their suitability for use as a cap block, and a poor choice can reduce the effectiveness of the hammer blow by 50 percent. Soft woods will act like a sponge in soaking up the energy of the hammer. Woods that are too hard will split and splinter and lose their effectiveness by disintegrating into a mass of wood chips. A good wood for a cap block should have the property of toughness. Eucalyptus, oak, maple, greenheart, and similar woods when properly seasoned have proved their adaptability for this use.

To be fully effective, the cushion block should be turned to the correct diameter to fit the driving head and cut to the correct length to fit in between the anvil blocks. The grain of the wood should run vertically. As soon as a cushion block has been compressed more than half its original thickness or heats up and begins to smoke, it should be discarded and a fresh cushion block used.

For the driving of precast-concrete piles, several layers of ¾- to 1-in. plywood placed on the head of the pile under the anvil block have proved effective as a cushion block. The required number of layers can be found only by trial and may be as few as three or four or as many as six to nine, depending on the relative size of pile and hammer and the ultimate resistance to which the pile is to be driven. The development of a "permanent" cushion block has lagged, although several companies have developed cushion blocks that are semipermanent and that can be calibrated with reasonable accuracy.

A *follower* is a member which is interposed between the pile and the hammer when it becomes necessary to drive the head of the pile below the reach of the hammer. Its use introduces an additional element of uncertainty if a dynamic formula is to be used to predict the bearing value of the pile. The follower should be sufficiently rugged to withstand repeated blows and to transmit the full impact of the hammer at refusal without undue deflection. In addition, it must be designed so that it can be held in positive alignment with the hammer and the pile. When driving H piling or pipe piling, the minimum follower requirement is a section of the pile being driven, reinforced at its head and base to prevent damage from repeated use.

Jets of water, or air and water, under pressure can be used to facilitate the driving of a pile. The jet serves to dislodge the soil particles or to increase the soil water content to a point of fluidity. Although a jetting pipe can be cast as an integral part of a concrete pile, the preferred practice is to use the jet as a separate tool during driving. It is frequently necessary to work the jet up and down in the hole or alongside the pile, which is impossible when the jet is part of the pile. Also, a jet which is a part of the pile can become plugged and made ineffective during driving, with no means of rectifying the trouble.

A jet should always be used with a nozzle which is designed for the type of soil being worked and for the desired result. A granular soil requires a "balanced" nozzle, which has holes drilled in the head so that the upward and downward thrusts at the nozzle are about equal. In this type of nozzle the downward jet action cuts the granular material away from the surrounding material and the upward jets tend to keep the grains in suspension and wash them toward the surface of the ground. However, the balanced nozzle is of no use whatsoever in a cohesive soil where straight pressure velocity does the cutting. In clays and similar soils, the hole cut by a jet will be only slightly larger than the diameter of the jet pipe. A large jet and a generous flow of water are needed to obtain large-diameter holes in these soils.

There are no hard and fast rules for the required size of jet or for the optimum volume and pressure of the fluid. If rule-of-thumb figures are used for these parameters, bear in mind that each jetting situation warrants its own study. For any but the lightest of jetting service, the minimum facility is a 2-in. jet at 100-psi pressure and 200-gpm flow; the jetting equipment requirements for large-diameter piles of great length may exceed 6 in. for the jet and involve 200-psi fluid pressures. As a rough guide, the diameter of the jet should be approximately one-quarter the diameter of the pile (average diameter for a tapered pile) and the pressure should be 50 psi plus an additional 1 psi for each foot of depth.

A water jet which is used to clean out an open-end pile should be combined with air. Both air and water should be at the same pressure, with the air slightly higher, if anything, but not less. The air and the water should come from the same nozzle and at the same time. The pressurized air will expand to break up the material being removed and will then carry water and soil with it as it rises to the top of the pile. The jet diameter should be about one-third that of the pile, but will seldom exceed 10 in. at the nozzle. The pressure of both water and air can be estimated at 50 psi plus $\frac{1}{2}$ psi for each foot of depth.

There are places where a jet is of no use in sinking a pile and may even increase the driving problems. In a sand and gravel mixture where sizes of 2 in. and larger predominate, the jet will merely wash away the fines (sand and small gravel). The large-sized gravel will then fall into the jetted hole, causing collapse of the hole as well as more difficult driving. Care must also be exercised in the use of jets in silty soils where there is a risk of loosening the soil around piles already driven. If this loosening occurs, retapping of each pile and sometimes retapping of adjacent piles is required to reseat them and reestablish their bearing values.

It sometimes happens that a dense layer of granular material overlies a soft or loose soil, which in turn overlies a denser layer. If piling is to be driven to end bearing on the lower firm stratum, a jet is a useful tool to reduce the hard driving through the upper granular stratum. The pile will ultimately function as an end-bearing one, once the softer material has yielded under the loads which will be transmitted through the upper crust layers; and the use of jetting through the crust layers will give a better indication of the ultimate bearing value of the pile.

A *solid-mandrel spud* is used where piles must be driven through an accumulation of debris and rubbish fill or where boulders are present above soft or loose materials. The spud can be driven at the proposed pile location to push aside these obstructions which might otherwise damage or deflect the pile. The diameter of the spud should be about 25 percent greater than the diameter of the pile.

A *preexcavator* is a form of spud which is used in cohesive soils where the displacement of a pile may cause lateral movement in adjacent structures or piles or, through ground heave, might cause previously driven piles to lift off their bearing strata. The preexcavator has an open end and removes a plug of soil as it is extracted from the ground. It is usually sufficient to remove 75 to 85 percent of the volume of the pile. Spuds of this type are fabricated from medium wall pipe, closed at the top and fitted with air or steam connections so that the plug of earth can be blown out.

A hollow pipe casing, containing a close-fitting *removable plug mandrel,* is still another type of spud. It is used either to remove material that could give a false bearing value for the pile, or to isolate the pile permanently from an undesirable soil, or to act as a form for a sand or concrete encasement for the pile. The first step is to drive the assembled casing and mandrel into the ground at the pile location, displacing all the soil. Then the inner mandrel is withdrawn, and the pile is driven through the open casing to the required resistance. After that, one of four things is generally done: the casing is withdrawn, allowing the ground to move back around the pile; or the casing is filled with sand to prevent a bond between the original soil and the pile and then withdrawn, or the casing is filled with concrete and then withdrawn, leaving a concrete-protected section of the pile in the ground; or the unfilled casing can be left permanently in the ground, thus completely isolating the pile from the cased portion of the soil.

Rotary preexcavation is used at depths of more than 50 ft, where plug spudding is impractical. The wet-rotary preexcavator is a rotating jet that is fitted with cutters at its end and with reamers along its sides. The jets provide a cutting action, the cutters help in breaking up cohesive materials, and the reamers size the hole to fit the shape of the pile. The excavated soil and water form a mixture with the consistency of driller's mud, and this seals off and coats over intermediate layers of granular material. The mud also keeps the hole open after preexcavation has been completed. The pile is inserted in the drilled hole, displacing the mud, and is driven to satisfactory bearing resistance in the undisturbed soil beneath the drilled hole.

The rotary preexcavation method is used where deep beds of clays and similar cohesive soils are present above hardpan, rock, or good granular materials. End-bearing high-load piles are then indicated, and displacement and heave can become problems. The method tends to decrease, by some indeterminate amount, the side-wall friction between the ground and the preexcavated length of the pile. It has been successfully used to depths of 120 ft, but it cannot be employed in a wholly granular soil.

DRIVING EFFECTS ON PILES

A thorough inspection of the pile should be made prior to pile driving. Piles which do not meet specification requirements should be rejected, since a pile which is bad to start with can only remain a bad pile after driving. If the inspection indicates that the pile is initially satisfactory, a number of factors in addition to the characteristics of the ground in which it is driven will determine its ultimate value as a structural member. All pile materials behave differently under the action of driving, even though the same hammer, etc., may be used. The pile should be driven to the bearing resistance which is predicated in the foundation design, but overdriving should be avoided. This can be as detrimental to the pile as underdriving is to the foundation.

Overdriving usually occurs when a pile either brings up abruptly or is driven excessively hard so as to develop a greater load capacity. Overdriving will break a wood pile, damage or break a concrete pile, cause section failure at the point or head of an H pile, and cause the walls of a pipe pile to "accordion" or fold up. In all cases the structural section of the pile is destroyed, and with it the value of the pile as a bearing member. The driving of a pile should be stopped when the hammer bounces on the anvil of the driving head. This is an indication that the pile has reached refusal and can be driven no further. At this point the total energy of the hammer is being dissipated in the material of the pile itself.

When a long, slender pile is driven, it has a tendency to whip or vibrate in the leads. This whipping absorbs hammer energy and can become great enough to cause permanent damage to the pile. To prevent its occurrence, the pile should be supported or braced against the leads at regular intervals. A single support at the midpoint of the pile is frequently sufficient, and only rarely must the supports be spaced closer than 40 ft on centers. However, when long, slender piles are driven on batters exceeding 1 on 3, the intermediate supports may be as close as 20 ft on centers.

Timber Piles The behavior of a timber pile during driving will depend to a large extent on the soil strata which it penetrates. If the soil properties are reasonably constant within the depth of interest, the resistance to driving will increase more or less uniformly as the pile goes into the ground. The driving energy is then dissipated into the ground along the length of the embedded pile, with only a small percentage of it actually reaching the point. If the hammer has not been oversized, the pile will be driven to final resistance without fracture or damage. On the other hand, when a wood pile is driven through a deep layer of soft ground into a hard layer, its tip end takes almost the full brunt of the hammer blow. There is then a danger that the pile will split or that its tip will be broken or "broomed" by the driving. When these soil conditions exist, the hammer should be sized for whatever portion of the pile is expected to be critical.

Relatively few wood piles are chalk-line straight, but this is normally of little consequence. Minor variations in the straightness of a pile will not reduce its structural capacity and, even though it will not drive exactly vertical, its head will generally end up at the place where its tip started. This is true even though the tip may now be displaced horizontally by a distance of several feet. For a pile that is only partially embedded (e.g., in water for a pier or on land for a trestle), slight misalignment can be corrected by pulling the pile into line. Large misalignment must be more carefully handled for fear of breaking the pile below the mud line, and piles too far out of line should be pulled and redriven or replaced.

The danger comes when a wood pile with a pronounced twist or bend is driven. Even though this pile is held straight in the leads, its tip will drive in the direction of the bend. As driving continues, the tip will continue to deviate until the pile is probably broken in the ground. Even if no significant deviation of the pile tip occurs, there is still the risk of an undetected pile fracture at the twist or bend as a result of the driving stresses. This type of failure is difficult to detect, since its only indications are a severe inclination of the pile and a tendency to pull out of the leads during driving.

To sum up, care and caution should guide the driving of a wood pile. The pile should be reasonably straight to start with, and care should be taken to ensure that it is not broken by overdriving under adverse conditions. Lastly, one should not expect to meet exacting tolerances as to line.

Concrete Piles Precast-concrete piles should be handled only at the pickup points. Assuming that this is properly done and that the pile is safely positioned under the hammer, there still remains the danger of breaking the pile during driving.

One limiting condition in the driving of a precast-concrete pile is its ultimate strength. The concrete should not be stressed to more than half its ultimate strength during driving and, at the same time, metal-to-concrete contact between hammer and pile should be avoided. The best form of cushion block on the head of a concrete pile is several layers of 1-in. plywood which have been bored to fit over the extended bars. No fewer than three layers are required, and more often five or six are desirable. The optimum number of layers can be determined only by trial.

A plain reinforced precast-concrete pile is stiff and brittle and cannot withstand twisting or bending during driving. It must also be driven to line and in position initially, since subsequent pulling into line can break it. Despite these words of caution, a precast-concrete pile with proper cushioning can take a lot of punishment.

The statements which just have been made for precast-concrete piles are not equally applicable to cast-in-place concrete piles. In almost every case, cast-in-place concrete piles of either the driven shell type or the dropped-in-place shell type are driven with an internal or external mandrel, and the concrete is poured into the pile after the shell is in place. As a result, cast-in-place piling can be driven to the stress limits of its metal components (core or mandrel).

In the driven shell type of piling, the danger during driving comes from three sources: (1) the shell may be torn by large gravel, boulders, or obstructions, thus permitting soil, sand, and water to enter and making it difficult or impossible to place concrete; (2) in certain types of soil, usually compact sands, the corrugations can be ironed out and the consequent loss in strength will cause the shell to collapse; (3) in incompressible clays, the back pressure developed by the driving of the pile itself or of adjacent piles can be sufficient to collapse the shell.

The solutions to these problems, listed in order, are (1) to use heavier-gauge shells, or even sections of pipe, in the areas of tearing shells; (2) to use heavier-gauge shells and/or pipe as in (1), supplemented by jetting as required; (3) to use either dry or wet rotary preexcavation, or jetting, or provide heavier shells or pipe of adequate strength to withstand the back pressure.

Steel Piles Steel H piles can be driven within close tolerances provided that proper equipment and adequate care are used, but they can be deflected from their intended positions if obstructions are encountered during driving. If a steel pile has its tip bent out of line by contact with a boulder, the remainder of the pile may continue the same bend or inclination with each successive blow of the hammer. The pile will then have the shape of a sweeping arc past the boulder. Although such extreme cases are not common, flanges are frequently bent or twisted to the extent that the piles twist or rotate during driving. This condition can cause the metal at the point of a bearing pile to be overstressed, possibly resulting in the failure of the pile. To prevent this kind of distortion, it is common practice to reinforce the point of the pile with plates welded to the flanges and sometimes to the web for a distance equal to 1½ to 2 times the pile section; i.e., a 12-in. H pile would have 12- by 18- to 24-in. plates welded to the flanges. The thickness of the plates should equal the flange thickness.

Pipe piles will behave much the same as steel H piles during driving and can also bend and deflect under adverse circumstances. However, they differ from H piles in that they can be inspected after driving, and damaged piles can be replaced. Where particularly hard driving is anticipated, the point and butt sections of pipe should have enough metal to withstand the driving conditions. Generally the lengths of these heavier sections need be only three to four times the diameter of the pipe.

A single pile can frequently be driven in a boulder soil to good bearing, with relatively little damage to the pile itself. However, the subsequent driving of adjacent piles can deflect boulders into the first pile, causing collapse of its wall. Where this condition exists, single or multiple spudding will be necessary. Multiple spudding requires a number of spuds, which are successively driven in each pile location and left in place. The first spud is then withdrawn and the first pile driven, the second spud is removed and the second pile driven, and so on, until all piles are installed.

DRIVING EFFECTS ON SOILS

Each type of soil reacts differently during the driving of piles, and, conversely, each type of pile and the sequence of pile driving will influence the behavior of the soil differently. As stated earlier, the compaction of granular soils can be increased by driving closely spaced piles. Under these circumstances, piles which are driven later may get less penetration than the earlier-driven piles. The nonuniform support conditions which result can be a matter of concern in large pile groups, such as the foundations for storage tanks. For this reason, in granular soils it is necessary to drive the first pile in the geometrical center of the proposed pile group and follow it progressively by those next in line and outward. This procedure will give a balanced foundation, with the center pile the longest and with the surrounding piles successively shorter as the soil is compacted uniformly outward.

In other types of soil, the pile-driving sequence should be different. In soft silts and clays, particularly along riverbanks or shorelines, the driving of piles will create soil pressures that will try to relieve themselves by movement toward the area of lesser pressure. This soil movement will affect the previously driven piles, bending or breaking them and moving them increasingly out of line as successive piles are driven. In such circumstances, the first piles to be driven should be those farthest away from the area of pressure relief. Subsequent pile driving should then proceed toward this area.

Most soils are neither all granular nor all silts or clays. Instead, they consist of stratified layers and of mixtures of several soil types in varying proportions. The borings should be studied carefully before pile driving begins, and a driving sequence should be set up to minimize the effects of any imbalance in the load-supporting capacities of the piles. It should also minimize any soil movements which may have detrimental effects on the piles or on adjacent structures.

Loose Sand Pile driving will improve the load-carrying capacity of the sand itself by compacting it from a relatively loose state to a more dense or compact state.

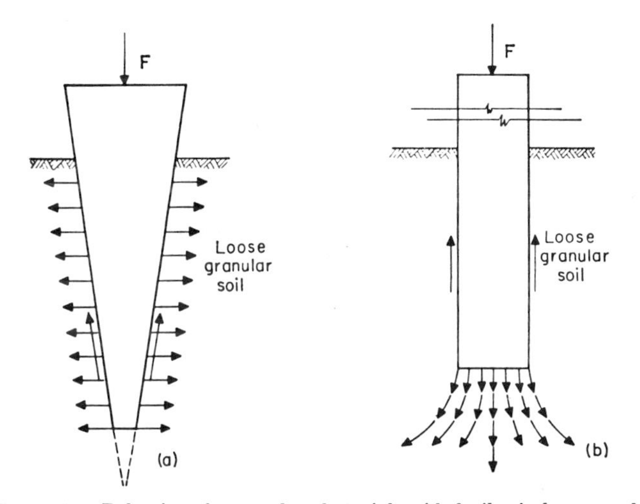

Fig. 27-14 Behavior of tapered and straight-sided piles in loose sand.

This results in an improved frictional resistance on the sides of the pile, plus a larger point resistance. While any type of pile will consolidate a loose sand as a result of driving, a tapered pile displaces the soil sideways and compacts it by the very nature of its wedge shape. Even though the tapered pile has a small point and only a small volume of soil is actually displaced in a lateral direction, the practical benefits of its wedging and compacting actions are most pronounced.

A straight-sided pile in a loose sand behaves in a somewhat different fashion. The flat point of the straight-sided pile compacts the material directly beneath it, and this compacted material is then pushed ahead and to the side as the pile penetrates further. The straight-sided pile is thus preceded by a cone and surrounded by a thin annular ring of compacted sand. Its bearing capacity is derived principally from the point compaction of the sand below it, although a small percentage also comes from the side-wall friction developed by the annular ring of compacted sand which surrounds it. For equal bearing capacity in loose sand, a straight-sided pile must be longer than a tapered pile. (See Fig. 27-14.)

An H pile transmits vibrations during driving and in this way increases the density of a loose sand. This compaction is analogous to that obtained by repeatedly tapping the side of a glass filled with loose sand and thus altering the loose sand to a dense state. The bearing capacity of an H pile in loose sands is derived principally from the improved friction values of the adjacent densified sands. Very little bearing is obtained at the point of the pile.

To sum up, each of the three principal types of piles derives its bearing in a loose sand in a different manner. The length of pile to obtain a required bearing capacity will be shortest for the tapered pile, followed by the straight-sided displacement pile, and will be longest for the nondisplacement H pile. Piles driven in loose gravels behave in much the same way as piles driven in loose sands.

Clay The driving of a pile into clay may change the character of the soil in the immediate area. A great deal of information regarding this "remolding" effect has been published in the journals of the engineering societies. The subject is technically involved but, briefly and simply, the lateral pressure as a pile is forced into the ground tends to squeeze the pore water out of the adjacent clay.[10] This has two effects: the released water acts as a lubricant on the sides of the pile, and the annular ring of clay around the pile experiences time-dependent alterations in its strength properties.

Because of the effects just described, it is sometimes difficult to get the pile to "take up" at the required bearing capacity within a reasonable length. More often, better results can be obtained by driving the pile to a predetermined length, regardless of the blow count, and then allowing the pile to set for 24 to 48 hr. This waiting period permits the lubricating pore water to return, at least in part, to the clay from which it originated. It may also permit a gain in the strength properties of the remolded soil. After this period of time has elapsed, it may be found that it is impossible to start the pile without increasing the blow count to a figure several times that predicted for the required bearing capacity. Consequently, many practical pile drivers believe that pile bearing-capacity formulas based on blow counts or driving energy are not applicable to piles driven in clays.

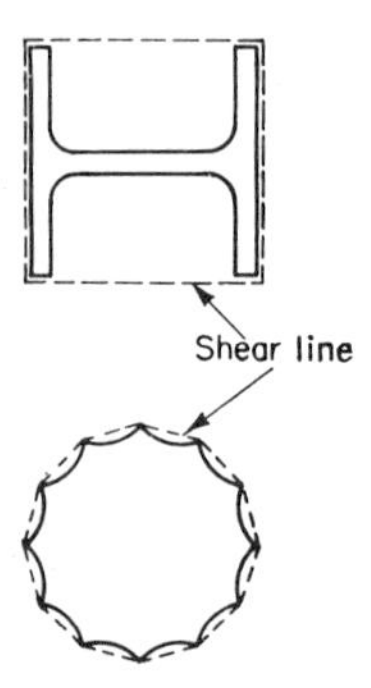

Fig. 27-15 Shear lines for pile sections in clay soils.

The limiting factor for the bearing capacity of a pile in clay is the shearing value of the clay itself. Since the adhesion of clay to the surface of the pile (the friction) is normally greater than its shearing value, the ultimate bearing capacity is equal to the surface area of the embedded pile multiplied by the shearing strength of the clay. The perimeter of the pile should be calculated by assuming that the extremities of the pile cross section are connected by straight lines, since the clay will fail along such planes rather than along the actual surface of the pile. This phenomenon can be observed by pulling an H pile. The pile will come out of the ground with the clay solidly filling the space between the flanges and also forming a layer on the surface of the flanges, showing conclusively that failure occurs by the shearing of the clay along the line of least resistance. For this reason, a fluted pile must be figured as a pile with multiple straight sides and an H pile must be figured as a square pile. (See Fig. 27-15.)

A phenomenon observed during the driving of a pile in or through clays is the heaving or lifting of adjacent piles and buildings. Clays are, to all intents, incompressible when subjected to a rapidly applied load. Nevertheless, something has to move when a pile is driven into a soil of this type. Such movement can only occur upward, and the clay will raise the previously driven piles and any adjacent buildings as it heaves.

Several steps can be taken to minimize the heaving of previously driven piles. A slip joint can be installed near the center of a cast-in-place shell pile, allowing the upper portion of the pile to heave upward without affecting its tip—or its bearing value should the pile be a point-bearing one. After driving a pile of the continuous-shell, pipe, or H-section type, the previously driven piles within a large area around it must be redriven to their original bearing values. Frequently, this redriving must be repeated several times to ensure a good foundation.

Another means of controlling the heave of the ground and the pile, which is only applicable in clay soils, is to remove a plug of clay from the ground before the pile is driven. However, the use of this method does not automatically mean that retapping of driven piles can be eliminated. The raising of adjacent structures can be controlled in much the same fashion, by removing a volume of clay which is approximately equal to the volume of the pile to be driven. This removal can be accomplished by driving an open-end pipe preexcavator until the desired amount of ground is removed or by use of the rotary preexcavator process. Both methods are applicable only to piles which are essentially end-bearing and not to friction piles.

Silt A driven pile tends to compress this soil and force out its pore water. Although pile driving does not alter the strength properties of a silt, in contrast to the remolding effects which were described for clays, it is still poor policy to rely on a silt to carry any significant load from a pile. Settlements of unpredictable magnitude can be expected, since all silts are in the process of slow consolidation.

Silt will tend to flow and move toward areas of lesser pressures as driving forces build up in the soil adjacent to the pile, in much the same manner as described for soft clays. Since this movement can (and does) bend previously driven piles, the order of driving piles should be arranged to minimize any adverse effects.

ALLOWABLE PILE LOADS

As noted earlier, a pile acts only as a medium for transmitting the load at its head to the ground around or below it. This load transmission, depending on the soil and its interaction with the loaded pile, can occur by direct end bearing, side friction, or a combination of these two mechanisms. Pile group action, in addition to the behavior of individual piles, must be carefully studied.

Analysis of Static Forces For an end-bearing pile, both the tip and the material on which it rests must be capable of sustaining the unit load with a reasonable factor of safety. These factors must be investigated before the selection of the pile is made, since failure can be expected in an end-bearing pile of insufficient tip area. Under conditions where it is known that the loads will remain constant and relatively uniform, a factor of safety of 1.5 against a failure in soil and/or pile bearing may be acceptable. This permits working stresses up to two-thirds of the ultimate.

The load carried by a friction pile is transmitted to the adjacent soil at a rate dependent upon the friction between the pile and the soil. The percentage of the total load that actually reaches the point may approach or equal zero. Samples of the various soil strata can be tested in the laboratory to evaluate their shearing strength. Subsequently, when calculating the allowable pile loads, friction values of one-half to one-third of the experimentally determined shear values are frequently employed. This provides a theoretical factor of safety of 2 to 3 to take care of the inaccuracies of laboratory procedure, job conditions, and temporary overload.

Typical values[8] for friction between pile and soil are reproduced in Table 27-5. When using this table, it is well to examine the ± ranges. Under no circumstances

TABLE 27-5 Friction between Pile and Soil

Material	*Ordinary range* of values, psf, of bounding area of pile*
Fine-grained soils:	
Mud	250 ± 200
Silt	300 ± 200
Soft clay	400 ± 200
Silty clay	600 ± 200
Sandy clay	600 ± 200
Firm clay	900 ± 200
Dense silty clay	1,200 ± 300
Hard (stiff) clay	1,500 ± 400
Coarse-grained soils:	
Silty sand	300 ± 200
Sand	1,200 ± 500†
Sand and gravel	2,000 ± 1,000
Gravel	2,500 ± 1,000

* The ± indicates that a range is governed by the character of the soil. Not all soils falling in the same general classification have equal properties.

† If not micaceous, muddy, or under hydrostatic pressure or vibration.

should a friction pile be driven to less than a factor of safety of 1.5, using the minimum value of shear (the tabulated value less the range value) for each stratum of soil through which the pile will penetrate. For piles which obviously will get their bearing from a combination of both friction and bearing, the friction component should be based on the minimum shear values of the strata which the pile penetrates and should include a factor of safety of not less than 2. The point condition should also be investigated for bearing adequacy.

While these recommendations may appear to be on the conservative side, one must bear in mind that in the majority of cases the foundations represent 10 to 15 percent of the total investment in the structure, and it is poor economy to risk a foundation failure and the loss of the total investment by taking chances unnecessarily. Inasmuch as the science of foundations is still in its infancy, we must, as engineers, be conservative.

TABLE 27-6 Classification of Supporting Soils[11]

Class	Material	Maximum allowable presumptive bearing in tons per square foot
1	Hard sound rock	60
2	Medium hard rock	40
3	Hardpan overlaying rock	12
4	Compact gravel and boulder-gravel formations; very compact sandy gravel	10
5	Soft rock	8
6	Loose gravel and sandy gravel; compact sand and gravelly sand; very compact sand-inorganic silt soils	6
7	Hard, dry, consolidated clay	5
8	Loose, coarse to medium sand; medium-compact fine sand	4
9	Compact sand-clay soils	3
10	Loose, fine sand; medium-compact sand-inorganic silt soils	2
11	Firm or stiff clay	1.5
12	Loose, saturated sand-clay soils; medium-soft clay	1

Presumptive Bearing Capacity The 1969 *New York Building Laws* manual gives the following criteria for bearing capacities to be used for soils and rock:[11]

1. Satisfactory bearing materials shall be ledge rock in its natural bed, natural deposits of gravel, sand, compact inorganic silt, clay, or any combination of these materials. These bearing materials shall not contain an appreciable amount of organic matter or other unsatisfactory material, nor shall they be underlaid by layers of such unsatisfactory materials of appreciable thickness.

2. Fill material, mud, muck, peat, organic silt, loose inorganic silt, and soft clay shall be considered as unsatisfactory bearing materials and shall be treated as having no presumptive bearing value.

3. The maximum allowable presumptive bearing values for satisfactory bearing materials shall, in the absence of satisfactory load tests or other evidence, be those established in the classification shown in Table 27-6. As further guidance, Table 27-7 correlates various states of soil compaction with representative blow counts.

Dynamic Pile Formulas It should be recognized that no dynamic formula will give the true static bearing value of a pile, since this value can be determined only by a load test. The dynamic formulas make the assumption that the ultimate carrying capacity of a pile is equal to its driving resistance. In essence, the weight of a ram

TABLE 27-7 Soil Compaction Related to Spoon Blows[11]

Descriptive term	Blows per ft	Remarks
		For sand
Loose Compact Very compact	15 or less 16–50 50 or more	These figures are approximate for medium sand, 2½-in. spoon, 300-lb hammer, 18-in. fall. Coarser soil requires more blows, finer material requires fewer blows.
Loose Med. compact Compact Very compact	15 or less 16–30 30–50 50 or more	These figures are approximate for 2-in. spoon, 148-lb hammer, 30-in. fall.
Loose Med. compact Compact Very compact	10 or less 11–25 26–45 45 or more	These figures are approximate for 2-in. spoon, 300-lb hammer, 18-in. fall.
Loose Med. compact Compact Very compact	25 or less 25–45 45–65 65 or more	These figures are approximate for 2-in. spoon, 150-lb hammer, 18-in. fall.
		For clay
Very soft	Push to 2	Molded with relatively slight finger pressure.
Soft	3–10	
Stiff	11–30	Molded with substantial finger pressure; might be removed by spading.
Hard	30 or more	Not molded by fingers, or with extreme difficulty; might require picking for removal.
Very soft Soft Stiff Hard	Push to 3 4–12 12–35 35 or more	These figures are approximate for 2-in. spoon, 148-lb hammer, 30-in. fall.
Very soft Soft Stiff Hard	Push to 2 3–10 10–25 25 or more	These figures are approximate for 2-in. spoon, 300-lb hammer, 18-in. fall.
Very soft Soft Stiff Hard	Push to 5 5–15 15–40 40 or more	These figures are approximate for 2-in. spoon, 150-lb hammer, 18-in. fall.

multiplied by its stroke is equated with the driving resistance multiplied by the net penetration per blow (set) for the pile tip.[8] The chief value of any dynamic pile-driving formula is to supply a standard of uniformity to which all piles can be driven, rather than as a measure of the allowable pile load.

The *Engineering News* formula and its various modifications, plus the West Coast and Navy formulas, are typical of this dynamic group. The first of these has the most widespread usage and appears in the following forms:

For drop hammers:

$$R = \frac{2W_r h}{s + 1.0} \tag{27-16}$$

For single-acting hammers:

$$R = \frac{2W_r h}{s + 0.1} \tag{27-17}$$

For differential or double-acting hammers:

$$R = \frac{2E}{s + 0.1} \tag{27-18}$$

where R = safe bearing value of pile, lb
W_r = weight of ram, lb
h = fall of ram, ft
s = set of pile in in. per blow, averaged for last few blows
E = rated energy per blow of hammer, ft-lb, as supplied by the manufacturer.

In one of the several modifications of the *Engineering News* formula, the constant in the denominators of Eqs. (27-17) and (27-18) is modified by the ratio of the pile weight W_p to the weight of the ram W_r. For single-acting steam hammers the formula then becomes:

$$R = \frac{2W_r h}{s + 0.1(W_p/W_r)} \tag{27-19}$$

When the weight of the pile W_p is the same as the weight of the hammer ram W_r, the safe load of the pile R by this modified *Engineering News* formula, Eq. (27-19), is the same as by the basic *Engineering News* formula, Eq. (27-17). The effect of the ratio W_p/W_r is to give a higher predicted capacity for the pile when the hammer weight is greater than the weight of the pile and a lower capacity when the pile weight is greater than the weight of the hammer. There is a great deal of merit in this approach since, for the same total expenditure of energy, a heavy hammer will drive a light pile deeper into the ground than will a lighter hammer driving a heavier pile. However, Eq. (27-19) can give misleading bearing values when a very lightweight pile is used with a heavy hammer, since the highest bearing value is theoretically obtained when the pile is weightless. Therefore, it is advisable to limit the use of the formula to cases where the pile weight is one-half to twice that of the hammer ram.

The Navy-McKay formula, which may be viewed as a further modification of the *Engineering News* formula, is expressed as follows:

$$R = \frac{2W_r h}{s[1 + 0.3(W_p/W_r)]} \tag{27-20}$$

The derivations for the *Engineering News* and Navy-McKay formulas suggest that the predicted safe bearing value R has a factor of safety of 6 with respect to the ultimate bearing resistance R_u of the pile. This is misleading, and load tests reveal wide variations in the ratio of R_u to R. In the lower ranges of blows per in. (2 to 3+), R_u/R may be only between 2 and 2½; in the medium range of 5 to 7 blows per in., R_u/R may range upward from around 3; and only in the higher ranges of 10 to 15 blows per in. does R_u/R approach 6.

The *Engineering News* formula and its modifications do not normally take the efficiency of the hammer into account. If this refinement is considered desirable, the energy term can be multiplied by the efficiency factor of the hammer. For drop hammers, an efficiency factor of 0.80 is used; for single-acting hammers, 0.90 is used. The manufacturer's rated energy E for double-acting hammers and differential hammers should be reported on a net basis, with efficiency losses already subtracted.

Load Tests The purpose of a load test is to determine the safe load that a pile can be expected to sustain in bearing, uplift, or bending. It is customary to load a pile to twice the design load, either in 5- to 10-ton increments or in regular percentage increments of the design load (i.e., one-quarter, one-half, etc.). It is assumed that a pile, as driven, will be satisfactory under its service load if it performs adequately under its test load. A load test is primarily a test of the ground into which a pile is driven rather than a test of the pile itself. Seldom is a load-test failure due to a structural failure of the pile; more often it is the ground in which the pile has been driven which is unable to sustain the load transmitted to it by the loaded pile.

To make a vertical load test, a rigid platform can be built over the driven pile and weighted to provide a reaction for jacking. Structural members are framed in the base to distribute the reaction of the loaded jack to the platform. The reaction load may be sand, sand fill, concrete blocks, pig iron, slabs of steel, or tanks of water. An alternative method, known as a "reaction test," utilizes the uplift resistance of four driven piles to counteract the downward reaction on the pile to be tested. This procedure also tests the anchor piles for uplift at loadings of up to one-half of their design bearing value. It is preferable to have all five piles in the same line when making the test, although it can also be performed by placing the piles in the form of a cross with the test pile in the center.

When the piles are in line, the four exterior piles are joined in pairs by spreader beams fastened at their centers by the reaction beam. The test jack is located at the center of the reaction beam and over the pile to be tested. This same basic procedure is used for a tension test except that only two adjacent bearing piles are used for reaction purposes. The uplift piles must be able to take tension throughout their lengths. Wood, steel, and precast-concrete piles are normally capable of meeting this requirement, as are cast-in-place concrete piles which have continuous longitudinal reinforcement.

To make a bending test on a pile, a deadman can be provided for reaction purposes or several piles can be framed together and braced. Under normal conditions, where the pile is to be tested to 50 or 100 percent overload rather than to destruction, it is far simpler and easier to place the reaction jack between two adjacent piles. The piles are then jacked apart, with measurements of strain and deflection taken for each pile.

To repeat, the purpose of a normal load test is to ascertain that the ground into which the pile is driven will take the design load on the pile, with a reasonable factor of safety for overload and uncertainties of design. It is both senseless and poor economy to test a pile (ground) to failure unless some engineering data that will benefit the project or science are obtained thereby.

COSTS OF PILING

Since the costs of materials, labor, equipment, insurance, and freight may vary widely in different areas, it is obvious that unit prices for the different types of piling will differ from place to place. Additional factors which affect the price of piling are the quantity involved, the job conditions under which the piling is to be driven, whether the piles can be driven at one time, the time of the year during which driving is to be done, etc.

Other considerations being equal, the pile should be selected that has the lowest dollar cost per ton of bearing capacity. Relatively speaking, piling costs per foot vary from the least expensive to the most expensive in approximately the following

Standard Pile Estimate

P.W. No ______

Type Pile	Load Per Pile	Tons Date
Prospect	Location	Ft
Estimated By At		Filled Shell Ft
Estimate For Piles @ Ft =		Unfilled Shell Ft
Soil Conditions		Wood / Pipe Ft
		Total Avg. Ft

Labor Costs Classification	No Men	Hrs /wk	Hourly Rate	Total w/o U.T.F.	U.T.F. Hr	U.T.F. Total
Foremen						
Engineer						
Firemen						
Oilers						
Pile Drivers						
Labor Foremen						
Laborers						
Standby Cost /Sh	Total For Sh Cost / Shift					

13. Concrete Type		Total
Cy / Pile x Piles =	Cy @	
Placing	Cy @	
Waste % Of Materials Cost		
Protection or Heat	Cy Piles @	
Cost / Ft Pile =	Total $	
14. Re Bars Cage Length	Ft	Total
Material	TN @	
Fabricate	TN @	
Handle & Place	@	
Cost / Ft Pile	$	

45 Insurance	Liab & Soc. Sec	Total
Prop. Dam. Lim. Broad	Conc. % x	
Limits / Yes	Other % x	
Work Inside Bldg No	P.D. % x	
Risk	Liab. % x	
Railroad /		
	$	

Schedule.	Rig No.1 Shifts	Rig No.1 Days	Rig No.2 Shifts	Rig No.2 Days	Rig No.3 Shifts	Rig No.3 Days
Load At						
Enroute						
Move On & Set Up						
Load Tests & Piles						
Driving						
Extra Moves						
Cribbing						
Delays - Weather						
Holidays						
Rig Repair						
Knock Down Move Off						
Enroute						
Unload & Store						
Total Work Shift						
Total Rental Time						
Total Job / Cal. Days						

Rig Availability
Time
Notes

A/C		Item	Total	Labor	√
1	On & Off	Load At Ds @			
2	On & Off	Freight Cwt @			
3	On & Off	Trucking @ /Cwt			
4	On & Off	Unload & Set Up Sh @			
5	On & Off	Knock Down & Load Sh @ /Sh			
6	On & Off	Unload At Ds @			
7	On & Off				
8	Materials	Shells			
9	Materials	Wood Piles & Acc			
10	Materials	Pipe Piles & Acc			
11	Materials	H. Piles & Acc			
12	Materials	Consumables @ $ /LF			
13	Materials	Concrete In Place			
14	Materials	Re. Bars in Place			
15	Materials	Forms			
16	Equipment	Co. Driver Rent Ds @			
17	Equipment	Outside Driver Rent Ds @			
18	Equipment	Other Rentals Ds @			
19	Equipment	Build Alter. Eq.			
20	Equipment	New Eq. Purch. % of $			
21	Equipment	Equip. Repairs			
22	Equipment	Fuel Sh @			
23	Equipment	Mats			
24	Equipment	Hose			
25	Equipment	Jet Plant Power Water			
26	Equipment				
27	Operating	Driving Sh @			
28	Operating	Moves Sh @			
29	Operating	Cribbing Sh @			
30	Operating	Delays - Weather Sh @			
31	Operating	Holidays Sh @			
32	Operating	Rig. Br. Dwn. Sh @			
33	Operating	Load Tests & Piles Sh @			
34	Operating				
35	Overhead	Supt. & G.S. Sal. Ds @			
36	Overhead	Clerk Sal. Ds @			
37	Overhead	Eng's. Sal. Ds @			
38	Overhead	Watchmen Ds @			
39	Overhead	Eng. Layout Ds @			
40	Overhead	Living Allow. No. $ /d x Ds			
41	Overhead	Trans. Men. No @ $			
42	Overhead	Office Exp. & Rent Ds @			
43	Overhead	Clean Up Ds @			
44	Overhead	U.T.F. Sh @			
45	Overhead	Insurance			
46	Overhead	Tool Loss %			
47		Sub. Contr.			
48					

		Tot. Unit Per Ft.	Opr. Cst. Per Ft.
Subtotal 1	$		
Less Wood, Pipe, Shell, H. Pile & Sub. C.			
Net Cost	$		
Overhead	%		
Subtotal 2	$		
Profit	%		
Add Wood, Pipe, Shell, H. Pile & S. Cont.			
Add % On Matl & Sub. Cont.			
Sales or Use Tax @	%		
Bond % Of			
Price	$		
Other Items			
Total Price	$		

Quote To
Price ______ Date ______
Add ______ /LF ______ / Pile
Ded. ______ /LF ______ / Pile

Fig. 27-16 Standard pile estimate.

Proj ____________________ Standard Pile Estimate. (cont'd) Sheet No. 2.

9 Wood Piles	Total	Labor	√
No. x Lf = Lf			
Lf @ $ /Lf			
Freight			
Unload Crane Ds @			
Labor Ds @			
Truck Ds @			
Handle on Job			
Heading & Boring No. @			
Cut off Points and/or Butts			
Waste % of			
Points No. @ $ /ea			
Total Cost $			
Cost/Lf $			

10 Pipe Piles	Total	Labor	√
No. x Lf =			
Lf @ $ /Lf			
Freight Cwt @ /Cwt			
Plates No @ /ea			
Freight Cwt @ /Cwt			
Backup Rings No. @ /ea			
Waste % of			
Unload - Crane Ds @ /d			
Labor Ds @ "			
Truck Ds @ "			
Handle Crane Ds @ "			
Labor Ds @ "			
Truck Ds @ "			
Cut off @ /ea			
Welding No. @ /Wld			
Total Cost $			
Cost/Lin Ft $			

11 H-Piles	Total	Labor	√
No. x Lf = Lf x	lb =	lb =	cwt
Cwt @ /Cwt			
Freight Cwt @ /Cwt			
Unload - Crane Ds @			
Labor Ds @			
Truck Ds @			
Handle - Crane Ds @			
Labor Ds @			
Truck Ds @			
Plate Cwt @			
Waste % of			
Welding No. @			
Cut Off No. @			
Total Cost $			
Cost/Lf $			

25 Jetting	Total	Labor	√
x x x Pump @ /			
Lf in Suct Hose @ /Lf			
Lf in Disc Hose @ /Lf			
Lf in Jet Hose @ /Lf			
Trucking			
Set Up Ds @ /d			
Operate Ds @ /d			
Fuel G/dx Dsx $ /Gal			
Lube %			
Dismantle Ds @ /d			
Trucking			
Water Piles G/p @ /m			
Total $			
Cost/Lf $			

8 Shells					Base Length Est @ Lf = Tlf		√
Ft sections stock length @ Lf = Tlf							
Sect	Wire	Ga	Price	Weight	Sets @ lb= Cwt	Total	
					Sets @ $		
					Frt Cwt @ $/Cwt		
					Trucking @		
					Labor Lf @/Lf		
					Subtotal		
					Waste %		
					Redrives No @		
					Cut off		
					Total $		
Totals					Cost/lf		

15 Forms - Tot. Sf =	Total	Labor	√
Level Area Sf @ /Sf			
Buy Lumber M @ /m			
" Acc. & Nails			
" Form Oil			
Build Bottoms Sf @ /Sf			
" Sides Sf @ /Sf			
Set Forms Sf @ /Sf			
Strip Clean Fms Sf @ /Sf			
Total $			
Cost/Sf $			

Additions	Total	Labor	√
Subtotal 1 Sheet 1			
Fixed Charges Deducted	——	——	
1 Load At			
2 Freight			
3 Trucking			
4 Unload & Set Up			
5 Knock Down & Load			
6 Unload At			
16 Co. Driver Rent Ds @			
17 Outside Rentals Ds @			
18 Other Rentals Ds @			
19 Build Alter Eq			
20 New Eq. Purch			
22 Fuel Ds @			
25 Jet Plant, Power, Water			
28 Moves Sh @			
29 Cribbing Sh @			
30 Delays - Weather Sh @			
31 - Holidays Sh @			
32 - Rig D.T. Sh @			
33 Load Tests & Piles			
35 Supt. Sal. Ds @			
36 Clerk " Ds @			
37 Engs " Ds @			
38 Watchmen Ds @			
39 Eng. Layout			
40 Living Allow Ds @			
41 Trans. Men			
42 Office Exp. Ds @			
43 Clean Up			
Insurance %			
Tools %			
Total Fixed Charges $			
Operating Subtotal			
Overhead %			
Subtotal $			
Profit %			
Operating Total $			
Oper. Tot./No. Piles =	Price/Add Ft Pile		
Oper. Tot/Lin Ft Pile	Price/Add/Ft Pile		

Fig. 27-16 (*Continued*).

order: plain wood, wood-concrete composite, creosoted wood, shell-less cast-in-place concrete, driven-shell cast-in-place concrete, precast concrete, steel pipe piles, and steel H piles. The order of the three last-mentioned piles can change, depending on size and load requirements.

When estimating the cost of piling, many contractors are prone to omit some of the important items of expense. For example, they may omit equipment charges for pile driving by reasoning that a crane will be needed for concrete work on the job in any event. Such reasoning is fallacious, since equipment-use charges have to be carried against a job for the time that the equipment is on the job and should be properly prorated against the classes of work that the equipment is to be used for. In the case mentioned, a proper allocation of equipment-use charges would increase the cost of the piling and decrease the cost of the concrete work. By so doing, the book costs for both classes of work would be brought nearer to their true values.

The normal crew for a pile-driving rig consists of a foreman, engineer, fireman, and four pile drivers. When a crane is used with an auxiliary boiler, union regulations may require an oiler on the crane in addition to the fireman on the boiler. If the rig is extremely large, a fifth pile-driving man will be needed, and a second fireman will be required if more than one boiler is used.

Manufacturers' tables will list the rated steam consumption of each hammer for normal operation, but it is advisable to supply excess boiler capacity to allow for the line loss and scaling of the boiler. As examples, nothing less than a 40-hp boiler working at 100-psi pressure should be used with a No. 1 Vulcan hammer (5,000-lb ram); or 50 hp with a 7,500-lb-ram hammer; or 60 hp with a 10,000-lb-ram hammer; or 70 hp with a 12,000-lb-ram hammer. It is not an uncommon practice to use compressed air in lieu of steam boilers, particularly where relatively few piles must be driven. The manufacturers' tables will also give the compressed-air requirements for hammers.

The Standard Pile Estimate form which is supplied as Fig. 27-16 has been developed as an aid in preparing pile estimates and covers nearly all the conditions and items occurring on a pile job. Each item of work therein is a definite item of cost and should not be passed over or discounted unless an analysis shows that it is not applicable to the job under consideration. Each contractor should enter his own cost units in the Standard Pile Estimate.

REFERENCES

1. *Pile Foundations and Pile Structures*, ASCE Manuals of Engineering Practice No. 27, American Society of Civil Engineers, New York, 1946.
2. W. W. Moore, Experiences with Predetermining Pile Lengths, *Trans. Am. Soc. Civil Engs.*, vol. 114, 1949.
3. Timber Engineering Co., *Timber Design and Construction Handbook*, McGraw-Hill Book Company, New York, 1956.
4. American Institute of Timber Construction, *Timber Construction Manual*, John Wiley & Sons, Inc., New York, 1966.
5. *Pressure Treated Timber Foundation Piles for Permanent Structures*, American Wood Preservers Institute, Washington, D.C., 1967.
6. *Pile Foundations Know-How*, American Wood Preservers Institute, Washington, D.C., February, 1969.
7. ACI 318-63, *Building Code Requirements for Reinforced Concrete*, American Concrete Institute, 1963.
8. Robert D. Chellis, *Pile Foundations*, 2d ed., McGraw-Hill Book Company, New York, 1961.
9. *Specification for the Design, Fabrication and Erection of Structural Steel for Buildings*, adopted Feb. 12, 1969, by the American Institute of Steel Construction.
10. K. Terzaghi and R. B. Peck, *Soil Mechanics in Engineering Practice*, 2d ed., John Wiley & Sons, Inc., New York, 1967.
11. *New York Building Laws*, manual published by New York Society of Architects, 1969.

Section **28**

Cofferdams and Caissons

BEN C. GERWICK, JR.

Executive Vice-president, Santa Fe-Pomeroy, Inc.,
San Francisco, Calif.

A. Introduction

Cofferdams and caissons are devices used for the construction of bridge piers and other structures which extend into water or unstable soils. Although they are but means to an end and often serve only a temporary purpose, some of these caissons and cofferdams rank as major construction achievements. The design of the larger caissons and cofferdams requires detailed engineering analysis and the utmost judgment and experience as to the loads from water, soils, ice, current, and other external sources, plus evaluations of such problems as erosive scour. The designer must be primarily a constructor, for the adverse conditions under which cofferdams and caissons are often built require that a thoroughly practicable method of construction be adopted. While some of the larger caissons have been designed by the bridge engineer, the usual practice is to make the contractor responsible for the design and construction of an adequate structure.

The specified requirements as to allowable tolerances or the methods of pouring concrete or the location of construction joints, etc., have frequently been the controlling factors in determining the type of cofferdam or caisson which can be used. Unrealistic requirements and arbitrary specifications have often made the construction far more costly and hazardous than it need have been. On the other hand, some specifications have been entirely too lax and indefinite. Where serious troubles have been encountered because of changed field conditions, improper selection of method and procedure by the contractor, accidents in construction, or unrealistic requirements, it has frequently been found necessary and expedient to reduce and modify the requirements of the specifications. Therefore, a great deal of thought should be given to setting forth the criteria upon which the contractor is to base his methods and designs. To rely entirely on the phrase "subject to the approval of the engineer" is neither sufficient nor equitable. The engineer should specify the limitations or prohibitions needed to safeguard his design and adjoining structures and then allow the maximum flexibility to the contractor. To protect against an inadequate or unsafe design, he may well specify that the design and procedure for the cofferdam or caisson be prepared and signed by a registered engineer.

Perhaps in no other branch of construction have there been so many failures. Some of these have been spectacular and disastrous, while others have been minor and either partially corrected or else compensated for by a design change. It is striking to note that serious troubles have been encountered by experienced contractors and engineers. This emphasizes that each bridge pier cofferdam or caisson must be treated individually to meet the particular conditions under which it is to be built and that the greatest care, skill, and alertness are required.

In retrospect, the cause of every failure has been determinable. Unanticipated characteristics, conditions, and behavior of the soil have caused the majority of the troubles. This indicates that a study of the soil should be undertaken at every important caisson or cofferdam site. The bridge engineer may have taken borings, but he is primarily interested in the supporting strata, whereas the contractor is interested in the overlying mantle.

A great number of factors must be considered by the designer or constructor of a large cofferdam or caisson. Most major cofferdams and caissons involve timber, steel, and concrete design in three-dimensional frames; continuity considerations; deflection analyses; column formulas; and determination of current pressures, hydrostatic pressures, pile supporting and uplift values, soil pressures, soil permeability, and fresh concrete pressures. Superimposed and accidental loads must be considered. The structure must be practicable to build. Interference of the various elements during construction, the method and quality of underwater connections, and the shape and sequence of constructing the pier itself will all introduce problems which must be considered.

Many of the serious failures have occurred in the course of attempts to correct minor deviations or to facilitate pier construction. While it seems hard to believe, there are numerous instances where a strut has been left out to expedite excavation or has been notched to accommodate prefabricated forms for concrete. To correct minor listing of caissons, extensive dredging and other radical steps have been taken in haste and without full engineering consideration. Symmetry of delicate caisson-sinking operations has been totally neglected in order that one crew might dredge more yardage than the previous shift. Cofferdams have been pumped down "just a few feet" without realization that a large percentage of the total head is acting, since the pressures vary as the square of the depths. Sheet-pile setting and driving records have been established, only leading the record setters to find one sheet pile driven out of the interlock. So the cofferdams have collapsed, and the bottoms have blown, and the caissons have tilted or shifted sidewise, but such results are always traceable to some error in design, procedure, or construction.

A cofferdam or caisson must not only lend itself to rapid construction but be economical as well. Critical path method (CPM) scheduling is particularly well adapted to cofferdam and caisson construction planning.* The confined area of operations and the numerous engineering considerations require a step-by-step planning procedure wherein temporary and permanent construction operations are integrated. The proper use of a critical path schedule may show, for example, that selected operations such as sheet-pile driving or excavation should be performed on an overtime or shift basis, or that all forms and reinforcing steel should be prefabricated in order to minimize their installation time. The removal of a cofferdam brings new problems of economic gains versus costs and the interference with and possible effect on the pier itself.

Finally, the human factor must be considered in all dimensions. Safety is paramount, as there have been far too many accidents and deaths in cofferdam and caisson construction. The desired design and construction procedures have to be transmitted to the actual builders, superintendents, foremen, and journeymen in such a way as to ensure compliance. During construction, the actual results must be continually checked with the design and appropriate steps taken to correct any new factors or conditions which arise. Good design and construction procedures must be made effective through constant supervision.

As opposed to major caissons and cofferdams, many hundreds of small cofferdams have been built in shallow water and in soil; these have been successfully used as building foundations and to support wharves and small bridge piers. The same principles apply to both large and small structures, but many of the complexities and problems are reduced in seriousness in the smaller structures.

* See Sec. 4, Construction Planning and Scheduling.

B. Cofferdams

GENERAL CONSIDERATIONS

Functions A cofferdam is a temporary structure designed to keep water and soil out of the excavation in which a bridge pier or other structure is to be built. Usually cofferdams are dewatered, in whole or part, so that the structures may be built substantially in the dry. However, there are many instances where the structure has been built under water, without dewatering the cofferdam. In the first case, the cofferdam must exclude both water and soil; in the second case, it has to exclude only the soil.

Cofferdams are much more expensive, in cost per cubic yard of material removed, than open-cut excavation. But where free-flowing water, or unstable soils, or heavy surcharge loads are encountered, open-cut excavation is generally impossible or impracticable and the use of cofferdams may be the best approach.

It is usually necessary to seal the bottom of the cofferdam so that it can be dewatered. To accomplish this, underwater concrete is placed in a sufficient thickness so that, by its own weight alone or in combination with piles, it can resist the uplift pressure. Other types of seal have been used in special cases, including clay blankets and concrete on the outside; however, the normal and usually economical method is to place a seal of underwater concrete.

Types Many successful shallow cofferdams for bridge piers have been built using earth dikes. In some shallow rivers, sand dikes have been used, with wellpoints to dewater both the dike and the excavation.

In the past, many major bridge piers were constructed using timber sheet piling for cofferdams. This piling was usually of the Wakefield type, in which three planks were laminated together to give a tongue-and-groove connection between adjoining piles. Although timber sheet piling has now been almost entirely replaced by steel sheet piling, its use remains a sound practice and should be kept in mind for unusual cases where the special properties of timber may be required. Timber sheet piling is open to attack by marine borers unless it has been treated, and one major cofferdam had to be abandoned after its timber sheet piling was riddled by teredo. Steel sheet piling combines fair watertightness, high strength in bending and shear, high interlock strength, ease of driving and removal, and high salvage or reuse value.

While some shallow cofferdams resist the pressure of water and soil entirely by cantilever action of the sheet piles, most cofferdams require additional bracing or similar support. The usual bridge pier requires a rectangular cofferdam, and for this a rectangular bracing system is normally used. (See Figs. 28-1 and 28-2.) Bracing systems usually consist of horizontal wales and struts, with as many levels as necessary to resist the external forces. Vertical soldier beams, with horizontal wales and struts, have occasionally been used to reduce the number of levels of cross bracing or to enable them to be placed at more favorable elevations. In other cases, timber cribs have been sunk by filling them with rock, then sheet piling is driven on the outside face. This method requires a much larger cofferdam in plan, but it is a good solution where there is bare rock bottom with little or no possibility of obtaining a toehold with the sheet piles and where there may be considerable current or swell from ocean waves.

Some of the largest and deepest cofferdams have been built in a circular or elliptical plan. In some instances, a double row of piling has been driven and the space between the walls filled with concrete. Thus the cofferdam acts as a ring, and no bracing is required. In other instances, a single wall of sheet piling is driven, with rings of steel, timber, or concrete taking the thrust around the cofferdam and thus eliminating cross bracing. Although a rectangular pier seems to be the most common shape, engineers should make more frequent use of circular or elliptical forms for deep bridge piers in order to take advantage of the safety and economy of circular or elliptical cofferdams.

Another type of cofferdam which is valuable in special cases is a composite box cofferdam and caisson. In this case, the entire bracing system, together with the walls and sometimes with the bottom, is floated or set in place on a previously prepared foundation. This type of cofferdam can be used only where no overburden is

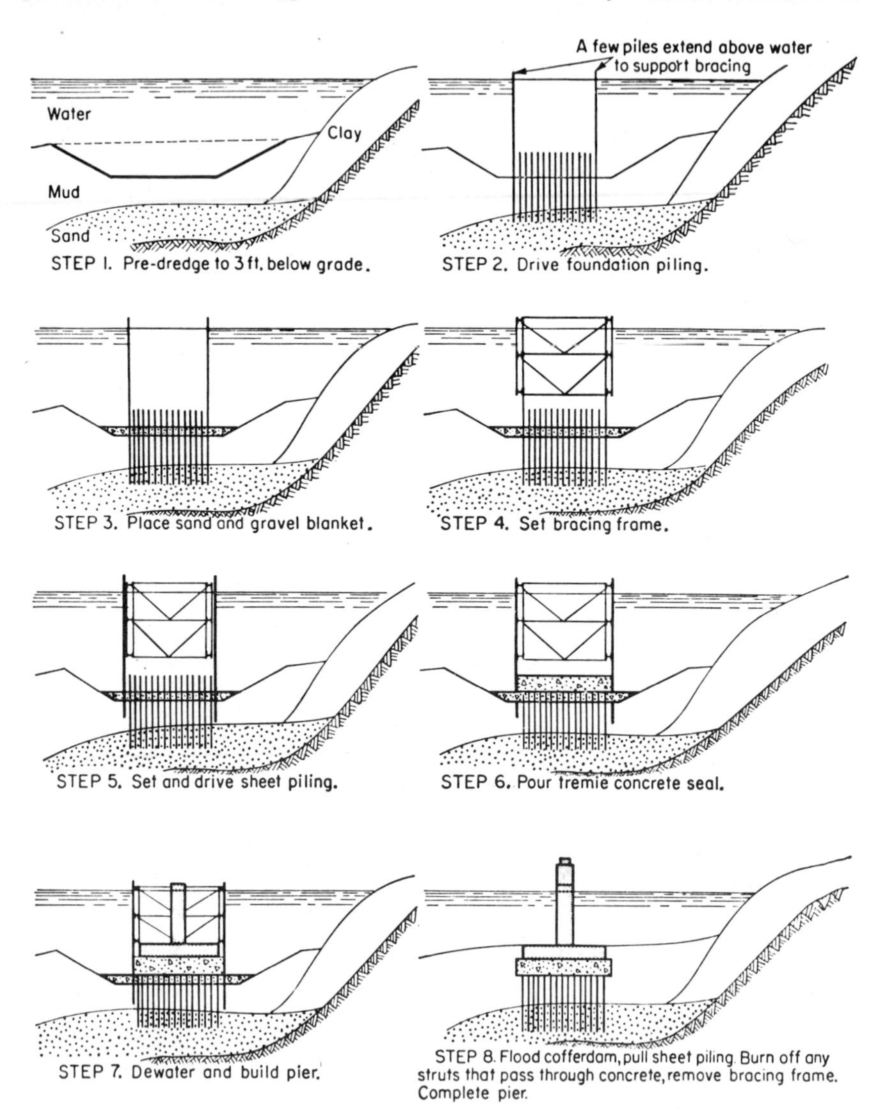

Fig. 28-1 Typical cofferdam construction sequence.

present or where it can be removed in advance. If the cofferdam is to be set on rock, its bottom is usually tailored to fit the rock contours closely. The prepared foundation may consist of underwater timber piles cut off to grade, or a dredged hole backfilled and leveled with crushed rock or sand, or hardpan carefully dredged to grade and leveled with sand backfill.

The box cofferdam may be constructed of timber, steel, or concrete and is frequently a combination of these materials. It may be built on ways or in drydock, launched and floated to place, and sunk. Similar "boxes" can sometimes be constructed in place by setting panels of timber, steel, or precast concrete around a bracing system. This method is well adapted to multiple use of the panels in successive cofferdams.

Cellular cofferdams of sheet-pile cells can be used for very large and deep bridge piers; however, their use is rare because they do not fit the plan shape of the usual pier and because they are relatively expensive. They may be a practicable solution in

Fig. 28-2 Large sheet-pile cofferdam. Note corner and intersection details and electric pump in foreground to maintain dry cofferdam. Steel strut was damaged by digging bucket, a possibility for which allowance must always be made in design.

very swift water or in exposed locations where progressive construction is necessary. In such cases, they would have the same general function as the circular cofferdams.

Some bridge piers for major structures have been constructed entirely under water, a method which has many advantages in time of construction, economy, and reduction of risk. High-quality underwater concrete can be obtained provided that the techniques which have been developed are followed carefully and properly. To realize the benefits of this method, design must be integrated with construction methods.

External Loads The external loads on cofferdams are principally those of water (hydrostatic) pressure and soil pressure. In addition, the cofferdam may have to resist current pressure, ice loads, log jams, and wind and wave loads. The soils may be higher on one side than the other, as on a steep bank, throwing an unbalanced load on the cofferdam. There can also be scour around and under the cofferdam due to currents.

Hydrostatic pressure is the fluid pressure due to the weight of water. In fresh water, this is 62.4 pcf; in ocean salt water, it is 64 pcf; and in the brackish water commonly encountered in tidal bays and estuaries, it runs somewhere in between. The variation in unit weight will assume real importance only in an inland salt sea, where the weight of water may run as high as 80 pcf.

Soil pressure against a structure is a highly technical and complex subject about which many good books[1-3]* and technical papers have been written. The characteristics and properties of the several strata surrounding and underlying the cofferdam must be determined first. The effective pressures can then be computed.

A rough approximation of the soil pressure can be obtained by the method of equivalent fluid pressure. This method assumes that the soil will act as a fluid, which is of course seldom true. For piers in water or saturated ground, full hydrostatic pressure is assumed to act from the water level down to the bottom of the excavation within the cofferdam. In addition, the submerged-soil pressure is assumed to be distributed and to act like a fluid. The equivalent fluid weight of the submerged soil will rarely exceed 26 to 30 pcf. The equivalent fluid pressure is determined from this assumed equivalent fluid weight, using the hydrostatic formulas $p = wh$ and $P = wh^2/2$. This pressure is assumed to act from the top of the soil down to the bottom of the excavation and is additive at every level to the water pressure at that level.

Having determined the total load from the soil on the basis of equivalent fluid pressure, experienced judgment may then be applied in predicting its variation with depth. Instead of a triangular (hydrostatic) pressure distribution, a rectangular, parabolic, or trapezoidal distribution may be postulated. This increases the design loads in the upper levels of the cofferdam and reduces them in the lower levels.

The great advances of soil mechanics, and particularly the dissemination of knowledge of this subject through the engineering profession, now make it practicable to make a more careful investigation of soil pressures on major cofferdams. The cost of a thorough investigation by a competent engineer will usually be repaid many times in economy of proper design and in safety.

In any method of determining soil pressure, consideration must be given to the effect that driving the sheet piles and building the cofferdam will have on the soils. Driving, and especially jetting, may open a passage through normally impervious strata. Driving will remold and disturb clays. If the bottom of the cofferdam blows, even without complete failure, the condition of the soils and the pressure exerted by them may be suddenly and radically changed.

A cofferdam is a flexible structure, and the deflection of the sheet piles serves to equalize pressures and distribute them. It can also cause the soils to yield and thus exert greater pressures. With a cofferdam in deep water (or in soils which behave like a fluid), considerable pressure is exerted on the full height of the sheet-pile walls just as soon as dewatering is under way. The total pressure varies as the square of the depth; and with an outside depth of, say, 50 ft, one-half of the total pressure is on the sheet-pile walls when the inside has been pumped down 15 ft. While this should be obvious, it is frequently proposed to "pump down in stages and install each bracing set as its level is reached." Again, when the structure inside is partially completed and it is necessary to remove a set of bracing, a proposal is erroneously made "to flood the cofferdam to that level and then take out that set of bracing."

Where deep cofferdams must be constructed through a great depth of soil (as opposed to water), it may be necessary to install each level of bracing as its elevation is reached in excavation. In this case, complete calculations have to be made for each stage of excavation. Additional sets of bracing may be required to take care of the pressures at the several stages as compared to those necessary to take care of the pressures at the final stage.

* Superior numbers refer to the list of references beginning on page 28-76.

One technique that has proved economical in such circumstances is to purposefully fill the excavation with water, pumped in if necessary. Then any remaining bracing sets are lowered through the water as the excavation proceeds and blocked to the sheet piles by a diver. By filling the sheeted excavation as high or even higher than the groundwater elevation outside, considerable control of pressures can be effected for the construction stage.

For small shafts or cofferdams, a bentonite slurry may be used to offset and balance the external pressures. Slurry densities of 70 to 80 pcf are used (higher ones are possible). Tremie concrete can be poured through this bentonite slurry just as it is through water; the major difference is a reduction of bond on the sheet piles. Brine densities of up to 75 pcf can be accomplished by saturated solutions of rock salt (NaCl) and even higher densities can be achieved, although at very high expense, with calcium chloride and sodium silicate.

Ice pressures, both static and impact, may impose extremely severe overloads on a cofferdam.[4] Solid sheet ice exerts pressures of 100 to 300 psi. This is so great that it is impracticable to design any cofferdam to resist more than a few inches of ice thickness, and the only feasible approach is to create open water around the cofferdam. Moored logs, through their ability to absorb and retain heat and their cushion effect, have been found extremely effective in protecting against crushing. Similarly, coal dust, etc., may absorb heat; salt and other chemicals may destroy the structure of the ice. Explosives are generally ineffective in breaking up ice and at the same time endanger the cofferdam.

Moving ice cakes, on the other hand, are dangerous to a cofferdam because of their tendency to raft and pile up. A log-boom shearwater can be effective in deflecting ice cakes around the cofferdam.

Moving *logs and debris* during runoff periods may endanger the cofferdam, both from local impact of a single large log and from rafting and piling up. A log boom, well anchored, can be utilized to keep floating logs and debris away from the cofferdam. During severe flood runoff stages, the cofferdam should be filled with water.

Current pressure is a relatively small loading factor except when the current is very swift. For example, in a river of 10-fps velocity (about 7 mph), the average unit pressure on the upstream side of a rectangular cofferdam would be about 130 psf. However, the unbalanced current load may distort the cofferdam or even overturn it. The total current load may be computed by the formula

$$P = A \frac{v^2}{2g} w c_s$$

where P = total current load, lb
A = area of cofferdam wall exposed to current, sq ft
v = velocity of current, fps
g = 32.2 fps per sec
w = unit weight of water, pcf
c_s = shape constant: 1.33 for rectangular, 0.75 for circular

For the usual rectangular cofferdam in salt water, this formula is reduced to

$$P = 1.3Av^2$$

Guide structures employing vertical and batter piles are used to hold the cofferdam in place during the construction operation. A floating boom or a barge moored upstream will break the current on the surface; this will help in setting sheet piles, although it provides little relief for total pressure against the cofferdam. In a swift river, special protection may have to be provided for the cofferdam to reduce the current pressure and to still the water during setting and driving operations.

Scour is a frightening phenomenon which may destroy a cofferdam or caisson in short order.[5] The very act of building the cofferdam places a block or dam in the river, and this blocking may be further aggravated by floating barges and moored derricks. Scour may occur where pier construction tends to block even relatively wide rivers, such as the Mississippi. It can also occur in tidal bays and estuaries where the tidal current may be locally quite high.

Scour is usually deepest in the eddy at the back corners of a rectangular cofferdam. The addition of a few sheet piles tailing back from the two downstream corners like streamlining fins can be of considerable help. This design was successfully used on the sheet-pile cofferdams for dams on the Mississippi.[5] Riprap, possibly combined with willow mattresses, should help; but on at least one caisson such mattresses proved ineffective. Providing a deflecting nose of sheet piles helps to prevent scour immediately behind the upstream corners, another vulnerable spot (Fig. 28-3). In many rivers or estuaries where currents are swift and scour is anticipated, the prior placing of willow mattresses or of 2 or 3 ft of small rock (3 in. minus) over the entire area, before setting the sheet piles, is effective in preventing the erosion from ever starting.

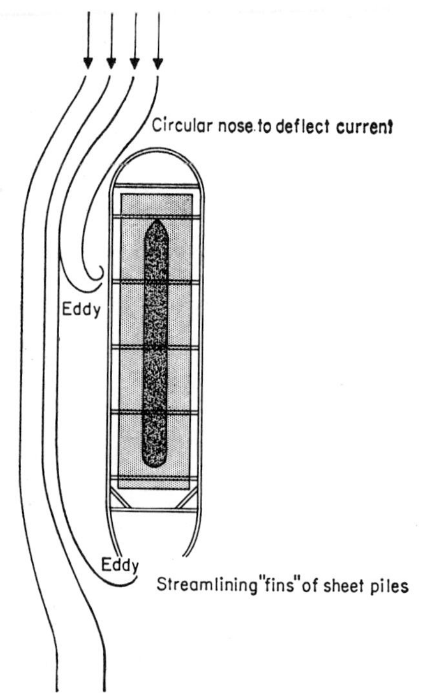

Fig. 28-3 Schematic drawing of cofferdam designed to reduce scour due to fast river current over sandy bottom.

SHEET PILING

Types and Sections Most cofferdams today are built using one of the standard rolled sections of steel sheet piling. Timber sheet piling has been widely used in the past, and fabricated-steel sheet-pile sections are occasionally needed and used on very deep or unusual cofferdams.

Timber sheet piling is of the Wakefield type, made up of three laminated timbers spiked together with the middle timber offset to form a tongue on one side and a groove on the other. The timbers should be either surfaced on four sides or surfaced on one edge. The toe of each makeup section is usually cut on a diagonal (with the point on the groove side) so that, in driving, the pile will be continuously wedged back against the previously driven pile. These piles are driven with the tongue leading.

Standard rolled-steel sheet-pile sections are manufactured by at least three large steel companies in the United States. Lighter sections for shallow cofferdams and trenches are manufactured by several other steel companies. There are also German, Belgian, and English standard sheet-pile sections which are coming into fairly wide use,[6] particularly in Canada. When ordering foreign sheet piles, a few points of difference in practice must be remembered. Most foreign sheet piles do not have as high interlock strength, nor are the interlock tolerances so close as in domestic sheets. Also, at least some of the foreign sheet piles have excessive sweep in the longer lengths. Both these matters can be taken care of by specifying exactly what tolerances are needed and acceptable for the particular job. In listing the section modulus for a sheet pile, the United States practice is to list the section modulus for an individual pile. Foreign practice is usually to list the section modulus for a linear foot of wall, assuming adjacent piles turned in opposite directions and full interlock friction. For the same deep-arch steel sheet-pile sections, therefore, the foreign catalog would show approximately twice the section modulus that is shown in the United States catalog.

The deep-arch type of sheet piling offers many advantages from an overall point of view. It combines reasonably high strength in bending with high interlock strength and ease of setting, driving, alignment, and removal. Two of these deep-arch sections are shown in Fig. 28-4. While these sheet piles can be set and driven with all the arches turned the same way, the highest-strength wall and greatest ease of setting, driving, and pulling can be obtained with each adjacent pile turned in the opposite direction. The lighter deep-arch section will interlock interchangeably with the heavier section. Similar sections of the several United States manufacturers will

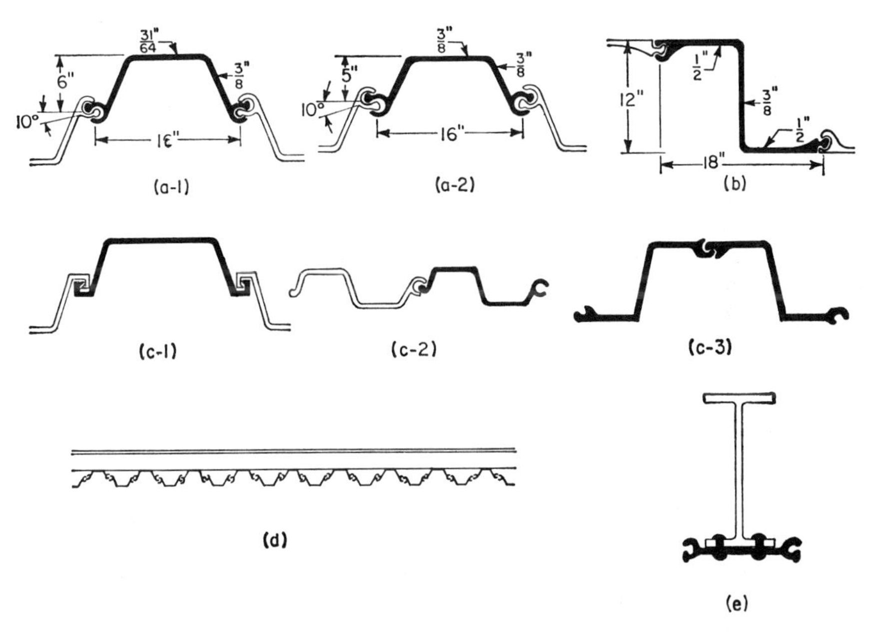

Fig. 28-4 Rolled-steel sheet piles. (*a*) Typical deep-arch sheet-pile sections, United States manufacture: (*a*-1) Weight per square foot of wall = 32.0 lb; section modulus per linear foot of wall = 15.3 in.^{3}; Bethlehem DP 1; U.S. Steel M-110, Inland I-32. (*a*-2) Weight per square foot of wall = 27.0 lb; section modulus per linear foot of wall = 10.7 in.^{3}; Bethlehem DP-2, U.S. Steel M-116, Inland I-27. (*b*) Typical vertical-web Z-pile section, United States manufacture: Weight per square foot of wall = 38.0 lb; section modulus per linear foot of wall = 46.8 in.^{3}; Bethlehem ZP38, U.S. Steel MZ 38. (*c*) Foreign manufactured steel sheet-pile sections distributed in the United States: (*c*-1) Larssen steel sheet piling—Larssen standard sections 1, 2, 3, 4, 5, 6, 7. (*c*-2) Belval sheet piling Type TR. (*c*-3) Belval sheet piling Type BZ. (*d*) Typical wall made up of deep-arch sheet piles alternately turned in opposite directions. (*e*) "Master" pile. Wide-flange beam riveted to flat-arch sheet pile.

also interlock with each other, although there may be an increased tendency to bind in the common interlock.

The Z type of steel sheet piling is being used more and more for deep cofferdams because of its very high strength in bending. The ratio of section modulus to weight (and cost) is more favorable for Z types than for the deep-arch types. However, in setting and driving there is a tendency for Z piles to roll and twist, and they may bind in the interlocks in both driving and pulling. There also seems to be a greater tendency for Z piles to drive out of the interlocks under very hard driving. Z piles are made with vertical webs and also in sections having inclined webs. The inclined-web piles are much more likely to twist, roll, and bind and therefore seem more suited for permanent bulkheads than for cofferdams. While a sheet pile of greater strength decreases the number of sets or levels of bracing, the total load to be carried across remains the same. Thus there may be a saving in the number of bracing levels but not necessarily a significant reduction in total weight of bracing.

It is generally accepted practice to use the deep-arch type of sheet piles wherever its strength is adequate. When a stronger sheet pile is required, the vertical-web Z pile, which is the heaviest and strongest Z-pile section, is preferred.

The manufacturers' catalogs set forth the properties of the several sections of steel sheet piles and also give a great deal of valuable information on their use. The catalogs of Bethlehem Steel, United States Steel, and Inland Steel are valuable and necessary reference books for a cofferdam designer.

Fabricated-steel sheet piling is sometimes used for very deep cofferdams or where it is necessary to have a very high-strength sheet pile to span between abnormally spread bracing levels. A very strong sheet pile can be made up by riveting an I beam to the back of a flat or shallow-arch sheet pile. Cofferdam walls have been built up using these "soldier beam" sheet piles spaced several feet apart, with connecting arcs of flat or shallow-arch sheet piles. These arcs act in tension through the interlocks. This scheme was used on the East Bay cofferdam piers for the San Francisco–Oakland Bay Bridge in order to eliminate all underwater bracing. The 36-in. wide-flange beams were riveted to a shallow-arch sheet pile on its inner face. Three shallow-arch sheet piles formed the arc between adjoining master piles.

Fabricated sheet piles have also been made up by riveting two deep-arch sections back to back, although this scheme is more adaptable to bulkheads where it is possible to fill the cells with reinforced concrete.

In special cases requiring extremely heavy bending resistance, a standard flat-section sheet pile has been split and the two halves welded to the sides of large-diameter pipe piles or to large, wide-flanged beams, which then act as master piles. Care must be taken to prevent distortion in welding, and this can best be accomplished by welding intermittently and using low-heat electrodes and techniques. Similarly, the split sheet-pile halves can be embedded in a large concrete pile. In this case, the sheet pile should be split in a wavy or corrugated pattern, which will give intermittently longer sections of the web for better anchorage in the concrete.

Corner sections, Y's, and T's are available from the manufacturers of the standard rolled-steel sheet-pile sections. For the arch-type sheet piles, corner sections are rolled. For Z-type sheet piles, corner sections are fabricated, using pieces cut from standard sections and riveted together. Corner sections should be selected with care from the manufacturer's catalog to be sure that the fingers of the interlock are turned in the correct direction. Where there is an odd number of sheet piles in a wall, a different corner piece is required than where there is an even number. Corner sections for arch-type sheet piles can also easily be bent from a section of shallow-arch sheet pile. Most shipyards are equipped to bend even a long sheet pile to any desired angle.

As shown on Fig. 28-4, the deep-arch sections may be rotated through 10 deg per pile to form an arc.

Alloy steel sheet piles have recently been made available by domestic and foreign manufacturers. These give significantly higher yield-point stresses, thus permitting higher design stresses in bending and longer spacing between bracing frames. This may be of major economic significance when the sheet piles are driven first and the bracing is placed as excavation proceeds. Again, it may permit better spacing of the frames to miss the concrete structure inside, and fewer penetrations. On one major water cofferdam, the use of alloy sheet piles would have allowed the bracing frame to be placed as a unit with the bottom frame at the groundline, instead of being lowered piecemeal and bolted in place by divers after excavation, as was actually done.

The increased yield point is not accompanied by any increase in ultimate strength. Deflections will be greater, which is not too important in water but which may cause greater movements or higher pressures in soils. The higher yield point reduces damage in driving.

Table 28-1 is a composite of current United States manufacturers' data. For more detailed information, the manufacturers' catalogs should be consulted.

These alloy steels are weldable, but E-70xx low-hydrogen electrodes are recommended to eliminate the need for preheating. The increased cost of alloy steels over standard grade is from 10 to 15 percent of mill price. Obviously, whenever their increased design stress can be utilized, alloy steels will show both direct and indirect economies. In addition to their higher yield point, some alloy steel sheet piles have a greater resistance to corrosion in the sea-water splash zone.

Setting and Driving Sheet piling can be ordered with one or two handling holes in one end for ease in lifting and in pulling. These handling holes must subsequently be welded or plugged if high tide or floods can reach them. If sheet piles are spliced, particular care should be taken to make sure that all handling holes are welded tight.

Sheet piling must be handled with care to prevent buckling or damage to the interlocks. A crimped flange on a steel H pile is of little concern; on a steel sheet pile it could, and probably would, cause excessive trouble and expense. Flat-arch and shallow-arch sheet piles have little strength in bending when lifted flat and must be picked up with equalized slings at several pickup points or with a strongback. Another method is to turn them and pick them up on edge instead of flat. Wind gusts can also cause buckling of long flat-arch piles while they are being picked up.

TABLE 28-1 Properties of Alloy Steel Sheet Piling

Grade	Minimum yield point, psi	Minimum tensile strength, psi	Recommended working stress, psi*
ASTM A328 (Standard).....	38,500	70,000	23,000–25,000
Grade 45....................	45,000	65,000–70,000	27,000–29,000
Grade 50....................	50,000	70,000	30,000–32,500
Grade 55....................	55,000	70,000	33,000–35,500

* In all cases the writer prefers the lower values listed.

Guide structures are required for the proper setting and alignment of sheet piles. Accurate, well-built guides will more than repay their cost in ease of setting, driving, and pulling and in the accuracy and safety of the cofferdam. The principal guide may be the top level of the bracing, which is usually located at or near the high-water mark. If the water is deep or the soil soft, so that the sheet piles when first set will run well down, leaving only short lengths sticking above the top bracing, then this may be all that is required. When the top bracing is to be placed after the sheet piles have been set and driven, then a temporary inside or outside guide frame may also be used. When the sheet piles will stand high after setting, then upper levels of guides must be built against which to set the sheet piles and to secure them against wind.

Wind loads on a high-standing wall of sheet piling can be very serious. One such wall of high-section-modulus sheet piles was bent over flat by a gust of wind, although it had been set against what was thought to be an adequate guide structure. If the guides are on the inside, which is the usual practice, then it may be necessary to provide temporary wire lines as ties. These can be fastened to the top of the sheet piles through the handling hole or to a clip welded onto the arch of an occasional sheet pile in such a way that the clip will pass the wale without hitting it. Ties and upper sets of guides are progressively removed as the piles are driven to grade.

The entire cofferdam should be set before any appreciable driving is begun. Setting should start at one corner, and every tenth pile or so may be driven a few feet to hold the wall. Particular care should be taken to get the corners exactly vertical and in the correct spot. Minor adjustments can then be made in the runs of standard sheet piles since there is about $\frac{1}{4}$-in. play per sheet pile in the interlock. Care must be taken not to gain too much, particularly with Z piles where a slight twisting will increase the space occupied by a single pile. Some experienced contractors allow about $\frac{1}{8}$-in. gain per pile in laying out the cofferdam design.

The piles must be set vertically and, should any tendency to lean be noted, setting must be stopped and the leaning piles reset. Some cofferdam builders set the four corners first and then set each way, making connections in the straight wall where there is considerable flexibility for adjustment. Steel sheet piles can often be set very quickly, and many times inexperienced crews will try to make a record in their setting. By so doing, they may waste far more time in subsequent driving and pulling due to friction and binding in the interlocks.

Driving should be done a few feet at a time, working around the cofferdam and bringing all sheet piles down approximately together. The corner piles are very rigid and will not yield. If the adjacent walls are driven down ahead of the corner piles and if there is a misalignment or if the piles are out of vertical in either direction, the corner will very likely be driven out of the interlock. (See Fig. 28-5.) If the corner is driven so that its toe is slightly ahead of the adjacent standard piles, then the latter piles will adjust to fit the corner. They can accomplish this by taking the tolerance in the interlock or by a slight spreading of the arch or twisting of the Z. Unfortunately, it is difficult to fit a standard sheet-pile hammer on a corner pile which does not have its head a foot or two above the adjacent piles. A solution is to order the corner piles 5 ft longer than the standard sheet piles.

Fig. 28-5 Steel sheet pile driven out of interlock because corner pile had not been driven ahead and because hammer used was too heavy.

Friction in the interlocks can contribute considerable resistance to both driving and pulling and may be greater than the resistance of the soils. In many cases, greasing the interlocks greatly aids setting, driving, and pulling. In some sands, however, the grease and sand may ball up. Before driving, pile interlocks should be thoroughly cleaned of wedged sand, etc., from previous uses and checked for defects. One major oil company (Standard Oil Co. of California) has developed a special zinc-and-grease lubricant for sheet-pile interlocks which has given very satisfactory results on a number of projects. One note of caution; greasing of interlocks should never be employed where friction in the interlocks is a design feature; e.g., sheet-pile cells.

It is particularly important to use a light hammer in initial driving when the piles are standing high, as a heavy hammer may cause the piles to buckle. It may also cause individual piles to run too far ahead and deflect the tips. Comparative tests have shown that a fairly light double-acting hammer will often penetrate obstructions better than a heavy hammer whose blow may damage the pile tip, turning it so that further penetration is impossible. The action of a light, fast hammer on a steel sheet pile is similar to that of a clay spade which penetrates slowly and steadily. After the piles are substantially down, and when driving has become very hard, then a heavier hammer may often be used to advantage. (See Fig. 28-6.)

The writer prefers a No. 6 McKiernan-Terry hammer or its equal for the light, fast hammer. This size of hammer can fit on a single pile, with "legs," so that the hammer rides the pile. It is easily handled, even high in the air. It is very fast acting, delivering some 275 blows per minute. When driving arch-type sheet piles through 50 ft of sand, comparative studies were made with light and medium hammers. The light hammer, although penetration per blow was very small, actually drove the piles faster and with less damage.

Fig. 28-6 Sheet pile driven down center of buried brick wall using light, fast hammer. When heavy hammer was put on pile, tip curled and further driving became impossible.

The McKiernan-Terry No. 65 double-acting hammer fits on a single pile and is very fast, but it has a heavier ram than the No. 6. It has proved extremely satisfactory in extensive use over several years. For the medium-heavy hammers, where driving resistance is hard, the Vulcan 50-C double-acting hammer and the McKiernan-Terry No. 11-B-3 have been used successfully.

A sheet-pile hammer must be properly lubricated. It is a fast-acting piece of machinery, working under adverse conditions and taking hard and repeated shocks. Insufficient oil or the use of the wrong kind of oil causes inefficient operation, excessive repairs, and excessive downtime. The manufacturer's instructions must be followed explicitly. Different oils are required for operation with air than for operation with steam. Table 28-2 lists some characteristics of hammers.

Where obstructions are expected, such as boulders, cemented gravel, buried logs, old foundations, or riprap, the thickest section sheet pile of the deep-arch series is preferable (Bethlehem DP-1, United States Steel M-110, or Inland I-32). The tips of these piles can be hard-faced, and each pile in succession should be carefully driven for not more than a foot or so. A light, fast hammer should be employed unless its inability to penetrate is definitely established. Alloy steel sheet piles are particularly desirable for such use; their higher yield strength resists local distortion and tearing.

A drop hammer is sometimes used in initial setting and driving, with built-in guides or legs that keep it aligned with the sheet pile. The height of fall must then be kept to a practical minimum to avoid damaging the sheet-pile head or tip or driving the pile out of interlock, etc.

Pile driving in sand can be aided by jetting, but this may produce undesirable side-effects. It may cut a pathway through semi-impervious strata so that the full hydrostatic head will act throughout the entire cofferdam height. It may also create a pathway for blowing during excavation and pumping, particularly where there is little penetration of the sheet piles or where it is planned to dewater the cofferdam without a seal. (See Fig. 28-7.)

TABLE 28-2 Characteristics of Steam (Air) Pile Hammers in Common Use for Driving Sheet Piles in Cofferdams

Make*	Model	Rated striking energy, ft-lb	Blows per min	Boiler hp required	Volume of air required, cfm
Vulcan	50-C	15,100	120	60	880
Vulcan	65-C	19,200	117	70	991
Vulcan	80-C	24,450	111	80	1,245
McKiernan-Terry	No. 6	2,500	275	25	500*
McKiernan-Terry	No. 7	4,150	225	35	500*
McKiernan-Terry	No. 9-B-3	8,750	145	45	600*
McKiernan-Terry	No. 11-B-3	19,150	95	60	900
McKiernan-Terry	No. 65	3,200	280	35	500–600†

* Both manufacturers make many pile hammers which are suitable for use on sheet piles. Their catalogs give complete descriptions and characteristics and are of great value in selecting, using, and maintaining sheet-pile hammers. All listed hammers are double acting, as is recommended by the writer for sheet-pile work.

† These air volumes are the writer's recommendations, and are greater than those recommended by the manufacturer. Presumably, the difference is due to pressure drop, etc., experienced in field operations.

Diesel hammers have been used in a number of cases for the driving of sheet piling. They have proved most useful in cases where continuous, moderately heavy driving resistance is encountered; e.g., clays.

A recent development is the use of vibratory hammers for driving (and pulling) sheet piling. These are extremely effective in cohesionless material—sands, silts, etc. They are relatively ineffective in clay, hardpan, sandy clay, gravel, etc. When driving through heterogeneous strata, it is frequently advantageous to use a vibratory hammer in the cohesionless soils and a steam hammer in the cohesive soils and gravel layers. Jetting assists the vibratory action; or in hard driving, just allowing water to flow freely at the surface will sometimes help. The water vibrating down along the sheet pile face serves to lubricate it.

The three types of vibratory hammer in common use are the following:

1. Vulcan VHD Vibro-hammer series (often called the "Japanese type"). This can be assembled in units of one, two, or three. Its frequency is in the range of 16 to 20 cycles per sec, and it is equipped with an excellent coil-spring type damper to reduce the transmission of vibration into the boom and rigging. Generators of 125, 150, and

200 kw capacity, respectively, are required as power sources for the VHD-1, VHD-2, and VHD-3 hammers.

2. Foster Vibro Driver-Extractor[6] (often called the "French type"). This also comes in three models or sizes, with vibration frequencies ranging from 12 to 20 cycles per sec. Power required is 34, 70, or 100 hp for model 2-17, 2-35, or 2-50, respectively.

3. Bodine Resonant Driver[7] (often called the "sonic type"). This vibratory driver oscillates at a speed corresponding to the natural resonance frequency of the mechanical assembly of which the pile is a part. The speed of the vibration is too fast for the soil to respond elastically, so skin friction is reduced and the pile punches

Fig. 28-7 Setting sheet piling for circular cofferdam. Note two levels of guides to ensure accuracy.

its way in numerous short vibrations through the soil. The applied power consists of two 500-hp gasoline engines. Frequency is 60 to 120 cycles per sec.

All three vibratory hammers are relatively quiet and thus are suitable for use in cities. All three are equipped with hydraulic clamping devices to clamp the sheet pile, thus minimizing damage to the sheet-pile head. Transmission of vibration to the rigging, boom, etc., is a potential source of fatigue failure of the handling equipment. Therefore, use of a damping device is necessary.

In some sands, where steam hammers develop excessive skin friction, the vibratory hammers will dramatically outperform their more conventional counterparts. The vibratory hammers transmit some vibration to the surrounding soil, and driving on one pair of piles will frequently cause other sheet piles in the cofferdam to sink simultaneously. The penetration is very rapid in cohesionless soils, ranging up to several feet per minute, and it is important not to run any pair of piles too far ahead of the adjoining piles. Frequent shifting of the hammer is necessary to keep sheet-pile alignment true.

Length and Penetration The determination of the proper length of sheet piling to use and the necessary penetration requires careful analysis. Generally, the top of the sheet piling must be above extreme high water. Spring tides in tidal bays and estuaries may be augmented by wind-driven waters and flooding tributaries. The extreme high tides may be 1 to 2 ft greater in one part of a tidal bay than at the recording station, and prolonged winds can often raise the water level appreciably. While the sheet piles do not have to keep out the splash of the waves under these extreme conditions, they must keep out the bulk of the water. It is common to place the top of the sheet piles 2 ft or more above the highest recorded tide at the site.

In seasonally fluctuating rivers, it is usually uneconomical to design a cofferdam to hold out flood levels. The cofferdam top is set at an elevation a few feet above the highest anticipated river level for the season, and work is scheduled for completion during that season. Although a cofferdam can always be flooded to meet an unexpected or unseasonal flood, damage to equipment and forms and the costs of pumping out again and cleaning out the silt and debris do not justify cutting the height too close.

The sheet piles must penetrate far enough to develop sufficient bearing capacity to hold up the weight of the pile, bracing frame, and tremie pipes, etc. They must also penetrate a sufficient distance below the lowest level of excavation to prevent the soil from running in underneath. There is a stage during the construction when the sheet piles have been driven and the excavation has been carried to its deepest level, yet the seal has not yet been poured. At this stage, when there is equal water pressure inside and out, the imbalance or tendency to run in underneath is due to the weight and characteristics of the soils. The stability of the soils around and below the tips of the sheet piles should be considered in this regard.

The sheet piles must also resist the unbalanced pressures caused by the greater height of the soils outside the cofferdam than inside it. In firm soils, such as hardpan or firm clays, this required lateral support may be developed in a few feet of pile penetration. In soft mud or unstable sands, it may take 15 to 20 ft or more. It can be destroyed, at least temporarily, by movement of soils under the tips of the sheet piles. In the writer's experience, more cofferdam failures have been due to inadequate sheet-pile penetration than to any other single cause. The usual reason given for using short sheet piles is to save costs, yet a few extra feet will provide a safeguard far beyond the small additional cost. The tendency of the soils to move underneath the piles can also be reduced or eliminated by predredging.

Substantial penetration into relatively firm materials will develop a degree of fixity at the lower reaction of the sheet piles. In many deep cofferdams, the greatest stress in the sheet piles exists after excavation but before the seal is poured. If a fixed reaction can be developed through adequate pile penetration, the critical bending stresses under this loading will be reduced. Similarly, in a later stage, when the seal has been poured and the cofferdam dewatered, adequate penetration to ensure a fixed end moment at the tremie seal will reduce critical bending stresses.

With deep-arch sheet piles turned in opposite directions, the effective section modulus will be greatly increased (1.5 to 1.75 times) if sliding along the interlock can be prevented. Adequate penetration into the bottom will help to develop this joint action. In some critical cofferdams, this phenomenon has been used to strengthen the wall. Adjoining sheet piles are welded for a foot or two at the top and are driven to a deep penetration in firm material so as to prevent sliding in the interlock. Rules of thumb for penetration are inadequate and dangerous, but usually, in sediments such as sands, silts, and soft clays, 8 to 20 ft of penetration will suffice.

If the cofferdam is to be dewatered without a seal, then a different and much more critical approach is involved. The penetration here must be adequate to prevent boiling, blowing, and piping due to the imbalance of water head and soil on the outside. In normal sediments, muds, soft and medium clays, and sands, the pile penetration required for the safe dewatering of even a moderately deep cofferdam without a seal is usually too great for economy. If the bottom does become quick and lateral resistance at the bottom is lost, the sheet pile will kick in and the cofferdam will collapse. Each such situation can be analyzed by means of flow nets, but normally the penetration below the lowest excavation must be at least one-half the total outside head. Therefore, a seal will usually be found cheaper, safer, and better.

When the cofferdam is founded on clay hardpan or on soft rock where a fair penetration of sheet piling can be obtained and where there is a fairly impervious stratum on the outside, it may be practicable and even desirable to eliminate the seal. A set of bracing should then be installed as low down in the cofferdam as possible as a safeguard against any softening of the bottom and a resulting tendency of the sheet piles to kick in. For example, serpentine is hard and sound until exposed to air, when it quickly slakes to a slimy mud. Sand may be hard packed and dense until continued pumping sucks the fines out and starts a piping action.

Splicing Sheet piles sometimes have to be spliced, although this should be avoided for major cofferdams whenever possible. Probably the best splice is a full-penetration butt weld which has been made in accordance with the best welding practice. Good workmanship is necessary, because a poor weld will fracture under the repeated hammer shocks. Use of low-hydrogen rods is very beneficial in reducing any tendency of welds to fracture under driving. Welding should be confined to the outside of the finger, so as not to deposit metal inside the interlock where it would impede the free movement of the adjoining pile.

It is very important to line up the interlocks perfectly. One jobsite method is to take each side of the two pieces which are to be spliced and thread it into the interlock of a single section. These single sections should extend 5 ft or more on each side of the joint. Splices should not be made at the point of maximum moment, as there is obviously a reduction in strength because of the incomplete welding of the interlocks. Splice plates are usually not required, since tests have shown that the full strength of the flange of a deep-arch section can be developed with the full-penetration butt weld. If necessary, plates can be added on the inside of the web just above the interlock to restore the full section of the pile.

On Z piles, if splices have to be made at points of maximum moment, splice plates on the flange can restore the section. These plates should be located on the inside of the flange so as not to hit the wale in driving. Experience has shown that well-made splices will stand up satisfactorily in both driving and pulling, but sheet-pile driving is a severe test of the quality of a weld.

Extraction Sheet piles are extracted either by pulling the entire pile or by cutting off the sheet pile at or above the mud line and then pulling the top portion. The second alternative, which leaves the lower portion of the sheet pile in place, provides excellent scour protection and prevents undue disturbance to the soils around and below the footing block. When a sheet pile is extracted from clay soils, it is usually found that a clay plug in the arch (or in the web of Z piles) is pulled along with the sheet pile itself. This removes an increment of soil from below the edge of the footing block. In sands, jetting will probably be necessary to accomplish the removal of the entire pile, and this also disturbs the soils below the edge of the footing block. Disturbance of soils in the vicinity of a footing block which rests directly on compact sand or on clay hardpan is undesirable.

Scour, either in a sudden flood or gradually through the years, may lower the bed of a river sufficiently to uncover or undermine the footing block. Dredging for channel deepening or for the installation of submarine pipelines, sewer siphons, or power cables or for the construction of future bridge piers or other waterfront installations may also uncover and undermine the footing block, exposing timber piles to the action of marine borers. For these reasons, leaving the lower portion of the sheet piling in place may often have engineering advantages that greatly outweigh the additional cost.

Pulling of sheet piles is performed with extractors, pulling frames, or vibratory hammers. Table 28-3 lists the extractors commonly used in the United States. These extractors are operated by steam or air. While a good strain must be kept on the line from which the extractor is suspended, an excessively large direct pull on the top of the extractor can cause the extractor casing to be distorted or even pulled apart.

In other words, the extractor itself should do the work. Common practice is to suspend the extractor from a derrick or crane boom and, maintaining a steady pull, let the extractor bring the pile clear of the bottom and of the tremie seal. The second line from the derrick boom can be shackled directly to the sheet pile. When the extractor has done its job, it is removed and the pile is lifted clear. Sometimes jetting will be necessary to aid the extractor.

The HD-10 and HD-15 extractors transmit considerable vibration and shock through the rigging to the boom of the crane or derrick: this may damage the boom or cause it to fail in fatigue. Use of an additional hydraulic damping device may be advisable. Even though these extractors are already equipped with a special shock-absorbing suspension unit, this becomes ineffective if the maximum crane pull is exceeded. These newer extractors are particularly effective in hard clay and on bent or distorted piles.

TABLE 28-3 Characteristics of Steam (Air) Pile Extractors in Common Use for Extraction of Sheet Piling

Make	Model	Energy per blow, ft-lb	Weight of ram, lb	Blows per min	Boiler hp required	Volume of air required, cfm	Max crane pull, tons
BSP-McKiernan-Terry	HD-10	Up to 8,000	1,650	150	60	600	20
BSP-McKiernan-Terry	HD-15	12,000 (at 100 psi) 17,000 (at 150 psi)	3,850	92 (at 100 psi) 130 (at 150 psi)	70	900	20
Nilens	A-4	9,525	5,654	150	50	700	50
Nilens	A-3	6,350	3,102	160	40	500	30
Nilens	A-2	3,175	1,617	160	22	300	25
Vulcan	1200-A	1,640	1,200	530	65	1,200	150
Vulcan	800-A	1,000	800	550	40	900	100
Vulcan	400-A	500	400	550	25	500	50
Vulcan	200-A	250	200	550	18	300	25
McKiernan-Terry	E-4	1,000	400	400	35	600	100
McKiernan-Terry	E-2	700	200	450	30	500	50

The use of an A frame or pulling beam is quite effective. The frame or beam is designed to rest on the next adjacent pile or on a completed portion of the pier, and it is held vertical by one line from the derrick. In some cases, the A frame rests on the bow of the derrick barge.* Heavy blocks are rigged to develop the necessary pull to raise the pile clear of the firm material in which it was embedded. After all the piles or a group of piles have been raised clear, the frame is removed and the piles are lifted out directly. The use of a frame transfers the reaction directly to the sheet piles, so that there is no heavy load on the boom itself. Many serious accidents have occurred in pulling directly off the boom, especially where many parts have been rigged. While there are exceptions, a good general rule is never to pull directly from the boom. In the case of the exceptions, the crane should be equipped with a boom stop, and an experienced rigger and operator must control the operation; even then it is a dangerous practice which has resulted in many accidents.

In some cases of unusually high resistance it has been necessary to use both an extractor and a direct pull from an A frame to the sheet pile (not through the extrac-

* See Sec. 18, Marine Equipment.

tor). In these cases, it is questionable whether the cost of removal justifies the salvage obtained, and it may be cheaper to burn off the pile at the mud line.

Vibratory hammers are extremely effective in extracting sheet piling driven in cohesionless materials. They will often extract piling readily from material through which the same vibratory hammer could not drive it. These vibrators require a hydraulic damping unit to prevent vibration fatigue and damage to the crane or derrick boom.

Removing the first pile is usually the hardest job. Often it is necessary to try several piles before finding one which can be pulled free. The center piles of a long wall are likely to be the easiest, although there are cases where a pile near a corner will come free. When a pulling frame is used, this first pile should be broken loose by driving it down a foot or two and then making an attempt to pull it. This same sequence can also be tried if any subsequent pile gives trouble. Pulling is often a longer and more costly operation than driving, and it is at this stage that care in setting, greasing of the interlocks, and similar foresightedness will reduce costs.

High removal costs can be experienced when sheet piles have been driven hard into firm material. It is often economical to abandon the lower portion, burning it off and salvaging only the easily removed upper section. In this case, it is likely that the extractor will only have to be used on the first pile; the rest can often be lifted clear. The burning must be performed by a diver. The electric-arc–oxygen method is the most efficient, although only a few marine divers are familiar with this equipment. With a good diver, the proper equipment, and an efficient top crew, sheet piles can be burned off under water at a reasonable cost.

One problem encountered in hard pulling, particularly when an extractor is used, is the tendency for the holes in the top of the sheet pile to shear out. Most extractors are equipped to pull with two pins, and both of these should be used in cases of hard removal. The holes in the sheet piles should be burned carefully, and in difficult cases should be burned undersize and reamed to exact size. As an alternative to the use of pins, with its consequent risk of shearing the pin holes during hard pulling, both mechanical and hydraulic clamps or tongs have been developed. Some extractors and vibrators come equipped with these clamps which, in comparison with pins, considerably reduce pulling damage to the tops of the piles. They are also much faster and safer in their operation, and their use is strongly recommended.

It is also possible to invert a double-acting hammer and use it as an extractor. However, it will be less efficient than the extractor which is designed especially for removal operations. The extractor is also handled more readily and expeditiously in the field.

Sealing to Existing Structures Occasionally a bridge pier must be built immediately adjacent and connected to an existing pier or structure. On one such occasion, the designer of the first bridge had been foresighted enough to leave the common wall of sheet piling in place, with a sheet-pile T at each corner. Such foresight is rare, and the problem often arises of sealing a new sheet-pile wall to an existing concrete pier (see Fig. 28-8).

One successful method used to seal against a 40-ft hydrostatic head was to form a cell against the concrete wall by using a Y-section sheet pile. As this cell was excavated, pumps kept an excess head of water inside it to prevent loose sand from flowing in. The cell sides were then jetted clean, and the cell was filled with tremie grout. The three sheet piles which form the cell should be longer than the rest so that cell excavation and seal can be carried well below the level of excavation planned for the main cofferdam.

On another occasion, the joint between the sheet piling and the concrete wall was carefully calked by a diver, using soft wood wedges which were placed as excavation proceeded inside the cofferdam. Although this worked successfully, it is not believed to be as safe a method. It does not seal below the excavation level against the possibility of material flowing in underneath.

Canvas can be used on the outside of the connection to seal the crack above the external mud line, but below this point it would not be effective. Both cement grout and chemical grouts have also been used effectively to seal. Cement grouts have good structural strength but will not penetrate silts and clays thoroughly. Chemical grouts penetrate well and may form a colloidal gel, but they generally have little structural strength. In some instances, the use of cement grout as an immediate structural barrier, followed by chemical grouting behind it, has proved to be the optimum solution.

To seal against a footing block of an existing pier, several methods are again possible. If the sheet piles have their tips hard-faced and if a light, fast hammer is used, the sheet piles can usually be driven up to a foot in depth into the concrete. Subsequent injection of grout into pervious materials just outside the pile tips and at the surface of the footing block should certainly help in making a seal.

The best method is to use a double row of sheet piles, formed as cells if the pressures require it, and to excavate underwater while keeping the water head slightly higher inside than out. The surface of the footing block should be cleaned with a high-pressure jet and all material removed with an air-lift pump. The cells are filled with at

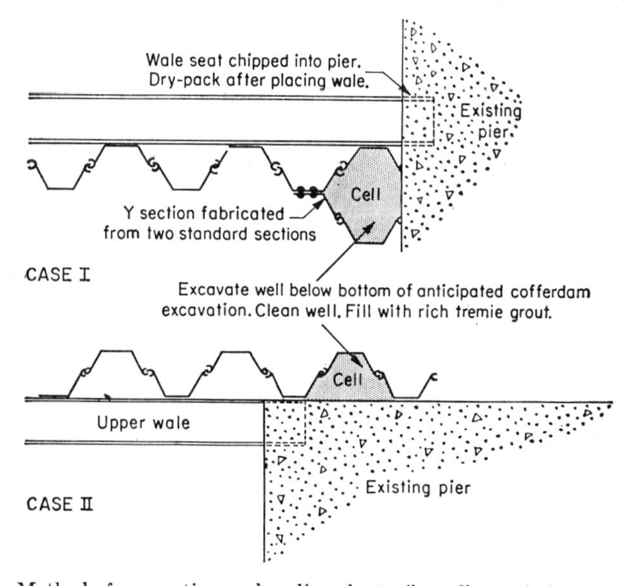

Fig. 28-8 Method of connecting and sealing sheet-pile wall to existing concrete pier.

least several feet of tremie grout and then backfilled. In this method, as in the others, use of a light, fast hammer to chip a little way into the existing footing block is most desirable.

BRACING SYSTEMS

Circular Cofferdams Several very important cofferdams have been built in a circular or nearly circular plan, so that the external pressures will be transmitted around the perimeter. In this way, all cross bracing can be eliminated. The method is an economical and comparatively safe one for constructing very deep cofferdams under difficult conditions, such as swift currents. The cofferdam may be constructed as a relatively rigid continuous wall, such as a heavy concrete cylinder, which is capable of transmitting pressures circumferentially; or it may be constructed as a flexible wall, such as a sheet-pile ring, which is braced by heavy circumferential wales. The condition of equal pressures and true ring action is best satisfied for a cofferdam in water. For a land cofferdam, an imbalance of pressure may result from the soils and from any surcharge, such as construction equipment. Unbalanced pressures will, of course, introduce bending moment into the ring.

Any variation from a true circle in the plan shape of a cofferdam also introduces bending moments in the wall or wales, as the case may be, but these moments can be resisted by the wall or wales if the shape approximates a circle. Thus equilateral polygons are sometimes used, or ellipses.

Circular cofferdams should be more frequently used for bridge piers. This is a fast, cheap, and safe method if the shape of the pier itself approximates a circle, but it is hard to build a rectangular pier inside a circular cofferdam without excessive waste of materials. Bridge designers should therefore keep the possibility of a circular cofferdam in mind when deciding the shape of the pier. The circular cofferdam is probably the only type that it is safe to build if the cofferdam is to be unwatered to very great depths. Because of the elimination of cross bracing, construction inside the pier is expedited. Corner sheet piles and corner connections are also eliminated.

One scheme of construction of a circular cofferdam is as follows. Guide piles are driven in the center, and from these is hung a bracing frame of steel rings to act as wales. Sheet piles are set around the frame, and after the last of these is entered they are driven to grade. Excavation is then completed to grade, and the foundation piles can be driven by placing a small floating pile driver (on pontoons) inside the cofferdam or by using a derrick or skid rig. In any event, there is no cross bracing to interfere with pile driving. Then the tremie seal is poured, the cofferdam unwatered, and the pier constructed. The cofferdam is flooded, the sheet piles are pulled, and the bracing frame is lifted intact for use on the next pier (Scheme 1, Fig. 28-9).

The wales need not be true circles but may instead be short chords rigidly connected at every intersection. Timber and steel have both been used as wale rings, but steel is preferable since it is easier to ensure a full moment connection at the chord intersections.

Another scheme for constructing circular cofferdams is to first drive the sheet piles around a guide frame and then lower a circular steel form to proper grade. This form, which may contain reinforcing steel, is filled with the highest-quality tremie concrete. Excavation proceeds until the next level is reached, and another tremie concrete wale is formed and poured. These concrete wales may be supported on steel H piles, or they may be suspended from the sheet piles. The method is best suited where a deep cofferdam must be built through a considerable depth of unstable sands or muds. In such a case, deep preexcavation is not possible, and bracing must be set as excavation proceeds step by step. As noted previously, pressures during construction stages may be greater than pressures in the completed cofferdam, and therefore more wales and different wale spacings may be required (see Scheme 2, Fig. 28-9).

Several schemes can be used to construct a circular cofferdam with a rigid wall. Forms may be set under water and a heavy cylinder or shell of tremie concrete poured, which may or may not contain reinforcing steel. On large cofferdams of this type, the shell may be poured in blocks or sections. The cofferdam is then excavated and a tremie seal is poured. This scheme can be used only where it is possible to predredge to grade (see Scheme 3, Fig. 28-10).

Another scheme for use in weaker soils is to drive a double row of steel sheet piling in a circle and fill this annular ring with tremie concrete. However, in this method there is a tendency for the sheet piles to bulge as the tremie concrete is poured. In a permanent cofferdam (e.g., where the cofferdam is to be used as a fender for the pier), anchor bars can be welded to the sheet piles before driving. These progressively anchor the sheet piles to the concrete as it is poured and takes its set. Anchors cannot be used where the sheet piles have to be removed.

If the concrete is poured very slowly or in small increments of height, bond may be sufficient to hold the sheet piles to the previous lift of concrete. However, there is always the danger that the new lift will bulge the sheet piles. This can allow fresh concrete to run down between the hardened concrete and the sheet piles, thus building up an increased hydrostatic head and the possibility of a progressive failure (see Scheme 5, Fig. 28-10).

In extremely exposed conditions it may not be practicable to construct the entire cofferdam at one time. In such a case, it may be constructed in units, with each one completed and made stable in itself before the next unit is started. Individual cells of large pipe to which sheet piles have been riveted, or of concrete cylinders with embedded sheet piles, are driven in an intermittent circle, excavated, and filled with concrete. These cells can then be connected with double walls of steel sheet piles, which in turn are excavated and filled. It is essential that the primary cells be properly positioned and driven vertically, although there is some "play" in the connecting arcs of sheet piles (see Schemes 4 and 6, Fig. 28-10).

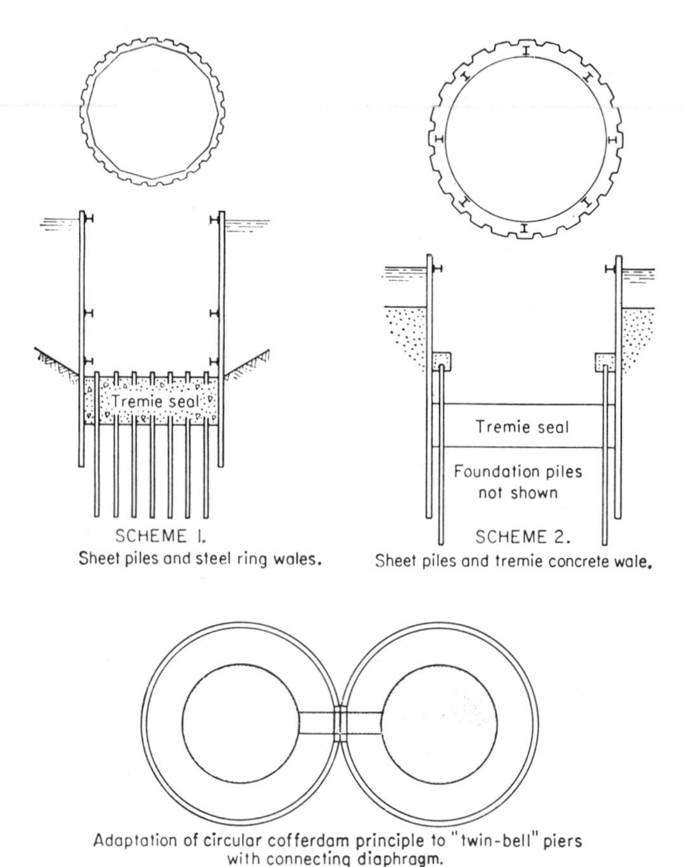

Fig. 28-9 Circular cofferdams.

Successive construction and completion of cells, progressing continuously around the ring, will eliminate the problem of close fit in making the connections. However, to function as a structural ring under high pressures, accurate construction of the cells is essential in any method.

Rectangular Cofferdams The rectangular cofferdam is the type most commonly constructed for bridge piers. Once the loads have been determined, it is necessary to select the section of sheet pile and the type of bracing system. Usually the selection of the sheet-pile section depends on several independent factors such as availability, driving conditions anticipated, length required and the accompanying handling problems, and strength before the seal is poured as well as after dewatering. Similarly,

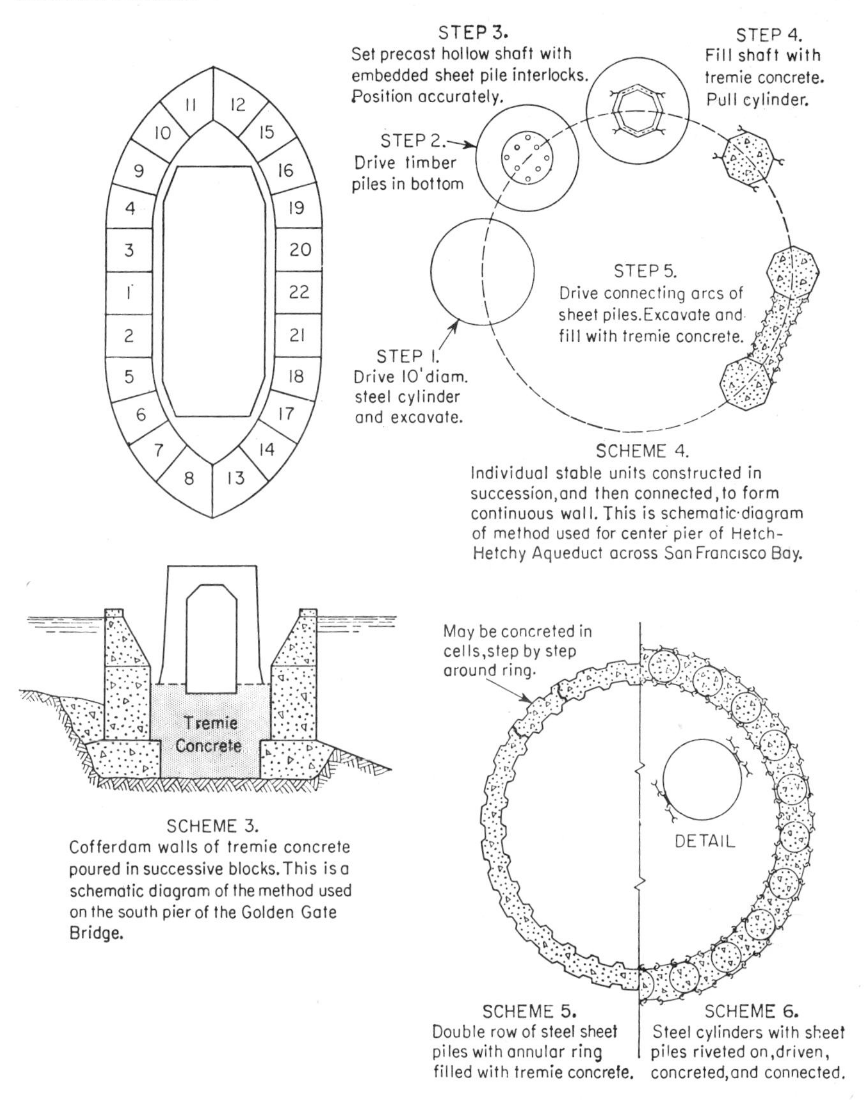

Fig. 28-10 Circular-type cofferdams.

the bracing system must meet other requirements besides initial economy and design strength.

The following factors must be considered in the selection of a bracing system:

1. Elevation of construction joints in finished pier
2. Elevation of foundation-pile cutoff
3. Elevation of tremie seal
4. Outside ground elevation
5. Extreme high-tide elevation
6. Plan dimensions of cofferdam
7. Outline of pier and shafts

8. Reuse required, salvage, removal
9. Weight of bracing for handling purposes

Many large and deep cofferdams have been built using numerous levels of heavily trussed timber wales and struts. Present-day labor costs dictate a simpler design with a minimum of labor in initial fabricating, setting, and removal. For this reason steel wales are normally used today, with timber or steel struts. The bracing system must be laid out so as to provide the least interference with the new structure inside the cofferdam. Consideration should also be given to building, setting, and removing the bracing.

One typical example is shown in Fig. 28-11, where two identical main piers were to be built in successive seasons. Each pier had two shafts, with a clear space between them above the distribution block. Most of the bracing and most of the trussing of the struts were placed in the open space between the shafts, leaving a few single-member struts and a few light angles going through the shafts. After the first pier was completed, the struts were burned off and left embedded in the shafts. Most of the bracing system was salvaged in large, reusable units.

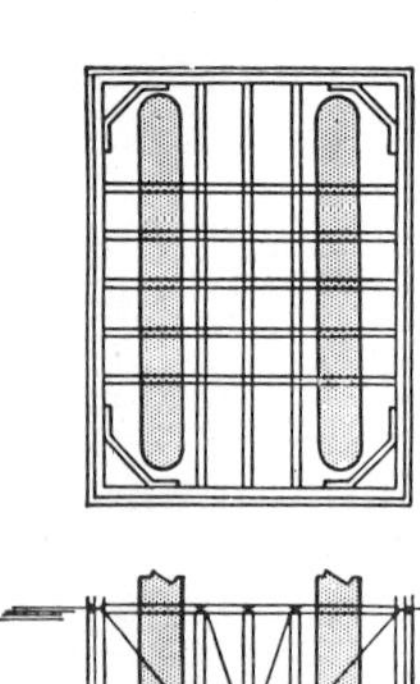

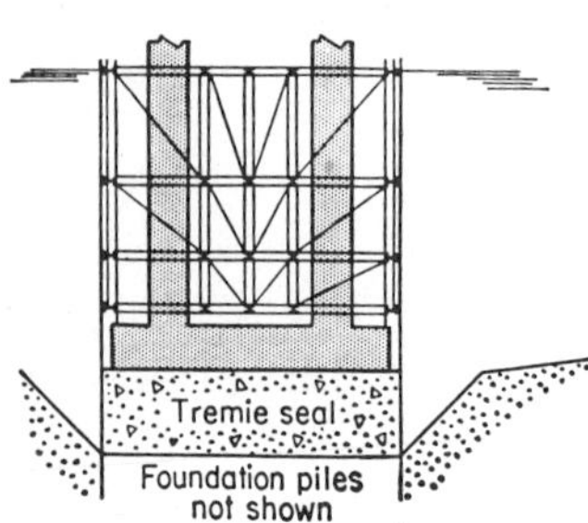

Fig. 28-11 Cofferdam bracing system.

Wales. The number and location of wales can be determined once the loads are known and the sheet-pile section is selected. The theoretical solution is given in many texts[8] and in the publications of the leading manufacturers of steel sheet piling.[9,10] The writer prefers to utilize simple approximations on a trial-and-error basis, since many considerations may influence wale location. A detailed computation is made only as a final check on very deep or unusual cofferdams.

An initial estimate for the required vertical spacing, d (feet), between wale levels at an average depth, h (feet), from the surface may be obtained by successive approximation, using the formula $d = 15\sqrt{S/h}$, where S = the section modulus, in.^{3} per foot of width, as taken from the sheet-pile catalog.

One set or level of bracing is desirable at or near the top, even though the sheet piles theoretically might carry the top hydrostatic loads as a cantilever. This top set is useful in aligning the cofferdam and as a guide against which to set the sheet piles. It provides the skeleton for supporting tremie pipes, pumps, compressors, and a pile template. Also, it resists lateral forces applied at the surface, such as the impact of the floating equipment working at the cofferdam. This bracing is therefore much heavier than would be required by the hydrostatic loading alone. One empirical rule is to use the same size of members that is determined for the second level of bracing, although this rule requires judgment in its application. If the struts are to carry any loads, such as equipment, then they must be sufficiently strong to carry both axial and bending loads and will frequently require trussing.

The usual procedure is to sketch the cofferdam, showing the elevations of high tide, outside soils, and bottom and top of tremie seal. A top set of bracing is tentatively established at the high-tide elevation. Additional levels of bracing are now tried at intermediate elevations, and the approximate bending moments are computed by assuming simple spans between wales. A fairly high bending stress (20,000 to 24,000 psi) can be allowed in steel sheet piling provided that all the loads used in the computations represent maximum conditions. Dividing the maximum bending moment by the allowable bending stress gives the required section modulus ($S = M/f_b$). By varying the trial levels of bracing, the section modulus for each of the spans can be made approximately equal to the tabulated section modulus for the sheet-pile section in question.

The sheet-pile section may have been previously determined by the availability of one particular section, or by driving conditions, etc. But in most cases it is wise to lay out tentative bracing levels for the light and heavy deep-arch sections and for the heavy Z-pile section.

Next, the stresses during construction must be analyzed. Where predredging to grade is performed, the external and internal pressures will be equal until the cofferdam is dewatered. Where predredging is not performed, it is necessary to check the stage of construction when all excavation is completed inside and before the seal is poured. There will then be considerable pressure from the soils outside the cofferdam, acting over the span from the lowest set of bracing to the bottom or below the bottom of the inside excavation. This condition usually requires that one set of bracing be placed at as low an elevation as possible; i.e. the groundline before final excavation is started inside the cofferdam. In many cases there will be a long span on which relatively small loads act, and this may produce a high bending moment and require a heavy section of sheet pile. Here is a condition for which alloy steel sheet piles may be well suited.

An estimate must be made as to how the reaction is taken from the sheet pile by the soil. In typical sands and clays, it may take 2 to 4 ft for the soil to take this reaction. In very soft mud and silt, the required distance may be 10 ft or more. The effective span is thus somewhat longer than the distance from the lower wale to the bottom of the excavation. A short distance below this reaction point, the sheet pile may develop fixity in the soil and thus reduce the bending moment in the lower sheet-pile span. Although more refined analyses may be made by applying the concepts of modern soil mechanics, in many cases a careful evaluation based on experience is adequate to indicate the points for use in computations. Since this is a temporary condition and one not involving dewatering—i.e., one not normally involving risk of life—higher allowable stresses may be used in the sheet piles at this stage (e.g., 70 percent of yield).

Since the bracing frames should preferably be set and tied together as a unit before the sheet piling is set, predredging will often be indicated. Predredging is cheap excavation, reduces soil pressures at all stages, makes pile driving easier, reduces or eliminates the need to jet and clean the arches of the sheet piles before pouring tremie concrete, and enables the bracing frame to be set at the best level to meet other criteria. Where predredging cannot be done or where it is limited by the proximity of structures which might be undermined or the stability of slopes, etc., the level of the bottom determines the lowest level to which the bracing frame can be set.

An alternative is to place the lower bracing under water as the cofferdam is excavated. This is difficult to perform satisfactorily, is hazardous and expensive, and has resulted in many cofferdam troubles and failures. It is usually preferable to use a heavier section of sheet piling to span from a prefabricated bracing frame which is set at or near the groundline. Vertical support must be provided by previously driven "pin" piles or trussing. Extreme care must be taken to ensure that the lowered bracing frame is truly level, well blocked to the sheet piles, and properly secured to the vertical supports and sheet piles.

The next criterion has to do with the structure inside the cofferdam. To facilitate removal of wales and struts and to obviate extensive shifting of bracing in the cofferdam while building the structure, bracing levels should be located just above construction joints wherever possible. Frequently, the lowest level can be located just above the footing block or distribution block. It is sometimes possible to raise the top level of bracing a little higher than would otherwise be required and thus keep it completely above the concrete which must be poured in the cofferdam. In some cohesive and impermeable soils, such as marine clays, it may be practicable to eliminate the tremie seal. In this case, the cofferdam may be kept full of water during excavation and a set of bracing may be lowered to the bottom, i.e., below the footing, before dewatering. In this case, of course, the lower set of bracing is not salvageable.

In designing the struts, their action as columns must be considered. Where vertical posts are not practicable, trussing of the several sets can supply the needed strength in the vertical plane. This may require pairs of bracing levels instead of single heavy sets, even though the sheet pile might take the bending with single sets. To use an extra level for this purpose is not too serious since the total load which must be carried is the same. The undesirable features are the extra fabricating cost and the interference during construction; the benefit is the provision for ample support of the struts.

Some early constructors preferred to use more sets of bracing than were required by the sheet piles, so as to equalize the load and to provide safety if one set should buckle or fail. However, even with the additional sets of bracing, it any one set is not properly designed or constructed there is the possibility of progressive failure. It is better practice to compute the loads properly and design and build each set of bracing adequately.

In computing bending moments, for selection of the proper section of sheet piling and for locating wale levels, an assumption is commonly made of simple spans between wales. This assumption is used by many engineering authorities and most cofferdam constructors. For example, see Jacoby and Davis, *Foundations of Bridges and Building.*[8] The *Steel Sheet Piling Design Manual, 1969,* which is issued by the United States Steel Company,[10] contains similar approximations. However, it uses $wl/10$ instead of $wl/8$ in computing sheet-pile bending moments in cofferdams having several sets of bracing.

Obviously the steel sheet piling is continuous over each wale and is partially or wholly fixed at the bottom. Before the seal is poured, the points of support and fixation are an indefinite distance below the surface; after the seal is poured, the point of support is at the surface of the seal, and the point of fixation may be only a short distance below. Even the top wale may not be a simple support unless it is above maximum high tide. Thus, questions arise as to the validity of the assumptions commonly made. Numerous comparisons between the approximate method and more detailed computations give roughly the same moments except that they are opposite in sign. That is, the maximum moment by the approximate method of assuming simple spans is computed as a positive moment between wales, whereas the more accurate computations show the maximum moment to be a negative moment over the supports. The numerical answers are roughly the same by either method. Since a steel sheet-pile wall has equal resisting moment either way, both methods lead to a proper selection of sheet-pile section and wale spacing.

The wale section should be a heavy, sizable, approximately square member which is adequate to resist the bending and shear between struts and the crushing and localized concentration at the struts. The reason for a roughly square section instead of a more efficient beam section is to prevent rolling since, as the sheet piles are driven past, there is a tendency to roll or twist a narrow wale section. The reasonably rugged wale section is specified in order to resist localized crimping and damage, such as a blow from a clamshell bucket in excavating. The allowable bending stress in wale design should be kept low (18,000 psi in steel) since the wale will probably carry some axial compressive stress also. The allowable shear stress in steel wales should be 16,000 psi. With steel wales, the beam must be adequately stiffened opposite the strut. Stiffeners can be welded between the flanges.

For timber wales, bending and horizontal shear must both be checked. Allowable stresses and methods of computation of horizontal shear should be those recommended by The American Institute of Timber Construction *Timber Construction Manual.*[11] The wales will be completely soaked at the time of dewatering and intermittently wet thereafter; so allowable stresses must be reduced accordingly. On the other hand, the time of loading is of short duration, a few months at the most; so if new, properly graded timbers are used, it is common practice to use the allowable stresses for timber "occasionally wet but quickly dried." Frequently cofferdams are built with old timbers, which seems penny-wise and pound-foolish, since relatively small quantities are involved and it is only wise to use the best timbers obtainable. If old timbers are used, allowable stresses must be reduced considerably.

Coast Douglas fir is the most widely used timber for cofferdam bracing. Final design should be based on the AITC *Timber Construction Manual* and the recommendation of Standard 102-62, National Design specifications for structural timber. Tentative design may be based on an allowable bending stress of 1,000 psi [for construction-grade timber (1350f)] and 72 psi horizontal shear. The horizontal shear may be computed at a distance from the centerline of the support equal to 1½ times the width of the strut.

In timber, crushing of the wale by the strut is a most frequent cause of failure. The allowable stress in bearing (compression perpendicular to the grain) for Coast Douglas fir, occasionally wet, is 240 psi, and where continuously damp or wet, 213 psi. These values apply to both select and common grades and, since experience has shown this bearing to be a source of trouble, 200 psi would be a good value to use. Usually a steel plate will be required, bolted to the wale, against which the strut may bear. The steel plate must be thick enough to resist bending and shear. Corbel blocks of hardwoods may also be used, since these have higher allowable bearing stresses. Oak, for example, has an allowable bearing stress of approximately 350 psi.

In computing the bending stress in wales, approximate methods are again usually satisfactory. The wale may be considered as a partially continuous ($wl/10$) beam between struts. The best way to compute the load per linear foot is to draw a pressure diagram, consisting of rectangles and triangles, from which the load coming on any particular wale can be quickly computed.

With steel wales, the intersection problems are greatly simplified. The corner must be designed to transfer thrust both ways since the wales will carry axial stress, acting as end struts. To make a rigid corner with a full-moment connection, a short, diagonal H beam with stiffeners between the flanges of the wales and the flanges of the diagonals may be used, supplemented by cover plates if necessary. This type of corner helps to reduce the bending stresses in the wale at the location of greatest axial stress.

In timber, the end connection may be more difficult. If the end wales are at the same elevation as the side wales, then a separate additional end strut is used. This is bolted to the end wale and bears against the side of the side wale. The side wale is cut short so as to bear against the full-length end wale. The end strut may also be shortened into a kicker block and bolted with shear connectors to the end wale. Bearing perpendicular to the grain must be carefully checked.

Where timber struts are used, and sometimes with steel struts, the longitudinal struts and end wales are placed just above the transverse struts and side wales, or vice versa. This arrangement eliminates strut intersections, which are particularly difficult in timber because of the limitation on bearing perpendicular to the grain. To make an end connection in this situation, kicker blocks are used. These may be steel beams, say 10 ft long, which are welded or bolted to the other wale; or timber beams which are bolted to the other wale with shear connectors.

Struts. These are essentially columns, and the main problem in design is to provide adequate lateral stiffness and support in both the vertical and horizontal planes. Lateral support can usually be provided by intersecting struts in the other direction, which makes the struts act as short columns in this plane. Vertical support is difficult to furnish and is also the most necessary because of the dead-load deflection of the member itself. When a single set of bracing is used and is above water, properly located piles can be used to support the struts. This vertical support cannot be removed until it has been replaced by a support on the structure itself or until the cofferdam is flooded again. This comment seems unnecessary, but during the construction of almost every cofferdam someone will propose the temporary removal of a strut or the notching of a wale or the burning of a post to facilitate forming or pouring concrete. Underwater struts can be supported by connecting several sets to a vertical pipe post. A steel pile can then be driven through the pipe and bolted or wedged to its top.

Where several sets of bracing are involved, common practice is to truss the struts of two or more levels in a vertical plane. With steel struts, these trussing members are usually angles. With timber struts, timber diagonals and steel tie-rod verticals make a Howe truss. Since the trussing must be provided for struts in both directions, quite a lot of truss bracing is required in large cofferdams. This makes forming and pouring more difficult and also increases the risk of accidental damage by a dredging or concrete bucket.

For these reasons, strut members should be selected with as large a radius of gyration as possible. In general, the depth of section should be approximately the same as the width of the wale against which the strut will bear. Steel H beams or wide-flange beams usually work out best. If the struts in one direction are detailed so as to lie just above the struts in the other direction, then the H beams can be installed with their webs vertical to give them maximum strength against dead-weight bending.

If, on the other hand, it is desired to have struts from both directions lie in the same plane, then the webs should be horizontal so as to simplify strut intersections. Stiffeners are essential at these intersections. (See Fig. 28-12.)

Fig. 28-12 Timber strut intersection with steel wale.

Pipe columns have excellent radii of gyration in relation to their weight. If welded tight, they can be used to provide buoyancy and thus reduce the weight of bracing once it is under water. The intersection details are difficult, since full steel area must be provided across the intersection. Square tubing is an excellent section for struts, although adequate stiffener details are required at intersections.

The design loads are those due to the external pressure which is transmitted through the wales, plus an allowance for possible impact from a dredging or concrete bucket. The most common way to take care of the impact loads is to use conservative stresses in the design of the struts and to provide more than adequate trussing, etc. Vertical sheathing of wood planks is sometimes hung on the struts, to guide buckets and protect the struts. In any event, care in bucket handling, not speed, must be emphasized. Some designers actually compute the stress arising from the impact of a bucket, but this is only a guide and not an exact answer.

Diagonal struts are frequently used at the cofferdam corners to reduce obstructions within the structure and to shorten the required length of strut. These diagonals carry an increased stress because of their inclination (1.4 times the normal load if on a 45° diagonal). Their reaction must be resisted by an adequate shear connection to each wale, and this is usually provided by a kicker block. This also increases the axial stresses in the wales.

Strut intersections must be carefully detailed to prevent crushing or buckling of flanges. Adequate stiffeners must be installed at all strut intersections and at all strut-wale intersections.

Ties. Frequently, a cofferdam must resist outward pressures during some stages of construction. One source of outward pressure is a high tide which hangs up longer inside the cofferdam than outside; this problem can be eliminated by a tide gate in the sheet piles. However, in a location exposed to long swells or surges, the tide gate may not act rapidly enough and external ties may be necessary. Above water level these ties can be simply provided by bolting every other sheet pile to the wale and making the strut also act as a tie. Below water, if ties are needed, special details may be necessary and can involve a considerable amount of diver work.

Outward pressures on a cofferdam are encountered when making a pour of underwater concrete that will extend well above the mud line. These pressures can be countered by backfilling against the sheet piles with sand prior to pouring the concrete, but this may give too great an inward pressure against the sheet piling. If backfill cannot be placed before pouring concrete, then it will have to be placed during concreting and coordinated with it. Large cofferdams are often tied with cables at the top to help maintain their alignment. Some small bridge piers have been constructed under water by using a sheet-pile cofferdam which is excavated and filled to its top with underwater concrete. The outward pressure of the fresh concrete on the sheet piles is resisted by outside wales.

Setting Bracing In large and deep piers, the bracing systems become very heavy. Wherever possible, the bracing frame should be completely fabricated and set as one unit. Two derricks can be used when the weight is beyond the capacity of one. The frame may be hung on temporary piles, as shown schematically in Fig. 28-13. After the sheet piles have been set and driven, the load of the bracing system can be transferred to the sheet piles.

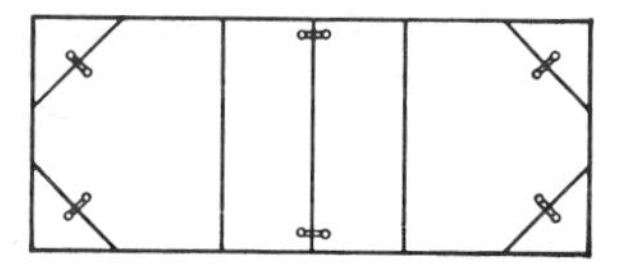

Fig. 28-13 Temporary support for prefabricated bracing frame.

When the bracing frame becomes too heavy to be handled with derricks or when heavy-lift derricks are not available, the bracing frame can be erected in units. Each level of wales and struts is bolted or welded together and hung from temporary piles. The uprights and trussing angles which will extend upward to the next lift are fastened on, and the unit is lowered until the uprights are just above water. The next unit is then erected, and again the completed portion is lowered. This procedure is continued until the complete frame is in place. Alternatively, the bracing can be fabricated into vertical sections (space frames) and bolted together under water by divers. For the anchorage cofferdam on the Mackinac Straits Bridge, which was 115 by 135 ft in plan and 88 ft deep, the bracing was fabricated into three sections, each 115 by 45 by 75 ft deep and weighing 107 tons.

The lowering is accomplished either by several heavy hand winches, reeved up in sufficient parts to carry the weight of the frame, or by long, threaded rods and nuts which allow the unit to be lowered by screwing. Each winch may be supported on a short cap over two falsework piles. Threaded rods, if used, must come up through the cap to a steel bearing plate. The nuts bear on this plate while they are turned. Figure 28-14 shows a well-designed bracing system.

Many ingenious schemes utilizing flotation can be employed. Certain sections of the bracing system can be completely boxed in and made watertight, although care must be taken that they do not unduly impede pile driving or concrete placement. Combinations of timber with steel bracing are effective in reducing the submerged weight. Hollow, watertight box girders can be used for wales, and pipe can be used for struts if proper details are employed at strut intersections. These means are most frequently employed when lowering the units in stages but, except in inaccessible locations or with extremely large bracing systems, elaborate methods of utilizing flotation are usually more costly than direct handling and setting.

The bracing system can also be erected between two properly spaced and anchored barges. It is then lowered in stages from winches on the barges. A floating timber raft can also be moored over the site and the bracing system erected thereon. The raft is allowed to sink under the bracing until it is supported from falsework piles, following which it is cut into sections and removed.

Timber and steel bracing cribs have been built on a launching ways or in dry dock and then launched and floated to the site. After positioning, they are sunk by weighting until they are hung from temporary supporting piles. The steel sheet piling is then set and driven.

Removing Bracing Where many similar piers are to be constructed in sequence, the bracing frame may be so laid out as to permit its being removed wholly or largely intact after the sheet piles have been pulled. In these cases, and depending on the bracing weight, it may be lifted up over the structure with the aid of one or more derricks.

Fig. 28-14 Well-designed bracing system for deep cofferdam, showing corner details and tie-rod supports for struts. Note exhaust from gasoline-engine-driven pump led up outside cofferdam to prevent concentration of carbon monoxide in cofferdam.

Most specifications permit steel struts (and usually timber struts also) to be left in the structure below the waterline. Such struts may be boxed by forms for the outside 6 to 12 in. and then burnt back after flooding the cofferdam. The boxed cavity is subsequently filled with an underwater-setting epoxy grout or hydraulic cement. Alternatively, and much cheaper, the exposed steel may be cut flush and thoroughly coated with epoxy. Where specifications require the removal of struts, they may be completely boxed out in the structure, burnt off under water, and pulled out sidewise after all the other bracing has been removed.

A method commonly used in the past, and one which is still necessary occasionally, is to remove the bracing piecemeal before flooding the cofferdam. In this case, after the footing or distribution block is poured, it is carefully blocked to the sheet piles and the next higher set of bracing is taken out. After the next pour, the pier is again blocked to the sheet piles and another level of bracing removed. This method requires a great deal more labor, consumes valuable time in a deep cofferdam, and increases the risk of carelessness or human error. For example, the bracing frame is usually rigid and husky enough to withstand the impact of a dropped beam, whereas the blocking may not be capable of taking a load except in its principal lateral direction. Also, since the full or substantially full pressures remain on the cofferdam until it is completely flooded, the necessity of placing the blocking below the level of the brace to to be removed means that the stress distribution will be different. Complete analysis and calculations are required for each loading condition, and additional bracing may be needed.

EXCAVATION

Predredging Whenever practicable, it is usually best to predredge the site to or just below the elevation of the bottom of the tremie-concrete seal. When the material at this bottom elevation is still soft, excavation is frequently carried a few feet further down to permit placing a gravel or sand-and-gravel blanket. This blanket provides a level base for the tremie-concrete seal and, if of adequate thickness, will also provide lateral support to the sheet piles in the critical stage before the seal is poured. Predredging is usually performed by a clamshell dredge, although hydraulic dredges have also been used.

Predredging quantities may run three times the "pay quantity" inside the cofferdam, but this type of excavation usually costs less than one-third as much as excavation through bracing. It facilitates other operations such as setting the bracing frame and pile driving, but it may raise a problem of outward pressures with deep tremie seals.

One common sequence is to predredge to an elevation 3 ft or so below grade. Using a floating driver, the foundation piling is then either driven to or cut off at approximate pile cutoff grade. A sand blanket is then placed, the bracing frame is set, sheet piles are set and driven, the tremie-concrete seal is poured, and the cofferdam is unwatered. This typical sequence is shown schematically in Fig. 28-1. Where it is impracticable to predredge to grade, it still is desirable to predredge down to an elevation below the bottom of the bracing frame.

The selection of type and size of clamshell bucket depends on the material encountered, and thus a change in bucket may frequently be required as different strata are reached.* In muds and silts, a bucket without teeth may suffice. In hard clay or sandy clay, teeth may be required and the bucket may have to be weighted. The most difficult material is densely packed sand, which is hard to dig and yet tends to run out of the bucket. On occasion it may be economical to loosen it by jetting ahead of the digging. Boulders and old riprap and logs are extremely difficult to remove and require patience. One good way is to dig one pocket considerably ahead of the remainder, then work back from this face.

Excavation Inside Cofferdam After predredging is completed, the bracing frame is put in place and the sheet piles are set and driven. This excavation can be performed with clamshell buckets, working through the frame. Care must be exercised to prevent damage to the frame by the bucket. Vertical sheathing on the bracing frame, struts, and wales may be of help in this regard.

The clamshell bucket cannot dig close to the sheet piles or under the wales and struts. A long chisel can be run up and down the sheet piling, and jets can be used to break down a wall of clay. Care should be exercised to avoid running chisels and jets below the bottom of the proposed seal. When excavating in the dry, a small crawler-mounted dozer or loader may be lowered into the bottom of the excavation. This equipment can clean beneath struts and work the material to the clamshell bucket (see Fig. 28-15).

A small hydraulic-dredge suction ladder may be hung over the side of a cofferdam and silts and sands worked to it by means of jets. In very small cofferdams, or when excavating nests of cobbles, an orange-peel bucket may be useful.

Dredging should be carried 1 to 2 ft below grade and the chiseling and jetting done thoroughly at the sheet piles and especially at the corners. Then a diver should

* See Sec. 10, Excavators.

usually be sent down with a jet to clean the arches of the sheet piles thoroughly, since many cofferdam blows have occurred where an arch was left full of clay. The bearing piles are then driven through the bracing, and a sand and gravel blanket is placed to bring the bottom to grade.

In discussing sheet-pile penetration and bracing design, it was pointed out that one of the critical stages occurring during the cofferdam construction is after excavation is completed but before the seal is poured. With a soft bottom, there is little lateral support for the sheet piles at that elevation. One solution has been to predredge a few feet below grade and place a sand or gravel blanket, which acts as a firm stratum.

Soft, soupy mud which is left in the bottom after the main clamshell dredging should be cleaned up, using an air-lift pump.* Success and economy in air-lift pumping can best be achieved by using an excess of air, and often a diver with a jet can wash material to the pump intake. Also, where piles have to be driven before final excavation, an air-lift pump assisted by a diver with a jet can excavate safely between the

Fig. 28-15 Excavating through bracing with clamshell bucket.

piles. The cost is considerably higher per cubic yard than other methods of excavation but, if the volume to be removed is small, this extra cost of air lifting may be more than offset by savings in other factors or by greater safety in the critical operation. Air-lift excavation may also be desirable where running or fine sand is encountered. The piles must be thoroughly jetted clean of all mud and clay in the zone of the tremie seal.

The air-lift pump is a relatively simple and effective tool for removing granular material of all shapes and sizes, from silts and sands up through gravel and cobbles. Its efficiency is not affected appreciably by depth and, in fact, may be somewhat better at greater depths. An air-lift pipe generally runs from 8 to 14 in. in diameter. It consists simply of a pipe with an elbow at the top to direct the discharge and an air fitting at the intake.

Compressed air is supplied by a smaller pipe, 1½ to 2½ in. in diameter, fastened to

* See Air Lifts in Sec. 29, Construction Dewatering.

and running down alongside the discharge pipe. At the intake, the air is fed into the discharge through an annular chamber and many small ports. These small ports break up the entering air into many small bubbles, well distributed throughout the column of water inside the pipe. The aerated water inside the pipe column flows upward due to its low unit weight and carries solids with it. As large pieces (e.g., cobbles) are sucked in, their tendency to plug the pipe increases the force moving them; consequently, rocks have actually been discharged that appeared to be the same diameter as the air lift. In protracted use, care must be taken to see that the intake holes or ports do not become plugged by small particles, etc.

Air requirements are large. While a 10- or 12-in. air lift will function with as little as 500 to 600 cfm, the optimum air supplies are:

10-in. dia	800 cfm
12-in. dia	1,200 cfm
14-in. dia	1,600 to 1,800 cfm

A common problem is keeping the material fed to the air-lift intake. Frequently, it is found that a diver using a jet is the most effective means of moving the material to the air lift.

PILE DRIVING WITHIN THE COFFERDAM

Where the pier site is predredged, the piling can be driven with an underwater hammer in telescopic leads or driven as above-water piles and cut off to grade. If the pier site cannot be predredged, it is almost always best to excavate the cofferdam before driving the piles. In the very unusual case where piles are driven before excavation is completed, excavation will have to be performed with an air-lift pump as noted above. Depending on the pile concentration and the type of soil, some overexcavation will be necessary to compensate for swell.

Several methods are possible for driving the piles within a constructed cofferdam:

1. When using a skid rig, the top level of bracing may be supported by timber falsework piles and the skid rig moved on these timbers, or external supports may be provided with girders spanning the cofferdam.

2. It has been practicable to use a floating pile driver inside very large cofferdams. In this case, the top level of cross bracing has to be placed later. The floating rig can be of the pontoon type, or one wall of sheet piling can be driven down to or cut off at low tide; the driver then floats in and out at high tide. After driving is completed, the low wall is extended up, top bracing placed, and work proceeds.

Great care must be taken where driving through bracing to make sure that it is not damaged. A careful inspection by a diver after completing the driving would be prudent.

3. A derrick or crane can set the piles through a template resting on the bracing. Then the template is removed and the piles are driven down, using an underwater hammer or using piles of sufficient extra length to keep their heads above water level.

In laying out the plan dimensions of the cofferdam and the bracing, consideration must be given to the pile-driving methods so as to eliminate interference and allow enough room for the hammer and leads to clear. This often means that the line of sheet piles must be laid out 6 in. to 1 ft wider, so that the external row of bearing piles can be driven. Where congestion is particularly serious, as it often is in relatively deep cofferdams containing both vertical and batter piles, it may be practicable and economical to drive the piles in lengths long enough to extend above water. Then, after the seal is poured, the piles are cut off. These relatively long sections can be used to splice piles for subsequent piers.

The outside row of bearing piles, if battered outward, may intersect the sheet piles. One solution is to make the outside row vertical and batter the second or third rows. Where this is not permitted, it may be necessary to move the sheet pile walls back a sufficient distance to clear. In some instances it may be practicable to leave the particular sheet piles involved a little high so as to clear, but this means less penetration and the possibility of a "blow" of material underneath.

Jetting of piling within a cofferdam can be very dangerous as it may cause sands to run underneath, destroying the lateral support at the bottom of the sheet piling. Where necessary, a set of bracing may be placed on the bottom after excavation just to hold the toe of the sheet piles. This set should be prefabricated a few inches smaller in plan dimension than the cofferdam and hung underneath the bottom of the bracing frame. After excavation, it can be lowered down to the bottom and wedged to the sheet piles by a diver.

BOTTOM SEAL

Functions The functions of the tremie concrete seal are:

1. To form the bottom of the cofferdam box with a seal that will withstand the uplift pressure from hydrostatic and soil forces
2. To provide a rigid support for the sheet piles at the bottom

In modern design, the seal is frequently used as a part of the permanent structure as a distribution or footing block, for weight in uplift, and for load transfer to the piles.

Excellent reinforced concrete can be constructed under water by controlled high-quality tremie methods. A great deal of the cost and the risk of deep cofferdams can be eliminated by a design that permits underwater tremie-concrete construction of the entire footing block. Provision must then be made for placing the reinforcing steel in welded cages hung from the bracing, and piling must be cut off to proper elevation under water. The savings on a deep cofferdam may then run 20 to 50 percent by allowing a lighter section of sheet piling, eliminating the lowest and heaviest set of bracing and, in many cases, facilitating salvage and reuse of the bracing system as a whole. Time of construction is reduced, and risk is materially lessened.

Resistance to Uplift The simplest case assumes that the uplift will be equal to the hydrostatic head; so a thickness of concrete seal is specified that will give a weight equal to the water head at the bottom of the seal. With sea water, this thickness of seal becomes $^{64}/_{150}$ or 43 percent of the water depth at the bottom of the seal. The sheet piles below the seal act as a cutoff wall and, if there is no completely impervious stratum cutting off water flow, full head will ultimately develop under a tight seal. However, the placement of a gravel layer and a few bleeder pipes under the seal will relieve much of this pressure.

The sheet piles themselves may have considerable frictional resistance to uplift. If their sides and arches have been cleaned before the seal is poured, they help to resist uplift on the seal. However, this assistance by the sheet piles is not normally considered except in a trench-type (narrow) cofferdam.

Foundation piles are very effective in resisting uplift. Timber piles taper, so that any tendency for the seal to lift is transferred directly to the pile by wedging as well as through bond. Peeled piles must be used for this purpose in order to develop skin-friction resistance to uplift in soil and to bond with the concrete. In very pervious soils, some leakage may occur through cracks or checks in the timber piles, but these are generally not serious.

With concrete piles or H piles, the uplift can be transferred through bond alone in a few feet. On a cleaned steel sheet pile, 80 psi has been observed without failure and 30 to 40 psi can be used for design. Unless it is thoroughly jetted clean by a diver, no bond at all can be allowed on a pile which has been driven before the zone of the tremie seal has been excavated. Algae and other marine growths can form quickly under proper conditions of salinity, warmth, and light, even at considerable depths in somewhat muddy water, and these reduce bond unless cleaned off.

The uplift value of the piles depends on the skin friction developed plus the dead weight of the pile itself. Common practice often allows uplift values of 50 to 80 percent of the allowable bearing capacities, but this should be more carefully investigated on any critical project. Often the limiting factor is the ability of the tremie concrete to span between piles, and this has been empirically set at about 4 to 6 ft in practice, although here again actual analysis should be made. Obviously a thicker seal will allow greater spacing of the piles.

Unless special conditions require otherwise, present practice on major bridge-pier cofferdams is to use a combination of the dead weight of the seal with uplift on the foundation piles and not to try to make the seals too thin. Thus, seals with a thickness of 15 to 20 percent of the pumped-out head are common.

Seal Construction by Bucket Method Bottom-dump buckets can be used only where no foundation piles project above the bottom. The bucket must be watertight and must be covered. It is lowered gently to the bottom and the gate is slowly opened, allowing the fresh concrete to flow out smoothly and uniformly.

This method has been widely used in the past, but it is not too satisfactory for cofferdam seals. As each load is deposited and the concrete flows out, some laitance is formed on the surface. The succeeding load may displace this upward, or it may trap it. In any event, experience has shown that an excessive amount of laitance is often formed by this method. In some cases, the depth of laitance above the concrete has actually been equal to the depth of seal itself.

Improved buckets have been developed by leading manufacturers, and the trend is to large buckets (up to 6 cu yd). The larger bucket and fewer loads result in less exposed surface and less laitance. If this method is used, it is believed that a slope might well be kept across the concrete surface and an air-lift pump operated at the low end of the slope to remove as much laitance as possible. It is especially important not to trap all the laitance at one end or one corner.

Seal Construction by Tremie Method The tremie method is currently the accepted standard for high-quality structural underwater concrete.[12] Tremie pipes are used so that the fresh concrete is always deposited under the surface of the previously deposited concrete. In this way only one surface is ever exposed to the water, and that surface is gradually raised to the top.

The tremie pipe is plugged at its lower end and, for depths of water up to 70 ft or so, the best plug is a sheet of 1-in. plywood with a rubber gasket. This plug is fastened to the end of the pipe by heavy twine. The plugged tremie pipe is lowered to the bottom and charged with tremie concrete to about half the water depth. It is then raised about 12 in. off the bottom, breaking the twine and allowing the concrete to flow. For deeper pours, where the buoyancy of the plugged tremie pipe presents problems, it is advisable to replace the plug by an inflated rubber ball which is wedged in the top of the pipe. The pipe is lowered to the bottom and slowly filled with concrete, the weight of which forces the ball down and out the bottom end of the pipe. The concrete flows out over the base, and the ball floats up and can be recovered and reused. From this time on, the lower end of the pipe is kept continually immersed in the concrete.

Experience has shown that tremie concrete sets in about 1½ to 2 hr in normal water temperatures and with normal cements. The pipe end must be kept in the freshly placed concrete which has not yet taken its set but, on the other hand, it must be kept embedded deeply enough so that the tremied concrete will not all run out and the seal be lost. The proper depth of pipe immersion will depend on the rate of pour and the mix, the water temperature, the time of set of the cement and, of course, the height of the concrete column in the tremie pipe. Rates of pour in standard practice generally run from a minimum raise of 18 in. per hr to a maximum of 6 ft per hr (to avoid excessive fluid head). The general average is 3 ft per hr.

Many specifications erroneously require that the tremie pipe be kept full to the top. With even moderately deep water, this might require the pipe to be embedded so far into the tremie concrete that the new concrete pushes up underneath concrete which has already taken its set. A proper balance of the concrete column in the pipe with depth of embedment will give better concrete and less laitance.

Sometimes a valve at the bottom end of the tremie pipe is specified; however, this seriously disturbs the flow of the concrete and causes segregation and laitance. A smooth and gradual restriction in the bottom of the pipe could be used on exceptionally deep pours, but the usual practice is to keep only a workable height of concrete in the pipe. New concrete dropped down the pipe will then remix before flowing out.

Plasticizing and retarding admixtures, such as Sika Plastiment, and air entrainment are very beneficial in preventing segregation, reducing laitance, ensuring a more level surface, reducing heat of hydration, and permitting a deeper embedment of the tremie pipe. Retarding admixtures increase the form pressures.

The loss of seal always results in the formation of laitance and should be avoided whenever possible. If the seal is lost, the pipe must be raised, replugged, lowered into the fresh concrete, recharged, and then slightly raised to start the flow. Use of the ball, while very convenient in that the tremie pipe does not have to be raised clear, does result in washing the fresh concrete by jet action as the ball is forced down the pipe.

Proper mix, proper workability, and gentle, substantially continuous placing are essential for obtaining the best results. Tremie seals can be and have been poured with practically no laitance whatsoever.

Gravel, not crushed rock, should be employed as the aggregate for this tremie concrete. The top size should be small, and it should be well graded. The United States Navy Bureau of Yards and Docks recommends $\frac{3}{4}$-in. maximum, while other agencies recommend $1\frac{1}{2}$-in. maximum. The writer has tried $2\frac{1}{2}$-in. rock, but some of the larger pieces wedged in the 12-in. tremie pipe, the seal was lost, and excessive laitance was formed.

Experience dictates the use of 42 to 45 percent sand, with 40 percent minimum. A rich mix is best, with a minimum cement content of $6\frac{1}{2}$ sacks per cu yd and a recommended amount of 7 to $7\frac{1}{2}$ sacks. Some even prefer 8 sacks. Slump should be 6 to 8 in.

Placing should be through good-sized tremie pipes, from 8 to 16 times the size of maximum aggregate. Pipes of 10- and 12-in. diameter have been very successfully used. Since the pipes must be raised gently and uniformly, some method of gradual, continuous hoisting should be provided. Pipes and hoppers can be raised by a derrick and blocked up or tied off, but this may tend to be jerky. The best way is by the use of air hoists.

On deep seals, the pipes may have to be raised so far as to present a problem in handling, supporting, and filling. This was solved on the tremie seal for a dry dock by having side gates, which could be unbolted so that concrete could be placed into the pipes at a relatively constant level. The most common solution is to have removable sections in the tremie pipes. Some means of holding the lower end of the pipe during this operation must be provided. Telescoping pipes have also been designed and used. Underwater joints must be gasketed and tight to prevent water leakage.

In placing tremie concrete, any and every disturbance produces laitance. Pipes must be raised slowly and evenly. Flow must be kept as uniform as possible. An ideal situation would be to have a continuous flow into the hopper and out the bottom. No disturbance of the surface may be permitted, and a diver should never be allowed to walk on the concrete surface until after it has taken its set.

Tremie-pipe spacing depends on depth of pour, practical limitations due to size and configuration, obstructions such as piles, slump, and whether or not admixtures are used. Formerly, specifications limited spacing to 10 to 18 ft but more modern practice is to use larger spacing, up to 20 and 30 ft on pours 6-ft thick and thicker. Excellent tremie concrete has been poured with one pipe in a circular form 34 ft in diameter. Some fairly large cofferdams with deep seals have even been poured with only one setting of the tremie; however, this can create a situation of overfalling and thus create pockets and segregation. For general practice, therefore, 15 to 30 ft would appear to be the normal range. In a large cofferdam it may be necessary to shift pipes forward as the pour progresses. In this case, the pipes should be resealed and set gently into the advancing slope. As with the bottom-dump bucket method, care must be exercised not to concentrate all the laitance in one corner or end. An air-lift pump may be beneficially used to remove it ahead of the advancing concrete.

Experience on the Chesapeake Bay piers indicated that most of the laitance is formed at the very start of the pour as the concrete flows out over the bottom. Experience has also shown that the amount of laitance is dependent more on the area of the seal than on its depth. Piles impede and disturb the flow and increase the amount of laitance. This is especially true of H piles, because of their shape.

The slope of the tremie-concrete surface will usually run 1:12 to 1:6. Piles will cause a steeper slope, as will a stiff or unworkable mix. For these reasons, thin tremie seals are not desirable. Even in shallower cofferdams, 3 ft should be the practicable minimum, and 5 ft is preferable. Because of slope, laitance, and inaccuracies, 6 in. to 1 ft of depth should be discounted in determining the weight and strength of seals. Thus, when a very thin seal is specified, good practice would be to start the seal 6 in. to 1 ft below the specified bottom grade.

In bringing the seal to grade where a footing block or distribution block is to be poured later over the entire surface, and especially if this block is heavily reinforced near the bottom, as is the usual case, the seal should be stopped 3 to 6 in. below the design grade. This cannot be safely done on very thin seals, but it is good practice on reasonably thick or noncritical seals. The reason for this is that it is very expensive and time consuming to chip out tremie concrete above grade. On some major cofferdams, 3 to 8 weeks have been consumed chipping out hard, sound concrete so that the reinforcing steel for the footing block could be placed. When stopped a few inches low, the deficiency can be made up while pouring the distribution block.

After the tremie seal is placed, which sometimes may involve thousands of cubic yards of concrete and many shifts, it is important to schedule one more operation before everyone goes home. This is to send down a diver 4 hr or so after the pour is completed and after the concrete has definitely set to jet off the laitance from the surface. At this stage it can readily be jetted off and pumped out by an air-lift pump. If left until the cofferdam is dewatered 7 to 10 days later, jackhammers usually will be required to remove it.

Tremie concrete which is properly made, mixed, and placed will develop compressive strengths of 4,000 to 7,000 psi. Tests have shown that it also develops excellent bond to steel or timber piles, or to other concrete, provided that the surfaces are clean. In actual practice, ultimate bond strengths of 80 psi have been developed.

Due to its continuous immersion, tremie concrete does not develop drying shrinkage. In normal, good tremie-concrete practice, the yield will run from 100 to 105 percent. Any greater yield is an indication of possible segregation and formation of excessive laitance.

Seal Construction by Grout Intrusion Method* This method has been used on several major cofferdams and caissons. A graded gravel without any fines smaller than ½ in. is placed in the location of the desired seal. Grout pipes which have been previously installed to the bottom of this seal are then slowly raised as grout is pumped through them. Volume and pressure are controlled so as to completely fill all voids in the gravel.

To ensure that the grout will not set before it has filled all the voids, a retarding agent is added. To ensure fluidity, a cement dispersing agent and a wetting agent are added. When properly placed, the resultant concrete will be strong and dense. Shrinkage is substantially prevented because the gravel particles are in contact before grouting and are already as close together as possible.

Core borings in concrete placed by this method have shown that the accidental presence of fines causes stratification and uncemented zones. Therefore, extreme care should be taken to keep fines out of the gravel. Also, if the gravel is just dumped through water, segregation may take place. One solution is to use all one size of gravel, say ¾ in., while another is to lower the gravel to or near the bottom in buckets. Placing the aggregate through pipes would also reduce its tendency to segregate.

The grout intrusion should be performed as soon as possible after the gravel is placed. Extreme care must be taken to prevent contamination of the gravel by fines, silt, or organic material. Even the deposition of silt from a muddy river can clog the aggregate and impede the grout flow.

Grout pipes should be spaced closely enough together to ensure that all voids can be filled. Also, the gravel should be placed a few feet higher than the grout to prevent it from being pumped out into water to form laitance.

* See Sec. 22, Special Concretes and Mortars.

DEWATERING*

Pumping Large, temporary pumps are used to pump out a cofferdam initially, and smaller automatic pumps are then used to keep it dry.

It is important to have a large pump capacity in initial dewatering, since leakage through the sheet piles will be excessive until a differential head is established and the sheet-pile interlocks tighten up under pressure. Several large pumps are normally required to give a satisfactory rate of dewatering. While electric pumps are fine if the power supply and connections are available, diesel- or gasoline-powered pumps are also satisfactory since they will be used for a short period only. The exhaust of gasoline- or diesel-powered pumps must be led over the sheet piles and clear of the cofferdam to prevent dangerous concentrations of carbon monoxide in the bottom of the cofferdam.

After the cofferdam has been dewatered, a few small electric float-operated pumps can be used to keep it dry. These pumps are economical in their operation and, being float operated, will automatically control the water level. An excess pump capacity of at least 100 percent should be installed in case a float valve sticks while unattended. The extra pump or pumps should be installed so as to go into operation automatically if required.

The maintenance pumps are usually installed in small sumps. If a tremie concrete seal is used, a sump can be formed before pouring the tremie seal or it can be jackhammered in a low spot in the tremie-concrete surface after dewatering. Where the distribution block is smaller in plan than the tremie-concrete seal and leakage occurs around its edges, sandbags can be used to channel this water to the sump.

It is often desirable to construct a tide gate in a cofferdam. This is usually done during the driving of the sheet piles by burning off one pile at the low-water line. The top section of this pile can then be raised a few feet to open the gate. By leaving this gate open during construction of the cofferdam, unbalanced pressures are avoided. Without the tide gate, the water will still fill up inside the cofferdam to the high-tide level but, as the tide starts to fall, the drop in the inside water level will occur at a slower rate. The unbalanced head produces an outwardly directed pressure for which few cofferdams are designed.

When the cofferdam is ready for dewatering, the tide gate is left open until low tide. As the tide starts to rise, the gate is closed and welded tightly shut. All available pumping capacity is then thrown into action. This has been found to be the best and sometimes the only way in which a start can be made on dewatering. Most well-built cofferdams will not leak excessively after this start in dewatering has been accomplished. The more rigid and heavily braced cofferdams generally leak more than flexible cofferdams, because in the latter case the interlocks are very tight under stress. This is a small evil and can be taken care of with a slightly larger pump capacity.

If excessive leakage through the interlocks is encountered, there are several techniques for reducing the flow. Often some material is dropped through the water just outside the sheet piles so that the inflowing water will suck it into the interlocks. Some of the mixtures which have been used are sand and sawdust; cement and sand; cinders, manure, and sand. On several cofferdams which leaked excessively, weighted canvas sheets were draped on the outside of the sheet piles. A diver can place sandbags at the bottom to hold the canvas against the sheet piles.

Serious leaks can occur where sheet piles have been driven out of interlock, and corrective measures are time consuming and expensive. Usually the fact that a sheet pile has been driven out of the interlocks is not known until efforts to pump down have been unsuccessful. A diver can then be sent down to inspect the interlocks but, since any opening is usually at or near the inside bottom, it may be hard to find. Under some conditions, a colored dye in the outside water might help to locate the bad spot. In some cases cofferdams have been successfully dewatered only to blow at the location of a gap.

Once a leaking spot has been definitely located, corrective measures can be started. One method is to drive a "blister" of two or three sheet piles on the outside. A gasket of some soft material is then used to seal the new piles against the old wall, or a diver can drive wedges. Where possible, the soil inside this blister is excavated down to the

* See Sec. 29, Construction Dewatering.

level of the bottom of the tremie seal, and the blister is then filled with tremie concrete. In some special cases the gap can be satisfactorily sealed by a diver driving wedges directly and placing canvas on the outside. This method is effective only where the tremie seal on the inside comes up above the mud line on the outside. A temporary patch may be made by a diver using an underwater quick-setting hydraulic cement.

Pressure grouting could be used below the mud line in granular, poorly graded materials to help seal areas which cannot otherwise be reached. One technique which has been successfully used in connection with the blister method described above is to keep the water level inside the blister higher than normal by pumping water in continuously. This excess head tends to keep the soils outside the blister from entering it during excavation.

Leaks through handling holes have occurred on several major projects and are due to nothing but carelessness. A routine procedure should be set up to check every pile where spliced or used piles are being driven. If a leak does occur it can be plugged by a diver, assuming that it is above the mud line and can be reached. Otherwise, a blister or similar method must be employed.

Splices will always leak since the interlock cannot be fully welded. Also, the repeated sharp blows in driving will cause poorly made splices to crack and leak. The individual leakage volume at an individual splice is usually small and is generally important only as another contribution to the total leakage. Occasionally the added leakage due to splices is enough to make corrective measures (such as canvas, etc.) necessary. For this, and for structural reasons, and for efficiency in removal afterward, first-quality welds are essential.

Other Dewatering Methods Wellpoints can often be effectively used as a substitute for pumping, particularly in very large cofferdams that are not too deep. Standard wellpoints are most effective in medium and coarse sands, and a vacuum wellpoint system should be used in fine sands. Electroosmosis can be used in conjunction with wellpoints, creating an additional head to help in moving water through rock flour and relatively impervious silts.

A clay blanket placed on the outside of a cofferdam has much the same effect as additional penetration of the sheet piles. It increases the flow path of the water, thus decreasing infiltration and reducing the tendency of the bottom to blow. Under proper conditions, the placing of several feet of gravel in the bottom of the cofferdam before starting the dewatering helps to stabilize the bottom. Gravel lets the water flow through but holds down the sand and other sediments, preventing a real blow. The gravel also acts as a strut to provide reaction for the sheet piles.

COFFERDAM DIFFICULTIES*

A few of the more interesting cofferdam difficulties are described in the following paragraphs.

Cofferdam Destroyed by Surge This cofferdam was built at the mouth of an estuary, where heavy ocean swells ran in past it almost continuously. During construction, when three walls of the cofferdam had been set, the swell inside would get out of phase with the swell outside. This caused the sheet piles to work back and forth until some particularly heavy swells broke one wall off as the result of fatigue. The cofferdam was rebuilt and tied securely with cables and turnbuckles to prevent working. More rigid bracing and guides were used, and the setting and driving of sheet piles were completed as rapidly as possible.[13]

Cofferdam Bottom Blows The project required precast-concrete piles to be driven as bearing piles for new bridge piers in a tidal lagoon. The soil consisted of medium and fine hard-packed sand. The cofferdam bracing was set, the sheet piles were set and driven, and some 20 ft of sand was excavated. The precas-concrtete piles were then set and driving was started, using swinging leads and an underwater hammer. Because it was impossible to get adequate penetration of the concrete piles, a jet was used. The bottom blew because of the unbalanced head of sand inside and out. The situation was aggravated by the lack of a tide gate, so that on a rising tide the water

* See also Problems in Building Cofferdams in Sec. 30, River Diversion.

level was slightly higher outside than inside. The blowing of the bottom caused the sheet piling to kick in so badly that the entire cofferdam had to be removed. It was finally rebuilt, using much longer sheet piles to get more penetration. Perhaps a better method would have been to predredge as far as possible and then drive the concrete piles, jetting as necessary. The bracing would then be set and sheet piles set and driven. Final excavation to grade could be performed by jet and air-lift pump and, of course, care would have to be taken to clean the piles thoroughly in the range of the tremie seal.

Improper Sheet-pile Section and Bracing In this case, the bridge pier was to be built in a river, with 30 ft of water overlying 30 ft of sand.[14] A 6-ft rise of water could be expected after rains, and a river current of as high as 10 knots had been experienced. The contractor built heavy timber pile guides to resist the current. He selected the lightest deep-arch section in 80-ft lengths which required six sets of timber bracing, two of which would be below the existing bottom. Deep predredging was impracticable because of the current.

The sheet piles were set and driven, using a jet, and the inside was excavated a few feet and leveled. The bottom set of timber bracing and frames was lowered to the sand, and the other five sets were erected above it.

As excavation proceeded, the bracing was forced down by weighting and even by driving down, using a spud set on the wales and driven on with a pile hammer. When about 10 ft of sand had been excavated, the sheet piles deflected inward so far that the bracing could not be forced down until additional driving and jacking forces were employed.

Excavation was completed, a 20-ft thick seal poured, and pumping started. As the water lowered, the third or middle level of bracing suddenly began to crush and give way. It was necessary to flood the cofferdam and have divers install an additional, heavier set of bracing before again commencing dewatering.

It is probable that the excessive pressures were due to the nature of the sands and their disturbance by jetting. However, the method that was employed is hazardous, whether successful or not. A heavier section of sheet pile such as a Z section would be preferable today, although Z piles were not manufactured when this particular cofferdam was built. The bottom could have been predredged some 3 ft, following which a complete bracing system could have been set as a unit. This bracing would have only two levels, one at the top and a very heavy set at the predredged sand line. The Z piling would then have been strong enough to take the unbalanced soil pressure during excavation and also the hydrostatic pressure after the seal had been poured and the cofferdam dewatered.

Sheet Piles Stopped by Boulders In this cofferdam,[15] a pier was to be carried approximately 60 ft to glacial till. Because of the excessive length of sheet piles, an auxiliary cofferdam of timber sheeting was used to allow the main cofferdam to be built with its top at low-tide level. A heavy, deep-arch sheet-pile section was used and 75-ft lengths were set and driven, using a heavy steam hammer to overcome the excessive driving resistance. Even so, many of the sheet piles stood high.

The excavation was started, and dewatering and bracing were carried out as the soil was removed. As the tip elevations of the high piles were reached, sand began to boil in underneath. Wellpoints were employed, both from on top and through holes in the sheet piles.

It was found that huge boulders had caused the trouble and that the sheet piles were bent, twisted, and split "beyond description." The boulders were drilled and shot, and those sheet piles which were not damaged too badly were driven farther. Nevertheless, conditions finally got too bad to proceed. The design of the pier was changed to employ foundation piling instead of excavating and pouring a concrete block on the glacial till.

This problem of boulders floating in sand or similar loose sediments is always a serious and costly one. Excavating, drilling, and shooting under water have sometimes been successfully employed. Line drilling down along a sheet pile which has hung up could be employed. Using a lighter, faster hammer and hard-facing the tips of the sheet piles would help. Alloy sheet piles, then unavailable, would minimize distortion.

Buckling of Long Struts One large, medium-depth cofferdam employed timber piles as long, unsupported struts for its single level of bracing. As the cofferdam was dewatered, these struts buckled and the cofferdam collapsed. While the timber-pile struts would have had sufficient column strength if they had been perfectly straight, their dead-load deflection made them curved columns before the loads ever came on; so they obviously had little effective column value to resist the water pressure. In this particular case, a few timber piles driven near the middle of the struts would have been a practicable and economical method of supporting the struts and reducing their unsupported lengths.

Improper Procedure in Unstable Soil Several piers had to be built in relatively shallow water, but excavation was to be carried a considerable distance into the underlying soils. The soil was glacial rock flour, very dense and impervious but unstable in that it would fail easily in shear on curved planes. It became much weaker as soon as it was disturbed. Foundation piles were specified, and a tremie seal was prohibited in order to keep foundation pressures low. The drawings indicated that a long penetration of sheet piles should be obtained.

The very long sheet piles required were not obtainable, so the contractor used the longest available. After setting and driving around a top frame, he proceeded to excavate, dewater, and place bracing progressively. The lower portion on two of the cofferdams collapsed suddenly; fortunately the failures occurred when no work was going on. Excessive pressures were noted in other cofferdams, and the bracing was doubled up and reinforced. The owner blamed the contractor for using shorter sheet piles, while the contractor made a claim against the owner for his failure to specify a tremie-concrete seal.

The steel bracing frames provided by the contractor were adequate to resist the final pressures in the completed cofferdam. But the procedure adopted, excavating and pumping down level by level and installing bracing as work progressed, permitted excessive deflection of the sheet piles. The remolding effect of driving the sheet piles made this unstable soil fail in shear, and the excessive deflection of the sheet piling caused shear planes to become well established and lubricated, thus multiplying the pressure. In at least one case, there had been slow plastic "squeezing" flow of the soil upward in the bottom. This reduced the toe support for the sheet piles, allowing them to collapse inward. In the other case, the plastic flow of the bottom was apparently simultaneous with the shear failures on the outside.

Based on previous experience in the glacial rock flour which is frequently encountered in Alaskan and Canadian lakes, the writer believes that the pier sites could have been predredged some 20 ft deep to within 5 ft of grade. Side slopes of about 1:5 would have been stable. Then the bracing frame could have been set as a unit, the sheet piles set and driven, the excavation completed, foundation piles driven under water, a few feet of gravel placed, and the cofferdam dewatered. The sheet piles would require a sufficient penetration below the bottom to cut off flow. In this impervious but weak material and with about a 40-ft external water head, the penetration should have been 20 ft at least. One heavy set of bracing would have been needed near the bottom to provide support. This extra level of bracing would be made up and set as part of the whole bracing unit at the predredged site and placed at as low a level as possible; i.e., at the predredged grade.

An alternative method of construction, if predredging were not possible, would have been to keep the cofferdam pumped full of water. Divers would then have been employed to lower and wedge the bracing as the excavation progressed.

Tying the sheet piling at the top would help to prevent rotation of the sheet piles about the additional bottom set of bracing. Heavier section sheet piles would have helped in any case, provided that there had been good penetration and tying at the top. A careful study of the soil characteristics, not necessarily highly technical, would have pointed up the dangerous conditions and probably have led to the selection of a safe procedure at the start.

Scour and Poor Tremie-concrete Procedure Pier 169 of the Columbia River Bridge at Astoria, Oregon,[16,17] was one of four twin cofferdam piers for the crossing of the main channel. The cofferdam construction procedure was to predredge to the bottom of the seals at −65, drive the piling with an underwater hammer, set and hang a bracing frame from falsework piles, set and drive sheet piles, then pour the tremie seal.

On Pier 169, after the H piles had been driven and the bracing frames set, the sheet piles were set and driven. Suddenly, on the occasion of a high ebb-tide runout which augmented the river's normal flow, the bottom scoured on the channel side to a depth of 20 ft. The sheet piles settled vertically and the cofferdam tilted. Attempts were made to deposit sand in the hole by using a hopper dredge, but to little avail. Brush mattresses were made up and sunk with rock, and then riprap, etc., was placed over them. This succeeded in stabilizing the bottom and restoring most of the material on the outside. The inside bottom had dropped about 5 ft, and the sheet pilings were extended by splicing them at low tide, then driving them down another 5 ft. Since the steel H piles for the foundation were battered closely under the original tips of the sheet piles, the additional driving forced the bottoms of the sheet piles out and possibly opened some interlocks.

The tremie seals were then poured and, as nearly as can be ascertained, only one pipe was used. This pipe was apparently charged and sealed correctly and the start of the pour was properly performed. According to verbal accounts, the tremie pipe was then raised vertically some 15 or 20 ft and the remainder of the concrete was poured. At one stage, therefore, all the concrete must have been falling through water. A mound eventually built up under the pipe and the tremie concrete was again being properly discharged. Rate of pour was only about 20 cu yd per hr and it was noted that only 64 to 76 percent of the theoretical volume of tremie concrete was required; that is, the swell was 150 percent. It has since been reported that the water surrounding the entire area was milky white, indicating excessive laitance.

After dewatering, chipping of high spots revealed generally good concrete on the surface. A few feet down, however, uncemented sand and gravel was encountered. The cofferdam began to leak excessively through the seal, and test probes showed that this loose sand and gravel extended to within a few feet of the bottom. The cofferdams were ordered flooded and the contractor was directed to remove and replace the seals. This led to considerable contractual argument and resulted in a shift to a new contractor.

The old seal was readily removed from one cofferdam by using high-pressure jets, chisels, and air-lift pumps. In the other cofferdam, a great deal of sound concrete was interbedded between the gravel layers. Its removal required extensive chipping by divers using air tools, so explosives were placed by divers and found to be effective. A much larger charge was then tried, and this damaged one side of the cofferdam. A steel blister was subsequently installed around the damaged area, and the work of removing the seal was completed by diver chipping.

The new contractor chose to use grout-intruded aggregate to replace the seals. After he had placed the gravel aggregate in one pier, it was discovered that excessive silt from the river had been deposited in and around this aggregate. It was therefore necessary to remove it and replace it with clean aggregate. After this had been done, the grout was intruded successfully, the cofferdam dewatered, and the pier completed.

Problems with Tremie-concrete Seal The East Anchorage of the Second Delaware Memorial Bridge was constructed in a cofferdam 110 by 250 ft in plan, excavated into firm clay at −65. The bracing frames were assembled just above high water and were supported on pin piles. Then the sheet piles were set and driven into the hard clay founding strata. The cofferdam was excavated progressively by clamshell and, as each particular level was reached, a bracing frame was lowered down to position and secured to the pin piles and sheet piles. The harder clay proved difficult to excavate near the sheet piles and required considerable jetting.

When the founding elevation was reached, a tremie-concrete footing block 30 ft thick was poured in one continuous operation lasting 13 days and nights, totaling 28,000 cu yd. The specifications required that it be poured in 8-ft layers, starting at one end and proceeding to the other before returning with the next lift. A 5-sack mix was specified in order to reduce the heat of hydration and as a means of economy. However, lean tremie mixes have never turned out to be economical, and this was certainly the case here. Further, a better means of reducing the heat of hydration would have been to allow each 8-ft lift to set several days before the next lift was poured.

This would have permitted thorough jet cleaning and laitance removal from the surface of each lift. It would have eliminated the logistic difficulties and costs of assembling supplies of aggregates and cement in the prodigious amounts needed to take care of a single 28,000-cu yd pour.

Fig. 28-16 Removal of laitance and segregated material. Second Delaware Memorial Bridge.

Fig. 28-17 West Anchorage of Delaware Bridge shows excellent results obtainable with proper procedures.

In addition to low cement content, the tremie mix had the further flaw of being undersanded (36 percent sand), so that flow could only be accomplished by addition of water. Such a mix was harsh, unworkable, and inherently subject to segregation. The result, shown in Fig. 28-16, was excessive zones and pockets of sand and gravel and laitance, sometimes partially covered by good concrete. It required several months of around-the-clock work to cut out the defective material preparatory to continuing work. (See Fig. 28-17.)

C. Box Caissons

GENERAL CONSIDERATIONS

A box caisson is a prefabricated box, with sides and a bottom, which is set on a prepared foundation and filled with concrete to form a bridge pier or similar structure. The excavation is completed before the box caisson is set. Box caissons and box cofferdams have many elements in common, the theoretical distinction being that the box caisson becomes part of the structure and the box cofferdam is only temporary. Most actual structures contain elements of both, and the question of terminology is unimportant.

Many variations of the box caisson have been developed and used on notable and important bridges. While some of these were not true box caissons, they have possessed many of the same elements: the prefabrication, the completion of excavation before setting, the setting on a prepared foundation, and their incorporation in the completed structure by filling with concrete.

Often the top section of a conventional box caisson is a box cofferdam. After the pier shaft is completed to an elevation above high water, the box cofferdam is flooded and the cofferdam units are removed. A representative box caisson of this type might extend up to 20 ft below water level, with the caisson and its fill of underwater concrete remaining in place as the permanent pier base structure. Above this elevation, the box might be a cofferdam and the pier might be formed as shafts poured in the dry. The sides of the box cofferdam would then be removed after completion of the shafts.

Box caissons and box cofferdams were much used in the past for comparatively small bridge piers. A box was built, using heavy timbers to form its sides and bottom, and its joints were well calked and sheathed. This box was towed to the site and sunk to rest on a previously excavated base by filling pockets with rock or concrete. The water was pumped out and the pier built in the dry. Where the bottom was sloping rock, it was usually excavated sufficiently to form a key and the box was tailored to fit the actual bottom.

As the method evolved, many piers were built from box caissons which were supported on foundations of timber piles. A basin was first predredged, following which timber piles were driven. Usually these piles extended above water so that a diver could measure down and accurately cut them off to grade. Guides were then constructed, and the box caisson was built, launched, towed to the site, and sunk to rest on the piles. A concrete seal was placed, the upper portion of the caisson dewatered, and the pier completed in the dry. The excavated hole was then backfilled to protect the timber piles and timber bottom of the box from marine borers, and riprap was placed to protect against scour.

Later box caissons of this type frequently had concrete bases and occasionally had concrete sides with removable timber cofferdam walls. The box could be floated to the site and sunk to grade by filling pockets with concrete. In another development, box caissons with timber bottoms were set on predredged and leveled bases. Timber piles with shoes or steel piles were driven, punching through the timber base. A tremie-concrete seal was placed and the caisson unwatered. Concrete-box caissons with sections of unreinforced concrete have been similarly employed, using steel or timber piles to break out the unreinforced bottom.

Where bedrock or hardpan can be exposed and cleaned by predredging, a steel or precast-concrete box can be sunk. Some structures of this type consist of bottomless boxes which are set in place by derricks. A concrete seal is poured and the pier constructed either by underwater methods or by dewatering. Precast-concrete or structural steel units have also been employed to form box caissons. These units are set in sequence and later filled with tremie concrete, which ties the entire structure together. A number of recent major bridges have been constructed by this method in water depths as great as 165 ft. [16–21]

Another variation is a box caisson which uses flotation permanently to support a part of the load. The complete underwater pier structure is built of concrete in a dry dock, floated to place, and sunk between guides on previously cut-off piles. The pier is cellular in construction, having many watertight compartments. Buoyancy then supports a large portion of the dead load of the pier, and the foundation piles support the live load and the remainder of the dead load. The compartments near the waterline may be left open to the outside water to eliminate change in buoyancy due to tide. Sinking is accomplished by filling specified cells with water. Extremely accurate control and very sturdy and effective guides are required if this method is to be successful.

Where bottom conditions are adequate, the piles may be omitted and the box sunk directly on the soil. The soil may be preconsolidated by flooding additional cells and later pumping them out as the steel is erected. Any voids under the box may be filled with grout under pressure after the structure is backfilled.

SITE AND FOUNDATION PREPARATION

Excavation at the site is performed by clamshell or dipper dredges or occasionally by hydraulic dredge. If the box caisson is to be set on bedrock, the site must first be stripped of all overburden. Then, if practicable, the rock should be leveled. Underwater drilling and shooting may be required for this purpose, and care should be taken to avoid breaking up too much of the rock below grade. If it is not practicable to level the rock at the site, it should be stepped and adequate keys should be provided.

Where the box is to be set on hardpan, predredging should go deep enough to ensure good support, and the site must be leveled accurately. A hydraulic dredge can be used to cut the final surface to a very small tolerance. If there is much current, the predredged bottom may scour considerably; this is especially likely to occur as the caisson is being sunk. In a few cases, a sheet-pile cofferdam has been built to hold back the soil. This has enabled predredging to grade despite unstable soil, steep banks, and considerable current.

Where the pier is to be supported by foundation piles, the site should be overexcavated to prevent interference with the caisson setting. If the box caisson is to be temporarily supported on the bottom, then a gravel or crushed-rock blanket can be placed and leveled off to grade. This rock should be of an adequate size (3 to 6 in. is frequently satisfactory) and should cover a sufficient area so that it will not erode or scour as the caisson is lowered. A Venturi effect can cause a substantial increase in bottom current velocity just before the caisson is landed.

Foundation piles are driven by the customary methods, using an underwater hammer riding in telescopic leads. These piles must be cut off accurately to grade if the box is to be set on them. A diver can cut off timber piles under water, using an air saw, but it is a slow and tedious process. It is necessary to first measure down on each pile and then clamp on girts so that the pile can be cut off accurately and squarely. For this reason, sizable projects will generally employ large underwater saws.

One method is to drive sufficient falsework piles on each side to support girders, which in turn support a gantry. The underwater saw rides on the gantry, which moves on tracks set on the girders. Piles can thus be cut off very quickly and accurately. A typical underwater saw will have a 48-in. diameter blade, driven by a large electric motor through a shaft. It has been found that the horsepower needed to power an underwater saw is about 50 percent more than for the same saw working above water. If some of the pile butts are "sinkers" and fall over, they may get wedged between piles and cause the saw to cut through them on a slant. This puts a great strain on the whole saw and is very likely to break the shaft. Timber pile butts may also jump out of the water and are dangerous to personnel and equipment unless proper precautions are taken.

An underwater saw can be powered hydraulically, using a diesel-driven pump to supply oil at a very high pressure through a flexible metal hose to a hydraulic motor. In one case, the hydraulic motor turned a 48-in. saw blade at 500 to 600 rpm and satisfactorily absorbed the shock when the saw hit the pile. The drive shaft can be short, since it can operate submerged.

An underwater saw can be rigidly attached to telescopic leads and mounted on a floating pile driver. This method requires smooth water, without swells, and a large, stable barge. Although the rig can operate only during calm periods, the piles can be cut off quickly and it is usually practicable to pick favorable times for the work. However, there is more likelihood of missing a pile when a floating rig is used. Grade adjustment is made continuously by an instrument man on a pile-supported platform nearby, sighting on a mark on the upper end of the telescopic leads. It may be necessary to check each pile, and a long pipe hung in the leads can be used as a level rod for this purpose. A diver is not necessary if the piles were accurately driven, since the instrument man can properly spot the rig for each pile. The floating rig should work uptide from the piles which are to be cut off, as the current will then prevent a pile butt from coming up under the barge and damaging it.

FABRICATION, LAUNCHING, AND TOWING

The conventional box caisson is designed to be floated to the site and sunk in place. Therefore, it is built much as a barge would be: in a dry dock, on a marine railway, or on a launching ways. A box caisson may also be built on a barge, which is then sunk to launch the caisson.

Since a box caisson is set on a previously excavated and prepared site, there will normally be no soil pressures against its walls. The caisson walls in the zones that are to be filled with underwater concrete must be designed to withstand the net fluid pressure of the concrete. The effective unit weight of underwater concrete will run 85 to 90 pcf, and the depth of concrete which has not yet taken its initial set must be considered to act as a fluid. Thus, there is a point of maximum lateral pressure which moves progressively up the walls; e.g., with a scheduled rate of pour of 3 ft per hr and a maximum permissible time of 2½ hr until initial set, there will be a fluid zone 7½ ft in depth which moves progressively up the wall. It exerts a maximum pressure at any point of $7\frac{1}{2} \times (150 - 64)$, or 645 psf. The walls of a circular caisson are designed to resist this load by hoop tension, just like a tank. Steel is well adapted to resist stress in this way, and reinforced concrete has also been used with success. The design tension in the reinforcing steel should be kept low, say 16,000 psi, in order to minimize "hair cracks" in the concrete. Prestressed concrete can also be used. Timber has been used in a manner similar to that used in wood-stave tanks, with reinforcing bands of steel to act in tension.

In rectangular caissons, the sides can be designed to take the load in bending horizontally between vertical soldier beams of steel, timber, or concrete. Alternatively, the sides may be designed to take the load in bending vertically, spanning between wales. The lateral load is once again caused by the fluid zone of concrete as it moves progressively up the walls. Steel dished plates may also be used to take the load in tension, even in a rectangular caisson.

Deflections during pouring must be watched, particularly with steel, to prevent the fluid concrete from pushing the whole wall outward and breaking it loose from the already set zone of concrete. Such movement would allow fresh concrete to run down the crack, thus building up a very high fluid head to further deflect the walls outward. Since this could lead to progressive wall failure, anchors should be provided which will tie the walls to the concrete as it sets. Time of set depends on the cement properties, the temperature of the cement, the temperature of the mix, and the temperature of the surrounding water. Accordingly, the rates of pour must be intelligently adjusted in the field to prevent excessive fluid heads from developing. A rough check can sometimes be made in the field with a rod to determine the depth of fluid concrete.

If any zone of the caisson is to be dewatered, its sides must be strong enough to resist the external hydrostatic pressure. This was also a factor in cofferdam design, but the load in a box caisson is usually taken in horizontal bending between vertical soldier beams or vertical solid walls. The walls in this zone must be relatively watertight and, if made of steel, should preferably be welded or else bolted or riveted and calked. Timber sheathing must be calked and sheathed. Several layers of Irish felt or canvas and asphalt protected with sheathing will make an extremely watertight wall.

A box caisson is normally constructed very rapidly, so there is usually no problem with marine borers in timber. If, however, the timber is an integral and essential part of the permanent structure, treated timber must be used. Irish felt or canvas and asphalt plus sheathing of redwood or copper will also prevent borer attack. Similarly, corrosion of steel is important only if the steel is to become an integral and essential part of the permanent structure.

The bottom of the box must be designed to distribute the load over the prepared foundation. If the box caisson is to be set on piles, it must resist the punching shear of the piles and support its loads while spanning between supports. Since one pile may be a little high and another a little low, the bottom should be designed to span across at least one missing pile and to resist two or more times the average punching shear. The bottom of the box must also resist upward hydrostatic pressure during the flotation and all stages of sinking. This force must be taken in bending between exterior and interior walls. Similarly, if the box is to be dewatered, the bottom which is reinforced with tremie-concrete fill must take the upward hydrostatic pressure. In addition to resisting the design stresses, each member and section must be rugged enough to take the stresses and impact blows of launching, towing to the site, and sinking.

One type of box caisson consists of several prefabricated units which are set under water. These are usually built in a yard, picked up by crane, set on a barge, towed to the site, and set with a derrick. To aid in lifting these heavy units, picking eyes or bolts are embedded in the concrete or welded or bolted to the steel. Picking eyes should be overdesigned to take care of impact and swinging of the units. Where bolts are used in tension, the net area at the root of the threads must be used in design. Welds must be well made and should be stressed in shear rather than in tension.

High-strength alloy steel bars, of the type used in post tensioning, have been adopted for heavy lifts on some recent bridge construction. It is important to have long lifting bolts and to suitably detail mild steel reinforcing so as to prevent the bolt and adjacent concrete from shearing out. A factor of safety suitable to rigging practice, i.e., 5 to 1 or 6 to 1, should be used in the design of lifting bolts. Loops of reinforcing steel are not safe to use as picking eyes because deformations cause local stress concentration. Loops of bundled prestressing strand, suitably splayed and well embedded, are quite satisfactory and are widely employed.

Very heavy caisson units are sometimes built directly on a barge, which is satisfactory as long as warping of the barge during pouring does not distort the unit. When two or more units are to be constructed in succession on the same barge, the pouring of one unit may warp the barge and thus distort another recently poured unit. Large caisson units have been constructed on girders spanning a basin; a barge has then been brought underneath at low tide and cribbed, so as to pick the unit off at high tide. A platform which is just awash at low tide has also been used to support a large caisson unit while it is being constructed. The unit is then lifted at high tide by a large derrick, taking advantage of the buoyancy of the water acting on the submerged lower portion. Portions of the caisson may be made watertight to help the flotation.

Oceangoing ships have been successfully launched sideways off a dock, and very large and long caissons have been launched in a similar manner. The caisson is usually allowed to slide from launching timbers into the water. A positive means must be used to ensure the simultaneous launching of both ends, since any tendency for one end to get ahead of the other would cause the caisson to rotate.

Several precautions must be taken in towing a floating caisson to its site. Splash boards may be needed to prevent waves from overtopping the sides or, if the caisson is towed on its side, a temporary end bulkhead is required. Baffles or compartmentation should be provided, particularly in long caissons, to prevent leakage from accumulating at one end and thus causing the caisson to list. In some totally enclosed box caissons where a temporary top bulkhead has been added, an inside air pressure equivalent to or slightly greater than the water head may be maintained to keep water out. Stability during towing may be improved by the use of temporary side pontoons bolted to the caisson.

SETTING

Guides or anchors are required to hold the caisson against current pressure and other lateral forces during sinking operations. Guides may consist of well-braced steel or timber-pile dolphins, often joined by heavy floating fenders or heavy trusses. They are built around three sides, the box is floated in, and the fourth side is then closed. Vertical guides are hung from the dolphins; these may be heavy timbers or steel beams which are suspended in exact position and are truly vertical. Some tolerance is necessary to prevent binding of the box in the guides, and 1 ft all around is common. The verticals must be strong enough to resist the load which results when one edge of the caisson is forced against the guide, either by the current or by tipping during sinking.

If anchors are used, they should be set off at a sufficient distance so that the change in horizontal length of the anchor cable due to extreme tides is well below the tolerance within the guides. The main cables are taken to points near the bottom of the box where blocks and tackle are reeved, with the load line being taken to a winch on top of the caisson. Large hand winches are usually used, with sufficient parts of line to enable a ready adjustment in length.

Several types of anchors may be used, including large concrete blocks jetted into place and heavy ship anchors with steel plates welded to the flukes. A steel H pile may be driven under water, so that its top is at the mud line, and used as an anchor. An anchor cable may be fastened on before driving the pile; this eliminates diver work and permits the cable to be fastened some distance down from the pile top. Another scheme is to use a cast-iron or cast-steel point, say 4 ft in diameter, with a cable fastened to it. The point is fitted on the tip of a timber pile and driven to refusal. The timber pile is then broken off at the mud line, leaving the point and attached cable deeply embedded in firm material.

Current pressure may be computed by the same formula given in the subsection on cofferdams, namely:

$$P = \frac{Av^2}{2g} wc_s$$

where P = total load, lb
A = area of caisson surface normal to current, sq ft
v = velocity of current, fps
g = 32.2 fps per sec
w = unit weight of water, pcf
c_s = shape factor: 0.75 for circular and 1.33 for rectangular

Another method of computing current pressure has been taken from ship-design practice. It considers the pull of the running tide to be made up of two components: the pull due to surface friction and the pull due to eddy action. The resistance due to surface friction is generally believed to be directly proportional to the entire area of the immersed surface and (approximately) to the square of the water velocity. For surface friction, the following formula can be employed:

$$P_f = A_t v^2 c \sin \theta$$

where P_f = resistance in pounds due to surface friction normal to the face of the caisson under consideration
A_t = total immersed surface area, sq ft
v = velocity of tide, knots
c = a constant, taken as 0.05
θ = angle that direction of flow makes with face, deg

The second component, that due to eddy action, can be calculated by the "rudder formula" given by Attwood in *Theoretical Naval Architecture*:[22]

$$P_e = 3.2Av^2 \sin \theta$$

where P_e = resistance in pounds normal to a surface due to eddy making
A = immersed area, in sq ft, of surface of face under consideration
v = velocity of tide, knots
θ = angle that direction of flow makes with face, deg

The total current pressure against a face can be found by combining the two formulas:

$$P = P_f + P_e = v^2 \sin \theta (A_t c + 3.2A)$$

Since the tidal current may make different angles with the faces of the caisson at different times and stages of the tide, the maximum pressure should be determined for each face. Thus, for the faces normal to the general direction of flow, the current pressure is computed for the maximum angle, where the current is 90° to the face and sin θ becomes 1.0. For the faces generally parallel to the current, the maximum possible angle of flow is used in computations, based on observations at the site. In many cases, the angle on the side has been taken as 30°, and sin θ becomes $\frac{1}{2}$.

The guides or anchors are designed to resist the current pressure, as determined by either of the two methods, acting through the caisson as it bears on the guides.

The caisson sinking is accomplished by weighting. The pouring of tremie concrete into the box adds weight at the bottom and increases stability. It also strengthens the bottom to resist upward pressure during sinking and again when the box is dewatered, if such is part of the construction plan. Sinking has also been accomplished by filling pockets with rock or by flooding with water. This latter may be very dangerous from a stability standpoint unless the free surface of the water is reduced by interior walls and bulkheads. After the box caisson is sunk to position, enough additional weight should be added to ensure stability and to resist current pressures. It is safest to support the box near its bottom as it is sunk, either by anchors or continuous guides. However, box caissons have been successfully sunk using guides which support the caisson only at the water surface. In the latter case, the box caisson must be quite stable in the stage just before it lands on the bottom.

In a few special cases, where the box has been sunk on a previously leveled base without piles, grout has been pumped in afterward to ensure uniform bearing. Care must be taken to keep the grout pressure low enough so as not to raise the whole caisson or overstress any part of its bottom.

Where timber piles form the prepared foundation, and sometimes with steel piles, the outside perimeter must be protected against scour by riprap or rock fill. After this is placed, a slurry of cement and sand or sandy loam may be pumped and jetted under the base. This filling of voids has been considered unnecessary in many box caissons—unless there was fear of scour during construction—as long as adequate backfill was provided and the backfill itself was protected from scour by rock.

In setting a box caisson which is made up of several units that are to be joined under water, it is especially important to get the first or lower units level and correctly positioned. If the base consists of more than one unit, these units should be joined by steel beams or trusses after setting to hold them together properly during subsequent setting and pouring operations. Guides should be provided, such as inclined steel lips or plates, which will tend to center the next higher unit as it is set. Bearing of one unit on another may be uneven and concentrated, and asphalt-impregnated fiber strips and other similar materials have been used to distribute the loads more evenly. Sufficient tolerance must be provided to enable minor corrections to be made as the upper units are set.

The weight of each unit aids its stability, but current pressure on upper units will tend to displace or overturn the lower units. Strong connections should be used to tie upper and lower units together.

CONCRETING

Underwater concrete may be placed in the box caisson by any of the three methods described for cofferdams: tremie, bottom-dump bucket, or intrusion grouting. The standard method is by the use of tremie. Bottom-dump buckets will produce a great deal more laitance and cannot be used at all if any piles protrude above the bottom. There is little if any shrinkage of underwater concrete placed by any of the three methods, and experience with tremie concrete indicates the possibility of a slight swell.

Good bond can be developed between precast concrete and tremie concrete and between steel and tremie concrete, although surfaces must be kept clean of algae. Corrugations and mechanical keys have been widely used, but these may partially defeat their purpose unless they are smoothly curved and gradually formed. The rising tremie-concrete surface tends to trap laitance and free water under any horizontal projection. "Windows" or openings in interior walls may be similarly undesirable; as the tremie concrete flows through the window, it drops on the far side and forms laitance.

Where precast units are set on top of one another, there must naturally be a seat and a change in the wall section. As far as possible, this seat should be formed on the outside, so that the interior may have an unchanged section at the joint. Then the concrete surface can rise past the joint without undue disturbance and without trapping water or laitance. Where a change in the interior section has to be made either at a joint or elsewhere, there should be a smooth, gradual transition.

Box-caisson piers have been built where the tremie concrete in a lower section must come to a true plane and elevation to support an upper section. In a relatively small unit, it is possible to fill the lower unit and overflow it, screeding off the top surface with a steel beam and cleaning the surface with a high-pressure jet handled by a diver several hours after the concrete has set. This joint will be clean but somewhat rough. The next precast unit can then be set down and wedged to exact grade. The joint may be sealed with sandbags placed by a diver around the second unit. Weighted canvas hung over the joint will also restrict the flow sufficiently to allow the grout to seal the joint. Tests show that a workable tremie concrete will run 4 to 8 in. into a horizontal slot 1 by 6 in. This will not be subject to wash or leaching other than by current action except where extremely turbulent water or waves act near the water surface.

Where the horizontal seat must be absolutely true, a seat of steel or precast concrete can be suspended on the bottom unit. This seat must have sides or bulkheads so that the tremie concrete can be brought up under and around it, and it must be smooth. A half-round pipe of large diameter would form a good seat. It must be rigidly held in position to prevent displacement by flotation or by accidental bumping with a tremie pipe.

Underwater construction joints require special attention. Because of the excessive laitance, they are not normally practicable where underwater concrete is placed by bottom-dump buckets. However, the so-called "cyclopean" quay wall construction which involves several such joints has been found to be successful. In this type of construction, the laitance runs down the back side of the sloping wall that is being built up, and thus leaves the concrete surface reasonably clean and sound. Large rocks are left sticking up through the concrete of one pour to form keys with the next pour. As far as is known, this method has never been applied to pier construction, but it is possible that in the future some scheme of letting the laitance flow off the surface and out of the caisson may be developed. Cyclopean construction can be performed with either tremie pipes or bottom-dump buckets. Because of the large volume occupied by the big rock and the fact that the pours can be intermittent instead of continuous, the method offers possibilities of economy and practicability for large masses of fill concrete, such as may be required in a caisson for an anchor pier or at inaccessible or remote construction sites.

With tremie concrete, the construction joint should be carefully constructed. The lower pour is made fluid and workable to get as nearly level a surface as possible. Several hours after it has set, the surface should be jetted clean to expose the coarse aggregate, and the laitance and scum are then pumped out. The laitance will gather in low spots around the perimeter and behind any foundation piles. Before starting the next pour, the surface should again be jetted clean of any silt. Unless the sediment is extremely heavy (e.g., sand), pumping will not be necessary as the jetting will carry silt into suspension and any semifluid sediments will be carried up on the surface of the concrete. As on dry pours, it is best to start the second pour with a rich tremie grout.

If there is a curtain of reinforcing steel around the perimeter, there will be a 3- to 12-in. drop in the concrete surface behind it. This can be reduced by spacing larger bars farther apart, by reducing the size of the coarse aggregate, and by keeping the steel a little farther away from the side walls. Regions where bars are spliced by lapping are particularly bad; the overlap of reinforcing steel bars should be kept 2 ft or more above the concrete construction joint.

Special effort and attention should be given to cleaning the pocket outside the reinforcing steel. It is here that laitance from the first pour gathers, along with other sediments. After cleaning, a diver can fill this pocket with tremie grout to a few inches above the general level of the concrete surface of the first pour. This scheme, while involving considerable expense, gives excellent results.

With intrusion grouting, the practice is to stop the mortar surface at least 5 ft below the aggregate surface. This protects the surface from any washing action of the water. Many individual pieces of aggregate will be half embedded in the lower pour and will key with the next pour. Any fines washed off the aggregate subsequently placed will be dispersed throughout the 5-ft or thicker layer so that they will not be concentrated at the construction joint. Particular care to avoid fines at all stages must be taken with this method of underwater concreting, and such care is absolutely essential in the zone of construction joints.

Problems may occur near the water surface when placing underwater concrete by any method. Tidal and river currents may cause turbulence and washing near the surface through even very small cracks or seams in the box caisson. If practicable, the caisson should be watertight in this zone. When underwater concrete has attained sufficient strength, the box caisson may be dewatered and the pier completed in the dry. Minor deviations in alignment and location can be corrected in the first dry pour.

EXAMPLES

Richmond–San Rafael Bridge, San Francisco Bay, California[19] This 4-mi bridge required 78 piers, of which 62 were of the so-called "bell-bottom" type. Water depths up to 60 ft, exposure to storm and waves, very soft mud bottom, tidal currents up to 5½ knots, and extreme distance to bedrock all contributed to the selection of the type of construction.

Precast-concrete units were assembled under water and filled with tremie concrete to form the piers. Figures 28-18 to 28-23 illustrate the steps in construction.

During the placement of the tremie concrete, some minor cracks were noted in the precast-concrete shafts, indicating fluid pressures of 800 psf. One diaphragm form failed, although designed for 600 psf. Other diaphragm forms were redesigned for 800 psf, which worked satisfactorily. It is believed that the high stresses were due to extreme variations in the rate of set of cement (up to 7 hr) of the particular type used, and to localized high-impact stresses in the diaphragm forms near the tremie-pipe discharge. These impact forces are apparently dissipated when pouring tremie concrete over a medium or large area and become significant only when, as in this case, the pour is made into a narrow wall.

Many mixes of tremie concrete were tried on this project, some as lean as 5½ sacks of cement with 1½-in. aggregate, but the most workable and satisfactory mix was a 7-sack mix with ¾-in. aggregate and 42 percent sand.

Many factors entered into the quality of tremie concrete on this job. The very strong tidal current caused turbulence inside the diaphragm forms, which was solved by making them as watertight as possible and by scheduling the pour for a period of slack or low current. Windows had been built in the shafts to permit the tremie concrete from the shafts and diaphragm to flow together to form a mechanical key. These windows were a cause of considerable laitance because the concrete would flow through from whichever side was higher and spill over.

The close spacing of reinforcing steel at splices at underwater construction joints, which were required in certain special piers, caused a drop of 12 in. or so in the tremie surface behind the steel. Further cleaning and filling of this depression with tremie grout before making the next pour was the best solution found, although it was an expensive and time-consuming operation. Increasing the clear spacing of bars to at least twice the maximum aggregate size and locating the splices above the construction joint would help prevent this depression. The shaft unit had an offset on its interior where it set on the cone. This offset trapped water in a small pocket, and the problem was corrected in later pours by flaring the offset in a smooth transition. Experience on this project showed that the best tremie concrete was obtained when the pour was made at a fairly rapid and continuous rate.

Certain of the larger piers had four very large bells forming a base. Dredging,

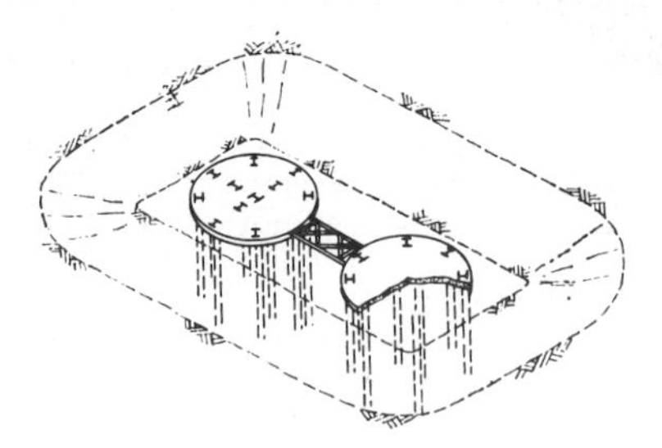

Step 1: Excavation of pier site, illustration of temporary support by timber piles and setting precast base grid.

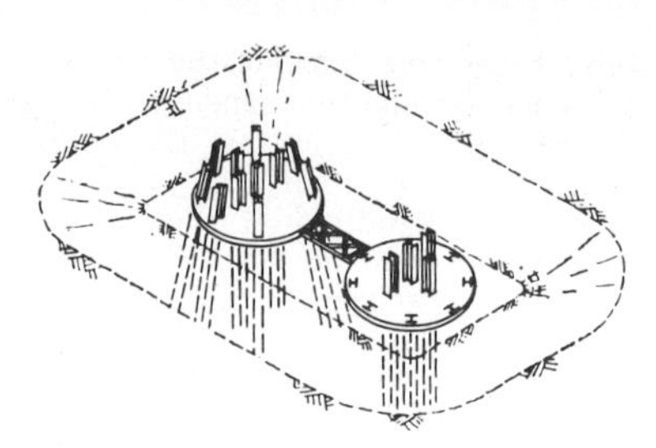

Step 2: Driving of steel H piles through grid. Grouting of H piles to base grid.

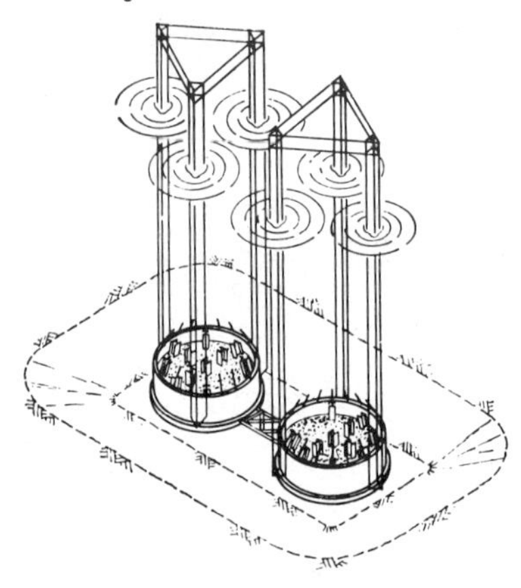

Step 3: Setting precast concrete bottom shells, with temporary steel towers bolted on. Tremie concrete seal poured.

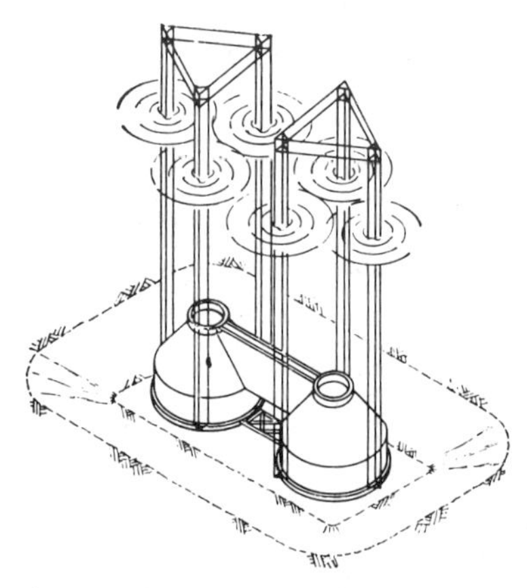

Step 4: Precast concrete cone-diaphragm set.

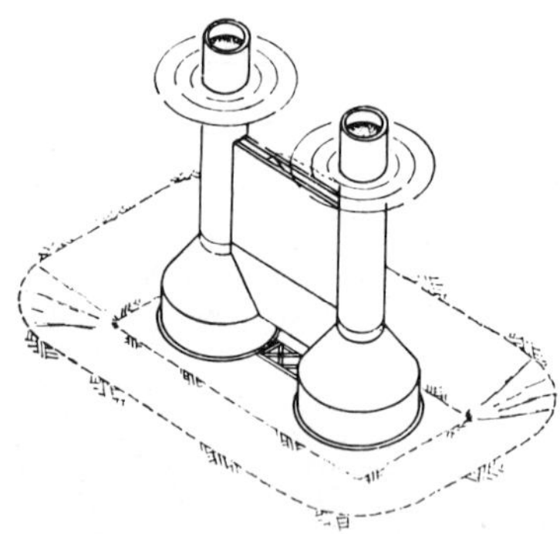

Step 5: Precast concrete shafts set. Prefabricated steel diaphragm set. Entire pier filled with tremie concrete.

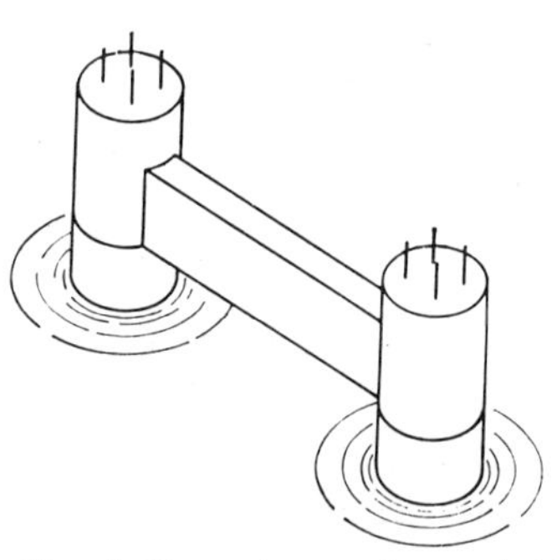

Step 6: Precast concrete spandrel beam set across tops of shafts. Anchor bolt assemblies set. Top of pier shafts concreted.

Fig. 28-18 Richmond–San Rafael Bridge.

Fig. 28-19 Richmond–San Rafael Bridge. Setting precast-concrete cone-diaphragm unit.

Fig. 28-20 Richmond–San Rafael Bridge. Setting precast-concrete shaft. Note temporary steel tower supports and spreader beam used in picking shaft.

Fig. 28-21 Richmond–San Rafael Bridge. Setting steel shell unit for large four-bell pier. Reinforcing steel cage supported inside shell.

setting the precast base grid, driving the steel piles, setting the precast bottom shell, and pouring the tremie-concrete seal proceeded as for the more typical piers. Then huge steel forms, designed for 500-psf fluid concrete pressure and 60-psf external tidal pressure, were set over the shells. Form plate was ¼ in. thick, with angle stiffeners. Truss units were provided on the inside face of the form to resist tidal pressure and to prevent any tendency of the diaphragms to work back and forth because of wave action.

Each prefabricated steel form unit had the reinforcing steel welded to it on the inside. Because of the great weights involved and the height of the main channel piers due to the depth of water, the forms were made in three sections with horizontal joints. They were set one on top of the other and bolted together. All lower sections of these steel units were anchored to the grids by cables and turnbuckles to prevent movement or overturning due to tidal pressure.

Fig. 28-22 Richmond–San Rafael Bridge. Pouring tremie concrete in one of the main piers. Note tugs holding floating equipment against swift tide, in addition to anchors.

Chesapeake Bay Bridge, Maryland[18] This bridge required several different types of piers, the most interesting and unusual of which were the "bell-bottom" piers used at 28 of the pier sites. These piers were similar in many respects to those described above for the Richmond–San Rafael Bridge, which was built after the Chesapeake Bay Bridge.

The basic steps of construction were:

1. A hole 12 ft deep was dredged into the bay bottom.
2. Timber falsework piles were driven and cut off underwater, just above the dredged bottom.

Fig. 28-23 Richmond–San Rafael Bridge. Pouring tremie concrete in main pier.

3. A large timber platform was floated to the site and sunk by placing precast-concrete blocks at the center. These blocks later served to support the diaphragm connecting the two bells. The timber mat had square holes in it, through which the H piles were later driven. Each hole was marked with raised Roman numerals, to enable the diver to verify his location in the muddy water.
4. Steel H piles up to 135 ft long were driven through the holes in the mat on radial lines and battered outward, using an underwater hammer and telescopic leads.
5. Steel "cans" of $\frac{1}{4}$-in. plate, suitably stiffened, were prefabricated at a shipyard with all reinforcing steel welded in place. They were transported to the site suspended from one or two derricks, depending on the weight, and set on the timber mat. On very large four-bell piers, the cans for the pier were made in three or four units which were set down successively, one on top of the other, and bolted together by divers.

6. A tremie-concrete seal 5 ft thick was poured which served to bond the steel H piles to support construction loads. Then the second tremie concrete pour was made, filling the can to just below the waterline.

7. The top 8 ft was then dewatered, the seal cleaned off, and the pier completed in the dry.

Considerable difficulty was reported as a result of the steel H piles damaging the timber mat during driving. Many of the H piles were driven on a batter outward, which gave them a tremendous wedging action against the mat. There were several cases in which the steel cans were badly damaged. The diaphragm ripped near its junction with the shaft, apparently due to fatigue from its constant working under wave action. Construction progress, other than delays caused by the above two problems, was very rapid. This was primarily because the piers were constructed on a production-line basis.

The method just described was similar to that used on the Potomac River Bridge at Dahlgren, Va., in 1940. There the largest of the cans were prefabricated complete, with temporary covers welded on. Compressed air was used to provide buoyancy, and a derrick was used to provide stability during the tow and sinking. The can was sunk at the site by releasing the air, meanwhile holding the caisson between guides with derricks since it was quite unstable during sinking.

San Mateo–Hayward Bridge, California[16] The bell piers for this bridge carried forward the development of the box-caisson concept in some interesting and significant ways. Pier construction was basically the same as for the Richmond–San Rafael Bridge, with these important changes:

1. The bottom precast element consisted of two bells connected with a diaphragm and set as one piece. This ensured that all upper units set subsequently were in proper relative position.

2. Elimination of cast-in-place underwater diaphragms. These had proved difficult to form and pour on the Richmond–San Rafael Bridge.

3. Reduction of the size of tremie pours to a maximum of about 358 cu yd. This in contrast to the large individual pours on the Richmond–San Rafael Bridge of up to 4,800 cu yd.

4. To ensure the highest possible quality of structural tremie concrete, the following steps were taken:

a. All pours were made by the same crew from start to finish, working in daylight hours as far as possible.

b. Pours were made as soon as practicable after the precast units had been set, to minimize the silting.

c. Extra cement was provided and a plasticizing admixture was used to ensure a workable, homogeneous mix.

d. All pours were under the direct supervision of a registered engineer who was qualified as an expert in tremie concrete.

e. Cores were taken to prove the soundness and homogeneity of each structural tremie pour. Excellent results were obtained. The cores showed good bond between the two tremie-concrete lifts and gave compressive strengths well above 4,000 psi.

Double-walled Caisson for Lighthouse at Kish Bank, Eire[23] This large, reinforced concrete caisson and tower was constructed in a protected harbor at Dublin, then towed to the site and sunk on a prepared bed. The outer caisson was 104 ft in diameter, 90 ft high, and included a complete reinforced-concrete flat slab base. Inside it was constructed a smaller-diameter caisson, also having a solid slab base. The entire assembly, weighing 6,700 tons, was then floated and towed to the site. Water was pumped into the outer caisson, sinking it on the prepared gravel base and raising the inner caisson tower, and sand was then pumped in. Piles were driven around the periphery of the outer caisson to give stability, and riprap was placed. Then a permanent connection was made between the two caissons by means of grout-intruded concrete.

Newport Bridge, Rhode Island[21] The main tower piers for the suspension bridge are located in water depths of 100 to 140 ft. The construction was complicated by the high tidal velocities and exposure to weather. A modified form of bell-pier box caisson was selected, with the piers supported on long steel H piles driven to shale and hard sand strata. The first step in construction was the predredging of 20 ft of hard-packed silt. Piles were then driven with an underwater hammer (McKiernan-Terry S-14) mounted on special leads capable of maintaining rigidity and alignment at a depth of 160 ft. These leads were 5 ft square, of heavy pipe framing. When they were lowered to the bottom, spuds penetrated about 5 ft into the bottom and secured them during driving. Although pile driving tolerances permitted piles to extend 2 ft above or 3 ft below cutoff, a great many required cutting. Divers, working in a Westinghouse saturation diving bell, made the cutoff with an electric oxygen arc. Eight 36-in.-diameter pipes, extending up to +10, were driven to act as guides or spuds for the box-caisson steel forms. These were secured by driving an H pile inside, connecting it to the 36-in. pipe by grout at the base, and welding guide cones at the top. Each box caisson was 68 by 134 by 30 to 32 ft, and, with its internal bracing and reinforcing, weighed 400 tons. It was set in place over the guides by a 500-ton derrick. The shaft forms consisted of four cylinders connected by diaphragm walls. They were floated on their sides to the site, upended with the assistance of the heavy-lift derrick, and sunk by flooding.

Reinforcing steel was detailed with particular care to ensure the free flow of tremie concrete. Internal bracing and strutting were incorporated in the steel forms so as to prevent distortion from their own weight, and the pressures of the current and the tremie concrete.

D. Open Caissons

GENERAL CONSIDERATIONS

Open-caisson construction is not usually economical in the range of depths at which cofferdam construction can be employed. It is also more costly than pile-supported bell piers and pile-supported and buoyant box caissons. However, the open caisson may be the only practicable solution where very heavy loadings, such as the main tower piers and anchorages for suspension bridges, etc., must be carried to deep strata.

An open caisson is sunk by dredging through its bottom while building up the walls. The lower end of the caisson is the cutting edge; this helps to shear through soils by concentrating the entire weight of the caisson on its perimeter. Sinking is often aided by jets and sometimes by adding weight to the caisson. Since the caisson is to become the basic structure of the pier, it is essential that the pier design and construction procedure be integrated and developed simultaneously.

A method of constructing an open caisson in the water is to build temporary false bottoms in the wells of the caisson. The entire caisson can then be floated to the site. The false bottoms can continue to supply the caisson with buoyancy until it has sunk into the mud a sufficient distance so that its weight is supported on its cutting edges and by skin friction on the outside. The temporary bottoms must therefore be designed to resist the upward hydrostatic pressure and possibly some mud pressure. This method of caisson construction is usually the lowest in cost but also the most hazardous. Guides and anchorages must be provided to hold the caisson while it is being sunk to the bottom, during the removal of the false bottoms, and throughout the initial stages of dredging. While these steps are being performed, the soft material will tend to run in under the cutting edges and cause the caisson to tip. The soil under the cutting edge may also be so overloaded that it shears on one side, moving the caisson sidewise. There is a strong tendency for bottom scour just before the caisson is landed, and this can possibly undermine the guides.

A scheme for false bottoms[24] which has been widely and successfully used on many deep caissons is to use heavy timber strongbacks which are held in place by diagonal kickers wedged tight. These strongbacks in turn hold the heavy timber planks which may be calked, sheathed, or covered with canvas. The arrangement is such that, by pulling a line attached to one of the kickers, the entire false bottom comes apart and floats or is raised to the surface. The wire line is usually attached to all kickers and strongback members, with a loop of line between each piece. This is to allow the false bottom to be pulled to the surface if it should fail to come up by flotation alone, because of wedging or jamming. To prevent such jamming, the design should permit the use of the shortest length individual members possible. If desired, steel beams could be used in lieu of the heavy timber strongbacks. It would seem to be good practice to have bolts secure the false bottom against accidental release until just before the time of flooding. Then the cables could be attached and the bolts freed.

The dredging well should always be flooded prior to removal of the false bottom. If excess mud pressure is suspected, it might be well to overfill the well in an attempt to balance the pressure. Other means of relieving excess mud pressure on the false bottom are by jetting or by removing mud with air-lift pumps. The pumps can operate either through small-diameter wells (holes) left in the interior walls or through pipe wells which penetrate the false bottoms. In very soft and unstable soils, steel domes can be placed over the dredging wells and compressed air used to control the stability and rate of sinking of the caissons.[25–27] This method is slower and much more expensive, but it has been successful in founding very large caissons to great depths in extremely unstable soils.

To protect the false bottoms from accidental tripping during the buoyant phase and to act as a safeguard for men working on forms and concreting the dredging wells high above the false bottoms, heavy safety nets should be hung in the dredging wells. These can be secured to the same inserts used to hold the forms on earlier lifts of concrete.

One basic type of open-caisson construction is used when the existing site is above water. The cutting edge is laid out in exact location and leveled, and the walls are built up progressively as the soils are excavated through the dredging wells. Difficulties may occur in the early stages when the weight of the caisson is insufficient to overcome the resistance of the soils without supplementary excavation below the cutting edge. Unstable material may then slough in, causing the caisson to tip. As a general rule, excavation should never be carried below the cutting edge. High-pressure jetting and further raising of the caisson walls are preferable methods of causing penetration. A stratum of cobbles or hardpan which is known to be present near the surface can be broken up, or even removed and replaced with sand, prior to starting caisson construction.

The sand-island method[28] envisages the construction of a temporary island upon which the cutting edge is built and through which the caisson will ultimately be sunk. The sand island serves as a working platform, enabling forming and concreting operations to be performed at or near ground level with conventional materials and equipment. It may consist of sheet-pile cells[29,30] or a large-diameter steel shell filled with sand; or it may be a naturally sloped sand island with its slopes protected by riprap. The initial sinking is uniform, since it is through selected sand, and the caisson is firmly held in position and alignment by the time that hard strata are encountered. The principal difficulties with this method are the tendency to scour around and under the sand island and the high cost of the sand island. Also, a sand island cannot be built in very deep water or where the underlying soils are incapable of supporting its weight.

The simplest open caisson is circular in plan and consists of a single heavy exterior wall. Larger open caissons are often rectangular in plan and have interior cross walls which divide the caisson into interior dredging wells. Some of the early, large open caissons had double walls and were floated to place and sunk by filling the space between the walls with concrete. This method requires more materials because the walls must be strong enough to resist the hydrostatic pressures. However, the caissons are considerably more stable and more easily controlled during sinking. Guides and anchorages are required to resist the current forces. This method is generally faster in construction and safer, but it does involve the costs of the strong, watertight double walls.[31]

Combinations of the double-wall and false-bottom methods have been used to reduce the great pressure and sudden changes of pressure which often occur with the false-bottom method. Another variation of the false-bottom method is to sink the caisson through the upper soft and unstable strata before removing the false bottoms.[32] In one such application, the soft mud was sucked from under the caisson by means of pipes extending through the false bottoms. Still another variation, employed where many sunken logs are embedded in the top strata of very unstable soil, is to land the caisson and sink it through the top stratum as a pneumatic caisson. The roofs in the wells are then removed, and open-caisson methods are employed thereafter.

Many caissons are surmounted by cofferdams to allow the upper portions of the pier to be completed by normal pours, thus permitting a reduction in cross section and weight at these upper levels. The cofferdam also permits minor adjustments in position and facilitates the installation of granite facing, fenders, and other protective devices which may be specified.

CUTTING EDGES

The cutting edge functions primarily as the bearing element of the caisson as it shears through the soil, boulders, buried timber, hardpan, and rock. Thus, the cutting edge must be extremely strong and rugged. It must be securely tied to the caisson so as to fully utilize the great stiffness against distortion or warping which is inherent in the caisson walls and boxlike structure.[33]

For a caisson which is started while floating in water, the cutting edge must also serve as a rigid floating frame or barge that will not warp or distort during the early stages of building up the caisson. The first construction loads and first concreting cannot be placed exactly symmetrically and simultaneously; so the entire cutting edge must act as the girders or trusses of a deep and heavy barge.

Cutting edges are almost always made of steel, often in combination with concrete. Heavily reinforced concrete has been used for smaller caissons, and it is well suited structurally and economically for large caissons if properly designed and detailed. Reinforced concrete, with steel-plate armor, has been used on some recent large caissons built by the sand-island method. The shoe or bearing element is usually roughly triangular in cross section and must meet the following requirements:

1. It must be strong and rugged to resist extremely high localized pressures, such as might be caused by a boulder.
2. It must be designed to resist twisting, shearing, crushing, and particularly the tendency to spread outward because of its sloping inner surface.
3. Its plates must be adequately stiffened, and outside plates must be heavy.
4. Connection and splice details must be rugged and strong.
5. There must be provision for ease of concrete placement inside the cutting edge and of placement of the seal so as to avoid voids.
6. There must be sufficient vertical diaphragms to make the cutting-edge cross section act as a whole.
7. All cutting edges must be tied together in a rigid frame so as to resist distortion.
8. The shoe must be designed to provide for excavation under the cutting edge in great depths of water.

The shoes for the interior cross walls are usually several feet above the exterior shoes and so do not have to shear through the soils nor resist the high localized pressures. They are often square or slightly beveled steel plate and angle boxes which can resist crushing and damage during excavation.

When a caisson is built on land or on a sand island, the concrete placed in the early stages can be used in conjunction with steel shoes to act as the girders of the cutting edge. Additional reinforcing may be required, and care must be taken to bond the shoes securely so that the entire cutting edge will act as a unit. Adequate support for the shoes must be provided in the early stages to hold the frames true.

In the normal floating caisson, the entire cutting edge is made of steel. The caisson may be watertight so as to float by itself or in combination with the false bottoms, if such are used. Heavy steel girders are used in the exterior and interior walls to hold the shoes rigidly and to act as a rigid barge. Sometimes the cutting edge is made slightly larger in plan than the caisson in order to reduce skin friction. This may be effective in cohesive soils such as hard clay, but it is probably ineffective in granular soils. It may be undesirable even in clays, since the skin friction on the sides is the guiding and supporting force during excavation. The scheme may be suitable only in heavy and hard cohesive soils where skin-friction forces are believed to be excessive in relation to the sinking forces available.

Some cutting-edge shoes[33] are made with flat bottoms, since these are said to give better alignment control when sinking a caisson in soft materials. This is a matter on which there is a divergence of opinion, and the choice must be based on a thorough analysis of all factors in the particular case.

SETTING (FLOATING CAISSONS)

A caisson must be held against unbalanced external lateral forces. These forces are principally due to current, but consideration must also be given to ice, debris, logs, waves, surge, and the effect of floating equipment tied alongside as well as accidental collisions. Formulas for determining current pressures and force are given in the subsection on box caissons. Because of the indefinite nature and amount of other forces, the best practice is to consider each of the elements as to its possible magnitude and probability and then to protect the caisson from the more serious ones by providing safeguards such as fenders, dolphins, icebreakers, camels, separate moorings, etc., and, lastly, to make an empirical but logical allowance for the combined effect of all forces.

Once a stable caisson is well embedded in the bottom, it develops considerable additional resistance to lateral movement. Eventually it is able to resist external forces without the aid of guides or anchorages. The most critical time is just before the caisson is landed, when its maximum surface is exposed to the current. At this same stage the current velocity underneath the caisson is usually increased, sometimes radically so. In a narrow channel, the entire body of water may be speeded up by the constriction caused by the caisson.

When anchors are used to hold the caisson and where differences in water elevation will occur, the scope must be long enough to make the change in horizontal length negligible. Thus scopes of 600 to 1,200 ft are commonly used. A maximum possible change in water elevation should produce a horizontal difference not to exceed the allowable tolerance, which will usually be of the order of magnitude of 3 to 6 in. With extremely stable caissons in shallow water and a long scope, it may not be necessary to adjust the length during sinking and landing on the bottom. The usual case, however, requires constant adjustment. Therefore, each anchor line is led to a heavy block near the bottom of the caisson. From this block, tackle is led to a pad eye on the cutting edge or caisson side, and the haul line is led up to a winch on the caisson top. Sufficient parts are used in the tackle so that the line pull can be kept within the capacity of a large, hand-operated winch. The winches must be shifted upward, one at a time during slack water, as the caisson sides are built up. In a strong tide, there will be excessive vibration in the lines and blocks, and this has caused many blocks to fail in fatigue. The blocks and shackles should be of the best quality and considerably oversize. Enough cables must be provided so that the failure or jamming of one or more will not endanger the caisson. After the caisson is landed, the lines may be progressively shifted upward to new pad eyes on the sides of the caisson until there is no further need for support.

The best features of this method of anchoring are the ability to control the location while landing to within a matter of inches, the freedom from interference with floating equipment, and usually the low initial cost. The disadvantages are the necessity of moving the winches up continually and the possibility of jamming or failure due to vibration.

Guide structures are essentially dolphins which form a cage around the caisson. Usually three sides are constructed, the caisson is floated in, and the fourth side is then closed. The several guide structures may then be joined at the top so as to develop multiple action. Since the guide itself must be truly vertical and in accurate position, the usual practice is to hang the guide on the face of the dolphin and adjust it accurately. The dolphin often consists of a combination of vertical and batter piles. On major structures, a frame may be set through which piles are driven. This makes a more rigid dolphin and provides bracing at the bottom. In a strong current, vibration is a source of trouble; therefore bracing members should be oversize and welds of the highest quality. The guide beam may extend all the way into the bottom, or it may provide lateral support near the waterline only. In the latter case, the caisson is expected to remain plumb by reason of its inherent stability. Anchors may be attached to the guides at or just below the surface to give additional support in the direction of greatest pressure. Heavy guide beams driven well into the bottom provide increased support against tipping and lateral movement during the most dangerous stages of sinking and dredging. Where more than one guide is used on a side, the guides must be located fairly near the center so that they will all act together to prevent the possibility of progressive failure.

The advantages of guide structures are the elimination of changes during sinking and landing, presumably greater reliability and strength, and the possibility of their ancillary use to support construction equipment or even to become a part of the permanent fender system, etc. The disadvantages are the cost, which is usually greater, and the fact that 1 to 2 ft of tolerance must ordinarily be provided to prevent binding. This latter problem may be easily solved by leading lines from the cutting edge to a sheave fastened to the guide structure near the bottom, then up to a winch on the top of the caisson walls. This allows exact adjustments to be made while landing.

Floating guides which are secured by anchor lines have been used to hold caissons in position.[34] The floating guide must be strong enough in bending and shear to hold the caisson should it bear on one end only. This method has the advantages of very low comparative cost and the elimination of constant adjustments. It has the disadvantages of lack of exact control at landing, of interference of anchor lines with floating equipment, and of being inherently less reliable than the other methods. A combination of this and other methods can sometimes be used to advantage; for example, using guide structures at the two ends where the current acts and using anchored floating guides on the sides.

When a floating caisson has been sunk through the water and is just about to land on the bottom, the work must be well programmed so that the landing can be completed as expeditiously as possible. It is at this stage that scour often takes place under the caisson because of the great increase in the velocity of the current. At times this scour has been so rapid and of such magnitude as to undermine the guide structures and endanger the whole caisson. In sand and sandy silt river bottoms, very large willow mattresses are often sunk by covering with rock, yet even these have been only partially successful. This problem of scour due to increased velocity must be carefully considered and kept to a minimum by positive steps. The number of piers under construction at one time, the amount of floating equipment moored in the river, the stage of the river or tide, and the character of the bottom are all important factors determining scour.

During landing, the exterior walls or cofferdam walls should be extended to a substantial freeboard so that the caisson can be landed promply even if some scour takes place during landing. At a tidal site, the caisson will probably be landed at low tide, but there must be sufficient freeboard so that water ballast can be added to keep it on the ground during the next high tide. Landing on low tide and then floating again on high tide is an invitation to scour and serious trouble.

SINKING

A double-walled caisson has walls of calked timber or steel. Concreting consists of filling the space between the double walls in stages, using buckets, elephant trunks, or intrusion grouting. With timber double-walled caissons, the presence and activity of marine borers such as teredo must be given consideration, and the use of treated timber has been necessary in some cases in the past.

With single-walled floating caissons, the outside form has usually been brought up in calked timber, steel, or precast-concrete panels[35] so as to act as a continuous shallow cofferdam as the caisson is sunk. This allows the center of gravity to be kept low and makes more freeboard available. The inside of the dredging wells can be formed with collapsible steel or panel forms which bolt at the bottom to inserts left in the previous pour. These forms preferably span the entire height of the lift to a crosstie above the top of the lift. They should be designed so that they can be raised directly to their next position without having to be lifted clear. On single-walled caissons built on land or sand islands, or in some cases of floating caissons, the outside walls can be formed with reusable panels. Care must be taken to ensure good construction joints so as to prevent leakage under hydrostatic head.

The removal of material through the dredging wells is usually accomplished by clamshell buckets. Where hard material is anticipated, the heaviest and largest bucket that will operate inside the wells should be used. Weighting of the bucket is often preferable and more effective than dropping it. Very soft materials and sands can be removed by air-lift pumps and other suction-type pumps. This method may be applied in the early sinking stages by sucking through pipes which pierce the interior walls or the false bottoms before they are removed.

When the caisson approaches the founding stratum or whenever digging becomes very hard, special rock teeth should be added to the bucket. Heavy chiseling can also be used to break up hard strata. The chisels may be long steel beams with tempered chisel points which are operated by raising and dropping or by using a steam hammer on them, like a gigantic clay spade. Churn-drill techniques can also be employed.

High-pressure jets can be used, both through holes in the interior walls and as free jets to help loosen the material. The most difficult material to dig is that which lies under interior and exterior walls. Jets and chisels can be operated through holes in the walls. Jets can be operated in the dredging wells, held in position by a cage which rides up and down in the well.[35] The direction of the jet nozzle can be controlled from above so as to knock down the clay that stands high under the walls. This jet may be equipped with a reaction nozzle so as to eliminate the high thrust. Special rotating nozzles have also been used.

Jet systems for cutting purposes are frequently built into the cutting edges, and sometimes into the interior walls as well. The systems are arranged with headers and valves so that any group of nozzles may be operated individually. While the installation of cutting jets in exterior cutting edges is good practice and is recommended, nevertheless it has been found that these jetting systems are seldom as efficient as planned. They tend to plug as the caisson is sunk, and the direction of the nozzle cannot be adjusted as needed. Also, it is not practicable to use as high a pressure and volume on groups of nozzles as with an individual free nozzle. The latter should also be provided and used to supplement the built-in cutting jet system.[36,37]

Jets are used in the exterior walls to reduce skin friction and must be built into the walls. High-velocity nozzles are not desired; the system should be designed to give a high-volume flow of water up the sides. The usual system consists of a row of nozzles around the perimeter of the caisson, several feet above the cutting edge. Additional rows may be added at 10- to 15-ft intervals up the side if excessive skin friction is expected and particularly if sinking through deep sand strata. The nozzles usually are pointed straight out. Turned-up nozzles have been tried but have become plugged by sand falling in when jetting stopped. When a nozzle is directed horizontally, a bead or other projection which partially blocks its lower edge will deflect the jet upward.

The material dredged from the cutting wells should be disposed of in such a way as to prevent any unbalanced load which might cause it to tie against the caisson. In many caisson jobs, the material is just dropped into the water alongside. This may be all right for a small amount of fine material in a river or tidal current, but the material must not be allowed to pile up against the caisson side. It may be necessary to use barges to haul away the excavated material.

The sinking of the caisson is a most delicate process and calls for extremely careful[26] control. Excessive dredging in wells along one side, dredging below the cutting edges, and other unsymmetrical operations may cause the caisson to tip. Dredging must be carried on symmetrically about the major axis of the caisson; that is, the axis parallel to the sides. On relatively long, narrow caissons, first one end and then the other may be dredged if it is not possible to work both simultaneously. Some authorities even recommend this alternating procedure, but the work at each end must be symmetrical with respect to the sides.

Dredging should never be carried below the exterior cutting edges during sinking operations until all other methods have been exhausted. Jetting and chiseling, combined with removal of material from under interior walls, should be used to lower the caisson, rather than digging under the walls. For this reason, the bottom of the interior walls should be built 5 ft or more above the exterior cutting-edge elevations. Dredging below the cutting edge initiates sudden and usually unsymmetrical failures of the soil in shear. Such failures may cause tipping of the caisson or a local blowout of material from under the cutting edge.

The removal of rock and boulders from under the cutting edge can be a very serious problem. Chiseling and jetting should be used wherever possible. Explosives have been used, both in holes drilled beneath the exterior walls and as shaped charges placed below the cutting edge. Experience has shown that blasting near the cutting edge is exceptionally risky as, in at least one case, the caisson wall was seriously damaged. Wherever it is known beforehand that rock will be encountered, the safest procedure and probably the cheapest in the long run is to drill and shoot before the caisson is set in position, or at least long before the cutting edge reaches the rock.

Some data are available on the actual skin-friction values experienced in open-caisson construction. R. F. Legget, writing in *Civil Engineering* in July, 1940[38] about experience in India on relatively small-diameter caissons across the river Hooghly, quotes values of 280 to 560 psf. At Willingdon, they experienced 1,000 psf. George L. Freeman, writing in *Civil Engineering* in September, 1940,[39] says that skin-friction values over 1,350 psf were experienced on the Huey P. Long Bridge across the Mississippi. E. S. Blaine, writing in *Engineering News-Record* in Feb. 6, 1947,[40] gives more detailed and explicit values for skin friction experienced on the Baton Rouge Bridge, as follows:

Material	*Skin friction, psf*
Watertight stiff clay	800
Tight clay grading to sandy clay	850
Same, lubricated by jets	647
30% sand and gravel, 25% clay and sand, 40% stiff clay, 5% sand	845
50% sand and clay, 50% sand	736
20% silt, 80% fine sand	1,120

Data plotted from experience on many open caissons gave an average value of 8 psf per foot of depth. Extreme values were generally within a range of 50 percent plus or minus of this value.

Explosives have been used to reduce skin friction while sinking, but this is not generally recommended. Vibration in the adjoining soil, caused by driving of sheet piling nearby, has been very effective in sinking a caisson through dense sand where jets alone were proving inadequate. Pumping down the dredging wells causes the material under the cutting edge to become quick and run or blow, thus aiding sinking. However, this method is very dangerous and may cause a sudden tipping. Weighting of caissons has been used to help sinking, but this is very time consuming and expensive. Also, it raises the center of gravity excessively if the weights are added above water. It is better and more economical to design the caisson to have sufficient weight to enable it to be sunk without additional weighting.

One phenomenon which may impede sinking is the formation of mud plugs—which act like a false bottom—in the dredging wells. These can be dangerous, since they may let go suddenly during dredging and cause tipping. Mud plugs can be broken up by high-pressure jets.

Injection of bentonite slurry through the embedded jets is a new development which has proved very successful. The slurry is injected through lubricating pipes into an area just above the enlarged cutting edge. This creates a thin (1- to 8-in.) annular ring of bentonite which greatly reduces skin friction. The Lorenz-Fehlmann method, which is extensively employed in Europe, makes use of a rubber membrane to prevent the bentonite from escaping downward. Careful control must be exercised over the mix composition and the injection pressure. After founding, the annular ring must be stabilized by injection of cement grout under high pressure, although it is generally assumed that skin friction against downward loads is permanently reduced by the bentonite.

Painting the exterior of a steel shell caisson with a paint having a low value of friction on sand has been tried, apparently without beneficial results. Another means of effectively increasing the sinking force has been proposed for the case where rock or similar firm material is available below the founding elevation. After the caisson is well seated in the overlying materials, holes are drilled down into the rock through wells in the caisson walls. Prestressing tendons are inserted and anchored to the rock by grouting. Forces of 1,000 tons per tendon are practicable, although 300 tons may be a more workable size. The caisson can be pulled down by means of jacks, which also provide an excellent means of control against tipping. If the caisson walls require further lifts to be added, the tendons must be extended by splicing. Splices are commercially available, but only in sizes of about 50 tons force per tendon. More separate tendons are thus required, or several tendons may be bundled.

During the floating stage, stability of the caisson is provided by the righting couple of the buoyancy and gravity, in much the same way as for a ship or a barge. The metacentric height is the measure of the stability. Since the caisson's center of buoyancy is generally quite high in relation to its center of gravity, the unit is very stable while it is afloat. The wider the caisson is in relation to its height, the greater is its stability. The several stages during construction and sinking prior to landing must be checked for freeboard and stability, and in some cases it may be necessary to carry the exterior walls up higher than the interior walls or to place a cofferdam on top of the caisson.

As soon as the caisson enters the mud, a definite determination of stability can no longer be made. The material at the cutting edge may act partly as a solid and partly as a fluid, and these states may change suddenly and irregularly as the material is disturbed. The stage at landing and for the next 15 to 25 ft of sinking is the most critical. The buoyant force of the water may be utilized as long as it is possible to keep some dredging wells dewatered. By filling the central wells first and keeping the perimeter buoyant, the metacentric height at the several stages is kept as large as possible.

TIPPING AND SLIDING[24,26,27,41]

The tipping of a caisson while it is being sunk is the major hazard and problem of open-caisson construction. During these early stages of sinking, the forces acting on the caisson which tend to counteract its tipping are the support of the soil under the cutting edge, the skin-friction support of the soil on the sides, the buoyancy of the water, and the lateral support of guide structures. By providing guides capable of resisting lateral forces at the level of the top of the caisson or water surface, a very large moment can be developed because of the long lever arm. Anchor lines can be used in lieu of the guides, but the latter have usually been found more practicable for this purpose. Few caisson guides or anchorages have been specifically designed to resist tipping, but many have acted that way and have saved their caissons from catastrophe. Well-braced guide structures, tied together to act as a unit, are therefore very desirable.

Many open caissons tend to slip sidewise in the early stages of sinking, after the caisson has been landed. This is due to overloading of the soil, such as may occur when flooding dredging wells prior to removal of the false bottom. Sidewise movement may also accompany tipping and the subsequent measures for its correction.

To provide external support against lateral slippage, several measures are possible. If anchor lines were originally used to provide lateral support to the caisson, the points of attachment could be shifted up the sides of the caisson during the early stages of sinking and a strain maintained. If guide structures were installed, the lateral support of these will tend to hold the caisson in position. This positioning can be supplemented by installing full-length guide beams, driven well into the bottom and held at the groundline or the top by the guide structure. Spuds driven through spud wells in the caisson have also been used. Guide beams and spuds should be driven to a sufficient penetration to eliminate the chance of their being undermined by the scour which frequently occurs just before the caisson lands.

As stated earlier, the most dangerous time is during the early stages of sinking. The time of removal of false bottoms is particularly hazardous. Once false bottoms have been removed and the caisson has been sunk deeper into firmer material, the danger is decreased. Nevertheless, serious tipping has occurred after considerable penetration into soils which offered little lateral support.

Caissons tip because support is lost under one side. Rectangular caissons seldom are in danger of tipping about the short axis because of the much greater righting moments which develop. But rectangular caissons, particularly long, narrow ones with a ratio of width to length in the neighborhood of 1 to 2, are very susceptible to tipping about the long axis.

Support is lost under one side because of failure of the soil under the cutting edge and sometimes the loss of skin-friction support on the side. When a dredging well suddenly blows, the soil under the cutting edge becomes momentarily quick. If the false bottoms are removed at too early a stage, the soil may fail in shear under the loading of the cutting edge. If dredging is carried below the cutting edge, sudden failure in shear may take place. If the dredging wells are flooded too early, the mud on which the caisson rests may be overloaded and flow out under the cutting edge.

A study of the history of open caissons leads to the following conclusions:

1. Most cases of initial tipping occur because of attempts to rush the work. For example, a night-shift dredging crew wants to dredge more yards than the day shift and it may therefore work too long in one pocket.
2. Most cases of serious tipping occur when too radical steps are taken to correct the minor initial tipping.
3. False-bottom caissons are the lowest in cost but also the most prone to tip.
4. Dome caissons, such as used on the San Francisco–Oakland Bay Bridge, are very expensive. While the domes are of theoretical value in preventing tipping, in practice they do not give positive assurance against its occurrence.
5. Modifications of the false-bottom type which permit substantial penetration in soft soils before removal of the false bottoms (for example, the caissons used on the Rappahannock River Bridge) are of major value in preventing tipping.
6. The double-walled caisson, such as that used on the Mackinac Straits Bridge, is expensive but is very unlikely to tip seriously because of its low center of gravity and high center of buoyancy.
7. Any caisson may, at times, act like a false-bottom type because of the formation of mud plugs in the dredging wells.

General rules to help prevent initial tipping of large, open caissons include the following:[26]

1. Keep the weight of the caisson and its center of gravity as low as possible. This can be done by keeping the interior concrete below the waterline during the dangerous stages.
2. Provide lateral support at as high a level as possible.
3. Maintain symmetry at the plane of the cutting edge. The removal of false bottoms and dredging in the wells should generally be such as to preserve symmetrical conditions, particularly about the long axis.
4. Avoid sudden movements of material. Dredge so as to remove mud from the center, maintaining bearing on the cutting edges.
5. Do not dredge below the exterior cutting edges until the caisson is founded. If the caisson will not settle satisfactorily, use jets on the sides to reduce skin friction.
6. In a large, rectangular caisson, sinking may be accomplished by rocking slightly about the short axis, dredging first one end and then the other. This must never be done about the other axis.
7. In unstable soils, avoid any action which might tend to make the soil under the cutting edge quick, particularly the use of blasting and pumping down the dredging wells. Jetting on the sides of the caisson, and in the dredge wells if mud plugs tend to form, is usually safe and desirable.

The correction of tipping is, as mentioned earlier, an undertaking which must be handled slowly and carefully. When the caisson first tips, it moves over until the righting forces equal the tipping forces. One of the main righting forces under these circumstances is supplied by the caisson's buoyancy or, more specifically, by the moment couple which acts through the center of gravity and the center of buoyancy.

When steps are taken to move the caisson back out of this tipped position, this buoyancy force gives it considerable rotational momentum which tends to tilt it in the other direction. If the corrective steps include overdredging and overweighting on the high side, it is obvious that the caisson will tip even more dangerously to the other side as it swings back from the initial tip. Therefore, certain general rules can be suggested for correcting initial tipping:

1. Do not take radical or drastic steps.
2. Provide means of restraining the caisson as it swings back, so as to reduce the momentum of the swing.
3. Keep the sides freed from skin friction by adequate jetting as righting measures are started, so that the caisson will not break free all at once.

4. Do not overdredge on the high side. As the caisson rights, it will rotate about its center of gravity and it may also sink for some distance.

5. Applied righting forces must be capable of rapid removal as the caisson swings back up to vertical. Lateral forces applied high up are much safer than weights. Buoyant forces on the low side are also safer than weights but not so safe as lateral forces.

In most cases the practical means of righting the caisson from a minor initial tip will include:

1. Applying a lateral righting pull as high up on the caisson as possible
2. Affixing lateral restraining cables on the low side to keep it from going back too fast
3. Dredging down to the grade that the cutting edge will come to when it rotates back to vertical

Lateral displacement can usually be corrected or reduced by dredging some material from outside the caisson on one side and dumping mud or gravel on the other side. Dredging slightly more in the wells on the one side than on the other will also help, provided that no tendency to tip is present. This eccentric dredging should probably be employed only when the direction of shift is along the long axis of a rectangular caisson; it would be extremely hazardous when attempting to shift the caisson in the direction of the short axis.

COMPLETING THE INSTALLATION

In the ideal case, the caisson is sunk until it penetrates a small distance into the rock stratum. The area enclosed by the cutting edges is excavated a sufficient additional distance into the bedrock to secure firm bearing over the entire area and to prevent slipping.

In actual cases, the conditions may vary considerably from this ideal. If the stratum which stops the cutting edge is decomposed or fractured rock, it may be necessary to excavate a substantial distance below the cutting edge. Often the most practicable way to perform this is to excavate down in sections, pouring the concrete seal in one section before excavating the adjoining ones. Thus the middle third of a rectangular caisson might be excavated and tremie concrete poured to support the exterior and interior cutting edges in this area. Then the end thirds could be completed in succession.

When the caisson comes to rest on sloping bedrock, serious problems may develop. One means of eliminating these problems, particularly where there is little overburden, is to tailor the caisson cutting edge to fit the rock. Another method is to remove the overburden and level the area by using submarine rock-removal techniques. However, the caisson may come to rest with the cutting edge bearing on only one side or corner, so that neither of these methods can be used. If the overburden is soft, there may be a considerable tendency to tip, and therefore the process of excavation into rock and immediate sealing with underwater concrete must be carried out in very small increments, possibly one dredging well at a time. In some cases, it has been necessary to drive short lengths of sheeting to keep out the mud and sand while digging down to rock. Blow-ins may occur under that part of the cutting edge which has not yet reached rock, and it may be necessary to inject grout to stabilize the material sufficiently to permit excavation to be continued.

Excavation in broken bedrock or other firm strata is particularly troublesome immediately under the interior and exterior cutting edges. High-pressure jets and chisels can be used, working through holes formed in the caisson walls or through the dredging wells. It is extremely difficult to control the jets at great depths, and guides may be affixed to the jet pipe to control its position and direction while working in the dredging well. Reaction jets handled by divers have been widely used, but at great depths this is a slow and costly method. Powder charges have been used in many cases with varying degrees of success. As stated earlier, it is probably sounder practice to break up the rock while the cutting edge is still well above the bottom. For example, with the caisson well embedded in overlying strata, jet probes through holes in the caisson walls could determine the profile of the rock. Then, through these same wells, holes could be drilled and the rock shot. Holes should be well stemmed between the powder and the cutting edge to prevent the blast effect from traveling up the holes.

Once excavation is completed, all loose material should be cleaned up. Specially designed buckets which scrape the bottom have been found more effective than standard clamshell buckets. Large air-lift pumps may be effectively employed. While it is generally necessary that clay and sand be removed from the bottom, the tremie concrete can incorporate loose fragments of rock and displace very light sediments which settle out of the water. Algae, other marine growth, and clay should be removed from the underside and sides of the interior and exterior cutting edges so that bond may be developed with the concrete seal.

The sealing of the caisson usually consists of an underwater-concrete pour which is carried up some distance above the interior cutting edges. Sometimes this seal pour is carried well up the caisson, as in an anchor pier for a suspension bridge. In any event, it must be deep enough to provide full transfer of load from the bedrock to the caisson walls, with an adequate factor of safety. It may be placed by any of the three methods described for cofferdam seals: tremie, bottom-dump bucket, or intrusion grouting.

The tremie process is somewhat complicated when the seal must be placed at great depths. Some restriction may then be necessary near the bottom of the tremie pipes in order to hold the concrete seal in the pipe. Particular care must be taken not to lose this seal when raising the pipes.

The bottom-dump bucket method has frequently been selected for caisson work. It will often prove the fastest if a large underwater-concrete bucket is used and the hoisting crane has a high line speed, but a great deal of laitance may be formed. If the dredging wells must be subsequently dewatered, the removal of this laitance can be a serious problem. To minimize this, care should be taken in concrete placement, the best possible buckets should be used, and the surface should be cleaned immediately after the pour by jetting and air-lift pumps. With both the tremie and bottom-dump bucket methods, it may be found advantageous to add a retardant admixture to the concrete.

The intrusion-grouting method is simple from a construction standpoint, but great care must be taken to avoid a stratum of fines in the aggregate which will prevent thorough infiltration of the grout. A thick blanket of aggregate must also be kept over the seal course to prevent loss of grout when operating at great depths.

Dewatering and completion of concreting in the dry were common in the past and may still be required on occasion. However, in many recent caissons, the dredging wells were not required to be dewatered. These wells can be left partially open, filled with water only, to reduce dead weight. Precast-concrete covers may be placed over them to serve as a base for the distribution block. The distribution block and upper portions of the caisson are then completed by normal pours. Cofferdam walls are frequently carried up as a continuation of the caisson exterior walls to permit pours in the dry. Minor corrections of position can be accomplished in the distribution block and subsequent pours.

EXAMPLES

Mackinac Straits Bridge, Michigan[31] The foundations for this great bridge were constructed under extremely adverse conditions of exposure to storm and ice. The construction was highly successful and was completed very rapidly.

The two main tower piers were constructed by means of double-walled steel caissons, circular in plan. They were 116 ft in diameter and depths to rock were 195 and 205 ft. One cable-support pier was built as a rectangular double-walled caisson containing 21 circular dredging wells. Water depth was about 100 ft, with some 30 ft of overburden above the bedrock.

Caissons were prefabricated near shore. The watertight steel double-walled shell provided flotation, thus eliminating the need for false bottoms or domes. As ballast, 8 ft of concrete was added, and the caisson was then towed to position and moored between heavy steel guide towers. The guide towers consisted of a frame of pipe piles and cross bracing, set on the bottom. Through the pipe piles, steel H piles were driven to rock. The H-pile spuds were locked to the pipe frames by filling the piles with concrete.

To sink a caisson, the space between the double walls was filled with concrete in progressive lifts of 4 to 10 ft in depth. As the caisson sank, the steel walls were built up to keep a safe freeboard. The concrete placement was performed by the intrusion-grouting method which permitted large yardages to be placed in a minimum of time, despite rough water.

The caissons were landed in 100 ft of water and penetrated about 16 ft because of the weight added by concreting between the double walls. Sinking through the harder material was aided by dredging through the dredging well or wells and by jetting. The concrete surface remained well below the water surface during sinking, thus ensuring a very low center of gravity. Absence of false bottoms, excellent ratios of least plan dimension to depth, symmetrical dredging, and sturdy lateral support by the guide towers all contributed to the relative sureness with which the caissons were sunk.

The maximum head to which the double-walled shell was subjected was 73 ft; that is, the concrete surface was 73 ft below the water surface at one stage. This meant that extremely heavy internal bracing had to be built into the walls. Heavy wide-flange soldier beams were placed vertically to support the steel shells, with structural-steel cross bracing closely spaced as struts between the soldier beams. After founding, the interior well or wells were cleaned out and filled with underwater concrete. The top portions of the piers were completed in the dry by dewatering the interior of the caisson, using the steel walls as a cofferdam. This permitted the wrought-iron protection plates to be placed in the dry.

The cutting edge was sturdily constructed, using ¾- and ⅜-in. plates supported by wide-flanged beams. The cutting edge was 20 ft in height. Heavy trussing 14 ft in height spanned between the cutting edges.

For the concrete filling operation, rock was placed by means of conveyor from self-unloading barges at rates up to 2,500 tons per hour. Simultaneously, mortar was pumped from the mixing barge through flexible hoses to vertical 1¼-in. diameter steel pipes spaced 10 to 20 ft on centers each way. These pipes were withdrawn gradually as the mortar height raised. Observation wells were used to keep track of the mortar surface and ensure that it remained between 20 and 5 ft below the top of the rock at all times.

Some difficulties were reported to have occurred due to thin layers of fines being inadvertently deposited as the rock was placed. Apparently, these fines initially accumulated on the decks of the barges during shipment and were then placed at the end of each unloading cycle by the self-unloading conveyors. These strata of fines were very difficult to fill with the intruded grout.

While this double-walled method of caisson construction involves large quantities of fabricated structural steel and thus high initial cost, it is undoubtedly one of the safest and surest methods of caisson construction. Its use enables completion in the shortest possible time under difficult conditions, such as prevailed on this project.

Rappahannock River Bridge, Virginia[32] This bridge made a notable innovation in the construction of caissons by the false-bottom method. Fifteen caissons were constructed, the largest being 62 by 43 ft in plan and sunk to a maximum depth of −153 ft. Each caisson was essentially a double-walled one, equipped with a false bottom for full flotation. The external walls were rectangular in plan, and the internal walls were formed by six 15-ft diameter steel dredging wells. As sections were added, they were filled with concrete, progressively sinking the caisson to the bottom of the estuary.

A stratum of very soft mud at the site is some 15 to 20 ft in thickness. If the conventional false-bottom method had been employed, difficulty might have been encountered because of overloading of this mud and consequent sideways slippage. Also, the removal of the false bottoms would have entailed considerable risk because of the character of the mud. For these reasons, the contractor designed and built removable domes which served as false bottoms for the dredging wells. A 12-in.-

diameter pipe was connected to the top of each dome and extended up to the top of the caisson walls as constructed at this stage. When the caisson had landed, the soft mud was removed from under the false bottoms by means of air-lift pumps. (These wells were also available for the operation of jets through them, had jetting been required.) By removing mud uniformly from under all the domes, by adding weight to the caisson as water was pumped into the closed dredging wells, and by adding more sections of the steel and concrete walls, the caisson was sunk through the soft mud and into firmer sediments under excellent control and with a minimum of risk.

The false-bottom domes were then removed. This was accomplished without any difficulty, since the caisson was well embedded at this stage, with no mud pressures under any false bottom. The caisson was then sunk to the bearing stratum of firm sand by open dredging, aided by jets operating through sounding wells.

Fraser River Bridge, British Columbia[42] This bridge was constructed across the Fraser River at Vancouver, British Columbia. The river at this site had a current velocity up to 8½ knots.

The false-bottom method was used for floating caissons constructed of timber with steel cutting edges. These were towed to the site, moored inside timber-pile enclosures, and sunk by filling the caisson walls with concrete. Considerable scour was experienced as the caissons were landed on the bottom. This phenomenon of underscour has occurred on many caissons in many different types of material and must always be reckoned with. In this case, scour was checked by dumping gravel around the caisson until it was well embedded. The caisson was then sunk to elevation −132 ft, timber foundation piles were driven in the bottom, and a tremie seal was poured.

During construction of pier 5, after the caisson had been founded and the tremie seal placed, a flood occurred in the river and the bottom was scoured out to below the elevation of the bottom of the tremie seal. There was no settlement of the pier, and remedial steps were based on the following requirements:

1. Restoration of bearing value of foundation piling
2. Permanent protection against deep scour
3. Streamlining of the pier to reduce scour

The scoured hole was first backfilled with sand and gravel to the elevation of the bottom of the tremie seal. Fine sharp sand was then jetted in under the tremie concrete seal, and steel sheet piling was driven around the outside of the pier to form a streamlined nose on each of its ends. Within the enclosures (the spaces between the sheet-pile noses and the pier itself), bearing piles were driven and a tremie-concrete slab was poured. The underlying gravel was pressure grouted, and the entire hole was then filled with gravel to the original elevation of the riverbed.

Second Carquinez Bridge, California[35] This bridge required three open caissons to be sunk in about 90 ft of water with currents up to 8 and 10 knots. A fender system of concrete-filled pipe piles was constructed first and used as an aid in anchoring. Sixteen concrete anchors were also used to hold each caisson. These weighed 25 tons each and were sunk into place by jets and air lifts. Anchor cables were then run through sheaves on the cutting edge and up to the top of the walls, thus taking their pull very low and resisting any tendency to tip.

The caisson walls were extended upward progressively by precast-concrete panels that served as forms and became an integral part of the walls. The concrete caissons were made buoyant by false bottoms in the shape of inverted timber domes with a 30-in. diameter steel-pipe dredging well in each cell. After the caisson reached the bottom, a hydraulic jet monitor and air-lift pump was lowered down each of these pipe wells. The jet was a 6-in. pipe fed by two 1,500-gpm pumps at 350 psi pressure. It could be both rotated and angled downward, and effectively loosened the sand and gravel for removal by the 10-in. air lift. This means of excavation was so effective that the false bottoms were left on until near final position.

After removal of the false bottoms, hard material was removed by clamshell bucket. In the caisson for the north pier, the rock contours sloped across the pier. As rock was reached at the high corner, shaped charges were placed by divers and used to break up the rock. Some damage was reportedly done to the lower walls and cutting edge, and progress, at best, was slow. It is believed that a better plan would have been to drill and shoot this rock before the caisson was even brought to the site or, at least, while the cutting edge was well above the rock and thus protected by a thickness of sand and gravel.

Tagus River Bridge, Lisbon, Portugal[25] The open caisson for one of the main piers of this bridge was founded at elevation −260 ft, the deepest caisson yet constructed. Because of the great depth to rock, the swift currents, open water, heavy overburden, and earthquake danger, an air-dome method similar to that used on the San Francisco–Oakland Bay Bridge was selected. Water depth was 90 ft, followed by 100 ft of mud and 60 ft of sand before bedrock was reached. Twenty-eight dredging wells were used, each 15.5 ft in diameter. The air domes were shifted from one well to another, and balance was maintained with compressed air at 18 to 20 psi. A computer was used to control the entire process. Excavation was performed by clamshell bucket working through the wells.

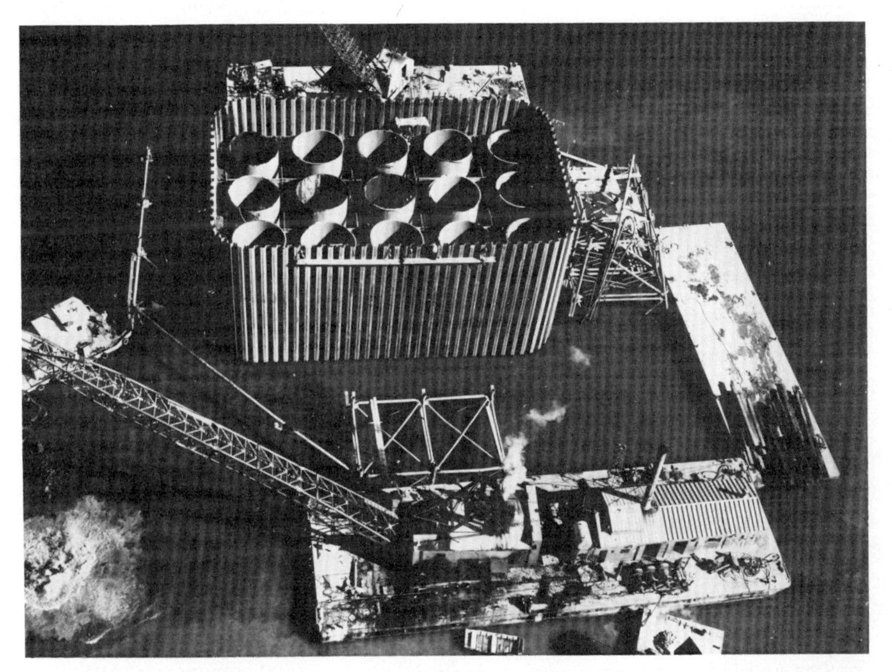

Fig. 28-24 Sunshine Bridge, Donaldsonville, La. Sheet-pile cofferdam extends exterior walls of open caisson.

Sunshine Bridge, Donaldsonville, Louisiana This bridge required four major open caissons. These were constructed in a shipyard, while at the site willow mattresses were sunk and weighted with rock. Three sides of the "pen" were built, using a tubular braced frame through which pin piles were driven, and the caisson was then towed in and the pen closed. The caisson wells had false bottoms which were bolted on and contained central pipes through which air lifting could be performed. After the caisson had been sunk through the mat and the soft overlying material, the bolts were removed by diver, a few wells at a time, and bucket excavation was used. When the caisson was well down, its top was extended by a sheet-pile cofferdam. (See Fig. 28-24.)

The first three piers went extremely well. The fourth and last was near the west levee bank. In earlier caissons, some difficulties had been experienced in removing the false bottom due to excessive upward soil pressure, so it was decided to remove some of the inshore false bottoms after only about 20 ft of penetration. After this removal, bucket dredging was started. Suddenly, the bank sheared and the caisson tilted some 15° toward the levee.

The corrective action taken was well planned and executed and was very successful. Heavy steel beams were bolted across the top of the caisson and made into a truss extending out through the structure of the pen. Weights consisting of steel piles were placed on top of the pen, and jacks were inserted to push downward on the beams with a force of several hundred tons. Dredging was then carried out in two of the central wells on the river side. Work was performed slowly and methodically, and the caisson gradually was righted. Extreme care was taken not to overdredge on the river side as the caisson returned to vertical. The caisson was ultimately sunk to founding elevation, less than two feet in plan from the originally designed location.

Verrazano-Narrows Bridge, New York[29] The two main piers for this, the longest-span bridge in the world, were constructed as open caissons inside sand islands. These islands were formed by sheet-pile cells filled with sand. They served as work platforms for construction and as docks for mooring sand and gravel barges. The caissons were 229 by 129 ft in plan, contained 66 cells, and were founded at depths of 105 to 170 ft. Each cell was 17 ft square at the base and 17 ft in diameter in the caisson shaft. The area inside the cells was filled with sand to an elevation of −12 ft and was kept dry by a wellpoint system during the assembly of the cutting edge and the start of the wall concrete. The 13-ft-high cutting edges were made of reinforced concrete, protected by a heavy "armor" of fabricated steel of ½-in. to ¾-in. thickness. To reduce skin friction, the cutting edge was stepped back 4 in. at its upper edge.

High-pressure water jets were installed in the outer walls, two tiers in the Staten Island caisson and three tiers in the Brooklyn caisson. Nozzles were 2½ in. to ¾ in. and discharged pointing upward 60° from the horizontal. Risers were 2½ in. in diameter, headers were 6 in. in diameter. Nozzles were spaced 13 ft 9 in. on centers and were connected and operated in groups of four or five. Operating water pressure was 300 to 350 psi. Provisions were made for the injection of bentonite slurry as a lubricant, but the water jets alone proved adequate.

The contractor took advantage of the great inherent stability of these caissons and the control made possible by the sand-island method to build the caisson walls up in lifts of approximately 40 ft. This procedure made adequate weight available for sinking, with the aid of jetting and clamshell excavation in the wells. Since the soils through which the caisson was to be sunk had high shear strengths, there was no danger in making high lifts; on the contrary, the contractor felt there was less risk. With the greater effective weight afforded by the high lift, he could obtain a satisfactory rate of caisson penetration without excessive disturbance of the surrounding soil by jetting, etc. Also, he would not have to excavate as close to the cutting edge.

Clamshells used were 1½ and 2 cu yd in size. To minimize stresses in the caisson walls and to ensure a condition of uniform support, jetting was employed during sinking and before the next stage was built up.

Bridge across the Garonne River at Bordeaux, France The main pier for this suspension bridge was constructed on a sand island formed by a 170-ft diameter steel sheet-pile ring. The concrete caisson was then constructed in the dry. It was sunk through sand, gravel, and hard shale and sandstone lenses by injecting an annular ring of bentonite just above the cutting edge. During excavation and sinking, the caisson was supported almost entirely on the cutting edge.

San Francisco–Oakland Bay Bridge, California The caissons for this bridge were both the largest and the deepest which had ever been sunk, and they were largely responsible for the selection of this bridge as one of the seven wonders of the civil engineering world.

The caissons for the West Bay crossing employed "domes" on top of the dredging wells so that air pressure could be used to regulate the flotation in the water and the semifluid muds. The caissons were founded on bedrock as deep as 240 ft below the surface. The caissons for the east crossing were sunk as false-bottom caissons, also to extreme depths.

The following points are selected from the excellent article written by Purcell,

Andrews, and Woodruff[26] about the difficult problems encountered and overcome in sinking these deep caissons.

The stability of a caisson can be computed while the caisson is floating but cannot be determined once the caisson enters the mud. The material at the bottom of the bay may be solid or a heavy fluid and may change from one state to the other suddenly on disturbance. The general precautions recommended are to keep the center of gravity of the caisson low and as near the cutting edge as possible, and prevent the mud from flowing in or out under the cutting edges. Side supports at a high level are very desirable. Maintain symmetry in all operations at the cutting edge. Avoid sudden movements of material. Use jets to cut away walls of clay instead of overdredging to undermine them.

Pier E-4 was a hollow-walled caisson (of timber) with a steel cutting edge. It was floated and sunk as a false-bottom caisson. It was designed so that during the flotation stage the concrete surface was well below the water surface. The necessary weight for landing might have been furnished by water ballast in the dredging wells, but since this would have greatly increased the head on the timber walls the necessary weight was obtained by pouring the concrete walls higher.

The caisson landed on a low tide, raised on the subsequent high tide, and alternated between landing and floating for several tide changes before finally entering the bottom. Then 5,000 tons of concrete was added to the walls, and the dredging wells were filled with 5,500 tons of water, giving a unit pressure on the semifluid mud of 2 tons per sq ft. The caisson sank 16 ft, pushed up a 4-ft mud wave outside, and rocked back and forth while the false bottoms were removed.

At pier E-3, constructed in similar fashion, the caisson landed on nonuniform material, with the mud under one corner being very soft. The caisson listed diagonally and skewed. Its motion was checked and corrected by piling mud outside the corner and by dredging at the opposite corner.

Extensive jetting through built-in jets greatly facilitated the sinking operations. Free jets, including reaction jets, and jets held in position by sliding frames working in the dredging wells were effective in cleaning off the bedrock some 240 ft below sea level.

The west piers were sunk as open caissons, with control by air pressure in the temporarily "domed" dredging wells. Domes were removed from one or more wells at a time for the extension of the well shaft and for dredging operations, and then replaced while other domes were being removed. There were 55 dredging wells with domes in the west anchorage caisson.

After pier W-6 had been landed and had penetrated 20 ft into the soft bottom, it hung up. The concrete walls were well above water. The caisson started to settle on the west side so that dredging was carried below the cutting edge on the east side in an attempt to correct the slight list. Suddenly it tilted 6° to the east, but fortunately was held at this inclination by the floating fender which was secured between two work platforms and guide towers.

Caisson W-6 was straightened up by dredging in the west wells, while using domes and compressed air to make sure it would not swing too far when it came back. Anchor lines were led to the west to help pull it back, and jetting was carried out along the west cutting edge. To provide adequate freeboard while tipped without unduly increasing the height of the center of gravity, the concrete in the exterior walls was carried up above the interior walls.

Carrying down alternately one end and then the other and jetting under the cross walls was successful from then on, although at 45-ft penetration the north end suddenly dropped 5 ft. This was corrected by dredging in the south end.

Pier W-3 was landed at -70 ft and sank to -79 ft. The concrete walls were built up to $+34$ ft, adding considerable weight and causing the caisson to sink to -94 ft while forcing mud up 30 ft into the cylinders. This mud formed a plug or mud false bottom, and material was forced out under the cutting edge. The air pressure was lowered in the dredging wells without any results. When the pressure was again lowered, the caisson dropped 6 ft and listed. This was corrected by varying the air pressure in the domed cylinders. Dredging was commenced in the dredging wells to break up these mud plugs. The mud suddenly rose 24 ft, and the caisson listed to the north.

Because the tipping was endwise instead of sidewise, it was not nearly so serious and was corrected without excessive difficulty. If the concrete walls had been kept lower, it is probable that the mud plugs could have been removed or broken up before bottom conditions became so bad. Also, the center of gravity would have been lower and the control by air would have been more effective.

At pier W-4, when there was 20-ft penetration into the bottom, the mud had risen 25 ft in the cylinders and had plugged, forming a false bottom of mud. The caisson then rocked back and forth for 3 weeks, with list up to 4 ft. At 30-ft penetration, with the caisson level, and while dredging the center wells, the mud was displaced sideways under the cutting edges, causing the pile-supported working platforms to heave. The caisson tipped slowly 8 ft to the north.

The exterior concrete walls were built up to give more freeboard. Domes were put on the north wells and air pressure was raised on the north and removed on the south, with no effect because of the restraint of the mud plugs. Extra concrete was added to the south walls, and the south wells were dredged. The caisson then slowly swung back and on over to a 9-ft list to the south. Mud rose in the south cylinders and was pushed out under the south cutting edge. The caisson was finally righted by pumping the water down in the north wells and by dredging in the north wells. The caisson was too heavy for the soil to support it at this stage. The concrete should have been kept lower to reduce weight and to lower the center of gravity.

By way of contrast, pier W-5 was the narrowest pier and in the softest mud. The concrete was kept low after landing. The anchor lines, after the cutting edge was embedded, were moved up above the then waterline on the caisson and tensioned to 30 tons per line. Symmetry was maintained in all operations, and the mud was kept excavated down near to, but not below, the cutting edge to eliminate mud plugs. No difficulty or tipping was encountered with this caisson.

The cleaning and sealing of the bottoms, after founding in bedrock, was carried out in three sections, one-third at a time. At pier W-4, the material tended to flow in under the cutting edge even at this great depth; so the caisson was sunk into rock, using "gads" or chisels and high-pressure jets. Where broken-rock fragments were flowing in, grout pipes were jetted down outside the caisson, and the outside material was thoroughly grouted.

References

1. D. P. Krynine, *Soil Mechanics, Its Principles and Structural Applications*, McGraw-Hill Book Company, New York, 1947.
2. Karl Terzaghi and Ralph B. Peck, *Soil Mechanics in Engineering Practice*, 2d ed., John Wiley & Sons, Inc., New York, 1967.
3. Gregory P. Tschebotarioff, *Soil Mechanics, Foundations, and Earth Structures*, McGraw-Hill Book Company, New York, 1951.
4. Krijn van Loenen Hoving, *Bridge Over the Eastern Scheldt* (in Dutch), *Cement (Amsterdam)*, November–December, 1964.
5. L. White and E. A. Prentis, *Cofferdams*, Columbia University Press, New York, 1950.
6. *Foster Steel Sheet Piling Catalog 400R*, L. B. Foster Company, New York and San Francisco, 1963.
7. *Bodine Resonant Driver*, The Resonant Pile Corporation, East Providence, R.I.
8. H. S. Jacoby and R. P. Davis, *Foundations for Bridges and Buildings*, 3d ed., McGraw-Hill Book Company, New York, 1941.
9. *Typical Installations of Bethlehem Sheet Piling*, Bethlehem Steel Company, Booklet 127-C, 1954.
10. United States Steel Company, *Steel Sheet Piling Design Manual, 1969.*

11. American Institute of Timber Construction, *Timber Construction Manual*, John Wiley & Sons, Inc., New York, 1966.
12. Ben C. Gerwick, Jr., Placement of Tremie Concrete, *Symp. Concrete Construction in Aqueous Environments*, American Concrete Institute Publication SP-8, 1964.
13. Edwin R. Albertson, Cofferdam Difficulties in a Narrow Tideway, *Eng. News-Record*, vol. 118, pp. 729–732, May 20, 1937.
14. Carl B. Jansen, The Toughest Part Is under the River, *Civil Eng.*, vol. 26, no. 2, pp. 82–96, February, 1956.
15. Sinking Open Cofferdams through Glacial Drift, *Eng. News-Record*, vol. 114, pp. 1–5, Jan. 3, 1935.
16. Ben C. Gerwick, Jr., Bell-Pier Construction, Recent Developments and Trends, *J. Am. Concrete Inst.*, October, 1965.
17. Jacking Barges Install Foundations (Columbia River Bridge at Astoria), *Eng. News-Record*, Oct. 17, 1963.
18. Six Contractors Tackle First Bridge over Chesapeake, *Eng. News-Record*, vol. 145, no. 17, pp. 32–34, Oct. 26, 1950.
19. Ben C. Gerwick, Jr., Hollow Precast Concrete Units of Great Size Form Bridge Substructure, *Civil Eng.*, vol. 24, no. 4, pp. 58–63, April, 1954.
20. The Chesapeake Bay Bridge–Tunnel, *Eng. News-Record*, Nov. 23, 1961.
21. A. F. Hedefine and L. G. Silano, Newport Bridge Foundations, *Civil Eng.*, vol. 38, no. 10, pp. 37–43, October, 1968.
22. E. L. Attwood, *Theoretical Naval Architecture*, 8th ed., Longmans, Green & Co., New York and London, 1953.
23. Prefabricated Telescopic Lighthouse for the Kish Bank, Eire, *International Construction*, January, 1965.
24. J. W. Rollins, Pier Construction for Mid-Hudson Bridge at Poughkeepsie, New York, *J. Boston Soc. Civil Eng.*, October, 1929.
25. Portugal Builds the Longest Suspension Bridge in Europe, *Eng. News-Record*, Dec. 12, 1963.
26. C. H. Purcell, C. E. Andrews, and Glenn B. Woodruff, Difficult Problems Overcome in Sinking Deep Caissons, *Eng. News-Record*, vol. 114, no. 7 (I), pp. 239–242, Feb. 14, 1935, and no. 8 (II), pp. 278–281, Feb. 21, 1935.
27. State of California, Department of Public Works, *First Annual Report, San Francisco–Oakland Bay Bridge*, California State Printing Office.
28. William L. Stone, Jr., Sand Islands Used to Sink Caissons for Piers of Philadelphia-Gloucester Bridge, *Civil Eng.*, vol. 25, no. 8, pp. 504–508, August, 1955.
29. Nomer Gray and George H. Decaneq, Verrazano-Narrows Bridge: Construction of Tower Foundations and Anchorages, *J. Constr. Div. CO2, Am. Soc. Civil Engrs.*, March, 1966.
30. Nomer Gray, Foundations for the Throgs Neck Bridge, *Civil Eng.*, vol. 29, no. 10, p. 706, October, 1959.
31. John C. King and O. H. Millikan, Mackinac Straits Bridge, Substructure Design and Construction, *Roads and Streets*, December, 1955, p. 90.
32. Albert C. Smith, Capping Bottoms of Dredge Wells Helps Control Caisson Buoyancy, *Construction Methods and Equipment*, vol. 37, no. 9, pp. 50–54, September, 1955.
33. C. W. Dunham, *Foundations of Structures*, pp. 522–527, McGraw-Hill Book Company, New York, 1950.
34. Long Piles and Floating Caissons for Potomac River Bridge Piers, *Eng. News-Record*, vol. 24, no. 9, pp. 306–310, Feb. 29, 1940.
35. Hydraulic Jets Sink Carquinez Caissons, *Eng. News-Record*, April 18, 1956.
36. Two Types of Caissons Support Piers for Louisiana Bridge (Baton Rouge), *Eng. News-Record*, September 3, 1964.
37. Deep Piers for Baton Rouge Bridge, *Eng. News-Record*, vol. 120, no. 3, pp. 106–108, Jan. 20, 1938.
38. R. F. Legget, Discussion of "Settlement Studies of Huey P. Long Bridge" by William P. Kimball, *Civil Eng.*, vol. 10, no. 7, p. 452, July, 1940.
39. George L. Freeman, Discussion of "Settlement Studies of Huey P. Long Bridge" by William P. Kimball, *Civil Eng.*, vol. 10, no. 9, p. 595, September, 1940.
40. E. S. Blaine, Practical Lessons in Caisson Sinking from the Baton Rouge Bridge, *Eng. News-Record*, vol. 138, no. 6, pp. 213–215, Feb. 6, 1947.
41. Glenn B. Woodruff, An Overturned 19,000-ton Caisson Successfully Salvaged, *Eng. News-Record*, vol. 106, no. 7, pp. 275–281, Feb. 12, 1931.
42. Some Recent Canadian Construction—Substructure of Fraser River Bridge, *Civil Eng.*. vol. 7, no. 12, pp. 831–832, December, 1937.

Section 29

Construction Dewatering

R. Y. BUSH

Consulting Engineer, Fullerton, Calif.

INTRODUCTION

Construction projects frequently involve deep excavations which extend below the groundwater table. In certain instances, excavation may be made "in the wet," such as in the cases of open caissons or bridge-pier cofferdams. Generally, however, it is preferable to remove the water from the excavation by the use of an appropriate dewatering method.

The dewatering method should effect the removal of the water from the excavation without causing instability in the side slopes or bottom of the excavation. The system may also be required to provide pressure relief in pervious strata underlying the bottom of the excavation.

The selection of the method of dewatering and the design of the system should be based on a careful study of all available soil and groundwater data. Frequently such information can be obtained from governmental agencies, water-supply well logs, previous construction records in the general area, etc. The design engineer will generally provide the contractor with drill logs on projects involving dewatering problems. Exploration holes should be sufficient in number to indicate continuity of strata. Samples should be taken for permeability evaluation. On projects involving major dewatering problems, pump tests may be required for the purpose of obtaining accurate permeability data.

All too frequently, exploration holes are not sufficiently deep to detect the presence of pervious strata requiring pressure relief. *Exploration should extend to a depth below the water table equal to at least twice the depth of the excavation bottom below the water table.* The importance of this rule cannot be overstressed.

Dewatering operations can be divided generally into two main classes: (1) surface pumping methods, and (2) predraining methods. In surface pumping, the water is pumped from open reservoirs or sumps. Predraining methods are characterized by pumping from below the ground surface from wells or wellpoints prior to excavation.

DEWATERING METHODS

Surface Pumping *Sumps and ditches* are frequently used for dewatering. The method simply involves pumping from sumps located inside the excavation area. It is applicable to soils whose stability is not significantly reduced by the action of seepage; for example, stiff clay or a soil having a high percentage of gravel or boulders.

The dewatering of an excavation using this method (frequently referred to as "sumping") usually consists of making an initial excavation to the water level, followed by the excavation of ditches and sumps around the perimeter of the inside of the excavation. Sumps are, if possible, excavated to an elevation below the deepest final grade within the excavation area. Excavations extending to appreciable depths below the water table require the use of a system of sumps progressively excavated to deeper elevations, with pumps relocated in "leap-frog" fashion until final excavation grade is reached.

The concurrent removal of water and soil from an excavation results in seepage entering the excavation through the bottom and the side slopes. In the case of fine-grained granular soils such as fine sands and silts, the bottom of the excavation frequently becomes "quick" and the side slopes unstable as a consequence of the seepage forces imparted to the soil grains by the water entering the excavation. The dewatering of such soils generally requires the application of predraining methods described later in this section.

The problem of instability due to seepage is frequently compounded by the contractor's desire to expedite the dewatering operation. Water is pumped from the excavation at a rapid rate, and the water level inside the excavation lowers considerably faster than the groundwater level in the adjacent ground. To minimize the head differential between the inside and outside of the excavation, it is important to pump at a relatively slow, *continuous* rate rather than at a rapid intermittent rate (the interruptions occur when pump capacities exceed the rate of inflow into the sumps).

Horizontal drainage systems consist of series of ditches leading to sumps. These ditches are frequently filled with gravel (the so-called "French drain") to prevent their caving and/or to permit later concreting or backfilling above the excavation level. Due to the coarseness of the gravel generally used, such drains are subject to clogging by removing fines from the adjacent natural material. Further, their hydraulic carrying capacity is quite low. A more trouble-free and efficient drain can be provided by utilizing either perforated or porous drainpipe surrounded by a properly graded sand filter.

To permit continuous pumping at later construction stages during backfilling, horizontal drains can be connected to vertical or sloping pipe "risers." These risers are extended as the backfilling progresses, thus making it possible to control the water level below the backfill. After the completion of construction, the drainpipe and risers may be grouted if this is considered necessary.

Cofferdam unwatering is utilized for structures, such as dams, powerhouses, and bridges, which involve construction in or immediately adjacent to "open" water. To perform this construction, it is first necessary to build a cofferdam around the work area.

Cofferdams such as those used for stream diversion to permit dam construction may be constructed of soil or rock materials.* The design of such cofferdams should be based on the principles of earth-dam design, with particular attention given to stability considerations inside the cofferdam.

Steel sheet piling is frequently used in cofferdam construction. It may be used in connection with an earth-dam-type embankment to reduce seepage; i.e., to act as a cutoff. Cellular-type cofferdams, consisting of circular cells of interlocking sheet piling filled with granular material, are used where it is necessary to enclose large work areas, particularly in swiftly flowing streams. Excavations of small areas, such as those required for bridge piers, are frequently made by using rectangular cofferdams consisting of sheet piling cross braced with a system of wales and struts. An extensive description of various types of cofferdams is contained in Section 28, Cofferdams and Caissons.

Excavation inside small, rectangular, sheet-pile cofferdams may be made "in the wet" by using a clamshell. By maintaining the same water level both inside and outside the cofferdam, the bottom of the excavation remains hydraulically stable; i.e., there is no upward flow of water. Excavation is followed by the underwater tremie placement of a concrete plug. The vertical thickness of this plug must be sufficient to withstand the hydrostatic uplift after pumping out the impounded water (correctly referred to as "unwatering") inside the cofferdam. In this case, the pumping rate selected will be dependent mainly on the desired duration of the pumping period. Leakage through sheet-pile interlocks and/or the concrete plug will generally be nominal if the cofferdam is constructed carefully. Occasionally, remedial measures are required (e.g., grouting in voids of the tremie plug, sealing interlocks with straw, etc.) to reduce the leakage into the cofferdam.

The unwatering of cofferdams which do not incorporate a concrete plug or a positive cutoff will require the removal of continuous infiltration in addition to the impounded water. In the case of an earth cofferdam, this seepage will occur through as well as under the embankment. The depth of sheet piling in a rectangular cofferdam may significantly affect the pumping rate as well as the stability of the bottom. Considering the frequency of cofferdam failure attributable to insufficient embedment of the sheet piling, it is prudent to use a conservatively large embedment. Analyses of seepage and stability may be made by utilizing flow-net techniques.[1]†

An extremely important consideration in the design of cofferdams, as well as in the design of dewatering systems in general, is that of bottom pressure relief. Figure 29-1 illustrates a cross section of a rectangular sheet-pile cofferdam which requires bottom pressure relief. To effect this relief, wells or wellpoints should be installed inside the cofferdam. If such relief is not provided in the example, it is apparent that the hydrostatic uplift pressure on the bottom of the silty clay layer will be twice that of the downward pressure due to the saturated soil weight, after excavation has reached subgrade! This obviously dangerous condition exists in spite of the fact that a reasonably large (what would normally be considered conservative) sheet-pile embedment is used; i.e., 10 ft for a 15-ft cofferdam depth.

If the underlying pervious soil in this example were a thin, continuous stratum, the

* See Sec. 30, River Diversion.

† Superior numbers refer to the list of references at the end of this section.

sheet piling should be driven sufficiently deep to cut off this pressure zone and thereby reduce the need for wells or wellpoints inside the cofferdam.

Figure 29-1 also serves to illustrate the recommendation that exploratory borings should extend to a depth below water equal to twice that of the excavation depth below water (for an "average" saturated-soil unit weight of 125 pcf, which is about twice the unit weight of water).

Well Method Predraining methods of dewatering generally require the use of wells of some type. These wells are installed around the excavation or adjacent to it, as in the case of trench excavation. The necessity for predraining is dictated by the presence of cohesionless soils which are subject to becoming unstable and quick if surface pumping methods are used.

Predraining systems range from large diameter (12 in. or more) deep wells to smaller-diameter wells or wellpoints which extend to relatively shallow depths of less than 30 ft. A discussion of the special case of wellpoint dewatering will be presented later in this section.

Ideal soil conditions for economical dewatering with wells are a high permeability and a reasonable degree of homogeneity below the water table. If these conditions extend to an appreciable depth below excavation level, widely spaced deep wells of relatively high capacity are applicable. The dewatering of stratified or finer-grained soils requires the use of smaller-capacity, more closely spaced wells. Wells are not

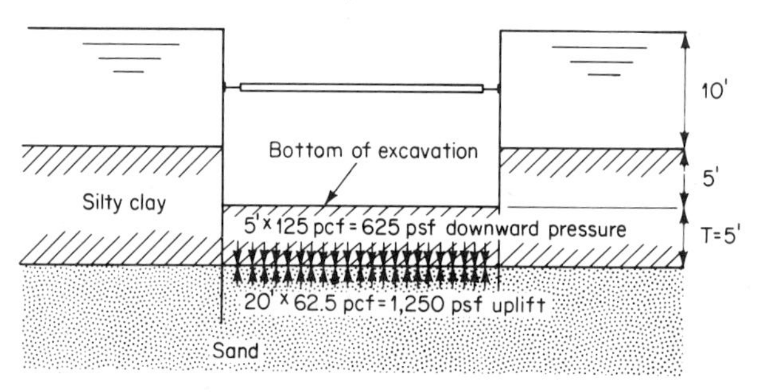

Fig. 29-1 Cross section of cofferdam requiring bottom pressure relief. Note that when $T = 10$ ft, downward pressure = 1,250 psf. Therefore, if depth to previous stratum is greater than or equal to twice excavation depth (both measured below water level), uplift pressure will be less than or equal to downward pressure.

suitable for dewatering down to the top of an impervious layer due to the reduction in the wells' collection capacity with the lowering of the water level. Either conventional or ejector-type wellpoints should be used in this situation.

Well drilling methods used in construction dewatering include cable-tool, rotary, and reverse-rotary methods. Support of the sides of the drilled hole may be provided by (1) a steel casing, as in the cable-tool method; (2) using "drilling mud" which is introduced into the drilled hole to obtain a fluid having a specific gravity slightly higher than that of water; or (3) maintaining a higher water level in the drilled hole than that of the groundwater. This excess head provides the necessary support, as in the case of the reverse-rotary method.

The cable-tool method is applicable in relatively coarse-grained materials which do not require gravel packing. The steel casing is installed progressively with the drilling of the well and provides permanent support to the sides of the hole after the completion of drilling. The casing is perforated either prior to installation or in place by using special perforating equipment.

The use of temporary steel casing is generally unsatisfactory for large-diameter gravel-packed wells which extend to appreciable depths (in the range of 100 ft). The necessity for installing and removing the casing in short sections, requiring welding and cutting, makes the method slow and frequently expensive. The pulling of the casing is particularly troublesome, necessitating the use of large jacks to overcome the friction between the casing and the natural ground and gravel pack.

The rotary method is frequently used with drilling mud to support the sides of the hole. Occasionally the "bucket-auger" method is used with drilling mud. By eliminating the necessity for casing support, these methods are generally quite efficient and relatively economical. The main disadvantage of methods utilizing drilling mud is the difficulty in removing the mud from the completed well. During drilling the mud forms a "cake" along the walls of the hole, and it is doubtful that this cake can ever be completely removed even though the well is treated with extensive bailing and surging. The mud that is not removed will reduce, in many cases appreciably, the collection capacity of the well.

The reverse-rotary method is used with an excess hydrostatic pressure of water in the hole to provide support. Clean water is used, and the problem of "mudding the hole" is reduced to that resulting from the suspension created by the drilling. Due to the size and expense of reverse-rotary drilling equipment, there are areas where it is not possible to find local drillers experienced with the method. Where equipment and qualified drillers are available, the method is one of the best to use for dewatering wells. Although the unit cost of a given size of well will generally be higher with this method, the greater productivity of the wells will frequently more than offset the additional drilling expense. In many cases the yield of reverse-rotary wells will be several times that of wells of comparable size drilled with mud.

Excellent dewatering wells have been installed using jetting methods. A specially designed casing is jetted in the ground with appropriate high-pressure pumping equipment. A crane of adequate size and boom length is used to handle the long casing during jetting and removal after gravel packing. Because of the fairly extensive amount of special equipment required, the jetting method is generally applicable only on projects requiring the installation of a large number of wells.

After the completion of drilling, the perforated casing or well screen is installed. The upper portion, above the final water level, is unperforated. The size of the perforations as well as the gradation of the gravel-pack material must be selected so that the well will perform efficiently and that, at the same time, the pumping of fines is prevented. Criteria for determining gravel-pack gradation and the size of well casing perforations are supplied in this section under Design of Dewatering Systems.

Deep-well turbine- or submersible-type pumps are generally used for pumping from the wells. Occasionally, horizontal-type pumps are used where the required drawdown in the well does not exceed the suction limit of the pump. Because of their high efficiencies and relatively low maintenance, deep-well turbine-type pumps are most frequently used for prolonged pumping on large-capacity wells.

Small clearances (less than 1 in.) between the inside diameter of the perforated casing and the maximum diameter of the pumping equipment installed inside it should be avoided. Larger clearances reduce the danger of sand or gravel particles lodging between the pump (or pump bowls) and the casing, which would make the ultimate removal of the pumping equipment quite difficult.

Wellpoint Method Wellpoint dewatering may be used for a wide variety of soil types and conditions. Wellpoints are installed on relatively close spacings, usually in the range of 2 to 8 ft, making them well suited to the dewatering of fine-grained or stratified soils. The broad range of application of the method is illustrated by Figs. 29-2 and 29-3, representing, respectively, a sewer-trench excavation dewatered with a single stage of wellpoints and a dam cutoff trench excavation dewatered with a multistage wellpoint system. Excavations extending to depths greater than about 18 ft below the water table require the use of a multistage wellpoint system.

A bottom-suction self-jetting wellpoint, similar to the type generally used on construction dewatering operations, is illustrated in Fig. 29-4. Such wellpoints are ordinarily approximately 3 ft long, having an outside screen diameter of about 3 in. The wellpoint is connected to a riser pipe, the length of which is governed by the amount of water lowering required. Generally, these riser lengths vary from 10 to 21 ft. The connection between the header manifold and the top of the riser is made by means of a "swing joint" which may be either of the flexible hose type or an assem-

Fig. 29-2 Single-stage wellpoints for sewer trench.

Fig. 29-3 Multistage wellpoints for dam core trench.

bly of pipe nipples and elbows. Each swing joint is provided with a valve which permits the regulation of the flow of water and air into the wellpoint. The water from the wellpoints is collected in the header manifold and carried to the wellpoint pump.

Wellpoint pumps are designed to handle air as well as water. All water entering the pump passes through an air-separation tank from which the air is removed by means of a supplementary vacuum pump. These features keep the system "primed"

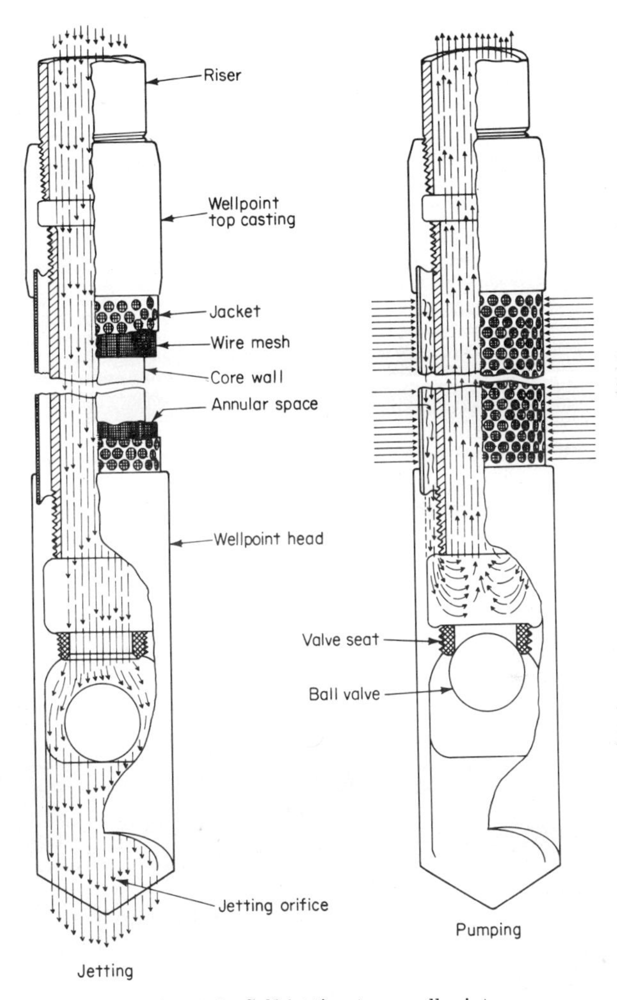

Fig. 29-4 Self-jetting-type wellpoint.

at all times by maintaining a continuous high vacuum. The water-handling pump of the unit is a horizontal-centrifugal type. (See Pumps for Dewatering in this section.)

Figure 29-5 illustrates a typical wellpoint installation and shows all the component parts assembled and operating. Installation of the wellpoint system is started by excavating to, or slightly above, groundwater level, and installing the header manifold at this elevation. In some cases, where the final depth of excavation below the ground surface is less than the practical suction lift, the header is installed on the ground surface. In either case, it is necessary to provide a level grade or one resulting in a

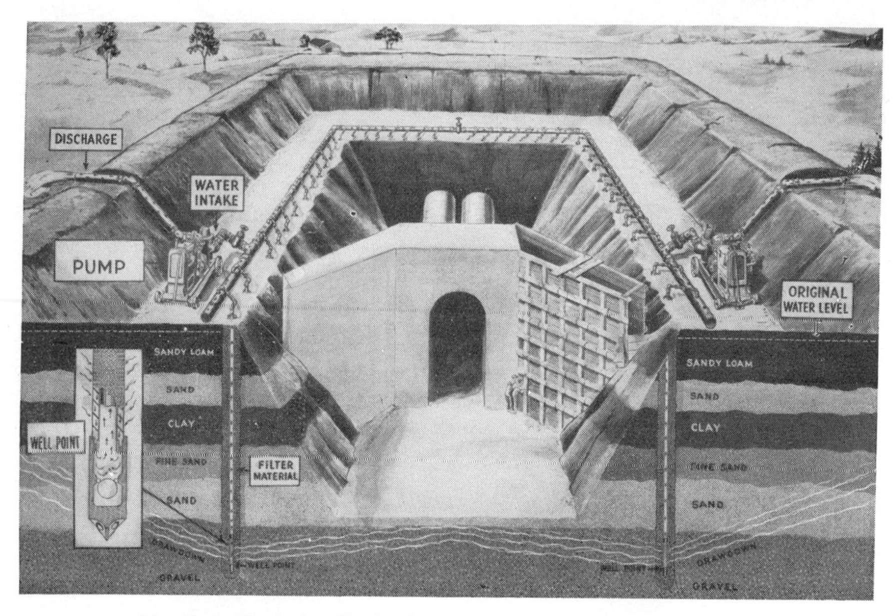

Fig. 29-5 Typical wellpoint installation. (*John W. Stang Corp.*)

slight slope of the header up to the pump. By eliminating "high" points in the header line, the accumulation of air at these points is prevented. Having installed the header with all the necessary elbows, gate valves, tees, and other fittings, the swing joints are connected to the threaded outlets at the required spacing. Unused outlets are plugged.

At this point the wellpoints, attached to the risers, are installed. If the soil below the water table is fairly coarse grained and homogeneous, wellpoints are usually self-jetted into the ground. This operation is illustrated in Fig. 29-6. Where a fine-grained and/or stratified soil profile exists, it is necessary to provide a sand filter

Fig. 29-6 Wellpoint being self-jetted into the ground.

around the wellpoint and riser. This filter increases the effective radius of the wellpoint and permits the dewatering of water-bearing strata which are separated by relatively impervious horizontal layers. The installation of filtered wellpoints is usually performed by utilizing a "sand casing." This casing is jetted into the ground as shown on Fig. 29-7. After jetting, the wellpoint and riser are placed inside the casing and the annular space is filled with the filter sand. The casing is then removed, leaving a filtered wellpoint such as those shown in Fig. 29-5.

The wellpoint pumps are carefully set on firm soil, or on timbers if it is necessary to place them on soft soil. To effect a maximum drawdown, the vertical distance between the header and the pump suction intake should be kept at a minimum. The discharge

Fig. 29-7 Jetting "sand casing" into ground.

line from the pump should extend a sufficient distance to minimize the reentry of water into the excavation.

All pipe connections on the suction side of the pumps must be airtight to maintain a maximum vacuum on the entire system. Couplings used on the header manifold must be provided with positive seal gaskets, and all threaded pipe connections on risers, swing joints, and wellpoints should be treated with pipe compound.

When pumping is started, it is frequently necessary to open the swing-joint valves gradually over a period of several hours to avoid excessively high initial entry velocities. Adjustment of these valves is also necessary during the later stages of drawdown to permit a uniform lowering of the water level and to prevent some of the wellpoints from "going dry." The regulation of these valves is extremely important to maintain a high vacuum on the line and also to prevent clogging the wellpoint screens.

With a properly designed wellpoint system, the desired drawdown is usually obtained in several hours when dewatering pervious soils. In less pervious materials, the rate of drawdown is slower, frequently requiring a period of several days or longer. After the water level is lowered below the deepest final elevation, the excavation is completed. In some cases, the excavation is performed as the water level is being lowered, but at no time should it be carried to a depth below the receding water level. Disregarding this rule will result in a wet "silting" condition in the bottom of the excavation which is frequently difficult to dewater satisfactorily. In the case of a multistage wellpoint operation, the excavation is continued to the lowered water level, at which point the next stage of wellpoints is installed. The wellpoint system is pumped continuously until excavation and construction including backfilling up to the original water level are completed, at which time the wellpoint system is removed.

On pipeline operations, sufficient wellpoint equipment is required to permit the predraining of the trench prior to excavation and also to allow for the reinstallation of a portion of the wellpoint equipment. Where only a short length of pipeline requires excavation below the groundwater level, it is generally more economical to install wellpoints along the entire length of the "wet" trench. Such an operation is quite efficient since the necessity for simultaneously reinstalling the wellpoint equipment along with the excavation and pipe-laying operation is eliminated.

Ejector Method Ejectors (also referred to as "eductors") can be used to lower the water table beyond the limits of a conventional wellpoint system. The ejectors are identical in principle to the so-called "jet pumps" used for domestic water supply where it is necessary to pump from depths beyond the suction limit. When water under pressure passes through the nozzle of the ejector at a high velocity, pressure is reduced below atmospheric. By connecting an ejector to a wellpoint screen, the "vacuum" is applied at this level rather than at the header manifold elevation as in a conventional wellpoint system.

An ejector system requires two manifolds—one, the high-pressure line which supplies the ejectors, and the other the collection manifold which carries the water discharged by the ejectors. This includes the groundwater collected by the screens. Each ejector is connected to each of the manifolds by separate small-diameter pipes, and water is circulated in the system by a horizontal centrifugal pump. The water picked up by the system is discharged at an overflow tank located on the suction side of the pump.

The design of an ejector system must be given careful consideration. It is particularly important to select properly sized ejectors for maximum efficiency at the anticipated head and discharge conditions. Maximum overall efficiencies of ejector systems are low—generally not more than about 20 percent—and their use should not be considered where large rates of pumping are anticipated.

Other Methods The more generally used dewatering methods, which incorporate some type of pumping system, have been given particular emphasis in this section. Less frequently used methods for controlling groundwater in connection with excavations will now be discussed briefly. For the most part these methods are not, strictly speaking, dewatering ones, since water removal is not a prime requisite. Generally, they would be more accurately described as either (1) stabilization methods, such as electroosmosis; (2) cutoff methods, such as chemical grout injection; or (3) a combination of stabilization and cutoff, such as the freezing method.

Electroosmosis[2] is applicable in extremely fine-grained, low-strength soils. The method is based upon the principle that, if a pair of electrodes is driven into water-bearing ground and connected to a source of direct current, the current passing through the ground actuates a flow of water from anode to cathode. Simultaneously, there is a reduction of "pore-water pressure" which changes the stress in the water contained in the interstices of the soil from compression to tension. This phenomenon results in the stabilizing of the soil since the water, being in a state of tension, actually tends to pull the soil particles together.

On a typical electroosmotic installation, the anodes and cathodes are placed alternately around the perimeter so that a wall of stabilized soil is formed by the action of the current flow. Ordinary wellpoints are frequently used as cathodes since they permit the collection of the water flowing to the cathodes. The amount of water collected is quite small as compared with the conventional dewatering methods. The anodes may be pipe or reinforcing steel of suitable length.

The spacing between anodes and cathodes, as well as the capacity and number of dc generators, can best be established by preliminary field testing to determine the conductivity of the soil. This conductivity varies over a wide range and appears to depend on water content, mineralogical composition of the soil, and the amount of chemicals in the water.

Particular emphasis should be placed on the fact that this method should be utilized only where excavation below the water table is required in *extremely fine-grained soils.* Soils that are coarser grained than fine silt can generally be more effectively dewatered by utilizing one of the methods of predraining discussed previously.

Cutoff methods should be considered in areas where pervious strata are underlain by a continuous, relatively impervious layer. Such methods are frequently used where permeabilities are too high to permit economical dewatering by pumping. The necessity for providing bottom pressure relief, as previously discussed, must be checked when cutoff methods are used.

Interlocking steel sheet piling is frequently used in cofferdam construction where driving conditions permit. A reasonably tight cutoff is provided by driving the sheet piling into an impervious stratum. Nominal pumping is generally required to remove seepage through the sheet-pile interlocks. This is a more serious problem when the sheet piling is unstressed, as in the case of a sheet-pile cutoff in an earth or rock cofferdam. Installation of the sheet piling through granular, pervious soils can frequently be performed effectively by using supplementary jetting with conventional dynamic pile driving equipment or by using vibratory pile driving equipment.

Cement grout cutoffs are used occasionally but are restricted to extremely pervious soils such as gravel. They are frequently required (too often overlooked, unfortunately) in gravel bedding under structures which extend below groundwater level where future extensions of the initial construction are contemplated.

The use of grouted cutoffs has significantly increased since the development of the relatively new chemical grouts. These grouts may be used in less pervious soils due to their relatively low viscosities. The newer chemical grouts, when used with wetting agents, have viscosities approximately equivalent to that of water. Only minor strength increase is generally obtained in the soil mass, but an impervious gel is formed which provides the cutoff.

Chemical grouting may be either "two shot" or "one shot." In two-shot grouting, successive injections of two chemicals are made. An example is the Joosten process, which utilizes sodium silicate followed by calcium chloride. In the one-shot process, a single mixture of the main gel-producing chemical with a catalyst is injected. The amount of the catalyst regulates the polymerization time—the time required for the formation of the gel.

Chemical grouts are generally quite expensive and, to use them effectively, it is important that soil properties (particularly permeability) be carefully investigated. In addition, the grouting should be done by specialists experienced in this field. To illustrate this point, if the polymerization time is estimated on the basis of a permeability that is lower than the correct amount, the injected solution will disperse over too wide an area and the result will be an ineffective cutoff.

Slurry-trench cutoffs have been widely used in recent years. The method basically consists of providing internal support to an excavated trench by introducing a bentonite slurry, similar to the method of supporting a rotary-drilled hole with drilling mud. The cutoff may be solely for temporary construction purposes, in which case the trench may be backfilled with a mixture of bentonite and natural material. This mixture will have a much lower permeability than the natural material. If the cutoff is to serve as a cofferdam wall and ultimately become a part of the permanent structure, the trench is usually excavated by special drilling or clamshell equipment. Excavation is followed by the placement of a structural or reinforced-concrete bulkhead. Extensive subway construction which is presently (1969) under way in the San Francisco Bay area involves this method of construction.[5] The wall consists of steel "soldier" piles and concrete and is referred to as an SPTC—soldier-pile tremie-concrete wall.

Freezing methods are also used to stabilize soil, thereby eliminating the necessity for dewatering. This method was originally used in the mining industry and was introduced in construction starting about 1960. Generally, it consists of freezing an ice wall which serves as a cutoff to water and provides support to the walls of the excavation or shaft due to the increase in soil strength resulting from freezing. This method has been used most frequently in connection with the construction of shafts, many extending to considerable depths. Frozen shafts of as much as 1,500 ft ± have been sunk in Saskatchewan, Canada, for the mining of potash.

Freezing pipes are installed in holes drilled on spacings which usually range from about 3 to 6 ft. A refrigeration unit circulates low-temperature brine solution through the freezing piping. The process may be considered as analogous to a dewatering problem to the extent that heat, rather than water, is "pumped out" of the ground. (The analogy between the principles of heat and water flow through soils is extremely helpful in making analyses of such problems.)

The freezing time required to obtain an ice wall of sufficient thickness to satisfy cutoff and structural considerations is usually in the range of from 2 to 4 months. This period may be shortened by reducing the spacing of freezing pipes. The design of the ice wall thickness is performed in a manner similar to the analysis of a comparable unreinforced-concrete structure. Ultimate compressive strengths of frozen soil range from about 500 psi for clay to about 2,000 psi for sand. For a given soil, strength increases with an increase in water content and with a decrease in temperature.

PUMPS FOR DEWATERING

A detailed discussion of pump types and their classification is beyond the scope of this section, but a brief review will be made of those pump types which are generally used for construction dewatering.

Pumps most frequently used for dewatering are of the centrifugal type. A centrifugal pump is basically an impeller or rotor with an intake at its center. Water entering the impeller is rotated and discharged by centrifugal force into the casing which surrounds the impeller. The head or pressure developed by the pump is a result of the velocity imparted to the water by the rotating impeller.

Centrifugal pumps may be single or multistage, the stages representing the number of impellers in the pump. Multistage pumps are required where discharge pressures are high. Centrifugal pumps are also classified as horizontal or vertical, depending on the position of the pump shaft. Centrifugal pumps are characterized by their ability to operate over relatively wide ranges of pumping rate and pressure. These ranges may be further increased by varying the pump speed.

Horizontal Centrifugal Pumps Many sizes and types of horizontal centrifugal pumps are available to the contractor. Some of these are the conventional self-priming "contractors' pumps," wellpoint pumps, and "trash" pumps. They are usually available either engine- or electric-motor-driven. Due to their relatively high efficiencies (70 percent ±) they are particularly well suited for large-capacity pumping requirements at moderate heads.

Horizontal centrifugal pumps have a suction lift limitation which is characteristic of suction-type pumps and is due to the restriction of atmospheric pressure. Generally, the maximum practical suction limit for these pumps is about 20 ft at sea level. This limit can be exceeded, in some cases up to a lift of 25 ft, at the expense of reduced pumping capacity and accelerated pump wear due to cavitation.

Horizontal pumps are not capable of "running dry" for extended periods without mechanical damage. In addition, if a continuous supply of water is not provided, the repriming of the pump becomes a problem. For this reason, suction strainers should always be provided with foot valves. Wellpoint pumps are occasionally used for surface-pumping dewatering because the supplementary vacuum pumps with which they are equipped eliminate the problem of priming.

In recent years, trash pumps have been developed. These pumps have special, open-type impellers, which are capable of handling considerable amounts of suspended solids without excessive wear. Because of this feature, the units are frequently used for the bypass pumping of sewer lines which are undergoing repair.

Vertical Pumps The most frequently used type of vertical pump is the so-called "deep-well turbine," which is a vertical centrifugal pump similar in performance characteristics to the horizontal centrifugal pump. Propeller and axial-flow vertical pumps are used occasionally for large-capacity pumping requirements at low heads.

A vertical pump installation consists of (1) a vertical electric motor, or gasoline or diesel engine with an "angle-gear drive"; (2) a discharge head which provides a mounting for the electric motor or angle-gear drive, and the connection for discharge piping; (3) water column, tubing, and shafting for connection between the pump "bowls" and discharge head; (4) pump impellers and bowl assemblies which are located at the bottom end of the column, tubing, and shafting. A strainer is usually provided at the bottom of the bowl assembly.

These features permit the use of vertical pumping equipment for a wide range of head requirements. Multiple impeller and bowl assemblies are used for high head requirements. The column, tubing, and shafting are usually manufactured in 10-ft lengths, which are combined to provide the necessary "pump setting"—the distance between discharge head and strainer.

Vertical pumps are ideally suited to pumping from deep wells and are also used for dewatering cofferdams. Due to their inability to "run dry" without excessive wear, vertical pumps are frequently installed with float or electrode switches which automatically control the pump motors with the rise and fall of the water at the pump intake.

Submersible Pumps The term "submersible" is used to describe centrifugal pumps which are driven by electric motors close-coupled to the impellers. The motor and impeller assembly is contained in one housing, which normally operates below the water surface. Submersible pumps used for construction dewatering are turbine type or contractors' type.

Submersible turbine pumps are similar to the previously described deep-well turbine except for the drive mechanism and their smaller size. They are frequently used as multistage pumps. Most manufacturers build the units to a maximum-size electric motor of about 5 hp. Due to the close impeller tolerances, these pumps will not handle suspended solids without excessive wear.

Contractors' submersible pumps are available in a wide range of pump sizes up to 8 in. with electric motors of a maximum size of about 65 hp. The smaller sizes, such as the 3-in. unit illustrated in Fig. 29-8, are quite mobile and can be moved by one or two men. Most contractors' submersible pumps can "run dry" for limited periods without mechanical damage and therefore require little attendance during their operation. Due to larger impeller clearances, they can handle dirty water, but large amounts of suspended particles will result in excessive maintenance. Contractors' submersibles are *not* dredge pumps.

Contractors' submersible pumps have certain disadvantages, among which are relatively low efficiencies (generally less than 60 percent), the need for engine-driven electrical generators on "moving jobs" such as pipeline construction, and their relatively high initial and maintenance costs.

Diaphragm Pumps Diaphragm pumps have been for many years the workhorses of pipeline contractors. Their mobility, simple construction, and low maintenance are some of the reasons for their popularity. They are usually engine driven, eliminating the need for generators. Diaphragm pumps are capable of handling dirty water with relatively little maintenance.

Their use is restricted to small pumping rates, due to their relatively low efficiencies and small available sizes (up to about 4 in.). Suction lift limits are similar to those of horizontal centrifugal pumps.

Air Lifts* Air-lift pumping requires the use of an air compressor rather than

* See also Excavation Inside Cofferdams in Sec. 28, Cofferdams and Caissons.

conventional pumping equipment. Although the method is relatively inefficient, it is often used because of its simplicity and its ability to handle large amounts of suspended materials. For the latter reason, an air lift is frequently used for well-development pumping. It may also be applied for limited dredging, such as the excavation of material inside a small bridge-pier cofferdam in a river.

Fig. 29-8 Pumps for dewatering. (*a*) 3-in. contractor's pump—gasoline-engine driven. (*Gorman-Rupp.*) (*b*) 3-in. contractor's submersible pump—electric-motor driven. (*Prosser Industries.*) (*c*) 3-in. diaphragm pump—gasoline-engine driven. (*Marlowe Pumps.*)

The method operates by injecting compressed air into the water inside the discharge pipe at a point below the water level. The unit weight of the created mixture of air and water is less than that of water alone and results in a flow of the air, soil, and water mixture from the top of the discharge pipe. This vertical discharge pipe, along with the smaller air line inside it, must be submerged for at least 40 percent of its length. Pipe diameters for discharge and air line, respectively, range from about 2 in. and ½ in. for 60 gpm up to 8 in. and 2½ in. for 700 gpm.[9]

Pump Selection An analysis of a dewatering problem will result in estimates of the required pumping rate and head, which are used for selecting an appropriate pumping plant.

The pumping conditions for which a given pump is applicable are graphically represented by its characteristic curves. These curves are compiled by the manufacturer on the basis of accurate tests and provide a picture of the relationships between capacity, head, brake horsepower, and efficiency.

Figure 29-9 illustrates a typical set of characteristic curves for a centrifugal pump. The procedure for using these curves can best be illustrated by an example.

Example: Provide a horizontal centrifugal pump capable of handling 2,000 gpm if the pump has 15 ft of suction lift and the discharge point is located 55 ft above the pump. Water is discharged from the pump through a 10-in. pipeline 300 ft long.

Item	*Calculated head, ft*
Suction lift, supply reservoir to inlet	15
Suction inlet to discharge outlet, vertical distance	1
Discharge outlet to discharge point, vertical distance	55
Friction loss, from tables, 3.59 × 3	11
Velocity head, $v^2/2g = 8.16^2/64.4$	1
Minor losses, entrance, exit, elbows, etc	2
Total dynamic head	85

The pump selected for this operation should be capable of efficiently pumping 2,000 gpm at a total dynamic head of 85 ft. Figure 29-9 indicates that this 10-in. pump fulfills these requirements very well since it would be operating in the range of maximum efficiency of about 72 percent. The brake-horsepower curve indicates that an electric motor or engine capable of providing 60 bhp would be required.

DESIGN OF DEWATERING SYSTEMS

The analysis of any dewatering problem requires a knowledge of (1) soil conditions, in particular, the permeability of the soil or the permeabilities of individual strata if the soil profile is variable; (2) the distance to open water, such as a river, lake, etc.; and (3) the depth to a continuous impervious stratum. On the basis of this information, the quantity of water to be pumped is calculated for the amount of water lowering required. The dewatering system selected must provide a collection capacity equal to or greater than the calculated rate of inflow. In other words, any dewatering problem must be considered from the standpoints of yield from the ground and collection capacity.

The nomenclature used in this section is defined below, along with the units used:

Q = rate of flow, gpm
Q_w = rate of flow from a single well, gpm
q = rate of flow per foot of perimeter, gpm
k = permeability; velocity for unit hydraulic gradient, fpm
i = hydraulic gradient (dimensionless)
H = vertical distance from undisturbed groundwater level to an impervious stratum, ft
h = vertical distance from lowered water level to an impervious stratum, ft
h_o = depth of immersion of well, ft
$H - h$ = drawdown, ft
m = aquifer thickness, artesian condition, ft
R = radius of influence of depressed water table, ft
r_1 = effective radius of a ring of wells or wellpoints, ft
r_w = effective radius of a well or wellpoint, ft
W = spacing between wells, ft
p = horizontal distance from center of ring of wells to shoreline of open water, ft

The flow of water through soils is governed by the basic relationship known as Darcy's law, which states that for a given soil the velocity of flow is directly proportional to the hydraulic gradient and the permeability of the soil:[10]

$$\text{Velocity} = ki$$

The permeability factor k can be considered as a coefficient representing the "hydraulic conductivity" of a given soil. Having the units of velocity (feet per minute), it can be considered as the rate of water movement through the soil at a hydraulic gradient of 1. The physical significance of Darcy's law can best be illustrated by Fig. 29-10, which shows a constant-head permeability test. The hydraulic gradient i is equal to the difference in piezometric head ΔH divided by the length of the sample.

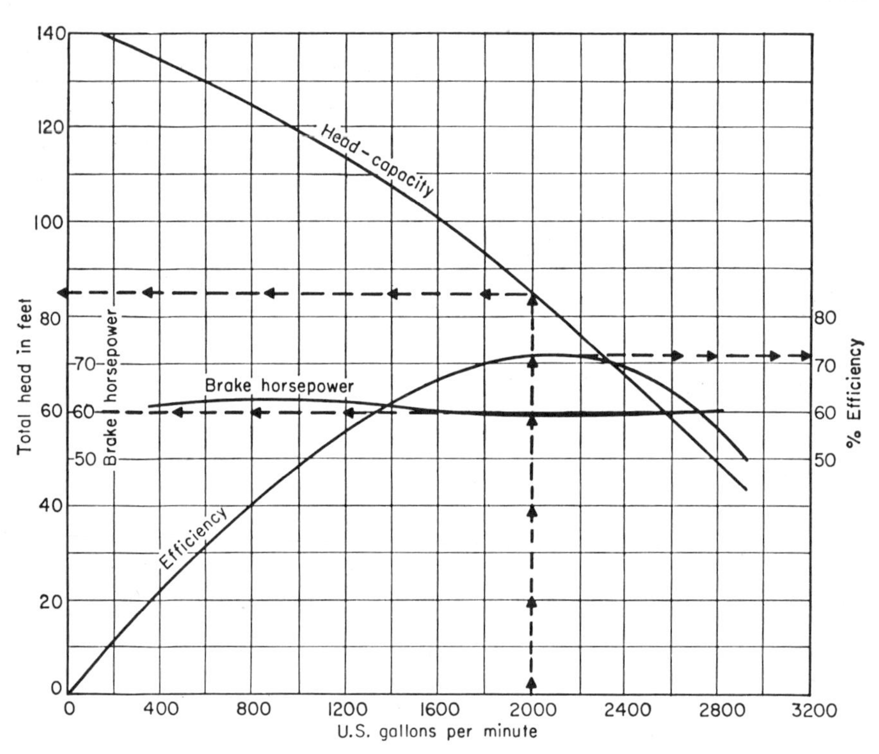

Fig. 29-9 Centrifugal-pump performance curve. (*John W. Stang Corp.*)

Determination of Permeability The design of a dewatering system requires a reasonably accurate evaluation of the permeability of the soils below the water table. Laboratory testing may be used, such as the constant-head test illustrated schematically in Fig. 29-10. However, due to the difficulty of obtaining relatively undisturbed samples of the granular pervious soils which are of principal concern in a dewatering analysis, permeability based on laboratory tests is frequently found to be misleading and significantly different from the in situ permeability.

Permeability is frequently estimated on the basis of grain size using a method originally proposed by Hazen.[11] Hazen found that, for uniformly graded filter sands, permeability (feet per minute) is roughly equal to $2(D_{10})^2$, where D_{10} is in millimeters. Here D_{10} is the so-called 10 percent size, which is defined as that grain size representing the division between the finer fraction which accounts for 10 percent of the total sample weight and the coarser fraction which accounts for 90 percent of the total sample. More recent studies indicate similar[12] relationships between in situ permeability and the D_{10} size. It must be emphasized that any method for estimating permeability based on grain size alone should be used cautiously, particularly if applied to fine-grained sands and silts.

The approximate relationship between permeability and the D_{10} size indicates that the fine fraction of a soil has a significant effect on its permeability. As an illustration of this important point, a relatively clean concrete sand will have a permeability of around 0.04 fpm. If it contains the maximum permissible amount of fines (10 percent passing a No. 100 sieve), its permeability can be reduced to the range of from 0.7×10^{-4} to 0.7×10^{-5} fpm.[13]

Permeability may be more accurately evaluated by performing pumping tests, and these are strongly recommended for projects involving large dewatering costs. Water is pumped from a well, preferably one which fully penetrates the aquifer and is located at a central point in the area to be dewatered. The rate of pumping should be kept constant during the test period since this permits the application of nonequilibrium well formulas relating permeability, pumping rate, and drawdown rate.[14,15] When

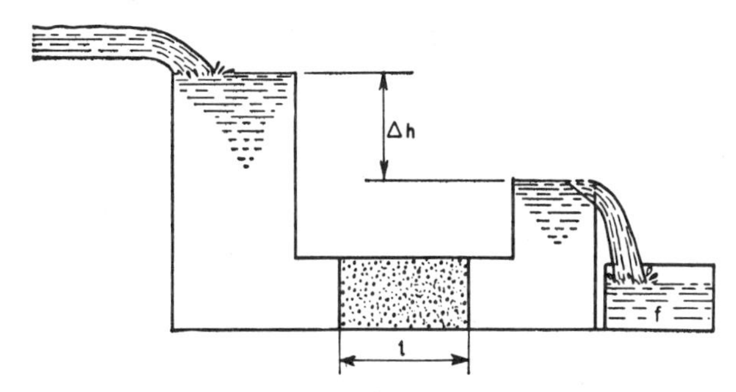

Fig. 29-10 Constant-head permeability test.

the drawdown in surrounding observation wells approaches a static condition, pumping can be stopped. Water elevations are measured in the observation wells during and following pumping. The pump test data should be evaluated on the basis of the nonequilibrium methods, both drawdown and "rebound," and using the classical equilibrium well formulas discussed in this section. A review of the nonequilibrium methods is beyond the scope of this section.

Well Formulas The basic well formula for a single gravity well, fully penetrating the pervious stratum, is[16]

$$Q_w = \frac{7.5\pi k(H^2 - h^2)}{\ln (R/r_w)}$$

The following expressions for multiple gravity wells can be derived by following Muscat's procedure:[17,18]

For two wells a distance W apart:

$$Q_w = \frac{7.5\pi k(H^2 - h^2)}{\ln (R^2/r_w W)}$$

For three wells in a pattern of an equilateral triangle:

$$Q_w = \frac{7.5\pi k(H^2 - h^2)}{\ln (R^3/r_w W^2)}$$

For three wells equally spaced in a straight line:

$$Q_w = \frac{7.5\pi k(H^2 - h^2) \ln (W/r_w)}{2 \ln R/W \ln W/r_w + \ln W/2r_w \ln R/r_w} \qquad \text{(Outer wells)}$$

$$Q_w = \frac{7.5\pi k(H^2 - h^2) \ln (W/2r_w)}{2 \ln (R/W) \ln (W/r_w) + \ln (W/2r_w) \ln (R/r_w)} \qquad \text{(Middle well)}$$

For four wells in a square pattern:

$$Q_w = \frac{7.5\pi k(H^2 - h^2)}{\ln (R^4/\sqrt{2}\, r_w W^3)}$$

The basic well formula can be applied for obtaining the flow into a ring consisting of an infinite number of wells:

$$Q = \frac{7.5\pi k(H^2 - h^2)}{\ln (R/r_1)}$$

The simplifying assumption that a continuous ring of wells exists permits the determination of the flow per foot of perimeter for estimating the spacing of wellpoints. For normal wellpoint spacings, the assumption does not introduce any appreciable error.

The above well formulas are all based on the assumption that a circular boundary exists having a radius of influence of R. Equations can be derived for various other boundary conditions. For example, the case of a ring of an infinite number of wells bounded by a line source (shore line) is[19]

$$Q = \frac{7.5\pi k(H^2 - h^2)}{\ln (2p/r_1)}$$

A high degree of precision in the estimation of the radius of influence R (or the distance to a line source p), and the effective radius of the ring of wells or wellpoints r_1, is not required because the ratio of these dimensions appears as a logarithmic term in the well formulas. Therefore, the formulas can be applied with reasonable accuracy to typical excavations which are square or rectangular in plan rather than circular. Approximate r_1 values can be readily determined on the basis of the dimensions of the excavation.

For excavations that are extremely long in plan, for example a pipeline trench with wellpoints on one or both sides of the trench, the selection of appropriate r_1 values is more difficult. The effective radius of a rectangular installation of wells or wellpoints having a length-to-width ratio greater than 10 is equal to approximately one-sixth of the length.

All the above well formulas are applicable to fully penetrating gravity wells. The mathematical treatment of the subject of partial penetration is quite complex and beyond the scope of this section.[17] It should be noted, however, that for a given soil profile and a required drawdown $(H - h)$, the pumping rate Q from partially penetrating wells will be less than that for wells that are fully penetrating. Therefore, well formulas which are strictly applicable only to fully penetrating wells can be used as a conservative basis for the design of dewatering systems which involve partially penetrating wells. The necessity for applying a correction for partial penetration is difficult to justify in many cases because of uncertainties regarding permeability and the depth to imperviousness.

The well formulas considered above are for *gravity flow* conditions; i.e., the drawdown effected by pumping occurs in an *unconfined* aquifer—the condition most frequently encountered in practical dewatering problems. Pressure relief problems, on the other hand, require pumping from a *confined* aquifer, such as the underlying layer shown in Fig. 29-1. Such conditions require the application of *artesian* well formulas.

Gravity well formulas may easily be adjusted for application to artesian conditions by considering the aquifer thickness through which seepage passes, in the following manner. The term $(H^2 - h^2)$ in the gravity well formulas may be written as $(H - h)(H + h)$, which could be considered as the drawdown $(H - h)$ times *roughly* twice the average wetted thickness of the aquifer $2 \times (H + h)/2$. In a confined aquifer, the wetted thickness of the aquifer is equal to the aquifer thickness except in the rare instance when the water level is lowered below the top of the aquifer. (For this last case the wetted thickness is reduced slightly.) The basic artesian well formula may be expressed as

$$Q_w = \frac{7.5\pi k(H - h)(2m)}{\ln (R/r_w)}$$

Again, $2m$ is twice the average (which in this case is a constant) wetted thickness of aquifer. From this it is apparent that the only difference between gravity and artesian well formulas is a factor which expresses (approximately) the variable of wetted aquifer thickness. All the previously discussed gravity well formulas may be expressed as corresponding artesian well formulas by replacing the term $(H^2 - h^2)$ by $(H - h)(2m)$.

Capacity of Wells The rate of flow into a pumped well or wellpoint depends upon the area and permeability of the ground immediately outside the well and upon the hydraulic gradient causing the flow. Applying Darcy's law, the following expression can be written:

$$Q = 15ki\pi r_w h_o$$

Sichardt found that the entry gradient into the well can be expressed in terms of the permeability of the soil.[21] His results are contained in the empirical formula

$$i = \frac{1}{1.07\sqrt{k}}$$

From this it follows that

$$Q \cong 14\sqrt{k}\,\pi r_w h_o$$

This equation permits the determination of the capacity of a well or wellpoint for any conditions. The effective radius of a well would be the radius of the perforated casing if not gravel packed and the radius of the filter envelope in the case of gravel-packed wells. For self-jetted wellpoints, the effective radius depends on the soil, since the cavity formed in the jetting process varies in size in different types of soil. In coarse-grained soils, a larger cavity is formed than is the case in fine-grained soils. During the jetting process, the coarser particles settle around the wellpoint screen, filling this cavity and forming a natural filter having a radius ranging up to approximately 6 in. Where wellpoints are used in fine-grained materials, it is generally advisable to use the sand casing discussed under Wellpoint Method. In this case, the effective radius of the wellpoint would be the radius of the sand casing. In estimating the capacity of a given well or wellpoint, it is important to allow for the reduced inflow area which results from the lowering of the water table; in terms of the above equation, this is a reduction in h_o.

Gravel or sand filters are used on wells or wellpoints for the purpose of preventing an excessive amount of fines from entering the screens or perforated casing, increasing the collection capacity, and providing vertical drainage in stratified soils. The gradation of the filter with regard to the natural soil is generally made on the basis of the Terzaghi criterion,[22] which is expressed in the following relationship:

$$\frac{D_{15} \text{ (of filter)}}{D_{85} \text{ (of natural soil)}} < 4 \quad \text{to} \quad 5 < \frac{D_{15} \text{ (of filter)}}{D_{15} \text{ (of natural soil)}}$$

The subscripts designate percent, as in the case of the previously defined D_{10} size. The slot width or hole diameter used in the well casing or screen should be equal to or smaller than the D_{70} size of the adjacent material.[23]

COST OF DEWATERING OPERATIONS

The following items must be considered in estimating the costs involved in any dewatering operation:

1. Cost or rental of pumping equipment, jetting equipment, pipe, valves, fittings, etc. Except on projects of unusually long duration, it is generally to the contractor's advantage to rent such equipment. The manufacturer generally provides free estimates covering equipment rental along with other cost estimates covering the additional items entering into the total cost of dewatering.

2. Transportation costs. This item includes freight charges plus an allowance for unloading and loading the equipment at the site.

3. Installation and removal costs. In the case of the well method, these include the cost of drilling the well, gravel packing and its installation, perforated casing and its installation, setting of the pump, and installation of all necessary pipe. In the case of a wellpoint installation, these costs include laying the header and discharge lines, installing the pumps, and jetting the wellpoints. The costs of the jetting operation will generally range from $\frac{1}{4}$ to 2 man-hours per wellpoint, depending on the nature of the soil and whether the self-jetting or sand-casing method is utilized. A light crane is generally required during the jetting operation. An allowance must be made for miscellaneous materials, such as filter sand and fuel to operate the jet pump.

4. Operational costs. (*a*) Fuel, grease, and oil. The cost of fuel or power can be estimated on the basis of the horsepower rating of the power unit and/or the performance rating provided by the manufacturer. On continuously operating pumping units with engine drives, oil changes are generally required at least once every 48 hr. (*b*) Pump operators are generally required around the clock on continuous pumping operations.

5. Miscellaneous. Allowance must be made for shortages, damages, maintenance, etc. These items are extremely indeterminate but can generally be satisfactorily estimated on the basis of the nature of the project and its anticipated duration.

REFERENCES

1. Harry R. Cedergran, *Seepage, Drainage, and Flow Nets*, chaps. 3 and 4, John Wiley & Sons, Inc., New York, 1967.
2. Leo Casagrande, *Review of Past and Current Work on Electro-osmotic Stabilization of Soils*, Harvard Soil Mechanics Series No. 45, Cambridge, Mass., 1953.
3. Chemical Grouting: Progress Report of the Task Committee on Chemical Grouting, *J. Soil Mech. Found. Div., Proc. Am. Soc. Civil Engrs.*, paper 1426, November 1957.
4. Reuben H. Karol: Chemical Grouting Technology, *J. Soil Mech. Found. Div., Proc. Am. Soc. Civil Engrs.*, paper 5748, January, 1968.
5. George J. Thon and Richard C. Harlan, *Slurry Wall Construction for BART Subway Stations*, meeting preprint 746, Am. Soc. Civil Engrs. Annual Meeting, Pittsburgh, September, 1968.
6. Frederick J. Sanger, Ground Freezing in Construction, *J. Soil Mech. Found. Div., Proc. Am. Soc. Civil Engrs.*, paper 5743, January, 1968.
7. Alfred R. Jumikis, *Thermal Soil Mechanics*, chaps. 10 and 11, Rutgers University Press, New Brunswick, N.J., 1966.
8. Mile High Deilmann Ltd., *Freezing Method and Its Application in Construction*, Wheatridge, Col., 1968.
9. Pumping Water by Air-Lift, *Johnson Drillers Journal*, St. Paul, Minn., November, 1960.
10. H. Darcy, *Les Fontaines Publiques de la Ville de Dijon*, 1856.
11. Allen Hazen, *Some Physical Properties of Sands and Gravels*, Massachusetts State Board of Health, 24th Annual Report for 1892, 1893.
12. *Investigation of Underseepage and Its Control, Lower Mississippi River Levees*, Waterways Experiment Station TM 3-424, Vicksburg, Miss., October, 1956.
13. H. R. Cedergren, Seepage Requirements of Filters and Pervious Bases, *Trans. Am. Soc. Civil Engrs.*, vol. 127, part I, pp. 1090–1113.
14. L. K. Wenzel, *Methods for Determining Permeability of Water Bearing Materials*, Geological Survey Water-Supply Paper 887.
15. C. E. Jacob, On The Flow of Water in an Elastic Artesian Aquifer, *Trans. American Geophysical Union*, vol. 21, pp. 574–586, 1940.

16. J. Dupuit, *Études Théoriques et Pratiques sur le Mouvement des Eaux*, 1863.
17. M. Muskat, *The Flow of Homogeneous Fluids through Porous Media*, chap. 9, J. W. Edwards, Publisher, Inc., Ann Arbor, Mich., 1946.
18. Harold E. Babbitt and David H. Caldwell, The Free Surface around, and Interference between, Gravity Wells, *Univ. Illinois Bull.*, vol. 45, no. 30, Jan. 7, 1948.
19. Stuart B. Avery, Jr., Analysis of Ground Water Lowering Adjacent to Open Water, *Proc. Am. Soc. Civil Engrs.*, vol. 77, December, 1951.
20. V. H. Collingridge, *Wells for Water Lowering*, unpublished pamphlet.
21. W. Sichardt, *Das Fassungsvermögen von Rohrbrunnen u. seine Bedeutung für die Grundwasserabsenkung, insbesondere für grössere Absenkungstiefen*, Springer-Verlag, Berlin, 1928.
22. Karl Terzaghi and Ralph Peck: *Soil Mechanics in Engineering Practice*, 1st ed., chap. 2, John Wiley & Sons, Inc., New York, 1948.
23. G. A. Leonards (Ed.), *Foundation Engineering*, chap. 3, McGraw-Hill Book Co., New York, 1962.

Section 30

River Diversion

LYMAN D. WILBUR

Vice-president, Morrison-Knudsen Company, Inc., Boise, Idaho; and Chairman of the Board, International Engineering Company, Inc., San Francisco, Calif.

INTRODUCTION

To construct a dam across a stream of flowing water, it is necessary to divert the stream from its natural bed to give access to the river bottom in an unwatered condition. Methods used for diverting streams for dam construction are also sometimes used for other structures, such as inverted siphons, but usually such diversions are much more simply accomplished by using only the first steps required for a dam.

FACTORS INFLUENCING METHOD OF DIVERSION

Characteristics of Stream Most streams have large variations between high and low flows. Low flows can be used to great advantage in the construction of cofferdams, particularly in making closures and turning the stream into diversion channels. The flow characteristics and the physical conditions at the site are the basic considerations in determining the method of diversion.

Some streams, such as the Zambezi in Africa, fluctuate with great regularity from a low flow of less than 7,000 to a high of over 570,000 cfs, the low-water season (flows less than 50,000 cfs) always occurring between July and January. Some years the low flow is somewhat higher than others and the low-flow period shorter, but it always occurs within the same months. The river stage may fluctuate as much as 70 ft.[1,2]*

California streams have long periods of low flow, with the main runoff coming from melting snows during a few months in the spring. However, these same streams have their greatest flows for short durations at any time from December until the end of the spring runoff, caused by warm rains on snow. Desert streams are frequently dry or have a negligible flow, except for flash floods. Such streams permit relatively simple and inexpensive diversion methods, as the work in the riverbeds below river level can frequently be accomplished in one season, requiring only the diversion of low-water flows. High water at the next season may then flow over low blocks in the finished work. For other types of streams, expensive works must be constructed to handle large flows of water, sometimes at high velocities.

Characteristics of Site The ideal site for a dam is in a narrow canyon with hard rock at or near the surface. Such a site allows little space or choice of natural material for diversion works, thus restricting the choice of method.

Many streams flow through wide valleys with gently sloping banks. Such a site for a dam provides sufficient space to permit the shifting of the river from its natural location to a portion of its bed or to a diversion channel during construction in the river bed. If the dam site is on a winding stream at the end of a narrow ridge, the ridge may permit construction of short diversion tunnels at low cost.

Exposed rock in the stream bed as compared to heavy earth overburden makes the driving of sheet piles difficult but tunneling easier, thus influencing the method of diversion. Available materials, earth or rock, and velocities in the stream during construction all affect the decision as to diversion method.

Characteristics of Method Many factors must be considered in determining the best method of diversion:

1. The cost of the diversion works and the effect the diversion will have on the entire cost of the project.

2. The time required to construct the project under various diversion schemes and its effect on interest charges during construction as well as overhead costs. If the time available for construction is established by other factors, only methods permitting the meeting of the schedule need be considered.

3. Safety of the diversion works, the permanent structure, and the public. Diversion works are rarely designed to prevent flooding of the works under the worst possible conditions of stream flow. In case the bypass channels are inadequate to handle the maximum possible flow, provision must be made so that the diversion works and permanent structure will not be damaged, or if damaged, that the possible cost of repairs will be less than the cost of preventing the damage. Provision must be made so that third parties and their property will not be adversely affected. Generally speaking, only concrete dams may be permitted to be overtopped.

4. Effect of overtopping of cofferdams and flooding of the permanent works. The cost of designing cofferdams against overtopping is frequently more than the value of the lost time for construction within the cofferdam area. Therefore, the works may be

* Superior numbers refer to the list of references at the end of this section.

permitted to flood during high water. The effect of the lost time on the construction schedule and costs must be studied and weighed against the cost of a higher cofferdam with no lost time or a lesser lost time because of flooding.

5. The necessity of preventing overtopping of cofferdams around earth-fill dams or overtopping of earth and rock-fill dams, or the cost of protecting earth and rock-fill dams if overtopped. Earth-fill dams will not stand overtopping except when the top has not yet been raised above the stream bed. They must not be permitted to be overtopped under any circumstance when raised to any appreciable height above the stream bed unless elaborate temporary paving of the surface is performed where overtopping might occur.[3] Rock-fill dams may be permitted to be overtopped only to the extent that eroding velocities will not develop. At Blakely Mountain Dam, a compacted earth-fill structure, two tunnels were available for diversion but were of insufficient capacity to take the maximum flood flow. Prior to the flood season, placing of earth embankment was stopped and a layer of dumped riprap 300 ft long and 3 ft thick was constructed on top of the fill to form a weir for a distance of 250 ft upstream from the downstream slope. The 1:3 downstream slope of the dam was paved with grouted riprap 5 ft deep. This weir was designed for a 20-ft depth of water,[4] but the dam was actually overtopped by only 6 or 7 in.

SCHEMES FOR ACCOMPLISHING DIVERSION

Tunnels, flumes, conduits, and canals may be used to divert a stream from its natural bed, or cofferdams may force a river to occupy only a portion of its bed. The most common schemes are described below.

Tunnel Bypass Tunnels driven through one or both abutments with a capacity sufficient to take the maximum flood flow permit uninterrupted construction of the dam for its entire length. From the standpoint of the dam construction itself, exclusive of the cost of diversion, this scheme will result in the lowest cost. At some sites this is the only practical solution, as at Hoover Dam.[5,6] This was constructed in a narrow canyon requiring deep excavation, and it took more than a year from the time unwatering of the foundation was started until concrete could reach river level. The project required four 50-ft-diameter concrete-lined tunnels, two through each abutment, and an upstream cofferdam 98 ft high of rolled earth fill, with a concrete slab on the upstream face and a rock-fill blanket on the downstream face. Diversion was accomplished during a low stage of the river by first removing temporary concrete-arch cofferdams left at the upstream portals of the diversion tunnels, then constructing a rock fill downstream from the dam. In relatively still water, the upstream portion of the upstream cofferdam and the downstream earth-fill cofferdam were constructed to river level. After unwatering the dam site, the remaining cofferdams were built in the dry to the elevations required for flood flows. The lower portions of the diversion tunnels were connected to the spillways by inclined tunnels to act as the spillways for the completed structures (see Fig. 30-1).

More frequently, for concrete dams where conditions permit, diversion tunnels are constructed to bypass only low river flows for a sufficient period of time to permit excavation and construction of the dam to stream bed during low-water periods. Some of the sections of the dam are left low, or openings are left to take high-river discharges, with construction of the balance of the dam continuing during high-river stages. The low section can be raised or openings closed during subsequent low-flow periods, when overtopping could not adversely affect the work (see Fig. 30-2). Openings should preferably be round or have peaked tops so as to facilitate later filling with concrete.

Diversion tunnels for earth-fill dams must be of sufficient size to prevent overtopping unless steps are taken to prevent erosion. They do not need to take the maximum flood flow in cases where the dam can be constructed to sufficient height during low water so that storage capacity above the dam will accommodate runoff in excess of tunnel capacity. The Sorensen graphical method may be used to determine the tunnel discharge capacity required.[7] In cases where an earth-fill dam cannot be economically completed in one low-water season, the portions of the dam outside the stream bed may be completed first, leaving only the river section to be completed in

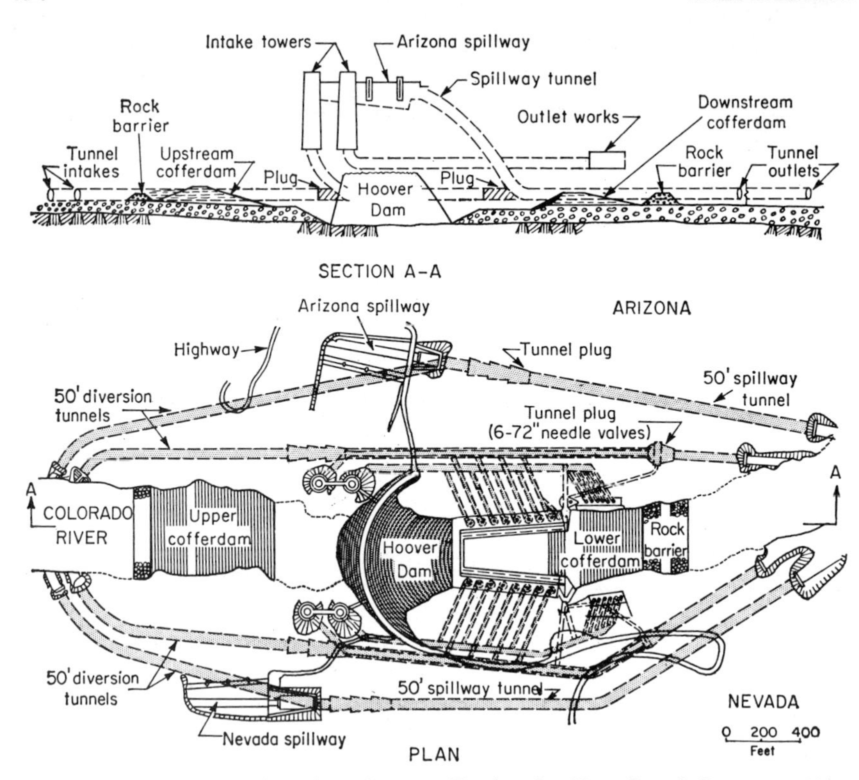

Fig. 30-1 Hoover Dam diversion scheme. (*Engineering News-Record, Dec. 15, 1932.*)

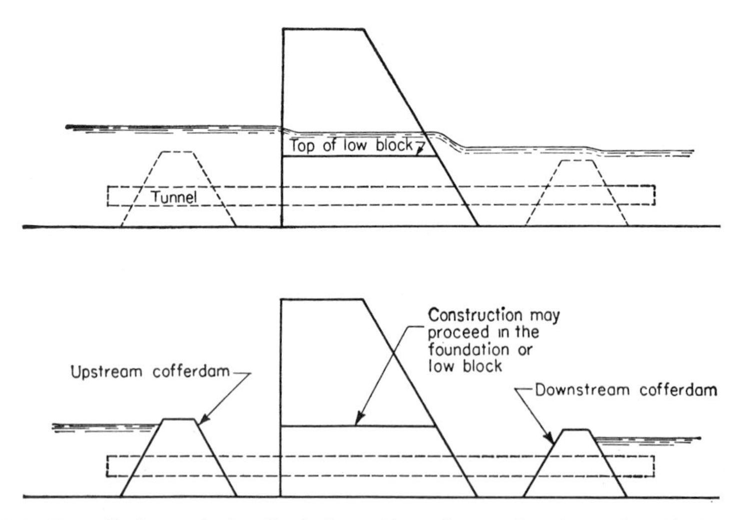

Fig. 30-2 Top: discharge during flood stages through tunnel and over low block in dam. Bottom: discharge during low-flow periods.

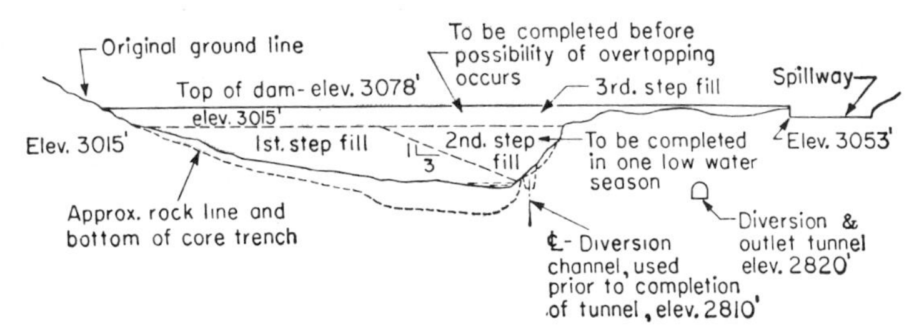

Fig. 30-3 Lucky Peak Dam fill sequence.

one season (see Fig. 30-3). It is sometimes possible to use the upstream and downstream portions of the main dam as upstream and downstream cofferdams to divert the river into diversion tunnels (Fig. 30-4). Sometimes the diversion tunnels can be incorporated in the finished work as the reservoir outlet or spillway outlet.

Diversion tunnels may be unlined in solid hard rock, but for earth, soft rock, or high velocities in seamy rock where objectionable erosion may occur, concrete lining is required.

After the dam has reached an elevation so that storage can begin without danger of overtopping or damage to the structure, the tunnel must be plugged. A permanent concrete plug is usually required unless the diversion tunnel is to be used as a permanent outlet, in which case permanent gates must be installed. The plug or gates can be installed only after flow through the tunnel has been stopped by a temporary closure at the upstream portal. The method of closure must be selected so as to be certain of success. Closure during a period of low-river flow is to be desired in order to minimize the head for which the temporary closure structure must be designed.

This type of diversion has been used at Hungry Horse Dam,[8] Anderson Ranch Dam,[9] Box Canyon Dam,[10] Boysen Dam,[11] Cabinet Gorge Dam,[12] Garrison Dam,[13] and many others.

Flume Bypass In some locations for concrete dams, a flume may be used for carrying the stream flow along one abutment, water being restrained from entering the dam foundation area by upstream and downstream cofferdams. Flumes may be constructed of wood, steel, or concrete as part of the permanent dam construction. A wood flume 25 ft high by 65 ft wide used at Canyon Ferry Dam near Helena, Mont. was designed to carry a discharge of 23,000 cfs[14] (see Fig. 30-5). Flow exceeds this 1 year out of 5, with a maximum of 38,000 cfs. If flows in excess of flume capacity are to occur, arrangements must be made to breach the cofferdam to allow flooding without damage. Care must be exercised to make a tight cutoff around the flume at the cofferdams and arrangements made for stop logs or other measures to close the flume. A concrete flume, later incorporated as part of the dam, may be used. While the stream is diverted through the flume, the dam is constructed in the river channel. Either temporary or permanent sluiceways near river-bed level may be constructed in the dam to take the stream flow while the dam within the area of the flume is constructed. Temporary sluiceways must be temporarily closed by gates and then plugged with concrete and the joints grouted. Instead of sluiceways, low blocks in the dam may be used to pass the stream flow. By using movable cofferdams at the up-

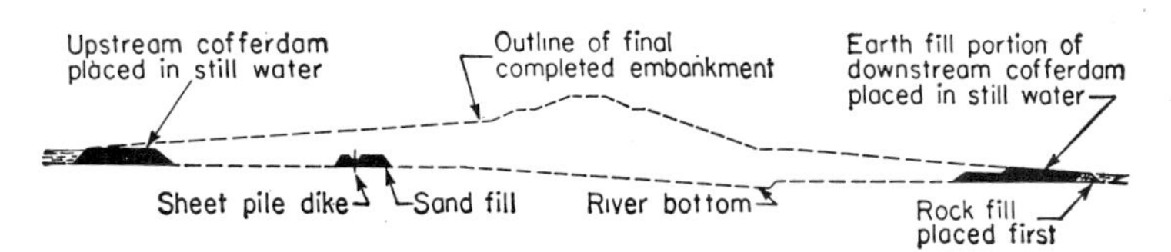

Fig. 30-4 Garrison Dam cofferdams.

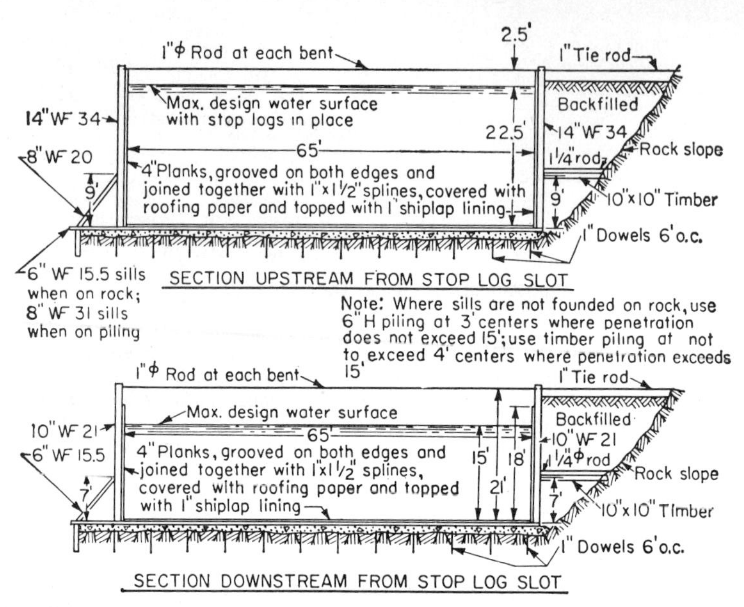

Fig. 30-5 Canyon Ferry Dam diversion flume. (*Engineering News-Record, Aug. 3, 1950.*)

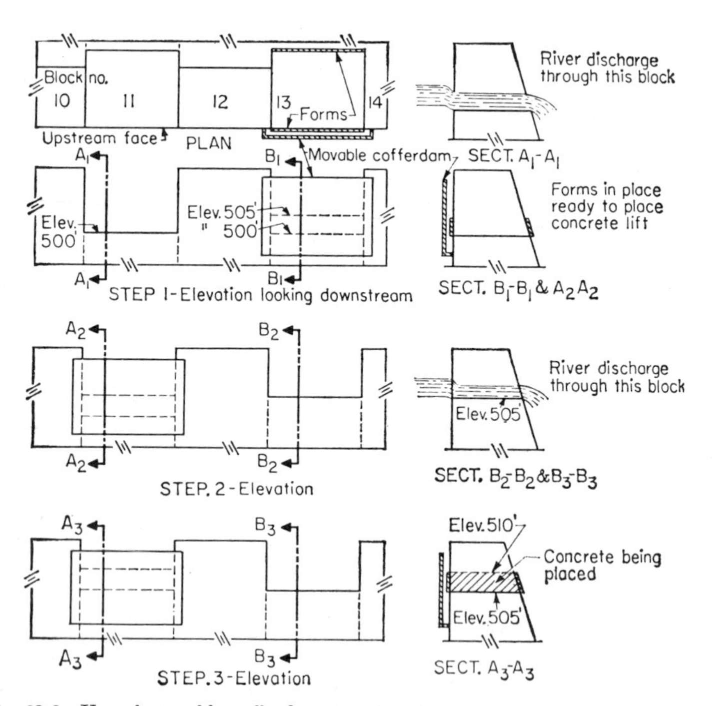

Fig. 30-6 Use of movable cofferdam to raise blocks below reservoir water level.

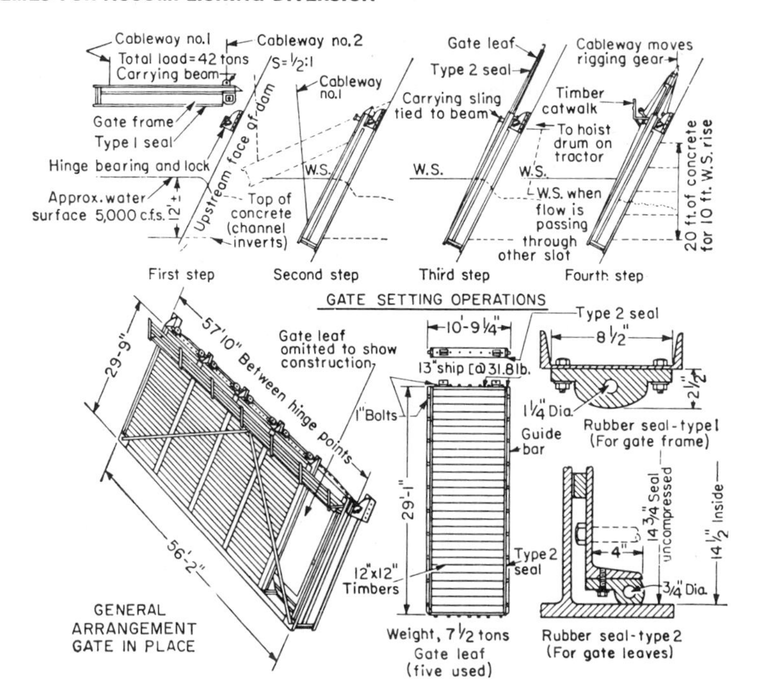

Fig. 30-7 Shasta Dam movable cofferdam. (*Engineering News-Record, Sept. 9, 1943.*)

stream face of the dam, concrete may be placed in one block while the river discharge is being carried by the alternate block. This type of cofferdam was used at Shasta Dam[15] (see Figs. 30-6 and 30-7).

Conduit Bypass A heavily reinforced cast-in-place concrete conduit placed as a permanent outlet may be used for diverting low flows through earth-fill dams. This method is suitable where the river section of the dam may be completed in one low-water season or where foundation excavation and fill construction to stream-bed level can be completed in one season and the remainder of the dam completed in the next season. Concrete-encased steel penstocks may also be used for diversion (see Fig. 30-8).

Two- or Three-stage Construction Where characteristics of site and stream, type of dam and its design, and the construction time schedule permit, it is frequently most economical to construct a first-stage cofferdam around a portion of the work. A portion of the dam is then constructed within this area, leaving permanent or temporary sluiceways to take the river flow during subsequent stages of construction.

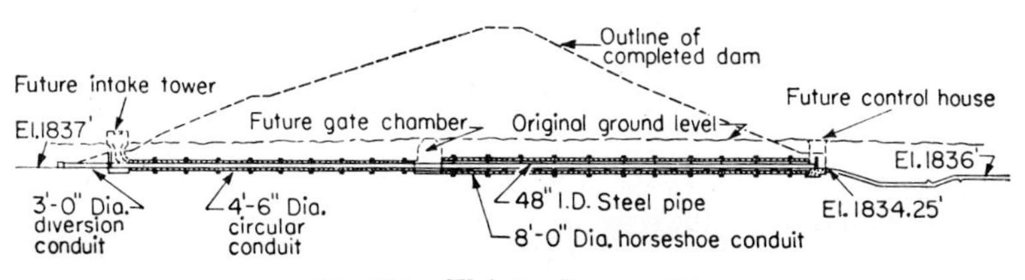

Fig. 30-8 Webster Dam outlet.

After the work within the first-stage cofferdam has progressed sufficiently to permit completion after flooding, the cofferdam is removed to allow diversion of the stream through the sluiceways and another section of the dam is enclosed within a second-stage cofferdam. The sequence may be repeated three or more times, or two stages may suffice. This scheme of diversion is usually applied to large rivers in wide valleys but may be employed to advantage on lesser streams. It has been used for the great dams on such rivers as the Columbia[16–21] and St. Lawrence, and for lesser dams on the Ohio. The diversion scheme at McNary Dam is typical of this type (see Fig. 30-9). In some cases permanent gated sluiceways can be constructed within the first-stage

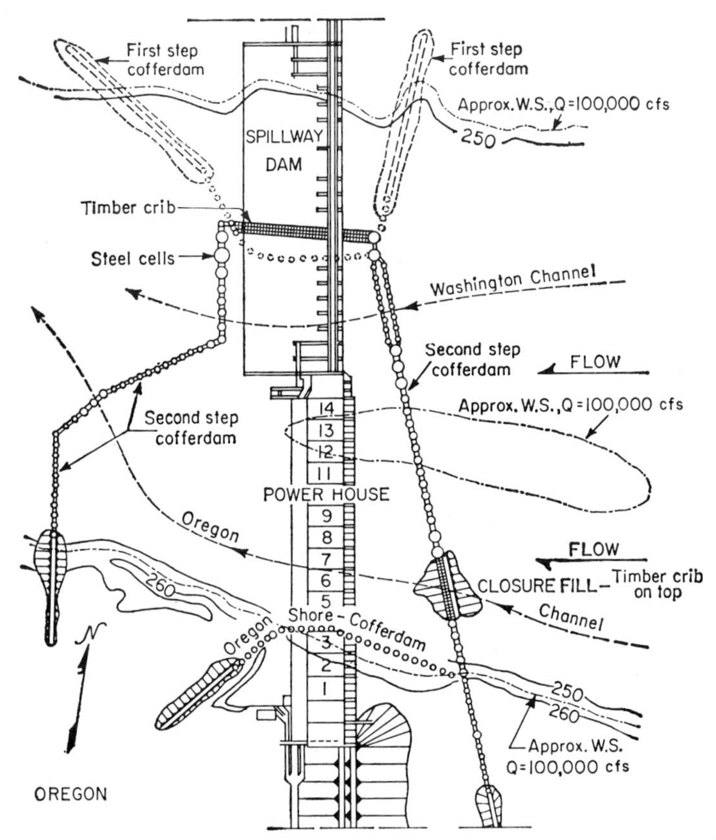

Fig. 30-9 McNary Dam river diversion. (*Corps of Engineers, Walla Walla District, Washington.*)

cofferdams, such as at Imperial Dam[22] and Morelos Dam,[23] and the stream can be diverted through these sluiceways by the second-stage cofferdam.

Permanent Outlets Perhaps the simplest diversion works consist of permanent outlets built in the river valley but outside the stream bed. Construction of a diversion channel from the river to the outlet works and cofferdams across the river will then permit unwatering of the riverbed. Where diversion-channel excavation is useful as fill in earth-fill dams and large-capacity permanent outlets are required at river level, this scheme can be accomplished at a small diversion cost.

CLOSURE PROBLEMS

In determining the diversion scheme and the type of cofferdam and materials to use, the most important consideration is how the final closure is to be made. The first steps of diversion are relatively easy, but at closure high velocities and high pressures are usually present and must be thoroughly understood and planned for. Closure must be scheduled to phase with river levels. Closure during periods of low water results in the lowest velocities and pressures and therefore the lowest cost. Cofferdams must be constructed of materials able to withstand velocities that will occur during closure.

Where diversion tunnels or temporary sluiceways are used, some method must be provided to close the diversion opening temporarily so that a permanent plug may be placed or the sluiceways filled. Where flows and hydraulic heads are small, closure may be effected by placing needle beams or stop logs, one by one, in prepared slots. At Detroit Dam, 14 concrete logs 30 ft long, coated with mastic and weighing 18 tons each, were used to close the diversion tunnel.[24] Where flows are large, the flow may best be stopped by dropping a heavy slide gate specially constructed of steel, concrete, and wood or other suitable material. Gate guides must be prepared in advance of diverting the stream into the tunnel or sluices. Gates may be hung from cables arranged so that they may be cut with a torch when time comes for closure. Gates must be designed with enough weight to overcome friction due to static as well as kinetic action of the water and with enough structural strength to withstand the loads during closure and the head that will build up as the reservoir fills.[4]

The weight of closure gate required to overcome friction due to water pressure may be expressed as follows:

$$W \geq fP$$

where W = weight of gate, lb
P = net force due to water pressure on gate, lb
f = coefficient of kinetic friction (see Table 30-1)

Because of uncertainties of the coefficient of friction, a factor of safety of 2 should be allowed.

The net force due to water pressure for a slowly closing gate may be expressed as follows:

$$P = 62.5A_1h + 62.5A_2\frac{v^2}{g}$$

where P = net static and dynamic forces due to head and flowing water, lb
A_1 = area of water face of gate, sq ft
h = average head of water against gate immediately after closure, ft
A_2 = effective area of gate subject to velocity, sq ft
v = velocity in fps of water through openings to be closed
g = acceleration due to gravity—32.2 fps per sec

TABLE 30-1 Values of f

Concrete on concrete	0.65
Steel on steel	0.30
Concrete on steel	0.40
Wood on concrete	0.50
Oak on steel	0.30
Oak on oak	0.25

Both v and A_2 are variables during closure. For small gates, A_2 may be considered equal to A, and the maximum velocity may be used. For large gates, detailed analysis involving differential equations and computer programs should be used for optimized design of the gate.

Controls can be designed for quick release, in order to utilize momentum of the gate as an aid to closure. In some cases, a tremendous down pull may be exerted when the gate is near closure. Damage to the gate and gate sill must then be prevented by the installation of crushing blocks, slowing down the gate when near closure, or by other means.

Where either side of the gate is a closed conduit and flowing full, the water pressure on that side of the gate due to quick closing of the gate is in the form of a pressure wave. The magnitude of the water pressure can be evaluated by the established method for water hammer in a closed conduit.

During a quick closure of a gate in an open-channel flow, a standing wave is created on either side of the gate (see Fig. 30-10). On the upstream side, a positive wave with a height of $d_2 - d_1$ travels upstream, and on the downstream side a negative wave with a height of $d_3 - d_4$ travels downstream. The quantities d_2 and d_4, which determine the pressure on the gate, are functions of the flow conditions before the closure operation and time of closure. A complete solution of d_2 and d_4 involves lengthy computations by a step method or tedious graphical method, both based on Massau's characteristic equations.[25] For a simplified case of a rectangular channel with instantaneous closure, d_2 and d_4 can be calculated from the following equations[26] by cut and try:

$$\frac{(d_1 + d_2)(d_2 - d_1)^2}{d_2} = \frac{2d_1v_1^2}{g}$$

$$\frac{(d_3 + d_4)(d_3 - d_4)^2}{d_4} = \frac{2d_3v_3^2}{g}$$

It is difficult to get a tight seal on closure gates, as closure may have to be made under a great depth of water. Heads may build up rapidly to 100 ft or more before a

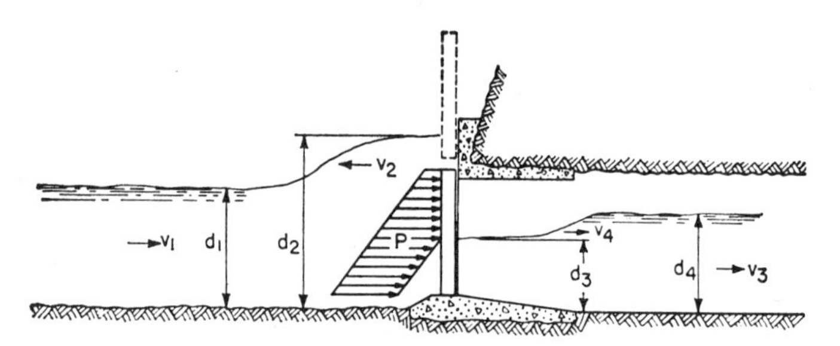

Fig. 30-10 Pressures developed during quick closure of diversion gate.

tight seal is accomplished. At Blakely Mountain Dam, where tunnel closure was made under an 88-ft head, it was necessary to place a steel bulkhead with rubber seals behind the closure gate that could not be calked tight.

In some cases wheel gates may be necessary to reduce friction. Caterpillar roller gates are not recommended because of the disastrous closure problem at Guri Dam, Venezuela.[27] Wheel gates may be designed so that axles shear as pressure builds up, permitting a simple water seal.

Frequently it is necessary to maintain a minimum flow in a stream below the dam site. Where a diversion tunnel is taking all the flow and outlets through the dam are at a higher elevation, some auxiliary outlet must be provided. This can be a small temporary opening through the dam or a small tunnel bypass around the tunnel closure gate. Closure arrangements for this smaller opening will be required. The opening should normally be designed small enough to allow flow without control and still permit the reservoir to rise until some of the permanent dam outlets are reached, at which time closure of the auxiliary opening may be made.

Where the stream is shifted back and forth by stage construction of cofferdams enclosing only part of the dam site at any one time, or during construction of cofferdams to divert the stream through tunnels, flumes, or conduits, it is usually necessary to increase the velocity of flow through the restricted area. This causes a head differential between the upstream and downstream sides of the cofferdam. The velocities around the end or over the top of an uncompleted cofferdam must be considered in determining the types of material for its construction.[28]

Velocities will be less in fill cofferdams which are constructed in lifts rather than by end dumping. Lift construction is more expensive, as it requires the construction of a bridge, cableway, or other device for spreading the lifts below water level. Sizes and shapes of individual particles of fill material must be sufficient to prevent erosion unless the fill can be placed faster than erosive action will remove it. At Cabinet Gorge Dam, a deep channel between rock walls with a large stream flowing at high velocities was closed by blasting the entire rock fill of the upstream cofferdam into place at one time. Extensive coyote holes were excavated into the canyon side well in advance, and advantage was taken of available upstream storage to reduce the flow of the stream at the time of the blast. At McNary Dam, where flows in excess of 100,000 cfs through a 240-ft gap 53 ft in depth had to be closed for the second-step cofferdam, 12-ton tetrahedrons constructed of concrete were successfully used to overcome velocities of 31 fps (although model tests indicated that 8-ton units would suffice). Large rock was used upstream from the tetrahedrons, but quarry run would have been satisfactory. The tetrahedron shape is considered to provide some mechanical interlocking to assist resistance to overturning and sliding. The pool-head differential was as much as 15 ft. Model tests indicated that velocities might be 28 fps over the crest and up to 38 fps for the submerged plunging jet near the toe of the slope.[29]

For fill-type cofferdams, stream velocity or erosion during construction will carry the material downstream from the point of dumping. Extra excavation for the dam may be avoided by placing the downstream cofferdam first, as any eroded material will be deposited outside the dam foundation. At McNary Dam, 12-ton tetrahedrons dropped in 25 ft of water flowing 13 fps drifted downstream 39 ft. Drops in 55 ft of water with velocities reducing from 12.5 to 6 fps drifted 40 ft downstream.[30]

At Box Canyon Dam, a rock-fill closure cofferdam was constructed by end dumping. High velocities as closure approached were met by anchoring with cables a large number of rocks placed by derrick. End dumping was unsuccessful at Ruby Dam. Attempts made to blast the canyon wall to fill the river channel failed until they were supplemented by rock tied with cable.[31] Similarly anchored steel shapes were used at Long Sault Dam.[32]

At Tres Marias Dam in Brazil, closure was made of a channel 550 ft wide with a flow of 10,000 cfs by placing lifts of rock fill in a wide weir with trucks hauling in water up to 18 in. deep. The weir was tilted in cross section from upstream to downstream so that a tractor could trim the upstream lip when the fill was part way across the channel, thereby reducing velocities in the uncompleted section and permitting that portion to be filled from the other bank. Larger rock was placed in the downstream portion of the weir where the velocities were greater. The discharge may be approximately computed from the broad crested weir formula:

$$Q = 5.67ld^{3/2}$$

where Q = discharge, cfs
l = length of weir, ft
d = depth of water over weir, ft

At Hell's Canyon Dam on the Snake River in Idaho, with closure to be made at a regulated flow of 5,000 cfs, the upstream cofferdam had to raise the water only 4 in. to put it through the diversion tunnel. However, the usual method of removing the cofferdam above the tunnel first and then constructing an end-dump fill across the river was not possible because of the steep gradient of the river and the high velocities that would have occurred due to the 10-ft head differential across the cofferdam as the flow decreased below the cofferdam. The solution was to build a submerged weir 350 ft long and 70 ft wide across the river before opening up the diversion tunnel. By this means, with a head differential of only 1 ft and a 2-ft depth over the submerged weir, the weir could be constructed by end dumping with trucks. The trucks operated in a maximum depth of 2 ft of water.

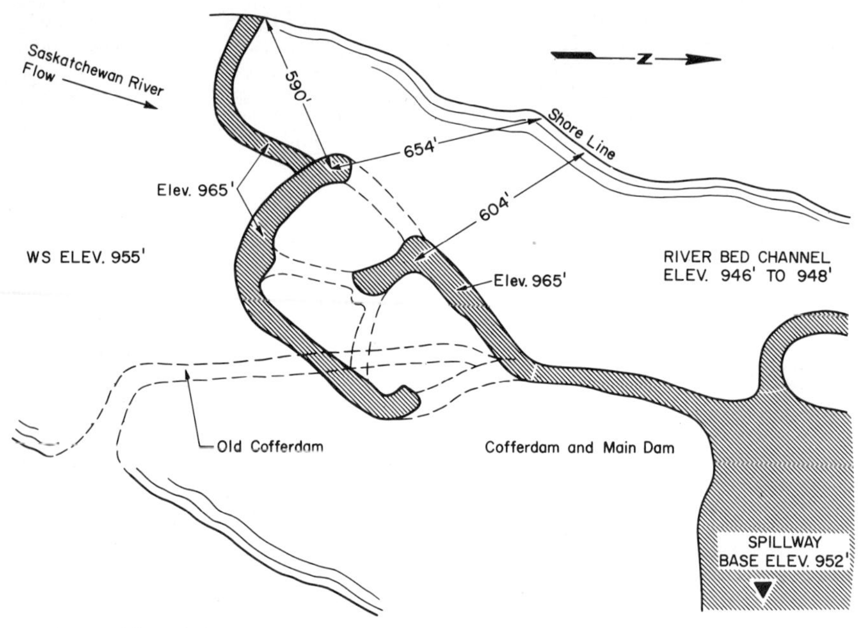

Fig. 30-11 Plan of Cascade closure at Squaw Rapids Dam. (*Adapted from Engineering and Contract Record.*)

At Squaw Rapids Dam in Saskatchewan, Canada, the waste material used for cofferdam dikes eroded at velocities of about 8 fps. Velocities were reduced within this limit by constructing a cascade channel that was closed by the simultaneous construction of four separate finger dikes (see Fig. 30-11).[33]

The cofferdams at Akosombo Dam in Ghana (see Fig. 30-12) were built of rock fill as part of the main dam. The dam foundation was first dredged to depths of 217 ft. Most of the rock fill was dumped under water. Transition and blanket materials were also dumped under water, first for the upstream cofferdam and finally for the downstream cofferdam. The cofferdams were designed for overtopping during early stages of construction.[34–37]

Provided that storage is available upstream so that the river flow may be temporarily reduced or cut off, low blocks left for diversion may be unwatered and bulkheaded

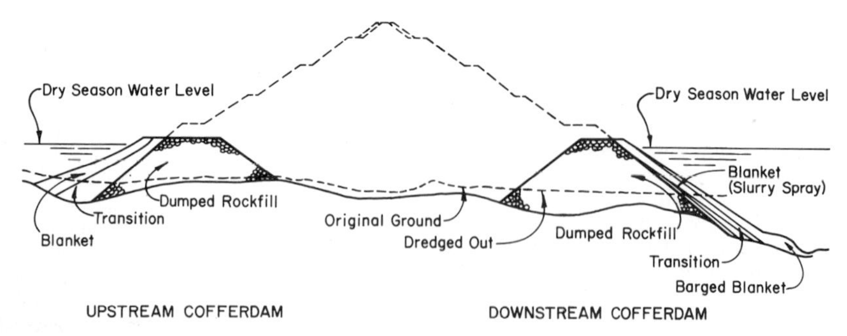

Fig. 30-12 Akosombo Dam cofferdams. (*Engineering News-Record, June 20, 1963.*)

off in the dry. At Canyon Ferry Dam this was accomplished by using reinforced-concrete bulkheads lowered into braced, vertical steel channel guides.[38]

Closure of rock-fill cofferdams sometimes requires emergency measures. The upstream cofferdam at Davis Dam was used for diverting the river into a bypass channel. Rock fill was dumped from a trestle, but as closure approached, the largest rock available (2 ton) was washed away by the current until frameworks of scrap steel were placed along the upstream side of the trestle to retain large rock dumped against the frames.[39]

Earth fills may be constructed partially across a stream by end-dump methods until erosive velocities develop around the ends of the fill. Earth can be stockpiled at the end of these fills for later rapid placement with draglines at a period of lower flow, raising the level of the fill uniformly across the remainder of the opening, as was done at Gavins Point Dam.[40]

At Oahe Dam on the Missouri, diversion into tunnels was accomplished by dumping and bulldozing 67,000 cu yd of shale in 21½ hr into a channel 900 ft long by 125 ft wide faster than it was eroded.[41]

At Chute-a-Caron Dam on the Saguenay River, diversion of the flow of 40,000 cfs at a velocity of 20 fps was accomplished by toppling into it a concrete monolith of 5,400 cu yd constructed on end on two piers. The riverward face of the concrete was contoured to fit the river bottom. When the pier in the channel side was blasted away, the precast dam toppled into place without breaking.[42]

At the Cuanza Development in Angola, an obelisk was dropped across the river to reduce the flow and permit a temporary rock cofferdam to be constructed above the permanent concrete cofferdam. A steel net, stretched across the center of the river, was used to prevent the rock from being swept downstream.[43]

Breaches in the dikes in Holland have been closed by floating in and sinking concrete caissons on willow mattresses where velocities ranged from 3 to 16 fps.[44]

At Chief Joseph Dam, a rock fill of 1- to 5-ton rocks was designed to make closure with a maximum head differential of 9 ft. For the final closure, 15- to 20-ton rocks with cable anchors were required.

At Long Sault Dam on the St. Lawrence, where a cellular steel sheet-pile cofferdam was specified for the stage-one cofferdam, cells were constructed until the stream was narrowed down to 265 ft, causing a velocity of 10 fps. Closure against the expected head differential of 16 ft, flow of 35,000 cfs, and final velocities up to 24 fps was impracticable by continuing cell construction. A 230-ton sled 115 ft long was constructed of heavy steel members, hinged in the center to fit the rough bottom better, with timber cribs on each end and a clear opening of approximately 60 ft. Braced posts at the upstream side formed guides into which heavy steel screens with 12-in. openings could be dropped. The sled was dragged through the water into its position near the center of the stream, a rock fill was built from one shore to the near crib by end dumping, and the cribs were filled. A rock fill with 2-ton maximum size was then constructed from the opposite shore to the other crib. Final closure required rock with a maximum size of 12 by 15 in. placed upstream of the screen.[45]

At Barkley Dam on the Cumberland River, two spans of a girder bridge were temporarily hung on their sides and sheet piling was driven against them to keep out the flow as the final closure cell of the cofferdam was constructed.[46]

For closing temporary sluiceways or low blocks in a concrete dam, it is sometimes possible to use, temporarily, permanent spillway gates both as cofferdam and forms.[47] Where tail water is above the elevation of the low block, a steel caisson with a pocket for a clay seal may be used for the downstream cofferdam.

In Sweden, semicircular arched steel beams with spans of 49 ft and flows of 238 cfs per ft of length have been used to divert or shut off water flows. Care must be used to ensure that their weight is sufficient to offset uplift forces.[48,49]

At the Manicouagan 5 Dam, water tightness of the upstream cofferdam foundation (200 ft deep) was accomplished by use of interlocking 24-in.-diameter concrete piles drilled in by the Icos patented method.[50–52]

At Wanapum and Wells Dams on the Columbia River, the cutoff to bedrock through gravel layers of cofferdam foundation up to 80 ft deep was accomplished by excavating vertical-sided narrow trenches to bedrock without unwatering. The trench sides were supported with rotary mud (bentonite slurry) until the trench could be refilled with clay and gravel. This, together with the bentonite slurry remaining in the trench, provided a watertight seal. Construction was similar to the cutoff for the permanent dam Wanapum except that no concrete was placed in the bottom of the trench.[53,54]

One of the most sensitive areas in a tunnel diversion scheme is the area immediately below the closure gate. Frequently the rock or tunnel lining is not competent to stand the head of water that will ultimately develop in the rock or on the outside of the lining. In such cases, careful grouting and drainage are required or the tunnel plug must be constructed quickly before excessive hydrostatic head develops. In the case of the Shihmen Dam in Taiwan, a temporary concrete plug was constructed within 6 days after dropping the tunnel stop logs in order to protect against possible excessive external pressures. At Oxbow Dam on the Snake River in Idaho, an old power tunnel was used for diversion. Because of the weak rock and unknown condition of the concrete lining, an elaborate system of grouting back of the lining and drain holes was provided, immediately below the closure gate, to reduce pressure on the lining after closure.

Gas explosions have occurred during diversion closures at Akosombo Dam, Ghana, and at Furnas Dam, Brazil.[55] Methane gas may have formed from decaying vegetable matter remaining in the uncleared reservoir. On reduction of pressure due to water leaking through the closure gate, the absorbed gas is released. Explosions at Furnas took place when concrete plug forms sealed the area upstream from the plug. Vents to reservoir level, upstream of plug, should remain in operation until plug concrete is completed, which should be as soon after closure as possible.

It is desirable that tops of diversion tunnels be above normal tailwater in the river so as to provide access to the gate in free air. The closure gate should be located upstream of overhangs that might prevent the lowering of supplemental devices for closure in case the primary scheme fails.

To avoid diverting the Mississippi at Chain of Rocks Dam, the rock fill was placed in layers under water. The smaller D stone (8-in. maximum) for the core was placed by means of a tremie after the outer shells of large rock and filter zone material had been dumped into place.[56]

COFFERDAM TYPES AND MATERIALS*

Cofferdams must be designed to withstand the dynamic effects of flowing water in addition to static water pressures. High velocities may produce scour and additional pressures, with the severest conditions usually occurring during closures.

Static water pressure may be computed by the following formula:

$$P = wh$$

where P = pressure, psf
w = unit weight of water, pcf
h = static head, ft

Pressure caused by the dynamic action of water flowing against a wall may be expressed by the following equation:

$$P = \frac{62.5v^2}{g}$$

where P = pressure, psf
v = velocity, fps
g = acceleration due to gravity—32.2 fps per sec

Earth Fills The simplest type of cofferdam is one constructed of earth. This type may be used only where low velocities will occur during or after construction. Sufficient space must also be available to accommodate its width. Where suitable

* See also Sec. 28, Cofferdams and Caissons.

earth can be found within economical haul distance, this type of cofferdam will be the cheapest. Velocities greater than those producing scour are permissible during construction provided that fill can be placed faster than it is carried away. Velocities after construction must be less than those producing scour. Scouring velocities are shown in Table 30-2.

TABLE 30-2 Scouring Velocities for Cofferdam Materials

Material	*Scouring velocity, fps*
Fine clay	0.25
Silt	0.50
Sand	1.0
Coarse gravel	4.0

Where scouring velocities will occur after construction of an earth-fill cofferdam, the fill may be protected from scour by a blanket of riprap. In most cases, a fill with the minimum top width required for construction equipment (usually 20 ft) and with side slopes at the natural angle of repose of the available material (approximately 1 vertical on 1½ horizontal for dry sections and flatter for fills placed under water) will prove adequate. For very high earth fills or where severe damage would result from failure, cofferdams must be designed and constructed according to the usual practice for earth-fill dams. Where coarse materials are used for the main body of the fill, it may be necessary to use a blanket of clay on the water face to reduce percolation through

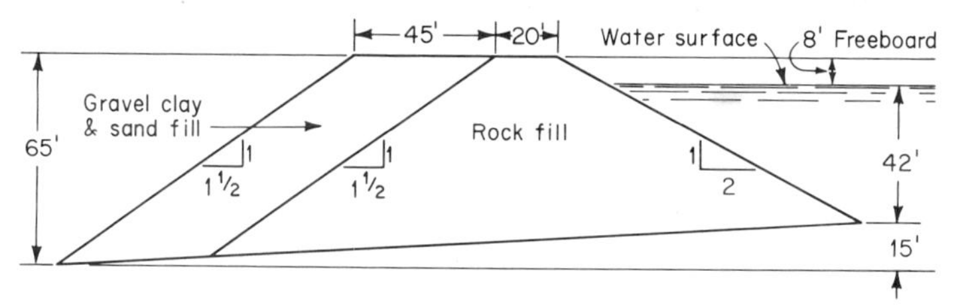

Fig. 30-13 Box Canyon cofferdam.

the fill. Earth-fill cofferdams must not be permitted to overtop. Earth fills are usually built by end dumping up to river level.

Rock Fills Where velocities are too great for earth fills but adequate space is available, rock fills sealed with blankets of finer material protected by riprap may be used. Velocities during construction must not cause erosion faster than the fill can be replaced. Usually the finer materials are placed on the water face, but if velocities are too great, the finer materials may be placed on the inside of the cofferdam, as at Box Canyon (see Fig. 30-13). Where fine materials are not available for sealing, a structural diaphragm of steel sheet piling or two layers of 2- by 12-in. dressed plank with 12-oz canvas between the layers may be used, as was done at Seminoe Dam.[57] Eroding velocities for various rock sizes have been determined from full-scale observations and model tests (see Fig. 30-14). The Dalles Dam model tests indicated that a 1,000-lb rock would withstand a velocity of 17 fps. Rock fills are usually built by end dumping, but they may be placed in layers where necessary to reduce velocities. Except for high structures, the minimum section possible to construct by dumping is adequate for the cofferdam.

Timber Cribs Where high velocities must be withstood or the area available for cofferdam construction is limited, cribs constructed of large timbers filled with rock and sealed on the water face may be used. This type of cofferdam is suitable for use where overtopping occurs. In order to prevent damage, means should be provided to flood the inside of the cofferdam before it is overtopped. Cribs may be constructed in deep and fast-flowing water by framing their lower portions (5 to 15 ft, more or less) on ways, launching and floating them into place under control of cables connected to hoists, and constructing the balance of each crib in place and sinking it by loading it with rock until it sinks to the bottom. Where the river bottom is uneven, soundings must be taken to permit accurate shaping of the crib to fit the contour of the foundation. At Grand Coulee,[58] a sounding barge consisting of a pile driver with 4-in., double, extra-heavy sounding pipe arranged to slide in leads was used to take soundings to 0.1 ft at 2-ft intervals in each direction. At Rock Island Dam,[16] soundings were taken at 4-ft intervals by a rod held in leads from a cable spanning the channel, the lower end of the rod being held by a bridle with cables reaching upstream. Where cribs are constructed in the dry, the foundations may be constructed of concrete, simplifying construction of the crib work and assisting in sealing.

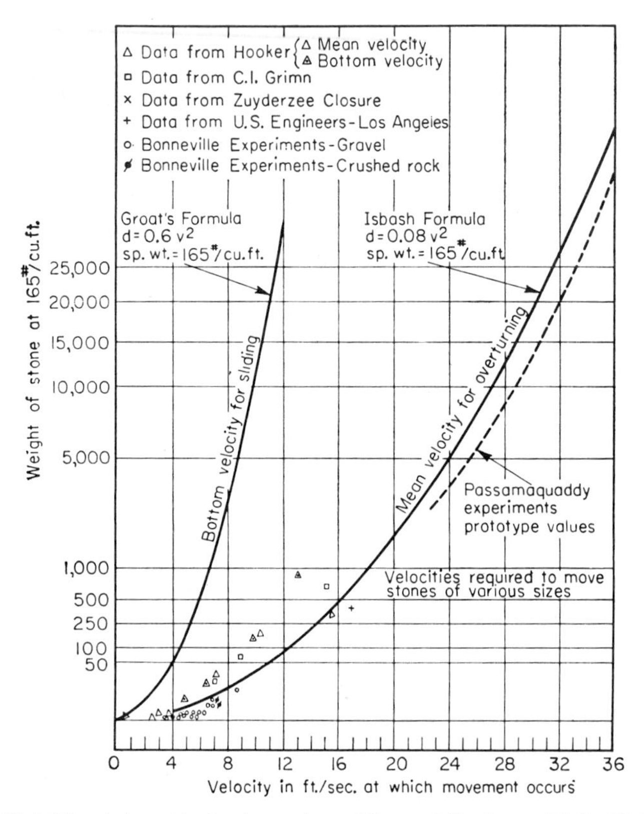

Fig. 30-14 Stability of stones in flowing water. *(Corps of Engineers, Walla Walla District, Washington.)*

Cribs are usually constructed of square timbers, bolted together, braced diagonally, and with a rock pocket at the base. However, logs may also be used. The water seal at the upstream face may be constructed of two layers of planks with roofing paper between or of a clay embankment on the water face. Where open-crib work is desirable during sinking operations in order to reduce the pressure on the crib or where cribs are founded on loose material requiring a cutoff below the bottom of the crib, a steel or Wakefield sheet-pile water seal may be used. At Bonneville Dam,[18] it was found from model studies that shore arms of cofferdams inclined in plan offered the minimum obstruction to river flow at stages above and below the top of the cofferdam and prevented eddies and crosscurrents from tending to scour the riverbank. Cribs

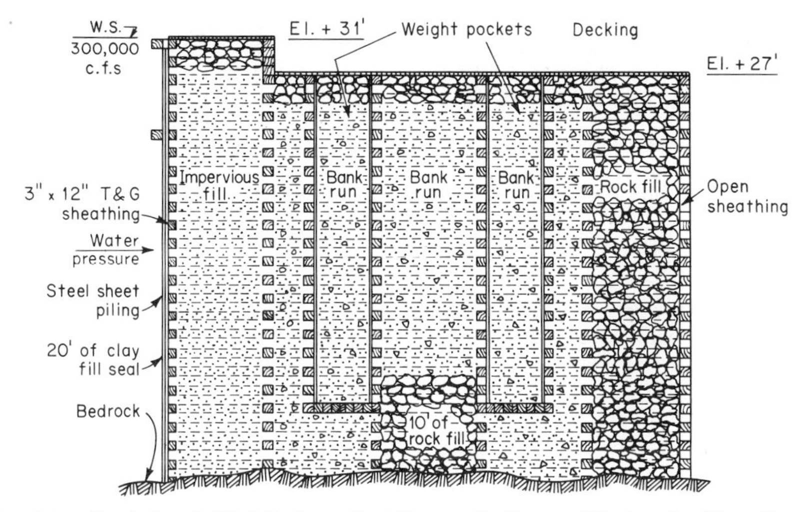

Fig. 30-15 Typical rock-filled timber crib at Bonneville Dam. (*Engineering News-Record, Sept. 5, 1935.*)

must be designed for safety against sliding and overturning and for bursting pressures exerted by the crib fill. A simple rule is to make the width of cofferdam equal to the height where not supported by a fill on the inside face (see Fig. 30-15).

Gravel, boulders, and rock may be used for crib-fill material. Loads up to 175 tons were exerted on cable controls at Bonneville Dam for cribs placed in 7-mph current.

Cribs at Rock Island Dam[16] with a height of 58 ft were placed in 15-mph current. These cribs, whose typical cross section is shown in Fig. 30-16, were constructed to excess width because their top elevation was undetermined at the start of construction. They were 16 and 32 ft long, with widths upstream and downstream ranging from 24 to 80 ft, and were framed of 12- by 12-in. timbers. Their bottoms were covered with 4- by 12-in. planks to form the ballast rock pockets. The cribs were floated into position when 5 or 6 ft high, being held with seven to ten $1\frac{1}{2}$-in. steel guy cables. They were designed to be overtopped 20 ft and were actually overtopped 11 ft 6 in.

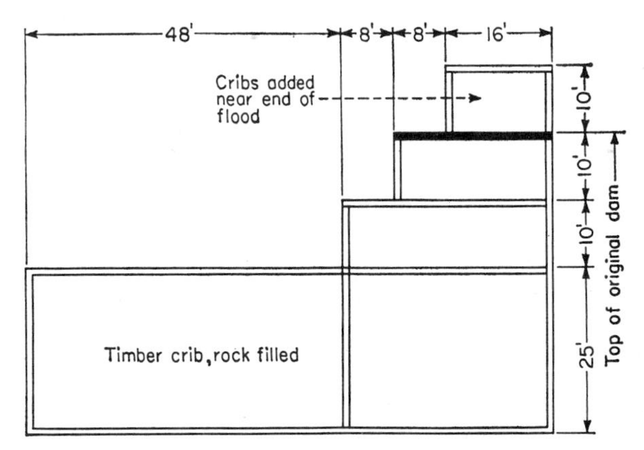

Fig. 30-16 Cross section of Rock Island Dam cribs. (*Engineering News-Record, Apr. 30, 1931.*)

The water seal was two layers of 2-in. planking with tarpaper filler except in the deepest sections, where wooden sheet piling was placed with the aid of a diver. The tops of the cribs were covered with timbers to prevent scour. The bottom on which the cribs were placed was clean, hard, basalt rock. Models of cribs were built to a scale of ½ in. to 1 ft to assist carpenters in framing the lower portions of the cribs. To facilitate the sealing of the cribs, 16,000 cu yd of clay, hay, sagebrush, and other loose binder was dumped along the water faces of cofferdams. Twenty-five acres were unwatered with one 10-in. pump operating intermittently.

In order to reduce the pressure during construction, some of the cribs at Grand Coulee Dam were floated into place when 40 ft high and constructed in a U shape to permit the stream to flow through the trough (see Fig. 30-17). Each crib was designed to sink when its 15-ft base section was filled with rock and the top constructed to 10 ft more than water depth. A head differential of not more than 2 ft developed until the trough was closed by stop logs constructed of three 12- by 18-in. timbers spanning the 32-ft trough. This cofferdam was sealed by driving sheet piles, before filling the cribs, along the downstream (water) face after the stop logs were in place. Piles were anchored to the cribs and the cribs filled by dumping from trucks.

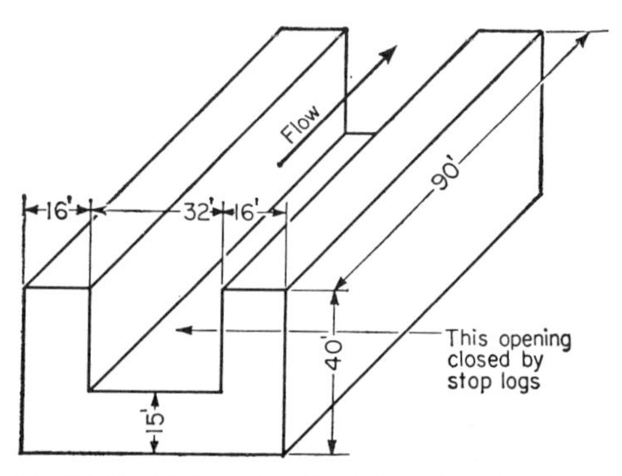

Fig. 30-17 U-shaped crib at Grand Coulee Dam.

The pressures against the cribs during floating into place and sinking may be computed by the formula[18] below.

$$P = Ah_v w c_s$$

where P = total force on crib, lb
c_s = constant depending on shape of crib (use c_s = 1.5)
w = unit weight of water, 62.5 pcf
A = gross area of crib presented to current, sq ft
h_v = velocity head of current, ft

Steel Sheet-pile Cellular Cofferdams Steel sheet-pile cellular cofferdams consist of interconnected cells constructed of interlocking steel sheet piles filled with earth, preferably a free-draining material such as sand or gravel. The more common types are shown in Fig. 30-18. Type A consists of circular cells of interlocking sheet piles driven closely together and interconnected with segments of circles formed of interlocking piles, the segments being attached to the circular units by means of T piles incorporated in the circles. Type B consists of two rows of segments of circles whose chords are perpendicular with diaphragms connecting the two rows of segments at the end of each segment. Type A has the advantage of permitting complete filling of each major cell independently of the others, whereas the fill in type B must be brought to approximately the same elevation on each side of the straight diaphragm. Cellular cofferdams must be designed for bursting pressures as well as stability against overturning. The height of this type of cofferdam is limited only by the strength of the interlock. Dimensions and piling required for type A cofferdams as prepared by the U. S. Steel Corporation are given in Table 30-3. Other sections of piling may also be used.

TABLE 30-3 Piling Sections MP-101 and MP-102 (15-in. driving distance)

Number of piles in cell*	D ft	Y ft	Number M piles	R ft	Number N piles	X ft	Z ft	E ft	Area, sq ft	
									Within circle	Between circles
48	19.10	24.83	11	8.01	9	1.09	5.73	16.33	286.5	137.5
52	20.69	25.97	12	8.01	9	1.66	5.28	17.57	336.5	141.3
56	22.28	28.22	13	8.80	10	1.66	5.94	18.94	390.6	169.6
60	23.87	29.35	14	8.80	10	2.23	5.48	20.20	448.6	173.1
64	25.46	31.61	15	9.60	11	2.23	6.15	21.52	510.7	204.5
68	27.06	32.74	16	9.60	11	2.79	5.68	22.82	576.8	207.8
72	28.65	33.87	17	9.60	11	3.36	5.22	24.12	646.9	210.4
76	30.24	35.01	18	9.60	11	3.93	4.77	25.36	721.1	212.6
80	31.83	37.19	19	10.39	12	3.90	5.36	26.64	794.2	248.1
84	33.42	38.32	20	10.39	12	4.46	4.90	27.90	876.2	249.7
88	35.01	39.45	21	10.39	12	5.03	4.44	29.20	962.1	250.7
92	36.61	41.71	22	11.19	13	5.03	5.10	30.60	1,052.1	289.9
96	38.20	42.84	23	11.19	13	5.59	4.64	31.89	1,146.1	290.6
100	39.79	43.97	24	11.19	13	6.16	4.18	33.15	1,244.1	290.7
104	41.38	45.10	25	11.19	13	6.72	3.72	34.10	1,346.1	290.0
108	42.97	47.36	26	11.99	14	6.73	4.39	35.79	1,452.2	333.7
112	44.56	48.49	27	11.99	14	7.30	3.93	37.08	1,562.3	332.8
116	46.15	49.62	28	11.99	14	7.86	3.47	38.37	1,676.4	331.2
120	47.75	51.81	29	12.78	15	7.83	4.06	39.66	1,787.0	378.3
124	49.34	52.94	30	12.78	15	8.39	3.60	40.99	1,908.9	376.5
128	50.93	54.07	31	12.78	15	8.96	3.14	42.33	2,034.8	374.0
132	52.52	56.32	32	13.58	16	8.96	3.80	43.66	2,164.8	424.5
136	54.11	57.45	33	13.58	16	9.53	3.34	44.98	2,298.7	421.8
140	55.70	58.59	34	13.58	16	10.10	2.89	46.30	2,436.7	418.9
144	57.30	60.83	35	14.37	17	10.10	3.53	47.62	2,578.7	471.8
148	58.89	61.85	36	14.37	17	10.66	2.96	48.94	2,724.7	469.4
152	60.48	64.23	37	15.17	18	10.67	3.75	50.27	2,874.8	526.6
156	62.07	65.36	38	15.17	18	11.23	3.29	51.72	3,028.8	522.8
160	63.66	67.62	39	15.96	19	11.23	3.96	52.90	3,186.9	583.1

* Including four T piles.

E is the theoretical width of a rectangular wall having a resistance against overturning equal to that of the cellular wall; i.e., rectangular wall and cellular wall have equal section moduli in cross section.

Dimensions refer to Fig. 30-18.

At Deadman Island Dam on the Ohio River, circular steel cells were used. These were 40 ft in diameter and 40 ft high with 2-ft gaps, driven to rock and filled with sand and gravel. Cells were constructed with a minimum distance of 10 ft between the inner walls and faces of the dam. For raising the "Maine," cofferdam cylinders 50 ft in diameter constructed with 75-ft piles driven 38 ft into the mud and filled with mud were used. An approximate rule for type A cofferdams is to make the diameter of the cells equal to the height. However, all cofferdams should be analyzed for stability against sliding, overturning, and bursting.[59] To reduce the diameter of the cells, an earth fill may be used to support the inside of the cofferdam. This type of cofferdam may be used for any velocity, where the cofferdam is to be overtopped and where space does not permit fill cofferdams. The advantage of this type of cofferdam over timber cribs, besides its economy where timber and labor costs are high, is that cells may be driven to rock or other impervious strata so as to cut off underground percolation into the cofferdammed area. Steel cells have been constructed to a height of

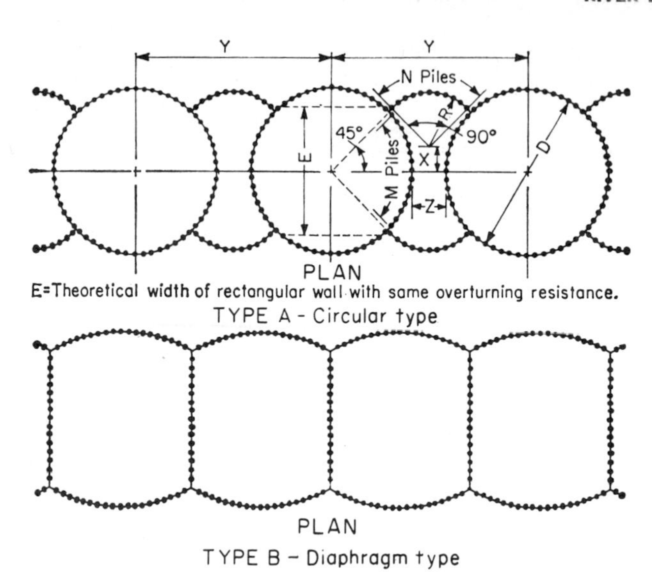

Fig. 30-18 Types of steel sheet-pile cellular cofferdams.

98 ft at the Kentucky River Dam (58.89 ft in diameter with a berm on the dry side to within 36 ft of the top of piling).[60–63]

Weep holes should be provided through the piles on the dry face of the cells to prevent the buildup of hydrostatic pressure inside the cells and thus adding to bursting stresses caused by the fill material.

Where overtopping is to occur, gated sluiceways or weirs with stop logs or needle beams should be constructed to provide for controlled flooding of the cofferdam area.

Precautions that may be needed for high cellular cofferdams include:

1. Grouting of rock below the cofferdam to ensure tightness against seepage and uplift
2. Adequate draining of berms where required for stability
3. Riprapping of berm slopes (using foundation excavation) to prevent erosion

Where the diameter of a single circular cell produces excessive stresses, a cloverleaf design may be used (see Fig. 30-19). For economy of the cofferdam, it is desirable to lay out the diversion scheme so as to get as much reuse of piling as possible.

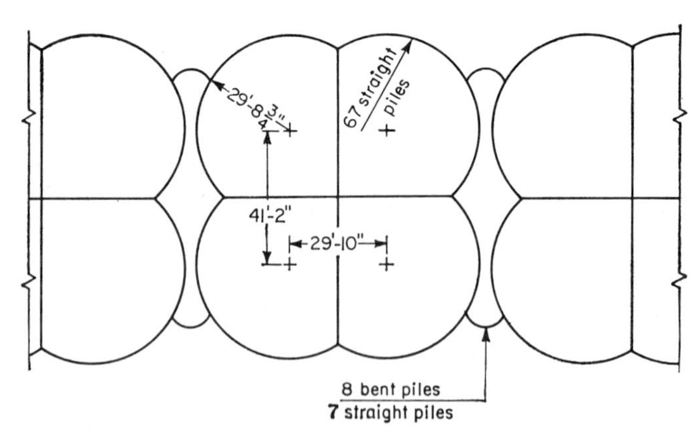

Fig. 30-19 Cloverleaf sheet-pile cell used at Kentucky Dam.

Parallel Rows of Sheet Piles For low heads, cofferdams may be constructed of sheet piles, either of tongue-and-groove wood or of interlocking steel. These are driven in parallel rows with tie rods and wales to tie the two rows together. The space between the rows is filled with earth, preferably sand and gravel (see Fig. 30-20). A single row of sheet piles may also be used, braced if necessary against the expected head.

Concrete Where cofferdams are to be constructed on rock in restricted areas and are subject to high velocities or overtopping, concrete gravity, buttress, or arch dams may be used if low water exposes the cofferdam site or if auxiliary cofferdams may be used to unwater for their construction. At the Kariba Gorge Dam site on the Zambezi River, a first-stage concrete-ring cofferdam about 60 ft high around one abutment

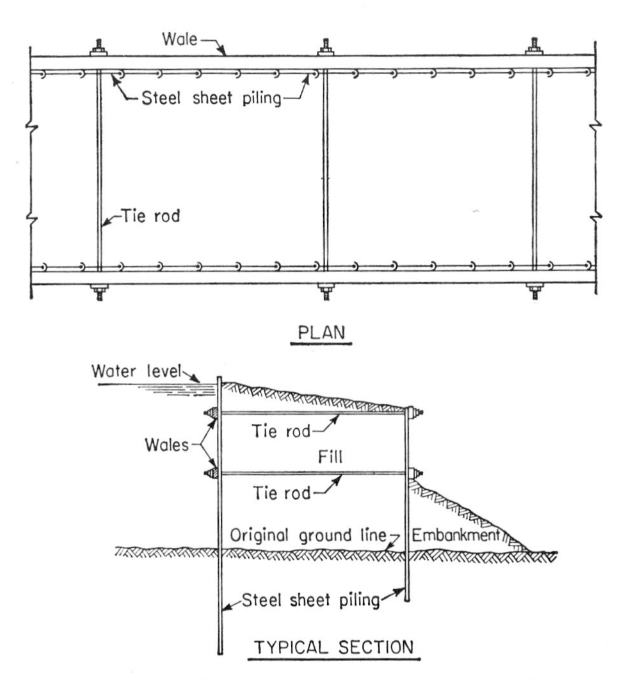

Fig. 30-20 Cofferdam of parallel rows of sheet piling.

was constructed.[64] The second-stage cofferdam consisted of a concrete ring with a radius of 175 ft and a maximum height of 140 ft. At Bull Shoals Dam, a section of the cofferdam was built with concrete buttresses in an excavated channel. The river was diverted into the channel during a low stage by low earth-fill cofferdams. Higher cofferdams of concrete were constructed within the earth fills for the first-stage operations. The buttresses were later closed with concrete stop logs when diversion to the next stage was made (Figs. 30-21 and 30-22).

To divert small streams such as canals from the area around head gates, a movable steel cofferdam may be used.[65]

PROBLEMS IN BUILDING COFFERDAMS*

In building fill cofferdams in water, it is sometimes very difficult to effect a watertight seal between the fill and the natural riverbed. This is particularly true where the riverbed materials are coarse, such as boulders, cobbles, and gravel. Where such

* See also Cofferdam Difficulties in Sec. 28, Cofferdams and Caissons.

materials are encountered, it is desirable, where practicable, to excavate a cutoff trench through the coarse material to an impervious stratum or to bedrock. Where this is not possible and excessive percolation through the fill prevents unwatering, steps must be taken to reduce the flow through the fill. A trench may be dug at the upstream toe of the cofferdam after the fill is constructed and backfilled with clay, or a blanket of clay may be placed upstream from the dam. Grouting with sawdust or mixtures of sawdust and cement has been used, but usually this is not too successful if flows are great. Frequently the best solution is to use more pumps. However, where flows through the cofferdam or its foundation cause erosion, steps must be taken to reduce

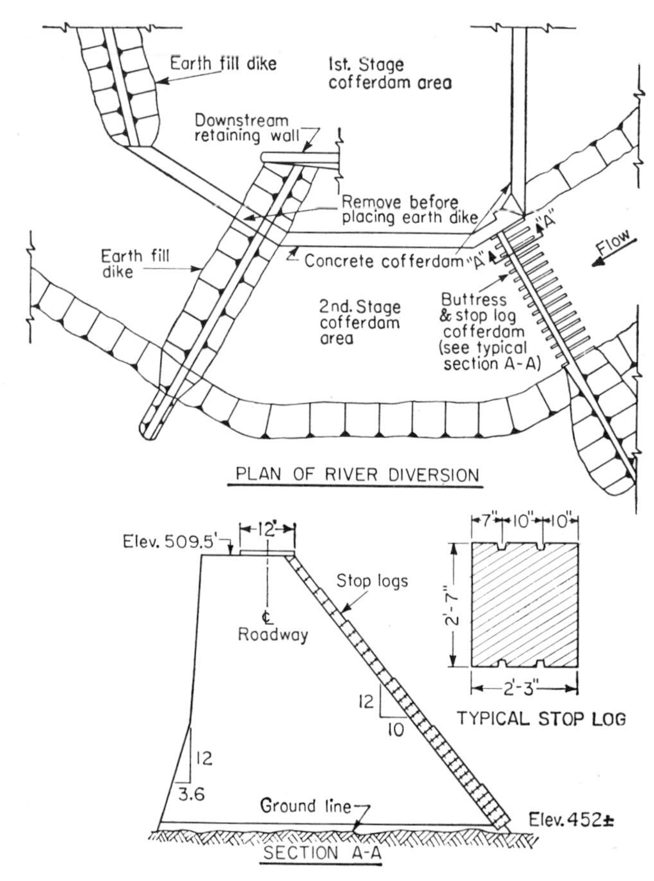

Fig. 30-21 Bull Shoals river diversion.

the flow below the eroding point. If the cost of pumping is too great, it may be necessary to construct a second cofferdam inside the original one.

Where steel cellular cofferdams are constructed on earth and are designed for a height above the original ground surface, it is necessary to leave a sufficient amount of earth behind the cofferdam to ensure support. It is desirable to drive sheet piles to rock at least on the water face. Where built on rock, steel cells must be supported by an elaborate template (Fig. 30-23). To simplify construction on rock and to assist in providing a watertight seal, an earth fill may sometimes be constructed during low water periods and the sheet piles driven through it, or alternative methods may be used.[66]

Sometimes hard driving will split open the pile interlocks. If it is impossible to drive sheet piles to bedrock or an impervious stratum, leakage may develop below the cells and cause trouble. If the leakage continues to increase, the indications would be that the foundation might blow out, and steps must be taken to seal off the leakage. At Grand Coulee Dam where driving of cells was stopped 15 or 20 ft above bedrock, an inflow of 35,000 gpm developed, with a vortex outside the cells. The sheet piling became distorted, some settled out of plumb, and riveted seams were ripped open.[67,68] Attempts to stop the flow by filling around the outside of the cofferdam were unsuccessful, making necessary the construction of a dike inside the cofferdam to provide a

Fig. 30-22 Bull Shoals cofferdam.

sump, further filling on the river side, the construction of a filter bed between cutoff and cells, the plugging of the filter bed, emptying and rebuilding of the cells, the construction of a cellular toe wall inside the cofferdam, the drilling of one hundred 4-in. holes to find the cavity, and the grouting of cavities with the following mix, using a concrete pump followed by air at 35 psi:

Mixed wet for 1½ min	Mixed dry for ½ min	Sawdust	8 cu ft
		Shavings, thin and wispy	10–16 cu ft
		Portland cement	2 cu ft
		Sand	6 cu ft
		Lump bentonite	50 lb
		Water	8–10 cu ft
		Pulverized bentonite	30 lb

The product of the above mix was a 14-cu-ft batch resembling wet cement plaster.

At Chief Joseph Dam on the Columbia River, high water of 40 to 50 ft with velocity of up to 35 fps sucked out fill from the bottom of the sheet-pile cells, which had been founded on practically bare, uneven rock. One cell lost 2,600 cu yd of its 6,000 cu yd of fill. Rocks as large as 9 in. could not seal the openings, so a blister wall or diaphragm of steel sheet piling was placed along the river side of the cells. The blister wall was held in place by heavy pipe spudded into rock and welded to the existing cells. Then the space between was sealed with bagged concrete and rock to prevent further scour and retain the cell backfill.

At Canyon Ferry Dam, water broke through the gravel and talus material under the floor of the flume into the dam foundation. This was attributed to inadequate cutoff under the flume at the upstream cofferdam.

Where cofferdams are to be placed on a porous rock formation, grouting may be necessary to prevent excessive flows under the cofferdams. Sanded intrusion grouting using the patented admixture Alfesil and Intrusion Aid was used successfully on a difficult cofferdam job at Jim Woodruff Dam.[69]

In high velocities above Niagara Falls, a metal shield held in place by cables from upstream and advanced a few feet at a time ahead of cofferdam construction provided

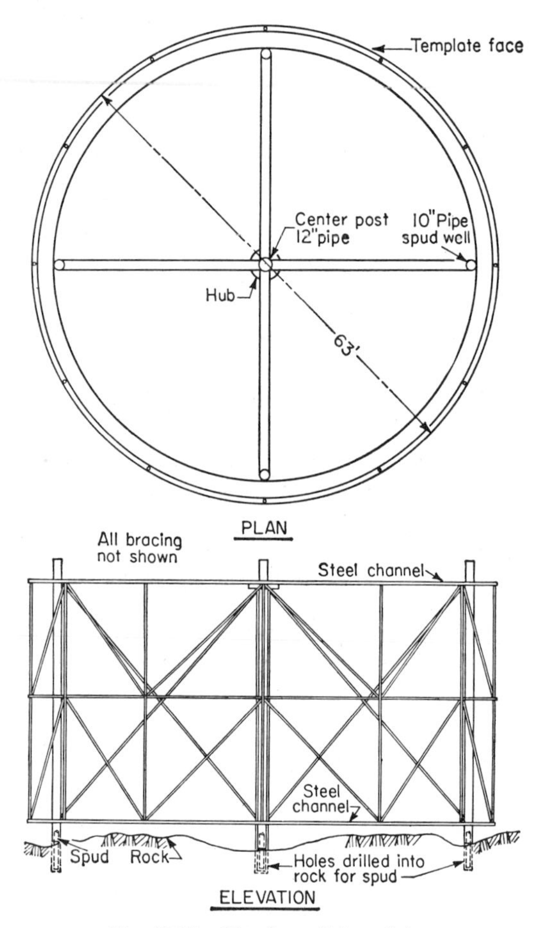

Fig. 30-23 Circular cell template.

relatively quiet water for the construction of a steel sheet-pile cofferdam with rock-fill backing.[70] A similar plan was used at Long Sault Dam.[71]

At Gorge Dam on the Skagit River, bedrock was covered by as much as 150 ft of porous gravel and boulders. Cutoff under the cofferdams was provided by drilling holes to bedrock, inserting cooling coils, and freezing a curtain across the valley under the cofferdams.[72]

At St. Anthony Falls on the Mississippi River, leakage was so great through the soft sandstone to which the cellular cofferdam had been driven that a sheet-pile cutoff inside the cells was driven 30 to 40 ft through the sandstone to a level 11 ft below the deepest excavation. Piles within the area of the dam required pulling, and the disturbed sandstone was solidified by a chemical process.[73]

A cellular cofferdam under construction was destroyed when water poured in the upper, open end and created a head differential that caused the downstream wall to overturn.

At Bonneville Dam, the second-stage cofferdam was to have been completed before high water, but a flood flow of 520,000 cfs overtopped the upstream arm of the cofferdam and permitted water to escape through the uncompleted downstream arm. Earth fill at the shore end of the upstream arm was carried away, as well as three nearly completed cribs on the downstream arm. The earth fill was replaced by rock up to 30-ton size dumped from a trestle constructed to make the repair.[19]

REFERENCES

1. Kariba Cofferdam Is in Business, *Eng. News-Record*, vol. 161, no. 11, p. 107, Sept. 11, 1958.
2. T. A. L. Paton, New Kariba Dam, Ready to Harness the Zambesi, *Eng. News-Record*, vol. 161, pp. 30–34, Nov. 20, 1958.
3. Andrew Weiss, Construction Technique of Passing Floods over Earth Dams, *Trans. Am. Soc. Civil Engrs.*, vol. 116, pp. 1158–1178, 1951.
4. J. R. Stephens, Divert Stream over Dam under Construction, *Construction Methods and Equipment*, vol. 35, pp. 92–99, September, 1953.
5. From Colorado River Compact to the Closure of Black Canyon, *Eng. News-Record*, vol. 109, pp. 701–728, Dec. 15, 1932.
6. Upstream Cofferdam for Hoover Dam, *Eng. News-Record*, vol. 110, p. 203, Feb. 9, 1933.
7. Kenneth E. Sorensen, Graphical Solution of Hydraulic Problems, *Trans. Am. Soc. Civil Engrs.*, vol. 118, pp. 61–77, 1953.
8. Hungry Horse Diversion Accomplished, *Eng. News-Record*, vol. 142, p. 33, Apr. 28, 1949.
9. Divert Water for Anderson Ranch Dam, *Eng. News-Record*, vol. 136, p. 679, Apr. 25, 1946.
10. Ralph Whitaker, Jr., Building with Boulders, *Western Construction*, pp. 63–67, October, 1954.
11. Shock Absorber for the Big Horn River, *Eng. News-Record*, vol. 149, pp. 32–34, Sept. 18, 1952.
12. Fast Schedule for Cabinet Gorge Dam, *Eng. News-Record*, vol. 147, pp. 36–38, Dec. 13, 1951.
13. Garrison Dam Event Has Minor Crisis, *Eng. News-Record*, vol. 150, p. 27, Apr. 23, 1953.
14. William P. Price, Jr., Diversion Is the Big Job at Canyon Ferry Dam, *Eng. News-Record*, vol. 145, pp. 32–34, Aug. 3, 1950.
15. David C. May, River Diversion at Shasta Dam, *Eng. News-Record*, vol. 131, pp. 397–399, Sept. 9, 1943.
16. Cofferdamming the Columbia at Rock Island, *Eng. News-Record*, vol. 106, pp. 716–719, Apr. 30, 1931.
17. Constructing the First Cofferdam, *Eng. News-Record*, vol. 115, pp. 148–151, Aug. 1, 1935.
18. C. I. Grimm, Cofferdams in Swift Water for Bonneville Dam, *Eng. News-Record*, vol. 115, pp. 315–318, Sept. 5, 1935.
19. Jetty Construction Methods Used to Repair Bonneville Cofferdam, *Eng. News-Record*, vol. 117, pp. 461, 463, Oct. 1, 1936.
20. J. R. Thatcher, Twelve-ton Tetrahedrons Close Cofferdam for McNary Powerhouse and Spillway, *Civil Eng.*, vol. 21, pp. 337–341, June, 1951.
21. Chief Joseph Arises, *Eng. News-Record*, vol. 150, pp. 40–44, June 18, 1953.
22. Progress at Imperial Dam, *Eng. News-Record*, vol. 118, pp. 92–94, Jan. 21, 1937.
23. Well-Points Dry Morelos Dam Excavation, *Eng. News-Record*, vol. 143, pp. 58–61, July 7, 1949.
24. Reinforced Concrete Logs Seal Detroit Dam Diversion Tunnel, *Eng. News-Record*, vol. 149, p. 53, Aug. 7, 1952.

25. J. Massau, Graphical Integration of Partial Differential Equations with Special Applications to Unsteady Flow in Open Channels, *Ann. de l'association des ingénieurs sortis des écoles spéciales de Gand*, vol. 23, pp. 95–214, 1900 (in French); English trans. by H. J. Putnam, Rocky Mountain Hydraulic Lab., Allenspark, Colo., 1948.
26. Robert W. Angus, *Hydraulics for Engineers*, p. 273, Sir Isaac Pittman & Sons, Ltd., London, 1943.
27. Kaiser Fights Guri's Leaky Gates, *Eng. News-Record*, vol. 182, pp. 83–84, Feb. 6, 1969.
28. S. V. Isbash and I. V. Lebeder, Change of Natural Streams During Construction of Hydraulic Structures, *Intern. Assoc. Hydraulic Res., Ninth Conv.*, pp. 1114–1121, Dubrovnik, Yugoslavia, 1961.
29. H. B. Elder, Diversion and Construction Techniques of the Corps of Engineers for Concrete Dams in Wide Valleys, *Trans. Seventh Intern. Cong. Large Dams*, pp. 431–451, 1961.
30. McNary Dam Diversion's Toughest Part, *Eng. News-Record*, vol. 145, p. 31, Dec. 14, 1950.
31. Anchored Rocks Aid River Diversion, *Eng. News-Record*, vol. 122, pp. 218–219, Feb. 16, 1939.
32. Sidney N. Smith, Diversion of Saint Lawrence River for Construction of Long Sault Dam, *Trans. Seventh Intern. Cong. Large Dams*, vol. 3, pp. 431–451, 1961.
33. Bruce Taylor, Introduction to Squaw Rapids, *Eng. and Contract Record*, vol. 74, no. 12, pp. 53–56, December, 1961.
34. Highest Underwater Cofferdams Keep Akosombo Dry, *Eng. News-Record*, vol. 170, no. 25, pp. 64–73, June 20, 1963.
35. Donald J. Bleifuss, Unwatering Akosombo Cofferdams, *J. Soil Mech. Found. Div., Proc. Am. Soc. Civil Engrs.*, vol. 90, no. SM 2, Part 1, p. 111, March, 1964.
36. Ghana's Akosombo Dam Has Volta River in Harness, *Eng. News-Record*, vol. 173, no. 5, pp. 26–28, July 30, 1964.
37. R. P. Ware and D. R. Hooper, Ghana's Akosombo Dam, *Civil Eng.*, vol. 34, no. 12, pp. 65–69, December, 1964.
38. Third-stage Diversion at Canyon Ferry Dam, *Eng. News-Record*, vol. 149, pp. 42–43, Oct. 30, 1952.
39. Rock Fill Diverts River at Davis Dam, *Eng. News-Record*, vol. 141, pp. 70–72, Aug. 5, 1948.
40. Closure Method Is Set at Gavins Point, *Eng. News-Record*, vol. 154, p. 25, Mar. 31, 1955.
41. Roland J. Carr, Men and Machines Beat a River, *Eng. News-Record*, vol. 161, no. 7, pp. 21–23, Aug. 14, 1958.
42. Saguenay River Diverted for New Water-power Plant, *Eng. News-Record*, vol. 105, pp. 187–188, July 31, 1930.
43. Goncalo Sarmento and P. C. Afouso, The Middle Cuanza Development, *Water Power*, vol. 14, no. 8, pp. 299–305, August, 1962.
44. Ira A. Hunt, Jr., Dutch Close Tidal Breach at Ouwerkek, *Civil Eng.*, vol. 24, pp. 647–651, October, 1954.
45. E. L. A. Weidner, Crib-sled Nicely Does the Trick, *Eng. News-Record*, vol. 156, pp. 47–50, Mar. 8, 1956.
46. Dual Function Bridge and Presplitting Methods Speed Barkley Dam Job, *Contractors and Engineers*, pp. 68–71, February, 1965.
47. Lee Warren, Cofferdams and Stream Control of Jordan Dam, Alabama, *Eng. News-Record*, vol. 106, pp. 804–808, May 14, 1931.
48. S. Angelin, The Use of Arched Beams as a Temporary Shut-off, *Proc. Sixth General Meeting, Intern. Assoc. Hydraulic Res.*, pp. D11-1–12.
49. H. L. Rundgren, Hydrodynamic Forces Against Steel Arch Cofferdam Segments Immersed in Streaming Water, *Proc. Seventh General Meeting, Intern. Assoc. Hydraulic Res.*, pp. D13-1–10.
50. I. V. Galbiati, Concrete Cutoff Wall Drilled Down 250 ft, *Civil Eng.*, vol. 33, no. 11, pp. 59–61, November, 1963.
51. The Manicouagan-Aux-Outardes Power Development, *Water Power*, vol. 16, no. 10, pp. 411–419, October, 1964.
52. William W. Jacobus, Jr., Hydro-Quebec's Big, Beautiful Manicouagan 5 Hides in the Bush, *Eng. News-Record*, vol. 171, no. 17, pp. 38–45, Oct. 24, 1963.
53. J. V. Engstrom, Innovations at Wanapum Dam, *Civil Eng.*, vol. 33, no. 10, pp. 43–47, October, 1963.

54. Bentonite Slurry Stabilizes Trench, Keeps Groundwater Out, *Eng. News-Record*, vol. 164, no. 6, pp. 42–46, Feb. 11, 1960.
55. F. H. Lyra and W. MacGregor, Furnas Hydro-Electric Scheme, Brazil: Closure of Diversion Tunnels, *Proc. Inst. Civil Engrs. (London)*, vol. 36, pp. 20–46, January, 1967.
56. Gordon R. Smith, Carefully Placed Rock Dams the Mississippi, Dam Built Without Channel Diversion, *Construction Methods & Equipment*, vol. 44, no. 11, p. 76, November, 1962.
57. J. H. Warner, Seminoe Dam Progress, *Eng. News-Record*, vol. 119, pp. 817–822, Nov. 18, 1937.
58. Program for Grand Coulee's Second Cofferdam, *Eng. News-Record*, vol. 117, pp. 464–466, Oct. 1, 1936.
59. Karl Terzaghi, Stability and Stiffness of Cellular Cofferdams, *Trans. Am. Soc. Civil Engrs.*, vol. 110, pp. 1083–1202, 1945.
60. A. F. Hedman, Cofferdam Design for Kentucky Dam, *Eng. News-Record*, vol. 128, pp. 50–54, Jan. 1, 1942.
61. Edwin Paul Swatek, Jr., Cellular Cofferdam Design and Practice, *J. Waterways Harbors Div., Proc. Am. Soc. Civil Engrs.* pp. 109–132, August, 1967.
62. O. R. Bengston, Template Takes Trouble Out of Driving 70 Foot Cofferdam Cells, *Eng. News-Record*, vol. 148, pp. 34–35, Jan. 17, 1952.
63. Byron J. Prugh, Miramar Cofferdam Dewatered by Well Points, *Civil Eng.*, vol. 23, pp. 452–453, July, 1953.
64. J. J. Martin, Kariba Left Bank Cofferdam, *J. Construction Div., Proc. Am. Soc. Civil Engrs.*, no. CO. 1, paper 3,454, pp. 34–54, March, 1963.
65. Dam Solves Canal Repair Problem, *Eng. News-Record*, vol. 149, p. 35, Nov. 27, 1952.
66. Divers Seal Cofferdam with Bags of Concrete, *Eng. News-Record*, vol. 162, no. 22, p. 32, June 4, 1959.
67. Serious Leak Checked in Cofferdam at Grand Coulee, *Eng. News-Record*, vol. 118, pp. 595–597, Apr. 22, 1937.
68. Contractors Win River Battle, *Eng. News-Record*, vol. 119, pp. 13–16, July 1, 1937.
69. Fred A. Robeson and William E. Webb, Cofferdam Grouting in Jim Woodruff Dam, *Eng. News-Record*, vol. 145, pp. 35–37, July 6, 1950.
70. Horseshoe Falls: On a Construction Anvil, *Eng. News-Record*, vol. 155, pp. 30–32, Aug. 25, 1955.
71. John Patterson, Taming the Long Sault Rapids, *Civil Eng.*, vol. 26, pp. 366–369, June, 1956.
72. River Bed to Be Frozen to Form Cofferdam Cut-Off, *Eng. News-Record*, vol. 152, p. 23, June 3, 1954.
73. Solving a Tricky Dewatering Problem, *Eng. News-Record*, vol. 147, pp. 39–43, Oct. 11, 1951.

Section **31**

Tunneling

JOHN R. TABOR

President, Mining Equipment Manufacturing Corporation, Racine, Wis. *(Tunnels in Earth)*

ROBERT S. MAYO

President, Mayo Tunnel & Mine Equipment, Inc., Lancaster, Pa. *(Tunnels in Rock)*

A. Introduction

Throughout history, man has used tunnels to turn his environment into a more comfortable, functional place in which to live. From earliest times, open areas below ground have been used for tombs, temples, and simple dwellings. Roman and Egyptian civilizations developed underground pipelines for fresh water supplies and sanitation facilities. Progressively, tunnel systems have evolved to a point where they are commonly used for city subway lines, power distribution, nuclear testing sites, and military installations. Today, as the supply of available surface land decreases, serious attention is being directed towards more intensive use of subsurface space. It is no longer unrealistic to think in terms of underground cities, intercity transportation lines, and material storage centers. The future demands for tunneling and for the utilization of subsurface space will be numerous. Indeed, many theorists hold that only through proper planning and efficient use of this space can man hope to survive.

B. Tunnels in Earth

For as long as man has been tunneling through the earth, he has been plagued by two basic problems: (1) economy, and (2) the variable parameters of the earth. Because of the competitive nature of soft-ground tunneling and the growing impact of mechanized techniques, the contractor's capabilities and the project's characteristics must be analyzed with discrimination. Economy will be realized through efficient financing, knowledgeable personnel, and updated facilities. The problems posed by the earth's unknowns will only be solved by continuous research and development and the application of space-age technologies.

CONTRACTOR EVALUATION

Financial Mechanized tunneling equipment is inherently a specially designed product which is custom manufactured. Because of this, its procurement requires large initial outlays of capital before the actual start of revenue-producing operations. Machinery manufacturing alone may take from 3 months to more than a year, depending on the individual job specifications.

Standardization is almost nonexistent in mechanized tunneling today. While basic component parts such as motors, pumps, and jacks may be somewhat similar, other features of a machine are generally unique. Contractors are seldom guaranteed another project on which their equipment can be used, and the entire equipment cost must therefore be written off by the time the job is complete; this is in contrast to the expenses for more common machinery such as trucks, cranes, and end-loaders, which can be spread over a number of jobs.

Because of the one-of-a-kind nature of some components of the machine, contractors must decide on their requirements for spare parts at the time of the original equipment purchase.* Complexity of the component plus manufacturing time will guide this decision, and the required duplicates should be on hand in advance of job start-up. These purchasing precautions will guard against costly downtime once work is underway and will act as insurance against paying premium prices at a later date. One rule of thumb: approximately 25 percent of the overall cost of the equipment should be allocated for spare-parts purchase.

Unlike other industries which benefit from government and foundation funding, the tunneling industry pays for its own research and development. Testing sites are, more often than not, the actual jobsites. Contractors, as a result, bear the financial burden of underwriting design and development of support equipment as well as the costs involved in acquiring updated technology and in training personnel.

Obviously, costs of financing overhead, equipment, spare parts, and research and development, along with the contractor's other early costs, must be included in the bid price of most tunnel-in-earth projects.

Personnel Modern tunneling places strenuous demands on contractors and personnel ranging from top management ranks to day laborers. Each member of the team must adapt to innovations and accept new criteria almost daily, for the face of tunneling is rapidly changing.

In a sense, tunneling can be compared to the missile program; both deal with similar unknowns of space and both are subject to new discoveries which can affect the outcome of an entire project. As laymen now understand the importance of split-second timing and functioning of fail-safe devices in missiles, so must tunneling personnel be fully aware of new technological discoveries and realize how each task, however insignificant it may seem, can become critical to the success of a project.

Equipment manufacturers often provide invaluable data and advice to help con-

* See Sec. 7, Equipment Economics.

tractors assess work-force needs. As concepts are finalized, plans for work-force training schedules must be adopted. Much of the needed training for specialized equipment will take place at manufacturing sites. Operators should be encouraged to familiarize themselves thoroughly with the multiplicity of components as they are assembled and installed in the system. Such firsthand observation and familiarity with new equipment is invaluable not only to operators but to supervisory personnel as well.

Whenever possible, and depending on the job, contractors may find it helpful to seek assistance from consultants in specialized but related fields such as geology, metallurgy, and mechanical and electrical engineering. The need for this type of outside expertise grows with every technological advance in the industry. Just as developments in the space program have forced rapid changes upon industry, so, too, is galloping technology replacing conventional techniques in tunneling. It is unwise to decide against a new idea or new approach arbitrarily, just because it is new. An open mind to fresh ideas is as important to technological progress as an open portal is to fresh air.

Fig. 31-1 A control panel for a rotary excavator. Sophisticated apparatus such as this requires updated maintenance facilities.

Facilities With money in hand and with well-designed equipment manned by a competent work force, there still remains one major consideration: can you service and maintain the new machinery adequately with your existing shop facilities? Like the first-time buyer of a foreign sports car, you may have the money to buy and the ability to drive, but you need to know where and how the car can be serviced.

In many cases, simple backup equipment used for normal repair and maintenance work will be insufficient to meet the requirements of high-powered equipment. Today's equipment systems require updated maintenance facilities. While there will be no attempt made here to itemize the tools for an up-to-date maintenance shop, the following is a suggested checklist of considerations:

1. Is your shop equipped to handle work with today's high-test alloy metals?
2. Is your shop equipped to repair high-pressure tubing, hose, and power units?
3. Is your shop equipped to repair or install sophisticated variable-speed electric controls?
4. Is your shop equipped with test meter devices to spot damage or failure in electrical systems?

In general, these problems in metallurgy and in hydraulic and electric systems require more than pliers and screw drivers (see Fig. 31-1). Know what jobs your shop can handle, and be prepared with troubleshooters for those which it cannot.

PROJECT EVALUATION

Technical aspects of construction methods and equipment, along with tunnel characteristics, form the basis of tunnel design. A properly designed tunnel takes into consideration both construction and end use; and by adjusting one to the other, maximum overall economy is attained. It is from tunnel characteristics such as shape, diameter, and soil conditions that tunneling techniques are established. While no extensive coverage will be given here, some major factors will be outlined.

Geological Conditions Usually geological conditions of the tunnel area are furnished as part of the information given to prospective bidders on tunnel projects. These data are derived from test boreholes and are vital to the contractor's analysis of the project. At best, however, such test borings give information which is minimal in content and general in nature. It is recommended that the contractor check for any inconsistencies in these reports and, whenever possible, perform his own supplemental field observations. A study of the records of previous tunneling or drilling work in the area is also beneficial.

For every separate soil characteristic, an ideal excavating method can be found. For example, sand is best excavated with a shovel, and clay is best excavated with a clay knife; however, when sand and clay are mixed, neither of these tools is efficient. Water added to this mixture creates still another property. Due to the multitude of possible soil combinations along the tunnel path and their bearing on equipment choice, it is advisable for both manufacturer and contractor to be fully informed.

Tunnel Supports The tunnel liner or support system is designed to satisfy three basic needs:

1. It must adequately sustain not only ground loads but also loads from within the tunnel itself.
2. It must be capable of efficiently transporting whatever material the tunnel is designed for.
3. It must withstand thrust loads produced by push jacks of the shield.

The tunnel lining is usually installed in two stages: first the primary lining, and then the secondary or final lining. The excavation of the tunnel will cause some disturbance of the surrounding ground, and a period of time will be required for redistribution of the loads and stress. Such shifting could cause movement of the lining, with adverse effects on alignment and grade. Additional misalignment could occur due to difficulty or carelessness in steering the shield. Once it is assured that movement of the adjacent ground has ceased and that the primary lining has assumed stability, then the secondary lining is cautiously placed in proper alignment for efficient transport of materials.

Because of the economic and time factors involved in the sequence just described, attempts have been made to place the lining in a single stage. Cast-iron segments have been used for one-step lining, but costs of manufacturing and installation proved to be prohibitive. Fabricated steel segments with a bituminous protective coating to prevent corrosion are being successfully used today to line tunnels constructed for vehicular traffic, subways, and transmission tubes for utility lines. However, the flanges and other appurtenances associated with this type of lining will inhibit efficient hydraulic flow. Concrete linings are extensively used where efficient flow characteristics become important, but plastics and other synthetic materials with smooth surfaces are also being researched and may someday be economically incorporated. Regardless of the type of support used, ease of handling and rapid installation coupled with economy of material are necessary in order to keep pace with the other tunneling operations.

Tunnel Shape End use of the tunnel has, in the past, been the prime factor in determining tunnel shape. Following is a list of shapes and corresponding economical uses:

1. *Egg shape:* raw water, sewage
2. *Circular:* internal pressure lines, gas, water
3. *Horseshoe:* vehicular, rail, and foot transport
4. *Rectangular:* multilane or multitrack transport

In determining these shapes, structural elements, hydraulic gradient, sludge deposit, vehicular characteristics, and even man's shape have all been considered. Such appraisals were necessary not only from a purely technical standpoint but also for construction considerations. Tunnels were planned to require minimal excavation while still providing a functional form with the least amount of wasted space. Historically, tunneling costs were closely tied to the amount of spoil removed. Today, however, circular and horseshoe-shaped tunnels are most common because of the efficiencies of modern tunneling equipment. Regardless of the amount of extraneous spoil removed, such equipment provides the most economically driven tunnels in these two shapes.

Tunnel Diameter Common sizes of tunnels driven today range from 4 to 40 ft in diameter; however, because of the increasing demand for use of underground space, it is conceivable that tunnels with diameters of 80 to 150 ft will be common in the near future. Finished diameter or size of tunnels is based on required capacity for the end use. Excavated sizes are determined by the thickness of lining and by tunneling techniques.

Generally, the cost of smaller tunnels can be reduced by slightly increasing the size to allow enough room for machinery and men to maneuver with ease. Small, cramped working areas in tunnels jeopardize personnel safety and hamper efficient production. Experience with modern tunneling techniques suggests that a tunnel of 8 to 9 ft in diameter is the smallest that can be constructed with acceptable speed and efficiency.

The equipment requirements for constructing large-size tunnels can impose problems in manufacturing, shipping, and handling (Fig. 31-2). The larger power units, bearings, and other components may be unique in their design, requiring special manufacturing processes and machinery. Road, rail, and air transport limitations will frequently influence the permissible maximum sizes for these components. Additional costs for special loading and handling of equipment, as well as on-site fabrication and erection facilities, must be considered.

If these considerations make large-diameter tunnels unfeasible, it may be necessary to consider the use of twin tunnels or a cluster of tunnels for the same job.

Tunnel Length The length of a tunnel project strongly influences the degree to which it is feasible to mechanize its construction. In the longer tunnels the cost of sophisticated mechanisms and devices can be spread out over a greater number of production units, thereby decreasing the equipment cost per unit of finished tunnel.

If the economics of the job and equipment have been belabored here, it is with good reason. When great outlays of capital are required for a single job, the contractor should continually be aware of opportunities for reuse of component parts in future tunnel projects or possible resale for other uses.

Tunnel Gradient This will depend on the surface terrain and on the end use of the tunnel. Tunneling equipment is affected by grades which are excessively steep or completely flat. If the grade is steep, for instance, additional horsepower in the power unit of the haulage system will be needed for braking and accelerating. A flat grade, on the other hand, will hamper the flow of the groundwater which is prevalent in most tunnels in earth. This condition could cause messy working areas and possible ground wash-in problems. Costly corrective measures such as sump excavations and pumping systems would then be required.* The minimum gradient for proper drainage should be not less than 0.25 percent.

The relationship of gradient to surface terrain along the tunnel route affects the amount of cover over the tunnel crown and also the nature of the tunnel access. Shallow tunnels demand exceptional tunneling techniques to avoid excessive settlement or even collapse. In some soils, insufficient depth does not allow for the arching of the ground. Without this arching effect, additional crown and face loads appear and normal tunneling procedures are hampered.

Access to the tunnel is most commonly achieved by either a shaft or a portal. The shaft method requires a vertical bore excavated on or near the tunnel line. The shaft

* See Sec. 29, Construction Dewatering.

excavation must continue below the invert of the tunnel to allow for sumps, material storage, and platform structures for haulage equipment. Portal entry requires temporary thrust or reaction structures from which a shield can be launched. A larger working area is also required to accommodate surface disposal equipment when using the portal method.

Tunnel Curvature When a curve is required, it should be made with a radius sufficiently large to accommodate the type of equipment to be used. Sharp curves increase construction costs and are often difficult to accomplish. Abrupt curves will not allow passage of the tunnel shield. The normal tunneling operations must then

Fig. 31-2 20-ft-diameter bearing weighing 75 tons manufactured for a 33-ft-diameter rotary excavator. Size of bearing was limited to allowable shipping clearances from Racine, Wis., to Osaka, Japan.

cease and a larger excavation, usually performed by hand, must be carried ahead of the shield and without its protection. In certain bad ground conditions this method is not desirable. Medium curves can be accomplished by shields but steering is, at best, difficult and should be executed with great care. In such cases, supplementary structural bracing is needed at the tunnel support to take additional eccentric loads during the steering operation.

The design of the muck haulage system may also be influenced by tunnel curvature. For example, it may be necessary to restrict the lengths of conveyors and switch platforms and thereby reduce the rate of production. Even if the tunnel curvature does not constrain the design of the hauling equipment, it may still interfere with the regular sequence of tunneling operations. For example, frequent dismantling and reassembly of the primary tunneling equipment may be required. During such periods, less efficient, temporary systems must be substituted, and it may even be necessary to remove the original gear from the tunnel in order to employ this temporary equipment.

Curves can also hamper the placement of the secondary lining. Steel forms used in this operation are commonly conveyed in long sections, but sharp curves may necessitate the use of shorter sections. The fillers or special liner forms which are required for each different curve are manufactured prior to tunneling and stockpiled in readiness for later use. Substantial inaccuracies in alignment of the curve can cause serious problems in installing these prefabricated fillers.

Tunnel Location This obviously affects the overall construction scheme, since it has a direct bearing on the availabilities of work forces, material resources, and power. By analyzing the actual conditions with respect to mobilization of and support for personnel and equipment, an economical selection of tunneling methods can be made.

Remote sites usually require greater numbers of imported personnel to handle mechanized equipment, although the capabilities of the local work force should be carefully evaluated in the overall scheme for the project. Thought should be given not only to the actual bodies available but also to wage and union factors, language problems, etc. If time will permit the training of relatively low-skilled local help, such training might prove less expensive than the importation of large numbers of skilled personnel from other locations. Local governmental laws and trade restrictions should be reviewed before a final determination is made.

On remote jobsites, even the simplest supplies are often scarce or nonexistent. It is recommended that a thorough logistic survey be conducted before the project begins and before importation commitments are made. Where importation of material is necessary, appropriate allowances should be provided in the project budget. While native or locally available materials can in some cases be substituted, testing of these materials is also recommended in advance of the job.

Obviously, many of the same considerations apply to power sources as well. Mechanized equipment today can use power supplies ranging from 50 hp to as much as 5,000 hp. Such requirements may necessitate on-site power generation if no adequate facilities are readily accessible.

FUNDAMENTALS OF TUNNELING IN EARTH

Operations Four major operations are performed during the construction of the tunnel proper, regardless of the type of equipment used. Each operation includes subordinate functions which are performed before and after each main function. It is from the main function that the operations are named. The construction sequence of the major operations is as follows:

1. Ground control
2. Excavation
3. Soil disposal
4. Tunnel support

When all four operations are completed in sequence, a complete work cycle has been performed and a unit of the tunnel has been produced. As the cycle is repeated, the tunnel advances.

Each operation is dependent upon the others for its progress. For instance, excavation cannot continue unless the ground is properly controlled, and erection of tunnel supports is impossible unless adequate space is provided and the spoil is removed. When some of the subordinate functions of one operation are performed at the same time as those of another, the work cycle flows smoothly through the operational transition periods. This overlapping of operations shortens the work cycle and thus increases the efficiency of tunnel production. Rapid tunneling is the main concern of the tunneling industry and can be achieved by increasing the degree of simultaneity of operations.

Soil Classification Geological conditions of tunnels in earth range from sloppy silt to an occasional full face of rock. Generally, two basic classifications exist: erratic soil conditions and relatively consistent soil conditions. Selection of equipment is based on the specific requirements of each classification.

Glacial till, which is a heterogeneous combination of silt, sand, clay, gravel, and boulders, best typifies the first category. Adequate ground-control equipment is needed to handle silt and sand but is detrimental if rock is encountered. Due to the irregular soil conditions in this category, a variety of tunneling appurtenances must be kept on hand. This equipment must be strategically placed for accessibility when needed and yet remain inconspicuous to avoid obstruction when not in use. Obviously, flexibility of equipment and methods is a prime factor in dealing with erratic soil conditions. The open-face shield is utilized most effectively in this category of soil.

The second classification includes soils where the conditions of interest remain relatively constant throughout the entire length of the tunnel line. An alluvial deposit is considered a consistent stratum, as is a uniform deposit of clay. Two or more layers of earth which maintain essentially the same pattern throughout the tunnel length are also placed in this classification; this latter condition is commonly referred

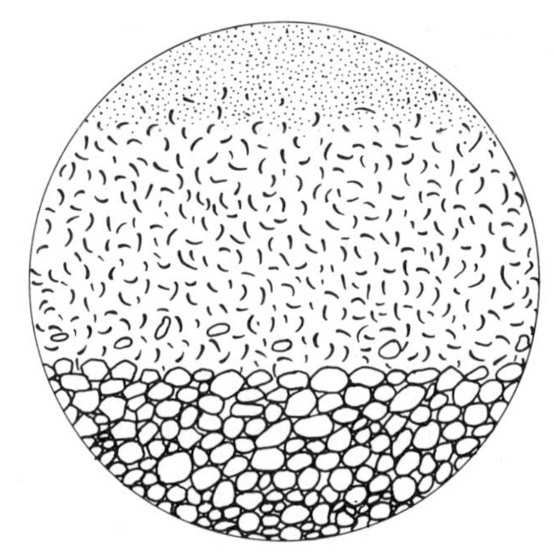

Fig. 31-3 Section through tunnel stratum illustrating a mixed face in the second classification of soils.

to as a "mixed face" (Fig. 31-3). A stable layer of earth at the crown of the tunnel is considered an ideal situation in a mixed face because it contributes to the stabilization of the whole face. It is feasible to provide equipment which is custom designed for the specific soil conditions when tunneling in consistent soils, and the rotary excavator has been found to be well suited to this application.

Mixed-face bores which consist of silt at the crown and rock at the bottom half of the tunnel are undesirable and are not classified as consistent soils.

OPEN-FACE SHIELD

The evolution of tunneling in soft ground has supplied today's industry with a basic tunneling unit, the open-face horseshoe shield. This unit performs the first operation, which is ground control. It supports the already excavated bore and also protects the tunnel face in preparation for the excavating operation. Fitted with auxiliary equipment, it allows the other operations to proceed safely and efficiently. Accompanied by the necessary excavating equipment, spoil disposal systems, and tunnel support techniques, the shield becomes a veritable workshop for tunneling in earth (Fig. 31-4).

Horseshoe Shield The basic shield has three significant structural areas: the hood, the body, and the tail section.

The hood is a visorlike extension on the top and front of the shield which gradually curves downward to the floor. It is designed to control soils with average angles of repose at the tunnel face and to provide safety to men and equipment in the working area.

The body or main structure houses the power elements: push jacks, motors, pumps, and control panels. Structurally, it is designed to withstand the thrust loads of the hydraulic jacks which advance the shield as excavation proceeds.

The primary steering device consists of the hydraulic push jacks which are positioned circumferentially within the body of the shield (Fig. 31-5). Manipulation of the jack thrusts can adjust direction of the shield. For instance, if the shield is off line on the right, only those jacks on the right side are used for advancement until such time as alignment is corrected.

The tail section furnishes temporary ground support at the rear of the shield where the tunnel supports are erected. It can be manufactured as a rigid structural element for heavy ground conditions or as a flexible unit in stable ground. This flexibility permits slight articulation with the shield and allows for greater accuracy and ease of steering.

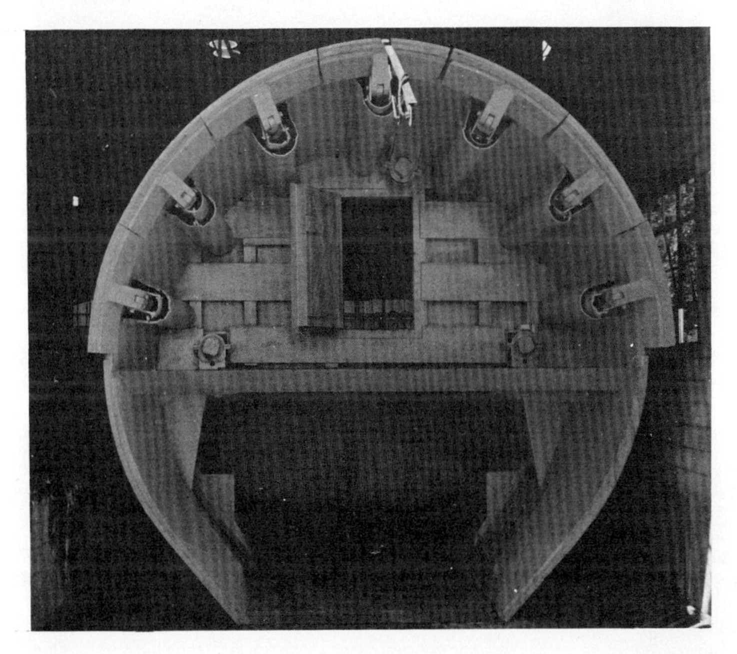

Fig. 31-4 Horseshoe shield with poling plates, breast jacks, platform, and blasting bulkhead and mat.

Because of its shape, the open-face horseshoe shield drives a flat floor which is ideal for both walking and working and for placing track for the spoil disposal system. It is also advantageous from the standpoint of steering: as the shield is thrust forward, there is little chance of ground displacement at its sides; thus the chance of rolling is held to a minimum.

Experience has shown that remining costs can be reduced by driving this type of shield within the top portion of the tolerance zone. If the shield then fluctuates off line or grade, tight areas will occur more frequently near the bottom where remining operations are less hazardous.

Poling Plates These provide a means of extension beyond the shield hood into the tunnel face. Soils with smaller angles of repose can be controlled more easily with poling plates than with the use of the shield hood alone (Fig. 31-6). Hydraulic jacks

Fig. 31-5 Rear view of horseshoe shield. The jack shoes about the circumference fit securely into a tunnel ring to distribute the thrust loads of the push jacks when the shield is advanced.

Fig. 31-6 Shield with poling plates extended. The tail section has been moved back to illustrate the push jacks in their fully extended position.

thrust the poling plates forward, independently of the shield, to supply the necessary range of extension. During this process, the plate is automatically furnished with an upward force component to support the overburden without deflecting the plates.

The operational sequence of the poling plates is as follows:

1. Plates are extended forward.
2. Soil is excavated as conditions permit.
3. Shield is advanced under the plates.
4. Soil is again excavated.
5. Plates are again extended and the cycle is repeated.

If the plates were extended and then allowed to advance along with the shield, undue stresses could cause buckling of the plates. It is for this reason that the poling plates operate with the shield in a leapfrog fashion. During the shield advancement, selective hydraulic controls supply sufficient pressures to prevent the plates from being forced back by the face loads.

In some instances, when not needed for ground-control measures, poling plates can also be used as supplemental excavation equipment. For this process, a conical hole is made at the center of the face by normal methods and poling plates are then thrust forward in a shearing or extruding action. This forces the surrounding ground to break up and fall into the prepared area, leaving only the mucking operation to clear the area of spoil.

Breasting Apparatus Breasting is essentially a face-supporting technique in which structural elements of the shield hold the tunnel face in the desired position. Soil conditions determine the need for breasting. Other methods of ground control include the following: clay pocket, clay blanket, chemical blanket, chemical hood, chemical solidification, electrochemical stabilization, grouting, freezing, wells, wellpoints, bulkhead control, and compressed air.

In its earliest applications, breasting was achieved by manually setting wood planks or so-called "breast boards" on edge against the face in a wall-like fashion. The wall, in turn, was secured by a "raker" or strut sloping back to the floor from a "soldier" or timber placed vertically at the center of the wall.

Advanced shield design frequently incorporates breasting by powered equipment as an integral part of the shield. Breast jacks, pads, and plates take the place of rakers, soldiers, and breasting boards. With mechanical breasting, the job is accomplished more quickly and effectively. The rapidity with which support is applied to the face significantly reduces the amount of ground disturbance and makes for smoother operation.

Partial breasting at the lower portion of the face shortens the height of the slope of the soil, thus allowing improved handling of runny soils or those with slight angles of repose. This technique is also applied to any part of the face to control pockets or strata of runny material.

Full breasting is required when the entire face is of a runny or uncontrollable consistency. Mining can be effective for short runs under these conditions, but other control measures may be required if adverse conditions persist.

It is mandatory that the face of the tunnel be fully breasted when operations cease temporarily, such as over weekends. This should be done regardless of the appearance of the tunnel face. Experienced tunnelers have discovered that atmospheric effects, internal but undetected water or gas pressures, and ground movements can all seriously endanger the work face if it is left uncontrolled.

Platforms The primary use of the work platform has been in larger shields, to enable the higher portions of the tunnel face to be reached. By dividing the larger area into several smaller working areas, the tunnel face could be handled more effectively. With the concentration of several structural elements and a larger work force at the tunnel face, ground control was more reliably accomplished and excavation production increased.

Mechanical devices can also be added to the standard platform to make it a more versatile piece of equipment which can be used in other phases of tunneling. Fitted with hydraulic jacks and properly hinged, the platform can be used as a dumping element (Fig. 31-7). Loose spoil placed on the platform by the miners can be loaded directly into the muck boxes below, thus eliminating duplication of effort by the muckers in rehandling the spoil. In addition, the platform can be adjusted in the longitudinal axis to enable it to move ahead and perform a breasting function.

Should larger boulders or a solid face of rock be encountered in a soft-ground tunnel, appropriate drilling and blasting accessories can be added to the shield. A work platform can accommodate the stationing of several drills. This type of platform can be hinged at both sides of the shield and folded back for other methods of excavation.

When considerable drilling and blasting is required, the platform can be fixed to the shield and a blasting bulkhead added. Bulkhead and blasting mats are used to confine the flying rock so that no damage or injury is inflicted on the shield controls or the men. Confinement of the flying debris also minimizes the amount of cleanup required following the blast.

General Excavation The cutting edge along the perimeter of the shield slices the soil and defines the bore area as the shield advances through the ground. As the shield continues, the soil about its perimeter is extruded, breaks up, and falls to the floor. This is possibly the extent of the excavation performed by the shield itself. In smaller tunnels a minimum of hand work or the like is required at the center of the tunnel face to complete the excavation. However, in larger tunnels a more definite means of excavation is required.

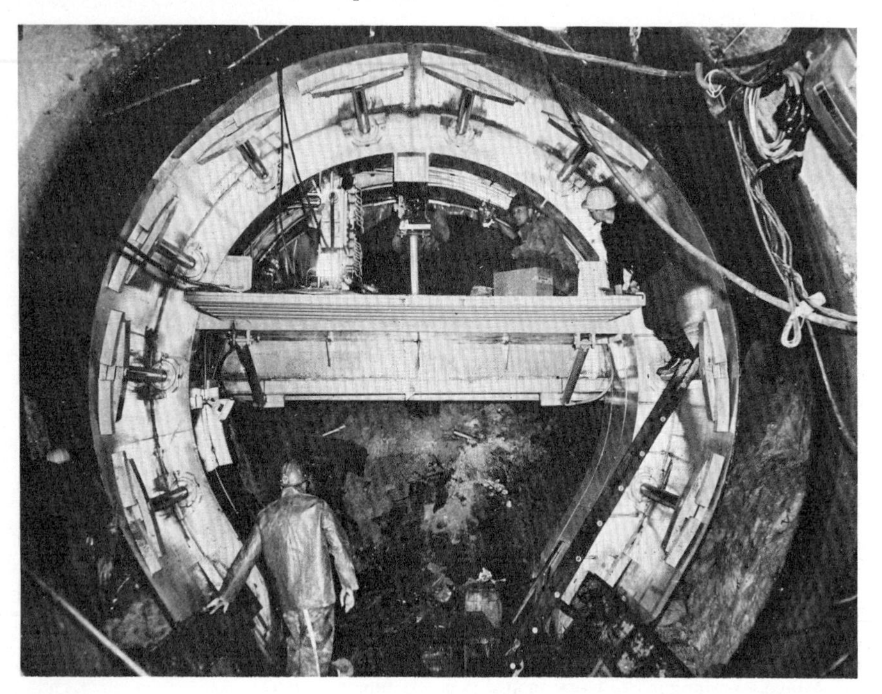

Fig. 31-7 The platform within this shield is composed of a fixed and an adjustable section. The fixed section is used as a work station from which the operator controls the shield. Notice the position of the transit which indicates the location of the shield. The movable portion of the platform is used for dumping and breasting.

In larger tunnels, where space permits, surface excavating equipment such as end-loaders and backhoes may be used. However, excavating units of this type are usually impractical for long tunnels because of problems relating to ventilation, equipment maneuverability, and equipment strength. These units have not provided an adequate solution to the excavation needs in earth tunneling.

Mechanical Bucket Excavator A mechanical bucket excavator has recently been developed and has advanced the progress of the tunneling industry immensely (Fig. 31-8). It is capable of performing excavation duties as an integral component of the shield and its other accessories, as described earlier. This excavator combines versatile maneuverability and tremendous power. It is fully adequate for tunneling in stable ground conditions, and its excavating capabilities can also be applied to rock tunnels.

Roller trucks mounted on each side of the excavator body permit the excavator to travel longitudinally in runways attached to the sides of the shield (Fig. 31-9). Large

Fig. 31-8 Mechanical bucket excavator being assembled for use in a 37-ft tunnel.

Fig. 31-9 A mechanical bucket excavator mounted to a 26-ft-diameter shield with poling plates.

travel jacks move the entire machine to or from the tunnel face. A rotary actuator allows the boom and bucket to approach the tunnel face from any desired circumferential position. Radial positioning is accomplished by the boom, and finer radial adjustments are provided by hydraulic jacks attached to the bucket. The bucket can be adjusted for capacity and function and, at the one extreme, it can be entirely closed like a clenched fist to apply crushing or ripping loads. At the other extreme, it can be completely opened to permit a large quantity of spoil to be removed in a scooping action.

The excavator model in current use contains a power unit supplying 4,000 hp. This power is distributed to the various components to produce the following work forces:

Traveling: 800 tons
Rotating: 1,000 tons
Boom thrust: 1,000 tons
Bucket movement: 800 tons
Bucket manipulation: 200 tons

This type of excavator has been successfully used in the excavation of weathered and fractured granite as well as other decomposed rock formations. Prior to its introduction, excavation in such ground conditions was dependent entirely upon prior drilling and blasting.

ROTARY TUNNEL EXCAVATOR

The rotary tunnel excavator is custom designed for use with a specific type of soil within the second or "consistent" classification. Since only one type of soil is to be excavated, the equipment can be custom designed to achieve a high level of performance. In recent tests, for example, it was found that the rate of advancement of an 11-ft-diameter rotary excavator in silt was 1 mi per day. However, due to slow liner erection and spoil disposal, the average rate of advance was substantially below this figure.

The rotary excavator consists of a shield, an excavating or cutting wheel, and some wheel-driving apparatus. In very stable ground, the shield may not be necessary. The concept of the rotary excavator shield is derived largely from the open-face horseshoe shield. It performs basically the same function although, of necessity, it is circular in shape. The hood of the rotary excavator shield controls overburden and protects the excavation area, the body houses the working mechanisms, and the tail furnishes ground support during the lining installation.

The cutting wheel is designed to perform three main functions: excavating, spoil removal, and breasting. Cutting tools such as teeth, scoops, blades, or combinations of these are selected and placed on the wheel face to form the desired excavating pattern. As the wheel rotates, excavated spoil falls within the wheel onto dump plates or other suitable apparatus where it is carried to a dumping position and dropped into the spoil disposal system.

Breasting is accomplished by selectively controlling wheel-face openings, adjusting the bite of the cutting tool, and supplying proper breasting pressures at the tunnel face. Breasting pressure is furnished by transferring the push-jack forces through the structure of the shield to the face of the wheel. The incorporation of the breasting elements in the rotary excavator has been exceptionally successful, to the extent that other methods of ground control such as chemical stabilization, freezing, or compressed air are considered unnecessary in many cases.

A reversible peripheral-drive system is incorporated to rotate the wheel. Transmission units are placed on a mounting diaphragm fixed within the shield, as indicated in Fig. 31-10. These provide power to the wheel, which rotates on a large bearing coupled to the diaphragm. The working zone is located at the center of the shield and the wheel drive mechanisms such as motors, pumps, and transmission units are placed circumferentially about the body of the shield (Fig. 31-11). They are compactly arranged in this fashion for easy access during repair or replacement without hindering other operational activities.

Fig. 31-10 A transmission unit with a double reduction specially designed into a compact unit for use in a rotary excavator.

Fig. 31-11 Rotary excavator with motors, pumps, and transmission units placed circumferentially within the shield. Each unit is capable of producing 312,000 ft-lb of torque.

Open-face Wheel A rotary excavator with an open-face wheel is generally employed in soils of reasonable stability. As a rough guide, such soils will stand unsupported at the tunnel face for two or three days without perceptible flaking or loss of tensile strength.

Many configurations of the open-face wheel have been manufactured and have been very successful. Figure 31-12 illustrates an open-face wheel. A combination of teeth, scoops, and cutting blades was used on this face to excavate a combined soil condition of clay and sand. The face was fitted with a steel ring at the outer edge of the wheel for partial breasting.

Fig. 31-12 Open-face wheel.

Fig. 31-13 Full-breasting face wheel with teeth placed to cut in two directions.

Full-breasting Face Wheel In conditions of runny soils, such as silt or sand with high water content, a rotary excavator with a full-breasting face wheel is ideal (Fig. 31-13). This particular face is composed of three cutting doors, a scoop, an inspection hole, and an access hole. The cutting doors and scoop are variably opened or closed as necessary during the excavation and breasting operations, while the other openings remain closed until visual inspection or the like is required at the tunnel face.

Fig. 31-14 Variable-face wheel with port plates overlapping the cutting blades.

Fig. 31-15 Variable-face wheel with port plates removed.

The amount of soil removed will indicate whether further breasting action is needed or not. For example, if more soil is removed than was calculated for a complete shove, more breasting pressure is required. The degree of breasting is achieved by simultaneously varying the door openings and the breasting pressures.

Variable-face Wheel A rotary excavator with a variable-face wheel (Fig. 31-14) possesses features of both the open-face and full-breasting face wheels.

The face is composed of radially located cutting blades and imbricated adjustable port plates. The openings between the spokes and port plates can be adjusted to meet the varying requirements of the soil or completely closed to provide a full-breasting face. The port plates are detachable, and when they are removed the face resembles an open-face wheel (Fig. 31-15). This type of wheel is used where one-half of the tunnel length requires full breasting and the other half requires a minimum of breasting.

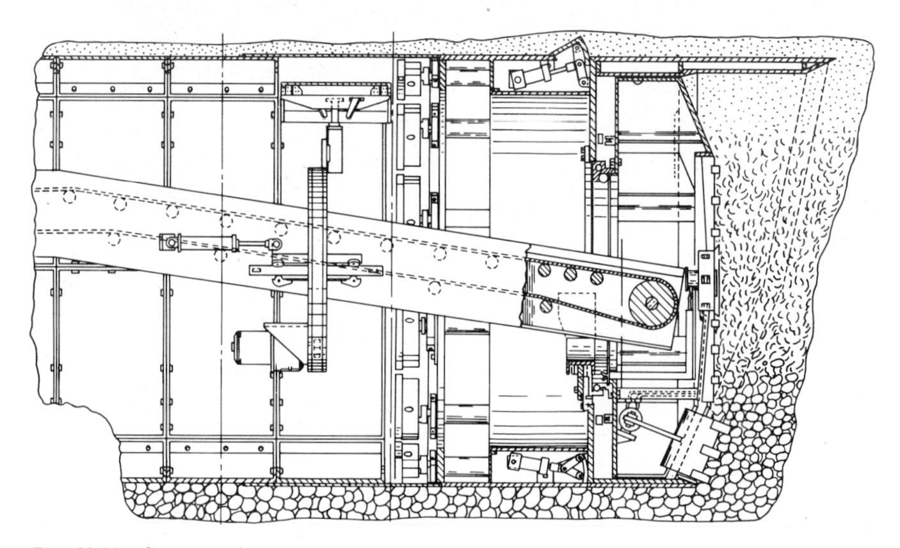

Fig. 31-16 Cross section of variable-face wheel with power-operated claw to excavate gravel. When at the top portion of the tunnel face, the opening at the claw can be closed for breasting the softer material.

The machine is also used where the tunnel stratum is of a mixed-face nature. For example, the bottom third of the face may consist of gravel and boulders 2 to 8 in. in diameter; the middle third may be sand, silt, and wet clay; and the top third may be loam (Fig. 31-16). The gravel and boulders can be excavated economically with a power-operated claw which can be worked in and out and side to side. Manipulation of this device, along with the rotation of the wheel, loosens any difficult area or large boulders to facilitate excavation.

SPOIL DISPOSAL

The spoil disposal system is involved in the complete handling of the spoil between the excavation face and the final surface disposal area. The system is divided into four separate functions: mucking, tunnel haulage, dumping or hoisting, and surface disposal and/or haulage.

Due to its close relationship to excavation, mucking is considered the critical function of the spoil disposal operation. The mucking operation transfers the spoil from the excavation area to the tunnel haulage equipment. The most successful method of mucking is the conveyor shown in Fig. 31-17, which works as an integral part of the open-face shield or the rotary excavator system. The complete unit is attached to the shield and travels with advancement of the shield.

The size of the conveyor is determined essentially from the number of cubic yards in one excavation cycle. The length of excavation in one cycle is determined by the length of the tunnel support, which is relatively standard at 4 ft. The volume of soil to be excavated in each cycle, including an appropriate allowance for bulking, will determine the muck-box dimensions.* One box or many boxes may be used to accept the spoil from one shove. The main point is to establish the length of train of muck boxes, liner car, and auxiliary cars along with necessary working tolerances. With these dimensions in mind, the height and length of the conveyor system can be finalized.

Fig. 31-17 Conveyor mounted on a travel gantry.

The conveyor system is composed of a chain flight conveyor and a belt conveyor.† The flight conveyor is set on the floor of the shield, near the tunnel face, sloping upward and back and high enough to dump onto the belt conveyor. It is fixed within the shield with automatic correction devices to permit self-leveling when the shield rolls. Surrounding the flight conveyor, on the floor of the shield, is a horizontal-type hopper that directs the spoil toward the flights. The hopper checks spillage and avoids unnecessary hand work.

The belt conveyor is set at a sufficient height to permit passage of the boxes underneath. It must also be long enough to fill each muck box in the train. It is supported by one or more gantry frames that are fitted with wheels and travel on rails set on the tunnel floor. Within or attached to the frames of the conveyor are electric panels, oil reservoirs, and compartments for spare parts and tools.

The prime function in tunnel haulage is the transferring of spoil from the mucking area to the intermediate handling area at the tunnel access. The secondary function is the transportation of men and materials back to the tunnel working area. Length of time of the haulage sequence must remain constant as the tunnel gets longer. In longer tunnels this becomes critical because the system must be designed for a maximum run.

Tunnel haulage in both earth and rock has been expediently performed by rail, as illustrated in Fig. 31-18. Systems of this type, which are used extensively, are described under Tunnels in Rock in this section. A haulage system consisting of several belt conveyors in tandem has also been used successfully in short tunnels.

* See Sec. 6, Engineering Fundamentals.

† See Sec. 15, Belt Conveyors.

In longer tunnels, however, an additional haulage system would be required to satisfy the secondary function of transporting men and materials back to the working area. Such a dual hauling system would probably not be economically feasible. It would also present problems of congestion within the tunnel. The concept of the dual haul-

Fig. 31-18 Train of muck boxes used for a tunnel haulage system.

Fig. 31-19 100-ton muck-box tipple.

age system does have potential, however, and with refinement it may be useful in the future.

Regardless of what method of tunnel haulage is incorporated, the spoil must be disposed of at the surface. Hoisting equipment such as crawler cranes or gantry cranes is used at shafts to lift the spoil from the tunnel and transfer it to surface haulage equipment. If the spoil must be removed from the work site, on-highway haulage equipment will normally be used.

If a portal is used as means for tunnel access, the tunnel haulage system will bring the spoil from the working face to a dump area near the portal (Fig. 31-19). If the project site is in a remote open area, the spoil can sometimes be disposed in fill areas near the portal. Earthmoving equipment will be used for this chore.

In the larger tunnels, trucks and other surface haulage equipment can be used for both the tunnel and surface haulage. Although this reduces the amount of material handling, surface haulage equipment is usually undesirable for use in the tunnel proper due to ventilation and maneuverability problems.

TUNNEL SUPPORTS

The rib-and-lagging method illustrated in Fig. 31-20 is the most popular for the primary tunnel support. It is extensively used because the materials—wood and steel beams—are readily available, relatively inexpensive, and easy to install. This method has a high performance record in a variety of ground conditions. Ribs can be rolled to form several different shapes, of which the horseshoe and circle are most common. The horseshoe shape is composed of arches, legs, and spreaders and can be made in sets of three, five, or more pieces; the circle is made of three equally sized sections. Joints consist of flat steel plates or angles welded to the ends of each section. The ring is firmly secured by welding or bolting these joints together. During installation, lagging is placed horizontally between two vertical rings and is held in place by the flanges of the rolled beams.

Fig. 31-20 Rib-and-lagging tunnel support in place.

The size of the rib (Fig. 31-21) is designed on the basis of the maximum ground loads acting on a 4-ft span. If the actual load is greater than this design load, the same rib can be used by simply reducing its spacing. On the other hand, if the ground is more stable than was anticipated, savings can be realized by placing partial lagging. When this is used, the lags should be placed so that they can withstand the thrust loads from the shield in addition to supporting the imposed ground loading. Because of the flexibility afforded by this method of ground support, it is widely used in the first classification of soil.

Fig. 31-21 Various sizes of horseshoe ribs.

A concrete segment liner, illustrated in Fig. 31-22, has recently been devised for use in an internal pressure tunnel. The results have been highly satisfactory due to the liner's simplicity of design and ease of installation. It is designed as a compact structural unit and thus requires the least amount of handling during erection. The full ring consists of three identical segments and is installed without key blocks or specials. These segments can be manufactured at the jobsite in assembly-line fashion (Fig. 31-23).

Each segment is made of wire mesh and two rolled ribs encased in concrete. A portion of each rib is left exposed at each end of the segment to allow for mechanical connection during installation. Small openings about the beams in the center of the segments are formed to accept the clamping device of the erector mechanism. On the outside diameter at the ends of the segment, a recess is furnished for installation of protection plates. These plates protect connecting joints from incoming soil after the complete ring is installed (Fig. 31-24).

Since tunnel supports are more frequently erected inside the tail section of the shield, it is obvious that the circumferential size of the completed support is smaller than the perimeter of the excavated area. Therefore, when the shield is advanced

Fig. 31-22 Concrete-segment tunnel support in place.

Fig. 31-23 Concrete segments manufactured in assembly-line fashion.

and the support is exposed to the surrounding ground, a void slightly larger than the thickness of the skin of the tail section is present. For the support to function properly, contact with the surrounding ground must be attained and the void filled. Normally, the void between surrounding ground and support is filled with cement

Fig. 31-24 Concrete segment with corrugated protection plate.

grout, pea gravel, or similar types of material. This method requires special equipment as well as extra room within the tunnel, and it is not always effective.

A second and more practical method of making this contact is by expanding the support itself. To do this, a joint of the rib or segment must be opened and spread apart to force the whole section against the surrounding ground. By applying additional hydraulic pressure, any soil that has sloughed can be recompacted into its original condition. Once the support is placed firmly against the bore surface, a structural filler is placed into the widened joint of the rib or segment to secure the complete ring. A "ring expander" device, shown in Fig. 31-25, is used to expand the ring and can be designed as an integral part of the machinery.

The "peg block" support is a self-expandable unit requiring no special filling, expanding operation, or equipment. The completed tunnel ring is composed of pre-

Fig. 31-25 Ring expander mounted in horseshoe shield.

Fig. 31-26 Concrete blocks of the "peg block" tunnel-support system.

fabricated, reinforced concrete blocks and wooden pegs, dowels, and strips. The blocks are formed (Fig. 31-26) so that the tapered pegs can be inserted at the longitudinal joints and the strips inserted at the circumferential joint. Each block is secured by doweling to the blocks of the previously set ring.

The ring is circumferentially expanded, making contact with the excavated wall, as the tapered pegs are forced deeper into the joints by the axial force of the push jacks. The system was designed to resist shock loads such as minor earth tremors, earthquakes, etc. The resiliency of the wood has proved adequate for this purpose, in addition to providing a seal for waterproofing.

CONTINUOUS TUNNELING SYSTEM

By combining the best methods and equipment into a tunneling system, the major operations are, to a high degree, performed at the same time. With additional modi-

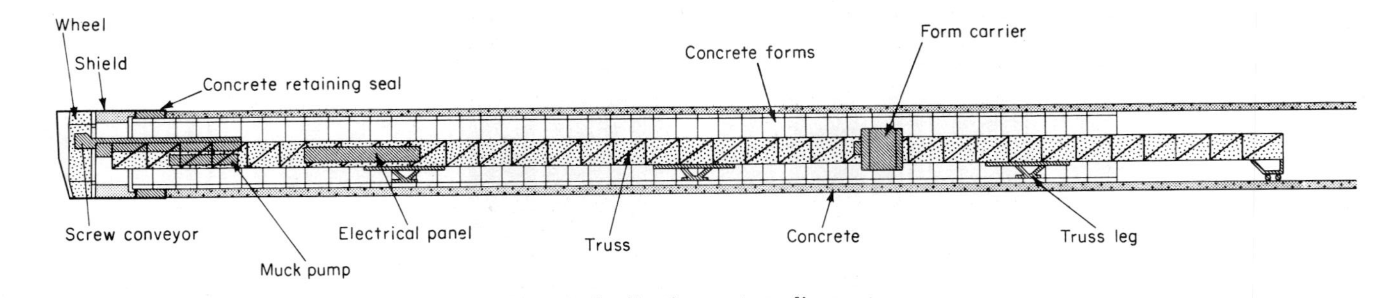

Fig. 31-27 Continuous tunneling system.

fication of this equipment, a completely integrated tunneling system can be developed whereby ground control, excavation, spoil disposal, supporting and lining of the bore are performed simultaneously in one operation to produce a complete tunnel.

Several techniques have already been developed in which the four major operations of tunneling have been integrated in one operation. Perhaps the most promising of these is a tunneling system which utilizes the rotary excavator for excavating, pumping equipment for disposal of spoil, and pumping equipment for delivery of lining material (Fig. 31-27). In this system, each component is synchronized with the others to ensure a continuous operation. In other words, excavating proceeds at a speed which delivers spoil at a predetermined rate which can be handled by the spoil disposal system and, at the same time, concrete is pumped into steel forms to produce

Fig. 31-28 Rear view of rotary excavator used in the continuous tunneling system.

a final lining. These operations are carried on simultaneously and at speeds which allow for considerable operating economy as well as simplified procedures. Some experts have estimated that the successful development of such a system can net overall savings of approximately 50 percent in both time and cost of the project. Further, these savings could be accomplished without increasing the danger to the men working underground. It even seems likely that safety conditions would be considerably improved under such a system.

The tunneling system just described is designed for ground consisting of silt and wet clay with a high water table. It is estimated that it could produce 60 ft of completed tunnel per day in such material. The basic unit from which the entire system is operated is a 16-ft-diameter rotary excavator (Fig. 31-28). It carries, as standard equipment, hydraulic and electrical systems that are connected to each other in such a way that if a motor, pump, or circuit in one system becomes inoperative, power can be supplied from other systems during its repair or replacement so that the tunneling operation is not completely stopped. Advancing thrust for the system is provided by a series of pumps and motors rated at 10,000 psi and developing 1,980 tons of thrust through 12 hydraulic pushing jacks. These push jacks advance the shield a maximum of 4 ft with each thrust, shoving off from reusable concrete forms.

Ground-control elements up front consist of the shield and a wheel face capable of 85 to 100 percent variable breasting action (Fig. 31-29). Directly behind the wheel, on the motor mounting diaphragm, a waterproof bulkhead is supplied to seal off the whole excavation area in case of excessive water inflow. A combination of concrete forms, concrete retaining bulkhead, and the final lining completes the support system at the rear. The concrete retaining bulkhead located between the tail section and concrete form keeps the freshly applied concrete in constant compression. This compressive force is also transmitted to the surrounding soil and causes compaction. When the soil and concrete reach a predetermined compression point greater than

Fig. 31-29 Rotary excavator with face of wheel capable of 85 to 100 percent variable breasting.

the input pressure, the pressure at the bulkhead is automatically relieved to permit additional concrete to enter. Because continuous pressure is exerted on the concrete and the soil, ground support is constant and settlement nonexistent (Fig. 31-30).

Trailing behind the shield is a 210-ft-long box truss attached to the motor mounting diaphragm. This support truss houses the spoil disposal apparatus, concrete line, and electric panels. The truss is also used as a means of transporting the concrete forms.

Supporting the truss in position are three support legs; these allow the truss to advance on a track-and-roller system. When the truss advances 15 or 20 ft, the legs are hydraulically raised, rolled ahead, and lowered once again to take the truss load. One leg is advanced at a time while the other two take the load of the truss. Movement of the legs does not interfere with any part of the operation.

The excavated spoil is directed to a screw conveyor through a hopper located within the wheel (Fig. 31-31). The conveyor feeds the muck pump, which forces the spoil through a 10-in.-diameter muck pipe housed in the truss, back through the completed tunnel, and ultimately up and out to a truck loading hopper located on the surface. Because of existing ground conditions and generally favorable strata for the process,

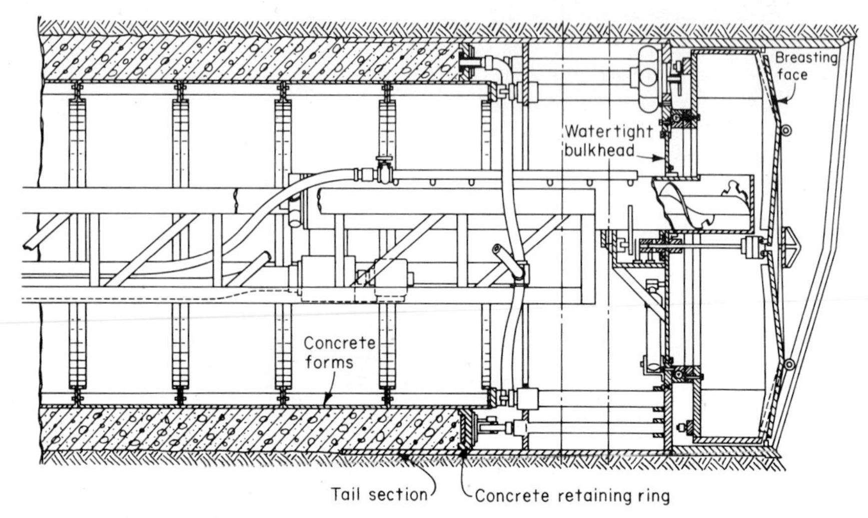

Fig. 31-30 Section illustrating ground-support elements of the continuous tunneling system.

Fig. 31-31 Screw conveyor with water ports to control consistency of the spoil.

slurry or breakdown of the spoil is not normally required, although provision to control the consistency of the spoil is included. The material is moved at a rate of 120 cu yd per hr through the muck pipe under 1,250 lb of pressure. The hydraulic device which operates the disposal system is rated at 5,000 psi and is located within the truss. As the excavation apparatus advances and pulls the truss ahead, a hydraulic telescopic pipe compensates for the increasing length of the muck pipe. After an advance of 20 ft, the telescopic pipe is hydraulically retracted and a new section of muck pipe is installed.

Concrete is delivered to the jobsite via ready-mix trucks from a nearby batch plant. The first-stage concrete pump, located on the surface, receives the concrete and sends it to a second pump which is located in the tunnel. The concrete is then fed at 350-psi pressure through a pipe system to a distribution center located at the forward end of the truss. From this center, a series of six hoses distributes the concrete evenly through the concrete retaining bulkhead to form the final lining of the tunnel.

Fig. 31-32 Carrier/erector apparatus for the continuous tunneling system.

The two concrete pumps work in conjunction with each other to provide a flow of the proper volume. However, should either of the pumps become inoperative, adjustment can rapidly be made to continue feeding the distribution center from one of the two pumps while the other is being repaired or replaced.

To maintain constant pressure and to guard against voids and/or cold joints, the concrete retaining bulkhead forms a continuous flexible seal around the concrete. It is held by a series of six hydraulic jacks attached to and operated from the tail of the shield. Each of these "concrete jacks," operating under a 5,000-psi system, has a maximum of 62 in. of stroke.

By utilizing an underground "leapfrog" system, forms are set to support the concrete for 72 hr while it is curing. They are then stripped, transported forward, and reused. Forty-five sections of concrete form are required, each of which is 4 ft in width. Two basic form units built of mild steel plate and angle comprise the entire 360° perimeter of the tunnel. As the machine bores and excavates its way through the face of the tunnel, exposing a new 4-ft section within the tail, the folding forms are carried forward by a cylindrical carrier/erector arm apparatus which rides on the 210-ft-long support truss.

The form carrier/erector illustrated in Fig. 31-32 is capable of traveling the length of the tunnel on wheels fitted into a track attached to the truss. Because these forms are variably positioned according to alignment requirements, the erector is designed to rotate 360° about the truss to place or strip the forms.

The carrier/erector apparatus is supplied with three hydraulic jacks placed approximately 90° apart. The center jack fits into the center or hinged part of the form, and the other two jacks are pinned on each side to secure the form into position. After unbolting the section to be moved from the successive ring, the jacks are retracted. This strips the form and folds it to size, ready to travel forward for erection.

Each concrete form is tapered to allow for corrective adjustments in its alignment (Fig. 31-33). If necessary, entire circular forms can be rotated to achieve the desired line and grade.

Fig. 31-33 Concrete forms with built-in adjustment fillers.

The following is a list of a few materials and operations eliminated by this concurrent excavating and lining technique:

1. Primary lining
2. Liner expanding apparatus
3. Special sets for squaring primary line
4. Grout and grouting apparatus
5. Muck cars and locomotives
6. Track, switches, and flooring
7. Air-compression equipment
8. Special vertical hauling equipment
9. Large, sophisticated water-pumping systems
10. Cleaning the tunnel prior to concreting
11. Remining
12. Time involved in performing the above operations

Although a total-system concept of continuous tunneling has been explored, problems peculiar to the system still remain to be solved. Nevertheless, it is readily apparent that any approach to tunneling which eliminates the step-by-step operations of the present techniques is intrinsically desirable. This is clearly the direction of tunnel technology of the future. It is even conceivable that, in the not too distant future, men will create below-ground open spaces simply by placing the appropriate equipment systems at the desired depth and then controlling the entire operation from a master control panel above.

C. Tunnels in Rock

The tunnel man's definition of rock is "anything which must be drilled and blasted." This is not a precise geological classification because it includes certain hardpans, hard clays, and other consolidated materials. In recent years, tunnel men have also begun to distinguish between "hard rock" and "soft rock." Hard rock is then defined as that which cannot be rapidly and economically mined by means of the present generation of mechanical tunnel excavators, or "moles."

There is no question but that within a few years some of the rock which is now considered too hard will be economically mined with a mechanical excavator. Basically, all that is needed is more usable power and a tooth which is harder than the present tungsten carbide. However, in spite of the continuing improvements in the mole, drilling and blasting will still be used even in the softer rock. Short tunnels cannot justify the high purchase cost of a mole if this cost must be written off on each job. Even when a mole of the proper diameter can be leased on a footage basis, the costs of incoming freight, lowering the mole into the tunnel, and shipping it back to the factory at the end of the job may more than offset any labor saving which can result from its use.

DRILLING

The Round Rock tunnels are always driven in a cycle which is known as "the round." The cycle elements are:

Move in drilling equipment and prepare to drill.
Drill holes of pattern.
Load holes with explosive.
Move out drilling equipment.
Blast, or "shoot."
Ventilate, known as "smoke time."
Scale down roof to remove any loose rock.
Load out broken rock, known as "mucking."
Place roof support if required.

These different operations follow one after another around the clock. A well-organized small- or medium-sized tunnel with proper equipment should be able to produce six rounds in 24 hr, and occasionally some jobs may hit seven or eight rounds per day. A large tunnel more than 20 ft in width will probably call for drill holes which are 12 ft or more in depth. These will take longer to drill and, when the round is fired, there will be an immense amount of muck to be loaded and hauled out. For this situation the contractor should plan on one round per 8- or 10-hr shift. Progress in a large tunnel should be referenced to cubic yards removed rather than to lineal feet advanced.

Actual progress is, of course, dependent on the depth the holes are drilled. The contractor may discover that he can complete more lineal feet of tunnel per day by drilling deeper holes, even though this may reduce the actual number of rounds. In small tunnels it may work the other way; a shorter drill hole will require more rounds, but it sometimes results in more progress. In either case, the secret of successful and economical tunneling is to have the same man do the same job at the same time every day.

Method of Attack Today, most tunnels are driven "full face," which means that the entire face of the tunnel is drilled and blasted at one time. Larger tunnels, such as those for highways, might be mined more economically by the "top heading" method illustrated in Fig. 31-34*a*. A top heading can be rapidly driven through until the other shaft or portal is reached. This means that ventilation will be simpli-

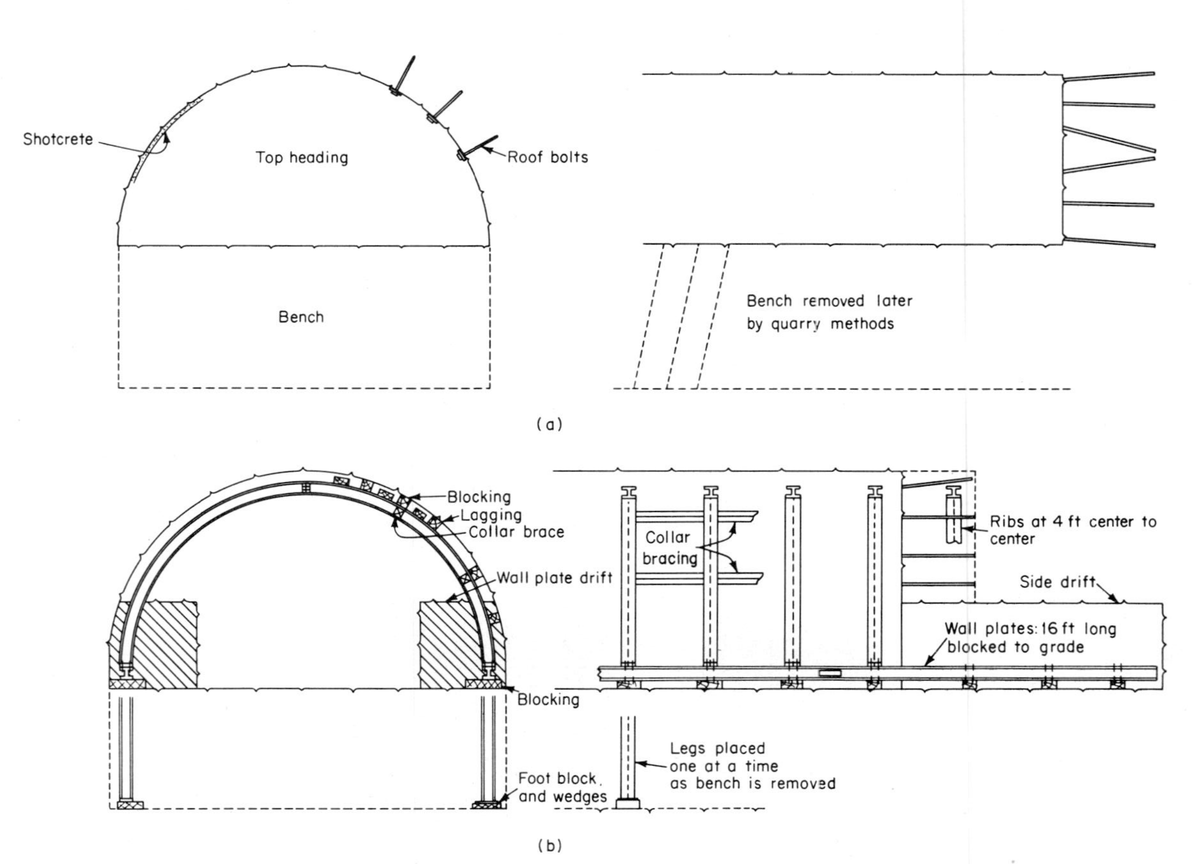

Fig. 31-34 Driving large tunnels by heading-and-bench method. (*a*) Heading and bench method—large tunnels in good ground; (*b*) drifting for wall plates—large tunnels in bad ground.

fied. The lower half of the tunnel, or "bench," is then taken out by quarry methods. These are a good deal cheaper and require less explosive.

Sometimes when the ground is very bad the tunnel can only be advanced by "drifting for wall plates" (Fig. 31-34*b*). In this method, small pilot drifts are driven along each side of the tunnel and wall plates are placed on temporary cribbing. The intervening ground is then shot down in very short sections, with steel ribs placed one at a time to furnish immediate support for the roof.

Rock Drills All tunnels driven today employ pneumatic drills. Small tunnels will use the air-leg drill (Fig. 31-35) which consists of a pneumatic drill supported on a telescoping pusher leg. Drills of this type should always be used in pairs since, although they can be operated by one man, it takes two men to get one started. Air-leg drills are sometimes used on larger tunnels, but a platform must then be supplied so that the operator can reach the upper holes. Platform decks should be 6 to 8 ft apart.

Fig. 31-35 Air-leg drills for small tunnels. (*Gardner-Denver.*)

The present method for most large tunnels is to mount a heavy pneumatic drill on a hydraulic boom (Fig. 31-36). These hydraulically controlled booms may be swung up, down, or sideways, so that the drills can be slanted in any desired direction. Some boom ends have a rotating device so that the drill may be turned from the upper to the lower position. Hydraulic booms are so big and heavy that they must always be mounted on some type of drill carriage or "jumbo." An alternative is the balanced pneumatic boom, better known as a "Long Tom," in which a pneumatic cylinder is used for holding the drill and for advancing it. Swinging and pointing the boom must be done manually. This boom is a good deal lighter than the hydraulic boom, but it still requires a jumbo.

"Ladder drilling" is used in Europe but has not been widely adopted in the United States. In this method an air-leg drill is mounted on top of a steel ladder whose rungs are about 12 in. apart. The drill will then push itself along the ladder, one rung at a time, with very little attention. In Europe one man will handle two or more ladder drills, but there is no big advantage in using the method in this country since there is a union rule which says, "one drill, one man."

Drill Jumbos Three types of drill jumbos are used in modern tunneling. The "main-line jumbo," Fig. 31-37*a*, is mounted on car wheels which run on the haulage track. The jumbo is advanced to the face to drill the round and, after the holes have been loaded, is pulled back to the nearest siding. It will have folding wings which are extended when it is at the face so that the men can reach the rib holes. When it is parked on the siding the wings are lowered to permit the passage of the mucking machine and muck trains.

The "straddle-type jumbo" shown in Fig. 31-37*b* is used on large- and medium-sized tunnels. It is designed to straddle the muck track and thus permit the passage of muck cars or trucks. It runs on its own wide-gauge track and, once the holes have been drilled and loaded, it is pulled back about 150 ft to prevent its damage by fly rock.

Fig. 31-36 Pneumatic drills on hydraulic booms. (*C. J. Langenfelder: Pennsylvania Turnpike.*)

If the tunnel is large enough for truck haulage, then the jumbo will be mounted on either a tractor or truck (Fig. 31-37*c*). Here again, once the round is drilled, the jumbo moves back by its own power and parks to one side to permit the passage of the mucker and the muck trucks.

Drill Steel Almost all drill steels today are tipped with tungsten carbide. Gone are the days when the blacksmith had to resharpen and retemper every drill rod after only 2 ft of usage. For air-leg drills, the holes are 1½ to 1⅝ in. in diameter, depending on the size of the drill and the type of ground. The bits are generally chisel-pointed except for ravelly ground. For boom-mounted drifters, the usual hole diameters are 1¼ to 2 in., and the holes are generally drilled with a four-pointed bit.

Integral steel, wherein the bit and the rod are one piece, is considered superior for air-leg and hand-held drills. The drill rods are then made of carburized alloy steel. If detachable bits are employed, the drill rod is generally made of alloy steel so that it can be rethreaded in the contractor's shop when battered. Carbide drill bits require resharpening on a silicon-carbide grinding wheel every 40 to 100 lin ft of drilling,

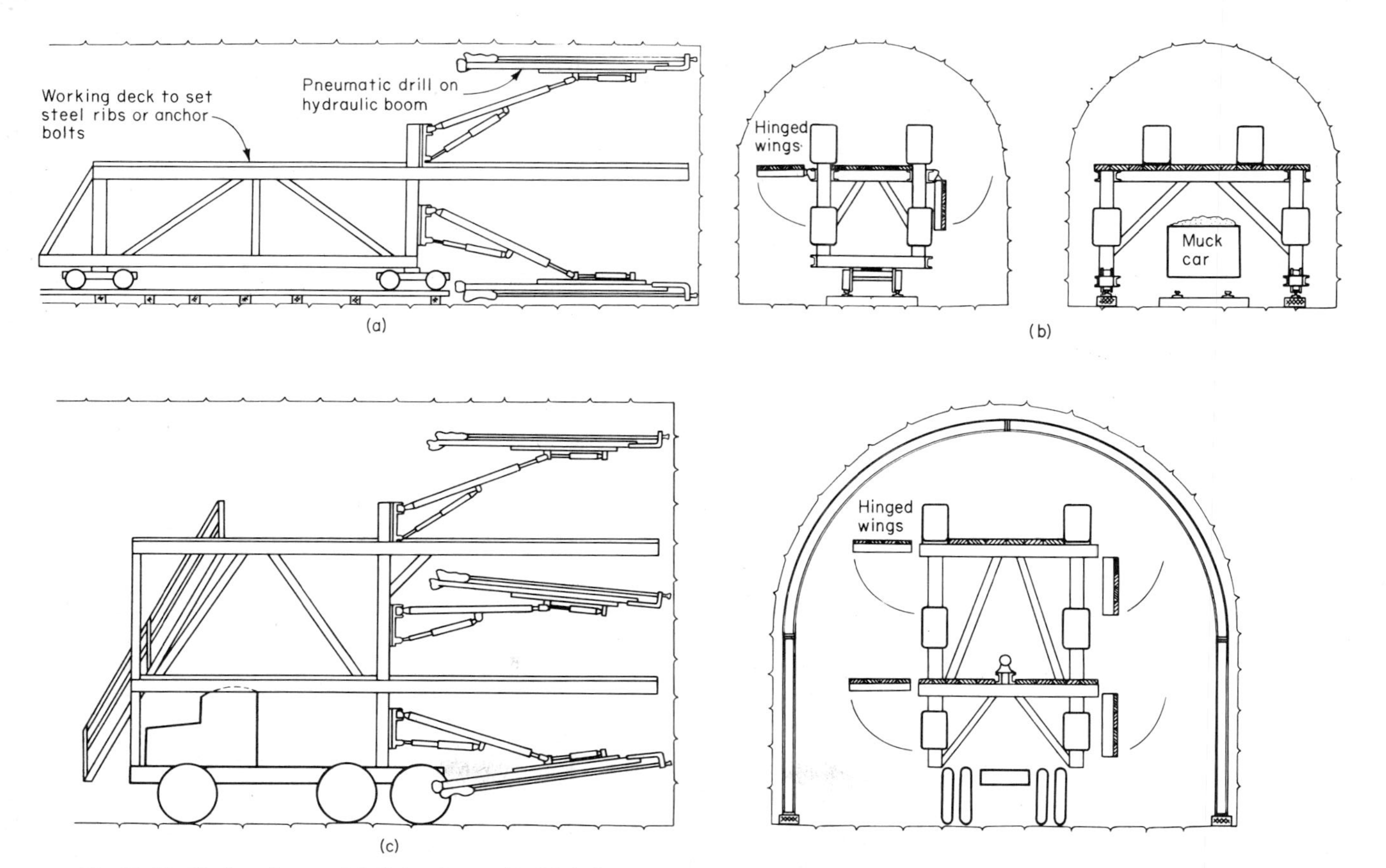

Fig. 31-37 Various types of drill jumbos. (*a*) Main-line drill jumbo; (*b*) straddle-type jumbo; (*c*) truck-mounted drill jumbo.

depending on the types of material being drilled. A bit may be resharpened 12 to 16 times before it is discarded.

Drill Patterns The arrangement of the holes in the face of the tunnel is known as the "drill pattern," and every tunnel superintendent has his favorite. These patterns fall into two basic classifications known as the "pyramid cut" and the "burn cut" (see Fig. 31-38).

The pyramid cut, as its name implies, is drilled so that the holes which are fired first will pull out a pyramid in the center of the tunnel face. The remaining holes in the round are then fired in sequence to break out the full dimensions of the tunnel. The holes at the roof of the tunnel are known as "back holes," those along the sides are "rib holes," and those at the floor are "lifters."

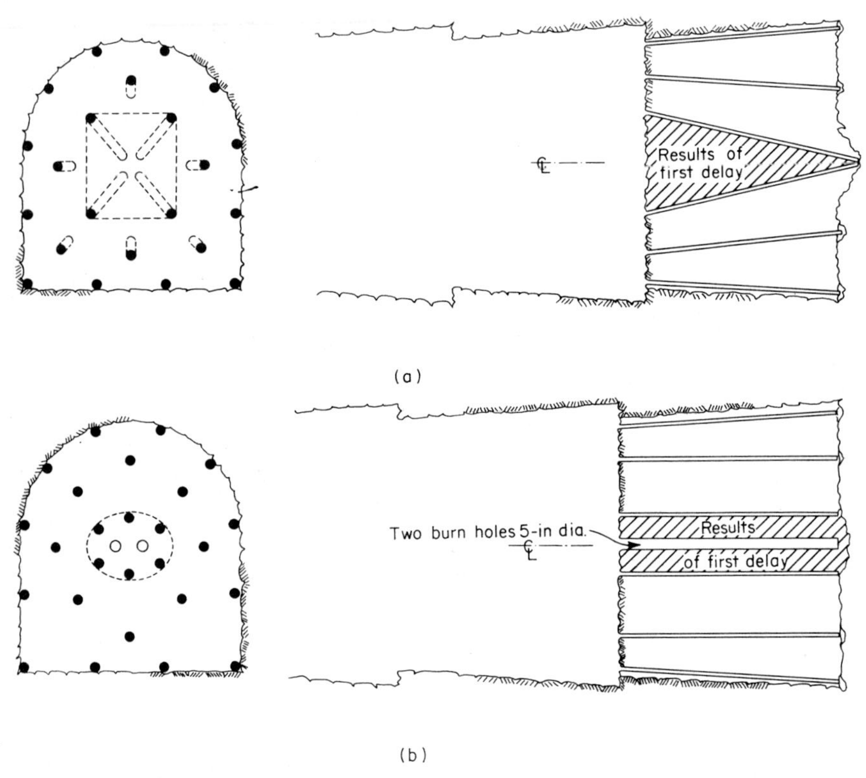

Fig. 31-38 Drill patterns for rock tunnels. (*a*) Conventional pyramid cut; (*b*) burn cut.

The burn cut consists of a centrally located hole, generally 5 in. in diameter, which is not loaded. This central hole is surrounded by four satellite holes, heavily loaded, which are fired first to blast a circular hole (about 18 in. in diameter) for the full length of the round. The remaining holes in the round are then fired in sequence to attain the shape of the tunnel. Proponents of the burn cut claim that it will break about 6 in. more tunnel on every round than the pyramid cut. It has the advantage that all holes are parallel with the longitudinal axis of the tunnel, thus requiring less skill in pointing the drill. This method of firing also reduces the amount of fly rock and gives a more compact muckpile.

The length of the round is dependent upon the type of rock and the width of the tunnel. In a tunnel 8 ft wide, the round would probably be 6 ft. A medium-sized tunnel would probably employ an 8-ft round and break 7 to 7½ ft of tunnel. A very large tunnel would drill a 12-ft round, which would probably pull 11½ ft. It is a general rule that the depth of the hole should be less than the width of the tunnel.

The spacing of the holes in the face will also vary with the type of rock. For a preliminary estimate, it can be assumed there will be one hole for every 4 to 5 sq ft of face. The peripheral holes of the rib and back will probably be spaced 2 ft on centers.

In all drilling there will be "overbreak" or "tights" due to irregular breaking of the rock. Removal of these tights by secondary blasting is very expensive, and contractors generally prefer to drive the tunnel 6 in. oversize all around. This is considered to be cheaper than going back through to remove the tights, even though it means that more muck must be hauled out of the tunnel and, if the tunnel is concrete-lined, more concrete must be placed. An overbreak of 10 percent may increase the amount of concrete required by 25 to 40 percent. The tunnel superintendent should constantly observe the way the round breaks so that, by rearranging the holes or by varying the quantity of dynamite, he can maintain the overbreak within acceptable limits. "Presplitting" is being widely used in open rock cuts to obtain a smooth, uniform surface, and perhaps it could be applied to tunneling to reduce overbreak.*

Explosives These are fully discussed in Section 19 and need not be elaborated on herein. The dynamite which is selected, always known as "powder," should be that which gives the least fumes and which breaks the rock to a size which can be loaded most conveniently. All explosive companies employ blasting experts who are familiar with the local types of rock and can make knowledgeable recommendations. For preliminary estimating, it can be assumed that a medium-sized tunnel will require 4½ to 5 lb of dynamite for each cubic yard of rock broken. A larger tunnel will require proportionately less dynamite, and this can be estimated at 3 to 4 lb per cu yd. A very hard rock, such as that around New York, may require 7 lb per cu yd.

A rather new development in tunnel blasting is ANFO, which is a mixture of ammonium nitrate (a fertilizer) with fuel oil. This explosive costs less than dynamite, and in some cases it is just as efficient. However, it cannot be used in a wet tunnel.

The round is always fired electrically, using electric blasting caps which are arranged by delays so that the center cut holes are fired first. The relievers are then fired, followed by the rib holes. The tunnel foreman should always count the delays so that he can detect a possible misfire. These misfires, which are known as "bootlegs," must be dealt with most cautiously before the mucking machine starts to work. A search should be made for the unexploded cap and the dynamite removed from the hole, either by air or by water. If this cannot be done, another hole should be drilled parallel to the bootleg and fired. A bootleg hole is more dangerous than a rattlesnake.

MUCKING

Muck Loading The material excavated in a tunnel is always known as "muck," and the loaders are always known as "mucking machines." And, of course, the men who follow this work are known as "muckers." The type of mucking machine selected is dependent on the type of haulage employed. If car haulage is being used, the mucking machine will run on the same rails. If truck haulage is employed, the mucker will be mounted on crawlers or rubber tires.

On small tunnels it is general practice to use an air-operated mucker, as shown in Fig. 31-39. A muck car is coupled to the rear of the mucking machine and is dragged back and forth as the mucker charges into the muckpile. Once the dipper is filled, it is raised and thrown back into the car. This machine has a limited side range and therefore it cannot be used in wide tunnels.

Larger tunnels which use rail haulage will have an electrically powered mucking machine of the type illustrated in Fig. 31-40. Of course, this means that high-voltage cable must be carried into the tunnel in addition to the ordinary lighting circuit. These machines have a much wider cleanup range, as can be seen in the illustration. Here again the muck car is attached to the rear of the machine and moves back and forth as the machine advances and retreats. This type of mucker

* See Sec. 19, Rock Excavation.

has a short belt conveyor and thus it can fill a longer car. It is general practice to use a two-position "jackknife" drawbar which permits filling the car more uniformly.

Truck haulage will be employed on larger tunnels, and in such cases the mucking machine will be mounted on crawlers and is usually diesel-powered. Figure 31-41

Fig. 31-39 Air-operated mucking machine for small tunnels. (*Eimco Corporation.*)

Fig. 31-40 Electrically powered mucking machine. (*Goodman Manufacturing Company.*)

shows the "overshot" loader which throws the muck from the front to the rear. In recent years contractors have adopted the side-discharge front-end loader shown in Fig. 31-42. This tractor can discharge to either side, and the operator has better control of cleanup and dumping.

Muck Haulage Muck is hauled out of the tunnel in narrow-gauge cars or diesel-powered trucks. There are other methods of muck transportation which may some-

Fig. 31-41 Overshot mucking machine, diesel-powered. (*Eimco Corporation.*)

Fig. 31-42 Side-discharge tunnel mucker. Note diesel-powered tunnel truck. (*Robertson & Fowler, Norfolk and Western Railway Company.*)

day be employed but are not today considered economical. One such method is to crush the muck to a uniform fineness, mix it with water, and pump it out as a slurry. Since a pumpable slurry can only contain about 15 percent of solids (by weight), a tremendous volume of water is required. Belt conveyors provide another method of moving muck but their cost, if assigned to only one job, is prohibitive.

Rail haulage is always employed in small tunnels and very often in larger ones, since it requires less labor and less ventilation than truck haulage. The muck is hauled in cars which run on narrow-gauge track. The most common gauges are 18, 24, 30, and 36 in. On rare occasions 42-in. gauge is selected, and on very small tunnels

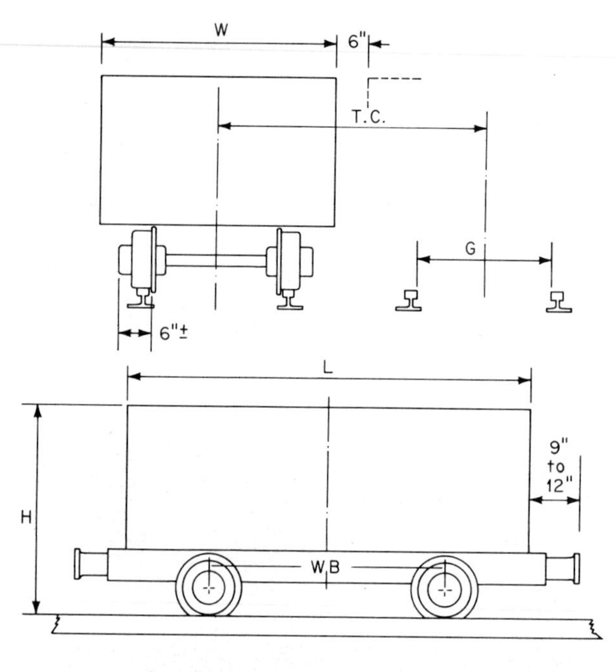

Fig. 31-43 Usual proportions of narrow-gauge mine cars.

14-in. gauge is sometimes used. Figure 31-43 shows the customary dimensions of narrow-gauge muck cars. It is a general rule that the cars must be able to pass within the tunnel; sometimes this requirement is based on the bore, and in other cases it is based on the finished inside dimension. Once this controlling dimension has been established, it is relatively simple to select the width of the car and the gauge of track.

The cars will either be "side-dump" or "lift-off boxes," depending on how the muck is being handled at the shaft or portal (see Fig. 31-44). It is always desirable to make the capacity of the car as large as possible, but some mucking machines cannot fill a car if it is too long or too high. This possibility must be carefully investigated.

Table 31-1 gives the weights of rails and minimum curvatures which are recommended for the different gauges of track. These values are based on the cars and locomotives which are normally used on these gauges. The minimum curve radius for a car or locomotive is taken as six times its wheelbase. The tabulated starting

TABLE 31-1 Design Data for Rail Haulage Systems

Track gauge, in.	Rail weight, lb/yd	Curve radius, ft	Starting friction, lb/ton	Rolling friction, lb/ton
18	20	25	30	25
24	30	30	30	25
30	40	40	25	20
36	60	60	20	15

and rolling frictions are for roller bearings and include an allowance for rough and dirty track.

Rails are always classified by their weight per lineal yard. It is not a good investment to buy too light a rail which may then become surface-bent and unfit for use on future jobs. Portable track can be obtained in 15-ft lengths, with five steel ties per length. However, one of these sections becomes too heavy for two men to handle if the rail weight exceeds 30 lb per yd.

Tunnel locomotives will be either battery- or diesel-powered, since safety engineers frown upon trolley locomotives. A battery locomotive will always have two batteries, so that one can be charged while the other is in service. The battery should be sized to give 6 to 8 hr of service, and changing batteries should not take longer than 10 min. Diesel locomotives require approximately 75 cfm of air per brake horsepower, and must be equipped with a "scrubber" to reduce the smoke and gases discharged into the tunnel.

Fig. 31-44 Tunnel car for 24-in.-gauge track with 2 cu yd lift-off box. This box is raised to the surface by a crane and dumped into a truck. (*Mayo.*)

Locomotives are always classified by weight. It can be assumed that the starting tractive effort (TE) is 25 percent of the weight of the locomotive, and a 3-ton locomotive would therefore have a starting TE of 1,500 lb. Table 31-1 supplies the starting friction and the rolling friction for mine cars, expressed in pounds per ton of weight. A known value of starting TE can be divided by the appropriate value of starting friction to find the weight in tons of the entire train (locomotive, cars, and muck) that can be started on level track by this particular locomotive. The critical design situation occurs when starting the train, despite the fact that the TE of a moving locomotive is only about 20 percent of the locomotive weight. This is because the rolling friction of the train is at least 20 percent less than its starting friction.

One kilowatthour is equivalent to 1,760,000 ft-lb and, knowing the total distance traveled in one shift and the required moving TE (loaded one way, empty returning) for a battery locomotive, the required size of its battery may be calculated. In a diesel locomotive, the engine must be large enough, in low gear, to develop the starting TE of 25 percent of the weight on drivers.

Grades in tunnel work are always given in percent, and a 1-percent grade will add (or subtract) 20 lb per ton to the TE required to move the train. Grades are classified as favorable or adverse, with a favorable grade defined as one that aids the locomotive in moving the loaded train. The TE should be calculated for both the loaded and empty trains. In some cases where a grade is too favorable, an overweight locomotive may be needed in order to provide the necessary braking effort. The following formula may be used to calculate the TE required on grades:

$$\text{TE (lb) required} = W(f \pm 20g)$$

where W = total weight of trains, tons
f = starting friction, lb per ton
g = percent of grade

The bottleneck in rail haulage is car changing. A mucking machine can load a car in 1 to 2 min, but it may take 5 min to remove the loaded car, shove in an empty car, and hook it onto the mucking machine. For small tunnels the car passer illustrated in Fig. 31-45 is generally employed. This is a small transfer table which rests on the main line. An empty car is shoved onto the table by a locomotive, and the car is

Fig. 31-45 Car passer for 24-in.-gauge track. This car passer can be used on either side of the main line without cutting rails and is readily portable. (*Mayo Tunnel & Mine Equipment.*)

then pushed sideways by one man or by an air-powered winch. With the main line thus cleared, the locomotive can advance to pick up the loaded car and then retreat far enough so that the empty car can be pushed back onto the main line. Once the line is clear, the locomotive pushes the empty up to the mucking machine.

The California switch which is shown in Fig. 31-46 consists of a short section of double track on steel ties with diamond turnouts on each end. The switch rests right on the main line and has a climber point on each of its ends so that the train can climb onto the passing track. The empty cars are kept on one track, and the locomotive uses the other track for the storage of loaded cars. An empty car is pulled up to the mucking machine by means of an air tugger. Once this car is filled, the locomotive advances and pulls it back onto the other track.

Both the California switch and the car passer are moved forward every day to keep them at the most efficient distance from the mucking machine. Wheel-mounted California switches are employed on some jobs because they can be moved more easily and quickly. These units are about 15 in. higher than the standard California switch.

The sliding platform or "magic carpet" is a platform of steel which is placed directly on the floor of the tunnel and includes the passing trucks, turnouts, and climber points. It is broken down into at least three sections which can be advanced 5 ft, one section at a time, by means of built-in hydraulic jacks. In this way, the sliding platform literally crawls along the tunnel. Its front end is kept tight up against the face of the tunnel and must be strong enough to resist the blast and support the muck-pile. New rails are added at the rear end, under the climber points, as the platform advances.

Truck haulage can be used in a flat-bottom tunnel when the width is about 20 ft, so that two trucks can pass. Occasionally trucks are used in narrower tunnels, but breakouts are then required in the wall to permit passing. This arrangement is of very doubtful economy.

The trucks should be double-enders so that they can go forward or back at the same speed and need not be turned around within the tunnel. Trucks have the advantage that they can run directly to the dump even when this is some distance from the portal. They have the disadvantage of requiring a tremendous amount of ventilation as compared to a diesel locomotive. Trucks with diesel engines must be fitted with scrubbers to reduce the pollutants in the exhaust.

Fig. 31-46 California switch, 30-in. gauge, used by New York Board of Water Supply. (*Midwest Steel.*)

The rolling friction on a rubber-tired truck is about 100 lb per ton as compared to 20 or 25 lb for mine cars. Thus, on level ground, the amount of power required for truck hauling is about four times that required for comparable rail hauling. However, trucks normally have better gradeability than locomotives and can be used on climbs as steep as 15 percent.

Muck Disposal This may be a real problem for tunnels driven in metropolitan areas. The contractor must find a dump area where the muck can be disposed of at all times, night and day. If muck is being hauled from the face by tunnel trucks, the same trucks can run directly to a dump which is not too far away. In most cases, however, the muck must be transferred to on-highway trucks.*

* See Sec. 13, Hauling Units.

When the muck is being hauled from the face by rail, the cars are run out of the tunnel and dumped from a trestle. In some cases, this muck can be chuted directly into a truck. Where this is not possible, the cars are unloaded onto the ground and the muck is reloaded into trucks with the aid of a front-end loader.* With sufficient storage capacity, all muck haulage can be performed during one shift. In very small tunnels, it is common practice to dig a pit at the portal into which the cars are dumped. This muck is then clammed out of the pit once a day.

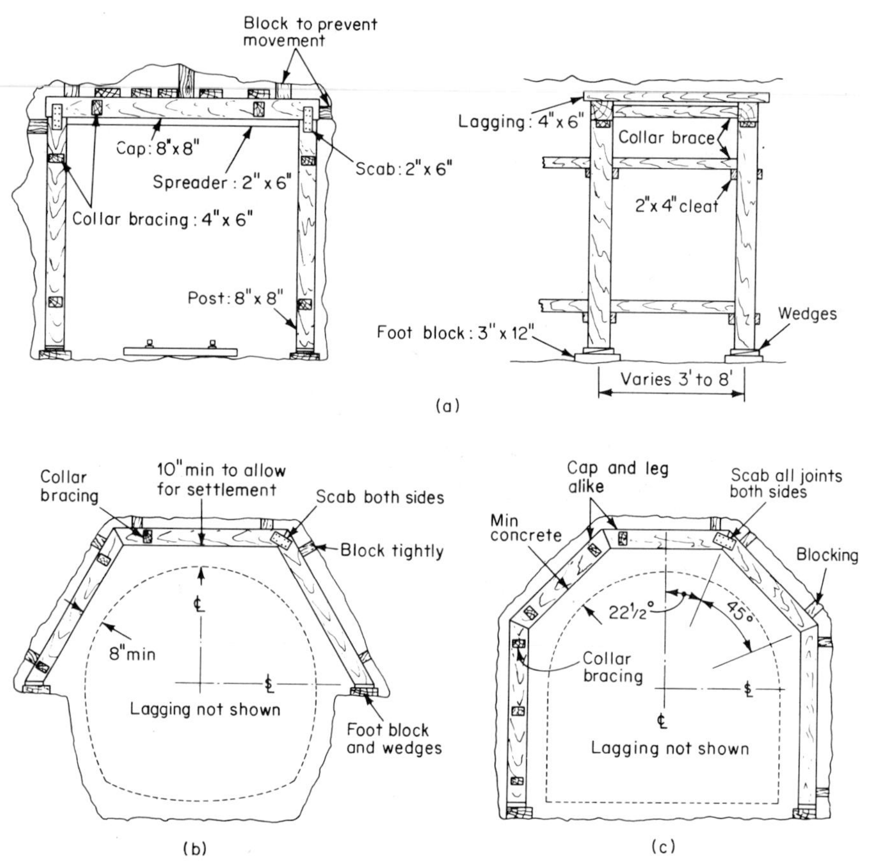

Fig. 31-47 Timbering for small and medium-sized tunnels. (*a*) Simple post-and-post set for small tunnels; (*b*) three-piece set in wall hitches; (*c*) three-piece set on plumb posts.

GROUND SUPPORT

Scaling Down One important item of safety in rock tunneling is "scaling down." As soon as the gang moves back into the heading after blasting, the most experienced miners should scale down the roof and sidewalls, using a long crowbar or a drill steel with a chisel point. Any loose rock must be pried down before the gang starts working on the muckpile. Even if there is no apparent loose rock, the roof should be "sounded" by tapping it with the scaling bar. An experienced miner can tell by the sound whether the rock is loose and liable to fall. Such rock must be scaled down or otherwise supported.

* See Sec. 11, Loaders.

Timbers Not too many years ago, if ground required support, the only type of support available was timber. It sometimes happens, even today, that there are short stretches of ground which require quick support. Timber is readily available, and the contractor may wish to resort to its use.* Figure 31-47*a* shows the "square set" which is used in small tunnels. The timbers will be 8 by 8 in. or 10 by 10 in., depending on the size of the tunnel. The longitudinal spacing will generally be 4 ft. Figure 31-47*b* shows a simple three-piece set which is used in circular tunnels. As shown in this illustration, the ends of the timbers set in a "hitch" at the spring line.

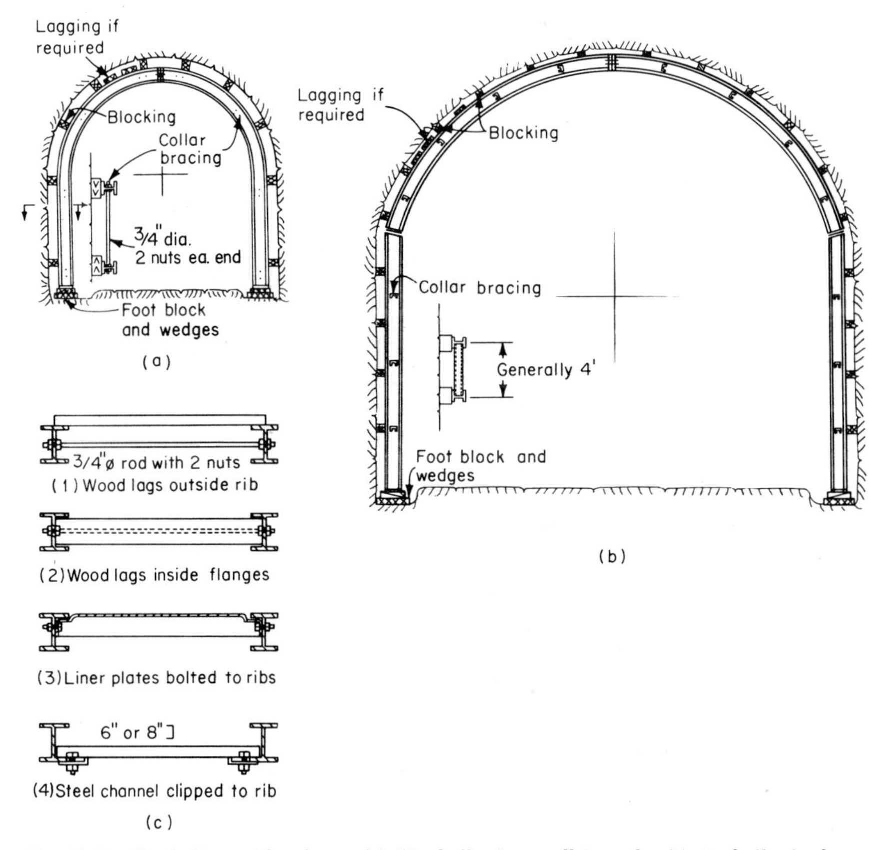

Fig. 31-48 Steel ribs and lagging. (*a*) Steel ribs for small tunnels; (*b*) steel ribs for large tunnels; (*c*) types of lagging.

The blocking at the ends of the cap and leg will develop arch action and prevent shifting of the timber set. Lagging should be laid on the outside of the timbers to prevent the fall of small pieces. Generally these lags are 4 by 6 in., spaced 8 to 12 in. apart. Figure 31-47*c* shows the three-piece set on legs, which is better adapted to many shapes of tunnels. In all cases, there must be longitudinal collar bracing between the posts and the caps to protect the set from being knocked out when blasting.

Steel Ribs These are the standard method of ground support and, in most cases, part of the rib is allowed to protrude into the concrete lining (see Fig. 31-48). Thus the tunnel need not be driven as large as in the case where timbers are employed.

* See Structural Timber in Sec. 24, Timber Construction.

Lagging is placed outside the ribs to support any loose ground. In most cases this lagging consists of 4- by 6-in. wood members, but in some cases steel lagging as illustrated in Fig. 31-48*c* will be specified by the engineers. This lagging, whether of steel or wood, is widely spaced unless the ground is very ravelly. As in timber sets, "blocking points" are most important to develop the strength of the steel rib. The closer the blocking points, the more support the rib gives. The usual practice is to set the rib up in its correction position on foot blocks, which are generally 4- by 12- by 24-in. timbers. After these blocks have been cut and set against the rock, wedges are driven between the foot block and the bottom of the rib to stress the rib. Collar bracing is placed about every 4 or 5 ft between the webs of the wide-flange ribs to prevent the rib being knocked out when blasting.

Fig. 31-49 Rib-setting carriage on top of a tractor-mounted drill jumbo. (*Norfolk and Western Railway Company.*)

In small tunnels these steel ribs are set by hand, one pair at a time. On large tunnels, such as those for turnpikes, the ribs are big and heavy. There should then be a system of chain hoists and air rams on the drill jumbo (Fig. 31-49) to raise and hold these ribs until the foot blocks are placed. Here again, blocking and lagging are laid over the ribs, and the wedges are driven on the foot block.

Roof Bolts This development has been borrowed from the coal mining industry. Roof bolts are particularly valuable in fastening weak strata together. There are several types of bolts, the most common of which is shown in Fig. 31-50. This bolt has a threaded rod which is inserted in a hole drilled in the roof. When the roof bolt is tightened, the shell expands and grips the side of the hole. Further tightening will pull the washer hard against the rock. Often a wire mesh is placed over the roof of the tunnel and held in place by roof bolts. This will prevent the falling of small rocks from between bolts.

Most rock bolts are ⅝ or ¾ in. in diameter, requiring a 1⅜- or 1¾-in. hole. Common lengths are 5 or 6 ft. They should be tightened with a torque wrench to prevent overloading of the bolt or the expansion shell. In mines these bolts are periodically rechecked for torque, since a loose bolt indicates slippage of the shell.

Shotcrete This development has come from Europe and is frequently referred to as the "new Austrian method." Basically the shotcrete gun is similar to a gunite machine except that, instead of spraying mortar, it can apply concrete with a small aggregate. The general practice is to apply shotcrete to the roof and walls immediately after they have been scaled down following the blast. Shotcrete is carried clear up to the face, as shown in Fig. 31-51. While some of the leading edge may be knocked off at the next blast, it still provides support for the rock and protection to the mucking crew.

Shotcrete is often used in conjunction with roof bolts and wire mesh. In other cases, where the ground is heavy, steel ribs are placed and blocked against the roof. Shotcrete is then applied between the ribs in place of lagging.

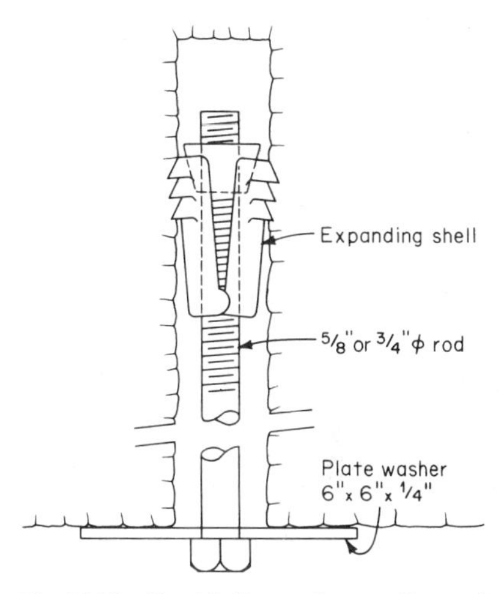

Fig. 31-50 Roof bolts are frequently used to tie weak strata back into solid rock.

VENTILATION

Adequate ventilation is specified by law in most states and is a necessity for rapid and efficient tunnel driving. Each blast releases vast quantities of smoke and dust, and until these fumes are removed or dissipated, tunneling cannot be resumed. Some-

Fig. 31-51 Shotcrete being applied on a Canadian National Railway tunnel. Note that it is carried right up to the face.

times the shot can be arranged to be fired during the gang's lunch period, but this can rarely be scheduled day after day. Three methods of ventilation will be outlined in the following paragraphs.

Blower System This consists of blowers or fans which force fresh air into the tunnel. The incoming air is carried through sheet-metal pipes to a point about 200 ft from the face. Beyond this point, a collapsible canvas tube is used to lead the air to the face. The canvas tube is removed immediately before each blast to prevent its damage by fly rock, and it is replaced as soon afterward as possible in order to bring fresh air into the area where blasting has taken place. The system has the advantage of bringing ventilation directly to the face through a removable and collapsible tube. It has the disadvantage that the fumes drift slowly back from the face so that the motormen and anybody else working back in the tunnel must be in the smoke for a long time.

Exhaust System This method employs suction to remove the smoke and foul air through a sheet-steel pipe, which must be strong enough to resist collapse due to the partial vacuum. The method has the advantage that smoke and fumes are quickly removed, permitting fresh air to drift into the tunnel for its whole length. However, because the end of the steel pipe must be kept about 200 ft from the face to prevent its damage by fly rock, a "dead end" will develop between the end of the ventilating pipe and the face of the tunnel. This difficulty may be corrected easily by employing a small auxiliary blower, set in the tunnel just back of the end of the main ventilating pipe, to force fresh air to the face through an 8-in. canvas tube. The canvas tube must then be rolled up and removed just before each shot.

Combined System A system designed for either blowing or suction is referred to as a "combined" system. Most of the time the blowers are used to force fresh air into the tunnel through a metal tube, to which is attached a canvas tube to lead the air to the face. Prior to the shot, the canvas tube is removed and the fans or blowers are set to exhaust. Air is exhausted for 20 min, which will remove most of the smoke. The fans are then set to blowing and the canvas tube is replaced to lead fresh air to where the men are working. This is the most popular system for ventilating long tunnels. Most fans can actually be reversed, but blowers cannot. When blowers are used for ventilating, as is the most common practice, some type of reversing valve such as that illustrated in Fig. 31-52 is required. This valve is set in the pipe, close to the blower, and is used to change the direction of flow.

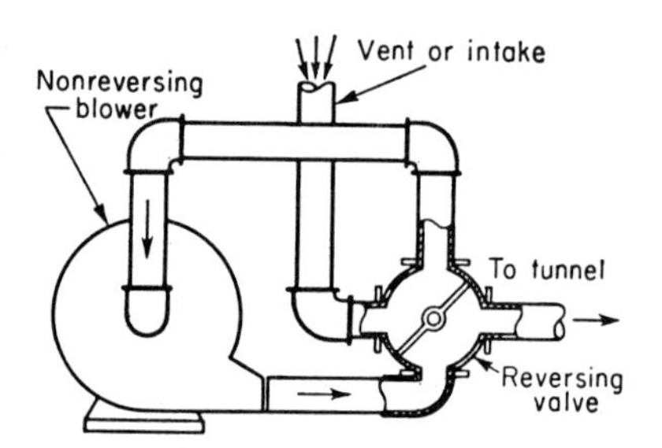

Fig. 31-52 Reversing valve for a nonreversing blower.

Quantity of Air The quantity of air required is based on the number of men employed underground. A minimum amount would be 100 cfm per man or 50 cfm per sq ft of tunnel, whichever is the larger. The high-pressure air which is exhausted from the drills should not be included in this quantity, as it may contain oil or moisture. If diesel locomotives or trucks are used for haulage, the U.S. Bureau of Mines recommends a minimum of 75 cfm of free air for each actual continuous horsepower. Knowing the quantity of air to be supplied and the maximum distance which it must travel, the horsepower necessary to force this air volume through a pipe of a specified diameter can be readily calculated. This presents an interesting problem in economics: i.e., to balance the continuing cost of power with the initial cost of the pipe.

Mechanical excavators require special ventilating procedures, and they always employ an exhaust system which will pick up the dust from immediately behind the cutting teeth or rollers. In order to control this dust, it is necessary to spray the rock face and/or the cuttings with water which contains a small amount of wetting agent or detergent. There should be a dust collector immediately adjacent to the mole to pick up this wet dust before it enters the ventilating system, since otherwise the dust will settle to the bottom of the pipe. It will then restrict the flow of air and, when it becomes too heavy, it is liable to pull out the supporting brackets.

MECHANICAL EXCAVATION

Tunnel Excavators A crude tunnel excavator was used successfully in 1882 on the pilot bore of a proposed tunnel under the English Channel. The formation was a uniform chalk and, since that time, inventors have spent a lot of time and money in trying to devise an excavator which could be used in harder rock. It was not until the development of bits tipped with tungsten carbide that real progress was observed along these lines. In the last 5 or 6 years, tremendous strides have been made in perfecting machines which will work in the harder rocks as well as in sandstone and shale. There have been some spectacular successes and, it should be added, there have also been some disappointing and expensive failures.

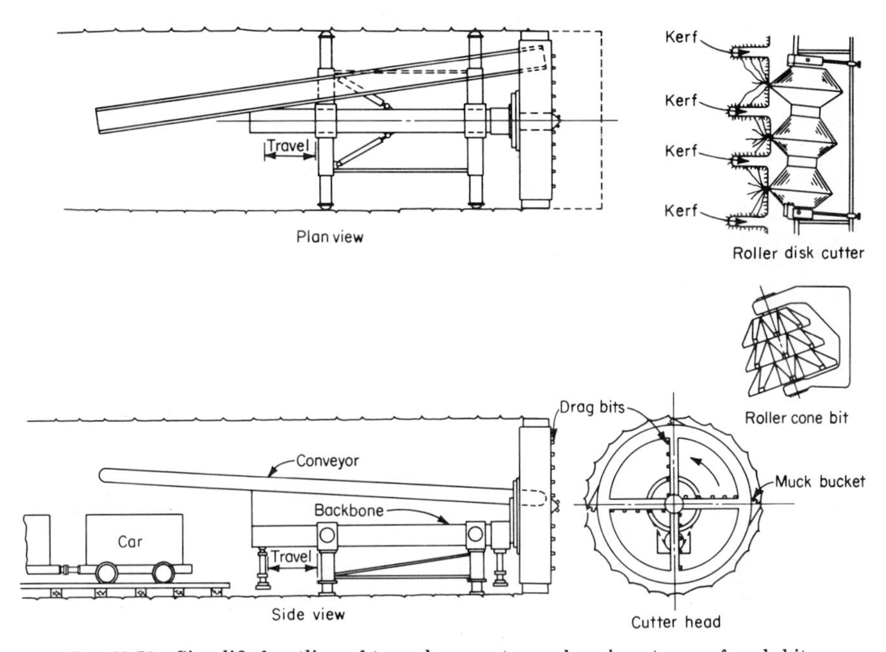

Fig. 31-53 Simplified outline of tunnel excavator and various types of rock bits.

There are several types of tunnel excavators, better known as "moles," which will work successfully and economically in various types of rock. Rocks are ranked as to hardness according to Mohs' scale, wherein 1 is talc and 10 is diamond. Limestone is rated at 3; sandstone will vary from 3 to 7; schist and gneiss are rated 6 to 7. The present generation of moles can handle rocks with a hardness of 6 to 7, but not 8.

Another basis of rock classification is compressive strength, which is expressed in pounds per square inch. The moles have no particular trouble in 6,000- to 10,000-psi rock, but they may experience trouble between 10,000 and 25,000 psi. Some rocks with a high psi will spall under pressure and thus may not be as difficult to mine as rocks of lower psi.

Figure 31-53 shows a simplified mole and various types of bits. The laser beam is essential for maintaining alignment.

Economics In preparing estimates for any rock tunnel, it is prudent to study the mole as an alternative to conventional drilling and blasting. However, in its present stage of development, it is seldom economical to use a mole on tunnels of less than 1,000 to 1,500 ft in length. This situation may be changed if a mole is available on a lease basis since, depending somewhat on its diameter, the original cost of a mole

(*a*)

(*c*)

Fig. 31-54 Various types of mechanical excavators for rock tunnels. (*a*) Jarva machine, diameter, showing muck conveyor and hydraulic pumps; (*c*) Robbins excavator, 12-ft diam-

(b)

(d)

20-ft-diameter, San Francisco subways (S & M Constructors); (b) Calweld mole, 13-ft 4-in. eter; (d) Alpine tunnel excavator. This machine will mine a horseshoe or rectangular tunnel.

will be between $250,000 and $1,000,000. This first cost must then be amortized over the total number of lineal feet of tunnel on which the mole can be used. Its use may reduce total labor by 25 to 40 percent, and it will obviously eliminate the drill jumbo and the cost of dynamite, but it will not eliminate the cars, locomotives, and hoisting and dumping equipment. Unless the tunnel is long, the savings in labor will not offset the original cost of the machine plus the cost of installation and operation.

Roughly, a mole will cost about $1,000 per horsepower. The horsepower required is approximately $4d^2$, where d is the diameter of the bore expressed in feet. The cost of the required power can then be estimated at $.02 per hp-hr. Bit cost will vary from $1 to $5 per cu yd excavated, depending on the type of rock and the type of bit. The manufacturer should be able to estimate this cost accurately and may be willing to guarantee it. Equipment downtime to change bits, grease the equipment, and make minor repairs will probably run 25 percent of the available working time.

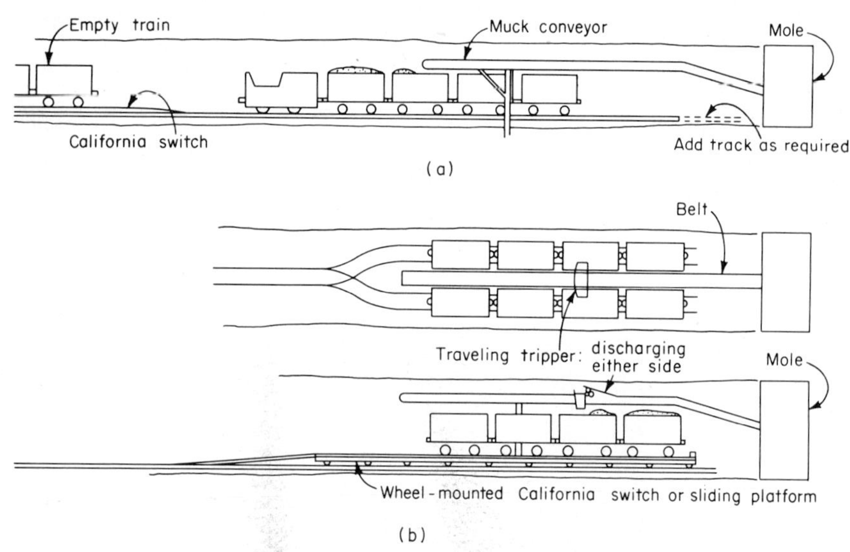

Fig. 31-55 Train-loading schemes for mole-driven tunnels. (*a*) Loading on single track; (*b*) loading on double track.

Knowing these items of cost, and estimating the progress and saving in the labor of the drilling and mucking crew, an economic comparison can be made between the alternative tunneling methods.

On all jobs the rock will vary, and progress may decrease if the rock becomes too hard or too soft. While the mole cuts a nice smooth tunnel roof, this method can be very treacherous in laminated rock. There will probably be as much rock support required in a tunnel which is mechanically excavated as in one which is driven by conventional means. Erecting steel ribs and setting lagging may seriously delay progress with a mole. The method does offer one big saving: overbreak is practically eliminated, and the total quantity of concrete required will be reduced by 25 to 35 percent. Figure 31-54 illustrates four types of mechanical excavators which have been used in rock tunneling. The Alpine tunnel excavator shown in Fig. 31-54*d* has its cutting head mounted on an adjustable boom. Unlike the other excavators illustrated in Fig. 31-54, it has the ability to mine tunnels with noncircular cross sections.

Mucking Tunnels driven with mechanical excavators have, up to now, used narrow-gauge cars for haulage. A circular tunnel is not adapted to truck haulage unless some muck is left in the bottom to give a level roadway which is wide enough so that two trucks can pass. Each of the present moles has a conveyor which projects back into the tunnel; this conveyor should be long enough so that an entire train can be loaded at one time. Generally the width and capacity of the cars are selected so that two trains can pass within the tunnel. The number of cars in the train is based on the capacity of the locomotive or the "shove" of the mole. As an example, if the mole can advance 4 ft on one setting, the train should have a capacity equal to this yardage. Then, while the mole retracts and advances its shove mechanism, the loaded train will leave and the next train will take its place.

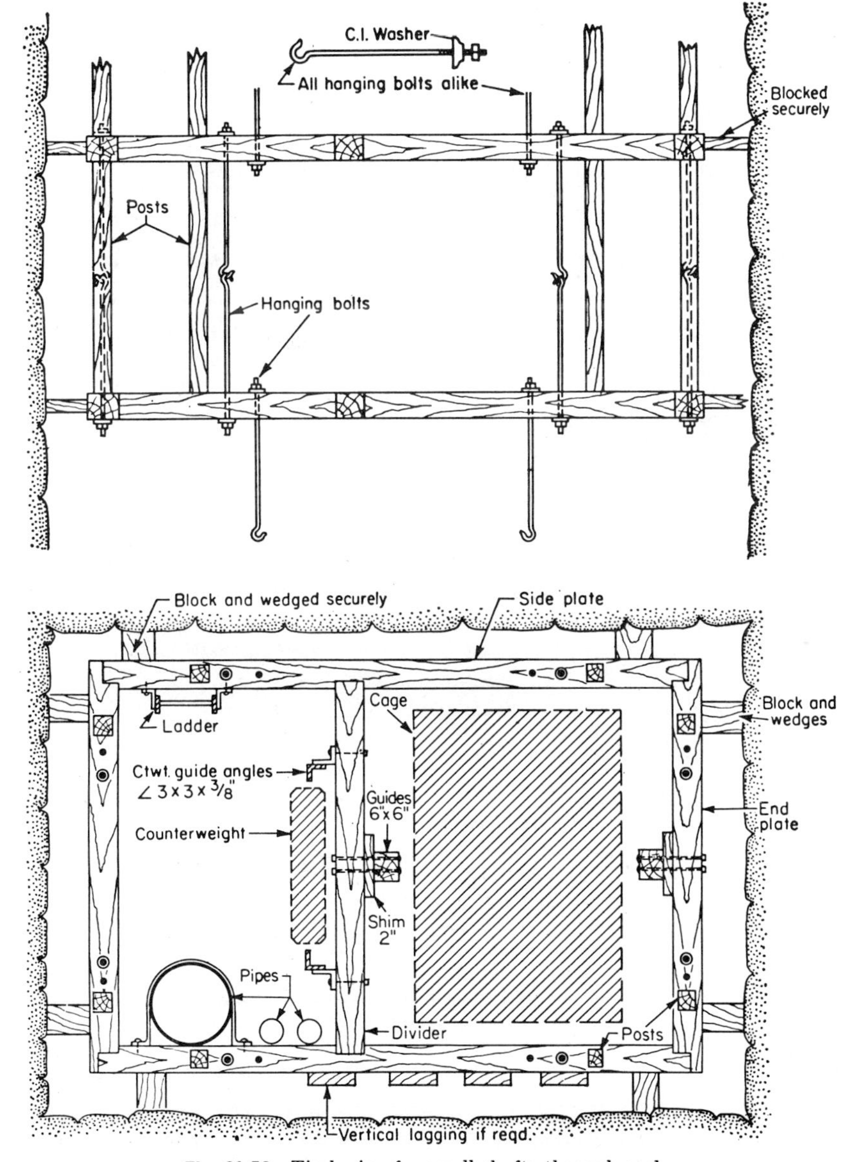

Fig. 31-56 Timbering for small shafts through rock.

There are two general schemes of loading cars. First as shown in Fig. 31-55*a*, the conveyor can load the entire train, one car at a time. The conveyor is in the center of the tunnel, and there is only one track at the loading point. Alternatively, where a larger tunnel is involved, there can be a double track right behind the mole as indicated in Fig. 31-55*b*. Two trains can then be in position for loading, and the belt has a "tripper" which can be moved back and forth so as to discharge either to the right or to the left. Once a train is loaded it is hauled to the portal. This method of loading may eliminate one locomotive.

Fig. 31-57 Steel liner plates, 16-ft diameter, for a shaft through overburden such as gravel or clay. (*Commercial Shearing and Stamping.*)

The muck produced by a mole is of more or less uniform size. Experiments have been made in grinding up this muck to a fineness equal to sand and then pumping it out of the tunnel as a slurry. A pumpable slurry will contain about 15 percent solids *by weight*, so that it could be said that one cubic yard of slurry must be pumped out of the tunnel for every cubic foot of rock. Belt conveyors would be another method of handling muck, but they would be too expensive in a long tunnel. In any case, tracks must be laid in the tunnel to carry men and material to the face.

SHAFTS

Diversion, highway, and railroad tunnels can always be driven from the portals. Other driving methods, or other types of tunnels, will require a shaft or shafts to get down to grade. Only "shallow" shafts will be considered here, with depths of less than 200 ft. Beyond this depth the sinking methods and the required high-speed hoisting equipment fall within the province of the mining engineer, and the reader is referred to a number of handbooks on mining practice.

Shaft Sinking Every shaft must pass through some overburden before rock is reached. Pressure from this overburden can generally be supported by timbering in the case of small rectangular shafts (see Fig. 31-56) or by liner plates in the case of circular shafts (see Fig. 31-57). For small-diameter circular shafts, the liner plates themselves will be sufficient, but for larger shafts a rib of rolled I beam must be set every two or three courses of plate (see Fig. 31-58). The excavation is accomplished with a clamshell bucket* on a crane boom. The miners set the liner plates, one at a

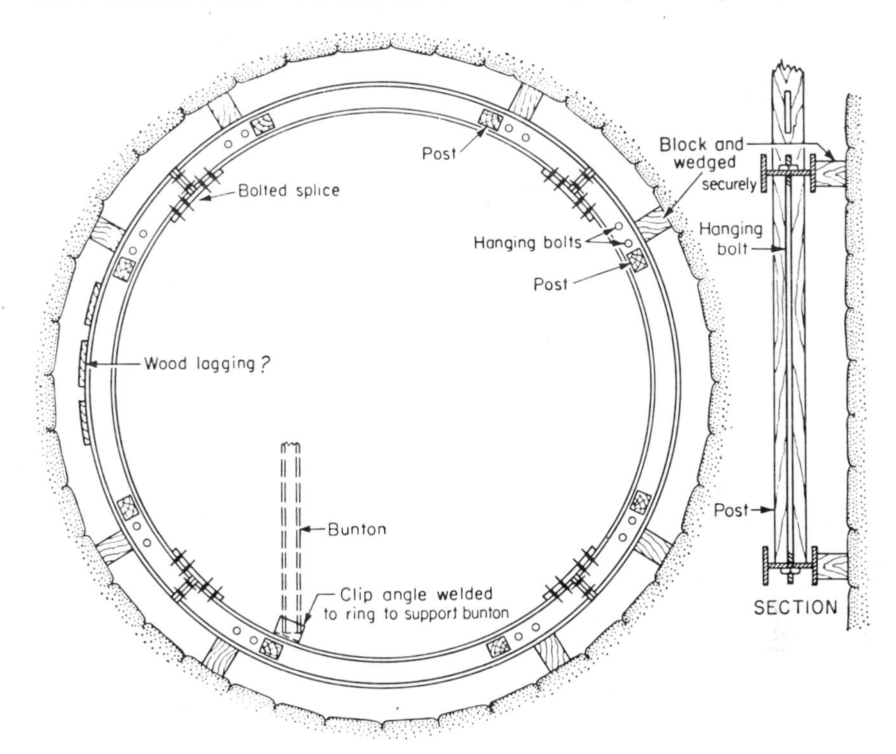

Fig. 31-58 Steel ribs for circular shafts in rock.

time, meanwhile shoveling the muck to the center of the shaft where the clamshell can reach it. It is most important that the shape of a circular shaft be maintained, because if it should become "egg shaped," the bending forces may far exceed the flexural strength of the liner. Any voids outside the liner plates should be promptly filled with sand or spalls.

Once rock is reached, the liner plates should be sealed off with grout to prevent, as much as possible, the infiltration of groundwater. Every gallon of water which falls into the shaft must be pumped out again, and besides, a wet shaft is a disagreeable place in which to work. On all rock shafts, if water or poor rock is anticipated, good practice is to drill a ring of holes about 10 ft away from the shaft and to the full depth and to grout them to refusal. This should eliminate all water problems during the sinking operations and speed up the whole process. If required, the excavated shaft can be lined as illustrated in Fig. 31-59.

* See Sec. 10, Excavators.

Shaft Drilling Drilling in shallow rock shafts will be done by hand-held air hammers with a modified pyramidal cut as shown in Fig. 31-60. It is general practice to pull only 4 to 6 ft on each round. If the shaft is deep, some type of shaft jumbo (Fig. 31-61) should be employed wherein boom-mounted drills are attached to a folding frame. This jumbo can be set down in the hole and leveled, and drilling can then commence simply by connecting the bull hose. Jumbos must be collapsible so that they will pass through the shaft door at the collar. Once at the surface, they must be stored out of the way until they are again required.

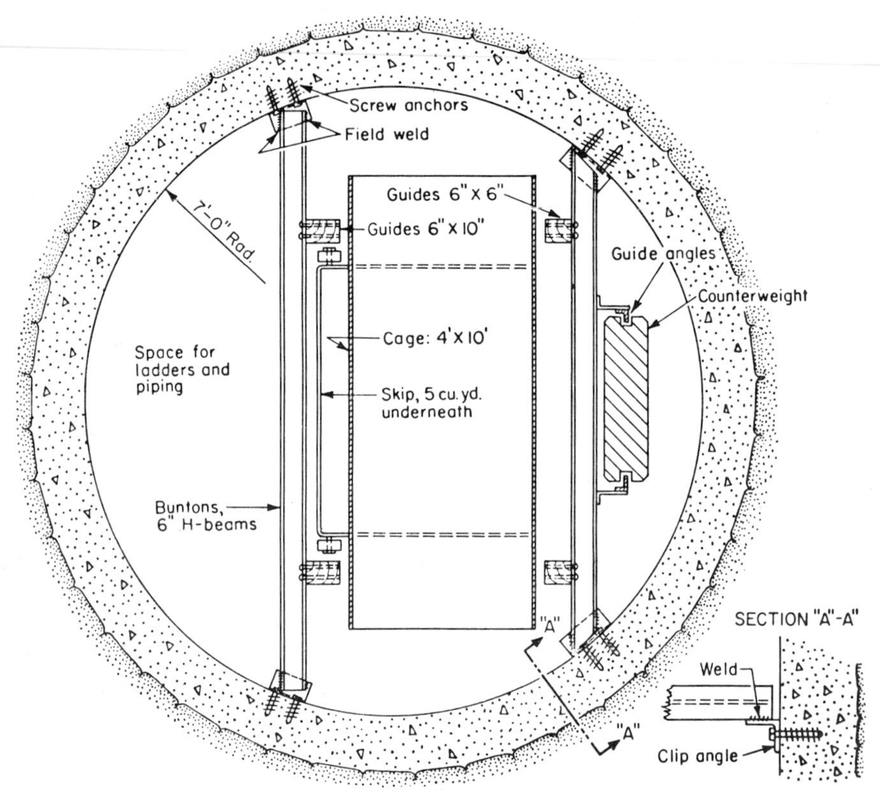

Fig. 31-59 Concrete-lined shaft through rock or soft ground. Provisions must be made in the forms for inserting screw anchors to support the buntons which carry the guides.

Shaft Mucking The first 60 to 100 ft of the shaft can be mucked with a surface crane which is fitted with a clamshell bucket. This depth is, in most cases, specified by the state mining laws. Beyond 100-ft depth it is good practice to set up a temporary headframe, as illustrated in Fig. 31-62, and hoist muck in "kibbles" or buckets. These kibbles, each with a capacity of 2 to 4 cu yd, are raised to the surface and dumped into a bin or onto the ground. The simplest method of dumping the buckets is to use a "bull line," which is suspended from a boom attached to the face of the sinking frame. A more sophisticated way is to use a "kibble dumper" which swings in under the kibble to engage a dingle ball attached to the bottom of the bucket. When the kibble is slowly lowered, the dingle ball engages a slot and automatically tips the kibble over.

On large shafts, of greater than 18-ft diameter, it is customary to use an air-operated crawler-mounted tunnel mucking machine of the type illustrated in Fig. 31-63. At the conclusion of the mucking cycle, this machine is raised to the surface and stored until again required. There are several types of small remote-controlled shaft muckers which are, basically, air-operated clamshell buckets. One of these is shown in Fig. 31-64.

A very small shaft, of less than 10-ft diameter, would have to be mucked by hand. Blasting should then be controlled to break the muck into large pieces which can be hand-loaded more easily.

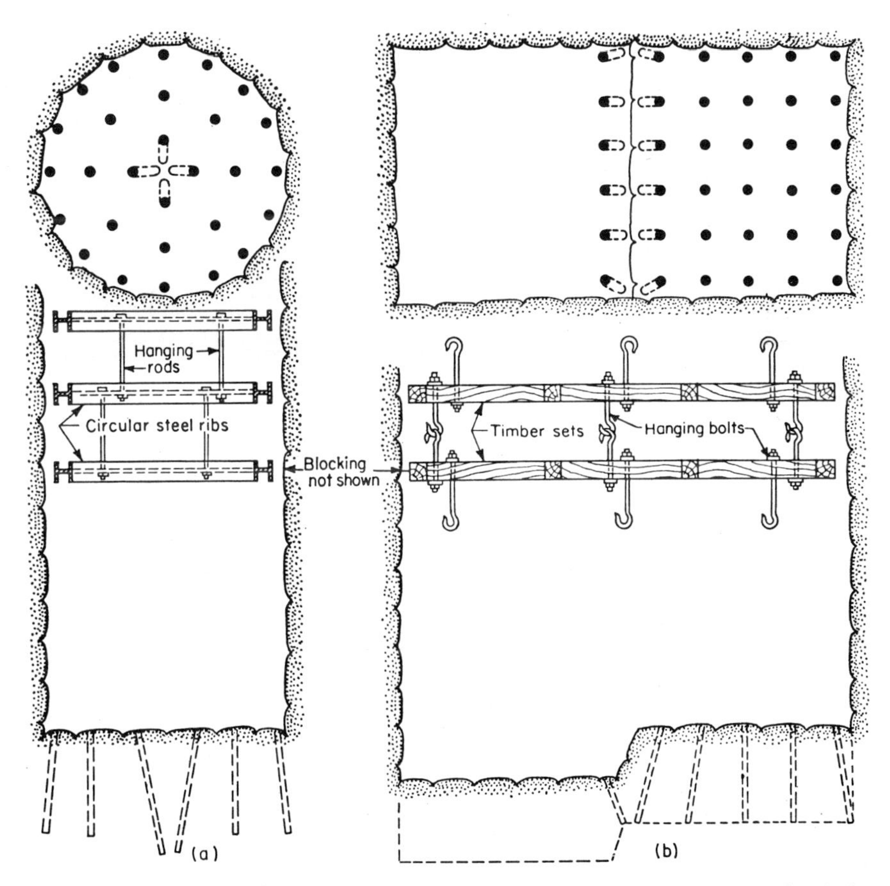

Fig. 31-60 Drill patterns for shaft sinking. (*a*) Circular shafts; (*b*) large rectangular shafts.

HOISTING

Tunnel Cars The simplest method of hoisting muck from tunnels is by the use of tunnel cars, Fig. 31-65. Each car has a removable box whose capacity will vary from 2 to 6 cu yd. A car is spotted at the foot of the shaft, and a crane at the surface hoists the box to the surface and dumps it either into a truck or onto the ground. Dumping is accomplished by means of a second line from the crane, which is hooked to a ring on the bottom of the box. The method is simple and economical for depths of 50 to 75 ft. The shaft must be large enough so that, if the muck box should swing around, it will not hit the shaft timbering or get hooked under the stairway.

Fig. 31-61 Shaft jumbo, mounting four jibs and drifters, for 26-ft 6-in.-diameter shaft, Muddy Run Project. (*Robert S. Mayo.*)

Fig. 31-62 Portable sinking frame for small shafts. (*Robert S. Mayo.*)

Fig. 31-63 Air-operated tunnel mucker, crawler-mounted, used for loading out shafts greater than 18-ft diameter. (*Eimco Corporation.*)

Substantial stairways should be built for the men. These are generally constructed of steel, much like fire escapes on apartment buildings. Often there is a piece of vertical ladder at the bottom to give room for a double track at the foot of the shaft.

Guide Frame Another hoisting arrangement for shallow shafts is shown in Fig. 31-66. This consists of a self-supporting tower, about 60 ft longer than the depth

Fig. 31-64 Air-operated shaft mucker for small inclined or vertical shafts. (*Cryderman.*)

of the shaft, which is erected within the shaft and attached to the walls. Inside this tower is a self-dumping skip large enough to carry one or two carloads. This skip is raised and lowered by means of a stiffleg derrick or a crawler crane. The method has the advantage that, when not hoisting muck, the crane can be disengaged and used for unloading trucks or lowering supplies into the shaft. One contractor in Pittsburgh used this type of hoist tower with a "self-service" passenger elevator

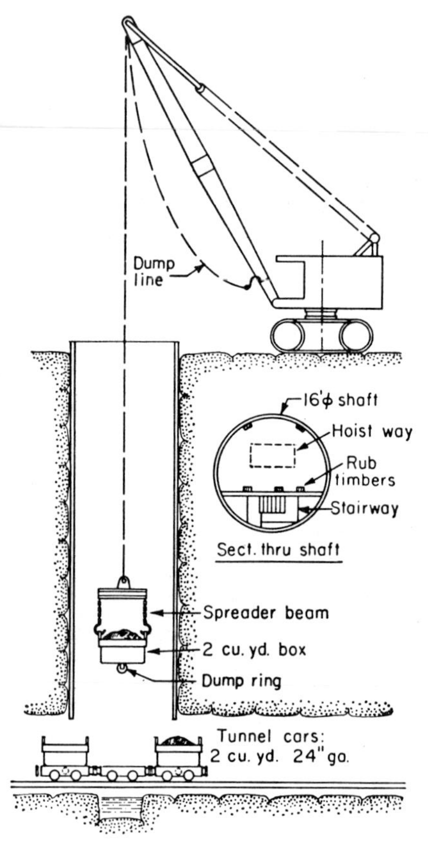

Fig. 31-65 Hoisting lift-off muck boxes in shallow shaft.

attached to one side. The elevator was pushbutton operated and had a capacity of 1,000 lb or four men. Another variation of this hoisting method is to run the tunnel car right under the guide frame. A "shroud" engages the stiffening channels on the outside of the muck box and hoists the box itself to the surface, where it is automatically dumped. On the return trip the box is landed right on the flatcar.

Headframe As the shafts get deeper, a cage such as that illustrated in Fig. 31-67 must be supplied to raise and lower the men and equipment. There will also be a need for a self-dumping skip (Fig. 31-68) to handle the muck. These requirements will necessitate a headframe. As shown in Fig. 31-69, the conventional headframe consists of a tower with an inclined backleg. This backleg should be adequate to take the breaking load of any one cable. At the top of the headframe are mounted the headsheaves, which usually have a diameter of forty times the rope diameter. On the face of the headframe are dumping plates, or scrolls, to engage and tip the skip. There should also be a muck bin to avoid any delays in waiting for trucks.

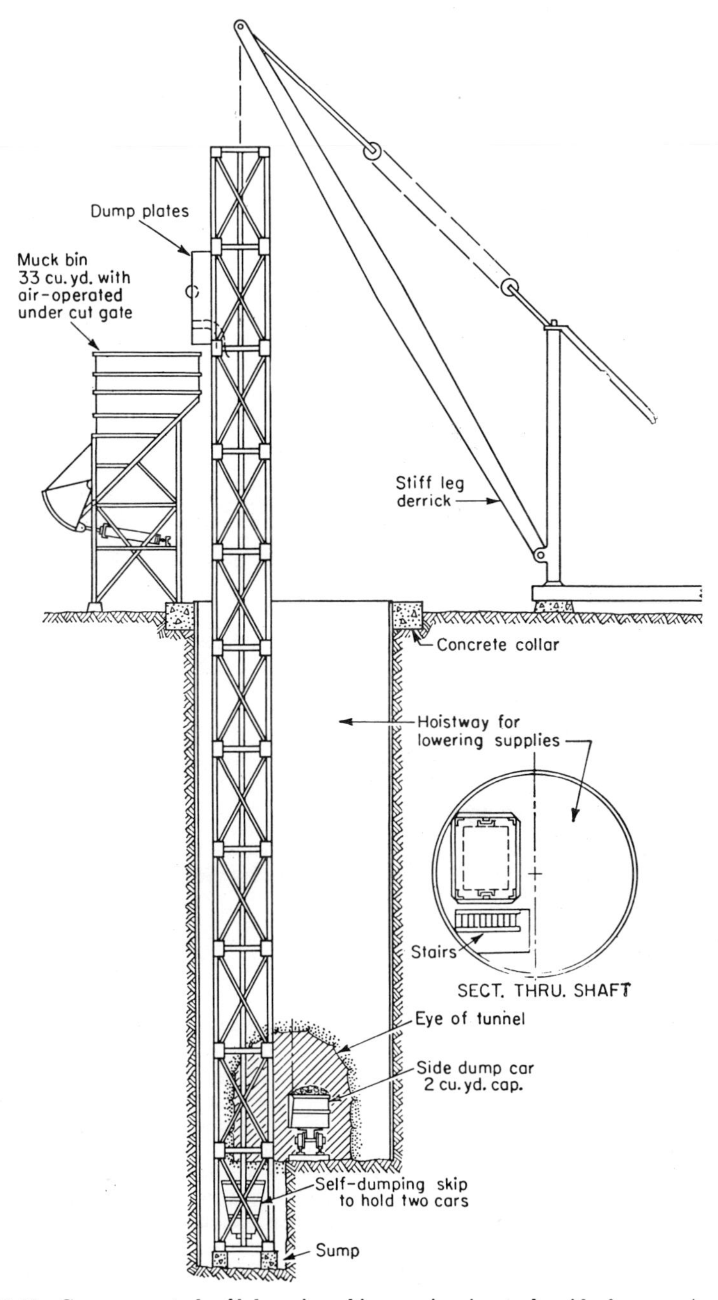

Fig. 31-66 Crane-operated self-dumping skip running in steel guide frame. A stairway should be built for the men.

The hoist will be electrically operated and must be equipped with suitable brakes and some type of depth indicator. In simple unbalanced hoisting, a single drum with sufficient capacity to wind the cable in several layers is all that is required. In balanced hoisting, either with a counterweight or with two skips, one large grooved drum is generally employed. This drum must be large enough to hold all the cable in one layer, the rope to the counterweight being unwound as the rope to the skip is being wound up. In mining practice, two drums are frequently employed; these drums are clutched together, and one drum winds the counterweight cable while the other

Fig. 31-67 Man cage for hoisting men, cars, or material. Note safety dogs for wood guides and bonnet over cage. (*Robert S. Mayo.*)

unwinds the cable to the skip. The line pull on the hoist will, of course, depend on the unbalanced load to be raised. The speed of the hoist will be governed by the depth of the shaft; roughly, the hoisting speed in fpm should be 60 to 80 percent of the depth of shaft in feet. If at all possible, the headframe and hoisting method should be laid out to use a counterweight which is equal to the dead weight of the cage and skip plus 50 percent of the muck load. Such a counterweight greatly reduces the unbalanced pull on the hoist and, if correctly designed, can cut the required horsepower in half. When it is difficult or undesirable to use a counterweight, it is recommended that mining practice be followed in which two skips (or one skip and one cage) are operated in balance.

Koepe Hoist The Koepe hoist (Fig. 31-70) is a rather new development in the United States although it has been widely used in Europe. It consists of a multiple-rope friction-drive hoist mounted directly on the top of the headframe. A two-speed ac motor is employed, as illustrated in Fig. 31-71, with pushbutton controls either from the cage or from an operator's booth at the surface. When the proper button is pushed, the skip will automatically accelerate, travel to the selected station, deceler-

Fig. 31-68 Self-dumping skip, 6-cu yd capacity. (*Robert S. Mayo.*)

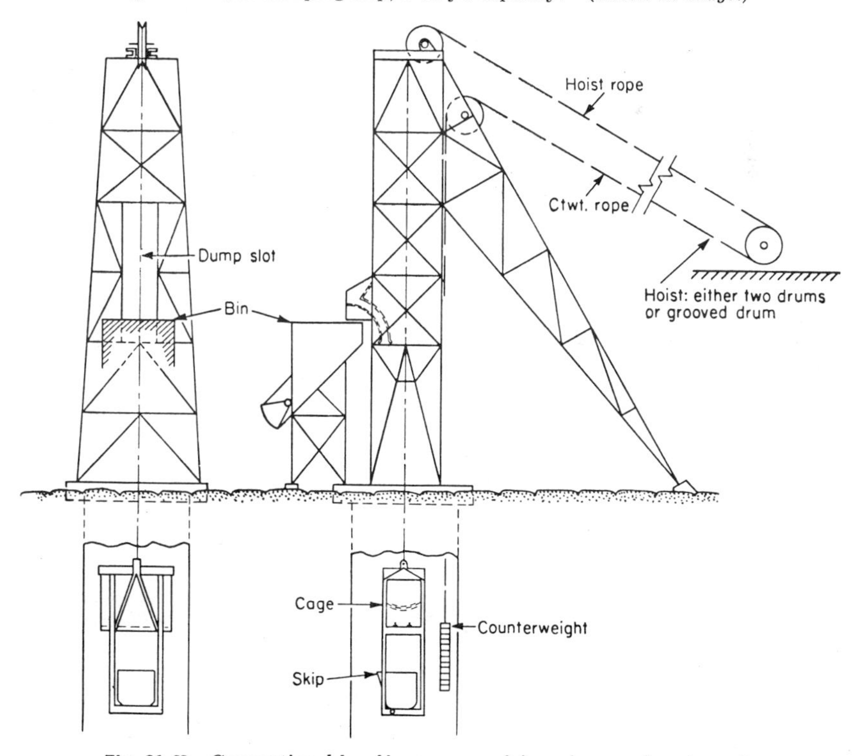

Fig. 31-69 Conventional headframe as used for mines or deep tunnels.

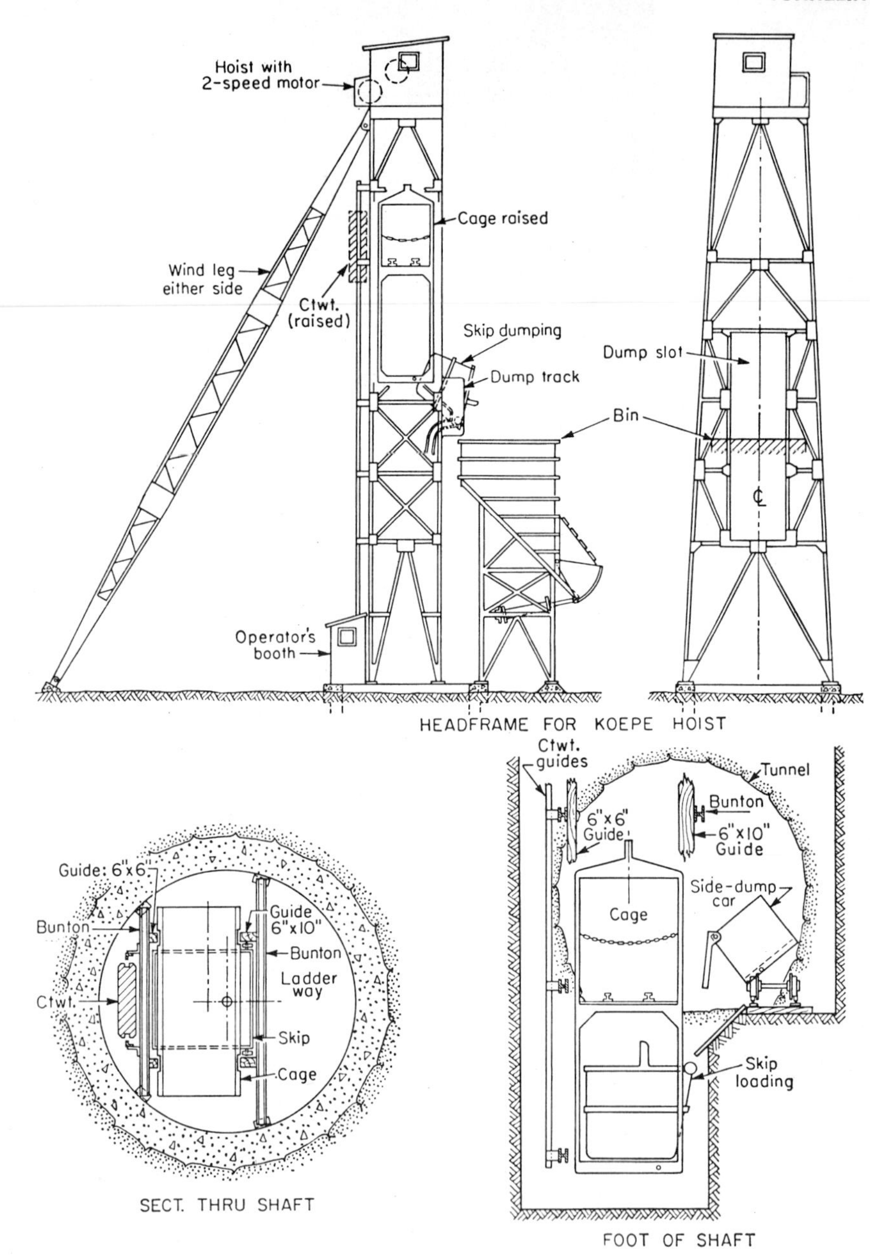

Fig. 31-70 Koepe hoist for tunnel work.

Fig. 31-71 Electric hoist for Koepe. This is a two-speed motor to give a landing speed of 60 fpm and a hoisting speed of 180 fpm.

ate, and stop. The brakes are spring-loaded and will engage when the power is shut off.

All cages that carry men must have safety devices as required by the state mining codes. In general, these devices will consist of a steel bonnet on the cage to protect the men from falling rock, safety dogs to engage the wood guides in the event the cable should part, and headsheaves of the proper diameter. Hoisting ropes should have a factor of safety of 5 when hoisting muck or equipment and 8 when hoisting men.

Index